The Enzyme Reference

A Comprehensive Guidebook to
Enzyme Nomenclature, Reactions, and Methods

The Enzyme Reference

A Comprehensive Guidebook to
Enzyme Nomenclature, Reactions, and Methods

Daniel L. Purich

R. Donald Allison

Department of Biochemistry
and Molecular Biology

University of Florida
College of Medicine

Gainesville, Florida

ACADEMIC PRESS
An imprint of Elsevier Science

Amsterdam Boston London New York Oxford
Paris San Diego San Francisco Singapore Sydney Tokyo

Academic Press
An imprint of Elsevier Science.
525 B Street, Suite 1900, San Diego, California 92101-4495, USA
http://www.academicpress.com

1003079986

Academic Press
84 Theobald's Road, London WC1X 8RR, UK
http://www.academicpress.com

Library of Congress Catalog Card Number: 2002113737

International Standard Book Number: 0-12-568041-4

PRINTED IN THE UNITED STATES OF AMERICA
02   03   04   05   06   07   9 8 7 6 5 4 3 2 1

Table of Contents

Preface

Few would quarrel with the assertion that the extraordinary catalytic power of enzymes is rivaled only by their vast diversity of reaction types and their ability to surprise and fascinate molecular life scientists and chemists alike. Contrary to predictions of a waning interest in intermediary metabolism, new enzymes continue to be discovered, especially in the fields of cell biology and neuroscience. Attesting to the vibrancy of modern enzyme research are the widely heralded recent discoveries of ribozymes, telomerases, proteasomes, NAD^+-dependent histone deacetylases, as well as hundreds of new proteases, protein kinases, and an ever-growing list of energases that includes novel molecular motors, ion/solute pumps, protein-folding chaperonins, and GTP-regulatory enzymes. One may even view the field of cell biology as a form of structural metabolism wherein the cell's supramolecular components are formed, remodeled, and degraded enzymatically. As more details emerge about the chemical foundations of cell biology, various processes (*e.g.*, membrane targeting, cell crawling, and cell division) are all taking on the appearance of biochemical pathways akin to those so frequently illustrated in metabolic pathway charts. Enzymes also remain as the principal targets for designing therapeutic agents—witness the burgeoning list of new inhibitors of angiotensinogen-converting enzymes, nitric oxide synthases, cyclooxygenases, and cholesterogenic enzymes. Given that entire genomes have been sequenced and that powerful combinatoric and array technologies have been perfected for detecting and quantifying enzyme interactions with substrates and inhibitors, expedited discovery of novel drugs and enzyme-based/directed therapies is especially likely.

Such considerations underscore the need for a single-volume reference offering molecular life scientists the name, class, EC number, reaction, essential cofactor(s), and other details about each of the enzyme-catalyzed reactions classified by the Enzyme Commission, as of January of 2002, as well as many presently unclassified enzymes for which we found sufficient documentation regarding their substrates, products, and reaction stoichiometry. We also included many alternative enzyme names, while ignoring those deemed to be arcane, abandoned, or obviously incorrect. This compendium lists over 6500 different enzyme-catalyzed reactions, often citing the original paper describing a particular enzyme and/or several authoritative reviews. When two different enzymes catalyze virtually identical or very similar reactions, we have attempted to include key distinguishing features, along with an italicized phrase directing the reader's attention to related enzyme-catalyzed reactions. We avoided distinctions made solely on a quantitative basis, because the identical reaction in different tissues and organisms is catalyzed by enzymes with different Michaelis constants, turnover numbers, molecular weights, etc. We also made a special effort to include reactions facilitated by plant enzymes, not only because these reactions tend to be overlooked, but because their reactions are both surprisingly complex and wonderfully fascinating. We also included the structures of nearly one thousand metabolites and cofactors to inform the reader about the chemical transformations occurring in many reactions.

Throughout the course of this effort, we were nagged by the question: *How should we handle those ATPase and GTPase reactions which, when written without reference to their noncovalent substratelike and productlike conformational states, are currently misclassified as hydrolases?* This problem arises from the Enzyme Commission's long-standing practice of describing enzyme reactions strictly in terms of transformations in covalent bonds. For example, myosin is classified as a hydrolase despite its catalysis of a force-generating contraction (*e.g.*, $Myosin_{Position-1} + ATP = Myosin_{Position-2} + ADP + Pi$, where Position-1 and Position-2 indicated the myosin's advancement along an actin filament). Thus, while the essence of myosin's function is the making and breaking of noncovalent bonds, the Enzyme Commission ignores such transformations when classifying and naming enzymes. In truth, biological catalysis need not be attended by any change in covalency, leading to the recent suggestion that an enzyme be redefined as a biological catalyst that accelerates the making/breaking of chemical bonds [Trends in Biochemical Sciences (2001) **26**, 417]. This new definition builds on Pauling's assertion in The Nature of the Chemical Bond that any long-lived, chemically distinct interaction constitutes a chemical bond. These considerations point to the need for a new enzyme class that would include ATPases and GTPases as specialized enzymes designed to transduce covalent bond energy into mechanical work. (*See text entry entitled* "Energases.") Even so, until inconsistencies in classifying so-called "ATPase" and "GTPase" reactions are resolved, the Enzyme Reference retains the current Enzyme Commission name, classification, and numbering system for these reactions.

The origins of the Enzyme Reference are rooted in the Handbook of Biochemical Kinetics, which we wrote several years ago to assist molecular life scientists in the design, execution, and evaluation of kinetic experiments. Because one of us (D.L.P.) has edited Methods in Enzymology volumes for nearly 25 years, we obtained permission to include within that handbook thousands of brief citations to Methods in Enzymology articles describing the preparation, assay, storage, and other properties for some 600 enzymes widely used in kinetic investigations. We now extend this practice by including these and similar citations for thousands of enzymes listed in the Enzyme Reference. Thus, while there are internet-accessible databases on enzyme-catalyzed reactions, none provides such extensive and convenient access to Methods in Enzymology.

Sharing an enduring fascination about the diversity of enzyme catalysis, we have consulted thousands of original research papers and reviews, not to mention countless abstracts, during the preparation of the Enzyme Reference. On many occasions, we questioned the wisdom of ever undertaking this task, and we often remarked that our parents were probably most responsible for our stubborn pursuit of the scholarship needed to complete the project. With an abiding reverence for their values, we dedicate this reference book to our parents, Violet and Edward Purich and Edna and Robert Allison.

We hope our readers will find the Enzyme Reference to be a trusted desk companion, always kept within reach. We painstakingly sought to achieve accuracy, and we would deeply appreciate reader comments, especially concerning required amendments and corrections. We would also welcome suggestions about new enzymes that should be included in later editions.

Daniel L. Purich

R. Donald Allison

Nonstandard Journal Abbreviations Used

ABB   Archives of Biochemistry and Biophysics

AE   Advances in Enzymology (book series)

BBA   Biochimica et Biophysica Acta

BBRC   Biochemistry & Biophysics Research Communications

BJ   Biochemical Journal

CBC   Comprehensive Biological Catalysis (book)

EJB   European Journal of Biochemistry

JACS   Journal of the American Chemical Society

JBC   Journal of Biological Chemistry

MIE   Methods in Enzymology (book series)

NAR   Nucleic Acids Research

PNAS   Proceedings of the National Academy of Sciences, U.S.A.

TiBS   Trends in Biochemical Sciences

*Aaa*I RESTRICTION ENDONUCLEASE

This type II restriction enzyme [EC 3.1.21.4], isolated from *Acetobacter aceti*, catalyzes the hydrolysis of both strands of DNA at 5′...C∇GGCCG...3′.[1]

[1] H. Tagami, K. Tayama, T. Tohyama, M. Fukaya, H. Okumura, Y. Kawamura, S. Horinouchi & T. Beppu (1988) *FEMS Microbiol. Lett.* **56**, 161.

*Aat*II RESTRICTION ENDONUCLEASE

This type II restriction endonuclease [EC 3.1.21.4], isolated from *Acetobacter aceti*, catalyzes the hydrolysis of both strands of DNA at 5′...GACGT∇C...3′.[1]

[1] H. Sugisaki, Y. Maekawa, S. Kanazawa & M. Takanami (1982) *NAR* **10**, 5747.

ABELSON NONRECEPTOR TYROSINE KINASE

This protein-tyrosine kinase (abbreviated Abl), originally identified from the transforming gene of Abelson murine leukemia virus, catalyzes the phosphorylation of a variety of proteins, including RNA polymerase II and c-Crk.[1–3] This kinase is an important factor in a number of signalling pathways and apoptosis.

[1] S. P. Goff, E. Gilboa, O. N. Witte & D. Baltimore (1980) *Cell* **22**, 777.
[2] J. G. Foulkes, M. Chow, C. Gorka, A. R. Frankelton & D. Baltimore (1985) *JBC* **260**, 8070.
[3] A. B. Raitano, Y. E. Whang & C. L. Sawyers (1997) *BBA* **1333**, F201.

ABEQUOSYLTRANSFERASE

This transferase[1,2] [EC 2.4.1.60] catalyzes the reaction of CDP-abequose (also referred to as CDP-3,6-dideoxy-D-galactose and CDP-3-deoxy-D-fucose as well as CDP-3,6-dideoxy-D-xylohexose) with D-mannosylrhamnosyl-D-galactose-1-diphospholipid to produce CDP and abequosyl-D-mannosylrhamnosyl-D-galactose-1-diphospholipid. The diastereoisomer CDP-tyvelose (*i.e.*, CDP-3,6-dideoxy-D-mannose) can also serve as an alternative, albeit weaker, substrate. Interestingly, the epimer CDP-paratose (*i.e.*, CDP-3,6-dideoxy-D-glucose) was not a substrate. One potential explanation for this interesting observation is that CDP-tyvelose may adopt a furanose or pyranose form similar to that of CDP-abequose whereas CDP-paratose would exist primarily in the other conformation.

The *Salmonella typhimurium* enzyme participates in the biosynthesis of the *O*-antigen of specific lipopolysaccharides. Since abequose-containing lipopolysaccharides have been observed in other organisms, an abequosyltransferase activity has been implicated in those species.[3]

pyranose form   furanose form

Note: Abequose is a dideoxyhexose, also referred to as 3,6-dideoxy-D-galactose, 3-deoxy-D-fucose, and 3,6-dideoxy-D-xylohexose. This dideoxy sugar rarely exists as the free hexose, and is primarily present as the CDP or glycophospholipid derivative.

[1] M. J. Osborn & I. M. Weiner (1968) *JBC* **243**, 2631.
[2] M. J. Osborn, M. A. Cynkin, J. M. Gilbert, L. Müller & M. Singh (1972) *MIE* **28**, 583.
[3] D. Liu, L. Lindqvist & P. R. Reeves (1995) *J. Bacteriol.* **177**, 4084.

Selected entries from ***Methods in Enzymology*** [vol, page(s)]:
General discussion: 28, 583
Other molecular properties: bacterial *O*-antigen biosynthesis and, **28**, 584; preparation, **28**, 597; properties, **28**, 589

*Aca*I RESTRICTION ENDONUCLEASE

This type II restriction endonuclease [EC 3.1.21.4] is obtained from *Anabaena catanula*. The enzyme has a DNA recognition sequence of $5' \ldots \text{TTCGAA} \ldots 3'$, but the site of cleavage is still unknown.

Selected entries from **Methods in Enzymology** [vol, page(s)]:
General discussion: 65, 2

AccI RESTRICTION ENDONUCLEASE

This type II restriction endonuclease [EC 3.1.21.4] is obtained from *Acinetobacter calcoaceticus* and acts on both strands of DNA at $5' \ldots \text{GT}\nabla\text{MKAC} \ldots 3'$,[1,2] where M refers to A or C and K refers to G or T. The active form of the enzyme appears to be tetrameric.[3]

[1] R. J. Roberts (1980) *Meth. Enzymol.* **65**, 1.
[2] S. C. Kang & O. J. Yoo (1985) *Korean J. Microbiol.* **23**, 13.
[3] B. Kawakami, C. Hilzheber, M. Nagatomo & M. Oka (1991) *Agric. Biol. Chem.* **55**, 1553.

Selected entries from **Methods in Enzymology** [vol, page(s)]:
Other molecular properties: recognition sequence, 65, 2

AccII RESTRICTION ENDONUCLEASE

This type II restriction endonuclease [EC 3.1.21.4] is obtained from *Acinetobacter calcoaceticus* and acts on both strands of DNA at $5' \ldots \text{CG}\nabla\text{CG} \ldots 3'$,[1,2] producing blunt-ended fragments. If the cytidine is 5-methylated, DNA cleavage is inhibited.[3]

[1] R. J. Roberts (1980) *MIE* **65**, 1.
[2] K. Kita, N. Hiraoka, F. Kimizuka & A. Obayashi (1984) *Agric. Biol. Chem.* **48**, 531.
[3] M. L. Gaido & J. S. Strobl (1987) *Arch. Microbiol.* **46**, 338.

Selected entries from **Methods in Enzymology** [vol, page(s)]:
Other molecular properties: recognition sequence, 65, 2

Acc65I RESTRICTION ENDONUCLEASE

This type II restriction endonuclease [EC 3.1.21.4], obtained from *Acinetobacter calcoaceticus* 65, catalyzes the hydrolysis of both strands of DNA at $5' \ldots \text{G}\nabla\text{GTACC} \ldots 3'$.[1] This enzyme is sensitive to overlapping *dcm* methylation.

[1] G. G. Prichodko, N. I. Rechnukova, V. E. Repin & S. K. Degtyarev (1991) *Sib. Biol. J.* **1**, 59.

trans-ACENAPHTHENE-1,2-DIOL DEHYDROGENASE

trans-acenaphthene-1,2-diol   acenaphthenequinone

This oxidoreductase [EC 1.10.1.1] catalyzes the reaction of *trans*-acenaphthene-1,2-diol with $NADP^+$ to produce acenaphthenequinone and NADPH.[1] Either stereoisomer of the diol serves as a substrate, and NAD^+ can be used as the coenzyme with the enzyme from some sources.

[1] R. P. Hopkins, E. C. Drummond & P. Callaghan (1973) *Biochem. Soc. Trans.* **1**, 989.

ACETALDEHYDE DEHYDROGENASE (ACETYLATING)

This oxidoreductase [EC 1.2.1.10] catalyzes the oxidation of acetaldehyde in the presence of NAD^+ and Coenzyme A to produce acetyl-CoA, NADH, and H^+. Other aldehyde substrates include glycolaldehyde, propanal, and butanal.

With the *Escherichia coli* enzyme, an essential thiol group probably attacks the hydrated aldehyde to form a thiohemiacetal that is oxidized to an *S*-acetyl enzyme species that undergoes thiolysis by Coenzyme A. Initial rate and competitive inhibition experiments support a Bi Uni Uni Uni ping pong kinetic mechanism in which NAD^+ binds to the free enzyme followed by acetaldehyde.[1,2] The product NADH is then released before Coenzyme A binds, and acetyl-CoA is the last product released. The enzyme appears to contain two distinct NAD^+ binding sites, an activator site and a catalytic site, based on pre-steady-state experiments with the NAD^+ analogues AMP and 3-pyridine-carboxaldehyde adenine dinucleotide.[2]

[1] C. C. Shone & H. J. Fromm (1981) *Biochemistry* **20**, 7494.
[2] F. B. Rudolph, D. L. Purich & H. J. Fromm (1968) *JBC* **243**, 5539.

Selected entries from **Methods in Enzymology** [vol, page(s)]:
General discussion: 1, 518
Assay: 1, 518, 520
Other molecular properties: kinetics of *Escherichia coli* enzyme, Fromm method, **63**, 13, 14; lag elimination, **63**, 9; properties, 1, 520; purification from *Clostridium kluyveri*, 1, 519; specificity, 1, 520; stereospecificity, **87**, 113; thiols, effect of, **63**, 9.

4-ACETAMIDOBUTYRATE DEACETYLASE

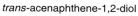

acetamidobutyrate

This deacetylase [EC 3.5.1.63] catalyzes the hydrolysis of 4-acetamidobutanoate to produce acetate and

4-aminobutanoate.[1,2] Alternative substrates include N-acetyl-β-alanine and 5-acetamidopentanoate.

[1]C. Gillyon, G. W. Haywood, P. J. Large, B. Nellen & A. Robertson (1987) J. Gen. Microbiol. 133, 2477.
[2]G. W. Haywood & P. J. Large (1986) J. Gen. Microbiol. 132, 7.

4-ACETAMIDOBUTYRYL-CoA DEACETYLASE

This enzyme [EC 3.5.1.51], also referred to as aminobutyryl-CoA thiolesterase, catalyzes the hydrolysis of 4-acetamidobutanoyl-CoA (CoAS-COCH$_2$CH$_2$NHCOCH$_3$) to produce acetate and 4-aminobutanoyl-CoA.[1]

[1]M. Ohsugi, J. Kahn, C. Hensley, S. Chew & H. A. Barker (1981) JBC 256, 7642.

2-(ACETAMIDOMETHYLENE)SUCCINATE HYDROLASE

This enzyme [EC 3.5.1.29], also called α-(N-acetylaminomethylene)succinate hydrolase, catalyzes the hydrolysis of 2-(acetamidomethylene)succinate to produce acetate, succinate semialdehyde, carbon dioxide, and ammonia. Oxaloacetate, itaconate, and other di- and tricarboxylic acids inhibit the enzyme, as does sulfite.[1]

[1]M. S. Huynh & E. E. Snell (1985) JBC 260, 2379.

Selected entries from **Methods in Enzymology** [vol, page(s)]:
Assay: 18A, 657
Other molecular properties: properties, 18A, 658; purification from Pseudomonas sp., 18A, 657

ACETATE CoA-TRANSFERASE

This CoA transferase [EC 2.8.3.8], also referred to as acetyl-CoA:acetoacetate CoA transferase and first isolated from Escherichia coli K12, catalyzes the reversible reaction of an acyl-CoA with acetate to produce a fatty acid anion and acetyl-CoA.[1] The acyl-donor substrates can be butanoyl-CoA, pentanoyl-CoA, and succinyl-CoA.

[1]E. Vanderwinkel, P. Furmanski, H. C. Reeves & S. J. Ajl (1968) BBRC 33, 902

ACETATE KINASE

This phosphotransferase [EC 2.7.2.1], also called acetokinase, catalyzes the thermodynamically favored reaction of ADP with acetyl phosphate to form ATP and acetate (i.e., K_{eq} = [ATP][acetate]/{[acetyl phosphate][ADP]} = 3000).[1–4] GDP is also a phosphoryl group acceptor. Phosphoryl transfer proceeds with inversion of configuration, indicating an in-line nucleophilic displacement reaction.

Incidental generation of a metaphosphate anion during catalysis may lead to the sub-stoichiometric formation of an enzyme-bound acyl-phosphate. Active site glutamyl residues have been identified.

Acetate kinase has a role in the acetate-stimulated change in the direction of flagellar rotation in Escherichia coli.[5] Acetate kinase is also ideally suited for the regeneration of ATP or GTP from ADP or GDP, respectively. A thermostable acetate kinase (optimum temperature of 90°C) was recently isolated from a hyperthermophilic bacterium.[5]

[1]P. A. Frey (1992) The Enzymes 20, 160.
[2]K. Singh-Wissmann, C. Ingram-Smith, R. D. Miles & J. G. Ferry (1998) J. Bacteriol. 180, 1129.
[3]Y. S. Kim & C. Park (1988) BBA 956, 103.
[4]R. Barak, W. N. Abouhamad & M. Eisenbach (1998) J. Bacteriol. 180, 985.
[5]A. K. Bock, J. Glasemacher, R. Schmidt & P. Schonheit (1999) J. Bacteriol. 181, 1861.

Selected entries from **Methods in Enzymology** [vol, page(s)]:
General discussion: 1, 591; 243, 94
Assay: 1, 591; activity assay, 44, 893, 894; Bacillus stearothermophilus, 90, 179; Veillonella alcalescens, (ATP formation assay, 71, 312; hydroxamate assay, 71, 311)
Other molecular properties: acetate assay, 3, 269; activation, 44, 889; activity assay, 44, 893, 894; acetyl phosphate and acetyl-CoA determination, 122, 44; alternative substrates, 87, 11; α-amino-δ-hydroxyvalerate, and, 87, 18; Bacillus stearothermophilus (assay, 90, 179; properties, 90, 183; purification, 90, 180); bridge-nonbridge transfer, 87, 19, 20, 226, 232; chiral methyl group analysis, 87, 130; chiral phosphoric monoester, 87, 300; chiral phosphoryl-ATP, 87, 18, 211, 231, 258, 300; cold denaturation, 63, 9; cysteine residues, 44, 887; equilibrium constant, 63, 5; 87, 16; exchange properties, 64, 9, 39; 87, 5, 6, 18, 656; hexokinase, and, in glucose 6-phosphate production, 136, 52; hydroxaminolysis, 87, 18; immobilized, 44, 891, 892 (dihydroxyacetone phosphate synthesis with, 136, 277; glucose 6-phosphate synthesis with, 136, 279; sn-glycerol 3-phosphate synthesis with, 136, 276); inhibitor, 63, 398; initial rate kinetics, 87, 18; intermediate, 87, 12, 18, 20; mechanism, 87, 8, 18, 19, 211, 656; mercuric ion inhibition, 87, 18; metal-ion binding, 63, 275; metaphosphate intermediate, and, 87, 12, 20; metaphosphate synthesis and, 6, 262; nucleoside diphosphokinase activity, 87, 18; phosphorothioates, 87, 200, 205, 226, 232, 258; phosphorylation potential, 55, 237; 87, 18, 656; pH stability profile, 87, 18; purification, (Bacillus stearothermophilus, 90, 180; Escherichia coli, 1, 593); purine nucleoside diphosphate kinase activity, 63, 8; pyruvic acid phosphoroclastic system, in, 243, 96, 99; ribulose-5-phosphate 4-epimerase and, 5, 253; specificity, 1, 594; 71, 316; synergism quotient, 87, 8; triple-displacement mechanism, 87, 19, 211; Veillonella alcalescens, from, 71, 311 (ATP formation assay, 71, 312; hydroxamate assay, 71, 311; properties, 71, 315; stability to heat, 71, 313; stimulation by succinate, 71, 316; substrate specificity, 71, 316); xylulose-5-phosphate 3-epimerase and, 5, 250; xylulose-5-phosphate phosphoketolase and, 5, 262

ACETATE KINASE (PYROPHOSPHATE)

This phosphotransferase [EC 2.7.2.12], also known as acetate kinase (diphosphate), catalyzes the conversion of acetate and pyrophosphate (or, diphosphate) to acetyl phosphate and orthophosphate.[1]

[1]R. Reeves & J. D. Guthrie (1975) *BBRC* **66**, 1389.

ACETOACETATE DECARBOXYLASE

This decarboxylase [EC 4.1.1.4], also referred to as acetoacetate carboxy-lyase, catalyzes the conversion of acetoacetate to acetone and carbon dioxide. The enzyme has an active-site lysyl residue with a highly perturbed ε-amino pK_a (about 5.9).[1–3] Decarboxylation of the imine intermediate produces an enamine which rearranges to a new imine intermediate; hydrolysis of this Schiff base releases acetone and regenerates the lysyl residue.

[1]F. C. Kokesh & F. H. Westheimer (1971) *JACS* **93**, 7270.
[2]P. A. Frey, F. C. Kokesh & F. H. Westheimer (1971) *JACS* **93**, 7266.
[3]D. E. Schmidt, Jr., & F. H. Westheimer (1971) *Biochemistry* **10**, 1249.

Selected entries from *Methods in Enzymology* [vol, page(s)]:
General discussion: 1, 624; 14, 231
Assay: 1, 624; 14, 231
Other molecular properties: active site, 14, 240; coenzyme for, 1, 626; crystallization, 14, 236; inhibitors, 1, 627; 14, 239; 63, 400; intermediate, 87, 84; kinetic isotope effect, 64, 94, 95; mechanism, 14, 240; perturbed lysine pK, 63, 210; properties, 1, 626; purification, *Clostridium acetobutylicum*, 1, 625; 14, 232; resolution, 1, 626; specificity, 14, 239; transition state and multisubstrate analogues, 249, 307

ACETOACETYL-CoA HYDROLASE

This enzyme [EC 3.1.2.11] catalyzes the hydrolysis of acetoacetyl-CoA to produce acetoacetate and Coenzyme A.[1]

[1]J. J. Aragon & J. M. Lowenstein (1983) *JBC* **258**, 4725.

ACETOACETYL-CoA REDUCTASE

This oxidoreductase [EC 1.1.1.36] catalyzes the reaction of acetoacetyl-CoA (*i.e.*, 3-oxobutyryl-CoA) with NADPH (stereospecificity: B-side) to produce (*R*)-3-hydroxybutyryl-CoA and NADP$^+$. The reductase acts on a number of 3-ketoacyl-CoA (*i.e.*, 3-oxoacyl-CoA), generating the corresponding 3-hydroxyacyl-CoA products.[1]

[1]O. Ploux, S. Masamune & C. T. Walsh (1988) *Eur. J. Biochem.* **174**, 177

Selected entries from *Methods in Enzymology* [vol, page(s)]:
Assay: 71, 96

ACETOACETYL-CoA SYNTHETASE

This enzyme [EC 6.2.1.16], also called acetoacetate:CoA ligase, catalyzes the reaction of ATP with acetoacetate and coenzyme A to produce AMP, pyrophosphate (or, diphosphate), and acetoacetyl-CoA.[1–3] L-3-Hydroxybutanoate is

a poor alternative substrate.[4] Many fatty acyl-CoA derivatives are inhibitors (*e.g.*, palmitoyl-CoA has a K_i value of 10 μM).[5]

[1]J. R. Stern & S. Ochoa (1951) *JBC* **191**, 161.
[2]J. R. Stern, M. J. Coon, & A. del Campillo (1953) *Nature* **171**, 28.
[3]J. R. Stern, M. J. Coon, A. del Campillo & M. C. Schneider (1956) *JBC* **221**, 15.
[4]T. Fukui, M. Ito & K. Tomita (1982) *Eur. J. Biochem.* **127**, 423.
[5]M. Ito, T. Fukui, T. Saito & K. Tomita (1987) *BBA* **922**, 287.

Selected entries from *Methods in Enzymology* [vol, page(s)]:
General discussion: 110, 3
Assay: 110, 3

S-ACETOACETYLHYDROLIPOATE HYDROLASE

This thiolester hydrolase activity, formerly classified as EC 3.1.2.9, is now a deleted Enzyme Commission entry.

ACETOIN DEHYDROGENASE

Acetoin dehydrogenase [EC 1.1.1.5], also known as diacetyl reductase, catalyzes the reversible reaction of acetoin (*i.e.*, 3-hydroxy-2-butanone) with NAD$^+$ to produce diacetyl (*i.e.*, 2,3-butanedione) and NADH. NADP$^+$ can also act as an alternative coenzyme substrate. Yeast produces two acetoin dehydrogenases that are stereospecific for two enantiomers of acetoin.[1] The *Staphylococcus aureus*[2] and hamster liver[3] enzymes were reported to have an ordered Bi Bi kinetic mechanism for the reverse reaction and a Theorell–Chance mechanism when 2,3-pentanedione was used as an alternative substrate.[2] The hamster enzyme is also reported to be B-side stereospecific with respect to NADPH.[3]

[1]J. Heidlas & R. Tressl (1990) *EJB* **188**, 165.
[2]J. Gonzalez, I. Vidal, A. Bernardo & R. Martin (1988) *Biochimie* **70**, 1791.
[3]H. Sawada, A. Hara, T. Nakayama & K. Seiriki (1985) *J. Biochem.* **98**, 1349.

Selected entries from *Methods in Enzymology* [vol, page(s)]:
Assay: 41, 529, 530; 89, 516
Other molecular properties: chromatography, 41, 531; disc gel electrophoresis, 89, 517; properties, 41, 532, 533; 89, 521; purification, 41, 530 (bovine liver, 89, 517; *Escherichia coli*, 89, 519; pigeon liver, 89, 518)

ACETOIN RACEMASE

This racemase [EC 5.1.2.4], formerly known as acetylmethylcarbinol racemase, catalyzes the interconversion of (*S*)-acetoin (*i.e.*, (*S*)-3-hydroxy-2-butanone) and (*R*)-acetoin.[1,2]

[1]M. B. Taylor & E. Juni (1960) *BBA* **39**, 448.
[2]M. Voloch, M. R. Ladisch, V. W. Rodwell & G. T. Tsao (1983) *Biotechnol. Bioeng.* **25**, 173.

ACETOIN:RIBOSE-5-PHOSPHATE TRANSALDOLASE

This thiamin pyrophosphate-dependent transaldolase [EC 2.2.1.4], also known as 1-deoxy-D-*altro*-heptulose-7-phosphate synthase, catalyzes the reversible reaction of acetoin (*i.e.*, 3-hydroxybutan-2-one) with D-ribose 5-phosphate to produce 1-deoxy-D-*altro*-heptulose 7-phosphate and acetaldehyde.[1] The enzyme can use methylacetoin as an alternative substrate, in this case generating acetone.

[1] A. Yokota & K. Sasajima (1983) *Agric. Biol. Chem.* **47**, 1545.

ACETOLACTATE DECARBOXYLASE

Acetolactate decarboxylase [EC 4.1.1.5] catalyzes the conversion of acetolactate (*i.e.*, (*S*)-2-hydroxy-2-methyl-3-oxobutanoate) to carbon dioxide and (*R*)-acetoin.[1]

[1] V. Phalip, C. Monnet, P. Schmitt, P. Renault, J. J. Godon & C. Divies (1994) *FEBS Lett.* **351**, 95.

Selected entries from *Methods in Enzymology* [**vol**, page(s)]:
General discussion: 1, 471
Assay: 1, 471; 41, 526, 527
Other molecular properties: *Aerobacter aerogenes*, from, **41**, 526 (assay, **41**, 526, 527; chromatography, **41**, 527; electrophoresis, **41**, 528; properties, **41**, 528, 529; purification, **41**, 527); preparation, 3, 283

2-ACETOLACTATE MUTASE

This ascorbate-dependent enzyme [EC 5.4.99.3] catalyzes the interconversion of 2-acetolactate (*i.e.*, 2-hydroxy-2-methyl-3-oxobutanoate) to 3-hydroxy-3-methyl-2-oxobutanoate as well as 2-aceto-2-hydroxybutanoate to 3-hydroxy-3-methyl-2-oxopentanoate.[1] *See also Ketol-Acid Reductoisomerase*

[1] H. S. Allaudeen & T. Ramakrishnan (1968) *ABB* **125**, 199.

ACETOLACTATE SYNTHASE

This thiamin-pyrophosphate-dependent enzyme [EC 4.1.3.18], also known as acetohydroxy acid synthetase (or synthase), catalyzes the reaction of two pyruvate ions to produce 2-acetolactate (*i.e.*, 2-hydroxy-2-methyl-3-oxobutanoate) and carbon dioxide.[1–7] Some isozymes also catalyze the (non-physiological) oxidative decarboxylation of pyruvate, resulting in peracetic acid. Although the reaction is not an oxidation–reduction process, FAD is required. The plant enzyme is a target for several classes of herbicides.

The reaction involves the decarboxylation of pyruvate, followed by condensation with either a second molecule of pyruvate or with α-ketobutyrate.[1–5] The enzyme has a ping pong Bi Bi kinetic mechanism. The first molecule of pyruvate binds to the enzyme and forms an α-lactylthiamin diphosphate intermediate. Decarboxylation yields an enamine intermediate that then reacts with the second molecule of pyruvate, such that 2-acetolactate is released. (The enzyme also exhibits a minor oxygenase side-reaction, in which the enamine intermediate reacts with dioxygen and peracetate (CH_3COOO^-) is generated.) The synthase will also produce 2-aceto-2-hydroxybutyrate from pyruvate and α-ketobutyrate.

[1] M. T. Tse & J. V. Schloss (1993) *Biochemistry* **32**, 10398.
[2] A. K. Chang & R. G. Duggleby (1997) *BJ* **327**, 161.
[3] J. H. Yang & S. S. Kim (1993) *BBA* **1157**, 178.
[4] S. S. Pang & R. G. Duggleby (1999) *Biochemistry* **38**, 5222.
[5] D. Chipman, Z. Barak & J. V. Schloss (1998) *BBA* **1385**, 401

Selected entries from *Methods in Enzymology* [**vol**, page(s)]:
General discussion: **166**, 101, 241, 436, 519; **324**, 95
Assay: **166**, 101, 241, 436; *Aerobacter aerogenes*, **41**, 519, 520; permeabilized bacteria, in, **166**, 230; reaction product assays, **166**, 234
Other molecular properties: *Aerobacter aerogenes*, from, **41**, 519 (assay, **41**, 519, 520; chromatography, **41**, 521, 522; crystallization, **41**, 522; inhibitors, **41**, 526; properties, **41**, 525, 526; purification, **41**, 520, 525); inhibitor sulfometuron methyl (hypersensitive microbial mutants, isolation, **166**, 105; isolation of resistant microbial and plant mutants, **166**, 103); isozyme I (assay, **166**, 101, 241, 436; properties, **166**, 440; purification from *Esherichia coli*, **166**, 438); isozyme II (assay, **166**, 101, 446; properties, **166**, 451; purification from *Salmonella typhimurium*, **166**, 448); isozyme III (assay, **166**, 101, 241, 455; properties, **166**, 456; purification from *Escherichia coli*, **166**, 456); production in *Salmonella typhimurium*, **22**, 89; reaction products, assay, **166**, 234; structural genes, isolation from microbes and plants, **166**, 101

ACETONE-CYANOHYDRIN LYASE

This lyase [EC 4.1.2.37] catalyzes the conversion of acetone-cyanohydrin ($H_3CC(OH)(CN)CH_3$) to produce cyanide and acetone. 2-Butanone cyanohydrin is an alternative substrate.[1,2] This enzyme is distinct from mandelonitrile lyase [EC 4.1.2.10] and hydroxymandelonitrile lyase [EC 4.1.2.11].

[1] H. Wajant & K. Pfizenmaier (1996) *JBC* **271**, 25830.
[2] U. Hanefeld, A. J. Straathof & J. J. Heijnen (1999) *BBA* **1432**, 185.

ACETOXYBUTYNYLBITHIOPHENE DEACETYLASE

5-(4-acetoxybut-1-ynyl)-2,2′-bithiophene

This highly specific deacetylase [EC 3.1.1.54] catalyzes the hydrolysis of 5-(4-acetoxybut-1-ynyl)-2,2′-bithiophene to produce 5-(4-hydroxybut-1-ynyl)-2,2′-bithiophene and acetate.[1]

[1]R. Sütfeld & G. H. N. Towers (1982) *Phytochemistry* **21**, 277.

N-ACETYL-β-ALANINE DEACETYLASE

This deacetylase [EC 3.5.1.21] catalyzes the hydrolysis of *N*-acetyl-β-alanine to produce acetate and β-alanine.[1] *N*-Acetyltaurine is a poor substrate.

[1]D. Fujimoto, T. Koyama & N. Tamiya (1968) *BBA* **167**, 407.

ACETYLALKYLGLYCEROL ACETYLHYDROLASE

This enzyme [EC 3.1.1.71], also known as alkylacetylglycerol acetylhydrolase, catalyzes the hydrolysis of a 2-acetyl-1-alkyl-*sn*-glycerol to produce a 1-alkyl-*sn*-glycerol and acetate.[1] An equally effective substrate is 1-alkyl-3-acetyl-*sn*-glycerol.

This enzyme is distinguishable from lipoprotein lipase [EC 3.1.1.34], because 1,2-diacetyl-*sn*-glycerols are not substrates. The hydrolase is also distinct from 1-acetyl-2-alkylglycerophosphocholine esterase [EC 3.1.1.47].

[1]M. L. Blank, Z. L. Smith, E. A. Cress & F. Snyder (1990) *BBA* **1042**, 153

2-ACETYL-1-ALKYLGLYCERO-PHOSPHOCHOLINE ESTERASE

This hydrolase [EC 3.1.1.47] (also known as 1-alkyl-2-acetylglycerophosphocholine esterase, platelet-activating factor acetylhydrolase, and lipoprotein-associated phospholipase A₂) catalyzes the hydrolysis of 2-acetyl-1-alkyl-*sn*-glycero-3-phosphocholine to produce 1-alkyl-*sn*-glycero-3-phosphocholine and acetate.[1–5]

Pre-steady-state kinetic analysis with the enzyme tightly bound to vesicles and utilizing a substrate that undergoes slow intervesicle exchange establishes that the esterase accesses its substrate from the aqueous phase and thus is not an interfacial enzyme.[2]

[1]M. L. Blank, T. Lee, V. Fitzgerald & F. Snyder (1981) *JBC* **256**, 175.
[2]J. H. Min, M. K. Jain, C. Wilder, L. Paul, R. Apitz-Castro, D. C. Aspleaf & M. H. Gelb (1999) *Biochemistry* **38**, 12935.
[3]D. M. Stafforini, T. M. McIntyre, G. A. Zimmerman & S. M. Prescott (1997) *JBC* **272**, 17895.
[4]E. A. Dennis (1997) *TiBS* **22**, 1.
[5]D. M. Stafforini, S. M. Prescott, G. A. Zimmerman & T. M. McIntyre (1996) *BBA* **1301**, 161.

Selected entries from *Methods in Enzymology* [vol, page(s)]:
General discussion: 187, 344; 197, 411
Assay: 141, 393; 187, 345; 197, 414

Other molecular properties: human erythrocyte (assay, 197, 414; biological role, 197, 425; comparison to other intracellular PAF acetylhydrolases, 197, 422; partial purification, 197, 418; properties, 197, 419); properties, 141, 393; 187, 348; purification, 187, 352; substrate specificity, 187, 355

ACETYLAMINODEOXYDEOXYGLUCOSE KINASE

This phosphotransferase activity, formerly classified as EC 2.7.1.9, is now a deleted Enzyme Commission entry.

ACETYLCHOLINESTERASE

This esterase [EC 3.1.1.7] (also known as true cholinesterase, choline esterase I, and cholinesterase) catalyzes the hydrolysis of acetylcholine to produce choline and acetate.[1,2] The enzyme, which also acts on other acetate esters, will in certain cases catalyze some transacetylation. Acetylcholinesterase has an ordered Uni Bi kinetic mechanism, proceeding by way of an *O*-acetylated enzyme intermediate. The active site contains the catalytic triad of seryl (which is acetylated), histidyl, and either aspartyl or glutamyl residues. Also present is an oxyanion hole similar to that of serine proteases. Low-energy hydrogen bonds may stabilize the transition state.

See also *Cholinesterase*

[1]D. Grisaru, M. Sternfeld, A. Eldor, D. Glick & H. Soreq (1999) *EJB* **264**, 672.
[2]I. Silman, C. B. Millard, A. Ordentlich, H. M. Greenblatt, M. Harel, D. Barak, A. Shafferman & J. L. Sussman (1999) *Chem. Biol. Interact.* **119**, 43.

Selected entries from *Methods in Enzymology* [vol, page(s)]:
General discussion: 1, 642; 34, 571; 46, 22, 25, 515; 54, 52; 71, 745
Assay: 1, 647; 32, 775; 71, 733; erythrocyte ghost sidedness assay, in, 31, 176, 177, 179; ESR, by, 251, 102; fluorometric, with 1-methylacetoxyquinolinium iodide, 17B, 783; microdetermination, 17B, 778; sarcolemmal vesicles, in, 157, 32; spectrophotometric, 17B, 782; titration, 17B, 783
Other molecular properties: acetylthiocholine as substrate, 251, 101; affinity labeling, 87, 482; affinity purification, technique (11S form, 34, 575; 14 S form, 34, 578; 18 S form, 34, 578); biological significance, 1, 642; brain mitochondria, 55, 58, 59; classification, criteria for, 1, 644; column mechanism and capacity, 34, 584, 585; correlation analysis, 87, 482; cultured cells, in, 32, 766, 768, 772, 773; *Electrophorus electricus* (intact tail subunits, isolation, 82, 335; pepsin-resistant fragments, isolation, 82, 332; purification, 82, 327); enzyme concentration effect, 64, 237, 238; erythrocyte ghost sidedness assay, in, 31, 176, 177, 179; gel electrophoresis, 32, 88; immobilization (activation energy, 64, 243, 245; enzyme concentration effect, 64, 237, 238; Michaelis constant, flow rate effect, 64, 238, 240; nylon tubing, on, 64, 240, 242; oscillatory phenomena, 135, 559; temperature effect, 64, 243, 244; Thiele function, 64, 232, 236); immunoassay of, in (eicosanoids, 187, 24; leukotrienes C₄ and E₄, 187, 82); inhibitors, 251, 103; interferon-treated cells, 79, 343; microdetermination, 17B, 778; modification by symmetrical disulfide radical, 251, 100; muscle (intracellular transport and fate, 96, 363; molecular forms, localization and assembly, 96, 365; subcellular distribution, 96, 354; synthesis, 96, 361); nerve impulse, in, 32, 310; neurobiology, in, 32, 765; other esterases and, 1, 643; purification, 34,

575 (covalent affinity technique, **34**, 581; *Electrophorus electricus*, from, **17B**, 782; 11 S, **34**, 575; 14 S, **34**, 578; 18 S, **34**, 578); quaternary ammonium salts, and, **87**, 482; release by phospholipase C, **71**, 733, 740; separation of 18 S and 14 S AChE, **34**, 580; sources, **1**, 644; spacers, **34**, 588; specific phosphorylation of impure enzyme with ^{32}P-diisopropylfluorophosphate, **11**, 694; spin-label studies, **49**, 445, 446; stability, **1**, 645; subunit, radioiodination, **70**, 232; thioester substrate, **248**, 16; transferrin effects, myotrophic assay, **147**, 296; transition state inhibitors and multisubstrate analogues, **249**, 305

ACETYL-CoA ACETYLTRANSFERASE

This acetyltransferase [EC 2.3.1.9], also known as thiolase and thiolase II, catalyzes the transfer of an acetyl group from one acetyl-CoA molecule to another to form free coenzyme A and acetoacetyl-CoA. The enzyme has a ping pong Bi Bi kinetic mechanism and forms an *S*-acetyl-enzyme intermediate.[1,2] **See also** *Acetyl-CoA Acyltransferase*

[1] T. Nishimura, T. Saito & K. Tomita (1978) *Arch. Microbiol.* **116**, 21

[2] F. Suzuki, W. L. Zahler & D. W. Emerich (1987) *ABB* **254**, 272.

Selected entries from *Methods in Enzymology* [vol, page(s)]:
General discussion: **1**, 574, 581; **71**, 398, 399
Assay: **1**, 581; **35**, 129, 167
Other molecular properties: bile acid biosynthesis, and, **15**, 581; chiral methyl groups, and, **87**, 147; *Clostridium kluyveri*, selenium incorporation, **107**, 621; crotonyl coenzyme A assay and, **1**, 562; cytoplasmic, **35**, 128, 167 (inhibitor, **35**, 134, 173; specificity, **35**, 129, 134); glyoxysomes, in, **31**, 569; **52**, 502; mitochondrial, **35**, 128 (activation by K^+ and NH_4^+, **35**, 129, 130, 135; inhibitor, **35**, 135; specificity, **35**, 129, 134); properties, **1**, 584; purification, **1**, 583

ACETYL-CoA ACYLTRANSFERASE

This acyltransferase [EC 2.3.1.16], also known as thiolase I and 3-ketoacyl-CoA thiolase, catalyzes acyl group transfer from an acyl-CoA to acetyl-CoA to form free coenzyme A and 3-ketoacyl-CoA (*i.e.*, 3-oxoacyl-CoA).[1]

The overall reaction can be described as a Claisen condensation of two thiol esters.

The rat liver peroxidomal enzyme exhibits a ping pong Bi Bi kinetic mechanism,[1] proceeding through an *S*-acyl-enzyme intermediate. The enzyme exhibits high activity with medium- and long-chain acyl-CoA derivatives; even so, there is significant activity with acetoacetyl-CoA. **See also** *Acetyl-CoA Acetyltransferase*

[1] S. Miyazawa, S. Furuta, T. Osumi, T. Hashimoto & N. Ui (1981) *J. Biochem.* **90**, 511.

Selected entries from *Methods in Enzymology* [vol, page(s)]:
General discussion: **71**, 378, 403, 404
Assay: **35**, 129, 130; peroxisomal, **148**, 524; pig heart, **71**, 399
Other molecular properties: effect of bromooctanoate, **72**, 568; glyoxysomes, from, **72**, 788; induction by clofibrate and di(2-ethylhexyl)phthalate, **72**, 507; inhibition by

(3-ketopent-4-enoyl-CoA, **72**, 610; 2,4-pentadienyl-CoA, **72**, 610); pig heart, from, **71**, 398 (antibodies, **71**, 403; assay, **71**, 399; inhibitors, **71**, 403; intracellular distribution, **71**, 398; properties, **71**, 402; specificity, **71**, 402); stability, **35**, 132; tissue contents of individual thiolases, **35**, 133

ACETYL-CoA CARBOXYLASE

This biotin-dependent carboxylase [EC 6.4.1.2] catalyzes the reaction of acetyl-CoA with bicarbonate and ATP to form malonyl-CoA, orthophosphate, and ADP.[1,2] The plant enzyme will also act on propionyl-CoA and butanoyl-CoA.

The biotin form of the enzyme first reacts with bicarbonate and ATP to produce ADP, orthophosphate, and the carboxybiotin form of the enzyme (forming a carboxyphosphate intermediate). This modified form of the protein then reacts with acetyl-CoA to produce malonyl-CoA and regenerate the original form of the coenzyme. The enzyme, which will also catalyze certain transcarboxylations, is highly regulated: at the level of gene expression, via allosteric regulation of the enzyme, and by reversible phosphorylation.[1]

[1] M. R. Munday & C. J. Hemingway (1999) *Adv. Enzyme Regul.* **39**, 205.

[2] R. W. Brownsey, R. Zhande & A. N. Boone (1997) *Biochem. Soc. Trans.* **25**, 1232.

Selected entries from *Methods in Enzymology* [vol, page(s)]:
Assay: **200**, 365; chloroplasts and cytosol of higher plants, from, **71**, 47; *Euglena gracilis*, **71**, 61; lactating rat mammary gland, **71**, 27; radioactive assay, **71**, 5, 37, 38, 55; rat liver, **35**, 3; **71**, 301 (radioactive assay, **71**, 5; spectrophotometric assay, **71**, 5, 6); rat mammary gland, **35**, 11; spectrophotometric assay, **71**, 5, 6, 37; yeast, **71**, 34
Other molecular properties: activation by hydroxycitrate, **72**, 489; activators and inhibitors, **71**, 12, 42, 58; biotin carboxyl carrier protein, **71**, 784; biotin content, **14**, 8, 16; **71**, 10, 42, 60; *Candida lipolytica*, from, **71**, 37 (kinetic properties, **71**, 42; pH optimum, **71**, 43; radioactive assay, **71**, 37, 38; regulation of cellular content and synthesis, **71**, 43 [mRNA, of, **71**, 43]; repression, **71**, 43; specificity, **71**, 43; spectrophotometric assay, **71**, 37; stability and storage, **71**, 42); cascade control, **64**, 325; carboxylation of acetyl-fatty acid synthase, **35**, 83; cell-free translation, **71**, 15, 43; chicken and rat liver, from, **14**, 9; chiral methyl groups, and, **87**, 148; chloroplasts and cytosol of higher plants, from, **71**, 44 (assay, **71**, 47; cytosolic form, **71**, 46; instability, **71**, 46; measurement of carboxybiotin, **71**, 48; prokaryotic form in chloroplasts, **71**, 46; transcarboxylation assay, **71**, 49); cooperativity, **64**, 218; *Escherichia coli*, from, **35**, 17 (biotin carboxylase, **35**, 25; biotin carboxyl carrier protein, **35**, 17; carboxyltransferase, **35**, 32); estimation of active and inactive forms by immunotitration, **71**, 292; *Euglena gracilis*, from, **71**, 60 (assay, **71**, 61; dissociation of complex, **71**, 70; properties, **71**, 72); hysteresis, **64**, 218; inhibition by 5-(tetradecyloxy)-2-furoyl-coenzyme A, **72**, 552; irreversible inhibitor, **72**, 580; lactating rabbit mammary gland, from, **71**, 26; lactating rat mammary gland, from, **71**, 16, 26 (assay, **71**, 27; biotin and phosphate content, **71**, 26; contamination with proteolytic activity, **71**, 25; effect of *S*-4-bromo-2,3-dioxobutyl-CoA, **71**, 25; heterogeneous population of filaments, **71**, 22; hypersharp peak, **71**, 24; immunochemical studies, **71**, 25; inhibition by biotin-binding antibodies, **71**, 26; polymeric form, **71**, 21; properties, **71**, 24, 32; protomers, **71**, 22; stability, **71**, 24);

molecular forms, **71**, 11; monovalent cations on activity, effect of, **71**, 58, 59; mRNA, **71**, 15; multifunctional polypeptide, **71**, 11, 42; mutant, **72**, 694; phosphorylation, **71**, 14; pigeon liver, from, **14**, 11; preparation, **6**, 540; rat adipose tissue, from, **14**, 11; rat liver, from, **14**, 9; **35**, 3 (activation of crude and pure enzyme, **35**, 8, 9; activator, **35**, 8, 9; assay, **35**, 3; **71**, 301; biotin content, **35**, 10; [^{14}C]biotin-labeled enzyme, **71**, 9; cell-free translation, **71**, 15; enzyme-inhibitor complex, **71**, 13; gel filtration of crude enzyme, **35**, 4; inhibition constants for long-chain acyl-CoA and analogues, **71**, 14; inhibitor, **35**, 8, 9; phosphate content, **35**, 10; radioactive assay, **71**, 5; spectrophotometric assay, **71**, 5, 6; stability, **35**, 8, 9; steady state kinetics, **71**, 12; subunit structure, **71**, 11); rat mammary gland, **35**, 11 (activator, **35**, 17; assay, **35**, 11; **71**, 27; inhibitor, **35**, 17; **71**, 26; stability, **35**, 16; **71**, 24); repression, **71**, 16; sedimentation coefficient, effects of activators and inhibitors, **14**, 16; stability, **14**, 6, 7, 12, 14, 16; synthesis and degradation, **71**, 15; *Turbatrix aceti*, **71**, 55 (carboxylation of propionyl-CoA and butyryl-CoA, **71**, 55, 59; molecular weight, **71**, 59; oligomeric structure, **71**, 60; radioactive assay, **71**, 55; subunit structure, **71**, 60); wheat germ, from, **14**, 11; yeast, from, **14**, 3; **71**, 34 (assay, **71**, 34; dissociation and reactivation, **71**, 37; properties, **71**, 36; subunit structure, **71**, 37)

[ACETYL-CoA CARBOXYLASE] KINASE

This phosphotransferase [EC 2.7.1.128; formerly classified as EC 2.7.1.111], catalyzes the reaction of ATP with the enzyme acetyl-CoA carboxylase [EC 6.4.1.2] to produce ADP and [acetyl-CoA carboxylase] phosphate, the inactive form of the carboxylase in which Ser77 and Ser1200 are phosphorylated.[1,2] Citrate inhibits the phosphorylation reaction.[1] The rat liver enzyme is identical to [3-hydroxy-3-methylglutaryl-CoA reductase] kinase [EC 2.7.1.109].[3]

[1] A. H. Mohamed, W. Y. Huang, W. Huang, K. V. Venkatachalam & S. J. Wakil (1994) *JBC* **269**, 6859.
[2] B. Lent & K. H. Kim (1982) *JBC* **257**, 1897.
[3] D. Carling, P. R. Clarke, V. A. Zammit & D. G. Hardie (1989) *EJB* **186**, 129.

[ACETYL-CoA CARBOXYLASE]-PHOSPHATASE

This phosphoprotein phosphatase [EC 3.1.3.44] catalyzes the hydrolysis of [acetyl-CoA carboxylase] phosphate (*i.e.*, the phosphorylated form of the carboxylase) to produce orthophosphate and the now active acetyl-CoA carboxylase [EC 6.4.1.2]. The enzyme will also dephosphorylate [3-hydroxy-3-methylglutaryl-CoA reductase] phosphate [EC 1.1.1.88], glycogen phosphorylase *a* [EC 2.4.1.1], [glycogen(starch) synthase] phosphate [EC 2.4.1.11], phosphoprotamine, and 4-nitrophenyl phosphate.[1,2] The enzyme is distinct from phosphorylase phosphatase [EC 3.1.3.17] and [pyruvate dehydrogenase (lipoamide)]-phosphatase [EC 3.1.3.43].

[1] G. R. Krakower & K. H. Kim (1981) *JBC* **256**, 2408.
[2] K. G. Thampy & S. J. Wakil (1985) *JBC* **260**, 6318.

ACETYL-CoA HYDROLASE

This hydrolase [EC 3.1.2.1], also called acetyl-CoA deacylase and acetyl-CoA acylase, catalyzes the hydrolysis of acetyl-CoA, producing free coenzyme A and acetate. Disulfide peptides activate the enzyme.[1] The rat liver enzyme is stimulated by ATP and inhibited by ADP.

[1] M. A. Namboodiri, J. T. Favilla & D. C. Klein (1982) *JBC* **257**, 10030.

Selected entries from *Methods in Enzymology* [**vol**, page(s)]:
General discussion: **1**, 606
Assay: **1**, 606
Other molecular properties: aceto-CoA kinase, absent in assay of, **5**, 464; brown adipose tissue mitochondria, **55**, 76; citrate-cleavage enzyme preparation, lack of presence in, **5**, 644; irreversible inhibitor, **72**, 580; malate synthetase preparation, assay for presence in, **5**, 634; preparation, **1**, 607

ACETYL-CoA:LONG-CHAIN BASE ACETYLTRANSFERASE

This acetyltransferase catalyzes the reaction of acetyl-CoA with a long-chain primary amine to produce the corresponding acetylated amine and free Coenzyme A. The specificity differs from that of arylamine acetyltransferase [EC 2.3.1.5] and aralkylamine acetyltransferase [EC 2.3.1.87]. The enzyme will also transfer an acetyl group from acetyl-CoA to sphingosine and dihydrosphingosine. The amino and hydroxyl groups of the base can be acetylated.

Selected entries from *Methods in Enzymology* [**vol**, page(s)]:
General discussion: **35**, 242
Assay: **35**, 242
Other molecular properties: inhibitors, **35**, 247; preparation, **35**, 245; properties, **35**, 246; specificity, **35**, 246, 247; stability, **35**, 246

ACETYL-CoA C-MYRISTOYLTRANSFERASE

This transferase [EC 2.3.1.155], also known as 3-oxopalmitoyl-CoA hydrolase and 3-oxopalmitoyl-CoA acetyltransferase, catalyzes the reversible reaction of myristoyl-CoA with acetyl-CoA to produce 3-oxopalmitoyl-CoA (*i.e.*, β-ketopalmitoyl-CoA) and Coenzyme A.[1]

myristoyl-CoA

3-oxopalmitoyl-CoA

This enzyme, which participates in the peroxisomal β-oxidation of branched-chain fatty acids, is distinct from propionyl-CoA C^2-trimethyldecanoyltransferase [EC 2.3.1.154].

[1] S. Miyazawa, S. Furuta, T. Osumi, T. Hashimoto & N. Ui (1981) *J. Biochem.* **90**, 511.

ACETYL-CoA SYNTHETASE

This enzyme [EC 6.2.1.1] (also known as acetate:CoA ligase, aceto-CoA kinase, acetyl-CoA kinase, acyl-activating enzyme, acetate thiokinase, acetothiokinase, short-chain fatty acyl-CoA synthetase, and acetyl-activating enzyme) catalyzes the reaction of acetate with Coenzyme A and ATP to form acetyl-CoA, AMP, and pyrophosphate (or, diphosphate). Certain isozymes will also utilize propanoate and propenoate as substrates. The reaction proceeds via an acetyl adenylate intermediate. **See also** Acetyl-Co Synthetase (ADP-Forming)

Selected entries from **Methods in Enzymology** [vol, page(s)]:
General discussion: 1, 585; 5, 461; 13, 375
Assay: 1, 536; 5, 461; 13, 375; 14, 625; coupled enzyme assay, 63, 33; rationale for choice, 13, 375
Other molecular properties: acetate assay, in, 35, 302; acetate kinase and, 1, 593; acetyl-AMP intermediate, 13, 375, 380; acylphosphatase preparation, not present in, 6, 327; adenosine 5'-*O*-(1-thiotriphosphate), and, 87, 224, 230, 251; bakers' yeast, from, 71, 317 (kinetic mechanism, 71, 321; molecular weight and sedimentation analysis, 71, 321; stability, 71, 321); brain mitochondria, 55, 58; bridge–nonbridge oxygens, 87, 251; cation requirements, 13, 380; choline acetylase assay and, 1, 623; glyoxysomes, low in, 31, 569; hydroxylamine, and, 1, 586; isotope exchange, 64, 9; kinetics, 87, 355; mechanism, 5, 461; 13, 375; 63, 158; 87, 251, 355; NMR, ^{31}P and ^{17}O, 87, 251; partial reactions, 5, 464; properties, 1, 590; 13, 379; purification, 1, 588; 5, 462; 13, 376; stereochemistry, 87, 206, 212, 224, 230, 251

ACETYL-CoA SYNTHETASE (ADP-FORMING)

This enzyme [EC 6.2.1.13], also known as acetate:CoA ligase (ADP-forming) and acetate thiokinase, catalyzes the reaction of ATP with acetate and Coenzyme A to produce ADP, orthophosphate, and acetyl-CoA. The enzyme can also use propanoate and, weakly, butanoate as alternative substrates.[1-4]

[1] R. E. Reeves, L. G. Warren, B. Susskind & H. S. Lo (1977) *JBC* **252**, 726.
[2] X. Mai & M. W. Adams (1996) *J. Bacteriol.* **178**, 5897.
[3] J. Glasemacher, A. K. Bock, R. Schmid & P. Schonheit (1997) *EJB* **244**, 561.
[4] L. B. Sanchez, H. G. Morrison, M. L. Sogin & M. Muller (1999) *Gene* **233**, 225.

N-ACETYLDIAMINOPIMELATE DEACETYLASE

This hydrolase [EC 3.5.1.47], first isolated from *Bacillus megaterium*, catalyzes the hydrolysis of *N*-acetyl-LL-2,6-diaminoheptanedioate to produce acetate and LL-2,6-diaminoheptanedioate.[1]

[1] G. Sundharadas & C. Gilvarg (1967) *JBC* **242**, 3983.

ACETYLENECARBOXYLATE HYDRATASE

This enzyme [EC 4.2.1.71], also known as acetylenemonocarboxylate hydratase and alkynoate hydratase, catalyzes the effectively irreversible addition of water to propynoate to form 3-hydroxypropenoate, which tautomerizes to 3-oxopropanoate. The enzyme also acts on 3-butynoate to form acetoacetate.[1] **See also** Acetylene Hydratase

[1] E. W. Yamada & W. B. Jakoby (1959) *JBC* **234**, 941.

ACETYLENEDICARBOXYLATE DECARBOXYLASE

This decarboxylase [EC 4.1.1.78], and previously classified as acetylenedicarboxylate hydratase [EC 4.2.1.72], catalyzes the reaction of acetylenedicarboxylate with water to produce pyruvate and carbon dioxide.[1] Hydration of the triple bond results in the formation of a proposed enzyme-bound oxaloacetate intermediate that undergoes decarboxylation.

[1] E. W. Yamada & W. B. Jakoby (1958) *JBC* **233**, 706.

ACETYLENE HYDRATASE

This tungsten-containing iron–sulfur hydratase has been isolated from *Peleobacter acetylenicus* and catalyzes the reaction of acetylene with water to produce acetaldehyde.[1,2] **See also** Acetylenecarboxylate Hydratase

[1] B. M. Rosner & B. Schink (1995) *J. Bacteriol.* **177**, 5767.
[2] R. U. Meckenstock, R. Krieger, S. Ensign, P. M. Kroneck & B. Schink (1999) *EJB* **264**, 176.

ACETYLESTERASE

This hydrolase [EC 3.1.1.6], also known as C-esterase in animal tissues, catalyzes the hydrolysis of an acetic ester to produce an alcohol and acetate. **See also** Cephalosporin Acetylesterase; specific acetylesterase

Selected entries from **Methods in Enzymology** [vol, page(s)]:
Other molecular properties: activity assay, 47, 78; gel electrophoresis, 32, 88

N-ACETYLGALACTOSAMINE DEACETYLASE

This clostridial enzyme catalyzes the deacetylation of the terminal non-reducing *N*-acetylgalactosamine residue of

the blood group A determinant, thereby destroying its serological activity as a blood group A substance.[1]

[1]D. M. Marcus (1972) *MIE* **28**, 967.

Selected entries from ***Methods in Enzymology*** **[vol**, page(s)]:
General discussion: 28, 967
Assay: 28, 967
Other molecular properties: blood group specificity and, **28**, 970; isolation from *Clostridium tertium*, **28**, 968; properties, **28**, 969; purification, **28**, 969

N-ACETYLGALACTOSAMINE-6-PHOSPHATE DEACETYLASE

This enzyme [EC 3.5.1.80] catalyzes the hydrolysis of *N*-acetyl-D-galactosamine 6-phosphate to produce D-galactosamine 6-phosphate and acetate.[1]

[1]J. Reizer, T. M. Ramseier, A. Reizer, A. Charbit & M. H. Saier, Jr., (1996) *Microbiol.* **142**, 231.

N-ACETYLGALACTOSAMINE-4-SULFATASE

This enzyme [EC 3.1.6.12], known variously as arylsulfatase B, chondroitinsulfatase, *N*-acetylgalactosamine-4-sulfate sulfatase, and chondroitinase, catalyzes the hydrolysis of the 4-sulfate groups of the *N*-acetylgalactosamine-4-sulfate moieties in chondroitin sulfate and dermatan sulfate. *N*-Acetylglucosamine 4-sulfate, methylumbelliferyl sulfate, and UDP-*N*-acetylgalactosamine-4-sulfate can act as alternative substrates.[1] The enzyme reportedly will also exhibit an exo-sulfatase activity in the degradation of glycosaminoglycans. ***See also*** *Arylsulfatase*

[1]D. A. Brooks, D. A. Robertson, C. Bindloss, T. Litjens, D. S. Anson, C. Peters, C. P. Morris & J. J. Hopwood (1995) *BJ* **307**, 457.

Selected entries from ***Methods in Enzymology*** **[vol**, page(s)]:
Assay: 50, 450, 451, 474, 475, 537
Other molecular properties: diagnostic assay for multiple sulfatase deficiency, **138**, 745; distribution, **50**, 450; human placental (effect on [spasmogenic mediators, **86**, 20; synthetic leukotrienes, **86**, 28]; inactivation of slow-reacting substance, **86**, 20; purification, **86**, 18); Maroteaux-Lamy syndrome, **50**, 451; multiple sulfatase deficiency, **50**, 453, 454, 474; properties, **50**, 546, 547; purification, **50**, 544, 545; specificity, **50**, 537

N-ACETYLGALACTOSAMINE-6-SULFATASE

This enzyme [EC 3.1.6.4], also known variously as chondroitinsulfatase, chondroitinase, *N*-acetylgalactosamine-6-sulfate sulfatase, and galactose-6-sulfate sulfatase, catalyzes the hydrolysis of the 6-sulfate groups of the *N*-acetylgalactosamine moieties in chondroitin sulfate and the galactose 6-sulfate groups in keratan sulfate.

Except with the human enzyme, *N*-acetylgalactosamine 6-sulfate is a substrate.[1] Other substrates include *N*-acetylgalactosamine 6-sulfate-(β,1-4)-glucuronic acid-(β,1-3)-*N*-acetylgalactosaminitol 6-sulfate and UDP-*N*-acetylgalactosamine-6-sulfate. ***See also*** *N-Acetylgalactosamine-4-Sulfatase; Chondro-6-Sulfatase*

[1]C. T. Lim & A. L. Horwitz (1981) *BBA* **657**, 344

Selected entries from ***Methods in Enzymology*** **[vol**, page(s)]:
General discussion: 8, 663
Assay: 8, 663
Other molecular properties: deficiency in mucopolysaccharidoses, assay, **83**, 563; properties, **8**, 669; purification, *Proteus vulgaris*, **8**, 665; radioactive substrates, preparation, **83**, 570

α-N-ACETYLGALACTOSAMINIDASE

This glycosidase [EC 3.2.1.49], also known as α-galactosidase B, catalyzes the hydrolysis of terminal non-reducing *N*-acetyl-D-galactosamine residues in *N*-acetyl-α-D-galactosaminides. The enzyme removes *N*-acetylgalactosaminyl groups from O-3 of seryl and threonyl residues. There are a number of isoforms of EC 3.2.1.49 and the specificity can vary considerably between isoforms and between sources. Substrates include *p*-nitrophenyl-α-*N*-acetylgalactosaminide, ovine submaxillary asialoglycoprotein, Forssman hapten, 4-methylumbelliferyl-α-galactopyranoside, *p*-nitrophenyl-2-deoxy-α-D-galactopyranoside, *o*-nitrophenyl-α-D-fucopyranoside, *o*-nitrophenyl-α-*N*-acetylgalactosaminide, and globopentaose.[1] ***See also*** *Glycopeptide α-N-Acetylgalactosaminidase*

[1]A. Zhu, C. Monahan & Z. K. Wang (1996) *BBA* **1297**, 99.

Selected entries from ***Methods in Enzymology*** **[vol**, page(s)]:
Assay: *Aspergillus niger*, **28**, 734; beef liver, **28**, 801; *Clostridium perfringens*, **28**, 756
Other molecular properties: *Aspergillus niger* (assay, **28**, 734; properties, **28**, 738; purification, **28**, 736); beef liver (assay, **28**, 801; properties, **28**, 803; purification, **28**, 802); blood-group A substance and, **8**, 705; carbohydrate structure and, **28**, 19; *Clostridium perfringens* (assay, **28**, 756; properties, **28**, 761; purification, **28**, 757); glycolipid structure studies, for, **32**, 363

β-N-ACETYLGALACTOSAMINIDASE

This glycosidase [EC 3.2.1.53] catalyzes the hydrolysis of terminal non-reducing *N*-acetyl-D-galactosamine residues in *N*-acetyl-β-D-galactosaminides.[1–3] A useful chromogenic substrate is 4-methylumbelliferyl-β-*N*-acetylgalactosaminide.

Note that β-*N*-acetylglucosaminidase [EC 3.2.1.30] also utilizes β-*N*-acetylgalactosaminides as alternative substrates. ***See also*** *β-N-Acetylglucosaminidase*

[1] Y. Z. Frohwein & S. Gatt (1967) *Biochemistry* **6**, 2775.
[2] T. Izumi & K. Suzuki (1983) *JBC* **258**, 6991.
[3] A. Tanaka & S. Ozaki (1997) *J. Biochem.* **122**, 330.

α-N-ACETYLGALACTOSAMINIDE α2 → 6-SIALYLTRANSFERASE

This transferase [EC 2.4.99.3], also called CMP-*N*-acetylgalactosaminide α-2,6-sialyltransferase and mucin sialyltransferase, catalyzes the reaction of a glycano-1,3-(*N*-acetylgalactosaminyl)-glycoprotein with CMP-*N*-acetylneuraminate to produce CMP and the glycano-(2,6-α-*N*-acetylneuraminyl)-(*N*-acetylgalactosaminyl)-glycoprotein.[1–3] The enzyme transfers sialic acid to the core region of *O*-linked glycans. An α-*N*-acetylgalactosamine residue linked to L-threonine or L-serine is also an acceptor, when substituted at the 3-position. The enzyme is distinct from (α-*N*-acetylneuraminyl-2,3-β-galactosyl-1,3)-*N*-acetylgalactosaminide α-2,6-sialyltransferase [EC 2.4.99.7]. Substrates include porcine and ovine submaxillary asialomucin, asialofetuin, asialoorosomucoid, as well as the antifreeze glycoprotein found in Antarctic fish. *See also Sialyltransferases; α-2,6-Sialyltransferases*

[1] J. E. Sadler, J. I. Rearick & R. L. Hill (1979) *JBC* **254**, 5934.
[2] H. H. Higa & J. C. Paulson (1985) *JBC* **260**, 8838.
[3] D. M. Carlson, E. J. McGuire, G. W. Jourdian & S. Roseman (1973) *JBC* **248**, 5763.

Selected entries from *Methods in Enzymology* [**vol**, page(s)]:
General discussion: 8, 361
Assay: 8, 361; 83, 480
Other molecular properties: erythrocyte resialylation with, 138, 163; ovine submaximillary gland (assay, 8, 361; properties, 8, 364; purification, 8, 362); porcine submaxillary gland (assay, 83, 480; properties, 83, 486; purification, 83, 481)

N-ACETYLGALACTOSAMINOGLYCAN DEACETYLASE

This deacetylase [EC 3.1.1.58], also known as poly(*N*-acetylgalactosamine) deacetylase, catalyzes the reaction of *N*-acetyl-D-galactosaminoglycan (possessing at least fourteen *N*-acetylgalactosamine units) with *n* molecules of water to produce a D-galactosaminoglycan with *n*-fewer acetyl groups and *n* molecules of acetate.[1,2]

[1] J. A. Jorge, S. G. Kinney & J. L. Reissig (1982) *Braz. J. Med. Biol. Res.* **15**, 29.
[2] Y. Araki, H. Takada, N. Fujii & E. Ito (1979) *EJB* **102**, 35.

Selected entries from *Methods in Enzymology* [**vol**, page(s)]:
General discussion: 161, 514
Assay: 161, 515
Other molecular properties: properties, 161, 517; purification from *Aspergillus parasiticus*, 161, 516

ACETYLGALACTOSAMINYL-O-GLYCOSYL-GLYCOPROTEIN β-1,3-N-ACETYLGLUCOSAMINYLTRANSFERASE

This glycosyltransferase [EC 2.4.1.147], also called *O*-glycosyl-oligosaccharide-glycoprotein *N*-acetylglucosaminyltransferase III and mucin core-3 β−3-*N*-acetylglucosaminyltransferase, catalyzes the reaction of UDP-*N*-acetyl-D-glucosamine and *N*-acetyl-D-galactosaminyl-R to produce UDP and *N*-acetyl-β-D-glucosaminyl-1,3-*N*-acetyl-D-galactosaminyl-R where R is the polypeptide derived from mucin or it is a phenyl or benzyl derivative. *See also β-1,3-Galactosyl-O-Glycosyl-Glycoprotein β-1,6-N-Acetylglucosaminyltransferase; β-1,3-Galactosyl-O-Glycosyl-Glycoprotein β-1,3-N-Acetylglucosaminyltransferase; Acetylgalactosaminyl-O-Glycosyl-Glycoprotein β-1,6-N-Acetylglucosaminyltransferase*

Selected entries from *Methods in Enzymology* [**vol**, page(s)]:
General discussion: 230, 303
Assay: 179, 386

ACETYLGALACTOSAMINYL-O-GLYCOSYL-GLYCOPROTEIN β-1,6-N-ACETYLGLUCOSAMINYLTRANSFERASE

This glycosyltransferase [EC 2.4.1.148], also called *O*-glycosyl-oligosaccharide-glycoprotein *N*-acetylglucosaminyltransferase IV, catalyzes the reaction of UDP-*N*-acetyl-D-glucosamine with *N*-acetyl-β-D-glucosaminyl-1,3-*N*-acetyl-D-galactosaminyl-R to produce UDP and *N*-acetyl-β-D-glucosaminyl-1,6-(*N*-acetyl-β-D-glucosaminyl-1,3)-*N*-acetyl-D-galactosaminyl-R, where R is either a polypeptide derived from mucin or a phenyl or benzyl derivative. The bovine enzyme reportedly has an ordered Bi Bi kinetic mechanism.[1] *See also β-1,3-Galactosyl-O-Glycosyl-Glycoprotein β-1,6-N-Acetylglucosaminyltransferase, β-1,3-Galactosyl-O-Glycosyl-Glycoprotein β-1,3-N-Acetylglucosaminyltransferase, and Acetylgalactosaminyl-O-Glycosyl-Glycoprotein β-1,3-N-Acetylglucosaminyltransferase*

[1] P. A. Ropp, M. R. Little & P. W. Cheng (1991) *JBC* **266**, 23863.

Selected entries from *Methods in Enzymology* [**vol**, page(s)]:
General discussion: 230, 303
Assay: 179, 389

N-ACETYLGLUCOSAMINE DEACETYLASE

This enzyme [EC 3.5.1.33] catalyzes the hydrolysis of *N*-acetyl-D-glucosamine to produce D-glucosamine and acetate.[1,2] *N*-Acetylgalactosamine can act as an alternative substrate, as can peptidoglycans containing *N*-acetylglucosamine residues.

[1] S. Roseman (1957) *JBC* **226**, 115.
[2] Y. Araki, S. Fukuoka, S. Oba & E. Ito (1971) *BBRC* **45**, 751.

Selected entries from *Methods in Enzymology* [vol, page(s)]:
Other molecular properties: *N*-acetyl-D-glucosamine kinase, potential contaminant in assay for, **9**, 421

N-ACETYLGLUCOSAMINE GALACTOSYLTRANSFERASE

This glycosyltransferase, previously classified as EC 2.4.1.98 but now a deleted EC entry, is listed under *N*-acetyllactosamine synthase [EC 2.4.1.90].

N-ACETYL-D-GLUCOSAMINE KINASE

This phosphotransferase [EC 2.7.1.59] catalyzes the phosphorylation by ATP of *N*-acetyl-D-glucosamine to generate ADP and *N*-acetyl-D-glucosamine 6-phosphate. The enzyme from some sources will also act on D-glucose.[1–3] The *Candida* and the rat enzymes are inhibited by sulfhydryl reagents, and substrates protect against inactivation by the sulfhydryl reagents.[3]

The rat liver and kidney enzymes show a significant pH-dependent lag phase before reaching a steady-state velocity.[2] Non-hyperbolic kinetics are observed with *N*-acetyl-D-glucosamine as an acceptor substrate, but not with *N*-acetyl-D-mannosamine or D-glucose.[2] There are also findings suggesting that negative cooperativity occurs when D-glucose is the substrate.[1]

[1] Y. P. Rai, B. Singh, N. Elango & A. Datta (1980) *BBA* **614**, 350.
[2] M. B. Allen & D. G. Walker (1980) *BJ* **185**, 577.
[3] S. Hinderlich, S. Nohring, C. Weise, P. Franke, R. Stasche & W. Reutter (1998) *EJB* **252**, 133.
[1] M. L. Vera, M. L. Cardenas & H. Niemeyer (1984) *ABB* **229**, 237.

Selected entries from *Methods in Enzymology* [vol, page(s)]:
General discussion: **9**, 415, 421; **42**, 58
Assay: **9**, 415, 421; **42**, 58
Other molecular properties: *Escherichia coli*, from (kinetic constants, **9**, 424; properties, **9**, 424; purification, **9**, 422; specificity, **9**, 424); hog spleen, from, **42**, 58 (chromatography, **42**, 61; inhibitors, **42**, 62; properties, **42**, 62; purification, **41**, 407; **42**, 60); *Streptococcus pyogenes*, from (inhibitors, **9**, 420; kinetic constants, **9**, 420; properties, **9**, 419; purification, **9**, 416; specificity, **9**, 419; stability, **9**, 419)

N-ACETYLGLUCOSAMINE-6-PHOSPHATE DEACETYLASE

This enzyme [EC 3.5.1.25] catalyzes the hydrolysis of *N*-acetyl-D-glucosamine 6-phosphate to produce D-glucosamine 6-phosphate and acetate.[1] The enzyme has an ordered Uni Bi kinetic mechanism.[2]

[1] P. Gopal, P. A. Sullivan & M. G. Shepherd (1982) *J. Gen. Microbiol.* **128**, 2319.
[2] J. M. Souza, J. A. Plumbridge & M. L. Calcagno (1997) *ABB* **340**, 338.

Selected entries from *Methods in Enzymology* [vol, page(s)]:
General discussion: **41**, 497
Assay: *Escherichia coli*, **41**, 498

ACETYLGLUCOSAMINE-PHOSPHATE ISOMERASE

This isomerase, previously classified as EC 5.3.1.11, is now a deleted EC entry.

N-ACETYLGLUCOSAMINE-1-PHOSPHODIESTER α-*N*-ACETYLGLUCOSAMINIDASE

This hydrolase [EC 3.1.4.45], also known as α-*N*-glucosaminyl phosphodiesterase, catalyzes the hydrolysis of a glycoprotein *N*-acetyl-D-glucosaminylphospho-D-mannose to produce *N*-acetyl-D-glucosamine and a glycoprotein phospho-D-mannose. This reaction is the second step in the synthesis of the mannose 6-phosphate determinant and is required for efficient intracellular targeting of newly synthesized lysosomal hydrolases to the lysosome. The enzyme utilizes a wide variety of compounds as substrates, provided that the *N*-acetyl-D-glucosamine residue is α-linked to a phosphate group. The enzyme cleaves the C–O bond rather than the O–P bond.[1]

[1] A. Varki, W. Sherman & S. Kornfeld (1983) *ABB* **222**, 145.

N-ACETYLGLUCOSAMINE-6-SULFATASE

This enzyme [EC 3.1.6.14], also known as glucosamine-6-sulfatase, *N*-acetylglucosamine-6-sulfate sulfatase, and chondroitinsulfatase, catalyzes the hydrolysis of the 6-sulfate moieties of the *N*-acetylglucosamine 6-sulfate subunits of heparan sulfate and keratan sulfate. In some organisms, this enzyme is identical to *N*-sulfoglucosamine-6-sulfatase. Alternative substrates include *O*-(α-glucosamine 6-sulfate)-(1→3)-L-idonic acid, glucose 6-sulfate, and *N*-acetylglucosamine 6-sulfate. Sulfate and phosphate ions are potent inhibitors.[1,2]

The bovine kidney enzyme exhibits an "endo" sulfatase activity, catalyzing the removal of sulfate from internally sulfated *N*-acetylglucosamine residues on glycopeptides and glycoproteins.[3]

[1] C. Freeman, P. R. Clements & J. J. Hopwood (1987) *BJ* **246**, 347.
[2] C. Freeman & J. J. Hopwood (1987) *BJ* **246**, 355.
[3] A. Shilatifard & R. D. Cummings (1994) *Biochemistry* **33**, 4273.

α-N-ACETYLGLUCOSAMINIDASE

This enzyme [EC 3.2.1.50] catalyzes the hydrolysis of terminal nonreducing N-acetylglucosamine residues in N-acetyl-α-glucosaminides. The enzyme will also hydrolyze UDP-N-acetylglucosamine to remove N-acetylglucosamine residues from heparin sulfate.

Selected entries from **Methods in Enzymology** [vol, page(s)]:
General discussion: 28, 796
Assay: 28, 796; **50**, 449, 450
Other molecular properties: distribution, **50**, 449; pig liver, from (assay, **28**, 796; properties, **28**, 799; purification, **28**, 798); Sanfilippo B syndrome, **50**, 450

β-N-ACETYLGLUCOSAMINIDASE

This enzyme activity, formerly classified as EC 3.2.1.30, is now a deleted EC entry.

N^4-(β-N-ACETYLGLUCOSAMINYL)-L-ASPARAGINASE

This hydrolase [EC 3.5.1.26], also known as aspartylglucosylamine deaspartylase, aspartylglucosylaminidase, and glycosylasparaginase, catalyzes the hydrolysis of N^4-(β-N-acetyl-D-glucosaminyl)-L-asparagine to produce N-acetyl-β-glucosaminylamine and L-aspartate. The enzyme is distinct from that of peptide-N^4-(N-acetyl-β-glucosaminyl) asparagine amidase [EC 3.5.1.52], which requires an asparaginyl residue as a part of an amino acid sequence. EC 3.5.1.26, which also has an asparaginase activity, requires an asparagine with a free amino group and a free carboxyl group. This enzyme also effectively catalyzes the synthesis of β-aspartyl peptides by transferring the β-aspartyl moiety from other β-aspartyl peptides or β-aspartylglycosylamine to a variety of amino acids and peptides akin to γ-glutamyltranspeptidase.[1]

This enzyme also is a member of the class of self-processing N-terminal nucleophile amidohydrolases. After the inactive precursor loses its signal peptide via the action of signal peptidase, the single chain proenzyme undergoes autoproteolysis (at an Asp–Thr bond) and the newly exposed N-terminal threonyl residue of the β-chain of the new heterodimer is the active-site nucleophile. Additional processing of the α-chain then takes place in the lysosome.[2,3]

[1] T. Noronkoski, I. B. Stoineva, I. P. Ivanov, D. D. Petkov & I. Mononen (1998) *JBC* **273**, 26295.
[2] C. Guan, T. Cui, V. Rao, W. Liao, J. Benner, C.-L. Lin & D. Comb (1996) *JBC* **271**, 1732.
[3] Y. Liu, G. Dunn & N. N. Aronson, Jr., (1996) *Glycobiology* **6**, 527.

Selected entries from **Methods in Enzymology** [vol, page(s)]:
General discussion: 28, 781, 787
Assay: 28, 781, 787
Other molecular properties: enzymatic properties, **28**, 785, 789; hen oviduct (properties, **28**, 785; purification, **28**, 783); isolation from pig kidney, **28**, 788; reaction, **28**, 781, 787

N-ACETYLGLUCOSAMINYLDIPHOSPHO-DOLICHOL N-ACETYLGLUCOSAMINYL-TRANSFERASE

This transferase [EC 2.4.1.141], which plays a role in the synthesis of N-linked glycoproteins, catalyzes the reaction of UDP-N-acetyl-D-glucosamine with N-acetyl-D-glucosaminyldiphosphodolichol to produce N,N″-diacetyl-chitobiosyldiphosphodolichol and UMP.[1]

[1] C. B. Sharma, L. Lehle & W. Tanner (1982) *EJB* **126**, 319.

N-ACETYLGLUCOSAMINYLDIPHOSPHO-UNDECAPRENOL N-ACETYL-β-D-MANNOSAMINYLTRANSFERASE

This glycosyltransferase [EC 2.4.1.187], which participates in the biosynthesis of teichoic acid linkage units in bacterial cell walls, catalyzes the reaction of UDP-N-acetyl-D-mannosamine and N-acetyl-D-glucosaminyldiphosphoundecaprenol to produce N-acetyl-β-D-mannosaminyl-1,4-N-acetyl-D-glucosaminyldiphosphoundecaprenol and UDP.[1]

[1] N. Murazumi, K. Kumita, Y. Araki & E. Ito (1988) *J. Biochem.* **104**, 980.

N-ACETYLGLUCOSAMINYL-DIPHOSPHOUNDECAPRENOL GLUCOSYLTRANSFERASE

This transferase [EC 2.4.1.188] catalyzes the reaction of UDP-D-glucose with N-acetyl-D-glucosaminyldiphosphoundecaprenol to produce β-D-glucosyl-1,4-N-acetyl-D-glucosaminyldiphosphoundecaprenol and UDP.[1]

[1] K. Kumita, N. Murazumi, Y. Araki & E. Ito (1988) *J. Biochem.* **104**, 985.

β-N-ACETYLGLUCOSAMINYL-GLYCOPEPTIDE β-1,4-GALACTOSYLTRANSFERASE

This glycosyltransferase [EC 2.4.1.38], also known as glycoprotein 4-β-galactosyltransferase and UDP-galactose:glycoprotein galactosyltransferase, catalyzes the reaction of UDP-D-galactose with an N-acetyl-β-D-glucosaminyl-glycopeptide to produce β-D-galactosyl-1,4-N-acetyl-β-D-glucosaminylglycopeptide and UDP. By means of this reaction, terminal N-acetyl-β-D-glucosaminyl residues in polysaccharides, glycoproteins, and glycopeptides become galactosylated. This transferase, which is the protein A component of lactose synthase [EC 2.4.1.22], is also reportedly identical to N-acetyllactosamine synthase [EC 2.4.1.90] in a number of organisms. *See N-Acetyllactosamine Synthase; Galactosyltransferases*

Selected entries from *Methods in Enzymology* [vol, page(s)]:
General discussion: 230, 302
Assay: avian salt gland, 96, 638; hepatic Golgi subfractions, in, 109, 222
Other molecular properties: Golgi apparatus, in, 22, 128; probe of O-GlcNAc on cytoplasmic and nuclear proteins, as, 230, 446; radiolabeling of glycoconjugates, in, 230, 37; reaction catalyzed by, 230, 306, 446

N-ACETYLGLUCOSAMINYL-PHOSPHATIDYLINOSITOL DEACETYLASE

This deacetylase [EC 3.1.1.69], which participates in the formation of the glycosylphosphatidylinositol membrane anchor of the trypanosome variant-surface glycoprotein, catalyzes the hydrolysis of N-acetyl-D-glucosaminylphosphatidylinositol to produce D-glucosaminylphosphatidylinositol and acetate.[1,2]

[1] T. L. Doering, W. J. Masterson, P. T. Englund & G. W. Hart (1989) *JBC* **264**, 11168.
[2] K. G. Milne, R. A. Field, W. J. Masterson, S. Cottaz, J. S. Brimacombe & M. A. Ferguson (1994) *JBC* **269**, 16403.

6-ACETYLGLUCOSE DEACETYLASE

This hydrolase [EC 3.1.1.33] catalyzes the hydrolysis of 6-acetyl-D-glucose to produce D-glucose and acetate.[1]

[1] R. B. Duff & D. M. Webley (1958) *BJ* **70**, 520.

N-ACETYLGLUTAMATE KINASE

This phosphotransferase [EC 2.7.2.8], also known as N-acetyl-γ-glutamokinase, catalyzes the reaction of ATP with N-acetyl-L-glutamate to produce N-acetyl-L-glutamate 5-phosphate and ADP. The pea (*Pisum sativum*) enzyme reportedly has a random Bi Bi kinetic mechanism as well as negative cooperativity.[1]

[1] G. McKay & P. D. Shargool (1981) *BJ* **195**, 71.

Selected entries from *Methods in Enzymology* [vol, page(s)]:
General discussion: 17A, 251, 269
Assay: 17A, 252, 269
Other molecular properties: properties, 17A, 255, 272; purification from (*Chlamydomonas reinhardti*, 17A, 270; *Escherichia coli*, 17A, 253)

N-ACETYL-γ-GLUTAMYL-PHOSPHATE REDUCTASE

This oxidoreductase [EC 1.2.1.38], also known as N-acetylglutamate semialdehyde dehydrogenase and

NAGSA dehydrogenase, catalyzes the reaction of N-acetyl-5-glutamyl phosphate with NADPH to generate N-acetylglutamate 5-semialdehyde and NADP$^+$.[1]

[1] A. U. Wandinger-Ness & R. L. Weiss (1987) *JBC* **262**, 5823

Selected entries from *Methods in Enzymology* [vol, page(s)]:
General discussion: 17A, 255
Assay: 17A, 255
Other molecular properties: properties, 17A, 259; purification from *Escherichia coli*, 17A, 257

N-ACETYLHEXOSAMINE 1-DEHYDROGENASE

This oxidoreductase [EC 1.1.1.240] catalyzes the reaction of N-acetyl-D-glucosamine with NAD$^+$ to produce N-acetyl-D-glucosaminate and NADH. Alternative substrates include N-acetyl-D-galactosamine and N-acetyl-D-mannosamine.[1] *See also N-Acylhexosamine Oxidase*

[1] T. Horiuchi & T. Kurokawa (1989) *Agric. Biol. Chem.* **53**, 1919.

β-N-ACETYLHEXOSAMINIDASE

This enzyme [EC 3.2.1.52], also known as β-hexosaminidase and N-acetyl-β-glucosaminidase, catalyzes the hydrolysis of terminal nonreducing N-acetylhexosamine residues in N-acetyl-β-hexosaminides.[1-4] N-Acetylglucosides and N-acetylgalactosides (including p-nitrophenyl-N-acetyl-β-D-glucosaminide, p-nitrophenyl-N-acetyl-β-D-galactosaminide, N,N-diacetylchitobiose, 4-methylumbelliferyl-β-D-galactosaminide, and 4-methylumbelliferyl-β-D-glucosaminide) are substrates. Many enzymes previously classified under β-N-acetylglucosaminidase [EC 3.2.1.30] are now listed in this EC entry.

Three major mammalian forms of this hydrolase are found in the lysosome in mammals. Form A, a trimer consisting of one α chain, one β-A chain, and one β-B chain, degrades the gangliosides G_{M2} and G_{M3}. Form B, a tetramer consisting of two β-A chains and two β-B chains, acts on the gangliosides G_{M1}, G_{M2}, and on globoside. Form S is a homodimer of two α chains. The human genetic disorders known as Tay-Sachs disease arises from defects in the α chain, whereas another genetic disorder known as Sandhoff's disease is associated with mutations in the β chain. *See also Mannosyl-Glycoprotein Endo-β-N-Acetylglucosaminidase; Chitinase*

[1] K. Zwierz, A. Zalewska & A. Zoch-Zwierz (1999) *Acta Biochim. Pol.* **46**, 739.
[2] G. Davies, M. L. Sinnott & S. G. Withers (1998) *CBC* **1**, 119.
[3] E. Conzelmann & K. Sandhoff (1987) *AE* **60**, 89.
[4] R. O. Brady (1983) *The Enzymes*, 3rd ed., **16**, 409.

Selected entries from *Methods in Enzymology* [**vol**, page(s)]:
General discussion: **8**, 577; **14**, 161; **28**, 720, 728, 756, 772; **34**, 3; **50**, 520; **161**, 462
Assay: **8**, 513, 580; **31**, 348; **28**, 857; **50**, 485, 486, 520, 521, 548; **161**, 463, 494, 524; *Clostridium perfringens*, **28**, 756; hen oviduct, **28**, 772; **50**, 580, 581; labeled trihexosyl ceramide, with, **14**, 162, 164; nitrophenyl derivatives of *N*-acetylglucosamine or galactosamine, with, **14**, 162, 163; *Phaseolus vulgaris*, **28**, 720; secretion from platelet, assay, **169**, 196, (lysosomes, from, **169**, 336); trihexosyl ceramide, with, **14**, 162, 163
Other molecular properties: *N*-acetylglucosaminyltransferase assay and, **8**, 426; β-*N*-acetylhexosaminidase B (pinocytosis measurements in Sandhoff disease fibroblasts, **98**, 292; purification from Tay-Sachs disease fibroblast secretions, **98**, 291); analysis in multiple systems, **22**, 18; *Aspergillus niger*, from (properties, **28**, 733; purification, **28**, 729); assay, for *O*-GlcNAc-bearing protein samples, **230**, 454; bovine testes (assay, **50**, 520, 521; properties, **50**, 522, 523; purification, **50**, 521, 522); brain fractions, in, **31**, 469; carbohydrate structure and, **28**, 702; ceramide glucosidase activity, **14**, 167; *Clostridium perfringens*, from (assay, **28**, 756; properties, **28**, 760; purification, **28**, 757); detergent removal from *Escherichia coli* cells, **161**, 526; diagnostic assay for G_{M2}-gangliosidosis, **138**, 755; *Diplococcus pneumoniae* (incubation conditions, **230**, 284; purification, **230**, 288; specificity, **230**, 284); effect of detergents, **14**, 166; as endosome marker, determination, **109**, 267; β-galactosidase activity, **14**, 167; glucocerebrosidase digestion before cell targeting, **149**, 35; glycolipid structure studies, for **32**, 363; hen oviduct, from (assay, **28**, 772; **50**, 580, 581; properties, **28**, 775; **50**, 583, 584; purification, **28**, 773; **50**, 581); β-hexosaminidase A, **28**, 857 (assay, **28**, 857; **50**, 485, 486, 548; biological significance in Tay-Sachs disease, **28**, 861, 866; isolation from human placenta, **28**, 858 [assay, **50**, 548; immunochemical determination, **50**, 554, 555; properties, **50**, 551; purification, **50**, 549]; leukocyte and fibroblast assay, **28**, 863; lysosomes, **50**, 490; mucolipidoses, **50**, 455; properties, **28**, 860, 866; Sandhoff's disease, **50**, 484; serum assay, **28**, 863; Tay-Sachs disease, **50**, 484); β-hexosaminidase B (assay, **50**, 485, 486, 548; human placenta, from [assay, **50**, 548; immunochemical determination, **50**, 554, 555; properties, **50**, 551; purification, **50**, 549]; lysosomes, **50**, 490; Sandhoff's disease, **50**, 484; Tay-Sachs disease, **50**, 484); hexosaminidase C, **50**, 490; hydrolysis of Tay-Sach ganglioside, **14**, 152; inhibition by aldonolactones, **8**, 573, 579; inhibitors, **14**, 167; jack bean (incubation conditions, **230**, 284; isolation from, **28**, 704, 708; for glycosphingolipid analysis, **230**, 385; purification, **230**, 288; specificity, **230**, 284); levels in sponge implants, measurement, **162**, 330; liver, in, **31**, 97; lysosomal marker, assay, **182**, 215; lysosomes, in, **8**, 513; marker for chromosome detection in somatic cell hybrids, as, **151**, 185; *Phaseolus vulgaris* (pinto beans), from (assay, **28**, 720; properties, **28**, 723; purification, **28**, 721; substrate specificity, **28**, 725); plasma membrane, in, **31**, 92, 100; PMN granules, in, **31**, 349; preparation, **8**, 578; properties, **8**, 577; **28**, 711; **161**, 467; *Pycnoporus cinnabarinus*, purification from, **161**, 464; removal of *N*-acetylgalactosamine during ganglioside hydrolysis in brain, **14**, 135, 136, 138, 139; secretion from (mast cells, **253**, 39; platelet lysosomes, assay, **169**, 196, 336, 436); substrate specificity, **14**, 167; **230**, 298; transition state and multisubstrate analogues, **249**, 306

N-ACETYLHISTIDINE DEACETYLASE

This hydrolase, previously classified as EC 3.5.1.34 but now a deleted entry, is listed under Xaa-methylhistidine dipeptidase [EC 3.4.13.5].

O-ACETYLHOMOSERINE (THIOL)-LYASE

This enzyme [EC 4.2.99.10], also known as *O*-acetyl-homoserine sulfhydrylase, catalyzes the reaction of *O*-acetylhomoserine with methanethiol to generate

methionine and acetate.[1,2] The enzyme also acts on other thiols or H_2S, producing homocysteine or thioethers. The enzyme from bakers yeast also catalyzes the same catalytic activity as *O*-acetylserine (thiol)-lyase [EC 4.2.99.8], albeit more slowly. The spinach chloroplast enzyme has two catalytically non-equivalent pyridoxal-5′-phosphate-containing active-sites and exhibits positive cooperativity with respect to *O*-acetylserine in the presence of sulfide, but negative cooperativity in the presence of another alternative cosubstrate.[1]

[1] N. Rolland, M. L. Ruffet, D. Job, R. Douce & M. Droux (1996) *EJB* **236**, 272.
[2] S. Yamagata (1989) *Biochimie* **71**, 1125.

Selected entries from *Methods in Enzymology* [**vol**, page(s)]:
General discussion: **17B**, 446; **143**, 478
Assay: **17B**, 446; **143**, 465, 470, 479
Other molecular properties: cystathionine γ-synthase, *Salmonella*, and, **17B**, 453; properties, **17B**, 449; **143**, 468, 473; purification from (*Brevibacterium flavum*, **143**, 471; *Neurospora crassa*, **17B**, 447; *Schizosaccharomyces pombe*, **143**, 466); pyridoxal phosphate and, **17B**, 449; yeast (properties, **143**, 482; purification, **143**, 480)

N-ACETYLINDOXYL OXIDASE

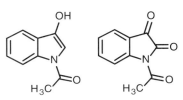

N-acetylindoxyl   *N*-acetylisatin

This oxidoreductase [EC 1.7.3.2] catalyzes the reaction of *N*-acetylindoxyl with dioxygen to produce *N*-acetylisatin and an uncharacterized coproduct.[1]

[1] H. Beevers & R. C. French (1954) *ABB* **50**, 427.

Selected entries from *Methods in Enzymology* [**vol**, page(s)]:
General discussion: **56**, 475

N-ACETYLLACTOSAMINE 3-α-GALACTOSYLTRANSFERASE

This transferase [EC 2.4.1.124] catalyzes the reaction of UDP-D-galactose with β-D-galactosyl-(1,4)-*N*-acetyl-D-glucosamine to produce UDP and α-D-galactosyl-(1,3)-β-D-galactosyl-(1,4)-*N*-acetyl-D-glucosamine. Non-reducing terminal *N*-acetyllactosamine residues of glycoproteins act as alternative substrate acceptors, including asialo-α_1-acid glycoprotein. The calf thymus enzyme requires manganese ion.[1]

[1] D. H. Van den Eijnden, W. M. Blanken, H. Winterwerp & W. E. Schiphorst (1983) *EJB* **134**, 523.

Selected entries from *Methods in Enzymology* [**vol**, page(s)]:
General discussion: **230**, 302

N-ACETYLLACTOSAMINE SYNTHASE

This glycosyltransferase [EC 2.4.1.90], also known as *N*-acetylglucosamine (β-1,4)galactosyltransferase and UDP-galactose:*N*-acetylglucosamine β-D-galactosyltransferase, catalyzes the reaction of UDP-D-galactose with *N*-acetyl-D-glucosamine to produce *N*-acetyllactosamine and UDP.[1,2] The transferase can utilize either free *N*-acetylglucosamine as the acceptor substrate or the non-reducing terminal monosaccharide of a carbohydrate chain of a glycoprotein or glycolipid. Acceptor substrates include asialoagalactotransferrin, ovomucoid, and desialated degalactosylated fetuin.

This enzyme activity is also catalyzed by the A protein component of lactose synthase [EC 2.4.1.22] in the absence of the B protein component (α-lactalbumin). This protein will also catalyze the reaction of β-*N*-acetylglucosaminylglycopeptide β-1,4-galactosyltransferase [EC 2.4.1.38]. In the murine F9 embryonal carcinoma cell line, EC 2.4.1.38 and EC 2.4.1.90 are identical, based on cDNA sequencing.[1] In fact, a number of investigators have classified the enzyme as EC 2.4.1.90/38. *See also* Lactose Synthase; β-*N*-Acetylglucosaminyl-Glycopeptide β-1,4-Galactosyltransferase

[1] K. Nakazawa, K. Furukawa, A. Kobata & H. Narimatsu (1991) *EJB* **196**, 363.
[2] L. N. Gastinel, C. Cambillau & Y. Bourne (1999) *EMBO J.* **18**, 3546.

Selected entries from *Methods in Enzymology* [vol, page(s)]:
Assay: 22, 139; 98, 122
Other molecular properties: bovine milk (assay, 98, 122; properties, 98, 124; purification, 98, 123); diagnostic enzyme, as, 31, 20, 24; Golgi apparatus, in, 22, 144; 31, 20, 24 (assay, 22, 139; reaction in Golgi apparatus, 22, 142); plasma membranes, in, 31, 180, 184

N-ACETYLLACTOSAMINIDE β-1,3-*N*-ACETYLGLUCOSAMINYLTRANSFERASE

This transferase [EC 2.4.1.149] catalyzes the reaction of UDP-*N*-acetyl-D-glucosamine with β-D-galactosyl-1,4-*N*-acetyl-D-glucosaminyl-R to produce UDP and *N*-acetyl-β-D-glucosaminyl-1,3-β-D-galactosyl-1,4-*N*-acetyl-D-glucosaminyl-R. The acceptor substrates are the β-galactosyl-1,4-*N*-acetylglucosaminyl termini on asialo-α₁-acid glycoprotein and other glycoproteins and oligosaccharides.

Selected entries from *Methods in Enzymology* [vol, page(s)]:
General discussion: 230, 303
Assay: 179, 393
Other molecular properties: blood group i (partial purification, 179, 394; properties, 179, 394); *N*- and *O*-glycan synthesis, in (HPLC assay, 179, 351; substrate preparation, 179, 360)

N-ACETYLLACTOSAMINIDE β-1,6-*N*-ACETYLGLUCOSAMINYLTRANSFERASE

This transferase [EC 2.4.1.150] catalyzes the reaction of UDP-*N*-acetyl-D-glucosamine with β-D-galactosyl-1,4-*N*-acetyl-D-glucosaminyl-R to produce UDP and *N*-acetyl-β-D-glucosaminyl-1,6-β-D-galactosyl-1,4-*N*-acetyl-D-glucosaminyl-R. Acceptor substrates are the β-galactosyl-1,4-*N*-acetylglucosaminyl termini on asialo-α₁-acid glycoproteins.

N-ACETYLLACTOSAMINIDE α-1,3-GALACTOSYLTRANSFERASE

This manganese-dependent transferase [EC 2.4.1.151] catalyzes the reaction of UDP-D-galactose with β-D-galactosyl-(1,4)-*N*-acetyl-D-glucosaminyl-R to produce UDP and α-D-galactosyl-(1,3)-β-D-galactosyl-(1,4)-*N*-acetyl-D-glucosaminyl-R. Acceptor substrates include the β-galactosyl-(1,4)-*N*-acetylglucosaminyl termini on asialo-α₁-acid glycoprotein and *N*-acetyllactosamine (*i.e.*, β-D-galactosyl-1,4-*N*-acetyl-β-D-glucosamine). However, the 2′-fucosylated derivative of *N*-acetyllactosamine is not a substrate.

Selected entries from *Methods in Enzymology* [vol, page(s)]:
General discussion: 230, 302

N-ACETYLLACTOSAMINIDE α-2,3-SIALYLTRANSFERASE

This sialyltransferase [EC 2.4.99.6], also known as CMP-*N*-acetylneuraminate:*N*-acetyllactosaminide α-2,3-sialyltransferase and CMP-*N*-acetylneuraminate:galactose-β—1,4-*N*-acetyl-D-glucosamine α—2,3-sialyltransferase, catalyzes the reaction of CMP-*N*-acetylneuraminate with a β-D-galactosyl-1,4-*N*-acetyl-D-glucosaminylglycoprotein to produce α-*N*-acetylneuraminyl-2,3-β-D-galactosyl-1,4-*N*-acetyl-D-glucosaminylglycoprotein and CMP, acting on the β-D-galactosyl-1,4-*N*-acetyl-D-glucosaminyl termini in glycoproteins.[1-3] This sialyltransferase does not act on either branch of complex type diantennary glycopeptides.[3]

[1] D. H. Van den Eijnden & W. E. C. M. Schiphorst (1981) *JBC* **256**, 3159.
[2] B. Bauvois, J. Montreuil & A. Verbert (1984) *BBA* **788**, 234.
[3] M. Nemansky & D. H. Van den Eijnden (1993) *Glycoconjugate J.* **10**, 99.

Selected entries from *Methods in Enzymology* [vol, page(s)]:
Assay: 98, 129

*N*⁶-ACETYL-β-LYSINE AMINOTRANSFERASE

This pyridoxal-phosphate-dependent aminotransferase [EC 2.6.1.65] catalyzes the reversible reaction of

3-amino-6-acetamidohexanoate (or, N^6-acetyl-β-lysine) with α-ketoglutarate (or, 2-oxoglutarate) to produce 3-oxo-6-acetamidohexanoate and L-glutamate.[1,2]

[1] G. Bozler, J. M. Robertson, M. Ohsugi, C. Hensley & H. A. Barker (1979) *ABB* **197**, 226.
[2] H. Schmidt & R. Bode (1992) *Antonie Van Leeuwenhoek* **62**, 285.

ACETYLMURAMOYL-L-ALANINE AMIDASE

This enzyme [EC 3.5.1.28; previously misclassified as EC 3.4.19.10], also known as murein hydrolase and autolysin, catalyzes the hydrolysis of the link between *N*-acetylmuramoyl residues and L-aminoacyl residues in certain bacterial cell-wall glycopeptides.[1] Amidases from different sources show different specificities and properties: for example, the enzyme from *Streptococcus pneumoniae* requires cell walls containing choline. **See also** *Autolysin*

[1] G. D. Shockman & J. V. Höltje (1994) in *New Comprehensive Biochemistry: Bacterial Cell Wall* (J. M. Ghuysen & R. Hakenbeck, eds.), Elsevier, Amsterdam, p. 131.

Selected entries from *Methods in Enzymology* [vol, page(s)]:
Other molecular properties: activity, **248**, 205; bacterial cell wall degradation and, **8**, 687, 690; properties, **248**, 191, 195, 205; synthesis, **248**, 205

N-ACETYLNEURAMINATE 4-*O*-ACETYLTRANSFERASE

This acetyltransferase [EC 2.3.1.44] catalyzes the reaction of acetyl-CoA with *N*-acetylneuraminate to produce free *N*-acetyl-4-*O*-acetylneuraminate and Coenzyme A. *N*-Acetyl- and *N*-glycolylneuraminate glycosides act as acceptor substrates.[1,2]

[1] R. Schauer (1987) *MIE* **138**, 611.
[2] M. Iwersen, V. Vandamme-Feldhaus & R. Schauer (1998) *Glycoconj. J.* **15**, 895.

Selected entries from *Methods in Enzymology* [vol, page(s)]:
General discussion: **50**, 386; **138**, 611
Other molecular properties: equine submandibular gland, from, reaction, **50**, 386

N-ACETYLNEURAMINATE 7-*O*-(OR 9-*O*)-ACETYLTRANSFERASE

This transferase [EC 2.3.1.45], also known as glycoprotein sialate 7(9)-*O*-acetyltransferase, catalyzes the reaction of acetyl-CoA with *N*-acetylneuraminate to produce free *N*-acetyl-7-*O*(or 9-*O*)-acetylneuraminate and Coenzyme A. Glycosidically bound *N*-acetyl- and *N*-glycolylneuraminates can act as alternative *O*-acetyl acceptor substrates.[1,2]

[1] R. Schauer (1987) *MIE* **138**, 611.

[2] S. Diaz, H. H. Higa & A. Varki (1989) *MIE* **179**, 416.

Selected entries from *Methods in Enzymology* [vol, page(s)]:
General discussion: **50**, 381; **138**, 611; **179**, 416
Assay: **50**, 381; **179**, 417
Other molecular properties: glycoprotein synthesis, **50**, 386; isolation from bovine submandibular gland, **50**, 384; occurrence, **50**, 385, 386; properties, **50**, 384, 385; **179**, 420; reaction, **50**, 381

N-ACETYLNEURAMINATE LYASE

This enzyme [EC 4.1.3.3], also known as *N*-acetylneuraminate aldolase and sialate aldolase, catalyzes the reversible cleavage of *N*-acetylneuraminate to *N*-acetylmannosamine and pyruvate.[1-4] The enzyme will also act on *N*-glycoloylneuraminate and on *O*-acetylated sialic acids, other than O^4-acetylated derivatives. The α-anomer of *N*-acetylneuraminate appears to be the form that binds to the enzyme.[3] The substrate forms a Schiff base intermediate with a lysyl residue. Pyruvate is a competitive inhibitor, suggesting it is the last product released. In the reverse direction, pyruvate forms a Schiff base but not *N*-acetylmannosamine. The reverse reaction is useful for the synthesis of sialic acid derivatives.

[1] S. Nees, R. Schauer & F. Mayer (1976) *Hoppe-Seyler's Z. Physiol. Chem.* **357**, 839.
[2] F. N. Kolisis (1985) *Arch. Int. Physiol. Biochim.* **93**, 89.
[3] C. M. Deijl & J. F. Vliegenthart (1983) *BBRC* **111**, 668.
[4] T. Izard, M. C. Lawrence, R. L. Malby, G. G. Lilley & P. M. Colman (1994) *Structure* **2**, 361.

Selected entries from *Methods in Enzymology* [vol, page(s)]:
General discussion: **5**, 391
Assay: **5**, 391
Other molecular properties: acylneuraminate cleavage, **50**, 81, 82; assay of sialic acids, **6**, 465; catalysis of *N*-acetyl-D-neuramine synthesis, **202**, 611; labeling, **46**, 139; properties, **5**, 393; purification, **5**, 392; **6**, 466; sialic acid determination, **50**, 75, 76; sialic acid synthesis and, **8**, 210; sialic acid removal from glycoconjugates, **230**, 194; sugar nucleotide regeneration system, **247**, 123

N-ACETYLNEURAMINATE MONOOXYGENASE

This iron-dependent oxidoreductase [EC 1.14.99.18], also known as CMP-*N*-acetylneuraminate monooxygenase or hydroxylase, catalyzes the reaction of *N*-acetylneuraminate with dioxygen and AH_2 to produce *N*-glycoloylneuraminate, water, and A where AH_2 represents the reduced acceptor substrate and A is the oxidized acceptor. The reducing equivalents can be provided by NADH, NADPH, or ascorbate. The enzyme is probably linked to an NADH:cytochrome b_5 reductase.[1,2]

While this enzyme can also use CMP-*N*-acetylneuraminate as a substrate, it should be noted that a separate CMP-*N*-acetylneuraminate monooxygenase is also listed as EC 1.14.13.45. **See also** *CMP-N-Acetylneuraminate Monooxygenase*

[1] L. Shaw, P. Schneckenburger, J. Carlsen, K. Christiansen & R. Schauer (1992) *EJB* **206**, 269.
[2] C. J. Mukuria, W. D. Mwangi, A. Noguchi, G. P. Waiyaki, T. Asano & M. Naiki (1995) *BJ* **305**, 459.

Selected entries from *Methods in Enzymology* [**vol**, page(s)]:
General discussion: 50, 374
Assay: 50, 375
Other molecular properties: glycoprotein synthesis, **50**, 380; isolation from porcine submandibular glands, **50**, 377; occurrence, **50**, 380; **52**, 12; properties, **50**, 378, 379; reaction, **50**, 375

N-ACETYLNEURAMINATE SYNTHASE

This synthase [EC 4.1.3.19] catalyzes the reaction of *N*-acetyl-D-mannosamine with phosphoenolpyruvate and water to produce *N*-acetylneuraminate and orthophosphate.[1]

[1] E. Komaki, Y. Ohta & Y. Tsukada (1997) *Biosci. Biotechnol. Biochem.* **61**, 2046.

α-*N*-ACETYLNEURAMINIDE α-2,8-SIALYLTRANSFERASE

This sialyltransferase [EC 2.4.99.8], also known as CMP-*N*-acetylneuraminate: α-*N*-acetylneuraminide α-2,8-sialyltransferase, catalyzes the reaction of CMP-*N*-acetylneuraminate with α-*N*-acetylneuraminyl-2,3-β-D-galactosyl-R to produce CMP and α-*N*-acetylneuraminyl-2,8-α-*N*-acetylneuraminyl-2,3-β-D-galactosyl-R.[1–3] Other sialyl acceptor substrates include the gangliosides G_{M3} and G_{D1a}. **See also** *Sialyltransferases*

[1] H. Higashi, M. Basu & S. Basu (1985) *JBC* **260**, 824.
[2] C. M. Eppler, D. J. Morré & T. W. Keenan (1980) *BBA* **619**, 318.
[3] K. Busam & K. Decker (1986) *EJB* **160**, 23.

(α-*N*-ACETYLNEURAMINYL-2,3-β-GALACTOSYL-1,3)-*N*-ACETYLGALACTOSAMINIDE α-2,6-SIALYLTRANSFERASE

This sialyltransferase [EC 2.4.99.7], also known as CMP-*N*-acetylneuraminate: (α-*N*-acetylneuraminyl-2,3-β-galactosyl-1,3)-*N*-acetylgalactosaminide α-2,6-sialyltransferase, catalyzes the reaction of CMP-*N*-acetylneuraminate with α-*N*-acetylneuraminyl-2,3-β-D-galactosyl-1,3-*N*-acetyl-D-galactosaminyl-R with CMP to produce α-*N*-acetylneuraminyl-2,3-β-D-galactosyl-1,3-(α-*N*-acetylneuraminyl-2,6)-*N*-acetyl-D-galactosaminyl-R.[1–4] The enzyme catalytically attaches an *N*-acetylneuraminate in an α-2,6-linkage to an *N*-acetylgalactosamine residue, only when present in the above stucture. (The R group may be a protein or *p*-nitrophenol.) This enzyme is distinct from α-*N*-acetylgalactosaminide α-2,6-sialyltransferase [EC 2.4.99.3]. **See also** *Sialyltransferases; α-2,6-Sialyltransferases; α-N-Acetylgalactosaminide α-2,6-Sialyltransferase*

[1] M. L. E. Bergh, G. J. M. Hooghwinkel & D. H. Van den Eijnden (1983) *JBC* **258**, 7430.
[2] M. L. E. Bergh & D. H. Van den Eijnden (1983) *EJB* **136**, 113.
[3] E. A. Higgins, K. A. Siminovitch, D. Zhuang, I. Brockhausen & J. W. Dennis (1991) *JBC* **266**, 6280.
[4] H. Baubichon-Cortay, P. Broquet, P. George & P. Louisot (1989) *Glycoconj. J.* **6**, 115.

(*N*-ACETYLNEURAMINYL) GALACTOSYL GLUCOSYLCERAMIDE *N*-ACETYL-GALACTOSAMINYLTRANSFERASE

This transferase [EC 2.4.1.92], also known as ganglioside G_{M2} synthase and ganglioside G_{M3} acetylgalactosaminyltransferase, catalyzes the reaction of UDP-*N*-acetyl-D-galactosamine with (*N*-acetylneuraminyl)-D-galactosyl-D-glucosylceramide (or, G_{M3}) to produce UDP and *N*-acetyl-D-galactosaminyl-(*N*-acetylneuraminyl)-D-galactosyl-D-glucosylceramide (or, G_{M2}). Initial-rate kinetic studies indicate the presence of cooperativity.[1] The enzyme also appears to be regulated by phosphorylation/dephosphorylation.[2]

[1] K. Yanagisawa, N. Taniguchi & A. Makita (1987) *BBA* **919**, 213.
[2] E. Bieberich, B. Freischutz, S. S. Liour & R. K. Yu (1998) *J. Neurochem.* **71**, 972.

Selected entries from *Methods in Enzymology* [**vol**, page(s)]:
Assay: 138, 594

(*N*-ACETYLNEURAMINYL)GALACTOSYL-GLUCOSYLCERAMIDE β-1,4-*N*-ACETYLGALACTOSAMINYLTRANSFERASE

This manganese-dependent transferase [EC 2.4.1.165] catalyzes the reaction of UDP-*N*-acetyl-D-galactosamine with *N*-acetylneuraminyl-2,3-α-D-galactosyl-1,4-β-D-glucosylceramide to form *N*-acetyl-β-D-galactosaminyl-1,4-(*N*-acetyl-α-neuraminyl-2,3)-β-D-galactosyl-1,4-β-D-glucosylceramide and UDP.[1] The enzyme is responsible for the addition of the immunodominant sugar of the so-called SDA-histodeterminant.[2] Only substances with sialic acid residues can act as acceptor substrates; these include bovine fetuin, 3′-sialyllactose, *N*-acetylneuraminyl-α-2,3-galactosyl-β-1,4-glucose, the urinary tract Tamm-Horsfall glycoprotein, $α_1$-acid glycoprotein, and human chorionic gonadotropin.

[1] A. Takeya, O. Hosomi & T. Kogure (1987) *J. Biochem.* **101**, 251
[2] N. Malagolini, F. Dall'Olio & F. Serafini-Cessi (1991) *BBRC* **180**, 681.

ACETYLORNITHINE δ-AMINOTRANSFERASE

This pyridoxal-phosphate-dependent aminotransferase [EC 2.6.1.11] catalyzes the reaction of N^2-acetyl-L-ornithine with α-ketoglutarate (or, 2-oxoglutarate) to produce N-acetyl-L-glutamate semialdehyde and L-glutamate. One *Klebsiella aerogenes* isozyme is induced by arginine, while a second isozyme is repressed by the very same amino acid.[1] Some sources also produce an enzyme possessing ornithine δ-aminotransferase activity.

The Enzyme Commission lists another enzyme catalyzing this reaction, namely N^2-acetylornithine 5-aminotransferase [EC 2.6.1.69].[2]

[1] B. Friedrich, C. G. Friedrich & B. Magasanik (1978) *J. Bacteriol.* **133**, 686.
[2] C. Vander Wauven & V. Stalon (1985) *J. Bacteriol.* **164**, 882.

Selected entries from *Methods in Enzymology* [vol, page(s)]:
General discussion: 17A, 260, 277
Assay: 17A, 261, 277
Other molecular properties: properties, 17A, 264, 280; purification from (*Chlamydomonas reinhardti*, 17A, 279; *Escherichia coli*, 17A, 262); pyridoxal phosphate and, 17A, 264, 281; synthesis of *N*-acetylglutamate γ-semialdehyde, use in, 17A, 256

ACETYLORNITHINE DEACETYLASE

This enzyme [EC 3.5.1.16], also known as acetylornithinase and *N*-acetylornithinase, catalyzes the reaction of water with N^2-acetyl-L-ornithine to produce acetate and L-ornithine.[1] *N*-Acetyl-L-methionine is an alternative substrate. The tobacco plant enzyme will also deacetylate *N*-acetyl-L-phosphinothricin.[2]

[1] H. J. Vogel & D. M. Bonner (1956) *JBC* **218**, 97.
[2] G. Kriete, K. Niehaus, A. M. Perlick, A. Puhler & I. Broer (1996) *Plant J.* **9**, 809.

Selected entries from *Methods in Enzymology* [vol, page(s)]:
General discussion: 17A, 265
Assay: 17A, 265
Other molecular properties: properties, 17A, 268; **248**, 223; purification from *Escherichia coli*, 17A, 266

(S)-N-ACETYL-1-PHENYLETHYLAMINE HYDROLASE

This enzyme [EC 3.5.1.85], which is inhibited by phenylmethanesulfonyl fluoride, catalyzes the hydrolysis of (*S*)-*N*-acetyl-1-phenylethylamine to produce (*S*)-1-phenylethylamine and acetate.[1] Related acetylated compounds can also be hydrolyzed.

[1] A. Brunella, M. Graf, M. Kittelmann, K. Lauma & O. Ghisalba (1997) *Appl. Microbiol. Biotechnol.* **47**, 515.

ACETYLPUTRESCINE DEACETYLASE

This enzyme [EC 3.5.1.62] catalyzes the hydrolysis of *N*-acetylputrescine to produce acetate and putrescine. Alternative substrates for the *Micrococcus luteus* enzyme include N^8-acetylspermidine and N^1-acetylcadaverine.[1]

[1] O. Suzuki, K. Ishikawa, K. Izu & T. Matsumoto (1986) *BBA* **882**, 140.

ACETYLPYRUVATE HYDROLASE

This enzyme [EC 3.7.1.6], first isolated from *Pseudomonas putida*, catalyzes the hydrolysis of acetylpyruvate ($CH_3COCH_2COCOO^-$) to produce acetate and pyruvate. Pyruvate, oxaloacetate, maleylpyruvate, fumarylpyruvate, and acetylacetone are not alternative substrates. Mn^{2+} and Mg^{2+} stimulate activity.[1]

[1] J. F. Davey & D. W. Ribbons (1975) *JBC* **250**, 3826.

ACETYLSALICYLATE DEACETYLASE

This deacetylase [EC 3.1.1.55], also known as aspirin esterase, catalyzes the hydrolysis of acetylsalicylate (*i.e.*, aspirin) to produce salicylate and acetate.[1–3] This particular deacetylase will act on a number of acetyl esters of aryl alcohols as well as thioesters. The microsomal enzyme hydrolyzes some other negatively charged esters.

The enzyme is distinct from carboxylesterase [EC 3.1.1.1], arylesterase [EC 3.1.1.2], acetylcholinesterase [EC 3.1.1.7], and cholinesterase [EC 3.1.1.8].

[1] D. H. Kim, Y. S. Yang & W. B. Jakoby (1990) *Biochem. Pharmacol.* **40**, 481.
[2] K. N. White & D. B. Hope (1984) *BBA* **785**, 138.
[3] K. N. White, V. L. Vale & D. B. Hope (1994) *Biochem. Soc. Trans.* **22**, 220S.

ACETYLSEROTONIN O-METHYLTRANSFERASE

This methyltransferase [EC 2.1.1.4], also known as hydroxyindole *O*-methyltransferase, catalyzes the reaction of *S*-adenosyl-L-methionine with *N*-acetylserotonin to produce *S*-adenosyl-L-homocysteine and *N*-acetyl-5-methoxytryptamine (or, melatonin). A number of other hydroxyindoles can act as acceptor substrates, albeit less efficiently. The kinetic mechanism is ordered Bi Bi.[1,3,4]

In the silkworm, the enzyme exhibits as a circadian rhythm of catalytic activity linked to environmental cycles of daylight and darkness.[2]

[1] D. Sugden, V. Cena & D. C. Klein (1987) *MIE* **142**, 590.
[2] M. T. Itoh, T. Nomura & Y. Sumi (1997) *Brain Res.* **765**, 61.

[3] D. J. Morton & N. Kock (1990) *J. Pineal Res.* **8**, 35.
[4] D. J. Morton & H. J. Forbes (1989) *J. Neural. Transm.* **75**, 65.

Selected entries from *Methods in Enzymology* [vol, page(s)]:
General discussion: 17B, 764; 142, 590
Assay: 17B, 764; 39, 391, 392; 142, 591
Other molecular properties: amino acid composition, 142, 593; melatonin biosynthesis, in, 39, 379, 380; properties, 17B, 766; 142, 594; purification from (pineal, 17B, 398, 765; 103, 490; 142, 592); radioenzymatic assay for serotonin, in, 103, 491

ACETYLSPERMIDINE DEACETYLASE

This enzyme [EC 3.5.1.48] catalyzes the hydrolysis of N^1-acetylspermidine to produce acetate and spermidine. Alternative substrates include N^8-acetylspermidine and N^1-acetylspermine. The enzyme is inhibited by metal chelators, several ω-amino-substituted carboxylates, and thiol reagents.[1]

[1] T. L. Huang, S. A. Dredar, V. A. Manneh, J. W. Blankenship & D. S. Fries (1992) *J. Med. Chem.* **35**, 2414.

Selected entries from *Methods in Enzymology* [vol, page(s)]:
Assay: rat liver, 94, 329
Other molecular properties: rat liver (assay, 94, 329; properties, 94, 331; purification, 94, 329)

ACETYLXYLAN ESTERASE

This esterase [EC 3.1.1.72] catalyzes the deacetylation of xylans and xylo-oligosaccharides. Alternative substrates include acetylated xylose, acetylated D-glucose, α-naphthyl acetate, and *p*-nitrophenyl acetate.[1]

[1] D. L. Blum, X. L. Li, H. Chen & L. G. Ljungdahl (1999) *Appl. Environ. Microbiol.* **65**, 3990.

Selected entries from *Methods in Enzymology* [vol, page(s)]:
Assay: 160, 701; *Streptomyces* species, in, 160, 551
Other molecular properties: detection in gels, 160, 702; properties, 160, 706; purification from *Schizophyllum commune*, 160, 705

ACID PHOSPHATASE

This enzyme [EC 3.1.3.2], also known as acid phosphomonoesterase, phosphomonoesterase, and glycerophosphatase, catalyzes the hydrolysis of an orthophosphoric monoester to generate an alcohol and orthophosphate. The enzyme displays a wide spectrum of activity and will even catalyze phosphotransfer reactions. *See also Phosphatase; Phosphomonoesterase; Alkaline Phosphatase; Glycerol-2-Phosphatase; Glycerol-1-Phosphatase*

Selected entries from *Methods in Enzymology* [vol, page(s)]:
General discussion: 2, 523; 46, 3; 71, 242
Assay: 31, 329, 361, 362; automated, 22, 18, 19; Fast Analyzer, by, 31, 816; histochemical assay, 4, 386; lysosomal, 18A, 68
Other molecular properties: adenosine deaminase and, 12A, 127; aleurone grains, in, 31, 576; amoebic phagosomes, in, 31, 697; analysis of, automated (continuous-flow method, 22, 11; multiple systems, in, 22, 18, 19, 22); assessment of phagocytic function, for, 132, 123, 163;

associated with cell fractions, 40, 86; bovine liver, extraction from, I, 33; brain fractions, in, 31, 464, 466, 467, 469; brush border, 31, 121; chiral phosphoric monoester, and, 87, 265, 300; cytochemistry, 39, 150, 151; diagnostic enzyme, as, 31, 20, 186, 187, 326, 327, 330, 743, 744 (lysosomes, for, 31, 330, 334, 337, 338, 344, 735); distribution, 2, 523; gelatin-immobilized, properties, 135, 297; glucose oxidase electrode, description, 137, 37; glucose-6-phosphatase, separation from, 9, 625; glycerophosphatase, acid (brain homogenates, in, 31, 464, 466; macrophage fractions, in, 31, 340; microsomal, 2, 542); group-specific, 2, 523; hepatic Golgi subfractions, in, 109, 224; histochemistry, 4, 386; hydrolysis of polyprenol pyrophosphates, in, 110, 153; intermediate, 64, 76; 87, 265, 300, 319; kinetics, 87, 321, 327; large-scale isolation utilizing osmotic shock, 22, 490; liver-cell fractions, in, 31, 209; lung lamellar bodies, in, 31, 423, 424; lysosomal, 10, 12; 18A, 68; 31, 23, 24; 55, 101, 103, 104 (assay, 18A, 68; 55, 102); macrophage fractions, in, 31, 340; marker enzyme, as, 32, 15 (marker for chromosome detection in somatic cell hybrids, as, 151, 177); osmotic shock and, 12B, 844; 22, 490; partial purification, 51, 273; peroxisomes, in, 31, 361, 362; phagocyte staining, in, 132, 133, 174; phosphoenolpyruvate, action on, 87, 93; phosphoenzyme intermediate and oxygen exchanges, 64, 76; potato acid phosphatase (preparation, 18A, 472; product inhibition, 249, 73; pyridoxal 5'-phosphate, in determination of, 18A, 472; use in hydrolysis of pyridoxal 5'-phosphate, 18A, 472, 473, 474); preparation, 2, 529; 18A, 472; product inhibition, 249, 73; progress curve analysis, 63, 180; prostate acid phosphatase, 2, 523 (phenyl phosphate hydrolysis, progress curve analysis, 249, 70); pyridoxine-phosphorylating activity, having, 18A, 630 (assay, 18A, 630; properties, 18A, 633; purification from *Escherichia freundii*, 18A, 631); quality index, 87, 299; ribonuclease T₁ and, 12A, 233; ribonuclease T₂ and, 12A, 234, 244; specificity, 2, 524; staining on gels, 22, 585, 591, 601; stereochemistry, 87, 263, 265, 299, 300; streptavidin, with, in detection of antigens on Western blots, 184, 441; tartrate-resistant, in detection of osteoclast activation, 198, 506; transphosphorylation by, 2, 556; transition state and multisubstrate analogues, 249, 305; viscosity barrier centrifugation, 55, 133, 134

ACID PROTEASE

Two proteases, designated I and II for those from the fungi *Sporotrichum pulverulentum*, exhibit acidic pH optima (5.0 and 5.2, respectively).[1] Protease I will hydrolyze the Val15–Arg16 peptide bond of purified human fibrinopeptide A, whereas protease II acts on Phe8–Leu9 and Leu9–Ala10. Both proteases play roles in the activation of fungal endo-1,4-β-glucanases.

[1] K. E. Eriksson & B. Pettersson (1988) *MIE* **160**, 500.

Selected entries from *Methods in Enzymology* [vol, page(s)]:
Assay: 160, 501
Other molecular properties: properties, 160, 506; purification from *Sporotrichum pulverulentum*, 160, 501

*Aci*I RESTRICTION ENDONUCLEASE

This type II restriction endonuclease [EC 3.1.21.4], obtained from *Arthrobacter citreus*, catalyzes the hydrolysis of both strands of DNA at 5′...C∇CGC...3′ and 3′...GGC∇G...5′.[1] Note that the enzyme has a non-palindromic recognition site.

[1] C. Polisson & R. D. Morgan (1990) *NAR* **18**, 5911.

*Acl*I RESTRICTION ENDONUCLEASE

This type II restriction endonuclease [EC 3.1.21.4], obtained from *Acinetobacter calcoaceticus* M4, catalyzes the hydrolysis of both strands of DNA at 5′...AA∇ CGTT...3′.[1]

[1]S. K. Degtyarev, M. A. Abdurashitov, A. A. Kolyhalov & N. I. Rechkunova (1992) *NAR* **20**, 3787.

ACONITASE

This [4Fe-4S] cluster-containing enzyme [EC 4.2.1.3], also known as citrate hydro-lyase and aconitate hydratase, catalyzes the conversion of citrate to *cis*-aconitate (*i.e.*, (*Z*)-prop-1-ene 1,2,3-tricarboxylate) and water.[1-17] The enzyme likewise catalyzes the conversion of isocitrate into *cis*-aconitate.

Upon complexation of the substrate with one of the iron ions in the cluster, the iron is converted into an octahedral site possessing a coordinated water molecule. An *aci*-acid intermediate, produced by β-proton abstraction, rearranges to expel the hydroxyl group to form *cis*-aconitate.[11]

The *Bacillus subtilis* aconitase is an RNA-binding protein and may have an important role in sporulation, particularly when the cells are iron-deprived.[2]

The iron regulatory proteins (IRPs) are major regulatory factors for iron uptake and storage in most mammalian cells.[3,4] IRP1 belongs to a family of proteins containing one [4Fe-4S] cluster and includes mitochondrial and cytoplasmic aconitase. Extensive disassembly of the cluster is required to allow the protein to interact with iron-responsive elements (IREs) on the mRNA of the target genes to repress translation or enhance mRNA stability.[5] The reciprocal control of both activities (the "iron-sulfur switch") appears as a key mechanism in the regulatory function of IRP1.

The (–)-*erythro* diastereomer of 2-fluorocitrate is reported to be a mechanism-based inhibitor. It is first converted to fluoro-*cis*-aconitate by aconitase followed by addition of hydroxide and the loss of fluoride to form 4-hydroxy-*trans*-aconitate which binds very tightly to the enzyme.[6]

[1]J. V. Schloss, M. H. Emptage & W. W. Cleland (1984) *Biochemistry* **23**, 4572.
[2]C. Alen & A. L. Sonenshein (1999) *PNAS* **96**, 10412.
[3]R. D. Klausner, T. A. Rouault & J. B. Harford (1993) *Cell* **72**, 19.
[4]M. W. Hentze & L. C. Kühn (1996) *PNAS* **93**, 8175.
[5]E. C. Theil (1994) *BJ* **304**, 1.
[6]H. Lauble, M. C. Kennedy, M. H. Emptage, H. Beinert & C. D. Stout (1996) *PNAS* **93**, 13699.

[7]M. J. Gruer, P. J. Artymiuk & J. R. Guest (1997) *Trends Biochem. Sci.* **22**, 3.
[8]T. Rouault & R. Klausner (1997) *Curr. Top. Cell. Regul.* **35**, 1.
[9]H. Beinert & M. C. Kennedy (1993) *FASEB J.* **7**, 1442.
[10]V. E. Anderson (1998) *CBC* **2**, 115.
[11]B. G. Fox (1998) *CBC* **3**, 261.
[12]H. Lauble, M. C. Kennedy, H. Beinert & D. C. Stout (1992) *Biochemistry* **31**, 2735.
[13]L. Zheng, M. C. Kennedy, H. Beinert & H. Zalkin (1992) *JBC* **267**, 7895.
[14]H. Kwack & R. L. Veech (1992) *Curr. Top. Cell. Reg.* **33**, 185.
[15]J. B. Howard & D. C. Rees (1991) *Adv. Protein Chem.* **42**, 199.
[16]N. E. Tolbert (1981) *Ann. Rev. Biochem.* **50**, 133.
[17]J. P. Glusker (1971) *The Enzymes*, 3rd ed., **5**, 413.

Selected entries from *Methods in Enzymology* [vol, page(s)]:
General discussion: 1, 695; 5, 614; 13, 26; **269**, 26
Assay: 1, 695; 5, 614; 13, 26; spectrophotometric, 269, 39
Other molecular properties: aconitate decarboxylase and, 5, 595; activation of, 13, 26, 27, 30, 448; activity in glutathione mutants of *Escherichia coli*, 252, 86; *Aspergillus niger*, in, 5, 614; citrate assay, 3, 426; 13, 446, 529; 55, 211; cytosolic ([Fe-S] center, model, 269, 28; IRE-binding activity, 269, 34; measurement, 269, 30); deuterium transfer by, 13, 588; electron nuclear double resonance spectroscopy, 246, 555, 583; equilibrium values, 3, 425; [Fe-S]-containing, 269, 26; glyoxysomes, in, 31, 565, 569; inactivation, 13, 449; inhibitors, 5, 595; 63, 400; IRE-binding activity, 269, 34; iron coordination, 246, 454; iron-responsive element, binding activity of aconitase, 269, 27, 34; latency, intact mitochondria, 55, 143; microbodies, in, 52, 496; mitochondrial, measurement, 269, 30; partition studies, 249, 334; preparations, treatment by nitric oxide donors, 269, 31; porcine heart, site-directed mutants, 249, 113; properties (*Aspergillus niger*, 5, 616; pig heart, 1, 698; 13, 29); purification, 1, 696; 5, 615; 13, 28; reactivation, rate constants, 269, 37; resonance Raman spectroscopy, 246, 455; specific radioactivity determination, use in, 13, 529; stability, 13, 26, 29; toluene-treated mitochondria, 56, 547; transition state and multisubstrate analogues, 87, 89; 249, 308

ACONITATE DECARBOXYLASE

This enzyme [EC 4.1.1.6] catalyzes the conversion of *cis*-aconitate to itaconate (*i.e.*, 2-methylenesuccinate) and carbon dioxide. It is a cytosolic enzyme in *Aspergillus terreus*.[1]

[1]W. M. Jaklitsch, C. P. Kubicek & M. C. Scrutton (1991) *J. Gen. Microbiol.* **137**, 533

Selected entries from *Methods in Enzymology* [vol, page(s)]:
General discussion: 5, 593
Assay: 5, 593
Other molecular properties: mechanism, 5, 596; properties, 5, 595; purification, 5, 594

ACONITATE Δ-ISOMERASE

This isomerase [EC 5.3.3.7] catalyzes the interconversion of *trans*-aconitate and *cis*-aconitate, the reaction reportedly to occur by an allelic rearrangement involving the (*pro-S*)-hydrogen in the methylene position.[1,2]

[1]J. P. Klinman & I. A. Rose (1971) *Biochemistry* **10**, 2253.
[2]J. P. Klinman & I. A. Rose (1971) *Biochemistry* **10**, 2259.

ACROCYLINDROPEPSIN

This aspartic proteinase [EC 3.4.23.28], also known as *Acrocylindrium* proteinase, is similar in structure to rhizopuspepsin. It was first isolated from a species of *Acro-cylindrium*, a plant pathogenic fungus. Studies indicated that the enzyme prefers tyrosyl, phenylalanyl, or leucyl residues at the P1′ position.[1] When casein is used as a substrate, a pH optimum of 2.0 is reported.

[1]S. Ichihara & F. Uchino (1975) *Agric. Biol. Chem.* **39**, 423.

ACROSIN

This endopeptidase [EC 3.4.21.10] catalyzes the hydrolysis of peptide bonds at Arg–Xaa and Lys–Xaa. The protein's name arises because this trypsin-like protease was isolated from the acrosome of the sperm head.[1] Acrosin is glycosylated and consists of a light and heavy chain held together by disulfide bonds. Acrosin is biosynthesized as an inactive precursor, proacrosin (which, in turn, is derived from preproacrosin), which can undergo autoactivation. This process converts proacrosin to α-acrosin, β-acrosin, and γ-acrosin, having reported molecular weights of 49, 34, and 25 kDa, respectively. ***See also*** *Acrolysin*

[1]U. A. Urch (1991) in *Elements of Mammalian Fertilization: Basic Concepts*, vol. I (P. M. Wassarman, ed.), CRC Press, Boston, p. 233.

Selected entries from ***Methods in Enzymology*** [**vol**, page(s)]:
General discussion: 34, 3; **46**, 206; **80**, 621

Assay: 80, 621; **45**, 331; active enzyme staining, **80**, 622, 623; active-site titration, with *p*-nitrophenyl-*p*′-guanidinobenzoate, **45**, 334; amidase activity, **45**, 334; **80**, 621, 622; combined N^α-benzoyl-L-arginine ethyl ester/alcohol dehydrogenase, **45**, 333; comparison and evaluation of assay techniques, **45**, 335; direct assay with N^α-benzoyl-L-arginine ethyl ester, **45**, 333; esterase activity, **45**, 332; proteolytic activity estimation, **45**, 331; sperm, **45**, 331; **80**, 261 (active enzyme staining, **80**, 622, 623; amidase activity, **80**, 621, 622; gelatin-film, **80**, 623)

Other molecular properties: inhibition, **45**, 337; **80**, 763, 801; radioiodination, **73**, 121; release during acrosome reaction, assays, **225**, 141; sperm, **45**, 330, 338; **80**, 621, 792 (assay, **45**, 331; **80**, 621 [active enzyme staining, **80**, 622, 623; active-site titration, with *p*-nitrophenyl-*p*′-guanidinobenzoate, **45**, 334; amidase activity, **45**, 334; **80**, 621, 622; combined N^α-benzoyl-L-arginine ethyl ester/alcohol dehydrogenase, **45**, 333; comparison and evaluation, **45**, 335; direct N^α-benzoyl-L-arginine ethyl ester, **45**, 333; esterase activity, **45**, 332; proteolytic activity estimation, **45**, 331]; boar sperm [amino acid composition, **80**, 630; inhibition, **80**, 801; molecular weight, **80**, 630; N-terminal amino acid sequence, **80**, 631, 632; purification, **80**, 625; specific activity, **80**, 632]; bovine, inhibitor, **80**, 801; composition, **45**, 341, 342; concentration, **80**, 632; human, amino acid composition, **80**, 630; function, **45**, 325; gelatin-film, **80**, 623; inhibitors, **45**, 337; kinetics, **45**, 340, 341; molecular weight, **45**, 340; properties, **45**, 330, 338; **80**, 629; purification, **45**, 336, 338; purity, **80**, 629; ram, amino acid composition, **80**, 630; specificity, **45**, 342; stability, **45**, 339; **80**, 629); stability, **45**, 339; **80**, 629, 630; trapped by α₂-macroglobulin, **80**, 748

ACTIN

Found in both muscle and nonmuscle cells, this major cytoskeletal protein readily undergoes head-to-tail polymerization in which monomeric globular actin (also known as G-actin) form filamentous actin (or F-actin).[1,2]

Assembly-induced hydrolysis of actin-bound ATP leads to the release of orthophosphate, but ADP remains firmly bound within the lattice of subunits comprising the actin filament.[3] Although the Enzyme Commission currently classifies proteins like actin as ATPases, the actual reaction is more correctly written as: Polymerization State₁ + ATP + H₂O = Polymerization State₁ + ADP + Pᵢ, where two affinity-modulated polymerization (or conformational) states of a filament are indicated by the subscripts.[4] In this respect, actin is an energase, a specialized mechanochemical enzyme that facilitates affinity-modulated reactions by coupling the $\Delta G_{\text{ATP-hydrolysis}}$ to the self-assembly and/or subsequent conformational maturation of the filament.[2] There is mounting evidence that ATP-hydrolysis during clamped-filament elongation is responsible for actin-based motility responsible for cell crawling. Dickinson and Purich[5] have formulated and analyzed the "Lock, Load & Fire" mechanism defining the fundamental force-producing step in actin motility. Polymerized actin also combines with myosin to form the actomyosin complex that also utilizes ATP hydrolysis in the sliding-filament contractile reactions taking place within both muscle and nonmuscle cells. ***See also*** *Energase*

[1]T. P. Stossel (1993) *Science* **260**, 1086.
[2]T. P. Stossel (1994) *Sci. Am.* **271**, 54.
[3]D. L. Purich & F. S. Southwick (1999) *MIE* **308**, 93.
[4]D. L. Purich (2001) *TiBS* **26**, 417.
[5]R. B. Dickinson & D. L. Purich (2001) *Biophys. J.* **82**, 605.

Selected entries from ***Methods in Enzymology*** [**vol**, page(s)]:
General discussion: I, 50; **46**, 294; **298**, 18, 26, 32, 114; **308**, 93

Other molecular properties: β-actin (assay by template titration, **218**, 465, 468; control sequence in nitric oxide synthase assay, **269**, 416); aggregation, dilatometry and, **26**, 405; amino-terminal acetylation (assays, **106**, 187, 189; inhibition, **106**, 181; reticulocyte lysate system, in **106**, 183); antibodies (isoform-specific [production, **134**, 461; purification, **134**, 462]; region-specific [production, **134**, 453; specificity, **134**, 458]); antibody binding, **50**, 55; assembly, **308**, 93; bilayers, effect on, **32**, 500, 535; binding protein, macrophage (purification, **85**, 483; sedimentation assay for crosslinking activity, **85**, 481); binding proteins, purification from intestinal brush border and properties, **139**, 137; circular dichroism, **12B**, 328; crosslinking in agonist-stimulated cells, assay, **196**, 486; cytoplasmic (assay, **85**, 371; properties, **85**, 372; purification from porcine brain, **85**, 371); depolymerization, **308**, 93 (actin-binding protein effects, **215**, 74; myosin effects, **215**, 74); dilution effect on platelet cytoskeleton, **215**, 76; disassembly, **308**, 93; disruption in yeast *cdc42* mutants, **256**, 282, 283; elongation, **308**, 96; energetics, **308**, 93; expression in *Escherichia coli*, **196**, 368 (growth conditions for, **196**, 378; solubility, effect of bacterial lysis, **196**, 380; vector-related variability, **196**, 370); F, dansylated (ATP addition to, effect on fluorescence polarization, **11**, 864; interaction with myosin, effect on fluorescence polarization, **11**, 865); filaments (decoration with myosin subfragment I, **236**, 480; equilibrium binding to [heavy meromyosin, **85**, 714; myosin single-headed fragment, **85**, 710]; fixation, **236**, 478; network characterization assays [falling-ball assay, **85**, 216; Fukada viscoelastometry, **85**, 231; gelmeter, **85**, 221; low-speed sedimentation, **85**, 215; micromagnetorheometry, **85**, 226; rotational viscometry, **85**,

223; test tube inversion, **85**, 213; Zaner viscoelastometry, **85**, 228]; properties, **85**, 182; role in ligand translocation, **98**, 368; steady-state binding to myosin single-headed fragment, **85**, 717); flow birefringence, **26**, 314 (electric field and, **26**, 322); fluorescein labeling, **196**, 50; fluorescently labeled (assay, **134**, 511; microinjection, **134**, 513; preparation, **134**, 508); gel overlays, assay for protein-protein interactions, **134**, 558; gels, extraction of fascin, **134**, 17; gelsolin complex (assay, **215**, 94; isolation from platelets, **215**, 89); globular (purification from *Dictyostelium discoideum*, **196**, 89; retroviral protease substrate, as, **241**, 290); growth factor-induced reorganization, **256**, 306; isoforms, separation, **196**, 105; labeling, **85**, 608 (anisotropic rotation and sample orientation, **85**, 616; restriction in rotation amplitude, **85**, 619); leukocytes, in (biochemistry, **162**, 246; intracellular behavior, **162**, 265; microfilament network, regulation, **162**, 247; purification, **162**, 250; role in locomotion, **162**, 246); mechanical instability after reaction with activated mixed disulfides, **251**, 362; modulating proteins, purification from macrophages, **162**, 254; monomeric columns with, preparation, **196**, 310; myosin ATPase and, **2**, 586; myosin interaction, **64**, 82 (EPR, **85**, 613); network in neutrophils, effect of C3 exoenzyme, **256**, 333; nonacetylated, preparation, **106**, 182; nucleotide exchange, **308**, 93; nucleation activity (agonist-stimulated, assay of inhibitor, **196**, 495; measurement, **196**, 493); oriented filaments, myosin movement (assays, **134**, 531, 538; inhibition by antibodies, **134**, 452); phalloidin binding site, **251**, 362; platelet (characterization, **215**, 58; purification, **215**, 58, 66; recombination with actin-binding protein and α-actinin, **215**, 73); polymerization, **46**, 293, 294; **308**, 93 (agonist-stimulated cells, in, assay, **196**, 486; assays [capillary viscometry, **85**, 185; bacterial invasion, during, **236**, 476; DNase I inhibition, **85**, 204; electron microscopy, **85**, 200; flow birefringence, **85**, 192; fluorescence, **85**, 195; light scattering, **85**, 198; Millipore filtration, **85**, 207; ultracentrifugation, **85**, 206; uv absorption, **85**, 190]; calmodulin effects, **139**, 852; mechanism, **85**, 182 [minimal, **308**, 93]; nucleotide exchange promotion, **308**, 93; properties [breaking and annealing, **85**, 209; critical concentration, **85**, 209; elongation rates, **85**, 208; filament length distribution, **85**, 210; nucleation rate, **85**, 208]; self-assembly models [ATP role {actin depolymerization factor control, **308**, 103, 105; bidirectional polymerization with hydrolysis, **308**, 99; delayed hydrolysis, **308**, 101; difficulty of study, **308**, 110; overview, **308**, 94; timing control, **308**, 102, 110; unidirectional polymerization with hydrolysis, **308**, 97}; condensation equilibrium model, **308**, 94; elongation without nucleotide hydrolysis, **308**, 96]); ponticulin, binding activity of, **196**, 58; preparation, **196**, 402; profilin complex, isolation from cell extracts, **196**, 97; properties in presence of modulating proteins, **162**, 260; purification, **1**, 50; **85**, 164; **134**, 20, 508; **196**, 50, 74 (human platelets, from, **139**, 847); pyrene labeling, **134**, 20; **196**, 138; radioiodination, **70**, 212; regulated (equilibrium binding to myosin single-headed fragment, **85**, 715; steady-state binding to myosin single-headed fragment, **85**, 722); removal from crude myosin preparation, **85**, 134; saturation-transfer studies, **49**, 495; **85**, 612; separation from profilin:actin, **196**, 115; Sepharose affinity resins, preparation, **134**, 463; skeletal muscle, labeling with pyrene, **196**, 138; sliding over myosin-coated surfaces, assay, **196**, 399; staining with fluorochrome-conjugated phalloidin, **194**, 729; stored, rejuvenation by recycling, **196**, 403; stress fibers, formation (regulation by Rho, **256**, 317; stimulation by growth factors, **256**, 311); structure-function relationship, optical activity measurements, **85**, 684; thiols, crosslinking, **251**, 363; 30-kDa protein crosslinking, preparation from *Dictyostelium discoideum*, **196**, 84; time-resolved X-ray scattering, **134**, 675; unfolding, monitoring, **251**, 362

ACTIN-HISTIDINE N-METHYLTRANSFERASE

This transferase, which is responsible for forming highly conserved 3-methylhistidine at position-73 in actin, catalyzes the reaction of *S*-adenosylmethionine with actin to produce methylated actin and *S*-adenosyl-L-homocysteine.[1,2]

[1] C. Vijayasarathy & B. S. Rao (1987) *BBA* **923**, 156.
[2] M. Raghavan, U. Lindberg & C. Schutt (1992) *EJB* **210**, 311.

ACTINIDAIN

This cysteine proteinase [EC 3.4.22.14], also known as actinidin and *Actinidia* anionic protease, catalyzes the hydrolysis of peptide bonds with a specificity similar to that of papain, via an acyl-enzyme intermediate. The enzyme is obtained from kiwi fruit or chinese gooseberry where it comprises a startling 50% of the soluble fruit protein. Differences from the action spectrum papain, particularly at the S2 subsite, can be observed when comparing k_{cat}/K_m values with a variety of peptide substrates.[1,2] Electrostatic probe kinetics have been used to further characterize the active site of actinidain.[3–5]

[1] M. J. Boland & M. J. Hardman (1973) *EJB* **36**, 575.
[2] E. N. Baker, M. J. Boland, P. C. Calder & M. J. Hardman (1980) *BBA* **616**, 30.
[3] K. Brocklehurst, M. O'Driscoll, D. Kowlessur, I. R. Phillips, W. Templeton, E. W. Thomas, C. M. Topham & C. W. Wharton (1989) *BJ* **257**, 309.
[4] D. Kowlessur, M. O'Driscoll, C. M. Topham, W. Templeton, E. W. Thomas, & K. Brocklehurst (1989) *BJ* **259**, 443.
[5] C. M. Topham, E. Salih, C. Frazao, D. Kowlessur, J. P. Overington, M. Thomas, S. M. Brocklehurst, M. Patel, E. W. Thomas, & K. Brocklehurst (1991) *BJ* **280**, 79.

ACTINOMADURA R39 D-Ala-D-Ala CARBOXYPEPTIDASE

This enzyme exhibits both carboxypeptidase and transpeptidase activities. The artificial substrate with the best activity observed so far is N^α-acetyl-L-Lys-D-Ala-D-Ala.[1–4] The carboxypeptidase activity of the enzyme results in the production of N^α-acetyl-L-Lys-D-Ala and D-alanine (with N^α-acetyl-L-Lys-D-Ala-D-Ala as the substrate, enzyme acylation is the rate-limiting step). In the presence of acceptor substrates such as glycine or a number of D-amino acids, transpeptidation will be observed (thus, N^α-acetyl-L-Lys-D-Ala-D-Ala reacts with R-NH$_2$ to produce N^α-acetyl-L-Lys-D-Ala-NH-R and D-alanine).[1,2]

[1] J.-M. Ghuysen, M. Leyh-Bouille, J. N. Campbell, R. Moreno, J.-M. Frère, C. Duez, M. Nieto & H. R. Perkins (1973) *Biochemistry* **12**, 1243.
[2] J.-M. Ghuysen, P. E. Reynolds, H. R. Perkins, J.-M. Frère & R. Moreno (1974) *Biochemistry* **13**, 2539.
[3] M. Jamin, J.-M. Wilkin & J.-M. Frère (1995) *Essays Biochem.* **29**, 1.
[4] J.-M. Frère, M. Leyh-Bouille, J.-M. Ghuysen, M. Nieto & H. R. Perkins (1976) *MIE* **45**, 610.

Selected entries from ***Methods in Enzymology*** [**vol**, page(s)]:
General discussion: 45, 610

ACTINOMYCIN LACTONASE

This enzyme [EC 3.1.1.39] catalyzes the hydrolysis of an actinomycin to produce an actinomycinic monolactone.[1,2]

[1] C. T. Hou & D. Perlman (1970) JBC **245**, 1289.
[2] R. J. Mehta, L. R. Fare, D. J. Newman & C. H. Nash (1978) Eur. J. Appl. Microbiol. Biotechnol. **5**, 165.

Selected entries from **Methods in Enzymology** [vol, page(s)]:
General discussion: 43, 763
Assay: 43, 764
Other molecular properties: Actinoplanes, **43**, 765, 766; ammonium sulfate fractionation, **43**, 765; calcium phosphate gel chromatography, **43**, 765, 766; crude extract preparation, **43**, 765; culture preparation, **43**, 764, 765; DEAE-cellulose column chromatography, **43**, 766; inactivation product, **43**, 767; properties, **43**, 767; purification, **43**, 764; reaction scheme, **43**, 763; Sephadex G-200 column chromatography, **43**, 766

ACTINOMYCIN SYNTHETASE II

This 4'-phosphopantetheine-containing enzyme catalyzes the ATP-dependent formation of the enzyme-bound actinomycin D precursor, 4-methyl-3-hydroxyanthraniloyl-L-threonyl-D-valine.[1,2] Interestingly, the enzyme uses L-valine as the substrate: hence, the enzyme also catalyzes an epimerization reaction of the thioester-bound peptide. The anthranilate component of the product is derived from the adenylate of 4-methyl-3-hydroxyanthranilate.

[1] A. Stindl & U. Keller (1993) JBC **268**, 10612.
[2] A. Stindl & U. Keller (1994) Biochemistry **33**, 9358.

ACULEACIN-A DEACYLASE

This enzyme [EC 3.5.1.70] catalyzes the hydrolysis of the amide bond in aculeacin A and related neutral lipopeptide antibiotics, releasing the long-chain fatty acid.[1]

[1] H. Takeshima, J. Inokoshi, Y. Takada, H. Tanaka & S. Omura (1989) J. Biochem. **105**, 606.

ACUTOLYSIN

This zinc metalloproteinase, often called Ac1-proteinase, is a hemorrhagic toxin isolated from the venom of the hundred-pace snake (Agkistrodon acutus) found in Taiwan and China.[1,2] It hydrolyzes the insulin B chain at Ala14–Leu15 and Tyr16–Leu17. The two other hemorrhagic enzymes in this snake venom are Ac2-proteinase and Ac3-proteinase. All three proteinases will act on casein. Ac3 will act on the insulin B chain at His10–Leu11, Ala14–Leu15, Tyr16–Leu17, and Phe24–Phe25.

[1] T. Nikai, H. Sugihara & T. Tanaka (1977) J. Pharm. Soc. Japan **97**, 507.
[2] T. Nikai, C. Kato, Y. Komori, H. Sugihara & M. Homma (1991) Int. J. Biochem. **23**, 311.

ACYL-[ACYL-CARRIER PROTEIN] DESATURASE

This oxidoreductase [EC 1.14.19.2; previously classified as EC 1.14.99.6], also known as stearyl-ACP desaturase or stearoyl-[acyl-carrier protein] Δ^9-desaturase, catalyzes the reaction of stearoyl-[acyl-carrier protein] with the reduced electron donor AH_2 and dioxygen to produce oleoyl-[acyl-carrier protein], the oxidized electron donor A, and two water molecules.[1–7]

The enzyme uses a diiron center to catalyze the desaturation reaction. The k_{cat}/K_m values for 18:0-ACP and 19:0-ACP for the Ricinus communis (castor) enzyme were similar, suggesting that the active-site channel past the diiron center can accommodate at least one more methylene group than is found in the natural substrate (note, however, that the V_{max} for palmitoyl-ACP is 100-fold lower than stearoyl-ACP while the K_m values are comparable).[2] The reaction appears to proceed via a μ-peroxo diferric intermediate.[3]

Other desaturases have been identified that catalyze the incorporation of a double bond at positions other than 9.

[1] T. A. McKeon & P. K. Stumpf (1982) JBC **257**, 12141.
[2] J. A. Haas & B. G. Fox (1999) Biochemistry **38**, 12833.
[3] J. A. Broadwater, C. Achim, E. Munck & B. G. Fox (1999) Biochemistry **38**, 12197.
[4] E. B. Cahoon, S. Shah, J. Shanklin & J. Browse (1998) Plant Physiol. **117**, 593.
[5] E. B. Cahoon, S. J. Coughlan & J. Shanklin (1997) J. Plant Mol. Biol. **33**, 1105.
[6] J. Browse & C. Somerville (1991) Ann. Rev. Plant Physiol. Plant Mol. Biol. **42**, 467.
[7] J. L. Harwood (1988) Ann. Rev. Plant Physiol. Plant Mol. Biol. **39**, 101.

Selected entries from **Methods in Enzymology** [vol, page(s)]:
Assay: safflower seeds, 71, 276
Other molecular properties: safflower seeds, from, 71, 275 (properties, **71**, 280; specificity, **71**, 280; stimulation by catalase, **71**, 280; synthesis of substrate, **71**, 277)

ACYL-[ACYL-CARRIER PROTEIN]: PHOSPHOLIPID O-ACYLTRANSFERASE

This acyltransferase [EC 2.3.1.40] catalyzes the reaction of an acyl-[acyl-carrier protein] (particularly, β-hydroxy-fatty acyl derivatives) with an O-(2-acyl-sn-glycero-3-phospho)ethanolamine to produce the free [acyl-carrier protein] and O-(1-β-acyl-2-acyl-sn-glycero-3-phospho)ethanolamine.[1,2] The enzyme has a role in lipid biosynthesis and in the incorporation of 3-hydroxymyristate into cell-wall lipids.

In *Escherichia coli*, the protein exhibiting an acyl-[acyl-carrier protein]:phospholipid *O*-acyltransferase activity is also reported to be an acyl-[acyl-carrier protein] synthetase.[3,4]

[1] S. S. Taylor & E. C. Heath (1969) *JBC* **244**, 6605.
[2] R. J. Heath & C. O. Rock (1998) *J. Bacteriol.* **180**, 1425.
[3] C. L. Cooper, L. Hsu, S. Jackowski & C. O. Rock (1989) *JBC* **264**, 7384.
[4] S. Jackowski, L. Hsu & C. O. Rock (1992) *MIE* **209**, 111.

Selected entries from *Methods in Enzymology* [vol, page(s)]:
Assay: radiochemical, **209**, 112

Other molecular properties: acyl-[acyl-carrier-protein] synthetase, with (properties, **209**, 116; purification, **209**, 114; radiochemical assay, **209**, 112)

ACYL-[ACYL-CARRIER PROTEIN] SYNTHETASE

This ligase [EC 6.2.1.20], also known as long-chain-fatty-acid:[acyl-carrier protein] ligase and long-chain-fatty-acyl-[acyl-carrier protein] synthetase, catalyzes the reaction of ATP with an acid and [acyl-carrier protein] to produce acyl-[acyl-carrier protein], AMP, and pyrophosphate (or, diphosphate). The enzyme has an apparent role in reintroducing free fatty acids into the biosynthetic pathway and in incorporating fatty acids into particular phospholipids. This ligase is distinct from long-chain-fatty-acyl-CoA synthetase [EC 6.2.1.3].

In *Escherichia coli*, the protein exhibiting an acyl-[acyl-carrier protein] synthetase activity is also reported to be a 2-acylglycerolphosphoethanolamine acyltransferase (or, acyl-[acyl-carrier protein]: phospholipid *O*-acyltransferase).[1,2] *See also Acyl-[Acyl-Carrier Protein]: Phospholipid O-Acyltransferase*

[1] C. L. Cooper, L. Hsu, S. Jackowski & C. O. Rock (1989) *JBC* **264**, 7384.
[2] S. Jackowski, L. Hsu & C. O. Rock (1992) *MIE* **209**, 111.

Selected entries from *Methods in Enzymology* [vol, page(s)]:
Assay: radiochemical, **209**, 112

Other molecular properties: 2-acylglycerophosphoethanolamine acyltransferase and (properties, **209**, 116; purification, **209**, 114; radiochemical assay, **209**, 112); assay of acyl carrier protein, **71**, 342; *Escherichia coli*, from, **71**, 163 (effect of chaotropic salts on activity, **71**, 168; properties, **71**, 167)

ACYL-[ACYL-CARRIER PROTEIN]: UDP-*N*-ACETYLGLUCOSAMINE *O*-ACYLTRANSFERASE

This acyltransferase [EC 2.3.1.129], also known as UDP-*N*-acetylglucosamine acyltransferase, catalyzes the reaction of (*R*)-3-hydroxytetradecanoyl-[acyl-carrier protein] with UDP-*N*-acetyl-D-glucosamine to produce the free acyl-carrier protein and UDP-3-*O*-(3-hydroxytetradecanoyl)-*N*-acetyl-D-glucosamine.[1–3]

UDP-*N*-acetylglucosamine

UDP-3-*O*-(3-hydroxytetradecanoyl)-*N*-acetylglucosamine

The enzyme participates with lipid-A-disaccharide synthase [EC 2.4.1.182] and tetraacyldisaccharide 4′-kinase [EC 2.7.1.130] in the biosynthesis of lipid A in the outer membrane of *Escherichia coli*.

[1] M. S. Anderson, C. E. Bulawa & C. R. Raetz (1985) *JBC* **260**, 15536.
[2] M. S. Anderson & C. R. Raetz (1987) *JBC* **262**, 5159.
[3] M. S. Anderson, H. G. Bull, S. M. Galloway, T. M. Kelly, S. Mohan, K. Radika & C. R. Raetz (1993) *JBC* **268**, 19858.

Selected entries from *Methods in Enzymology* [vol, page(s)]:
General discussion: 209, 449
Assay: 209, 449
Other molecular properties: *Escherichia coli* (assay, **209**, 449; properties, **209**, 454; purification, **209**, 452)

ACYLAGMATINE AMIDASE

This enzyme [EC 3.5.1.40] catalyzes the hydrolysis of benzoylagmatine to produce benzoate and agmatine. Alternative substrates include acetylagmatine (producing acetate and agmatine), propanoylagmatine (producing propanoate and agmatine), and bleomycin B2 (producing bleomycinate and agmatine).[1]

[1] Y. Takahashi, T. Shirai & S. Ishii (1975) *J. Biochem.* **77**, 823.

N-ACYL-D-AMINO-ACID DEACYLASE

This zinc-dependent enzyme [EC 3.5.1.81], also known as D-aminoacylase, catalyzes the hydrolysis of an *N*-acyl-D-amino acid to produce a carboxylate anion and a D-amino acid.[1] The enzyme exhibits broad substrate specificity. *N*-Acetyl-D-methionine, *N*-acetyl-D-phenylalanine, *N*-chloroacetyl-D-valine, and *N*-acetyl-D-leucine are good substrates whereas *N*-acetyl-D-valine is a weak substrate. L-Amino acid derivatives are not substrates. The enzyme has been used in the separation of D- and L-amino acids.

[1] Y.-C. Tsai, C.-P. Tseng, K.-M. Hsiao & L.-Y. Chen (1988) *App. Environ. Microbiol.* **54**, 984.

N-ACYLAMINO ACID RACEMASE

This racemase catalyzes the interconversion of an *N*-acyl-L-amino acid and an *N*-acyl-D-amino acid.[1–3]

Substrates include *N*-acetyl-L-methionine, *N*-chloroacetyl-L-phenylalanine, *N*-acetyl-L-tyrosine, and *N*-acetyl-L-valine.

[1]S. Tokuyama, K. Hatano & T. Takahashi (1994) *Biosci. Biotech. Biochem.* **58**, 24.
[2]S. Tokuyama, H. Miya, K. Hatano & T. Takahashi (1994) *Appl. Microbiol. Biotechnol.* **40**, 835.
[3]S. Tokuyama & K. Hatano (1995) *Appl. Microbiol. Biotechnol.* **42**, 853.

ACYLAMINOACYL-PEPTIDASE

This peptidase [EC 3.4.19.1], also known as acylamino-acid releasing enzyme and *N*-acylpeptide hydrolase, catalyzes the hydrolysis of an acylaminoacyl-peptide to generate an acylamino acid and the free peptide. Interestingly, N-terminally blocked proteins are not substrates for this peptidase whereas the blocked thirteen-mer α-melanocyte stimulating hormone is a substrate. Catalysis is most efficient when the P1 site is occupied by a seryl, alanyl, or methionyl residue. The enzyme is a poor catalyst if P1 is a glycyl, tyrosyl, aspartyl, asparaginyl, or prolyl residue.

A catalytic mechanism involving an acyl-enzyme intermediate is supported by the formation of acetyl-alanyl hydroxamate during the hydrolysis of acetylalanine *p*-nitroanilide in the presence of hydroxylamine. Hydrolysis of the acyl-enzyme intermediate is rate-limiting.[1]

[1]A. Scaloni, D. Barra, W. M. Jones & J. M. Manning (1994) *JBC* **269**, 15076.

Selected entries from *Methods in Enzymology* [vol, page(s)]:
Assay: **244**, 288; ninhydrin, by, **45**, 552
Other molecular properties: activators, **45**, 559; composition, **45**, 558, 560; distribution, **45**, 559; function, **45**, 552 (peptide sequence determination, for, **45**, 561); inhibitors, **45**, 559; **244**, 230; isoelectric point, **45**, 557; **244**, 230; kinetics, **45**, 557, 558; molecular weight, **45**, 557; pH optimum, **45**, 557; properties, **45**, 557; residues, **45**, 556, 557; **244**, 228; reaction catalyzed by, **244**, 227; reaction mechanism, **244**, 230; residues, terminal, **45**, 558; size, **244**, 230; specificity, **45**, 557; stability, **45**, 557; structure, subunit, **45**, 557; substrate specificity, **244**, 227, 230; synthetic substrate, **45**, 553

N-ACYL-D-ASPARTATE DEACYLASE

This zinc-dependent enzyme [EC 3.5.1.83], also known as *N*-acyl-D-aspartate amidohydrolase, catalyzes the hydrolysis of an *N*-acyl-D-aspartate to produce a carboxylate anion and D-aspartate.[1] *See also Aspartoacylase*

[1]M. Moriguchi, K. Sakai, Y. Katsuno, T. Maki & M. Wakayama (1993) *Biosci. Biotechnol. Biochem.* **57**, 1145.

ACYLCARNITINE HYDROLASE

This enzyme [EC 3.1.1.28] catalyzes the hydrolysis of an *O*-acylcarnitine to produce a free fatty acid anion and L-carnitine. The *O*-acylcarnitine substrates are six-to-eighteen carbon esters of L-carnitine, with the highest activity observed with *O*-decanoyl-L-carnitine. The enzyme's role appears to prevent the accumulation of cytotoxic esters. The rat liver enzyme is mainly associated with the endoplasmic reticulum.[1]

[1]R. Mentlein, H. Rix-Matzen & E. Heymann (1988) *BBA* **964**, 319.

[ACYL-CARRIER PROTEIN] ACETYLTRANSFERASE

This transferase [EC 2.3.1.38], also known as acetyl-CoA:ACP transacylase and acetyl-CoA:[acyl-carrier protein] *S*-acetyltransferase, catalyzes the transfer of an acetyl group from one acetyl-CoA to an acyl-carrier protein (ACP) to produce free Coenzyme A and the acetyl-[acyl-carrier protein]. Pantetheine can replace acyl-carrier protein. The *Escherichia coli* transferase exhibits a ping pong Bi Bi kinetic mechanism with an acetyl-enzyme intermediate.[1] **See also** *Fatty Acid Synthase*

[1]P. N. Lowe & S. Rhodes (1988) *BJ* **250**, 789.

Selected entries from *Methods in Enzymology* [vol, page(s)]:
General discussion: **14**, 50
Assay: **14**, 50; spinach leaves, **122**, 53
Other molecular properties: involvement in bacterial fatty acid synthesis, **14**, 41; spinach leaves (assay, **122**, 53; function, **122**, 58; properties, **122**, 56; purification, **122**, 54); properties, **14**, 52; purification, **14**, 51

[ACYL-CARRIER PROTEIN] S-MALONYLTRANSFERASE

This transferase [EC 2.3.1.39], also known as malonyl-CoA:[acyl-carrier protein] transacylase, catalyzes the reaction of malonyl-CoA with [acyl-carrier protein] to produce coenzyme A and malonyl-[acyl-carrier protein].

Mammalian fatty acid synthase is a dimer of identical multifunctional polypeptides that includes a malonyl-transferase activity. Complementation analysis indicates that the malonyltransferase domain of the complex can cooperate with the acyl-carrier protein domain of either subunit.[1]

[1]A. K. Joshi, A. Witkowski & S. Smith (1998) *Biochemistry* **37**, 2515.

Selected entries from *Methods in Enzymology* [vol, page(s)]:
General discussion: **14**, 53
Assay: **14**, 53
Other molecular properties: assay for acyl carrier protein, **14**, 44; involvement in bacterial fatty acid synthesis, **14**, 41; mutant, **72**, 694; properties, **14**, 56; purification, **14**, 54

[ACYL-CARRIER PROTEIN] PHOSPHODIESTERASE

This phosphodiesterase [EC 3.1.4.14] catalyzes the hydrolysis of [acyl-carrier protein] (*i.e.*, holo-[acyl-carrier protein]) to produce 4'-phosphopantetheine and the apoprotein.[1,2]

[1] P. R. Vagelos & A. R. Larrabes (1967) *JBC* **242**, 1776.
[2] A. S. Fischl & E. P. Kennedy (1990) *J. Bacteriol.* **172**, 5445.

Selected entries from *Methods in Enzymology* [vol, page(s)]:
General discussion: 14, 81
Assay: 14, 81
Other molecular properties: metal ion requirement, 14, 83; properties, 14, 83; 62, 256; purification, 14, 82 (rat liver, 62, 255, 256)

ACYL-CoA:6-AMINOPENICILLANATE ACYLTRANSFERASE

This acyltransferase, also known as acyl-CoA:6-aminopenicillanic acid acyltransferase and 6-aminopenicillanate acyltransferase, catalyzes the reaction of 6-aminopenicillanate with an acyl-CoA to produce a penicillin and free Coenzyme A. This is the last step in the biosynthesis of penicillins. Because of the somewhat broad specificity for the acyl-CoA substrate, a number of penicillins can be prepared. The kinetic mechanism is likely ping pong Bi Bi.[1–3]

[1] S. Gatenbeck (1975) *MIE* **43**, 474.
[2] M. J. Alonzo, F. Bermejo, A. Reglero, J. M. Fernández-Cañón, G. González de Buitrago & J. M. Luengo (1988) *J. Antibiot.* **41**, 1074.
[3] J. M. Luengo (1995) *J. Antibiot.* **48**, 1195.

Selected entries from *Methods in Enzymology* [vol, page(s)]:
General discussion: 43, 474, 476, 485
Assay: 43, 474, 475
Other molecular properties: activators, 43, 476; ammonium sulfate precipitation, 43, 475; hydroxyapatite gel adsorption and elution, 43, 475; *Penicillium chrysogenum* preparation, 43, 475; pH effect, 43, 476; phenylacetyl-6-aminopenicillanate hydrolase, and, 43, 485; purification, 43, 475, 476; Sephadex G-200 column chromatography, 43, 475; specific activity, 43, 474; specificity, 43, 476; unit definition, 43, 474

ACYL-CoA DEHYDROGENASES

A number of enzymes with this name catalyze the reversible formation of a 2,3-dehydroacyl-CoA derivative from the acyl-CoA substrate.

Acyl-CoA dehydrogenase (NADP+) [EC 1.3.1.8], also known as 2-enoyl-CoA reductase and crotonyl-CoA reductase, uses NADP+ as the coenzyme. The liver protein acts on enoyl-CoA substrates having a carbon chain length of four-to-sixteen carbon atoms, with an optimal activity with 2-hexenoyl-CoA. In *Escherichia coli*, both *cis*-specific and *trans*-specific enzymes exist (*cis*-2-enoyl-CoA reductase (NADPH) [EC 1.3.1.37] and *trans*-2-enoyl-CoA reductase (NADPH) [EC 1.3.1.38]). In other sources, EC 1.3.1.8 acts on both *cis* and *trans* substrates, *trans* more active than *cis*. An active-site cysteinyl residue is present.[1] *See also trans-2-Enoyl-CoA Reductase (NADPH); cis-2-Enoyl-CoA Reductase (NADPH)*

Medium-chain acyl-CoA dehydrogenase [EC 1.3.99.3; formerly classified as EC 1.3.2.2], often known as simple acyl-CoA dehydrogenase, is FAD-dependent and utilizes an electron-transferring flavoprotein instead of NADP+. This enzyme is a component of a system to reduce ubiquinone and other acceptors. Binding of the acyl-CoA substrate results in production of the reduced form of the enzyme which is reoxidized by the electron-transferring flavoprotein. The enzyme, which has an ordered kinetic mechanism, abstracts the α-proton from certain weakly acidic acyl-CoA derivatives, without the concerted transfer of a hydride equivalent to the flavin.[2,3]

A short-chain acyl-CoA dehydrogenase (butyryl-CoA dehydrogenase, [EC 1.3.99.2]) and a long-chain enzyme [EC 1.3.99.13] have also been identified.[3] *See also Long-Chain Acyl-CoA Dehydrogenase; Butyryl-CoA Dehydrogenase*

[1] K. A. Strom, W. L. Galeos, L. A. Davidson & S. Kumar (1979) *JBC* **254**, 8153.
[2] S. M. Lau, R. K. Brantley & C. Thorpe (1988) *Biochemistry* **27**, 5089.
[3] B. A. Palfey & V. Massey (1998) *CBC* **3**, 83.

Selected entries from *Methods in Enzymology* [vol, page(s)]:
General discussion: 5, 546; 53, 502; 54, 143; 72, 308
Assay: 5, 546; pig liver mitochondria, 71, 376
Other molecular properties: activity, 53, 502; acyl dehydrogenase (reduced), electron-transferring flavoprotein as electron acceptor, 14, 115; beef heart, 5, 554; 53, 506, 507, 510, 518; beef liver, 5, 552; bromooctanoate, effect of, 72, 568; brown adipose tissue mitochondria, 55, 76; catalytic properties, 53, 506, 507; deficiency, in *Escherichia coli* mutants, 32, 859; electrophoretic properties, 53, 508, 509, 512; flavin content, 53, 508; glyoxysomes, in, 52, 502; long-chain, 53, 402 (mutant, 72, 695); medium chain, 5, 546; 71, 375; octanoyl-CoA dehydrogenase (assay, 5, 549; electrophoresis of, 5, 551); optical properties, 53, 504, 505, 508, 510, 511; pig kidney, from, 71, 366 (electron-transfer flavoprotein, 71, 366; flavoprotein, 71, 366; inhibitors, 71, 372; properties, 71, 372; specificity, 71, 374); pig liver, 5, 549; 53, 506 (mitochondria, 71, 375, 378, 386 [acyl-CoA-reducible flavin, 71, 379; assay, 71, 376; DCPIP assay, 71, 382, 386; other sources, 71, 379; PMS mediated and ETF mediated DCPIP reduction, 71, 382; properties, 71, 384; substrate-product inhibition, 71, 385; thioesterase activity, 71, 384; turnover numbers, 71, 385]); short chain, 71, 375 (mutant, 72, 695); stability, 53, 518; stereochemistry, 87, 110; substrate specificities, 53, 502, 503

ACYL-CoA HYDROLASE

This enzyme [EC 3.1.2.20] catalyzes the hydrolysis of an acyl-CoA to produce free coenzyme A and a carboxylate anion, with broad specificity for medium- to long-chain acyl-CoA derivatives. Depending on the source, this hydrolase acts on octadecanoyl-CoA, hexadecanoyl-CoA, tetradecanoyl-CoA, dodecanoyl-CoA, *n*-decanoyl-CoA, octanoyl-CoA, hexanoyl-CoA, *n*-butyryl-CoA, acetoacetyl-CoA, acetyl-CoA, and succinyl-CoA. Acyl-CoA hydrolase is insensitive to NAD^+. *See ADP-Dependent Medium-Chain Acyl-CoA Hydrolase*

Selected entries from *Methods in Enzymology* [**vol**, page(s)]:
Other molecular properties: induction by clofibrate and di(2-ethylhexyl)phthalate, **72**, 507; propionyl-CoA carboxylase preparation, not present in, **5**, 575

ACYL-CoA OXIDASE

This FAD-dependent oxidoreductase [EC 1.3.3.6] catalyzes the reaction of an acyl-CoA with dioxygen to produce *trans*-2,3-dehydroacyl-CoA and hydrogen peroxide. This enzyme, which catalyzes the first step in peroxisomal β-oxidation, can act on acyl-CoA derivatives with a chain length of eight-to-eighteen carbon atoms. The plant glyoxisomal enzyme appears to be more selective in its substrate profile.[1] In fact, there appear to be a number of different acyl-CoA oxidases in both peroxisomes and glyoxisomes. In the rat liver and *Candida lipolytica* enzymes, the reaction proceeds via *anti*-elimination of the *pro*-2*R* and *pro*-3*R* hydrogens of the acyl-CoA derivative.[2]

[1] T. Kirsch, H. G. Loffler & H. Kindl (1986) *JBC* **261**, 8570.
[2] A. Kawaguchi, S. Tsubotani, Y. Seyama, T. Yamakawa, T. Osumi, T. Hashimoto, T. Kikuchi, M. Ando & S. Okuda (1980) *J. Biochem.* **88**, 1481.

Selected entries from *Methods in Enzymology* [**vol**, page(s)]:
Assay: peroxisomal, assay, **148**, 523
Other molecular properties: coupled assay for acyl-CoA synthetase, **250**, 423, 425; glyoxysomes, in, **31**, 369; induction by clofibrate and di(2-ethylhexyl)phthalate, **72**, 507; substrate specificity, **250**, 425

ACYL-CoA SYNTHETASE (GDP-FORMING)

This ligase [EC 6.2.1.10], also known as acid:CoA ligase (GDP-forming), catalyzes the reaction of GTP with an acid and Coenzyme A to produce GDP, orthophosphate, and an acyl-CoA.[1,2] *See also Long-Chain Fatty-Acyl-CoA Synthetase; Butyryl-CoA Synthetase*

[1] J. C. Londesborough & L. T. Webster (1974) *The Enzymes*, 3rd. ed., **10**, 469.
[2] M. V. Park (1978) *Biochem. Soc. Trans.* **6**, 70.

Selected entries from *Methods in Enzymology* [**vol**, page(s)]:
General discussion: **14**, 91, 622
Assay: **14**, 91, 622 (nitroprusside assay, **14**, 92)
Other molecular properties: activators and inhibitors, **14**, 95; assay methods, summary, **14**, 91; nitroprusside assay, **14**, 92; specificity, **14**, 94

N-ACYLGLUCOSAMINE 2-EPIMERASE

This epimerase [EC 5.1.3.8], also known as *N*-acetylglucosamine 2-epimerase, catalyzes the interconversion of *N*-acylglucosamine and *N*-acylmannosamine. ATP play an essential allosteric activator role rather than participating directly in catalysis.[1]

[1] I. Maru, J. Ohnishi, Y. Ohta, W. Hashimoto, Y. Tsukada, K. Murata & B. Mikami (1996) *J. Biochem.* **120**, 481

Selected entries from *Methods in Enzymology* [**vol**, page(s)]:
General discussion: **8**, 191; **41**, 407
Assay: **8**, 191; **41**, 408
Other molecular properties: ATP dependence, **8**, 191; **41**, 407; properties, **8**, 194; **41**, 411; purification, **8**, 192; **41**, 409

N-ACYLGLUCOSAMINE-6-PHOSPHATE 2-EPIMERASE

This epimerase [EC 5.1.3.9], also known as *N*-acetylglucosamine-6-phosphate 2-epimerase, catalyzes the interconversion of *N*-acylglucosamine 6-phosphate and *N*-acylmannosamine 6-phosphate.

Selected entries from *Methods in Enzymology* [**vol**, page(s)]:
General discussion: **8**, 185
Assay: **8**, 185, 186

N-ACYL-D-GLUTAMATE DEACYLASE

This zinc-dependent enzyme [EC 3.5.1.82], also known as *N*-acyl-D-glutamate amidohydrolase, catalyzes the hydrolysis of an *N*-acyl-D-glutamate to produce a carboxylate anion and D-glutamate.[1]

[1] M. Wakayama, T. Ashika, Y. Miyamoto, T. Yoshikawa, Y. Sonoda, K. Sakai & M. Moriguchi (1995) *J. Biochem.* **118**, 204.

2-ACYLGLYCEROL O-ACYLTRANSFERASE

This transferase [EC 2.3.1.22], also known as acylglycerol palmitoyltransferase and monoglyceride acyltransferase, catalyzes the reaction of an acyl-CoA with a 2-acylglycerol to produce free coenzyme A and an *sn*-1,2-diacylglycerol.[1,2] The stereospecificity depends on the source. The 1-monoacylglycerol can act as a substrate; however, the 2-adduct is preferred. The specificity for the acyl-CoA substrate is rather broad.

The dependence on palmitoyl-CoA with the liver enzyme was highly cooperative (Hill constant > 2.4). An altered K_m value for palmitoyl-CoA in the presence of fatty acid or anionic phospholipid suggests that both long-chain fatty acids and phospholipid cofactors may induce a conformational change in the transferase.[3]

[1] R. M. Bell & R. A. Coleman (1983) *The Enzymes*, 3rd ed., **16**, 87.
[2] R. A. Coleman (1992) *MIE* **209**, 98.
[3] R. A. Coleman, P. Wang & B. G. Bhat (1996) *Biochemistry* **35**, 9576.

Selected entries from *Methods in Enzymology* [vol, page(s)]:
General discussion: 209, 98
Assay: 209, 102
Other molecular properties: activity in intestine and liver, **209**, 98; activity of phospholipase A₁, **71**, 684

ACYLGLYCEROL KINASE

This phosphotransferase [EC 2.7.1.94], also known as monoacylglycerol kinase, catalyzes the reaction of ATP with an acylglycerol to produce ADP and an acyl-*sn*-glycerol 3-phosphate. Both 1- and 2-monoacylglycerols will act as substrates. The bovine brain enzyme shows a preference for substrates with unsaturated fatty acylglycerols except for 1- or 2-monostearin.[1]

[1] Y. H. Shim, C. H. Lin & K. P. Strickland (1989) *Biochem. Cell Biol.* **67**, 233.

ACYLGLYCEROL LIPASE

This enzyme [EC 3.1.1.23], also known as monoacylglycerol lipase and monoglyceride lipase, catalyzes the hydrolysis of glycerol monoesters of long-chain fatty acids to produce glycerol and the free fatty acid anion.[1–4] The human erythrocyte enzyme exhibits kinetics typical of an interfacial lipolytic enzyme, with optimal activity on emulsified substrate particles.[5]

[1] P. Belfrage, G. Fredrikson, P. Stralfors & H. Tornqvist (1984) in *Lipases* (B. Borgström & H. L. Brockman, eds.), p. 365, Elsevier, Amsterdam.
[2] J. C. Khoo & D. Steinberg (1981) *MIE* **71**, 627.
[3] H. Tornqvist & P. Belfrage (1981) *MIE* **71**, 646.
[4] B. A. Wolf (1998) *Meth. Mol. Biol.* **105**, 167.
[5] C. Somma-Delpero, A. Valette, J. Lepetit-Thevenin, O. Nobili, J. Boyer & A. Verine (1995) *BJ* **312** (Pt 2), 519.

Selected entries from *Methods in Enzymology* [vol, page(s)]:
General discussion: 71, 627, 646; **77**, 334
Assay: 71, 646; **197**, 317
Other molecular properties: chicken adipose tissue, **71**, 627, 633, 634; enzymatic activity, **31**, 522; rat adipose tissue, **71**, 646 (activation by lipid–water interphases, **71**, 652; assay, **71**, 646; properties, **71**, 651; specificity, **71**, 651; true lipase, **71**, 652); role in arachidonate liberation from phospholipids, **191**, 677

1-ACYLGLYCEROL-3-PHOSPHATE O-ACYLTRANSFERASE

This acyltransferase [EC 2.3.1.51] catalyzes the reaction of an acyl-CoA derivative with 1-acyl-*sn*-glycerol 3-phosphate to generate coenzyme A and 1,2-diacyl-*sn*-glycerol 3-phosphate.[1,2] The enzyme is very specific for the 1-acyl derivative. The animal enzyme is also very specific for the transfer of unsaturated fatty acyl groups. Interestingly, the acylated acyl-carrier protein can act as the acyl donor (particularly in phospholipid biosynthesis in chloroplasts). Inhibition by 1,2-diacylglycerol of the rat liver enzyme was essentially uncompetitive with respect to the acyl-CoA substrate.[3]

[1] R. M. Bell & R. A. Coleman (1983) *The Enzymes*, 3rd ed., **16**, 87.
[2] S. Yamashita, K. Hosaka, Y. Miki & S. Numa (1981) *MIE* **71**, 528.
[3] J. J. Mukherjee, P. G. Tardi & P. C. Choy (1992) *BBA* **1123**, 27.

Selected entries from *Methods in Enzymology* [vol, page(s)]:
General discussion: 71, 528, 550, 551
Other molecular properties: acyl-acceptor specificities, **71**, 533; acyl-donor specificities, **71**, 533; *Escherichia coli* auxotrophs, in, **32**, 864; phosphatidic acid biosynthesis, **69**, 299, 301; product of glycerophosphate acyltransferase, **71**, 550; properties, **71**, 533; rat liver, from, **71**, 528; stabilization with ethylene glycol, **71**, 531; substrate for 1-acylglycerophosphate acyltransferase, **71**, 529

2-ACYLGLYCEROL-3-PHOSPHATE O-ACYLTRANSFERASE

This acyltransferase [EC 2.3.1.52] catalyzes the reaction of an acyl-CoA with a 2-acyl-*sn*-glycerol 3-phosphate to produce free coenzyme A and a 1,2-diacyl-*sn*-glycerol 3-phosphate. Saturated acyl-CoA derivatives are the most effective acyl donors; however, specificities vary with the source.[1]

[1] S. M. Cooper & M. R. Grigor (1980) *BJ* **187**, 289.

Selected entries from *Methods in Enzymology* [vol, page(s)]:
General discussion: 71, 536

ACYLGLYCERONE-PHOSPHATE REDUCTASE

This oxidoreductase [EC 1.1.1.101], also known as palmitoyldihydroxyacetone-phosphate reductase, catalyzes the reaction of palmitoylglycerone phosphate with NADPH to produce 1-palmitoylglycerol 3-phosphate and NADP⁺. Most of the rat and guinea pig liver enzymes are located in the peroxisomes.[1] The guinea pig enzyme will also catalyze the reduction of hexadecyldihydroxyacetone phosphate, indicating that the enzyme is actually an acyl/alkyl dihydroxyacetone-phosphate (*i.e.*, acyl/alkylglycerone-phosphate) reductase.[2,3]

[1] M. K. Ghosh & A. K. Hajra (1986) *ABB* **245**, 523.
[2] S. C. Datta, M. K. Ghosh & A. K. Hajra (1990) *JBC* **265**, 8268.
[3] A. K. Hajra, S. C. Datta & M. K. Ghosh (1992) *MIE* **209**, 402.

Selected entries from *Methods in Enzymology* [vol, page(s)]:
General discussion: 209, 402.
Assay: 209, 402
Other molecular properties: hepatic peroxisomal, guinea pig (assay, **209**, 402; properties, **209**, 406; purification, **209**, 404); stereospecificity, **87**, 111

1-ACYLGLYCEROPHOSPHOCHOLINE O-ACYLTRANSFERASE

This acyltransferase [EC 2.3.1.23], also called lysolecithin acyltransferase and lysophosphatidylcholine acyltransferase, catalyzes the reaction of an acyl-CoA derivative with 1-acyl-sn-glycero-3-phosphocholine to produce coenzyme A and 1,2-diacyl-sn-glycero-3-phosphocholine. The enzyme preferentially acts on unsaturated acyl-CoA donor substrates. 1-Acyl-sn-glycero-3-phosphoinositol can also act as the acceptor substrate.

Selected entries from **Methods in Enzymology** [vol, page(s)]:
General discussion: 71, 528
Assay: 129, 790; radioassay, 209, 81
Other molecular properties: acyl-acceptor specificities, 71, 533; acyl specificity, 71, 533; 209, 84; human plasma (assays, 129, 790; properties, 129, 794, 795); kinetics, 209, 85; properties, 71, 533; purification, 209, 83; radioassay, 209, 81; rat liver, from, 71, 528; regulation, 209, 85; stabilization with ethylene glycol, 71, 531 subcellular localization, 209, 83

2-ACYLGLYCEROPHOSPHOCHOLINE O-ACYLTRANSFERASE

This transferase [EC 2.3.1.62] catalyzes the reaction of an acyl-CoA with 2-acyl-sn-glycero-3-phosphocholine to produce free coenzyme A and a phosphatidylcholine. The guinea pig heart enzyme did not exhibit a distinct preference for saturated fatty acids.[1]

[1] G. Arthur (1989) BJ 261, 575.

N-ACYLHEXOSAMINE OXIDASE

This oxidoreductase [EC 1.1.3.29] catalyzes the reaction of N-acetyl-D-glucosamine with dioxygen to produce N-acetyl-D-glucosaminate and hydrogen peroxide. Alternative substrates include N-glycolyl-D-glucosamine, N-acetyl-D-galactosamine, and N-acetyl-D-mannosamine (which is a relatively poor substrate). The enzyme is specific for the β-anomer of N-acetyl-D-glucosamine.[1] **See also** N-Acetylhexosamine Dehydrogenase

[1] T. Horiuchi (1989) Agric. Biol. Chem. 53, 361.

ACYL-LYSINE DEACYLASE

This deacylase [EC 3.5.1.17] catalyzes the hydrolysis of N^6-acyl-L-lysine to generate a carboxylate anion and L-lysine.[1,2]

[1] I. Chibata, T. Ishikawa & T. Tosa (1970) MIE 19, 756.
[2] J. Leclerc & L. Benoiton (1968) Can. J. Biochem. 46, 471.

Selected entries from **Methods in Enzymology** [vol, page(s)]:
General discussion: 19, 756
Assay: Achromobacter pestifer, 19, 756
Other molecular properties: Achromobacter pestifer, from, 19, 756 (assay method for, 19, 756; distribution, 19, 762; inhibitors and

activators, 19, 760; kinetics, 19, 761; pH optimum, 19, 760; physical properties and purity, 19, 760; properties, 19, 759; purification procedure, 19, 757, 760; specificity, 19, 761; stability, 19, 759); various sources, from, 19, 762

N-ACYLMANNOSAMINE 1-DEHYDROGENASE

This oxidoreductase [EC 1.1.1.233] catalyzes the reaction of an N-acyl-D-mannosamine with NAD^+ to produce an N-acyl-D-mannosaminolactone and NADH. Acylated substrates include N-acetyl-D-mannosamine and N-glycolyl-D-mannosamine.[1]

[1] T. Horiuchi & T. Kurokawa (1988) J. Biochem. 104, 466.

N-ACYLMANNOSAMINE KINASE

This phosphotransferase [EC 2.7.1.60], also known as N-acetylmannosamine kinase, catalyzes the ATP-dependent phosphorylation of N-acyl-D-mannosamine to produce ADP and N-acyl-D-mannosamine 6-phosphate. The enzyme can act on both the acetyl and the glycolyl derivatives.

The mouse, rat, and human enzyme is bifunctional (UDP-N-acetylglucosamine-2-epimerase/N-acetylmannosamine kinase), participating in neuraminate biosynthesis.[1–3]

[1] L. Lucka, M. Krause, K. Danker, W. Reutter & R. Horstkorte (1999) FEBS Lett. 454, 341.
[2] R. Horstkorte, S. Nohring, N. Wiechens, M. Schwarzkopf, K. Danker, W. Reutter & L. Lucka (1999) EJB 260, 923.
[3] S. Hinderlich, S. Nohring, C. Weise, P. Franke, R. Stasche & W. Reutter (1998) EJB 252, 133.

Selected entries from **Methods in Enzymology** [vol, page(s)]:
General discussion: 8, 195; 42, 53
Assay: 8, 195; 42, 54
Other molecular properties: chromatography, 42, 57; properties, 8, 200; 42, 57; purification, 8, 199; 42, 56

ACYLMURAMOYL-ALANINE CARBOXYPEPTIDASE

This enzyme, previously been classified as EC 3.4.17.7 and EC 3.4.19.10, is now listed under N-acetylmuramoyl-L-alanine amidase [EC 3.5.1.28].

ACYLNEURAMINATE CYTIDYLYLTRANSFERASE

This transferase [EC 2.7.7.43], also known as CMP-N-acetylneuraminate synthetase and CMP-sialate synthase, catalyzes the conversion of CTP and N-acylneuraminate to produce pyrophosphate (or, diphosphate) and CMP-N-acylneuraminate. The protein will act on both the N-acetyl and N-glycolyl derivatives. It has a requirement for

Mg^{2+} or Mn^{2+} ions. The *Haemophilus ducreyi* enzyme has an ordered Bi Bi kinetic mechanism.[1] The rate-limiting step appears to be the desorption of the final product, CMP-*N*-acylneuraminate.

[1] N. M. Samuels, B. W. Gibson & S. M. Miller (1999) *Biochemistry* **38**, 6195.

Selected entries from *Methods in Enzymology* [vol, page(s)]:
General discussion: 8, 208; 28, 413
Assay: 8, 208; 28, 413
Other molecular properties: chromatography, 28, 420; distribution, 8, 214; isolation from liver, 28, 415; nucleotide specificity, 28, 419; properties, 8, 213; purification, 8, 211; reaction, 28, 413; sources, 28, 413, 418; subcellular distribution, 28, 420; synthesis of CMP-activated analogs, 247, 163; synthesis of sialic acid analogues, 247, 163, 180; synthetic sialylation system, 247, 114

N-ACYLNEURAMINATE 9-PHOSPHATASE

This enzyme [EC 3.1.3.29] catalyzes the hydrolysis of *N*-acylneuraminate 9-phosphate to produce *N*-acylneuraminate and orthophosphate.[1]

[1] J. Van Rinsum, W. Van Dijk, G. J. Hooghwinkel & W. Ferwerda (1984) *BJ* **223**, 323

Selected entries from *Methods in Enzymology* [vol, page(s)]:
General discussion: 8, 205
Assay: 8, 205

N-ACYLNEURAMINATE-9-PHOSPHATE SYNTHASE

This enzyme [EC 4.1.3.20] catalyzes the reaction of *N*-acyl-D-mannosamine 6-phosphate with phosphoenolpyruvate and water to produce *N*-acylneuraminate 9-phosphate and orthophosphate. The reverse reaction has not been observed.

Selected entries from *Methods in Enzymology* [vol, page(s)]:
General discussion: 8, 201
Assay: 8, 201

ACYLPHOSPHATASE

This enzyme [EC 3.6.1.7], also known as acetylphosphatase, catalyzes the irreversible hydrolysis of a wide spectrum of acyl phosphates to yield the free acid anion and orthophosphate. This enzyme will not catalyze transphosphorylation between substrate and glycerol and methanol.

Acylphosphatase fails to catalyze an [^{18}O]water–orthophosphate exchange, suggesting that no phosphoenzyme intermediate forms. Enzymic hydrolysis of benzoyl phosphate in the presence of [^{18}O]water leads to isotope incorporation into orthophosphate (and not benzoate),

indicating P–O, and not C–O, bond cleavage. The enzyme is competitively inhibited by orthophosphate, but is not inhibited by the carboxylate anions, suggesting the last step of catalysis may be the release of orthophosphate.[1]

[1] P. Paoli, P. Cirri, L. Camici, G. Manao, G. Cappugi, G. Moneti, G. Pieraccini, G. Camici & G. Ramponi (1997) *BJ* **327**, 177.

Selected entries from *Methods in Enzymology* [vol, page(s)]:
General discussion: 2, 555; 6, 324
Assay: 2, 556; 6, 324
Other molecular properties: acetyl phosphate assay and, 2, 650; distribution, 2, 555; 6, 327; kinetic parameters, 6, 327; phosphorus oxygen bond and, 2, 556; plasma membranes, in, 31, 88; preparation, 2, 555; 6, 325; properties, 2, 555; 6, 326; skeletal muscle, 6, 324; substrate specificity, 2, 554; 6, 326

ACYL PHOSPHATE:HEXOSE PHOSPHOTRANSFERASE

This phosphotransferase [EC 2.7.1.61], also known as hexose phosphate:hexose phosphotransferase, catalyzes the reaction of an acyl phosphate with a hexose to produce a carboxylate anion and a hexose phosphate. If the sugar is D-glucose or D-mannose, phosphorylation is on O6. If the sugar is D-fructose, phosphorylation is on O1 or O6. The enzyme has a random Bi Bi kinetic mechanism.[1]

[1] J. P. Casazza & H. J. Fromm (1977) *Biochemistry* **16**, 3091.

Selected entries from *Methods in Enzymology* [vol, page(s)]:
General discussion: 9, 392
Assay: 9, 392
Other molecular properties: *Aerobacter* (now *Klebsiella*) *aerogenes*, from, (kinetic constants, 9, 395; pH optimum, 9, 395; properties, 9, 395; purification, 9, 393; specificity, 9, 395; stability, 9, 396); mechanism, 87, 324

5′-ACYLPHOSPHOADENOSINE HYDROLASE

This enzyme [EC 3.6.1.20] catalyzes the hydrolysis of a 5′-acylphosphoadenosine to produce AMP and a fatty acid anion. Alternative substrates include acylphosphoinosine and acylphosphouridine.[1]

[1] G. M. Kellerman (1959) *BBA* **33**, 101.

ACYLPYRUVATE HYDROLASE

This enzyme [EC 3.7.1.5] catalyzes the hydrolysis of a 3-acylpyruvate to produce a fatty acid anion and pyruvate. Alternative substrates include formylpyruvate, 2,4-dioxopentanoate, 2,4-dioxohexanoate, and 2,4-dioxoheptanoate.[1]

[1] G. K. Watson, C. Houghton & R. B. Cain (1974) *BJ* **140**, 277.

N-ACYLSPHINGOSINE GALACTOSYLTRANSFERASE

This transferase [EC 2.4.1.47], also known as ceramide galactosyltransferase, catalyzes the reaction of UDP-D-galactose with an N-acylsphingosine (i.e., a ceramide) to produce a D-galactosylceramide and UDP.[1]

[1]Y. Fujino & M. Nakano (1969) BJ 113, 573.

Selected entries from **Methods in Enzymology** [vol, page(s)]:
General discussion: 311, 73
Assay: 138, 553; 311, 73
Other molecular properties: glycosphingolipid-synthesizing, assays, 138, 581; isolation from chick embryo brain, 28, 492; testosterone effects on, 28, 493

Acyl RESTRICTION ENDONUCLEASE

This type II restriction endonuclease [EC 3.1.21.4] is obtained from Anabaena cylindrica and acts on both strands of DNA at 5′ . . . GR∇CGYC . . . 3′, where R represents A or G and Y represents T or C.[1,2]

[1]A. de Waard, J. Korsuize, C. P. van Beveren & J. Maat (1978) FEBS Lett. 96, 106.
[2]R. J. Roberts (1980) MIE 65, 1.

Selected entries from **Methods in Enzymology** [vol, page(s)]:
Other molecular properties: recognition sequence, 65, 2

ADAMALYSIN

This zinc metalloproteinase [EC 3.4.24.46] was first isolated from the venom of the eastern diamondback rattlesnake, Crotalus adamanteus. Two forms were initially identified and known as proteinase I and proteinase II.[1] Proteinase II is the form usually regarded as adamalysin (or adamalysin II). Neither enzyme elicits a hemorrhagic response in mice or rabbits (such a response obtained from the snake venom is due to proteinase H). Interestingly, adamalysin inactivates human serpins with a single peptide bond cleavage (in α_1-proteinase inhibitor, at Ala350–Met351, and in antithrombin III, at Ala375–Ser376).[2] The insulin B chain is hydrolyzed at Phe1–Val2, His5–Leu6, Ala14–Leu15, Leu15–Tyr16, and Tyr16–Leu17.[1]

[1]T. Kurecki, M. Laskowski & L. F. Kress (1978) JBC 253, 8340.
[2]L. F. Kress (1986) J. Cell. Biochem. 32, 51.

Selected entries from **Methods in Enzymology** [vol, page(s)]:
Other molecular properties: adamalysin II, structure, 248, 448, 633; calcium binding site, 248, 357; cleavage specificity on insulin B chain, 248, 361; isozymes, 248, 349; properties, 248, 194, 200, 351, 356, 370; structure, 248, 191, 198, 346, 361, 366

ADENINE DEAMINASE

This enzyme [EC 3.5.4.2] (also called adenase, adenine aminohydrolase, and adenine aminase) catalyzes the hydrolysis of adenine to produce hypoxanthine and ammonia.

In Bacillus subtilis, adenine deaminase is the only enzyme that can deaminate adenine. The uptake of adenine is strictly coupled to its further metabolism. The concentration of adenine deaminase decreases when exogenous guanosine serves as the purine source or when L-glutamine is the nitrogen source.[1]

[1]P. Nygaard, P. Duckert & H. H. Saxild (1996) J. Bacteriol. 178, 846.

Selected entries from **Methods in Enzymology** [vol, page(s)]:
General discussion: 6, 203
Assay: 6, 203
Other molecular properties: adenylic acid ribosidase and, 6, 118, 121; bacteria, in, 51, 263; properties, 6, 207; purification, 6, 205

ADENINE PHOSPHORIBOSYLTRANSFERASE

This transferase [EC 2.4.2.7], also known as AMP pyrophosphorylase and transphosphoribosidase, catalyzes the reaction of adenine with 5-phospho-α-ribose 1-diphosphate to produce AMP and pyrophosphate (or, diphosphate). 5-Amino-4-imidazolecarboxamide can act as an alternative acceptor substrate. The equilibrium lies significantly in favor of nucleotide formation.[1–4] The Escherichia coli[1] and Saccharomyces cerevisiae[5] enzymes catalyze a ping pong Bi Bi kinetic mechanism.

[1]J. Hochstadt (1978) MIE 51, 558.
[2]W. J. Arnold & W. N. Kelley (1978) MIE 51, 568.
[3]D. P. Groth, L. G. Young & J. G. Kenimer (1978) MIE 51, 574.
[4]J. G. Flaks (1963) MIE 6, 136.
[5]J. D. Alfonzo, A. Sahota & M. W. Taylor (1997) BBA 1341, 173.

Selected entries from **Methods in Enzymology** [vol, page(s)]:
General discussion: 6, 136; 51, 558, 568, 574
Assay: 6, 136; 51, 559, 568, 574, 575
Other molecular properties: activity, 51, 558, 559; amino acid analysis, 51, 580; 5-amino-4-imidazolecarboxamide ribotide synthesis and, 6, 696; 5′-amino-1-ribosyl-4-imidazolecarboxamide 5′-phosphate synthesis, in, 51, 189; cation requirement, 51, 556, 573; cellular localization, 51, 559, 568; Escherichia coli, from, 51, 558; effectors, 51, 566, 573; human erythrocytes, from, 51, 568; isoelectric point, 51, 573, 579, 580; kinetic properties, 51, 565, 573, 580; marker for chromosome detection in somatic cell hybrids, as, 151, 178; mechanism, 63, 51; molecular weight, 51, 565, 573, 579; monkey liver, from, 51, 574; mouse cells deficient in, in cell hybridization studies, 32, 578; pH optimum, 51, 566, 574, 580; physiological function, 51, 566, 567; preparation, 6, 139, 697; properties, 6, 143; 51, 567; PRPP synthetase assay, in, 51, 5; purification, 6, 139; 51, 554, 561, 570, 575; purity, 51, 565, 579; rat liver, from, 51, 574; reaction stoichiometry, 51, 565; regulation, 51, 558, 559; sedimentation coefficient, 51, 573; specific activity, 51, 565; stability, 51, 564, 565, 579; Stokes radius, 51, 573; substrate specificity, 51, 565, 572, 573; sulfhydryl content, 51, 580; tissue distribution, 51, 568; turnover number, 51, 565

ADENOSINE DEAMINASE

This deaminase [EC 3.5.4.4], also known as adenosine aminohydrolase, catalyzes the hydrolysis of adenosine to produce inosine and ammonia.[1–7] The deaminase is strongly inhibited by transition-state analogues that mimic the unstable hydrated intermediate, including coformycin, 2′-deoxycoformycin and 1,6-hydroxymethyl-1,6-dihydropurine ribonucleoside.

[1] D. K. Wilson & F. A. Quiocho (1994) *Nature Struct. Biol.* **1**, 691.
[2] D. K. Wilson & F. A. Quiocho (1993) *Biochemistry* **32**, 1689.
[3] J. J. Villafranca & T. Nowak (1992) *The Enzymes*, 3rd ed., **20**, 63.
[4] D. K. Wilson, F. B. Rudolph & F. A. Quiocho (1991) *Science* **252**, 1278.
[5] L. F. Thompson & J. E. Seegmiller (1980) *AE* **51**, 167.
[6] R. Wolfenden (1976) *Ann. Rev. Biophys. Bioeng.* **5**, 271.
[7] C. L. Zielke & C. H. Suelter (1971) *The Enzymes*, 3rd ed., **4**, 47.

Selected entries from *Methods in Enzymology* [vol, page(s)]:
General discussion: 2, 473, 475; **12A**, 126; **34**, 3; **46**, 19, 25, 327; **51**, 502
Assay: modified assay, **51**, 502, 503, 508, 509; nonspecific, **2**, 475; **12A**, 126

Other molecular properties: activity, **51**, 502, 508; antibody production, **74**, 351; assay of mung bean nuclease, **65**, 266; ATP-regeneration system, **238**, 35; auxiliary enzyme, **63**, 32, 36; binding energy analysis; **308**, 404, 407, 409; binding protein, competitive idiotype-anti-idiotype enzyme immunoassay, **178**, 171; clinical significance, **74**, 351; complexes, group contributions and role of solvent water, **249**, 301; congenital immunodeficiency, **51**, 506; coupled assay, **63**, 32; difference spectrum with inhibitors, **46**, 333; *Escherichia coli*, from, **51**, 508; friction ratio, **51**, 505; 6-halopurine ribosides as substrates, **46**, 328; hemolysate sample preparation, **74**, 355; heterologous gene selection and co-amplification in mammalian cells, in, **185**, 552; human erythrocytes, from, **51**, 502; immunogen preparation, **74**, 351; immunoprecipitation with protein A, **74**, 356 (precision, **74**, 356, 358; procedure, **74**, 355; protein A preparation, **74**, 352; radioiodination of immunogen, **74**, 352; sensitivity, **74**, 356; specificity, **74**, 357; standard curve, **74**, 357; specific activity, **74**, 358); inactivation kinetics, **46**, 334; inhibitors, **51**, 506, 507, 511 (slow-binding, **63**, 464; tight-binding, **63**, 462); kinetic properties, **51**, 506, 511; marker for chromosome detection in somatic cell hybrids, as, **151**, 178; mechanistic probes, **249**, 296; molecular weight, **51**, 505, 512; nonspecific, **2**, 475 (assay method, **2**, 475; **12A**, 126; properties, **2**, 477; **12A**, 128; purification, **2**, 476; **12A**, 127); 3′-nucleotidase assay, **2**, 551; nucleotidase assay, in, **32**, 127, 369; partial specific volume, **51**, 505; polyethylene glycol modification, **242**, 90; properties, **2**, 477; **12A**, 128; **51**, 505, 506, 511, 512; purification, **2**, 476; **12A**, 127; **51**, 503, 509; **74**, 351; purity, **51**, 511; radioimmunoassay, **74**, 351 (antibody production, **74**, 351; data analysis, **74**, 356; hemolysate sample preparation, **74**, 355; immunogen preparation, **74**, 351; immunoprecipitation with protein A, **74**, 356; precision, **74**, 356, 358; procedure, **74**, 355; protein A preparation, **74**, 352; radioiodination of immunogen, **74**, 352; sensitivity, **74**, 356; specificity, **74**, 357; standard curve, **74**, 357); reaction, **46**, 332, 333; sedimentation coefficient, **51**, 505; slow-binding inhibitor, **63**, 464; specific, **2**, 473 (3′-nucleotidase assay, **2**, 551); specific activity, **74**, 358; stability, **51**, 505, 512; Stokes radius, **51**, 505; substrate specificity, **51**, 505, 511; subunit structure, **51**, 505; tight-binding inhibitor, **63**, 462; transition state analogues, **249**, 307 (complexes, group contributions and role of solvent water, **249**, 301; mechanistic probes, as, **249**, 296)

ADENOSINE KINASE

This phosphotransferase [EC 2.7.1.20] catalyzes the ATP-dependent phosphorylation of adenosine to generate AMP and ADP. 2-Aminoadenosine can act as an alternative substrate.

The structure of the human enzyme reveals a Mg^{2+} ion binding site in a trough between the two adenosine sites. In addition, an active-site aspartyl residue is possibly involved in the deprotonation of the 5′-hydroxyl during the phosphate transfer.[1] The rat liver enzyme catalyzes an ordered Bi Bi kinetic mechanism.[2] Interestingly, the enzyme reportedly catalyzes an adenosine/AMP exchange in the absence of ATP. Even so, no phosphorylenzyme intermediate could be identified. The observed exchange is enhanced significantly with ADP and the basal exchange rates could be accounted for by ADP impurities in the AMP.[2]

[1] I. I. Mathews, M. D. Erion & S. E. Ealick (1998) *Biochemistry* **37**, 15607.
[2] M. Mimouni, F. Bontemps & G. Van den Berghe (1994) *JBC* **269**, 17820.

Selected entries from *Methods in Enzymology* [vol, page(s)]:
Assay: coupled, **63**, 32
Other molecular properties: marker for chromosome detection in somatic cell hybrids, as, **151**, 179; preparation of 3′-amino-3′-deoxy-ATP, **59**, 135, 136; stereochemical studies, **87**, 206, 231, 300; transition state and multisubstrate analogues, **249**, 305

ADENOSINE MONOPHOSPHATASE

This term, often abbreviated AMPase, describes a catalytic activity in which AMP is hydrolyzed to adenosine and orthophosphate. However, care should be exercised when encountering this term since a number of enzymes are known to catalyze this reaction, including alkaline phosphatase [EC 3.1.3.1], acid phosphatase [EC 3.1.3.2], and 5′-nucleotidase [EC 3.1.3.5].

Selected entries from *Methods in Enzymology* [vol, page(s)]:
Other molecular properties: microsomal fractions, in, **52**, 83; myokinase preparation, not present in, **6**, 230

ADENOSINE NUCLEOSIDASE

This enzyme [EC 3.2.2.7], also known as adenosinase, catalyzes the hydrolysis of adenosine to produce adenine and D-ribose. The enzyme has an acidic pH_{opt} and adenosine N-oxide, 7-methyladenosine, and 2′-deoxyadenosine can act as alternative substrates.

ADENOSINE-PHOSPHATE DEAMINASE

This deaminase [EC 3.5.4.17] catalyzes the hydrolysis of 5′-AMP to produce 5′-IMP and ammonia. Alternative substrates include ADP, ATP, NAD^+, and adenosine. The bacterial enzyme also acts on the deoxy derivatives. The enzyme is distinct from AMP deaminase [EC 3.5.4.6], which is much more specific.[1] *See also AMP Deaminase*

[1] J.-C. Su, C.-C. Li & C. C. Ting (1966) *Biochemistry* **5**, 536.

Selected entries from *Methods in Enzymology* [vol, page(s)]:
General discussion: **46**, 305

ADENOSINE TETRAPHOSPHATASE

This enzyme [EC 3.6.1.14] catalyzes the hydrolysis of adenosine 5'-tetraphosphate to produce ATP and orthophosphate. Inosine tetraphosphate and tripolyphosphate are alternative substrates.[1] The yellow lupine (*Lupinus luteus*) seed enzyme[2] appears to be less specific than the rabbit hydrolase. *Saccharomyces cerevisiae* exopolyphosphatase [EC 3.6.1.11] also has an adenosine 5'-tetraphosphatase activity.[3]

[1] G. D. Small & C. Cooper (1966) *Biochemistry* **5**, 14.
[2] A. Guranowski, E. Starzynska, P. Brown & G. M. Blackburn (1997) *BJ* **328**, 257.
[3] A. Guranowski, E. Starzynska, L. D. Barnes, A. K. Robinson & S. Liu (1998) *BBA* **1380**, 232.

ADENOSINE TRIPHOSPHATASE (Mg^{2+}-ACTIVATED)

This enzymatic activity, formerly classified EC 3.6.1.4 and now deleted, is listed under ATPase [EC 3.6.1.3].

S-ADENOSYLHOMOCYSTEINASE

This enzyme [EC 3.3.1.1], also known as adenosylhomocysteine hydrolase, catalyzes the hydrolysis of S-adenosyl-L-homocysteine to produce adenosine and L-homocysteine.[1–5]

This enzyme contains tightly bound NAD$^+$. In the hydrolytic direction, the first step of the enzyme mechanism involves the oxidation of the 3'-hydroxyl group of S-adenosyl-L-homocysteine by the enzyme-bound NAD$^+$. L-Homocysteine is then eliminated, generating 3-keto-4,5-didehydro-5-deoxyadenosine. Michael addition of water to the 5-position of this tightly bound intermediate produces 3-ketoadenosine. This is then reduced by the enzyme-bound NADH to generate adenosine. Note that the 5-hydrolytic activity depends upon the 3-oxidative activity.[4]

This enzyme is an attractive target for the design of antiviral, antiparasitic, and antiarthritic agents due to its important role in regulating biological methylation reactions.

[1] A. R. Clarke & T. R. Dafforn (1998) *CBC* **3**, 1.
[2] P. K. Chiang (1987) *MIE* **143**, 377.
[3] A. Guranowski & H. Jakubowski (1987) *MIE* **143**, 430.
[4] J. L. Palmer & R. H. Abeles (1979) *JBC* **254**, 1217.
[5] M. A. Ator & P. R. Ortez de Montellano (1990) *The Enzymes*, 3rd ed., **19**, 213.

Selected entries from *Methods in Enzymology* [vol, page(s)]:
General discussion: 5, 752; 71, 134; 143, 377, 430
Assay: 5, 752; 143, 378, 430

Other molecular properties: coupled enzyme assay, use in a, **63**, 36; properties, **5**, 755; **6**, 573; **143**, 382, 432; purification (bovine liver, **143**, 380; rat liver, **5**, 753; yellow lupine, **143**, 431)

S-ADENOSYLHOMOCYSTEINE DEAMINASE

This enzyme [EC 3.5.4.28] catalyzes the hydrolysis of S-adenosyl-L-homocysteine to produce S-inosyl-L-homocysteine and ammonia. This enzyme is the first step in the main pathway for S-adenosyl-L-homocysteine catabolism in *Streptomyces flocculus*.[1]

[1] J. J. Zulty & M. K. Speedie (1989) *J. Bacteriol.* **171**, 6840.

S-ADENOSYLHOMOCYSTEINE NUCLEOSIDASE

This enzyme [EC 3.2.2.9] catalyzes the hydrolysis of S-adenosyl-L-homocysteine to produce adenine and S-ribosyl-L-homocysteine. The enzyme will also act on 5'-methylthioadenosine to yield adenine and 5-methylthioribose: however, the *Alcaligenes faecalis* and *Sarcina lutea* enzymes do not act on 5'-methylthioadenosine.

A key step in the mechanism appears to be the protonation of N-7 of the purine ring.[1] Evidence has also been presented against a possible general-base catalysis involving the anchimeric assistance of the 2'-α-hydroxy group and the formation of a 1,2-epoxide as an intermediate in the catalytic process.[2] The transition state appears to have substantial oxocarbenium character.[2] **See also** *Methylthioadenosine Nucleosidase*

[1] F. Della Ragione, M. Porcelli, M. Carteni-Farina, V. Zappia & A. E. Pegg (1985) *BJ* **232**, 335.
[2] B. Allart, M. Gatel, D. Guillerm & G. Guillerm (1998) *EJB* **256**, 155.

Selected entries from *Methods in Enzymology* [vol, page(s)]:
General discussion: 17B, 411
Assay: 17B, 412
Other molecular properties: properties, **17B**, 415; purification from *Escherichia coli*, **17B**, 413

ADENOSYLMETHIONINE:8-AMINO-7-OXONONANOATE AMINOTRANSFERASE

This pyridoxal-phosphate-dependent enzyme [EC 2.6.1.62] (also known as adenosylmethionine:8-amino-7-ketononanoate aminotransferase, 7,8-diamino-pelargonate aminotransferase, 7,8-diaminononanoate aminotransferase, and DAPA aminotransferase) catalyzes the reaction of S-adenosyl-L-methionine with 8-amino-7-oxononanoate (*i.e.*, 7-keto-8-aminopelargonate; $^+H_3NCH(CH_3)$ $CO(CH_2)_5COO^-$) to produce *S*-adenosyl-4-methylthio-2-oxobutanoate and 7,8-diaminononanoate (*i.e.*, 7,8-diaminopelargonate; $^+H_3NCH(CH_3)CH(NH_3^+)(CH_2)_5$-$COO^-$).[1,2] Note that the product *S*-adenosyl-4-methylthio-2-oxobutanoate can decompose nonenzymatically to form 5′-methylthioadenosine and 2-oxo-3-butenoate.[3] *S*-Adenosyl-L-homocysteine can also act as an alternative donor substrate. The enzyme, which participates in biotin biosynthesis in microorgansims, has a ping pong Bi Bi kinetic mechanism with strong substrate inhibition by 7-keto-8-aminopelargonate.[3]

[1]Y. Izumi, Y. Tani & K. Ogata (1979) *MIE* **62**, 326.
[2]M. A. Eisenberg & G. L. Stoner (1979) *MIE* **62**, 342.
[3]G. L. Stoner & M. A. Eisenberg (1975) *JBC* **250**, 4029.

Selected entries from *Methods in Enzymology* [**vol**, page(s)]:
General discussion: 62, 326, 342
Assay: 62, 333, 334, 342, 343
Other molecular properties: distribution, 62, 336; properties, 62, 335, 336, 347; purification, from (*Brevibacterium divaricatum*, **62**, 334, 335; *Escherichia coli* regulatory mutants, **62**, 343)

ADENOSYLMETHIONINE CYCLOTRANSFERASE

This alkyltransferase [EC 2.5.1.4] catalyzes the conversion of *S*-adenosyl-L-methionine to produce 5′-methylthio-adenosine and 2-aminobutan-4-olide (or, homoserine lactone).[1] In some systems, this enzyme may be identical to adenosylmethionine hydrolase [EC 3.3.1.2]. **See also** *Adenosylmethionine Hydrolase*

[1]S. H. Mudd (1959) *JBC* **234**, 87.

Selected entries from *Methods in Enzymology* [**vol**, page(s)]:
General discussion: 17B, 406
Assay: 17B, 407
Other molecular properties: properties, 17B, 410; purification (*Escherichia coli* infected with bacteriophage T3, **17B**, 408; *Escherichia coli* infected with UV-T3, **17B**, 409)

S-ADENOSYLMETHIONINE DECARBOXYLASE

This enzyme [EC 4.1.1.50] catalyzes the pyruvate-dependent conversion of *S*-adenosyl-L-methionine to carbon dioxide and the (5-deoxy-5-adenosyl)-(3-aminopropyl)-methylsulfonium salt.

During catalysis, a Schiff base is formed between the substrate, *S*-adenosyl-L-methionine, and a pyruvoyl moiety of the protein. Carbon dioxide is then released from this intermediate, followed by protonation of the α carbon. Subsequent hydrolysis results in the release of the

final product.[1] Putrescine activates the mammalian enzyme by lowering the K_m value for the substrate.

[1]H. Xiong, B. A. Stanley & A. E. Pegg (1999) *Biochemistry* **38**, 2462.

Selected entries from *Methods in Enzymology* [**vol**, page(s)]:
General discussion: 5, 756; 17B, 647
Assay: 5, 756; 17B, 647; 79, 347; 94, 219, 228, 231, 234
Other molecular properties: *Escherichia coli* (deficient mutants, screening, **94**, 84; gene cloning, **94**, 121; properties, **5**, 759; **17B**, 650; **94**, 230; purification, **5**, 757; **17B**, 648; **94**, 229); induction, **79**, 345; inhibition by interferon, **79**, 343, 344; inhibition assay for methylglyoxal bis(guanylhydrazone), in, **94**, 247; inhibitors, synthesis and characterization, **94**, 239; preparation, **94**, 219; rat liver (antibody preparation, **94**, 238; assay, **94**, 234; properties, **94**, 237; purification, **94**, 235); *Saccharomyces cerevisiae* (deficient mutants, mass screening, **94**, 107; properties, **94**, 233; purification, **94**, 232)

ADENOSYLMETHIONINE HYDROLASE

This enzyme [EC 3.3.1.2], also known as *S*-adenosylmethionine cleaving enzyme and methylmethionine-sulfonium salt hydrolase, catalyzes the hydrolysis of *S*-adenosyl-L-methionine to produce methylthioadenosine and L-homoserine. The enzyme will also convert *S*-methyl-L-methionine sulfonium salt to dimethyl sulfide and L-homoserine. In some systems, this enzyme may be identical to adenosylmethionine cyclotransferase [EC 2.5.1.4].
See also *Adenosylmethionine Cyclotransferase*

Selected entries from *Methods in Enzymology* [**vol**, page(s)]:
General discussion: 17B, 406
Assay: 17B, 407; 30, 661
Other molecular properties: properties, 17B, 410; purification from *Escherichia coli* (after bacteriophage T3 infection, **17B**, 408; after UV-T3 infection, **17B**, 409)

ADENOVIRUS PROTEASE

This cysteine endopeptidase [EC 3.4.22.39], officially called adenain, has been observed in a wide variety of adenoviruses (for example, avian adenovirus 1, dog mastadenovirus c1, human adenovirus type 41, mastadenovirus mus1, and pig adenovirus type 3). Most of the biochemical investigations have been carried out on the human adenovirus type 2 enzyme, often abbreviated as AVP.[1] The enzyme catalyzes the hydrolysis of peptide bonds at (Met,Ile,Leu)–Xaa–Gly–Gly-∇-Xaa and at (Met,Ile,Leu)–Xaa–Gly–Xaa-∇-Gly, where Xaa can be any amino acid. (Note: ∇ indicates the cleavage site.)

[1]J. M. Weber (1995) in *Molecular Repertoire of the Adenoviruses* (W. Doerfler & P. Bohm, eds.), *Curr. Top. Microbiol. Immunol* **199/I**, 227.

Selected entries from *Methods in Enzymology* [**vol**, page(s)]:
General discussion: 244, 595
Assay: 244, 596
Other molecular properties: activation, 244, 599; affinity labeling, 244, 598; catalytic residues, 244, 474, 599; discovery, 244, 595; family, 244, 474; inhibitors, 244, 601; pH optimum, 244, 601; purification, 244, 597;

recombinant, expression, **244**, 596; role in virus infection, **244**, 602; sequence homology, **244**, 598; size, **244**, 595, 601; substrate specificity, **244**, 599; Western blotting, **244**, 598

ADENYLATE CYCLASE

This enzyme [EC 4.6.1.1], also known as adenylyl cyclase and 3′,5′-cyclic AMP synthase, catalyzes the conversion of ATP to 3′,5′-cyclic AMP and pyrophosphate (or, diphosphate). The enzyme requires pyruvate as a cofactor and will also utilize dATP as a substrate (thereby producing 3′,5′-cyclic dAMP). In the presence of NAD(P)$^+$-arginine ADP-ribosyltransferase, this enzyme is activated by covalent modification.

The adenylate cyclases comprise a family of membrane-bound enzymes catalyzing cAMP formation from 5′-ATP. The catalytic site exhibits specificity for the adenine moiety, enhanced binding via substrate and inhibitor phosphate groups, and tolerance of modifications to the ribose portion of the molecule.

Recent reports indicated that the active site of adenylate cyclase is similar to the sites of DNA polymerases, suggesting that adenylate cyclase uses a two-metal-ion mechanism.[1,2] The catalytic mechanism involves deprotonation and activation of the ATP 3′-hydroxyl for nucleophilic attack as well as stabilization of the transition state at the α-phosphate. In addition, the increased negative charge on the pyrophosphate leaving group has to be stabilized as well.

The adenylate cyclases appear to be regulated by free metal ion concentrations, P-site inhibitors (inhibitors containing a purine ring and are either non- or uncompetitive with respect to the forward reaction but competitive with respect to cAMP in the reverse reaction), forskolin (activates all adenylate cyclases except type IX), G-protein subunits (for example, all types of mammalian adenylate cyclases are activated by the GTP-bound G-protein α-subunit $G_s\alpha$ whereas $G_i\alpha$ inhibits adenylyl cyclase types V and VI), Ca^{2+}/calmodulin (activates type I adenylyl cyclase), and protein phosphorylation (the type II enzyme is activated by protein kinase C-dependent phosphorylation of Thr1057 whereas type III adenylate cyclase is inhibited by CaM kinase II-dependent phosphorylation of Ser1056).[2]

[1] J. J. Tesmer, R. K. Sunahara, R. A. Johnson, G. Gosselin, A. G. Gilman & S. R. Sprang (1999) *Science* **285**, 756.
[2] J. H. Hurley (1999) *JBC* **274**, 7599.

Selected entries from *Methods in Enzymology* [**vol**, page(s)]:
General discussion: 46, 591, 598; 79, 162, 163
Assay: 38, 42, 72; 39, 480; 74, 464; 79, 164; 195, 66, 94, 124, 130, 138, 165; automated chromatographic assay, 38, 125, 126, 134, 135 (cyclase preparation, 38, 128, 129; identification of product, 38, 132; materials, 38, 127, 128; nucleotide separation, 38, 126, 127; procedure, 38, 129, 130; sensitivity and reproducibility, 38, 131); chromatographic methods, 195, 15; 238, 45, 92; contaminating enzyme activities, 195, 4 (inhibition, 238, 33); cyanobacteria, in, 167, 588; data analysis, 195, 19; 238, 53, 93; enzyme concentration, 238, 36; general principles (determination of residual substrate, 38, 124, 125; divalent cations, 38, 120; enzyme concentration, 38, 117; 238, 36; labeled nucleoside triphosphates, 38, 118, 119; linearity of cyclic nucleotide production, 38, 122; nucleotide triphosphate concentration and volume, 38, 115; protection of cyclase activity, 38, 121, 122; purification, proof of identity, determination of product, 38, 123, 124; reduction of [product degradation, 38, 120, 121; substrate degradation, 38, 119, 120]; termination of cyclase reaction, 38, 122, 123); G protein βγ subunits, for, 237, 451, 455; incubation temperature, 238, 37; inhibition assays, 238, 90; methods, 31, 103 (procedure, 31, 111, 112; stopping solution, 31, 112); permeabilized adipocytes, 159, 200; α-^{32}P-labeled nucleotide preparation for, 195, 29; preparation of activated $G_s\alpha$ and G_i, in, 237, 116; radioimmunoassay, 238, 31; reaction conditions for, establishment, 195, 3; reaction stopping (ATP/sodium dodecyl sulfate/cAMP, 238, 43; hydrochloric acid, 238, 43, 50; zinc acetate/sodium carbonate/cAMP, 238, 41, 45, 52); reaction time, 238, 36; skeletal muscle triads, in assay, 157, 68; transfected cells, 238, 111; yeast cAMP cascade system mutants, in, assay, 159, 35
Other molecular properties: activated, isolation from rabbit myocardium, 195, 84; activation, 39, 253, 256 (bacterial toxins, by, 235, 624, 626; cholera toxin, by, role of soluble guanine nucleotide-dependent ADP-ribosylation factor, 195, 243; gastric parietal cells, in, 191, 649; iodinated glucagon, by, in glucagon receptor assay, 109, 11; stimulatory G protein, by, 237, 437); activity, 79, 167 (characterization of β-adrenergic receptor antibodies, 74, 464); adenosine 5′-O-(1-thiotriphosphate), and, 87, 203, 230; adenylyliminodiphosphate, 38, 420, 421; β-adrenergic receptor-adenylate cyclase complex, 38, 187; amiloride, characterization with, 191, 746; anthrax toxin assay, in, 165, 113; associated coupling proteins N_s and N_i (ADP-ribosylation by pertussis and cholera toxins, 109, 566; analytical procedures and assays, 109, 450; purification, 109, 446, 454); ATP requirements, 238, 32; *Bacillus anthracis* (cell culture assay, 195, 168; characterization, 195, 153; expression systems, 195, 162; purification, 195, 153); binding subunit, 79, 167; bone cells, in, effects of hormones, 145, 330; bovine brain, preparation, 38, 128, 129; *Brevibacterium liquefaciens*, 38, 160 (forward reaction assay, 38, 161; properties, 38, 167; purification, 38, 163; reverse reaction assay, 38, 162, 163); brush borders, in, 31, 122; calcium ion-sensitive/-insensitive, separation, 195, 127; calcium-sensitive (assay, 238, 74, 101; calcium concentration effects, 238, 72, 80; calmodulin effects, 238, 77, 95; hormone effects, 238, 79; immunoblot analysis, 238, 99; ionomycin effects, 238, 79; magnesium effects, 238, 76; manganese effects, 238, 76); calmodulin-depleted, from equine sperm, chromatography, 195, 101; calmodulin-mediated, from equine sperm, 195, 91; calmodulin-sensitive (bovine brain, from, [characterization, 195, 65; molecular weight, 195, 72; polyclonal antibodies, isolation, 195, 110; purification, 139, 777; 195, 65]; catalytic subunit [elution from electrophoretic gel, 195, 149; gel electrophoresis, 195, 147; interaction with calmodulin, 195, 76; interaction with wheat germ agglutinin, 195, 78; purification, 195, 74]; cell-invasive, from *Bordetella pertussis* [assay, 195, 137; purification, 195, 137]; distribution in rat tissues, 195, 120; immunoprecipitation, 195, 112; iodination, 195, 111; murine polyclonal antibodies, isolation, 195, 111; partially purified, immunoprecipitation, 195, 119; purification, 195, 80, 111; reconstitution with [β-adrenergic receptors, 139, 788; calmodulin, 139, 787; multilamellar liposomes, 139, 780]); cAMP inhibitor formation, 38, 274, 275; catalytic reactions, termination (ATP/SDS/cAMP with or without [^{3}H]cAMP, by, 195, 14; zinc acetate/Na$_2$CO$_3$/cAMP, with, 195, 421; β-adrenergic receptor-adenylate

cyclase complex, **38**, 187; amiloride, characterization with, **191**, 746; anthrax toxin assay, in, **165**, 113; associated coupling proteins N_s and N_i (ADP-ribosylation by pertussis and cholera toxins, **109**, 566; analytical procedures and assays, **109**, 450; purification, **109**, 446, 454); ATP requirements, **238**, 32; *Bacillus anthracis* (cell culture assay, **195**, 168; characterization, **195**, 153; expression systems, **195**, 162; purification, **195**, 153); binding subunit, **79**, 167; bone cells, in, effects of hormones, **145**, 330; bovine brain, preparation, **38**, 128, 129; *Brevibacterium liquefaciens*, **38**, 160 (forward reaction assay, **38**, 161; properties, **38**, 167; purification, **38**, 163; reverse reaction assay, **38**, 162, 163); brush borders, in, **31**, 122; calcium ion-sensitive/-insensitive, separation, **195**, 127; calcium-sensitive (assay, **238**, 74, 101; calcium concentration effects, **238**, 72, 80; calmodulin effects, **238**, 77, 95; hormone effects, **238**, 79; immunoblot analysis, **238**, 99; ionomycin effects, **238**, 79; magnesium effects, **238**, 76; manganese effects, **238**, 76); calmodulin-depleted, from equine sperm, chromatography, **195**, 101; calmodulin-mediated, from equine sperm, **195**, 91; calmodulin-sensitive (bovine brain, from, [characterization, **195**, 65; molecular weight, **195**, 72; polyclonal antibodies, isolation, **195**, 110; purification, **139**, 777; **195**, 65]; catalytic subunit [elution from electrophoretic gel, **195**, 149; gel electrophoresis, **195**, 147; interaction with calmodulin, **195**, 76; interaction with wheat germ agglutinin, **195**, 78; purification, **195**, 74]; cell-invasive, from *Bordetella pertussis* [assay, **195**, 137; purification, **195**, 137]; distribution in rat tissues, **195**, 120; immunoprecipitation, **195**, 112; iodination, **195**, 111; murine polyclonal antibodies, isolation, **195**, 111; partially purified, immunoprecipitation, **195**, 119; purification, **195**, 80, 111; reconstitution with [β-adrenergic receptors, **139**, 788; calmodulin, **139**, 787; multilamellar liposomes, **139**, 780]); cAMP inhibitor formation, **38**, 274, 275; catalytic reactions, termination (ATP/SDS/cAMP with or without [^{3}H]cAMP, by, **195**, 14; zinc acetate/Na_2CO_3/cAMP, with, **195**, 12); catalytic subunit, **79**, 167; cell-invasive properties, assay, **195**, 143; cerebral (basal form [isolation, **195**, 85; solubilization, **195**, 86]; purification, **195**, 83, 122); chemokine assay, **288**, 326; cloning, **238**, 108, 116, 124; cooperativity, **64**, 218; cytochemical localization, **39**, 480; degenerative primers, **238**, 118; 2'-deoxyadenosine 3',5'-cyclic phosphate, and, **87**, 208, 209, 254, 263; divalent metal cation requirements, **238**, 32; effects of toxin ADP-ribosyltransferases, **109**, 411; endosome marker, as, determination, **109**, 267; erythrocyte (assay, **38**, 170, 171; properties, **38**, 173; purification, **38**, 171, 172); *Escherichia coli*, **38**, 155 (assay, **38**, 156; properties, **38**, 159, 160; purification, **38**, 159); fat-cell plasma membrane, in, **31**, 65, 67 (hormonal regulation, **31**, 68, 70); follicle stimulating hormone effects, **39**, 253; functional analysis with pertussis toxin as probe, **109**, 563; fusion to hepatic membranes, in analysis of glucagon receptor function, **109**, 346; gel electrophoresis, **195**, 124; gonadotropin regulation, **39**, 237; G protein regulation, **238**, 37, 81, 116; GTPase stimulation, **237**, 18; GTP modulation, **237**, 108; in GTPγS binding assay, **237**, 35; heart and skeletal muscle (assay, **38**, 143, 144; preparation, **38**, 144; properties, **38**, 146; solubilization, **38**, 149); hormonal regulation, assay, **109**, 333; hormone antagonist assay, in, **37**, 435; hormone-sensitive (assay, **38**, 190; binding to isoproterenol on glass beads, **38**, 189; preparation, **38**, 188); hysteresis, **64**, 218; identification from SDS-polyacrylamide gels, **195**, 148; identification with PCR, **238**, 121; immunoblotting, **195**, 124; immunoprecipitation, **195**, 126; inhibition, **46**, 600; **63**, 395 (adenosine, **195**, 80); invasive, isolation, **195**, 140; isoforms (cloning, **238**, 124; distribution, **238**, 96; forskolin response, **238**, 114; functions, **238**, 126; G_s response, **238**, 127; immunoblot analysis, **238**, 99; regulation, **238**, 96; sizes, **238**, 96, 106; types, **238**, 95, 116, 124); isoproterenol-stimulated, **237**, 434; kidney, preparation, **38**, 150; leukocyte, assays, **132**, 422; mammalian (expression in Sf9 cells, **238**, 95; transient expression assays, **238**, 108); membrane topology, **238**, 118; monoclonal antibodies, characteristics, **195**, 121; myocardial (activation, **195**, 84; purification, **195**, 83; solubilization and role of phospholipids, **38**, 174; **195**, 84); neonatal bone, **38**, 152, 153; norepinephrine activation, **39**, 376, 380; oxidation, **238**, 68; particulate from brain, **38**, 135, 136 (assay, **38**, 136; characteristics, **38**, 140; dispersion, **38**, 138; preparation, **38**, 136);

phosphoenolpyruvate effects, **238**, 34; pineal, in, **39**, 396; plasma membranes, in, **31**, 88, 90, 143, 145; "P"-site (affinity ligands, preparation, **238**, 56; inhibition, **238**, 34, 37); purification, **74**, 469 (recombinant histidine-tagged enzyme, **238**, 102); receptor, **79**, 167; reconstituted cyc^- membranes, in, **237**, 252; regulation, **237**, 269, 452; related mutations in Y1 adrenal tumor cells, characterization, **109**, 355; retinal, characterization and localization in (microdissected sample, **81**, 518; tissue homogenates, **81**, 517); rod outer segments, in, **38**, 154, 155; sensitivity to Mn^{2+}, NaF, guanylyl imidodiphosphate, and forskolin, **195**, 72; sequence homology between species, **238**, 117, 124; skeletal muscle triads, in, assay, **157**, 68; solubilized from brain, characteristics, **38**, 140; soluble, purification from testis, **195**, 130; sperm, equine (activity loss, restoration by calmodulin, **195**, 101; effects of proteases, **195**, 107; extracts, HPLC, **195**, 104; inhibition by anti-calmodulin agents, **195**, 101; mediation by calmodulin, **195**, 100; preparation, **195**, 97; properties, **195**, 97); stereochemistry, **87**, 161, 203, 208, 209, 212, 230, 254, 263, 288; stimulation by GTPγS *in vitro*, **237**, 404; substrate affinity, **238**, 32; testis mitochondria, in, **55**, 11; TSH effects, **37**, 262; ultraviolet irradiation, **238**, 68; vasopressin-sensitive, target analysis studies, **191**, 566; water flow, **37**, 255, 256; yeast (activation by Ras, **255**, 468, 474; assay, **255**, 474; cAMP cascade system mutants, in, assay, **159**, 35; epitope-tagged, purification, **255**, 473; extract preparation, **255**, 471)

ADENYLATE ISOPENTENYLTRANSFERASE

This magnesium-dependent transferase [EC 2.5.1.27], also called 2-isopentenyl-diphosphate:AMP Δ^2-isopentenyl-transferase and cytokinin synthase, catalyzes the reaction of AMP with Δ^2-isopentenyl diphosphate to yield pyrophosphate (or, diphosphate) and N^6-(Δ^2-isopentenyl)adenosine 5'-monophosphate.[1] ADP is also a substrate for the slime mold enzyme (*Dictyostelium discoideum*).[2] Note that the substrate Δ^2-isopentenyl diphosphate is commonly known as dimethylallyl diphosphate.

[1] C. M. Chen & D. K. Melitz (1979) *FEBS Lett.* **107**, 15.
[2] M. Ihara, Y. Taya, S. Nishimura & Y. Tanaka (1984) *ABB* **230**, 652.

Selected entries from **Methods in Enzymology** [vol, page(s)]:
Other molecular properties: adenylate dimethylallyltransferase, **110**, 340 (assay, tobacco crown gall tumor, **110**, 344, properties, **110**, 346; purification, **110**, 343, 346)

ADENYLATE KINASE

This phosphotransferase [EC 2.7.4.3], also known as myokinase, catalyzes the freely reversible reaction ($K_{eq} \sim 1$) of MgATP with AMP to produce MgADP and ADP.[1–15] The kinetic mechanism is of the random Bi Bi sequential type,[13] leading to an enzyme · MgATP^{2-} · AMP ternary complex and subsequent in-line phosphoryl transfer[1,2] (a later study suggests an Iso random Bi Bi kinetic mechanism[15]). Inorganic triphosphate can also act as substrate with this enzyme. Other nucleotide substrates include dATP and dAMP.

The geometrical analogues AppppA ($K_i \sim 10^{-4}$ M) and ApppppA ($K_i \sim 10^{-5}$ M) are potent inhibitors that are anchored simultaneously on the enzyme surface.[11,12] The relationship of enzyme-bound ApppppA or AppppppA to the transition state is less obvious, because metal ion binding may not mimic the enzyme $\cdot$ MgATP^{2-} $\cdot$ AMP ternary complex.

[1] H. Yan & M.-D. Tsai (1999) AE **73**, 103.
[2] M.-D. Tsai, R.-T. Jiang, T. Dahnke & Z. Shi (1995) MIE **249**, 425.
[3] L. A. Kleczkowski (1994) Ann. Rev. Plant Physiol. Plant Mol. Biol. **45**, 339.
[4] J. J. Villafranca & T. Nowak (1992) The Enzymes, 3rd ed. **20**, 63.
[5] G. E. Schulz (1992) Faraday Discuss. **1992** (93), 85
[6] M.-D. Tsai & H. G. Yan (1991) Biochemistry **30**, 6806.
[7] M. Hamada, H. Takenaka, M. Sumida & S. A. Kuby (1991) A Study of Enzymes **2**, 403.
[8] P. A. Frey (1989) AE **62**, 119.
[9] A. S. Mildvan & D. C. Fry (1987) AE **59**, 241.
[10] L. Noda (1973) The Enzymes, 3rd ed., **8**, 279.
[11] D. L. Purich & H. J. Fromm (1972) BBA **276**, 563.
[12] G. E. Lienhard & I. I. Secemski (1973) JBC **248**, 1121.
[13] D. G. Rhoads & J. M. Lowenstein (1968) JBC **243**, 3963.
[14] L. Noda (1962) The Enzymes, 2nd ed., **6**, 139.
[15] X. R. Sheng, X. Li & X. M. Pan (1999) JBC **274**, 22238.

Selected entries from **Methods in Enzymology** [**vol**, page(s)]:
General discussion: 2, 598; 6, 223; 46, 21, 25, 298, 299; 51, 459, 467
Assay: 2, 598; 6, 223; 10, 457; 51, 459, 460, 468, 469; 55, 285, 286, 288; 57, 30; activity assay, 44, 892, 893; coupled assay, 63, 32; kinetic properties, 51, 473; luciferase and, 3, 873; luminescent determination of ATP, 56, 536, 537, 543; marker enzymes, as, 56, 210
Other molecular properties: aceto-CoA kinase and, 5, 464; activation, 2, 603; 6, 230; 44, 889; activators, 51, 464; activity, 51, 459, 473; 57, 74, 90, 108; acylphosphatase and, 6, 327; adenosine kinase assay and, 2, 497; adenosine phosphotransferase, assay of, 63, 32; adenylic acid ribosidase and, 6, 121; ADP assay and, 2, 497; affinity for calcium phosphate gel, 51, 584; amino acid analysis, 51, 463, 464; amino acid-incorporation enzymes and, 5, 734, 735; AMP assay, 55, 208; 63, 32; AMP phosphorylation and, 29, 105, 111; 238, 34; analytical use of, 3, 869; arginine condensing system and, 2, 362; assay of (cyclic nucleotide phosphodiesterase, 57, 95; modulator protein, 57, 108); associated with cell fractions, 40, 86; ATP exchange and, 6, 319, 323; ATP-sulfurylase, not present in preparation of, 5, 972; bacterial photophosphorylation and, 6, 317; bifurcation analysis, 240, 809; bound reactants and products, equilibrium constant determination, 177, 365; cAMP assay, 38, 62, 63, 65; carbamoyl phosphate synthetase and, 5, 921; cellular function, 51, 467; chiral-phosphoryl-AMP, and, 87, 209, 218, 225, 300; chromium-nucleotide specificity, 87, 176, 177; citrate-cleavage enzyme and, 5, 644; contaminant in (glycine reductase preparations, 53, 381; isocitrate dehydrogenase, 13, 42; luciferase preparation, 57, 7, 29, 35, 38, 56, 62; phosphotransferases, 63, 7); cysteine residues, 44, 887; distribution, 2, 602; 51, 459; energy charge, 55, 231; enzyme contamination, 64, 24; Escherichia coli and, 2, 619; exchange reactions, 55, 258; extraction of, 10, 6; FAD forming enzyme and, 6, 344; flavokinase and, 2, 645; fluorometric assay for AMP, 13, 494; N^{10}-formyltetrahydrofolate synthetase and, 6, 379; geometric analogue, 87, 16; glycinamide ribotide synthetase and, 6, 64; GMP synthetase, assay of, 51, 219; 63, 32; hepatoma mitochondria, 55, 88; human erythrocytes, from, 51, 467; immobilization, in polyacrylamide gel, 44, 890, 891; immobilized, 5-phospho-α-D-ribosyl pyrophosphate synthesis with, 136, 278; immunological properties, 51, 473; inhibition, 2, 603; 6, 230; 63, 7, 398, 401, 483; 87, 16, 309; 249, 305; insect muscle mitochondria, 55, 23; interference by, in metabolite extraction, 13, 438; interference in (GTP determinations, 57, 93, 94; luciferase assay, 57, 11); isoelectric point, 51, 464, 472; isolated nuclei and, 12A, 432; isotope exchange properties, 64, 8; kinetics, 51, 473; 63, 18; 240, 809, 812; labeled ATP and, 4, 914; luminescent determination of ATP, 56, 536, 537, 543; marker enzymes,

as, 56, 210 (marker for chromosome detection in somatic cell hybrids, as, 151, 179); mass action ratio, 63, 18; mechanism, 63, 18; microbial phosphorylation and, 6, 292; microbial supernatant and, 6, 291; mitochondrial location, 10, 26, 457, 467; mitochondrial subfractions, 55, 94; molecular weight, 51, 463, 464, 472; multiple forms, 51, 467 (electrophoretic patterns, 51, 469, 470); multisubstrate analogue inhibitor, 63, 7; 249, 292; muscle relaxation and, 2, 602; mutant, 72, 694; NAD$^+$ synthetase and, 6, 352; nomenclature of, 2, 602; NMR spectroscopy, nonresonant effects, 239, 77; nucleoli and, 12A, 457; nucleoside diphosphates and, 3, 874; nucleoside diphosphokinase and, 6, 163; partial specific volume, 51, 464; perchloric acid, 55, 201; ^{31}P exchanges in, 176, 303; phosphofructokinase, problems in assay of, 9, 437f; phosphoglyceraldehyde dehydrogenase and, 2, 601; phosphorothioates, and, 87, 200, 258, 301; phosphorus stereospecificity (adenosine 5'-monothiophosphate, toward, 249, 426 [enhancement with R97M mutant, 249, 432; reversal with R44M mutant, 249, 431; wild-type, conformation, 249, 429]; adenosine 5-(1-thiotriphosphate), toward, 249, 433 [perturbation with T23A mutant, 249, 437; relaxation with R128A mutant, 249, 435]; AMP site, at, 249, 426; ATP site, at, 249, 426; demonstration, 249, 426; manipulation by site-directed mutagenesis, 249, 425 [active site conformations and, 249, 441; kinetic experiments, 249, 428; methods, 249, 428; microscopic rates and, 249, 440; procedures, 249, 429; results interpretation, 249, 440]; P$_\alpha$ of MgATP, at, 249, 433; wild-type and site-directed mutant enzymes, comparison, 249, 440); photoactivatable probe synthesis, in, 237, 88, 93; photophosphorylation, 69, 656; properties, 2, 602; 6, 230; propionyl-CoA carboxylase and, 5, 572, 575; prostaglandins, effect of, 51, 466; purification, 2, 601; 6, 225; 51, 460; purity, 6, 230; 51, 463, 472; pyruvate kinase and, 5, 367, 368; quenching problem, 9, 16; rabbit muscle, 2, 601; 6, 225; rat liver, from, 51, 459; related enzyme, 3, 788; removal from myosin, 2, 585; sedimentation coefficients, 51, 464; site-directed mutants, 249, 428; soluble coupling enzyme and, 6, 283; stability, 2, 602; 6, 230; 51, 463, 472; 57, 109; stereochemistry, 87, 200, 258, 300; substrate binding, quadrupolar NMR study, 177, 331; substrate, metal–ion complex, 63, 259; substrates, 240, 810; substrate specificity, 2, 602; 6, 230; 51, 464, 473; 57, 82; subunit structure, 51, 463, 464, 473; supernatant fraction and, 5, 56; thiol-disulfide interchange in, thiol pK$_a$ values, 143, 134; transition state and multisubstrate analogues, 249, 305; tyrosine-activating enzyme and, 5, 725

ADENYLOSUCCINATE LYASE

This lyase [EC 4.3.2.2], also known as adenylosuccinase, catalyzes the conversion of N^6-(1,2-dicarboxyethyl)AMP (i.e., adenylosuccinate) to fumarate and AMP.[1–4] The enzyme will also convert 1-(5-phosphoribosyl)-4-(N-succinocarboxamide)-5-aminoimidazole to 5'-phosphoribosyl-5-amino-4-imidazolecarboxamide and fumarate as well as similar reactions with 2'-deoxysuccino-AMP and 8-azasuccino-AMP.

Stone et al.[1] demonstrated that human enzyme exhibits an ordered Uni Bi kinetic mechanism, with fumarate as the first product released. The k_{cat} for cleavage of 5-amino-4-imidazole-N-succinocarboxamide ribotide (SAICAR) was 90 s^{-1} with a K_m of 2.35 μM, whereas the k_{cat} for adenylosuccinate (SAMP) cleavage was 97 s^{-1} with a K_m of 1.79 μM. The catalytic mechanism appears to involve a general base catalyst (pK$_a$ = 6.4) and a general acid catalyst ($pK_a = 7.5$). AICAR and AMP were competitive with SAICAR and SAMP, whereas fumarate inhibited non-competitively. Competitive inhibition by AICAR and AMP supported a single active site that binds both SAICAR and SAMP.

[1]R. L. Stone, H. Zalkin & J. E. Dixon (1993) *JBC* **268**, 19710.
[2]R. M. Pinto, A. Faraldo, A. Fernandez, J. Canales, A. Sillero & M. A. Sillero (1983) *JBC* **258**, 12513.
[3]S. Ratner (1972) *The Enzymes*, 3rd ed., **7**, 167.
[4]W. A. Bridger & L. H. Cohen (1968) *JBC* **243**, 644.

Selected entries from *Methods in Enzymology* [vol, page(s)]:
General discussion: 6, 85, 790; 46, 302, 305, 306; 51, 202
Assay: 6, 85, 790; 51, 202, 203
Other molecular properties: activity, 51, 202, 203; adenylosuccinate synthetase, 6, 100, 101; aggregated forms, 51, 206; 5-amino-4-imidazole-*N*-succinocarboxamide ribotide and, 6, 88, 692, 695; intragenic complementation, 51, 206, 207; mechanism, 63, 51; molecular weight, 51, 206; *Neurospora crassa*, from, 51, 202; pH optimum, 6, 88; 51, 206; product inhibition, 63, 432; properties, 6, 88, 791; 51, 206, 207; purification, 6, 86, 790; 51, 190; specificity, 6, 88; succino-AICAR synthesis, in, 51, 189; succino-AICAR synthetase assay, in, 51, 186, 187; transition state and multisubstrate analogues, 249, 308; yeast, preparation, 6, 86, 790; 51, 190

ADENYLOSUCCINATE SYNTHETASE

This enzyme [EC 6.3.4.4], also known as IMP:aspartate ligase and succinoadenylic kinosynthetase, catalyzes the reaction of IMP with L-aspartate and MgGTP^{2-} to produce AMP-succinate (*i.e.*, adenylosuccinate), MgGDP^{1-}, and orthophosphate.[1–6] Aside from the natural nucleophile, L-aspartate, the only other agents that can attack the 6-carbon of IMP in a highly specific manner are hydroxylamine and L-alanine 3-nitronate.

As the first committed step in the biosynthesis of AMP from IMP, adenylosuccinate synthetase plays a central role in *de novo* purine nucleotide biosynthesis. A 6-phosphoryl-IMP intermediate appears to be formed during catalysis, and kinetic studies of the *Escherichia coli* enzyme demonstrated that the substrates bind to the enzyme active sites randomly.[1] With mammalian synthetases, L-aspartate exhibits preferred binding to the E·GTP·IMP complex rather than to the free enzyme as is seen in equilibrium exchange[3] and isotope scrambling[4] investigations of the rat liver enzyme. Other kinetic data support the inference that Mg·L-aspartate complex formation occurs within the adenylosuccinate synthetase active site and that such a complex may be an important factor in the activation of the protonated amino group of L-aspartate, enhancement of the enzyme's binding affinity, and its specificity for L-aspartate.

[1]R. B. Honzatko, M. M. Stayton & H. J. Fromm (1999) *AE* **73**, 57.
[2]R. B. Honzatko & H. J. Fromm (1999) *ABB* **370**, 1.

[3]B. F. Cooper, H. J. Fromm & F. B. Rudolph (1986) *Biochemistry* **25**, 7323.
[4]M. B. Bass, H. J. Fromm & F. B. Rudolph (1984) *JBC* **259**, 12330.
[5]F. M. Raushel & J. J. Villafranca (1988) *Crit. Rev. Biochem.* **23**, 1.
[6]M. M. Stayton, F. B. Rudolph & H. J. Fromm (1983) *Curr. Top. Cell. Regul.* **22**, 103.

Selected entries from *Methods in Enzymology* [vol, page(s)]:
General discussion: 6, 100; 51, 207
Assay: 6, 100; 51, 207, 208, 212, 213; coupled enzyme, 63, 32
Other molecular properties: adenylosuccinate (product of adenylosuccinate synthetase, 51, 207); buffers, 51, 212; crystallization, 51, 210, 211; *Escherichia coli*, from (kinetic mechanism, equilibrium isotope exchange study, 249, 466; positional isotope exchange studies, 249, 423; product inhibition studies, three substrates:three products reactions, 249, 207; properties, 6, 102; purification, 6, 101; 104, 109; site-directed mutagenesis, 249, 93); inhibitors, 51, 212; kinetic mechanism, equilibrium isotope exchange study, 249, 466; mechanism, 63, 51, 157; 87, 357; 249, 466; molecular weight, 51, 211; pH optimum, 6, 102; 51, 211; positional isotope exchange studies, 249, 423; product inhibition studies, three substrates:three products reactions, 249, 207; properties, 6, 102; 51, 211; purification, 6, 101; 51, 208; purity, 51, 211; rabbit tissues, from, 51, 207; reaction mechanism, 63, 51, 157; sources, 51, 212; stability, 51, 211; structural formula, 51, 203; substrate, of adenylosuccinate AMP-lyase, 51, 202

ADENYLYL-[GLUTAMINE SYNTHETASE] HYDROLASE

This enzyme [EC 3.1.4.15], also known as adenylyl-[glutamate:ammonia ligase] hydrolase and [glutamine synthetase] deadenylylating enzyme, catalyzes the hydrolysis of adenylyl-[glutamine synthetase] (*i.e.*, a subunit of glutamine synthetase that has an adenylyl group covalently linked to a tyrosyl residue) to produce adenylate (*i.e.*, AMP) and the deadenylylated glutamine synthetase.[1]

O-adenylyltyrosyl residue

tyrosyl residue

In *Escherichia coli*, the enzyme is part of a cascade system that also catalyzes an adenylyltransferase activity ([glutamine-synthetase] adenylyltransferase, EC 2.7.7.42): however, in this case, orthophosphate is used in phosphorolysis to produce ADP. ***See also** [Glutamine-Synthetase] Adenylyltransferase*

[1]B. M. Shapiro (1969) *Biochemistry* **8**, 659.

Selected entries from *Methods in Enzymology* [vol, page(s)]:
Other molecular properties: activation, 113, 227; identity with Holzer's inactivation enzyme, 182, 798; role in adenylylation and deadenylylation of glutamine synthetase, 113, 213; 182, 797

ADENYLYLSULFATASE

This enzyme [EC 3.6.2.1], also called adenosine 5'-phosphosulfate sulfohydrolase, catalyzes the hydrolysis of adenylylsulfate to produce AMP and sulfate.[1-5] Product inhibition studies suggest that product released is ordered.[3] The enzyme also exhibits an ATPase and a pyrophosphatase activity as well.

[1] G. F. White, M. G. Rowlands, K. S. Dodgson & W. J. Payne (1979) *FEMS Microbiol. Lett.* **5**, 267.
[2] K. M. Rogers, G. F. White & K. S. Dodgson (1978) *BBA* **527**, 70.
[3] A. M. Stokes, W. H. B. Denner & K. S. Dodgston (1973) *BBA* **315**, 402.
[4] A. Schmidt & K. Jäger (1992) *Ann. Rev. Plant Physiol. Plant Mol. Biol.* **43**, 325.
[5] A. B. Roy (1971) *The Enzymes*, 3rd ed., **5**, 1.

Selected entries from *Methods in Enzymology* [**vol**, page(s)]:
Other molecular properties: product inhibition, **63**, 432

ADENYLYLSULFATE:AMMONIA ADENYLYLTRANSFERASE

This transferase [EC 2.7.7.51] catalyzes the reaction of adenylylsulfate with ammonia to produce adenosine 5'-phosphoramidate and sulfate.[1-3]

[1] A. Schmidt & K. Jäger (1992) *Ann. Rev. Plant Physiol. Plant Mol. Biol.* **43**, 325.
[2] H. Frankhauser, J. A. Schiff & L. J. Garber (1981) *BJ* **195**, 545.
[3] H. Frankhauser & J. A. Schiff (1979) *Plant Physiol.* **65S**, 17.

Selected entries from *Methods in Enzymology* [**vol**, page(s)]:
General discussion: **143**, 354
Assay: **143**, 354
Other molecular properties: properties, **143**, 359; purification from *Chlorella pyrenoidosa*, **143**, 357

ADENYLYLSULFATE KINASE

This phosphotransferase [EC 2.7.1.25], also known as APS kinase, catalyzes the reaction of ATP with adenylylsulfate to generate ADP and 3'-phosphoadenylylsulfate.[1-8] The ATP sulfurylase:adenylylsulfate complex does not serve as a substrate for adenylylsulfate kinase. Hence, there is no "channeling" between the two enzymes in the metabolic pathway.[1]

The *Escherichia coli* enzyme proceeds through a phospho-enzyme intermediate.[2,3] Interestingly, the reverse reaction is reported to be rapid equilibrium ordered Bi Bi for both the *Penicillium chrysogenum*[4] and *Escherichia coli*[2] enzymes. In the forward reaction, adenylylsulfate is a potent substrate inhibitor.[2,4] This inhibition is uncompetitive due to the formation of an E·adenylylsulfate·MgADP abortive complex.[5]

[1] F. Renosto, R. L. Martin & I. H. Segel (1989) *JBC* **264**, 9433.
[2] C. Satishchandran & G. D. Markham (1989) *JBC* **264**, 15012.
[3] C. Satishchandran, Y. N. Hickman & G. D. Markham (1992) *Biochemistry* **31**, 11684.

[4] F. Renosto, P. A. Seubert & I. H. Segel (1984) *JBC* **259**, 2113.
[5] I. J. MacRae & I. H. Segel (1999) *ABB* **361**, 277.
[6] T. S. Leyh (1993) *Crit. Rev. Biochem. Mol. Biol.* **28**, 515.
[7] A. Schmidt & K. Jäger (1992) *Ann. Rev. Plant Physiol. Plant Mol. Biol.* **43**, 325.
[8] H. D. Peck, Jr. (1974) *The Enzymes*, 3rd ed., **10**, 651.

Selected entries from *Methods in Enzymology* [**vol**, page(s)]:
General discussion: **5**, 974; **143**, 329
Assay: **5**, 974; **63**, 32; **143**, 342
Other molecular properties: absence in sulfate-reducing bacteria, **243**, 241; activity, **57**, 245; adenosine phosphosulfate assay and, **6**, 769; ATP-sulfurylase preparation free of, **5**, 972; catalyzed reaction, **143**, 334; properties, **5**, 975; purification (bakers' yeast, **5**, 974; *Penicillium chrysogenum*, **143**, 344); separation from ATP-sulfurylase, **5**, 971

ADENYLYLSULFATE REDUCTASE

This iron- and FAD-dependent oxidoreductase [EC 1.8.99.2] catalyzes the reversible reaction of adenylylsulfate with a reduced donor substrate to produce AMP, sulfite, and the oxidized acceptor.[1-6] Reduced methyl viologen can act as the donor substrate as can ferricyanide and reduced cytochrome *c*. This is the key enzyme of dissimilatory sulfate respiration. Free flavin coenzyme may be the electron carrier in the *Desulfovibrio vulgaris Miyazaki* enzyme.[2] Recent studies of the *Arabidopsis thaliana* protein identified SO_3^{2-} bound to a cysteinyl residue.[1] The intermediate could be release upon treatment of the enzyme with thiols, inorganic sulfide, and sulfite.

[1] M. Weber, M. Suter, C. Brunold & S. Kopriva (2000) *EJB* **267**, 3647.
[2] T. Yagi & M. Ogata (1996) *Biochimie* **78**, 838.
[3] K. Adachi & I. Suzuki (1977) *Can. J. Microbiol.* **55**, 91.
[4] Y. Hatefi & D. L. Stiggall (1976) *The Enzymes*, 3rd ed., **13**, 175.
[5] G. Palmer (1975) *The Enzymes*, 3rd ed., **12**, 1.
[6] H. D. Peck, R. Bramlett & D. V. DerVartanian (1972) *Z. Naturforsch.* **27B**, 1084.

Selected entries from *Methods in Enzymology* [**vol**, page(s)]:
General discussion: **243**, 241, 331, 400
Assay: **243**, 333; adenylylsulfate formation method, **243**, 406; *Archaeoglobus fulgidus* enzyme, **243**, 333; cytochrome *c*-dependent assay, **243**, 406; ferricyanide-dependent assay, **243**, 406; lithotrophs and heterotrophs, in, assay, **243**, 508; *Thiobacillus*, **243**, 393; *Thiocapsa roseopersicina* (BBS, **243**, 406; DSM 219, **243**, 406; MI, **243**, 406)
Other molecular properties: *Archaeoglobus fulgidus* enzyme (assay, **243**, 333; in dissimilatory sulfate reduction, **243**, 331; gene, cloning, **243**, 340, 346; properties, **243**, 336); *Chlorobium limicola* f. *thiosulfatophilum*, **243**, 417; *Chromatium minutissimum*, **243**, 406; *Chromatium vinosum*, **243**, 406; distribution, **243**, 243, 393; lithotrophs and heterotrophs, in, assay, **243**, 508; phototrophic sulfur metabolism, in, **243**, 401; purification, **243**, 611 (*Archaeoglobus fulgidus*, **243**, 334; *Desulfovibrio gigas*, **243**, 335; *Desulfovibrio vulgaris*, **243**, 335; sulfate-reducing bacteria, **243**, 244; *Thermodesulfobacterium commune*, **243**, 334; *Thermodesulfobacterium mobilis*, **243**, 335; *Thiobacillus denitrificans*, **243**, 395; *Thiobacillus thioparus*, **243**, 397; *Thiocapsa roseopersicina* (DSM 219), **243**, 415; *Thiocapsa roseopersicina* MI, **243**, 417); reaction catalyzed by, **243**, 393; sulfate-reducing bacteria, from, (activity studies, **243**, 245; assimilatory sulfate reduction pathway, **243**, 241; cellular localization, **243**, 246; comparison, **243**, 253; dissimilatory sulfate reduction pathway, **243**, 243; distribution, **243**, 244; EPR study, **243**, 247, 257; mechanism, **243**, 257;

molecular mass, **243**, 246; Mössbauer spectroscopy, **243**, 250; physicochemical characterization, **243**, 244; properties, **243**, 253; redox properties, **243**, 251, 259; stability, **243**, 245; subunits, **243**, 246; ultraviolet–visible spectroscopy, **243**, 246, 256); *Thiobacillus* (assay, **243**, 393; concentration, **243**, 422; electron acceptors, **243**, 399; inhibitors, **243**, 400; molecular weight, **243**, 398; pH optimum, **243**, 399; purification, **243**, 394; purity, **243**, 397; stability, **243**, 398; substrate specificity, **243**, 399; unit of enzyme activity, **243**, 394); *Thiocapsa roseopersicina* (BBS, assay, **243**, 406; DSM 219 [assay, **243**, 406; preparation and properties, **243**, 415]; MI [assay, **243**, 406; purification and properties, **243**, 417]); *Thiocystis violacea*, properties, **243**, 416

ADENYLYLSULFATE REDUCTASE (GLUTATHIONE)

This oxidoreductase [EC 1.8.4.9], also known as plant-type 5′-adenylylsulfate reductase and 5′-adenylylsulfate reductase (a term also used for EC 1.8.99.2), catalyzes the reaction of adenylyl sulfate (*i.e.*, adenosine-5′-phosphosulfate) with two molecules of glutathione (*i.e.*, L-γ-glutamyl-L-cysteinylglycine) to produce AMP, sulfite, and glutathione disulfide.[1–3] This enzyme is distinct from adenylyl-sulfate reductase (acceptor) [EC 1.8.99.2] in that glutathione is the reductant: however, glutathione can be replaced by γ-L-glutamyl-L-cysteine or dithiothreitol.

[1] J. F. Gutierrez-Marcos, M. A. Roberts, E. I. Campbell & J. L. Wray (1996) *PNAS* **93**, 13377.
[2] A. Setya, M. Murillo & T. Leustek (1996) *PNAS* **93**, 13383.
[3] J. A. Bick, F. Aslund, Y. Cen & T. Leustek (1998) *PNAS* **95**, 8404.

ADP DEAMINASE

This deaminase [EC 3.5.4.7] catalyzes the hydrolysis of ADP to produce IDP and ammonia. Other substrates include ATP, AMP, dADP, cAMP, and 3′-AMP.[1,2]

[1] S.-T. Chung, K. Aida & T. Uemura (1967) *J. Gen. Appl. Microbiol.* **13**, 335.
[2] S.-T. Chung, S. Ito, K. Aida & T. Uemura (1968) *J. Gen. Appl. Microbiol.* **14**, 111.

ADP-DEPENDENT MEDIUM-CHAIN-ACYL-CoA HYDROLASE

This hydrolase [EC 3.1.2.19], which requires the presence of ADP (ATP and AMP are not effectors), catalyzes the hydrolysis of a medium-chain acyl-CoA to produce free coenzyme A and a carboxylate anion. The best substrate reported is nonanoyl-CoA. The hydrolase is also inhibited by β-NADH, with no effect observed with α-NADH and NAD+.[1]

[1] S. E. Alexson & J. Nedergaard (1988) *JBC* **263**, 13564.

ADP-DEPENDENT SHORT-CHAIN-ACYL-CoA HYDROLASE

This hydrolase [EC 3.1.2.18], which requires the essential activator ADP for activity (ATP and AMP are not effectors), catalyzes the hydrolysis of a short-chain acyl-CoA to produce free coenzyme A and a carboxylate anion. The best substrate reported is propanoyl-CoA. The hydrolase is also inhibited by β-NADH (α-NADH and NAD+ have no effect).[1]

[1] S. E. Alexson & J. Nedergaard (1988) *JBC* **263**, 13564.

ADP-*GLYCEROMANNO*-HEPTOSE 6-EPIMERASE

This epimerase [EC 5.1.3.20], also known as ADP-L-*glycero*-D-*manno*-heptose 6-epimerase, catalyzes the interconversion of ADP-D-*glycero*-D-*manno*-heptose and ADP-L-*glycero*-D-*manno*-heptose. The *Escherichia coli* enzyme contains a tightly bound NAD+.[1]

[1] L. Ding, B. L. Seto, S. A. Ahmed & W. G. Coleman, Jr. (1994) *JBC* **269**, 24384.

ADP-PHOSPHOGLYCERATE PHOSPHATASE

This enzyme [EC 3.1.3.28] catalyzes the hydrolysis of 3-(ADP)-2-phosphoglycerate to produce 3-(ADP)glycerate and orthophosphate.[1] Other substrates include 2,3-bisphosphoglycerate.

[1] G. T. Zancan, E. F. Recondo & L. F. Leloir (1964) *BBA* **92**, 125.

ADP-RIBOSE PYROPHOSPHATASE

This enzyme [EC 3.6.1.13], also known as ADP-ribose diphosphatase and ADP-ribose phosphohydrolase, catalyzes the hydrolysis of ADP-D-ribose to produce AMP and D-ribose 5-phosphate.[1–3]

[1] J. M. Wu, M. B. Lennon & R. J. Suhadolnik (1978) *BBA* **520**, 588.
[2] M. D. Doherty & J. F. Morrison (1962) *BBA, Short Comm.* **65**, 364.
[3] A. Miro, M. T. Hernandez, M. J. Costas & J. C. Cameselle (1991) *J. Biochem. Biophys. Meth.* **22**, 177.

ADP-RIBOSYLARGININE HYDROLASE

This enzyme [EC 3.2.2.19], also known as ADP-ribose-L-arginine cleaving enzyme, catalyzes the hydrolysis of N^2-(ADP-D-ribosyl)-L-arginine to produce L-arginine and ADP-D-ribose.[1,2] ADP-D-ribose is a potent inhibitor of the hydrolase.[2–3]

[1] J. Moss, S. C. Tsai, R. Adamik, H. C. Chen & S. J. Stanley (1988) *Biochemistry* **27**, 5819.
[2] J. Moss, N. J. Oppenheimer, R. E. West & S. J. Stanley (1986) *Biochemistry* **25**, 5408.
[3] P. Konczalik & J. Moss (1999) *JBC* **274**, 16736.

ADP-RIBOSYL-[DINITROGEN REDUCTASE] HYDROLASE

This enzyme [EC 3.2.2.24], also called dinitrogenase reductase-activating glycohydrolase (DRAG), catalyzes the hydrolysis of ADP-D-ribosyl-[dinitrogen reductase] to produce ADP-D-ribose and [dinitrogen reductase].[1,2] Dinitrogen reductase is a component of the nitrogenase complex. This enzyme, along with NAD$^+$:[dinitrogen reductase] ADP-D-ribosyltransferase [EC 2.4.2.37], regulates the activity of nitrogenase [EC 1.18.6.1]. The hydrolase had been reported to be rapidly inactivated in air, owing to the presence of sodium dithionite and Mn^{2+} throughout the purification of the enzyme.[3]

The redox state of nitrogenase Fe protein plays a significant role in regulation of the activities of the hydrolase *in vivo*.[4]

[1] L. L. Saari, E. W. Triplett & P. W. Ludden (1984) *JBC* **259**, 15502.
[2] M. R. Pope, L. L. Saari & P. W. Ludden (1986) *JBC* **261**, 10104.
[3] G. M. Nielsen, Y. Bao, G. P. Roberts & P. W. Ludden (1994) *BJ* **302**, 801.
[4] C. M. Halbleib, Y. Zhang & P. W. Ludden (2000) *JBC* **275**, 3493.

ADP-SUGAR PYROPHOSPHATASE

This enzyme [EC 3.6.1.21], also known as ADP-sugar diphosphatase, catalyzes the hydrolysis of an ADP-sugar to produce AMP and the corresponding sugar 1-phosphate.[1–3] Substrates include ADP-D-glucose, ADP-D-mannose, ADP-D-ribose, and ADP-riboflavin. This substrate specificity is distinct from that of UDP-sugar pyrophosphatase [EC 3.6.1.45].

[1] R. Providenzia, S. T. Bass & R. G. Hansen (1968) *BBA* **167**, 199.
[2] L. Glaser, A. Melo & R. Paul (1967) *JBC* **242**, 1944.
[3] A. Melo & L. Glaser (1966) *Biochem. Biophys. Res. Commun.* **22**, 524.

Selected entries from *Methods in Enzymology* [vol, page(s)]:
Other molecular properties: properties, **28**, 975; purification from *Escherichia coli*, **28**, 973

ADP:THYMIDINE KINASE

This phosphotransferase [EC 2.7.1.118] catalyzes the reaction of ADP with thymidine to produce AMP and thymidine 5'-phosphate (dTMP). This reaction is also catalyzed by the deoxypyrimidine kinase complex induced by herpes simplex virus.[1] **See also** *Thymidine Kinase; AMP:Thymidine Kinase; dTMP Kinase*

[1] D. Falke, J. Labenz, D. Brauer & W. E. Müller (1982) *BBA* **708**, 99.

[β-ADRENERGIC RECEPTOR] KINASE

This phosphotransferase [EC 2.7.1.126] catalyzes the reaction of ATP with the β-adrenergic receptor (*i.e.*, the agonist-occupied form of the receptor) to produce ADP and [β-adrenergic receptor] phosphate.[1,2] Rhodopsin is a somewhat poorer alternative substrate, and the enzyme will not act on casein or histones. This enzyme is inhibited by ammonia, heparin, and digitonin but is unaffected by cyclic-AMP. The enzyme exhibits a marked preference for acidic aminoacyl residues at the amino-terminal side of the seryl or threonyl residue being phosphorylated.[3] **See also** *Rhodopsin Kinase*

[1] J. L. Benovic, R. Strasser, M. G. Caron & R. J. Lefkowitz (1986) *PNAS* **83**, 2797.
[2] J. L. Benovic, J. W. Regan, H. Matsui, F. Mayor, S. Cotecchia, L. M. Leeb-Lundberg, M. G. Caron & R. J. Lefkowitz (1987) *JBC* **262**, 17251.
[3] J. J. Onorato, K. Palczewski, J. W. Regan, M. G. Caron, R. J. Lefkowitz & J. L. Benovic (1991) *Biochemistry* **30**, 5118.

Selected entries from *Methods in Enzymology* [vol, page(s)]:
General discussion: 200, 351
Assay: 200, 353
Other molecular properties: β-adrenergic receptor kinase 1 (expression systems, **250**, 150; G protein association, **250**, 150; polyisoprenoid analysis by HPLC, **250**, 153; radiolabeling with mevalonic acid, **250**, 152; translocation, **250**, 149 [assay, **250**, 155; effect of prenylation, **250**, 149, 157]); characterization, **200**, 351; inhibition, **200**, 360; purification from cerebral cortex, **200**, 355; structure, **200**, 362; substrate specificity, **200**, 359

ADRENODOXIN REDUCTASE

This FAD-dependent oxidoreductase, currently regarded as a variant of ferredoxin:NADP$^+$ reductase [EC 1.18.1.2], catalyzes the reversible reaction of two molecules of oxidized adrenodoxin with NADPH to produce two molecules of reduced adrenodoxin and NADP$^+$. **See** *Ferredoxin:NADP$^+$ Reductase*

AEROLYSIN

This archaebacterial protease, a toxin that causes eukaryotic cell death by producing channels in cell membranes, from the hyperthermophile *Pyrobaculum aerophilum* is optimally active at 100°C. The enzyme resembles subtilisin in its action.[1] The term is confusing, because aerolysin also is used to refer to the soluble toxin protein produced by *Aeromonas hydrophila*.[2]

[1] P. Völkl, P. Markiewicz, K. O. Stetter & J. I. Miller (1994) *Prot. Sci.* **3**, 1329.
[2] J. T. Buckley & S. P. Howard (1988) *MIE* **165**, 193.

AERUGINOLYSIN

This zinc-dependent metallopeptidase [EC 3.4.24.40] was first isolated from the growth medium of *Pseudomonas aeruginosa* (earlier described as *Pseudomonas myxogenes*) and initially called *Pseudomonas aeroginosa* alkaline proteinase.[1,2] It is closely related to serralysin (the sequence is 55% identical to the serralysin of *Serratia* sp. E-15) and is listed under serralysin's EC classification number. The short peptides Z–Gly–Leu-∇-Gly–Gly–Ala, Z–Ala–Gly-∇-Gly–Leu–Ala, and Abz–Gly–Phe–Arg-∇-Leu–Leu–4-nitrobenzylamide are all effective artificial substrates. (Note: ∇ indicates the cleavage site.)

[1] K. Morihara (1957) *Bull. Agric. Chem. Soc. Japan* **21**, 11.
[2] H. Maeda & K. Morihara (1993) *MIE* **248**, 395.

*Afe*I RESTRICTION ENDONUCLEASE

This type II restriction endonuclease [EC 3.1.21.4], obtained from *Alcaligenes faecalis* T2774, catalyzes the hydrolysis of both strands of DNA at 5′ . . . AGC∇GCT . . . 3′, producing blunt-ended fragments.[1]

[1] P. R. Whitehead & N. L. Brown (1985) *J. Gen. Microbiol.* **131**, 951.

*Afl*II RESTRICTION ENDONUCLEASE

This type II restriction endonuclease [EC 3.1.21.4], obtained from *Anabaena flos-aquae*, catalyzes the hydrolysis of both strands of DNA at 5′ . . . C∇TTAAG . . . 3′.[1]

[1] P. R. Whitehead & N. L. Brown (1985) *J. Gen. Microbiol.* **131**, 951.

*Afl*III RESTRICTION ENDONUCLEASE

This type II restriction endonuclease [EC 3.1.21.4], obtained from *Anabaena flos-aquae*, catalyzes the hydrolysis of both strands of DNA at 5′ . . . A∇CRYGT . . . 3′, where R refers to either A or G and Y refers to either T or C.[1]

[1] P. R. Whitehead & N. L. Brown (1985) *J. Gen. Microbiol.* **131**, 951.

AGARASE

This enzyme [EC 3.2.1.81] catalyzes the hydrolysis of D-galactosidic linkages in agarose, producing the tetramer as the predominant product (*i.e.*, neoagarotetraose). Other substrates include porphyran. β-Agarases act on the β-linkages in agarose whereas α-agarases act on the α1,3-linkages. Agarases from certain sources will also act on neoagarotetraoses.

AGARITINE γ-GLUTAMYLTRANSFERASE

This transferase [EC 2.3.2.9] catalyzes the reaction of agaritine with an acceptor substrate to generate 4-hydroxymethylphenylhydrazine and the γ-glutamyl acceptor.[1] Substrate acceptors include 4-hydroxyaniline, cyclohexylamine, 1-naphthylhydrazine, and similar compounds. The enzyme will also catalyze the hydrolysis of agaritine. **See also** *γ-Glutamyl Transpeptidase*

[1] H. J. Gigliotti & B. Levenberg (1964) *JBC* **239**, 2274.
Selected entries from ***Methods in Enzymology*** [vol, page(s)]:
General discussion: 17A, 877

*Age*I RESTRICTION ENDONUCLEASE

This type II restriction endonuclease [EC 3.1.21.4], obtained from *Agrobacterium gelatinovorum*, catalyzes the hydrolysis of both strands of DNA at 5′ . . . A∇CCGGT . . . 3′.[1,2]

[1] Y. Yamada, H. Mizuno, H. Sato, M. Akagawa & K. Yamasato (1989) *Agric. Biol. Chem.* **53**, 1747.
[2] H. Mizuno, T. Suzuki, M. Akagawa, K. Yamasato & Y. Yamada (1990) *Agric. Biol. Chem.* **54**, 1797.

Ag⁺-EXPORTING ATPase

This so-called P-type ATPase [EC 3.6.3.53], also known as silver-transporting ATPase, catalyzes the ATP-dependent (ADP- and orthophosphate-producing) transport of Ag^+_{in} to produce Ag^+_{out}:[1,2] thus, it directs the ATP-dependent transport of silver ions out of certain microorganisms and animals.

The idea that a transporter is an enzyme is in keeping with a new definition of enzyme catalysis as the facilitated making/breaking of chemical bonds, not just covalent bonds.[3] This idea builds on Pauling's assertion that any long-lived, chemically distinct interaction (in this case, the persistent location of a solute with respect to the faces of a membrane) can be regarded as a chemical bond. Note also that the equilibrium constant ($K_{eq} = [Ag^+_{out}][ADP][P_i]/[Ag^+_{in}][ATP]$) does not conform to that expected for an ATPase (*i.e.*, $K_{eq} = [ADP][P_i]/[ATP]$). Thus, although the overall reaction yields ADP and orthophosphate, the enzyme is misclassified as a hydrolase, and instead should be regarded as an energase-type reaction. Energases facilitate affinity-modulated reactions by coupling the $\Delta G_{ATP-hydrolysis}$ to a force-generating or work-producing step.[3] In this case, P-O-P bond-scission supplies the energy to drive silver ion transport. **See** *ATPase; Energase*

[1] A. Gupta, K. Matsui, J. F. Lo & S. Silver (1999) *Nature Med.* **5**, 183.
[2] N. R. Bury, M. Grosell, A. K. Grover & C. M. Wood (1999) *Toxicol. Appl. Pharmacol.* **159**, 1.
[3] D. L. Purich (2001) *TiBS* **26**, 417.

AGAVAIN

This protease, formerly classified as EC 3.4.99.2, is now a deleted Enzyme Commission entry.

AGMATINASE

This manganese-dependent enzyme [EC 3.5.3.11], also known as agmatine ureohydrolase, catalyzes the hydrolysis of agmatine to produce putrescine and urea.[1–3]

[1] C. Satishchandran & S. M. Boyle (1986) *J. Bacteriol.* **165**, 843.
[2] C. Vicente & M. E. Legaz (1982) *Physiol. Plant.* **55**, 335.
[3] N. Carvajal, J. Olate, M. Salas, V. Lopez, J. Cerpa, P. Herrera & E. Uribe (1999) *BBRC* **264**, 196.

Selected entries from *Methods in Enzymology* [vol, page(s)]:
General discussion: 94, 117
Other molecular properties: *Escherichia coli* (deficient mutants, screening, **94**, 90; gene cloning, **94**, 118)

AGMATINE 4-COUMAROYLTRANSFERASE

4-coumaroyl-CoA   *N*-(4-guanidinobutyl)-4-hydroxycinnamide

This transferase [EC 2.3.1.64], also known as *p*-coumaroyl-CoA:agmatine *N-p*-coumaroyltransferase, catalyzes the reaction of 4-coumaroyl-CoA with agmatine to produce coenzyme A and *N*-(4-guanidinobutyl)-4-hydroxycinnamamide.[1] Other acceptor substrates include feruloyl-CoA, caffeoyl-CoA, sinapoyl-CoA, and cinnamoyl-CoA.

[1] C. R. Bird & T. A. Smith (1981) *Phytochemistry* **20**, 2345.

Selected entries from *Methods in Enzymology* [vol, page(s)]:
General discussion: 94, 344
Assay: barley seedling, 94, 345
Other molecular properties: barley seedling (properties, **94**, 347; purification, **94**, 346; substrate synthesis, **94**, 344)

AGMATINE DEIMINASE

This enzyme [EC 3.5.3.12], also known as agmatine iminohydrolase, catalyzes the hydrolysis of agmatine to produce *N*-carbamoylputrescine and ammonia.[1–4]

agmatine   *N*-carbamoylputrescine

The enzyme from several plants also catalyzes the reactions of EC 2.1.3.3 (ornithine carbamoyltransferase), EC 2.1.3.6 (putrescine carbamoyltransferase), and EC 2.7.2.2 (carbamate kinase), thereby functioning as a putrescine synthase that converts agmatine and ornithine into putrescine and citrulline, respectively.

[1] T. A. Smith (1985) *Ann. Rev. Plant Physiol.* **36**, 117.
[2] C. W. Tabor & H. Tabor (1984) *Ann. Rev. Biochem.* **53**, 749.
[3] R. K. Sindhu & H. V. Desai (1979) *Phytochemistry* **18**, 1937.
[4] H. Yanagisawa & Y. Suzuki (1981) *Plant Physiol.* **67**, 697.

AGMATINE KINASE

This phosphotransferase [EC 2.7.3.10], also known as phosphagen phosphokinase, catalyzes the reaction of ATP with agmatine to produce ADP and N^4-phosphoagmatine.[1]

N^4-phosphoagmatine

Other acceptor substrates include L-arginine, albeit not as effectively. N^4-Phosphoagmatine acts as a phosphagen in a number of invertebrates (for example, the sponge *Geodia gigas*).

[1] E. Piccinni & O. Coppellotti (1979) *Comp. Biochem. Physiol.* **62B**, 287.

*Aha*III RESTRICTION ENDONUCLEASE

This type II restriction endonuclease [EC 3.1.21.4], obtained from *Aphanothece halophytica*, catalyzes the hydrolysis of both strands of DNA at 5′ . . . TTT∇AAA . . . 3′.[1] Note that the recognition site contains only A and T.

[1] P. R. Whitehead & N. L. Brown (1982) *FEBS Lett.* **143**, 296.

*Ahd*I RESTRICTION ENDONUCLEASE

This type II restriction endonuclease [EC 3.1.21.4], obtained from *Aeromonas hydrophila*, catalyzes the hydrolysis of both strands of DNA at 5′ . . . GACNNN∇NNGTC . . . 3′, where N refers to any base.[1] The resulting DNA fragments have a single-base 3′ extension that is more difficult to ligate than blunt-ended fragments.

[1] D. Mernagh, P. Marks & G. Kneale (1999) *Biochem. Soc. Trans.* **27**, A126.

D-ALANINE:ALANYL-POLY(GLYCEROLPHOSPHATE) LIGASE

This enzyme [EC 6.3.2.16], also known as D-alanyl-alanyl-poly(glycerolphosphate) synthetase and D-alanine:membrane-acceptor ligase, catalyzes the reaction of ATP with D-alanine and alanyl-poly(glycerolphosphate) to produce D-alanyl-alanyl-poly(glycerolphosphate), ADP, and orthophosphate.[1,2] This enzyme participates in teichoic acid biosynthesis.

[1] F. Neuhaus, R. Linzer & V. M. Reusch (1974) *Ann. N. Y. Acad. Sci.* **235**, 502.
[2] V. M. Reusch & F. C. Neuhaus (1971) *JBC* **246**, 6136.

D-ALANINE AMINOTRANSFERASE

This pyridoxal-phosphate-dependent aminotransferase [EC 2.6.1.21] (also known as D-aspartate aminotransferase, D-amino acid aminotransferase, and D-amino acid transaminase) catalyzes the reversible reaction of D-alanine with α-ketoglutarate (or, 2-oxoglutarate) to produce pyruvate and D-glutamate.[1–8] Other substrates include the D-stereoisomers of leucine, aspartate, glutamate, 2-aminobutyrate, norvaline, norleucine, and asparagine. The kinetic mechanism is ping pong Bi Bi.[6]

The coenzyme is initially bound as an internal aldimine form (i.e., a Schiff base linkage with an active site lysyl residue). A transaldimination occurs subsequent to binding of the D-amino acid substrate, forming an external aldimine. The lysyl residue now acts as a catalytic base and a 1,3 prototropic shift converts the internal aldimine into a ketimine. Hydrolysis of the ketimine generates the pyridoxamine phosphate form of the coenzyme and the new α-keto acid. The second half-reaction is the reversal of these steps with a different keto acid.

D-Alanine aminotransferase's stereospecificity is determined in large part by a group of three aminoacyl residues (tyrosyl, arginyl, and histidyl) forming four hydrogen bonds to the substrate's α-carboxyl group.[2] Interestingly, the pyridoxamine phosphate form of the coenzyme is bound to the enzyme with its re side facing the protein and the catalytic lysyl residue. In L-aspartate aminotransferase [EC 2.6.1.1] and other L-amino acid transferases, the si side of the coenzyme faces the protein.

Because the enzyme is crucial in providing D-alanine, D-glutamate, and other D-amino acids for bacterial cell wall synthesis, it represents a target enzyme for the development of antimicrobial agents.

[1] P. K. Mehta & P. Christen (2000) AE 74, 129.
[2] D. Peisach, D. M. Chipman, P. W. Van Ophem, J. M. Manning & D. Ringe (1998) Biochemistry 37, 4958.
[3] R. A. John (1998) CBC 2, 173.
[4] H. Hayashi, H. Wada, T. Yoshimura, N. Esaki & K. Soda (1990) Ann. Rev. Biochem. 59, 87.
[5] C. T. Walsh (1984) Ann. Rev. Biochem. 53, 493.
[6] J. P. Gosling & P. F. Fottrell (1973) Biochem. Soc. Trans. 1, 252.
[7] A. E. Braunstein (1973) The Enzymes, 3rd ed., 9, 379.
[8] M. Martinez-Carrion & W. T. Jenkins (1965) JBC 240, 3538.

Selected entries from **Methods in Enzymology** [vol, page(s)]:
General discussion: 17A, 167; 113, 108
Assay: 113, 109, 110
Other molecular properties: properties, 113, 112; purification from (Bacillus sphaericus, 113, 110; Bacillus subtilis, 17A, 167); pyridoxal phosphate and, 17A, 171; site-directed mutants, acid-base catalysis, 249, 117.

L-ALANINE AMINOTRANSFERASE

This pyridoxal-phosphate-dependent aminotransferase [EC 2.6.1.2], also known as glutamic–pyruvic transaminase and glutamic–alanine transaminase, catalyzes the reversible reaction of L-alanine with α-ketoglutarate (or, 2-oxoglutarate) producing pyruvate and L-glutamate.[1–6] 2-Aminobutanoate will also act as the amino donor, albeit slowly. Some isozymes from certain sources are also very active with β-alanine. The enzyme has a ping pong Bi Bi kinetic mechanism (for more on mechanism, **See** D-Alanine Aminotransferase).[1–4,6]

Alanine:oxo-acid aminotransferase [EC 2.6.1.12] also catalyzes the pyridoxal-phosphate-dependent reaction of L-alanine with an α-keto acid to generate pyruvate and an α-amino acid. **See also** Alanine:Glyoxylate Aminotransferase; Alanine:Oxo-Acid Aminotransferase; D-Alanine Aminotransferase

[1] B. Lain-Guelbenzu, J. Cardenas & J. Munoz-Blanco (1991) EJB 202, 881.
[2] N. Tamaki, H. Aoyama, K. Kubo, T. Ikeda & T. Hama (1982) J. Biochem. 92, 1009.
[3] G. De Rosa, T. L. Burk & R. W. Swick (1979) BBA 567, 116.
[4] J. Rech & J. Crouzet (1974) BBA 350, 392.
[5] A. E. Braunstein (1973) The Enzymes, 3rd ed., 9, 379.
[6] B. Bulos & P. Handler (1965) JBC 240, 3283.

Selected entries from **Methods in Enzymology** [vol, page(s)]:
General discussion: 17A, 153; 113, 69
Assay: 113, 70; dry reagent chemistries, with, 137, 414; Fast Analyzer, by, 31, 816; serum, in, 17B, 868
Other molecular properties: affinity partitioning, 228, 126, 128; L-alanine determination, 8, 698, 699; clinical aspects, 17B, 866; glucocorticoid effect on, 17A, 155, 159; immobilization and assay, 136, 488, 490; mechanism, 63, 51; multienzymatic NADH recycling, in, 136, 71; natural occurrence, 113, 69; nuclear labels, indirect detection with spin echoes, 176, 347; purification from (porcine heart, 17A, 159, 805; rat liver, 17A, 153); pyridoxal phosphate and, 17A, 158, 163; pyridoxal phosphate-binding site, isolation and sequence determination, 106, 123; serum, in, assay of, 17B, 868

ALANINE CARBOXYPEPTIDASE

A zinc-dependent carboxypeptidase [EC 3.4.17.6] catalyzes the hydrolysis of peptidyl-L-alanine to produce a peptide and L-alanine. There appears to be a preference for substrates having an aromatic aminoacyl residue at the penultimate position, and the enzyme will hydrolyze folate analogues such as 4-amino-4-deoxypteroylalanine and 4-aminobenzoylalanine.[1,2] A similar enzyme isolated from the soil microorganism *Corynebacterium equi* H-7 will not act on Ac-Ala or Z-Ala. Although the Enzyme Commission lists both enzymes under EC 3.4.17.6, they are clearly distinct, both in their action and properties.

[1] C. C. Levy & P. Goldman (1969) *JBC* **244**, 4467.
[2] E. Miyagawa, T. Harada & Y. Motoki (1986) *Agric. Biol. Chem.* **50**, 1527.

L-ALANINE DEHYDROGENASE

This oxidoreductase [EC 1.4.1.1] catalyzes the reaction of L-alanine with water and NAD^+ to produce pyruvate, ammonia, and NADH.[1–6] The *Bacillus subtilis* enzyme has a predominantly steady-state ordered kinetic mechanism, in which NAD^+ binds prior to the amino acid; however, above pH 10.9, the mechanism becomes rapid equilibrium ordered.[5,6]

[1] A. R. Clarke & T. R. Dafforn (1998) *CBC* **3**, 1.
[2] N. M. W. Brunhuber & J. S. Blanchard (1994) *Crit. Rev. Biochem. Mol. Biol.* **29**, 415.
[3] M. T. Smith & D. W. Emerich (1993) *JBC* **268**, 10746.
[4] P. M. Weiss, C.-Y. Chen, W. W. Cleland & P. F. Cook (1988) *Biochemistry* **27**, 4814.
[5] C. E. Grimshaw & W. W. Cleland (1981) *Biochemistry* **20**, 5650.
[6] C. E. Grimshaw, P. F. Cook & W. W. Cleland (1981) *Biochemistry* **20**, 5655.

Selected entries from *Methods in Enzymology* [**vol**, page(s)]:
General discussion: 5, 673; **17A**, 176
Assay: 5, 673
Other molecular properties: binding, to agarose-bound nucleotides, **66**, 196; commitment, **87**, 631; distribution, **5**, 676; enzyme reactor, in, **44**, 880; *Escherichia coli* and, **1**, 136; glutamate dehydrogenase, and, **87**, 405, 653; histidine, active site, **87**, 405; intermediate, **87**, 405; isomerization, **87**, 397; isotope effect, **87**, 395, 397, 631; isotope exchange study, fingerprint kinetic patterns for modifier action, **87**, 652; pH effects, **87**, 395, 397, 405; properties, **5**, 675; purification from *Bacillus subtilis*, **5**, 674; **17A**, 177; stereochemistry, **87**, 116, 405; sticky substrates, **87**, 395, 631; synthesis in *Chlamydomonas* cultures, **23**, 73; transition-state and multisubstrate analogues, **249**, 304

D-ALANINE γ-GLUTAMYLTRANSFERASE

This transferase [EC 2.3.2.14] catalyzes the reaction of L-glutamine with D-alanine to produce γ-L-glutamyl-D-alanine and ammonia. In addition to D-alanine, D-phenylalanine and D-2-aminobutyrate can also act as γ-glutamyl acceptor substrates.[1]

[1] Y. Kawasaki, T. Ogawa & K. Sasaoka (1982) *BBA* **716**, 194.

ALANINE:GLYOXYLATE AMINOTRANSFERASE

This pyridoxal-phosphate-dependent aminotransferase [EC 2.6.1.44] catalyzes the reversible reaction of L-alanine with glyoxylate to produce pyruvate and glycine.[1,2] One component of the animal enzyme can utilize 2-oxobutanoate (*i.e.*, α-ketobutyrate) as a substrate instead of glyoxylate. Alternative substrates for L-alanine include L-phenylalanine, L-glutamine, L-tyrosine, and L-leucine. A second component of the enzyme can also catalyze the reaction of L-alanine with 3-hydroxypyruvate to produce pyruvate and L-serine. The enzyme has a ping pong Bi Bi kinetic mechanism. ***See also*** *Serine:Pyruvate Aminotransferase*

[1] Y. Nakamura & N. E. Tolbert (1983) *JBC* **258**, 7631.
[2] A. E. Braunstein (1973) *The Enzymes*, 3rd ed., **9**, 379.

Selected entries from *Methods in Enzymology* [**vol**, page(s)]:
General discussion: 17A, 163
Other molecular properties: leucine:glyoxylate aminotransferase, microbodies, in, **52**, 496; purification from human liver, **17A**, 164; pyridoxal phosphate and, **17A**, 166

D-ALANINE HYDROXYMETHYLTRANSFERASE

This transferase [EC 2.1.2.7], also called 2-methylserine hydroxymethyltransferase, catalyzes the reversible reaction of D-alanine with water and 5,10-methylenetetrahydrofolate to produce tetrahydrofolate and 2-methylserine.[1] The enzyme can also use 2-hydroxymethylserine as a substrate in the reverse reaction to generate D-serine.

[1] E. M. Wilson & E. E. Snell (1962) *JBC* **237**, 3171 and 3180.

Selected entries from *Methods in Enzymology* [**vol**, page(s)]:
General discussion: 17B, 341
Assay: 17B, 341

ALANINE:OXO-ACID AMINOTRANSFERASE

This pyridoxal-phosphate-dependent aminotransferase [EC 2.6.1.12] catalyzes the reversible reaction of L-alanine with a 2-oxo acid anion (α-keto acid anion) to produce pyruvate and an L-α-amino acid.[1,2] Keto substrates include 2-oxobutanoate (producing 2-aminobutanoate), 2-oxohexanoate (producing norleucine), and 2-oxopentanoate (producing norvaline).

[1] R. A. Altenbern & R. D. Housewright (1953) *JBC* **204**, 159.
[2] Y. Koide, M. Honma & T. Shimomura (1977) *Agric. Biol. Chem.* **41**, 781.

ALANINE:OXOMALONATE AMINOTRANSFERASE

This pyridoxal-phosphate-dependent aminotransferase [EC 2.6.1.47], also known as L-alanine:ketomalonate transaminase, catalyzes the reversible reaction of L-alanine with oxomalonate (*i.e.*, ketomalonate) to produce pyruvate and aminomalonate.[1] In addition to L-alanine, other amino donors include L-glutamate, L-2-aminobutyrate, and L-aspartate.

[1] H. Nagayama, M. Muramatsu & K. Shimura (1958) *Nature* **181**, 417.

D-ALANINE:POLY(PHOSPHORIBITOL) LIGASE

This ligase [EC 6.1.1.13], also known as D-alanyl-poly(phosphoribitol) synthetase, catalyzes the reaction of ATP with D-alanine and poly(ribitol phosphate) to produce AMP, pyrophosphate (or, diphosphate), and O-D-alanyl-poly(ribitol phosphate).[1–3] The enzyme, a participant in the biosynthesis of teichoic acids, reportedly catalyzes the reaction by means of a D-alanyl-AMP intermediate.[1]

[1]M. Perego, P. Glaser, A. Minutello, M. A. Strauch, K. Leopold & W. Fischer (1995) *JBC* **270**, 15598.
[2]V. M. Reusch & F. C. Neuhaus (1971) *JBC* **246**, 6136.
[3]J. Baddiley & F. C. Neuhaus (1960) *BJ* **75**, 579.

β-ALANINE:PYRUVATE AMINOTRANSFERASE

This pyridoxal-phosphate-dependent aminotransferase [EC 2.6.1.18], also known as β-alanine aminotransferase and ω-amino acid aminotransferase, catalyzes the reversible reaction of β-alanine with pyruvate to produce 3-oxopropanoate (*i.e.*, malonate semialdehyde) and L-alanine.[1–5] Other amino donor substrates include 4-aminobutanoate and 3-aminoisobutanoate.

[1]O. W. Griffith (1986) *Ann. Rev. Biochem.* **55**, 855.
[2]Y. Nakano, H. Tokunaga & S. Kitaoka (1977) *J. Biochem.* **81**, 1375.
[3]A. E. Braunstein (1973) *The Enzymes*, 3rd ed., **9**, 379.
[4]R. A. Stinson & M. S. Spencer (1969) *BBRC* **34**, 120.
[5]O. Hayashi, Y. Nishizuka, M. Tatibana, M. Takeshita & S. Kuno (1961) *JBC* **236**, 781.

Selected entries from *Methods in Enzymology* [vol, page(s)]:
General discussion: 5, 698; 143, 500
Assay: 5, 698; 143, 500

ALANINE RACEMASE

This pyridoxal-phosphate-dependent racemase [EC 5.1.1.1] catalyzes the interconversion of D-alanine and L-alanine.[1–11] A tyrosyl and lysyl residue have been proposed to be general acid/base catalysts in a two-base racemization mechanism.[1–4] The lysyl residue abstracts the proton from the C2 position of the D-alanine aldimine intermediate. The resulting quinonoid structure can now be protonated by the tyrosyl residue to produce the L-alanine aldimine intermediate.

[1]S. Sun & M. D. Toney (1999) *Biochemistry* **38**, 4058.
[2]M. E. Tanner & G. L. Kenyon (1998) *CBC* **2**, 7.
[3]R. A. John (1998) *CBC* **2**, 173.
[4]J. P. Shaw, G. A. Petsko & D. Ringe (1997) *Biochemistry* **36**, 1329.
[5]K. Soda & N. Esaki (1994) *Pure Appl. Chem.* **66**, 709.
[6]T. Yoshimura, N. Esaki & K. Soda (1992) *Bull. Inst. Chem. Res.* **70**, 378.
[7]M. A. Ator & P. R. Ortez de Montellano (1990) *The Enzymes*, 3rd ed., **19**, 213.

[8]K. Soda & K. Tanizawa (1990) *Ann. N. Y. Acad. Sci.* **585**, 386.
[9]A. E. Martell (1982) *AE* **53**, 163.
[10]E. Adams (1976) *AE* **44**, 69.
[11]E. Adams (1972) *The Enzymes*, 3rd ed., **6**, 479.

Selected entries from *Methods in Enzymology* [vol, page(s)]:
General discussion: 2, 212; **17A**, 171
Assay: 2, 212
Other molecular properties: L-alanine dehydrogenase, presence in preparation of, **5**, 676; properties, **2**, 215; purification from (*Pseudomonas*, **17A**, 173; *Streptococcus faecalis*, **2**, 213); transition-state and multisubstrate analogues, **249**, 308

ALANOPINE DEHYDROGENASE

This oxidoreductase [EC 1.5.1.17] catalyzes the reversible reaction of 2,2′-iminodipropanoate (or, alanopine) with water and NAD$^+$ to produce L-alanine, pyruvate, and NADH.[1–4] In the reverse reaction, L-alanine can be replaced as a substrate by L-cysteine, L-serine, L-2-aminobutyrate, or L-threonine. Glycine acts very slowly as a substrate.

[1]J. Thompson & S. P. F. Miller (1991) *AE* **64**, 317.
[2]W. C. Plaxton & K. B. Storey (1982) *Can. J. Zool.* **60**, 1568.
[3]W. C. Plaxton & K. B. Storey (1982) *J. Comp. Physiol.* **149**, 57.
[4]J. H. A. Fields & P. W. Hochachka (1981) *EJB* **114**, 615.

β-ALANOPINE DEHYDROGENASE

This oxidoreductase [EC 1.5.1.26] catalyzes the reaction of β-alanopine with NAD$^+$ and water to produce β-alanine, pyruvate, and NADH.[1–2]

[1]C. S. Hammen & R. C. Bullock (1991) *Biochem. Syst. Ecol.* **19**, 263.
[2]M. Sato, M. Takahara, N. Kanno, Y. Sato & W. R. Ellington (1987) *Comp. Biochem. Physiol.* **88B**, 803.

D-ALANYL-D-ALANINE CARBOXYPEPTIDASE VanY

This carboxypeptidase catalyzes the hydrolysis of D-alanine from *N,N*-diacetyl-L-Lys-D-Ala-D-Ala and from some peptidoglycan precursors such as undecaprenylpyrophosphatidyl-*N*-acetylmuramoyl-L-Ala-D-Glu-mDAP-D-Ala-D-Ala.[1] The carboxypeptidase appears to contribute to drug resistance.[2,3] The enzyme is not identical to the zinc-type or serine-type D-Ala-D-Ala carboxypeptidases.

See also *Serine-Type D-Ala-D-Ala Carboxypeptidase; Zinc-Type D-Ala-D-Ala Carboxypeptidase*

[1]G. G. Wright, C. Molinas, M. Arthur, P. Courvalin & C. C. Walsh (1992) *Antimicrob. Agents Chemother.* **36**, 1514.
[2]C. C. Walsh, S. S. Fisher, I.-S. Park, M. Prohalad & Z. Wu (1996) *Chem. Biol.* **3**, 21.
[3]M. Arthur & P. Courvalin (1993) *Antimicrob. Agents Chemother.* **37**, 1563.

D-ALANYL-D-ALANINE SYNTHETASE

This enzyme [EC 6.3.2.4], also known as D-alanine:D-alanine ligase, catalyzes an essential reaction for growth of certain enterococci, and failure to synthesize this dipeptide prevents the cross-linking of the underlying peptidoglycan structure of the bacterial cell wall.[1–6] The resulting aminoacyl-D-Ala-D-Ala strand is the target of vancomycin binding, such that this antibiotic arrests peptidoglycan cross-linking and blocks cell-wall synthesis.

The enzyme catalyzes the reaction of MgATP with two molecules of D-alanine to produce D-alanyl-D-alanine, MgADP, and orthophosphate. The kinetic mechanism appears to be random, and for the reaction to proceed, all substrates must reside as an E·D-Ala·D-Ala·MgATP quaternary complex. Except for its activation of an α-carboxylate to form a peptide bond, the enzyme's mechanism appears to be completely analogous to that catalyzed by glutamine synthetase, which forms a γ-glutamyl-phosphate intermediate. There is strong evidence for the occurrence of an acyl-phosphate intermediate and subsequent attack by the amino group of a second D-alanine molecule. Indeed, a phosphinate dipeptide analogue[1] is converted by enzymatic phosphorylation from low-affinity inhibitor to extremely tightly bound analogue of the ligase's reaction intermediates. [1(S)-Aminoethyl][2-carboxy-2(R)-methyl-1-ethyl]phosphinate is an ATP-dependent, slow-binding inhibitor of the D-Ala:D-Ala ligase from *Salmonella typhimurium*, and the enzyme-inhibitor complex (after ATP-dependent phosphorylation) has a half-life of 17 days at 37°C. The mechanism of inhibition is analogous to that of glutamine synthetase by methionine sulfoximine and phosphinothricin.

Vancomycin-resistant enterococci produce a mutant enzyme[4] (designated, VanA) that substitutes α-hydroxy acids in place of D-alanine. Because these so-called depsipeptides fail to bind vancomycin, but still function in the essential cell wall cross-linking reactions, these mutant bacteria are resistant to vancomycin antibiosis.

[1]K. Duncan & C. T. Walsh (1988) *Biochemistry* **27**, 3709.

[2]A. E. McDermott, F. Creuzet, R. G. Griffin, L. E. Zawadzke, Q. Z. Ye & C. T. Walsh (1990) *Biochemistry* **29**, 5767.

[3]C. T. Walsh (1993) *Science* **261**, 308.

[4]T. D. Bugg, S. Dutka-Malen, M. Arthur, P. Courvalin & C. T. Walsh (1991) *Biochemistry* **30**, 2017.

[5]C. Fan, P. C. Moews, C. T. Walsh & J. R. Knox (1994) *Science* **266**, 439.

[6]S. Evers, B. Casadewall, M. Charles, S. Dutka-Malen, M. Galimand & P. Courvalin (1996) *Mol. Evol.* **42**, 706.

Selected entries from *Methods in Enzymology* [**vol**, page(s)]:
General discussion: 8, 333
Assay: 8, 333

β-ALANYL-CoA AMMONIA-LYASE

This lyase [EC 4.3.1.6], also known as acrylyl-CoA aminase, catalyzes the conversion of β-alanyl-CoA to acrylyl-CoA (CoAS–COCH=CH_2) and ammonia.[1] When the enzyme uses acrylyl pantetheine as a substrate, and the equilibrium constant favors the formation of the β-alanyl product.

[1]E. R. Stadtman (1955) *JACS* **77**, 5765.

Selected entries from *Methods in Enzymology* [**vol**, page(s)]:
General discussion: 5, 587
Assay: 5, 587

O-ALANYLPHOSPHATIDYLGLYCEROL SYNTHASE

This enzyme [EC 2.3.2.11], also known as alanyl-tRNA:phosphatidylglycerol alanyltransferase, catalyzes the reaction of L-alanyl-tRNA and phosphatidylglycerol to generate tRNA and O^3-L-alanyl-O^1-phosphatidylglycerol.[1,2]

[1]R. L. Soffer (1974) *AE* **40**, 91.

[2]R. M. Gould, M. P. Thornton, V. Liepkalns & W. J. Lennarz (1968) *JBC* **243**, 3096.

D-ALANYL-sRNA SYNTHETASE

This enzyme, formerly classified as EC 6.1.1.8, is now a deleted Enzyme Commission entry.

ALANYL-tRNA SYNTHETASE

This enzyme [EC 6.1.1.7], also known as alanine:tRNA ligase and alanine translase, catalyzes the reaction of ATP with L-alanine and tRNA^Ala to produce AMP, pyrophosphate (or, diphosphate), and L-alanyl-tRNA^Ala.[1] Alanyl-tRNA synthetase is the only class II aminoacyl-tRNA synthetase with an α_4 quaternary structure. As with other members of the category of enzymes, the reaction proceeds via the formation of adenyl-adenylate intermediate. Interestingly, in binding of tRNA^Ala to the synthetase, no contact is made with the anticodon. Acceptor stem interactions have been preserved in widely distributed species. ***See also*** *Aminoacyl-tRNA Synthetases*

[1]P. Schimmel (1990) *AE* **63**, 233.

Selected entries from *Methods in Enzymology* [**vol**, page(s)]:
General discussion: 29, 547; 34, 170
Other molecular properties: purification, 29, 570; separation, 29, 562; subcellular distribution, 59, 233, 234

ALBENDAZOLE MONOOXYGENASE

albendazole

albendazole S-oxide

This FAD-dependent oxidoreductase [EC 1.14.13.32], also known as albendazole sulfoxidase, catalyzes the reaction of albendazole with NADPH and dioxygen to produce albendazole S-oxide, NADP$^+$, and water.[1]

[1]X. Fargetton, P. Galtier & P. Delatour (1986) Vet. Res. Commun. **10**, 317.

ALCOHOL O-ACETYLTRANSFERASE

This acetyltransferase [EC 2.3.1.84] catalyzes the reaction of acetyl-CoA with an alcohol to produce Coenzyme A and an acetyl ester.[1] Acceptor substrates include a range of short-chain aliphatic alcohols, including methanol, ethanol, isobutanol, and isoamyl alcohol.[2]

[1]M. Harada, Y. Ueda & T. Iwata (1985) Plant Cell Physiol. **26**, 1067.
[2]T. Minetoki, T. Bogaki, A. Iwamatsu, T. Fujii & M. Hamachi (1993) Biosci. Biotechnol. Biochem. **57**, 2094.

ALCOHOL O-CINNAMOYLTRANSFERASE

This acyltransferase [EC 2.3.1.152] catalyzes the reaction of 1-O-trans-cinnamoyl-β-D-glucopyranose with an alcohol to produce an alkyl cinnamate and D-glucose. The alcohol substrate can be methanol, ethanol, or 1-propanol. 1-O-trans-Cinnamoyl-β-D-gentobiose can also act as the acyl donor, albeit less effectively.

ALCOHOL DEHYDROGENASES

These enzymes catalyze the reversible oxidation of a broad spectrum of alcohol substrates, producing aldehydes and ketones, usually with NAD$^+$ or NADP$^+$ as the coenzyme.

The most widely studied alcohol dehydrogenase is a zinc- or iron-dependent oxidoreductase [EC 1.1.1.1], also known as aldehyde reductase, that catalyzes the reversible reaction of an alcohol with NAD$^+$ to produce an aldehyde or ketone and NADH.[1-11] Substrates include a wide variety of primary or secondary alcohols or hemiacetals: examples include ethanol, 1-propanol, 2-propanol, 1-butanol, 2-butanol, cyclohexanol, retinol, and benzyl alcohol. The enzyme has an ordered Bi Bi kinetic mechanism (the coenzyme binds first) with abortive complexes. The animal, but not the yeast, enzyme acts also on cyclic secondary alcohols. The yeast and horse liver enzymes are probably the most extensively characterized oxidoreductases. Only one of two zinc ions is(are) catalytically important, and the general mechanistic properties of the yeast and liver enzymes are similar but not identical (the yeast enzyme is less ordered than the liver protein). Alcohol dehydrogenase can be regarded as a model enzyme system for the exploration of hydrogen kinetic isotope effects and hydrogen tunneling.

Alcohol dehydrogenase (NADP$^+$) [EC 1.1.1.2], also known as aldehyde reductase (NADPH), catalyzes the reversible zinc-dependent reaction of an alcohol with NADP$^+$ to produce an aldehyde and NADPH (with A-side stereospecificity).[12] Some members of this group oxidize only primary alcohols while others can act on secondary alcohols as well. In some organisms this enzyme may be identical with glucuronate reductase [EC 1.1.1.19], mevaldate reductase (NADPH) [EC 1.1.1.33], or lactaldehyde reductase (NADPH) [EC 1.1.1.55]. The pig liver enzyme has an ordered Bi Bi kinetic mechanism that is partially random with certain substrates at elevated concentrations.[13]

Alcohol dehydrogenase (NAD(P)$^+$) [EC 1.1.1.71], a variant of which is often called retinal reductase, catalyzes the reversible reaction of an alcohol with NAD(P)$^+$ to produce an aldehyde and NAD(P)H.[14,15] In the reverse direction, the aldehyde substrates contain two-to-fourteen carbon atoms with highest activity observed with butanal, hexanal, and octanal. Retinal can also be reduced to retinol.

Alcohol dehydrogenase (acceptor) [EC 1.1.99.8], also known as methanol dehydrogenase, catalyzes the pyrroloquinoline-quinone-dependent reversible reaction of a primary alcohol with an acceptor substrate to produce an aldehyde and the reduced acceptor. Substrates include a wide range of primary alcohols. The *Hyphomicrobium* enzyme reportedly has a ping pong Bi Bi kinetic mechanism. The acceptor substrate varies with enzyme source: examples include benzyl viologen, phenosaffarin, ferricyanide, phenazine methosulfate, nitro blue tetrazolium, 2,6-dichlorophenolindophenol, cytochrome c, and cytochrome c_L. **See Methanol Dehydrogenase**

Note that alcohol oxidases will also produce aldehydes from primary alcohols.

[1]A. R. Clarke & T. R. Dafforn (1998) CBC **3**, 1.
[2]M. W. W. Adams & A. Kletzin (1996) Adv. Protein Chem. **48**, 101.
[3]N. J. Oppenheimer & A. L. Handlon (1992) The Enzymes, 3rd ed., **20**, 453.

[4]H. Jörnvall, B. Persson, M. Krook & R. Kaiser (1988) *Biochem. Soc. Trans.* **18**, 169.

[5]H. Jörnvall, J.-O. Höög, M. Von Bahr-Lindström, J. Johansson, R. Kaiser & B. Persson(1988) *Biochem. Soc. Trans.* **16**, 223.

[6]G. Pettersson (1987) *CRC Crit. Rev. Biochem.* **21**, 349.

[7]E. Herrera, A. Zorzano & V. Fresneda (1983) *Biochem. Soc. Trans.* **11**, 729.

[8]C.-I. Brändén, H. Jörnvall, H. Eklund & B. Furugren (1975) *The Enzymes*, 3rd ed., **11**, 103.

[9]M. G. Rossman, A. Liljas, C.-I. Brändén & L. J. Banaszak (1975) *The Enzymes*, 3rd ed., **11**, 61.

[10]K. Dalziel (1975) *The Enzymes*, 3rd ed., **11**, 1.

[11]M. Feraudi, M. Kohlmeier & G. Schmolz (1975) *Biochem. Soc. Trans.* **3**, 1063.

[12]T. G. Flynn (1982) *Biochem. Pharmacol.* **31**, 2705.

[13]A. Magnien & G. Branlant (1983) *EJB* **131**, 375.

[14]N. H. Fridge & D. S. Goodman (1968) *JBC* **243**, 4372.

[15]A. Hatanaka, T. Kajiwara & S. Tomohiro (1974) *Agric. Biol. Chem.* **38**, 1819.

Selected entries from *Methods in Enzymology* [vol, page(s)]:

General discussion: 1, 495, 500, 504; 9, 346, 357; 27, 66, 262, 263; 31, 576; 34, 3, 122, 238; 41, 369; 46, 21, 33, 83, 145, 249; 77, 18; 89, 429, 435, 445, 450, 507; 188, 14, 21, 26, 33, 202, 210; 189, 436; 243, 17

Assay: 1, 495, 500; 9, 347, 350; 52, 361; 57, 210, 214; 188, 29; acceptor-dependent, 89, 450; 188, 33, 202, 210; *Acetobacter aceti* membrane-bound (acceptor-dependent), 89, 450; *Drosophila melanogaster*, 41, 376; 89, 446; equine liver, 1, 495; 89, 436; Fast Analyzer, by, 31, 3; *Gluconobacter suboxydans* membrane-bound, 9, 350; 89, 450; human liver, 41, 369; isoenzymes in retinoid metabolism, 189, 437; long-chain, 188, 172; NADP⁺-dependent, 1, 504; 89, 507; NAD(P)⁺-dependent, soluble form, 188, 15; particulate, 9, 350; propane-specific, 188, 21; soluble, 9, 347

Other molecular properties: abortive complex, 63, 420; acceptor-dependent (See *Methanol Dehydrogenase*); *Acetobacter aceti* membrane-bound (assay, 89, 450; properties, 89, 455; purification, 89, 454); activation, 87, 657; active-center overlapping subsites, analysis by Yonetani-Theorell graphical method, 87, 504; active site, 47, 66; 249, 94; adenosine 5'-diphosphoribose, 87, 504, 509; ADP, and, 87, 504, 509; affinity chromatography, of NAD⁺, 66, 43; affinity partitioning with dye ligands, 228, 132; alcohol oxidation, 44, 838; aldehyde assay and, 3, 295; aldehyde dehydrogenase assay and, 1, 518; AMP, and, 87, 504, 509; analysis with cobalt probe, 226, 67; assay of (acetaldehyde, 17B, 818; glucose, 57, 212; NAD⁺, 66, 41, 42, 80); associated with cell fractions, 40, 86; atomic emission studies, 54, 458; atomic fluorescence studies, 54, 469; bakers' yeast (properties, 1, 502; purification, 1, 501); binding, to agarose-bound nucleotides, 66, 196; binding site, 46, 75; bromoacetamide, and, 87, 491; 4-(3-bromoacetylpyridino)butyl diphosphoadenosine, 87, 475; carboxymethylated, 87, 470; chromogenic substrates, 246, 177, 189; cloning from bacteria, mutant complementation, 258, 217; cobalt probe, 226, 67; coenzyme oxidation, 44, 863; coimmobilized with NAD⁺ (application to ethanol analyzer, 136, 31; assay, 136, 28; preparation, 136, 26, 56); commitment to catalysis, 87, 630, 631, 634; conjugate (activity, 44, 271; kinetics, 44, 406, 407; stability, 44, 841); coupled assay, use in, 63, 34 (plasmalogen-metabolizing enzymes, for, 197, 79); coupling to 1,1'-carbonyldiimidazole-activated support, 135, 115; cryoenzymology, 63, 338; crystallization, 22, 261; cysteine residue, 87, 470, 475, 485, 491; cytosolic, differentiation from retinol dehydrogenase, 189, 478; deoxyribose phosphate aldolase, assay of, 1, 384; 9, 550; *Desulfovibrio gigas* (NAD⁺-dependent) (assay, 243, 17; inhibitors, 243, 20; Michaelis constants, 243, 21; properties, 243, 20; purification, 243, 18; structure, 243, 20; unit of enzyme activity, 243, 18); diacetylmethyl carbinol and, 6, 492; *Drosophila melanogaster*, from, 41, 374 (assay, 41, 376; 89, 446; electrophoresis, 41, 379; molecular weight, 41, 378; properties, 41, 378, 379; purification, 41, 376; 89, 446; sequencing at nanomole level, 91, 470; thin-layer peptide mapping, 91, 466); enzymic activity, 46, 254; equilibrium perturbation, 64, 120, 123; 87, 643, 645; equine liver (active site, 249, 94; affinity HPLC, 104, 218; assay, 89, 436; configuration, stability after coupling to sulfonated supports, 104, 67; coupling to sulfonated support, 104, 61; NAD⁺ complexes [assays, 89, 461; properties, 89, 464; synthesis, 89, 459]; oxidation of benzyl alcohol, isotope effects, 249, 383; purification, 89, 436; rapid scanning stopped-flow spectroscopy [alcohol oxidation, 246, 184, 186, 189; aldehyde reduction, 246, 184; cobalt substitution for zinc, 246, 181; experimental design, 246, 183; isobutyramide binding, 246, 183; mechanism elucidation, 246, 191, 193; pyrazole inhibition, 246, 183; structural assignment of intermediates, 246, 190]; reaction, hydrogen tunneling in, 249, 390, 396; silica-bound, in reversed affinity chromatography, 104, 220; site-directed mutants [altered pH dependencies, 249, 110; catalytic efficiency, 249, 104; steady-state kinetic analysis, 249, 101; transient kinetic analysis, 249, 108]; steroid-active SS isozyme [assays, 89, 430; catalytic properties, 89, 431; gel electrophoresis, 89, 431; purification, 89, 432, 439]; structure, 249, 94; ethanol assay and, 3, 253; ethanol metabolism, in, 52, 355; N-ethylmaleimide, and, 87, 491; fluorescence and, 26, 503; fluorometric assay for NAD⁺, 13, 481; free sulfhydryl groups, determination by ESR, 251, 97; *Gluconobacter suboxydans* membrane-bound (assay, 89, 450; properties, 89, 455; purification, 89, 452); glutathione homocystine transhydrogenase and, 6, 449; half-site reactivity, 64, 184; histidine residue, 87, 405; histochemistry, 6, 892; horse liver (properties, 1, 498; purification, 1, 496); human liver, from, 41, 369 (assay, 41, 369; 89, 507; chromatography, 41, 371; inhibitors, 41, 373; properties, 41, 372; 89, 510; purification, 41, 370; 89, 508; site-directed mutants, acid-base catalysis, 249, 117); hydrogen tunneling, 249, 383, 396; hydrophobicity studies, 44, 63, 64; hyperconjugation, 87, 637; imidazole, effect of, 87, 657; immobilization, 44, 836, 837; 57, 212, 214 (hollow-fiber membrane-device, in, 44, 307; inert protein, on, 44, 903, 908; multistep enzyme system, in, 44, 315, 316; nylon tube, on, 44, 120; water-insoluble carriers, on, 44, 108); inactivation rates, 87, 491; inhibition, 63, 398, 408; 87, 504; interaction constant, 87, 506; intermediate, 87, 405, 627, 630; iodoacetamide, and, 87, 491; iodoacetate, and, 87, 491; isobutyramide, and, 87, 509; isoenzymes in retinoid metabolism (assay, 189, 437; purification from rat, 189, 438; starch gel electrophoresis, 189, 438); isotope effect, 63, 110; 87, 397, 626, 643, 645, 718; 249, 383; isotope exchange, 64, 8, 28, 29; 87, 657; keto-6-phosphogluconate reductase and, 5, 298; kinetics by continuous-flow studies (yeast), 22, 14; lactate dehydrogenase site-to-site complex, with (assays, 136, 108; characterization, 136, 110; immobilized system, 136, 104; soluble system, 136, 106); ligand interactions, evaluation, 202, 520; liquid–liquid partition chromatography, 228, 195, 197; liver, from (histidine content, 47, 431; reductive alkylation, 47, 474; use in assay of retinal, 18C, 639); long-chain (assay, 188, 172; isolation from *Candida*, 188, 173; properties, 188, 174); lysine residue, 87, 470; marker enzyme, as, 32, 733; mechanism, 63, 51; 246, 178, 191, 193; membrane-bound, microdetermination of alcohol, 89, 28; membrane osmometry, 48, 75; metal-depleted, preparation, 158, 32; metal-hybrid, preparation, 158, 87; metal ligands, 246, 182; methionine residue, 87, 485; modification by symmetrical disulfide radical, 251, 97; multienzymatic NADH recycling, in, 136, 70, 74, 77; NAD⁺ assay and, 2, 660, 670; 3, 891, 897; NADP⁺-dependent (human liver [assay, 89, 507; properties, 89, 510; purification, 89, 508]; *Leuconostoc mesenteroides*, 1, 504); NAD(P)⁺-dependent (soluble form [assay, 188, 15; properties, 188, 16]; stereospecificity, 87, 109); NAD⁺ photoaffinity labeling, 237, 72; NAD⁺ reduction by, 2, 694, 707; nicotinamide 5-bromoacetyl-4-methylimidazole dinucleotide, and, 87, 475; nitroaryl reductase assay and, 2, 407, 411; oxaloacetate assay, 55, 207; particulate (assay, 9, 350; *Gluconobacter oxydans*, of [coenzyme, 9, 354; inhibitors, 9, 354; properties, 9, 353; purification, 9, 351; specificity, 9, 353]); partition analysis, 249, 323; pH effects, 87, 397, 405, 406, 631, 634; o-phenanthroline, and, 87, 504, 509; phosphopyridoxylated, 87, 470; photosynthetic pyridine nucleotide reductase and, 6, 445; polarography, 56, 478; preparation of deuterated NAD⁺, in, 54, 226, 228; primary and secondary, from *Gluconobacter*, 9, 346; product inhibition, 63, 436;

propane-specific (assay, **188**, 21; properties, **188**, 25; purification from *Rhodococcus rhodochrous* PNKb1, **188**, 23); pulse-chase experiments, **249**, 320; pyridine nucleotide and, **4**, 282, 284, 840 (assay, **55**, 264, 265, 267); rate-limiting step, **87**, 406, 657; reaction intermediates, stopped-flow absorbance studies, **61**, 319; reaction with (arylazido-β-alanine, **46**, 282, 284; arylazido-β-alanine NAD$^+$, **46**, 282, 284); reactor (ethanol assay with, **137**, 276; preparation, **137**, 276); reduction of NAD$^+$ and NADP$^+$ analogues, **44**, 873; removal from liver microsomal components, **52**, 359; *Rhodotorula*, from, **41**, 361 (assay, **41**, 361; chromatography, **41**, 363; electrofocusing, **41**, 363; inhibitors, **41**, 364; properties, **41**, 364; purification, **41**, 361); secondary structure analysis, **246**, 514; Sepharose-bound (absorption spectrophotometry, **135**, 547; fluorometry, **135**, 552); site-directed mutants, **249**, 101 (catalytic contributions of particular interactions or residues, **249**, 94; catalytic efficiency, **249**, 104); soluble (assay, **9**, 347; *Gluconobacter oxydans*, from properties, **9**, 349; purification, **9**, 348); specificity, **3**, 296; stabilization factor, **87**, 509; stain for, **224**, 106; staining on gels, **22**, 594; starch gel electrophoresis, **6**, 968; stereochemistry, **5**, 430; **87**, 102, 109; stereospecific reduction of ketone, **44**, 837, 838; **77**, 18; substrate effect, **63**, 494; substrate inhibition, **63**, 507; sulfate-reducing bacteria, in, **243**, 41; threonine assay and, **5**, 952; transition state and multisubstrate analogues, **249**, 304; tresyl-silica-coupled (affinity chromatography with, **135**, 77; characterization, **135**, 73; preparation, **135**, 69); use in spectrophotometric recycling assay for NAD$^+$, **17B**, 26, 27, 29, 30; yeast (affinity labeling, **47**, 423, 424, 428; histidine modification, **47**, 433; peptide separation, **47**, 205; reaction [hydrogen tunneling in, **249**, 383, 396; isotope effects, **249**, 383]; site-directed mutants [altered pH dependencies, **249**, 110; catalytic efficiency, **249**, 104; steady-state kinetic analysis, **249**, 101]; uneicosapeptide containing CM-cysteine from, **11**, 396; urea denaturation of, effect on fluorescence quantum yield, **11**, 830); Yonetani-Theorell plot, **87**, 504; zinc, **87**, 475, 485, 505, 627

ALCOHOL OXIDASE

This FAD-dependent oxidoreductase [EC 1.1.3.13], also known as methanol oxidase and alkyl alcohol oxidase, catalyzes the reaction of a primary alcohol with dioxygen to produce an aldehyde and hydrogen peroxide. Substrates include lower primary alcohols and unsaturated alcohols (for example, methanol, ethanol, *n*-propanol, and 2-propyn-1-ol). Branched-chain and secondary alcohols are not substrates. In some organisms, this enzyme may be identical to methanol oxidase. A ping pong kinetic mechanism has been reported.[1–3] *See also Methanol Oxidase*

[1] J. Geissler, S. Ghisla & P. M. Kroneck (1986) *EJB* **160**, 93.
[2] F. M. Dickinson & C. Wadforth (1992) *BJ* **282**, 325.
[3] R. Teschke, Y. Hasumura, J.-G. Joly, H. Ishii & C. S. Lieber (1972) *BBRC* **49**, 1187.

Selected entries from *Methods in Enzymology* [vol, page(s)]:
General discussion: 41, 364; 52, 17, 18; 89, 424; 161, 322; 188, 420
Assay: 41, 365; 89, 424; 188, 421
Other molecular properties: basidiomycetes, from, **41**, 364 (assay, **41**, 365; production, **41**, 366, 367; properties, **41**, 368; purification, **41**, 366); *Candida boidinii* (assay, **89**, 424; production, **89**, 425; properties, **89**, 427; purification, **89**, 426); cytochrome P450-dependent, ethanol oxidase, **52**, 16; enzyme electrode, in, **44**, 590; immobilization, **56**, 483, 489; immunocytochemical detection, **188**, 417; localization in *Hansenula polymorpha*, **188**, 414; properties, **188**, 424; purification from *Hansenula polymorpha* CBS 4732, **188**, 422; purification using nonionic polymers, **22**, 240

ALCOHOL SULFOTRANSFERASE

This transferase [EC 2.8.2.2], also known as hydroxysteroid sulfotransferase, catalyzes the reaction of 3′-phosphoadenylylsulfate with an alcohol to produce adenosine 3′,5′-bisphosphate and an alkyl sulfate.[1] The alcohols that can act as substrates include aliphatic alcohols, ascorbate, chloramphenicol, ephedrine, hydroxysteroids, and other primary and secondary alcohols. However, phenolic steroids will not serve as substrates (such alcohols can be acted upon by steroid sulfotransferases). *See also Aryl Sulfotransferase; Steroid Sulfotransferase*

[1] E. S. Lyon & W. B. Jakoby (1980) *ABB* **202**, 474.

Selected entries from *Methods in Enzymology* [vol, page(s)]:
General discussion: 77, 206
Assay: 77, 206 (chromatographic, **77**, 207; filter, **77**, 206); cholesterol sulfotransferase, **190**, 48
Other molecular properties: affinity chromatography, **77**, 211; cholesterol sulfotransferase (assay, **190**, 48; down-regulation by retinoids, **190**, 42); definition of units, **77**, 208; isoelectric point, **77**, 212; kinetic constants, **77**, 212; molecular weight, **77**, 212; pH optimum, **77**, 212; preparation, **77**, 206; properties, **77**, 211; purification procedures, **77**, 208 (hydroxysteroid sulfotransferase 1, **77**, 209, 210; hydroxysteroid sulfotransferase 2, **77**, 209, 210; hydroxysteroid sulfotransferase 3, **77**, 209); stability, **77**, 211; substrates, **77**, 212, 213

ALDEHYDE DEHYDROGENASES

These oxidoreductases catalyze the conversion of an aldehyde to a carboxylate anion. A number of enzymes have been known as aldehyde dehydrogenase. Class 1 aldehyde dehydrogenases are tetrameric cytosolic enzymes, class 2 represents tetrameric mitochondrial enzymes, and class 3 are the dimeric cytosolic aldehyde dehydrogenases. The dehydrogenases from several sources exhibit substrate inhibition. *See also Acetaldehyde Dehydrogenase (Acetylating); Aldehyde Ferredoxin Oxidoreductase*

Aldehyde dehydrogenase (NAD$^+$) [EC 1.2.1.3] catalyzes the reaction of an aldehyde with NAD$^+$ and water to form a carboxylate anion and NADH.[1–4] The enzyme is capable of acting on a wide variety of aldehydes (for example, acetaldehyde, cinnamaldehyde, crotonaldehyde, hexanal, glyoxal, propanal, and phenylacetaldehyde). It will also oxidize D-glucuronolactone to form D-glucarate (formerly called D-glucuronolactone dehydrogenase and previously classified as EC 1.1.1.70).

Aldehyde dehydrogenase (NADP$^+$) [EC 1.2.1.4] catalyzes the reaction of an aldehyde with NADP$^+$ to produce a carboxylate anion and NADPH.[4–6] Substrates include acetaldehyde and dodecanal.

Aldehyde dehydrogenase $(NAD(P)^+)$ [EC 1.2.1.5] catalyzes the conversion of an aldehyde to a carboxylate anion using either NAD^+ or $NADP^+$ as the coenzyme.[7–10] Substrates include acetaldehyde, glyceraldehyde, phenylacetaldehyde, propanal, butanal, glycolaldehyde, and cinnamaldehyde. The bovine lens and sheep liver enzymes have ordered Bi Bi kinetic mechanisms.[9,10]

Aldehyde dehydrogenase (pyrroloquinoline-quinone) [EC 1.2.99.3], formerly known as aldehyde dehydrogenase (acceptor), catalyzes the reaction of an aldehyde with an acceptor substrate and water to produce a carboxylate anion and the reduced acceptor using pyrroloquinoline-quinone (PQQ).[11–13] This enzyme can utilize a wide variety of aldehydes as substrates including straight-chain aldehydes containing as many as ten carbon atoms (*e.g.*, acetaldehyde, heptanal, crotonaldehyde, benzaldehyde, glyoxylate, and glyceraldehyde). Acceptor substrates include ferricyanide, phenazine methosulfate, phenazine ethosulfate, and 2,6-dichlorophenol indophenol. *See also* Aldehyde Ferredoxin Oxidoreductase

[1] N. J. Oppenheimer & A. L. Handlon (1992) *The Enzymes*, 3rd ed., **20**, 453.
[2] K. Inoue, H. Nishimukai & K. Yamasawa (1979) *BBA* **569**, 117.
[3] T.-K. Li (1977) *AE* **45**, 427.
[4] L. Guerrillot & J. P. Vandecasteele (1977) *EJB* **81**, 185.
[5] H. Aurich, H. Sorger, R. Bergmann & J. Lasch (1987) *Biol. Chem. Hoppe-Seyler* **368**, 101.
[6] W. F. Bosron & R. L. Prairie (1972) *JBC* **247**, 4480.
[7] F. M. Dickinson (1989) *Biochem. Soc. Trans.* **17**, 299.
[8] J. P. Hill & F. M. Dickinson (1988) *Biochem. Soc. Trans.* **16**, 856 and 857.
[9] H. H. Ting & M. J. Crabbe (1983) *BJ* **215**, 351 and 361.
[10] G. J. Hart & F. M. Dickinson (1982) *BJ* **203**, 617.
[11] H. Muraoka, Y. Watanabe, N. Ogasawara & H. Takahashi (1981) *J. Ferment. Technol.* **59**, 247.
[12] R. N. Patel, C. T. Hou, P. Derelanko & A. Felix (1980) *ABB* **203**, 654.
[13] R. Patel, C. T. Hou & A. I. Laskin (1982) *Dev. Ind. Microbiol.* **23**, 187.

Selected entries from *Methods in Enzymology* [**vol**, page(s)]:
General discussion: 1, 508; 77, 18; 89, 480, 484, 491; 188, 26, 323
Assay: 1, 508; 188, 29; *Acetobacter aceti* membrane-bound, 89, 491; baker's yeast, 41, 354, 355; *Gluconobacter suboxydans* membrane-bound, 89, 491; liver, 71, 778; long-chain, 188, 176; NAD^+-dependent (bovine liver, 1, 514; 89, 497; *Clostridium kluyveri*, 1, 518; *Hyphomicrobium* X, 188, 328; isozymes, equine liver, 89, 474; *Pseudomonas aeruginosa*, 89, 484); $NADP^+$-dependent (*Proteus vulgaris*, 89, 480; *Pseudomonas aeruginosa*, 89, 489; yeast, 1, 511); $NAD(P)^+$-dependent, 188, 19 (yeast, 1, 508; 89, 469); *Pseudomonas aeruginosa*, 41, 348; 89, 489; *Pseudomonas oleovorans*, 42, 313, 314; pyrroloquinoline-quinone-dependent, 188, 323
Other properties: acceptor-dependent, (See *pyrroloquinoline-quinone-dependent*); *Acetobacter aceti* membrane-bound (properties, 89, 495; purification, 89, 494); aldehyde assay and, 3, 295; baker's yeast, from, 1, 508; 41, 354 (assay, 1, 508; 41, 354, 355; inhibition, 41, 360; properties, 41, 359, 360; purification, 41, 355); crystallization, 22, 250; *Desulfovibrio gigas*, from (activity, 243, 28; CD spectra, 243, 30; extended XAFS, 243, 35, 38; functionality, 243, 38; genetic advances, 243, 38; iron-sulfur centers [EPR studies, 243, 35; Mössbauer studies, 243, 36]; molybdenum center, EPR studies, 243, 31; molybdenum cofactor extrusion, 243, 37; optical absorption spectra, 243, 30, 32; physicochemical characterization, 243, 26; physiological significance, 243, 40; physiological studies, 243, 39; related molybdenum-containing proteins, 243, 40; X-ray crystallography, 243, 39); electron acceptors and, 4, 330; *Gluconobacter suboxydans* membrane-bound (assay, 89, 491; properties, 89, 495; purification, 89, 493); immobilization, in enzyme system, 44, 315, 316; inhibition, slow-binding, 63, 450, 465; liver, from, 71, 772 (affinity chromatography, 71, 775; assay, 71, 778; isozymes, 71, 772; mitochondrial and cytoplasmic forms, 71, 773; properties of horse, sheep, and human liver isozymes, 71, 773, 774; purification, 71, 773; sulfhydryl group sensitivity, 71, 773; susceptibility to disulfiram, 71, 773); long-chain (assay, 188, 176; properties, 188, 177); membrane-bound, microdetermination of aldehydes, 89, 27; NAD^+-dependent (bovine liver [properties, 1, 517; 89, 499; purification, 1, 515; 89, 498]; *Clostridium kluyveri* [properties, 1, 520; purification, 1, 519]; *Hyphomicrobium* X [properties, 188, 329; purification, 188, 329]; isozymes, equine liver [properties, 89, 477; purification, 89, 475]; *Pseudomonas aeruginosa* [properties, 89, 488; purification, 89, 485]); $NADP^+$-dependent (*Proteus vulgaris* [properties, 89, 483; purification, 89, 480]; *Pseudomonas aeruginosa* [properties, 89, 490; purification, 89, 489]; yeast [properties, 1, 513; purification, 1, 512]); $NAD(P)^+$-dependent (properties, 188, 20; yeast [properties, 1, 510; 89, 472; purification, 1, 509; 89, 469]); partition analysis, 249, 323 (pulse-chase experiments, 249, 320); *Pseudomonas aeruginosa*, from, 41, 348 (amino acids, 41, 353; assay, 41, 348; 89, 484; chromatography, 41, 351; properties, 41, 352; 89, 488; purification, 41, 349; 89, 485); *Pseudomonas oleovorans*, from, 42, 313 (assay, 42, 313, 314; chromatography, 42, 315; properties, 42, 315; purification, 42, 314); pyrroloquinoline-quinone-dependent (assay, 188, 323; properties, 188, 326; purification from *Hyphomicrobium* X, 188, 325); specificity, 3, 296; staining on gels, 22, 594; stereochemistry, 87, 113

ALDEHYDE OXIDASES

These heme- and molybdenum-dependent oxidoreductases [EC 1.2.3.1] catalyze the reaction of aldehydes with water and dioxygen to produce carboxylate anions and superoxide (or hydrogen peroxide).[1–8] In addition, the enzyme can also catalyze the oxidation of quinoline and pyridine derivatives. In some systems, this enzyme may be identical with xanthine oxidase [EC 1.1.3.22];[8] however, the substrate specificity is broader for aldehyde oxidase. *See also* Glyoxal Oxidase

[1] P. E. Baugh, D. Collison, C. D. Garner & J. A. Joule (1998) *CBC* **3**, 377.
[2] J. G. Ferry (1992) *Crit. Rev. Biochem. Mol. Biol.* **27**, 473.
[3] S. Yoshihara & K. Tatsumi (1986) *ABB* **249**, 8.
[4] R. C. Bray (1980) *AE* **51**, 107.
[5] R. C. Bray (1975) *The Enzymes*, 3rd ed., **12**, 299.
[6] G. Palmer (1975) *The Enzymes*, 3rd ed., **12**, 1.
[7] V. Massey (1973) in *Iron-Sulfur Proteins* (W. Lovenberg, ed.), **1**, 301, Academic Press, New York.
[8] T. A. Krenitsky, S. M. Neil, G. B. Elion & G. H. Hitchings (1972) *ABB* **150**, 585.

Selected entries from *Methods in Enzymology* [**vol**, page(s)]:
General discussion: 1, 523; 9, 364; 34, 3; 52, 17, 18; 53, 401
Assay: 1, 523; 9, 364
Other molecular properties: dinitrobenzene reduction by, 2, 409; N^1-methylnicotinamide oxidase activity, 18B, 216 (assay, 18B, 216; properties, 18B, 221, 222; purification from rabbit liver, 18B, 219); pig liver, from (inhibitors, 1, 527; properties, 1, 526; purification, 1, 525); rabbit liver, from (inhibitors, 9, 368; molecular weight, 9, 367;

properties, **9**, 367; purification, **9**, 366; specificity, **9**, 367); reversible dissociation, **53**, 435; sulfate-reducing bacteria, **243**, 41

ALDEHYDE REDUCTASE

This oxidoreductase [EC 1.1.1.21], also known as aldose reductase and polyol dehydrogenase (NADP⁺), catalyzes the reversible reaction of an aldose (with wide specificity) with NAD(P)H to produce an alditol and NAD(P)⁺.[1–5] Sugar substrates include D-galactose, D-erythrose, D-glucose, and D-ribose. In addition, many non-sugar aldehydes (*e.g.*, hexanal, propanal, and benzaldehyde) are substrates. The coenzyme stereochemistry is A-side. *See also Polyol Dehydrogenase; Sorbitol Dehydrogenase; Alcohol Dehydrogenases; Carbonyl Reductase (NADPH); Glucuronate Reductase; Xylose Reductase*

[1] B. Das & S. K. Srivastava (1985) *ABB* **238**, 670.
[2] F. B. Negm (1986) *Plant Physiol.* **80**, 972.
[3] B. Wermuth, H. Bürgisser, K. Bohren & J.-P. von Wartburg (1982) *EEJB* **127**, 279.
[4] M. A. Attwood & C. C. Doughty (1974) *BBA* **370**, 358.
[5] S. Hayman & J. H. Kinoshita (1965) *JBC* **240**, 877.

Selected entries from *Methods in Enzymology* [vol, page(s)]:
General discussion: 5, 333; 9, 163, 167; 89, 181
Assay: 5, 333; 9, 171, 193 (NAD⁺-linked, 9, 163; NADP⁺-linked, 9, 167; particulate, 9, 178); 41, 159, 165, 166; 71, 269; 89, 182
Other molecular properties: aldose reductase (mammalian tissues, from, 5, 333; 41, 159 [activators and inhibitors, 41, 164; assay, 41, 159; chromatography, 41, 162, 163; immunological data, 41, 165; properties, 5, 335; 41, 164, 165; purification, 41, 161; sources, 41, 160, 165]; seminal vesicle and placenta of ruminants, from, 41, 165 [activators and inhibitors, 41, 168; assay, 41, 165, 166; chromatography, 41, 167; properties, 41, 168; purification, 41, 166, 167, 169]); *Brassica oleracea*, from, 71, 269 (specificity, 71, 272); *Candida utilis*, of, 9, 163 (activators, 9, 165; inhibitors, 9, 165; NAD⁺-linked dehydrogenase, 9, 163; NADP⁺-linked dehydrogenase, 9, 166; pH optimum, 9, 166, 169; properties, 9, 165, 169; purification, 9, 164, 167; specificity, 9, 165, 169; stability, 9, 165, 169); crystals, 277, 40; *Gluconobacter*, soluble, of, 9, 170 (activators, 9, 177; inhibitors, 9, 177; pH optima, 9, 177; properties, 9, 174; purification, 9, 171; specificity, 9, 174); human brain (assay, 89, 182; properties, 89, 184; purification, 89, 183); low resolution phasing, 276, 643, 646; specificity of, Bertrand-Hudson rule, 9, 170; stereospecificity, 87, 102, 103

ALDONOLACTONASE

This enzyme, formerly classified as EC 3.1.1.18, is now listed under gluconolactonase [EC 3.1.1.17].

ALDOSE 1-DEHYDROGENASE

This oxidoreductase [EC 1.1.1.121] catalyzes the reaction of a D-aldose with NAD⁺ to produce a D-aldonolactone and NADH.[1,2] Aldose substrates include D-glucose, 2-deoxy-D-glucose, 6-deoxy-D-glucose, D-galactose, 6-deoxy-D-galactose (*i.e.*, D-fucose), 2-deoxy-L-arabinose, D-allose, D-mannose, and D-xylose.

[1] A. S. Dahms & R. L. Anderson (1972) *JBC* **247**, 2222.
[2] A. L. Cline & A. S. L. Hu (1965) *JBC* **240**, 4488, 4493, and 4498.

D-*threo*-ALDOSE 1-DEHYDROGENASE

This oxidoreductase [EC 1.1.1.122], also known as L-fucose dehydrogenase and (2*S*,3*R*)-aldose dehydrogenase, catalyzes the reaction of a D-*threo*-aldose with NAD⁺ to produce a D-*threo*-aldono-1,5-lactone and NADH.[1–3] Substrates include L-fucose, D-arabinose, and L-xylose. The *Pseudomonas caryophylii* dehydrogenase will also use L-glucose as a substrate.[2] Notice that in all of the substrates mentioned, the configuration at C2 and C3 is the same. The Enzyme Commission mentions that the mammalian enzyme will also act on L-arabinose: however, the configurations at C2 and C3 are opposite of those substrates listed above.

[1] P. W. Mobley & R. Metzger (1978) *ABB* **186**, 184.
[2] K.-I. Sasajima & A. J. Sinskey (1979) *BBA* **571**, 120.
[3] H. Schachter, J. Sarney, E. J. McGuire & S. Roseman (1969) *JBC* **244**, 4785.

Selected entries from *Methods in Enzymology* [vol, page(s)]:
General discussion: 41, 173
Assay: sheep liver, 41, 174
Other molecular properties: L-fucose microassay, in, 41, 5; sheep liver, from, 41, 173 (assay, 41, 174; chromatography, 41, 176; molecular weight, 41, 176; properties, 41, 176; purification, 41, 174); staining on gels, 22, 594; stereospecificity, 87, 111

ALDOSE 1-EPIMERASE

This epimerase [EC 5.1.3.3], also known as mutarotase and aldose mutarotase, catalyzes the reversible epimerization of the hemiacetal carbon atom of aldoses (thus, anomerization):[1–4] hence, α-D-glucose is reversibly interconverted with β-D-glucose. Other sugars can act as substrates, including L-arabinose, D-xylose, D-galactose, maltose, and lactose. Interestingly, the ketose D-fructose is reported to be an excellent substrate for the *Aspergillus niger* enzyme.[4]

[1] M. E. Tanner (1998) *CBC* **1**, 208.
[2] P. W. Kuchel, B. T. Bulliman & B. E. Chapman (1988) *Biophys. Chem.* **32**, 89.
[3] E. Adams (1976) *AE* **44**, 69.
[4] S. Kinoshita, K. Kadota & H. Taguchi (1981) *BBA* **662**, 285.

Selected entries from *Methods in Enzymology* [vol, page(s)]:
General discussion: 5, 219; 9, 608; 41, 471, 484
Assay: 5, 219; 9, 116, 608; 41, 473; coupled enzyme assay, 63, 33; higher plants, 41, 484; kidney cortex, 5, 219; 41, 473
Other molecular properties: anomerase activity observed with glucosephosphate isomerase from baker's yeast and other sources, 41, 57; bovine kidney, from, substrate specificities, 41, 487; *Capsicum frutescens*, from, 41, 484 (inhibitors, 41, 487; purification, 41, 485, 486; substrate specificities, 41, 487); coupling enzyme, use as, 63, 32; distribution and abundance, 41, 478; *Escherichia coli*, from (molecular weight, 9, 611; properties, 9, 611; purification, 9, 609; substrate specificities, 41, 487); galactokinase, absence in preparations of, 9, 411; glucose aerodehydrogenase and, 1, 341, 342; higher plants, from, 41, 484

(assay, **41**, 484; chromatography, **41**, 485; inhibitors, **41**, 486; properties, **41**, 486, 487; purification, **41**, 485, 486); kidney cortex, from, **5**, 222; **41**, 471 (assay, **41**, 473; catalytic coefficients, **41**, 484; chromatography, **41**, 479; inhibitors, **41**, 481, 482; molecular weights, **41**, 483; properties, **41**, 480; purification, **41**, 478); inhibitors, **5**, 224; liquid–liquid partition chromatography, **228**, 199; *Penicillium notatum*, from, substrate specificities, **41**, 487; pig kidney (properties, **5**, 223; purification, **5**, 222); sources, **41**, 471

ALDOSE β-D-FRUCTOSYLTRANSFERASE

This transferase [EC 2.4.1.162] catalyzes the reaction of an α-D-aldosyl$_a$ β-D-fructoside with a D-aldose$_b$ to produce a D-aldose$_a$ and an α-D-aldosyl$_b$ β-D-fructoside, where subscripts designate the different aldose units involved in the reaction. The fructoside substrate must be a fructosyl-containing disaccharide in which the β-fructosyl ring is linked to the anomeric carbon of an aldose: examples include sucrose, raffinose, and stachyose. The aldose donor substrate can be monomeric or oligomeric hexoses or pentoses: for example, glucose, ribose, sucrose, lyxose, arabinose, xylose, galactose, sorbose, mannose, glucose 6-phosphate, 6-deoxyglucose, rhamnose, mellibiose, lactose, isomaltose, cellobiose, maltose, maltopentaose, maltohexaose, gluconic acid, and xylitol.[1]

[1]P. S. J. Cheetham, A. J. Hacking & M. Vlitos (1989) *Enz. Microbial. Technol.* **11**, 212.

ALDOSE-1-PHOSPHATE ADENYLYLTRANSFERASE

This transferase [EC 2.7.7.36], also known as sugar-1-phosphate adenylyltransferase and ADP-aldose phosphorylase, catalyzes the reaction of orthophosphate with an ADP-D-aldose (such as ADP-D-glucose) to produce D-aldose 1-phosphate and ADP.[1] *See also Aldose-1-Phosphate Nucleotidyltransferase*

[1]M. Dankert, I. R. J. Goncalves & E. Recondo (1964) *BBA* **81**, 78.

ALDOSE-1-PHOSPHATE NUCLEOTIDYLTRANSFERASE

This transferase [EC 2.7.7.37], also known as sugar-1-phosphate nucleotidyltransferase, sugar nucleotide phosphorylase, and NDP-sugar phosphorylase, catalyzes the reaction of orthophosphate with an NDP-aldose to produce an NDP and an aldose 1-phosphate.[1] NDP-sugar substrates include UDP-D-mannose, GDP-D-mannose, ADP-D-mannose, dTDP-D-mannose, and UDP-D-glucose. *See also Aldose-1-Phosphate Adenylyltransferase*

[1]E. Cabib, H. Carminatti & N. M. Woyskovsky (1965) *JBC* **240**, 2114.

Selected entries from *Methods in Enzymology* [**vol**, page(s)]:

General discussion: 8, 224

Assay: wheat germ, 8, 228; yeast, 8, 224

Other molecular properties: wheat germ, from, 8, 228 (properties, 8, 229; purification, 8, 228); yeast, from, 8, 224 (properties, 8, 228; purification, 8, 225)

ALDOSE-6-PHOSPHATE REDUCTASE (NADPH)

This oxidoreductase [EC 1.1.1.200], also known as $NADP^+$-dependent D-sorbitol-6-phosphate dehydrogenase, catalyzes the reversible reaction of D-glucose 6-phosphate with NADPH to produce D-sorbitol 6-phosphate and $NADP^+$.[1,2] Alternative substrates also include D-galactose 6-phosphate, D-mannose 6-phosphate (weakly), and 2-deoxy-D-glucose 6-phosphate (also weakly).

[1]W. H. Loescher, G. C. Marlow & R. A. Kennedey (1982) *Plant Physiol.* **70**, 335.

[2]F. B. Negm & W. H. Loescher (1981) *Plant Physiol.* **67**, 139.

ALEURAIN

This cysteine aminopeptidase, similar in action to cathepsin H, was first observed from barley aleurone layers and is related to oryzain from rice, CCP2 from maize, and PeTh3 from petunia. Good artificial substrates are Arg–NHNap, Arg–NHMec, and Leu–NHMec.[1] The enzyme is derived from an inactive precursor, proaleurain.

[1]B. C. Holwerda & J. C. Rogers (1993) *J. Exp. Bot.* **44** (suppl.), 321.

ALGINASE

This glycosidase activity, formerly classified as EC 3.2.1.16, is now a deleted Enzyme Commission entry.

ALGINATE LYASE

This lyase [EC 4.2.2.3], also known as poly(β-D-mannuronate) lyase and poly(manna)alginate lyase, catalyzes the eliminative cleavage of polysaccharides containing β-D-mannuronate residues (for example, alginate) to give oligosaccharides with 4-deoxy-α-L-*erythro*-hex-4-enopyranuronosyl groups at their ends.[1–5] The enzyme has at least three subsites.[5] *See also Polymannuronate Hydrolase*

[1]A. Nakagawa, T. Ozaki, K. Chubachi & T. Hosoyama (1998) *J. Appl. Microbiol.* **84**, 328.

[2]H. Ertesvag, F. Erlien, G. Skjak-Braek, B. H. A. Rehm & S. Valla (1998) *J. Bacteriol.* **180**, 3779.

[3]I. W. Davidson, C. J. Lawson & I. W. Sutherland (1977) *J. Gen. Microbiol.* **98**, 223.

[4]L. A. Elyakova & V. V. Favorov (1974) *BBA* **358**, 341.

[5]F. Chavagnat, A. Heyraud, P. Colin-Morel, M. Guinand & J. Wallach (1998) *J. Carbohydr. Res.* **308**, 409.

ALGINATE SYNTHASE

This enzyme [EC 2.4.1.33], also known as mannuronosyl-transferase, catalyzes the reaction of GDP-D-mannuronate with [alginate]$_n$ to produce GDP and [alginate]$_{n+1}$ (*i.e.*, alginate with an additional β-D-mannuronate residue).[1]

[1] T.-Y. Lin & W. Z. Hassid (1966) *JBC* **241**, 5284.

ALIPHATIC NITRILASE

This enzyme [EC 3.5.5.7] catalyzes the hydrolysis of aliphatic nitriles to produce a carboxylate anion and ammonia. Substrates include crotononitrile, acrylonitrile, and glutaronitrile.[1] Some substrates for this enzyme are not substrates for nitrilase [EC 3.5.5.1].

[1] J. Dhillon, S. Chhatre, R. Shanker & N. Shivaraman (1999) *Can. J. Microbiol.* **45**, 811.

ALIZARIN 2-β-GLUCOSYLTRANSFERASE

This glucosyltransferase [EC 2.4.1.103] catalyzes the reaction of UDP-D-glucose with alizarin (*i.e.*, 1,2-dihydroxy-9,10-anthraquinone) to produce UDP and 1-hydroxy-2-(β-D-glucosyloxy)-9,10-anthraquinone.[1,2] Other acceptor substrates include other hydroxy- and dihydroxy-derivatives of 9,10-anthraquinone: for example, 1,3-dihydroxy-9,10-anthraquinone, 3-hydroxy-1-methoxy-9,10-anthraquinone, 1-hydroxy-9,10-anthraquinone, and 2-hydroxy-9,10- anthraquinone.

[1] J. Mateju, J. Cudlín, N. Steinerov, M. Blumauerov & Z. Vanek (1979) *Folia Microbiol.* **24**, 205.
[2] J. Mateju & M. Nohynek (1991) *Folia Microbiol.* **36**, 314.

ALKALINE D-PEPTIDASE

This extracellular endopeptidase, with its characteristic pH$_{optimum}$ of around 10, catalyzes the hydrolysis of peptides composed of aromatic D-amino acids. It is distinct from DD-carboxypeptidase or D-aminopeptidase. The enzyme is assayed by observing the production of D-phenylalanyl-D-phenylalanine from (D-Phe)$_4$.[1]

[1] Y. Asano, H. Ito, T. Dairi & Y. Kato (1996) *JBC* **271**, 30256.

ALKALINE PHOSPHATASE

This zinc- and magnesium-dependent enzyme [EC 3.1.3.1] (also known as alkaline phosphomonoesterase, phosphomonoesterase, and glycerophosphatase), which has a high pH optimum, catalyzes the hydrolysis of many orthophosphoric monoesters (the substrate specificity is quite wide) to generate an alcohol and orthophosphate.[1-14] In addition to this phosphatase activity, the enzyme can also catalyze certain transphosphorylations. In some systems, the enzyme can also act on pyrophosphate (or, diphosphate) (hence, a pyrophosphatase activity). The reaction proceeds via a phosphoenzyme intermediate (at a seryl residue). The rate-limiting step at pH > 7 is the release of orthophosphate. There is a net retention of configuration at the phosphorus atom.

Phosphoester substrates include 5′-AMP, 5′-UMP, β-glycerophosphate, phenyl phosphate, D-glucose 1-phosphate, D-glucose 6-phosphate, phosphocholine, phosphoethanolamine, and *p*-nitrophenylphosphate.

[1] A. C. Hengge (1998) *CBC* **1**, 517.
[2] J. A. Gerlt (1992) *The Enzymes*, 3rd ed., **20**, 95.
[3] J. E. Coleman (1992) *Ann. Rev. Biophys. Biomol. Struct.* **21**, 441.
[4] W. Meyer-Sabellek, P. Sinha & E. Köttgen (1988) *J. Chromatogr.* **429**, 419.
[5] F. Herz (1985) *Experientia* **41**, 1357.
[6] B. L. Vallee & A. Galdes (1984) *AE* **56**, 283.
[7] H. W. Wyckoff, M. Handschumacher, K. Murthy & J. M. Sowadski (1983) *AE* **55**, 453.
[8] J. E. Coleman & P. Gettins (1983) *AE* **55**, 381.
[9] M. Cohn & G. H. Reed (1982) *Ann. Rev. Biochem.* **51**, 365.
[10] R. B. McComb, G. N. Bowers & S. Posen (1979) *Alkaline Phosphatase*, Plenum Press, New York.
[11] T. W. Reid & I. B. Wilson (1971) *The Enzymes*, 3rd ed., **4**, 373.
[12] H. N. Fernley (1971) *The Enzymes*, 3rd ed., **4**, 417.
[13] T. C. Stadtman (1961) *The Enzymes*, 2nd ed., **5**, 55.
[14] J. Roche (1950) *The Enzymes*, 1st ed., **1**, 473.

44, 51; enzyme production, **12B**, 214; *Hph*I endonuclease, of, **65**, 164; post-Fenton reaction digestion of DNA, **234**, 54; properties, **12B**, 217; purification procedure, **12B**, 215; ribonucleic acid hydrolysis and, **12A**, 306; source, **12A**, 156; strains used, **12B**, 213; substrate specificity, **65**, 190; tritium derivative method, **65**, 640, 641, 645, 647, 648, 653, 658, 660; tRNA sequence analysis, in, **59**, 61, 75, 99, 102); bacterial membrane, in, **22**, 113; biotinylation, **184**, 149, 474; bone, **2**, 539 (preparation, **2**, 540); brush-border marker, as, **31**, 120, 125; butanol, effect of, **1**, 49, 50; calf intestinal, **68**, 86 (catalytic properties, **65**, 79; digestion procedure, **65**, 79, 80, 665; tRNA sequence analysis, in, **59**, 61); characterization of (2′P-ADP, **66**, 117, 118; tRNA digestion products, **59**, 171); chiral phosphoryl groups, **87**, 257, 288, 300; cobalt and, **87**, 247; cobalt-substituted, magnetic circular dichroism, **49**, 168; conjugate (kinetics, **44**, 426; optimal effective pH, **44**, 429; preparation, for immunoassays, **70**, 432, 433; stability, **73**, 162, 163); conjugated secondary antibodies, indirect labeling of blots, **196**, 345; conjugation, with antibody, **44**, 711; **73**, 150, 165, 166, 409, 495, 659, 660 (anti-Ig antibodies for Ig detection on blots, to, **121**, 499; applications, **73**, 282); coupled assay, use in, **63**, 36; cryoenzymology, **63**, 338; cyanoethyl tRNA and, **20**, 164; cytochemistry, **39**, 151; delivery systems, polymeric and copolymeric (kinetic analysis, **112**, 414; preparation, **112**, 400); deoxyribonuclease in, **21**, 334; deoxyribonucleic acid joining enzyme assay and, **21**, 312, 314, 333, 334; deoxyribonucleic acid molecular weight and, **12B**, 416; deoxyribonucleic acid sequence analysis and, **29**, 247, 283, 284, 288, 291, 292; dephosphorylation of (DNA, **65**, 505, 520, 521; insert DNA, **217**, 371; phosphorylated PDGF β receptor, **200**, 375; 7-α-D-ribofuranosyladenine 5′-phosphate by *Escherichia coli* enzyme, **18C**, 98; suppressor-containing plasmid, **217**, 367); detection, **235**, 429; determination of deuterium content of RNA, **59**, 646; differential extraction of tissue, in, **1**, 33; digestion of tRNA 3′-terminus, **59**, 175, 186, 343; endogenous, blocking in frozen tissue sections, **245**, 329; enzyme-linked immunosorbent assay, **74**, 591, 592; *Escherichia coli*, of, **9**, 639; **12B**, 212; **27**, 66 (crystallization, **9**, 641; GDP-mannose synthesis and, **28**, 283; properties, **9**, 642; **12B**, 217; purification, **9**, 640; **12B**, 215; stains, **12B**, 213); *Escherichia coli* enzyme used in hydrolysis of ribonucleotides of nicotinate and nicotinamide, **18B**, 60; exchanges, **87**, 247; extraction, **1**, 33, 38; **22**, 205 (butanol, and, **1**, 49); fingerprinting procedure and, **12A**, 362, 370, 377; flip-flop mechanism, **64**, 185; fluorine nuclear magnetic resonance, **49**, 273, 274 (chemical shift, **49**, 292, 293; spectra, **49**, 276, 279, 280, 286, 287); functional, recovery from SDS gels, **91**, 274; fusions, **326**, 35, 159; β-galactosidase, and, double staining of retrovirus-infected tissues for, **225**, 947; gastrointestinal, targeting for prodrug absorption, **112**, 367; gel electrophoresis, **32**, 83, 90; gene expression in Rhodospirillaceae, **204**, 482; gene fusions, **326**, 35, 159; glucose-1,6-bisphosphate preparation, **1**, 355; human placental (glycophosphatidylinositol modification [cDNA truncation for analysis on gels, **250**, 540, 547; cotranslational processing assay, **250**, 542; evidence for incorporation, **250**, 545; fusions, **326**, 159; kinetics of incorporation, **250**, 546; translation-independent processing assay, **250**, 543, 545]; ω site, determination [antibody production, **250**, 574; HPLC of peptides, **250**, 574; immunoprecipitation, **250**, 576; proteolysis of protein, **250**, 573; radioimmunoassay, **250**, 574; sequencing, **250**, 573; site-directed mutagenesis, **250**, 575, 581]; ω site sequence, **250**, 572; precursor forms, identification, **250**, 539; site-directed antipeptide antibody generation, **250**, 541); human semen, nearest-neighbor base sequences and, **6**, 742; hybrid proteins, detection, **235**, 430; hydrolysis of DNA, **234**, 6, 21; hydrolysis of farnesyl pyrophosphate and, **15**, 448; immobilization (collodion membrane, in, **44**, 904, 908; hollow-fiber membrane, in, **44**, 308, 309; microencapsulation, by, **44**, 214; water-insoluble supports, on, **44**, 108); immobilized subunits (catalytic activity, **135**, 492; preparation, **135**, 493; properties, **135**, 499; structural analysis, **135**, 494); immunoassay, **73**, 384; inactivation procedure, **65**, 267; induced cell-free synthesis, **12B**, 797, 803, 807, 817; inhibitors, **65**, 75 (phosphate, **6**, 244); inositol-1-phosphate synthase, assay of, **9**, 699; intermediate, **87**, 300, 319; intestine (properties, **2**, 533; purification, **2**, 530); intestinal mucosa,

in, **32**, 667; isolated nuclei and, **12A**, 440, 445; isozyme conversion protein, *Escherichia coli*, **248**, 117, 118; kinetics, **63**, 238; **87**, 321; labeled DNA preparation and, **21**, 245; labeled oligodeoxynucleotides and, **29**, 331, 333, 335, 337; labeling of blunt-ended restriction fragments, in, **212**, 251; large-scale isolation, **22**, 490, 536; leukemic cell phosphodiesterase and, **6**, 254; linked antibodies, in protein detection on nitrocellulose, **121**, 848; linked immunoassay for prostaglandin F$_{2\alpha}$, **86**, 272; lipoprotein complexes, dissociation from, **1**, 42, 49; lung lamellar bodies, in, **31**, 423, 424; measurement of activity, **12A**, 196; membrane fragments and, **12B**, 797; membrane ribonucleic acid gradient sedimentation and, **12B**, 807; metal-hybrid, preparation, **158**, 89; metal studies, **87**, 544; microvillous membrane, in, **31**, 130, 132; micromethods, **4**, 371; milk (properties, **2**, 538; purification, **2**, 535); nearest-neighbor method, in, **6**, 744; negative cooperativity, **64**, 185; nitration and reduction of, **25**, 520; NMR, ^{31}P and ^{17}O, **87**, 542; nucleoside-2′,3′-cyclic phosphodiesterase and, **6**, 255; nucleoside diphosphates and, **12B**, 222, 223; nucleotide sugar degradation, **14**, 663; oligonucleotide conjugates, applications (chemiluminescent detection of DNA, **217**, 411; genomic sequencing, **217**, 409; genomic Southern analysis, **217**, 407); oligosaccharide cyclic phosphate esters and, **27**, 607; oligonucleotide analysis and, **20**, 121; oligonucleotide chain length and, **12A**, 320; oxidation with NBS, **11**, 521; phagocyte staining, in, **132**, 133, 173; pH effects, **63**, 215 (lipoprotein complexes, and, **1**, 42); phenylphosphate, and chirality, **87**, 288; phosphate–water exchange, **64**, 83; phosphate-protein bond cleavage and, **30**, 582; phosphatidic acid, **69**, 301; phosphodiesterase assay, **38**, 260 (completeness of digestion, assay of, **6**, 744; inhibition with phosphate, **6**, 244); phosphodiesterase preparation, removal from, **2**, 569; phosphoenolpyruvate, **87**, 93; phosphohydrolase assay, in, **32**, 131; phosphorus-31 magnetic resonance, **64**, 83; **87**, 542; phosphoryl-alkaline phosphatase, **16**, 247 (dephosphorylation, **16**, 247); phosphorothioates, and, **87**, 223, 257; placental (histochemical marker gene, as, **225**, 946; purification, **197**, 568; secreted form [assay, **216**, 366; eukaryotic reporter gene, as, **216**, 362]); plasma membranes, in, **22**, 128; PMN granules, in, **31**, 345; ^{31}P NMR studies, **107**, 73; production in *Escherichia coli*, **22**, 88; proline racemase, effect on, **5**, 878; proline reductase, inactivation of, **5**, 875; prostatic phosphomonoesterase (use in hydrolysis of ribonucleotides of nicotinate and nicotinamide, **18B**, 60, 137); proteases, purification by addition of, **1**, 39; purification, fusion proteins, **326**, 35, 301; purine oligonucleotides and, **12A**, 218; quantitation, **235**, 429; radioiodination, **73**, 121 (precursor form, **73**, 120); reaction catalyzed, progress curve analysis, **249**, 87; reaction pathway, **64**, 76; recombinant engineering for immunoassay, **270**, 90; recovery from SDS gels, **91**, 274; reductive denaturation studies, **44**, 517; regulation, **12B**, 818; release by phospholipase C, **71**, 732; release, osmotic shock and, **12B**, 844; removal of 5′-phosphates from DNA, **152**, 99; reporter activity in HepG2 cells, **216**, 390; reporter gene, in analysis of bacterial cell-surface virulence determinants, **235**, 426; reporter, **326**, 35, 159 (assays, **326**, 163, 175, 202); reporter molecule for immunohistochemical reactions, as, **245**, 322, 337; ribonuclease T$_1$ and, **12A**, 233; ribonuclease T$_2$ and, **12A**, 234, 244; ribonucleotide diphosphate reductases, assay of, **12A**, 160; ribosubstituted DNA fragments and, **29**, 302, 303, 315, 320; SDS complex, properties, **48**, 6; site-specific reagent, **47**, 484; spectrokinetic probe, **61**, 323; stain for detection, **12B**, 213; **44**, 713; staining (gels, on, **22**, 585, 590, 601; primordial germ cells for, of, **225**, 46, 65, 73); stepwise amplified immunoenzymatic staining, **121**, 855; stereochemistry, **87**, 288, 300; structure, pressure effects, **259**, 406; substrate-specific, **2**, 541; sucrose gradient sedimentation, **12B**, 807; tagging agent, **326**, 35; terminal phosphorus determination and, **12B**, 218, 219; transphosphorylation by, **2**, 556; transition state and multisubstrate analogues, **249**, 305; treatment of linearized vector DNA, **152**, 203; tRNA and, **29**, 474 (hydrolysis, **29**, 495; preparation for nucleotidyltransferase, **29**, 708; stepwise degradation, **29**, 657); trypsin, purification by addition of, **1**, 39; ultraviolet-endonuclease assay and, **21**, 245; unit of activity, **235**, 429

ALKANAL MONOOXYGENASE (FMN-LINKED)

This FMN-dependent oxidoreductase [EC 1.14.14.3], commonly known as bacterial luciferase, catalyzes the reaction of an aldehyde (with eight or more carbon atoms) with $FMNH_2$ and dioxygen to produce a carboxylate anion, FMN, water, and light.[1–5] $FMNH_2$ binds to the enzyme followed by dioxygen and a reaction purportedly results in a C4a-hydroperoxy adduct of $FMNH_2$. The subsequent reaction between the aldehyde and the hydroperoxy adduct forms a peroxyhemiacetal intermediate. An activated complex, possibly a radical of the C4a-hydroxyflavin, breaks down with the emission of light. Examples of aldehyde substrates include tetradecanal, dodecanal, decanal, and octanal.

[1] P. Angell, D. Langley & A. H. L. Chamberlain (1989) FEMS Microbiol. Lett. **65**, 177.
[2] S.-C. Tu (1982) JBC **257**, 3719.
[3] M. Kurfürst, S. Ghisla, R. Presswood & J. W. Hastings (1982) EJB **123**, 355.
[4] E. M. Kosower (1980) BBRC **92**, 356.
[5] J. W. Hastings (1978) CRC Crit. Rev. Biochem. **5**, 163.

Selected entries from **Methods in Enzymology** [vol, page(s)]:
General discussion: 2, 857; **18B**, 381; **52**, 17; **53**, 403, 558; **57**, 135, 171, 174; **133**, 98, 109; **305**, 135, 152, 157, 164, 279
Assay: 2, 857; **53**, 564; **133**, 99, 113, 200; coupled, **57**, 140; photometer, **57**, 137; principles, **57**, 135, 136; using immobilized enzyme rods, **57**, 205
Other molecular properties: absorbance coefficient, **57**, 172; active center-based enzyme immunoassays, in, **133**, 248; active site, **57**, 179; activity, **53**, 558; activity units, **57**, 150; affinity methods, **133**, 98; anion stabilizers, **57**, 199; assay (aldehyde, **57**, 138, 139; **133**, 247; bile acids, **133**, 215, 246; coupled, **57**, 140; ethanol, **133**, 246; FMN, **53**, 422; glycerol, **133**, 247; malate, **57**, 187; NADH, **133**, 235; NAD(P)H, **133**, 244; oxaloacetate, **57**, 187; 6-phosphogluconate, **133**, 235; photometer, **57**, 137; principles, **57**, 135, 136; using immobilized enzyme rods, **57**, 205); assay in reporter gene systems, **273**, 323; Beneckea harveyi (now Vibrio harveyi), from, **53**, 566; **57**, 143; binding to 2,2-diphenylpropylamine-Sepharose, **133**, 102; bioluminescence quantum yields, **53**, 570; bioluminescent response to aldehydes, **133**, 191; chemical modification, **57**, 132, 133; classification, **57**, 136; coimmobilized with other enzymes, bioluminescent assays with, **136**, 82; comparative molecular weights, **53**, 560, 567; cryoenzymology, **63**, 338, 366; diaphorase, assay, **133**, 200; dithionite assay for aldehydes, in, **133**, 190; effect of pH, **57**, 220; extinction coefficient, **57**, 149; flavin intermediates (formation and properties, **133**, 141, 146; reaction with superoxide, **133**, 144); flavin specificity, **53**, 569; generic definition, **53**, 558; glycoprotein, **53**, 562; **57**, 132, 149; hybrid molecules, preparation, **57**, 178, 179; 4a-hydroperoxyflavin intermediates (isolation, **133**, 131, 133; properties, **133**, 135; reaction products, **133**, 138; stabilization, **133**, 130); immobilization, **57**, 204, 205; **133**, 201, 218, 233; immobilized, assay, **57**, 205; inhibitors, 2, 860; **57**, 151, 152; intermediate, **63**, 366; kinetics, for identification of luminous species, **57**, 161; kinetic properties, 2, 861; **53**, 570; ligand binding studies, **57**, 132, 133; light measurement, **53**, 564; luciferase-luciferin enzyme preparations (commercial, sensitivity comparison, **57**, 559); membranes, **53**, 562; molecular weight, **57**, 149; oxidoreductase system, stability, **133**, 203; oxygenated-flavin intermediate, **53**, 570; photoexcitable, **53**, 569; **57**, 140, 141; preparation of acetone-powder extract from Photobacterium fischeri (now Vibrio fischeri), **18B**, 382, 383; properties, 2, 860; protease

sensitivity, **57**, 198; purification, 2, 858; **53**, 566; **57**, 143; **133**, 98, 106, 109, 119 (Beneckea harveyi [now Vibrio harveyi], **57**, 143; Photobacterium fischeri [now Vibrio fischeri], **57**, 147, 148; Photobacterium phosphoreum, **57**, 149); quantum yield, **57**, 151; reaction, **53**, 562, 563; reaction intermediate, **57**, 127, 133, 152, 195 (absorption spectra, **57**, 197; assay, **57**, 196; fluorescence spectra, **57**, 197; isolation, **57**, 195, 196; quantum yield, **57**, 197; stability, **57**, 196, 197); reaction sequence, **57**, 126, 133; reconstitution, **53**, 567, 569; **57**, 174; reporter gene, (application in cyanobacteria, **167**, 746; assay, **217**, 55); specific activities, **57**, 150, 172; stability, 2, 860; **57**, 149, 150; structural alterations, mutation, **57**, 168; structure, **57**, 258; substrate specificity, **57**, 151; subunit, **53**, 567 (absorbance coefficients, **57**, 173; chemical modification, **57**, 174; hydrolysis by proteases, **57**, 201; molecular weights, **57**, 171; preparation, **57**, 172, 173; properties, **57**, 173, 174; role in activity, **57**, 132, 179; structures, **57**, 132, 149, 171); α-subunit, **57**, 132, 133; β-subunit, function, **57**, 132; thermal stability, tests, **57**, 168, 169; transient kinetic analysis, with site-directed mutants, **249**, 109; turnover rate, **57**, 126, 127, 132, 136, 195, 223; use in assay for FMN and FAD, **18B**, 381

ALKANE 1-MONOOXYGENASE

This iron-dependent oxidoreductase [EC 1.14.15.3] (also known as alkane 1-monooxygenase, laurate ω-hydroxylase, ω-hydroxylase, and fatty acid ω-hydroxylase) catalyzes the reaction of n-octane with dioxygen and reduced rubredoxin to produce 1-octanol, water, and oxidized rubredoxin.[1,2] The enzyme can also hydroxylate fatty acids at the ω-position.

In many organisms this enzyme is a heme-thiolate-dependent protein (i.e., a cytochrome P450 system) and there may be a requirement for NAD(P)H. **See also CYP4 Family**

[1] B. G. Fox (1998) CBC **3**, 261.
[2] E. J. McKenna & M. J. Coon (1970) JBC **245**, 3882.

Selected entries from **Methods in Enzymology** [vol, page(s)]:
General discussion: 52, 14, 20, 310, 318; **53**, 356; **188**, 3
Assay: 188, 4
Other molecular properties: cell-free production of oxygenated products, **188**, 7; composition, **53**, 360; inhibitor, **53**, 360; isolation from Pseudomonas oleovorans, **188**, 7; molecular weight, **53**, 359; oxygenation activity, assay, **188**, 4; prosthetic group, **53**, 360; purification, **53**, 357; purity, **53**, 359; reductase component, isolation, **188**, 6; rubredoxin component (isolation, **188**, 5; properties, **188**, 8); stability, **53**, 359; substrate specificity, **53**, 360

ALKANE 2-MONOOXYGENASE

This oxidoreductase, also known as fatty acid (ω–2) hydroxylase, catalyzes the reaction of n-heptane with dioxygen and an acceptor substrate to produce 2-heptanol, water, and the oxidized acceptor.[1] The enzyme reportedly catalyzes the hydroxylation of fatty acids at the ω–2 position.[1,2] **See also CYP4 Family**

[1] U. Frommer, V. Ullrich, H. Staudinger & S. Orrenius (1972) *BBA* **280**, 487.
[2] R. S. Hare & A. J. Fulco (1975) *BBRC* **65**, 665.

Selected entries from *Methods in Enzymology* [vol, page(s)]:
General discussion: 52, 14, 20

ALKAN-1-OL DEHYDROGENASE (ACCEPTOR)

This pyrroloquinoline-quinone-dependent oxidoreductase [EC 1.1.99.20], also known as polyethylene glycol dehydrogenase, catalyzes the reaction of a primary alcohol with an acceptor substrate to produce an aldehyde and the reduced acceptor.[1–5] The enzyme acts on three-to-sixteen carbon linear-chain saturated primary alcohols and non-ionic surfactants containing polyethylene glycol residues. 2,6-Dichloroindophenol can act as the acceptor substrate.

[1] F. Kawai, H. Yamanaka, M. Ameyama, E. Shinagawa, K. Matsushita & O. Adachi (1985) *Agric. Biol. Chem.* **49**, 1071.
[2] F. Kawai, T. Kimura, Y. Tani, H. Yamada, T. Ueno & H. Fukami (1983) *Agric. Biol. Chem.* **47**, 1669.
[3] F. Kawai, T. Kimura, T. Yoshiki, H. Yamada & K. Mamoru (1980) *Appl. Environ. Microbiol.* **40**, 701.
[4] H. Yamanaka & F. Kawai (1989) *J. Ferment. Bioeng.* **67**, 300 and 324.
[5] M. Yasuda, A. Cherepanov & A. Duine (1996) *FEMS Microbiol. Lett.* **138**, 23.

1-ALKENYL-2-ACYLGLYCEROL CHOLINEPHOSPHOTRANSFERASE

This transferase [EC 2.7.8.22] catalyzes the reaction of CDP-choline with a 1-alkenyl-2-acylglycerol to produce CMP and a plasmenylcholine.[1]

[1] M. Wientzek, R. Y. Man & P. C. Choy (1987) *Biochem. Cell Biol.* **65**, 860.

1-ALKENYLGLYCEROPHOSPHOCHOLINE O-ACYLTRANSFERASE

This acyltransferase [EC 2.3.1.104] catalyzes the reaction of an acyl-CoA with a 1-alkenylglycerophosphocholine to produce Coenzyme A and a 1-alkenyl-2-acylglycerophosphocholine.[1] Unsaturated acyl-CoA substrates are preferred: *e.g.*, linoleoyl-CoA, oleoyl-CoA, linolenoyl-CoA, and arachidonoyl-CoA. The enzyme is distinct from 1-alkenylglycerophosphoethanolamine *O*-acyltransferase [EC 2.3.1.121].

[1] G. Arthur & P. C. Choy (1986) *BJ* **236**, 481.

Selected entries from *Methods in Enzymology* [vol, page(s)]:
General discussion: 209, 89
Assay: 209, 89; spectrophotometric, **209**, 90
Other molecular properties: properties, **209**, 91

ALKENYLGLYCEROPHOSPHOCHOLINE HYDROLASE

This enzyme [EC 3.3.2.2], also known as lysoplasmalogenase, catalyzes the hydrolysis of a 1-(1-alkenyl)-*sn*-glycero-3-phosphocholine to produce an aldehyde and *sn*-glycero-3-phosphocholine.[1–3] **See also** *Alkenylglycerophosphoethanolamine Hydrolase*

[1] Y. Hirashima, M. S. Jurkowitz-Alexander, A. A. Farooqui & L. A. Horrocks (1989) *Anal. Biochem.* **176**, 180.
[2] G. Arthur, L. Page, T. Mock & P. C. Choy (1986) *BJ* **236**, 475.
[3] H. R. Warner & W. E. M. Lands (1961) *JBC* **236**, 2404.

Selected entries from *Methods in Enzymology* [vol, page(s)]:
General discussion: 197, 483
Assay: coupled enzyme assays, **197**, 79
Other molecular properties: hepatic microsomal, solubilization and purification, **197**, 483

1-ALKENYLGLYCEROPHOSPHOETHANOL-AMINE O-ACYLTRANSFERASE

This acyltransferase [EC 2.3.1.121] catalyzes the reaction of an acyl-CoA with a 1-alkenylglycerophosphoethanolamine to produce coenzyme A and a 1-alkenyl-2-acyl-glycerophosphoethanolamine. The best donor substrates are long-chain unsaturated acyl-CoA derivatives (such as linoleoyl-CoA and arachidonoyl-CoA). The enzyme is distinct from 1-alkenylglycerophosphocholine *O*-acyltransferase [EC 2.3.1.104].[1]

[1] G. Arthur, L. Page & P. C. Choy (1987) *BBA* **921**, 259.

ALKENYLGLYCEROPHOSPHOETHANOL-AMINE HYDROLASE

This enzyme [EC 3.3.2.5] catalyzes the hydrolysis of a 1-(1-alkenyl)-*sn*-glycero-3-phosphoethanolamine to produce an aldehyde and *sn*-glycero-3-phosphoethanolamine.[1] **See also** *Alkenylglycerophosphocholine Hydrolase*

[1] J. Gunawan & H. Debuch (1981) *Hoppe-Seyler's Z. Physiol. Chem.* **362**, 445.

1-ALKYL-2-ACETYLGLYCEROL O-ACYLTRANSFERASE

This acyltransferase [EC 2.3.1.125], also known as 1-hexadecyl-2-acetylglycerol acyltransferase, catalyzes the reaction of an acyl-CoA and a 1-*O*-alkyl-2-acetyl-*sn*-glycerol (such as 1-hexadecyl-2-acetyl-*sn*-glycerol) to produce coenzyme A and a 1-*O*-alkyl-2-acetyl-3-acyl-*sn*-glycerol.[1] Several acyl-CoA derivatives can act as acyl donors, the best reported being linoleoyl-CoA. The enzyme is distinct from diacylglycerol acyltransferase [EC 2.3.1.20].

[1] T. Kawasaki & F. Snyder (1988) *JBC* **263**, 2593.

1-ALKYL-2-ACETYLGLYCEROLCHOLINE PHOSPHOTRANSFERASE

This transferase activity, formerly classified as EC 2.7.8.16 but now a deleted entry, is listed under diacylglycerol cholinephosphotransferase [EC 2.7.8.2].

ALKYLACETYLGLYCEROPHOSPHATASE

This phosphatase [EC 3.1.3.59], also known as 1-alkyl-2-acetyl-*sn*-glycero-3-phosphate phosphatase, catalyzes the hydrolysis of a 1-alkyl-2-acetyl-*sn*-glycerol 3-phosphate to produce a 1-alkyl-2-acetyl-*sn*-glycerol and orthophosphate.[1,2] It participates in the biosynthesis of thrombocyte activating factor in animal tissues. There appears to be relatively little substrate selectivity with respect to the alkyl chain length at the *sn*-1 position. The acetyl group at the *sn*-2 position can be replaced by four-to-six carbon acyl groups.

[1] T. C. Lee, B. Malone & F. Snyder (1986) *JBC* **261**, 5373.
[2] T. C. Lee, B. Malone & F. Snyder (1988) *JBC* **263**, 1755.

Selected entries from *Methods in Enzymology* [vol, page(s)]:
General discussion: 209, 230
Assay: 209, 231
Other molecular properties: product identification, **209**, 232; properties, **209**, 232

ALKYLAMIDASE

This enzyme [EC 3.5.1.39] catalyzes the hydrolysis of *N*-methylhexanamide to produce hexanoate and methylamine.[1] Substrates include *N*-monosubstituted and *N*,*N*-disubstituted amides, with some activity towards primary amides, in which the chain length is not short, including *N*,*N*-dimethylhexanamide, *N*-butylhexanamide, *N*-methylheptanamide, *N*-phenylhexanamide, and *N*-methyloctanamide.

[1] P. R. S. Chen & W. C. Dauterman (1971) *BBA* **250**, 216.

S-ALKYLCYSTEINE LYASE

This pyridoxal-phosphate-dependent lyase [EC 4.4.1.6], also known as *S*-alkyl-L-cysteinase, catalyzes the reaction of an *S*-alkyl-L-cysteine with water to produce an alkyl thiol, ammonia, and pyruvate.[1-3] Substrates include *S*-propyl-L-cysteine, *S*-benzyl-L-cysteine, and djenkolate. While the enzyme reportedly acts on *S*-alkyl-L-cysteine sulfoxide, it should not be confused with alliin lyase. The reaction proceeds via an α,β-elimination. In yeast, the enzyme may be identical with cystathionine β-lyase [EC 4.4.1.8]. *See also Alliin Lyase*

[1] A. J. L. Cooper (1994) *Adv. Pharmacol.* **27**, 71.
[2] H. Kamitani, N. Esaki, H. Tanaka & K. Soda (1991) *J. Biochem.* **109**, 645.

[3] J. Nomura, Y. Nishizuta & O. Hayaishi (1963) *JBC* **238**, 1441.

Selected entries from *Methods in Enzymology* [vol, page(s)]:
General discussion: 17B, 470
Assay: 17B, 471
Other molecular properties: properties, **17B**, 474; purification from *Pseudomonas cruciviae*, **17B**, 472; stereochemistry, **87**, 149

ALKYLDIHYDROXYACETONE KINASE

This phosphotransferase [EC 2.7.1.84], officially known as alkylglycerone kinase, catalyzes the reaction of ATP with an *O*-alkyldihydroxyacetone (*i.e.*, *O*-alkylglycerone) to produce ADP and an *O*-alkylglycerone phosphate.[1]

[1] K. Chae, C. Piantadosi & F. Snyder (1973) *JBC* **248**, 6718.

ALKYLDIHYDROXYACETONE-PHOSPHATE SYNTHASE

This enzyme [EC 2.5.1.26], officially known as alkylglycerone-phosphate synthase, catalyzes the reaction of 1-acyldihydroxyacetone 3-phosphate (*i.e.*, 1-acylglycerone 3-phosphate) with a long-chain alcohol to produce 1-alkyldihydroxyacetone 3-phosphate (*i.e.*, 1-alkylglycerone 3-phosphate) and a long-chain acid anion.[1-4]

In this ping pong Bi Bi reaction,[1,3] the ester-linked fatty acid of the substrate is removed and replaced with a long-chain alcohol in an ether linkage. The enzyme will also catalyze exchange reactions, including the reaction of a 1-acyl$_a$dihydroxyacetone 3-phosphate with a long-chain acid anion$_b$ to produce the new 1-acyl$_b$dihydroxyacetone 3-phosphate and new acid anion$_a$, where the subscripts are used to designate the different acyl groups. In addition, the enzyme also catalyzes the reaction of a 1-alkyl$_a$dihydroxyacetone 3-phosphate with a long-chain primary alcohol$_b$ to produce the new 1-alkyl$_b$dihydroxyacetone 3-phosphate and new alcohol$_a$. Although catalysis is not attended by a net change in redox state, the enzyme does contain FAD, which is reduced upon incubation with the substrate palmitoyldihydroxyacetone 3-phosphate. This redox step suggests that the latter undergoes oxidation, and de Vet *et al.*[2] presented a hypothetical mechanism explaining the role played by the flavin.

[1] A. J. Brown & F. Snyder (1982) *JBC* **257**, 8835.
[2] E. C. J. M. de Vet, Y. H. A. Hilkes, M. W. Fraaije & H. van den Bosch (2000) *JBC* **275**, 6276.
[3] E. C. de Vet & H. van den Bosch (1999) *BBA* **1436**, 299.
[4] H. van den Bosch & E. C. de Vet (1997) *BBA* **1348**, 35.

Selected entries from *Methods in Enzymology* [vol, page(s)]:
General discussion: 209, 377, 385
Assay: 209, 379, 385

Other molecular properties: delipidation, **209**, 380; hepatic peroxisomal (assay, **209**, 385; properties, **209**, 389; purification, **209**, 388; solubilization, **209**, 388); product identification, **209**, 382; properties, **209**, 383; purification, **209**, 380; solubilization, **209**, 380; transition-state and multisubstrate analogues, **249**, 305

ALKYLGLYCEROL KINASE

This phosphotransferase [EC 2.7.1.93] catalyzes the reaction of ATP with a 1-*O*-alkyl-*sn*-glycerol to generate ADP and a 1-*O*-alkyl-*sn*-glycerol 3-phosphate.[1] Other substrates include 1-octadecyl-*sn*-glycerol and 1-hexadecyl-*sn*-glycerol.

[1]C. O. Rock & F. Snyder (1974) *JBC* **249**, 5382.

Selected entries from *Methods in Enzymology* [vol, page(s)]:
General discussion: 209, 211
Assay: radiochemical, **209**, 213
Other molecular properties: product identification, **209**, 214; properties, **209**, 214

ALKYLGLYCEROPHOSPHATE 2-O-ACETYLTRANSFERASE

This acetyltransferase [EC 2.3.1.105] catalyzes the reaction of acetyl-CoA with a 1-alkyl-*sn*-glycerol 3-phosphate to produce coenzyme A and a 1-alkyl-2-acetyl-*sn*-glycerol 3-phosphate.[1] The enzyme participates in the biosynthesis of thrombocyte activating factor in animal tissues. The enzyme is distinct from 1-alkylglycerophosphocholine acetyltransferase [EC 2.3.1.67]. MgATP has been reported to inhibit the rabbit brain enzyme.[2]

[1]T. C. Lee, B. Malone & F. Snyder (1986) *JBC* **261**, 5373.
[2]R. R. Baker & H. Y. Chang (1994) *BBA* **1213**, 27.

Selected entries from *Methods in Enzymology* [vol, page(s)]:
Assay: 141, 385
Other molecular properties: properties, **141**, 386

1-ALKYLGLYCEROPHOSPHOCHOLINE O-ACETYLTRANSFERASE

This acetyltransferase [EC 2.3.1.67] catalyzes the reaction of acetyl-CoA with a 1-alkyl-*sn*-glycero-3-phosphocholine to produce coenzyme A and the 1-alkyl-2-acetyl-*sn*-glycero-3-phosphocholine.[1-3] Acceptor substrates include 1-hexadecenyl-*sn*-glycero-3-phosphocholine, 1-octadecyl-*sn*-glycero-3-phosphocholine, and 1-octadecadienyl-*sn*-glycero-3-phosphocholine. *See also* 1-*Alkylglycerophosphocholine O-Acyltransferase*

[1]J. Gomez-Cambronero, S. Velasco, M. Sanchez-Crespo, F. Vivanco & J. M. Mato (1986) *BJ* **237**, 439.
[2]J. Gomez-Cambronero, M. L. Nieto, J. M. Mato & M. Sanchez-Crespo (1985) *BBA* **845**, 511.
[3]P. W. Schindler & E. Ninio (1991) *Lipids* **26**, 1004.

Selected entries from *Methods in Enzymology* [vol, page(s)]:
General discussion: 209, 396
Assay: 141, 382; **209**, 397, 408
Other molecular properties: product identification, **209**, 399, 410; properties, **141**, 384; **209**, 399, 410

1-ALKYLGLYCEROPHOSPHOCHOLINE O-ACYLTRANSFERASE

This acyltransferase [EC 2.3.1.63] catalyzes the reaction of an acyl-CoA and a 1-alkyl-*sn*-glycero-3-phosphocholine to yield coenzyme A and the 1-alkyl-2-acyl-*sn*-glycero-3-phosphocholine.[1] Unsaturated acyl donors were preferred over saturated donors: for example, arachidonyl-CoA and linolenoyl-CoA. In some systems, this enzyme may be identical to 1-acylglycerophosphocholine acyltransferase [EC 2.3.1.23]. *See also* 1-*Alkylglycerophosphocholine O-Acetyltransferase; 1-Acylglycerophosphocholine Acyltransferase*

[1]K. Waku & Y. Nakazawa (1977) *J. Biochem.* **82**, 1779.

Selected entries from *Methods in Enzymology* [vol, page(s)]:
General discussion: 209, 86
Assay: 209, 87; arachidonyltransferase, **141**, 391
Other molecular properties: properties, **141**, 392; **209**, 91

ALKYLGLYCEROPHOSPHOETHANOLAMINE PHOSPHODIESTERASE

This phosphodiesterase [EC 3.1.4.39], also known as lysophospholipase D, catalyzes the hydrolysis of a 1-alkyl-*sn*-glycero-3-phosphoethanolamine to produce 1-alkyl-*sn*-glycerol 3-phosphate and ethanolamine.[1-3] The enzyme will also act on acyl and choline analogues.

[1]T. Kawasaki & F. Snyder (1987) *BBA* **920**, 85.
[2]R. L. Wykle, W. F. Kraemer & J. M. Schremmer (1980) *BBA* **619**, 58.
[3]R. L. Wykle, W. F. Kraemer & J. M. Schremmer (1977) *ABB* **184**, 149.

Selected entries from *Methods in Enzymology* [vol, page(s)]:
General discussion: 197, 583
Assay: 197, 583

ALKYLHALIDASE

This enzyme [EC 3.8.1.1], also known as halogenase and haloalkane dehalogenase, catalyzes the hydrolysis of bromochloromethane ($BrClCH_2$) to produce formaldehyde, bromide, and chloride.[1] In a number of organisms, this activity may be identical with haloalkane dehalogenase. *See also Haloalkane Dehalogenase*

[1]L. A. Heppel & V. T. Porterfield (1948) *JBC* **176**, 763.

ALKYLHYDRAZINE OXIDASE

This enzyme catalyzes the $NADP^+$-dependent formation of methane from methylhydrazine and ethane from ethylhydrazine.

Selected entries from *Methods in Enzymology* [vol, page(s)]:
General discussion: 52, 21

ALKYL-HYDROPEROXIDE REDUCTASE

This protein complex catalyzes the reaction of NADH and H^+ with an alkylhydroperoxide to produce an alcohol, water, and NAD^+.[1-3] The complex consists of a disulfide-containing flavoprotein which binds the pyridine coenzyme and a smaller protein component which also contains a redox-active disulfide. Reaction with NADH generates $FADH_2$ which reduces a disulfide in the large subunit. A thiol-disulfide exchange then occurs with the smaller subunit. This newly formed dithiol reacts with either hydrogen peroxide or an alkylhydroperoxide to generate an alcohol and convert one of the thiols to a sulfenate. The other thiol (or, thiolate) reacts to displace water and reform the disulfide.

[1] L. B. Poole & H. R. Ellis (1996) *Biochemistry* **35**, 56.
[2] L. B. Poole (1996) *Biochemistry* **35**, 65.
[3] Y. Niimura & V. Massey (1996) *JBC* **271**, 30459.

ALKYLMERCURY LYASE

This lyase [EC 4.99.1.2], also known as organomercury lyase and organomercurial lyase, catalyzes the reaction of $R–Hg^+$ with a proton to produce Hg^{2+} and $R–H$.[1,2] Substrates include alkylmercury halides such as $CH_3Hg^+ \ Cl^-$ and vinylmercury bromide, and a number of arylmercury compounds (for example, phenylmercuric acetate). Substrates also include organotin compounds such as triethylvinyltin.[2] The mechanism of this protonolytic carbon-mercury bond cleavage reaction is thought to be of the S_E2 type.[1]

[1] T. P. Begley, A. E. Walts & C. T. Walsh (1986) *Biochemistry* **25**, 7186.
[2] A. Walts & C. T. Walsh (1988) *JACS* **110**, 1950.

ALKYLSULFATASE

Alkylsulfatases are enzymes that catalyze the hydrolysis of aliphatic sulfate esters resulting in the release of sulfate and the parent alcohol. Several alkylsulfatases have been isolated from bacterial cell extracts. The tetrameric *Pseudomonas* C12B enzyme acts on primary alkyl sulfates having a chain length of six-to-thirteen carbon atoms (it also has a weak arylsulfatase activity).[1] The dimeric coryneform B1a enzyme acts on primary alkyl sulfates having chain lengths of between three and seven carbon atoms.[2] The tetrameric enzyme from *Comamonas terrigena* acts on (+)-secondary sulfate esters in which the sulfate is on the C2 position and the chain length is between six and ten carbon atoms.[3]

[1] T. J. Bateman, K. S. Dodgson & G. F. White (1986) *BJ* **236**, 401.
[2] P. J. Matts, G. F. White & W. J. Payne (1994) *BJ* **304**, 937.
[3] G. W. Matcham & K. S. Dodgson (1977) *BJ* **167**, 723.

Selected entries from *Methods in Enzymology* [vol, page(s)]:
General discussion: 2, 324
Assay: 2, 330

2-ALKYN-1-OL DEHYDROGENASE

This oxidoreductase [EC 1.1.1.165] catalyzes the reaction of 2-butyne-1,4-diol with NAD^+ to produce 4-hydroxy-2-butynal and NADH. Other 2-alkyn-1-ols (such as 2-propyn-1-ol, 2-butyn-1-ol, and 2,4-hexadiyne-1,6-diol) can act as alternative substrates. Other non-alkyne substrates include *trans*-2-butene-1,4-diol and 1,4-butanediol. $NADP^+$ can act as the coenzyme, albeit weakly.[1]

[1] T. Miyoshi, H. Sato & T. Harada (1974) *BBA* **358**, 231.

ALLANTOATE DEIMINASE

This deiminase [EC 3.5.3.9] catalyzes the hydrolysis of allantoate to produce ureidoglycine, ammonia, and carbon dioxide.[1,2]

[1] G. D. Vogels (1966) *BBA* **113**, 277.
[2] C. van der Drift, F. E. de Windt & G. D. Vogels (1970) *ABB* **136**, 273.

ALLANTOICASE

This enzyme [EC 3.5.3.4], also called allantoate amidinohydrolase, catalyzes the hydrolysis of allantoate to produce (−)-ureidoglycolate and urea.[1-4] The enzyme can also catalyze the hydrolysis of (+)-ureidoglycolate to generate glyoxylate and urea.

[1] P. Piedras, A. Munoz, M. Aguilar & M. Pineda (2000) *ABB* **378**, 340.
[2] T. G. Cooper (1984) *AE* **56**, 91.
[3] N. E. Tolbert (1981) *Ann. Rev. Biochem.* **50**, 133.
[4] F. Tribjels & G. D. Vogels (1966) *BBA* **118**, 387.

ALLANTOINASE

This enzyme [EC 3.5.2.5] catalyzes the hydrolysis of allantoin to produce allantoate.[1-3]

[1] V. Romanov, M. T. Merski & R. P. Hausinger (1999) *Anal. Biochem.* **268**, 49.
[2] T. G. Cooper (1984) *AE* **56**, 91.
[3] N. E. Tolbert (1981) *Ann. Rev. Biochem.* **50**, 133.

Selected entries from *Methods in Enzymology* [vol, page(s)]:
Other molecular properties: microbodies, in, **52**, 496, 500

ALLANTOIN RACEMASE

This racemase [EC 5.1.99.3] catalyzes the interconversion of (*S*)-(+)-allantoin and (*R*)-(−)-allantoin.[1,2] Studies in D_2O with the *Candida utilis* enzyme demonstrate that racemization proceeds in parallel with release of the hydrogen atom attached to the chiral C5 position of allantoin.[1] This was not the case for nonenzymatic racemization.

[1] I. Okumura & T. Yamamoto (1978) *J. Biochem.* **84**, 891.
[2] L. van der Drift, G. D. Vogels & C. van der Drift (1975) *BBA* **391**, 240.

ALLENE-OXIDE CYCLASE

This enzyme [EC 5.3.99.6] catalyzes the conversion of (9Z)-(13S)-12,13-epoxyoctadeca-9,11,15-trienoate to (15Z)-12-oxophyto-10,15-dienoate.[1,2]

(9Z)-(13S)-12,13-epoxyoctadeca-9,11,15-trienoate   (15Z)-12-oxophyto-10,15-dienoate

Allene oxides formed by the action of hydroperoxide dehydratase [EC 4.2.1.92] are converted into cyclopentenone derivatives. Allene oxides will cyclize nonenzymatically provided one pair of conjugated double bonds are present and an additional isolated double bond in the β,γ position relative to the epoxide group (in other words, having the structural entity 4,5-epoxy-1,3,7-octatriene present within the molecule). Enzymatic cyclization takes place when this structural element is present in a fatty acid chain with its epoxide group in the ω–6,7 position (numbering from the methyl group) and the isolated double bond in the ω–3 position.[1]

[1] J. Ziegler, C. Wasternack & M. Hamberg (1999) *Lipids* **34**, 1005.
[2] G. Sembdner & B. Parthier (1993) *Ann. Rev. Plant Physiol. Plant Mol. Biol.* **44**, 569.

ALLIIN LYASE

This pyridoxal-phosphate-dependent lyase [EC 4.4.1.4], also known as alliinase and cysteine sulfoxide lyase, catalyzes the conversion of an *S*-alkyl-L-cysteine *S*-oxide to an alkyl sulfenate and 2-aminoacrylate.[1–4] **See also** *S-Alkylcysteine Lyase*

[1] A. E. Braunstein & E. V. Goryachenkova (1984) *AE* **56**, 1.
[2] J. E. Lancaster & H. A. Collin (1981) *Plant Sci. Lett.* **22**, 169.
[3] K. Iwami & K. Yasumoto (1980) *Agric. Biol. Chem.* **44**, 3003.
[4] L. Davis & D. E. Metzler (1972) *The Enzymes*, 3rd ed., **7**, 33.

Selected entries from *Methods in Enzymology* [vol, page(s)]:
General discussion: 17B, 475
Assay: 17B, 475; 143, 435
Other molecular properties: *S*-substituted (properties, 17B, 477; 143, 437; purification from [onion, 17B, 476; shiitake mushroom, 143, 436]; specificity, 17B, 476)

ALLOHYDROXY-D-PROLINE OXIDASE

This oxidoreductase reportedly catalyzes the reaction of allohydroxy-D-proline (*i.e.*, *cis*-4-hydroxy-D-proline)

with dioxygen to produce Δ^1-pyrroline-4-hydroxy-2-carboxylate and water. Proline dehydrogenase may also catalyze the conversion of allohydroxy-D-proline to Δ^1-pyrroline-3-hydroxy-5-carboxylate. **See also** *Proline Dehydrogenase*

Selected entries from *Methods in Enzymology* [vol, page(s)]:
General discussion: 17B, 266, 290; 52, 21
Assay: 17B, 290
Other molecular properties: phenazine methosulfate and, 17B, 293; properties, 17B, 293; purification from *Pseudomonas putida*, 17B, 292; solubilization, 17B, 292

ALLOPHANATE HYDROLASE

This enzyme [EC 3.5.1.54] catalyzes the hydrolysis of urea 1-carboxylate (*i.e.*, allophanate) to produce two carbon dioxide molecules and two molecules of ammonia.[1] The yeast (*Saccharomyces cerevisiae*) enzyme, but not that from green algae, also catalyzes the reaction of urea carboxylase (hydrolyzing) [EC 6.3.4.6] wherein ATP reacts with urea and bicarbonate to form ADP, orthophosphate, two NH_3 molecules and two CO_2 molecules:[2] in this way, allophanate hydrolase can hydrolyze urea to carbon dioxide and ammonia in the presence of ATP and bicarbonate.

[1] G. S. Maitz, E. M. Haas & P. A. Castric (1982) *BBA* **714**, 486.
[2] R. A. Sumrada & T. G. Cooper (1982) *JBC* **257**, 9119.

ALLOSE KINASE

D-allose

This phosphotransferase [EC 2.7.1.55], also known as allokinase, catalyzes the reaction of ATP with D-allose to produce ADP and D-allose 6-phosphate.[1] D-Glucose is weakly phosphorylated.

[1] L. N. Gibbins & F. J. Simpson (1963) *Can. J. Biochem.* **9**, 769.

Selected entries from *Methods in Enzymology* [vol, page(s)]:
General discussion: 9, 412
Assay: 9, 412
Other molecular properties: *Aerobacter aerogenes*, from (properties, 9, 415; purification, 9, 413)

ALLOTHREONINE ALDOLASE

This enzyme activity, once classified under the number EC 4.1.2.6 but now a deleted EC entry, probably arises from serine hydroxymethyltransferase [EC 2.1.2.1].

Selected entries from *Methods in Enzymology* [vol, page(s)]:
General discussion: 5, 931
Assay: 5, 931

ALLYL-ALCOHOL DEHYDROGENASE

This oxidoreductase [EC 1.1.1.54] catalyzes the reaction of allyl alcohol with $NADP^+$ to produce acrolein and NADPH.[1] Other substrates include 2,3-butylene glycol.

[1] K. Otsuka (1958) *J. Gen. Appl. Microbiol.* **4**, 211.

ALTERNANSUCRASE

This enzyme [EC 2.4.1.140] catalyzes the transfer of an α-D-glucosyl residue alternately to the 6-position or the 3-position of the nonreducing terminal position of an α-D-glucan, thus producing a glucan having alternating α-1,6 and α-1,3 linkages.[1,2] The product is often called alternan.

[1] H. Tsumori, A. Shimamura & H. Mukasa (1985) *J. Gen. Microbiol.* **131**, 3347.
[2] G. L. Cote & J. F. Robyt (1982) *Carbohydr. Res.* **101**, 57.

ALTERNATIVE-COMPLEMENT-PATHWAY C3/C5 CONVERTASE

This serine endoproteinase [EC 3.4.21.47] (also known as complement factor B, factor B, complement component C3/C5 convertase (alternative), properdin factor B, glycine-rich β-glycoprotein, C3 proactivator, C3 convertase, and C5 convertase) participates in a pathway that is activated by antibody-antigen aggregates and by direct contact with the surfaces of a number of microorganisms and viruses. The enzyme catalyzes the cleavage of complement component C3 (185 kDa) at Arg77–Ser78 to yield C3a (Arg77 becoming the C-terminal residue of C3a) and C3b.[1-7] This is the same bond that is hydrolyzed in the classical pathway. In the presence of elevated C3b, C3b binds noncovalently to Bb and the convertase will also act on complement component C5 at Arg74–Leu75 to produce C5a and C5b. No other natural substrates have been identified for this convertase. However, Bb does have an esterase activity (for example, with Ac-Gly-Lys-∇-OMe and Z-Gly-Leu-Ala-Arg-∇-SBzl). (Note: ∇ indicates the cleavage site.)

The alternative convertase is a complex of complement fragment Bb containing the active site of the convertase, a size of 63 kDa, and formed by the action of complement factor D on factor B with either C3b or cobra venom factor (cobra C3b). Note that C3b, a product of the convertase, is also a component of the endopeptidase. Hence, there is a rapid amplification of the reaction with time. The reaction can begin because C3 undergoes a slow, nonenzymatic hydrolysis of an internal thioester to generate an analogue of C3b, producing a fluid-phase C3 convertase.

The alternative convertase is extremely unstable in solution, having a half-life of only a few minutes. This half-life can be increased or decreased by the presence of other naturally occurring factors. The convertase is stabilized by properdin and nephritic factors.

[1] M. A. Kerr (1981) *MIE* **80**, 102.
[2] M. K. Pangburn & H. J. Müller-Eberhard (1984) *Springer Semin. Immunopathol.* **7**, 163.
[3] D. R. Bentley & R. D. Campbell (1986) *Biochem. Soc. Symp.* **51**, 7.
[4] B. P. Morgan (1990) *Complement: Clinical Aspects and Relevance*, Academic Press, Orlando.
[5] K. Whaley, M. Loos & J. Weiler (1993) *Complement in Health and Disease*, 2nd ed., Kluwer Publ., Amsterdam.
[6] M. A. Kerr (1994) in *LABFAX Immunochemistry* (M. A. Kerr & R. Thorpe, eds.), Bios Scientific, Oxford, p. 211.
[7] S. K. A. Law & K. B. M. Reid (1995) *Complement: In Focus*, 2nd ed., IRL Press, Oxford.

Selected entries from *Methods in Enzymology* [vol, page(s)]:
General discussion: 34, 731; 80, 63, 102, 103
Assay: 80, 104; 93, 381
Other molecular properties: activation, 80, 4, 5; alternative pathway, assembly, 80, 102, 103; assembly, 80, 54, 55; classical pathway, effect of C4-binding protein, 80, 124, 133; factor Ba (human, 80, 102, 103, 134 [amino acid composition, 80, 108; molecular weight, 80, 108]); factor Bb (human, 80, 3, 4, 102, 103, 134 [amino acid composition, 80, 108; molecular weight, 80, 108]; thioester substrates, 248, 13); genetic polymorphism, 80, 110, 111; human, 80, 74, 85, 102, 134 (activation, 80, 4, 5; amino acid composition, 80, 109; amino acid sequence, 80, 111, 112; ammonium sulfate precipitation, 80, 106); human, in alternative complement pathway (hemolytic assay, 93, 381; immunoquantitation, 93, 381; purification, 93, 379); inhibition, 80, 112; molecular properties, 80, 108; molecular weight, 80, 108; proteolytic enzyme, as, 80, 111, 112; purification, 80, 103, 105; separation of C2, 80, 61; similarity to C2, 80, 62, 63; stability, 80, 55; thioester substrates, kinetic constants, 248, 13

ALTRONATE DEHYDRATASE

This enzyme [EC 4.2.1.7] catalyzes the conversion of D-altronate to 2-dehydro-3-deoxy-D-gluconate and water.[1-3] The *Escherichia coli* enzyme requires Fe^{2+} but not in an iron–sulfur cluster.[1]

[1] J.-L. Dreyer (1987) *EJB* **166**, 623.
[2] W. A. Wood (1971) *The Enzymes*, 3rd ed., **5**, 573.
[3] J. D. Smiley & G. Ashwell (1960) *JBC* **235**, 1571.

Selected entries from *Methods in Enzymology* [vol, page(s)]:
General discussion: 5, 199; 90, 288
Assay: *Escherichia coli*, 5, 199; 90, 288
Other molecular properties: *Escherichia coli* (properties, 5, 202; 90, 291; purification, 5, 201; 90, 289)

*Alu*I RESTRICTION ENDONUCLEASE

This type II restriction endonuclease [EC 3.1.21.4], obtained from *Arthrobacter luteus*, catalyzes the hydrolysis of both strands of DNA at 5′ . . . AG∇CT . . . 3′, producing blunt-ended fragments.[1] The presence of flanking bases around the recognition sequence also appears to be of importance.[2]

[1]R. J. Roberts, P. A. Myers, A. Morrison & K. Murray (1976) *J. Mol. Biol.* **102**, 157.
[2]K. Majumder (1989) *Biophys. J.* **55**, 48a.

Selected entries from *Methods in Enzymology* [vol, page(s)]:
General discussion: 65, 21, 27, 363, 394, 481, 570, 573, 799
Other molecular properties: recognition sequence, **65**, 3, 555, 572

*Alw*NI RESTRICTION ENDONUCLEASE

This type II restriction endonuclease [EC 3.1.21.4], obtained from *Acinetobacter lwoffii* N, catalyzes the hydrolysis of both strands of DNA at 5′ . . . CAGNNN∇CTG . . . 3′, where N refers to any base.[1] The enzyme is blocked by overlapping *dcm* methylation.[2,3]

[1]R. D. Morgan, M. Dalton & R. Stote (1987) *NAR* **15**, 7201.
[2]A. Rieger & M. Nassal (1993) *NAR* **21**, 4148.
[3]M. Bourbonniere & J. Nalbantoglu (1991) *NAR* **19**, 4774.

*Alw*I RESTRICTION ENDONUCLEASE

This type II restriction endonuclease (subtype s) [EC 3.1.21.4], obtained from *Acinetobacter lwoffii*, catalyzes the hydrolysis of DNA at 5′ . . . GGATCNNNN∇ . . . 3′ and 3′ . . . CCTAGNNNNN∇ . . . 5′, where N refers to any base. The enzyme is blocked by overlapping *dam* methylation. Note that this endonuclease produces DNA fragments that have a single-base 5′ extension that is more difficult to ligate than blunt-ended fragments.

AMELOPROTEASE

This serine protease is responsible for the degradation of amelogenin during enamel maturation.[1]

[1]J. Moradian-Oldak, W. Leung, J. P. Simmer, M. Zeichner-David & A. G. Fincham (1996) *BJ* **318**, 1015.

AMIDASE

This enzyme [EC 3.5.1.4], also known as acylamidase, catalyzes the hydrolysis of a monocarboxylic acid amide to produce a monocarboxylate anion and ammonia.[1–4] Substrates include acetamide, acetanilide, propionamide, chloroacetamide, and 2-chloropropionamide. The *Pseudomonas aeruginosa* enzyme has a ping pong kinetic mechanism and proceeds through an acyl enzyme

intermediate.[3] The enzyme from many sources also has esterase activity.[5] *See also* specific amidase; ω-*Amidase*; *4-L-Aspartylglycosylamine Amidohydrolase*

[1]A. Thiery, M. Maestracci, A. Arnaud & P. Galzy (1986) *J. Gen. Microbiol.* **132**, 2205.
[2]R. C. Wyndham & J. H. Slater (1986) *J. Gen. Microbiol.* **132**, 2195.
[3]M. J. Woods, J. D. Findlater & B. A. Orsi (1979) *BBA* **567**, 225.
[4]J. Alt, E. Heyman & K. Krisch (1975) *EJB* **53**, 357.
[5]M. Kobayashi, M. Goda & S. Shimizu (1998) *FEBS Lett.* **439**, 325.

Selected entries from *Methods in Enzymology* [vol, page(s)]:
General discussion: 43, 208; 46, 22; 77, 333
Assay: 77, 405; acetanilide, with, 77, 406; aniline derivatives, with, 77, 405; butanilicaine, with, 77, 407; diazotization, 77, 405; fluorometric, 77, 408, 409; 4-nitroacetanilide, with, 77, 407; paracetamol, with, 77, 408, 409; phenacetin, with, 77, 408, 409; *p*-phenetidine, with, 77, 405; spectrophotometric, 77, 407
Other molecular properties: amino acid amides and, 2, 397; amino acid resolution, and, 3, 554, 559, 569; bacterial cell wall degradation and, 8, 690; chymotrypsin as, 2, 22; diaminopimelate isomers and, 6, 627, 629; induction in microorganisms, 22, 87, 88, 91; properties, 77, 342; purification, 77, 337; solubilization, 77, 337; transition-state and multisubstrate analogues, 249, 307; trypsin as, 2, 34

ω-AMIDASE

This enzyme [EC 3.5.1.3] catalyzes the hydrolysis of a monoamide of a dicarboxylate to generate the dicarboxylate and ammonia.[1–3] Substrates include glutaramate, succinamate, and their corresponding α-keto derivatives. The reaction proceeds through an acyl-enzyme intermediate.[3] The enzyme exhibits an esterase activity and catalyzes hydroxaminolysis and transamidation reactions.

[1]A. J. L. Cooper, T. E. Duffy & A. Meister (1985) *MIE* **113**, 350.
[2]A. J. L. Cooper & A. Meister (1976) *Crit. Rev. Biochem.* **4**, 281.
[3]L. B. Hersh (1972) *Biochemistry* **11**, 2251.

Selected entries from *Methods in Enzymology* [vol, page(s)]:
General discussion: 2, 384; 113, 350
Assay: 2, 384; 113, 352, 353
Other molecular properties: α-ketoglutaramate, assay for, 113, 356; properties, 2, 385; 113, 355; purification, rat liver, 2, 385; 113, 353, 357; rat liver, in, 2, 384; 17A, 1017; 113, 353, 357

AMIDINOASPARTASE

This enzyme [EC 3.5.3.14] catalyzes the hydrolysis of *N*-amidino-L-aspartate to produce L-aspartate and urea.[1,2] Other substrates include *N*-amidino-L-glutamate.

[1]S. Milstien & P. Goldman (1972) *JBC* **247**, 6280.
[2]T. Yorifuji & S. Furuyoshi (1986) *Agric. Biol. Chem.* **50**, 1327.

AMIDOPHOSPHORIBOSYLTRANSFERASE

This transferase [EC 2.4.2.14], also known as glutamine phosphoribosyl-pyrophosphate amidotransferase and phosphoribosyldiphosphate 5-amidotransferase, catalyzes the reaction of L-glutamine with 5-phospho-α-D-ribose

1-diphosphate and water to produce 5-phospho-β-D-ribosylamine, pyrophosphate (or, diphosphate), and L-glutamate.[1-6]

[1]G. Davies, M. L. Sinnott & S. G. Withers (1998) *CBC* **1**, 119.
[2]J. L. Smith (1998) *Curr. Opin. Struct. Biol.* **8**, 686.
[3]H. Zalkin (1993) *AE* **66**, 203.
[4]W. D. L. Musick (1981) *Crit. Rev. Biochem.* **11**, 1.
[5]W. N. Keley & J. B. Wyngaarden (1974) *AE* **41**, 1.
[6]D. E. Koshland, Jr. & A. Levitzki (1974) *The Enzymes*, 3rd ed. **10**, 539.

Selected entries from **Methods in Enzymology** [vol, page(s)]:
General discussion: 6, 56; **46**, 420, 424; **113**, 264
Assay: 6, 56; **51**, 171; **113**, 265
Other molecular properties: chicken liver, from, **6**, 58; **51**, 171; *Escherichia coli* (properties, **113**, 272; purification, **113**, 270); hormone induction of biorhythms, **36**, 480; inhibitors, **6**, 60; **51**, 178; **113**, 273; kinetic properties, **6**, 61; **51**, 178; **113**, 272; molecular weight, **51**, 177; properties, **6**, 59; **51**, 177, 178; **113**, 272; purification, **6**, 58; **51**, 173, 183; **113**, 270 (low ionic strength precipitation, **51**, 174); purity, **51**, 177; thiol, active-site, **87**, 83

AMINE DEHYDROGENASE

This oxidoreductase [EC 1.4.99.3], also known as methyl-amine dehydrogenase and primary-amine dehydrogenase, catalyzes the reaction of a primary amine with water and an acceptor substrate to produce an aldehyde, ammonia, and the reduced acceptor.[1-3]

Tryptophan tryptophylquinone (TTQ) is the cofactor for this enzyme. Amine substrates include methyl-amine, *n*-propylamine, *n*-butylamine, benzylamine, and *n*-nonylamine. The acceptor substrate is usually a specific blue copper protein (the acceptor is amicyanin in methylo-tropic bacteria); artificial acceptors include phenazine methosulfate. The TTQ-containing amine dehydrogenases catalyze a ping pong Bi Bi kinetic mechanism in which the aldehyde is released before the acceptor substrate binds. The first half-reaction involves the formation of an iminoquinone Schiff base between the amine-containing substrate and the cofactor. Proton abstraction from the α-carbon produces a transient carbanion. Reduction of the TTQ ring system produces the tautomeric form of the Schiff base and subsequent hydrolysis generates the aldehyde product and the aminoquinol form of the cofactor. A large deuterium kinetic isotope effect demonstrates that proton abstraction is the rate-determining step. Binding of an electron acceptor results in the formation of the iminoquinone form of the cofactor which is hydrolyzed to release ammonia and regenerate the TTQ quinone (note that, since amicyanin is a single electron acceptor, a semiquinone intermediate may also form in the second half reaction). **See also** *Aralkylamine*

Dehydrogenase, Trimethylamine Dehydrogenase, Dimethyl-amine Dehydrogenase

[1]C. Anthony (1998) *CBC* **3**, 155.
[2]J. P. Klinman & D. Mu (1994) *Ann. Rev. Biochem.* **63**, 299.
[3]J. A. Duine, J. Frank & J. A. Jongejan (1987) *AE* **59**, 169.

Selected entries from **Methods in Enzymology** [vol, page(s)]:
General discussion: **188**, 241; **258**, 149, 176; **280**, 98
Assay: **188**, 236, 242, 247; **258**, 152
Other molecular properties: absorption spectra, **258**, 152, 155, 163; adducts (ammonium, **258**, 185; hydroxide, **258**, 186; structures, **258**, 179); blue copper proteins as electron acceptors for, assay, **188**, 285; carbonyl reagents (effect on absorption spectra, **258**, 186; phenylhydrazine derivatization, **258**, 135; sensitivity, **258**, 156); cloning from bacteria (mutant complementation, **258**, 217; oligonucleotide probing, **258**, 220); electron acceptors, **258**, 151, 164, 192; electron spin-echo envelope modulation, **246**, 583; genes, bacterial (expression systems, **258**, 221; mutant construction, **258**, 226; regulation, **258**, 226; sequencing, **258**, 222); *Methylobacillus extorquens*, cloning of small subunit, **258**, 162; *Methylobacillus flagellatum* (assay, **188**, 247; properties, **188**, 249; purification, **188**, 248); methylotrophic bacterial (assay, **188**, 242; properties, **188**, 245; purification, **188**, 242; subunit resolution and reconstruction, **188**, 245); reaction mechanism, **258**, 178; redox forms (absorption spectra, **258**, 180; aminoquinol formation, **258**, 182; aminosemiquinone formation, **258**, 184; electron paramagnetic resonance, **258**, 184; fluorescence spectra, **258**, 183; reduction with dithionite, **258**, 181; structures, **258**, 179); resonance Raman spectroscopy, **258**, 138; staining on gels, **22**, 594; stopped-flow spectroscopy (amicyanin oxidation, **258**, 190; apparatus, **258**, 187, 189; methylamine reduction, **258**, 189); subunit structure, **258**, 151, 178, 191; *Thiobacillus versutus* (assay, **188**, 236; properties, **188**, 239; purification, **188**, 236); tryptophan tryptophylquinone isolation (amino acid composition, **258**, 159; carboxypeptidase-Y treatment, **258**, 158; cofactor stabilization, **258**, 156; leucine aminopeptidase treatment, **258**, 158; pronase treatment, **258**, 157; sequence analysis, **258**, 159; subunit isolation, **258**, 157; sulfhydryl group protection, **258**, 156); X-ray crystallography, **258**, 163, 168, 178 (amino acid sequence analysis, **258**, 199; cofactor identification, **258**, 200; crystallization conditions, **258**, 194; heavy subunit, **258**, 204; light subunit, **258**, 205; *Paracoccus denitrificans* enzyme [amicyanin complex, **258**, 193, 208; cytochrome *c*-amicyanin ternary complex, **258**, 210]; structure analysis and refinement, **258**, 194, 196; superbarrel motif, **258**, 213, 215; *Thiobacillus versutus* enzyme, hydrazine complex, **258**, 208; tryptophan tryptophylquinone orientation, **258**, 168, 201, 203)

AMINE *N*-METHYLTRANSFERASE

This methyltransferase [EC 2.1.1.49] (also known as nicotine *N*-methyltransferase, tryptamine *N*-methyltrans-ferase, indoleamine *N*-methyltransferase, and arylamine *N*-methyltransferase) catalyzes the reaction of *S*-adenosyl-L-methionine with an amine to produce *S*-adenosyl-L-homocysteine and a methylated amine.[1-4] Acceptor substrates include a wide range of primary, secondary, and tertiary amines: for example, tryptamine, aniline, nicotine, β-phenylethylamine, anisidine, pyrazole, and *N*-methylcyclohexylamine.

[1]P. A. Crooks, C. S. Godin, C. G. Nwosu, S. S. Ansher & W. B. Jakoby (1986) *Biochem. Pharmacol.* **35**, 1600.
[2]S. S. Ansher & W. B. Jakoby (1986) *JBC* **261**, 3996.

[3]L. L. Hsu & A. J. Mandell (1973) *Life Sci.* **13**, 847.
[4]A. H. Bahnmaier, B. Woesle & H. Thomas (1999) *Chirality* **11**, 160.

Selected entries from **Methods in Enzymology** [vol, page(s)]:
General discussion: 77, 263;142, 660
Assay: 142, 661, 669
Other molecular properties: indoleamine N-methyltransferase (assay, 142, 669; properties, 142, 671; purification from rabbit lung, 142, 670); properties, 142, 666; purification from rabbit liver (type A, 142, 662; type B, 142, 665)

AMINE OXIDASES

This group of enzymes catalyzes the reaction of an amine with dioxygen to produce an aldehyde, ammonia, and hydrogen peroxide.[1-17]

Amine oxidase (flavin-containing) [EC 1.4.3.4] (also known as monoamine oxidase, tyramine oxidase, tyraminase, and adrenalin oxidase) catalyzes the FAD-dependent reaction of an organic primary amine with dioxygen and water to produce an aldehyde, ammonia, and hydrogen peroxide. Secondary and tertiary amines can also serve as substrates. Examples of substrates include phenylethylamine, n-butylamine, ethylamine, amylamine, benzylamine, tyramine, dopamine, and kynuramine. Monoamine oxidase A has a ping pong Bi Bi kinetic mechanism which is distinct from the scheme seen with monoamine oxidase B.[9]

Amine oxidase (copper-containing) [EC 1.4.3.6] (also known as diamine oxidase, diamino oxhydrase, and histaminase) catalyzes the trihydroxyphenylalanyl- and copper-dependent reaction of a primary amine with water and dioxygen to produce an aldehyde, ammonia, and hydrogen peroxide. This classification represents a group of enzymes which can also act on diamines and histamine. Substrates include not only histamine but also putrescine, benzylamine, cadaverine, dopamine, tryptamine, phenylethylamine, spermine, and spermidine. The enzyme has a ping pong Bi Bi kinetic mechanism. The reaction mechanism involves the nucleophilic attack of the amine-containing substrate on the 5'-position of the 2',5'-dioxo-4'-hydroxyphenylalanyl residue (*i.e.*, TOPA quinone) of the protein to produce a carbinolamine intermediate. This transient intermediate is readily converted to the Schiff base. Release of the aldehyde product generates the aminoquinol form of the prosthetic group. The copper-dependent reaction of the aminoquinol with dioxygen and water results in the release of ammonia and hydrogen peroxide and the regeneration of TOPA quinone. **See also** *Methylamine Oxidase; Lysyl Oxidase*

[1]C. W. Abell & S. W. Kwan (2000) *Prog. Nucl. Acid Res. Mol. Biol.* **65**, 129.
[2]C. Anthony (1998) *CBC* **3**, 155.
[3]A. Messerschmidt (1998) *CBC* **3**, 401.
[4]C. Hartmann & W. S. McIntire (1997) *MIE* **280**, 98.
[5]J. P. Klinman (1996) *JBC* **271**, 27189.
[6]J. M. Janes & J. P. Klinman (1995) *MIE* **258**, 20.
[7]B. J. Bahnson & J. P. Klinman (1995) *MIE* **249**, 373.
[8]E. J. Brush & J. W. Kozarich (1992) *The Enzymes*, 3rd ed., **20**, 317.
[9]R. R. Ramsay (1991) *Biochemistry* **30**, 4624.
[10]J. F. Powell (1991) *Biochem. Soc. Trans.* **19**, 199.
[11]R. B. Silverman (1991) *Biochem. Soc. Trans.* **19**, 201.
[12]R. R. Ramsay & T. P. Singer (1991) *Biochem. Soc. Trans.* **19**, 219.
[13]M. A. Ator & P. R. Ortez de Montellano (1990) *The Enzymes*, 3rd ed., **19**, 213.
[14]J. A. Duine, J. Frank & J. A. Jongejan (1987) *AE* **59**, 169.
[15]V. Massey & P. Hemmerich (1975) *The Enzymes*, 3rd ed., **12**, 191.
[16]B. G. Malmström, L.-E. Andréasson & B. Reinhammer (1975) *The Enzymes*, 3rd ed., **12**, 507.
[17]W. G. Bardsley, M. J. C. Crabbe & J. S. Shindler (1973) *BJ* **131**, 459.

Selected entries from **Methods in Enzymology** [vol, page(s)]:
General discussion: 2, 390, 394; 17B, 682, 686, 692, 698, 705, 709, 717, 722, 726, 730, 735, 741; 46, 38, 160; 52, 10, 18; 56, 474; 142, 617; 249, 373; 258, 20; 280, 98 (methylamine oxidase, 52, 21)
Assay: 2, 390, 394; 17B, 682, 686, 689, 692, 698, 705, 709, 717, 722, 726, 730, 736, 741, 745; 39, 388; 53, 496; 55, 101, 102; 94, 315; 142, 619, 628, 647; activity assay, 54, 492, 497; 63, 34; *Arthrobacter* P1, 188, 227; bovine plasma, 17B, 698; 94, 315; *Candida boidinii*, 188, 427; coupled assay, 63, 34; fluorometric, 17B, 741; oxygen electrode assay, 258, 71; pea seedlings, 17B, 730; *Pichia pastoris*, 188, 427; pig kidney, 2, 394; 17B, 736; putrescine oxidase, 17B, 726; 188, 427; radiochemical assay, 17B, 744; 258, 84; radiometric, 17B, 745; spectrophotometric assay, 17B, 741; 258, 71, 74
Other molecular properties: absorption spectra, 56, 696; antibody (affinity chromatography, 56, 700; conjugation to ferritin, 56, 711, 712; specificity, 56, 690); beef liver mitochondria, from, 53, 495; benzylamine oxidation, 77, 20 (benzylamine oxidase, 52, 10 [assay, 188, 427; properties, 188, 433; purification from *Candida boidinii*, 188, 429]; benzylamine/putrescine oxidase [assay, 188, 427; properties, 188, 433; purification from *Pichia pastoris*, 188, 429]); bovine plasma (assays, 94, 315; polyamine system, structure-activity relationship, 94, 416; properties, 94, 316; purification, 94, 315); clinical aspects, 17B, 697; copper-containing (ammonia [production, radiochemical analysis, 258, 84; release, freeze-quench kinetics, 258, 84; retention on enzyme, 258, 83]; *Arthrobacter* P1 [properties, 188, 232; purification, 188, 229]; bovine serum [amino acid sequence, 258, 113; cloning {library screening, 258, 115, 119; mixed oligonucleotide primed amplification of cDNA, 258, 116; N-terminal sequencing, 258, 108; primer design, 258, 117, 121}; hydrogen tunneling in, 249, 393, 397; resonance Raman spectroscopy, 258, 137; sequence analysis {HPLC/electrospray ionization mass spectrometry, 258, 112; tandem mass spectrometry, 258, 92, 97, 102, 105, 110}; site of synthesis, 258, 115]; *Candida boidinii* [properties, 188, 433; purification, 188, 429]; carbonyl reagent sensitivity, 258, 20, 122; copper role in catalysis, 258, 53, 70; electron transfer, temperature-jump detection, 258, 89; inhibition assay for methylglyoxal bis(guanylhydrazone), in, 94, 250; multiple regression analysis in Hammett value determination, 258, 79; oxidative half reaction, 258, 70, 73; pea seedlings, in, 17B, 730 [copper and, 17B, 735; purification, 17B, 731; spectra, 17B, 735]; peptide isolation [thermolytic digest, 258, 25, 30; tryptic digest, 258, 26, 91, 100]; phenylhydrazine reaction, 258, 21, 23, 33, 133; pig kidney, in, 17B, 735 [copper and, 17B, 739; properties, 2, 396; purification, 2, 395; 17B, 737; pyridoxal phosphate, and, 17B, 739]; properties, 2, 396; 17B, 727, 734, 739; proteolytic digestion, 258, 24, 30; purification [*Micrococcus rubens*, 17B, 727; pea seedling, 17B, 731; pig kidney, 2, 395; 17B, 737]; putrescine oxidase, 17B, 726 [flavin adenine dinucleotide and, 17B, 729; purification from *Micrococcus rubens*,

17B, 727; spectrum, 17B, 729]; pyridoxamine-5-phosphate oxidase and, 6, 333; reductive half reaction, 258, 70, 72; Schiff base intermediates [product, 258, 83; substrate, 258, 74; sodium cyanoborohydride effects [enzyme-substrate complex inactivation, 258, 74; free enzyme inactivation, 258, 75]; steady-state kinetics, 258, 78; stopped-flow spectrometry [anaerobic conditions, 258, 77; data analysis, 258, 77, 81; deuterium isotope effects, 258, 82; rapid-scanning analysis, 258, 80; relaxation characteristics, 258, 81]; subunit structure, 258, 69; topa quinone chromophore [accessibility to modifying reagents, 258, 35; identification attempts, history, 258, 20; peptide analysis mass spectrometry, 258, 28, 95, 97; nuclear magnetic resonance, 258, 29; resonance Raman spectroscopy, 258, 31, 34, 139; visible absorbance spectroscopy, 258, 31, 34; pyrroloquinoline quinone studies, 258, 21, 32, 34; semiquinone intermediate {absorption spectroscopy, 258, 88; circular dichroism, 258, 87; electron paramagnetic resonance, 258, 85}]; transamination mechanism, 258, 39, 53); coupling to Sepharose 4B, 56, 698, 699, 700; digitonin, 56, 692; distribution (inner surface of outer membrane, on, 56, 686; mitochondrial outer membrane, on, 56, 684, 691; outer surface of outer membrane vesicles, on, 56, 688); flavin content, 17B, 721, 725; 53, 500, 501; flavin linkage, 53, 450; formaldehyde generation, 52, 297; hepatoma mitochondria, 55, 81, 82, 88; human serum, in, 17B, 692 (cirrhosis, in, 17B, 697; congestive heart failure, in, 17B, 697; pregnancy, in, 17B, 697; sarcoidosis, in, 17B, 697); 5-hydroxytryptamine and, 6, 598, 605; immunocytochemical localization in brain, 142, 649; inactivation, 53, 441; inhibitors, 53, 501; intestinal epithelium, in, 77, 156; isolation of flavin peptide, 53, 454; liver, 2, 393; location, 10, 457, 461; mechanism, 249, 273; mechanism-based inactivation, 249, 272; melatonin biosynthesis, in, 39, 379; methylamine oxidase, absorption of phenylhydrazine derivatives, 258, 133; mitochondrial membranes, in, 32, 91; molecular weight, 53, 500; monoamine oxidase B (hydrogen tunneling in, 249, 395); multiple forms, 142, 618; outer mitochondrial membranes, in, 31, 319; 55, 101, 103, 104; 56, 684, 686; pea seedling, 17B, 731; pH optimum, 53, 501; pineal, in, 39, 396; plasma membrane, 31, 90; production of aromatic aldehydes, ultraviolet extinction of, 17B, 689; preparation and properties, 56, 694, 695; properties, 2, 393, 399; 17B, 685, 690, 695, 701, 707, 714, 721, 725, 729, 734, 739; 142, 618, 636, 638, 648; purification, 53, 495, 496, 498 (Aspergillus niger, 17B, 705; beef liver, mitochondria, 17B, 710; beef plasma, 2, 391; 17B, 699; human serum, 17B, 694; mammalian liver, 142, 624; pig brain, mitochondria, 17B, 719; Micrococcus rubens, 17B, 727; pea seedling, 17B, 731; pig kidney, 2, 395; 17B, 737; pig plasma, 17B, 683; rabbit serum, 17B, 687; Sarcina lutea, 17B, 723); purity, 53, 500; putrescine oxidase 17B, 726; pyridoxamine-5-phosphate oxidase and, 6, 333; release from outer membrane, 55, 91, 93, 94; spectrum of, 17B, 702, 708, 725; staining on gels, 22, 601; storage, 53, 500; type A, purification from (human liver, 142, 648; human placenta, 142, 631); type B (antibody complex, characterization, 142, 647; purification from [bovine liver, 142, 633; human liver, 142, 645]); tyramine oxidase, 17B, 722

AMINO-ACID ACETYLTRANSFERASE

This transferase [EC 2.3.1.1], also known as *N*-acetyl-glutamate synthase and acetyl-CoA:glutamate acetyl-transferase, catalyzes the reaction of acetyl-CoA with L-glutamate to form coenzyme A and *N*-acetyl-L-glutamate. The enzyme can also act on L-aspartate and, less efficiently, on some other amino acids. The mammalian enzyme is activated by L-arginine, whereas the *Neurospora crassa* enzyme and the enzyme from a number of microorganisms is inhibited by L-arginine. *See also Glutamate Acetyltransferase; specific acyltransferases*

Selected entries from *Methods in Enzymology* [vol, page(s)]:
General discussion: 1, 616; 113, 27
Assay: 1, 616; 113, 28
Other molecular properties: *Clostridium kluyveri* (properties, 1, 618; purification, 1, 617); rat liver (assays, 113, 28; properties, 113, 32; purification, 113, 29)

D-AMINO-ACID N-ACETYLTRANSFERASE

This acetyltransferase [EC 2.3.1.36] catalyzes the reaction of acetyl-CoA with a D-amino acid to produce Coenzyme A and an *N*-acetyl-D-amino acid.[1]

[1] M. H. Zenk & J. H. Schmitt (1965) *Biochem. Z.* **342**, 54.

D-AMINO-ACID DEHYDROGENASE

This FAD-dependent oxidoreductase [EC 1.4.99.1] catalyzes the reaction of a D-amino acid with an acceptor substrate and water to produce an α-keto acid (*i.e.*, a 2-oxo acid), ammonia, and the reduced acceptor.[1] Acceptor substrates include dichlorophenolindophenol, coenzyme Q, methylene blue, ferricyanide, phenazine methosulfate, and pyocyanine. Most D-amino acids, with the noted exceptions of D-aspartate and D-glutamate, are substrates.

[1] P. J. Olsiewski, J. Kaczorowski & C. Walsh (1980) *JBC* **255**, 4487.

Selected entries from *Methods in Enzymology* [vol, page(s)]:
General discussion: 17B, 623
Assay: 17B, 623
Other molecular properties: properties, 17B, 627; purification from *Pseudomonas fluorescens*, 17B, 624; spectrum, 17B, 627

L-AMINO-ACID DEHYDROGENASE

This oxidoreductase [EC 1.4.1.5] catalyzes the reaction of an L-amino acid with NAD^+ and water to produce an α-keto acid (*i.e.*, a 2-oxo acid), ammonia, and NADH.[1,2] The best substrates are L-valine and L-isoleucine.[2]

[1] N. M. W. Brunhuber & J. S. Blanchard (1994) *Crit. Rev. Biochem. Mol. Biol.* **29**, 415.
[2] Y. Nitta, Y. Yasuda, K. Tochikubo & Y. Hachisuka (1974) *J. Bacteriol.* **117**, 588.

α-AMINO-ACID ESTERASE

This esterase [EC 3.1.1.43], also known as α-amino-acid ester hydrolase, catalyzes the hydrolysis of an α-amino acid ester to produce an α-amino acid and an alcohol.[1–3] Substrate examples include L-arginine methyl ester, L-tyrosine propyl ester, L-leucine ethyl ester, and D-alanine methyl ester. The enzyme will also catalyze an α-aminoacyl transfer to a number of amine nucleophiles.

[1] K. Kato, K. Kawahara, T. Takahashi & A. Kakinuma (1980) *Agric. Biol. Chem.* **44**, 1069 and 1075.
[2] K. Kato (1980) *Agric. Biol. Chem.* **44**, 1083.
[3] T. Takahashi, Y. Yamazaki & K. Kato (1974) *BJ* **137**, 497.

D-AMINO-ACID OXIDASE

This FAD-dependent oxidoreductase [EC 1.4.3.3] catalyzes the reaction of a D-amino acid with dioxygen and water to produce an α-keto acid (*i.e.*, 2-oxo acid), ammonia, and hydrogen peroxide.[1-8] Substrate specificity is broad and includes glycine (the best substrate is D-proline in most organisms studied).[7]

In the first half-reaction, the D-amino acid substrate reduces the enzyme-bound FAD cofactor and generates the imino acid intermediate. When the imino acid dissociates from the enzyme, solvolysis produces the corresponding α-keto acid and ammonia. In the second half-reaction, the reduced coenzyme · enzyme complex reacts with dioxygen to reform the FAD and hydrogen peroxide. With neutral D-amino acids, the dissociation of the resultant imino acid is slow and hydrogen peroxide is formed prior to the release of the imino acid. With good substrates, imino acid release is the rate-determining step. The first step in the mechanism has been suggested to involve a hydride transfer from the α-carbon of the substrate to the flavin N5 position, resulting in the formation of a transient carbanion intermediate.

[1] B. A. Palfey & V. Massey (1998) *CBC* 3, 83.
[2] A. Mattevi, M. A. Vanoni & B. Curti (1997) *Curr. Opin. Struct. Biol.* 7, 804.
[3] K. Yagi (1991) *A Study of Enzymes* 2, 271.
[4] L. Pollegioni, A. Falbo & M. S. Pilone (1992) *BBA* 1120, 11.
[5] G. A. Hamilton (1985) *AE* 57, 85.
[6] H. J. Bright & D. J. T. Porter (1975) *The Enzymes*, 3rd ed., 12, 421.
[7] L. A. Lichtenberg & D. Wellner (1968) *Anal. Biochem.* 26, 313.
[8] M. Dixon & K. Kleppe (1965) *BBA* 96, 357, 368, and 383.

Selected entries from *Methods in Enzymology* [vol, page(s)]:
General discussion: 2, 199; 17B, 593, 608; 46, 37; 52, 18; 53, 402; 56, 474
Assay: 2, 199; 17B, 595, 596, 609; ammonia determination, by, 17B, 609; 63, 34; coupled enzyme assay, 63, 34; fluorometric, 17B, 596; oxygen consumption, 2, 199; 17B, 609; peroxide formation, 17B, 609; spectrophotometric, 17B, 595
Other molecular properties: D-alanine determination and, 2, 171; 8, 698; alanine racemase assay and, 2, 212; 17A, 171; alanine transamination and, 2, 176; amino acid identification and, 3, 526; amino acid purity and, 2, 171; 3, 567; 5, 691; anaerobic photoreduction, 18B, 509; apoenzyme (assay of FAD, 53, 422; porcine kidney, in luminometric FAD assay [inactivation during assay, 122, 190; preparation, 122, 187]); assay of (D-alanine, 2, 171; 8, 698; 17B, 630; D-leucine, 17B, 630); borohydride reduction, 18B, 474, 475, 478, 479; D-β-chloroalanine, 87, 88; cofactor requirement of, 2, 227; cryoenzymology, 63, 338; dissociation kinetics, 53, 430; electrophoresis convection of, 5, 40; enzyme electrode, in, 44, 589; extraction of, 1, 33; flavin adenine dinucleotide and, 17B, 614, 615; flavin adenine dinucleotide assay and, 2, 200, 673, 698; 3, 955; fluorescence of, 17B, 620; formation of N^5-acyl leuco FAD during illumination with products, 18B, 512; gluconate dehydrogenase preparation, in, 5, 291; glycine oxidase and, 2, 227; holoenzyme reconstitution, 53, 436; immobilization, in gel, 44, 908; inactivation, 53, 442, 446, 448; inhibition, 17B, 619; 87, 493, 500, 509 (riboflavin 5'-monosulfate, by,

18B, 465); intermediate, 17B, 615; 87, 88; keto acid preparation and, 3, 412; large-scale isolation of, 22, 536; microbodies, in, 52, 496, 498; peroxisomes, in, 31, 364, 367; D-propargylglycine, 87, 493; properties, 17B, 618; purification from (hog kidney, 17B, 610; sheep kidney, 2, 200); purification of calcium phosphate gel-cellulose, 22, 342; pyridoxamine-5-phosphate oxidase and, 6, 333; reaction with benzoate, 17B, 609; redox potential, 53, 400; reversible dissociation, 53, 429, 433; spectrum of, 17B, 618; staining on gels, 22, 594; substrates, 17B, 619; 44, 615; suicide inhibition, 87, 493; sulfite complex, 18B, 473; Yagi-Ozawa plots, 87, 500, 509

L-AMINO-ACID OXIDASE

This FAD-dependent oxidoreductase [EC 1.4.3.2], also known as ophio-amino-acid oxidase (specifically referring to the enzyme from the king cobra, *Ophiophagus hannah*), catalyzes the reaction of an L-amino acid with dioxygen and water to generate an α-keto acid (*i.e.*, 2-oxoacid), ammonia, and hydrogen peroxide.[1-3] Glycine is a substrate for the enzyme isolated from some sources. **See** *D-Amino-Acid Oxidase*

[1] B. A. Palfey & V. Massey (1998) *CBC* 3, 83.
[2] H. J. Bright & D. J. T. Porter (1975) *The Enzymes*, 3rd ed., 12, 421.
[3] A. Meister & D. Wellner (1963) *The Enzymes*, 2nd ed., 7, 609.

Selected entries from *Methods in Enzymology* [vol, page(s)]:
General discussion: 2, 205, 209, 211; 17B, 593, 597, 601; 46, 37; 52, 18; 56, 474
Assay: 2, 204, 205, 209; 17B, 593, 595, 601; coupled enzyme, 63, 34; fluorometric, 17B, 595; 3-hydrazinoquinoline, with, 17B, 601; spectrophotometric, 17B, 593
Other molecular properties: amino acid purity and, 3, 566; anaerobic photoreduction, 18B, 510, 511; borohydride reduction, 18B, 474, 475, 478, 479; conjugate, activity, 44, 271; detection of, 2, 179; diaminopimelate assay and, 6, 633; enzyme electrode, in, 44, 589; immobilization, 56, 484, 485 (gel, in, 44, 908; inert protein, on, 44, 908); inactivation, 53, 443, 445; α-keto acid preparation and, 3, 406; microbodies, in, 52, 496, 498; *Neurospora crassa*, 2, 211; properties, 2, 207, 210; 17B, 599, 604; purification from (rat kidney, 2, 209; 17B, 602; snake venom, 2, 205; 17B, 597); racemase assay and, 2, 213; sources, 2, 211; spectrum of, 17B, 599, 604; substrates, 44, 615; sulfite complex, 18B, 473; transaminases and, 5, 686; use for synthesis of D-Δ¹-piperideine-6-carboxylate, 17B, 190

AMINO ACID RACEMASE

This pyridoxal-phosphate-dependent racemase [EC 5.1.1.10] catalyzes the interconversion of L- and D-amino acids.[1-3] The rates of racemization are not the same for all amino acids. **See also** *specific racemase*

[1] M. E. Tanner & G. L. Kenyon (1998) *CBC* 2, 7.
[2] E. Adams (1972) *The Enzymes*, 3rd ed., 6, 479.
[3] K. Soda & T. Osumi (1969) *Biochem. Biophys. Res. Commun.* 35, 363.

Selected entries from *Methods in Enzymology* [vol, page(s)]:
General discussion: 2, 212; 17B, 629
Assay: 17B, 629
Other molecular properties: properties, 17B, 635; purification from *Pseudomonas striata*, 17B, 633

AMINOACYLASE

This enzyme [EC 3.5.1.14] (also known as histozyme, hippuricase, benzamidase, dehydropeptidase II, aminoacylase I, and acylase I) catalyzes the hydrolysis of an N-acyl-L-amino acid to produce a carboxylate anion and an L-amino acid.[1-4] The enzyme has a wide specificity for the amino acid derivative. It will also catalyze the hydrolysis of dehydropeptides. At pH values above neutrality, the enzyme will also catalyze the exchange of L-alanine oxygens with those of water.[2] The enzyme has been used to resolve racemic mixtures of amino acids. *See also Hippurate Hydrolase; Aspartoacylase; Membrane Dipeptidase*

[1]M. W. Anders & W. Dekant (1994) *Adv. Pharmacol.* **27**, 431.
[2]K. H. Rohm & R. L. Van Etten (1986) *EJB* **160**, 327.
[3]J. Gentzen, H.-G. Loeffler & F. Schneider (1980) *Z. Naturforsch.* **35c**, 544.
[4]J. Matsumoto & S. Nagai (1972) *J. Biochem.* **72**, 269.

Selected entries from *Methods in Enzymology* [vol, page(s)]:
General discussion: 2, 109, 115
Assay: 2, 109, 115; 44, 199, 747, 748; dehydropeptidase activity, 2, 109
Other molecular properties: amino acid resolution by, 2, 119; 3, 554, 558, 568; *Bacillus stearothermophilus*, 248, 224; commercial use, 44, 165; dehydropeptidase activity of, 2, 109; entrapment, in polyacrylamide gel, 44, 750; exocystine desulfhydrase, present in preparation of, 2, 319; fiber entrapment, industrial application, 44, 241; identity of, 2, 115; immobilization on DEAE-Sephadex, 44, 748, 749 (iodoacetylcellulose, on, 44, 749, 750; protein copolymerization, by, 44, 198; regeneration of activity during, 44, 164, 165); immobilized, kinetic properties, 44, 750, 751; industrial application, 44, 746; marker for chromosome detection in somatic cell hybrids, as, 151, 180; native, preparation, 44, 748; properties, 2, 113, 117; 248, 224; purification, pig kidney, 2, 112, 116; source, 44, 196; specificity, 2, 117; stability, 2, 117; supports, 44, 759; thermal stability, 44, 751, 752; vinylation, effect on pH, 44, 199

AMINOACYL-tRNA HYDROLASE

This enzyme [EC 3.1.1.29] (also known as N-acylaminoacyl-tRNA hydrolase, peptidyl-tRNA hydrolase, and N-acetylaminoacyl-tRNA hydrolase) catalyzes the hydrolysis of an N-substituted aminoacyl-tRNA to produce an N-substituted amino acid and tRNA.[1-4] The protein from some sources can act on substrates that are not substituted at the amino group.

Whereas the bacterial and yeast enzymes act by deacylation, the rabbit reticulocyte enzyme is reported[3] to hydrolyze an N-acylaminoacyl-tRNA to N-acylaminoacyl-AMP, upon which an aminoacyl-AMP deacylase acts.

The bacterial enzyme recycles aborted peptidyl-tRNA molecules resulting from premature termination. Any N-blocked aminoacyl-tRNA can serve as a substrate (except for formylmethionyl-tRNA$_f$^{Met} of eubacterial origin).

[1]J.-P. Jost & R. M. Bock (1969) *JBC* **244**, 5866.

[2]C. Ferreiro & C. F. Heredia (1986) *BBA* **882**, 410.
[3]M. Gross, P. Crow & J. White (1992) *JBC* **267**, 2080.
[4]P. R. Schimmel (1980) *Crit. Rev. Biochem.* **9**, 207.

Selected entries from *Methods in Enzymology* [vol, page(s)]:
General discussion: 20, 194
Assay: 20, 198
Other molecular properties: esterification, assaying extent of, 20, 92; methionyl-tRNA hydrolase, 29, 726 (assay, 29, 729; characterization, 29, 729; fractionation, 29, 733; materials and solutions, 29, 728); preparation, 20, 197; removal from ribosomes, 30, 307; specificity, 20, 196

2-AMINOADIPATE AMINOTRANSFERASE

This pyridoxal-phosphate-dependent aminotransferase [EC 2.6.1.39] catalyzes the reversible reaction of L-2-aminoadipate with α-ketoglutarate (or, 2-oxoglutarate) to generate α-ketoadipate (*i.e.*, 2-oxoadipate) and L-glutamate.[1,2] Certain isozymes of aminoadipate aminotransferase also exhibit a kynurenine aminotransferase activity.

[1]M. R. Mawal & D. R. Deshmukh (1991) *Prep. Biochem.* **21**, 63.
[2]D. R. Deshmukh & S. M. Mungre (1989) *BJ* **261**, 761.

Selected entries from *Methods in Enzymology* [vol, page(s)]:
General discussion: 17B, 119
Assay: 113, 666
Other molecular properties: *Neurospora crassa*, in, 17B, 119; rat kidney (properties, 113, 672; purification, 113, 669); yeast, in bakers', 17B, 119

L-2-AMINOADIPATE-6-SEMIALDEHYDE DEHYDROGENASE

This oxidoreductase [EC 1.2.1.31], also known as α-aminoadipate reductase, catalyzes the reaction of L-2-aminoadipate 6-semialdehyde with NADP$^+$ and water to generate L-2-aminoadipate and NADPH.[1-3]

[1]H. Schmidt, R. Bode & D. Birnbaum (1990) *FEMS Microbiol. Lett.* **70**, 41.
[2]V.-F. Chang, P. Ghosh & V. V. Rao (1990) *BBA* **1038**, 300.
[3]D. E. Ehmann, A. M. Gehring & C. T. Walsh (1999) *Biochemistry* **38**, 6171.

Selected entries from *Methods in Enzymology* [vol, page(s)]:
General discussion: 17B, 120, 195
Assay: 17B, 120, 196

δ-(L-α-AMINOADIPOYL)-L-CYSTEINYL-D-VALINE SYNTHETASE

This phosphopantothenate-containing enzyme system catalyzes the formation of δ-(L-α-aminoadipoyl)-L-cysteinyl-D-valine from ATP, L-α-aminoadipate, L-cysteine, and L-valine.[1,2] An ATP exchange with pyrophosphate is observed in the presence of all three amino acids, suggesting that each amino acid forms an amino acyl-adenylate intermediate. At some point in the synthesis, either a racemization occurs (*i.e.*, L-valine to D-valine) or an epimerization (for example, with a di- or tripeptide or with the valyl-adenylate).

[1] J. E. Baldwin, M. F. Byford, R. A. Field, C.-Y. Shiau, W. J. Sobey & C. J. Schofield (1993) *Tetrahedron* **49**, 3221.
[2] C.-Y. Shiau, J. E. Baldwin, M. F. Byford, W. J. Sobey & C. J. Schofield (1995) *FEBS Lett.* **358**, 97.

Selected entries from *Methods in Enzymology* [vol, page(s)]:
General discussion: 43, 471
Assay: 43, 471
Other molecular properties: *Cephalosporium acremonium* preparation, 43, 473; specificity, 43, 473

2-AMINOBENZENESULFONATE 2,3-DIOXYGENASE

This oxidoreductase [EC 1.14.12.14], also known as 2-aminosulfobenzene 2,3-dioxygenase, catalyzes the reaction of 2-aminobenzenesulfonate with NADH, H^+, and dioxygen to produce 3-sulfocatechol, ammonia, and NAD^+. This dioxygenase is a multicomponent protein complex that includes a flavin-dependent reductase component and an iron-dependent oxygenase component.[1]

[1] J. Mampel, J. Ruff, F. Junker & A. M. Cook (1999) *Microbiology* **145**, 3255.

AMINOBENZOATE DECARBOXYLASE

This pyridoxal-phosphate-dependent decarboxylase [EC 4.1.1.24] catalyzes the conversion of 4-aminobenzoate to aniline and carbon dioxide.[1] Other substrates include 2-aminobenzoate.[1]

[1] W. G. McCullough, J. T. Piligian & I. J. Daniel (1957) *JACS* **79**, 628.

4-AMINOBENZOATE 1-MONOOXYGENASE

This FAD-dependent oxidoreductase [EC 1.14.13.27], also known as 4-aminobenzoate hydroxylase and 4-aminobenzoate dehydrogenase, catalyzes the reaction of 4-aminobenzoate with NAD(P)H and dioxygen to produce 4-hydroxyaniline, $NAD(P)^+$ (with A-side stereospecificity), water, and carbon dioxide.[1] Alternative substrates include anthranilate, 4-amino-2-chlorobenzoate, 2-amino-5-chlorobenzoate, and 4-aminosalicylate; however, salicylate is not a substrate, instead reacting with salicylate 1-monooxygenase [EC 1.14.13.1].

Studies with labeled dioxygen demonstrate that one atom of molecular oxygen is incorporated into the 4-hydroxyaniline formed from 4-aminobenzoate.[1]

[1] H. Tsuji, T. Ogawa, N. Bando & K. Sasaoka (1986) *JBC* **261**, 13203.

p-AMINOBENZOATE SYNTHASE

This enzyme, which is similar in action to anthranilate synthase [EC 4.1.3.27], catalyzes the reaction of chorismate with L-glutamine to produce p-aminobenzoate, pyruvate, and L-glutamate.[1] Ammonia can act as a nitrogen donor as well. The *Bacillus subtilis* enzyme is activated by guanosine. *See also* Anthranilate Synthase

[1] B. Roux & C. T. Walsh (1992) *Biochemistry* **31**, 6904.

Selected entries from *Methods in Enzymology* [vol, page(s)]:
General discussion: 113, 293
Assay: 113, 293

4-AMINOBUTYRALDEHYDE DEHYDROGENASE

This oxidoreductase [EC 1.2.1.19], also known as 4-aminobutanal dehydrogenase and γ-guanidinobutyraldehyde dehydrogenase, catalyzes the reaction of 4-aminobutanal with NAD^+ and water to produce 4-aminobutanoate and NADH.[1,2] The rat liver enzyme is also regulated by the coenzyme.[1] Enzymes from *Pseudomonas putida* and *Vicia faba* also act on 4-guanidinobutanal.

[1] G. Testore, S. Colombatto, F. Silvagno & S. Bedino (1995) *Int. J. Biochem. Cell Biol.* **27**, 1201.
[2] M. I. Prieto, J. Martin, R. Balana-Fouce & A. Garrido-Pertierra (1987) *Biochimie* **69**, 1161.

Selected entries from *Methods in Enzymology* [vol, page(s)]:
General discussion: 5, 767
Assay: 5, 767
Other molecular properties: formation of, 5, 765; properties, 5, 770; purification, *Pseudomonas fluorescens*, 5, 768; Δ[1]-pyrroline assay, 5, 767; stereospecificity determination, 54, 229, 231; 87, 114; substrate preparation, 5, 767

4-AMINOBUTYRATE AMINOTRANSFERASE

This pyridoxal-phosphate-dependent aminotransferase [EC 2.6.1.19], also known as γ-amino-N-butyrate transaminase and GABA transaminase, catalyzes the reversible reaction of γ-aminobutyrate with α-ketoglutarate (or, 2-oxoglutarate) to produce succinate semialdehyde and L-glutamate.[1-3] A number of enzyme preparations have been reported to also use β-alanine, 5-aminopentanoate, and (R,S)-3-amino-2-methylpropanoate as substrates.

The enzyme has a ping pong Bi Bi kinetic mechanism typical of aminotransferases.[2] The coenzyme is initially bound as an internal aldimine (*i.e.*, a Schiff base linkage with an active site lysyl residue). A transaldimination occurs subsequent to binding of the γ-aminobutyrate, forming an external aldimine. A catalytic base removes a proton from the C4 position and a prototropic shift converts the aldimine into a ketimine. Hydrolysis of the ketimine generates the pyridoxamine phosphate form of the coenzyme and the new succinate semialdehyde. The second half-reaction is the reversal of these steps with α-ketoglutarate.

[1] M. A. Ator & P. R. Ortez de Montellano (1990) *The Enzymes*, 3rd ed., **9**, 213.
[2] M. Maitre, L. Ciesielski, C. Cash & P. Mandel (1975) *EJB* **52**, 157.
[3] A. E. Braunstein (1973) *The Enzymes*, 3rd ed., **9**, 379.

Selected entries from *Methods in Enzymology* [vol, page(s)]:
General discussion: 5, 771; 46, 33, 37; 113, 80
Assay: 5, 771; 113, 81
Other molecular properties: brain mitochondria, 55, 58; mechanism, 249, 263; mechanism-based inactivation, 249, 263; natural occurrence, 113, 80; properties, 5, 773; purification, *Pseudomonas fluorescens*, 5, 771; sources, 113, 81

3-AMINOBUTYRYL-CoA AMMONIA-LYASE

This lyase [EC 4.3.1.14], also known as L-3-aminobutyryl-CoA deaminase, catalyzes the conversion of L-3-aminobutyryl-CoA to crotonoyl-CoA and ammonia.[1]

[1] I.-M. Jeng & H. A. Barker (1974) *JBC* **249**, 6578.

AMINOCARBOXYMUCONATE-SEMIALDEHYDE DECARBOXYLASE

This decarboxylase [EC 4.1.1.45], also called picolinate decarboxylase, catalyzes the conversion of 2-amino-3-(3-oxoprop-2-enyl)but-2-enedioate to 2-aminomuconate semialdehyde and carbon dioxide.[1] The product then rearranges nonenzymically to picolinate.

[1] Y. Egashira, H. Kouhashi, T. Ohta & H. Sanada (1996) *J. Nutr. Sci. Vitaminol.* **42**, 173.

Selected entries from *Methods in Enzymology* [vol, page(s)]:
General discussion: 17A, 471; 18B, 162, 175
Assay: *in vitro*, 18B, 162, 177, 178; *in vivo*, 18B, 172
Other molecular properties: properties, 18B, 179, 180; purification from (liver, 18B, 178, 179; cat liver, 17A, 474)

L-2-AMINO-4-CHLOROPENT-4-ENOATE DEHYDROCHLORINASE

This enzyme [EC 4.5.1.4], isolated from *Proteus mirabilis*, catalyzes the hydrolysis of L-2-amino-4-chloropent-4-enoate to produce 2-oxopent-4-enoate, HCl, and ammonia.[1]

[1] M. Moriguchi, S. Hoshino & S.-I. Hatanaka (1987) *Agric. Biol. Chem.* **51**, 3295.

1-AMINOCYCLOPROPANE-1-CARBOXYLATE DEAMINASE

This deaminase [EC 4.1.99.4] catalyzes the reaction of 1-aminocyclopropane-1-carboxylate with water to produce α-ketobutyrate (or, 2-oxobutanoate) and ammonia.

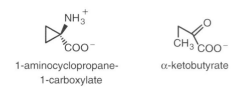

1-aminocyclopropane-1-carboxylate   α-ketobutyrate

Note that ring cleavage occurs between the *pro-S* and the α-carbon of the substrate. The pseudomonad enzyme is pyridoxal-phosphate-dependent and incubation in D_2O results in one deuterium located in the methyl carbon of the product as well as in one of the methylene protons at C3.[1] The α-anion equivalent of the vinylglycyl-PLP aldimine is a key intermediate in the reaction pathway which is formed after nucleophilic addition to open the ring, followed by β-proton abstraction.[2]

[1] C. Walsh, R. A. Pascal, M. Johnston, R. Raines, D. Dikshit, A. Krantz & M. Honma (1981) *Biochemistry* **20**, 7509.
[2] K. Li, W. Du, N. L. S. Que & H.-W. Liu (1996) *JACS* **118**, 8763.

1-AMINOCYCLOPROPANE-1-CARBOXYLATE OXIDASE

This ascorbate- and iron-dependent oxidoreductase catalyzes the reaction of 1-aminocyclopropane-1-carboxylate with dioxygen to produce ethene, carbon dioxide, cyanide, and two molecules of water.

Early studies suggested that the ascorbate is consumed stoichiometrically as well.[1] Carbon dioxide has also been shown to be catalytically essential and the formation of a lysine carbamate residue coordinated with Fe^{2+} has been suggested.[2] The reaction has been proposed to first form an *N*-hydroxy intermediate that releases the ethene and produces cyanoformate which decomposes to cyanide and carbon dioxide.[1]

[1] J. G. Dong, J. C. Fernández-Maculet & S. F. Yang (1992) *PNAS* **89**, 9789.
[2] J. C. Fernández-Maculet, J. G. Dong & S. F. Yang (1993) *BBRC* **193**, 1168.

1-AMINOCYCLOPROPANE-1-CARBOXYLATE SYNTHASE

S-adenosyl-L-methionine

1-aminocyclopropane-1-carboxylate

This pyridoxal-phosphate-dependent enzyme [EC 4.4.1.14] catalyzes the conversion of S-adenosyl-L-methionine to 1-aminocyclopropane-1-carboxylate and methylthio-adenosine via an α,γ-elimination.[1–3]

[1] M. F. White, J. Vasquez, S. F. Yang & J. F. Kirsch (1994) PNAS **91**, 12428.
[2] H. Kende (1993) Ann. Rev. Plant Physiol. Plant Mol. Biol. **44**, 283.
[3] Y.-B. Yu, D. O. Adams & S. F. Yang (1979) ABB **198**, 280.

Selected entries from **Methods in Enzymology** [**vol**, page(s)]:
General discussion: 143, 426
Assay: 143, 426
Other molecular properties: properties, 143, 429; purification from tomato, 143, 428; transition state and multisubstrate analogues, 249, 307

AMINODEOXYGLUCONATE DEHYDRATASE

This pyridoxal-phosphate-dependent dehydratase [EC 4.2.1.26], also known as glucosaminate dehydratase, catalyzes the reaction of 2-amino-2-deoxy-D-gluconate (*i.e.*, D-glucosaminate) with water to produce 2-dehydro-3-deoxy-D-gluconate, ammonia, and water (an α,β-elimination).[1,2] Other substrates, albeit weaker, include D-mannosaminate and D-galactosaminate. The enzyme can also exhibit a slight D-glucosaminate aldolase activity, about 25% of the rate of the main reaction.[1]

The rate-determining step in the reaction is the abstraction of the α-proton. This enzyme may be identical with glucosaminate ammonia-lyase [EC 4.3.1.9]. **See** *Glucosaminate Ammonia-Lyase*

[1] R. Iwamoto, H. Taniki, J. Koishi & S. Nakura (1995) Biosci. Biotechnol. Biochem. **59**, 408.
[2] R. Iwamoto, Y. Imanaga, S. Sawada, & K. Soda (1983) FEBS Lett. **156**, 33.

(2-AMINOETHYL)PHOSPHONATE: PYRUVATE AMINOTRANSFERASE

This pyridoxal-phosphate-dependent aminotransferase [EC 2.6.1.37] catalyzes the reversible reaction of (2-aminoethyl) phosphonate (also called ciliatine) with pyruvate to produce 2-phosphonoacetaldehyde and L-alanine. Other substrates include 2-aminoethylarsonate.

The enzyme has a ping pong Bi Bi kinetic mechanism.[1] The *pro-S* hydrogen atom at the C2 position of the substrate is the hydrogen that is abstracted following formation of the aldimine intermediate.[2]

[1] C. Dumora, A.-M. Lacoste & A. Cassaigne (1983) EJB **133**, 119.
[2] A. M. Lacoste, C. Dumora, L. Balas, F. Hammerschmidt & J. Vercauteren (1993) EJB **215**, 841.

AMINOGLYCOSIDE N³'-ACETYLTRANSFERASE

This transferase [EC 2.3.1.81], also known as gentamicin 3'-N-acetyltransferase, catalyzes the reaction of acetyl-CoA with a 2-deoxystreptamine antibiotic to produce coenzyme A and the N³'-acetyl-2-deoxystreptamine antibiotic.[1,2] Compounds that contain a 2-deoxystreptamine ring can act as acceptor substrates: examples include gentamicin, kanamycin, tobramycin, neomycin, and apramycin. This acetyltransferase is distinct from gentamicin 3'-N-acetyltransferase [EC 2.3.1.60]. **See also** *Gentamicin 3'-N-Acetyltransferase*

[1] J. Davies & S. O'Connor (1978) Antimicrob. Agents Chemother. **14**, 69.
[2] E. Wolf, A. Vassilev, Y. Makino, A. Sali, Y. Nakatani & S. K. Burley (1998) Cell **94**, 439.

2-AMINOHEXANOATE AMINOTRANSFERASE

This pyridoxal-phosphate-dependent aminotransferase [EC 2.6.1.67], also called norleucine aminotransferase, catalyzes the reaction of L-2-aminohexanoate with α-ketoglutarate (or, 2-oxoglutarate) to produce α-ketohexanoate (*i.e.*, 2-oxohexanoate) and L-glutamate. Alternative substrates include L-leucine, L-isoleucine, L-2-aminopentanoate, and 6-N-acetyllysine. The kinetic reaction mechanism is ping pong Bi Bi.[1,2]

[1] P. A. Der Garabedian & J. J. Vermeersch (1987) EJB **167**, 141.
[2] P. A. Der Garabedian & J. J. Vermeersch (1989) Biochimie **71**, 497.

6-AMINOHEXANOATE-CYCLIC DIMER HYDROLASE

This enzyme [EC 3.5.2.12] catalyzes the hydrolysis of 1,8-diazacyclotetradecane-2,9-dione to produce N-(6-aminohexanoyl)-6-aminohexanoate; that is, the cyclic dimer of 6-aminohexanoate is converted to the linear dimer.[1,2]

[1] K. Tsuchiya, S. Fukuyama, N. Kanzaki, K. Kanagawa, S. Negoro & H. Okada (1989) J. Bacteriol. **171**, 3187.
[2] S. Kinoshita, S. Negoro, M. Muramatsu, V. S. Bisaria, S. Sawada & H. Okada (1977) EJB **80**, 489.

6-AMINOHEXANOATE-DIMER HYDROLASE

This enzyme [EC 3.5.1.46], also known as 6-aminohexanoate oligomer hydrolase, catalyzes the hydrolysis of N-(6-aminohexanoyl)-6-aminohexanoate to produce two molecules of 6-aminohexanoate.[1,2] Oligomers of 6-aminohexanoate, containing up to six residues, are also hydrolyzed (removed sequentially from the N-terminus), albeit more slowly.

[1] K. Kanagawa, M. Oishi, S. Negoro, I. Urabe & H. Okada (1993) *J. Gen. Microbiol.* **139**, 787.
[2] S. Kinoshita, T. Terada, T. Taniguchi, Y. Takene, S. Masuda, N. Matsunaga & H. Okada (1981) *EJB* **116**, 547.

2-AMINOHEXANO-6-LACTAM RACEMASE

This pyridoxal-phosphate-dependent racemase [EC 5.1.1.15], also known as 2-amino-*hexano*-6-lactam racemase and α-amino-ε-caprolactam racemase, catalyzes the interconversion of L-2-aminohexano-6-lactam and D-2-aminohexano-6-lactam.[1–3] Other substrates include the enantiomers of 2-aminopentano-5-lactam (α-amino-δ-valerolactam) and 2-amino-4-thiahexano-6-lactam. The enzyme also exhibits a minor aminotransferase activity with certain α-amino acids.

This enzyme catalyzes α-hydrogen exchange during racemization of the substrate in deuterium oxide.[1] An overshoot of the optical rotation observed during the racemization in D_2O, but not in H_2O, is attributed to a primary kinetic isotope effect.[1] Tritium-labeling experiments showed that the enzyme catalyzes both retention and inversion of configuration of the substrate with a similar probability in each turnover. Conversion of [α-^{2}H]-D-α-amino-ε-caprolactam in water and unlabeled D-α-amino-ε-caprolactam in deuterium oxide into the L-isomer under nearly single turnover conditions with the enzyme showed significant internal return of the α-hydrogen. A single base mechanism best describes this enzyme-catalyzed racemization reaction.

[1] S. A. Ahmed, N. Esaki, H. Tanaka & K. Soda (1986) *Biochemistry* **28**, 385.
[2] M. E. Tanner & G. L. Kenyon (1998) *CBC* **2**, 7.
[3] H. Hayashi, H. Wada, T. Yoshimura, N. Esaki & K. Soda (1990) *Ann. Rev. Biochem.* **59**, 87.

2-AMINO-4-HYDROXY-6-HYDROXY-METHYLDIHYDROPTERIDINE PYROPHOSPHOKINASE

This pyrophosphotransferase [EC 2.7.6.3], also known as 6-hydroxymethyl-7,8-dihydropterin pyrophosphokinase and 7,8-dihydro-6-hydroxymethylpterin pyrophosphokinase, catalyzes the reaction of ATP with 2-amino-4-hydroxy-6-hydroxymethyl-7,8-dihydropteridine to generate AMP and 2-amino-7,8-dihydro-4-hydroxy-6-(diphosphooxymethyl)-pteridine.[1–4]

[1] T. L. Talarico, I.K. Dev, W. S. Dallas, R. Ferone & P. H. Ray (1991) *J. Bacteriol.* **173**, 7029.
[2] R. L. Switzer (1974) *The Enzymes*, 3rd ed., **10**, 607.

[3] T. Shiota, C. M. Baugh, R. Jackson & R. Dillard (1969) *Biochemistry* **8**, 5022.
[4] D. P. Richey & G. M. Brown (1969) *JBC* **244**, 1582.

Selected entries from *Methods in Enzymology* [vol, page(s)]:
General discussion: 18B, 765; **66**, 564
Assay: 18B, 766
Other molecular properties: properties, **18B**, 770; purification, **18B**, 768

AMINOIMIDAZOLASE

This iron-dependent enzyme [EC 3.5.4.8] catalyzes the hydrolysis of 4-aminoimidazole to produce an unidentified product and ammonia.[1,2] The unidentified product rapidly converts to formiminoglycine.

[1] J. C. Rabinowitz & W. E. Pricer (1956) *JBC* **222**, 537.
[2] J. C. Rabinowitz (1956) *JBC* **218**, 175.

β-AMINOISOBUTYRATE AMINOTRANSFERASE

This pyridoxal-phosphate-dependent aminotransferase [EC 2.6.1.40], also known as (R)-3-amino-2-methylpropionate: pyruvate aminotransferase, catalyzes the reversible reaction of (R)-3-amino-2-methylpropanoate with pyruvate to produce 2-methyl-3-oxopropanoate and L-alanine.[1–4] In rat liver, this enzyme and alanine:glyoxylate aminotransferase 2 [EC 2.6.1.44] are identical.[4] *See also (S)-3-Amino-2-Methylpropanoate Aminotransferase; (R)-3-Amino-2-Methylpropanoate Aminotransferase*

[1] S.-I. Ueno, H. Morino, A. Sano & Y. Kakimoto (1990) *BBA* **1033**, 169.
[2] N. Tamaki, M. Kaneko, C. Mizota, M. Kikugawa & S. Fujimoto (1990) *EJB* **189**, 39.
[3] O. W. Griffith (1986) *Ann. Rev. Biochem.* **55**, 855.
[4] Y. Kontani, M. Kaneko, M. Kikugawa, S. Fujimoto & N. Tamaki (1993) *BBA* **1156**, 161.

Selected entries from *Methods in Enzymology* [vol, page(s)]:
General discussion: 324, 376

5-AMINOLEVULINATE AMINOTRANSFERASE

This pyridoxal phosphate-dependent aminotransferase [EC 2.6.1.43], also known as L-alanine:4,5-dioxovalerate aminotransferase, catalyzes the reversible reaction of 5-aminolevulinate with pyruvate to produce 4,5-dioxopentanoate and L-alanine. The enzyme has a ping pong Bi Bi kinetic mechanism.[1–4] It may be identical to L-alanine:glyoxylate aminotransferase [EC 2.6.1.44] in some organisms.

[1] H.-i. Rhee, K. Murata & A. Kimura (1988) *J. Biochem.* **103**, 1045.
[2] Y. Shioi, M. Nagamine & T. Sasa (1984) *ABB* **234**, 117.
[3] Y. Shioi, M. Doi & T. Sasa (1984) *Plant Cell Physiol.* **25**, 1487.
[4] A. E. Braunstein (1973) *The Enzymes*, 3rd ed., **9**, 379.

Selected entries from *Methods in Enzymology* [vol, page(s)]:
General discussion: 17A, 188

5-AMINOLEVULINATE SYNTHASE

This pyridoxal-phosphate-dependent enzyme [EC 2.3.1.37], also known as δ-aminolevulinate synthase, catalyzes the reaction of succinyl-CoA with glycine to yield 5-aminolevulinate, coenzyme A, and carbon dioxide.[1–5] In mammals, the enzyme isolated from erythrocytes is genetically distinct from that in other tissues. The *Rhodopseudomonas sphaeroides* enzyme has an ordered Bi Bi kinetic mechanism.[4]

[1] P. M. Shoolingin-Jordan, J. E. LeLean & A. J. Lloyd (1997) *MIE* **281**, 309.
[2] G. C. Ferreira & J. Gong (1995) *J. Bioenerg. Biomembr.* **27**, 151.
[3] S. Granick & S. I. Beale (1978) *AE* **46**, 33.
[4] M. Fanica-Gaignier & J. Clement-Metral (1973) *EJB* **40**, 19.
[5] P. M. Jordan & D. Shemin (1972) *The Enzymes*, 3rd ed., **7**, 339.

Selected entries from **Methods in Enzymology** [vol, page(s)]:
General discussion: 17A, 195; **123**, 395, 435; **281**, 309, 336
Assay: 17A, 195, 201; **52**, 351; **123**, 395, 435; radiochemical assay, 17A, 201
Other molecular properties: chicken embryo hepatic mitochondria (assay, **123**, 395; purification, **123**, 397); *Chlorella* mutants, in, **23**, 168; deficiency, **56**, 118, 119; induction, in liver, 17A, 203; inhibition by hemin, 17A, 204; inhibitor, **63**, 400; liver, in, 17A, 201, 204; pyridoxal phosphate and, 17A, 204; rat hepatic mitochondria (assay, **123**, 395; purification, **123**, 401); *Rhodopseudomonas sphaeroides* (applications, **123**, 439; assay, **123**, 435; properties, **123**, 441; purification, 17A, 197; **123**, 437); tissue levels, problems in determination, **52**, 353, 354

AMINOMALONATE DECARBOXYLASE

This enzymatic activity, previously classified as EC 4.1.1.10, is now listed under aspartate β-decarboxylase [EC 4.1.1.12].

(R)-3-AMINO-2-METHYLPROPANOATE AMINOTRANSFERASE

This aminotransferase [EC 2.6.1.61] catalyzes the reversible reaction of (R)-3-amino-2-methylpropanoate with α-ketoglutarate (or, 2-oxoglutarate) to produce 2-methyl-3-oxopropanoate and L-glutamate.[1] The enzyme is clearly distinct from (S)-3-amino-2-methylpropionate aminotransferase [EC 2.6.1.22]. **See** *(S)-3-Amino-2-Methylpropanoate; β-Aminoisobutyrate Aminotransferase*

[1] K. Kakimoto, K. Taniguchi & I. Sano (1969) *JBC* **244**, 335.

(S)-3-AMINO-2-METHYLPROPIONATE AMINOTRANSFERASE

This aminotransferase [EC 2.6.1.22], also known as L-3-aminoisobutyrate aminotransferase, catalyzes the reversible reaction of (S)-3-amino-2-methylpropanoate with α-ketoglutarate (or, 2-oxoglutarate) to produce 2-methyl-3-oxopropanoate and L-glutamate. Other donor substrates include β-alanine and other ω-amino acids having carbon chains between two and five carbon atoms in length (for example, 4-aminobutanoate and 5-aminopentanoate). This aminotransferase is distinct from (R)-3-amino-2-methylpropanoate aminotransferase [EC 2.6.1.61].[1,2] **See** *(S)-3-Amino-2-Methylpropanoate Aminotransferase; β-Aminoisobutyrate Aminotransferase*

[1] Y. Kakimoto, A. Kanazawa, K. Taniguchi & I. Sano (1968) *BBA* **156**, 374.
[2] N. Tamaki, S. Fujimoto, C. Mizota & M. Kikugawa (1987) *BBA* **925**, 238.

AMINOMETHYLTRANSFERASE

This transferase [EC 2.1.2.10], also known as the glycine-cleavage system T-protein and tetrahydrofolate aminomethyltransferase, catalyzes the reversible reaction of (6S)-tetrahydrofolate with S-aminomethyldihydrolipoyl-protein to produce (6R)-5,10-methylenetetrahydrofolate, ammonia, and dihydrolipoylprotein. It is a component of the glycine cleavage system (*i.e.*, glycine synthase).[1] The decarboxylated substrate is attached to the H-protein of the complex. The reverse reaction has an ordered Ter Bi kinetic mechanism.[2] **See** *Glycine Synthase*

[1] K. Okamura-Ikeda, K. Fujiwara & Y. Motokawa (1982) *JBC* **257**, 135.
[2] K. Okamura-Ikeda, K. Fujiwara & Y. Motokawa (1987) *JBC* **262**, 6746.

2-AMINOMUCONATE DEAMINASE

2-aminomuconate   4-oxalocrotonate

This deaminase [EC 3.5.99.5] catalyzes the hydrolysis of 2-aminomuconate to produce 4-oxalocrotonate and ammonia. Deamination is spontaneous under acid conditions. The enzyme has been reported to function in the degradation of nitrobenzene by *Pseudomonas pseudocaligenes* JS45, and apparently proceeds via an imine intermediate.[1]

[1] Z. He & J. C. Spain (1998) *J. Bacteriol.* **180**, 2502.

α-AMINOMUCONATE REDUCTASE

This oxidoreductase reportedly catalyzes the reaction of 2-aminomuconate with NAD(P)H to produce α-ketoadipate, ammonia, and NAD(P)$^+$. This overall reaction in tryptophan degradation is now thought to be due to a sequence of enzyme-catalyzed reactions.

Selected entries from **Methods in Enzymology** [vol, page(s)]:
General discussion: 17A, 483
Assay: 17A, 483

AMINOMUCONATE-SEMIALDEHYDE DEHYDROGENASE

This oxidoreductase [EC 1.2.1.32] catalyzes the reaction of 2-aminomuconate 6-semialdehyde with NAD^+ and water to produce 2-aminomuconate and NADH.[1] Other substrates include 2-hydroxymuconate semialdehyde.

[1] A. Ichiyama, S. Nakamura, H. Kawai, T. Honjo, Y. Nishizuka, O. Hayaishi & S. Senoh (1965) *JBC* **240**, 740.

Selected entries from *Methods in Enzymology* [**vol**, page(s)]:
General discussion: **17A**, 476

8-AMINO-7-OXONONANOATE SYNTHASE

This pyridoxal-phosphate-dependent enzyme [EC 2.3.1.47], also known as 8-amino-7-ketopelargonate synthase and 7-keto-8-aminopelargonate synthase, catalyzes the reaction of 6-carboxyhexanoyl-CoA (*i.e.*, pimeloyl-CoA) with L-alanine to produce 8-amino-7-oxononanoate (also called 8-amino-7-ketopelargonate), coenzyme A, and carbon dioxide.[1–4]

L-Alanine binds rapidly to form an external aldimine with the enzyme-bound cofactor. Significant quinonoid formation occurs upon addition of pimeloyl-CoA to the aldimine-containing complex. A lysyl residue abstracts the proton from the α-carbon position in a step that is partially rate-limiting.[3] The quinonoid intermediate subsequently reacts with the second substrate in a Claisen-type condensation to form a new aldimine intermediate. Decarboxylation then follows to form the product.

[1] S. P. Webster, D. Alexeev, D. J. Campopiano, R. M. Watt, M. Alexeeva, L. Sawyer & R. L. Baxter (2000) *Biochemistry* **39**, 516.
[2] O. Ploux, O. Breyne, S. Carillon & A. Marquet (1999) *EJB* **259**, 63.
[3] O. Ploux & A. Marquet (1996) *EJB* **236**, 301.
[4] O. Ploux & A. Marquet (1992) *BJ* **283**, 327.

Selected entries from *Methods in Enzymology* [**vol**, page(s)]:
General discussion: **62**, 334, 335
Assay: **62**, 330, 331

5-AMINOPENTANAMIDASE

This enzyme [EC 3.5.1.30], also known as 5-amino-valeramidase, catalyzes the hydrolysis of 5-aminopenta-namide (also known as 5-aminovaleramide) to produce 5-aminopentanoate (*i.e.*, 5-aminovalerate) and ammonia.[1] The *Pseudomonas putida* enzyme also acts on 4-aminobutanamide and 6-aminohexanamide.

[1] M. S. Reitz & V. M. Rodwell (1970) *JBC* **245**, 3091.

Selected entries from *Methods in Enzymology* [**vol**, page(s)]:
General discussion: **17B**, 158
Assay: **17B**, 160

AMINOPEPTIDASE I

This zinc-dependent aminopeptidase [EC 3.4.11.22], also known as vacuolar aminopeptidase I, aminopeptidase III, leucine aminopeptidase IV, and aminopeptidase yscI, catalyzes peptide-bond scission in protein substrates containing a N-terminal leucine or other hydrophobic amino acid.[1,2] $Leu-NHPhNO_2$ is a commonly used chromogenic substrate. Aminoacylarylamides are relatively poor substrates, even though they are often used in enzyme assays.

[1] P. Matile, A. Wiemken & W. Guyer (1971) *Planta* **96**, 43.
[2] G. Metz & K. H. Röhm (1976) *BBA* **429**, 933.

AMINOPEPTIDASE B

This metallopeptidase [EC 3.4.11.6], also known as arginyl aminopeptidase, cytosol aminopeptidase IV, arginine aminopeptidase, and arylamidase II, catalyzes the release of N-terminal arginine and lysine from oligopeptides containing a prolyl residue in the P1′ position.[1–3] Non-basic aminoacyl residues located at the N-terminus are not released by aminopeptidase B. The 2-naphthylamides of arginine and lysine serve as synthetic chromogenic substrates.

Interestingly, aminopeptidase also catalyzes the hydrolysis of leukotriene A_4 to LTB_4. At one time, this aminopeptidase was thought to be identical to leukotriene A_4 hydrolase [EC 3.3.2.6] and/or bleomycin hydrolase. Both of these hydrolases exhibit an arginyl aminopeptidase activity; however, leukotriene A_4 hydrolase is not specific for basic amino acids, and bleomycin hydrolase is a cysteine aminopeptidase.

[1] V. K. Hopsu, K. K. Mäkinen & G. G. Glenner (1966) *ABB* **114**, 557.
[2] V. K. Hopsu, K. K. Mäkinen & G. G. Glenner (1966) *ABB* **114**, 567.
[3] S. Cadel, A. R. Pierotti, T. Foulon, C. Créminon, N. Barrè, D. Segrétain & P. Cohen (1995) *Mol. Cell. Endocrinol.* **110**, 149.

Selected entries from *Methods in Enzymology* [**vol**, page(s)]:
Assay: fluorimetric, **248**, 601
Other molecular properties: alkaline phosphatase isozyme conversion protein, *Escherichia coli*, **248**, 117, 118

AMINOPEPTIDASE C (PepC) OF LACTIC ACID BACTERIA

This cysteine peptidase, first isolated from *Lactococcus lactis* and commonly known as PepC, is a strict aminopeptidase, absolutely requiring the presence of a free α-amino group. While the enzyme's specificity is broad, no peptide-bond scission occurs when a prolyl residue is present in the P1 or P1′ position.[1,2]

[1] E. Neviani, C.-Y. Boquien, V. Monnet, L. Phan Thanh & J.-C. Gripon (1989) *Appl. Environ. Microbiol.* **55**, 2308.

[2] M.-Y. Mistou, P. Rigolet, M.-P. Chapot-Chartier, M. Nardi, J.-C. Gripon & S. Brunie (1994) *J. Mol. Biol.* **237**, 160.

AMINOPEPTIDASE DmpA

This threonine-dependent aminopeptidase, isolated from *Ochrobactrum anthropi*, will catalyze the hydrolysis of peptides, preferring N-termini residues with the L-configuration.[1] The enzyme also exhibits a D-esterase activity. Aminopeptidase DmpB (also called D-stereospecific aminopeptidase) of *O. anthropi* is a serine-dependent enzyme. Chromogenic substrates include glycyl-*p*-nitroanilide and D-alanyl-*p*-nitroanilide. **See also** D-*Stereospecific Aminopeptidase*

[1] L. Fanuel, C. Goffin, A. Cheggour, B. Devreese, G. Van Driessche, B. Joris, J. Van Beeumen & J. M. Frère (1999) *BJ* **341**, 147.

AMINOPEPTIDASE Ey

This zinc-dependent metalloaminopeptidase [EC 3.4.11.20], first isolated from hen's egg yolk, exhibits a broad specificity for substrates with N-terminal hydrophobic, basic, and acidic amino acids, as well as those containing proline in this position.[1–3]

[1] E. Ichishima, Y. Yamagata, H. Chiba, K. Sawaguchi & T. Tanaka (1989) *Agric. Biol. Chem.* **53**, 1867.

[2] T. Tanaka & E. Ichishima (1993) *Comp. Biochem. Physiol.* **105B**, 105.

[3] T. Tanaka & E. Ichishima (1993) *Int. J. Biochem.* **25**, 1681.

AMINOPEPTIDASE T

This metallopeptidase, AP-T, first isolated from the thermophile *Thermus aquaticus* YT-1, catalyzes the hydrolysis of N-terminal amino acids and exhibits a broad specificity. Even so, hydrolysis does not occur if a prolyl residue in the penultimate residue. The more effective N-terminal amino acids are Leu, Val, Phe, and Tyr.[1] **See also** *Aminopeptidase II; Bacillus Aminopeptidase I*

[1] E. Minagawa, S. Kaminogawa, H. Matsuzawa, T. Ohta & K. Yamauchi (1988) *Agric. Biol. Chem.* **52**, 1755.

AMINOPEPTIDASE Y

This zinc-dependent metalloaminopeptidase [EC 3.4.11.15], first isolated from yeast, catalyzes peptide-bond scission in various polypeptide and protein substrates.[1] The best N-terminal amino acid of aminoacyl arylamides are L-arginine and L-lysine, followed by L-leucine, L-methionine, and L-alanine.[2] Co^{2+} activates the enzyme at concentrations above 0.25 mM. **See** *Aminopeptidase yscCo II*

[1] T. Achstetter, C. Ehmann & D. H. Wolf (1982) *BBRC* **109**, 341.

[2] T. Yasuhara, T. Nakai & A. Ohashi (1994) *JBC* **269**, 13644.

2-AMINOPHENOL OXIDASE

This flavin- and manganese-dependent oxidoreductase [EC 1.10.3.4], also known as *o*-aminophenol oxidase and isophenoxazine synthase, catalyzes the reaction of two molecules of 2-aminophenol with three molecules of dioxygen to produce two molecules of isophenoxazine and six molecules of water.[1,2] *o*-Quinoneimine is suggested to be an intermediate and isophenoxazine may be formed by a secondary condensation from this initial oxidation product.

[1] P. M. Nair & I. C. Vining (1965) *BBA* **96**, 318.

[2] P. V. S. Rao & C. S. Vaidyanathan (1967) *ABB* **118**, 388.

Selected entries from ***Methods in Enzymology*** [vol, page(s)]: **General discussion: 56**, 475

AMINOPHOSPHOLIPID-TRANSPORTING ATPase

This so-called P-type ATPase [EC 3.6.3.13], which undergoes covalent phosphorylation during the transport cycle, catalyzes the ATP-dependent (ADP- and orthophosphate-producing) export of aminophospholipids.[1–3] It resembles a flippase in function, catalyzing the movement of phospholipids from one lipid bilayer face to the other. Phospholipid substrates include phosphatidyl-L-serine and phosphatidylethanolamine. Optimal activity was observed with substrates containing one saturated and one unsaturated fatty acyl group.

The idea that a transporter is an enzyme is in keeping with a new definition of enzyme catalysis as the facilitated making/breaking of chemical bonds, not just covalent bonds.[4] This idea builds on Pauling's assertion that any long-lived, chemically distinct interaction (in this case, the persistent location of a solute with respect to the faces of a membrane) can be regarded as a chemical bond. Note also that the equilibrium constant (K_{eq} = [Phospholipid$_{out}$][ADP][P_i]/[Phospholipid$_{in}$][ATP]) does not conform to that expected for an ATPase (*i.e.*, K_{eq} = [ADP][P_i]/[ATP]). Thus, although the overall reaction yields ADP and orthophosphate, the enzyme is misclassified as a hydrolase, and instead should be regarded as an energase-type reaction. Energases facilitate affinity-modulated reactions by coupling the $\Delta G_{ATP-hydrolysis}$ to a force-generating or work-producing step.[4] In this case, P-O-P bond-scission supplies the energy to drive aminophospholipid transport. **See** *ATPase; Energase*

[1] M. L. Zimmerman & D. L. Daleke (1993) *Biochemistry* **32**, 12257.

[2] Z. Beleznay, A. Zachowski, P. F. Devaux, M. P. Navazo & P. Ott (1993) *Biochemistry* **32**, 3146.

[3] X. Tang, M. S. Halleck, R. A. Schlegel & P. Williamson (1996) *Science* **272**, 1495.

[4] D. L. Purich (2001) *TiBS* **26**, 417.

5-AMINO-6-(5-PHOSPHORIBOSYLAMINO) URACIL REDUCTASE

This oxidoreductase [EC 1.1.1.193], also called aminodioxyphosphoribosylaminopyrimidine reductase, catalyzes the reaction of 5-amino-6-(5-phosphoribosylamino)uracil with NADPH to produce 5-amino-6-(5-phosphoribitylamino) uracil and $NADP^+$.[1]

[1] R. B. Burrows & G. M. Brown (1978) *J. Bacteriol.* **136**, 657.

D-AMINOPROPANOL DEHYDROGENASE

This NAD^+-dependent oxidoreductase activity, formerly classified as EC 1.1.1.74, is now listed under (*R,R*)-butanediol dehydrogenase [EC 1.1.1.4].

(R)-AMINOPROPANOL DEHYDROGENASE

This potassium-dependent oxidoreductase [EC 1.1.1.75] catalyzes the reversible reaction of (*R*)-1-aminopropan-2-ol with NAD^+ to produce aminoacetone and NADH.[1,2] Other substrates include 1,3-diaminopropan-2-ol.

[1] N. Cox, J. M. Turner & A. J. Willetts (1968) *BBA* **170**, 438.

[2] M. A. Pickard, I. J. Higgins & J. M. Turner (1968) *J. Gen. Microbiol.* **54**, 115.

5-AMINOVALERATE AMINOTRANSFERASE

This pyridoxal-phosphate-dependent aminotransferase [EC 2.6.1.48], also known as δ-aminopentanoate aminotransferase and δ-aminovalerate transaminase, catalyzes the reversible reaction of 5-aminopentanoate (*i.e.*, 5-aminovalerate) with α-ketoglutarate (or, 2-oxoglutarate) to produce glutarate semialdehyde (*i.e.*, 5-oxopentanoate) and L-glutamate.[1–3] Other donor substrates include 6-aminohexanoate and 4-aminobutanoate.

[1] H. A. Barker, L. D'Ari & J. Kahn (1987) *JBC* **262**, 8994.

[2] P. A. Der Garabedian (1986) *Biochemistry* **25**, 5507.

[3] A. Ichihara, E. A. Ichihara & M. Suda (1960) *J. Biochem.* **48**, 412.

AMMONIA KINASE

phosphoramidate

This phosphotransferase [EC 2.7.3.8] catalyzes the reversible reaction of ADP with phosphoramide to produce ATP and ammonia. There is broad specificity for the donor substrate: for example, substrates include *N*-phosphoglycine and *N*-phospho-L-histidine.[1] Acceptor substrates include ADP, dADP, GDP, CDP, dTDP, dCDP, IDP, and UDP (in decreasing order of activity). Evidence has been presented that the reaction proceeds via a phospho-enzyme intermediate.[1]

[1] M. J. Dowler & H. I. Nakada (1968) *JBC* **243**, 1434.

AMOEBAPAIN

This cysteine endopeptidase first isolated from lysosome-like vesicles of *Entamoeba histolytica*, is regarded by some as a variant of histolysain [EC 3.4.22.35] and classified under the same EC number. Amoebapain catalyzes peptide-bond hydrolysis in a variety of proteins, including components of the extracellular matrix. Peptide specificity studies indicated a preference for basic aminoacyl residues in the P2 position and a broad specificity at P1. The synthetic substrate Arg-Arg-∇-NHMec is hydrolyzed as is Glu-Arg-Gly-∇-Phe-Phe (a pentapeptide fragment of the insulin B chain).[1,2] (Note: ∇ indicates the cleavage site.) ***See also*** *Histolysain*

[1] H. Scholze, U. Löhden-Bendinger, G. Müller & T. Bakker-Grunwald (1992) *Arch. Med. Res.* **23**, 105.

[2] H. Scholze & E. Tannich (1994) *MIE* **244**, 512.

Selected entries from ***Methods in Enzymology*** [**vol**, page(s)]:
General discussion: 244, 512
Other molecular properties: cleavage site specificity, **244**, 519; isolation from pathogenic *Entamoeba histolytica*, **244**, 515; sequence, **244**, 521; substrate specificity, **244**, 518

AMP DEAMINASE

This deaminase [EC 3.5.4.6] (also known as AMP aminohydrolase, adenylate deaminase, and myoadenylate deaminase) catalyzes the hydrolysis of AMP to produce IMP and ammonia.[1–4] The muscle enzyme is activated by monovalent cations such as K^+, apparently by promoting oligomerization.[3] The yeast enzyme displays cooperativity, contains two ATP regulatory sites per polypeptide, is zinc-dependent, has a rapid equilibrium random Uni Bi kinetic mechanism, and uses a Zn^{2+}-activated water molecule to attack the C6 position of AMP to displace ammonia.[1] ***See also*** *Adenosine-Phosphate Deaminase*

[1] D. J. Merkler & V. L. Schramm (1993) *Biochemistry* **32**, 5792.

[2] K. Kaletha & G. Nowak (1988) *BJ* **249**, 255.

[3] C. L. Zielke & C. H. Suelter (1971) *The Enzymes*, 3rd ed., **4**, 47.

[4] Y. Lee (1957) *JBC* **227**, 987, 993, and 999.

AMP:THYMIDINE KINASE

This phosphotransferase [EC 2.7.1.114], also known as adenylate:nucleoside phosphotransferase, catalyzes the reaction of AMP with thymidine to produce adenosine and thymidine 5′-phosphate. Other acceptor substrates for the plant enzyme include guanosine, deoxyuridine, adenosine, cytidine, and uridine.[1]

This reaction is also catalyzed by the deoxypyrimidine kinase complex induced by herpes simplex virus (which also has a thymidine kinase activity [EC 2.7.1.21], an [acetyl-CoA carboxylase] kinase activity [EC 2.7.1.128], and a thymidylate kinase activity [EC 2.7.4.9]).[2,3]

[1] A. R. Grivell & J. F. Jackson (1976) BJ 155, 571.
[2] D. Falke, W. Nehrbass, D. Brauer & W. E. G. Müller (1981) J. Gen. Virol. 53, 247.
[3] J. Labenz, W. E. G. Müller & D. Falke (1984) Arch. Virol. 81, 205.

AMYGDALIN β-GLUCOSIDASE

amygdalin

This highly specific glycosidase [EC 3.2.1.117] catalyzes the hydrolysis of (R)-amygdalin to produce (R)-prunasin and D-glucose.[1–4] Prunasin, linamarin, gentiobiose, and cellobiose are not substrates. The enzyme can be assayed with the chromogenic substrate 2-nitrophenyl-β-D-glucopyranoside. **See also** *β-Glucosidase*

[1] G. Kuroki, P. A. Lizotte & J. E. Poulton (1983) Z. Naturforsch. 39c, 232.
[2] E. Swain, C. P. Li & J. E. Poulton (1992) Plant Physiol. 100, 291.
[3] C. P. Li, E. Swain & J. E. Poulton (1992) Plant Physiol. 100, 282.
[4] G. W. Kuroki & J. E. Poulton (1986) ABB 247, 433.

AMYLASES

These glycosidases catalyze the hydrolysis of O-glucosyl bonds in various glucan substrates.[1–16]

α-Amylase [EC 3.2.1.1], also known as 1,4-α-D-glucan glucanohydrolase, catalyzes the endohydrolysis of a

AMP NUCLEOSIDASE

This enzyme [EC 3.2.2.4] catalyzes the hydrolysis of AMP to produce adenine and D-ribose 5-phosphate.[1–6] The *Azotobacter vinelandii* enzyme has a rapid-equilibrium random Uni Bi kinetic mechanism in the presence of MgATP, the allosteric activator.[6]

Kinetic isotope effect data[2,5] suggest that the transition state has a weak reaction-coordinate bond to C1′ with substantial carbenium ion character in the ribose ring. The N9–C1′ bond to the leaving group is nearly broken and the adenine ring is protonated at the transition state.[6] The substrate binds in a *syn* configuration and, presumably, the transition state also has a *syn* configuration.[4]

[1] G. Davies, M. L. Sinnott & S. G. Withers (1998) CBC 1, 119.
[2] D. W. Parkin, F. Mentch, G. A. Banks, B. A. Horenstein, & V. L. Schramm (1991) Biochemistry 30, 4586.
[3] M. H. O'Leary (1989) Ann. Rev. Biochem. 58, 377.
[4] V. L. Giranda, H. M. Berman & V. L. Schramm (1988) Biochemistry 27, 5813.
[5] F. Mentch, D. W. Parkin & V. L. Schramm (1987) Biochemistry 26, 921.
[6] W. E. DeWolf, F. A. Emig & V. L. Schramm (1986) Biochemistry 25, 4132.

1,4-α-D-glucosidic bond in a polysaccharide or oligosaccharide (requiring at least three glucosyl units in the oligosaccharide). Starch, glycogen, and related polysaccharides and oligosaccharides will serve as substrates. The endoglycosidic linkages are hydrolyzed in a random manner. The newly generated glycoside has a anomeric carbon in the α-configuration. Incubation of amylose and limit dextrins with α-amylase ultimately yields mainly maltose. Most α-amylases have five subsites, with cleavage taking place between subsites 2 and 3.

β-Amylase [EC 3.2.1.2], also known as saccharogen amylase and 1,4-α-D-glucan maltohydrolase, catalyzes the hydrolysis of 1,4-α-glucosidic linkages in polysaccharides near the nonreducing of the polysaccharide chain such that successive maltose units are released from that nonreducing end. Starch, glycogen, and related polysaccharides and oligosaccharides can serve as substrates, all generating β-maltose by an inversion reaction. Lengthy incubation of amylopectin with β-amylase results in the formation of a β-amylase limit dextrin since β-amylase will not act near branch points.

Various amylases include *Bacillus macerans* amylase (**See** *Cyclomaltodextrin Glucanotransferase*), amylase III (**See** *Oligosaccharide 4-α-D-Glucosyltransferase*), γ-amylase (**See** *Glucan 1,4-α-Glucosidase*), maltase-glucoamylase (**See** *α-Glucosidase*), G$_4$-amylase (**See** *Glucan 1,4-α-Maltotetrahydrolase*), maltotetraose-forming amylase (**See** *Glucan 1,4-α-Maltotetrahydrolase*), isoamylase (**See** *Isoamylase*), G$_6$-amylase (**See** *Glucan 1,4-α-Maltohexaosidase*), maltohexaose-producing amylase (**See** *Glucan 1,4-α-Maltohexaosidase*), and maltogenic α-amylase (**See** *Glucan 1,4-α-Maltohydrolase*).

[1] A. Pandey, P. Nigam, C. R. Soccol, V. T. Soccol, D. Singh & R. Mohan (2000) *Biotechnol. Appl. Biochem.* **31**, 135.

[2] G. Davies, M. L. Sinnott & S. G. Withers (1998) *CBC* **1**, 119.

[3] S. Janecek (1997) *Prog. Biophys. Mol. Biol.* **67**, 67.

[4] M. W. Bauer, S. B. Halio & R. M. Kelly (1996) *Adv. Protein Chem.* **48**, 271.

[5] R. R. Ray & G. Nanda (1996) *Crit. Rev. Microbiol.* **22**, 181.

[6] E. H. Van Beers, H. A. Büller, R. J. Grand, A. W. C. Einerhand & J. Dekker (1995) *Crit. Rev. Biochem. Mol. Biol.* **30**, 197.

[7] J. Lehmann & M. Schmidt-Schuchardt (1994) *MIE* **247**, 265.

[8] G. Mooser (1992) *The Enzymes*, 3rd ed., **20**, 187.

[9] I. S. Pretorius, M. G. Lambrechts & J. Marmur (1991) *Crit. Rev. Biochem. Mol. Biol.* **26**, 53.

[10] M. Vihinen & P. Mantsala (1989) *Crit. Rev. Biochem. Mol. Biol.* **24**, 329.

[11] J. D. Allen (1980) *MIE* **64**, 248.

[12] M. B. Ingle & R. J. Erikson (1978) *Adv. Appl. Microbiol.* **24**, 257.

[13] J. A. Thoma, J. E. Spradlin & S. Dygert (1971) *The Enzymes*, 3rd ed., **5**, 115.

[14] T. Takagi, H. Toda & T. Isemura (1971) *The Enzymes*, 3rd ed., **5**, 235.

[15] E. H. Fischer & E. A. Stein (1960) *The Enzymes*, 2nd. ed. **4**, 313.

[16] D. French (1960) *The Enzymes*, 2nd ed., **4**, 345.

Selected entries from *Methods in Enzymology* [**vol**, page(s)]:
General discussion: 1, 149, 154; 28, 925; 34, 592; 44, 63; 330, 269, 354
Assay: 1, 149; 28, 927; α-amylase, 1, 149; 8, 533; 137, 691 (biosensor, with, 137, 64; dye-labeled CM-amylose, with, 160, 81; spectrophotometric, 44, 98); β-amylase, 1, 149 (spectrophotometric, 44, 98)
Other molecular properties: α-amylase, 1, 149; 8, 533 (amylo-1,6-glucosidase, and, starch degradation in aqueous two-phase system, 137, 661; amylopectin and, 1, 152; amylosucrase and, 1, 185; barrier, subsite, 64, 272, 273; binding energy, 64, 260, 267; bond cleavage frequency, 64, 257, 269, 270; branching enzyme and, 1, 223, 224; cellular DNA isolation, in, 36, 296; computer model, 64, 265, 266; conjugate [activity, 44, 100, 271; stability, 44, 100, 101; storage, 44, 98]; cross-partitioning, determination of isoelectric point, 228, 228; cyclic dextrin, 64, 256; diazo binding, 44, 98; fluorine nuclear magnetic resonance, 49, 274; glycogen complexes and, 8, 533; immobilization [adsorption, by, effects on activity, 44, 44; Enzacryl AA, on, 44, 98, 99; Enzacryl AH, on, 44, 99, 100; polyacrylamide gel, in, 44, 902, 909; silk, on, 44, 909]; isothiocyanato coupling, 44, 99; kinetics, nonclassical, 64, 269; labeled maltose and, 4, 507; liquefying and saccharifying, multiple sequence comparisons, 183, 452; malt, 9, 13; nonproductive binding, 64, 269; oligosaccharides, TLC analysis, 160, 175; oxidation with NBS, effect on tryptophan in, 11, 516; phosphorylase phosphatase purification and, 8, 549; plant phosphorylase and, 1, 193; porcine pancreatic [HPLC of peptides, 247, 285, 287; inhibition constant determination, 247, 280; oligosaccharide binding site, 247, 270; photoaffinity labeling, 247, 271 {detection of labeled peptides, 247, 283, 288; kinetics, 247, 282}; structure, 247, 270; trypsinization, 247, 285]; powder, source of S$_1$ nuclease, 65, 249; production by *Aspergillus*, 137, 689, 693; properties, 1, 152; purification, 8, 534 [glycogen complex, 8, 534; salivary, human, 1, 150; starch, with, 34, 164]; salivary, human, 1, 150; secretion by pancreas, stimulation by secretagogues, 192, 248; separation from glucanosyltransferase and, 28, 686; source, 8, 535; 44, 98; starch hydrolysis, 44, 784, 788, 789; subsite mapping, 64, 260; substrate, multimolecular reaction, 64, 273, 274; transglycosylase activity, 64, 269; transition state and multisubstrate analogues, 249, 306; urease conjugate, 87, 466); β-amylase (amylopectin and, 1, 156, 157; conjugate [activity, 44, 100; stability, 44, 100, 101; storage, 44, 99]; conjugation with oxo-agarose, procedure, 44, 40; diazo binding, 44, 98; entrapment, in N,N'-methylenebisacrylamide, 44, 171; immobilization on [Enzacryl AA, 44, 98, 99; Enzacryl AH, 44, 99, 100; hexyl agarose, 44, 44; polyacrylamide gel, in, 44, 902]; immobilized, relative activity, 44, 40; isothiocyanate coupling, 44, 99; malt, 9, 13; phosphorylase limit dextrins and, 3, 53; plant phosphorylase and, 1, 193; properties, 1, 156; purification [mycobacterial polysaccharide MMP, in, 35, 91; sweet potato, 1, 154]; repetitive attack, 64, 274, 276; source, 44, 98; subsite mapping, 64, 258); amylase digestion and gel filtration, purification of glycogen synthetase, 28, 535; amylomaltase assay and, 5, 142; effector changes, X-ray scattering studies, 61, 227; enzymatic properties, 28, 928; glycogen synthase and, 5, 147; 28, 541; hyperthermophilic, 330, 269, 354; identification of luminous bacteria, 57, 161, 162; isolated nuclei and, 12A, 445; isolation by foaming, 22, 526; isolation from *Pseudomonas stutzeri*, 28, 926; liver catalase and, 2, 776; pancreas, in, 31, 53 (clinical significance, 74, 290; heat stability in pancreatic juice, 74, 296, 297; isoenzymes, 74, 291, 294, 295; normal serum level, 74, 296; purification, 74, 291; radioimmunoassay, 74, 290 [antibody production, 74, 291; application, 74, 291, 296; precision, 74, 293; procedure, 74, 292; results, comparative, 74, 295; second antibody stage, 74, 292; solutions, 74, 292; specificity, 74, 294, 295; standard, 74, 291; standard curve, 74, 293]; radioiodination, 74, 292); preparative acrylamide gel electrophoresis, 22, 432; purification, 9, 13; secretion by parotid slices, 39, 463, 464; stain for, 224, 106; staining on gels, 22, 600; tissue culture, 58, 125

AMYLO-1,6-GLUCOSIDASE

This enzyme [EC 3.2.1.33], also known as dextrin 6-α-D-glucosidase, catalyzes the endohydrolysis of 1,6-α-D-glucoside linkages at points of branching in chains of 1,4-linked α-D-glucose residues. The enzyme acts on the branch points of amylopectin and glycogen and removes unsubstituted glucose residues linked α1,6 to the main polysaccharide chain.[1-4] In mammals and yeast, amylo-1,6-glucosidase is linked to a glycosyltransferase resembling 4-α-glucanotransferase [EC 2.4.1.25]; together these two activities constitute the glycogen debranching system. *See Glycogen Debranching Enzyme*

[1] E. Y. C. Lee & J. H. Carter (1973) *ABB* **154**, 636.
[2] D. H. Brown, R. B. Gordon & B. I. Brown (1973) *Ann. N. Y. Acad. Sci.* **210**, 238.
[3] T. N. Palmer & B. E. Ryman (1971) *FEBS Lett.* **18**, 277.
[4] T. E. Nelson & J. Larner (1970) *BBA* **198**, 538.

Selected entries from *Methods in Enzymology* [vol, page(s)]:
General discussion: 1, 211; **8**, 515; **90**, 479
Assay: 1, 211; **8**, 516; biopsy material, in, **8**, 527; hexokinase and, **3**, 110
Other molecular properties: α-amylase, and, starch degradation in aqueous two-phase system, **137**, 661; contaminants of, **3**, 51, 52; glycogen and, **8**, 516; polysaccharide analysis and, **3**, 51; properties, 1, 214; **8**, 522; purification, 1, 212; **8**, 519, 527; **90**, 479

AMYLOMALTASE

This enzyme, previously classified as EC 2.4.1.3 but now a deleted Enzyme Commission entry, is now listed under 4-α-D-glucanotransferase [EC 2.4.1.25].

AMYLOPECTIN 6-GLUCANOHYDROLASE

This enzyme, previously classified as EC 3.2.1.69 but now a deleted Enzyme Commission entry, is now listed under α-dextrin endo-1,6-α-glucosidase [EC 3.2.1.41].

AMYLOPECTIN-1,6-GLUCOSIDASE

This enzyme activity, previously classified as EC 3.2.1.9, is now a deleted Enzyme Commission entry.

AMYLOSUCRASE

This enzyme [EC 2.4.1.4], also known as sucrose:glucan glucosyltransferase, catalyzes the reaction of sucrose with a glucan (specifically, $[(1,4)\text{-}\alpha\text{-}D\text{-glucosyl}]_n$) to yield D-fructose and a longer glucan (*i.e.*, $[(1,4)\text{-}\alpha\text{-}D\text{-glucosyl}]_{n+1}$).[1,2] Contrary to earlier reports on amylosucrases, the *Neisseria polysacchare* enzyme only acts on sucrose. In addition to polymer synthesis, amylosucrase exhibits a sucrase activity (*i.e.*, the ability to produce maltose) and maltotriose synthesis by means of successive transfers of the D-glucosyl moiety of sucrose onto the released glucose derived by the sucrase activity. Likewise, the enzyme synthesizes turanose and trehalulose synthesis through the transfer of glucosyl units to fructose.[3] Glycogen activates this enzyme system.

[1] B . Y. Tao, P. J. Reilly & J. F. Robyt (1988) *Carbohydr. Res.* **181**, 163.
[2] G. Okada & E. J. Hehre (1974) *JBC* **249**, 126.
[3] G. Potocki de Montalk, M. Remaud-Simeon, R. M. Willemot, P. Sarcabal, V. Planchot & P. Monsan (2000) *FEBS Lett.* **471**, 219.

Selected entries from *Methods in Enzymology* [vol, page(s)]:
General discussion: 1, 184; **230**, 304
Assay: 1, 184
Other molecular properties: properties, 1, 185; purification, 1, 184

ANANAIN

This hydrolytic enzyme [EC 3.4.22.31] catalyzes peptide-bond hydrolysis of proteins with broad specificity.[1-3] The enzyme prefers substrates having polar amino acids at the P1' subsite, exercizes broad specificity at P1, and prefers a hydrophobic side chain at the P2 position. The best reported low-molecular-weight substrate is Bz-Phe-Val-Arg-NHMec. *See also Bromelain, Fruit; Bromelain, Stem; Comosain*

[1] A. D. Napper, S. P. Bennett, M. Borowski, M. B. Holdridge, M. J. C. Leonard, E. E. Rogers, Y. Duan, R. A. Laursen, B. Reinhold & S. L. Shames (1994) *BJ* **301**, 727.
[2] A. D. Rowan & D. J. Buttle (1994) *MIE* **244**, 555.
[3] A. D. Rowan, D. J. Buttle & A. J. Barrett (1988) *ABB* **267**, 262.

Selected entries from *Methods in Enzymology* [vol, page(s)]:
Assay: **244**, 560

ANANDAMIDE AMIDOHYDROLASE

This enzyme catalyzes the reversible hydrolysis of anandamide (or, arachidonylethanolamide) to produce arachidonate and ethanolamine. Anandamide is an endogenous ligand for the cannabinoid receptor.[1,2] Porcine brain anandamide amidohydrolase catalyzes both anandamide hydrolysis and its net synthesis; however, the reaction in the ligating direction may not be physiologically relevant.[3,4]

Highly efficient amide synthesis requires carboxyl-group activation, usually achieved by acyl-group phosphorylation or adenylylation. Anandamide synthesis is reportedly independent of ATP and coenzyme A.[5] *See also Fatty Acid Amide-Synthesizing Enzymes*

[1] C. J. Hillard (2000) *Prosta. Other Lipid Mediat.* **61**, 3.
[2] D. G. Deutsch & S. A. Chin (1993) *Biochem. Pharmacol.* **46**, 791.
[3] Y. Kurahashi, N. Ueda, H. Suzuki, M. Suzuki & S. Yamamoto (1997) *BBRC* **237**, 512.
[4] N. Ueda, Y. Kurahashi, S. Yamamoto & T. Tokunaga (1995) *JBC* **270**, 23823.
[5] K. K. Kruszka & R. W. Gross (1994) *JBC* **269**, 14345.

ANCROD

This venom endopeptidase from the Malayan pit viper (*Calloselasma rhodostoma*, formerly *Agkistrodon rhodostoma*) blocks blood coagulation and is classified as a variant of venombin A [EC 3.4.21.74].[1-3] The enzyme resembles thrombin in its action and catalyzes the hydrolysis of the Arg16–Gly17 bond in fibrinogen Aα. Ancrod contains more carbohydrate than batroxobin or crotalase. *See also Venombin A; Batroxobin; Crotalase*

[1] M. P. Esnouf & G. W. Tunnah (1967) *Brit. J. Haematol.* **13**, 581.
[2] C. Nolan, L. S. Hall & G. W. Barlow (1976) *MIE* **45**, 205.
[3] H. Pirkle & I. Theodor (1998) in *The Enzymology of Snake Venoms* (G. S. Bailey, ed.), Alaken, Fort Collins.

Selected entries from *Methods in Enzymology* [vol, page(s)]:
General discussion: 45, 205
Assay: 45, 205; coagulant activity, 45, 205; esterolytic activity, 45, 206
Other molecular properties: composition, 45, 213, 232; inhibitors, 45, 210, 211; properties, 45, 210; physical, 45, 212; purification, 45, 207, 208; purity, 45, 212; specificity, 45, 210; stability, 45, 212

ANDROSTENOLONE SULFOTRANSFERASE

This sulfotransferase, a variant of steroid sulfotransferase [EC 2.8.2.15], catalyzes the reaction of androstenolone with 3′-phosphoadenylylsulfate to produce adenosine 3′,5′-bisphosphate or the corresponding steroid *O*-sulfate.[1,2] *See Steroid Sulfotransferase*

[1] J. B. Adams & D. McDonald (1980) *BBA* **615**, 275 and (1979) **567**, 144.
[2] C. N. Falany, M. E. Vasquez & J. M. Kalb (1989) *BJ* **260**, 641.

Selected entries from *Methods in Enzymology* [vol, page(s)]:
General discussion: 15, 732

ANGIOTENSIN CONVERTING ENZYME 2

This zinc-dependent metallopeptidase catalyzes the conversion of angiotensin I to angiotensin 1-9, acting as a carboxypeptidase, hydrolyzing the terminal His-Leu peptide bond. This enzyme is insensitive to classical angiotensin converting enzyme inhibitors such as captopril, lisinopril, and enalaprilat.[1] The enzyme also acts on des-Arg bradykinin and neurotensin.

[1] S. R. Tipnis, N. M. Hooper, R. Hyde, E. Karran, G. Christie & A. J. Turner (2000) *JBC* **275**, 33238.

1,5-ANHYDRO-D-FRUCTOSE REDUCTASE

This oxidoreductase [EC 1.1.1.263] catalyzes the reaction of 1,5-anhydro-D-fructose with NADPH and H^+ to produce 1,5-anhydro-D-glucitol and $NADP^+$.[1] Other substrates include pyridine-3-aldehyde and 2,3-butanedione. Acetaldehyde, 2-dehydroglucose (*i.e.*, glucosone), and glucuronate are weak substrates.

[1] M. Sakuma, S. Kametani & H. Akanuma (1998) *J. Biochem.* **123**, 198.

ANHYDROSIALIDASE

This hydrolase [EC 3.2.1.138], also known as anhydroneuraminidase and sialidase L, catalyzes the hydrolysis of α-sialosyl linkages in *N*-acetylneuraminic acid glycosides, releasing 2,7-anhydro-α-*N*-acetylneuraminate. The enzyme also acts on *N*-glycolylneuraminic acid glycosides.[1] Anhydrosialidase is the first sialidase found to exhibit a strict specificity toward the hydrolysis of the NeuAc α2 → 3Gal linkage.[2] *See also Exo-α-Sialidase; Endo-α-Sialidase*

[1] Y. T. Li, H. Nakagawa, S. A. Ross, G. C. Hansson & S. C. Li (1990) *JBC* **265**, 21629.
[2] M. Y. Chou, S. C. Li, M. Kiso, A. Hasegawa & Y. T. Li (1994) *JBC* **269**, 18821.

ANHYDROTETRACYCLINE MONOOXYGENASE

This oxidoreductase [EC 1.14.13.38], also known as anhydrotetracycline oxygenase (ATC oxygenase), catalyzes the reaction of anhydrotetracycline with NADPH and dioxygen to produce 12-dehydrotetracycline, $NADP^+$, and water.[1-3]

[1] V. Behal (1987) *CRC Crit. Rev. Biotech.* **5**, 275.
[2] I. Vancurova, M. Flieger, J. Volc, M. J. Benes, J. Novotna, J. Neuzil & V. Behal (1987) *Basic Microbiol.* **27**, 529.
[3] I. Vancurova, J. Volc, M. Flieger, J. Neuzil, J. Novotna, J. Vlach & V. Behal (1988) *BJ* **253**, 263.

ANTHOCYANIDIN 3-*O*-GLUCOSYLTRANSFERASE

This transferase [EC 2.4.1.115] catalyzes the reaction of UDP-D-glucose with anthocyanidin to produce UDP and anthocyanidin-3-*O*-D-glucoside.[1,2] Other acceptor substrates include pelargonidin and delphinidin. The enzyme will not catalyze 5-*O*-glucosylation of cyanidin; nor does it act on such flavonols as quercetin and kaempferol which are substrates for flavonol 3-*O*-glucosyltransferase [EC 2.4.1.91].

[1] J. Kamsteeg, J. van Brederode & G. van Nigtevecht (1978) *Biochem. Genet.* **16**, 1045.
[2] M. Teusch, G. Forkmann & W. Seyffert (1986) *Z. Naturforsch.* **41c**, 699.

ANTHOCYANIN 5-AROMATIC ACYLTRANSFERASE

This acyltransferase [EC 2.3.1.153] catalyzes the reaction of hydroxycinnamoyl-CoA with anthocyanidin-3,5-diglucoside to produce Coenzyme A and anthocyanidin 3-glucoside-5-hydroxycinnamoylglucoside.[1] Only the C-5 glucoside of the anthocyanin substrate is modified.

[1] H. Fujiwara, Y. Tanaka, Y. Fukui, M. Nakao, T. Ashikari & T. Kusumi (1997) *EJB* **249**, 45.

ANTHRANILATE ADENYLYLTRANSFERASE

This transferase [EC 2.7.7.55] catalyzes the reaction of ATP with anthranilate to produce pyrophosphate (or, diphosphate) and N-adenylylanthranilate.[1,2]

[1] W. Lerbs & M. Luckner (1985) J. Basic Microbiol. **25**, 387.
[2] M. Gerlach, N. Schwelle, W. Lerbs & M. Luckner (1985) Phytochemistry **24**, 1935.

ANTHRANILATE N-BENZOYLTRANSFERASE

This acyltransferase [EC 2.3.1.144], which participates in the biosynthesis of phytoalexins, catalyzes the reaction of benzoyl-CoA with anthranilate to produce Coenzyme A and N-benzoylanthranilate.[1] Other acyl-CoA alternative substrates include cinnamoyl-CoA, 4-coumaroyl-CoA, and salicyloyl-CoA.

[1] K. Reinhard & U. Matern (1989) ABB **275**, 295.

ANTHRANILATE 1,2-DIOXYGENASE

This iron-dependent oxidoreductase [EC 1.14.12.1], also known as anthranilate hydroxylase (deaminating, decarboxylating), catalyzes the reaction of anthranilate with NAD(P)H, dioxygen, and two water molecules to produce catechol, carbon dioxide, NAD(P)$^+$, and ammonia.[1–3] The reaction probably proceeds by forming a cyclic peroxide intermediate.[4]

[1] J. V. Schloss & M. S. Hixon (1998) CBC **2**, 43.
[2] O. Hayaishi, M. Nozaki & M. T. Abbott (1975) The Enzymes, 3rd ed., **12**, 119.
[3] O. Salcher & F. Lingens (1980) J. Gen. Microbiol. **121**, 465.
[4] S. Kobayashi, S. Kino, N. Hada & O. Hayaishi (1964) BBRC **16**, 556.

Selected entries from **Methods in Enzymology** [vol, page(s)]:
General discussion: 52, 12

ANTHRANILATE N-MALONYLTRANSFERASE

This malonyltransferase [EC 2.3.1.113] catalyzes the reaction of malonyl-CoA with anthranilate to produce coenzyme A and N-malonylanthranilate.[1]

[1] U. Matern, C. Feser & W. Heller (1984) ABB **235**, 218.

ANTHRANILATE N-METHYLTRANSFERASE

anthranilate   N-methylanthranilate

This methyltransferase [EC 2.1.1.111] catalyzes the reaction of S-adenosyl-L-methionine with anthranilate to produce S-adenosyl-L-homocysteine and N-methylanthranilate. The protein participates in plant biosynthesis of acridine alkaloids.[1–3]

[1] U. Eilert & B. Wolters (1989) Plant Cell Tissue Organ Cult. **18**, 1.
[2] A. Baumert, J. Crèche, M. Rideau, J.-C. Chénieux & D. Gröger (1990) Plant Physiol. Biochem. **28**, 587.
[3] A. Baumert, W. Maier, B. Schumann & D. Gröger (1991) J. Plant Physiol. **139**, 224.

ANTHRANILATE 3-MONOOXYGENASE

This iron-dependent oxidoreductase [EC 1.14.16.3], also known as anthranilate 3-hydroxylase, catalyzes the reaction of anthranilate with tetrahydropteridine and dioxygen to produce 3-hydroxyanthranilate, dihydropteridine, and water.[1]

[1] P. M. Nair & C. S. Vaidyanathan (1965) BBA **110**, 521.

ANTHRANILATE 3-MONOOXYGENASE (DEAMINATING)

This oxidoreductase [EC 1.14.13.35; formerly classified as 1.14.12.2], also known as anthranilate hydroxylase and anthranilate 2,3-hydroxylase (deaminating), catalyzes the reaction of anthranilate with NADPH and dioxygen to produce 2,3-dihydroxybenzoate (i.e., o-pyrocatechuate), NADP$^+$, and ammonia.[1–4] The enzyme from *Aspergillus niger* is an iron protein[4] whereas that from the yeast *Trichosporon cutaneum* is a flavoprotein (FAD).[1–3]

Oxygen-labeling experiments with the flavoenzyme of *T. cutaneum* with labeled dioxygen results in the appearance of the label at the 3-position of the product whereas the oxygen at the 2-position is derived from water; a similar observation is made with the *A. niger* enzyme.[4] A mechanism has been proposed for the flavoenzyme involving imine formation and hydrolysis during the reaction with a C4a-hydroperoxyflavin intermediate formed from reduced flavin and molecular oxygen.[1–3] In addition, the yeast enzyme undergoes slow conformational changes that occur on binding of the aromatic ligand as well as on reduction of the enzyme. These changes are apparently important for rapid anthranilate binding to occur in turnovers subsequent to the first. Bound anthranilate is also required for rapid reduction of enzyme-bound FAD by NADPH.[5]

[1] J. B. Powlowski, S. Dagley, V. Massey & D. B. Ballou (1987) JBC **262**, 69.
[2] J. Powlowski, V. Massey & D. P. Ballou (1989) JBC **264**, 5606.
[3] J. Powlowski, D. P. Ballou & V. Massey (1990) JBC **265**, 4969.
[4] V. Subramanian & C. S. Vaidyanathan (1984) J. Bacteriol. **160**, 651.

[5]J. Powlowski, D. Ballou & V. Massey (1989) JBC **264**, 16008.

Selected entries from *Methods in Enzymology* [vol, page(s)]:
General discussion: 17A, 510

ANTHRANILATE PHOSPHORIBOSYLTRANSFERASE

This enzyme [EC 2.4.2.18], also known as phosphoribosyl-anthranilate pyrophosphorylase, catalyzes the reaction of anthranilate with phosphoribosylpyrophosphate to produce N-5′-phosphoribosylanthranilate and pyrophosphate (or, diphosphate).[1–3] In certain species, this enzyme is part of a multifunctional protein, together with one or more other components of the system for the biosynthesis of tryptophan (*i.e.*, indole-3-glycerol-phosphate synthase, anthranilate synthase, tryptophan synthase, and phosphoribosylanthranilate isomerase).

[1]R. Bentley (1990) Crit. Rev. Biochem. Mol. Biol. **25**, 307.
[2]U. Hommel, A. Lustig & K. Kirschner (1989) EJB **180**, 33.
[3]W. D. L. Musick (1981) Crit. Rev. Biochem. **11**, 1.

Selected entries from *Methods in Enzymology* [vol, page(s)]:
General discussion: 17A, 368, 380, 393; 34, 377, 389; 142, 366
Assay: 17A, 368
Other molecular properties: complex with anthranilate synthase in (A. aerogenes, 17A, 380, 386; Escherichia coli, 17A, 371); inactivation on column, 34, 392; product, 17A, 368, 382, 387 (synthesis of, 17A, 365); purification from (A. aerogenes, 17A, 383; Neurospora crassa, 17A, 393)

ANTHRANILATE SYNTHASE

This enzyme [EC 4.1.3.27] catalyzes the reaction of chorismate with L-glutamine to generate anthranilate, pyruvate, and L-glutamate.[1–8] In certain species, this enzyme is part of a multifunctional protein together with one or more other components of the system for the biosynthesis of tryptophan (*i.e.*, indole-3-glycerol-phosphate synthase, anthranilate phosphoribosyltransferase, tryptophan synthase, and phosphoribosylanthranilate isomerase). The anthranilate synthase that is present in these complexes has been reported to be able to utilize either L-glutamine or ammonia as the nitrogen source. However, it has also been reported that when anthranilate synthase is separated from this complex, only ammonia can serve as a substrate. The pH_{opt} for the ammonia- and glutamine-dependent activities are different.

See also p-Aminobenzoate Synthase

[1]F. Massiere & M. A. Badet-Denisot (1998) Cell Mol. Life Sci. **54**, 205.
[2]B. Bartel (1997) Ann. Rev. Plant Physiol. Plant Mol. Biol. **48**, 51.
[3]R. M. Romero, M. F. Roberts & J. D. Phillipson (1995) Phytochemistry **39**, 263.
[4]H. Zalkin (1993) AE **66**, 203.
[5]R. Bentley (1990) Crit. Rev. Biochem. Mol. Biol. **25**, 307.
[6]D. E. Koshland, Jr., & A. Levitzki (1974) The Enzymes, 3rd ed., **10**, 539.
[7]J. M. Buchanan (1973) AE **39**, 91.
[8]H. Zalkin (1973) AE **38**, 1.

Selected entries from *Methods in Enzymology* [vol, page(s)]:
General discussion: 17A, 371, 380, 393, 401; 34, 377, 389; 46, 420; 113, 287; 142, 300
Assay: 17A, 368, 391; 113, 288; 142, 301
Other molecular properties: affinity labeling, **87**, 474; anthranilate synthase-phosphoribosyltransferase (assays, 142, 368; catalyzed reactions, 142, 368; properties, 142, 370; purification from Salmonella typhimurium [AS partial complex, 142, 377; component I subunit, 113, 290; 142, 380; component II subunit, 142, 381; whole complex, 142, 376]); complex, 34, 377, 389 (A. aerogenes, 17A, 380; Escherichia coli, 17A, 371; feedback-inhibitor as ligand, 34, 390; inactivation on column, 34, 392; tryptophan-agarose, 34, 390); coupling enzyme in arom multifunctional enzyme assay, as, preparation, 142, 331; properties, 113, 292; purification, 34, 389 (Aerobacter (Klebsiella) aerogenes, 17A, 383; Neurospora crassa, 17A, 393; Pseudomonas putida, 142, 302; Salmonella typhimurium, 17A, 403; Serratia marcescens, 113, 289); Serratia marcescens (component ASI, purification, 113, 290; properties, 113, 292; purification, 113, 289)

ANTHRANILOYL-CoA MONOOXYGENASE

This FAD-dependent oxidoreductase [EC 1.14.13.40], also known as 2-aminobenzoyl-CoA monooxygenase/reductase, catalyzes the reaction of 2-aminobenzoyl-CoA with dioxygen and two molecules of NAD(P)H to produce 2-amino-5-oxocyclohex-1-enecarboxyl-CoA, water, and two molecules of $NAD(P)^+$. Upon hydrolysis of the thiolester, the product nonenzymatically releases carbon dioxide and ammonia and generates 1,4-cyclohexanedione.[1–4]

The enzyme initially catalyzes the flavin- and NAD(P)H-dependent hydroxylation reaction, via a flavin hydroperoxide, to form 2-amino-5-hydroxybenzoyl-CoA as an intermediate. In a second flavin- and NAD(P)H-dependent step, 2-amino-5-oxocyclohex-1-enecarboxyl-CoA is produced. The enzyme will also catalyze an additional NAD(P)H-dependent reduction to form 2-amino-5-hydroxycyclohex-1-enecarboxyl-CoA. Note that three different products can be observed depending on reaction conditions.

[1]R. Buder & G. Fuchs (1989) EJB **185**, 629.
[2]R. Buder, K. Ziegler, G. Fuchs, B. Langkau & S. Ghisla (1989) EJB **185**, 637.
[3]B. Langkau, S. Ghisla, R. Buder, K. Ziegler & G. Fuchs (1990) EJB **191**, 365.
[4]B. Langkau & S. Ghisla (1995) EJB **230**, 686.

ANTHRANILYL-CoA SYNTHETASE

This ligase [EC 6.2.1.32], also known as anthranilate:CoA ligase, catalyzes the reaction of ATP with anthranilate and coenzyme A to produce AMP, pyrophosphate (or, diphosphate), and anthranilyl-CoA. Alternative substrates include benzoate, 2-fluorobenzoate, and 4-fluorobenzoate.[1]

[1]U. Altenschmidt, B. Oswald & G. Fuchs (1991) J. Bacteriol. **173**, 5494.

*Apa*LI RESTRICTION ENDONUCLEASE

This type II restriction endonuclease [EC 3.1.21.4], obtained from *Acetobacter pasteurianus*, catalyzes the hydrolysis of both strands of DNA at 5′ . . . G∇TGCAC . . . 3′.[1,2]

[1]Y. Yamada & M. Murakami (1985) *Agric. Biol. Chem.* **49**, 3627.
[2]M. Murakami & Y. Yamada (1990) *Agric. Biol. Chem.* **54**, 1791.

*Apa*I RESTRICTION ENDONUCLEASE

This type II restriction endonuclease [EC 3.1.21.4] catalyzes the hydrolysis of both strands of DNA at 5′ . . . GGGCC∇C . . . 3′.[1] The enzyme is inhibited by salt concentrations larger than 50 mM. It is also inhibited by overlapping *dcm* methylation.[2]

[1]J. Seurinck, A. Van de Voorde & M. Van Montagu (1983) *NAR* **11**, 4409.
[2]F. W. Larimer (1987) *NAR* **15**, 9087.

APIGENIN 4′-*O*-METHYLTRANSFERASE

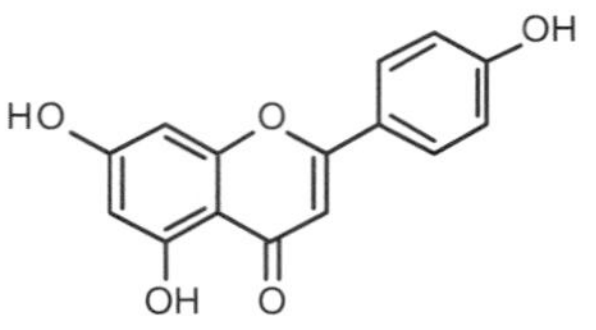

apigenin

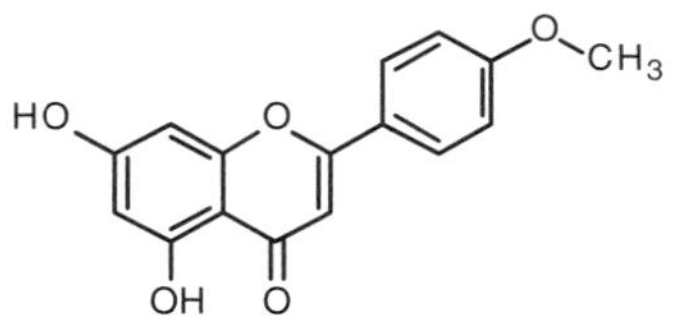

acacetin

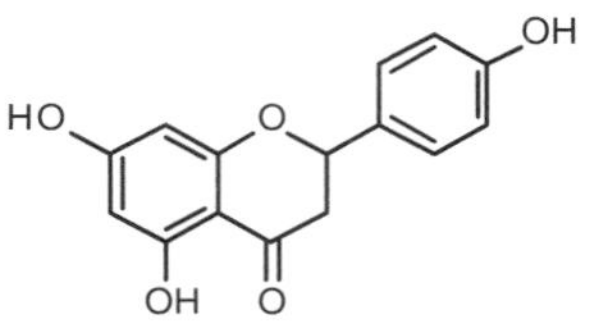

naringenin

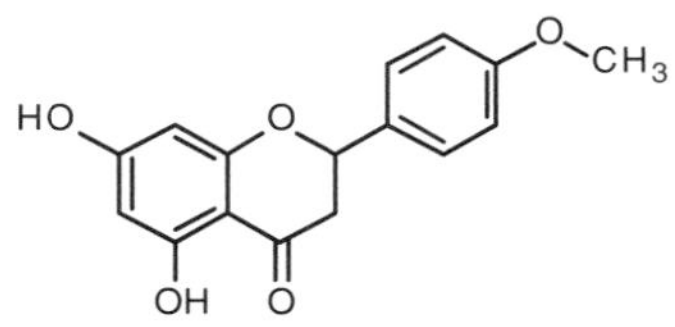

4′-methoxy-5,7-dihydroxyflavonone

This methyltransferase [EC 2.1.1.75], also known as flavonoid *O*-methyltransferase, catalyzes the reaction of *S*-adenosyl-L-methionine with 5,7,4′-trihydroxyflavone (or, apigenin) to produce *S*-adenosyl-L-homocysteine and 4′-methoxy-5,7-dihydroxyflavone (or, acacetin). Naringenin (5,7,4′-trihydroxyflavonone) can also act as a weaker alternative substrate.[1,2]

[1]G. Kuroki & J. E. Poulton (1981) *Z. Naturforsch.* **36C**, 916.
[2]V. De Luca & R. K. Ibrahim (1985) *ABB* **238**, 606.

APIOSE 1-REDUCTASE

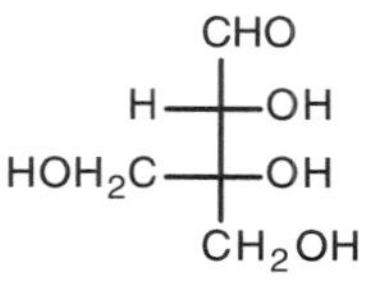

D-apiose

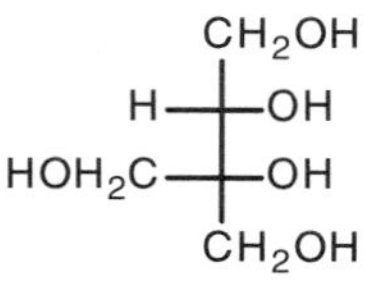

D-apiitol

This oxidoreductase [EC 1.1.1.114], also known as D-apiose reductase and D-apiitol reductase, catalyzes the reversible reaction of D-apiose with NADH to produce D-apiitol and NAD+.[1,2]

[1]R. Hanna, M. Picken & J. Mendicino (1973) *BBA* **315**, 259.
[2]D. L. Neal & P. K. Kindel (1970) *J. Bacteriol.* **101**, 910.

Selected entries from *Methods in Enzymology* [vol, page(s)]:
General discussion: 89, 228
Assay: 89, 229

Apo-β-CAROTENOID 14′,13′-DIOXYGENASE

This thiol-dependent oxidoreductase [EC 1.13.12.12] catalyzes the reaction of 8′-apo-β-carotenol (but not β-carotene) with dioxygen to produce 14′-apo-β-carotenal and water.[1] The enzyme is distinct from β-carotene 15,15′-dioxygenase [EC 1.13.11.21].

[1]A. A. Dmitrovskii, N. N. Gessler, S. B. Gomboeva, Y. V. Ershov & V. Y. Bykhovsky (1997) *Biochemistry (Moscow)* **62**, 787.

*Apo*I RESTRICTION ENDONUCLEASE

This type II restriction endonuclease [EC 3.1.21.4], obtained from *Arthrobacter protophormiae*, catalyzes the hydrolysis of both strands of DNA at 5′ . . . R∇AATTY . . . 3′, where R refers to either A or G and Y refers to either T or C.[1] The resulting DNA fragments have a 5′ AATT extension that can be ligated to DNA fragments generated by *Eco*RI digestion. The enzyme can also exhibit star activity. *See "Star" Activity*

[1]C. Polisson & D. Robinson (1992) *NAR* **20**, 2888.

APYRASE

This calcium ion-dependent enzyme [EC 3.6.1.5], also known as ATP-diphosphatase, adenosine diphosphatase, ADPase, and ATP-diphosphohydrolase, catalyzes the hydrolysis of ATP to generate AMP and two orthophosphate ions.[1] The enzyme also utilizes ADP as a substrate as well as other nucleoside triphosphates and diphosphates. *See also ATPase*

[1]M. Komoszynski & A. Wojtczak (1996) *BBA* **1310**, 233.

Selected entries from *Methods in Enzymology* [vol, page(s)]:
General discussion: 2, 591
Assay: 2, 591; luciferase and, **3**, 873
Other molecular properties: adenylate kinase assay and, **2**, 600; **6**, 230; analytical use of, **3**, 869; ATP removal by, **2**, 643; hydrolysis of ATP, **57**, 69, 70, 71, 72; insect, **2**, 590, 595; myokinase and, **6**, 230; plasma membranes, in, **31**, 88; potato, **2**, 591; **65**, 668, 669, 677 (preparation, **2**, 592; **12A**, 157; properties, **2**, 593; ribonucleoside diphosphate

reductases and, **12A**, 160); properties, **2**, 593; purification, **2**, 592; temperature and, **2**, 593

*Apy*I RESTRICTION ENDONUCLEASE

This type II restriction endonuclease [EC 3.1.21.4] is obtained from *Arthrobacter pyridinolis*. It acts on both strands of DNA at 5′ . . . CCⱯWGG . . . 3′, where W refers to either A or T.[1] The enzyme will act at this sequence when the internal cytosine is methylated, but is inactive on hemimethylated DNA in which both cytosines are methylated.[2]

[1] R. J. Roberts (1980) *Meth. Enzymol.* **65**, 1.
[2] Y. Gruenbaum, H. Cedar & A. Razin (1981) *Nucleic Acids Res.* **9**, 2509.

Selected entries from **Methods in Enzymology** [vol, page(s)]:
Other molecular properties: recognition sequence, 65, 3.

AQUACOBALAMIN REDUCTASES

Aquacobalamin reductase (NADH) [EC 1.6.99.8] catalyzes the flavin-dependent reaction of two molecules of aquacob(III)alamin with NADH to form two molecules of cob(II)alamin and NAD$^+$.[1,2]

Aquacobalamin reductase (NADPH) [EC 1.6.99.11] catalyzes the flavin-dependent reaction of NADPH with two molecules of aquacob(III)alamin to produce NADP$^+$ and two molecules of cob(II)alamin.[1,3] The enzyme will also act on hydroxycobalamin, but not on cyanocobalamin. In rat liver, the microsomal NADPH-linked aquacobalamin reductase appears to be identical to the NADPH-cytochrome *c* reductase.[4]

[1] F. Watanabe, Y. Nakano, N. Tachikake, Y. Tamura, H. Yamanaka & S. Kitaoka (1990) *J. Nutr. Sci. Vitaminol.* **36**, 349.
[2] G. A. Walker, S. Murphy & F. M. Huennekens (1969) *ABB* **134**, 95.
[3] F. Watanabe, Y. Oki, Y. Nakano & S. Kitaoka (1987) *JBC* **262**, 11514.
[4] F. Watanabe, Y. Nakano, H. Saido, Y. Tamura & H. Yamanaka (1992) *BBA* **1119**, 175.

Selected entries from **Methods in Enzymology** [vol, page(s)]:
General discussion: 281, 289, 295
Other molecular properties: purification (*Euglena gracilis*, **281**, 289; mammals, **281**, 295)

ARABINAN ENDO-1,5-α-L-ARABINOSIDASE

This enzyme [EC 3.2.1.99], also known as endo-1,5-α-L-arabinanase, catalyzes the endohydrolysis of 1,5-α-L-arabinofuranosidic linkages in 1,5-arabinans.[1,2]

[1] A. Kaji & T. Saheki (1975) *BBA* **410**, 354.
[2] S. M. Pitson, A. G. Voragen, J. P. Vincken & G. Beldman (1997) *Carbohydr. Res.* **303**, 207.

D-ARABINITOL 2-DEHYDROGENASE

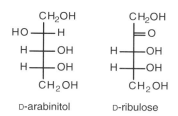

D-arabinitol   D-ribulose

This oxidoreductase [EC 1.1.1.250], also known as D-arabinitol 2-dehydrogenase (ribulose-forming), catalyzes the reversible reaction of D-ribulose with NADH to produce D-arabinitol and NAD$^+$, thus acting on and producing the enantiomers of the substrate and product of L-arabinitol 2-dehydrogenase [EC 1.1.1.13].[1–5]

[1] B. Wong, J. S. Murray, M. Castellanos & K. D. Croen (1993) *J. Bacteriol.* **175**, 6314.
[2] J. S. Murray, M. L. Wong, C. G. Miyada, A. C. Switchenko, T. C. Goodman & B. Wong (1995) *Gene* **155**, 123.
[3] M. W. Quong, C. G. Miyada, A. C. Switchenko & T. C. Goodman (1993) *BBRC* **196**, 1323.
[4] J. Hallborn, M. Walfridsson, S. Penttila, S. Keranen & B. Hahn-Haegerdal (1995) *Yeast* **11**, 839.
[5] J. M. Ingram & W. A. Wood (1966) *MIE* **9**, 186.

Selected entries from **Methods in Enzymology** [vol, page(s)]:
General discussion: 9, 186
Assay: 9, 186
Other molecular properties: *Saccharomyces rouxii* (properties, **9**, 188; purification, **9**, 187)

D-ARABINITOL 4-DEHYDROGENASE

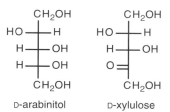

D-arabinitol   D-xylulose

This oxidoreductase [EC 1.1.1.11], also known as D-arabitol 4-dehydrogenase, catalyzes the reaction of D-arabinitol with NAD$^+$ to yield D-xylulose and NADH.[1–3]

[1] W. A. Wood, M. J. McDonough, L. B. Jacobs (1961) *JBC* **236**, 2190.
[2] E. C. C. Lin (1961) *JBC* **236**, 31.
[3] M. S. Neuberger, R. A. Patterson & B. S. Hartley (1979) *BJ* **183**, 31.

Selected entries from **Methods in Enzymology** [vol, page(s)]:
General discussion: 9, 184
Assay: 9, 184

L-ARABINITOL 4-DEHYDROGENASE

This oxidoreductase [EC 1.1.1.12], also known as L-arabitol 4-dehydrogenase, catalyzes the reaction of

L-arabinitol with NAD$^+$ to yield L-xylulose and NADH.[1,2]

[1] C. Chiang & S. G. Knight (1960) *BBRC* **3**, 554.
[2] C. F. B. Witteveen, R. Busnik, P. Van de Vondervoort, C. Dijkema, K. Swart & J. Visser (1989) *J. Gen. Microbiol.* **135**, 2163.

Selected entries from *Methods in Enzymology* [**vol**, page(s)]:
General discussion: 9, 180
Assay: 9, 180
Other molecular properties: *Aerobacter aerogenes* (properties, 9, 183)

L-ARABINITOL 2-DEHYDROGENASE (RIBULOSE FORMING)

This oxidoreductase [EC 1.1.1.13], also called L-arabitol 2-dehydrogenase, catalyzes the reaction of L-arabinitol with NAD$^+$ to yield L-ribulose and NADH.[1]

[1] C. Chiang & S. G. Knight (1961) *BBA* **46**, 271.

α-L-ARABINOFURANOSIDASE

This enzyme [EC 3.2.1.55], also called arabinosidase, catalyzes the hydrolysis of terminal nonreducing α-L-arabinofuranoside residues in α-L-arabinosides.[1–4] Other substrates include α-L-arabinofuranosides, α-L-arabinans containing (1,3)- and/or (1,5)-linkages, arabinoxylans, and arabinogalactans. Some β-galactosidases and β-D-fucosidases will also hydrolyze α-L-arabinosides.

Initial-rate investigations of the *Monilinia fructigena* enzyme indicate that bond-breaking is rate-limiting and that proton donation to the leaving group is far advanced at the transition state.[2] The reaction proceeds with retention of configuration.[5]

[1] G. Davies, M. L. Sinnott & S. G. Withers (1998) *CBC* **1**, 119.
[2] M. A. Kelly, M. L. Sinnott & M. Herrchen (1987) *BJ* **245**, 843.
[3] J. R. Whitacker (1984) *Enzyme Microb. Technol.* **6**, 341.
[4] P. M. Dey & E. Del Campillo (1984) *AE* **56**, 141.
[5] S. M. Pitson, A. G. Voragen & G. Beldman (1996) *FEBS Lett.* **398**, 7.

Selected entries from *Methods in Enzymology* [**vol**, page(s)]:
General discussion: 160, 707, 712
Assay: 160, 707, 712
Other molecular properties: properties, 160, 711, 717; purification from (*Aspergillus niger*, 160, 708; *Scopolia japonica*, 160, 713); transition state and multisubstrate analogues, 249, 306

α-L-ARABINOFURANOSIDE HYDROLASE

This enzyme activity, previously classified as EC 3.2.1.79 but now a deleted Enzyme Commission entry, is listed under α-L-arabinofuranosidase [EC 3.2.1.55].

ARABINOGALACTAN ENDO-1,3-β-GALACTOSIDASE

This galactosidase [EC 3.2.1.90], also known as endo-1,3-β-galactanase, galactanase, and arabinogalactanase, catalyzes the endohydrolysis of 1,3-β-D-galactosidic linkages in arabinogalactans.[1,2]

[1] R. F. H. Dekker (1985) *Biosynthesis and Biodegradation of Wood Components* (T. Higuchi, ed.), p. 505.
[2] Y. Hashimoto (1971) *J. Agric. Chem. Soc. Jpn.* **45**, 147.

Selected entries from *Methods in Enzymology* [**vol**, page(s)]:
Assay: 160, 720

ARABINOGALACTAN ENDO-1,4-β-GALACTOSIDASE

This galactosidase [EC 3.2.1.89], also known as endo-1,4-β-galactanase, galactanase, and arabinogalactanase, catalyzes the endohydrolysis of 1,4-β-D-galactosidic linkages in arabinogalactans.[1]

[1] R. F. H. Dekker (1985) *Biosynthesis and Biodegradation of Wood Components* (T. Higuchi, ed.), p. 505.

Selected entries from *Methods in Enzymology* [**vol**, page(s)]:
Assay: 160, 720

D-ARABINOKINASE

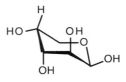

β-D-arabinose

This phosphotransferase [EC 2.7.1.54] catalyzes the ATP-dependent phosphorylation of D-arabinose to generate D-arabinose 5-phosphate and ADP.[1]

[1] W. A. Volk (1962) *JBC* **237**, 19.

Selected entries from *Methods in Enzymology* [**vol**, page(s)]:
General discussion: 9, 442
Assay: 9, 442

L-ARABINOKINASE

This phosphotransferase [EC 2.7.1.46] catalyzes the ATP-dependent phosphorylation of D-arabinose to generate L-arabinose 5-phosphate and ADP.[1,2]

[1] E. F. Neufeld, D. S. Feingold & W. Z. Hassid (1960) *JBC* **235**, 906.
[2] P. H. Chan & W. Z. Hassid (1975) *Anal. Biochem.* **64**, 372.

ARABINONATE DEHYDRATASE

Arabinonate dehydratase [EC 4.2.1.5] catalyzes the elimination of water from D-arabinonate to yield 2-dehydro-3-deoxy-D-arabinonate.[1,2]

[1] W. A. Wood (1971) *The Enzymes*, 3rd ed., **5**, 573.
[2] R. Weinberg & M. Doudoroff (1955) *JBC* **217**, 607.

L-ARABINONATE DEHYDRATASE

Arabinonate dehydratase [EC 4.2.1.25] catalyzes the elimination of water from L-arabinonate to yield 2-dehydro-3-deoxy-L-arabinonate.[1,2]

[1] F. O. Pedrosa & G. T. Zancan (1974) *J. Bacteriol.* **119**, 336.
[2] N. J. Novick & M. E. Tyler (1982) *J. Bacteriol.* **149**, 364.

D-ARABINONOLACTONASE

This enzyme [EC 3.1.1.30] catalyzes the hydrolysis of D-arabinono-1,4-lactone to produce D-arabinonate.[1] *See also 1,4-Lactonase; Gluconolactonase; L-Arabinonolactonase*

[1] N. J. Palleroni & M. Doudoroff (1957) *J. Bacteriol.* **74**, 180.

L-ARABINONOLACTONASE

This enzyme [EC 3.1.1.15] catalyzes the hydrolysis of L-arabinono-1,4-lactone to produce L-arabinonate.[1] *See also 1,4-Lactonase; Gluconolactonase; D-Arabinonolactonase*

[1] M. J. Dilworth, R. Arwas, I. A. McKay, S. Saroso & A. R. Glenn (1986) *J. Gen. Microbiol.* **132**, 2733.

Selected entries from *Methods in Enzymology* [**vol**, page(s)]:
Other molecular properties: arabinose dehydrogenase, presence in preparation of, **5**, 344

D-ARABINONO-1,4-LACTONE OXIDASE

This FAD-dependent oxidoreductase [EC 1.1.3.37] catalyzes the reaction of D-arabinono-1,4-lactone with dioxygen to produce D-*erythro*-ascorbate and hydrogen peroxide.[1]

[1] W.-K. Huh, S.-T. Kim, K.-S. Yang, Y.-J. Seok, Y. C. Hah & S.-O. Kang (1994) *EJB* **225**, 1073

D-ARABINOSE 1-DEHYDROGENASE

This oxidoreductase [EC 1.1.1.116] catalyzes the reaction of D-arabinose with NAD$^+$ to produce D-arabinono-1,4-lactone and NADH. The pig liver enzyme, which has an ordered Bi Bi kinetic mechanism, also catalyzes the oxidation of L-fucose.[1]

[1] W. R. Carper, K. W. Chang, W. G. Thorpe, M. A. Carper & C. M. Buess (1974) *BBA* **358**, 49.

L-ARABINOSE 1-DEHYDROGENASE

This oxidoreductase [EC 1.1.1.46] catalyzes the reaction of L-arabinose with NAD$^+$ to produce L-arabinono-1,4-lactone and NADH.[1]

[1] N. J. Novick & M. E. Tyler (1983) *Can. J. Microbiol.* **29**, 242.

Selected entries from *Methods in Enzymology* [**vol**, page(s)]:
General discussion: 5, 342
Assay: 5, 342; 41, 151
Other molecular properties: galactose dehydrogenase preparation, not present in, **5**, 341; pseudomonad, from, **41**, 150 (chromatography, **41**, 152; properties, **5**, 343; **41**, 153; purification, **5**, 342; **41**, 151, 152)

D-ARABINOSE 1-DEHYDROGENASE (NAD(P)$^+$)

This oxidoreductase [EC 1.1.1.117] catalyzes the reaction of D-arabinose with NAD(P)$^+$ to produce D-arabinono-1,4-lactone and NAD(P)H. Other substrates include L-galactose, 6-deoxy-L-galactose (*i.e.*, L-fucose), and 3,6-dideoxy-L-galactose (*i.e.*, L-colitose).[1]

[1] A. L. Cline & A. S. L. Hu (1965) *JBC* **240**, 4488.

D-ARABINOSE ISOMERASE

This isomerase [EC 5.3.1.3] catalyzes the interconversion of D-arabinose and D-ribulose.[1-3] The enzyme also interconverts L-fucose and L-fuculose as well as L-xylose and L-xylulose. Although this enzyme reportedly uses L-fucose as an alternative substrate, a distinct L-fucose isomerase [EC 5.2.1.25] has been isolated and characterized.

[1] E. J. Oliver & R. P. Mortlock (1971) *J. Bacteriol.* **108**, 293.
[2] J. R. Boulter & W. O. Gielow (1973) *J. Bacteriol.* **113**, 687.
[3] K. Izumoti & K. Yamanaka (1974) *Agric. Biol. Chem.* **38**, 267.

Selected entries from *Methods in Enzymology* [**vol**, page(s)]:
General discussion: 1, 366; 9, 583;.41, 462
Assay: 1, 367; 9, 583; 41, 462, 463
Other molecular properties: *Aerobacter aerogenes*, from, **41**, 462 (assay, **41**, 462, 463; inhibitors, **41**, 465; molecular weight, **41**, 465; properties, **9**, 585; **41**, 465; purification, **9**, 584; **41**, 463); *Escherichia coli*, from, in D-ribulose preparation, **9**, 40; properties, **1**, 370; **9**, 585; **41**, 465; purification, **1**, 369; **9**, 584; **41**, 465

L-ARABINOSE ISOMERASE

This isomerase [EC 5.3.1.4] catalyzes the interconversion of L-arabinose and L-ribulose.[1-3]

[1] E. C. Heath, B. L. Horecker, P. Z. Smyrniotis & Y. Takagi (1958) *JBC* **231**, 1031.
[2] T. Nakamatu & K. Yamanaka (1969) *BBA* **178**, 156.
[3] K. Izumori, K. Yamanaka & A. D. Elbein (1976) *J. Bacteriol.* **128**, 587.

Selected entries from *Methods in Enzymology* [**vol**, page(s)]:
General discussion: 5, 344; 9, 596; 41, 453, 458
Assay: 5, 344; 9, 597; *Escherichia coli*, 41, 453, 454; *Lactobacillus gayonii*, 41, 458
Other molecular properties: *Aerobacter aerogenes*, from (inhibitors, **9**, 602; metal requirement, **9**, 602; properties, **9**, 601; purification, **9**, 599; specificity, **9**, 601); bifunctional imidoester and, **25**, 648; *Escherichia coli*, from, **41**, 453 (assay, **41**, 453, 454; chromatography, **41**, 455; inhibition, **41**, 457; properties, **41**, 456; purification, **41**, 454); *Escherichia coli* B/r, in, isolation of mutants, **9**, 15; labeled cyanate and, **25**, 584; *Lactobacillus gayonii*, from, **9**, 601; **41**, 458 (assay, **41**, 458; chromatography, **41**, 459; inhibition constants of pentitols, **41**, 461; metal requirement, **9**, 602;

molecular weight, **41**, 460; properties, **9**, 601; **41**, 460, 461; purification, **9**, 601; **41**, 458); *Lactobacillus plantarum* (properties, **5**, 346; purification, **5**, 345); ribulokinase free of, **5**, 301; sources, **9**, 598

ARABINOSE-5-PHOSPHATE ISOMERASE

This isomerase [EC 5.3.1.13] catalyzes the interconversion of D-arabinose 5-phosphate and D-ribulose 5-phosphate.[1,2] This enzyme catalyzes a key step in the biosynthesis of lipopolysaccharide, an essential component of the outer membrane of Gram-negative bacteria. The reaction mechanism probably involves an enediol intermediate.

[1]E. A. Noltmann (1972) *The Enzymes*, 3rd ed., **6**, 271.
[2]E. C. Bigham, C. E. Gragg, W. R. Hall, J. E. Kelsey, W. R. Mallory, C. B. Richardson, C. Benedict & P. H. Ray (1984) *J. Med. Chem.* **27**, 717.

Selected entries from *Methods in Enzymology* [vol, page(s)]:
General discussion: 9, 585
Assay: 9, 585
Other molecular properties: D-arabinokinase preparation, presence in, **9**, 445; heavy metal as inhibitor, **9**, 588; *Propionibacterium pentosaceum*, from (pH optimum, **9**, 588; properties, **9**, 588; purification, **9**, 586; specificity, **9**, 588); transition-state and multisubstrate analogues, **249**, 308

β-L-ARABINOSIDASE

This enzyme [EC 3.2.1.88], also known as vicianosidase, catalyzes the hydrolysis of a β-L-arabinoside to produce an alcohol and L-arabinose.[1,2] The hydrolase can be assayed with the chromogenic substrate *p*-nitrophenyl β-L-arabinoside.

[1]P. M. Dey (1983) *BBA* **746**, 8.
[2]P. M. Dey (1973) *BBA* **302**, 393.

ARACHIDONATE 5-LIPOXYGENASE

This iron-dependent enzyme [EC 1.13.11.34], also known as 5-lipoxygenase and occasionally known as leukotriene A$_4$ synthase, catalyzes the reaction of arachidonate (*i.e.*, 5Z,8Z,11Z,14Z-eicosatetraenoate) with dioxygen to produce (6E,8Z,11Z,14Z)-(5S)-5-hydroperoxyeicosa-6,8,11,14-tetraenoate, which rapidly converts to leukotriene A$_4$.[1–10]

arachidonate

(6*E*,8*Z*,11*Z*,14*Z*)-(5*S*)-5-hydroperoxy-
eicosa-6,8,11,14-tetraenoate

The enzyme will also catalyze the oxidation of 5,8,11,14,17-eicosapentaenoate to 5-hydroperoxy-6,8,11,14,17-eicosapentaenoate and 5,8,11-eicosatrienoate to 5-hydroperoxy-6,8,11-eicosatrienoate.

In mammals, this lipoxygenase participates in the biosynthesis of the leukotrienes which have important roles in bronchoconstriction and inflammation. Upon cell stimulation, the mammalian leukocyte 5-lipoxygenase is translocated to the nuclear membrane and becomes associated with FLAP, an integral membrane protein essential for leukotriene biosynthesis in intact cells by acting as an arachidonate transfer protein. *See also* Lipoxygenase

[1]J. Z. Haeggstrom (2000) *Amer. J. Respir. Crit. Care Med.* **161**, S25.
[2]B. G. Fox (1998) *CBC* **3**, 261.
[3]A. W. Ford-Hutchinson, M. Gresser & R. N. Young (1994) *Ann. Rev. Biochem.* **63**, 383.
[4]S. Yamamoto & Y. Ishimura (1991) *A Study of Enzymes* **2**, 315.
[5]W. E. DeWolf (1991) in *Lipoxygenases and Their Products* (S. T. Crooke, ed.), p. 105, Academic Press, New York.
[6]M. A. Ator & P. R. Ortez de Montellano (1990) *The Enzymes*, 3rd ed., **19**, 213.
[7]T. Schewe, S. M. Rapoport & H. Kühn (1986) *AE* **58**, 191.
[8]H. Kühn, T. Schewe & S. M. Rapoport (1986) *AE* **58**, 273.
[9]F. J. Papatheofanis & W. E. M. Lands (1985) in *Biochemistry of Arachidonic Acid Metabolism* (W. E. M. Lands, ed.) p. 9, Martinus Nijhoff Publ., Boston.
[10]C. R. Pace-Asciak & W. L. Smith (1983) *The Enzymes*, 3rd ed., **16**, 543.

Selected entries from *Methods in Enzymology* [vol, page(s)]:
General discussion: 163, 344; 187, 268, 296, 312, 338
Assay: 163, 345; potato, 187, 297; potato tuber, 187, 269
Other molecular properties: activation in bone marrow-derived mast cells, differentiation from exocytosis, **187**, 518; expression in insect cells with baculovirus vector, **187**, 496; human leukocyte (isolation and characterization, **187**, 312; stimulatory factors, preparation, **187**, 316); molecular biology and cloning, **187**, 491; monoclonal antibodies, preparation, **187**, 341; peptide sequence analysis, **187**, 492; properties, **163**, 347; polyclonal antibody preparation, **187**, 493; porcine leukocyte (immunoaffinity purification, **187**, 338; properties, **187**, 342); potato (assay, **187**, 297; properties, **187**, 301; purification, **187**, 300); potato tuber (assay, **187**, 269; properties, **187**, 273; purification, **187**, 268); purification from human neutrophils, **163**, 346; recombinant (characterization, **187**, 500; transfer vector, construction, **187**, 497)

ARACHIDONATE 8-LIPOXYGENASE

arachidonate

(5*Z*,9*E*,11*Z*,14*Z*)-(8*R*)-8-hydroperoxy-
eicosa-5,9,11,14-tetraenoate

This iron-dependent oxidoreductase [EC 1.13.11.40], also known as 8-lipoxygenase, catalyzes the reaction of arachidonate (*i.e.*, 5*Z*,8*Z*,11*Z*,14*Z*-eicosatetraenoate) with dioxygen to produce (5*Z*,9*E*,11*Z*,14*Z*)-(8*R*)-8-hydroperoxyeicosa-5,9,11,14-tetraenoate.[1–5] The 8*S* enantiomer is produced by the mouse enzyme.[3,4]

[1]G. L. Bundy, E. G. Nidy, D. E. Epps, S. A. Mizsak & R. J. Wnuk (1986) *JBC* **261**, 747.

[2]A. R. Brash, S. W. Baertschi, C. D. Ingram & T. M. Harris (1987) *JBC* **262**, 15829.

[3]G. Fürstenberger, H. Hagedorn, T. Jacobi, E. Besemfelder, M. Stephan, W.-D. Lehman & F. Marks (1991) *JBC* **266**, 15738.

[4]A. R. Brash, W. E. Boeglin, M. S. Chang & B. H. Shieh (1996) *JBC* **271**, 20949.

[5]C. R. Pace-Asciak & W. L. Smith (1983) *The Enzymes*, 3rd ed., **16**, 543.

ARACHIDONATE 12-LIPOXYGENASE

This iron-dependent oxidoreductase [EC 1.13.11.31], better known as 12-lipoxygenase, catalyzes the reaction of arachidonate (*i.e.*, 5*Z*,8*Z*,11*Z*,14*Z*-eicosatetraenoate) with dioxygen to produce (5*Z*,8*Z*,10*E*,14*Z*)-(12*S*)-12-hydroperoxyeicosa-5,8,10,14-tetraenoate, which then converts rapidly to the corresponding 12*S*-hydroxy compound (12-HETE).[1–8]

arachidonate

(5*Z*,8*Z*,10*E*,14*Z*)-(12*S*)-12-hydroperoxy-eicosa-5,8,10,14-tetraenoate

There are distinct mammalian arachidonate 12-lipoxygenases. For example, the platelet, leukocyte, and epidermal 12-lipoxygenases have distinct sequences, catalytic activities, and functions. Some 12-lipoxygenases catalyze the formation of a mixture of products, and the bovine, porcine, rat, and mouse leukocyte enzymes (but not the platelet protein) also catalyze oxygenation at the C15 position.

[1]S. Yamamoto, H. Suzuki, M. Nakamura & K. Ishimura (1999) *Adv. Exp. Med. Biol.* **447**, 37.

[2]B. G. Fox (1998) *CBC* **3**, 261.

[3]S. Yamamoto, H. Suzuki & N. Ueda (1997) *Prog. Lipid Res.* **36**, 23.

[4]T. Yoshimoto & S. Yamamoto (1995) *J. Lipid Mediat. Cell Signal.* **12**, 195.

[5]S. Yamamoto & Y. Ishimura (1991) *A Study of Enzymes* **2**, 315.

[6]T. Schewe, S. M. Rapoport & H. Kühn (1986) *AE* **58**, 191.

[7]H. Kühn, T. Schewe & S. M. Rapoport (1986) *AE* **58**, 273.

[8]C. R. Pace-Asciak & W. L. Smith (1983) *The Enzymes*, 3rd ed., **16**, 543.

Selected entries from *Methods in Enzymology* [vol, page(s)]:
General discussion: 86, 49
Assay: bovine platelet, **86**, 50; rabbit peritoneal polymorphonuclear leukocyte, **86**, 46
Other molecular properties: bovine platelet (assays, **86**, 50; partial purification, **86**, 51; properties, **86**, 53); rabbit peritoneal

polymorphonuclear leukocyte (assay, **86**, 46; properties, **86**, 48; purification, **86**, 46)

ARACHIDONATE 15-LIPOXYGENASE

This iron-dependent oxidoreductase [EC 1.13.11.33], also known as 15-lipoxygenase or arachidonate ω–6 lipoxygenase, catalyzes the reaction of arachidonate (*i.e.*, 5*Z*,8*Z*,11*Z*,14*Z*-eicosatetraenoate) with dioxygen to produce (5*Z*,8*Z*,11*Z*,13*E*)-(15*S*)-15-hydroperoxyeicosa-5,8,11,13-tetraenoate, which rapidly converts to the corresponding 15*S*-hydroxy compound.[1–7]

arachidonate

(5*Z*,8*Z*,11*Z*,13*E*)-(15*S*)-15-hydroperoxy-eicosa-5,8,11,13-tetraenoate

Plant 13-lipoxygenase, which acts on the 18-carbon acids linoleate and linolenate, corresponds to mammalian 15-lipoxygenase since these enzymes "count" the substrate carbons from the methyl end: they both react at the ω–6 oxygen.

Some mammalian 15-lipoxygenases generate a mixture of products, and the rabbit and human reticulocyte exhibit a minor oxygenation activity at the C12 position.

[1]H. Kuhn & S. Borngraber (1999) *Adv. Exp. Med. Biol.* **447**, 5.

[2]B. G. Fox (1998) *CBC* **3**, 261.

[3]H. Kuhn & B. J. Thiele (1995) *J. Lipid Mediat. Cell Signal.* **12**, 157.

[4]S. Yamamoto & Y. Ishimura (1991) *A Study of Enzymes* **2**, 315.

[5]A. W. Ford-Hutchinson (1991) *Eicosanoids* **4**, 65.

[6]M. A. Ator & P. R. Ortez de Montellano (1990) *The Enzymes*, 3rd ed., **19**, 213.

[7]C. R. Pace-Asciak & W. L. Smith (1983) *The Enzymes*, 3rd ed., **16**, 543.

Selected entries from *Methods in Enzymology* [vol, page(s)]:
General discussion: 86, 45; **163**, 345
Assay: 163, 345

ARACHIDONOYL-CoA SYNTHETASE

This enzyme [EC 6.2.1.15], also known as arachidonate:CoA ligase, catalyzes the reaction of ATP with arachidonate and coenzyme A to produce AMP, pyrophosphate (or, diphosphate), and arachidonoyl-CoA.[1,2] Other substrates include 8,11,14-eicosatrienoate and 5,8,11,14,17-eicosapentaenoate, but not most other long-chain fatty acids. The enzyme is distinct from long-chain-fatty-acyl-CoA synthetase [EC 6.2.1.3].

[1] T. S. Reddy & N. G. Bazan (1983) *ABB* **226**, 125.
[2] D. B. Wilson, S. M. Prescott & P. W. Majerus (1982) *JBC* **257**, 3510.

Selected entries from *Methods in Enzymology* [vol, page(s)]:
General discussion: 187, 237
Assay: 141, 351; 187, 238
Other molecular properties: defective mutant cell line, production, 187, 241; separation from long-chain-fatty-acid:CoA ligase, 187, 240; solubilization, 187, 238

ARALKYLAMINE *N*-ACETYLTRANSFERASE

This acetyltransferase [EC 2.3.1.87], also known as serotonin acetyltransferase and serotonin acetylase, catalyzes the reaction of acetyl-CoA and an aralkylamine to produce coenzyme A and an *N*-acetylaralkylamine.[1] There is a narrow specificity with respect to the acceptor substrate (which includes tryptamine, serotonin, 5-methoxytryptamine, 6-fluorotryptamine, and phenylethylamine).

The enzyme is distinct from arylamine acetyltransferase [EC 2.3.1.5]. This enzyme, which is reported to have essential cysteinyl and histidyl residues at the active site,[2] catalyzes the first step in the synthesis of melatonin from serotonin.[1]

Structural studies indicate that this acetyltransferase catalyzes an ordered Bi Bi reaction.[3] The binding of acetyl-CoA leads to a large conformational change that facilitates serotonin binding. In addition, a water-filled channel leading from the active site to the surface has been identified, suggesting the presence of a specific pathway for proton removal following amine deprotonation. An active-site tyrosyl residue has also been proposed to have a role in the catalytic process.[3] By measuring the enzymatic rates versus increasing buffer microviscosity, investigators have demonstrated that diffusional release of product is most likely the principal rate-determining step.[4] The nucleophilicity of the amine in the substrate is also important in the enzyme-catalyzed acetyl transfer.[4]

[1] P. Voisin, M. A. Namboodiri & D. C. Klein (1984) *JBC* **259**, 10913.
[2] X. Zhan-Poe & C. M. Craft (1999) *J. Pineal Res.* **27**, 49.
[3] A. B. Hickman, M. A. Namboodiri, D. C. Klein & F. Dyda (1999) *Cell* **97**, 361.
[4] E. M. Khalil, J. De Angelis & P. A. Cole (1998) *JBC* **273**, 30321.

Selected entries from *Methods in Enzymology* [vol, page(s)]:
General discussion: 142, 583
Assay: 39, 389

ARALKYLAMINE DEHYDROGENASE

This oxidoreductase [EC 1.4.99.4], also known as aromatic amine dehydrogenase, catalyzes the reaction of a primary amine with water and an acceptor substrate to produce an aldehyde, ammonia, and the reduced acceptor.[1,2] The R group can be an aromatic moiety (for example, dopamine, serotonin, tryptamine, and tyramine) or a long-chain aliphatic group (for example, *n*-octylamine and *n*-hexylamine), which is not as effective a substrate. However, methylamine and ethylamine are not substrates (thus, this enzyme is distinct from amine dehydrogenase [EC 1.4.99.3]). The acceptor substrate can be phenazine methosulfate.

[1] S. M. Cuskey, V. Peccoraro & R. H. Olsen (1987) *J. Bacteriol.* **169**, 2398.
[2] M. Iwaki, T. Yagi, K. Horiike, Y. Saeki, T. Ushijima & M. Nozaki (1983) *ABB* **220**, 253.

Selected entries from *Methods in Enzymology* [vol, page(s)]:
General discussion: 142, 650
Assay: 142, 650
Other molecular properties: absorption properties, 258, 182; aminoquinol formation, 258, 183; dithionite reduction, 258, 182; electron acceptors, 258, 177; properties, 142, 653; purification from *Alcaligenes faecalis*, 142, 651; subunit structure, 258, 178

ARGINASE

This enzyme [EC 3.5.3.1], also known as arginine amidinase and canavanase, catalyzes the hydrolysis of L-arginine resulting in the formation of urea and L-ornithine.[1–7]

The enzyme, which has tightly bound manganese ions as cofactors, also utilizes α-*N*-substituted L-arginines and canavanine as substrates. The rat liver enzyme, which contains a dimanganese(II,II) center per subunit, has a water or hydroxide ion that bridges the two metal ions. This ligand has been proposed to act as the nucleophile in the hydrolysis of L-arginine.[2]

[1] D. E. Ash, J. D. Cox & D. W. Christianson (2000) *Met. Ions Biol. Syst.* **37**, 407.
[2] J. E. Penner-Hahn (1998) *CBC* **3**, 439.
[3] S. V. Khangulov, T. M. Sossong, D. E. Ash & G. C. Dismukes (1998) *Biochemistry* **37**, 8539.
[4] C. P. Jenkinson, W. W. Grody & S. D. Cederbaum (1996) *Comp. Biochem. Physiol. B Biochem. Mol. Biol.* **114**, 107.
[5] C. Reyero & F. Dorner (1975) *EJB* **56**, 137.
[6] S. Ratner (1973) *AE* **39**, 1.
[7] D. M. Greenberg (1960) *The Enzymes*, 2nd ed., **4**, 257.

Selected entries from *Methods in Enzymology* [vol, page(s)]:
General discussion: 2, 368; 17A, 313
Assay: 2, 368; microassay for, 17A, 328; serum, in, 17B, 858
Other molecular properties: aminoacyl tRNAs and, 20, 45; arginine assay, 2, 356, 359, 360, 364; 3, 1044; 17B, 632; clinical aspects, 17B, 857; crude, preparation of, 3, 1045; deficient mutants, *Neurospora crassa*, in analysis of polyamine metabolism, 94, 112; heat and, 2, 370; isolated nuclei and, 12A, 445; liver, 1, 33; 2, 357, 358; microassay for, 17A, 328;

properties, **2**, 373; purification from (horse liver, **2**, 371; rat liver, **17A**, 314); purity, tests for, **2**, 373; reactor, preparation from dialyzer, **137**, 485; renal mitochondria, **55**, 12; seasonal variation of, **2**, 371; serum, in, assay of, **17B**, 858; transamidinase assay and, **5**, 844, 845

D-ARGINASE

This enzyme [EC 3.5.3.10] catalyzes the hydrolysis of D-arginine to produce D-ornithine and urea.[1]

[1] Y. Nadai (1958) *J. Bacteriol.* **45**, 1011.

ARGININE DECARBOXYLASE

This pyridoxal-phosphate-dependent decarboxylase [EC 4.1.1.19] catalyzes the conversion of L-arginine to carbon dioxide and agmatine.[1,2] L-Canavanine is also decarboxylated. There are two distinct isoforms in plants, one of which functions in arginine degradation while the other has a role in the biosynthesis of agmatine and putrescine.

Upon binding of L-arginine, an external Schiff base is formed between the coenzyme and the substrate. Decarboxylation then occurs and agmatine is released. At room temperature, both Schiff base formation and decarboxylation are rate-contributing. As the temperature is lowered, decarboxylation becomes rate-limiting.

[1] T. Hashimoto & Y. Yamada (1994) *Ann. Rev. Plant Physiol. Plant Mol. Biol.* **45**, 257.
[2] E. A. Boeker & E. E. Snell (1972) *The Enzymes*, 3rd ed., **6**, 217.

Selected entries from ***Methods in Enzymology*** [vol, page(s)]:
General discussion: 2, 187; **17B**, 657; **94**, 125, 176
Assay: 17B, 657; **94**, 125
Other molecular properties: L-arginine assay, **3**, 465; conjugate, activity, **44**, 271; end-group analysis, **25**, 120; *Escherichia coli* (biosynthetic and biodegradative forms [assay, **94**, 125; properties, **17B**, 660; **94**, 133; purification, **2**, 187; **3**, 464; **17B**, 658; **94**, 127]; deficient mutants, screening, **94**, 89; gene cloning, **94**, 118); guanidine hydrochloride and, **3**, 1049; isotope effect, **64**, 100; oat seedlings (assays, isotopic and spectrophotometric, **94**, 177; properties, **94**, 180; purification, **94**, 178); properties, **17B**, 660; **94**, 133, 180; purification, from *Escherichia coli*, **2**, 187; **3**, 464; **17B**, 658; **94**, 127; pyridoxal phosphate and, **17B**, 660; resolution of, **2**, 189; spectrum, **17B**, 660; transaminase reactions and, **2**, 171

ARGININE DEIMINASE

This enzyme [EC 3.5.3.6], also known as arginine dihydrolase and arginine desimidase, catalyzes the hydrolysis of L-arginine to produce L-citrulline and ammonia.[1,2] L-Canavanine is also a substrate.

[1] J. Thompson & S. P. F. Miller (1991) *AE* **64**, 317.
[2] M. C. Manca de Nadra, A. A. Pesce de Ruiz Holgado & G. Oliver (1988) *Biochimie* **70**, 367.

Selected entries from ***Methods in Enzymology*** [vol, page(s)]:
General discussion: 2, 374, 375; **17A**, 310; **107**, 624
Assay: 2, 375; **107**, 628

Other molecular properties: citrullinase of, **2**, 374, 376; citrulline formation by, **2**, 374; **3**, 642; distribution, **2**, 376; immune response, effect on, **44**, 705; properties, **2**, 376; **107**, 631; purification, **2**, 375 (*Mycoplasma*, **17A**, 311)

ARGININE KINASE

This phosphotransferase [EC 2.7.3.3] catalyzes the reaction of L-arginine with ATP to produce ADP and *N*-phospho-L-arginine. Other substrates include L-canavanine and L-homoarginine.[1] The enzyme has a random Bi Bi kinetic mechanism. While widely distributed in invertebrates and many lower chordates, arginine kinases are absent in vertebrates.[1]

[1] J. F. Morrison (1973) *The Enzymes*, 3rd ed., **8**, 457.

Selected entries from ***Methods in Enzymology*** [vol, page(s)]:
General discussion: 17A, 330; **46**, 21
Other molecular properties: adenosine 5'-*O*-(1-thiodiphosphate), **87**, 216; adenosine 5'-*O*-(1-thiotriphosphate), **87**, 224; bound reactants and products, equilibrium constant determination, **177**, 362; chromium-nucleotide specificity, **87**, 177; interconversion rates, NMR analysis, **176**, 297; isotope exchange, **64**, 8; MgATP as the substrate, **63**, 258, 292; ^{31}P NMR, simulated and experimental spectra, comparison, **176**, 299; purification from (arthropod muscle, **17A**, 331; *Limulus*, **17A**, 333); screw sense, **87**, 178; specificity, **87**, 216, 224; transition-state and multisubstrate analogues, **249**, 305

ARGININE 2-MONOOXYGENASE

This flavin-dependent oxidoreductase [EC 1.13.12.1] catalyzes the reaction of L-arginine with dioxygen to generate 4-guanidobutanamide, carbon dioxide, and water.[1–4] Other substrates include L-canavanine and L-homoarginine.

[1] S. Yamamoto & Y. Ishimura (1991) *A Study of Enzymes* **2**, 315.
[2] V. Massey & P. Hemmerich (1975) *The Enzymes*, 3rd ed., **12**, 191.
[3] M. S. Flashner & V. Massey (1974) *Mol. Mech. Oxygen Activ.* (O. Hayaishi, ed.), p. 283, Academic Press, New York.
[4] N. Van Thoai & A. Olomucki (1962) *BBA* **59**, 533 and 545.

Selected entries from ***Methods in Enzymology*** [vol, page(s)]:
General discussion: 17A, 335

ARGININE *N*-SUCCINYLTRANSFERASE

argininosuccinate

This transferase [EC 2.3.1.109] catalyzes the reaction of succinyl-CoA with L-arginine to produce Coenzyme A and N^2-succinyl-L-arginine. An alternative substrate is L-ornithine (producing N^2-succinyl-L-ornithine).[1–3]

[1]C. Vander Wauven, A. Jann, D. Haas, T. Leisinger & V. Stalon (1988) *Arch. Microbiol.* **150**, 400.
[2]C. Tricot, V. Stalon & C. Legrain (1991) *J. Gen. Microbiol.* **137**, 2911.
[3]C. V. Wauven & V. Stalon (1985) *J. Bacteriol.* **11**, 882.

ARGININOSUCCINATE LYASE

This lyase [EC 4.3.2.1], also known as arginosuccinase, catalyzes the conversion of *N*-(L-arginino)succinate to fumarate and L-arginine.[1–8] The enzyme has a random Uni Bi kinetic mechanism.[5] A stepwise E1cB mechanism has been proposed in which a transient carbanion intermediate forms.[4]

[1]V. E. Anderson (1998) *CBC* **2**, 115.
[2]J. V. Schloss & M. S. Hixon (1998) *CBC* **2**, 43.
[3]L. S. Mullins & F. M. Raushel (1995) *MIE* **249**, 398.
[4]S. C. Kim & F. M. Raushel (1986) *Biochemistry* **25**, 4744.
[5]F. M. Raushel & R. Nygaard (1983) *ABB* **221**, 143.
[6]H. J. Vogel & R. H. Vogel (1974) *AE* **40**, 65.
[7]S. Ratner (1973) *AE* **39**, 1.
[8]S. Ratner (1972) *The Enzymes*, 3rd ed., **7**, 167.

Selected entries from *Methods in Enzymology* [**vol**, page(s)]:
General discussion: 17A, 304
Assay: coupling assay, **63**, 33; microassay, **17A**, 326
Other molecular properties: cold lability of, **17A**, 309; localization in gels by simultaneous capture technique, **104**, 436; microassay, **17A**, 326; positional isotope exchange studies, **249**, 413; purification from steer liver, **17A**, 306; synchronous cultures and, **21**, 469

ARGININOSUCCINATE SYNTHETASE

This enzyme [EC 6.3.4.5], also known as citrulline:aspartate ligase, catalyzes the reaction of L-citrulline with L-aspartate and ATP to produce *N*-(L-arginino)succinate, AMP, and pyrophosphate (or, diphosphate).[1–6]

[1]F. M. Raushel & J. J. Villafranca (1988) *Crit. Rev. Biochem.* **23**, 1.
[2]C. Ghose & F. M. Raushel (1985) *Biochemistry* **24**, 5894.
[3]F. M. Raushel & J. L. Seiglie (1983) *ABB* **225**, 979.
[4]F. Hilger, J. P. Simon & V. Stalon (1979) *EJB* **94**, 153.
[5]H. J. Vogel & R. H. Vogel (1974) *AE* **40**, 65.
[6]S. Ratner (1973) *AE* **39**, 1.

Selected entries from *Methods in Enzymology* [**vol**, page(s)]:
General discussion: 17A, 299
Assay: coupling assay, **63**, 34; microassay, **17A**, 325
Other molecular properties: purification from steer liver, **17A**, 300

ARGINYLTRANSFERASE

This transferase [EC 2.3.2.8], also known as arginyl-tRNA:protein transferase, catalyzes the reaction of L-arginyl-tRNA with a protein to produce tRNA and an L-arginyl-protein.[1,2] The protein or polypeptide acceptor substrate must have an N-terminal L-glutamyl, L-aspartyl, or L-cystinyl residue (di- and tripeptides can also serve as acceptor substrates provided they have these same N-terminal specificities). Examples of protein acceptor substrates include bovine serum albumin and bovine thyroglobulin. The reaction requires the presence of a sulfhydryl reagent such as 2-mercaptoethanol as well as a univalent cation.

[1]R. L. Soffer (1973) *JBC* **248**, 2918.
[2]A. Ciechanover, S. Ferber, D. Ganoth, S. Elias, H. Avram & S. Arfin (1988) *JBC* **263**, 11155.

Selected entries from *Methods in Enzymology* [**vol**, page(s)]:
General discussion: 106, 198
Assay: 106, 198
Other molecular properties: deficient mutants, characterization, **106**, 204; purification from rabbit liver, **106**, 201; substrate identification, **106**, 202

ARGINYL-tRNA SYNTHETASE

This enzyme [EC 6.1.1.19], also known as arginine:tRNA ligase and arginine translase, catalyzes the reaction of ATP with L-arginine and tRNA^{Arg} to produce AMP, pyrophosphate (or, diphosphate), and L-arginyl-tRNA^{Arg}. This class I aminoacyl-tRNA synthetase has a random Ter Ter kinetic mechanism,[1–4] different than that of most other aminoacyl-tRNA synthetases. ***See also*** *Aminoacyl-tRNA Synthetases*

[1]S. Char & K. P. Gopinathan (1986) *J. Biochem.* **100**, 349.
[2]H. Y. Wang & F. Pan (1984) *Int. J. Biochem.* **16**, 1379.
[3]R. Thiebe (1983) *EJB* **130**, 517.
[4]J. Charlier & E. Gerlo (1979) *Biochemistry* **18**, 3171.

Selected entries from *Methods in Enzymology* [**vol**, page(s)]:
General discussion: 34, 170
Other molecular properties: binding constants, **29**, 640; isotope exchange, **64**, 9; mechanism, **29**, 630, 633; subcellular distribution, **59**, 233, 234

ARISTOLOCHENE SYNTHASE

This enzyme [EC 4.2.3.9; previously classified as EC 4.1.99.7 and EC 2.5.1.40], also known as sesquiterpene cyclase, catalyzes the conversion of *trans,trans*-farnesyl diphosphate to aristolochene and pyrophosphate (or, diphosphate).[1,2]

farnesyl diphosphate　　　germacrene A　　　aristolochene

The initial internal cyclization of the farnesyl group at the C10 position produces the monocyclic intermediate germacrene A; further cyclization and methyl transfer steps converts the intermediate into aristolochene. The rate-limiting step in the reaction occurs after the initial chemical step involving rupture of the carbon–oxygen bond in farnesyl diphosphate in the formation of germacrene A.[1]

[1]J. R. Mathis, K. Back, C. Starks, J. Noel, C. D. Poulter & J. Chappell (1997) *Biochemistry* **36**, 8340.

[2]D. E. Cane, P. C. Prabhakaran, J. S. Oliver & D. B. McIlwaine (1990) *JACS* **112**, 3209.

AROMATIC-AMINO-ACID AMINOTRANSFERASE

This pyridoxal-phosphate-dependent aminotransferase [EC 2.6.1.57], also known as aromatic amino acid transferase, catalyzes the reversible reaction of an aromatic amino acid with α-ketoglutarate (*i.e.*, 2-oxoglutarate) to generate an aromatic oxo acid and L-glutamate.[1–5] Donor substrates include L-phenylalanine, L-tryptophan, L-tyrosine, 5-hydroxy-L-tryptophan, 3-iodo-L-tyrosine, L-histidine, and 1-carboxy-4-hydroxy-2,5-cyclohexadien-1-alanine. L-Methionine can also act as a weak donor substrate. Oxaloacetate can substitute for α-ketoglutarate. Partial proteolysis results in a protein exhibiting the activities of aspartate aminotransferase [EC 2.6.1.1].

The *Pyrococcus horikoshii* enzyme, which is remarkably thermostable with a melting temperature of 120°C, has unusual specificities in that L-glutamate is an effective amino donor.[5] *See also* Aromatic-Amino-Acid:Glyoxylate Aminotransferase

[1]J. DiRuggiero & F. T. Robb (1996) *Adv. Protein Chem.* **48**, 311.

[2]G. Andreotti, M. V. Cubellis, G. Nitti, G. Sannia, X. Mai, G. Marino & M. W. W. Adams (1994) *EJB* **220**, 543.

[3]H. G. Beschle, R. Süssmuth & F. Lingens (1982) *Hoppe-Seyler's Z. Physiol. Chem.* **363**, 1365.

[4]D. A. Weigent & E. W. Nester (1976) *JBC* **251**, 6974.

[5]I. Matsui, E. Matsui, Y. Sakai, H. Kikuchi, Y. Kawarabayasi, H. Ura, S. Kawaguchi, S. Kuramitsu & K. Harata (2000) *JBC* **275**, 4871.

Selected entries from *Methods in Enzymology* [vol, page(s)]:
General discussion: 113, 73; 142, 253, 267
Assay: 113, 74; 142, 256, 267
Other molecular properties: catalyzed reactions, 113, 73; 142, 255; limited proteolysis, 142, 264; natural occurrence, 113, 73; properties, 142, 261, 265, 270; purification from (*Escherichia coli*, 142, 258; rat kidney, 113, 74; rat small intestine, 142, 268)

AROMATIC-L-AMINO-ACID DECARBOXYLASE

This pyridoxal-phosphate-dependent decarboxylase [EC 4.1.1.28] (also known as DOPA decarboxylase, tryptophan decarboxylase, and hydroxytryptophan decarboxylase) catalyzes the conversion of L-tryptophan to tryptamine and carbon dioxide.[1–7] The enzyme can also utilize 5-hydroxy-tryptophan and dihydroxyphenylalanine as substrates.

Transient and steady-state investigations have demonstrated that this decarboxylase preferentially binds of DOPA with its amino group unprotonated.[1] This is due, in part, to the high pK_a value of the internal aldimine between the cofactor and the ε-amino group of a lysyl residue. The pK_a in aspartate aminotransferase is low at 6.8 and the form of aspartate that binds has a protonated amino group. However, the pK_a of this aldimine is high in aromatic-amino-acid aminotransferase. In addition, the pK_a of the amino group of DOPA is decreased upon formation of the Michaelis complex, promoting generation of the species that undergoes transaldimination to form the external aldimine complex.[1] *See also* Tyrosine Decarboxylase; Phenylalanine Decarboxylase; Tryptophan 2-Monooxygenase

[1]H. Hayashi, F. Tsukiyama, S. Ishii, H. Mizuguchi & H. Kagamiyam (1999) *Biochemistry* 38, 15615.

[2]M. D. Berry, A. V. Juorio, X. M. Li & A. A. Boulton (1996) *Neurochem. Res.* **21**, 1075.

[3]M. Y. Zhu & A. V. Juorio (1995) *Gen. Pharmacol.* **26**, 681.

[4]T. Hashimoto & Y. Yamada (1994) *Ann. Rev. Plant Physiol. Plant Mol. Biol.* **45**, 257.

[5]M. A. Ator & P. R. Ortez de Montellano (1990) *The Enzymes*, 3rd ed., **19**, 213.

[6]M. J. Jung (1986) *Bioorg. Chem.* **14**, 429.

[7]E. A. Boeker & E. E. Snell (1972) *The Enzymes*, 3rd ed., **6**, 217.

Selected entries from *Methods in Enzymology* [vol, page(s)]:
General discussion: 2, 195; 17B, 652; 142, 170, 179
Assay: 17B, 652; 142, 171, 180; 3,4-dihydroxyphenylalanine, 2, 195
Other molecular properties: fetal tissues, in, 17B, 673; histidine decarboxylase and, 17B, 676; inhibition by NSD-1039, 17A, 610; inhibitor, 63, 397, 400; mastocytoma, in, 17B, 670; neurobiological studies, in, 32, 787, 788; preparation from bovine brainstem, 17A, 453; properties, 2, 197; 17B, 654; 142, 177, 183; purification from (guinea pig kidney, 2, 196; 17B, 653; porcine kidney, 142, 176, 182); pyridoxal phosphate and, 17A, 455; rat tissues, in, 17B, 673; sources, 2, 195, 196, 199

AROMATIC-AMINO-ACID:GLYOXYLATE AMINOTRANSFERASE

This pyridoxal-phosphate-dependent aminotransferase [EC 2.6.1.60] catalyzes the reversible reaction of an aromatic amino acid with glyoxylate to produce an aromatic oxo acid and glycine.[1] Aromatic amino acids that can act as substrates include L-phenylalanine, L-kynurenine, L-tyrosine, and L-histidine. Pyruvate and hydroxypyruvate can also act as acceptor substrates. *See also* Aromatic-Amino-Acid Aminotransferase

[1]I. Harada, T. Noguchi & R. Kido (1978) *Hoppe-Seyler's Z. Physiol. Chem.* **359**, 481.

Selected entries from *Methods in Enzymology* [vol, page(s)]:
General discussion: 142, 273
Assay: 142, 274
Other molecular properties: properties, 142, 275; purification from rat liver, 142, 275

AROMATIC-HYDROXYLAMINE *O*-ACETYLTRANSFERASE

This acetyltransferase [EC 2.3.1.56] catalyzes the reaction of *N*-hydroxy-4-acetylaminobiphenyl with *N*-hydroxy-4-aminobiphenyl to produce *N*-hydroxy-4-aminobiphenyl and *N*-acetoxy-4-aminobiphenyl: hence, it transfers the *N*-acetyl group of some aromatic acethydroxamates to the *O*-position of some aromatic hydroxylamines.[1]

[1] W. T. Allaben & C. M. King (1984) *JBC* **259**, 12128.

arom MULTIFUNCTIONAL ENZYME

This multifunctional enzyme complex, found in *Neurospora crassa* and some other fungi and in *Euglena gracilis*, catalyzes several steps in the biosynthesis of aromatic amino acids.[1] The five enzyme activities present in this polypeptide, often known as pentafunctional *arom* polypeptide, are 3-dehydroquinate synthase [EC 4.6.1.3], 3-dehydroquinate dehydratase [EC 4.2.1.10], shikimate dehydrogenase [EC 1.1.1.25], shikimate kinase [EC 2.7.1.71], and 5-enolpyruvoylshikimate phosphate synthase [EC 2.5.1.19; also called 3-phosphoshikimate 1-carboxyvinyltransferase]. *See individual enzyme*

[1] J. W. Jacobson, B. A. Hart, C. H. Doy & N. H. Giles (1972) *BBA* **289**, 1.

Selected entries from *Methods in Enzymology* [vol, page(s)]:
Assay: 142, 333
Other molecular properties: catalyzed reactions, 142, 327; coupling enzymes, 142, 328; properties, 142, 338; purification from *Neurospora crassa*, 142, 337; substrates, 142, 328

ARSENATE REDUCTASE (DONOR)

This oxidoreductase [EC 1.97.1.6] catalyzes the reaction of arsenate (As(V) or $H_2AsO_4^-$) with a donor substrate to produce arsenite (As(III) or $H_2AsO_3^-$) and the oxidized donor.[1] Reduced benzylviologen can serve as the reducing agent: however, reduced glutaredoxin cannot (thus distinguishing this enzyme from arsenate reductase (glutaredoxin) [EC 1.97.1.5]).[2]

[1] T. Krafft & J. M. Macy (1998) *EJB* **255**, 647.
[2] T. R. Radabaugh & H. V. Aposhian (2000) *Chem. Res. Toxicol.* **13**, 26.

ARSENATE REDUCTASE (GLUTAREDOXIN)

This molybdenum-dependent oxidoreductase [EC 1.97.1.5] catalyzes the reaction of arsenate with reduced glutaredoxin to yield arsenite and oxidized glutaredoxin.[1–7] Glutathione is required to regenerate reduced glutaredoxin. Arsenate reductase participates in pathways for removal of toxic arsenate (even though arsenite is more toxic than arsenate, some microorganisms utilize an arsenite-transporting ATPase [EC 3.6.3.16] for the release of the toxic metabolite) or in methylation reactions in the biosynthesis of organoarsenicals. *See also Arsenate Reductase (Donor); Glutaredoxin*

[1] T. Gladysheva, J. Y. Liu & B. P. Rosen (1996) *JBC* **271**, 33256.
[2] T. B. Gladysheva, K. L. Oden & B. P. Rosen (1994) *Biochemistry* **33**, 7288.
[3] G. Y. Ji, E. A. E. Garber, L. G. Armes, C. M. Chen, J. A. Fuchs & S. Silver (1994) *Biochemistry* **33**, 7294.
[4] T. Krafft & J. M. Macy (1998) *EJB* **255**, 647.
[5] J. Messens, G. Hayburn, A. Desmyter, G. Laus & L. Wyns (1999) *Biochemistry* **38**, 16857.
[6] T. R. Radabaugh & H. V. Aposhian (2000) *Chem. Res. Toxicol.* **13**, 26.
[7] J. Shi, V. Vlamis-Gardikas, F. Aslund, A. Holmgren & B. P. Rosen (1999) *JBC* **274**, 36039.

ARSENITE METHYLTRANSFERASE

This methyltransferase [EC 2.1.1.137] catalyzes the reaction of *S*-adenosyl-L-methionine with arsenite ($H_2AsO_3^-$) to produce *S*-adenosyl-L-homocysteine and methylarsonate ($CH_3AsO_3H^-$).[1–3]

[1] R. A. Zakharyan & H. V. Aposhian (1999) *Chem. Res. Toxicol.* **121**, 1278.
[2] R. A. Zakharyan, F. Ayala-Fierro, W. R. Cullen, D. M. Carter & H. V. Aposhian (1999) *Toxicol. Appl. Pharmacol.* **158**, 9.
[3] R. A. Zakharyan, Y. Wu, G. M. Bogdan & H. V. Aposhian (1995) *Chem. Res. Toxicol.* **8**, 1029.

ARSENITE-TRANSPORTING ATPase

This so-called ATPase [EC 3.6.3.16] catalyzes the ATP-dependent (ADP- and orthophosphate-producing) transport of $arsenite_{in}$ to produce $arsenite_{out}$.[1,2] The bacterial enzyme often contains two large subunits where one forms the channel in the membrane and the other contains the ATP-binding site. In addition to arsenite (AsO_2^-), the *Escherichia coli* system will also transport antimonite (SbO_2^-).

The idea that a transporter is an enzyme is in keeping with a new definition of enzyme catalysis as the facilitated making/breaking of chemical bonds, not just covalent bonds.[3] This idea builds on Pauling's assertion that any long-lived, chemically distinct interaction (in this case, the persistent location of a solute with respect to the faces of a membrane) can be regarded as a chemical bond. Note also that the equilibrium constant ($K_{eq} = [Arsenite_{out}][ADP][P_i]/[Arsenite_{in}][ATP]$) does not conform to that expected for an ATPase (*i.e.*, $K_{eq} = [ADP][P_i]/[ATP]$). Thus, although the overall reaction yields ADP and orthophosphate, the enzyme is misclassified as a hydrolase, and instead should be regarded as an energase-type reaction. Energases facilitate affinity-modulated reactions by coupling the $\Delta G_{ATP\text{-hydrolysis}}$ to a force-generating or work-producing step.[3] In this case, P-O-P bond scission supplies the energy to drive ion transport. *See ATPase; Energase*

[1] B. P. Rosen, U. Weigel, R. A. Monticello & B. P. Edwards (1991) *Arch. Biochem. Biophys.* **284**, 381.
[2] T. Zhou, B. P. Rosen & D. L. Gatti (1999) *Acta Crystallogr. D Biol. Crystallogr.* **55**, 921.
[3] D. L. Purich (2001) *TiBS* **26**, 417.

ARYLACETONITRILASE

This thiol-dependent enzyme [EC 3.5.5.5] catalyzes the hydrolysis of 4-chlorophenylacetonitrile with two water molecules to produce 4-chlorophenylacetate and ammonia. Other substrates include 4-substituted phenylacetonitriles (*e.g.*, 4-fluorobenzylcyanide, 4-bromobenzylcyanide, 4-iodobenzylcyanide), thien-2-ylacetonitrile (*e.g.*, 2-thiopheneacetonitrile), and tolylacetonitriles (*e.g.*, *p*-tolylacetonitrile).[1] *See also Nitrilase*

[1] T. Nagasawa, J. Mauger & H. Yamada (1990) *EJB* **194**, 765.

ARYL ACYLAMIDASE

This enzyme [EC 3.5.1.13] catalyzes the hydrolysis of an anilide (for example, *N*-acetyl-*o*-toluidine) to produce a fatty acid anion and aniline.[1–4] Other substrates include 4-substituted anilides. Cholinesterases can exhibit aryl acylamidase activity. Multiple forms of the acylamidase have been identified in brain that can be differentiated by the effects of several classes of drugs such as serotonin, tetrahydro-β-carbolines, and lysergic acid diethylamide.[1]

[1] L. L. Hsu (1982) *Int. J. Biochem.* **14**, 1037.
[2] R. E. Hoagland (1975) *Phytochemistry* **14**, 383.
[3] R. E. Hoagland & G. Graf (1974) *Can. J. Biochem.* **52**, 903.
[4] G. Engelhardt & P. R. Wallnöfer (1973) *Appl. Microbiol.* **26**, 709.

ARYL-ALCOHOL DEHYDROGENASES

The aryl-alcohol dehydrogenases are oxidoreductases that catalyze the conversion of an aromatic alcohol to an aromatic aldehyde.

The NAD$^+$-dependent enzyme [EC 1.1.1.90], also known as *p*-hydroxybenzyl alcohol dehydrogenase and benzyl alcohol dehydrogenase, catalyzes the reversible reaction of an aromatic alcohol with NAD$^+$ to produce an aromatic aldehyde and NADH.[1–3] This classification actually represents an entire group of enzymes with broad specificity toward primary alcohols with an aromatic or cyclohex-1-ene ring, but with low or no activity toward short-chain aliphatic alcohols. Substrates include benzyl alcohol, 3,4-dimethoxybenzyl alcohol, 4-hydroxybenzyl alcohol, cinnamyl alcohol, 4-isopropylbenzyl alcohol, and 2-aminobenzyl alcohol.

Aryl-alcohol dehydrogenase (NADP$^+$) [EC 1.1.1.91] catalyzes the reversible reaction of an aromatic alcohol with NADP$^+$ to produce an aromatic aldehyde and NADPH.[4,5] Other substrates include some aliphatic alcohols (for example, 1-butanol and 1-pentanol); nevertheless, cinnamyl alcohol was the best substrate identified. Other substrates include benzyl alcohol, 4-methoxybenzyl alcohol, 4-hydroxybenzyl alcohol, and 3,4-dimethoxybenzyl alcohol.

[1] J. P. Shaw & S. Harayama (1990) *EJB* **191**, 705.
[2] R. M. Chalmers, A. J. Scott & C. A. Fewson (1990) *J. Gen. Microbiol.* **136**, 637.
[3] R. W. MacKintosh & C. A. Fewson (1988) *BJ* **255**, 653 and **250**, 743.
[4] D. D. Davies, E. N. Ugochukwu, K. D. Patil & G. H. N. Towers (1973) *Phytochemistry* **12**, 531.
[5] G. G. Gross & M. H. Zenk (1969) *EJB* **8**, 420.

ARYL-ALCOHOL OXIDASE

This oxidoreductase [EC 1.1.3.7], also known as veratryl alcohol oxidase and β-naphthylcarbinyl oxidase, catalyzes the reaction of an aromatic primary alcohol with dioxygen to produce an aromatic aldehyde and hydrogen peroxide.[1–3] Alcohol substrates include (2-naphthyl)-methanol, 3-methoxybenzyl alcohol, veratryl alcohol (or, 3,4-dimethoxybenzyl alcohol), benzyl alcohol, cinnamyl alcohol (or, 3-phenyl-2-propen-1-ol), and anisyl alcohol (or, 4-methoxybenzyl alcohol).

[1] F. Guillen, A. T. Martinez & M. J. Martinez (1992) *EJB* **209**, 603.
[2] F. Guillen, A. T. Martinez & M. J. Martinez (1990) *Appl. Microbiol. Biotechnol.* **32**, 465.
[3] R. Bourbonnais & M. G. Paice (1988) *BJ* **255**, 445.

Selected entries from *Methods in Enzymology* [**vol**, page(s)]: **General discussion: 52**, 18; **56**, 474

ARYL-ALDEHYDE DEHYDROGENASES

The NAD$^+$-dependent enzyme [EC 1.2.1.29] catalyzes the reaction of an aromatic aldehyde with NAD$^+$ and water to produce an aromatic acid and NADH.[1] Substrates include a number of aromatic aldehydes; however, aliphatic aldehydes are not substrates.

Aryl-aldehyde dehydrogenase (NADP$^+$) [EC 1.2.1.30] catalyzes the reversible reaction of an aromatic acid with NADPH and ATP to produce an aromatic aldehyde, NADP$^+$, AMP, pyrophosphate (or, diphosphate), and water.[2,3] An acyl-adenylate intermediate has been suggested.[2]

[1] J. K. Raison, G. Henson & K. G. Rienits (1966) *BBA* **118**, 285.
[2] G. G. Gross (1972) *EJB* **31**, 585 and (1971) *FEBS Lett.* **17**, 309.
[3] G. G. Gross & M. H. Zenk (1969) *EJB* **8**, 413.

ARYL-ALDEHYDE OXIDASE

This oxidoreductase [EC 1.2.3.9], first isolated from *Streptomyces viridosporus*, catalyzes the reaction of an aromatic aldehyde with dioxygen to produce an aromatic acid anion and hydrogen peroxide.[1] Substrates include benzaldehyde, vanillin, terephthalaldehyde, veratraldehyde, salicylaldehyde, isophthalaldehyde, and *m*-hydroxybenzaldehyde. Aliphatic aldehydes and aldoses are not substrates.

[1] D. L. Crawford, J. B. Sutherland, A. L. Pometto, III, & J. M. Miller (1982) *Arch. Microbiol.* **131**, 351.

ARYLALKYL ACYLAMIDASE

This enzyme [EC 3.5.1.76], also known as aralkyl acylamidase, catalyzes the hydrolysis of an *N*-acetylarylalkylamine to produce an arylalkylamine and acetate.[1] Substrates include *N*-acetyl-2-phenylethylamine, *N*-acetyl-3-phenylpropylamine, *N*-acetyldopamine, *N*-acetyl-serotonin, and melatonin. This enzyme will not act on acetanilide derivatives, which are substrates for aryl acylamidase [EC 3.5.1.13].

This enzyme can be used to prepare optically active α-methyl arylalkylamines and α-methyl arylalkylalcohols through enantioselective hydrolysis of their racemic amides and esters.[2]

[1] S. Shimizu, J. Ogawa, M. C. Chung & H. Yamada (1992) *EJB* **209**, 375.
[2] J. Ogawa, S. Shimizu & H. Yamada (1994) *Bioorg. Med. Chem.* **2**, 429.

ARYLALKYLAMINE N-ACETYLTRANSFERASE

This enzyme [EC 2.3.1.87], also known as serotonin acetyltransferase and serotonin acetylase, catalyzes the reaction of acetyl-CoA and an arylalkylamine to produce coenzyme A and an *N*-acetylarylalkylamine.[1-4] The enzyme, which is distinct from arylamine acetyltransferase [EC 2.3.1.5], exhibits a rather narrow specificity toward other arylalkylamines. Substrates include serotonin, phenylethylamine, 6-fluorotryptamine, tryptamine, and 3-methoxytryptamine.

[1] N. Fajardo, P. Abreu & R. Alonso (1992) *J. Pineal Res.* **13**, 80.
[2] T. J. Smith (1990) *Bioessays* **12**, 30.
[3] P. Voisin, M. A. A. Namboodiri & D. C. Klein (1984) *JBC* **259**, 10913.

[4] J. J. Morrissey, S. B. Edwards & W. Lovenberg (1977) *BBRC* **77**, 118.
Selected entries from *Methods in Enzymology* [vol, page(s)]:
General discussion: 142, 583
Assay: 142, 584

ARYLAMINE N-ACETYLTRANSFERASE

This acetyltransferase [EC 2.3.1.5], also known as arylamine acetylase and acetyl-CoA:amine acetyltransferase, catalyzes the reaction of acetyl-CoA with an arylamine to produce free coenzyme A and an *N*-acetylarylamine.[1-9] The enzyme has a wide specificity for aromatic amines, including serotonin (other substrates include 2-aminobenzoate, 3-aminobenzoate, and 4-methylaniline). In fact, the *Hansenula ciferri* enzyme can utilize water-soluble alkyl amines (for example, D-glucosamine, histamine, and tryptamine). Interestingly, the enzyme also reportedly catalyzes the exchange of an acetyl group between different arylamines. **See also** *Aralkylamine N-Acetyltransferase*

[1] I. Cascorbi, J. Brockmoller, P. M. Mrozikiewicz, A. Muller & I. Roots (1999) *Drug Metab. Rev.* **31**, 489.
[2] L. W. Wormhoudt, J. N. Commandeur & N. P. Vermeulen (1999) *Crit. Rev. Toxicol.* **29**, 59.
[3] C. M. King, S. J. Land, R. F. Jones, M. Debiec-Rychter, M. S. Lee & C. Y. Wang (1997) *Mutat. Res.* **376**, 123.
[4] M. Watanabe, T. Igarashi, T. Kaminuma, T. Sofuni & T. Nohmi (1994) *Environ. Health Perspect.* **102**, Suppl. 6, 83.
[5] E. Sim, D. Hickman, E. Coroneos & S. L. Kelly (1992) *Biochem. Soc. Trans.* **20**, 304
[6] D. W. Hein (1988) *BBA* **948**, 37.
[7] H. H. Andres, A. J. Klem, L. M. Schopfer, J. K. Harrison & W. W. Weber (1988) *JBC* **263**, 7521.
[8] H. H. Andres, H. J. Kolb, R. J. Schreiber & L. Weiss (1983) *BBA* **746**, 193.
[9] J. J. Morrissey, S. B. Edwards & W. Lovenberg (1977) *BBRC* **77**, 118.
Selected entries from *Methods in Enzymology* [vol, page(s)]:
General discussion: 17B, 805; 35, 247; 77, 263, 272
Assay: 17B, 805; 35, 247; 77, 263, 264, 273, 274, 275; caffeine assay, **272**, 124, 130
Other molecular properties: acetylation, **77**, 279; affinity chromatography, **77**, 265; assay of (acetyl-CoA, **13**, 546; **63**, 33; acetyl-CoA synthetase, **63**, 33; ATP-citrate lyase, **63**, 33; coenzyme A, **13**, 539); coupling enzyme, as a, **63**, 33; definition of unit, **77**, 264; distribution, **77**, 279; *Drosophila melanogaster*, purification, **103**, 489; genetics, **77**, 279; glucosamine-6-phosphate acetylase, presence in preparation of, **9**, 707; inhibitor, **17B**, 811; **35**, 253; **77**, 280; intestinal epithelium, in, **77**, 156; isoelectric point, **77**, 266; molecular weight, **77**, 266; mutagen formation, **77**, 280; pH optimum, **17B**, 811; **35**, 252; **77**, 266, 280; properties, **17B**, 810; **35**, 252; **77**, 266, 279, 280; purification, **17B**, 807; **35**, 250; **77**, 264, 277 (rabbit liver, from, **17B**, 807); rat liver, purification, **103**, 489; serotonin, radioenzymatic assay for, **103**, 491; size, **77**, 266; solubilization, **77**, 277; specificity, **17B**, 810; **35**, 253; **77**, 266; stability, **17B**, 810; **35**, 252; substrates, **77**, 266, 279, 280

ARYLAMINE GLUCOSYLTRANSFERASE

This glucosyltransferase [EC 2.4.1.71], also known as UDP-glucose:arylamine glucosyltransferase, catalyzes the reaction of UDP-D-glucose with an arylamine to produce

UDP and an N-D-glucosylarylamine.[1] Acceptor substrates include 2,3-dichloroaniline, 3-amino-2,5-dichlorobenzoate, 5-amino-2,3-dichlorobenzoate, 3-aminobenzoate, 2,5-dibromoaniline, and especially 3,4-dichloroaniline. Donor substrates also include TDP-D-glucose.

[1] D. S. Frear (1968) *Phytochemistry* **7**, 381.

ARYLAMINE SULFOTRANSFERASE

This sulfotransferase [EC 2.8.2.3], also known as amine sulfotransferase, catalyzes the reaction of 3′-phospho-adenylylsulfate with an arylamine to produce adenosine 3′,5′-bisphosphate and an arylsulfamate.[1–3] Amine substrates include aniline, 2-naphthylamine, cyclohexylamine, and octylamine. The human brain enzyme acts on dopamine, m-tyramine, norepinephrine, and p-tyramine.[2]

[1] S. G. Ramaswamy & W. B. Jakoby (1987) *JBC* **262**, 10039.
[2] P. H. Yu, B. Rozdilsky & A. A. Boulton (1985) *J. Neurochem.* **45**, 836.
[3] J. S. Hernandez, S. P. Powers & R. M. Weinshilboum (1991) *Drug Metab. Dispos.* **19**, 1071.

Selected entries from *Methods in Enzymology* [vol, page(s)]:
General discussion: 143, 201
Assay: radiochemical assay, 143, 205
Other molecular properties: ion-pair extraction, 143, 203; products of, 5, 983; thin-layer chromatography, 143, 203

ARYLDIALKYLPHOSPHATASE

This divalent-cation-dependent metalloenzyme [EC 3.1.8.1; it was previously regarded as a variant of arylesterase[1], EC 3.1.1.2] (also known as paraoxonase, A-esterase, aryl-triphosphatase, phosphotriesterase, and paraoxon hydrolase) catalyzes the hydrolysis of an aryl dialkyl phosphate to produce a dialkyl phosphate and an aryl alcohol.[1–5] Substrates include a number of organophosphorus compounds including esters of phosphonic and phosphinic acids: such as, paraoxon (O,O-diethyl-p-nitrophenylphosphate), organophosphorus insecticides, pirimiphos-methyloxon, Soman, Tabun, diazoxon, and diisopropylfluorophosphate. It is inhibited by chelating agents. A tryptophanyl and either an aspartyl or a glutamyl residue are at the active site of the human plasma isozyme.[4]

paraoxon

The active site of *Pseudomonas diminuta* phosphotri-esterase contains two zinc ions, the two cations being bridged by a carbamoyl group formed from a lysyl residue (the metal ions can be replaced with Cd^{2+}, Co^{2+}, Ni^{2+}, or Mn^{2+}). In addition, a hydroxide ion is either bonded to one zinc ion or bridges both metal ions. The best known substrate is the pesticide paraoxon, with rate accelerations approaching 10^{12}. The reaction proceeds with stereochemical inversion at the phosphorus. Brønsted plots and isotope-effect investigations shows that hydrolysis of the substrate is almost diffusion limited. The phosphoryl bond order in the transition state is almost that of the product.[5]

[1] M. I. Mackness, H. M. Thompson, A. R. Hardy & C. H. Walker (1987) *BJ* **245**, 293.
[2] F. Gil, A. Pla, M. C. Gonzalvo, A. F. Hernandez & E. Villanueva (1993) *Chem. Biol. Interact.* **87**, 149.
[3] F. Gil, M. C. Gonzalvo, A. F. Hernandez, E. Villanueva & A. Pla (1994) *Biochem. Pharmacol.* **48**, 1559.
[4] D. Josse, W. Xie, F. Renault, D. Rochu, L. M. Schopfer, P. Masson & O. Lockridge (1999) *Biochemistry* **38**, 2816.
[5] N. H. Williams (1998) *CBC* **1**, 557.

ARYLESTERASE

This class of esterases [EC 3.1.1.2] catalyzes the hydrolysis of a phenyl acetate to produce a phenol and acetate. The enzyme acts on many phenolic esters: examples include 2-naphthyl acetate, phenyl acetate, and p-nitrophenyl butyrate. Note that aryldialkylphosphatase, also called paraoxonase, used to be listed as a variant of arylesterase.[1,2]

See also *Aryldialkylphosphatase*

[1] J. Debord, T. Dantoine, J.-C. Bollinger, M. H. Abraham, B. Verneuil & L. Merle (1998) *Chem.-Biol. Interact.* **113**, 105.
[2] H. B. Bosmann (1972) *BBA* **276**, 180.

Selected entries from *Methods in Enzymology* [vol, page(s)]:
General discussion: 160, 551
Other molecular properties: gel electrophoresis, 32, 88; intestinal epithelium, in, 77, 157

ARYLFORMAMIDASE

This enzyme [EC 3.5.1.9], also called kynurenine formamidase and formylkynureninase, catalyzes the hydrolysis of N-formyl-L-kynurenine to produce formate and L-kynurenine.[1–3] The enzyme will also use other aromatic formylamines as substrates.

[1] D. Brown, M. J. Hitchcock & E. Katz (1986) *Can. J. Microbiol.* **32**, 465.
[2] E. M. Gal & A. D. Sherman (1980) *Neurochem. Res.* **5**, 223.
[3] C. G. Bailey & C. Wagner (1974) *JBC* **249**, 4439.

Selected entries from *Methods in Enzymology* [vol, page(s)]:
General discussion: 2, 246; 142, 225
Assay: 2, 246; 142, 226
Other molecular properties: assay of L-tryptophan 2,3-dioxygenase, 17A, 416, 418, 421; properties, 2, 248; 142, 229; *Pseudomonas fluorescens*, in, 17A, 430; purification from (rat liver, 2, 247; *Streptomyces parvulus*, 142, 227); rabbit intestine, in, 17A, 435; sources, 2, 253

ARYLMALONATE DECARBOXYLASE

This decarboxylase [EC 4.1.1.76] catalyzes the conversion of a 2-aryl-2-methylmalonate to a 2-arylpropionate and carbon dioxide.[1] This enzyme is not inhibited by avidin.

[1]K. Miyamoto & H. Ohta (1992) *EJB* **210**, 475.

2-ARYLPROPIONYL-CoA EPIMERASE

This epimerase catalyzes the interconversion of (2*R*)-arylpropionyl-CoA and (2*S*)-arylpropionyl-CoA. Both cytosolic and mitochondrial variants of this enzyme have been identified.[1,2] Substrates include the CoA-derivative of ibuprofen. In the presence of D_2O, a deuterium atom is incorporated in the C2 position.

[1]C.-S. Chen, W.-R. Shieh, P.-H. Lu, S. Harriman & C.-Y. Chen (1991) *BBA* **1078**, 411.
[2]W.-R. Shieh & C.-S. Chen (1993) *JBC* **268**, 3487.

ARYLPYRUVATE KETO-ENOL TAUTOMERASE

See *Phenylpyruvate Tautomerase*

ARYLSULFATASES

Arylsulfatases are a subset of enzymes that catalyze the hydrolysis of a phenol sulfate to produce a phenol and sulfate.

Arylsulfatase [EC 3.1.6.1], also known simply as sulfatase and aryl-sulfate sulfohydrolase, is the general enzyme in this category and also represents a group of similar enzymes.[1–8] Typically, enzymes in this classification will also catalyze the reaction of at least one of the following enzymes: steryl-sulfatase [EC 3.1.6.2], cerebroside-sulfatase [EC 3.1.6.8], or, *N*-acetylgalactosamine-4-sulfatase [EC 3.1.6.12]. Examples of substrates include *p*-nitrophenyl sulfate, nitrocatechol sulfate, tyrosine sulfate, and phenol sulfate. In the presence of tyramine, aryl sulfatase II of *Aspergillus oryzae* acts as an arylsulfate sulfotransferase. **See also** *Arylsulfate Sulfotransferase*

Steryl-sulfatase [EC 3.1.6.2], also known as arylsulfatase C and steroid sulfatase, catalyzes the hydrolysis of 3-β-hydroxyandrost-5-en-17-one 3-sulfate to produce 3-β-hydroxyandrost-5-en-17-one and sulfate. Other substrates include androstenediol-3-sulfate, dehydroandrosterone sulfate, 4-methylumbelliferyl sulfate, estrone sulfate, dehydroisoandrosterone sulfate, 5α-androstan-17-one-3α-sulfate, 5α-androstan-17-one-3β-sulfate, and β-estradiol-3-sulfate. **See** *Steryl-Sulfatase*

Cerebroside-sulfatase [EC 3.1.6.8], also known as arylsulfatase A, catalyzes the hydrolysis of a cerebroside 3-sulfate to yield a cerebroside and sulfate. The enzyme will also hydrolyze the galactose 3-sulfate bond present in a number of lipids. In addition, the enzyme will also hydrolyze ascorbate 2-sulfate, other phenol sulfates, and 1-*O*-alkyl-2-*O*-acyl-3-*O*-(β-D-galactopyranoside-3′-sulfate)-glycerol. **See also** *Cerebroside-Sulfatase; Chondrosulfatase; Glycosulfatase*

N-Acetylgalactosamine-4-sulfatase [EC 3.1.6.12], also known as arylsulfatase B and chondroitin sulfatase, catalyzes the hydrolysis of the 4-sulfate groups of the *N*-acetyl-D-galactosamine 4-sulfate units of chondroitin sulfate and dermatan sulfate. *N*-Acetylglucosamine 4-sulfate is also a substrates. **See also** *N-Acetylgalactosamine-4-Sulfatase; Chondrosulfatase; Glycosulfatase*

[1]G. Lowe (1998) *CBC* **1**, 627.
[2]E. Conzelmann & K. Sandhoff (1987) *AE* **60**, 89.
[3]Y.-T. Li & S.-C. Li (1983) *The Enzymes*, 3rd ed., **16**, 427.
[4]M. Rasburn & C. H. Wynn (1973) *BBA* **293**, 191.
[5]G. J. Delisle & F. H. Milazzo (1972) *Can. J. Microbiol.* **18**, 561.
[6]A. B. Roy (1971) *The Enzymes*, 3rd ed., **5**, 1.
[7]R. G. Nicholls & A. B. Roy (1971) *The Enzymes*, 3rd ed., **5**, 21.
[8]A. B. Roy (1960) *Adv. Enzymol.* **22**, 205.

Selected entries from *Methods in Enzymology* [vol, page(s)]:
General discussion: **2**, 324, 330; **15**, 684, 689; **28**, 837; **143**, 361
Assay: **143**, 207, 361; bioluminescence assay, **57**, 257
Other molecular properties: brain fractions, in, **31**, 469; diagnostic enzyme, as, **31**, 20, **31**, 186, 187; limpet (effect on spasmogenic mediators, **86**, 20; inactivation of slow-reacting substance of anaphylaxis, **86**, 20; synthetic leukotrienes, **86**, 28); macrophage fractions, in, **31**, 340; phenol sulfokinase assay and, **5**, 978, 979; 3′-phosphoadenosine 5′-phosphate assay and, **5**, 979; properties, **143**, 365; purification from *Helix pomatia*, **143**, 363; synthesis of benzyl luciferyl sulfate, **57**, 252; Taka-diastase, **2**, 328; transition-state and multisubstrate analogues, **249**, 305

ARYLSULFATE SULFOTRANSFERASE

This sulfotransferase [EC 2.8.2.22] catalyzes the reaction of an aryl sulfate with a phenol to produce a new phenol and a new aryl sulfate. The hydroxyl groups of tyrosyl residues in peptides (*e.g.*, in angiotensin) can act as acceptor substrates. Other phenol acceptor substrates include phenol, tyramine, salicylamide, phenolphthalein, naphthol, estradiol, tyrosine methyl ester, epinephrine, 9-phenanthrol, lavine, chalcone, xanthone, and 3-chlorophenol. Other peptides sulfated include enkephalin, vasopressin, proctorin, cholecys-tokinin octapeptide, and phyllocerulein.[1–3]

[1] D. H. Kim, L. Konishi & K. Kobashi (1986) *BBA* **872**, 33.
[2] K. Kobashi, D. H. Kim & T. Morikawa (1987) *J. Protein Chem.* **6**, 237.
[3] N. S. Lee, B. T. Kim, D. H. Kim & K. Kobashi (1995) *J. Biochem.* **118**, 796

ARYL SULFOTRANSFERASE

This sulfotransferase [EC 2.8.2.1], also known as sulfok-inase and phenol sulfotransferase, catalyzes the reaction of 3′-phosphoadenylylsulfate with a phenol to produce adenosine 3′,5′-bisphosphate and an aryl sulfate.[1–4] Aryl substrates include phenol, dopamine, 3-nitrophenol, 4-nitrophenol, 2-hydroxyestrone, 2-hydroxyestradiol, epinephrine, norepinephrine, 1-naphthol, and 2-naphthol. Organic hydroxylamines are not substrates (these can be acted upon by tyrosine-ester sulfotransferase [EC 2.8.2.9]).

See also Arylamine Sulfotransferase; Tyrosine-Ester Sulfotransferase; Arylsulfate Sulfotransferase

[1] R. M. Weinshilboum (1986) *Fed. Proc.* **45**, 2223.
[2] M. E. Veronese, W. Burgess, X. Zhu & M. E. McManus (1994) *BJ* **302**, 497.
[3] R. M. Whittemore, L. B. Pearce & J. A. Roth (1986) *ABB* **249**, 464.
[4] R. M. Whittemore, L. B. Pearce & J. A. Roth (1985) *Biochemistry* **24**, 2477.

Selected entries from *Methods in Enzymology* [vol, page(s)]:
General discussion: 5, 977; **17B**, 825; **77**, 197
Assay: 5, 977; **17B**, 825; **77**, 198, 199; chromatographic, **77**, 198, 199; **143**, 204, 205
Other molecular properties: affinity chromatography, **77**, 202, 203; assay of PAP, **57**, 254, 256, 257; definition of units, **77**, 199; differentiation, **77**, 198; Ecteola-cellulose chromatography, **143**, 204; effect of salt, **77**, 199, 204, 205; inhibitors, **77**, 206; intestinal epithelium, in, **77**, 156; ion-pair extraction, **143**, 202; isoelectric focusing, **77**, 203; isoelectric point, **77**, 204; kinetic constants, **77**, 205; molecular weight, **77**, 204; pH optima, **5**, 980; **77**, 204, 205; phosphoadenosine phosphosulfate, assay of, **6**, 773; properties, **5**, 980; **17B**, 828; **77**, 204; purification, **77**, 200 (liver, from rabbit, **5**, 979; **17B**, 827); separation from sulfate-activating system, **5**, 966; serotonin sulfation, **17B**, 825 (assay, **17B**, 825; properties, **17B**, 828; purification from rabbit liver, **17B**, 827); substrates, **77**, 205, 206; sulfate activation, assay of, **5**, 965

ASCLEPAIN

This protease [EC 3.4.22.7] catalyzes the hydrolysis of proteins having a specificity even broader than that of papain.[1–4]

[1] T. Winnick, A. R. Davies & D. M. Greenberg (1940) *J. Gen. Physiol.* **23**, 275.
[2] W. J. Brockbank & K. R. Lynn (1979) *BBA* **578**, 13.
[3] B. E. Barragán, A. Hernández-Arana, M. Oliver, M. Castañeda-Agulló & L. M. Del Castillo (1986) *Rev. Latinoamer. Quim.* **16**, 158.
[4] M. Tablero, R. Arreguín, B. Arreguín, M. Soriano, R. I. Sanchéz, A. Rodríguez-Romero & A. Hernández-Arana (1991) *Plant Sci.* **74**, 7.

Selected entries from *Methods in Enzymology* [vol, page(s)]:
General discussion: 2, 56

L-ASCORBATE:CYTOCHROME-b_5 REDUCTASE

This oxidoreductase [EC 1.10.2.1] catalyzes the reaction of L-ascorbate with ferricytochrome b_5 to produce monodehydroascorbate and ferrocytochrome b_5.[1–4]

[1] J. V. Schloss & M. S. Hixon (1998) *CBC* **2**, 43.
[2] G. Scherer & W. Weis (1978) *Hoppe-Seyler's Z. Physiol. Chem.* **359**, 1527.
[3] H. Weber, W. Weis, W. Schaeg & H. Staudinger (1973) *Hoppe-Seyler's Z. Physiol. Chem.* **354**, 1277.
[4] F. B. Everling, W. Weis & H. Staudinger (1969) *Hoppe-Seyler's Z. Physiol. Chem.* **350**, 1485.

ASCORBATE 2,3-DIOXYGENASE

This iron-dependent oxidoreductase [EC 1.13.11.13] catalyzes the reaction of ascorbate with dioxygen to produce oxalate and threonate.[1]

[1] G. A. White & R. M. Krupka (1965) *Arch. Biochem. Biophys.* **110**, 448.

Selected entries from *Methods in Enzymology* [vol, page(s)]:
General discussion: 52, 11

ASCORBATE OXIDASE

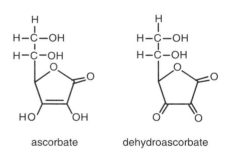

ascorbate   dehydroascorbate

This copper-dependent oxidoreductase [EC 1.10.3.3], also called ascorbase, catalyzes the reaction of two molecules of L-ascorbate with dioxygen to produce two dehydroascorbate molecules and two water molecules.[1–9] Other substrates include D-isoascorbate and D-glucoascorbate. Ceruloplasmin is also reported to have an ascorbate oxidase activity.

The enzyme contains a type-1 copper site and a trinuclear copper center. Electron transfer must occur between the two sites. The enzyme has a two-site ping pong Bi Bi kinetic mechanism.[1] All copper ions in the resting form of ascorbate oxidase are Cu(II). Ascorbate binds and reacts producing Cu(I) at site 1 and semihydroascorbate (which then undergoes dismutation to form ascorbate and dehydroascorbate). Electrons are then transferred to the trinuclear site. When four electrons have been transferred, dioxygen binds at a Cu(I) in site 2, forming the hydroperoxide intermediate. Ultimately, a second water molecule is produced.[1]

[1] A. Messerschmidt (1998) *CBC* **3**, 401.
[2] J. V. Schloss & M. S. Hixon (1998) *CBC* **2**, 43.
[3] E. T. Adman (1991) *Adv. Protein Chem.* **42**, 145.
[4] A. Messerschmidt & R. Huber (1990) *EJB* **187**, 341.
[5] J. Dayan & C. R. Dawson (1976) *BBRC* **73**, 451.
[6] R. A. Holwerda, S. Wherland & H. B. Gray (1976) *Ann. Rev. Biophys. Bioeng.* **5**, 363.
[7] B. G. Malmström, L.-E. Andréasson & B. Reinhammer (1975) *The Enzymes*, 3rd ed., **12**, 507.
[8] C. R. Dawson, K. G. Strothkamp & K. G. Krul (1975) *Ann. N. Y. Acad. Sci.* **258**, 209.
[9] G. R. Stark & C. R. Dawson (1963) *The Enzymes*, 2nd ed., **8**, 297.

Selected entries from *Methods in Enzymology* [**vol**, page(s)]:
General discussion: 2, 831; 52, 9; 62, 30
Assay: 2, 831; 62, 30, 31
Other molecular properties: apo and reconstituted oxidase, 62, 37; ascorbate-TMPD oxidase, effect of DABS or PMPS labeling, 56, 621; holoascorbate oxidase, 62, 31; properties, 2, 834; 62, 34; prosthetic group, 52, 4; purification, 2, 832; 62, 35; Raman frequencies and assignments, 49, 144; reaction mechanism, 52, 39; reactor (ascorbic acid assay with, 137, 281; preparation, 137, 282); resonance Raman spectrum, 49, 142

ASCORBATE PEROXIDASE

This iron-dependent oxidoreductase [EC 1.11.1.11] catalyzes the reaction of L-ascorbate with hydrogen peroxide to produce dehydroascorbate and two water molecules.[1-4] A ferryl porphyrin π-cation radical forms as an intermediate.[1,2]

[1] H. B. Dunford (1998) *CBC* **3**, 195.
[2] L. A. Marquez, M. Quitoriano, B. A. Zilinskas & H. B. Dunford (1996) *FEBS Lett.* **389**, 153.
[3] T. L. Poulos, W. R. Patterson & M. Sundaramoorthy (1990) *Biochem. Soc. Trans.* **23**, 228.
[4] D. A. Dalton, F. J. Hanus, S. A. Russell & H. J. Evans (1987) *Plant Physiol.* **83**, 789.

AscI RESTRICTION ENDONUCLEASE

This type II restriction endonuclease [EC 3.1.21.4], obtained from *Arthrobacter* species, catalyzes the hydrolysis of both strands of DNA at 5′ . . . GG∇CGCGCC . . . 3′.[1]

[1] C. Polisson & R. D. Morgan (1988) *NAR* **16**, 10365.

AseI RESTRICTION ENDONUCLEASE

This type II restriction endonuclease [EC 3.1.21.4], obtained from *Aquaspirillum serpens*, catalyzes the hydrolysis of both strands of DNA at 5′ . . . AT∇TAAT . . . 3′.[1] Note that the recognition site consists of only A and T bases. The enzyme can also exhibit star activity. **See** *"Star" Activity*

[1] C. Polisson & R. D. Morgan (1988) *NAR* **16**, 10365.

ASPARAGINASE

This enzyme [EC 3.5.1.1] catalyzes the hydrolysis of L-asparagine to produce ammonia and L-aspartate.[1-3] Other substrates include β-cyano-L-alanine and β-aspartylhydroxamate; in addition, there is a slight glutaminase activity.

[1] J. C. Wriston, Jr. & T. O. Yellin (1973) *AE* **39**, 185.
[2] J. C. Wriston, Jr. (1971) *The Enzymes*, 3rd ed., **4**, 101.
[3] G. R. Guy & R. M. Daniel (1982) *BJ* **203**, 787.

Selected entries from *Methods in Enzymology* [**vol**, page(s)]:
General discussion: 2, 383; 17A, 732; 17B, 861; 34, 405; 46, 22, 35, 432; 113, 608
Assay: 2, 383; 34, 407; 113, 609; 242, 86; automated assay, 17B, 862; 22, 8; coupled enzyme, 63, 33, 34; serum, in, 17B, 862; urine, in, 17B, 862
Other molecular properties: *Acinetobacter*, of, 27, 401; 113, 612; active site, 46, 432; activity (assay procedure, native protein, 44, 690, 691; relative, conjugate, 44, 271; specific conjugate, 44, 255); albumin-IgG conjugate, preparation, 137, 570; analysis of (automated from manual, 22, 8; standards for, 22, 10); antilymphoma activity of, 17A, 736, 738; asparaginase II, large-scale isolation of, 22, 490; bacterial (assay, 113, 609; modifications, 113, 616; properties, 113, 615; purification, 113, 610); clinical aspects, 17B, 861; crude δ-aminovaleramidase, in, 17B, 164; crystallization, 22, 262; encapsulated, exposed antigenic sites, 44, 692; entrapment in (polyacrylamide gel, 44, 692; poly-2-hydroxyethylmethacrylate, 44, 176, 692; fiber entrapment (biomedical application, 44, 242; cellulose triacetate fiber, in, 44, 693); guinea pig serum, from (preparation, 2, 383; 11, 78; properties, 2, 383; 11, 80; 113, 612); half-life, 242, 87; immobilization on (collagen, 44, 255, 260, 693); Dacron, 44, 693; glass plates, 44, 693; inert protein, 44, 909; nylon tube, 44, 120, 693; polymethylmethacrylate, 44, 693); immobilized enzyme kinetics, 64, 237; immune response, effect on, 44, 705; inhibition of tumor growth by, 17B, 861; mammalian, occurrence and properties, 113, 618; microencapsulation, 44, 212; microparticle-entrapped (hemofilter with, 137, 493; preparation, 137, 492); peptide active site, 46, 434; polyethylene glycol modification (activated PEG₂ modification, 242, 86; effect on [clearance time, 242, 87; immunoreactivity, 242, 85; kinetic constants, 242, 87, 89]; PM modification, 242, 88); preparative acrylamide gel electrophoresis, 22, 432; properties, 113, 612, 615, 618; purification, 34, 409, 410 (*Escherichia coli*, 17A, 736; 113, 612, 614; guinea pig serum, 2, 383; 11, 78; 17A, 733; 113, 612; *Proteus vulgaris*, 113, 611; *Serratia marcescens*, 17A, 740; *Vibrio succinogenes*, 113, 610; yeast, 2, 384); quantitative precipitin reaction, 242, 86; reaction with DONV, 46, 434, 435; reactor, preparation (dialyzer, from, 137, 483, 487; fibrin, with, 137, 482); sedimentation coefficient, 61, 108; serum, in, assay of, 17B, 862; storage in solution, 44, 212, 213; therapeutic application, 17B, 861; 44, 679, 689 (leukemia, 242, 85; lymphosarcoma, 242, 85); transition state and multisubstrate analogues, 249, 307; urine, in, assay of, 17B, 862

D-ASPARAGINASE

This enzyme catalyzes the hydrolysis of D-asparagine to produce ammonia and D-aspartate.[1]

[1] G. R. Guy & R. M. Daniel (1982) *BJ* **203**, 787.

ASPARAGINE:α-KETO-ACID AMINOTRANSFERASE

This pyridoxal-phosphate-dependent aminotransferase [EC 2.6.1.14], also known as asparagine:2-oxo-acid aminotransferase, catalyzes the reversible reaction of L-asparagine and an α-keto acid (or, 2-oxo acid) to produce α-ketosuccinamate (*i.e.*, 2-oxosuccinamate) and an α-L-amino acid.[1,2]

α-ketosuccinamate

Acceptor substrates include α-ketosuccinamate, oxomalonate, glyoxylate, and pyruvate. Both cytosolic and mitochondrial isotypes of the aminotransferase have been reported, each with unique aspects in the specificity profiles.[2]

[1] A. J. L. Cooper (1977) *JBC* **252**, 2032.
[2] D. M. Maul & S. M. Schuster (1986) *ABB* **251**, 577.

Selected entries from *Methods in Enzymology* [vol, page(s)]:
General discussion: 113, 602
Assay: 113, 603
Other molecular properties: asparaginase, distinguishing from, **2**, 382; liver extracts, in, **17A**, 1023; rat liver (purification, **113**, 605; substrate specificity, **113**, 606)

ASPARAGINE SYNTHETASES

The Enzyme Commission has classified three enzyme activities that catalyze the ATP-dependent formation of L-asparagine from L-aspartate.[1–13]

Asparagine synthetase (glutamine-hydrolyzing) [EC 6.3.5.4] catalyzes the reaction of L-aspartate with ATP and L-glutamine to produce L-asparagine, AMP, pyrophosphate (or, diphosphate), and L-glutamate. The enzyme will also hydrolyze L-glutamine to L-glutamate and ammonia in the absence of the other substrates. The enzyme from many sources can also exhibit an ammonia-dependent activity (*i.e.*, EC 6.3.1.1). Incubation of the enzyme with 6-diazo-5-oxo-L-norleucine prior to reaction will remove the glutamine-dependent activity but not the ammonia-dependent activity. The enzyme forms a transient γ-glutamyl thioester with an active-site cysteinyl residue.[1,2,5]

It has been proposed that the enzyme has a Uni Uni ordered Bi Ter ping pong kinetic mechanism which proceeds via the formation of a β-aspartyl-AMP intermediate and possibly including a Theorell-Chance step in the overall scheme.[8–10] In this mechanism, L-glutamine is the first substrate to bind followed by the release of L-glutamate. More recent studies utilizing product inhibition and isotopic probes suggest that the β-aspartyl-AMP intermediate forms prior to L-glutamine binding.[3]

Asparagine synthetase [EC 6.3.1.1], also known as aspartate:ammonia ligase, catalyzes the reaction of L-aspartate with ammonia and ATP to produce L-asparagine, AMP, and pyrophosphate (or, diphosphate). Hydroxylamine can also act as the nitrogen-donor.

Asparagine synthetase (ADP-forming) [EC 6.3.1.4] reportedly catalyzes the reaction of L-aspartate with ammonia and ATP to produce L-asparagine, ADP, and orthophosphate. Hydroxylamine can also act as a nitrogen donor.[13]

[1] H. G. Schnizer, S. K. Boehlein, J. D. Stewart, N. G. J. Richards & S. M. Schuster (1999) *Biochemistry* **38**, 3677.
[2] N. G. J. Richards & S. M. Schuster (1998) *AE* **72**, 145.
[3] S. K. Boehlein, J. D. Stewart, E. S. Walworth, R. Thirumoorthy, N. G. J. Richards & S. M. Schuster (1998) *Biochemistry* **37**, 13230.
[4] H.-M. Lam, K. T. Coschigamo, I. C. Oliveira, R. Melo-Oliveira & G. M. Coruzzi (1996) *Ann. Rev. Plant Physiol. Plant Mol. Biol.* **47**, 569.
[5] S. Sheng, D. A. Moraga-Amador, G. van Heeke, R. D. Allison, N. G. J. Richards & S. M. Schuster (1993) *JBC* **268**, 16771.
[6] H. Zalkin (1993) *AE* **66**, 203.
[7] P. M. Mehlhaff, C. A. Luehr & S. M. Schuster (1985) *Biochemistry* **24**, 1104.
[8] T. A. Huber & J. G. Streeter (1985) *Plant Sci.* **42**, 9.
[9] S. Hongo & T. Sato (1985) *ABB* **238**, 410.
[10] H. A. Milman, D. A. Cooney & C. Y. Huang (1980) *JBC* **255**, 1862.
[11] A. Meister (1974) *The Enzymes*, 3rd ed., **10**, 561.
[12] J. M. Buchanan (1973) *AE* **39**, 91.
[13] P. M. Nair (1969) *ABB* **133**, 208.

Selected entries from *Methods in Enzymology* [vol, page(s)]:
General discussion: **17A**, 722; **46**, 420
Other molecular properties: isotope exchange, **64**, 9; mechanism, **63**, 52; **87**, 355; purification from (*Lactobacillus arabinosus*, **17A**, 724; mouse leukemia cells, **17A**, 729); tumor concentration of, **17A**, 731

ASPARAGINYL-tRNA SYNTHETASE

This enzyme [EC 6.1.1.22], also known as asparagine:tRNA ligase and asparagine translase, catalyzes the reaction of ATP with L-asparagine and tRNA^Asn to produce AMP, pyrophosphate (or, diphosphate), and L-asparaginyl-tRNA^Asn.[1–3] Other substrates include isoasparagine and aspartate-3-hydroxamate. This class II aminoacyl-tRNA synthetase catalyzes the reaction via the formation of an asparaginyl-adenylate intermediate. ***See also*** *Aminoacyl-tRNA Synthetases*

[1] C. Berthet-Colominas, L. Seignovert, M. Hartlein, M. Grotli, S. Cusack & R. Leberman (1998) *EMBO J.* **17**, 2947.
[2] J. Anselme & M. Härtlein (1991) *FEBS Lett.* **280**, 163.
[3] P. J. Lea & L. Fowden (1973) *Phytochemistry* **12**, 1903.

Selected entries from *Methods in Enzymology* [vol, page(s)]:
Other molecular properties: subcellular distribution, **59**, 233, 234

ASPARAGUSATE REDUCTASE (NADH)

This oxidoreductase [EC 1.6.4.7], also known as asparagusate dehydrogenase, catalyzes the reaction of NADH with asparagusate to produce NAD^+ and 3-mercapto-2-mercaptomethylpropanoate (or, dihydroasparagusate).[1,2] The enzyme acts on lipoate to produce dihydrolipoate.

[1]H. Yanagawa & F. Egami (1976) JBC **251**, 3637.
[2]H. Yanagawa & F. Egami (1975) BBA **384**, 342.
Selected entries from **Methods in Enzymology** [**vol**, page(s)]:
General discussion: **62**, 172; **143**, 516
Assay: **62**, 172, 173; **143**, 517
Other molecular properties: distribution, **62**, 181; properties, **62**, 176; **143**, 519; purification from asparagus, **62**, 173; **143**, 518

ASPARTATE N-ACETYLTRANSFERASE

This transferase [EC 2.3.1.17] catalyzes the reaction of acetyl-CoA with L-aspartate to produce coenzyme A and N-acetyl-L-aspartate.[1-3]

[1]H. Knizley, Jr., (1967) JBC **242**, 4619.
[2]F. B. Goldstein (1959) JBC **234**, 2702.
[3]M. E. Truckenmiller, M. A. Namboodiri, M. J. Brownstein & J. H. Neale (1985) J. Neurochem. **45**, 1658.
Selected entries from **Methods in Enzymology** [**vol**, page(s)]:
Other molecular properties: occurrence, **6**, 562; pH activity profile, **44**, 741, 742; thermal stability, **44**, 741

ASPARTATE AMINOTRANSFERASE

This pyridoxal-phosphate-dependent aminotransferase [EC 2.6.1.1] (also known as transaminase A, glutamate: oxaloacetate transaminase, and glutamic:aspartic transaminase) catalyzes the reversible reaction of L-aspartate with α-ketoglutarate (or, 2-oxoglutarate) to produce oxaloacetate and L-glutamate.[1-8] The enzyme has a relatively broad specificity: L-tyrosine, L-phenylalanine, L-tryptophan, L-glutamate, L-methionine, and α-L-aminoadipate can all serve as amino donor substrates.

The enzyme has a ping pong Bi Bi kinetic mechanism and is the prototypic aminotransferase. The pyridoxal phosphate is covalently linked to a lysyl residue as an internal Schiff base. A transaldimination occurs upon binding of L-aspartate, forming an external aldimine. The lysyl residue then acts as a catalytic base and a 1,3 prototropic shift converts the external aldimine into a ketimine. Hydrolysis of the ketimine generates the pyridoxamine phosphate form of the coenzyme and oxaloacetate. The second half-reaction is the reversal of these steps with a different α-keto acid.

[1]R. A. John (1998) CBC **2**, 173.

[2]J. DiRuggiero & F. T. Robb (1996) Adv. Protein Chem. **48**, 311.
[3]H.-M. Lam, K. T. Coschigamo, I. C. Oliveira, R. Melo-Oliveira & G. M. Coruzzi (1996) Ann. Rev. Plant Physiol. Plant Mol. Biol. **47**, 569.
[4]M. A. Ator & P. R. Ortez de Montellano (1990) The Enzymes, 3rd ed., **19**, 213.
[5]D. J. Creighton & N. S. R. K. Murthy (1990) The Enzymes, 3rd ed., **19**, 323.
[6]A. E. Martell (1982) AE **53**, 163.
[7]D. E. Metzler (1979) AE **50**, 1.
[8]A. E. Braunstein (1973) The Enzymes, 3rd ed., **9**, 379.

Selected entries from **Methods in Enzymology** [**vol**, page(s)]:
General discussion: **5**, 677; **17A**, 167; **34**, 4; **46**, 31, 32, 37, 52, 163, 432, 445; **62**, 551; **113**, 66, 83; **142**, 253; **259**, 590
Assay: **5**, 677; **113**, 67, 83; **142**, 256; clinical, **17B**, 868; dry reagent chemistries, with, **137**, 414; Fast Analyzer, by, **31**, 816
Other molecular properties: affinity labeling, **87**, 474; amino-terminal mutants, **259**, 593 (expression in Escherichia coli, **259**, 594; free energy of unfolding [differential scanning calorimetry, **259**, 601; fluorescence monitoring, **259**, 598; guanidine hydrochloride perturbation, **259**, 596; reactivation curves, **259**, 598]; lipid vesicle interactions, **259**, 604); analysis of (continuous-flow method, **22**, 7, 8, 10; fluorometric and spectrophotometric methods compared, **22**, 10, 11); L-asparaginase assay, **44**, 690; assay for (aspartate, **13**, 473; **63**, 33; α-ketoglutarate, **13**, 458); binding of alkyl-substituted pyridoxal 5-phosphate analogs, **18A**, 587, 590; brain mitochondria, **55**, 58; catalyzed reactions, **142**, 256; 3-chloro-L-alanine-labeled active site (effect on catalytic mechanism, **106**, 133; isolation, **106**, 131; structural analysis, **106**, 133); clinical aspects, **9**, 569; **17B**, 866; coimmobilization, **44**, 472; complex formation with oligosaccharide, **138**, 421; coupled enzyme assay, in, **63**, 33; cyanylation, **47**, 131; cysteinesulfinic acid assay with, **143**, 160; elimination from mitochondria, **56**, 423; Escherichia coli (properties, **113**, 87; purification, **113**, 84, 86, 89; **142**, 258; site-directed mutants, acid-base catalysis, **249**, 116); fluorescence spectroscopy (1-anilino-8-naphthalene sulfonate, **259**, 598; covalent probes, **259**, 602; tryptophan, **259**, 598); folding pathway, **259**, 600; gluconeogenic catalytic activity, **37**, 281; glyoxysomes, in, **31**. 569; hydroxylamine, reaction with, **87**, 435; immobilization (assay, **136**, 486, 488, 490; collagen film, on, **44**, 908); inactivation, **46**, 45 (α-ketoglutarate assay, **55**, 206); inhibitors, **5**, 683; interaction with spin-labeled derivatives of vitamin B_6, **62**, 495; marker enzyme for (chromosome detection in somatic cell hybrids, **151**, 183; peroxisomes, as, **23**, 681); microbodies, in, **52**, 496; mitochondrial (amino-terminal sequences, **259**, 592; homology with cytosolic enzyme, **259**, 591; liposome interactions, **259**, 603; presequence contribution to stability, **259**, 607); natural occurrence, **113**, 66; nitration, substrates and, **25**, 518; open-closed conformational transition, **259**, 592, 603; oxaloacetate assay, **55**, 207; photooxidation of, **25**, 405; plasma membrane, in, **31**, 89; progress curve, **63**, 181 (nonlinear regression analysis, **249**, 76); properties, **5**, 682; **113**, 87; **142**, 261, 265; purification, **55**, 208 (Escherichia coli, from, **113**, 84, 89; **142**, 258; pig heart, **5**, 679); reactions, **87**, 434 (enzyme progress curves, nonlinear regression analysis, **249**, 76; proton transfer in, **249**, 501); serum, in, assay of, **17B**, 868; simulated annealing, **277**, 260; spectral properties, **18A**, 467, 468; spinach, in, **31**, 741; staining on gels, **22**, 598; stain for, **224**, 107; sulfhydryl peptide alkylated with N-ethylmaleimide, isolation by diagonal electrophoresis, **91**, 394; syncatalytic modification, **46**, 42; synthesis of deuterated malate, **13**, 582

D-ASPARTATE AMINOTRANSFERASE

This transferase, previously classified as EC 2.6.1.10 but now a deleted Enzyme Commission entry, is listed under D-alanine aminotransferase [EC 2.6.1.21].

ASPARTATE AMMONIA-LYASE

This lyase [EC 4.3.1.1], also known as aspartase and fumaric aminase, catalyzes the conversion of L-aspartate to fumarate and ammonia, using a carbanion mechanism (an *anti* elimination reaction).[1-6]

[1] R. E. Viola (2000) *AE* **74**, 295.
[2] V. E. Anderson (1998) *CBC* **2**, 115.
[3] J. V. Schloss & M. S. Hixon (1998) *CBC* **2**, 43.
[4] W. Shi, J. Dunbar, M. M. K. Jayasekera, R. E. Viola & G. K. Farber (1997) *Biochemistry* **36**, 9136.
[5] M. Y. Yoon, K. A. Thayer-Cook, A. J. Berdis, W. E. Karsten, K. D. Schnackerz & P. F. Cook (1995) *ABB* **320**, 115.
[6] K. R. Hanson & E. A. Havir (1972) *The Enzymes*, 3rd ed., **7**, 75.

Selected entries from *Methods in Enzymology* [vol, page(s)]:
General discussion: 2, 386; 34, 405, 407, 410, 411; 113, 618; 136, 463; 228, 590
Assay: 2, 386; 113, 619, 624; 136, 465; 228, 591; ammonia determination, by, 13, 355; spectrophotometric, 13, 354
Other molecular properties: L-aspartate, assay for, 2, 389; catalyzed reaction, 228, 590; *Escherichia coli* (properties, 113, 622; purification, 113, 620; 228, 590); extractive purification from *Escherichia coli* K12, 228, 590; heterologous gene selection and coamplification in mammalian cells, in, 185, 557; heterotropic interactions with nucleotides, 13, 360; homotropic interaction of substrate, 13, 361; industrial L-aspartate production with, 136, 466; *Lactobacillus helveticus*, purification, 2, 386; occurrence, 2, 388; 13, 361; 113, 627; partitioning, 228, 592, 597; preparation of labeled aspartate, 13, 578; properties, 2, 387; 113, 622, 626; *Pseudomonas fluorescens* (properties, 2, 387; 113, 626; purification, 2, 387; 113, 624); SDS-PAGE, 228, 592; subunit structure, 13, 358; transition state and multisubstrate analogues, 249, 308

ASPARTATE CARBAMOYLTRANSFERASE

Aspartate carbamoyltransferase [EC 2.1.3.2], also known as aspartate transcarbamylase and carbamylaspartotranskinase, catalyzes the first committed step (*i.e.*, the transfer of a carbamoyl group from carbamoyl phosphate to the α-amino group of L-aspartate to form *N*-carbamoyl-L-aspartate with the liberation of orthophosphate) in pyrimidine biosynthesis.[1-7] The *Escherichia coli* enzyme (molecular weight 310,000) is the prototypical allosteric regulatory protein. This enzyme (ATCase) can be dissociated by mercurials into two catalytic trimers and three regulatory dimers. ATCase is feedback-inhibited by the pyrimidine nucleotide biosynthetic end-product CTP and is activated by ATP.

Although this enzyme operates by an ordered kinetic mechanism with carbamoyl phosphate binding first and orthophosphate dissociating last, the transition-state (or, geometric multisubstrate) analogue known as *N*-(phosphonoacetyl)-L-aspartate readily binds with very high affinity. Structural studies have demonstrated the likelihood of (a) general base catalysis in increasing the nucleophilicity of the α-amino group of L-aspartate and (b) general acid catalysis in making the carbonyl group of carbamoyl phosphate more susceptible to attack.

[1] W. N. Lipscomb (1994) *AE* **68**, 67.
[2] E. Lolis & G. A. Petsko (1990) *Ann. Rev. Biochem.* **59**, 597.
[3] E. R. Kantrowitz & W. N. Lipscomb (1990) *TiBS* **15**, 53.
[4] N. M. Allewell (1989) *Ann. Rev. Biophys. Biophys. Chem.* **18**, 71.
[5] E. R. Kantrowitz & W. N. Lipscomb (1988) *Science* **241**, 669.
[6] B. L. Vallee & A. Galdes (1984) *AE* **56**, 283.
[7] G. R. Jacobson & G. R. Stark (1973) *The Enzymes*, 3rd ed., **9**, 225.

Selected entries from *Methods in Enzymology* [vol, page(s)]:
General discussion: 5, 904, 913, 925; 27, 13, 15, 16, 19; 46, 21, 90; 51, 35, 41, 51, 105, 111, 121; 113, 627; 135, 569; 259, 608
Assay: 5, 913, 925; 51, 41, 42, 106, 107, 122, 123; 113, 628
Other molecular properties: acid fractionation, 5, 906; activation, 64, 177; activators, 51, 48, 50; active site titration, with tritiated *N*-phosphonacetyl-L-aspartate, 51, 126; activity, 51, 35; allosteric activator site, 51, 50; allosteric, generation by site-directed mutagenesis *in vitro*, 202, 717, 725; allosteric interaction, 48, 305, 306; allosteric mechanism, 259, 614, 627; assay of carbamoyl-phosphate synthetase, 51, 105; *Bacillus subtilis* (properties, 113, 632; purification, 113, 629); bifunctional imidoester and, 25, 648; binding site number, 259, 615; binding studies, 44, 557; carbamoyl-phosphate synthetase activity, 51, 57, 105; carbamoyl-phosphate synthetase preparation, in, 5, 909; carbamoyl-phosphate synthetase (glutamine):aspartate carbamoyltransferase:dihydroorotase enzyme complex, 51, 111 (aggregation studies, 51, 126; assay, 51, 112, 113, 122, 123; dissociation, 51, 120, 121; gene selection and coamplification, in, 185, 550; hamster cells, from, 51, 121; isoelectric point, 51, 118; kinetic properties, 51, 118, 119; molecular weight, 51, 118, 126, 133; primary structure, 51, 133, 134; properties, 51, 118, 126; purification from [mutant hamster cells, 51, 123; rat ascites hepatoma cells, 51, 114; rat liver, 51, 116]; purity, 51, 126; sedimentation coefficient, 51, 118; sources, 51, 119, 120, 121; stability, 51, 134; N-terminus, 51, 133; tissue distribution, 51, 119, 120); carbamoyl-phosphate synthetase (glutamine):aspartate carbamoyltransferase enzyme complex, 51, 105 (assay, 51, 105; inhibitors, 51, 111; kinetic properties, 51, 111; molecular weight, 51, 111; *Neurospora*, from, 51, 105; purification, 51, 107; purity, 51, 110; stability, 51, 110); carboxyl-terminal amino acids, 25, 377; catalytic subunits, 27, 14, 15; chimeric, formation, 202, 701; concentration, 259, 615; concerted model, 64, 173; conformational states, chemical stabilization, 135, 569; C::R interface residue, site-directed alterations, 202, 705; crosslinked subunits (preparation, 135, 574; reassociation, 135, 575; separation, 135, 573); crosslinking with tartryl diazide, 135, 571; cyanylation, 47, 131; dicarboxylic acids and, 5, 915; differential scanning calorimetry (assembly effects, 259, 624; buffer sensitivity, 259, 625; ligation effects, 259, 625; mutation effects, 259, 626; electrostatic interactions, 259, 626; end-group analysis, 25, 120; energy transfer studies, 48, 378; enzyme complexes, in, 51, 105; enzyme production and, 5, 926; 22, 89; *Escherichia coli*, from, 5, 925; 51, 35 (cooperativity in [allosteric structures and model testing, 249, 554; experimental evaluation, 249, 548; heterotropic, 249, 552; homotropic, 249, 551; mutational analysis, 249, 554; structural model, 249, 549]; intersubunit ligand binding sites, 249, 559; site-directed mutants, acid-base catalysis, 249, 115); hybrid formation, 202, 697; inhibitors, 51, 48; isotope effects, 249, 363; isotope exchange, 64, 10 (fingerprint kinetic patterns for modifier action, 87, 651); kinetic properties, 51, 48, 57, 58; 249, 362; large-scale, 22, 547; ligand binding (assays, 259, 615; cooperativity, detection, 259, 616; dissociation constants, 259, 616; enthalpy, 259, 618; entropy, 259, 618; free energy determination, 259, 623; nucleotide triphosphates, 259, 619, 624; sites, interaction distances, 249, 561; subunit interactions, 259, 610, 622); linkage analysis, 259, 614, 621, 627; localization in gels, simultaneous capture and postincubation capture techniques, 104, 437; mechanism, 249, 362 (contribution to protein

structure knowledge, **259**, 612; equilibrium isotope exchange investigation, **87**, 651; **249**, 470; molecular modeling, **259**, 612; thermodynamic definition by state functions, **259**, 611, 628); modifier action, equilibrium isotope exchange investigation, **87**, 651; **249**, 471; molecular weight, **51**, 46, 55, 121; negative cooperativity, **64**, 189; ornithine carbamoyltransferase preparation, in, **5**, 912; p*K* values, sensitivity to side-chain positions, **259**, 627; properties, **5**, 915, 930;**51**, 46, 120; **113**, 632; *Pseudomonas fluorescens*, from, **51**, 51; purification, **51**, 43, 52; **113**, 629; rapid relaxation measurement, **64**, 188, 189; rat cells, from, **51**, 111; reaction catalyzed, **5**, 904; **259**, 609; regulation, **51**, 35; **259**, 609; renaturation, **51**, 55, 56; reversible denaturation, **259**, 626; sedimentation velocity pattern, **27**, 13, 15, 16, 19; separation of subunits from, **11**, 191; sequential model, **64**, 173; site-directed mutants, equilibrium isotope exchange investigations, **249**, 474; stability, **51**, 46, 47; **113**, 632; *Streptococcus faecalis*, from, **5**, 914; **51**, 41; structure, **259**, 609; structure-function analysis, **202**, 694; subunits (dissociation, **51**, 36 [identification, **51**, 36; isolation, **51**, 37; purity, **51**, 39, 40; structure, **51**, 55]; hybrids and, **25**, 538; subunit interactions, free energy, **259**, 613, 617, 620, 623); succinylation of, **25**, 539; supersecondary structures, genetic exchange within domains, **202**, 704; synthesis in *Chlamydomonas* cultures, **23**, 73; time-resolved X-ray scattering, **134**, 674; transition state and multisubstrate analogues, **249**, 304

ASPARTATE CARBOXYPEPTIDASE

This metallocarboxypeptidase, formerly classified as EC 3.4.17.5, is now a deleted Enzyme Commission entry.

ASPARTATE 1-DECARBOXYLASE

This decarboxylase [EC 4.1.1.11] catalyzes the conversion of L-aspartate to β-alanine and carbon dioxide.[1-4] Pyruvate is a cofactor for the *Escherichia coli* enzyme.

The crystal structure of the tetrameric *E. coli* enzyme shows catalytic pyruvoyl groups at three active sites and an ester at the fourth. The ester is an intermediate in the autocatalytic self-processing leading to formation of the pyruvoyl group.[2] L-Aspartate initially binds to the enzyme and forms a Schiff base intermediate with the pyruvoyl group.

[1] M. L. Hackert & A. E. Pegg (1998) *CBC* **2**, 201.
[2] A. Albert, V. Dhanaraj, U. Genschel, G. Khan, M. K. Ramjee, R. Pulido, B. L. Sibanda, F. von Delft, M. Witty, T. L. Blundell, A. G. Smith & C. Abell (1998) *Nat. Struct. Biol.* **5**, 289.
[3] P. D. van Poelje & E. E. Snell (1990) *Ann. Rev. Biochem.* **59**, 29.
[4] J. M. Williamson & G. M. Brown (1979) *JBC* **254**, 8074.

Selected entries from *Methods in Enzymology* [vol, page(s)]:
General discussion: 2, 188; 113, 589
Assay: 113, 589
Other molecular properties: aspartate aminotransferase, assay of, **2**, 171; *Clostridium welchii*, **2**, 182 (properties, **2**, 188; purification, **2**, 188;**3**, 465); *Escherichia coli* (properties, **113**, 594; purification, **113**, 591)

ASPARTATE 4-DECARBOXYLASE

This pyridoxal-phosphate-dependent decarboxylase [EC 4.1.1.12], also known as desulfinase, catalyzes the conversion of L-aspartate to L-alanine and carbon dioxide.[1,2]

The enzyme will also catalyze the decarboxylation of aminomalonate (producing glycine and CO_2) as well as the desulfination of 3-sulfinoalanine to sulfite and alanine.

Kinetic isotope effect studies of the *Alcaligenes faecalis* enzyme have allowed investigators to be able to estimate the relative free energies of the transition states and to demonstrate that the decarboxylation step is not entirely rate limiting.[1]

[1] R. M. Rosenberg & M. H. O'Leary (1985) *Biochemistry* **24**, 1598.
[2] S. S. Tate & A. Meister (1971) *AE* **35**, 503.

Selected entries from *Methods in Enzymology* [vol, page(s)]:
General discussion: 17A, 681; 34, 407; 46, 37, 427
Assay: desulfinase activity, 2, 333
Other molecular properties: activation by α-keto acid, **17A**, 682; activity stabilization by glutaraldehyde, **136**, 476; aminotransferase activity, **17A**, 681, 693; aspartate degradation and, **4**, 722; desulfinase activity, **2**, 333; labeling, **46**, 429; purification from (*Achromobacter*, **17A**, 691; *Alcaligenes faecalis*, **17A**, 684); pyridoxal 5'-phosphate and, **17A**, 687, 693; stereochemistry, **87**, 148

ASPARTATE KINASE

This phosphotransferase [EC 2.7.2.4], also known as aspartokinase, catalyzes the reaction of L-aspartate with ATP to produce 4-phospho-L-aspartate (*i.e.*, β-L-aspartyl phosphate) and ADP.[1,2] There are three *Escherichia coli* isozymes: (1) aspartate kinase I (repressed by L-threonine and L-isoleucine; inhibited by L-threonine) is a multifunctional enzyme possessing L-homoserine dehydrogenase activity [EC 1.1.1.3]; (2) aspartate kinase II (repressed by L-methionine) also has a homoserine dehydrogenase activity; (3) aspartate kinase III (repressible by L-lysine; inhibited by L-lysine) lacks homoserine dehydrogenase activity.

[1] I. Saint-Gerons, C. Parsot, M. M. Zakin, O. Bârzy & G. N. Cohen (1988) *Crit. Rev. Biochem.* **23**, S1.
[2] P. Truffa-Bachi (1973) *The Enzymes*, 3rd ed., **8**, 509.

Selected entries from *Methods in Enzymology* [vol, page(s)]:
General discussion: 5, 821; 113, 596
Assay: 5, 821; 113, 596
Other molecular properties: aspartokinase III (properties, **113**, 599; purification, **113**, 597); aspartokinase I-homoserine dehydrogenase I, **17A**, 694 (kinetic mechanism, equilibrium isotope exchange investigation, **249**, 468; purification from *Escherichia coli*, **17A**, 696); aspartokinase II-homoserine dehydrogenase II, **17A**, 699 (purification from *Escherichia coli*, **17A**, 700); aspartate kinase-homoserine dehydrogenase complex (activation energetics, **64**, 225, 226; cooperativity, **64**, 191, 192, 218, 222, 223; feedback inhibition, **64**, 191, 192; hysteresis, **64**, 196, 218, 222, 223); β-aspartyl phosphate assay and, **6**, 623; chromium-nucleotides, and, **87**, 655; *Escherichia coli* (properties, **113**, 599; purification, **17A**, 696, 700; **113**, 597); isotope exchange, **64**, 9, 25; **87**, 655; mechanism, **87**, 655; product inhibition studies, **249**, 188; propertries, **5**, 822; **113**, 599; synchronous cultures and, **21**, 469; yeast, purification, **5**, 821

D-ASPARTATE OXIDASE

This FAD-dependent oxidoreductase [EC 1.4.3.1] catalyzes the reaction of D-aspartate with water and dioxygen to generate oxaloacetate, ammonia, and hydrogen peroxide.[1,2] Other substrates include *meso*-2,3-diaminosuccinate (producing 2-amino-3-oxosuccinate) and 4-substituted thiazolidine-2-carboxylates.

A mechanism has been proposed in which the enzyme is reduced by the substrate, the resulting product then dissociates from the reduced enzyme, a second molecule of substrate binds, and reoxidation of the reduced enzyme-substrate complex occurs.[1]

[1] A. Negri, V. Massey, C. H. Williams & L. M. Schopfer (1988) *JBC* **263**, 13557.
[2] G. A. Hamilton (1985) *AE* **57**, 85.

Selected entries from **Methods in Enzymology** [vol, page(s)]:
General discussion: 17A, 713; **52**, 18

L-ASPARTATE OXIDASE

This FAD-dependent oxidoreductase [EC 1.4.3.16] catalyzes the reaction of L-aspartate with water and dioxygen to produce oxaloacetate, ammonia, and hydrogen peroxide.[1,2]

[1] J. Seifert, N. Kunz, R. Flachmann, A. Laüfer, K.-D. Jany & H. G. Gassen (1990) *Biol. Chem. Hoppe-Seyler* **371**, 239.
[2] R. Flachmann, N. Kunz, J. Seifert, M. Gütlich, F.-J. Wientjes, A. Läufer & H. G. Gassen (1988) *EJB* **175**, 221.

ASPARTATE:PHENYLPYRUVATE AMINOTRANSFERASE

This pyridoxal-phosphate-dependent aminotransferase [EC 2.6.1.70] catalyzes the reversible reaction of L-aspartate with phenylpyruvate to produce oxaloacetate and L-phenylalanine. The enzyme from *Pseudomonas putida* will also act on 4-hydroxyphenylpyruvate. L-Glutamate and L-histidine are weak alternative substrates in place of L-aspartate.[1,2]

[1] H. Ziehr & R.-M. Kula (1985) *J. Biotechnol.* **3**, 19.
[2] H. Ziehr, R.-M. Kula, E. Schmidt, C. Wandrey & J. Klein (1987) *Biotechnol. Bioeng.* **24**, 482.

ASPARTATE RACEMASE

This racemase [EC 5.1.1.13] catalyzes the interconversion of L- and D-aspartate. The enzyme will also utilize alanine as a substrate, albeit at only half the reported rates for aspartate.[1–3]

Other substrates include cysteate and cysteine sulfinate. In addition, this racemase will catalyze the exchange of the α-hydrogen of the substrate with the solvent hydrogen. If L-aspartate is incubated with the enzyme in D_2O, an overshoot in the optical rotation is observed before the substrate is fully racemized (due to the kinetic isotope effect). Abstraction of the α-hydrogen is at least partially rate-determining.[2]

[1] M. E. Tanner & G. L. Kenyon (1998) *CCBC* **2**, 7.
[2] T. Yamaguchi, S.-Y. Choi, H. Okada, M. Yohda, H. Kumagai, N. Esaki & K. Soda (1992) *JBC* **267**, 18361.
[3] E. Adams (1972) *The Enzymes*, 3rd ed., **6**, 479.

ASPARTATE-SEMIALDEHYDE DEHYDROGENASE

This oxidoreductase [EC 1.2.1.11] catalyzes the reversible reaction of aspartate-4-semialdehyde with orthophosphate and $NADP^+$ to produce 4-aspartyl phosphate and NADPH.[1,2] Other substrates for the reverse reaction include β-3-methylaspartyl phosphate and β-2-methylaspartyl phosphate.

The *Escherichia coli* enzyme has a random kinetic mechanism with deadend complexes, substrate inhibition, and a preference for L-aspartate-4-semialdehyde binding to the $E \cdot NADP^+ \cdot P_i$ ternary complex. A catalytically essential cysteinyl residue acts on the carbonyl carbon of aspartate-β-semialdehyde, with general acid assistance by a lysyl residue. The thiohemiacetal intermediate is oxidized by $NADP^+$ to a thioester which is subsequently attacked by orthophosphate. The transient tetrahedral intermediate produces the acyl-phosphate product and regenerates the original unmodified enzyme.[1]

[1] W. E. Karsten & R. E. Viola (1991) *BBA* **1077**, 209.
[2] I. Saint-Gerons, C. Parsot, M. M. Zakin, O. Bârzy & G. N. Cohen (1988) *Crit. Rev. Biochem.* **23**, S1.

Selected entries from **Methods in Enzymology** [vol, page(s)]:
General discussion: 5, 823; 17A, 708; 113, 600
Assay: 5, 823; 113, 600
Other molecular properties: affinity labeling, 87, 475; *Escherichia coli* (properties, 113, 601; purification, 17A, 709; 113, 600); hysteresis, **64**, 215; stereochemistry, 87, 113; yeast (properties, 5, 824; purification, 5, 823)

ASPARTOACYLASE

This enzyme [EC 3.5.1.15], also known as aminoacylase II, catalyzes the hydrolysis of an *N*-acyl-L-aspartate to produce a fatty acid anion and L-aspartate.[1,2] Substrates include *N*-acetyl-L-aspartate, *N*-formyl-L-aspartate, and glycyl-L-aspartate. **See also** *N-Acyl-D-Aspartate Deacylase; Aminoacylase*

[1]A. F. D'Adamo, E. Wertman, F. Foster & H. Schneider (1978) *Life Sci.* **23**, 791.
[2]F. B. Goldstein (1976) *J. Neurochem.* **26**, 45.

Selected entries from *Methods in Enzymology* [vol, page(s)]:
General discussion: 2, 115, 117
Assay: 2, 115
Other molecular properties: amino acid resolution by, 2, 119; 3, 554, 558, 568; preparation, 2, 117; properties, 2, 117; specificity, 2, 119

β-ASPARTYL-N-ACETYLGLUCOSAMINIDASE

This enzyme [EC 3.2.2.11], also known as 1-aspartamido-β-N-acetylglucosamine amidohydrolase, catalyzes the hydrolysis of 1-β-aspartyl-N-acetyl-D-glucosaminylamine to produce N-acetyl-D-glucosamine and L-asparagine.[1]

[1]E. Conzelmann & K. Sandhoff (1987) *AE* **60**, 89.

Selected entries from *Methods in Enzymology* [vol, page(s)]:
General discussion: 8, 597
Assay: 8, 597

β-ASPARTYL DIPEPTIDASE

This dipeptidase, also known as isoaspartyl dipeptidase, was first identified in *Escherichia coli*[1,2] and is most active with β-aspartyl-L-leucine followed by the corresponding β-aspartyl analogues of L-serine, L-methionine, and L-valine. The enzyme is distinct from the mammalian β-aspartyl peptidase [EC 3.4.19.5] which has a different specificity profile and can act on tripeptides. *See also β-Aspartyl Peptidase*

[1]E. E. Haley (1968) *JBC* **243**, 5748.
[2]J. D. Gary & S. Clarke (1995) *JBC* **270**, 4076.

4-L-ASPARTYLGLYCOSYLAMINE AMIDOHYDROLASE

This activity, previously classified as EC 3.5.1.37 but now a deleted Enzyme Commission entry, is listed under N^4-(β-N-acetylglucosaminyl)-L-asparaginase [EC 3.5.1.26].

β-ASPARTYL PEPTIDASE

This mammalian, cytosolic enzyme [EC 3.4.19.5; previously classified as EC 3.4.13.10] catalyzes the cleavage of a β-linked aspartic residue from the amino terminus of a peptide. The best dipeptide substrate is β-aspartylglycine, followed by the methionine, leucine, and serine analogues.[1] In addition, certain tripeptides can serve as substrates, including β-Asp–Gly–Gly and β-Asp–Gly–Ala. Other isopeptide bonds, such as γ-glutamyl and β-alanyl, were not hydrolyzed. The enzyme is distinct from the bacterial β-aspartyl dipeptidase.[2] *See also β-Aspartyl Dipeptidase*

[1]F. E. Dorer, E. E. Haley & D. L. Buchanan (1968) *ABB* **127**, 490.
[2]E. E. Haley (1970) *MIE* **19**, 737.

Selected entries from *Methods in Enzymology* [vol, page(s)]:
General discussion: 19, 730, 737
Assay: 19, 730, 737
Other molecular properties: activators and inhibitors, 19, 735, 739; distribution, 19, 736, 741; peptides hydrolyzed by, 19, 740; pH optimum and buffer effects, 19, 735, 741; physical properties and purity, 19, 735; properties, 19, 735, 739; purification, 19, 732 (rat liver, from, 19, 738); purity, 19, 739; specificity, 19, 735, 739; stability, 19, 735, 739; substrate affinity, 19, 736; substrates for, 19, 736

ASPARTYLTRANSFERASE

This transferase [EC 2.3.2.7] catalyzes the reversible reaction of L-asparagine with hydroxylamine to produce L-aspartylhydroxamate and ammonia.[1] D-Asparagine is also an effective substrate.

[1]H. N. Jayaram, T. Ramakrishnan & C. S. Vaidyanathan (1969) *Indian J. Biochem.* **6**, 106.

ASPARTYL-tRNA SYNTHETASE

This enzyme [EC 6.1.1.12], also known as aspartate:tRNA ligase and aspartate translase, catalyzes the reaction of L-aspartate with ATP and $tRNA^{Asp}$ to generate L-aspartyl-$tRNA^{Asp}$, AMP, and pyrophosphate (or, diphosphate).[1–4]

This class II aminoacyl-tRNA synthetase will also exhibit an aspartate-dependent ATP↔pyrophosphate exchange, the reaction proceeding via the formation of an aspartyl-AMP intermediate. The order of addition of the first two substrates, ATP first followed by L-aspartate, is opposite of that seen with class I synthetases. The α-carboxyl group is positioned by hydrogen-binding interactions for nucleophilic in-line attack on the α-phosphate of ATP. Following the release of pyrophosphate, $tRNA^{Asp}$ binds such that the 3'-hydroxyl group is positioned for nucleophilic attack on the aspartyl-adenylate intermediate (the free oxygen atom on the α-phosphate group of the intermediate may serve as a general base to abstract a proton from the attacking hydroxyl group). A tetrahedral transition state is stabilized by hydrogen bonds and the final products are released.[1] *See also Aminoacyl-RNA Synthetases*

[1]E. A. First (1998) *CBC* **1**, 573.
[2]B. Rees, J. Cavarelli & D. Moras (1996) *Biochimie* **78**, 624.
[3]J. Cavarelli, G. Eriani, B. Rees, M. Ruff, M. Boeglin, A. Mitschler, F. Martin, J. Gangloff, J.-C. Thierry & D. Moras (1994) *EMBO J.* **13**, 327.
[4]P. Schimmel (1987) *Ann. Rev. Biochem.* **56**, 125.

Selected entries from *Methods in Enzymology* [vol, page(s)]:
General discussion: 34, 170

Other molecular properties: crystallization, **276**, 643, 648, 654; preparation, **12B**, 697; subcellular distribution, **59**, 233, 234

ASPERGILLOPEPSIN I

This aspartic proteinase [EC 3.4.23.18] catalyzes peptide-bond hydrolysis in proteins with a broad specificity. When the insulin B chain is used as a substrate, two peptide bonds are readily hydrolyzed, Leu15–Tyr16 and Phe24–Phe25, as well as a number of minor cleavages.[1] In general, the enzyme from *A. saitoi* (now reclassified as *A. phoenicis*) prefers hydrophobic aminoacyl residues at P1 and P1'. Interestingly, Lys at P1 is also accepted and this can lead to the activation of trypsinogen at acidic pH values.

[1] N. Tanaka, M. Takeuchi & E. Ichishima (1977) *BBA* **485**, 406.

Selected entries from *Methods in Enzymology* [vol, page(s)]:
General discussion: 19, 373, 386
Assay: 19, 397
Other molecular properties: activators and inactivators, **19**, 405; amino acid composition, **19**, 405; distribution, **19**, 406; nomenclature, **19**, 373, 375; physical properties, **19**, 404; purification, **19**, 399; specificity, **19**, 405; stability, **19**, 403; trypsinogen activation by, **19**, 374

ASPERGILLOPEPSIN II

This aspartic proteinase [EC 3.4.23.19; previously classified as EC 3.4.23.6] (also known as proteinase A, proctase A, and *Aspergillus niger* var. *macrosporus* aspartic proteinase) catalyzes the hydrolysis of specific peptide bonds in proteins. With the A chain of insulin, Asn3–Gln4, Gly13–Ala14, and Tyr26–Thr27 bonds are cleaved. The P1 subsite appears to require Tyr, Phe, His, Asn, Asp, Gln, or Glu. The enzyme is less susceptible than aspergillopepsin I to pepstatin inhibition, and acts on collagen and gelatin with greater facility than aspergillopepsin I.[1]

[1] K. Takahashi (1995) *MIE* **248**, 146.

Selected entries from *Methods in Enzymology* [vol, page(s)]:
General discussion: 248, 115, 146
Assay: 248, 146
Other molecular properties: amino acid composition, **248**, 149; heavy chain, **248**, 149 (amino acid sequence, **248**, 149); inhibitors, **248**, 155; light chain, **248**, 149 (amino acid sequence, **248**, 149); molecular weight, **248**, 148; pH profile, **248**, 152; properties, **248**, 109, 116, 146, 148 (physical, **248**, 148); purification, **248**, 147; purity, **248**, 148; stability, **248**, 152; structure (core, **248**, 152; primary, **248**, 150; secondary, **248**, 152); substrate specificity, **248**, 153

ASPERGILLUS DEOXYRIBONUCLEASE K₁

This enzyme [EC 3.1.22.2], also known as *Aspergillus* DNase K₁, catalyzes the endonucleolytic cleavage of DNA (preferring single-stranded DNA) to produce nucleoside 3'-phosphate and 3'-phosphooligonucleotide

end-products.[1,2] A preference is also reported for bonds between dG and dG and between dG and dA.

[1] M. Kato & Y. Ikeda (1968) *J. Biochem.* **64**, 321.
[2] M. Kato, T. Ando & Y. Ikeda (1968) *J. Biochem.* **64**, 329.

ASPERGILLUS NUCLEASE S₁

This zinc-dependent endonuclease [EC 3.1.30.1] (also known as endonuclease S₁, single-stranded-nucleate endonuclease, nuclease S₁, and deoxyribonuclease S₁) catalyzes the hydrolysis of ribo- or deoxyribonucleic acids (with single strand breaks in double-stranded DNA), with a preference for single-stranded substrates.[1–5] This *Aspergillus oryzae* enzyme, which utilizes three bound zinc ions in the reaction, catalyzes the endonucleolytic cleavage of those nucleic acids to generate 5'-phosphomononucleotides and 5'-phosphooligonucleotide end-products with no apparent base specificity. Similar enzymes are *Neurospora crassa* nuclease, mung bean nuclease, and *Penicillium citrinum* nuclease P₁. **See also** *Penicillium* Nuclease P₁; *Neurospora crassa* Nuclease

[1] S. A. Martin, R. C. Ullrich & W. L. Meyer (1986) *BBA* **876**, 67.
[2] K. Shishido & T. Ando (1982) *Cold Spring Harbor Monogr. Ser.* **14**, 155.
[3] I. R. Lehman (1981) *The Enzymes*, 3rd ed., **14**, 193.
[4] A. E. Oleson & E. D. Hoganson (1981) *ABB* **211**, 478.
[5] A. E. Oleson & M. Sasakuma (1980) *ABB* **204**, 361.

Selected entries from *Methods in Enzymology* [vol, page(s)]:
General discussion: 65, 248, 263, 275, 418, 660, 729, 730, 743, 751; **68**, 18, 81, 249, 475; **79**, 321
Assay: 65, 249, 250
Other molecular properties: activity, **65**, 275, 751, 754; assay of psoralen binding to DNA, in, **212**, 249; cation requirement, **65**, 248; cDNA library construction, in, **225**, 595; characterization of pro-α2(I) collagen gene mutation, in, **145**, 213; characterization of RNA, in, **180**, 334; commercial source, **65**, 729; cutting of hairpin loop in second-strand cDNA synthesis, **152**, 328; digestion, **65**, 730, 754 (cDNA for analysis of residual hairpin loops, of, **152**, 333; DNA, of, **170**, 287; effect, of temperature, **65**, 743; haloacetaldehyde-modified DNA in eukaryotic cells, **212**, 171; mRNA·DNA-B hybrid, **216**, 189; oligonucleotide 3'-dialcohol, of, **65**, 660; RNA–DNA hybridization reaction products, of, **180**, 344); double-stranded complementary DNA preparation, in, **79**, 604; endo–exonuclease, *Neurospora crassa*, **65**, 255 (assay, **65**, 255, 256; cation requirements, **65**, 261, 262; commercial preparation, properties, **65**, 263; conversion, to single strand-specific endonuclease, **65**, 260; inhibitors, **65**, 262; ionic requirements, **65**, 261; molecular size, **65**, 262, 263; properties, **65**, 261; purification, **65**, 256; substrate specificity, **65**, 261); hybridization of single-copy DNA sequences, in, **224**, 239; limited digestion of 5S RNA, **164**, 726; mapping gene transcripts, in, **152**, 623; mapping of haloacetaldehyde-modified DNA bases, in, **212**, 157, 169; measurement of DNA–DNA hybridization, in, **224**, 343; *Neurospora crassa*, **65**, 255 (assay, **65**, 255, 256; cation requirements, **65**, 261, 262; commercial preparation, properties, **65**, 263; conversion, to single strand-specific endonuclease, **65**, 260; inhibitors, **65**, 262; ionic requirements, **65**, 261; molecular size, **65**, 262, 263; properties, **65**, 261; purification, **65**, 256; substrate specificity, **65**, 261); nucleic acid hybridization, in, **79**, 321; pH optimum, **65**, 252; purification, **65**, 250; RNA structure (analysis, **261**, 334, 339; probing, specificity and properties, for, **180**, 197); stability, **65**, 251;

structural probing of DNA *in vitro*, **212**, 296, 300; structural probing of native snRNA, in, **180**, 214, 221; synthetic oligonucleotides, in RNA mapping, and, **154**, 87; tritium derivative method, **65**, 647, 648; two-dimensional gel mapping, of RNA transcripts, **65**, 418; uses, **65**, 253

*Asp*101 RESTRICTION ENDONUCLEASE

This type II restriction enzyme [EC 3.1.21.4], isolated from *Aeromonas* species, catalyzes the hydrolysis of DNA; however, the recognition sequence and cleavage site are unknown.[1]

[1] A. M. Kul'ba & A. A. El'gammudi (1997) *Vestn. Beloruss. Gos. Univ.* **2**, 47.

*Asp*281 RESTRICTION ENDONUCLEASE

This type II restriction enzyme [EC 3.1.21.4], isolated from *Aeromonas* species, catalyzes the hydrolysis of DNA; however, the recognition sequence and cleavage site are unknown.[1]

[1] A. M. Kul'ba & A. A. El'gammudi (1997) *Vestn. Beloruss. Gos. Univ.* **2**, 47.

ASPULVINONE DIMETHYLALLYLTRANSFERASE

This transferase [EC 2.5.1.35] catalyzes the reaction of two molecules of dimethylallyl diphosphate with aspulvinone E to produce two molecules of pyrophosphate (or, diphosphate) and aspulvinone H.[1] Other acceptor substrates include aspulvinone G and aspulvinone I.

[1] I. Takahashi, N. Ojima, K. Ogura & S. Seto (1978) *Biochemistry* **17**, 2696.

Selected entries from **Methods in Enzymology** [**vol**, page(s)]:
General discussion: **110**, 320
Assay: **110**, 320

ASTACIN

This zinc-dependent endopeptidase [EC 3.4.24.21; formerly classified as EC 3.4.99.6] catalyzes the hydrolysis of peptide bonds in substrates having five or more amino acids.[1–4] Astacin has an extended binding domain, exhibiting a preference for an alanyl residue in P1′, prolyl in P2′ and P3′, hydrophobic residues in P3′ and P4′, and lysyl, arginyl, asparaginyl, or tyrosyl residues in P1 and P2. Studies with a transition-state analogue suggest the participation of a tyrosyl side-chain in the catalytic reaction (the "tyrosine switch").[2]

[1] W. L. Mock (1998) *CBC* **1**, 425.
[2] F. Grams, V. Dive, A. Yiotakis, I. Yiallouros, S. Vassilou, R. Zwilling, W. Bode & W. Stöcker (1996) *Nature Struct. Biol.* **3**, 671.
[3] J. S. Bond & R. B. Beynon (1995) *Protein Sci.* **4**, 1247.
[4] W. Stöcker, F. Grams, U. Baumann, P. Reinemer, F.-X. Gomis-Rüth, D. B. McKay & W. Bode (1995) *Protein Sci.* **4**, 823.

Selected entries from **Methods in Enzymology** [**vol**, page(s)]:
General discussion: **80**, 633, **248**, 305

Assay: **248**, 307; dansylated substrates for, **248**, 308; chromatographic method, **248**, 310; fluorescent oligopeptide substrate method, **248**, 308; quenched fluorescent substrates, with, **248**, 308; spectrofluorimetric, with N-dansylated substrates, **248**, 308; succinylalanylalanylalanyl-4-nitroanilide method, **248**, 307

Other molecular properties: absorbance coefficient, **248**, 310; activation mechanism, **248**, 198; amino acid sequence, **248**, 411; apoastacin (enzymatic activity, **248**, 319; preparation, **248**, 319); astacin subfamily, **248**, 192, 196; chelator inhibition, time dependence, **248**, 234; cleavage reactions (dansylated oligopeptide substrates, **248**, 313; peptide nitroanilides, **248**, 312; specificity, **248**, 312; type I collagen, **248**, 314); crystallization, **248**, 321; discovery, **248**, 306; disulfide bridges, **248**, 199; domain structure, **248**, 335; EGF-like domains, **248**, 198, 200, 336; inhibitors, **248**, 343 (metal-directed, **248**, 318; protein, **248**, 317; synthetic, **248**, 317; transition state analog, **248**, 316); isoelectric point, **248**, 311; kinetic parameters, determination, **248**, 307; kinetic properties, **248**, 310, 312; MAM domains, **248**, 198, 337; metal content, **248**, 311; metal removal, kinetics, **248**, 318; metal-substituted derivatives (enzymatic activity, **248**, 319; properties, **248**, 319; spectroscopy, **248**, 320; structure, **248**, 324); molecular weight, **248**, 311; pH optimum, **248**, 310; phylogenetic tree, **248**, 196; polypeptide chain structure, **248**, 198; preparation, **248**, 310; proenzyme, **248**, 307; properties, **248**, 196, 306, 310; purification, **248**, 310; reprolysin subfamily, **248**, 192, 196, 200, 345; solubility, **248**, 310; source, **248**, 306; species distribution, **248**, 305; stability, **248**, 310; structure, **248**, 191, 198, 306, 312, 321, 343, 448, 633; synthesis, **248**, 306; zinc dissociated from, half-lives for, **248**, 236; zinc replacement with other metals, **248**, 240; zymogen, **248**, 307

*Asu*I RESTRICTION ENDONUCLEASE

This type II restriction endonuclease [EC 3.1.21.4] is obtained from *Anabaena subcylindrica*, is Mg^{2+}-dependent, and acts at 5′-G∇GNCC-3′ (where N represents any base)[1] to generate staggered ends.

[1] S. G. Hughes, T. Bruce & K. Murray (1980) *Biochem. J.* **185**, 59.

Selected entries from **Methods in Enzymology** [**vol**, page(s)]:
Other molecular properties: recognition sequence, **65**, 2

ATP ADENYLYLTRANSFERASE

This transferase [EC 2.7.7.53], also known as bis(5′-nucleosyl)-tetraphosphate phosphorylase (NDP-forming) and diadenosinetetraphosphate α,β-phosphorylase, catalyzes the reaction of ADP with ATP to produce orthophosphate and P^1,P^4-bis(5′-adenosyl)tetraphosphate.[1–4] Both GTP and adenosine tetraphosphate can act as alternative adenylyl acceptor substrates. The enzyme requires the presence of divalent cations such as Mg^{2+} or Mn^{2+}. It also has been shown to catalyze an exchange reaction between the β-phosphate of nucleoside diphosphate and orthophosphate in the medium.[3]

[1] A. Guranowski & S. Blanquet (1985) *JBC* **260**, 3542.
[2] A. Brevet, H. Coste, M. Fromant, P. Plateau & S. Blanquet (1987) *Biochemistry* **26**, 4763.
[3] A. Guranowski & S. Blanquet (1986) *JBC* **261**, 5943.
[4] A. K. Robinson & L. D. Barnes (1991) *BJ* **279**, 135.

ATPase

This enzymatic activity [EC 3.6.1.3], also known as adenosinetriphosphatase and adenylpyrophosphatase, catalyzes the hydrolysis of ATP to produce ADP and orthophosphate. Whether any enzyme actually has a sole function of catalyzing ATP hydrolysis is still disputed.[1] Nearly every overt ATPase activity appears to be the consequence of functional uncoupling of ATP hydrolysis and changes in noncovalent bonding associated with mechanical work.

In a few cases, a low-level ATPase activity is observed for kinase-type reactions, especially when a phosphoryl acceptor substrate is scarce. An excellent example is six-times recrystallized yeast hexokinase, which in the absence of glucose or another hexose substrate, catalyzes ATP hydrolysis.[2] The observation that competitive inhibitors of glucose in the hexokinase reaction (e.g., N-acetylglucosamine) inhibited the ATPase reaction helped to demonstrate that the ATPase activity was an intrinsic property of hexokinase. Zewe et al.[3] first suggested that a water molecule resides in place of the $-CH_2OH$ of glucose, and this suggestion is in harmony with the observation that the pentoses lyxose and xylose, which lack this functional group in their preferred pyranose forms, strongly activate the ATPase activity.[4]

Another type of ATPase activity arises when two enzymes catalyzing opposing metabolic reactions are simultaneously active within the same cellular compartment. Their combined action results in a futile cycle that hydrolyzes ATP. Consider the net effect of simultaneous catalysis of the phosphofructokinase and fructose-1,6-bisphosphatase reactions:

$$Fructose\text{-}6\text{-}P + ATP = Fructose\ 1,6\text{-}bisP + ADP$$

$$Fructose\text{-}1,6\text{-}bisP + H_2O = Fructose\ 6\text{-}P + P_i$$

$$Net Reaction:\ ATP + H_2O = ADP + P_i$$

Hoffmann et al.[5] examined the dynamic and functional organization of the fructose 6-phosphate/fructose 1,6-bisphosphate cycle as an open, homogeneous reconstituted enzyme system using a model based on the kinetic properties of the individual enzymes. At low maximum activities of phosphofructokinase, they observed a domain of multiple stationary states, in which stable stationary states coexist with a stable oscillatory or with an alternate stable stationary state. Oscillations and the emergence of alternate stationary motions are caused mainly by the reciprocal

effect of the allosteric effectors AMP and fructose-2,6-bisP. Newsholme[6] also considered the role of the fructose 6-phosphate/fructose 1,6-bisphosphate cycle in metabolic regulation and heat generation.

Discoveries of the past two decades have convincingly demonstrated the pervasiveness of mechanochemical proteins that transduce the Gibbs free energy of nucleotide hydrolysis into some form of useful work. The product of these reactions can be described as some form of translational movement, rotation, or solute gradient. Under normal physiologic conditions, nucleotide hydrolysis is stoichiometrically coupled to the production of an increment of useful work. Purich[1] has coined the term energase to describe these mechanochemical systems as a distinct enzyme class. The present Enzyme Commission classification does not explicitly account for the noncovalent "work steps", thereby treating energases as hydrolases. That energases constitute a separate class becomes obvious by considering the following ATP-dependent reactions, written here without specifying how metal ions are involved or how protons are released:

Adenosine 5′-Triphosphatase:

$$ATP + H_2O = ADP + P_i$$

Glutamine Synthetase:

$$ATP + Glutamate + NH_3 = Glutamine + ADP + P_i$$

Energase:

$$ATP + Work\text{-}State_1 = Work\text{-}State_2 + ADP + P_i$$

The corresponding reaction equilibrium constants are:

Adenosine 5′-Triphosphatase:

$$K_{eq} = [ADP][P_i]/[ATP]$$

Glutamine Synthetase:

$$K_{eq} = [Glutamine][ADP][P_i]/$$
$$[Glutamate][NH_3][ATP]$$

Energase:

$$K_{eq} = [Work\text{-}State_2][ADP][P_i]/$$
$$[Work\text{-}State_1][ATP]$$

Note that the fundamental nature of the mechanical work step in energase reactions is illustrated by treating Work-State$_1$ as a substrate-like species and Work-State$_2$ as a product-like species. The relative abundance of these conformational states (i.e., [State$_2$]/[State$_2$]) must be explicitly indicated when writing the chemical reaction and the equilibrium constant. While the quotient [ADP][P$_i$]/[ATP] is common to all three reactions, in many respects, energases share more in common with synthases than with hydrolases. This argument shows that one simply cannot ignore the substrate-like and product-like species that differ only in the energetics of their noncovalent interactions.

Ironically, the hydrolytic activities of energases often become exaggerated when steps in an energase reaction are uncoupled by various treatments or agents. A good example is the tubulin GTPase. Strict stoichiometric coupling of tubulin incorporation and GTP hydrolysis occurs under normal physiologic conditions, but is lost in the presence of the mitotic spindle poison known as colchicine.[7] Only in the presence of colchicine, which blocks microtubule self-assembly, does tubulin exhibit an enhanced capacity to hydrolyze GTP.[8] Similar statements apply to the ability of other uncoupling agents that increase the hydrolase activities of ATP synthase and other ATP-dependent transporters. *See Energase*

Finally, the idea that all transporters are enzymes is in keeping with a new definition of enzyme catalysis as the facilitated making/breaking of chemical bonds, not just covalent bonds.[8] This idea builds on Pauling's assertion that any long-lived, chemically distinct interaction (including the persistent location of a solute with respect to the faces of a membrane) can be regarded as a chemical bond. Consideration of the overall transport reaction and the equilibrium constant ($K_{eq.transport} =$ [Solute$_{in}$][ADP][P$_i$]/[Solute$_{out}$][ATP]) does not conform to that of an ATPase, *i.e.*, $K_{eq} = $ [ADP][P$_i$]/[ATP]) confirms that the so-called ATP-dependent transporters are not ATPases.

[1]D. L. Purich (2001) *TiBS* **26**, 417.
[2]K. A. Trayser & S. P. Colowick (1961) *ABB* 94, **161**, 169.
[3]V. Zewe, H. J. Fromm & R. Fabiano (1964) *JBC* **239**, 1625.
[4]F. B. Rudolph & H. J. Fromm (1964) *JBC* **246**, 2104.
[5]E. Hofmann, K. Eschrich & W. Schellenberger (1985) *Adv. Enzyme Regul.* **23**, 331
[6]E. A. Newsholme (1976) *Biochem. Soc. Trans.* **4**, 978.
[7]R. K. MacNeal and D. L. Purich (1978) *JBC* **253**, 4683.
[8]D. L. Purich and J. M. Angelastro (1994) *AE* **69**, 121.

ATP:CITRATE (*pro-3S*)-LYASE

This lyase [EC 4.1.3.8], also known as ATP–citrate (*pro-S*-)-lyase and citrate cleavage enzyme, catalyzes the reaction of citrate with ATP and coenzyme A to yield oxaloacetate, acetyl-CoA, ADP, and orthophosphate.[1–6] The multiprotein enzyme complex can be dissociated into smaller components of which two are identical with citryl-CoA lyase and citrate:CoA ligase. The reaction proceeds through the formation of a phosphoenzyme intermediate[2] and has a random BC, random PQ kinetic mechanism.[6] *See also Citrate (pro-3S)-Lyase*

[1]D. J. Creighton & N. S. R. K. Murthy (1990) *The Enzymes*, 3rd ed., **19**, 323.
[2]B. Houston & H. G. Nimmo (1984) *BJ* **224**, 437.
[3]P. A. Srere (1975) *AE* **43**, 57.
[4]L. B. Spector (1972) *The Enzymes*, 3rd ed., **7**, 357.
[5]P. A. Srere (1972) *Curr. Top. Cell Regul.* **5**, 229.
[6]K. M. Plowman & W. W. Cleland (1967) *JBC* **242**, 4239.

Selected entries from *Methods in Enzymology* [vol, page(s)]:
General discussion: 5, 641; 13, 153
Assay: 5, 641; 13, 153; coupled enzyme, 63, 33; hydroxamate, 13, 153; radioactive, 13, 154; spectrophotometric, 5, 641; 13, 154
Other molecular properties: assay for coenzyme A, 13, 540; immunology, 13, 160; induction in rats, 13, 155; inhibition, 13, 159; 63, 400; isotope effect, 64, 10; mechanism, 63, 51, 158; 87, 358; organ distribution, 5, 644; 13, 159; parenchyma cells, in, 32, 706; properties, 5, 644; 13, 158; purification (chicken liver, 5, 642; rat liver, 13, 155); stereochemistry, 87, 147

ATP DEAMINASE

This deaminase [EC 3.5.4.18] catalyzes the hydrolysis of ATP to produce ITP and ammonia.[1,2] The enzyme from certain sources is relatively nonspecific, acting on adenosine, AMP, and ADP

[1]C. L. Zielke & C. H. Suelter (1971) *The Enzymes*, 3rd ed., **4**, 47.
[2]S.-T. Chung, K. Aida & T. Uemura (1967) *J. Gen. Appl. Microbiol.* **13**, 237.

ATP-DEPENDENT H$_4$ NAD(P)OH DEHYDRATASE

This dehydratase [EC 4.2.1.93] catalyzes the reaction of ATP with (6S)-6-β-hydroxy-1,4,5,6-tetrahydronicotinamideadenine dinucleotide to produce ADP, orthophosphate, and NADH. Hydrated NADPH will also act as a substrate. The 6S epimer is the true substrate; however, the epimers undergo rapid spontaneous equilibration.[1]

[1]S. A. Acheson, H. N. Kirkman & R. Wolfenden (1988) *Biochemistry* **27**, 7371.

ATP PHOSPHORIBOSYLTRANSFERASE

This transferase [EC 2.4.2.17], also known as phosphoribosyl-ATP pyrophosphorylase, catalyzes the reaction of

ATP with 5-phospho-α-D-ribose 1-diphosphate to produce 1-(5-phospho-D-ribosyl)-ATP and pyrophosphate (or, diphosphate).[1-5] The enzyme has an ordered Bi Bi kinetic mechanism.[2,3]

[1] W. D. L. Musick (1981) *Crit. Rev. Biochem.* **11**, 1.
[2] D. P. Morton & S. M. Parsons (1976) *ABB* **175**, 677.
[3] J. E. Kleeman & S. M. Parsons (1976) *ABB* **175**, 687.
[4] R. G. Martin (1963) *JBC* **238**, 257.
[5] B. N. Ames, R. G. Martin & B. J. Garry (1961) *JBC* **236**, 2019.

Selected entries from *Methods in Enzymology* [vol, page(s)]:
General discussion: 17B, 18, 40, 41; **46**, 294
Assay: 17B, 18
Other molecular properties: exchanges, **87**, 5; feedback inhibition, **17B**, 21; intermediate, possible, **87**, 6, 17; molecular weight, **17B**, 40, 41; protein-tRNA interaction studies, **59**, 323, 324, 328, 329; properties, **17B**, 40, 41; purification from *Salmonella typhimurium*, **17B**, 19; stereochemistry, **87**, 17; substrate contamination, **87**, 7, 17; subunits, **17B**, 40, 41

ATP PYROPHOSPHATASE

This enzyme [EC 3.6.1.8], also known as ATP diphosphatase and, confusingly, ATPase, catalyzes the hydrolysis of ATP to produce AMP and pyrophosphate (or, diphosphate).[1-3] Other substrates include ITP, GTP, CTP, UTP, β,γ-methyleneadenosine 5′-triphosphate (or, p(CH$_2$)ppA), and β,γ-imidoadenosine 5′-triphosphate (or, p(NH)ppA).[3] *See also* Apyrase

[1] H. T. Hsu (1983) *JBC* **258**, 3463.
[2] C. Torp-Pedersen, H. Flodgaard & T. Saermark (1979) *BBA* **571**, 94.
[3] H. Flodgaard & C. Torp-Pedersen (1978) *BJ* **171**, 817.

ATP SYNTHASE, H$^+$-TRANSPORTING

This enzyme [EC 3.6.3.14; previously classified as EC 3.6.1.34] also known as mitochondrial ATPase, chloroplast ATPase, and F$_o$ ATP synthase, utilizes the cell's transmembrane proton motive force to catalyze ATP synthesis from ADP and orthophosphate (P$_i$). In the absence of a proton gradient or in the presence of an uncoupling proton ionophore, the enzyme catalyzes the hydrolysis of ATP.[1-7] The mitochondrial complex consists of an oligomycin-sensitive membrane-bound F$_o$ component and the multisubunit α$_3$β$_3$γδε rotary motor complex that retains the ATPase activity, but lacks the proton-transport activity. Phosphoryl transfer is attended by inversion of stereochemical configuration at the phosphorus atom indicating that no phosphorylated enzyme intermediate is formed. The passage of proton through the channels within the F$_o$ subunit drives rotation and ATP synthesis.

Vacuolar ATP synthase and its Archebacter counterpart structurally resemble the mitochondrial ATP synthase, they are designed to extrude protons, rather than synthesize ATP, under typical physiological conditions. *See also CF$_1$CF$_o$ATP Synthase*

[1] P. D. Boyer (2000) *BBA* **1458**, 252.
[2] H. S. Penefsky & R. L. Cross (1991) *AE* **64**, 173.
[3] P. Mitchell (1979) *Science* **206**, 1148.
[4] A. Abrams & J. B. Smith (1974) *The Enzymes*, 3rd ed., **10**, 395.
[5] H. S. Penefsky (1974) *The Enzymes*, 3rd ed., **10**, 375.
[6] R. W. Estabrook & M. E. Maynard, eds. (1967) *MIE*, vol. **10**, Academic Press, New York.
[7] W. W. Kielley (1961) *The Enzymes*, 2nd ed., **5**, 149.

Selected entries from *Methods in Enzymology* [vol, page(s)]:
General discussion: 2, 593; 46, 83, 277, 278, 287; **55**, 302, 320, 333, 372; **56**, 584; **69**, 313; **97**, 510; **126**, 417, 545; **260**, 133, 163
Assay: 2, 593; 10, 506; 260, 168, 174; ATPase activity, **55**, 313, 314, 320, 324, 325; **260**, 140, 156, 168, 174; ATP-P$_i$ exchange reaction, for (principle, **55**, 310, 311; procedure, **55**, 312, 313; reagents, **55**, 311, 312); ATPase inhibitor protein, **260**, 184; ATP synthesis, **260**, 157; cellular respiration parameters, **260**, 149, 159; chloroplast, **97**, 513; **126**, 518; chromaffin granule-associated, **157**, 622; clathrin-coated vesicles, **157**, 634; endocytotic vesicles, in, **191**, 509, 517; endoplasmic reticulum-associated, **157**, 614; *Escherichia coli*, **126**, 548; Golgi membrane-associated, **157**, 614; inorganic phosphate exchange with ATP, **260**, 174; mutants, **56**, 113; oligomycin sensitivity conferral protein, **260**, 183; plasma membrane-associated, **157**, 513, 528, 534; proton translocation in *Neurospora crassa*, **174**, 668, 674; respiration substrate selection, **260**, 157; tonoplast-associated, **148**, 126; **157**, 584; vacuolar membrane-associated, **157**, 548; yeast (mitochondrial, assay, **194**, 658; plasma membrane, in, assay, **194**, 658; vacuolar, assay, **194**, 656)
Other molecular properties: alternate site participation, **64**, 82; ATPase TF$_o$ ^{o}F$_1$ (reconstitution, **56**, 600, 601; subunits [properties, **55**, 785; purification, **55**, 784, 785; TF$_1$ reconstitution, **55**, 786; TF$_o$ ^{o}F$_1$ reconstitution, **55**, 786, 787]; thermophilic bacterium, from, crystallization, **55**, 372); bacterial, F$_1$ portion (inactivating modifications, **126**, 713; noninactivating modifications, **126**, 729); bacterial, soluble, **6**, 279; blue-native gel electrophoresis, **260**, 195, 200; **264**, 560; bovine heart (differential solubilization, **260**, 195; F$_o$ [detergent extraction, **260**, 177; ion-exchange chromatography, **260**, 177; reconstitution, **260**, 179; removal from mitochondrial inner membranes, **260**, 175; subunit composition, **260**, 178]; F$_1$-ATPase [crystallization conditions, **260**, 168; crystal properties, **260**, 169; interaction with bacterially expressed F$_1$F$_o$-ATPase subunits, **260**, 185; nucleotide removal, **260**, 168; oligomycin sensitivity conferral protein interactions, **260**, 163, 185; purification, **260**, 166; subunit composition, **260**, 163, 167; X-ray crystallographic structural determination, **260**, 164]; F$_1$-ATPase stalk complex [activity, **260**, 189; purification, **260**, 186; reconstitution, **260**, 185, 188; stoichiometry, **260**, 189; structure, **260**, 163]; F$_1$F$_o$-ATPase [ATPase inhibitor protein, purification, **260**, 181; F$_6$, purification, **260**, 182; fragment bC, purification, **260**, 182; fragment d′, purification, **260**, 183; gel filtration, **260**, 172; ion-exchange chromatography, **260**, 173; oligomycin sensitivity conferral protein, purification, **260**, 181; subunit b, purification, **260**, 183; subunit d, purification, **260**, 183; subunit expression in *Escherichia coli*, **260**, 179; subunit structure, **260**, 175]); carbon dioxide-induced exocytotic insertion, regulation of urinary epithelial H$^+$ transport, **172**, 54; chloroplast (assay, **97**, 513; **126**, 518; coupling factor 1, **23**, 547 [activated enzyme, assay of, **23**, 549; assay, **23**, 253; assay of Ca^{2+}-ATPase activity of, **23**, 548; assay of coupling activity of, **23**, 547; chloroplast preparations deficient in, **23**, 251; photophosphorylating system of *Rhodopseudomonas capsulata*, in, **23**, 556; preparation, **23**, 550; properties, **23**, 555; purification by sucrose density gradient centrifugation, **23**, 553]; F$_1$ portion [inactivating modifications, **126**, 713; noninactivating modifications, **126**, 729; synthesis of enzyme-bound ATP, **126**, 643]; individual subunits [antibody preparation, **118**, 358; isolation, **118**, 358; quantitative estimation, **118**,

plasma membrane, in, assay, **194**, 658; properties of F$_I$ ATPase, **55**, 353, 354 [F$_I$ subunits, of, **55**, 356; oligomycin-sensitive ATPase, of, **55**, 356]; purification of F$_I$ ATPase, **55**, 352, 353 [F$_I$ subunits, of, **55**, 354, 355; sensitive ATPase, of, **55**, 356]; subunits, **55**, 358; subunit 6 [immunoprecipitation, **97**, 302; one-dimensional gel electrophoresis, **97**, 297; source, **97**, 304; two-dimensional gel electrophoresis, **97**, 299]; vacuolar, assay, **194**, 656); venturicidin, **55**, 506

ATRAZINE CHLOROHYDROLASE

This enzyme [EC 3.8.1.8] catalyzes the hydrolysis of the herbazide atrazine (*i.e.*, 2-chloro-4-(ethylamino)-6-(isopropylamino)-1,3,5-triazine) to produce 4-(ethylamino)-2-hydroxy-6-(isopropylamino)-1,3,5-triazine and hydrochloric acid.[1] Other substrates include simazine and desethylatrazine.

[1] M. L. de Souza, M. J. Sadowsky & L. P. Wackett (1996) *J. Bacteriol.* **178**, 4894.

ATROLYSIN A

This zinc metalloproteinase [EC 3.4.24.1], a highly potent hemorrhagic toxin (HT-a) from the venom of the diamondback rattlesnake, catalyzes peptide-bond hydrolysis in blood-clotting enzymes and can be assayed with the quenched-fluorescence substrate Abz–Ala–Gly-∇-Leu–Ala-nitrobenzylamide.[1-4] (Note: ∇ indicates the cleavage site.) Atrolysin A is weakly inhibited by α_2-macroglobulin.

See also *Atrolysin; Atrolysin B*

[1] J. B. Bjarnason & A. T. Tu (1978) *Biochemistry* **17**, 3395.
[2] J. B. Bjarnason & J. W. Fox (1995) *MIE* **248**, 345.
[3] J. W. Fox & J. B. Bjarnason (1995) *MIE* **248**, 369.
[4] J. W. Fox & J. B. Bjarnason (1996) in *Zinc Metalloproteinases in Health and Disease* (N. M. Hooper, ed.) Taylor & Francis, London, p. 47.

Selected entries from *Methods in Enzymology* [**vol**, page(s)]:
Other molecular properties: cDNA sequence, **248**, 363, 374; hemorrhagic activity, **248**, 352, 355; properties, **248**, 194, 360, 362, 368, 371, 376; purification, **248**, 369; structure, **248**, 363

ATROLYSIN B

This zinc metalloproteinase [EC 3.4.24.41], a hemorrhagic toxin (HT-b) from the venom of the diamondback rattlesnake, catalyzes peptide-bond hydrolysis in Abz–Ala–Gly–Leu-∇-Ala-nitrobenzylamide and protein substrates.[1,2] (Note: ∇ indicates the cleavage site.) Insulin B-chain is cleaved at His5–Leu6, His10–Leu11, Ala14–Leu15, Tyr16–Leu17, and Gly23–Phe24. When injected into muscle, atrolysin B causes localized necrosis of the tissue, a phenomena not often observed with other atrolysins.[3]

See also *Atrolysin; Atrolysin A*

[1] J. B. Bjarnason & A. T. Tu (1978) *Biochemistry* **17**, 3395.
[2] J. W. Fox & J. B. Bjarnason (1996) in *Zinc Metalloproteinases in Health and Disease* (N. M. Hooper, ed.) Taylor & Francis, London, p. 47.

[3] C. L. Ownby, J. B. Bjarnason & A. T. Tu (1978) *Amer. J. Pathol.* **93**, 201.

Selected entries from *Methods in Enzymology* [**vol**, page(s)]:
Other molecular properties: amino acid sequence, **248**, 349; cDNA sequences, **248**, 374; cleavage specificity on insulin B chain, **248**, 361; properties, **248**, 351, 356, 360, 368, 371, 373; purification, **248**, 369

ATROLYSIN C

This zinc metalloproteinase [EC 3.4.24.42], a hemorrhagic toxin (HT-c) from the venom of the western diamondback rattlesnake, catalyzes peptide-bond hydrolysis in Abz–Ala–Gly–Leu-∇-Ala-nitrobenzylamide and protein substrates.[1,2] (Note: ∇ indicates the cleavage site.) The insulin B chain is cleaved at His5–Leu6, His10–Leu11, Ala14–Leu15, Tyr16–Leu17, and Gly23–Phe24 (identical to the sites of atrolysin B). With small molecule substrates, atrolysin C appears to prefer proteins possessing a hydrophobic aminoacyl residue at P2′ and small aminoacyl residues such as alanyl or glycyl at P1. Atrolysin C action on fibrinogen produces nonclottable peptides, and this same metalloproteinase will digest type IV collagen.[3,4] ***See also*** *Atrolysin; Atrolysin B*

[1] J. B. Bjarnason & A. T. Tu (1978) *Biochemistry* **17**, 3395.
[2] L. A. Hite, L.-G. Jia, J. B. Bjarnason & J. W. Fox (1994) *ABB* **308**, 182.
[3] J. W. Fox & J. B. Bjarnason (1996) in *Zinc Metalloproteinases in Health and Disease* (N. M. Hooper, ed.) Taylor & Francis, London, p. 47.
[4] J. B. Bjarnason & J. W. Fox (1994) *J. Pharmacol. Therapeut.* **62**, 325.

Selected entries from *Methods in Enzymology* [**vol**, page(s)]:
Other molecular properties: amino acid sequence, **248**, 349; atrolysin Cc (cDNA sequences, **248**, 374; properties, **248**, 368, 371; purification, **248**, 369); atrolysin Cd (cDNA sequences, **248**, 374; properties, **248**, 194, 368, 371; purification, **248**, 369); calcium binding site, **248**, 357; cleavage specificity on insulin B chain, **248**, 361; isozymes, **248**, 349; properties, **248**, 194, 200, 351, 356, 360, 373; structure, **248**, 346, 361, 366

ATROLYSIN E

This zinc metalloproteinase [EC 3.4.24.44], a hemorrhagic toxin (HT-e) from the venom of the western diamondback rattlesnake, catalyzes peptide-bond hydrolysis in Abz–Ala–Gly–Leu-∇-Ala-nitrobenzylamide and protein substrates.[1-3] (Note: ∇ indicates the cleavage site.) The first site cleaved in the insulin B-chain is Ala14–Leu15, followed by Ser9–His10 and Asn3–Gln4.[2] Atrolysin E cleaves type IV collagen, laminin, and nidogen at specific sites.[4] The basement membrane surrounding endothelial cells is degraded or lost.

[1] J. B. Bjarnason & A. T. Tu (1978) *Biochemistry* **17**, 3395.
[2] J. W. Fox & J. B. Bjarnason (1995) *MIE* **248**, 369.
[3] J. W. Fox & J. B. Bjarnason (1996) in *Zinc Metalloproteinases in Health and Disease* (N. M. Hooper, ed.) Taylor & Francis, London, p. 47.

[4]E. N. Baramova, J. D. Shannon, J. B. Bjarnason, S. L. Gonias & J. W. Fox (1990) *Biochemistry* **29**, 1158.

Selected entries from *Methods in Enzymology* [vol, page(s)]:
Other molecular properties: cDNA sequences, **248**, 374; cleavage of oxidized insulin B chain, **248**, 370; hemorrhagic activity, **248**, 370; inhibitors, **248**, 372; molecular mass, **248**, 370; properties, **248**, 194, 199, 351, 356, 360, 368; purification, **248**, 369; specificity, **248**, 370; substrates, **248**, 370; synthesis, **248**, 362

ATROLYSIN F

This zinc metalloproteinase [EC 3.4.24.45], a hemorrhagic toxin (HT-f) from the venom of the western diamondback rattlesnake, cleaves the insulin B-chain at Val2–Asn3, Gln3–His5, Leu6–Cys7, His10–Leu11, Ala14–Leu15, and Tyr16–Leu17 bonds. Although this cleavage pattern closely resembles that of atrolysin A, antibodies generated against atrolysin F fail to cross-react with other atrolysins.[1-3]

[1]T. Nikai, N. Mori, M. Kishida, H. Sugihara & A. T. Tu (1984) *ABB* **231**, 309.
[2]J. W. Fox & J. B. Bjarnason (1996) in *Zinc Metalloproteinases in Health and Disease* (N. M. Hooper, ed.) Taylor & Francis, London, p. 47.
[3]J. B. Bjarnason & J. W. Fox (1994) *J. Pharmacol. Therapeut.* **62**, 325.

Selected entries from *Methods in Enzymology* [vol, page(s)]:
Other molecular properties: properties, **248**, 351, 368, 371, 377; purification, **248**, 369

ATROXASE

This zinc-dependent nonhemorrhagic endopeptidase [EC 3.4.24.43], a nonhemorrhagic metalloendopeptidase from the venom of the western diamondback rattlesnake, catalyzes the hydrolysis of peptide bonds in proteins and oligopeptides. Action on the insulin B chain results in bond-scission at His5–Leu6, Ser9–His10, His10–Leu11, Ala14–Leu15, and Tyr16–Leu17. The fibrinogen γ chain resists atroxase action.[1-3]

[1]T. W. Willis & A. T. Tu (1988) *Biochemistry* **27**, 4769.
[2]B. J. Baker & A. T. Tu (1996) in *Natural Toxins II* (B. R. Singh & A. T. Tu, eds.), Plenum Press, New York, p. 203.
[3]N. Marsh (1994) *Thromb. Haemost.* **71**, 793.

Selected entries from *Methods in Enzymology* [vol, page(s)]:
Other molecular properties: cleavage specificity on insulin B chain, **248**, 361; properties, **248**, 352, 358, 378

AUREOLYSIN

This zinc-dependent, extracellular metalloproteinase [EC 3.4.24.29] (also known as protease III, *Staphylococcus aureus* neutral proteinase, and staphylococcal metalloprotease) catalyzes peptide-bond hydrolysis with a selectivity resembling thermolysin. The insulin B-chain is hydrolyzed at His5–Leu6, His10–Leu11, Ala14–Leu15, Tyr16–Leu17, Gly23–Phe24, and Phe25–Tyr26. Hydrophobic aminoacyl residues are apparently preferred at P1'.[1,2] While lacking elastinolytic activity, aureolysin activates *S. aureus* glutamyl endopeptidase [EC 3.4.21.19].

[1]S. Arvidson, T. Holme & B. Lindholm (1972) *Acta Pathol. Microbiol. Scand. Sect. B* **80**, 835.
[2]C. C. Häse & R. A. Finkelstein (1993) *Microbiol. Rev.* **57**, 823.

AZOBENZENE REDUCTASE

This oxidoreductase [EC 1.6.6.7], also called dimethylaminoazobenzene reductase, catalyzes the NADPH-dependent reductive cleavage of 4-(dimethylamino)-azobenzene to produce $NADP^+$, *N,N*-dimethyl-1,4-phenylenediamine, and aniline.[1]

4-(dimethylamino)azobenzene

aniline   *N,N*-dimethyl-1,4-phenylenediamine

The enzyme also acts on methyl red (*i.e.*, 2′-carboxy-4-*N,N*-dimethylazobenzene) and orange II (4-[(2-hydroxy-1-naphthalenyl)azo]benzenesulfonic acid). P450 type cytochromes are responsible for the azoreductase activity in liver microsomes.[2] The antipsychotic agent chlorpromazine reportedly inhibits the reductase.[2]

[1]S. Fujita & J. Peisach (1978) *JBC* **253**, 4512.
[2]M. H. Mostafa & E. K. Weisburger (1980) *J. Natl. Cancer. Inst.* **64**, 925.

B

BACILLOLYSIN

This extracellular zinc metalloendopeptidase [EC 3.4.24.28], also referred to as megateriopeptidase, catalyzes an activity very similar to that of thermolysin.[1,2] Substrate specificity studies suggest that a hydrophobic residue is needed at the P1' position. However, bacillolysin does not appear to be as thermally stable as thermolysin.

[1] W. Stark, R. A. Pauptit, K. S Wilson & J. N. Jansonius (1992) *EJB* **207**, 781.
[2] C. Häse & R. A. Finkelstein (1993) *Microbiol. Rev.* **57**, 823.

BACILLUS SUBTILIS RIBONUCLEASE

This enzyme [EC 3.1.27.2] catalyzes the endonucleolytic cleavage of ribonucleic acids to produce 2',3'-cyclic nucleoside monophosphates.[1]

[1] M. Yamasaki & K. Arima (1970) *BBA* **209**, 475.

BACITRACIN SYNTHETASE

This complex catalyzes the ATP-dependent synthesis of bacitracins, the branched cyclic dodecylpeptide antibiotics containing both L- and D-amino acyl residues produced by *Bacillus subtilis* and *B. licheniformis*. Several aminoacyladenylate intermediates are formed during the course of the multistep reaction.

This high-molecular-weight multienzyme complex activates and incorporates the constituent amino acids by a thiotemplate mechanism in which aminoacyl-AMP intermediates are transferred to a covalently enzyme-bound 4'-phosphopantetheinyl-group as thiolesters. The product is generated by the sequential incorporation of the thiolesterified amino acids in a series of transpeptidation reactions. The thiazoline ring within the bacitracins is formed from an isoleucylcysteine dipeptide intermediate early in the reaction scheme.

Selected entries from *Methods in Enzymology* [vol, page(s)]:
General discussion: 43, 548
Assay: 43, 551
Other molecular properties: ammonium sulfate precipitation, **43**, 558; ATP-^{32}PP$_i$ exchange measurement, **43**, 551, 553, 555, 558, 559; *Bacillus licheniformis* preparation, **43**, 198; cell, harvesting and storage, **43**, 556; cell lysis, **43**, 556; DEAE-cellulose chromatography, **43**, 557; diafiltration and concentration, **43**, 557; disc gel electrophoresis, **43**, 554, 558; fermentation, **43**, 554; *Micrococcus flavus* preparation, **43**, 551; Millipore filter test, **43**, 552, 553, 558; molecular weight, **43**, 555; protein determination, **43**, 554; purification, **43**, 556; radio thin-layer chromatography, **43**, 552; reaction scheme, **43**, 549; SDS disc gel electrophoresis, **43**, 554; Sephadex G-50 chromatography, **43**, 556; Sephadex G-200 chromatography, **43**, 557; sucrose density gradient centrifugation, **43**, 554; unit definition, **43**, 549

BACTERIAL *O*-ANTIGEN LIGASE

This ligase catalyzes the reaction of (-D-mannosyl[-abequosyl]-rhamnosyl)$_n$-D-mannosyl[-abequosyl]-rhamnosyl-D-galactose-1-diphospholipid with a core polysaccharide to produce the diphospho lipid and the modified core polysaccharide.

Selected entries from *Methods in Enzymology* [vol, page(s)]:
Assay: 28, 588, 591, 600
Other molecular properties: bacterial *O*-antigen biosynthesis and, **28**, 583, 591; isolation from *Salmonella typhimurium*, **28**, 597; properties, **28**, 601; *Salmonella typhimurium* mutations and, **28**, 591, 601

BACTERIAL *O*-ANTIGEN POLYMERASE

This enzyme catalyzes the polymerization of n molecules of D-mannosyl[-abequosyl]-rhamnosyl-D-galactose-1-diphospholipid to produce $n - 1$ molecules of the diphospholipid and -(D-mannosyl[-abequosyl]-rhamnosyl)$_{n-1}$-D-mannosyl[-abequosyl]-rhamnosyl-D-galactose-1-diphospholipid.

Selected entries from *Methods in Enzymology* [vol, page(s)]:
Assay: 28, 588, 600
Other molecular properties: biosynthesis of bacterial *O*-antigen and, **28**, 583, 599; isolation from *Salmonella typhimurium*, **28**, 586, 598; *Salmonella typhimurium* mutations and, **28**, 599

BACTERIAL PROTEASOME

This high-molecular-weight, barrel-like multi-subunit protease [EC 3.4.25.1] exhibits a specificity that is similar to that of chymotrypsin, with good activity with

Suc-Leu-Leu-Val-Tyr-∇-NHMec.[1,2] (Note: ∇ indicates the cleavage site.) **See also** *Archaean Proteasome; Multicatalytic Endopeptidase Complex*

[1] T. Tamura, I. Nagy, A. Lupas, F. Lottspeich, Z. Cejka, G. Schoofs, K. Tanaka, R. De Mot & W. Baumeister (1995) *Curr. Biol.* **5**, 766.

[2] F. Zühl, T. Tamura, I. Dolene, Z. Cejka, I. Nagy, R. De Mot & W. Baumeister (1997) *FEBS Lett.* **400**, 83.

BACTERIOPHAGE T4 PROHEAD ENDOPEPTIDASE

This protease, also referred to as T4 prohead proteinase, prohead endopeptidase, and gp21, is a key component in the morphopoietic pathway of the bacteriophage prohead.[1–3] Its only substrates known are the T4 prohead proteins with cleavage occurring most often at Glu–Ala bonds.[1,2]

[1] M. K. Showe, E. Isobe & L. Onorato (1976) *J. Mol. Biol.* **107**, 55.

[2] J. M. Mullaney & L. W. Black (1996) *J. Mol. Biol.* **261**, 372.

*Bae*I RESTRICTION ENDONUCLEASE

This type II restriction endonuclease (subtype s) [EC 3.1.21.4], obtained from *Bacillus sphaericus*, catalyzes the hydrolysis of both strands of DNA at $5' \ldots \nabla(N)_{15}AC(N)_4GTAYC(N)_{12}\nabla \ldots 3'$ and $3' \ldots \nabla(N)_{15}TG(N)_4CATRG(N)_7\nabla \ldots 5'$, where R refers to either A or G, Y refers to either T or C, and N refers to any base. Note that this endonuclease cleaves DNA twice to excise the recognition site, releasing a 28-base fragment with five-base 3' extensions.[1,2] The enzyme requires the presence of *S*-adenosyl-L-methionine for optimal activity.

[1] L. E. Sears, B. Zhou, J. M. Aliotta, R. D. Morgan & H. Kong (1996) *NAR* **24**, 3590.

[2] S. E. Halford, D. T. Bilcock, N. P. Stanford, S. A. Williams, S. E. Milsom, N. A. Gormley, M. A. Watson, A. J. Bath, M. L. Embleton, D. M. Gowers, L. E. Daniels, S. H. Parry & M. D. Szczelkun (1999) *Biochem. Soc. Trans.* **27**, 696.

*Bal*I RESTRICTION ENDONUCLEASE

This type II restriction endonuclease [EC 3.1.21.4] is obtained from *Brevibacterium albidum*. It acts on both strands of DNA at $5' \ldots TGG\nabla CCA \ldots 3'$, producing blunt-ended fragments.[1]

[1] R. E. Gelinas, P. A. Myers, G. H. Weiss, R. J. Roberts & K. Murray (1977) *J. Mol. Biol.* **114**, 433.

Selected entries from **Methods in Enzymology** [vol, page(s)]:
General discussion: **65**, 446
Other molecular properties: recognition sequence, **65**, 5

*Bam*FI RESTRICTION ENDONUCLEASE

This type II restriction endonuclease [EC 3.1.21.4] is obtained from *Bacillus amyloliquefaciens* F. The enzyme recognizes the DNA sequence $5' \ldots GGATCC \ldots 3'$, but the site of action is unclear.[1]

[1] T. Shibata, S. Ikawa, C. Kim & T. Ando (1976) *J. Bacteriol.* **128**, 473.

Selected entries from **Methods in Enzymology** [vol, page(s)]:
Other molecular properties: recognition sequence, **65**, 3

*Bam*HI RESTRICTION ENDONUCLEASE

This type II restriction endonuclease [EC 3.1.21.4] is Mg^{2+}-dependent, is obtained from *Bacillus amyloliquefaciens* H, and acts on both strands of DNA at $5' \ldots G\nabla GATCC \ldots 3'$.[1–3] The presence of hydrophobic agents such as glycerol or dimethylsulfoxide will alter the recognition sequence.[4] The reaction mechanism is reported to act in two steps.[5] Interestingly, both strands of the DNA duplex appear to be cleaved almost simultaneously.[6] Dimeric and tetrameric forms of the enzyme have been observed[5], and the enzyme has been found to form large aggregates.[7] The K_m value for pJC 80 DNA was determined to be 0.36 nM.[7] An active-site arginyl residue is crucial in sequence recognition.[8] A kinetic model of *Bam* HI has been presented,[9,11,] the reaction rates being faster for longer DNA substrates.[10] In addition, osmotic pressure has been reported to alter the specificity for the substrate, implicating a role for bound water.[12,13]

[1] G. A. Wilson & F. E. Young (1975) *Abstr. Gen. Meet. Am. Soc. Microbiol.* **75**, 103.

[2] G. A. Wilson & F. E. Young (1975) *J. Mol. Biol.* **97**, 123.

[3] R. J. Roberts, G. A. Wilson & F. E. Young (1977) *Nature* **265**, 82.

[4] J. George, R. W. Blakesley & J. G. Chirikjian (1980) *JBC* **255**, 6521.

[5] L. A. Smith & J. G. Chirikjian (1979) *JBC* **254**, 1003.

[6] S. E. Halford, N. P. Johnson & J. Grinsted (1979) *BJ* **179**, 353.

[7] B. Hinsch & M.-R. Kula (1980) *Nucleic Acids Res.* **8**, 623.

[8] J. George, G. Nardone & J. G. Chirikjian (1985) *JBC* **260**, 14387.

[9] G. Nardone, M. Wastney, P. Hensely & J. G. Chirikjian (1986) *Fed. Proc.* **45**, 1504.

[10] G. Nardone, J. George & J. G. Chirikjian (1986) *JBC* **261**, 12128.

[11] P. Hensley, G. Nardone, J. G. Chirikjian & M. E. Wastney (1990) *JBC* **265**, 15300.

[12] C. R. Robinson & S. G. Sligar (1994) *Biophys. J.* **66**, A34.

[13] C. R. Robinson & S. G. Sligar (1995) *PNAS* **92**, 3444.

Selected entries from **Methods in Enzymology** [vol, page(s)]:
General discussion: **65**, 23, 34, 35, 147, 361, 363, 385, 432, 450, 462, 493, 566, 787, 789, 793, 799, 823

Other molecular properties: cation requirement, **65**, 150; cleavage sites, of SV40 DNA, **65**, 713; fragment end structure produced, **65**, 511; immobilization, **65**, 174 (codigestion studies, **65**, 181, 182; reaction kinetics, **65**, 178; stability, **65**, 178); isoschizomers, **65**, 137, 170; molecular weight, **65**, 150; properties, **65**, 150, 151; recognition sequence, **65**, 3, 147, 151, 511; stability, **65**, 178; use in linkage studies, **76**, 816

*Bam*N$_X$I RESTRICTION ENDONUCLEASE

This type II restriction endonuclease [EC 3.1.21.4] requires Mg^{2+} and is obtained from *Bacillus amyloliquefaciens* N. The enzyme acts on both strands of DNA at 5′ ... G∇GWCC... 3′, where W refers to either A or T.[1,2]

[1] T. Shibata & T. Ando (1976) *BBA* **442**, 184.
[2] S. Ikawa, T. Shibata & T. Ando (1979) *Agric. Biol. Chem.* **43**, 873.

Selected entries from *Methods in Enzymology* [vol, page(s)]:
General discussion: 65, 3

*Ban*I RESTRICTION ENDONUCLEASE

This type II restriction endonuclease [EC 3.1.21.4], obtained from *Bacillus aneurinolyticus*, catalyzes the hydrolysis of both strands of DNA at 5′ ...G∇GYRCC ... 3′, where Y refers to either T or C and R refers to either A or G.[1,2] *Ban*I binds in the major groove of DNA and the central pyrimidines, particularly thymines, make important contacts with the protein.[3] This endonuclease is inhibited by salt concentrations larger than 150 mM. It can also exhibit star activity. ***See "Star" Activity***

[1] H. Sugisaki, Y. Maekawa, S. Kanazawa & M. Takanami (1982) *NAR* **10**, 5747.
[2] I. Schildkraut, J. Lynch & R. Morgan (1987) *NAR* **15**, 5492.
[3] S. Advani & K. B. Roy (2000) *BBRC* **269**, 35.

*Ban*II RESTRICTION ENDONUCLEASE

This type II restriction endonuclease [EC 3.1.21.4], obtained from *Bacillus aneurinolyticus*, catalyzes the hydrolysis of both strands of DNA at 5′ ...GRGCY∇C ... 3′, where Y refers to either T or C and R refers to either A or G.[1]

[1] H. Sugisaki, Y. Maekawa, S. Kanazawa & M. Takanami (1982) *NAR* **10**, 5747.

BARBITURASE

barbituric acid   urea   malonate

This cytosolic enzyme [EC 3.5.2.1], which participates in pyrimidine catabolism in certain microorganisms, catalyzes the hydrolysis of barbiturate to produce malonate and urea.[1]

[1] O. Hayaishi & A. Kornberg (1952) *JBC* **197**, 717.

Selected entries from *Methods in Enzymology* [vol, page(s)]:
General discussion: 2, 492
Assay: 2, 492

BARRIERPEPSIN

This heavily glycosylated aspartic peptidase [EC 3.4.23.35], also known as extracellular "barrier" protein and bar proteinase, catalyzes the selective hydrolysis of the Leu6–Lys7 peptide bond in a diffusible peptide hormone (also known as α factor) of the budding yeast *Saccharomyces cerevisiae*.[1]

[1] V. L. MacKay, J. Armstrong, C. Yip, S. Welch, K. Walker, S. Osborn, P. Sheppard & J. Forstrom (1991) in *Proc. of the Aspartic Proteinase Conference on Structure and Function of the Aspartic Proteinases* (B. M. Dunn, ed.), Plenum Publ., New York, p. 161.

BASILYSIN

This zinc-dependent proteinase, formerly called basilase, catalyzes the hydrolysis of the α and β chains of fibrin and fibrinogen and is nonhemorrhagic.[1] Specificity studies suggest that Ala–Leu, Tyr–Leu, and Lys–Leu bonds are preferentially hydrolyzed.

[1] G. Datta, A. Dong, J. Witt & A. T. Tu (1995) *ABB* **317**, 365.

*Bbr*I RESTRICTION ENDONUCLEASE

This type II restriction endonuclease [EC 3.1.21.4] is obtained from *Bordetella bronchiseptica* and the enzyme acts on both strands of DNA at 5′ ...A∇AGCTT... 3′.[1]

[1] V. N. Kalinin, I. A. Lapaeva, V. G. Lunin, E. A. Skripkin, V. D. Smirnov & T. I. Tikchonenko (1986) *Mol. Gen. Mikrobiol. Virusol.* **9**, 16.

Selected entries from *Methods in Enzymology* [vol, page(s)]:
Other molecular properties: recognition sequence, **65**, 5

*Bbs*I RESTRICTION ENDONUCLEASE

This type II restriction endonuclease, subtype s, [EC 3.1.21.4] from *Bacillus laterosporus* acts on DNA at 5′ ... GAAGACNN∇ ... 3′ and 3′ ...CTTCTGNNNNNNN∇ ...5′ where N refers to any base. *Bbs*I is unable to hydrolyze an internally thiomodified phosphodiester.[1]

[1] J. A. Schenk, S. Heymann & B. Micheel (1995) *Biochem. Mol. Biol. Int.* **36**, 1037.

*Bbv*I RESTRICTION ENDONUCLEASE

This type II restriction endonuclease (subtype s) [EC 3.1.21.4] is obtained from *Bacillus brevis* and it recognizes and acts at the DNA sequence 5′ ... GCACGNNNNNNNN ∇NNNN ... 3′ and 3′ ... CGTCGNNNNNNNNNNNNN∇ ... 5′, where N represents any base.[1]

[1] T. R. Gingeras, J. P. Milazzo & R. J. Roberts (1978) *Nucleic Acids Res.* **5**, 4105.

Selected entries from *Methods in Enzymology* [vol, page(s)]:
Other molecular properties: recognition sequence, **65**, 3

*Bbv*CI RESTRICTION ENDONUCLEASE

This type II restriction endonuclease (subtype s) [EC 3.1.21.4] is obtained from *Bacillus brevis* and recognizes and acts at the DNA sequence 5′ . . . CC∇TCAGC . . . 3′ and 3′ . . . GGAGT∇CG . . . 5′.

*Bcg*I RESTRICTION ENDONUCLEASE

This type II restriction endonuclease (subtype s) [EC 3.1.21.4] is obtained from *Bacillus coagulans* and recognizes and acts on DNA at the sequence 5′ . . . ∇(N)$_{10}$CGA(N)$_6$TCG(N)$_{12}$∇ . . . 3′ and 3′ . . . ∇ (N)$_{12}$GCT(N)$_6$ACG(N)$_{10}$∇ . . . 5′, where N refers to any base.[1,2] Note that the endonuclease cleaves DNA twice and excises the recognition site. A thirty-two base-pair fragment is generated containing two-base 3′ extensions. This endonuclease requires the presence of *S*-adenosyl-L-methionine for maximal activity.

[1]H. Kong, R. D. Morgan, R. E. Maunus & I. Schildkraut (1993) *NAR* **21**, 987.
[2]H. Kong, S. E. Roemer, P. A. Waite-Rees, J. S. Benner, G. G. Wilson & D. O. Nwankwo (1994) *JBC* **269**, 683.

*Bci*VI RESTRICTION ENDONUCLEASE

This type II restriction endonuclease (subtype s) [EC 3.1.21.4], obtained from *Bacillus circulans*, acts on both strands of DNA at 5′ . . . GTATCCNNNNNN∇ . . . 3′ and 3′ . . . CATAGGNNNNN∇ . . . 5′, where N refers to any base.[1] Note that this restriction endonuclease produces DNA fragments with single-base 3′-extensions which are more difficult to ligate than blunt-ended fragments.

[1]R. J. Roberts & D. Macelis (1997) *NAR* **25**, 248.

*Bcl*I RESTRICTION ENDONUCLEASE

This type II restriction endonuclease [EC 3.1.21.4] is obtained from *Bacillus caldolyticus* and it acts on both strands of DNA at the sequence 5′ . . . T∇GATCA . . . 3′.[1] Notice that the stop triplet TAG is present in the 3′-5′ strand.[2] The enzyme is blocked by *dam* methylation. The product fragments have a 5′ GATC extension and this extension can be readily ligated to DNA fragments generated by *Bam*HI, *Bgl*II, *Bst*YI, *Mbo*I, and *Sau*3AI.

[1]A. H. A. Bingham, T. Atkinson, D. Sciaky & R. J. Roberts (1978) *Nucleic Acids Res.* **5**, 3457.
[2]C. C. Rowland, P. P. Lim & R. E. Glass (1992) *Gene* **116**, 21.

Selected entries from ***Methods in Enzymology*** [**vol**, page(s)]:
General discussion: **65**, 3, 151

BENZALDEHYDE DEHYDROGENASES

Benzaldehyde dehydrogenase (NAD$^+$) [EC 1.2.1.28; formerly classified as EC 1.2.1.6] catalyzes the reaction of benzaldehyde with NAD$^+$ and water to produce benzoate and NADH.[1–4] A number of substituted benzaldehydes are also substrates (for example, 2-methylbenzaldehyde, 3-methylbenzaldehyde, and 4-methylbenzaldehyde). Interestingly, substitution at the *o*-position with electron-donating groups essentially abolished reactivity.[1]

Benzaldehyde dehydrogenase (NADP$^+$) [EC 1.2.1.7] catalyzes the reaction of benzaldehyde with NADP$^+$ and water to produce benzoate and NADPH.[5,6] Certain substituted benzaldehydes are also substrates.

[1]J. P. Shaw, F. Schwager & S. Harayama (1992) *BJ* **283**, 789.
[2]J. P. Shaw & S. Harayama (1990) *EJB* **191**, 705.
[3]R. M. Chalmers, A. J. Scott & C. A. Fewson (1990) *J. Gen. Microbiol.* **136**, 637.
[4]R. W. MacIntosh & C. A. Fewson (1988) *BJ* **250**, 743 and **255**, 653.
[5]U. Altenschmidt & G. Fuchs (1991) *Arch. Microbiol.* **156**, 152.
[6]C. S. Stachow, I. L. Stevenson & D. Day (1967) *JBC* **242**, 5294.

Selected entries from ***Methods in Enzymology*** [**vol**, page(s)]:
General discussion: **2**, 273, 280
Assay: **2**, 280

BENZENE 1,2-DIOXYGENASE

This FAD- and iron-dependent oxidoreductase [EC 1.14.12.3], also known as benzene hydroxylase, catalyzes the reaction of benzene with NADH and dioxygen to produce *cis*-1,2-dihydrobenzene-1,2-diol (*i.e.*, *cis*-1,2-dihydroxycyclohexa-3,5-diene) and NAD$^+$.[1–3]

cis-1,2-dihydrobenzene 1,2-diol

This multiprotein complex contains an iron–sulfur flavoprotein, an iron–sulfur oxygenase, and ferredoxin. Other substrates include toluene, ethylbenzene, *n*-propylbenzene, fluorobenzene, chlorobenzene, and bromobenzene. ***See also*** *Phthalate Dioxygenase*

[1]B. G. Fox (1998) *CBC* **3**, 261.
[2]S. E. Crutcher & P. J. Geary (1979) *BJ* **177**, 393.
[3]O. Hayaishi, M. Nozaki & M. T. Abbott (1975) *The Enzymes*, 3rd ed., **12**, 119.

Selected entries from ***Methods in Enzymology*** [**vol**, page(s)]:
General discussion: **52**, 11, 12; **188**, 52
Assay: oxygen electrode assay, **188**, 53; spectrophotometric assay, **188**, 54

Other molecular properties: detection by antibody probing, **188**, 55; ferredoxin component, purification, **188**, 59; oxygen electrode assay, **188**, 53; purification, **188**, 56; reductase component, purification, **188**, 58; spectrophotometric assay, **188**, 54; terminal dioxygenase (properties, **188**, 59; purification, **188**, 58; subunit preparation, **188**, 59)

BENZOATE:CoA LIGASE

This ligase [EC 6.2.1.25], also known as benzoyl-CoA synthetase, catalyzes the reaction of ATP with benzoate and coenzyme A to produce AMP, pyrophosphate (or, diphosphate), and benzoyl-CoA.[1] Alternative substrates include 2-, 3-, and 4-fluorobenzoate as well as 2-aminobenzoate. The corresponding chloro- analogues are weak substrates.

[1] J. Gibson, J. F. Geissler & C. S. Harwood (1990) *Meth. Enzymol.* **188**, 154.

Selected entries from **Methods in Enzymology** [vol, page(s)]:
General discussion: 188, 154
Assay: 188, 154
Other molecular properties: properties, **188**, 157; purification from *Rhodopseudomonas palustris*, **188**, 156

BENZOATE 1,2-DIOXYGENASE

This iron-dependent oxidoreductase [EC 1.14.12.10; formerly listed as EC 1.13.99.2], also referred to as benzoate hydroxylase and benzoic hydroxylase, catalyzes the reaction of benzoate with NADH and dioxygen to produce catechol, carbon dioxide, and NAD^+.[1,2] 1-Carboxy-*cis*-1,2-dihydroxy-3,5-cyclohexadiene is a transiently formed intermediate.

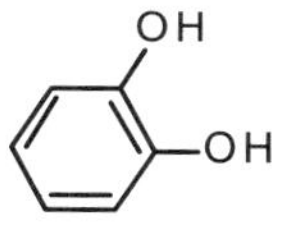

catechol

Benzoate 1,2-dioxygenase is a protein complex containing an iron–sulfur FAD-dependent reductase and an iron–sulfur oxygenase. Other substrates include 3-fluorobenzoate, 3-chlorobenzoate, 3-aminobenzoate, 4-fluorobenzoate, and 3-methylbenzoate.

[1] M. Yamaguchi & H. Fujisawa (1980) *JBC* **255**, 5058.
[2] M. Yamaguchi & H. Fujisawa (1982) *JBC* **257**, 12497.

Selected entries from **Methods in Enzymology** [vol, page(s)]:
General discussion: 52, 12

BENZOATE 4-MONOOXYGENASE

This iron- and tetrahydropteridine-dependent oxidoreductase [EC 1.14.13.12], also referred to as benzoate 4-hydroxylase, catalyzes the reaction of benzoate with NADPH and dioxygen to produce 4-hydroxybenzoate,

$NADP^+$, and water.[1–3] Other substrates include 2-chlorobenzoate and 3-chlorobenzoate. A cytochrome P450 system from *Rhodotorula minuta* also catalyzes this reaction.[3]

[1] C. McNamee & D. R. Durham (1985) *BBRC* **129**, 485.
[2] C. C. Reddy & C. S. Vaidyanathan (1976) *ABB* **177**, 488.
[3] H. Fukuda, K. Nakamura, E. Sukita, T. Ogawa & T. Fujii (1996) *J. Biochem.* **119**, 314.

Selected entries from **Methods in Enzymology** [vol, page(s)]:
General discussion: 52, 12

BENZOIN ALDOLASE

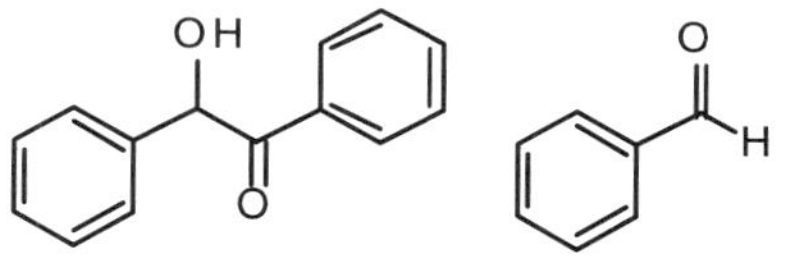

benzoin   benzaldehyde

This aldolase [EC 4.1.2.38], also called benzaldehyde lyase, catalyzes the thiamin-pyrophosphate-dependent conversion of benzoin to two molecules of benzaldehyde. Anisoin (*i.e.*, 4,4'-dimethoxybenzoin) is an alternative substrate.[1]

[1] B. Gonzalez & R. Vicuna (1989) *J. Bacteriol.* **171**, 2401.

BENZOPHENONE SYNTHASE

This enzyme [EC 2.3.1.151] reportedly catalyzes a multistep reaction in which three molecules of malonyl-CoA react with 3-hydroxybenzoyl-CoA to produce four molecules of coenzyme A, three molecules of carbon dioxide, and 2,3',4,6-tetrahydroxybenzophenone.[1]

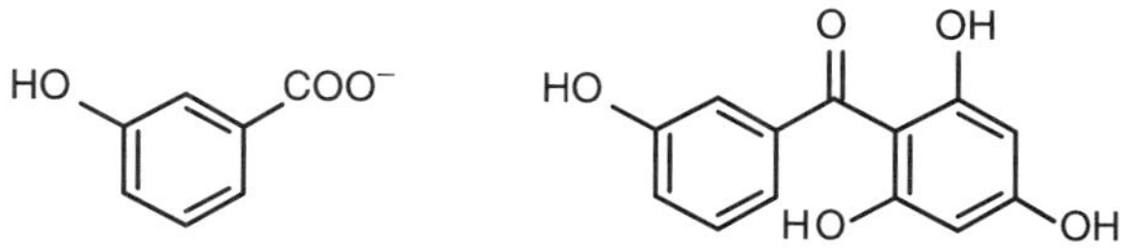

3-hydroxybenzoate   2,3',4,6-tetrahydroxybenzophenone

The enzyme participates in the biosynthesis of plant xanthones. Benzoyl-CoA can replace 3-hydroxybenzoyl-CoA as an alternative substrate.

[1] L. Beerhues (1996) *FEBS Lett.* **383**, 264.

BENZO[a]PYRENE 3-MONOOXYGENASE

This oxidoreductase, formerly classified as EC 1.14.14.2 and a deleted entry, is now listed with other unspecific monooxygenases [EC 1.14.14.1].

p-BENZOQUINONE REDUCTASE (NADPH)

1,4-benzoquinone   1,4-hydroquinone

This oxidoreductase [EC 1.6.5.6] catalyzes the reaction of *p*-benzoquinone with NADPH and a proton to produce hydroquinone (also called 1,4-benzenediol) and NADP⁺.[1] This enzyme participates in the bacterial degradation of *p*-nitrophenol.

[1]J. C. Spain & D. T. Gibson (1991) *Appl. Environ. Microbiol.* **57**, 812.

D-BENZOYLARGININE-4-NITROANILIDE AMIDASE

N-benzoyl-D-arginine 4-nitroanilide   *N*-benzoyl-D-arginine

This stereospecific amidase [EC 3.5.1.72], first isolated from *Bacillus* sp., catalyzes the hydrolysis of *N*-benzoyl-D-arginine-4-nitroanilide to produce *N*-benzoyl-D-arginine and 4-nitroaniline.[1]

[1]L. V. Gofshtein-Gandman, A. Keynan & Y. Milner (1988) *J. Bacteriol.* **170**, 5895.

BENZOYLCHOLINESTERASE

This enzyme activity, formerly classified as EC 3.1.1.8 and now a deleted Enzyme Commission entry, is regarded as a side reaction of cholinesterase [EC 3.1.1.7].

BENZOYL-CoA 3-MONOOXYGENASE

This FAD- or FMN-dependent oxidoreductase [EC 1.14.13.58], also known as benzoyl-CoA 3-hydroxylase, catalyzes the reaction of benzoyl-CoA with NADPH and dioxygen to produce 3-hydroxybenzoyl-CoA, NADP⁺, and water.[1]

[1]R. Niemetz, U. Altenschmidt, S. Brucker & G. Fuchs (1995) *EJB* **227**, 161.

BENZOYL-CoA REDUCTASE

benzoyl-CoA   cyclohexa-1,5-diene-1-carbonyl-CoA

This iron-sulfur-dependent oxidoreductase system [EC 1.3.99.15], also referred to as benzoyl-CoA reductase (dearomatizing), catalyzes the reaction of benzoyl-CoA with a reduced donor substrate and two molecules of ATP to produce cyclohexa-1,5-diene-1-carbonyl-CoA, the oxidized donor, two molecules of ADP, and two molecules of orthophosphate.[1,2]

The enzymes benzoyl-CoA reductase and cyclohexa-1,5-diene-1-carbonyl-CoA hydratase catalyze the first steps of benzoyl-CoA conversion under anoxic conditions in the denitrifying bacterium, *Thauera aromatica*. Reaction products obtained with [ring-¹³C6]benzoyl-CoA and [ring-¹⁴C]benzoyl-CoA as substrates were analyzed by NMR spectroscopy and high performance liquid chromatography.[2] The main product obtained with titanium(III) citrate or with reduced [8Fe-8S]-ferredoxin was identified as cyclohexa-1,5-diene-1-carbonyl-CoA. The cyclic diene was converted into 6-hydroxycyclohex-1-ene-1-carbonyl-CoA by the hydratase. Assay mixtures containing reductase, hydratase, and sodium dithionite or a mixture of sulfite and titanium(III) citrate as reducing agent yielded cyclohex-2-ene-1-carbonyl-CoA and 6-hydroxycylohex-2-ene-1-carbonyl-CoA. The potential required for the first electron transfer to the model compound *S*-ethyl-thiobenzoate yielding a radical anion was determined by cyclic voltammetry as 1.9 V versus a standard hydrogen electrode.

[1]M. Boll & G. Fuchs (1995) *EJB* **234**, 921.
[2]M Boll, D. Laempe, W. Eisenreich, A. Bacher, T. Mittelberger, J. Heinze & G. Fuchs (2000) *JBC* **275**, 21889.

BENZOYLFORMATE DECARBOXYLASE

This thiamin-pyrophosphate-dependent enzyme [EC 4.1.1.7] catalyzes the conversion of benzoylformate to benzaldehyde and carbon dioxide.[1,2] Other substrates include (*p*-methylbenzoyl)formate, (*p*-methoxybenzoyl)formate, (*p*-fluoromethylbenzoyl)formate, and (*p*-hydroxybenzoyl)formate.

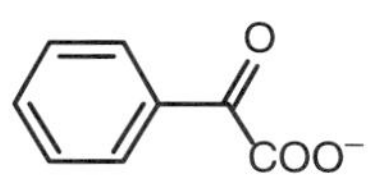

benzoylformate

The decarboxylase is a component in the mandelate pathway of pseudomonads and associates with other enzymes in that pathway to form a multienzyme complex. This pathway was the first pathway whose genes were recognized as being coordinately regulated.

[1] L. J. Reynolds, G. A. Garcia, J. W. Kozarich & G. L. Kenyon (1988) *Biochemistry* **27**, 5530.
[2] P. M. Weiss, G. A. Garcia, G. L. Kenyon, W. W. Cleland & P. F. Cook (1988) *Biochemistry* **27**, 2197.

Selected entries from *Methods in Enzymology* [**vol**, page(s)]:
General discussion: **2**, 273, 278; **17A**, 674
Assay: **2**, 278; **17A**, 674

N-BENZOYL-4-HYDROXYANTHRANILATE 4-*O*-METHYLTRANSFERASE

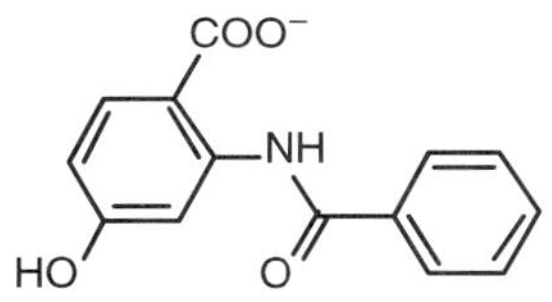

N-benzoyl-4-hydroxyanthranilate

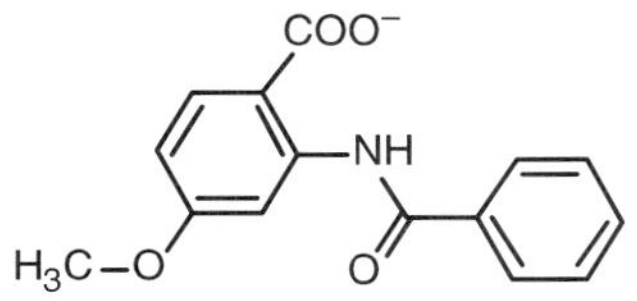

N-benzoyl-4-methoxyanthranilate

This methyltransferase [EC 2.1.1.105], which participates in phytoalexin biosynthesis, catalyzes the reaction of *S*-adenosyl-L-methionine with *N*-benzoyl-4-hydroxyanthranilate to produce *S*-adenosyl-L-homocysteine and *N*-benzoyl-4-methoxyanthranilate (the phytoalexin methoxydianthramide B).[1]

[1] K. Reinhard & U. Matern (1989) *ABB* **275**, 295.

BENZPHETAMINE *N*-DEMETHYLASE

This cytochrome P450 monooxygenase catalyzes the NADPH- and dioxygen-dependent *N*-demethylation of benzphetamine to produce formaldehyde and α-methyl-*N*-(phenylmethyl)benzeneethaneamine. ***See*** *specific cytochrome P450*

Selected entries from *Methods in Enzymology* [**vol**, page(s)]:
Assay: 52, 362
Other molecular properties: Ah locus, **52**, 232; isolated MEOS fraction, in, **52**, 362, 363

BENZYLAMINE/PUTRESCINE OXIDASE

This PQQ-dependent (pyrroloquinoline quinone-dependent) amine oxidase of *Pichia pastoris* catalyzes the reaction of benzylamine with dioxygen and water to produce benzaldehyde, ammonia, and hydrogen peroxide. The enzyme is also an α,ω-diamine oxidase and will act on putrescine. ***See also*** *Amine Oxidases*

Selected entries from *Methods in Enzymology* [**vol**, page(s)]:
Assay: 188, 427

BENZYL-2-METHYL-HYDROXYBUTYRATE DEHYDROGENASE

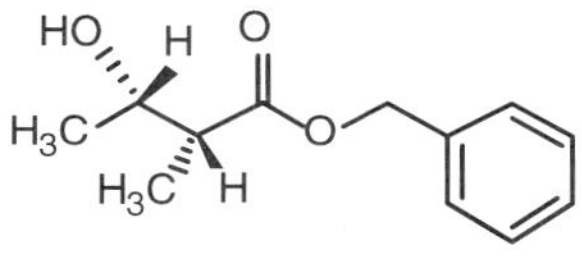

benzyl (2*R*,3*S*)-2-methyl-3-hydroxybutanoate

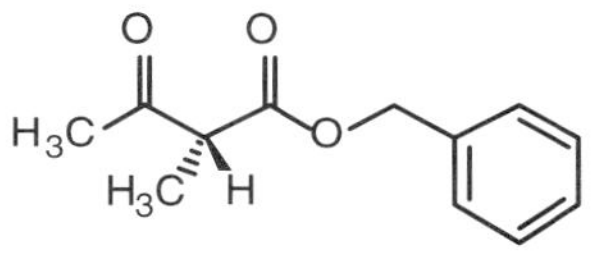

benzyl (2*R*)-2-methyl-3-oxobutanoate

This oxidoreductase [EC 1.1.1.217] catalyzes the reaction of benzyl (2*R*,3*S*)-2-methyl-3-hydroxybutanoate with NADP$^+$ to produce benzyl 2-methyl-3-oxobutanoate and NADPH. Other substrates include benzyl (2*S*,3*S*)-2-methyl-3-hydroxybutanoate.[1]

[1] A. Furuichi, H. Akita, H. Matsukura, T. Oishi & K. Horikoshi (1985) *Agric. Biol. Chem.* **49**, 2563.

N-BENZYLOXYCARBONYLGLYCINE HYDROLASE

This zinc/cobalt-dependent enzyme [EC 3.5.1.58] catalyzes the hydrolysis of *N*-benzyloxycarbonylglycine to produce benzyl alcohol, carbon dioxide, and glycine.[1]

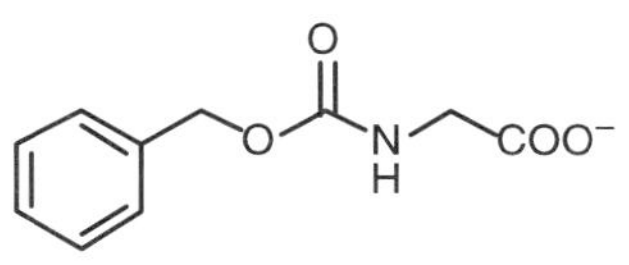

N-benzyloxycarbonylglycine

An alternative substrate is *N*-benzyloxycarbonylalanine; however, other corresponding derivatives of other amino acids are not hydrolyzed. **See also** N^{α}-*Benzyloxycarbonyl-Leucine Hydrolase*

[1]S. Murao, E. Matsumura & T. Kawano (1985) *Agric. Biol. Chem.* **49**, 967.

N^{α}-BENZYLOXYCARBONYL-LEUCINE HYDROLASE

This enzyme [EC 3.5.1.64] catalyzes the hydrolysis of N^{α}-benzyloxycarbonyl-L-leucine to produce benzyl alcohol, carbon dioxide, and L-leucine.

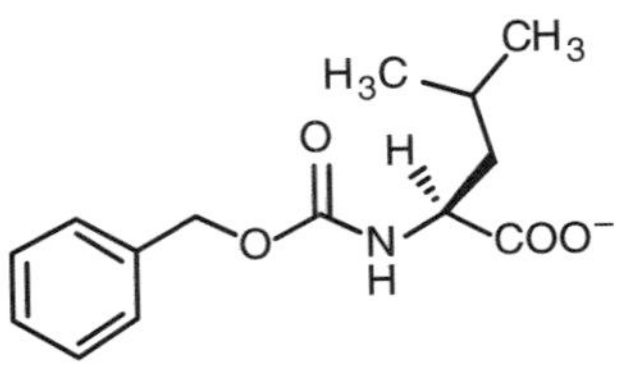

N-benzyloxycarbonyl-L-leucine 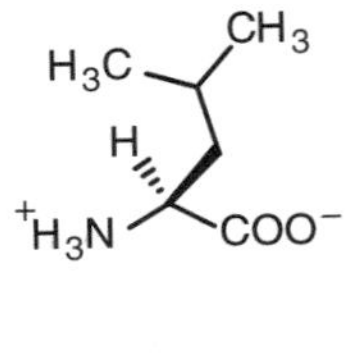   L-leucine

Alternative substrates include N^{α}-*t*-butoxycarbonyl-L-leucine and the corresponding derivatives of L-aspartate, L-methionine, L-glutamate, and L-alanine.[1] **See also** *N-Benzyloxycarbonylglycine Hydrolase*

[1]E. Matsumura, T. Shin, S. Murao, M. Sakaguchi & T. Kawano (1985) *Agric. Biol. Chem.* **49**, 3643.

BENZYLSUCCINATE SYNTHASE

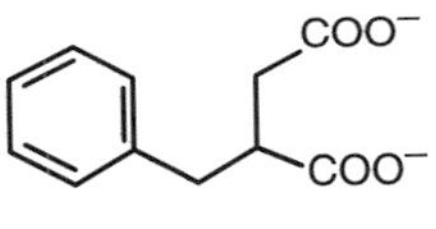

benzylsuccinate

This enzyme [EC 4.1.99.11] catalyzes the reaction of toluene with fumarate to produce benzylsuccinate.[1,2] During catalysis, the enzyme employs a glycyl radical that is inactivated by dioxygen.

[1]H. R. Beller & A. M. Spormann (1998) *J. Bacteriol.* **180**, 5454.
[2]B. Leuthner, C. Leutwein, H. Schultz, P. Hörth, W. Haehnel, E. Schiltz, H. Schägger & J. Heider (1998) *Mol. Microbiol.* **28**, 615.

(R,S)-1-BENZYL-1,2,3,4-TETRAHYDROISOQUINOLINE N-METHYLTRANSFERASE

This methyltransferase [EC 2.1.1.115], also referred to as (*R,S*)-tetrahydrobenzylisoquinoline *N*-methyltransferase and norreticuline *N*-methyltransferase, catalyzes the reaction of *S*-adenosyl-L-methionine with (*R,S*)-1-benzyl-1,2,3,4-tetrahydroisoquinoline to produce *S*-adenosyl-L-homocysteine and *N*-methyl-(*R,S*)-1-benzyl-1,2,3,4-tetrahydroisoquinoline (both *R*- and *S*-enantiomers are

substrates).[1] Other acceptor substrates include coclaurine (a probable physiological substrate), norcoclaurine, isococlaurine, norarmepavine, norreticuline, and tetrahydropapaverine. The enzyme, which functions in plant benzylisoquinoline alkaloid biosynthesis, had been called norreticuline *N*-methyltransferase, but norreticuline has not been identified in any organism and use of that name is discouraged.

[1]T. Frenzel & M. H. Zenk (1990) *Phytochemistry* **29**, 3491.

BERBAMUNINE SYNTHASE

This cytochrome P450-dependent oxidoreductase [EC 1.1.3.34], also referred to as (*S*)-*N*-methylcoclaurine oxidase (C-O phenol-coupling) and CYP80, catalyzes the reaction of (*S*)-*N*-methylcoclaurine with (*R*)-*N*-methylcoclaurine, NADPH, and dioxygen to produce the bisbenzylisoquinoline alkaloid berbamunine, NADP$^+$, and two water molecules. Note that no oxygen is incorporated in the alkaloid product.

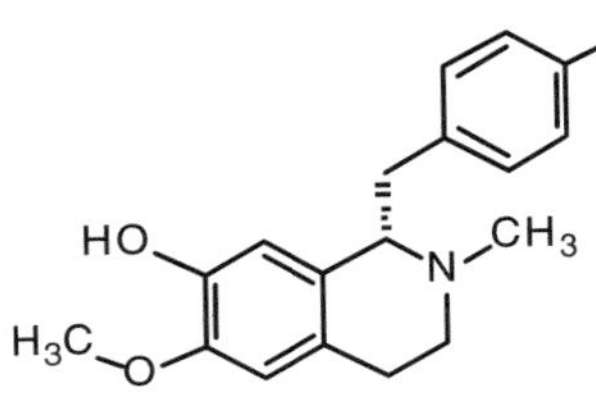

(S)-N-methylcoclaurine 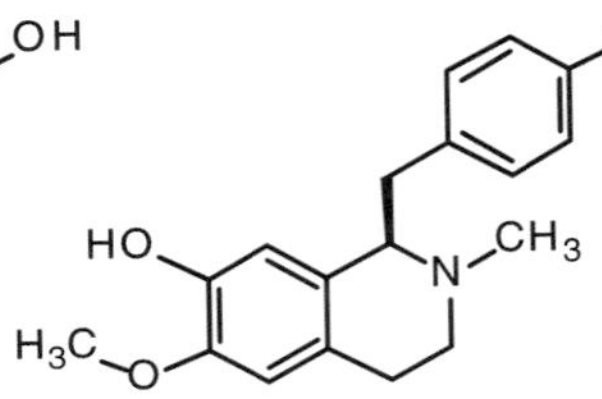   (R)-N-methylcoclaurine

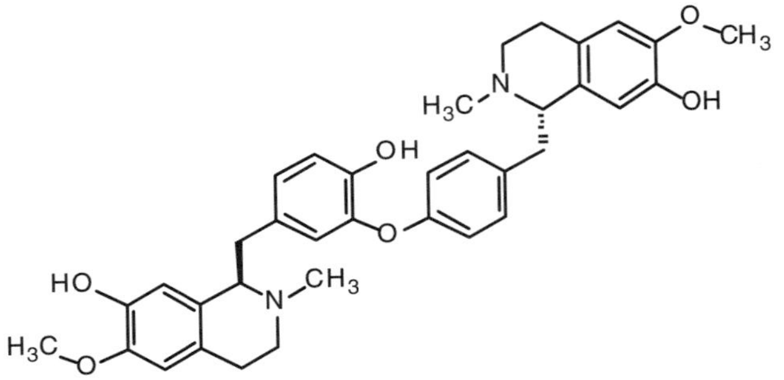

berbamunine

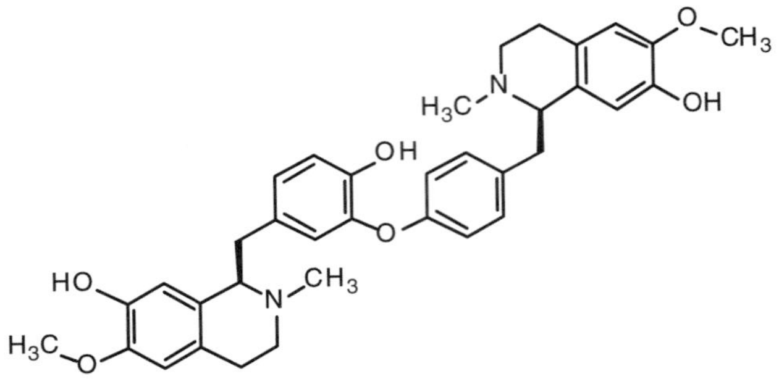

guattagaumerine

When two molecules of (*R*)-*N*-methylcoclaurine are used as substrates, the alternative product is guattagaumerine. The relative amounts of the two possible bisbenzylisoquinoline products formed could be altered by the reductase source or by varying the enantiomeric composition of the substrates.[1]

[1]P. F. Kraus & T. M. Kutchan (1995) *PNAS* **92**, 2071.

BERGAPTOL *O*-METHYLTRANSFERASE

This methyltransferase [EC 2.1.1.92], first identified in parsley (*Petroselinum crispum*), catalyzes the reaction of *S*-adenosyl-L-methionine with bergaptol (*i.e.*, 5-hydroxyfuranocoumarin) to produce *S*-adenosyl-L-homocysteine and *O*-methylbergaptol (also called bergapten).[1,2] Another excellent substrate is 5-hydroxyxanthotoxin (generating isopimpinellin).

[1]K. D. Hauffe, K. Hahlbrock & D. Scheel (1986) *Z. Naturforsch.* **41C**, 228.
[2]W. Knogge, E. Kombrink, E. Schmelzer & K. Hahlbrock (1987) *Planta* **171**, 279.

BETAINE-ALDEHYDE DEHYDROGENASE

This oxidoreductase [EC 1.2.1.8] catalyzes the reaction of betaine aldehyde with NAD^+ and water to produce betaine and NADH.[1–3]

This porcine kidney enzyme and the protein from the leaves of *Amaranthus hypochondriacus* have an Iso ordered Bi Bi kinetic mechanism. In some organisms, the enzyme is substrate inhibited by betaine aldehyde.

[1]C. G. Figueroa-Soto & E. M. Valenzuela-Soto (2000) *BBRC* **269**, 596.
[2]E. M. Valenzuela-Soto & R. A. Munoz-Clares (1993) *JBC* **268**, 23818.
[3]P. Falkenberg & A. R. Strom (1990) *BBA* **1034**, 253.

Selected entries from *Methods in Enzymology* [vol, page(s)]:
Assay: 4, 331

BETAINE:HOMOCYSTEINE METHYLTRANSFERASE

This zinc-dependent methyltransferase [EC 2.1.1.5] catalyzes the reaction of trimethylammonioacetate (*i.e.*, glycine betaine) with L-homocysteine to produce *N,N*-dimethylglycine and L-methionine, in which the methyl transfer appears to occur directly from one substrate to the other.[1] The enzyme has an ordered Bi Bi kinetic mechanism.[2]

[1]W. M. Awad, P. L. Whitney, W. E. Skiba, J. H. Mangum & M. S. Wells (1983) *JBC* **258**, 12790.
[2]J. D. Finkelstein, W. E. Kyle & B. J. Harris (1974) *ABB* **165**, 774.

Selected entries from *Methods in Enzymology* [vol, page(s)]:
General discussion: 143, 384
Assay: 143, 384

*Bfa*I RESTRICTION ENDONUCLEASE

This type II restriction endonuclease [EC 3.1.21.4], obtained from *Bacteroides fragilis*, acts on both strands of DNA at 5′ . . . C∇TAG . . . 3′.[1]

[1]S. N. Reinecke & R. N. Morgan (1991) *NAR* **19**, 1152.

*Bgl*I RESTRICTION ENDONUCLEASE

This type II restriction endonuclease [EC 3.1.21.4] is Mg^{2+}-dependent and is obtained from *Bacillus globigii*. It acts on both strands of DNA at the sequence 5′ . . . GCCNNNN∇NGGC . . . 3′,[1–5] where N represents any base. The enzyme has active-site lysyl and arginyl residues.[6] A crystal structure has been determined.[7]

[1]V. Pirrotta (1976) *Nucleic Acids Res.* **3**, 1747.
[2]C. H. Duncan, G. A. Wilson & F. E. Young (1978) *J. Bacteriol.* **134**, 338.
[3]H. Van Heuverswyn & W. Fiers (1980) *Gene* **9**, 195.
[4]J. A. Lautenberger, C. T. White, N. L. Haigwood, M. H. Edgell & C. A. Hutchison (1980) *Gene* **9**, 213.
[5]T. A. Bickle & K. Ineichen (1980) *Gene* **9**, 205.
[6]Y. H. Lee & J. G. Chirikjian (1979) *JBC* **254**, 6838.
[7]M. Newman, K. Lunnen, G. Wilson, J. Greci, I. Schildkraut & S. E. V. Phillips (1998) *EMBO J.* **17**, 5466.

Selected entries from *Methods in Enzymology* [vol, page(s)]:
General discussion: 65, 132, 147, 150, 739, 741, 813, 824, 825
Assay: 65, 133, 134
Other molecular properties: cleavage sites, of SV40 DNA, **65**, 713; optimum reaction conditions, **65**, 136; properties, **65**, 136; purification, **65**, 94, 133; recognition sequence, **65**, 3, 136; storage, **65**, 135; termini produced, **65**, 392

BglII RESTRICTION ENDONUCLEASE

This type II restriction endonuclease [EC 3.1.21.4] is obtained from *Bacillus globigii* and acts on both strands of DNA at the sequence 5′ ... A∇GATCT... 3′.[1,2] The enzyme is not blocked by *dam* methylation.

[1] V. Pirrotta (1976) *Nucleic Acids Res.* **3**, 1747.
[2] C. H. Duncan, G. A. Wilson & F. E. Young (1976) *J. Bacteriol.* **134**, 338.

Selected entries from *Methods in Enzymology* [**vol**, page(s)]:
General discussion: 65, 132, 147, 150, 787, 789, 791
Assay: 65, 133, 134
Other molecular properties: fragment end structure produced, **65**, 511; isoschizomers, **65**, 137; molecular weight, **65**, 137; properties, **65**, 137; purification, **65**, 94, 133; recognition sequence, **65**, 3, 137, 151, 152, 511; storage, **65**, 135

BILE-SALT SULFOTRANSFERASE

This transferase [EC 2.8.2.14], which exists as a number of isozymes, catalyzes the reaction of 3′-phosphoadenylylsulfate with taurolithocholate to produce adenosine 3′,5′-bisphosphate and taurolithocholate sulfate.[1,2]

taurolithocholate   taurolithocholate sulfate

Acceptor substrates include both conjugated and unconjugated bile salts: for example, 5α-cholanoate-3β-ol, 5-cholenate-3β-ol, glycolithocholate, lithocholate, chenodeoxycholate, glycochenodeoxycholate, and taurochenodeoxycholate.

[1] S. Barnes, R. Waldrop, J. Crenshaw, R. J. King & K. B. Taylor (1986) *J. Lipid Res.* **27**, 1111.
[2] L. J. Chen & I. H. Segel (1985) *ABB* **241**, 371.

Selected entries from *Methods in Enzymology* [**vol**, page(s)]:
General discussion: 77, 213
Assay: 77, 213
Other molecular properties: affinity chromatography, **77**, 216, 217; distribution, **77**, 218; inhibition, **77**, 217; isoelectric focusing, **77**, 216, 217; isoelectric point, **77**, 217; properties, **77**, 217, 218; purification, **77**, 215; size, **77**, 217; specificity, **77**, 217

BILIRUBIN-GLUCURONOSIDE GLUCURONOSYLTRANSFERASE

This transferase [EC 2.4.1.95], also known as bilirubin monoglucuronide transglucuronidase, catalyzes the reaction of two molecules of bilirubin-glucuronoside to produce bilirubin and bilirubin-bisglucuronoside.[1,2]

[1] P. L. M. Jansen, J. R. Chowdhury, E. B. Fischberg & I. M. Arias (1977) *JBC* **252**, 2710.
[2] J. R. Chowdhury, N. R. Chowdhury, M. M. Bhargava & I. M. Arias (1979) *JBC* **254**, 8336.

Selected entries from *Methods in Enzymology* [**vol**, page(s)]:
General discussion: 77, 192

BILIRUBIN OXIDASE

This oxidoreductase [EC 1.3.3.5] catalyzes the reaction of bilirubin with dioxygen to produce biliverdin and water, a four electron reduction of bilirubin.[1–4]

bilirubin IXα   biliverdin IXα

The *Myrothecium verrucaria* enzyme is a monomeric protein that contains catalytically essential copper.[1]

[1] A. Shimizu, J.-H. Kwon, T. Sasaki, T. Satoh, N. Sakurai, T. Sakurai, S. Yamaguchi & T. Samejima (1999) *Biochemistry* **38**, 3034.
[2] O. Yokosuka & B. Billing (1987) *BBA* **923**, 268.
[3] R. Cardenas-Vazquez, O. Yokosuka & B. H. Billing (1986) *BJ* **236**, 625.
[4] N. Tanaka & S. Murao (1982) *Agric. Biol. Chem.* **46**, 2499.

BILIVERDIN REDUCTASE

This oxidoreductase [EC 1.3.1.24] catalyzes the reaction of biliverdin and NAD(P)H to produce bilirubin and NAD(P)$^+$.

biliverdin IXα   bilirubin IXα

The enzyme has been reported to have an ordered Bi Bi kinetic mechanism[1,4] or a random Bi Bi scheme with an abortive complex.[2,3] Certain variants of this enzyme have been reported that differ in the specificity of the biliverdin substrate.

[1] O. Cunningham, M. G. Gore & T. J. Mantle (2000) *BJ* **345**, 393.
[2] B. F. Cooper & F. B. Rudolph (1995) *Meth. Enzymol.* **249**, 188.
[3] J. E. Bell & M. D. Maines (1988) *ABB* **263**, 1.
[4] E. Rigney & T. J. Mantle (1988) *BBA* **957**, 237.

Selected entries from *Methods in Enzymology* [vol, page(s)]:
Other molecular properties: biliverdin, assay of, **63**, 36; heme oxygenase assay, in, **52**, 368; **63**, 36; product inhibition studies, two substrates:two products reactions, **249**, 197

BIOCHANIN-A REDUCTASE

This oxidoreductase [EC 1.3.1.46] catalyzes the reaction of biochanin A (*i.e.*, 6,7-dihydroxy-4'-methoxyisoflavone) with NADPH to produce dihydrobiochanin A and NADP+.[1,2] The NADPH-stereospecificity is A-side. A number of other isoflavones are reduced to the corresponding isoflavonones: for example, pratensein (*i.e.*, 5,7,3'-trihydroxy-4'-methoxyisoflavone) to 2-dihydro pratensein and genistein (*i.e.*, 5,7,3'-trihydroxyisoflavone) to 2-dihydrogenistein.

[1] D. Schlieper & W. Barz (1987) *Phytochemistry* **26**, 2495.
[2] D. Schlieper, K. Tiemann & W. Barz (1990) *Phytochemistry* **29**, 1519.

BIOTIN:[ACETYL-CoA-CARBOXYLASE] LIGASE

This enzyme [EC 6.3.4.15] (also referred to as biotin:[acetyl-CoA carboxylase] synthetase, biotin:protein ligase, and acetyl-CoA carboxylase biotin holoenzyme synthetase) catalyzes the reaction of ATP with biotin and apo-[acetyl-CoA carboxylase] to produce AMP, pyrophosphate (or, diphosphate) and acetyl-CoA carboxylase (*i.e.*, the holo-enzyme: EC 6.4.1.2).[1]

[1] D. F. Barker & A. M. Campell (1981) *J. Mol. Biol.* **146**, 451 and 469.

Selected entries from *Methods in Enzymology* [vol, page(s)]:
Assay: **107**, 265
Other molecular properties: properties, **107**, 262, 274; purification, **107**, 271; reaction mechanism, **107**, 275; sequence homology with other synthetases, **107**, 264; roles for, **107**, 276; yeast, **14**, 8

BIOTIN CARBOXYLASE

This enzyme [EC 6.3.4.14] catalyzes the reaction of ATP with a biotin-carboxyl-carrier protein and carbon dioxide to produce ADP, orthophosphate, and the carboxybiotin-carboxyl-carrier protein. The *Escherichia coli* enzyme can also act on free biotin.[1–3]

This enzyme is a component in all biotin-dependent carboxylation reactions: for example the acetyl-CoA carboxylase multienzyme complex. Characterization of the bicarbonate-dependent biotin-independent ATPase activity as well as isotope-effect investigations suggest carboxylation proceeds by forming a carboxyphosphate intermediate.[1] The first step in this reaction involves the deprotonation of the 1' nitrogen of biotin. A number of investigations have suggested that an active-site lysyl residue removes the proton from the thiol group of a cysteinyl residue. The thiolate anion then abstracts the N1' proton of biotin while the protonated lysyl residue stabilizes the negative charge on the ureido oxygen, although recent studies dispute this.[2]

[1] P. A. Tipton & W. W. Cleland (1988) *Biochemistry* **27**, 4317 and 4325.
[2] K. L. Levert, R. B. Lloyd & G. L. Waldrop (2000) *Biochemistry* **39**, 4122.
[3] C. Z. Blanchard, D. Amspacher, R. Strongin & G. L. Waldrop (1999) *BBRC* **266**, 466.

Selected entries from *Methods in Enzymology* [vol, page(s)]:
General discussion: **35**, 25
Assay: **35**, 26

BIOTIN:CoA LIGASE

This enzyme [EC 6.2.1.11], also known as biotinyl-CoA synthetase, catalyzes the reaction of ATP with biotin and Coenzyme A to produce AMP, pyrophosphate (or, diphosphate), and biotinyl-CoA.[1,2] The *Mycoplana* enzyme can also act on dethiobiotin and actithiazic acid.[2]

[1] J. E. Christner, M. J. Schlesinger & M. J. Coon (1964) *JBC* **239**, 3997.
[2] M. Tanaka, H. Yamamoto, Y. Izumi & H. Yamada (1986) *ABB* **251**, 479.

BIOTINIDASE

This enzyme [EC 3.5.1.12] catalyzes the hydrolysis of biotin amide to produce biotin and ammonia.[1–3] Other substrates include biotin esters and biocytin (*i.e.*, *N*-(+)-biotinyl-L-lysine).

[1] D. V. Craft, N. H. Goss, N. Chandramouli & H. G. Wood (1985) *Biochemistry* **24**, 2471.
[2] J. Moss & M. D. Lane (1971) *AE* **35**, 321.
[3] J. Knappe, W. Brümmer & K. Biederbick (1963) *Biochem. Z.* **338**, 599.

Selected entries from *Methods in Enzymology* [vol, page(s)]:
General discussion: **184**, 103; **279**, 422, 435, 442
Assay: **184**, 97, 104; **279**, 435, 442
Other molecular properties: biotin-binding assay, **184**, 97; catalytic properties, **184**, 108; inhibition, **184**, 110; overview, **184**, 95; physical properties, **184**, 108; specificity, **184**, 109; stability, **184**, 111

BIOTIN:[METHYLCROTONOYL-CoA-CARBOXYLASE] LIGASE

This enzyme [EC 6.3.4.11], also referred to as biotin-[methylcrotonoyl-CoA-carboxylase] synthetase, catalyzes the reaction of ATP with biotin and apo-[3-methylcrotonoyl-CoA carboxylase] (*i.e.*, the apo form of EC 6.4.1.4) to produce AMP, pyrophosphate (or, diphosphate), and 3-methyl-crotonoyl-CoA carboxylase (*i.e.*, the holo-enzyme).[1]

[1] T. Höpner & J. Knappe (1965) *Biochem. Z.* **342**, 190.

Selected entries from *Methods in Enzymology* [vol, page(s)]:
Assay: 107, 265
Other molecular properties: properties, **107**, 262, 274; purification, **107**, 271; reaction mechanism, **107**, 275; roles for, **107**, 276; sequence homology with other synthetases, **107**, 264

BIOTIN:[METHYLMALONYL-CoA-CARBOXYLTRANSFERASE] LIGASE

This enzyme [EC 6.3.4.9], also known as biotin-[methylmalonyl-CoA-carboxyltransferase] synthetase and biotin-transcarboxylase synthetase, catalyzes the reaction of ATP with biotin and apo-[methylmalonyl-CoA carboxyltransferase] (*i.e.*, the apo form of EC 2.1.3.1) to produce AMP, pyrophosphate (or, diphosphate), and methylmalonyl-CoA carboxyltransferase (*i.e.*, the holo-enzyme).[1–3]

[1] B. C. Shenoy & H. G. Wood (1988) *FASEB J.* **2**, 2396.
[2] H. G. Wood, F. R. Harmon, B. Wühr, K. Hübner & F. Lynen (1980) *JBC* **255**, 7397.
[3] M. D. Lane, D. L. Young & F. Lynen (1964) *JBC* **239**, 2858.

Selected entries from *Methods in Enzymology* [vol, page(s)]:
Assay: 107, 265
Other molecular properties: properties, **107**, 262, 274; purification, **107**, 271; reaction mechanism, **107**, 275; sequence homology with other synthetases, **107**, 264

BIOTIN-[PROPIONYL-CoA-CARBOXYLASE (ATP-HYDROLYZING)] SYNTHETASE

This enzyme [EC 6.3.4.10], also known as biotin:[propionyl-CoA-carboxylase] ligase and holocarboxylase synthetase, catalyzes the reaction of biotin with ATP and apo-[propanoyl-CoA carboxylase] (*i.e.*, the apo-form of EC 6.4.1.3) to produce AMP, pyrophosphate (or, diphosphate), and propanoyl-CoA carboxylase (*i.e.*, the holo-enzyme).[1]

[1] L. Siegel, J. L. Foote & M. J. Coon (1965) *JBC* **240**, 1025.

Selected entries from *Methods in Enzymology* [vol, page(s)]:
General discussion: 107, 261; 279, 386
Assay: 107, 265; 279, 386
Other molecular properties: overview, **184**, 94; properties, **107**, 262, 274; **279**, 397; purification, **107**, 271; **279**, 390; reaction mechanism, **107**, 275; sequence homology with other synthetases, **107**, 264

BIOTIN-SULFOXIDE REDUCTASE

This molybdenum-dependent oxidoreductase catalyzes the reaction of biotin sulfoxide with a reducing agent to produce biotin, water, and the oxidized agent. A marked preference for NADPH as the reducing agent has been observed for the *Rhodobacter sphaeroides* enzyme, which has a ping pong Bi Bi kinetic mechanism.[1] Other substrates include nicotinamide-*N*-oxide, dimethyl sulfoxide, methionine sulfoxide, and trimethylamine-*N*-oxide.

[1] V. V. Pollock & M. J. Barber (2001) *Biochemistry* **40**, 1430.

BIOTIN SYNTHASE

This *S*-adenosylmethionine-dependent, iron-sulfur enzyme [EC 2.8.1.6] catalyzes the reaction of dethiobiotin with sulfur to produce biotin.[1] The iron-sulfur cluster functions as the immediate sulfur donor for biotin formation. Radical intermediates are generated by hydrogen atom transfer from dethiobiotin to the adenosyl radical. This radical is formed by the reductive cleavage of *S*-adenosyl-L-methionine by the reduced iron-sulfur cluster.[2–4]

[1] I. Sanyal, G. Cohen & D. H. Flint (1994) *Biochemistry* **33**, 3625.
[2] T. P. Begley, J. Xi, C. Kinsland, S. Taylor & F. McLafferty (1999) *Curr. Opin. Chem. Biol.* **3**, 623.
[3] N. M. Shaw, O. M. Birch, A. Tinschert, V. Venetz, R. Dietrich & L. A. Savoy (1998) *BJ* **330**, 1079.
[4] D. Guianvarc'h, D. Florentin, B. Tse Sum Bui, F. Nunzi & A. Marquet (1997) *BBRC* **236**, 402 and **240**, 246.

Selected entries from *Methods in Enzymology* [vol, page(s)]:
General discussion: 279, 349, 356

BIPHENYL-2,3-DIOL 1,2-DIOXYGENASE

This iron-dependent oxidoreductase [EC 1.13.11. 39], also referred to as 2,3-dihydroxybiphenyl dioxygenase, catalyzes the reaction of biphenyl-2,3-diol with dioxygen to produce 2-hydroxy-6-oxo-6-phenylhexa-2,4-dienoate and water.[1–4] 3-Isopropylcatechol is an alternative substrate, producing 7-methyl-2-hydroxy-6-oxoocta-2,4-dienoate: however, the enzyme is not identical with catechol 2,3-dioxygenase [EC 1.13.11.2].

The *Sphingomonas* enzyme can also oxidize 3-chloro-catechol to 3-chloro-2-hydroxymuconic semialdehyde by a distal (1,6) cleavage mechanism.[2,3] 3-Methylcatechol and 2,3-dihydroxybiphenyl are oxidized by this enzyme *via* proximal (2,3) cleavage.[3]

[1] H. Ishigooka, Y. Yoshida, T. Omori & Y. Minoda (1986) *Agric. Biol. Chem.* **50**, 1045.
[2] U. Riegert, G. Heiss, A. E. Kuhm, C. Muller, M. Contzen, H. J. Knackmuss & A. Stolz (1999) *J. Bacteriol.* **181**, 4812.
[3] U. Riegert, G. Heiss, P. Fischer & A. Stolz (1998) *J. Bacteriol.* **180**, 2849.
[4] B. G. Fox (1998) *CBC* **3**, 261.

BIPHENYL 2,3-DIOXYGENASE

This iron-dependent oxidoreductase [EC 1.14.12.18] catalyzes the reaction of biphenyl with NADH, a proton, and dioxygen to produce (2R,3S)-3-phenylcyclohexa-3,5-diene-1,2-diol and NAD$^+$.[1] The *Pseudomonas* dioxygenase is part of a multicomponent protein complex composed of an NADH:ferredoxin oxidoreductase (FAD cofactor), a [2Fe-2S] Rieske-type ferredoxin, and the terminal oxygenase. Other substrates include chlorine-substituted biphenyls.

[1] J. D. Haddock & D. T. Gibson (1995) *J. Bacteriol.* **177**, 5834.

BIS(5′-ADENOSYL)-TRIPHOSPHATASE

This enzyme [EC 3.6.1.29] (also referred to as dinucleosidetriphosphatase, diadenosine 5′,5′′′-P^1,P^3-triphosphate hydrolase, and Ap₃A hydrolase) catalyzes the hydrolysis of P^1,P^3-bis(5′-adenosyl)triphosphate to produce ADP and AMP.[1] Other substrates include Gp₃G, Up₃U, Ap₄A, and Ap₅A.[2] In humans, there are distinct enzymes that act on Ap₃A and Ap₄A.[3] ***See also*** *Bis(5′-Nucleosyl)-Tetraphosphatase (Symmetrical); Bis(5′-Nucleosyl)-Tetraphosphatase (Asymmetrical)*

[1] M. A. G. Sillero, R. Villalba, A. Moreno, M. Quintanilla, C. D. Lobatón & A. Sillero (1977) *EJB* **76**, 331.
[2] H. Jakobowski & A. Guranowski (1983) *JBC* **258**, 9982.
[3] A. Guranowski, M. Galbas, R. Hartmann & J. Justesen (2000) *ABB* **373**, 218.

BIS(2-ETHYLHEXYL)PHTHALATE ESTERASE

This esterase [EC 3.1.1.60] catalyzes the hydrolysis of bis(2-ethylhexyl)phthalate (a plasticizer chemical) to produce 2-ethylhexylphthalate and 2-ethylhexan-1-ol.[1] The enzyme also acts on 4-nitrophenyl esters, with optimum acyl chain length of six-to-eight carbon atoms. While hydrolysis of bis(2-ethylhexyl)phthalate is considerably slower than cleavage of 4-nitrophenyl octanoate, the esterase is the rate-limiting step in the metabolism of bis(2-ethylhexyl)phthalate.[1]

[1] H. W. Krell & H. Sandermann (1984) *EJB* **143**, 57.

BIS-γ-GLUTAMYLCYSTINE REDUCTASE (NADPH)

This FAD-dependent oxidoreductase [EC 1.6.4.9] catalyzes the reaction of NADPH with bis-γ-glutamylcystine to produce NADP$^+$ and two molecules of γ-glutamyl-cysteine.[1,2] The enzyme is distinct from glutathione reductase (NAD(P)H) [EC 1.6.4.2] or CoA-disulfide reductase [EC 1.6.4.10].

[1] A. R. Sundquist & R. C. Fahey (1988) *J. Bacteriol.* **170**, 3459.
[2] A. R. Sundquist & R. C. Fahey (1989) *JBC* **264**, 719.

BIS(5′-NUCLEOSYL)-TETRAPHOSPHATASE (ASYMMETRICAL)

This enzyme [EC 3.6.1.17] (also referred to as bis(5′-guanosyl)-tetraphosphatase, bis(5′-adenosyl)-tetraphosphatase, dinucleosidetetraphosphatase (asymmetrical), and AP₄A hydrolase) catalyzes the hydrolysis of P^1,P^4-bis(5′-guanosyl)tetraphosphate to produce GTP and GMP.[1,2] Other substrates include bis(5′-xanthosyl)-tetraphosphate, bis(5′-adenosyl)-tetraphosphate, and bis(5′-uridyl)-tetraphosphate.

The catalyzed reaction proceeds with inversion of configuration at phosphorus, indicating that displacement by water occurs by an in-line mechanism.[3] ***See also*** *Bis(5′-Nucleosyl)-Tetraphosphatase (Symmetrical); Bis(5′-Adenosyl)-Triphosphatase*

[1] C. G. Vallejo, M. A. G. Sillero & A. Sillero (1974) *BBA* **358**, 117.
[2] C. D. Lobatón, C. G. Vallejo, A. Sillero & M. A. G. Sillero (1975) *EJB* **50**, 495.
[3] R. M. Dixon & G. Lowe (1989) *JBC* **264**, 2069.

BIS(5′-NUCLEOSYL)-TETRAPHOSPHATASE (SYMMETRICAL)

This enzyme [EC 3.6.1.41], also called dinucleoside tetraphosphatase (symmetrical) and diadenosine tetraphosphatase (symmetrical), catalyzes the hydrolysis of P^1,P^4-bis(5′-adenosyl)tetraphosphate to produce two molecules of ADP. Alternative substrates include bis(5′-guanosyl) tetraphosphate and bis(5′-adenosyl) pentaphosphate (producing ADP and ATP), and, more slowly, on some other polyphosphates, forming a nucleoside bis-phosphate as one product in all cases.

The enzyme cleaves the polyphosphate chain at the second phosphate from the bound adenosine moiety. The enzyme exhibits biphasic kinetics with ADP acting as a potent inhibitor.[1] The firefly (*Photinus pyralis*) enzyme is competitively inhibited by adenosine 5′-tetraphosphate ($K_i = 7.5$ nM), as compared to a 1.9 µM K_m for AppppA with the firefly enzyme. ***See also*** *Bis(5′-Nucleosyl)-Tetraphosphatase (Asymmetrical); Bis(5′-Adenosyl)-Triphosphatase*

[1] L. D. Barnes & C. A. Culver (1982) *Biochemistry* **21**, 6123.

3′(2′),5′-BISPHOSPHATE NUCLEOTIDASE

This nucleotidase [EC 3.1.3.7] (also referred to as phosphoadenylate 3′-nucleotidase, 3′(2′),5′-bisphosphonucleoside 3′(2′)-phosphohydrolase, and 3′-phosphoadenylylsulfate 3′-phosphatase) catalyzes the hydrolysis of adenosine 3′,5′-bisphosphate to produce adenosine 5′-monophosphate (AMP) and orthophosphate.[1,2] Other substrates include 3′-phosphoadenylylsulfate, adenosine 3′-phosphate 5′-phosphosulfate, and the corresponding 2′-phosphates. The enzyme is inhibited by sodium and lithium ions, but not potassium ion.[3]

[1] M. Lik-Tsang & J. A. Schiff (1976) EJB 65, 113.
[2] E. G. Brunngraber (1958) JBC 233, 472.
[3] S. G. Ramaswamy & W. B. Jakoby (1987) JBC 262, 10044.

BISPHOSPHOGLYCERATE MUTASE

This mutase [EC 5.4.2.4] (also known as bisphosphoglycerate synthase, diphosphoglycerate mutase, and glycerate phosphomutase) catalyzes the interconversion of 3-phospho-D-glyceroyl phosphate and 2,3-bisphospho-D-glycerate.[1–3] In erythrocytes, the main function of this enzyme is the synthesis of 2,3-bisphosphoglycerate, the allosteric modulator of hemoglobin oxygen affinity.

The enzyme is first phosphorylated at a histidyl residue by 3-phosphoglyceroyl phosphate to produce a phosphoenzyme and 3-phosphoglycerate. The intermediate is subsequently phosphorylated to yield 2,3-bisphosphoglycerate; however, the 3-phosphoglycerate intermediate can also dissociate from the enzyme, resulting in reduced reaction rate. The mutase is considerably more active in the presence of added 3-phosphoglycerate. The mutase will also catalyze the activities of bisphosphoglycerate phosphatase [EC 3.1.3.13] and phosphoglycerate mutase [EC 5.4.2.1], albeit slowly.

[1] L. A. Fothergill-Gilmore & H. C. Watson (1989) Adv. Enzymol. 62, 227.
[2] H. Chiba & R. Sasaki (1978) Curr. Top. Cell Regul. 14, 75.
[3] W. J. Ray, Jr., & E. J. Peck, Jr. (1972) The Enzymes, 3rd ed., 6, 407.

Selected entries from **Methods in Enzymology** [vol, page(s)]:
General discussion: 1, 425; **42**, 450; **87**, 42
Assay: 1, 425; **87**, 50
Other molecular properties: mechanism, **87**, 42; phosphorylated intermediates (kinetic competence, **87**, 47; phosphoprotein bond characterization, **87**, 46; preparation and assay, **87**, 50; primary structure at active site, homology with phosphoglycerate mutase, **87**, 49; role in enzyme reaction, **87**, 43); properties, 1, 427; **87**, 42, 47; purification, rabbit erythrocytes, 1, 426

BISPHOSPHOGLYCERATE PHOSPHATASE

This enzyme [EC 3.1.3.13] catalyzes the hydrolysis of 2,3-diphosphoglycerate to produce 3-phosphoglycerate and orthophosphate. Note that bisphosphoglycerate mutase [EC 5.4.2.4] also exhibits bisphosphoglycerate phosphatase activity.

Selected entries from **Methods in Enzymology** [vol, page(s)]:
General discussion: 5, 243; 9, 644; **42**, 410
Assay: 5, 243; **42**, 410
Other molecular properties: amino acids, **42**, 422; bisphosphoglycerate mutase, presence in assays of, 1, 426; bovine brain, from, preparation and purification, **42**, 415, 417; distribution, 5, 247; heart muscle, from, preparation and purification, **42**, 418, 419; human erythrocyte, from, preparation and purification, **42**, 419, 420; inhibition, **42**, 424, 425; molecular weight, **42**, 421, 422; preparation, from horse liver, **42**, 413, 415, 416; properties, 5, 246; **42**, 420; purification from (bovine brain, **42**, 415; chicken breast muscle, 5, 243; horse liver, **42**, 413; horse muscle, **42**, 412; human erythrocyte, **42**, 419; yeast, 5, 244); sources, **42**, 409

BIURET AMIDOHYDROLASE

This enzyme [EC 3.5.1.84] catalyzes the hydrolysis of biuret to produce urea, carbon dioxide, and ammonia.[1]

biuret   urea

The enzyme participates in the bacterial degradation of cyanuric acid and of the herbicide atrazine (2-chloro-4-(ethylamino)-6-(isopropylamino)-1,3,5-triazine).

[1] A. M. Cook, P. Beilstein, H. Grossenbacher & R. Hutter (1985) BJ 231, 25.

BLASTICIDIN-S DEAMINASE

This zinc-dependent deaminase [EC 3.5.4.23] catalyzes the hydrolysis of the antibiotic blasticidin S to produce deaminohydroxyblasticidin S and ammonia.[1,2] Other substrates include cytomycin and acetylblasticidin S.

[1] I. Yamaguchi, H. Shibata, H. Seto & T. Misato (1975) J. Antibiot. 28, 7.
[2] M. Kimura, S. Sekido, Y. Isogai & I. Yamaguchi (2000) J. Biochem. 127, 955.

BLEOMYCIN HYDROLASE

This bleomycin-inactivating enzyme [EC 3.4.22.40] catalyzes the hydrolysis of the carboxyamide bond of the β-alanine moiety of bleomycins, a family of related antibiotics that form complexes with Fe(III).[1–4] Bleomycins, which are potent antitumor agents that cleave single-stranded DNA by hydroxyl free radicals, often block cells in the G_2 phase of the cell cycle.

The mammalian and yeast bleomycin hydrolases have pH optima of 7.0–7.5 whereas the bacterial enzyme has values between 6.5 and 7.0. The enzyme will cleave small peptides with an unblocked N-terminus and is inhibited by cysteine protease inhibitors. Interestingly, the yeast and rat enzymes will bind nucleic acids, preferably single-stranded DNA or RNA. The enzyme is expressed in every tissue in mammals. However, the only known natural substrate identified so far is bleomycin.

[1] H. Umezawa, T. Takeuchi, S. Hori, T. Sawa & M. Ishizuka (1972) *J. Antibiot.* **25**, 409.

[2] D. Brömme, A. B. Rossi, S. P. Smeekens, D. C. Anderson & D. G. Payan (1996) *Biochemistry* **35**, 6706.

[3] L. Joshua-Tor, E. H. Xu, S. A. Johnston & D. C. Rees (1995) *Science* **268**, 533.

[4] H. E. Xu & S. A. Johnston (1994) *JBC* **269**, 21177.

BLOOD-GROUP-SUBSTANCE ENDO-1,4-β-GALACTOSIDASE

This glycosidase [EC 3.2.1.102], also referred to simply as endo-β-galactosidase, catalyzes the endohydrolysis of 1,4-β-D-galactosidic linkages in blood group A and B substances.[1] The bond hydrolyzed is the 1,4-β-D-galactosyl linkages adjacent to a 1,3-α-D-galactosyl or N-acetylgalactosaminyl residues and a 1,2-α-D-fucosyl residue. *See also Keratan-Sulfate Endo-1,4-β-Galactosidase*

[1] S. Takasaki & A. Kobata (1976) *JBC* **251**, 3603.

Selected entries from *Methods in Enzymology* [**vol**, page(s)]:
General discussion: 50, 560
Assay: releasing Gal(α1 →3)Gal, 179, 496
Other molecular properties: *Escherichia freundii* (glycosphingolipid analysis, for, **230**, 385; specificity, **230**, 414); modification of erythrocytes, **236**, 226; releasing Gal(α1 →3)Gal (assay, **179**, 496; properties, **179**, 499; purification from *Clostridium perfringens*, **179**, 498)

*Blp*I RESTRICTION ENDONUCLEASE

This type II restriction endonuclease [EC 3.1.21.4], obtained from *Arcanobacterium haemolyticum* (formerly *Bacillus lentus* and *Bacillus* species lp), acts on both strands of DNA at the sequence 5′ . . . GC∇TNAGC . . . 3′, where N refers to any base.

*Blu*I RESTRICTION ENDONUCLEASE

This type II restriction endonuclease [EC 3.1.21.4] is obtained from *Brevibacterium luteum* and acts on both strands of DNA at the sequence 5′ . . . C∇TCGAG . . . 3′.[1]

[1] T. R. Gingeras, P. A. Myers, J. A. Olson, F. A. Hanberg & R. J. Roberts (1978) *J. Mol. Biol.* **118**, 113.

Selected entries from *Methods in Enzymology* [**vol**, page(s)]:
Other molecular properties: fragment end structure produced, **65**, 511; recognition sequence, **65**, 5, 511

*Bmr*I RESTRICTION ENDONUCLEASE

This type II restriction endonuclease (subtype s) [EC 3.1.21.4], obtained from *Bacillus megaterium*, acts on DNA at the sequences 5′ . . . ACTGGGNNNNN∇ . . . 3′ and 3′ . . . TGACCCNNNN∇ . . . 5′, where N refers to any base. Note that the fragments generated contain a single-base 3′ extension which is more difficult to ligate than blunt-ended fragments.

BONTOXILYSINS

These zinc metalloproteinases [EC 3.4.24.69], also known as botulinum neurotoxins, act on rat synaptobrevin at Gln–Phe, Lys–Leu, Gln–Lys, and Ala–Ala, respectively.[1-3] Bontoxilysins A, C, and E hydrolyze rat SNAP-25 (25-kDa synaptosomal-associated protein) at Gln–Arg, Arg–Ala, and Arg–Ile, respectively. Bontoxilysin C also acts on syntaxin at Lys–Ala.

[1] C. Montecucco (ed.) (1995) *Clostridial Neurotoxins. Current Topics in Microbiology and Immunology*, vol. **195**, Springer-Verlag, Heidelberg.

[2] C. Montecucco & G. Schiavo (1995) *Q. Rev. Biophys.* **28**, 423.

[3] D. M. Gill (1982) *Microbiol. Rev.* **46**, 86.

(+)-BORNEOL DEHYDROGENASE

This oxidoreductase [EC 1.1.1.198] catalyzes the reaction of (+)-borneol with NAD+ to produce (+)-camphor and NADH. NADP+ is a weaker coenzyme substrate.[1]

[1] S. S. Dehal & R. Croteau (1987) *ABB* **258**, 287.

(−)-BORNEOL DEHYDROGENASE

This oxidoreductase [EC 1.1.1.227] catalyzes the reaction of (−)-borneol with NAD+ to produce (−)-camphor and NADH. NADP+ is a weaker coenzyme substrate.[1]

[1] S. S. Dehal & R. Croteau (1987) *ABB* **258**, 287.

BOTHROLYSIN

This zinc metalloproteinase [EC 3.4.24.50], also referred to as *Bothrops* metalloendopeptidase J and J protease, catalyzes the hydrolysis of the Pro7–Phe8 bond in angiotensin I and Gln4–His5, Ser9–His10, and Ala14–Leu15 of the insulin B chain.[1]

[1] M. M. Tanizaki, R. B. Zingali, H. Kawazaki, S. Imajoh, S. Yamazaki & K. Suzuki (1989) *Toxicon* **27**, 747.

BOTHROPASIN

This zinc metalloendopeptidase [EC 3.4.24.49] catalyzes the hydrolysis of casein by cleaving insulin B chain at His5–Leu6, His10–Leu11, Ala14–Leu15, Tyr16–Leu17, and Phe24–Phe25.[1,2] A related metalloendopeptidase, called MPB, has been isolated from the venom of *Bothrops moojeni*.[3]

[1]F. R. Mandelbaum, A. P. Reichl & M. T. Assakura (1982) *Toxicon* **20**, 955.
[2]J. B. Bjarnason & J. W. Fox (1995) *Meth. Enzymol.* **248**, 345.
[3]S. M. T. Serrano, C. A. M. Sampaio & F. R. Mandelbaum (1993) *Toxicon* **31**, 483.

*Bp*ll RESTRICTION ENDONUCLEASE

This type II restriction endonuclease [EC 3.1.21.4], obtained from *Bacillus pumilus*, catalyzes the hydrolysis of both strands of DNA at 5′ . . . ∇(N)$_8$GAG(N)$_5$CTC (N)$_{13}\nabla$ . . . 3′, where N refers to any base. Note that this endonuclease cleaves DNA twice to excise the recognition site.[1] The enzyme can also exhibit star activity. *See "Star" Activity*

[1]J. Vitkute, Z. Maneliene, M. Petrusyte & A. Janulaitis (1997) *NAR* **25**, 4444.

*Bpm*l RESTRICTION ENDONUCLEASE

This type II restriction endonuclease (subtype α) [EC 3.1.21.4], obtained from *Bacillus pumilus*, acts on DNA at the sequences 5′. . . CTGGAG(N)$_{16}\nabla$ . . . 3′ and 3′. . . GACCTC(N)$_{14}\nabla$ . . . 5′, where N refers to any base. The enzyme is blocked by overlapping *dcm* methylation.

*Bpu*l RESTRICTION ENDONUCLEASE

This type II restriction endonuclease [EC 3.1.21.4], obtained from *Bacillus pumilus* AHU1387, catalyzes the hydrolysis of both strands of DNA at 5′ . . . GRGCY∇C . . . 3′,[1] where R represents A or G and Y represents T or C.[1,2]

[1]S. Ikawa, T. Shibata & T. Ando (1976) *J. Biochem.* **80**, 1457.
[2]M. Iwabuchi, S. Tajima, T. Inoue, T. Shibata & T. Ando (1992) *Nucleic Acids Res.* **20**, 5850.

Selected entries from *Methods in Enzymology* [vol, page(s)]:
General discussion: **65**, 3

*Bpu*10I RESTRICTION ENDONUCLEASE

This type II restriction endonuclease [EC 3.1.21.4], obtained from *Bacillus pumilus* 10, catalyzes the hydrolysis of both strands of DNA at 5′ . . . CC∇TNAGC . . . 3′ and 3′ . . . GGANT^CG . . . 5′.[1,2] Note that the recognition site is quasipalindromic.[2] The protein, which can exhibit star activity, is a heterodimer. *See "Star" Activity*

[1]S. K. Degtyarev, P. A. Zilkin, G. G. Prihodko, V. E. Repin & N. I. Rechkunova (1989) *Mol. Biol. (Mosk)* **23**, 1051.

[2]S. K. Degtyarev, N. I. Rechkunova, A. A. Kolyhalov, V. S. Dedkov & P. A. Zhilkin (1990) *NAR* **18**, 5807.

BRANCHED-CHAIN AMINO ACID AMINOTRANSFERASE

This pyridoxal-phosphate-dependent aminotransferase [EC 2.6.1.42], also referred to as transaminase B, catalyzes the reversible reaction of L-leucine with α-ketoglutarate (or, 2-oxoglutarate) to produce 4-methyl-2-oxopentanoate (*i.e.*, δ-methyl-α-ketopentanoate) and L-glutamate.[1] Other substrates include L-isoleucine (producing 3-methyl-2-oxopentanoate) and L-valine (producing 3-methyl-2-oxobutanoate). Nevertheless, this enzyme is distinct from that of valine:pyruvate aminotransferase [EC 2.6.1.66]. *See also Leucine Aminotransferase; Valine:Pyruvate Aminotransferase*

[1]A. E. Braunstein (1973) *The Enzymes*, 3rd ed., **9**, 379.

Selected entries from *Methods in Enzymology* [vol, page(s)]:
General discussion: 17A, 802; **113**, 71; **166**, 269, 275
Assay: 113, 72; **166**, 270, 275; **324**, 23, 103
Other molecular properties: natural occurrence, **113**, 71; properties, **166**, 277, 279, 280; purification from (canine pancreas, **166**, 280; hog brain, supernatant, **17A**, 813; human pancreas, **166**, 278; pig heart, soluble, **17A**, 804; **166**, 272; pig heart mitochondria, **17A**, 808; rat pancreas, **166**, 276)

BRANCHED-CHAIN-FATTY-ACID KINASE

This phosphotransferase [EC 2.7.2.14], also called isobutyrate kinase and first isolated from a spirochete, catalyzes the reaction of ATP with 2-methylpropanoate (*i.e.*, isobutyrate) to produce ADP and 2-methylpropanoyl phosphate. Alternative substrates include 3-methylbutanoate, 2-methylbutanoate, pentanoate, butanoate, and propanoate.[1] *See also Butyrate Kinase*

[1]C. S. Harwood & E. Canale-Parola (1982) *J. Bacteriol.* **152**, 246.

BRANCHED-CHAIN α-KETO ACID DECARBOXYLASE

This decarboxylase [EC 4.1.1.72], also called branched-chain-2-oxoacid decarboxylase, catalyzes the conversion of (3*S*)-3-methyl-2-oxopentanoate to 2-methylbutanal and carbon dioxide. Substrates include a number of α-keto acids (or, 2-oxoacids), with a high affinity towards branched-chain substrates. The aldehyde generated may stay enzyme-bound and may be an intermediate in the bacterial system for biosynthesis of branched-chain fatty acids.[1] *See also Branched-Chain α-Keto Acid Dehydrogenase Complex*

[1]H. Oku & T. Kaneda (1988) *JBC* **263**, 18386.

BRANCHED-CHAIN α-KETO ACID DEHYDROGENASE COMPLEX

This multienzyme system, also referred to as 3-methyl-2-ketobutanoate dehydrogenase complex and 3-methyl-2-oxobutanoate dehydrogenase complex, catalyzes the overall reaction of 3-methyl-2-oxobutanoate (*i.e.*, α-keto-isovalerate) with NAD^+ and coenzyme A to produce isobutyryl-CoA, NADH, and carbon dioxide.[1,2] The complex, which uses five cofactors (NAD^+, coenzyme A, thiamin pyrophosphate, lipoamide, and FAD), contains three distinct enzyme activities. 2-Ketoisovalerate dehydrogenase (lipoamide) [EC 1.2.4.2], also referred to as 2-methyl-2-ketobutanoate dehydrogenase (lipoamide) and 2-methyl-2-oxobutanoate dehydrogenase (lipoamide), is the E_1 component of the complex. This component is thiamin-pyrophosphate-dependent and catalyzes the reaction of α-ketoisovalerate with lipoamide to produce S(2-methylpropanoyl)dihydrolipoamide and carbon dioxide. Dihydrolipoamide S-(2-methylpropanoyl)transferase, often referred to as E_2, catalyzes the reaction of S-(2-methylpropanoyl)dihydrolipoamide with coenzyme A to produce dihydrolipoamide and isobutyryl-CoA. Dihydrolipoamide dehydrogenase [EC 1.8.1.4] is an FAD-dependent oxidoreductase that catalyzes the reaction of dihydrolipoamide with NAD^+ to produce lipoamide and NADH. Other substrates for the complex include (*S*)-3-methyl-2-ketopentanoate (*i.e.*, (*S*)-2-keto-3-methylvalerate) and 4-methyl-2-ketopentanoate (*i.e.*, 2-ketoisocaproate). **See also** *specific enzyme*

The complex is regulated by [3-methyl-2-ketobutanoate dehydrogenase (lipoamide)] kinase [EC 2.7.1.115] and [3-methyl-2-ketobutanoate dehydrogenase (lipoamide)] phosphatase [EC 3.1.3.52].

[1] P. J. Randle, P. A. Patston & J. Espinal (1987) *The Enzymes*, 3rd ed., **18**, 97.
[2] L. J. Reed, Z. Damuni & M. L. Merryfield (1985) *Curr. Top. Cell. Regul.* **27**, 41.

Selected entries from **Methods in Enzymology** [vol, page(s)]:
General discussion: 17A, 818; 166, 114, 146, 189, 298, 303, 309, 313, 342, 350; 324, 129, 179, 192, 200, 329, 336, 355, 389, 365, 376, 453, 465, 479
Assay: 166, 144, 147, 175, 189, 298, 303, 309, 313; 324, 329, 336, 389 (maple syrup urine disease fibroblasts, with, 166, 135; potential interfering substances, 166, 196; potential sources of error, 166, 176); liver, in, 166, 204; muscle, in, 166, 209; *Pseudomonas*, from, 166, 345
Other molecular properties: activation by α-chloroisocaproate, 166, 118; activator protein, assay, 166, 183; antibodies to constituent proteins (analysis of mutations, in, 166, 109; preparation, 166, 108; screening of expression libraries, in, 166, 110); component E_1 (assay, 166, 147; properties, 166, 150); component E_2 (assay, 166, 152; properties, 166, 154); flux rates in perfused heart and liver,

measurement, 166, 484; lipoyl domain, in *Bacillus stearothermophilus* (lipoylation [recognition site, 251, 446; role in catalysis, 251, 436, 445]; purification, 251, 438; reductive acetylation, assay, 251, 439; size, 251, 444); phosphorylation sites, analysis, 166, 10; properties, 166, 150, 301, 312; *Pseudomonas*, from (assay, 166, 345; gene cloning, 166, 350; purification, 166, 346); purification (bovine kidney, 166, 304, 310; bovine liver, 17A, 820; 166, 299; kinase assays, for, 166, 174; rabbit liver, 166, 316); pyruvate dehydrogenase complex (assay, 166, 331; properties, 166, 337; purification from *Bacillus subtilis*, 166, 332); reconstitution of components, 166, 308; resolution into active components, 166, 306; staining on gels, 22, 595

BRANCHED-DEXTRAN EXO-1,2-α-GLUCOSIDASE

This glycosidase [EC 3.2.1.115], also referred to as dextran α-(1,2)-debranching enzyme, catalyzes the hydrolysis of 1,2-α-D-glucosidic linkages at the branch points of dextrans and related polysaccharides, producing free D-glucose.[1,2] The enzyme does not act on 1,2-disaccharides or oligosaccharides containing linear 1,2-α-glucosidic linkages.

[1] Y. Mitsuishi & M. Kobayashi & K. Matsuda (1979) *Agric. Biol. Chem.* **43**, 2283.
[2] Y. Mitsuishi, M. Kobayashi & K. Matsuda (1980) *Carbohydr. Res.* **83**, 303.

BROMELAIN, FRUIT

Fruit bromelain [EC 3.4.22.33] is the major proteolytic enzyme of the juice of the pineapple (*Ananas comosus*): almost 90% of the proteolytic activity of the fruit is due to EC 3.4.22.33. This cysteine endopeptidase catalyzes the hydrolysis of peptide bonds in proteins with a broad specificity.[1] The synthetic substrate Bz-Phe-Val-Arg-∇-NHMec is a good substrate for fruit bromelain but a poor substrate for stem bromelain. Interestingly, fruit bromelain is weakly inhibited by chicken cystatin.

The cysteine endopeptidase pinguinain, formerly classified under EC 3.4.99.18, is now listed as a fruit bromelain, EC 3.4.22.33. **See also** *Bromelain, Stem; Ananain; Comosain; Pinguinain*

[1] A. D. Rowan & D. J. Buttle (1994) *Meth. Enzymol.* 244, 555.

Selected entries from **Methods in Enzymology** [vol, page(s)]:
General discussion: 2, 56, 62; 19, 273, 283; 244, 555
Assay: 19, 283; 244, 559
Other molecular properties: amino acid composition, 19, 280; 45, 480; amino acid sequence, 45, 484; properties, 19, 284; 45, 484; 244, 565; meat tenderizing by, 2, 55; nomenclature, 45, 475; 244, 555; purification, 19, 283; 45, 483; 244, 562, 564; storage, 244, 565

BROMELAIN, STEM

Stem bromelain [EC 3.4.22.32], also known simply as bromelain, is the major endopeptidase present in extracts of the stem of the pineapple plant (*Ananas comosus*).[1,2] Originally, all of the endopeptidases from the pineapple plant were called the bromelains. The stem and fruit enzymes were assigned different EC numbers (originally EC 3.4.4.24 and EC 3.4.4.25, respectively and now classified as 3.4.22.4). This cysteine endopeptidase shows broad specificity and high activity with protein substrates; with synthetic substrates, the enzyme exhibits a strong preference for Z-Arg-Arg-∇-NHMec, which is a poor substrate for fruit bromelain and ananain. (Note: ∇ indicates the cleavage site.) *See also Bromelain, Fruit; Ananain; Comosain*

[1] A. D. Rowan, D. J. Buttle & A. J. Barrett (1990) *BJ* **266**, 869.
[2] A. D. Rowan & D. J. Buttle (1994) *MIE* **244**, 555.

Selected entries from *Methods in Enzymology* [vol, page(s)]:
General discussion: 2, 56, 62; **19**, 273, 274; **34**, 4, 418, 532; **244**, 555
Assay: **19**, 274 (amidase activity, **19**, 275; caseinolytic activity, **19**, 274; esterolytic activity, **19**, 275); **244**, 557; p-nitrophenyl-N^α-benzyloxycarbonyl-L-lysine, by, **45**, 740
Other molecular properties: activators, **19**, 282; **45**, 482; active site residues, **87**, 461; amino acid composition, **19**, 280; **45**, 480; amino acid sequence, **45**, 481; chemical modifications, **45**, 482; chemical properties, **19**, 281; CM-cellulose derivative of (preparation, **19**, 970; properties, **19**, 971); conjugate, kinetics, **44**, 406, 437; enzymatic mechanism, **45**, 482; γ-glutamyl transpeptidase, purification of, **77**, 243; immobilization ((carboxymethyl)cellulose, on, **44**, 52; property changes, **44**, 53); inhibitors, **19**, 282; **45**, 482 (pineapple stem [amino acid composition, **45**, 747; sequence, **45**, 749; assay, by p-nitrophenyl-N^α-benzyloxycarbonyl-L-lysine, **45**, 742; properties, **45**, 748; protein chemistry, **45**, 748; purification, **45**, 745]); kinetics, **19**, 282; **45**, 482; meat tenderizing by, **2**, 55; nomenclature, **45**, 475; **244**, 555; properties, **19**, 280, 281; **45**, 479; **244**, 565; purification, **19**, 276; **45**, 476, 743; **244**, 562; purity, **19**, 279; **45**, 478; specificity, **19**, 282; **45**, 482; storage, **244**, 565; titration, **45**, 9 (burst, **45**, 10; sulfhydryl group, **45**, 10); trapped by α₂-macroglobulin, **80**, 748

BROMOPEROXIDASE

This haloperoxidase catalyzes the oxidation of a bromide anion with hydrogen peroxide, resulting in the concomitant halogenation of some organic substrate.[1] A common substrate in assay protocols is monochlorodimedone (*i.e.*, 2-chloro-5,5-dimethyl-1,3-dimedone), forming bromochlorodimedone. Bromoperoxidase will preferentially oxidize Br^- but will also act on iodide and chloride *See Chloroperoxidase*; however, in the case of I^-, triiodide is produced (I_3^-). There are a number of different haloperoxidases. These include iron-heme proteins, vanadium-dependent enzymes, amd metal-free systems. Lactoperoxidase has a bromoperoxidase activity.

Vanadium-dependent bromoperoxidase displays a ping pong kinetic mechanism. Hydrogen peroxide first binds to vanadium(V) to produce a peroxyvanadium(V) species.

Bromide binds and a two-electron-oxidized intermediate is produced (such as Br^+, HOBr, Br_3^- etc.). This intermediate can react with another molecule of hydrogen peroxide, generating dioxygen in the singlet excited state ($^1\Delta_g$); or, it can react with an organic substrate to produce the brominated product.

Iron-heme bromoperoxidases also function as two-electron redox catalysts. *See also Peroxidase; Chloroperoxidase; Lactoperoxidase*

[1] A. Butler (1998) *CBC* **3**, 427.

Selected entries from *Methods in Enzymology* [vol, page(s)]:
Assay: **107**, 442
Other molecular properties: brown algal, extraction in aqueous two-phase systems, **228**, 667; protein bromination, **107**, 441, 444; purification from *Penicillus capitatus*, **107**, 442

BROMOXYNIL NITRILASE

This highly specific bacterial hydrolase [EC 3.5.5.6] catalyzes the reaction of two molecules of water with 3,5-dibromo-4-hydroxybenzonitrile, the herbicide bromoxynil, to produce 3,5-dibromo-4-hydroxy-benzoate and ammonia. Other substrates include 3,5-dichloro-4-hydroxybenzonitrile (*i.e.*, chloroxynil) and 3,5-diiodo-4-hydroxybenzonitrile (*i.e.*, ioxynil).[1] *See also Nitrilase*

[1] D. M. Stalker, L. D. Malyj & K. E. McBride (1988) *JBC* **263**, 6310.

*Bsa*I RESTRICTION ENDONUCLEASE

This type II restriction endonuclease (subtype s) [EC 3.1.21.4], obtained from *Bacillus stearothermophilus*, catalyzes the hydrolysis of DNA at 5′ ... GGTCTCN∇ ... 3′ and 3′ ... CCAGAGNNNNN∇ ... 5′, where N refers to any base.

*Bsa*AI RESTRICTION ENDONUCLEASE

This type II restriction endonuclease [EC 3.1.21.4], obtained from *Bacillus stearothermophilus* G668, catalyzes the hydrolysis of both strands of DNA at 5′ ... YAC∇GTR ... 3′, where Y refers to T or C and R refers to A or G.[1]

[1] H. Kong, R. D. Morgan & Z. Chen (1990) *NAR* **18**, 2832.

*Bsa*BI RESTRICTION ENDONUCLEASE

This type II restriction endonuclease [EC 3.1.21.4], obtained from *Bacillus stearothermophilus* B674, catalyzes the hydrolysis of both strands of DNA at 5′ ... GATNN∇ ... 3′ (where N refers to any base), producing blunt-ended fragments. The enzyme is blocked by overlapping *dam* methylation.

*Bsa*HI RESTRICTION ENDONUCLEASE

This type II restriction endonuclease [EC 3.1.21.4], obtained from *Bacillus stearothermophilus* CPW11, catalyzes the hydrolysis of both strands of DNA at $5' \dots GR\nabla CGYC \dots 3'$, where R refers to A or G and Y refers to T or C.

*Bsa*JI RESTRICTION ENDONUCLEASE

This type II restriction endonuclease [EC 3.1.21.4], obtained from *Bacillus stearothermophilus* J695, catalyzes the hydrolysis of both strands of DNA at $5' \dots C^{\wedge}CNNGG \dots 3'$, where N refers to any base.

*Bsa*WI RESTRICTION ENDONUCLEASE

This type II restriction endonuclease [EC 3.1.21.4], obtained from *Bacillus stearothermophilus* W1718, catalyzes the hydrolysis of both strands of DNA at $5' \dots W\nabla CCGGW \dots 3'$, where W refers to A or T.

*Bsa*XI RESTRICTION ENDONUCLEASE

This type II restriction endonuclease (subtype s) [EC 3.1.21.4], obtained from *Bacillus stearothermophilus* Cpw230, catalyzes the hydrolysis of DNA at $5' \dots \nabla(N)_9AC(N)_5CTCC(N)_{10}\nabla \dots 3'$ and $3' \dots \nabla(N)_{12}\text{-}TG(N)_5GAGG(N)_7\nabla \dots 5'$, where N refers to any base.

*Bse*MII RESTRICTION ENDONUCLEASE

This type II restriction endonuclease (subtype s) [EC 3.1.21.4], obtained from *Bacillus stearothermophilus* Isl 15-111, catalyzes the hydrolysis of both strands of DNA at $5' \dots CTCAG(N)_{10}\nabla \dots 3'$ and $3' \dots GAGTC(N)_8\nabla NN \dots 5'$, where N refers to any base.

*Bse*RI RESTRICTION ENDONUCLEASE

This type II restriction endonuclease (subtype s) [EC 3.1.21.4], obtained from *Bacillus* species R, catalyzes the hydrolysis of DNA at $5' \dots GAGGAG(N)_{10}\nabla \dots 3'$ and $3' \dots CTCCTC(N)_8\nabla \dots 5'$, where N refers to any base.[1]

[1]R. Mushtaq, S. Naeem, A. Sohail & S. Riazuddin (1993) *NAR* **21**, 3585.

*Bse*SI RESTRICTION ENDONUCLEASE

This type II restriction endonuclease [EC 3.1.21.4], obtained from *Bacillus stearothermophilus* Jo-553,

catalyzes the hydrolysis of both strands of DNA at $5' \dots GKGCM\nabla C \dots 3'$, where K refers to G or T and M refers to A or C.[1]

[1]D. Steponaviciene, Z. Maneliene, M. Petrusyte & A. Janulaitis (1999) *NAR* **27**, 2644.

*Bsg*I RESTRICTION ENDONUCLEASE

This type II restriction endonuclease (subtype α) [EC 3.1.21.4], obtained from *Bacillus sphaericus* B922, catalyzes the hydrolysis of DNA at $5' \dots GTGCAG(N)_{16}\nabla \dots 3'$ and $3' \dots CACGTC(N)_{14}\nabla \dots 5'$, where N refers to any base.[1] This restriction endonuclease requires the presence of 80 μM *S*-adenosyl-L-methionine for maximal activity.

[1]S. E. Halford, D. T. Bilcock, N. P. Stanford, S. A. Williams, S. E. Milsom, N. A. Gormley, M. A. Watson, A. J. Bath, M. L. Embleton, D. M. Gowers, L. E. Daniels, S. H. Parry & M. D. Szczelkun (1999) *Biochem. Soc. Trans.* **27**, 696.

*Bsi*EI RESTRICTION ENDONUCLEASE

This type II restriction endonuclease [EC 3.1.21.4], obtained from *Bacillus* species, catalyzes the hydrolysis of both strands of DNA at $5' \dots CGRY\nabla CG \dots 3'$, where R refers to A or G and Y refers to T or C.[1]

[1]Y. K. Mok, D. R. Clark, K. M. Kam & P. C. Shaw (1990) *NAR* **18**, 4954.

*Bsi*HKAI RESTRICTION ENDONUCLEASE

This type II restriction endonuclease [EC 3.1.21.4], obtained from *Bacillus stearothermophilus*, catalyzes the hydrolysis of both strands of DNA at $5' \dots GWGCW\nabla C \dots 3'$, where W refers to A or T.[1] The optimum reaction temperature is 60 °C.[1]

[1]K.-F. Lee, S.-D. Shi, K.-M. Kam & P.-C. Shaw (1992) *NAR* **20**, 921.

*Bsi*WI RESTRICTION ENDONUCLEASE

This type II restriction endonuclease [EC 3.1.21.4], obtained from *Bacillus* species, catalyzes the hydrolysis of both strands of DNA at $5' \dots C\nabla GTACG \dots 3'$.

*Bsl*I RESTRICTION ENDONUCLEASE

This type II restriction endonuclease [EC 3.1.21.4], obtained from *Bacillus* species, catalyzes the hydrolysis of both strands of DNA at $5' \dots CCNNNNN\nabla NNGG \dots 3'$, where N refers to any base.[1]

[1]P.-C. Hsieh, J.-P. Xiao, D. O'Loane & S.-Y. Xu (2000) *J. Bacteriol.* **182**, 949.

*Bsm*I RESTRICTION ENDONUCLEASE

This type II restriction endonuclease [EC 3.1.21.4], obtained from *Bacillus stearothermophilus* NUB36, catalyzes the hydrolysis of both strands of DNA at the nonpalindromic sequence 5′ … GAATGCN∇ … 3′ and 3′ … CTTAC∇GN … 5′, where N refers to any base.

*Bsm*AI RESTRICTION ENDONUCLEASE

This type II restriction endonuclease (subtype s) [EC 3.1.21.4], obtained from *Bacillus stearothermophilus* A664, catalyzes the hydrolysis of both strands of DNA at 5′ … GTCTCN∇ … 3′ and 3′ … CAGAGNNNNN∇ … 5′, where N refers to any base.[1] *Bsm*AI can cleave one strand in a three-way DNA junction.[2]

[1]H. Kong, R. D. Morgan & Z. Chen (1990) *NAR* **18**, 686.
[2]D. M. Wong & J. G. Wetmur (1994) *Gene* **150**, 63.

*Bsm*BI RESTRICTION ENDONUCLEASE

This type II restriction endonuclease (subtype s) [EC 3.1.21.4], obtained from *Bacillus stearothermophilus* B61, catalyzes the hydrolysis of both strands of DNA at 5′ … CGTCTCN∇ … 3′ and 3′ … GCAGAGNNNNN∇ … 5′, where N refers to any base.

*Bsm*FI RESTRICTION ENDONUCLEASE

This type II restriction endonuclease (subtype s) [EC 3.1.21.4], obtained from *Bacillus stearothermophilus* F, catalyzes the hydrolysis of both strands of DNA at 5′ … GGGAC(N)$_p$∇ … 3′ and 3′ … CCCTG(N)$_q$∇ … 5′, where N refers to any base, $p = 10$, and $q = 14$. Occasionally, *Bsm*FI will cleave both strands of DNA when $p = 9$ and $q = 13$. The frequency of this alternative cleavage pattern is unknown.

*Bso*BI RESTRICTION ENDONUCLEASE

This type II restriction endonuclease [EC 3.1.21.4], obtained from *Bacillus stearothermophilus* JN2091, catalyzes the hydrolysis of both strands of DNA at 5′ … C∇YCGRG … 3′, where Y refers to C or T and R refers to G or A.[1,2] This is a thermophilic enzyme; however, the incubation temperature is not 60 °C but 37 °C in order to minimize star activity. (**See** *"Star" Activity*) Structural studies have led to a suggestion that the catalytic mechanism is mediated by a histidyl residue (His253) and two metal ions. This would make *Bso*BI the first restriction endonuclease known to utilize such a mechanism.[3]

[1]H. Ruan, K. D. Lunnen, M. E. Scott, L. S. Moran, B. E. Slatko, J. J. Pelletier, E. J. Hess, J. Benner, G. G. Wilson & S.-Y. Xu (1996) *Mol. Gen. Genet.* **252**, 695.
[2]H. Ruan, K. D. Lunnen, J. J. Pelletier & S.-Y. Xu (1997) *Gene* **188**, 35.
[3]M. J. van der Woerd, J. J. Pelletier, S.-Y. Xu & A. M. Friedman (2001) *Structure* **9**, 133.

*Bsp*1286I RESTRICTION ENDONUCLEASE

This type II restriction endonuclease [EC 3.1.21.4] is obtained from *Bacillus sphaericus* IAM 1286 and acts on both strands of DNA at the sequence 5′ … GDGCH∇C … 3′[1] where D represents any base other than C (*i.e.*, A, T, or G) and H represents any base other than G (*i.e.*, A, T, or C).[1,2]

[1]T. Shibata, S. Ikawa, C. Kim & T. Ando (1976) *J. Bacteriol.* **128**, 473.
[2]M. Nelson, C. Christ & I. Schildkraut (1984) *Nucleic Acids Res.* **12**, 5165.

Selected entries from *Methods in Enzymology* [vol, page(s)]: **General discussion: 65**, 3

*Bsp*DI RESTRICTION ENDONUCLEASE

This type II restriction endonuclease [EC 3.1.21.4], obtained from *Bacillus* species, catalyzes the hydrolysis of both strands of DNA at 5′ … AT∇CGAT … 3′. The enzyme is blocked by overlapping *dam* methylation.

*Bsp*EI RESTRICTION ENDONUCLEASE

This type II restriction endonuclease [EC 3.1.21.4], obtained from *Bacillus* species, catalyzes the hydrolysis of both strands of DNA at 5′ … T∇CCGGA … 3′.[1] The enzyme is blocked by overlapping *dam* methylation.

[1]D. O. Nwankwo (1995) *Gene* **157**, 31.

*Bsp*HI RESTRICTION ENDONUCLEASE

This type II restriction endonuclease [EC 3.1.21.4], obtained from *Bacillus* species H, catalyzes the hydrolysis of both strands of DNA at 5′ … T∇CATGA … 3′. The enzyme is blocked by overlapping *dam* methylation.[1]

[1]S. Servos, C. Silva, G. Dougan & I. G. Charles (1991) *NAR* **19**, 183.

*Bsp*MI RESTRICTION ENDONUCLEASE

This type II restriction endonuclease (subtype s) [EC 3.1.21.4], obtained from *Bacillus* species M, catalyzes the hydrolysis of both strands of DNA at 5′ … ACCTGCNNNN∇ … 3′ and 3′ … TGGACGNNNN NNNN∇ … 5′, where N refers to any base.[1] The enzyme also exhibits star activity. **See** *"Star" Activity*

[1] S. E. Halford, D. T. Bilcock, N. P. Stanford, S. A. Williams, S. E. Milsom, N. A. Gormley, M. A. Watson, A. J. Bath, M. L. Embleton, D. M. Gowers, L. E. Daniels, S. H. Parry & M. D. Szczelkun (1999) *Biochem. Soc. Trans.* **27**, 696.

*Bsp*RI RESTRICTION ENDONUCLEASE

This type II restriction endonuclease [EC 3.1.21.4] is Mg^{2+}-dependent, obtained from *Bacillus sphaericus* R, and acts on both strands of DNA at the sequence 5′ ... GG∇CC ... 3′, producing blunt-ended fragments.[1,2]

[1] A. Kiss, B. Sain, E. Csordas-Toth & P. Venetianer (1977) *Gene* **1**, 323.
[2] C. Koncz, A. Kiss & P. Venetianer (1978) *EJB* **89**, 523.
Selected entries from *Methods in Enzymology* [vol, page(s)]:
General discussion: 65, 109
Assay: 65, 109
Other molecular properties: cation requirements, **65**, 111; inhibitor, **65**, 111; isoschizomers, **65**, 109; molecular weight, **65**, 111; pH optimum, **65**, 111; recognition sequence, **65**, 4, 109; stability, **65**, 111; substrate specificity, **65**, 111, 112; turnover number, **65**, 111

*Bsr*I RESTRICTION ENDONUCLEASE

This type II restriction endonuclease (subtype s) [EC 3.1.21.4], obtained from *Bacillus stearothermophilus*, catalyzes the hydrolysis of both strands of DNA at 5′ ... ACTGGN∇ ... 3′ and 3′...TGAC∇CN ... 5′, where N refers to any base.[1]

[1] C. Polisson & R. D. Morgan (1988) *NAR* **16**, 5205.

*Bsr*BI RESTRICTION ENDONUCLEASE

This type II restriction endonuclease [EC 3.1.21.4], obtained from *Bacillus stearothermophilus* CPW193, catalyzes the hydrolysis of both strands of DNA at 5′ ... GAG∇CGG ... 3′ and 3′...CTC∇GCC... 5′ (a nonpalindromic recognition sequence), producing blunt-ended fragments. DNA cleaved by *Bsr*BI and then ligated will contain ligated sites only half of which are recleavable by *Bsr*BI (due to the nonpalindromic nature of the sequence). Those sites not recleavable by *Bsr*BI can still be cleaved by *Sac*I or *Sac*II.

*Bsr*DI RESTRICTION ENDONUCLEASE

This type II restriction endonuclease (subtype s) [EC 3.1.21.4], obtained from *Bacillus stearothermophilus* D70, catalyzes the hydrolysis of both strands of DNA at 5′ ... GCAATGNN∇ ... 3′ and 3′...CGTTAC∇NN ... 5′, where N refers to any base.

*Bsr*FI RESTRICTION ENDONUCLEASE

This type II restriction endonuclease [EC 3.1.21.4], obtained from *Bacillus stearothermophilus* CPW16, catalyzes the hydrolysis of both strands of DNA at 5′ ... R∇CCGGY ... 3′, where R refers to A or G and Y refers to T or C. The enzyme may exhibit star activity. *See "Star" Activity*

*Bsr*GI RESTRICTION ENDONUCLEASE

This type II restriction endonuclease [EC 3.1.21.4], obtained from *Bacillus stearothermophilus* GR75, catalyzes the hydrolysis of both strands of DNA at 5′ ... T∇GTACA ... 3′.[1]

[1] Z. F. Chen & X. S. Pan (1994) *Chinese Sci. Bull.* **39**, 526.

*Bss*HII RESTRICTION ENDONUCLEASE

This type II restriction endonuclease [EC 3.1.21.4], obtained from *Bacillus stearothermophilus* H3, catalyzes the hydrolysis of both strands of DNA at 5′ ... G∇CGCGC ... 3′. The enzyme also exhibits star activity. (*See "Star" Activity*) Kinetic investigations confirm that *Bss*HII slides and scans along DNA until it encounters the recognition sequence.[1]

[1] B. Berkhout & J. van Wamel (1996) *JBC* **271**, 1837.

*Bss*KI RESTRICTION ENDONUCLEASE

This type II restriction endonuclease [EC 3.1.21.4], obtained from *Bacillus stearothermophilus* TBI, catalyzes the hydrolysis of both strands of DNA at 5′ ... ∇CCNGG ... 3′. The enzyme is blocked by overlapping *dcm* methylation.

*Bss*SI RESTRICTION ENDONUCLEASE

This type II restriction endonuclease (subtype s) [EC 3.1.21.4], obtained from *Bacillus stearothermophilus* 27S, catalyzes the hydrolysis of both strands of DNA at 5′ ... C∇TCGTG ... 3′ and 3′...GAGCA∇C ... 5′. The enzyme can also exhibit star activity. *See "Star" Activity*

*Bst*API RESTRICTION ENDONUCLEASE

This type II restriction endonuclease [EC 3.1.21.4], obtained from *Bacillus stearothermophilus* AP, catalyzes the hydrolysis of both strands of DNA at 5′ ... GCANNNN∇NTGC ... 3′, where N refers to any base.[1]

[1] M. A. Abdurashitov, O. A. Belichenko, A. V. Shevchenko, V. S. Dedkov & S. K. Degtyarev (1997) *NAR* **25**, 2301.

*Bst*BI RESTRICTION ENDONUCLEASE

This type II restriction endonuclease [EC 3.1.21.4], obtained from *Bacillus stearothermophilus* B225, catalyzes the hydrolysis of both strands of DNA at 5′ . . . TT∇CGAA . . . 3′.

*Bst*4CI RESTRICTION ENDONUCLEASE

This type II restriction endonuclease [EC 3.1.21.4], obtained from *Bacillus stearothermophilus* 4C, catalyzes the hydrolysis of both strands of DNA at 5′ . . . ACN∇GT . . . 3′, where N refers to any base.

*Bst*DSI RESTRICTION ENDONUCLEASE

This type II restriction endonuclease [EC 3.1.21.4], obtained from *Bacillus stearothermophilus* DS, catalyzes the hydrolysis of both strands of DNA at 5′ . . . CC∇RYGG . . . 3′, where R refers A or G and Y refers to T or C.

*Bst*EII RESTRICTION ENDONUCLEASE

This type II restriction endonuclease [EC 3.1.21.4] is obtained from *Bacillus stearothermophilus* ET and acts on both strands of DNA at 5′ . . . G∇GTNACC . . . 3′[1], where N represents any base. Note that the fragments generated contain five base extensions.

[1] J. A. Lautenberger, M. H. Edgell & C. A. Hutchison III (1980) *Gene* **12**, 171.

*Bst*F5I RESTRICTION ENDONUCLEASE

This type II restriction endonuclease [EC 3.1.21.4], obtained from *Bacillus stearothermophilus* F5, catalyzes the hydrolysis of both strands of DNA at 5′ . . . GGATGNN∇ . . . 3′ and 3′ . . . CCTAC∇NN . . . 5′, where N refers to any base.[1]

[1] M. A. Abdurashitov, E. V. Kileva, N. M. Shinkarenko, A. V. Shevchenko, V. S. Dedkov & S. K. Degtyarev (1996) *Gene* **172**, 49.

*Bst*NI RESTRICTION ENDONUCLEASE

This type II restriction endonuclease [EC 3.1.21.4], obtained from *Bacillus stearothermophilus* N, catalyzes the hydrolysis of both strands of DNA at 5′ . . . CC∇WGG . . . 3′, where W refers to A or T.[1]

[1] M. M. Baryshev, Y. I. Buryanov, V. G. Kosykh & A. A. Bayev (1989) *Biokhimiia* **54**, 1894.

*Bst*I RESTRICTION ENDONUCLEASE

This type II restriction endonuclease [EC 3.1.21.4] is Mg^{2+}-dependent, is obtained from *Bacillus stearothermophilus*

1503-4R, and acts on both strands of DNA at 5′ . . . G∇GATCC . . . 3′.[1,2]

[1] J. F. Catterall & N. E. Welker (1977) *J. Bacteriol.* **129**, 1110.
[2] C. M. Clarke & B. S. Hartley (1979) *BJ* **177**, 49.

Selected entries from *Methods in Enzymology* [vol, page(s)]:
General discussion: **65**, 147, 150, 167
Assay: **65**, 167, 168
Other molecular properties: cation requirement, **65**, 169; isoelectric point, **65**, 169; isoschizomer, **65**, 170; molecular weight, **65**, 151, 169, 170; optimum reaction conditions, **65**, 169; properties, **65**, 151, 169, 170; purification, **65**, 168, 169; recognition sequence, **65**, 4, 170; stability, **65**, 167, 169; subunit composition, **65**, 169

*Bst*UI RESTRICTION ENDONUCLEASE

This type II restriction endonuclease [EC 3.1.21.4], obtained from *Bacillus stearothermophilus* U458, catalyzes the hydrolysis of both strands of DNA at 5′ . . . CG∇CG . . . 3′, producing blunt-ended fragments.

*Bst*XI RESTRICTION ENDONUCLEASE

This type II restriction endonuclease [EC 3.1.21.4], obtained from *Bacillus stearothermophilus* X1, catalyzes the hydrolysis of both strands of DNA at 5′ . . . C CANNNNN∇NTGG . . . 3′, where N refers to any base. The enzyme also exhibits star activity. *See "Star" Activity*

*Bst*YI RESTRICTION ENDONUCLEASE

This type II restriction endonuclease [EC 3.1.21.4], obtained from *Bacillus stearothermophilus* Y406, catalyzes the hydrolysis of both strands of DNA at 5′ . . . R∇GATCY . . . 3′, where R refers A or G and Y refers to T or C.[1] The enzyme is not blocked by *dam* methylation.

[1] Z. Chen & H. Kong (1988) *FEBS Lett.* **234**, 169.

*Bst*Z17I RESTRICTION ENDONUCLEASE

This type II restriction endonuclease [EC 3.1.21.4], obtained from *Bacillus stearothermophilus* 38M, catalyzes the hydrolysis of both strands of DNA at 5′ . . . GTA∇TAC . . . 3′, producing blunt-ended fragments. The enzyme may exhibit star activity. *See "Star" Activity*

*Bsu*36I RESTRICTION ENDONUCLEASE

This type II restriction endonuclease [EC 3.1.21.4], obtained from *Bacillus subtilis* 36, catalyzes the hydrolysis of both strands of DNA at 5′ . . . CC∇TNAGG . . . 3′, where N refers to any base.[1]

[1] B. Zhou & Q. Li (1987) *Acta Biochim. Biophys. Sin.* **19**, 537.

*Bsu*BI RESTRICTION ENDONUCLEASE

This type II restriction endonuclease [EC 3.1.21.4], also referred to as *Bsu* 1247I, is obtained from *Bacillus subtilis* IAM 1247. It acts on both strands of DNA at 5′ ... CTGCA∇G ... 3′.[1,2]

[1] T. Shibata, S. Ikawa, C. Kim & T. Ando (1976) *J. Bacteriol.* **128**, 473.
[2] G. L. Xu, W. Kapfer, J. Walter & T. A. Trautner (1992) *Nucleic Acids Res.* **20**, 6517.

*Bsu*FI RESTRICTION ENDONUCLEASE

This type II restriction endonuclease [EC 3.1.21.4], once referred to as *Bsu* 1231, is obtained from *Bacillus subtilis* IAM 1231 and acts on both strands of DNA at 5′ ... C∇CGG ... 3′.[1,2]

[1] T. Shibata, S. Ikawa, C. Kim & T. Ando (1976) *J. Bacteriol.* **128**, 473.
[2] S. Jentsch (1983) *J. Bacteriol.* **156**, 800.

*Bsu*RI RESTRICTION ENDONUCLEASE

This type II restriction endonuclease [EC 3.1.21.4] is Mg^{2+}-dependent, is obtained from *Bacillus subtilis* strain R, and acts on both strands of DNA at 5′ ... GG∇CC ... 3′.[1–3]

[1] T. A. Trautner, B. Pawlek, S. Bron & C. Anagnostopoulos (1974) *Mol. Gen. Genet.* **131**, 181.
[2] S. Bron, K. Murray & T. A. Trautner (1975) *Mol. Gen. Genet.* **143**, 13.
[3] S. Bron & K. Murray (1975) *Mol. Gen. Genet.* **143**, 25.

Selected entries from ***Methods in Enzymology*** [vol, page(s)]:
General discussion: 65, 112, 147
Assay: using gel electrophoresis of DNA fragments, **65**, 117, 118; using transfection, **65**, 113
Other molecular properties: cation requirements, **65**, 125; contaminating nonspecific nuclease activities, **65**, 124, 125; ionic requirements, **65**, 125, 126; isoschizomers, **65**, 109, 126; molecular weight, **65**, 126; purification, **65**, 118, 124; reaction conditions (optimum, **65**, 130; standard, **65**, 125, 126); recognition sequence, **65**, 4, 113, 126; specificity, **65**, 126; storage and stability, **65**, 125; substrate specificity, **65**, 126

*Btg*I RESTRICTION ENDONUCLEASE

This type II restriction endonuclease [EC 3.1.21.4], obtained from *Bacillus thermoglucosidasius*, catalyzes the hydrolysis of both strands of DNA at 5′ ... C∇CRYGG ... 3′, where R refers to A or G and Y refers to T or C.

*Btr*I RESTRICTION ENDONUCLEASE

This type II restriction endonuclease (subtype s) [EC 3.1.21.4], obtained from *Bacillus stearothermophilus* SE-U62, catalyzes the hydrolysis of both strands of DNA at 5′ ... CAC∇GTC ... 3′ and 3′...GTG∇CAG ... 5′ (a nonpalindromic sequence), producing blunt-ended fragments.[1]

[1] S. K. Degtyarev, O. A. Belichenko, N. A. Lebedeva, V. S. Dedkov & M. A. Abdurashitov (2000) *NAR* **28**, e56.

*Bts*I RESTRICTION ENDONUCLEASE

This type II restriction endonuclease (subtype s) [EC 3.1.21.4], obtained from *Bacillus thermoglucosidasius*, catalyzes the hydrolysis of both strands of DNA at 5′ ... GCAGTGNN∇ ... 3′ and 3′ ... CGTCAC^NN ... 5′, where N refers to any base.[1]

[1] R. J. Roberts & D. Macelis (1999) *NAR* **27**, 312.

BUTANAL DEHYDROGENASE

This oxidoreductase [EC 1.2.1.57] catalyzes the reaction of butanal with coenzyme A and $NAD(P)^+$ to produce butanoyl-CoA and NAD(P)H.[1] Acetaldehyde and propanal are weaker alternative substrates.[2]

[1] D. T. Jones & D. R. Woods (1986) *Microbiol. Rev.* **50**, 484.
[2] N. R. Palasaari & P. Rogers (1988) *J. Bacteriol.* **170**, 2971.

meso-BUTANEDIOL DEHYDROGENASE

This oxidoreductase, first reported in *Escherichia coli*, catalyzes the reaction of *meso*-butane-2,3-diol with NAD^+ to produce (R)-acetoin and NADH.[1]

meso-butane-2,3-diol   (R)-acetoin

Note that both (R,R)-butanediol dehydrogenase [EC 1.1.1.4] and (S,S)-butanediol dehydrogenase [EC 1.1.1.76] have been reported to act on *meso*-butane-2,3-diol.

[1] S. Ui, A. Mimura, M. Ohkuma & T. Kudo (1999) *Lett. Appl. Microbiol.* **28**, 457.

(R,R)-BUTANEDIOL DEHYDROGENASE

This oxidoreductase [EC 1.1.1.4], also referred to as butyleneglycol dehydrogenase, catalyzes the reversible reaction of (R,R)-butane-2,3-diol with NAD^+ to produce (R)-acetoin and NADH.[1,2]

(R,R)-butane-2,3-diol   (R)-acetoin

Interestingly, the enzyme also catalyzes the reaction of diacetyl (*i.e.*, 2,3-butanedione) with NADH to produce acetoin and NAD^+ (*i.e.*, a diacetyl reductase activity; this reaction heavily favors acetoin formation). In addition, the enzyme also catalyzes the reaction of D-1-amino-2-propanol with NAD^+ to produce 1-amino-2-propanone and NADH (this latter activity had been previously classified as EC 1.1.1.74, but is now a deleted EC entry).

[1] J. J. Kelley & E. E. Dekker (1984) *JBC* **259**, 2124.
[2] H. Höhn-Bentz & F. Radler (1978) *Arch. Microbiol.* **116**, 197.

(S,S)-BUTANEDIOL DEHYDROGENASE

This oxidoreductase [EC 1.1.1.76] catalyzes the reversible reaction of (S,S)-butane-2,3-diol with NAD^+ to produce (S)-acetoin and NADH.[1] The *Bacillus stearothermophilus* enzyme will also act on *meso*-butane-2,3-diol[2]; nevertheless, a *meso*-butane-2,3-diol dehydrogenase has been reported as well.[3]

[1] S. Ui, N. Matsuyama, H. Masuda & H. Muraki (1984) *J. Ferment. Technol.* **62**, 551.
[2] P. P. Giovannini, A. Medici, C. M. Bergamini & M. Rippa (1996) *Bioorg. Med. Chem.* **4**, 1197.
[3] S. Ui, A. Mimura, M. Ohkuma & T. Kudo (1999) *Lett. Appl. Microbiol.* **28**, 457.

(S)-1,3-BUTANEDIOL DEHYDROGENASE

(S)-1,3-butanediol   4-hydroxy-2-butanone

This oxidoreductase, first isolated from *Candida parapsilosis*, catalyzes the reaction of (S)-1,3-butanediol with NAD^+ to produce 4-hydroxy-2-butanone and NADH.[1]

[1] H. Yamamoto, A. Matsuyama, Y. Kobayashi & N. Kawada (1995) *Biosci. Biotechnol. Biochem.* **59**, 1769.

BUTYRATE:ACETOACETATE CoA-TRANSFERASE

This transferase [EC 2.8.3.9] catalyzes the reversible reaction of butanoyl-CoA with acetoacetate to produce butanoate and acetoacetyl-CoA.[1,2] The enzyme also acts on other monocarboxylates (and their CoA derivatives) containing between two and six carbon atoms. Succinyl-CoA is not a substrate.

[1] H. A. Barker, I.-M. Jemg, N. Neff, J. M. Robertson, F. K. Tam & S. Hosaka (1978) *JBC* **253**, 1219.
[2] S. J. Sramek & F. E. Frerman (1975) *ABB* **171**, 14.

BUTYRATE CoA-TRANSFERASE

This transferase, formerly classified as EC 2.8.3.4, is now a deleted Enzyme Commission entry.

BUTYRATE KINASE

This phosphotransferase [EC 2.7.2.7] catalyzes the reaction of ATP with butanoate to produce ADP and butanoyl phosphate.[1,2]

butyryl phosphate

The *Clostridium acetobutylicum* enzyme also acts on pentanoate, propanoate, and some branched-chain fatty acids, albeit slowly. A second butyrate kinase was recently isolated from the same organism.[3]

[1] M. G. N. Hartmanis (1987) *JBC* **262**, 617.
[2] R. Twarog & R. S. Wolfe (1962) *JBC* **237**, 2474.
[3] K. X. Huang, S. Huang, F. B. Rudolph & G. N. Bennett (2000) *J. Mol. Microbiol. Biotechnol.* **2**, 33.

γ-BUTYROBETAINE:2-OXOGLUTARATE DIOXYGENASE

This iron- and ascorbate-dependent oxidoreductase [EC 1.14.11.1], also known as γ-butyrobetaine hydroxylase, catalyzes the reaction of 4-trimethylammoniobutanoate (*i.e.*, γ-butyrobetaine) with α-ketoglutarate (or, 2-oxoglutarate) and dioxygen to yield 3-hydroxy-4-trimethylammoniobutanoate (*i.e.*, L-carnitine), succinate, and carbon dioxide.[1–7]

γ-butyrobetaine   L-carnitine

Potassium ions stimulate the activity.[4] The enzyme reportedly has an ordered Ter Ter kinetic mechanism, and catalysis is likely to proceed via radical intermediates.[3]

[1] F. M. Vaz, S. van Gool, R. Ofman, L. IJlst & R. J. Wanders (1999) *Adv. Exp. Med. Biol.* **466**, 117.
[2] S. Galland, F. Le Borgne, D. Guyonnet, P. Clouet & J. Demarquoy (1998) *Mol. Cell. Biochem.* **178**, 163.
[3] B. G. Fox (1998) *CBC* **3**, 261.
[4] R. S. Wehbie, N. S. Punekar & H. A. Lardy (1988) *Biochemistry* **27**, 2222.
[5] E. Holme, S. Lindstedt & I. Nordin (1982) *BBRC* **107**, 518.
[6] O. Hayaishi, M. Nozaki & M. T. Abbott (1975) *The Enzymes*, 3rd ed., **12**, 119.
[7] G. Lindstedt & S. Lindstedt (1970) *JBC* **245**, 4178.

Selected entries from *Methods in Enzymology* [vol, page(s)]:
General discussion: 52, 12

BUTYRYL-CoA DEHYDROGENASE

This FAD-dependent oxidoreductase [EC 1.3.99.2], also referred to as short-chain acyl-CoA dehydrogenase and unsaturated acyl-CoA reductase, catalyzes the reaction of butanoyl-CoA and an electron-transferring flavoprotein to produce 2-butenoyl-CoA and the reduced electron-transferring flavoprotein.[1-4]

This enzyme forms a complex with the electron-transferring flavoprotein and electron-transferring-flavoprotein dehydrogenase [EC 1.5.5.1] that will reduce ubiquinone and other acceptors. Binding of the acyl-CoA substrate results in production of the reduced form of the enzyme which is reoxidized by the electron-transferring flavoprotein. The *Megasphera elsdenii* enzyme catalyzes the exchange of the α- and β-hydrogens of the substrate with the solvent. The β-hydrogen is transferred to the flavin N^5-position as a hydride.[1,3] *See also Electron-Transferring Flavoprotein; Acyl-CoA Dehydrogenases; Long-Chain Acyl-CoA Dehydrogenase*

[1] B. A. Palfey & V. Massey (1998) *CBC* **3**, 83.
[2] M. A. Ator & P. R. Ortez de Montellano (1990) *The Enzymes*, 3rd ed., **19**, 213.
[3] S. Ghisla, C. Thorpe & V. Massey (1984) *Biochemistry* **23**, 3154.
[4] H. Beinert (1963) *The Enzymes*, 2nd. ed., **7**, 447.

Selected entries from *Methods in Enzymology* [**vol**, page(s)]:
General discussion: 1, 553; **14**, 110; **53**, 402; **71**, 359
Assay: 1, 553; **5**, 549
Other molecular properties: adsorbents and, 1, 96; electron-transferring flavoprotein and, **5**, 547; electrophoresis, **5**, 551; ethanol precipitation, **5**, 554; inhibition by (hypoglycin, **72**, 611, 616; palmitoyl-CoA, **5**, 549); lack of sulfite complexing with pig liver enzyme, **18B**, 473; *Megasphaera elsdenii*, from, **71**, 359 (absorption spectrum, **71**, 363; alicyclic acyl-CoA substrates, **71**, 365; anaerobic dialysis, **71**, 365; charge-transfer interaction, **71**, 364; deazaflavin, **71**, 365; extinction coefficient, **71**, 365; green form, **71**, 363, 365; β-hydroxybutyryl-CoA as substrate, **71**, 365; long-wavelength band, **71**, 364; prosthetic group, **71**, 363; sodium borohydride, effect, **71**, 365; specificity, **71**, 365; stability, **71**, 365; yellow form, **71**, 365); properties, 1, 567; prosthetic group, FAD, **14**, 110, 114; purification, beef liver, 1, 555; reduced, substrate for electrontransferring flavoprotein, **14**, 118; reduction of, **5**, 555; requirement for electron-transferring flavoprotein, **14**, 110, 114

BUTYRYL-CoA SYNTHETASE

This ligase [EC 6.2.1.2] (also referred to as butyrate:CoA ligase, acyl-activating enzyme, and fatty acid thiokinase (medium chain)) catalyzes the reaction of ATP with an acid and Coenzyme A to produce AMP, pyrophosphate (or, diphosphate) and an acyl-CoA.[1-5] Acceptor substrates include carboxylic acids containing between four and eleven carbon atoms as well as the corresponding 3-hydroxy- and 2,3- or 3,4-unsaturated acids. *See also Long-Chain Fatty-Acyl-CoA Synthetase; Acyl-CoA Synthetase (GDP-Forming)*

[1] W. D. Reed & P. T. Ozand (1980) *ABB* **205**, 94.
[2] J. R. Scaife & J. Z. Tichivangana (1980) *BBA* **619**, 445.
[3] J. C. Londesborough & L. T. Webster (1974) *The Enzymes*, 3rd. ed., **10**, 469.
[4] L. T. Webster, L. D. Gerowin & L. Rakita (1965) *JBC* **240**, 29.
[5] H. R. Mahler & S. J. Wakil (1953) *JBC* **204**, 453.

Selected entries from *Methods in Enzymology* [**vol**, page(s)]:
Assay: **14**, 622

C

*Cac*8I RESTRICTION ENDONUCLEASE

This type II restriction endonuclease [EC 3.1.21.4], isolated from *Clostridium acetobutylicum* ABKn8, catalyzes the cleavage of both strands of DNA at 5′ . . . GCN∇NGC . . . 3′, where N refers any base, generating blunt-ended fragments and ∇ is the cleavage site.[1]

[1]H. Azeddoug & G. Reysset (1991) *FEMS Microbiol. Lett.* **78**, 153 and **83**, 121.

(+)-δ-CADINENE SYNTHASE

(+)-δ-cadinene

This magnesium-ion-dependent lyase [EC 4.2.3.13; previously classified as EC 4.6.1.11], also called D-cadinene synthase, catalyzes the conversion of 2-*trans*,6-*trans*-farnesyl diphosphate to (+)-δ-cadinene and pyrophosphate (or, diphosphate).[1–3]

[1]G. D. Davis & M. Essenberg (1995) *Phytochemistry* **39**, 553.
[2]E. M. Davis, J. Tsuji, G. D. Davis, M. L. Pierce & M. Essenberg (1996) *Phytochemistry* **41**, 1047.
[3]X. Y. Chen, Y. Chen, P. Heinstein & V. J. Davisson (1995) *ABB* **324**, 255.

C3 ADP-RIBOSYLTRANSFERASE

This transferase, also called C3 exoenzyme and C3 transferase, catalyzes the reaction of NAD$^+$ with an asparaginyl residue on RhoA, B, and C to produce nicotinamide and ADP-D-ribosylated protein.

Selected entries from ***Methods in Enzymology*** [**vol**, page(s)]:
Assay: catalyzed ADP-ribosylation, **256**, 188, 197
Other molecular properties: applications in biological systems, **256**, 202; catalyzed ADP-ribosylation (applications, **256**, 195; assay, **256**, 188,

197; detection in intact cells, **256**, 193); diphtheria toxin fragment B (cytopathic effect on Vero cells, **256**, 303; expression from *Escherichia coli*, **256**, 300; genetic construction, **256**, 299; immunoblotting, **256**, 303; purification, **256**, 301, 302; SDS-PAGE, **256**, 303); effect on actin network of neutrophils, **256**, 333; effect on cell morphology, **256**, 333; glutathione *S*-transferase (cleavage from glutathione beads by thrombin, **256**, 178; expression in *Escherichia coli*, **256**, 175; precipitation on glutathione beads, **256**, 177; purification from bacterial cell lysate, **256**, 176); inhibitory effects (JY cell aggregation, **256**, 295; lymphocyte-mediated cytotoxicity, **256**, 320; neutrophil motility, **256**, 331); loading into neutrophils, **256**, 329; lymphocyte treatment (electropermeabilized cytotoxic cells, **256**, 322; JY cells, **256**, 291); PC-12 cell treatment with, **256**, 206; properties, **256**, 185, 201; purification from *Clostridium botulinum* culture filtrate, **256**, 200; reaction in broken cell extracts, inhibitors, **256**, 181; recombinant (activity on Rho proteins, **256**, 179; assay, **256**, 179; ion-exchange chromatography, **256**, 178; probe for Rho proteins, as, **256**, 179; purification from *Escherichia coli*, **256**, 201); related exoenzymes, **256**, 187 (acceptor amino acid, testing, **256**, 192); Swiss 3T3 cell treatment with, **256**, 205; uptake by electroporation, control, **256**, 324

CAFFEATE 3,4-DIOXYGENASE

This oxidoreductase [EC 1.13.11.22] catalyzes the reaction of 3,4-dihydroxy-*trans*-cinnamate (*i.e.*, caffeate) with dioxygen to produce 3-(2-carboxyethenyl)-*cis*,*cis*-muconate.[1]

[1]M. M. Seidman, A. Toms & J. M. Wood (1969) *J. Bacteriol.* **97**, 1192.

Selected entries from ***Methods in Enzymology*** [**vol**, page(s)]:
General discussion: 52, 19

CAFFEATE *O*-METHYLTRANSFERASE

This methyltransferase [EC 2.1.1.68] catalyzes the reaction of *S*-adenosyl-L-methionine with 3,4-dihydroxy-*trans*-cinnamate (*i.e.*, caffeate) to produce *S*-adenosyl-L-homocysteine and 3-methoxy-4-hydroxy-*trans*-cinnamate (*i.e.*, ferulate). Other acceptor substrates include 5-hydroxyferulate, 3,4-dihydroxybenzaldehyde, and catechol.[1–4]

[1]J. E. Poulton, K. Hahlbrock & H. Grisebach (1976) *ABB* **176**, 449.
[2]C. P. Vance & J. W. Bryan (1981) *Phytochemistry* **20**, 41.
[3]K. Inoue, K. Parvathi & R. A. Dixon (2000) *ABB* **375**, 175.
[4]S. Maury, P. Geoffroy & M. Legrand (1999) *Plant Physiol.* **121**, 215.

CAFFEOYL-CoA *O*-METHYLTRANSFERASE

This methyltransferase [EC 2.1.1.104], also referred to as *trans*-caffeoyl-CoA 3-*O*-methyltransferase, catalyzes the reaction of *S*-adenosyl-L-methionine with caffeoyl-CoA to produce *S*-adenosyl-L-homocysteine and feruloyl-CoA. Alternative substrates include several *trans*-caffeic acid esters. The enzyme is reported to have an ordered Bi Bi kinetic mechanism.[1–4]

[1] A.-E. Pakusch, U. Matern & E. Schiltz (1991) *Plant Physiol.* **95**, 137.
[2] D. Schmidt, A.-E. Pakusch & U. Matern (1991) *JBC* **266**, 17416.
[3] A.-E. Pakusch, R. E. Kneusel & U. Matern (1989) *ABB* **271**, 488.
[4] S. Maury, P. Geoffroy & M. Legrand (1999) *Plant Physiol.* **121**, 215.

CALCINEURIN

This calmodulin-binding phosphoprotein phosphatase [EC 3.1.3.16], also known as protein phosphatase 2B, catalyzes the removal of phosphoryl groups from phosphoseryl and phosphothreonyl residues.[1–5] The enzyme consists of two subunits: the catalytic subunit (calcineurin A) which contains an active-site dinuclear metal center, and the myristoylated Ca^{2+}-binding subunit (calcineurin B). *See Serine/Threonine Specific Protein Phosphatase*

[1] F. Rusnak & P. Mertz (2000) *Physiol. Rev.* **80**, 1483.
[2] J. Aramburu, A. Rao & C. B. Klee (2000) *Curr. Top. Cell Regul.* **36**, 237.
[3] E. N. Olson & R. S. Williams (2000) *Cell* **101**, 689.
[4] B. G. Fox (1998) *CBC* **3**, 261.
[5] C. B. Klee, G. F. Draetta & M. J. Hubbard (1988) *AE* **61**, 149.

Selected entries from ***Methods in Enzymology*** [vol, page(s)]:
General discussion: 102, 227

CALCIUM/CALMODULIN-DEPENDENT PROTEIN KINASE

This phosphotransferase [EC 2.7.1.123], with variant forms referred to as calcium/calmodulin-dependent protein kinase type II and microtubule-associated protein 2 kinase (MAP2 kinase), catalyzes the calcium- and calmodulin-dependent reaction of ATP with a protein to produce ADP and the *O*-phosphoprotein. Protein substrates include vimentin, synapsin, glycogen synthase, myosin light-chains, MAP2, myelin basic protein, ribosomal protein S6, histone H1, estrogen receptor, casein, and the microtubule-associated tau protein.[1–5]

This protein kinase is distinct from [myosin light-chain] kinase [EC 2.7.1.117], caldesmon kinase [EC 2.7.1.120], or [tau protein] kinase [EC 2.7.1.135].

[1] H. Schulman, J. Kuret, A. B. Jefferson, P. S. Nose & K. H. Spitzer (1985) *Biochemistry* **24**, 5320.
[2] J. Baudier & R. D. Cole (1987) *JBC* **262**, 17577.
[3] N. C. Schanen & G. Landreth (1992) *Mol. Brain Res.* **14**, 43.
[4] T. G. Boulton, J. S. Gregory & M. H. Cobb (1991) *Biochemistry* **30**, 278.
[5] M. Hoshi, E. Nishida & H. Sakai (1989) *EJB* **184**, 477.

CALDECRIN

A calcium-decreasing factor that exhibits a chymotrypsin-like protease activity.[1,2] *See Chymotrypsin C*

[1] A. Tomomura, M. Tomomura, T. Fukushiga, M. Akiyama, N. Kubota, K. Kumaki, Y. Nishii, T. Noikura & T. Saheki (1995) *JBC* **270**, 30315.
[2] A. Tomomura, T. Fukushiga, M. Tomomura, T. Noikura, Y. Nishii & T. Saheki (1993) *FEBS Lett.* **335**, 213.

CALDESMON KINASE

This calcium-dependent phosphotransferase [EC 2.7.1.120] catalyzes the reaction of ATP with caldesmon to produce ADP and [caldesmon] phosphate (*i.e.*, the phosphorylated seryl residue in caldesmon, which now cannot bind to actin and thus cannot inhibit the ATPase activity of actinomyosin).[1–4] The enzyme requires calmodulin unless it has undergone autophosphorylation.[3]

Although once considered an isozyme of [myosin light-chain] kinase [EC 2.7.1.117][3] or casein kinase II, recent reports suggest that this enzyme is a more specific protein kinase.[5]

[1] G. C. Scott-Woo & M. P. Walsh (1988) *BJ* **252**, 463.
[2] P. K. Ngai & M. P. Walsh (1984) *JBC* **259**, 13656.
[3] M. Ikebe, S. Reardon, G. C. Scott-Woo, Z. Zhou & Y. Koda (1990) *Biochemistry* **29**, 11242.
[4] A. V. Vorotnikov, N. B. Gusev, S. Hua, J. H. Collins, C. S. Redwood & S. B. Marston (1993) *FEBS Lett.* **334**, 18.
[5] M. A. Krymsky, M. V. Chibalina, V. P. Shirinsky, S. B. Marston & A. V. Vorotnikov (1999) *FEBS Lett.* **452**, 254.

CALDESMON-PHOSPHATASE

This phosphatase [EC 3.1.3.55] catalyzes the hydrolysis of phosphorylated caldesmon to produce caldesmon and orthophosphate.[1,2] Upon dephosphorylation, the calmodulin- and actin-binding ability of caldesmon are activated. The enzyme will also catalyze the dephosphorylation of phosphorylated myosin LC20 and phosphorylated calponin.[2]

[1] P. K. Ngai & M. P. Walsh (1984) *JBC* **259**, 13656.
[2] M. D. Pato, C. Sutherland, S. J. Winder & M. P. Walsh (1993) *BJ* **293**, 35.

CALMODULIN-LYSINE *N*-METHYLTRANSFERASE

This methyltransferase [EC 2.1.1.60] catalyzes the reaction of *S*-adenosyl-L-methionine with calmodulin-L-lysine (*i.e.*, a lysyl residue in desmethyl-calmodulin) to produce *S*-adenosyl-L-homocysteine and calmodulin containing an N^6-methyl-L-lysyl residue.[1] Upon further methylation, an N^6-trimethyllysyl residue is produced.[1]

[1] A. Sitaramayya, L. S. Wright & F. L. Siegel (1980) *JBC* **255**, 8894.

Selected entries from *Methods in Enzymology* [vol, page(s)]:
General discussion: 102, 158
Assay: radiometric, 102, 160
Other molecular properties: calmodulin *N*-methylation assay with, 139, 674; partial purification, 102, 161

CALPAIN

This Ca^{2+}-dependent cysteine protease [EC 3.4.22.17], also called CANP (for calcium-activated neutral protease), catalyzes peptide-bond hydrolysis in various proteins, although casein is most often used in activity assays.[1–3] These enzymes appear to have a preference for hydrophobic (*i.e.*, Tyr, Met, Leu, or Val) or arginyl residues in the P2 position of the substrate. There are no apparent differences in substrate specificity for μ- and m-calpain. For μ-calpain, half-maximal activity is observed at about $50\,\mu M$ calcium ion, and the $[Ca^{2+}]_{0.5}$ is lowered to the $0.1–1\,\mu M$ range in the presence of such phosphoinositides as phosphatidylinositol 4,5-bisphosphate.

An additional type of mammalian calpain, initially referred to as nCANP (for novel Ca^{2+}-activated neutral protease), is now more widely known as p94, based on its apparent molecular weight. Substrates include itself, myotonin protein kinase, and fodrin. Interestingly, proteolysis does not require calcium ions, although there is a Ca^{2+}-binding domain.[4–6] Muscle calpain p94 undergoes rapid autolysis.

[1] T. Murachi (1989) *Biochem. Int.* 18, 263.
[2] R. L. Mellgren & T. Murachi (1990) *Intracellular Calcium-Dependent Proteolysis*, CRC Press, Boston.
[3] T. C. Saido, H. Sorimachi & K. Suzuki (1994) *FASEB J.* 8, 814.
[4] H. Sorimachi, S. Kimura, K. Kinbara, J. Kazama, M. Takahashi, H. Yajima, S. Ishiura, N. Sasagawa, I. Nonaka, H. Sugita & K. Suzuki (1996) *Adv. Biophys.* 33, 101.
[5] H. Sorimachi & K. Suzuki (1992) *BBA* 1160, 55.
[6] K. Suzuki, H. Sorimachi, T. Yoshizawa, K. Kinbara & S. Ishiura (1995) *Biol. Chem. Hoppe-Seyler* 376, 523.

Selected entries from *Methods in Enzymology* [vol, page(s)]:
General discussion: 102, 279; 139, 363
Assay: types I and II, 169, 444
Other molecular properties: affinity labeling, 244, 655; cDNA cloning, 139, 372; evolution, 139, 377; family, 244, 468; mRNA distribution and complexity, 139, 373; structure, 139, 374; 244, 468; types I and II (assay, 169, 444; natural substrates, 169, 454; properties, 169, 449; purification from human platelets, 169, 445)

CAMPHOR 1,2-MONOOXYGENASE

This FMN-dependent oxidoreductase [EC 1.14.15.2], also known as 2,5-diketocamphane 1,2-monooxygenase, 2,5-diketocamphane lactonizing enzyme, and camphor ketolactonase I, catalyzes the reaction of (+)-bornane-2,5-dione (also known as 2,5-diketocamphane) with reduced

rubredoxin and dioxygen to produce 5-oxo-1,2-campholide, oxidized rubredoxin, and water.[1–3] Camphor is also converted to 1,2-campholide.

[1] D. G. Taylor & P. W. Trudgill (1986) *J. Bacteriol.* 165, 489.
[2] C. A. Yu & I. C. Gunsalus (1969) *JBC* 244, 6149.
[3] P. W. Trudgill, R. DuBus & I. C. Gunsalus (1966) *JBC* 241, 1194.

CAMPHOR 5-MONOOXYGENASE

This heme-thiolate-dependent oxidoreductase [EC 1.14.15.1], also known as camphor 5-*exo*-methylene hydroxylase, cytochrome P450$_{cam}$, and CYP101, catalyzes the reaction of (+)-camphor with putidaredoxin and dioxygen to generate (+)-*exo*-5-hydroxycamphor, oxidized putidaredoxin, and water.[1–7] The enzyme can also utilize (–)-camphor as a substrate, and 1,2a-campholide will result in the formation of 5-*exo*-hydroxy-1,2-campholide.

During the course of reaction, camphor first binds to the Fe^{3+}-species of the enzyme to form a five-coordinated, high-spin species with one vacant coordination site. The reduction potential increases, and a one-electron reduction takes place to form the Fe^{2+}-enzyme-substrate complex. In this step, NADH is the source of the electrons which are transferred through a flavoprotein dehydrogenase and an iron-sulfur redoxin. Dioxygen then binds to generate a ternary complex. Addition of a second electron and two protons produces water and an active enzyme intermediate (probably a heme oxyferryl species: $Fe^V{=}O$). The oxygen atom is then stereospecifically inserted into the substrate, and the ferric enzyme is regenerated. Note that this step involves both hydrogen abstraction as well as oxygen addition. The oxyferryl species abstracts the hydrogen atom to form a hydroxyl radical ([$Fe^{IV}{-}O{-}H$]·) and a carbon radical. The radicals react to form the hydroxylated product which is released.[1–3]

[1] M. J. Honeychurch, A. O. Hill & L. L. Wong (1999) *FEBS Lett.* 451, 351.
[2] L. L. Wong (1998) *Curr. Opin. Chem. Biol.* 2, 263.
[3] H. B. Dunford (1998) *CBC* 3, 195.
[4] T. L. Poulos & R. Raag (1992) *FASEB J.* 6, 674.
[5] P. V. Gould, M. H. Gelb & S. G. Sligar (1981) *JBC* 256, 6686.
[6] V. Ullrich & W. Duppel (1975) *The Enzymes*, 3rd ed., 12, 253.
[7] C. T. Tyson, J. D. Lipscomb & I. C. Gunsalus (1972) *JBC* 247, 5777.

Selected entries from *Methods in Enzymology* [vol, page(s)]:
General discussion: 52, 14, 166; 206, 31; 272, 358
Assay: 52, 152, 153
Other molecular properties: active site analysis, 206, 44; bacterial, purification, 52, 152; carbon monoxide binding, IR studies, 54, 310; CO difference spectrum, 52, 152; crystallization (crystal recovery, 272, 360; precipitant, 272, 360; substrate addition and removal, 272, 359; thiol protection, 272, 359); crystalline spectra, 52, 158; cytochrome b_5 complex formation, 206, 41; cytochrome P420$_{cam}$, carbon monoxide binding, IR studies, 54, 310; expression in *Escherichia coli*, 206, 35;

functional background, **206**, 32; gene isolation in *Pseudomonas putida*, **206**, 35; interaction with camphor, role of Tyr96, **206**, 45; molecular weight, **52**, 156; monooxygenase components, **52**, 187; Mössbauer spectrum, **54**, 353, 357; mutagenesis, **206**, 31, 38; prosthetic group, **52**, 157; *Pseudomonas putida*, of (difference spectra, **52**, 259; electron paramagnetic spectra, **52**, 252, 253; spin states, **52**, 259, 267); putidaredoxin complex, formation, **206**, 41; role of (heme axial ligand, **206**, 47; Thr252, **206**, 46); reaction cycle, **272**, 337; specific activity determination, **52**, 156; spectral properties, **52**, 131, 155; structural alignment with P450s (algorithms, **272**, 317; core structure, **272**, 316; C-terminal alignment, **272**, 319, 321, 325; N-terminal alignment, **272**, 323); substrate access, **206**, 48; substrate recognition by hydrophobic contacts, **206**, 46; substrate specificity (active site flexibility, **272**, 350, 355; DOCK screening, **272**, 339, 342; reaction rate prediction, **272**, 328, 330; stages of substrate interaction, **272**, 337); system, **52**, 167; X-ray structure, in modeling of mammalian P450, **206**, 11

(S)-CANADINE SYNTHASE

(S)-tetrahydrocolumbamine

(S)-canadine

This oxidoreductase [EC 1.1.3.36], also known as (S)-tetrahydrocolumbamine oxidase (methylenedioxy-bridge-forming) and (S)-tetrahydroberberine synthase, catalyzes the reaction of (S)-tetrahydrocolumbamine with NADPH and dioxygen to produce (S)-canadine, NADP⁺, and two water molecules. This heme-thiolate-dependent system does not incorporate oxygen into the canadine product.

CANCER PROCOAGULANT

This calcium-dependent cysteine endopeptidase [EC 3.4.22.26], often abbreviated CP, catalyzes the hydrolysis of particular peptide bonds with a $[Ca^{2+}]_{opt}$ of 7 mM. The only known physiological substrate is coagulation factor X.[1,2] The bond cleaved in the heavy chain of factor X is Pro-Tyr-∇-Asp22, activating the coagulation factor to factor Xa. (Note: ∇ indicates the cleavage site.) Cancer procoagulant, first identified in extracts of neoplastic human

and animal tissues, accelerates blood coagulation in a host with cancer.

[1] S. G. Gordon (1994) *Meth. Enzymol.* **244**, 568.
[2] W. P. Mielicki, E. Mielicka & S. G. Gordon (1997) *Thromb. Res.* **87**, 251.
Selected entries from *Methods in Enzymology* [vol, page(s)]:
General discussion: 244, 568
Assay: activity assays (factor X activation [chromogenic assay, **244**, 570; coupled assays, **244**, 571]; recalcified clotting time assay, **244**, 569, 572; sensitivity, **244**, 572)
Other molecular properties: active site (configuration, **244**, 582; oxidation, **244**, 573, 580); amino acid composition, **244**, 580; antigen, immunoassay, **244**, 574; discovery, **244**, 568; glycosylation, **244**, 582; hydrophobicity, **244**, 579; inhibitors, **244**, 569, 582, 573; metal ion dependence, **244**, 572; purification, **244**, 577; size, **244**, 568; species conservation, **244**, 569; tissue distribution, **244**, 569

CANDIDAPEPSIN

This aspartic endopeptidases [EC 3.4.23.24; formerly classified as EC 3.4.23.6], also known as candialbicin and *Candida albicans* aspartic proteinase, comprise a family of seven secretory aspartic proteinases (SAP).[1,2] The SAP endopeptidases have a broad specificity; however, SAP2 exhibits a preference for hydrophobic aminoacyl residues at the P1 and P1′ positions during short incubation times.[1–3] The oxidized insulin B chain is hydrolyzed at Val2–Asn3, Leu11–Val12, Val12–Glu13, Ala14–Leu15, and Phe25–Tyr26 (however, the enzyme fails to cleave Leu15–Tyr16, Tyr16–Leu17, and Phe24–Phe25). SAP2 activates trypsinogen and degrades keratin, the Fc portion of IgG, IgA, and complement factor C3 (all contributing to fungal virulence). The endopeptidase is inhibited by pepstatin A ($K_i = 6$ nM). **See also** *Canditropsin; Candiparapsin*

[1] M. Fusek, E. Smith & S. I. Foundling (1995) in *Aspartic Proteinases: Structure, Function, Biology, and Biomedical Implications* (K. Takahashi, ed.), Plenum Press, New York, p. 489.
[2] R. Rüchel, F. De Bernardis, T. L. Ray, P. A. Sullivan & G. T. Cole (1992) *J. Med. Vet. Mycol.* **30** (suppl. 1), p. 123.
[3] M. Fusek, E. A. Smith, M. Monod, B. M. Dunn & S. I. Foundling (1994) *Biochemistry* **33**, 9791.
Selected entries from *Methods in Enzymology* [vol, page(s)]:
General discussion: 19, 373
Assay: 19, 387

CAPSULAR-POLYSACCHARIDE ENDO-1,3-α-GALACTOSIDASE

This enzyme [EC 3.2.1.87], also known as polysaccharide depolymerase, catalyzes the random hydrolytic scission of 1,3-α-D-galactosidic linkages in the *Aerobacter aerogenes* capsular polysaccharide, resulting in depolymerization.[1–3]

[1] E. C. Yurewicz, M. A. Ghalambor & E. C. Heath (1971) *JBC* **246**, 5596.
[2] E. C. Yurewicz, M. A. Ghalambor, D. H. Duckworth & E. C. Heath (1971) *JBC* **246**, 5607.
[3] A. Tsugita (1971) *The Enzymes*, 3rd ed., **5**, 343.

Selected entries from *Methods in Enzymology* [vol, page(s)]:
General discussion: 28, 990
Assay: 28, 993
Other molecular properties: enzymatic properties, 28, 995; homogeneity, 28, 995; isolation from *Aerobacter aerogenes*, 28, 994; reaction, 28, 990

CAPSULAR-POLYSACCHARIDE-TRANSPORTING ATPase

This so-called ATPase [EC 3.6.3.38] catalyzes the ATP-dependent (ADP- and orthophosphate-producing) transport of [capsular polysaccharide]$_{in}$ to produce [capsular polysaccharide]$_{out}$ in Gram-negative bacteria.[1-3] This enzyme is an ABC-type (ATP-binding cassette-type) ATPase that has two similar ATP-binding domains and does not undergo phosphorylation during the transport process.

The idea that a transporter is an enzyme is in keeping with a new definition of enzyme catalysis as the facilitated making/breaking of chemical bonds, not just covalent bonds.[4] This idea builds on Pauling's assertion that any long-lived, chemically distinct interaction (in this case, the persistent location of a solute with respect to the faces of a membrane) can be regarded as a chemical bond. Note also that the equilibrium constant (K_{eq} = [Capsular Polysaccharide$_{out}$][ADP][P$_i$]/[Capsular Polysaccharide$_{in}$][ATP]) does not conform to that expected for an ATPase (*i.e.*, K_{eq} = [ADP][P$_i$]/[ATP]). Thus, although the overall reaction yields ADP and orthophosphate, the enzyme is misclassified as a hydrolase, and instead should be regarded as an energase-type reaction. Energases facilitate affinity-modulated reactions by coupling the $\Delta G_{ATP\text{-}hydrolysis}$ to a force-generating or work-producing step.[4] In this case, P-O-P bond-scission supplies the energy to drive capsular polysaccharide transport.

See ATPase; Energase

[1] M. J. Fath & R. Kolter (1993) *Microbiol. Rev.* 57, 995.
[2] I. T. Paulsen, A. M. Beness & M. H. Saier, Jr. (1997) *Microbiology* 143, 2685.
[3] R. P. Pigeon & R. P. Silver (1997) *FEMS. Microbiol. Lett.* 156, 217.
[4] D. L. Purich (2001) *TiBS* 26, 417.

CARBAMATE KINASE

This phosphotransferase [EC 2.7.2.2], also known as carbamyl phosphokinase, catalyzes the reaction of ATP with ammonia and carbon dioxide to produce ADP and carbamoyl phosphate.[1-5] The equilibrium constant is 0.027 at 25°C, suggesting that the enzyme functions physiologically in the direction of ATP synthesis, using the carbamoyl phosphate generated during fermentative catabolism of arginine.

[1] A. Marina, P. M. Alzari, J. Bravo, M. Uriarte, B. Barcelona, I. Fita & V. Rubio (1999) *Protein Sci.* 8, 934.
[2] M. C. Manca de Nadra, A. A. Pesce de Ruiz Holgado & G. Oliver (1987) *Biotechnol. Appl. Biochem.* 9, 141.
[3] V. N. Pandey & D. S. Pradhan (1981) *BBA* 660, 284.
[4] L. Raijman & M. E. Jones (1973) *The Enzymes*, 3rd ed., 9, 97.
[5] M. Marshall & P. P. Cohen (1966) *JBC* 241, 4197.

Selected entries from *Methods in Enzymology* [vol, page(s)]:
General discussion: 5, 904; 17A, 229
Assay: 5, 906
Other molecular properties: aspartate carbamoyltransferase, in preparation of, 5, 915; mechanism, 63, 51; ornithine carbamoyltransferase, 5, 911, 912; product inhibition, 63, 434, 435; properties, 5, 909; purification from *Streptococcus faecalis*, 5, 908; 17A, 230

N-CARBAMOYL-D-AMINO ACID HYDROLASE

This enzyme [EC 3.5.1.77] catalyzes the hydrolysis of an N-carbamoyl-D-amino acid to produce the D-amino acid, ammonia, and carbon dioxide.[1,2] Note that L-amino acid derivatives, N-formyl amino acids, N-carbamoylsarcosine, N-carbamoylcitrulline, N-carbamoylallantoin, and N-carbamoylureidopropionate are not substrates.

[1] Y. Ikenaka, H. Nanba, K. Yajima, Y. Yamada, M. Takano & S. Takahashi (1999) *Biosci. Biotechnol. Biochem.* 63, 91.
[2] J. Ogawa, S. Shimizu & H. Yamada (1993) *EJB* 212, 685.

CARBAMOYLASPARTATE DECARBOXYLASE

This decarboxylase activity, formerly classified as EC 4.1.1.13, is now a deleted Enzyme Commission entry.

CARBAMOYLPHOSPHATE SYNTHETASES

Two enzymes catalyze the ATP-/bicarbonate-dependent synthesis of carbamoyl phosphate, depending on the nitrogen source.

Carbamoyl-phosphate synthetase (ammonia) [EC 6.3.4.16], also known as carbamoyl-phosphate synthetase I and carbon-dioxide:ammonia ligase, catalyzes the reaction of two molecules of ATP with carbon dioxide, ammonia, and water to produce two molecules of ADP, orthophosphate, and carbamoyl phosphate.[1-9] The synthetase has two distinct ATP binding sites and the carboxyphosphate intermediate has been trapped and identified.[5-7] The mammalian mitochondrial enzyme requires the essential activator N-acetyl-L-glutamate.

Carbamoyl-phosphate synthetase (glutamine-hydrolyzing) [EC 6.3.5.5], also known as carbamoyl-phosphate synthetase II, catalyzes the reaction of two molecules of ATP with carbon dioxide, L-glutamine, and water to produce two molecules of ADP, orthophosphate, L-glutamate, and carbamoyl phosphate.[10–14] There are also two distinct ATP binding sites and the reaction proceeds via a carboxyphosphate intermediate.[5–7,12] In addition, L-glutamine reacts with a cysteinyl residue to form a transient γ-glutamyl thioester intermediate (hence, the enzyme will also exhibit a glutaminase activity). It has been proposed that an tunnel connects the L-glutamine binding site in the *Escherichia coli* enzyme to the other active site within the larger subunit and that ammonia generated from L-glutamine transports to that other site via this tunnel.[10] Ammonia can act as a substrate at concentrations in excess of about 100 mM (K_m for ammonia for the *Escherichia coli* enzyme is 93 mM[12]).

[1] J. N. Earnhardt & D. N. Silverman (1998) *CBC* **1**, 495.
[2] V. Rubio & J. Cervera (1995) *Biochem. Soc. Trans.* **23**, 879.
[3] A. Meister (1989) *AE* **62**, 315.
[4] P. P. Cohen (1981) *Curr. Top. Cell Regul.* **18**, 1.
[5] S. G. Powers & A. Meister (1977) *JBC* **253**, 1258.
[6] S. G. Powers, O. W. Griffith & A. Meister (1977) *JBC* **252**, 3558.
[7] S. G. Powers & A. Meister (1976) *PNAS* **73**, 3020.
[8] S. Ratner (1973) *AE* **39**, 1.
[9] M. Tatibana & K. Shigesada (1972) *Adv.Enzyme Regul.* **10**, 249.
[10] X. Huang & F. M. Raushel (2000) *JBC* **275**, 26233.
[11] H. Zalkin (1993) *AE* **66**, 203.
[12] D. S. Kaseman & A. Meister (1985) *MIE* **113**, 305.
[13] M. E. Jones (1980) *Ann. Rev. Biochem.* **49**, 253.
[14] J. M. Buchanan (1973) *Adv. Enzymol.* **39**, 91.

Selected entries from **Methods in Enzymology** [vol, page(s)]:
General discussion: 5, 904, 916; **17A**, 235; **46**, 152, 420, 425, 426; **51**, 21, 105; **113**, 305
Assay: 51, 30, 31, 105, 122; *Escherichia coli*, **113**, 306; frog liver, **5**, 917
Other molecular properties: activity, 51, 21, 29, 30; **308**, 140; carbamoyl-phosphate synthetase (glutamine):aspartate carbamoyltransferase:dihydroorotase enzyme complex, **51**, 111 (aggregation studies, **51**, 126; assay, **51**, 112, 113, 122, 123; dissociation, **51**, 120, 121; gene selection and coamplification, in, **185**, 550; hamster cells, from, **51**, 121; isoelectric point, **51**, 118; kinetic properties, **51**, 118, 119; molecular weight, **51**, 118, 126, 133; primary structure, **51**, 133, 134; properties, **51**, 118, 126; purification from [mutant hamster cells, **51**, 123; rat ascites hepatoma cells, **51**, 114; rat liver, **51**, 116]; purity, **51**, 126; sedimentation coefficient, **51**, 118; sources, **51**, 119, 120, 121; stability, **51**, 134; N-terminus, **51**, 133; tissue distribution, **51**, 119, 120); carbamoyl-phosphate synthetase (glutamine):aspartate carbamoyltransferase enzyme complex, **51**, 105 (assay, **51**, 105; inhibitors, **51**, 111; kinetic properties, **51**, 111; molecular weight, **51**, 111; *Neurospora*, from, **51**, 105; purification, **51**, 107; purity, **51**, 110; stability, **51**, 110); catalytic activities and subunits, **308**, 140; channeling, substrate, **308**, 140 (kinetics, **308**, 142); cobalt-ATP, and, **87**, 192, 193; dissociation, **51**, 34, 35; *Escherichia coli*, from, **51**, 21; **113**, 305 (assays, **113**, 306; heavy and light subunits [interactions, **113**, 318; separation, **113**, 314, 315]; inhibition by oxidation, **113**, 320; mechanism of action, **113**, 322; properties, **113**, 316, 317; purification, **113**, 309); feedback inhibition by uridine 5′-monophosphate, **17A**, 236; feedback repression by arginine, **17A**, 235; frog (*Rana catesbiana*) liver **5**, 916 (properties, **5**, 921; purification, **5**, 918); γ-glutamylhydroxamatase activity (of carbamoyl-phosphate synthetase (glutamine-hydrolyzing) subunits, **51**, 26, 28; hydrolysis of ATP by, bicarbonate dependent, **17A**, 242 (glutamine by, of, **17A**, 242); isotope trapping, **64**, 58; kinetics, **51**, 33, 34; ligand binding constants from anisotropy measurements, **246**, 288; mapping, active site, **87**, 192; mitochondrial (intracellular transport, **97**, 402; precursor posttranslational uptake and processing *in vitro*, **97**, 405; synthesis, **97**, 399); molecular weight, **51**, 23, 33, 121; mutant lacking, **5**, 927; *Neurospora*, from, **51**, 105; NMR, ³¹P, **87**, 193; glutamine:aspartate carbamoyltransferase:dihydroorotase enzyme complex, **51**, 111 (aggregation studies, **51**, 126; assay, **51**, 112, 113, 122, 123; dissociation, **51**, 120, 121; gene selection and coamplification, in, **185**, 550; hamster cells, from, **51**, 121; isoelectric point, **51**, 118; kinetic properties, **51**, 118, 119; molecular weight, **51**, 118, 126, 133; primary structure, **51**, 133, 134; properties, **51**, 118, 126; purification from [mutant hamster cells, **51**, 123; rat ascites hepatoma cells, **51**, 114; rat liver, **51**, 116]; purity, **51**, 126; sedimentation coefficient, **51**, 118; sources, **51**, 119, 120, 121; stability, **51**, 134; N-terminus, **51**, 133; tissue distribution, **51**, 119, 120); carbamoyl-phosphate synthetase (glutamine):aspartate carbamoyltransferase enzyme complex, **51**, 105 (assay, **51**, 105; inhibitors, **51**, 111; kinetic properties, **51**, 111; molecular weight, **51**, 111; *Neurospora*, from, **51**, 105; purification, **51**, 107; purity, **51**, 110; stability, **51**, 110); catalytic activities and subunits, **308**, 140; channeling, substrate, **308**, 140 (kinetics, **308**, 142); cobalt-ATP, and, **87**, 192, 193; dissociation, **51**, 34, 35; *Escherichia coli*, from, **51**, 21; **113**, 305 (assays, **113**, 306; heavy and light subunits [interactions, **113**, 318; separation, **113**, 314, 315]; inhibition by oxidation, **113**, 320; mechanism of action, **113**, 322; properties, **113**, 316, 317; purification, **113**, 309); feedback inhibition by uridine 5′-monophosphate, **17A**, 236; feedback repression by arginine, **17A**, 235; frog (*Rana catesbiana*) liver **5**, 916 (properties, **5**, 921; purification, **5**, 918); γ-glutamylhydroxamatase activity (of carbamoyl-phosphate synthetase (glutamine-hydrolyzing) subunits, **51**, 26, 28; hydrolysis of ATP by, bicarbonate dependent, **17A**, 242 (glutamine by, of, **17A**, 242); isotope trapping, **64**, 58; kinetics, **51**, 33, 34; ligand binding constants from anisotropy measurements, **246**, 288; mapping, active site, **87**, 192; mitochondrial (intracellular transport, **97**, 402; precursor posttranslational uptake and processing *in vitro*, **97**, 405; synthesis, **97**, 399); molecular weight, **51**, 23, 33, 121; mutant lacking, **5**, 927; *Neurospora*, from, **51**, 105; NMR, ³¹P, **87**, 193; oxygen exchanges, **87**, 247; positional isotope exchange studies, **249**, 413, 418; preparation, **51**, 23; product inhibition, **63**, 429; properties, **51**, 25, 33, 34; purification, **51**, 31 (*Agaricus bisporus*, **17A**, 245; *Escherichia coli*, **17A**, 237; **113**, 309); rat cells, from, **51**, 111; reconstituted, enzymic activities, **51**, 26; regulation, **51**, 21, 30, 32; *Salmonella typhimurium*, from, **51**, 29; stability, **51**, 118; substrate channelling, **308**, 140; subunit interactions, **51**, 27; subunit properties, enzymic, **51**, 35; sulfur, active site, **87**, 83; synthesis of ATP by, **17A**, 242; thiol group, active site, **87**, 83

N-CARBAMOYLPUTRESCINE AMIDOHYDROLASE

This enzyme [EC 3.5.1.53], also known as *N*-carbamoylputrescine amidase, catalyzes the hydrolysis of *N*-carbamoylputrescine to produce putrescine, carbon dioxide, and ammonia.[1,2] *N*-Carbamoylagmatine is also a substrate.

[1] H. Yanagisawa & Y. Suzuki (1982) *Phytochemistry* **21**, 2201.
[2] A. Mercenier, J. P. Simon, D. Haas & V. Stalon (1980) *J. Gen. Microbiol.* **116**, 381.

N-CARBAMOYLSARCOSINE AMIDASE

This amidase [EC 3.5.1.59], also known as *N*-carbamoylsarcosine amidohydrolase and CSHase, catalyzes the hydrolysis of *N*-carbamoylsarcosine to produce sarcosine, carbon dioxide, and ammonia.[1–3] This enzyme, which is involved in the microbial degradation of creatinine, is not related to N-carbamoyl-D-amino acid hydrolase even though it will hydrolyze a number of N-carbamoyl-D-amino acids.[1] X-ray crystallographic analysis of an enzyme-inhibitor complex suggests that the catalytic nucleophile is the Cys117 thiol.[2]

[1] J. M. Kim, S. Shimizu & H. Yamada (1986) JBC **261**, 11832.
[2] M. J. Romao, D. Turk, F. X. Gomis-Ruth, R. Huber, G. Schumacher, H. Mollering & L. Russmann (1992) J. Mol. Biol. **226**, 1111.
[3] M. Hermann, H.-J. Knerr, N. Mai, A. Groß & H. Kaltwasser (1992) Arch. Microbiol. **157**, 395.

CARBAMOYL-SERINE AMMONIA-LYASE

This pyridoxal-phosphate-dependent lyase [EC 4.3.1.13], also known as O-carbamoyl-L-serine deaminase, catalyzes the reaction of O-carbamoyl-L-serine with water to produce pyruvate, carbon dioxide, and two molecules of ammonia.[1]

[1] A. J. L. Cooper & A. Meister (1973) BBRC **55**, 780.

CARBONIC ANHYDRASE

This zinc-dependent enzyme [EC 4.2.1.1] catalyzes the reversible reaction of carbon dioxide with water to form bicarbonate and a proton (it was the first zinc enzyme identified).[1–8] The zinc ion acts as a Lewis acid to activate the water molecule, resulting in a zinc-bound hydroxide, and this metal ion also appears to delocalize the negative charge developing in the transition state after carbon dioxide binds. Nucleophilic attack then takes place, and bicarbonate is liberated, leaving the active site with zinc-bound water. During the second and rate-limiting stage of this catalytic cycle, a proton is transferred from the active site to solution. Brønsted analysis of the intramolecular proton transfer and Marcus theory treatments suggest that the most energy-requiring process in proton transfer is the thermodynamic restructuring of hydrogen bonds in the active site.[2]

[1] J. Kivela, S. Parkkila, A. K. Parkkila, J. Leinonen & H. Rajaniemi (1999) J. Physiol. **520**, pt. 2, 315.
[2] J. N. Earnhardt & D. N. Silverman (1998) CBC **1**, 483.
[3] D. W. Christianson (1991) Adv. Protein Chem. **42**, 281.
[4] D. N. Silverman & S. Lindskog (1988) Acc. Chem. Res. **21**, 30.
[5] B. L. Vallee & A. Galdes (1984) Adv. Enzymol. **56**, 283.
[6] Y. Pocker & S. Sarkanen (1978) Adv. Enzymol. **47**, 149.
[7] S. Lindskog, L. E. Henderson, K. K. Kannan, A. Liljas, P. O. Nyman & B. Strandberg (1971) The Enzymes, 3rd ed., **5**, 587.
[8] R. P. Davis (1961) The Enzymes, 2nd ed., **5**, 454.

Selected entries from **Methods in Enzymology** [**vol**, page(s)]:
General discussion: 2, 836; 28, 549; 34, 4; 46, 87, 155; 87, 732; 249, 479
Assay: 2, 836
Other molecular properties: activation, 2, 846; active sites, structures, 249, 481; affinity labeling, 87, 473; amidinated, 25, 591; amino acid sequence, 249, 480; atomic emission studies, 54, 458; atomic fluorescence spectroscopy, 54, 469; bovine, phosphorescence studies, 49, 246; Brønsted plots, 308, 291; buffer concentration, effect of, 63, 206; 87, 746; carbamino compound formation, 76, 488, 489, 498; carbon dioxide transport, 44, 920, 921; carbonic anhydrase I, Mn(II)-, chemical exchange reactions, magnetization transfer NMR study, 176, 320; carbonic anhydrase B (histidine residues, NMR spectra, 26, 650); carboxymethylation of, 25, 433, 438; catalysis (active-site residue effects, 249, 489; measurement at steady state, 249, 486; proton transfer in, 249, 484 [active-site residue effects, 249, 489; application of Marcus rate theory, 249, 495, 502; 308, 289; Brønsted plots for, 249, 491; 308, 291; histidine-64 as proton shuttle in, 249, 487; 308, 291; intrinsic energy barrier, 249, 495, 502; rate-equilibria relationships, 249, 491; rates, factors affecting, 249, 498; site-directed mutagenesis studies, 249, 485]); catalytic mechanism, 249, 483; catalytic stages, 308, 286; catalyzed CO_2 hydration, ^{18}O exchange analysis, 87, 746 (kinetic expressions, 87, 739); chelator inhibition, mechanism, 248, 233; cobalt-substituted, magnetic circular dichroism, 49, 168; conjugate, activity, 44, 271; cyanuration of, 25, 511, 512; exchanges, 87, 732; free energy profile, 308, 13; fluorescence and, 26, 503, 528; gel electrophoresis, 32, 101; His-64 in proton shuttling, 308, 286; human, 249, 480 (isozymes, steady-state properties, 249, 480, 489; physiological functions, 249, 480; reaction catalyzed, 249, 480; tissue distribution, 249, 480); immobilization (on membrane, 44, 905, 909, 920, 921; by microencapsulation, 44, 215); inhibition, 2, 845; isotope effects, 87, 748, 750; 308, 292; isotope exchange, 87, 732; 249, 479; isozyme, 34, 595; isozyme I (capillary electrophoresis-mass spectrometry, 271, 475, 477, 481, 483; steady-state properties, 249, 480, 489; tissue distribution, 249, 480); isozyme II (catalysis, proton transfer in, 249, 501; cloning, 249, 485; expression, 249, 485; site-specific mutagenesis, 249, 485; steady-state properties, 249, 480, 489; tissue distribution, 249, 480); isozyme III (assay to determine protein S-thiolation in cells, 251, 425; cloning, 249, 485; expression, 249, 485; gel electrofocusing, 251, 425; model system, 308, 287; site-specific mutagenesis, 249, 485; steady-state properties, 249, 480, 489; tissue distribution, 249, 480); isozyme IV, tissue distribution, 249, 480; isozyme V, tissue distribution, 249, 480; isozyme VI, tissue distribution, 249, 480; isozyme VII, tissue distribution, 249, 480; kinetics, 87, 732 (expressions, 87, 735); labeling, 46, 139; ligand interactions, evaluation, 202, 515; Marcus parameters for proton transfer, 308, 289, 291, 293, 295; mechanism, 87, 732; oxygen-18 exchange kinetics, 87, 732; 249, 486; oxyntic cells, in, 32, 716, 717; pH effect, 2, 844; 87, 734, 748; plant, 2, 842; 249, 480; progestin assay, 36, 457; properties, 2, 843; purification, ox red cells, 2, 841; rate-limiting steps, 308, 285; spin-label studies, 49, 463; stain, 22, 585, 593; 224, 106; structural classification, 249, 480; sulfonamide binding, stopped flow CD, 61, 320; therapeutic use, 44, 697; triplet excitation transfer and, 21, 41; zinc dissociated from, half-lives for, 248, 236

CARBON-MONOXIDE DEHYDROGENASE

This oxidoreductase [EC 1.2.99.2], also known as acetyl-CoA synthase and carbon-monoxide/acetyl-CoA synthase, catalyzes the reversible reaction of carbon-monoxide with water and an electron acceptor substrate to produce carbon dioxide and the reduced acceptor.[1–4] Aerobic microbes utilize a Mo-Fe flavin enzyme while phototrophic anaerobes have a dehydrogenase possessing nickel and iron-sulfur clusters. The enzyme from *Hydrogenophaga pseudoflava* is a seleno-molybdo-iron protein. Methyl viologen can serve as the acceptor substrate as well as ferredoxins, rubredoxin, and flavodoxin. In the presence of the dehydrogenase, an exchange reaction of carbon between C1 of acetyl-CoA and carbon monoxide (*i.e.*, $CH_3-^{14}COSCoA \leftrightarrow {}^{14}CO$) can be observed.

This enzyme participates in the synthesis of acetyl-CoA in these organisms (for example, acetogenic bacteria such as *Clostridium thermoacetium*). The enzyme first reduces carbon dioxide to carbon monoxide, followed by the reaction of coenzyme A with the carbon monoxide and a methyl source (for example, methylcobalamin) to produce acetyl-CoA (a metal-methyl intermediate is formed in the mechanism). The carbon monoxide dehydrogenase activity of *C. thermoaceticum* has a ping pong Bi Bi kinetic mechanism with an enzyme-bound carboxyl intermediate.[2]

[1] B. T. Golding & W. Buckel (1998) *CBC* **3**, 239.
[2] P. E. Baugh, D. Collison, C. D. Garner & J. A. Joule (1998) *CBC* **3**, 377.
[3] J. Seravalli, M. Kumar, W. P. Lu & S. W. Ragsdale (1995) *Biochemistry* **34**, 7879.
[4] G. Fuchs (1986) *FEMS Microbiol. Rev.* **39**, 181.

Selected entries from *Methods in Enzymology* [vol, page(s)]:
Other molecular properties: exchanges, **249**, 466; kinetic mechanism, equilibrium isotope exchange study, **249**, 466; *Rhodospirillum rubrum*, direct electrochemical studies, **227**, 521

CARBON-MONOXIDE OXIDASE

This iron-sulfur-, molybdenum-, and flavin-dependent oxidoreductase [EC 1.2.3.10] catalyzes the reaction of carbon monoxide with water and dioxygen to produce carbon dioxide and hydrogen peroxide.[1,2] Methylene blue can serve as an electron acceptor substrate.

[1] R. C. Bray, G. N. George, R. Lange & O. Meyer (1983) *BJ* **211**, 687.
[2] O. Meyer & K. V. Rajagopalan (1984) *JBC* **259**, 5612.

CARBON MONOXIDE OXYGENASE (CYTOCHROME *b*-561)

This FAD-, iron-, and molybdenum-dependent oxidoreductase [EC 1.2.2.4] catalyzes the reaction of carbon monoxide with water and ferrocytochrome *b*-561 to produce carbon dioxide, two protons, and ferricytochrome *b*-561.[1,2] Methylene blue and iodonitrotetrazolium chloride can serve as artificial electron acceptors.

[1] S. Jacobitz & O. Meyer (1989) *J. Bacteriol.* **171**, 6294.
[2] M. Kraut, I. Hugendieck, S. Herwig & O. Meyer (1989) *Arch. Microbiol.* **152**, 335.

CARBONYL REDUCTASE (NADPH)

This oxidoreductase [EC 1.1.1.184], also known as aldehyde reductase I, xenobiotic ketone reductase, and prostaglandin 9-ketoreductase, catalyzes the reaction of a ketone with NADPH (B-side stereospecificity) to produce a secondary alcohol and $NADP^+$.[1-4] The enzyme acts on a wide range of carbonyl compounds, including quinones (*e.g.*, methyl-1,4-benzoquinone, ubiquinone, and tocopherolquinone), aromatic aldehydes (*e.g.*, 4-nitrobenzaldehyde), ketoaldehydes, daunorubicin, and prostaglandins E_2 and 15-ketoprostaglandin $F_{2\alpha}$. *See also Prostaglandin-E_2 9-Reductase*

[1] N. Iwata, N. Inazu, S. Takeo & T. Satoh (1990) *EJB* **193**, 75.
[2] K. M. Bohren, J.-P. von Wartburg & B. Wermuth (1987) *BJ* **244**, 165.
[3] S. Usui, A. Hara, T. Nakayama & H. Sawada (1984) *BJ* **223**, 697.
[4] B. Wermuth (1981) *JBC* **256**, 1206.

Selected entries from *Methods in Enzymology* [vol, page(s)]:
Assay: 86, 143
Other molecular properties: porcine kidney (assay, **86**, 143; properties, **86**, 145; purification, **86**, 144); stereospecificity, **87**, 112

2-CARBOXY-D-ARABINITOL-1-PHOSPHATASE

This phosphatase [EC 3.1.3.63] catalyzes the hydrolysis of the rubisco inhibitor 2-carboxy-D-arabinitol 1-phosphate to produce 2-carboxy-D-arabinitol and orthophosphate.[1-4] Alternative substrates include 2-carboxy-D-arabinitol 1, 5-bisphosphate and 2-carboxy-D-ribitol 1,5-bisphosphate.[4]

[1] M. E. Salvucci & G. P. Holbrook (1989) *Plant Physiol.* **90**, 679.
[2] S. Gutteridge & B. Julien (1989) *FEBS Lett.* **254**, 225.
[3] G. P. Holbrook, G. Bowes & M. E. Salvucci (1989) *Plant Physiol.* **90**, 673.
[4] A. H. Kingston-Smith, I. Major, M. A. Parry & A. J. Keys (1992) *BJ* **287**, 821.

CARBOXYCYCLOHEXADIENYL DEHYDRATASE

This dehydratase [EC 4.2.1.91], also called arogenate dehydratase, catalyzes the conversion of L-arogenate to L-phenylalanine, water, and carbon dioxide.[1,2] The enzyme from some sources will also use prephenate as a substrate. L-Phenylalanine is a strong product inhibitor ($K_i \approx 24\,\mu M$).[1] *See also Prephenate Dehydratase*

[1] D. L. Siehl & E. E. Conn (1988) *ABB* **260**, 822.
[2] G. S. Zhao, T. H. Xia, R. S. Fischer & R. A. Jensen (1992) *JBC* **267**, 2487.

Selected entries from *Methods in Enzymology* [vol, page(s)]:
General discussion: 142, 495
Assay: 142, 488, 495, 513
Other molecular properties: amino acid composition, **142**, 518; properties, **142**, 494, 502, 517; purification from (*Phenylobacterium immobile*, **142**, 516; *Pseudomonas diminuta*, **142**, 501; tobacco, **142**, 490); substrate preparation, **142**, 515

3-CARBOXYETHYLCATECHOL 2,3-DIOXYGENASE

This iron-dependent oxidoreductase [EC 1.13.11.16], also called 2,3-dihydroxyphenylpropionate 1,2-dioxygenase, catalyzes the reaction of 3-(2,3-dihydroxyphenyl)propa-noate (*i.e.*, 3-carboxyethylcatechol) with dioxygen to produce 2-hydroxy-6-oxonona-2,4-diene-1,9-dioate.[1,2]

3-(2,3-dihydroxyphenyl)propionate

2-hydroxy-6-oxonona-2,4-diene-1,9-dioate

During catalysis, the substrate's two hydroxyl groups bind the Fe(II) ion to form a bidentate complex. Dioxygen then binds to the iron and the distal oxygen attacks the ring of the aryl substrate. A transient semiquinone-like intermediate ultimately produces the final product. Other substrates include *cis*- and *trans*-2-(2,3-dihydroxyphenyl)cyclopro-pane-1-carboxylate (both geometric isomers produce *trans*-2-(1-oxo-5-carboxy-5-hydroxy-2,4-pentadienyl) cyclopropane-1-carboxylate, suggesting the formation of a radical intermediate that allows rearrangement of the cyclopropane ring).[3]

[1]P. J. Geary & S. Dagley (1968) *BBA* **167**, 459.
[2]S. Dagley, P. J. Chapman & D. T. Gibson (1965) *BJ* **97**, 643.
[3]F. Spence, G. J. Langley & T. D. H. Bugg (1996) *JACS* **118**, 8336.

Selected entries from *Methods in Enzymology* [**vol**, page(s)]:
General discussion: **52**, 11

N^5-(CARBOXYETHYL)ORNITHINE SYNTHASE

This oxidoreductase [EC 1.5.1.24] catalyzes the reaction of L-ornithine with pyruvate and NADPH to produce N^5-(L-1-carboxyethyl)-L-ornithine, NADP$^+$, and water.[1–4] L-Lysine is an alternative amino acid substrate.

[1]J. Thompson (1989) *JBC* **264**, 9592.
[2]J. Thompson, N. Y. Nguyen, D. L. Sackett & J. A. Donkersloot (1991) *JBC* **266**, 14573.
[3]J. A. Donkersloot & J. Thompson (1995) *JBC* **270**, 12226.
[4]J. Thompson & S. P. F. Miller (1991) *Adv. Enzymol.* **64**, 317.

6-CARBOXYHEXANOYL-CoA SYNTHETASE

This enzyme [EC 6.2.1.14], also known as 6-carboxy-hexanoate:CoA ligase and pimeloyl-CoA synthetase,

catalyzes the reaction of ATP with 6-carboxyhexanoate (*i.e.*, pimelate or heptanedioate) and Coenzyme A to produce AMP, pyrophosphate (or, diphosphate), and 6-carboxyhexanoyl-CoA.[1–3]

[1]Y. Izumi, H. Morita, K. Sato, Y. Tani & K. Ogata (1972) *BBBA* **264**, 210.
[2]Y. Izumi, H. Morita, Y. Tani & K. Ogata (1974) *Agric. Biol. Chem.* **38**, 2257.
[3]O. Ploux, P. Soularue, A. Marquet, R. Gloeckler & Y. Lemoine (1992) *BJ* **287**, 685.

Selected entries from *Methods in Enzymology* [**vol**, page(s)]:
General discussion: **62**, 327
Assay: **62**, 327, 328

3-CARBOXY-2-HYDROXYADIPATE DEHYDROGENASE

This oxidoreductase [EC 1.1.1.87] catalyzes the reaction of 3-carboxy-2-hydroxyadipate with NAD$^+$ to produce α-ketoadipate (or, 2-oxoadipate), carbon dioxide, and NADH.[1]

[1]M. Strassman & L. N. Ceci (1965) *JBC* **240**, 4357.

4-CARBOXY-2-HYDROXYMUCONATE-6-SEMIALDEHYDE DEHYDROGENASE

This oxidoreductase [EC 1.2.1.45] catalyzes the reaction of 4-carboxy-2-hydroxy-*cis*,*cis*-muconate 6-semialdehyde with NADP$^+$ to produce 4-carboxy-2-hydroxy-*cis*,*cis*-muconate and NADPH.[1,2] NAD$^+$ can act as the coenzyme substrate, albeit less effectively. Unsubstituted aliphatic aldehydes, aromatic aldehydes, or D-glucose are not substrates. The enzyme has an ordered Bi Bi kinetic mechanism.[1]

[1]K. Maruyama (1988) *J. Biochem.* **103**, 714 and (1979) **86**, 1671.
[2]K. Maruyama, N. Ariga, M. Tsuda & K. Deguchi (1978) *J. Biochem.* **83**, 1125.

CARBOXYLATE REDUCTASE

This tungsten-dependent oxidoreductase [EC 1.2.99.6] catalyzes the reaction of a carboxylate anion with a reduced donor substrate to produce an aldehyde, the oxidized donor, and water.[1] Reduced methyl viologen can act as the donor substrate.

[1]H. White, R. Feicht, C. Huber, F. Lottspeich & H. Simon (1991) *Biol. Chem. Hoppe Seyler* **372**, 999.

S-CARBOXYMETHYLCYSTEINE SYNTHASE

This pyridoxal-phosphate-dependent enzyme [EC 4.5.1.5], first isolated from *Escherichia coli*, catalyzes the reaction of 3-chloro-L-alanine with thioglycolate (HSCH$_2$COO$^-$)

to produce *S*-carboxymethyl-L-cysteine and chloride ion.[1] Alternative thiol-containing substrates include methylmercaptan and ethylmercaptan, respectively producing *S*-methyl-L-cysteine and *S*-ethyl-L-cysteine.

[1] H. Kumagai, H. Suzuki, H. Shigematsu & T. Tochikura (1989) *Agric. Biol. Chem.* **53**, 2481.

CARBOXYMETHYLENEBUTENOLIDASE

This enzyme [EC 3.1.1.45], also known as dienelactone hydrolase and maleylacetate enol-lactonase, catalyzes the hydrolysis of 4-carboxymethylenebut-2-en-4-olide to produce 4 oxohex-2-enedioate.[1–3]

[1] K.-L. Ngai, M. Schlomann, H.-J. Knackmus & L. N. Ornsten (1987) *J. Bacteriol.* **169**, 699.
[2] E. Schmidt & H.-J. Knackmus (1980) *BJ* **192**, 339.
[3] A. L. Beveridge & D. L. Ollis (1995) *Protein Eng.* **8**, 135.

CARBOXYMETHYLHYDANTOINASE

carboxymethylhydantoin   *N*-carbamoyl-L-aspartate

This enzyme [EC 3.5.2.4], also known as 5-(acetic acid)-hydantoinase, catalyzes the reversible hydrolysis of L-5-carboxymethylhydantoin to produce *N*-carbamoyl-L-aspartate.[1]

[1] J. Lieberman & A. Kornberg (1954) *JBC* **207**, 911

Selected entries from *Methods in Enzymology* [**vol**, page(s)]:
Other molecular properties: cyclization of ureidosuccinate by, **2**, 496; dihydroörotase and, **6**, 180

4-CARBOXYMETHYL-4-HYDROXYISOCROTONOLACTONASE

This lactonase activity, originally classified as EC 3.1.1.16, but now deleted, is attributed to the combined action of muconolactone Δ-isomerase [EC 5.3.3.4] and β-ketoadipate-enol-lactonase [EC 3.1.1.24].

5-CARBOXYMETHYL-2-HYDROXYMUCONATE Δ-ISOMERASE

This isomerase [EC 5.3.3.10] catalyzes the interconversion of 5-carboxymethyl-2-hydroxymuconate and 5-carboxy-2-oxohept-3-enedioate.[1–3]

[1] M. L. Martin & A. Garrido-Pertierra (1985) *Z. Naturforsch.* **40c**, 503.
[2] A. Garrido-Pertierra & R. A. Cooper (1981) *EJB* **117**, 581.
[3] D. I. Roper & R. A. Cooper (1990) *FEBS Lett.* **266**, 63.

5-CARBOXYMETHYL-2-HYDROXYMUCONATE-SEMIALDEHYDE DEHYDROGENASE

This oxidoreductase [EC 1.2.1.60] catalyzes the reaction of 5-carboxymethyl-2-hydroxymuconate semialdehyde with water and NAD^+ to produce 5-carboxymethyl-2-hydroxymuconate, NADH, and a proton.[1–3]

[1] E. R. Blakley (1977) *Can. J. Microbiol.* **23**, 1128.
[2] J. M. Alonso & A. Garrido-Pertierra (1982) *BBA* **719**, 165.
[3] R. A. Cooper & M. A. Skinner (1980) *J. Bacteriol.* **143**, 302.

4-CARBOXYMETHYL-4-METHYLBUTENOLIDE MUTASE

4-carboxymethyl-4-
methylbut-2-en-1,4-olide

4-carboxymethyl-3-
methylbut-2-en-1,4-olide

This mutase [EC 5.4.99.14], also called 4-methylmuconolactone methylisomerase and 4-methyl-3-enelactone methylisomerase, catalyzes the interconversion of 4-carboxymethyl-4-methylbut-2-en-1,4-olide (*i.e.*, (+)-(4*S*)-4-methylmuconolactone) and 4-carboxymethyl-3-methylbut-2-en-1,4-olide (*i.e.*, (−)-(4*S*)-3-methylmuconolactone).[1,2]

[1] N. C. Bruce, R. B. Cain, D. H. Pieper & K. H. Engesser (1989) *BJ* **262**, 303.
[2] D. H. Pieper, K. Stadler-Fritzsche, H. J. Knackmuss, K. H. Engesser, N. C. Bruce & R. B. Cain (1990) *BJ* **271**, 529.

CARBOXYMETHYLOXYSUCCINATE LYASE

This lyase [EC 4.2.99.12] catalyzes the conversion of carboxymethyloxysuccinate to fumarate and glycolate.[1]

[1] D. Peterson & J. Llaneza (1971) *ABB* **162**, 135.

CARBOXY-*cis*,*cis*-MUCONATE CYCLASE

This cyclase [EC 5.5.1.5], also known as 3-carboxy-*cis*,*cis*-muconate lactonizing enzyme and 3-carboxymuconate cyclase, catalyzes the interconversion of 3-carboxy-*cis*,*cis*-muconate and 3-carboxy-2,5-dihydro-5-oxofuran-2-acetate (*i.e.*, 3-carboxymuconolactone).[1] This is a *syn*-1,2 addition-elimination reaction.[2] The *Neurospora crassa* enzyme also exhibits a minor muconate cycloisomerase [EC 5.5.1.1] activity.[2]

[1]D. R. Thatcher & R. B. Cain (1974) *EJB* **56**, 193.
[2]P. Mazur, W. J. Henzel, S. Mattoo & J. W. Kozarich (1994) *J. Bacteriol.* **176**, 1718.

3-CARBOXY-*cis,cis*-MUCONATE CYCLOISOMERASE

This isomerase [EC 5.5.1.2], also known as 3-carboxy-muconate lactonizing enzyme, catalyzes the interconversion of *cis,cis*-butadiene-1,2,4-tricarboxylate (*i.e.*, 3-carboxy-*cis,cis*-muconate) and 2-carboxy-5-oxo-2,5-dihydrofuran-2-acetate.[1,2]

[1]D. R. Thatcher & R. B. Cain (1972) *BJ* **127**, 32P and 33P.
[2]L. N. Ornston (1966) *JBC* **241**, 3787.

Selected entries from *Methods in Enzymology* [vol, page(s)]:
General discussion: 17A, 529

γ-CARBOXYMUCONOLACTONE DECARBOXYLASE

This decarboxylase [EC 4.1.1.44] catalyzes the conversion of 2-carboxy-2,5-dihydro-5-oxofuran-2-acetate to 4,5-dihydro-5-oxofuran-2-acetate and carbon dioxide.[1]

[1]R. V. J. Chari, C. P. Whitman & J. W. Kozarich (1987) *JACS* **109**, 5520.

Selected entries from *Methods in Enzymology* [vol, page(s)]:
General discussion: 17A, 529, 543
Other molecular properties: β-carboxymuconate decarboxylase (protocatechuate oxidase, in preparation of, **2**, 286); purification from *Pseudomonas putida*, **17A**, 531, 544

CARBOXYPEPTIDASE

Members of this hydrolase subclass catalyze peptide-bond hydrolysis at the C-terminal amino acid of a protein or oligopeptide. *See specific carboxypeptidase*

CARBOXYPEPTIDASE II

This member of the carboxypeptidase D subset [EC 3.4.16.6] of serine-type carboxypeptidases catalyzes the hydrolysis of C-terminal amino acids from oligopeptides and exhibits a significant preference for protein substrates containing basic amino acids in either the P1′ or P1 positions (for example, furylacryloyl-Ala-∇-Lys can be used to assay the enzyme, where ∇ indicates the cleavage site).[1-3] Nevertheless, peptides containing hydrophobic residues at the C-terminus can be hydrolyzed (*e.g.*, furylacryloyl-Phe-∇-Leu). Carboxypeptidase II also exhibits an amidase activity (releasing ammonia from peptide amides) and an esterase activity.

Carboxypeptidase II has a catalytic triad consisting of Ser146, His397, and Asp338. An oxyanion hole has also been identified.[4] *See also Carboxypeptidase D*

[1]K. Breddam (1986) *Carlsberg Res. Commun.* **51**, 83.
[2]S. J. Remington (1993) *Curr. Opin. Biotechnol.* **4**, 462.
[3]S. J. Remington & K. Breddam (1994) *Meth. Enzymol.* **244**, 231.
[4]D.-I. Liao, K. Breddam, R. M. Sweet, T. Bullock & S. J. Remington (1992) *Biochemistry* **31**, 9796.

Selected entries from *Methods in Enzymology* [vol, page(s)]:
General discussion: 244, 231
Other molecular properties: cleavage site specificity, **244**, 243; glycosylation, **244**, 237; substrate specificity, **244**, 235; three-dimensional structure, wheat enzyme (carboxylate-binding site, **244**, 241; catalytic residues, **244**, 240; disulfide bridges, **244**, 244; evolution, **244**, 245; overall fold, **244**, 238)

CARBOXYPEPTIDASE A

This zinc-dependent enzyme [EC 3.4.17.1] catalyzes C-terminal peptide-bond hydrolysis, preferring peptide or protein substrates with a C-terminal Phe, Tyr, Trp, Leu, or Ile.[1-8] The "A" in the name refers to the observation that the enzyme prefers substrates with C-terminal aromatic or branched-chain amino acids. (Because of the later identification of another carboxypeptidase (carboxypeptidase A2), some workers suggest that the original enzyme now be known as carboxypeptidase A1.)

Mast cell carboxypeptidase, also called carboxypeptidase A3, is currently regarded as a variant of carboxypeptidase A and classified as EC 3.4.17.1. However, its specificity, while similar, is distinct from that of pancreatic carboxypeptidase A. *See also Carboxypeptidase; Carboxypeptidase B; Carboxypeptidase A2, Carboxypeptidase A, Lysosomal; Procarboxypeptidase A*

[1]J. A. Hartsuck & W. N. Lipscomb (1971) *The Enzymes*, 3rd ed., **3**, 1.
[2]D. W. Christianson & W. N. Lipscomb (1989) *Acc. Chem. Res.* **22**, 62.
[3]W. L. Mock (1998) *CBC* **1**, 425.
[4]F. A. Quiocho & W. N. Lipscomb (1971) *Adv. Protein Chem.* **25**, 1.
[5]J. F. Riordan & B. Holmquist (1984) *Meth. Enzymatic Analysis*, 3rd ed., **5**, 43.
[6]B. L. Vallee, A. Galdes, D. S. Auld & J. F. Riordan (1983) in *Zinc Enzymes* (T. G. Spiro, ed.), Wiley, New York, p. 26.
[7]D. S. Auld & B. L. Vallee (1987) in *Hydrolytic Enzymes* (A. Neuberger & K. Brocklehurst, eds.), Elsevier, Amsterdam, p. 201.
[8]D. S. Auld & B. L. Vallee (1970) *Biochemistry* **9**, 4352.

Selected entries from *Methods in Enzymology* [vol, page(s)]:
General discussion: 2, 77; **19**, 460; **25**, 262; **27**, 414, 439, 497, 702, 705, 729, 730, 732; **34**, 4, 418; **45**, 728; **46**, 22, 225; **76**, 167
Assay: 2, 79; **19**, 475 (esterase activity, **19**, 477; peptidase activity, **19**, 475); crayfish, **80**, 660; mast cell, peritoneal, **80**, 604, 605
Other molecular properties: acetylation of, **25**, 501, 502, 504; acetylation of with *N*-acetylimidazole, **11**, 575; acid-base catalysis, site-directed mutagenesis studies, **249**, 114, 117; active site of, **19**, 502; active site probes, **61**, 322, 323; activity in aqueous salt cryosolvents, **226**, 559; affinity labeling, **87**, 472; aldolase B, action on, **9**, 498; amidated, **25**, 591; amidation of, **11**, 597; antibody production,

CARBOXYPEPTIDASE A2

This zinc-dependent metallopeptidase [EC 3.4.17.15], found in rat and human but not in cattle, is similar to carboxypeptidase A [EC 3.4.17.1] since it prefers to catalyze the hydrolysis of the peptide bond at the C-terminus of polypeptides and proteins in which the C-terminal amino acid is an aromatic or branched-chain amino acid. However, carboxypeptidase A2 prefers substrates with bulky C-terminal residues.[1] *See also Carboxypeptidase A*

[1] S. J. Gardell, C. S. Craik, E. Clauser, E. J. Goldsmith, C.-B. Stewart, M. Graff & W. J. Rutter (1988) *JBC* **263**, 17828.

CARBOXYPEPTIDASE B

This zinc-dependent peptidase [EC 3.4.17.2] catalyzes the hydrolysis of the peptide bond at the C-terminus of a polypeptide in which the C-terminus is lysine or arginine. The "B" in the name refers to the preferential release of basic amino acids from the C-terminus.

Synthetic substrates such as Bz-Gly-∇-Arg and FA-Ala-∇-Arg are very useful in enzyme assays. (Note: ∇ indicates the cleavage site.) The inactive precursor of this enzyme, procarboxypeptidase B, is generated by the action of trypsin: interestingly, trypsin is the only pancreatic enzyme identified so far that processes carboxypeptidase B. *See also Carboxypeptidase; Carboxypeptidase A; Lysine(Arginine) Carboxypeptidase*

Selected entries from *Methods in Enzymology* [vol, page(s)]:
General discussion: 19, 504; 25, 262; 34, 411, 418; 46, 225; 76, 167
Assay: 19, 504
Other molecular properties: activators and inhibitors, 19, 508; activity resembling, in beta granules, 31, 378; affinity labeling, 87, 472; alkylated, 34, 414; amino acid composition, 19, 507; antibody production, 74, 278; bovine (amino acid composition, 80, 662 [pancreatic, 80, 607, 608]; amino-terminal sequence, 80, 663, 664; inhibition, 80, 786, 787); calmodulin-converting enzyme, 102, 167 (assay, 102, 169; rat anterior pituitary, properties, 102, 170); carboxypeptidase A activity of, 11, 651; chicken muscle, 80, 788; distribution, 19, 508; human, amino-terminal sequence, 80, 663, 664; hydrolysis of peptide substrates, kinetics, 248, 679; inhibition, by amino group alkylation, 47, 474; inhibition by argininic acid, 11, 651; iodination of, 25, 441; membrane-bound, 248, 663; modified proteins, 34, 414; paired-iodination, 25, 445; pancreatic, 248, 663; pancreatic, types A and B, molecular replacement study, 115, 70; phosphorescence studies, 49, 239, 248; porcine (inhibition, 80, 787, 788; purification, 80, 791); preparation, 25, 146; primary specificity (S1')

pocket, **248**, 678; properties, **248**, 213, 216; purification, **19**, 505; purity and physicochemical properties, **19**, 507; radioimmunoassay, equilibration step, **74**, 284; removal of C-terminal lysine residues from tryptic peptides with, **11**, 444; separation of fragments, **47**, 99; transition-state and multisubstrate analogues, **249**, 306; zinc replacement with other metals, **248**, 240

CARBOXYPEPTIDASE C

The subclass of serine-type carboxypeptidases [classified as EC 3.4.16.5] prefers substrates with hydrophobic residues at the P1' position.[1–4] *See also Carboxypeptidase; Carboxypeptidase Y; Carboxypeptidase A, Lysosomal*

[1] K. Breddam (1986) *Carlsberg Res. Commun.* **51**, 83.
[2] S. J. Remington (1993) *Curr. Opin. Biotechnol.* **4**, 462.
[3] R. Hayashi (1977) *Meth. Enzymol.* **47**, 84.
[4] S. J. Remington & K. Breddam (1994) *Meth. Enzymol.* **244**, 231.

Selected entries from *Methods in Enzymology* [vol, page(s)]:
General discussion: **11**, 155, 437; **19**, 292; **45**, 561, 568; **46**, 206; **47**, 73, 84; **244**, 231
Assay: **19**, 286, 291; **47**, 76, 77 (by Cbo-Leu-Phe or Cbo-Glu-Tyr, **45**, 562, 563; colorimetric, **47**, 77; spectrophotometric, **47**, 76, 77); **244**, 232
Other molecular properties: activators and inhibitors, **19**, 295; commercial preparations of, contaminants, **47**, 74; bovine spleen, **80**, 788; diisopropylfluorophosphate-inhibited, isolation, sources, **45**, 599; distribution, **19**, 296; family (active site residues, **244**, 44, 231, 240; biological role, **244**, 44, 232; members, **244**, 43; pH optimum, **244**, 44; processing, **244**, 44; sequence homology, **244**, 44; tertiary structure, **244**, 44); function, **45**, 561; hydrolase assay, **19**, 291; inhibitors, **46**, 206; **244**, 231; kinetic properties, **19**, 296; preparation, **47**, 75, 76; pretreatment with DFP, **47**, 78, 79; processing, **244**, 237; properties, **45**, 566, 567; proteinase assay, **19**, 286; purification, **19**, 292; **45**, 563, 565; **244**, 233; reaction conditions, **47**, 79, 80; reaction mixture analysis, **47**, 80, 81; sample preparation, **47**, 79; specificity, **11**, 437; **19**, 285, 295; **45**, 566; **47**, 74, 75; **244**, 235, 243; stability, **19**, 295; **47**, 78; units, **45**, 563

CARBOXYPEPTIDASE D

These serine carboxypeptidases [EC 3.4.16.6], which catalyze the preferential release of C-terminal amino acids, were originally called acid carboxypeptidases in view of their acidic pH optimum.[1–3] The best characterized member is wheat carboxypeptidase II. Other members include carboxypeptidase Kex1 from *Saccharomyces cerevisiae*, carboxypeptidase S1 from the fungus *Penicillium janthinellum*, barley carboxypeptidase MI, and carboxypeptidase AII from *Aspergillus niger*. These enzymes are readily assayed with the use of synthetic furylacryloyl-dipeptide substrates (for example, furylacryloyl-Ala-∇-Lys, where ∇ indicates the cleavage site).

The substrate specificity for carboxypeptidase S1 of *Penicillium janthinellum* is pH dependent. At pH 6.0, the pH$_{opt}$ for the esterase activity, Bz-Arg-OMe and Bz-Lys-OMe are readily hydrolyzed.[4] *See also Carboxypeptidase II; Carboxypeptidase Kex I*

[1] K. Breddam (1986) *Carlsberg Res. Commun.* **51**, 83.
[2] S. J. Remington (1993) *Curr. Opin. Biotechnol.* **4**, 462.
[3] S. J. Remington & K. Breddam (1994) *Meth. Enzymol.* **244**, 231.
[4] K. Breddam (1988) *Carlsberg Res. Commun.* **53**, 309.

Selected entries from *Methods in Enzymology* [vol, page(s)]:
General discussion: **244**, 231
Assay: **244**, 232
Other molecular properties: active site residues, **244**, 231, 240; inhibitors, **244**, 231; penicillocarboxypeptidase (or, carboxypeptidases S1 and S2 from *Penicillium janthinellum*) (assay, **19**, 384 [mixture, in, **45**, 592; ninhydrin, with, **45**, 591; 2,4,6-trinitrobenzenesulfonic acid, with, **45**, 591]; definition of, **19**, 376; *Penicillium janthinellum* [assay, **19**, 384; nomenclature, **19**, 373; properties, **19**, 385]; S-1 enzyme [amino acid composition, **45**, 594; assay {mixture, in, **45**, 592; ninhydrin, with, **45**, 591; 2,4,6-trinitrobenzenesulfonic acid, with, **45**, 591}; endopeptidase activity, **45**, 589; inhibitors, **45**, 595; kinetic, **45**, 596; pH dependence, **45**, 597; properties, **45**, 588, 593, 595; purification, **45**, 587, 590; specificity, **45**, 596; stability, **45**, 593; unit definition, **45**, 592]; S-2 enzyme [amino acid composition, **45**, 594; assay {mixtures, from, **45**, 592; ninhydrin, with, **45**, 591; 2,4,6-trinitrobenzene sulfonic acid, **45**, 591}; inhibitors, **45**, 595; kinetic, **45**, 596; pH dependence, **45**, 597; properties, **45**, 589 {enzymic, **45**, 595, 596; molecular, **45**, 593}; purification, **45**, 587, 590; specificity, **45**, 596; stability, **45**, 593; units, **45**, 592]); processing, **244**, 237; purification, **244**, 233; substrate specificity, **244**, 235, 243

CARBOXYPEPTIDASE G$_3$

This variant of glutamate carboxypeptidase [EC 3.4.17.11] catalyzes the hydrolysis of *N*-fatty acyl-glutamate or *N*-fatty acyl-aspartate derivatives as well as the release of C-terminal amino acids from polypeptides.[1] Because it will act on both the L- and D-acidic amino acids, this enzyme differs from both carboxypeptidase G$_1$ and G$_2$. The enzyme also catalyzes the hydrolysis of *N*-octanoyl-DL-Glu and -DL-Asp. The best fatty acyl moiety appears to be decanoyl (*N*-benzoylglutamate is a poor substrate). This carboxypeptidase will also act on glutamylglutamate, *N*-octanoyl-Glu-∇-Glu, and *N*-octanoyl-Tyr-∇-Glu. (Note: ∇ indicates the cleavage site.)

[1] N. Yasuda, M. Kaneko & Y. Kimura (1992) *Biosci. Biotechnol. Biochem.* **56**, 1536.

CARBOXYPEPTIDASE H

This zinc-dependent metalloproteinase [EC 3.4.17.10], also called carboxypeptidase E (first discovered in enkephalin-containing chromaffin vesicles) and enkephalin convertase, catalyzes the hydrolysis of C-terminal basic amino acids from peptides and proteins.[1,2] The enzyme appears to prefer substrates with an alanyl residue in the P1 position, although all residues are tolerated in that position. Carboxypeptidase H has a pH$_{opt}$ of 5.0–5.5 with activity falling rapidly at higher pH values (this decrease is due to changes in the K_m value). The enzyme is activated by millimolar concentrations of Co^{2+} and inhibited by 1,10-phenanthroline and thiol-directed reagents. Carboxypeptidase H plays an important role in the processing of protein hormones and bioactive peptides. The enzyme is distinct from that of carboxypeptidase B [EC 3.4.17.2] and lysine(arginine) carboxypeptidase [EC 3.4.17.3].

[1] L. D. Fricker (1991) in *Peptide Biosynthesis and Processing* (L. D. Fricker, ed.), CRC Press, Boca Raton, p. 199.
[2] L. D. Fricker (1995) *Meth. Neurosci.* **23**, 237.

CARBOXYPEPTIDASE M

This zinc-dependent metalloproteinase [EC 3.4.17.12] catalyzes the release of C-terminal arginine or lysine residues, and only those residues, from polypeptides or proteins.[1,2] optimally active at neutral pH. The enzyme is activated by Co^{2+} and inhibited by 1,10-phenanthroline. Its membrane association (hence, the descriptor "M") distinguishes this enzyme from lysine(arginine) carboxypeptidase [EC 3.4.17.3], carboxypeptidase H [EC 3.4.17.10], and other carboxypeptidases. Release from membranes occurs upon treatment with bacterial phosphatidylinositol-specific phospholipase C.

[1] F. Tan, P. A. Deddish & R. A. Skidgel (1995) *MIE* **248**, 663.
[2] R. A. Skidgel (1996) in *Zinc Metalloproteases in Health and Disease* (N. M. Hooper, ed.), Taylor & Francis, London, p. 241.

Selected entries from *Methods in Enzymology* [vol, page(s)]:
General discussion: 248, 663.
Assay: 248, 664 (dansyl-Ala-Arg substrate, 248, 665; procedure, 248, 666)
Other molecular properties: human, 248, 663 (activator, 248, 672; activity units, calculation, 248, 667; cleavage specificity, 248, 672; cloning, 248, 674; distribution, 248, 675; inhibitors, 248, 672; membrane anchoring, 248, 674; molecular weight, 248, 672; physiological functions, 248, 675; properties, 248, 672 [enzymatic, 248, 672]; purification, 248, 667; sequencing, 248, 674; solubilization, 248, 668 [detergent, with, 248, 669; phosphatidylinositol-specific phospholipase C, with, 248, 669]; storage, 248, 672; substrates, 248, 664, 673; transferrin contaminant, removal, 248, 668); properties, 248, 213, 216

CARBOXYPEPTIDASE T

This zinc-dependent, bacterial metalloproteinase [EC 3.4.17.18] catalyzes the release of a C-terminal amino acid residue from a polypeptide or protein in which the residue may be hydrophobic or positively charged. The enzyme will release C-terminal Arg, Lys, His, Phe, Tyr, Trp, Leu, Met, and Ala. Aminoacyl residues in the P1 position also influence the rate, and hydrolysis does not occur if a prolyl residue is located at P1.

Selected entries from *Methods in Enzymology* [vol, page(s)]:
General discussion: 248, 675.

Assay: 248, 680
Other molecular properties: catalytic mechanism, functional groups involved in, 248, 679; homologues, 248, 676; hydrolysis of peptide substrates, kinetics, 248, 679; isoelectric point, 248, 676; molecular characteristics, 248, 676; molecular mass, 248, 676; primary specificity (S1') pocket, 248, 678; primary structure, 248, 677; properties, 248, 676; purification, 248, 682; secretion, 248, 676; substrate binding site, 248, 678; substrate specificity, 248, 676, 678; tertiary structure, 248, 677; *Thermoactinomyces*, 248, 213, 216

CARBOXYPEPTIDASE TAQ

This zinc-dependent metallopeptidase [EC 3.4.17.19], also called thermostable carboxypeptidase 1, catalyzes the release of a C-terminal amino acid from polypeptides with a broad specificity; however, a C-terminal proline is not released. Prolyl residues at the P1 and P2 positions are tolerated. The optimal pH is about 8, and the optimal temperature is 80°C. Activity is maximal with 1 mM Co^{2+}.[1]

[1] S.-H. Lee, E. Minagawa, H. Taguchi, H. Matsuzawa, T. Ohta, S. Kaminogawa & K. Yamauchi (1992) *Biosci. Biotechnol. Biochem.* **56**, 1839.

CARBOXYPEPTIDASE U

This zinc-dependent metallopeptidase [EC 3.4.17.20] (also known as arginine carboxypeptidase, carboxypeptidase R, plasma carboxypeptidase B, and thrombin-activatable fibrinolysis inhibitor [TAFI]) catalyzes the release of a C-terminal arginine or lysine from a polypeptide or protein.[1,2] The enzyme will also act on synthetic substrates such as Bz-Gly-∇-Lys, FA-Ala-∇-Lys, and FA-Ala-∇-Arg, where ∇ indicates the cleavage site. The precursor procarboxypeptidase U in plasma is activated by thrombin or plasmin during clotting, and the enzyme is unstable (hence, the descriptor "U"). *See also Lysine Carboxypeptidase*

[1] D. Hendriks, W. Wang, S. Scharpé, M. P. Lommaert & M. van Sande (1990) *BBA* **1034**, 86.
[2] A. K. Tan & D. L. Eaton (1995) *Biochemistry* **34**, 5811.

CARBOXYPEPTIDASE Y

Carboxypeptidase Y, originally called proteinase C, was first isolated from the yeast *Saccharomyces cerevisiae* and has since been identified in many organisms. It is a serine-type carboxypeptidase preferring hydrophobic residues in the P1' position: thus, it is now regarded as a variant of carboxypeptidase C and is classified under that EC number [EC 3.4.16.5].[1-3] *See also Carboxypeptidase C*

[1] K. Breddam (1986) *Carlsberg Res. Commun.* **51**, 83.
[2] S. J. Remington (1993) *Curr. Opin. Biotechnol.* **4**, 462.
[3] U. H. Mortensen, H. R. Stennicke, M. Raaschou-Nielsen & K. Breddam (1994) *JACS* **116**, 34.

Selected entries from *Methods in Enzymology* [vol, page(s)]:
General discussion: 46, 206; **47**, 84
Assay: 45, 568 (anilidase activity, **45**, 571; esterase activity, **45**, 570, 584; **47**, 86, 87; peptidase activity, **45**, 569; **47**, 86); enzymatic assay, **185**, 384; **194**, 448; filter assay, **194**, 659
Other molecular properties: activators, **45**, 579; active site residues, **244**, 231; activity, potential, **45**, 575; amidase activity, **45**, 584, 585; characteristics, **45**, 568; complete hydrolysis, **47**, 41, 43; composition, **45**, 578; coupling to 1,1'-carbonyldiimidazole-activated support, **135**, 116; C-terminal protein sequence determination, **244**, 246; DFP sensitivity, **47**, 89; distribution, **45**, 586; esterase activity, **45**, 584; glycosylation, **244**, 237; hydrolysis, limited, **45**, 585; immobilized, deblocking in peptide synthesis, **136**, 157; inhibitors, **45**, 579, 580; **46**, 206; **47**, 89, 90; kinetics, **45**, 581, 585; metal ion sensitivity, **47**, 89; nomenclature, **45**, 586; oxidation, **135**, 144; peptidase activity, **45**, 581; pH dependence, **244**, 234; processing, **244**, 233; properties, **45**, 576, 586; **47**, 84, 85, 88, 90; purification, **45**, 572; **47**, 87, 88; purity, **45**, 576; reaction conditions, **47**, 91, 92; specificity, **45**, 581; **47**, 90; **244**, 237; stability, **45**, 576; **47**, 88; synthesis of peptide bonds, **244**, 247; three-dimensional structure, **244**, 241

4-(2-CARBOXYPHENYL)-2-OXOBUT-3-ENOATE ALDOLASE

(3*E*)-4-(2-carboxyphenyl)-2-oxobut-3-enoate

2-carboxybenzaldehyde

This aldolase [EC 4.1.2.34], also called 2'-carboxybenzalpyruvate aldolase, catalyzes the conversion of (3*E*)-4-(2-carboxyphenyl)-2-oxobut-3-enoate to 2-carboxybenzaldehyde and pyruvate.[1]

[1] E. A. Barnsley (1983) *J. Bacteriol.* **154**, 113.

CARBOXYVINYL-CARBOXYPHOS-PHONATE PHOSPHORYLMUTASE

This mutase [EC 2.7.8.23], also known as carboxyphosphonoenolpyruvate phosphonomutase and CPEP phosphonomutase, catalyzes the conversion of 1-carboxyvinyl carboxyphosphonate to 3-(hydrohydroxyphosphoryl)pyruvate and carbon dioxide:[1]

[1] S. J. Pollack, S. Freeman, D. L. Pompliano & J. R. Knowles (1992) *EJB* **209**, 735.

CARICAIN

This cysteine endopeptidase [EC 3.4.22.30], also known as papaya peptidase A, papaya proteinase Ω, and papaya proteinase II, catalyzes the hydrolysis of peptide bonds with an activity very similar to papain [EC 3.4.22.2] and chymopapain [EC 3.4.22.6].[1,2] The highly basic nature of caricain (pI $\approx$ 11.7) readily distinguishes the enzyme from the other cysteine endopeptidases. The enzyme also exhibits charge heterogeneity, as do many other cysteine endopeptidases. pH-rate studies demonstrate that multiple ionizations affect the acylation step.

Caricain prefers hydrophobic aminoacyl residues in both the S2 and S3 subsites; however, other amino acids are accepted. The synthetic substrates Bz-Arg-∇-NHPhNO$_2$ or Z-Phe-Arg-∇-NHMec (∇ = cleavage site) can be used to assay the enzyme. *See also Papain; Chymopapain*

[1] S. Zucker, D. J. Buttle, M. J. H. Nicklin & A. J. Barrett (1985) *BBA* **828**, 196.
[2] R. W. Pickersgill, P. Rizkallah, G. W. Harris & P. W. Goodenough (1991) *Acta Crystallogr. B* **47**, 766.

CARNITINAMIDASE

This stereospecific hydrolase [EC 3.5.1.73] catalyzes the hydrolysis of L-carnitinamide to produce L-carnitine and ammonia.

L-carnitinamide   L-carnitine

The high Michaelis constant (12 mM) for L-carnitinamide raises serious questions regarding the nature of the natural substrate. A search for the physiologic substrate demonstrated that the enzyme failed to hydrolyze any of sixty different aliphatic/aromatic amides, nitriles, amino acid amides, and dipeptide amides.[1]

[1] U. Joeres & M. R. Kula (1994) *Appl. Microbiol. Biotechnol.* **40**, 606.

CARNITINE *O*-ACETYLTRANSFERASE

This acetyltransferase [EC 2.3.1.7], also known as carnitine acetylase, catalyzes the freely reversible reaction of acetyl-CoA with L-carnitine to yield coenzyme A and *O*-acetyl-L-carnitine ($K_{eq} \approx 1.7$).[1-5] Other acyl-donor substrates include propanoyl-CoA and butanoyl-CoA.

L-carnitine   *O*-acetyl-L-carnitine

The β-oxidation metabolite 3-keto-4-pentenoyl-CoA is an effective inhibitor of the enzyme, but only in the presence of L-carnitine. In fact, 3-keto-4-pentenoyl-CoA is a mechanism-based inhibitor (or suicide substrate) that irreversibly inactivates the enzyme.[5] The enzyme reportedly has a random Bi Bi kinetic mechanism.[6] **See also** *Carnitine O-Octanoyltransferase; Carnitine O-Palmitoyltransferase*

[1] R. R. Ramsay (2000) *Biochem. Soc. Trans.* **28**, 182.
[2] V. A. Zammit (1999) *Prog. Lipid Res.* **38**, 19.
[3] S. O. Farrell, C. J. Fiol, J. K. Reddy & L. L. Bieber (1984) *JBC* **259**, 13089.
[4] L. L. Bieber & S. Farrell (1983) *The Enzymes*, 3rd ed., **16**, 627.
[5] J. Zhong, J. C. Fong & H. Schulz (1985) *ABB* **240**, 524.

Selected entries from *Methods in Enzymology* [vol, page(s)]:
General discussion: 13, 387; 71, 351; 72, 277; 123, 276
Assay: 13, 387; bovine cardiac mitochondria, 123, 281; coupled with citrate synthase and malate dehydrogenase, 13, 388; fluorometric with (acetylcarnitine, 13, 509; L-carnitine, 13, 505); hydroxamate, 13, 389; isotope exchange, 13, 389; murine hepatic peroxisomes, 123, 281; peroxisomes and microsomes, 71, 351; thioester formation, 13, 387; thiol release, 13, 389
Other molecular properties: assays for (acetylcarnitine, 13, 509; 14, 695; carnitine, 13, 505; 55, 211, 212; long chain fatty acyl-CoA, in, 35, 273; short-chain acyl-CoA, 13, 536, 544, 546, 549); bovine cardiac mitochondria, 123, 280; chemical modification, 63, 229; enzymatic synthesis of acetylcarnitine, 14, 691, 695; induction by clofibrate and di(2-ethylhexyl)phthalate, 72, 507; microbodies, in, 52, 496, 500, 501; peroxisomes and microsomes, from, 71, 351; pH kinetics, 63, 227; properties, 13, 392; purification, pigeon breast muscle, 13, 389; transitionstate and multisubstrate analogues, 249, 304

CARNITINE DECARBOXYLASE

L-carnitine   L-2-methylcholine

This ATP-dependent decarboxylase [EC 4.1.1.42] catalyzes the conversion of L-carnitine to 2-methylcholine and carbon dioxide.[1,2]

[1] M. Habibulla & R. W. Newburgh (1972) *J. Insect Physiol.* **18**, 1929.
[2] E. A. Khairallah & G. Wolf (1967) *JBC* **242**, 32.

L-CARNITINE DEHYDRATASE

4-(trimethylammonio)but-2-enoate

This dehydratase [EC 4.2.1.89] catalyzes the reversible conversion of L-carnitine to 4-(trimethylammonio)but-2-enoate (*i.e.*, crotonobetaine) and water.[1]

[1] H. Jung, K. Jung & H. P. Kleber (1989) *BBA* **1003**, 270.

CARNITINE 3-DEHYDROGENASE

L-carnitine   3-dehydrocarnitine

This oxidoreductase [EC 1.1.1.108] catalyzes the conversion of (*R*)-carnitine (*i.e.*, L-carnitine) and NAD$^+$ to produce 3-dehydrocarnitine and NADH. (*S*)-Carnitine is a competitive inhibitor of the *Agrobacterium* enzyme.[1,2] **See also** *(S)-Carnitine 3-Dehydrogenase*

[1] H. Hanschmann, R. Ehricht & H. P. Kleber (1996) *BBA* **1290**, 177
[2] P. Goulas (1988) *BBA* **957**, 335

(S)-CARNITINE 3-DEHYDROGENASE

This oxidoreductase [EC 1.1.1.254], also known as D-carnitine dehydrogenase, catalyzes the reaction of (*S*)-carnitine with NAD$^+$ to produce 3-dehydrocarnitine and NADH.[1,2]

D-carnitine   3-dehydrocarnitine

Note that the carnitine substrate is the enantiomer of the substrate acted upon by carnitine 3-dehydrogenase [EC 1.1.1.108]. Interestingly, the pH optima for the forward and reverse reactions are quite different. (*R*)-Carnitine and choline, are competitive inhibitors. **See also** *Carnitine 3-Dehydrogenase*

[1] H. Hanschmann & H. P. Kleber (1997) *BBA* **1337**, 133
[2] S. Setyahadi, T. Ueyama, T. Arimoto, N. Mori & Y. Kitamoto (1997) *Biosci., Biotech., and Biochem.* **61**, 1055

CARNITINE O-OCTANOYLTRANSFERASE

O-octanoyl-L-carnitine

This transferase [EC 2.3.1.137] catalyzes the reversible reaction of octanoyl-CoA with L-carnitine to produce coenzyme A and octanoyl-L-carnitine.[1-3] **See also** *Carnitine O-Acetyltransferase; Carnitine O-Palmitoyltransferase*

[1] R. R. Ramsay (2000) *Biochem. Soc. Trans.* **28**, 182.
[2] V. A. Zammit (1999) *Prog. Lipid Res.* **38**, 19.
[3] S. O. Farrell, C. J. Fiol, J. K. Reddy & L. L. Bieber (1984) *JBC* **259**, 13089.

Selected entries from *Methods in Enzymology* [vol, page(s)]:
General discussion: 123, 276
Assay: 123, 281
Other molecular properties: induction by clofibrate and
   di(2-ethylhexyl)phthalate, 72, 507; microbodies, in, 52, 496, 500,
   501; murine hepatic peroxisomes (assays, 123, 281; purification,
   123, 278)

CARNITINE *O*-PALMITOYLTRANSFERASE

This transferase [EC 2.3.1.21] catalyzes the reversible reaction of palmitoyl-CoA with L-carnitine to produce coenzyme A and *O*-palmitoyl-L-carnitine.[1–6]

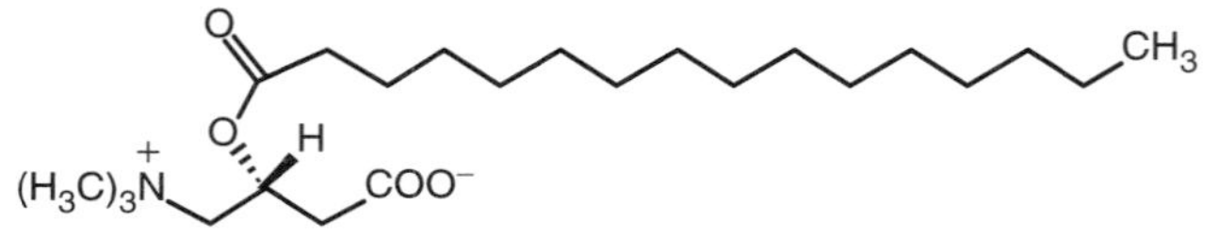

O-palmitoyl-L-carnitine

The enzyme exhibits a broad specificity toward acyl-CoA substrates, covering a range from C_8 to C_{18}, but optimal activity is observed with palmitoyl-CoA. At least two isozymes of this transferase have been identified: in the inner and outer mitochondrial membranes. The enzyme has a random Bi Bi kinetic mechanism.[4,5] *See also* Carnitine *O*-Acetyltransferase; Carnitine O-Octanoyltransferase

[1] R. R. Ramsay (2000) *Biochem. Soc. Trans.* 28, 182.
[2] G. Woldegiorgis, H. Shi, H. Zhu & D. N. Arvidson (2000) *J. Nutr.* 130
   (2S Suppl), 310S.
[3] V. A. Zammit (1999) *Prog. Lipid Res.* 38, 19.
[4] R. R. Ramsay, J. P. Derrick, A. S. Friend & P. K. Tubbs (1987) *BJ*
   244, 271.
[5] J. P. Derrick, P. K. Tubbs & R. R. Ramsay (1986) *Biochem. Soc. Trans.*
   14, 698.
[6] L. L. Bieber & S. Farrell (1983) *The Enzymes*, 3rd ed., 16, 627.

Selected entries from *Methods in Enzymology* [vol, page(s)]:
General discussion: 56, 368; 71, 242; 123, 276
Assay: 56, 368, 369; bovine cardiac mitochondria, 123, 281; CoASH
   released, of, 56, 371, 372; hydroxamate, 56, 372; isotope exchange, 56,
   370, 371; radioactive forward method, 56, 372; spectophotometric, 56,
   369, 370
Other molecular properties: assay for carnitine and its *O*-acyl
   derivatives, 14, 612; bovine cardiac mitochondria, 123, 281; brown
   adipose tissue mitochondria, 55, 76; distribution, methods of evaluation,
   56, 372; induced cooperativity in, 202, 716; induction by clofibrate and
   di(2-ethylhexyl)phthalate, 72, 507; palmitoyl-CoA assay, 13, 544, 551;
   properties, 56, 374; purification, 56, 373, 374; synthesis of
   acylcarnitines, 14, 690

CARNOSINASE

This glycosylated enzyme [EC 3.4.13.20] (also known as serum carnosinase, β-alanylhistidine dipeptidase, and β-Ala-His dipeptidase) catalyzes the preferential hydrolysis of the peptide bond in β-alanyl-L-histidine (*i.e.*, carnosine), as well as anserine (β-alanyl-1-methylhistidine),

Xaa–His, and other dipeptides including homocarnosine (γ-aminobutyrylhistidine). Interestingly, the affinity for homocarnosine is about ten times tighter than that for carnosine. It has been suggested by some investigators that homocarnosine is the true physiological substrate.[1,2] β-Alanylalanine is hydrolyzed only about a tenth of the rate of carnosine.

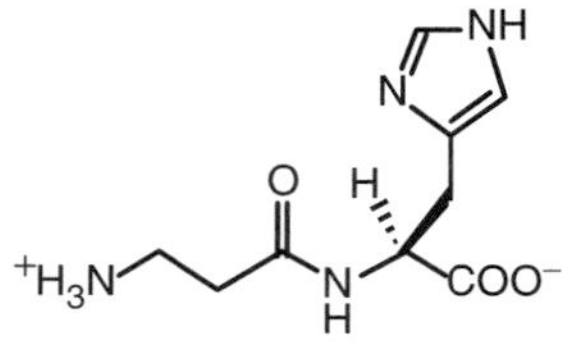

carnosine

There is considerable confusion in the literature with respect to serum carnosinase, Xaa-His dipeptidase [EC 3.4.13.3; tissue carnosinase], and cytosol nonspecific dipeptidase [EC 3.4.13.18]. Serum carnosinase hydrolyzes homocarnosine and anserine (but not prolylleucine), is inhibited by dithiothreitol, and has a different pH optimum (8.0–8.5). *See also Xaa-His Dipeptidase; Cytosol Nonspecific Dipeptidase*

[1] M. C. Jackson, C. M. Kucera & J. F. Lenney (1991) *Clin. Chim. Acta* 196, 193.
[2] L. R. Gjessing, H. A. Lunde, L. Mokrid, J. F. Lenney & O. Sjaastad (1990)
   J. Neural. Transm. Suppl. 29, 91.

CARNOSINE *N*-METHYLTRANSFERASE

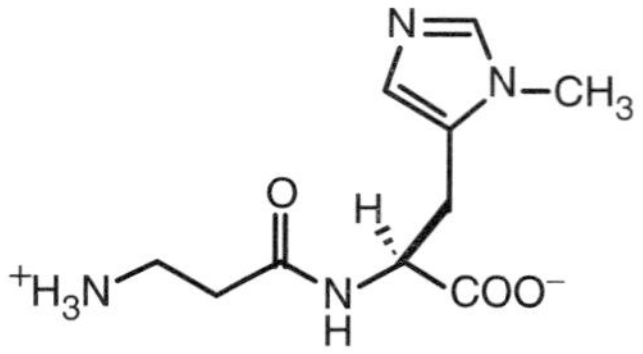

anserine

This methyltransferase [EC 2.1.1.22] catalyzes the reaction of *S*-adenosyl-L-methionine with carnosine to produce *S*-adenosyl-L-homocysteine and anserine.[1,2] This enzyme also methylates specific histidyl residues in certain actin-derived peptides.[2]

[1] R. McManus (1962) *JBC* 237, 1207.
[2] M. Raghavan, U. Lindberg & C. Schutt (1992) *EJB* 210, 311.

CARNOSINE SYNTHETASE

This enzyme [EC 6.3.2.11], also called carnosine synthase, catalyzes the reaction of L-histidine with β-alanine and ATP to produce carnosine, AMP, and pyrophosphate (or, diphosphate).[1-3] Other alternative substrates for histidine include 3-methyl-L-histidine, 1-methyl-L-histidine, and 2-methyl-L-histidine. γ-Aminobutyrate can substitute for β-alanine. It has been suggested that the enzyme mechanism utilizes an acyladenylate intermediate.

[1] S. J. Kish, T. J. Perry & S. Hansen (1979) *J. Neurochem.* **32**, 1629.
[2] H. Horinishi, M. Grillo & F. L. Margolis (1978) *J. Neurochem.* **31**, 909.
[3] R. H. Ng & F. D. Marshall (1978) *J. Neurochem.* **30**, 187.

Selected entries from *Methods in Enzymology* [**vol**, page(s)]:
General discussion: 17B, 102
Assay: 17B, 102

CAROTENE 7,8-DESATURASE

This oxidoreductase [EC 1.14.99.30], also known as ζ-carotene desaturase, catalyzes the reaction of neurosporene with AH_2 and dioxygen to produce lycopene, A, and two water molecules (where AH_2 is the reduced acceptor substrate and A is the oxidized product).[1]

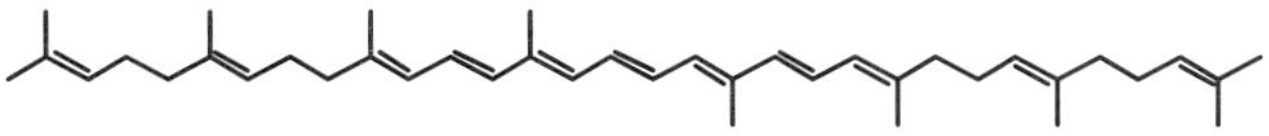

ζ-carotene

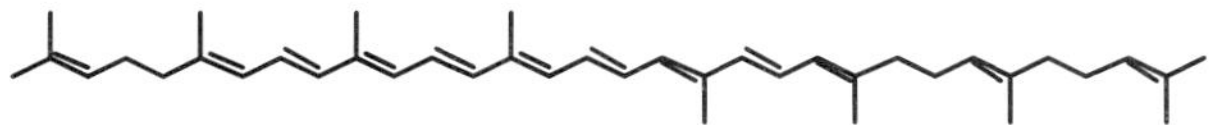

neurosporene

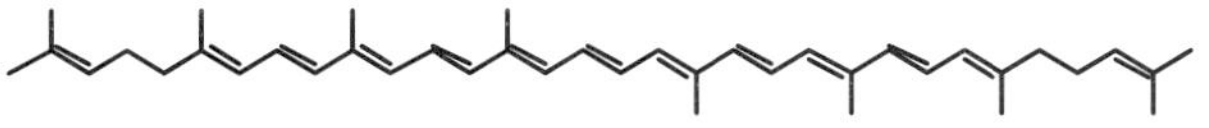

lycopene

The enzyme will also act on ζ-carotene twice to generate lycopene, on pro-ζ-carotene (producing prolycopene), and on β-zeacarotene (producing γ-carotene).

[1] M. Albrecht, H. Linden & G. Sandmann (1996) *EJB* **236**, 115.

β-CAROTENE 15,15′-DIOXYGENASE

This iron-dependent oxidoreductase [EC 1.13.11.21] catalyzes the reaction of β-carotene with dioxygen to produce two molecules of retinal.[1-4] Bile salts are essential activators.

[1] O. Hayaishi, M. Nozaki & M. T. Abbott (1975) *The Enzymes*, 3rd ed., **12**, 119.
[2] H. Singh & H. R. Cama (1974) *BBA* **370**, 49.
[3] M. R. Lakshamanan, H. Chansang & J. A. Olson (1972) *J. Lipid Res.* **13**, 477.
[4] D. S. Goodman, H. S. Huang, M. Kanai & T. Shiratori (1967) *JBC* **242**, 3543.

Selected entries from *Methods in Enzymology* [**vol**, page(s)]:
General discussion: 15, 462; **52**, 11; **189**, 425, 433; **214**, 168, 256
Assay: 15, 462; **189**, 426; homogenates of rat intestinal mucosal scraping, in, **214**, 168
Other molecular properties: effect of (chelators, **15**, 474; detergents on, **15**, 471); extract preparation, **214**, 170; mechanism, **15**, 474; pH optimum, **15**, 473; physiological implication, **15**, 474; preparation and purification from rat liver and rat intestinal mucosa, **15**, 470; properties, **15**, 471; **189**, 431; stability, **15**, 473

β-CAROTENE MONOOXYGENASE

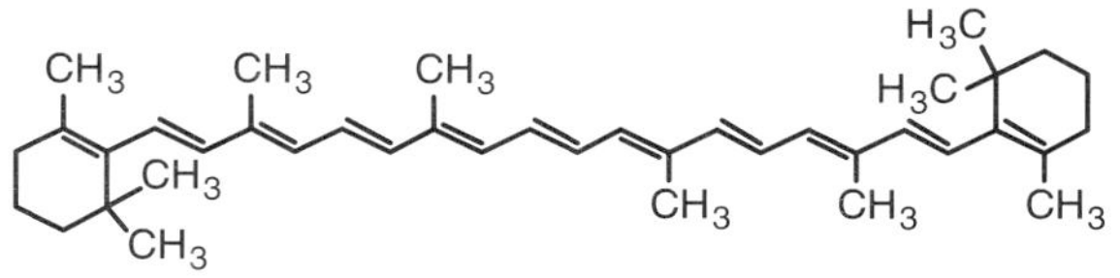

β-carotene

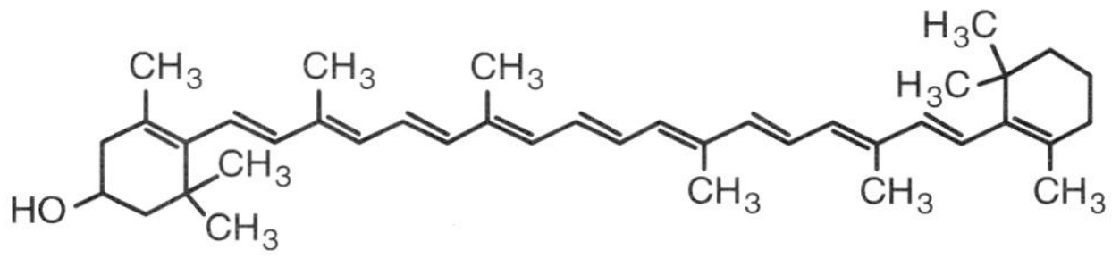

β-cryptoxanthin

This oxidoreductase catalyzes the dioxygen-dependent conversion of β-carotene to β-cryptoxanthin.[1]

[1] J. C. B. McDermott, D. J. Brown, G. Britton & T. W. Goodwin (1974) *BJ* **144**, 231.

Selected entries from *Methods in Enzymology* [**vol**, page(s)]:
General discussion: 52, 20

CAROTENE OXYGENASE

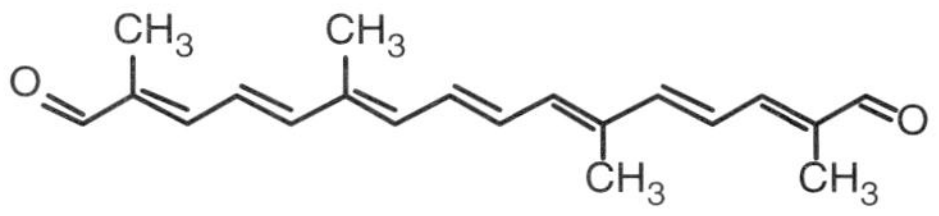

crocetindial

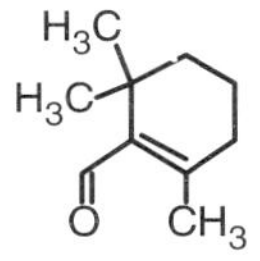

β-cyclocitral

This oxidoreductase catalyzes the reaction of β-carotene with two molecules of dioxygen to produce crocetindial (*i.e.*, 8,8′-diapocarotene-8,8′-dial) and two molecules of β-cyclocitral (*i.e.*, 2,6,6-trimethylcyclohex-1-enecarboxaldehyde).

Selected entries from *Methods in Enzymology* [**vol**, page(s)]:
General discussion: 167, 336
Assay: 167, 337

CARVEOL DEHYDROGENASE

(−)-*trans*-carveol   (−)-carvone

This oxidoreductase [EC 1.1.1.243] catalyzes the reaction of (−)-*trans*-carveol with $NADP^+$ to produce (−)-carvone and NADPH. It has a role in monoterpene biosynthesis in *Mentha spicata* (spearmint).[1,2]

[1]J. Gershenzon, M. Maffei & R. Croteau (1989) *Plant Physiol.* **89**, 1351.
[2]M. J. van der Werf, C. van der Ven, F. Barbirato, M. H. Eppink, J. A. de Bont & W. J. van Berkel (1999) *JBC* **274**, 26296.

CASBENE SYNTHASE

This enzyme [EC 4.2.3.8; previously classified as EC 4.6.1.7] catalyzes the conversion of geranylgeranyl diphosphate to casbene and pyrophosphate (or, diphosphate).[1–4]

geranylgeranyl diphosphate   casbene

Casbene is a diterpene phytoalexin with antibacterial and antifungal activity that is produced by seedlings of castor bean (*Ricinus communis* L.) in response to fungal attack. Two active-site aspartyl residues participate in the reaction.[3]

[1]R. A. Gibbs (1998) *CBC* **1**, 31.
[2]P. Moesta & C. A. West (1985) *ABB* **238**, 325.
[3]K. Huang, Q. I. Huang & A. I. Scott (1998) *ABB* **352**, 144.
[4]A. M. Hill, D. E. Cane, C. J. D. Mau & C. A. West (1996) *ABB* **336**, 283.

CASEIN KINASE

These phosphotransferases catalyze the ATP-dependent phosphorylation of the milk protein casein to produce phosphorylated casein and ADP. **See** *Protein Kinase*

CASPASE

This family of cytosolic endopeptidases catalyze hydrolysis of peptide bonds at aspartyl residues (the name is derived from cysteinyl aspartate-specific proteinase).

CASPASE-1

This cysteine endopeptidase [EC 3.4.22.36], also known as interleukin-1β converting enzyme and interleukin-1β convertase precursor, catalyzes the hydrolysis of the human prointerleukin-1β at Asp116–Ala117 to produce the bioactive cytokine. The enzyme exhibits an almost absolute specificity for an aspartyl residue at the P1 position of the substrate.[1,2] Hydrophobic aminoacyl residues are preferred at P4. The enzyme will also hydrolyze the small peptide, Ac-TyrValAlaAsp–NHMEC.

Caspase-1 can be distinguished from caspase-3 by its ability to hydrolyze prointerleukin-1β, its inhibition by the cowpox serpin CrmA ($K_i = 4$ pM), and inhibition by Ac-Tyr-Val-Ala-Asp-CHO ($K_i = 0.76$ nM).

[1]D. W. Nicholson & N. A. Thornberry (1997) *TiBS* **22**, 299.
[2]N. A. Thornberry & S. M. Molineaux (1995) *Protein Sci.* **4**, 3.

Selected entries from ***Methods in Enzymology*** [**vol**, page(s)]:
General discussion: 244, 615; **322**, 91, 100, 110, 143, 154, 162, 177
Assay: 244, 618; **322**, 91
Other molecular properties: apoptosis role, **244**, 616; **322**, 91, 100, 110, 143, 154, 162, 177; catalytic residues, **244**, 631; cellular distribution, **244**, 616; concentration dependence of activity, **244**, 621; inhibitors (CrmA, **244**, 629, 631; metal chelators, **244**, 628; peptide (acyloxy)methyl ketones, **244**, 629; peptide aldehydes, **244**, 629; peptide diazomethanes, **244**, 629); processing, **244**, 616, 620; purification, **244**, 623; sequence homology between human and mouse, **244**, 630; size, **244**, 626; specific activity of pure enzyme, **244**, 626; subcellular localization, **244**, 617; substrate, fluorigenic, **248**, 34; substrate specificity, **244**, 616; subunit structure, **244**, 616, 620

CASPASE-2

This Golgi-associated cysteine endopeptidase has a similar tetrapeptide specificity to that of caspase-3, preferring substrates with an aspartyl residue at the P1 position.[1,2] The enzyme cleaves golgin-160 at a unique site.[1]

[1]M. Mancini, C. E. Machamer, S. Roy, D. W. Nicholson, N. A. Thornberry, L. A. Casciola-Rosen & A. Rosen (2000) *J. Cell Biol.* **149**, 603.
[2]K. McCall & H. Steller (1998) *Science* **279**, 230.

CASPASE-3

This cysteine endopeptidase, initially named CPP32 (for cysteine proteinase of protein size 32 kDa), hydrolyzes poly(ADPribose) polymerase, has an almost absolute preference for substrates with an aspartyl residue at the P1 position of the substrate. Poly(ADPribose) polymerase is hydrolyzed at Asp-Glu-Val-Asp216-∇-Gly217.[1] (Note: ∇ indicates the cleavage site.) Specificity studies suggest that there is a strict preference for an aspartyl residue at the P4 position. Unlike caspase-1, this enzyme is not significantly inhibited by cowpox serpin CrmA.

[1] D. W. Nicholson, A. Ali, N. A. Thornberry, J. P. Vaillancourt, C. K. Ding, M. Gallant, Y. Gareau, P. R. Griffin, M. Labelle, Y. A. Lazebnik, N. A. Munday, A. M. Raju, M. E. Smulson, T.-T. Yamin, V. L. Yu & D. K. Miller (1995) *Nature* **376**, 37.

CASPASE-4

Caspase-4, originally called ICH-2, is a cysteine endopeptidase that catalyzes the production of interleukin-1β cytokine from its inactive precursor. Overproduction of caspase-4 in insect cells induces apoptosis.[1,2] As is true of the other caspases, the enzyme has a preference for substrates with an aspartyl residue at position P1.

[1] J. Kamens, M. Paskind, M. Hugunin, R. V. Talanian, H. Allen, D. Banach, N. Bump, M. Hackett, C. G. Johnston, P. Li, J. A. Mankovich, M. Terranova & T. Ghayur (1995) *JBC* **270**, 15250.
[2] M. Garcia-Calvo, E. P. Peterson, D. M. Rasper, J. P. Vaillancourt, R. Zamboni, D. W. Nicholson & N. A. Thornberry (1999) *Cell Death Differ.* **6**, 362.

CASPASE-5

Caspase-5, originally called ICH-3, is a cysteine endopeptidase that catalyzes the hydrolysis of peptide bonds with a preference for substrates with an aspartyl residue at position P1.[1] Its gene is regulated by lipopolysaccharides and interferon-γ.[2]

[1] M. Garcia-Calvo, E. P. Peterson, D. M. Rasper, J. P. Vaillancourt, R. Zamboni, D. W. Nicholson & N. A. Thornberry (1999) *Cell Death Differ.* **6**, 362.
[2] X. Y. Lin, M. S. K. Choi & A. G. Porter (2000) *JBC* **275**, 39920.

CASPASE-6

Caspase-6, originally called Mch2, is a cysteine endopeptidase that catalyzes the hydrolysis of peptide bonds with a preference for substrates with an aspartyl residue at position P1.[1,2]

[1] M. Garcia-Calvo, E. P. Peterson, D. M. Rasper, J. P. Vaillancourt, R. Zamboni, D. W. Nicholson & N. A. Thornberry (1999) *Cell Death Differ.* **6**, 362.
[2] H. R. Stennicke & G. S. Salvesen (1997) *JBC* **272**, 25719.

CASPASE-7

Caspase-7, originally called Mch3, is a cysteine endopeptidase that catalyzes the hydrolysis of peptide bonds with a preference for substrates with an aspartyl residue at position P1.[1,2] Nonpeptide inhibitors of caspase-7 inhibit apoptosis.[3]

[1] M. Garcia-Calvo, E. P. Peterson, D. M. Rasper, J. P. Vaillancourt, R. Zamboni, D. W. Nicholson & N. A. Thornberry (1999) *Cell Death Differ.* **6**, 362.
[2] H. R. Stennicke & G. S. Salvesen (1997) *JBC* **272**, 25719.
[3] D. Lee, S. A. Long, J. L. Adams, G. Chan, K. S. Vaidya, T. A. Francis, K. Kikly, J. D. Winkler, C. M. Sung, C. Debouck, S. Richardson, M. A. Levy, W. E. DeWolf, Jr., P. M. Keller, T. Tomaszek, M. S. Head, M. D. Ryan, R. C. Haltiwanger, P. H. Liang, C. A. Janson, P. J. McDevitt, K. Johanson, N. O. Concha & W. Chan (2000) *JBC* **275**, 16007.

CASPASE-8

This cysteine endopeptidase catalyzes the hydrolysis of peptide bonds with a preference for an aspartyl residue at position P1.[1,2] Caspase-8 also exhibits a preference for substrates with a small hydrophobic aminoacyl residue at the P4 position.

[1] M. Garcia-Calvo, E. P. Peterson, D. M. Rasper, J. P. Vaillancourt, R. Zamboni, D. W. Nicholson & N. A. Thornberry (1999) *Cell Death Differ.* **6**, 362.
[2] H. R. Stennicke & G. S. Salvesen (1997) *JBC* **272**, 25719.

CASPASE-9

This cysteine endopeptidase catalyzes the hydrolysis of peptide bonds with a preference for substrates with an aspartyl residue at position P1.[1] The proenzyme form of caspase-9 is found in the intermembrane space of mitochondria.[2]

[1] M. Garcia-Calvo, E. P. Peterson, D. M. Rasper, J. P. Vaillancourt, R. Zamboni, D. W. Nicholson & N. A. Thornberry (1999) *Cell Death Differ.* **6**, 362.
[2] P. M. Ritter, A. Marti, C. Blanc, A. Baltzer, S. Krajewski, J. C. Reed & R. Jaggi (2000) *Eur. J. Cell Biol.* **79**, 358.

CASPASE-10

This cysteine endopeptidase, originally called Mch4 and FLICE2, catalyzes the hydrolysis of peptide bonds with a preference for substrates with an aspartyl residue at position P1.[1]

[1] M. Garcia-Calvo, E. P. Peterson, D. M. Rasper, J. P. Vaillancourt, R. Zamboni, D. W. Nicholson & N. A. Thornberry (1999) *Cell Death Differ.* **6**, 362.

CASPASE-11

This cysteine endopeptidase catalyzes peptide-bond hydrolysis, preferring substrates with an aspartyl residue at position P1.[1] Because this enzyme participates in the activation of caspase-1 and caspase-3, it is believed to assist in regulating cytokine maturation[2]

[1] M. Garcia-Calvo, E. P. Peterson, D. M. Rasper, J. P. Vaillancourt, R. Zamboni, D. W. Nicholson & N. A. Thornberry (1999) *Cell Death Differ.* **6**, 362.

[2]S. J. Kang, S. Wang, H. Hara, E. P. Peterson, S. Namura, S. Amin-Hanjani, Z. Huang, A. Srinivasan, K. J. Tomaselli, N. A. Thornberry, M. A. Moskowitz & J. Yuan (2000) *J. Cell Biol.* **149**, 613.

CASPASE-12

This cysteine endopeptidase catalyzes peptide-bond hydrolysis, preferring substrates with an aspartyl residue at position P1.[1] This member of the murine caspase family is localized in the endoplasmic reticulum and is activated by endoplasmic reticulum stress, but not by membrane- or mitochondrial-targeted apoptotic signals.[2] m-Calpain has been reported to catalyze the conversion of procaspase-12 to caspase-12.[3]

[1]M. Garcia-Calvo, E. P. Peterson, D. M. Rasper, J. P. Vaillancourt, R. Zamboni, D. W. Nicholson & N. A. Thornberry (1999) *Cell Death Differ.* **6**, 362.
[2]T. Nakagawa, H. Zhu, N. Morishima, E. Li, J. Xu, B. A. Yankner & J. Yuan (2000) *Nature* **403**, 98.
[3]T. Nakagawa & J. Yuan (2000) *J. Cell Biol.* **150**, 887.

CASPASE-13

This cysteine endopeptidase, also called ERICE (for evolutionarily related interleukin-1β converting enzyme), catalyzes peptide-bond hydrolysis, preferring substrates with an aspartyl residue at position P1. Caspase-13 can be activated by caspase-8: hence, it is postulated that this caspase participates in the receptor-initiated death pathway.[1]

[1]E. W. Humke, J. Ni & V. M. Dixit (1998) *JBC* **273**, 15702.

CASPASE-14

This cysteine endopeptidase, also called MICE, catalyzes peptide-bond hydrolysis, preferring substrates with an aspartyl residue at position P1. The enzyme plays a role in death receptor and granzyme B-induced apoptosis.[1,2] Caspase-14 is regulated during keratinocyte differentiation.[3]

[1]M. Ahmad, S. M. Srinivasula, R. Hegde, R. Mukattash, T. Fernandes-Alnemri & E. S. Alnemri (1998) *Cancer Res.* **58**, 5201.
[2]M. Van de Craen, G. Van Loo, S. Pype, W. Van Criekinge, I. Van den Brande, F. Molemans, W. Fiers, W. Declercq & P. Vandenabeele (1998) *Cell Death Differ.* **5**, 838.
[3]L. Eckhart, J. Ban, H. Fischer & E. Tschachler (2000) *BBRC* **277**, 655.

CATALASE

This oxidoreductase [EC 1.11.1.6] catalyzes the conversion of two molecules of hydrogen peroxide to dioxygen and two water molecules.[1–6] Most catalases use heme or manganese as cofactors. Several organic substances, such as ethanol, can act as the hydrogen donor. Heme-dependent catalases form an $Fe^{IV}=O$ porphyrin π-cation radical as an intermediate.

The protein containing two Mn(III) ions in the resting state is often called pseudocatalase. These heme-independent catalases cycle between Mn(II)/Mn(II) and Mn(III)/Mn(III) oxidation states. An oxo-bridge between the two manganese(III) ions is proposed to form in the reaction mechanism and promote the binding of the second molecule of hydrogen peroxide.

In several organisms, the enzyme also acts as a peroxidase [EC 1.11.1.7] for which several organic substances, especially ethanol, can act as a hydrogen donor.

[1]D. W. Yoder, J. Hwang & J. E. Penner-Hahn (2000) *Met. Ions Biol. Syst.* **37**, 527.
[2]M. Zamocky & F. Koller (1999) *Prog. Biophys. Mol. Biol.* **72**, 19.
[3]H. B. Dunford (1998) *CBC* **3**, 195.
[4]J. E. Penner-Hahn (1998) *CBC* **3**, 439.
[5]D. C. Rees & D. Farrelly (1990) *The Enzymes*, 3rd ed., **19**, 37.
[6]G. R. Schonbaum & B. Chance (1976) *The Enzymes*, 3rd ed., **13**, 363.

Selected entries from *Methods in Enzymology* [vol, page(s)]:
General discussion: 2, 764, 775, 781, 784, 789, 791; 18A, 59; 27, 609; 31, 40; 77, 16; 105, 121; 188, 463
Assay: 2, 764, 769, 781, 785, 789, 791; 4, 293; 18A, 60; 31, 361, 743; 74, 344; 105, 122, 126; 137, 601; 188, 463; luminol, with, 57, 404; perborate method, 44, 685; 74, 344; soluble, assay, 44, 479, 482
Other molecular properties: activation energy, 63, 241; activity, 233, 523, 602; activity studies with salicylate hydroxylase, 53, 539; affinity for calcium phosphate gel, 51, 584; aggregation procedure, in, 44, 268; alanine racemase assay and, 2, 212, 213; alcohol-chloroform treatment of, 2, 782; amino acid oxidases and, 2, 199, 200, 209; 56, 484; amino groups, assay, 137, 603; antibody production (immune complexes [active, 74, 344; inactive, 74, 345]; immunogen preparation, 74, 344); assay of (hydrogen peroxide, 77, 16; phenylalanine hydroxylase, 53, 279; superoxide dismutase, 53, 385); associated with cell fractions, 40, 86; atebrin fluorescence quenching, 56, 114; avidin labeling with, 137, 115; blood, 2, 781; catalase fluoride, EPR studies, 76, 321; catalytic reaction, 44, 478, 479; circular dichroism studies, 54, 283; C_1 metabolism, 77, 18; cobalt-nucleotide complexes, 87, 160; coelomic cell homogenization buffer, in, 57, 376; commercial source, 44, 481; conformational composition, 54, 269, 271; conjugate, activity, 44, 156, 200, 271, 272, 482, 483; Cotton effect and, 6, 941; coupled enzyme, use in, 63, 32, 33; crosslinked (characterization, 137, 609; gel electrophoresis, 137, 603; immobilization, 137, 604; preparation, 137, 602; thermostability, 137, 613); cryoenzymology, 63, 338; cyanide and, 4, 277; delivery systems, polymeric and copolymeric (kinetic analysis, 112, 414; preparation, 112, 400); detection of, in (drug-induced H_2O_2 formation, 105, 508; soluble immune response suppressor, 116, 397); diamine oxidase assay and, 2, 394; diffusional control, 63, 253; dopamine β-hydroxylase, and, 87, 717, 720, 723; effects of azide, nitrite, and nitrate, 269, 319; effects on (endotoxin-induced circulatory shock, 186, 661; endotoxin-induced disseminated intravascular coagulation, 186, 657); electron microscopic tracer, as, 39, 147; entrapment with photo-crosslinkable resin prepolymers, 135, 234; enzyme thermistor, in, 44, 674, 675; enzymic activity in immune complex, 74, 343 (absence, 74, 349); EPR studies, 76, 321; erythrocyte, 2, 781; formaldehyde generation, 52, 297, 344; formation of, 1, 136; galactosidase assay and, 1, 242; gel electrophoresis, 32, 101; gene encoding, analysis, 236, 201; glucose aerodehydrogenase and, 1, 341; glucose electrode, 56, 482, 483; glucose oxidase assay, in, 9, 82; 44, 348, 349; glucosone effects,

137, 606; glycine oxidase assay and, **2**, 225; **3**, 108; glyoxysomes, in, **31**, 565, 569; **72**, 788; heme-containing protein, as, in chemiluminescent oxidation system, **57**, 445; heme-deficient mutants, **56**, 559; Hill reaction and, **4**, 354; hydrogen peroxide assay, in, **52**, 344; immobilization on collagen, **44**, 153, 255; immobilized, effect of inhibitor, **44**, 916, 917; immunoassay, **74**, 346; inactivation-reactivation studies on, **22**, 505, 506, 531; induction in *Rhodopseudomonas sphaeroides*, **22**, 90; inhibition by ascorbate, **18A**, 59; inhibitor (Dextran 500, **55**, 131; phagocyte chemiluminescence, of, **57**, 492); interaction parameters, preferential, **61**, 48; interference (aromatic hydroxylation assays, in, **52**, 412; hydrogen peroxide assay, in, **52**, 344, 349, 350); iodination interferent, as, **32**, 108, 109; isoelectric point, **228**, 227; isolated nuclei and, **12A**, 445; isolation (foaming, by, **22**, 526; large-scale, **22**, 508); α-keto acid preparation and, **3**, 406; kidney cells, in, **77**, 142; kinetic parameters, **63**, 240; kinetic properties *in vitro*, **105**, 121; kinetics, **44**, 483, 686; liver, **2**, 775, 791; liver-cell fractions, in, **31**, 209; localization in *Hansenula polymorpha*, **188**, 413; luciferase-mediated oxygen assay, **57**, 225; marker enzyme, as, **23**, 680; **31**, 20, 46, 496, 735, 741; **32**, 15, 29, 35; microbodies, in, **52**, 495; *Micrococcus lysodeikticus*, **2**, 784; molecular weight, **53**, 350, 351; mutarotase assay and, **5**, 220; NADP$^+$ regenerating system, **13**, 593; nicotinic acid receptors, **32**, 320; nucleoli and, **12A**, 457; oxidation state of Fe, **76**, 315; oxygen-binding in blood, **76**, 429, 512; oxygen consumption studies, in, **54**, 487, 492, 493; oxygen determination, in, **54**, 501; oxygen electrode, and, sulfhydryl oxidase assay with, **143**, 120; oxygen scavenging system, in, **52**, 223; partial specific volumes, **61**, 48; peptide, Edman degradation, **25**, 312, 364; peroxisomes, in, **10**, 13; **31**, 361, 364, 367 (peroxisomal marker, assay, **182**, 219); plant, **2**, 789 (microbodies, in, **31**, 489); polarographic measurement, **56**, 459; preparation, **18A**, 60; prepolymer-entrapped, applications, **135**, 245; preservation, **22**, 530; properties, **2**, 784, 788, 790; **44**, 479; **188**, 466; protection of hemoglobin from peroxides, **76**, 156; protective effects in ischemia-reperfusion, **233**, 603; *Pseudomonas fluorescens*, **2**, 764; purification from (beef liver, **2**, 776, 792; *Candida boidinii*, **188**, 465; horse blood, **2**, 782; *Micrococcus lysodeikticus*, **2**, 786; pig blood, **2**, 782; spinach leaves, **2**, 789); radioiodination, **70**, 212; reactions, **77**, 18; recombination of hemoglobin chains, in, **76**, 117, 118; release of O$_2$, in measurement of H$_2$O$_2$, **105**, 397; removal from liver microsomal components, **52**, 359, 364; resonance Raman studies, **226**, 362; ribosomal subunits, preparation of cross-linked, **59**, 537; role in ethanol metabolism, **52**, 355; role in myocardial preservation during heart transplantation, **186**, 742; SDS complex, properties, **48**, 6; soluble, assay, **44**, 479, 482; sources, **2**, 209, 765, 775; specific activity, calculation of, **2**, 767, 785; specific gravity, **44**, 255; spin diameter, **44**, 160; spinach, **2**, 789; *Spirillum lipoferum* groups, **69**, 738; staining on gels, **22**, 594; steroid hydroxylase and, **5**, 512; stimulation of desaturase reaction, **71**, 280; stopped-flow studies, **4**, 306; storage in solution, **44**, 212, 213; substrates, **77**, 18; synthesis, free and membrane-bound polysomes and, **30**, 326; temperature-jump studies, **76**, 691; test, **52**, 361; therapeutic application, **44**, 679, 681, 684; thermodenaturation, **137**, 605; thymol-free, lipid peroxidation, **52**, 304; tryptophan peroxidase assay and, **2**, 244, 246; unit cell dimensions, **44**, 159; viscosity barrier centrifugation, **55**, 133, 134, 135; visual spectroscopy, **4**, 311; xanthine oxidase assay and, **2**, 483; Zwischenferment and, **2**, 712, 718

CATECHOL 1,2-DIOXYGENASE

This iron-dependent oxidoreductase [EC 1.13.11.1], also known as pyrocatechase, catalyzes the reaction of catechol with dioxygen to produce *cis,cis*-muconate.[1–8]

catechol   *cis,cis*-muconate

Among its physiological roles, catechol 1,2-dioxygenase participates in the metabolism of nitro-aromatic compounds. Other substrates include 4-methylcatechol and 3-methylcatechol.

During catalysis, catechol binds to the enzyme such that the two oxygen atoms displace two ligands from the pentacoordinated Fe(II). Dioxygen binds to the iron and the electron rearrangement effectively generates a Fe(III)-superoxide intermediate. Attack of the distal oxygen on the ring structure results in the transient formation of a semiquinone-like structure. Insertion of an oxygen into the ring followed by ring cleavage generates the product. A hydroperoxy-intermediate forms in the course of the reaction. Structure:activity studies with a number of 4- and 5-substituted catechols indicate that the rate-limiting step of the reaction is dependent on the nucleophilic reactivity of the substrate.[3] ***See also*** *Catechol 2,3-Dioxygenase; Chlorocatechol 1,2-Dioxygenase*

[1] B. G. Fox (1998) *CBC* **3**, 261.
[2] J. V. Schloss & M. S. Hixon (1998) *CBC* **2**, 43.
[3] L. Ridder, F. Briganti, M. G. Boersma, S. Boeren, E. H. Vis, A. Scozzafava, C. Veeger & I. M. Rietjens (1998) *EJB* **257**, 92.
[4] L. Que & R. Y. N. Ho (1996) *Chem. Rev.* **96**, 2607.
[5] S. Yamamoto & Y. Ishimura (1991) *A Study of Enzymes* **2**, 315.
[6] T. A. Walsh, D. P. Ballou, R. Mayer & L. Que, Jr., (1983) *JBC* **258**, 14422.
[7] O. Hayaishi, M. Nozaki & M. T. Abbott (1975) *The Enzymes*, 3rd ed., **12**, 119.
[8] O. Hayaishi (1963) *The Enzymes*, 2nd. ed., **8**, 353.

Selected entries from ***Methods in Enzymology*** [vol, page(s)]:
General discussion: 2, 273, 281; 17A, 518; 52, 6, 11, 38; 188, 122
Assay: 2, 282; 188, 123; 204, 508
Other molecular properties: adaptive cells, presence in, 1, 134; electronic absorption spectroscopy, 226, 48; ferric ion in, 17A, 522; purification from (*Acinetobacter calcoaceticus*, 188, 124; *Pseudomonas*, 2, 282; 17A, 519); tyrosinase and, 2, 817, 819; yield, storage, stability, and properties, 188, 126

CATECHOL 2,3-DIOXYGENASE

This iron-dependent oxidoreductase [EC 1.13.11.2], also called metapyrocatechase, catalyzes the reaction of catechol with dioxygen to produce 2-hydroxymuconate semialdehyde.[1–8]

catechol   2-hydroxy-*cis,cis*-muconate
                    semialdehyde

The enzyme has an ordered Bi Uni kinetic mechanism.[7] The dioxygenase from *Alcaligenes* sp. strain O-1 reportedly catalyzes the reaction of 3-sulfocatechol with dioxygen and water to produce (2*E*,4*Z*)-2-hydroxymuconate and bisulfite. **See also** *Catechol 1,2-Dioxygenase*

[1] B. G. Fox (1998) *CBC* **3**, 261.
[2] T. D. Bugg, J. Sanvoisin & E. L. Spence (1997) *Biochem. Soc. Trans.* **25**, 81.
[3] L. Shu, Y. M. Chiou, A. M. Orville, M. A. Miller, J. D. Lipscomb & L. Que, Jr., (1995) *Biochemistry* **34**, 6649.
[4] S. Yamamoto & Y. Ishimura (1991) *A Study of Enzymes* **2**, 315.
[5] M. A. Ator & P. R. Ortez de Montellano (1990) *The Enzymes*, 3rd ed., **19**, 213.
[6] O. Hayaishi, M. Nozaki & M. T. Abbott (1975) *The Enzymes*, 3rd ed., **12**, 119.
[7] K. Hori, T. Hashimoto & M. Nozaki (1973) *J. Biochem.* **74**, 375.
[8] D. T. Gibson (1971) *Meth. Microbiol.* **6A**, 463.

Selected entries from **Methods in Enzymology** [**vol**, page(s)]:
General discussion: 17A, 522; 52, 11, 19; **188**, 115
Assay: 188, 116
Other molecular properties: amino acid composition, **188**, 119; coordination geometry of iron, **246**, 96; *d* orbital energy level diagram, **246**, 108; electronic absorption spectroscopy, **226**, 48; induction, **188**, 120; inhibition by azide, **246**, 106; iron content, **17A**, 525; magnetic circular dichroism, **246**, 98, 105; physiological role, **188**, 120; properties, **188**, 118; purification from (*Pseudomonas*, **17A**, 523; *Pseudomonas aeruginosa*, **188**, 116); reaction catalyzed by, **246**, 104; saturation magnetization curve, **246**, 98, 107

CATECHOL *O*-METHYLTRANSFERASE

This methyltransferase [EC 2.1.1.6], which exists in both soluble and membrane-bound forms, catalyzes the reaction of catechol with *S*-adenosyl-L-methionine to produce *S*-adenosyl-L-homocysteine and guaiacol (*i.e.*, *o*-methoxyphenol).[1–5]

catechol   guaiacol

Actually, catecholamines are better substrates: for example, epinephrine and norepinephrine. The enzyme has an ordered Bi Bi kinetic mechanism and isotope-effect investigations clearly demonstrate that the methyl transfer event is rate limiting.[2,5] In addition, the transition state is compressed structurally compared to the transition state of the uncatalyzed reaction.

The enzyme is not strictly regiospecific. With a number of catechol substrate analogues, methylation occurs at the 3- and 4-hydroxyl groups, albeit more rapidly at the 3-position. For example, with dopamine as the acceptor substrate, both the 3- and the 4-methoxy products are generated.[1]

[1] F. Takusagawa, M. Fujioka, A. Spies & R. L. Schowen (1998) *CBC* **1**, 1.

[2] T. Lotta, J. Vidgren, C. Tilgmann, I. Ulmanen, K. Melén, I. Julkunen & J. Taskinen (1995) *Biochemistry* **34**, 4202.
[3] J. Vidgren, L. A. Svenson & A. Liljas (1994) *Nature* **368**, 354.
[4] A. J. Rivett & J. A. Roth (1982) *Biochemistry* **21**, 1740.
[5] J. K. Coward, E. P. Slisz & F. Y.-H. Wu (1973) *Biochemistry* **12**, 2291.

Selected entries from **Methods in Enzymology** [**vol**, page(s)]:
General discussion: 5, 748; 15, 721, 722; 34, 700; 46, 554; 74, 469, 470; 77, 267
Assay: 5, 748; 32, 777; 63, 32; 77, 267, 268; 2-hydroxy-17β-estradiol, with, 15, 721
Other molecular properties: affinity chromatography, **77**, 270, 271; affinity labeling, **46**, 559; catecholamine radioimmunoassay with, **103**, 486; **142**, 559; inhibition, **5**, 751; **15**, 723; **77**, 272; isolation, **46**, 559; isotope effects, **87**, 587, 639; molecular weight, **77**, 272; neurobiology, in, **32**, 765; pineal, in, **39**, 396; properties, **5**, 750; **15**, 722; **77**, 272; purification, **77**, 268 (rat liver, from, **5**, 750; **15**, 722; **103**, 485; **142**, 554); size, **77**, 272; solvent isotope effects, **87**, 587; specificity, **5**, 751; **15**, 723; **77**, 272; stability, **15**, 723; **77**, 271, 272; stereochemistry, **87**, 149

CATECHOL OXIDASE

These copper ion-dependent oxidoreductases [EC 1.10.3.1] (also known as diphenol oxidases, *O*-diphenolase, phenolases, polyphenol oxidases, or tyrosinases) catalyze the reaction of two molecules of catechol with dioxygen to produce two molecules of 1,2-benzoquinone and two molecules of water.[1–7]

catechol   1,2-benzoquinone

A variety of substituted catechols can act as substrates: other substrates include dopamine, dopa, gallate, 4-methylcatechol, and epinephrine. The quinone products are highly reactive and tend to polymerize, resulting in the brown discoloration of fruits and vegetables during storage.

The proposed reaction mechanism proceeds by binding of the *o*-diphenol substrate to the copper site, with formation of the cuprous form of the enzyme and release of 1,2-benzoquinone. The second molecule of catechol binds with dioxygen, and the second molecule of 1,2-benzoquinone is produced, with regeneration of the dicupric form of the enzyme.[1] **See also** *Monophenol Monooxygenase; Tyrosine Monooxygenase; Laccase*

[1] C. Eicken, B. Krebs & J. C. Sacchettini (1999) *Curr. Opin. Struct. Biol.* **9**, 677.
[2] A. M. Mayer (1987) *Phytochemistry* **26**, 11.
[3] T. Yoshimoto, K. Yamamoto & D. Tsuru (1985) *J. Biochem.* **97**, 1747.
[4] E. O. Anosike & A. O. Ayaebene (1981) *Phytochemistry* **20**, 2625.
[5] A. Mayer & E. Harel (1979) *Phytochemistry* **18**, 193.
[6] T. C. Wong, B. S. Luh & J. R. Whitaker (1971) *Plant Physiol.* **48**, 19.
[7] S. Motoda (1979) *J. Ferment. Technol.* **57**, 71 and 79.

Selected entries from *Methods in Enzymology* [**vol**, page(s)]:
General discussion: 17A, 615; **52**, 10, 19
Other molecular properties: dopa-containing protein, and, cosecretion, **258**, 1; inhibitors, **31**, 539, 540; luminescence, **226**, 535; plant tissues, from, **13**, 559 (extraction in aqueous two-phase systems, **228**, 666); plastocyanin preparation, **69**, 226; thiooxidase, **52**, 10; tyrosinase and, **2**, 817, 819

CATECHOL OXIDASE (DIMERIZING)

This oxidoreductase [EC 1.1.3.14] catalyzes the reaction of four molecules of catechol with three molecules of dioxygen to produce two molecules of dibenzo[1,4]dioxin-2,3-dione and six molecules of hydrogen peroxide.[1,2]

[1] S. Uchiyama, M. Tamata, Y. Tofuku & S. Suzuki (1988) *Anal. Chim. Acta* **208**, 287.
[2] P. M. Nair & L. C. Vining (1964) *ABB* **106**, 422.

CATHEPSIN B

This lysosomal cysteine peptidase [EC 3.4.22.1], also known as cathepsin B_1, catalyzes peptide-bond hydrolysis with a broad specificity.[1-6] Cathepsin B appears to prefer large, hydrophobic aminoacyl residues at the S2 subsite. In addition, cathepsin B will also accept arginyl residues at this subsite (due to the presence of Glu245).[1] To assist in distinguishing this particular cathepsin from other lysosomal cathepsins, small synthetic substrates such as Z-Arg-Arg-∇-NHPhNO$_2$ or Z-Arg-Arg-∇-Mec have proved very useful. (Note: ∇ indicates the cleavage site.) Cathepsin B has both endopeptidase and exopeptidase activities. Under acidic conditions, the exopeptidase activity is considerably greater of the two.

[1] S. Hasnain, T. Hirama, C. P. Huber, P. Mason & J. S. Mort (1993) *JBC* **268**, 235.
[2] I. M. Berquin & B. F. Sloane (1996) *Adv. Exp. Med. Biol.* **389**, 281.
[3] J. S. Mort & D. J. Buttle (1997) *Int. J. Biochem. Cell Biol.* **29**, 715.
[4] K. Brocklehurst, F. Willenbrock & E. Salih (1988) *New Compr. Biochem.* **16**, 39.
[5] J. S. Fruton (1982) *AE* **53**, 239.
[6] L. M. Greenbaum (1971) *The Enzymes*, 3rd. ed., **3**, 475.

Selected entries from *Methods in Enzymology* [**vol**, page(s)]:
General discussion: **19**, 285; **46**, 206; **80**, 535, 820; **244**, 500, 512
Assay: **19**, 286; **80**, 56; amidase assay, **19**, 288; esterase assay, **19**, 289; **80**, 825; fluorimetric, **248**, 20; hydrolase assay, **19**, 291; proteinase assay, **19**, 286; proteoglycan degradation method, **248**, 52; transferase assay, **19**, 290; Z-Arg-Arg-NNap, with, **80**, 552, 553
Other molecular properties: action on oxidized B chain of insulin, **80**, 553, 555; activators and inhibitors, **19**, 300; active-site titration, **80**, 543, 552; activity, staining for, **80**, 544; affinity chromatography, **244**, 642, 645; affinity labeling, **244**, 655; bovine, inhibition by cystatin, **80**, 772, 777; comparison to papain, **19**, 300; 2,2'-dipyridyl disulfide, and, **87**, 461; distribution, **19**, 301; gene locus, **244**, 511; human liver (amino acid composition, **80**, 552; purification, **80**, 546, 547); inactivation, **80**, 823, 824 (aldolase, of, **80**, 553, 554; α_1-proteinase inhibitor, of, **80**, 763); inhibited by (antipain, **45**, 683; leupeptin, **45**, 679); inhibition assay, **251**, 394; inhibitors, **80**, 538, 558; **244**, 510; kinetic properties and constants,

19, 301; **80**, 552, 553; molecular weight, **80**, 551; peptidyldipeptidase activity, **80**, 553, 554; pH optima, **80**, 552; pig, amino acid composition, **80**, 552; processing, **244**, 506; properties, **80**, 536, 551; purification, **19**, 297; **244**, 505, 641; purity and physical properties, **19**, 299; rat liver (amino acid composition, **80**, 552; purification, **80**, 549); specificity, **19**, 285, 300; **244**, 500, 682; splice variants, **244**, 511; stability, **19**, 299; storage, **80**, 551; tertiary structure, **244**, 510; trapped by α_2-macroglobulin, **80**, 748

CATHEPSIN D

This lysosomal aspartic endopeptidase [EC 3.4.23.5], often assayed with hemoglobin as the substrate and similar in action to pepsin A, catalyzes peptide-bond hydrolysis, with preference for substrates with hydrophobic residues in the P1 and P1′ positions.[1-5] Cathepsin D is greatly inhibited by pepstatin ($K_i \approx 4$ pM).

[1] P. E. Scarborough, G. R. Richo, J. Kay, G. E. Conner & B. M. Dunn (1991) *Adv. Exp. Med. Biol.* **306**, 343.
[2] P. E. Scarborough, K. Guruprasad, C. Topham, G. R. Richo, G. E. Conner, T. L. Blundell & B. M. Dunn (1993) *Protein Sci.* **2**, 264.
[3] P. E. Scarborough & B. M. Dunn (1994) *Protein Eng.* **7**, 495.
[4] A. J. Barrett (1977) in *Proteinases in Mammalian Cells and Tissues* (A. J. Barrett, ed.), North-Holland Biomed. Press, Amsterdam, p. 209.
[5] M. Fusek & V. Vetvicka (1995) *Aspartic Proteinases: Physiology and Pathology*, CRC Press, Boca Raton.

Selected entries from *Methods in Enzymology* [**vol**, page(s)]:
General discussion: **19**, 285, 309; **80**, 535, 565
Assay: **80**, 566; hydrolase assay, **19**, 291; lysosomal, **96**, 767; proteinase assay, **19**, 286
Other molecular properties: activators and inhibitors, **19**, 31; active-site titration with tight-binding inhibitors, **248**, 96; bovine spleen (amino acid composition, **80**, 577; isoelectric point, **80**, 575; light chain, structure, **80**, 578; molecular weight, **80**, 575; properties, **80**, 575, 581; purification, **80**, 579; structure, **80**, 577, 578, 581); brain homogenates, in, **31**, 464, 466, 467, 469; distribution, **19**, 312; disulfide loops, **248**, 110; effect on purified NADH-cytochrome b_5 reductase, **52**, 108; hydrolysis of peptide substrates, **248**, 130; immunostaining, **236**, 150; inhibition assay, **251**, 394; kinetic properties, **19**, 312; light chains, homology to gastric and microbial proteases, **80**, 578; lysosomal, **10**, 12 (antibody preparation, **96**, 770; assay, **96**, 767; immunoprecipitation, **96**, 773; purification, **96**, 767; synthesis *in vitro*, **96**, 770); lysosome marker, as, **31**, 406, 408; macrophage fractions, in, **31**, 340, 344, 355; porcine spleen (amino acid composition, **80**, 577; enzymatic properties, **80**, 572, 573; heavy and light chains [separation, **80**, 573, 574; structure, **80**, 575, 576, 578]; high-molecular-weight component, isolation, **80**, 571; isoelectric point, **80**, 575; molecular weight, **80**, 575; properties, **80**, 575; purification, **80**, 568; structure, **80**, 574); procathepsin D, maltose-binding protein fusion, **326**, 314, 317; purification, **19**, 293, 309; purity and physical properties, **19**, 311; specificity, **19**, 312; stability, **19**, 311; structure, **248**, 108, 110; trapped by α_2-macroglobulin, **80**, 748; two-chain molecule, **248**, 111

CATHEPSIN E

This aspartic proteinase [EC 3.4.23.34], also known as slow-moving proteinase and erythrocyte membrane aspartic proteinase, catalyzes peptide-bond hydrolysis with a broad specificity. The most commonly used substrate is hemoglobin. Although the specificity range is broad for the

P1 and P1′ positions, cathepsin E prefers substrates with hydrophobic residues in both positions.[1,2] The P1′ position can contain branched aminoacyl residues but not the P1 position. Cathepsin E hydrolyzes the oxidized insulin B chain at Glu13–Ala14, Ala14–Leu15, Glu21–Arg22, Phe24–Phe25, and Phe25–Tyr26.

[1] C. Rao-Naik, K. Guruprasad, B. Batley, S. Rapundalo, J. Hill, T. L. Blundell, J. Kay & B. M. Dunn (1995) *Proteins: Struct. Funct. Genet.* **22**, 168.
[2] M. Fusek & V. Vetvicka (1995) *Aspartic Proteinases: Physiology and Pathology*, CRC Press, Boca Raton, p. 207.

Selected entries from *Methods in Enzymology* [**vol**, page(s)]:
General discussion: 19, 285, 313; 248, 120
Assay: hemoglobin-digestion method, 248, 122; hydrolase assay, 19, 291; proteinase assay, 19, 286; substance P hydrolysis method, 248, 124
Other molecular properties: activators and inhibitors, 19, 314; 248, 132; dimeric form, 248, 110, 128 (monomeric form, and, interconversion, 248, 128); discovery, 248, 121; distribution, 19, 315; hydrolysis of peptide substrates, 248, 129 (cleavage site specificity, 248, 130; kinetics, 248, 130; pH effects, 248, 129); hydrolysis of protein substrates, 248, 129 (cleavage site specificity, 248, 130; pH optimum, 248, 129); inhibitors, 248, 132; isolation, 248, 127; molecular mass, 248, 128; monomeric form, 248, 110; precursor, 248, 120 (purification, 248, 125; structure, 248, 133); properties, 19, 314; 248, 120, 128; purification, 19, 313; purity, 19, 314; 248, 128; specificity, 19, 314; stability, 19, 314; 248, 128; structure, 248, 108, 110, 121; synthesis, 248, 121; tissue distribution, 248, 135; units of activity, 248, 123

CATHEPSIN F

This cysteine endopeptidase [EC 3.4.22.41], a lysosomal enzyme, catalyzes peptide-bond hydrolysis in artificial substrates containing phenylalanyl or leucyl residues (better than valyl) in the P2 position.[1-3] Its specificity is similar to that of cathepsin L. Another cathepsin F, now known as cathepsin F (old), had been reported earlier and is suggested to degrade proteoglycans.[4]

[1] I. Santamaría, G. Velasco, A. M. Pendás, A. Paz & C. López-Otín (1999) *JBC* **274**, 13800.
[2] D. K. Nägler, T. Sulea & R. Ménard (1999) *BBRC* **257**, 313.
[3] T. Wex, B. Levy, H. Wex & D. Brömme (1999) *BBRC* **259**, 401.
[4] J. T. Dingle, A. J. Barrett, A. M. J. Blow & P. E. N. Martin (1977) *BJ* **167**, 775.

CATHEPSIN G

This serine endopeptidase [EC 3.4.21.20] is a single-chain glycoprotein having a pI value of about 12 that catalyzes the hydrolysis of peptide bonds with a specificity similar to chymotrypsin C. The best artificial substrates contain Phe at position P1.[1,2] Interestingly, Glu226 is found at the P1 site suggesting that basic residues might also be accommodated at P1.[3] This was later confirmed with lysyl-containing substrates, making cathepsin G rather unusual in the chymotrypsin family. However, arginyl-containing substrates are poor substrates.

[1] P. M. Starkey & A. J. Barrett (1976) *BJ* **155**, 273.
[2] G. C. Szabó, J. Tõzsér, L. Aurell & P. Elõdi (1986) *Acta Biochim. Biophys. Hung.* **21**, 349.
[3] G. Salvesen, D. Farley, J. Shuman, A. Pryzbyla, C. Reilly & J. Travis (1987) *Biochemistry* **26**, 2289.

Selected entries from *Methods in Enzymology* [**vol**, page(s)]:
General discussion: 46, 201; 80, 561, 765
Assay: 80, 562, 563; 163, 324
Other molecular properties: active-site titration, 80, 563, 564; 248, 87, 100; activity, staining, 80, 564; amino-terminal sequence, 80, 595, 602, 603; distribution, 80, 565; human, source, 248, 7; inactivation, 80, 771; inhibition, 46, 206; 80, 565, 763, 801 (eglin, by, 80, 804, 812); molecular weight, 80, 564; N-terminal sequence, homology with chymotrypsin, 80, 564; properties, 80, 564, 565; 163, 314; purification, 80, 564 (human neutrophils, from, 163, 313); trapped by α₂-macroglobulin, 80, 748

CATHEPSIN H

This mammalian endopeptidase [EC 3.4.22.16] is the only cysteine peptidase from the lysosome also possessing a strong aminopeptidase activity that is not inhibited by puromycin or bestatin (potent inhibitors of cytosolic aminopeptidases).[1-4] The enzyme (also known as aleurain, cathepsin B₃, cathepsin Ba, and benzoylarginine:naphthylamide hydrolase) catalyzes the hydrolysis of peptide bonds. Unblocked synthetic substrates such as Arg-∇-NHMec are readily hydrolyzed. However, care should always be exercised since several noncysteine peptidases will also hydrolyze these substrates. Artificial endopeptidase substrates that have been used include Bz-Arg-∇-NHNap, Bz-Arg-∇-NHMec, Bz-Phe-Val-Arg-∇-NHMec, and Pro-Gly-∇-Phe.[1-4] (Note: ∇ indicates the cleavage site.) Collagen and laminin are not degraded by cathepsin H.

[1] A. J. Barrett & H. Kirschke (1981) *Meth. Enzymol.* **80**, 535.
[2] H. Kirschke & A. J. Barrett (1987) in *Lysosomes: Their Role in Protein Breakdown* (H. Glaumann & F. J. Ballard, eds.) Academic Press, London, p. 193.
[3] M. Roth & J. Dodt (1992) *EJB* **210**, 759.
[4] X. Q. Xin, B. Gunesekera & R. W. Mason (1992) *ABB* **299**, 334.

Selected entries from *Methods in Enzymology* [**vol**, page(s)]:
General discussion: 80, 535
Assay: 80, 539, 540
Other molecular properties: activity, staining, 80, 544; affinity chromatography, 244, 642, 646; aminopeptidase activity, 80, 554; human (kinetic constants, 80, 556; liver [isoelectric forms, 80, 554; purification, 80, 546]; inhibition assay, 251, 394; inhibitors, 80, 558; molecular weight, 80, 554; processing, 244, 506; properties, 80, 536, 554; proteolytic activity, 80, 554; rabbit lung, isoelectric forms, 80, 554; rat liver (inactivation, 80, 824; isoelectric point, 80, 554; purification, 80, 549; stability, 80, 554); storage, 80, 551; substrates, 80, 554, 556; substrate specificity, 244, 500; trapped by α₂-macroglobulin, 80, 748

CATHEPSIN K

This cysteine endopeptidase [EC 3.4.22.38], also known as cathepsin O, cathepsin X, and cathepsin O2, catalyzes the hydrolysis of peptide bonds with a broad specificity. The enzyme is abundant in osteoclasts and is believed to play a role in bone resorption. The S2 subsite exhibits a preference for hydrophobic, albeit not aromatic, aminoacyl residues (for example, leucyl, norleucyl, or methionyl).[1,2] In contrast to cathepsin B, arginyl residues are poor substituents at P2.

[1] D. Brömme, J. L. Klaus, K. Okamoto, D. Rasnick & J. T. Palmer (1996) *BJ* **315**, 85.
[2] D. Brömme, K. Okamoto, B. B. Wang & S. Biroc (1996) *JBC* **271**, 2126.

CATHEPSIN L

This cysteine endopeptidase [EC 3.4.22.15] is a lysosomal endopeptidase with a specificity similar to that of papain. Cathepsin L, which lacks exopeptidase activity, will not act on Z-Arg-Arg-NHMec.[1–3] The enzyme acts on protein substrates with a higher activity than cathepsin B, preferring substrates with hydrophobic aminoacyl residues in positions P2 and P3. The synthetic substrate Z-Phe-Arg-∇-NHMec has frequently been used to assay the enzyme. (Note: ∇ indicates the cleavage site.) However, this substrate is also hydrolyzed by cathepsins B and S; hence, use of cathepsin L-specific inhibitors such as Z-Phe-Phe-CHN$_2$ or Z-Phe-Tyr(tBu)-CHN$_2$ are needed to kinetically distinguish cathepsin L from other family members. The use of monoclonal antibodies has also helped to distinguish cathepsin L.

[1] C. Le Boulay, A. Van Wormhoudt & D. Sellos (1995) *Comp. Biochem. Physiol. [B]* **111B**, 353.
[2] M. V. Laycock, R. M. MacKay, M. Di Fruscio & J. W. Gallant (1991) *FEBS Lett.* **292**, 115.
[3] S. J. Hawthorne, M. Pagano, D. W. Halton & B. Walker (2000) *BBRC* **277**, 79.

Selected entries from *Methods in Enzymology* [vol, page(s)]:
General discussion: 80, 535
Assay: 80, 540; esterase, 80, 825
Other molecular properties: action on oxidized B chain of insulin, 80, 555, 558; active-site titration, 80, 543, 544; affinity chromatography, 244, 646; affinity labeling, 244, 655; immunostaining, 236, 150; inhibition, 80, 558, 820, 823, 824; 244, 511; inhibition assay, 251, 394; pH optima, 80, 558; processing, 244, 506; properties, 80, 536, 556; rat (liver [isoelectric point, 80, 556; molecular weight, 80, 556; purification, 80, 549]; substrate specificity, 80, 557, 558); storage, 80, 551; substrate specificity, 244, 500, 509; trapped by α_2-macroglobulin, 80, 748

CATHEPSIN M

This cysteine endopeptidase, also called cathepsin M (new), was isolated from mice and is a placenta-specific lysosomal protease related to both cathepsin P and cathepsin L.[1]

[1] K. Sol-Church, J. Frenck & R. W. Mason (2000) *BBA* **1491**, 289.

CATHEPSIN N

This cysteine endopeptidase, isolated from spleen and liver, has collagenolytic activity.[1,2]

[1] P. Evans & D. J. Etherington (1979) *FEBS Lett.* **99**, 55.
[2] R. A. Maciewicz & D. J. Etherington (1988) *BJ* **256**, 433.

CATHEPSIN O

This cysteine endopeptidase [EC 3.4.22.42] catalyzes the hydrolysis of peptide bonds in Z-Phe-Arg-NHMec and Z-Arg-Arg-NHMec. It is a lysosomal peptidase that is active at pH 6.0.[1,2] *See also Cathepsin K*

[1] I. Santamaría, A. M. Pendás & C. López-Otín (1998) *Genomics* **53**, 231.
[2] G. Velasco, A. A. Ferrando, X. S. Puente, L. M. Sanchez & C. López-Otín (1994) *JBC* **269**, 27136.

CATHEPSIN P

This cysteine endopeptidase, also called cathepsin J, was obtained from mouse placenta and suggested to play a role during implantation and fetal development.[1,2]

[1] K. Sol-Church, J. Frenck, D. Troeber & R. W. Mason (1999) *BJ* **343**, 307.
[2] K. Tisljar, J. Deussing & C. Peters (1999) *FEBS Lett.* **459**, 299.

CATHEPSIN Q

This cysteine endopeptidase is a lysosomal protein expressed in placenta.[1]

[1] K. Sol-Church, J. Frenck & R. W. Mason (2000) *BBRC* **267**, 791.

CATHEPSIN R

This newly recognized cysteine proteinase is expressed only in placenta.[1] A serine endopeptidase activity, also called cathepsin R, ribosomal cathepsin and ribosomal serine endopeptidase, was formerly classified as EC 3.4.21.52, but is now a deleted Enzyme Commission entry.

[1] K. Sol-Church, J. Frenck, G. Bertenshaw & R. W. Mason (2000) *BBA* **1492**, 488.

CATHEPSIN S

This lysosomal cysteine peptidase [EC 3.4.22.27] catalyzes the hydrolysis of peptide bonds with a substrate specificity similar to cathepsin L. However, cathepsin S acts on the synthetic substrate Z-Phe-Arg-∇-NHMec less readily than cathepsin L. (Note: ∇ indicates the cleavage site.) Cathepsin S appears to prefer substrates with hydrophobic aminoacyl residues in the P2 position.[1] Synthetic substrates include Bz-Phe-Val-Arg-∇-NHMec and Z-Val-Val-Arg-∇-NHMec. Cathepsin S is also stable and active at pH 7.5 (its pH$_{opt}$ is 6.5).

[1]P. Locnikar, T. Popovic, T. Lah, I. Kregar, J. Babnik, M. Kopitar & V. Turk (1981) in *Proteinases and Their Inhibitors: Structure, Function, and Applied Aspects* (V. Turk & L. Vitale, eds.), Pergamon Press, Oxford, p. 109.

Selected entries from *Methods in Enzymology* [vol, page(s)]:
Assay: 244, 502
Other molecular properties: bovine, purification, 244, 504; inhibitors, 244, 511; processing, 244, 506; sequence homology with cathepsins, 244, 506; sources of pure enzyme, 244, 507; storage, 244, 505; substrate specificity, 244, 500, 509; tissue distribution, 244, 500, 508

CATHEPSIN T

This protease [EC 3.4.22.24] acts on the N-terminal region of the 53 kDa monomer of rat tyrosine aminotransferase (form I of the aminotransferase is composed of two 53 kDa subunits) to generate a 49 kDa monomer and a 4.5 kDa peptide.[1–3] Cathepsin T will also act on azocasein and denatured hemoglobin; however, tyrosine aminotransferase appears to be the only naturally-occurring substrate.

[1]H. C. Pitot & E. Gohda (1987) *Meth. Enzymol.* 142, 279.
[2]J. B. Dietrich, G. Genot & G. Beck (1988) *Biochimie* 70, 673.
[3]J. L. Hargrove, H. A. Scoble, W. R. Mathews, B. R. Baumstark & K. Biemann (1989) *JBC* 264, 45.

Selected entries from *Methods in Enzymology* [vol, page(s)]:
General discussion: 142, 279
Assay: 142, 280
Other molecular properties: properties, 142, 287; purification from rat liver, 142, 285

CATHEPSIN V

This cysteine endopeptidase [EC 3.4.22.43], also known as cathepsin L2 and cathepsin U, catalyzes the hydrolysis of peptide bonds in proteins (such as serum albumin and collagen) as well as in artificial substrates such as Z-Phe-Arg-NHMec, Z-Leu-Arg-NHMec, and Z-Val-Arg-NHMec.[1–3] The human lysosomal enzyme has a optimal pH at 5.7 and is unstable at neutral pH.

[1]D. Brömme, Z. Li, M. Barnes & E. Mehler (1999) *Biochemistry* 38, 2377.
[2]W. Adachi, S. Kawamoto, I. Ohno, K. Nishida, S. Kinoshita, K. Matsubara & K. Okubo (1998) *Invest. Ophthalmol. Vis. Sci.* 39, 1789.
[3]I. Santamaría, G. Velasco, M. Cazorla, A. Fueyo, E. Campo & C. López-Otín (1998) *Cancer Res.* 58, 1624.

CATHEPSIN X

This lysosomal carboxypeptidase [EC 3.4.18.1] (also called lysosomal carboxypeptidase B, cathepsin B2, cathepsin IV, catheptic carboxypeptidase, carboxypeptidase LB, cysteine-type carboxypeptidase, and acid carboxypeptidase) catalyzes the release of a C-terminal amino acid from proteins, polypeptides, and *N*-acylated dipeptides (unacylated dipeptides are not hydrolyzed).[1,2] The enzyme has broad specificity but will not act on a C-terminal proline. It will deaminate small synthetic substrates such as Z-Arg-NH₂ but it has a weak endopeptidase activity and will

not catalyze the hydrolysis of arylamide derivatives. Interestingly, peptides with an amidated C-terminus (such as substance P) are not substrates.

[1]J. K. McDonald & C. Schwabe (1977) in *Proteinases in Mammalian Cells and Tissues* (A. J. Barrett, ed.), North Holland Publ., Amsterdam, p. 311.
[2]J. K. McDonald & A. J. Barrett (1986) *Mammalian Proteases: A Glossary and Bibliography*, vol. 2, Academic Press, London.

Selected entries from *Methods in Enzymology* [vol, page(s)]:
General discussion: 19, 315
Other molecular properties: inhibition, 46, 206; specificity, 19, 285

Ca²⁺-TRANSPORTING ATPase

This class of so-called ATPases [EC 3.6.3.8] (previously classified as [EC 3.6.1.38], also known as calcium pumps, the sarcoplasmic reticulum ATPase, and the sarco(endo)plasmic reticulum Ca²⁺-ATPase), catalyzes the ATP-dependent (ADP- and orthophosphate-producing) transport of Ca^{2+}_{cis} to produce Ca^{2+}_{trans}.[1–9] This enzyme is a P-type ATPase that undergoes covalent phosphorylation during the transport cycle and undergoes conformational changes during the course of ATP hydrolysis. There are three distinct types of calcium-transporting protein complexes: those found in the plasma membrane, the sarcoplasmic reticulum, and in yeast. The sarcoplasmic reticulum system transports two cations per ATP hydrolyzed whereas the other two subclasses transport only one.

The idea that a transporter is an enzyme is in keeping with a new definition of enzyme catalysis as the facilitated making/breaking of chemical bonds, not just covalent bonds.[10] This idea builds on Pauling's assertion that any long-lived, chemically distinct interaction (in this case, the persistent location of a solute with respect to the faces of a membrane) can be regarded as a chemical bond. Note also that the equilibrium constant ($K_{eq} = [Ca^{2+}_{out}][ADP][P_i]/[Ca^{2+}_{in}][ATP]$) does not conform to that expected for an ATPase (*i.e.*, $K_{eq} = [ADP][P_i]/[ATP]$). Thus, although the overall reaction yields ADP and orthophosphate, the enzyme is misclassified as a hydrolase, and instead should be regarded as an energase-type reaction. Energases facilitate affinity-modulated reactions by coupling the $\Delta G_{ATP\text{-hydrolysis}}$ to a forcegenerating or work-producing step.[10] In this case, P-O-P bond scission supplies the energy to drive Ca²⁺ transport. ***See*** *ATPase; Energase*

[1]D. H. MacLennan, W. J. Rice & N. M. Green (1997) *JBC* 272, 28815.
[2]E. Carafoli (1992) *JBC* 267, 2115.
[3]F. Wuytack & L. Raeymaekers (1992) *J. Bioenerg. Biomembr.* 24, 285.
[4]W. P. Jencks (1992) *Biochem. Soc. Trans.* 20, 555.
[5]G. Inesi, D. Lewis, D. Nikic, A. Hussain & M. E. Kirtley (1992) *AE* 65, 185.

[6]G. Inesi, T. Watanabe, C. Coan & A. Murphy (1982) *Ann. NY Acad. Sci.* **402**, 515.

[7]W. P. Jencks (1980) *AE* **51**, 75.

[8]W. Hasselbach (1974) *The Enzymes*, 3rd ed., **10**, 431.

[9]H. J. Schatzmann & F. F. Vicenzi (1969) *J. Physiol.* **201**, 369.

[10]D. L. Purich (2001) *TiBS* **26**, 417.

Selected entries from *Methods in Enzymology* [vol, page(s)]:
Assay: 32, 303; 139, 794; one-step inside-out vesicles, in, 173, 373; sarcoplasmic reticulum fractions, in, 157, 43

Other molecular properties: aerobic photosensitivity, 54, 52; analysis by controlled proteolysis, 139, 801; ATP binding during ATP hydrolysis, assay, 157, 238; calcium binding during ATP hydrolysis, analysis, 157, 236; cardiac sarcoplasmic reticulum-associated (assay, 157, 117; calmodulin-dependent stimulation, 157, 134; cAMP-dependent stimulation, 157, 128; functional reconstitution, 157, 314; role in Ca^{2+} transport, 157, 120); coupling rules, 171, 156; crystals (freeze-fracture-etch electron microscopy, 157, 279; induction in sarcoplasmic reticulum, 157, 272; negative staining electron microscopy, 157, 273; structural analysis by image reconstruction, 157, 283; unstained, electron microscopic observation, 157, 278); isolation, from kidney cells, 32, 303; labeling with maleimides and iodoacetamide derivative, 157, 251; monomeric, preparation with (deoxycholate, 157, 267; octaethylene glycol dodecyl monoether, 157, 268); occluded Ca^{2+}, assay, 157, 229; oligomeric, preparation with Tween 80, 157, 265; pancreatic islet plasma membrane, assay, 98, 192; phospholamban kinase-phospholamban complex, purification, 102, 269; phosphorylation, assay, 157, 230; plasma membrane-associated (assays, 157, 350, 353; isolation, 157, 345; reconstitution into liposomes, 157, 347; solubilization, purification, and reconstitution, 157, 355); probe binding, 54, 51; properties, 32, 305, 306; purification by transport-specific fractionation, 172, 34; purification from human erythrocytes, 139, 792; rabbit muscle sarcoplasmic reticulum, crosslinking, 172, 606; reaction pathway, 64, 76; reconstitution, 139, 795; 172, 35; rotation studies, 54, 58; sarcoplasmic reticulum (active cation pumping, induction by electric field, 157, 240; assay with N,N'-dicyclohexylcarbodiimide, 125, 96; associated ATP or ITP hydrolysis, 157, 200; associated exchange reactions, 157, 182, 197; ATP binding, fluorescence assay, 157, 214; ATP synthesis, 157, 180, 201, 220; calcium binding, measurement, 157, 163, 210; catalytic cycle, 157, 191; crosslinking, 172, 606; magnesium-induced changes, fluorescence assay, 157, 217; phosphoenzyme assay, 157, 195; phosphoenzyme functional characterization, 157, 174; phosphoenzyme hydrolysis and reversal, 157, 176; phosphorylation, 157, 168, 192, 215; role in Ca^{2+} transport, 157, 156; soluble, inactivation, 157, 196)

*Cau*I RESTRICTION ENDONUCLEASE

This type II restriction endonuclease [EC 3.1.21.4] is a dimer, Mg^{2+}-dependent, and is obtained from *Chloroflexus aurantiacus*. It acts on both strands of DNA at 5′...G∇GWCC...3′ where W represents either A or T.[1,2] In the presence of dimethyl sulfoxide and high pH, the specificity is decreased.[3]

[1]F. Molemans, J. van Emmelo & W. Fiers (1982) *Gene* **18**, 93.

[2]A. H. A. Bingham & J. Darbyshire (1982) *Gene* **18**, 87.

[3]S. P. Bennett & S. E. Halford (1987) in *DNA-Ligand Interactions: From Drugs To Proteins* (W. Guschlbauer & W. Saenger, eds.), pp 239–250, Plenum Press, New York

*Cau*II RESTRICTION ENDONUCLEASE

This type II restriction endonuclease [EC 3.1.21.4] is a heterogeneous dimer, Mg^{2+}-dependent, and is obtained from *Chloroflexus aurantiacus*. It acts on both strands of DNA at 5′...CC∇SGG...3′ where S represents either G or C.[1,2] In the presence of dimethyl sulfoxide and high pH, the specificity is decreased.[3]

[1]F. Molemans, J. van Emmelo & W. Fiers (1982) *Gene* **18**, 93.

[2]A. H. A. Bingham & J. Darbyshire (1982) *Gene* **18**, 87.

[3]S. P. Bennett & S. E. Halford (1987) in *DNA-Ligand Interactions: From Drugs To Proteins* (W. Guschlbauer & W. Saenger, eds.), pp. 239–250, Plenum Press, New York

CC-PREFERRING ENDODEOXYRIBONUCLEASE

This magnesium-activated hydrolase [EC 3.1.21.6], also known as 5′-CC-3′-preferring endodeoxyribonuclease and *Streptomyces glaucescens* exocytoplasmic endodeoxyribonuclease, catalyzes endonucleolytic cleavage in CC sequences within circular double-stranded DNA (ds-DNA) as well as linear ds-DNA, yielding 5′-phosphooligonucleotide end-products. By introducing nicks in ds-DNA, the enzyme creates ds-fragments with 5′- and/or 3′-protruding single-stranded tails. Catalysis is unaffected by cytosine methylation in hemimethylated DNA.[1,2]

[1]S. Cal, J. F. Aparicio, C. G. De Los Reyes-Gavilan, R. G. Nicieza & J. Sanchez (1995) *BJ* **306**, 93.

[2]J. F. Aparicio, J. M. Freije, C. Lopez-Otin, S. Cal, & J. Sanchez (1992) *EJB* **205**, 695.

Cd^{2+}-EXPORTING ATPase

This so-called P-type ATPase [EC 3.6.3.3], which undergoes covalent phosphorylation[2] during the transport cycle in protozoa, fungi, and plants, catalyzes the ATP-dependent (ADP- and orthophosphate-producing) transport of Cd$^{2+}_{in}$ to produce Cd$^{2+}_{out}$.[1,2] *See also* ATPase, Cd^{2+}-Transporting

The idea that a transporter is an enzyme is in keeping with a new definition of enzyme catalysis as the facilitated making/breaking of chemical bonds, not just covalent bonds.[3] This idea builds on Pauling's assertion that any long-lived, chemically distinct interaction (in this case, the persistent location of a solute with respect to the faces of a membrane) can be regarded as a chemical bond. Note also that the equilibrium constant ($K_{eq} = [Cd^{2+}_{out}][ADP][P_i]/[Cd^{2+}_{in}][ATP]$) does not conform to that expected for an ATPase (*i.e.*, $K_{eq} = [ADP][P_i]/[ATP]$). Thus, although the overall reaction yields ADP and orthophosphate, the enzyme is misclassified as a hydrolase, and instead should be regarded as an energase-type reaction. Energases facilitate affinity-modulated reactions by coupling the $\Delta G_{\text{ATP-hydrolysis}}$ to a force-generating or work-producing step.[3] In this case, P-O-P bond scission supplies the energy to drive Cd^{2+} transport. **See ATPase; Energase**

[1] S. Silver & G. Ji (1994) *Environ. Health Perspect.* **102**, Suppl. 3, 107.
[2] K. J. Tsai & A. L. Linet (1993) *ABB* **305**, 267.
[3] D. L. Purich (2001) *TiBS* **26**, 417.

Cd^{2+}-TRANSPORTING ATPase

This so-called ATPase [EC 3.6.3.46] (also known as yeast cadmium factor and heavy-metal-exporting ATPase) catalyzes the ATP-dependent (forming ADP and orthophosphate) export of heavy metal ions from the cytosol into a membrane-enclosed vacuole. The enzyme, which is an ABC-type (ATP-binding cassette-type) ATPase having two similar ATP-binding domains[1], does not undergo phosphorylation during the transport process. **See also ATPase, Cd^{2+}-Exporting**

The idea that a transporter is an enzyme is in keeping with a new definition of enzyme catalysis as the facilitated making/breaking of chemical bonds, not just covalent bonds.[2] This idea builds on Pauling's assertion that any long-lived, chemically distinct interaction (in this case, the persistent location of a solute with respect to the faces of a membrane) can be regarded as a chemical bond. Note also that the equilibrium constant ($K_{\text{eq}} = [\text{Cd}^{2+}_{\text{out}}][\text{ADP}][\text{P}_i]/[\text{Cd}^{2+}_{\text{in}}][\text{ATP}]$) does not conform to that expected for an ATPase (*i.e.*, $K_{\text{eq}} = [\text{ADP}][\text{P}_i]/[\text{ATP}]$). Thus, although the overall reaction yields ADP and orthophosphate, the enzyme is misclassified as a hydrolase, and instead should be regarded as an energase-type reaction. Energases facilitate affinity-modulated reactions by coupling the $\Delta G_{\text{ATP-hydrolysis}}$ to a force-generating or work-producing step.[2] In this case, P-O-P bond scission supplies the energy to drive cadmium transport. **See ATPase; Energase**

[1] Z. S. Li, M. Szczypka, Y. P. Lu, D. J. Thiele & P. A. Rea (1996) *JBC* **271**, 6509.
[2] D. L. Purich (2001) *TiBS* **26**, 417.

CDP-ABEQUOSE EPIMERASE

This NAD$^+$-dependent epimerase [EC 5.1.3.10], also known as CDP-paratose epimerase, catalyzes the interconversion of CDP-3,6-dideoxy-D-glucose (*i.e.*, CDP-paratose) and CDP-3,6-dideoxy-D-mannose.[1]

[1] S. Matsuhashi (1966) *JBC* **241**, 4275.

Selected entries from *Methods in Enzymology* [vol, page(s)]:
General discussion: 8, 310, 315
Assay: 8, 315

CDP-ACYLGLYCEROL O-ARACHIDONYLTRANSFERASE

This transferase [EC 2.3.1.70] catalyzes the reaction of arachidonyl-CoA with a CDP-acylglycerol to produce coenzyme A and the corresponding CDP-diacylglycerol.[1] The enzyme is highly specific for both donor and acceptor substrates.

[1] W. Thomson & G. MacDonald (1979) *JBC* **254**, 3311.

CDP-4-DEHYDRO-6-DEOXYGLUCOSE REDUCTASE

This oxidoreductase [EC 1.17.1.1], also known as CDP-4-keto-6-deoxyglucose reductase, catalyzes the reversible reaction of CDP-4-dehydro-3,6-dideoxy-D-glucose with NAD(P)$^+$ and water to produce CDP-4-dehydro-6-deoxy-D-glucose and NAD(P)H.[1-3]

[1] P. Gonzalez-Porque (1986) *Coenzymes Cofactors*, **1**, 391.
[2] H. Pape & J. L. Strominger (1969) *JBC* **244**, 3598.
[3] P. Gonzalez-Porque & J. L. Strominger (1972) *JBC* **247**, 6748.

Selected entries from *Methods in Enzymology* [vol, page(s)]:
General discussion: 8, 310

CDP-DIACYLGLYCEROL: sn-GLYCEROL-3-PHOSPHATE 3-PHOSPHATIDYLTRANSFERASE

This transferase [EC 2.7.8.5] (also known as phosphatidylglycerophosphate synthase, glycerophosphate phosphatidyltransferase, and 3-phosphatidyl-1′-glycerol-3′-phosphate synthase) catalyzes the reaction of CDP-diacylglycerol with glycerol 3-phosphate to produce CMP and 3-(3-phosphatidyl)glycerol 1-phosphate.[1-5] The enzyme has an ordered Bi Bi kinetic mechanism.[5]

[1] S. A. Minskoff & M. L. Greenberg (1997) *BBA* **1348**, 187.
[2] R. A. Pieringer (1983) *The Enzymes*, 3rd ed., **16**, 255.
[3] J. D. Esko & C. R. H. Raetz (1983) *The Enzymes*, 3rd ed., **16**, 207.
[4] T. S. Moore, Jr. (1982) *Ann. Rev. Plant Physiol.* **33**, 235.
[5] T. Hirabayashi, T. J. Larson & W. Dowhan (1976) *Biochemistry* **15**, 5205.

Selected entries from *Methods in Enzymology* [vol, page(s)]:
General discussion: 71, 555; 209, 313
Assay: castor bean endosperm, 71, 611; plant endoplasmic reticulum, 148, 593
Other molecular properties: castor bean endosperm, from, 71, 611 (assay, 71, 611; detergent effects, 71, 613; intracellular distribution, 71, 613; metal ion specificity, 71, 612; properties, 71, 612); *Escherichia coli*, from, 71, 555 (affinity chromatography, 71, 558; preparation of membrane extract, 71, 558; properties, 71, 560); mutant, 72, 694; plant endoplasmic reticulum, in (assay, 148, 593; properties, 148, 593)

CDP-DIACYLGLYCEROL:INOSITOL 3-PHOSPHATIDYLTRANSFERASE

This transferase [EC 2.7.8.11], also known as phosphatidylinositol synthase, catalyzes the reaction of CDP-diacylglycerol with *myo*-inositol to produce CMP and phosphatidyl-1D-*myo*-inositol.[1–5]

[1] J. E. Bleasdale, P. Wallis, P. C. MacDonald & J. M. Johnston (1979) *BBA* **575**, 135.
[2] P. P. N. Murthy & B. W. Agranoff (1982) *BBA* **712**, 473.
[3] M. E. Monaco, M. Feldman & D. L. Kleinberg (1994) *BJ* **304**, 301.
[4] B. E. Antonsson (1994) *BJ* **297**, 517.
[5] J. D. Esko & C. R. H. Raetz (1983) *The Enzymes*, 3rd ed., **16**, 207.

Selected entries from ***Methods in Enzymology*** [vol, page(s)]:
General discussion: **209**, 305
Assay: **209**, 307; castor bean endosperm, **71**, 605; plant endoplasmic reticulum, **148**, 594; rat liver, **109**, 474
Other molecular properties: castor bean endosperm, from, **71**, 605 (assay, **71**, 605; detergent effects, **71**, 606; intracellular distribution, **71**, 606; metal ion specificity, **71**, 606; properties, **71**, 605); electroblotting, **209**, 308; plant endoplasmic reticulum, in (assay, **148**, 594; properties, **148**, 595); properties, **209**, 311; purification, **209**, 308; reconstitution, **209**, 310; synthetic and analytical uses, **209**, 312

CDP-DIACYLGLYCEROL PYROPHOSPHATASE

This pyrophosphatase [EC 3.6.1.26], also known as CDP-diacylglycerol diphosphatase and CDP-diacylglycerol phosphatidylhydrolase, catalyzes the hydrolysis of a CDP-diacylglycerol to produce CMP and a phosphatidate.[1–3]

[1] C. R. H. Raetz, C. B. Hirschberg, W. Dowhan, W. T. Wickner & E. P. Kenney (1972) *JBC* **247**, 2245.
[2] P. P. N. Murthy & B. W. Agranoff (1982) *BBA* **712**, 473.
[3] A. Mok, T. Wong, O. Filgueiras, P. G. Casola, D. W. Nicholson, W. C. McMurray, P. G. Harding & F. Possmayer (1988) *Biochem. Cell Biol.* **66**, 425.

CDP-DIACYLGLYCEROL:SERINE *O*-PHOSPHATIDYLTRANSFERASE

This transferase [EC 2.7.8.8], also known as phosphatidylserine synthase and CDP-diglyceride:serine *O*-phosphatidyltransferase, catalyzes the reaction of CDP-diacylglycerol with L-serine to produce CMP and *O-sn*-phosphatidyl-L-serine.[1–5]

[1] T. J. Larson & W. Dowhan (1976) *Biochemistry* **15**, 5212.
[2] M. S. Bae-Lee & G. M. Carman (1984) *JBC* **259**, 10857.
[3] A. Dutt & W. Dowhan (1985) *Biochemistry* **24**, 1073.
[4] K. Matsumoto (1997) *BBA* **1348**, 214.
[5] R. A. Pieringer (1983) *The Enzymes*, 3rd ed., **16**, 255.

Selected entries from ***Methods in Enzymology*** [vol, page(s)]:
General discussion: **71**, 561; **209**, 287, 298
Assay: yeast, **209**, 299
Other molecular properties: *Escherichia coli*, from, **71**, 561 (amino acid sequence, **209**, 296; assay, **209**, 287; CMP-CDPdiacylglycerol exchange reaction, **71**, 571; detergent-substrate mixed micelles, **71**, 570;

properties, **209**, 291; purification, **209**, 289); gene (hybrid plasmids, **71**, 562; structural, **71**, 562); mutant, **72**, 694; stereospecificity at phosphorus, **197**, 258; yeast (assay, **209**, 299; electroblotting, **209**, 300; properties, **209**, 303; purification, **209**, 301; reconstitution, **209**, 303; synthetic and analytical uses, **209**, 304)

CDP-GLUCOSE 4,6-DEHYDRATASE

This NAD$^+$-dependent dehydratase [EC 4.2.1.45], also known as CDP-glucose oxidoreductase, catalyzes the conversion of CDP-D-glucose to CDP-4-dehydro-6-deoxy-D-glucose and water, with NAD$^+$ functioning as an intermediate hydride carrier.[1–4]

[1] X. He, J. S. Thorson & H. W. Liu (1996) *Biochemistry* **35**, 4721.
[2] L. Glaser & H. Zarkowsky (1971) *The Enzymes*, 3rd ed., **5**, 465.
[3] A. E. Hey & A. D. Elbein (1966) *JBC* **241**, 5473.
[4] S. Matsuhashi, M. Matsuhashi, J. G. Brown & J. L. Strominger (1966) *JBC* **241**, 4283.

Selected entries from ***Methods in Enzymology*** [vol, page(s)]:
General discussion: **8**, 310, 313; **28**, 461
Assay: **8**, 313; **28**, 462

CDP-GLYCEROL GLYCEROPHOSPHOTRANSFERASE

This transferase [EC 2.7.8.12], also known as teichoic-acid synthase and poly(glycerol phosphate) polymerase, catalyzes the reaction of CDP-glycerol with (glycerophosphate)$_n$ to produce CMP and (glycerophosphate)$_{n+1}$.[1,2]
See also CDP-Ribitol Ribitolphosphotransferase

[1] M. M. Burger & L. Glaser (1964) *JBC* **239**, 3168.
[2] H. M. Pooley, F.-X. Abellan & D. Karamata (1992) *J. Bacteriol.* **174**, 646.

Selected entries from ***Methods in Enzymology*** [vol, page(s)]:
General discussion: **8**, 430; **50**, 394
Assay: **8**, 430

CDP-GLYCEROL PYROPHOSPHATASE

This pyrophosphatase [EC 3.6.1.16], also known as CDP-glycerol diphosphatase, catalyzes the hydrolysis of CDP-glycerol to produce CMP and *sn*-glycerol 3-phosphate.[1]

[1] L. Glaser (1965) *BBA* **101**, 6.

CDP-RIBITOL RIBITOLPHOSPHOTRANSFERASE

This transferase [EC 2.7.8.14], also known as teichoic-acid synthase and poly(ribitol phosphate) polymerase, catalyzes the reaction of CDP-ribitol with (ribitol phosphate)$_n$ to produce CMP and (ribitol phosphate)$_{n+1}$.[1–3] ***See also*** CDP-Glycerol Glycerophosphotransferase

[1] N. Ishimoto & J. L. Strominger (1966) *JBC* **241**, 639.
[2] F. Fiedler & L. Glaser (1974) *JBC* **249**, 2684.

[3]W. Fischer, H. U. Koch, P. Rösel, F. Fiedler & L. Schmuck (1980) *JBC* **255**, 4550.

Selected entries from *Methods in Enzymology* [vol, page(s)]:
General discussion: 8, 423
Assay: 8, 423; 50, 389
Other molecular properties: preparation, 8, 425; properties, 8, 426; 50, 392; purification, 50, 390, 391

CELLOBIOSE DEHYDROGENASES

These oxidoreductases catalyze the conversion of cellobiose (*i.e.*, 4-*O*-β-D-glucopyranosyl-D-glucose) to cellobiono-1,5-lactone.

cellobiose

Cellobiose dehydrogenase (quinone) [EC 1.1.5.1], also known as cellobiose:quinone oxidoreductase, catalyzes the FAD-dependent reaction of cellobiose with a quinone to produce cellobiono-1,5-lactone and a phenol.[1,2] Other substrates include cello-oligosaccharides, lactose, and D-glucosyl-1,4-β-D-mannose. Cellulose is not a substrate.

Cellobiose dehydrogenase (acceptor) [EC 1.1.99.18] catalyzes the reaction of cellobiose with an acceptor substrate to produce cellobiono-1,5-lactone and the reduced acceptor.[3–5] Acceptor substrates include 2,6-dichloroindophenol, phenol blue, and cytochrome *c*. Other carbohydrate substrates include cello-oligosaccharides, lactose, and D-glucosyl-(1,4)-β-D-mannose.

[1]F. F. Morpeth & G. D. Jones (1986) *BJ* **236**, 221.
[2]K.-E. Eriksson (1978) *Biotechnol. Bioeng.* **20**, 317.
[3]J. C. Sadana & R. V. Patil (1985) *J. Gen. Microbiol.* **131**, 1917.
[4]M.-R. Coudray, G. Canevascini & H. Meier (1982) *BJ* **203**, 277.
[5]R. F. H. Dekker (1980) *J. Gen. Microbiol.* **120**, 309.

Selected entries from *Methods in Enzymology* [vol, page(s)]:
General discussion: acceptor-dependent, 160, 443, 448, 454; quinone-dependent, 160, 463
Assay: acceptor-dependent, 160, 444, 448, 457; quinone-dependent, 160, 464

CELLOBIOSE 2'-EPIMERASE

This epimerase [EC 5.1.3.11] catalyzes the interconversion of cellobiose (*i.e.*, 4-*O*-β-D-glucopyranosyl-D-glucose) and β(1,4)-D-glucosyl-D-mannose (*i.e.*, 4-*O*-β-D-glucopyranosyl-D-mannose).[1–4]

[1]M. E. Tanner & G. L. Kenyon (1998) *CBC* **2**, 7.
[2]E. Adams (1976) *Adv. Enzymol.* **44**, 69.
[3]L. Glaser (1972) *The Enzymes*, 3rd ed., **6**, 355.
[4]T. R. Tyler & J. M. Leatherwood (1967) *ABB* **119**, 363.

CELLOBIOSE OXIDASE

This FAD- and cytochrome *b*-dependent oxidoreductase [EC 1.1.3.25] catalyzes the reaction of cellobiose (*i.e.*, 4-*O*-β-D-glucopyranosyl-D-glucose) with dioxygen to produce cellobiono-1,5-lactone and hydrogen peroxide.[1,2] Other substrates include the cellodextrins, lactose, maltose, and, more slowly, 4-β-D-glucosyl-D-mannose. Strong substrate inhibition is observed at elevated concentrations of cellobiose.[1]

[1]F. F. Morpeth (1985) *BJ* **228**, 557.
[2]G. D. Jones & M. T. Wilson (1988) *BJ* **256**, 713.

Selected entries from *Methods in Enzymology* [vol, page(s)]:
General discussion: 89, 129
Assay: *Sporotrichum pulverulentum*, 89, 129

CELLOBIOSE PHOSPHORYLASE

This phosphorylase [EC 2.4.1.20] catalyzes the reversible reaction of cellobiose (*i.e.*, 4-*O*-β-D-glucopyranosyl-D-glucose) with orthophosphate to produce α-D-glucose 1-phosphate and D-glucose.[1–4] β-Cellobiose undergoes phosphorolysis faster than the α-anomer.[1]

[1]M. Kitaoka, T. Sasaki & H. Taniguchi (1992) *J. Biochem.* **112**, 40.
[2]J. J. Mieyal & R. H. Abeles (1972) *The Enzymes*, 3rd ed., **7**, 515.
[3]M. Cohn (1961) *The Enzymes*, 2nd ed., **5**, 179.
[4]M. Doudoroff (1961) *The Enzymes*, 2nd ed., **5**, 229.

Selected entries from *Methods in Enzymology* [vol, page(s)]:
General discussion: 28, 944; 160, 468
Assay: 28, 944; 160, 392, 468
Other molecular properties: enzymatic properties, 28, 947; isolation from *Clostridium thermocellum*, 28, 945; properties, 160, 470; purification from *Cellvibrio gilvus*, 160, 469; reaction, 28, 944, 947; staining on gels, 22, 598

CELLODEXTRIN PHOSPHORYLASE

This phosphorylase [EC 2.4.1.49] catalyzes the reversible reaction of $[(1,4)\text{-}\beta\text{-}D\text{-glucosyl}]_n$ with orthophosphate to produce $[(1,4)\text{-}\beta\text{-}D\text{-glucosyl}]_{n-1}$ and α-D-glucose 1-phosphate.[1] Glucosyl-containing substrates include cellotriose, cellotetraose, cellopentaose, and cellohexaose.[1]

[1]K. Sheth & J. K. Alexander (1969) *JBC* **244**, 457.

Selected entries from *Methods in Enzymology* [vol, page(s)]:
General discussion: 28, 948
Assay: 28, 948

CELLULASE

This enzyme [EC 3.2.1.4], also known as endo-1,4-β-glucanase and carboxymethyl cellulase, catalyzes random, endohydrolytic scission of 1,4-β-D-glucosidic linkages in cellulose.[1–6] The enzyme also catalyzes the hydrolysis of β-1,4-linkages in β-D-glucans also containing 1,3-linkages. Many organisms contain an entire set of cellulases, each having its own distinct specificities, regulatory properties, and structure. Some cellulases catalyze their reaction with inversion of anomeric configuration, while others exhibit retention.

Note that the term "cellulase" is often used indiscriminantly to refer to any lytic enzyme acting on cellulose (*e.g.*, endo-1,3(4)-β-glucanase [EC 3.2.1.6], β-glucosidase [EC 3.2.1.21], and cellulose 1,4-β-cellobiosidase [EC 3.2.1.91]).

[1] G. Davies, M. L. Sinnott & S. G. Withers (1998) *CBC* 1, 119.
[2] M. W. Bauer, S. B. Halio & R. M. Kelly (1996) *Adv. Protein Chem.* 48, 271.
[3] A. A. Klyosov (1990) *Biochemistry* 29, 10577.
[4] J.-P. Aubert, P. Beguin & J. Millet, eds., (1988) *Biochemistry and Genetics of Cellulose Degradation*, Academic Press, New York.
[5] D. R. Whitaker (1971) *The Enzymes*, 3rd ed., 5, 273.
[6] J. Larner (1960) *The Enzymes*, 2nd ed., 4, 369.

Selected entries from **Methods in Enzymology** [vol, page(s)]:
General discussion: 1, 173; 8, 603; 44, 108, 165, 166; 160, 19, 45, 74, 87, 112, 117, 135, 200, 216, 221, 234, 243, 259, 264, 274, 300, 314, 323, 338, 342, 351, 355, 363, 368, 378, 382, 472; 330, 290, 346.
Assay: 1, 173; 8, 603; 63, 32; 160, 87, 100, 130, 352, 364, 377; cellobiose dehydrogenase, based on, 160, 112; nephelometric and turbidometric assay, 160, 117; type D from *Clostridium thermocellum*, 160, 356; viscosimetric assay, 160, 130
Other molecular properties: activation by acid proteases, 160, 507; amorphous, crystalline, and dyed substrates, preparation, 160, 19; carboxymethylcellulase activity (assay, 160, 377 [viscosimetric assay, 160, 130]; properties, 160, 381; purification from *Sclerotium rolfsii*, 160, 378); complexes in *Clostridium thermocellum* (assays, 160, 486; fractionation, 160, 487; properties, 160, 490); detection in (gels, 160, 139; prokaryotes, 160, 182); (1 → 6)-α-D-glucopyranan hydrolysis, 242, 253; *Helix pomatia*, purification, 1, 173, 174; hemicellulase complex (properties, 160, 335; purification from *Phoma hibernica*, 160, 333); hyperthermophilic, 330, 290, 346; nephelometric and turbidometric assay, 160, 117; Onozuka cellulase (source, 69, 78); plant cell (isolation, 69, 59, 65, 67; protoplasts, 69, 78); plant plasma membrane marker, as, assay, 148, 545; production *in vitro*, effect of culture conditions, 160, 207; properties, 160, 354, 367, 375, 381, 385, 387 (*Aspergillus* enzyme, 160, 262, 293; *Cellulomonas* enzymes, 160, 214; *Eupenicillium* enzymes, 160, 256; *Helix pomatia*, 1, 175; *Humicola* enzymes, 160, 330; kidney bean enzymes, 160, 350; mutant *Trichoderma* enzymes, 160, 248; *Penicillium notatum*, 8, 607; *Pseudomonas* enzymes, 160, 206; *Ruminococcus* enzyme, 160, 221; *Sporocytophaga* enzymes, 160, 341; *Thermoascus* enzymes, 160, 304; *Thermomonospora* enzyme, 160, 316; *Trichoderma* enzymes, 160, 233, 241); protoplast preparation, use in, 23, 200, 201; 31, 582, 583; purification, 160, 384 (*Aspergillus aculeatus*, 160, 279; *Aspergillus fumigatus*, 160, 268; *Aspergillus niger*, 160, 261; *Cellulomonas uda*, 160, 212; *Clostridium thermocellum*, 160, 352; *Eupenicillium javanicum*, 160, 254; *Helix pomatia*, 1, 174; 8, 415f; *Humicola*, 160, 325, 327, 329; kidney bean, 160, 347; *Penicillium notatum*, 8, 604; *Pseudomonas fluorescens*, 160, 202; *Ruminococcus albus*, 160, 219; *Sclerotium rolfsii*, 160, 378; *Sporocytophaga myxococcoides*, 160, 341; *Sporotrichum pulverulentum*, 160, 370; *Talaromyces emersonii*, 160, 365;

Thermoascus aurantiacus, 160, 303; *Thermomonospora fusca*, 160, 316; *Trichoderma koningii*, 160, 224, 227; *Trichoderma reesei*, 160, 237; *Trichoderma viride* QM 9414 mutant strain, 160, 245); snails, preparation from, 1, 173; 8, 414; sources, 1, 177; 69, 78; *Trichoderma* cellulase (use in protoplast preparation, 23, 203); *Trichoderma viride*, from, 72, 776 (chloroplast isolation, 69, 123; making protoplasts, for, 72, 776); type B, *in vitro* conversion to type A, 160, 210; type C, purification from *Escherichia coli*, 160, 192; type D from *Clostridium thermocellum* (assay, 160, 356; crystallization, 160, 360; properties, 160, 360; purification from *Escherichia coli*, 160, 357)

CELLULOSE 1,4-β-CELLOBIOSIDASE

This enzyme [EC 3.2.1.91], also known as exoglucanase, exocellobiohydrolase, and 1,4-β-cellobiohydrolase, catalyzes the hydrolysis of 1,4-β-D-glucosidic linkages in cellulose and cellotetraose, thereby releasing cellobiose from the non-reducing ends of the chains.[1–3] *See also Glucan 1,4-β-Glucosidase*

[1] D. Fracheboud & G. Canevascini (1989) *Enzyme Microb. Technol.* 11, 220.
[2] L. Huang, C. W. Forsberg & D. Y. Thomas (1988) *J. Bacteriol.* 170, 2923.
[3] R. M. Gardner, K. C. Doerner & B. A. White (1987) *J. Bacteriol.* 169, 4581.

Selected entries from **Methods in Enzymology** [vol, page(s)]:
General discussion: 160, 211, 221, 224, 243, 307, 323, 398, 403
Assay: 160, 98, 126, 403
Other molecular properties: cellobiosidase (assay, 160, 392; properties, 160, 395; purification from *Ruminococcus albus*, 160, 393); detection in gels, 160, 142; exocellulase activity from *Irpex lacteus* (assay, 160, 403; properties, 160, 406; purification, 160, 404); function, 160, 313; properties, 160, 311, 402, 406; purification from (*Humicola*, 160, 326; *Irpex lacteus*, 160, 404; *Penicillium pinophilum*, 160, 188, 400; *Sclerotium rolfsii*, 160, 310; *Trichoderma koningii*, 160, 231; *Trichoderma reesei*, 160, 188)

CELLULOSE POLYSULFATASE

This enzyme [EC 3.1.6.7] catalyzes the hydrolysis of the 2- and 3-sulfate groups of the polysulfates of cellulose and charonin.[1,2] Dextran polysulfate and chondroitin sulfate are hydrolyzed very slowly.

[1] A. B. Roy (1971) *The Enzymes*, 3rd ed., 5, 1.
[2] N. Takahashi & F. Egami (1961) *BJ* 80, 384.

CELLULOSE SYNTHASE (GDP-FORMING)

This enzyme [EC 2.4.1.29] catalyzes the reaction of GDP-D-glucose with [(1,4)-β-D-glucosyl]$_n$ to produce GDP and [(1,4)-β-D-glucosyl]$_{n+1}$.[1–3]

[1] G. Davies, M. L. Sinnott & S. G. Withers (1998) *CBC* 1, 119.
[2] J. Chambers & A. D. Elbein (1970) *ABB* 138, 620.
[3] H. M. Flowers, K. K. Batra, J. Kemp & W. Z. Hassid (1969) *JBC* 244, 4969.

Selected entries from **Methods in Enzymology** [vol, page(s)]:
General discussion: 8, 416; 28, 581

Assay: 8, 416; **28**, 581
Other molecular properties: enzymatic and physical properties, **28**, 582; isolation from *Acanthamoeba castellanii*, **28**, 582; occurrence, 8, 418; preparation, mung bean, 8, 417; properties, 8, 418

CELLULOSE SYNTHASE (UDP-FORMING)

This enzyme [EC 2.4.1.12], also known as UDP-glucose-β-D-glucan glucosyltransferase and UDP-glucose-cellulose glucosyltransferase, catalyzes the reaction of UDP-D-glucose with $[(1,4)-\beta-D-glucosyl]_n$ to produce UDP and $[(1,4)-\beta-D-glucosyl]_{n+1}$.[1-4]

[1]G. Davies, M. L. Sinnott & S. G. Withers (1998) *CBC* **1**, 119.
[2]I. Kuribayashi, S. Kimura, T. Morita & I. Igaue (1992) *Biosci. Biotechnol. Biochem.* **56**, 388.
[3]D. Haass, G. Hackspacher & G. Franz (1985) *Plant Sci.* **41**, 1.
[4]L. Glaser (1958) *JBC* **232**, 627.

CEPHALOSPORINASE

This enzyme activity, formerly classified as EC 3.5.2.8, is now listed under β-lactamase [EC 3.5.2.6].

CEPHALOSPORIN-C AMINOTRANSFERASE

This pyridoxal 5′-phosphate-dependent aminotransferase [EC 2.6.1.74] catalyzes the reaction of cephalosporin C with pyruvate to produce 7-(5-carboxyl-5-oxopentanyl) aminocephalosporinate and L-alanine.[1]

[1]W. Aretz & K. Sauber (1988) *Ann. N. Y. Acad. Sci.* **542**, 366.

CEPHALOSPORIN-C DEACETYLASE

This esterase [EC 3.1.1.41] catalyzes the hydrolysis of cephalosporin C to produce deacetylcephalosporin C and acetate.[1,2] The *Rhodosporidium toruloides* enzyme is a glycoprotein that hydrolyzes short-chain *p*-nitrophenyl esters and will also acetylate deacetylcephalosporins.[3]

[1]A. Hinnem & J. Nüesch (1976) *Antimicrob. Agents Chemother.* **9**, 824.
[2]B. J. Abbott & D. Fukuda (1975) *Appl. Microbiol.* **30**, 413.
[3]M. Politino, S. M. Tonzi, W. V. Burnett, G. Romancik & J. J. Usher (1997) *Appl. Environ. Microbiol.* **63**, 4807.

Selected entries from ***Methods in Enzymology*** [**vol**, page(s)]:
General discussion: 43, 728, 731
Assay: *Bacillus subtilis*, **43**, 732, 733; citrus, **43**, 728 (manometric, **43**, 728, 729; potentiometric, **43**, 729)

CERAMIDASE

This enzyme [EC 3.5.1.23], also known as acylsphingosine deacylase, catalyzes the hydrolysis of an *N*-acyl-sphingosine (*i.e.*, a ceramide) to produce a fatty acid anion and sphingosine.[1-5] At least three types of ceramidases exist (acid, neutral, and alkaline) depending on their

pH optima. Saposine D activates the acid enzyme but inhibits the alkaline protein.

[1]A. Nilsson & R. D. Duan (1999) *Chem. Phys. Lipids* **102**, 97.
[2]D. F. Hassler & R. M. Bell (1993) *Adv. Lipid Res.* **26**, 49.
[3]J. D. Esko & C. R. H. Raetz (1983) *The Enzymes*, 3rd ed., **16**, 207.
[4]Y. Kishimoto (1983) *The Enzymes*, 3rd ed., **16**, 357.
[5]S. Gatt (1966) *JBC* **241**, 3724.

Selected entries from ***Methods in Enzymology*** [**vol**, page(s)]:
General discussion: 14, 139; **28**, 837; **311**, 194, 201, 276
Assay: 14, 139; **28**, 488, 490; **50**, 463
Other molecular properties: cerebroside-β-glucosidase and, **28**, 830; cholate activation, **14**, 144; diagnostic assay for Farber disease, **138**, 734; galactosyl ceramide synthetase and, **28**, 488; glucosyl ceramide synthetase and, **28**, 486; hydrolysis of brain ganglioside, in, **14**, 135, 136, 138; isolation from rat brain, **14**, 142; **28**, 488; properties, **14**, 143; separation from sphingomyelinase, **14**, 148; sphingomyelinase and, **28**, 874

CERAMIDE CHOLINEPHOSPHOTRANSFERASE

This transferase [EC 2.7.8.3], also known as phosphorylcholine:ceramide transferase, catalyzes the reaction of CDP-choline with an *N*-acyl-*threo-trans*-sphingosine to produce CMP and a sphingomyelin.[1,2]

[1]M. Sribney & E. P. Kennedy (1958) *JBC* **233**, 1315.
[2]W. D. Marggraf & J. N. Kanfer (1987) *BBA* **897**, 57.

Selected entries from ***Methods in Enzymology*** [**vol**, page(s)]:
General discussion: 5, 486
Assay: 5, 486

CERAMIDE GLUCOSYLTRANSFERASE

This transferase [EC 2.4.1.80], also known as glucosyl-ceramide synthase and UDP-glucose:ceramide glucosyltransferase, catalyzes the reaction of UDP-D-glucose with an *N*-acylsphingosine (a ceramide) to produce UDP and D-glucosyl-*N*-acylsphingosine (a cerebroside).[1-4] Other acceptor substrates include sphingosine and dihydrosphingosine. CDP-D-glucose is also a donor substrate.

[1]S. Ichikawa & Y. Hirabayashi (1998) *Trends Cell Biol.* **8**, 198.
[2]N. Matsuo, T. Nomura & G. Imokawa (1992) *BBA* **1116**, 97.
[3]S. Basu, B. Kaufman & S. Roseman (1973) *JBC* **248**, 1388.
[4]S. N. Shah (1973) *ABB* **159**, 143.

Selected entries from ***Methods in Enzymology*** [**vol**, page(s)]:
General discussion: 28, 486; **311**, 42, 50, 373
Assay: 28, 486; **72**, 384; **138**, 578
Other molecular properties: glycosphingolipid-synthesizing, assay, **138**, 578; inhibition, **72**, 677 (DL-2-decanoylamino-3-morpholino-1-phenylpropanol, **72**, 679; DL-2-decanoylamino-3-morpholinopropio-phenone, **72**, 677); properties, **28**, 486

CERAMIDE KINASE

This divalent cation-stimulated phosphotransferase [EC 2.7.1.138], also known as acylsphingosine kinase, catalyzes the reaction of ATP with a ceramide to produce ADP and ceramide 1-phosphate.[1,2]

[1]S. M. Bajjalieh, T. F. Martin & E. Floor (1989) *JBC* **264**, 14354.
[2]R. N. Kolesnick & M. R. Hemer (1990) *JBC* **265**, 18803.

Selected entries from *Methods in Enzymology* [vol, page(s)]:
General discussion: 311, 207

CEREBROSIDE-SULFATASE

This sulfatase [EC 3.1.6.8], also known as arylsulfatase A, catalyzes the hydrolysis of a cerebroside 3-sulfate to produce a cerebroside and sulfate. Other substrates include galactose 3-sulfate residues in a number of lipids, ascorbate 2-sulfate, nitrocatechol sulfate, and many phenol sulfates (hence, cerebroside-sulfatase exhibits an arylsulfatase-like [EC 3.1.6.1] activity). A key catalytic residue in sulfate ester cleavage is a Cα-formylglycyl residue that is post-translationally generated from a cysteinyl residue in the endoplasmic reticulum.

In the first catalytic half-reaction, one of the geminal oxygens of the hydrated formyl group attacks the sulfur atom of the sulfate ester; the resulting transesterification of the sulfate group to the aldehyde hydrate is attended by release of the alcohol product. In the second half-reaction, sulfate is eliminated from the enzyme-sulfate intermediate by an intramolecular rearrangement induced by the second oxygen, thereby regenerating the catalytic aldehyde group.[1]

[1]A. Waldow, B. Schmidt, T. Dierks, R. von Bülow & K. von Figura (1999) *JBC* **274**, 12284.

Selected entries from *Methods in Enzymology* [vol, page(s)]:
General discussion: 28, 880; 34, 4; 50, 537; 62, 42; 311, 255
Assay: 28, 880; 50, 537
Other molecular properties: activity assay, 73, 551, 553, 554; analysis in multiple systems, 22, 18, 20; antiserum preparation, 73, 553; L-ascorbate-2-sulfate, hydrolysis of, 62, 42; barium ion inactivation, 50, 450; cross reactivity between species, 73, 563, 564, 577; diagnostic assay (metachromatic leukodystrophy, 138, 742; multiple sulfatase deficiency, 138, 745); human placental (effect on [slow-reacting substance, 86, 20; spasmogenic mediators, 86, 20]; purification, 86, 18); isolation from urine, 28, 882; metachromatic leukodystrophy and, 28, 884; 50, 471; mucolipidoses II and III, 50, 455; multiple sulfatase deficiency, 50, 453, 454, 474; properties, 28, 884; 50, 546, 547; purification, 50, 543, 544; 73, 551, 559, 562, 571, 578; specificity, 50, 537; stability, 73, 562; substrates, 50, 472, 473

CEREVISIN

This serine proteinase [EC 3.4.21.48], also known as yeast proteinase B and proteinase yscB, catalyzes peptide-bond hydrolysis with a broad specificity. The enzyme also exhibits esterase activity, with Z-Tyr-OPhNO₂ serving as a chromogenic substrate.[1,2] The proteinase ycaB from *Candida albicans* is similar to cerevisin.[3]

IB2, a 74-residue cytoplasmic inhibitor of cerevisin, is a noncompetitive inhibitor.[1] The enzyme is also inhibited by chymostatin, mercurials, and DFB.

[1]E. W. Jones & D. G. Murdock (1994) in *Cellular Proteolytic Systems* (A. J. Ciechanover & A. L. Schwartz, eds.), Wiley-Liss, New York, p. 115.
[2]H. B. van den Hazel, M. C. Kielland-Brandt & J. R. Winter (1996) *Yeast* **12**, 1.
[3]P. C. Farley, M. G. Shepherd & P. A. Sullivan (1986) *BJ* **236**, 177.

Selected entries from *Methods in Enzymology* [vol, page(s)]:
General discussion: 194, 428
Assay: 194, 447; Golgi apparatus, in, 194, 658.
Other molecular properties: overlay test, 194, 435; vacuolar (biochemical analyses, 185, 379; 194, 444; genetic analyses, 185, 372; 194, 435; interference with yeast studies, prevention, 194, 428).

CETRAXATE BENZYLESTERASE

This esterase [EC 3.1.1.70], first isolated from *Aspergillus niger*, catalyzes the hydrolysis of cetraxate benzyl ester, generating cetraxate and benzyl alcohol.

cetraxate benzyl ester

Benzyl esters of substituted phenyl propanoates as well as the benzyl esters of phenylalanine and tyrosine are also substrates.[1]

[1]H. Kuroda, A. Miyadera, A. Imura & A. Suzuki (1989) *Chem. Pharm. Bull.* **37**, 2929.

CF₁CF₀ ATP SYNTHASE (AND CF₁CF₀-ATPase)

This plant H⁺-transporting ATP-synthesizing enzyme [EC 3.6.1.34], also known as chloroplast ATPase and CF₁CF₀-ATPase, catalyzes ATP synthesis from ADP and orthophosphate when coupled to a transmembrane proton gradient. In its uncoupled state, the enzyme catalyzes ATP hydrolysis to produce ADP, orthophosphate, and a proton.[1–3] (The subscript "o" refers to the oligomycin-sensitive component.) The multisubunit complex known as CF₁, when isolated from the complete membrane-bound complex, retains the ATPase activity but not the proton-translocating activity. *See also ATP Synthase, H⁺-Transporting; Energase*

[1] G. E. Tusnady, E. Bakos, A. Varadi & B. Sarkadi (1997) *FEBS Lett.* **402**, 1.
[2] D. N. Sheppard & M. J. Welsh (1999) *Physiol. Rev.* **79**, S23.
[3] H. S. Penefsky (1974) *The Enzymes*, 3rd ed., **10**, 375.

Selected entries from **Methods in Enzymology** [vol, page(s)]:
General discussion: 97, 513; 126, 518; 297, 139
Assay: 97, 513; 126, 518
Other molecular properties: chloroplast (assay, 97, 513; 126, 518; coupling factor I, 23, 547 [activated enzyme, assay of, 23, 549; assay, 23, 253; assay of Ca^{2+}-ATPase activity of, 23, 548; assay of coupling activity of, 23, 547; chloroplast preparations deficient in, 23, 251; photophosphorylating system of *Rhodopseudomonas capsulata*, in, 23, 556; preparation, 23, 550; properties, 23, 555; purification by sucrose density gradient centrifugation, 23, 553]; extraction from chloroplasts, 22, 206; F$_1$ portion [inactivating modifications, 126, 713; noninactivating modifications, 126, 729; synthesis of enzyme-bound ATP, 126, 643]; individual subunits [antibody preparation, 118, 358; isolation, 118, 358; quantitative estimation, 118, 360; structure and function, 118, 355]; labeling *in vivo* in *Chlamydomonas reinhardi* cells, 97, 520; polypeptide electrotransfer and immunodecoration, 97, 514; purification, 97, 510; 126, 513; reconstitution, 126, 516; subunit-specific antibodies, preparation, 97, 515; synthesis assay, 97, 518; synthesis during development, analytical techniques, 118, 367)

*Cfr*10I RESTRICTION ENDONUCLEASE

This type II restriction endonuclease [EC 3.1.21.4], isolated from *Citrobacter freundii* RFL10, catalyzes the cleavage of both strands of DNA at 5′ . . . R∇CCGGY . . . 3′, where R refers to A or G and Y refers to T or C.[1] The enzyme can also exhibit star activity. The tetrameric form of *Cfr*10I is enzymatically active.[2]

[1] A. A. Janulaitis, P. S. Stakenas & Y. A. Berlin (1983) *FEBS Lett.* **161**, 210.
[2] V. Siksnys, R. Skirgaila, G. Sasnauskas, C. Urbanke, D. Cherny, S. Grazulis & R. Huber (1999) *J. Mol. Biol.* **291**, 1105.

*Cfr*AI RESTRICTION ENDONUCLEASE

This type I site-specific deoxyribonuclease [EC 3.1.21.3] is subtype β, isolated from *Citrobacter freundii*, and recognizes 5′ . . . GCANNNNNNNGTGG . . . 3′, where N refers to any base. However, the cleavage sites are random.[1] The type I site-specific methyltransferase activity of this system is also referred to specifically as M.*Cfr*AI.

[1] A. S. Daniel, F. V. Fuller-Pace, D. M. Legge & N. E. Murray (1988) *J. Bacteriol.* **170**, 1775.

*Cla*I RESTRICTION ENDONUCLEASE

This type II restriction endonuclease [EC 3.1.21.4], obtained from *Caryophanon latum* L., catalyzes the hydrolysis of both strands of DNA at 5′ . . . AT∇CGAT . . . 3′.[1] The rate-limiting step in the catalyzed reaction is the very slow dissociation of the enzyme-product complex.[2] The hydrolysis is blocked by overlapping *dam* methylation.

[1] H. Mayer, R. Grosschedl, H. Schutte & G. Hobom (1981) *Nucleic Acids Res.* **9**, 4833.
[2] B. Hinsch & M.-R. Kula (1981) *Nucleic Acids Res.* **9**, 3159.

cGMP PHOSPHODIESTERASE

This membrane-bound, homodimeric phosphodiesterase [EC 3.1.4.35] catalyzes the hydrolysis of 3′,5′-cyclic-GMP to form 5′-GMP.[1-3] **See also** *3′,5′-Cyclic-Nucleotide Phosphodiesterase*

[1] S. H. Francis, I. V. Turko & J. D. Corbin (2000) *Prog. Nucl. Acid Res. Mol. Biol.* **65**, 1.
[2] W. K. Sonnenburg & J. A. Beavo (1994) *Adv. Pharmacol.* **26**, 87.
[3] N. D. Goldberg & M. K. Haddox (1977) *Ann. Rev. Biochem.* **46**, 823.

Selected entries from **Methods in Enzymology** [vol, page(s)]:
General discussion: 81, 526, 532, 542; 159, 702, 722, 730; 315, 635, 646, 742
Assay: 57, 94; 81, 532; 238, 9, 11, 19
Other molecular properties: activation by (rhodopsin, 238, 3, 11; transducin, 238, 3, 11, 21; trypsin, 238, 9, 11, 13, 19); activity, 57, 96; bovine retina (assay, 238, 9, 11, 19; chromatography, 238, 6; extraction from rod outer segments, 238, 4, 19; structure, 238, 3, 13); inhibitory subunit, 238, 3, 13, 19 (factor Xa fusion protein, 238, 23; fluorescence labeling, 238, 24; transducin α subunit binding [affinity, 238, 24; assay, 238, 24; site, 238, 23, 25); noncatalytic binding sites, photoaffinity labeling, 159, 730; purification from rod outer segments, 159, 707; retinal rod outer segment (bovine [activation by {transducin α-subunit, 96, 623; trypsin, 96, 626}; controller enzymes, and, real-time assay, 81, 532; properties, 81, 537; 96, 621; purification, 96, 625 {light-regulated protein binding, by, 81, 564}; γ-subunit, purification, 96, 626]; inhibitors [assay, 81, 545; isolation, 81, 543; preparation from purified enzyme, 81, 543; role in visual transduction, 81, 546]; *Rana catesbiana* [properties, 81, 530; purification, 81, 527]); stimulation by calmodulin in intact tissue, assay, 159, 601

CHALCONE ISOMERASE

This isomerase [EC 5.5.1.6], also known as chalcone:flavonone isomerase, catalyzes the interconversion of a chalcone, such as 2′,4′,4-trihydroxychalcone, to a flavonone, in this case (2S)-4′,7-dihydroxyflavanone.[1-4]

[1] J. M. Jez, M. E. Bowman, R. A. Dixon & J. P. Noel (2000) *Nat. Struct. Biol.* **7**, 786.
[2] R. A. Bednar & J. R. Hadcock (1988) *JBC* **263**, 9583.
[3] R. A. Dixon, P. M. Dey & C. J. Lamb (1983) *Adv. Enzymol.* **55**, 1.
[4] M. J. Boland & E. Wong (1975) *EJB* **50**, 383.

CHANNEL-CONDUCTANCE-CONTROLLING ATPase

This so-called ATPase [EC 3.6.3.49], also known as cystic-fibrosis membrane-conductance-regulating protein, catalyzes the ATP-dependent (ADP- and orthophosphate-producing) transport of certain ions.[1,2] A mammalian member of this class of ATPases is active in forming a chloride channel. It participates in the functioning of other transmembrane channels. **See also** *ATPase, Cl⁻- Transporting*

The idea that a transporter is an enzyme is in keeping with a new definition of enzyme catalysis as the facilitated making/breaking of chemical bonds, not just covalent bonds.[3] This idea builds on Pauling's assertion that any long-lived, chemically distinct interaction (in this case, the persistent location of a solute with respect to the faces of a membrane) can be regarded as a chemical bond. Note also that the equilibrium constant ($K_{eq} = [Ion_{out}][ADP][P_i]/[Ion_{in}][ATP]$) does not conform to that expected for an ATPase (*i.e.*, $K_{eq} = [ADP][P_i]/[ATP]$). Thus, although the overall reaction yields ADP and orthophosphate, the enzyme is misclassified as a hydrolase, and instead should be regarded as an energase-type reaction. Energases facilitate affinity-modulated reactions by coupling the $\Delta G_{ATP\text{-hydrolysis}}$ to a force-generating or work-producing step.[3] In this case, P-O-P bond scission supplies the energy to drive ion transport. *See ATPase; Energase*

[1] G. E. Tusnady, E. Bakos, A. Varadi & B. Sarkadi (1997) *FEBS Lett.* **402**, 1.
[2] D. N. Sheppard & M. J. Welsh (1999) *Physiol. Rev.* **79**, S23.
[3] D. L. Purich (2001) *TiBS* **26**, 417.

Selected entries from *Methods in Enzymology* [vol, page(s)]:
General discussion: 294, 227

CHAPERONIN ATPases

This multisubunit enzymes [EC 3.6.4.9] catalyze the hydrolysis of ATP to produce ADP and orthophosphate in the maintainence of an unfolded polypeptide chain prior to folding or entry into an organelle such as the mitochondria or the chloroplast.[1–3] Examples include the *Escherichia coli* GroEL.

Although the overall reaction yields ADP and orthophosphate, the enzyme is misclassified as a hydrolase, and instead should be regarded as an energase-type reaction. Energases facilitate affinity-modulated reactions by coupling the $\Delta G_{ATP\text{-hydrolysis}}$ to a force-generating or work-producing step. In this case, P-O-P bond scission supplies the energy to drive protein folding.[4] The overall enzyme-catalyzed reaction is: $State_1 + ATP = State_1 + ADP + P_i$, where two affinity-modulated conformational states are indicated.[4] **See also** *Non-Chaperonin Molecular Chaperon ATPases; Chaperones (Chaperonins); Energase*

[1] N. A. Ranson, H. E. White & H. R. Saibil (1998) *BJ* **333**, 233.
[2] R. J. Ellis, ed., (1996) *The Chaperonins*, Academic Press, San Diego.

[3] S. M. Hemmingsen, C. Woolford, S. M. van der Vies, K. Tilly, D. T. Dennis, G. C. Georgopoulos, R. W. Hendrix & R. J. Ellis (1988) *Nature* **333**, 330.
[4] D. L. Purich (2001) *TiBS* **26**, 417.

Selected entries from *Methods in Enzymology* [vol, page(s)]:
General discussion: vol. 290

(S)-CHEILANTHIFOLINE SYNTHASE

This heme-thiolate-dependent (*i.e.*, cytochrome P450-dependent) oxidoreductase [EC 1.1.3.33], also known as (*S*)-scoulerine oxidase (methylenedioxy-bridge-forming), catalyzes the reaction of (*S*)-scoulerine with NADPH and dioxygen to produce (*S*)-cheilanthifoline, NADP⁺, and two water molecules.[1] Note that oxygen is not incorporated into the alkaloid product. A methylenedioxy bridge is formed by oxidative ring closure of adjacent phenolic and methoxy groups of scoulerine.

[1] W. Bauer & M. H. Zenk (1991) *Phytochemistry* **30**, 2953.

CHENODEOXYCHOLOYLTAURINE HYDROLASE

This enzyme [EC 3.5.1.74] catalyzes the hydrolysis of chenodeoxycholoyltaurine to produce chenodeoxycholate and taurine. Neither glycine conjugates nor choloyltaurine are substrates, indicating that the enzyme is highly specific.[1]

[1] K. Kawamoto, I. Horibe & K. Uchida (1989) *J. Biochem.* **106**, 1049.

CHITINASE

This enzyme [EC 3.2.1.14] (also known as chitodextrinase, 1,4-β-poly-*N*-acetylglucosaminidase, endochitinase, and poly-β-glucosaminidase) catalyzes the random hydrolytic scission of 1,4-β-linkages in *N*-acetyl-D-glucosamine polymers of chitin.[1–5] Some chitinases will also display lysozyme [EC 3.2.1.17] activity.

There are several different chitanases. Family 18 chitinases are found in bacteria, fungi, and higher plants. These chitinases act by a retaining mechanism in which the β-linked polymer is cleaved to release a β-anomeric product. Family 19 chitinases are found primarily in plants and some microorganisms. They are structurally unrelated to the family 18 proteins and operate through an stereo-inverting mechanism. Family 18 chitinases may form an oxazoline ion intermediate that is stabilized by the neighboring C2′ acetamido group. This distinguishes family 18 chitinase from family 19 chitinases which have two acidic residues near the active site. Family 19 barley chitinase has a glutamyl residue (Glu67) which protonates the anomeric oxygen linking two sugar residues in the substrate. An asparaginyl residue (Asn199) hydrogen bonds with the C2' N-acetyl group of one of these two sugar residues, and prevents the formation of an oxazoline ion intermediate. A hydrolysis product with inversion of the anomeric configuration occurs because of nucleophilic attack by a water molecule that is coordinated by Glu89 and Ser120.

[1] J. D. Robertus & A. F. Monzingo (1999) EXS **87**, 125.
[2] D. Koga, M. Mitsutomi, M. Kono & M. Matsumiya (1999) EXS **87**, 111.
[3] G. Davies, M. L. Sinnott & S. G. Withers (1998) CBC **1**, 119.
[4] E. Cabib (1987) Adv. Enzymol. **59**, 59.
[5] T. Imoto, L. N. Johnson, A. C. T. North, D. C. Phillips & J. A. Rupley (1972) The Enzymes, 3rd ed., **7**, 665.

Selected entries from **Methods in Enzymology** [vol, page(s)]:
General discussion: 8, 644; 161, 424, 426, 430, 457, 460, 462, 471, 474, 479, 484, 490, 498; 330, 319
Assay: 8, 645; 161, 430, 460, 467, 471, 474, 480, 485, 493, 496, 498; radiochemical assay, 161, 424; viscosimetric assay, 161, 426
Other molecular properties: hyperthermophilic, 330, 319; immunoprecipitation, 194, 694; O-linked mannooligosaccharides from, analysis, 194, 691; O-mannosylated, mobility shifts, 194, 687; occurrence, 8, 645; properties, 8, 649; 161, 500 (kidney bean, 161, 482; Neurospora, 161, 473; Pycnoporus, 161, 470; Serratia, 161, 461; Streptomyces antibioticus, 8, 649; tomato, 161, 489; Verticillium, 161, 477; wheat germ, 161, 500); purification, 8, 647 (kidney bean leaves, 161, 481; Neurospora crassa, 161, 472; puffballs, 161, 494; Pycnoporus cinnabarinus, 161, 468; Serratia marcescens, 161, 460; soybean seeds, 161, 492; Streptomyces antibioticus, 8, 648; tomato, 161, 487; Verticillium albo-atrum, 161, 475; wheat germ, 161, 498); secreted, isolation by chitin binding, 194, 696; separation from melibiase, 1, 250

CHITIN DEACETYLASE

This deacetylase [EC 3.5.1.41] catalyzes the hydrolysis of chitin (or, more specifically, the N-acetamido groups of N-acetyl-D-glucosamine residues in chitin) to produce chitosan and acetate.[1–4] The length of the oligomeric substrate is important. The *Mucor rouxii* enzyme cannot effectively deacetylate chitin oligomers with a degree of oligomerization lower than three.[1] In addition, for longer oligomers, the first acetyl group removed is from the nonreducing-end residue and hydrolysis continues in a processive fashion.[1]

[1] I. Tsigos, N. Zydowicz, A. Martinou, A. Domard & V. Bouriotis (1999) EJB **261**, 698.
[2] E. Cabib (1987) AE **59**, 59.
[3] L. L. Davis & S. Bartnicki-Garcia (1984) Biochemistry **23**, 1065.
[4] Y. Araki & E. Ito (1975) EJB **55**, 71.

Selected entries from **Methods in Enzymology** [vol, page(s)]:
Assay: 161, 511, 520
Other molecular properties: biological implications, 161, 523; properties, 161, 513, 522; protein determination, 161, 522; purification from (Colletotrichum lindemuthianum, 161, 521; Mucor rouxii, 161, 512)

CHITIN SYNTHASE

This enzyme [EC 2.4.1.16], also known as chitin-UDP N-acetylglucosaminyltransferase, catalyzes the reaction of UDP-N-acetyl-D-glucosamine with [(1,4)-(N-acetyl-β-D-glucosaminyl)]$_n$ to produce UDP and [(1,4)-(N-acetyl-β-D-glucosaminyl)]$_{n+1}$.[1–3] Interestingly, the synthase can also initiate the synthesis of chitin chains without a primer.[2]

[1] E. Cabib (1987) Adv. Enzymol. **59**, 59.
[2] M. S. Kang, N. Elango, E. Mattia, J. Au-Young, P. W. Robbins & E. Cabib (1984) JBC **259**, 14966.
[3] M. Fähnrich & J. Ahlers (1981) EJB **121**, 113.

Selected entries from **Methods in Enzymology** [vol, page(s)]:
General discussion: 28, 572; 138, 643
Assay: 28, 575; 138, 643
Other molecular properties: activator (assay, 28, 576; enzymatic properties, 28, 577; inhibitor [assay, 28, 578; isolation from Saccharomyces carlsbergensis, 28, 579; isolation from Saccharomyces cerevisiae, 28, 580; properties, 28, 580]; isolation from Saccharomyces cerevisiae, 28, 577); preparation from Saccharomyces cerevisiae, 28, 574; properties, 28, 575; 138, 648; purification from yeast, 138, 645; zymogen (activation of, 28, 573; assay, 28, 572; isolation from Saccharomyces cerevisiae, 28, 574)

CHITOBIASE

This enzyme activity, once classified as EC 3.2.1.29 but now a deleted entry, is now included with N-acetyl-β-glucosaminidase [EC 3.2.1.30].

CHITOBIOSYLDIPHOSPHODOLICHOL α-MANNOSYLTRANSFERASE

This transferase [EC 2.4.1.142] catalyzes the reaction of GDP-D-mannose with chitobiosyldiphosphodolichol to produce GDP and α-D-mannosylchitobiosyldiphosphodolichol.[1–3]

[1] G. P. Kaushal & A. D. Elbein (1987) Biochemistry **26**, 7953.
[2] G. P. Kaushal & A. D. Elbein (1986) ABB **250**, 38.
[3] C. B. Sharma, L. Lehle & W. Tanner (1982) EJB **126**, 319.

CHITOSANASE

This glycosidase [EC 3.2.1.132] catalyzes the endohydrolysis of β-1,4-linkages between N-acetyl-D-glucosamine and D-glucosamine residues in a partly acetylated (that is, 30% to 60% acetylation) chitosan.[1] The *Bacillus cereus* S1 enzyme reportedly recognizes more than two glucosamine residues posited in both sides against scissile bond for the glucosamine polymer.[2] There are six subsites in the *Streptomyces* enzyme.[1] It appears that a glutamyl residue acts as an acid and an aspartyl residue acts as a general base to activate a water molecule for an S_N2 attack on the glycosidic bond.[3] The enzyme was found to produce only the α-anomeric form of the product; hence, *Streptomyces chitosanase* is an inverting enzyme.

[1]T. Fukamizo & R. Brzezinski (1997) *Biochem. Cell Biol.* **75**, 687.

[2]M. Kurakake, S. Yo-u, K. Nakagawa, M. Sugihara & T. Komaki (2000) *Curr. Microbiol.* **40**, 6.

[3]E. M. Marcotte, A. F. Monzingo, S. R. Ernst, R. Brzezinski & J. D. Robertus (1996) *Nature Struct. Biol.* **3**, 155.

Selected entries from *Methods in Enzymology* [vol, page(s)]:
General discussion: 161, 501, 505
Assay: 161, 502, 506
Other molecular properties: properties, 161, 503, 508; purification from (*Bacillus*, 161, 502; *Streptomyces griseus*, 161, 507)

CHLORAMPHENICOL ACETYLTRANSFERASE

chloramphenicol

This transferase [EC 2.3.1.28] catalyzes the reaction of acetyl-CoA with chloramphenicol to produce Coenzyme A and chloramphenicol 3-acetate.[1-5] This acetyltransferase forms the basis for acquired chloramphenicol resistance in certain bacterial strains. The gene for the enzyme is also frequently used as reporter gene-product, because the level of its expression can be quantified by a sensitive activity assay.

[1]I. A. Murray & W. V. Shaw (1997) *Antimicrob. Agents Chemother.* **41**, 1.

[2]W. V. Shaw & A. G. W. Leslie (1991) *Ann. Rev. Biophys. Biophys. Chem.* **20**, 363.

[3]J. Ellis, C. R. Bagshaw & W. V. Shaw (1991) *Biochemistry* **30**, 10806.

[4]W. V. Shaw, P. Day, A. Lewendon & I. A. Murray (1988) *Biochem. Soc. Trans.* **16**, 939.

[5]W. V. Shaw (1983) *Crit. Rev. Biochem.* **14**, 1.

Selected entries from *Methods in Enzymology* [vol, page(s)]:
General discussion: 43, 737, 739
Assay: 43, 740; 204, 628; 217, 604; 238, 274; 255, 419 (reaction sequence, 43, 740); 189, 269; assay in reporter gene systems, 273, 324; radioactive assay, 43, 739, 743 ([^{14}C]acetyl-CoA, 43, 744; culture preparation, 43, 745, 746); fluorescent chloramphenicol derivative as substrate, 216, 369; spectrophotometric assay, 43, 739, 742, 743
Other molecular properties: acetylation techniques, 43, 739; [^{14}C]acetyl-chloramphenicol direct measurement, 43, 743; activators, 43, 755; acyl acceptor, specificity, 43, 753; acyl donor, specificity, 43, 753; affinity chromatography, 43, 750; alumina gel procedure, 43, 750; ammonium sulfate precipitation, 43, 748; assay (identification of gene regulatory elements, in, 152, 717; plant transformants, in, 153, 303; reporter gene systems, in, 273, 324); bacterial growth, 43, 747, 748; bidirectional vector construction, 273, 320; catabolite repression of synthesis, 43, 738, 739; chromatographic detection of acetylation, 43, 740, 741; coenzyme A complex (ligand conformation, 239, 667; NOESY experiment, 239, 668; coenzyme A sulfhydryl group, 43, 742; crude extract preparation, 43, 748; DEAE-cellulose

chromatography, 43, 749; diacetylchloramphenicol complex, protein-ligand interaction study, 239, 683; encoding gene, in collagen gene expression analysis, 144, 67; extracts, preparation, 216, 371; fluorescent chloramphenicol derivative as substrate, 216, 369; fusions, 326, 202 (*lac*, with, 326, 86); gel filtration, 43, 749; hepatoma cells, in, assay, 206, 413, 415; HIV protease fusion protein, 241, 33; hybrid formation, 43, 755; inhibitors, 43, 755; K_m values, 43, 754; molecular weight, 43, 752; native tetrameric, 43, 755; pH optimum, 43, 752; properties, 43, 752; protein fusions, 326, 202; purification, 43, 746; pyruvate-kinase system, 43, 744; reporter plasmids, 255, 414, 433; 326, 202; sensitivity as reporter gene, 273, 324; 326, 202; site-directed mutagenesis studies, 249, 107; specificity, 43, 753; stability, 43, 755; 255, 421; *Staphylococcus*, from, 43, 751, 752; structural gene, 43, 738; thin-layer chromatographic activity assay, 43, 739

CHLORAMPHENICOL HYDROLASE

p-nitrophenylserinol

This enzyme catalyzes the hydrolysis of chloramphenicol to produce dichloroacetate and *p*-nitrophenylserinol.

Selected entries from *Methods in Enzymology* [vol, page(s)]:
General discussion: 43, 734
Assay: *Streptomyces*, 43, 735
Other molecular properties: *Streptomyces*, 43, 734 (acetone powder, 43, 736; culture preparation, 43, 736; properties, 43, 736; purification, 43, 736; reaction scheme, 43, 734)

CHLORATE REDUCTASE

This oxidoreductase [EC 1.97.1.1] catalyzes the reaction of a reducing agent (AH$_2$) with chlorate (ClO$_3^-$) to produce the oxidized agent (A), water, and chlorite (ClO$_2^-$).[1-4] Electron acceptor substrates include flavins and benzylviologen. The enzyme from the chlorate-respiring bacterial strain GR-1 is an iron-molybdenum-selenium enyme that will also act on perchlorate.[4]

[1]L. F. Oltmann, V. P. Claassen, P. Kastelein, W. N. M. Reijnders & A. H. Stouthamer (1979) *FEBS Lett.* **106**, 43.

[2]L. F. Oltmann, W. N. M. Reijnders & A. H. Stouthamer (1976) *Arch. Microbiol.* **1111**, 25.

[3]E. Azoulay, S. Mutaftschiev & M. L. M. Rosado de Sousa (1971) *BBA* **237**, 579.

[4]S. W. Kengen, G. B. Rikken, W. R. Hagen, C. G. van Ginkel & A. J. Stams (1999) *J. Bacteriol.* **181**, 6706.

CHLORDECONE REDUCTASE

This oxidoreductase [EC 1.1.1.225] catalyzes the reaction of chlordecone (*i.e.*, 1,1a,3,3a,4,5,5a,5b,6-decachloroocta-hydro-1,3,4-metheno-2*H*-cyclobuta[*cd*]pentalen-2-one) with NADPH to produce chlordecone alcohol and NADP$^+$. Chlordecone is an organochlorine insecticide and fungicide.

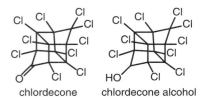

chlordecone   chlordecone alcohol

The bioreduction of the pesticide has been observed in human, gerbil, and rabbit liver but not in mouse, rat, or hamster.[1,2]

[1] D. T. Molowa, A. G. Shayne & P. S. Guzelian (1986) JBC **261**, 12624.
[2] C. J. Winters, D. T. Molowa & P. S. Guzelian (1990) Biochemistry **29**, 1080.

CHLORIDAZON-CATECHOL DIOXYGENASE

This iron-dependent oxidoreductase [EC 1.13.11.36] catalyzes the reaction of 5-amino-4-chloro-2-(2,3-dihydroxyphenyl)-3(2H)-pyridazinone with dioxygen to produce 5-amino-4-chloro-2-(2-hydroxymuconoyl)-3(2H)- pyridazinone.[1,2] The enzyme is distinct from catechol 1,2-dioxygenase [EC 1.13.11.1], catechol 2,3-dioxygenase [EC 1.13.11.2], or homogentisate 1,2-dioxygenase [EC 1.13.11.5].

[1] S. Schmitt, R. Müller & F. Lingens (1984) J. Immunol. Meth. **68**, 263.
[2] R. Müller, S. Haug, J. Eberspächer & F. Lingens (1977) Hoppe-Seyler's Z. Physiol. Chem. **358**, 797.

CHLORIDE-TRANSPORTING ATPase

This so-called ATPase [EC 3.6.3.11], also known as Cl$^-$-translocating ATPase, Cl$^-$-motive ATPase, and Cl$^-$ pump, catalyzes the ATP-dependent (ADP- and orthophosphate-producing) transport of Cl$^-_{out}$ to produce Cl$^-_{in}$.[1–3] This is a P-type ATPase that undergoes covalent phosphorylation during the transport cycle. **See also** ATPase, Channel-Conductance-Controlling

The idea that a transporter is an enzyme is in keeping with a new definition of enzyme catalysis as the facilitated making/breaking of chemical bonds, not just covalent bonds.[4] This idea builds on Pauling's assertion that any long-lived, chemically distinct interaction (in this case, the persistent location of a solute with respect to the faces of a membrane) can be regarded as a chemical bond. Note also that the equilibrium constant ($K_{eq} = $ [Cl$^-_{out}$][ADP][P$_i$]/[Cl$^-_{in}$][ATP]) does not conform to that expected for an ATPase

(*i.e.*, $K_{eq} = $ [ADP][P$_i$]/[ATP]). Thus, although the overall reaction yields ADP and orthophosphate, the enzyme is misclassified as a hydrolase, and instead should be regarded as an energase-type reaction. Energases facilitate affinity-modulated reactions by coupling the $\Delta G_{ATP\text{-hydrolysis}}$ to a force-generating or work-producing step.[4] In this case, P-O-P bond scission supplies the energy to drive chloride transport. **See** ATPase; Energase

[1] T. Ohhashi, T. Katsu & M. Ikeda (1992) BBA **1106**, 165.
[2] G. A. Gerencser & K. R. Purushotham (1996) J. Bioenerget. Biomembr. **28**, 459.
[3] C. Inagaki, M. Hara & X. T. Zeng (1996) J. Exp. Zool. **275**, 262.
[4] D. L. Purich (2001) TiBS **26**, 417.

Selected entries from **Methods in Enzymology** [vol, page(s)]:
General discussion: 294, 227

3-CHLORO-D-ALANINE DEHYDROCHLORINASE

This pyridoxal-phosphate-dependent enzyme [EC 4.5.1.2] catalyzes the α,β-elimination reaction of 3-chloro-D-alanine with water to produce pyruvate, chloride ion, and ammonium ion.[1,2] Hydrogen sulfide can replace water in a β-replacement reaction to generate D-cysteine, chloride ion, and a proton.[2]

[1] H. Yamada, T. Nagasawa, H. Ohkishi, B. Kawakami & Y. Tani (1981) BBRC **100**, 1104.
[2] T. Nagasawa, H. Ohkishi, B. Kawakami, H. Yamano, Y. Hosono, Y. Tani & H. Yamada (1982) JBC **257**, 13749.

4-CHLOROBENZOATE DEHALOGENASE

This enzyme [EC 3.8.1.6] catalyzes the hydrolysis of 4-chlorobenzoate to produce 4-hydroxybenzoate and chloride ion. The 4-bromo- and 4-iodo-analogues are also substrates. In many microorganisms (for example, *Pseudomonas* sp. and *Arthrobacter* sp.), this activity comprises three separate enzymes: 4-chlorobenzoate-CoA ligase [EC 6.2.1.33], 4-chlorobenzoyl-CoA dehalogenase [EC 3.8.1.7], and 4-hydroxybenzoyl-CoA thioesterase [EC 3.1.2.23].[1] **See also** 4-Chlorobenzoyl-CoA Dehalogenase

[1] K.-H. Chang, P.-H. Liang, W. Beck, J. D. Scholten & D. Dunaway-Mariano (1992) Biochemistry **31**, 5605.

2-CHLOROBENZOATE 1,2-DIOXYGENASE

This iron-dependent oxidoreductase [EC 1.14.12.13], also known as 2-halobenzoate 1,2-dioxygenase, catalyzes the reaction of 2-chlorobenzoate with NADH and dioxygen to produce catechol, chloride ion, carbon dioxide, and NAD$^+$. Other substrates include the 2-fluoro-, 2-bromo-, 2-iodo-, 2-methoxy-, and 2-aminobenzoates. Two components have been identified in the *Burkholderia* (*Pseudomonas*) *cepacia* enzyme: an iron-sulfur NADH-acceptor reductase and an iron-sulfur flavoprotein.[1]

[1] S. Fetzner, R. Muller & F. Lingens (1992) *J. Bacteriol.* **174**, 279.

4-CHLOROBENZOYL-CoA DEHALOGENASE

This enzyme [EC 3.8.1.7] catalyzes the hydrolysis of 4-chlorobenzoyl-CoA to produce 4-hydroxybenzoyl-CoA and chloride ion.[1–5] The dehalogenase is specific for the 4-position. Alternative substrates include the 4-fluoro-, 4-bromo-, and 4-iodo-derivatives (however, the rate with the fluoro derivative is quite slow: in fact, there is a report that the *Pseudomonas* enzyme will not act at all on 4-fluorobenzoyl-CoA[3]).

The reaction proceeds by attack of an active-site aspartyl residue (Asp145) at the 4-position of the benzoyl ring to form a Meisenheimer complex. Loss of chloride ion from this intermediate results in the formation of an arylated enzyme intermediate. Finally, the arylated enzyme is hydrolyzed to free enzyme plus the product.[4,5] A histidyl residue serves as the general base in the hydrolysis step.

[1] K.-H. Chang, P.-H. Liang, W. Beck, J. D. Scholten & D. Dunaway-Mariano (1992) *Biochemistry* **31**, 5605.
[2] G. P. Crooks & S. D. Copley (1994) *Biochemistry* **33**, 11645.
[3] F. Loffler, F. Lingens & R. Muller (1995) *Biodegradation* **6**, 203.
[4] G. Yang, P.-H. Liang & D. Dunaway-Mariano (1994) *Biochemistry* **33**, 8527.
[5] K. L. Taylor, R. Q. Liu, P. H. Liang, J. Price, D. Dunaway-Mariano, P. J. Tonge, J. Clarkson & P. R. Carey (1995) *Biochemistry* **34**, 13881.

4-CHLOROBENZOYL-CoA SYNTHETASE

This ligase [EC 6.2.1.33], also called 4-chlorobenzoate: CoA ligase, catalyzes the reaction of 4-chlorobenzoate with coenzyme A and ATP to produce 4-chlorobenzoyl-CoA, AMP, and pyrophosphate (or, diphosphate). This magnesium-dependent enzyme is a component in the bacterial 2,4-dichlorobenzoate degradation pathway. Alternative substrates include benzoate, 4-bromobenzoate, 4-iodobenzoate, and 4-methylbenzoate.[1]

[1] K.-H. Chang, P.-H. Liang, W. Beck, J. D. Scholten & D. Dunaway-Mariano (1992) *Biochemistry* **31**, 5605.

2-CHLORO-4-CARBOXYMETHYLENEBUT-2-EN-1,4-OLIDE ISOMERASE

This isomerase [EC 5.2.1.10] catalyzes the interconversion of *cis*-2-chloro-4-carboxymethylenebut-2-en-1,4-olide and *trans*-2-chloro-4-carboxymethylenebut-2-en-1,4-olide.[1]

[1] U. Schwien, E. Schmidt, H.-J. Knackmuss & W. Reinecke (1988) *Arch. Microbiol.* **150**, 78.

CHLOROCATECHOL 1,2-DIOXYGENASE

This iron-dependent oxidoreductase [EC 1.13.11.1] catalyzes the reaction of 4-chlorocatechol (*i.e.*, 1,2-dihydroxy-4-chlorobenzene) with dioxygen to produce 3-chloro-*cis*,*cis*-muconate. The enzyme has the same EC number as catechol 1,2-dioxygenase and is currently regarded as a variant of that enzyme; even so, catechol 1,2-dioxygenase does not act on 4-chlorocatechol.

Selected entries from *Methods in Enzymology* [vol, page(s)]:
Assay: 188, 123
Other molecular properties: 5-chloro-3-methylcatechol dioxygenase, 18C, 396; 52, 11; purification from *Pseudomonas* B13, 188, 125; yield, storage, stability, and properties, 188, 126

CHLOROGENATE:GLUCARATE O-HYDROXYCINNAMOYLTRANSFERASE

This acyltransferase [EC 2.3.1.98] catalyzes the reaction of chlorogenate with glucarate to produce quinate and 2-O-caffeoylglucarate.[1] Galactarate can act as an acceptor substrates, albeit not as effectively as glucarate. This enzyme participates with quinate O-hydroxycinnamoyltransferase [EC 2.3.1.99] in the biosynthesis of caffeoylglucarate in tomatoes.

[1] D. Strack & W. Gross (1990) *Plant Physiol.* **92**, 41.

CHLOROGENATE HYDROLASE

This enzyme [EC 3.1.1.42], also known as chlorogenase, catalyzes the hydrolysis of chlorogenate to produce caffeate and quinate.[1,2] Other substrates include isochlorogenate and 3,5-di-*O*-caffeylquinate.

[1] B. Schöbel (1980) *Z. Naturforsch.* **35c**, 209.
[2] B. Schöbel & W. Pollmann (1980) *Z. Naturforsch.* **35c**, 699.

CHLOROMUCONATE CYCLOISOMERASE

This manganese-dependent enzyme [EC 5.5.1.7], also known as muconate cycloisomerase II, catalyzes the interconversion of 3-chloro-*cis*,*cis*-muconate and 2-chloro-2,5-dihydro-5-oxofuran-2-acetate.[1] The product undergoes spontaneous elimination of HCl and forms *cis*-4-carboxymethylenebut-2-en-4-olide. Other substrates include 2-chloro-*cis*,*cis*-muconate. The enzyme is distinct from muconate cycloisomerase [EC 5.5.1.1] or dichloromuconate cycloisomerase [EC 5.5.1.11].

[1] E. Schmidt & H. J. Knackmuss (1980) *BJ* **192**, 339.

CHLOROPEROXIDASE

This oxidoreductase [EC 1.11.1.10], also called chloride peroxidase, catalyzes the reaction of hydrogen peroxide with two RH and two chloride anions to produce two R–Cl and two water molecules.[1–5] An HOCl intermediate has been detected spectrophotometrically. This haloperoxidase can also catalyze bromination and iodination, but not fluorination; however, there is a preference for Cl⁻. A common substrate in assay protocols is monochlorodimedone (*i.e.*, 2-chloro-5,5-dimethyl-1,3-dimedone), resulting in the formation of dichlorodimedone. There are a number of different haloperoxidases, including iron-heme enzymes, vanadium-dependent enzymes, and metal-free systems. The iron-heme enzymes also have a catalase activity. Myeloperoxidase also exhibits chloroperoxidase activity.

See also *Peroxidase; Bromoperoxidase; Iodide Peroxidase; Myeloperoxidase*

[1]H. B. Dunford (1998) *CBC* **3**, 195.
[2]A. Butler (1998) *CBC* **3**, 427.
[3]M. A. Ator & P. R. Ortez de Montellano (1990) *The Enzymes*, 3rd ed., **19**, 213.
[4]A.-M. Lambeir & H. B. Dunford (1983) *JBC* **258**, 13558.
[5]B. N. Campbell, T. A. Araiso, A. L. Reinisch, K. T. Yue & L. P. Hager (1982) *Biochemistry* **21**, 4343.

Selected entries from *Methods in Enzymology* [vol, page(s)]:
General discussion: 17A, 648; 52, 521
Assay: 52, 522
Other molecular properties: activity, 52, 521, 522; *Caldariomyces fumago*, in, 52, 521; chemical properties, 52, 528; extinction coefficients, 17A, 652; inhibitors, 52, 529; iodination of tyrosine by, 17A, 649; magnetic circular dichroism, 49, 170; Mössbauer studies, 54, 355, 378; compound I, 54, 379; molecular weight, 52, 528; purification, 52, 525 (from *Caldariomyces fumago*, 17A, 651); purity determination, 52, 524; specificity, 52, 529; spectral properties, 52, 528, 529; stability, 52, 528

CHLOROPHENOL O-METHYLTRANSFERASE

This methyltransferase [EC 2.1.1.136], also known as halogenated phenol O-methyltransferase and trichlorophenol O-methyltransferase, catalyzes the reaction of S-adenosyl-L-methionine with trichlorophenol to produce S-adenosyl-L-homocysteine and trichloroanisole.

4-CHLOROPHENYLACETATE 3,4-DIOXYGENASE

This oxidoreductase [EC 1.14.12.9] catalyzes the reaction of 4-chlorophenylacetate with NADH, dioxygen, and water to produce 3,4-dihydroxyphenylacetate, chloride ion, and NAD⁺.[1,2] The *Pseudomonas* dioxygenase is a protein complex containing an iron-sulfur reductase and an iron-sulfur FMN-dependent oxygenase but

no independent ferredoxin. Alternative substrates include 4-bromophenylacetate and 4-fluorophenylacetate.

[1]A. Markus, D. Krekel & F. Lingens (1986) *JBC* **261**, 12883.
[2]D. Schweizer, A. Markus, M. Seez, H. H. Ruf & F. Lingens (1987) *JBC* **262**, 9340.

CHLOROPHYLLASE

This enzyme [EC 3.1.1.14] catalyzes the hydrolysis of chlorophyll to produce phytol and chlorophyllide.[1–4] The enzyme also reportedly catalyzes chlorophyllide transfer reactions, including the conversion of chlorophyll to methylchlorophyllide. The enzyme interacts preferentially with substrates having the isocyclic carbomethoxy and the C17 propionate groups facing opposite sides of the porphyrin macrocycle.[5]

[1]K. Tanaka, T. Kakuno, J. Yamashita & T. Horio (1982) *J. Biochem.* **92**, 1763.
[2]K. Shimokawa (1982) *Phytochemistry* **21**, 543.
[3]S. Granick & S. I. Beale (1978) *Adv. Enzymol.* **46**, 33.
[4]A. O. Klein & W. Vishniac (1961) *JBC* **236**, 2544.
[5]L. Fiedor, V. Rosenbach-Belkin & A. Scherz (1992) *JBC* **267**, 22043.

Selected entries from *Methods in Enzymology* [vol, page(s)]:
General discussion: 123, 421
Assay: *Chlorella prototothecoides*, 123, 422
Other molecular properties: action on chlorophyll, 23, 421, 470; *Chlorella prototothecoides* (assays, 123, 422; properties, 123, 426; purification, 123, 424); cytochrome photooxidase and, 6, 409

CHLOROPLAST PROTEIN-TRANSPORTING ATPase

This set of so-called ATPase [EC 3.6.3.52] catalyze ATP-dependent (forming ADP and orthophosphate) transport of proteins and preproteins into the chloroplast stroma.[1,2] There are a number of variants of this system that vary considerably in their specificity and kinetics. Included among the proteins translocated is the light-harvesting chlorophyll a/b protein.

The idea that a transporter is an enzyme is in keeping with a new definition of enzyme catalysis as the facilitated making/breaking of chemical bonds, not just covalent bonds.[3] This idea builds on Pauling's assertion that any long-lived, chemically distinct interaction (in this case, the persistent location of a solute with respect to the faces of a membrane) can be regarded as a chemical bond. Note also that the equilibrium constant ($K_{eq} = [\text{Protein}_{out}][\text{ADP}][P_i]/[\text{Protein}_{in}][\text{ATP}]$) does not conform to that expected for an ATPase (*i.e.*, $K_{eq} = [\text{ADP}][P_i]/[\text{ATP}]$). Thus, although the overall reaction yields ADP and orthophosphate, the enzyme is misclassified as a hydrolase, and instead should be regarded as an energase-type reaction. Energases facilitate affinity-modulated reactions by coupling the $\Delta G_{ATP\text{-hydrolysis}}$ to a force-generating or work-producing step.[3] In this case, P-O-P bond scission supplies the energy to drive protein transport. *See ATPase; Energase*

[1]K. Cline, N. F. Ettinger & S. M. Theg (1992) *JBC* **267**, 2688.
[2]S. V. Scott & S. M. Theg (1996) *J. Cell Biol.* **132**, 63.
[3]D. L. Purich (2001) *TiBS* **26**, 417.

CHOLERA TOXIN

Cholera toxin, also known as choleragen, is an enterotoxin produced by *Vibrio cholerae* that has an ADP-ribosyltransferase activity which permanently activates the G_S regulatory protein in the adenylate cyclase pathway.[1-5]

[1]J. Moss, R. S. Hawn, S.-C. Tsai, C. F. Welsh, F.-J. S. Lee, S. R. Price & M. Vaughan (1994) *Meth. Enzymol.* **237**, 44.
[2]J. Moss, S.-C. Tsai & M. Vaughan (1994) *Meth. Enzymol.* **235**, 640.
[3]N. J. Oppenheimer & A. L. Handlon (1992) *The Enzymes*, 3rd ed., **20**, 453.
[4]G. S. Kopf & M. J. Woolkalis (1991) *Meth. Enzymol.* **195**, 257.
[5]J. Moss & M. Vaughan (1988) *Adv. Enzymol.* **61**, 303.

Selected entries from *Methods in Enzymology* [vol, page(s)]:
General discussion: 34, 613; **195**, 257; **235**, 640; **237**, 44
Assay: 165, 170; **195**, 253, 255, 279; **235**, 524, 624; ADP-ribosyltransferase, **106**, 414; **165**, 245; **235**, 642
Other molecular properties: activation (ADP-ribosylation factors, by [ARF characteristics, **235**, 641; assay, **235**, 644; **237**, 48; effects of assay components, **235**, 642; overview, **237**, 44]; GTP, by, **237**, 46); activation of adenylyl cyclase, role of soluble guanine-nucleotide-dependent ADP-ribosylation factor, **195**, 243; ADP-ribosylation of membrane N_s and N_i components, **109**, 566; ADP-ribosyltransferase activity (assays, **106**, 414; **165**, 245; comparison with pertussis toxin enzymatic activity, **165**, 242; competing reactions and artifactual results, **165**, 243; demonstration, **106**, 413; effect on adenylate cyclase system, **106**, 411; presence of high NADase activity, in, **165**, 241; procedures and analyses *in vitro*, **165**, 236; required cytosolic factor, assay and purification, **165**, 246); antibodies (purification, **84**, 239, 241; radioimmunoassay, **84**, 251; radioiodination, **84**, 241; solid phase-bound, preparation, **84**, 241); antiserum, preparation, **84**, 238; A1 protein, **237**, 44; A2 protein, **237**, 44; auto-ADP-ribosylation, assay, **235**, 646; B subunit, ligand-binding domain, analysis with anti-Id antibody to G_{M1}, **178**, 165; cAMP, **38**, 3; catalysis of ADP-ribosylation (activity, **235**, 632; assay, **235**, 642; G proteins, **237**, 24 [α_s subunits, **237**, 243, 268]; structural analysis of heterotrimeric G proteins, in, **237**, 71); competition with interferon, **79**, 459, 460; continuous epitopes in, location, prediction, **203**, 190; crude, affinity chromatography, **34**, 616, 618, 619; cytopathogenic effects, **235**, 682; effects of ADP-ribosylation factor, mechanism, **195**, 249; effects on (GTP hydrolysis, **237**, 14; neutrophil chemotaxis, **236**, 66); enzyme-linked immunosorbent assay, **235**, 525; expression (environmental growth conditions affecting, **235**, 521; strain differences, **235**, 520); fragment A_1, generation, **195**, 268; ganglioside-agarose, **34**, 610; genes, genetic regulation, **235**, 519; G protein sensitivity, **238**, 373; G_s, reconstitution assay for ADP-ribosylation factor, **195**, 236; hydrolysis, **237**, 45; ^{125}I-labeled, **34**, 613 (assay method, **34**, 616); immunoassay, **73**, 391; mechanism of action, **235**, 680; NAD$^+$:cholera toxin A1 auto-ADP-ribosyltransferase, assay,**195**, 253; NAD$^+$:$G_{S\alpha}$ ADP-ribosyltransferase, assay, **195**, 255; polyfunctional spacer arms, **34**, 94; production, **165**, 171; properties, **165**, 175; purification, **165**, 171; purified, affinity chromatography of, **34**, 616, 617; radioimmunoassay, **84**, 243; radiolabeled (binding to thin-layer chromatograms, glycolipid

ligand detection, **83**, 238; preparation, **84**, 240; purification, **84**, 241); receptor binding, **235**, 623; receptor, gangliosides, **50**, 250; regulation by environmental factors, **235**, 517; structure, **235**, 518, 640; substrates, **237**, 45; subunit, radioiodination, **70**, 237; subunits, **235**, 623; synthesis, **235**, 623; vaccine preparation with glutaraldehyde, **93**, 36

CHOLESTANETETRAOL 26-DEHYDROGENASE

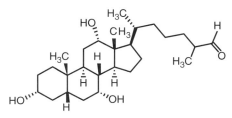

5β-cholestan-3α,7α,12α,26-tetraol

3α,7α,12α-trihydroxy-5β-cholestan-26-al

This oxidoreductase [EC 1.1.1.161] catalyzes the reversible reaction of 5β-cholestane-3α,7α,12α,26-tetraol with NAD$^+$ to produce 3α,7α,12α-trihydroxy-5β-cholestan-26-al and NADH.[1,2] In some organisms, this activity may be catalyzed by an isozyme of alcohol dehydrogenase [EC 1.1.1.1].[1]

[1]A. Okuda & K. Okuda (1983) *JBC* **258**, 2899.
[2]T. Masui, R. Herman & E. Staple (1966) *BBA* **117**, 266.

CHOLESTANETRIOL 26-MONOOXYGENASE

This heme-thiolate-dependent oxidoreductase [EC 1.14.13.15], also known as cholestanetriol 26-hydroxylase and 5β-cholestane-3α,7α,12α-triol 26-hydroxylase, catalyzes the reaction of 5β-cholestane-3α,7α,12α-triol with NADPH and dioxygen to produce 5β-cholestane-3α,7α, 12α,26-tetraol, NADP$^+$, and water.[1,2]

Note that both the 25S and 25R stereoisomers have been observed in several vertebrates; hence, there may be more than one 26-monooxygenase.[3] The rat liver enzyme produces the 25R diastereoisomer.[1]

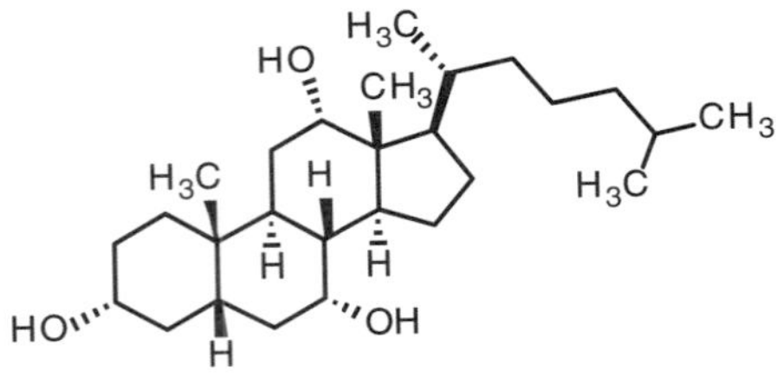

5β-cholestan-3α,7α,12α-triol

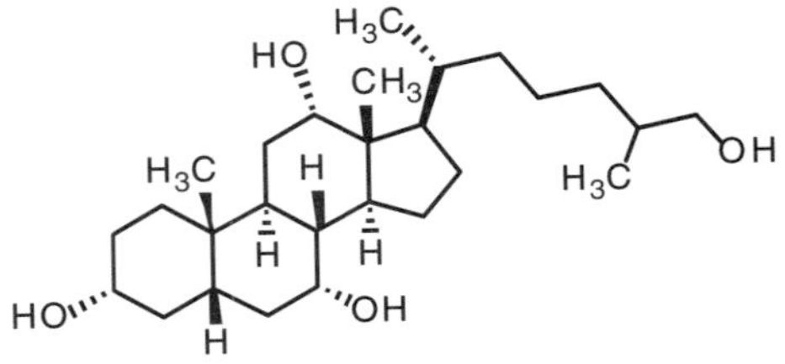

5β-cholestan-3α,7α,12α,26-tetraol

[1] Y. Atsuta & K. Okuda (1982) *J. Lipid Res.* **23**, 345.
[2] S. Taniguchi, N. Hoshita & K. Okuda (1973) *EJB* **40**, 607.
[3] M. Une, H. G. Kim, M. Yoshii, T. Kuramoto & T. Hoshita (1997) *Steroids* **62**, 458.

CHOLEST-5-ENE-3β,7α-DIOL 3β-DEHYDROGENASE

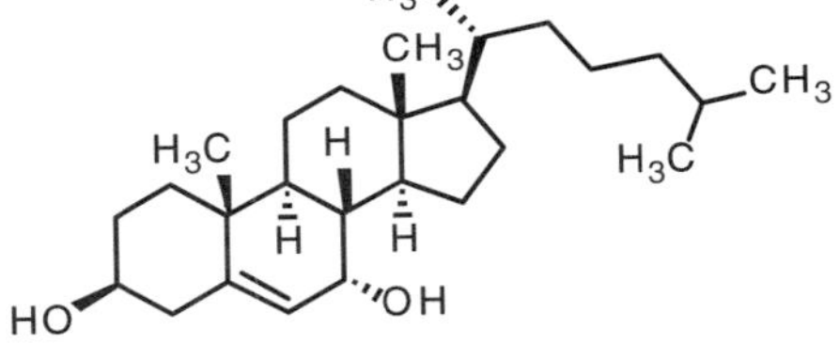

cholest-5-ene-3β,7α-diol

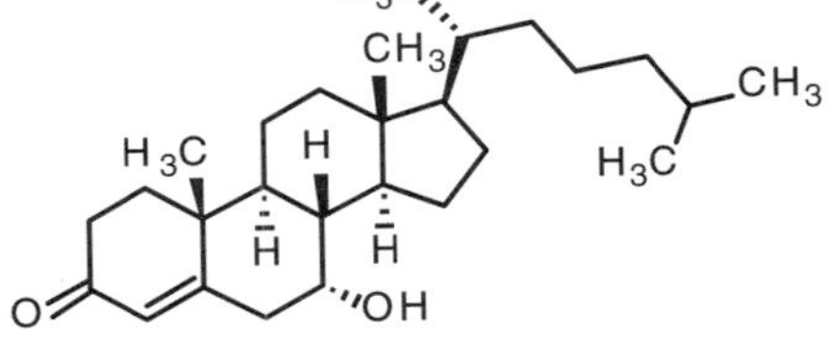

7α-hydroxycholest-4-en-3-one

This oxidoreductase [EC 1.1.1.181], also known as 3β-hydroxy-Δ^5-C_{27}-steroid oxidoreductase, catalyzes the reaction of cholest-5-ene-3β,7α-diol with NAD^+ to produce 7α-hydroxycholest-4-en-3-one and NADH.[1] The enzyme is very specific for 3β-hydroxy-C_{27}-steroids with a Δ^5-double bond.

[1] K. Wikvall (1981) *JBC* **256**, 3376.

CHOLESTENOL Δ-ISOMERASE

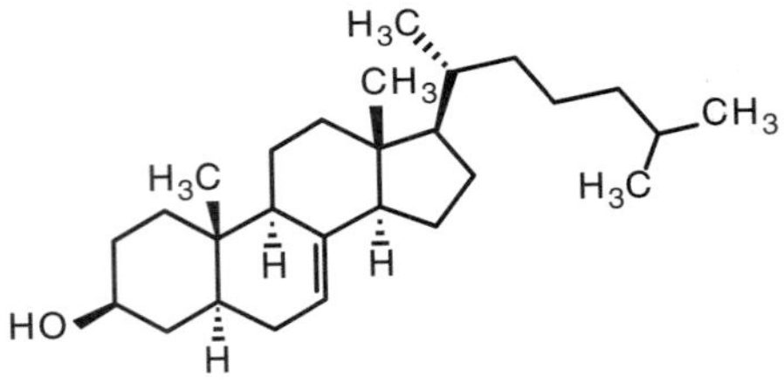

5α-cholest-7-en-3β-ol

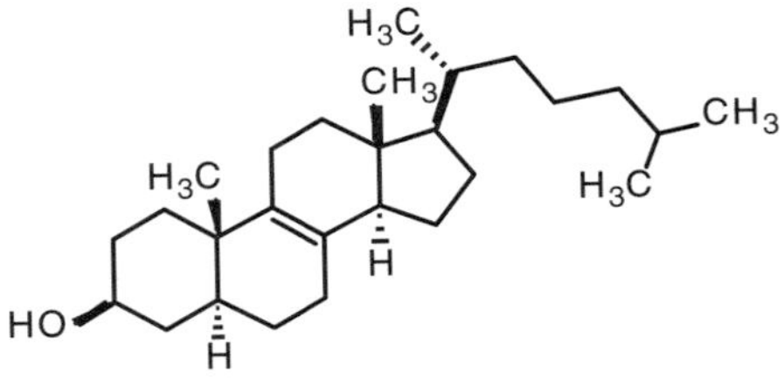

5α-cholest-8-en-3β-ol

This isomerase [EC 5.3.3.5] catalyzes the interconversion of 5α-cholest-7-en-3β-ol and 5α-cholest-8-en-3β-ol.[1-3]

[1] A. Svala, M. Galli-Kienle, M. Anastasia & G. Galli (1974) *EJB* **48**, 263.
[2] W. H. Lee, R. Kammereck, B. N. Lutsky, J. A. Closkey & G. J. Schroepfer, Jr., (1969) *JBC* **244**, 2033.
[3] D. C. Wilton, A. D. Rahimtula & M. Akhtar (1969) *BJ* **114**, 71.

CHOLESTENONE 5α-REDUCTASE

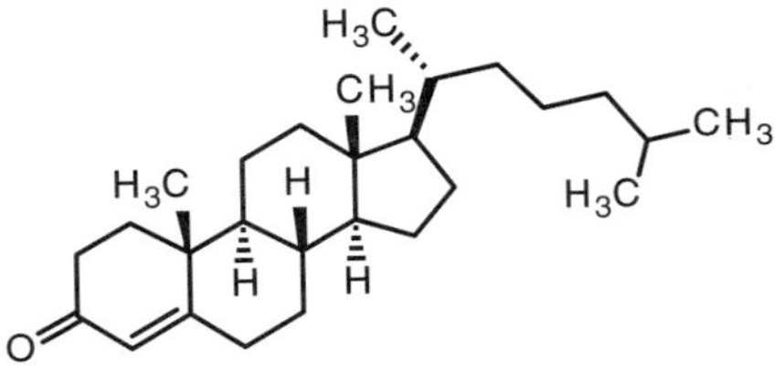

cholest-4-en-3-one

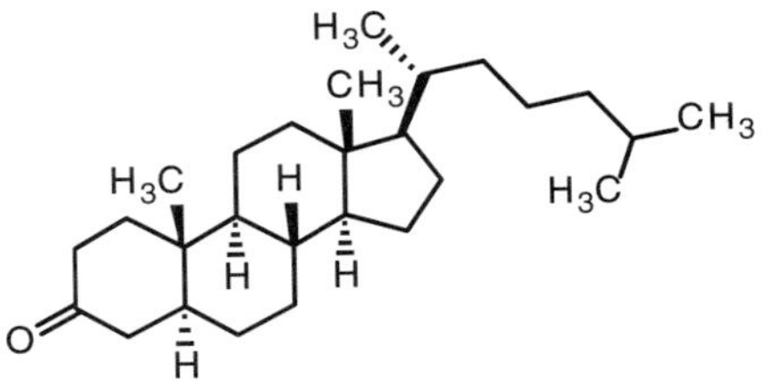

5α-cholestan-3-one

This oxidoreductase [EC 1.3.1.22] catalyzes the reaction of cholest-4-en-3-one with NADPH to produce 5α-cholestan-3-one and $NADP^+$.[1-4] Other substrates include testosterone, androstenedione, and progesterone.

[1] H. Scheer & B. Robaire (1983) *BJ* **211**, 65.
[2] D. Thouvenot & R. F. Morfin (1980) *J. Steroid Biochem.* **13**, 1337.

[3]B. Houston, G. D. Chisholm & F. K. Habib (1987) *Steroids* **49**, 355.
[4]G. M. Cooke & B. Robaire (1987) *J. Steroid Biochem.* **26**, 361.

Selected entries from *Methods in Enzymology* [**vol**. page(s)]:
General discussion: **36**, 466

CHOLESTENONE 5β-REDUCTASE

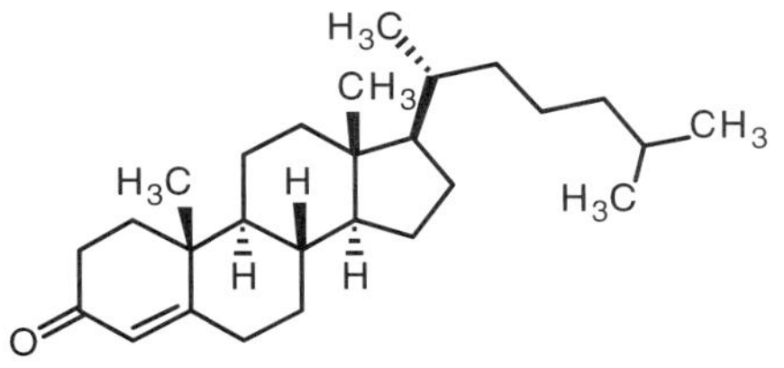

cholest-4-en-3-one

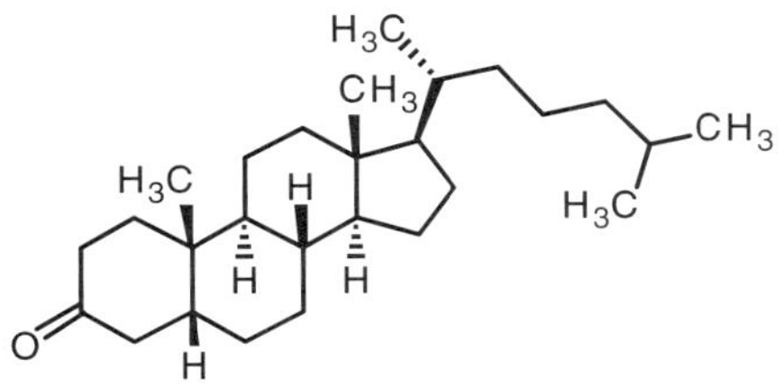

5β-cholestan-3-one

This oxidoreductase [EC 1.3.1.23], also known as 5β-steroid reductase, catalyzes the reaction of cholest-4-en-3-one with NADPH to produce 5β-cholestan-3-one and $NADP^+$.[1-4] Other steroid substrates include 4-androstene-3,17-dione, 7α-hydroxy-4-cholesten-3-one, testosterone, and 7α,12α-dihydroxy-4-cholesten-3-one. The rat liver enzyme is identical to cortisone β-reductase [EC 1.3.1.3].[1]

See also Cortisone β-Reductase

[1]M. Furuebisu, S. Deguchi & K. Okuda (1987) *BBA* **912**, 110.
[2]A. Okuda & K. Okuda (1984) *JBC* **259**, 7519.
[3]W. Collins & E. H. D. Cameron (1975) *BJ* **147**, 165.
[4]E. J. Van Doorn, J. C. Nduaguba & A. F. Clark (1973) *Can. J. Biochem.* **51**, 1661.

Selected entries from *Methods in Enzymology* [**vol**. page(s)]:
General discussion: **15**, 557
Assay: **15**, 557
Other molecular properties: purification, **15**, 559; relative rates of reaction with different steroids, **15**, 560

CHOLESTEROL ACYLTRANSFERASE

This acyl transferase [EC 2.3.1.26] (also known as sterol *O*-acyltransferase, sterol-ester synthase, acyl-CoA:cholesterol *O*-acyltransferase, and ACAT) catalyzes the reaction of an acyl-CoA with cholesterol to produce coenzyme A and the cholesterol ester.[1-4]

The mammalian enzyme prefers acyl groups with a single *cis* double bond at position C9. Acyl donor substrates include oleoyl-CoA, palmitoleoyl-CoA, palmitoyl-CoA, and linoleoyl-CoA. Other acceptor substrates include ergosterol, zymosterol, episterol, and fecosterol. A number of isozymes of ACAT have been identified. *See also* Phosphatidylcholine:Cholesterol Acyltransferase

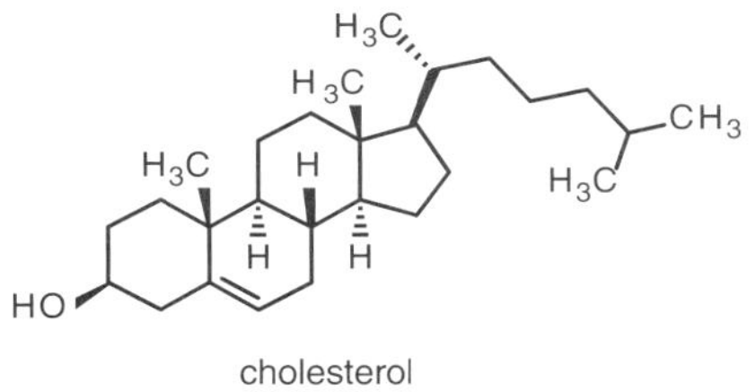

cholesterol

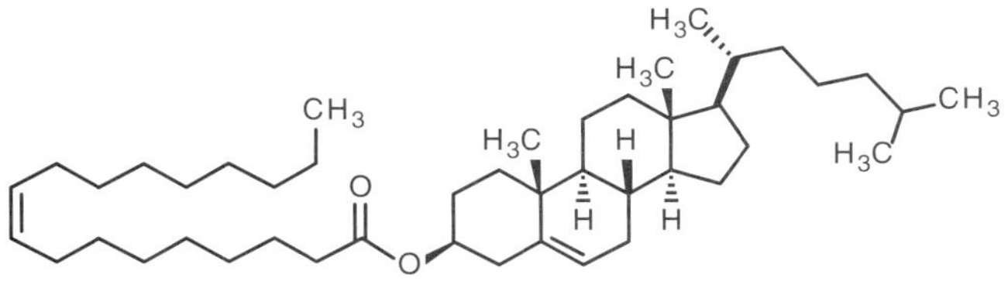

cholesteroyl oleate

[1]T.-Y. Chang, C. C. Y. Chang & D. Cheng (1997) *Ann. Rev. Biochem.* **66**, 613.
[2]T.-Y. Chang & G. M. Doolittle (1983) *The Enzymes*, 3rd ed., **16**, 523.
[3]E. R. Simpson & M. F. Burkhart (1980) *ABB* **200**, 79.
[4]S. Taketani, T. Nishino & H. Katsuki (1979) *BBA* **575**, 148.

Selected entries from *Methods in Enzymology* [**vol**, page(s)]:
General discussion: **111**, 286
Assay: rat liver, **111**, 286
Other molecular properties: metabolic function, **128**, 17; rat liver (activation, **111**, 293; assay, **111**, 286; inhibitors, **111**, 291; substrate specificity, **111**, 290)

CHOLESTEROL ESTERASE

This enzyme [EC 3.1.1.13] (also known as cholesterol ester hydrolase, sterol esterase, cholesterol ester synthase, and triterpenol esterase) catalyzes the hydrolysis of a steryl ester to produce a sterol and a fatty acid anion.[1-4] This class represents a group of enzymes acting with broad specificity on esterol esters. These enzymes are typically activated by bile salts. Substrates include cholesteryl esters where the acyl group can be vary from acetyl to arachidonyl. A number of lipases also catalyze this reaction. *See also* Esterases; Lipase

An esterase has been reported that will catalyze the hydrolysis of cholesteryl oleate and dolichyl oleate but not cholesteryl palmitate. *See also* Dolichyl Esterase

[1]D. P. Hajjar (1994) *Adv. Enzymol.* **69**, 45.
[2]S. D. Fowler & W. J. Brown (1984) in *Lipases* (B. Borgström & H. L. Brockman, eds.) p. 329.

[3]E. A. Rudd & H. L. Brockman (1984) in *Lipases* (B. Borgström & H. L. Brockman, eds.) p. 185.

[4]M. A. Wells & N. A. DiRenzo (1983)*The Enzymes*, 3rd ed., **16**, 113.

Selected entries from *Methods in Enzymology* [vol, page(s)]:

General discussion: 15, 528, 537; **44**, 633; **71**, 664; **284**, 340; **286**, 116, 231

Assay: 15, 528, 537, 676; **71**, 664, 665

Other molecular properties: enzyme replacement therapy with, **137**, 574; inhibitors, **286**, 231; microbial sterol esterases, **15**, 675 (*Cylindrocarpon radicicola*, **15**, 679 [properties, **15**, 683; purification, **15**, 681]; *Nocardia restrictus*, **15**, 677 [properties, **15**, 679; purification, **15**, 678]); microvillous membrane, in, **31**, 130; myelin marker, as, **31**, 444; pancreatic, **15**, 537; **71**, 664; polarography, **56**, 471; preparation and purification, **15**, 529; properties, **15**, 531, 542; rat liver, purification, **15**, 529; sources, **56**, 471; transition-state and multisubstrate analogues, **249**, 305

CHOLESTEROL 20-HYDROXYLASE

This oxidoreductase activity, formerly classified as EC 1.14.1.9, is now a deleted Enzyme Commission entry. **See** *Steroid 20-Monooxygenase; Steroid 20α,22R-Monooxygenase*

CHOLESTEROL 7α-MONOOXYGENASE

cholesterol

7α-hydroxycholesterol

This heme-thiolate-dependent oxidoreductase [EC 1.14.13.17], also known as cholesterol 7α-hydroxylase and CYP7, catalyzes the reaction of cholesterol with NADPH and dioxygen to produce 7α-hydroxycholesterol, NADP$^+$, and water.[1–4]

[1]J. Y. L. Chiang, W. F. Miller & G.-M. Lin (1990) *JBC* **265**, 3889.

[2]T. Ogishima, S. Deguchi & K. Okuda (1987) *JBC* **262**, 7646.

[3]G. S. Boyd, A. M. Grimwade & M. E. Lawson (1973) *EJB* **37**, 334.

[4]L. B. Nguyen, S. Shefer, G. Salen, J. Y. Chiang & M. Patel (1996) *Hepatology* **24**, 1468.

Selected entries from *Methods in Enzymology* [vol, page(s)]:

General discussion: 111, 364

Assay: microsomal, **206**, 487; reversed-phase HPLC assay, **206**, 483

Other molecular properties: kinetic characterization, **206**, 490; metabolites, chromatography, **206**, 487; microsomal (activity, **206**, 489; assay, **206**, 487); purification, **206**, 486; reconstitution, **206**, 487; regulation in liver, **206**, 490; reversed-phase HPLC assay, **206**, 483

CHOLESTEROL MONOOXYGENASE (SIDE-CHAIN-CLEAVING)

cholesterol

pregnenolone

This heme-thiolate-dependent (*i.e.*, cytochrome P450-dependent) oxidoreductase [EC 1.14.15.6] (also known as cholesterol desmolase, cholesterol side-chain cleaving enzyme, cytochrome P450$_{scc}$, CYP11A1, steroid 20-22 desmolase, and steroid 20-22-lyase) catalyzes the reaction of cholesterol with reduced adrenal ferredoxin and dioxygen to produce pregnenolone, 4-methylpentanal, oxidized adrenal ferredoxin, and water.[1–3]

[1]I. Hanukoglu, V. Spitsberg, J. A. Bumpus, K. M. Dus & C. R. Jefcoate (1981) *JBC* **256**, 4321.

[2]I. Hanukoglu, C. T. Privalle & C. R. Jefcoate (1981) *JBC* **256**, 4329.

[3]M. Shikita & P. F. Hall (1973) *JBC* **248**, 5598.

Selected entries from *Methods in Enzymology* [vol, page(s)]:

General discussion: 15, 585, 591; **272**, 358

Assay: 15, 587, 591, 593; **52**, 125, 126, 139

Other molecular properties: activity, **52**, 14, 124, 131; adrenal, **15**, 591 (adrenal cortex mitochondria, in, **52**, 14); bile acid biosynthesis, in, **15**, 562; bovine corpus luteum preparation, **15**, 587; complexes, g value, **52**, 254; crystallization, **272**, 361; double isotope method, **15**, 595; enzyme preparation, **15**, 570; gonadal tissue, **15**, 585; immunologic properties, **52**, 124, 131; inhibition by 22-amino-23,24-bisnor-5-cholen-3β-ol, **206**, 550; inhibitors, **15**, 589; **52**, 132; **206**, 550; interaction with adrenodoxin, **206**, 51; intermediates, **15**, 585; kinetics, **52**, 131; lysine residues, **206**, 53; microassay, **15**, 591; molecular weight, **52**, 130; nucleotide specificity, **15**, 590; preparation, **15**, 586; properties of gonadal enzyme systems, **15**, 588; rat adrenal preparation, **15**, 592; rat testis preparation, **15**, 586; spectral properties, **52**, 131; spin states, **52**, 268; stability, **52**, 131; structural alignment with P450s, **272**, 318; substrate specificity, **52**, 132

CHOLESTEROL OXIDASE

This FAD-dependent oxidoreductase [EC 1.1.3.6], also known as cholesterol:O_2 oxidoreductase and 3β-hydroxysteroid oxidase, catalyzes the reaction of cholesterol with dioxygen to produce cholest-4-en-3-one and hydrogen peroxide.[1-4] Note that cholest-5-ene-3-one is an intermediate.

cholesterol

cholest-4-en-3-one

The oxidase catalyzes two distinct reactions: the oxidation of Δ^5-ene-3β-hydroxysteroids to the corresponding Δ^5-3-ketosteroid and the isomerization to the Δ^4-3-ketosteroid. The isomerization catalyzed by the *Brevibacterium sterolicum* enzyme occurs by means of a *cis*-diaxial intramolecular transfer of protons from the C4β to C6β positions. Other steroid substrates include β-sitosterol, fucosterol, dihydroepiandrosterone, and pregnenolone.

[1] J. MacLachlan, A. T. Wotherspoon, R. O. Ansell & C. J. Brooks (2000) *J. Steroid Biochem. Mol. Biol.* **72**, 169.
[2] I. J. Kass & N. S. Sampson (1995) *BBRC* **206**, 688.
[3] T. Abe, T. Nihira, A. Tanaka & S. Fukui (1983) *BBA* **749**, 69.
[4] A. G. Smith & C. J. W. Brooks (1976) *J. Steroid Biochem.* **7**, 705.

Selected entries from *Methods in Enzymology* [vol, page(s)]:
General discussion: I, 678; **52**, 18
Assay: I, 678
Other molecular properties: assay of cholesterol, **71**, 665; electrode system, **56**, 485; enzyme thermistor, in, **44**, 674; macrophage fractions, in, **31**, 340, 344; properties, I, 681; purification, *Mycobacterium*, I, 680; sources, **56**, 471

CHOLINE *O*-ACETYLTRANSFERASE

This transferase [EC 2.3.1.6], also known as choline acetylase, catalyzes the reversible reaction of acetyl-CoA with choline to produce coenzyme A and *O*-acetylcholine.[1-3]

acetylcholine

Other substrates, albeit weaker, include propionyl-CoA. The enzyme has a Theorell-Chance Bi Bi kinetic mechanism, although isotope exchange at equilibrium studies suggest a random pathway.[4]

[1] D. Wu & L. B. Hersh (1995) *JBC* **270**, 29111.
[2] S. Tucek (1988) in *Handbook of Experimental Pharmacology* (V. P. Whittaker, ed.) **86**, 129, Springer-Verlag, Berlin.
[3] D. R. Haubrich (1976) in *Biology of Cholinergic Function* (A. M. Goldberg & I. Hanin, eds.), pp. 239-266, Raven Press, New York.
[4] D. S. Sigman & G. Mooser (1975) *Ann. Rev. Biochem.* **44**, 889.

Selected entries from *Methods in Enzymology* [vol, page(s)]:
General discussion: I, 619; **17B**, 778, 780, 788, 798; **34**, 4
Assay: I, 622; **17B**, 780, 790, 798; **32**, 781
Other molecular properties: affinity labeling, **47**, 428; cultured cells, in, **32**, 766, 772, 773; fly heads, in, **17B**, 794; *Lactobacillus plantarum*, in, I, 622; **17B**, 794; microdetermination, **17B**, 780; neurobiology, in, **32**, 765; properties, **17B**, 794, 800; purification from (brain, I, 622; **17B**, 798; *Lactobacillus plantarum*, I, 622; **17B**, 794; placenta, **17B**, 793; squid ganglia, I, 620; **17B**, 793); sources, I, 620

CHOLINE DEHYDROGENASE

This PQQ-dependent (*i.e.*, pyrroloquinoline quinone-dependent) oxidoreductase [EC 1.1.99.1], also called choline oxidase, catalyzes the reaction of choline with an acceptor substrate to produce betaine aldehyde and the reduced acceptor.[1-3]

betaine aldehyde

The acceptor substrate can be phenazine methosulfate, dioxygen, coenzyme Q, cytochrome *c*, and ferricyanide.

[1] D. R. Haubrich & N. H. Gerber (1981) *Biochem. Pharmacol.* **30**, 2993.
[2] Y. Hatefi & D. L. Stiggall (1976) *The Enzymes*, 3rd ed., **13**, 175.
[3] M. C. Barrett & A. P. Dawson (1975) *BJ* **151**, 677.

Selected entries from *Methods in Enzymology* [vol, page(s)]:
General discussion: **5**, 562; **53**, 405
Assay: **4**, 330; **5**, 562
Other molecular properties: choline oxidase and, I, 675; extraction from mitochondria, **22**, 205; extraction, I, 37 (effect of extraction methods, **53**, 409); ferricyanide and, I, 675; hydroxyapatite columns, on, **5**, 437, 568; properties, **5**, 569; purification, difficulty in removing impurities, **5**, 434, 437; purification, rat liver, **5**, 564, 567; soluble, **5**, 437, 567; solubilized by treatment with phospholipase A, **5**, 437

CHOLINE KINASE

This phosphotransferase [EC 2.7.1.32] catalyzes the reaction of ATP with choline to produce ADP and *O*-phosphocholine.[1-6]

$$(H_3C)_3\overset{+}{N}\diagdown\diagup O\diagdown PO_3H^-$$

O-phosphocholine

Other substrates include ethanolamine, *N*-methylethanol-amine, *N,N*-dimethylethanolamine, and *N,N*-diethyl-ethanolamine.

[1] K. Ishidate (1997) *BBA* **1348**, 70.
[2] S. Yamashita & K. Hosaka (1997) *BBA* **1348**, 63.
[3] A. J. Kinney & T. S. Moore, Jr. (1988) *ABB* **260**, 102.
[4] R. R. Reinhardt, L. Wecker & P. F. Cook (1984) *JBC* **259**, 7446.
[5] J. D. Esko & C. R. H. Raetz (1983) *The Enzymes*, 3rd ed., **16**, 207.
[6] M. A. Brostrom & E. T. Browning (1973) *JBC* **248**, 2364.

Selected entries from *Methods in Enzymology* [**vol**, page(s)]:
General discussion: 5, 560; 209, 121, 134, 147
Assay: 5, 560; cerebral, rat, radiochemical assay, 209, 147; hepatic, rat, radiochemical assay, 209, 136; renal, rat, radiochemical assay, 209, 121
Other molecular properties: cerebral, rat (properties, 209, 151; purification, 209, 149; radiochemical assay, 209, 147); hepatic, rat (properties, 209, 143; purification, 209, 136; radiochemical assay, 209, 136); magnesium activation, 63, 270, 292; mechanism, 63, 292; renal, rat (properties, 209, 128; purification, 209, 124; radiochemical assay, 209, 121; subcellular localization, 209, 128); yeast (properties, 5, 561; purification, 5, 560)

CHOLINE MONOOXYGENASE

This magnesium-dependent oxidoreductase [currently EC 1.14.14.4; may be reclassified as EC 1.14.15.7] catalyzes the reaction of choline with dioxygen, two molecules of reduced ferredoxin, and two protons to produce betaine aldehyde hydrate, water, and two molecules of oxidized ferredoxin.[1,2]

$$(H_3C)_3\overset{+}{N}\diagdown\diagup\overset{\displaystyle O}{\underset{\displaystyle H}{\diagdown}}$$

betaine aldehyde

$$(H_3C)_3\overset{+}{N}\diagdown\diagup\overset{\displaystyle OH}{\underset{\displaystyle H}{\diagdown OH}}$$

betaine aldehyde hydrate

The spinach chloroplast enzyme contains a Rieske-type iron-sulfur cluster. Alternative substrates containing larger alkyl groups on the nitrogen have significantly weaker affinity. *See also Choline Dehydrogenase; Choline Oxidase*

[1] M. Burnet, P. J. Lafontaine & A. D. Hanson (1995) *Plant Physiol.* **108**, 581.
[2] B. Rathinasabapathi, M. Burnet, B. L. Russell, D. A. Gage, P. Liao, G. J. Nye, P. Scott, J. H. Golbeck & A. D. Hanson (1997) *PNAS* **94**, 3454.

CHOLINE OXIDASE

This FAD-dependent oxidoreductase [EC 1.1.3.17] catalyzes the reaction of choline with dioxygen to produce betaine aldehyde and hydrogen peroxide.[1–3] *See also Choline Dehydrogenase; Choline Monooxygenase*

[1] M. Ohta-Fukuyama, Y. Miyake, S. Emi & T. Yamano (1980) *J. Biochem.* **88**, 197.
[2] H. Yamada, N. Mori & Y. Tani (1979) *Agric. Biol. Chem.* **43**, 2173.
[3] S. Ikuta, S. Imamura, H. Misaki & Y. Horiuti (1977) *J. Biochem.* **82**, 1741.

Selected entries from *Methods in Enzymology* [**vol**, page(s)]:
General discussion: 1, 674
Assay: 1, 676

CHOLINE-PHOSPHATE CYTIDYLYLTRANSFERASE

This lipid-activated transferase [EC 2.7.7.15], also known as phosphorylcholine transferase and CTP:phosphocholine cytidylyltransferase, catalyzes the reversible reaction of CTP with choline phosphate to produce pyrophosphate (or, diphosphate) and CDP-choline.[1–4]

$$CDP\diagdown\overset{\displaystyle O}{\diagup}\diagdown\diagup\overset{+}{N}(CH_3)_3$$

CDPcholine

The rat brain enzyme, which catalyzes the rate-limiting step in phosphatidylcholine biosynthesis, has a random Bi Bi kinetic mechanism. The mammalian enzyme exists in both soluble and membrane-associated forms.

[1] J. M. Clement & C. Kent (1999) *BBRC* **257**, 643.
[2] C. Kent (1997) *BBA* **1348**, 79.
[3] H. Jamil & D. E. Vance (1991) *BBA* **1086**, 335.
[4] Y. Kishimoto (1983) *The Enzymes*, 3rd ed., **16**, 357.

Selected entries from *Methods in Enzymology* [**vol**, page(s)]:
General discussion: 5, 480; 14, 121; 71, 576; 209, 248
Assay: 5, 480; 14, 121; 209, 248 (alternate assay, 14, 122; contamination with enzymes crude tissue preparations, 14, 123; phospholipid addition for choline enzyme, 14, 123, 124)
Other molecular properties: activation on storage, 14, 125; lipid activators, 14, 124; 71, 579, 581 (phospholipid, 71, 577; properties, 71, 581); onion stem dictyosomes, in, 22, 147; phospholipid activation of choline enzyme from liver, 14, 123, 124; preparation of enzymes, 14, 123 (guinea pig liver, 5, 481); preparation of labeled CDPcholine, 14, 684; properties, 5, 482; 14, 124; 209, 256; purification, 5, 481; 14, 123; 209, 251; rat liver, from, 71, 576; substrate specificity, 5, 482; 14, 124

CHOLINESTERASE

This serine-esterase [EC 3.1.1.8] (also known as pseudo-cholinesterase, acylcholine acylhydrolase, non-specific cholinesterase, and benzoylcholinesterase) catalyzes the hydrolysis of an acylcholine to produce choline and a carboxylate anion.[1–6] Hydrolysis occurs by way of an acyl-enzyme intermediate. Substrates include a wide variety of acylcholines and other esters: for example, acetylcholine, butyrylcholine (highest activity), (butyrylthio)choline, benzoylcholine, succinoylcholine, and propionylthiocholine.

Hydrolysis of cocaine (ecgonine methyl ester benzoate) to ecgonine methyl ester by human cholinesterase is an important detoxification step leading to pharmacologically inactive metabolites of cocaine. **See Acetylcholinesterase**

[1]P. Taylor (1991) *JBC* **266**, 4025.
[2]S. S. Brown, W. Kalow, W. Pilz, M. Whittaker & C. L. Woronick (1981) *Adv. Clin. Chem.* **22**, 1.
[3]A. R. Main, E. Tarkan, J. L. Aull & W. G. Soucie (1972) *JBC* **247**, 566.
[4]D. Nachmansohn & I. B. Wilson (1951) *Adv. Enzymol.* **12**, 259.
[5]D. M. Quinn & S. R. Feaster (1998) *CBC* **1**, 455.
[6]H. Okuda (1991) *A Study of Enzymes* **2**, 563.

Selected entries from **Methods in Enzymology** [vol, page(s)]:
General discussion: 44, 647
Assay: Fast Analyzer, by, **31**, 816; fluorometric, **44**, 626
Other molecular properties: affinity labeling, **87**, 488; brain, **1**, 36, 43, 47, 50; calcium phosphate columns and, **5**, 32; definition of, **1**, 644; extraction from plasma membrane, **22**, 206; gel electrophoresis, **32**, 88; histochemical method for, **4**, 390; immobilized (activity assay, **44**, 653; entrapment, by, **44**, 173, 178; monitoring system, in, **44**, 647; starch pad, in, **44**, 652, 653); immunocytochemical detection, **207**, 254; inactivation, **46**, 85; inhibitor, **63**, 439; intestine, **1**, 50; pesticide inhibitors, **44**, 637; red cell, **1**, 37, 47, 50; serum, **1**, 644, 647; staining on gels, **22**, 600; transition-state and multisubstrate analogues, **249**, 305; turnover number, **44**, 647

CHOLINE SULFATASE

This enzyme [EC 3.1.6.6] catalyzes the hydrolysis of choline sulfate to produce choline and the sulfate anion.[1-3] The enzyme has an ordered Uni Bi kinetic mechanism.[1]

[1]J. J. Lucas, S. W. Burchiel & I. H. Segel (1972) *ABB* **153**, 664.
[2]A. B. Roy (1971) *The Enzymes*, 3rd ed., **5**, 1.
[3]J. M. Scott & B. Spencer (1968) *BJ* **106**, 471.

CHOLINE SULFOTRANSFERASE

This transferase [EC 2.8.2.6], also called choline sulfokinase, catalyzes the reaction of 3'-phosphoadenylylsulfate with choline to produce adenosine 3',5'-bisphosphate and choline sulfate.[1-3] *N,N*-Dimethylaminoethanol also a substrate.

[1]J. Rivoal & A. D. Hanson (1994) *Plant Physiol.* **106**, 1187.
[2]F. Renosto & I. H. Segel (1977) *ABB* **180**, 416.
[3]B. A. Orsi & B. Spencer (1964) *J. Biochem.* **56**, 81.

Selected entries from **Methods in Enzymology** [vol, page(s)]:
General discussion: 5, 982
Assay: 5, 983

CHOLOYL-CoA SYNTHETASE

This enzyme [EC 6.2.1.7], also known as cholate:CoA ligase, catalyzes the reaction of ATP with cholate and Coenzyme A to produce AMP, pyrophosphate (or, diphosphate), and choloyl-CoA.[1-3]

[1]F. A. Simion, B. Fleischer & S. Fleischer (1983) *Biochemistry* **22**, 5029.
[2]D. A. Vessey & D. Zakim (1977) *BJ* **163**, 357.
[3]M. A. Polokoff & R. M. Bell (1977) *JBC* **252**, 1167.

Selected entries from **Methods in Enzymology** [vol, page(s)]:
General discussion: 5, 473
Assay: 5, 473

CHOLOYLGLYCINE HYDROLASE

This enzyme [EC 3.5.1.24, also known as glycocholase and bile salt hydrolase, catalyzes the hydrolysis of 3α,7α,12α-trihydroxy-5β-cholan-24-oylglycine (*i.e.*, choloylglycine) to produce 3α,7α,12α-trihydroxy-5β-cholanate (*i.e.*, cholate) and glycine.[1-4] Other substrates include 3α,12α-dihydroxy-5β-cholan-24-oylglycine, 3α,7α,12α-trihydroxy-5β-cholan-24-oyltaurine, and the glycine and taurine conjugates of chenodeoxycholate.

[1]S. M. Huijghebaert & A. F. Hofmann (1986) *J. Lipid Res.* **27**, 742.
[2]E. J. Stellwag & P. B. Hylemon (1976) *BBA* **452**, 165.
[3]P. P. Nair, M. Gordon & J. Reback (1967) *JBC* **242**, 7.
[4]S. G. Lundeen & D. C. Savage (1992) *J. Bacteriol.* **174**, 7217.

Selected entries from **Methods in Enzymology** [vol, page(s)]:
General discussion: 77, 308

CHONDROITIN ABC LYASE

This lyase [EC 4.2.2.4], also known as chondroitinase and chondroitin ABC eliminase, catalyzes the eliminative degradation of polysaccharides containing 1,4-β-D-hexosaminyl and 1,3-β-D-glucuronosyl or 1,3-α-L-iduronosyl linkages to produce disaccharides containing 4-deoxy-β-D-gluc-4-enuronosyl groups.[1,2] Substrates include chondroitin 4-sulfate, chondroitin 6-sulfate, and dermatan sulfate, and even hyaluronate, albeit slowly.

[1]R. J. Linhardt, P. M. Galliher & C. L. Cooney (1986) *Appl. Biochem. Biotechnol.* **12**, 135.
[2]S. Linn, T. Chan, L. Lipeski & A. A. Salyers (1983) *J. Bacteriol.* **156**, 859.

Selected entries from **Methods in Enzymology** [vol, page(s)]:
General discussion: 28, 911
Assay: 28, 912

CHONDROITIN AC LYASE

This lyase [EC 4.2.2.5], also known as chondroitinase, chondroitin sulfate lyase, and chondroitin AC eliminase, catalyzes the eliminative degradation of polysaccharides containing 1,4-β-D-hexosaminyl and 1,3-β-D-glucuronosyl linkages to produce disaccharides containing 4-deoxy-β-D-gluc-4-enuronosyl groups.[1,2] Substrates include chondroitin 4-sulfate and chondroitin 6-sulfate. Hyaluronate is a weak substrate.

[1]D. S. P. Q. Horton & Y. M. Michelacci (1986) *EJB* **161**, 139.
[2]K. Hiyama & S. Okada (1977) *J. Biochem.* **82**, 429.

Selected entries from **Methods in Enzymology** [vol, page(s)]:
General dfiscussion: 28, 911
Assay: 28, 912; proteoglycan degradation method, **248**, 50

CHONDROITIN-GLUCURONATE 5-EPIMERASE

This epimerase [EC 5.1.3.19], also called polyglucuronate 5-epimerase and dermatan-sulfate 5-epimerase, catalyzes the interconversion of chondroitin D-glucuronate and dermatan L-iduronate.[1,2] The enzyme is distinct from heparosan-*N*-sulfate-glucuronate 5-epimerase [EC 5.1.3.17].

[1]A. Malmstrom & L. Aberg (1982) *BJ* **201**, 489.
[2]A. Malmstrom (1984) *JBC* **259**, 161.

CHONDROITIN SULFATE LYASE

This enzyme activity, formerly classified as EC 4.2.99.6 but now a deleted entry, is thought to arise from the combined action of chondroitin ABC lyase [EC 4.2.2.4] and chondroitin AC lyase [EC 4.2.2.5].

CHONDROITIN 4-SULFOTRANSFERASE

This transferase [EC 2.8.2.5] catalyzes the reaction of 3'-phosphoadenylylsulfate with chondroitin to produce adenosine 3',5'-bisphosphate and chondroitin 4'-sulfate.[1–4] The sulfation occurs at the 4-position of the *N*-acetylgalactosamine residues of chondroitin. *See also Chondroitin 6-Sulfotransferase; Heparitin Sulfotransferase; Heparin-Glucosamine 3-O-Sulfotransferase*

[1]O. Habuchi (2000) *BBA* **1474**, 115.
[2]O. Habuchi & N. Miyashita (1982) *BBA* **717**, 414.
[3]O. Habuchi & K. Miyata (1980) *BBA* **616**, 208.
[4]D. M. Delfert & H. E. Conrad (1985) *JBC* **260**, 14446.

Selected entries from *Methods in Enzymology* [vol, page(s)]:
General discussion: **8**, 496
Assay: 8, 496

CHONDROITIN 6-SULFOTRANSFERASE

This sulfotransferase [EC 2.8.2.17] catalyzes the reaction of 3'-phosphoadenylylsulfate with chondroitin to produce adenosine 3',5'-bisphosphate and chondroitin 6-sulfate.[1–3] *See also Chondroitin 4-Sulfotransferase; Heparitin Sulfotransferase; Heparin-Glucosamine 3-O-Sulfotransferase*

[1]O. Habuchi & N. Miyashita (1982) *BBA* **717**, 414.
[2]G. Sugumaran & J. E. Silbert (1988) *JBC* **263**, 4673.
[3]O. Habuchi, Y. Matsui, Y. Kotoya, Y. Aoyama, Y. Yasuda & M. Noda (1993) *JBC* **268**, 21968.

CHONDRO-4-SULFATASE

This enzyme [EC 3.1.6.9] catalyzes the hydrolysis of 4-deoxy-β-D-gluc-4-enuronosyl-(1,4)-*N*-acetyl-D-galactosamine 4-sulfate to produce 4-deoxy-β-D-gluc-4-enuronosyl-(1,4)-*N*-acetyl-D-galactosamine and sulfate.[1]

Other substrates include the saturated analogue. Higher oligosaccharides are not substrates and neither is the 6-sulfate analogue.[2]

[1]T. Yamagata, H. Saito, O. Habuchi & S. Suzuki (1968) *JBC* **243**, 1523.
[2]K. Sugahara & T. Kojima (1996) *EJB* **239**, 865.

Selected entries from *Methods in Enzymology* [vol, page(s)]:
General discussion: **2**, 324; **8**, 663
Assay: 8, 663

CHONDRO-6-SULFATASE

This enzyme [EC 3.1.6.10] catalyzes the hydrolysis of 4-deoxy-β-D-gluc-4-enuronosyl-(1,4)-*N*-acetyl-D-galactosamine 6-sulfate to produce 4-deoxy-β-D-gluc-4-enuronosyl-(1,4)-*N*-acetyl-D-galactosamine and sulfate.[1] Other substrates include the saturated analogue and *N*-acetyl-D-galactosamine 4,6-bissulfate.[2] *See also N-Acetylgalactosamine-6-Sulfatase; Chondro-4-Sulfatase*

[1]T. Yamagata, H. Saito, O. Habuchi & S. Suzuki (1968) *JBC* **243**, 1523.
[2]K. Sugahara & T. Kojima (1996) *EJB* **239**, 865.

Selected entries from *Methods in Enzymology* [vol, page(s)]:
General discussion: **2**, 324; **8**, 663
Assay: 8, 663; **28**, 918
Other molecular properties: enzymatic properties, **8**, 671; **28**, 920; isolation from *Proteus vulgaris*, **8**, 665; **28**, 919; properties, **8**, 671; separation from chondro-4-sulfatase, **28**, 919

CHORISMATE MUTASE

chorismate

This mutase [EC 5.4.99.5] catalyzes the interconversion of chorismate and prephenate.[1–4] The [3,3]-pericyclic process is formally analogous to a Claisen rearrangement. In a number of organisms (for example, *Escherichia coli*) this activity is part of a bifunctional enzyme: chorismate mutase and prephenate dehydrogenase (but not for all organisms; *e.g.*, the *Bacillus subtilis* enzyme is monofunctional).

[1]J. V. Schloss & M. S. Hixon (1998) *CBC* **2**, 43.
[2]A. Y. Lee, J. D. Stewart, J. Clardy & B. Ganem (1995) *Chem. Biol.* **2**, 195.
[3]R. Bentley (1990) *Crit. Rev. Biochem. Mol. Biol.* **25**, 307.
[4]S. D. Copley & J. R. Knowles (1987) *J. Amer. Chem. Soc.* **109**, 5008.

Selected entries from *Methods in Enzymology* [vol, page(s)]:
General discussion: **34**, 4; **142**, 432, 440, 450, 463
Assay: 142, 451, 463
Other molecular properties: prephenate dehydratase, and (assays, **142**, 432; properties, **142**, 435; purification from *Escherichia coli*,

142, 433); prephenate dehydrogenase, and (assays, **142**, 442; properties, **142**, 448; purification from *Escherichia coli*, **142**, 444; substrate preparation, **142**, 440); properties (mung bean isozymes, **142**, 457; sorghum isozymes, **142**, 462; *Streptomyces* enzyme, **142**, 470); purification from (*A. aerogenes*, **17A**, 567, 571; mung bean, **142**, 455; peas, **17A**, 572; sorghum, **142**, 459; *Streptomyces aureofaciens*, **142**, 466); transition state and multisubstrate analogues, **249**, 308

CHORISMATE SYNTHASE

This enzyme [EC 4.2.3.5; previously classified as EC 4.6.1.4], also known as 5-enolpyruvylshikimate-3-phosphate phospholyase, catalyzes the conversion of 5-*O*-(1-carboxyvinyl)-3-phosphoshikimate to chorismate and orthophosphate.[1–5] This reaction is a 1,4-elimination of orthophosphate and the C-($6proR$) hydrogen is lost from the substrate with overall *anti* stereochemistry via a nonconcerted mechanism (studies of the nonenzymatic reaction with model substrates exhibit *syn* elimination). Reduced FMN is required for the enzymatic reaction, despite the fact that there is no overall redox change in the conversion of substrate to chorismate. A flavin intermediate forms, possibly a flavin semiquinone, and possibly a substrate-derived radical as well, before the C($6proR$)-H bond is cleaved.

[1] P. Macheroux, J. Schmid, N. Amrhein & A. Schaller (1999) *Planta* **207**, 325.
[2] B. A. Palfey & V. Massey (1998) *CBC* **3**, 83.
[3] S. Bornemann, D. J. Lowe & R. N. Thorneley (1996) *Biochemistry* **35**, 9907.
[4] R. Bentley (1990) *Crit. Rev. Biochem. Mol. Biol.* **25**, 307.
[5] G. R. Welch, K. W. Cole & F. H. Gaertner (1974) *ABB* **165**, 505.

Selected entries from *Methods in Enzymology* [vol, page(s)]:
General discussion: 142, 362
Assay: 17A, 390; 142, 362
Other molecular properties: coupling enzyme in *arom* multifunctional enzyme assay, preparation, as, **142**, 331; hysteresis, **64**, 218; properties, **142**, 365; purification from *Neurospora crassa*, **17A**, 393; **142**, 363

CHYMASE

This serine endopeptidase [EC 3.4.21.39], also known as mast cell protease I and skeletal muscle (SK) protease, has been isolated from mast cell granules and catalyzes the hydrolysis of peptide bonds, preferring Phe–Xaa > Tyr–Xaa > Trp–Xaa > Leu–Xaa.[1–3] There are at least five chymases in mice (chymase 2 is also known as mouse mast cell protease 1 and rat mast cell protease 2). Whereas all chymases have a chymotrypsin-like activity, substrate specificity and efficiency can vary. Nevertheless, all prefer an aromatic residue at the P1 position. *See also Tryptase*

Note that in the older literature the term "chymase" was occasionally used to refer to gastricsin [EC 3.4.23.3].

[1] G. H. Caughey, ed., (1995) *Mast Cell Proteases in Immunology and Biology*, Marcel Dekker, New York.
[2] L. B. Schwartz (1990) *Neutral Proteases of Mast Cells*, S. Karger AG, Basel.
[3] J. S. Bond & P. E. Butler (1987) *Ann. Rev. Biochem.* **56**, 333.

Selected entries from *Methods in Enzymology* [vol, page(s)]:
General discussion: 80, 588, 592, 765
Assay: 80, 589; 163, 327
Other molecular properties: active-site titration, **248**, 100; chymase I, **80**, 565; mast cell, inactivation, **80**, 771; properties, **163**, 314; purification from human mast cells, **163**, 323; rat, **80**, 589 (amino acid composition, **80**, 593; amino-terminal sequence, **80**, 594, 595; inhibitors, **80**, 596; isoelectric point, **80**, 594; molecular weight, **80**, 593; physiological role, **80**, 596, 597; properties, **80**, 592; purification, **80**, 589; purity, **80**, 592; specific activity, **80**, 592; stability, **80**, 592; substrate specificity, **80**, 594, 603; unit of activity, **80**, 589)

CHYMOPAPAIN

This cysteine endopeptidase [EC 3.4.22.6], also known as papaya proteinase II, has a specificity similar to that of papain.[1–5] The enzyme exhibits a preference for substrates with hydrophobic residues in the S2 and S3 subsites; however, other residues can still be accepted. The other subsites have a very broad specificity; however, prolyl residues are not accepted at S1′. The enzyme will also act on synthetic substrates such as Bz-Arg-∇-NHPhNO$_2$ and Z-Phe-Arg-∇-NHMec, where ∇ indicates the cleavage site. *See also Papain; Caricain*

[1] K. Brocklehurst, F. Willenbrock & E. Salih (1988) *New Compr. Biochem.* **16**, 39.
[2] K. Brocklehurst, F. Willenbrock & E. Salih (1987) in *Hydrolytic Enzymes* (A. Neuberger & K. Brocklehurst, eds.), p. 39, Elsevier, Amsterdam.
[3] B. S. Baines & K. Brocklehurst (1982) *J. Protein Chem.* **1**, 119.
[4] K. Brocklehurst (1982) *Meth. Enzymol.* **87**, 427.
[5] K. Brocklehurst, B. S. Baines & M. P. L. Kierstan (1981) *Top. Enzyme Ferment. Biotechnol.* **5**, 262.

Selected entries from *Methods in Enzymology* [vol, page(s)]:
General discussion: 2, 61; 19, 244; 34, 532, 542, 546; 244, 639
Assay: chymopapain B, 19, 244
Other molecular properties: chymopapain A, physicochemical properties, **19**, 250; chymopapain B, **19**, 244 (activation and inhibition, **19**, 251; amino acid composition, **19**, 250; assay method, **19**, 244; homogeneity, **19**, 249; pH optimum, **19**, 251; physicochemical properties, **19**, 249; purification procedure, **19**, 246; specificity, **19**, 249; stability, **19**, 251); comparison to papain, **19**, 243; detection in preparations of papain, **87**, 460; 2,2′-dipyridyl disulfide, reaction with, **87**, 460, 461; isolation, **244**, 540; purification, **2**, 61; **19**, 246

CHYMOSIN

This aspartic endopeptidase [EC 3.4.23.4], also known as rennin, catalyzes the hydrolysis of peptide bonds, exhibiting a broad specificity similar to that of pepsin A.[1,2] This neonatal gastric enzyme exhibits high milk-clotting activity. Despite an exhaustive literature on chymosin's use in cheese-making, relatively little is known about the physiological roles of chymosin. *See also Mucoropepsin*

[1] B. Foltmann (1966) *C. R. Trav. Lab. Carlsberg* **35**, 143.
[2] T. Beppu (1983) *Trends Biotechnol.* **1**, 85.
Selected entries from ***Methods in Enzymology*** [vol, page(s)]:
General discussion: 2, 69; **19**, 421
Assay: 2, 69; **19**, 422
Other molecular properties: amino acid composition, **19**, 431, 434,
456; amino acid sequence, **241**, 216; bovine, homology to cathepsin D,
80, 578; chymosin B, β hairpin loop mutations, accommodation, **202**, 76,
81; comparison with pepsin, **19**, 434, 435; crystal (immobilization, **44**,
548; X-ray analysis, **44**, 555, 556); immobilization, with collagen, **44**, 244;
lactoperoxidase and, **2**, 813; molecular weights and structure, **19**, 432;
Mucor species, from, [**See Mucor Rennin**]; properties, **2**, 76;
propionibacteria growth medium and, **2**, 388; purification, **2**, 72; **19**,
427; trapped by α$_2$-macroglobulin, **80**, 748

CHYMOTRYPSIN

This serine endopeptidase [EC 3.4.21.1], also known as
α-chymotrypsin, catalyzes the hydrolysis of peptide bonds
with the preference for Tyr–Xaa, Trp–Xaa, Phe–Xaa, and
Leu–Xaa bonds.[1–12] The enzyme also possesses esterase
and amidase activities, which share features of the peptide
hydrolysis mechanism. The active catalyst is formed by
proteolysis of its precursor chymotrypsinogen (see below).

The enzyme operates by a three-step mechanism (*i.e.*, formation of an enzyme-substrate complex, conversion to an
O-acyl-enzyme species, and hydrolysis of this intermediate). The catalytic triad of His57, Asp102, and Ser195
is essential for catalysis. The seryl residue is hydrogen
bonded to the histidyl residue which, in turn is hydrogen
bonded to the β-carboxyl group of Asp102. An oxyanion hole is formed by the amide nitrogens of Gly193 and
Ser195. Upon binding of the substrate, the hydroxyl group
of the seryl residue attacks the carbonyl group of the peptide bond to generate a tetrahedral intermediate. In this
transient structure, the oxygen atom of the substrate now
occupies the oxyanion hole. The acyl-enzyme intermediate now forms, assisted by proton donation from His57.
The N-terminal portion of the peptide is now released
and replaced by water. The acyl enzyme subsequently
undergoes hydrolysis and the enzyme is regenerated.

Early evidence for acyl-enzyme formation and product
release was provided by so-called burst-phase kinetics, detecting liberated *p*-nitrophenol when *p*-nitrophenyl
acetate or *p*-nitrophenyl ethyl carbonate was the
substrate.[12] Later studies on formation of stable acyl/tosyl
intermediates, X-ray crystallography of enzyme-bound
reactants, and detection of unstable transients by
cryoenzymology[5] and fast kinetic techniques are all consistent with this mechanism.

Two pathways exist for the activation of chymotrypsinogen
to form chymotrypsin. The final products of each pathway,
α- and γ-chymotrypsin, have identical primary structures:
three polypeptide chains held together by disulfide bonds
(Cys1 . . . Leu13, Ile16 . . . Tyr146, and Ala149 . . . Asn246).
However, these two forms have slightly different enzymatic properties, stabilities, and X-ray structure.[1,7,8,11]

[1] C. W. Wharton (1998) *CBC1*, 345.
[2] J. J. Perona & C. S. Craik (1995) *Protein Sci.* **4**, 337.
[3] M. A. Ator & P. R. Ortez de Montellano (1990) *The Enzymes*, 3rd ed.,
19, 213.
[4] E. T. Kaiser, D. S. Scott & S. E. Rokita (1985) *Ann. Rev. Biochem.* **54**, 565.
[5] R. J. Coll, P. D. Compton & A. L. Fink (1982) *MIE* **87**, 66.
[6] J. S. Fruton (1982) *AE* **53**, 239.
[7] G. H. Cohen, E. W. Silverton & D. R. Davies (1981) *J. Mol. Biol.* **148**, 449.
[8] D. M. Blow (1976) *Acc. Chem. Res.* **9**, 145.
[9] D. M. Blow (1971) *The Enzymes*, 3rd ed., **3**, 185.
[10] G. P. Hess (1971) *The Enzymes*, 3rd ed., **3**, 213.
[11] P. Desnuelle (1960) *The Enzymes*, 2nd ed., **4**, 93.
[12] B. S. Hartley & B. A. Kilby (1954) *BJ* **56**, 288.

Selected entries from ***Methods in Enzymology*** [vol, page(s)]:
General discussion: 2, 8; **19**, 38, 64-108; **34**, 7, 415, 585; **46**, 22, 39, 75,
84, 89, 90, 96, 156, 197, 201, 206, 207, 215, 431, 478, 537, 538; **308**, 201
Assay: 2, 19; α-chymotrypsin, **11**, 235; chymotrypsin A, **19**, 73
(*N*-acetyl-L-tyrosine ethyl ester, **19**, 73); chymotrypsin A$_\alpha$,
spectrophotometric assay, **19**, 37, 43; fluorimetric assay, **248**, 20;
hornet, **80**, 644
Other molecular properties: abortive complex, **63**, 205; acetophenone
phosphorescence and, **21**, 41; acetylchymotrypsin, **87**, 66, 73;
N-acetyl-3-nitrotyrosinyl-α-chymotrypsin, isolation, **87**, 67; activation
energy, **63**, 243; **64**, 226; active site, **36**, 390; **44**, 410; **46**, 198; active
site residues, **87**, 470; active-site titration, **248**, 87, 100; activity, **44**, 30,
31, 34, 36, 200, 266, 267, 271, 446, 447, 530, 561; acylation, **44**, 447,
562; acylchymotrypsin, **87**, 68; adsorption, onto glass, **61**, 77;
amidinated, **25**, 592; anhydrochymotrypsin, **34**, 4;
4-azidocinnamoyl-α-chymotrypsin, **46**, 84; bifunctional reagents and, **25**,
634; burst kinetics, **64**, 199; charge transfer complexes and, **26**, 601;
chromatography of, **1**, 109; α-chymotrypsin, **2**, 8; **28**, 206, 278, 318, 359,
364, 413, 419, 473, 514, 608, 611, 612, 702, 705; **34**, 514;
γ-chymotrypsin, **2**, 14; δ-chymotrypsin, **2**, 8, 15; π-chymotrypsin, **2**, 8;
chymotrypsin A, **34**, 418; **46**, 213, 214 (assay methods, **19**, 73; bovine,
80, 641, 651; heavy metal inhibitors, **19**, 80; pancreatic, **80**, 644; physical
and chemical properties, **19**, 79, 104; preparation and purification, **19**,
102; specific activity, **19**, 73, 103; specificity, **19**, 88; stability, **19**, 79);
chymotrypsin A$_\alpha$ (formation, **19**, 64, 65; physical properties, **19**, 80;
purification, **19**, 75, 77; specific activity, **19**, 74; spectrophotometric
assay, **19**, 37, 43; stability, **19**, 79); chymotrypsin B, **2**, 18; **20**, 81; **34**,
418 (bovine, homology to crab collagenase, **80**, 729; α-chymotrypsin
and, **2**, 9; fin whale, from, **19**, 104; formation, **2**, 9, 18; pancreatic, **80**,
644; preparation, **2**, 18; specificity characterization, **19**, 36, 87; stability,
19, 87); chymotrypsin B$_\alpha$, **19**, 86 (definition of, **19**, 81; formation, **19**,
81; stability, **19**, 87); chymotrypsin B$_\pi$, **19**, 81 (crystallization and
properties, **19**, 86; stability, **19**, 87); chymotrypsin II, **73**, 121;
chymotrypsinogen, preparation, **2**, 9, 16; coupling to tresyl-agarose, **135**,
69; cryoenzymology, **63**, 338; crystallographic analysis of mechanism,
308, 201; cyanate and, **25**, 583, 584; cyanuration of, **25**, 510, 514;
derivatives of, **87**, 66, 470, 598, 627; dielectric constant, effects, **63**, 212;
dye binding, **44**, 535; ester hydrolysis, **87**, 38, 490; family (biological role,
244, 23; catalytic serine codon, **244**, 30; domains, **244**, 25, 29;
evolution, **244**, 30; members, **244**, 22, 26; species distribution, **244**, 26,
31); fluorescence and, **26**, 503, 528; **87**, 73; fluorimetric assay, **248**, 20;
forms of, **19**, 64; gel-entrapped (catalytic activity, **135**, 593; preparation,

135, 593; reactivation, **135**, 594, 595; thermoinactivation, **135**, 593); glass-immobilized, **135**, 544; hemiketal intermediate, **87**, 480; hormone receptor studies, in, **37**, 212; hysteresis, **64**, 217, 220; immobilized (analysis of protein digestibility changes, in, **135**, 601; peptide synthesis with, **136**, 182, 184); immobilization (activated agarose, on, **44**, 33; activated Spheron, on, **44**, 72, 83; acyl azide intermediate method, using, **44**, 446; adsorption, by, effects on activity, **44**, 44; aggregation, by, **44**, 265; anhydride derivative of hydroxyalkylmethacrylate gel, on, **44**, 83; cellophane membrane, in, **44**, 905, 909; cellulose, on, **44**, 52; glutaraldehyde, using, **44**, 265; hollow-fiber membrane device, in, **44**, 310, 311; imidoester-containing polyacrylonitrile, on, **44**, 323, 324; inert protein, on, **44**, 909; polyacrylamide beads, on, **44**, 446; polyacrylamide gel, in, **44**, 902; Sephadex, on, **44**, 30, 529); inactivation of, **4**, 267; inhibition, **19**, 131; **63**, 180; **80**, 771 (eglin, **80**, 812; α$_1$-proteinase inhibitor, **80**, 763); interaction parameters, preferential, **61**, 48; intermediate, **87**, 480; **308**, 201; isotope effects, **87**, 598, 627; iterative template refinement, **266**, 333, 335, 337; kinetic parameters, **63**, 237, 238, 244; kinetics, calorimetric determination, **61**, 264; low-temperature dynamics, **61**, 325, 326; mapping, active-site, **87**, 471, 495; mechanism overview, **308**, 207; microcalorimetric studies, of binding, **61**, 303; modification, **44**, 310 (2-bromoacetamido-4-nitrophenol and, **26**, 582; 4-bromoacetamido-2-nitrophenol and, **26**, 584; effect on activity, **44**, 409, 448); naphthacyl-α-chymotrypsin, **46**, 89; negative cooperativity, **64**, 220; nitrogen isotope effect, **64**, 95; nuclear magnetic resonance, **63**, 212 (^{13}C and ^{19}F, **87**, 72); oxidation, **4**, 264; oxidation with hydrogen peroxide, **25**, 397; partial specific volumes, **61**, 48; pH effect, **44**, 404; **87**, 401, 479, 497; phosphonate inhibitors, **244**, 434; pK values, perturbed, **63**, 210, 220; preparation, **19**, 105, 106; **235**, 566; pressure-jump studies, **48**, 314; progress curve analysis, **63**, 180; protonation, calorimetry, **61**, 265; purification of flavin peptides, **53**, 453; radioimmunoassay, **74**, 273; removal from trypsin, **19**, 57; reovirus subviral particles and, **30**, 715; resolution of organic acids, **44**, 833; Sepharose-immobilized (EPR studies, **135**, 504, 510; preparation, **135**, 504; spin labeling, **135**, 504); similarity to (cathepsin G, **80**, 564; thrombin, **80**, 294, 295); site-directed mutants, acid-base catalysis, **249**, 114; solvent perturbation, **87**, 401; solvent trap, **308**, 201; specificity, **87**, 496; spectrokinetic probe, **61**, 323; structure, **240**, 703, 705; structure-function, **80**, 733; substrates (fluorogenic, **248**, 19; thioester, **248**, 10); temperature effects, **63**, 244; terminal groups of, **2**, 8; tissue culture, **58**, 125; transglutaminase digestion, **87**, 41; transition-state and multisubstrate analogues, **249**, 306; trapped by α$_2$-macroglobulin, **80**, 748; water-insoluble derivatives of, **19**, 955 (EMA-chymotrypsin, **19**, 955; miscellaneous types, **19**, 956); zymogen conformational change, **64**, 225

CHYMOTRYPSIN C

This serine endopeptidase [EC 3.4.21.2] catalyzes the hydrolysis of peptide bonds, exhibiting a preference for Leu–Xaa, Tyr–Xaa, Phe–Xaa, Met–Xaa, Trp–Xaa, Gln–Xaa, and Asn–Xaa.[1–3] The enzyme reacts more readily with Tos–Leu–CH$_2$Cl than Tos–Phe–CH$_2$Cl in contrast to chymotrypsin. Note that there is high activity for the hydrolysis of leucyl-containing peptide bonds. Three leucyl bonds in the insulin B chain that are unaffected by chymotrypsin are hydrolyzed by chymotrypsin C.

Chymotrypsin A and chymotypsin B were later found to be homologous and now are listed under chymotrypsin [EC 3.4.21.1]. **See also** *Caldecrin; Chymotrypsin*

[1] A. Thomson & I. S. Denniss (1976) *BBA* **429**, 581.
[2] V. Keil-Dlouha, A. Puigserver, A. Marie & B. Keil (1972) *BBA* **276**, 531.
[3] J. E. Folk & E. W. Schirmer (1965) *JBC* **240**, 181.

Selected entries from *Methods in Enzymology* [**vol**, page(s)]:
General discussion: 19, 109
Assay: 19, 97, 109
Other molecular properties: activators and inhibitors, **19**, 112; amino acid composition, **19**, 112; amino acid sequence, **19**, 99; esteroproteolytic enzyme as, **19**, 91, 99, 112; molecular weight, **19**, 98; pig, from, **19**, 90, 97, 109; purification, **19**, 97, 109; reaction with *N*-tosyl-L-phenylalanine chloromethyl ketone, **19**, 99; specificity, **19**, 99, 112; stability, **19**, 98, 111

CINNAMATE GLUCOSYLTRANSFERASE

This glucosyltransferase [EC 2.4.1.177] catalyzes the reaction of UDP-D-glucose with *trans*-cinnamate to produce UDP and *trans*-cinnamoyl-β-D-glucose. Alternative acceptor substrates include 4-methoxycinnamate, 4-coumarate, 2-coumarate, benzoate, feruloate, and caffeate.[1]

[1] T. Shimizu & M. Kojima (1984) *J. Biochem.* **95**, 205.

trans-CINNAMATE 2-MONOOXYGENASE

trans-cinnamate   *o*-coumarate

This oxidoreductase [EC 1.14.13.14], also known as cinnamate 2-hydroxylase, catalyzes the reaction of *trans*-cinnamate with NADPH and dioxygen to produce 2-hydroxycinnamate (*i.e.*, *o*-coumarate), NADP$^+$, and water.[1]

[1] B. Gestetner & E. E. Conn (1974) *ABB* **163**, 617.

Selected entries from *Methods in Enzymology* [**vol**, page(s)]:
General discussion: 52, 20

trans-CINNAMATE 4-MONOOXYGENASE

This heme-thiolate-dependent oxidoreductase [EC 1.14.13.11], also known as cinnamate 4-hydroxylase and CYP73A, catalyzes the reaction of *trans*-cinnamate with NADPH and dioxygen to generate 4-hydroxycinnamate (*i.e.*, *p*-coumarate), NADP$^+$, and water.[1–5]

trans-cinnamate   *p*-coumarate

The enzyme, which is a cytochrome P450 system (CYP73A), can also replace NADPH with NADH (however, the reaction will proceed slower). CYP73A1 of *Helianthus tuberosus* will act on a number of small planar molecules with dipolar character: alternate substrates include *p*-chloro-*N*-methylaniline, methoxycoumarin, ethoxycoumarin, and the herbicide chortoluron.

[1] I. Raskin (1992) *Ann. Rev. Plant Physiol. Plant Mol. Biol.* **43**, 439.
[2] R. A. Dixon, P. M. Dey & C. J. Lamb (1983) *Adv. Enzymol.* **55**, 1.
[3] I. Beneviste, J.-P. Salaun & F. Durst (1977) *Phytochemistry* **16**, 69.
[4] R. Pfändler, D. Scheel, H. Sandermann & H. Grisebach (1977) *ABB* **178**, 315.
[5] J. R. M. Potts, R. Weklych & E. E. Conn (1974) *JBC* **249**, 5019.

Selected entries from *Methods in Enzymology* [vol, page(s)]:
General discussion: 52, 15; 272, 51, 259
Assay: 2-naphthoate hydroxylation and fluorescence detection, 272, 266; radiolabeled substrate and thin-layer chromatography, 272, 265
Other molecular properties: distribution in plant species, 272, 261; immunoblotting, 272, 267; induction (chemicals, 272, 263; light, 272, 262; wounding, 272, 262); microsome preparation (seedlings, 272, 261, 264; tuber slices, 272, 261); Northern blot analysis, 272, 268; phenylpropanoid synthesis, 272, 260

CINNAMOYL-CoA REDUCTASE

This oxidoreductase [EC 1.2.1.44] catalyzes the reaction of cinnamoyl-CoA with NADPH to produce cinnamaldehyde, coenzyme A, and $NADP^+$.[1,2] Other substrates include feruloyl-CoA, *p*-coumaroyl-CoA, sinapoyl-CoA, and caffeoyl-CoA.

[1] F. Sarni, C. Grand & A. M. Boudet (1984) *EJB* **139**, 259.
[2] H. Wengenmayer, J. Ebel & H. Grisebach (1976) *EJB* **65**, 529.

CINNAMYL-ALCOHOL DEHYDROGENASE

This oxidoreductase [EC 1.1.1.195] catalyzes the reversible reaction of cinnamyl alcohol with $NADP^+$ to produce cinnamaldehyde and NADPH.[1–4]

trans-cinnamyl alcohol   *trans*-cinnamaldehyde

Other substrates include coniferyl alcohol, sinapyl alcohol, 4-coumaryl alcohol, and 3,4-dimethoxycinnamyl alcohol.
See also *Coniferyl-Alcohol Dehydrogenase*

[1] F. Sarni, C. Grand & A. M. Boudet (1984) *EJB* **139**, 259.
[2] H. Kutsuki, M. Shimada & T. Higuchi (1982) *Phytochemistry* **21**, 19.
[3] D. Wyrambik & H. Grisebach (1979) *EJB* **97**, 503.
[4] A. M. Boudet, J. Grima-Pettenati & K. T. Douglas (1995) *Biochemistry* **34**, 12426.

CITRAMALATE CoA-TRANSFERASE

This transferase [EC 2.8.3.11], a component of citramalate lyase [EC 4.1.3.22], catalyzes the reaction of acetyl-CoA with citramalate to produce acetate and (3*S*)-citramalyl-CoA.[1] The enzyme will also catalyze the transfer of the thioacyl-carrier protein from its acetylate form to citramalate. ***See also*** *Citramalate Lyase; Citramalyl-CoA Lyase*

[1] P. Dimroth, W. Buckel, R. Loyal & H. Eggerer (1977) *EJB* **80**, 469.

CITRAMALATE LYASE

This lyase [EC 4.1.3.22] catalyzes the reversible conversion of (3*S*)-citramalate to acetate and pyruvate.[1,2] The enzyme complex can be dissociated into components, two of which are identical with citamalate CoA transferase [EC 2.8.3.11] and citramalyl-CoA lyase [EC 4.1.3.25]. ***See also*** *Citramalyl-CoA Lyase; Citramalate CoA-Transferase*

[1] H. A. Barker (1967) *Arch. Mikrobiol.* **59**, 4.
[2] W. Buckel & S. L. Miller (1987) *EJB* **164**, 565.

Selected entries from *Methods in Enzymology* [vol, page(s)]:
General discussion: 13, 344
Assay: 13, 344
Other molecular properties: extraction, 13, 345; metal ion requirement, 13, 346; properties, 13, 346; stereochemistry, 87, 147; storage, 13, 345

CITRAMALYL-CoA LYASE

This lyase [EC 4.1.3.25] catalyzes the conversion of (3*S*)-citramalyl-CoA to acetyl-CoA and pyruvate.[1,2] The (3*S*)-citramalyl thioacyl-carrier protein can also be utilized as a substrate. This activity is a component of citramalate lyase [EC 4.1.3.22]. ***See also*** *Citramalate Lyase; Citramalate CoA-Transferase*

[1] K. Sasaki & H. Katsuki (1973) *J. Biochem.* **73**, 599.
[2] P. Dimroth, W. Buckel, R. Loyal & H. Eggerer (1977) *EJB* **80**, 469.

Selected entries from *Methods in Enzymology* [vol, page(s)]:
General discussion: 13, 316, 317

CITRATE CoA-TRANSFERASE

This transferase [EC 2.8.3.10] catalyzes the reaction of acetyl-CoA with citrate to produce acetate and (3*S*)-citryl-CoA.[1]

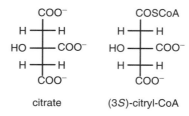

citrate   (3*S*)-citryl-CoA

This activity is a component of citrate (*pro-3S*)-lyase [EC 4.1.3.6]. Note that the transferase activity in the isolated state is greater than that observed in the citrate (*pro-3S*)-lyase complex. The enzyme will also catalyze the transfer of the thioacyl-carrier protein from the acetylated protein to citrate.

[1] P. Dimroth, R. Loyal & H. Eggerer (1977) *EJB* **80**, 479.

CITRATE DEHYDRATASE

citrate   *cis*-aconitate

This enzyme [EC 4.2.1.4] catalyzes the conversion of citrate to *cis*-aconitate (*i.e.*, (*Z*)-prop-1-ene-1,2,3-tricarboxylate) and water.[1] Isocitrate is not a substrate.

[1] N. E. Neilson (1955) *BBA* **17**, 139.

CITRATE (*pro-3S*)-LYASE

This lyase [EC 4.1.3.6] (also known as citrate (*pro-3S*)-lyase, citrase, citratase, citritase, citridesmolase, and citrate aldolase) catalyzes the conversion of citrate to acetate and oxaloacetate.[1–5]

The seryl residue of one of the three subunits of the *Klebsiella pneumoniae* enzyme acts as an acyl-carrier protein (the γ subunit) and contains a prosthetic group attached via a phosphodiester bond: 2′-(5″-phosphoribosyl)-3′-dephospho-CoA. The thiol moiety of this prosthetic group must be acetylated (catalyzed by [citrate (*pro-3S*)-lyase] ligase [EC 6.2.1.22]), generating an enzyme-bound acetyl thioester, in order for the overall catalytic activity to be expressed. The α subunit (citrate CoA-transferase [EC 2.8.3.10]) catalyzes the exchange of the acetyl group with a citryl group, producing acetate and citryl-[acyl-carrier protein]. The β subunit (citryl-CoA lyase [EC 4.1.3.34]) then catalyzes the regeneration of acetyl-[acyl-carrier protein] via the release of oxaloacetate from citryl-[acyl-carrier protein]. [Citrate (*pro-3S*)-lyase] thiolesterase [EC 3.1.2.16] catalyzes the deacetylation of the lyase and inactivates the enzyme. **See also** ATP:Citrate (*pro-3S*)-Lyase

[1] D. J. Creighton & N. S. R. K. Murthy (1990) *The Enzymes*, 3rd ed., **19**, 323.
[2] S. Subramanian & C. Sivaraman (1984) *J. Biosci.* **6**, 379.
[3] H. Kulla & G. Gottschalk (1977) *J. Bacteriol.* **132**, 764.
[4] P. A. Srere (1975) *Adv. Enzymol.* **43**, 57.
[5] L. B. Spector (1972) *The Enzymes*, 3rd ed., **7**, 357.

Selected entries from *Methods in Enzymology* [vol, page(s)]:
General discussion: 5, 623; 13, 160
Assay: 5, 623; 13, 161
Other molecular properties: aconitic hydrase preparation, in, **5**, 615; citrate, assay of, **3**, 426; **13**, 450, 517; **55**, 211; equilibrium constant, **5**, 627; **13**, 161; inhibition by oxaloacetate, **13**, 160; properties, **5**, 626; purification, *Aerobacter aerogenes* (now called *Klebsella pneumoniae*), **5**, 625; **13**, 161, 518; sources, **5**, 622, 625; stereochemistry, **87**, 147

[CITRATE (*pro-3S*)-LYASE] LIGASE

This ligase [EC 6.2.1.22], also known as citrate lyase ligase and citrate lyase synthetase, catalyzes the reaction of ATP with acetate and the thiol form of citrate (*pro-3S*)-lyase to produce AMP, pyrophosphate (or, diphosphate), and the acetyl form of citrate (*pro-3S*)-lyase.[1] The thiol form of citrate (*pro-3S*)-lyase [EC 4.1.3.6] is the inactive form of the enzyme.

[1] G. Antranikian, C. Herzberg & G. Gottschalk (1985) *EJB* **153**, 413.

CITRATE (*pro-3S*)-LYASE THIOLESTERASE

This thiolesterase [EC 3.1.2.16], also known as citrate lyase deacetylase, catalyzes the hydrolysis of the acetyl form of citrate (*pro-3S*)-lyase to produce acetate and the thiol form of citrate (*pro-3S*)-lyase, thus inactivating that lyase [EC 4.1.3.6].[1,2] The *Rhodopseudomonas gelatinosa* enzyme is strongly inhibited by L-glutamate.

[1] F. Giffhorn & G. Gottschalk (1975) *J. Bacteriol.* **124**, 1052.
[2] F. Giffhorn, H. Rode, A. Kuhn & G. Gottschalk (1980) *EJB* **111**, 461.

CITRATE (*re*)-SYNTHASE

This enzyme [EC 4.1.3.28] catalyzes the reversible reaction of acetyl-CoA with oxaloacetate to produce citrate and coenzyme A.[1,2]

CH₃COSCoA   citrate

This activity, found in a number of microorganisms, has the opposite stereochemistry of citrate (*si*)-synthase [EC 4.1.3.7]. The acetyl group is transferred to the *re*-face of the carbonyl of oxaloacetate.

[1] P. Wunderwald, W. Buckel, H. Lenz, V. Buschmeier, H. Eggerer, G. Gottschalk, J. W. Cornforth, J. W. Redmond & R. Mallaby (1971) *EJB* **24**, 216.

[2] G. Gottschalk & S. Dittbrenner (1970) *Hoppe-Seyler's Z. Physiol. Chem.* **351**, 1183.

CITRATE (*si*)-SYNTHASE

This enzyme [EC 4.1.3.7] (also known as citrate condensing enzyme, oxaloacetate transacetase, and citrogenase) catalyzes the reversible reaction of acetyl-CoA with oxaloacetate and water to produce citrate and coenzyme A.[1-7]

CH3COSCoA

citrate

Citrate (*re*)-synthase [EC 4.1.3.28] catalyzes the same reaction, albeit with the opposite stereochemistry. The addition of the acetyl group is to the *si*-face of the carbonyl of oxaloacetate. The reaction is believed to utilize low-barrier hydrogen bonds. The enzyme also catalyzes the enolization of acetyl-CoA.

[1] J. V. Schloss & M. S. Hixon (1998) *CBC* **2**, 43.

[2] S. J. Remington (1992) *Curr. Top. Cell. Reg.* **33**, 209.

[3] R. F. Colman (1990) *The Enzymes*, 3rd ed., **19**, 283.

[4] D. J. Creighton & N. S. R. K. Murthy (1990) *The Enzymes*, 3rd ed., **19**, 323.

[5] P. A. Srere (1975) *AE* **43**, 57.

[6] L. B. Spector (1972) *The Enzymes*, 3rd ed., **7**, 357.

[7] J. R. Stern (1961) *The Enzymes*, 2nd ed., **5**, 367.

Selected entries from *Methods in Enzymology* [vol, page(s)]:
General discussion: 1, 685; 13, 3, 11, 16, 19, 22; 71, 151; 188, 350
Assay: 1, 685; 13, 3, 12, 16, 19, 22, 365; methylotrophs, 188, 350; polarographic assay for, 13, 365; spectrophotometric assay, 1, 687; 13, 3, 12, 16, 19, 22; 264, 496; ultramicroassay, 191, 548
Other molecular properties: activity assay, 44, 458, 474; acyl CoA-binding site photoaffinity labeling, 91, 639; allosteric effectors, 13, 11, 25; amino acid composition of pig heart enzyme, 13, 10, 11; chiral methyl groups, and, 87, 129; citryl-CoA as substrate for, 13, 8; fluorocitrate synthesis, use for, 13, 655; glyoxysomes, from, 31, 565, 569; 72, 788; immobilization in multistep enzyme system, 44, 181, 457, 473; inhibitors, 13, 9, 19, 25; irreversible inhibitor of, 72, 580; kinetics, in multistep enzyme system, 44, 434, 466; labeled citrate, preparation, use for, 13, 584, 588; malyl-CoA as substrate for, 13, 9; marker enzyme, as, 56, 217, 218; mechanism, 63, 51; microbodies, in, 52, 496; modifiers, 13, 11, 25; mutant, 72, 695; occurrence, 13, 11; porcine heart, acyl CoA-binding site photoaffinity labeling, 91, 639; preparation from (bakers' yeast, 13, 17; *Escherichia coli*, 13, 23; lemon fruit, 13, 20; methylotrophs, 188, 351; moth flight muscle, 13, 8; pigeon breast muscle, 13, 7; pig heart, 1, 689; 13, 5; rat liver, 13, 13; yeast, 13, 16); properties, 1, 693; 13, 8, 15, 19, 25; 188, 352; stereochemistry, 87, 129, 147, 156, 157; substrate specificity, 1, 693; 13, 8; succinate dehydrogenase, and, 87, 157; synthesis of

deuterated citrate, 13, 584; transition-state and multisubstrate analogues, 249, 307; tritium exchange reaction, 13, 9; ultramicroassay, 191, 548

CITRULLINASE

L-citrulline   L-ornithine

This enzyme [EC 3.5.1.20] catalyzes the hydrolysis of L-citrulline to produce L-ornithine, carbon dioxide, and ammonia.[1,2]

[1] B.-S. Park, A. Hirotani, Y. Nakano & S. Kitaoka (1985) *Agric. Biol. Chem.* **49**, 2205.

[2] D. L. Hill & P. Chambers (1967) *BBA* **148**, 435.

Selected entries from *Methods in Enzymology* [vol, (s)]:
General discussion: 2, 374, 376
Assay: 2, 376

CITRYL-CoA LYASE

This enzyme [EC 4.1.3.34] catalyzes the conversion of (3*S*)-citryl-CoA to acetyl-CoA and oxaloacetate.[1] The lyase, which is a component of citrate (*pro-3S*)-lyase [EC 4.1.3.6] and ATP:citrate (*pro-3S*)-lyase [EC 4.1.3.8], will also act on the (3*S*)-citryl thioacyl-carrier protein.

[1] S. Nilekani & C. SivaRaman (1983) *Biochemistry* **22**, 4657.

CITRYL-CoA SYNTHETASE

This ligase [EC 6.2.1.18], also known as citrate:CoA ligase, catalyzes the reaction of ATP with citrate and coenzyme A to produce ADP, orthophosphate, and (3*S*)-citrylCoA. It is a component of ATP:citrate (*pro-3S*)-lyase [EC 4.1.3.8].[1]

[1] U. Lill, A. Schreil & H. Eggerer (1982) *EJB* **125**, 645.

c-JUN KINASE

This protein kinase catalyzes the ATP-dependent phosphorylation of c-Jun, a transcription factor. *See also* Protein Kinase

Selected entries from *Methods in Enzymology* [vol, page(s)]:
Assay: cell extract preparation, 255, 351; immune complex kinase assay, 255, 349, 353; *in vitro* kinase assay, 255, 349, 352; transcription assay, 255, 355
Other molecular properties: domains, 255, 342; glutathione S-transferase fusion protein, 255, 350, 352; immunoprecipitation, 255, 346; phosphorylation-induced activation, 255, 342, 348 (pattern mapping of kinase in cultured cells, 255, 343; substrate specificity, 255, 343); radiolabeling in cells, 255, 346; SDS-PAGE, 255, 344, 347; tryptic digestion, 255, 344; Western blot analysis, 255, 347

CLOSTRIDIAL AMINOPEPTIDASE

This aminopeptidase [EC 3.4.11.13] catalyzes the release of any N-terminal aminoacyl residue, including prolyl and hydroxyprolyl residues, from a polypeptide or protein (charged aminoacyl residues are released slowly).[1,2] The enzyme will not act when the second aminoacyl residue (*i.e.*, that occupying the P1' position) is a prolyl residue. Accordingly, the enzyme cannot act on polyproline. **See also** *Aminopeptidase*

[1] G. Flaminger & A. Yaron (1984) *BBA* **789**, 245 and (1983) **743**, 437.
[2] E. Kessler & A. Yaron (1976) *EJB* **63**, 271.

Selected entries from **Methods in Enzymology** [vol, page(s)]:
General discussion: 45, 544; 136, 170
Assay: 45, 544; 136, 171, 173

CLOSTRIPAIN

This cysteine endopeptidase [EC 3.4.22.8], also known as clostridiopeptidase B, has been isolated from *Clostridium histolyticum* and catalyzes the hydrolysis of peptide bonds with a preference for the Arg–Xaa bond (including the Arg–Pro bond): the Lys–Xaa peptide bond is hydrolyzed at a significantly lower rate.[1–7] Small synthetic substrates such as Bz-Arg-OEt and Z-Phe-Arg-∇-NHMec can be used to assay the enzyme. Clostripain can also act as a transpeptidase.[5–7]

[1] E. Shaw (1990) *AE* **63**, 271.
[2] K. Morihara (1974) *AE* **41**, 179.
[3] W. M. Mitchell & W. F. Harrington (1971)*The Enzymes*, 3rd ed., **3**, 699.
[4] A. A. Kembhavi, D. J. Buttle, P. Rauber & A. J. Barrett (1991) *FEBS Lett.* **283**, 277.
[5] G. Fortier & J. Gagnon (1990) *ABB* **276**, 317.
[6] V. Schellenberger, C. W. Turck, L. Hedstrom & W. J. Rutter (1993) *Biochemistry* **32**, 4349.
[7] D. Ullmann & H.-D. Jakubke (1994) *EJB* **223**, 865.

Selected entries from **Methods in Enzymology** [vol, page(s)]:
General discussion: 19, 635; 46, 206, 229; 47, 165
Assay: esterase assay, 80, 826
Other molecular properties: active site properties, 47, 165; *Clostridium histolyticum*, from, specificity, 59, 600; esterase and amidase activities, 19, 641; family, 244, 479; inactivation, 80, 824; inhibition, 46, 234; inhibitors, 47, 167; pH optimum, 19, 639; properties, 19, 639; protein cleavage, 47, 165; purification, 19, 637; sequence tool, utility as, 47, 169, 170; substrate specificity, 19, 640; sulfhydryl requirement, 19, 639; trapped by α2-macroglobulin, 80, 748; tubulin, proteolysis of, 134, 182; zymography in nondissociating gels with copolymerized substrates, 235, 588

CMP-N-ACETYLNEURAMINATE MONOOXYGENASE

This iron-dependent monooxygenase [EC 1.14.13.45], commonly known as *N*-acetylneuraminate monooxygenase or hydroxylase, catalyzes the reaction of CMP-*N*-acetylneuraminate with dioxygen and NAD(P)H to produce CMP-*N*-glycoloylneuraminate, water, and NAD(P)$^+$.[1–4] Free *N*-acetylneuraminate is not a substrate.[2] The Enzyme Commission lists a related *N*-acetylneuraminate monooxygenase [EC 1.14.99.18]. Ascorbate and 6,7-dimethyl-5,6,7,8-tetrahydrobiopterin can also support the reaction in place of NAD(P)H.[3] **See also** *N-Acetylneuraminate Monooxygenase*

[1] L. Shaw, P. Schneckenburger, J. Carlsen, K. Christiansen & R. Schauer (1992) *EJB* **206**, 269.
[2] E. A. Muchmore, M. Milewski, A. Varki & S. Diaz (1989) *JBC* **264**, 20216.
[3] L. Shaw & R. Schauer (1988) *Biol. Chem. Hoppe Seyler* **369**, 477.
[4] L. Shaw & R. Schauer (1989) *BJ* **263**, 355.

Selected entries from **Methods in Enzymology** [vol, page(s)]:
General discussion: 50, 374
Assay: 50, 375

CMP-N-ACYLNEURAMINATE PHOSPHODIESTERASE

This enzyme [EC 3.1.4.40], also known as CMP-sialate hydrolase, catalyzes the hydrolysis of CMP-*N*-acyl-neuraminate to produce CMP and *N*-acylneuraminate.[1–3]

[1] L. Masson & B. E. Holbein (1983) *J. Bacteriol.* **154**, 728.
[2] W. Van Dijk, H. Maier & D. H. Van den Eijnden (1976) *BBA* **444**, 816.
[3] E. L. Kean & K. J. Bighouse (1974) *JBC* **249**, 7813.

Selected entries from **Methods in Enzymology** [vol, page(s)]:
General discussion: 28, 983
Assay: electrophoresis, 28, 986; ion exchange, 28, 983

CoA-DISULFIDE REDUCTASE (NADH)

This oxidoreductase [EC 1.6.4.10] catalyzes the reaction of NADH with coenzyme A disulfide to produce NAD$^+$ and two coenzyme A molecules.[1] The enzyme is distinct from cystine reductase (NADH) [EC 1.6.4.1], glutathione reductase (NADPH) [EC 1.6.4.2], and bis-γ-glutamylcystine reductase (NADPH) [EC 1.6.4.9].

[1] B. Setlow & P. Setlow (1977) *J. Bacteriol.* **132**, 444.

CoA:GLUTATHIONE REDUCTASE (NADPH)

This flavin-dependent oxidoreductase [EC 1.6.4.6] catalyzes the reaction of NADPH with CoA-glutathione (*i.e.*, the mixed disulfide) to produce NADP$^+$, coenzyme A, and glutathione (*i.e.*, γ-L-glutamyl-L-cysteinylglycine).[1] **See also** *Glutathione Reductase; CoA-Disulfide Reductase*

[1] R. N. Ondarza, E. Escamilla, J. Gutiérrez & G. de la Chica (1974) *BBA* **341**, 162.

Selected entries from **Methods in Enzymology** [vol, page(s)]:
General discussion: 18A, 318
Assay: 18A, 318

COAGULATION FACTOR VIIA

This serine endoproteinase [EC 3.4.21.21], catalyzing the first enzyme in the blood coagulation cascade, converts coagulation factor X to produce factor Xa by the hydrolysis of a single peptide bind (Arg152–Ile153).[1,2]

[1]R. Bach (1988) CRC Crit. Rev. Biochem. **23**, 339.
[2]L. Luchtman-Jones & G. J. Broze, Jr., (1995) Ann. Med. **27**, 47.

Selected entries from *Methods in Enzymology* [**vol**, page(s)]:
General discussion: 2, 155; **80**, 299, 325
Other molecular properties: activation by factor XIIa, **163**, 68; activity measurements, **222**, 192; clotting mechanism and, **2**, 139, 150, 155, 156; complexes of, **2**, 151; composition, **45**, 54; human, **80**, 229; inhibition (antithrombin III, by, with heparin, **80**, 237; DFP, by, **80**, 237); purification, **2**, 155; radioligand assay of tissue factor mutant receptor function, in, **222**, 217; synthetic substrates for, **222**, 189; thioester substrates, **248**, 13

COAGULATION FACTOR IXA

This serine endopeptidase [EC 3.4.21.22], also known as activated Christmas factor, catalyzes the conversion of coagulation factor X to Xa by the hydrolysis of a single peptide bond (Arg–Ile).[1-4] This conversion requires the presence of three cofactors: Ca^{2+}, phospholipid, and factor VIIIa.

[1]K. Fujikawa & E. W. Davie (1976) Meth. Enzymol. **45**, 74.
[2]J. P. Miletich, G. J. Broze, Jr., & P. W. Majerus (1981) Meth. Enzymol. **80**, 221.
[3]S. P. Bajaj & J. J. Birktoft (1993) Meth. Enzymol. **222**, 96.
[4]H. Brandstetter, M. Bauer, R. Huber, P. Lollar & W. Bode (1995) PNAS **92**, 9796.

Selected entries from *Methods in Enzymology* [**vol**, page(s)]:
General discussion: 45, 74; **80**, 221, 299, 325; **222**, 96
Assay: peptide chromogenic substrates, with, **80**, 354
Other molecular properties: activation process, **222**, 185 (assays, **222**, 186; discontinuous measurements, **222**, 186); activator (factor VII, of, **80**, 236; factor X, of, with factor VIII and phospholipids, **80**, 245, 246); bovine (activator of factor X, **80**, 247; assay, with peptide chromogenic substrates, **80**, 354; composition, **45**, 82; conversion from factor IX, **45**, 81, 82; hirudin, inactivated, **45**, 678; homology to crab collagenase, **80**, 729; interaction with other factors, **45**, 83); factor IXαβ, active site labeling, **222**, 189; human, overview, **222**, 96; serine protease, as, **80**, 156; thioester substrates, **248**, 13 (subsite mapping studies, **248**, 17)

COAGULATION FACTOR XA

This serine endopeptidase [EC 3.4.21.6] (also known as Stuart's factor, Prower's factor, thrombokinase, and pro-thrombase) catalyzes the conversion of prothrombin to thrombin with the hydrolysis of two specific peptide bonds, requiring the presence of calcium ions and a coagulation cofactor, factor Va.[1,2] Factor Va binds to factor Xa extremely tightly, having a K_d value of about 1 nM. However, factor Va does not appear to bind to the precursor factor X.

Scutelarin [EC 3.4.21.60], a toxin from the venom of the taipan snake, has a very similar specificity but does not require factor Va.

[1]H. L. James (1994) in *Haemostasis and Thrombosis*, 3rd ed. (A. L. Bloom, C. D. Forbes, D. P. Thomas & E. G. D. Tuddenham, eds.), Churchill Livingstone, Edinburgh, p. 439.
[2]K. G. Mann, R. J. Jenny & S. Krishnaswamy (1988) Ann. Rev. Biochem. **57**, 915.

Selected entries from *Methods in Enzymology* [**vol**, page(s)]:
General discussion: 45, 97; **80**, 325
Assay: bovine, **45**, 101; human, **80**, 345, 350, 351; peptide chromogenic and fluorogenic substrates, with, **80**, 345, 350, 351, 359
Other molecular properties: activation, measurement in vitro, **222**, 522; activation (proteolysis) of prothrombin, **19**, 169; **45**, 124, 155; **80**, 287; activator of factor VII, **80**, 236; active-site titration, benzyl p-guanidinothiobenzoate in, **248**, 14; activity in prothrombinase complex, **223**, 293; affinity labeling, **80**, 834, 835, 838; assay of erythrocyte fusion, in, **220**, 171; assay, with peptide chromogenic and fluorogenic substrates, **80**, 345, 350, 351, 359; binary macromolecular interactions in, **222**, 270; binding (kinetics, **215**, 319; thrombin-activated human platelets, with, kinetics, **215**, 354); binding to unstimulated platelets, **215**, 336; bovine, **80**, 357 (activation, cleavage sites, **45**, 106; amino-terminal sequence, **80**, 641; assay, **45**, 101; **80**, 345, 350, 351; binding to platelets, **215**, 331; chromatography of tissue factor pathway inhibitor, in, **222**, 204; fluorescent derivatives, preparation, **222**, 262; homology to crab collagenase, **80**, 729; preparation, **222**, 261; proteolytic action, **45**, 105; prothrombin activation, **45**, 155); cleavage at monocyte surface, visualization, **222**, 292; complex, **80**, 250, 272, 273; **222**, 260 (platelet surface, at, mediation by component E, **215**, 347); CYP52A3, factor Xa recognition site in amino terminus (factor Xa proteolysis [buffer, **272**, 70; comparison to complementary DNA truncation, **272**, 74; cytosolic domain liberation from intact microsomes, **272**, 73; efficiency, **272**, 70, 72; reaction conditions, **272**, 71; specificity, **272**, 72]) digestion of fusion proteins produced in *Escherichia coli*, **153**, 475; effect on proteolysis of platelet-bound factor Va by activated protein C, **215**, 345; elastase and, **5**, 672; factor Va, and, coordinate binding to unstimulated platelets, **215**, 338; factor Va solution-phase interaction, measurement, **222**, 276; factor Xaα (bovine [composition, **45**, 94, 95, 98; preparation, **45**, 103]); factor Xaβ (bovine [composition, **45**, 94, 98; preparation, **45**, 104]); formation, affected by factor X, via extrinsic pathway, **80**, 245; human (activator of factor VII, **80**, 229; assay, **80**, 345, 350, 351); inhibition by tissue factor pathway inhibitor, **222**, 198, 206; inhibitor (assay, **223**, 292; characterization, **223**, 291; purification from hematophagous organisms, **223**, 291); interaction with phospholipid-bound factor Va, **80**, 270; mathematical simulation, **215**, 316; model, **80**, 272; platelet binding, **80**, 250, 251, 272; platelet-factor Va binding interactions, stoichiometry, **215**, 355; platelet interaction, mathematical expression for, **215**, 340; protection of factor Va from inactivation, **80**, 270; reaction with α2-antiplasmin, **80**, 404; receptors (platelet [binding studies, **215**, 329; bovine, component E in, **215**, 351; human, binding parameters, **215**, 351]); serine protease, as, **80**, 156; substrate, and (binding distributions, calculation, **215**, 323; bulk and local concentrations, calculation, **215**, 323); ternary interactions in, **222**, 266

COAGULATION FACTOR XIA

This serine endopeptidase [EC 3.4.21.27], formerly known as plasma thromboplastin antecedent (PTA), catalyzes the conversion of coagulation factor IX (57 kDa) to factor IXa by the selective hydrolysis of the Arg145–Ala146 and Arg180–Val181 peptide bonds in factor IX.[1] Factor IXa will also hydrolyze H-kininogen slowly and it is unclear whether this activity is physiologically significant. In addition, the enzyme will catalyze the hydrolysis of a number of short peptides (for example, Glp-Pro-Arg-NHPhNO$_2$) and peptide thioesters.

[1] P. N. Walsh (1992) *Semin. Hematol.* **29**, 189.

Selected entries from *Methods in Enzymology* [vol, page(s)]:
General discussion: **45**, 65; **80**, 155, 299, 325
Assay: **80**, 214
Other molecular properties: activation, **80**, 196, 197, 208; **163**, 67; binding sites on platelet, **215**, 368; binding to (human platelet receptors, **215**, 361; platelets [assay, **215**, 363; characteristics, **215**, 365]); bovine, **80**, 357 (activator of factor IX, **80**, 248, 249); conversion of factor IX to factor IXa, **80**, 212; human (enzyme activity, **80**, 220; inhibition, **80**, 220; structure, **80**, 220); inhibition, **80**, 51, 220; platelet-bound, characterization, **215**, 368; role, **45**, 65; substrate, fluorogenic, **248**, 24

COAGULATION FACTOR XIIA

This serine endopeptidase [EC 3.4.21.38], also known as Hageman factor, is the first component of the intrinsic coagulation pathway. The precursor to XIIa undergoes autoactivation (at Arg353–Val354) under those physiological conditions that initiate coagulation or is activated by plasma kallikrein. Factor XIIa catalyzes the conversion of factor XI to factor XIa as well as convert factor VII to factor VIIa, at Arg–Ile peptide bonds. Other substrates include plasma prekallikrein, plasminogen, and complement component C1.[1-3]

[1] M. Silverberg & A. P. Kaplan (1982) *Blood* **60**, 64.
[2] Y. Hojima, J. V. Pierce & J. J. Pisano (1982) *Biochemistry* **21**, 3741.
[3] E. W. Davie, K. Fujikawa, K. Kurachi & W. Kisiel (1979) *AE* (1979) **48**, 277.

Selected entries from *Methods in Enzymology* [vol, page(s)]:
General discussion: **34**, 435; **80**, 155, 325; **163**, 54, 68, 170; **222**, 51
Assay: human, **45**, 57; peptide chromogenic and fluorogenic substrates, with, **80**, 354, 359
Other molecular properties: activation, **163**, 75; **222**, 64 (plasma kallikrein, by, **163**, 66); activator (factor VII, of, **80**, 236; factor XI, of, **80**, 212, 220; prekallikrein, of, **80**, 157, 196, 199); active-site titration, benzyl *p*-guanidinothiobenzoate in, **248**, 14; assay, with peptide chromogenic and fluorogenic substrates, **80**, 354, 359; digestion into fragments, by plasmin, **80**, 365; enzymatic properties, **163**, 79; factor VII activation, **163**, 68; factor XI and plasma prekallikrein activation, **163**, 67; formation, **80**, 209, 210; fragment, **80**, 210, 211; human (assay, **45**, 57; chromatography, **45**, 61; composition, **45**, 64; enzyme activity, **80**, 208; inhibitors, **45**, 65; properties, **45**, 62, 63 [immunochemical, **45**, 65]; purification, **45**, 60; purity, **45**, 63; separation from factor XII, **80**, 207, 208; stability, **45**, 62; zymogen form, **45**, 56); inhibition, **45**, 65; **80**, 51; polypeptide chains, physical properties, **80**, 209, 210; purification, **45**, 60; **163**, 77; rabbit, amino acid composition, **45**, 64

COB(I)ALAMIN ADENOSYLTRANSFERASE

This manganese-dependent transferase [EC 2.5.1.17], also known as aquacob(I)alamin adenosyltransferase, catalyzes the reaction of ATP with cob(I)alamin and water to produce orthophosphate, pyrophosphate (or, diphosphate), and adenosylcobalamin. The reaction proceeds with overall inversion of configuration at the C5' carbon of the adenosyl moiety.[1]

[1] R. J. Parry, J. M. Ostrander & I. Y. Arzu (1985) *J. Amer. Chem. Soc.* **107**, 2190.

Selected entries from *Methods in Enzymology* [vol, page(s)]:
General discussion: **67**, 41
Assay: **67**, 42

COB(II)ALAMIN REDUCTASE

This flavin-dependent oxidoreductase [EC 1.6.99.9], also known as vitamin B$_{12r}$ reductase, catalyzes the reaction of NADH with two molecules of cob(II)alamin to produce NAD$^+$ and two molecules of cob(I)alamin.[1,2] *See also* Aquacobalamin Reductase; Aquacobalamin Reductase (NADPH)

[1] G. A. Walker, S. Murphy & F. M. Huennekens (1969) *ABB* **134**, 95.
[2] F. Watanabe, Y. Nakano, N. Tachikake, S. Kitaoka, Y. Tamura, H. Yamanaka, S. Haga, S. Imai & H. Saido (1991) *Int. J. Biochem.* **23**, 531.

COCCOLYSIN

This zinc-dependent endopeptidase [EC 3.4.24.30], also known as streptococcal gelatinase and *Streptococcus thermophilus* intracellular proteinase, is a single-chain protein of 31.5 kDa from *Enterococcus thermophilus* and *Enterococcus faecalis*. The enzyme catalyzes the hydrolysis of peptide bonds in substrates that have at least six aminoacyl residues. The enzyme prefers peptide bonds in which the P1' position is Leu, Phe, Ile, or Ala. The oxidized insulin B chain is rapidly hydrolyzed at Phe24–Phe25 followed by His5–Leu6 and then His10–Leu11, Ala14–Leu15, and Phe25–Tyr26. Coccolysin will also act on bradykinin, substance P, human endothelin I, angiotensin I, and angiotensin II.[1,2]

[1] P.-L. Mäkinen, D. B. Clewell, F. An & K. K. Mäkinen (1989) *JBC* **264**, 3325.
[2] P.-L. Mäkinen & K. K. Mäkinen (1994) *BBRC* **200**, 981.

CODEINONE REDUCTASE (NADPH)

This oxidoreductase [EC 1.1.1.247] catalyzes the reaction of codeinone with NADPH to produce codeine and NADP$^+$.

Codeine is the immediate precursor of morphine in the poppy plant, *Papaver somniferum*, which displays a number of isoforms of the enzyme.[1,2]

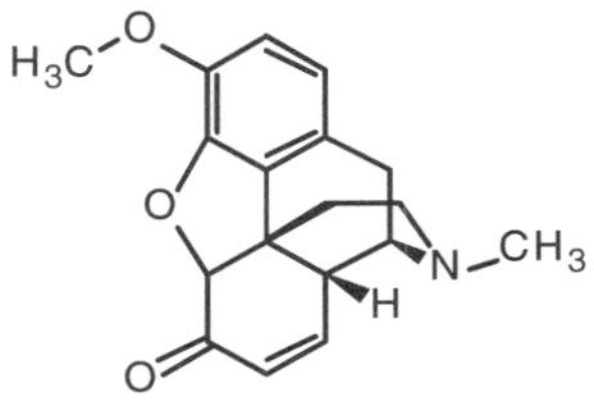

codeinone 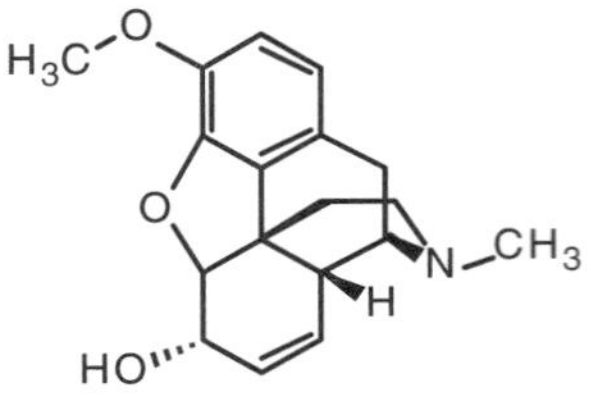   codeine

[1] R. Lenz & M. H. Zenk (1995) *EJB* **233**, 132.
[2] B. Unterlinner, R. Lenz & T. M. Kutchan (1999) *Plant J.* **18**, 465.

COENZYME F420-DEPENDENT N^5,N^{10}-METHYLENETETRAHYDROMETHANO-PTERIN REDUCTASE

This oxidoreductase [EC 1.5.99.11] catalyzes the reaction of N^5,N^{10}-methylenetetrahydromethanopterin with reduced coenzyme F420 to produce 5-methyl-5,6,7,8-tetrahydromethanopterin and coenzyme F420.[1,2]

[1] K. Ma & R. K. Thauer (1990) *EJB* **191**, 187.
[2] B. W. te Brommelstroet, W. J. Geerts, J. T. Keltjens, C. van der Drift & G. D. Vogels (1991) *BBA* **1079**, 293.

COENZYME F420-DEPENDENT NADP$^+$ REDUCTASE

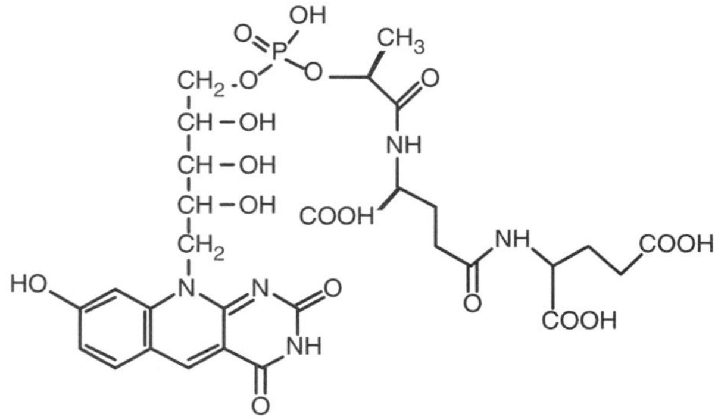

oxidized F$_{420}$

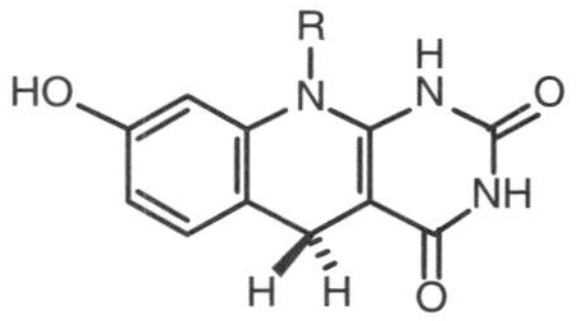

reduced F$_{420}$

This oxidoreductase, also known as 8-hydroxy-5-deazaflavin-dependent NADP$^+$ reductase, catalyzes the reaction of reduced coenzyme F420 with NADP$^+$ to produce the oxidized coenzyme and NADPH.[1]

[1] S. Yamazaki, L. Tsai, T. C. Stadtman, F. S. Jacobson & C. Walsh (1980) *JBC* **255**, 9025.

COENZYME F420 HYDROGENASE

This iron- and nickel-dependent oxidoreductase [EC 1.12.99.1], also known as 8-hydroxy-5-deazaflavin-reducing hydrogenase, catalyzes the reaction of molecular hydrogen with coenzyme F420 (*i.e.*, 7,8-didemethyl-8-hydroxy-5-deazariboflavin 5'-phosphoryllactylglutamyl-glutamate) to produce reduced coenzyme F420.

The enzyme has been reported to have a two-site hybrid ping-pong kinetic mechanism.[1] When D$_2$ is used, no deuterium is found at the C5 position of the reduced coenzyme F420; however, in the presence of D$_2$O, a 5-monodeuterio reduced product is formed. This finding suggests that there is complete exchange of hydrogens from H$_2$ with the solvent before the final transfer of a hydride ion.

[1] D. J. Livingston, J. A. Fox, W. H. Orme-Johnson & C. T. Walsh (1987) *Biochemistry* **26**, 4228.

COENZYME-M-7-MERCAPTOHEPTANOYL-THREONINE-PHOSPHATE-HETERODISULFIDE HYDROGENASE

This iron-/nickel-, and FAD-dependent oxidoreductase [EC 1.12.99.2] catalyzes the reaction of coenzyme M 7-mercaptoheptanoylthreonine-phosphate-heterodisulfide with molecular hydrogen to produce coenzyme M and N-(7-mercaptoheptanoyl)threonine O^3-phosphate.[1–5]

[1] R. Hedderich & R. K. Thauer (1988) *FEBS Lett.* **234**, 223.
[2] R. Hedderich, A. Berkessel & R. K. Thauer (1989) *FEBS Lett.* **255**, 67.
[3] R. Hedderich, A. Berkessel & R. K. Thauer (1990) *EJB* **193**, 255.
[4] M. C. Brenner, L. Ma, M. K. Johnson & R. A. Scott (1992) *BBA* **1120**, 160.
[5] E. Setzke, R. Hedderich, S. Heiden & R. K. Thauer (1994) *EJB* **220**, 139.

COLLAGENASE 3

This protease, also known as matrix metalloproteinase 13 (MMP-13), catalyzes the hydrolysis of type I collagen at the same site as interstitial collagenase (collagenase 1, EC 3.4.24.7), namely Gly775–Ile776.[1–3]

[1] V. Knäuper, C. López-Otin, B. Smith, G. Knight & G. Murphy (1996) *JBC* **271**, 1544.
[2] J. M. P. Freije, I. Díez-Itza, M. Balbín, L. M. Sánchez, R. Blasco, J. Tolivia & C. López-Otin (1994) *JBC* **269**, 16766.
[3] W. T. Roswit, J. Halme & J. J. Jeffrey (1983) *ABB* **225**, 285.

COLLAGENASE 4

This protease, also known as matrix metalloproteinase 18 (MMP-18), catalyzes the hydrolysis of the same Gly775–Ile776 site in the α1(I) chain as interstitial collagenase (EC 3.4.24.7).[1] Collagenase 4 will also act on types II and III collagens, on human α_1-antitrypsin, and human α_2-macroglobulin.

[1] M. A. Stolow, D. D. Bauzon, J. Li, T. Sedgwick, V. C.-T. Liang, Q. A. Sang & Y.-B. Shi (1996) *Mol. Biol. Cell* **7**, 1471.

COLLAGENASE, INTERSTITIAL

This zinc-dependent endopeptidase [EC 3.4.24.7] (also known as vertebrate collagenase, matrix metalloproteinase 1 [MMP-1], and collagenase 1) was first identified and described from tadpole tail. The enzyme was given the descriptor "interstitial" since it catalyzed the hydrolysis of the native collagen triple helix at a single site on all three α chains.[1–3] The active-site cleft of the collagenase can only bind one chain at a time. The human α1(I) is hydrolyzed at Gly-Pro-Gln-Gly775-∇-Ile776-Ala-Gly-Gln, ∇ indicates the cleavage site. The enzyme will also act on collagen types VII and X. Interstitial collagenase also exhibits an autoprocessing activity. The human enzyme will act at Pro269–Ile270.

Interstitial collagenase will also hydrolyze α_2-macroglobulin at Gly-Pro-Glu-Gly679-∇-Leu680-Arg-Val-Gly. In fact, α_2-macroglobulin is cleaved more rapidly than collagen. The collagenase will also act on α_1-proteinase inhibitor, α_1-antichymotrypsin, and serum amyloid A. *See also Collagenase, Neutrophil; Collagenase 3; Collagenase 4*

[1] A. J. Barrett & J. K. McDonald (1980) in *Mammalian Proteases: A Glossary and Bibliography*, vol. 1, Academic Press, New York, p. 359.
[2] T. E. Cawston & G. Murphy (1981) *MIE* **80**, 711.
[3] M. Dioszegi, P. Cannon & H. E. Van Wart (1995) *MIE* **248**, 413.

Selected entries from *Methods in Enzymology* [vol, page(s)]:
General discussion: **80**, 711; **248**, 413
Assay: **80**, 711, 712; **82**, 428; fibroblast-type, **248**, 419 (collagen-based, **248**, 419; peptide-based, **248**, 422); mammalian, **80**, 711, 712 (fibril, **80**, 714; radiolabeling of substrate, **80**, 713, 714; substrate preparation, **80**, 712, 713)
Other molecular properties: activation, **82**, 443; **248**, 424; activity, **235**, 595; **248**, 414; cleavage specificity, **248**, 414; fibroblast-type, **248**, 415 (activation, **248**, 424; assay, **248**, 419 [collagen-based, **248**, 419; peptide-based, **248**, 422]; molecular mass, **248**, 417; proenzyme, **248**, 415; properties, **248**, 415; purification, **248**, 428; recombinant, **248**, 426; substrate specificity, **248**, 418, 433; synthesis, **248**, 415); human, polypeptide chain structure, **248**, 204; isolation, **82**, 439; mammalian, **80**, 711 (assay, **80**, 711, 712 [fibril, **80**, 714; radiolabeling of substrate, **80**, 713, 714; substrate preparation, **80**, 712, 713]; culture media, **80**, 715, 716; inhibitors, **80**, 719, 720; latency, **80**, 715, 720; molecular charge, **80**, 721; molecular size, **80**, 722; properties, **80**, 719; purification, **80**, 715; sources, **80**, 715, 716; specific activity, **80**, 720, 721; stability, **80**, 720; substrate specificity, **80**, 721; tissue extracts, **80**, 716); matrixin subfamily, **248**, 192, 200, 415, 485, 511; phylogenetic tree, **248**, 201; polypeptide chain structure, **248**, 203; proenzyme, **248**, 432; properties, **248**, 191, 511; purification, **82**, 444; **248**, 428; recombinant, **248**, 425; serralysin subfamily, **248**, 192, 201, 395; sources, **248**, 425; species distribution, **248**, 201; structure, **248**, 191; substrates, **248**, 414; synthetic substrates, **82**, 427; tadpole, from, **19**, 616, 621 (zymogen, **40**, 351); type IV, assay, **82**, 452

COLLAGENASE, MICROBIAL

The microbial collagenases [EC 3.4.24.3] are metalloendopeptidases that catalyze hydrolysis of peptide bonds in native collagen at Xaa–Gly bonds. *See Vibrio Collagenase*

Selected entries from *Methods in Enzymology* [vol, page(s)]:
Assay: **19**, 616; **235**, 600; bacterial, type I, assays (collagen fibril substrate, **235**, 597; with collagen synthetic peptide analogs, **235**, 599; with dinitrophenyl coupled to collagen-like peptides, **235**, 601; insoluble tendon collagen as substrate, **235**, 599; solution, **235**, 595)
Other molecular properties: *Achromobacter iophagus* (activity, **235**, 595; assay, **235**, 600; preparation for assay, **235**, 566); amino acid composition, **19**, 630; bacterial (digestion of collagen, **224**, 127; radiolabeled collagen determination, **82**, 394); gelatinase activity (activities, **235**, 602; affinity chromatography, **235**, 602; assays, **235**, 605; identification of luminous bacteria, **57**, 161, 162; purification, **235**, 602); properties, **19**, 616, 628; **82**, 453; possible of, subunits of, **19**, 629; preparation and purification, **19**, 626; specificity, **19**, 634; transition state and multisubstrate analogues, **249**, 307; zymographic characterization, **235**, 580

COLLAGENASE, NEUTROPHIL

This zinc-/calcium-dependent endopeptidase [EC 3.4.24.34], also known as matrix metalloproteinase 8 (MMP-8) and collagenase-2, is stored in specific granules of polymorphonuclear leukocytes and articular chondrocytes. The enzyme catalyzes hydrolysis of interstitial collagens in the triple helical domain.[1] Unlike interstitial collagenase [EC 3.4.24.7], however, this enzyme cleaves type III collagen more slowly than type I. In fact, neutrophil collagenase prefers type I collagen. The Gly775–Leu776 and Gly775–Ile776 bonds in the different α chains are cleaved. Other naturally occurring substrates include fibronectin, human C1-inhibitor, α_1-antitrypsin, β-casein, angiotensin, and substance P. Neutrophil collagenase exhibits a preference for an alanyl residue at position P1 and a tyrosyl or phenylalanyl residue at P1'. The best synthetic peptide substrate identified to date is Gly-Pro-Gln-Gly-∇-Ile-Trp-Gly-Gln, where ∇ indicates the cleavage site. *See also Collagenase, Interstitial; Collagenase 3; Collagenase 4*

[1] H. E. Van Wart (1992) *Matrix*, suppl. 1, 31.

Selected entries from *Methods in Enzymology* [vol, page(s)]:
General discussion: **248**, 431

Assay: 163, 325; **248**, 419, 435 (collagen-based, **248**, 419; peptide-based, **248**, 422); gelatinase activity, human neutrophil, **132**, 273

Other molecular properties: activation, **248**, 424, 432; amino acid sequence, **248**, 444; catalytic activity, **248**, 445; catalytic domain, **248**, 445 (refolding and purification, **248**, 442); cellular localization, **248**, 431; C-terminally truncated, refolding and purification, **248**, 442; gene, **248**, 432; leukocyte (polymorphonuclear, **80**, 711, 716, 719; trapped by α_2-macroglobulin, **80**, 748); metal content, **248**, 447; molecular mass, **248**, 417, 433; proenzyme, **248**, 415, 432 (activation, **248**, 443; purification, **248**, 434; truncation by trypsin to latent enzyme, **248**, 444); properties, **163**, 314; **248**, 192, 415, 445, 511; proteolytic properties, **248**, 433; purification, **248**, 428 (from human neutrophils, **163**, 319); purity, **248**, 439; recombinant, **248**, 427, 440 (expression, **248**, 441; extraction, **248**, 442; preparation, **248**, 440; purification, **248**, 442); structure, **248**, 191, 445; substrate specificity, **248**, 418, 433; superactivation, **248**, 445, 449; synthesis, **248**, 415, 432; variants (cDNA coding for, cloning, **248**, 440; expression, **248**, 441; extraction, **248**, 442; purification, **248**, 442)

COLUMBAMINE *O*-METHYLTRANSFERASE

columbamine

palmatine

This methyltransferase [EC 2.1.1.118] catalyzes the reaction of *S*-adenosyl-L-methionine with columbamine to produce *S*-adenosyl-L-homocysteine and palmatine, which is a protoberberine alkaloid widely found in the plant kingdom.[1] The specificity of EC 2.1.1.118 differs from that of flavonol 8-*O*-methyltransferase (8-hydroxyquercetin 8-*O*-methyltransferase, EC 2.1.1.88).

[1] M. Rueffer, M. Amann & M. H. Zenk (1986) *Plant Cell Reports* **5**, 182

COLUMBAMINE OXIDASE

This iron-dependent oxidoreductase [EC 1.1.3.26], also known as berberine synthase, catalyzes the reaction of two molecules of columbamine with dioxygen to produce two molecules of berberine and two water molecules.[1]

[1] M. Rueffer & M. H. Zenk (1985) *Tetrahedron Lett.* **26**, 201.

COMPLEMENT COMPONENT C5 CONVERTASE

This protease activity, formerly classified as EC 3.4.21.44, is now included with classical-complement-pathway C3/C5 convertase [EC 3.1.21.43].

COMPLEMENT COMPONENT C$\overline{1r}$

The serine protease [EC 3.4.21.41] that activates complement subcomponent C1s by the selective hydrolysis of a single Arg–Ile bond is due to complement component C1r (note that the active form of C1r is usually designated by an overbar, C$\overline{1r}$).[1] The only two natural substrates are C1s and a corresponding Arg–Ile bond in C1r (thus, autoactivation). Interestingly, purified C$\overline{1r}$) will undergo further autolytic cleavage at two Arg–Gly peptide bonds.[2] This endopeptidase also has an esterase activity, acting on Ac-Gly-Lys-∇-OMe and Z-Gly-Arg-∇-SBzl, where an arginyl or a lysyl residue is at position P1, and ∇ is the cleavage site. The only known physiological inhibitor is C1 inhibitor.[3]

[1] G. J. Arlaud & N. M. Thielens (1993) *MIE* **223**, 61.
[2] G. J. Arlaud, A. C. Willis & J. Gagnon (1987) *BJ* **241**, 711.
[3] A. E. Davis, K. S. Aulak, K. Zahedi, J. J. Bissler & R. A. Harrison (1993) *MIE* **223**, 97.

Selected entries from *Methods in Enzymology* [**vol**, page(s)]:
General discussion: 80, 3, 26; **223**, 61
Assay: C1s activation rate, by, **80**, 29; esterolytic, **80**, 28; hemolytic, **80**, 27, 28
Other molecular properties: A chain (amino acid composition, **80**, 37, 38; carbohydrate composition, **80**, 36, 37); activation, **80**, 41, 42; amino acid composition, **80**, 37, 38; B chain, **80**, 37, 38; carbohydrate composition, **80**, 36, 37; dimer, dissociation, **80**, 52; esterolytic activity, **80**, 40 (pH optima, **80**, 40); inhibition, **80**, 40, 51; isoelectric point, **80**, 37; nonhuman, rabbit, **80**, 33, 34; partial specific volume, **80**, 34; properties (chemical, **80**, 37, 38; enzymic, **80**, 38; structural, **80**, 34); purification, **80**, 32 (yield, **80**, 33); stability, **80**, 42

COMPLEMENT COMPONENT C$\overline{1s}$

This serine endopeptidase [EC 3.4.21.42], the active form of which was formerly known as C1 esterase, catalyzes the conversion of component C2 to C2a and C2b (by hydrolyzing the peptide bond at Ser-Leu-Gly-Arg-∇-Lys-Ile-Gln-Ile) as well as the conversion of C4 to C4a and C4b (by hydrolyzing the peptide bond at Gly-Leu-Gln-Arg-∇-Ala-Leu-Glu-Ile).[1] (Note: ∇ indicates the cleavage site.) Artificial substrates require an arginyl, a lysyl, or, surprisingly, a tyrosyl residue in the P1 position. The proteinase also has an esterase activity (for example, Tos-Arg-∇-OMe, Z-Tyr-∇-OPhNO$_2$, and Z-Gly-Arg-∇-Sbzl, where ∇ indicates the cleavage site.).[1] The only known physiological inhibitor is C1 inhibitor.[2]

[1] G. J. Arlaud & N. M. Thielens (1993) *MIE* **223**, 61.

[2]A. E. Davis, K. S. Aulak, K. Zahedi, J. J. Bissler & R. A. Harrison (1993) *MIE* **223**, 97.

Selected entries from *Methods in Enzymology* [vol, page(s)]:
General discussion: 34, 731, 735, 738; **80**, 3, 26; **223**, 61
Assay: destruction of C4, by, **80**, 29; esterolytic, **80**, 28, 29; hemolytic, **80**, 27, 28
Other molecular properties: A chain (amino acid composition, **80**, 37, 38; amino acid sequence, **80**, 37, 39; carbohydrate composition, **80**, 36, 37); activation, **80**, 41; amino acid composition, **80**, 37, 38; B chain (amino acid composition, **80**, 37, 38; amino acid sequence, **80**, 37, 39; carbohydrate composition, **80**, 36, 37); carbohydrate composition, **80**, 36, 37; esterolytic activity, **80**, 40 (inhibition, **80**, 53; pH optimum, **80**, 40); inhibition, **80**, 40, 51; inhibitors, **46**, 115, 116, 120; **80**, 69, 74; nonhuman (bovine, **80**, 34; rabbit, **80**, 33, 34); partial specific volume, **80**, 34; properties (chemical, **80**, 37, 38; enzymic, **80**, 38; structural, **80**, 34); purification, **80**, 32 (yield, **80**, 33); stability, **80**, 41; thioester substrates, kinetic constants, **248**, 13

COMPLEMENT COMPONENT C2 AND THE CLASSICAL PATHWAY C3/C5 CONVERTASE

This serine endopeptidase [EC 3.4.21.43] (also known as classical-complement-pathway C3/C5 convertase, complement C2, C3 convertase, and C5 convertase) catalyzes the cleavage of component C3, the central protein of the complement system, at Arg77–Ser78 to yield a large fragment, C3b (185 kDa), and a small fragment, C3a (9 kDa).[1–4]

[1]G. D. Ross (1986) *Immunobiology of the Complement System*, Academic Press, Orlando.
[2]B. P. Morgan (1990) *Complement: Clinical Aspects and Relevance*, Academic Press, Orlando.
[3]K. Whaley, M. Loos & J. Weiler (1992) *Complement in Health and Disease*, 2nd ed., Kluwer Publ., Amsterdam.
[4]M. A. Kerr (1981) *MIE* **80**, 54.

Selected entries from *Methods in Enzymology* [vol, page(s)]:
General discussion: 34, 732
Assay: hemolytic, **80**, 55, 56
Other molecular properties: complement system, in, control, **223**, 13, 17; human, **80**, 6, 54, 74, 75, 103, 109 (activation, **80**, 4, 62, 63; amino acid composition, **80**, 62; amino acid sequence, N-terminal, **80**, 62, 63; assay, hemolytic, **80**, 55, 56; esterolytic activity, **80**, 64; inhibition, by DFP, **80**, 64; molecular properties, **80**, 61; plasma concentration, **80**, 54; proteolytic enzyme, as, **80**, 63, 64; purification, **80**, 56 [precautions against inactivation, **80**, 61; preparation of chromatography resins, **80**, 56, 57; yield, **80**, 60]; stability, **80**, 60); nonhuman, guinea pig (assay of C1, **80**, 8; purification, **80**, 60); purification, **74**, 168; thioester substrates, **248**, 13

COMPLEMENT FACTOR D

This serine endopeptidase [EC 3.4.21.46] (also known as C3 convertase activator, C3 proactivator convertase (C3PAse), glycine-rich β-glycoprotein convertase (GBGase), properdin factor D, and adipsin) catalyzes the cleavage of component factor B (at Arg233–Lys234) when in complex with C3b, with C3 with a hydrolyzed thioester bond, or with cobra venom factor (CoVF): the complexes often known as complement factor $\overline{D}$.[1]

[1]J. E. Volanakis & S. V. L. Narayana (1996) *Protein Sci.* **5**, 553.

Selected entries from *Methods in Enzymology* [vol, page(s)]:
General discussion: 34, 731; **80**, 134; **223**, 82
Assay: **223**, 83; human, **80**, 104; **93**, 393; complement factor $\overline{D}$, **80**, 135, 136
Other molecular properties: complement factor $\overline{D}$, **80**, 3, 134 (amino acid composition, **80**, 141; amino acid sequence, **80**, 142, 143; isoelectric points, **80**, 142; physical properties, **80**, 141; purification, **80**, 136 [yield, **80**, 140, 141]; serum concentration, **80**, 135; stability, **80**, 141; typical serine esterase, as, **80**, 143); esterolytic activity, **223**, 95; human, **80**, 85 (assay, **80**, 104; cleavage of Factor B, **80**, 102, 103, 108, 109; elution, **80**, 106; plasmin as substitute for, in alternate pathway, **80**, 365); human, in alternative complement pathway (assay, **93**, 393; purification, **93**, 390); properties, **223**, 82; purification from, **223**, 82 (peritoneal dialysis fluid, **223**, 92; urine, **223**, 87); serum concentration, in health and disease, **223**, 96; structure, **223**, 92; thiolester substrates, **248**, 13

COMPLEMENT FACTOR I

This serine endopeptidase [EC 3.4.21.45] (also known as complement component C3b inactivator, conglutinogen-activating factor (KAF), C3b inactivator, and C3b/C4b inactivator) catalyzes the inactivation of complement subcomponent C3b by the hydrolysis at Arg1303–Ser1304 and Arg1320–Ser1321, producing a small, seventeen amino acid fragment (C3f) and iC3b (formerly called C3bi).[1–3] The hydrolysis of C3b requires the presence of one of the following cofactors: complement factor H, membrane cofactor protein (MCP), or complement receptor type 1 (CR1).

[1]R. B. Sim, A. J. Day, B. E. Moffatt &. M. Fontaine (1993) *Meth. Enzymol.* **223**, 13.
[2]T. J. Vyse, P. Spath, K. A. Davies, B. J. Morley, P. Philippe, P. Athanassiou, C. M. Giles & M. J. Walport (1994) *Q. J. Med.* **87**, 385.
[3]L. G. Crossley (1981) *Meth. Enzymol.* **80**, 112.

Selected entries from *Methods in Enzymology* [vol, page(s)]:
General discussion: 80, 69, 74, 112; **223**, 13
Assay: 80, 113
Other molecular properties: amino acid composition, **80**, 118, 119; antisera, **80**, 118; chemical composition, **80**, 117, 118; cofactors, and, in control of complement system convertase enzymes, **223**, 13; effect of protease inhibitors, **80**, 121; enzymatic properties, **223**, 34; inhibition, **80**, 121; isoelectric point, **80**, 118; isolation, **223**, 27; molecular weight, **80**, 117; properties, **80**, 117; protein and gene structure, **223**, 18; purification, **80**, 115 (yield, **80**, 117, 118); purity, **80**, 117; stability, **80**, 118; substrate formation in activated system, **223**, 13; substrate specificity, **80**, 119; thioester substrates, **248**, 13

CONIFERIN β-GLUCOSIDASE

This plant cell wall glycosidase [EC 3.2.1.126] catalyzes the hydrolysis of coniferin to produce coniferol (*i.e.*, coniferyl alcohol) and D-glucose. Alternative substrates include syringin and 4-cinnamyl alcohol β-glucoside.[1–3]

coniferin

[1] W. Hösel, E. Surholt & E. Borgmann (1978) *EJB* **84**, 487.
[2] S. Marcinowski & H. Grisebach (1978) *EJB* **87**, 37.
[3] D. P. Dharmawardhana, B. E. Ellis & J. E. Carlson (1995) *Plant Physiol.* **107**, 331.

CONIFERYL-ALCOHOL DEHYDROGENASE

This oxidoreductase [EC 1.1.1.194] catalyzes the reaction of coniferyl alcohol with $NADP^+$ to produce coniferyl aldehyde and NADPH.[1] The enzyme is highly specific for coniferyl alcohol.

[1] D. Wyrambik & H. Grisebach (1975) *EJB* **59**, 9.

Selected entries from *Methods in Enzymology* [**vol**, page(s)]:
Assay: 161, 302

CONIFERYL-ALCOHOL GLUCOSYLTRANSFERASE

This transferase [EC 2.4.1.111] catalyzes the reaction of UDP-D-glucose with coniferyl alcohol to produce UDP and coniferin.[1] Sinapyl alcohol is also an acceptor substrate.

[1] G. Schmid & H. Grisebach (1982) *EJB* **123**, 363.

CONIFERYL-ALDEHYDE DEHYDROGENASE

coniferyl aldehyde   ferulate

This oxidoreductase [EC 1.2.1.68] catalyzes the reaction of coniferyl aldehyde with water and $NAD(P)^+$ to produce ferulate, NAD(P)H, and a proton.[1] Other aromatic aldehydes will also act as substrates.

[1] S. Achterholt, H. Priefert & A. Steinbuchel (1998) *J. Bacteriol.* **180**, 4387.

COPROPORPHYRINOGEN OXIDASE

This iron-dependent oxidoreductase [EC 1.3.3.3] (also known as coproporphyrinogenase, coproporphyrinogen-III oxidase, and coprogen oxidase) catalyzes the reaction of coproporphyrinogen-III with dioxygen to generate protoporphyrinogen-IX and two molecules of carbon dioxide: hence, the propionate side chains of coproporphyrinogen III are converted into the vinyl groups of protoporphyrinogen IX.[1-5]

coproporphyrinogen III   protoporphyrinogen IX

The reaction occurs stepwise starting from position 2 to 4 via a possible β-hydroxypropionate porphyrinogen as an intermediate.[1,2,5] A catalytically essential tyrosyl residue has been identified.[5]

Note that two coproporphyrinogen oxidase systems have been identified: one in animals and other aerobic organisms requiring dioxygen, and another type found in anaerobic systems linked to an alternative electron acceptor (*e.g.*, a cytochrome chain).

[1] M. Akhtar (1994) *Ciba Found Symp.* **180**, 131.
[2] J. S. Seehra, P. M. Jordan & M. Akhtar (1983) *BJ* **209**, 709.
[3] T. Yoshinaga & S. Sano (1980) *JBC* **255**, 4727.
[4] S. Granick & S. I. Beale (1978) *AE* **46**, 33.
[5] G. H. Elder, J. O. Evans, J. R. Jackson & A. H. Jackson (1978) *BJ* **169**, 215.

Selected entries from *Methods in Enzymology* [**vol**, page(s)]:
General discussion: 52, 21; 281, 355, 367

CORTICOSTEROID SIDE-CHAIN-ISOMERASE

This isomerase [EC 5.3.1.21] catalyzes the interconversion of 11-deoxycorticosterone and 20-hydroxy-3-oxopregn-4-en-21-al, a ketol-aldol interconversion.[1-3]

11-deoxycorticosterone   20-hydroxy-3-oxopregn-4-en-21-al

Other substrates include corticosterone and 17-deoxycortisol. The enzyme also catalyzes an epimerization reaction at C20 and C21. Isotope exchange with water and $(21S)$-$[21-^3H]$-substrate is a pseudo-first-order reaction.

The corresponding exchange from the 21R isomer proceeds with first-order kinetics after a lag associated with the epimerization to the 21S form.[3]

[1] K. O. Martin, S.-W. Oh, H. J. Lee & C. Monder (1977) *Biochemistry* **16**, 3803.
[2] F. Iohan & C. Monder (1984) *ABB* **230**, 440.
[3] A. Marandici & C. Monder (1990) *Biochemistry* **29**, 1147.

CORTICOSTERONE 18-MONOOXYGENASE

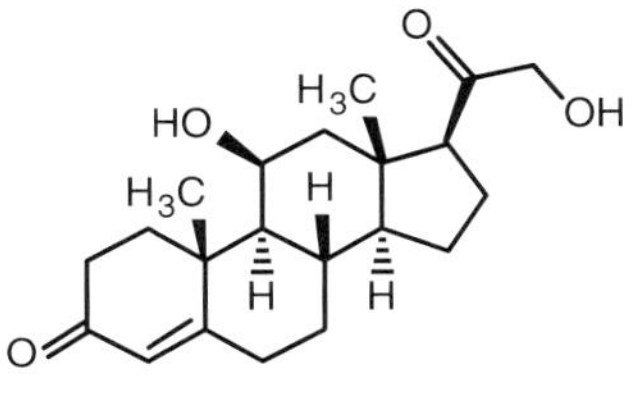

corticosterone

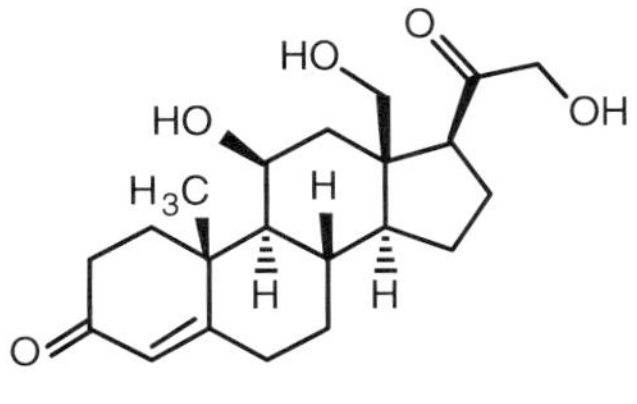

18-hydroxycorticosterone

This heme-thiolate-dependent oxidoreductase [EC 1.14.15.5], also known as corticosterone 18-hydroxylase and corticosterone methyloxidase, catalyzes the reaction of corticosterone with reduced adrenal ferredoxin and dioxygen to produce 18-hydroxycorticosterone, oxidized adrenal ferredoxin, and water.[1]

[1] P. B. Raman, D. C. Sharma & R. I. Dorfman (1966) *Biochemistry* **5**, 1795.

CORTISOL O-ACETYLTRANSFERASE

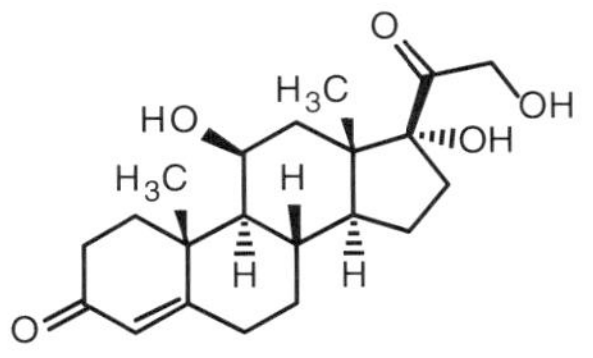

cortisol

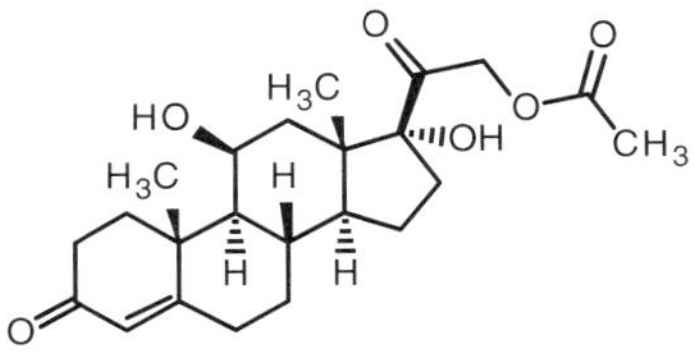

cortisol 21-acetate

This transferase [EC 2.3.1.27] catalyzes the reaction of acetyl-CoA with cortisol to produce coenzyme A and cortisol 21-acetate.[1]

[1] P. J. Thomas (1968) *BJ* **109**, 695.

CORTISOL SULFOTRANSFERASE

This sulfotransferase [EC 2.8.2.18], also called glucocorticosteroid sulfotransferase, catalyzes the reaction of 3'-phosphoadenylylsulfate with cortisol to produce adenosine 3',5'-bisphosphate and cortisol 21-sulfate.[1,2] Alternative substrates include deoxycorticosterone, dehydroepiandrosterone, 17β-estradiol, and testosterone.

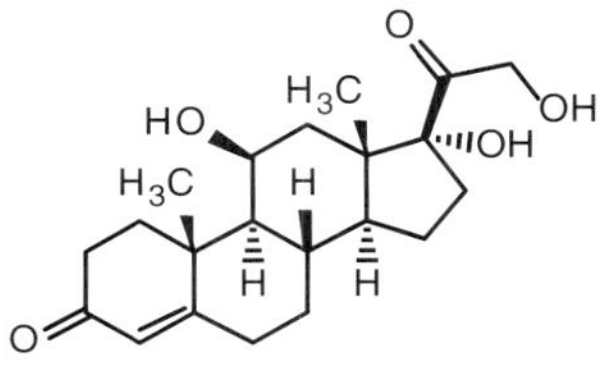

cortisol

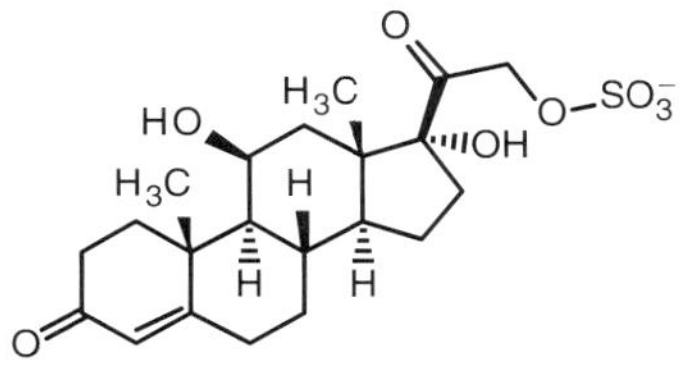

cortisol 21-sulfate

[1] S. S. Singer (1984) *Biochem. Soc. Trans.* **12**, 35.
[2] S. S. Singer & L. Bruns (1980) *Can. J. Biochem.* **58**, 660.

Selected entries from *Methods in Enzymology* [**vol**, page(s)]:
General discussion: 77, 206

CORTISONE α-REDUCTASE

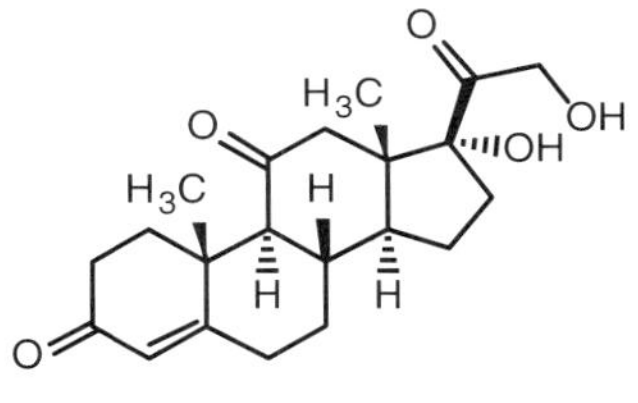

cortisone

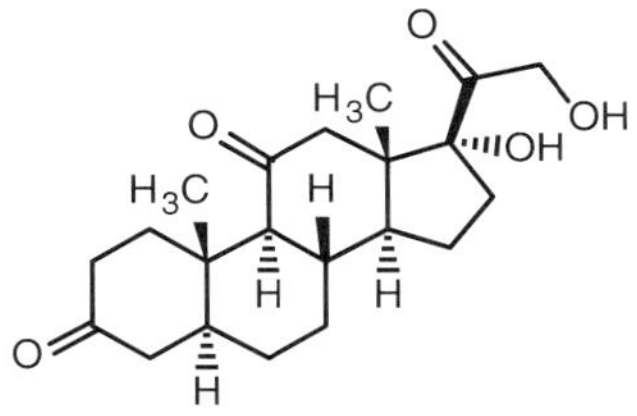

4,5α-dihydrocortisone

This oxidoreductase [EC 1.3.1.4] catalyzes the reaction of cortisone with NADPH to produce 4,5α-dihydrocortisone and NADP$^+$.[1] Other Δ^4-3-ketosteroid substrates include testosterone and 17α,21-dihydroxypregn-4-ene-3,20-dione.

[1] J. S. McGuire, V. W. Hollis & G. M. Tomkins (1960) *JBC* **235**, 3112.

CORTISONE β-REDUCTASE

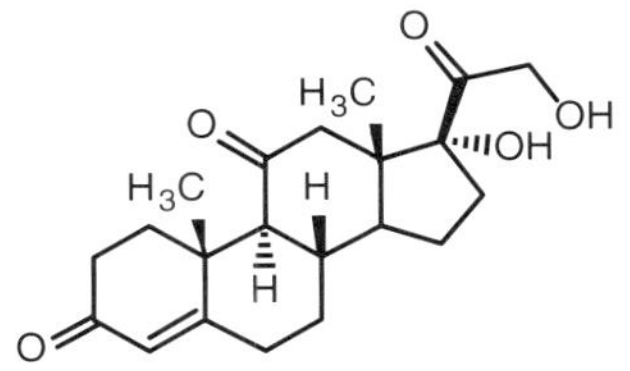

cortisone

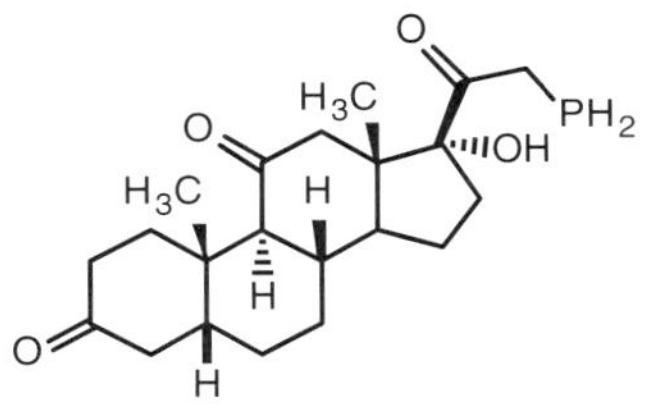

4,5β-dihydrocortisone

This oxidoreductase [EC 1.3.1.3], also known as steroid 5β-reductase, catalyzes the reaction of cortisone with NADPH to produce 4,5β-dihydrocortisone and NADP$^+$. Testosterone is an alternative substrate.[1–3] In some sources (for example, rat liver), this enzyme and cholestenone 5β-reductase [EC 1.3.1.23] are identical.[2] *See also 20-Hydroxysteroid Dehydrogenase; Cholestenone 5β-Reductase*

[1] G. M. Tomkins (1957) *JBC* **225**, 13.
[2] M. Furuebisu, S. Deguchi & K. Okuda (1987) *BBA* **912**, 110.
[3] R. Ozon, C. Fouchet & F. Perin (1974) *BBA* **348**, 425.

Selected entries from *Methods in Enzymology* [vol, page(s)]:
General discussion: 5, 499
Assay: 5, 499

2-COUMARATE O-β-GLUCOSYLTRANSFERASE

This transferase [EC 2.4.1.114] catalyzes the reaction of UDP-D-glucose with *trans*-2-hydroxycinnamate to produce UDP and *trans*-β-D-glucosyl-2-hydroxycinnamate.[1,2] Coumarinate (*i.e.*, *cis*-2-hydroxycinnamate) is not an acceptor substrate.

[1] J. E. Poulton, D. E. McRee & E. E. Conn (1980) *Plant Physiol.* **65**, 171.
[2] A. Kleinhofs, F. A. Haskins & H. J. Gorz (1967) *Phytochemistry* **6**, 1313.

cis-4-COUMARATE GLUCOSYLTRANSFERASE

This transferase [EC 2.4.1.209] catalyzes the reaction of UDP-D-glucose with *cis-p*-coumarate to produce 4'-O-β-D-glucosyl-*cis-p*-coumarate and UDP.[1]

[1] S. Rasmussen & H. Rudolph (1997) *Phytochemistry* **46**, 449.

4-COUMARATE 3-MONOOXYGENASE

This oxidoreductase activity, formerly classified as EC 1.14.17.2 and now a deleted EC entry, is now listed with monophenol monooxygenase [EC 1.14.18.1].

COUMARATE REDUCTASE

This oxidoreductase [EC 1.3.1.11], also known as melilotate dehydrogenase, catalyzes the reaction of 2-coumarate with NADH to produce 3-(2-hydroxyphenyl)propanoate (*i.e.*, melilotate) and NAD$^+$.[1]

[1] C. C. Levy & G. D. Weinstein (1964) *Biochemistry* **3**, 1944.

4-COUMAROYL-CoA SYNTHETASE

trans-4-coumarate   *trans*-4-coumaroyl-CoA

This enzyme [EC 6.2.1.12], also known as hydroxycinnamoyl-CoA ligase and 4-coumarate:CoA ligase, catalyzes the reaction of 4-coumarate with ATP and coenzyme A to yield 4-coumaroyl-CoA, AMP, and pyrophosphate (or, diphosphate).[1–4]

[1] W. J. Hu, A. Kawaoka, C. J. Tsai, J. Lung, K. Osakabe, H. Ebinuma & V. L. Chiang (1998) *PNAS* **95**, 5407.

[2] R. A. Dixon, P. M. Dey & C. J. Lamb (1983) *AE* **55**, 1.
[3] K. H. Knobloch & K. Hahlbrock (1977) *ABB* **184**, 237.
[4] T. Lüderitz, G. Schatz & H. Grisebach (1982) *EJB* **123**, 583.

5-O-(4-COUMAROYL)-D-QUINATE 3'-MONOOXYGENASE

This oxidoreductase [EC 1.14.13.36], also known as 5-O-(4-coumaroyl)-D-quinate/shikimate 3'-hydroxylase, catalyzes the reaction of *trans*-5-O-(4-coumaroyl)-D-quinate with NADPH and dioxygen to produce *trans*-5-O-caffeoyl-D-quinate (chlorogenate), NADP$^+$, and water.[1] Other coumaroyl substrates include *trans*-5-O-(4-coumaroyl) shikimate, producing *trans*-5-O-(4-caffeoyl)-D-shikimate.

[1] T. Kühnl, U. Koch, W. Heller & E. Wellmann (1987) *ABB* **285**, 226.

CREATINASE

This enzyme [EC 3.5.3.3], also known as creatine amidinohydrolase, catalyzes the hydrolysis of creatine to produce sarcosine and urea.[1]

[1] M. Coll, S. H. Knof, Y. Ohga, A. Messerschmidt, R. Huber, H. Moellering, L. Russmann & G. Schumacher (1990) *J. Mol. Biol.* **214**, 597.

CREATINE KINASE

This muscle phosphotransferase [EC 2.7.3.2] catalyzes the reversible reaction of creatine with ATP to produce phosphocreatine and ADP (the magnesium complexes of the nucleotides are the true substrates and products).[1–5] N-Ethylglycocyamine can also act as a substrate.

creatine   phosphocreatine

The equilibrium constant actually favors the rephosphorylation of ADP to form ATP (K_{eq} = [MgATP][creatine]/ ([phosphocreatine][MgADP]) = 30). In resting muscle, creatine phosphate is synthesized at the expense of abundant stores of ATP; intracellular creatine phosphate stores often reach 50–60 mM. If ATP is suddenly depleted by muscle contraction, its product ADP is immediately converted back into ATP by the reverse of the creatine kinase reaction. Depending on the pH at which the enzyme is studied, the kinetic mechanism can be either rapid equilibrium random or rapid equilibrium ordered Bi Bi.

[1] J. J. Villafranca & T. Nowak (1992) *The Enzymes*, 3rd ed., **20**, 63.
[2] M. Wyss, J. Smeitink, R. A. Wevers & T. Wallimann (1992) *BBA* **1102**, 119.
[3] G. L. Kenyon & G. H. Reed (1983) *Adv. Enzymol.* **54**, 367.

[4]D. C. Watts (1973) *The Enzymes*, 3rd ed., **8**, 384.

[5]S. A. Kuby & E. A. Noltmann (1962) *The Enzymes*, 2nd ed., **6**, 515.

Selected entries from *Methods in Enzymology* [**vol**, page(s)]:

General discussion: 2, 605; 17A, 995; 34, 246, 532; 46, 21; 90, 185

Assay: 2, 605; bioluminescent, **57**, 63; canine isozymes, **90**, 186; chicken skeletal muscle, spectrophotometric assay, **90**, 490; Fast Analyzer, **31**, 816; human isozymes, **90**, 186; isozyme (assay, using firefly luciferase, **57**, 56 [spectrophotometric, **57**, 59, 60]); thymidine kinase, of, **51**, 366

Other molecular properties: abortive complex formation, **63**, 432; activators, **2**, 610; activity, **57**, 58; adenylate cyclase assay, **79**, 165; affinity labeling, **47**, 427; analysis by continuous-flow method, **22**, 7, 8, 12; aspartokinase assay and, **5**, 821; assay (cGMP, **38**, 73, 78, 79; creatine, **44**, 630, 631; cyclase, **38**, 119, 154; fluorometric, **44**, 628, 629); assay to determine protein S-thiolation, **251**, 424, 454; associated with cell fractions, **40**, 86; ATP-regeneration system, **238**, 34; bound reactants and products, equilibrium constant determination, **177**, 364; calmodulin contamination, **238**, 76; canine isozymes (assay, **90**, 186; purification, **90**, 195); chicken skeletal muscle (isolation, **90**, 491; M-line isoprotein [immunoassays, **85**, 142; properties, **85**, 142; purification, **85**, 140]; spectrophotometric assay, **90**, 490); chiral phosphoric monoester, and, **87**, 300; chromium nucleotides, and, **87**, 161, 177, 178; cobalt nucleotides, and, **87**, 161, 177, 178; coupled enzyme assay, **63**, 32; crystallization, **2**, 608; dimethylformamide, and, **87**, 631; electron paramagnetic resonance, ^{17}O effect, **87**, 276; FDNB, reaction with, **11**, 548; heart, **1**, 31; heart mitochondria, **55**, 14; histidine, active-site, **87**, 403; human isozymes (assay, **90**, 186; purification, **90**, 187); hybrid enzymes, **17A**, 999; inhibition, **63**, 463; **87**, 161, 178 (competitive, **63**, 292; determination of metal ion-nucleotide complex dissociation constants, in, **249**, 183; tight-binding, **63**, 463); interconversion rates, NMR analysis, **176**, 297; isolation of mitochondria, **55**, 4; isotope partitioning, and, **87**, 395; isozyme (assay, using firefly luciferase, **57**, 56 [spectrophotometric, **57**, 59, 60]; double-antibody immunoprecipitation assay, **182**, 712; clinical significance, **74**, 198; hybridization, **74**, 201; normal plasma levels, **74**, 199; properties, **74**, 198; purification, **74**, 200; radioimmunoassay, **74**, 198 [antisera, **74**, 199, 202; assay requirements, **74**, 205; immunogen preparation, **74**, 200; procedure, **74**, 205; protocol, **74**, 205; results, **74**, 209; second antibody technique, **74**, 206; sensitivity, **74**, 199, 209; sources of error, **74**, 201, 202, 204, 206; standard inhibition curve, **74**, 207; unknown sample analysis, **74**, 208]; radioiodination, **70**, 231; **73**, 121; **74**, 202; separation, **57**, 56 [ion-exchange chromatography, by, **57**, 62]; storage, **74**, 202); kinetics, **2**, 606, 610; **64**, 7, 23, 32, 33, 39; **249**, 467 (mechanism, equilibrium isotope exchange study, **249**, 467); marker for chromosome detection in somatic cell hybrids, as, **151**, 180; measurement of interferon effect on cellular differentiation, in, **119**, 619; measurement of mixed disulfide bond reduction, **251**, 172; mechanism, **64**, 7, 23, 32, 33, 39; **249**, 467; metal ion activation, **63**, 258, 263, 264, 275, 292, 327; myokinase (assay of, **6**, 223; not present in preparations of, **6**, 230; stabilization of, **6**, 224); nuclear labels, indirect detection with spin echoes, **176**, 351; (2′-5′)-oligoadenylate synthetase assay, **79**, 157, 158; partition analysis, **249**, 323; pH effect, **2**, 610; **63**, 280; **87**, 395, 403; [$^{16}O,^{17}O,^{18}O$]phosphopropanediol, and, **87**, 265; phosphoprotein phosphatase assay, **79**, 180; phosphorothioates, **87**, 178, 200, 224; photoactivatable probe synthesis, in, **237**, 88, 93; photooxidation, **25**, 405; ^{31}P NMR, simulated and experimental spectra, comparison, **176**, 299; properties, **2**, 609; protein synthesis and, **20**, 436; purification **1**, 50 (chicken heart, **17A**, 997; chicken skeletal muscle, **17A**, 996; rabbit brain, **17A**, 999; rabbit muscle, **2**, 607); radioiodination, **70**, 231; ribosomal proteins, in *in vitro* labeling of, **59**, 517, 527; screw sense, **87**, 178; soluble ribonucleic acid-enzyme complexes and, **6**, 30; solvent perturbations, **87**, 396, 631; specificity, **2**, 609; **87**, 161, 177, 200, 398; staining on gels, **22**, 598; stereochemistry, **87**, 200, 224, 262, 265, 300; sticky substrates, **87**, 395, 396, 631; temperature effect, **87**, 396; thiocyanate, and, **87**, 276; transition state, **87**, 276; transition state and multisubstrate analogues, **249**, 305; tritium-labeled glucose 6-phosphate and, **6**, 877

CREATININASE

creatinine   creatine

This enzyme [EC 3.5.2.10], also known as creatinine amidohydrolase, catalyzes the reversible hydrolysis of creatinine to produce creatine.[1]

[1]A. Kaplan & L. L. Szabo (1974) *Mol. Cell. Biochem.* **3**, 17.

CREATININE DEIMINASE

creatinine   *N*-methylhydantoin

This deiminase [EC 3.5.4.21] catalyzes the hydrolysis of creatinine to produce *N*-methylhydantoin and ammonia. The enzyme will also act on cytosine, producing uracil and ammonia (a cytosine deiminase activity [EC 3.5.4.1]).

4-CRESOL DEHYDROGENASE (HYDROXYLATING)

This FAD-dependent oxidoreductase [EC 1.17.99.1], also known as *p*-cresol methylhydroxylase, catalyzes the reaction of 4-cresol with water and an acceptor substrate to produce 4-hydroxybenzaldehyde and the reduced acceptor, probably via a quinone methide intermediate.[1–4]

4-cresol   4-hydroxybenzaldehyde

The enzyme first catalyzes the oxidation of *p*-cresol to *p*-hydroxybenzyl alcohol, utilizing one atom of oxygen derived from water. This product is then further oxidized to produce 4-hydroxybenzaldehyde.

[1]L. M. Cunane, Z. W. Chen, N. Shamala, F. S. Mathews, C. N. Cronin & W. S. McIntire (2000) *J. Mol. Biol.* **295**, 357.

[2]I. D. Bossert, G. Whited, D. T. Gibson & L. Y. Young (1989) *J. Bacteriol.* **171**, 2956.

[3]M. J. Keat & D. J. Hopper (1978) *BJ* **175**, 649.

[4]D. J. Hopper & D. G. Taylor (1977) *BJ* **167**, 155.

CROSSOVER JUNCTION ENDORIBONUCLEASE

This endodeoxyribonuclease [EC 3.1.22.4], also called Holliday junction nuclease, catalyzes the endonucleolytic cleavage at a junction, leading to the reciprocal single- stranded crossover between two homologous DNA duplexes.[1] The enzyme will not act on single-stranded or double-stranded DNA lacking Holliday junctions.

[1]L. S. Symington & R. Kolodner (1985) *PNAS* **82**, 7247.

CROTONOYL-[ACYL-CARRIER PROTEIN] HYDRATASE

This hydratase [EC 4.2.1.58] catalyzes the reversible conversion of (3R)-3-hydroxybutanoyl-[acyl-carrier protein] to but-2-enoyl-[acyl-carrier protein] and water.[1] The acyl portion of the substrate can be between four and eight carbons in length.

[1]P. W. Majerus, A. W. Alberts & P. R. Vagelos (1965) *JBC* **240**, 618.

Selected entries from *Methods in Enzymology* [vol, page(s)]:
General discussion: 14, 64
Assay: 14, 64

CTP SYNTHETASE

This enzyme [EC 6.3.4.2], also known as UTP:ammonia ligase, catalyzes the reaction of UTP with ATP and ammonia (or L-glutamine) to produce CTP, ADP, and orthophosphate (and L-glutamate if L-glutamine had been used as the nitrogen source).[1–5] Rapid-quench studies, isotope partitioning experiments, and positional isotope exchange investigations suggest that CTP formation involves phosphorylation of UTP followed by attack of ammonia (or the nitrogen donor).[2,3]

[1]H. Zalkin (1993) *Adv. Enzymol.* **66**, 203.

[2]D. A. Lewis & J. J. Villafranca (1989) *Biochemistry* **28**, 8454.

[3]W. von der Saal, P. M. Anderson & J. J. Villafranca (1985) *JBC* **260**, 14997.

[4]D. E. Koshland, Jr., & A. Levitzki (1974) *The Enzymes*, 3rd ed., **10**, 539.

[5]J. M. Buchanan (1973) *AE* **39**, 91.

Selected entries from *Methods in Enzymology* [vol, page(s)]:
General discussion: 46, 420; 51, 79, 84; 113, 282
Assay: 51, 79, 80, 84; 113, 282
Other molecular properties: activators, 51, 82; activity, 51, 79, 84; bovine calf liver, from, 51, 84; dimer-tetramer interconversion, 113, 287; *Escherichia coli*, from, 51, 79 (properties, 113, 286; purification, 113, 283); inhibitors, 51, 89; kinetics, 51, 83, 84, 89, 90; molecular weight, 51, 82, 90; partition analysis, 249, 323; pH optimum, 113, 286; positional isotope exchange studies, 249, 423; properties, 51, 82, 83, 89, 90; 113, 286; purification, 51, 80, 86; 113, 283; sedimentation coefficients, 51, 90; self-association studies, 48, 306; storage, 51, 82; structure, 51, 82, 83

CUCUMISIN

This serine endopeptidase [EC 3.4.21.25] was first isolated from the sarcocarp of the musk melon (*Cucumis melo*). The enzyme exhibits a broad specificity and can be assayed with synthetic substrates such as Glt-Ala-Ala-Pro-Leu-∇-NHPhNO$_2$ and Suc-Ala-Ala-Pro-Phe-∇-NHPhNO$_2$.[1,2] (Note: ∇ indicates the cleavage site.)

[1]T. Uchikoba, H. Yonezawa & M. Kaneda (1995) *J. Biochem.* **117**, 1126.

[2]M. Kaneda, H. Yonezawa & T. Uchikoba (1995) *Biotechnol. Appl. Biochem.* **22**, 215.

CUCURBITACIN Δ^{23}-REDUCTASE

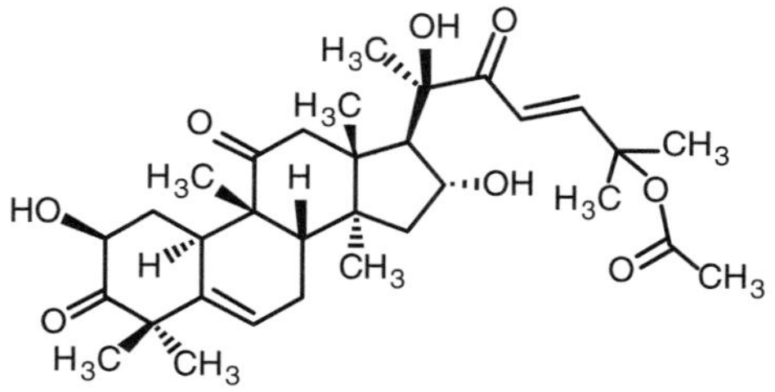

cucurbitacin B

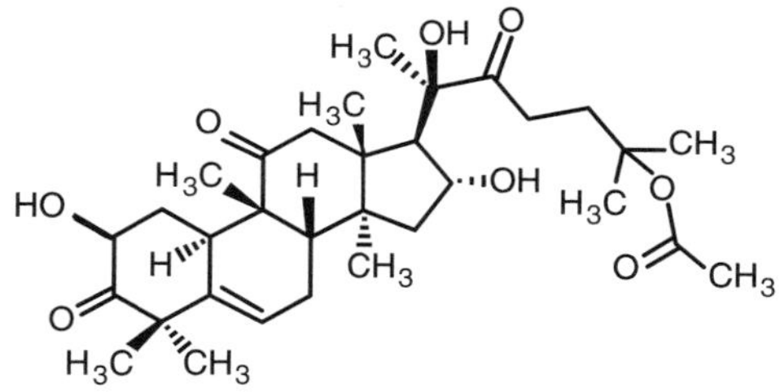

23,24-dihydrocucurbitacin B

This manganese-dependent oxidoreductase [EC 1.3.1.5] catalyzes the reaction of cucurbitacin with NAD(P)H to produce 23,24-dihydrocucurbitacin and NAD(P)$^+$. Iron or zinc can replace manganese to some extent.[1]

[1]J. C. Schabort, D. J. J. Potgieter & V. De Villiers (1968) *BBA* **151**, 33.

Cu^{2+}-EXPORTING ATPase

This so-called ATPase [EC 3.6.3.4] catalyzes the ATP-dependent (ADP- and orthophosphate-producing) transport of Cu$^{2+}_{in}$ to produce Cu$^{2+}_{out}$.[1,2] This enzyme is called a P-type ATPase because it undergoes covalent phosphorylation during the transport cycle.

The idea that a transporter is an enzyme is in keeping with a new definition of enzyme catalysis as the facilitated making/breaking of chemical bonds, not just covalent bonds.[3] This idea builds on Pauling's assertion that any long-lived, chemically distinct interaction (in this case, the persistent location of a solute with respect to the faces of a membrane) can be regarded as a chemical bond. Note also that the equilibrium constant ($K_{eq} = [Cu_{out}^{2+}][ADP][P_i]/[Cu_{in}^{2+}][ATP]$) does not conform to that expected for an ATPase (i.e., $K_{eq} = [ADP][P_i]/[ATP]$). Thus, although the overall reaction yields ADP and orthophosphate, the enzyme is misclassified as a hydrolase, and instead should be regarded as an energase-type reaction. Energases facilitate affinity-modulated reactions by coupling the $\Delta G_{ATP\text{-}hydrolysis}$ to a force-generating or work-producing step.[3] In this case, P-O-P bond scission supplies the energy to drive Cu^{2+} transport. **See Energase**

[1]C. Vulpe, B. Levinson, S. Whitney, S. Packman & J. Gitschier (1993) Nat. Genet. **3**, 7.
[2]K. Petrukhin, S. Lutsenko, I. Chernov, B. M. Ross, J. H. Kaplan & T. C. Gilliam (1994) Hum. Mol. Genet. **3**, 1647.
[3]D. L. Purich (2001) TiBS **26**, 417.

CUTINASE

This plant lipase [EC 3.1.1.74] catalyzes the hydrolysis of cutin to produce cutin monomers.[1,2] Other substrates include synthetic polymers such as polycaprolactone. The cutinase from *Fusarium solani pisi* will also act on certain water-soluble substrates such as *p*-nitrophenyl butyrate.

[1]R. Garcia-Lepe, O. M. Nuero, F. Reyes & F. Santamaria (1997) Lett. Appl. Microbiol. **25**, 127.
[2]R. E. Purdy & P. E. Kolattukudy (1975) Biochemistry **14**, 2824 and 2832.

Selected entries from **Methods in Enzymology** [**vol**, page(s)]:
General discussion: 284, 130
Assay: fungi, 71, 656
Other molecular properties: diffraction, 276, 282; fungi, from, 71, 653 (amino acid and carbohydrate composition, 71, 658; assay, 71, 656; chain length specificity, 71, 661; inhibitors, 71, 660; properties, 71, 658); *Fusarium solani pisi* (amino-terminal glucuronamide linkage [occurrence, 106, 216; detection, 106, 210]; amino-terminal region, structure, 106, 217; biosynthesis in vitro, 106, 216); pollen, from, 71, 660 (chain length specificity, 71, 660; inhibitors, 71, 663; properties, 71, 663; substrate specificity, 71, 663)

*Cvi*JI RESTRICTION ENDONUCLEASE

This type II restriction endonuclease [EC 3.1.21.4], obtained from *Chlorella* strain NC64A (IL-3A), catalyzes the hydrolysis of both strands of DNA at 5′...RG∇CY...3′, where R refers to A or G and Y refers to T or C.[1] This enzyme produces blunt-ended fragments.

[1]Y. Xia, D. E. Burbank, L. Uher, D. Rabussay & J. L. Van Etten (1987) NAR **15**, 6075.

*Cvi*TI RESTRICTION ENDONUCLEASE

This type II restriction endonuclease [EC 3.1.21.4], obtained from *Chlorella* strain NC64A (CA-1A), catalyzes the hydrolysis of both strands of DNA at 5′...RG∇CY...3′, where R refers to A or G and Y refers to T or C, producing blunt-ended fragments.

CYANAMIDE HYDRATASE

This enzyme [EC 4.2.1.69] catalyzes the reaction of cyanamide ($H_2N-C\equiv N$) with water to produce urea.[1]

[1]H. Stransky & A. Amberger (1973) Z. Pflanzenphysiol. **70**, 74.

CYANATE LYASE

This lyase [EC 4.3.99.1], formerly called cyanate hydrolase (and formerly classified as EC 3.5.5.3) and cyanase, catalyzes the reaction of cyanate (NCO^-) with bicarbonate to produce carbon dioxide and carbamate (H_2NCOO^-). Carbamate will then generate ammonia and CO_2.

The kinetic mechanism has been reported to be rapid equilibrium random Bi Bi.[1,2] ^{18}O from labeled water is not incorporated into the CO_2 formed from either bicarbonate or cyanate. However, oxygen from ^{18}O-labeled bicarbonate is incorporated into the CO_2 produced from cyanate.[1]

[1]W. V. Johnson & P. M. Anderson (1987) JBC **262**, 9021.
[2]P. M. Anderson & R. M. Little (1986) Biochemistry **25**, 1621.

CYANIDE HYDRATASE

This enzyme [EC 4.2.1.66], also known as formamide hydrolyase, catalyzes the reaction of cyanide with water to produce formamide ($HCONH_2$).[1,2]

[1]W. E. Fry & R. L. Millar (1972) ABB **151**, 468.
[2]N. Nazly & C. J. Knowles (1981) Biotechnol. Lett. **3**, 363.

CYANIDIN-3-RHAMNOSYLGLUCOSIDE 5-*O*-GLUCOSYLTRANSFERASE

This transferase [EC 2.4.1.116] catalyzes the reaction of UDP-D-glucose with cyanidin-3-*O*-D-rhamnosyl-(1,6)-D-glucoside to produce UDP and cyanidin-3-*O*-[D-rhamnosyl-(1,6)-D-glucoside]-5-*O*-D-glucoside.[1,2] Other acceptor substrates include pelargonidin-3-*O*-rhamnosyl glucoside and cyanidin 3-*O*-glucoside.

[1]J. Kamsteeg, J. van Brederode & G. van Nigtevecht (1978) *Biochem. Genet.* **16**, 1059.

[2]J. Kamsteeg, J. van Brederode & G. van Nigtevecht (1980) *Z. Pflanzenphysiol.* **96**, 87.

3-CYANOALANINE HYDRATASE

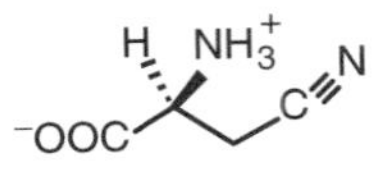

3-cyano-L-alanine 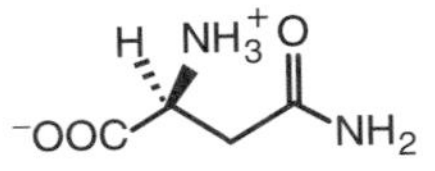   L-asparagine

This enzyme [EC 4.2.1.65] catalyzes the reaction of 3-cyano-L-alanine with water to produce L-asparagine.[1,2]

[1]H. Yanase, T. Sakai & K. Tonomura (1983) *Agric. Biol. Chem.* **47**, 473.

[2]P. A. Castric, J. F. Farnden & E. E. Conn (1972) *ABB* **152**, 62.

3-CYANOALANINE NITRILASE

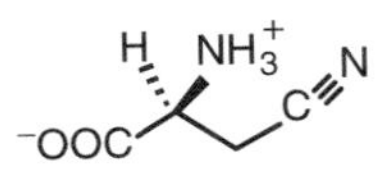

3-cyano-L-alanine 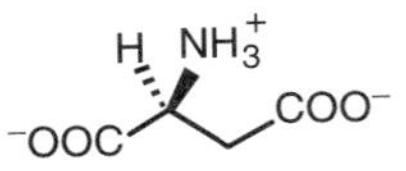   L-aspartate

This enzyme [EC 3.5.5.4] catalyzes the hydrolysis of 3-cyano-L-alanine with two water molecules to produce L-aspartate and ammonia, with L-asparagine formed as an intermediate.[1,2] **See also** *Nitrilase*

[1]H. Yanase, T. Sakai & K. Tonomura (1983) *Agric. Biol. Chem.* **47**, 473.

[2]W. G. Robinson & R. H. Hook (1971) *MIE* **17B**, 244.

3-CYANOALANINE SYNTHASE

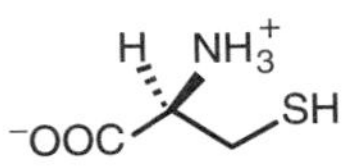

L-cysteine 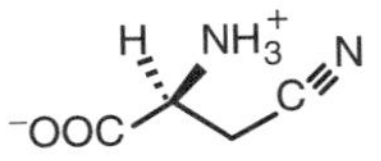   3-cyano-L-alanine

This enzyme [EC 4.4.1.9] catalyzes the reaction of L-cysteine with cyanide to produce hydrogen sulfide and 3-cyano-L-alanine.[1,2] The enzyme from a number of organisms also has a cysteine synthase (*i.e.*, *O*-acetylserine (thiol)-lyase: EC 4.2.99.8) activity.[3]

[1]F. Ikegami, K. Takayama & I. Murakoshi (1988) *Phytochemistry* **27**, 3385.

[2]A. E. Braunstein & E. V. Goryachenkova (1984) *Adv. Enzymol.* **56**, 1.

[3]A. Maruyama, K. Ishizawa & T. Takagi (2000) *Plant Cell Physiol.* **41**, 200.

Selected entries from *Methods in Enzymology* [**vol**, page(s)]:
General discussion: 17B, 233
Assay: 17B, 233

CYANOCOBALAMIN REDUCTASE (NADPH, CYANIDE-ELIMINATING)

This FAD/FMN-dependent oxidoreductase [EC 1.6.99.12] catalyzes the reaction of NADPH with cyanocob(III)alamin to produce $NADP^+$, cob(I)alamin, and cyanide.[1]

[1]F. Watanabe, Y. Oki, Y. Nakano & S. Kitaoka (1988) *J. Nutr. Sci. Vitaminol.* **34**, 1.

CYANOHYDRIN β-GLUCOSYLTRANSFERASE

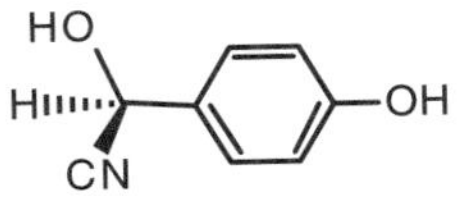

(*S*)-4-hydroxymandelonitrile

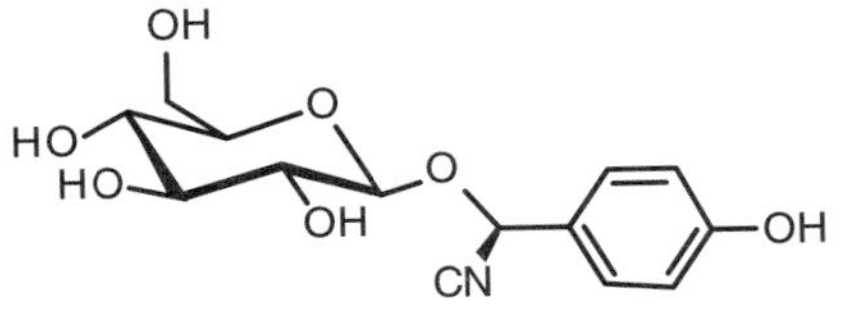

dhurrin

This transferase [EC 2.4.1.85], also known as UDP-glucose: *p*-hydroxymandelonitrile glucosyltransferase, catalyzes the reaction of UDP-D-glucose with (*S*)-4-hydroxymandelonitrile to produce UDP and (*S*)-4-hydroxymandelonitrile β-D-glucoside (commonly known as dhurrin).[1,2] Other acceptor substrates include (*S*)-mandelonitrile, hydroquinone, and 4-hydroxybenzyl alcohol.

[1]P. F. Reay & E. E. Conn (1974) *JBC* **249**, 5826.

[2]M. Kojima, J. E. Poulton, S. S. Thayer & E. E. Conn (1979) *Plant Physiol.* **63**, 1022.

CYANURIC ACID AMIDOHYDROLASE

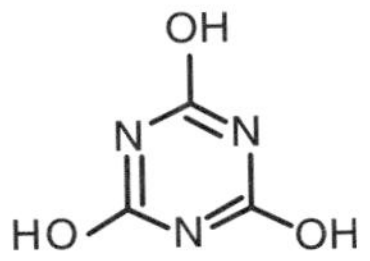

cyanuric acid 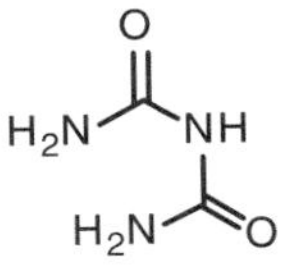   biuret

This enzyme [EC 3.5.2.15], which participates in the bacterial degradation of the herbacide atrazine (*i.e.*, 2-chloro-4-(ethylamino)-6-(isopropylamino)-1,3,5-triazine), catalyzes the hydrolysis of cyanuric acid to produce biuret and carbon dioxide.[1]

[1]R. W. Eaton & J. S. Karns (1991) *J. Bacteriol.* **173**, 1215 and 1363.

CYCLAMATE SULFOHYDROLASE

cyclohexylsulfamate   cyclohexylamine

This enzyme [EC 3.10.1.2], also known as cyclamate sulfamatase and cyclamate sulfamidase, catalyzes the hydrolysis of cyclohexylsulfamate (*i.e.*, a cyclamate) to produce cyclohexylamine and sulfate.[1] Other substrates include aliphatic sulfamates containing three to eight carbon atoms.

[1]T. Niimura, T. Tokiedo & T. Yamaha (1974) *J. Biochem.* **75**, 407.

1,2-CYCLIC-INOSITOL-PHOSPHATE PHOSPHODIESTERASE

This phosphodiesterase [EC 3.1.4.36] catalyzes the reaction of D-*myo*-inositol 1,2-cyclic phosphate with water to produce D-*myo*-inositol 1-phosphate.[1,2]

[1]T. S. Ross & P. W. Majerus (1986) *JBC* **261**, 11119.
[2]R. M. C. Dawson & N. G. Clarke (1973) *BJ* **134**, 59.

3′,5′-CYCLIC-NUCLEOTIDE PHOSPHODIESTERASE

This phosphodiesterase [EC 3.1.4.17], often known as cAMP phosphodiesterase, catalyzes the hydrolysis of 3′,5′-cAMP (or other 3′,5′-cyclic nucleotides, cNMP) to form 5′-AMP (or, 5′-NMP).[1-7] The mammalian enzyme is activated by calcium ions in the presence of calmodulin. Substrates include 3′,5′-cyclic dAMP, 3′,5′-cyclic IMP, 3′,5′-cyclic GMP, and 3′,5′-cyclic CMP. Note that at least eleven families of phosphodiesterases with varying selectivities for cAMP or cGMP have been identified in mammalian tissues. In addition, there are low-affinity (type 1) and high-affinity (type 2) enzymes. *See also* cGMP *Phosphodiesterase*

[1]S. H. Francis, I. V. Turko & J. D. Corbin (2000) *Prog. Nucl. Acid Res. Mol. Biol.* **65**, 1.
[2]E. Degerman, P. Belfrage & V. Manganiello (1997) *JBC* **272**, 6823.
[3]J. A. Gerlt (1992) *The Enzymes*, 3rd ed., **20**, 95.
[4]J. A. Beavo (1988) *Adv. Second Messenger Phosphoprotein Res.* **22**, 1.
[5]Y. M. Lin & W. Y. Cheung (1980) in *Calcium Cell Funct.* (W. Y. Cheung, ed.) **1**, 79, Academic Press, New York.
[6]M. Chasin & D. N. Harris (1972) *Methods Mol. Biol.* **3**, 169.
[7]G. I. Drummond & M. Yamamoto (1971) *The Enzymes*, 3rd ed., **4**, 355.

Selected entries from ***Methods in Enzymology*** [**vol**, page(s)]:
General discussion: **31**, 111; **38**, 218, 223, 240, 244, 257; **159**, 457, 543, 557, 582, 660, 675, 777

Assay: **38**, 62, 63; **39**, 480; **57**, 94; hormone-sensitive low-K_m, **159**, 737; modulator-deficient, preparation, **57**, 109; insulin-sensitive, **159**, 745, 752; low-K_m, **159**, 767; modulator protein, of, **57**, 108, 109

Other molecular properties: activation by *Acanthamoeba* calmodulin, assay, **139**, 62; activity, **57**, 96, 108; acylpeptide inhibitors from *Bacillus subtilis*, **159**, 497; adenylyl cyclase preparation contamination, **238**, 33; assay (cyanobacteria, in, **167**, 589; α-^{32}P-labeled nucleotide preparation for, **195**, 29; radiolabeled and fluorescent substrates, with, **159**, 457); authenticity, **38**, 26; calmodulin, interaction with, **87**, 524; calmodulin-sensitive (catalytic site, analysis, **159**, 526; purification, **159**, 522); calmodulin-stimulated (analysis with calmodulin derivatives, **159**, 605; antibodies, preparation, characterization, and applications, **159**, 627; assay, **159**, 548, 664; association with calmodulin in intact cells, assay, **159**, 594; heat-sensitive inhibitor [assay, **159**, 665; preparation, **159**, 662; properties, **159**, 666]; high-affinity testicular isoforms [age effects and distribution, **159**, 679; assay, **159**, 678]; isozymes [properties, **159**, 593; purification from bovine brain, **159**, 588]; properties, **159**, 566; purification from [bovine brain, **159**, 573; mammalian brain, **159**, 557; mammalian tissues, **159**, 543]; selective inhibitors, description, **159**, 652); calspermin assay with, **139**, 127; cGMP-stimulated (allosteric site, analysis, **159**, 528; binding of cyclic nucleotide analogs, **159**, 529; catalytic site, analysis, **159**, 526; purification, **159**, 523); contamination of adenylyl cyclase assay, **195**, 5; [^{17}O,^{18}O]-2′-deoxyadenosine-3′,5′-cyclic phosphate, **87**, 208; diffusion; **87**, 524; fat cells, in, **31**, 67, 68; histo- and cytochemical assays, **159**, 477; hormone-sensitive low-K_m (analysis with cAMP analogs, **159**, 531; assay, **159**, 737; hormone effects, observation conditions, **159**, 738; particulate, solubilization, **159**, 740; properties, **159**, 744; purification, **159**, 740; subcellular distribution, **159**, 742); hydrolysis rate, **32**, 128; inhibitors, **38**, 275, 276, 282; **238**, 33 (acylpeptide inhibitors from *Bacillus subtilis*, **159**, 497); insulin-sensitive (assay, **159**, 745, 752; properties, **159**, 750, 757; purification from rat liver, **159**, 753; stimulation by insulin, **159**, 748); Job plot, **87**, 524; low-K_m (assay, **159**, 767; properties, **159**, 770; purification from [human platelets, **159**, 772; rat brain, **159**, 767]; subcellular distribution and regulation, **159**, 771); methylxanthine inhibitors, **159**, 489; mutational analysis in *Drosophila*, **159**, 786; NMR, ^{31}P, **87**, 263, 265; osmotic shock and, **12B**, 844; preparation from porcine brain, **139**, 128; purification from canine kidney, **159**, 760; quality index, **87**, 299; reaction mixtures, HPLC analysis, **159**, 471; relationship to cAMP turnover, **159**, 52; specific inhibitors, effects on adipocytes, **159**, 504; stereochemistry, **87**, 206, 209, 234, 254, 263, 288, 299, 303; yeast cAMP cascade system mutants, in, assay, **159**, 38; zinc-containing, purification from yeast, **159**, 777

2′,3′-CYCLIC-NUCLEOTIDE 2′-PHOSPHODIESTERASE

This phosphodiesterase [EC 3.1.4.16] catalyzes the hydrolysis of a nucleoside 2′,3′-cyclic phosphate to produce nucleoside 3′-phosphate.[1-4] Other substrates include 3′-nucleoside monophosphates and bis-4-nitrophenyl phosphate. Similar reactions are carried out by ribonuclease T$_1$ [EC 3.1.27.3] and pancreatic ribonuclease [EC 3.1.27.5].

[1]G. I. Drummond & M. Yamamoto (1971) *The Enzymes*, 3rd ed., **4**, 355.
[2]T. Unemoto & M. Hayashi (1969) *BBA* **171**, 89.
[3]J.-F. Coqil, N. Virmaux, P. Mandel & C. Goridis (1975) *Biochim. Biophys. Acta* **403**, 425.
[4]B. M. Anderson, D. W. Kahn & C. D. Anderson (1985) *J. Gen. Microbiol.* **131**, 2041.

2',3'-CYCLIC-NUCLEOTIDE 3'-PHOSPHODIESTERASE

This phosphodiesterase [EC 3.1.4.37] catalyzes the hydrolysis of a nucleoside 2',3'-cyclic phosphate to produce a nucleoside 2'-phosphate.[1-6] The enzyme liver acts on 2',3'-cCMP more readily than on purine analogues and is often known as cCMP phosphodiesterase (it also acts on 3',5'-cyclic nucleotides, albeit slowly): however, the brain enzyme acts on 2',3'-cyclic AMP more rapidly than on the cUMP or cCMP analogues.

[1] T. J. Sprinkle (1989) CRC Crit. Rev. Clin. Neurobiol. 4, 235.
[2] U. S. Vogel & R. J. Thompson (1988) J. Neurochem. 50, 1667.
[3] K. Tyc, C. Kellenberger & W. Filipowicz (1987) JBC 262, 12994.
[4] Y. Tsukada & H. Suda (1980) Cell. Mol. Biol. 26, 493.
[5] B. W. Agranoff & M. H. Aprison (1978) Adv. Neurochem. 3, 1.
[6] G. I. Drummond & M. Yamamoto (1971) The Enzymes, 3rd ed., 4, 355.

Selected entries from Methods in Enzymology [vol, page(s)]:
Assay: 32, 124

CYCLOARTENOL SYNTHASE

This enzyme [EC 5.4.99.8], also known as 2,3-epoxysqualene:cycloartenol cyclase, catalyzes the conversion of (S)-2,3-epoxysqualene to cycloartenol, the first triterpene intermediate in the biosynthesis of sterols in photosynthetic organisms.[1-5]

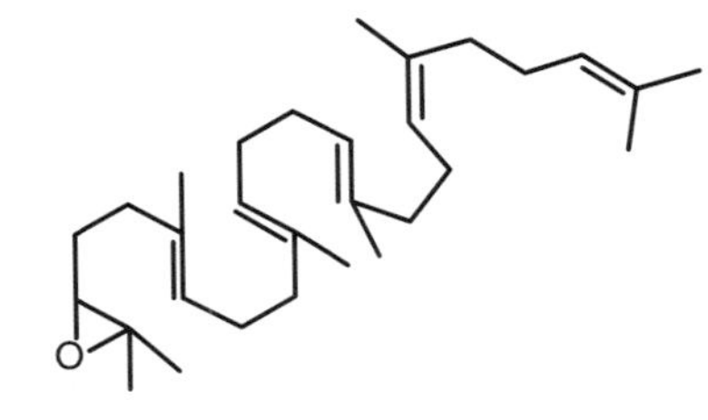

squalene 2,3-epoxide

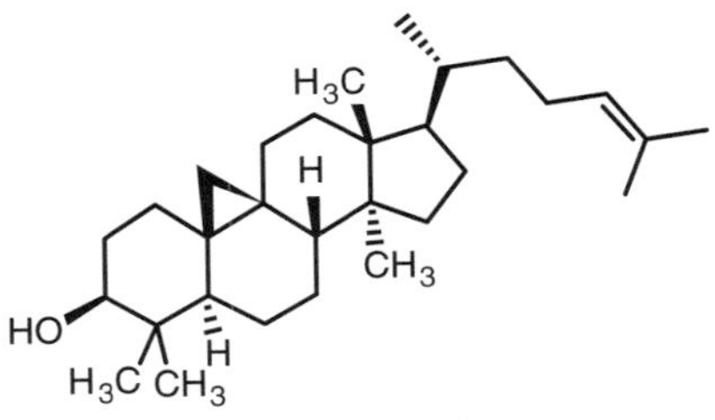

cycloartenol

The reaction proceeds via a number of Wagner-Meerwein shifts. Squalene 2,3-oxide is bound to the active site in a chair-boat-chair-boat-unfolded conformation. Electrophilic attack by H$^+$ occurs on the epoxy oxygen, followed by a series of cyclization reactions. Transient carbonium ion formation at C20 is followed by Wagner-Meerwein rearrangements.

[1] G. H. Beastall, H. H. Rees & T. W. Goodwin (1971) FEBS Lett. 18, 175.
[2] I. Shechter, F. W. Sweat & K. Bloch (1970) BBA 220, 463.
[3] H. H. Rees, L. J. Goad & T. W. Goodwin (1969) BBA 176, 892.
[4] S. Yamamoto, K. Lin & K. Bloch (1969) PNAS 63, 110.
[5] E. J. Corey, P. R. Ortiz de Montellano, K. Lin & P. D. G. Dean (1967) JACS 89, 2797.

CYCLOEUCALENOL CYCLOISOMERASE

This isomerase [EC 5.5.1.9], also known as cycloeucalenol:obtusifoliol isomerase, catalyzes the interconversion of cycloeucalenol and obtusifoliol, in plant sterol biosynthesis.

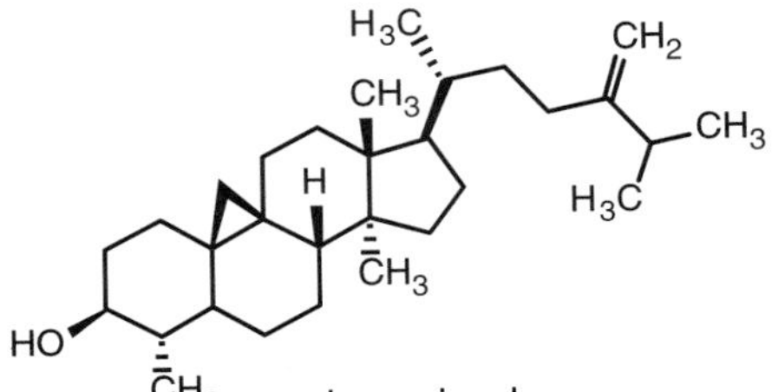

cycloeucalenol

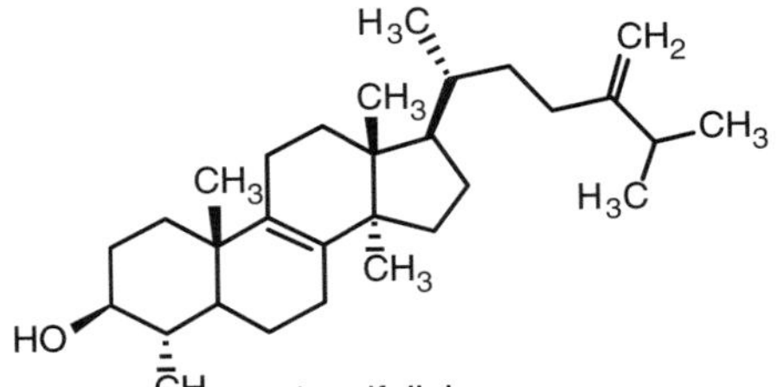

obtusifoliol

A number of alternative substrates include 4α-methyl-9β,19-cyclosterols. The 4β derivatives are not substrates. Specificity studies indicate that the 3β-hydroxyl group appears to be required for substrate binding.[1]

[1] A. Rahier, M. Taton & P. Benveniste (1989) EJB 181, 615.

Selected entries from Methods in Enzymology [vol, page(s)]:
Other molecular properties: plant, active cell-free preparations, isolation, 111, 323

CYCLOHEPTAGLUCANASE

This enzyme activity, formerly classified as EC 3.2.1.12 and now a deleted entry, is included under cyclomaltodextrinase [EC 3.2.1.54].

CYCLOHEXA-1,5-DIENECARBONYL-CoA HYDRATASE

This enzyme [EC 4.2.1.100], also known as dienoyl-CoA hydratase and cyclohexa-1,5-diene-1-carbonyl-CoA hydratase, catalyzes the reaction of cyclohexa-1,5-dienecarbonyl-CoA with water to produce 6-hydroxy-cyclohex-1-enecarbonyl-CoA.[1]

[1] D. Laempe, W. Eisenreich, A. Bacher & G. Fuchs (1998) *EJB* **255**, 618.

CYCLOHEXADIENYL DEHYDROGENASE

This oxidoreductase [EC 1.3.1.43], also known as pre-tyrosine dehydrogenase and arogenate dehydrogenase, catalyzes the reaction of L-arogenate (*i.e.*, 3-(1-carboxy-4-hydroxycyclohexa-2,5-dien-1-yl)-L-alanine) with NAD⁺ to produce L-tyrosine, NADH, and carbon dioxide. In some organisms, NADP⁺ is reported to be a better substrate.[1,2]

[1] T. H. Xia & R. A. Jensen (1990) *JBC* **265**, 20033.
[2] G. Zhao, T. Xia, L. O. Ingram & R. A. Jensen (1993) *EJB* **212**, 157.
Selected entries from *Methods in Enzymology* [vol, page(s)]:
General discussion: 142, 488, 513

CYCLOHEXAGLUCANASE

This enzyme activity, formerly classified as EC 3.2.1.13 and now a deleted entry, is included under cyclomaltodextrinase [EC 3.2.1.54].

CYCLOHEXANE-1,2-DIOL DEHYDROGENASE

This oxidoreductase [EC 1.1.1.174] catalyzes the reaction of *trans*-cyclohexane-1,2-diol with NAD⁺ to produce 2-hydroxycyclohexan-1-one and NADH.[1] Other substrates include the *cis*-cyclohexane-1,2-diol and 2-hydroxycyclohexanone.

[1] J. F. Davey & P. W. Trudgill (1977) *EJB* **74**, 115.

CYCLOHEXANE-1,3-DIONE HYDROLASE

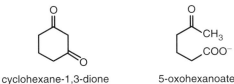

cyclohexane-1,3-dione   5-oxohexanoate

This enzyme [EC 3.7.1.10] catalyzes the hydrolysis of cyclohexane-1,3-dione to produce 5-oxohexanoate. The enzyme will not act on other dione derivatives of cyclohexane, cyclopentane, or cycloheptane.[1]

[1] C. Kluge, A. Tschech & G. Fuchs (1990) *Arch. Microbiol.* **155**, 68.

CYCLOHEXANOL DEHYDROGENASE

This oxidoreductase [EC 1.1.1.245] catalyzes the reaction of cyclohexanol with NAD⁺ to generate cyclohexanone and NADH.[1–3]

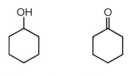

cyclohexanol   cyclohexanone

The enzyme will also use a number of other alicyclic alcohols and diols as alternative substrates (for example, cyclopentanol, cyclooctanol, 2-cyclohexenol, 3-methylcyclohexanol, and cyclohexane-1,4-diol). Straight-chain alcohols are very poor substrates or are not substrates at all with a number of exceptions: for example, 3-hexanol is a good substrate for the *Nocardia* enzyme.[3]

[1] M. K. Trower, R. M. Buckland, R. Higgins & M. Griffin (1985) *Appl. Environ. Microbiol.* **49**, 1282.
[2] W. Dangel, A. Czech & G. Fuchs (1989) *Arch. Microbiol.* **152**, 273.
[3] L. A. Stirling & J. J. Perry (1980) *Curr. Microbiol.* **4**, 37.

CYCLOHEXANONE DEHYDROGENASE

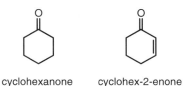

cyclohexanone   cyclohex-2-enone

This oxidoreductase [EC 1.3.99.14] catalyzes the reaction of cyclohexanone with an acceptor substrate to produce cyclohex-2-enone and the reduced acceptor.[1] Cyclopentanone is not a substrate. 2,6-Dichloroindophenol can act as the acceptor substrate.

[1] W. Dangel, A. Tschech & G. Fuchs (1989) *Arch. Microbiol.* **153**, 273.

CYCLOHEXANONE MONOOXYGENASE

This FAD-dependent oxidoreductase [EC 1.14.13.22] catalyzes the reaction of cyclohexanone with NADPH and dioxygen to produce 6-hexanolide (an ε-caprolactone), NADP⁺, and water.[1–3] A number of other cyclic ketones can serve as substrates.

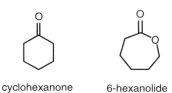

cyclohexanone   6-hexanolide

The reaction is a Baeyer-Villiger oxidation in which a flavin hydroperoxide acts as a nucleophile, presumably generating a transient 4a-flavin-hydroxide intermediate. In the absence of a cyclic ketone, the enzyme exhibits an NADPH oxidase activity. The enzyme from *Xanthobacter* utilizes FMN instead of FAD.

[1]B. A. Palfey & V. Massey (1998) *CBC* **3**, 83.
[2]M. A. Ator & P. R. Ortez de Montellano (1990) *The Enzymes*, 3rd ed., **19**, 213.
[3]C. Walsh, B. Branchaud, B. Fox & J. Latham (1983) *Colloq. Ges. Biol. Chem. Mosbach, 34th*, 140.

Selected entries from **Methods in Enzymology** [**vol**, page(s)]:
General discussion: 188, 70
Assay: 188, 71
Other molecular properties: properties, **188**, 74; purification from *Acinebacter* NCIMB 9871, **188**, 72

CYCLOHEXYLAMINE OXIDASE

cyclohexylamine   cyclohexanone

This FAD-dependent oxidoreductase [EC 1.4.3.12] catalyzes the reaction of cyclohexylamine with dioxygen and water to produce cyclohexanone, ammonia, and hydrogen peroxide.[1] Other substrates include some other cyclic amines.

[1]T. Tokieda, T. Niimura, F. Takamura & T. Yamaha (1977) *J. Biochem.* **81**, 851.

CYCLOMALTODEXTRINASE

This enzyme [EC 3.2.1.54] catalyzes the hydrolysis of a cyclomaltodextrin to produce a linear maltodextrin.[1-4] Substrates include α-, β-, and γ-cyclodextrins. It will also act on linear maltodextrins (*e.g.*, maltotetraose, maltopentaose, and maltohexaose), producing maltose. Pullulan is also hydrolyzed, producing panose. The cyclodextrins are the more effective substrates with ring-opening being the rate-limiting step.

Note that α- and β-cyclodextrins are also hydrolyzed, albeit slowly, by human salivary α-amylase as well as the α-amylases from human and porcine pancreas. Most α-amylases will act on γ-cyclodextrin.[1]

[1]K. H. Park, T. J. Kim, T. K. Cheong, J. W. Kim, B. H. Oh & B. Svensson (2000) *BBA* **1478**, 165.

[2]S. Kitahata & S. Okada (1985) *Carbohydr. Res.* **137**, 217.
[3]S. Kitahata, M. Taniguchi, S. D. Beltran, T. Sugimoto & S. Okada (1983) *Agric. Biol. Chem.* **47**, 1441.
[4]J. A. DePinto & L. L. Campbell (1968) *Biochemistry* **7**, 121.

CYCLOMALTODEXTRIN GLUCANOTRANSFERASE

Cyclodextrins are formed in the degradation of starch and dextran by the action of cyclodextrin/glucanosyltransferases [EC 2.4.1.19], also known as cyclodextrin glycosyltransferase and cyclodextrin glucanotransferase: note that this enzyme catalyzes the formation of a new α1,4-glycosidic linkage. The cyclic products formed are the Schardinger dextrins of various sizes (for example, cyclomaltohexaose, cyclomaltoheptaose, and cyclomaltooctaose: also known as α-, β-, and γ-cyclodextrins, respectively).[1-5] These enzymes also catalyze so-called acceptor reactions in which the cyclodextrin ring is opened and an acceptor molecule (*e.g.*, D-glucose) is added to the reducing end of the maltodextrin chain (a 4-α-glucanotransferase activity). The kinetics of three transglucosylation reactions catalyzed by the *Bacillus circulans* enzyme reveal three distinctions in mechanism, each proceeding via a covalent enzyme substrate intermediate. The disproportionation reaction (*i.e.*, cleavage of an α-glycosidic bond of a linear malto-oligosaccharide and transfer of one part to an acceptor substrate) proceeds via a ping-pong Bi Bi kinetic mechanism utilizing a glycosyl-enzyme intermediate. Cyclodextrin formation is a single substrate reaction and coupling (*i.e.*, cleavage of an α-glycosidic bond in a cyclodextrin ring and transfer of the resulting linear malto-oligosaccharide to an acceptor substrate) reportedly proceeds according to a random Bi Bi kinetic mechanism.[5] X-ray crystal investigations show that cyclodextrins bind to the active site differently than do the corresponding linear dextrins.[6]

[1]J. F. Robyt (1998) *Essentials of Carbohydrate Chemistry*, Springer, New York.
[2]I. S. Pretorius, M. G. Lambrechts & J. Marmur (1991) *Crit. Rev. Biochem. Mol. Biol.* **26**, 53.
[3]M. Vihinen & P. Mäntsälä (1989) *Crit. Rev. Biochem. Mol. Biol.* **24**, 329.
[4]J. A. DePinto & L. L. Campbell (1968) *Biochemistry* **7**, 114.
[5]B. A. van der Veen, G. J. van Alebeek, J. C. Uitdehaag, B. W. Dijkstra & L. Dijkhuizen (2000) *EJB* **267**, 658.
[6]J. C. Uitdehaag, K. H. Kalk, B. A. van Der Veen, L. Dijkhuizen & B. W. Dijkstra (1999) *JBC* **274**, 34868.

Selected entries from **Methods in Enzymology** [**vol**, page(s)]:
General discussion: 5, 148
Assay: 5, 148

CYCLOPENTANOL DEHYDROGENASE

cyclopentanol   cyclopentanone

This oxidoreductase [EC 1.1.1.163] catalyzes the reaction of cyclopentanol with NAD^+ to produce cyclopentanone and NADH.[1]

[1] M. Griffin & P. W. Trudgill (1972) *BJ* **129**, 595.

CYCLOPENTANONE MONOOXYGENASE

cyclopentanone   5-valerolactone

This oxidoreductase [EC 1.14.13.16] catalyzes the reaction of cyclopentanone with NADPH and dioxygen to produce 5-valerolactone, $NADP^+$, and water.[1] Other substrates include cyclobutanone, cyclohexanone, and 2-methylcyclohexanone.

[1] M. Griffin & P. W. Trudgill (1976) *EJB* **63**, 199.

Selected entries from *Methods in Enzymology* [**vol**, page(s)]:
General discussion: 188, 77
Assay: 188, 77

β-CYCLOPIAZONATE DEHYDROGENASE

This FAD-dependent oxidoreductase [EC 1.3.99.9], also known as β-cyclopiazonate oxidocyclase, catalyzes the reaction of β-cyclopiazonate with an acceptor substrate to produce α-cyclopiazonate and the reduced acceptor.[1,2] Acceptor substrates include cytochrome *c*, 2,6-dichlorophenolindophenol, and phenazine methosulfate.

[1] D. J. Steenkamp, J. Schabort & N. P. Ferreira (1973) *BBA* **309**, 440.
[2] J. C. Schabort & D. J. J. Potgieter (1971) *BBA* **250**, 329.

CYCLOPROPANE-FATTY-ACYL-PHOSPHOLIPID SYNTHASE

This methyltransferase [EC 2.1.1.79] (also known as cyclopropane synthetase, unsaturated-phospholipid methyltransferase, cyclopropane fatty acid synthase, and CFA synthase) catalyzes the reaction of *S*-adenosyl-L-methionine with a phospholipid olefinic fatty acid to produce *S*-adenosyl-L-homocysteine and a phospholipid cyclopropane fatty acid.[1–5]

The enzyme catalyzes the addition of a methylene group across the 9,10 position of a Δ^9-olefinic acyl chain in phosphatidylethanolamine, thus producing a cyclopropane derivative. Other substrates include phosphatidylglycerol and cardiolipin (however, they are converted more slowly than phosphatidylethanolamine). Phosphatidylcholine is a weak substrate. Note that this methyltransferase has a different specificity compared to methylene-fatty-acyl-phospholipid synthase [EC 2.1.1.16]. *See also Methylene-Fatty-Acyl-Phospholipid Synthase*

[1] A. E. Chung & J. H. Law (1964) *Biochemistry* **3**, 967.
[2] H. Zalkin, J. H. Law & H. Goldfine (1963) *JBC* **238**, 1242.
[3] A. Y. Wang, D. W. Grogan & J. E. Cronan, Jr. (1992) *Biochemistry* **31**, 11020.
[4] F. R. Taylor & J. E. Cronan, Jr. (1979) *Biochemistry* **18**, 3292.
[5] D. D. Smith, Jr., & S. J. Norton (1980) *ABB* **205**, 564.

Selected entries from *Methods in Enzymology* [**vol**, page(s)]:
General discussion: 71, 133
Assay: radioactive S-adenosylmethionine, with, **71**, 134

CYPIAI

This cytochrome P450 system, also called cytochrome P450c and cytochrome $P450_{\beta NF-B}$, catalyzes the NADPH- and dioxygen-dependent hydroxylation of numerous polycyclic aromatic hydrocarbons and participates in the metabolism of xenobiotics. Substrates include benzo[*a*]pyrene and α-naphthoflavin. The enzyme will also catalyze the demethylation of caffeine as well as the *O*-deethylation of phenacetin, 7-ethoxyresorufin, and 7-ethoxycoumarin. *See also Cytochrome P450*

Selected entries from *Methods in Enzymology* [**vol**, page(s)]:
Other molecular properties: alleles (cancer association, **272**, 226, 232; *Ile-Val*, **272**, 228, 230; *MspI*, **272**, 227, 232; polymerase chain reaction analysis, **272**, 228, 230; single-strand conformational polymorphism, **272**, 230; Southern blot analysis, **272**, 227); analysis with antipeptide antibodies, **206**, 220; antibody affinity for, **206**, 229; antipeptide antibodies, effect on enzyme activity, **206**, 232; cigarette smoke detoxification, **272**, 226, 232; encoding gene, upstream protein-DNA interactions, gel retardation analysis, **206**, 403; family members, proof of identity, **206**, 430; functional activity measurements, **206**, 426; gene locus, **272**, 227; induction by aromatic hydrocarbons, **272**, 396; *in vitro* induction, **206**, 425; *in vivo* induction, **206**, 423; messenger RNA quantitation with polymerase chain reaction, **272**, 411; photoaffinity labeling with 4′-azidowarfarin, **206**, 61; role in transcriptional activation of genes with responsive element sequences, **206**, 417; stable expression in V79 cells, **206**, 118; tamoxifen metabolism, **272**, 162

CYPIA2

This cytochrome P450 system, also known as cytochrome $P450_{LM4}$, catalyzes the NADPH- and dioxygen-dependent oxidation of numerous hydrocarbons, resulting in the activation of several promutagens and procarcinogens: for example, aflatoxin B_1, phenacetin, and heterocyclic arylamines. It will also catalyze the demethylation of caffeine to produce paraxanthine, theophylline, and theobromine (the enzyme has a preference for 3-demethylation), convert imipramine to desipramine, catalyze the *O*-deethylation of ethoxyresorufin and phenacetin, hydroxylate cyclohexane, and *p*-hydroxylate acetanilide. **See also** *Cytochrome P450*

Selected entries from **Methods in Enzymology** [vol, page(s)]:
Assay: caffeine assay (blood [high-performance liquid chromatography, **272**, 126; sampling, **272**, 126]; systemic clearance estimation, **272**, 125, 131; urine [high-performance liquid chromatography, **272**, 128, 130; sampling, **272**, 128])

Other molecular properties: absorption spectrum, **52**, 117; analysis with antipeptide antibodies, **206**, 220; antibody affinity for, **206**, 229; antipeptide antibodies, effect on enzyme activity, **206**, 232; carbohydrate content, **52**, 117; family members, proof of identity, **206**, 430; functional activity measurements, **206**, 426; heme content, **52**, 116; imipramine metabolism, **272**, 177; *in vitro* induction, **206**, 425; *in vivo* induction, **206**, 423; isolation and purification, **52**, 114, 115; molecular weight, **52**, 117; propafenone (dealkylation assay, **272**, 102; metabolism, **272**, 100); role in autoimmune drug-induced hepatitis, **272**, 85; solubility, **52**, 116; stable expression in V79 cells, **206**, 118; substrate specificity, **52**, 117

CYP2A SUBFAMILY

Enzymes in this subfamily of cytochrome P450 systems catalyze the oxidation of a number of xenobiotics and steroids.

CYP2A1

This cytochrome P450 system, also called cytochrome $P450_{UT-F}$, catalyzes the NADPH-/dioxygen-dependent conversion of testosterone to 7α-hydroxytestosterone.[1]
See also *Cytochrome P450*

[1] N. Hanioka, F. J. Gonzalez, N. A. Lindberg, G. Liu, H. V. Gelboin & K. R. Korzekwa (1992) *Biochemistry* **31**, 3364.

Selected entries from **Methods in Enzymology** [vol, page(s)]:
Assay: testosterone hydroxylase activity in microsomes, **206**, 252

CYP2A2

This cytochrome P450 system, also called cytochrome $P450_{RLM2}$, catalyzes the NADPH-/dioxygen-dependent conversion of testosterone to 15α-hydroxytestosterone.[1]
See also *Cytochrome P450*

[1] N. Hanioka, F. J. Gonzalez, N. A. Lindberg, G. Liu, H. V. Gelboin & K. R. Korzekwa (1992) *Biochemistry* **31**, 3364.

Selected entries from **Methods in Enzymology** [vol, page(s)]:
Assay: testosterone hydroxylase activity in microsomes, **206**, 252
Other molecular properties: catalysis of progesterone hydroxylation, **206**, 474

CYP2A5

This mouse cytochrome P450 system, also called cytochrome $P450_{coh}$, catalyzes the oxidation of a number of xenometabolites and steroids, including the NADPH-/dioxygen-dependent 7-hydroxylation of coumarin.[1] **See also** *Cytochrome P450*

[1] M. Iwasaki, T. A. Darden, L. G. Pedersen, D. G. Davis, R. O. Juvonen, T. Sueyoshi & M. Negishi (1993) *JBC* **268**, 759.

CYP2A6

This human cytochrome P450 system, also called cytochrome $P450_{IIA3}$, catalyzes the NADPH- and dioxygen-dependent conversion of coumarin to 7-hydroxycoumarin. Other substrates include diethylnitrosamine. The enzyme will also activate some procarcinogens (*e.g.*, 4-methylnitrosoamino-1-(3-pyridyl)-1-butanone) and catalyze a *C*-oxidase activity on nicotine. **See also** *Cytochrome P450*

CYP2A10

This cytochrome P450 system catalyzes the NADPH- and dioxygen-dependent hydroxylation of coumarin and 17-hydroxylation of testosterone and the *gem*-diol intermediate forms of androstendione. Other substrates include ethanol, *N*-nitrosodiethylamine, and phenacetin.[1]

Cytochrome P450NMa is a mixture of CYP2A10 and CYP2A11 (the descriptor NMa refers to "nasal microsomal P450 form a"). **See** *CYP2A11; Cytochrome P450NMa; Cytochrome P450*

[1] H. M. Peng, X. Ding & M. J. Coon (1993) *JBC* **268**, 17253.

CYP2A11

This rabbit cytochrome P450 system catalyzes the NADPH- and dioxygen-dependent hydroxylation of coumarin. Other substrates include ethanol, *N*-nitrosodiethylamine, and phenacetin.[1] The enzyme will also catalyze the 17-hydroxylation of testosterone, but not as actively as CYP2A10.

Cytochrome P450Nma is a mixture of CYP2A10 and CYP2A11 (the descriptor NMa refers to "nasal microsomal P450 form a"). **See** *CYP2A10; Cytochrome P450NMa; Cytochrome P450*

[1] H. M. Peng, X. Ding & M. J. Coon (1993) *JBC* **268**, 17253.

CYP2A13

This human cytochrome P450 system catalyzes the NADPH- and dioxygen-dependent oxidation of hexamethylphosphoramide, *N,N*-dimethylaniline, 2'-methoxyacetophenone, and *N*-nitrosomethylphenylamine. This enzyme is not as effective as CYP2A6 on coumarin 7-hydroxylation; however, it is very active with the tobacco-specific carcinogen, 4-(methylnitrosamino)-1-(3-pyridyl)-1-butanone.[1] *See also Cytochrome P450*

[1] T. Su, Z. Bao, Q. Y. Zhang, T. J. Smith, J. Y. Hong & X. Ding (2000) *Cancer Res.* **60**, 5074.

CYP2B1

This phenobarbital-inducible cytochrome P450 system, also called cytochrome $P450_{PB-4}$ and cytochrome P450b, catalyzes the NADPH- and dioxygen-dependent *O*-deethylation of 7-ethoxy-4-(trifluoromethyl)coumarin. *See also Cytochrome P450*

CYP2B4

This rabbit cytochrome P450 system, also known as cytochrome $P450_{LM2}$, catalyzes the oxidation of a large number of xenometabolites and steroids: for example, oxidation of hexobarbital and the *N*-demethylation of benzphetamine. Other substrates include cyclohexane, ethylmorphine, and *p*-nitroanisole. CYP2B4 will also catalyze the 4-hydroxylation of biphenyl and the 16α-hydroxylation of testosterone. *See also Cytochrome P450; CYP2B6*

Selected entries from *Methods in Enzymology* [vol, page(s)]:
Assay: 52, 205; **272**, 31
Other molecular properties: absorption spectrum, **52**, 117; carbohydrate content, **52**, 117; heme content, **52**, 116; isolation and purification, **52**, 111; molecular weight, **52**, 117; N-terminal modification (effects [catalytic activity, **272**, 27, 34; subcellular localization, **272**, 26, 32, 34]; expression in *Escherichia coli* [culture conditions, **272**, 28; harvesting, **272**, 28; vector construction, **272**, 27]; immunoblot analysis for subcellular localization, **272**, 29; insertion of positive charges, **272**, 26; purification of glutathione *S*-transferase fusion protein, **272**, 30; truncation, **272**, 25); reconstituted mixed-function oxidase system, in, **52**, 200; solubility, **52**, 116; substrate specificity, **52**, 117

CYP2B6

This human cytochrome P450 system, also known as cytochrome $P450_{LM2}$, catalyzes the oxidation of a large number of xenometabolites and steroids: for example, the *N*-demethylation of benzphetamine, and the *O*-deethylation of 7-ethoxycoumarin. Other substrates include cyclohexane, cyclophosphamide, bupropion, ethylmorphine, and *p*-nitroanisole. CYP2B6 will also catalyze the 4-hydroxylation of biphenyl and the 16α-hydroxylation of testosterone. *See also Cytochrome P450; CYP2B4*

CYP2C SUBFAMILY

This subfamily of cytochrome P450 systems catalyzes the oxidation of many xenobiotics and steroids. Substrates include mephenytoin, tolbutamide, progesterone, and testosterone. *See also Cytochrome P450*

Selected entries from *Methods in Enzymology* [vol, page(s)]:
Other molecular properties: hepatic, rat, expression analysis, **206**, 249; human isozymes, characterization with yeast expression system, **206**, 183; mRNA, analysis with gene-specific oligonucleotide probes, **206**, 260; steroid hydroxylase activity, assay, **206**, 251; testosterone hydroxylase activity in microsomes, assay, **206**, 252; Western blotting analysis, **206**, 255

CYP2C2

This phenobarbital-inducible cytochrome P450 system, also called cytochrome $P450_{PBc2}$, catalyzes the NADPH- and dioxygen-dependent oxidation of a number of steroids and xenometabolites. *See also Cytochrome P450*

Selected entries from *Methods in Enzymology* [vol, page(s)]:
Other molecular properties: membrane integration assay, **206**, 72; processing in membrane, analysis, **206**, 71; targeting to microsomal membranes, cell-free analysis, **206**, 64; translation systems, *in vitro*, **206**, 70; translocated, protease protection, **206**, 73

CYP2C3

This cytochrome P450 system catalyzes the NADPH- and dioxygen-dependent 16α-hydroxylation of progesterone. A variant of CYP2C3 called CYP2C3v, differing in only five amino acids out of 489, catalyzes both 6-hydroxylation and 16α-hydroxylation of progesterone with a higher catalytic efficiency. *See also Cytochrome P450*

CYP2C5

This cytochrome P450 system catalyzes the NADPH- and dioxygen-dependent 21-hydroxylation of progesterone. *See also Cytochrome P450*

Selected entries from *Methods in Enzymology* [vol, page(s)]:
Other molecular properties: steroid hydroxylase activity, inhibition by 22-amino-23,24-bisnor-5-cholen-3β-ol, **206**, 552

CYP2C6

This cytochrome P450 system, also called cytochrome $P450_{PB1}$, catalyzes the NADPH- and dioxygen-dependent conversion of progesterone to 21-hydroxyprogesterone. *See also Cytochrome P450*

Selected entries from *Methods in Enzymology* [vol, page(s)]:
Assay: steroid hydroxylase activity, **206**, 251

CYP2C8

This cytochrome P450 system catalyzes the NADPH- and dioxygen-dependent conversion of taxol to 6-hydroxytaxol as well as torsemide tolylmethylhydroxylation. It will also catalyze the *N*-deethylation of amiodarone, (*R*)-mephenytoin 4'-hydroxylation, and act on tolbutamide. *See also Cytochrome P450*

Selected entries from *Methods in Enzymology* [vol, page(s)]:
Assay: taxol hydroxylation assay (fecal samples, **272**, 148; high-performance liquid chromatography, **272**, 147; interference, **272**, 149, 151; microsomes, **272**, 149; plasma samples, **272**, 147; quantitation, **272**, 149)
Other molecular properties: characterization, **206**, 583; purification from human liver, **206**, 578

CYP2C9

This cytochrome P450 system catalyzes the NADPH- and dioxygen-dependent 4'-hydroxylation of diclofenac, the conversion of tolbutamide to hydroxytolbutamine, (*R*)-mephenytoin 4'-hydroxylation, and the *O*-demethylation of 7-methoxy-4-trifluoromethylcoumarin. It has been estimated that roughly 15% of currently used therapeutics are acted upon by CYP2C9. CYP2C9 specifically 7-hydroxylates (*S*)-warfarin. *See also Cytochrome P450*

Selected entries from *Methods in Enzymology* [vol, page(s)]:
Assay: tolbutamide hydroxylation assay (extraction, **272**, 141; high-performance liquid chromatography, **272**, 142; incubation conditions, **272**, 141; microsome preparation, **272**, 141; reagents, **272**, 141; validation, **272**, 142)
Other molecular properties: characterization, **206**, 583; cDNA expression, plasmids for, construction, **206**, 184; purification from human liver, **206**, 578; role in autoimmune drug-induced hepatitis, **272**, 81, 85; yeast, crude extract preparation, **206**, 186

CYP2C11

This cytochrome P450 system catalyzes the oxidation of a variety of drugs and metabolites, including the hydroxylation of testosterone. *See also Cytochrome P450*

Selected entries from *Methods in Enzymology* [vol, page(s)]:
Other molecular properties: expression in cultured hepatocytes (cell culture, **272**, 381; ceramide effects [ceramide-activated protein phosphatase, **272**, 384; delivery of ceramide to cells, **272**, 382; exogenous sphingomyelinase and ceramide elevation, **272**, 383; RNA analysis of gene expression, **272**, 384; Western blotting, **272**, 385])

CYP2C19

This cytochrome P450 system catalyzes the oxidation of a variety of drugs and xenometabolites: for example, the NADPH- and dioxygen-dependent 5-hydroxylation of omeprezole and the 4'-hydroxylation of *S*-mephenytoin. The enzyme will also catalyze the conversion of flunitrazepam to both desmethylflunitrazepam and 3-hydroxyflunitrazepam as well as catalyze the

4-hydroxylation of (*R*)-mephobarbital. *See also Cytochrome P450*

Selected entries from *Methods in Enzymology* [vol, page(s)]:
Other molecular properties: alleles (nomenclature, **272**, 212; polymerase chain reaction analysis [digestion of products, **272**, 211, 214; interpretation, **272**, 214; primers, **272**, 211; reaction conditions, **272**, 213; sensitivity, **272**, 217; solutions, **272**, 212; specificity, **272**, 217; troubleshooting, **272**, 215]; types, **272**, 210); imipramine metabolism, **272**, 177; mephenytoin (assay, *in vitro*, **272**, 113; assay, *in vivo* [drug administration, **272**, 107; measurement of urinary 4'-hydroxymephenytoin, **272**, 108; measurement of urinary enantiomeric ratio, **272**, 109; phenotyping, **272**, 108, 111; principle, **272**, 107]; hydroxylation phenotypes, **272**, 105, 210); omeprazole (hydroxylation, **272**, 106, 112; phenotyping, **272**, 112)

CYP2D SUBFAMILY

Enzymes in this subfamily of cytochrome P450 systems catalyze the oxidation of a number of steroids and xenometabolites. *See also Cytochrome P450*

CYP2D6

This cytochrome P450 system, also called cytochrome $P450_{buf}$, catalyzes the oxidation of a variety of drugs and xenometabolites: for example, the NADPH- and dioxygen-dependent *O*-demethylation of dextromethorphan to produce dextrorphen and formaldehyde. This enzyme will also catalyze the 4-hydroxylation of debrisoquine, the 5-hydroxylation of propafenone, the 1'-hydroxylation of bufuralol, and act on sparteine. *See also Cytochrome P450*

Selected entries from *Methods in Enzymology* [vol, page(s)]:
Assay: [*O-methyl*-^{14}C]dextromethorphan assay (extraction, **272**, 190; inhibitor analysis, **272**, 195; microsome [incubation, **272**, 189; preparation, **272**, 189]; phenotyping, **272**, 191, 193; principle, **272**, 187; reagents, **272**, 189; recombinant protein, **272**, 191; substrate synthesis, **272**, 188)
Other molecular properties: alleles (detection [*CYP2D6A*, **272**, 203; *CYP2D6B*, **272**, 203; *CYP2D6D*, **272**, 204; *CYP2D67*, **272**, 206; DNA preparation, **272**, 200; polymerase chain reaction assay, **272**, 201; single-strand conformational polymorphism analysis, **272**, 207]; effects on activity, **272**, 199, 202; types, **272**, 200, 202); gene nomenclature, **206**, 182; human (immunoisolation from hepatic microsomes, **206**, 206; prototype substrates, **206**, 509; recombinant, as diagnostic reagent, **206**, 216); mutant alleles (restriction fragment analysis, **206**, 173; specific amplification, **206**, 173); phenotypes, **272**, 186, 200; propafenone (hydroxylation assay, **272**, 101; metabolism, **272**, 99); substrates, **272**, 186, 349

CYP2E1

This cytochrome P450 system will catalyze the NADPH- and dioxygen-dependent oxidation of more than one hundred pharmaceuticals, xenometabolites, and natural metabolites: for example, this enzyme will act on ethanol, acetaldehyde, acetone, carbon tetrachloride,

N-nitrosoalkylamines (for example, the demethylation of *N*-nitrosodimethylamine), and fatty acids.

chlorzoxazone   6-hydroxychlorzoxazone

This enzyme catalyzes the reaction of chlorzoxazone with dioxygen and NADPH to produce 6-hydroxychlorzoxazone, water, and NADP$^+$. It will also catalyze a demethylation of caffeine to produce theophylline and theobromine. Humans, rats, and mice express only one member of the CYP2E subfamily. **See also** *Cytochrome P450*

Selected entries from **Methods in Enzymology** [vol, page(s)]:
Assay: 206, 112, 600; 272, 31; *in vitro* (cell culture, 272, 118; high-performance liquid chromatography, 272, 119; microsome incubation, 272, 117; reagents, 272, 117; thin-layer chromatography, 272, 119); *in vivo* (alcoholics, 272, 122; drug administration, 272, 121; high-performance liquid chromatography, 272, 121; interference, 272, 120, 122; reagents, 272, 121; sample preparation, 272, 121)
Other molecular properties: alleles (disease association, 272, 220; polymerase chain reaction detection [DNA purification, 272, 222; 5'-flanking region, 272, 223; intron 6, 272, 224; primers, 272, 223; reagents, 272, 221; restriction analysis, 272, 223]; types, 272, 219); characterization, 206, 114, 583, 595; chlorzoxazone 6-hydroxylation (assay, *in vitro* [cell culture, 272, 118; high-performance liquid chromatography, 272, 119; microsome incubation, 272, 117; reagents, 272, 117; thin-layer chromatography, 272, 119]; assay, *in vivo* [alcoholics, 272, 122; drug administration, 272, 121; high-performance liquid chromatography, 272, 121; interference, 272, 120, 122; reagents, 272, 121; sample preparation, 272, 121]; evidence, 272, 115; phenotypes, 272, 120, 122); distribution, 206, 595; evidence, 272, 115; induction, 206, 595; 272, 396; *N*-nitrosodiethylamine deethylase activity of cytochrome P450 2E1, assay, 206, 112; *N*-nitrosodiethylamine deethylase activity, assay, 206, 112; *N*-nitrosodimethylamine demethylase activity, assay, 206, 600; N-terminal modification (effects [catalytic activity, 272, 27, 34; subcellular localization, 272, 26, 32, 34]; expression in *Escherichia coli* [culture conditions, 272, 28; harvesting, 272, 28; vector construction, 272, 27]; immunoblot analysis for subcellular localization, 272, 29; insertion of positive charges, 272, 26; purification of glutathione *S*-transferase fusion protein, 272, 30; truncation, 272, 25); partial purification, 206, 114; phenotypes, 272, 120, 122; purification, 206, 595 (acetone-induced rat hepatic microsomes, from, 206, 598; human liver, from, 206, 578); spectral properties, 206, 603; structural alignment with P450s, 272, 318; structure, 206, 603; substrates, 272, 115, 218

CYP2G1

This cytochrome P450 system, also known as cytochrome P450 NMb (the descriptor NMb refers to "nasal microsomal P450 form b"), catalyzes the NADPH- and dioxygen-dependent 15α- and 15β-hydroxylation of testosterone and 7-hydroxylation of coumarin. The mouse enzyme, cyp2g1, has been shown to catalyze the 3-hydroxylation of acetaminophen. **See also** *Cytochrome P450*

Selected entries from **Methods in Enzymology** [vol, page(s)]:
Other molecular properties: properties, 206, 610; purification from rabbit nasal microsomes, 206, 603; separation from cytochrome P450NMa, 206, 605

CYP2J SUBFAMILY

Several members of this cytochrome P450 subfamily act on arachidonate and produce a number of epoxide derivatives: for example, the NADPH and dioxygen-dependent production of 5,6-, 8,9-, 11,12-, and 14,15-epoxyeicosatrienoate. **See also** *Cytochrome P450*

CYP2J1

This rabbit cytochrome P450 system catalyzes the NADPH- and dioxygen-dependent *N*-demethylation of benzamphetamine.[1] **See also** *Cytochrome P450*

[1] Y. Kikuta, K. Sogawa, M. Haniu, M. Kinosaki, E. Kusunose, Y. Nojima, S. Yamamoto, K. Ichihara, M. Kusunose & Y. Fujii-Kuriyama (1991) *JBC* 266, 17821.

CYP2J2

This human cytochrome P450 system catalyzes the NADPH- and dioxygen-dependent formation of all four *cis*-epoxy-derivatives of arachidonate: *cis*-5,6-, 8,9-, 11,12-, and 14-15-epoxyeicosatrienoate.[1] **See also** *Cytochrome P450*

[1] S. Wu, C. R. Moomaw, K. B. Tomer, J. R. Falck & D. C. Zeldin (1996) *JBC* 271, 3460.

CYP2J3

This rat cytochrome P450 system catalyzes the NADPH- and dioxygen-dependent conversion of arachidonate to 14,15-, 11,12-, and 8,9-epoxyeicosatrienoates as well as formation of 19-hydroxyeicosatetraenoate.[1] **See also** *Cytochrome P450*

[1] S. Wu, W. Chen, E. Murphy, S. Gabel, K. B. Tomer, J. Foley, C. Steenbergen, J. R. Falck, C. R. Moomaw & D. C. Zeldin (1997) *JBC* 272, 12551.

CYP2J4

This rat cytochrome P450 system catalyzes the NADPH- and dioxygen-dependent conversion of arachidonate to a number of epoxyeicosatrienoate and hydroxyeicosatrienoate derivatives.[1] In addition, CYP2J4 is active toward all-*trans*- and 9-*cis*-retinal, generating the corresponding retinoates.[2] *See also Cytochrome P450*

[1] Q. Y. Zhang, X. Ding & L. S. Kaminsky (1997) *ABB* **340**, 270.
[2] Q. Y. Zhang, G. Raner, X. Ding, D. Dunbar, M. J. Coon & L. S. Kaminsky (1998) *ABB* **353**, 257.

CYP2J5

This mouse cytochrome P450 system is expressed in kidney and liver but not in other tissues.[1] It catalyzes the NADPH- and dioxygen-dependent conversion of arachidonate to 14,15-, 11,12-, and 8,9-epoxyeicosatrienoates as well as 11- and 15-hydroxyeicosatetraenoates. *See also Cytochrome P450*

[1] J. Ma, W. Qu, P. E. Scarborough, K. B. Tomer, C. R. Moomaw, R. Maronpot, L. S. Davis, M. D. Breyer & D. C. Zeldin (1999) *JBC* **274**, 17777.

CYP3A SUBFAMILY

This subfamily of cytochrome P450 systems includes members that function in xenobiotic and steroid metabolism. Included are proteins that exhibit the ability to demethylate tamoxifen, to oxidize nifedipine, and to catalyze the *N*-dealkylation of haloperidol and bromperidol. *See also Cytochrome P450*

Selected entries from *Methods in Enzymology* [vol, page(s)]:
Other molecular properties: induction of gene expression (ethanol, **272**, 396; reporter genes, **272**, 399; species specificity of hepatocyte response, **272**, 389, 395; steroids, **272**, 388, 394, 397); *in vitro* induction in hepatocytes, **206**, 345; *in vivo* induction, **206**, 345; tamoxifen metabolism, **272**, 161

CYP3AI

This cytochrome P450 system catalyzes the oxidation of a variety of drugs and metabolites: for example, the NADPH- and dioxygen-dependent *N*-demethylation of tamoxiphen. *See also Cytochrome P450*

CYP3A3

This cytochrome P450 system catalyzes the oxidation of a variety of drugs and xenometabolites: for example, the NADPH- and dioxygen-dependent oxidation of nifedipine, cyclosporin, aflatoxin B_1, and testosterone. CYP3A3 and CYP3A4 have very similar substrate specificities, although the site of oxidation may vary. *See also Cytochrome P450*

CYP3A4

This cytochrome P450 system catalyzes the oxidation of a variety of drugs and xenometabolites: for example, the NADPH- and dioxygen-dependent conversion of taxol to 3'-hydroxytaxol, the *N*-depropylation of propafenone, the mono-*N*-dealkylation of diisopyramide, and the hydroxylation and demethylation of flunitrazepam. Other substrates include nifedipine, cyclosporin, aflatoxin B_1, and testosterone. *See also Cytochrome P450*

Selected entries from *Methods in Enzymology* [vol, page(s)]:
Assay: taxol hydroxylation assays (fecal samples, **272**, 148; high-performance liquid chromatography, **272**, 147; interference, **272**, 149, 151; microsomes, **272**, 149; plasma samples, **272**, 147; quantitation, **272**, 149)
Other molecular properties: coexpression with NADPH-P450 reductase in baculovirus-insect cell system (cell culture, **272**, 88, 91; heme supplementation, **272**, 88; host cells, **272**, 87, 92; microsomal fraction isolation, **272**, 93; plaque assay, **272**, 90; promoters, **272**, 87; recombinant virus formation, **272**, 87; scaleup, **272**, 94; vectors, **272**, 86; virus amplification, **272**, 91); imipramine metabolism, **272**, 177; propafenone (dealkylation assay, **272**, 102; metabolism, **272**, 100)

CYP3A6

This cytochrome P450 system, also called cytochrome P450$_{3c}$, catalyzes NADPH- and dioxygen-dependent 6β-hydroxylation of progesterone. *See also Cytochrome P450*

Selected entries from *Methods in Enzymology* [vol, page(s)]:
Other molecular properties: steroid hydroxylase activity, inhibition by 22-amino-23,24-bisnor-5-cholen-3β-ol, **206**, 553

CYP4 FAMILY

The CYP4 family of cytochrome P450 systems participates in NADPH- and dioxygen-dependent fatty acid ω and ω − 1 hydroxylations. *See also Alkane 1-Monooxygenase; Alkane 2-Monooxygenase; Cytochrome P450*

Selected entries from *Methods in Enzymology* [vol, page(s)]:
Other molecular properties: arachidonate oxygenase activity (assay, **187**, 386; microsomal [biochemical characterization, **187**, 385; reconstitution, **187**, 392]); gene isolation from cockroach using polymerase chain reaction (amplification, **272**, 309; cloning, **272**, 310; primer design, **272**, 308; sequence analysis, **272**, 310; strategy, **272**, 307; template isolation, **272**, 309; validation of method, **272**, 311)

CYP4A SUBFAMILY

This subfamily of cytochrome P450 systems catalyzes the ω-hydroxylation of a number of fatty acids, including prostaglandins with certain members of the subfamily (*e.g.*, CYP4A7). While other cytochrome P450s can catalyze the hydroxylation of methylene carbons in fatty acids, only CYP4 members preferentially hydroxylate the terminal methyl carbon. *See also Alkane 1-Monooxygenase; Cytochrome P450*

Selected entries from *Methods in Enzymology* [vol, page(s)]:
Other molecular properties: functional analysis, **206**, 273; induction in primary hepatocyte cultures, **206**, 353, 360; induction, *in vivo*, **206**, 353, 356

CYP4A1

This cytochrome P450 system catalyzes the NADPH- and dioxygen-dependent ω-hydroxylation of laurate [EC 1.14.15.3]. The rat enzyme is present in liver and is induced by clofibrate. **See also** *Alkane 1-Monooxygenase*

Selected entries from **Methods in Enzymology** [vol, page(s)]:
Other molecular properties: induction in primary hepatocyte cultures, **206**, 353, 360; induction, *in vivo*, **206**, 353, 356

CYP4A4

This cytochrome P450 system catalyzes the NADPH- and dioxygen-dependent ω-hydroxylation of arachidonate but not laurate [EC 1.14.15.3].[1] In rabbits, this cytochrome P450 is elevated during pregnancy. **See also** *Alkane 1-Monooxygenase; Cytochrome P450*

[1] L. J. Roman, C. N. Palmer, J. E. Clark, A. S. Muerhoff, K. J. Griffin, E. F. Johnson & B. S. Masters (1993) *ABB* **307**, 57.

CYP4A5

This cytochrome P450 system catalyzes the NADPH- and dioxygen-dependent ω-hydroxylation of laurate and exhibits little activity with respect to arachidonate or palmitate [EC 1.14.15.3].[1] **See also** *Alkane 1-Monooxygenase; Cytochrome P450*

[1] L. J. Roman, C. N. Palmer, J. E. Clark, A. S. Muerhoff, K. J. Griffin, E. F. Johnson & B. S. Masters (1993) *ABB* **307**, 57.

CYP4A6

This cytochrome P450 system catalyzes the NADPH- and dioxygen-dependent ω-hydroxylation of laurate and half of that activity with respect to arachidonate [EC 1.14.15.3].[1] CYP4A6 is highly induced by clofibrate. **See also** *Alkane 1-Monooxygenase; Cytochrome P450*

[1] L. J. Roman, C. N. Palmer, J. E. Clark, A. S. Muerhoff, K. J. Griffin, E. F. Johnson & B. S. Masters (1993) *ABB* **307**, 57.

CYP4A7

This cytochrome P450 system catalyzes the NADPH- and dioxygen-dependent ω-hydroxylation of laurate, arachidonate, and prostaglandins (CYP4A7 is the only member of the CYP4A subfamily from rabbit kidney that can act on prostaglandins) [EC 1.14.15.3].[1] **See also** *Alkane 1-Monooxygenase; Cytochrome P450*

[1] L. J. Roman, C. N. Palmer, J. E. Clark, A. S. Muerhoff, K. J. Griffin, E. F. Johnson & B. S. Masters (1993) *ABB* **307**, 57.

Selected entries from **Methods in Enzymology** [vol, page(s)]:
Assay: 187, 260

CYP6D1

This insect cytochrome P450 system catalyzes the NADPH- and dioxygen-dependent hydroxylation of polyaromatic hydrocarbons such as phenanthrene. It is the system that is responsible for resistance to pyrethroid insecticides in the house fly. **See also** *Cytochrome P450*

Selected entries from **Methods in Enzymology** [vol, page(s)]:
Other molecular properties: purification from house fly (anion-exchange chromatography, **272**, 291; buffers, **272**, 287; hydrophobic interaction chromatography, **272**, 290; microsome preparation, **272**, 287; yield, **272**, 292)

CYP14 FAMILY

These cytochrome P450 systems include those proteins that can catalyze the NADPH- and dioxygen-dependent 14α-demethylation of lanosterol. **See also** *Cytochrome P450*

CYP19

This cytochrome P450 system, also known as cytochrome P450$_{arom}$ and aromatase, catalyzes the NADPH- and dioxygen-dependent conversion of testosterone to 17β-estradiol and androstenedione into estrone. The enzyme catalyzes three consecutive hydroxylation reactions converting C19 androgens to aromatic C18 estrogenic steroids. The first two hydroxylations take place on position C19 (a methyl carbon) of the androgens (thus, androstendione is converted first to the 19-hydroxyandrostendione intermediate and then to the 19-aldehyde). The third oxygen activation leads to aromatization of ring A and loss of carbon-19.

These and other conversions play a crucial role in controlling the influence of testosterone on many behavioral and physiological processes (*e.g.*, the activation of male sexual behavior, sexual differentiation in different tissues, and steroid-hormone feedback on the secretion of gonadotropic hormones). **See also** *Cytochrome P450*

Selected entries from **Methods in Enzymology** [vol, page(s)]:
General discussion: 15, 703
Assay: 15, 703; **206**, 477
Other molecular properties: granulosa cell, in bioassay for follicle-stimulating hormone, **168**, 414; inhibition, **206**, 551; properties, **15**, 704; structural alignment with P450s, **272**, 318

CYP52A3

This alkane-inducible cytochrome P450 system, also known as cytochrome P450$_{Alk-1}$ and cytochrome P450$_{Cm1}$, catalyzes the NADPH and dioxygen-dependent conversion of hexadecane to 1-hexadecanol. Further oxidations can occur to produce hexadecanal, hexadecanoate, 1, 16-hexadecanediol, 16-hydroxyhexadecanoate, and 1, 16-hexadecanedioate.[1] *See also Alkane 1-Monooxygenase; Cytochrome P450*

[1]U. Scheller, T. Zimmer, D. Becher, F. Schauer & W. H. Schunck (1998) *JBC* **273**, 32528.

Selected entries from *Methods in Enzymology* [vol, page(s)]:
Other molecular properties: factor Xa recognition site in amino terminus (cytosolic domain purification [detergent removal, **272**, 73; hydrophobic interaction chromatography, **272**, 72]; expression in *Saccharomyces cerevisiae* [culture conditions, **272**, 67, 75; factor Xa recognition site, insertion at membrane anchor region, **272**, 66, 74; strain selection, **272**, 66; vector, **272**, 66]; factor Xa proteolysis [buffer, **272**, 70; comparison to complementary DNA truncation, **272**, 74; cytosolic domain liberation from intact microsomes, **272**, 73; efficiency, **272**, 70, 72; reaction conditions, **272**, 71; specificity, **272**, 72]; purification [hydrophobic interaction chromatography, **272**, 69; hydroxyapatite chromatography, **272**, 70; microsome preparation, **272**, 68; solubilization, **272**, 69])

CYP102

This cytochrome P450 system, also known as cytochrome P450$_{BM-3}$ and isolated from *Bacillus megaterium*, catalyzes the NADPH- and dioxygen-dependent subterminal hydroxylation, preferably, of long-chain fatty acids (containing twelve to twenty carbon atoms: note that short- and medium-chain acids can still be hydroxylated).[1] This enzyme catalyzes the reaction of chlorzoxazone with dioxygen and NADPH to produce 6-hydroxychlorzoxazone, water, and NADP$^+$. It will also catalyze a demethylation of caffeine to produce theophylline and theobromine. Humans, rats, and mice express only one member of the CYP2E subfamily. *See also Cytochrome P450*

[1]G. Truan, M. R. Komandla, J. R. Falck & J. A. Peterson (1999) *ABB* **366**, 192.

Selected entries from *Methods in Enzymology* [vol, page(s)]:
Other molecular properties: structural alignment with P450s (algorithms, **272**, 317; core structure, **272**, 316; C-terminal alignment, **272**, 319, 321, 325; N-terminal alignment, **272**, 323); substrate specificity (active site flexibility, **272**, 350, 355; reaction rate prediction, **272**, 328, 330)

CYP108

This cytochrome P450 system, also known as cytochrome P450$_{terp}$ (since it acts on α-terpineol), catalyzes the NADPH- and dioxygen-dependent hydroxylation of a wide variety of substrates: for example, methyl hydroxylation of 4-methylthioanisole and 4-methylstyrene (note that thioanisole undergoes sulfoxidation and styrene is epoxidized). *See also Cytochrome P450*

Selected entries from *Methods in Enzymology* [vol, page(s)]:
Other molecular properties: crystallization, **272**, 362; structural alignment with P450s (algorithms, **272**, 317; core structure, **272**, 316; C-terminal alignment, **272**, 319, 321, 325; N-terminal alignment, **272**, 323)

CYPRIDINA-LUCIFERIN 2-MONOOXYGENASE

Cypridina luciferin

Cypridina oxyluciferin

This oxidoreductase [EC 1.13.12.6], also known as *Cypridina*-type luciferase, catalyzes the reaction of *Cypridina* luciferin with dioxygen to produce the oxidized *Cypridina* luciferin, carbon dioxide, and light.[1] It was first studied in detail from the marine ostracod crustacean *Cypridina hilgendorfii* (note that not all species of *Cypridina* are luminous: e.g., *C. bairdii* is not. In addition, note that the original crustacean has recently been reclassified as *Vargula hilgendorfii*). *Cypridina* luciferin is [3-[3,7-dihydro-6-(1H-indol-3-yl)-2-[(S)-1-methyl-6-propyl]-3-oxoimidazo-[1,2a]pyrazin-8-yl]propyl]guanidine. The luciferins and the luciferases of a number of luminous fish (for example, *Apogon ellioti*, *Parapriacanthus beryciformes*, *Porichthys porosissimus*) are either identical or closely similar.

[1]F. I. Tsuji, R. V. Lynch & C. L. Stevens (1974) *Biochemistry* **13**, 5204.

Selected entries from *Methods in Enzymology* [vol, page(s)]:
General discussion: 2, 851; 52, 20; 57, 331, 364; 326, 172
Assay: 2, 851; 57, 371; 326, 172, 174
Other molecular properties: cation requirement, 57, 371; content, in living *Cypridina*, 57, 337; diffusion constant, 57, 352, 353; expression, 326, 167; isoelectric point, 57, 352; kinetics, 57, 353, 355, 356; luciferase-oxyluciferin complex, light-emission, 57, 360, 361; luciferin, 57, 338, 366; 326, 166; mechanism, 57, 345; molecular weight, 57, 350,

371; pH optimum, **2**, 853; **57**, 352, 353; properties, **2**, 853; **57**, 352; purification, **2**, 852; **57**, 350, 366, 367, 369; reported gene, **326**, 172; salt concentration, **57**, 350, 351; sedimentation constant, **57**, 352, 353

CYSTATHIONINE β-LYASE

This pyridoxal-phosphate-dependent enzyme [EC 4.4.1.8] (also known as β-cystathionase, cysteine lyase, and cystine lyase) catalyzes the reaction of cystathionine with water to produce L-homocysteine, pyruvate, and ammonia.[1–5] The enzyme from a number of sources will also act on L-cystine (producing pyruvate, ammonia, L-cysteine persulfide), *meso*-lanthionine, and djenkolate. The yeast enzyme may be identical with *S*-alkylcysteine lyase [EC 4.4.1.6].

cystathionine

L-homocysteine   pyruvate

After binding of cystathionine to the enzyme, an external aldimine is generated with the enzyme-bound pyridoxal phosphate. This step is followed by α-proton abstraction, yielding an α-carbanion equivalent stabilized as a ketimine quinonoid intermediate. L-Homocysteine is then eliminated and the pyridoxal phosphate derivative of aminoacrylate undergoes protonation and reverse transaldimination to form iminopropionate and regenerate the lyase-bound coenzyme. Iminopropionate is then hydrolyzed to pyruvate and ammonia presumably after product release.[1,5]

[1] T. Clausen, B. Laber & A. Messerschmidt (1997) *Biol. Chem.* **378**, 321.
[2] C. M. Dwivedi, R. C. Ragin & J. R. Uren (1982) *Biochemistry* **21**, 3064.
[3] N. W. Anderson & J. F. Thompson (1979) *Phytochemistry* **18**, 1953.
[4] J. N. Burnell & F. R. Whatley (1977) *BBA* **481**, 246.
[5] L. Davis & D. E. Metzler (1972) *The Enzymes*, 3rd ed., **7**, 33.

Selected entries from *Methods in Enzymology* [**vol**, page(s)]:
General discussion: 2, 314; **17B**, 439, 450; **143**, 443, 483
Assay: 17B, 439, 451; **143**, 439, 444, 484
Other molecular properties: derepression in *Salmonella typhimurium*, **17B**, 440; inhibition by *N*-ethylmaleimide, **17B**, 451; properties (cabbage enzyme, **143**, 441; *Escherichia coli* enzyme, **143**, 486; *Salmonella typhimurium* enzyme, **17B**, 441; spinach enzyme, **143**, 447); purification from (cabbage, **143**, 440; *Escherichia coli*, **143**, 484; *Neurospora*, **17B**, 452; *Proteus morganii*, **2**, 314; spinach, **143**, 446; *Salmonella typhimurium*, **17B**, 440)

CYSTATHIONINE γ-LYASE

This pyridoxal-phosphate-dependent enzyme [EC 4.4.1.1] (also known as homoserine deaminase, homoserine dehydratase, γ-cystathionase, cystine desulfhydrase, and cysteine desulfhydrase) catalyzes the reaction of cystathionine with water to produce L-cysteine, ammonia, and α-ketobutanoate (or, 2-oxobutanoate).[1–4]

cystathionine

α-ketobutyrate   L-cysteine

The enzyme will also catalyze elimination reactions with L-homoserine (forming water, ammonia, and α-ketobutanoate), L-cystine (producing *S*-thio-L-cysteine, pyruvate, and ammonia), and L-cysteine (producing pyruvate, ammonia, and hydrogen sulfide). Note that the yeast and rat liver enzymes also exhibit a cystathionine β-lyase activity whereas the human enzyme does not.[1] The human enzyme will also not act on L-cystine.[1]

[1] C. Steegborn, T. Clausen, P. Sondermann, U. Jacob, M. Worbs, S. Marinkovic, R. Huber & M. C. Wahl (1999) *JBC* **274**, 12675.
[2] A. E. Braunstein & E. V. Goryachenkova (1984) *AE* **56**, 1.
[3] A. E. Martell (1982) *AE* **53**, 163.
[4] L. Davis & D. E. Metzler (1972) *The Enzymes*, 3rd ed., **7**, 33.

Selected entries from *Methods in Enzymology* [**vol**, page(s)]:
General discussion: 2, 311; **5**, 936; **17B**, 433; **46**, 31, 32, 163, 445; **143**, 486
Assay: 2, 311; **5**, 936; **17B**, 433, 451; **143**, 487
Other molecular properties: borohydride treatment, **5**, 682; derepression in *Neurospora*, **17B**, 435; inactivation by propargylglycine, **252**, 100; mercaptopyruvate transsulfurase, not present in preparation of, **5**, 989; precipitating immunoassay, **121**, 704; properties, **2**, 313; **5**, 941; **17B**, 438; **143**, 491; purification from (*Neurospora*, **17B**, 435; pig liver, **2**, 312; rat liver, **5**, 938; *Streptomyces phaeochromogenes*, **143**, 488)

CYSTATHIONINE β-SYNTHASE

This pyridoxal-phosphate-dependent enzyme [EC 4.2.1.22] (also known as serine sulfhydrase, β-thionase, and methylcysteine synthase) catalyzes the reaction of L-homocysteine with L-serine to produce cystathionine and water.[1–4]

Other β-replacement reactions can be catalyzed using a number of β-substituted L-α-amino acids reacting with a variety of thiol-containing substrates. The rat and human, but not the yeast, enzymes have been reported to be both pyridoxal-phosphate- and heme-containing proteins; however, it is not clear what is the role of the heme group as the yeast and human proteins are very similar.[1,2]

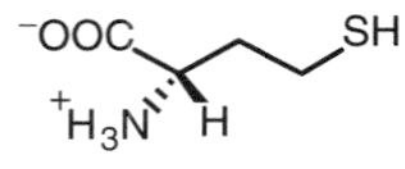

L-homocysteine 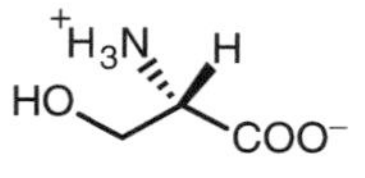   L-serine

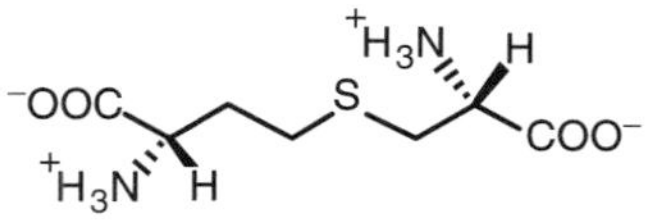

cystathionine

L-Serine binds to the enzyme and reacts with the enzyme-bound pyridoxal phosphate to generate the aldimine of L-serine. Dehydration subsequently occurs to produce the aldimine of aminoacrylate which reacts with the incoming L-homocysteine. Hydrolysis of the aldimine of cystathionine releases the product and regenerates the internal aldimine of the coenzyme.[2]

[1] K. N. Maclean, M. Janosik, J. Oliveriusova, V. Kery & J. P. Kraus (2000) *J. Inorg. Biochem.* **81**, 161.
[2] K. H. Jhee, P. McPhie & E. W. Miles (2000) *JBC* **275**, 11541.
[3] A. E. Braunstein & E. V. Goryachenkova (1984) *AE* **56**, 1.
[4] L. Davis & D. E. Metzler (1972) *The Enzymes*, 3rd ed. **7**, 33.

Selected entries from *Methods in Enzymology* [vol, page(s)]:
General discussion: 17B, 450, 454; 143, 388
Assay: 5, 943; 17B, 454; 143, 389
Other molecular properties: *Neurospora*, in, 17B, 450; properties, 143, 393; purification from (liver, human, 143, 391; liver, rat, 5, 946; 17B, 456); pyridoxal phosphate and, 17B, 458; separation from serine dehydratase, 5, 946; 17B, 350, 458

CYSTEAMINE DEHYDROGENASE

This oxidoreductase, formerly classified as EC 1.8.1.1, is now a deleted Enzyme Commission entry.

CYSTEAMINE DIOXYGENASE

This iron-dependent oxidoreductase [EC 1.13.11.19], also known as persulfurase, catalyzes the reaction of cysteamine with dioxygen to produce hypotaurine.[1-4]

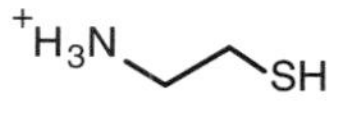

cysteamine 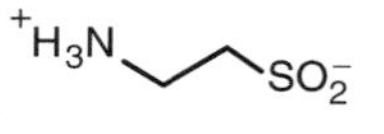   hypotaurine

Artificial substrates include piperazinylcysteamine, *N*-acetylcysteamine, *N,N*-dimethylcysteamine, trimethyl (2-mercaptoethyl)ammonium chloride, 2-mercaptoethanol, and cysteine methyl ester. It has been suggested that

superoxide ions and thiyl radicals may be produced at the active site during catalysis.[1]

[1] G. Ricci, S. Dupre, G. Federici, M. Nardini, G. Spoto & D. Cavallini (1987) *Free Radic. Res. Commun.* **3**, 365.
[2] O. Hayaishi, M. Nozaki & M. T. Abbott (1975) *The Enzymes*, 3rd ed., **12**, 119.
[3] D. Cavallini, G. Federici, G. Ricci, S. Dupre, A. Antonucci & C. De Marco (1975) *FEBS Lett.* **56**, 348.
[4] M. Nozaki (1974) in *Mol. Mech. Oxygen Activ.* (O. Hayaishi, ed.), p. 135.

Selected entries from *Methods in Enzymology* [vol, page(s)]:
General discussion: 17B, 479; 52, 11; 143, 148, 410
Assay: 17B, 479; 143, 411
Other molecular properties: cystamine and cysteamine assay with, 143, 150; properties, 17B, 482; 143, 413; purification from (horse kidney, 17B, 480; porcine liver, 143, 411)

CYSTEINE AMINOTRANSFERASE

This pyridoxal-phosphate-dependent aminotransferase [EC 2.6.1.3] catalyzes the reversible reaction of L-cysteine with α-ketoglutarate (or, 2-oxoglutarate) to produce mercaptopyruvate and L-glutamate.[1]

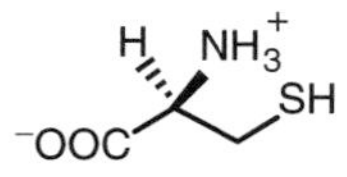

L-cysteine 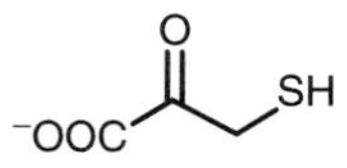   mercaptopyruvate

In a number of organisms, cytosolic cysteine aminotransferase may be identical with cytosolic aspartate aminotransferase [EC 2.6.1.1].

[1] R. J. Thibert & D. E. Schmidt (1977) *Can. J. Biochem.* **55**, 958.

CYSTEINE-*S*-CONJUGATE *N*-ACETYLTRANSFERASE

This transferase [EC 2.3.1.80] catalyzes the reaction of acetyl-CoA with an *S*-substituted L-cysteine to produce coenzyme A and an *S*-substituted *N*-acetyl-L-cysteine.

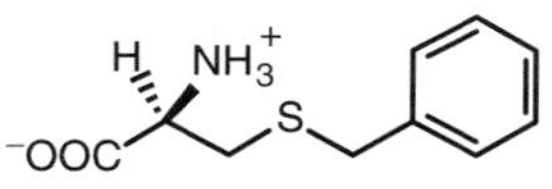

S-benzyl-L-cysteine 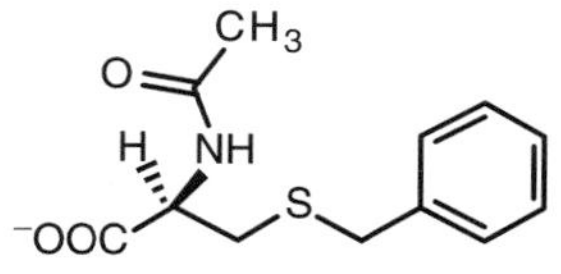   *N*-acetyl-*S*-benzyl-L-cysteine

Acceptor substrates include *S*-benzyl-L-cysteine, *S*-butyl-L-cysteine, *S*-propyl-L-cysteine, *O*-benzyl-L-serine, and *S*-ethyl-L-cysteine. Leukotriene E₄ *N*-acetyltransferase is currently regarded as a variant of this enzyme that is specific for leukotriene E₄ as the acceptor substrate.[1,2]

[1] A. Aigner, M. Jager, R. Pasternack, P. Weber, D. Wienke & S. Wolf (1996) *BJ* **317**, 213.
[2] M. W. Duffel & W. B. Jakoby (1982) *Mol. Pharmacol.* **21**, 444.

Selected entries from *Methods in Enzymology* [vol, page(s)]:
General discussion: 113, 516
Assay: 113, 516

CYSTEINE-CONJUGATE AMINOTRANSFERASE

This pyridoxal-phosphate-dependent aminotransferase [EC 2.6.1.75], also known as cysteine-conjugate transaminase, catalyzes the reversible reaction of S-(4-bromophenyl)-L-cysteine with α-ketoglutarate (or, 2-oxoglutarate) to produce S-(4-bromophenyl)mercaptopyruvate and L-glutamate.

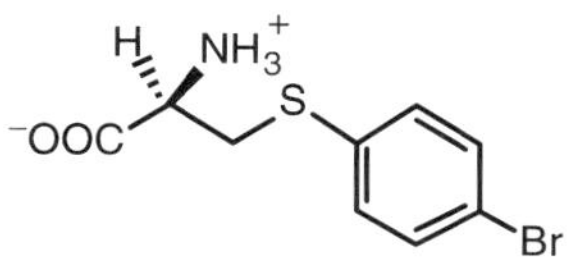

S-(4-bromophenyl)-L-cysteine

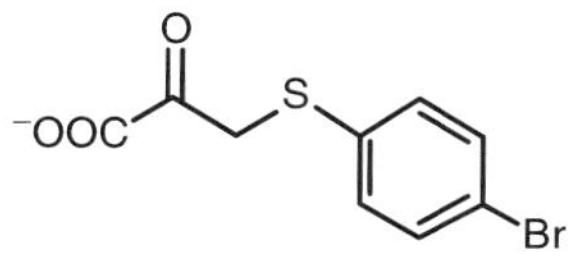

S-(4-bromophenyl)mercaptopyruvate

A number of cysteine conjugates can also act as substrates (*e.g.*, S-phenyl-L-cysteine, S-benzyl-L-cysteine, S-(1-butyl)-L-cysteine, and S-(1-propyl)-L-cysteine). Three isozymes have been identified in rat liver.[1]

[1]H. Tomisawa, N. Ichimoto, Y. Takanohashi, S. Ichihara, H. Fukazawa & M. Tateishi (1988) *Xenobiotica* 18, 1015.

CYSTEINE S-CONJUGATE β-LYASE

This pyridoxal-phosphate-dependent lyase [EC 4.4.1.13] catalyzes the conversion of an S-substituted L-cysteine to RSH, ammonia, and pyruvate.[1]

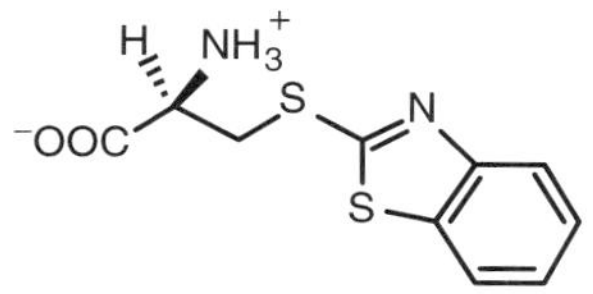

S-(2-benzothiazolyl)-L-cysteine

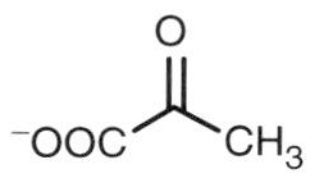

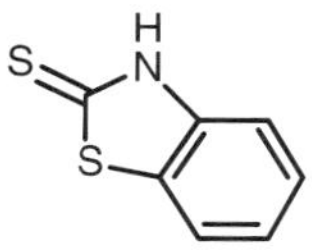

pyruvate   2-mercaptobenzothiazole

Substrates include S-(4-bromophenyl)-L-cysteine, S-(2-benzothiazolyl)-L-cysteine, S-(1,2-dichlorovinyl)-L-cysteine, and S-(2,4-dinitrophenyl)-L-cysteine. **See also** *S-Substituted Cysteine Sulfoxide Lyase*

[1]A. J. L. Cooper (1998) *Adv. Enzymol.* 72, 199.

Selected entries from *Methods in Enzymology* [vol, page(s)]:
General discussion: 77, 253; 113, 510
Assay: 77, 253; 113, 511
Other molecular properties: activators, 77, 256; distribution, 77, 256; inhibitors, 77, 256; pH optima, 77, 256; 113, 515; properties, 77, 256; 113, 514; purification, 77, 254; 113, 512; pyridoxal phosphate requirement, 77, 256; 113, 515; rat liver (properties, 113, 514; purification, 113, 512); substrate specificity, 77, 256; 113, 514

D-CYSTEINE DESULFHYDRASE

This enzyme [EC 4.4.1.15] catalyzes the hydrolysis of D-cysteine to produce hydrogen sulfide, ammonia, and pyruvate.

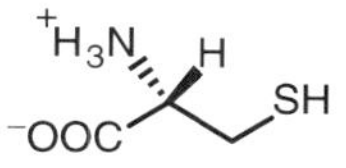

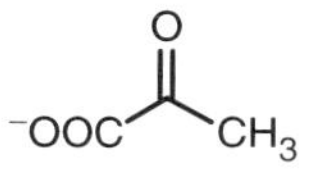

D-cysteine   pyruvate

The *Escherichia coli* enzyme is pyridoxal-phosphate-dependent and can also react 3-chloro-D-alanine with thioglycolate to generate S-carboxymethyl-D-cysteine.[1]

[1]T. Nagasawa, T. Ishii, H. Kumagai & H. Yamada (1985) *EJB* 153, 541.

Selected entries from *Methods in Enzymology* [vol, page(s)]:
General discussion: 143, 449
Assay: 143, 450

CYSTEINE DIOXYGENASE

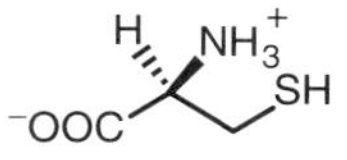

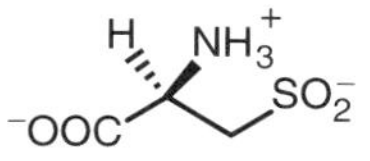

L-cysteine   3-sulfino-L-alanine

This iron- and NAD(P)H-dependent oxidoreductase [EC 1.13.11.20] catalyzes the reaction of L-cysteine with dioxygen to yield 3-sulfino-L-alanine.[1-4]

[1]V. Kumar, B. Maresca, M. Sacco, R. Goewert, G. S. Kobayashi & G. Medoff (1983) *Biochemistry* 22, 762.
[2]K. Yamaguchi, Y. Hosokawa, N. Kohashi, Y. Kori, S. Sakakibara & I. Ueda (1978) *J. Biochem.* 83, 479.
[3]O. Hayaishi, M. Nozaki & M. T. Abbott (1975) *The Enzymes*, 3rd ed. 12, 119.
[4]J. B. Lombardini, T. P. Singer & P. D. Boyer (1969) *JBC* 244, 1172.

Selected entries from *Methods in Enzymology* [vol, page(s)]:
General discussion: 52, 11; 143, 395
Assay: 143, 397, 403

CYSTEINE LYASE

This pyridoxal-phosphate-dependent lyase [EC 4.4.1.10] catalyzes the reaction of L-cysteine with sulfite (SO_3^{2-}) to produce L-cysteate and hydrogen sulfide.[1-3]

L-cysteine   L-cysteate

The enzyme can also catalyze the reaction of two molecules of L-cysteine to producing lanthionine. Other alkyl thiols can act as substrates to react with L-cysteine.

[1]A. E. Braunstein & E. V. Goryachenkova (1984) *AE* **56**, 1.
[2]A. E. Braunstein & E. V. Goryachenkova (1976) *Biochimie* **58**, 5.
[3]L. Davis & D. E. Metzler (1972) *The Enzymes*, 3rd ed., **7**, 33.

CYSTEINE SYNTHASE

This pyridoxal-phosphate-dependent enzyme [EC 4.2.99.8], also known as *O*-acetylserine (thiol)-lyase and *O*-acetylserine sulfhydrylase, catalyzes the reaction of hydrogen sulfide with O^3-acetyl-L-serine to produce L-cysteine and acetate.[1-5]

O^3-acetyl-L-serine   L-cysteine

Other acceptor substrates include alkyl thiols, cyanide, pyrazole, and other heterocyclic compounds. The enzyme is distinct from β-pyrazolylalanine synthase (acetylserine) [EC 4.2.99.14], L-mimosine synthase [EC 4.2.99.15], and uracilylalanine synthase [EC 4.2.99.16]. It also catalyzes the synthesis of L-selenocysteine from *O*-acetyl-L-serine and selenide. ***See also*** *O-Acetylhomoserine (Thiol)-Lyase*

[1]C. H. Tai & P. F. Cook (2000) *AE* **74**, 185.
[2]S. Benci, S. Vaccari, A. Mozzarelli & P. F. Cook (1997) *Biochemistry* **36**, 15419.
[3]C. C. Hwang, E. U. Woehl, D. E. Minter, M. F. Dunn & P. F. Cook (1996) *Biochemistry* **35**, 6358.
[4]C. R. Kuske, L. O. Ticknor, E. Guzman, L. R. Gurley, J. G. Valdez, M. E. Thompson & P. J. Jackson (1994) *JBC* **269**, 6223.
[5]C. H. Tai, S. R. Nalabolu, T. M. Jacobson, D. E. Minter & P. F. Cook (1993) *Biochemistry* **32**, 6433

Selected entries from ***Methods in Enzymology*** [**vol**, page(s)]:
General discussion: **17B**, 459
Assay: **17B**, 460; **143**, 474
Other molecular properties: blue lupine (*Lupinus angustifolius*), in, **17B**, 239; cysteine synthase and, **17B**, 467; properties, **17B**, 463; **143**, 477; purification from (*Bacillus sphaericus*, **143**, 475; *Salmonella typhimurium*, **17B**, 461, 465, 467; yeast, **143**, 480); pyridoxal phosphate and, **17B**, 463, 470; serine transacetylase and, **17B**, 459; spectrum, **17B**, 464

CYSTEINYLGLYCINE DIPEPTIDASE

This manganese-dependent dipeptidase [EC 3.4.13.6], also called Cys-Gly dipeptidase and cysteinylglycinase, catalyzes the hydrolysis of L-cysteinylglycine, producing L-cysteine and glycine. Cobalt and iron divalent cations can substitute for Mn^{2+}.[1,2] The enzyme is present in virtually all tissues other than nervous tissue. It should be pointed out that L-cysteinylglycine can be hydrolyzed by a number of other enzymes, most notably membrane alanyl aminopeptidase (or, aminopeptidase N; EC 3.4.11.2). Evidence has been presented that the dipeptidase in the brush border of the kidney tubule is identical to EC 3.4.11.2.[3] However, cells lacking the aminopeptidase are reported still able to hydrolyze L-cysteinylglycine as well as *S*-adducts.[4] ***See also*** *Membrane Alanyl Aminopeptidase*

[1]C. K. Olson & F. Binkley (1950) *JBC* **186**, 731.
[2]F. Binkley (1952) *Exp. Cell Res.* **2** (suppl.), 145.
[3]B. B. Rankin, T. M. McIntyre & N. P. Curthoys (1980) *BBRC* **96**, 991.
[4]S. S. Tate (1985) *MIE* **113**, 471.

Selected entries from ***Methods in Enzymology*** [**vol**, page(s)]:
Other molecular properties: kidney cells, in, **77**, 144; rat kidney (assays, **113**, 472; properties, **113**, 480; purification, **113**, 475; substrate specificity, **113**, 482)

CYSTEINYL-tRNA SYNTHETASE

This enzyme [EC 6.1.1.16], also known as cysteine:tRNA ligase and cysteine translase, catalyzes the reaction of ATP with L-cysteine and tRNA^{Cys} to produce AMP, pyrophosphate (or, diphosphate), and L-cysteinyl-tRNA^{Cys}. L-Selenocysteine is also a substrate. The enzyme from certain archaea is also a prolyl-tRNA synthetase.[1,2] ***See also*** *Aminoacyl-tRNA Synthetases*

[1]C. Stathopoulos, T. Li, R. Longman, U. C. Vothknecht, H. D. Becker, M. Ibba & D. Soll (2000) *Science* **287**, 479.
[2]Y. Motorin & J. P. Waller (1998) *Biochimie* **80**, 579.

Selected entries from ***Methods in Enzymology*** [**vol**, page(s)]:
Other molecular properties: subcellular distribution, **59**, 233, 234

CYSTINE REDUCTASE (NADH)

This oxidoreductase [EC 1.6.4.1] catalyzes the reaction of NADH with L-cystine to produce NAD^+ and two molecules of L-cysteine.[1,2]

[1]J. E. Carroll, G. W. Kosicki & R. J. Thibert (1970) *BBA* **198**, 601.
[2]A. H. Romano & W. J. Nickerson (1954) *JBC* **208**, 409.

CYSTINYL AMINOPEPTIDASE

This zinc-dependent aminopeptidase [EC 3.4.11.3] (also known as cystyl-aminopeptidase, oxytocinase, vasopressinase, and cystine aminopeptidase) catalyzes the hydrolysis of an N-terminal aminoacyl residue, that is Cys-Xaa-..., in which Cys is a half-cystinyl residue involved in a disulfide loop. Naturally occurring substrates include oxytocin and vasopressin. Interestingly, the hydrolysis of synthetic aminoacyl arylamides such as Leu-NHNap, Arg-NHNap, and Ala-NHNap exceed the rates for the corresponding cystinyl derivative. In fact, all aminoacyl arylamides are hydrolyzed except the aspartyl and glutamyl derivatives.[1,2] Because L-methionine inhibits common leucyl arylamidase activity but not the activity expressed by cystinyl aminopeptidase, this enzyme is often assayed with Leu-NHNap in the presence of 20 mM L-methionine.

[1]H. Sakura, T. Y. Lin, M. Doi, S. Mizutani & Y. Kawashima (1981) *Biochem. Int.* **2**, 173.
[2]S. Mizutani & Y. Tomoda (1992) *Semin. Reprod. Endocrinol.* **10**, 146.

Selected entries from **Methods in Enzymology** [vol, page(s)]:
General discussion: 168, 385
Assay: 168, 385

CYTIDINE DEAMINASE

This deaminase [EC 3.5.4.5], also called cytidine aminohydrolase, catalyzes the hydrolysis of cytidine to produce uridine and ammonia.[1–4]

The *Escherichia coli* enzyme is a dimeric protein containing a single zinc ion required for catalytic activity, in which the Zn^{2+} is coordinated to a histidyl residue and two cysteinyl thiols. The *Bacillus subtilis* and human enzyme are tetrameric with a single zinc ion that is coordinated to three cysteinyl residues.[1] A bound water molecule is activated by the zinc and the resulting hydroxide ion attacks the carbon atom in cytidine to generate the corresponding tetrahedral intermediate. The subsequent proton transfer steps are different than those observed with adenosine deaminase.

[1]D. C. Carlow, C. W. Carter, Jr., N. Mejlhede, J. Neuhard & R. Wolfenden (1999) *Biochemistry* **38**, 12258.
[2]C. W. Carter (1995) *Biochimie* **77**, 92.
[3]A. Vita, T. Cacciamani, P. Natalini, S. Ruggieri, N. Raffaelli & G. Magni (1989) *Comp. Biochem. Physiol. B* **93**, 591.
[4]R. M. Cohen & R. Wolfenden (1971) *JBC* **246**, 7561.

Selected entries from **Methods in Enzymology** [vol, page(s)]:
General discussion:2, 478; **51**, 394, 401, 408
Assay: 2, 478; **51**, 395, 401, 402, 405, 406, 408, 409
Other molecular properties: activity, **51**, 401, 408; binding energy analysis, **308**, 405, 409, 414, 424, 426; crystal growth, **276**, 89, 97; *Escherichia coli*, from, **2**, 479; **51**, 401; effects of deuterium oxide, **51**, 404, 407; genetic polymorphism, **51**, 412; heat sensitivity, **51**, 401; human liver, from, **51**, 405; inhibitors, **51**, 405, 411; kinetic properties, **51**, 400; leukemic mouse spleen, from, **51**, 408; molecular weight, **51**, 400, 404, 411; normal mouse spleen, from, **51**, 411, 412; partial specific volume, **51**, 404; pH effects, **51**, 404, 407, 412; product specificity, **51**, 411; properties, **2**, 479; **51**, 404, 405, 407, 411, 412; purification, **2**, 479; **51**, 397, 406, 409, 410; sedimentation coefficient, **51**, 404; sources, **51**,

408, 412; stability, **51**, 404, 411; Stokes radius, apparent, **51**, 404; substrate specificity, **51**, 404, 407, 411; transition state analogues, **249**, 307 (complexes [characterization, **249**, 294; group contributions and role of solvent water, **249**, 301]; mechanistic probes, as, **249**, 298); yeast, from, **51**, 394

CYTIDYLATE CYCLASE

This enzyme [EC 4.6.1.6] (also known as cytidylyl cyclase, cytidyl cyclase, 3′,5′-cyclic-CMP synthase, and cCMP synthase) catalyzes the conversion of CTP to 3′,5′-cyclic CMP and pyrophosphate (or, diphosphate).[1,2]

[1]R. P. Newton, B. J. Salvage & N. A. Hakeem (1990) *BJ* **265**, 581.
[2]N. Muto, M. Kanoh & I. Yamamoto (1993) *Life Sci.* **52**, 13.

CYTIDYLATE KINASE

This phosphotransferase [EC 2.7.4.14] (also called deoxycytidylate kinase, pyrimidine nucleoside monophosphate kinase, uridylate kinase, and uridylate-cytidylate kinase) catalyzes the reversible reaction of CMP with ATP to form CDP and ADP, as well as similar reactions with dCMP, UMP, and dUMP.[1–5] Interestingly, CTP can substitute for ATP; thus, the enzyme can also catalyze the formation of two molecules of CDP from CTP and CMP.

[1]J. Seagrave & P. Reyes (1987) *ABB* **254**, 518.
[2]E. M. Scott & R. C. Wright (1979) *BBA* **571**, 45.
[3]E. P. Anderson (1973) *The Enzymes*, 3rd ed., **9**, 49.
[4]T. Q. Gravey, F. K. Millar & E. P. Anderson (1973) *BBA* **302**, 38.
[5]L. Noda (1962) *The Enzymes*, 2nd ed., **6**, 139.

Selected entries from **Methods in Enzymology** [vol, page(s)]:
General discussion: 6, 167; **51**, 321, 331
Assay: 6, 167; **51**, 322, 323, 332
Other molecular properties: activators, **6**, 170; **51**, 331; activity, **51**, 321, 322, 331, 332; assay of phosphatidylserine synthase, **71**, 562; bacterial extracts, in, **51**, 332; calf thymus, from, **51**, 332; cation requirements, **51**, 329; inhibitors, **6**, 170; **51**, 331; kinetic properties, **51**, 329, 330, 337; Novikoff ascites cells, from, **51**, 326, 330, 331; pH optimum, **6**, 170; **51**, 336; purification, **6**, 168; **51**, 323, 334; rat liver, from, **51**, 323, 329, 330; reaction mechanism, **51**, 337; regulation, **51**, 331; stability, **51**, 336; substrate specificity, **6**, 170; **51**, 329, 330, 336; sulfhydryl reducing agent requirement, **51**, 330, 331; *Tetrahymena pyriformis*, from, **51**, 331

CYTOCHROME b_6f COMPLEX

This is a multiprotein complex found in chloroplasts containing cytochrome b_6, cytochrome f, and the Rieske protein. **See** *Plastoquinol:Plastocyanin Reductase*

CYTOCHROME b_5 REDUCTASE

This FAD-dependent oxidoreductase [EC 1.6.2.2] catalyzes the reaction of NADH with two molecules of ferricytochrome b_5 to produce NAD^+ and two molecules of ferrocytochrome b_5 (cytochrome b_5 is an electron carrier for many oxygenases).[1-7] The reaction mechanism proceeds in two successive one-electron steps.

[1] T. Guray & E. Arinc (1991) *Int. J. Biochem.* **23**, 1315.
[2] M. Tamura, S. Yoshida, T. Tamura, T. Saitoh & M. Takeshita (1990) *Arch. Biochem. Biophys.* **280**, 313.
[3] R. E. Utecht & D. M. Kurtz, Jr., (1988) *BBA* **953**, 164.
[4] T. Yubisui & M. Takeshita (1982) *J. Biochem.* **91**, 1467.
[5] P. Strittmatter (1963) *The Enzymes*, 2nd. ed., **8**, 113.
[6] P. W. Holloway (1983) *The Enzymes*, 3rd ed., **16**, 63.
[7] C. H. Williams, Jr. (1976) *The Enzymes*, 3rd ed., **13**, 89.

Selected entries from *Methods in Enzymology* [vol, page(s)]:
General discussion: 10, 561; 52, 102, 463; 71, 258
Assay: 10, 561; 52, 103, 207
Other molecular properties: absorption spectrum, 52, 107; activity, 52, 107; alternative names, 52, 464; amidination of, 25, 593, 596; antibody against, purification, 52, 241, 242; binding to liposomes, 52, 209, 210; contaminant, of cytochrome P450$_{LM}$ forms, 52, 117; cytochrome b_5 assay, in, 52, 97; deficiency, methemoglobinemia, 76, 717, 720, 729; dispersion state, 52, 106; distribution, 52, 102; effect of extraction methods on properties, 53, 408, 409 (of protease treatment, 52, 108); erythrocyte, alternative names, 52, 464; extraction from microsomes, 22, 205, 227; ferricyanide reductase activity, 53, 103; flavin content, 52, 107; inhibition, by phosphate anions, 52, 244; lipid peroxidation, 52, 305; membrane binding, 52, 108; microsomal fractions, in, 22, 205; 52, 88; microsomal, purification, 74, 264; molecular weight, 52, 106; outer membrane, in, 10, 448; properties, 10, 564; purification, 10, 562; 52, 104, 105 (partial purification, 52, 99); reactivity, 52, 101; requirement for regeneration of stearyl-CoA desaturase activity, 35, 258, 259; stability, 52, 107; stearyl-CoA desaturase activity assay, in, 52, 189; stereospecificity, 10, 459; 87, 118; storage, 52, 105

[CYTOCHROME-*c*]-ARGININE *N*-METHYLTRANSFERASE

This methyltransferase [EC 2.1.1.124] catalyzes the reaction of *S*-adenosyl-L-methionine with a single L-arginyl residue of cytochrome *c* to produce *S*-adenosyl-L-homocysteine and an N^ω-methyl-L-arginyl residue in cytochrome *c*.[1]

[1] J. Z. Farooqui, M. Tuck & W. K. Paik (1985) *JBC* **260**, 537.

CYTOCHROME c_3 HYDROGENASE

This iron-dependent oxidoreductase [EC 1.12.2.1], also called hydrogenase, catalyzes the reversible reaction of molecular hydrogen with two molecules of ferricytochrome c_3 to produce two protons and two molecules of ferrocytochrome c_3.[1-3] Methylene blue and other acceptors can be reduced with H_2 with this enzyme, the H_2 undergoing heterolytic cleavage. The enzyme from many sources also contains nickel ions. ***See also Hydrogenase; Hydrogen Dehydrogenase***

[1] M. A. Halcrow (1998) *CBC* **3**, 359.
[2] M. W. Adams (1990) *BBA* **1020**, 115.
[3] I. Okura, K. Nakamura & S. Nakamura (1981) *J. Inorg. Biochem.* **14**, 15.

[CYTOCHROME-*c*]-LYSINE *N*-METHYLTRANSFERASE

This methyltransferase [EC 2.1.1.59] catalyzes the reaction of *S*-adenosyl-L-methionine with [cytochrome *c*]-L-lysine (that is, a lysyl residue in cytochrome *c*) to produce *S*-adenosyl-L-homocysteine and cytochrome *c* containing an N^6-methyl-L-lysyl residue.[1] The enzyme has a hybrid ping pong kinetic mechanism.[2]

[1] W. K. Paik, Y. B. Cho, B. Frost & S. Kim (1989) *Biochem. Cell Biol.* **67**, 602.
[2] E. Durban, S. Kim, G. J. Jun & W. K. Paik (1983) *Korean J. Biochem.* **15**, 19.

Selected entries from *Methods in Enzymology* [vol, page(s)]:
General discussion: 106, 274

[CYTOCHROME-*c*]-METHIONINE *S*-METHYLTRANSFERASE

This methyltransferase [EC 2.1.1.123] catalyzes the reaction of *S*-adenosyl-Lmethionine with an L-methionyl residue in cytochrome *c* to produce *S*-adenosyl-L-homocysteine and an *S*-methyl-L-methionyl residue in the cytochrome *c*.[1]

[1] J. Z. Farooqui, M. Tuck & W. K. Paik (1985) *JBC* **260**, 537.

CYTOCHROME *c* OXIDASE

This heme- and copper-dependent oxidoreductase [EC 1.9.3.1] (also known as cytochrome aa_3, cytochrome oxidase, and complex IV) catalyzes the reaction of four molecules of ferrocytochrome *c* with dioxygen to produce four molecules of ferricytochrome *c* and two water molecules.[1-7]

This enzyme-catalyzed reaction is the final step in mitochondrial electron transport. Four protons are translocated across the inner mitochondrial membrane with the reduction of dioxygen. There appear to be two distinct channels to facilitate proton translocation within cytochrome *c* oxidase and they are important at different parts of the catalytic cycle. A number of analogous oxidoreductases have been reported in many microorganisms.

An electron is transferred from cytochrome *c* to the Cu_A site, which contains two copper ions. The electron is then transferred to the heme *a* site and then to the heme a_3-Cu_B couple. The active site for the reduction of dioxygen is at this second site. The iron-peroxy adduct leads to the formation of a ferryl Fe^{4+}=O and $(H_2O)Cu^{2+}$ intermediate. Addition of electrons and protons results in the release of water and the regeneration of Fe^{3+} and Cu^{2+} at the cytochrome a_3 site. **See also** *Pseudomonas Cytochrome Oxidase*

[1] R. J. Maier (1996) *Adv. Protein Chem.* **48**, 35.

[2] S. M. Musser, M. H. B. Stowell & S. I. Chan (1995) *AE* **71**, 79.

[3] E. T. Adman (1991) *Adv. Protein Chem.* **42**, 145.

[4] A. Azzi & M. Müller (1990) *ABB* **280**, 242.

[5] M. Wikström, K. Krab & M. Saraste (1981) *Ann. Rev. Biochem.* **50**, 623.

[6] M. Wikström, K. Krab & M. Saraste (1981) *Cytochrome Oxidase: A Synthesis*, Academic Press, New York.

[7] W. S. Caughey, W. J. Wallace, J. A. Volpe & S. Yoshikawa (1976) *The Enzymes*, 3rd ed., **13**, 299.

Selected entries from **Methods in Enzymology** [vol, page(s)]:

General discussion: 16, 180, 224; 44, 371, 372, 545; 46, 86; 52, 15; 53, 40, 54-79, 155; 126, 13, 22, 32, 45, 64, 72, 78, 123, 138, 145, 153, 159; 148, 491; 167, 437; 280, 97, 117

Assay: 10, 245, 332; 32, 398, 400; 53, 45, 73, 74; 56, 101, 102; chicken cardiac mitochondria, 125, 18; content in yeast, 260, 152; gel electrophoresis, 260, 100; heme content, 260, 100, 119; *in situ*, 260, 114; mitochondrial, 54, 492, 493; polarographic assay, 260, 101, 121; protein determination, 260, 99; sea urchin embryo, 134, 215; spectrophotometric assay, 260, 101, 121; 264, 495; staining, 260, 101, 113; thermophilic bacterium PS3, 126, 131, 146

Other molecular properties: abnormal composition, 56, 565; absorbance coefficients, 53, 127; absorption spectra, 56, 693; activity, 10, 246; 53, 40, 54, 73; activity staining, 264, 514, 520; amino acid composition, 53, 71, 73; antibody, 56, 227 (affinity chromatography, 56, 690, 700; conjugation to ferritin, 56, 710, 711; specificity, 56, 690); assembly, inhibition, 56, 55; associated proton-motive force, measurement, 221, 405; binding of CO, IR spectroscopic studies, 226, 273; bovine heart, 53, 54 (activation and dispersion with alkyl glycoside detergents, 125, 33; cross-reaction with rat liver enzyme, 56, 706, 707; crystal preparation, 126, 22, 23; gross structure, 126, 24; high-affinity site for cytochrome c, structure, 126, 30; H⁺ pumping activity, measurement, 126, 15, 20; monomeric and dimeric forms [centrifugation, 126, 53, 61; detergent binding assay, 126, 51; gel filtration, 126, 47; sedimentation equilibrium measurements, 126, 56; sedimentation velocity measurements, 126, 57]; polypeptide components, topography, 126, 25; prosthetic groups, locus, 126, 29; purification, 126, 64; reaction with N,N'-dicyclohexylcarbodiimide, 125, 98; reconstitution, 126, 14, 78; redox Bohr effects, measurement, 126, 338; subcomplex splitting, 126, 73; subunit III depletion, 126, 74); carbohydrate, 53, 76; carbon monoxide binding, IR spectroscopic studies, 226, 273; carbon monoxide, reaction with, 4, 273; centrifugation of heart microsomes, in, 5, 64; chicken cardiac mitochondria, assay, 125, 18; choline dehydrogenase and, 5, 570; complex IV, 56, 580, 583, 584; complex V, 56, 580; components, 56, 12, 40, 580, 583, 584; composition, 53, 44, 64, 76; concentration determination, 53, 127; conformational composition, 54, 269, 271; coordination geometries of metals, 246, 491; copper of, 10, 250; coupling to Sepharose 4B, 56, 698, 699; cross-linking, 56, 637, 641; cryoenzymology, 63, 338, 368; Cu K-edge spectra, 54, 335, 336; Cu K-edge transitions, 54, 339; cyanide inhibition, 4, 277; 10, 48; cytochrome a, 732; 4, 277 (amytal and, 4, 280, 304; anaerobiosis and, 4, 286; antimycin A and, 4, 304; assay of, 2, 733; bacterial cells and, 4, 302, 303; complex IV, in, 53, 44; content of membrane fractions, 10, 263, 440; cytochrome a_3 and, 2, 733; cytochrome c oxidase and, 2, 739; derivative spectrum, 24, 20; determination, 10, 489; electron transport particles and, 6, 421; enzyme sequence and, 4, 303; extinction coefficient, 2, 733; 10, 256, 490; 53, 71; heart, 2, 737, 739; hematin peptide and, 2, 169; hemochromogen of, 10, 493; kidney cells, in, 77, 142; ligand binding, 53, 191, 196, 197; midpoint redox potential, 53, 197; *Nitrosomonas europaea*, of, 53, 207; pentomeric form of, 10, 670; photosynthetic bacteria, of, properties, 23, 348, 349; prosthetic group, 53, 203; redox titration, 54, 430; resonance Raman studies, 54, 245; sarcosomes and, 4, 300; yeast and, 4, 296); cytochrome

a_3, 4, 311; 52, 15 (amytal and, 4, 280, 304; anaerobiosis and, 4, 286; antimycin A and, 4, 304; assay, 2, 735; azide binding, 53, 197; *Bacillus subtilis*, from, 53, 207; bacterial, 2, 734, 735; bacterial hemochromogen and, 2, 699; blocked, 54, 76; carbon monoxide binding, 2, 734; 4, 273; 53, 197; complex IV, in, 53, 44; cyanide binding, 53, 192, 193; cytochrome b and, 2, 740; cytochrome c and, 2, 732, 750; distribution, 2, 732; enzyme sequence and, 4, 303; extinction coefficients for, 2, 734; flash photolysis experiments, in, 54, 101; fluoride binding, 53, 200, 201; heart, 2, 734, 737; ligand binding, 53, 191; 54, 531, 532; MCD studies, 54, 296, 301; midpoint redox potential, 53, 197; muscle, 4, 299; *Mycobacterium phlei*, from, 53, 207, 208; nitrosyl binding, 53, 199, 200; oxygen and, 2, 732; prosthetic group, 52, 5; reaction kinetics, 4, 306; redox titrations, 54, 429; sarcosomes and, 4, 300; terminal respiration and, 2, 732; yeast and, 4, 296); cytochrome c complex, dissociation (hydrostatic pressure effects, 259, 414; osmotic pressure effects, 259, 414); cytochrome photooxidase, 6, 405 (assay, 6, 405; properties, 6, 409; purification, chloroplasts, 6, 407); deficiency and disease, 260, 118; determination of cytochrome c activity, 53, 158; determination of nitric oxide concentration, 269, 5; diagnostic enzyme, as, 31, 334, 337, 338, 361, 393; distribution on inner membrane matrix particle, 56, 686 (inverted inner membrane vesicles, on, 56, 689); effects of azide, nitrite, and nitrate, 269, 319; electron microscopy (activity staining, 264, 541, 544; immunostaining [antibody visualization, 264, 550; antigen localization, 264, 549; fixation, 264, 546, 551; immunolabeling, 264, 551; resin embedding and polymerization, 264, 546, 549]); electron spin resonance and, 6, 914; 49, 405; EPR signal properties, 54, 135, 136; 76, 321; 246, 552; error in protein determination, 53, 4; extinction coefficient of, 10, 332; 53, 71; extraction from mitochondria, 22, 220, 222, 227, 228, 229; ferrocytochrome a_3, complex with nitric oxide, 269, 5 (percentage of molecules containing, 269, 11); flash photolysis experiments, in, 54, 100, 102, 105; fluorescent immunostaining, 264, 516, 521; free subunit, labeling kinetics, 56, 54; freeze-quench experiments, in, 54, 92; gel electrophoresis of, 10, 679; heart, 1, 29, 36; 5, 65; heart EP, of, 10, 258; HeLa cell, 56, 78; heme a in, 10, 250, 335; 53, 44; heme absorbance spectra, 260, 119; heme stoichiometries, 54, 145; hemopocket relaxation, nanosecond time-resolved resonance Raman spectroscopy, 226, 427; hepatoma mitochondria, 55, 88; histochemical method for, 4, 382; immunoprecipitation of human enzyme (antibody incubation, 260, 208; antisera preparation, 260, 206; controls, 260, 210; gel electrophoresis, 260, 208; immunoadsorbent preparation, 260, 207; mitochondrial lysis, 260, 208; protein radiolabeling, 260, 205); indirect coulometric titration, 246, 715; indirect electrochemical studies, 227, 510; inhibitors, 10, 250; 53, 44, 45; inner membrane, 55, 101, 103; interference in (cytochrome P-450 difference spectrum measurements, 52, 213; α-ketoglutarate dehydrogenase determination, 13, 53); interaction with (azide, 53, 194, 196, 197; carbon monoxide, 53, 194, 197; cyanide, 53, 192; fluoride, 53, 194, 200, 201; formate, 53, 201; hydrogen sulfide, 53, 194, 201; hydroxylamine, 53, 199; isonitriles, 53, 201); interference in superoxide dismutase assay, 53, 385; intestinal mucosa, in, 32, 667; IR studies, 54, 309, 310, 312; isolated nuclei and, 12A, 432; isolation of, 6, 422; Keilin-Hartree preparation, 55, 126; kinetic properties, 53, 44; kinetic studies, using monospecific antibodies, 74, 260; labeled immunoglobulin, 56, 228; D-lactate:cytochrome c reductase, in assay of, 9, 303; ligand, 53, 191 (binding studies, 54, 531, 532; inhibition mechanism, 53, 193); lipid of, 10, 250; liver-cell fractions, in, 31, 97, 208, 209, 262; localization, 56, 707; *Locusta migratoria*, from, 53, 66, 72, 73; luciferase-mediated oxygen assay, 57, 226; lung lamellar bodies, in, 31, 423; magnetic circular dichroism, 49, 170; 54, 295, 301; mammalian, 4, 473 (activity in various tissues, 260, 121, 123; extraction from tissue, 260, 118; immunoaffinity chromatography, 260, 132; reduction with sodium dithionite, 260, 122; substrate specificity, 260, 122; subunits [composition, 260, 123; gel electrophoresis, 260, 123; isoforms, 260, 123; monoclonal antibody, species cross-reaction, 260, 127; sequencing, 260, 125; switching and fetal development, 260, 126; Western blot analysis, 260, 129]); manganese dioxide and, 4, 334; marker enzyme, as, 31, 735, 739; 32, 15,

393; **55**, 5; **182**, 214; mechanism, **246**, 491; membrane potential, **55**, 595; mit⁻ mutants, **56**, 15; mitochondrial, **1**, 29; **31**, 23, 24, 52, 326, 327 (assay, **54**, 492, 493; marker, assay, **182**, 214; properties, **148**, 501; purification from sweet potato, **148**, 497; tissue-specific isozymes [amino-terminal sequence analysis, **126**, 44; carboxylic groups, reactivity, **126**, 39; immunological characterization, **126**, 41; isolation, **126**, 34; properties, **126**, 36; subunit isolation, **126**, 43]); mitochondrial DNA depletion assay in rho⁰ mutants, **264**, 301; mitochondrial mRNA, **56**, 10, 11; mitochondrial subfractions, **55**, 94, 104; monomeric complex, **56**, 635; mutants, **56**, 105 (polypeptides, screening, **56**, 602, 603); NADH-cytochrome c reductase and, **6**, 410, 413; nanosecond magnetic CD spectroscopy, **226**, 176; nanosecond time-resolved resonance Raman spectroscopy, **226**, 425; Neurospora crassa, from, **53**, 66; **55**, 26, 148; nitric oxide interactions, **269**, 3, 151; nuclear mutations, **56**, 13; organ absorbance spectrophotometry, **52**, 56; oxidase reaction, kinetic constants, determination, **53**, 46, 47; oxidation kinetics, **54**, 108; oxidation of o-aminophenols by, **17A**, 557; oxidation, proton release, **55**, 624; oxidation-reduction midpoint potentials, **53**, 65, 66; oxy form, spectrum, **52**, 36; oxygen affinity, **57**, 223; oxygen complex, **54**, 282; oxygen intermediates (detection, methodological approach, **105**, 32; general features, **105**, 34); Paracoccus denitrificans, of, **53**, 207 (purification, **126**, 156); partial volume, **26**, 118; phospholipid-deficient, preparation, **53**, 60, 61; phospholipid-sufficient, preparation, **53**, 61; phosphorylation with, **10**, 38; planar membranes containing, **55**, 755, 756; plasma membrane, **31**, 90, 100, 148, 249; polypeptides, **56**, 592, 637 (molecular weights, **56**, 595, 606); Polytoma mirum, from, **53**, 66, 72, 73; potassium cyanide inhibition, **264**, 477, 500; preparation, **53**, 3, 4, 6, 40; **269**, 4 ; properties, **10**, 249, 334; **56**, 693, 694; **105**, 31; protein rotation studies, in, **54**, 58; proteoliposomes, **55**, 761, 762, 765, 771; Pseudomonas fluorescens and P. quercito-pyrogallica, not in, **9**, 97; purification, **10**, 246, 333; **53**, 55, 74 (chromatographic, **53**, 67, 75); purity, **53**, 76; purity index of, **10**, 332, 334; Rattus norwegicus, from, **53**, 66, 72, 73; reaction mechanism, **52**, 39, 40; reconstituted, H⁺ and charge stoichiometry, **172**, 154; reconstitution, **55**, 699, 703, 706 (into liposomes, **221**, 396); redox centers, **53**, 126, 127, 191; reductive titration, **54**, 124; resonance Raman spectroscopy, **54**, 240; Rhodopseudomonas sphaeroides aa₃-type (characterization, **126**, 141; purification, **126**, 138); Saccharomyces cerevisiae, from, **53**, 73 (essential subunits, **260**, 98; genes, **260**, 98 [assembly-related genes, **260**, 111; mutagenesis, **260**, 113; regulatory genes, **260**, 111; structural genes, **260**, 111]; holoenzyme purification [ammonium sulfate fractionation, **260**, 103; detergent exchange chromatography, **260**, 104; microisolation, **260**, 106; submitochondrial particle preparation, **260**, 102, 105]; infrared spectroscopy [carbon monoxide ligand, **260**, 399, 401, 405; enzyme preparation, **260**, 400; isozyme analysis, **260**, 403, 405; nitric oxide ligand, **260**, 399, 401, 405; spectra acquisition, **260**, 401]; mutants [nondenaturing gel electrophoresis, **260**, 115; screening, **260**, 112; turnover number, in situ determination, **260**, 114; Western blotting, **260**, 114]; subunit purification [gel filtration, **260**, 109; holoenzyme fractionation, **260**, 107, 110; reversed-phase HPLC, **260**, 107; sequencing, **260**, 125]); sea urchin embryo, assay, **134**, 215; sensitive to nitric oxide, **269**, 151; smooth muscle mitochondria, of, **55**, 62; solubilization, **1**, 29; Soret CD spectrum, **54**, 260, 262; spectrally invisible redox components, **54**, 398; spectral properties, **10**, 250, 335; **53**, 65, 66, 69, 72, 126, 127; spectrophotometric assay, **264**, 495; stability, **10**, 249; **53**, 63, 76; storage, **53**, 56; structure, **246**, 491 (structure in humans, **264**, 514, 553); submitochondrial particles, **55**, 109; subunit II (gene, COX2, deletion repair by biolistic transformation, **264**, 271; immunohistochemistry, **264**, 227; polyacrylamide gel electrophoresis, **264**, 208); subunit III-deficient mutants, as source for mitochondrial ATPase subunit 6, **97**, 304; subunits (assembly, **260**, 118; composition, **53**, 54, 70, 76, 77; labeling kinetics, **56**, 52; molecular weights, **53**, 70, 72, 73; properties, **53**, 79; purification, **53**, 76; resolution, **56**, 602; structure, **260**, 97, 99, 117, 123, 200); sulfide inhibition, **10**, 49; temperature-jump relaxation kinetics, **54**, 76, 77; temperature-sensitive mutants, **56**, 139; thermophilic bacterium PS3

(assays, **126**, 131, 146; properties, **126**, 149; purification, **126**, 146; reconstitution, **126**, 124, 151); time-resolved resonance Raman spectroscopy (difference spectra, **246**, 493; intermediates in oxygen binding, **246**, 493; isotopic substitution, **246**, 493, 498); turnover number, **53**, 64, 71; **57**, 223; turnover rate, **10**, 332; vesicle (activity, **55**, 703, 705, 709; preparation, **55**, 749); viscosity barrier centrifugation, **55**, 133, 134; visible CD spectra, **54**, 279; yeast (components, **56**, 41; isolation, **56**, 44, 45, 49, 50); Xenopus muelleri, from, **53**, 66, 72, 73

CYTOCHROME c PEROXIDASE

This heme-dependent oxidoreductase [EC 1.11.1.5] catalyzes the reaction of two molecules of ferrocytochrome c with hydrogen peroxide to produce two molecules of ferricytochrome c and two water molecules.[1–6] Other cytochrome substrates include cytochrome c-551 and cytochrome c_4.

Hydrogen peroxide first reacts with the heme group to produce a stable oxy-ferryl ($Fe^{IV}=O$) heme and an indolyl cation radical at a tryptophanyl residue. The two oxidized sites are then reduced by two molecules of ferrocytochrome c to regenerate the ferric enzyme.[1,2]

[1] H. B. Dunford (1998) CBC **3**, 195.
[2] M. A. Miller (1996) Biochemistry **35**, 15791.
[3] J. S. Zhou & B. M. Hoffman (1994) Science **265**, 1693.
[4] H. Pelletier & J. Kraut (1992) Science **258**, 1748.
[5] T. E. Meyer & M. D. Kamen (1982) Adv. Protein Chem. **35**, 105.
[6] T. Yonetani (1976) The Enzymes, 3rd ed., **13**, 345.

Selected entries from **Methods in Enzymology** [vol, page(s)]:
General discussion: 2, 761; **53**, 156
Assay: 2, 761
Other molecular properties: assay for hydrogen peroxide production by mitochondria, in, **105**, 433; crystal intensity, multiwire area detector diffractometry, **114**, 470; crystallization, **22**, 262; cytochrome a₃ assay and, **2**, 737; direct electrochemical studies, **227**, 516; distribution, **2**, 764; formation, **1**, 136; interference in superoxide dismutase assay, **53**, 385; magnetic susceptibility measurements, **26**, 695, 698; oxidation state of Fe, **76**, 315; properties, **2**, 763; purification, Pseudomonas fluorescens, **2**, 762; redox potentiometry, in, **54**, 419; resonance Raman studies, **226**, 362; site-directed mutants, nonadditivity effects, **249**, 118; two-dimensional NMR spectra, **239**, 501

CYTOCHROME P450

Cytochrome P450, formerly called cytochrome m, is a superfamily of heme-dependent monooxygenases that utilize dioxygen and NADPH.[1–10] These enzymes (b-type cytochromes in which a sulfur of a cysteinyl residue is ligated to the iron: hence, often called heme-thiolate proteins) are localized in the endoplasmic reticulum, and are often used as a marker for microsomal fractions obtained upon homogenization of cells. Especially abundant in liver, cytochrome P450 enzymes play a major role in detoxification. They also catalyze the first step in ω-oxidation of medium- and long-chain fatty acids. In general, the reaction catalyzed is RH with dioxygen and the reduced flavoprotein to produce ROH, water, and the oxidized flavoprotein. *See specific CYP*

[1]H. B. Dunford (1998) *CBC* **3**, 195.
[2]Y. Watanabe & J. T. Groves (1992) *The Enzymes*, 3rd ed., **20**, 405.
[3]F. P. Guengerich (1991) *JBC* **266**, 10019.
[4]D. C. Rees & D. Farrelly (1990) *The Enzymes*, 3rd ed., **19**, 37.
[5]M. A. Ator & P. R. Ortez de Montellano (1990) *The Enzymes*, 3rd ed., **19**, 213.
[6]S. D. Black & M. J. Coon (1987) *AE* **60**, 35.
[7]P. Ortiz de Montellano, ed., (1986) *Cytochrome P-450: Structure, Mechanism, and Biochemistry*, Plenum, New York.
[8]M. J. Coon & D. R. Koop (1983) *The Enzymes*, 3rd ed., **16**, 645.
[9]I. C. Gunsalus & S. G. Sligar (1978) *AE* **47**, 1.
[10]V. Ullrich & W. Duppel (1975) *The Enzymes*, 3rd ed., **12**, 253.

CYTOCHROME P450LM

This is a set made up of a number of cytochrome P450 systems that catalyze the NADPH- and dioxygen-dependent oxidation of a number of xenobiotics (for example, the hydroxylation of cyclohexane). *See specific enzyme*

Selected entries from *Methods in Enzymology* [vol, page(s)]:
Other molecular properties: carbon monoxide binding, IR studies, **54**, 310; multiple forms, **52**, 203

CYTOCHROME P450NMa

Cytochrome P450NMa, first identified in rabbit nasal microsomes, is a mixture of CYP2A10 and CYP2A11 (the descriptor NMa refers to "nasal microsomal P450 form a").[1] *See CYP2A10; CYP2A11*

[1]X. Ding, H. M. Peng & M. J. Coon (1994) *BBRC* **203**, 373.

Selected entries from *Methods in Enzymology* [vol, page(s)]:
Other molecular properties: properties, **206**, 610; purification from rabbit nasal microsomes, **206**, 603; separation from cytochrome P450NMb, **206**, 605

CYTOCHROME REDUCTASE (NADPH)

This oxidoreductase, formerly classified as EC 1.6.2.3, is now a deleted Enzyme Commission entry.

β-(9-CYTOKININ)-ALANINE SYNTHASE

This enzyme [EC 4.2.99.13], also known as lupinate synthase, catalyzes the reaction of *O*-acetyl-L-serine with zeatin (*i.e.*, N^6-(4-hydroxy-3-methyl-but-*trans*-2-enylamino)purine) to produce lupinate and acetate.[1,2] A number of other N^6-substituted purines can function as acceptor substrates.

[1]C. W. Parker, B. Entsch & D. S. Letham (1986) *Phytochemistry* **25**, 303.
[2]B. Entsch, C. W. Parker & D. S. Letham (1983) *Phytochemistry* **22**, 375.

CYTOKININ 7β-GLUCOSYLTRANSFERASE

This transferase [EC 2.4.1.118], also known as UDP-glucose:zeatin 7-glucosyltransferase, catalyzes the reaction of UDP-D-glucose with an N^6-alkylaminopurine to produce UDP and an N^6-alkylaminopurine-7-β-D-glucoside.[1,2] The enzyme can act on a number of N^6-substituted adenine derivatives (*e.g.*, zeatin and N^6-benzylaminopurine): however, N^6-benzyladenine does not appear to be a substrate. Depending on the nature of the substrate and the source of the enzyme, a 9-β-D-glucoside may be the product formed.

[1]B. Entsch & D. S. Letham (1979) *Plant Sci. Lett.* **14**, 205.
[2]B. Entsch, C. W. Parker, D. S. Letham & R. E. Summons (1979) *BBA* **570**, 124.

CYTOKININ OXIDASE

This FAD-dependent oxidoreductase [EC 1.4.3.18] catalyzes the reaction of N^6-(3-methylbut-2-enyl)adenine with water and dioxygen to produce adenine, 3-methylbut-2-enal, and hydrogen peroxide.[1,2] Other isoprenylated adenines can act as substrates.

[1]N. Houba-Hérin, C. Pether, J. d'Alayer & M. Laloue (1999) *Plant J.* **17**, 615.
[2]R. O. Morris, K. D. Bilyeu, J. G. Laskey & N. N. Cheikh (1999) *BBRC* **255**, 328.

CYTOMEGALOVIRUS ASSEMBLIN

This serine protease [EC 3.4.21.97], from the cytomegalovirus of the Herpesviridae family, catalyzes the cleavage of the assembly protein precursor. This activity is required in packaging of the viral genome during capsid maturation and dissolution of the capsid scaffold.[1,2]

[1]W. Gibson, A. R. Welch & M. R. T. Hall (1994) *Perspect. Drug Discovery Design* **2**, 413.
[2]M. C. Smith, J. Giordano, J. A. Cook, M. Wakulchik, E. C. Villarreal, G. W. Becker, K. Bemis, J. Labus & J. S. Manetta (1994) *MIE* **244**, 412.

Selected entries from *Methods in Enzymology* [vol, page(s)]:
General discussion: **244**, 412.

CYTOSINE DEAMINASE

This deaminase [EC 3.5.4.1], also known as cytosine aminohydrolase, catalyzes the hydrolysis of cytosine to produce uracil and ammonia.[1,2] Other substrates include 5-methylcytosine and the 5-halocytosines.

[1]T. Yu, T. Sakai & S. Omata (1976) *Agric. Biol. Chem.* **40**, 543 and 551.
[2]P. L. Ipata, F. Marmocchi, G. Magni, R. Felicioli & G. Polidoro (1971) *Biochemistry* **10**, 4270.

Selected entries from *Methods in Enzymology* [vol, page(s)]:
General discussion: 51, 394
Assay: 51, 395

CYTOSOL ALANYL AMINOPEPTIDASE

This metallopeptidase [EC 3.4.11.14] (also known as arylamidase, puromycin-sensitive aminopeptidase, and cytosol aminopeptidase III) catalyzes the release of an N-terminal amino acid from a wide range of peptides, amides, and arylamides.[1,2]

[1] P. M. Dando, N. E. Young & A. J. Barrett (1997) in *Proteolysis in Cell Functions: Proceedings of the 11th Intern. Conf. on Proteolysis and Protein Turnover* (V. K. Hopsu-Havu, ed.), IOS Press, Amsterdam, p. 88.
[2] D. B. Constam, A. R. Tobler, A. Rensing-Ehl, I. Kemler, L. B. Hersh & A. Fontana (1995) *JBC* **270**, 26931.

Selected entries from *Methods in Enzymology* [vol, page(s)]:
Assay: aminoacyl-*p*-nitroanilide substrate, with, **45**, 498; amino-β-naphthylamide substrate, with, **45**, 496, 498; fluorimetric, **248**, 602
Other molecular properties: action, **45**, 495; commercial, **248**, 533; composition, **45**, 503; human liver aminopeptidase (action, **45**, 495; assay, aminoacyl-*p*-nitroanilide substrate, with, **45**, 498; assay, amino-β-naphthylamide substrate, with, **45**, 496, 498; composition, **45**, 503; properties, **45**, 501; physical, **45**, 502; purification, **45**, 500; specificity, **45**, 495); membrane, properties, **248**, 185, 189; properties, **45**, 501; physical, **45**, 502; purification, **45**, 500; **248**, 533; specificity, **45**, 495; structure, **248**, 186

CYTOSOL NONSPECIFIC DIPEPTIDASE

This zinc-dependent dipeptidase [EC 3.4.13.18] (also known as glycylglycine dipeptidase, glycylleucine dipeptidase, and N^2-β-alanylarginine dipeptidase) catalyzes the hydrolysis of dipeptides.[1–3] Hydrophobic dipeptides are cleaved preferentially (glycyl-L-leucine provides the highest activity), including prolyl-amino acid dipeptides such as L-prolyl-L-leucine (which provides the highest activity of all the Pro-Xaa dipeptides; however, this enzyme is distinct from Xaa-Pro dipeptidase, EC 3.4.13.9). In addition, EC 3.4.13.18 exhibits a carnosinase activity; however, the enzyme appears to be distinct from that of either serum carnosinase [EC 3.4.13.20] or Xaa-His dipeptidase [EC 3.4.13.3]. The specificity of EC 3.4.13.18 varies somewhat from species to species. ***See also*** Carnosinase; Xaa-His *Dipeptidase; Xaa-Pro Dipeptidase*

[1] E. L. Smith (1948) *JBC* **176**, 9.
[2] D. A. Priestman & J. Butterworth (1985) *BJ* **231**, 689.
[3] N. Kunze, H. Kleinkauf & K. Bauer (1986) *EJB* **160**, 605.

Selected entries from *Methods in Enzymology* [vol, page(s)]:
General discussion: **2**, 105, 107; **19**, 748
Assay: **2**, 105, 107
Other molecular properties: iminodipeptidase and, **2**, 99; properties, **2**, 109; **19**, 756; purification, **2**, 108

D

DACTYLYSIN

This endopeptidase [EC 3.4.24.60], also known as peptide hormone inactivating endopeptidase (PHIE), catalyzes peptide-bond hydrolysis in peptides of at least six residues. The enzyme shows preference for substrates containing bulky, hydrophobic residues (*e.g.*, Trp, Tyr, Phe, Ile, Leu, and Val) at the P1' position. Specificity studies also indicate that a hydrophobic residue at P1 improves substrate binding. Dactylysin will hydrolyze Phe–Phe and Gly–Leu in substance P, Phe–Leu in [Leu^5,Arg^6]enkephalin, and Gly–Leu in neurokinin A.[1,2] The action of dactylysin resembles that of neprilysin and meprin, except dactylysin is insensitive to thiorphan and unable to digest either [Met^5]enkephalin or [Leu^5]enkephalin.

[1]K. de M. Carvalho, C. Joudiou, H. Boussetta, A. M. Leseney & P. Cohen (1992) *PNAS* **89**, 84.
[2]C. Joudiou, K. de M. Carvalho, G. Camarao, H. Boussetta & P. Cohen (1993) *Biochemistry* **32**, 5959.

dATP(dGTP):DNA PURINE TRANSFERASE

This transferase [EC 2.6.99.1] catalyzes the reaction of dATP (or dGTP) with depurinated DNA to produce deoxyribose triphosphate and DNA: hence, the purine base of the deoxynucleotide is transferred to the apurinic site on the DNA and forms a normal *N*-glycosidic linkage.[1,2] When the substrate is depurinated poly(dA/dT), dGTP is not active as the donor substrate; likewise, when depurinated poly(dG/dC) is the acceptor substrate, dATP cannot act as the donor.

[1]W. A. Deutsch & S. Linn (1979) *PNAS* **76**, 141.
[2]Z. Livneh, D. Elad & J. Sperling (1979) *PNAS* **76**, 1089.

DAUNORUBICIN REDUCTASE

This oxidoreductase, which catalyzes the reaction of daunorubicin (also known as daunomycin) with NADPH (A-side) to produce daunorubicinol and NADP$^+$, is a variant of alcohol dehydrogenase (NADP$^+$) [EC 1.1.1.2]. Another daunorubicin reductase from rabbit liver exhibits B-side stereospecificity. **See also** Carbonyl Reductase

dCMP DEAMINASE

This deaminase [EC 3.5.4.12], also known as deoxycytidylate deaminase, catalyzes the hydrolysis of dCMP to produce dUMP and ammonia.[1–3] Other 5-substituted dCMPs can also serve as substrates.

[1]F. Maley & G. F. Maley (1990) *Prog. Nucl. Acid Res. Mol. Biol.* **39**, 49.
[2]P. Reichard (1988) *Ann. Rev. Biochem.* **57**, 349.
[3]F. Maley, M. Belfort & G. Maley (1984) *Adv. Enzyme Regul.* **22**, 413.

Selected entries from ***Methods in Enzymology*** [**vol**, page(s)]:
General discussion: **12A**, 170; **51**, 412
Assay: **12A**, 170; **51**, 413; principle, **12A**, 170; procedure, **12A**, 171; reagents, **12A**, 171
Other molecular properties: activated R conformation (crosslinked, properties, **135**, 580; crosslinking, **135**, 579); amino acid analysis, **51**, 417; conformational states, chemical stabilization, **135**, 577; feedback regulation, **12A**, 182; inhibited T conformation (crosslinked, properties, **135**, 583; crosslinking, **135**, 581); *in vitro* synthesis, assay for, **20**, 540; kinetic properties, **51**, 417; molecular weight, **51**, 417; occurrence, **12A**, 170; properties, **12A**, 181; purification, **51**, 413 (chick embryo, **12A**, 173; monkey and rabbit liver, **12A**, 175); purity, **51**, 417; regulation, **51**, 412, 417, 418; stability, **51**, 417; substrate specificity, **51**, 417; subunit structure, **51**, 417; T2-infected *Escherichia coli*, from, **51**, 412; transition-state and multisubstrate analogues, **249**, 307

dCTP DEAMINASE

This deaminase [EC 3.5.4.13], also known as deoxycytidine triphosphate deaminase, catalyzes the hydrolysis of dCTP to produce dUTP and ammonia.[1,2] In certain organisms, this deaminase reportedly converts 5-methyl-dCTP to dTTP.

[1]C. F. Beck, A. R. Eisenhardt & J. Neuhard (1975) *JBC* **250**, 609.
[2]A. R. Price (1974) *J. Virol.* **14**, 1314.

Selected entries from ***Methods in Enzymology*** [**vol**, page(s)]:
General discussion: **51**, 418
Assay: **51**, 418
Other molecular properties: activity, **51**, 418; cation requirement, **51**, 422; inhibitors, **51**, 422; kinetic properties, **51**, 423; molecular weight, **51**, 423; pH optimum, **51**, 422; purity, **51**, 422; *Salmonella typhimurium*, from, **51**, 418; stability, **51**, 422; substrate specificity, **51**, 422

dCTP PYROPHOSPHATASE

This enzyme [EC 3.6.1.12], also known as dCTP diphosphatase, catalyzes the hydrolysis of dCTP to produce dCMP and pyrophosphate (or, diphosphate).[1] Other substrates include dCDP, forming dCMP and orthophosphate. The alternative names deoxycytidine-triphosphatase, dCTPase, and deoxy-CTPase can be easily confused with reactions forming dCDP and orthophosphate.

[1]S. B. Zimmerman & A. Kornberg (1961) *JBC* **236**, 1480.

*Dde*I RESTRICTION ENDONUCLEASE

This type II restriction endonuclease [EC 3.4.21.4], isolated from *Desulfovibrio desulfuricans Norway* strain, catalyzes the hydrolysis of both strands of DNA at 5′ . . . C∇TNAG . . . 3′, where N refers to any base.[1] The enzyme also cleaves single-stranded DNA slowly.

[1]R. A. Makula & R. B. Meagher (1980) *NAR* **8**, 3125.

DDT-DEHYDROCHLORINASE

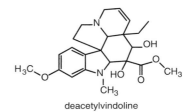

DDT   DDE

This glutathione-dependent lyase [EC 4.5.1.1] catalyzes the conversion of 1,1,1,-trichloro-2,2-bis(4-chlorophenyl)-ethane, the long-lived pesticide known widely as DDT, to 1,1-dichloro-2,2-bis(4-chlorophenyl)ethylene (DDE) and hydrogen chloride.[1] ***See also*** *Glutathione S-Transferase; Dichloromethane Dehalogenase*

[1]A. H. Tang & C. P. Tu (1994) *JBC* **269**, 27876.

DEACETYL-[CITRATE-(*pro-3S*)-LYASE] S-ACETYLTRANSFERASE

This acetyltransferase [EC 2.3.1.49], also known as *S*-acetyl-phosphopantetheine:deacetyl citrate lyase *S*-acetyltransferase and deacetyl-[citrate-(*pro-3S*)-lyase] acetyltransferase, catalyzes the reaction of *S*-acetylphosphopantotheine with deacetyl-[citrate-oxaloacetate-lyase-((*pro-3S*)-CH$_2$COO$^-$ → acetate)] to produce phosphopantotheine and [citrate-oxaloacetate-lyase((*pro-3S*)-CH$_2$COO$^-$ → acetate)].[1]

[1]M. Singh, B. Böttger, C. Stewart, G. C. Brooks & P. A. Srere (1973) *BBRC* **53**, 1.

DEACETYLIPECOSIDE SYNTHASE

This enzyme [EC 4.3.3.4] catalyzes the reaction of dopamine with secologanin to produce deacetylipecoside and water.[1] The enantiomeric specificity of the *Alangium lamarckii* leaf enzyme differs from deacetylisoipecoside synthase [EC 4.3.3.3].

[1]W. DeEknamkul, A. Ounaroon, T. Tanahashi, T. Kutchan & M. H. Zenk (1997) *Phytochemistry* **45**, 477.

DEACETYLISOIPECOSIDE SYNTHASE

This enzyme [EC 4.3.3.3] catalyzes the reaction of dopamine with secologanin to produce deacetylisoipecoside and water.[1] The *Alangium* leaf enzyme differs in enantiomeric specificity from deacetylipecoside synthase [EC 4.3.3.4].

[1]W. DeEknamkul, A. Ounaroon, T. Tanahashi, T. Kutchan & M. H. Zenk (1997) *Phytochemistry* **45**, 477.

17-*O*-DEACETYLVINDOLINE *O*-ACETYLTRANSFERASE

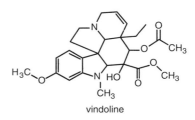

deacetylvindoline

vindoline

This acetyltransferase [EC 2.3.1.107] catalyzes the reaction of acetyl-CoA with 17-*O*-deacetylvindoline to produce coenzyme A and vindoline.[1–4] In the periwinkle, this reaction is the last step in vindoline biosynthesis from tabersonine.

[1]R. Power, W. G. Kurz & V. De Luca (1990) *ABB* **279**, 370.
[2]W. Fahn, H. Gundlach, B. Deus-Neumann & J. Stöckigt (1985) *Plant Cell Rep.* **4**, 333.
[3]V. De Luca & A. J. Cutler (1987) *Plant Physiol.* **85**, 1099.
[4]B. St-Pierre, P. Laflamme, A. M. Alarco & V. De Luca (1998) *Plant J.* **14**, 703.

DEBRANCHING ENZYME

A debranching enzyme catalyzes the cleavage and/or removal of branched points in biopolymers, particularly in branched polysaccharides. **See** *specific enzyme; for example, Glycogen Debranching Enzyme*

DECYLCITRATE SYNTHASE

This enzyme [EC 4.1.3.23] catalyzes the reaction of lauroyl-CoA ($CH_3(CH_2)_{10}COSCoA$) with water and oxaloacetate to produce (2S,3S)-2-hydroxytridecane-1,2,3-tricarboxylate (*i.e.*, 2-decylcitrate) and Coenzyme A.[1,2]

[1] A. Måhlén (1971) *EJB* **22**, 104.
[2] A. Måhlén & S. Gatenbeck (1968) *Acta Chem. Scand.* **22**, 2617.

DECYLHOMOCITRATE SYNTHASE

This enzyme [EC 4.1.3.29] catalyzes the reaction of lauroyl-CoA with water and α-ketoglutarate (or, 2-oxoglutarate) to produce 3-hydroxytetradecane-1,3,4-tricarboxylate (*i.e.*, decylhomocitrate) and Coenzyme A.[1] Other donor substrates include decanoyl-CoA.

[1] A. Måhlén (1973) *EJB* **38**, 32.

DEDIMETHYLAMINO-4-AMINOANHYDROTETRACYCLINE *N*-METHYLTRANSFERASE

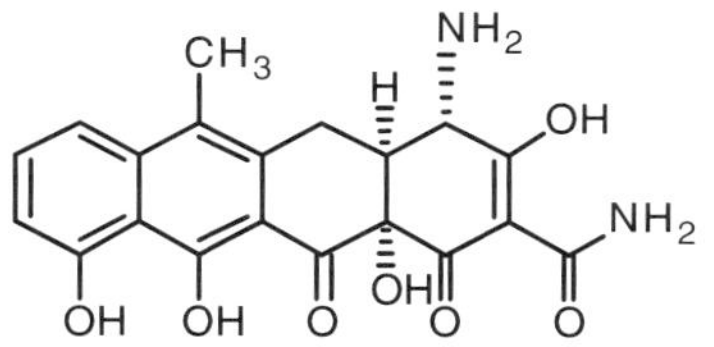

dedimethylamino-4 amino-
anhydrotetracycline

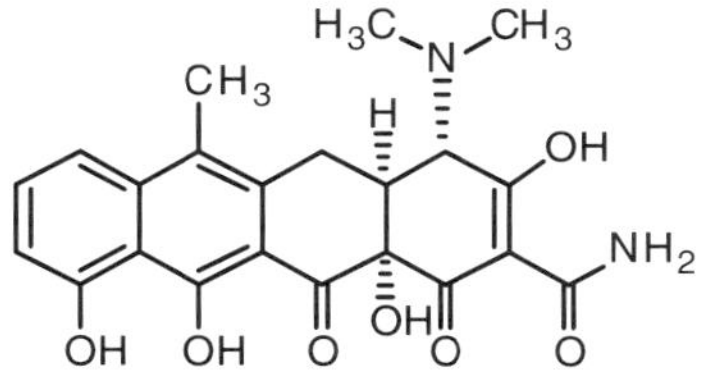

anhydrotetracycline

This methyltransferase catalyzes the reaction of dedimethyl-amino-4-aminoanhydrotetracycline with two molecules of *S*-adenosyl-L-methionine to produce anhydrotetracycline and two molecules of *S*-adenosyl-L-homocysteine.

Selected entries from **Methods in Enzymology** [**vol**, page(s)]:
General discussion: 43, 603
Assay: 43, 603
Other molecular properties: properties, **43**, 605, 606; reaction scheme, **43**, 603; *Streptomyces aureofaciens* preparation, **43**, 605; *Streptomyces rimosus* preparation, **43**, 604

DegP

This high-molecular-weight serine endopeptidase, also known as protease Do, is a heat-shock proteinase required for growth of *E. coli* at elevated temperatures.[1–4] It plays a major role in the degradation of cell envelope proteins, cleaving at Val–Xaa and Ile–Xaa bonds are cleaved.[1,2]

[1] B. Lipinska, M. Zylicz & C. Georgopoulos (1990) *J. Bacteriol.* **172**, 1791.
[2] K. H. S. Swamy, C. H. Chung & A. Goldberg (1983) *ABB* **224**, 543.
[3] C. H. Chung (1993) *Science* **262**, 372.
[4] H. Kolmar, P. R. Waller & R. T. Sauer (1996) *J. Bacteriol.* **178**, 5925.

7-DEHYDROCHOLESTEROL REDUCTASE

This oxidoreductase [EC 1.3.1.21] catalyzes the reaction of cholesta-5,7-dien-3β-ol (*i.e.*, 7-dehydrocholesterol) with NADPH to produce cholesterol and $NADP^+$.[1–3]

[1] A. A. Kandutsch (1962) *JBC* **237**, 358.
[2] R. Niemiro & R. Fumagalli (1965) *BBA* **98**, 624.
[3] I. Björkhem & I. Holmberg (1973) *EJB* **33**, 364.

Selected entries from **Methods in Enzymology** [**vol**, page(s)]:
General discussion: 15, 506
Assay: 15, 502, 506, 508
Other molecular properties: preparation, **15**, 509; properties, **15**, 512; stereospecificity, **87**, 116

2-DEHYDRO-3-DEOXY-L-ARABONATE DEHYDRATASE

This dehydratase [EC 4.2.1.43], also known as 2-keto-3-deoxy-L-arabonate dehydratase, catalyzes the conversion of 2-dehydro-3-deoxy-L-arabonate to 2,5-dioxopentanoate (*i.e.*, α-ketoglutarate semialdehyde) and water.[1–4]

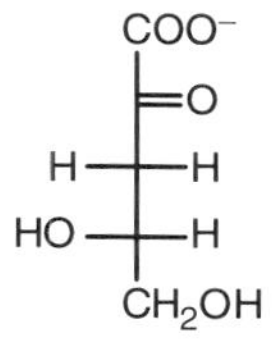

2-dehydro-3-deoxy-
D-arabonate

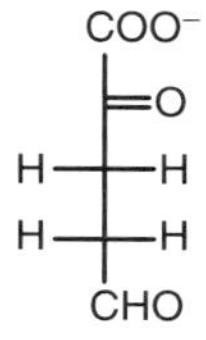

α-ketoglutarate
semialdehyde

The reaction proceeds by formation of a Schiff base intermediate that can be trapped by borohydride reduction. The dehydratase catalyzes the exchange of protons between the solvent and the α-hydrogen adjacent to the carbonyl group.

[1] K. N. Allen (1998) *CBC* **2**, 135.
[2] W. A. Wood (1971) *The Enzymes*, 3rd ed., **5**, 573.
[3] D. Portsmouth, A. C. Stoolmiller & R. H. Abeles (1967) *JBC* **242**, 2751.
[4] A. C. Stoolmiller & R. H. Abeles (1966) *JBC* **241**, 5764.

Selected entries from **Methods in Enzymology** [**vol**, page(s)]:
General discussion: 42, 308
Assay: 42, 308, 309

2-DEHYDRO-3-DEOXYGALACTONOKINASE

This phosphotransferase [EC 2.7.1.58], also known as 2-keto-3-deoxy-galactonokinase and 2-oxo-3-deoxygalactonate kinase, catalyzes the reaction of ATP with 2-dehydro-3-deoxy-D-galactonate to produce ADP and 2-dehydro-3-deoxy-D-galactonate 6-phosphate.[1-3]

[1]T. Szumilo (1981) *J. Bacteriol.* **148**, 368.
[2]J. Deacon & R. A. Cooper (1977) *FEBS Lett.* **77**, 201.
[3]A. H. Stouthamer (1961) *BBA* **48**, 484.

2-DEHYDRO-3-DEOXYGLUCARATE ALDOLASE

This aldolase [EC 4.1.2.20], also known as 2-keto-3-deoxy-D-glucarate aldolase, catalyzes the reversible conversion of 2-dehydro-3-deoxy-D-glucarate to pyruvate and tartronate semialdehyde.[1] The enzyme also acts on 5-keto-4-deoxy-D-glucarate.

[1]W. A. Wood (1972) *The Enzymes*, 3rd ed., **7**, 281.

5-DEHYDRO-4-DEOXYGLUCARATE DEHYDRATASE

This dehydratase [EC 4.2.1.41], also known as 5-keto-4-deoxyglucarate dehydratase, catalyzes the conversion of 5-dehydro-4-deoxy-D-glucarate to 2,5-dioxopentanoate, water, and carbon dioxide.[1-3] The reaction mechanism proceeds *via* a Schiff's base intermediate that can be trapped upon sodium borohydride treatment.[3]

[1]W. A. Wood (1971) *The Enzymes*, 3rd ed., **5**, 573.
[2]R. Jeffcoat (1975) *BJ* **145**, 305.
[3]R. Jeffcoat, H. Hassall & S. Dagley (1969) *BJ* **115**, 977.

2-DEHYDRO-3-DEOXY-D-GLUCONATE 5-DEHYDROGENASE

This oxidoreductase [EC 1.1.1.127], also known as 2-keto-3-deoxygluconate dehydrogenase, catalyzes the reaction of (4S)-4,6-dihydroxy-2,5-dioxohexanoate with NADH to produce 2-dehydro-3-deoxy-D-gluconate and NAD+ (reversible at elevated pH).[1,2]

(4S)-4,6-dihydroxy-2,5-dioxohexanoate

2-dehydro-3-deoxy-D-gluconate

The enzyme isolated from *Pseudomonas* can utilize either NADPH or NADH as the coenzyme whereas the enzymes isolated from *Erwinia chrysanthemi* and *Escherichia coli* are more specific for NADH.

[1]J. Preiss & G. Ashwell (1963) *JBC* **238**, 1577.
[2]G. Condemine, N. Hugouvieux-Cotte-Pattat & J. Robert-Baudouy (1984) *J. Gen. Microbiol.* **130**, 2839.

2-DEHYDRO-3-DEOXY-D-GLUCONATE 6-DEHYDROGENASE

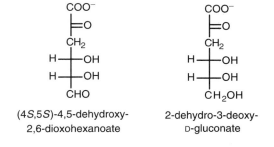

(4S,5S)-4,5-dehydroxy-2,6-dioxohexanoate

2-dehydro-3-deoxy-D-gluconate

This oxidoreductase [EC 1.1.1.126], also known as 2-keto-3-deoxy-D-gluconate dehydrogenase and 4-deoxy-L-erythro-5-hexosulose-uronate reductase, catalyzes the reversible reaction of (4S,5S)-4,5-dehydroxy-2,6-dioxohexanoate with NADPH to produce 2-dehydro-3-deoxy-D-gluconate and NADP+.[1]

[1]J. Preiss & G. Ashwell (1962) *JBC* **237**, 309, 317.

2-DEHYDRO-3-DEOXYGLUCONOKINASE

```
    COO⁻              COO⁻
    |=O               |=O
    CH₂               CH₂
 H——OH            H——OH
 H——OH            H——OH
    CH₂OH             CH₂OPO₃H⁻

2-dehydro-3-deoxy-    6-phospho-2-dehydro-3-
   D-gluconate          deoxy-D-gluconate
```

This phosphotransferase [EC 2.7.1.45], also known as 2-keto-3-deoxygluconokinase and 3-deoxy-2-oxo-D-gluconate kinase, catalyzes the ATP-dependent phosphorylation of 2-dehydro-3-deoxy-D-gluconate to produce 6-phospho-2-dehydro-3-deoxy-D-gluconate and ADP.[1,2] **See** *5-Dehydro-2-Deoxygluconokinase; Dehydrogluconokinase*

[1] M. A. Cynkin & G. Ashwell (1969) *JBC* **235**, 1576.
[2] J. Pouyssegur & F. Stoeber (1971) *Biochimie* **53**, 771.

Selected entries from *Methods in Enzymology* [vol, page(s)]:
General discussion: 5, 190, 205
Assay: 5, 205

5-DEHYDRO-2-DEOXYGLUCONOKINASE

```
    COO⁻              COO⁻
    CH₂               CH₂
 HO——H            HO——OH
 H——OH            H——OH
    |=O               |=O
    CH₂OH             CH₂OPO₃H⁻

5-dehydro-2-deoxy-    6-phospho-5-dehydro-2-
   D-gluconate          deoxy-D-gluconate
```

This phosphotransferase [EC 2.7.1.92], also known as 5-keto-2-deoxygluconokinase, catalyzes the ATP-dependent phosphorylation of 5-dehydro-2-deoxy-D-gluconate to produce ADP and 6-phospho-5-dehydro-2-deoxy-D-gluconate.[1] **See also** *2-Dehydro-3-Deoxygluconokinase; Dehydrogluconokinase*

[1] W. A. Anderson & B. Magasanik (1971) *JBC* **246**, 5662.

2-DEHYDRO-3-DEOXY-D-PENTONATE ALDOLASE

This aldolase [EC 4.1.2.28], also known as 2-keto-3-deoxy-D-pentonate aldolase, catalyzes the reversible cleavage of 2-dehydro-3-deoxy-D-pentonate to pyruvate and glycolaldehyde.[1]

[1] A. S. Dahms (1974) *BBRC* **60**, 1433.

2-DEHYDRO-3-DEOXY-L-PENTONATE ALDOLASE

This aldolase [EC 4.1.2.18], also known as 2-keto-3-deoxy-L-pentonate aldolase and 2-keto-3-deoxyarabonate aldolase, catalyzes the reversible cleavage of 2-dehydro-3-deoxy-L-pentonate to pyruvate and glycolaldehyde.[1,2] The enzyme from some organisms will also act on 2-keto-3-deoxy-D-fuconate, 2-keto-5-deoxy-L-arabonate, and 2-keto-3-deoxy-D-xylonate.

[1] W. A. Wood (1972) *The Enzymes*, 3rd ed., **7**, 281.
[2] A. S. Dahms & R. L. Anderson (1972) *JBC* **247**, 2238.

Selected entries from *Methods in Enzymology* [vol, page(s)]:
General discussion: 42, 269; **90**, 269
Assay: pseudomonad MSU-1, **42**, 269, 270; **90**, 269
Other molecular properties: pseudomonad MSU-1, from, **42**, 269 (assay, **42**, 269, 270; **90**, 269; chromatography, **42**, 271; properties, **42**, 271, 272; **90**, 271; purification, **42**, 270, 271; **90**, 270)

DEHYDROGLUCONATE DEHYDROGENASE

This flavin-dependent oxidoreductase [EC 1.1.99.4], also known as ketogluconate dehydrogenase, catalyzes the reaction of 2-dehydro-D-gluconate with an acceptor substrate (*i.e.*, ferricyanide, phenazine methosulfate, and 2,6-dichlorophenolindophenol) to produce 2,5-didehydro-D-gluconate and the corresponding reduced acceptor.[1] The cofactor is covalently attached as 8α-(N^3-histidyl)-riboflavin.[2]

[1] E. Shinagawa, K. Matsushita, O. Adachi & M. Ameyama (1981) *Agric. Biol. Chem.* **45**, 1079.
[2] W. McIntire, T. P. Singer, M. Ameyama, O. Adachi, K. Matsushita & E. Shinagawa (1985) *BJ* **231**, 651.

Selected entries from *Methods in Enzymology* [vol, page(s)]:
Assay: *Gluconobacter melanogenus* membrane-bound, **89**, 194
Other molecular properties: *Gluconobacter melanogenus*, from (assay, **89**, 194; properties, **89**, 197; purification, **89**, 195); 2-keto-D-gluconate microdetermination, **89**, 26

DEHYDROGLUCONOKINASE

This phosphotransferase [EC 2.7.1.13], also known as ketogluconokinase, catalyzes the reaction of 2-dehydro-D-gluconate (*i.e.*, 2-keto-D-gluconate) with ATP to produce ADP and 6-phospho-2-dehydro-D-gluconate.[1]

[1] E. W. Frampton & W. A. Wood (1961) *JBC* **236**, 2578.

Selected entries from *Methods in Enzymology* [vol, page(s)]:
General discussion: 5, 291
Assay: 5, 291

DEHYDRO-L-GULONATE DECARBOXYLASE

3-keto-L-gulonate   L-xylulose

This decarboxylase [EC 4.1.1.34], also known as keto-L-gulonate decarboxylase, catalyzes the conversion of 3-dehydro-L-gulonate to L-xylulose and carbon dioxide.[1] The enzyme also catalyzes the decarboxylation of 2,3-diketo-L-gulonate to produce L-xylonate in L-ascorbate catabolism.

[1]D. Mukherjee, N. C. Kar, N. Sasmal & G. C. Chatterjee (1968) *BJ* **106**, 627.

Selected entries from **Methods in Enzymology** [**vol**, page(s)]:
General discussion: **18A**, 47
Assay: 18A, 47

3-DEHYDRO-L-GULONATE 2-DEHYDROGENASE

3-keto-L-gulonate   (4R,5S)-4,5,6-trihydroxy-
                            2,3-dioxohexanoate

This oxidoreductase [EC 1.1.1.130], also known as 3-keto-L-gulonate dehydrogenase, catalyzes the reversible reaction of 3-dehydro-L-gulonate (*i.e.*, 3-keto-L-gulonate) with NAD(P)$^+$ to produce (4R,5S)-4,5,6-trihydroxy-2,3-dioxohexanoate and NAD(P)H.[1]

[1]W. A. Volk & J. L. Larsen (1962) *JBC* **237**, 2454.

(R)-DEHYDROPANTOATE DEHYDROGENASE

This oxidoreductase [EC 1.2.1.33], also known as D-aldopantoate dehydrogenase, catalyzes the reversible reaction of (R)-4-dehydropantoate with NAD$^+$ and water to produce (R)-3,3-dimethylmalate and NADH.[1]

[1]P. T. Magee & E. E. Snell (1966) *Biochemistry* **5**, 409.

2-DEHYDROPANTOATE 2-REDUCTASE

This oxidoreductase [EC 1.1.1.169], also known as ketopantoate reductase, catalyzes the reversible reaction of 2-dehydropantoate with NADPH to produce (R)-pantoate and NADP$^+$.[1] The *Escherichia coli* enzyme has an ordered Bi Bi kinetic mechanism, and hydride transfer is not rate-limiting.[2]

[1]S. Shimizu, M. Kataoka, M. C.-M. Chung & H. Yamada (1988) *JBC* **263**, 12077.
[2]R. Zheng & J. S. Blanchard (2000) *Biochemistry* **39**, 3708.

2-DEHYDROPANTOYLLACTONE REDUCTASE (A-SPECIFIC)

This oxidoreductase [EC 1.1.1.168], also known as ketopantoyl lactone reductase, catalyzes the reaction of 2-dehydropantolactone with NADPH to produce (R)-pantolactone and NADP$^+$.[1,2] The yeast enzyme is specific for the coenzyme's A-face, whereas the *Escherichia coli* protein [EC 1.1.1.214] is specific for the B-face. 3-Hydroxycyclohexanone dehydrogenase [EC 1.1.99.26] also catalyzes this reaction as well but with different receptor requirements.

[1]H. Hata, S. Shimizu, S. Hattori & H. Yamada (1989) *FEMS Microbiol. Lett.* **58**, 87.
[2]H. Hata, S. Shimizu, S. Hattori & H. Yamada (1989) *BBA* **990**, 175.

Selected entries from **Methods in Enzymology** [**vol**, page(s)]:
General discussion: **62**, 209
Assay: 62, 210, 211
Other molecular properties: inhibitors, **62**, 213, 214; properties, **62**, 213; purification from *Saccharomyces cerevisiae*, **62**, 211; stereospecificity, **87**, 112

2-DEHYDROPANTOYLLACTONE REDUCTASE (B-SPECIFIC)

This oxidoreductase [EC 1.1.1.214], also known as 2-dehydropantoyl-lactone reductase (B-specific), catalyzes the reaction of 2-dehydropantolactone with NADPH to produce (R)-pantolactone and NADP$^+$.

(R)-pantoyllactone   2-dehydropantolactone

This enzyme, first isolated from *Escherichia coli*, is distinct from that of yeast (*Saccharomyces cerevisiae*) [EC 1.1.1.168], which is specific for the A-face of $NADP^+$.[1] The enzyme also differs in acceptor requirements from 3-hydroxycyclohexanone dehydrogenase [EC 1.1.99.26].

[1]D. R. Wilken, H. L. King & R. E. Dyar (1975) *JBC* **250**, 2311.

16-DEHYDROPROGESTERONE HYDRATASE

This enzyme [EC 4.2.1.86], also known as 16α-hydroxyprogesterone dehydratase, catalyzes the reversible reaction of 16-dehydroprogesterone with water to produce 16α-hydroxyprogesterone.[1,2]

[1]T. L. Glass & C. Z. Burley (1984) *J. Steroid Biochem.* **21**, 65.
[2]W. E. Watkins & T. L. Glass (1991) *J. Steroid Biochem. Mol. Biol.* **38**, 257.

3-DEHYDROQUINATE DEHYDRATASE

This enzyme [EC 4.2.1.10], also known as 3-dehydroquinase and 5-dehydroquinase, catalyzes the conversion of 3-dehydroquinate to 3-dehydroshikimate and water.[1-3]

3-dehydroquinate   3-dehydroshikimate

Two distinct types of dehydroquinate dehydratase have been identified: Type I enzymes are heat labile dimers that catalyze dehydroquinate dehydration by a Schiff base mechanism; and Type II enzymes catalyze the reaction by forming an enediolate intermediate. The stereochemistry of the type I elimination reaction is *syn*, but *anti* for the type II enzyme.[1]

[1]K. N. Allen (1998) *CBC* **2**, 135.
[2]R. Bentley (1990) *Crit. Rev. Biochem. Mol. Biol.* **25**, 307.
[3]S. Chaudhuri, J. M. Lambert, L. A. McColl & J. R. Coggins (1986) *BJ* **239**, 699.

Selected entries from ***Methods in Enzymology*** [**vol**, page(s)]:
General discussion: 2, 305; **142**, 320
Assay: 2, 305; **142**, 321
Other molecular properties: coupling enzyme in *arom* multifunctional enzyme assay, as, preparation, **142**, 330; 3-deoxy-D-*arabino*-heptulosonate 7-phosphate synthase preparation, not present in, **5**, 398; dehydroquinate synthase, not present in preparation of, **5**, 401; 5-dehydroshikimate reductase and, **2**, 306; distribution, **2**, 307; properties, **2**, 307; **142**, 323; purification from (*Escherichia coli*, **2**, 306; **5**, 400f; **142**, 322; *Neurospora crassa*, **17A**, 387); quinate dehydrogenase assay and, **2**, 308

3-DEHYDROQUINATE SYNTHASE

This cobalt-dependent enzyme containing a tightly bound NAD^+ [EC 4.2.3.4; previously classified as EC 4.6.1.3],

also known as 5-dehydroquinate synthase, catalyzes the conversion of 3-deoxy-*arabino*-heptulosonate 7-phosphate to 3-dehydroquinate and orthophosphate.[1-5] (Note that the numbering system has changed, such that 5-dehydroquinate is now 3-dehydroquinate.)

The hydrogen atoms located on C7 of the substrate are retained on C2 of the product. The crystal structure illustrates how a single active site can catalyze a series of consecutive reactions without the release of reaction intermediates and unwanted products.[2] The substrate is oxidized at C5 by the enzyme-bound NAD^+, producing NADH and the tightly bound 5-keto form of the substrate. Elimination of orthophosphate produces an α,β-unsaturated ketone. The C5 position is then reduced to regenerate NAD^+ and C5 hydroxy form of the intermediate that proceeds to 3-dehydroquinate.[1]

[1]A. R. Clarke & T. R. Dafforn (1998) *CBC* **3**, 1.
[2]E. P. Carpenter, A. R. Hawkins, J. W. Frost & K. A. Brown (1998) *Nature* **394**, 299.
[3]R. Bentley (1990) *Crit. Rev. Biochem. Mol. Biol.* **25**, 307.
[4]P. A. Bartlett & K. Satake (1988) *JACS* **110**, 1628.
[5]S. L. Rotenberg & D. B. Sprinson (1970) *PNAS* **67**, 1669.

Selected entries from ***Methods in Enzymology*** [**vol**, page(s)]:
General discussion: 5, 398; **142**, 306
Assay: 5, 399; **142**, 308
Other molecular properties: action of, **5**, 398; 3-deoxy-D-*arabino*-heptulosonate 7-phosphate synthase preparation, absent in, **5**, 397; properties, **5**, 401; **142**, 313; purification from (*Escherichia coli*, **5**, 399; **142**, 309; *Neurospora crassa*, **17A**, 393); substrate isolation from bacterial growth medium, **142**, 312

1,2-DEHYDRORETICULINIUM REDUCTASE (NADPH)

This oxidoreductase [EC 1.5.1.27], also known as 1,2-dehydroreticulinium ion reductase, catalyzes the reaction of a 1,2-dehydroreticulinium ion with NADPH to produce (*R*)-reticuline and $NADP^+$.[1]

[1]W. De-Eknamkul & M. H. Zenk (1992) *Phytochemistry* **31**, 813.

3-DEHYDROSPHINGANINE REDUCTASE

This endoplasmic reticulum oxidoreductase [EC 1.1.1.102] catalyzes the reversible reaction of 3-dehydrosphinganine with NADPH to produce sphinganine and $NADP^+$.[1,2]

[1]W. Stoffel, D. LeKim & G. Sticht (1968) *Hoppe-Seyler's Z. Physiol. Chem.* **349**, 664 and 1637.
[2]E. C. Mandon, I. Ehses, J. Rother, G. van Echten & K. Sandhoff (1992) *JBC* **267**, 11144.

Selected entries from ***Methods in Enzymology*** [**vol**, page(s)]:
General discussion: 209, 427
Assay: 209, 431

11-O-DEMETHYL-17-O-DEACETYLVINDOLINE O-METHYLTRANSFERASE

This methyltransferase [EC 2.1.1.94], which participates in vindoline biosynthesis in the periwinkle, catalyzes the reaction of S-adenosyl-L-methionine with 11-O-demethyl-17-O-deacetylvindoline to produce S-adenosyl-L-homocysteine and 17-O-deacetylvindoline.[1]

[1]W. Fahn, E. Laußermair, B. Deus-Neumann & J. Stöckigt (1985) *Plant Cell Rep.* **4**, 337.

DEMETHYLMACROCIN O-METHYLTRANSFERASE

This enzyme [EC 2.1.1.102] catalyzes the penultimate step in tylosin biosynthesis, namely the reaction of S-adenosyl-L-methionine with demethylmacrocin to produce S-adenosyl-L-homocysteine and macrocine.[1] This methyltransferase will also convert demethyllactenocin to lactenocin.

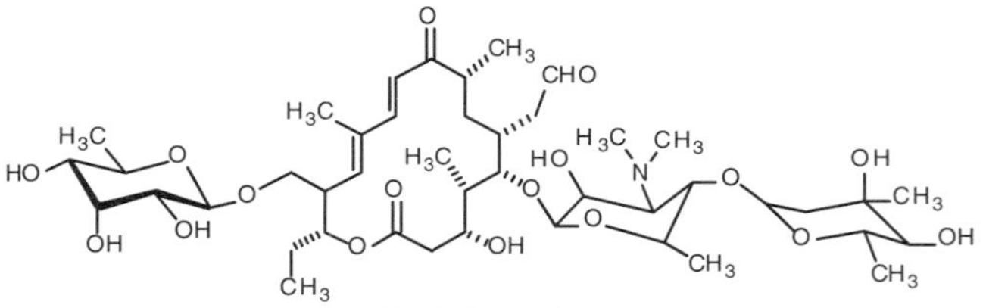

demethylmacrocin

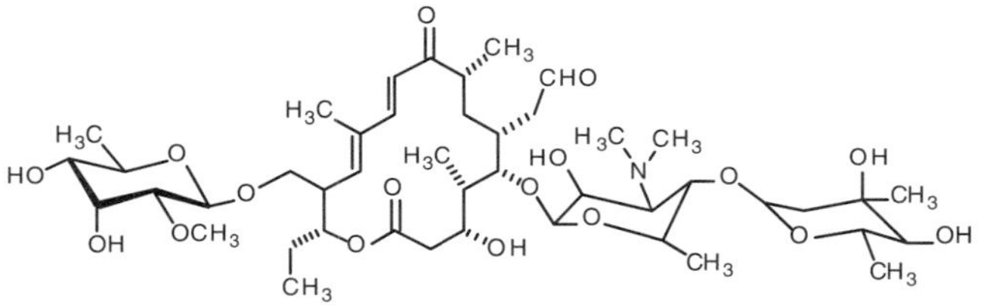

macrocin

The enzyme is distinct from macrocin O-methyltransferase [EC 2.1.1.101] which produces tylosin, a growth promoting agent in swine.

[1]A. J. Kreuzman, J. R. Turner & W.-K. Yeh (1988) *JBC* **263**, 15626.

O-DEMETHYLPUROMYCIN O-METHYLTRANSFERASE

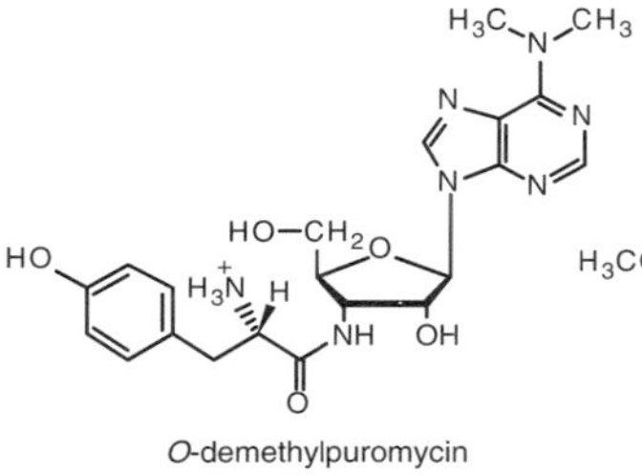

O-demethylpuromycin 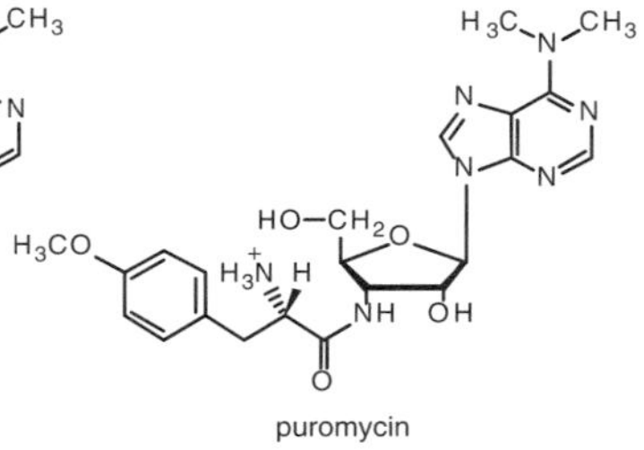   puromycin

This methyltransferase [EC 2.1.1.38] catalyzes the reaction of S-adenosyl-L-methionine with O-demethylpuromycin to produce S-adenosyl-L-homocysteine and puromycin.[1,2]

[1]M. M. Rao, P. F. Rebello & B. M. Pogell (1969) *JBC* **244**, 112.
[2]L. Sankaran & B. M. Pogell (1975) *Antimicrob. Agents Chemother.* **8**, 721.

Selected entries from *Methods in Enzymology* [**vol**, page(s)]:
General discussion: 43, 508
Assay: 43, 509 (cell-free extract, 43, 512)
Other molecular properties: S-adenosylmethionine, 43, 509; DEAE-cellulose column chromatography, 43, 512; O-demethylpuromycin, 43, 509; pH optimum, 43, 514; properties, 43, 513, 514; purification, 43, 511; puromycin biosynthesis, scheme, 43, 508; salt fractionation, 43, 512; Sephadex G-200 gel filtration, 43, 513; specific activity, 43, 510; specificity, 43, 514; stability, 43, 514; unit definition, 43, 510

3′-DEMETHYLSTAUROSPORINE O-METHYLTRANSFERASE

This methyltransferase [EC 2.1.1.139], also known as 3′-demethoxy-3′-hydroxystaurosporine O-methyltransferase and staurosporine synthase, catalyzes the reaction of S-adenosyl-L-methionine with 3′-demethylstaurosporine to produce S-adenosyl-L-homocysteine and staurosporine, a potent protein kinase C inhibitor.[1]

[1]S. Weidner, M. Kittelmann, K. Goeke, O. Ghisalba & H. Zahner (1998) *J. Antibiot.* **51**, 679.

DEMETHYLSTERIGMATOCYSTIN 6-O-METHYLTRANSFERASE

This methyltransferase [EC 2.1.1.109], which participates in fungal biosynthesis of aflatoxins, catalyzes the reaction of S-adenosyl-L-methionine with 6-demethylsterigmatocystin to produce S-adenosyl-L-homocysteine and sterigmatocystin.[1,2] Dihydrodemethylsterigmatocystin can also be methylated by this enzyme.

[1]K. Yabe, Y. Ando, J. Hashimoto & T. Hamasaki (1989) *Appl. Environ. Microbiol.* **55**, 2172.
[2]K. Yabe, K.-I. Matsushima, T. Koyama & T. Hamasaki (1998) *Appl. Environ. Microbiol.* **64**, 166.

3-DEMETHYLUBIQUINONE-9 3-O-METHYLTRANSFERASE

This methyltransferase [EC 2.1.1.64] catalyzes the reaction of S-adenosyl-L-methionine with 3-demethylubiquinone-9 to produce S-adenosyl-L-homocysteine and ubiquinone-9.[1]

[1]R. M. Houser & E. Olson (1977) *JBC* **252**, 4017.

DEOXYADENOSINE KINASE

This phosphotransferase [EC 2.7.1.76] catalyzes the reaction of ATP with deoxyadenosine to produce ADP and dAMP.[1-3] Other acceptor substrates include deoxyguanosine. In several organisms, this enzyme is identical with deoxycytidine kinase. The *Lactobacillus acidophilus* system contains two separate but interacting active sites responsible for deoxyadenosine kinase and deoxycytidine kinase activities. *See also Deoxycytidine Kinase; Deoxyguanosine Kinase; (Deoxy)Adenylate Kinase; Deoxynucleoside Kinase*

[1]M. R. Deibel, R. B. Reznik & D. H. Ives (1977) JBC **252**, 8240.
[2]B. Kierdaszuk & S. Eriksson (1990) Biochemistry **29**, 4109.
[3]E. P. Anderson (1973) The Enzymes, 3rd ed., **9**, 49.

Selected entries from *Methods in Enzymology* [vol, page(s)]:
Other molecular properties: deoxynucleoside kinase, **51**, 346 (assay, **51**, 346, 347; kinetic properties, **51**, 353, 354; *Lactobacillus acidophilus*, from, **51**, 346; properties, **51**, 352, 353; purification, **51**, 347; stability, **51**, 352, 353; substrate specificity, **51**, 352)

(DEOXY)ADENYLATE KINASE

This phosphotransferase [EC 2.7.4.11] catalyzes the reaction of ATP with dAMP to produce ADP and dADP. The enzyme can also use AMP as the acceptor substrate to producing a second molecule ADP, akin to adenylate kinase reaction.[1,2] *See also Adenylate Kinase; Deoxyadenosine Kinase*

[1]E. P. Anderson (1973) The Enzymes, 3rd ed., **9**, 49.
[2]T. J. Griffith & C. W. Helleiner (1965) BBA **108**, 114.

DEOXYCYTIDINE DEAMINASE

This deaminase [EC 3.5.4.14] catalyzes the hydrolysis of deoxycytidine to produce deoxyuridine and ammonia.[1] Other substrates include cytidine, 5-methyldeoxycytidine, 5-bromodeoxycytidine, and arabinosylcytosine.

[1]F. le Floc'h & A. Guillot (1974) Phytochemistry **13**, 2503.

DEOXYCYTIDINE KINASE

This phosphotransferase [EC 2.7.1.74] catalyzes the reaction of a nucleoside triphosphate with deoxycytidine to produce a nucleoside diphosphate and dCMP.[1-4] This enzyme can use any nucleoside triphosphate (except dCTP) as the phosphate-donor substrate and it can phosphorylate cytosine arabinoside as well. The human enzyme, which is feedback inhibited by dCTP, has an ordered Bi Bi kinetic mechanism.[3,4] *See also Deoxyadenosine Kinase; Deoxyguanosine Kinase; Deoxynucleoside Kinase*

[1]B. Turk, R. Awad, E. V. Usova, I. Bjork & S. Eriksson (1999) Biochemistry **38**, 8555.
[2]M. Y. Kim & D. H. Ives (1989) Biochemistry **28**, 9043.
[3]T. L. Hughes, T. M. Hahn, K. K. Reynolds & D. S. Shewach (1997) Biochemistry **36**, 7540.
[4]E. P. Anderson (1973) The Enzymes, 3rd ed., **9**, 49.

Selected entries from *Methods in Enzymology* [vol, page(s)]:
General discussion: **51**, 337, 346
Assay: **51**, 337, 346, 347

Other molecular properties: activators, **51**, 345; activity, **51**, 337; calf thymus, from, **51**, 337; cation requirement, **51**, 345; isoelectric point, **51**, 345, 353; kinetic properties, **51**, 345, 353, 354; molecular weight, **51**, 345, 353; pH optimum, **51**, 353; purification by affinity chromatography, **51**, 348; regulation, **51**, 345; stability, **51**, 345; substrate specificity, **51**, 344, 345

DEOXYCYTIDYLATE HYDROXYMETHYLTRANSFERASE

This transferase [EC 2.1.2.8], also known as deoxycytidylate hydroxymethylase and dCMP hydroxymethylase, catalyzes the reaction of 5,10-methylenetetrahydrofolate with water and deoxycytidylate to produce tetrahydrofolate and 5-hydroxymethyldeoxycytidylate.[1,2]

The results of kinetic isotope effect measurements suggest the reaction proceeds via sequential $sp^2 \to sp^3 \to sp^2$ hybridization changes at C6 of the substrate nucleotide, a finding that is consistent with a transient covalent linkage of C6 to the thiol of a cysteinyl residue.[1] This intermediate forms a C5 carbanion, which nucleophilically attacks the *exo*-methylene of the 5-iminium cation generated from 5,10-methylenetetrahydrofolate. This reaction produces an enzyme-linked dCMP covalently bonded to the N5 position of tetrahydrofolate by means of CH_2. An elimination reaction then generates an enzyme-linked C5-methylene-dCMP. Addition of water to the *exo*-CH_2 leads to the elimination of the cysteinyl residue as well as 5-hydroxymethyl-dCMP release.

[1]K. L. Graves & L. W. Hardy (1994) Biochemistry **33**, 13049.
[2]M. H. Lee, M. Gautam-Basak, C. Wooley & E. G. Sander (1988) Biochemistry **27**, 1367.

Selected entries from *Methods in Enzymology* [vol, page(s)]:
General discussion: **6**, 131
Assay: **6**, 131
Other molecular properties: mechanism, **64**, 127; occurrence, **6**, 131; properties, **6**, 135; purification, *Escherichia coli*, **6**, 133; site-directed mutagenesis studies, **249**, 106

DEOXYCYTIDYLATE KINASE

This enzyme activity, previously classified as EC 2.7.4.5, is now included with cytidylate kinase [EC 2.7.4.14].

DEOXYCYTIDYLATE C-METHYLTRANSFERASE

This methyltransferase [EC 2.1.1.54] catalyzes the reaction of 5,10-methylenetetrahydrofolate with dCMP to produce dihydrofolate and deoxy-5-methylcytidylate.[1] Other acceptor substrates, albeit weaker, include CMP, dCTP, and CTP.

[1]T.-Y. Feng, J. Tu & T.-T. Kuo (1978) EJB **87**, 29.

2-DEOXY-D-GLUCONATE 3-DEHYDROGENASE

This oxidoreductase [EC 1.1.1.125], also known as 2-keto-3-deoxygluconate oxidoreductase, catalyzes the reaction of 2-deoxy-D-gluconate with NAD^+ to produce 3-dehydro-2-deoxy-D-gluconate and NADH.[1,2]

[1]M. M. Eichhorn & M. A Cynkin (1965) *Biochemistry*, **4**, 159.
[2]P. Allenza & T. G. Lessie (1982) *J. Bacteriol.* **150**, 1340.

2-DEOXYGLUCOSE-6-PHOSPHATASE

This phosphatase [EC 3.1.3.68], also known as 2-deoxyglucose-6-phosphate phosphatase, catalyzes the hydrolysis of 2-deoxy-D-glucose 6-phosphate to produce 2-deoxy-D-glucose and orthophosphate.[1,2] An alternative substrate is fructose 1-phosphate.

[1]P. Sanz, F. Randez-Gil & J. A. Prieto (1994) *Yeast* **10**, 1195.
[2]F. Randez-Gil, A. Blasco, J. A. Prieto & P. Sanz (1995) *Yeast* **11**, 1233.

2-DEOXYGLUCOSIDASE

This glucosidase [EC 3.2.1.112] catalyzes the hydrolysis of a 2-deoxy-D-glucoside to produce an alcohol and 2-deoxy-D-glucose. The enzyme can be easily assayed by using 4-nitrophenyl 2-deoxy-α-D-glucopyranoside as a chromogenic substrate.[1]

[1]Z. N. Canellakis, P. K. Bondy, J. A. May, Jr., M. K. Myers-Robfogel & A. C. Sartorelli (1984) *EJB* **143**, 159.

DEOXYGUANOSINE KINASE

This phosphotransferase [EC 2.7.1.113] catalyzes the reaction of ATP with 2′-deoxyguanosine (or 2′-deoxyinosine) to produce ADP and 2′-deoxyguanosine 5′-phosphate (or 2′-deoxyinosine 5′-phosphate).[1-4] This enzyme reaction is the rate-limiting step in the so-called alternative dNTP biosynthetic pathway which commences with phosphorylation of deoxynucleosides derived from endogenous dNTP breakdown or uptake. The human enzyme phosphorylates both the β-D and β-L stereoisomers.[4] The beef liver enzyme has a random Bi Bi kinetic mechanism. *See also Deoxyadenosine Kinase; Deoxycytidine Kinase; Deoxynucleoside Kinase*

[1]J. Barker & R. A. Lewis (1981) *BBA* **658**, 111.
[2]I. Park & D. H. Ives (1988) *ABB* **266**, 51.
[3]T. G. Petrakis, E. Ktistaki, L. Wang, S. Eriksson & I. Talianidis (1999) *JBC* **274**, 24726.
[4]I. Park & D. H. Ives (1995) *J. Biochem.* **117**, 1058.

Selected entries from *Methods in Enzymology* [vol, page(s)]:
General discussion: 51, 346
Assay: 51, 346, 347

Other molecular properties: isoelectric point, **51**, 353; molecular weight, **51**, 353; pH optimum, **51**, 353; purification by affinity chromatography, **51**, 348

DEOXYGUANOSINE-TRIPHOSPHATE TRIPHOSPHOHYDROLASE

This enzyme [EC 3.1.5.1], also known as dGTPase and deoxy-GTPase, catalyzes the hydrolysis of dGTP to produce deoxyguanosine and triphosphate.[1-3] Other substrates include GTP. Kinetic analyses demonstrate a requirement for two divalent cations.[1]

[1]D. N. Frick, D. J. Weber, J. R. Gillespie, M. J. Bessman & A. S. Mildvan (1994) *JBC* **269**, 1794.
[2]D. Seto, S. K. Bhatnagar & M. J. Bessman (1988) *JBC* **263**, 1494.
[3]S. R. Kornberg, I. R. Lehman, M. J. Bessman, E. S. Simms & A. Kornberg (1957) *JBC* **233**, 159.

4-DEOXY-L-*threo*-5-HEXOSULOSE-URONATE KETOL-ISOMERASE

This isomerase [EC 5.3.1.17], also known as 5-keto-4-deoxyuronate isomerase, catalyzes the interconversion of 4-deoxy-L-*threo*-5-hexosulose uronate and 3-deoxy-D-*glycero*-2,5-hexodiulosonate.[1]

[1]J. Preiss & G. Ashwell (1963) *JBC* **238**, 1577.

Selected entries from *Methods in Enzymology* [vol, page(s)]:
General discussion: 9, 602
Assay: 9, 602
Other molecular properties: properties, **9**, 604; purification, *Pseudomonas*, **9**, 603; stability, **9**, 604

DEOXYHYPUSINE MONOOXYGENASE

This oxidoreductase [EC 1.14.99.29], also known as deoxyhypusine hydroxylase, catalyzes the reaction of AH_2 with dioxygen and an N^6-(4-aminobutyl)-L-lysyl residue (*i.e.*, a deoxyhypusyl residue) in a protein to produce A, water, and an N^6-[(R)-4-amino-2-hydroxybutyl]-L-lysyl residue (*i.e.*, a hypusyl residue) in the protein (where AH_2 is the reduced acceptor substrate and A is the oxidized acceptor).[1-3] A ferric ion appears to be essential for enzymatic activity.[3]

[1]M. H. Park, H. L. Cooper & J. E. Folk (1982) *JBC* **257**, 7217.
[2]A. Abbruzzese, M. H. Park & J. E. Folk (1986) *JBC* **261**, 3085.
[3]R. Csonga, P. Ettmayer, M. Auer, C. Eckerskorn, J. Eder & H. Klier (1996) *FEBS Lett.* **380**, 209.

DEOXYHYPUSINE SYNTHASE

This highly specific enzyme [EC 2.5.1.46; formerly classified as EC 1.1.1.249], also known as spermidine dehydrogenase and (4-aminobutyl)lysine synthase, catalyzes the NAD^+-dependent reaction of spermidine with [precursor-eIF-5A]-lysine to produce 1,3-diaminopropane and [precursor-eIF-5A]-deoxyhypusine.

This unique aminoacyl residue, also known as N^{ε}-(4-aminobutyl)lysine, is the immediate precursor to hypusine (N^{ε}-(4-amino-2-hydroxybutyl)lysine) found solely in the eukaryotic initiation factor 5A (eIF-5A). The product of this synthase is the substrate of deoxyhypusine monooxygenase [EC 1.14.99.29]. This modified aminoacyl residue has a crucial role in cell proliferation and viability.[1–3]

[1] E. C. Wolff, J. E. Folk & M. H. Park (1997) *JBC* **272**, 15865
[2] Y. A. Joe, E. C. Wolff, Y. B. Lee & M. H. Park (1997) *JBC* **272**, 32679.
[3] D. Ober & T. Hartmann (1999) *JBC* **274**, 32040.

Selected entries from *Methods in Enzymology* [vol, page(s)]:
Other molecular properties: transition state and multisubstrate analogue, **249**, 304

DEOXYLIMONATE A-RING-LACTONASE

This lactonase [EC 3.1.1.46] catalyzes the hydrolysis of deoxylimonate to produce deoxylimonic acid D-ring-lactone (*i.e.*, the A ring is opened while the D-ring-lactone remains intact).[1]

[1] S. Hasegawa, R. D. Bennett & C. P. Verdon (1980) *Phytochemistry* **19**, 1445.

DEOXYNUCLEOSIDE KINASE

This phosphotransferase [EC 2.7.1.145], also known as multispecific deoxynucleoside kinase, multisubstrate deoxyribonucleoside kinase, and multifunctional deoxynucleoside kinase, catalyzes the reaction of ATP with a 2'-deoxynucleoside to produce ADP and the corresponding 2'-deoxynucleoside 5'-phosphate.[1,2] *See also Deoxyadenosine Kinase; Deoxyguanosine Kinase*

[1] B. Munch-Petersen, J. Piskur & L. Søndergaard (1998) *JBC* **273**, 3926.
[2] B. Munch-Petersen, W. Knecht, C. Lenz, L. Søndergaard & J. Piskur (2000) *JBC* **275**, 6673.

(DEOXY)NUCLEOSIDE-PHOSPHATE KINASE

This phosphotransferase [EC 2.7.4.13] catalyzes the reaction of ATP with a deoxynucleoside monophosphate to produce ADP and a deoxynucleoside diphosphate.[1] The enzyme can also utilize dATP as the phosphoryl-donor substrate as well. Acceptor substrates include dAMP, dTMP, dGMP, dCMP, and 5-bromo-dUMP.[2] *See also specific enzymes; for example, Deoxycytidylate Kinase*

[1] M. J. Bessman, S. T. Herriott & M. J. Van Bibber Orr (1965) *JBC* **240**, 439.
[2] A. Teplyakov, P. Sebastiao, G. Obmolova, A. Perrakis, G. S. Brush, M. J. Bessman & K. S. Wilson (1996) *EMBO J.* **15**, 3487.

Selected entries from *Methods in Enzymology* [vol, page(s)]:
General discussion: **6**, 166, 170, 173
Assay: **6**, 171, 173
Other molecular properties: deoxynucleoside triphosphate synthesis and, **29**, 239; properties, **6**, 173, 176; purification (*Escherichia coli*, **6**, 172; T2-infected *E. coli*, **6**, 173)

DEOXYNUCLEOTIDE 3'-PHOSPHATASE

This enzyme [EC 3.1.3.34], also known as 3'-deoxynucleotidase, catalyzes the hydrolysis of a deoxynucleoside 3'-phosphate to produce a deoxynucleoside and orthophosphate.[1–4]

[1] A. Becker & J. Hurwitz (1967) *JBC* **242**, 936.
[2] C. C. Richardson (1969) *Ann. Rev. Biochem.* **38**, 795.
[3] G. Magnusson (1971) *EJB* **20**, 225.
[4] Y. Habraken & W. G. Verly (1988) *EJB* **171**, 59.

3-DEOXYOCTULOSONASE

This enzyme [EC 3.2.1.144] catalyzes the hydrolysis of a 3-deoxyoctulosonyl-lipopolysaccharide to produce 3-deoxyoctulosonate and the lipopolysaccharide, releasing β-linked 3-deoxy-D-*manno*-octulosonate from different lipopolysaccharides.[1]

[1] Y. T. Li, L. X. Wang, N. V. Pavlova, S. C. Li & Y. C. Lee (1997) *JBC* **272**, 26419.

3-DEOXY-D-*manno*-OCTULOSONATE ALDOLASE

This aldolase [EC 4.1.2.23] catalyzes the reversible conversion of 3-deoxy-D-*manno*-octulosonate to pyruvate and D-arabinose.[1,2] Other substrates include 3-deoxy-D-*arabino*-heptulosonate and 3-deoxy-D-*erythro*-hexulosonate.

[1] W. A. Wood (1972) *The Enzymes*, 3rd ed., **7**, 281.
[2] M. A. Ghalambor & E. C. Heath (1966) *JBC* **241**, 3222.

Selected entries from *Methods in Enzymology* [vol, page(s)]:
General discussion: **8**, 134; **9**, 534
Assay: **9**, 534
Other molecular properties: 3-deoxyoctulosonate synthesis, **8**, 134; properties, **9**, 537; purification, *Aerobacter cloacae*, **9**, 535; specificity, **9**, 537; stability, **9**, 537

3-DEOXY-*manno*-OCTULOSONATE CYTIDYLYLTRANSFERASE

This transferase [EC 2.7.7.38], also known as CMP-3-deoxy-D-*manno*-octulosonate pyrophosphorylase and CMP-2-keto-3-deoxyoctulosonate synthase, catalyzes the reaction of CTP with 3-deoxy-D-*manno*-octulosonate (the β-anomer) to produce pyrophosphate (or, diphosphate) and CMP-3-deoxy-D-*manno*-octulosonate.[1,2] The product is relatively unstable (half-life = 34 min); under neutral conditions, breakdown is thought to proceed by means of glycosidic cleavage.[3] An enzyme with similar properties has been reported in maize.[4]

[1]M. A. Ghalambor & E. H. Heath (1966) *JBC* **241**, 3216.
[2]P. H. Ray, C. D. Benedict & H. Grasmuk (1981) *J. Bacteriol.* **145**, 1273.
[3]C. H. Lin, B. W. Murray, I. R. Ollmann & C. H. Wong (1997) *Biochemistry* **36**, 780.
[4]J. Royo, E. Gomez & G. Hueros (2000) *JBC* **275**, 24993.
Selected entries from ***Methods in Enzymology*** [**vol**, page(s)]:
General discussion: 8, 221; 83, 535
Assay: 8, 221; **83**, 535

3-DEOXY-*manno*-OCTULOSONATE-8-PHOSPHATASE

This enzyme [EC 3.1.3.45] catalyzes the hydrolysis of 3-deoxy-D-*manno*-octulosonate 8-phosphate to produce 3-deoxy-D-*manno*-octulosonate and orthophosphate.[1]

[1]P. H. Ray & C. D. Benedict (1980) *J. Bacteriol.* **142**, 60.
Selected entries from ***Methods in Enzymology*** [**vol**, page(s)]:
General discussion: 83, 531
Assay: 83, 531

3-DEOXY-2-OCTULOSONIDASE

This glycosidase [EC 3.2.1.124] catalyzes the endohydrolysis of the β-ketopyranosidic linkages of 3-deoxy-D-*manno*-2-octulosonate in capsular polysaccharides (*i.e.*, it facilitates the depolymerization of capsular polysaccharides containing 3-deoxy-2-octulosonide in the cell walls of organisms such as *Escherichia coli*).[1]

[1]F. Altmann, B. Kwiatkowski, S. Stirm, L. März & F. M. Unger (1986) *BBRC* **136**, 329.

DEOXYRIBODIPYRIMIDINE ENDONUCLEOSIDASE

This nucleosidase [EC 3.2.2.17], also known as pyrimidine dimer DNA-glycosylase, repairs pyrimidine dimer-containing DNA by catalyzing the cleavage of the *N*-glycosidic bond between the 5′-pyrimidine residue in cyclobutadipyrimidine (in DNA) and the corresponding deoxy-D-ribose residue.[1-7] Because the bacteriophage-T4 and *Micrococcus luteus* enzymes catalyze β-elimination reactions, they are not true endonucleases.[4]

[1]S. McMillan, H. J. Edenberg, E. H. Radany, R. C. Friedberg & E. C. Friedberg (1981) *J. Virol.* **40**, 211.
[2]R. H. Grafstrom, L. Park & L. Grossman (1982) *JBC* **257**, 13465.
[3]E. H. Radany & E. C. Friedberg (1982) *J. Virol.* **41**, 88.
[4]V. Bailly, B. Sente & W. G. Verly (1989) *BJ* **259**, 751.
[5]J. F. Garvish & R. S. Lloyd (1999) *JBC* **274**, 9786.
[6]T. Lindahl (1982) *Ann. Rev. Biochem.* **51**, 61.
[7]B. K. Duncan (1981) *The Enzymes*, 3rd ed., **14**, 565.

DEOXYRIBODIPYRIMIDINE PHOTOLYASE

This flavoenzyme [EC 4.1.99.3], also known as photo-reactivating enzyme and DNA photolyase, catalyzes the light-dependent conversion of a cyclobutadipyrimidine in DNA to two pyrimidine residues in the DNA (*i.e.*, the reactivation by light of the irradiated DNA).[1-7] A second chromophore, such as pterin or F420, serves as a light-harvesting antenna.

While the optimal wavelength varies from one organism to another, it is typically between 300 and 600 nm. Energy transfer from the antenna to the reduced flavin takes place as a Förster-type dipole–dipole interaction. The phosphodiester backbone in the strand containing the pyrimidine dimer and the cyclobutane ring of the dimer itself are the important substrate-recognition determinants.[1] The excited flavin transfers an electron to the pyrimidine dimer, forming a ketyl radical that decomposes, transferring the electron back to the flavin. The repaired DNA is then released.[5] A similar reaction is observed with RNA, but is probably the consequence of catalysis by another enzyme.

[1]I. Husain, G. B. Sancar, S. R. Holbrooks & A. Sancar (1987) *JBC* **262**, 13188.
[2]G. B. Sancar & A. Sancar (1987) *Trends Biochem. Sci.* **12**, 259.
[3]B. M. Sutherland (1981) *The Enzymes*, 3rd ed., **14**, 481.
[4]H. Werbin (1977) *Photochem. Photobiol.* **26**, 675.
[5]B. A. Palfey & V. Massey (1998) *CBC* **3**, 83.
[6]A. Sancar (1994) *Biochemistry* **33**, 2.
[7]E. J. Brush & J.W. Kozarich (1992) *The Enzymes*, 3rd ed., **20**, 317.
Selected entries from ***Methods in Enzymology*** [**vol**, page(s)]:
General discussion: 258, 319
Assay: coupled enzyme assay, **258**, 325; DNA repair quantum yield measurement, **258**, 326; gel retardation, **258**, 325; spectrophotometric assay, **258**, 325
Other molecular properties: apoenzyme preparation, **258**, 323; electron paramagnetic resonance (instrumentation, **258**, 327; isotope effects, **258**, 342; time-resolved spectra, **258**, 337); flash photolysis, **258**, 326 (camera flash photolysis, **258**, 332; nanosecond laser flash photolysis, **258**, 332; picosecond laser flash photolysis, **258**, 335); mapping of cyclobutane-pyrimidine dimers, **304**, 447; photosensitizing cofactors, **258**, 319, 329; purification from *Escherichia coli* (affinity chromatography, **258**, 322; ammonium sulfate precipitation, **258**, 322; cell growth, **258**, 321; cell lysis, **258**, 322; gel filtration, **258**, 323; hydroxylapatite chromatography, **258**, 323); reconstitution, **258**, 323; substrate synthesis, **258**, 324; thymine dimers and, **12A**, 30; tryptophans (binding contribution, **258**, 320, 327; electron transfer, **258**, 320, 343; flavin radical photoreduction, **258**, 332; photosensitized repair of pyrimidine dimers, **258**, 329; site-directed mutagenesis, **258**, 320, 328, 331, 334)

DEOXYRIBONUCLEASE

These enzymes, which catalyze the hydrolytic scission of phosphodiester bonds in DNA, are broadly classified as endodeoxyribonucleases generating 5′-phosphomonoesters [EC 3.1.21.x], endodeoxyribonucleases yielding products other than 5′-phosphomonoesters [EC 3.1.22.x], site-specific endodeoxyribonucleases acting on altered bases [EC 3.1.25.x], and exodeoxyribonucleases producing 5′-phosphomonoesters [EC 3.1.11.x].[1–3] **See** *specific enzyme*

(1) Deoxyribonuclease I [EC 3.1.21.1], also known as pancreatic DNase and thymonuclease, catalyzes the endonucleolytic cleavage of DNA, preferring dsDNA, to generate a 5′-phosphodinucleotide and 5′-phosphooligonucleotide end products.

(2) Deoxyribonuclease IV (phage T_4-induced) [EC 3.1.21.2], also known as endodeoxyribonuclease IV (phage T_4-induced) and endonuclease IV, catalyzes the endonucleolytic cleavage of DNA (preferring single-stranded DNA) to generate 5′-phosphooligonucleotide end products.

(3) Type I site-specific deoxyribonuclease [EC 3.1.21.3], also known as type I restriction enzyme, catalyzes the ATP- (or dATP-) and *S*-adenosyl-L-methionine-dependent endonucleolytic cleavage of DNA to give random, double-stranded fragments with terminal 5′-phosphates. ATP (or dATP) is simultaneously hydrolyzed to ADP and orthophosphate. This large group of enzymes recognizes specific short DNA sequences and hydrolyzes bonds at sites remote from the recognition sequence. The $\Delta G_{\text{ATP-hydrolysis}}$ is probably converted into mechanical work needed to promote the endonucleolytic cleavage reaction. **See** *Energase*

(4) Type II site-specific deoxyribonuclease [EC 3.1.21.4], also known as type II restriction enzyme, catalyzes the magnesium-dependent endonucleolytic cleavage of DNA to give specific, double-stranded fragments with terminal 5′-phosphates.

(5) Type III site-specific deoxyribonuclease [EC 3.1.21.5], also known as type III restriction enzyme, catalyzes the ATP-dependent endonucleolytic cleavage of DNA to give specific, double-stranded fragments with terminal 5′-phosphates. Note that the enzyme does not hydrolyze the ATP. While *S*-adenosyl-L-methionine stimulates the reaction, it is not absolutely required. These enzymes recognize specific short DNA sequences and hydrolyze bonds a short distance away from the recognition sequence.

(6) Deoxyribonuclease II [EC 3.1.22.1], also known as pancreatic DNase II, catalyzes the endonucleolytic hydrolysis of DNA (preferring double-stranded DNA) to produce nucleoside 3′-phosphate and 3′-phosphooligonucleotide end products.

(7) Deoxyribonuclease V [EC 3.1.22.3], also known as endodeoxyribonuclease V, catalyzes the endonucleolytic cleavage of DNA at apurinic or apyrimidinic sites to generate products with a 3′-phosphate.

(8) Crossover junction endoribonuclease [EC 3.1.22.4], also known as Holliday junction nuclease, catalyzes the endonucleolytic hydrolysis of a bond(s) in DNA at a junction such as a reciprocal single-stranded crossover between two homologous DNA duplexes (*i.e.*, Holliday junction).

(9) Deoxyribonuclease X [EC 3.1.22.5] catalyzes the endonucleolytic cleavage of supercoiled plasma DNA to produce linear DNA duplexes. The enzyme exhibits a preference for supercoiled DNA.

(10) Deoxyribonuclease (pyrimidine dimer) [EC 3.1.25.1] catalyzes the endonucleolytic hydrolysis of a bond in DNA near pyrimidine dimers to generate products with 5′-phosphates. The enzyme acts on damaged strands of DNA, 5′ from the damaged site.

(11) DNA-(apurinic or apyrimidinic site) lyase [EC 4.2.99.18, formerly listed with EC 3.1.25.2] acts on the C–O–P bond 3′ to the apurinic or apyrimidinic site in DNA. This bond is broken by a β-elimination reaction, leaving a 3′-terminal unsaturated sugar and a product with a terminal 5′-phosphate. Note that this "nicking" of the phosphodiester bond is a lyase-type reaction, not hydrolysis.

(12) Exodeoxyribonuclease I [EC 3.1.11.1] catalyzes the hydrolysis of bonds in DNA (preferably single-stranded), acting progressively in a 3′- to 5′-direction, releasing 5′-phosphomononucleotides.

(13) Exodeoxyribonuclease III [EC 3.1.11.2] catalyzes the hydrolysis of bonds in DNA (preferably double-stranded),

acting progressively in a 3'- to 5'-direction, releasing 5'-phosphomononucleotides.

(14) Exodeoxyribonuclease (λ-induced) [EC 3.1.11.3], also known as lambda exonuclease and λ exonuclease, catalyzes the hydrolysis of double-stranded DNA, acting progressively in a 5'- to 3'-direction and releasing 5'-phosphomononucleotides. This enzyme does not attack single-strand breaks.

(15) Exodeoxyribonuclease (phage SP3-induced) [EC 3.1.11.4] catalyzes exonucleolytic cleavage of DNA (preferring single-stranded DNA) in the 5'- to 3'-direction to produce 5'-phosphodinucleotides.

(16) Exodeoxyribonuclease V [EC 3.1.11.5] catalyzes the exonucleolytic cleavage of double-stranded DNA in the presence of ATP, in either the 5'- to 3'- or the 3'- to 5'-direction to produce 5'-phosphooligonucleotides. Note that this enzyme exhibits a DNA-dependent ATP hydrolysis activity. The $\Delta G_{\text{ATP-hydrolysis}}$ is probably converted into mechanical work needed to promote the endonucleolytic cleavage reaction. *See Energase*

(17) Exodeoxyribonuclease VII [EC 3.1.11.6] catalyzes the exonucleolytic cleavage of DNA (preferring single-stranded DNA) in either the 5'- to 3'- or the 3'- to 5'-direction to produce 5'-phosphomononucleotides.

[1] P. D. Boyer, ed., *The Enzymes*, 3rd ed., vols. 14 and 21.
[2] I. R. Lehman (1971) *The Enzymes*, 3rd ed., 4, 251.
[3] G. Bernardi (1971) *The Enzymes*, 3rd ed., 4, 271.

Selected entries from *Methods in Enzymology* [vol, page(s)]:
General discussion: 27, 140; 34, 4, 468; 68, 40
Other molecular properties: adenylate cyclase purification, 38, 159, 172; aminoacyl-tRNA synthetase preparation and, 20, 76; 29, 537; arabinose dehydrogenase, preparation of, 5, 343; bacterial particles, in, 5, 61; bacteriophage lysates and, 21, 451; bacteriophage preparation and, 30, 718; cAMP-binding protein preparation, 38, 369; cell-adhesion studies, in, 32, 601; chain elongation factor extraction and, 20, 285, 296, 304, 308; chymotrypsinogen B samples, in, estimation of, 19, 83; deoxyribonuclease-agarose, 34, 517; deoxyribonucleoprotein extraction and, 12B, 68; determination of tryptophan in, 25, 50, 51; digestion of chromosomes and nuclei, 40, 87; digest mix, 68, 136; disrupted cells and, 5, 58; DNA-cellulose and, 12, 201; DNA fragmentation and, 29, 258, 272, 380, 445; DNA-free crude extract and, 21, 202; DNA polymerase assays and, 29, 56, 57; DNA polymerases and, 29, 42, 43, 60, 64, 69; DNA–RNA hybridization and, 21, 362; dogfish liver (immunological assay of, 12A, 270; purification, 12A, 270); electrodecantation, 5, 39; elongation factor Tu–GDP complex and, 30, 225; extraction of, for (FAD-linked malate dehydrogenase, 13, 142; nuclear RNA, 65, 770, 772; oxalyl-CoA decarboxylase, 13, 370; phosphoenolpyruvate carboxylase, 13, 295); formate dehydrogenase, purification of, 9, 362; fragments, sequence analysis, 29, 282 (dinucleotides, 29, 286; labeling of 5'-terminus, 29, 284; mononucleotides, 29, 284; reagents, 29, 283;

tetranucleotides, 29, 290; trinucleotides, 29, 287); functional, recovery in SDS gels, 91, 268; galactose dehydrogenase, preparation of, 5, 340; glyoxylate dehydrogenase, purification of, 9, 344; hybrid nucleic acid and, 12B, 655; immunodiffusion and, 6, 851; induced enzyme synthesis and, 12B, 808, 811, 817, 818; infrared spectrum, 26, 468; initiation factor isolation and, 20, 239, 252; initiation factor purification and, 30, 8, 12, 26, 35, 42, 47; isolation of acidic chromatin proteins, 40, 174; *lac* repressor purification and, 21, 484; mass screening of mutants (exonuclease I assay, 29, 189; exonuclease III assay, 29, 190; general procedures, 29, 185; microassays in genetic experiments, 29, 191; outline of method, 29, 180; possibilities and limitations of method, 29, 192; special equipment, 29, 183); membrane fractions and, 12B, 796; membrane nucleic acids and, 12B, 803; membrane preparation, 55, 782, 803; membrane vesicle preparation, 56, 381, 383, 402; metals and, 3, 696; micrococcal, nearest-neighbor base sequences and, 6, 741, 743; mitochondrial DNA, 56, 5, 6, 186; *Neurospora*, 27, 77; nuclear RNA isolation and, 12A, 611, 612; 65, 770, 772; osmotic shock and, 12B, 844; petite mutant DNA, 56, 156, 162, 163; phage preparation and, 29, 233; phage SP3, methylation sites and, 29, 291; phosphodiesterase purification, 38, 251; plant nucleic acid preparation and, 12B, 108, 111; polyribosome preparation and, 12A, 504, 511, 512, 517, 523; 30, 471, 472; preparation of aminoacyl-tRNA synthetase, 59, 314 (*Bacillus* ribosomes, of, 59, 439; CTP(ATP):tRNA nucleotidyltransferase, of, 59, 123; *Escherichia coli* ribosomes, of, 59, 444, 445, 451, 554; hydrogenase/thiosulfate-sulfite reductases extract, of, 53, 618; MS2 RNA, of, 59, 299; polysomes, of, 59, 355, 365); preparation, of cellular RNA, 65, 582; protein digestion, concentration and incubation period, 40, 38; protein isolation, in, 52, 153, 160, 178; protein synthesis and, 6, 17, 21 (*in vitro*, 20, 531); protoplast ghost preparation, 55, 181; purification, 65, 719; purification of (adenylate nucleosidase, 51, 268; dihydroorotate dehydrogenase, 51, 61; double-stranded RNA, 60, 552; ω-hydroxylase, 53, 358; *p*-hydroxybenzoate hydroxylase, 53, 546; mitochondria, 12A, 468; nitrate reductase, 53, 348; nitrite reductase, 53, 643; nitrogenase, 53, 316, 318, 322, 327; *Nitrosomonas ubiquinone*, 53, 639; *Pseudomonas* cytochrome oxidase, 53, 650; rubredoxin, 53, 342; rubredoxin:NAD⁺ oxidoreductase, 53, 619; uridine:cytidine kinase, 51, 311); purple-membrane isolation, for, 31, 670; recipient cell purification and, 21, 160; recovery in SDS gels, 91, 268; release factor preparation and, 20, 372; replicative form phage RNA and, 12A, 614; reticulocyte RNA and, 12A, 613; reverse transcriptase and, 29, 148, 156; ribonuclease T₁, and, 12A, 234; ribonuclease T₂, presence in preparation of, 12A, 245; ribosome preparation and, 20, 392, 450, 458, 466; 30, 454, 555; ribosome wash protein preparation and, 30, 56, 71; RNA and, 12A, 196; RNA isolation and, 30, 659, 660 (messenger, 30, 651; nuclear, 30, 609); RNA polymerase, presence in assays of, 12B, 556, 558, 576; RNA polymerase preparation and, 21, 510, 514; RNA preparation and, 21, 476; RNA transcription and, 29, 266; seminal fluid, 3, 698; sequence analysis and, 29, 282, 283, 284, 287; single strand breaks and, 21, 414; single-stranded phage RNA, in assay for, 12B, 885; snail, 27, 473, 475 (hyperchromic shift and, 29, 344; preparation, 29, 343); specificity, 29, 349; spheroplasts, 56, 175; spore preparation and, 21, 432; spore purification and, 29, 503; terminal 5'-phosphomonoester formation by, 3, 769; thymus, 3, 697; tissue culture, 58, 125, 126; transcription inhibitor and, 29, 210; transformation assay and, 12B, 868; use in bacterial membrane preparation, 22, 106; virus protein purification by, 1, 40; yeast, 2, 445; yeast cell extract and, 29, 717

DEOXYRIBONUCLEASE I

This enzyme [EC 3.1.21.1], also known as pancreatic DNase, DNase, and thymonuclease, catalyzes the endonucleolytic cleavage of DNA (preferably double-stranded) to produce a 5'-phosphodinucleotide and a 5'-phosphooligonucleotide.[1-5] Similar enzymes include streptococcal DNase (streptodornase), T4 endonuclease II, T7 endonuclease II, *Escherichia coli* endonuclease I, nicking nuclease of calf thymus, and colicin E2 and E3.

The enzyme binds tightly in the minor groove of dsDNA, and to the sugar-phosphate backbones of both strands. Binding induces a widening of the minor groove and a bending toward the major groove. Cleavage rates appear to be dependent on the DNA base sequences.[5] *See also Deoxyribonuclease*

[1] S. Moore (1981) *The Enzymes*, 3rd ed., **14**, 281.
[2] J. D. Love & R. R. Hewitt (1979) *JBC* **254**, 12588.
[3] M. Laskowski, Sr., (1971) *The Enzymes*, 3rd ed., **4**, 289.
[4] I. R. Lehman (1971) *The Enzymes*, 3rd ed., **4**, 251.
[5] D. Suck (1994) *J. Mol. Recognit.* **7**, 65.

Selected entries from *Methods in Enzymology* [vol, page(s)]:
General discussion: **2**, 437; **170**, 264
Assay: **2**, 437; fluorometric assay and, **21**, 267
Other molecular properties: actin polymerization assay, in, **85**, 204; affinity labeling, **87**, 485, 486; DNaseI-agarose, in isolation of gelsolin–actin complexes, **215**, 91; antisera and, **2**, 447; assay for (actin, **162**, 261, 270; actin-depolymerizing factor, **196**, 136; actin filaments, **215**, 54; protein–DNA interactions *in vitro*, **152**, 731); bacterial spores and, **12B**, 99; binding to microfilaments, **79**, 466; borate and, **2**, 443; **12A**, 101; CCAAAT/enhancer-binding protein, footprinting assay, **274**, 102, 107; chymotrypsinogen B and, **2**, 10, 17; cleavage of bent DNA, **208**, 351; degradation of nucleic acids, **3**, 757; digestion of (nuclei, **170**, 326; nucleosomes, **170**, 264); disaggregation of murine embryonic gonads, in, **225**, 71; DNA joining enzyme and, **21**, 315, 329; DNA modification with, **152**, 110; DNA polymerase and, **12B**, 590; DNA preparation and, **29**, 23, 39, 82, 87, 90, 100, 132; DNA sequencing, in, **155**, 93; footprinting, in analysis of reconstituted nucleosome core particle authenticity, **170**, 580; footprinting, in mapping of OxyR binding, **234**, 222; footprinting of protein–DNA contacts, in, (comparison with hydroxyl radical method, **155**, 551; procedure, **155**, 542); footprinting reactions, **208**, 359; footprint titrations, in analysis of protein–DNA interactions, **130**, 132; GA-binding protein, footprinting assay, **274**, 102, 108, 110; GAGA factor, footprinting assay, **274**, 293; GAL4-VP16, footprinting assay, **274**, 140, 142; genomic footprinting with, **225**, 573; hydrolysis of DNA, **234**, 6; hypersensitive sites in chromatin, analysis, **170**, 269; inhibitor, **2**, 443; **12A**, 101 (deoxyribonuclease preparation and, **2**, 446); inhibitor protein, **34**, 517; interferon-β enhancer complex assembly assay, **274**, 168, 173; leukemic tissue, **2**, 447; mitochondrial DNA and, **12A**, 533, 536; mitochondrial transcription termination factor, footprinting assay, **264**, 166, 173; oligonucleotide identification and, **12A**, 283, 290, 320; oligonucleotide preparation and, **12B**, 608; pancreatic **27**, 473, 475; **34**, 464, 468 (activated DNA and, **29**, 4, 17, 28, 54, 55; assay, of *Bsu* endonuclease, **65**, 114, 116; digestion, partial, of short DNA fragments, **65**, 627, 634; DNA duplex labeling studies, **65**, 44, 53; DNA ligase substrate and, **21**, 334; DNA–protein interactions, **65**, 854; DNA ring form, **12B**, 428; fluorometric assay, **21**, 263, 267; hyperchromic shift and, **29**, 344 [linkages cleaved, **29**, 288; 3′-penultimate nucleotide determination and, **29**, 348, 349; specificity, **29**, 352; 5′-terminal nucleotide determination and, **29**, 347]; oligonucleotide preparation and, **21**, 179; preparation, of polyoma virus-induced cellular RNA, **65**, 727; primed RNA synthesis, **65**, 498; properties, **2**, 441; purification, **2**, 438; sequence determination, **47**, 91; stop solution, of cell-free protein synthesis system, **65**, 806; stop solution, of transformation reaction, **65**, 143; synthesis of cRNA, **65**, 725); plasmolyzed cells and, **29**, 227; post-Fenton reaction digestion of DNA, **234**, 53; removal of contaminating nuclear DNA, **264**, 444;

ribonuclease I and, **6**, 252; ribonuclease-free, preparation of *Escherichia coli* crude extract, **59**, 194; sensitivity of nucleosome core particles, internal structure, **170**, 107; streptococcal, **2**, 446; structural probing of DNA *in vitro*, **212**, 299; thiol-disulfide interchange in, thiol pKₐ values, **143**, 134; thymus, **3**, 697, 698 (purification, **2**, 443); transcription factor binding to nucleosomes, footprinting assay, **274**, 280

DEOXYRIBONUCLEASE II

This enzyme [EC 3.1.22.1], also known as DNase II and lysosomal DNase II, catalyzes the endonucleolytic cleavage of DNA (preferring double-stranded DNA) to produce a nucleoside 3′-phosphate and a 3′-phosphooligonucleotide.[1–4] Similar enzymes include crab testis DNase, snail DNase, salmon testis DNase, liver acid DNase, and human acid DNases of gastric mucosa and cervix. Studies of the human lymphoblast enzyme indicate that every available site along the duplexed portion of a paired oligonucleotide substrate is cleaved with the exception of the last four nucleotides in the sequence.[4] *See also Deoxyribonuclease*

[1] W. Hörz, F. Miller, G. Klobeck & H. G. Zachau (1980) *J. Mol. Biol.* **144**, 329.
[2] G. Bernardi (1971) *The Enzymes*, 3rd ed., **4**, 271.
[3] A. Maly (1971) *Collect. Czech. Chem. Commun.* **36**, 2966 and 2980.
[4] I. Harosh, D. M. Binninger, P. V. Harris, M. Mezzina & J. B. Boyd (1991) *EJB* **202**, 479.

Selected entries from *Methods in Enzymology* [vol, page(s)]:
General discussion: **170**, 264
Other molecular properties: brain homogenates, in, **31**, 464, 466, 467; chromaffin granule, in, **31**, 381; digestion of nucleosomes, **170**, 264; lysosomes, of, **10**, 12; **31**, 330; macrophage fractions, in, **31**, 340; removal of, **21**, 151; snail, **27**, 473, 475; spleen, **2**, 444 (hyperchromic shift and, **29**, 344; preparation, **29**, 343)

DEOXYRIBONUCLEASE IV (PHAGE T₄-INDUCED)

This enzyme [EC 3.1.21.2], also known as endodeoxyribonuclease IV (phage T4-induced) and endonuclease IV, catalyzes the endonucleolytic cleavage of DNA (preferring single-stranded DNA) to produce 5′-phosphooligonucleotide end-products.[1] Similar enzymes include deoxyribonuclease V (mammalian), *Aspergillus sojae* DNase, *Bacillus subtilis* endonuclease, T7 endonuclease I, *Aspergillus* DNase K2, vaccinia virus DNase VI, T4 endonuclease III, yeast DNase, and *Chlorella* DNase.

[1] P. D. Sadowski & J. Hurwitz (1969) *JBC* **244**, 6192.

Selected entries from *Methods in Enzymology* [vol, page(s)]:
Assay: *Escherichia coli*, **65**, 212, 213
Other molecular properties: *Escherichia coli*, **65**, 212 (molecular weight, **65**, 216; optimum reaction conditions, **65**, 216; properties, **65**,

215, 216; purification, **65**, 213; stability, **65**, 215, 216; substrate specificity, **65**, 216); yeast, **2**, 445; yeast cell extract and, **29**, 717

DEOXYRIBONUCLEASE V

This enzyme [EC 3.1.22.3], also known as endodeoxyribonuclease V, catalyzes the endonucleolytic cleavage at apurinic or apyrimidinic sites of DNA to generate products with a 3′-phosphate. DNase V recognizes some peculiarities of abnormal DNA structure. However, these peculiarities are not simply distortions, since some lesions, including pyrimidine dimers, are not substrates. Similar enzymes include *Escherichia coli* endonuclease II, thymus endonuclease, and human placenta endonuclease.[1] *See also DNA-(Apurinic or Apyrimidinic Site) Lyase*

[1]V. Bailly & W. G. Verly (1989) *NAR* **17**, 3617.

Selected entries from ***Methods in Enzymology*** [**vol**, page(s)]:
General discussion: 65, 224

DEOXYRIBONUCLEASE X

This endodeoxyribonuclease [EC 3.1.22.5], first identified in *Escherichia coli*, catalyzes the endonucleolytic cleavage of supercoiled plasma DNA to form linear DNA duplexes.[1] The enzyme exhibits a preference for supercoiled DNA; there is little action on linear, double-stranded DNA. ATP, AMP, and single-stranded DNA are inhibitors.

[1]S. Ghosh & U. DasGupta (1984) *Curr. Trends Life Sci.* **12**, 79.

DEOXYRIBONUCLEASE (PYRIMIDINE DIMER)

This enzyme [EC 3.1.25.1], also known as endodeoxyribonuclease (pyrimidine dimer), catalyzes the endonucleolytic cleavage of DNA near pyrimidine dimers to generate products with 5′-phosphates (hence, the enzyme acts on a damaged DNA strand, 5′ from the damaged site).[1–5] *See also DNA-(Apurinic or Apyrimidinic Site) Lyase*

[1]L. H. Pearl & R. Savva (1995) *Trends Biochem. Sci.* **20**, 421.
[2]P. W. Doetsch & R. P. Cunningham (1990) *Mutat. Res.* **236**, 173.
[3]E. A. Gruskin & R. S. Lloyd (1988) *JBC* **263**, 12728.
[4]T. Inaoka, M. Ishida & E. Ohtsuka (1989) *JBC* **264**, 2609.
[5]S. Riazuddin & L. Grossman (1977) *JBC* **252**, 6280, 6287, and 6294.

Selected entries from ***Methods in Enzymology*** [**vol**, page(s)]:
General discussion: 65, 185
Assay: bacteriophage T$_4$(coupled nuclease assay, **65**, 192, 193; DNA nicking assay, **65**, 194, 195; filter binding assay, **65**, 193, 194); *Escherichia coli*, **65**, 224
Other molecular properties: bacteriophage T$_4$, **65**, 191 (assay by [coupled nuclease assay, **65**, 192, 193; DNA nicking assay, **65**, 194, 195;

filter binding assay, **65**, 193, 194]; cation requirements, **65**, 201; inhibitors, **65**, 201; molecular weight, **65**, 200; properties, **65**, 200, 201; purification, **65**, 196; storage, **65**, 198; substrate specificity, **65**, 200); *Escherichia coli*, **65**, 224 (assay, **65**, 224; cation requirement, **65**, 230; inhibitor, **65**, 230; molecular weight, **65**, 230, 231; pH optimum, **65**, 230; properties, **65**, 230, 231; purification, **65**, 227; substrate specificity, **65**, 230); pyrimidine dimer correndonuclease I, **65**, 185 (assay, **65**, 185; homogeneity, **65**, 188; properties, **65**, 188; purification, **65**, 186; reaction conditions, optimum, **65**, 188, 190; reaction products, **65**, 190; recognition sequence, **65**, 245, 247; role, in cellular repair, **65**, 191; stability, **65**, 188; substrate specificity, **65**, 190); pyrimidine dimer correndonuclease II, **65**, 185 (assay, **65**, 185; homogeneity, **65**, 188; properties, **65**, 188; purification, **65**, 186; reaction conditions, optimum, **65**, 188, 190; reaction products, **65**, 190; role, in cellular repair, **65**, 191; stability, **65**, 188; substrate specificity, **65**, 190); yeast, detection, **234**, 38

2-DEOXYRIBOSE-5-PHOSPHATE ALDOLASE

This class I aldolase [EC 4.1.2.4], also known as phosphodeoxyriboaldolase and deoxyriboaldolase, catalyzes the reversible conversion of 2-deoxy-D-ribose 5-phosphate to D-glyceraldehyde 3-phosphate and acetaldehyde.[1–4] Other substrates include a number of 2-methyl-2-deoxypentose phosphates (for example, 2-methyl-2-deoxy-L-xylose 5-phosphate). The reaction proceeds via the formation of a Schiff base with a lysyl residue. In the reverse direction, acetaldehyde is the compound that forms the Schiff base with the lysyl residue (Lys167 in the *Escherichia coli* enzyme).

In the reverse direction, propanal can substitute for acetaldehyde in a reaction with L-glyceraldehyde 3-phosphate to produce 2-methyl-2-deoxy-L-xylose 5-phosphate. When D-glyceraldehyde 3-phosphate is used in the reaction with propanal, the reaction is significantly less favored, presumably because the product, 2-methyl-2-deoxy-D-ribose 5-phosphate, has a destabilized ring structure attributable to an axial methyl group and an axial carbinol phosphate group.[1]

[1]K. N. Allen (1998) *CBC* **2**, 135.
[2]D. S. Feingold & P. A. Hoffee (1972) *The Enzymes*, 3rd ed., **7**, 303.
[3]O. M. Rosen, P. Hoffee & B. L. Horecker (1965) *JBC* **240**, 1517.
[4]E. Racker (1951) *JBC* **196**, 347.

Selected entries from ***Methods in Enzymology*** [**vol**, page(s)]:
General discussion: 1, 384; **9**, 545, 549; **42**, 276
Assay: 1, 384; **9**, 545, 549; **42**, 276, 277
Other molecular properties: action of, **3**, 187; deoxyribokinase, not present in preparation of; **5**, 309; D-2-deoxyribose-5-phosphate, preparation of, **3**, 186; *Escherichia coli* (properties, **1**, 386; purification, **1**, 385); *Lactobacillus plantarum* (properties, **9**, 548; purification, **9**, 546; specificity, **9**, 548); liver (kinetic properties, **9**, 554; molecular weight, **9**, 553; properties, **9**, 553; purification, **9**, 551); regulation, **51**, 438, 517; *Salmonella typhimurium*, from, **42**, 276 (molecular weight, **42**, 279; properties, **42**, 279; purification, **42**, 277); sources, **9**, 549; stereochemistry, **87**, 148

DEPHOSPHO-CoA KINASE

This phosphotransferase [EC 2.7.1.24] catalyzes the reaction of ATP with 3'-hydroxyl of the ribosyl group on dephospho-CoA to produce ADP and coenzyme A.[1-4] The pig and rat enzymes form a bifunctional system with pantetheine-phosphate adenylyltransferase [EC 2.7.7.3].[4]

[1]S. Skrede & O. Halvorsen (1983) *EJB* **131**, 57.
[2]S. Skrede & O. Halvorsen (1979) *BBRC* **91**, 1536.
[3]T. P. Wang & N. O. Kaplan (1954) *JBC* **206**, 311.
[4]D. M. Worrall & P. K. Tubbs (1983) *BJ* **215**, 153.

Selected entries from *Methods in Enzymology* [vol, page(s)]:
General discussion: 2, 649; **18A**, 358
Assay: 2, 649; **18A**, 359
Other molecular properties: dephospho-CoA phosphorylase preparation, presence in, **2**, 667, 669; bifunctional nature, **18A**, 358; properties, **2**, 651; **18A**, 361; purification, **2**, 650; **18A**, 360; role in CoA biosynthesis, **62**, 237

DEPHOSPHO-[REDUCTASE KINASE] KINASE

This phosphotransferase [EC 2.7.1.110], also known as reductase kinase kinase, catalyzes the reaction of ATP with dephospho-[[3-hydroxy-3-methylglutaryl-CoA reductase (NADPH)] kinase] to produce ADP and [[3-hydroxy-3-methylglutaryl-CoA reductase (NADPH)] kinase] (*i.e.*, this enzyme phosphorylates and activates [hydroxymethylglutaryl-CoA reductase (NADPH)] kinase [EC 2.7.1.109] which, in turn, phosphorylates and inactivates hydroxymethylglutaryl-CoA reductase (NADPH) [EC 1.1.1.34]. Serine/threonine-specific protein phosphatase [EC 3.1.3.16] catalyzes the dephosphorylation of the product formed by dephospho-[reductase kinase] kinase. This protein kinase also catalyzes casein phosphorylation, and the mammalian enzyme is activated by AMP.[1]

[1]J. Weekes, S. A. Hawley, J. Corton, D. Shugar & D. G. Hardie (1994) *EJB* **219**, 751.

DESULFOGLUCOSINOLATE SULFOTRANSFERASE

This highly specific sulfotransferase [EC 2.8.2.24] catalyzes the reaction of 3'-phosphoadenylylsulfate with desulfoglucotropeolin to produce adenosine 3',5'-bisphosphate and glucotropeolin.[1,2] The enzyme participates with thiohydroxamate β-D-glucosyltransferase [EC 2.4.1.195] in the biosynthesis of thioglycosides in cruciferous plants.

[1]T. M. Glendening & J. E. Poulton (1990) *Plant Physiol.* **94**, 811.
[2]J. C. Jain, J. W. D. Groot Wassink, A. D. Kolenovsky & E. W. Underhill (1990) *Phytochemistry* **29**, 1425.

DESULFOHEPARIN SULFOTRANSFERASE

This transferase [EC 2.8.2.8] catalyzes the reaction of 3'-phosphoadenylylsulfate with *N*-desulfoheparin to produce adenosine 3',5'-bisphosphate and heparin.[1-4] The enzyme will also catalyze the sulfation of heparan sulfate, heparitin, chondroitin 4-sulfate, and dermatan sulfate (albeit, to a more limited extent with these later substrates). In some organisms, this enzyme may be identical with heparitin sulfotransferase [EC 2.8.2.12].

[1]A. Orellana, C. B. Hirschberg, Z. Wei, S. J. Swiedler & M. Ishihara (1994) *JBC* **269**, 2270.
[2]E. Brandan & C. B. Hirschberg (1988) *JBC* **263**, 2417.
[3]D. Göhler, R. Niemann & E. Buddecke (1984) *EJB* **138**, 301.
[4]L. Jansson, M. Höök, A. Wasteson & U. Lindahl (1975) *BJ* **149**, 49.

DETHIOBIOTIN SYNTHETASE

This metal ion-dependent enzyme [EC 6.3.3.3], also known as dethiobiotin synthase, catalyzes the reaction of ATP with (7*R*,8*S*)-7,8-diaminononanoate and carbon dioxide to produce ADP, orthophosphate, and dethiobiotin.[1-3]

(7*R*,8*S*)-7,8-diaminononanoate   dethiobiotin

CTP is not quite as effective as ATP. The first intermediate in the reaction pathway is the 7-carbamate of 7,8-diaminononanoate; a second mixed carbamic-phosphoric anhydride intermediate is formed when the carbamate is phosphorylated by ATP.[1] This intermediate has been observed via kinetic crystallography.[2]

[1]K. J. Gibson (1997) *Biochemistry* **36**, 8474.
[2]H. Kack, K. J. Gibson, Y. Lindqvist & G. Schneider (1998) *PNAS* **95**, 5495.
[3]K. Krell & M. A. Eisenberg (1970) *JBC* **245**, 6558.

Selected entries from *Methods in Enzymology* [vol, page(s)]:
General discussion: 62, 336, 348; **279**, 376
Assay: 62, 336, 348, 349

DEUTEROLYSIN

This zinc-dependent metalloproteinase [EC 3.4.24.39], also known as microbial neutral proteinase II (NpII), and acid metalloproteinase, catalyzes the hydrolysis of basic proteins, including histones and protamine.[1,2] The enzyme exhibits low activity with casein (the pH$_{opt}$ with casein is 5.5–6.0), hemoglobin, albumin, and gelatin. The sites of cleavage are dependent somewhat on the source of the enzyme. For example, the *Aspergillus sojae* enzyme hydrolyzes the insulin B chain at Phe1–Val2, Asn3–Gln4, Gln4–His5, Gly8–Ser9, Ser9–His10, His10–Leu11, Leu11–Val12, Glu13–Ala14, Ala14–Leu15, Tyr16–Leu17, Arg22–Gly23, and Tyr26–Thr27. The *Penicillium roqueforti* enzyme will catalyze the hydrolysis of all of these peptide bonds except Ser9–His10, His10–Leu11, Leu11–Val12, and Arg22–Gly23. In addition, it will cleave Val12–Glu13.[1] Deuterolysin is not inhibited by sulfhydryl reagents or phosphoramidon. Interestingly, EDTA only inhibits at acidic pH values.

[1] J. C. Gripon, B. Auberger & J. Lenoir (1980) *Int. J. Biochem.* **12**, 451.
[2] H. Sekine (1976) *Agric. Biol. Chem.* **40**, 703.

DEXTRANASE

This enzyme [EC 3.2.1.11], also known as α-1,6-glucan-6-glucanohydrolase, catalyzes the endohydrolysis of 1,6-α-D-glucosidic linkages in dextran.[1]

[1] E. H. Fischer & E. A. Stein (1960) *The Enzymes*, 2nd ed., **4**, 301.

Selected entries from *Methods in Enzymology* [vol, page(s)]:
General discussion: 8, 615; 34, 720
Assay: 8, 615
Other molecular properties: immobilization, by adsorption, effects on activity, 44, 44; properties, 8, 619; purification, 8, 616; source, 44, 30; *Streptococcus mutans* enzyme, 64, 260, 271; subsite mapping, 64, 271; substrate, 64, 255

DEXTRAN 1,6-α-ISOMALTOTRIOSIDASE

This enzyme [EC 3.2.1.95], also known as exo-isomaltotriohydrolase, catalyzes the hydrolysis of 1,6-α-D-glucosidic linkages in dextrans, thereby removing successive isomaltotriose units from the nonreducing ends of the chains.[1]

[1] M. Sugiura, A. Ito & T. Yamaguchi (1974) *BBA* **350**, 61.

DEXTRANSUCRASE

This enzyme [EC 2.4.1.5], also known as sucrose 6-glucosyltransferase, catalyzes the reaction of sucrose with $[(1,6)\text{-}\alpha\text{-}D\text{-glucosyl}]_n$ to produce fructose and $[(1,6)\text{-}\alpha\text{-}D\text{-glucosyl}]_{(n+1)}$.[1–7] Other donor substrates include α-D-glucopyranosyl fluoride and *p*-nitrophenyl α-D-glucopyranoside. A glycosyl-enzyme intermediate has been trapped in the *Streptococcus sobrinus* enzyme (at an aspartyl residue).[4] The *Streptococcus mutans* enzyme has a random Bi Bi kinetic mechanism.[6] Note that dextransucrases are glucosyltransferases whereas levansucrases (which catalyze a similar reaction) are fructosyltransferases.

An alternative two-site mechanism has been proposed in which two sucrose molecules bind to the enzyme. The enzyme thus becomes doubly glucosylated. An isomaltosyl-enzyme intermediate is formed, and another sucrose molecule can then bind to promote elongation.[2,7]

There are at least three distinct dextransucrases in *Streptococcus mutans*. Dextransucrase I produces a water-soluble product containing α(1 → 6) linkages, dextransucrase S generates a water-insoluble product containing both α(1 → 6) and α(1 → 3) linkages, and dextransucrase SI produces both types of glucans.

[1] V. Monchois, R. M. Willemot & P. Monsan (1999) *FEMS Microbiol. Rev.* **23**, 131.
[2] G. Davies, M. L. Sinnott & S. G. Withers (1998) *CBC* **1**, 119.
[3] G. Mooser (1992) *The Enzymes*, 3rd ed., **20**, 187.
[4] G. Mooser, S. A. Hefta, R. J. Paxton, J. F. Snively & T. D. Lee (1991) *JBC* **266**, 8916.
[5] A. W. Miller, S. H. Eklund & J. F. Robyt (1986) *Carbohydr. Res.* **147**, 119.
[6] G. Mooser, D. Shur, M. Lyou & C. Watanabe (1985) *JBC* **260**, 6907.
[7] D. Su & J. F. Robyt (1994) *ABB* **308**, 471.

Selected entries from *Methods in Enzymology* [vol, page(s)]:
General discussion: 1, 179
Assay: 1, 179; 136, 241; 138, 650
Other molecular properties: dextran synthesis with, 136, 251; immobilization, 136, 245; production in *Leuconostoc* fed-batch culture, 136, 242; properties, 1, 182; 138, 655; purification from (*Leuconostoc mesenteroides*, 1, 181; 136, 242; streptococcus, 138, 652)

DEXTRIN DEXTRANASE

This enzyme [EC 2.4.1.2], also known as dextrin 6-glucosyltransferase, catalyzes the reaction of $[(1,4)\text{-}\alpha\text{-}D\text{-glucosyl}]_n$ with $[(1,6)\text{-}\alpha\text{-}D\text{-glucosyl}]_m$ to produce $[(1,4)\text{-}\alpha\text{-}D\text{-glucosyl}]_{n-1}$ and $[(1,6)\text{-}\alpha\text{-}D\text{-glucosyl}]_{m+1}$.[1,2]

[1] K. Yamamoto, K. Yoshikawa & S. Okada (1993) *Biosci. Biotechnol. Biochem.* **57**, 47 and 136.
[2] E. J. Hehre (1953) *JBC* **192**, 161.

α-DEXTRIN ENDO-1,6-α-GLUCOSIDASE

This enzyme [EC 3.2.1.41] (also known as pullulanase, pullulan 6-glucanohydrolase, limit dextrinase, debranching enzyme, and amylopectin 6-glucanohydrolase) is the debranching enzyme for starch and pullulan, catalyzing the hydrolysis of α(1 → 6)-glucosidic linkages to produce maltotriose.[1–3] Another substrate is glycogen. ***See also* *Isoamylase; Isopullulanase; Neopullulanase***

[1] M. Vihinen & P. Mantsala (1989) *Crit. Rev. Biochem. Mol. Biol.* **24**, 329.
[2] E. Y. C. Lee & W. J. Whelan (1972) *The Enzymes*, 3rd ed., **5**, 191.
[3] C. J. Brandt, B. J. Catley & W. M. Awad, Jr., (1976) *J. Bacteriol.* **125**, 501.

5-(3,4-DIACETOXYBUT-1-YNYL)-2,2'-BITHIOPHENE DEACETYLASE

5-(3,4-diacetoxybut-1-ynyl)-
2,2'-bithiophene

5-(3-hydroxy-4-acetoxybut-1-ynyl)-
2,2'-bithiophene

This highly specific deacetylase [EC 3.1.1.66], first isolated from the roots of the French marigold (*Tagetes patula*), catalyzes the hydrolysis of 5-(3,4-diacetoxybut-1-ynyl)-2,2'-bithiophene to produce 5-(3-hydroxy-4-acetoxybut-1-ynyl)-2,2'-bithiophene and acetate.[1]

[1]R. Pensl & R. Sütfeld (1985) *Z. Naturforsch.* **40c**, 3.

DIACYLGLYCEROL *O*-ACYLTRANSFERASE

This transferase [EC 2.3.1.20], also known as diglyceride acyltransferase, catalyzes the reaction of an acyl-CoA with *sn*-1,2-diacylglycerol to produce a triacylglycerol and Coenzyme A.[1,2]

sn-1,2-diacylglycerol   triacylglycerol

The acyl-CoA derivative can be palmitoyl-CoA or another long-chain acyl-CoA compound. In many organisms, 1,3-diacylglycerols do not serve as acceptor substrates. *sn*-2,3-Diacylglycerols are typically weak substrates.

[1]R. V. Farese, Jr., S. Cases & S. J. Smith (2000) *Curr. Opin. Lipidol.* **11**, 229.
[2]R. M. Bell & R. A. Coleman (1983) *The Enzymes*, 3rd ed., **16**, 87.

Selected entries from *Methods in Enzymology* [vol, page(s)]:
General discussion: 71, 528; **209**, 98
Assay: 209, 99
Other molecular properties: acyl-acceptor specificities, 71, 533; acyl-donor specificities, 71, 533; activity in intestine and liver, **209**, 98; properties, 71, 533; rat liver, from, 71, 528; stabilization with ethylene glycol, 71, 531

DIACYLGLYCEROL CHOLINEPHOSPHOTRANSFERASE

This transferase [EC 2.7.8.2] (also known as CDP-choline:1,2-diacylglycerol cholinephosphotransferase, cholinephosphotransferase, phosphorylcholine:glyceride transferase, alkylacylglycerol cholinephosphotransferase, and 1-alkyl-2-acetylglycerol cholinephosphotransferase) catalyzes the reaction of CDP-choline with an *sn*-1,2-diacylglycerol to produce CMP and a phosphatidyl-choline.[1,2] 1-Alkyl-2-acylglycerol derivatives can also serve as acceptor substrates; action on this alternative substrate was previously designated as 1-alkyl-2-acetylglycerolcholine phosphotransferase [EC 2.7.8.16].

[1]C. R. McMaster & R. M. Bell (1997) *BBA* **1348**, 100.
[2]C. R. Mantel, A. R. Schulz, K. Miyazawa & H. E. Broxmeyer (1993) *BJ* **289**, 815.

Selected entries from *Methods in Enzymology* [vol, page(s)]:
General discussion: 5, 484; 71, 536; **209**, 267, 272, 279
Assay: 5, 484; **141**, 387; **209**, 281; castor bean endosperm, 71, 600; mammalian, **209**, 267; microsomes, 55, 102; myocardial, **209**, 422; plant endoplasmic reticulum, **148**, 586; yeast, **209**, 273
Other molecular properties: castor bean endosperm, from, 71, 600 (assay, 71, 600; inhibitors, 71, 602; metal ion specificity, 71, 602; properties, 71, 602); chicken liver, from, 5, 485; diagnostic enzyme, as, 31, 24, 303; mammalian (assay, **209**, 267; partial purification, **209**, 269; properties, **209**, 271; solubilization, **209**, 269); microsomes, 55, 101, 104 (assay, 55, 102); myocardial, assay, **209**, 422; plant endoplasmic reticulum, in (assay, **148**, 586; properties, **148**, 587); preparation, 5, 485; product identification, **209**, 281; properties, 5, 486; **141**, 388; **209**, 282; rat liver, from, 71, 536 (activators and inhibitors, 71, 545; acyl chain length specificity, 71, 544; back-reaction, 71, 540; interference by microsomal lipase, 71, 540; localization in microsomal membranes, 71, 546; membrane-bound substrates, 71, 539; phospholipase A₂ treatment of membrane-bound enzyme, 71, 545; reversibility, 71, 544; separation from ethanolaminephosphotransferase, 71, 545; solubilization, 71, 541; substrate specificity, 71, 542); reconstitution after detergent inactivation, **209**, 270; yeast, assay, **209**, 273

1,2-DIACYLGLYCEROL 3-β-GALACTOSYLTRANSFERASE

sn-1,2-diacylglycerol   3-β-D-galactosyl-1,2-diacylglycerol

This transferase [EC 2.4.1.46], also known as UDP-galactose:1,2-diacylglycerol 3-β-galactosyltransferase, catalyzes the reaction of UDP-D-galactose with an *sn*-1,2-diacylglycerol to produce UDP and a 3-β-D-galactosyl-1,2-diacylglycerol. Initial-rate kinetics demonstrate that the spinach chloroplast enzyme does not operate by a ping-pong kinetic mechanism.[1]

[1] E. Marechal, M. A. Block, J. Joyard & R. Douce (1994) *JBC* **269**, 5788.

1,2-DIACYLGLYCEROL 3-GLUCOSYLTRANSFERASE

This glucosyltransferase [EC 2.4.1.157] catalyzes the reaction of UDP-D-glucose with an *sn*-1,2-diacylglycerol (including several with long-chain acyl groups) to produce UDP and a 3-D-glucosyl-1,2-diacylglycerol.[1,2] Cooperativity is observed with respect to the diacylglycerol substrate.[3]

[1] N. Sato & N. Murata (1982) *Plant Cell Physiol.* **23**, 1115.
[2] T. Omata & N. Murata (1986) *Plant Cell Physiol.* **27**, 485.
[3] O. P. Karlsson, A. Dahlqvist, S. Vikstrom & A. Wieslander (1997) *JBC* **272**, 929.

DIACYLGLYCEROL KINASE

sn-1,2-diacylglycerol   phosphatidate

This phosphotransferase [EC 2.7.1.107], also known as diglyceride kinase, catalyzes the reaction of ATP with an *sn*-1,2-diacylglycerol to produce ADP and an *sn*-1,2-diacylglycerol 3-phosphate (*i.e.*, phosphatidate).[1–6] This enzyme is a key participant in the regulation of signal transduction.[1–3] Several isozymes have been identified. With many, activity is measurably higher if the acceptor substrate contains an arachidonyl group in the *sn*-2 position. The *Escherichia coli* enzyme has a random Bi Bi kinetic mechanism (note that the *E. coli* enzyme is the smallest known kinase at 13 kDa).

[1] M. K. Topham & S. M. Prescott (1999) *JBC* **274**, 1144.
[2] W. J. van Blitterswijk & B. Houssa (1999) *Chem. Phys. Lipids* **98**, 95.
[3] F. Sakane & H. Kanoh (1997) *Int. J. Biochem. Cell Biol.* **29**, 1139.
[4] P. Badola & C. R. Sanders II (1997) *JBC* **272**, 24176.
[5] J. D. Esko & C. R. H. Raetz (1983) *The Enzymes*, 3rd ed., **16**, 207.
[6] R. A. Pieringer (1983) *The Enzymes*, 3rd ed., **16**, 255.

Selected entries from *Methods in Enzymology* [vol, page(s)]:
General discussion: **209**, 153, 162, 173
Assay: arachidonoyl-specific, radiochemical assay, **209**, 174; cerebral, **209**, 420; cerebral isozymes, suspension assay, **209**, 164; *Escherichia coli*, radiochemical assay, **209**, 154; lymphoid tissue isozymes, suspension assay, **209**, 164; rat liver, **109**, 475
Other molecular properties: arachidonoyl-specific (radiochemical assay, **209**, 174; separation from nonspecific diacylglycerol kinases, **209**, 177; substrate specificity, **209**, 180); ceramides, assay of, **312**, 22; cerebral isozymes, suspension assay, **209**, 164; 80-kDa isozyme (immunological studies, **209**, 170; intracellular translocation, **209**, 171; 150-kDa isozyme, and, separation, **209**, 166; phosphorylation, **209**, 171; primary structure, **209**, 169; purification, **209**, 165); *Escherichia coli* (properties, **209**, 159; radiochemical assay, **209**, 154; lymphoid tissue isozymes, suspension assay, **209**, 164; mutant, **72**, 694; non-specific, separation from arachidonoyl-diacylglycerol kinase, **209**, 177; 150-kDa isozyme (and 80-kDa isozyme, separation, **209**, 166; purification, **209**, 165); refolded, phorbol ester binding, effect of zinc, **256**, 119

DIACYLGLYCEROL:STEROL O-ACYLTRANSFERASE

cholesterol ester

This transferase [EC 2.3.1.73] catalyzes the reaction of a 1,2-diacyl-*sn*-glycerol with a sterol to produce a mono-acylglycerol and a sterol ester.[1] Acceptor substrates include cholesterol, sitosterol, campesterol, and even diacylglycerol. A wide variety of acyl groups can be transferred.

[1] R. E. Garcia & J. B. Mudd (1978) *ABB* **191**, 487.

Selected entries from *Methods in Enzymology* [vol, page(s)]:
General discussion: **71**, 768
Assay: spinach leaves, **71**, 768
Other molecular properties: spinach leaves, from, **71**, 768 (activators and inhibitors, **71**, 771; assay, **71**, 768; properties, **71**, 771; specificity, **71**, 771; stability, **71**, 771)

2,2-DIALKYLGLYCINE DECARBOXYLASE (PYRUVATE)

This pyridoxal-phosphate- and potassium-dependent enzyme [EC 4.1.1.64], also known as L-alanine:α-keto-butyrate aminotransferase and α,α-dialkylamino acid aminotransferase, catalyzes the reaction of 2,2-dialkyl-glycine with pyruvate to produce a dialkyl ketone, carbon dioxide, and L-alanine.[1–3] Substrates include 2-amino-2-methylpropanoate (*i.e.*, 2-methylalanine; producing acetone), 2-amino-2-methylbutanoate (generating butanone), and 1-aminocyclopentane carboxylate (producing cyclopentanone). The enzyme has also been observed to catalyze an amino transfer reaction between L-alanine and pyruvate. There are two alkali metal binding sites: one at the active site, and another at the carboxyl terminus of an α-helix.

During catalysis, a Schiff base is first formed between the substrate and the coenzyme. pH-rate kinetics indicate that an aminoacyl residue with a pK_a value of about 7.4 influences the ratio between the ketoenamine and enolimine tautomers in the reaction pathway.[3] The kinetic mechanism is ping pong; in the first half-reaction oxidative decarboxylation produces pyridoxamine phosphate, and in the second half-reaction, one observes classical transamination of pyruvate to L-alanine.

[1] G. B. Bailey & W. B. Dempsey (1967) Biochemistry **6**, 1526.
[2] Y. Koide, M. Honma & T. Shimomura (1981) Agric. Biol. Chem. **45**, 775.
[3] X. Zhou & M. D. Toney (1999) Biochemistry **38**, 311.

Selected entries from **Methods in Enzymology** [vol, page(s)]:
General discussion: 17A, 829
Assay: 17A, 829

DIAMINE N-ACETYLTRANSFERASE

This acetyltransferase [EC 2.3.1.57], also known as spermidine acetyltransferase and putrescine acetyltransferase, catalyzes the reaction of an alkane-α,ω-diamine with acetyl-CoA to produce coenzyme A and an N-acetyl-diamine.[1–4] When spermidine is the substrate, both N^1- and N^8-acetylspermidine are produced. Other substrates include 1,3-diaminopropane, 1,5-diaminopentane, putrescine, spermine, N^1-acetylspermidine, and N^8-acetylspermidine. The *Bacillus subtilis* enzyme reportedly has a random Bi Bi kinetic mechanism.[4] The enzyme will also produce N^1- and N^{12}-spermine from spermine.

[1] G. W. Haywood & P. J. Large (1985) EJB **148**, 277.
[2] F. Della Ragione & A. E. Pegg (1982) Biochemistry **21**, 6152.
[3] R.-M. Wittich & R. D. Walter (1989) BJ **260**, 265.
[4] D. P. Woolridge, J. D. Martinez, D. E. Stringer & E. W. Gerner (1999) BJ **340**, 753.

Selected entries from **Methods in Enzymology** [vol, page(s)]:
General discussion: 94, 321
Assay: bovine liver, 94, 325; rat liver, 94, 321
Other molecular properties: bovine liver (properties, 94, 328; purification, 94, 326); rat liver (properties, 94, 324; purification, 94, 322)

DIAMINE AMINOTRANSFERASE

This pyridoxal-phosphate-dependent aminotransferase [EC 2.6.1.29] catalyzes the reaction of an α,ω-diamine with α-ketoglutarate (or, 2-oxoglutarate) to generate an ω-aminoaldehyde and L-glutamate.[1–3]

putrescine   4-aminobutanal

Diamine substrates include putrescine, cadaverine, 1,7-diaminoheptane, and 1,4-diaminobutan-2-ol.

[1] A. E. Braunstein (1973) The Enzymes, 3rd ed., **9**, 379.
[2] K.-H. Kim (1964) JBC **239**, 783.
[3] K. Madduri, C. Stuttard & L. C. Vining (1989) J. Bacteriol. **171**, 299.

Selected entries from **Methods in Enzymology** [vol, page(s)]:
General discussion: 17B, 812
Assay: 17B, 812

L-2,4-DIAMINOBUTYRATE ACTIVATING ENZYME

This enzyme, also known as L-2,4-diaminobutyrate adenylyltransferase, participates in the biosynthesis of the polymyxins by catalyzing the reaction of L-2,4-diaminobutyrate with ATP to produce L-2,4-diaminobutyryl-AMP and pyrophosphate (or, diphosphate). **See also** Polymyxin Synthetase

Selected entries from **Methods in Enzymology** [vol, page(s)]:
General discussion: 43, 579
Assay: 43, 579

2,4-DIAMINOBUTYRATE: α-KETOGLUTARATE AMINOTRANSFERASE

This aminotransferase [EC 2.6.1.76] (also known as 2,4-diaminobutyrate 4-transaminase) catalyzes the reversible reaction of L-2,4-diaminobutyrate with α-ketoglutarate (or, 2-oxoglutarate) to produce L-glutamate and L-aspartate 4-semialdehyde.[1] The enzyme participates in the biosynthesis of 1,3-diaminopropane in *Haemophilus influenzae* and *Acinetobacter baumannii*. It is distinct from diaminobutyrate:pyruvate aminotransferase [EC 2.4.1.46].

[1] H. Ikai & S. Yamamoto (1997) J. Bacteriol. **179**, 5118.

DIAMINOBUTYRATE:PYRUVATE AMINOTRANSFERASE

This pyridoxal-phosphate-dependent aminotransferase [EC 2.6.1.46], also known as L-diaminobutyrate transaminase, catalyzes the reversible reaction of L-2,4-diaminobutanoate with pyruvate to produce L-aspartate

L-2,4-diaminobutyrate   L-aspartate 4-semialdehyde

4-semialdehyde and L-alanine.[1,2] The enzyme is distinct from 2,4-diaminobutyrate:α-ketoglutarate aminotransferase [EC 2.6.1.76].

[1]D. R. Rao, K. Hariharan & K. R. Vijayalakshmi (1969) *BJ* **114**, 107.
[2]P. Peters, E. A. Galinski & H. G. Trüper (1990) *FEMS Microbiol. Lett.* **71**, 157.

DIAMINOHEXANOATE DEHYDROGENASE

This enzyme [EC 1.4.1.11], also known as L-*erythro*-3,5-diaminohexanoate dehydrogenase, catalyzes the reversible reaction of L-*erythro*-3,5-diaminohexanoate with NAD^+ and water to produce (*S*)-5-amino-3-oxohexanoate, ammonia, and NADH.[1-4] Some organisms (for example, *Clostridium sticklandii*)[3] may prefer $NADP^+$.

[1]N. M. W. Brunhuber & J. S. Blanchard (1994) *Crit. Rev. Biochem. Mol. Biol.* **29**, 415.
[2]S. C. L. Hong & H. A. Barker (1973) *JBC* **248**, 41.
[3]T. C. Stadtman (1973) *AE* **38**, 413.
[4]J. J. Baker, I. Jeng & H. A. Barker (1972) *JBC* **247**, 7724.

DIAMINOHYDROXYPHOSPHORIBOSYL-AMINOPYRIMIDINE DEAMINASE

This deaminase [EC 3.5.4.26] catalyzes the hydrolysis of 2,5-diamino-6-hydroxy-4-(5-phosphoribosylamino)pyrimidine to produce 5-amino-6-(5-phosphoribosylamino)-uracil and ammonia.[1]

[1]R. B. Burrows & G. M. Brown (1978) *J. Bacteriol.* **136**, 657.

2,4-DIAMINOPENTANOATE DEHYDROGENASE

This oxidoreductase [EC 1.4.1.12] catalyzes the reaction of 2,4-diaminopentanoate with $NAD(P)^+$ and water to produce 2-amino-4-oxopentanoate, ammonia, and $NAD(P)H$.[1-4] The enzyme can also utilize 2,5-diaminohexanoate as a substrate, although not as effectively as diaminopentanoate, to form 2-amino-5-oxohexanoate; the latter then cyclizes nonenzymatically to form 1-pyrroline-2-methyl-5-carboxylate.

[1]N. M. W. Brunhuber & J. S. Blanchard (1994) *Crit. Rev. Biochem. Mol. Biol.* **29**, 415.
[2]R. Somack & R. N. Costilow (1973) *JBC* **247**, 385.
[3]T. C. Stadtman (1973) *AE* **38**, 413.
[4]Y. Tsuda & H. C. Friedmann (1970) *JBC* **245**, 5914.

DIAMINOPIMELATE DECARBOXYLASE

This pyridoxal-phosphate-dependent decarboxylase [EC 4.1.1.20] catalyzes the conversion of *meso*-2,6-diaminoheptanedioate to L-lysine and carbon dioxide.[1-4] Lanthionine is a weak alternative substrate. Both the wheat germ and *Bacillus sphaericus* enzymes catalyze the decarboxylation with inversion of configuration, unusual for pyridoxal-phosphate-dependent α-decarboxylases.[2,3]

[1]J. G. Kelland, L. D. Arnold, M. M. Palcic, M. A. Pickard & J. C. Vederas (1986) *JBC* **261**, 13216.
[2]J. G. Kelland, M. M. Palcic, M. A. Pickard & J. C. Vederas (1985) *Biochemistry* **24**, 3263.
[3]Y. Asada, K. Tanizawa, S. Sawada, T. Suzuki, H. Misono & K. Soda (1981) *Biochemistry* **20**, 6881.
[4]E. A. Boeker & E. E. Snell (1972) *The Enzymes*, 3rd ed., **6**, 217.

Selected entries from *Methods in Enzymology* [vol, page(s)]:
General discussion: 5, 864; **17B**, 140, 146
Assay: 5, 864; **17B**, 140, 147
Other molecular properties: diaminopimelate assay and, **6**, 632; diaminopimelate racemase, assay of, **5**, 858, 861, 862; diaminopimelate synthesis, not present in, **5**, 849; diaminopimelate transamination, assay of, **5**, 854; *Escherichia coli* mutant, not present in, **5**, 849; lysine, if present in spectrophotomeric determination of, **17B**, 228f; properties, **5**, 869; **17B**, 145, 150; purification from (*Aerobacter aerogenes*, **5**, 867; *Escherichia coli*, **17B**, 142; *Spirodeia oligorhiza*, **17B**, 148); pyridoxal phosphate and, **5**, 864; **17B**, 145, 147; racemase and, **5**, 858, 861, 862; *N*-succinyl-L-diaminopimelate:glutamate transaminase, assay of, **5**, 854

DIAMINOPIMELATE DEHYDROGENASE

This oxidoreductase [EC 1.4.1.16], also known as *meso*-diaminopimelate D-dehydrogenase, catalyzes the reaction of *meso*-2,6-diaminoheptanedioate with $NADP^+$ and water to produce L-2-amino-6-oxoheptanedioate, ammonia, and NADPH.[1-4] The enzyme is absolutely specific for the *meso* isomer of the amino acid substrate and must be able to distinguish between the two chiral centers (the D-center is the site of oxidation). The substrate binds in a single orientation in an elongated cavity, such that the D-stereocenter is positioned near the coenzyme.[1] The *Bacillus sphaericus* enzyme has an ordered Bi Bi kinetic mechanism.[4]

[1]G. Scapin, M. Cirilli, S. G. Reddy, Y. Gao, J. C. Vederas & J. S. Blanchard (1998) *Biochemistry* **37**, 3278.
[2]N. M. W. Brunhuber & J. S. Blanchard (1994) *Crit. Rev. Biochem. Mol. Biol.* **29**, 415.
[3]H. Misono, M. Ogasawara & S. Nagasaki (1986) *Agric. Biol. Chem.* **50**, 1329.
[4]H. Misono & K. Soda (1980) *JBC* **255**, 10599.

DIAMINOPIMELATE EPIMERASE

This pyridoxal-phosphate-independent epimerase [EC 5.1.1.7] catalyzes the interconversion of LL-2,6-diaminoheptanedioate and *meso*-diaminoheptanedioate.[1-5] This enzyme utilizes two aminoacyl residues as a general acid and general base in the chemical mechanism (at least one is a cysteinyl residue). When the LL-stereoisomer is incubated with the enzyme in the presence of D_2O, an equilibrium "overshoot" is observed, similar to that observed with proline racemase; however, if the *meso*-stereoisomer is incubated with the enzyme with D_2O, a double "overshoot" is seen.[1] The DD-stereoisomer is not a substrate.

[1]C. W. Koo & J. S. Blanchard (1999) *Biochemistry* **38**, 4416.
[2]M. E. Tanner & G. L. Kenyon (1998) *CBC* **2**, 7.
[3]J. S. Wiseman & J. S. Nichols (1984) *JBC* **259**, 8907.
[4]E. Adams (1972) *The Enzymes*, 3rd ed., **6**, 479.
[5]P. J. Withe (1969) *BJ* **113**, 589.

Selected entries from *Methods in Enzymology* [vol, page(s)]:
General discussion: **5**, 858
Assay: **5**, 858
Other molecular properties: *N*-succinyl-L-diaminopimelate transaminase, assay of, **5**, 854; diaminopimelate assay and, **6**, 632; diaminopimelate transamination and, **5**, 854; diaminopimelate decarboxylase and, **5**, 867, 868; properties, **5**, 863; purification, *Escherichia coli*, **5**, 862; transition-state and multisubstrate analogues, **249**, 308

DIAMINOPROPIONATE AMMONIA-LYASE

This pyridoxal-phosphate-dependent lyase [EC 4.3.1.15], also known as diaminopropionatase and α,β-diamino-propionate ammonia-lyase, catalyzes the conversion of either stereoisomer of 2,3-diaminopropionate ($^+H_3NCH_2CH(NH_3^+)COO^-$) to pyruvate and two molecules of ammonia. D- and L-Serine are weaker alternative substrates.[1]

[1]T. Nagasawa, K. Tanizawa, T. Satoda & H. Yamada (1988) *JBC* **263**, 958.

2,3-DIAMINOPROPIONATE N-OXALYLTRANSFERASE

L-2,3-diaminopropionate

N^3-oxalyl-L-2,3-diaminopropionate

This transferase [EC 2.3.1.58], also known as oxalyl-diaminopropionate synthase, catalyzes the reaction of oxalyl-CoA with L-2,3-diaminopropanoate to produce coenzyme A and N^3-oxalyl-L-2,3-diaminopropanoate.[1] L-2,4-Diaminobutyrate is also an acceptor substrate, albeit weaker.

[1]K. Malathi, G. Padmanaban & P. S. Sarma (1970) *Phytochemistry* **9**, 1603.

2,5-DIAMINO-6-RIBOSYLAMINO-4(3H)-PYRIMIDINONE 5'-PHOSPHATE DEAMINASE

This deaminase catalyzes the hydrolysis of 2,5-diamino-6-ribosylamino-4(3H)-pyrimidinone to produce ammonia and 5-amino-6-ribosylamino-4(3H)-pyrimidinone. 2,5-Diamino-6-ribitylamino-4(3H)-pyrimidinone is reportedly also a substrate.

Selected entries from *Methods in Enzymology* [vol, page(s)]:
General discussion: **280**, 382

2,5-DIAMINO-6-RIBOSYLAMINO-4(3H)-PYRIMIDINONE 5'-PHOSPHATE REDUCTASE

This oxidoreductase catalyzes the reaction of 2,5-diamino-6-ribosylamino-4(3H)-pyrimidinone with NAD(P)H to produce $NAD(P)^+$ and 2,5-diamino-6-ribitylamino-4(3H)-pyrimidinone. 5-Amino-6-ribosylamino-4(3H)-pyrimidinone is also a substrate.

Selected entries from *Methods in Enzymology* [vol, page(s)]:
General discussion: **280**, 382

2,5-DIAMINOVALERATE AMINOTRANSFERASE

This pyridoxal-phosphate-dependent aminotransferase [EC 2.6.1.8], also known as diamino-acid amino-transferase, catalyzes the reversible reaction of 2,5-diaminopentanoate (also known as 2,5-diaminovalerate) with α-ketoglutarate (or, 2-oxoglutarate) to produce 5-amino-α-ketopentanoate (*i.e.*, 5-amino-2-oxopentanoate) and L-glutamate.[1] Other amino donor substrates include 2,5-diaminoglutarate.

[1]E. Roberts (1954) *ABB* **48**, 395.

DIAPHORASE

This archaic term refers to oxidoreductases catalyzing hydride transfer from NADH and/or NADPH to an electron acceptor other than dioxygen. Electron acceptors include cytochromes, 2,6-dichlorophenolindolphenol, ferricyanide, methylene blue, and certain quinones. This term should be avoided, especially when the natural electron acceptor has been identified. Dihydrolipoamide dehydrogenase [EC 1.8.1.4] is often termed diaphorase. **See Dihydrolipoamide Dehydrogenase; NADPH Dehydrogenase; NAD(P)H Dehydrogenase (Quinone); NADH Dehydrogenase**

Selected entries from *Methods in Enzymology* [vol, page(s)]:
General discussion: **34**, 288

Other molecular properties: assay in membrane fragments, **8**, 165; coenzyme A transferase, in preparation of, **1**, 580; heart, **1**, 32, 49, 50; **2**, 707; **3**, 291; hydroxysteroid dehydrogenase and, **5**, 520; malic dehydrogenase and, **1**, 739; methemoglobin reduction, **76**, 12; NAD:cytochrome c reductase and, **2**, 692, 697; NADP$^+$ and, **2**, 711; plastocyanin preparation, **69**, 227; *Pseudomonas*, **1**, 137; sarcosine dehydrogenase and, **5**, 742; succinate dehydrogenase and, **5**, 613; use in isocitrate dehydrogenase assay, **13**, 43

DIASTASE

This now archaic term was used to describe α-amylase [EC 3.2.1.1] as well as crude preparations containing amylases. *See* α-*Amylase*

DIBENZOTHIOPHENE DIHYDRODIOL DEHYDROGENASE

This oxidoreductase [EC 1.3.1.60] catalyzes the reaction of *cis*-1,2-dihydroxy-1,2-dihydrodibenzothiophene with NAD$^+$ to produce 1,2-dihydroxydibenzothiophene, NADH, and a proton.[1,2]

[1] A. L. Laborde & D. T. Gibson (1977) *Appl. Environ. Microbiol.* **34**, 783.
[2] S. S. Denome, D. C. Stanley, E. S. Olson & K. D. Young (1993) *J. Bacteriol.* **175**, 6890.

DICARBOXYLATE:CoA LIGASE

This ligase [EC 6.2.1.23], also known as dicarboxylyl-CoA synthetase, catalyzes the reaction of ATP with coenzyme A and an ω-dicarboxylic acid to produce AMP, pyrophosphate (or, diphosphate), and an ω-carboxyacyl-CoA. The ligase, first identified in rat liver, acts on dicarboxylic acids having a chain length of five to sixteen carbons: the best substrate reported is dodecanedioate.[1]

[1] J. Vamecq, E. de Hoffmann & F. Van Hoof (1985) *BJ* **230**, 683.

3,4-DICHLOROANILINE N-MALONYLTRANSFERASE

This acyltransferase [EC 2.3.1.114] catalyzes the reaction of malonyl-CoA with 3,4-dichloroaniline to produce coenzyme A and N-(3,4-dichlorophenyl)malonate.[1,2] 4-Chloroaniline is an alternative substrate.

[1] U. Matern, C. Feser & W. Heller (1984) *ABB* **235**, 218.
[2] H. Sandermann, R. Schmitt, H. Eckey & T. Bauknecht (1991) *ABB* **287**, 341.

2,4-DICHLOROBENZOYL-CoA REDUCTASE

This oxidoreductase [EC 1.3.1.63] catalyzes the reaction of 2,4-dichlorobenzoyl-CoA with NADPH and a proton to produce 4-chlorobenzoyl-CoA, NADP$^+$, and HCl.[1]

[1] V. Romanov & R. P. Hausinger (1996) *J. Bacteriol.* **178**, 2656.

DICHLOROMETHANE DEHALOGENASE

This dihalomethane-specific glutathione *S*-transferase [EC 4.5.1.3] catalyzes the hydrolysis of dichloromethane (CH_2Cl_2) to produce formaldehyde and two molecules of HCl.[1,2] *Hyphomicrobium*, *Methylobacterium*, and *Methylophilus* dehalogenases reportedly form a halomethylthioether intermediate, using glutathione as a cofactor. Alternative substrates include CH_2Br_2, CH_2ClBr, and CH_2I_2. *See also Glutathione S-Transferase; DDT-Dehydrochlorinase*

[1] R. Scholtz, L. P. Wackett, C. Egli, A. M. Cook & T. Leisinger (1988) *J. Bacteriol.* **170**, 5698.
[2] F. A. Blocki, M. S. Logan, C. Baoli & L. P. Wackett (1994) *JBC* **269**, 8826.
Selected entries from *Methods in Enzymology* [vol, page(s)]:
General discussion: **188**, 355
Assay: **188**, 355

DICHLOROMUCONATE CYCLOISOMERASE

This enzyme [EC 5.5.1.11] catalyzes the manganese ion-dependent interconversion of 2,4-dichloro-*cis,cis*-muconate to 2,4-dichloro-2,5-dihydro-5-oxofuran-2-acetate: the product spontaneously eliminates chloride to produce *cis*-4-carboxymethylene-3-chlorobut-2-en-4-olide. Alternative substrates include *cis,cis*-muconate and its monochloro-derivatives.[1] This cycloisomerase is distinct from muconate cycloisomerase [EC 5.5.1.1] or chloromuconate cycloisomerase [EC 5.5.1.7].

[1] A. E. Kuhm, M. Schlomann, H. J. Knackmuss & D. H. Pieper (1990) *BJ* **266**, 877.

2,4-DICHLOROPHENOL 6-MONOOXYGENASE

This FAD-dependent oxidoreductase [EC 1.14.13.20], also known as 2,4-dichlorophenol hydroxylase, catalyzes the reaction of 2,4-dichlorophenol with NADPH and dioxygen to produce 3,5-dichlorocatechol, NADP$^+$, and water.[1,2] Other substrates include 4-chlorophenol, 4-chloro-2-nitrophenol, and 4-chloro-2-methylphenol. NADH reacts more slowly.

[1] C. A. Beadle & A. R. W. Smith (1982) *EJB* **123**, 323.
[2] T. Liu & P. J. Chapman (1984) *FEBS Lett.* **173**, 314.

2,4-DICHLOROPHENOXYACETATE DIOXYGENASE

This iron-dependent oxidoreductase, isolated from *Alcaligenes eutrophus* JMP134, catalyzes the reaction of 2,4-dichlorophenoxyacetate with dioxygen and α-ketoglutarate (or, 2-oxoglutarate) to produce 2,4-dichlorophenol, carbon dioxide, succinate, and glyoxylate.[1,2] Ascorbate is required to preserve enzyme activity by reducing the rate of metal ion-mediated inactivation. Substrates include a wide range of phenyloxyacetates.

[1] F. Fukumori & R. P. Hausinger (1993) *J. Bacteriol.* **175**, 2083.
[2] F. Fukumori & R. P. Hausinger (1993) *JBC* **268**, 24311.

2,4-DIENOYL-CoA REDUCTASE (NADPH)

This oxidoreductase [EC 1.3.1.34], also known as 4-enoyl-CoA reductase (NADPH), catalyzes the reaction of *trans,trans*-2,3,4,5-tetradehydroacyl-CoA with NADPH to produce *trans*-2,3-didehydroacyl-CoA and NADP[+].[1] The best substrates contain a 2,4-diene structure with chain length of eight or ten carbon atoms. Other substrates include the *trans,cis*-2,3-analogues. The *Escherichia coli* enzyme, which is required for the β-oxidation of unsaturated fatty acids with even-numbered double bonds, is an iron-sulfur flavoprotein.[2]

[1] V. Dommes, C. Baumgart & W.-H. Kunau (1981) *JBC* **256**, 8259.
[2] X. Liang, C. Thorpe & H. Schulz (2000) *ABB* **380**, 373.

DIETHYL 2-METHYL-3-OXOSUCCINATE REDUCTASE

This oxidoreductase [EC 1.1.1.229] catalyzes the reaction of diethyl-2-methyl-3-oxosuccinate with NADPH to produce diethyl (2*R*,3*R*)-2-methyl-3-hydroxysuccinate and NADP[+].[1] The reaction represents an asymmetric reduction, with the product being a mixture of the (2*R*,3*R*)-*syn*- and (2*S*,3*R*)-*anti*-β-hydroxyester. The dimethyl esters can act as substrates, albeit more weakly.

[1] A. Furuichi, H. Akita, H. Matsukura, T. Oishi & K. Horikoshi (1987) *Agric. Biol. Chem.* **51**, 293

DIFERRIC-TRANSFERRIN REDUCTASE

This oxidoreductase [EC 1.16.1.2], also known as transferrin reductase, catalyzes the reaction of transferrin-[Fe(III)]$_2$ with NADH to produce transferrin-[Fe(II)]$_2$ and NAD[+]. The enzyme has been identified in rat liver plasma membranes and it has been suggested that the reduction of diferric transferrin at the cell surface may be an important function in the stimulation of cell growth and in iron transport.[1–3] The reductase also stimulates the Na[+]/H[+] antiport of HeLa cells.[3]

[1] I. L. Sun, P. Navas, F. L. Crane, D. J. Morre & H. Löw (1987) *JBC* **262**, 15915.
[2] M. Fry (1989) *BBRC* **158**, 469.

[3] I. L. Sun, W. Toole-Simms, F. L. Crane, E. S. Golub, T. Diaz de Pagan, D. J. Morre & H. Löw (1987) *BBRC* **146**, 976.

DIFRUCTOSE-ANHYDRIDE SYNTHASE

This enzyme [EC 3.2.1.134] catalyzes the conversion of inulobiose to bis-D-fructose 2′,1:2,1′-dianhydride and water.[1]

[1] T. Matsuyama, K. Tanaka, M. Mashiko & M. Kanamoto (1982) *J. Biochem.* **92**, 1325.

DIGLUCOSYL DIACYLGLYCEROL SYNTHASE

This magnesium-dependent glucosyltransferase [EC 2.4.1.208] (also known as monoglucosyl diacylglycerol (1 → 2)glucosyltransferase, MGlcDAG(1 → 2) glucosyltransferase, and DGlcDAG synthase) catalyzes the reaction of UDP-D-glucose with a 1,2-diacyl-3-*O*-(α-D-glucopyranosyl)-*sn*-glycerol to produce a 1,2-diacyl-3-*O*-(α-D-glucopyranosyl(1 → 2)-*O*-α-D-glucopyranosyl)-*sn*-glycerol and UDP.[1]

[1] O. P. Karlsson, M. Rytomaa, A. Dahlqvist, P. K. Kinnunen & A. Wieslander (1996) *Biochemistry* **35**, 10094.

DIGUANIDINOBUTANASE

This enzyme [EC 3.5.3.20] catalyzes the hydrolysis of 1,4-diguanidinobutane (*i.e.*, *N,N*′-diamidinoputrescine or areanine) to produce agmatine and urea. Alternative substrates include other diguanidinoalkanes with three to ten methylene groups: for example, 1,5-diguanidinopentane (*N,N*′-diamidinocadaverine or andonine) and 1,6-diguanidinohexane.[1]

[1] T. Yorifuji, M. Kaneoke, E. Shimizu, K. Shiota & R. Matsuo (1989) *Agric. Biol. Chem.* **53**, 3003.

cis-1,2-DIHYDROBENZENE-1,2-DIOL DEHYDROGENASE

This oxidoreductase [EC 1.3.1.19], also known as *cis*-benzene glycol dehydrogenase and *cis*-1,2-dihydroxy-cyclohexa-3,5-diene dehydrogenase, catalyzes the reaction of *cis*-1,2-dihydrobenzene-1,2-diol with NAD[+] to produce catechol and NADH.[1–3]

[1] P. J. Artymiuk, C. C. F. Blake & P. J. Geary (1977) *J. Mol. Biol.* **111**, 203.
[2] B. C. Axcell & P. J. Geary (1973) *BJ* **136**, 927.
[3] K. P. Fong & H. M. Tan (1999) *FEBS Lett.* **451**, 5.

Selected entries from *Methods in Enzymology* [vol, page(s)]:
General discussion: 188, 134
Assay: 188, 134
Other molecular properties: properties, **188**, 137; purification from *Pseudomonas putida* NCIB 12190, **188**, 136

trans-1,2-DIHYDROBENZENE-1,2-DIOL DEHYDROGENASE

This oxidoreductase [EC 1.3.1.20], also known as *trans*-benzene glycol dehydrogenase and *trans*-1,2-dihydroxycyclohexa-3,5-diene dehydrogenase, catalyzes the reaction of *trans*-1,2-dihydrobenzene-1,2-diol with NADP$^+$ to produce catechol and NADPH.[1–3] Alternative substrates include naphthalene dihydrodiol. The liver enzyme has an ordered Bi Bi kinetic mechanism.[1,2] It remains unclear whether both stereoisomers of the *trans*-diol are substrates of this dehydrogenase.

[1] H. Nanjo, H. Adachi, S. Morihana, T. Mizoguchi, T. Nishihara & T. Terada (1995) *BBA* **1244**, 53.
[2] K. Sato, M. Nakanishi, Y. Deyashiki, A. Hara, K. Matsuura & I. Ohya (1994) *J. Biochem.* **116**, 711.
[3] L. E. Bolcsak & D. E. Nerland (1983) *JBC* **258**, 7252.

DIHYDROBENZOPHENANTHRIDINE OXIDASE

This copper-dependent oxidoreductase [EC 1.5.3.12] catalyzes the reaction of dihydrosanguinarine with dioxygen to produce sanguinarine and hydrogen peroxide.[1] Other conversions for this plant enzyme include dihydrochelirubine to chelirubine and dihydromacarpine to macarpine.

[1] H. Arakawa, W. G. Clark, M. Psenak & C. J. Coscia (1992) *ABB* **299**, 1.

cis-2,3-DIHYDROBIPHENYL-2,3-DIOL DEHYDROGENASE

This oxidoreductase [EC 1.3.1.56], also known as 2,3-dihydro-2,3-dihydroxybiphenyl dehydrogenase, catalyzes the reaction of *cis*-3-phenylcyclohexa-3,5-diene-1,2-diol with NAD$^+$ to produce biphenyl-2,3-diol, NADH, and a proton.[1] The enzyme participates in the bacterial degradation of biphenyl. A catalytic triad comprised of seryl, tyrosyl, and lysyl residues has been identified.[2]

[1] M. Sylvestre, Y. Hurtubise, D. Barriault, J. Bergeron & D. Ahmad (1996) *Appl. Environ. Microbiol.* **62**, 2710.
[2] M. Vedadi, D. Barriault, M. Sylvestre & J. Powlowski (2000) *Biochemistry* **39**, 5028.

DIHYDROBUNOLOL DEHYDROGENASE

This oxidoreductase [EC 1.1.1.160], also known as bunolol reductase, catalyzes the reversible reaction of (±)-5-[(*tert*-butylaminol)-2′-hydroxypropoxy]-1,2,3,4-tetrahydro-1-naphthol (*i.e.*, dihydrobunolol) with NADP$^+$ to produce (±)-5-[(*tert*-butylaminol)-2′-hydroxypropoxy]-3,4-dihydro-1(2*H*)-naphthalenone (*i.e.*, bunolol) and

NADPH.[1] The enzyme also utilizes NAD$^+$, albeit more slowly.

[1] F.-J. Leinweber, R. C. Greenough, C. F. Schwender, H. R. Kaplan & F. J. di Carlo (1972) *Xenobiotica* **2**, 191.

DIHYDROCERAMIDE DESATURASE

This oxidoreductase catalyzes the dioxygen- and NAD(P)H-dependent conversion of dihydroceramide to ceramide.[1]

[1] C. Michel, G. van Echten-Deckert, J. Rother, K. Sandhoff, E. Wang & A. H. Merrill, Jr., (1997) *JBC* **272**, 22432.

Selected entries from *Methods in Enzymology* [vol, page(s)]:
General discussion: 311, 22
Assay: 311, 23

DIHYDROCHELIRUBINE 12-MONOOXYGENASE

This heme-thiolate-dependent oxidoreductase [EC 1.14.13.57], also known as dihydrochelirubine 12-hydroxylase, catalyzes the reaction of dihydrochelirubine with NADPH and dioxygen to produce 12-hydroxy-dihydrochelirubine, NADP$^+$, and water.[1] The cytochrome P450-dependent system catalyzes a step in the biosynthesis of macarpine.

[1] L. Kammerer, W. De-Eknamkul & M. H. Zenk (1994) *Phytochemistry* **36**, 1409.

DIHYDROCOUMARIN LIPASE

This enzyme [EC 3.1.1.35], also known as dihydrocoumarin hydrolase, catalyzes the hydrolysis of dihydrocoumarin to produce melilotate.[1] Other substrates include benzenoid 1,4-lactones, including homogentisic acid lactone, 7-hydrodihydrocoumarin, and *o*-hydroxyphenylacetic acid lactone.

[1] T. Kosuge & E. E. Conn (1962) *JBC* **237**, 1653.

2,3-DIHYDRO-2,3-DIHYDROXYBENZOATE DEHYDROGENASE

This oxidoreductase [EC 1.3.1.28], also known as 2,3-dihydroxycyclohexa-4,6-diene-1-carboxylate dehydrogenase, catalyzes the reaction of 2,3-dihydro-2,3-dihydroxybenzoate with NAD$^+$ to produce 2,3-dihydroxybenzoate and NADH.[1]

2,3-dihydro-2,3-hydroxybenzoate   2,3-dihydroxybenzoate

A transient intermediate in the reaction pathway is 2-hydroxy-3-oxo-4,6-cyclohexadiene-1-carboxylate.[1] It remains unclear if all geometric and stereoisomers of this diol are substrates of this dehydrogenase.

[1] M. Sakaitani, F. Rusnak, N. R. Quinn, C. Tu, R. B. Frigo, G. A. Berchtold & C. T. Walsh (1990) *Biochemistry* **29**, 6789.

cis-1,2-DIHYDRO-1,2-DIHYDROXYNAPHTHALENE DEHYDROGENASE

This oxidoreductase [EC 1.3.1.29], also known as *cis*-naphthalene dihydrodiol dehydrogenase and *cis*-dihydrodiol naphthalene dehydrogenase, catalyzes the reaction of *cis*-1,2-dihydronaphthalene-1,2-diol with NAD$^+$ to produce naphthalene-1,2-diol and NADH.[1,2]

cis-1,2-dihydro-naphthalene-1,2-diol   naphthalene-1,2-diol

It remains unclear whether both enantiomers of the *cis*-diol are substrates. Other substrates include *cis*-anthracene dihydrodiol, *cis*-phenanthrene dihydrodiol, and *cis*-biphenyl dihydrodiol.

[1] C. M. Serdar & D. T. Gibson (1989) *BBRC* **164**, 764.
[2] T. R. Patel & D. T. Gibson (1976) *J. Bacteriol.* **128**, 842 and (1974) **119**, 879.

DIHYDRODIPICOLINATE REDUCTASE

This oxidoreductase [EC 1.3.1.26] catalyzes the reaction of NAD(P)H with 2,3-dihydrodipicolinate to produce NAD(P)$^+$ and 2,3,4,5-tetrahydrodipicolinate.[1–3]

2,3-dihydrodipicolinate   tetrahydrodipicolinate

Kinetic isotope effects and structural investigations suggest that the reaction exhibits rapid hydride transfer step, followed by a slower, kinetically significant, proton transfer step. The C3 carbanion intermediate or its resonance tautomer, the enamine, must be stabilized at the active site prior to the proton transfer.[1] The *Escherichia coli* enzyme has an ordered Bi Bi kinetic mechanism.[2]

[1] G. Scapin, S. G. Reddy, R. Zheng & J. S. Blanchard (1997) *Biochemistry* **36**, 15081.

[2] S. G. Reddy, J. C. Sacchettini & J. S. Blanchard (1995) *Biochemistry* **34**, 3492.
[3] K. Kimura & T. Goto (1977) *J. Biochem.* **81**, 1367.
Selected entries from *Methods in Enzymology* [vol, page(s)]:
General discussion: 17B, 134
Assay: 17B, 134

DIHYDRODIPICOLINATE SYNTHASE

This enzyme [EC 4.2.1.52], also known as pyruvate: aspartate semialdehyde condensing enzyme, catalyzes the reaction of L-aspartate 4-semialdehyde with pyruvate to produce 2,3-dihydrodipicolinate and two water molecules.

The *Escherichia coli* enzyme reportedly has an ordered Bi Uni kinetic mechanism under acidic conditions or ping pong mechanism at pH 8 in which a proton is released prior to the binding of the second substrate. Pyruvate first binds to the enzyme and forms of a Schiff base with a lysyl residue (Lys161 of the *E. coli* enzyme) as well as releasing the first water molecule. The enamine form of the intermediate reacts with the aldehyde carbon of L-aspartate 4-semialdehyde resulting in the formation of the 4-hydroxy-2,3,4,5-tetrahydrodipicolinate Schiff base.[1] This intermediate is then converted enzymatically or nonenzymatically to 2,3-dihydrodipicolinate and the second water molecule.

[1] W. E. Karsten (1997) *Biochemistry* **36**, 1730.
Selected entries from *Methods in Enzymology* [vol, page(s)]:
General discussion: 17B, 129
Assay: 17B, 129

cis-DIHYDROETHYLCATECHOL DEHYDROGENASE

This oxidoreductase [EC 1.3.1.66], which participates in the bacterial degradation of ethylbenzene, catalyzes the reaction of *cis*-1,2-dihydro-3-ethylcatechol with NAD$^+$ to produce 3-ethylcatechol, NADH, and a proton.[1]

[1] D. T. Gibson, B. Gschwendt, W. K. Yeh & V. M. Kobal (1973) *Biochemistry* **12**, 1520.

DIHYDROFOLATE DEHYDROGENASE

This oxidoreductase, formerly classified as EC 1.5.1.4, is now included under dihydrofolate reductase [EC 1.5.1.3].

DIHYDROFOLATE REDUCTASE

This oxidoreductase [EC 1.5.1.3], also known as tetrahydrofolate dehydrogenase and folate reductase, catalyzes the reversible reaction of 7,8-dihydrofolate with NADPH to produce 5,6,7,8-tetrahydrofolate and NADP$^+$.[1–11]

7,8-dihydrofolate

5,6,7,8-tetrahydrofolate

The enzyme, which has an ordered Bi Bi kinetic mechanism with NADPH binding first (there are also reports of a random scheme), is isolated from mammals and some microorganisms. It can also slowly catalyze, at pH 4, the reduction of folate to 5,6,7,8-tetrahydrofolate and 8-methylpterin to 8-methyl-7,8-dihydropterin. The binding of the pyridine coenzyme is facilitated by protonation of a carboxyl group having a perturbed pK_a value of 8.09. The rate-limiting step at elevated pH is the hydride transfer whereas, at low or neutral pH values, the rate-limiting step is product release. **See also** *Dihydrofolate Reductase-Thymidylate Synthase*

[1] A. R. Clarke & T. R. Dafforn (1998) *CBC* **3**, 1.
[2] R. L. Blakley (1995) *Adv. Enzymol.* **70**, 23.
[3] H. Eisenberg, M. Mevarech & G. Zaccai (1992) *Adv. Protein Chem.* **43**, 1.
[4] K. A. Johnson & S. J. Benkovic (1990) *The Enzymes*, 3rd ed., **19**, 159.
[5] R. F. Colman (1990) *The Enzymes*, 3rd ed., **19**, 283.
[6] J. L. Hamlin & C. Ma (1990) *BBA* **1087**, 107.
[7] J. Kraut & D. A. Matthews (1987) in *Biological Macromolecules & Assemblies* (F. A. Jurnak & A. McPherson, eds.) **3**, p. 1, Wiley, New York.
[8] J. H. Freisheim & D. A. Matthews (1984) in *Folate Antagonists Ther. Agents* (F. M. Sirotnak, ed.) **1**, 69.
[9] B. Roth, E. Bliss & C. R. Beddell (1983) *Top. Mol. Struct. Biol.* **3**, 363.
[10] F. M. Huennekens, K. S. Vitols, J. M. Whiteley & V. G. Neef (1976) *Methods Cancer Res.* **13**, 199.
[11] B. R. Baker (1969) *Acc. Chem. Res.* **2**, 129.

Selected entries from **Methods in Enzymology** [vol, page(s)]:
General discussion: 6, 364; 18B, 775; 32, 576; 34, 48, 272, 281; 46, 68; 122, 360
Assay: 6, 364; 18B, 775; radiometric, 122, 346, 360
Other molecular properties: affinity labeling, 87, 494; bacterial, crystal intensity, multiwire area detector diffractometry, 114, 468; buffer mixture, and, 87, 426; catalysis, transient kinetics, 249, 23; copurification, with thymidylate synthase, 51, 94; cryoenzymology, 63, 338; *Escherichia coli* (hysteresis, structural basis, 249, 545; kinetic mechanism, derivation, 249, 25; ligand binding, association and dissociation rate constants, 249, 24; site-directed mutants, altered pH dependencies, 249, 111; transient kinetics, 249, 23 [with site-directed mutant, 249, 108]); expression plasmid, calcium phosphate

cotransfection of mammalian cells, 185, 563; gene amplification (FACS analysis, 108, 237; methotrexate resistance selection for, 185, 544); gene from mutant CHO cells, repetitive cloning, 151, 451; HIV protease fusion protein, 241, 33; hysteresis, 240, 318; 249, 545; inhibitors, 18B, 779; 63, 396, 399 (inhibitor binding, transient kinetics, 249, 23; tight-binding, 63, 437, 462, 466); *Lactobacillus casei*, deuteration, for protein–ligand interaction study, 239, 684; ligand interactions, evaluation, 202, 532; mechanism, 240, 317; methotrexate-agarose purification, 34, 272 (aminoethyl cellulose attachment, 34, 272; ω-aminohexylagarose, 34, 274, 275; assay, 34, 277; formylaminopterin acrylamide, 34, 274; methotrexate-AH-agarose, 34, 276; spacer effect, 34, 273); mitochondrial import of fusion protein, 260, 243, 248, 271; NADPH off-rate, 240, 318; peptide separation, 47, 206; pK_a determination, 308, 193; preparation from (Bakers' yeast, 18B, 610, 611; chicken liver, 6, 365; 18B, 610); pre-steady-state burst, 240, 318; properties, 6, 368; pteroyllysine-agarose in purification, 34, 281 (assay, 34, 282, 283; coupling of agarose, 34, 284; folate as eluent, 34, 286; pteroic acid, 34, 283; N^α-pteroyl-L-lysine, 34, 283; purification, 34, 284; stability, 34, 288); Raman spectroscopy, difference, 308, 193; reaction catalyzed, 249, 23; selectable marker in eukaryotic expression system, as, 245, 303, 311; staining on gels, 22, 594; stereospecificity, 87, 117; synthesis (*Escherichia coli* system, in, 217, 133, 136; wheat germ system, in, 217, 136); tetrahydrofolate preparation and, 6, 805; transition state and multisubstrate analogues, 249, 304; use in determination of (dihydrobiopterin, 18B, 615; dihydrofolate, 18B, 609); unfolding, reaction coordinate diagram, 202, 117; use in induction of enzymes, 22, 92; use of sheep liver enzyme to recycle phenylalanine hydroxylase cofactor, 18B, 601

DIHYDROFOLATE SYNTHETASE

This enzyme [EC 6.3.2.12], also known as dihydrofolate synthase, catalyzes the reaction of ATP with dihydropteroate and L-glutamate to generate ADP, orthophosphate, and dihydrofolate.[1,2]

7,8-dihydropteroate

7,8-dihydrofolate

Oxygen exchange data are in keeping with the formation of an acyl phosphate intermediate during catalysis.[1]

[1] R. V. Banerjee, B. Shane, J. J. McGuire & J. K. Coward (1988) *Biochemistry* **27**, 9062.
[2] G. M. Brown & J. M. Williamson (1982) *Adv. Biochem.* **53**, 345.

DIHYDROKAEMPFEROL 4-REDUCTASE

This oxidoreductase [EC 1.1.1.219], which participates in the biosynthesis of plant anthocyanidins, catalyzes the reaction of (+)-dihydrokaempferol with NADPH (or, more weakly, NADH) to produce *cis*-3,4-leucopelargonidin and $NADP^+$.

(+)-dihydrokaempferol   *cis*-3,4-leucopelargonidin

Alternative substrates include (+)-dihydroquercetin, which is actually a better substrate (producing *cis*-3,4-leucocyanidin) than dihydrokaempferol, and (+)-dihydromyricetin (producing *cis*-3,4-leucodelphinidin).

DIHYDROLIPOAMIDE S-ACETYLTRANSFERASE

This lipoyl-dependent transferase, [EC 2.3.1.12] (also known as lipoate acetyltransferase, thioltransacetylase A, dihydrolipoamide transacetylase, and E_2 of the pyruvate dehydrogenase complex), catalyzes the reversible reaction of coenzyme A with *S*-acetyldihydrolipoamide to produce acetyl-CoA and dihydrolipoamide.[1-5] The physiological substrate is a lipoyl group (or an *S*-acetyldihydrolipoyl group) attached to the ε-amino group of a lysyl residue of the protein.

The enzyme consists of one-to-three lipoyl-containing domains, a dihydrolipoyl-dehydrogenase-binding domain, and a catalytic domain, the three domains being connected via highly flexible prolyl- and alanyl-rich linkers. X-ray structure of the catalytic domain from *Azotobacter vinelandii* demonstrate that this acetyltransferase consists of eight trimers arranged at the corners of a cube. Coenzyme A and dihydrolipoamide bind at opposite ends of a long channel. *See also Pyruvate Dehydrogenase Complex*

[1] A. de Kok, A. F. Hengeveld, A. Martin & A. H. Westphal (1998) *BBA* 1385, 353.
[2] L. J. Reed & M. L. Hackert (1990) *JBC* 265, 8971.
[3] L. J. Reed & S. J. Yeaman (1987) *The Enzymes*, 3rd ed., 18, 77.
[4] S. J. Yeaman (1986) *Trends Biochem. Sci.* 11, 293.
[5] P. J. Butterworth, S. Tsai, M. H. Eley, T. E. Roche & L. J. Reed (1975) *JBC* 250, 1921.

DIHYDROLIPOAMIDE DEHYDROGENASE

This FAD-dependent oxidoreductase [EC 1.8.1.4] (also known as lipoamide reductase (NADH), the E_3 component of the pyruvate dehydrogenase complex, lipoyl dehydrogenase, dihydrolipoyl dehydrogenase, and diaphorase) catalyzes the reversible reaction of dihydrolipoamide with NAD^+ to produce lipoamide and NADH.[1-8] The protein is a component (E_3) of the pyruvate dehydrogenase, the α-ketoglutarate (or, 2-oxoglutarate) dehydrogenase, and the branched-chain α-keto acid dehydrogenase complexes. The physiological substrate is a dihydrolipoyl group covalently linked to the ε-amino group of a lysyl residue in E_2 of the α-keto acid dehydrogenase complexes. If the dehydrogenase is removed from its multienzyme complex, the enzyme will act on free dihydrolipoamide.

dihydrolipoyl group   lipoyl group

Dihydrolipoamide dehydrogenase contains a reactive disulfide. When the dihydrolipoyl group binds, a disulfide exchange occurs and the oxidized lipoyl group is released. The dihydrolipoamide dehydrogenase now contains two thiols (*i.e.*, reduced E_3). These two cysteinyl residues are reoxidized by the tightly bound FAD (a transient cysteine-flavin C4a adduct has been observed) and the resultant $FADH_2$ reacts with NAD^+ to produce FAD and $NADH + H^+$.

Two different kinetic mechanisms have been reported: a ping pong Bi Bi scheme, having significantly different pH_{opt} for the forward and reverse reactions, and an ordered Bi Bi mechanism under conditions of high NAD^+ concentrations.[1] *See also Pyruvate Dehydrogenase Complex; α-Ketoglutarate Dehydrogenase Complex*

[1] A. R. Clarke & T. R. Dafforn (1998) *CBC* 3, 1.
[2] B. A. Palfey & V. Massey (1998) *CBC* 3, 83.
[3] A. de Kok, A. F. Hengeveld, A. Martin & A. H. Westphal (1998) *BBA* 1385, 353.
[4] R. N. Perham & L. C. Packman (1989) *Ann. N.Y. Acad. Sci.* 573, 1.
[5] R. N. Perham, L. C. Pakman & S. E. Radford (1987) *Biochem. Soc. Symp.* 54, 67.
[6] S. J. Yeaman (1986) *Trends Biochem. Sci.* 11, 293.
[7] C. H. Williams, Jr. (1976) *The Enzymes*, 3rd ed., 13, 89.
[8] V. Massey (1963) *The Enzymes*, 2nd ed., 7, 275.

Selected entries from *Methods in Enzymology* [vol, page(s)]:
General discussion: 9, 247, 272, 273; 18A, 298; 18B, 582; 34, 288; 89, 408, 414, 420; 166, 330, 342; 252, 186; 324, 192, 453, 465
Assay: 9, 253, 272, 273; 18B, 582, 583; 52, 244; 166, 144, 332; acetylpyridine adenine dinucleotide as substrate, 252, 191; *Bacillus*, 89, 401; bovine heart, 89, 376; bovine kidney, 89, 376; broccoli mitochondrial, 89, 408; cauliflower mitochondrial, 89, 408; diaphorase activity, 252, 192; *Escherichia coli*, 89, 392; ferricyanide reductase activity, 9, 273; *Hansenula miso*, 89, 420; luciferase luminescent assay, 133, 200; NADH production, 252, 187, 190; *Neurospora crassa*, 89, 387; pigeon breast muscle, 89, 414; reverse reaction, 252, 191; spectrophotometric assay, 133, 201
Other molecular properties: absorption spectrum, 252, 193; apoenzyme and recombination with coenzyme, 18B, 585; *Bacillus* (assay, 89, 401; properties, 89, 405; purification, 89, 402); binding, to agarose-bound nucleotides, 66, 196; borohydride reduction, 18B, 475; bovine (heart [assay, 89, 376; properties, 89, 386; purification, 89, 383]; kidney [assay, 89, 376; properties, 89, 382; purification, 89, 377]); broccoli mitochondrial (assay, 89, 408; properties, 89, 412; purification, 89, 411); cauliflower mitochondrial (assay, 89, 408; properties, 89, 410; purification, 89, 409); catalyzed reactions, role of thiyl radicals, 186, 328; coimmobilized with NAD^+ (application to ethanol analyzer, 136, 31; assay, 136, 28; preparation, 136, 26); component of α-ketoglutarate dehydrogenase, 13, 55; crude extract preparation (bacteria, 252, 187; cultured skin fibroblasts, 252, 189; eukaryotic cells, 252, 188); dihydrolipoate formation with, 234, 456; *Escherichia coli*, from, 9, 253; 17B, 504 (assay, 89, 392; behavior on ethanol-Sepharose 2B, 89, 398; inhibitors, 9, 257; physical constants, 9, 257; properties, 9, 257; purification, 89, 393; resolution from pyruvate dehydrogenase complex, 9, 254; site-directed mutagenesis studies, 249, 107; specificity, 9, 257); FAD cofactor, 252, 186, 191; flavoprotein classification, 53, 397; *Hansenula miso* (assay, 89, 420; properties, 89, 423; purification, 89, 421); holoenzyme reconstitution, 53, 436; immobilized, bioluminescent assays with, 133, 198; inhibition by BCNU, 251, 182; isocitrate dehydrogenase assay, 13, 43; isolation, 9, 274; kinetic mechanism, 252, 187, 193; kinetic parameters, 252, 193; lack of sulfite complexing, 18B, 473; lipoamide dehydrogenase-valine (assay, 166, 345; purification from *Pseudomonas aeruginosa*, 166, 349); luciferase-coimmobilized

(bioluminescent assays with, 136, 88; preparation, 136, 85); luciferase luminescent assay, 133, 200; marker for chromosome detection in somatic cell hybrids, as, visualization, 151, 181; mechanism, 63, 51; metal sensitivity, 252, 193; NAD^+ (assay, 62, 172, 173; distribution, 62, 181; immobilization, in enzyme system, 44, 315, 316; properties, 62, 176; purification, from asparagus, 62, 173; radial diffusion assay, 62, 166; use in spectrophotometric titrations, 62, 190); *Neurospora crassa* (assay, 89, 387; properties, 89, 390; purification, 89, 389); pH optimum, 252, 193; photosynthetic pyridine nucleotide reductase and, 6, 445; pigeon breast muscle (assay, 89, 414; properties, 89, 418; purification, 89, 416); pig heart, 9, 272; 18B, 583 (catalytic mechanism, 9, 277; kinetic constants, 9, 277; molecular weight, 9, 277; pH optima, 9, 278; properties, 9, 276; purification, 9, 274; 18B, 583); preparation of deuterated NADH, in, 54, 226; primary structure, 252, 193; properties, 9, 257, 276; purification, 252, 189 (*Bacillus subtilis*, 166, 335; *Escherichia coli*, 9, 253; 17B, 504; pig heart, 9, 272; 18B, 583); purification on (calcium phosphate gel-cellulose, 22, 342; lipoamide glass, 34, 292); pyruvate dehydrogenase complex, 61, 215; pyruvic oxidase and, 1, 494; redox potential, 251, 15; reversible dissociation, 53, 433; scavenger enzyme in assay for site-to-site enzyme complex, as, 136, 109, 112; sensitivity to arsenite and Cd^{2+}, 13, 55, 56; spectrophotometric assay, 133, 201; stability, 252, 193; staining on gels, 22, 595; stereospecificity, 87, 119; substrates, 252, 186

DIHYDROLIPOAMIDE S-(2-METHYLPROPANOYL)TRANSFERASE

This lipoyl-dependent transferase, also known as E_2 of the branched-chain α-keto acid dehydrogenase complex, catalyzes the reversible reaction of coenzyme A with *S*-(2-methylpropanoyl)dihydrolipoamide to produce 2-methylpropanyl-CoA and dihydrolipoamide. *See also Branched Chain α-Keto Acid Dehydrogenase Complex*

Selected entries from *Methods in Enzymology* [vol, page(s)]:
General discussion: 324, 192
Assay: 166, 144, 152

DIHYDROLIPOAMIDE S-SUCCINYLTRANSFERASE

This lipoyl-dependent transferase [EC 2.3.1.61], also known as E_2 of α-ketoglutarate dehydrogenase complex, catalyzes the reversible reaction of coenzyme A with *S*-succinyldihydrolipoamide to produce succinyl-CoA and dihydrolipoamide.[1,2] This transferase is the structural and catalytic core of the α-ketoglutarate dehydrogenase complex. *See also α-Ketoglutarate Dehydrogenase Complex*

[1] J. E. Knapp, D. Carroll, J. E. Lawson, S. R. Ernst, L. J. Reed & M. L. Hackert (2000) *Protein Sci.* 9, 37.
[2] L. J. Reed & D. J. Cox (1970) *The Enzymes*, 3rd ed., 1, 213.

Selected entries from *Methods in Enzymology* [vol, page(s)]:
General discussion: 5, 651, 655
Assay: 5, 655
Other molecular properties: beef kidney, of, 27, 633; component of α-ketoglutarate dehydrogenase, 13, 55; distribution, 5, 651; *Escherichia coli*, of, 27, 621, 622, 625, 632, 644, 646 (properties, 5, 656; purification, 5, 656)

DIHYDRONEOPTERIN ALDOLASE

This aldolase [EC 4.1.2.25] catalyzes the conversion of 2-amino-4-hydroxy-6-(D-*erythro*-1,2,3-trihydroxypropyl)-7,8-dihydropteridine to 2-amino-4-hydroxy-6-hydroxy-methyl-7,8-dihydropteridine and glycolaldehyde.[1–4]

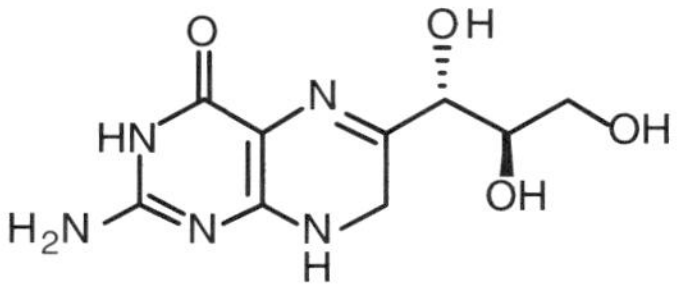

7,8-dihydro-D-neopterin

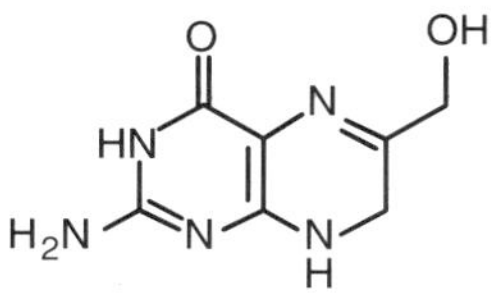

6-hydroxymethyl-
7,8-dihydropterin

Other substrates include the L-*threo* diastereoisomer. The enzyme can also catalyze the epimerization of carbon 2′ of dihydroneopterin and dihydromonopterin. Polarization of the 2′-hydroxy group of the substrate may serve as the initial reaction step for the aldolase.[1]

[1] C. Haußmann, F. Rohdich, E. Schmidt, A. Bacher & G. Richter (1998) *JBC* **273**, 17418.
[2] M. Hennig, A. D'Arcy, I. C. Hampele, M. G. Page, C. Oefner & G. E. Dale (1998) *Nat. Struct. Biol.* **5**, 357.
[3] G. M. Brown & J. M. Williamson (1982) *AE* **53**, 345.
[4] J. B. Mathis & G. M. Brown (1970) *JBC* **245**, 3015.

Selected entries from *Methods in Enzymology* [vol, page(s)]:
General discussion: **66**, 556
Assay: **66**, 557, 558

DIHYDRONEOPTERIN MONOPHOSPHATE PHOSPHOHYDROLASE

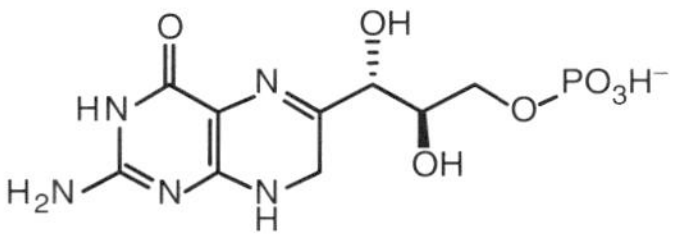

7,8-dihydro-D-neopterin 3′-phosphate

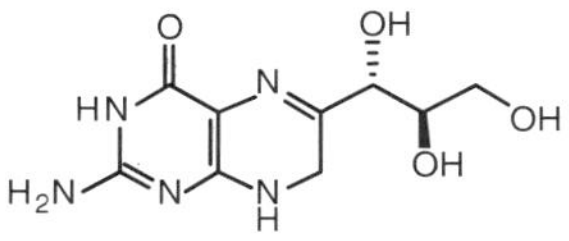

7,8-dihydro-D-neopterin

This enzyme has been proposed to catalyze the hydrolysis of 7,8-dihydro-D-neopterin 3′-monophosphate to 7,8-dihydro-D-neopterin and orthophosphate in the biosynthetic pathway for tetrahydrofolate.[1]

[1] Y. Suzuki & G. M. Brown (1974) *JBC* **249**, 2405.

DIHYDRONEOPTERIN TRIPHOSPHATE EPIMERASE

This epimerase catalyzes the interconversion of 7,8-dihydro-D-neopterin 3′-triphosphate and 7,8-dihydro-L-monopterin 3′-triphosphate has been observed in

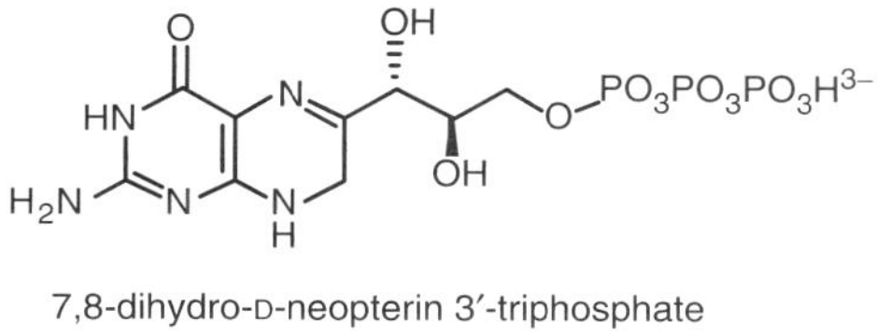

7,8-dihydro-D-neopterin 3′-triphosphate

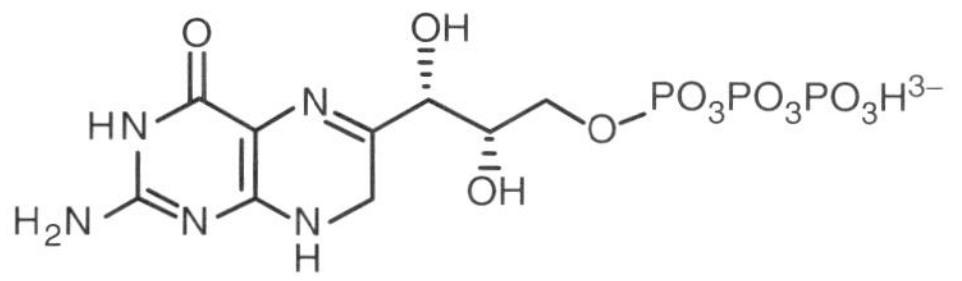

7,8-dihydro-L-monopterin 3′-triphosphate

Escherichia coli.[1,2] Dihydroneopterin aldolase has also been shown to catalyze this reaction.

[1] M. C. Heine & G. M. Brown (1975) *BBA* **411**, 236.
[2] C. Haußmann, F. Rohdich, E. Schmidt, A. Bacher & G. Richter (1998) *JBC* **273**, 17418.

DIHYDRONEOPTERIN TRIPHOSPHATE PYROPHOSPHOHYDROLASE

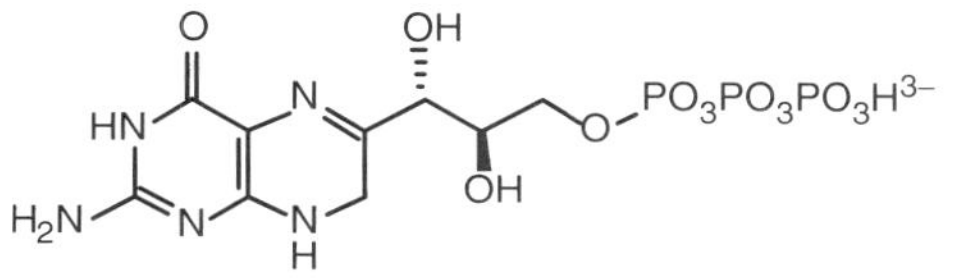

7,8-dihydro-D-neopterin 3′-triphosphate

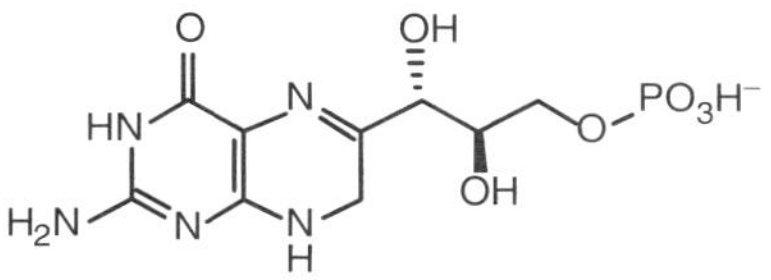

7,8-dihydro-D-neopterin 3′-phosphate

This enzyme has been proposed to catalyze the hydrolysis of 7,8-dihydro-D-neopterin 3′-triphosphate to produce 7,8-dihydro-D-neopterin 3′-monophosphate and pyrophosphate (or, diphosphate) in the biosynthetic pathway for tetrahydrofolate.[1,2] It has also been suggested that this pyrophosphohydrolase activity is nonenzymatic and is due to the presence of Ca^{2+} and Mg^{2+} ions.[3]

[1] G. M. Brown & J. M. Williamson (1982) *AE* **53**, 345.
[2] Y. Suzuki & G. M. Brown (1974) *JBC* **249**, 2405.
[3] A. De Saizieu, P. Vankan & A. P. van Loon (1995) *BJ* **306**, 371.

DIHYDROOROTASE

This enzyme [EC 3.5.2.3], also known as carbamoyl-aspartic dehydrase, catalyzes the reversible hydrolysis of (*S*)-dihydroorotate to yield *N*-carbamoyl-L-aspartate.[1–3]

N-carbamoyl-L-aspartate   dihydroorotate

The CAD protein of many eukaryotes has an dihydroorotase activity (CAD contains carbamoyl phosphate synthetase, aspartate carbamoyltransferase, and dihydroorotase activities).

[1] R. R. Bidigare, E. G. Sander & D. W. Pettigrew (1985) *BBA* **831**, 159.
[2] M. W. Washabaugh & K. D. Collins (1984) *JBC* **259**, 3293.
[3] J. E. Scheffler, J. Ma & E. G. Sander (1979) *BBRC* **91**, 563.

Selected entries from *Methods in Enzymology* [vol, page(s)]:
General discussion: 6, 180; 51, 111, 121, 135
Assay: 6, 180; 51, 123
Other molecular properties: association with OPRTase:OMPdecase complex activity, 51, 135, 138, 140; carbamoyl-phosphate synthetase-aspartate transcarbamoylase, and, in gene selection and coamplification, 185, 550; carbamoyl-phosphate synthetase (glutamine):aspartate carbamoyltransferase:dihydroorotase enzyme complex, 51, 111 (aggregation studies, 51, 126; assay, 51, 112, 113, 122, 123; dissociation, 51, 120, 121; hamster cells, from, 51, 121; isoelectric point, 51, 118; kinetic properties, 51, 118, 119; molecular weight, 51, 118, 126, 133; primary structure, 51, 133, 134; properties, 51, 118, 126; purification from [mutant hamster cells, 51, 123; rat ascites hepatoma cells, 51, 114; rat liver, 51, 116]; purity, 51, 126; sedimentation coefficient, 51, 118; sources, 51, 119, 120, 121; stability, 51, 134; N-terminus, 51, 133; tissue distribution, 51, 119, 120); molecular weight, 51, 121; properties, 6, 181; 51, 120; purification (hamster cells, 51, 123; rat cells, 51, 114, 116; *Zymobacterium oroticum*, 6, 180); rat cells, from, 51, 111; transition-state and multisubstrate analogues, 249, 307

DIHYDROOROTATE DEHYDROGENASE

This oxidoreductase [EC 1.3.99.11] catalyzes the reaction of (S)-dihydroorotate with an acceptor substrate to produce orotate and the reduced acceptor. Both iron and zinc ions are needed as cofactors.[1–3]

dihydroorotate   orotate

Acceptor substrates include soluble quinones, 2,6-dichloroindophenol, 1,10-phenanthroline, and dioxygen (although dioxygen isn't as effective as the others). The enzyme has a two-site ping pong Bi Bi kinetic mechanism.[1]

Isotope-effect investigations suggest a concerted mechanism for the liver protein.[1] **See also** *Orotate Reductase; Dihydroorotate Oxidase*

[1] V. Hines & M. Johnston (1989) *Biochemistry* **28**, 1222 and 1227.
[2] H. J. Forman & J. Kennedy (1978) *ABB* **191**, 23.
[3] G. Palmer (1975) *The Enzymes*, 3rd ed., **12**, 1.

Selected entries from *Methods in Enzymology* [vol, page(s)]:
General discussion: 51, 58, 63
Assay: 51, 58, 64; oxidase, as, 51, 59
Other molecular properties: associated cofactors, 51, 68; cytochrome, 56, 172; *Escherichia coli*, from, 51, 58; electron transport and, 51, 68, 69; inhibitors, 51, 68; kinetics, 51, 62; molecular weight, 51, 62, 67; *Neurospora*, from, 51, 63; oxidase activity, 51, 69; properties, 51, 62, 63, 67; purification, 51, 60, 65; purity, 51, 67, 68; stability, 51, 62

DIHYDROOROTATE OXIDASE

This FMN- and FAD-dependent oxidoreductase [EC 1.3.3.1], also known as dihydroorotate dehydrogenase, catalyzes the reaction of (S)-dihydroorotate with dioxygen to produce orotate and hydrogen peroxide.[1,2] Ferricyanide can also serve as a substrate as well. Isotope-effect investigations of the *Crithidia fasciculata* enzyme suggest that the reaction proceeds in a stepwise rather than concerted fashion and the both C–H bond-breaking steps are partially rate-limiting.[2] **See also** *Dihydroorotate Dehydrogenase; Orotate Reductase*

[1] R. A. Pascal & C. T. Walsh (1984) *Biochemistry* **23**, 2745.
[2] R. A. Pascal, N. L. Trang, A. Cerami & C. T. Walsh (1983) *Biochemistry* **22**, 171.

Selected entries from *Methods in Enzymology* [vol, page(s)]:
General discussion: 52, 17, 19; 56, 474

cis-3,4-DIHYDROPHENANTHRENE-3,4-DIOL DEHYDROGENASE

This oxidoreductase [EC 1.3.1.49] catalyzes the reaction of (+)-cis-3,4-dihydrophenanthrene-3,4-diol with NAD[+] to produce phenanthrene-3,4-diol and NADH.[1]

[1] K. Nagao, N. Takizawa & H. Kiyahara (1988) *Agri. Biol. Chem.* **52**, 2621.

DIHYDROPTERIDINE REDUCTASE

This oxidoreductase [EC 1.6.99.7] catalyzes the reaction of NAD(P)H with 6,7-dihydropteridine (*i.e.*, the quinoid form of dihydropteridine) to produce NAD(P)[+] and 5,6,7,8-tetrahydropteridine.[1–4]

The enzyme is not identical with dihydrofolate reductase. Note that although the reaction listed by the Enzyme Commission suggests that this classification can utilize either NADPH or NADH, NADPH-specific systems have been reported as have systems exhibiting a preference for

NADH. A number of studies suggest that the enzyme has an ordered Bi Bi kinetic mechanism in which the C–H bond breaking step is not rate-limiting. Hydride transfer is believed to be to the C5 position of the 6,7-dihydropteridine ring and a proton adding to either N3 or to the carbonyl oxygen.[1]

[1] A. R. Clarke & T. R. Dafforn (1998) CBC **3**, 1.
[2] A. Niederwieser & H. C. Curtius (1997) Enzyme **38**, 302.
[3] D. Randles (1986) EJB **155**, 301.
[4] S. Poddar & J. Henkin (1984) Biochemistry **23**, 3143.

Selected entries from *Methods in Enzymology* [vol, page(s)]:
General discussion: **142**, 97, 103, 111, 116, 127
Assay: **142**, 97, 104, 118, 127; NADPH-specific, **142**, 111
Other molecular properties: activity, **53**, 278; antiserum (assay, **66**, 724, 725; preparation, of antigen and immunization, **66**, 725; properties, of antiserum, **66**, 725; purification, **66**, 724, 725); assay of phenylalanine hydroxylase, **53**, 278, 279; function and use in assay for 7,8-dihydrobiopterin, **18B**, 602, 615; NADPH-specific (assay, **142**, 111; properties, **142**, 115; purification from bovine liver, **142**, 112; staining, **142**, 113); partition analysis, **249**, 323; properties, **142**, 101, 110, 125, 131; purification from (bovine liver, **142**, 107, 112; human liver, **142**, 120; ovine brain, **142**, 128; ovine liver, **142**, 99); site-directed mutagenesis studies, **249**, 106; stereospecificity, **87**, 120

DIHYDROPTERIDINE REDUCTASE (NADH)

This oxidoreductase activity, originally classified as EC 1.6.99.10, is now included with dihydropteridine reductase [EC 1.6.99.7].

DIHYDROPTEROATE SYNTHASE

This enzyme [EC 2.5.1.15], also known as dihydropteroate pyrophosphorylase, catalyzes the reaction of 2-amino-4-hydroxy-6-hydroxymethyl-7,8-dihydropteridine diphosphate with 4-aminobenzoate to produce pyrophosphate (or, diphosphate) and dihydropteroate.[1]

[1] G. M. Brown & J. M. Williamson (1982) Adv. Enzymol. **53**, 345.

Selected entries from *Methods in Enzymology* [vol, page(s)]:
General discussion: **18B**, 765; **66**, 560, 564
Assay: **18B**, 766; plant enzyme, **66**, 561; *Plasmodium chabaudi*, **66**, 565, 566
Other molecular properties: inhibition (p-aminobenzoate derivatives, by, **66**, 570; *Escherichia coli* system, **66**, 576); plant enzyme, **66**, 560 (assay, **66**, 561; distribution, **66**, 562; preparation, **66**, 562, 563; properties, **66**, 563); *Plasmodium chabaudi*, **66**, 564 (assay, **66**, 565, 566; properties, **66**, 568; purification, **66**, 566); properties, **18B**, 771; purification from *Escherichia coli*, **18B**, 770, 771

DIHYDROPYRIMIDINASE

This enzyme [EC 3.5.2.2], also known as hydantoinase and hydropyrimidine hydrase, catalyzes the hydrolysis of 5,6-dihydrouracil to produce 3-ureidopropionate.[1] Other substrates include dihydrothymine (producing carbamoyl-β-aminoisobutyrate), hydrouracil (generating carbamoyl-β-alanine), and hydantoin (producing carbamoylglycine).

pH kinetic investigations of the mammalian liver enzyme suggest that the reaction proceeds with a general base mechanism in which the enzyme has a water molecule bound to the active-site zinc ion. An amino-acyl residue with a pK_a value of 7.5–8 is required to activate the water for nucleophilic attack on the pyrimidine substrate, which is coordinated to the active-site zinc.[1]

[1] K. Jahnke, B. Podschun, K. D. Schnackerz, J. Kautz & P. F. Cook (1993) Biochemistry **32**, 5160.

Selected entries from *Methods in Enzymology* [vol, page(s)]:
General discussion: **6**, 184; **12A**, 57, 58
Assay: **6**, 184

DIHYDROSANGUINARINE 10-MONOOXYGENASE

This heme-thiolate-dependent oxidoreductase [EC 1.14.13.56], also known as dihydrosanguinarine 10-hydroxylase, catalyzes the reaction of dihydrosanguinarine with NADPH and dioxygen to produce 10-hydroxydihydrosanguinarine, NADP$^+$, and water.[1] This cytochrome P450-dependent enzyme is a component in benzophenanthridine alkaloid synthesis in higher plants.

[1] W. De-Eknamkul, T. Tanahashi & M. H. Zenk (1992) Phytochemistry **31**, 2713.

DIHYDROSTREPTOMYCIN-6-PHOSPHATE 3′α-KINASE

This phosphotransferase [EC 2.7.1.88] catalyzes the reaction of ATP with dihydrostreptomycin 6-phosphate to produce ADP and dihydrostreptomycin 3′α-6-bisphosphate.[1] Other acceptor substrates include 3′-deoxydihydrostreptomycin 6-phosphate.

[1] J. B. Walker & M. Skorvaga (1973) JBC **248**, 2435.

Selected entries from *Methods in Enzymology* [vol, page(s)]:
General discussion: **43**, 634
Assay: **43**, 635
Other molecular properties: biological distribution, **43**, 636, 637; properties, **43**, 636, 637; reaction scheme, **43**, 634; specificity, **43**, 637; stability, **43**, 636; *Streptomyces bikiniensis*, preparation, **43**, 636

DIHYDROURACIL DEHYDROGENASES

Dihydrouracil dehydrogenase (NAD$^+$) [EC 1.3.1.1] catalyzes the reversible reaction of 5,6-dihydrouracil with NAD$^+$ to produce uracil and NADH. The *Clostridium uracilicum* enzyme is very specific for NAD$^+$: NADP$^+$ is inactive as a substrate.[1]

5,6-dihydrouracil   uracil

Dihydrouracil dehydrogenase (NADP$^+$) [EC 1.3.1.2], also known as dihydropyrimidine dehydrogenase (NADP$^+$) and dihydrothymine dehydrogenase, catalyzes the same reaction, albeit with NADP$^+$.[2–7] Other substrates include dihydrothymine. Liver dehydrogenase activities are controlled by circadian rhythms.

[1] L. L. Campbell (1957) *JBC* **227**, 693.
[2] U. Schmitt, K. Jahnke, K. Rosenbaum, P. F. Cook & K. D. Schnackerz (1996) *ABB* **332**, 175.
[3] F. N. Naguib, S. J. Soong & M. H. el Kouni (1993) *Biochem. Pharmacol.* **45**, 667.
[4] D. J. T. Porter & T. Spector (1993) *JBC* **268**, 19321.
[5] B. Podschun, P. F. Cook & K. D. Schnackerz (1990) *JBC* **265**, 12966.
[6] O. W. Griffith (1986) *Ann. Rev. Biochem.* **55**, 855.
[7] T. Shiotani & G. Weber (1981) *JBC* **256**, 219.

Selected entries from *Methods in Enzymology* [vol, page(s)]:
General discussion: 6, 181; **12A**, 50
Assay: bacterial, **12A**, 55; beef liver, 6, 181; **12A**, 51; rat liver, **12A**, 54
Other molecular properties: bacterial (properties, 6, 184; **12A**, 57, 58; purification, 6, 184; **12A**, 56); beef liver (properties, 6, 183; **12A**, 53, 58; purification, 6, 182; **12A**, 52); occurrence, **12A**, 50; rat liver (properties, **12A**, 55, 58; purification, 6, 184; **12A**, 54); related enzymes, **12A**, 57

DIHYDROURACIL OXIDASE

This FMN-dependent oxidoreductase [EC 1.3.3.7] catalyzes the reaction of 5,6-dihydrouracil with dioxygen to produce uracil and hydrogen peroxide.[1] The enzyme will also catalyze the oxidation of dihydrothymine to thymine.

[1] J. Owaki, K. Uzura, Z. Minami & K. Kuai (1986) *J. Ferment. Technol.* **64**, 205.

2,4'-DIHYDROXYACETOPHENONE DIOXYGENASE

This iron-dependent oxidoreductase [EC 1.13.11.41] catalyzes the reaction of 2,4'-dihydroxyacetophenone with dioxygen to produce 4-hydroxybenzoate and formate.[1] The *Alcaligenes* enzyme is a homotetramer.[2]

[1] D. J. Hopper (1986) *BJ* **239**, 469.
[2] D. J. Hopper & M. A. Kaderbhai (1999) *BJ* **344**, 397.

DIHYDROXY-ACID DEHYDRATASE

This iron–sulfur-containing dehydratase [EC 4.2.1.9] catalyzes the conversion of 2,3-dihydroxy-3-methylbutanoate (*i.e.*, α,β-dihydroxyisovalerate) to 3-methyl-2-ketobutanoate (*i.e.*, α-ketoisovalerate) and water.

2,3-dihydroxy-3-methylbutanoate   3-methyl-2-ketobutanoate

Other substrates include 2,3-dihydroxy-3-methyl-pentanoate.[1–3] The enzyme from most sources has an *R* stereochemical requirement at the C2 position. The 3-hydroxy group of the dihydroxy substrate interacts with one of the iron ions (*i.e.*, a Lewis acid) of the iron–sulfur cluster either by displacing a cluster ligand or by expansion of the iron's coordination sphere.[1,2]

Interestingly, spinach 2,3-dihydroxy-acid dehydratase, which contains a [2Fe:2S] cluster, is stable in the presence of dioxygen.[2] However, those variants containing [4Fe:4S] clusters, such as from *Escherichia coli*, are inactivated by either dioxygen or superoxide.

[1] B. G. Fox (1998) *CBC* **3**, 261.
[2] D. H. Flint & M. H. Emptage (1988) *JBC* **263**, 3558.
[3] D. H. Altmiller (1972) *BBRC* **49**, 1000.

Selected entries from *Methods in Enzymology* [vol, page(s)]:
General discussion: **17A**, 755

2,3-DIHYDROXYBENZOATE 2,3-DIOXYGENASE

2,3-dihydroxybenzoate   2-carboxy-*cis,cis*-muconate

This oxidoreductase [EC 1.13.11.28] catalyzes the reaction of 2,3-dihydroxybenzoate with dioxygen to produce 2-carboxy-*cis,cis*-muconate.[1,2] Other, albeit weaker, substrates include 2,3-dihydroxy-4-toluate and 2,3-dihydroxy-4-cumate.

[1] H. K. Sharma & C. S. Vaidyanathan (1975) *EJB* **56**, 163.
[2] H. K. Sharma & C. S. Vaidyanathan (1975) *Phytochemistry* **14**, 2135.

Selected entries from *Methods in Enzymology* [vol, page(s)]:
General discussion: 52, 9, 19

2,3-DIHYDROXYBENZOATE 3,4-DIOXYGENASE

2,3-dihydroxybenzoate   3-carboxy-2-hydroxymuconate
semialdehyde

This oxidoreductase [EC 1.13.11.14], also known as
o-pyrocatechuate oxygenase and 2,3-dihydroxybenzoic
oxygenase, catalyzes the reaction of 2,3-dihydroxybenzoate
with dioxygen to produce 3-carboxy-2-hydroxymuconate
semialdehyde.[1,2] Other substrates include 2,3-dihydroxy-
p-toluate.

[1] H. Rettenmaier & F. Lingens (1985) Biol. Chem. Hoppe-Seyler 366, 637.
[2] D. W. Ribbons & J. J. Senior (1970) ABB 138, 557.

(2,3-DIHYDROXYBENZOYL)ADENYLATE SYNTHASE

This enzyme [EC 2.7.7.58], also known as 2,3-
dihydroxybenzoate:AMP ligase, catalyzes the reaction of
ATP with 2,3-dihydroxybenzoate to produce pyrophosphate (or, diphosphate) and (2,3-dihydroxybenzoyl)
adenylate.[1] The thousandfold difference in the rates
observed between the ATP ↔ pyrophosphate exchange
and the initial rates of pyrophosphate release suggest that
intermediates are accumulating on the enzyme surface. The
product remains enzyme-bound for further reaction in the
overall biosynthesis of enterobactin.

[1] F. Rusnak, W. S. Faraci & C. T. Walsh (1989) Biochemistry 28, 6827.

N-(2,3-DIHYDROXYBENZOYL)SERINE SYNTHETASE

This enzyme [EC 6.3.2.14], also known as 2,3-
dihydroxybenzoate:serine ligase, catalyzes the reaction
of ATP with 2,3-dihydroxybenzoate and L-serine to produce AMP, pyrophosphate (or, diphosphate), and N-(2,3-
dihydroxybenzoyl)-L-serine.[1–3]

2,3-dihydroxybenzoate   N-(2,3-dihydroxybenzoyl)-L-serine

This enzyme participates in the biosynthesis of enterobactin; a 2,3-dihydroxybenzoyl-AMP formation is followed by the transfer of the aryl group to a thiol. Similarly,
the serine is carried by a thiol group.[1]

[1] D. E. Ehmann, C. A. Shaw-Reid, H. C. Losey & C. T. Walsh (2000) PNAS 97, 2509.
[2] G. F. Bryce & N. Brot (1971) ABB 142, 399.
[3] N. Brot & J. Goodwin (1968) JBC 243, 510.

3α,7α-DIHYDROXY-5β-CHOLESTANATE:CoA LIGASE

3α,7α-dihydroxy-5β-cholestanoate

This enzyme [EC 6.2.1.28], also known as DHCA:CoA
ligase and 3α,7α-dihydroxy-5β-cholestanoyl-CoA synthetase, catalyzes the reaction of ATP with 3α,7α-dihydroxy-
5β-cholestanate and Coenzyme A to produce AMP,
pyrophosphate (or, diphosphate), and 3α,7α-dihydroxy-
5β-cholestanoyl-CoA. It has been identified in rat liver
endoplasmic reticulum.[1]

[1] K. Prydz, B. F. Kase, I. Bjorkhem & J. I. Pedersen (1988) J. Lipid Res. 29, 997.

DIHYDROXYCOUMARIN 7-O-GLUCOSYLTRANSFERASE

daphnetin   β-glucose   daphnin

This transferase [EC 2.4.1.104] catalyzes the reaction of
UDP-D-glucose with 7,8-dihydroxycoumarin (i.e., daphnetin) to produce UDP and daphnin.[1,2] The enzyme will
also convert esculetin into cichorin, umbelliferone into
skimmin, hydrangetin into hydrangin, and scopoletin into
scopolin, albeit more slowly.

[1] R. K. Ibrahim & B. Boulay (1980) Plant Sci. Lett. 18, 177.
[2] R. K. Ibrahim (1980) Phytochemistry 19, 2459.

1,6-DIHYDROXYCYCLOHEXA-2,4-DIENE-1-CARBOXYLATE DEHYDROGENASE

This oxidoreductase [EC 1.3.1.25] catalyzes the reaction of 1,6-dihydroxycyclohexa-2,4-diene-1-carboxylate with NAD$^+$ to produce catechol, carbon dioxide, and NADH.[1,2] **See also** *(3S,4R)-3,4-Dihydroxycyclohexa-1,5-Diene-1,4-Dicarboxylate Dehydrogenase; cis-1,2-Dihydroxycyclohexa-3,5-Diene-1-Carboxylate Dehydrogenase*

[1] W. Reineke & H.-J. Knackmuss (1978) *BBA* **542**, 424.
[2] A. M. Reiner (1972) *JBC* **247**, 4960.

(3S,4R)-3,4-DIHYDROXYCYCLOHEXA-1,5-DIENE-1,4-DICARBOXYLATE DEHYDROGENASE

This iron-dependent oxidoreductase [EC 1.3.1.53], also known as (1R,2S)-dihydroxy-3,5-cyclohexadiene-1,4-dicarboxylate dehydrogenase and dihydroxy-3,5-cyclohexadiene-1,4-dicarboxylate dehydrogenase, catalyzes the reaction of (3S,4R)-3,4-dihydroxycyclohexa-1,5-diene-1,4-dicarboxylate with NAD$^+$ to produce 3,4-dihydroxybenzoate, carbon dioxide, and NADH.[1]

[1] E. Saller, H. Laue, H. R. Schläfli-Oppenberg & A. M. Cook (1995) *FEMS Microbiol. Lett.* **130**, 97.

cis-1,2-DIHYDROXYCYCLOHEXA-3,5-DIENE-1-CARBOXYLATE DEHYDROGENASE

This oxidoreductase [EC 1.3.1.55], also known as *cis*-1,2-dihydroxy-3,4-cyclohexadiene-1-carboxylate dehydrogenase and 2-hydro-1,2-dihydroxybenzoate dehydrogenase, catalyzes the reaction of *cis*-1,2-dihydroxycyclohexa-3,5-diene-1-carboxylate with NAD$^+$ to produce catechol, NADH and a proton.[1,2] **See also** *1,6-Dihydroxycyclohexa-2,4-Diene-1-Carboxylate Dehydrogenase*

[1] A. M. Reiner (1972) *JBC* **246**, 4960.
[2] E. L. Neidle, C. Hartnett, N. L. Ornston, A. Bairoch, M. Rekik & S. Harayama (1992) *EJB* **204**, 113.

2,3-DIHYDROXY-2,3-DIHYDRO-*p*-CUMATE DEHYDROGENASE

This oxidoreductase [EC 1.3.1.58], which participates in the degradation of *p*-cumate in *Pseudomonas putida*, catalyzes the reaction of *cis*-5,6-dihydroxy-4-isopropylcyclohexa-1,3-dienecarboxylate (*i.e.*, 2,3-dihydroxy-2,3-dihydro-*p*-cumate) with NAD$^+$ to produce 2,3-dihydroxy-*p*-cumate, NADH, and a proton.[1] It is not clear if both stereoisomers of the *cis*-5,6-dihydroxy acid are substrates.

[1] R. W. Eaton (1996) *J. Bacteriol.* **178**, 1351.

DIHYDROXYFUMARATE DECARBOXYLASE

This decarboxylase [EC 4.1.1.54] catalyzes the conversion of dihydroxyfumarate to tartronate semialdehyde and carbon dioxide.[1,2] Note that the substrate exists primarily as an ene-diol dicarboxylate rather than an α-hydroxy keto dicarboxylate.[1]

[1] J. V. Schloss & M. S. Hixon (1998) *CBC* **2**, 43.
[2] K. Fukunaga (1960) *J. Biochem.* **47**, 741.

2,3-DIHYDROXYINDOLE 2,3-DIOXYGENASE

This oxidoreductase [EC 1.13.11.23] catalyzes the reaction of 2,3-dihydroxyindole with dioxygen to produce anthranilate and carbon dioxide.[1,2]

[1] O. Hayaishi, M. Nozaki & M. T. Abbott (1975) *The Enzymes*, 3rd ed., **12**, 119.
[2] M. Fujioka & H. Wada (1968) *BBA* **158**, 70.

Selected entries from *Methods in Enzymology* [vol, page(s)]:
General discussion: **52**, 19

DIHYDROXYISOVALERATE DEHYDROGENASE (ISOMERIZING)

This oxidoreductase, formerly classified as EC 1.1.1.89, catalyzes the reaction of 2,3-dihydroxy-3-methylbutanoate (*i.e.*, 2,3-dihydroxyisovalerate) with NADP$^+$ to produce 2-hydroxy-2-methyl-3-ketobutanoate and NADPH. This activity is now included with ketol-acid reductoisomerase [EC 1.1.1.86]. **See** *Ketol-Acid Reductoisomerase*

7,8-DIHYDROXYKYNURENATE 8,8a-DIOXYGENASE

This iron-dependent oxidoreductase [EC 1.13.11.10], also known as 7,8-dihydroxykynurenate oxygenase, catalyzes the reaction of 7,8-dihydroxykynurenate with dioxygen to produce 5-(3-carboxy-3-oxopropenyl)-4,6-dihydroxypyridine-2-carboxylate.[1]

[1] S. Kuno, M. Tashiro, H. Taniuchi, K. Horibata, O. Hayaishi, S. Seno, T. Tokuyama & T. Sakan (1961) *Fed. Proc.* **20**, 3.

Selected entries from *Methods in Enzymology* [vol, page(s)]:
General discussion: **52**, 11

2,4-DIHYDROXY-7-METHOXY-2H-1,4-BENZOXAZIN-3(4H)-ONE 2-D-GLUCOSYLTRANSFERASE

This glucosyltransferase [EC 2.4.1.202], first isolated from corn (*Zea mays*), catalyzes the reaction of UDP-D-glucose with 2,4-dihydroxy-7-methoxy-2H-1,4-benzoxazin-3(4H)-one to produce UDP and

2,4-dihydroxy-7-methoxy-2H-1,4-benzoxazin-3(4H)-one 2-D-glucoside.[1]

[1] B. A. Bailey & R. L. Larson (1989) Plant Physiol. **90**, 1071.

cis-1,2-DIHYDROXY-4-METHYLCYCLOHEXA-3,5-DIENE-1-CARBOXYLATE DEHYDROGENASE

This oxidoreductase [EC 1.3.1.67], which participates in the bacterial degradation of p-xylene, catalyzes the reaction of cis-1,2-dihydroxy-4-methylcyclohexa-3,5-diene-1-carboxylate with NAD(P)$^+$ to produce 4-methylcatechol, carbon dioxide, NAD(P)H, and a proton.[1]

[1] G. M. Whited, W. R. McCombie, L. D. Kwart & D. T. Gibson (1986) J. Bacteriol. **166**, 1028.

1,2-DIHYDROXY-6-METHYLCYCLOHEXA-3,5-DIENECARBOXYLATE DEHYDROGENASE

This oxidoreductase [EC 1.3.1.68], which participates in the bacterial degradation of o-xylene, catalyzes the reaction of 1,2-dihydroxy-6-methylcyclohexa-3,5-diene-carboxylate with NAD$^+$ to produce 3-methylcatechol, carbon dioxide, NADH, and a proton.[1]

[1] F. K. Higson & D. D. Focht (1992) Appl. Environ. Microbiol. **58**, 194.

1,6-DIHYDROXY-5-METHYLCYCLOHEXA-2,4-DIENECARBOXYLATE DEHYDROGENASE

This oxidoreductase [EC 1.3.1.59], which participates in bacterial degradation of m-xylene, catalyzes the reaction of 1,6-dihydroxy-5-methylcyclohexa-2,4-diene-carboxylate with NAD$^+$ to produce 3-methylcatechol, carbon dioxide, NADH, and a proton.[1]

[1] E. Neidle, C. Hartnett, L. N. Ornston, A. Bairoch, M. Rekik & S. Harayama (1992) EJB **204**, 113.

5,6-DIHYDROXY-3-METHYL-2-OXO-1,2,5,6-TETRAHYDROQUINOLINE DEHYDROGENASE

This oxidoreductase [EC 1.3.1.65] catalyzes the reaction of 5,6-dihydroxy-3-methyl-2-oxo-1,2,5,6-tetrahydroquinoline with NAD$^+$ to produce 5,6-dihydroxy-3-methyl-2-oxo-1,2-dihydroquinoline, NADH, and a proton.[1]

[1] S. Schach, G. Schwarz, S. Fetzner & F. Lingens (1993) Biol. Chem. Hoppe Seyler **374**, 175.

3,4-DIHYDROXYPHENYLACETATE 2,3-DIOXYGENASE

This iron-dependent oxidoreductase [EC 1.13.11.15], also known as homoprotocatechuate 2,3-dioxygenase, catalyzes the reaction of 3,4-dihydroxyphenylacetate with dioxygen to produce 2-hydroxy-5-carboxymethylmuconate semialdehyde.[1–5] Other substrates include 3,4-dihydroxymandelate.

3,4-dihydroxy-phenylacetate   2-hydroxy-5-carboxy-methylmuconate semialdehyde

There are also a few reports of manganese-dependent dioxy-genases: two from the genus Arthrobacter and one from Bacillus brevis. The enzyme from Bacillus brevis, which contains two tightly bound Mn(II), has an ordered Bi Uni kinetic mechanism.[3] The Fe(II)-dependent enzyme from the Gram-positive bacterium Brevibacterium fuscum also exhibits a catalase activity.[2]

The two hydroxyl groups of the substrate initially form bidentate coordination with the iron. This is followed by binding of dioxygen. Partial charge transfer from the iron makes the oxygen more nucleophilic.[1]

[1] J. E. Penner-Hahn (1998) CBC **3**, 439.
[2] M. A. Miller & J. D. Lipscomb (1996) JBC **271**, 5524.
[3] L. Que, Jr., J. Widom & R. L. Crawford (1981) JBC **256**, 10941.
[4] O. Hayaishi, M. Nozaki & M. T. Abbott (1975) The Enzymes, 3rd ed., **12**, 119.
[5] M. Nozaki (1974) in Mol. Mech. Oxygen Activ. (O. Hayaishi, ed.), pp. 135–165, Academic Press, New York.

Selected entries from *Methods in Enzymology* [vol, page(s)]:
General discussion: 17A, 645; **52**, 11

3,4-DIHYDROXYPHENYLACETATE 3,4-DIOXYGENASE

This iron-dependent oxidoreductase, formerly classified as EC 1.13.11.7, is now a deleted Enzyme Commission entry.

DIHYDROXYPHENYLALANINE AMINOTRANSFERASE

This pyridoxal-phosphate-dependent aminotransferase [EC 2.6.1.49], also known as DOPA aminotransferase, catalyzes the reversible reaction of 3,4-dihydroxy-L-phenylalanine with α-ketoglutarate (or, 2-oxoglutarate) to produce 3,4-dihydroxyphenylpyruvate and L-glutamate.[1]

[1] F. Fonnum & K. Larsen (1965) J. Neurochem. **12**, 589.

DIHYDROXYPHENYLALANINE AMMONIA-LYASE

This lyase [EC 4.3.1.11] catalyzes the conversion of 3,4-dihydroxy-L-phenylalanine to *trans*-caffeate and ammonia.[1]

[1] N. J. Macleod & J. B. Pridham (1963) BJ **88**, 45P.

DIHYDROXYPHENYLALANINE DECARBOXYLASE

This pyridoxal-phosphate-dependent decarboxylase, formerly classified as EC 4.1.1.26, catalyzes the conversion of 3,4-dihydroxyphenylalanine to 3,4-dihydroxyphenylethylamine and carbon dioxide. This activity is now included with aromatic-L-amino-acid decarboxylase [EC 4.1.1.28]. Note that tyrosine decarboxylase [EC 4.1.1.25] will also catalyze this reaction. **See** *Aromatic-L-Amino-Acid Decarboxylase; Tyrosine Decarboxylase*

3,4-DIHYDROXYPHTHALATE DECARBOXYLASE

This decarboxylase [EC 4.1.1.69] catalyzes the conversion of 3,4-dihydroxyphthalate to 3,4-dihydroxybenzoate and carbon dioxide.[1]

[1] R. W. Eaton & D. W. Ribbons (1982) J. Bacteriol. **151**, 48.

4,5-DIHYDROXYPHTHALATE DECARBOXYLASE

This decarboxylase [EC 4.1.1.55] catalyzes the conversion of 4,5-dihydroxyphthalate to 3,4-dihydroxybenzoate and carbon dioxide.[1–3] Other substrates include 4-hydroxyphthalate.

[1] J. V. Schloss & M. S. Hixon (1998) CBC **2**, 43.
[2] B. G. Pujar & D. W. Ribbons (1985) Appl. Environ. Microbiol. **49**, 374.
[3] T. Nakazawa & E. Hayashi (1978) Appl. Environ. Microbiol. **36**, 264.

3,9-DIHYDROXYPTEROCARPAN 6a-MONOOXYGENASE

This heme-thiolate-dependent oxidoreductase [EC 1.14.13.28], also known as 3,9-dihydroxypterocarpan 6a-hydroxylase, catalyzes the reaction of (6aR,11aR)-3,9-dihydroxypterocarpan with NADPH and dioxygen to produce (6aS,11aS)-3,6a,9-trihydroxypterocarpan

(a precursor of the phytoalexin glyceollin), NADP+, and water.[1]

[1] M. L. Hagmann, W. Heller & H. Grisebach (1984) EJB **142**, 127.

2,5-DIHYDROXYPYRIDINE 5,6-DIOXYGENASE

This iron-dependent oxidoreductase [EC 1.13.11.9], also known as 2,5-dihydroxypyridine oxygenase and pyridine-2,5-diol dioxygenase, catalyzes the reaction of 2,5-dihydroxypyridine with dioxygen to produce maleamate and formate.[1–3]

[1] O. Hayaishi, M. Nozaki & M. T. Abbott (1975) The Enzymes, 3rd ed., **12**, 119.
[2] M. Nozaki (1974) in Mol. Mech. Oxygen Activ. (O. Hayaishi, ed.), pp. 135–165, Academic Press, New York.
[3] R. B. Cain, C. Houghton & K. A. Wright (1974) BJ **140**, 293.

Selected entries from **Methods in Enzymology** [vol, page(s)]:
General discussion: **52**, 11

3,4-DIHYDROXYPYRIDINE 2,3-DIOXYGENASE

This oxidoreductase catalyzes the reaction of 3,4-dihydroxypyridine with dioxygen to produce 3-formiminopyruvate and formate.[1]

[1] G. K. Watson, C. Houghton & R. B. Cain (1974) BJ **140**, 277.

Selected entries from **Methods in Enzymology** [vol, page(s)]:
General discussion: **52**, 19

2,6-DIHYDROXYPYRIDINE 3-MONOOXYGENASE

This flavin-dependent oxidoreductase [EC 1.14.13.10] catalyzes the reaction of 2,6-dihydroxypyridine with NADH and dioxygen to produce 2,3,6-trihydroxypyridine, NAD+, and water.[1]

[1] P. E. Holmes & S. C. Rittenberg (1972) JBC **247**, 7622 and 7628.

Selected entries from **Methods in Enzymology** [vol, page(s)]:
General discussion: **52**, 16

3,4-DIHYDROXYQUINOLINE 2,4-DIOXYGENASE

This oxidoreductase [EC 1.13.99.5] (variously known as: 1H-3-hydroxy-4-oxoquinoline 2,4-dioxygenase; 3-hydroxy-4-oxo-1,4-dihydroquinoline 2,4-dioxygenase; 3-hydroxy-4(1H)-one, 2,4-dioxygenase; and quinoline-3,4-diol 2,4-dioxygenase) catalyzes the reaction of quinoline-3,4-diol with dioxygen to produce N-formylanthranilate and carbon monoxide.[1] This reaction is a 2,4-dioxygenolytic cleavage of two C–C bonds with the concomitant release of CO. Note that quinolin-4-ols exist primarily as their quinolin-4(1*H*)-one tautomers: thus, quinoline-2,4-diol can be called 3-hydroxy-4-oxo-1,4-dihydroquinoline. The enzyme from *Pseudomonas putida* is highly specific for the substrate, whereas the enzyme from *Arthrobacter* also acts on 2-methylquinoline-3,4-diol to produce *N*-acetylanthranilate.

[1] I. Bauer, N. Max, S. Fetzner & F. Lingens (1996) *EJB* **240**, 576.

2,3-DIHYDROXY-9,10-SECOANDROSTA-1,3,5(10)-TRIENE-9,17-DIONE 4,5-DIOXYGENASE

This iron-dependent oxidoreductase [EC 1.13.11.25], also known as steroid 4,5-dioxygenase and 3-alkylcatechol 2,3-dioxygenase, catalyzes the reaction of 3,4-dihydroxy-9,10-secoandrosta-1,3,5(10)-triene-9,17-dione with dioxygen to produce 3-hydroxy-5,9,17-trioxo-4,5:9,10-disecoandrosta-1(10),2-dien-4-oate.[1-3] Other substrates include 3-isopropylcatechol and 3-*tert*-butyl-5-methylcatechol.

[1] O. Hayaishi, M. Nozaki & M. T. Abbott (1975) *The Enzymes*, 3rd ed., **12**, 119.
[2] M. Nozaki (1974) in *Mol. Mech. Oxygen Activ.* (O. Hayaishi, ed.), pp. 135–165, Academic Press, New York.
[3] H. H. Tai & C. J. Sih (1970) *JBC* **245**, 5062.

DIIODOPHENYLPYRUVATE REDUCTASE

This oxidoreductase [EC 1.1.1.96], also known as aromatic α-keto acid reductase, catalyzes the reversible reaction of β-(3,5-diiodo-4-hydroxyphenyl)pyruvate with NADH to produce β-(3,5-diiodo-4-hydroxyphenyl)lactate and NAD$^+$.[1,2] The substrates for this enzyme must contain an aromatic ring with a pyruvate side chain (or lactate), the most active substrates being halogenated derivatives. Potential substrates with hydroxyl or amino groups in the 3 or 5 position are inactive.

[1] Y. Takada, T. Noguchi & R. Kido (1977) *Life Sci.* **20**, 609.
[2] V. G. Zannoni & W. W. Weber (1966) *JBC* **241**, 1340.

Selected entries from *Methods in Enzymology* [vol, page(s)]:
General discussion: 17A, 665

3,5-DIIODOTYROSINE AMINOTRANSFERASE

This pyridoxal-phosphate-dependent aminotransferase [EC 2.6.1.24] catalyzes the reversible reaction of 3,5-diiodo-L-tyrosine with α-ketoglutarate (or, 2-oxoglutarate) to generate 3,5-diiodo-4-hydroxyphenylpyruvate and L-glutamate.[1,2] Other substrates include the 3,5-dichloro,

3,5-dibromo, and the 3-iodo derivatives of L-tyrosine, as well as thyroxine and triiodothyronine. ***See also*** Thyroid Hormone Aminotransferase

[1] A. E. Braunstein (1973) *The Enzymes*, 3rd ed., **9**, 379.
[2] M. Nakano (1967) *JBC* **242**, 73.

Selected entries from *Methods in Enzymology* [vol, page(s)]:
General discussion: 17A, 660
Assay: rat kidney, 113, 668

DIISOPROPYL-FLUOROPHOSPHATASE

diisopropylfluorophosphate

This divalent cation-dependent enzyme [EC 3.1.8.2; formerly classified as EC 3.8.2.1], also known as DFPase, tabunase, somanase, dialkylphosphofluoridase, and organophosphorus acid anhydrolase, catalyzes the hydrolysis of diisopropyl fluorophosphate (K_m value of 4.3 mM) to produce diisopropyl phosphate and fluoride anion.[1-4] Substrates include organophosphorus compounds containing phosphorus anhydride bonds (such as phosphorus-halide and phosphorus-cyanide. Nerve gases such as Soman (*i.e.*, methylphosphonofluoridic 1,2,2-trimethylpropyl ester; K_m value of 33 mM) and Tabun (*i.e.*, dimethylphosphoramidocyanidic acid) are also substrates. The enzyme, which can be inhibited by chelating agents, is related to aryldialkylphosphatase [EC 3.1.8.1].

[1] J. A. Cohen & M. G. P. J. Warringa (1957) *BBA* **26**, 29.
[2] F. C. Hoskin & R. J. Long (1972) *ABB* **150**, 548.
[3] J. J. DeFrank & T. C. Cheng (1991) *J. Bacteriol.* **173**, 1938.
[4] W. G. Landis, M. V. Haley & D. W. Johnson (1986) *J. Protozool.* **33**, 216.

Selected entries from *Methods in Enzymology* [vol, page(s)]:
General discussion: 1, 651
Assay: 1, 651

β-DIKETONE HYDROLASE

This enzyme [EC 3.7.1.7], also known as oxidized PVA hydrolase (since the enzyme acts on the products generated by the action of secondary-alcohol oxidase [EC 1.1.3.18] on polyvinyl alcohols), catalyzes the hydrolysis of nonane-4,6-dione to produce pentan-2-one and butanoate. Other β-diketone substrates include 2,4-hexanedione, acetylacetone, 6-methyl-2,4-heptanedione, 6,8-tridecanedione, and 1-phenyl-2,4-pentanedione.[1]

[1] K. Sakai, N. Hamada & Y. Watanabe (1986) *Agric. Biol. Chem.* **50**, 989.

DIMETHYLALLYL*cis*TRANSFERASE

This enzyme [EC 2.5.1.28], also known as neryl-diphosphate synthase, catalyzes the reaction of dimethylallyl diphosphate with isopentenyl diphosphate to produce pyrophosphate (or, diphosphate) and neryl diphosphate.[1,2] Larger prenyl diphosphates are not efficient donor substrates. The product is a precursor for several cyclic monoterpenes. **See also** Dimethylallyltranstransferase

[1] D. V. Banthorpe, G. A. Bucknall, H. J. Doonan, S. Doonan & M. G. Rowan (1976) *Phytochemistry* **15**, 91.
[2] E. Beytia, P. Valenzuela & O. Cori (1969) *ABB* **129**, 346.

DIMETHYLALLYL*trans*TRANSFERASE

This enzyme [EC 2.5.1.1], also known as prenyltransferase and geranyl-diphosphate synthase, catalyzes the reaction of dimethylallyl diphosphate and isopentenyl diphosphate to produce geranyl diphosphate and pyrophosphate (or, diphosphate).[1,2] The enzyme will not accept larger prenyl diphosphates as substrates. **See also** Dimethylallylcistransferase; Geranyltranstransferase; Farnesyltranstransferase; specific prenyltransferase

[1] H. Kleinig (1989) *Ann. Rev. Plant Physiol. Plant Mol. Biol.* **40**, 39.
[2] L. Heide & U. Berger (1989) *ABB* **273**, 331.

Selected entries from **Methods in Enzymology** [**vol**, page(s)]:
General discussion: 15, 425, 426; 110, 188
Assay: 15, 428, 433; chicken liver, 110, 179; eukaryotic, 110, 145; human liver, 110, 156; *Micrococcus luteus*, 110, 189; pumpkin fruit, 110, 167; tomato fruit chromoplasts, 110, 213; yeast, 110, 179
Other molecular properties: analysis of reaction products, 15, 433; characterization, 214, 357; chicken liver (assays, 110, 179; purification, 110, 180); distribution in ammonium sulfate fractions, 15, 429; domains in, detection in CrtB and CrtE, 214, 303; eukaryotic (assay, 110, 145; gene cloning, 214, 417; interconvertible forms, 110, 149; *Neurospora crassa*, from, conserved domains, 214, 423; photoaffinity labeling, 110, 128; photolabile substrate analogues [preparation, 110, 125; properties, 110, 128]; properties, 110, 151; purification, 110, 146, 150); human liver (assay, 110, 156; properties, 110, 161; purification, 110, 159); *Micrococcus luteus* (assay, 110, 189; properties, 110, 191; purification, 110, 189); partition analysis, 249, 323; pig liver (preparation, 15, 430, 434; properties, 15, 437); *trans*-prenyl transferase reaction, stereochemistry of, 15, 427; pumpkin fruit (assays, 110, 167; properties, 110, 170; purification, 110, 169); separation from *Capsicum* phytoene synthase complex, 214, 353; tomato fruit chromoplasts (assay, 110, 213; extraction and partial purification, 110, 217; properties, 110, 219); yeast (assays, 110, 179; purification, 110, 183)

DIMETHYLAMINE DEHYDROGENASE

This FMN-dependent oxidoreductase [EC 1.5.99.10] catalyzes the reaction of dimethylamine with water and an acceptor substrate to produce methylamine, formaldehyde, and the reduced acceptor. Electron transfer to the [4Fe-4S] cluster is an obligatory step in catalysis.[1]

[1] D. J. Steenkamp & H. Beinert (1982) *BJ* **207**, 241.

DIMETHYLANILINE MONOOXYGENASE (*N*-OXIDE-FORMING)

This FAD-dependent oxidoreductase [EC 1.14.13.8], also known as microsomal flavin-containing monooxygenase, dialkyl arylamine *N*-oxidase, and dimethylaniline oxidase, catalyzes the reaction of *N*,*N*-dimethylaniline with NADPH and dioxygen to produce *N*,*N*-dimethylaniline *N*-oxide, $NADP^+$, and water.[1–6] The enzyme has a broad specificity toward relatively nonpolar substrates containing a heteroatom, acting on many dialkylarylamines as well as 1,1-dimethylhydrazine.[5]

The pig liver enzyme has an ordered kinetic mechanism.[6] NADPH binds first, reducing the flavin cofactor. Dioxygen subsequently binds to form an enzyme · flavin hydroperoxide · $NADP^+$ complex. The organic substrate finally binds and is oxidized. While the precise order of product release is unclear, $NADP^+$ is known to dissociate last. In the absence of organic substrate, hydrogen peroxide is slowly released from the flavin-hydroperoxide form of the enzyme, the rate increasing with an increase in pH.

[1] B. A. Palfey & V. Massey (1998) *CBC* **3**, 83.
[2] J. R. Cashman (1995) *Chem. Res. Toxicol.* **8**, 166.
[3] L. L. Poulsen & D. M. Ziegler (1995) *Chem. Biol. Interact.* **96**, 57.
[4] F. P. Guengerich (1990) *Crit. Rev. Biochem. Mol. Biol.* **25**, 97.
[5] R. A. Prough, P. C. Freeman & R. N. Hines (1981) *JBC* **256**, 4178.
[6] L. L. Poulsen & D. M. Ziegler (1979) *J. Biol. Chem.* **254**, 6449.

Selected entries from **Methods in Enzymology** [**vol**, page(s)]:
General discussion: 44, 849; 52, 17, 142
Assay: 52, 142, 143; *in vitro*, 52, 144
Other molecular properties: activators and inhibitors, 52, 148, 149; activity, 52, 142; conjugate (activity assay, 44, 852, 853; concentration determination, 44, 852; properties, 44, 853, 854); distribution, 52, 142; flavin content, 52, 148; immobilization, on glass beads, 44, 851, 852; molecular weight, 52, 148; oxygen sensitivity, 44, 853; postmortem inactivation, 52, 143; purification procedure, 52, 143; soluble (properties, 44, 849; substrate specificity, 44, 849, 850); specificity, 52, 149, 150

DIMETHYLANILINE-*N*-OXIDE ALDOLASE

This aldolase [EC 4.1.2.24] catalyzes the conversion of *N*,*N*-dimethylaniline *N*-oxide to *N*-methylaniline and formaldehyde.[1] Other substrates include such *N*,*N*-dialkylarylamine *N*-oxides as *N*,*N*-dimethyl-*p*-toluidine *N*-oxide and *N*,*N*-dimethylnaphthylamine *N*-oxide.

[1] J. M. Machinist, W. H. Orme-Johnson & D. M. Ziegler (1966) *Biochemistry* **5**, 2939.

DIMETHYLARGININASE

This enzyme [EC 3.5.3.18], also known as dimethylarginine dimethylaminohydrolase and N^γ, N^γ-dimethylarginine dimethylaminohydrolase, catalyzes the hydrolysis of

N^γ, N^γ-dimethyl-L-arginine to produce dimethylamine and L-citrulline. N^γ-Monomethyl-L-arginine is an alternative substrate. The rat kidney enzyme is strongly inhibited by p-chloromercuribenzoate and mercuric chloride.[1] Zinc is present in the bovine brain enzyme as a stabilizing factor.[2]

[1] T. Ogawa, M. Kimoto & K. Sasaoka (1989) JBC 264, 10205.
[2] R. Bogumil, M. Knipp, S. M. Fundel & M. Vasak (1998) Biochemistry 37, 4791.

N,N-DIMETHYLFORMAMIDASE

This iron-dependent enzyme [EC 3.5.1.56] catalyzes the hydrolysis of N,N-dimethylformamide to produce dimethylamine and formate. Alternative substrates include N-ethylformamide, N-methylformamide, and, much more weakly, N,N-diethylformamide, N,N-dimethylacetamide, and unsubstituted acyl amides.[1]

[1] H. P. Schar, W. Holzmann, G. M. Ramos Tombo & O. Ghisalba (1986) EJB 158, 469.

DIMETHYLGLYCINE DEHYDROGENASE

This oxidoreductase [EC 1.5.99.2], containing covalently bound FAD, catalyzes the reaction of N,N-dimethylglycine with water and an acceptor substrate to produce sarcosine, formaldehyde, and the reduced acceptor.[1] Acceptor substrates include 2,6-dichlorophenolindophenol and phenazine methosulfate. The enzyme will act on other methylated substrates, including sarcosine, N-methyl-L-alanine, and N^ε-methyl-L-lysine. When tetrahydrofolate is present, N^5, N^{10}-methylenetetrahydrofolate is produced rather than formaldehyde. **See also Dimethylglycine Oxidase**

[1] D. H. Porter, R. J. Cook & C. Wagner (1985) ABB 243, 396.

Selected entries from **Methods in Enzymology** [vol, page(s)]:
General discussion: 17A, 976; 122, 255
Assay: rat liver, 122, 256
Other molecular properties: flavin linkage, 53, 450; purification from rat liver, 17A, 977; 122, 257; rat liver (properties, 122, 259; purification, 17A, 977; 122, 257)

DIMETHYLGLYCINE OXIDASE

This FAD-dependent oxidoreductase [EC 1.5.3.10] catalyzes the reaction of N,N-dimethylglycine with water and dioxygen to produce sarcosine, formaldehyde, and hydrogen peroxide.[1,2] Sarcosine and choline are not substrates. **See also Dimethylglycine Dehydrogenase**

[1] N. Mori, B. Kawakami, Y. Tani & H. Yamada (1980) Agric. Biol. Chem. 44, 1383.
[2] P. R. Levering, D. J. Binnema, J. P. Van Dijken & W. Harder (1981) FEMS Microbiol. Lett. 12, 19.

DIMETHYLHISTIDINE N-METHYLTRANSFERASE

This methyltransferase [EC 2.1.1.44] catalyzes the reaction of S-adenosyl-L-methionine with N^α, N^α-dimethyl-L-histidine to produce S-adenosyl-L-homocysteine and $N^\alpha, N^\alpha, N^\alpha$-trimethyl-L-histidine.[1] Other substrates include N^α-methyl-L-histidine and L-histidine.

[1] Y. Ishikawa & D. B. Melville (1970) JBC 245, 5967.

DIMETHYLMALEATE HYDRATASE

This oxygen-sensitive, iron-dependent enzyme [EC 4.2.1.85] catalyzes the reversible reaction of dimethylmaleate with water to produce (2R,3S)-2,3-dimethylmalate.[1]

[1] A. Kollmann-Koch & H. Eggerer (1984) Hoppe-Seyler's Z. Physiol. Chem. 365, 847.

DIMETHYLPROPIOTHETIN DETHIOMETHYLASE

This enzyme [EC 4.4.1.3] catalyzes the conversion of S,S-dimethyl-β-propiothetin to dimethyl-sulfide and acrylate.[1] Other substrates include dimethylacetothetin.

[1] G. L. Cantoni & D. G. Anderson (1956) JBC 222, 171.

3,7-DIMETHYLQUERCETIN 4'-O-METHYLTRANSFERASE

This methyltransferase [EC 2.1.1.83], also known as flavonol 4'-O-methyltransferase, catalyzes the reaction of S-adenosyl-L-methionine with 3',4',5-trihydroxy-3,7-dimethoxyflavone (i.e., 3,7-dimethylquercetin) to produce S-adenosyl-L-homocysteine and 3',5-dihydroxy-3,4',7-trimethoxyflavone (i.e., 3,7,4'-trimethylquercetin).[1]

[1] V. de Luca & R. K. Ibrahim (1985) ABB 238, 596.

DIMETHYL SULFOXIDE REDUCTASE

This oxidoreductase catalyzes the reaction of dimethyl sulfoxide with a reducing agent to produce dimethyl sulfide and the oxidized agent. The molybdenum- and iron-dependent enzyme from Escherichia coli uses menaquinol as the reducing agent. The enzyme contains three subunits: a catalytic subunit that is the site of dimethyl sulfoxide reduction (containing the molybdenum coordinated to two molybdopterin guanine dinucleotides), the electron-transfer subunit containing four iron–sulfur clusters, and a subunit located within the membrane and containing the menaquinol binding site. Recent studies suggest that the crucial stage in catalysis is facilitated when the active site is in the intermediate Mo(V) oxidation state.[1] **See also** *Methionine S-Oxide Reductase*

[1] K. Heffron, C. Léger, R. A. Rothery, J. H. Weiner & F. A. Armstrong (2001) *Biochemistry* **40**, 3117.

2,6-DIOXO-6-PHENYLHEXA-3-ENOATE HYDROLASE

This enzyme [EC 3.7.1.8] catalyzes the hydrolysis of 2,6-dioxo-6-phenylhexa-3-enoate to produce benzoate and 2-oxopent-4-enoate.[1,2]

[1] T. Omori, K. Sugimura, H. Ishigooka & Y. Minoda (1986) *Agric. Biol. Chem.* **50**, 931.
[2] P. V. Bünz, R. Falchetto & A. M. Cook (1993) *Biodegradation* **4**, 171.

2,5-DIOXOPIPERAZINE HYDROLASE

This highly specific enzyme [EC 3.5.2.13], also known as cyclo(Gly-Gly) hydrolase and 2,5-diketopiperazine hydrolase, catalyzes the hydrolysis of 2,5-dioxopiperazine to produce glycylglycine.[1]

[1] T. Muro, Y. Tominaga & S. Okada (1985) *Agric. Biol. Chem.* **49**, 1567.

DIOXOTETRAHYDROPYRIMIDINE PHOSPHORIBOSYLTRANSFERASE

This transferase [EC 2.4.2.20], also known as dioxotetrahydropyrimidine-ribonucleotide pyrophosphorylase and dioxotetrahydropyrimidine-ribonucleotide diphosphorylase, catalyzes the reaction of a 2,4-dioxotetrahydropyrimidine with 5-phospho-α-D-ribose 1-diphosphate to produce a 2,4-dioxotetrahydropyrimidine D-ribonucleotide and pyrophosphate (or, diphosphate).[1] Substrates include uracil and other pyrimidines and pteridines containing a 2,4-diketo structure (for example, orotate, thymine, 6-azathymine, and 5-fluorouracil).

[1] D. Hatfield & J. B. Wyngaarden (1964) *JBC* **239**, 2580.

2,5-DIOXOVALERATE DEHYDROGENASE

This oxidoreductase [EC 1.2.1.26], also known as α-ketoglutarate semialdehyde dehydrogenase, catalyzes the reaction of 2,5-dioxopentanoate (*i.e.*, α-ketoglutatate semialdehyde) with NADP$^+$ and water to produce α-ketoglutarate (*i.e.*, 2-oxoglutarate) and NADPH.[1–4] The enzyme will also catalyze the NADP$^+$-dependent conversion of glutarate semialdehyde to glutarate. **See also** *Glutarate Semialdehyde Dehydrogenase*

[1] M. J. Dilworth, R. Arwas, I. A. McKay, S. Saroso & A. R. Glenn (1986) *J. Gen. Microbiol.* **132**, 2733.
[2] K. L. Schimz & G. Kurz (1975) *Biochem. Soc. Trans.* **3**, 1087.

[3] E. Adams & G. Rosso (1967) *JBC* **242**, 1802.
[4] E. Adams & L. Frank (1980) *Ann. Rev. Biochem.* **49**, 1005.
Selected entries from **Methods in Enzymology [vol, page(s)]**:
General discussion: 17B, 303; **22**, 595
Assay: 17B, 303

DIOXYGENASES

These dioxygen-utilizing oxidoreductases catalyze reactions in which both oxygen atoms from dioxygen are incorporated in one of the products. **See specific enzyme;** *Lipoxygenase*

DIPEPTIDASE

These enzymes, which catalyze the hydrolysis of a dipeptide to produce two amino acids, were formerly classified under EC 3.4.13.11; most are now reclassified.

LD-DIPEPTIDASE

A peptidase has been isolated from *Bacillus sphaericus* that catalyzes the hydrolysis of dipeptides containing an unblocked N-terminal L-aminoacyl residue and a C-terminal D-amino acid (or a C-terminal glycine). The best substrate reported is L-lysylglycine.[1]

[1] F. Ciroussel, M. J. Vacheron, M. Guinand & G. Michel (1990) *Int. J. Biochem.* **22**, 525.

DIPEPTIDASE DA

Dipeptidase DA (also known as dipeptidase A, pepDA and PepD) has been reported in *Lactobacillus helveticus* and *Lactobacillus sake*. It catalyzes the hydrolysis of dipeptides that do not contain proline (L-leucyl-L-leucine is readily hydrolyzed).[1–3] Specificity studies indicate that the enzyme requires a free carboxyl group on the dipeptide. The enzyme is not inhibited with EDTA or with 1,10-phenanthroline.

[1] E. G. Dudley, A. C. Husgen, W. He & J. L. Steele (1996) *J. Bacteriol.* **178**, 701.
[2] E. R. S. Kunji, I. Mierau, A. Hagting, B. Poolman & W. N. Konings (1996) *Antoine van Leeuwenhoek* **70**, 187.
[3] E. Vesanto, K. Peltoniemi, T. Purtsi, J. L. Steele & A. Palva (1996) *Appl. Microbiol. Biotechnol.* **45**, 638.

DIPEPTIDASE E

Dipeptidase E [EC 3.4.11.23], also known as aspartyl dipeptidase and often abbreviated PepE, catalyzes the hydrolysis of Asp–Xaa dipeptides. The enzyme will not hydrolyze Glu–Xaa, Gln–Xaa, or Asn–Xaa nor will it act on β-aspartyl dipeptides or tripeptides.[1,2] Artificial substrates include Asp-Phe-OMe, Asp-Phe-NH$_2$, and Asp-NHPhNO$_2$, the latter of which can be used in a straightforward assay.

[1]T. H. Carter & C. G. Miller (1984) *J. Bacteriol.* **159**, 453.
[2]C. A. Conlin, K. Håkensson, A. Liljas & C. G. Miller (1994) *J. Bacteriol.* **176**, 166.

DIPEPTIDASE, LYSOSOMAL

This metallodipeptidase (also known as LDP, serylmethionine dipeptidase, and leucylglycyl dipeptidase) catalyzes the hydrolysis of dipeptides containing free, unsubstituted N- and C-termini and has an acidic pH optimum. Specificity is fairly broad. It is distinguished from other dipeptidases by its cellular location, its EDTA-sensitive hydrolysis of L-seryl-L-methionine at acidic pH, and lack of susceptibility to inhibitors of cysteine-, serine-, or aspartic-dependent enzymes.[1,2]

[1]J. K. McDonald & A. J. Barrett (1986) in *Proteases: A Glossary and Bibliography*, vol. **2**, Academic Press, New York, p. 298.
[2]J. K. McDonald, P. X. Callahan & S. Ellis (1972) *MIE* **25B**, 272.

DIPEPTIDYL AMINOPEPTIDASE B

The fungal dipeptidyl aminopeptidase B is a serine-dependent enzyme that is an integral membrane glycoprotein of the yeast vacuole.[1]

[1]C. J. Roberts, G. Pohlig, J. H. Rothman & T. H. Stevens (1989) *J. Cell Biol.* **108**, 1363.

DIPEPTIDYL DIPEPTIDASE

This cysteine peptidase [EC 3.4.14.6], also known as dipeptidyl tetrapeptide hydrolase and tetrapeptide dipeptidase, catalyzes the hydrolysis of tetrapeptides to form two dipeptides. The hydrolysis of Ala-Gly-∇-Ala-Gly (where ∇ indicates the cleavage site) proceeds twice as rapidly as that of tetraglycine. Triglycine and pentaglycine are not substrates. The reaction is readily reversible (hence the third alternative name: dipeptide ligase), and tetrapeptides can be synthesized from dipeptides without the prior activation of a carboxyl group of one of the dipeptides. The highest rate of peptide bond formation is observed with L-alanyl-L-alanine.[1]

[1]F. W. H. T. Eng (1984) *Can. J. Biochem. Cell Biol.* **62**, 516.

DIPEPTIDYL-PEPTIDASE I

This enzyme [EC 3.4.14.1] (also known as cathepsin C, cathepsin J, dipeptidyl aminopeptidase I, and DPP I) catalyzes the hydrolysis of a peptide bond resulting in the release of an N-terminal dipeptide: that is XaaXbb from XaaXbb–Xcc… where Xaa is not an arginyl or lysyl residue and neither Xbb nor Xcc is a prolyl residue. The enzyme also catalyzes a transferase activity. The enzyme is often assayed with derivatives of Gly-Phe- or Gly-Arg-.[1–4] Surprisingly, dipeptidyl-peptidase I also catalyzes the hydrolysis of Z-Phe-Arg-∇-NHMec where ∇ indicates the cleavage site.

See also *Dipeptidase, Lysosomal*

[1]J. K. McDonald, P. X. Callahan, S. Ellis & R. E. Smith (1971) in *Tissue Proteinases* (A. J. Barrett & J. T. Dingle, eds.), p. 69, North-Holland Publ., Amsterdam.
[2]J. K. McDonald & C. Schwabe (1977) in *Proteinases in Mammalian Cells and Tissues* (A. J. Barrett, ed.), p. 311, North Holland Publ., Amsterdam.
[3]J. K. McDonald & A. J. Barrett (1986) in *Mammalian Proteases: A Glossary and Bibliography*, vol. **2**, p. 111, Academic Press, New York.
[4]H. Kirschke, A. J. Barrett & N. D. Rawlings (1995) *Protein Profile* **2**, 1587.

Selected entries from ***Methods in Enzymology*** [vol, page(s)]:
General discussion: 2, 64; 19, 302; 25, 272; 77, 333
Assay: 2, 64; 25, 273; 47, 393; amidase assay, 19, 288; esterase assay, 19, 289; hydrolase assay, 19, 291; transferase assay, 19, 290
Other molecular properties: activators and inhibitors, 19, 306; amidase assay, 19, 288; bovine spleen, preparation, 2, 66; 91, 512; distribution, 19, 309; esterase assay, 19, 289; hydrolase assay, 19, 291; inactivation, 80, 824; inhibition by cystatin, 80, 772, 777; kinetic properties, 19, 309; large-scale isolation, 22, 481, 536, 553; partial specific volume, 26, 117, 118; physical properties and purity, 2, 68; 19, 305; polypeptide sequencing, 91, 516; preparation, 2, 66; 47, 392, 393; preparation suitable for sequence studies, 25, 274; production of natural sequence growth hormone from derivatives, in, 153, 399; purification, 2, 66; 19, 303; sequence studies, 91, 516 (dipeptide identification, 25, 290; ordering of dipeptide fragments, 25, 294; other means of sequencing and, 25, 295; peptide digestion procedure, 25, 283; product dipeptide separation, 25, 285); specificity, 2, 68; 19, 285, 306; 25, 278; stability, 19, 305; transferase assay, 19, 290

DIPEPTIDYL-PEPTIDASE II

This lysosomal serine peptidase [EC 3.4.14.2], also known as dipeptidyl aminopeptidase II, dipeptidyl arylamidase II, and carboxytripeptidase, catalyzes the release of an N-terminal dipeptide: at Xaa-Xbb-$^\wedge$-Xcc, preferentially when Xbb (*i.e.*, position P1) is an alanyl or prolyl residue.[1–3] Actually, the P1 position can be any aminoacyl residue, although the acidic aminoacyl residues are the least favored. Tripeptides are the best substrates, the highest rates being observed with L-alanyl-L-alanyl-L-alanine. The enzyme also exhibits amidase and esterase activities. High activity with synthetic substrates is seen with Lys-Ala-NHNap and Lys-Pro-NHNap.

[1]J. K. McDonald, F. H. Leibach, R. E. Grindeland & S. Ellis (1968) *JBC* **243**, 4143.
[2]D. A. Eisenhauer & J. K. McDonald (1986) *JBC* **261**, 8859.
[3]K. Huang, M. Takagaki, K. Kani & I. Ohkubo (1996) *BBA* **1290**, 149.

Selected entries from ***Methods in Enzymology*** [vol, page(s)]:
Assay: 248, 602

DIPEPTIDYL-PEPTIDASE III

This broadly specific, cytosolic metallopeptidase [EC 3.4.14.4] (also known as dipeptidyl aminopeptidase III, dipeptidyl arylamidase III, red cell angiotensinase, and enkephalinase B) catalyzes the release of an N-terminal dipeptide from a peptide composed of four or more aminoacyl residues. Most tripeptides are inactive as substrates. Also resistant to hydrolysis are large polypeptides and proteins (for example, glucagon, albumin, ribonuclease, proinsulin, and corticotropin). Glu_4 and Gly_4 are similarly not hydrolyzed. If a prolyl residue is at positions P1 or P1', no hydrolysis will occur. Acceptable substrates include Ala^4, Ala^6, Lys^4, Val^5-angiotensin II, Leu-enkephalin, and angiotensin III. The enzyme is typically assayed with Arg-Arg-NHNap at pH 9.0.[1,2]

[1] S. Ellis & J. M. Nuenke (1967) JBC **242**, 4623.
[2] J. K. McDonald & A. J. Barrett (1986) in *Mammalian Proteases: A Glossary and Bibliography*, vol. **2**, p. 127, Academic Press, New York.

DIPEPTIDYL-PEPTIDASE IV

This serine peptidase [EC 3.4.14.5] (also known as dipeptidyl aminopeptidase IV, Xaa-Pro-dipeptidylaminopeptidase, Gly-Pro naphthylamidase, postproline dipeptidyl aminopeptidase IV, and postproline dipeptidylpeptidase) catalyzes the hydrolysis of a peptide bond in a protein such that there is a release of an N-terminal dipeptide (*i.e.*, XaaXbb is released from XaaXbb–Xcc–...), preferentially when Xbb is a prolyl residue and provided Xcc is neither a prolyl nor a hydroxyprolyl residue.[1] This peptidase is typically assayed with $Gly-Pro-\nabla-NHPhNO_2$, where ∇ indicates the cleavage site.

[1] Y. Ikehara, S. Ogata & Y. Misumi (1994) *Meth. Enzymol.* **244**, 215
Selected entries from *Methods in Enzymology* [vol, page(s)]:
General discussion: 77, 333; 244, 215
Assay: fluorimetric, 248, 602; rat liver, 244, 216
Other molecular properties: isolation by high-performance liquid chromatography, 271, 122; polypeptide sequencing, 91, 516; porcine kidney, preparation, 91, 513; preparation, 47, 394; rat liver (assay, 244, 216; homology between rat and human, 244, 224; membrane form, purification, 244, 220; mutants, 244, 226; properties, 244, 221; radioiodination, 73, 120; sequence, 244, 224; soluble form, purification, 244, 218); stability, 47, 394

DIPEPTIDYL-PEPTIDASE VI

This cysteine peptidase, also known as γ-D-glutamyl-L-diamino acid endopeptidase, catalyzes the hydrolysis of the peptide sequence of peptidoglycans of chemotype $A_1\gamma$. The substrates have the general structure L-Ala-γ-D-Glu-$^\wedge$-Zaa-Y where Zaa is a diaminoacyl residue (for example, lysyl or *meso*-diaminopimelyl) and Y is either D-Ala or D-Ala-D-Ala. The N-terminal L-alanyl residue must have a free amino group. Note: Few enzymes catalyze the hydrolysis of γ-glutamyl isopeptide bonds.[1,2]

[1] T. Bourgogne, M. J. Vacheron, M. Guinand & G. Michel (1992) *Int. J. Biochem.* **24**, 471.
[2] M. J. Vacheron, M. Guinand, A. Françon & G. Michel (1979) *EJB* **100**, 189.

DIPHOSPHATE-PURINE NUCLEOSIDE KINASE

This phosphotransferase [EC 2.7.1.143], also known as pyrophosphate-purine nucleoside kinase and pyrophosphate-dependent nucleoside kinase, catalyzes the reaction of pyrophosphate (or, diphosphate) with a purine nucleoside (such as adenosine, guanosine, or inosine) to produce orthophosphate and a purine mononucleotide (such as AMP, GMP, or IMP, respectively).[1] ATP is not a phosphoryl donor.

[1] V. V. Tryon & J. D. Pollack (1984) *J. Bacteriol.* **159**, 265.

DIPHTHAMIDE SYNTHETASE

diphthamide

This ligase [EC 6.3.2.22], also known as diphthine:ammonia ligase, catalyzes the reaction of ATP with diphthine and ammonia to produce ADP, orthophosphate, and diphthamide.[1] Diphthamide is a product of posttranslational modification of histidine in eukaryotic and archaebacterial elongation factor 2.

[1] T. J. Moehring, D. E. Danley & J. M. Moehring (1984) *Mol. Cell Biol.* **4**, 642.

DIPHTHERIA TOXIN

This ADP-ribosyltransferase produced by strains of *Corynebacterium diphtheriae* inactivates host cell ribosomal elongation factor eEF-2 by NAD^+-dependent covalent modification of diphthamide, an acceptor-modified histidyl residue (2-[3-carboxyamido-3-(trimethylammonio)propyl] histidine): thus, NAD^+ reacts with the peptide-diphthamide to produce nicotinamide and peptide-(ADP-D-ribosyl) diphthamide (NAD^+:diphthamide ADP-ribosyltransferase; EC 2.4.2.36).[1–3] **See** *NAD$^+$:Diphthamide ADP-Ribosyltransferase*

[1] L. Passador & W. Iglewski (1994) *Meth. Enzymol.* **235**, 617.
[2] N. J. Oppenheimer & A. L. Handlon (1992) *The Enzymes*, 3rd ed., **20**, 453.
[3] C.-Y. Lai (1986) *AE* **58**, 99.

DIPHTHINE SYNTHASE

This methyltransferase [EC 2.1.1.98] catalyzes the reaction of S-adenosyl-L-methionine with 2-(3-carboxy-3-aminopropyl)-L-histidine to produce S-adenosyl-L-homocysteine and 2-[3-carboxy-3-(methylammonio)propyl]-L-histidine.[1,2] The product and the corresponding dimethyl derivative can also act as methyl acceptor substrates in this system.

[1]J. M. Moehring & T. J. Moehring (1988) JBC **263**, 3840.
[2]J.-Y. C. Chen & J. W. Bodley (1988) JBC **263**, 11692.

DIPLOCARDIA LUCIFERASE

This earthworm monooxygenase catalyzes the reaction of *Diplocardia* luciferin (*i.e.*, N-isovaleryl-3-aminopropanal) with dioxygen.

Selected entries from **Methods in Enzymology** [vol, page(s)]:
General discussion: 57, 375
Assay: 57, 377, 381
Other molecular properties: assay of hydrogen peroxide, 57, 378; copper stimulation, 57, 380, 381; purification, 57, 375; quantum yield, 57, 376

DISCADENINE SYNTHASE

This transferase [EC 2.5.1.24] catalyzes the reaction of S-adenosyl-L-methionine with N^6-(Δ^2-isopentenyl)adenine to produce 5′-methylthioadenosine and the potent cytokinin discadenine (*i.e.*, 3-(3-amino-3-carboxypropyl)-N^6-(Δ^2-isopentenyl)adenine.[1,2] Other substrates include N^6-benzyladenine.

[1]Y. Taya, Y. Tanaka & S. Nishimura (1978) FEBS Lett. **89**, 326.
[2]M. Ihara, Y. Tanaka, K. Yanagisawa, Y. Taya & S. Nishimura (1986) BBA **881**, 135.

DISULFOGLUCOSAMINE-6-SULFATASE

This sulfatase [EC 3.1.6.11], also known as N-sulfoglucosamine-6-sulfatase, catalyzes the hydrolysis of N,6-O-disulfo-D-glucosamine to produce N-sulfo-D-glucosamine and sulfate.[1,2] In some organisms, the enzyme may be identical with N-acetylglucosamine-6-sulfatase [EC 3.1.6.14].

[1]B. Weissmann, H. Chao & P. Chow (1980) BBRC **97**, 827.
[2]J. S. Bruce, M. W. McLean, F. B. Williamson & W. F. Long (1985) EJB **152**, 75.

DNA-(APURINIC OR APYRIMIDINIC SITE) LYASE

This lyase [EC 4.2.99.18] (also known as AP lyase, AP endonuclease class I, endodeoxyribonuclease [apurinic or apyrimidinic], deoxyribonuclease [apurinic or apyrimidinic], E. coli endonuclease III, phage-T4 UV endonuclease, and *Micrococcus luteus* UV endonuclease) catalyzes a reaction in which the C–O–P bond 3′ to the apurinic or apyrimidinic site in DNA is broken by β-elimination, leaving a 3′-terminal unsaturated sugar and a product with a terminal 5′-phosphate. The phosphodiester bond is "nicked," not hydrolyzed.[1,2] This group of enzymes was previously listed as hydrolyzing endonucleases, under the number EC 3.1.25.2.

The calf thymus enzyme removes UV-/oxidation-damaged bases by forming an N-glycosylase activity followed by a 3′ apurinic/apyrimidinic endonuclease activity.[3,4] The human lyase is quasi-processive,[5] cleaving 7 or 8 abasic sites while traveling at least 200 nucleotides before dissociating from the DNA substrate.

The T4 endonuclease V and E. coli endonuclease III catalyze an N-glycosylase action (*i.e.*, cleavage of the bond between the modified base and the sugar) and a apyrimidinic lyase reaction. The reaction proceeds with the formation of a Schiff base between an enzyme amino group (the α-amino group in the T4 protein and the ε-amino group of Lys120 of the E. coli endonuclease III) and the aldehyde group of a ring-opened apyrimidinic deoxyribose or the imine group of a ring-opened thymine glycol.[4] *See also Deoxyribonuclease (Pyrimidine Dimer)*

[1]V. Bailly & W. G. Verly (1987) BJ **242**, 565.
[2]J. Kim & S. Linn (1988) Nucleic Acids Res. **16**, 1135.
[3]P. W. Doetsch, D. E. Helland & W. A. Haseltine (1986) Biochemistry **25**, 2212.
[4]Y. W. Kow & S. S. Wallace (1987) Biochemistry **26**, 8200.
[5]D. C. Carey & P. R. Strauss (1999) Biochemistry **38**, 16553.

Selected entries from **Methods in Enzymology** [vol, page(s)]:
General discussion: 21, 244
Assay: 21, 244; human placenta, 65, 203, 217
Other molecular properties: activity of yeast redoxyendonuclease, 234, 111; crystals, stabilization, 276, 136; DNA damage testing protocols, 269, 251; human placenta, of, 65, 216 (assay, 65, 203, 217; cation requirements, 65, 223, 224; isoelectric point, 65, 222; molecular weight, 65, 223; pH optimum, 65, 223; properties, 65, 223, 224; purification, 65, 219; stability, 65, 224; substrate specificity, 65, 224); properties, 21, 248; purification, 21, 246

DNA (CYTOSINE-5-)-METHYLTRANSFERASE

This transferase [EC 2.1.1.37] catalyzes the reaction of S-adenosyl-L-methionine with DNA to produce S-adenosyl-L-homocysteine and DNA containing a 5-methylcytosine residue.[1-4] Site-specific DNA methyltransferase (cytosine-specific) [EC 2.1.1.73] catalyzes the reaction of

S-adenosyl-L-methionine with DNA containing a cytosine at a specific site to produce *S*-adenosyl-L-homocysteine and DNA containing a 5-methylcytosine. The *Hha*I methyltransferase has an ordered Bi Bi kinetic mechanism. **See also** Site-Specific DNA Methyltransferase (Cytosine-Specific); specific transferase

[1]I. Ahmad & D. N. Rao (1996) *Crit. Rev. Biochem. Mol. Biol.* **31**, 361.
[2]S. Kumar, X. Cheng, S. Klimasauskas, S. Mi, J. Posfai, R. J. Roberts & G. G. Wilson (1994) *Nucleic Acid Res.* **22**, 1.
[3]S. Hattman (1981) *The Enzymes*, 3rd ed., **14**, 517.
[4]S. J. Kerr & E. Borek (1973) *The Enzymes*, 3rd ed., **9**, 167.

Selected entries from **Methods in Enzymology** [vol, page(s)]:
General discussion: **155**, 32; **216**, 244

DNA-DEOXYINOSINE GLYCOSYLASE

This enzyme [EC 3.2.2.15], also known as DNA-deoxyinosine glycosidase and DNA(hypoxanthine) glycohydrolase, catalyzes the hydrolysis of DNA and polynucleotides containing deoxyinosine, resulting in the release of hypoxanthine.[1–3] This glycosylase prefers double-stranded DNA and polynucleotides. Hypoxanthine is removed fifteen-to-twenty times more readily from an I–T base pair than from an I–C base pair.[3]

[1]P. Karran & T. Lindahl (1980) *Biochemistry* **19**, 6005.
[2]P. Karran & T. Lindahl (1978) *JBC* **253**, 5877.
[3]G. Dianov & T. Lindahl (1991) *NAR* **19**, 3829.

DNA-DIRECTED RNA POLYMERASE

This polymerase [EC 2.7.7.6], also known as RNA nucleotidyltransferase (DNA-directed), catalyzes the DNA-template-directed reaction on *n* molecules of nucleoside triphosphates to produce *n* molecules of pyrophosphate (or, diphosphate) and RNA containing *n* bases: *i.e.*, the extension of the 3'-end of an RNA strand by one nucleotide at a time.[1,2] The polymerase can initiate the synthesis of an RNA chain *de novo*. Three forms of the eukaryotic enzyme are distinguished on the basis of their sensitivity of α-amanitin and the type of RNA synthesized: RNA polymerase I (not inhibited by α-amanitin) catalyzes the synthesis of most precursors of ribosomal RNA; RNA polymerase II (strongly inhibited by α-amanitin) catalyzes the synthesis of mRNA precursors; and RNA polymerase III (inhibition by α-amanitin varies with the species) catalyzes the synthesis of precursors of 5S rRNA, tRNAs, and other small RNAs. **See** *Bacteriophage T7, RNA Polymerase/Promoter System; Coronavirus RNA Polymerase; NS5B; Poliovirus RNA Polymerase 3D^Pol; RNA Polymerase I; RNA Polymerase II; RNA Polymerase III; RNA-Directed RNA Polymerase; Polynucleotide Adenylyltransferase*

[1]S. T. Jacob (1973) *Prog. Nucl. Acid Res. Mol. Biol.* **13**, 93.
[2]V. S. Sethi (1971) *Progr. Biophys. Mol. Biol.* **23**, 67.

Selected entries from **Methods in Enzymology** [vol, page(s)]:
Monographs: *Meth. Enzymol.* (1996) vols. **273** and **274** (S. Adhya, ed.)
General discussion: 12B, 555; **34**, 5, 464, 469; **46**, 89, 90; **60**, 323; **167**, 592
Assay: 12B, 555, 566; **21**, 507, 508; **31**, 270; **36**, 322; **46**, 350; **79**, 455; cyanobacterial, DNA-dependent, **167**, 592; filter paper disk assay for, 12B, 170; micrococcal, 12B, 559; rat liver, 12B, 556; bacterial and bacteriophage, **101**, 549, 559; abortive probability determination, **273**, 70; gel electrophoresis, **273**, 60, 62; gel electrophoresis of transcripts, **273**, 95; mitochondrial transcription factor A-dependent activity, **264**, 156; nuclear extract transcription system, **273**, 87; promoter binding, **273**, 5, 46, 55; reconstituted transcription system, **273**, 82, 93; regulatory protein assay, **273**, 55, 89, 102; reiterative transcription, **273**, 71, 77, 81; ribonuclease elimination, **273**, 74, 87; template preparation, **273**, 89; in vitro assay correlation with in vivo assay, **273**, 98
Other molecular properties: acrylamide gel electrophoresis, **21**, 513, 516, 520 (in situ assay, **21**, 188); *Azotobacter* enzyme, **6**, 11; 12B, 566; *Bacillus subtilis* enzyme, **273**, 155; bacterial and bacteriophage (application for extensive transcription in vitro, **101**, 562; assays, **101**, 549, 559; purification, **101**, 541, 556); bacteriophage T4 deoxyribonucleic acid and, **20**, 263, 267; β/β' subunits, **273**, 309; bacteriophage T7 (in vitro transcription [heterogeneity of products, **261**, 330; isotope-enriched nucleotides, **261**, 317, 319, 567; product purification, **261**, 332, 568; reaction conditions, **261**, 328; omptin substrate, **244**, 387; purification, **261**, 326); binding sites (primer, **273**, 310; rifampicin, **273**, 303; sorangicin A, **273**, 308; streptolydigin, **273**, 309); Brij 58 lysate and, 12A, 515, 516; cascade control, **64**, 325; cellulose-bound polynucleotides and, **21**, 191; chromatin, in, **31**, 279; (DNA-dependent, in nuclei, **31**, 259, 261, 262); core enzyme (assay, **21**, 508; elution, **21**, 504; isolation, **21**, 512; properties, **21**, 519); cross-linked to DNA, radioiodination, **73**, 120; CTD kinase phosphorylating, purification, **200**, 301; cyanobacterial, DNA-dependent (assay, **167**, 592; properties, **167**, 596; purification, **167**, 593; σ factor, purification, **167**, 595); DNA-dependent, **46**, 346; **64**, 281 (affinity labeling, **46**, 355, 356; assay, 12B, 566; atomic fluorescence studies, **54**, 469; chromatin, in, **31**, 259; daily rhythms, **36**, 477, 479; metal-hybrid, preparation, **158**, 89; preparation, 12A, 527, 528; **29**, 257; properties, 12B, 570; purification, 12B, 567; stereochemistry, **87**, 203, 212, 224, 231); DNA–RNA complex, partition analysis, **249**, 324; DNA transcription assay, **38**, 372; editing mechanism, **64**, 296, 297; energy transfer studies, **48**, 378; *Escherichia*, **6**, 23; **64**, 291 (CAP binding, **65**, 866; in vitro RNA synthesis, in, **59**, 847; primed RNA synthesis, **65**, 498; synthesis, of cRNA to polyoma virus DNA, **65**, 725); *Escherichia coli* (elongation arrest suppressors [GreA, **274**, 315, 318, 320; GreB, **274**, 318, 320, 323; N antiterminator, **274**, 364, 375, 386, 388, 390]; elongation rate, **274**, 334; epitope mapping of subunits, **274**, 511; *Escherichia coli* Eσ^70 [-DNA complex, conformational changes, **185**, 40; escape from promoters, **185**, 41; open complex with promoter {association and dissociation, activation free energy analysis, **208**, 338; association and dissociation, temperature and salt effects, **208**, 330; formation mechanism, **208**, 328}; promoter recognition and mRNA initiation, **185**, 37]; fluorescence spectroscopy [kinetic assay of RNA polymerases, **274**, 475; labeling of subunits, **274**, 476; nucleotide probes {binding to *Escherichia coli* RNA polymerase, **274**, 463, 469, 474; resonance energy transfer studies with rifampicin as acceptor, **274**, 465; synthesis, **274**, 458, 472, 477}; polarization studies of protein binding, **274**, 500; tryptophan {intrinsic protein fluorescence, **274**, 456; time-resolved emission in transcription factors, **274**, 457}];

guanosine 5′-diphosphate 3′-diphosphate binding and regulatory effects, **274**, 471; histidine-tagged protein [immobilization on solid phase, **274**, 326, 332; purification, **274**, 328; reconstitution, **274**, 331; solid-phase transcription, **274**, 331; tagging methods, **274**, 328]; immunoprinting of N complex [antibody probing, **274**, 367; elongation reaction, **274**, 367; principle, **274**, 366; template walking with affinity tags, **274**, 373]; initiation proteins, **273**, 99; pausing assays [average dwell time determination, **274**, 349; detection of pausing {gel electrophoresis, **274**, 340; radiolabeling of transcripts, **274**, 336, 341; reaction conditions, **274**, 340; reagents, **274**, 337; sampling times, **274**, 337; templates, **274**, 337}; half-life determination, **274**, 345; kinetic simulation and rate constant determination, **274**, 350; mapping of pause site [marker comparison method, **274**, 342; RNA fingerprinting, **274**, 342; RNA sequence ladder comparison, **274**, 343}; mechanisms of pausing, **274**, 344; pause efficiency, **274**, 348]; promoter binding, **273**, 5; purification of reconstituted enzyme, **273**, 128; subunits, **274**, 403, 476, 503 [composition, **273**, 5, 14, 121, 130, 149; genes, **273**, 301; reconstitution, **273**, 121, 128, 133]); *Escherichia coli* Eσ[70] (-DNA complex, conformational changes, **185**, 40; escape from promoters, **185**, 41; open complex with promoter [association and dissociation {activation free energy analysis, **208**, 338; temperature and salt effects, **208**, 330}]; promoter recognition and mRNA initiation, **185**, 37); fidelity, **64**, 294; filter paper disk assay for, **12B**, 170; holoenzyme, **21**, 513; hormone induced biorhythms, **36**, 480; interaction with chromatin, **12B**, 62; initiation of transcription, global steps, **273**, 109; labeled ribonucleic acid preparation and, **21**, 476; localization, **12A**, 425; membrane fractions and, **12B**, 797, 804, 805, 812, 814; metal-hybrid, preparation, **158**, 89; micrococcal enzyme, **12B**, 559; mitochondrial, **12A**, 465, 476; motifs of sequence homology, **275**, 3; N4 enzyme (early promoters, **274**, 14; purification, **274**, 11; subunits, **274**, 9; transcription assay *in vitro*, **274**, 12); nuclei, in, **31**, 263, 722 (assay, **31**, 270; preservation, **31**, 271); nucleic acid-free, **12A**, 579; nucleoli and, **12A**, 457; phage, assay of synthesis, **30**, 661; plant cell, **12B**, 106; polyguanylic acid formation by, **12B**, 523; processivity, **64**, 291, 296; product, ribonuclease and, **12B**, 564; promoter binding, **273**, 4, 30, 45, 47, 51, 101; promoter clearance, **273**, 60, 70, 108; promoter complexes, kinetic studies, **208**, 236 (stable intermediates, **208**, 238; transient intermediates, **208**, 251); promoter sequences for, identification in chloroplast extract, **118**, 267; purification, **12B**, 58, 560; purification of recombinant SP6 enzyme for *in vitro* transcription, **275**, 383; chromatography; rate expression, **64**, 282; rat liver (assay, **12B**, 556; preparation, **12B**, 557; properties, **12B**, 558); response regulator interactions, **273**, 281, 285; Rho-dependent termination, **273**, 317; RNA polymerase I (activation by transcriptional activators, **273**, 25; assay, **273**, 236; comparison to prokaryotic holoenzyme, **273**, 28; gene specificity, **273**, 14, 165, 170; promoter binding, **273**, 21; purification from mouse, **273**, 236; RNA polymerase II (abortive initiation assay, **273**, 106; activation by transcriptional activators, **273**, 23; assay [distinguishing initiation from elongation effects 3′-extended template assays, **274**, 423; promoter-specific initiation, **274**, 427]; elongation rate determination, **274**, 429; elongation stimulation by factors, **274**, 421, 439; mapping of 3′ ends, **274**, 430; pausing, **274**, 431; promoter clearance, **274**, 428; readthrough, **274**, 432; RNA cleavage by factors, **274**, 433, 436; template preparation, **274**, 60; termination, **274**, 431]; closed complex formation [holoenzyme binding, **273**, 18; minimal promoter sequence, **273**, 15; recognition proteins, **273**, 15, 28, 101, 110, 168; Sarkosyl effects, **273**, 102]; comparison to prokaryotic holoenzyme, **273**, 28; epitope tagging, **194**, 517; fractionation with Polymin P, **194**, 517; immunoaffinity purification [wheat germ enzyme, **274**, 521; yeast enzyme and associated factors, **274**, 523]; open-complex formation [assay, **273**, 104; nucleotide triphosphate requirement, **273**, 101, 103]; permanganate probing, **273**, 103, 106]; plant, *in vitro* transcription [efficiency, **273**, 273; primer extension assay of transcripts, **273**, 272; reaction conditions, **273**, 271; template preparation, **273**, 270]; preinitiation complex [assembly, **273**, 16, 110; gel filtration, **273**, 116;

purification using immobilized DNA template {DNA immobilization, **273**, 113; immunoblot analysis, **273**, 114; principle, **273**, 112; transcriptionally competent complex formation on DNA, **273**, 114; troubleshooting, **273**, 116}]; promoter detection by GRAIL, **266**, 272, 277; protein affinity chromatography with initiation factors, **274**, 120; purification from HeLa cells [core enzyme purification {heparin affinity chromatography, **274**, 98; immunoaffinity chromatography, **274**, 99; ion-exchange chromatography, **274**, 98; yield, **274**, 99}; holoenzyme purification, **274**, 87; heparin affinity chromatography, **274**, 62; nuclear extraction, **274**, 58, 61; solution preparation, **274**, 59; yield, **274**, 62}]; reconstitution system [activity, **274**, 57; preparation, **274**, 69]; regulation of transcription elongation, **274**, 419; subunits, **274**, 97, 520; yeast enzyme [assay of holoenzyme transcription, **273**, 175, 183; core enzyme, **273**, 172; holoenzyme purification, **273**, 173; initiation assay, **194**, 548; transcription *in vitro*, **194**, 545]); RNA polymerase III (activation by transcriptional activators, **273**, 24; assay [gel-retardation assay, **273**, 252; nonspecific transcription assay, **273**, 251; single-round transcription assay, **273**, 251]; plant, *in vitro* transcription, **273**, 277; promoter binding [preinitiation complex formation, **273**, 19; promoter elements {class 1, **273**, 19; class 2, **273**, 20; class 3, **273**, 20}; recognition proteins, **273**, 19, 28, 167, 249]; purification from yeast, **273**, 255; subunit structure, **273**, 257; transcription from chromatin and cloned gene templates *in vitro*, analysis, in, **170**, 347); *Saccharomyces cerevisiae* mitochondrial enzyme (core polymerase, nonselective transcription assay, **264**, 58; dilution, **264**, 63; holoenzyme, selective transcription assay, **264**, 59; promoter consensus sequence, **264**, 62; recombinant core polymerase purification, **264**, 60; recombinant specificity factor purification, **264**, 61; selectivity, **64**, 294; sigma factor, **21**, 500, 506; stereochemistry, **87**, 203, 212, 224, 231; subunits, **21**, 500, 517; T4-modified enzyme, T7 (cell clones carrying, selection, **217**, 58; delivery to cell by bacteriophage CE6, **185**, 65; directed expression of cloned genes in *Escherichia coli*, **185**, 60; induction by IPTG: expression of target DNA, **185**, 79; nucleoside triphosphate substrate specificity, **180**, 61; preparation of M1 RNA, in, **181**, 578; purification, **202**, 307; recombinant vaccinia virus system, Rab protein overexpression, **219**, 403; stable high-level expression of genes in mammalian cells, in, **217**, 47; synthesis of [long capped transcripts *in vitro*, **180**, 42; pre-tRNA, **180**, 67; small RNAs, **180**, 51]; transcription accuracy, **180**, 62); termination-altered mutants, **273**, 312, 317; vaccinia virus enzyme, **275**, 211, 212, 216; vertebrate mitochondrial enzyme isolation (assay, **264**, 144; buffers, **264**, 143; DEAE chromatography, **264**, 144; extraction, **264**, 143; heparin-Sepharose chromatography, **264**, 144)

DNA α-GLUCOSYLTRANSFERASE

This transferase [EC 2.4.1.26], first identified in *Escherichia coli* infected with bacteriophages T2, T4, or T6, catalyzes the transfer of an α-D-glucosyl residue from UDP-D-glucose to an hydroxymethylcytosyl residue in DNA.[1,2] *See also DNA β-Glucosyltransferase; Glucosyl-DNA β-Glucosyltransferase*

[1] S. R. Kornberg, S. B. Zimmerman & A. Kornberg (1961) *JBC* **236**, 1487.
[2] J. Tomaschewski, H. Gram, J. W. Crabb & W. Rüger (1985) *NAR* **13**, 7551.

Selected entries from *Methods in Enzymology* [vol, page(s)]:
General discussion: 12B, 496
Assay: 12B, 498; **20**, 540
Other molecular properties: *in vitro* synthesis, assay for, **20**, 540 properties, **12B**, 507; purification procedures, **12B**, 500, 502, 504

DNA β-GLUCOSYLTRANSFERASE

This transferase [EC 2.4.1.27], first identified in *Escherichia coli* infected with bacteriophages T4, catalyzes the transfer of an β-D-glucosyl residue from UDP-D-glucose to an hydroxymethylcytosyl residue in DNA.[1,2] The crystal structure reveals a deep cleft in which UDP-D-glucose binds with extensive hydrogen bonding interactions.[3] **See also** *DNA α-Glucosyltransferase; Glucosyl-DNA β-Glucosyltransferase*

[1] S. R. Kornberg, S. B. Zimmerman & A. Kornberg (1961) *JBC* **236**, 1487.

[2] J. Tomaschewski, H. Gram, J. W. Crabb & W. Rüger (1985) *NAR* **13**, 7551.

[3] A. Vrielink, W. Rüger, H. P. Driessen & P. S. Freemont (1994) *EMBO J.* **13**, 3413.

Selected entries from **Methods in Enzymology** [vol, page(s)]:
General discussion: 12B, 496
Assay: 12B, 498, 499
Other molecular properties: properties, 12B, 507; purification procedures, 12B, 500, 504

DNA HELICASE

DNA helicase is an ATP-dependent enzyme that catalyzes the unwinding of DNA double helices during DNA replication and repair. Helicases separate dsDNA into single-stranded DNA intermediates that are required during replication and recombination, using the Gibbs free energy available from the ATPase activity to unwind the duplex and translocate along the nucleic acid lattice. (**See** *Energase*) Most helicases require a ssDNA tail adjacent to a dsDNA region in order to initiate unwinding of the duplex. **See also** *DNA Topoisomerases*

Selected entries from **Methods in Enzymology** [vol, page(s)]:
Assay: ATPase assay, 262, 395, 402; DNA–RNA substrates, 262, 397; DNA substrate preparation, 262, 393, 396, 403; gel electrophoresis, 262, 395; magnesium dependence, 262, 395; nucleotide preference, 262, 391; polarity, 262, 398; primer design, 262, 394; quantitation, 262, 396; synthetic minifork as substrate, 262, 460, 464
Other molecular properties: ATP hydrolysis, 262, 389, 391, 395; macroscopic reaction mechanism analysis (distributive mechanism, 262, 400; limited unwinding reaction, 262, 401; processivity, 262, 400); oligomerization, 262, 402; polarity of reaction, 262, 389, 398; purification (assay monitoring, 262, 402; bacteriophage T4 enzyme, 262, 577; chromatography, 262, 403, 578; ultracentrifugation, 262, 404); quaternary structure, 262, 405; substrate specificity, 262, 391, 396; types in *Escherichia coli*, 262, 389; unwinding reaction, 262, 389, 459

DNA LIGASES

These enzymes catalyze reactions that lead to the reformation of phosphodiester linkages in variously damaged forms of DNA.[1-8]

DNA ligase (ATP) [EC 6.5.1.1] (also known as polydeoxyribonucleotide synthase (ATP), polynucleotide ligase (ATP), sealase, DNA repair enzyme, breakage–reunion enzyme, and DNA joinase) catalyzes the reaction of ATP with (deoxyribonucleotide)$_n$ and (deoxyribonucleotide)$_m$ (*i.e.*, duplex DNA with a single-strand break) to produce AMP, pyrophosphate (or, diphosphate), and (deoxyribonucleotide)$_{n+m}$. RNA can also act as substrate, albeit not as effectively.

DNA ligase (NAD$^+$) [EC 6.5.1.2] (also known as polydeoxyribonucleotide synthase (NAD$^+$), polynucleotide ligase (NAD$^+$), DNA repair enzyme, and DNA joinase) catalyzes the reaction of NAD$^+$ with (deoxyribonucleotide)$_n$ and (deoxyribonucleotide)$_m$ (*i.e.*, duplex DNA with a single-strand break) to produce AMP, nicotinamide nucleotide (NMN), and (deoxyribonucleotide)$_{n+m}$: thus, a phosphodiester bond is formed between a 3'-hydroxyl group and an adjacent 5'-phosphate terminus in a polynucleotide hydrogen-bonded to a complementary strand. As with EC 6.5.1.1, RNA is also a substrate. In the ping pong kinetic mechanism, NAD$^+$ first binds and forms an adenylyl-enzyme intermediate at a lysyl residue (phosphonamide linkage) with the release of NMN.[7] Upon binding the break-containing dsDNA, the adenylate group is transferred to the 5'-phosphate, forming a pyrophosphate linkage to the adenosine (a DNA-adenylate intermediate). The adjacent 3'-hydroxyl then attacks this new linkage, AMP is released, and the break is now sealed.[7]

[1] S. P. Jackson (1999) *Biochem. Soc. Trans.* **27**, 1.

[2] A. E. Tomkinson & Z. B. Mackey (1998) *Mutat. Res.* **407**, 1.

[3] A. E. Tomkinson & D. S. Levin (1997) *Bioessays* **19**, 893.

[4] S. Shuman (1996) *Structure* **4**, 653.

[5] M. J. Engler & C. C. Richardson (1982) *The Enzymes*, 3rd ed., **15**, 3.

[6] S. Söderhäll & T. Lindahl (1976) *FEBS Lett.* **67**, 1.

[7] I. R. Lehman (1974) *Science* **186**, 790.

[8] I. R. Lehman (1974) *The Enzymes*, 3rd ed., **10**, 237.

Selected entries from **Methods in Enzymology** [vol, page(s)]:
General discussion: 34, 473; 65, 190; 68, 41, 50, 51, 65, 180, 248, 346
Assay: 21, 311; 262, 545; assay by covalent joining of bacteriophage λ DNA molecules (circle formation, 21, 327; dimer formation, 21, 330; principle, 21, 326); bacteriophage T4 induced, 21, 319; rabbit tissue, 21, 333
Other molecular properties: bacteriophage T4 induced (preparation, 21, 322; properties, 21, 324; synthesis of synthetic *lac* operator, 65, 883; uses of, 21, 326); cohesive end ligation, 68, 58, 485; contaminating exonuclease, 68, 106; conversion of joint to recombinant molecules and, 21, 294, 296; DNA polymerases and, 29, 11, 51; large-scale isolation of, 22, 535; phage, assay of synthesis, 30, 662; product inhibition studies, 249, 203; properties, 21, 317; purification, 21, 315; rabbit tissue (cellular distribution, 21, 337;

properties, **21**, 337; purification procedure, **21**, 335; subcellular localization, **21**, 337); role in SV40 DNA replication, **262**, 545; T4, **68**, 16, 54, 104, 133, 248 (blunt-end ligation, **68**, 18, 57, 104, 249, 475, 485 [of *Eco*RI 8-mer, **68**, 104; of *Hin*dIII 10-mer, **68**, 104]; circularization experiments, in, **212**, 5; DNA modification with, **152**, 108; specific elution by substrate, **68**, 133; terminal crosslinking, **68**, 56; terminal mismatch, **68**, 56)

DNA-3-METHYLADENINE GLYCOSYLASE I

This glycosylase [EC 3.2.2.20] (also known as DNA-3-methyladenine glycosidase I, 3-methyladenine-DNA glycosylase I, and DNA glycosidase I) catalyzes the hydrolysis of methylated or alkylated DNA, resulting in the release of 3-methyladenine.[1-3]

This catalytic activity protects cells against the mutagenicity and/or cytotoxicity of certain alkylating agents. One mechanistic view is that the enzyme intercalates into the minor groove of DNA, causing the abasic pyrrolidine nucleotide to flip into the glycosylase active site, where an active-site water molecule is poised for nucleophilic attack.

[1]M. D. Wyatt, J. M. Allan, A. Y. Lau, T. E. Ellenberger & L. D. Samson (1999) *BioEssays* **21**, 668.
[2]L. Thomas, C. H. Yang & D. A. Goldthwait (1982) *Biochemistry* **21**, 1162.
[3]B. K. Duncan (1981) *The Enzymes*, 3rd ed., **14**, 565.

Selected entries from *Methods in Enzymology* [vol, page(s)]:
Assay: 65, 290, 291
Other molecular properties: *Escherichia coli*, **65**, 290 (inhibitors, **65**, 294, 295; kinetic parameters, **65**, 294; mechanism of action, **65**, 294, 295; molecular weight, **65**, 293; occurrence, **65**, 295; physical parameters, **65**, 293; properties, **65**, 293; purification, **65**, 291; substrate specificity, **65**, 294)

DNA-3-METHYLADENINE GLYCOSYLASE II

This glycosylase [EC 3.2.2.21] (also known as DNA-3-methyladenine glycosidase II, 3-methyladenine-DNA glycosylase II, and DNA glycosidase II) catalyzes the hydrolysis of alkylated DNA (*i.e.*, DNA methylated or alkylated at certain positions), yielding 3-methyladenine, 3-methylguanine, 7-methylguanine, or 7-methyladenine.[1,2]

[1]M. D. Wyatt, J. M. Allan, A. Y. Lau, T. E. Ellenberger & L. D. Samson (1999) *BioEssays* **21**, 668.
[2]L. Thomas, C. H. Yang & D. A. Goldthwait (1982) *Biochemistry* **21**, 1162.

DNA NUCLEOTIDYLEXOTRANSFERASE

This transferase [EC 2.7.7.31] (also known as terminal addition enzyme, terminal transferase, and terminal deoxyribonucleotidyltransferase) catalyzes the reaction of n molecules of deoxynucleoside triphosphates (*i.e.*, n dNTP) with [deoxynucleotide]$_m$ (*i.e.*, DNA containing m deoxynucleotides) to produce n molecules of pyrophosphate (or, diphosphate) and [deoxynucleotide]$_{m+n}$. In this manner, the enzyme catalyzes the template-independent extension of the $3'$-end of a DNA strand by one nucleotide at a time.[1-7] This transferase will not initiate *de novo* synthesis of DNA oligomers. Ribonucleotides can also be donor substrates. The enzyme has a random Bi Bi kinetic mechanism.[3,5]

[1]M. J. Modak & V. N. Pandey (1991) in *A Study of Enzymes* (S. A. Kuby, ed.), vol. **2**, 391, CRC Press, Boca Raton.
[2]L. M. S. Chang & F. J. Bollum (1986) *Crit. Rev. Biochem.* **21**, 27.
[3]B. L. Vallee & A. Galdes (1984) *Adv. Enzymol.* **56**, 283.
[4]R. L. Ratliff (1981) *The Enzymes*, 3rd ed., **14**, 105.
[5]M. R. Deibel & M. S. Coleman (1980) *JBC* **255**, 4206.
[6]F. J. Bollum (1978) *Adv. Enzymol.* **47**, 347.
[7]F. J. Bollum (1974) *The Enzymes*, 3rd ed., **10**, 145.

Selected entries from *Methods in Enzymology* [vol, page(s)]:
General discussion: 68, 17, 41, 51, 58, 346, 399
Assay: calf thymus, **29**, 72; filter paper disk assay for, **12B**, 170
Other molecular properties: assay for, in (polypeptide β₁, **116**, 247; thymosin α₁, **116**, 247; thymosin β₄, **116**, 253; thymosin fraction 5 and 5A, **116**, 231); biotinylation of ssDNA, **216**, 59; calf thymus (assay, **29**, 72; homogeneous preparation, **29**, 77; properties, **29**, 79; purification, **29**, 73, 77); cellulose-bound polynucleotides and, **21**, 191; DNA end-labeling, **65**, 43, 508, 510, 512, 523; **152**, 104; double-stranded cDNA tailing, **79**, 606, 607; endonuclease contamination, **68**, 49; filter paper disk assay for, **12B**, 170; oligodeoxynucleotide labeling and, **29**, 322, 340, 357 (micro scale, **29**, 334, 337; semimicro scale, **29**, 330, 333); **152**, 342; poly(dA) tailing, procedure, **65**, 489; positive lymphoid precursor cells, generation *in vitro*, **150**, 363; preparation, **29**, 326, 356; tailing of DNA and for *in vitro* mutagenesis, in, **100**, 96; tailing of plasmid, **79**, 610

DNA POLYMERASE

These polymerases [EC 2.7.7.7], also known as DNA nucleotidyltransferases (DNA-directed), catalyze DNA template-directed extension of the $3'$-end of a nucleic acid strand one nucleotide at a time. In this elongation reaction, n deoxynucleoside triphosphates produce n pyrophosphate (or, diphosphate) ions and DNA$_n$ containing x bases.[1-8] RNA can also serve as the primer. Enzymes within at least five distinct DNA polymerase families all catalyze the transfer reaction by a mechanism in which two metal ions act to stabilize a pentacoordinated transition state. One of the metal ions activates the primer's $3'$-hydroxyl group for attack on the α-phosphate of the dNTP; the other chelates the β- and γ-phosphates and stabilizes the negative charge that builds up on the leaving oxygen. *See specific polymerase*

RNA-directed DNA polymerases [EC 2.7.7.49], also known as DNA nucleotidyltransferases (RNA-directed) and reverse transcriptases (or revertases), catalyze the RNA template-directed elongation of the 3′-end of a nucleic acid strand one nucleotide at a time. In this primer extension process, n deoxynucleoside triphosphates produce n pyrophosphate (or, diphosphate) ions and DNA$_n$. As was the case above, this enzyme cannot initiate the synthesis of a polymeric chain *de novo*; it requires a primer which may be DNA or RNA. **See** *Reverse Transcriptase*

[1]N. H. Williams (1998) *CBC* **1**, 543.
[2]F. B. Perler, S. Kumar & H. Kong (1996) *Adv. Protein Chem.* **48**, 377.
[3]P. A. Fisher (1994) *Prog. Nucl. Acid Res. Mol. Biol.* **47**, 371.
[4]V. Mizrahi & S. J. Benkovic (1988) *Adv. Enzymol.* **61**, 437.
[5]I. R. Lehman (1981) *The Enzymes*, 3rd ed., **14**, 15 and 51.
[6]C. McHenry & A. Kornberg (1981) *The Enzymes*, 3rd ed., **14**, 39.
[7]A. Weissbach (1981) *The Enzymes*, 3rd ed., **14**, 67.
[8]T. Kornberg & A. Kornberg (1974) *The Enzymes*, 3rd ed., **10**, 119.

Selected entries from **Methods in Enzymology** [vol, page(s)]:
General discussion: 6, 34; 12B, 586, 591; 27, 93; 29, 3, 13, 22, 27, 38, 46, 53, 70, 89; 34, 4, 468; 46, 90, 171; 262, 3, 13, 22, 35, 42, 49, 62, 77, 84, 93, 98, 108, 147, 189, 202, 217, 232, 257, 270, 283, 274, 303, 323, 331, 363, 442
Assay: 6, 34; *Bacillus subtilis*, 29, 27; bacteriophage T4, 29, 46; calf thymus, 12B, 592; 29, 71; chloroplast, 118, 187; KB cells, 29, 90; low molecular weight of mammalian cells, 29, 82; *Micrococcus luteus*, 29, 39; mitogen-treated lymphocytes, in, 150, 63; polymerase assay (activity gel assays [DNA-dependent DNA polymerase, 275, 335, 340; principle, 275, 331; reagents, 275, 340; renaturation conditions, 275, 339; reverse transcriptase, 275, 336, 340; sample preparation, 275, 339; sensitivity optimization, 275, 339; substrates, 275, 339]; nonisotopic labeling assay, 275, 258, 275; polymerization with radiolabeled primer, 275, 246; primer selection, 275, 242; product detection by hybridization, 275, 259; radiolabeled nucleotide incorporation [acid precipitation, 275, 245; DE81 filter-binding assay, 275, 244; gel electrophoresis, 275, 246; isotope selection, 275, 243; reaction conditions, 275, 243; stopping of reaction, 275, 244]; template selection, 275, 242); sea urchin nucleus, 29, 54; thymus gland, 12B, 592

DNA POLYMERASE I

This prokaryotic DNA-directed polymerase [EC 2.7.7.7] catalyzes the attack of the 3′-hydroxy of the growing DNA chain on the α-phosphoryl group of an incoming dATP, dCTP, dGTP or dTTP to elongate the chain by one base by producing a new phosphodiester bond as well as pyrophosphate (or, diphosphate). The enzyme also acts as a $3′ \rightarrow 5′$ exonuclease and as a $5′ \rightarrow 3′$ exonuclease: hence to edit out noncomplementary bases erroneously inserted during polymerization.[1] DNA polymerase I plays central roles in DNA repair, nick translation, and RNA-primer excision. **See also** *DNA Polymerases; DNA Polymerase II; DNA Polymerase III*

[1]N. H. Williams (1998) *CBC* **1**, 543.

Selected entries from **Methods in Enzymology** [vol, page(s)]:
General discussion: 6, 34; 34, 469; 262, 3

Assay: 6, 34; DNA synthesis, 29, 3; $3′ \rightarrow 5′$ exonuclease, 29, 5; 262, 12; $5′ \rightarrow 3′$ exonuclease, 29, 6; 262, 12; polymerase activity, 262, 11
Other molecular properties: adenosine 5′-O-(1-thiotriphosphate), and, 87, 203, 213; 2′-deoxyadenosine 5′-O-(1-thiotriphosphate), and, 87, 213, 224, 230; DNA-dependent, 46, 357; double-stranded complementary DNA synthesis, 79, 604; editing mechanism, 64, 295 (change, 64, 288, 292; product distribution, 64, 289); *Escherichia coli*, 6, 34; 65, 190, 234, 719 (DNA extension reaction, 65, 397; labeling viral DNA, by nick translation, 65, 471; ^{32}P labeling, of polyoma virus DNA, 65, 723, 724; processivity, 64, 292; properties, 6, 39; purification, 6, 36; 262, 3 [historical account, 182, 783]; primed synthesis, 65, 560, 564; 3′ terminus blocking reaction, with ddTTP, 65, 567); exonuclease activity, 262, 3, 13; expression system (host strains, 262, 4; induction [cell growth, 262, 5; heat induction, 262, 6; monitoring, 262, 7; nalidixic acid induction, 262, 6]; plasmids, 262, 4); fidelity, 64, 273, 294; gene cloning in λ vectors, 262, 3; general properties, 29, 11; kinetic pathway, 64, 282; Klenow fragment, 262, 3, 11, 148, 203, 231, 257, 265, 385; 3′ labeling of DNA primers, in, 180, 358; mechanism, 64, 279; 240, 318; nucleotide binding, 64, 282; peptide I, NMR structural analysis, 262, 161; polymerization cycle, 262, 203; processivity, 64, 288, 292; proofreading, 64, 295; purification, 29, 8 (*Escherichia coli*, from, [ammonium sulfate fractionation, 262, 9; cell lysis, 262, 8; gel filtration, 262, 9; historical account, 182, 783; hydrophobic interaction chromatography, 262, 9; ion-exchange chromatography, 262, 9; ion-filtration chromatography, 262, 33; rationale, 262, 10]; purification strategy, 275, 6); purified, other enzyme activities present, 29, 11; reactions catalyzed, 29, 3; ribonucleotide incorporation and, 29, 253; selectivity, 64, 293, 294; specificity, 29, 4, 6, 8; stereochemistry, 87, 203, 211, 224, 227, 230; 262, 194; subtilisin digestion, 262, 344; triphosphate-metal configuration, determination, 262, 199

DNA POLYMERASE II

This prokaryotic DNA-directed polymerase [EC 2.7.7.7] catalyzes the attack of the 3′-hydroxy of the growing DNA chain on the α-phosphoryl group of an incoming dATP, dCTP, dGTP or dTTP to elongate the chain by one base by producing a new phosphodiester bond as well as pyrophosphate (or, diphosphate). While this enzyme possesses $3′ \rightarrow 5′$ exonuclease activity, it lacks the $5′ \rightarrow 3′$ exonuclease activity observed with DNA polymerase I. DNA polymerase II probably participates in DNA repair during the so-called SOS response. **See also** *DNA Polymerases; DNA Polymerase I; DNA Polymerase III*

Selected entries from **Methods in Enzymology** [vol, page(s)]:
General discussion: 34, 469; 262, 13
Assay: 29, 13; $3′ \rightarrow 5′$ exonuclease activity, 262, 15; polymerase activity, 262, 14
Other molecular properties: exonuclease activity, 262, 14; fidelity, 64, 294; gene regulation, 262, 13; pathway, 64, 295; processivity, 64, 291, 292; 262, 14; properties, 29, 16 (extent of reaction, 29, 20; other reactions, 29, 21; template requirements, 29, 17); purification, 29, 14; 262, 13 (*Escherichia coli* enzyme [ammonium sulfate fractionation, 262, 16; cell growth, 262, 16; cell lysis, 262, 16; exonuclease-deficient mutant, 262, 19; gel electrophoresis, 262, 20; ion-exchange chromatography, 262, 18; phosphocellulose chromatography, 262, 17; plasmid construction, 262, 21; recovery, 262, 20; storage, 262, 21]); selectivity, 64, 294

DNA POLYMERASE III

This prokaryotic DNA-directed polymerase [EC 2.7.7.7] catalyzes the attack of the 3'-hydroxy of the growing DNA chain on the α-phosphoryl group of an incoming dATP, dCTP, dGTP or dTTP to elongate the chain by one base by producing a new phosphodiester bond as well as pyrophosphate (or, diphosphate). This enzyme also possesses a 3' → 5' exonuclease activity, but not the 5' → 3' exonuclease activity observed with DNA polymerase I.[1,2] DNA polymerase III functions as the replicase responsible for replicating the *Escherichia coli* genome. Note that the exonuclease activity only occurs on single-stranded DNA: thus, while this polymerase can edit, it cannot catalyze nick translation. *See also DNA Polymerases; DNA Polymerase I; DNA Polymerase II*

[1] Z. Kelman & M. O'Donnell (1995) *Ann. Rev. Biochem.* **64**, 171.
[2] C. S. McHenry (1991) *JBC* **266**, 19127.

Selected entries from *Methods in Enzymology* [vol, page(s)]:
General discussion: **262**, 22, 35, 442
Assay: **29**, 22; exonuclease activity, **262**, 42; polymerase activity, **262**, 26, 41; solution preparation, **262**, 25
Other molecular properties: *Bacillus subtilis* (abundance in cell, **262**, 40; exonuclease activity, **262**, 36; nucleotide inhibitors, **262**, 36, 40; purification [cloned gene product, **262**, 38; Gram-positive bacteria extracts, **262**, 36]); crude extract preparation, **262**, 40; cycling and lagging strand synthesis (assay [acceptor complex construction, **262**, 444; agarose gel electrophoresis, **262**, 447; DNA purification, **262**, 447; donor complex construction, **262**, 444; mixing of reaction, **262**, 446]; mechanism, **262**, 444); holoenzyme (processivity, **262**, 22; structure, **262**, 22); kinase protection assay, **262**, 441; properties, **29**, 25; purification, **29**, 24; **262**, 22, 35 (*Escherichia coli* holoenzyme [ammonium sulfate fractionation, **262**, 27; cell growth, **262**, 24; cell lysis, **262**, 27; hydrophobic interaction chromatography, **262**, 31; ion-exchange chromatography, **262**, 28]); radiolabeling of γ complex, **262**, 437, 439; sliding clamp activity of β dimer, **262**, 442, 444; subunit (exchange with radiolabeled proteins, **262**, 440; structure, **262**, 13, 22); topological binding of β subunit to DNA, **262**, 439

DNA POLYMERASE α

This eukaryotic DNA-directed polymerase [EC 2.7.7.7] catalyzes the attack of the 3'-hydroxy of the growing DNA chain on the α-phosphoryl group of an incoming dATP, dCTP, dGTP or dTTP to elongate the chain by one base by producing a new phosphodiester bond as well as pyrophosphate (or, diphosphate).[1] This nuclear enzyme participates in the replication of chromosomal DNA, and its activity changes with the rate of cellular proliferation. The polymerase has a DNA primase activity, but not an exonuclease activity. The enzyme also exhibits moderate processivity and is strongly inhibited by aphidicolin and *N*-ethylmaleimide. Dideoxynucleoside triphosphates are not inhibitory. *See also DNA Polymerases; DNA Primase*

[1] I. R. Lehman & L. S. Kaguni (1989) *JBC* **264**, 4265.

Selected entries from *Methods in Enzymology* [vol, page(s)]:
General discussion: **262**, 62, 77
Assay: **275**, 140; exonuclease activity, **262**, 58; inhibitors, **262**, 57; polymerase activity, **262**, 57, 64, 94, 537; primase activity, **262**, 64, 94; spinach, **118**, 98; substrates, **262**, 57, 60
Other molecular properties: domains, **275**, 152; *Drosophila melanogaster*, purification (182-kDa subunit [exonuclease activity, **262**, 70; isolation, **262**, 70; overexpression in Sf9 cells, **262**, 70]; embryo culture, **262**, 63; extraction, **262**, 65; glycerol gradient sedimentation, **262**, 66; hydroxylapatite chromatography, **262**, 65; inhibitor sensitivity, **262**, 67, 69; ion-exchange chromatography, **262**, 67; phosphocellulose chromatography, **262**, 65; primase separation from polymerase, **262**, 69; proteolysis minimization, **262**, 62; single-stranded DNA cellulose chromatography, **262**, 63, 66); exonuclease-like domain, **275**, 150; expression during cell cycle, **275**, 154, 166; gene (cloning, **275**, 151; locus, **275**, 152; structure, **275**, 151); human, immunoaffinity purification (antibody preparation, **262**, 78, 83; baculovirus expression system, **262**, 78; cell extraction [cultured human cells, **262**, 79; recombinant insect cells, **262**, 79, 83]; immunoaffinity chromatography, **262**, 80, 84; inhibitor sensitivity, **262**, 82; kinetic properties of purified protein, **262**, 81; storage, **262**, 81; time required, **262**, 82; yield, **262**, 81, 83); immunoaffinity chromatography, **262**, 537; immunological characterization, **275**, 156; inhibitors, **275**, 157; phosphorylation, **275**, 154; primase complex (purification from HeLa cells [extraction, **275**, 137; hydroxyapatite chromatography, **275**, 139; immunoaffinity chromatography, **275**, 137; phosphocellulose chromatography, **275**, 139]; purification from mouse, **275**, 140; resolution of activities, **275**, 140; subunits and activities, **275**, 136, 149; role in SV40 replication, **275**, 135; *Saccharomyces cerevisiae* (genes, **262**, 50, 108; homology with mammalian enzyme, **262**, 52; inhibitor sensitivity, **262**, 53, 60, 126; purification [anion-exchange chromatography, **262**, 52, 57, 59; cation-exchange chromatography, **262**, 62; cell growth, **262**, 55, 58; desalting column, **262**, 56, 59; extraction, **262**, 55, 59; phosphocellulose batch adsorption, **262**, 56, 61; proteolysis minimization, **262**, 53]; telomere synthesis, **262**, 126); sequence alignment with other polymerases, **262**, 295; sequence homology between species, **275**, 153; site-directed mutagenesis (conserved regions, **262**, 296, 298; fidelity effects, **262**, 302; metal-binding domain, **262**, 302; nucleotide-binding region, **262**, 302; polymerization domain, **262**, 302; rationale of substitutions, **262**, 298); spinach (assay, **118**, 98; properties, **118**, 102; purification, **118**, 99); substrate-binding order, **275**, 151; template specificity, **275**, 150, 158

DNA POLYMERASE β

This eukaryotic DNA-directed polymerase [EC 2.7.7.7] catalyzes the attack of the 3'-hydroxy of the growing DNA chain on the α-phosphoryl group of an incoming dATP, dCTP, dGTP or dTTP to elongate the chain by one base by producing a new phosphodiester bond as well as pyrophosphate (or, diphosphate). This relatively small nuclear enzyme probably participates in DNA repair, and its activity does not vary with the rate of cellular proliferation. The enzyme is strongly inhibited by dideoxynucleoside triphosphates, but not by aphidicolin or *N*-ethylmaleimide. *See also DNA Polymerases*

Selected entries from *Methods in Enzymology* [vol, page(s)]:
General discussion: **262**, 98, 108

Assay: gap synthesis, **262**, 106; polymerase activity, **262**, 113

Other molecular properties: domains (8-kDa domain [NMR structural analysis, **262**, 105, 107, 161, 163, 166; purification, **262**, 105]; 31-kDa domain [NMR structural analysis, **262**, 161, 167; nucleotidyltransferase, **262**, 102, 107; purification, **262**, 106]; chemical cleavage, **262**, 104; DNA-binding domains, **262**, 101, 107; overexpression and purification of recombinant domains, **262**, 104; proteolytic mapping, **262**, 103, 107; structure, **262**, 101; substrate binding conformation, **262**, 169, 171); N-ethylmaleimide sensitivity, **262**, 116, 127; isotopic labeling for NMR, **262**, 105, 158; mammalian, recombinant (expression in *Escherichia coli*, **262**, 100; gene cloning, **262**, 99; purification, **262**, 100); phosphorylation, **262**, 125; physiological functions, **262**, 109, 119, 123, 128; *Saccharomyces cerevisiae* polymerase β-70 (DNA repair role, **262**, 119, 127; gene disruption, **262**, 110, 112, 119, 125, 127; induction during meiosis, **262**, 110, 114, 121, 125, 128; inhibitor sensitivity, **262**, 116, 126; methylmethane sulfonate sensitivity, **262**, 112, 120; nucleotide substrate specificity, **262**, 117, 126; processivity, **262**, 117; purification [cell lysis, **262**, 114; hemagglutinin epitope tagging, **262**, 110, 115; immunoaffinity purification, **262**, 114; yeast strains in overexpression, **262**, 110, 115]; reverse transcriptase activity, **262**, 117, 126; sequence homology with mammalian proteins, **262**, 125; sporulation role, **262**, 123, 128; ultraviolet light sensitivity, **262**, 112, 119; X-ray sensitivity, **262**, 112, 119); tissue distribution, mammals, **262**, 99

DNA POLYMERASE γ

This eukaryotic DNA-directed polymerase [EC 2.7.7.7] catalyzes the attack of the 3′-hydroxy of the growing DNA chain on the α-phosphoryl group of an incoming dATP, dCTP, dGTP or dTTP to elongate the chain by one base by producing a new phosphodiester bond as well as pyrophosphate (or, diphosphate). This mitochondrial (or chloroplast) enzyme participates in the replication of mitochondrial (or chloroplast) DNA. The enzyme is strongly inhibited by dideoxynucleoside triphosphates and N-ethylmaleimide, but not by aphidicolin. **See also** DNA Polymerases

Selected entries from **Methods in Enzymology** [vol, page(s)]:
General discussion: 262, 49
Assay: inhibitors, **262**, 57; polymerase activity, **262**, 57; substrates, **262**, 57, 60

Other molecular properties: *Saccharomyces cerevisiae* (genes, **262**, 50; homology with mammalian enzyme, **262**, 52; inhibitor sensitivity, **262**, 53, 60; purification [anion-exchange chromatography, **262**, 52, 57, 59; cation-exchange chromatography, **262**, 62; cell growth, **262**, 55, 58; desalting column, **262**, 56, 59; extraction, **262**, 55, 59; phosphocellulose batch adsorption, **262**, 56, 61; proteolysis minimization, **262**, 53])

DNA POLYMERASE δ

This eukaryotic DNA-directed polymerase [EC 2.7.7.7] catalyzes the attack of the 3′-hydroxy of the growing DNA chain on the α-phosphoryl group of an incoming dATP, dCTP, dGTP or dTTP to elongate the chain by one base by producing a new phosphodiester bond as

well as pyrophosphate (or, diphosphate).[1] This nuclear polymerase, which exhibits a 3′ → 5′ exonuclease activity but lacks DNA primase activity, participates in replication of chromosomal DNA. The highly processive enzyme is strongly inhibited by aphidicolin and N-ethylmaleimide, but only weakly inhibited by dideoxynucleoside triphosphates. **See also** DNA Polymerases

[1] R. A. Bambara & C. B. Jessee (1991) *BBA* **1088**, 11.

Selected entries from **Methods in Enzymology** [**vol**, page(s)]:
General discussion: 262, 62, 84
Assay: exonuclease activity, **262**, 72, 85, 88, 92; polymerase activity, **262**, 72, 89

Other molecular properties: abundance in human tissue, **275**, 156; accessory proteins, **275**, 145, 155; accessory protein proliferating cell nuclear antigen (effect on Pol δ processivity, **262**, 77, 85, 89, 92, 543; purification, **262**, 85, 88); domains, **275**, 159; exonuclease activity, **275**, 150, 154, 157; expression during cell cycle, **275**, 160, 166; gene (cloning, **275**, 158; locus, **275**, 160; sequence homology with other polymerases, **275**, 159; structure, **275**, 158); immunological characterization, **275**, 156; inhibitors, **275**, 154, 157; purification from calf thymus (crude extract, **262**, 86, 543; hydrophobic interaction chromatography, **262**, 86; hydroxylapatite chromatography, **262**, 87; inhibitor sensitivity, **262**, 91; ion-exchange chromatography, **262**, 86, 543; kinetic parameters of pure enzyme, **262**, 90; phosphocellulose chromatography, **262**, 87; physical properties of pure enzyme, **262**, 90; storage, **262**, 88); purification from *Drosophila melanogaster* (embryo culture, **262**, 63, 71; extraction, **262**, 72; heparin-Sepharose chromatography, **262**, 73; hydroxylapatite chromatography, **262**, 75; inhibitor sensitivity, **262**, 77; ion-exchange chromatography, **262**, 75; phosphocellulose chromatography, **262**, 72; proteolysis minimization, **262**, 62; single-stranded DNA cellulose chromatography, **262**, 63, 76); purification of human enzyme (cultured cells [anion-exchange chromatography, **275**, 144; cation-exchange chromatography, **275**, 144; cell lysis, **275**, 143; gel filtration, **275**, 144; glycerol gradient sedimentation, **275**, 145; immunoaffinity chromatography, **275**, 145; phosphocellulose chromatography, **275**, 144]; tissue preparations [anion-exchange chromatography, **275**, 142; DNA affinity chromatography, **275**, 143; extraction, **275**, 141; immunoaffinity chromatography, **275**, 142; phenyl-agarose chromatography, **275**, 142]); role in SV40 replication, **275**, 135; substrate specificity, **275**, 157; subunits, **262**, 85; **275**, 155, 157

DNA POLYMERASE ε

This eukaryotic DNA-directed polymerase [EC 2.7.7.7] catalyzes the attack of the 3′-hydroxy of the growing DNA chain on the α-phosphoryl group of an incoming dATP, dCTP, dGTP or dTTP to elongate the chain by one base by producing a new phosphodiester bond as well as pyrophosphate (or, diphosphate).[1] A nuclear enzyme lacking DNA primase activity, DNA polymerase ε participates in DNA repair and quite possibly replication. The enzyme also exhibits a 3′ → 5′ exonuclease activity with single-stranded DNA, producing six- or seven-residue oligonucleotides. It is high degree of processivity, even in the absence of the proliferating cell nuclear antigen. The enzyme is strongly inhibited by aphidicolin and *N*-ethylmaleimide, but only weakly by dideoxynucleoside triphosphates. **See also** *DNA Polymerases*

[1]R. A. Bambara & C. B. Jessee (1991) *BBA* **1088**, 11.

Selected entries from *Methods in Enzymology* [vol, page(s)]:
General discussion: **262**, 93
Assay: exonuclease activity, **262**, 93; fidelity of replication, **262**, 231; inhibitors, **262**, 57; polymerase activity, **262**, 57, 94; substrate preparation, **262**, 57, 60, 93
Other molecular properties: copurification with other polymerases, **275**, 149; immunological characterization, **275**, 156; processivity, **262**, 98; purification from HeLa cells (ammonium sulfate fractionation, **262**, 96; crude cell lysate, **262**, 95; hydroxyapatite chromatography, **262**, 96; ion-exchange chromatography, **262**, 96; phosphocellulose chromatography, **262**, 96; storage, **262**, 98); role in viral DNA replication, **275**, 155; *Saccharomyces cerevisiae* (genes, **262**, 50, 108; homology with mammalian enzyme, **262**, 52; inhibitor sensitivity, **262**, 53, 60; purification [anion-exchange chromatography, **262**, 52, 57, 59; cation-exchange chromatography, **262**, 62; cell growth, **262**, 55, 58; desalting column, **262**, 56, 59; extraction, **262**, 55, 59; phosphocellulose batch adsorption, **262**, 56, 61; proteolysis minimization, **262**, 53]; temperature-sensitive mutants, **262**, 93); subunit structure, **262**, 97; template specificity, **275**, 158

DNA PRIMASE

This magnesium-dependent single-stranded DNA-directed RNA polymerase catalyzes the synthesis of short RNA primers for DNA replication, and in so doing initiates chain synthesis on single-stranded DNA (or ssDNA).[1-3]
See also *DNA Polymerase α*

[1]M. Foiani, G. Lucchini & P. Plevani (1997) *TiBS* **22**, 424.
[2]H. Masai & K. Arai (1996) *Biochimie* **78**, 1109.
[3]M. A. Griep (1995) *Indian J. Biochem. Biophys.* **32**, 171.

Selected entries from *Methods in Enzymology* [vol, page(s)]:
Assay: coupled assay (eukaryotic primase activity, **262**, 408; prokaryotic primase activity, **262**, 406, 566); direct assay, **262**, 409; oligoribonucleotide synthesis assay (bacteriophage T7 primase, **262**, 410; calf thymus primase, **262**, 410; principle, **262**, 409); rate measurement, **275**, 140, 253; RNA-primed DNA synthesis assay (bacteriophage T7 primase, **262**, 407; *Drosophila melanogaster* primase, **262**, 408); synthetic minifork as substrate, **262**, 460; total primer synthesis assay, **262**, 411; template location of primer synthesis initiation, **275**, 256; 5'-terminal nucleotide of primer, identification, **275**, 256
Other molecular properties: active site characterization, **262**, 414; associated proteins in eukaryotes, **262**, 406; DNA polymerase α complex (purification from HeLa cells [extraction, **275**, 137; hydroxyapatite chromatography, **275**, 139; immunoaffinity chromatography, **275**, 137; phosphocellulose chromatography, **275**, 139]; purification from mouse, **275**, 140; resolution of activities, **275**, 140; subunits and activities, **275**, 136, 149); leading strand synthesis, initiation, **262**, 406; primer (identification of 5' ribonucleotide, **262**, 413; sequencing, **262**, 413); purification of bacteriophage T4 enzyme (extraction, **262**, 579; hydroxylapatite chromatography, **262**, 581; phosphocellulose chromatography, **262**, 580); recognition site (characterization, **262**, 412; T7 primase, **262**, 405); role in SV40 replication, **275**, 135; zinc motif characterization, **262**, 414

DNA TERMINASE

This enzyme, consisting of a complex of the phage proteins gp*A* and gp*Nu1*, participates in the packaging of bacteriophage λ DNA by making twelve-nucleotide staggered cuts in the two DNA strands at the *cos* site.

Selected entries from *Methods in Enzymology* [vol, page(s)]:
Assay: bacteriophage λ, **100**, 183
Other molecular properties: bacteriophage λ (assay, **100**, 183; host factor, **100**, 189 [purification, **100**, 190]; properties, **100**, 189; purification, **100**, 185)

DNA TOPOISOMERASES

Several enzymes catalyze the topological modification of supercoiled DNA, thereby altering the linking number.[1-11]

DNA topoisomerase [EC 5.99.1.2], also known as DNA topoisomerase I, relaxing enzyme, untwisting enzyme, swivelase, and type I DNA topoisomerase, catalyzes the ATP-independent breakage of single-strand of DNA, followed by the passage of the other strand through that nick and the rejoining of the DNA. The enzyme catalyzes the conversion of one topological isomer of DNA into another, thereby relaxing DNA superhelicity, interconverting simple and knotted rings of a single DNA strand, and modifying the intertwisting of single-stranded rings of complementary sequences. The prokaryotic enzyme catalyzes the reaction via the formation of a transient phosphotyrosyl diester linkage to the 5'-end of the DNA at the nick. Eukaryotic enzymes are linked at the 3'-end.

DNA topoisomerase (ATP-hydrolyzing) [EC 5.99.1.3] (also known as DNA topoisomerase II, DNA gyrase, and Type II DNA topoisomerase) catalyzes the ATP-dependent transient breakage of both strands of the DNA helix (often termed the gated helix) via the formation of a transient covalent linkage between a tyrosyl residue and a phosphate group in one of the strands. Both strands are ultimately cleaved resulting in the formation of overhangs (an equilibrium has been proposed to be established between the cleaved and ligated DNA in the gated helix). Binding of ATP stimulates a conformational change and passage of the transported helix through the break in the gated helix. The topoisomerase catalyzes a religation and the double-stranded DNA in the gated helix is rejoined. ATP hydrolysis occurs to regain the original conformation. The enzyme can introduce negative superhelical turns into double-stranded circular DNA. One unit of the protein has a nicking-closing activity, whereas another unit catalyzes the supertwisting that is attended by hydrolysis of ATP. (**See** *Energase*) There are at least two isoforms of mammalian DNA topoisomerase II, α and β. **See also** *DNA Helicase*

[1] J. M. Berger (1998) *BBA* **1400**, 3.
[2] Y. Pommier, P. Pourquier, Y. Fan & D. Strumberg (1998) *BBA* **1400**, 83.
[3] D. A. Burden & N. Osheroff (1998) *BBA* **1400**, 139.
[4] J. C. Wang (1998) *Q. Rev. Biophys.* **31**, 107.
[5] R. A. Grayling, K. Sandman & J. N. Reeve (1996) *Adv. Protein Chem.* **48**, 437.
[6] R. J. Reese & A. Maxwell (1991) *Crit. Rev. Biochem. Mol. Biol.* **26**, 335.
[7] A. Maxwell & M. Gellert (1986) *Adv. Prot. Chem.* **38**, 69.
[8] S. A. Wasserman & N. R. Cozzarelli (1986) *Science* **232**, 951.
[9] J. C. Wang (1981) *The Enzymes*, 3rd ed., **14**, 331.
[10] M. Gellert (1981) *The Enzymes*, 3rd ed., **14**, 345.
[11] H. A. Nash (1981) *The Enzymes*, 3rd ed., **14**, 471.

Selected entries from *Methods in Enzymology* [vol, page(s)]:
General discussion: 100, 133, 137, 144, 161
Assay: 29, 197; **100**, 133, 137; **150**, 64; DNA topoisomerase (ATP-hydrolyzing)**100**, 144, 161 (*Drosophila melanogaster* embryo, **100**, 162; *Escherichia coli*, **100**, 173; *Escherichia coli* phage T4, **100**, 150); *Escherichia coli*, **29**, 197; HeLa cell, **100**, 133; mitogen-treated lymphocytes, in, **150**, 64; rat liver, **100**, 137
Other molecular properties: activity, **228**, 207; ATP coupling, **308**, 65; classification, **308**, 63; complexed to DNA, partitioning (methods, **228**, 208; modification, **228**, 218; principle, **228**, 208); DNA topoisomerase (ATP-hydrolyzing)**100**, 144, 161 (assay [*Drosophila melanogaster* embryo, **100**, 162; *Escherichia coli*, **100**, 173; *Escherichia coli* phage T4, **100**, 150]; *Drosophila melanogaster* embryo [properties, **100**, 167; purification, **100**, 164]; *Escherichia coli* [purification, **100**, 175; reactions and uses, **100**, 178]; *Escherichia coli* phage T4 [comparison with other type II topoisomerases, **100**, 159; properties, **100**, 154; purification, **100**, 145]; *Escherichia coli* (properties, **29**, 203; purification, **29**, 201; reaction catalyzed, **29**, 197); HeLa cell (properties, **100**, 136; purification, **100**, 134); inhibitor, assay, **228**, 219; purification, **228**, 207; rat liver, multiple forms (characterization, **100**, 141); regulatory functions, **228**, 207; requirement for, **228**, 207; subunits, **308**, 65

DODECANOYL-[ACYL-CARRIER PROTEIN] HYDROLASE

This enzyme [EC 3.1.2.21], also known as dodecyl-[acyl-carrier protein] hydrolase and lauryl-[acyl-carrier protein] hydrolase, catalyzes the hydrolysis of dodecanoyl-[acyl-carrier protein] to produce dodecanoate and the free acyl-carrier protein.[1] The fourteen-carbon fatty-acylated protein is a significantly weaker substrate.

[1] H. M. Davies, L. Anderson, C. Fan & D. J. Hawkins (1991) *ABB* **290**, 37.

DODECENOYL-CoA Δ-ISOMERASE

This isomerase [EC 5.3.3.8], also known as Δ³-*cis*-Δ²-*trans*-enoyl-CoA isomerase and acetylene:allene isomerase, catalyzes the interconversion of 3-*cis*-dodecenoyl-CoA and 2-*trans*-dodecenoyl-CoA.[1–3]

3-*cis*-dodecenoyl-CoA

2-*trans*-dodecenoyl-CoA

The enzyme will also interconvert 3-acetylenic fatty acyl thioesters and (+)-2,3-dienoyl fatty acyl thiolesters, with fatty acid chain lengths between six and twelve carbon atoms.

[1] D. K. Novikov, K. T. Koivuranta, H. M. Helander, S. A. Filppula, A. I. Yagi, Y. M. Qin & K. J. Hiltunen (1999) *Adv. Exp. Med. Biol.* **466**, 301.
[2] F. M. Miscowicz & K. Bloch (1979) *JBC* **254**, 5868.
[3] W. Stoffel, R. Ditzer & H. Caesar (1964) *Hoppe-Seyler's Z. Physiol. Chem.* **339**, 167.

Selected entries from *Methods in Enzymology* [vol, page(s)]:
General discussion: 14, 99; **71**, 409
Assay: radio gas chromatography, **14**, 99, 100; spectrophotometry, **14**, 99, 100; thin-layer chromatography, **14**, 99, 100
Other molecular properties: β-oxidation of fatty acids, in, **14**, 99; **71**, 403; properties, **14**, 104; purification, rat liver, **14**, 103; substrate specificity, **14**, 105

DOLICHOL O-ACYLTRANSFERASE

This acyltransferase [EC 2.3.1.123] catalyzes the reaction of palmitoyl-CoA with dolichol to produce coenzyme A and dolichyl palmitate.[1] Alternative acyl-CoA substrates include stearoyl-CoA, myristoyl-CoA, oleoyl-CoA, acetyl-CoA, and arachidonoyl-CoA.

[1] O. Tollbom, C. Valtersson, T. Chojnacki & G. Dallner (1988) *JBC* **263**, 1347.

DOLICHOL KINASE

This phosphotransferase [EC 2.7.1.108] catalyzes the reaction of CTP with dolichol to produce CDP and dolichyl phosphate.[1] dCTP can also act as the donor substrate. Phospholipids enhance the interaction between the kinase and dolichol.[2]

[1] C. M. Allen, Jr., J. R. Kalin, J. Sack & D. Verizzo (1978) *Biochemistry* **17**, 5020.
[2] C. P. Genain & C. J. Waechter (1990) *J. Neurochem.* **54**, 855.

Selected entries from *Methods in Enzymology* [vol, page(s)]:
General discussion: 111, 471
Assay: bovine brain, **111**, 471

DOLICHYLDIPHOSPHATASE

This enzyme [EC 3.6.1.43] catalyzes the hydrolysis of dolichyl diphosphate to produce dolichyl phosphate and orthophosphate.[1,2] Dolichol is an inhibitor of this enzyme.

[1]M. G. Scher & C. J. Waechter (1984) JBC **259**, 14580.
[2]E. Belocopitow & D. Boscoboinik (1982) EJB **125**, 167.

Selected entries from *Methods in Enzymology* [vol, page(s)]:
General discussion: 111, 547
Assay: bovine brain, 111, 548

DOLICHYL-DIPHOSPHATE: POLYPHOSPHATE PHOSPHOTRANSFERASE

This transferase [EC 2.7.4.20] catalyzes the reaction of dolichyl diphosphate (*i.e.*, dolichyl pyrophosphate) with (phosphate)$_n$ (*i.e.*, polyphosphate) to produce dolichyl phosphate and (phosphate)$_{n+1}$.[1,2]

[1]A. V. Naumov, Y. A. Shabalin, V, M. Vagabov & I. S. Kulaev (1985) Biokhimiya **50**, 652.
[2]Y. A. Shabalin & I. S. Kulaev (1989) Biokhimiya **54**, 68.

DOLICHYL-DIPHOSPHOOLIGO-SACCHARIDE:PROTEIN GLYCOSYLTRANSFERASE

This transferase [EC 2.4.1.119], also known as oligosaccharidyltransferase, catalyzes the reaction of dolichyl diphosphooligosaccharide with a protein-L-asparagine (*i.e.*, an L-asparaginyl residue in a protein) to produce dolichyl diphosphate and a glycoprotein with the oligosaccharide chain attached by glycosylamine linkage to the asparaginyl residue in the protein.[1–5] The asparaginyl residue is a member of the sequence ...–Asn–Xaa–Ser/Thr– ... (where Xaa is not a prolyl residue).

[1]Q. Yan & W. J. Lennarz (1999) BBRC **266**, 684.
[2]T. Xu & J. K. Coward (1997) Biochemistry **36**, 14683.
[3]H. A. Kaplan, J. K. Welply & W. J. Lennarz (1987) BBA **906**, 161.
[4]J. K. Welply, P. Shenbagamurthi, W. J. Lennarz & F. Naider (1983) JBC **258**, 11856.
[5]S. Silberstein & R. Gilmore (1996) FASEB J. **10**, 849.

DOLICHYL-PHOSPHATASE

This enzyme [EC 3.1.3.51], also known as dolichol phosphate phosphatase and polyisoprenyl phosphate phosphatase, catalyzes the hydrolysis of dolichyl phosphate to produce dolichol and orthophosphate.[1–4] Other substrates include phytanyl phosphate and solanesyl phosphate. The mammalian brain enzyme is an intrinsic membrane glycoprotein with a dual function, catalyzing: (a) the hydrolysis of dolichyl phosphate, and (b) the hydrolysis of phosphatidate.[5]

[1]R. K. Keller, W. L. Adair, N. Cafmeyer, F. A. Simion, B. Fleischer & S. Fleischer (1986) ABB **249**, 207.

[2]J. W. Rip, C. A. Rupar, N. Chaudhary & K. K. Carroll (1981) JBC **256**, 1929.
[3]J. F. Wedgwood & J. L. Strominger (1980) JBC **255**, 1120.
[4]G. S. Adrian & R. W. Keenan (1979) BBA **575**, 431.
[5]D. W. Frank & C. J. Waechter (1998) JBC **273**, 11791.

Selected entries from *Methods in Enzymology* [vol, page(s)]:
General discussion: 111, 471
Assay: bovine brain, 111, 477
Other molecular properties: properties, bovine brain, 111, 478

DOLICHYL-PHOSPHATE α-N-ACETYLGLUCOSAMINYLTRANSFERASE

This transferase [EC 2.4.1.153] catalyzes the reaction of UDP-N-acetyl-D-glucosamine with dolichyl phosphate to produce UDP and dolichyl N-acetyl-α-D-glucosaminyl phosphate.[1,2]

[1]W. E. Riedell & J. A. Miernyk (1988) Plant Physiol. **87**, 420.
[2]H. Arakawa & S. Mookerjea (1984) EJB **140**, 297.

DOLICHYL-PHOSPHATE-GLUCOSE PHOSPHODIESTERASE

This phosphodiesterase [EC 3.1.4.48] catalyzes the hydrolysis of dolichyl β-D-glucosyl phosphate to produce dolichyl phosphate and D-glucose.[1]

[1]E. V. Crean (1984) BBA **792**, 149.

DOLICHYL-PHOSPHATE β-GLUCOSYLTRANSFERASE

This transferase [EC 2.4.1.117] catalyzes the reversible reaction of UDP-D-glucose with dolichyl phosphate to produce UDP and dolichyl β-D-glucosyl phosphate.[1–3] Other acceptor substrates, albeit weaker, include solanesyl phosphate and ficaprenyl phosphate.

[1]H. Matern & S. Matern (1989) BBA **1004**, 67.
[2]N. H. Behrens & L. F. Leloir (1970) PNAS **66**, 153.
[3]J. Rodriguez-Bonilla, L. Vargas-Rodriguez, C. Calvo-Mendez, A. Flores-Carreon & E. Lopez-Romero (1998) Antonie Van Leeuwenhoek **73**, 373.

Selected entries from *Methods in Enzymology* [vol, page(s)]:
Other molecular properties: photoaffinity labeling with azido-nucleoside diphosphate sugar analogues, **230**, 335; preparative gel electrophoresis, **230**, 336

DOLICHYL-PHOSPHATE-MANNOSE: GLYCOLIPID α-MANNOSYLTRANSFERASE

This transferase [EC 2.4.1.130] catalyzes the transfer of an α-D-mannosyl residue from dolichyl-phosphate D-mannose into a membrane lipid-linked oligosaccharide.[1] Note that four of the nine mannosyl residues in the main membrane lipid-linked oligosaccharide of the structure Glc$_3$Man$_9$GlcNAc$_2$ are reportedly provided by this enzyme.[1] The enzyme was fairly specific for dolichyl-phosphate mannose as the mannosyl donor substrate and $Man_5GlcNAc_2$-diphosphodolichol as the initial mannosyl acceptor.[2]

[1]J. I. Rearick, K. Fujimoto & S. Kornfeld (1981) JBC **256**, 3762.
[2]C. B. Sharma, G. P. Kaushal, Y. T. Pan & A. D. Elbein (1990) Biochemistry **29**, 8901.

DOLICHYL-PHOSPHATE-MANNOSE PHOSPHODIESTERASE

This enzyme [EC 3.1.4.49], also known as mannosylphosphodolichol phosphodiesterase, catalyzes the hydrolysis of dolichyl β-D-mannosyl phosphate to produce dolichyl phosphate and D-mannose. Dolichyl phosphate and dolichol are competitive inhibitors.[1]

[1]Y. Tomita & Y. Motokawa (1987) EJB **170**, 363.

DOLICHYL-PHOSPHATE-MANNOSE: PROTEIN MANNOSYLTRANSFERASE

This transferase [EC 2.4.1.109] catalyzes the reaction of dolichyl phosphate D-mannose with a seryl or threonyl residue in a protein to produce dolichyl phosphate and the O-D-mannosyl-protein (e.g., a cell-wall mannoprotein).[1–5] The donor dolichyl substrate has to be a long-chain α-dihydropolyprenyl derivative containing at least thirty-five carbon atoms. At least seven distinct mannosyltransferases have been identified in Saccharomyces cerevisiae. All are integral membrane glycoproteins and utilize dolichyl phosphate D-mannose as the donor substrate; however, there are differences in protein specificities.[4]

[1]S. Strahl-Bolsinger, M. Gentzsch & W. Tanner (1999) BBA **1426**, 297.
[2]A. Weston, P. M. Nassau, C. Henley & M. S. Marriott (1993) EJB **215**, 845.
[3]S. Strahl-Bolsinger & W. Tanner (1991) EJB **196**, 185.
[4]M. Gentzsch & W. Tanner (1997) Glycobiology **7**, 481.
[5]V. Girrbach, T. Zeller, M. Priesmeier & S. Strahl-Bolsinger (2000) JBC **275**, 19288.

DOLICHYL-PHOSPHATE β-D-MANNOSYLTRANSFERASE

This transferase [EC 2.4.1.83], also known as GDP-mannose:dolichyl-phosphate O-β-D-mannosyltransferase and dolichol-phosphate mannose synthase, catalyzes the reaction of GDP-D-mannose with dolichyl phosphate to produce GDP and dolichyl-D-mannosyl phosphate.[1–3] The only reported acceptor substrates are long-chain polyprenyl phosphates and α-dihydropolyprenyl phosphates larger than thirty-five carbon atoms in length.

This enzyme is associated with the endoplasmic reticulum membrane.

[1]P. Orlean (1992) Biochem. Cell Biol. **70**, 438.
[2]Y. Tomita & Y. Motokawa (1985) BBA **842**, 176.
[3]K. A. Presper & E. C. Heath (1983) The Enzymes, 3rd ed., **16**, 449.

DOLICHYL-PHOSPHATE D-XYLOSYLTRANSFERASE

This transferase [EC 2.4.2.32] catalyzes the reaction of UDP-D-xylose with dolichyl phosphate to produce UDP and dolichyl D-xylosyl phosphate.[1]

[1]C. J. Waechter, J. J. Lucas & W. J. Lennarz (1974) BBRC **56**, 343.

DOLICHYL-XYLOSYL-PHOSPHATE: PROTEIN XYLOSYLTRANSFERASE

This transferase [EC 2.4.2.33] catalyzes the reaction of dolichyl D-xylosyl phosphate with a protein to produce dolichyl phosphate and a D-xylosylprotein.[1]

[1]C. J. Waechter, J. J. Lucas & W. J. Lennarz (1974) BBRC **56**, 343.

DOPACHROME ISOMERASE

L-dopachrome   5,6-dihydroxyindole-2-carboxylate

This zinc-dependent enzyme [EC 5.3.3.12] (also known as dopachrome Δ-isomerase, dopachrome tautomerase, dopachrome conversion factor, and dopachrome D-isomerase) catalyzes the conversion of L-dopachrome (i.e., 5,6-dioxo-2,3-5,6-tetrahydroindole-2-carboxylate) to 5,6-dihydroxy-indole-2-carboxylate in the melanin biosynthetic pathway.[1] Other substrates include dopachrome methyl ester.

[1]J. M. Pawelek (1990) BBRC **166**, 1328.

DOPA DECARBOXYLASE

This pyridoxal-phosphate-dependent decarboxylase, formerly classified as EC 4.1.1.26, is now included with aromatic-L-amino-acid decarboxylase [EC 4.1.1.28]. Tyrosine decarboxylase [EC 4.1.1.25] also catalyzes this reaction.

DOPAMINE β-MONOOXYGENASE

This copper-dependent oxidoreductase [EC 1.14.17.1], also known as dopamine β-hydroxylase, catalyzes the reaction of 3,4-dihydroxyphenethylamine (*i.e.*, dopamine) with dioxygen and two molecules of ascorbate to produce norepinephrine, water, and two molecules of semidehydroascorbate.[1–7] The enzyme, which is stimulated by fumarate, can utilize other substrates such as 1-phenyl-1-aminomethylethene and 1-(4-hydroxybenzyl) imidazole. The monooxygenase has a Uni Uni random Bi Uni ping-pong kinetic mechanism in which ascorbate is the first substrate to bind (resulting in the conversion of Cu^{2+} to Cu^{1+}).[6,7] Upon reduction by ascorbate, dopamine and dioxygen bind randomly.

The overall reaction involves the formation of a substrate-derived benzylic radical via a hydrogen atom abstraction mechanism.[5] Detailed studies of intrinsic O^{18} isotope effects indicate that O–O cleavage occurs prior to proton abstraction from the substrate.[1] Hydrogen transfer from an active-site tyrosyl residue to Cu(II)–OOH results in the formation of water, Cu(II)–O·, and the tyrosyl free radical. The Cu(II)–O· radical acts as the hydrogen abstracting agent producing the substrate radical (a benzylic radical) and a hydrogen-bonded intermediate which equilibrates between Cu(II)–OH/[tyrosyl radical] and Cu(II)–O·/[tyrosyl residue]. The benzylic radical that had been generated now reacts with the oxygen, producing an alkoxide ligated to the copper.[1] A second electron is transferred from a second copper, cleaving the Cu–O bond and releasing the product which obtains a proton from the solvent.

[1] G. Tian, J. A. Berry & J. P. Klinman (1994) *Biochemistry* **33**, 226.
[2] M. A. Ator & P. R. Ortez de Montellano (1990) *The Enzymes*, 3rd ed., **19**, 213.
[3] F. P. Guengerich (1990) *Crit. Rev. Biochem. Mol. Biol.* **25**, 97.
[4] S. M. Miller & J. P. Klinman (1985) *Biochemistry* **24**, 2114.
[5] S. M. Miller & J. P. Klinman (1982) *MIE* **87**, 711.
[6] R. A. Long, R. M. Weppelman, J. E. Taylor, R. L. Tolman & G. Olson (1981) *Biochemistry* **20**, 7423.
[7] V. Ullrich & W. Duppel (1975) *The Enzymes*, 3rd ed., **12**, 253.

Selected entries from *Methods in Enzymology* [vol, page(s)]:
General discussion: 52, 9; 87, 711; 142, 596, 603, 608
Assay: 74, 374; 142, 597, 603, 608; coupled enzyme assay, 63, 36
Other molecular properties: antibody production, 74, 372; chromaffin granules, in, 31, 383; function and regulation, 74, 370; genetic regulation, 74, 375, 378; immunotitration, 74, 371 (applications, 74, 375; copper sulfate addition, 74, 374; equivalence point analysis, 74, 376; interference by endogenous inhibitors, 74, 374; precision, 74, 375; principle, 74, 371; procedure, 74, 373; results, 74, 374; second antibody stage, 74, 374); isotope effects, 87, 711; kinetic mechanisms, 87, 711; mechanism, 87, 711; properties, 142, 600, 606, 613; purification, 74, 372 (adrenal gland, bovine, 17B, 754; 142, 599; adrenal gland, chicken, 142, 610; human plasma, 142, 604); radioimmunoassay, 74, 371, 377 (applications, 74, 378; immunogen preparation, 74, 372; principles, 74, 371; procedure, 74,

377; radioiodination procedure, 74, 373; second antibody stage, 74, 377; species specificity, 74, 376, 380; standard curve, 74, 377, 378); radioiodination, 70, 232; transition state and multisubstrate analogues, 249, 304

*Dpn*I RESTRICTION ENDONUCLEASE

This type II methyl-directed restriction endonuclease [EC 3.1.21.4], obtained from *Diplococcus pneumoniae*, catalyzes the hydrolysis of both strands of DNA at 5′...GA∇TC...3′ only when the adenine in the sequence is methylated (thus, 6-methylaminopurine: 5′...GA*^TC...3′), producing blunt-ended fragments.[1–3]

[1] S. Lacks & B. Greenberg (1975) *JBC* **250**, 4060.
[2] G. F. Vovis & S. Lacks (1977) *J. Mol. Biol.* **115**, 525.
[3] S. Lacks & B. Greenberg (1977) *J. Mol. Biol.* **114**, 153.

Selected entries from *Methods in Enzymology* [vol, page(s)]:
General discussion: 65, 138
Assay: 65, 141
Other molecular properties: applications, 65, 146; cation requirements, 65, 145; distribution, 65, 146; molecular weight, 65, 145; properties, 65, 145, 146; purification, 65, 139; recognition sequence, 65, 5, 138; requirement, for methylated adenine, 65, 138; stability, 65, 141

*Dpn*II RESTRICTION ENDONUCLEASE

This type II restriction endonuclease [EC 3.1.21.4], obtained from *Diplococcus pneumoniae*, catalyzes the hydrolysis of both strands of DNA at 5′...∇GATC...3′.[1–3] The enzyme, which can exhibit star activity under conditions in which pH > 6.5, is blocked by *dam* methylation. The DNA fragments generated have a 5′ GATC extension which can be ligated to other fragments generated by *Bam*HI, *Bcl*I, *Bgl*II, *Mbo*I, *Sau*3AI, and *Bst*YI.

[1] S. Lacks & B. Greenberg (1975) *JBC* **250**, 4060.
[2] G. F. Vovis & S. Lacks (1977) *J. Mol. Biol.* **115**, 525.
[3] S. Lacks & B. Greenberg (1977) *J. Mol. Biol.* **114**, 153.

Selected entries from *Methods in Enzymology* [vol, page(s)]:
General discussion: 65, 138
Assay: 65, 141
Other molecular properties: applications, 65, 146; cation requirements, 65, 145; distribution, 65, 146; molecular weight, 65, 145; properties, 65, 145, 146; purification, 65, 139; recognition sequence, 65, 5, 138; stability, 65, 141

*Dra*I RESTRICTION ENDONUCLEASE

This type II restriction endonuclease [EC 3.1.21.4], obtained from *Deinococcus radiophilus*, catalyzes the hydrolysis of both strands of DNA at 5′...TTT∇AAA ...3′, producing blunt-ended fragments.[1]

[1] I. J. Purvis & B. E. B. Moseley (1983) *NAR* **11**, 5467.

*Dra*II RESTRICTION ENDONUCLEASE

This type II restriction endonuclease [EC 3.1.21.4], obtained from *Deinococcus radiophilus*, catalyzes the hydrolysis of both strands of DNA at 5′...RG∇GNCCY ...3′, where R refers to either A or G, Y refers to either T or C, and N refers to any base.[1,2] The enzyme is inhibited by overlapping *dcm* methylation.[3]

[1]R. Grosskopf, W. Wolf & C. Kessler (1985) *NAR* **13**, 1517.
[2]C. M. de Wit, B. M. M. Dekker, A. C. Neele & A. de Waard (1985) *FEBS Lett.* **180**, 219.
[3]K. Schnetz & B. Rak (1988) *NAR* **16**, 1623.

*Dra*III RESTRICTION ENDONUCLEASE

This type II restriction endonuclease [EC 3.1.21.4], obtained from *Deinococcus radiophilus*, catalyzes the hydrolysis of both strands of DNA at 5′...CACNNN∇ GTG...3′, where N refers to any base, and ∇ indicates the cleavage site.[1,2]

[1]R. Grosskopf, W. Wolf & C. Kessler (1985) *NAR* **13**, 1517.
[2]C. M. de Wit, B. M. M. Dekker, A. C. Neele & A. de Waard (1985) *FEBS Lett.* **180**, 219.

*Drd*I RESTRICTION ENDONUCLEASE

This type II restriction endonuclease [EC 3.1.21.4], obtained from *Deinococcus radiodurans*, catalyzes the hydrolysis of both strands of DNA at 5′...GACNNNN∇ NNGTC...3′, where N refers to any base, and ∇ indicates the cleavage site.[1]

[1]C. Polisson & R. D. Morgan (1989) *NAR* **17**, 3316.

dTDP-4-AMINO-4,6-DIDEOXYGALACTOSE AMINOTRANSFERASE

This pyridoxal-phosphate-dependent aminotransferase [EC 2.6.1.59] catalyzes the reversible reaction of dTDP-4-amino-4,6-dideoxy-D-galactose with α-ketoglutarate (or, 2-oxoglutarate) to produce dTDP-4-dehydro-6-deoxy-D-galactose (also known as dTDP-4-dehydro-6-deoxy-D-glucose) and L-glutamate.[1] *See also dTDP-4-Amino-4,6-Dideoxy-D-Glucose Aminotransferase*

[1]H. Ohashi, M. Matsuhashi & S. Matsuhashi (1971) *JBC* **246**, 2325.

Selected entries from *Methods in Enzymology* [vol, page(s)]:
General discussion: 8, 317
Assay: 8, 320

dTDP-4-AMINO-4,6-DIDEOXYGLUCOSE ACETYLTRANSFERASE

This acetyltransferase, also known as TDP-4-amino-4,6-dideoxyglucose:acetyl-CoA transacetylase, catalyzes the reaction of dTDP-4-amino-4,6-dideoxyglucose with acetyl-CoA to produce dTDP-4-acetamido-4,6-dideoxy-glucose and free coenzyme A. *See also dTDP-4-Amino-4,6-Dideoxygalactose Acetyltransferase*

Selected entries from *Methods in Enzymology* [vol, page(s)]:
General discussion: 8, 323
Assay: 8, 323

dTDP-4-AMINO-4,6-DIDEOXY-D-GLUCOSE AMINOTRANSFERASE

This pyridoxal-phosphate-dependent aminotransferase [EC 2.6.1.33], also known as thymidine diphospho-4-keto-6-deoxy-D-glucose transaminase, catalyzes the reversible reaction of dTDP-4-amino-4,6-dideoxy-D-glucose with α-ketoglutarate (or, 2-oxoglutarate) to produce dTDP-4-dehydro-6-deoxy-D-glucose (also known as dTDP-4-dehydro-6-deoxy-D-galactose) and L-glutamate.[1] *See also dTDP-4-Amino-4,6-Dideoxygalactose Aminotransferase*

[1]M. Matsuhashi & J. L. Strominger (1966) *JBC* **241**, 4738.

Selected entries from *Methods in Enzymology* [vol, page(s)]:
General discussion: 8, 317, 320
Assay: 8, 320

dTDP-4-DEHYDRO-6-DEOXYGLUCOSE REDUCTASE

This oxidoreductase [EC 1.1.1.266] catalyzes the reaction of dTDP-4-dehydro-6-deoxy-D-glucose with NADPH and a proton to produce dTDP-D-fucose and NADP+.[1]

[1]Y. Yoshida, Y. Nakano, T. Nezu, Y. Yamashita & T. Koga (1999) *JBC* **274**, 16933.

dTDP-4-DEHYDRORHAMNOSE 3,5-EPIMERASE

This epimerase [EC 5.1.3.13], also known as dTDP-4-keto-6-deoxyglucose 3,5-epimerase, dTDP-4-dehydro-L-rhamnose synthase, and thymidine diphospho-4-keto-rhamnose 3,5-epimerase, catalyzes the interconversion of dTDP-4-dehydro-6-deoxy-D-glucose (*i.e.*, dTDP-4-dehydro-D-quinovose) and dTDP-4-dehydro-6-deoxy-L-mannose (*i.e.*, dTDP-4-dehydro-L-rhamnose).[1–3]

The enzyme occurs in a complex with dTDP-4-dehydro-rhamnose reductase [EC 1.1.1.133]. Note that the reaction involves the epimerization of two chiral centers. The process, probably involving two sequential deprotonation/reprotonation reactions (the C3 epimerization occurring first), has been suggested to proceed via two enol intermediates. Hydrogen exchanges are observed at both C3 and C5. *See also dTDP-4-Dehydrorhamnose Reductase*

[1]H. P. Wahl, U. Matern & H. Grisebach (1975) *BBRC* **64**, 1041.
[2]R. W. Gaugler & O. Gabriel (1973) *JBC* **248**, 6041.
[3]D. Christendat, V. Saridakis, A. Dharamsi, A. Bochkarev, E. F. Pai, C. H. Arrowsmith & A. M. Edwards (2000) *JBC* **275**, 24608.

Selected entries from *Methods in Enzymology* [vol, page(s)]:
General discussion: 28, 451

dTDP-4-DEHYDRORHAMNOSE REDUCTASE

This oxidoreductase [EC 1.1.1.133], also known as dTDP-4-keto-L-rhamnose reductase and dTDP-6-deoxy-L-mannose dehydrogenase, catalyzes the reaction of dTDP-4-dehydro-6-deoxy-L-mannose (*i.e.*, dTDP-4-dehydro-L-rhamnose) with NADPH to produce dTDP-6-deoxy-L-mannose (*i.e.*, dTDP-L-rhamnose) and NADP[+].[1,2] Note that the reduction only occurs when the substrate is bound to another enzyme (dTDP-4-dehydrorhamnose 3,5-epimerase): the entire complex has often been referred to as dTDP-L-rhamnose synthase. The reductase contains a seryl/tyrosyl/lysyl catalytic triad.[2]

[1]A. Melo & L. Glaser (1968) *JBC* **243**, 1475.
[2]M. Graninger, B. Nidetzky, D. E. Heinrichs, C. Whitfield & P. Messner (1999) *JBC* **274**, 25069.

Selected entries from *Methods in Enzymology* [vol, page(s)]:
General discussion: 28, 451

dTDP-6-DEOXY-L-TALOSE 4-DEHYDROGENASE

This oxidoreductase [EC 1.1.1.134] catalyzes the reaction of dTDP-6-deoxy-L-talose (*i.e.*, dTDP-L-quinovose) with NADP[+] to produce dTDP-4-dehydro-6-deoxy-L-mannose (*i.e.*, dTDP-4-dehydro-L-rhamnose) and NADPH.[1] Note that the oxidation reaction at the C4 position only occurs while the substrate is bound to another enzyme (a 3,5-epimerase).

[1]R. W. Gaugler & O. Gabriel (1973) *JBC* **248**, 6041.

dTDP-DIHYDROSTREPTOSE: STREPTIDINE-6-PHOSPHATE DIHYDROSTREPTOSYLTRANSFERASE

This transferase [EC 2.4.2.27], which participates in streptomycin biosynthesis, catalyzes the reaction of dTDP-L-dihydrostreptose with streptidine 6-phosphate to produce dTDP and *O*-1,4-α-L-dihydrostreptosylstreptidine 6-phosphate.[1]

[1]B. Kniep & H. Grisebach (1980) *EJB* **105**, 139.

dTDP-GALACTOSE 6-DEHYDROGENASE

This oxidoreductase [EC 1.1.1.186], also known as thymidine-diphosphate-galactose dehydrogenase, catalyzes the reaction of dTDP-D-galactose with water and two molecules of NAD[+] to produce dTDP-D-galacturonate, two molecules of NADH, and two protons.[1]

[1]R. Katan & G. Avigad (1966) *BBRC* **24**, 18.

dTDP-GLUCOSE 4,6-DEHYDRATASE

This NAD[+]-dependent dehydratase [EC 4.2.1.46], also known as deoxythymidine diphosphoglucose oxidoreductase, catalyzes the conversion of dTDP-D-glucose to dTDP-4-dehydro-6-deoxy-D-glucose (*i.e.*, dTDP-4-dehydro-D-quinovose) and water.[1-4] Other substrates include dTDP-3-deoxy-D-glucose and dTDP-3-azido-3-deoxy-D-glucose.[4] Upon binding of the substrate, reaction with NAD[+] produces the dTDP-4-dehydro-D-glucose intermediate. Dehydration then occurs to generate the 5,6-glucoseen and reduction by NADH yields the final product.

[1]L. Glaser & H. Zarkowsky (1971) *The Enzymes*, 3rd ed., **5**, 465.
[2]J. A. Vara & C. R. Hutchinson (1988) *JBC* **263**, 14992.
[3]H. Matern, G. U. Brillinger & H. Pape (1973) *Arch. Mikrobiol.* **88**, 37.
[4]A. Naundorf & W. Klaffke (1996) *Carbohydr. Res.* **285**, 141.

Selected entries from *Methods in Enzymology* [vol, page(s)]:
General discussion: 8, 317, 319
Other molecular properties: *Escherichia coli*, from, **8**, 319; **28**, 446, 454; properties, **8**, 319; staining on gels, **22**, 584, 586, 595; stereospecificity, **87**, 122, 152, 156

dTMP KINASE

This phosphotransferase [EC 2.7.4.9], also known as thymidylate kinase, catalyzes the reaction of ATP with thymidine 5′-phosphate (dTMP) to produce ADP and thymidine 5′-diphosphate (dTDP).[1-4] Other acceptor substrates include TMP, UMP, dUMP, and 3′-azidothymidine monophosphate (AZTMP).

[1]N. Ostermann, I. Schlichting, R. Brundiers, M. Konrad, J. Reinstein, T. Veit, R. S. Goody & A. Lavie (2000) *Structure Fold Des.* **8**, 629.
[2]N. Tamiya, T. Yusa, Y. Yamaguchi, R. Tsukifuji, N. Kuroiwa, Y. Moriyama & S. Fujimura (1989) *BBA* **995**, 28.
[3]R. Bone, Y.-C. Cheng & R. Wolfenden (1986) *JBC* **261**, 16410.
[4]A. Y. S. Jong & J. L. Campbell (1984) *JBC* **259**, 14394.

DUODENASE AND SHEEP MAST CELL PROTEINASE 1, CATTLE

A serine endopeptidase, similar to chymotrypsin, was isolated from sheep gastric mucosa mucosal mast cells in the mid-1980s and was named sheep mast cell proteinase-1 (sMCP-1).[1,2] A few years later a serine proteinase having specificities similar to both chymotrypsin and trypsin was isolated from cattle duodenal mucosa and named duodenase.[3,4] Later specificity studies suggest that the two enzymes may be species variants.[5]

[1] J. F. Huntley, S. Gibson, D. Knox & H. R. P. Miller (1986) *Int. J. Biochem.* **18**, 673.
[2] D. P. Knox & J. F. Huntley (1988) *Int. J. Biochem.* **20**, 193.
[3] T. S. Zamolodchikova, T. I. Vorotyntseva & V. K. Antonov (1995) *EJB* **227**, 866.
[4] T. S. Zamolodchikova, T. I. Vorotyntseva, I. V. Nazimov & G. A. Grishina (1995) *EJB* **227**, 873.
[5] A. D. Pemberton, J. F. Hundley & H. R. P. Miller (1997) *BJ* **321**, 665.

dUTP PYROPHOSPHATASE

This enzyme [EC 3.6.1.23], also known as dUTP diphosphatase, dUTPase, desoxyuridine 5′-triphosphate nucleotidohydrolase, and deoxyuridine 5′-triphosphatase, catalyzes the hydrolysis of dUTP to produce dUMP and pyrophosphate (or, diphosphate).[1–4] Removal of dUTP by this pyrophosphatase prevents misincorporation of dUMP units into DNA.

[1] L. E. Bertani, A. Häggmark & P. Reichard (1963) *JBC* **238**, 3407.
[2] A. R. Price & J. Frato (1962) *JBC* **250**, 8804.
[3] J. Shlomai & A. Kornberg (1978) *JBC* **253**, 3305.
[4] G. Larsson, P. O. Nyman & J. O. Kvassman (1996) *JBC* **271**, 24010.

DYNAMIN AND DYNAMIN GTPase

This enzyme [EC 3.6.1.50], which catalyzes the hydrolysis of GTP to produce GDP and orthophosphate, participates in endocytosis and is instrumental in pinching off membrane vesicles.[1–5] Purified dynamin readily self-assembles into rings or spirals, suggesting that it wraps around the necks of budding vesicles in its role in membrane fission. This enzyme is currently misclassified as a hydrolase, because P-O-P bond scission supplies the Gibbs free energy required to drive the pinching-off of membrane vesicles.[4] Energases transduce chemical energy into mechanical work, and the overall enzyme-catalyzed reaction should be written as: Membrane-State$_1$ + GTP = Membrane-State$_1$ + GDP + P$_i$, where two affinity-modulated membranal states are indicated.[5]

[1] D. E. Warnock & S. L. Schmid (1996) *Bioessays* **18**, 885.
[2] S. J. McClure & P. J. Robinson (1996) *Mol. Membr. Biol.* **13**, 189.
[3] P. Oh, D. P. McIntosh & J. E. Schnitzer (1998) *J. Cell Biol.* **141**, 101.
[4] M. A. McNiven, H. Cao, K. R. Pitts & Y. Yoon (2000) *TiBS* **25**, 115.
[5] D. L. Purich (2001) *TiBS* **26**, 417.

Selected entries from *Methods in Enzymology* [vol, page(s)]:
General discussion: **196**, 192; **329**, 447
Assay: functional assays, **196**, 198
Other molecular properties: cells expressing, growth and selection, **257**, 215; characterization, **196**, 192; expression activation (effect of

tetracycline concentration, **257**, 220; time course of induction, **257**, 218); expression with tetracycline-responsive promoter, **257**, 210; purification from bovine brain, **196**, 192; role in endocytosis in mammalian cells, **257**, 209; wild-type and mutant, inducible expression (characterization, **257**, 217; stable cell lines with [expression vectors, **257**, 214; general considerations, **257**, 213; selectable marker selection, **257**, 214; transfer and screening of selected clones, **257**, 216])

DYNEIN AND DYNEIN ATPases

This family of ATP-dependent motors [EC 3.6.4.2; previously classified as EC 3.6.1.33] forms ADP and orthophosphate from ATP during its energy-dependent translocation along the surface of microtubules.[1–4] Unique "cargo" sites on dynein molecules allow for the specific transport of cellular organelles and other macromolecular components. Dyneins also are responsible for the motile properties of eukaryotic flagella and cilia.

The idea that a molecular motor is an enzyme is in keeping with a new definition of enzyme catalysis as the facilitated making/breaking of chemical bonds, not just covalent bonds.[5] This idea builds on Pauling's assertion that any long-lived, chemically distinct interaction can be regarded as a chemical bond. Note also that the equilibrium constant (K_{eq} = [Microtubule-Position$_2$][ADP][P$_i$]/[Microtubule-Position$_1$][ATP]) does not conform to that expected for an ATPase (*i.e.*, K_{eq} = [ADP][P$_i$]/[ATP]). Thus, although the overall reaction yields ADP and orthophosphate, the enzyme is misclassified as a hydrolase, and instead should be regarded as an energase-type reaction. Energases facilitate affinity-modulated reactions by coupling the $\Delta G_{\text{ATP-hydrolysis}}$ to a force-generating or work-producing step.[5] In this case, P-O-P bond scission supplies the energy to drive motility. *See ATPase; Energase; Kinesin and Kinesin ATPase; Tubulin*

[1] S. M. King (2000) *BBA* **1496**, 60.
[2] K. K. Pfister (2000) *Mol Neurobiol.* **20**, 81.
[3] K. A. Johnson (1992) *The Enzymes*, 3rd ed., **20**, 1.
[4] I. R. Gibbons (1988) *JBC* **263**, 15837.
[5] D. L. Purich (2001) *TiBS* **26**, 417.

Selected entries from *Methods in Enzymology* [vol, page(s)]:
General discussion: 85, 450; **134**, 291, 306, 318, 325, 337; **291**, 307; **298**, 171
Assay: **134**, 301, 321, 327, 339; bovine sperm, **134**, 321; cytoplasmic, **196**, 188; sea urchin egg cytoplasm, **134**, 327, 339
Other molecular properties: bovine sperm (assay, **134**, 321; characterization, **134**, 322; extraction, **134**, 321); *Chlamydomonas* flagella (analytical techniques, **134**, 297; purification, **134**, 293); *Chlamydomonas*, mutant selection, **196**, 348; cytoplasmic (assay, **196**, 188; properties, *in vitro*, **196**, 181; purification from brain, **196**, 181); heavy chains (iron(III)-mediated photocleavage, **196**, 438; UV

irradiation, **196**, 431; vanadate-mediated photocleavage, **196**, 431 [at VI site, **196**, 432; at V2 site, **196**, 436]); inhibition by vanadate, **134**, 485; kinetic analysis (chemical quench-flow methods, **134**, 692; stopped-flow methods, **134**, 678); latent form (preparation from sea urchin sperm flagella, **85**, 453; properties, **85**, 463; subunits, properties, **85**, 465); outer arm, rainbow trout (ATPase activity, **196**, 219; characterization, **196**, 201; isolation, **196**, 212; polypeptide composition, **196**, 219; purification, **196**, 201; subfractionation, **196**, 221); photochemical analysis, **196**, 460; preparation, **196**, 431 (*Chlamydomonas* flagella, **85**, 467; lamellibranch gill cilia, **85**, 470; sea urchin sperm, **85**, 472; starfish sperm, **85**, 471; *Tetrahymena* cilia, **85**, 466; trout sperm, **85**, 473); properties, **85**, 468; sea urchin egg cytoplasm (assays, **134**, 327, 339; properties, **134**, 334, 350; purification, **134**, 329, 344); *Tetrahymena thermophila* cilia (characterization, **134**, 313; isolation, **134**, 309; -microtubule complex, formation, **134**, 316); tracheal cilia, porcine (assay, **196**, 225; characterization, **196**, 232; purification, **196**, 223); trout sperm outer arm dynein, **196**, 219

DYNORPHIN-CONVERTING ENZYME

This endopeptidase, often abbreviated DCE, catalyzes the conversion of dynorphin B-29 to dynorphin B-13, both of which are κ-opioid receptor active peptides. The DCE catalyzes the cleavage of the Thr13–Arg14 bond of dynorphin B-29. The enzyme is sensitive to thiol reagents and is distinct in its monobasic cleavage site specificity.[1]

[1]Y. L. Berman, L. Juliano & L. Devi (1995) *JBC* **270**, 23845.

E

*Eae*I RESTRICTION ENDONUCLEASE

This type II restriction endonuclease [EC 3.1.21.4], obtained from *Enterobacter aerogenes*, acts on both strands of DNA at 5′ . . . Y∇GGCCR . . . 3′, where Y represents C or T and R refers to A or G, and ∇ indicates the cleavage site.[1] The enzyme is blocked by overlapping *dcm* methylation.

[1] P. R. Whitehead & N. L. Brown (1983) *FEBS Lett.* **155**, 97.

*Eag*I RESTRICTION ENDONUCLEASE

This type II restriction endonuclease [EC 3.1.21.4], obtained from *Enterobacter agglomerans*, cleaves both strands of DNA at 5′ . . . C∇GGCCG . . . 3′.[1] The symbol ∇ indicates the cleavage site.

[1] L. Kauc & K. Leszczynska (1986) *Acta Microbiol. Pol.* **35**, 317.

*Ear*I RESTRICTION ENDONUCLEASE

This type II restriction endonuclease (subtype s) [EC 3.1.21.4], obtained from *Enterobacter aerogenes*, acts on DNA at 5′ . . . CTCTTCN∇ . . . 3′ and 3′ . . . GAGAAGNNNN∇ . . . 5′, where N refers to any base, and ∇ indicates the cleavage site.[1]

[1] C. Polisson & R. D. Morgan (1988) *NAR* **16**, 9872.

*Eca*I RESTRICTION ENDONUCLEASE

This type II restriction endonuclease [EC 3.1.21.4], obtained from *Enterobacter cloacae*, acts on both strands of DNA at 5′ . . . G∇GTNACC . . . 3′, where N represents A or G or T or C, and ∇ indicates the cleavage site.[1]

[1] G. Hobom, E. Schwarz, M. Melzer & H. Mayer (1981) *NAR* **9**, 4823.

ECARIN

This metalloproteinase, also referred to as prothrombin-activating principle, specifically catalyzes the conversion of prothrombin to α-thrombin via a meizothrombin

intermediate. In cattle prothrombin, the cleavage site is at Arg–Ile; in human prothrombin, there are two cleavage sites, Gly158–Ser159 and Arg232–Ile233.[1,2]

[1] T. Morita & S. Iwanaga (1978) *J. Biochem.* **83**, 559.
[2] S. Nishida, T. Fujita, N. Kohno, H. Atoda, T. Morita, H. Takeya, I. Kido, M. J. Paine, S. Kawabata & S. Iwanaga (1995) *Biochemistry* **34**, 1771.

ECDYSONE *O*-ACYLTRANSFERASE

ecdysone

This transferase [EC 2.3.1.139], first isolated from the ovaries of *Periplaneta americana* (the American cockroach), catalyzes the reaction of palmitoyl-CoA with ecdysone to produce Coenzyme A and ecdysone palmitate.[1]

[1] A. J. Slinger & R. E. Isaac (1988) *Insect Biochem.* **18**, 779.

ECDYSONE 3-EPIMERASE

This epimerase, also called 3-hydroxysteroid 3-epimerase, catalyzes the interconversion of 3β-ecdysteroids and 3α-ecdysteroids and requires dioxygen and NAD(P)H. **See also** *Ecdysone Oxidase*

Selected entries from **Methods in Enzymology** [vol, page(s)]:
General discussion: **111**, 437
Assay: *Manduca sexta*, **111**, 439
Other molecular properties: *Manduca sexta* (isolation, **111**, 438; properties, **111**, 440; substrate specificity, **111**, 440)

ECDYSONE 20-MONOOXYGENASE

This heme-thiolate-dependent (*i.e.*, a cytochrome P450 system) oxidoreductase [EC 1.14.99.22] catalyzes the reaction of α-ecdysone (*i.e.*, (22R)-2β,3β,14α,22,25-pentahydroxy-5β-cholest-7-en-6-one) with a reduced acceptor substrate (AH$_2$) and dioxygen to produce

20-hydroxyecdysone, the oxidized acceptor (A), and water. AH_2 can be $NADPH + H^+$.

Selected entries from **Methods in Enzymology** [vol, page(s)]:
General discussion: 111, 454
Assay: *Manduca sexta*, 111, 455

ECDYSONE OXIDASE

This oxidoreductase [EC 1.1.3.16] catalyzes the reaction of ecdysone with dioxygen to produce 3-dehydroecdysone and hydrogen peroxide.[1,2] 20-Hydroxyecdysone is an alternative substrate.

[1] G. F. Weinrich, M. J. Thompson & J. A. Svoboda (1989) *Arch. Insect Biochem. Physiol.* **12**, 201.
[2] J. Koolman & P. Karlson (1978) *EJB* **89**, 453.

Selected entries from **Methods in Enzymology** [vol, page(s)]:
General discussion: 52, 21; 111, 419
Assay: blowfly, 111, 420

*Eci*I RESTRICTION ENDONUCLEASE

This type II restriction endonuclease (subtype s) [EC 3.1.21.4], obtained from *Escherichia coli* T-8, catalyzes the hydrolysis of DNA at 5′ . . . GGCGGANNNNNNNNNNN ∇ . . . 3′ and 3′ . . . CCGCCTNNNNNNNNN∇ . . . 5′, where N refers to any base, and ∇ indicates the cleavage site.

*Ecl*I RESTRICTION ENDONUCLEASE

This type II restriction endonuclease [EC 3.1.21.4], obtained from *Enterobacter cloacae*, acts on both strands of DNA at 5′ . . . CAG∇CTG . . . 3′, producing blunt-ended fragments.[1] The symbol ∇ indicates the cleavage site.

[1] H. Hartmann & W. Goebel (1977) *FEBS Lett.* **80**, 285.

*Ecl*II RESTRICTION ENDONUCLEASE

This type II restriction endonuclease [EC 3.1.21.4], obtained from *Enterobacter cloacae*, recognizes the sequence 5′ . . . CCWGG . . . 3′ where W refers to either A or T.[1] However, the cleavage site is unknown.

[1] H. Hartmann & W. Goebel (1977) *FEBS Lett.* **80**, 285.

*Eco*57I RESTRICTION ENDONUCLEASE

This type II restriction endonuclease (subtype s) [EC 3.1.21.4], obtained from *Escherichia coli* RFL57, catalyzes the hydrolysis of DNA at 5′ . . . CTGAAG(N)$_{16}$∇ . . . 3′ and 3′ . . . GACTTC(N)$_{14}$∇NN . . . 5′, where N refers to any base, and the symbol ∇ indicates the cleavage site.[1] The enzyme also exhibits star activity. **See "Star"** *Activity*

[1] A. Janulaitis, M. Petrusyte, Z. Maneliene, S. Klimasauskas & V. Butkus (1992) *NAR* **20**, 6043.

*Eco*BI RESTRICTION ENDONUCLEASE

This type I restriction endonuclease [EC 3.1.21.3], obtained from *Escherichia coli* B, recognizes the sequence 5′ . . . TGANNNNNNNNTGCT . . . 3′ where N refers to either A, G, C, or T. The cleavage site is random, ATP is hydrolyzed in the reaction, and there is a requirement for *S*-adenosyl-L-methionine.[1–7]

[1] S. Linn & W. Arber (1968) *PNAS* **59**, 1300.
[2] J. Schell (1969) *Virology* **39**, 66.
[3] J. A. Lautenberger & S. Linn (1972) *JBC* **247**, 6176.
[4] B. Eskin & S. Linn (1972) *JBC* **247**, 6183.
[5] J. V. Ravetch, K. Horiuchi & N. D. Zinder (1978) *PNAS* **75**, 2266.
[6] J. A. Lautenberger, N. C. Kan, D. Lackey, S. Linn, M. H. Edgell & C. A. Hutchison III (1978) *PNAS* **75**, 2271.
[7] J. Rosamond, B. Endlich & S. Linn (1979) *J. Mol. Biol.* **129**, 619.

*Eco*KI RESTRICTION ENDONUCLEASE

This type I restriction endonuclease (subtype A) [EC 3.1.21.3], obtained from *Escherichia coli* K, recognizes the sequence 5′ . . . AACNNNNNNGTGC . . . 3′ where N refers to either A, G, C, or T. The cleavage site is random, ATP is hydrolyzed in the reaction, and there is a requirement for *S*-adenosyl-L-methionine.[1,2] Type I DNA restriction/modification enzymes protect the bacterial cell from viral infection by cleaving foreign DNA which lacks N^6-adenine methylation within a target sequence and maintaining the methylation of the targets on the host chromosome.

[1] M. Meselson & R. Yuan (1968) *Nature* **217**, 1110.
[2] N. C. Kan, J. A. Lautenberger, M. H. Edgell & C. A. Hutchison III (1978) *J. Mol. Biol.* **130**, 191.

*Eco*NI RESTRICTION ENDONUCLEASE

This type II restriction endonuclease [EC 3.1.21.4], obtained from *Escherichia coli* CDC A-193, acts on both strands of DNA at 5′ . . . CCTNN∇NNNAGG . . . 3′, where N refers to any base. The DNA fragments generated have a single-base 5′-extension that is more difficult to ligate than blunt-ended fragments.

*Eco*O109I RESTRICTION ENDONUCLEASE

This type II restriction endonuclease [EC 3.1.21.4], obtained from *Escherichia coli* H709c, acts on both strands of DNA at 5′ . . . RG∇GNCCY . . . 3′, where R refers to A or G, Y refers to C or T, and N refers to any base.[1] The enzyme is blocked by *dcm* methylation.[2]

[1] K. Mise & K. Nakajima (1985) *Gene* **36**, 363.
[2] K. Schnetz & B. Rak (1988) *Nucl. Acids Res.* **16**, 1623.

*Eco*P1 RESTRICTION ENDONUCLEASE

This type III restriction endonuclease [EC 3.1.21.5], obtained from *Escherichia coli* (P1), recognizes the sequence 5′ . . . AGACC . . . 3′.[1-4] *S*-Adenosyl-L-methionine is not required for the reaction but it will stimulate cleavage.[2] It has been reported that ATP is hydrolyzed in the cleavage reaction and that the endonuclease has an intrinsic ATPase activity.[4]

[1] J. Schell & S. W. Glover (1966) *Genet. Res.* **7**, 273 and 277.
[2] A. Haberman (1974) *J. Mol. Biol.* **89**, 545.
[3] B. Bachi, J. Reiser & V. Pirrotta (1979) *J. Mol. Biol.* **128**, 143.
[4] S. Saha & D. N. Rao (1995) *J. Mol. Biol.* **247**, 559.

*Eco*P15I RESTRICTION ENDONUCLEASE

This type III restriction endonuclease [EC 3.1.21.5], obtained from *Escherichia coli* P15, acts at the sequence 5′ . . . CAGCAG(N)$_{23}$NN∇NN . . . 3′ and 3′ . . . GTCGTC (N)$_{23}$NNNN∇ . . . 5′ where N refers to either A, T, G, or C.[1,2] *S*-Adenosyl-L-methionine is not required for the reaction but it will stimulate cleavage. *Eco*P15I possesses an intrinsic ATPase activity, the potential driving force of DNA translocation.[3]

[1] J. Reiser & R. Yuan (1977) *JBC* **252**, 451.
[2] S. M. Hadi, B. Bachi, J. C. W. Shepherd, R. Yuan, K. Ineichen & T. A. Bickle (1979) *J. Mol. Biol.* **134**, 655.
[3] A. Meisel, P. Mackeldanz, T. A. Bickle, D. H. Kruger & C. Schroeder (1995) *EMBO J.* **14**, 2958.

*Eco*RI RESTRICTION ENDONUCLEASE

This type II restriction enzyme [EC 3.4.21.4], isolated from *Escherichia coli* RY13, catalyzes the hydrolysis of both strands of DNA at 5′ . . . G∇AATTC . . . 3′, where ∇ indicates the cleavage site.[1-4] The rate-limiting step for endonuclease turnover occurs after double-strand cleavage.

[1] J. Hedgpeth, H. M. Goodman & H. W. Boyer (1972) *PNAS* **69**, 3448.
[2] D. J. Wright, K. King & P. Modrich (1989) *JBC* **264**, 11816.
[3] K. King, S. J. Benkovic & P. Modrich (1989) *JBC* **264**, 11807.
[4] D. J. Wright, W. E. Jack & P. Modrich (1999) *JBC* **274**, 31896.

Selected entries from ***Methods in Enzymology*** [**vol**, page(s)]:
General discussion: 54, 96; **65**, 23, 27, 29, 40, 64, 96, 147, 150, 173, 315, 317, 324, 325, 339, 342, 350, 361, 363, 407, 418, 421, 422, 432, 433, 440, 442, 446, 483, 566, 739, 762, 778, 780, 783, 784, 787, 789, 791; **68**, 264; **208**, 433
Assay: 65, 96, 177
Other molecular properties: activities, **65**, 96; catalytic requirements, **65**, 101; cleavage (double-strand, mechanism, **65**, 102, 103; sites on

[adenovirus 2 DNA, **65**, 769; SV40 DNA **65**, 713]); construction, of plasmid pLJ3, **65**, 237, 238; DNA binding (hydrogen-bonding network, **259**, 339, 343; specificity, **259**, 457; structural adaptations, **259**, 332); DNA footprinting at high pressure, **259**, 420; fragment end structure produced, **65**, 511; immobilization, **65**, 174, 175 (codigestion studies, **65**, 181, 182 [reaction kinetics, **65**, 178; stability, **65**, 178]); molecular weight, **65**, 101; optimal temperature, **65**, 416; pH optimum, **65**, 101; properties, **65**, 100; purification, **65**, 98; recognition sequence, **65**, 5, 96, 511; restriction, of pLJ3 plasmid, **65**, 240; sequence specificity, reduction, **65**, 103, 104; spectral properties, **65**, 101; stability, **65**, 178; star site cleavage, **259**, 418; substrate specificity (hydrostatic pressure effects, **259**, 418; osmotic pressure effects, **259**, 418); subunit composition, **65**, 101; turnover number, **65**, 111; volume change calculation, **259**, 419

*Eco*RII RESTRICTION ENDONUCLEASE

This type II restriction enzyme [EC 3.4.21.4], isolated from *Escherichia coli* R245, catalyzes the hydrolysis of both strands of DNA at 5′ . . . ∇CCWGG . . . 3′, where W refers to either A or T.[1,2]

[1] R. Yoshimori, D. Roulland-Dussoix & H. W. Boyer (1972) *J. Bacteriol.* **112**, 1275.
[2] H. W. Boyer, L. T. Chow, A. Dugaiczyk, J. Hedgpeth & H. M. Goodman (1973) *Nature New Biol.* **244**, 40.

Selected entries from ***Methods in Enzymology*** [**vol**, page(s)]:
General discussion: 65, 21, 64, 147; **208**, 433
Other molecular properties: fragment end structure produced, **65**, 511; recognition sequence, **65**, 5, 511

*Eco*RV RESTRICTION ENDONUCLEASE

This type II restriction endonuclease [EC 3.4.21.4], isolated from *Escherichia coli* J62 pLG74, catalyzes the hydrolysis of both strands of DNA at 5′ . . . GAT∇ATC . . . 3′, producing blunt-ended fragments.[1,2] In the magnesium-dependent reaction, the reaction proceeds with stereochemical inversion at phosphorus. A three-metal ion mechanism has been proposed, based upon metal ion binding sites and pH-rate profiles. One divalent cation increases the electrophilicity of the scissile O–P bond by direct ligation to the *proS* oxygen.[3] The metal ions also precisely position both the substrate and the attacking water molecule.[4] The kinetics of linear diffusion of *Eco*RV on DNA have also been investigated. At 50 mM Na$^+$ and 10 mM Mg^{2+} concentrations up to 10 mM, *Eco*RV accurately scans 2×10^6 base pairs per binding event with a velocity of about 1.7×10^6 base-pairs per second. The enzyme tends to overlook recognition sites at low magnesium ion concentration (1 mM).[5]

[1] G. V. Kholmina, B. A. Rebentish, Y. S. Skoblov, A. A. Mironov, N. Y. Yankovskii, Y. I. Kozlov, L. I. Glatmann, A. F. Moroz & V. G. Debabov (1980) *Dokl. Akad. Nauk.* **253**, 495.

[2]I. Schildkraut, C. D. B. Banner, C. S. Rhodes & S. Parekh (1984) *Gene* **27**, 327.

[3]N. C. Horton, M. D. Sam, B. A. Connolly & J. J. Perona (2000) *Biophys. J.* **78**, 417A.

[4]G. S. Baldwin, R. B. Sessions, S. G. Erskine & S. E. Halford (1999) *J. Mol. Biol.* **288**, 87.

[5]A. Jeltsch & A. Pingoud (1998) *Biochemistry* **37**, 2160.

EICOSANOYL-CoA SYNTHASE

This acyltransferase [EC 2.3.1.119], also known as acyl-CoA elongase and eicosanoyl-CoA synthase, catalyzes the reaction of stearoyl-CoA with malonyl-CoA and two molecules of NAD(P)H to produce eicosanoyl-CoA (also called icosanoyl-CoA and arachidoyl-CoA), carbon dioxide, and two molecules of $NAD(P)^+$.[1] The enzyme activity is strongly stimulated by the presence of lipids.

[1]J. J. Bessoule, R. Lessire & C. Cassagne (1989) *ABB* **268**, 475.

ELASTASE, LEUKOCYTE

This serine endopeptidase [EC 3.4.21.37], also called lysosomal elastase, neutrophil elastase, bone marrow serine protease, and medullasin, catalyzes peptide-bond hydrolysis in many proteins, with highly effective action on elastin. Catalysis proceeds through an acyl-enzyme intermediate.[1,2] The enzyme prefers a valyl residue in the P1 position of the substrate. Other residues accepted are alanyl, seryl, and cysteinyl; however, Val–Xaa is clearly preferred over Ala–Xaa. In addition, the activity increases with chain length. There are at lease eight subsites on the human enzyme. The specificity for synthetic substrates differs from the other elastases: MeOSuc-Ala-Ala-Pro-Val-∇-NHPhNO$_2$, MeOSucAla-Ala-Pro-Val-∇-SBzl, and Suc-Ala-Ala-Ala-∇-NHPhNO$_2$ have all been used to assay the enzyme. (Note: ∇ indicates the cleavage site.)

This enzyme is not found in all white blood cells, and the name is somewhat misleading. Its main source is the polymorphonuclear leukocyte. *See also Elastase, Pancreatic; Pancreatic Endopeptidase E; Pseudolysin; Elastase, Macrophage*

[1]J. G. Bieth (1986) in *Biology of Extra-Cellular Matrix*, vol. I (R. D. Mecham, ed.), Academic Press, New York, p. 217.

[2]W. Bode, E. Meyer, Jr. & J. C. Powers (1989) *Biochemistry* **28**, 1951.

Selected entries from *Methods in Enzymology* [vol, page(s)]:
General discussion: 80, 564, 581; 235, 554
Assay: 80, 582 (fluorometric, 80, 583, 584; spectrophotometric, 80, 582, 583); 163, 324; 235, 554
Other molecular properties: activation, 80, 749; active-site titration, 80, 584; 248, 87, 100; affinity chromatography at subzero temperatures,
135, 535; distribution, 80, 587, 588; human (active-site titration, 248, 87; ^{1}H NMR spectroscopy, 230, 133, 152, 157; similarity to myeloblastin, 244, 63, 66; source, 248, 7; thioester substrates, 248, 10, 16 [kinetic constants, 248, 10, 12]); inhibition, 80, 587 (polymorphonuclear, by α_1-proteinase inhibitor, 80, 755; by α_1-proteinase inhibitor, 80, 763); isoelectric points, 80, 586; molecular weight, 80, 586; N-terminal amino acid sequence, 80, 586; properties, 80, 586; 163, 314; purification, 80, 585, 586; 163, 313; storage, 80, 586; trapped by α_2-macroglobulin, 80, 748; units of activity, 80, 584

ELASTASE, MACROPHAGE

This zinc-dependent metalloendopeptidase [EC 3.4.24.65], also called macrophage metalloelastase and matrix metalloproteinase 12 (MMP-12), catalyzes peptide-bond hydrolysis in soluble and insoluble forms of elastin. Calcium ion (typically 5 mM) is required for activity. The enzyme catalyzes the hydrolysis of the Ala14–Leu15 and Tyr16–Leu17 bonds in the insulin B chain, suggesting a preference for substrates possessing a leucyl residue at the P1' position.[1,2] Interestingly, Pro357–Met358 of α_1-proteinase inhibitor is hydrolyzed; when the methionyl residue of α_1-proteinase inhibitor is oxidized, macrophage elastase catalyzes the hydrolysis of Phe352–Leu353. *See also Elastase, Pancreatic; Elastase, Leukocyte; Pancreatic Endopeptidase E; Pseudolysin*

[1]C. Kettner, E. Shaw, R. White & A. Janoff (1981) *BJ* **195**, 369.
[2]M. J. Banda, E. J. Clark, S. Sinha & J. Travis (1987) *J. Clin. Invest.* **79**, 1314.

Selected entries from *Methods in Enzymology* [vol, page(s)]:
General discussion: 144, 288; 235, 554
Other molecular properties: properties, 248, 192, 511

ELASTASE, PANCREATIC

This serine endopeptidase [EC 3.4.21.36], also referred to as pancreatopeptidase E and pancreatic elastase I, catalyzes peptide-bond hydrolysis in proteins, including elastin.[1–4] The enzyme reaction proceeds by forming an acyl-enzyme intermediate. This elastase exhibits a distinct preference for substrates containing a nonbulky aminoacyl residue in the P1 position; in such cases, Ala–Xaa bonds are readily cleaved. Longer peptides have proven to be better substrates, and with porcine pancreatic elastase, Suc-Ala-Ala-Ala-∇-*p*-nitroaniline is a 3000-fold better substrate than Suc-Ala-Ala-∇-*p*-nitroaniline. (Note: ∇ indicates the cleavage site.)

Trypsin activates the elastin precursor, proelastin, in the duodenum. The precursors are stored in the acinar cells of the pancreas.

Note: The human protein often referred to as human pancreatic elastase I is actually pancreatic endopeptidase E [EC 3.4.21.70]. *See also Elastase, Leukocyte; Endopeptidase E, Pancreatic; Pseudolysin; Elastase, Macrophage; Elastase II, Pancreatic*

[1]X. Ding, B. F. Rasmussen, G. A. Petsko & D. Ringe (1994) *Biochemistry* **33**, 9285.
[2]M. A. Ator & P. R. Ortez de Montellano (1990) *The Enzymes*, 3rd ed., **19**, 213.
[3]L. Robert & W. Hornebeck, eds. (1989) *Elastin and Elastases*, CRC Press, Boca Raton.
[4]B. S. Hartley & D. M. Shotton (1971) *The Enzymes*, 3rd ed., **3**, 323.

Selected entries from *Methods in Enzymology* [vol, page(s)]:
General discussion: 5, 665; 19, 113; 34, 4; 46, 20, 22, 198, 201, 206, 207, 214, 215; 80, 582, 587, 733; 235, 554
Assay: 5, 665; 19, 113; 82, 593; 144, 289; 235, 554; Congo red-elastin procedure, 19, 113; coupled enzyme assay, 63, 33; fluorimetric assay, 248, 20; synthetic substrates, with, 19, 117
Other molecular properties: activators and inactivators, 19, 131; active center sequence, 19, 133; active halogen compounds and, 25, 607, 608; active-site titration, 80, 424, 584; acylelastase, 87, 69; affinity chromatography at subzero temperatures, 135, 531; amino acid composition, 19, 128; 45, 423; amino acid sequence determination and, 6, 838; amino acid sequences, 19, 130, 133; carboxy terminal sequence analysis, 240, 705, 708, 710, 712; CBZ-alanyl-*p*-nitrophenyl ester, and, 87, 70; chemical modification, 63, 218, 219; α-chymotrpysin and, 19, 129; classification, 144, 299; cleavage site specificity, 244, 114; competitive labeling, 63, 218, 219; conformation related to chymotrypsin, 19, 58, 133; cryoenzymology, 63, 338; 87, 70; cryosolvent for, 87, 70; crystal intensity, multiwire area detector diffractometry, 114, 470; dimethylaminocinnamoylelastase, 87, 73; dissociation of rat liver, enzyme complex, 51, 121; distribution, 19, 139; elastase 2 (antibody production, 74, 278; molecular form in normal serum, 74, 273, 274; radioimmunoassay [applications, 74, 273; equilibration conditions, 74, 284; interference by protease inhibitor complexes, 74, 282; standard, 74, 278]); electrophoresis, 5, 43; enzymatic activity and substrate specificity, 19, 132; enzymatic properties compared to leukocyte elastase, 80, 586; hexokinase, preparation of, 9, 372; inhibition, 80, 587, 801 (assay, 45, 861, 870; bronchial mucus inhibitor, human, 45, 870; eglin, by, 80, 804, 812; elastatinal, by, 45, 687; functional, measurement, 82, 604; garden bean elastase inhibitor, reactive site, amino acid composition, 80, 764; α₁-proteinase, 80, 763; submandibular gland proteinase inhibitor, 45, 860, 865); iodinated proteins and, 25, 446; kinetic properties, 19, 135; modulation by (elastin ligands, 82, 601, 603; physical-chemical environment, 82, 600); mRNA sequences, cloning, 144, 304; N-terminal analysis, 19, 127; peptide separation, 47, 206; pH study, 63, 218, 219; physical properties, 19, 128, 131; porcine (active-site titration, 80, 424; amino-terminal sequence, 80, 641, 651; covalent incorporation in thiol ester-containing α-macroglobulin, 223, 135; homology to crab collagenase, 80, 729; hydrolysis of low molecular weight synthetic substrate, 80, 731; generation of C3 activation fragments, 80, 90, 91; similarity to thrombin, 80, 294; source, 248, 7; thioester substrates, 248, 10, 16 [kinetic constants, 248, 10, 12]; trapped by α₂-macroglobulin, 80, 748); product inhibition studies, 249, 194; proenzyme, 80, 586; properties, 5, 671; 19, 128, 131; proteolysis of tubulin, 134, 182; purification, 5, 668; 19, 120 (summary, 19, 126); Ser¹⁹⁵-peptide sequence, 80, 651; site-specific reagents and, 25, 657; solubilization of hexokinase, in, 9, 372; specific activity, 19, 115; spin-label studies, 49, 445, 446; stability, 19, 125, 127; structure, 240, 703, 705; subzero temperature, 87, 70; tetrahedral intermediate, 87, 71; thioester substrates, 248, 10, 16; tissue culture, 58, 125; TNBS inactivation, 19, 128; tosylelastase (β strand mutations,

accommodation, 202, 79); transition state analogue complexes, characterization, 249, 294; transition state and multisubstrate analogues, 249, 306; trimethylacetyl elastase, deacylation of, 19, 132

ELASTASE II, PANCREATIC

This serine endopeptidase [EC 3.4.21.71], PE II, has been isolated from only pig and human pancreas. It catalyzes the hydrolysis of peptide bonds, proceeding via an acyl-enzyme intermediate. PE II exhibits a preference for protein substrates bearing a medium to large side-chain residue (for example, Leu, Met, Phe, or Tyr) in the P1 position. The synthetic substrates glutaryl-Ala-Ala-Pro-Leu-∇-NHPhNO₂ (where ∇ indicates the cleavage site) is often used to assay the enzyme. Pancreatic elastin II can act on elastin, complement C3, kininogen, and fibronectin.[1]

[1]J. G. Bieth (1986) in *Biology of Extra-Cellular Matrix*, vol. I (R. D. Mecham, ed.), Academic Press, New York, p. 217.

Selected entries from *Methods in Enzymology* [vol, page(s)]:
General discussion: 5, 665
Assay: 5, 665
Other molecular properties: elastase 2 (antibody production, 74, 278; molecular form in normal serum, 74, 273, 274; radioimmunoassay [applications, 74, 273; equilibration conditions, 74, 284; interference by protease inhibitor complexes, 74, 282; standard, 74, 278]); properties, 5, 671; purification, hog, 5, 668

ELECTRON-TRANSFERRING FLAVOPROTEIN AND ELECTRON-TRANSFERRING-FLAVOPROTEIN DEHYDROGENASE

This iron-sulfur- and FAD-dependent oxidoreductase [EC 1.5.5.1], also known as electron transfer flavoprotein:ubiquinone oxidoreductase, ETF:ubiquinone oxidoreductase, and ETF dehydrogenase, catalyzes the reaction of reduced ETF with ubiquinone to produce ETF and ubiquinol by means of a ping-pong kinetic mechanism.[1,2]

The electron-transferring flavoprotein is an FAD-containing protein that acts with a number of acyl-CoA dehydrogenases to form a system that will oxidize an acyl-CoA and reduce ubiquinone. *See also Butyryl-CoA Dehydrogenase; Acyl-CoA Dehydrogenase*

[1]F. Frerman (1988) *Biochem. Soc. Trans.* **16**, 416.
[2]J. D. Beckmann & F. E. Frerman (1985) *Biochemistry* **24**, 3913.

Selected entries from *Methods in Enzymology* [vol, page(s)]:
General discussion: 6, 424; 53, 502
Assay: 5, 547; 53, 503, 504
Other molecular properties: activity, 53, 502; acyl dehydrogenases and, 5, 546, 552, 554, 555; assay for butyryl-CoA dehydrogenase, 14, 110; catalytic properties, 53, 506, 507; dehydrogenase, 53, 402, 404, 406; diaphorase activity, 14, 118; electrophoretic properties, 5, 551; 53, 508, 509, 512; flavin content, 53, 508; function, 14, 110, 114, 115;

monkey liver, from, **14**, 115; natural acceptor, **14**, 118; pig liver, from, **53**, 506; requirement for butyryl-CoA dehydrogenase, **14**, 110, 114; role in electron-transport, **53**, 402; sources, **14**, 115; spectral properties, **53**, 508, 510, 511; stability, **14**, 117; **53**, 518; substrate specificities, **53**, 502, 503

ELONGATION FACTORS

These enzymes promote polypeptide chain elongation during mRNA-dependent translation on ribosomes.[1–5] The Enzyme Commission classifies all GTP hydrolysis-dependent elongation factors (abbreviated EF) as protein-synthesizing GTPases [EC 3.6.1.48]. Despite the fact that the overall reaction yields GDP and orthophosphate, the enzyme is misclassified as a hydrolase, because P-O-P bond scission supplies the energy to drive peptidyl-tRNA translocation. Therefore, this enzyme should be regarded as an energase, a special class of mechanochemical enzymes that transduce chemical energy into mechanical work (*i.e.*, exerting a force over a distance).[6] The reaction for transcription factors should be written as: Positional State$_1$ + GTP = Positional State$_2$ + GDP + P$_i$, where transfer between the two positional states requires mechanical work. **See** *Energases*

In prokaryotes, the GTP-hydrolyzing elongation factors are EF-Tu and EF-G. EF-Tu (where Tu stands for transfer factor, heat unstable) promotes the GTP-dependent binding of the aminoacyl-tRNA to the ribosome's acceptor site. Prokaryotic EF-G, also simply called translocase, catalyzes GTP-dependent peptidyl-tRNA translocation from the acceptor site to the peptide site on the ribosome.

In eukaryotes, elongation factors EF-1α and EF-2 both possess GTPase activities. EF-1α plays a role similar to EF-Tu, and EF-2 by catalyzing GTP-dependent peptidyl-tRNA translocation from the acceptor site to the peptide site.

[1] I. M. Krab & A. Parmeggiani (1998) *BBA* **1443**, 1.
[2] J. Nyborg & A. Liljas (1998) *FEBS Lett.* **430**, 95.
[3] B. S. Negrutskii, A. V. El'skaya (1998) *Prog. Nucl. Acid. Res. Mol. Biol.* **60**, 47.
[4] M. V. Rodnina, A. Savelsberg, V. I. Katunin & W. Wintermeyer (1997) *Nature* **385**, 37.
[5] B. F. Clark & J. Nyborg (1997) *Curr. Opin. Struct. Biol.* **7**, 110.
[6] D. L. Purich (2001) *TiBS* **26**, 417.

Selected entries from *Methods in Enzymology* [vol, page(s)]:
Assay: 20, 292, 310, 323
Other molecular properties: assay for amino acid polymerization (definition of unit and specific activity, **20**, 294; principle, **20**, 292; procedure, **20**, 293, 324; reagents, **20**, 292); bacterial (isolation of T and G factors, **20**, 281; microscale isolation, **20**, 288; ribosome isolation and, **20**, 288; subfractionation of T factor, **20**, 289); complex II formation and, **20**, 312; discovery, resolution, purification, and function, **182**, 809;

eEF-1 (affinity for guanosine nucleotides, **60**, 583, 584, 589; assay by polyphenylalanine synthesis, **60**, 115, 142, 640, 677, 689, 690; function, **60**, 162, 163, 578, 579, 638, 657, 676, 695; nomenclature, **60**, 686, 687; polyacrylamide gel electrophoresis, **60**, 176, 177; preparation from *Artemia salina* cysts, **60**, 690 [from ascites cells, **60**, 92; from reticulocyte supernatant, **60**, 140, 582, 583, 638, 639, 641, 647, 648; from yeast, **60**, 678]; properties, **60**, 658, 686, 694; separation from elongation factor eEF-2, **60**, 640, 641; size classes, **60**, 694, 695); eEF-2 (adenosine diphosphate ribosyl derivative [assay, **60**, 677, 678, 704, 706; dansylated, **60**, 718; preparation, **60**, 710]; assay [binding of guanosine nucleotides, by, **60**, 704, 705; polyphenylalanine synthesis, by, **60**, 115, 142, 583, 650, 651, 662, 664, 704, 705; *Pseudomonas* toxin A, for, **60**, 788, 789; ribosome-dependent GTP hydrolysis, by, **60**, 704, 705, 708, 710]; binding of guanosine nucleotides, **60**, 583, 584, 708, 710, 711; contaminant of 80S ribosomes, **60**, 580; fluorescence polarization, **60**, 712, 716; functional role, **60**, 162, 613, 638, 649, 657, 676, 702, 703; labeling with dansyl chloride, **60**, 715; molecular shape, **60**, 718, 719; polyacrylamide gel electrophoresis, **60**, 176, 177; preparation from [Krebs ascites cells, **60**, 652, 656; pig liver, **60**, 667, 668; reticulocyte supernatant, **60**, 140, 639, 645; wheat germ, **60**, 706]; properties, **60**, 658, 668, 681, 682, 708, 710; separation from other elongation factors [eEF-1, **60**, 640, 641; EF-1a, **60**, 666]; stimulation by elongation factor EF-1b, **60**, 656); effect on structure of ribosomal proteins L7/L12, **164**, 154; elongation factor EF-1 (competition with elongation factor 2 for common ribosomal site, **30**, 243; isolation, **30**, 240; plants [assays, **118**, 144; properties, **118**, 152; purification, **118**, 143, 145]; ribosomal binding [materials, **30**, 239; procedure, **30**, 241]); elongation factor EF-1α (assay, **60**, 650, 662; complex with GDP, **60**, 663 [GTP and aminoacyl-tRNA, with, **60**, 663]; functional role, **60**, 649, 658; nomenclature, **60**, 686, 687; preparation, **60**, 654, 656, 668, 670; properties, **60**, 654, 658, 670; separation from elongation factor EF-2, **60**, 666, 667; stimulation by elongation factor EF-1b, **60**, 656; varieties, **60**, 670); elongation factor EF-1αβγ, complex with elongation factor EF-Tu (composition, **60**, 675, 676; high-molecular-weight form of EF-1, **60**, 659; preparation from pig liver mitochondria, **60**, 659; purification, **60**, 674, 675); elongation factor EF-1β (assay, **60**, 651, 662; elution, **60**, 655, 656; functional role, **60**, 649, 656, 658, 673, 674; isolation, **60**, 672; molecular weight, **60**, 657, 671, 672; nomenclature, **60**, 686, 687); elongation factor EF-1βγ (assay, **60**, 662; gel electrophoresis, **60**, 671, 672; molecular weight, **60**, 671; nomenclature, **60**, 686, 687; purification, **60**, 670, 671; separation into EF-1β and EF-1γ, **60**, 659, 672, 673); elongation factor EF-1γ (isolation, **60**, 672; molecular weight, **60**, 671, 672; nomenclature, **60**, 686, 687); elongation factor EF-2 (ADP-ribosylation, **235**, 617, 631, 648; competition for elongation factor 1 for common ribosomal site, **30**, 243; cytoplasmic, separation from mitochondrial factor, **30**, 250; diphthamide in [ADP-ribosylation by diphtheria toxin, **106**, 379, 392; biosynthesis *in vitro*, **106**, 388; demonstration, **106**, 395; properties, **106**, 385; purification from yeast, **106**, 381]; heat and, **30**, 257; isolation, **30**, 240; polyphenylalanine synthesis and, **30**, 241; preparation of ribosomal subunits, **59**, 404; ribosomal binding [materials, **30**, 239; procedure, **30**, 241]); elongation factor EF-3 (assay, **60**, 679, 682; functional role, **60**, 676, 677; separation from elongation factor EF-2, **60**, 681, 682); elongation factor EF-1$_H$, **60**, 649, 656, 658, 686, 687; elongation factor G (activity in protein synthesis, **60**, 593, 594, 605, 606, 761, 762, 775; adjacency of sulfhydryl group and binding center, **60**, 744, 745; assay [GTP hydrolysis, of, **59**, 361; polypeptide chain elongation, of, **59**, 358; poly(U)-dependent phenylalanine incorporation assay, **264**, 252; principle, **20**, 303; procedure, **20**, 303; protein synthesis from aminoacyl-tRNA, by, **60**, 598, 599; quaternary complex, as, **60**, 607; reagents, **20**, 303; stimulation of GTPase, by, **60**, 595 {of polyphenylalanine synthesis, **60**, 597, 598}; translocation, for, **59**, 359]; binding to ribosomal subparticles, **60**, 719, 720; binding to ribosome [comment, **30**, 238; materials, **30**, 236; principle, **30**, 235; procedure, **30**, 237; reconstituted subunits and, **30**, 562]; crystallization, **20**, 297, 305; detachment from ribosomes as requirement for further elongation, **60**, 775; enzymic homogeneity, **60**,

614; free *vs.* ribosome-bound, binding of guanosine nucleotides, **60**, 737; functional role, **60**, 719, 720; fusidic acid and, **30**, 281; GTP, and, binding to ribosomal pretranslocation complex, kinetics, **164**, 594; labeling with iodoacetamide, **60**, 743; limited trypsinolysis, **60**, 743, 744; mitochondrial [further purification, **30**, 253; separation from cytoplasmic factor, **30**, 250]; molecular weight, **60**, 614; photoactivated analogues [complex with GTP and ribosomes, **60**, 722, 723; preparation, **60**, 720, 742; use in proof of interaction with ribosomal subunits, **60**, 723]; primary structure, **60**, 742; purification, **20**, 304 [bovine mitochondria {extract preparation, **264**, 254; high-performance liquid chromatography, **264**, 250, 257, 259; mitochondria isolation, **264**, 253; yield, **264**, 261}]; radiolabeled, preparation, **59**, 790, 791; ribosome release, **59**, 862; tests of purity, **60**, 613, 614; thiostrepton and, **30**, 279; titrations, in analysis of translational kinetics, **164**, 616; translation in ribosomes induced by attachment, **60**, 766; translocation and, **30**, 462); elongation factor II (tRNA binding factor and, **30**, 213); elongation factor T (assay of polypeptide chain elongation, **59**, 358; crystallization, **20**, 298; mitochondrial, purification of, **30**, 250; preparation, **30**, 233; separation into factors Ts and Tu, **20**, 298; tosylphenylalanyl chloromethane and, **30**, 281); elongation factor T-I (rabbit reticulocytes, **20**, 316 [assays, **20**, 323; isolation procedures, **20**, 324; solutions, **20**, 317; special procedures, **20**, 318]); elongation factor T-II (inactivation, **20**, 318, 331; rabbit reticulocyte, **20**, 330, 336 [assay procedures, **20**, 332; biological materials, **20**, 331; isolation, **20**, 334]); elongation factor T2 (diphtheria toxin and, **30**, 282); elongation factor Ts (activity in protein synthesis, **60**, 593, 594, 605, 606, 702, 703; assay, **30**, 221 [binding of aminoacyl-tRNA to 80S ribosomes, **60**, 700; exchange of GDP in its complex with elongation factor EF-Tu, **60**, 636, 637, 700; guanine nucleotide exchange assay, **264**, 252; protein synthesis from aminoacyl-tRNA, **60**, 598, 599; stimulation of polyphenylalanine synthesis, **60**, 597, 598, 700]; complex with elongation factor EF-Tu, **60**, 602, 616, 626, 627, 636 [Tu-Ts complex, purification from bovine mitochondria {extract preparation, **264**, 254; high-performance liquid chromatography, **264**, 250, 257; mitochondria isolation, **264**, 253; yield, **264**, 261}]; constituent of initiation factor EF-1, **60**, 686, 687, 694, 703; elongation factor Tu assay and, **30**, 220, 221, 227; function of, **30**, 219; **60**, 700; isolation, **20**, 289; **60**, 602, 700, 701; nomenclature, **60**, 687; properties, **60**, 701; removal of, **30**, 225; role in *in vitro* translation, **164**, 620); elongation factor Tu, **46**, 151 (activity in protein synthesis, **60**, 593, 594, 605, 615; assay, **30**, 220 [GDP binding, **60**, 594, 636; poly(U)-dependent phenylalanine incorporation assay, **264**, 251; protein synthesis from aminoacyl-tRNA, **60**, 598, 599; stimulation of polyphenylalanine synthesis, **60**, 597, 598]; cGMP assay, **38**, 85; complex with elongation factor EF-Ts, **60**, 602, 616, 626, 627, 636, 703 [GDP, with, **60**, 616, 622; GTP, with, **60**, 627; GTP and aminoacyl-tRNA, with, **60**, 615, 616; guanylyl imidodiphosphate and aminoacyl-tRNA, **60**, 622, 624, 625; Tu-Ts complex, purification from bovine mitochondria {extract preparation, **264**, 254; high-performance liquid chromatography, **264**, 250, 257; mitochondria isolation, **264**, 253; yield, **264**, 261}]; constituent of EF-1, **60**, 194, 197, 686, 687, 703; crystallization, **20**, 298; effector region, **238**, 15; function of, **30**, 219; GDP, and [crystallization of, **30**, 226; properties, **30**, 227; purification and isolation of, **30**, 224; removal of GDP, **30**, 228]; GTP, and [preparation, **30**, 229; substances reacting with, **30**, 224]; isolation, **20**, 290 [Sephadex A-50, using, **60**, 602; Sephadex C-50, using, **60**, 696, 697, 699; Sephadex G-100, using, **60**, 599]; nomenclature, **60**, 687; properties, **60**, 697, 698; protein-tRNA interaction studies, **59**, 324; radiolabeled, preparation, **59**, 790, 791; regulation of binding of aminoacyl-tRNA to 70S ribosomes, **60**, 615, 616; removal from elongation factor EF-G, **60**, 610, 611, 613; thiostrepton and, **30**, 279; titrations, in analysis of translational kinetics, **164**, 616); elution from phosphocellulose, **30**, 139; *Euglena gracilis* chloroplast (assay, **118**, 297; properties, **118**, 308; purification, **118**, 296, 305); globin synthesis and, **30**, 121; human tonsil, preparation of, **30**, 284; initial dipeptide synthesis and, **30**, 118; isolation, **20**, 281 (ammonium sulfate fractionation, **20**, 286; assays, **20**, 282; crude extract preparation, **20**, 285; diethylaminoethyl cellulose and, **20**, 287; electrophoresis, **20**,

287; growth conditions and, **20**, 284; hydroxylapatite and, **20**, 287; materials, **20**, 283; polyethylene glycol-dextran and, **20**, 285); liver, separation of, **30**, 183; methionyl puromycin synthesis and, **30**, 113, 119; mitochondrial, isolation of, **30**, 248; nomenclature, **20**, 306; polyphenylalanine synthesis and, **30**, 110; preparation of, **30**, 80; properties (electrophoresis, **20**, 301; pH optimum, **20**, 302; physical, **20**, 302; stability, **20**, 302); purification (ammonium sulfate fractionation, **20**, 296; cell extract supernatant, **20**, 295; crystallization of factors G, T, and Tu, **20**, 297; growth and harvest of bacteria, **20**, 295; hydroxylapatite and, **20**, 297; isopropanol and, **20**, 296; notes, **20**, 300; polyacrylamide gel electrophoresis, **20**, 297; separation of factor T into Ts and Tu, **20**, 298; Sephadex G-200 and, **20**, 296); rat liver, **20**, 337 (assay, **20**, 339; crude preparation, **20**, 339; partial purification of T-I, **20**, 341; purification of T-II, **20**, 342; reagents, **20**, 338); requirement, circumvention of, **30**, 464; ribosomal subunit reassociation and, **30**, 192; separation of, **20**, 311; spinach chloroplast (assays, **118**, 297; properties, **118**, 300, 303; purification, **118**, 296, 298); wheat germ (assays, **118**, 114, 135; functional properties, **118**, 127; physical properties, **118**, 124; purification, **118**, 110, 116, 130); yeast, **20**, 349 (assays, **20**, 351; general preparative procedures, **20**, 353; isolation, **20**, 355; properties, **20**, 357; reagents, **20**, 350)

EMULSIN

This name refers to an almond extract possessing β-glucosidase activity. Nineteenth century investigations of emulsin-catalyzed hydrolysis of amygdalin provided some of the earliest evidence for the phenomenon of chemical catalysis.

ENAMELYSIN

This metalloproteinase, also known as matrix metalloproteinase 20 (MMP-20), catalyzes the hydrolytic degradation of amelogenin, the major protein component of the enamel matrix.[1] The enzyme is expressed by ameloblasts and odontoblasts at the onset of dentin mineralization.[2]

[1] E. Llano, A. M. Pendás, V. Knäuper, T. Sorsa, T. Salo, E. Salido, G. Murphy, J. P. Simmer, J. D. Bartlett & C. López-Otín (1997) *Biochemistry* **36**, 15101.
[2] O. H. Ryu, A. G. Fincham, C. C. Hu, C. Zhang, Q. Qian, J. D. Bartlett & J. P. Simmer (1999) *J. Dent. Res.* **78**, 743.

ENDOGALACTOSAMINIDASE

This enzyme [EC 3.2.1.109] catalyzes the endohydrolysis of galactosaminidic linkages in oligo- and poly(D-galactosamine).[1]

[1] J. L. Reissig, H.-H. Lai & J. E. Glasgow (1975) *Can. J. Biochem.* **53**, 1237.

ENDO-β-GALACTOSIDASE

This is a family of enzymes that catalyzes the endohydrolysis of β-galactosides.[1,2] They include the endogalactosidases [EC 3.2.1.102] that hydrolyze the 1,4-β-D-galactosidic linkages in blood groups A and B, the endohydrolases [EC 3.2.1.103] that act on the 1,4-β-D-galactosidic linkages in keratan sulfate, arabinogalactan endo-1,4-β-galactosidase [EC 3.2.1.89], and arabinogalactan endo-1,3-β-galactosidase [EC 3.2.1.90]. *See Keratan-Sulfate Endo-1,4-β-Galactosidase; Blood-Group-Substance Endo-1,4-β-Galactosidase; Arabinogalactan Endo-1,4-β-Galactosidase; Arabinogalactan Endo-1,3-β-Galactosidase*

[1] R. J. Staneloni & L. F. Leloir (1982) *Crit. Rev. Biochem.* **12**, 289.
[2] H. M. Flowers & N. Sharon (1979) *AE* **48**, 29.

ENDO-1,3(4)-β-D-GLUCANASE

This enzyme [EC 3.2.1.6], also known as endo-1,4-β-glucanase, endo-1,3-β-glucanase, and laminarinase, catalyzes the endohydrolysis of 1,3- or 1,4-linkages involving the reducing group of glucosyl residues (substituted at C3) in such β-D-glucans as laminarin, lichenin, and cereal glucans.[1-3] Plant endo-1,3(4)-glucanases act with retention of configuration, but are inhibited with (3R)-epoxybutyl-β-cellobiosides, which exert little effect on bacterial endo-1,3(4)-glucanases.[4]

This enzyme should not be confused with licheninase [EC 3.2.1.73] which is also referred to as an endo-β-1,3-1,4 glucanase.

[1] G. Davies, M. L. Sinnott & S. G. Withers (1998) *CBC* **1**, 119.
[2] A. Sharma & J. P. Nakas (1987) *Enzyme Microb. Technol.* **9**, 89.
[3] A. Totsuka & T. Usui (1986) *Agric. Biol. Chem.* **50**, 543.
[4] P. B. Høj, E. B. Rodriguez, J. R. Iser, R. V. Stick & B. A. Stone (1991) *JBC* **266**, 11628.

ENDOGLYCOSYLCERAMIDASES

These enzymes [EC 3.2.1.123], also known as endoglucosylceramidase, ceramide glycanase, and endoglycoceramidase, catalyze the hydrolysis of an oligosaccharyl-ceramide (*e.g.*, oligoglycosylglucosylceramide) to produce an oligosaccharide and a ceramide.[1-3] The enzyme from *Rhodococcus* sp. degrades various acidic and neutral glycosphingolipids to oligosaccharides and ceramides.[1] It does not act on monoglycosylceramides (as does glycosylceramidase [EC 3.2.1.62]). The *Corynebacterium* enzyme also exhibits a transglycosylation activity.[3]

[1] M. Ito & T. Yamagata (1986) *JBC* **261**, 14278.
[2] H. Ashida, K. Yamamoto, H. Kumagai & T. Tochikura (1992) *EJB* **205**, 729.
[3] H. Ashida, Y. Tsuji, K. Yamamoto, H. Kumagai & T. Tochikura (1993) *ABB* **305**, 559.

Selected entries from *Methods in Enzymology* [vol, page(s)]:
General discussion: **179**, 479; **311**, 287; **312**, 196
Assay: **179**, 479, 489; **311**, 289
Other molecular properties: glycosphingolipid analysis, for, **230**, 385; isolation from (earthworm, **179**, 482; leech, **179**, 481);

oligosaccharide-transferring activity (acceptors in neoglycoconjugate synthesis, **242**, 156; acceptor specificity, **242**, 155; assay, **242**, 146); properties, **179**, 482, 484, 493; **242**, 150; **311**, 290; purification from leech, **242**, 148; purification from *Rhodococcus* sp. G-74-2, **179**, 490; release of oligosaccharides from glycosphingolipids, **230**, 292; sphingolipids, assay of, **312**, 196; synthesis of octyl-II$_3$ NeuAc-GgOse$_4$, in, **242**, 150

ENDOPEPTIDASE Clp

This serine endopeptidase [EC 3.4.21.92], also known as endopeptidase Ti, caseinolytic protease, and ClpAB, catalyzes peptide-bond hydrolysis in small peptides and proteins in the presence of ATP and magnesium ion. Activity assays are often carried out with α-casein serving as the substrate. The acronym "Clp" has multiple meanings: (a) indicating that the complex <u>cl</u>ipped proteins into a number of small peptides without generating free amino acids; (b) designating <u>c</u>aseino<u>l</u>ytic <u>p</u>rotease; and (c) standing for <u>c</u>haperone-<u>l</u>inked <u>p</u>rotease, since ClpA partially unfolds proteins to promote degradation. The proteolytic activity of the complex is found in ClpP. Solo ClpP will act on short peptides of ten or fewer amino acids. ClpAP will act on polypeptides and proteins. The enzyme prefers nonpolar aminoacyl residues in the P1 position but significant hydrolysis rates are observed with polar and charged residues as well. The final products are typically peptides with five to fifteen aminoacyl residues.

Short peptides (less than forty amino acids) are acted upon by the endopeptidase without hydrolysis of ATP (however, ATP promotes binding). Larger polypeptides and proteins are degraded with continuous ATP hydrolysis. ClpA has a basal ATPase activity.[1,2] Although the overall reaction yields ADP and orthophosphate, P-O-P bond scission supplies the energy to drive protein unfolding.[3] The overall enzyme-catalyzed reaction is of the energase type (*e.g.*, State$_1$ + ATP = State$_2$ + ADP + P$_i$, where two affinity-modulated conformational states of the protein substrate are indicated.[3] *See Energase*

[1] M. R. Maurizi, M. W. Thompson, S. K. Singh & S.-H. Kim (1994) *MIE* **244**, 314.
[2] S. Gottesman, S. Wickner & M. R. Maurizi (1997) *Genes Dev.* **11**, 815.
[3] D. L. Purich (2001) *TiBS* **26**, 417.

Selected entries from *Methods in Enzymology* [vol, page(s)]:
General discussion: **244**, 314
Assay: ClpA subunit (ATPase assay, **244**, 323); ClpP subunit (assay, **244**, 321, 325)
Other molecular properties: affinity of complex components, **244**, 320; ATPase activity, **244**, 314, 327; ATP hydrolysis activity,

stoichiometry, **244**, 327; biological role, **244**, 330; ClpA subunit (ATPase assay, **244**, 323; ATP binding, **244**, 318, 326; purification, **244**, 315; size, **244**, 318; solubility, **244**, 317; stability, **244**, 317; thiol dependence, **244**, 330); ClpP subunit (assay, **244**, 321, 325; cleavage site specificity, **244**, 324; ClpX complex activity, **244**, 331; purification, **244**, 318; self-association, **244**, 320; size, **244**, 320; stability, **244**, 320); family (ATP-binding domains, **244**, 54, 56; sequence homology, **244**, 56; subunit structure, **244**, 54); inhibitors (diisopropyl fluorophosphate, **244**, 329; peptide, **244**, 325); ion inhibition, **244**, 328; kinetic parameters, **244**, 325; magnesium dependence, **244**, 328; peptide bond cleavage, stoichiometry, **244**, 327; pH optimum, **244**, 329; substrates, **244**, 314, 324; substrate specificity (nucleotide, **244**, 326; peptide, **244**, 321, 323)

ENDOPEPTIDASE E, PANCREATIC

This serine endopeptidase [EC 3.4.21.70] is also known as protease E, and cholesterol-binding pancreatic proteinase, preferentially catalyzes peptide-bond hydrolysis in substrates possessing nonbulky aminoacyl residues in the P1 position of the substrate: for example, Ala–Xaa or Val–Xaa.[1] It will hydrolyze synthetic substrates of pancreatic elastase but will not hydrolyze elastin to any significant extent. The enzyme also binds cholesterol and deoxycholate.

The human enzyme has often been misleadingly referred to as human pancreatic elastase I, because the enzyme will not hydrolyze elastin.

[1]P. A. Mallory & J. Travis (1975) *Biochemistry* **14**, 722.

ENDOPEPTIDASE La

This broadly distributed serine endopeptidase [EC 3.4.21.53], originally isolated as an *Escherichia coli* heat-shock protein, is also called ATP-dependent serine proteinase, protease La, as well as Lon protease, since it was found to be encoded by the *lon* gene. Endopeptidase La catalyzes peptide-bond hydrolysis in relatively large proteins, such as globin, casein, and denatured serum albumin. Specificity studies suggest that endopeptidase La has a preference for substrates possessing hydrophobic residues in the P1 position. Synthetic substrates such as glutaryl-Ala-Ala-Phe-^-NHNapOMe and Suc-Phe-Leu-Phe-^-NHNapOMe, but not the ester analogues, have been used to assay the enzyme.

Typically, two ATP molecules are hydrolyzed per peptide bond cleaved. Certain protein substrates (for example, casein or denatured albumin) cooperatively stimulate the ATPase activity two- to fourfold; interestingly, other proteins have no effect on the ATPase activity (*e.g.*, native albumin or hemoglobin).[1-4] This enzyme has also been incorrectly termed a protein-activated ATPase. Although the overall reaction yields ADP and orthophosphate, P-O-P bond scission supplies the energy to drive unfolding of the protein substrate and/or expulsion of peptide-bond cleavage products.[5] The overall enzyme-catalyzed reaction is of the energase type (*e.g.*, $State_1 + ATP = State_2 + ADP + P_i$, where two affinity-modulated conformational states of the protein substrate are indicated.) Vanadate inhibits both peptide-bond hydrolysis and ATP cleavage. **See Energase**

Endopeptidase La has a high affinity for DNA. ssDNA activates the proteolytic activity whereas mRNA and tRNA have no effect.

[1]A. L. Goldberg (1992) *Eur. J. Biochem.* **203**, 9.
[2]A. L. Goldberg, R. P. Marshall, C. H. Chung & M. R. Maurizi (1994) *MIE* **244**, 350.
[3]S. Kuzela & A. L. Goldberg (1994) *MIE* **244**, 376.
[4]S. Gottesman (1996) *Ann. Rev. Genet.* **30**, 465.
[5]D. L. Purich (2001) *TiBS* **26**, 417.

Selected entries from *Methods in Enzymology* [vol, page(s)]:
General discussion: 244, 350, 376
Assay: protease activity, assay (amino group generation, **244**, 357; fluorogenic peptide cleavage, **244**, 357; radiolabeled casein, **244**, 356; reverse-phase HPLC, **244**, 358); yeast mitochondrial, **260**, 487, 493
Other molecular properties: activation (allosteric, by protein substrates, **244**, 373; ATP analogs, by, **244**, 365, 372); ADP release, **244**, 370, 374; ATP affinity, **244**, 366, 370; ATPase activity, **244**, 366 (colorimetric assay, **244**, 359; inhibitors, **244**, 370; magnesium dependence, **244**, 366; peptide effects, **244**, 372; pH optimum, **244**, 372; radioassay, **244**, 358; substrate specificity, **244**, 370); ATP binding (site, **244**, 363; stoichiometry, **244**, 370, 372); ATP-binding domains, **244**, 54, 56; ATP dependence, **244**, 350, 365; biological role, **244**, 330, 352; chaperone proteins, **244**, 374; cloning, **244**, 356; deg mutants, **244**, 353; DNA (binding affinity, **244**, 369; effect on activity, **244**, 369); endopeptidase La-like protein (ATP-binding domains, **244**, 54, 56); gene, **244**, 350; heat shock (enzyme expression response, **244**, 353; regulation of enzyme by proteins, **244**, 374); homology with mitochondrial ATP-dependent protease, **244**, 376, 383; inhibitors, **244**, 368, 375; magnesium dependence, **244**, 358; peptide substrate specificity, **244**, 367; protease activity, assay (amino group generation, **244**, 357; fluorogenic peptide cleavage, **244**, 357; radiolabeled casein, **244**, 356; reverse-phase HPLC, **244**, 358); proteolytic action, role of ATP, **244**, 371; purification, **244**, 359; regulation, **244**, 351; sequence homology among species, **244**, 352, 354, 363; size, **244**, 362; species distribution, **244**, 352; stability, **244**, 362; storage, **244**, 357, 362; structure, **244**, 54, 363, 365; substrates (nucleotide, **244**, 370; peptide, **244**, 350, 367); thiol dependence, **244**, 369; yeast mitochondrial (assay, **260**, 487, 493; ATP dependence, **260**, 486, 495; inhibitors, **260**, 486; purification [hydroxyapatite chromatography, **260**, 491; ion-exchange chromatography, **260**, 491; mitochondria preparation, **260**, 487, 490]; sequence homology between species, **260**, 487; substrate specificity, **260**, 493)

ENDOPEPTIDASE So

This serine endopeptidase [EC 3.4.21.67], also called *E. coli* cytoplasmic proteinase and protease So, catalyzes peptide-bond hydrolysis in such proteins as casein, denatured serum albumin, glucagon, and globin. This peptidase also acts on *N*-acetyl-(Ala)$_4$, but not *N*-acetyl-(Ala)$_3$, Bz-Tyr-OEt, or Ac-Phe-β-naphthyl ester.[1,2]

[1] A. L. Goldberg, K. H. S. Swamy, C. H. Chung & F. S. Larimore (1981) *MIE* **80**, 680.
[2] Y. S. Lee, S. C. Park, A. L. Goldberg & C. H. Chung (1988) *JBC* **263**, 6643.

Selected entries from *Methods in Enzymology* [**vol**, page(s)]:
General discussion: **80**, 680

ENDOPOLYPHOSPHATASE

This enzyme [EC 3.6.1.10], also known as polyphosphate depolymerase, metaphosphatase, and polyphosphatase, catalyzes the hydrolysis of polyphosphate to produce oligophosphate products containing four or five phosphoryl residues.[1–4] The enzyme also acts on polymetaphosphate.

See also *Exopolyphosphatase; Trimetaphosphatase*

[1] M. A. Nesmeyanova (2000) *Biochemistry (Mosc)* **65**, 309.
[2] K. D. Kumble & A. Kornberg (1996) *JBC* **271**, 27146.
[3] H. Malmgren (1952) *Acta Chem. Scand.* **6**, 16.
[4] B. Ingelman & H. Malmgren (1949) *Acta Chem. Scand.* **3**, 157.

Selected entries from *Methods in Enzymology* [**vol**, page(s)]:
General discussion: **2**, 577
Assay: **2**, 577

ENDORIBONUCLEASE III

This ribonuclease activity, previously classified as EC 3.1.4.24, is now a deleted EC entry.

ENDO-α-SIALIDASE

This enzyme [EC 3.2.1.129], also called endo-*N*-acetylneuraminidase, endo-*N*-acylneuraminidase, endoneuraminidase, endo N, and poly(α-2,8-sialosyl)-endo-*N*-acetylneuraminidase, catalyzes the random endohydrolysis of (2→8)-α-sialosyl linkages in oligo- or poly(sialic) acids.[1–3] The enzyme requires a minimum of five sialyl residues for activity.[3] **See also** *Anhydrosialidase; Exo-α-Sialidase*

[1] K. Kitajima, S. Inoue, Y. Inoue & F. A. Troy (1988) *JBC* **263**, 18269.
[2] J. Finne & P. H. Mäkelä (1985) *JBC* **260**, 1265.
[3] P. C. Hallenbeck, E. R. Vimr, F. Yu, B. Bassler & F. A. Troy (1987) *JBC* **262**, 3553.

Selected entries from *Methods in Enzymology* [**vol**, page(s)]:
Assay: **230**, 467, 481; bacteriophage-derived, **138**, 171
Other molecular properties: bacteriophage-derived (polysialosyl glycoprotein detection with, **138**, 183; properties, **138**, 177; purification, **138**, 174); effects on (cerebral membranes, **138**, 634; polysialic acid electrophoresis, **138**, 631); properties, **230**, 466; release of sialyl oligomers from glycoconjugates, **230**, 473

ENDOTHELIN-CONVERTING ENZYME 1

This zinc-dependent metalloproteinase [EC 3.4.24.71], abbreviated ECE-1, catalyzes endothelin 1 formation by cleaving the Trp21–Val22 bond in the precursor (the so-called big endothelin 1). Rat lung big endothelins-2 and -3 are alternative substrates. The narrow pH *versus* rate profile (optimal at pH 7.0) of this protease contrasts with the acidic optimum of endothelin-converting enzyme 2.[1] **See also** *Endothelin-Converting Enzyme 2*

[1] A. J. Turner & L. J. Murphy (1996) *Biochem. Pharmacol.* **51**, 91.

ENDOTHELIN-CONVERTING ENZYME 2

This metalloproteinase, ECE-2, catalyzes the conversion of big endothelins (*i.e.*, endothelin precursors) to endothelin. Bovine ECE-2 actually exhibits a strong preference for big endothelin-1, as compared to big endothelin-2 and big endothelin-3. ECE-2 has optimum activity near pH 5.5.[1] **See also** *Endothelin-Converting Enzyme 1*

[1] N. Emoto & M. Yanagisawa (1995) *JBC* **270**, 15262.

ENDOTHIAPEPSIN

This aspartic proteinase [EC 3.4.23.22], named for its source *Endothia parasitica*, catalyzes peptide-bond hydrolysis in proteins, exhibits a broad specificity resembling that of pepsin A. This hydrolase preferentially acts on substrates containing hydrophobic residues at positions P1 and P1′. The relative rates of hydrolysis of the insulin B chain are: Phe24–Phe25 > Tyr16–Leu17 > Gln4–His5 ≫ Leu17–Val18 > Asn3–Gln4. Endothiapepsin's ability to clot milk results from the hydrolysis of Ser104–Phe105 of κ-casein, whereas calf chymosin and other aspartic proteinases that clot milk hydrolyze the Phe105–Met106 bond.[1–3]

[1] D. C. Williams, J. R. Whitaker & P. V. Caldwell (1972) *ABB* **149**, 52.
[2] H. B. Drohse & B. Foltmann (1989) *BBA* **995**, 221.
[3] D. Bailey & J. B. Cooper (1994) *Protein Sci.* **3**, 2129.

Selected entries from *Methods in Enzymology* [**vol**, page(s)]:
General discussion: **19**, 373, 436, 447
Assay: **19**, 436
Other molecular properties: activators and inhibitors, **19**, 444; amino acid composition, **19**, 444; distribution, **19**, 445; β hairpin loop mutations, accommodation, **202**, 76, 81; physical and chemical properties, **19**, 442; potential use in cheese making, **19**, 375; production and purification, **19**, 439; specificity, **19**, 444; stability, **19**, 442

ENERGASE

This presently unofficial term treats mechanochemical and affinity-modifying enzyme reactions as a distinct enzyme class.[1] Rooted in the word "energy", the term reinforces the idea that these enzymes transduce the ΔG of nucleoside 5'-triphosphate hydrolysis (or other similar sources of Gibbs free energy) into some form of translation, rotation or twist, or as solute and/or electrogenic gradient. Under normal physiologic conditions, energases couple nucleotide hydrolysis to force production. Energases participate in what may be regarded as a long-lived catalytic cycle. Overt ATPase or GTPase activity only occurs when an energy-transducing step of an energase process is disrupted by nonphysiologic treatments or uncoupling agents. In this respect, the current Enzyme Commission designation of energase-type reactions as ATPases is erroneous.

For nearly 50 years, systematic classification of enzyme reactions has been based on the organic chemistry (and more recently the metallo-organic chemistry) of covalent bond formation, scission, and/or rearrangement. Contemporary biochemistry has shown that metabolism necessarily includes many other reactions characterized by changes in the strength of noncovalent bonding interactions. Molecular motors power contractile processes, intracellular organelle trafficking, and cell crawling, thereby forming a branch of metabolism. Likewise, the building up and tearing down of macromolecular and supramolecular structures depends on ATP-driven chaperonins and proteasomes.

A New Definition of Enzyme Catalysis Accounting for Substrate-like and Product-like Noncovalent States— In the face of so many instances where biological catalysis is not attended by changes in covalent bonding, Purich[1] offered the following new definition: *Enzymes are biological catalysts that enhance the rate of chemical bond making/breaking.* While appearing to be no more encompassing than existing definitions of enzyme catalysis, the crucial difference lies in the use of *chemical* bond in place of *covalent* bond. In his book "The Nature of The Chemical Bond," Linus Pauling[2] offered the following guiding comment: "We shall say that there is a chemical bond between two atoms or groups of atoms in case that the forces acting between them are such as to lead to the formation of an aggregate with sufficient stability to make it convenient for the chemist to consider it as an independent molecular species." Because many protein conformational states and numerous protein-ligand

complexes are sufficiently long-lived to exhibit chemically definable properties, their formation and/or transformation must be considered as chemical reactions. This new definition of enzyme catalysis even includes factors that catalyzed nucleotide exchange, as exemplified by profilin-catalyzed ATP exchange with actin-bound ADP, a process that is attended only by changes in noncovalent bonding interactions.[3,4] Other examples are the agents that catalyze noncovalent self-assembly of macromolecular and supramolecular structures. In this vein, catalysts of protein folding/refolding, cytoskeletal filament assembly, as well as chromatin condensation should be regarded as enzymes. ATP-driven nucleic acid helicases also transduce P-O-P bond-scission energy into the mechanical work of separating/rearranging complementary strands of double-helical nucleic acids.[5]

While recognizing that changes in a metabolite's position with respect to a membrane define substrate-like and product-like states, the Enzyme Commission continues to categorize transporters as ATPases. However, with modest tinkering, Pauling's prescient definition of a chemical bond can be extended to include the persistent, chemically definable position of a solute relative to the faces of a membrane. In this way, the proposed definition of enzyme catalysis also treats membrane transporters as specialized enzymes.

Stoichiometry of Energase Reactions— The present EC classification does not account for the noncovalent "work steps" occurring in many reactions presently categorized as ATPases. Failure to name the reactions in terms of the complete reaction, as well as the current practice of merely indicating ATP or GTP hydrolysis, deters wider appreciation of the inherent force-producing biophysics of these processes. That energases comprise a separate class becomes obvious, especially when one explicitly accounts for the noncovalent forms, written here as $State_1$ and $State_2$:

$$State_1 + ATP \text{ (or GTP)} = State_2 + ADP \text{ (or GDP)} + P_i$$

The relative abundance of these noncovalent states (*i.e.*, $[State_2]/[State_2]$) must likewise be indicated in the equilibrium constant ($K_{eq} = [State_2][(A/G)DP][P_i]/[State_1] \times [(A/G)TP]$). Ignoring these noncovalent substrate-like and product-like species is erroneous.

Energases that Modify Enzyme Performance—The free energy of nucleoside 5′-triphosphate hydrolysis can also produce substrate-like and product-like conformational states that alter an enzyme's ability to catalyze a reaction. Two notable examples are: (a) ATP sulfurylase[6] ($ATP + SO_4^{2-} = AMP$-Sulfate $+ PP_i$) which also transduces the energy of GTP hydrolysis to modulate both substrate binding and catalytic performance; and nicotinate phosphoribosyl-transferase (Nicotinate + PRPP = Nicotinate-D-ribonucleotide $+ PP_i$) which also hydrolyzes ATP for similar reasons.[7] Any change in substrate affinity alters the Michaelis constant, which alters one or more of the other parameters in the Haldane relationship (*i.e.*, $K_{eq} = V_f K_{mr}/V_r K_{mf}$). Eventually, these and related enzymes should be assigned two separate EC numbers (one for the overall chemical reaction, and another for the energase activity), much as other bifunctional enzymes are currently classified with two EC numbers.

Uncoupled NTPase Activity is a Nonphysiologic Phenomenon—As noted above, explicit ATPase or GTPase activity is only observed when energy-transducing steps within an energase process are disrupted by nonphysiologic treatments or uncoupling agents. A good example is the tubulin GTPase. Strict stoichiometric coupling of tubulin incorporation and GTP hydrolysis occurs under normal physiologic conditions, but is lost in the presence of the mitotic spindle poison known as colchicine.[3] Only in the presence of colchicine, which blocks microtubule self-assembly, does tubulin exhibit a enhanced capacity to hydrolyze GTP. Similar statements apply to the ability of other uncoupling agents that increase the hydrolase activities of ATP synthase and other ATP-dependent transporters.

[1] D. L. Purich (2001) *TiBS* **26**, 417.
[2] L. Pauling (1945) *The Nature of the Chemical Bond*, p. 3, Cornell University Press, Ithaca, New York.
[3] D. L. Purich & R. D. Allison (2000) *Handbook of Biochemical Kinetics*, Academic Press, New York.
[4] H. J. Kinosian, L. A. Selden, L. C. Gershman, & J. E. Estes (2000) *Biochemistry* **39**, 13176.
[5] P. H. von Hippel & E. Delagoutte (2001) *Cell* **104**, 177.
[6] T. S. Leyh (1999) *MIE* **308**, 48.
[7] C. T. Grubmeyer, J. W. Gross & M. Rajavel (1999) *MIE* **308**, 28.

2-ENOATE REDUCTASE

This FAD- and iron-sulfur-dependent oxidoreductase [EC 1.3.1.31] catalyzes the reaction of 2-butenoate with NADH to produce butanoate and NAD^+. Substrates include a wide range of alkyl and aryl α,β-unsaturated carboxylate anions: the best substrate examined was (*E*)-2-butenoate. The *Clostridium kluyveri* enzyme exhibits a ping pong Bi Bi kinetic mechanism.[1]

[1] M. Buhler & H. Simon (1982) *Hoppe Seyler's Z. Physiol. Chem.* **363**, 609.

Selected entries from *Methods in Enzymology* [vol, page(s)]:
General discussion: **136**, 302
Assay: **136**, 312
Other molecular properties: immobilization, **136**, 313; purification from *Clostridium* La 1, **136**, 309; repeated use after immobilization, **136**, 315; substrate specificity and kinetic data, **136**, 305

ENOLASE

Magnesium-dependent enolase [EC 4.2.1.11], also referred to as phosphopyruvate hydratase and 2-phosphoglycerate dehydratase, catalyzes the interconversion of 2-phospho-D-glycerate to phosphoenolpyruvate and water in a stepwise mechanism.[1-5] Other substrates include 3-phospho-D-erythronate. Investigations suggest that the reaction proceeds via a carbanion intermediate.[6]

The enzyme also catalyzes the relatively rapid exchange of the C2 proton with bulk solvent water. Proton abstraction and at least one additional step (perhaps loss of hydroxide ion or release of product) appear to limit the overall reaction rate.

[1] R. Jaenicke, H. Schurig, N. Beaucamp & R. Ostendorp (1996) *Adv. Protein Chem.* **48**, 181.
[2] J. J. Villafranca & T. Nowak (1992) *The Enzymes*, 3rd ed., **20**, 63.
[3] L. Leboida & B. Stec (1991) *Biochemistry* **30**, 2817.
[4] F. Wold (1971) *The Enzymes*, 3rd ed., **5**, 499.
[5] B. G. Malmström (1961) *The Enzymes*, 2nd ed., **5**, 471.
[6] S. R. Anderson, V. E. Anderson & J. R. Knowles (1994) *Biochemistry* **33**, 10545.

Selected entries from *Methods in Enzymology* [vol, page(s)]:
General discussion: 1, 427; 9, 670; 27, 264; 42, 323, 329, 335; 46, 23, 381, 382; 90, 479
Assay: 1, 427; 9, 670; chicken muscle, 90, 490; *Escherichia coli*, 42, 323; fish muscle, 42, 330; human muscle, 42, 335
Other molecular properties: affinity partitioning with dye ligands, **228**, 135; bifunctional aryl halides and, **25**, 637; 2,3-bisphosphoglycerate assay and, **6**, 484, 485; brewers' yeast, **1**, 427 (properties, **1**, 433; purification, **1**, 430); chicken muscle (assay, **90**, 490; isolation, **90**, 492); chiral methyl groups, synthesis of, **87**, 140, 142; conformation studies, **49**, 357, 358; electron spin resonance and, **6**, 913; elution of, **5**, 30; energy for catalysis, **308**, 21; equilibrium isotope effect, **64**, 110; *Escherichia coli*, from, **42**, 323 (activation and inhibition, **42**, 327; amino acid composition, **42**, 327; assay, **42**, 323; chromatography, **42**, 326; molecular weight, **42**, 327; preparation and purification, **42**, 324; properties, **42**, 327, 329); fish muscle, from, **42**, 329 (assay, **42**, 330; molecular weight, **42**, 334; preparation and purification, **42**, 329; properties, **42**, 334); gluconeogenic catalytic activity, **37**, 281; D-glycerate-2-[^{18}O]phosphorothioate, **87**, 220, 233; human muscle, from, **42**, 335 (assay, **42**, 335; molecular weight, **42**, 338; preparation and purification, **42**, 335; properties, **42**, 337, 338); inhibitor, **42**, 327; **63**, 399; inhibitor complex, dissociation constants and stoichiometry, **41**, 122; isolated nuclei and, **12A**, 445; magnesium ion and,

87, 394; marker for chromosome detection in somatic cell hybrids, as, 151, 181; membrane osmometry, 48, 75; mercuric ions and, 1, 85, 432; molecular forms, 9, 678; paper electrophoresis, 4, 28; pH effect, 87, 394; 3-phosphoglycerate, assay of, 63, 33; phosphoglycerate kinase, assay of, 63, 33; phosphoglycerate mutase assay and, 5, 236; photooxidation of, 25, 404, 405; preparation, 1, 430; 5, 237; pyruvate kinase, 87, 220, 233 (contamination, 63, 7); rabbit muscle, from (crystallization, 9, 677; magnesium requirements, 9, 679; molecular weight, 9, 678; preparation, 9, 676); site-specific reagent, 47, 483; thermophilic bacteria, from, properties, 42, 329; transition-state analogues, and active site-specific reagents, 41, 120; yeast, from, 1, 427 (magnesium requirement, 9, 679; molecular weight, 9, 678; preparation, 9, 673; properties, 1, 433; purification, 1, 430 reaction with *p,p'*-difluoro-*m,m'*-dinitrodiphenylsulfone, 11, 629); zone electrophoresis, 4, 16

ENOYL-[ACYL-CARRIER PROTEIN] REDUCTASES

The several oxidoreductases acting on enoyl derivatives of acyl-carrier proteins are distinguished on the basis of substrate specificity and coenzyme stereochemistry.[1–7]

Enoyl-[acyl-carrier protein] reductase (NADH) [EC 1.3.1.9], also referred to as enoyl-ACP reductase and NADH:enoyl acyl-carrier protein reductase, catalyzes the reaction of a *trans*-2,3-dehydroacyl-[acyl-carrier protein] with NADH to produce an acyl-[acyl-carrier protein] and NAD^+. The acyl chain length can range between four to sixteen carbon atoms.

Enoyl-[acyl-carrier protein] reductase (NADPH, B-specific) [EC 1.3.1.10] (also referred to as acyl-ACP dehydrogenase, NADPH 2-enoyl CoA reductase, enoyl acyl-carrier protein reductase, and enoyl-ACP reductase) catalyzes the reaction of a *trans*-2,3-dehydroacyl-[acyl-carrier protein] (*i.e.*, a *trans*-2-enoyl [acyl-carrier protein]) with NADPH to produce an acyl-[acyl-carrier protein] and $NADP^+$. The enoyl-acyl moiety of the substrate is a carbon chain of between four and sixteen carbon atoms. The yeast and the *Escherichia coli* enzymes are B-specific with respect to the coenzyme (that is, *pro*-4S). However, the yeast enzyme catalyzes an *anti* addition to the double bond whereas the *E. coli* enzyme catalyzes a *syn* addition.[3,4] Note that the early literature does not distinguish between the A-side and B-side NADPH-dependent reductases. **See also Fatty Acid Synthetase**

Enoyl-[acyl-carrier protein] reductase (NADPH, A-specific) [EC 1.3.1.39], also referred to as acyl-ACP dehydrogenase, catalyzes the reaction of a *trans*-2,3-dehydroacyl-[acyl-carrier protein] (*i.e.*, a *trans*-2-enoyl-[acyl-carrier protein]) with NADPH to produce an

acyl-[acyl-carrier protein] and $NADP^+$. The enzyme is A-side specific with respect to the coenzyme (that is, the *pro*-4R hydrogen of NADPH).[5–7]

[1] J. Browse & C. Somerville (1991) *Ann. Rev. Plant Physiol. Plant Mol. Biol.* **42**, 467.
[2] S. J. Wakil & J. K. Stoops (1983) *The Enzymes*, 3rd ed., **16**, 3.
[3] K. Saito, A. Kawaguchi, Y. Seyama, T. Yamakawa & S. Okuda (1981) *EJB* **116**, 581.
[4] B. Sedgwick & C. Morris (1980) *J. Chem. Soc. Chem. Commun.* **1980**, 96.
[5] R. E. Dugan, L. L. Slakey & J. W. Porter (1970) *JBC* **245**, 6312.
[6] G. F. Leanz & G. G. Hammes (1986) *Biochemistry* **25**, 5617.
[7] Y. Seyama, H. Otsuka, A. Kawaguchi & T. Yamakawa (1981) *J. Biochem.* **90**, 789.

Selected entries from *Methods in Enzymology* [vol, page(s)]:
General discussion: 14, 66
Other molecular properties: activity of fatty acid synthase, 71, 96; effect of pH, 14, 71; effect of thiol inhibitors, 14, 73; involvement in bacterial fatty acid synthesis, 14, 42; separation of NADH-specific and NADPH-specific enzymes, 14, 69; stereospecificity, 87, 115; substrate specificity, 14, 72

ENOYL-CoA HYDRATASE

This enzyme [EC 4.2.1.17] (also referred to as enoyl hydrase, crotonase, and unsaturated acyl-CoA hydratase) catalyzes the reversible addition of water to *trans*-2(or 3)-enoyl-CoA to produce (3S)-3-hydroxyacyl-CoA (*i.e.*, the L-stereoisomer).[1,2] The hydratase also acts on *cis* derivatives to generate (3R)-3-hydroxyacyl-CoA (the D-stereoisomer) and on pantetheine analogues, including *trans*-2-butenoyl-pantetheine. Kinetic isotope effect data suggest that the reaction mechanism is concerted.[3] **See also Long-Chain-Enoyl-CoA Hydratase; 3-Hydroxybutyryl-CoA Dehydratase**

[1] R. L. Hill & J. W. Teipel (1971) *The Enzymes*, 3rd ed., **5**, 539.
[2] J. R. Stern (1961) *The Enzymes*, 2nd ed., **5**, 511.
[3] B. J. Bahnson & V. E. Anderson (1991) *Biochemistry* **30**, 5894.

Selected entries from *Methods in Enzymology* [vol, page(s)]:
General discussion: 1, 559; 35, 136; 71, 247, 390, 403, 421
Assay: 1, 559; 35, 137; 71, 247, 248, 378, 403, 404; *Clostridium acetobutylicum*, 71, 421; peroxisomal, 148, 523
Other molecular properties: acyl dehydrogenases and, 5, 555; affinity labeling, 47, 422; bile acid biosynthesis, and, 15, 580; bromooctanoate, effect of, 72, 568; *Clostridium acetobutylicum*, from, 71, 421, 423 (assay, 71, 421; hydration of crotonylpantetheine, 71, 426; properties, 71, 425; substrate complementation, 71, 426); crotonyl-CoA assay and, 1, 561; effect of chain length on activity, 35, 149, 150; *Escherichia coli*, from 71, 421, 426 (properties, 71, 428; size heterogeneity, 71, 429; stability, 71, 428; substrate specificities, 71, 428); glyoxysomes, in, 31, 569; 52, 502; 72, 788; β-hydroxybutyryl-CoA racemase and, 5, 557; β-hydroxy-β-methylglutaryl coenzyme A cleavage enzyme and, 5, 902; induction by clofibrate and di(2-ethylhexyl)phthalate, 72, 507; inhibition by sulfhydryl reagents, 35, 150; inhibitors, 35, 149, 150; lactyl-CoA dehydrase preparation, present in, 9, 683; methacrylyl-CoA and, 5, 452; β-methylcrotonyl carboxylase and, 5, 900; β-oxidation, 35, 136; ox liver (properties, 1, 564; purification, 1, 562); *trans*-2-pentenoyl coenzyme A and, 1, 564; properties, 1, 564; purification, ox liver, 1, 562; rat liver, from, 71, 247

(effect of [deoxycholate, **71**, 251; Triton X-100, **71**, 250]; kinetics, **71**, 251; radioactive assay, **71**, 248; reversibility, **71**, 252; spectrophotometric assay, **71**, 247; substrate specificity, **71**, 251); regulation of β-oxidation of fatty acids, **35**, 149, 150; short-chain, from pig heart, **71**, 390 (activity per gram of tissue, **71**, 398; preparation of substrates, **71**, 391; properties, **71**, 396; substrate specificity, **71**, 397); stability, **35**, 138, 145; stereospecificity, **35**, 148; tiglyl-CoA and, **5**, 452; transition-state and multisubstrate analogues, **249**, 308

trans-2-ENOYL-CoA REDUCTASE (NADH)

This oxidoreductase [EC 1.3.1.44] catalyzes the reaction of a *trans*-2,3-didehydroacyl-CoA (*i.e.*, the *trans*-2-enoyl-CoA) with NADH to produce an acyl-CoA and NAD$^+$.[1-4] Enoyl-CoA substrates include crotonoyl-CoA, *trans*-2-octenoyl-CoA, *trans*-2-decenoyl-CoA, *trans*-2-dodecenoyl-CoA, and *trans*-2-hexadecenoyl-CoA. The rat liver enzyme can use NADPH as the coenzyme substrate, depending on the chain length of the enoyl-CoA.

[1] H. Inui, K. Miyatake, Y. Nakano & S. Kitaoka (1986) *J. Biochem.* **100**, 995.
[2] T. Shimakata & T. Kasaka (1981) *J. Biochem.* **89**, 1075.
[3] T. Shimakata, Y. Fujita & T. Kusaka (1980) *J. Biochem.* **88**, 1051.
[4] M. N. Nagi, R. M. Prasad, L. Cook & D. L. Cinti (1983) *ABB* **226**, 50.

cis-2-ENOYL-CoA REDUCTASE (NADPH)

This oxidoreductase [EC 1.3.1.37] catalyzes the reaction of a *cis*-2,3-dehydroacyl-CoA (*i.e.*, a *cis*-2-enoyl-CoA) with NADPH to produce an acyl-CoA and NADP$^+$.[1,2] The *Escherichia coli* enzyme catalyzes an *anti* addition to the double bond.[1] The enzyme is distinct from *trans*-2-enoyl-CoA reductase (NADPH) [EC 1.3.1.38], which acts on both *cis* and *trans* derivatives.

[1] M. Mizugaki, T. Unuma, T. Nishimaki & T. Shiraishi, A. Kawaguchi, K. Saito, S. Okuda & H. Yamanaka (1982) *Chem. Pharm. Bull.* **30**, 2155.
[2] M. Mizugaki, T. Unuma & H. Yamanaka (1979) *Chem. Pharm. Bull.* **27**, 2334.

trans-2-ENOYL-CoA REDUCTASE (NADPH)

This oxidoreductase [EC 1.3.1.38] catalyzes the reaction of a *trans*-2,3-dehydroacyl-CoA (*i.e.*, a *trans*-2-enoyl-CoA) with NADPH to produce an acyl-CoA and NADP$^+$.[1-3]

[1] M. Mizugaki, T. Nishimaki, T. Shiraishi, A. Kawaguchi, S. Okuda & H. Yamanaka (1982) *J. Biochem.* **92**, 1649.
[2] T. Nishimaki, H. Yamanaka & M. Mizugaki (1984) *J. Biochem.* **95**, 1315.
[3] M. Mizugaki, T. Nishimaki, T. Shiraishi & H. Yamanaka (1982) *Chem. Pharm. Bull.* **30**, 2503.

ENTEROPEPTIDASE

This serine endopeptidase [EC 3.4.21.9], also known as enterokinase, forms trypsin by catalyzing the cleavage of the Lys^6–Ile^7 bond in trypsinogen. While the enzyme is fairly specific with respect to side-chains occupying positions P1-P5, there is very little specificity with respect to the C-terminal side of the scissile bond. The preferred substrate is the Asp-Asp-Asp-Asp-Lys-∇-Ile sequence (where ∇ indicates the cleavage site) occurring in trypsinogens from many sources.[1]

[1] A. Light & H. Janska (1989) *Trends Biochem. Sci.* **14**, 110.

Selected entries from *Methods in Enzymology* [vol, page(s)]:
General discussion: 2, 31; **34**, 449

ENTOMOPHTHORA COLLAGENOLYTIC PROTEINASE

This endopeptidase activity, formerly classified as EC 3.4.21.33, is now a deleted EC entry.

ENVELYSIN

This calcium- and zinc-dependent metalloproteinase [EC 3.4.24.12], also called sea urchin-hatching proteinase and hatching enzyme, catalyzes peptide-bond hydrolysis in proteins forming the fertilization envelope, preferring substrates with bulky, hydrophobic aminoacyl residues (for example, Leu, Ile, Phe, Tyr) at the P1′ position.[1,2] The oxidized insulin B chain is hydrolyzed at Gln4–His5, Ala14–Leu15, Tyr16–Leu17, and Phe25–Tyr26. *See also Choriolysin L; Choriolysin H*

[1] D. Barrett & B. F. Edwards (1976) *MIE* **45**, 354.
[2] K. Nomura, H. Tanaka, Y. Kikkawa, M. Yamaguchi & N. Suzuki (1991) *Biochemistry* **30**, 6115.

Selected entries from *Methods in Enzymology* [vol, page(s)]:
General discussion: 45, 354, 371
Other molecular properties: casein, digestion of modified, **45**, 356; chromatography, **45**, 364; dialysis, **45**, 363; fertilization envelope dissolution, **45**, 355; gametes shedding, **45**, 360; inhibitors, **45**, 368; kinetics, **45**, 367; metals, dependence, **45**, 367; molecular weight, **45**, 367; pH, **45**, 367; polypeptide chain structure, **248**, 204; preparation, **29**, 65; **45**, 359 (crude enzyme, **45**, 360; incubating embryos, **45**, 362; yield, **45**, 363); properties, **45**, 366; **248**, 192; purification, **45**, 363 (urea treatment, **45**, 363); purity, **45**, 367; sea urchin egg protease (activators, **45**, 352; assay, **45**, 343 [bioassays, **45**, 343; chemical, **45**, 345]; esterase activity, **45**, 345, 346 sperm receptor hydrolase, **45**, 344; vitelline delaminase, **45**, 343]; cortical granule, in, **45**, 343 [exudate preparation, **45**, 347]; distribution, **45**, 353; gametes, shedding, **45**, 346; inhibitors, **45**, 352; isoelectric focusing, **45**, 349; isoelectric precipitation, **45**, 347; kinetics, **45**, 353; preparation, **45**, 346; properties, **45**, 351; purification, **45**, 346; specificity, **45**, 353; stability, **45**, 351); specificity, **45**, 369; stability, **45**, 366; temperature, **45**, 367

ENZYME-THIOL TRANSHYDROGENASE (GLUTATHIONE DISULFIDE)

This oxidoreductase [EC 1.8.4.7], also referred to as glutathione-dependent thiol:disulfide oxidoreductase, catalyzes the reversible reaction of xanthine dehydrogenase [EC 1.1.1.204] with glutathione disulfide to produce xanthine oxidase [EC 1.1.3.22] and glutathione.[1] The transhydrogenase will also act on the disulfide bond of ricin.

[1] M. G. Battelli & E. Lorenzoni (1982) *BJ* **207**, 133.

EPHEDRINE DEHYDROGENASE

ephedrine

This oxidoreductase [EC 1.5.1.18] catalyzes the reaction of (–)-ephedrine with NAD^+ to produce (*R*)-2-methylimino-1-phenylpropan-1-ol and NADH.[1] The product immediately hydrolyzes to methylamine and 1-hydroxy-1-phenylpropan-2-one. Other substrates include (−)-sympatol, (+)-pseudoephedrine, and (+)-norephedrine.

[1] E. Klamann & F. Lingens (1980) *Z. Naturforsch.* **35c**, 80.

EPIDERMAL GROWTH FACTOR-BINDING PROTEIN

This poorly named serine endopeptidase of the kallikrein family, is often referred to as high-molecular-weight epidermal growth factor (HMW-EGF), epidermal growth factor urogastrone, and mGK-9. The enzyme, which proteolytically processes the immediate precursor to produce EGF and also acts on mouse L-kininogen, can be assayed with the following synthetic substrates: Tos-Arg-OMe, Bz-Arg-OEt, and D-Val-Leu-Arg-NHPhNO₂.[1,2]

[1] P. Frey, R. Forand, T. Maciag & E. M. Shooter (1979) *PNAS* **76**, 6294.

[2] Y. Hirata & D. N. Orth (1979) *J. Clin. Endocrinol. Metab.* **49**, 481.

EPOXIDE HYDROLASE

This enzyme [EC 3.3.2.3], also known as epoxide hydratase and arene-oxide hydratase, catalyzes the hydrolysis of various epoxides and arene oxides to yield the corresponding glycol.[1,2] **See also** *specific enzyme*

[1] R. N. Armstrong (1987) *Crit. Rev. Biochem.* **22**, 39.

[2] F. P. Guengerich (1982) *Rev. Biochem. Toxicol.* **4**, 5.

Selected entries from *Methods in Enzymology* [**vol**. page(s)]:

General discussion: 52, 193; 77, 344

Assay: 52, 193, 194, 416; 77, 345; vertebrate and lepidopteran, assays (with cholesterol epoxide, 111, 309; with juvenile hormone, 111, 305; with *trans*- and *cis*-stilbene oxide, 111, 308)

Other molecular properties: activators and inhibitors, 52, 200; affinity chromatography, 77, 349; *Ah* locus, 52, 232; alternate activity, 52, 416; benzo[*a*]pyrene and benzo[*a*]pyrene epoxide metabolism, 52, 284, 416 (specificity, alternate activity, 52, 416); biosynthesis and processing in endoplasmic reticulum membrane, 96, 537; contaminant of cytochrome P450_LM preparations, 52, 116, 121; cytosolic, 77, 344; 16α,17α-epoxyestra-1,3,5(10)-trien-3-ol hydrolase, 15, 725, 726; hepatic (purification from cytosol, 187, 328; subcellular distribution, 187, 327); hydrolysis of 2,2-dimethyloxirane, mechanistic studies, 177, 387; induction, 77, 346; kinetics, 187, 331; microsomal, 77, 344 (purification, 74, 264); molecular weight, 52, 199; 77, 349; purification, 52, 195; 77, 347; purity, 52, 199; size, 77, 349; solubilization, 77, 347; stability, 52, 199, 417; 77, 349; styrene oxide hydrase (alternate activity, 52, 416); substrate specificity, 52, 199, 200; 187, 327

EPOXYFARNESOATE METHYLTRANSFERASE

This transferase catalyzes the reaction of *S*-adenosyl-L-methionine with 7-ethyl-9-(3-ethyl-3-methyloxiranyl)-3-methyl-2,6-nonadienoate to produce *S*-adenosyl-L-homocysteine and 7-ethyl-9-(3-ethyl-3-methyloxiranyl)-3-methyl-2,6-nonadienoic acid methyl ester, also known as juvenile hormone C-18. Other substrates include 7-methyl-9-(3-ethyl-3-methyloxiranyl)-3-methyl-2,6-nonadienoate, in this case producing juvenile hormone C-17.

Selected entries from *Methods in Enzymology* [**vol**, page(s)]:

Other molecular properties: *Hyalophora cecropia* (preparation, 111, 541; product isolation and analysis, 111, 543)

trans-EPOXYSUCCINATE HYDROLASE

This enzyme [EC 3.3.2.4], also known as *trans*-epoxysuccinate hydratase and tartrate epoxidase, catalyzes the hydrolysis of both stereoisomers of *trans*-2,3-epoxysuccinate to produce *meso*-tartrate.[1,2]

[1] W. B. Jakoby & T. A. Fjellstedt (1972) *The Enzymes*, 3rd ed., **7**, 199.

[2] R. H. Allen & W. B. Jakoby (1969) *JBC* **244**, 2078.

EPRALYSIN

This extracellular metalloproteinase, also known as AprX, catalyzes peptide-bond hydrolysis in histones and casein. The best synthetic substrate reported is *t*-butyloxycarbonyl-Arg-Val-Arg-Arg-^-4-methylcoumaryl-7-amide.[1]

[1] H. J. Kim, Y. Tamanoue, G. H. Jeohn, A. Iwamatsu, A. Yokota, Y. T. Kim, T. Takahashi & K. Takahashi (1997) *J. Biochem.* **121**, 82.

EQUILIN DEHYDROGENASE

This oxidoreductase catalyzes the reaction of equilin (*i.e.*, 3-hydroxy-$\Delta^{1,3,5(10),7}$-estratetraen-17-one) with NAD(P)⁺ to produce equilenin (*i.e.*, 3-hydroxy-$\Delta^{1,3,5(10),6,8}$-estrapentaen-17-one), NAD(P)H, and a proton.

Selected entries from *Methods in Enzymology* [**vol**, page(s)]:

Assay: 15, 720

ERGOSTEROL 5,8-DIOXYGENASE

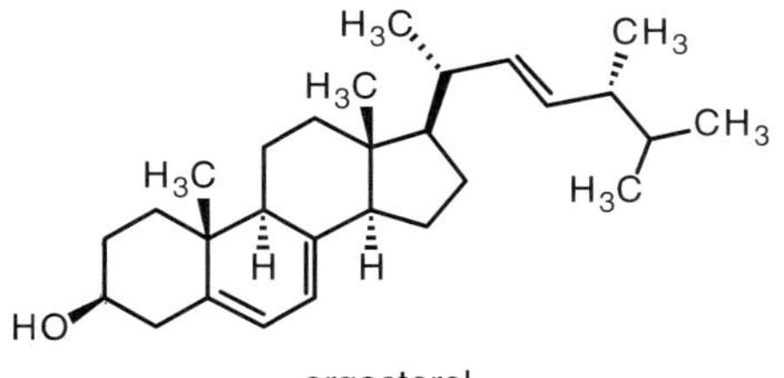

ergosterol

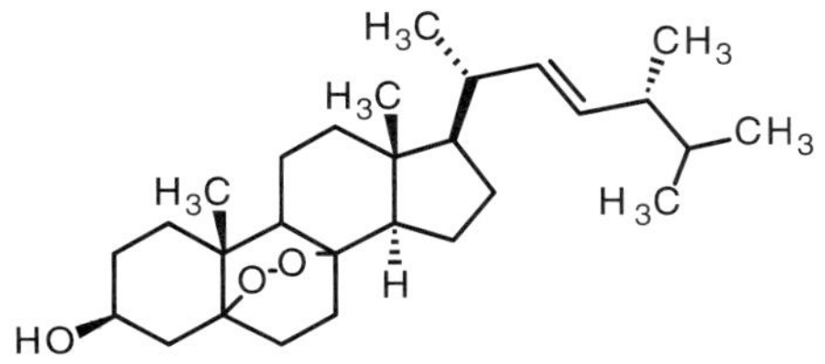

ergosterol 5,8-epidioxide

This oxidoreductase catalyzes the reaction of ergosterol with dioxygen to produce ergosterol 5,8-epidioxide.[1]

[1] J. D. White, D. W. Perkins & S. I. Taylor (1973) *Bioorg. Chem.* **2**, 163.

Selected entries from *Methods in Enzymology* [vol, page(s)]:
General discussion: **52**, 19

ERGOTHIONASE

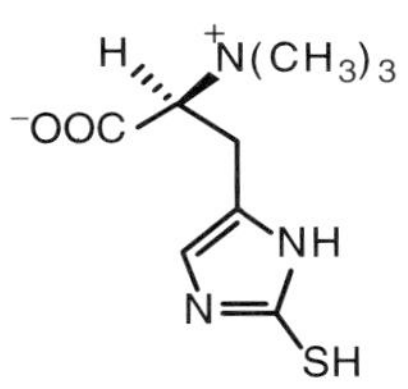

L-ergothioneine

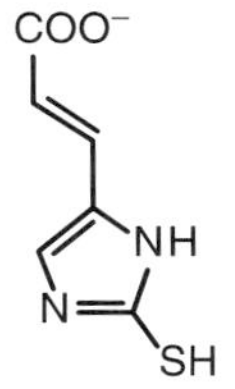

2-thiolurocanate

This lyase catalyzes the conversion of L-ergothioneine to 2-thiolurocanate and trimethylamine.

Selected entries from *Methods in Enzymology* [vol, page(s)]:
General discussion: **17B**, 105
Assay: **17B**, 105
Other molecular properties: induction in *Escherichia coli*, **17B**, 106; properties, **17B**, 107; purification, **17B**, 106

ERYTHRITOL KINASE

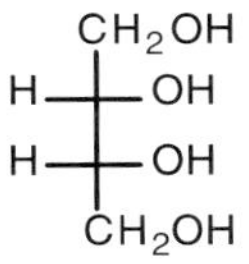

erythritol

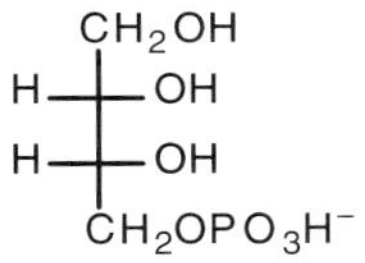

D-erythritol 4-phosphate

This phosphotransferase [EC 2.7.1.27] catalyzes the reaction of ATP with erythritol, a *meso* compound, to produce ADP and D-erythritol 4-phosphate (also known as L-erythritol 1-phosphate).[1]

[1] D. Holten & H. J. Fromm (1961) *JBC* **236**, 2581.

ERYTHROMYCIN C O-METHYLTRANSFERASE

This methyltransferase, also called *S*-adenosylmethionine: erythromycin C O-methyltransferase, catalyzes the reaction of *S*-adenosyl-L-methionine with erythromycin C to produce *S*-adenosyl-L-homocysteine and erythromycin A.[1]

[1] J. W. Corcoran (1975) *MIE* **43**, 487.

Selected entries from *Methods in Enzymology* [vol, page(s)]:
General discussion: **43**, 487
Assay: **43**, 488
Other molecular properties: *S*-adenosylmethionine, **43**, 490, 491; ammonium sulfate precipitation, **43**, 492, 494; cell-free extract, **43**, 492; cellular location, **43**, 495, 496; L-cladinose moiety, radioactivity, **43**, 489; erythromycin A (derivatization, **43**, 494; measurement of formation from erythromycin C, **43**, 489; recrystallization, **43**, 490); erythromycin C and spiroketal (6→9:12→9), **43**, 490; incubation, **43**, 488, 489; inhibitors, **43**, 496; kinetic properties, **43**, 496, 497; partition chromatography, **43**, 492, 493; pH effect, **43**, 496; purification, **43**, 491, 492; radioactivity measurement, **43**, 495; scintillation system, **43**, 495; specificity, **43**, 497; *Streptomyces erythreus* preparation, **43**, 491; temperature dependence, **43**, 496; thin-layer chromatography, **43**, 493, 494; transmethylase reaction product, **43**, 495; transmethylase reaction reversibility, **43**, 495

ERYTHRONATE-4-PHOSPHATE 2-DEHYDROGENASE

This oxidoreductase catalyzes the reaction of D-erythronate 4-phosphate (*i.e.*, the aldonic acid of D-erythrose 4-phosphate) with NAD^+ to produce D-erythrulonate 4-phosphate and NADH.[1]

[1] S. Kochhar, P. E. Hunziker, P. Leong-Morgenthaler & H. Hottinger (1992) *BBRC* **184**, 60.

ERYTHRONOLIDE SYNTHASE

This enzyme [EC 2.3.1.94], also known as erythronolide condensing enzyme, reportedly catalyzes the reaction of six molecules of malonyl-CoA with propionyl-CoA to produce seven coenzyme A molecules and 6-deoxyerythronolide B, an intermediate in the biosynthetic pathway leading to erythromycin antibiotics.[1]

[1] G. Roberts & P. L. Leadly (1984) *Biochem. Soc. Trans.* **12**, 642.

ERYTHROSE ISOMERASE

This isomerase activity, originally classified EC 5.3.1.2, is now a deleted Enzyme Commission entry.

ERYTHRULOSE REDUCTASE

This oxidoreductase [EC 1.1.1.162] catalyzes the reaction of D-erythrulose with NADPH to produce D-threitol and NADP$^+$. NADH can also serve as the reducing agent, albeit more slowly. Later studies[2] indicated that the original investigation[1] misidentified the product as erythritol. Other substrates include diketones, including diacetyl.

[1]K. Uehara, T. Tanimoto & H. Sato (1974) *J. Biochem.* **75**, 333.
[2]M. Maeda, S. Hosomi, T. Mizoguchi & T. Nishihara (1998) *J. Biochem.* **123**, 602.

Selected entries from *Methods in Enzymology* [vol, page(s)]:
General discussion: 89, 232
Assay: bovine liver, 89, 232
Other molecular properties: bovine liver (assay, **89**, 232; properties, **89**, 236; purification, **89**, 233)

ESTRADIOL 17α-DEHYDROGENASE

estradiol-17α   estrone

This oxidoreductase [EC 1.1.1.148] catalyzes the reaction of estradiol-17α with NAD(P)$^+$ to produce estrone and NAD(P)H.[1–3]

[1]L. H. LaRhee & J. C. Warren (1984) *Biochemistry* **23**, 486.
[2]J. Johnston & A. G. C. Renwick (1984) *BJ* **222**, 761.
[3]S. Hasnain & D. G. Williamson (1977) *BJ* **161**, 279 and (1975) **147**, 457.

Selected entries from *Methods in Enzymology* [vol, page(s)]:
Other molecular properties: staining on gels, **22**, 594; stereospecificity, **87**, 111

ESTRADIOL 17β-DEHYDROGENASE

This oxidoreductase [EC 1.1.1.62], also referred to as 17β-hydroxysteroid dehydrogenase, oestradiol 17β-dehydrogenase, and estrogen 17-oxidoreductase, catalyzes the reversible reaction of 17β-estradiol with NAD(P)$^+$ to produce estrone and NAD(P)H (with B-side stereospecificity).

Other substrates include testosterone, 3-*O*-methyl-17β-estradiol, and etiocholanediol. Structural studies indicate that seryl, tyrosyl, and lysyl residues are all critical for hydride transfer.[1]

17β-estradiol   estrone

(S)-20-hydroxypregn-4-en-3-one   progesterone

The enzyme also catalyzes the oxidation of the 20α-hydroxyl group (*i.e.*, (S)-20-hydroxy) on certain steroids, including (S)-20-hydroxypregn-4-en-3-one to produce progesterone. ***See also*** *3α(or 17β)-Hydroxysteroid Dehydrogenase; Testosterone 17β-Dehydrogenase; 3(or 17β)-Hydroxysteroid Dehydrogenase*

[1]T. Puranen, M. Poutanen, D. Ghosh, P. Vihko & R. Vihko (1997) *Mol. Endocrinol.* **11**, 77.

Selected entries from *Methods in Enzymology* [vol, page(s)]:
General discussion: 5, 523; **15**, 746; **27**, 66; **34**, 3, 4, 552, 555, 558, 564; **46**, 145, 148, 150, 451, 456, 457; 20α-dehydrogenase activity, **15**, 638
Assay: 5, 523; **15**, 747; 20α-dehydrogenase activity, **15**, 638
Other molecular properties: affinity labeling, **87**, 478; cold inactivation, **15**, 751; dehydrogenation assay, **15**, 747; interleukin 3 assay, in, **116**, 541; isotope exchange, **64**, 8; ovarian, purification of, **15**, 638; properties, **5**, 525; **15**, 751 (20α-dehydrogenase activity, **15**, 641); purification, human placenta, **5**, 524; **15**, 748 (20α-dehydrogenase activity, **15**, 639); staining on gels, **22**, 594; stereospecificity, **87**, 109; transhydrogenation assay, **15**, 747

ESTRADIOL 6β-MONOOXYGENASE

estradiol-17β   6β-hydroxyestradiol-17β

This oxidoreductase [EC 1.14.99.11], also referred to as estradiol 6β-hydroxylase, catalyzes the reaction of estradiol-17β with dioxygen and an electron donor substrate (AH$_2$) to produce 6β-hydroxyestradiol-17β, water, and the oxidized donor (A).[1]

[1]G. C. Müller & G. Rumney (1957) *JACS* **79**, 1004.

ESTRADIOL RECEPTOR-KINASE

This protein kinase catalyzes the reaction of ATP with the estradiol receptor to produce the phosphorylated receptor and ADP.

Selected entries from *Methods in Enzymology* [vol, page(s)]:
Assay: 139, 735

ESTROGEN ETHER-CLEAVING ENZYME

This oxidoreductase catalyzes the reaction of 2-methoxy-17β-estradiol with NADPH and dioxygen to produce 2-hydroxy-17β-estradiol, NADP+, formaldehyde, and water.[1] Other substrates include 17β-estradiol 3-methyl ether.

2-methoxy-17β-estradiol   2-hydroxy-17β-estradiol

Demethylation of 2-methoxyestrogens was fivefold faster than that of 4-methoxy-analogues in hamster liver microsomes.[2] Intestinal bacteria have also been observed to catalyze demethylation.[3]

[1] H. Breuer, M. Knuppen, D. Gross & C. Mittermayer (1964) *Acta Endocrinol.* 46, 361.
[2] B. T. Zhu, E. N. Evaristus, S. K. Antoniak, S. F. Sarabia, M. J. Ricci & J. G. Liehr (1996) *Toxicol. Appl. Pharmacol.* 136, 186.
[3] M. Axelson & J. Sjovall (1983) *BBA* 751, 162.

Selected entries from *Methods in Enzymology* [vol, page(s)]:
General discussion: 15, 723, 724
Assay: 15, 723
Other molecular properties: properties, 15, 724

ESTRONE SULFOTRANSFERASE

This transferase [EC 2.8.2.4] catalyzes the reaction of 3′-phosphoadenylylsulfate with estrone to produce adenosine 3′,5′-bisphosphate and estrone 3-sulfate.[1,2] Other substrates include 17β-estradiol, 17-deoxyestrone, estriol, and 16-epiestriol. The human enzyme has a random Bi Bi kinetic mechanism with two abortive complexes.[3]

[1] B. Chatterjee, C. S. Song, J. M. Kim & A. K. Roy (1994) *Chem. Biol. Interact.* 92, 273.
[2] J. B. Adams, R. K. Ellyard & J. Low (1974) *BBA* 370, 160.
[3] H. Zhang, O. Varlamova, F. M. Vargas, C. N. Falany & T. S. Leyh (1998) *JBC* 273, 10888.

Selected entries from *Methods in Enzymology* [vol, page(s)]:
General discussion: 15, 732, 733
Assay: 15, 732, 734

Other molecular properties: properties, 15, 733, 735; purification procedure, 15, 732, 734

ETHANOLAMINE AMMONIA-LYASE

This orange-colored, cobalamin-dependent lyase [EC 4.3.1.7], also known as ethanolamine deaminase, catalyzes the conversion of ethanolamine to acetaldehyde and ammonia.[1-4]

This reaction proceeds via cobalt(II)-radical intermediates and is accompanied by the migration of a hydrogen atom from the carbinol carbon of ethanolamine to the methyl carbon of acetaldehyde.[3] Coenzyme-mediated abstraction of a hydrogen followed by loss of a proton from the hydroxyl group results in the formation of an anion radical intermediate. Elimination of ammonia produces the acetaldehyde radical. Addition of ammonia and transfer of a hydrogen from the coenzyme generates the 1-aminoethanol. This intermediate releases ammonia to produce the final product.

[1] R. J. O'Brien, J. A. Fox, M. G. Kopczynski & B. M. Babior (1985) *JBC* 260, 16131.
[2] B. M. Babior (1982) in B_{12} (D. Dolphin, ed.) 2, p. 263, Wiley, New York.
[3] T. C. Stadtman (1972) *The Enzymes*, 3rd ed., 6, 539.
[4] T. T. Harkins & C. B. Grissom (1994) *Science* 263, 958.

Selected entries from *Methods in Enzymology* [vol, page(s)]:
General discussion: 17B, 818; 281, 235
Assay: 17B, 818
Other molecular properties: activators, 17B, 822; color, 17B, 823; inhibition, 17B, 822; 63, 399; properties, 17B, 821; purification from *Clostridium* sp., 17B, 820; specificity, 17B, 822; stereochemistry, 87, 153

ETHANOLAMINE KINASE

This phosphotransferase [EC 2.7.1.82] catalyzes the ATP-dependent phosphorylation of ethanolamine to produce *O*-phosphoethanolamine and ADP.[1]

ethanolamine   *O*-phosphoethanolamine

Many isozymes of this kinase catalyze the phosphorylation of choline kinase. ***See also Choline Kinase***

[1] J. D. Esko & C. R. H. Raetz (1983) *The Enzymes*, 3rd ed., 16, 207.

Selected entries from *Methods in Enzymology* [vol, page(s)]:
General discussion: 209, 121, 134, 147
Assay: cerebral, radiochemical assay, 209, 147; hepatic, radiochemical assay, 209, 136; renal, radiochemical assay, 209, 123
Other molecular properties: renal (properties, 209, 128; purification, 209, 124; radiochemical assay, 209, 123; subcellular localization, 209, 128)

ETHANOLAMINE OXIDASE

This oxidoreductase [EC 1.4.3.8] catalyzes the reaction of ethanolamine with water and dioxygen to produce glycolaldehyde, ammonia, and hydrogen peroxide.[1,2] Other substrates include 3-amino-1-propanol and 1-amino-2-propanol.

[1] S. A. Narrod & W. B. Jakoby (1964) JBC 239, 2189.
[2] A. P. Kulkarni & E. Hodgson (1973) Comp. Biochem. Physiol. 44B, 407.

Selected entries from Methods in Enzymology [vol, page(s)]:
General discussion: 9, 354; 52, 18; 56, 474
Assay: 9, 354

ETHANOLAMINE-PHOSPHATE CYTIDYLYLTRANSFERASE

This nucleotidyltransferase [EC 2.7.7.14], also referred to as phosphorylethanolamine transferase, catalyzes the reaction of CTP with ethanolamine phosphate to produce CDP-ethanolamine and pyrophosphate (or, diphosphate).[1,2] The rat liver enzyme has an ordered Bi Bi kinetic mechanism.[2]

[1] B. A. Bladergroen & L. M. van Golde (1997) BBA 1348, 91.
[2] R. Sundler (1975) JBC 250, 8585.

Selected entries from Methods in Enzymology [vol, page(s)]:
General discussion: 14, 121; 209, 258
Assay: 14, 121 (alternate assay, 14, 122; contamination with other enzymes in crude tissue preparations, 14, 123; phospholipid addition for choline enzyme, 14, 123, 124); 209, 258
Other molecular properties: activation on storage, 14, 125; phospholipid activation of choline enzyme from liver, 14, 123, 124; preparation of enzymes, 14, 123; preparation of labeled CDPethanolamine, 14, 684; properties, 14, 124; 209, 263; purification from rat liver, 209, 260; substrate specificity, 14, 124

ETHANOLAMINE-PHOSPHATE PHOSPHO-LYASE

This pyridoxal-phosphate-dependent enzyme [EC 4.2.3.2; previously classified as EC 4.2.99.7] catalyzes the reaction of ethanolamine phosphate with water to produce acetaldehyde, ammonia, and orthophosphate.[1–3] Other substrates include D-(or L-)-1-aminopropan-2-ol O-phosphate.

[1] H. L. Fleshood & H. C. Pitot (1970) JBC 245, 4414.
[2] A. Jones, A. Faulkner & J. M. Turner (1973) BJ 134, 959.
[3] A. Faulkner & J. M. Turner (1974) Biochem. Soc. Trans. 2, 133.

ETHANOLAMINEPHOSPHOTRANSFERASE

This enzyme [EC 2.7.8.1] converts CDP-ethanolamine and a 1,2-diacylglycerol into CMP and phosphatidylethanolamine.[1] The hamster liver enzyme also acts on CDP-choline.

[1] C. R. McMaster & R. M. Bell (1997) BBA 1348, 117.

Selected entries from Methods in Enzymology [vol, page(s)]:
General discussion: 71, 536; 209, 272

Assay: castor bean endosperm, 71, 603; myocardial, 209, 422; plant endoplasmic reticulum, in, 148, 588; yeast, 209, 273
Other molecular properties: castor bean endosperm, from, 71, 603 (assay, 71, 603; inhibitors, 71, 604; intracellular distribution, 71, 604; metal ion specificity, 71, 604; properties, 71, 604); plant endoplasmic reticulum, in (assay, 148, 588; properties, 148, 588); rat liver, from, 71, 536 (activators and inhibitors, 71, 545; acyl chain length specificity, 71, 544; localization in microsomal membranes, 71, 546; membrane-bound substrates, 71, 539; phospholipase A2 treatment of membrane-bound enzyme, 71, 545; reversibility, 71, 544; separation from cholinephosphotransferase, 71, 545; solubilization, 71, 541; substrate specificity, 71, 542)

2-ETHYLMALATE SYNTHASE

This enzyme [EC 4.1.3.33] catalyzes the reversible reaction of acetyl-CoA with water and α-ketobutyrate (or, 2-oxobutanoate) to produce (R)-2-ethylmalate and coenzyme A.[1,2] The enzyme also catalyzes the reaction of acetyl-CoA with water and α-ketovalerate to produce (R)-2-(n-propyl)malate.

[1] M. Strassman & L. N. Ceci (1967) ABB 119, 420.
[2] R. Rabin, I. I. Salamon, A. S. Bleiweis, J. Carlin & S. J. Ajl (1968) Biochemistry 7, 377.

3-ETHYLMALATE SYNTHASE

This enzyme [EC 4.1.3.10], also known as 2-ethyl-3-hydroxybutanedioate synthase, catalyzes the reaction of butanoyl-CoA with water and glyoxylate to produce 3-ethylmalate and coenzyme A.[1,2]

[1] R. Rabin, H. C. Reeves & S. J. Ajl (1963) J. Bacteriol. 86, 937.
[2] W. S. Wegener, P. Furmanski & S. J. Ajl (1967) BBA 144, 34.

Selected entries from Methods in Enzymology [vol, page(s)]:
General discussion: 13, 362
Assay: 13, 362

EUKARYOTIC 20S PROTEASOME

This protein complex [EC 3.4.25.1; previously classified as EC 3.4.99.46], also referred to as proteasome endopeptidase complex, is a high-molecular-weight cylinder/barrel-shaped structure possessing a number of proteolytic activities. The enzyme system is known by over thirty different names, including multicatalytic endopeptidase complex, multicatalytic proteinase (complex), ingensin, macropain, prosome, lens neutral proteinase, cylindrin, and protease YscE. Proteasomes from different organisms often differ in their catalytic properties. Even so, there are typically three main cleavage sites: (1) a trypsin-like preference for substrates having basic aminoacyl residues (Lys,Arg)–Xaa at the P1 position, (2) a chymotrypsin-like preference for protein substrates containing large hydrophobic aminoacyl residues at P1, and (3) a preference for substrates possessing acidic aminoacyl residues at P1.[1–3] Mammalian proteosomes have been shown to also catalyze cleavages where the P1 position is occupied by branched-chain aminoacyl residues and when the scissile bond is flanked by small neutral amino acids. Synthetic substrates used to assay proteasome activity include Suc-Leu-Leu-Val-Tyr-∇-NHMec, Suc-Ala-Ala-Phe-∇-NHMec, Boc-Leu-Ser-Thr-Arg-∇-NHMec, Boc-Leu-Arg-Arg-∇-NHMec, Z-Gly-Gly-Arg-∇-NHMec, and Z-Leu-Leu-Glu-∇-NHNap, where the symbol ∇ indicates the cleavage site. However, cleavage preferences determined from synthetic substrates frequently are not predictive of cleavage patterns with such proteins as the insulin B chain, casein, human histone H3, ovalbumin, bradykinin, or gonadotropin-releasing hormone.

The proteasome is processive and degrades protein substrate to small peptides, typically yielding short peptides of four to ten residues. There appears to be cooperative interactions between multiple active sites within a single protein complex.

Eukaryotic proteasomes consist of twenty-eight subunits, arranged in four rings of seven. The entire structure consists of a cylinder of about 15 nm in length and 11 nm in diameter and a overall molecular weight of about 700 kDa. The cylinder contains a channel which allows an unfolded polypeptide chain to pass. *See also* Archaean Proteasome; Bacterial Proteasome; Eukaryotic 26S Proteasome

[1] A. Lupas, P. Zwickl, T. Wenzel, E. Seemüller & W. Baumeister (1995) *Cold Spring Harbor Symp. Quant. Biol.* **60**, 515.
[2] J.-M. Peters (1994) *Trends Biochem. Sci.* **19**, 377.
[3] A. J. Rivett, P. J. Savory & H. Djaballah (1994) *MIE* **244**, 331.

Selected entries from *Methods in Enzymology* [vol, page(s)]:
General discussion: 244, 331
Assay: casein radiolabeled substrate, **244**, 337, 339; fluorimetric, **244**, 335; reverse-phase HPLC, **244**, 335; substrates, **244**, 333, 335
Other molecular properties: active sites (broad specificity, **244**, 332; catalytic mechanisms, **244**, 345; nomenclature, **244**, 347); biological role, **244**, 331, 349; effectors, **244**, 346; family (active site residues, **244**, 58; ATP-binding domains, **244**, 54; cleavage site specificity, **244**, 57; inhibitors, **244**, 57; processing, **244**, 58; sequence homology, **244**, 57; subunit structure, **244**, 54, 57); genes, **244**, 343; inhibitors, **244**, 345; kinetic properties, **244**, 332, 334; macroxyproteinase, characteristics, **186**, 493; pure enzyme sources, **244**, 338; purification (latent enzyme forms, **244**, 340; rat liver enzyme, **244**, 341; ubiquitin-conjugate degrading proteinase, **244**, 342; yield, **244**, 339); sequence homology between subunits, **244**, 345; size, **244**, 331; structure, **244**, 343; substrate specificity, **244**, 332

EUKARYOTIC 26S PROTEASOME

Also known as ubiquitin-conjugate degrading enzyme, this massive (2100 kDa) protein complex contains the 20S proteasome and at least one other multisubunit protein known as PA700. All of the proteolytic activity of this complex is accounted for by the 20S complex. The PA700 regulatory complex (also referred to as 19S cap, μ-particle, ball, and ATPase complex) possesses additional enzymatic activities (*e.g.*, ATPase and isopeptidase activities).[1,2]

ATP-dependent channel-gating and polypeptide unfolding facilitate proteolytic action. Although the overall reaction yields ADP and orthophosphate, the reaction is currently misclassified as an ATPase [EC 3.6.4.8]. However, P-O-P bond-scission supplies the Gibbs free energy to drive protein unfolding.[3] The overall enzyme-catalyzed reaction is of the energase type (*e.g.*, $State_1 + ATP = State_2 + ADP + P_i$, where two affinity-modulated conformational states of the protein substrate are indicated.) *See* Archaean Proteasome; Bacterial Proteasome; Eukaryotic 20S Proteasome; Energase

[1] O. Coux, K. Tanaka & A. L. Goldberg (1996) *Ann. Rev. Biochem.* **65**, 801.
[2] D. M. Rubin & D. Finley (1995) *Curr. Biol.* **5**, 854.
[3] D. L. Purich (2001) *TiBS* 26, 417.

Selected entries from *Methods in Enzymology* [vol, page(s)]:
Assay: 244, 343
Other molecular properties: biological role, 244, 349; inhibition by, **300**, 345; purification, **244**, 342

EXODEOXYRIBONUCLEASE I

This enzyme [EC 3.1.11.1], also known as exonuclease I and *E. coli* exonuclease I, catalyzes the degradation of single-stranded DNA, acting progressively in a 3′-to-5′ direction, releasing nucleoside 5′-phosphates.[1–4] The enzyme has a preference for single-stranded DNA. The nuclease from *Escherichia coli* will also act on glucosylated DNA. Similar enzymes include mammalian deoxyribonuclease III, exonuclease IV, and T2- and T4-induced exodeoxyribonucleases.

[1] R. S. Brody, K. G. Doherty & P. D. Zimmerman (1986) *JBC* **261**, 7136.
[2] B. Weiss (1981) *The Enzymes*, 3rd ed., **14**, 203.
[3] I. R. Lehman (1971) *The Enzymes*, 3rd ed., **4**, 251.
[4] I. R. Lehman (1966) *Procedures in Nucleic Acid Research* (G. L. Cantoni & D. R. Davies, eds.) p. 203.

Selected entries from *Methods in Enzymology* [vol, page(s)]:
General discussion: 6, 40
Assay: *Escherichia coli*, **6**, 40; **29**, 189; spectrophotometric assay, **12B**, 471
Other molecular properties: assay of *Escherichia coli* endonuclease V, **65**, 224; DNA characterization and, **21**, 311; DNA polymerase and, **12B**, 590; hydroxymethylated nucleic acid and, **12B**, 477; properties, **6**, 43; purification (*Escherichia coli*, **6**, 42); removal of deoxyribonuclease III, **21**, 151; sequence analysis and, **29**, 282, 283, 286

EXODEOXYRIBONUCLEASE II

This enzymatic activity, formerly classified as EC 3.1.4.26, is now a deleted Enzyme Commission entry.

EXODEOXYRIBONUCLEASE III

This enzyme [EC 3.1.11.2], also referred to as exonuclease III and *E. coli* exonuclease III, catalyzes the degradation of double-stranded DNA acting progressively in a 3′-to-5′ direction, releasing nucleoside 5′-phosphates.[1-6] Hydrolysis ceases when the bihelical structure is lost as result of the exonucleolytic attack. The enzyme is distinct from exodeoxyribonuclease I [EC 3.1.11.1] in that there is a preference for double-stranded DNA and the enzyme also has an endonucleolytic activity near apurinic sites on DNA. Similar enzymes include *Haemophilus influenzae* exonuclease. *See also Exodeoxyribonuclease V*

[1]C. F. Kuo, C. D. Mol, M. M. Thayer, R. P. Cunningham & J. A. Tainer (1994) *Ann. N. Y. Acad. Sci.* **726**, 223.
[2]J. D. Hoheisel (1993) *Anal. Biochem.* **209**, 238.
[3]P. W. Doetsch & R. P. Cunningham (1990) *Mutat. Res.* **236**, 173.
[4]Y. W. Know (1989) *Biochemistry* **28**, 3280.
[5]B. Weiss (1981) *The Enzymes*, 3rd ed., **14**, 203.
[6]I. R. Lehman (1971) *The Enzymes*, 3rd ed., **4**, 251.

Selected entries from *Methods in Enzymology* [**vol**, page(s)]:
General discussion: 34, 469; 68, 475
Assay: *Escherichia coli*, 29, 190; 65, 201, 203
Other molecular properties: analysis of nucleosome positions, in, **170**, 639; assay of Z-DNA, **212**, 261; chromatin structure, and, **304**, 584; detection of psoralen monoadducts and crosslinks in DNA, **212**, 250; direct sequencing of PCR products, in (selection for, influencing factors, **218**, 90; sequencing reactions, **218**, 89); DNA ligase assay and, **30**, 662; DNA polymerase primer and, **29**, 18, 19, 20, 23, 39, 51, 68, 101, 133; DNA resection by, **29**, 445; DNA sequence analysis and, **29**, 249 (specific deletion of nucleotides, and, **100**, 60); DNA synthesis and, **29**, 227; *Escherichia coli* K12, **65**, 275, 562 (activities, **65**, 201, 202; AP endonuclease activity, **65**, 201, 208; assay, **65**, 201; cation requirement, **65**, 208; dilution, **65**, 207, 208; enzymatic reagent, as, **65**, 211; molecular weight, **65**, 208; overproduction technique, **65**, 204, 205; pH optimum, **65**, 208; preparation of primed synthesis template, **65**, 566, 567; properties, **65**, 208; purification, **65**, 206; reaction mechanism, **65**, 209, 210; sequence analysis, of short DNA fragments, **65**, 624; stability, **65**, 207; storage, **65**, 207, 208; substrate specificities, **65**, 208); gapping of (nicked DNA strand, **217**, 203; nicked plasmid, **217**, 211); influenza virus endonuclease (inhibitor screening assay [extension reaction, **275**, 323; filtration, **275**, 323; principle, **275**, 320; specificity, **275**, 322; template-primer design, **275**, 324, 326; validation, **275**, 324]; virus processing, **275**, 320); mapping of hypersensitive sites in chromatin, in, **170**, 275; measurement of Z-DNA supercoiling, in, **212**, 331; restriction mapping with, **216**, 574; separation from DNA polymerase I, **29**, 11; unidirectional deletions of inserts in pBluescriptII vector, **216**, 489; unidirectional DNA digestion, in DNA sequencing, **155**, 156

EXODEOXYRIBONUCLEASE V

This enzyme [EC 3.1.11.5], also referred to as exonuclease V, *Escherichia coli* exonuclease V, and *Rec*BC DNase, catalyzes the ATP-dependent exonucleolytic cleavage of linear DNA (preferably double-stranded DNA) in either the 5′-to-3′ or the 3′-to-5′ direction to generate 5′-phosphooligonucleotides.[1-5] The enzyme will also act endonucleolytically on single-stranded circular DNA. In addition, it has a DNA-dependent ATPase activity.

The *Rec*B subunit catalyzes ATP hydrolysis in the presence of either single- or double-stranded DNA while *Rec*C stimulates that activity of *Rec*B, particularly with double-stranded DNA. Although the overall reaction yields ADP and orthophosphate, the reaction is currently misclassified as an ATPase. Nevertheless, P-O-P bond-scission supplies the Gibbs free energy to drive enzyme-assisted DNA degradation.[3] The overall enzyme-catalyzed reaction is of the energase type (*e.g.*, $State_1 + ATP = State_2 + ADP + P_i$, where two affinity-modulated conformational states of the enzyme-substrate interaction are indicated.)[6] *Haemophilus influenzae* ATP-dependent DNase is a similar enzyme. *See Energase*

[1]M. M. Cox & I. R. Lehman (1987) *Ann. Rev. Biochem.* **56**, 229.
[2]K. M. T. Muskavitch & S. Linn (1981) *The Enzymes*, 3rd ed., **14**, 233.
[3]I. R. Lehman (1971) *The Enzymes*, 3rd ed., **4**, 251.
[4]K. M. T. Muskavitch & S. Linn (1982) *JBC* **257**, 2641.
[5]F. Korangy & D. A. Julin (1993) *Biochemistry* **32**, 4873.
[6]D. L. Purich (2001) *TiBS* 26, 417.

Selected entries from *Methods in Enzymology* [**vol**, page(s)]:
Other molecular properties: *Escherichia coli*, assay, of restriction endonucleases **65**, 26 (*Eco*RI, **65**, 97; *Hpa*I, **65**, 154)

EXODEOXYRIBONUCLEASE VII

This hydrolase [EC 3.1.11.6], also referred to as exonuclease VII and *E. coli* exonuclease VII, catalyzes the exonucleolytic cleavage in either the 5′-to-3′ or 3′-to-5′ direction to yield nucleoside 5′-phosphates.[1] The enzyme will act on both irradiated and unirradiated DNA, preferring single-stranded forms. *Micrococcus luteus* exonuclease is a similar enzyme. Also related is human placental DNase VII. This variant initiates exonucleolytic hydrolysis from the 3′-end of DNA in a nonprocessive, or distributive, manner. The mononucleotide deoxyadenylic acid product is a noncompetitive inhibitor.[2]

[1]L. D. Vales, B. A. Rabin & J. W. Chase (1982) *JBC* **257**, 8799.
[2]G. L. Chen & L. Grossman (1985) *JBC* **260**, 5073.

Selected entries from *Methods in Enzymology* [**vol**, page(s)]:
General discussion: 21, 249; 68, 248
Assay: 21, 250
Other molecular properties: *Escherichia coli*, 65, 190 (activity, 65, 752, 756; digestion procedure, 65, 756; viral transcriptional RNA mapping, 65, 738); *Micrococcus luteus* (preparation, 21, 250; properties, 21, 253; removal from endonuclease, 21, 247, 248)

EXODEOXYRIBONUCLEASE (LAMBDA-INDUCED)

This DNase [EC 3.1.11.3] (also referred to as exodeoxyribonuclease (λ-induced), lambda exonuclease, and λ-exonuclease) catalyzes the degradation of double-stranded DNA, acting progressively in a 5′-to-3′ direction, releasing nucleoside 5′-phosphates.[1–5] There is a distinct preference for double-stranded DNA; the enzyme does not act at single-strand breaks. Similar enzymes include the T4, T5, and T7 exonucleases as well as mammalian deoxyribonuclease IV.

This exonuclease binds to the end of the DNA polymer, whether or not the end has an extension; however, the k_{cat} for the hydrolytic reaction decreases in the order 5′ recessed > blunt ≫ 5′ overhang. Nevertheless, the k_{cat}/K_m ratio is approximately the same for all three forms.[1]

This torroidal exonuclease consists of three subunits, with a funnel-shaped central channel; the wider end has an inner diameter of about 30 Å, whereas the narrow end is half that diameter.[2] **See also** *Exonuclease*

[1]P. G. Mitsis & J. G. Kwagh (1999) *Nucl. Acids Res.* **27**, 3057.
[2]R. Kovall & B. W. Matthews (1997) *Science* **277**, 1824.
[3]J. W. Little (1981) *Gene Amplif. Anal.* **2**, 135.
[4]B. Weiss (1981) *The Enzymes*, 3rd ed., **14**, 203.
[5]I. R. Lehman (1971) *The Enzymes*, 3rd ed., **4**, 251.

Selected entries from *Methods in Enzymology* [vol, page(s)]:
General discussion: 12A, 263; 21, 148; 68, 22, 398
Assay: 12A, 263; 21, 148; 34, 466
Other molecular properties: crystallization, 12A, 267; direct sequencing of PCR-amplified DNA, in, 218, 532; generation of ssDNA, 218, 7; interfering enzymes, absence of, 21, 151; preparation, 12A, 264; 21, 149; production of free 3′ termini, 65, 488, 489; properties, 12A, 267; 21, 152; resection of DNA fragment ends, 216, 43; specificity, 29, 445

EXODEOXYRIBONUCLEASE (PHAGE SP₃-INDUCED)

This DNase [EC 3.1.11.4], also known as phage SP₃ DNase and DNA 5′-dinucleotidohydrolase, catalyzes the exonucleolytic cleavage of DNA in the 5′-to-3′ direction to yield nucleoside 5′-phosphates, preferring single-stranded DNA.[1–3] A minor fraction of the products released are trinucleotides.

[1]N. W. Y. Ho (1979) *BBA* **563**, 393.
[2]I. R. Lehman (1971) *The Enzymes*, 3rd ed., **4**, 251.
[3]D. M. Trilling & H. V. Aposhian (1968) *PNAS* **60**, 214.

EXO-(1,4)-α-D-GLUCAN LYASE

This lyase, EC 4.2.2.13 (also called exo-α-1,4-glucan lyase, α-1,4-glucan lyase, α-1,4-glucan exo-lyase, and α-1,4-glucan 1,5-anhydro-D-fructose eliminase), catalyzes the release of 1,5-anhydro-D-fructose from the non-reducing end of linear (1→4)-α-D-glucans:[1,2] hence, maltose, maltosaccharides, and amylose are all completely degraded. The enzyme, first isolated from red seaweed and red alga, will not act on (1→6)-α-glucosidic bonds. Action amylopectin or glycogen results in the formation of a limit dextrin.

[1]S. Yu & M. Pedersén (1993) *BBA* **1156**, 313.
[2]K. Yoshinaga, M. Fujisue, J. Abe, I. Hanashiro, Y. Takeda, K. Muroya & S. Hizukuri (1999) *BBA* **1472**, 447.

EXOPOLYGALACTURONATE LYASE

This lyase [EC 4.2.2.9] catalyzes the eliminative cleavage of 4-(4-deoxy-α-D-galact-4-enuronosyl)-D-galacturonate from the reducing end of de-esterified pectin.[1–3]

[1]R. J. Linhardt, P. M. Galliher & C. L. Cooney (1986) *Appl. Biochem. Biotechnol.* **12**, 135.
[2]J. D. Macmillan & R. H. Vaughn (1964) *Biochemistry* **3**, 564.
[3]M. Sato & A. Kaji (1979) *Agric. Biol. Chem.* **43**, 1547.

Selected entries from *Methods in Enzymology* [vol, page(s)]:
General discussion: 8, 632
Assay: 8, 632

EXOPOLYGALACTURONOSIDASE

This enzyme [EC 3.2.1.82] catalyzes the hydrolysis of pectic acid from the nonreducing end, thereby releasing digalacturonate.[1–3]

[1]K. Heinrichova, M. Dzurova & L. Rexova-Benkova (1992) *Carbohydr. Res.* **235**, 269.
[2]K. Heinrichova, M. Wojciechowicz & A. Ziolecki (1989) *J. Appl. Bacteriol.* **66**, 169.
[3]A. Collmer, C. H. Whalen, S. V. Beer & D. F. Bateman (1982) *J. Bacteriol.* **149**, 626.

EXOPOLYPHOSPHATASE

This enzyme [EC 3.6.1.11] catalyzes the sequential hydrolysis of polyphosphate$_n$ to produce orthophosphate and polyphosphate$_{n-1}$.[1–3] Tetrametaphosphate is a poor substrate.

[1]H. Wurst & A. Kornberg (1994) *JBC* **269**, 10996.
[2]L. C. M. Capaccio & J. A. Callow (1982) *New Phytol.* **91**, 81.
[3]T. P. Afansieva & I. S. Kulaev (1973) *BBA* **321**, 336.

EXORIBONUCLEASE II

This enzyme [EC 3.1.13.1], also referred to as ribonuclease II, catalyzes the exonucleolytic cleavage of RNA, preferring single-stranded RNA, in the 3′- to 5′-direction to yield nucleoside 5′-phosphates. The enzyme processes the 3′-terminal extranucleotides of monomeric tRNA precursors, following the action of ribonuclease P

[EC 3.1.26.5]. Polyuridylate is hydrolyzed to 5′-UMP and an oligonucleotide and polyadenylate is converted to 5′-AMP and an oligonucleotide. The *Escherichia coli* enzyme has two essential binding sites. The catalytic site binds the first few 3′-nucleotides of the RNA substrate. The other site is an anchor site which binds the RNA approximately fifteen to twenty-five bases from the 3′-end. As exoribonuclease II acts on the single-stranded RNA substrate, the enzyme-substrate complex dissociates at discrete intervals of twelve nucleotides. The enzyme-substrate complex weakens progressively.[1]

Variants of exoribonuclease II can be quite specific for the nucleotide released (for example, ribonuclease BN). Similar enzymes include ribonuclease BN, ribonuclease Q, ribonuclease PIII, and ribonuclease Y. **See also** Ribonuclease BN; Ribonuclease T; Ribonuclease D

[1]V. J. Cannistraro & D. Kennell (1999) *BBA* **1433**, 170.

EXORIBONUCLEASE H

This viral enzyme [EC 3.1.13.2] catalyzes the exonucleolytic cleavage of RNA in duplex with a DNA strand to generate 5′-phosphomonoester oligonucleotides in both the 5-to-3′ and 3′-to-5′ directions.[1,2] Retroviral RNase H and transcriptase activities are present in a single polypeptide.

[1]R. J. Crouch & M.-L. Dirksen (1982) *Cold Spring Harbor Monogr. Ser.* **14**, 211.
[2]G. F. Gerard (1981) *J. Virol.* **37**, 748.

F

α-FACTOR-TRANSPORTING ATPase

This so-called ATPase [EC 3.6.3.48] catalyzes the ATP-dependent (ADP- and orthophosphate-producing) transport of α-factor$_{in}$ to produce α-factor$_{out}$ (α-factor, or a-factor, is the yeast mating pheromone and is an isoprenylated and methylated oligopeptide signaling molecule).[1] The enzyme is an ABC-type (ATP-binding cassette-type) ATPase having two similar ATP-binding domains and does not undergo phosphorylation during the transport process.

The idea that a transporter is an enzyme is in keeping with a new definition of enzyme catalysis as the facilitated making/breaking of chemical bonds, not just covalent bonds.[2] This idea builds on Pauling's assertion that any long-lived, chemically distinct interaction (in this case, the persistent location of a solute with respect to the faces of a membrane) can be regarded as a chemical bond. Note also that the equilibrium constant ($K_{eq} = [$α-Factor$_{out}][$ADP$][P_i]/[$α-Factor$_{in}][$ATP$]$) does not conform to that expected for an ATPase (i.e., $K_{eq} = [$ADP$][P_i]/[$ATP$]$). Thus, although the overall reaction yields ADP and orthophosphate, the enzyme is misclassified as a hydrolase, and instead should be regarded as an energase-type reaction. Energases facilitate affinity-modulated reactions by coupling the $\Delta G_{ATP\text{-}hydrolysis}$ to a force-generating or work-producing step.[2] In this case, P-O-P bond scission supplies the energy to drive α-factor transport. *See ATPase; Energase*

[1] S. Michaelis (1993) *Semin. Cell Biol.* **4**, 17.
[2] D. L. Purich (2001) *TiBS* **26**, 417.

FAD PYROPHOSPHATASE

This enzyme [EC 3.6.1.18], also known as FAD diphosphatase, catalyzes the hydrolysis of FAD to produce AMP and FMN.[1-4] The mung bean and spinach enzymes will also catalyze NAD$^+$ and NADH hydrolysis, and the rat liver enzyme also acts on coenzyme A.[3] In a number of organisms, FAD pyrophosphatase may be identical with nucleotide pyrophosphatase [EC 3.6.1.9]. Human 5′-nucleotidase exhibits significant FAD pyrophosphatase activity.[4]

[1] M. H. Ragab, R. Brightwell & A. L. Tappel (1968) *ABB* **123**, 179.
[2] S. D. Ravindranath & N. A. Rao (1969) *ABB* **133**, 54.
[3] H. J. Shin & J. L. Mego (1988) *ABB* **267**, 95.
[4] R. S. Lee & H. C. Ford (1988) *JBC* **263**, 14878.

Selected entries from *Methods in Enzymology* [**vol**, page(s)]: **General discussion: 280**, 424

FARNESOL DEHYDROGENASE

This oxidoreductase [EC 1.1.1.216] catalyzes the reaction of 2-*trans*,6-*trans*-farnesol with NADP$^+$ to produce NADPH and 2-*trans*,6-*trans*-farnesal.

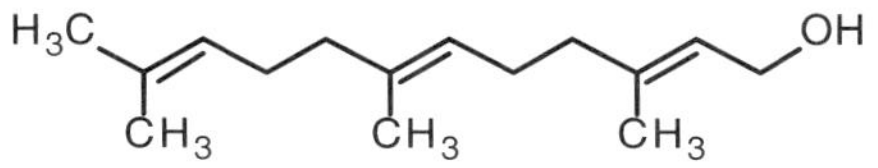

(*E,E*)-2,6-farnesol

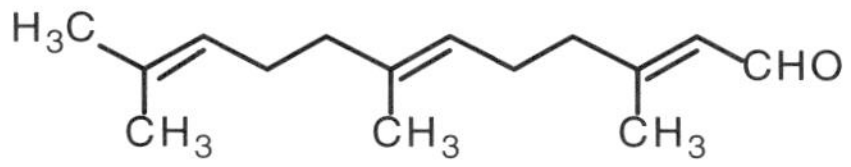

(*E,E*)-2,6-farnesal

Weaker alternative substrates include 2-*cis*,6-*trans*-farnesol, geraniol, citronerol, and nerol. The enzyme was isolated from sweet potato roots (*Ipomoea batatas*). It appears to be biosynthesized in response to infection by the black rot fungus *Ceratocystis fimbriata*.[1]

[1] H. Inoue, H. Tsuji & I. Uritani (1984) *Agric. Biol. Chem.* **48**, 733.

FARNESOL 2-ISOMERASE

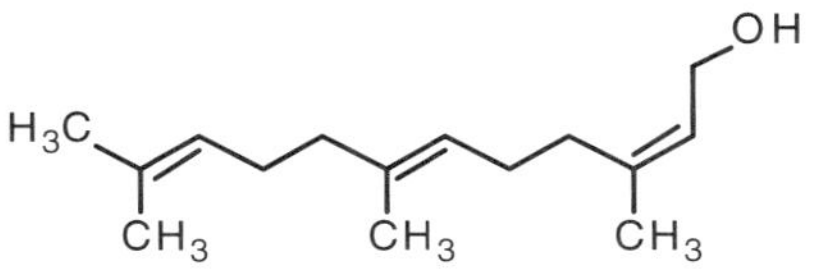

2(*Z*),6(*E*)-farnesol

This isomerase [EC 5.2.1.9] catalyzes the interconversion of 2-*trans*,6-*trans*-farnesol and 2-*cis*,6-*trans*-farnesol.[1]

[1]P. Anastasis, I. Freer, C. Gilmore, H. Mackie, K. Overton, D. Picken & S. Swanson (1984) *Can. J. Chem.* **62**, 2079.

FARNESYL-DIPHOSPHATE KINASE

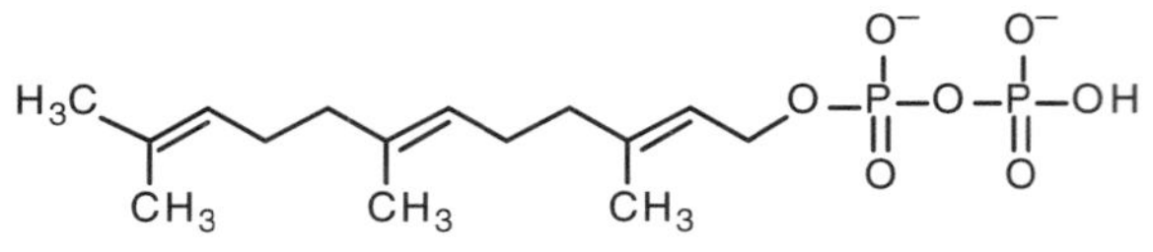

all-*trans*-farnesyl diphosphate

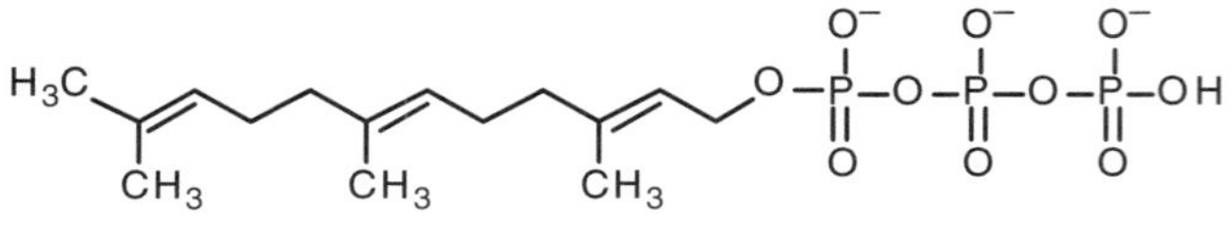

all-*trans*-farnesyl triphosphate

This phosphotransferase [EC 2.7.4.18], also known as farnesyl-pyrophosphate kinase, catalyzes the reversible reaction of ATP with farnesyl diphosphate to produce ADP and farnesyl triphosphate.[1] Interestingly, ADP can act as the phosphoryl donor substrate.

[1]I. Shechter (1974) *BBA* **362**, 233 and (1973) **316**, 222.

FARNESYL*trans*TRANSFERASE

This transferase [EC 2.5.1.29], also known as geranyl-geranyl-diphosphate synthase, catalyzes the condensation of *trans,trans*-farnesyl diphosphate (*i.e.*, (*E,E*)-farnesyl diphosphate) and isopentenyl diphosphate to yield *trans,trans,trans*-geranylgeranyl diphosphate and pyrophosphate (or, diphosphate).[1–3]

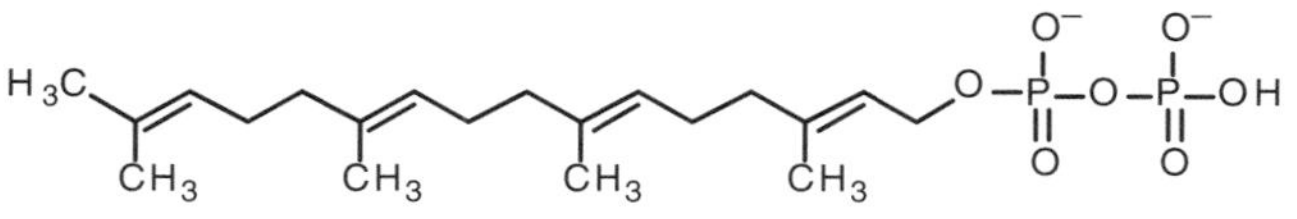

all-*trans*-geranylgeranyl diphosphate

Some variants of this enzyme also use geranyl diphosphate or dimethylallyl diphosphate as donor substrates, producing farnesyl diphosphate and geranyl diphosphate, respectively. In these organisms, geranylgeranyl diphosphate can be synthesized from dimethylallyl diphosphate and three molecules of isopentenyl diphosphate.

[1]K. Ogura, T. Koyama & H. Sagami (1997) *Subcell. Biochem.* **28**, 57.
[2]A. Laferriere & P. Beyer (1991) *BBA* **1077**, 167.
[3]O. Dogbo & B. Camara (1987) *BBA* **920**, 140.

Selected entries from *Methods in Enzymology* [vol, page(s)]:
General discussion:15, 490; 110, 167, 184, 188

Assay: *Micrococcus lysodeikticus*, 15, 490; porcine liver, 110, 185; pumpkin fruit, 110, 167
Other molecular properties: *Micrococcus lysodeikticus* (properties, 15, 494; purification, 15, 492); porcine liver (assay, 110, 185; properties, 110, 187; purification, 110, 185); pumpkin fruit (assays, 110, 167; properties, 110, 170; purification, 110, 169); transition state and multisubstrate analogues, 249, 305

Δ[12]-FATTY ACID DEHYDROGENASE

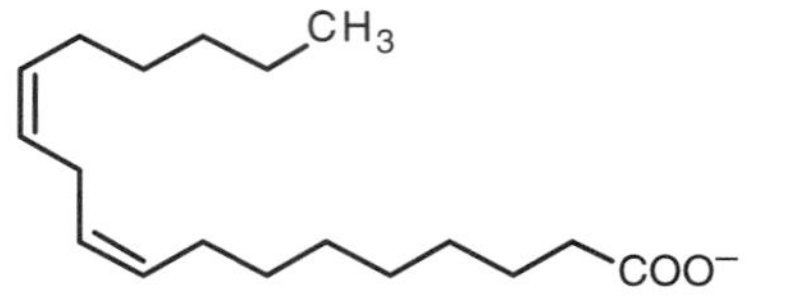

linoleate

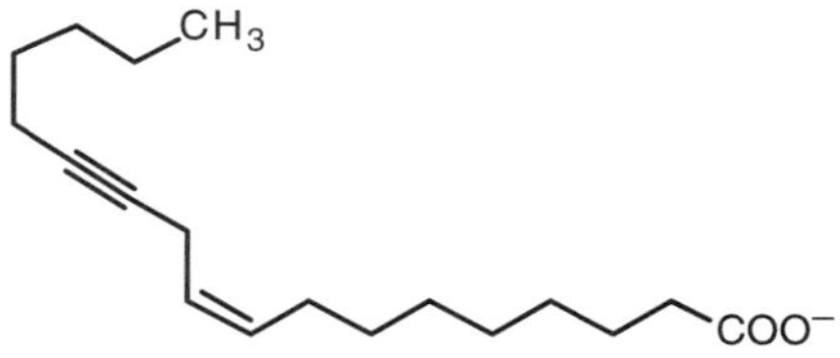

crepenynate

This iron-dependent oxidoreductase [EC 1.14.99.33], also known as crepenynate synthase and linoleate Δ[12]-fatty acid acetylenase (desaturase), catalyzes the reaction of linoleate (*i.e.*, (9Z,12Z)-octadeca-9,12-dienoate) with a reduced cosubstrate (AH_2) and dioxygen to produce crepenynate (*i.e.*, (9Z)-octadec-9-en-12-ynoate), the oxidized cosubstrate (A), and water.[1,2]

[1]A. Banas, M. Bafor, E. Wiberg, M. Lenman, U. Staahl & S. Stymne (1997) *Physiol. Biochem. Mol. Biol. Plant Lipids [Proc. Int. Symp. Plant Lipids] 12th*, 57.
[2]M. Lee, M. Lenman, A. Banas, M. Bafor, S. Singh, M. Schweizer, R. Nilsson, C. Liljenberg, A. Dahlqvist, P. O. Gummeson, S. Sjodahl, A. Green & S. Stymne (1998) *Science* **280**, 915.

FATTY ACID HYDROPEROXIDE LYASE

This heme-dependent enzyme catalyzes the cleavage of hydroperoxides of fatty acids such as linoleate.[1] Isozymes from sunflowers act on both 13-hydroperoxylinoleate and 13-hydroperoxylinolenate.[2] The lyase has been identified as a member of the CYP74 family of cytochrome P450s.

[1]Y. Shibata, K. Matsui, T. Kajiwara & A. Hatanaka (1995) *BBRC* **207**, 438.
[2]A. Itoh & B. A. Vick (1999) *BBA* **1436**, 531.

FATTY-ACID *O*-METHYLTRANSFERASE

This methyltransferase [EC 2.1.1.15] catalyzes the reaction of *S*-adenosyl-L-methionine with a fatty acid anion to produce *S*-adenosyl-L-homocysteine and a fatty acid methyl ester.[1,2] The best acceptor substrate is oleate.

[1] Y. Akamatsu & J. H. Law (1970) *JBC* **245**, 709.
[2] J. Orpiszewski, C. Hebda, J. Szykula, R. Powls, S. Clasper & H. J. Rees (1991) *FEMS Microbiol. Lett.* **82**, 233.

FATTY ACID PEROXIDASE

This oxidoreductase [EC 1.11.1.3] catalyzes the reaction of palmitate with two molecules of hydrogen peroxide to produce pentadecanal, carbon dioxide, and three molecules of water.[1,2] Other substrates include other long-chain fatty acids from laurate (*i.e.*, dodecanoate) to stearate (*i.e.*, octadecanoate).

[1] R. O. Martin & P. K. Stumpf (1959) *JBC* **234**, 2548.
[2] Z. A. Placer, L. L. Cushman & B. C. Johnson (1966) *Anal. Biochem.* **16**, 359.

FATTY ACID SYNTHASE

This enzyme complex [EC 2.3.1.85] catalyzes the conversion of acetyl-CoA, n molecules of malonyl-CoA, and $2n$ molecules of NADPH to produce a long-chain fatty acid anion, $(n + 1)$ molecules of coenzyme A, n molecules of carbon dioxide, and $2n$ molecules of $NADP^+$.[1-4] In mammals, the complex is a multifunctional protein, and in *Escherichia coli* and many other microorganisms, it is a multiprotein complex. The system includes the following activities: [acyl-carrier protein] *S*-acetyltransferase [EC 2.3.1.38], [acyl-carrier protein] *S*-malonyltransferase [EC 2.3.1.39], 3-ketoacyl-[acyl-carrier protein] synthase [EC 2.3.1.41], 3-ketoacyl-[acyl-carrier protein] reductase [EC 1.1.1.100], 3-hydroxypalmitoyl-[acyl-carrier protein] dehydratase [EC 4.2.1.61], enoyl-[acyl-carrier protein] reductase (NADPH) [EC 1.3.1.10] (in liver, this enzyme is enoyl-[acyl-carrier protein] reductase (NADPH, A-specific) [EC 1.3.1.39]), oleoyl-[acyl-carrier protein] hydrolase [EC 3.1.2.14], and the acyl-carrier protein. A fatty acyl-CoA [acyl-carrier protein] transacylase is also thought to be present. In most mammalian systems, the fatty acid product is palmitate. In yeast, the final product is a fatty acyl-CoA (*See Fatty-Acyl-CoA Synthase*, EC 2.3.1.86).

See also specific component enzymes

[1] S.-I. Chang & G. G. Hammes (1990) *Acc. Chem. Res.* **23**, 363.
[2] S. J. Wakil & J. K. Stoops (1983) *The Enzymes*, 3rd ed., **16**, 3.
[3] K. Bloch (1977) *AE* **45**, I.
[4] R. P. Vagelos (1973) *The Enzymes*, 3rd ed., **8**, 155.

Selected entries from *Methods in Enzymology* [vol, page(s)]:
General discussion: 14, 33; 27, 110; 35, 45; 71, 73, 103, 109, 127, 133, 140
Assay: 14, 34; bicyclic dione substrate, using, 72, 303 (comparison with *S*-acetoacetyl-*N*-cysteamine, 72, 306 [with other assays, 72, 305]; method, 72, 304); bovine mammary gland, lactating, 71, 86; *Brevibacterium ammoniagenes*, 71, 120; *Cephalosporium caerulens*, 71, 117; *Ceratitis capitata* (radiochemical assay, 71, 128; spectrophotometric assay, 71, 128); chicken liver, 35, 60, 61; human (radioactive assay, 71, 73, 74; spectrophotometric assay, 71, 73); *Mycobacterium smegmatis*, 71,

110; pigeon liver, 14, 34; rabbit mammary gland, 35, 75, 76; rat liver, 35, 37, 38; rat mammary gland, 35, 65; 71, 28; red blood cells, 71, 97; uropygial gland (radioactive, 71, 104; spectrophotometric, 71, 105)
Other molecular properties: absence of flavin compounds in pigeon liver enzyme, 14, 38; activation by mercaptans, 14, 22, 38; activation of pigeon liver enzyme by phosphate compounds, 14, 38, 39; amino acid composition, 71, 108, 131; antibodies, 71, 109, 144; apofatty acid synthase (assays, 62, 257; immunochemical determination, 62, 260; interconversion, 62, 249 [assay, for conversion of apofatty to holofatty acid synthetase, 62, 257 {conversion of holofatty to apofatty acid synthetase, for, 62, 259, 260}; separation, of apo and holo forms, of {pigeon liver enzyme, 62, 251; rat liver enzyme, 62, 251}]; preparation, from [pigeon liver, 62, 251; rat liver, 62, 250]); apparent lack of multienzyme complex in *Escherichia coli*, 14, 40; assay of (acetyl-CoA carboxylase, 14, 3, 4, 9; 35, 9; biotin carboxyl carrier protein, 35, 18, 19; long-chain fatty acyl-CoA, of, 35, 273, 274); bacterial membrane, in, 22, 113; binding of NADPH, first-order rate constant, 249, 8; *Brevibacterium ammoniagenes*, from, 71, 120 (assay, 71, 120; cerulenin inhibition, 71, 125; 3-decynoyl-*N*-acetylcysteamine inhibition, 71, 125; β,γ-dehydration, 71, 120; 3-hydroxyacyl thioester intermediate, 71, 120; mass fragmentographic assay, 71, 122; properties, 71, 124; specificity, 71, 125; stereospecificity of NADPH and NADH, 71, 126); cascade control, 64, 325; *Cephalosporium caerulens*, from, 71, 117 (assay, 71, 117; cerulenin resistance, 71, 119); *Ceratitis capitata*, from, 71, 127 (conformational properties, 71, 132; development and metamorphosis, 71, 127; kinetic properties, 71, 131; larvae, 71, 127; lipid content, 71, 132; partial specific volume, 71, 131; products, 71, 129, 131; radiochemical assay, 71, 128; spectrophotometric assay, 71, 128; stability, 71, 132); chemical modification, 71, 108; chicken liver, from, 35, 59 (acetyl-CoA binding sites, 35, 65; assay, 35, 60, 61; dissociation and reassociation, 35, 63; malonyl-CoA binding site, 35, 65; products of reaction, 35, 63; stability, 35, 63; sulfhydryl content, 35, 65); classes of reactive SH-groups in, 14, 22, 23; copurification with acetyl-CoA carboxylase, 71, 22; cow mammary gland, from, 71, 227; cyclopropane fatty acids, for, 71, 133; dependence of level of pigeon-liver enzyme on nutritional state, 14, 35; enoyl-CoA reductase site, 71, 108; *Escherichia coli*, from, 14, 39 (apparent lack of multienzyme complex in *Escherichia coli*, 14, 40); estimation of active and inactive forms by immunotitration, 71, 292; flavin mononucleotide content, 14, 23; fructose 1,6-bisphosphate as allosteric activator of pigeon liver enzyme, 14, 39; goat mammary gland, from, 71, 227; human liver, from, 71, 73 (activation and inhibition, 71, 79; dissociation, 71, 78; methylmalonyl-CoA as inhibitor, 71, 79; molecular weight, 71, 76; 4'-phosphopantetheine group, 71, 78; phosphorylated sugars as activators, 71, 79; primer substrate specificity, 71, 78; products, 71, 76, 79; prosthetic group, 71, 76; radioactive assay, 71, 73, 74; specificity, 71, 78; spectrophotometric assay, 71, 73; stability, 71, 76); hydrolase component, 71, 181; immunological comparisons, 71, 109; immunoprecipitation, 71, 144; inhibition by (cerulenin, 72, 520; sulfhydryl reagents, 14, 22, 39 [pH dependence, 14, 22]); inhibition of pigeon liver enzyme by arsenite, 14, 39; irreversible inhibitor of, 72, 580; ketoacyl reductase site, 71, 108; lactating bovine mammary gland, from, 71, 86 (analysis of products, 71, 89; assay, 71, 86; partial reactions, 71, 95; properties, 71, 95); lactating rabbit mammary gland, from, 71, 26; lactating rat mammary gland, from, 71, 16, 26 (assay, 71, 28; properties, 71, 32); mallard, 71, 106, 155, 234; mechanism, 14, 23, 39, 39; 249, 37; methylmalonyl-CoA stereospecificity, 71, 107; model systems for partial reactions (acetyl transfer, 14, 26, 27; condensation, 14, 28; dehydration, 14, 28, 29; malonyl transfer, 14, 26; palmityl transfer, 14, 31; "second" reduction, 14, 29, 30); molecular weight, 71, 107, 131; mRNA, 71, 140, 148; multienzyme complex, 35, 37, 45, 59, 84 (pigeon liver, 14, 33; yeast, 14, 17); *Mycobacterium phlei*, from, 35, 84 (activators, 35, 84, 85, 87; assay for de novo synthesis, 35, 85, 86; effect of long-chain acyl-CoA thioesterase on elongation activity, 35, 89, 90 [primer length on elongation activity, of, 35, 89]; elongation of acyl-CoA derivatives, 35, 89, 90; inhibitors, 35, 88, 89; methylated polysaccharide activators, 35,

84, 87, 95; phosphopantetheine content, **35**, 87; primer specificity, **35**, 88; products of *de novo* synthesis, **35**, 84, 89; type I, **35**, 84; type II, **35**, 84, 90); *Mycobacterium smegmatis*, **71**, 110 (assays, **71**, 110; 2,6-di-O-methylcycloheptaamylose, **71**, 115; fatty acid distribution in products, **71**, 115; mycobacterial polysaccharide, **71**, 110; properties, **71**, 115); nicotinamide adenine dinucleotide phosphate as cofactor, 14, 23, 38; palmitate synthesis, **249**, 31 (transient kinetic studies, **249**, 33); pantothenate-^{14}C labeled fatty acid synthase (pigeon liver, from, **18A**, 364 [assay, **18A**, 368, 369; purification, **18A**, 366; release of 4′-phosphopantetheine-^{14}C, **18A**, 369; synthesis, *in vivo*, **18A**, 365, 366]); Peking duck, from, **71**, 106, 234; phosphopantetheine carrier in, 14, 22; 4′-phosphopantetheine content, **35**, 42, 43, 58, 59, 87; physical properties, from (pigeon liver, **14**, 38; yeast, **14**, 23); pigeon liver, from, **14**, 33; **35**, 45 (absence of flavin compounds in pigeon liver enzyme, **14**, 38; acetyl-CoA binding sites, **35**, 59; activation of pigeon-liver enzyme by phosphate compounds, **14**, 38, 39; assay of component reaction activities, **35**, 49 [covalent binding sites of acetyl and malonyl groups, of, **35**, 51, 52; overall activity, of, **14**, 34; **35**, 46, 57]; effect of nutritional state, **35**, 52, 54; inactivation and dissociation, **35**, 48, 55; malonyl-CoA binding sites, **35**, 59; properties, **14**, 38; purification, **14**, 35; reaction mechanism, **14**, 39; **35**, 45, 46; reactivation and reassociation, **35**, 57; stability, **14**, 37; **35**, 47, 48, 53, 59; sulfhydryl content, **35**, 58); pig liver, from, **71**, 79 (absorption and fluorescence spectra, **71**, 85; decalone reductase partial reaction assay, **71**, 80; NADPH specificity, **71**, 80; properties, **71**, 84; reduction of alicyclic ketones, **71**, 80; stability, **71**, 85); *Podiceps caspicus*, from, **71**, 109; preparation of model substrates (S-crotonyl-N-acetylcysteamine, **14**, 30; D,L-S-β-hydroxybutyryl-N-acetylcysteamine, **14**, 29); primer chain length, effect on reaction rate, **14**, 22; product inhibition, **63**, 435; product specificity, **71**, 188; purification from (*Euglena gracilis*, **35**, 111; pigeon liver, **14**, 35); rabbit mammary gland, from, **35**, 74 (acetylation with acetic anhydride, **35**, 83; assay, **35**, 75, 76; chain termination factor, **35**, 83; inhibitors, **35**, 80; products of reaction, **35**, 81, 82; stability, **35**, 79; substrate control of chain length, **35**, 81, 82; sulfhydryl content, **35**, 81); rat adipose tissue, from, **14**, 39; rat liver, from, **14**, 39; **35**, 37; **71**, 293 (assay, **35**, 37, 38; concentration, **71**, 297; control of rate of synthesis, **35**, 44; immunochemical assay, **35**, 43, 44; preparation of antibody, **71**, 293, 294; stability, **35**, 41, 42; sulfhydryl content, **35**, 42); rat mammary gland, from, **35**, 65 (acetyl-CoA binding sites, **35**, 73; assay, **35**, 65; **71**, 28; dissociation and reassociation, **35**, 69, 70; effect of nutritional state, **35**, 67; inhibitors, **35**, 73; malonyl-CoA binding sites, **35**, 73; products of the reaction, **35**, 73; properties, **71**, 32; stability, **35**, 70; substrate control of chain length, **35**, 73; sulfhydryl content, **35**, 73); reaction catalyzed, **249**, 31; red blood cells, from, **71**, 97 (assays, **71**, 97; immunochemical cross-reactivity, **71**, 102; products, **71**, 102; properties, **71**, 101; separation of blood cells, **71**, 99; specific activities of components, **71**, 101); reduction of enzyme-bound acetoacetate to butyrate, **249**, 32, 35; stability of enzyme from (pigeon liver, **14**, 36, 37; **35**, 47, 48, 53, 59; rat liver, **35**, 41, 42; rat mammary gland, **35**, 70; yeast, **14**, 19, 20); thioesterase activity, **71**, 108; transient kinetics studies, **249**, 31; uropygial gland, from, **71**, 103, 140, 230 (amino acid composition, **71**, 233; assay [radioactive, **71**, 104; spectrophotometric, **71**, 105]; branched fatty acids produced from methylmalonyl-CoA, **71**, 106; effect of bovine serum albumin, **71**, 233; immunological comparison, **71**, 233; inhibitors, **71**, 233; molecular weight, **71**, 107, 233; mRNA, **71**, 148; products, **71**, 103, 106; purity, **71**, 106; substrates, **71**, 103, 106; substrate specificity, **71**, 107, 233; subunits, **71**, 107; synthesis, *in vitro*, **71**, 140; trypsin treatment, **71**, 231); variation of activity with preparation, **14**, 32; yeast, from, **14**, 17

FATTY ACYL-CoA [ACYL-CARRIER PROTEIN] TRANSACYLASE

This enzyme participates in the fatty acid synthase complex by catalyzing the transfer of a short acyl group from the phosphopantetheinoyl residue of the acyl-carrier protein to the sulfhydryl of a cysteinyl residue of the 3-ketoacyl-[acyl-carrier protein] synthase. At this point in the fatty acid synthase reaction sequence, the deacylated sulfhydryl of the phosphopantetheinoyl group on the acyl-carrier protein is free to react with a new malonyl-CoA (via [acyl-carrier protein] *S*-malonyltransferase) and fatty acyl elongation can continue. In this way, the transacetylase catalyzes the transfer of several acyl groups (*e.g.*, acetyl, butyryl, caproyl, capryloyl, capryl, lauroyl, and myristoyl) between two distinct thiols in the fatty acid synthase complex.

FATTY-ACYL-CoA SYNTHASE

This protein complex [EC 2.3.1.86], also known as yeast fatty acid synthase, catalyzes the overall reaction of acetyl-CoA with n molecules of malonyl-CoA with n molecules of NADH and n molecules of NADPH to produce a long-chain acyl-CoA, n molecules of coenzyme A, n molecules of carbon dioxide, n molecules of NAD^+, and n molecules of $NADP^+$.[1-4]

The yeast system is a multifunctional protein catalyzing the activities of [acyl-carrier protein] acetyltransferase [EC 2.3.1.38], [acyl-carrier protein] malonyltransferase [EC 2.3.1.39], 3-ketoacyl-[acyl-carrier protein] synthase [EC 2.3.1.41], 3-ketoacyl-[acyl-carrier protein] reductase [EC 1.1.1.100], 3-hydroxypalmitoyl-[acyl-carrier protein] dehydratase [EC 4.2.1.61], enoyl-[acyl-carrier protein] reductase (NADH) [EC 1.3.1.9], and the acyl-carrier protein. One distinctive characteristic of this complex compared to the mammalian system is that the final product is a long-chain fatty acyl-CoA (palmitoyl-CoA in *Saccharomyces cerevisiae*; thus, $n = 7$) whereas palmitate is released from fatty acid synthase [EC 2.3.1.85]. Hence, a palmitoyltransferase activity must also be present.

[1] J. K. Stoops & S. J. Wakil (1980) *PNAS* **77**, 4544.
[2] H. Engeser, K. Hübner, J. Straub & F. Lynen (1979) *EJB* **101**, 407.
[3] D. E. Vance, O. Mitsuhashi & K. Bloch (1973) *JBC* **248**, 2303.
[4] R. P. Vagelos (1973) *The Enzymes*, 3rd ed., **8**, 155.

Selected entries from ***Methods in Enzymology*** [vol, page(s)]:
General discussion: 14, 17
Assay: 14, 17

FATTY-ACYL-CoA-TRANSPORTING ATPase

This so-called ATPase [EC 3.6.3.47] catalyzes the ATP-dependent (ADP- and orthophosphate-producing) transport of [fatty acyl-CoA]$_{cis}$ to produce [fatty acyl-CoA]$_{trans}$.[1,2] The acyl substrates are long-chain fatty-acyl-CoA derivatives. This ABC-type (ATP-binding cassette-type) ATPase has two similar ATP-binding domains. This transport system functions in peroxisomes and does not undergo phosphorylation during the transport process.

The idea that a transporter is an enzyme is in keeping with a new definition of enzyme catalysis as the facilitated making/breaking of chemical bonds, not just covalent bonds.[3] This idea builds on Pauling's assertion that any long-lived, chemically distinct interaction (in this case, the persistent location of a solute with respect to the faces of a membrane) can be regarded as a chemical bond. Note also that the equilibrium constant (K_{eq} = [Fatty Acyl-CoA$_{trans}$][ADP][P$_i$]/[Fatty Acyl-CoA$_{cis}$][ATP]) does not conform to that expected for an ATPase (i.e., K_{eq} = [ADP][P$_i$]/[ATP]). Thus, although the overall reaction yields ADP and orthophosphate, the enzyme is misclassified as a hydrolase, and instead should be regarded as an energase-type reaction. Energases facilitate affinity-modulated reactions by coupling the $\Delta G_{ATP\text{-hydrolysis}}$ to a force-generating or work-producing step.[3] In this case, P-O-P bond scission supplies the energy to drive fatty acyl-CoA transport. **See ATPase; Energase**

[1]K. Kamijo, S. Taketani, S. Yokota, T. Osumi & T. Hashimoto (1990) *JBC* **265**, 4534.
[2]E. H. Hettema, C. W. T. van Roermund, B. Distel, M. van den Berg, C. Vilela, C. Rodrigues-Posada, R. J. A. Wanders & H. F. Tabak (1996) *EMBO J.* **15**, 3813.
[3]D. L. Purich (2001) *TiBS* **26**, 417.

FATTY-ACYL-ETHYL-ESTER SYNTHASE

This enzyme [EC 3.1.1.67] catalyzes the reaction of a long-chain carboxylate with ethanol to produce a long-chain-acyl ethyl ester and water.[1–4] The best acceptor substrates are unsaturated octadecanoates. The enzyme acts on other substrates, including palmitate, stearate, and arachidonate, albeit at slower rates. Methanol, 1-propanol, and 1-butanol can substitute for ethanol.

[1]S. Mogelson & L. G. Lange (1984) *Biochemistry* **23**, 4075.
[2]S. Mogelson, S. J. Pieper, P. M. Kinnunen & L. G. Lange (1984) *BBA* **798**, 144.
[3]P. S. Bora, C. A. Spilburg & L. G. Lange (1989) *FEBS Lett.* **258**, 236.
[4]D. J. Riley, E. M. Kyger, C. A. Spilburg & L. G. Lange (1990) *Biochemistry* **29**, 3848.

*Fau*I RESTRICTION ENDONUCLEASE

This type II restriction endonuclease (subtype s) [EC 3.4.21.4], isolated from *Flavobacterium aquatile*, catalyzes the hydrolysis of DNA at 5'...CCCGCNNNN∇...3'

and 3'...GGGCGNNNNNN∇...5', where the symbol ∇ indicates the cleavage site.[1]

[1]S. K. Degtyarev, A. A. Kolyhalov, N. I. Rechkunova & V. S. Dedkov (1989) *Bioorg. Khim.* **15**, 130.

FENCHOL DEHYDROGENASE

This oxidoreductase activity, previously classified as EC 1.1.1.182 but now a deleted Enzyme Commission entry, is listed under (+)-borneol dehydrogenase [EC 1.1.1.198], (−)-borneol dehydrogenase [EC 1.1.1.227], and (+)-sabinol dehydrogenase [EC 1.1.1.228].

(−)-*endo*-FENCHOL SYNTHASE

This enzyme [EC 4.2.3.10; previously classified as EC 4.6.1.8] catalyzes the conversion of geranyl diphosphate to (−)-*endo*-fenchol and pyrophosphate (or, diphosphate). (3R)-Linalyl diphosphate is an intermediate in the reaction.[1]

(−)-*endo*-fenchol

Isotope tracer experiments confirm that water is the sole source of the carbinol oxygen atom of the product.[1]

[1]R. Croteau, J. H. Miyazaki & C. J. Wheeler (1989) *ABB* **269**, 507.

FERREDOXIN:NADP⁺ REDUCTASE

This FAD-dependent oxidoreductase [EC 1.18.1.2], also known as adrenodoxin reductase, catalyzes a centrally important reaction in photosynthesis, namely the reversible reaction of two molecules of reduced ferredoxin with NADP⁺ to produce two molecules of oxidized ferredoxin and NADPH.[1–6] Other substrates include adrenodoxin.

Rapid reaction studies demonstrate the formation of a ternary complex (i.e., E·ferredoxin·NADP⁺) in the reaction pathway.[4] NADP⁺ binds to the enzyme followed by the binding of the first molecule of reduced ferredoxin, forming the ternary complex. A one-electron transfer reaction occurs, producing the semiquinone form of the flavoenzyme and the release of the oxidized ferredoxin. The second molecule of reduced ferredoxin binds, and the fully reduced form of the flavoenzyme is generated. Hydride transfer now occurs and NADPH is formed.

[1]A. K. Arakaki, E. A. Ceccarelli & N. Carrillo (1997) *FASEB J.* **11**, 133.
[2]P. A. Karplus & C. M. Bruns (1994) *J. Bioenerg. Biomembr.* **26**, 89.

[3] G. Zanetti & A. Aliverti (1991) in *Chemistry and Biochemistry of Flavoenzymes* (F. Müller, ed.) **2**, 305, CRC Press, Boca Raton.
[4] C. J. Batie & H. Kamin (1984) *JBC* **259**, 11976.
[5] H. Koike & S. Katoh (1980) *Photosynth. Res.* **1**, 163.
[6] G. Palmer (1975) *The Enzymes*, 3rd ed., **12**, 1.

Selected entries from **Methods in Enzymology** [vol, page(s)]:
General discussion: 23, 440; **69**, 250
Assay: 23, 441; **69**, 250, 251; cytochrome *c* reduction, **69**, 251; ferricyanide reduction, **69**, 251
Other molecular properties: adrenal cortex mitochondria, **55**, 10; adrenal gland, in, **52**, 124; apoadrenodoxin reductase, preparation, **52**, 134, 135; assay of (ω-hydroxylase, **53**, 356, 357; rubredoxin, **53**, 341); borohydride reduction, **18B**, 475; catalytic roles, **69**, 250; cytochrome P-450 activity assay, in, **52**, 125, 126, 131, 139; ferredoxin assay, in, **23**, 417; isolation from spinach, **23**, 443; kinetics, **52**, 141; lack of sulfite complexing with spinach enzyme, **18B**, 473; molecular weight, **52**, 141; nature of, **6**, 312; occurrence, **69**, 250; optical properties, **52**, 141; plastocyanin, preparation, **69**, 226; properties, **23**, 445; prosthetic group, **52**, 141; purification, **52**, 133, 134; **69**, 251 (adrenodoxin-substituted Sepharose column, **52**, 134); renal, solubilization, **67**, 433; solubilization, **52**, 141; specificity, **23**, 446; spectral properties, **52**, 134; spinach, properties, **69**, 253; stereospecificity, **87**, 121; synthesis in *Chlamydomonas* cultures, **23**, 73; temperature-jump studies, use in, **76**, 691

FERREDOXIN:NAD$^+$ REDUCTASE

This oxidoreductase [EC 1.18.1.3] catalyzes the reversible reaction of two molecules of reduced ferredoxin with NAD$^+$ to produce two molecules of oxidized ferredoxin and NADH.[1–4] This enzymre has also been identified as a component of a number of dioxygenases, including biphenyl 2,3-dioxygenase[1] and naphthalene dioxygenase.[2] The NADH oxidase of the thermoacidophilic archaea *Acidianus ambivalens* also exhibits a ferredoxin: NAD$^+$ reductase activity.[5]

[1] R. M. Broadus & J. D. Haddock (1998) *Arch. Microbiol.* **170**, 106.
[2] B. E. Haigler & D. T. Gibson (1990) *J. Bacteriol.* **172**, 457.
[3] Y.-P. Chen & D. C. Yoch (1989) *J. Bacteriol.* **171**, 5012.
[4] V. Subramanian, T.-N. Liu, W.-K. Yeh, M. Narro & D. T. Gibson (1981) *JBC* **256**, 2723.
[5] C. M. Gomes & M. Teixeira (1998) *BBRC* **243**, 412.

FERREDOXIN:NITRATE REDUCTASE

This molybdenum- and iron-sulfur-dependent oxidoreductase [EC 1.7.7.2], also known as assimilatory nitrate reductase, catalyzes the reaction of nitrate with two molecules of reduced ferredoxin to produce nitrite, water, and two molecules of oxidized ferredoxin.[1–3]

[1] I. Yamamoto, H. Shimizu, T. Tsuji & M. Ishimoto (1986) *J. Biochem.* **99**, 961.
[2] B. Mikami & S. Ida (1986) *Plant Cell Physiol.* **27**, 1013.
[3] B. Mikami & S. Ida (1984) *BBA* **791**, 294.

FERREDOXIN:NITRITE REDUCTASE

This iron- and heme-dependent oxidoreductase [EC 1.7.7.1] catalyzes the reaction of nitrite with three molecules of reduced ferredoxin to produce ammonia, water, hydroxide ion, and three molecules of oxidized ferredoxin.[1]

[1] B. Mikami & S. Ida (1989) *J. Biochem.* **105**, 47.

Selected entries from **Methods in Enzymology** [vol, page(s)]:
General discussion: 69, 255
Assay: methods (dithionite, **69**, 256; NADPH, **69**, 257, 258; reduced methyl viologen, **69**, 257)
Other molecular properties: iron-sulfur center, **69**, 269, 270; properties (absorption spectrum, **69**, 256, 264, 265; composition, **69**, 267; enzyme complexes, **69**, 268; EPR spectra, **69**, 265, 267; inhibitors, **69**, 263, 264; intracellular location, **69**, 264; mechanism of catalysis, **69**, 268, 269; oxidation-reduction, **69**, 267, 268; pH optimum, **69**, 264; specificity, **69**, 262, 263; stability, **69**, 260, 262; substrate affinity, **69**, 262); purification, from spinach, **69**, 258

FERRIC-CHELATE REDUCTASE

This oxidoreductase [EC 1.6.99.13] catalyzes the reaction of NADH with two ferric ions to produce NAD$^+$ and two ferrous ions. This protein participates in the transport of iron ions across plant plasma membranes. The iron ions are chelated, often by citrate or oxalate.[1–4]

[1] P. Askerlund, C. Larsson & S. Widell (1988) *FEBS Lett.* **239**, 23.
[2] W. Brüggemann, P. Moog, H. Nakagawa, P. Janiesch & P. J. C. Kuiper (1990) *Physiol. Plant.* **79**, 339.
[3] M. J. Holden, D. G. Luster, R. L. Chaney. T. J. Buckhout & C. Robinson (1991) *Plant Physiol.* **97**, 537.
[4] A. Bérczi, K. M. Fredlund & I. M. Møller (1995) *ABB* **320**, 65.

FERROCHELATASE

This [2Fe-2S] cluster-containing enzyme [EC 4.99.1.1], also known as protoheme ferro-lyase and heme synthase, catalyzes the reaction of a protoporphyrin with Fe^{2+} to yield protoheme and two protons.[1–3] Substrates include protoporphyrin IX, mesoporphyrin, deuteroporphyrin, hematoporphyrin, and 2,4-diacetyldeuteroporphyrin. *Bacillus subtilis* ferrochelatase is the only known "water-soluble" ferrochelatase. In eukaryotic cells, ferrochelatase is found in the inner mitochondrial membrane.

[1] G. C. Ferreira (1999) *Int. J. Biochem. Cell Biol.* **31**, 995.
[2] S. Taketani & R. Tokunaga (1981) *JBC* **256**, 12748.
[3] S. Granick & S. I. Beale (1978) *AE* **46**, 33.

Selected entries from **Methods in Enzymology** [vol, page(s)]:
General discussion: 123, 401, 408; **281**, 378
Assay: bacterial, **123**, 409; bovine liver, **123**, 402; chicken erythrocyte, **123**, 402; murine liver, **123**, 402
Other molecular properties: bacterial (assay, **123**, 409; properties, **123**, 413; purification, **123**, 410); bovine liver (assay, **123**, 402; properties, **123**, 405; purification, **123**, 403); chicken erythrocyte (assay, **123**, 402; purification, **123**, 403); murine liver (assay, **123**, 402; purification, **123**, 403)

FERROXIDASE

This copper-dependent oxidoreductase [EC 1.16.3.1], also known as ceruloplasmin, catalyzes the reaction of four ferrous ions with four protons and dioxygen to produce four ferric ions and two water molecules.[1–5] The mammalian enzyme is often known as ceruloplasmin whereas the enzyme from *Thiobacillus ferroxidans* is known as rusticyanin. Iron(II) oxidation also occurs within ferritin. **See also** *Ferritin*

[1] E. Frieden & H. S. Hsieh (1976) *AE* **44**, 187.
[2] B. G. Malmström, L.-E. Andréasson & B. Reinhammer (1975) *The Enzymes*, 3rd ed., **12**, 507.
[3] C. T. Huber & E. Frieden (1970) *JBC* **245**, 3973.
[4] S. Osaki & O. Walaas (1967) *JBC* **242**, 2653.
[5] H. Mukasa, T. Kaya & T. Sato (1967) *J. Biochem.* **61**, 485.

Selected entries from **Methods in Enzymology** [vol, page(s)]:
General discussion: **52**, 10; **80**, 73, 74; **163**, 441
Other molecular properties: calcium phosphate columns and, **5**, 32; characterization, **163**, 448; crystallization, **22**, 262; electron paramagnetic resonance studies, **49**, 514; genetic polymorphism, **163**, 450; human serum, countercurrent distribution, **228**, 165; isolation, **80**, 69; oxy form, spectrum, **52**, 36; partition coefficients, **12A**, 580; prosthetic group, **52**, 4; purification (ion-exchange chromatography, by, **22**, 275; human plasma, **163**, 446; nonionic polymers, using, **22**, 239); Raman frequencies and assignments, **49**, 144; reaction mechanism, **52**, 39; resonance Raman spectrum, **49**, 142; sialic acid removal (asialoceruloplasmin [oxidation of terminal galactose, **28**, 207; preparation, **28**, 205; tritiation of terminal galactose, **28**, 207]); sialic acid transferase and, **8**, 372; staining on gels, **22**, 595; structure, **50**, 273; tritiation of sialic acid in, **28**, 209, 211

trans-FERULOYL-CoA HYDRATASE

This enzyme [EC 4.2.1.101] catalyzes the reaction of *trans*-feruloyl-CoA with water to produce 4-hydroxy-3-methoxyphenyl-β-hydroxypropionyl-CoA.[1]

[1] A. Narbad & M. J. Gasson (1998) *Microbiology* **144**, 1397.

trans-FERULOYL-CoA SYNTHETASE

trans-ferulate   *trans*-feruloyl-CoA

This magnesium-dependent enzyme [EC 6.2.1.34], also known as *trans*-feruloyl-CoA synthase, catalyzes the reaction of ferulate with Coenzyme A and ATP to produce *trans*-feruloyl-CoA and the products of ATP breakdown (*i.e.*, either ADP and orthophosphate or AMP and pyrophosphate).[1]

[1] A. Narbad & M. J. Gasson (1998) *Microbiology* **144**, 1397.

FERULOYL ESTERASE

This esterase [EC 3.1.1.73], also known as hydroxycinnamoyl esterase and hemicellulase accessory enzyme, catalyzes the hydrolysis of a feruloyl-polysaccharide to produce ferulate and the polysaccharide.[1–3] This esterase is occasionally known as an hemicellulase accessory enzyme, because it assists xylanases and pectinases in the breakdown of hemicellulose in plant cell walls.

[1] C. B. Faulds & G. Williamson (1994) *Microbiology* **140**, 779.
[2] P. A. Kroon, C. B. Faulds & G. Williamson (1996) *Biotechnol. Appl. Biochem.* **23**, 255.
[3] A. Castanares, S. I. McCrae & T. M. Wood (1992) *Enzyme Microbiol. Technol.* **14**, 875.

N-FERULOYLGLYCINE DEACYLASE

This deacylase [EC 3.5.1.71] catalyzes the hydrolysis of *N*-feruloylglycine to produce feruloate (*i.e.*, 3-methoxy-4-hydroxy-*trans*-cinnamate) and glycine. Other substrates include a wide range of *N*-cinnamoyl-L-amino acids and substituted *N*-cinnamoyl-L-amino acids.[1] The deacylase is distinct from aminoacylase [EC 3.5.1.14].

[1] M. Martens, M. Cottenie-Ruysschaert, R. Hanselaer, L. De Cooman, K. Van de Casteele & C. F. van Sumere (1988) *Phytochemistry* **27**, 2457 and 2465.

Fe^{3+}-TRANSPORTING ATPase

This so-called ATPase [EC 3.6.3.30] catalyzes the ATP-dependent (ADP- and orthophosphate-producing) transport of Fe$^{3+}_{out}$ to produce Fe$^{3+}_{in}$.[1,2] Note that the uptake of ferric ions is in the absence of solubilizing siderophores. This ABC-type (ATP-binding cassette-type) ATPase has two similar ATP-binding domains and does not undergo phosphorylation during the transport process.

The idea that a transporter is an enzyme is in keeping with a new definition of enzyme catalysis as the facilitated making/breaking of chemical bonds, not just covalent bonds.[3] This idea builds on Pauling's assertion that any long-lived, chemically distinct interaction (in this case, the persistent location of a solute with respect to the faces of a membrane) can be regarded as a chemical bond. Note also that the equilibrium constant ($K_{eq} = $ [Fe$^{3+}_{in}$][ADP][P$_i$]/[Fe$^{3+}_{out}$][ATP]) does not conform to that expected for an ATPase (*i.e.*, $K_{eq} = $ [ADP][P$_i$]/[ATP]). Thus, although the overall reaction yields ADP and orthophosphate, the enzyme is misclassified as a hydrolase, and instead should be regarded as an energase-type reaction. Energases facilitate affinity-modulated reactions by coupling the $\Delta G_{\text{ATP-hydrolysis}}$ to a force-generating or work-producing step.[3] In this case, P-O-P bond scission supplies the energy to drive Fe^{3+} transport. **See ATPase; Energase**

[1] A. Angerer, B. Klupp & V. Braun (1992) *J. Bacteriol.* **174**, 1378.
[2] H. H. Khun, S. D. Kirby & B. C. Lee (1998) *Infect. Immun.* **66**, 2330.
[3] D. L. Purich (2001) *TiBS* **26**, 417.

FIBROBLAST ACTIVATION PROTEIN α

This serine endopeptidase, FAPα, is a cell-surface antigen of cultured normal fibroblasts and is typically assayed with the synthetic substrate Gly-Pro-∇-7-amino-4-trifluoromethylcoumarin. (Note: ∇ indicates the cleavage site.) Its natural substrate in different antigen-expressing tissues has yet to be described.[1]

[1] W. J. Rettig, S. L. Su, S. R. Fortunato, M. J. Scanlan, B. K. Mohan Raj, P. Garin-Chesa, J. H. Healey & L. J. Old (1994) *Int. J. Cancer* **58**, 385.

FIBROLASE

This zinc-dependent metalloproteinase [EC 3.4.24.72], also known as fibrinolytic proteinase, is obtained from the venom of the southern copperhead snake and has fibrinolytic activity. This 23 kDa enzyme catalyzes the hydrolysis of peptide bonds in proteins, acting initially on the Lys413–Leu414 bond in the Aα chain of fibrinogen. The hydrolase acts more slowly on the Bβ chain, but not on the γ chain.[1]

[1] F. S. Markland (1996) in *Natural Toxins II: Adv. Exp. Med. Biol.* **391**, (B. R. Singh & A. T. Tu, eds.), Marcel Dekker, New York, p. 173.

Selected entries from *Methods in Enzymology* [vol, page(s)]:
Other molecular properties: Southern copperhead snake, properties, **248**, 192

FICAIN

This cysteine endopeptidase [EC 3.4.22.3], also called ficin, catalyzes the hydrolysis of peptide bonds with broad specificity (similar to that of papain; however, there are differences: the insulin B chain is hydrolyzed at six positions by ficain and ten by papain). As with papain, the reaction proceeds via an acylated enzyme intermediate. The P1 position in the substrate can be occupied by Gly, Ser, Glu, Tyr, or Phe and the P2 position can be occupied by Leu, Val, Gly, Phe, or cysteic acid. The synthetic substrates Bz-Arg-∇-NHPhNO₂, Z-Phe-Arg-∇-NHPhNO₂, and Bz-Phe-Val-Arg-∇-NHPhNO₂ have all been used to assay the enzyme.[1,2] (Note: ∇ indicates the cleavage site.)

Ficain is the main proteolytic component of the latex of the fig *Ficus glabrata*. Cysteine endopeptidases with similar properties are present in other members of the large genus *Ficus*. **See also Ficin**

[1] D. J. Buttle (1994) *Meth. Enzymol.* **244**, 639.
[2] P. T. Englund, T. P. King, L. C. Craig & A. Walti (1968) *Biochemistry* **7**, 163.

Selected entries from *Methods in Enzymology* [vol, page(s)]:
General discussion: 2, 56; 19, 261; 34, 532; 244, 639
Assay: 19, 261
Other molecular properties: activators and inhibitors, 19, 268, 270; active site, 19, 271; affinity labeling, 87, 435, 461, 463, 491; amino acid composition, 19, 270; azide binding, 44, 52; carboxymethylation of, 25, 430; CM-cellulose derivative of, preparation and properties, 19, 970; conjugate, assay, 44, 338, 339; 2,2′-dipyridyl disulfide, and, 87, 435, 441, 461; distribution, 19, 272; hormone receptor studies, in, 37, 212; immobilization ((carboxymethyl)cellulose, on, 44, 52, 338, 339; collagen, with, 44, 244; property changes, 44, 53); inhibition by cystatin, 80, 771, 777; inhibitor from egg white, 19, 891; kinetics, of substrate diffusion, 44, 437; pH effects, 19, 268; 87, 435, 441, 463; physical properties, 2, 56; 19, 268; 2-PROD, 87, 463; purification, 2, 61; 19, 264; 87, 467; purity, 19, 268; specificity, 19, 272; stability, 19, 265; synthetic substrates for, 19, 261; transition state and multisubstrate analogues, 249, 307; trapped by α₂-macroglobulin, 80, 748

FICIN

A generic term for proteases collectively present in the latex. **See Ficain**

FLAP ENDONUCLEASE-1

This magnesium-dependent flap endonuclease-1, also known as FEN-1, recognizes and cleaves 5′-flap DNA. The 5′-flap structure results from the release of an upstream primer in DNA replication, during double-stranded break repair, homologous recombination, and in excision repair. The enzyme first binds to the 5′ free arm of the DNA, slides along the arm until it reaches duplex DNA, and at the junction region, it releases the unannealed single-strand region via an endonucleolytic cleavage.[1–4]

Other roles for this endonuclease have been identified. The calf enzyme can act as a $5' \rightarrow 3'$ exonuclease during nick translation. SV40 FEN-1 removes the last base of RNA primer attached to Okazaki fragments. *Xenopus laevis* FEN-1 participates in base excision repair. The yeast homologue of FEN-1, known as RAD27, acts to reduce nonhomologous DNA end-joining, thereby preventing trinucleotide repeat expansion and contraction.

[1] R. S. Murante, L. Rust & R. A. Bambara (1995) *JBC* **270**, 30377.
[2] J. J. Harrington & M. R. Lieber (1994) *EMBO J.* **13**, 1235.
[3] J. J. Harrington & M. R. Lieber (1994) *Genes Dev.* **8**, 1344.
[4] C.-Y. Kim, M. S. Park & R. B. Dyer (2001) *Biochemistry* **40**, 3208.

Selected entries from *Methods in Enzymology* [vol, page(s)]:
Assay: 262, 546

FLAVASTACIN

This zinc-dependent metalloendopeptidase [EC 3.4.24.76], also known as P40, catalyzes the hydrolysis of peptide bonds in proteins and has a preference for an aspartyl residue at the P1′ position (*i.e.*, at Xaa-∇-Asp, where ∇ indicates the cleavage site). If a glutamyl residue is at P1′, hydrolysis is slow. Peptide bonds containing an asparaginyl or a glutaminyl residue at P1′ are not hydrolyzed.[1]

[1] A. L. Tarentino, G. Quinones, B. G. Grimwood, C. R. Hauer & T. H. Plummer, Jr. (1995) *ABB* **319**, 281.

FLAVIRIN

This serine endopeptidase [EC 3.4.21.91], also known as yellow fever virus protease and NS2B-3 endopeptidase, catalyzes a trypsin-like activity that forms a complex with cofactor encoded into the polyprotein. Flavirin mediates the hydrolytic scission of a number of proteins in the viral polyprotein. The enzyme prefers hydrolysis at Xaa-Xbb-∇-Xcc (∇ = cleavage site) in which Xaa and Xbb (*i.e.*, the P2 and P1 positions, respectively) are usually arginyl or lysyl residues (sometimes, a glutaminyl residue can be at P2) and Xcc (*i.e.*, the P1′ position) is typically a seryl or glycyl residue (and, less commonly, alanyl or threonyl).[1,2]

[1] J. F. Bazan & R. J. Fletterick (1990) *Semin. Virol.* **1**, 311.
[2] C. M. Rice (1996) in *Fields Virology* (B. N. Fields, D. M. Knipe & P. M. Howley, eds.), Raven Press, New York, p. 931.

FLAVONE APIOSYLTRANSFERASE

7-*O*-β-D-glucosyl-5,7,4′-trihydroxyflavone

7-*O*-(β-D-apiofuranosyl-1,2-β-D-glucosyl)-
5,7,4′-trihydroxyflavone

This transferase [EC 2.4.2.25] catalyzes the reaction of UDP-apiose with 7-*O*-β-D-glucosyl-5,7,4′-trihydroxyflavone to produce UDP and 7-*O*-(β-D-apiofuranosyl-1,2-β-D-glucosyl)-5,7,4′-trihydroxyflavone (also known as apiin).[1]

Other substrates include 7-*O*-β-D-glucosides of a number of flavonoids (for example, biochanin A) as well as 4-substituted phenols.

[1] R. Ortmann, A. Sutter & H. Grisebach (1972) *BBA* **289**, 293.

Selected entries from *Methods in Enzymology* [vol, page(s)]:
General discussion: 28, 473
Assay: 28, 473

FLAVONE *O*⁷-β-GLUCOSYLTRANSFERASE

This transferase [EC 2.4.1.81], also known as UDP-glucose:apigenin β-glucosyltransferase and UDP-glucose:luteolin β-D-glucosyltransferase, catalyzes the reaction of UDP-D-glucose with 5,7,3′,4′-tetrahydroxyflavone (*i.e.*, luteolin) to produce UDP and 7-*O*-β-D-glucosyl-5,7,3′,4′-tetrahydroxyflavone.[1]

luteolin

7-*O*-β-D-glucosyl-5,7,3′,4′-tetrahydroxyflavone

Other acceptor substrates include a number of flavones (*e.g.*, luteolin, apigenin, and chrysoeriol), flavanones (*e.g.*, naringenin), and flavonols: however, the enzyme is distinct from flavonol *O*³-glucosyltransferase [EC 2.4.1.91].

[1] A. Sutter, R. Ortmann & H. Grisebach (1972) *BBA* **258**, 71.

Selected entries from *Methods in Enzymology* [vol, page(s)]:
Other molecular properties: preparation of, **28**, 475; purification of, **28**, 475

FLAVONOID 3′-MONOOXYGENASE

This oxidoreductase [EC 1.14.13.21], also known as flavonoid 3′-hydroxylase, catalyzes the reaction of a flavonoid with NADPH and dioxygen to produce a 3'-hydroxyflavonoid, NADP[+], and water.[1]

apigenin   luteolin

Substrates include a number of flavonoids: for example, naringenin, dihydrokaempferol, and apigenin. Substrates do not include 4-coumarate or 4-coumaroyl-CoA.

[1]K. Stich & G. Forkmann (1988) Z. Naturforsch. **43C**, 311.

FLAVONOL-3-O-β-GLUCOSIDE O-MALONYLTRANSFERASE

This acyltransferase [EC 2.3.1.116] catalyzes the reaction of malonyl-CoA with a flavonol 3-O-β-D-glucoside to produce coenzyme A and a malonyl-flavonol 3-O-β-D-glucoside.[1] Flavonol glucoside substrates include kaempferol 3-O-glucoside, isorhamnetin 3-O-glucoside, and quercetin 3-O-glucoside. Acylation usually takes place at the C6 position of the sugar moiety.

[1]U. Matern, C. Feser & D. Hammer (1983) ABB **226**, 206.

FLAVONOL-3-O-GLUCOSIDE L-RHAMNOSYLTRANSFERASE

This glycosyltransferase [EC 2.4.1.159] catalyzes the reaction of UDP-L-rhamnose with a flavonol 3-O-D-glucoside to produce UDP and a flavonol 3-O-L-rhamnosylglucoside. Flavonol glucoside substrates include quercetin 3-O-glucoside, kaempferol 3-O-glucoside, and rutin 3-O-glucoside.[1]

[1]G. Kleinehollenhorst, H. Behrens, G. Pegels, N. Srunk & R. Wiermann (1982) Z. Naturforsch. **37c**, 587.

FLAVONOL-3-O-GLYCOSIDE XYLOSYLTRANSFERASE

This transferase [EC 2.4.2.35], first isolated from the tulip, catalyzes the reaction of UDP-D-xylose with a flavonol 3-O-glycoside to produce UDP and a flavonol 3-O-D-xylosylglycoside. Substrates include kaempferol 3-O-glucoside, quercetin 3-O-galactoside, quercetin 3-O-glucoside, and isorhamnetin 3-O-glucoside.[1]

[1]G. Kleinehollenhorst, H. Behrens, G. Pegels, N. Srunk & R. Wiermann (1982) Z. Naturforsch. **37c**, 587.

FLAVONOL O[3]-GLUCOSYLTRANSFERASE

This transferase [EC 2.4.1.91], also known as UDP-glucose flavonol 3,O-glucosyltransferase, catalyzes the reaction of UDP-D-glucose with a flavonol to produce UDP and a flavonol 3-O-D-glucoside.[1–4]

quercetin   quercetin 3-O-glucoside

Acceptor substrates include quercetin, quercetin 7-O-glucoside, kaempferol, dihydrokaempferol, kaempferid, fisetin, and isorhamnetin. The enzyme is distinct from flavone O[7]-β-glucosyltransferase [EC 2.4.1.81].

[1]A. Sutter & H. Grisebach (1973) BBA **309**, 289.
[2]G. Kleinehollenhorst, H. Behrens, G. Pegels, N. Srunk & R. Wiermann (1982) Z. Naturforsch. **37C**, 587.
[3]N. Ishikura & M. Mato (1993) Plant Cell Physiol. **34**, 329.
[4]N. Ishikura & Y. Kazumi (1990) Plant Cell Physiol. **31**, 1109.

FLAVONOL 3-SULFOTRANSFERASE

This sulfotransferase [EC 2.8.2.25] catalyzes the reaction of 3'-phosphoadenylylsulfate with quercetin to produce adenosine 3',5'-bisphosphate and quercetin 3-sulfate.[1–4] A number of other flavonol aglycones can act as substrates: e.g., isorhamnetin, rhamnetin, patuletin, kaempferol, eupatin, and tamarixetin. Quercetin 3-sulfate is a non-competitive inhibitor with respect to either substrate and the enzyme is reported to have an ordered Bi Bi kinetic mechanism.[5] Lysyl and arginyl residues have been identified as crucial for enzyme activity.[6]

[1]L. Varin (1988) Bull. Liaison-Groupe Polyphenols **14**, 248.
[2]L. Varin, P. Gulick & R. Ibrahim (1994) Plant Physiol. **106**, 485.
[3]L. Varin & R. K. Ibrahim (1989) Plant Physiol. **90**, 977.
[4]L. Varin, V. DeLuca, R. K. Ibrahim & N. Brisson (1992) PNAS **89**, 1286.
[5]L. Varin & R. K. Ibrahim (1992) JBC **267**, 1858.
[6]F. Marsolais & L. Varin (1995) JBC **270**, 30458.

FLAVONONE 7-O-β-GLUCOSYLTRANSFERASE

This glucosyltransferase [EC 2.4.1.185] catalyzes the reaction of UDP-D-glucose and a flavonone to produce UDP and a flavonone 7-O-β-D-glucoside. Flavonone substrates include naringenin (i.e., 4',5',7-trihydroxyflavanone) and hesperetin (i.e., 3',5,7-trihydroxy-4-methoxyflavanone). Flavones and flavonols are not glucosyl acceptors for this enzyme.[1,2]

[1]C. A. McIntosh, L. Latchinian & R. L. Mansell (1990) *ABB* **282**, 50.
[2]C. A. McIntosh & R. L. Mansell (1990) *Phytochemistry* **29**, 1533.

FLAVONONE 4-REDUCTASE

This oxidoreductase [EC 1.1.1.234] catalyzes the reaction of (2S)-flavonone with NADPH to produce (2S)-flavon-4-ol and NADP$^+$. Flavonones such as (2S)-naringenin and (2S)-eriodictyol can serve as substrates, thus producing apiforol and luteoforol as products, respectively.[1] This reductase participates in the biosynthesis of 3-deoxyanthocyanidins.

[1]K. Stich & G. Forkmann (1988) *Biochemistry* **27**, 785.

FLUOREN-9-OL DEHYDROGENASE

This *Arthrobacter* oxidoreductase [EC 1.1.1.256] catalyzes the reaction of fluoren-9-ol with two molecules of NAD(P)$^+$ to produce fluoren-9-one, two molecules of NAD(P)H, and two protons.[1]

[1]M. Casellas, M. Grifoll, J. M. Bayona & A. M. Solanas (1997) *Appl. Environ. Microbiol.* **63**, 819.

FMN ADENYLYLTRANSFERASE

This enzyme [EC 2.7.7.2], also known as FAD synthase and FAD pyrophosphorylase, catalyzes the ATP-dependent transfer of an adenylyl group to FMN to yield FAD and pyrophosphate (or, diphosphate). The rat liver enzyme has an ordered Bi Bi kinetic mechanism.[1]

The *Corynebacterium ammoniagenes* enzyme is bifunctional, also exhibiting a riboflavin kinase activity. Interestingly, the FMN product of the phosphotransferase activity has to come off the enzyme surface and than rebind prior to the adenylyltransferase activity.[2]

[1]Y. Yamada, A. H. Merrill, Jr., & D. B. McCormick (1990) *ABB* **278**, 125.
[2]I. Efimov, V. Kuusk, X. Zhang & W. S. McIntire (1998) *Biochemistry* **37**, 9716.

Selected entries from ***Methods in Enzymology*** [vol, page(s)]:
General discussion: 2, 673; 18B, 555; 280, 407
Assay: 2, 673; 18B, 555, 556
Other molecular properties: properties, 2, 675; purification, beer yeast, 2, 673; radiosubstrate HPLC assay, 122, 239; specificity, 18B, 556, 557

FMN CYCLASE

This enzyme (also known as FAD-AMP lyase and FAD lyase) catalyzes the conversion of FAD to AMP and 4′,5′-cyclic FMN (*i.e.*, riboflavin 4′,5′-cyclic phosphate).[1] **See also** *Flavin Cyclic Nucleotide Synthetase*

[1]F. J. Fraiz, R. M. Pinto, M. J. Costas, J. Aavalos, J. Canales, A. Cabezas & J. C. Cameselle (1998) *BJ* **330**, 881.

*Fnu*AI RESTRICTION ENDONUCLEASE

This type II restriction enzyme [EC 3.4.21.4], isolated from *Fusobacterium nucleatum* A, catalyzes the hydrolysis of both strands of DNA at 5′...G∇ANTC...3′, where N refers to any base, and the symbol ∇ indicates the cleavage site.[1]

[1]D. W. Leung, A. C. P. Lui, H. Merilees, B. C. McBride & M. Smith (1979) *NAR* **6**, 17.

*Fnu*CI RESTRICTION ENDONUCLEASE

This type II restriction enzyme [EC 3.4.21.4], isolated from *Fusobacterium nucleatum* C, catalyzes the hydrolysis of both strands of DNA at 5′... ∇GATC...3′, where the symbol ∇ indicates the cleavage site.[1]

[1]D. W. Leung, A. C. P. Lui, H. Merilees, B. C. McBride & M. Smith (1979) *NAR* **6**, 17.

*Fnu*DI RESTRICTION ENDONUCLEASE

This type II restriction enzyme [EC 3.4.21.4], isolated from *Fusobacterium nucleatum* D, catalyzes the hydrolysis of both strands of DNA at 5′...GG∇CC...3′, where the symbol ∇ indicates the cleavage site.[1] Structurally, the enzyme is very similar to *Ngo*PII and *Mth*TI restriction endonucleases.

[1]D. W. Leung, A. C. P. Lui, H. Merilees, B. C. McBride & M. Smith (1979) *NAR* **6**, 17.

*Fnu*DII RESTRICTION ENDONUCLEASE

This type II restriction enzyme [EC 3.4.21.4], isolated from *Fusobacterium nucleatum* D, catalyzes the hydrolysis of both strands of DNA at 5′...CG∇CG...3′, where the symbol ∇ indicates the cleavage site.[1] The presence of 5-methylcytidine at either cytidine position in this recognition sequence inhibits DNA cleavage.[2]

[1]D. W. Leung, A. C. P. Lui, H. Merilees, B. C. McBride & M. Smith (1979) *NAR* **6**, 17.
[2]M. L. Gaido & J. S. Strobl (1987) *Arch. Microbiol.* **46**, 338.

*Fnu*DIII RESTRICTION ENDONUCLEASE

This type II restriction enzyme [EC 3.4.21.4], isolated from *Fusobacterium nucleatum* D, catalyzes the hydrolysis of both strands of DNA at 5′...GCG∇C...3′, where the symbol ∇ indicates the cleavage site.[1]

[1]D. W. Leung, A. C. P. Lui, H. Merilees, B. C. McBride & M. Smith (1979) *NAR* **6**, 17.

*Fnu*EI RESTRICTION ENDONUCLEASE

This type II restriction enzyme [EC 3.4.21.4], isolated from *Fusobacterium nucleatum* E, catalyzes the hydrolysis of both strands of DNA at $5' \ldots \nabla$GATC$\ldots 3'$, where the symbol ∇ indicates the cleavage site.[1]

[1] D. W. Leung, A. C. P. Lui, H. Merilees, B. C. McBride & M. Smith (1979) *NAR* **6**, 17.

*Fnu*4HI RESTRICTION ENDONUCLEASE

This type II restriction endonuclease [EC 3.4.21.4], isolated from *Fusobacterium nucleatum* 4H, catalyzes the hydrolysis of both strands of DNA at $5' \ldots$ GC∇NGC$\ldots 3'$, where N refers to any base, and the symbol ∇ indicates the cleavage site.[1] Note that the DNA fragments generated contain a single-base 5'-extension that is more difficult to ligate than blunt-ended fragments. *Fnu*4HI is inhibited if either the internal cytosine at position 2 or the external cytosine at position 5 is methylated.

[1] D. W. Leung, A. C. P. Lui, H. Merilees, B. C. McBride & M. Smith (1979) *NAR* **6**, 17.

*Fok*I RESTRICTION ENDONUCLEASE

This type II restriction endonuclease (subtype s) [EC 3.4.21.4], isolated from *Flavobacterium okeanokoites*, catalyzes the hydrolysis of DNA at $5' \ldots$ GGATGNNNNNN NNN$\nabla \ldots 3'$ and $3' \ldots$ CCTACNNNNNNNNNNNNN $\nabla \ldots 5'$ where N refers to any base, and the symbol ∇ indicates the cleavage site.[1] *Fok*I can cleave DNA between virtually any two bases by constructing a complementary oligonucleotide to the sequence to be cleaved.[2] Interestingly, early reports suggested that *Fok*I functions as a monomer, using a single catalytic center to cleave both strands of DNA.[3] Crystal studies show that the cleavage domain is sequestered in a "piggyback" fashion by the recognition domain.[4] More recently, a model has been proposed that requires the dimerization of *Fok*I on DNA to cleave both DNA strands.[5,6]

[1] H. Sugisaki & S. Kanazawa (1981) *Gene* **16**, 73.
[2] A. J. Podhajska & W. Szybalski (1985) *Gene* **40**, 175.
[3] D. S. Waugh & R. T. Sauer (1993) *PNAS* **90**, 9596.
[4] D. A. Wah, J. A. Hirsch, L. F. Dorner, I. Schildkraut & A. K. Aggarwal (1997) *Nature* **388**, 97.
[5] D. A. Wah, J. Bitinaite, I. Schildkraut & A. K. Aggarwal (1998) *PNAS* **95**, 10564.
[6] J. Bitinaite, D. A. Wah, A. K. Aggarwal & I. Schildkraut (1998) *PNAS* **95**, 10570.

FOLATE OLIGOGLUTAMATE:AMINO ACID TRANSPEPTIDASE

This enzyme, also called pteroyl oligoglutamate:amino acid transpeptidase, catalyzes the exchange of the terminal γ-glutamyl residue in an oligoglutamyl form of tetrahydrofolate (for example, pteroyl-Glu-γ-Glu) with a free amino acid (such as L-glutamate, L-methionine, L-glutamine, or glycine) to produce L-glutamate and a folate derivative containing a new terminal amino acid.[1] *See also* γ-*Glutamyl Transpeptidase*

[1] T. Brody & E. L. R. Stokstad (1982) *JBC* **257**, 14271.

Selected entries from ***Methods in Enzymology*** [**vol**, page(s)]:
Assay: 122, 367

FOLYLPOLYGLUTAMATE ENDOPEPTIDASE

This endopeptidase, also known as pteroyl oligo-γ-L-glutamyl endopeptidase, catalyzes the hydrolysis of internal isopeptide bonds in oligo-γ-L-glutamylated pteridines.[1-3] The enzyme from certain sources will also exhibit an exopeptidase activity. *See also* *Folylpolyglutamate Hydrolase*; γ-*Glutamyl Hydrolase*

[1] Y. Wang, Z. Nimec, T. J. Ryan, J. A. Dias & J. Galivan (1993) *BBA* **1164**, 227.
[2] S. D. Bhandari, J. F. Gregory, 3rd, D. R. Renuart & A. M. Merritt (1990) *J. Nutr.* **120**, 467.
[3] L. L. Samuels, L. J. Goutas, D. G. Priest, J. R. Piper & F. M. Sirotnak (1986) *Cancer Res.* **46**, 2230.

Selected entries from ***Methods in Enzymology*** [**vol**, page(s)]:
Other molecular properties: affinity chromatography, 34, 5; preparation, **66**, 667, 668; properties, **66**, 669, 670

FOLYLPOLYGLUTAMATE HYDROLASE

This enzyme [EC 3.4.19.8] (also known as pteroylpolyglutamate hydroxylase, pteroylpoly-γ-glutamate carboxypeptidase, and microsomal γ-glutamyl carboxypeptidase) catalyzes the hydrolysis of a γ-glutamyl residue from the unsubstituted C-terminus of pteroylpoly-γ-glutamate.[1-5] In this way, the exoisopeptidase sequentially removes γ-glutamyl residues from pteroylpoly-γ-glutamate to generate pteroyl-α-glutamate (folate) and free glutamate. The enzyme from certain sources reportedly also exhibits endopeptidase activity. *See also* *Folylpolyglutamate Endopeptidase*; γ-*Glutamyl Hydrolase*

[1] S. Lin, S. Rogiers & E. A. Cossins (1993) *Phytochemistry* **32**, 1109.
[2] T. T. Y. Wang, C. J. Chandler & C. H. Halsted (1986) *JBC* **261**, 13551.
[3] A. M. Reisenauer & C. H. Halsted (1981) *BBA* **659**, 62.
[4] K. N. Rao & J. M. Noronha (1977) *BBA* **481**, 594 and 608.
[5] M. Silink, R. Reddel, M. Bethel & P. B. Rowe (1975) *JBC* **250**, 5982.

FOLYLPOLYGLUTAMATE SYNTHETASE

This enzyme [EC 6.3.2.17], also known as folylpoly-γ-glutamate synthetase, catalyzes the sequential ATP-dependent addition of glutamyl groups onto tetrahydrofolyl-[Glu]$_n$ to yield tetrahydrofolyl-[Glu]$_{n+1}$, ADP, and orthophosphate.[1–5] The enzymes from different sources exhibit different substrate specificities (for example, some are active with N^5 and/or N^{10} one-carbon tetrahydrofolate derivatives, and the activity of others depends on the number of glutamyl groups).

[1] D. J. Cichowicz & B. Shane (1987) *Biochemistry* 26, 504 and 512.
[2] B. Shane (1986) *Chem. Biol. Pteridines, Proc. Int. Symp. Pteridines, Folic Acid Deriv., Chem. Biol. Clin. Aspects*, p. 719.
[3] B. Shane & D. J. Cichowicz (1983) *Adv. Exp. Med. Biol.* 163, 149.
[4] G. M. Brown & J. M. Williamson (1982) *AE* 53, 345.
[5] J. J. McGuire & J. R. Bertino (1981) *Mol. Cell. Biochem.* 38, 19.

FORMALDEHYDE DEHYDROGENASE

This oxidoreductase [EC 1.2.1.46] catalyzes the reaction of formaldehyde with NAD$^+$ and water to produce formate and NADH. While a similarly named enzyme [EC 1.2.1.1] converts formaldehyde and glutathione in the presence of NAD$^+$ into S-formylglutathione plus NADH, EC 1.2.1.46 does not require glutathione. Other substrates include propanal, glyoxal, and pyruvaldehyde. The *Pseudomonas putida* enzyme, which contains an active-site zinc ion, reportedly has a ping pong Bi Bi kinetic mechanism.[1,2] *See also Aldehyde Dehydrogenase; Formaldehyde Dehydrogenase (Glutathione)*

[1] K. T. Douglas (1987) *AE* 59, 103.
[2] M. Ando, T. Yoshimoto, S. Ogushi, K. Rikitake, S. Shibata & D. Tsuru (1979) *J. Biochem.* 85, 1165.

FORMALDEHYDE DEHYDROGENASE (GLUTATHIONE)

This oxidoreductase [EC 1.2.1.1], also known as formic dehydrogenase and NAD$^+$-linked formaldehyde dehydrogenase, catalyzes the reversible reaction of formaldehyde with glutathione and NAD$^+$ to produce S-formylglutathione and NADH. (Note that the product can undergo hydrolysis to produce glutathione and formate: in such cases, the overall reaction would be a glutathione-dependent conversion of formaldehyde to formate.) In the forward reaction, the actual substrate is the hemimercaptal adduct of formaldehyde and glutathione.

S-(hydroxymethyl)glutathione

S-formylglutathione

The human enzyme has a random Bi Bi kinetic mechanism.[1,2] Early studies suggested that free glutathione also might bind at an allosteric site[1]; however, detailed kinetic and equilibrium binding experiments demonstrated that this was not the case.[2] The yeast *Candida boidinii* reportedly has an ordered kinetic scheme.[3]

[1] L. Uotila & B. Mannervik (1979) *BJ* 177, 869.
[2] P. C. Sanghani, C. L. Stone, B. D. Ray, E. V. Pindel, T. D. Hurley & W. F. Bosron (2000) *Biochemistry* 39, 10720.
[3] N. Kato, H. Sahm & F. Wagner (1979) *BBA* 566, 12.

FORMALDEHYDE DEHYDROGENASE, MYCOTHIOL-DEPENDENT

This oxidoreductase [EC 1.2.1.66], also known as NAD^+/factor-dependent formaldehyde dehydrogenase, catalyzes the reaction of formaldehyde with mycothiol (*i.e.*, 1-*O*-[2-(*N*-acetyl-L-cysteinamido)-2-deoxy-α-D-glucopyranosyl]-D-*myo*-inositol) and NAD^+ to produce *S*-formylmycothiol, NADH, and a proton.[1,2] The formyl product hydrolyzes to mycothiol and formate. The activity is similar to that of formaldehyde dehydrogenase (glutathione) [EC 1.2.1.1].

[1]M. Misset-Smits, P. W. Van Ophem, S. Sakuda & J. A. Duine (1997) *FEBS Lett.* **409**, 221.

[2]A. Norin, P. W. Van Ophem, S. R. Piersma, B. Person, J. A. Duine & H. Jornvall (1997) *EJB* **248**, 282.

FORMALDEHYDE DISMUTASE

This oxidoreductase [EC 1.2.99.4], also called cannizzanase since it catalyzes a Cannizzaro-type reaction, catalyzes the reaction of two molecules of formaldehyde with water to produce formate and methanol.[1–4] Acetaldehyde can act as a donor or acceptor substrate, whereas propanal, acrolein, or butanal can act as an acceptor substrate.

[1]N. Kato, H, Kobayashi, M. Shimano & C. Sakazawa (1984) *Agric. Biol. Chem.* **48**, 2017.

[2]N. Kato, T. Yamagami, M. Shimao & C. Sakazawa (1986) *EJB* **156**, 59.

[3]N. Kato, S. Mizuno, Y. Imada, M. Shimao & C. Sakazawa (1988) *Appl. Microbiol. Biotechnol.* **27**, 567.

[4]R. P. Mason & J. K. M. Sanders (1989) *Biochemistry* **28**, 2160.

FORMALDEHYDE TRANSKETOLASE

This thiamin-pyrophosphate-dependent enzyme [EC 2.2.1.3], also known as dihydroxyacetone synthase and glycerone synthase, catalyzes the reaction of D-xylulose 5-phosphate with formaldehyde to produce D-glyceraldehyde 3-phosphate and dihydroxyacetone (*i.e.*, glycerone).[1,2]

D-xylulose 5-phosphate   D-glyceraldehyde 3-phosphate   glycerone

Formaldehyde can also react with hydroxypyruvate to produce dihydroxyacetone and carbon dioxide. The enzyme is not identical with transketolase [EC 2.2.1.1].

[1]M. J. Waites & J. R. Quayle (1981) *J. Gen. Microbiol.* **124**, 309.

[2]N. Kato, T. Higuchi, C. Sakazawa, T. Nishizawa, Y. Tani & H. Yamada (1982) *BBA* **715**, 143.

Selected entries from *Methods in Enzymology* [**vol**, page(s)]:
General discussion: 188, 435
Assay: 188, 435

FORMAMIDASE

This enzyme [EC 3.5.1.49], also known as formamide amidohydrolase, catalyzes the hydrolysis of formamide to produce formate and ammonia.[1] Other substrates include acetamide, butanamide, and propanamide.

[1]P. H. Clarke (1970) *Adv. Microb. Physiol.* **4**, 179.

FORMAMIDOPYRIMIDINE-DNA GLYCOSYLASE

This glycosylase [EC 3.2.2.23], often abbreviated FAPY-DNA glycosylase and Fpg protein and also known as DNA-formamidopyrimidine glycosylase, catalyzes the hydrolysis of the deoxyribosyl bond N6 of either 2,6-diamino-4-hydroxy-5(*N*-methyl) formamidopyrimidine (FAPY; *i.e.*, DNA containing ring-opened N^7-methylguanine residues) or 4,6-diamino-5-formamidopyrimidine to release the pyrimidine.

The enzyme appears to play a significant role in processes leading to recovery from mutagenesis and/or cell death by alkylating agents. It participates in the GO system responsible for removing an oxidatively damaged form of guanine (7,8-dihydro-8-oxoguanine) from DNA. The enzyme was also found to have a small but relevant activity on 5,6-dihydrothymine-containing oligonucleotides, thus broadening the substrate specificity of this enzyme to a new modified pyrimidine.[1]

The reaction mechanism involves protonation of 8-oxoguanine at O6 and nucleophilic attack of the deoxyribose moiety at C1′ leading to the formation of an enzyme-substrate Schiff base intermediate.[2,3]

[1]C. D'Ham, A. Romieu, M. Jaquinod, D. Gasparutto & J. Cadet (1999) *Biochemistry* **38**, 3335.

[2]J. Tchou, V. Bodepudi, S. Shibutani, I. Antoshechkin, J. Miller, A. P. Grollman & F. Johnson (1994) *JBC* **269**, 15318.

[3]J. Tchou & A. P. Grollman (1995) *JBC* **270**, 11671.

FORMATE *C*-ACETYLTRANSFERASE

This key transferase [EC 2.3.1.54], also known as pyruvate formate-lyase, participates in anaerobic glucose fermentation in *Escherichia coli* and other microorganisms by catalyzing the reversible CoA-dependent cleavage of pyruvate to acetyl-CoA and formate.[1–6] The *E. coli* enzyme has a ping pong Bi Bi kinetic mechanism.[5] In its active form, this enzyme contains a stable glycyl radical (Gly734) required for catalysis. Studies of mutants and substrate analogues propose that on substrate binding, the spin is transferred from Gly734 to the reaction center (Cys418/Cys419), where a thiyl radical initiates the homolytic cleavage of the pyruvate C–C bond. Acetylphosphinate is a mechanism-based inhibitor.[6]

Initial generation of the radical in the enzyme only occurs anaerobically via an iron–sulfur protein that uses S-adenosyl-L-methionine as a cofactor and reduced flavodoxin as the reductant (formate acetyltransferase activating enzyme, EC 1.97.1.4). The radical form of the lyase is inactivated by dioxygen; hence, the enzyme is protected during the transition to aerobiosis by conversion back to the radical-free, oxygen-stable form.[2]

[1] H. Eklund & M. Fontecave (1999) Structure Fold. Des. **7**, R257.
[2] G. Sawers & G. Watson (1998) Mol. Microbiol. **29**, 945.
[3] J. Knappe, S. Elbert, M. Frey & A. F. Wagner (1993) Biochem. Soc. Trans. **21**, 731.
[4] E. J. Brush & J. W. Kozarich (1992) The Enzymes, 3rd ed., **20**, 317.
[5] J. Knappe, H. P. Balschkowski, P. Gröbner & T. Schmitt (1974) EJB **50**, 253.
[6] L. Ulissi-DeMario, E. J. Brush & J. W. Kozarich (1991) JACS **113**, 4341.

Selected entries from **Methods in Enzymology** [vol, page(s)]:
General discussion: I, 476, 479; **41**, 508
Assay: **41**, 510; **258**, 348, 354
Other molecular properties: activating enzyme, **258**, 344 (conversion of purified enzyme, **258**, 347; purification, **258**, 346); catalytic parameters, **258**, 354; clostridiae, in, **41**, 518; covalent intermediates (acetyl-enzyme, **258**, 356; 1-hydroxyethylphosphoryl thioester, **258**, 358; pyruvate thiohemiketal-enzyme, **258**, 357); electron donors, **258**, 346; Escherichia coli, from (activator system, **41**, 508; assay, **41**, 510; chromatography, **41**, 515; molecular weight, **41**, 517; properties, **41**, 517; purification, **41**, 513); glycyl radical (activation, **258**, 344; AdhE protein-catalyzed reduction, **258**, 353; catalytic role, **258**, 345, 355, 361; electron paramagnetic resonance [data collection, **258**, 349; spectra, **258**, 350]; hydrogen abstraction mechanism, **258**, 348; hydrogen exchange, **258**, 351; oxygen fragmentation, **258**, 352; stereochemistry, **258**, 348); induction in bacteria, **258**, 344; purification, recombinant Escherichia coli enzyme (cell growth, **258**, 345; chromatography, **258**, 345); reaction catalyzed, **258**, 344; reaction mechanism, **258**, 355; stability, **258**, 348, 353; Streptococcus faecalis, in, **41**, 518; subunit structure, **258**, 346

[FORMATE ACETYLTRANSFERASE] ACTIVATING ENZYME

This iron-sulfur-cluster-dependent oxidoreductase [EC 1.97.1.4], also known as [pyruvate formate-lyase]-activating enzyme and PFL activase, catalyzes the reaction of S-adenosyl-L-methionine with dihydroflavodoxin and a glycyl residue in formate acetyltransferase [EC 2.3.1.54; Gly734 of the Escherichia coli enzyme] to produce 5′-deoxyadenosine, L-methionine, flavodoxin, and [formate acetyltransferase]-glycine-2-yl radical (i.e., a radical formed by the loss of the pro-S hydrogen from

the methylene carbon of the glycyl residue).[1] The flavodoxin may be in the partially reduced, radical form. The abstracted hydrogen atom is recovered in the methyl group of 5′-deoxyadenosine, suggesting that a 5′-deoxyadenosyl radical intermediate is the actual hydrogen abstracting species in this system.[2]

[1] M. Frey, M. Rothe, A. F. V. Wagner & J. Knappe (1994) JBC **269**, 12432.
[2] R. Külzer, T. Pils, R. Kappl, J. Hüttermann & J. Knappe (1998) JBC **273**, 4897.

FORMATE DEHYDROGENASES

The four enzymes that catalyze the oxidation of formate to carbon dioxide are distinguished by their requirement for NAD^+, cytochrome b_1, cytochrome c-553, or $NADP^+$.

Formate dehydrogenase [EC 1.2.1.2] catalyzes the reaction of formate with NAD^+ to produce carbon dioxide and NADH.[1–16] In the presence of hydrogen dehydrogenase [EC 1.12.1.2], this enzyme forms a system previously known as formate hydrogen-lyase. The yeast enzyme has an ordered Bi Bi kinetic mechanism[6] and the Pseudomonas protein has a random scheme.[16] The Escherichia coli enzyme reportedly has a ping pong Bi Bi mechanism.[3] Kinetic isotope effect data suggest that the pyridine ring remains planar during the course of hydride transfer.[4]

Formate dehydrogenase (cytochrome) [EC 1.2.2.1] catalyzes the reaction of formate with ferricytochrome b_1 to produce carbon dioxide and ferrocytochrome b_1.[1,8–10] Other electon acceptors include ferricyanide, methylene blue, and coenzyme Q_6.

Formate dehydrogenase (cytochrome c-553) [EC 1.2.2.3] catalyzes the reaction of formate with ferricytochrome c-553 to produce carbon dioxide and ferrocytochrome c-553.[11,12] Other electon acceptors include yeast cytochrome c, ferricyanide, and phenazine methosulfate. The Desulfovibrio desulfuricans enzyme has a ferredoxin-like domain that interacts with cytochrome c-553.[13]

Formate dehydrogenase ($NADP^+$) [EC 1.2.1.43] catalyzes the reversible reaction of formate with $NADP^+$ to produce carbon dioxide and NADPH.[14,15] The Clostridium thermoaceticum enzyme requires the presence of selenium, iron, and tungsten.

[1] T. C. Stadtman (1996) Ann. Rev. Biochem. **65**, 83.
[2] V. O. Popov & V. S. Lamzin (1994) BJ **301**, 625.
[3] M. J. Axley & D. A. Grahame (1991) JBC **266**, 13731.
[4] N. S. Rotberg & W. W. Cleland (1991) Biochemistry **30**, 4068.

[5] J. G. Ferry (1990) *FEMS Microbiol. Rev.* **87**, 377.
[6] J. S. Blanchard & W. W. Cleland (1980) *Biochemistry* **19**, 3543.
[7] W. Babel & G. Mothes (1980) *Z. Allg. Mikrobiol.* **20**, 167.
[8] R. C. Bray (1980) *AE* **51**, 107.
[9] T. C. Stadtman (1979) *AE* **48**, 1.
[10] K. Dalziel (1975) *The Enzymes*, 3rd ed., **11**, 1.
[11] T. Yagi (1979) *BBA* **548**, 96.
[12] T. Yagi (1969) *J. Biochem.* **66**, 473.
[13] X. Morelli & F. Guerlesquin (1999) *FEBS Lett.* **460**, 77.
[14] I. Yamamoto, T. Saiki, S.-M. Liu & L. G. Ljungdahl (1983) *JBC* **258**, 1826.
[15] J. R. Andreesen & L. G. Ljungdahl (1973) *J. Bacteriol.* **116**, 867 and (1974) **120**, 6.
[16] V. I. Tishkov, A. G. Galkin & A. M. Egorov (1989) *Biochimie* **71**, 551.

Selected entries from *Methods in Enzymology* [vol, page(s)]:
General discussion: 9, 360; 53, 360; 89, 531, 537; 188, 331
Assay: 9, 360; 53, 363; 228, 602; *Escherichia coli*, 89, 538; *Methylosinus trichosporium* OB3b, 188, 331; methylotrophic yeast, 188, 459; *Pseudomonas oxalaticus*, 89, 531; *Wolinella succinogenes*, 126, 397
Other molecular properties: acetyl-NAD$^+$, 87, 638, 640; activators, 53, 371; activity, 53, 360, 370; affinity purification (*Candida boidinii*, from, 228, 600; dye ligands, with, 228, 132-135; scaleup, 228, 600, 602); assay of formic acid formation from plasmalogens, in, 234, 613; azide, and, 87, 641; *Clostridium acidurici*, from, 53, 372; *Clostridium formicoaceticum*, from, 53, 372; *Clostridium sticklandii*, from, 53, 375, 376; *Clostridium thermoaceticum*, from, 53, 360; commitment-to-catalysis, 87, 635; cytochrome *b* complex, extraction, 22, 229; *Escherichia coli*, from, 53, 372 (assay, 89, 538; properties, 89, 542; purification, 89, 540); formate assay and, 3, 290; fumarate reductase, and, anaerobic electron transfer, 56, 386; inhibitors, 53, 371; 63, 398; isotope effect, 87, 635, 638; 249, 353; kinetic properties, 53, 370; 249, 353; liquid–liquid partition chromatography, 228, 195, 197; metal content, 53, 371, 372, 375, 376; methanogens, from, 243, 24; *Methylosinus trichosporium* OB3b (assay, 188, 331; properties, 188, 333; purification, 188, 332); methylotrophic yeast (assay, 188, 459; properties, 188, 461; purification, 188, 460); molecular weight, 53, 371; NAD-coimmobilized (assay, 136, 28; L-malate production with, 136, 29; preparation, 136, 26); nitrate reductase, and, anaerobic electron transfer, 56, 386; pH optimum, 53, 370; preparation, 3, 290; *Pseudomonas oxalaticus*, from, 9, 360 (assay, 9, 360; 89, 531; flavin removal and reactivation, 89, 536; inhibitors, 9, 363; properties, 9, 363; 89, 530; purification, 9, 362; 89, 528; specificity, 9, 363); purification, 9, 362; 53, 365; 89, 536; 126, 389; 188, 332; radiolabeling, 53, 369; reactions catalyzed (hydride transfer in, 249, 353; hydrogen tunneling in, 249, 396); reversible dissociation, 53, 436; selenium in, 243, 78; sources, 9, 364; stereospecificity, 87, 113; substrate specificity, 53, 369; temperature effects, 53, 367, 370, 371; thermophiles, from, 243, 24; transition state, 249, 353; transition-state analogue, 87, 641; turnover number, 53, 369; *Wolinella succinogenes* (assays, 126, 397; properties, 126, 391; purification, 126, 389; reconstitution, 126, 392)

FORMATE HYDROGEN LYASE

This multienzyme system is comprised of formate dehydrogenase [EC 1.2.1.2] and hydrogen dehydrogenase [EC 1.12.1.2].

FORMATE KINASE

This phosphotransferase [EC 2.7.2.6] catalyzes the reaction of ATP with formate to produce ADP and formyl phosphate.[1] Formate kinase II may be identical with acetate kinase [EC 2.7.2.1].

[1] W. S. Sly & E. R. Stadtman (1963) *JBC* **238**, 2639.

FORMIMIDOYLASPARTATE DEIMINASE

This deiminase [EC 3.5.3.5], also called forminoaspartate deiminase, catalyzes the hydrolysis of *N*-formimidoyl-L-aspartate to produce *N*-formyl-L-aspartate and ammonia.[1]
See also *Formimidoylglutamate Deiminase*

[1] P. Elödi & E. Szörenyi (1956) *Acta Physiol. Acad. Sci. Hung.* **9**, 367.

FORMIMIDOYLGLUTAMASE

N-formimino-L-glutamate

This enzyme [EC 3.5.3.8], also known as formimino-glutamate hydrolase and formiminoglutamase, catalyzes the hydrolysis of *N*-formimino-L-glutamate to produce formamide and L-glutamate.[1,2]

[1] P. Lund & B. Magasanik (1965) *JBC* **240**, 4316.
[2] E. Kaminskas, Y. Kimhi & B. Magasanik (1970) *JBC* **245**, 3536.

Selected entries from *Methods in Enzymology* [vol, page(s)]:
General discussion: 17B, 57
Assay: *Bacillus subtilis*, 17B, 57

FORMIMIDOYLGLUTAMATE DEIMINASE

This deiminase [EC 3.5.3.13], also called forminoglutamate deiminase, catalyzes the hydrolytic deamination of *N*-formimidoyl-L-glutamate to produce *N*-formyl-L-glutamate and ammonia.[1] **See also** *Formimidoylaspartate Deiminase*

[1] R. B. Wickner & H. Tabor (1972) *JBC* **347**, 1605.

Selected entries from *Methods in Enzymology* [vol, page(s)]:
General discussion: 17B, 80
Assay: *Pseudomonas*, 17B, 80

FORMIMIDOYLTETRAHYDROFOLATE CYCLODEAMINASE

This enzyme [EC 4.3.1.4], also called formiminotetrahydrofolate cyclodeaminase, catalyzes the conversion of 5-formimidoyltetrahydrofolate to 5,10-methenyltetrahydrofolate and ammonia.[1–3] In eukaryotes, this enzyme activity is a component of a bifunctional enzyme exhibiting glutamate formiminotransferase [EC 2.1.2.5] activity. Channeling has been proposed to occur between the two active sites.[2] **See also** *Glutamate Formiminotransferase*

[1]W. A. Findlay & R. E. MacKenzie (1988) *Biochemistry* **27**, 3404.

[2]J. Paquin, C. M. Baugh & R. E. MacKenzie (1985) *JBC* **260**, 14925.

[3]J. I. Rader & F. M. Huennekens (1973) *The Enzymes*, 3rd ed., **9**, 197.

Selected entries from **Methods in Enzymology** [vol, page(s)]:
General discussion: 5, 789; 6, 380; 66, 626
Assay: 5, 789; 6, 380; 66, 626
Other molecular properties: formiminotetrahydrofolate assay and, **6**, 814; formiminotransferase and, **5**, 787; glycine formiminotransferase and, **6**, 96, 98; properties, **5**, 790; **6**, 382; **66**, 630; purification from (*Clostridium cylindrosporum*, **6**, 381; pig liver, **5**, 790; **66**, 628)

FORMYLASPARTATE DEFORMYLASE

This enzyme [EC 3.5.1.8] catalyzes the hydrolysis of *N*-formyl-L-aspartate to produce formate and L-aspartate.[1] Deformylase II will also act on *N*-formyl-L-glutamate. **See also** *N-Formylglutamate Deformylase*

[1]O. Einosuke & O. Hayaishi (1957) *JBC* **227**, 181.

4-FORMYLBENZENESULFONATE DEHYDROGENASE

This oxidoreductase [EC 1.2.1.62], which participates in the degradation of toluene-4-sulfonate, catalyzes the reaction of 4-formylbenzenesulfonate with NAD^+ and water to produce 4-sulfobenzoate, NADH, and a proton.[1,2]

[1]F. Junker, E. Saller, H. R. Schläfli Oppenberg, P. M. Kroneck, T. Leisinger & A. M. Cook (1996) *Microbiology* **142**, 2419.

[2]F. Junker, R. Kiewitz & A. M. Cook (1997) *J. Bacteriol.* **179**, 919.

FORMYL-CoA HYDROLASE

This enzyme [EC 3.1.2.10] catalyzes the hydrolysis of formyl-CoA to produce coenzyme A and formate.[1] Other substrates include formylpantetheine.

[1]W. S. Sly & E. R. Stadtman (1963) *JBC* **238**, 2632.

10-FORMYLDIHYDROFOLATE SYNTHETASE

This ligase [EC 6.3.4.17], also known as formate:dihydrofolate ligase, catalyzes the reaction of ATP with formate and dihydrofolate to produce ADP, orthophosphate, and 10-formyldihydrofolate.[1] The enzyme is distinct from 10-formyltetrahydrofolate synthetase [EC 6.3.4.3].

[1]J. C. Drake, J. Baram & C. J. Allegra (1990) *Biochem. Pharmacol.* **39**, 615.

N-FORMYLGLUTAMATE DEFORMYLASE

This enzyme [EC 3.5.1.68], also called *N*-formylglutamate hydrolase and β-citryl-L-glutamate hydrolase, catalyzes the hydrolysis of *N*-formyl-L-glutamate to produce formate and L-glutamate.[1,2] Other substrates for the rat enzyme include β-citryl-L-glutamate and β-citryl-L-glutamine.[3] **See also** *Formylaspartate Deformylase*

[1]R. N. Wickner & H. Tabor (1971) *Meth. Enzymol.* **17B**, 80 (footnote, p. 83)

[2]E. Ohmura & O. Hayaishi (1957) *JBC* **227**, 181.

[3]M. Asakura, Y. Nagahashi, M. Hamada, M. Kawai, K. Kadobayashi, M. Narahara, S. Nakagawa, Y. Kawai, T. Hama & M. Miyake (1995) *BBA* **1250**, 35.

S-FORMYLGLUTATHIONE HYDROLASE

This enzyme [EC 3.1.2.12] catalyzes the hydrolysis of *S*-formylglutathione to produce glutathione (*i.e.*, γ-L-glutamyl-L-cysteinylglycine) and formate.[1–4] Other substrates include *S*-acetylglutathione, albeit more slowly.

[1]M. Koivusalo, R. Lapatto & L. Uotila (1995) *Adv. Exp. Med. Biol.* **372**, 42.

[2]L. Uotila (1989) *Coenzymes Cofactors* **3**(*Glutathione, Chem. Biochem. Med. Aspects, Part A*), 767.

[3]N. Kato, C. Sakazawa, T. Nishizawa, Y. Tani & H. Yamada (1980) *BBA* **611**, 323.

[4]I. Neben, H. Sahm & M. R. Kula (1980) *BBA* **614**, 81.

Selected entries from **Methods in Enzymology** [vol, page(s)]:
General discussion: 77, 314, 320, 424
Assay: 77, 321, 322

FORMYLMETHANOFURAN DEHYDROGENASE

This oxidoreductase [EC 1.2.99.5] catalyzes the reaction of formylmethanofuran with water and an acceptor substrate (*e.g.*, methyl viologen) to produce carbon dioxide, methanofuran, and the reduced acceptor. *N*-Furfurylformamide is an alternative substrate.[1] Several isozyme forms display their own metal ion dependencies: for example, (a) molybdenum formylmethanofuran dehydrogenase in thermophilic and mesophilic methanogens,[2,3] (b) a selenium-dependent enzyme in the hyperthermophilic *Methanopyrus kandleri*,[4] and (c) a tungsten-containing enzyme in the thermophilic archaeon *Methanobacterium wolfei*.[5]

[1]J. Breitung, G. Borner, M. Karrasch, A. Berkessel & R. K. Thauer (1990) *FEBS Lett.* **268**, 257.

[2]M. Karrasch, G. Borner, M. Enssle & R. K. Thauer (1989) *FEBS Lett.* **253**, 226.

[3]M. Karrasch, G. Borner & R. K. Thauer (1990) *FEBS Lett.* **274**, 48.

[4]J. A. Vorholt, M. Vaupel & R. K. Thauer (1997) *Mol. Microbiol.* **23**, 1033.

[5]R. A. Schmitz, M. Richter, D. Linder & R. K. Thauer (1992) *EJB* **207**, 559.

FORMYLMETHANOFURAN: TETRAHYDROMETHANOPTERIN N-FORMYLTRANSFERASE

This formyltransferase [EC 2.3.1.101] catalyzes the reaction of *N*-formylmethanofuran with 5,6,7,8-tetrahydromethanopterin to produce methanofuran and 5-formyl-5,6,7,8-tetrahydromethanopterin.[1–3] The enzyme isolated from the thermophilic *Methanopyrus kandleri* is absolutely dependent on the presence of phosphate or sulfate salts for activity.[1]

[1]J. Breitung, G. Borner, S. Scholz, D. Linder, K. O. Stetter & R. K. Thauer (1992) *EJB* **210**, 971.
[2]B. Mukhopadhyay, E. Purwantini & L. Daniels (1993) *Arch. Microbiol.* **159**, 141.
[3]B. Schwörer, J. Breitung, A. R. Klein, K. O. Stetter & R. K. Thauer (1993) *Arch. Microbiol.* **159**, 225.

N-FORMYLMETHIONINE DEFORMYLASE

This enzyme [EC 3.5.1.31] catalyzes the hydrolysis of *N*-formyl-L-methionine to produce formate and L-methionine.[1–3]

[1]S. K. Ackerman & S. D. Douglas (1979) *BJ* **182**, 885.
[2]S. Grisolia, A. Reglero & J. Rivas (1977) *BBRC* **77**, 237.
[3]J. N. Aronson & J. C. Lugay (1969) *BBRC* **34**, 311.

N-FORMYLMETHIONYLAMINOACYL-tRNA DEFORMYLASE

This enzyme [EC 3.5.1.27] catalyzes the hydrolysis of an *N*-formyl-L-methionylaminoacyl-tRNA to produce formate and L-methionylaminoacyl-tRNA.[1] Other substrates include *N*-formyl-L-methionylpuromycin and peptides containing an N-terminal *N*-formyl-L-methionyl residue.

[1]D. M. Livingston & P. Leder (1969) *Biochemistry* **8**, 435.

N-FORMYLMETHIONYL PEPTIDASE

This peptidase [EC 3.4.19.7], also known as (fMet)-releasing enzyme, catalyzes the release of an N-terminal *N*-formyl-L-methionyl residue from a polypeptide. It has been isolated from rat intestinal mucosal homogenates and rat liver. The enzyme is very specific for *N*-formylmethionyl-peptides (*i.e.*, for fMet in position P1; but not Met). It will hydrolyze *N*-formylmethionyl-β-naphthylamide (or fMet-NHNap) but not the corresponding analogues of fAla, fVal, fLeu, fAsp, fSer, fArg, or fPhe. Both acetyl and formyl di- and tripeptides are also hydrolyzed.[1–3]

[1]R. M. Sherriff, M. F. Broom & V. S. Chadwick (1992) *BBA* **1119**, 275.
[2]H. Suda, K. Yamamoto, T. Aoyagi & H. Umezawa (1980) *BBA* **616**, 60.
[3]G. Radhakrishna & F. Wold (1986) *JBC* **261**, 9572.

FORMYLTETRAHYDROFOLATE DEFORMYLASE

This enzyme [EC 3.5.1.10], also known as formyltetrahydrofolate amidohydrolase and formyl-FH$_4$ hydrolase, catalyzes the hydrolysis of 10-formyltetrahydrofolate to produce formate and tetrahydrofolate.[1,2] The *Escherichia*

coli enzyme is activated by L-methionine and inhibited by glycine.[1] Note that liver 10-formyltetrahydrofolate dehydrogenase [EC 1.5.1.6] has an NADP$^+$-independent hydrolase activity. ***See also*** *Formyltetrahydrofolate Synthetase*

[1]P. L. Nagy, A. Marolewski, S. J. Benkovic & H. Zalkin (1995) *J. Bacteriol.* **177**, 1292.
[2]J. I. Rader & F. M. Huennekens (1973) *The Enzymes*, 3rd ed., **9**, 197.
Selected entries from ***Methods in Enzymology*** [vol, page(s)]:
General discussion: 6, 373; 18B, 607; 281, 214
Other molecular properties: *Clostridium cylindrosporum*, 18B, 607, 608; oxalate decarboxylation and, 5, 638; use in assay for tetrahydrofolate, 18B, 608, 609

N^{10}-FORMYLTETRAHYDROFOLATE DEHYDROGENASE

This oxidoreductase [EC 1.5.1.6] catalyzes the reaction of 10-formyltetrahydrofolate with NADP$^+$ and water to produce tetrahydrofolate, NADPH, and carbon dioxide. Interestingly, the liver cytosolic enzyme will also catalyze the hydrolysis of 10-formyltetrahydrofolate to produce tetrahydrofolate and formate (however, this reaction is probably not physiologically relevant).[1] The enzyme also catalyzes the reaction of propanal with NADP$^+$ to produce propanoate and NADPH. A transient adduct has been proposed to form in the reaction mechanism between a cysteinyl residue and the formyl group of the substrate.[2]

[1]S. A. Krupenko & C. Wagner (1999) *JBC* **274**, 35777.
[2]V. Schirch (1998) *CBC* **1**, 211.
Selected entries from ***Methods in Enzymology*** [vol, page(s)]:
General discussion: 18B, 793; 77, 18; 281, 129, 146
Assay: 18B, 793

FORMYLTETRAHYDROFOLATE SYNTHETASE

This enzyme [EC 6.3.4.3], which has the recommended systematic name of formate:tetrahydrofolate ligase, catalyzes the reaction of formate with MgATP and tetrahydrofolate to produce MgADP, orthophosphate, and N^{10}-formyltetrahydrofolate.[1–4]

The reaction mechanism involves the initial formation of formyl phosphate as an intermediate which subsequently formylates tetrahydrofolate at the N^{10} position. Interestingly, formyl phosphate will formylate tetrahydrofolate nonenzymatically, but at the N^5 position. Oxygen-18 tracer studies demonstrate that this ligase formylates directly at N^{10}.[1] The eukaryotic enzyme is a trifunctional protein also having methylenetetrahydrofolate dehydrogenase [EC 1.5.1.5] and methenyl-tetrahydrofolate cyclohydrolase [EC 3.5.4.9] activities. The enzyme has a random

Ter Ter kinetic mechanism.[4] **See also** *Formyltetrahydrofolate Deformylase*

[1] S. Song, H. Jahansouz & R. H. Himes (1993) *FEBS Lett.* **332**, 150.
[2] J. I. Rader & F. M. Huennekens (1973) *The Enzymes*, 3rd ed., **9**, 197.
[3] R. H. Himes & J. A. K. Harmony (1973) *CRC Crit. Rev. Biochem.* **1**, 501.
[4] B. K. Joyce & R. H. Himes (1966) *JBC* **241**, 5716.

Selected entries from **Methods in Enzymology** [vol, page(s)]:
General discussion: 6, 375; **66**, 585, 616; **77**, 18; **281**, 146, 171
Assay: 6, 375; **66**, 585, 601, 609, 616
Other molecular properties: mechanism, **63**, 51; properties, **6**, 379; **66**, 595, 607, 615, 624; purification from (*Clostridium cylindrosporum*, **6**, 377; **66**, 591; *Clostridium thermoaceticum*, **66**, 604; pig liver, **66**, 611; sheep liver, **66**, 618; yeast, **66**, 621); solvent isotope effect, **87**, 579

FRAGILYSIN

This zinc-dependent metalloproteinase [EC 3.4.24.74], also known as *Bacteroides fragilis* enterotoxin, catalyzes the hydrolysis of peptide bonds in proteins with a broad specificity. It acts on proteins in tight junctions and basement membranes. Proteins hydrolyzed include collagen type IV, fibrinogen, actin, myosin, human complement C3, and gelatin. It will act on the synthetic substrate Pz-Pro-Leu-∇-Gly-Pro-D-Arg. (Note: ∇ indicates the cleavage site.) Other bonds observed hydrolyzed in proteins include Gly–Leu, Met–Leu, Cys–Leu, Ser–Leu, Thr–Leu, and Leu–Gly. Fragilysin is a major factor in the diarrhea caused by *B. fragilis* in a number of animals.[1,2]

[1] W. L. Stöcker, F. Grams, U. Bauman, P. Reinemer, F. X. Gomis-Rüth, D. B. McKay & W. Bode (1995) *Protein Sci.* **4**, 823.
[2] C. C. Häse & R. A. Finkelstein (1993) *Microbiol. Rev.* **57**, 823.

FRUCTAN β-FRUCTOSIDASE

This glycosidase [EC 3.2.1.80], also known as exo-β-D-fructosidase and fructanase, catalyzes the hydrolysis of terminal nonreducing 2,1- and 2,6-linked β-D-fructofuranose residues in fructans.[1,2] Substrates include inulin, levan, raffinose, and sucrose. **See also** *Inulinase; Levanase; 2,6-β-Fructan 6-Levanbiohydrolase*

[1] R. A. Burne, K. Schilling, W. H. Bowen & R. E. Yasbin (1987) *J. Bacteriol.* **169**, 4507.
[2] M. Frehner, F. Keller & A. Wiemken (1984) *J. Plant Physiol.* **116**, 197.

Selected entries from **Methods in Enzymology** [vol, page(s)]:
General discussion: 8, 621, 627
Assay: 8, 627

1,2-β-FRUCTAN 1^F-FRUCTOSYLTRANSFERASE

This transferase [EC 2.4.1.100] catalyzes fructosyl transfer between $(1,2)$-β-D-fructosyl$_m$ and $(1,2)$-β-D-fructosyl$_n$ to yield $(1,2)$-β-D-fructosyl$_{m-1}$ plus $(1,2)$-β-D-fructosyl$_{n+1}$.[1]

Substrates include different oligofructans of the inulin series, 1-kestose, nystoses, 1,1,1-logose, and sucrose.

[1] M. Luescher, M. Frehner & J. Noesberger (1993) *New Phytol.* **123**, 437 and 717.

2,6-β-FRUCTAN 6-LEVANBIOHYDROLASE

This glycosidase [EC 3.2.1.64] catalyzes the hydrolysis of 2,6-β-D-fructan so as to remove successive levanbiose residues from the end of the chain.[1] The *Streptomyces exfoliatus* enzyme produces levanbiose from levan in an exo-acting manner. This enzyme also acts on β-2,1-linkages of such fructooligosaccharides as 1-kestose, nystose, and 1-fructosylnystose.[2]

[1] G. Avigad & R. Zelikson (1963) *Bull. Res. Counc. Isr.* **11**, 253.
[2] K. Saito, K. Kondo, I. Kojima, A. Yokota & F. Tomita (2000) *Appl. Environ. Microbiol.* **66**, 252.

β-FRUCTOFURANOSIDASE

This glycosidase [EC 3.2.1.26] is also known as invertase (referring to the enzyme's action in inverting the optical activity of a sucrose solution) as well as saccharase, and β-fructosidase. The enzyme catalyzes the hydrolysis of terminal nonreducing β-D-fructofuranoside residues in β-D-fructofuranosides, resulting in the release of β-D-fructofuranose.[1–6] The reaction proceeds with retention of configuration at the anomeric carbon.

This enzyme also exhibits an exoinulinase activity (from the nonreducing end) as well as fructotransferase reactions. Substrates include sucrose, raffinose, gentianose, levan, inulin, 1-kestose, and nystose. It has been suggested that the reaction proceeds via an α-fructofuranosyl-enzyme intermediate (at an active-site carboxyl group). **See also** *Sucrose*

[1] A. Sturm (1999) *Plant Physiol.* **121**, 1.
[2] G. Davies, M. L. Sinnott & S. G. Withers (1998) *CBC* **1**, 119.
[3] P. M. Dey & E. Del Campillo (1984) *AE* **56**, 141.
[4] J. O. Lampen (1971) *The Enzymes*, 3rd ed., **5**, 291.
[5] K. Myrbäck (1960) *The Enzymes*, 2nd ed., **4**, 379.
[6] L. Michaelis & M. L. Menten (1913) *Biochem. Z.* **49**, 333.

Selected entries from **Methods in Enzymology** [vol, page(s)]:
General discussion: 1, 231, 251; **42**, 504
Assay: 1, 251; **3**, 109; exoinulinase activity, **8**, 625
Other molecular properties: accumulated forms in yeast secretory mutant, electrophoretic analysis, **96**, 812; aldose 1-epimerase assay, **63**, 33; coimmobilization, **44**, 472; conjugate, kinetics, **44**, 408; conjugation to protein A, **73**, 179; efficiency, effect of substrate concentration, **44**, 238; entrapment with urethane prepolymers, **135**, 241; exoinulinase activity, **8**, 625; fiber entrapment, **44**, 230, 231 (activity, **44**, 237; industrial application, **44**, 241); stability, **44**, 241); gelatin-immobilized, properties, **135**, 297; β-D-glucose assay; **63**, 33; glucose oxidase electrode, description, **137**, 38, 42; N-glycosylated, mobility shifts, **194**,

685; immobilization (bead polymerization, by, **44**, 194; collagen, on, **44**, 255, 260; entrapment, by, **44**, 176, 178; liposomes, in, **44**, 224, 702; magnetic carrier, with, **44**, 278, 279; microencapsulation by, **44**, 214; photochemical, **44**, 286, 287; radiation polymerization, by, **135**, 150); immobilized (preparation, **137**, 585, 593; stability, **137**, 593); immunoprecipitation, **194**, 694; induction in microorganisms, **22**, 93; industrial saccharification, in, (immobilized enzyme system, **136**, 377; liquid enzyme system, **136**, 374); intestinal mucosa, in, **32**, 667; invertase, transfer activity of, **1**, 255; invertases and transglycosidase activity, **1**, 251, 255, 258; iodine and, **25**, 400; K_m dependence on pH, **63**, 207, 208; methionine residues, regeneration of, **25**, 396; pH optimum, **1**, 259; preparation from intestine, **8**, 585; prepolymer-entrapped, applications, **135**, 245; prepro human serum albumin fusion cassette, construction, **185**, 476; purification, **1**, 252; **8**, 585; specific activity, **44**, 255; staining on gels, **22**, 601; transglycosylase activity, **1**, 258; yeast, from (external [assay, **1**, 251; **42**, 504, 505; properties, **1**, 255; **42**, 508; purification, **1**, 252; **42**, 506]; internal [activators and inhibitors, **42**, 511; assay, **42**, 509; properties, **42**, 511; purification, **42**, 509]; *mnn* mutant, gel electrophoresis, **185**, 452; secretion signals for directing proteins into secretory pathway, **194**, 497)

FRUCTOKINASE

This phosphotransferase [EC 2.7.1.4] catalyzes the ATP-dependent phosphorylation of D-fructose to form D-fructose 6-phosphate and ADP.[1–3] Other phosphoryl acceptor substrates include D-mannose.

The term fructokinase has also been used as an alternative name for ketohexokinase [EC 2.7.1.3] which produces D-fructose 1-phosphate. **See also** Ketohexokinase

[1] A. Gardner, H. V. Davies & L. R. Burch (1992) *Plant Physiol.* **100**, 178.
[2] J. Thompson, D. L. Sackett & J. A. Donkersloot (1991) *JBC* **266**, 22626.
[3] W. W. Cleland (1977) *AE* **45**, 373.

Selected entries from *Methods in Enzymology* [vol, page(s)]:
General discussion: **42**, 39
Assay: **63**, 32; *Leuconostoc mesenteroides*, **42**, 39, 40
Other molecular properties: chromium-ATP, **87**, 179; *Leuconostoc mesenteroides*, from, **42**, 39 (assay, **42**, 39, 40; chromatography, **42**, 42; molecular weight, **42**, 43; properties, **42**, 42, 43; purification, **42**, 40); myokinase, not present in preparations of, **6**, 230

FRUCTOSE-1,6-BISPHOSPHATASE

This hydrolase [EC 3.1.3.11] catalyzes the conversion of D-fructose 1,6-bisphosphate to D-fructose 6-phosphate and orthophosphate.[1–7] The mammalian enzyme also acts on sedoheptulose 1,7-bisphosphate. The bovine liver enzyme exhibits a steady-state ordered Uni Bi kinetic mechanism.

[1] S. J. Pilkis, T. H. Claus, P. D. Kountz & M. R. El-Maghrabi (1987) *The Enzymes*, 3rd ed., **18**, 3.
[2] E. Van Schaffingen (1987) *AE* **59**, 315.
[3] N. J. Ganson & H. J. Fromm (1984) *Curr. Top. Cell. Regul.* **24**, 197.
[4] G. A. Tejwani (1983) *AE* **54**, 121.
[5] J. P. Casazza, S. R. Stone & H. J. Fromm (1979) *JBC* **254**, 4661.
[6] B. L. Horecker, E. Melloni & S. Pontremoli (1975) *AE* **42**, 193.
[7] S. Pontremoli & B. L. Horecker (1971) *The Enzymes*, 3rd ed., **4**, 611.

Selected entries from *Methods in Enzymology* [vol, page(s)]:
General discussion: **2**, 543; **5**, 272; **9**, 625, 632, 636; **27**, 641; **34**, 164, 165; **42**, 347, 354, 360, 363, 369, 375, 397; **90**, 327, 330, 334, 340, 345, 349, 352, 357, 366, 371, 378, 384
Assay: **2**, 543; **5**, 272; **9**, 625, 632, 636; **22**, 380; *Bacillus licheniformis*, **90**, 385; bovine liver, **42**, 363, 364; **90**, 330; bumblebee flight muscle, **90**, 366; *Candida utilis*, **9**, 632; **42**, 347; chicken muscle, **90**, 340; chloroplast enzyme, **252**, 223; *Euglena gracilis*, **9**, 636; murine intestinal mucosa, **90**, 358; *Polysphondylium pallidum*, **42**, 360; rabbit intestinal mucosa, **90**, 358; rabbit liver, **2**, 543; **42**, 354, 355, 369, 370; **90**, 327, 345; rabbit muscle, **90**, 340; rat liver, **90**, 352; *Rhodopseudomonas palustris*, **90**, 379; snake muscle, **90**, 349; spinach, **5**, 272; **42**, 398; **90**, 372; swine kidney, **42**, 385, 386; turkey liver, **90**, 334
Other molecular properties: activation, thioredoxin *f* assay, **69**, 383, 384, 388; aldolase and, **5**, 311, 316; anomeric specificity, **63**, 374; *Bacillus licheniformis* (assay, **90**, 385; properties, **90**, 388; purification, **90**, 386); binding to AMP, assay by equilibrium gel penetration, **117**, 345; bovine liver, from, **42**, 363 (activators and inhibitors, **42**, 368; assay, **42**, 363, 364; **90**, 330; molecular weight, **42**, 367; properties, **42**, 367, 368; **90**, 333; purification, **42**, 364; **90**, 331); bumblebee flight muscle (assay, **90**, 366; properties, **90**, 370; purification, **90**, 367); *Candida utilis*, from, **9**, 632; **42**, 347 (activators and inhibitors, **42**, 352; assay, **9**, 632; **42**, 347; crystallization, **9**, 634; metal requirements, **9**, 635; molecular weight, **9**, 635; **42**, 352; pH optimum, **9**, 635; properties, **9**, 635; **42**, 351; purification, **9**, 633; **42**, 348); chicken muscle (assay, **90**, 340; properties, **90**, 344; purification, **90**, 341); chloroplast enzyme (assay, **252**, 223; light-dependent activation, **252**, 220; purification, **252**, 222; redox titration (curve fitting of data, **252**, 227; determination of E_h, **252**, 228; establishment of equilibrium, **252**, 225; monobromobimane labeling of thiols, **252**, 226; stability of ambient redox potential, **252**, 225); subunit structure, **252**, 221); cooperativity in, allosteric structures and model testing, **249**, 555; elution from carboxymethyl-cellulose, **25**, 229; *Euglena gracilis*, from, (assay, **9**, 636; pH optimum, **9**, 638; properties, **9**, 638; purification, **9**, 637); ferredoxin activation of, **23**, 415; ferredoxin-thioredoxin reductase assay, **69**, 389; gluconeogenic catalytic activity, **37**, 281; ligand binding sites (interaction distances, **249**, 561; intersubunit, **249**, 559); liver, from, **2**, 543; **9**, 10, 625; **42**, 354, 369 (activators and inhibitors, **2**, 546; **9**, 629; **42**, 358, 374; assay, **2**, 543; **9**, 10, 625; **42**, 354, 355, 363, 364, 369, 370; **90**, 327, 330, 345; chromatography, **9**, 12; **42**, 356; crystallization, **9**, 629; distribution, **42**, 359; molecular weight, **9**, 631; **42**, 357; pH optimum, **2**, 546; **9**, 630; properties, **2**, 545; **9**, 629; **42**, 357, 373, 374; **90**, 347; purification, **2**, 544; **9**, 10, 627; **42**, 355, 356, 364, 370; **74**, 211, 212; **90**, 327, 331, 346; specificity, **2**, 545; **9**, 629; subunits, **9**, 631; zinc-free, preparation, **90**, 328); maleylation, **25**, 533; murine intestinal mucosa (assay, **90**, 358; properties, **90**, 365; purification, **90**, 360); partition analysis, **249**, 322; *Polysphondylium pallidum*, from, **42**, 360 (assay, **42**, 360; chromatography, **42**, 362; properties, **42**, 362, 363; purification, **42**, 362); purification by elution with substrate, **22**, 379; rabbit intestinal mucosa (assay, **90**, 358; properties, **90**, 365; purification, **90**, 362); rabbit liver, from, **2**, 543; **9**, 625; **42**, 354, 369 (activators and inhibitors, **2**, 546; **9**, 629; **42**, 358, 374; assay, **2**, 543; **9**, 625; **42**, 354, 355, 369, 370; **90**, 327, 345; chromatography, **9**, 12; **42**, 356; crystallization, **9**, 629; distribution, **42**, 359; molecular weight, **9**, 631; **42**, 357; pH optimum, **2**, 546; **9**, 630; properties, **2**, 545; **9**, 629; **42**, 357, 373, 374; **90**, 347; purification, **2**, 544; **9**, 10, 627; **42**, 355, 356, 370; **74**, 211, 212; **90**, 327, 346; specificity, **2**, 545; **9**, 629; subunits, **9**, 631; zinc-free, preparation, **90**, 328); rabbit muscle (assay, **90**, 340; properties, **90**, 344; purification, **90**, 343); radioimmunoassay, **74**, 210 (antibody preparation, **74**, 212; application to tissue extracts, **74**, 219; chromatography columns, **74**, 211; fluorescamine assay for enzyme protein, **74**, 213; immune reaction, **74**, 216; immunogen iodination, **74**, 214; immunogen purification, **74**, 211; solutions, **74**, 210; standard competitive binding curve, **74**, 217); rat liver (assay, **90**, 352; properties, **90**, 355; purification, **90**, 353); reaction with FDNB, **11**, 548; reversible inhibitors as mechanistic probes, **249**, 141; *Rhodopseudomonas palustris* (assay, **90**, 379; properties, **90**, 383;

purification, **90**, 379); site-specific reagent, **47**, 481, 483; snake muscle (assay, **90**, 349; properties, **90**, 351; purification, **90**, 350); spinach, isolation from, **5**, 273; **23**, 691; spinach chloroplasts, in, **42**, 397 (activation and inhibition, **42**, 397, 402; amino acids, **42**, 402; assay, **42**, 398; chromatography, **42**, 401; molecular weight, **42**, 402; properties, **42**, 402, 404, 405; protein factor component from spinach chloroplasts [assay, **42**, 403; chromatography, **42**, 404; properties, **42**, 404; purification, **42**, 404]; purification, **42**, 400; **90**, 374); spinach leaf (assay, **5**, 272; **90**, 372; properties, **5**, 275; **90**, 377; purification from [chloroplast, **90**, 374; cytoplasm, **5**, 273; **90**, 375]); storage, **74**, 211; swine kidney, from, **42**, 375, 385 (assay, **42**, 385, 386; inhibition, **42**, 389; molecular weight, **42**, 388; properties, **42**, 388, 389; purification, **42**, 386); total liver activity, **74**, 229, 230; turkey liver (assay, **90**, 334 properties, **90**, 337; purification, **90**, 335)

FRUCTOSE-2,6-BISPHOSPHATASE

This hydrolase [EC 3.1.3.46], also known as fructose-2,6-bisphosphate 2-phosphatase, catalyzes the conversion of D-fructose 2,6-bisphosphate to yield D-fructose 6-phosphate and orthophosphate.[1–4] In many organisms, this enzymatic activity is associated with 6-phosphofructo-2-kinase. The bisphosphatase catalyzes D-fructose 2,6-bisphosphate hydrolysis via a covalent $N^{3'}$-phosphohistidine intermediate by an ordered, sequential mechanism. D-Fructose 6-phosphate dissociation is rate-limiting.[1]

See *6-Phosphofructo-2-Kinase/Fructose-2,6-Bisphosphatase; Fructose-2,6-Bisphosphate 6-Phosphatase*

[1] D. A. Okar, D. H. Live, M. H. Devany & A. J. Lange (2000) *Biochemistry* **39**, 9754.
[2] K. Uyeda (1991) *A Study of Enzymes* **2**, 445.
[3] S. J. Pilkis, T. H. Claus, P. D. Kountz & M. R. El-Maghrabi (1987) *The Enzymes*, 3rd ed., **18**, 3.
[4] E. Van Schaffingen (1987) *AE* **59**, 315.

FRUCTOSE-1,6-BISPHOSPHATE ALDOLASE

This aldolase [EC 4.1.2.13], often simply known simply as aldolase, catalyzes the reversible cleavage of D-fructose 1,6-bisphosphate to dihydroxyacetone phosphate (*i.e.*, glycerone phosphate) and D-glyceraldehyde 3-phosphate.[1–13] The enzyme, which has an ordered Uni Bi kinetic mechanism, will also act on (3*S*,4*R*)-ketose 1-phosphates (for example, D-fructose 1-phosphate). The ΔG for this reaction is +23.9 kJ/mol, indicating that the reaction is highly endergonic. Although this thermodynamic property favors gluconeogenesis, the reaction still mediates C–C bond cleavage at intracellular concentrations of reactants and products, in which case, the ΔG is approximately −1.3 kJ/mol.

Aldolases are grouped into two different classes. Class I aldolases are not inhibited by chelating agents such as EDTA and an intermediate can be trapped with borohydride treatment. This class of aldolases proceeds by covalent catalysis (*i.e.*, through the formation of a protonated imine intermediate). After the class I fructose-1,6-bisphosphate aldolase binds the substrate, a transient carbinolamine intermediate forms by reaction with a lysyl residue. A Schiff base is produced from this intermediate. Release of D-glyceraldehyde 3-phosphate generates an enamine which is subsequently converted to a new Schiff base. Subsequent hydrolysis produces dihydroxyacetone phosphate. Class II aldolases (commonly found in bacteria, fungi, and algae) require Zn^{2+} or Fe^{2+} and are inhibited by EDTA; no covalent intermediate is formed. In Class II aldolases, the divalent cation stabilizes the enolate intermediate or forms a metal ion–sugar complex that acts as an electrophile.

[1] K. N. Allen (1998) *CBC* **2**, 135.
[2] J. V. Schloss & M. S. Hixon (1998) *CBC* **2**, 43.
[3] J. A. Littlechild & H. C. Watson (1993) *Trends Biochem. Sci.* **18**, 36.
[4] R. Kluger (1992) *The Enzymes*, 3rd ed., **20**, 271.
[5] S. J. Gamblin, G. J. Davies, J. M. Grimes, R. M. Jackson, J. A. Littlechild & H. C. Watson (1991) *J. Mol. Biol.* **219**, 573.
[6] D. J. Hupe (1991) in *A Study of Enzymes* (S. A. Kuby, ed.), vol. **2**, p. 485, CRC Press, Boca Raton.
[7] D. J. Creighton & N. S. R. K. Murthy (1990) *The Enzymes*, 3rd ed., **19**, 323.
[8] A. S. Mildvan & D. C. Fry (1987) *AE* **59**, 241.
[9] B. L. Vallee & A. Galdes (1984) *AE* **56**, 283.
[10] A. E. Martell (1982) *AE* **53**, 163.
[11] C. Y. Lai & B. L. Horecker (1972) *Essays in Biochem.* **8**, 149.
[12] B. L. Horecker, O. Tsolas & C. Y. Lai (1972) *The Enzymes*, 3rd ed., **7**, 213.
[13] D. E. Morse & B. L. Horecker (1968) *AE* **31**, 125.

Selected entries from **Methods in Enzymology** [vol, page(s)]:
General discussion: **1**, 310, 315, 320; **5**, 310; **9**, 479; **27**, 90, 262; **34**, 165; **42**, 223, 228, 234, 240, 249; **46**, 23, 25, 49, 132, 383; **90**, 235, 241, 251, 254, 259

Assay: **1**, 310, 315; **3**, 202; **4**, 375; **5**, 310; **9**, 481, 486, 492; *Ascaris suum*, **90**, 255; *Bacillus subtilis*, **90**, 235; blue-green algae, **42**, 228; bovine liver, **5**, 310; chicken muscle, **90**, 490; *Clostridium perfringens*, **9**, 486; Fast Analyzer, by, **31**, 816; fructose 1-phosphate, with, **1**, 320; *Helix pomatia*, **90**, 259; human erythrocyte, **90**, 251; lobster muscle, **42**, 223, 224; mammalian tissues, **1**, 310; **42**, 240; **9**, 492; **90**, 251; *Mycobacterium smegmatis*, **90**, 242; *Mycobacterium tuberculosis*, **90**, 242; *Peptococcus aerogenes*, **42**, 250, 251; rabbit liver, **9**, 492; rabbit muscle, **1**, 310; spinach, **42**, 235, 236; yeast, **1**, 315; **9**, 481

Other molecular properties: *N*-acetylneuraminate and, **6**, 458, 465; activity determination, **47**, 495; affinity labeling, **47**, 483, 484, 491; **87**, 473; aldolase A, **9**, 480; aldolase B, **9**, 480 (distribution, **9**, 498; rabbit liver, from [molecular weight, **9**, 498; pH optimum, **9**, 497; properties, **9**, 496; purification, **9**, 494]); aldolase C, **9**, 480; anomeric specificity, **63**, 374 (direct kinetic determination, **63**, 378); *Ascaris suum* (assay, **90**, 255; properties, **90**, 257; purification, **90**, 256); *Bacillus subtilis* (assay, **90**, 235; properties, **90**, 240; purification, **90**, 238); bifunctional reagents and (alkyl halides, **25**, 635; imidoesters, **25**, 648); bisphosphoglycerate mutase assay and, **1**, 425; blue-green algae, from, **42**, 228 (assay, **42**, 228; chromatography, **42**, 232; molecular weight, **42**, 232; properties, **42**, 232; purification, **42**, 230); bovine liver, **5**, 310 (properties, **5**, 316; purification, **5**, 313); bovine muscle, simultaneous purification with glyceraldehyde-3-phosphate dehydrogenase, phosphoglycerate kinase, and phosphoglycerate mutase, **90**, 509; *N*-bromoacetylethanolamine phosphate, and, **87**, 473; chicken muscle (assay, **90**, 490; purification, **90**,

492); chiral acetate, and synthetsis of, **87**, 141; chiral methyl groups, and, **87**, 147; chiral thiophosphate, and analysis of, **87**, 314; *Chlamydomonas mundana*, from, **9**, 13; citraconylated, **25**, 551; class I (*Mycobacterium smegmatis* [assay, **90**, 242; properties, **90**, 245; purification, **90**, 243]; *Mycobacterium tuberculosis* [assay, **90**, 242; properties, **90**, 248; purification, **90**, 249]); class II, *Mycobacterium tuberculosis* (assay, **90**, 242; properties, **90**, 248; purification, **90**, 247); *Clostridium perfringens*, from (assay, **9**, 486; inhibitors, **9**, 490; molecular weight, **9**, 491; pH optimum, **9**, 490; properties, **9**, 490; purification, **9**, 487; specificity, **9**, 490); conjugate (assay, **44**, 336, 337 [monomeric derivatives, of, **44**, 495]; tetramer dissociation, **44**, 491); coupled assay, use in, **63**, 34; crystallization, **1**, 313, 394; **22**, 262; cytoplasmic inhibition, **23**, 227; dehydrogenase localization and, **6**, 969; denaturation studies, **44**, 498; density perturbation ultracentrifugation, **27**, 91; deoxyheptulosonate synthetase and, **5**, 397; detection, by tetrazolium dye reduction, **41**, 66, 67; diffusion coefficient, **27**, 351; *Drosophila melanogaster*, from, **42**, 223; EDTA-resistant fructose-1,6-bisphosphate aldolase, synthesis in *Chlamydomonas* cultures, **23**, 73; electrophoresis, **41**, 67; eneamine, **87**, 89; enediol, **87**, 88, 96; enzymic detection, **41**, 68, 70; epoxide synthesis, **47**, 490, 491 (site-specific reagent, **47**, 483); erythrose 4-phosphate and, **6**, 484; exchanges, **87**, 87, 96; ferricyanide, **87**, 88; fructose bisphosphate, **3**, 198; **9**, 479 (acid-induced difference spectra, **11**, 765; assay, **11**, 668; class I, **9**, 479; class II, **9**, 479; β-glycerophosphate derivative, **11**, 667 [preparation, **11**, 668]; reaction with CIDNB, **11**, 548, 549; solvent perturbation difference spectra, **11**, 756, 757; sources, **9**, 480); D-fructose 1-phosphate and, **1**, 320, 322 (properties, **1**, 322; purification, **1**, 321); gluconeogenic catalytic activity of, **37**, 281, 286; *Helix pomatia* (assay, **90**, 259; properties, **90**, 262; purification, **90**, 259); histidine residue, **87**, 96, 474; hollow-fiber retention data, **44**, 296; human erythrocyte (assay, **90**, 251; properties, **90**, 253; purification, **90**, 251); immobilization (polyacrylamide gel, in, **44**, 902, 909; Sepharose 4B, on, **44**, 493); indole 3-glycerolphosphate and, **6**, 594; interference by, in metabolite extraction, **13**, 438; intermediates, **87**, 6, 84, 91, 95, 96 (tetranitromethane and, **25**, 518); introduction, **9**, 479; isolated nuclei and, **12A**, 445; isotope exchange, **64**, 10, 20; *Leuconostoc mesenteroides* and, **1**, 330; lobster muscle, from, **42**, 223 (amino acids, **42**, 226, 227; assay, **42**, 223, 224; chromatography, **42**, 225; molecular weight, **42**, 226; properties, **42**, 226, 239; purification, **42**, 224); lysine residue, **87**, 474; maize, catalytic efficiency, site-directed mutagenesis studies, **249**, 105; maleylation, **25**, 533, 535 (reaggregation and, **25**, 534); mammalian tissues, from, **1**, 310, 320; **5**, 310; **9**, 491; **42**, 240 (assay, **1**, 310, 320; **5**, 310; **9**, 492; **42**, 240; chromatography, **42**, 243, 244, 247; electrophoresis, **42**, 241, 242; purification, **1**, 312, 321; **5**, 313; **9**, 13, 494; **42**, 242); membrane osmometry, **48**, 75; metal ion, **87**, 96, 97; methylglyoxal formation, **87**, 91, 95, 96; microcalorimetric studies of binding, **61**, 302; molecular weight standard, **28**, 55, 57; muscle, from, **1**, 310; **9**, 13 (purification, **1**, 312; **9**, 13); *Mycobacterium smegmatis* (assay, **90**, 242; properties, **90**, 245; purification, **90**, 243); *Mycobacterium tuberculosis* (assay, **90**, 242; properties, **90**, 248, 250; purification, **90**, 247, 249); myokinase and, **6**, 230; oxidation by, **87**, 88; partial volume, **26**, 109, 110, 118; *Peptococcus aerogenes*, of, **42**, 249 (amino acids, **42**, 256, 257; assay, **42**, 250, 251; chromatography, **42**, 253; properties, **42**, 255; purification, **42**, 252); pH effect, **87**, 474; phosphate assay and formation, **55**, 212; **87**, 91; phosphofructokinase, assay of, **9**, 425, 430, 437; phosphohexokinase assay and, **1**, 307; phosphoketotetrose aldolase preparation, presence in, **5**, 286; potential site-specific reagent, **25**, 666; preferential solvent interaction, **61**, 33; pyruvate kinase and, **5**, 368; rabbit liver, **9**, 491 (assay, **9**, 492; properties, **9**, 496; purification, **9**, 13, 494); rabbit muscle, of, **1**, 310; **27**, 108, 505 (crosslinking, **172**, 623; **215**, 406, 408; properties, **1**, 313; purification, **1**, 312; **9**, 13); radioiodination, **70**, 212; rat liver, fructose-1-phosphate aldolase activity (assay, **1**, 320; properties, **1**, 322; purification, **1**, 321); reaction quenching, **64**, 57; reactions of, **5**, 310; recrystallization, **5**, 315; rhamnulose-1-phosphate aldolase, similarity to, **9**, 545; sedimentation coefficient, **61**, 108; sedoheptulose and, **3**, 195; **4**, 890; sedoheptulose-1,7-bisphosphate preparation, in, **41**, 77; sorbose 1-phosphate and, **3**, 179; spinach, from,

42, 234 (assay, **42**, 235, 236; chromatography, **42**, 237; molecular weight, **42**, 238; properties, **42**, 238, 239; purification, **42**, 236); staining on gels, **22**, 585, 592; structural homology, **87**, 86; succinylation, **25**, 539, 540; tetranitromethane, and, **25**, 518; **87**, 88, 96; thiol group accessibility, **251**, 235; transaldolase assay and, **1**, 381; transition-state and multisubstrate analogues, **249**, 307; triokinase and, **5**, 362; triosephosphate isomerase and, **87**, 96; tryptophan content, **49**, 162; tryptophan synthetase assay and, **5**, 802; xylulose-5-phosphate phosphoketolase and, **5**, 263; yeast, from, **1**, 315; **9**, 480; **27**, 351, 384, 399, 541, 612 (assay, **1**, 315; **9**, 481; inhibitors, **1**, 317; **9**, 484; kinetic properties, **9**, 483; molecular weight, **9**, 485; pH optimum, **9**, 484; properties, **1**, 317; **9**, 483; purification, **1**, 316; **9**, 482; specificity, **9**, 483); zinc, exchangeable, **87**, 96

FRUCTOSE-2,6-BISPHOSPHATE 6-PHOSPHATASE

This phosphatase [EC 3.1.3.54], first isolated from *Saccharomyces cerevisiae*, catalyzes the hydrolysis of D-fructose 2,6-bisphosphate to produce D-fructofuranose 2-phosphate and orthophosphate.[1-3] ***See also*** *Fructose-2,6-Bisphosphatase*

[1] C. Purwin, M. Laux & H. Holzer (1987) *EJB* **164**, 27.
[2] C. Purwin, M. Laux & H. Holzer (1987) *EJB* **165**, 543.
[3] U. Plankert, C. Purwin & H. Holzer (1988) *FEBS Lett.* **239**, 69.

FRUCTOSE 5-DEHYDROGENASE

This oxidoreductase [EC 1.1.99.11], a quinohemoprotein, catalyzes the reaction of D-fructose with an acceptor substrate to produce 5-dehydro-D-fructose and the reduced acceptor.[1,2] Acceptor substrates include 2,6-dichloroindophenol, ferricyanide, and phenazine methosulfate. The *Gluconobacter* enzyme has a ping pong Bi Bi kinetic mechanism.[3]

[1] Y. Yamada, K. Aida & T. Uemura (1967) *J. Biochem.* **61**, 636.
[2] M. Ameyama, E. Shinagawa, K. Matsushita & O. Adachi (1981) *J. Bacteriol.* **145**, 814.
[3] J. Marcinkeviciene & G. Johansson (1993) *FEBS Lett.* **318**, 23.

Selected entries from ***Methods in Enzymology*** [vol, page(s)]:
General discussion: **89**, 154
Assay: *Gluconobacter industrius* membrane-bound, **89**, 154
Other molecular properties: *Gluconobacter industrius* membrane-bound (microdetermination of D-fructose, **89**, 24; properties, **89**, 158; purification, **89**, 156)

FRUCTOSE 5-DEHYDROGENASE (NADP⁺)

This oxidoreductase [EC 1.1.1.124], also known as 5-ketofructose reductase (NADP⁺), catalyzes the reaction of 5-dehydro-D-fructose with NADPH to produce D-fructose and NADP⁺.[1-3] The *Erwinia* enzyme also uses NADH.[1] ***See also*** *Sorbose Dehydrogenase (NADP⁺)*

[1] J. L. Schrimsher, P. T. Wingfield, A. Bernard, R. Mattaliano & M. A. Payton (1988) *BJ* **253**, 511.
[2] G. Avigad, S. Englard & S. Pifko (1966) *JBC* **241**, 373.

[3]M. Ameyama, K. Matsushita & E. Shinagawa (1981) *Agric. Biol. Chem.* **45**, 863.

Selected entries from *Methods in Enzymology* [vol, page(s)]:
General discussion: **41**, 127
Assay: *Gluconobacter cerinus*, **41**, 127

FRUCTOSE-6-PHOSPHATE PHOSPHOKETOLASE

This enzyme [EC 4.1.2.22] catalyzes the reaction of D-fructose 6-phosphate with orthophosphate to produce acetyl phosphate, D-erythrose 4-phosphate, and water.[1–3] Other substrates include D-xylulose 5-phosphate, producing acetyl phosphate, D-glyceraldehyde 3-phosphate, and water.

[1]D. A. Whitworth & C. Ratledge (1977) *J. Gen. Microbiol.* **102**, 397.
[2]B. Sgorbath, G. Lenaz & F. Casalicchio (1976) *Antonie Leeuwenhoek* **42**, 49.
[3]M. Schramm, V. Klybas & E. Racker (1958) *JBC* **233**, 1283.

Selected entries from *Methods in Enzymology* [vol, page(s)]:
General discussion: **5**, 276
Assay: **5**, 276

FRUCTURONATE REDUCTASE

This oxidoreductase [EC 1.1.1.57], also known as D-mannonate oxidoreductase and D-mannonate dehydrogenase, catalyzes the reversible reaction of D-fructuronate with NADH to produce D-mannonate and NAD^+. Other substrates include D-tagaturonate (producing D-altronate). The *Escherichia coli* enzyme has a random Bi Bi kinetic mechanism.[1,2]

[1]R. C. Portalier & F. R. Stoeber (1972) *EJB* **26**, 290.
[2]R. C. Portalier (1972) *EJB* **30**, 220.

Selected entries from *Methods in Enzymology* [vol, page(s)]:
General discussion: **89**, 210
Assay: *Escherichia coli*, **89**, 210

*Fse*I RESTRICTION ENDONUCLEASE

This type II restriction endonuclease [EC 3.4.21.4], isolated from *Frankia* species Eu11b, catalyzes the hydrolysis of both strands of DNA at $5' \ldots GGCCGG\nabla CC \ldots 3'$, where the symbol ∇ indicates the cleavage site.[1]

[1]J. M. Nelson, S. M. Miceli, M. P. Lechevalier & R. J. Roberts (1990) *NAR* **18**, 2061.

*Fsp*I RESTRICTION ENDONUCLEASE

This type II restriction endonuclease [EC 3.4.21.4], isolated from *Fischerella* species, catalyzes the hydrolysis of both strands of DNA at $5' \ldots TGC\nabla GCA \ldots 3'$, where the symbol ∇ indicates the cleavage site, to produce blunt-ended fragments.

FtsH PROTEASE

This ATP-dependent protease was originally identified in a cell-division-defective mutation of *Escherichia coli* ("fts" is the descriptor for filamentation temperature-sensitive). This protease has also been referred to as HflB protease (for high frequency lysogenation for bacteriophage λ). FtsH protease is a membrane-bound protein that is essential for growth. The enzyme has a basal ATPase activity and protein hydrolysis, against σ^{32} and SecY proteins (SecY is a subunit of the SecYEG protein translocase in the cytosolic membrane and σ^{32} is a heat-shock factor), requires concomitant ATP hydrolysis. The λ cII protein has also been suggested to be a substrate for the FtsH protease (this protein is critical for the lysogenation of λ).[1]

Although ATP hydrolysis is considered to be an ATPase reaction, the actual reaction is more correctly written as: Binding State$_1$ + ATP + H_2O = Binding State$_1$ + ADP + P_i, where two affinity-modulated binding states of a filament are indicated by the subscripts. In this respect, FtsH protease possesses the properties of an energase, a specialized mechanochemical enzyme that facilitates affinity-modulated reactions by coupling the $\Delta G_{\text{ATP-hydrolysis}}$ to protease action.[2] **See Energase**

[1]T. Tomoyasu, J. Gamer, J. Bakau, M. Kanemori, H. Mori, A. J. Rutman, A. B. Oppenheim, T. Yura, K. Yamanaka, H. Niki, S. Hiraga & T. Ogura (1995) *EMBO J.* **14**, 2551.
[2]D. L. Purich (2001) *TiBS* **26**, 417.

FtsZ GTPase

This tubulin-like bacterial protein, which is responsible for the ring-like septation furrow (or cell-wall and cell-membrane constriction) that forms two daughter cells, catalyzes the hydrolysis of GTP.[1–3] Although the Enzyme Commission currently classifies proteins like FtsZ as GTPases, the actual reaction is more correctly written as: Polymerization State$_1$ + GTP + H_2O = Polymerization State$_1$ + GDP + P_i, where two affinity-modulated polymerization (or conformational) states of a filament are indicated by the subscripts. In this respect, FtsZ is an energase, a specialized mechanochemical enzyme that facilitates affinity-modulated reactions by coupling the $\Delta G_{\text{GTP-hydrolysis}}$ to the self-assembly and/or subsequent conformational maturation of the FtsZ protofilament sheets and minirings.[4] **See Energase**

[1]P. de Boer, R. Crossley & L. Rothfield (1992) *Nature* **359**, 254.
[2] H. P. Erickson, D. W. Taylor, K. A. Taylor & C. Bramhill (1996) *PNAS* **93**, 519.

[3] J. Lowe & L. A. Amos (1998) *Nature* **391**, 121.
[4] D. L. Purich (2001) *TiBS* **26**, 417.

FUCOIDANASE

This enzyme [EC 3.2.1.44] catalyzes the endohydrolysis of 1,2-α-L-fucoside linkages in fucoidan without the release of sulfate.[1,2] Fucoidans are branched, sulfated heteropolysaccharides containing mostly L-fucosyl residues (many of which are sulfated).

[1] G. A. Levvy & A. McAllan (1967) *BJ* **80**, 435.
[2] N. M. Thanassi & H. I. Nakada (1967) *ABB* **118**, 172.

FUCOKINASE

This phosphotransferase [EC 2.7.1.52] catalyzes the reaction of ATP with 6-deoxy-L-galactose (*i.e.*, β-L-fucose) to produce ADP and 6-deoxy-L-galactose 1-phosphate (*i.e.*, L-fucose 1β-phosphate) with retention of configuration at the anomeric carbon.[1]

[1] W. Butler & G. S. Serif (1985) *BBA* **829**, 238.
Selected entries from *Methods in Enzymology* [vol, page(s)]:
General discussion: **28**, 399
Assay: pig liver, **28**, 399
Other molecular properties: pig liver, from (assay, **28**, 399; GDP-L-fucose synthesis and, **28**, 285; properties, **28**, 402; purification, **28**, 400); 32P-labeling of fucose, **50**, 191; quantitation of fucose, **50**, 200

D-FUCONATE DEHYDRATASE

This dehydratase [EC 4.2.1.67] catalyzes the conversion of D-fuconate to 2-dehydro-3-deoxy-D-fuconate and water.[1,2] Other substrates include L-arabinonate, producing 2-keto-3-deoxy-L-arabonate. *See also Galactonate Dehydratase*

[1] A. S. Dahms & R. L. Anderson (1972) *JBC* **247**, 2233.
[2] W. A. Wood (1971) *The Enzymes*, 3rd ed., **5**, 573.
Selected entries from *Methods in Enzymology* [vol, page(s)]:
General discussion: **42**, 305; **90**, 299
Assay: pseudomonad MSU-1, **42**, 305, 306
Other molecular properties: pseudomonad MSU-1, from, **42**, 305 (assay, **42**, 305, 306; chromatography, **42**, 306; properties, **42**, 307, 308; purification, **42**, 306, 307)

L-FUCONATE DEHYDRATASE

This enzyme [EC 4.2.1.68] catalyzes the conversion of L-fuconate to 2-dehydro-3-deoxy-L-fuconate and water.[1,2] Other substrates include D-arabinonate, producing 2-keto-3-deoxy-D-arabonate.

[1] R. Yuen & H. Schachter (1972) *Can. J. Biochem.* **50**, 798.
[2] J. Y. Chan, N. A. Nwokoro & H. Schachter (1979) *JBC* **254**, 7060.

L-FUCOSE ISOMERASE

This isomerase [EC 5.3.1.25] catalyzes the interconversion of L-fucose and L-fuculose. The *Escherichia coli* enzyme uses Mn^{2+} as a cofactor and the reaction probably belongs to the ene-diol type.[1] One should note that D-arabinose isomerase can also utilize L-fucose as an alternative substrate (in fact, D-arabinose isomerase has been known as L-fucose isomerase in a number of publications). *See also D-Arabinose Isomerase*

[1] J. E. Seemann & G. E. Schulz (1997) *J. Mol. Biol.* **273**, 256.

FUCOSE-1-PHOSPHATE GUANYLYLTRANSFERASE

This enzyme [EC 2.7.7.30] catalyzes the reaction of GTP with L-fucose 1-phosphate to produce GDP-L-fucose and pyrophosphate (or, diphosphate).[1,2]

[1] E. Adams (1976) *AE* **44**, 69.
[2] H. Ishihara & E. C. Heath (1968) *JBC* **243**, 1110.
Selected entries from *Methods in Enzymology* [vol, page(s)]:
General discussion: **28**, 403
Assay: **28**, 403
Other molecular properties: GDP-L-fucose synthesis and, **28**, 285; purification, **28**, 404

β-D-FUCOSIDASE

This enzyme [EC 3.2.1.38] catalyzes the hydrolysis of terminal non-reducing β-D-fucose residues in β-D-fucosides. Enzymes from some sources also hydrolyze β-D-galactosides and/or β-D-glucosides and/or α-L-arabinosides.[1,2] The synthetic substrate *p*-nitrophenyl β-D-fucopyranoside can be used to assay this fucosidase. Some xylan 1,4-β-xylosidases [EC 3.2.1.37] also exhibit this activity.

[1] R. Giordani & G. Noat (1988) *EJB* **175**, 619.
[2] M. J. Melgar, J. A. Cabezas & P. Calvo (1985) *Comp. Biochem. Physiol.* **80B**, 149.

α-L-FUCOSIDASE

This set of enzymes [EC 3.2.1.51] catalyzes the hydrolysis of an α-L-fucoside to produce α-L-fucose and the free alcohol.[1–5] Synthetic substrates include *p*-nitrophenyl α-L-fucoside and methylumbelliferyl α-L-fucopyranoside. Mammalian fucosidases are sialoglycoproteins.

[1] S. C. Fry (1995) *Ann. Rev. Plant Physiol. Plant Mol. Biol.* **46**, 497.
[2] S. W. Johnson & J. A. Alhadeff (1991) *Comp. Biochem. Physiol. B* **99**, 479.
[3] E. Conzelmann & K. Sandhoff (1987) *AE* **60**, 89.
[4] P. M. Dey & E. del Campillo (1984) *AE* **56**, 141.
[5] H. M. Flowers & N. Sharon (1979) *AE* **48**, 29.
Selected entries from *Methods in Enzymology* [vol, page(s)]:
General discussion: **8**, 584; **50**, 505; **83**, 625
Assay: **50**, 453; almond emulsin, **83**, 626; rat liver lysosomes, **50**, 506
Other molecular properties: almond emulsin (assay, **83**, 626; properties, **83**, 630; purification, **83**, 628); analysis in multiple enzyme systems, **22**, 18, 20; blood-group substances and, **8**, 709, 711; bovine

kidney, for glycosphingolipid analysis, **230**, 385; carbohydrate structure, **28**, 19; *Charonia lampas* (digestion conditions, **230**, 284; purification, **230**, 288; specificity, **230**, 284); distribution, **50**, 453; α-fucosidase I (almond [digestion conditions, **230**, 285; specificity, **230**, 285]); α-fucosidase III (almond [digestion conditions, **230**, 285; specificity, **230**, 285, 290]); fucosidosis, **50**, 453; isolation of 6′-galactosyllactose, **50**, 219; mucolipidoses II and III, **50**, 453; rat liver lysosomes, from (assay, **50**, 506; properties, **50**, 508; purification, **50**, 506); staining on gels, **22**, 600

1,2-α-L-FUCOSIDASE

This enzyme [EC 3.2.1.63], also known as almond emulsin fucosidase II, catalyzes the hydrolysis of methyl-2-α-L-fucopyranosyl-β-D-galactoside to produce L-fucose and methyl-β-D-galactoside.[1,2] Other substrates are non-reducing terminal L-fucosyl residues linked to D-galactosyl residues by an 1,2-α-linkage. The enzyme is distinct from 1,3-α-L-fucosidase [EC 3.2.1.111].

[1] M. Ogata-Arakawa, T. Muramatsu & A. Kobata (1977) *ABB* **181**, 353.
[2] O. P. Bahl (1970) *JBC* **245**, 299.

Selected entries from *Methods in Enzymology* [**vol**, page(s)]:
Assay: *Aspergillus niger*, **28**, 738; *Clostridium perfringens*, **28**, 764
Other molecular properties: *Aspergillus niger*, from (assay, **28**, 738; properties, **28**, 742; purification, **28**, 740; reaction, **28**, 738); *Clostridium perfringens*, from (assay, **28**, 764; enzymatic properties, **28**, 767; other sources, **28**, 763; purification, **28**, 765)

1,3-α-L-FUCOSIDASE

This glycosidase [EC 3.2.1.111], also known as almond emulsin fucosidase I, catalyzes the hydrolysis of 1,3-linkages between α-L-fucose and *N*-acetylglucosamine residues in glycoproteins.[1,2] This enzyme is distinct from 1,2-α-L-fucosidase [EC 3.2.1.63]. Substrates include lacto-*N*-fucopentaitol II and III, asialoorosomucoid, and lacto-ferrin.

[1] M. Ogata-Arakawa, T. Muramatsu & A. Kobata (1977) *ABB* **181**, 353.
[2] M. J. Imber, L. R. Glasgow & S. V. Pizzo (1982) *JBC* **257**, 8205.

1,6-α-L-FUCOSIDASE

This glycosidase [EC 3.2.1.127] catalyzes the hydrolysis of 1,6-linkages between α-L-fucose and *N*-acetyl-D-glucosamine in glycopeptides such as bovine immunoglobulin G glycopeptide and fucosylasialoagalactofetuin. The enzyme from *Aspergillus niger* does not act on 1,2-, 1,3- or 1,4-L-fucosyl linkages.[1]

[1] S. Yazawa, R. Madiyalakan, R. P. Chawda & K. L. Matta (1986) *BBRC* **136**, 563.

FUCOSTEROL-EPOXIDE LYASE

This enzyme [EC 4.1.2.33], isolated from the tobacco hornworm (*Manduca sexta*), catalyzes the conversion of (24R,24′R)-fucosterol epoxide to desmosterol and

acetaldehyde.[1] The enzyme participates in the conversion of sitosterol into cholesterol in insects. Intriguingly, there appears to be a low degree of stereospecificity during this conversion.[2]

[1] G. D. Prestwich, M. Angelastro, A. De Palma & M. A. Perino (1985) *Anal. Biochem.* **151**, 315.
[2] Y. Fujimoto, M. Morisaki & N. Ikekawa (1980) *Biochemistry* **19**, 1065.

L-FUCULOKINASE

This phosphotransferase [EC 2.7.1.51] catalyzes the reaction of ATP with L-fuculose to produce ADP and L-fuculose 1-phosphate.[1] Other substrates include D-ribulose, D-xylulose, and D-fructose.

[1] E. C. Heath & M. A. Ghalambor (1962) *JBC* **237**, 2423.

Selected entries from *Methods in Enzymology* [**vol**, page(s)]:
General discussion: **9**, 461
Assay: **9**, 461

L-FUCULOSE-1-PHOSPHATE ALDOLASE

This zinc-dependent class II aldolase [EC 4.1.2.17] catalyzes the reversible cleavage of L-fuculose 1-phosphate to produce glycerone phosphate (*i.e.*, dihydroxyacetone phosphate) and (S)-lactaldehyde.[1,2] The metal ion is thought to polarize the substrate, with subsequent steps mediated by general acid catalysis.[3]

[1] M. A. Ghalambor & E. C. Heath (1962) *JBC* **237**, 2427.
[2] D. S. Feingold & P. A. Hoffee (1972) *The Enzymes*, 3rd ed., **7**, 303.
[3] A. C. Joerger, C. Gosse, W.-D. Fessner & G. E. Schulz (2000) *Biochemistry* **39**, 6033.

Selected entries from *Methods in Enzymology* [**vol**, page(s)]:
General discussion: **9**, 538
Assay: **9**, 538
Other molecular properties: enediol-3-phosphate, and, **87**, 97; *Escherichia coli*, from (pH optimum, **9**, 542; properties, **9**, 541; purification, **9**, 540; specificity, **9**, 541)

FUMARATE EPOXIDASE

This oxidoreductase, first isolated from *Aspergillus fumigatus*, catalyzes the reaction of fumarate with dioxygen to produce *trans*-epoxysuccinate and water.[1]

[1] L. J. Wilkoff & W. R. Martin (1963) *JBC* **238**, 843.

FUMARATE HYDRATASE

This hydratase [EC 4.2.1.2] catalyzes the reversible hydration of fumarate to produce (S)-malate.[1–4]

fumarate   (S)-malate

When written in the direction of fumarate formation, the enzyme binds malate to form a carbanionic intermediate that eliminates OH$^-$ to produce fumarate (a β-elimination). The electron-withdrawing properties of the nitro group in 3-nitro-2(S)-hydroxypyruvate increases the acidity of the C3 proton (pK_a around 10) such that this analogue becomes a transition state mimic.[2]

[1] D. J. Creighton & N. S. R. K. Murthy (1990) *The Enzymes*, 3rd ed., **19**, 323.
[2] D. J. T. Porter & H. J. Bright (1980) *JBC* **255**, 4772.
[3] R. L. Hill & J. W. Teipel (1971) *The Enzymes*, 3rd ed., **5**, 539.
[4] R. A. Alberty (1961) *The Enzymes*, 2nd ed., **5**, 531.

Selected entries from *Methods in Enzymology* [vol, page(s)]:
General discussion: **1**, 729; **13**, 91; **34**, 410, 411; **44**, 832
Assay: **1**, 729; **13**, 91; **31**, 742; **136**, 456; coupled enzyme assay, **63**, 33
Other molecular properties: acetylene dicarboxylate, and, **87**, 90; activity, **4**, 374; adenylosuccinase and, **6**, 790; arginine synthesis and, **2**, 364, 367; brain mitochondria, **55**, 58; carboxymethylation of, **25**, 430; chiral methyl groups, and, **87**, 129, 137, 142, 159; chromaffin granule, in, **31**, 381; continuous crosscurrent extraction, **228**, 576, 582; cooperativity and slow transitions in, **249**, 545; counterflow experiments with, **249**, 331; dansylated, fluorescence polarization of, **11**, 865; determination of specific radioactivity of fumarate, **13**, 528; deuterium transfer by, **13**, 588; effect of anions on activity, **13**, 91, 96, 97; enol product, **87**, 90; equilibrium isotope effect, **64**, 110; equilibrium perturbation, **64**, 112; exchangeable proton, **63**, 206, 229, 230; extinction coefficient, **13**, 91; fluorometric assay of fumarate, **13**, 463; histidine, active site, **87**, 398, 400; horse heart (properties, **13**, 96; purification, **13**, 92); hydrogen bonding, **87**, 398; induced transport catalyzed by, **249**, 233; inhibition, **13**, 99; **63**, 229; **87**, 627; intermediate, **87**, 627; Iso mechanism, **249**, 212; isotope effects, **87**, 627, 633; isotope exchange, **64**, 20; latency, intact mitochondria, **55**, 143; ligand-independent recycling, **249**, 328; malate degradation and, **4**, 603; mechanism, **64**, 124, 125; L-malic acid production by immobilized bacteria, in, **136**, 458, 459; mitochondria marker, as, **31**, 406, 408, 735, 740, 742; nonliganded forms, partition studies, **249**, 331; pH effects, **63**, 229; **87**, 398, 400; pig heart (properties, **1**, 734; purification, **1**, 730); preparation of deuterated malate, **13**, 579, 580; progess curve analysis, **63**, 180; reaction (hydrogen transfer in, **249**, 336; progress curve analysis, Selwyn's test, **249**, 81); reversible reactions, progress curve analysis, **249**, 77; specific radioactivity determination with, **13**, 529, 531, 533; stability, **13**, 91, 96; substrate activation, **249**, 330; subunit structure, **13**, 98; succinate dehydrogenase and, **5**, 599; toluene-treated mitochondria, **56**, 547; transition state and multisubstrate analogues, **87**, 627; **249**, 308

FUMARATE REDUCTASE (NADH)

This oxidoreductase [EC 1.3.1.6] catalyzes the reaction of fumarate with NADH and a proton to produce succinate and NAD$^+$.[1–3] Other substrates include maleate and mesaconate. NADH can be replaced with FADH$_2$.

Confusion arises when this enzyme name is used to designate succinate dehydrogenase [EC 1.3.99.1] or succinate dehydrogenase (ubiquinone) [EC 1.3.5.1].

[1] K. M. Noll (1995) *Meth. Enzymol.* **251**, 470.
[2] B. J. Aue & R. H. Deibel (1967) *J. Bacteriol.* **93**, 1770.
[3] H.-G. Wetzstein & G. Gottschalk (1985) *Arch. Microbiol.* **143**, 157.

Selected entries from *Methods in Enzymology* [vol, page(s)]:
General discussion: **10**, 729; **251**, 470
Assay: **10**, 733

FUMARYLACETOACETASE

This enzyme [EC 3.7.1.2], also known as β-diketonase and fumarylacetoacetate hydrolase, catalyzes the hydrolysis of 4-fumarylacetoacetate to produce acetoacetate and fumarate (the reaction is essentially the reverse of a Claisen condensation).

The mechanism appears to involve a Glu/His/water catalytic triad in which the imidazole group functions as a general base.[1]

[1] D. E. Timm, H. A. Mueller, P. Bhanumoorthy, J. M. Harp & G. J. Bunick (1999) *Structure Fold Des.* **7**, 1023.

Selected entries from *Methods in Enzymology* [vol, page(s)]:
General discussion: **2**, 298
Assay: **2**, 298
Other molecular properties: acylpyruvase and, **2**, 298; homogentisate oxidase and, **2**, 293, 294; liver, **2**, 293, 294, 297, 298, 300; maleylacetoacetate isomerase and, **2**, 296; properties, **2**, 299; purification, **2**, 299

FURIN

This calcium-activated serine endopeptidase [EC 3.4.21.75], also called PACE and SPC1, catalyzes peptide bond hydrolysis during the processing of various proproteins in the *trans*-Golgi network, including growth factors, receptors, as well as serum factors.[1–3] Hydrolysis results in the release of mature proteins from their proproteins by hydrolysis of ArgXaaYaaArg–Zaa bonds, where Xaa can be any aminoacyl residue and Yaa is an arginyl or a lysyl residue. Albumin, complement component C3, and von Willebrand factor are thus released from their respective precursors.

[1] C. W. Wharton (1998) *CBC* **1**, 345.
[2] P. A. Halban & J.-C. Irminger (1994) *BJ* **299**, 1.
[3] W. J. M. Van de Ven, J. L. P. Van Duijnhoven & A. J. M. Roebroek (1993) *Crit. Rev. Oncol.* **4**, 115.

Selected entries from *Methods in Enzymology* [vol, page(s)]:
Assay: **244**, 171, 174
Other molecular properties: activation, **244**, 178; active site residues, **244**, 176; biological role, **244**, 167, 174; cleavage site specificity, **244**, 167; gene loci, **244**, 185; homology with Kex2 protease, **244**, 168; substrates, **244**, 174; tissue distribution, **244**, 168, 184; truncated soluble enzyme (expression in CHO cells, **244**, 170; plasmid construction, **244**, 169; preparation of conditioned medium, **244**, 170; processing, **244**, 172; purification steps, **244**, 171; storage, **244**, 173)

2-FUROYL-CoA DEHYDROGENASE

This copper- and molybdenum-dependent oxidoreductase [EC 1.3.99.8], also known as furoyl-CoA hydroxylase, catalyzes the reaction of 2-furoyl-CoA with water and an acceptor substrate to produce S-(5-hydroxy-2-furoyl)-CoA and the reduced acceptor.[1] The product's hydroxyl group is derived from water, not dioxygen. Electron acceptor substrates include methylene blue and nitro blue tetrazolium.

[1] J. P. Kitcher, P. W. Trudgill & J. S. Rees (1972) BJ **130**, 121.

2-FUROYL-CoA SYNTHETASE

2-furoate

This ligase [EC 6.2.1.31], also known as 2-furoate:CoA ligase, catalyzes the reaction of ATP with 2-furoate and coenzyme A to produce AMP, pyrophosphate (or, diphosphate), and 2-furoyl-CoA.[1]

[1] K. Koenig & J. R. Andreesen (1989) Appl. Environ. Microbiol. **55**, 1829.

FURYLFURAMIDE ISOMERASE

This NAD(P)H-dependent isomerase [EC 5.2.1.6] catalyzes the interconversion of (E)-2-(2-furyl)-3-(5-nitro-2-furyl)acrylamide and (Z)-2-(2-furyl)-3-(5-nitro-2-furyl)-acrylamide.[1,2]

[1] B. Kalyanaraman, R. P. Mason, R. Rowlett & L. D. Kispert (1981) BBA **660**, 102.
[2] M. Tomoeda & R. Kitamura (1977) BBA **480**, 315.

FUSARININE-C ORNITHINESTERASE

This enzyme [EC 3.1.1.48], also known as ornithine esterase, catalyzes the hydrolysis of an N^5-acyl-L-ornithine ester to produce an N^5-acyl-L-ornithine and an alcohol. The esterase acts on the three ornithine ester bonds in fusarinine C as well as on N^5-dinitrophenyl-L-ornithine methyl ester.[1]

[1] T. Emery (1976) Biochemistry **15**, 2723.

G

GALACTAN 1,3-β-GALACTOSIDASE

This galactosidase [EC 3.2.1.145], also called exo-(1 → 3)-D-galactanase, catalyzes the hydrolytic scission and release of terminal, non-reducing β-D-galactose residues in (1 → 3)-β-D-galactopyranans.[1,2] The enzyme will also act on terminal galactosyl residues of oligosaccharides containing other oligosaccharide chains on the O6 position (for example, the release of branches in [arabino-galacto-(1 → 6)]-(1 → 3)-β-D-galactans and release of (1 → 6)-β-D-galactobiose from [galacto-(1 → 6)]-(1 → 3)-β-D-galactans).

[1]Y. Tsumuraya, N. Mochizuki, Y. Hashimoto & P. Kovac (1990) *JBC* **265**, 7207.
[2]P. Pellerin & J. M. Brillouet (1994) *Carbohydr. Res.* **264**, 281.

GALACTARATE DEHYDRATASE

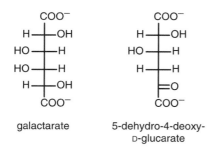

galactarate

5-dehydro-4-deoxy-D-glucarate

This enzyme [EC 4.2.1.42] catalyzes the conversion of the *meso*-compound galactarate to 5-dehydro-4-deoxy-D-glucarate and water.[1,2]

[1]B. S. Sharma & H. J. Blumenthal (1973) *J. Bacteriol.* **116**, 1346.
[2]W. A. Wood (1971) *The Enzymes*, 3rd ed., **5**, 573.

Selected entries from *Methods in Enzymology* [**vol**, page(s)]:
General discussion: **9**, 665
Assay: **9**, 665
Other molecular properties: *Escherichia coli*, from (pH optimum, **9**, 669; properties, **9**, 668; purification, **9**, 666; specificity, **9**, 669); D-glucarate dehydrase, presence in assay of, **9**, 661; 2-keto-3-deoxy-D-glucarate aldolase, action on product, **9**, 533; α-keto-β-deoxy-D-glucarate, preparation of, **9**, 56; sources, **9**, 669

GALACTARATE O-HYDROXYCINNAMOYL-TRANSFERASE

This acyltransferase [EC 2.3.1.130] catalyzes the reaction of feruloyl-CoA with galactarate to produce coenzyme A and *O*-feruloylgalactarate.[1] Alternative substrates include sinapoyl-CoA and 4-coumaroyl-CoA.

[1]D. Strack, H. Keller & G. Weissenböck (1987) *J. Plant Physiol.* **131**, 61.

GALACTINOL:RAFFINOSE GALACTOSYLTRANSFERASE

This transferase [EC 2.4.1.67], also known as stachyose synthase, catalyzes the reaction of 1α-D-galactosyl-*myo*-inositol (*i.e.*, galactinol) with raffinose to produce *myo*-inositol and stachyose, the tetrasaccharide *O*-α-D-galactopyranosyl-(1 → 6)-*O*-α-D-galactopyranosyl-(1 → 6)-*O*-α-D-glucopyranosyl-β-D-fructofuranoside.[1–3] Other galactosyl-acceptor substrates include melibiose (producing manninotriose), D-pinitol, and D-ononitol. ***See also*** *Inositol 1α-Galactosyltransferase; Galactinol:Sucrose Galactosyltransferase*

[1]G. Hoch, T. Peterbauer & A. Richter (1999) *ABB* **366**, 75.
[2]T. Peterbauer & A. Richter (1998) *Plant Physiol.* **117**, 165.
[3]P.-R. Gaudreault & J. A. Webb (1981) *Phytochemistry* **20**, 2629.

Selected entries from *Methods in Enzymology* [**vol**, page(s)]:
General discussion: **28**, 526
Assay: **28**, 526

GALACTINOL:SUCROSE GALACTOSYLTRANSFERASE

This galactosyltransferase [EC 2.4.1.82], also known as raffinose synthase, catalyzes the reaction of 1α-D-galactosyl-*myo*-inositol (*i.e.*, galactinol) with sucrose to produce *myo*-inositol and raffinose, the trisaccharide *O*-α-D-galactopyranosyl-(1→6)-*O*-α-D-glucopyranosyl-β-D-fructofuranoside.[1] Other galactosyl-donor substrates include 4-nitrophenyl-α-D-galactopyranoside. The enzyme also catalyzes an exchange reaction between raffinose and sucrose. ***See also*** *Inositol 1α-Galactosyltransferase; Galactinol:Raffinose Galactosyltransferase*

[1]L. Lehle & W. Tanner (1973) *EJB* **38**, 103.

Selected entries from *Methods in Enzymology* [vol, page(s)]:
General discussion: 28, 522
Assay: 28, 523

GALACTITOL DEHYDROGENASE

This oxidoreductase [EC 1.1.1.16], also called dulcitol dehydrogenase, catalyzes the reaction of galactitol with NAD^+ to produce D-tagatose and NADH.[1,2] The enzyme also acts on other alditols possessing a L-*threo* configuration adjacent to a primary alcohol group (*e.g.*, sorbitol, L-iditol, and L-arabitol).

[1]P. Allenza, Y. N. Lee & T. G. Lessie (1982) *J. Bacteriol.* **150**, 1348.
[2]S. B. Primrose & C. W. Ronson (1980) *J. Bacteriol.* **141**, 1109.

Selected entries from *Methods in Enzymology* [vol, page(s)]:
General discussion: 5, 323
Assay: 5, 323

GALACTITOL-1-PHOSPHATE 5-DEHYDROGENASE

This zinc-dependent oxidoreductase [EC 1.1.1.251] catalyzes the reaction of galactitol 1-phosphate with NAD^+ to produce L-tagatose 6-phosphate and NADH.[1]

[1]B. Nobelmann & J. W. Lengeler (1995) *BBA* **1262**, 69.

GALACTITOL-6-PHOSPHATE DEHYDROGENASE

This oxidoreductase reportedly catalyzes the reaction of D-galactitol 6-phosphate with NAD^+ to produce D-tagatose 6-phosphate and NADH. *See also Galactitol-1-Phosphate 5-Dehydrogenase; Galactitol 2-Dehydrogenase*

Selected entries from *Methods in Enzymology* [vol, page(s)]:
General discussion: 89, 275
Assay: *Klebsiella pneumoniae*, **89**, 275
Other molecular properties: *Klebsiella pneumoniae* (properties, **89**, 277; purification, **89**, 276)

β-GALACTOFURANOSIDASE

This enzyme [EC 3.2.1.146], also called exo-β-galactofuranosidase, catalyzes the hydrolysis of terminal non-reducing β-D-galactofuranosides, releasing D-galactose.[1,2] The *Helminthosporium sacchari* enzyme acts to detoxify helminthosporoside, fungal bis(digalactosyl) terpene through the hydrolytic release of four molecules of galactosyl groups.[2]

[1]M. Rietschel-Berst, N. H. Jentoft, P. D. Rick, C. Pletcher, F. Fang & J. E. Gander (1977) *JBC* **252**, 3219.
[2]L. S. Daley & G. A. Strobel (1983) *Plant Sci. Lett.* **30**, 145.

GALACTOGEN 6-β-GALACTOSYLTRANSFERASE

This transferase [EC 2.4.1.205], also known as 1,6-D-galactosyltransferase, catalyzes the reaction of UDP-D-galactose with galactogen to produce UDP and 1,6-β-D-galactosylgalactogen.[1]

[1]E. M. Goudsmit, P. A. Ketchum, M. K. Grossens & D. A. Blake (1989) *BBA* **992**, 289.

GALACTOKINASE

β-D-galactopyranose 1-P

This phosphotransferase [EC 2.7.1.6] catalyzes the reaction of ATP with D-galactose to produce ADP and D-galactose 1-phosphate.[1,2] Other substrates are D-galactosamine and 2-deoxy-D-galactose. The *Escherichia coli* enzyme displays a random Bi Bi kinetic mechanism.[2]

[1]P. A. Frey (1996) *FASEB J.* **10**, 461.
[2]J. S. Gulbinsky & W. W. Cleland (1968) *Biochemistry* **7**, 566.

Selected entries from *Methods in Enzymology* [vol, page(s)]:
General discussion: 1, 290; **8**, 229; **9**, 407; **42**, 43, 47; **44**, 698; **90**, 30
Assay: 1, 290; **8**, 230; **9**, 407; broken cell preparations, in, **5**, 174; *Escherichia coli*, **8**, 230; human erythrocytes, **42**, 48; pig liver, **42**, 43, 44; principle of, **5**, 174; **9**, 407; *Saccharomyces cerevisiae*, **90**, 30; *Saccharomyces fragilis*, **1**, 290; **9**, 407; **28**, 275
Other molecular properties: activation, **63**, 258, 292; bentonite and, **5**, 233; broken cell preparations, in, **5**, 174; carbon dioxide formation, **77**, 4, 5; *Escherichia coli*, **8**, 229 (properties, **8**, 235; purification, **8**, 231); human erythrocytes, from, **42**, 47 (assay, **42**, 48; chromatography, **42**, 51, 52; clinical significance, **42**, 47, 48; molecular weight, **42**, 53; properties, **42**, 52, 53; purification, **42**, 51, 52); isotope exchange, **64**, 8; marker for chromosome detection in somatic cell hybrids, as, **151**, 182; pig liver, from, **42**, 43 (assay, **42**, 43, 44; chromatography, **42**, 45; properties, **42**, 45, 47; purification, **42**, 44); properties, **8**, 235; purification, **8**, 231; *Saccharomyces cerevisiae* (assay, **90**, 30; properties, **90**, 33; purification, **90**, 31); Leloir pathway, part of, **1**, 290; **5**, 174; **87**, 20, 21; *Saccharomyces fragilis* (preparation, **1**, 292; **9**, 408; **28**, 275; properties, **1**, 292; **9**, 410; purification, **1**, 292; **9**, 409; reaction, **1**, 290; **9**, 407; **28**, 274, 278; specificity, **1**, 292; **9**, 411; stability, **9**, 411; steric specificity, **28**, 278); sources, **9**, 408; staining on gels, **22**, 598; sugar nucleotide regeneration system, **247**, 110; synthetic galactosylation of oligosaccharides, **247**, 110; yeast, from, antibody binding, **50**, 55

GALACTOLIPASE

This lipase [EC 3.1.1.26] catalyzes the hydrolysis of a 1,2-diacyl-3-β-D-galactosyl-*sn*-glycerol to produce 3-β-D-galactosyl-*sn*-glycerol and two fatty acid molecules.[1-3] The enzyme also exhibits lipase activity with monoacyl-3-β-D-galactosyl-*sn*-glycerols, 2,3-di-*O*-acyl-O^1-(O^6-α-D-galactosyl-β-D-galactosyl)-D-glycerol, phosphatidylcholine, and other phospholipids. Because galactolipids are a major component of the total acylated lipids in chloroplast membranes, galactolipase is thought to play an important role in a plant's response to chilling.[4]

[1] H. Matsuda & O. Hiayama (1979) *BBA* **573**, 155.
[2] O. Hirayama, H. Matsuda, H. Takeda, K. Maenaka & H. Takatsuka (1975) *BBA* **384**, 127.
[3] P. J. Helmsing (1969) *BBA* **178**, 519 and (1967) **144**, 470.
[4] Z. Kaniuga (1997) *Acta Biochim. Pol.* **44**, 21.

Selected entries from *Methods in Enzymology* [vol, page(s)]:
General discussion: 14, 204
Assay: 14, 204
Other molecular properties: chloroplast inactivation, **69**, 614, 615; distribution, **14**, 208; enzymatic activity, **31**, 522, 591; properties, **14**, 207; purification, runner bean (*Phaseolus multiflorus*), **14**, 206

GALACTOLIPID *O*-ACYLTRANSFERASE

This acyltransferase [EC 2.3.1.134], also known as galactolipid:galactolipid acyltransferase, catalyzes the reaction of two molecules of a mono-β-D-galactosyldiacylglycerol to produce an acylmono-β-D-galactosyldiacylglycerol and a mono-β-D-galactosylacylglycerol.[1,2] Di-D-galactosyldiacylglycerol is also a substrate.

[1] J. W. M. Heemskerk, J. F. G. M. Wintermans, J. Joyard, M. A. Block, A.-J. Dorne & R. Douce (1986) *BBA* **877**, 281.
[2] E. Heinz (1973) *Z. Pflanzenphysiol.* **69**, 359.

GALACTOLIPID GALACTOSYLTRANSFERASE

This transferase [EC 2.4.1.184], also known as galactolipid:galactolipid galactosyltransferase, catalyzes the reaction of two molecules of mono-β-D-galactosyldiacylglycerol to produce an α-D-galactosyl-β-D-galactosyldiacylglycerol and a diacylglycerol.[1–3] Trigalactosyldiacylglycerols and tetragalactosyldiacylglycerols can also be synthesized by subsequent transfer of D-galactosyl residues.[1–3]

[1] A.-J. Dorne, M. A. Block, J. Joyard & R. Douce (1982) *FEBS Lett.* **145**, 30.
[2] J. W. Heemskerk, G. Bögemann & J. F. G. M. Wintermans (1983) *BBA* **754**, 181.
[3] J. W. M. Heemskerk, F. H. H. Jacobs, M. A. M. Scheijen, J. P. F. G. Helsper & J. F. G. M. Wintermans (1987) *BBA* **918**, 189.

GALACTONATE DEHYDRATASE

This enzyme [EC 4.2.1.6] catalyzes the conversion of D-galactonate to 2-dehydro-3-deoxy-D-galactonate and water.[1–3] D-Fuconate is also dehydrated to form 2-keto-3-deoxy-D-fuconate and water in the presence of the *Klebsiella pneumoniae* enzyme. **See also** *D-Fuconate Dehydratase*

[1] W. A. Wood (1971) *The Enzymes*, 3rd ed., **5**, 573.
[2] A. Donald, D. Sibley, D. E. Lyons & A. S. Dahms (1979) *JBC* **254**, 2132.
[3] T. Szumilo (1981) *BBA* **661**, 240.

Selected entries from *Methods in Enzymology* [vol, page(s)]:
General discussion: 90, 294, 299
Assay: *Klebsiella pneumoniae*, **90**, 299; pseudomonad MSU-1, **90**, 294
Other molecular properties: *Klebsiella pneumoniae* (assay, **90**, 299; properties, **90**, 302; purification, **90**, 300); pseudomonad MSU-1 (assay, **90**, 294; properties, **90**, 297; purification, **90**, 295)

GALACTONOLACTONE DEHYDROGENASE

This oxidoreductase [EC 1.3.2.3], also known as L-galactono-γ-lactone dehydrogenase, catalyzes the reaction of L-galactono-1,4-lactone with two molecules of ferricytochrome *c* to produce L-ascorbate and two molecules of ferrocytochrome *c*.[1,2] **See also** *L-Galactonolactone Oxidase*

[1] L. W. Mapson & E. Breslow (1957) *BJ* **65**, 29P.
[2] K. Oba, S. Ishikawa, M. Nishikawa, H. Mizuno & T. Yamamoto (1995) *J. Biochem.* **117**, 120.

L-GALACTONOLACTONE OXIDASE

This FAD-dependent oxidoreductase [EC 1.1.3.24], also known as L-xylono-1,4-lactone oxidase, catalyzes the reaction of L-galactono-1,4-lactone with dioxygen to produce L-ascorbate and hydrogen peroxide. Other substrates include D-altrono-1,4-lactone, L-fucono-1,4-lactone, D-arabinono-1,4-lactone, and D-threono-1,4-lactone. The *Saccharomyces cerevisiae* oxidase contains an FAD covalently linked to a histidyl residue. The enzyme is distinct from L-gulonolactone oxidase [EC 1.1.3.8]; however, the sweet potato enzyme appears to be identical to galactonolactone dehydrogenase [EC 1.3.2.3].[1,2] **See also** *Galactonolactone Dehydrogenase*

[1] T. Imai, S. Karita, G. Shiratori, M. Hattori, T. Nunome, K. Oba & M. Hirai (1998) *Plant Cell Physiol.* **39**, 1350.
[2] H. S. Bleeg & F. Christensen (1982) *EJB* **127**, 391.

β-GALACTOSAMIDE α-2,3-SIALYLTRANSFERASE

This glycosyltransferase [EC 2.4.99.4], also known as CMP-*N*-acetylneuraminate-β-galactosamide-α-2,3-sialyltransferase, catalyzes the reaction of CMP-*N*-acetylneuraminate with a β-D-galactosyl-1,3-*N*-acetyl-α-D-galactosaminyl-R to produce α-*N*-acetylneuraminyl-2,3-β-D-galactosyl-1,3-*N*-acetyl-α-D-galactosaminyl-R (where R is H, a threonyl residue, or seryl residue in a glycoprotein, or a glycolipid) and CMP.[1,2] Lactose is also an acceptor substrate. Other substrates include *p*-nitrophenyl-β-D-galactoside and β-D-galactosyl-1,3-*N*-acetyl-α-D-glucosamine. The enzyme may be identical with monosialoganglioside sialyltransferase [EC 2.4.99.2] in some organisms. *See also Sialyltransferases; α-2,3-Sialyltransferases*

[1]J. I. Rearick, J. E. Sadler, J. C. Paulson & R. L. Hill (1979) *JBC* **254**, 4444.
[2]K. R. Westcott & R. L. Hill (1985) *JBC* **260**, 13116.

Selected entries from *Methods in Enzymology* [vol, page(s)]:
General discussion: 8, 358
Assay: 8, 358; 230, 189; 242, 135; fluorometric assay, 247, 178, 182, 191; radioassay, 247, 178
Other molecular properties: affinity chromatography, 247, 238; erythrocyte resialylation with, 138, 163; ganglioside acceptor G_MI sialylation, 247, 167; porcine liver, 230, 301; properties, 8, 359; purification, rat mammary gland, 8, 358; rat liver (preparation, 230, 311; synthesis of NeuAcα2→3Galβ1→4[Fucα1→3]GlcNAcβ-OR, in, 230, 310); sialylation efficiency, 247, 168; substrate specificity, 247, 154, 168, 176, 183, 238; synthetic sialylation systems, 247, 171, 238

β-GALACTOSAMIDE α-2,6-SIALYLTRANSFERASE

This sialyltransferase [EC 2.4.99.1], also known as β-galactoside α-2,6-sialyltransferase, CMP-*N*-acetylneuraminate:β-galactosamide α-2,6-sialyltransferase, and CMP-*N*-acetylneuraminate: galactose-β-1,4-*N*-acetyl-D-glucosamine α-2,6-sialyltransferase, catalyzes the reaction of CMP-*N*-acetylneuraminate with β-D-galactosyl-1,4-acetyl-β-D-glucosamine to produce α-*N*-acetylneuraminyl-2,6-β-D-galactosyl-1,4-*N*-acetyl-β-D-glucosamine and CMP.[1-4] The enzyme will also catalyze the transfer of a sialyl group to the terminal β-D-galactosyl residue of the oligosaccharide of glycoproteins as well as to *N*-acetyllactosamine. The bovine enzyme exhibits a random Bi Bi kinetic mechanism.[1] *See also Sialyltransferases; α-2,6-Sialyltransferases*

[1]J. C. Paulson, J. I. Rearick & R. L. Hill (1977) *JBC* **252**, 2363.
[2]J. Weinstein, U. de Souza-e-Silva & J. C. Paulson (1982) *JBC* **257** 13835.
[3]H. B. Bosmann (1973) *J. Neurochem.* **20**, 1037.
[4]B. A. Bartholomew, G. W. Jourdian & S. Roseman (1973) *JBC* **248**, 5751.

Selected entries from *Methods in Enzymology* [vol, page(s)]:
General discussion: 8, 368
Assay: 8, 368; 83, 495; 98, 127; 247, 191; fluorometric assay, 247, 178, 182, 191; hepatic Golgi subfractions, assay, in, 109, 223; porcine colostrum, 83, 495; radioassay, 247, 178
Other molecular properties: affinity chromatography, 247, 238; based assay, in control of secretory protein cell-free formation, 219, 92; colostrum, of bovine (properties, 98, 129; purification, 98, 128); colostrum, of goat, 8, 368 (properties, 8, 371; purification, 8, 369); erythrocyte resialylation with, 138, 163; Golgi apparatus, in, 22, 128; hepatic Golgi subfractions, assay, in, 109, 223; particulate *N*-acetylneuraminate monooxygenase, 50, 380; plasma membrane, in,

31, 102; porcine colostrum (assay, 83, 495; properties, 83, 499; purification, 83, 496); properties, 8, 371; 83, 499; 98, 129; rat liver, 230, 301; substrate specificity, 247, 183, 238; synthetic sialylation systems, 247, 166, 171, 238

GALACTOSE DEHYDROGENASES

This oxidoreductase [EC 1.1.1.48] catalyzes the reaction of D-galactose with NAD+ to produce D-galactono-1,4-lactone and NADH.[1-4] Other substrates include L-arabinose, 2-deoxy-D-galactose, and 6-deoxy-D-galactose. The *Pseudomonas saccharophila* enzyme is specific for the β-anomer of D-hexose.

Galactose 1-dehydrogenase (NADP+) [EC 1.1.1.120] catalyzes the same reaction, but uses NADP+ as the redox cofactor.

[1]E.-O. Blachnitzky, F. Wengenmayer & G. Kurz (1974) *EJB* **47**, 235.
[2]F. Wengenmayer, K.-H. Ueberschär & G. Kurz (1974) *EJB* **43**, 49.
[3]A. S. L. Hu & A. L. Cline (1964) *BBA* **93**, 237.
[4]J. De Ley & M. Doudoroff (1957) *JBC* **227**, 745.
[5]A. L. Cline & A. S. L. Hu (1965) *JBC* **240**, 4488.

Selected entries from *Methods in Enzymology* [vol, page(s)]:
General discussion: 5, 339; 9, 112; 89, 176
Assay: 5, 339; 9, 112; *Pseudomonas fluorescens*, 89, 177
Other molecular properties: aldose-1-epimerase, assay of, 9, 608; arabinose dehydrogenase and, 5, 344; binding, to agarose-bound nucleotides, 66, 196; enzyme reactor, in, 44, 880; D-galactose determination, in, 9, 114; 63, 33; D-galactose microassay, in, 41, 4; β-D-galactosidase and, 28, 821; lactase, assay of, 63, 33; molecular weight, 9, 114; mutarotase assay, for, 9, 116, 608; *Pseudomonas fluorescens* (assay, 89, 177; properties, 89, 181; purification, 89, 178); *Pseudomonas saccharophila*, 5, 339; 9, 112 (properties, 5, 340; 9, 114; purification, 5, 339; 9, 112); stereospecificity, 87, 106; subunits, 9, 114; uses, 9, 114

GALACTOSE OXIDASE

This copper-dependent oxidoreductase [EC 1.1.3.9] catalyzes the reaction of D-galactose with dioxygen to produce D-*galacto*-hexodialdose and hydrogen peroxide.[1-5] Other substrates include dihydroxyacetone, glycerol, raffinose, and polysaccharides containing a D-galactosyl residue at the nonreducing terminus; however, neither D-glucose nor L-galactose are substrates.

The catalytic mechanism utilizes a protein radical intermediate. The Cu^{2+} ion is coordinated to two histidyl residues, a tyrosinate, and a covalently modified tyrosine crosslinked to a cysteinyl residue (a thioether linkage at the 3' position). This metal ion is antiferromagnetically coupled to a tyrosyl radical. D-Galactose binds to the active site with O6 directly coordinated to the metal ion.[3-5]

[1] J. W. Whittaker (1999) *Essays Biochem.* **34**, 155
[2] A. Messerschmidt (1998) *CBC* **3**, 401.
[3] M. J. McPherson, C. Stevens, A. J. Baron, Z. B. Ogel, K. Seneviratne, C. Wilmot, N. Ito, I. Brocklebank, S. E. Phillips & P. F. Knowles (1993) *Biochem. Soc. Trans.* **21**, 752.
[4] E. T. Adman (1991) *Adv. Protein Chem.* **42**, 145.
[5] B. G. Malmström, L.-E. Andréasson & B. Reinhammer (1975) *The Enzymes*, 3rd ed., **12**, 507.

Selected entries from *Methods in Enzymology* [vol, page(s)]:
General discussion: 9, 87; **52**, 10; **89**, 163; **258**, 235, 262
Assay: 9, 87; commercial samples, from, **89**, 174; *Dactylium dendroides*, **89**, 167; *Diplocardia* bioluminescence, with, **57**, 381
Other molecular properties: active site structure, **258**, 264; assay with, for β-galactosidase, **14**, 157; circular dichroism, **258**, 274; commercial samples, from (assay, **89**, 174; properties, **89**, 175; separation from contaminating proteins, **89**, 172); copper(II) complex with, **9**, 91; copper role, **258**, 236, 257, 272; *Dactylium dendroides* (assay, **89**, 167; properties, **89**, 170; purification, **89**, 168; **258**, 237); electron nuclear double resonance spectroscopy, **246**, 577; EPR studies, **9**, 91; **258**, 236, 254 (apoenzyme, **258**, 267; double integration, **258**, 265; nitric oxide complex, **258**, 269; silent copper complex, **258**, 267; spin standards, **258**, 265); human immune interferon production, in, **78**, 161, 162; immobilization, by entrapment, **44**, 176; interferon inducer, as, **78**, 295; magnetic circular dichroism, **258**, 276; molecular weight, **9**, 90; neuraminidase, and, stimulation of T cells, **150**, 105; optical absorption (azide effects, **258**, 272; free radical identification, **258**, 273; ligand-to-metal charge transfer, **258**, 271; spectra, **258**, 270; titration curves, **258**, 272); *Polyporus circinatus*, of, **9**, 87 (preparation, **9**, 88; properties, **9**, 90; purification, **9**, 89; specificity, **9**, 91; stability, **9**, 92); purification from *Dactylium dendroides*, **258**, 237; radiolabeling studies (cell surface glycoconjugates, **230**, 33, 42; *N,N*-dilactitol-tyramine, **242**, 9); reaction catalyzed, **56**, 465; **258**, 235, 263; reaction specificity, **242**, 5; reaction with triantennary glycopeptides, **247**, 37; sources, **56**, 465; specificity, **9**, 91; **242**, 5; **258**, 236, 260; spectroscopic techniques, **258**, 262; stability, **9**, 92; substrate specificity, **258**, 236, 260; thioether bond between tyrosinyl and cysteinyl residues (biosynthesis, **258**, 257; catalysis role, **258**, 257, 264; confirmatory evidence, **258**, 251; mutation effects, **258**, 256; stacking interaction of tryptophan, **258**, 254; X-ray crystallography, **258**, 251); tritiated borohydride, and, in labeling of lymphocyte receptors, **150**, 412; tritiation of (carbohydrates and, **28**, 205, 207, 298, 302, 304; cell-surface glycoproteins, **50**, 204; glycolipids, **50**, 204); UDP-*N*-acetyl-D-glucosamine 4-epimerase assay and, **8**, 277; uronic acid analysis, **50**, 31, 32; X-ray crystallography (accuracy of model, **258**, 244; acetate ion, **258**, 243; copper coordination, **258**, 247; copper site geometry, **258**, 249; crystallization, **258**, 239; data collection, **258**, 238; heavy atom derivatives, **258**, 239; initial model building, **258**, 241; phase determination, **258**, 238; refinement, **258**, 243; sequence determination, **258**, 241; stacking tryptophan mutant, **258**, 255; substrate binding, **258**, 260; superbarrel motif, **258**, 213, 215, 245; thioether cysteine mutant, **258**, 256)

D-GALACTOSE-6-PHOSPHATE ISOMERASE

This isomerase [EC 5.3.1.26], a component of the tagatose 6-phosphate pathway of lactose catabolism in microorganisms, catalyzes the interconversion of D-galactose 6-phosphate and D-tagatose 6-phosphate.

Selected entries from *Methods in Enzymology* [vol, page(s)]:
General discussion: 89, 562
Assay: *Staphylococcus aureus*, **89**, 562

GALACTOSE-1-PHOSPHATE THYMIDYLYLTRANSFERASE

This transferase [EC 2.7.7.32], also known as dTDP-galactose pyrophosphorylase and dTDP-galactose diphosphorylase, catalyzes the reaction of dTTP with α-D-galactose 1-phosphate to produce pyrophosphate (or, diphosphate) and dTDP-D-galactose.[1] The bacterial enzyme participates in cell wall biosynthesis.

[1] J. H. Pazur & J. S. Anderson (1963) *JBC* **238**, 3155.

GALACTOSE-6-SULFURYLASE

This enzyme [EC 2.5.1.5], also known as porphyran sulfatase and galactose-6-sulfatase, catalyzes the elimination of sulfur from the galactose 6-sulfate residues of porphyran, thereby generating the corresponding 3,6-anhydrogalactose residues.[1]

[1] D. A. Rees (1961) *BJ* **80**, 449 and **81**, 347.

α-GALACTOSIDASE

This magnesium- and NAD$^+$-dependent enzyme [EC 3.2.1.22], also known as melibiase, catalyzes the hydrolysis of melibiose to yield D-galactose and D-glucose.[1–6] The enzyme also acts on terminal, non-reducing α-D-galactose residues in α-D-galactosides (for example, *p*-nitrophenyl-α-D-galactoside, raffinose, and stachyose as well as oligo- and polysaccharides). An α-galactosidase catalyzes the release of the terminal D-galactose from Galα1 → 4Galβ1 → 4Gluβ1 → 1′-ceramide. Also hydrolyzed are many α-D-fucosides, including *p*-nitrophenyl-α-D-fucoside.

[1] E. Conzelmann & K. Sandhoff (1987) *AE* **60**, 89.
[2] P. M. Dey & E. del Campillo (1984) *AE* **56**, 141.
[3] A. Crueger & W. Crueger (1984) in *Biotechnology* (K. Kieslich, ed.) **6A**, p. 421, Verlag Chemie, Weinheim.
[4] R. O. Brady (1983) *The Enzymes*, 3rd ed., **16**, 409.
[5] A. W. Schram & J. M. Tager (1981) *TiBS* **6**, 328.
[6] H. M. Flowers & N. Sharon (1979) *AE* **48**, 29.

Selected entries from *Methods in Enzymology* [vol, page(s)]:
General discussion: 1, 231, 249; **8**, 565; **34**, 347; **160**, 627; **330**, 246
Assay: 1, 249; **160**, 628; activity assay, **247**, 72; coffee beans, of, **8**, 566; ficin, from, **28**, 714; α-galactosidase A, **50**, 481; *Phaseolus vulgaris*, from, **28**, 720; placenta, from, **28**, 849
Other molecular properties: activity assay, **247**, 72; analysis of in multiple systems, **22**, 18, 20, 21; *Aspergillus niger*, from (properties, **28**, 733; purification, **28**, 729); blood-group substances and, **8**, 707; ceramide trihexosidase, **28**, 849; **50**, 494 (assay, **28**, 849; **50**, 533, 534; biological significance in Fabry's disease, **28**, 855; enzymatic properties, **28**, 853, 855; isolation from human placenta, **28**, 851 [assay, **50**, 533, 534; properties, **50**, 535, 536; purification, **50**, 534, 535; treatment of Fabry's disease, **50**, 536, 537]; reaction, **28**, 849); coffee beans, of, **8**, 565 (digestion conditions, **230**, 285; glycosphingolipid analysis, for, **230**, 385; preparation, **8**, 566; properties, **8**, 568; purification, **8**, 567; **230**, 283,

287; specificity, **8**, 569; **230**, 285); conjugate, assay, **44**, 350; cyclodextrin analysis with, **247**, 78; cyclodextrin modification by, **247**, 69; ficin, from (assay, **28**, 714; carbohydrate structure, **28**, 19, 153; isolation, **28**, 715; properties, **28**, 718); α-galactosidase A (assay, **50**, 481; distribution, **50**, 480; Fabry's disease, **50**, 480); α-galactosidase B (Fabry's disease, **50**, 481); hyperthermophilic, **330**, 246; inhibition by aldonolactones, **8**, 573; *Phaseolus vulgaris*, from (properties, **28**, 723; purification, **28**, 721; substrate specificity, **28**, 726); placenta, from (assay, **28**, 849; properties, **28**, 853, 855; purification, **28**, 851); properties, **1**, 251; **160**, 629, 631; purification, **1**, 250; **8**, 567; **34**, 349, 350 (guar seeds, from, **160**, 628; lucerne seeds, from, **160**, 630); reaction specificity, **247**, 69; sweet almonds (properties, **1**, 251; purification, **1**, 250); transition-state and multisubstrate analogues, **249**, 306

β-GALACTOSIDASE

This enzyme [EC 3.2.1.23], known also as LacZ and lactase, catalyzes the hydrolysis of terminal (nonreducing) β-D-galactosyl residues of β-D-galactosides.[1-6] Typical substrates are lactose, *p*-nitrophenyl-β-D-galactoside, and galactooligosaccharides containing β-1,6 or β-1,3 linkages. Some β-galactosidases also catalyze hydrolysis of α-L-arabinosides. Animal β-galactosidases often hydrolyze β-D-fucosides and β-D-glucosides as well.

The enzyme catalyzes a two-step mechanism that forms a covalent intermediate, and this step is rate-determining with alkyl-D-galactopyranoside or *p*-nitrophenyl-D-galactopyranoside substrates. There is broad specificity with regard to the structure the leaving group.

The LacZ β-galactosidase isozyme of *Escherichia coli* catalyzes a ping pong Bi Bi mechanism via a glycosyl-enzyme intermediate, a glutamyl residue at the bottom of an active-site pocket acting as the nucleophile. In addition, the enzyme will also catalyze the conversion of lactose to allolactose (that is, the β(1,6) isomer). The ebg isozyme of *E. coli*, which has a different quaternary structure, also has a double displacement mechanism. ***See also* Lactase**

Isopropyl-β-D-thiogalactoside, a weak competitive inhibitor of LacZ, is a potent and widely used inducer of the *lac* operon.

[1] G. Mooser (1992) *The Enzymes*, 3rd ed., **20**, 187.
[2] M. L. Sinnott (1990) *Chem. Rev.* **90**, 1171.
[3] E. Conzelmann & K. Sandhoff (1987) *AE* **60**, 89.
[4] H. Kresse & J. Glössl (1987) *AE* **60**, 217.
[5] K. Wallenfels & R. Weil (1972) *The Enzymes*, 3rd ed., **7**, 617.
[6] K. Wallenfels & O. M. Malhotra (1960) *The Enzymes*, 2nd ed., **4**, 409.

Selected entries from *Methods in Enzymology* [vol, page(s)]:
General discussion: **1**, 231, 234, 241; **5**, 212; **8**, 582; **14**, 156; **28**, 844; **34**, 105, 110, 118, 119, 350, 731; **42**, 497; **46**, 26, 36, 156, 364, 365, 398; **50**, 514; **89**, 59; **136**, 230, 411; **330**, 201

Assay: **1**, 234, 241; **5**, 212; **8**, 583; **9**, 116; **12B**, 805; **14**, 156; **28**, 701, 844; **34**, 351, 352; **63**, 33; **65**, 884, 885; **189**, 267; **217**, 639; **235**, 499; activity assay, **44**, 793; **73**, 476; **247**, 69; bovine testes, **50**, 515; *Clostridium perfringens*, from, **28**, 757; diagnosis of genetic mucopolysaccharide storage disorders, in, **83**, 565; fluorescent derivation of galactosyl cerebroside, with, **72**, 362; glucose-6-phosphate dehydrogenase, **5**, 212; human liver, from, **28**, 820; *Neurospora crassa*, from, **42**, 498, 502; *Phaseolus vulgaris*, from, **28**, 720; reporter gene assays, **326**, 177; trinitrophenyl-ω-aminolauroylgalactosylsphingosine, with, **72**, 360

Other molecular properties: *N*-acetylglucosamine binding protein, **50**, 291; activation energy, **44**, 796, 797; activity assay, **44**, 793; **73**, 476; **247**, 69; activity of β-*N*-acetylhexosaminidase, **14**, 167; affinity labeling, **87**, 473; agarose derivatives, **34**, 353; alkaline phosphatase, and, double staining of retrovirus-infected tissues for, **225**, 947; analysis in multiple systems, **22**, 18; antibody coupling for nerve growth factor immunoassay, **147**, 176; *Aspergillus niger*, from (properties, **28**, 733; purification, **28**, 729); assay (identification of gene regulatory elements, in, **152**, 717; measurement of tryptophan repressor activity, in, **208**, 646); assay with (ceramide hexosides, **14**, 156, 157; labeled ceramide hexosides, **14**, 156, 158; *o*- or *p*-nitrophenyl-β-D-galactopyranoside, **14**, 156); avidin, and, preparation as endocytic probe, **219**, 13; bacterial membrane, in, **22**, 113; biotinylation, **184**, 154, 474; bovine testes, from (assay, **50**, 515; properties, **50**, 518; purification, **50**, 515); bradykinin, enzyme immunoassay of, **80**, 175; cAMP binding, **38**, 367; carbohydrate analysis of glycoproteins and, **8**, 41; cells positive for (electron microscopy, **225**, 467; *in vivo* staining in suspension, **225**, 465); chemical glycosylation, **242**, 40; chromatography on DEAE-cellulose, **5**, 24; *Clostridium perfringens*, from (assay, **28**, 757; properties, **28**, 762; purification, **28**, 757); conjugation to antibody, **73**, 157, 475; conjugate (activity assay, **44**, 269, 343, 352; kinetics, **44**, 426); coupled to cellulose acetate beads (applications, **135**, 292; assay, **135**, 291); coupling efficiencies, **44**, 795; coupling with glycosyltransferase reactions, **247**, 125; cross-reacting protein, **34**, 351; cryoenzymology, **63**, 338; cyclodextrin substrates, **247**, 66; diagnostic assay (galactosialidosis, **138**, 760; G_{M1}-gangliosidosis, **138**, 752); *Diplococcus pneumoniae* (digestion conditions, **230**, 284; purification, **230**, 288; specificity, **230**, 284); encapsulated in crystallized carbohydrate microspheres (biological activity, assay, **112**, 121; preparation, **112**, 120; release studies, *in vitro*, **112**, 122); encapsulation of yeast cells with, **137**, 646; endosome marker, as, determination, **109**, 267; entrapment in cellulose triacetate fibers, **44**, 824; enzyme thermistor, in, **44**, 674, 675 (polyacrylamide gels, in, **44**, 176); *Escherichia coli*, from (antibody binding, **50**, 55, 56; assay in SDS gels, **91**, 274; (2*R*)-glycerol-*o*-β-D-galactopyranoside synthesis, **89**, 59; properties, **1**, 247; **5**, 216, 219; purification, **1**, 243; **5**, 213); expression in yeast, assays, **185**, 356; fiber entrapment, industrial application, **44**, 241; fibronectin fusion proteins (preparation, **144**, 459; purification from λgt11 lysogens, **144**, 461; segment-specific antibodies, **144**, 462); fluorescence immunoassay, reagent in, **74**, 80, 81; fluorine nuclear magnetic resonance, **49**, 273, 274; fusion proteins (analysis in bacterial transformants, **154**, 142; partition coefficients, **228**, 628, 636; partitioning, **228**, 627, 631; production, **228**, 631 [of antibody probes, in, **154**, 144]; protein stabilization, in, **182**, 109; purification, **182**, 110; **228**, 639; recovery, **228**, 639; solubility, **228**, 637; stability, **228**, 639; staphylococcal protein A and streptococcal protein G, with, [production, **228**, 633; purification, **228**, 627]; structure, **228**, 628); β-galactosidase chymotrypsin, **46**, 155; ganglioside G_{M1} β-galactosidase, **28**, 868 (assay, **28**, 868; **50**, 482, 483; enzymatic properties, **28**, 871; galactosylceramidase, **50**, 483; G_{M1}-gangliosidosis and, **28**, 872; histochemical assay for subcellular localization, **28**, 872; isolation from human liver, **28**, 871; lactosylceramide, **50**, 471; reaction, **28**, 868, 872); gene expression in *Rhodospirillaceae*, monitoring, **204**, 481; gene fusions (analyzing gene expression in *Escherichia coli* and yeast, for, **100**, 293; construction and application in yeast, **100**, 167); glucocerebrosidase digestion before cell targeting, **149**, 32; glutaraldehyde binding, **44**, 52; glycolipid structure studies, for, **32**, 363; HIV protease fusion protein, **241**, 33; hyperthermophilic, **330**, 201; human liver, from (assay, **28**, 820;

enzymatic properties, **28**, 823; preparation and purification, **28**, 821); immobilization, **64**, 236, 237, 239 ((aminoethyl)cellulose, on, **44**, 52; chemical aggregation, by, **44**, 269, 270; collagen, on, **44**, 153, 255, 260; hollow-fiber membrane device, in, **44**, 313, 314; inert protein, on, **44**, 909; microencapsulation, by, **44**, 212, 214; multistep enzyme system, in, **44**, 456, 457; porous ceramics, on, **44**, 793); immobilized composites (industrial applications [developments, **136**, 422; engineering considerations, **136**, 414; microbial contamination problems, **136**, 417; operating strategy, **136**, 419]; preparation, **136**, 413; properties, **136**, 413); immobilized, galactosyl-*N*-acetylgalactosamine synthesis with, **136**, 230; immunoassay of S-100 protein, in, **102**, 259; inactivation, **46**, 401, 402; induced, cell-free synthesis of, **12B**, 797, 805, 811, 819, 820; induction, DCCD, **55**, 498; induction, permeability and, **12B**, 842; induction of in microorganisms, **22**, 87, 90, 91, 94, 95; industrial saccharification, in, (immobilized enzyme system, **136**, 377; liquid enzyme system, **136**, 374); inhibition, **8**, 583; **14**, 161; **34**, 7; **63**, 401; **72**, 680; **73**, 541, 542 (aldonolactones, by, **8**, 572, 573, 583; DL-*erythro*-2-(2′-hydroxydodecanoyl)amino-1-phenyl-1,3-propanediol, by, **72**, 680; metal ions, by, **44**, 261; slow-binding, **63**, 450, 465); internal control plasmid in transfection assays, **255**, 416; isopropyl-β-D-thiogalactoside as inducer for, **22**, 87; jack bean, from (carbohydrate structure, **28**, 19, 152, 702; digestion conditions, **230**, 284; glycosphingolipid analysis, for, **230**, 385; isolation, **28**, 704, 709; properties, **28**, 713; purification, **230**, 282, 286; specificity, **230**, 284, 298); kinetics, in multistep enzyme system, **44**, 465, 466; label release from plasma membrane glycoconjugates, in, **98**, 419; lactose reduction of milk, for, **44**, 822; lactosylceramide β-galactosidase, **14**, 156; **28**, 844 (assay, **14**, 156; **28**, 844; isolation from rat brain, **14**, 159; **28**, 845; lactosylceramidosis and, **28**, 848; pathological assay in lactosylceramidosis patients, **28**, 847; properties, **14**, 160; **28**, 847); *lacZ* gene fusions, in analysis of transport proteins, **125**, 150; linked antibodies, in protein detection on nitrocellulose, **121**, 855; linked immunoassay for prostaglandin F$_{2\alpha}$, **86**, 269; lysosome, **8**, 513; **10**, 12 (lysosomal marker, assay, **182**, 215); macrophage fractions, in, **31**, 340; magnesium requirement, **44**, 827; magnetically enhanced phase separation in multistage separator, **228**, 114; membrane fractions and, **12B**, 795, 797; 4-methylumbelliferyl-β-D-galactoside assay, **225**, 449; microvillous membrane, in, **31**, 130; molecular weight, **53**, 350, 351, 353 (molecular weight calibration standard, as, **237**, 92); mouse, in, screening for, **225**, 468; mucolipidoses II and III, **50**, 455; *Neurospora crassa*, from, **42**, 497 (amino acid, **42**, 501, 502; assay, **42**, 498, 502; chromatography, **42**, 500, 501; properties, **42**, 501; purification, **42**, 498, 502); nicotinic acid receptors, **32**, 320; osmotic shock and, **12B**, 845; pH activity profile, **44**, 793, 795; *Phaseolus vulgaris*, from (assay, **28**, 720; purification, **28**, 721; substrate specificity, **28**, 727); poly(*N*-acetyl-β-lactosaminide-carrying acrylamide), in, **242**, 228; polyacrylamide derivatives, **34**, 352, 353; preparation from intestine, **8**, 585; preparation of trihexosyl ceramide, in, **14**, 162; production by yeast two-hybrid system, **256**, 236; properties, **1**, 237, 247; **5**, 216, 219; **8**, 582; **14**, 160; **28**, 727; **34**, 358; purification, **1**, 236, 243; **5**, 213; **14**, 159; **28**, 721; **34**, 355; **73**, 541, 542 (affinity chromatography, by, **22**, 367, 368 [elution with substrate, by, **22**, 385; large-scale, yield, **22**, 536]; *Streptococcus pneumoniae*, from, **149**, 34); radioiodination, **70**, 212; rat brain, **14**, 156 (properties, **14**, 160; purification, **14**, 159); rate equation, **44**, 798, 799; reaction specificity, **247**, 66; removal of galactose during ganglioside hydrolysis in brain, **14**, 135, 136, 138, 139; reporter, **326**, 177; secretion from platelet lysosomes, assay, **169**, 336; sedimentation coefficient, **61**, 108; soluble aggregate, activity assay, **44**, 474; sources, **44**, 793; **46**, 401; spacers, **34**, 356, 357; specific activity, **44**, 255; sphingolipidoses, **50**, 457; stability of antibody conjugate, **73**, 162; staining on gels, **22**, 600; substrate specificity, **14**, 161; sweet almond emulsin, from (assay, **28**, 701; properties, **1**, 237; **28**, 702; purification, **1**, 236; **28**, 701); synthesis, stimulation, **65**, 857, 865, 884; transition state and multisubstrate analogues, **249**, 306; transport of, in bacterial membranes, **22**, 120

GALACTOSIDE *O*-ACETYLTRANSFERASE

This acetyltransferase [EC 2.3.1.18], also known as thiogalactoside acetyltransferase and thiogalactoside transacetylase, catalyzes the reaction of acetyl-CoA with a β-D-galactoside (such as β-D-methylgalactoside) to produce coenzyme A and a 6-acetyl-β-D-galactoside.[1,2]

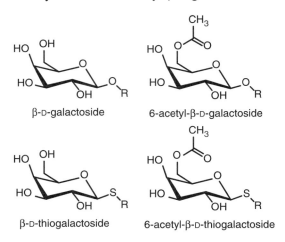

β-D-galactoside 6-acetyl-β-D-galactoside

β-D-thiogalactoside 6-acetyl-β-D-thiogalactoside

The enzyme also acts on thiogalactosides (for example, isopropyl-β-D-thiogalactoside and β-D-thiomethylgalactoside) and phenylgalactosides (for example, *p*-nitrophenyl-β-D-galactoside). The *Escherichia coli* enzyme reportedly has an ordered Bi Bi kinetic mechanism.[1] His115 has an importent catalytic role in the catalytic mechanism.[2]

[1] R. E. Musso & I. Zabin (1973) *Biochemistry* **12**, 553.
[2] A. Lewendon, J. Ellis & W. V. Shaw (1995) *JBC* **270**, 26326.

Selected entries from *Methods in Enzymology* [vol, page(s)]:
Assay: 12B, 806
Other molecular properties: end-group analysis, **25**, 120; induced, cell-free synthesis, **12B**, 797, 805; membrane fractions and, **12B**, 795, 797

GALACTOSIDE 2-L-FUCOSYLTRANSFERASE

This transferase [EC 2.4.1.69], also known as blood group H α-2-fucosyltransferase and α-(1→2)-L-fucosyltransferase, catalyzes the reaction of GDP-L-fucose with β-D-galactosyl-R to produce GDP and α-L-fucosyl-1,2-β-D-galactosyl-R, where the R group represents the rest of the glycopeptide, glycoprotein, or glycolipid. Lactose can act as an alternative acceptor substrate.

The porcine enzyme displays either a rapid-equilibrium random Bi Bi or a steady-state ordered Bi Bi mechanism.[1]

The enzymatic activity, formerly listed as EC 2.4.1.89, was thought to be that of galactosylglucosaminylgalactosylglucosylceramide α-L-fucosyltransferase. This EC entry was deleted and is now classified as EC 2.4.1.69. *See also Galactoside 3-L-Fucosyltransferase; Galactoside 3(4)-L-Fucosyltransferase*

[1] T. A. Beyer & R. L. Hill (1980) *JBC* **255**, 5373

Selected entries from *Methods in Enzymology* [vol, page(s)]:
General discussion: 8, 351
Assay: 8, 351; 138, 599; porcine submaxillary gland, 83, 460
Other molecular properties: glycosphingolipid-synthesizing, assays, 138, 599; porcine submaxillary gland (assay, 83, 460; properties, 83, 469; purification, 83, 462); properties, 8, 353; 98, 132; purification, 8, 352; transition state and multisubstrate analogues, 249, 305

GALACTOSIDE 3-L-FUCOSYLTRANSFERASE

This transferase [EC 2.4.1.152], also known as Lewis-negative α-3-fucosyltransferase, catalyzes the reaction of GDP-L-fucose with 1,4-β-D-galactosyl-*N*-acetyl-D-glucosaminyl-R to produce GDP and 1,4-β-D-galactosyl-(α-1,3-L-fucosyl)-*N*-acetyl-D-glucosaminyl-R, where R represents the rest of the glycopeptide, glycoprotein, or glycolipid.[1] A general base mechanism is supported by a pH-rate profile which suggests a catalytic residue with a pK_a value of 4.1.[2] Solvent kinetic isotope effect studies indicate that only one-proton transfer is involved in the catalytic step leading to the formation of the transition state.[2]

This enzyme is distinct from that of galactoside 3(4)-L-fucosyltransferase [EC 2.4.1.65] since it has no action on the corresponding 1,3-galactosyl derivative. *See also Galactoside 2-L-Fucosyltransferase; Galactoside 3(4)-L-Fucosyltransferase*

Note that five human α-1,3-fucosyltransferases have been identified, each having distinct specificities for the acceptor sugar. α-1,3-Fucosyltransferase V, which exhibits an ordered Bi Bi kinetic mechanism[3] with inversion of anomeric configuration, is responsible for the terminal step in the biosynthesis of Lewis X and sialyl Lewis X.

[1] C. J. Britten & M. I. Bird (1997) *BBA* **1334**, 57.
[2] B. W. Murray, S. Takayama, J. Schultz & C. H. Wong (1996) *Biochemistry* **35**, 11183.
[3] L. Qiao, B. W. Murray, M. Shimazaki, J. Schultz & C. H. Wong (1996) *JACS* **118**, 7653.

Selected entries from *Methods in Enzymology* [vol, page(s)]:
General discussion: 98, 132; 230, 303
Assay: 138, 600
Other molecular properties: glycosphingolipid-synthesizing, assays, 138, 600; properties, 98, 132; synthetic fucosylation system, 247, 116

GALACTOSIDE 3(4)-L-FUCOSYLTRANSFERASE

This transferase [EC 2.4.1.65] (also known as blood group Lewis α-4-fucosyltransferase, *N*-acetylglucosaminide α1 → 3-fucosyltransferase, *N*-acetylglucosaminide α1 → 4-fucosyltransferase and α-3/4-fucosyltransferase) catalyzes the reaction of GDP-L-fucose with 1,3-β-D-galactosyl-*N*-acetyl-D-glucosaminyl-R to produce GDP and 1,3-β-D-galactosyl-(α-1,4-L-fucosyl)-*N*-acetyl-D-glucosaminyl-R, where R represents the rest of the glycopeptide, glycoprotein, or glycolipid. The enzyme also acts on the corresponding 1,4-galactosyl derivative, forming 1,3-L-fucosyl links.[1] *See also Galactoside 2-L-Fucosyltransferase; Galactoside 3-L-Fucosyltransferase*

[1] F. Dupuy, J. M. Petit, R. Mollicone, R. Oriol, R. Julien & A. Maftah (1999) *JBC* **274**, 12257.

Selected entries from *Methods in Enzymology* [vol, page(s)]:
General discussion: 83, 501; 230, 303
Assay: human milk, 83, 501
Other molecular properties: human milk (properties, 83, 505; purification, 83, 503; 230, 310); properties, 83, 505; 98, 132; synthesis of NeuAcα2 →3Galβ1 →4[Fucα1 →3]GlcNAcβ-OR, in, 230, 310

β-GALACTOSYL-*N*-ACETYLGLUCOS-AMINYLGALACTOSYLGLUCOSYL-CERAMIDE β-1,3-ACETYLGLUCOS-AMINYLTRANSFERASE

This manganese-dependent transferase [EC 2.4.1.163], also known as poly-*N*-acetyllactosamine extension enzyme, catalyzes the reaction of UDP-*N*-acetyl-D-glucosamine with β-D-galactosyl-1,4-*N*-acetyl-β-D-glucosaminyl-1,3-β-D-galactosyl-1,4-β-D-glucosylceramide to produce *N*-acetyl-D-glucosaminyl-1,3-β-D-galactosyl-1,4-*N*-acetyl-β-D-glucosaminyl-1,3-β-D-galactosyl-1,4-β-D-glucosylceramide and UDP.[1]

[1] M. Basu & S. Basu (1984) *JBC* **259**, 12557.

GALACTOSYL-*N*-ACETYLGLUCOS-AMINYLGALACTOSYLGLUCOSYL-CERAMIDE β-1,6-*N*-ACETYLGLUCOS-AMINYLTRANSFERASE

This manganese-dependent transferase [EC 2.4.1.164] catalyzes the reaction of UDP-*N*-acetyl-D-glucosamine with D-galactosyl-1,4-*N*-acetyl-β-D-glucosaminyl-1,3-β-D-galactosyl-1,4-β-D-glucosylceramide to produce

N-acetyl-D-glucosaminyl-1,6-β-D-galactosyl-1,4-N-acetyl-β-D-glucosaminyl-1,3-β-D-galactosyl-1,4-β-D-glucosylceramide and UDP.[1]

[1]M. Basu & S. Basu (1984) JBC **259**, 12557.

β-GALACTOSYL-N-ACETYLGLUCOSAMINYLGLYCOPEPTIDE α-1,3-GALACTOSYLTRANSFERASE

This glycosyltransferase [EC 2.4.1.87] catalyzes the reaction of UDP-D-galactose with β-D-galactosyl-β-1,4-N-acetyl-D-glucosaminylglycopeptide to produce UDP and α-D-galactosyl-β-1,3-D-galactosyl-β-1,4-N-acetyl-D-glucosaminylglycopeptide.[1] Both membrane-bound and soluble forms of α1,3-galactosyltransferases have been identified.[2] *See also* β-N-Acetylglucosaminyl-Glycopeptide β-1,4-Galactosyltransferase

[1]M. Basu & S. Basu (1973) JBC **248**, 1700.
[2]S. K. Cho & R. D. Cummings (1997) JBC **272**, 13622.

1,3-β-GALACTOSYL-N-ACETYLHEXOSAMINE PHOSPHORYLASE

This enzyme [EC 2.4.1.211] catalyzes the reaction of β-D-galactopyranosyl-(1 → 3)-N-acetyl-D-glucosamine with orthophosphate to produce α-D-galactopyranose 1-phosphate and N-acetyl-D-glucosamine.[1] Other substrates include β-D-galactopyranosyl-(1 → 3)-N-acetyl-D-galactosamine, yielding α-D-galactopyranose 1-phosphate and N-acetyl-D-galactosamine.

[1]D. Derensy-Dron, F. Krzewinski, C. Brassart & S. Bouquelet (1999) Biotechnol. Appl. Biochem. **29**, 3.

GALACTOSYLACYLGLYCEROL O-ACYLTRANSFERASE

This acyltransferase [EC 2.3.1.141] catalyzes the reaction of an acyl-[acyl-carrier protein] with an sn-3-D-galactosyl-sn-2-acylglycerol to produce the acyl-carrier protein and a D-galactosyldiacylglycerol.[1] Palmitoyl, stearoyl, and oleoyl groups are transferred to the sn-1 position of the glycerol moiety.

[1]H. H. Chen, A. Wickrema & J. G. Jaworski (1988) BBA **963**, 493.

GALACTOSYLCERAMIDASE

This enzyme [EC 3.2.1.46], also known as galactocerebrosidase, GalCerase, and galactosylceramide β-galactosidase, catalyzes the hydrolysis of a D-galactosyl-N-acylsphingosine to produce D-galactose and

an N-acylsphingosine (*i.e.*, a ceramide).[1,2] *See also* Glycosylceramidase; Glucosylceramidase

[1]S. Goda, T. Kobayashi & I. Goto (1987) BBA **920**, 259.
[2]E. Hanada & K. Suzuki (1979) BBA **575**, 410.
Selected entries from *Methods in Enzymology* [**vol**, page(s)]:
General discussion: 8, 595; 28, 301, 834
Assay: 8, 595; 28, 834
Other molecular properties: enzymatic properties, 28, 838; intestine, from rat, 8, 595 (properties, 8, 596; purification, 8, 595); isolation from rat brain, 28, 837; pathological material assay in globoid cell leukodystrophy, 28, 839

GALACTOSYLCERAMIDE SULFOTRANSFERASE

This transferase [EC 2.8.2.11] catalyzes the reaction of 3′-phosphoadenylylsulfate with a galactosylceramide to produce adenosine 3′,5′-bisphosphate and a galactosylceramide sulfate.[1-3] Other substrates include lactosylceramides and galactocerebrosides. The human renal enzyme fails to act on substrates containing an α-galactoside at the non-reducing terminus.[4]

[1]K. S. Sundaram & M. Lev (1992) JBC **267**, 24041.
[2]G. Tennekoon, S. Aitchinson & M. Zabura (1985) ABB **240**, 932.
[3]G. Tennekoon & G. M. McKhann (1978) J. Neurochem. **31**, 329.
[4]K. Honke, M. Yamane, A. Ishii, T. Kobayashi & A. Makita (1996) J. Biochem. **119**, 421.
Selected entries from *Methods in Enzymology* [**vol**, page(s)]:
General discussion: 311, 94

GALACTOSYLDIACYLGLYCEROL α-2,3-SIALYLTRANSFERASE

This sialyltransferase [EC 2.4.99.5], also known as CMP-N-acetylneuraminate:galactosyldiacylglycerol α-2,3-sialyltransferase, catalyzes the reaction of CMP-N-acetylneuraminate with a 1,2-diacyl-3-β-D-galactosyl-sn-glycerol to produce CMP and a 1,2-diacyl-3-[3-(α-D-N-acetylneuraminyl)-β-D-galactosyl]-sn-glycerol.[1] The β-D-galactosyl residue in glycoproteins may also act as the acceptor substrate. *See also* Sialyltransferases; α-2,3-Sialyltransferases

[1]J. Pieringer, S. Keech & R. A. Pieringer (1981) JBC **256**, 12306.

GALACTOSYLGALACTOSYLGLUCOSYL-CERAMIDASE

This enzyme [EC 3.2.1.47] catalyzes the hydrolysis of a D-galactosyl-D-galactosyl-D-glucosyl-N-acylsphingosine to produce a lactosyl-N-acylsphingosine and D-galactose.[1,2]

[1]M. W. Ho (1973) BJ **133**, 1.
[2]R. O. Brady, A. E. Gal, R. M. Bradley & E. Martensson (1967) JBC **242**, 1021.
Selected entries from *Methods in Enzymology* [**vol**, page(s)]:
General discussion: 50, 533; **311**, 255

GALACTOSYLGALACTOSYLGLUCOSYL-CERAMIDE β-D-ACETYLGALACTOS-AMINYLTRANSFERASE

This transferase [EC 2.4.1.79], also known as globoside synthase, catalyzes the reaction of UDP-N-acetyl-D-galactosamine with a D-galactosyl-(1,4)-D-galactosyl-(1,4)-D-glucosylceramide to produce UDP and an N-acetyl-D-galactosaminyl-(1,3)-D-galactosyl-(1,4)-D-galactosyl-(1,4)-D-glucosylceramide.[1–3]

[1] T. Ishibashi, S. Kijimoto & A. Makita (1974) BBA 337, 92.
[2] J.-L. Chien, T. Williams & S. Basu (1973) JBC 248, 1778.
[3] N. Taniguchi & A. Makita (1984) JBC 259, 5637.

GALACTOSYLGALACTOSYLXYLOSYL-PROTEIN 3-β-GLUCURONOSYL-TRANSFERASE

This manganese-dependent transferase [EC 2.4.1.135], also known as glucuronosyltransferase I, catalyzes the reaction of UDP-D-glucuronate with a 3-β-D-galactosyl-4-β-D-galactosyl-O-β-D-xylosyl protein to produce UDP and the 3-β-D-glucuronosyl-3-β-D-galactosyl-4-β-D-galactosyl-O-β-D-xylosyl protein.[1,2] The enzyme participates in the biosynthesis of the heparin-polypeptide linkage region and of the proteochondroitin sulfate linkage region of certain glycoproteins. Glucuronosyl transfers to the nonreducing terminus can also be achieved with such acceptor substrates as 3-β-D-galactosyl-D-galactose, 3-β-D-galactosyl-D-galactosyl-4-xylose, and 3-β-D-galactosyl-D-galactosyl-4-xylosyl-L-serine. This transferase is very specific for this oligosaccharide substrate and does not act on other galactosides.[3]

[1] T. Helting & L. Rodén (1969) JBC 244, 2799.
[2] T. Helting (1972) JBC 247, 4327.
[3] Y. Tone, H. Kitagawa, K. Imiya, S. Oka, T. Kawasaki & K. Sugahara (1999) FEBS Lett. 459, 415.

Selected entries from **Methods in Enzymology** [vol, page(s)]:
General discussion: 28, 653
Assay: 28, 653
Other molecular properties: chondroitin sulfate biosynthesis and, 28, 639, 644; properties, 28, 654; reaction, 28, 653; sources, 28, 654; specificity and, 28, 642, 655; substrate preparation, 28, 670

GALACTOSYLGLUCOSAMINYL GALACTOSYLGLUCOSYLCERAMIDE α-L-FUCOSYLTRANSFERASE

This fucosyltransferase activity, formerly classified as EC 2.4.1.89 and now a deleted entry, is listed under galactoside 2-L-fucosyltransferase [EC 2.4.1.69].

β-1,3-GALACTOSYL-O-GLYCOSYL-GLYCOPROTEIN β-1,3-N-ACETYLGLUCOSAMINYLTRANSFERASE

This transferase [EC 2.4.1.146], also known as O-glycosyloligosaccharide-glycoprotein N-acetylglucosaminyltransferase II and elongation 3-β-GalNAc-transferase, catalyzes the reaction of UDP-N-acetyl-D-glucosamine with β-D-galactosyl-1,3-(N-acetyl-D-glucosaminyl-1,6)-N-acetyl-D-galactosaminyl-R to produce UDP and an N-acetyl-β-D-glucosaminyl-1,3-β-D-galactosyl-1,3-(N-acetyl-β-D-glucosaminyl-1,6)-N-acetyl-D-galactosaminyl-R, where R can be (a) the polypeptide chain of mucin or an antifreeze glycoprotein, (b) a benzyl group, or (c) a o-nitrophenyl group.[1,2] **See also** β-1,3-Galactosyl-O-Glycosyl-Glycoprotein β-1,6-N-Acetylglucosaminyltransferase; Acetylgalactosaminyl-O-Glycosyl-Glycoprotein β-1,3-N-Acetylglucosaminyltransferase; Acetylgalactosaminyl-O-Glycosyl-Glycoprotein β-1,6-N-Acetylglucosaminyltransferase

[1] I. Brockhausen, E. S. Rachaman, K. L. Matta & H. Schachter (1983) Carbohydr. Res. 120, 3.
[2] H. Schachter (1991) Glycobiology 1, 453.

Selected entries from **Methods in Enzymology** [vol, page(s)]:
General discussion: 179, 391; 230, 303
Assay: 179, 391

β-1,3-GALACTOSYL-O-GLYCOSYL-GLYCOPROTEIN β-1,6-N-ACETYLGLUCOSAMINYLTRANSFERASE

This transferase [EC 2.4.1.102], also known as β^6-N-acetylglucosaminyltransferase and O-glycosyloligosaccharide-glycoprotein N-acetylglucosaminyltransferase I, catalyzes the reaction of UDP-N-acetyl-D-glucosamine with β-D-galactosyl-1,3-N-acetyl-D-galactosaminyl-R to produce UDP and β-D-galactosyl-1,3-(N-acetyl-β-D-glucosaminyl-1,6)-N-acetyl-D-galactosaminyl-R where R can be (a) the polypeptide chain of mucin or an antifreeze glycoprotein, (b) a benzyl group, or (c) a o-nitrophenyl group.[1,2] **See also** β-1,3-Galactosyl-O-Glycosyl-Glycoprotein β-1,3-N-Acetylglucosaminyltransferase; Acetylgalactosaminyl-O-Glycosyl-Glycoprotein β-1,3-N-Acetylglucosaminyltransferase; Acetylgalactosaminyl-O-Glycosyl-Glycoprotein β-1,6-N-Acetylglucosaminyltransferase

[1] D. Williams, G. Longmore, K. L. Matta & H. Schachter (1980) JBC 255, 11253.
[2] T. Szumilo, G. P. Kaushal & A. D. Elbein (1987) Biochemistry 26, 5498.

Selected entries from **Methods in Enzymology** [vol, page(s)]:
General discussion: 179, 383; 230, 303
Assay: 179, 383

GALACTOSYL-PYROPHOSPHATE-[GLYCOYL-CARRIER LIPID] L-RHAMNOSYLTRANSFERASE

This glycosyltransferase catalyzes the reaction of galactosyl-pyrophosphate-[glycolyl-carrier lipid] with dTDP-L-rhamnose to produce rhamnosylgalactosyl-pyrophosphate-[glycolyl carrier lipid] and dTDP.

Selected entries from **Methods in Enzymology** [vol, page(s)]:
General discussion: 28, 583
Assay: 28, 588, 591

GALACTOSYLXYLOSYLPROTEIN 3-β-GALACTOSYLTRANSFERASE

This manganese-dependent glycosyltransferase [EC 2.4.1.134], also known as galactosyltransferase II, catalyzes the reaction of UDP-D-galactose with a 4-β-D-galactosyl-O-β-D-xylosylprotein to produce UDP and a 3-β-D-galactosyl-4-β-D-galactosyl-O-β-D-xylosylprotein.[1,2] Alternative acceptor substrates include 4-β-D-galactosyl-D-xylose and 4-β-D-galactosyl-O-β-D-xylosyl-L-serine. Phenyl and pyridine 3-O-β-galactosides can also serve as substrates.[2]

[1] N. B. Schwartz & L. Rodén (1975) JBC 250, 5200.
[2] J. A. Robinson & H. C. Robinson (1985) BJ 227, 805.

Selected entries from **Methods in Enzymology** [vol, page(s)]:
General discussion: 28, 651, 652
Assay: 28, 651, 652

GALACTURAN 1,4-α-GALACTURONIDASE

This enzyme [EC 3.2.1.67], also known as exopolygalacturonase and poly(galacturonate) hydrolase, catalyzes the hydrolysis of [(1,4)-α-D-galacturonide]$_n$ to produce [(1,4)-α-D-galacturonide]$_{n-1}$ and D-galacturonate.[1] In *Aspergillus*, this enzyme reaction proceeds with inversion of anomeric configuration, and β-D-galacturonate is initially released from the non-reducing end of the polymeric substrate.[2] The *Aspergillus* hydrolase also acts on xylogalacturonans.[3,4]

[1] R. Pressey & J. K. Avants (1975) Phytochemistry 14, 957.
[2] S. M. Pitson, M. Mutter, L. A. van den Broek, A. G. Voragen & G. Beldman (1998) BBRC 242, 552.
[3] H. C. Kester, J. A. Benen & J. Visser (1999) Biotechnol. Appl. Biochem. 30, 53.
[4] H. C. Kester, M. A. Kusters-van Someren, Y. Muller & J. Visser (1996) EJB 240, 738.

Selected entries from **Methods in Enzymology** [vol, page(s)]:
General discussion: 161, 373
Assay: 161, 373
Other molecular properties: chromatography with Spheron ion exchangers, 161, 388; 4-deoxy-L-*threo*-5-hexulose uronate, preparation of, 9, 603; physiological function, 161, 380; properties, 161, 376, 379; purification from (carrot, 161, 374; liverwort, 161, 377)

GALACTURONOKINASE

This phosphotransferase [EC 2.7.1.44] catalyzes the reaction of ATP with D-galacturonate to produce ADP and 1-phospho-α-D-galacturonate.[1]

[1] E. F. Neufeld, D. S. Feingold, S. M. Ilves, G. Kessler & W. Z. Hassid (1961) JBC 236, 3102.

GALLATE DECARBOXYLASE

This magnesium-dependent decarboxylase [EC 4.1.1.59] catalyzes the conversion of 3,4,5-trihydroxybenzoate (*i.e.*, gallate) to produce pyrogallol (*i.e.*, 1,2,3-benzenetriol) and carbon dioxide.[1,2]

[1] H. Yoshida, Y. Tani & H. Yamada (1982) Agric. Biol. Chem. 46, 2539.
[2] L. R. Krumholz, R. L. Crawford, M. E. Hemling & M. P. Bryant (1987) J. Bacteriol. 169, 1886.

GALLATE 1-β-GLUCOSYLTRANSFERASE

This transferase [EC 2.4.1.136], also known as UDP-glucose:vanillate 1-glucosyltransferase, catalyzes the reaction of UDP-D-glucose with gallate to produce UDP and 1-galloyl-β-D-glucose (also known as β-glucogallin).[1] Acceptor substrates include *p*-anisate, benzoate, protocatechuate, vanillate, and veratrate.[2]

[1] G. Gross (1983) Phytochemistry 22, 2179.
[2] G. Gross (1982) FEBS Lett. 148, 67.

GAMETOLYSIN

This metalloproteinase [EC 3.4.24.38], a green algal enzyme responsible for lysis of gamete cell walls, catalyzes the hydrolysis of peptide bonds in certain proline- and hydroxyproline-rich proteins.[1,2] Other peptides hydrolyzed include α-neoendorphin (at Phe4–Leu5), dynorphin (at Phe4–Leu5), neurotensin (at Pro–Tyr), mastoparan (at Ala–Ala and Ala–Leu), azocasein, gelatin, and Leu-Trp-Met-∇-Arg-Phe-Ala. (Note: ∇ indicates the cleavage site.) **See also** Autolysin; Sporangin

[1] Y. Matsuda (1988) Jpn. J. Phycol. 36, 246.
[2] M. J. Buchanan & W. J. Snell (1988) Exp. Cell Res. 179, 181.

GANGLIOSIDE G_{M1} β-GALACTOSIDASE

This enzyme activity, currently classified under β-galactosidase [EC 3.2.1.23], catalyzes the hydrolysis of the terminal galactosyl residue in ganglioside G_{M1} to produce D-galactose and ganglioside G_{M2}. **See** *β-Galactosidase*

Selected entries from *Methods in Enzymology* [vol, page(s)]:
General discussion: 28, 868
Assay: 28, 868; **50**, 482, 483
Other molecular properties: enzymatic properties, **28**, 871; galactosylceramidase, **50**, 483; G_{M1}-gangliosidosis and, **28**, 872; histochemical assay for subcellular localization, **28**, 872; isolation from human liver, **28**, 871; lactosylceramide, **50**, 471; reaction, **28**, 868, 872

GANGLIOSIDE GALACTOSYLTRANSFERASE

This transferase [EC 2.4.1.62] catalyzes the reaction of UDP-D-galactose with *N*-acetyl-D-galactosaminyl-(*N*-acetylneuraminyl)-D-galactosyl-D-glucosyl-*N*-acylsphingosine (*i.e.*, ganglioside G_{M2}) to produce UDP and D-galactosyl-*N*-acetyl-D-galactosaminyl-(*N*-acetylneuraminyl)-D-galactosyl-D-glucosyl-*N*-acylsphingosine (*i.e.*, the ganglioside G_{M1}).[1–4] Note that gangliosides G_{M3}, G_{M1}, G_{D1a}, and G_{T1} are not substrates.

[1] M. C. M. Yip & J. A. Dain (1970) *BJ* **118**, 247.
[2] G. B. Yip & J. A. Dain (1970) *BBA* **206**, 252.
[3] H.-J. Senn, M. Wagner & K. Decker (1983) *EJB* **135**, 231.
[4] S. Ghosh, J. W. Kyle, S. Dastgheib, F. Daussin, Z. Li & S. Basu (1995) *Glycoconj. J.* **12**, 838.

Selected entries from *Methods in Enzymology* [vol, page(s)]:
General discussion: 7., 521; **311**, 59
Assay: rat brain, **71**, 521, 523

GASTRICSIN

This aspartic endopeptidase [EC 3.4.23.3], also called pepsin C, parapepsin II, and pepsin II, catalyzes the hydrolysis of peptide bonds in proteins, preferring substrates with a tyrosyl residue at position P1 (*i.e.*, at Tyr–Xaa). However, Ac-Phe-∇-Leu-Val-His is a threefold better substrate than Ac-Tyr-∇-Leu-Val-His.[1] (Note: ∇ indicates the cleavage site.) The enzyme is found in the gastric juices of most vertebrates (including fish) and is formed from progastricsin, which has an additional 43 aminoacyl residues at the N-terminus (progastricsin has also been found in human seminal fluid). Gastricsin has slightly higher activity toward hemoglobin than pepsin A and is more specific. With hemoglobin as a substrate, the pH optimum is near 3, and two active-site aspartyl residues (pK_a values of 1.42 and 4.88) comprise the enzyme's catalytic center.[2]

[1] J. Tang (1970) *Meth. Enzymol.* **19**, 406.
[2] C. A. Auffret & A. P. Ryle (1979) *BJ* **179**, 239.

Selected entries from *Methods in Enzymology* [vol, page(s)]:
General discussion: 19, 336, 406
Assay: 19, 316, 317, 320, 406

GDP-4-DEHYDRO-D-RHAMNOSE REDUCTASE

This oxidoreductase [EC 1.1.1.187], also known as GDP-4-keto-6-deoxy-D-mannose reductase and GDP-4-keto-D-rhamnose reductase, catalyzes the reaction of GDP-4-dehydro-6-deoxy-D-mannose (*i.e.*, GDP-4-dehydro-D-rhamnose) with NAD(P)H to produce GDP-6-deoxy-D-mannose (*i.e.*, GDP-D-rhamnose) and NAD(P)+.[1,2]

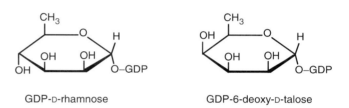

GDP-4-keto-D-rhamnose

GDP-D-rhamnose   GDP-6-deoxy-D-talose

The enzyme also produces GDP-6-deoxy-D-talose, the C4-epimer.

[1] N. W. Winkler & A. Markovitz (1971) *JBC* **246**, 5868.
[2] G. A. Barber (1968) *BBA* **165**, 68.

Selected entries from *Methods in Enzymology* [vol, page(s)]:
General discussion: 8, 296
Assay: 8, 297

GDP-6-DEOXY-D-TALOSE DEHYDROGENASE

This oxidoreductase [EC 1.1.1.135] catalyzes the reaction of GDP-6-deoxy-D-talose with NAD(P)+ to produce GDP-4-dehydro-6-deoxy-D-talose and NAD(P)H.[1]

[1] A. Markovitz (1964) *JBC* **239**, 2091.

GDP-GLUCOSIDASE

This enzyme [EC 3.2.1.42] catalyzes the hydrolysis of GDP-D-glucose to produce GDP and D-glucose.[1]

[1] S. Sonnino, H. Carminatti & E. Cabib (1966) *ABB* **116**, 26.

Selected entries from *Methods in Enzymology* [vol, page(s)]:
General discussion: 28, 979
Assay: 28, 979

GDP-MANNOSE 4,6-DEHYDRATASE

This NAD$^+$/NADP$^+$-dependent enzyme [EC 4.2.1.47] catalyzes the dehydration of GDP-D-mannose to yield GDP-4-dehydro-6-deoxy-D-mannose (also known as GDP-4-keto-D-rhamnose and GDP-4-keto-6-deoxy-D-mannose) and water.[1–6]

Analysis of the stereochemistry indicates: (a) the oxidoreductase reaction involves transfer of H4 to C6; (b) transfer is predominantly intramolecular; and (c) the transfer is stereospecific, such that H4 replaces the C6 hydroxyl group with inversion of configuration.[6] The C5 proton will also exchange with a solvent proton in the course of the reaction. The initial step of the reaction mechanism is the oxidation by the tightly bound coenzyme to produce a GDP-4-keto-D-mannose intermediate. Dehydration then occurs to generate an α,β-unsaturated ketone which is subsequently reduced by the coenzyme with transfer of the proton between C4 and C6 occurring concomitantly.[6]

[1] L. Glaser & H. Zarkowsky (1971) The Enzymes, 3rd ed., 5, 465.
[2] K. O. Broschat, S. Chang & G. Serif (1985) EJB 153, 397.
[3] T. H. Liao & G. A. Barber (1972) BBA 276, 85.
[4] A. D. Elbein & E. C. Heath (1965) JBC 240, 1926.
[5] A. Bisso, L. Sturla, D. Zanardi, A. De Flora & M. Tonetti (1999) FEBS Lett. 456, 370.
[6] P. J. Oths, R. M. Mayer & H. G. Floss (1990) Carbohydr. Res. 198, 91.

Selected entries from Methods in Enzymology [vol, page(s)]:
General discussion: 8, 296
Other molecular properties: preparation, 8, 298; properties, 8, 299

GDP-MANNOSE DEHYDROGENASE

This oxidoreductase [EC 1.1.1.132] catalyzes the reaction of GDP-D-mannose with water and two molecules of NAD$^+$ (or NADP$^+$) to produce GDP-D-mannuronate, two molecules of NADH (or NADPH), and two protons.[1–3] dGDP-D-mannose is also a substrate.

[1] S. Shankar, R. W. Ye, D. Schlictman & A. M. Chakrabarty (1995) AE 70, 221.
[2] K. Yamamoto, I. Katayama, Y. Onoda, M. Inami, H. Kumagai & T. Tochikura (1995) ABB 300, 694.
[3] S. Roychoudhury, T. B. May, J. F. Gill, S. K. Singh, D. S. Feingold & A. M. Chakrabarty (1989) JBC 264, 9380.

Selected entries from Methods in Enzymology [vol, page(s)]:
General discussion: 8, 285
Assay: 8, 285

GDP-MANNOSE 3,5-EPIMERASE

This epimerase [EC 5.1.3.18] catalyzes the interconversion of GDP-D-mannose and GDP-L-galactose.[1] The enzyme incorporates tritium from tritium-labeled water into carbon

atoms 3 and 5 of the hexose, suggesting that the reaction proceeds via an ene-diol(ate) intermediate.

[1] G. A. Barber (1979) J. Biol. Chem. 254, 7600.

Selected entries from Methods in Enzymology [vol, page(s)]:
General discussion: 83, 522
Assay: 83, 522

GEISSOSCHIZINE DEHYDROGENASE

This oxidoreductase [EC 1.3.1.36], which participates in the interconversion of heteroyohimbine alkaloids in the periwinkle, catalyzes the reaction of geissoschizine with NADP$^+$ to produce 4,21-dehydrogeissoschizine and NADPH.[1]

[1] A. Pfitzner & J. Stöckigt (1982) Phytochemistry 21, 1585.

GELATINASE A

This zinc-dependent metalloproteinase [EC 3.4.24.24] (also known as 72-kDa gelatinase, matrix metalloproteinase 2 (MMP-2), and type IV collagenase) catalyzes the hydrolysis of peptide bonds in gelatin type I and collagen types I, IV, V, VII, and X. (Note: The enzyme's action on type IV collagen depends on the extent of exposure of the protein substrate to reducing agents). The enzyme prefers to hydrolyze bonds between a glycyl residue and a hydrophobic residue; however, it also will cleave Gly–Glu, Gly–Asn, and Gly–Ser. Gelatinase A also has a requirement for hydroxyprolyl residues at the P5′ position (and usually at P5 as well). Other protein substrates include elastin, laminin, and fibronectin.[1] See also Gelatinase B

[1] G. Murphy & T. Crabbe (1995) Meth. Enzymol. 248, 470.

Selected entries from Methods in Enzymology [vol, page(s)]:
General discussion: 248, 470
Assay: 248, 471 (fluorometric, 248, 473; gelatinolytic, 248, 473; proteoglycan degradation method, 248, 52)
Other molecular properties: activation, 248, 480; active-site titration, 248, 100, 474, 502; activity, 248, 470; distribution, 248, 471; domain structure, 248, 477; Drosophila, gelatinase A-like, 245, 273; gene, 248, 204; inhibition, 248, 484; molecular weight, 248, 471; peptide thioester substrate, 248, 15; polypeptide chain structure, 248, 204; proenzyme (activation, 248, 472, 479; amino acid sequence, 248, 477, 479; binding to TIMP-2, 248, 471, 474, 477, 484, 504; membrane binding, 248, 479; properties, 248, 477; storage, 248, 476); properties, 248, 192, 480, 511; purification, 248, 475; substrate specificity, 248, 482; tissue extracts, in, measurement, 248, 523; zymography, 248, 474, 525

GELATINASE B

This zinc-dependent metalloproteinase [EC 3.4.24.35] (also known as 92-kDa gelatinase, matrix metalloproteinase 9 (MMP-9), type IV or V collagenase, and macrophage gelatinase) catalyzes the hydrolysis of peptide bonds in gelatin types I and V and collagen types III, IV, and V.[1,2] The enzyme's active site only tolerates alanyl and glycyl residues in the P1 position, and hydrophobic residues are preferred in P1′. For gelatinase B, the most effective synthetic substrate is Gly-Pro-Glu-Gly-^-Ile-Phe-Gly-Gln. **See also** *Gelatinase A*

[1]G. Murphy & T. Crabbe (1995) *Meth. Enzymol.* **248**, 470.
[2]H. Birkedal-Hansen (1995) *Curr. Opin. Cell Biol.* **7**, 728.

Selected entries from **Methods in Enzymology** [**vol**, page(s)]:
General discussion: **248**, 470
Assay: **163**, 326; **248**, 471 (fluorometric, **248**, 473; gelatinolytic, **248**, 473; proteoglycan degradation method, **248**, 52)
Other molecular properties: activation, **248**, 480; active-site titration, **248**, 100, 474, 502 (with tight-binding inhibitors, **248**, 97); activity, **248**, 470; distribution, **248**, 471; domain structure, **248**, 477; gene, **248**, 204; inhibition, **248**, 484; molecular weight, **248**, 471; peptide thioester substrate, **248**, 15; polypeptide chain structure, **248**, 204; proenzyme (activation, **248**, 472; activation by stromelysin 1, **248**, 468, 473, 481; amino acid sequence, **248**, 477, 479; binding to TIMP-1, **248**, 471, 474, 477, 484, 504; properties, **248**, 477; storage, **248**, 476); properties, **163**, 314; **248**, 192, 480, 511; purification, **248**, 475 (from human neutrophils, **163**, 320); substrate specificity, **248**, 482; tissue extracts, in, measurement, **248**, 523; zymography, **248**, 474, 525

GENTAMICIN 2′-ACETYLTRANSFERASE

This acetyltransferase [EC 2.3.1.59], also known as gentamicin acetyltransferase II, catalyzes the reaction of acetyl-CoA with the 2′-amino group of the hexose moiety of gentamicin C_{1a} to produce Coenzyme A and $N^{2'}$-acetylgentamicin C_{1a}.[1,2] Other acceptor substrates include butirosin, dideoxykanamycin, gentamicin A, gentamicin C_1, gentamicin C_2, kanamycin B, kanamycin C, neomycin B, paromomycin, sisomicin, tobramycin, and certain peptidoglycans.[3]

[1]M. Chevereau, P. J. L. Daniels, J. Davies & F. LeGoffic (1974) *Biochemistry* **13**, 598.
[2]R. Benveniste & J. Davies (1973) *PNAS* **70**, 2276.
[3]K. G. Payie & A. J. Clarke (1997) *J. Bacteriol.* **179**, 4106.

Selected entries from **Methods in Enzymology** [**vol**, page(s)]:
General discussion: **34**, 4; **43**, 611, 616
Assay: **43**, 612
Other molecular properties: *Escherichia coli*, from, **43**, 624; *Providencia*, from, **43**, 624; *Pseudomonas aeruginosa*, from, **43**, 624; transition state and multisubstrate analogues, **249**, 304

GENTAMICIN 3′-ACETYLTRANSFERASE

This acetyltransferase [EC 2.3.1.60], also known as gentamicin acetyltransferase I and aminoglycoside acetyltransferase AAC(3)-I, catalyzes the reaction of acetyl-CoA with the 3-*N*-position of the deoxystreptamine ring of gentamicin C to produce coenzyme A and $N^{3'}$-acetylgentamicin C. Other substrates include gentamicin B,

gentamicin C_1, gentamicin C_{1a}, gentamicin C_2, sisomicin, and tobramycin. The enzyme exhibits a steady-state random Bi Bi kinetic mechanism with synergism between substrate binding sites; with poor substrates such as tobramycin, the kinetic mechanism becomes rapid equilibrium random Bi Bi.[1,2] This transferase is distinct from that of aminoglycoside $N^{3'}$-acetyltransferase [EC 2.3.1.81]. **See also** *Aminoglycoside $N^{3'}$-Acetyltransferase*

[1]J. W. Williams & D. B. Northrop (1976) *Biochemistry* **15**, 125.
[2]J. W. Williams & D. B. Northrop (1978) *JBC* **253**, 5902 and 5908.

Selected entries from **Methods in Enzymology** [**vol**, page(s)]:
General discussion: **43**, 616, 625
Assay: **43**, 612

GENTAMICIN 2″-NUCLEOTIDYLTRANSFERASE

This transferase [EC 2.7.7.46] catalyzes the reaction of a nucleoside triphosphate with the 2-hydroxyl group of the 3-amino-3-deoxy-D-glucose moiety in gentamicin to produce a 2″-nucleotidylgentamicin and pyrophosphate (or, diphosphate).[1–3] The nucleoside triphosphate can be ATP, dATP, CTP, ITP, or GTP. Dibekacin, kanamycin, sisomicin and tobramycin can also act as acceptor substrates. The enzyme exhibits a Theorell-Chance Bi Bi kinetic mechanism with product release being the rate-determining step.[4]

[1]P. A. Frey (1989) *AE* **62**, 119.
[2]F. Angelatou, S. B. Litsas & P. Kontomichalou (1982) *J. Antibiot.* **35**, 235.
[3]J. E. Van Pelt & D. B. Northrop (1984) *ABB* **230**, 250.
[4]C. A. Gates & D. B. Northrop (1988) *Biochemistry* **27**, 3820, 3826, and 3834.

Selected entries from **Methods in Enzymology** [**vol**, page(s)]:
General discussion: **43**, 616, 625

GENTISATE DECARBOXYLASE

gentisate   hydroquinone

This decarboxylase [EC 4.1.1.62], also known as 2,5-dihydroxybenzoate decarboxylase, catalyzes the conversion of 2,5-dihydroxybenzoate (*i.e.*, gentisate) to hydroquinone and carbon dioxide.[1]

[1]D. J. W. Grant & J. C. Patel (1969) *Antonie Leeuwenhoek* **35**, 325.

GENTISATE 1,2-DIOXYGENASE

This iron-dependent oxidoreductase [EC 1.13.11.4] catalyzes the reaction of 2,5-dihydroxybenzoate with dioxygen to produce maleylpyruvate.[1,2]

gentisate   maleylpyruvate

Other substrates include 3-ethylgentisate, 4-fluorogentisate, and 3-methylgentisate. The substrate binds directly to the Fe^{2+} by means of its carboxylate and the C2 hydroxyl groups to correctly position the site of ring scission.[2]

[1] R. I. Crawford, S. W. Hutton & P. J. Chapman (1975) *J. Bacteriol.* **121**, 794.
[2] M. R. Harpel & J. D. Lipscomb (1990) *JBC* **265**, 6301 and 22187.

Selected entries from *Methods in Enzymology* [**vol**, page(s)]:
General discussion: 52, 11; 188, 101
Assay: 188, 102

GERANIOL DEHYDROGENASE

This oxidoreductase [EC 1.1.1.183] catalyzes the reaction of geraniol (*i.e.*, 3,7-dimethyl-2,6-octadien-1-ol) with $NADP^+$ to produce geranial (*i.e.*, 3,7-dimethyl-2,6-octadienal) and NADPH. Citronellol, farnesol, and nerol are weak alternative substrates.[1,2]

[1] V. H. Potty & J. H. Bruemmer (1970) *Phytochemistry* **9**, 1003.
[2] R. S. Sangwan, N. Singhsangwan & R. Luthra (1993) *J. Plant Physiol.* **142**, 129.

GERANIOL MONOOXYGENASE

geraniol   10-hydroxygeraniol

This heme-thiolate-dependent oxidoreductase catalyzes the reaction of geraniol with dioxygen and NADPH to produce 10-hydroxygeraniol, water, and $NADP^+$.[1]

[1] T. D. Meehan & C. J. Coscia (1973) *BBRC* **53**, 1043.

Selected entries from *Methods in Enzymology* [**vol**, page(s)]:
General discussion: 52, 14

GERANOYL-CoA CARBOXYLASE

This biotin-dependent enzyme [EC 6.4.1.5], also known as geranyl-CoA carboxylase, catalyzes the reaction of ATP and bicarbonate with geranoyl-CoA ($CoAS-COCH=C(CH_3)CH_2CH_2CH=C(CH_3)_2$) to produce ADP, orthophosphate, and 3-(4-methylpent-3-en-1-yl)pent-2-enedioyl-CoA.[1,2] The initial partial reaction utilizes ATP and bicarbonate to generate the carboxybiotin intermediate which is used to carboxylate the substrate.

[1] R. R. Fall (1981) *MIE* **71**, 791.
[2] X. Guan, T. Diez, T. K. Prasad, B. J. Nikolau & E. S. Wurtele (1999) *ABB* **362**, 12.

Selected entries from *Methods in Enzymology* [**vol**, page(s)]:
General discussion: 71, 791
Assay: *Pseudomonas citronellolis*, **71**, 792

GERANYL-DIPHOSPHATE CYCLASE

This enzyme [EC 5.5.1.8] (also known as geranyl-pyrophosphate cyclase, bornyl pyrophosphate synthase, and bornyl diphosphate synthase) catalyzes the conversion of geranyl diphosphate to (+)-bornyl diphosphate.[1–4] Other substrates include neranyl diphosphate and (3S)-linalyl diphosphate. Antipodal cyclization is thought to proceed after initial isomerization of the substrate to the bound allylic intermediate, (−)-(3R)-linalyl diphosphate.[3]

See also *Monoterpene Cyclases; Limonene Synthase*

[1] R. A. Gibbs (1998) *CBC* **1**, 31.
[2] F. M. Raushel & J. J. Villafranca (1988) *Crit. Rev. Biochem.* **23**, 1.
[3] R. Croteau, D. M. Satterwhite, D. E. Cane & C. C. Chang (1986) *JBC* **261**, 13438.
[4] R. Croteau (1986) *ABB* **251**, 777.

GERANYLGERANYLGLYCEROL-PHOSPHATE GERANYLGERANYLTRANSFERASE

This transferase [EC 2.5.1.42] catalyzes the reaction of geranylgeranyl diphosphate (*i.e.*, geranylgeranyl pyrophosphate) with *sn*-3-*O*-(geranylgeranyl)glycerol 1-phosphate to produce pyrophosphate (or, diphosphate) and 2,3-bis-*O*-(geranylgeranyl)glycerol 1-phosphate.[1] [1]

[1] D.-L. Zhang, L. Daniels & C. D. Poulter (1990) *JACS* **112**, 1264.

GERANYLGERANYL-PYROPHOSPHATE GERANYLGERANYLTRANSFERASE

This transferase [EC 2.5.1.32], also known as geranyl-geranyl-diphosphate geranylgeranyltransferase and prephytoene-diphosphate synthase, catalyzes the reaction of two geranylgeranyl pyrophosphate molecules to produce prephytoene pyrophosphate and pyrophosphate (or, diphosphate).[1] Prephytoene pyrophosphate is also converted to *trans*-phytoene. [1]

[1] O. Dogbo, A. Laferriere, A. D'Harlingue & B. Camara (1988) *PNAS* **85**, 7054.

Selected entries from *Methods in Enzymology* [vol, page(s)]:
General discussion: 110, 209; **214**, 352

GERANYL*trans*TRANSFERASE

This transferase [EC 2.5.1.10], also known as farnesyl-diphosphate synthase, catalyzes the reaction of geranyl diphosphate with isopentenyl diphosphate to produce *trans,trans*-farnesyl diphosphate and pyrophosphate (or, diphosphate).[1–7] Many sources of the enzyme also catalyze the reaction of dimethylallyl diphosphate with isopentenyl diphosphate to produce geranyl diphosphate and pyrophosphate. Those enzyme variants that can utilize both dimethylallyl diphosphate and geranyl diphosphate in reacting with isopentenyl diphosphate will not accept larger prenyl diphosphates as effective substitutes for geranyl diphosphate.

The kinetic mechanism is ordered Bi Bi with substrate inhibition by isopentenyl diphosphate. The reaction mechanism exhibits inversion of configuration at the diphosphate-bearing carbon. The *pro-R* hydrogen from the C2 of isopentenyl diphosphate is removed in the course of the reaction.[1,7] Most of the experimental evidence strongly suggests that the reaction proceeds via a stabilized allylic carbocation.[1] *See also* Dimethylallyltranstransferase

[1] R. A. Gibbs (1998) *CBC* **1**, 31.
[2] K. Ogura, T. Koyama & H. Sagami (1997) *Subcell. Biochem.* **28**, 57.
[3] B. A. Kellogg & C. D. Poulter (1997) *Curr. Opin. Chem. Biol.* **1**, 570.
[4] P. A. Edwards, M. N. Ashby, D. H. Spear, P. F. Marrero, A. Joly & G. Popjak (1992) *Biochem. Soc. Trans.* **20**, 475.
[5] H. Kleinig (1989) *Ann. Rev. Plant Physiol. Plant Mol. Biol.* **40**, 39.
[6] F. M. Raushel & J. J. Villafranca (1988) *Crit. Rev. Biochem.* **23**, 1.
[7] A. E. Leyes, J. A. Baker & C. D. Poulter (1999) *Org. Lett.* **1**, 1071.

GERMINATION CYSTEINE PROTEINASES

These plant cysteine endopeptidases, abbreviated GerCP, are isolated from germinating seeds. They catalyze peptide bond hydrolysis most likely in storage proteins. Most of these proteases are monomeric and have an acidic pH optimum (although some have a neutral optima), and several are hormonally induced by gibberellins.[1]

[1] A. Granell, M. Cercós & J. Carbonell (1998) in *Handbook of Proteolytic Enzymes* (A. J. Barrett, N. D. Rawlings & J. F. Woessner, eds.), Academic Press, New York, p. 578.

GIARDIA CYSTEINE ENDOPEPTIDASES

Multiple cysteine endopeptidase activities from *Giardia* sp. hydrolyze a wide variety of proteins, including hemoglobin, immunoglobulins, and extracellular matrix proteins.[1–4] These peptidases often prefer an arginyl residue in the P2 position. One such endopeptidase, giardain, reportedly facilitates excystation of the organism.

[1] D. F. Hare, E. L. Jarroll & D. G. Lindmark (1989) *Exp. Parasitol.* **68**, 168.
[2] M. J. North (1991) in *Biochemical Protozoology* (G. H. Coombs & M. J. North, eds.), Taylor & Francis, London, p. 234.
[3] A. G. Williams & G. H. Coombs (1995) *Int. J. Parasitol.* **25**, 771.
[4] E. Werries, A. Franz, H. Hippe & Y. Acil (1991) *J. Protozool.* **38**, 378.
[5] W. Ward, L. Alvarado, N. D. Rawlings, J. Engel, C. Franklin & J. H. McKerrow (1997) *Cell* **89**, 437.

GIBBERELLIN 2β-DIOXYGENASE

This oxidoreductase [EC 1.14.11.13], also known as gibberellin 2β-hydroxylase, catalyzes the reaction of gibberellin 1 with α-ketoglutarate (*i.e.*, 2-oxoglutarate) and dioxygen to produce 2β-hydroxygibberellin 1, succinate, and carbon dioxide.[1–3]

[1] V. A. Smith & J. MacMillan (1986) *Planta* **167**, 9.
[2] D. L. Griggs, P. Hedden & C. M. Lazarus (1991) *Phytochemistry* **30**, 2507.
[3] D. L. Griggs, P. Hedden, K. E. Temple-Smith & W. Rademacher (1991) *Phytochemistry* **30**, 2513.

GIBBERELLIN 3β-DIOXYGENASE

This ascorbate- and iron-dependent oxidoreductase [EC 1.14.11.15], also known as gibberellin 3β-hydroxylase, catalyzes the reaction of gibberellin 20 with α-ketoglutarate (*i.e.*, 2-oxoglutarate) and dioxygen to produce gibberellin 1, succinate, and carbon dioxide.[1,2] The enzyme also catalyzes the conversion of gibberellin 9 to gibberellin 4.

[1] S. S. Kwak, Y. Kamiya, A. Sakurai, N. Takahashi & J. Graebe (1988) *Plant Cell Physiol.* **29**, 935.
[2] T. Saito, S. S. Kwak, Y. Kamiya, H. Yamane, A. Sakurai, N. Murofushi & N. Takahashi (1991) *Plant Cell Physiol.* **32**, 239.

GIBBERELLIN-44 DIOXYGENASE

This iron-dependent oxidoreductase [EC 1.14.11.12], also known as gibberellin A44 oxidase, catalyzes the reaction of gibberellin 44 with α-ketoglutarate (*i.e.*, 2-oxoglutarate) and dioxygen to produce gibberellin 19, succinate, and carbon dioxide.[1]

[1] S. J. Gilmour, A. B. Bleecker & J. A. D. Zeevaart (1987) *Plant Physiol.* **85**, 87.

GIBBERELLIN β-GLUCOSYLTRANSFERASE

This transferase [EC 2.4.1.176] catalyzes the reaction of UDP-D-glucose with a gibberellin to produce UDP

and a gibberellin 2-O-β-D-glucoside. Acceptor substrates include gibberellins GA$_3$, GA$_9$, and GA$_7$.[1]

[1]G. Sembdner, H.-D. Knöfel, E. Schwarzkopf & H. W. Liebisch (1985) *Biol. Plant.* **27**, 231.

GINGIPAIN K

This cysteine endopeptidase (obtained from the oral pathogen *Porphyromonas gingivalis*, hence the name) catalyzes the hydrolysis of peptide bonds in proteins, including collagens and immunoglobulins. The descriptor K in the name indicates a preference for a lysyl residue in the P1 position (*i.e.*, Lys–Xaa). The enzyme also exhibits a preference for hydrophobic aminoacyl residues in positions P2 and P3. Gingipain K is readily assayed with the synthetic substrates Z-Lys-NHPhNO$_2$ and Tos-Gly-Pro-Lys-NHPhNO$_2$.[1–3] *See also Gingipain R*

[1]J. Travis, J. Potempa & H. Maeda (1995) *Trends Microbiol.* **3**, 405.
[2]J. Potempa & J. Travis (1996) *Acta Biochem. Pol.* **43**, 455.
[3]R. Pike, W. McGraw, J. Potempa & J. Travis (1994) *JBC* **269**, 406.

GINGIPAIN R

This cysteine endopeptidase [EC 3.4.22.37], also known as gingivain, Arg-gingipain, and gingipain 1, catalyzes the peptide-bond hydrolysis in various proteins, including collagens and immunoglobulins. The descriptor R in the name refers to the preference for substrates containing an arginyl residue in the P1 position (*i.e.*, Arg–Xaa). The enzyme also has a preference for hydrophobic aminoacyl residues in positions P2 and P3. Gingipain R is easily assayed with the synthetic substrate Bz-Arg-NHPhNO$_2$. Interestingly, this activity is greatly enhanced in the presence of glycine-containing compounds such as glycylglycine.[1–3] *See Gingipain K*

[1]J. Travis, J. Potempa & H. Maeda (1995) *Trends Microbiol.* **3**, 405.
[2]J. Potempa & J. Travis (1996) *Acta Biochem. Pol.* **43**, 455.
[3]Z. Chen, J. Potempa, A. Polanowski, M. Wilström & J. Travis (1992) *JBC* **267**, 18896.

GLOBOSIDE α-N-ACETYLGALACTOSAMINYLTRANSFERASE

This manganese-dependent transferase [EC 2.4.1.88] catalyzes the reaction of UDP-N-acetyl-D-galactosamine with an N-acetyl-D-galactosaminyl-(1,3)-D-galactosyl-(1,4)-D-galactosyl-(1,4)-D-glucosylceramide (known also as globoside) to produce UDP and an N-acetyl-D-galactosaminyl-N-acetyl-D-galactosaminyl-(1,3)-D-galactosyl-(1,4)-D-galactosyl-(1,4)-D-glucosylceramide (*i.e.*, Forssman hapten). UDP is a competitive inhibitor with respect to UDP-N-acetylgalactosamine and a noncompetitive inhibitor with respect to the globoside.[1] *See also Globotriosylceramide β-1,6-N-Acetylgalactosaminyltransferase*

[1]N. Taniguchi, N. Yokosawa, S. Gasa & A. Makita (1982) *JBC* **257**, 10631.

Selected entries from *Methods in Enzymology* [**vol**, page(s)]:
Assay: **138**, 595

GLOBOTRIOSYLCERAMIDE β-1,6-N-ACETYLGALACTOSAMINYLTRANSFERASE

This manganese-dependent transferase [EC 2.4.1.154], also called globoside N-acetylgalactosaminyltransferase, catalyzes the reaction of UDP-N-acetyl-D-galactosamine with a globotriosylceramide to produce UDP and a globotetraosylceramide containing a terminal β-N-acetylgalactosamine residue. Globotetraosylceramides and lactosylceramides are not acceptor substrates.[1] *See also Globoside α-N-Acetylgalactosaminyltransferase*

[1]M. W. Lockney & C. C. Sweeley (1982) *BBA* **712**, 234.

Selected entries from *Methods in Enzymology* [**vol**, page(s)]:
Assay: **138**, 594
Other molecular properties: properties, **138**, 594

1,4-α-GLUCAN BRANCHING ENZYME

This transferase [EC 2.4.1.18], which catalyzes the formation of 1,6-glucosidic linkages of amylopectin and glycogen,[1,2] is often named on the basis of its glucan substrate, hence glycogen branching enzyme, amylo-(1,4 to 1,6)-transglucosidase, branching enzyme, and amylo-(1,4-1,6)-transglycosylase. The minimal chain transferred by the plant enzyme contains five-to-seven glucosyl units. The number of glucosyl units between branch points in the final product is relatively constant per source of branching enzyme.[2] Specific aspartyl and glutamyl residues appear to be essential for catalytic activity.[2]

[1]D. S. Tolmasky, C. Labriola & C. R. Krisman (1998) *Cell Mol. Biol. (Noisy-le-grand)* **44**, 455.
[2]H. Takata, T. Takaha, T. Kuriki, S. Okada, M. Takagi & T. Imanaka (1994) *Appl. Environ. Microbiol.* **60**, 3096.

Selected entries from *Methods in Enzymology* [**vol**, page(s)]:
General discussion: **1**, 222; **8**, 395
Assay: **1**, 222; **8**, 395
Other molecular properties: conjugate, pH optimum shift, **44**, 468, 469; immobilization multistep enzyme system, **44**, 460, 461; phosphorylase, effect on action of plant, **1**, 193; properties, **8**, 402; purification (rabbit muscle, **8**, 398; rat liver, **1**, 223)

GLUCAN ENDO-1,2-β-GLUCOSIDASE

This enzyme [EC 3.2.1.71], also known as endo-1,2-β-glucanase, catalyzes the random hydrolytic scission of 1,2-glucosidic linkages in 1,2-β-D-glucans.[1,2]

[1] E. T. Reese, F. W. Parrish & M. Mandels (1961) *Can. J. Microbiol.* **7**, 309.
[2] S. Kitahata & S. Edagawa (1987) *Agric. Biol. Chem.* **51**, 2701.

Selected entries from *Methods in Enzymology* [vol, page(s)]:
General discussion: 8, 607, 611
Assay: 8, 611

GLUCAN ENDO-1,3-α-GLUCOSIDASE

This enzyme [EC 3.2.1.59], also known as endo-1,3-α-glucanase, catalyzes the endohydrolysis of 1,3-α-D-glucosidic linkages in pseudonigeran (in this case forming the products nigerose and D-glucose), isolichenin, and nigeran.[1-4]

[1] G. Davies, M. L. Sinnott & S. G. Withers (1998) *CBC* **1**, 119.
[2] L. G. Simonson, R. W. Gaugler, B. L. Lamberts & D. A. Reiher (1982) *BBA* **715**, 189.
[3] T. Takehara, M. Inoue, T. Morioka & K. Yokogawa (1981) *J. Bacteriol.* **145**, 729.
[4] M. T. Meyer & H. J. Phaff (1980) *J. Gen. Microbiol.* **118**, 197.

GLUCAN ENDO-1,3-β-GLUCOSIDASE

This hydrolase [EC 3.2.1.39], also known as $(1 \to 3)$-β-glucan endohydrolase, endo-1,3-β-glucanase, and laminarinase, catalyzes the hydrolysis of 1,3-β-D-glucosidic linkages in 1,3-β-D-glucans.[1-3] Despite limited action on mixed-link (1,3/1,4)-β-D-glucans, the enzyme catalyzes the hydrolysis of laminarin, paramylon, and pachyman. The enzymes from the marine mollusc and barley catalyze the reaction with retention of the anomeric configuration.[2,3] Note: This enzyme is distinct from endo-1,3(4)-β-glucanase [EC 3.2.1.6].

[1] G. Davies, M. L. Sinnott & S. G. Withers (1998) *CBC* **1**, 119.
[2] V. Lepagnol-Descamps, C. Richard, M. Lahaye, P. Potin, J. C. Yvin & B. Kloareg (1998) *Carbohydr. Res.* **310**, 283.
[3] L. Chen, M. Sadek, B. A. Stone, R. T. Brownlee, G. B. Fincher & P. B. Hoj (1995) *BBA* **1253**, 112.

Selected entries from *Methods in Enzymology* [vol, page(s)]:
General discussion: 8, 608, 609
Assay: 8, 609

GLUCAN ENDO-1,6-β-GLUCOSIDASE

This enzyme [EC 3.2.1.75], also known as endo-1,6-β-glucanase, catalyzes the random hydrolysis of 1,6-linkages in 1,6-β-D-glucans.[1-4] The enzyme will act on lutean, luteose, pustulan, and 1,6-oligo-β-D-glucosides. The *Acremonium persicinum* enzyme acts with retention of configuration. The *Gibberella fujikuroi* enzyme differs from other members of this class in that, while acting on $1 \to 6$ linkages, it will also catalyze the hydrolysis of $1 \to 3$ linkages adjacent to 6-substituted glucosyl residues.

[1] N. Hiura, T. Nakajima & K. Matsuda (1987) *Agric. Biol. Chem.* **51**, 3315.

[2] V. Y. Rudakova, N. M. Shevchenko & L. A. Elyakova (1985) *Comp. Biochem. Physiol.* **81B**, 677.
[3] G. P. Schep, M. G. Shepherd & P. Sullivan (1984) *BJ* **233**, 707.
[4] S. M. Pitson, R. J. Seviour, B. M. McDougall, B. A. Stone & M. Sadek (1996) *BJ* **316**, 841.

Selected entries from *Methods in Enzymology* [vol, page(s)]:
General discussion: 8, 608, 612
Assay: 8, 613

GLUCAN 1,3-α-GLUCOSIDASE

This glucosidase [EC 3.2.1.84], also known as exo-1,3-α-glucanase, catalyzes the hydrolysis of terminal 1,3-α-D-glucosidic linkages in 1,3-α-D-glucans.[1-3] Nigeran is not a substrate.

[1] S. Saxena, K. Shailubhai, B. Dong-Yu & I. K. Vijay (1987) *BJ* **247**, 563.
[2] D. Brada & U. C. Dubach (1984) *EJB* **141**, 149.
[3] D. M. Burns & O. Touster (1982) *JBC* **257**, 9991.

Selected entries from *Methods in Enzymology* [vol, page(s)]:
General discussion: 83, 416

GLUCAN 1,3-β-GLUCOSIDASE

This enzyme [EC 3.2.1.58], also known as exo-1,3-β-glucanase and exo-1,3-β-glucosidase, catalyzes the successive hydrolysis of β-D-glucose units from the non-reducing ends of 1,3-β-D-glucans, resulting in the release of D-glucose.[1-5] The enzyme will also act on oligosaccharides. Other substrates include laminarin, paramylon, and pachyman: however, laminaribiose is a weak substrate.

[1] S. M. Cutfield, G. J. Davies, G. Murshudov, B. F. Anderson, P. C. Moody, P. A. Sullivan & J. F. Cutfield (1999) *J. Mol. Biol.* **294**, 771.
[2] S. M. Pitson, R. J. Seviour & B. M. McDougall (1993) *Enzyme Microb. Technol.* **15**, 178.
[3] C. Nombela, M. Molina, R. Cenamor & M. Sanchez (1988) *Microbiol. Sci.* **5**, 328.
[4] A. Sanchez, J. R. Villanueva & T. G. Villa (1982) *J. Gen. Microbiol.* **128**, 3051.
[5] C. F. Talbot & V. D. Vacquier (1982) *JBC* **257**, 742.

Selected entries from *Methods in Enzymology* [vol, page(s)]:
General discussion: 8, 607, 609
Assay: 8, 609

GLUCAN 1,4-α-GLUCOSIDASE

This enzyme [EC 3.2.1.3] (known variously as glucoamylase, 1,4-α-D-glucan glucohydrolase, amyloglucosidase, γ-amylase, exoamylase, lysosomal α-glucosidase, and exo-1,4-α-glucosidase) catalyzes the successive hydrolytic scission of terminal 1,4-linked α-D-glucosyl residues from non-reducing ends of the glucan chains, resulting in the release of β-D-glucose, with inversion of anomeric configuration.[1-4] Substrates include amylose, glycogen, amylopectin, and dextrin. Most variants of this enzyme will catalyze the hydrolysis of α1 $\to$ 6 linkages as well, albeit with considerable less efficiency. All enzymes listed under EC 3.2.1.3 require at least six glucosyl residues to be present in the substrate and act more rapidly on polysaccharides than oligosaccharide substrates. Even so, low-molecular-weight glucosides, such as *p*-nitrophenyl α-glucoside, possessing exceptionally good leaving groups can be used to assay the enzyme.

Two carboxyl groups appear to be essential for enzyme activity, one acting as a general catalyst and the other as a catalytic base that guides nucleophilic attack by water on the C1 carbon of the aglycone moiety. Kinetic isotope effect investigations suggest that the transition state has an advanced oxocarbenium ion character.

Mammalian intestine α-glucosidase [EC 3.2.1.20] catalyzes similar reactions. **See also** α-Amylase; α-Glucosidase

[1] A. Pandey, P. Nigam, C. R. Soccol, V. T. Soccol, D. Singh & R. Mohan (2000) *Biotechnol. Appl. Biochem.* **31**, 135.
[2] S. Chiba (1997) *Biosci. Biotechnol. Biochem.* **61**, 1233.
[3] M. Vihinen & P. Mantsala (1989) *Crit. Rev. Biochem. Mol. Biol.* **24**, 329.
[4] P. Manjunath, B. C. Shenoy & M. R. R. Roa (1983) *J. Appl. Biochem.* **5**, 235.

Selected entries from *Methods in Enzymology* **[vol, page(s)]:**
General discussion: 28, 931
Assay: 28, 931; 137, 691
Other molecular properties: anomeric specificity, 64, 275; binding energetics, 64, 260, 261; brush border, in, 31, 130; commercial source, 44, 777; conjugate, activity assay, 44, 778; coupling to derivatized titanium(IV)-activated supports, 135, 127; enlarged, preparation for coentrapment, 137, 646; ethylenediamine adducts, coupling to carboxyalkyl agarose, 135, 145; exoamylase, *Pseudomonas saccharophila*, 9, 13; fiber entrapment, industrial application, 44, 241; glucose oxidase electrode, description, 137, 36; immobilization, 44, 42 (bridge formation, by, 44, 167; cellulose, on, 44, 52, 167; entrapment, by, 44, 178, 180; liposomes, in, 44, 224; porous ceramics, on, 44, 136, 137, 778; radiation polymerization, by, 135, 151; scale-up data, 44, 781); industrial production of glucose, 44, 776; industrial saccharification, in (immobilized enzyme system, 136, 377; liquid enzyme system, 136, 374); isolation from *Aspergillus niger*, 28, 931; 137, 689, 693; kinetics, in enzyme system, 44, 433, 434; production by *Aspergillus niger*, 137, 689, 693; properties, 28, 933; *Pseudomonas saccharophila*, 9, 13; subsite mapping, 64, 258; support for, choice, 44, 776, 777, 779; titanium chloride binding, 44, 52; transition metal-chelated, glutaraldehyde crosslinking, 135, 126

GLUCAN 1,4-β-GLUCOSIDASE

This enzyme [EC 3.2.1.74], also known as exo-1,4-β-glucosidase, catalyzes the successive hydrolysis of 1,4-linkages in 1,4-β-D-glucan polymers.[1,2] Substrates not only include 1,4-β-D-glucans, but a number of related oligosaccharides. Cellobiose is also hydrolyzed, albeit quite slowly.

A barley isozyme exhibits characteristics of both glucan 1,4-β-glucosidase and β-glucosidases [EC 3.2.1.21],

as exemplified by its ability to catalyze the hydrolysis of β(1,4)-oligoglucosides much more efficiently than 4-nitrophenyl-β-D-glucopyranoside.[3] **See also** β-Glucosidase; Cellulose 1,4-β-Cellobiosidase

[1] M. E. Himmel, M. P. Tucker, S. M. Lastick, K. K. Oh, J. W. Fox, D. D. Spindler & K. Grohmann (1986) *JBC* **261**, 12948.
[2] T. M. Wood & S. I. McCrae (1982) *Carbohydr. Res.* **110**, 291.
[3] M. Hrmova, E. A. MacGregor, P. Biely, R. J. Stewart & G. B. Fincher (1998) *JBC* **273**, 11134.

Selected entries from *Methods in Enzymology* **[vol, page(s)]:**
General discussion: 160, 408
Other molecular properties: cellobiosidase (assay, 160, 392; properties, 160, 395; purification from *Ruminococcus albus*, 160, 393); detection in gels, 160, 143

GLUCAN 1,6-α-GLUCOSIDASE

This enzyme [EC 3.2.1.70], also known as exo-1,6-α-glucosidase and glucodextranase, catalyzes the successive hydrolysis of glucosyl residues from 1,6-α-D-glucans and derived oligosaccharides, producing β-D-glucose by inversion.[1] Substrates include dextrans, isomaltosaccharides, and isomaltose, albeit weakly. Some variants of this glucosidase also act on 1,3-α-D-glucosidic bonds in dextrans.

[1] A. Ramos & I. Spencer-Martins (1983) *Antonie Van Leeuwenhoek* **49**, 183.

Selected entries from *Methods in Enzymology* **[vol, page(s)]:**
Other molecular properties: substrate preparation, 64, 255

(1,4)-α-D-GLUCAN 1-α-D-GLUCOSYLMUTASE

This mutase [EC 5.4.99.15], also called malto-oligosyl-trehalose synthase and maltodextrin α-D-glucosyltransferase, catalyzes the interconversion of 4-[(1,4)-α-D-glucosyl]$_{n-1}$-D-glucose to 1-α-D-[(1,4)-α-D-glucosyl]$_{n-1}$-α-D-glucopyranoside, and this reaction can be regarded as an intramolecular transglucosylation. The enzyme uses as substrates (1,4)-α-D-glucans containing three or more (1,4)-α-linked D-glucose units: hence, it is not active toward maltose.[1]

[1] T. Nakada, K. Maruta, K. Tsusaki, M. Kubota, H. Chaen, T. Sugimoto, M. Kurimoto & Y. Tsujisaka (1995) *Biosci. Biotech. Biochem.* **59**, 2210.

1,4-α-GLUCAN 6-α-GLUCOSYLTRANSFERASE

This glucosyltransferase [EC 2.4.1.24], also known as oligoglucan-branching glycosyltransferase, catalyzes the transfer of an α-D-glucosyl residue in a (1,4)-α-D-glucan or oligosaccharide (also in maltose) to the C6 hydroxyl group of D-glucose or C6-OH of a glucosyl residue in a 1,4-α-D-glucan.[1,2] **See also** 1,4-α-Glucan Branching Enzyme

[1]D. Kobrehel & R. Deponte (1982) *Enzyme Microb. Technol.* **4**, 185.
[2]S. A. Barker & T. R. Carrington (1953) *J. Chem. Soc.* **1953**, 3588.

GLUCAN 1,6-α-ISOMALTOSIDASE

This enzyme [EC 3.2.1.94], also known as exo-isomaltohydrolase and isomalto-dextranase, catalyzes the hydrolysis of 1,6-α-D-glucosidic linkages in polysaccharides so as to remove successive isomaltose units from the non-reducing ends of the chains.[1] The best substrates observed are 1,6-α-D-glucans containing six, seven, or eight glucosyl units.

[1]G. Okada, T. Takayanagi, S. Miyahara & T. Sawai (1988) *Agric. Biol. Chem.* **52**, 829.

GLUCAN 1,4-α-MALTOHEXAOSIDASE

This enzyme [EC 3.2.1.98], also known as exo-maltohexaohydrolase, G6-amylase, and maltohexaose-producing amylase, catalyzes the successive hydrolysis of 1,4-α-D-glucosidic linkages located at the nonreducing ends of amylosaccharides, producing maltohexaose residues of α anomeric configuration.[1]

Note: The enzyme is distinct from other glucan glucosidases: glucan 1,4-α-glucosidase [EC 3.2.1.3], catalyzing the removal of successive glucosyl residues from the nonreducing end; β-amylose [EC 3.2.1.2] and glucan 1,4-α-maltohydrolase [EC 3.2.1.133], catalyzing the removal of successive maltose residues; glucan 1,4-α-maltotriohydrolase [EC 3.2.1.116], catalyzing the removal of successive maltotriose residues; and glucan 1,4-α-maltotetraohydrolase [EC 3.2.1.60], catalyzing the removal of successive maltotetraose residues.

[1]M. Vihinen & P. Mäntsälä (1989) *Crit. Rev. Biochem. Mol. Biol.* **24**, 329.

GLUCAN 1,4-α-MALTOHYDROLASE

This glycosidase [EC 3.2.1.133], also known as maltogenic α-amylase, catalyzes the hydrolysis of (1→4)-α-D-glucosidic linkages in polysaccharides so as to remove successive α-maltose residues from the non-reducing ends of the chains. Substrates include starch and related polysaccharides and oligosaccharides.[1] The enzyme is similar in action to β-amylase [EC 3.2.1.2]; however, with that enzyme the product is β-maltose.

[1]I.-C. Kim, J.-H. Cha, J.-R. Kim, S.-Y. Jang, B.-C. Seo, T.-K. Cheong, D.-S. Lee, Y.-D. Choi & K.-H. Park (1992) *JBC* **267**, 2210.

GLUCAN 1,4-α-MALTOTETRAHYDROLASE

This glucosidase [EC 3.2.1.60], also known as exo-maltotetraohydrolase, G4-amylase, and maltotetraose-forming amylase, catalyzes the hydrolysis of 1,4-α-D-glucosidic linkages in amylaceous polysaccharides resulting in the removal of successive maltotetraose residues from the non-reducing ends of the oligomer or polymer.[1,2]

[1]Y. Sakano, E. Kashiyama & T. Kobayashi (1983) *Agric. Biol. Chem.* **47**, 1761.
[2]M. Vihinen & P. Mäntsälä (1989) *Crit. Rev. Biochem. Mol. Biol.* **24**, 329.

GLUCAN 1,4-α-MALTOTRIOHYDROLASE

This glycosidase [EC 3.2.1.116], also known as exo-maltotriohydrolase, catalyzes the hydrolysis of 1,4-α-D-glucosidic linkages in amylaceous polysaccharides by successive release of maltotriose residues from the non-reducing chain ends of the polymer.[1]

[1]T. Nakakuki, K. Azuma & K. Kainuma (1984) *Carbohydr. Res.* **128**, 297.

4-α-GLUCANOTRANSFERASE

This transferase [EC 2.4.1.25], also known as disproportionating enzyme, D-enzyme, dextrin glycosyltransferase, and oligo-1,4-1,4-glucantransferase, catalyzes the transfer of a segment of a (1,4)-α-D-glucan to a new 4-position in acceptor substrates (including D-glucose, maltose, and α(1,4)-D-glucan), with overall retention of anomeric configuration.[1-3] ***See also*** *Glycogen Debranching Enzyme*

[1]T. N. Palmer, B. E. Ryman & W. J. Whelan (1976) *EJB* **69**, 105.
[2]T. Takaha, M. Yanase, H. Takata, S. Okada & S. M. Smith (1996) *JBC* **271**, 2902.
[3]G. Davies, M. L. Sinnott & S. G. Withers (1998) *CBC* **1**, 119.
Selected entries from ***Methods in Enzymology*** [vol, page(s)]:
General discussion: **8**, 515, 517
Assay: **8**, 517
Other molecular properties: amylomaltase, **1**, 189; **5**, 141; **230**, 304 (induction in microorganisms, **22**, 94); glycogen and, **8**, 515, 517; properties, **8**, 522; purification, **8**, 519

4-α-D-[(1 → 4)-α-D-GLUCANO]TREHALOSE TREHALOHYDROLASE

This enzyme [EC 3.2.1.141], also known as malto-oligosyltrehalose trehalohydrolase, catalyzes the hydrolysis of α-(1 → 4)-D-glucosidic linkage in 4-α-D-[(1 → 4)-α-D-glucanosyl]$_n$-trehalose to produce trehalose and the α-(1 → 4)-D-glucan.[1]

[1]T. Nakada, K. Maruta, H. Mitsuzumi, M. Kubota, H. Chaen, T. Sugimoto, M. Kurimoto & Y. Tsujisaka (1995) *Biosci, Biotech. Biochem.* **59**, 2215.

1,3-β-GLUCAN PHOSPHORYLASE

This phosphorylase [EC 2.4.1.97], also known as laminarin phosphorylase, catalyzes the reaction of a 1,3-β-glucan (*i.e.*, [(1,3)-β-D-glucosyl]$_n$) with orthophosphate to produce [(1,3)-β-D-glucosyl]$_{n-1}$ and α-D-glucose 1-phosphate.[1,2] Substrates include a range of β-(1,3)-oligoglucans as well as laminarin. The enzyme is distinct from 1,3-β-oligoglucan phosphorylase [EC 2.4.1.30], which cannot act on laminarin, and laminaribiose phosphorylase [EC 2.4.1.31].

[1] Y. Lienart, J. Comtat & F. Barnoud (1988) *Plant Sci.* **58**, 165.
[2] G. J. Albrecht & H. Kauss (1971) *Phytochemistry* **10**, 1293.

α-1,4-GLUCAN-PROTEIN SYNTHASE (ADP-FORMING)

This enzyme [EC 2.4.1.113] catalyzes the reaction of ADP-D-glucose with a protein to produce ADP and the α-D-glucosyl-protein.[1]

[1] R. Barengo & C. R. Krisman (1978) *BBA* **540**, 190.

α-1,4-GLUCAN-PROTEIN SYNTHASE (UDP-FORMING)

This enzyme [EC 2.4.1.112] catalyzes the reaction of UDP-D-glucose with a protein to produce UDP and the α-D-glucosyl-protein.[1–3]

[1] R. Barengo & C. R. Krisman (1978) *BBA* **540**, 190.
[2] C. R. Krisman & R. Barengo (1975) *EJB* **52**, 117.
[3] N. Lavintman, J. S. Tandecarz, M. Carceller, S. Mendiara & C. E. Cardini (1974) *EJB* **50**, 145.

α-1,3-GLUCAN SYNTHASE

This transferase [EC 2.4.1.183] catalyzes the reaction of UDP-D-glucose with [α-D-glucosyl-(1,3)]$_n$ (*i.e.*, an α-1,3-glucosyl polymer) to produce UDP and [α-D-glucosyl-(1,3)]$_{n+1}$. A glucan primer is required to initiate the reaction.[1] The *Streptococcus mutans* enzyme reportedly catalyzes the hydrolysis of sucrose.[2]

[1] N. Hanada & T. Takehara (1987) *Carbohydr. Res.* **168**, 120.
[2] Y. Yamashita, N. Hanada, M. Itoh-Andoh & T. Takehara (1989) *FEBS Lett.* **243**, 343.

1,3-β-GLUCAN SYNTHASE

This transferase [EC 2.4.1.34], also known as 1,3-β-D-glucan:UDP-glucosyltransferase, UDP-glucose:1,3-β-D-glucan glucosyltransferase, and callose synthase, catalyzes the reaction of UDP-D-glucose with [(1,3)-β-D-glucosyl]$_n$

to produce UDP and [(1,3)-β-D-glucosyl]$_{(n+1)}$, where n is 60–80 for the *Saccharomyces cerevisiae* enzyme.[1–4]

[1] K. Fredrikson & C. Larsson (1992) *Biochem. Soc. Trans.* **20**, 710.
[2] J. Ruiz-Herrera (1991) *Antonie Van Leeuwenhoek* **60**, 72.
[3] E. M. Shematek, J. A. Braatz & E. Cabib (1980) *JBC* **255**, 888.
[4] L. R. Marechal & S. H. Goldemberg (1964) *JBC* **239**, 3163.

Selected entries from *Methods in Enzymology* [vol, page(s)]:
General discussion: 8, 404; 138, 637
Assay: 8, 404; 138, 637
Other molecular properties: membrane-associated, catalytic subunits, identification, **230**, 335; preparation (fungi, **138**, 638; mung beans [*Phaseolus aureus*] **8**, 405); properties, **8**, 405; **138**, 641

β-GLUCAN-TRANSPORTING ATPase

This so-called ATPase [EC 3.6.3.42] catalyzes the ATP-dependent (ADP- and orthophosphate-producing) transport of β-glucan$_{in}$ to produce β-glucan$_{out}$.[1,2] It has been identified in Gram-negative bacteria. This ABC-type (ATP-binding cassette-type) ATPase contains two similar ATP-binding domains and does not undergo phosphorylation during the transport process.

The idea that a transporter is an enzyme is in keeping with a new definition of enzyme catalysis as the facilitated making/breaking of chemical bonds, not just covalent bonds.[3] This idea builds on Pauling's assertion that any long-lived, chemically distinct interaction (in this case, the persistent location of a solute with respect to the faces of a membrane) can be regarded as a chemical bond. Note also that the equilibrium constant ($K_{eq} = $ [β-glucan$_{out}$][ADP][P$_i$]/[β-glucan$_{in}$][ATP]) does not conform to that expected for an ATPase (*i.e.*, $K_{eq} = $ [ADP][P$_i$]/[ATP]). Thus, although the overall reaction yields ADP and orthophosphate, the enzyme is misclassified as a hydrolase, and instead should be regarded as an energase-type reaction. Energases facilitate affinity-modulated reactions by coupling the $\Delta G_{ATP\text{-hydrolysis}}$ to a force-generating or work-producing step.[3] In this case, P-O-P bond-scission supplies the energy to drive β-glucan transport. **See** *ATPase; Energase*

[1] M. J. Fath & R. Kolter (1993) *Microbiol. Rev.* **57**, 995.
[2] J. K. Griffiths & C. E. Sansom (1998) *The Transporter Factsbook*, Academic Press, San Diego.
[3] D. L. Purich (2001) *TiBS* **26**, 417.

GLUCARATE DEHYDRATASE

This dehydratase [EC 4.2.1.40], also known as glucarate hydro-lyase, catalyzes the conversion of D-glucarate to 5-dehydro-4-deoxy-D-glucarate (*i.e.*, 5-keto-4-deoxy-D-glucarate) and water. The *Escherichia coli* enzyme, which utilizes an enol intermediate in the reaction pathway, also catalyzes the dehydration of L-idarate resulting in the epimerization reaction interconverting D-glucarate and L-idarate.[1–3]

[1] W. A. Wood (1971) *The Enzymes*, 3rd ed., **5**, 573.
[2] R. Jeffcoat (1972) *EJB* **25**, 515.
[3] A. M. Gulick, B. K. Hubbard, J. A. Gerlt & I. Rayment (2000) *Biochemistry* **39**, 4590.

Selected entries from *Methods in Enzymology* [vol, page(s)]:
General discussion: 9, 660
Assay: 9, 660

GLUCARATE *O*-HYDROXYCINNAMOYLTRANSFERASE

This acyltransferase [EC 2.3.1.131] catalyzes the reaction of sinapoyl-CoA with glucarate to produce coenzyme A and *O*-sinapoylglucarate.[1] Alternative acyl-CoA substrates include 4-coumaroyl-CoA, feruloyl-CoA, and caffeoyl-CoA. *See also Glucarolactone O-Hydroxycinnamoyltransferase*

[1] D. Strack, H. Keller & G. Weissenböck (1987) *J. Plant Physiol.* **131**, 61.

GLUCAROLACTONE *O*-HYDROXYCINNAMOYLTRANSFERASE

This acyltransferase [EC 2.3.1.132] catalyzes the reaction of sinapoyl-CoA with glucarolactone to produce coenzyme A and *O*-sinapoylglucarolactone.[1] Alternative acyl-CoA substrates include 4-coumaroyl-CoA, feruloyl-CoA, and caffeoyl-CoA. *See also Glucarate O-Hydroxycinnamoyltransferase*

[1] D. Strack, H. Keller & G. Weissenböck (1987) *J. Plant Physiol.* **131**, 61.

β-GLUCOGALLIN *O*-GALLOYLTRANSFERASE

This acyltransferase [EC 2.3.1.90] catalyzes the reaction of two molecules of 1-*O*-galloyl-β-D-glucoside (*i.e.*, β-glucogallin) to produce D-glucose and 1-*O*,6-*O*-digalloyl-β-D-glucoside.[1] Digalloylglucose can also act as the acceptor substrate (producing 1-*O*,2-*O*,6-*O*-trigalloyl-β-D-glucoside) as can 1-*O*-protocatechuoyl-β-D-glucoside. This enzyme participates in gallotannin biosynthesis in higher plants.

[1] G. G. Gross & K. Denzel (1990) *Z. Naturforsch.* **45c**, 37.

β-GLUCOGALLIN: TETRAKISGALLOYLGLUCOSE *O*-GALLOYLTRANSFERASE

This acyltransferase [EC 2.3.1.143] catalyzes the reaction of 1-*O*-galloyl-β-D-glucoside (*i.e.*, β-glucogallin)

with 1,2,3,6-tetrakis-*O*-galloyl-β-D-glucoside to produce D-glucose and 1,2,3,4,6-pentakis-*O*-galloyl-β-D-glucoside, the common precursor of gallotannins and the related ellagitannins.[1]

[1] J. Cammann, K. Denzel, G. Schilling & G. Gross (1989) *ABB* **273**, 58.

GLUCOKINASE

This phosphotransferase [EC 2.7.1.2] catalyzes the ATP-dependent phosphorylation of D-glucose to yield D-glucose 6-phosphate and ADP.[1–9] The Michaelis constant of the inducible liver enzyme for D-glucose is nearly 100 times greater than the corresponding value for hexokinase. Although originally thought to be specific for D-glucose, this enzyme also phosphorylates other hexoses: D-fructose, for example, binds some 30 to 50 times more weakly.[5,7] Bacterial and rat liver enzymes have ordered Bi Bi kinetic mechanisms.[6,9] *See also Hexokinase*

[1] J. H. Hurley (1996) *Ann. Rev. Biophys. & Biomol. Struct.* **25**, 137.
[2] P. B. Iynedjian (1993) *BJ* **293**, 1.
[3] R. L. Printz, M. A. Magnuson & D. K. Granner (1993) *Ann. Rev. Nutr.* **13**, 463.
[4] A. Cornish-Bowden & M. L. Cárdenas (1991) *Trends Biochem. Sci.* **16**, 281.
[5] R. J. Middleton (1990) *Biochem. Soc. Trans.* **18**, 180.
[6] D. Pollard-Knight & A. Cornish-Bowden (1982) *Mol. Cell. Biochem.* **44**, 71.
[7] D. L. Purich, H. J. Fromm & F. B. Rudolph (1973) *AE* **39**, 249.
[8] S. P. Colowick (1973) *The Enzymes*, 3rd ed., **9**, 1.
[9] E. V. Porter, B. M. Chassy & C. E. Holmlund (1982) *BBA* **709**, 178.

Selected entries from *Methods in Enzymology* [vol, page(s)]:
General discussion: 9, 381, 388; 34, 328, 490; 90, 25
Assay: 9, 382, 388; 63, 37, 38; liver, from rat, 42, 31, 32; *Streptococcus mutans*, 90, 25; yeast, from, 42, 25, 26
Other molecular properties: N-acetyl-D-glucosamine kinase, and, 9, 425; *Aerobacter aerogenes*, from (pH optimum, 9, 391; properties, 9, 391; purification, 9, 389; specificity, 9, 391; stability, 9, 392); cooperativity, 64, 218, 221; 249, 545; hepatic (cooperativity and slow transitions in, 249, 545; reaction sequence, 249, 316); hormonal control, 37, 293; hysteresis, 64, 215, 218, 221; 249, 545; liver, from rat, 42, 31 (assay, 42, 31, 32; chromatography, 42, 34; electrophoresis, 42, 32, 35, 36; inhibition, 42, 38; molecular weight, 42, 38; properties, 9, 387; 42, 38, 39; purification, 9, 383; 42, 33; specificity, 9, 387); marker enzyme, as, 32, 733; metal ion effects, 63, 325; Michaelis constant, 63, 38; plasma membrane, 31, 90; *Streptococcus mutans* (assay, 90, 25; properties, 90, 28; purification, 90, 27); vertebrate, kinetic cooperativity in, 249, 555; yeast, from, 42, 25 (assay, 42, 25, 26; chromatography, 42, 27, 28; inhibitors, 42, 30; molecular weight, 42, 30; properties, 42, 30; purification, 42, 26)

GLUCOMANNAN 4-β-MANNOSYLTRANSFERASE

This transferase [EC 2.4.1.32], also known as glucomannan-synthase, catalyzes the reaction of GDP-D-mannose with (glucomannan)$_n$ to produce GDP and (glucomannan)$_{n+1}$, such that a β(1 → 4) linkage is formed.[1–3]

[1] G. Dalessandro, G. Piro & D. H. Northcote (1986) *Planta* **169**, 564.

[2]M. M. Smith, M. Axelos & C. Peaud-Lenoel (1976) *Biochimie* **58**, 1195.
[3]A. D. Elbein (1969) *JBC* **244**, 1608.

GLUCONATE DEHYDRATASE

This enzyme [EC 4.2.1.39] catalyzes the elimination of water from D-gluconate to produce 2-dehydro-3-deoxy-D-glucarate.[1–3]

[1]N. Budgen & M. Danson (1986) *FEBS Lett.* **196**, 207.
[2]R. Bender & G. Gottschalk (1973) *EJB* **40**, 309.
[3]W. A. Wood (1971) *The Enzymes*, 3rd ed., **5**, 573.

Selected entries from ***Methods in Enzymology*** [**vol**, page(s)]:
General discussion: 42, 301; **90**, 283
Assay: *Alcaligenes*, from, **42**, 301; *Clostridium pasteurianum*, **90**, 283
Other molecular properties: *Alcaligenes*, from, **42**, 301 (assay, **42**, 301; distribution, **42**, 304; electrophoresis, **42**, 302, 303; inhibitors, **42**, 304; properties, **42**, 303, 304; purification, **42**, 302, 303); *Clostridium pasteurianum* (assay, **90**, 283; properties, **90**, 286; purification, **90**, 284)

GLUCONATE 2-DEHYDROGENASE

Two enzymes have the designation gluconate 2-dehydrogenase. The FAD-dependent enzyme [EC 1.1.99.3] catalyzes the reversible reaction of D-gluconate with an acceptor substrate to produce 2-dehydro-D-gluconate and the reduced acceptor.[1–3] The reverse reaction proceeds more readily. If the acceptor substrate is NADP$^+$, as is the case with the *Gluconobacter oxydans* enzyme, the protein is gluconate 2-dehydrogenase (NADP$^+$). The second gluconate 2-dehydrogenase [EC 1.1.1.215], also known as 2-keto-D-gluconate reductase, catalyzes the reversible reaction of D-gluconate with NADP$^+$ to produce 2-dehydro-D-gluconate and NADPH.[2,4] Alternative substrates include L-idonate, D-galactonate, and D-xylonate.

[1]K. Matsushita, E. Shinagawa, O. Adachi & M. Ameyama (1979) *J. Biochem.* **86**, 249.
[2]T. Chiyonobu, E. Shinagawa, O. Adachi & M. Ameyama (1976) *Agric. Biol. Chem.* **40**, 175.
[3]E. Shinagawa, T. Chiyonobu, O. Adachi & M. Ameyama (1976) *Agric. Biol. Chem.* **40**, 475.
[4]O. Adachi, T. Chiyonobu, E. Shinagawa, K. Matsushita & M. Ameyama (1978) *Agric. Biol. Chem.* **42**, 2057.

Selected entries from ***Methods in Enzymology*** [**vol**, page(s)]:
General discussion: 5, 287; **9**, 196; **89**, 187, 203
Assay: 5, 287; **9**, 196; acetic acid bacteria, **89**, 203; bacterial membrane-bound, **89**, 187
Other molecular properties: acetic acid bacteria, from (assay, **89**, 203; properties, **89**, 208; purification, **89**, 204); bacterial membrane-bound (assay, **89**, 187; microdetermination of D-gluconate, **89**, 25; properties, **89**, 191; purification, **89**, 188); *Brevibacterium helvolum*, from, **9**, 196 (pH optimum, **9**, 199; physical constants, **9**, 199; physiological function, **9**, 199; properties, **9**, 198; purification, **9**, 197; specificity, **9**, 198); FAD-dependent (*Pseudomonas fluorescens*, electrochemical studies, **227**, 521); *Gluconobacter oxydans*, from, **9**, 197, 198 (pH optimum, **9**, 199; physical constants, **9**, 199; physiological function, **9**, 199; properties, **9**, 197; purification, **9**, 198; specificity, **9**, 198); *Pseudomonas fluorescens* (electrochemical studies, **227**, 521; properties, **5**, 290; purification, **5**, 288); sources, **9**, 197

GLUCONATE 5-DEHYDROGENASE

This oxidoreductase [EC 1.1.1.69], also known as 5-keto-D-gluconate 5-reductase and 5-ketogluconate reductase, catalyzes the reversible reaction of D-gluconate with NAD(P)$^+$ to produce 5-dehydro-D-gluconate and NAD(P)H.[1,2] **See also** *L-Idonate Dehydrogenase*

[1]O. Adachi, E. Shinagawa, K. Matsushita & M. Ameyama (1979) *Agric. Biol. Chem.* **43**, 75.
[2]M. Ameyama, T. Chiyonobu & O. Adachi (1974) *Agric. Biol. Chem.* **38**, 1377.

Selected entries from ***Methods in Enzymology*** [**vol**, page(s)]:
General discussion: 9, 200; **89**, 198
Assay: 9, 200; **89**, 199
Other molecular properties: *Acetobacter*, from, **9**, 201; *Escherichia*, from, **9**, 201; *Gluconobacter oxydans*, from, **9**, 201; *Gluconobacter suboxydans* (pH optimum, **9**, 203; physical constants, **9**, 203; properties, **9**, 202; **89**, 201; purification, **9**, 201; **89**, 200; specificity, **9**, 202); 2-ketogluconate reductase, separation from, **9**, 198; *Klebsiella*, from, **9**, 201; polyol dehydrogenase, separation from, **9**, 174; sources, **9**, 201

GLUCONOKINASE

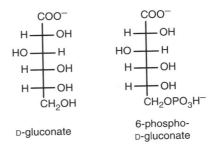

D-gluconate   6-phospho-D-gluconate

This phosphotransferase [EC 2.7.1.12], also known as gluconate kinase, catalyzes the reaction of ATP with D-gluconate to produce ADP and 6-phospho-D-gluconate.[1–3] The yeast enzyme reportedly has a rapid-equilibrium random Bi Bi kinetic mechanism.[4]

[1]M. Zachariou & R. K. Scopes (1985) *Biochem. Int.* **10**, 367.
[2]C. J. Coffee & A. S. L. Hu (1972) *ABB* **149**, 549.
[3]I. G. Leder (1957) *JBC* **225**, 125.
[4]C. S. Tsai, J. L. Shi & H. G. Ye (1995) *ABB* **316**, 163.

Selected entries from ***Methods in Enzymology*** [**vol**, page(s)]:
General discussion: 1, 350
Assay: 1, 350
Other molecular properties: brewers' yeast, purification, **1**, 353; *Escherichia coli*, **1**, 350 (properties, **1**, 354; purification, **1**, 353); ethylgluconate 6-phosphate preparation, **1**, 334; D-gluconate assay, **3**, 233; 2-ketogluconokinase, not present in preparation of, **5**, 294; *Leuconostoc mesenteroides*, **1**, 334; role in microorganisms, possible, **9**, 200

GLUCONOLACTONASE

This enzyme [EC 3.1.1.17], also known as lactonase and aldonolactonase, catalyzes the hydrolysis of

D-glucono-1,5-lactone to produce D-gluconate.[1–6] Other substrates include a wide range of hexono-1,5-lactones.

CH₂OH — O ... OH ... H ... OH ... OH ... H ... OH

D-gluconolactone   D-gluconate

Aldonolactonase, previously classified as EC 3.1.1.18, catalyzes the hydrolysis of L-gulono-γ-lactone to produce L-gulonate and that activity is now included with gluconolactonase. **See also** *Actinomycin Lactonase; Peptide Antibiotic Lactonase*

[1] M. Zachariou & R. K. Scopes (1986) *J. Bacteriol.* **167**, 863.
[2] W. R. Carper, A. S. Mehra, D. P. Campbell & J. A. Levisky (1982) *Experientia* **38**, 1046.
[3] G. D. Bailey, B. D. Roberts, C. M. Buess & W. R. Carper (1979) *ABB* **192**, 482.
[4] B. D. Roberts, G. D. Bailey, C. M. Buess & W. R. Carper (1978) *BBRC* **84**, 322.
[5] S. H. Grossman & B. Axelrod (1973) *JBC* **248**, 4846.
[6] A. F. Brodie & F. Lipmann (1955) *JBC* **212**, 677.

Selected entries from **Methods in Enzymology** [vol, page(s)]:
General discussion: 6, 337; **18A**, 50
Assay: 6, 337
Other molecular properties: actinomycin, **43**, 763 (*Actinoplanes*, **43**, 765, 766; assay, **43**, 764; inactivation product, **43**, 767; properties, **43**, 767; purification, **43**, 764; reaction scheme, **43**, 763); bovine liver (properties, **18A**, 52; purification, **18A**, 51); D(L)-gulono-γ-lactone hydrolase (preparation, **18A**, 50, 51; properties, **18A**, 52); peptide antibiotic, **43**, 767 (assay, **43**, 770, 771; inhibitors, **43**, 772; kinetic properties, **43**, 772; pH optimum, **43**, 772; properties, **43**, 772, 773; purification, **43**, 771, 772; specific activity, **43**, 772, 773; specificity, **43**, 772); rat liver (properties, **6**, 339; purification, **6**, 338)

GLUCOSAMINATE AMMONIA-LYASE

COO⁻ — NH₃⁺ ... HO — H ... H — OH ... H — OH ... CH₂OH

COO⁻ — O ... H — H ... H — OH ... H — OH ... CH₂OH

D-glucosaminate   2-keto-3-deoxy-
D-gluconate

This pyridoxal-phosphate-dependent lyase [EC 4.3.1.9], also known as D-glucosaminate dehydrase, catalyzes the

conversion of D-glucosaminate to 2-dehydro-3-deoxy-D-gluconate and ammonia.[1,2] **See *Aminodeoxygluconate Dehydratase***

[1] W. A. Wood (1971) *The Enzymes*, 3rd ed., **5**, 573.
[2] L. Davis & D. E. Metzler (1972) *The Enzymes*, 3rd ed., **7**, 33.

Selected entries from **Methods in Enzymology** [vol, page(s)]:
General discussion: 9, 657
Assay: 9, 657

GLUCOSAMINE ACETYLTRANSFERASE

This transferase [EC 2.3.1.3] catalyzes the reaction of acetyl-CoA with D-glucosamine to produce coenzyme A and *N*-acetyl-D-glucosamine.[1,2]

A human acetyltransferase, which also acetylates glucosaminyl residues at the nonreducing end of heparan sulfate, has an random Bi Bi kinetic mechanism[3] and operates via an acetyl-enzyme intermediate.[4] **See *Heparan-α-Glucosaminide N-Acetyltransferase***

[1] H. Kresse & J. Glössl (1987) *AE* **60**, 217.
[2] T. C. Chou & M. Soodak (1957) *JBC* **196**, 105.
[3] P. J. Meikle, A. M. Whittle & J. J. Hopwood (1995) *BJ* **308**, 327.
[4] K. J. Bame & L. H. Rome (1986) *JBC* **261**, 10127.

Selected entries from **Methods in Enzymology** [vol, page(s)]:
General discussion: 138, 607
Other molecular properties: diagnosis of genetic mucopolysaccharide storage disorders, in, **83**, 561; radioactive substrates, preparation, **83**, 566

GLUCOSAMINE KINASE

This phosphotransferase [EC 2.7.1.8] catalyzes ATP-dependent phosphorylation of D-glucosamine to produce ADP and D-glucosamine 6-phosphate.[1,2]

[1] J. L. Trujillo, W. J. Horng & J. C. Gan (1971) *BBA* **252**, 443.
[2] E. Bueding & J. A. MacKinnon (1955) *JBC* **215**, 495.

GLUCOSAMINE-1-PHOSPHATE *N*-ACETYLTRANSFERASE

This acetyltransferase [EC 2.3.1.157] catalyzes the reaction of acetyl-CoA with D-glucosamine 1-phosphate to produce coenzyme A and *N*-acetyl-D-glucosamine 1-phosphate.[1]

[1] D. Mengin-Lecreulx & J. van Heijenoort (1994) *J. Bacteriol.* **176**, 5788.

GLUCOSAMINE-6-PHOSPHATE *N*-ACETYLTRANSFERASE

This transferase [EC 2.3.1.4], also known as phosphoglucosamine transacetylase, phosphoglucosamine acetylase, and hexosamine 6-phosphate *N*-acetylase, catalyzes the reaction of acetyl-CoA (or propionyl-CoA) with D-glucosamine 6-phosphate to produce coenzyme A and N-acetyl-D-glucosamine 6-phosphate (or N-propionyl-D-glucosamine 6-phosphate). The rat liver acetyltransferase exhibits a ping pong Bi Bi kinetic mechanism.[1]

[1] A. P. Corfield, D. J. Mortimer & P. J. Winterburn (1984) *Biochem. Soc. Trans.* **12**, 565.

Selected entries from *Methods in Enzymology* [vol, page(s)]:
General discussion: 9, 704
Assay: 9, 704

GLUCOSAMINE-6-PHOSPHATE DEAMINASE

This isomerase [EC 3.5.99.6; previously classified as EC 5.3.1.10] catalyzes the reaction of D-glucosamine 6-phosphate with water to produce D-fructose 6-phosphate and ammonia.[1–5]

Porcine and *Escherichia coli* isomerases are allosteric enzymes activated by N-acetyl-D-glucosamine 6-phosphate. The reaction involves an aldose-ketose isomerization in which the ... $-CH(NH_2)CHO$ group of the acyclic glucosamine 6-phosphate is converted into ... $-C(=NH)-CH_2OH$ forming 2-deoxy-2-imino-D-*arabino*-hexitol, which then hydrolyzes to yield the final products.[1] Interestingly, 2-amino-2-deoxyglucitol 6-phosphate is strong competitive inhibitor with a K_i value of $0.2\,\mu M$.

[1] C. F. Midelfort & I. A. Rose (1977) *Biochemistry* **16**, 1590.
[2] I. A. Rose (1975) *AE* **43**, 491.
[3] J. M. Buchanan (1973) *AE* **39**, 91.
[4] E. A. Noltmann (1972) *The Enzymes*, 3rd ed., **6**, 271.
[5] D. G. Comb & S. Roseman (1958) *JBC* **232**, 807.

Selected entries from *Methods in Enzymology* [vol, page(s)]:
General discussion: 5, 418, 422; 9, 575; 41, 400, 497
Assay: 5, 418, 422; 9, 576; *Escherichia coli*, 5, 422; 41, 498; house flies, from, 41, 400
Other molecular properties: *Escherichia coli*, 5, 422; 41, 497 (activators, 5, 425; 41, 502; chromatography, 41, 499; properties, 5, 425; 41, 501, 502; purification, 5, 423; 41, 498); house flies, from, 41, 400 (activators and inhibitors, 41, 406; assay, 41, 400; properties, 41, 406; purification, 41, 403); intermediate, 87, 84; pig kidney, in, 5, 418; 41, 497 (properties, 5, 421; purification, 5, 419); *Proteus vulgaris*, from, 9, 575; 41, 497 (pH optimum, 9, 579; properties, 9, 578; purification, 9, 577); sources, 9, 575

GLUCOSAMINYLGALACTOSYL-GLUCOSYLCERAMIDE β-GALACTOSYLTRANSFERASE

This transferase [EC 2.4.1.86], also known as paragloboside synthase, catalyzes the reaction of UDP-D-galactose with N-acetyl-D-glucosaminyl-(1,3)-D-galactosyl-(1,4)-D-glucosylceramide to produce UDP and D-galactosyl-N-acetyl-D-glucosaminyl-(1,3)-D-galactosyl-(1,4)-D-glucosylceramide (a paragloboside).[1,2]

[1] M. Basu, K. A. Presper, S. Basu, L. M. Hoffmann & S. E. Brooks (1979) *PNAS* **76**, 4270.
[2] M. Basu & S. Basu (1972) *JBC* **247**, 1489.

GLUCOSE-1,6-BISPHOSPHATE SYNTHASE

This enzyme [EC 2.7.1.106] catalyzes the reaction of 3-phospho-D-glyceroyl phosphate (*i.e.*, 1,3-bisphosphoglycerate) with D-glucose 1-phosphate to produce 3-phospho-D-glycerate and D-glucose 1,6-bisphosphate.[1–3] Other acceptor substrates include D-glucose 6-phosphate (producing D-glucose 1,6-bisphosphate) and D-mannose 1-phosphate.

[1] I. A. Rose, J. V. B. Warms & G. Kaklij (1975) *JBC* **250**, 3466.
[2] M. Ueda, M. Hirose, R. Sasaki & H. Chiba (1978) *J. Biochem.* **83**, 1721.
[3] I. A. Rose, J. V. B. Warms & L.-J. Wong (1977) *JBC* **252**, 4262.

GLUCOSE DEHYDROGENASES

There are several different glucose dehydrogenases, and their EC classification depends on the nature of the electron acceptor.

Glucose 1-dehydrogenase [EC 1.1.1.47] catalyzes the reaction of β-D-glucose with $NAD(P)^+$ to produce D-glucono-1,5-lactone and $NAD(P)H$.[1–3] Other substrates include D-xylose, D-glucose 6-phosphate, D-idose, D-mannose, D-galactose, and 2-deoxy-D-glucose. The *Halobacterium salinarum* enzyme has an ordered Bi Bi kinetic mechanism.[3]

Glucose 1-dehydrogenase (NAD$^+$) [EC 1.1.1.118], also referred to as aldohexose dehydrogenase, catalyzes the same reaction but is specific for NAD$^+$.[4,5] Other substrates include D-galactose, D-mannose, 2-deoxy-D-glucose, 6-deoxy-D-galactose, 2-deoxy-D-galactose, D-altrose, 6-deoxy-D-glucose, and D-allose.

Glucose 1-dehydrogenase (NADP$^+$) [EC 1.1.1.119], also referred to as aldohexose dehydrogenase (NADP$^+$), catalyzes the same reaction but is specific for NADP$^+$.[6–8] Other substrates include D-mannose, 2-deoxy-D-glucose, and 2-amino-2-deoxy-D-mannose. The *Schizosaccharomyces pombe* enzyme has an ordered Bi Bi kinetic mechanism.[8]

Glucose dehydrogenase (acceptor) [EC 1.1.99.10], also referred to as glucose dehydrogenase (*Aspergillus*), catalyzes the FAD-dependent reaction of D-glucose with an acceptor substrate to produce D-glucono-1,5-lactone and the reduced acceptor.[9–12] Acceptor substrates include 2,6-dichloroindophenol, phenazine methosulfate, and ferricyanide.

Glucose dehydrogenase (pyrroloquinoline-quinone) [EC 1.1.99.17], also referred to as quinoprotein glucose dehydrogenase, catalyzes the PQQ-dependent reaction of D-glucose (the β-anomer) with an acceptor substrate to produce D-glucono-1,5-lactone and the reduced acceptor. The enzyme is distinct from glucose dehydrogenase (acceptor) [EC 1.1.99.10].[13–16] Other substrates include D-xylose, D-arabinose, lactose, maltose, and D-galactose. The reaction mechanism involves a general base-catalyzed hydride transfer (*i.e.*, a hydride transfer from glucose to PQQ, followed by tautomerization of C5-reduced PQQ to $PQQH_2$).[13] Catalysis by the enzyme proceeds according to a ping pong kinetic mechanism in which substrate inhibition by glucose and a negative cooperativity effect are exhibited.[14,15] The membrane-bound variant of EC 1.1.99.7 uses ubiquinone as the acceptor substrate.

[1] W. R. Carper, W. C. Groutas & D. B. Coffin (1988) *Experientia* **44**, 29.

[2] R. F. Ramaley & N. Vasantha (1983) *JBC* **258**, 12558.

[3] S. R. Bhaumik & H. M. Sonawat (1999) *Indian J. Biochem. Biophys.* **36**, 143.

[4] A. S. L. Hu & A. L. Cline (1964) *BBA* **93**, 237.

[5] Y. Kobayashi & K. Horikoshi (1989) *Agric. Biol. Chem.* **44**, 41.

[6] G. Avigad, Y. Alroy & S. Englard (1968) *JBC* **243**, 1936.

[7] O. Adachi, K. Matsushita, E. Shinagawa & M. Ameyama (1980) *Agric. Biol. Chem.* **44**, 301.

[8] C. S. Tsai, J. L. Shi & H. G. Ye (1995) *ABB* **316**, 163.

[9] K. Matsushita, Y. Ohno, E. Shinagawa, O. Adachi & M. Ameyama (1980) *Agric. Biol. Chem.* **44**, 1505.

[10] H.-M. Müller (1977) *Zentralbl. Bakteriol. Parasitenkd. Infektionskr. Hyg.* **132**, 14.

[11] T.-G. Bak (1967) *BBA* **139**, 277.

[12] J. B. Hauge (1964) *JBC* **239**, 3630.

[13] A. Oubrie, H. J. Rozeboom, K. H. Kalk, A. J. Olsthoorn, J. A. Duine & B. W. Dijkstra (1999) *EMBO J.* **18**, 5187.

[14] A. J. Olsthoorn, T. Otsuki & J. A. Duine (1998) *EJB* **255**, 255.

[15] A. R. Dewanti & J. A. Duine (2000) *Biochemistry* **39**, 9384.

[16] C. Anthony (1992) *Int. J. Biochem.* **24**, 29.

Selected entries from *Methods in Enzymology* [vol, page(s)]:
General discussion: 1, 335; 89, 149, 159; 280, 89; NAD$^+$-specific, 41, 147; NADP$^+$-specific, 41, 142
Assay: 9, 93, 98, 103, 107; *Gluconobacter cerinus*, 41, 142; *Gluconobacter suboxydans*, 89, 159; *Pseudomonas fluorescens* membrane-bound, 89, 149; pseudomonad MSU-1, 41, 147, 148
Other molecular properties: *Acetobacter suboxydans*, from, 9, 98; 41, 147; aryl-4-hydroxylase and, 5, 817, 818; assay of phenylalanine dehydrogenase, 53, 278, 279; azide inhibition, 9, 103; cloning from bacteria (mutant complementation, 258, 217; oligonucleotide probing, 258, 220); continuous flow analyzer, in, 44, 639, 646; cyanide inhibition, 9, 96, 103; *Gluconobacter cerinus*, from, 41, 142 (assay, 41, 142; inhibitors, 41, 145, 146; occurrence, 41, 146; properties, 41, 145; purification, 41, 143, 144); *Gluconobacter suboxydans* (assay, 89, 159; properties, 89, 162;

purification, 89, 160); *Gluconobacter suboxydans* membrane-bound, microdetermination of D-glucose, 89, 23; glucose dehydrogenase (acceptor) (assay, *Pseudomonas fluorescens* membrane-bound, 89, 149; *Gluconobacter suboxydans* membrane-bound, microdetermination of D-glucose, 89, 23; *Pseudomonas fluorescens* membrane-bound [assay, 89, 149; microdetermination of D-glucose, 89, 23; properties, 89, 152; purification, 89, 150]); glucose oxidase electrode, description, 137, 39, 40; hydropyrimidine dehydrogenase and, 12A, 55; immobilization, on nylon tube, 44, 640, 646; multienzymatic NADH recycling, in, 136, 71; particulate, 9, 92 (*Acetobacter* species, 9, 98; *Acetobacter suboxydans*, 9, 98; 41, 147; assay, 9, 93, 98; *Bacterium anitratum*, 9, 92; properties, 9, 94, 102; *Pseudomonas* species, 9, 92; purification, 9, 92, 99; specificity, 9, 102; *Xanthomonas phaseoli*, in, 9, 93); phenylalanine hydroxylase and, 5, 810; prosthetic group, 9, 98; pseudomonad MSU-1, from, 41, 147 (assay, 41, 147, 148; chromatography, 41, 149; properties, 41, 150; purification, 41, 148); *Pseudomonas fluorescens* membrane-bound (assay, 89, 149; microdetermination of D-glucose, 89, 23; properties, 89, 152; purification, 89, 150); pyridine nucleotide and, 2, 493, 494; 6, 107; soluble, 9, 103 (activators, 9, 106; assay, 9, 103, 107; *Bacillus cereus*, in, 9, 103; *Bacterium anitratum*, in, 9, 107; heat resistance, 9, 107; inhibitors, 9, 106; kinetic properties, 9, 106; molecular weight, 9, 111; pH optimum, 9, 110; properties, 9, 106, 110; prosthetic group, 9, 111; purification, 9, 104; specificity, 9, 106, 110; turnover number, 9, 111); stain for, 224, 107; stereospecificity, 87, 106; use of beef liver enzyme in assay for phenylalanine hydroxylase cofactor, 18B, 601, 615

GLUCOSE:FRUCTOSE OXIDOREDUCTASE

This oxidoreductase [EC 1.1.99.28] catalyzes the reaction of D-glucose with D-fructose to produce D-gluconolactone and D-glucitol. Other aldose substrates, although weaker than D-glucose, include D-mannose, D-xylose, D-galactose, 2-deoxy-D-*erythro*-hexose, and L-arabinose. Interestingly, the ketose substrate must be in the open-chain form. Other ketose substrates, albeit weaker, are D-xylulose and dihydroxyacetone (*i.e.*, glycerone). The *Zymomonas mobilis* enzyme, which contains tightly bound NADP$^+$, exhibits a ping pong Bi Bi kinetic mechanism.[1]

[1] M. Zachariou & R. K. Scopes (1986) *J. Bacteriol.* **167**, 863.

GLUCOSE ISOMERASE

This enzyme activity, formerly classified as EC 5.3.1.18, is attributable to glucose-6-phosphate isomerase [EC 5.3.1.9] in the presence of arsenate or to xylose isomerase [EC 5.3.1.5] for which D-glucose is an alternative substrate. *See Glucose-6-Phosphate Isomerase; Xylose Isomerase*

GLUCOSE OXIDASE

This FAD-dependent oxidoreductase [EC 1.1.3.4] catalyzes the reaction of β-D-glucose with dioxygen to produce D-glucono-1,5-lactone and hydrogen peroxide.[1–3] The enzyme has a ping pong Bi Bi kinetic mechanism in which β-D-glucose binds near the *re*-face of FAD. Reduction of the coenzyme is very fast. Reoxidation of the flavin by dioxygen is also rapid with the probable formation of a flavin C4a-hydroperoxide intermediate. Other sugar substrates, albeit weaker, include D-xylose, D-mannose, and D-galactose. Surprisingly, glucose oxidase also catalyzes the oxidation of nitroalkanes.[2]

[1]B. A. Palfrey & V. Massey (1998) *CBC* **3**, 83.
[2]H. J. Bright & D. J. T. Porter (1975) *The Enzymes*, 3rd ed., **12**, 421.
[3]R. Bentley (1963) *The Enzymes*, 2nd ed., **7**, 567.

Selected entries from *Methods in Enzymology* [vol, page(s)]:
General discussion: 1, 340; 9, 82; **34**, 338; **52**, 6, 18; **56**, 474; **161**, 307
Assay: 1, 340; 9, 82; **161**, 308
Other molecular properties: activity assay of immobilized enzyme, **44**, 198, 347, 482, 483 (insoluble aggregate, of, **44**, 268; microcalorimetry, by, **44**, 663, 664; native protein, of, **44**, 479; soluble aggregate, of, **44**, 474); aldose-1-epimerase, assay of, **9**, 608; **63**, 33; anaerobic photoreduction, **18B**, 510, 511; *Aspergillus niger*, **9**, 82 (properties, **9**, 85; purification, **9**, 83); assay (cellobiose, of, **63**, 32; β-D-glucose, of, **57**, 403, 452; **63**, 33; with luminol, **57**, 403, 446); biotinylation, **184**, 474; borohydride reduction, **18B**, 475; catalytic reaction, **44**, 478; cellulase, assay of, **63**, 32; cobalt-nucleotide conversions, **87**, 160; commercial source, **52**, 222; controlled-pore glass columns, **56**, 487, 488; cryoenzymology, **63**, 338; denaturation, **5**, 222; enzyme electrode, in, **44**, 585, 597; electrodes (long-term peritoneal implantation, **137**, 68; preparation, **137**, 25); enzyme channeling immunoassays with, **136**, 98; ethylenediamine adducts, coupling to carboxyalkyl agarose, **135**, 145; β-galactosidase assay and, **5**, 212; glucose assay by, **3**, 107; **57**, 403, 452; **63**, 33; glucose determination and, **8**, 9, 528, 589, 672; β-glucosidase assay and, **1**, 233, 242; **14**, 153; holoenzyme reconstitution, **53**, 436; horseradish peroxidase, and, cellobiase assay with, **160**, 110; hydrogenase assay and, **2**, 866; hydrogen production, **69**, 628; immobilization, **56**, 482, 483, 489 (carbohydrate residue activation, by, **44**, 319, 320; cellophane membrane, in, **44**, 904, 905, 908; **56**, 487; co-cross-linking, by, **44**, 268, 908; collagen, on, **44**, 260; controlled-pore ceramics, on, **44**, 152, 153, 156; copolymerization, by, **44**, 198, 200; entrapment, by, **44**, 173, 908; glass beads, on, **44**, 347, 671; magnetic carrier, with, **44**, 278, 279; microencapsulation, by, **44**, 214; multistep enzyme systems, in, **44**, 181, 460, 461, 472, 917; nylon tube, on, **44**, 124; **56**, 490); immobilized, assay of glucose, **57**, 454; kinetics (conjugate, on, **44**, 201, 425, 426, 483; multistep enzyme system, in, **44**, 433, 434); lactoperoxidase, and, in ^{125}I labeling of interferons, **119**, 263; loading capacity on collagen membrane, **44**, 255; mechanism, **63**, 51; methionine residues, modified, **25**, 396; multienzyme electrodes (amplification principle, **137**, 39; antiinterference principle, **137**, 42; competition principle, **137**, 40; sequence principle, **137**, 36); mutarotase assay and, **5**, 219, 224, 225; **9**, 608; **63**, 33; needle-type glucose sensor, in, **137**, 326; nitrate and, **2**, 579; oxidation, periodate, **44**, 318; oxygen scavenging system, in, **52**, 223; *Penicillum amagsakiense*, from, **9**, 85; *Penicillium notatum*, **1**, 343; peptide radiolabeling, in, **124**, 24; peroxidase inhibition, **44**, 160; peroxide and, **2**, 244; **32**, 108; photometry or fluorimetry, **56**, 461; polyacrylamide adduct, **56**, 486; portable continuous blood glucose analyzer, in, **137**, 319; properties, **1**, 344; proton resonances, of aromatic amino acid residues, **66**, 409; properties, **9**, 85; **44**, 478, 479; **161**, 311; purification, **89**, 46 (*Aspergillus niger*, **9**, 83; *Penicillium notatum*, **1**, 343; *Phanerochaete chrysosporium*, **161**, 309); quantitative immunodiffusion, use in, **44**, 716, 717; quantitization, **44**, 389; radioiodination, lactoperoxidase-catalyzed, **70**, 330; reactors (glucose assay with, **137**, 263, 279, 295; preparation, **137**, 261, 280, 289); reporter molecule for immunohistochemical reactions, as, **245**, 322; sources, **56**, 461; specific activity, **44**, 255; specificity, **9**, 86; **44**, 583; **63**, 371; spontaneous structuration studies, **44**, 927, 928; stain for detection, **44**, 713, 905; staining on gels, **22**, 594; substrates, **44**, 615; sulfite complex, **18B**, 473; temperature-activity profile, **44**, 259; transistor (glucose assay with, **137**, 259; preparation, **137**, 257); triazole treatment, **56**, 490; unit cell dimensions, **44**, 159

GLUCOSE-1-PHOSPHATASE

This enzyme [EC 3.1.3.10] catalyzes the hydrolysis of D-glucose 1-phosphate to produce D-glucose and orthophosphate. Other substrates include D-galactose 1-phosphate.[1,2] The enzyme from *Pholiota nameko* has a very high specificity for D-glucose 1-phosphate.[3] Note: This same reaction is also catalyzed by acid phosphatase and alkaline phosphatase.

[1]E. Pradel & P. L. Boquet (1988) *J. Bacteriol.* **170**, 4916.
[2]P. Faulkner (1955) *BJ* **60**, 590.
[3]T. Joh, J. Yazaki, K. Suzuki & T. Hayakawa (1998) *Biosci. Biotechnol. Biochem.* **62**, 2251.

Selected entries from *Methods in Enzymology* [vol, page(s)]:
Other molecular properties: *Azotobacter vinelandii* RNA protein particles, not present in, **5**, 61

GLUCOSE-6-PHOSPHATASE

This enzyme [EC 3.1.3.9] catalyzed the hydrolysis of D-glucose 6-phosphate to yield D-glucose and orthophosphate.[1-4] Other phosphorylated compounds (for example, carbamoyl phosphate and pyrophosphate) can also be hydrolyzed. Many glucose-6-phosphatases also catalyze transphosphorylation reactions from carbamoyl phosphate, hexose phosphates, pyrophosphate, and phosphoenolpyruvate using D-glucose, D-mannose, 3-methyl-D-glucose, or 2-deoxy-D-glucose as phosphoryl acceptors. The enzyme has a branched, ping pong Bi Bi kinetic mechanism involving a phosphoenzyme intermediate, and the phosphotransferase activity approaches the efficiency of the hydrolytic activity.[3,4]

[1]J. D. Foster, B. A. Pederson & R. C. Nordlie (1997) *Proc. Soc. Exp. Biol. Med.* **215**, 314.
[2]K. A. Sukalski & R. C. Nordlie (1989) *AE* **62**, 93.
[3]R. C. Nordlie (1982) *Meth. Enzymol.* **87**, 319.
[4]R. C. Nordlie (1971) *The Enzymes*, 3rd ed., **4**, 543.

Selected entries from *Methods in Enzymology* [vol, page(s)]:
General discussion: 2, 541; 9, 619; **87**, 319; **89**, 44; **90**, 396
Assay: 2, 541; 9, 619; **31**, 20, 91, 93, 94; **55**, 102; biopsy material, in, **8**, 530; endoplasmic reticulum, in, **22**, 138; glucose oxidase, using, **3**, 108; spectrophotometric assay, **89**, 44
Other molecular properties: bacterial particles and, **5**, 61; brush borders, in, **31**, 121; chromaffin granule, in, **31**, 381; diagnostic enzyme, as, **31**, 20, 31, 33, 35, 84, 262, 303, 327, 338, 361, 726, 729, 743; endoplasmic reticulum, in, **10**, 15; **22**, 135, 140, 141; **31**, 23, 24 (assay, **22**, 138); gluconeogenic catalytic activity, **37**, 281; glucose 6-phosphate and, **2**, 560; glucose-6-phosphate dehydrogenase, not present in preparations of, **9**, 371; glycogen synthase and, **5**, 147; inhibition, **87**, 346; isolated nuclei and, **12A**, 432; isotope exchange, **64**, 9, 36, 37; kinetic characteristics, **87**, 319, 327; lipid peroxidation, **52**, 304; liver, **2**, 542; **31**, 97, 195 (rat, preparation, **9**, 623); mannose-6-phosphatase (assay, **174**, 60; latency, in assessment of intactness of endoplasmic reticulum, **174**, 59); mechanism, **63**, 51; **87**, 319; microsomal fractions, in, **52**, 83; microsomal membranes and, **5**, 68; nuclear membrane, in, **31**, 290; nucleoli and, **12A**, 457; pH effects, **87**, 345; plasma membranes, in,

31, 88, 91, 100, 185, 186; properties, **2**, 542; **87**, 319; purification, **2**, 542; **9**, 623; rat (brain [assays, **90**, 398; comparison with hepatic form, **90**, 401; extraction and purification, **90**, 400]; liver, **9**, 623; **87**, 319 [differential effects of copper, **87**, 350; inhibitor effects, **87**, 346; kinetic mechanism, **87**, 327; pH kinetics, **87**, 345; rate constants {applicable equations, **87**, 334; experimental determination, **87**, 340; interrelationships, **87**, 335, 344; metabolic implications, **87**, 340; representative values, **87**, 343}]); rate equations, **87**, 334; ribonucleoprotein particles and, **5**, 66; rough microsomal subfractions, in, **52**, 75, 76; site-specific reagent, **47**, 483

GLUCOSE-1-PHOSPHATE ADENYLYLTRANSFERASE

This adenylyltransferase [EC 2.7.7.27], also referred to as ADP-glucose synthase and ADP-glucose pyrophosphorylase, catalyzes the reversible reaction of ATP with α-D-glucose 1-phosphate to produce pyrophosphate (or, diphosphate) and ADP-D-glucose.[1-5] This reaction is the first committed and highly regulated step of starch synthesis in all plant tissues. The enzyme from photosynthetic organisms typically exhibits cooperativity and is activated by 3-phosphoglycerate. Organisms which use glycolysis for carbon assimilation have glucose-1-phosphate adenylyltransferases that utilize D-fructose 1,6-bisphosphate as the primary activating effector. The barley (*Hordeum vulgare*) leaf enzyme has an Iso ordered Bi Bi kinetic mechanism.[2]

[1] J. Preiss (1996) *Biotechnol. Ann. Rev.* **2**, 259.
[2] L. A. Kleczkowski, P. Villand, J. Preiss & O. A. Olsen (1993) *JBC* **268**, 6228.
[3] L. A. Kleczkowski, P. Villand, A. Lönneborg, O.-A. Olsen & E. Lüthi (1991) *Z. Naturforsch.* **46C**, 605.
[4] J. Preiss (1978) *AE* **46**, 317.
[5] J. Preiss (1973) *The Enzymes*, 3rd ed., **8**, 73.

Selected entries from *Methods in Enzymology* [vol, page(s)]:
General discussion: **8**, 259; **23**, 618; **28**, 406
Assay: **23**, 618; **28**, 406; *Arthrobacter*, **8**, 262; corn grain, **8**, 259
Other molecular properties: ADP-glucose synthesis and, **28**, 279; *Arthrobacter*, from, **8**, 262 (assay, **8**, 262; properties, **8**, 264; purification, **8**, 263); bacteria, in, **8**, 266; corn grain, from, **8**, 259 (properties, **8**, 261; purification, **8**, 260); homogeneity and structural determinations, **28**, 410; isolation from *Escherichia coli*, **28**, 408; kinetics, **28**, 412; properties, **23**, 621; purification from spinach leaves, **23**, 619

GLUCOSE-1-PHOSPHATE CYTIDYLYLTRANSFERASE

This transferase [EC 2.7.7.33], also referred to as CDP-glucose pyrophosphorylase and CDP-glucose diphosphorylase, catalyzes the reversible reaction of CTP with D-glucose 1-phosphate to produce pyrophosphate (or, diphosphate) and CDP-D-glucose.[1-4] Other acceptor substrates include D-glucosamine 1-phosphate. The enzyme from *Salmonella enterica* has a ping pong Bi Bi kinetic mechanism.[4]

[1] R. M. Mayer & V. Ginsburg (1965) *JBC* **240**, 1900.
[2] K. Kimata & S. Suzuki (1966) *JBC* **241**, 1099.

[3] P. A. Rubenstein & J. L. Strominger (1974) *JBC* **249**, 3789.
[4] L. Lindqvist, R. Kaiser, P. R. Reeves & A. A. Lindberg (1994) *JBC* **269**, 122.

Selected entries from *Methods in Enzymology* [vol, page(s)]:
General discussion: **8**, 256
Assay: **8**, 256
Other molecular properties: 3,6-dideoxy sugar biosynthesis and, **28**, 461; properties, **8**, 258; purification, *Salmonella paratyphi*, **8**, 257

GLUCOSE-6-PHOSPHATE DEHYDROGENASE

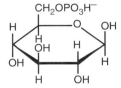 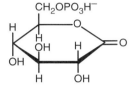

β-D-glucose 6-phosphate   6-phospho-D-glucono-δ-lactone

This oxidoreductase [EC 1.1.1.49], also referred to as glucose-6-phosphate 1-dehydrogenase and Zwischenferment, catalyzes the NADP+-dependent reduction of D-glucose 6-phosphate to produce 6-phospho-D-gluconolactone, NADPH, and a proton.[1-4] β-D-Glucose is a poor substrate.

[1] G. Martini & M. V. Ursini (1996) *Bioessays* **18**, 631.
[2] J. Chayen, D. W. Howat & L. Bitensky (1986) *Cell. Biochem. Funct.* **4**, 249.
[3] H. R. Levy (1979) *AE* **48**, 97.
[4] H. J. Engel, W. Domschke, M. Alberti & G. F. Domagk (1969) *BBA* **191**, 509.

Selected entries from *Methods in Enzymology* [vol, page(s)]:
General discussion: **1**, 323, 328; **9**, 116, 126, 131; **34**, 164, 252, 478; **41**, 177, 183, 188, 196, 201, 205, 208; **44**, 342, 629; **46**, 453; **89**, 252, 258, 261, 271; **188**, 335, 339
Assay: **1**, 323, 330; **9**, 116, 126, 132; **58**, 167, 168; *Arthrobacter globiformis*, **188**, 339; *Bacillus subtilis* vegetative and spore, **89**, 258; bovine adrenal cortex, from, **41**, 188; bovine mammary gland, from, **41**, 183, 184; *Candida utilis*, from, **41**, 205, 206; Fast Analyzer, by, **31**, 816; human erythrocytes, from, **9**, 126, 132; **41**, 209; immobilized enzyme rods, using, **57**, 210; fluorescence immunoassay, **74**, 48; lag time, **63**, 37; *Leuconostoc mesenteroides*, from, **41**, 196, 197; *Methylobacillus flagellatum*, **188**, 335; *Methylomonas* M15, **89**, 271; murine, **89**, 252; *Neurospora crassa*, from, **41**, 177; *Penicillium duponti*, from, **41**, 201, 202; *Pseudomonas fluorescens*, **89**, 263; *Pseudomonas* W6, **188**, 346; yeast cytoplasm, **194**, 657
Other Molecular Properties: adenosine 5′-O-(3-thiotriphosphate), synthesis of, **87**, 309; ADPglucose pyrophosphorylase assay and, **8**, 259; aldose reductase and, **5**, 334; amylomaltase assay and, **5**, 141, 142; *Arthrobacter globiformis* (assay, **188**, 339; properties, **188**, 342; purification, **188**, 341); assay for (acetyl-CoA carboxylase, **14**, 4; acyl-phosphate:hexose phosphotransferase, **9**, 392; adenylate cyclase, **38**, 162; adenylate kinase, **44**, 892; **51**, 468; α-amylase, **63**, 33; cytochrome P450, **52**, 126, 139; dimethylaniline monooxygenase, **52**, 143; fatty acid synthase, **14**, 17; fructokinase, **63**, 32; fructose-1,6-bisphosphatase, **9**, 626, 632; fructose 6-phosphate, **63**, 32; galactose-1-phosphate uridylyltransferase, **9**, 708; **63**, 33; glucokinase, **9**, 382, 388; **63**, 38; glucose, **3**, 52, 107, 109 [D-glucose microassay, **41**, 4]; glucose-6-phosphatase, **9**, 622; glucose 1-phosphate, **63**, 33; glucose 6-phosphate, **3**, 153; **9**, 622; hexokinase, **3**, 110; **9**, 371; inorganic

phosphate, **55**, 212; maltose, **63**, 33; phosphofructokinase, **9**, 437; phosphoglucosamine isomerase, **9**, 576; phosphoglucose isomerase, **9**, 557; pyridine nucleotide, **55**, 267); associated with cell fractions, **40**, 86; ATP assay and, **2**, 497; **13**, 488; *Bacillus subtilis* vegetative and spore (assay, **89**, 258; properties, **89**, 261; purification, **89**, 259); binding, to agarose-bound nucleotides, **66**, 196; bovine adrenal cortex, from, **41**, 188 (assay, **41**, 188; chromatography, **41**, 191, 192; inhibitors, **41**, 195; properties, **41**, 195; purification, **41**, 190); bovine mammary gland, from, **41**, 183 (assay, **41**, 183, 184; chromatography, **41**, 186, 187; purification, **41**, 184); brewer's yeast, from (crystallization, **9**, 123; properties, **1**, 325; **9**, 125; purification, **1**, 324; **9**, 118); *Candida utilis*, from, **41**, 205 (activators and inhibitors, **41**, 208; assay, **41**, 205, 206; chromatography, **41**, 207; properties, **41**, 207; purification, **41**, 206); CDPglucose pyrophosphorylase assay and, **8**, 256; CDP reduction and, **12A**, 158; cell monitoring, **58**, 167, 168; cell permeabilization test with, **137**, 640; cholesterol formation and, **5**, 498; chromatography, DEAE-cellulose, **5**, 24; clinical significance, **9**, 131; conjugate, activity assay, **44**, 200, 271, 340, 341; cortexolone formation, in, **37**, 310; cortisone 5β-reductase and, **5**, 499; creatine kinase, assay of, **17A**, 995; cytochrome reductase-free, **2**, 699; determination of guanosine nucleotides, **57**, 86, 89, 93; erythrocytes, from (amino terminal amino acids, **9**, 131; coenzyme, **9**, 130; kinetic properties, **9**, 131; molecular weight, **9**, 130; properties, **9**, 130; purification, **9**, 127; stability, **9**, 131; turnover number, **9**, 131); fatty acid synthesis and, **5**, 446; fluorescence immunoassay, **74**, 48; fluorometric assays with for (ATP, **13**, 488; NADP+, **13**, 483); fructose-1,6-bisphosphatase and, **5**, 272; galactokinase assay and, **5**, 175; galactose-1-phosphate uridylyltransferase, and, **9**, 708; **63**, 33; **87**, 22, 31; β-galactosidase assay, **5**, 212; glucose assay and, **3**, 52, 107, 109 (D-glucose microassay, **41**, 4); glucose effect, **63**, 40; glucose 1-phosphate assay and, **2**, 675; **5**, 176, 179; glucose 6-phosphate assay and, **3**, 153; **9**, 622; glutamine:fructose-6-phosphate transamidase and, **5**, 417; human erythrocytes, from, **41**, 208 (amino acids, **41**, 214; assay, **9**, 126, 132; **41**, 209; properties, **9**, 130; **41**, 212, 213; purification, **9**, 127, 133; **41**, 210); hydropyrimidine dehydrogenase and, **12A**, 54; immobilization in (collodion membrane, **44**, 906, 908; multistep enzyme system, **44**, 454); interference in malonyl-CoA assay, **35**, 313; intestinal mucosa, in, **32**, 667; ischemia-reperfusion, in, **233**, 604; isoenzyme, **58**, 167, 168; isolated nuclei and, **12A**, 432, 445; isotope effect, **87**, 398; kinetics, in enzyme system, **44**, 432, 433, 463 (phosphoribosylpyrophosphate synthetase, of, **51**, 6); lag time, **63**, 37; *Leuconostoc mesenteroides*, from, **1**, 328; **41**, 196 (assay, **1**, 330; **41**, 196, 197; chromatography, **41**, 198; inhibitors, **1**, 332; **41**, 201; properties, **1**, 331; **41**, 199; purification, **1**, 331; **41**, 197); mammary uptake, **39**, 454; marker enzyme, as, **31**, 735 (marker for chromosome detection in somatic cell hybrids, as, **151**, 182); methemoglobin reduction, **76**, 12; methionine sulfoxide reduction and, **5**, 993; *Methylobacillus flagellatum* (assay, **188**, 335; partial purification, **188**, 338; properties, **188**, 336); *Methylomonas* M15 (assay, **89**, 271; production, **89**, 272; properties, **89**, 274; purification, **89**, 272); molecular weight, **57**, 210; multienzymatic NADPH recycling, in, **136**, 72; murine (assay, **89**, 252; properties, **89**, 257; purification, **89**, 255); NADP+ and, **2**, 699, 712, 715, 719; **3**, 879; **8**, 792; NADPH-generating system, in, **44**, 850, 851, 865; *Neurospora crassa*, from, **41**, 177 (assay, **41**, 177; chromatography, **41**, 180; genetics, **41**, 182; properties, **41**, 181, 182; purification, **41**, 179); nonequilibrium exchange tool, as, **64**, 37, 38; oxidative stress in deficient cells, **251**, 286; parenchyma cells, in, **32**, 706; *Penicillium duponti*, from, **41**, 201 (assay, **41**, 201, 202; chromatography, **41**, 203; properties, **188**, 204, 205; purification, **41**, 202); pentose shunt, and **9**, 506; phagocytic function, assay for assessment of, **132**, 121, 161; pH effect, **87**, 398; 6-phosphogluconate dehydrogenase and, **6**, 799, 800; phosphohexoisomerase assay and, **1**, 299, 300, 305; P/O ratio, **55**, 226; preparation, **9**, 131; propanediol dehydrogenase preparation, absent from, **9**, 336; propionyl-CoA carboxylase and, **5**, 575; *Pseudomonas* **188**, 347 (assay, **188**, 346; properties, **188**, 348; purification, **188**, 347); purification by elution with substrate, **22**, 380; reduction of NADP+ analogues, **44**, 874; removal, **2**, 677; ribose 1,5-bisphosphate formation

and, **2**, 504; screening assay of, **9**, 135; specificity, **63**, 378; squalene oxidocyclase and, **5**, 498; stain, **58**, 168, 169; staining on gels, **22**, 594; stereospecificity, **54**, 229, 230; **87**, 106; steroid hydroxylase and, **5**, 508, 510; studies of glucose-marked liposomes, in, **32**, 512; substrate inhibition, alternative, **63**, 494; substrate specificity, **57**, 93; temperature-jump studies, use in, **76**, 691; thioredoxin *m* assay with, **167**, 411; thymidine diphosphate glucose pyrophosphorylase assay and, **8**, 254; transaldolase assay and, **1**, 381; UDPglucose:fructose transglucosylase assay and, **8**, 341; UDPglucose pyrophosphorylase and, **6**, 356 (assay, **8**, 248); yeast cytoplasm, assay, **194**, 657

GLUCOSE-6-PHOSPHATE 1-EPIMERASE

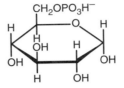

α-D-glucose 6-phosphate

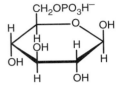

β-D-glucose 6-phosphate

This enzyme [EC 5.1.3.15] catalyzes the anomeric interconversion of α-D-glucose 6-phosphate and β-D-glucose 6-phosphate.[1] The *Saccharomyces cerevisiae* enzyme will also act on D-glucose. **See also** *Glucose-6-Phosphate Isomerase; Aldose 1-Epimerase*

[1] E. M. Chance, B. Hess, T. Plesser & B. Wurster (1975) *EJB* **50**, 419.

Selected entries from **Methods in Enzymology** [vol, page(s)]:
General discussion: **41**, 488
Assay: Baker's yeast, **41**, 488
Other molecular properties: anomerization kinetics, **63**, 370; baker's yeast, from, **41**, 488 (chromatography, **41**, 492; inhibitors, **41**, 493; properties, **41**, 493; purification, **41**, 490); *Escherichia coli*, from, molecular weight, **41**, 493; potato tubers, from, molecular weight, **41**, 493; *Rhodotorula gracilis*, from, molecular weight, **41**, 493

GLUCOSE-1-PHOSPHATE GUANYLYLTRANSFERASE

This transferase [EC 2.7.7.34], also referred to as GDP-glucose pyrophosphorylase and GDP-glucose diphosphorylase, catalyzes the reversible reaction of GTP with α-D-glucose 1-phosphate to produce pyrophosphate (or, diphosphate) and GDP-D-glucose.[1,2] Other substrates include D-mannose 1-phosphate, albeit reacting more slowly.

[1] I. Danishefsky & O. Heritier-Watkins (1967) *BBA* **139**, 349.
[2] K. Kawaguchi, S. Tanida, K. Matsuda, Y. Tani & K. Ogata (1973) *Agric. Biol. Chem.* **37**, 75.

Selected entries from **Methods in Enzymology** [vol, page(s)]:
General discussion: **8**, 266
Assay: **8**, 266

GLUCOSE-6-PHOSPHATE ISOMERASE

This isomerase [EC 5.3.1.9] (also known as phosphoglucose isomerase, phosphohexose isomerase, phosphohexomutase, oxoisomerase, and phosphoglucoisomerase) catalyzes the interconversion of D-glucose 6-phosphate and D-fructose 6-phosphate.[1–6] The enzyme also catalyzes the anomerization of D-glucose 6-phosphate (thus, a glucose-6-phosphate 1-epimerase activity, EC 5.1.3.15). In addition, glucose-6-phosphate isomerase can also generate, albeit slowly, the production of mannose 6-phosphate. It has been suggested that *cis*-enediol intermediate can rotate about its C2-C3 bond, followed by protonation at C2 (but not at C1) from the *si* face.[1] *See also Glucose-6-Phosphate 1-Epimerase*

[1] S. H. Seeholzer (1993) *PNAS* **90**, 1237.
[2] D. J. Creighton & N. S. R. K. Murthy (1990) *The Enzymes*, 3rd ed., **19**, 323.
[3] A. Achari, S. E. Marshall, H. Muirhead, R. H. Palmieri & E. A. Noltmann (1981) *Philos. Trans. R. Soc. Lond. B Biol. Sci.* **293**, 145.
[4] E. A. Noltmann (1972) *The Enzymes*, 3rd ed., **6**, 271.
[5] K. U. Yüksel & R. W. Gracy (1991) in *A Study of Enzymes* (S. A. Kuby, ed.), vol. 2, p. 457, CRC Press, Boca Raton.
[6] D. J. Creighton & N. S. R. K. Murthy (1990) *The Enzymes*, 3rd ed., **19**, 323.

Selected entries from *Methods in Enzymology* [vol, page(s)]:
General discussion: I, 304; **9**, 557, 565, 568; **27**, 541; **41**, 383, 388, 392; **46**, 23, 381, 382; **89**, 559
Assay: I, 304; **9**, 557, 565; *Bacillus stearothermophilus*, **41**, 383, 384; catfish liver and muscle, **89**, 550; clinical assay, **9**, 569; *Drosophila melanogaster*, **89**, 559; human erythrocytes and cardiac tissues, **41**, 392; mammalian tissue, I, 304; **9**, 557, 565; **89**, 550; murine muscle, **89**, 559; peas, **41**, 388, 389
Other molecular properties: affinity for calcium phosphate gel, **9**, 415; **51**, 584; affinity labeling, **47**, 493, 494; **87**, 473; anomerase activity, from Baker's yeast, **41**, 57; anomeric specificity, **63**, 377; *Bacillus stearothermophilus*, from, **41**, 383 (assay, **41**, 383, 384; chromatography, **41**, 385; inhibitors, **41**, 387; properties, **41**, 387; purification, **41**, 384); bovine mammary gland, of, **9**, 565 (molecular weight, **9**, 568; pH optimum, **9**, 568; properties, **9**, 567; purification, **9**, 566); catfish liver and muscle (assay, **89**, 550; purification, **89**, 555); chiral acetate, synthesis, **87**, 141; clinical aspects, **9**, 568; conjugate (activity, **44**, 271; kinetics, **44**, 917); determination of anomerase activity, from baker's yeast, **41**, 57; *Drosophila melanogaster* (assay, **89**, 559; properties, **89**, 562; purification, **89**, 560); equilibrium constant, I, 304; **63**, 378; fructokinase, assay of, **63**, 32; fructose-1,6-bisphosphatase, assay of, **5**, 272; **9**, 626, 632; fructose 6-phosphate, assay of, **63**, 32; glucosamine-6-phosphate deaminase, removal of in preparation of, **5**, 421, 425; glucosamine 6-phosphate formation and, **5**, 412, 414, 417; glucose 6-phosphate, preparation of, **3**, 152; glucosidase, in preparation of, **3**, 51, 52; glutamine:fructose-6-phosphate aminotransferase, in preparations of, **5**, 417; hexokinase, in preparations of, **5**, 232, 234; human erythrocytes and cardiac tissues, from, **41**, 392 (assay, **41**, 392; isolation, **41**, 394; properties, **41**, 399); immobilization, on inert protein, **44**, 909; inositol-1-phosphate synthetase, in preparations of, **9**, 703; isozyme electrophoresis for detection of lymphocyte hybridomas, **92**, 237; mammalian tissue (allozyme resolution, **89**, 555; assay, **89**, 550; isozymes and allozymes, **89**, 551; purification, **89**, 552); mammary gland, **9**, 565 (properties, **9**, 567; purification, **9**, 566); mannose isomerase and, **5**, 338; marker for chromosome detection in somatic cell hybrids, as, **151**, 182; microsomes, removal from, **2**, 542; murine muscle (assay, **89**, 559; properties, **89**, 562; purification, **89**, 560); mutant, **72**, 695; peas, from, **41**, 388 (assay, **41**, 388, 389; chromatography, **41**, 391; properties, **41**, 391, 392; purification, **41**, 389); phosphofructokinase, assay of, **9**, 437; phosphoglucosamine isomerase, assay of, **9**, 575; pH stability (as allose-6-phosphate isomerase), **9**, 415; plants, in, **31**, 740; 196; rabbit

muscle, I, 304; **9**, 557 (crystallization, **9**, 563; molecular weight, **9**, 565; properties, I, 306; **9**, 565; purification, I, 305; **9**, 560); rat muscle, crude preparation, **3**, 153; site-specific reagent, **47**, 482, 483; source, **44**, 811; stain for, **224**, 107; staining on gels, **22**, 585, 593; transaldolase, assay of, I, 381; transition state and multisubstrate analogues, **249**, 308; trehalose-forming enzyme and, **5**, 167; tritium-labeled glucose 6-phosphate and, **6**, 878; UDPglucose:fructose transglucosylase assay and, **8**, 341

GLUCOSE-1-PHOSPHATE PHOSPHODISMUTASE

This enzyme [EC 2.7.1.41], also referred to as glucose-1-phosphate transphosphorylase, catalyzes the reversible reaction of two molecules of D-glucose 1-phosphate to produce D-glucose and D-glucose 1,6-bisphosphate.[1,2]

[1] J. B. Sidbury, L. L. Rosenberg & V. A. Najjar (1956) *JBC* **222**, 89.
[2] F. Climent, M. Carreras & J. Carreras (1985) *Comp. Biochem. Physiol. B Comp. Biochem.* **81B**, 737.

GLUCOSE-1-PHOSPHATE THYMIDYLYLTRANSFERASE

This transferase [EC 2.7.7.24], also referred to as dTDP-glucose synthase, dTDP-glucose pyrophosphorylase, and dTDP-glucose diphosphorylase, catalyzes the reversible reaction of dTTP with α-D-glucose 1-phosphate to produce pyrophosphate (or, diphosphate) and dTDP-D-glucose.[1,2] The *Salmonella enterica* enzyme has a ping pong Bi Bi kinetic mechanism.[3]

[1] J. H. Pazur & E. W. Shuey (1961) *JBC* **236**, 1780.
[2] S. Kornfeld & L. Glaser (1961) *JBC* **236**, 1791.
[3] L. Lindquist, R. Kaiser, P. R. Reeves & A. A. Lindberg (1993) *EJB* **211**, 763.

Selected entries from *Methods in Enzymology* [vol, page(s)]:
General discussion: 8, 253
Assay: 8, 253
Other molecular properties: occurrence, **8**, 256; properties, **8**, 255; purification, *Escherichia coli*, **8**, 255; staining on gels, **22**, 584, 588; thymidine diphosphate glucose synthesis and, **28**, 290, 295

GLUCOSE-1-PHOSPHO-D-MANNOSYLGLYCOPROTEIN PHOSPHODIESTERASE

This phosphodiesterase [EC 3.1.4.51], also known as α-glucose-1-phosphate phosphodiesterase, catalyzes the hydrolysis of a 6-(D-glucose-1-phospho)-D-mannosylglycoprotein to produce D-glucose α-1-phosphate and the D-mannosylglycoprotein.[1]

[1] C. Srisomsap, K. L. Richardson, J. C. Jay & R. B. Marchase (1989) *JBC* **264**, 20540.

α-GLUCOSIDASE

This exoglycosidase [EC 3.2.1.20], also known as maltase, glucosidosucrase, maltase-glucoamylase, and acid maltase, catalyzes the hydrolysis of terminal, non-reducing 1,4-linked D-glucose residues on oligosaccharides with the release of D-glucose.[1-4] The enzyme also exhibits disaccharidase activity, catalyzing the hydrolysis of maltose to produce two molecules of D-glucose; other similar substrates include maltotriose, maltotetraose, maltopentaose, and *p*-nitrophenyl-α-D-glucoside.

The α-glucosidases actually comprise a family of enzymes whose specificity is directed mainly toward the exohydrolysis of 1,4-α-glucosidic linkages, such that oligosaccharides are hydrolyzed more rapidly than polysaccharides, which are hydrolyzed slowly, if at all. This hydrolytic reaction mechanism proceeds by means of an intermediate displaying substantial oxocarbenium character, even though the usual transition-state analogues such as the gluconolactones are relatively poor inhibitors.

The intestinal enzyme also catalyzes the hydrolysis of polysaccharides (a glucan 1,4-α-glucosidase [EC 3.2.1.3] activity) and, more slowly, hydrolyzes 1,6-α-D-glucose linkages. *See also Glucan 1,4-α-Glucosidase; α-Dextrin Endo-1,6-α-Glucosidase; Sucrose α-Glucosidase; Glucan Endo-1,3-α-Glucosidase; Glucan 1,6-α-Glucosidase; Glucan 1,3-α-Glucosidase; Branched-Dextran Exo-1,2-α-Glucosidase; Maltose-6′-Phosphate Glucosidase*

[1] G. Davies, M. L. Sinnott & S. G. Withers (1998) *CBC* 1, 119.
[2] T. P. Frandsen & B. Svensson (1998) *Plant Mol. Biol.* 37, 1.
[3] S. Chiba (1997) *Biosci. Biotechnol. Biochem.* 61, 1233.
[4] G. Mooser (1992) *The Enzymes*, 3rd ed., 20, 187.

Selected entries from *Methods in Enzymology* [vol, page(s)]:
General discussion: 8, 531, 559, 584; 46, 368; 330, 260
Assay: biopsy material, in, 8, 531; glucose oxidase, coupled enzyme assay with, 3, 109; 8, 513; rat liver, 28, 806; yeast, 8, 559
Other molecular properties: acid maltase, 8, 513, 531; albumin conjugate, preparation, 137, 569; albumin-insulin conjugate, preparation, 137, 569; amylomaltase assay and, 5, 142; analysis in multiple systems, 22, 18, 20; brush borders, in, 31, 122, 129, 130; exoamylase, *Pseudomonas saccharophila*, 9, 13; gelatin-immobilized, properties, 135, 297; glucose-6-phosphate dehydrogenase preparation, presence in, 5, 142; hexokinase preparations, presence in, 5, 142; hyperthermophilic, 330, 260; human jejunum, activity in, 8, 585; immobilization, by microencapsulation, 44, 214; α-glucosidase I (porcine liver, 230, 285); α-glucosidase II (porcine liver [incubation conditions, 230, 285]; specificity, 230, 285]); inhibition, 46, 369 (bromconduritol, with, 46, 381; gluconolactones, by, 8, 573); localization in gels by autochromic method with sandwich-type incubation, 104, 439; lysosomes, in, 8, 513; marker for chromosome detection in somatic cell hybrids, as, 151, 183; preparation from intestine, 8, 585; rat liver, from (assay, 28, 806; biological importance, 28, 805; properties, 28, 811; purification, 28, 808); reactions, 46, 369, 370; staining on gels, 22, 600; synchronous cultures and, 21, 469, 470; yeast, of, 8, 559 (properties, 8, 561; purification, 8, 560)

β-GLUCOSIDASE

This enzyme [EC 3.2.1.21], also referred to as cellobiase and amygdalase, catalyzes the hydrolysis of terminal, non-reducing β-D-glucosyl residues from β-D-glucosides, including oligosaccharides, with the resultant release of β-D-glucose (substrates include cellobiose, cellotriose, cellotetraose, cellopentaose, and *p*-nitrophenyl-β-D-glucopyranoside). The enzyme will also act on β-D-galactosides, α-L-arabinosides, β-D-xylosides, and β-D-fucosides. Variants of the enzyme have been reported to also act on β1 → 6 or β1 → 3 linkages as well. A covalent α-glucosyl enzyme intermediate was first isolated for the *Agrobacterium faecalis* enzyme.[1] *See specific enzyme*

[1] S. G. Withers & I. P. Street (1988) *J. Amer. Chem. Soc.* 110, 8551.

Selected entries from *Methods in Enzymology* [vol, page(s)]:
General discussion: 1, 177, 231, 234; 14, 152; 46, 368; 160, 221, 243, 274, 300, 323, 415, 424, 432, 437, 575; 330, 364
Assay: 1, 234; 160, 109, 409, 416, 424, 432, 438, 576; glucosyl ceramide, with, 14, 153, 155; glucose oxidasae, with, 3, 109; *p*-nitrophenyl-glucoside, with, 14, 152, 155; *Phaseolus vulgaris*, 28, 720; spectrophotometric, 44, 101
Other molecular properties: acyl azide coupling, 44, 103, 104; amebic phagosomes, in, 31, 697; amygdalin electrode, in, 44, 592, 604; analysis in multiple systems, 22, 18, 22; assay for cellobiose, 63, 32; assay for cellulase, 63, 32; conjugate (activity, 44, 103, 104, 271, 272; stability, 44, 104; storage, 44, 103); detection in (gels, 160, 138; prokaryotes, 160, 184); effect of detergents, 14, 155; enlarged, coentrapment with yeast cells, 137, 642; function, 160, 423; β-glucosidase A₃, 46, 379, 380; glycolipid structure studies, for, 32, 363; hyperthermophilic, 330, 364; immobilization in multistep enzyme system, 44, 473, 917 (polyacrylic type copolymers, on, 44, 103, 104; porous matrix, in, 44, 270); inhibition, 46, 369 (aldonolactones, 8, 572; bromoconduritol, with, 46, 381); inhibitors, 1, 240; 14, 155; kinetics, in enzyme system, 44, 433; melibiase, separation from, 1, 250; *Phaseolus vulgaris*, from (assay, 28, 720; purification, 28, 721; substrate specificity, 28, 727); pH optimum, 44, 607; properties (*Aspergillus* enzyme, 160, 579; bovine brain enzymes, 14, 155; *Ruminococcus* enzyme, 160, 412; *Sclerotium* enzymes, 160, 428; *Sporotrichum* enzymes, 160, 421; sweet almond enzyme, 1, 237; 28, 702; *Talaromyces* enzymes, 160, 442); purification from (*Aspergillus niger*, 160, 577; *Aspergillus oryzae*, 1, 178; bovine brain, 14, 154; *Humicola*, 160, 328; *Ruminococcus albus*, 160, 410; *Schizophyllum commune*, 160, 435; *Sclerotium rolfsii*, 160, 426; *Sporotrichum pulverulentum*, 160, 419; sweet almonds, 1, 236; 28, 701; *Talaromyces emersonii*, 160, 439; *Trichoderma koningii*, 160, 224); reactions, 46, 369, 370, 377; separation from cellulase, 1, 175; source, 44, 101; staining on gels, 22, 600; substrate specificity, 14, 155; sweet almond emulsin, 28, 701 (properties, 28, 702; purification, 28, 701); transition-state and multisubstrate analogues, 249, 306

GLUCOSIDE 3-DEHYDROGENASE

This FAD-dependent oxidoreductase [EC 1.1.99.13], also referred to as D-aldohexopyranoside dehydrogenase, catalyzes the reaction of sucrose with an acceptor substrate to produce 3-dehydro-α-D-glucosyl-β-D-fructofuranoside and the reduced acceptor.[1,2] Other substrates include D-glucose (producing 3-keto-D-glucose),

D-galactose, D-glucosides (for example, validoxylamine A and cellobiose), and D-galactosides (D-glucosides are more effective substrates than D-galactosides). Acceptor substrates include phenazine methosulfate, 2,6-dichlorophenolindophenol, ferricyanide, and certain cytochromes.

[1] K. Hayano & S. Fukui (1967) JBC **242**, 3665.
[2] M. Takeuchi, K. Ninomiya, K. Kawabata, N. Asano, Y. Kameda & K. Matsui (1986) J. Biochem. **100**, 1049.

Selected entries from *Methods in Enzymology* [vol, page(s)]:
General discussion: 9, 78; 41, 153
Assay: *Agrobacterium*, from, 9, 78; 41, 153
Other molecular properties: *Agrobacterium*, from, 9, 78; 41, 153 (chromatography, 41, 155; electrophoresis, 41, 156; pH optimum, 9, 81; properties, 9, 81; 41, 156; purification, 9, 79; 41, 155; 3-ulose determination, in, 41, 25, 26); specificity, 9, 81; stability, 9, 81; synthesis of sugar nucleotides, 28, 296

β-GLUCOSIDE KINASE

This phosphotransferase [EC 2.7.1.85] catalyzes the reaction of ATP with cellobiose to produce ADP and 6-phospho-β-D-glucosyl-(1,4)-D-glucose.[1] A number of β-D-glucosides can act as acceptor substrates, including cellotriose, gentiobiose, arbutin, sophorose, and amygdalin. Other less effective phosphoryl donors include GTP, CTP, ITP, and UTP.

[1] R. E. Palmer & R. L. Anderson (1972) JBC **247**, 3415.

Selected entries from *Methods in Enzymology* [vol, page(s)]:
General discussion: 42, 3
Assay: 42, 3, 4

α-GLUCOSIDURONASE

This enzyme [EC 3.2.1.139], also called α-glucuronidase, catalyzes the hydrolysis of an α-D-glucuronoside to produce an alcohol and D-glucuronate.[1]

[1] H. Uchida, T. Nanri, Y. Kawabata, I. Kusakabe & K. Murakami (1992) Biosci. Biotech. Biochem. **56**, 1608.

Selected entries from *Methods in Enzymology* [vol, page(s)]:
General discussion: 160, 560
Assay: 160, 561

GLUCOSINOLATE SULFATASE

This enzyme catalyzes the release of sulfate from glucosinolate sulfate, a component of mustard oil glycosides, and other molecules containing $R_1R_2C{=}NOSO_3^-$.

Selected entries from *Methods in Enzymology* [vol, page(s)]:
General discussion: 143, 361
Assay: 143, 361

GLUCOSYLCERAMIDASE

This enzyme [EC 3.2.1.45] (also known as β-glucocerebrosidase, acid β-glucosidase, and D-glucosyl-N-acylsphingosine glucohydrolase) catalyzes the hydrolysis of a D-glucosyl-N-acylsphingosine (i.e., a D-glucosylceramide) to produce D-glucose and an N-acylsphingosine (i.e., a ceramide).[1-5] Other substrates include D-glucosylsphingosine and 4-methylumbelliferyl-β-D-glucopyranoside. The human enzyme is inactivated by 2-deoxy-2-fluoro-β-D-glucopyranosyl fluoride which forms a stable pyranosyl-enzyme intermediate with an active-site glutamyl residue.[5] *See also Glycosylceramidase; Galactosylceramidase*

[1] G. A. Grabowski, S. Gatt & M. Horowitz (1990) Crit. Rev. Biochem. Mol. Biol. **25**, 385.
[2] E. Conzelmann & K. Sandhoff (1987) AE **60**, 89.
[3] A. W. Schram, J. M. F. G. Aerts, S. van Weely, J. A. Barranger & J. M. Tager (1987) Methodol. Surv. Biochem. Anal. **17**, 113.
[4] R. O. Brady (1983) The Enzymes, 3rd ed., **16**, 409.
[5] S. Miao, J. D. McCarter, M. E. Grace, G. A. Grabowski, R. Aebersold & S. G. Withers (1994) JBC **269**, 10975.

Selected entries from *Methods in Enzymology* [vol, page(s)]:
General discussion: 8, 591; 44, 702; 50, 529
Assay: 8, 591; 28, 830; fluorescent derivatives of glucocerebroside, with, 72, 367; glucosylceramide, with, 50, 476; human placenta, 50, 529, 530; 4-methylumbelliferyl β-glucoside, with, 50, 478, 479; trinitrophenyl-ω-aminolauroylglucosylsphingosine, with, 72, 364
Other molecular properties: beef spleen, isolation, 28, 831; deglycosylated, targeting to hepatic cells, 149, 38; diagnostic assay for Gaucher disease, 138, 747; digestion before cell targeting (N-acetylglucosamine removal, 149, 35; galactose removal, 149, 32; sialic acid removal, 149, 31); Gaucher's disease, 28, 830; 50, 476, 532; 138, 747; human placenta, from (assay, 50, 529, 530; properties, 50, 532; purification, 50, 530; 149, 29); human spleen, from, 8, 591 (properties, 8, 593; purification, 8, 593); inhibition, 72, 673 (conduritol β-epoxide, by, 72, 674; N-hexylglucosylsphingosine, by, 72, 673); reaction, 28, 830; specificity, 28, 833; transition-state and multisubstrate analogues, 249, 306

GLUCOSYL-DNA β-GLUCOSYLTRANSFERASE

This transferase [EC 2.4.1.28], first isolated from *Escherichia coli* infected with bacteriophage T6, catalyzes the transfer of a β-D-glucosyl residue from UDP-D-glucose to a glucosylhydroxymethylcytosyl residue in the DNA.[1] *See also DNA α-Glucosyltransferase; DNA β-Glucosyltransferase*

[1] S. R. Kornberg, S. B. Zimmerman & A. Kornberg (1961) JBC **236**, 1487.

Selected entries from *Methods in Enzymology* [vol, page(s)]:
General discussion: 12B, 496
Assay: 12B, 498
Other molecular properties: properties, 12B, 507; purification, 12B, 500, 506

GLUCURONAN LYASE

This lyase [EC 4.2.2.14] catalyzes the eliminative cleavage of (1,4)-β-D-glucuronans to produce oligosaccharides having 4-deoxy-β-D-gluc-4-enuronosyl groups at their non-reducing ends (note: glucuronans are polymers and oligomers of glucuronate).[1] Complete degradation of glucuronans results in the ultimate generation of tetrasaccharides.

[1]P. Michaud, P. Pheulpin, E. Petit, J. P. Seguin, J. N. Barbotin, A. Heyraud, B. Courtois & J. Courtois (1997) *Int. J. Biol. Macromol.* **21**, 3.

GLUCURONATE ISOMERASE

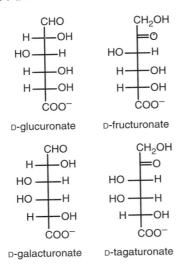

D-glucuronate   D-fructuronate

D-galacturonate   D-tagaturonate

This isomerase [EC 5.3.1.12], also referred to as uronate isomerase, catalyzes the interconversion of D-glucuronate and D-fructuronate as well as the interconversion of D-galacturonate and D-tagaturonate.[1,2]

[1]W. W. Kilgore & M. P. Starr (1959) *JBC* **234**, 2227.
[2]G. Ashwell, A. J. Wahba & J. Hickman (1960) *JBC* **235**, 1559.

Selected entries from *Methods in Enzymology* [**vol**, page(s)]:
General discussion: 5, 190
Assay: 5, 190

GLUCURONATE-1-PHOSPHATE URIDYLYLTRANSFERASE

This transferase [EC 2.7.7.44] catalyzes the reversible reaction of UTP with 1-phospho-α-D-glucuronate to produce UDP-D-glucuronate and pyrophosphate (or, diphosphate).[1-3] CTP can also serve as the nucleotide substrate, albeit less effectively. The *Typha latifolia* enzyme has a Theorell-Chance Bi Bi kinetic mechanism.[3]

[1]F. A. Loewus & M. W. Loewus (1983) *Ann. Rev. Plant Physiol.* **34**, 137.
[2]R. M. Roberts (1971) *JBC* **246**, 4995.
[3]H. Toshinobu, H. Akira & F. Tooru (1983) *Plant Cell Physiol.* **24**, 1535.

GLUCURONATE REDUCTASE

This oxidoreductase [EC 1.1.1.19], also known as glucuronate dehydrogenase and L-hexonate dehydrogenase, catalyzes the reaction of D-glucuronate with NADPH to produce L-gulonate and NADP+.[1-3] D-Galacturonate is also reduced by this enzyme, which in some organisms may be identical to alcohol dehydrogenase (NADP+) [EC 1.1.1.2]. The porcine kidney enzyme, which exhibits A-side stereospecificity in hydride transfer, has an ordered Bi Bi kinetic mechanism. **See also** *Glucuronolactone Reductase*

[1]E. E. Kaufman & T. Nelson (1981) *JBC* **256**, 6890.
[2]D. R. P. Tulsiani & O. Touster (1977) *JBC* **252**, 2545.
[3]J. L. York, A. P. Grollman & C. Bublitz (1961) *BBA* **47**, 298.

Selected entries from *Methods in Enzymology* [**vol**, page(s)]:
General discussion: 6, 334; 18A, 55; 89, 501
Assay: 6, 334; 18A, 30, 56; 89, 501
Other molecular properties: mammalian tissues, in, 41, 159; porcine kidney (assay, 6, 334; 89, 501; properties, 6, 336; 89, 505; purification, 6, 335; 89, 502); properties, 6, 336; 18A, 31; 89, 505; purification, 6, 335; 18A, 57, 58; 89, 502; source, 18A, 29

GLUCURONATE-2-SULFATASE

This sulfatase [EC 3.1.6.18], also called chondro-2-sulfatase, catalyzes the hydrolysis of the 2-sulfate groups of the 2-*O*-sulfo-D-glucuronate residues of chondroitin sulfate, heparin, and heparitin sulfate.[1,2] The enzyme does not act on iduronate 2-sulfate residues, a reaction catalyzed by iduronate 2-sulfatase [EC 3.1.6.13]. Sulfate and phosphate are potent inhibitors of the human enzyme.[2]

[1]P. N. Shaklee, J. H. Glaser & H. E. Conrad (1985) *JBC* **260**, 9146.
[2]C. Freeman & J. J. Hopwood (1991) *BJ* **279**, 399.

β-GLUCURONIDASE

This enzyme [EC 3.2.1.31] catalyzes the hydrolysis of a β-D-glucuronoside to produce an alcohol and D-glucuronate.[1-5] β-Glucuronidases have been identified that act on *p*-nitrophenyl-β-D-glucoronide, luteolin 7-*O*-diglucuronide 4'-*O*-glucuronide, phenolphthalein glucuronide, estriol-3-glucuronide, and 4-methyl-umbelliferyl-β-D-glucuronide.

[1]K. Paigen (1989) *Prog. Nucl. Acid Res. Mol. Biol.* **37**, 155.
[2]H. Kresse & J. Glössl (1987) *AE* **60**, 217.
[3]P. M. Dey & E. del Campillo (1984) *AE* **56**, 141.
[4]G. A. Levy & C. A. Marsh (1960) *The Enzymes*, 2nd ed., **4**, 397.
[5]W. H. Fishman (1955) *AE* **16**, 361.

Selected entries from *Methods in Enzymology* [**vol**, page(s)]:
General discussion: 1, 262; 3, 55; 8, 575; 10, 575; 34, 4; 50, 510
Assay: 1, 262; 8, 576; 50, 451, 452; human neutrophil, 132, 274; lysosomal, 96, 559, 765; microsomal, 96, 559; neutrophil granule-associated, 162, 546; plant tissue culture media, in, nondestructive assay, 216, 357; rat liver, 28, 814; rat preputial gland, 50, 510, 511; 77, 390; reporter gene assays, 326, 185

Other molecular properties: analysis in multiple systems, **22**, 18; anionic detergent, effect in isolation with, **1**, 37; assay for (correction of luciferase assays, **216**, 408; estrogens, **15**, 194; steroids, **15**, 231); brain homogenates, in, **31**, 464, 466, 467, 469; capsular polysaccharide and, **28**, 608; distribution, **50**, 451; compartmentalized expression, **217**, 548; effect on urine metabolites, **77**, 29; expression in (anther-derived albino plants, **217**, 548; callus derived from immature embryo, **217**, 549; plant chloroplasts, assay, **217**, 544); gel electrophoresis, **32**, 83, 89; glucuronide formation and, **5**, 162, 163; hyaluronate degradation by, **1**, 167; immunostaining, **236**, 150; inhibition by (aldonolactones, **8**, 572, 576; intestinal epithelium, in, **77**, 157; glucaro-(1 → 4)-lactone, **8**, 576); limpet, for glycosphingolipid analysis, **230**, 385; lysosomal, **8**, 513; **10**, 12; **31**, 330, 340, 344 (antibody preparation, **96**, 770; assay, **96**, 559, 765; immunoprecipitation, **96**, 773; purification, **96**, 560, 765; synthesis *in vitro*, **96**, 770); macrosome fraction, **31**, 340; marker enzyme, as, **31**, 753 (marker for chromosome detection in somatic cell hybrids, as, **151**, 183); microsomal (assay, **96**, 559; purification, **96**, 561); mucolipidosis II, **50**, 452, 455; mucolipidosis III, **50**, 452, 455; mucopolysaccharidosis VII, **50**, 452; phagocytic function, assay for assessment of, **132**, 119, 160; plant tissue culture media, in, nondestructive assay, **216**, 357; preparation, **8**, 575; **50**, 147; preparation of fungal mitochondria and plasma and vacuolar membranes, in, **157**, 563; properties, **1**, 267; **8**, 575; purification from (calf liver, **1**, 264; rat liver, **28**, 815; rat preputial glands, **98**, 302); radioiodination, **70**, 232; rat liver, from, **50**, 489 (assay, **28**, 814; properties, **28**, 817; purification, **28**, 815); rat preputial gland (assay, **50**, 510, 511; properties, **50**, 513, 514; purification, **50**, 511); reporter, **326**, 185; secretion from platelet, assay, **169**, 196, (lysosomes, from, **169**, 336); snail gut juice, in, **31**, 612; test for β-glucuronide, **50**, 147; uptake and binding assays in macrophages, **98**, 303

GLUCURONOARABINOXYLAN ENDO-1,4-β-XYLANASE

This enzyme [EC 3.2.1.136] catalyzes the endohydrolysis of 1,4-β-D-xylosyl links in some gluconoarabinoxylans, exhibiting high activity toward feruloylated arabinoxylans (*i.e.*, a feraxan) from cereal plant cell walls. This hydrolase recognizes glucuronosyl moieties inserted as monomeric side chains along the xylan backbone and mediates the hydrolysis of the β(1 → 4)-xylosyl linkage of the adjacent unsubstituted xylosyl residue in heteroxylans.[1]

[1]K. Nishitani & D. J. Nevins (1991) *JBC* **266**, 6539.

GLUCURONOKINASE

This phosphotransferase [EC 2.7.1.43] catalyzes the reaction of ATP with D-glucuronate to produce ADP and 1-phospho-α-D-glucuronate.[1]

[1]M. D. Leibowitz, D. B. Dickinson, F. A. Loewus & M. Loewus (1977) *ABB* **179**, 559.

D-GLUCURONOLACTONE DEHYDROGENASE

This oxidoreductase activity, formerly classified as EC 1.1.1.70, is now listed under aldehyde dehydrogenase (NAD[+]) [EC 1.2.1.3].

GLUCURONOLACTONE REDUCTASE

This enzyme [EC 1.1.1.20] catalyzes the reversible reaction of D-glucurono-3,6-lactone with NADPH to produce L-gulono-1,4-lactone and NADP[+].[1] *See also Glucuronate Reductase*

[1]S. Hayashi, M. Watanabe & A. Kimura (1984) *J. Biochem.* **95**, 223.

Selected entries from *Methods in Enzymology* [vol, page(s)]:
General discussion: 18A, 55
Assay: 18A, 56; L-gluconolactone oxidase, associated with, **18A**, 33
Other molecular properties: activators, 18A, 58; L-gluconolactone oxidase, associated with, **18A**, 29, 33; kinetic properties, **18A**, 58; properties, **18A**, 31; purification, **18A**, 57, 58; specificity, **18A**, 58; stability, **18A**, 58

GLUCURONOSYL-DISULFOGLUCOSAMINE GLUCURONIDASE

This enzyme [EC 3.2.1.56] catalyzes the hydrolysis of 3-D-glucuronosyl-N^2,6-disulfo-β-D-glucosamine to produce N^2,6-disulfo-D-glucosamine and D-glucuronate.[1,2]

[1]C. P. Dietrich (1969) *Biochemistry* **8**, 2089.
[2]C. P. Dietrich, M. E. Silva & Y. M. Michelacci (1973) *JBC* **248**, 6408.

GLUCURONOSYLTRANSFERASE

This family of enzymes [EC 2.4.1.17], also known as UDP-glucuronosyltransferases, catalyzes the reaction of UDP-D-glucuronate with an acceptor substrate to produce UDP and the corresponding β-D-glucuronoside.[1,2] They act on a wide range of substrates, including phenols, alcohols, amines, and fatty acids. Reported acceptor substrates include *p*-nitrophenol, 1-naphthol, morphine, bilirubin, testosterone, 17β-estradiol, phenolphthalein, estriol, deoxycholate, and 2-aminophenol. *See also Bilirubin-Glucuronoside Glucuronosyltransferase; Galactosyl-galactosylxylosylprotein 3-β-Glucuronosyltransferase; Luteolin-7-O-Glucuronide 7-O-Glucuronosyltransferase; Luteolin-7-O-Diglucuronide 4'-O-Glucuronosyltransferase*

[1]R. N. Armstrong (1987) *CRC Crit. Rev. Biochem.* **22**, 39.
[2]K. W. Bock, B. Burchell, G. J. Dutton, O. Hänninen, G. J. Mulder, I. S. Owens, G. Siest & T. R. Tephly (1983) *Biochem. Pharmacol.* **32**, 953.

Selected entries from *Methods in Enzymology* [vol, page(s)]:
General discussion: 5, 159; **15**, 726, 728, 730; **28**, 657, 679; **77**, 169, 177, 188
Assay: 5, 159; **15**, 726, 728, 729, 731; **28**, 615, 657, 679; **77**, 383 (2-aminophenol, **77**, 384, 385; bilirubin, **77**, 188; cells, in, **77**, 388; estradiol, **77**, 388; estrone, **15**, 726;**77**, 177; harmalol, **77**, 388; harmol, **77**, 388; high-performance liquid chromatography, **77**, 391; 3-hydroxybenzo[*a*]pyrene, **77**, 387; menthol, **77**, 388; 4-methylumbelliferone, **77**, 387; morphine, **77**, 388; 1-naphthol, **77**, 386; nitrophenol, **77**, 169; 4-nitrophenol, **77**, 177; 4-nitrothiophenol, **77**, 388; phenolphthalein, **77**, 388; resorufin, **77**, 387; testosterone, **77**, 387; umbelliferone, **77**, 387, 388)
Other molecular properties: activation, 77, 190; activators, **77**, 177; affinity chromatography, **77**, 183, 184; Ah^b allele, **52**, 231; capsular

GLUCURONOXYLAN 4-O-METHYLTRANSFERASE

This methyltransferase [EC 2.1.1.112] catalyzes the reaction of S-adenosyl-L-methionine with a D-glucuronate residue in a glucuronoxylan to produce S-adenosyl-L-homocysteine and a 4-O-methyl-D-glucuronate residue in the glucuronoxylan.[1]

[1] E. A.-H. Baydoun J. A.-R. Usta, K. W. Waldron & C. T. Brett (1989) *J. Plant Physiol.* **135**, 81.

GLUCURONYLGALACTOSYL-PROTEOGLYCAN β-1,4-N-ACETYLGALACTOSAMINYLTRANSFERASE

This manganese-dependent transferase [EC 2.4.1.174], also known as N-acetylgalactosaminyltransferase I, catalyzes the reaction of UDP-N-acetyl-D-galactosamine with a D-glucuronyl-1,3-β-D-galactosylproteoglycan to produce UDP and N-acetyl-D-galactosaminyl-1,4-β-D-glucuronyl-1,3-β-D-galactosylproteoglycan, participating in the biosynthesis of chondroitin sulfate.[1] Acceptor substrates require a terminal glucuronyl residue and a galactosyl residue in the penultimate position. β-D-Glucuronyl-1,3-D-galactosyl-1,3-D-galactose, for example, is a good substrate. ***See also*** *Glucuronyl-N-Acetylgalactosaminylproteoglycan β-1,4-N-Acetylgalactosaminyl-transferase*

[1] K. Rohrmann, R. Niemann & E. Buddecke (1985) *EJB* **148**, 463.

Selected entries from ***Methods in Enzymology*** [**vol**, page(s)]:
General discussion: **28**, 655, 677
Assay: **28**, 655
Other molecular properties: chondroitin sulfate biosynthesis and, **28**, 639, 644; heparin biosynthesis and, **28**, 676; isolation from mast-cell tumors, **28**, 678, 679; properties, **28**, 656

GLUCURONYL-N-ACETYLGALACTOS-AMINYLPROTEOGLYCAN β-1,4-N-ACETYLGALACTOSAMINYLTRANSFERASE

This enzyme [EC 2.4.1.175], also known as N-acetyl-galactosaminyltransferase II, catalyzes the reaction of UDP-N-acetyl-D-galactosamine with a D-glucuronyl-N-acetyl-1,3-β-D-galactosaminylproteoglycan to produce UDP and N-acetyl-D-galactosaminyl-1,4-β-D-glucuronyl-N-acetyl-1,3-β-D-galactosaminylproteoglycan, participating in the biosynthesis of chondroitin sulfate.[1] Acceptor substrates are even-numbered chondroitin oligosaccharides having a terminal glucuronyl-N-acetyl-1,3-β-D-galactosamine sequence at the non-reducing end: for example, (glucuronyl-N-acetyl-1,3-β-D-galactosamine)$_5$ is an excellent substrate. ***See also*** *Glucuronylgalactosylproteoglycan β-1,4-N-Acetylgalactosaminyltransferase*

[1] K. Rohrmann, R. Niemann & E. Buddecke (1985) *EJB* **148**, 463.

GLUSULASE

Glusulase is a proprietory name for the snail β-glucuronidase preparation (also containing arylsulfatase activity) used in the analysis of steroid conjugates and in the digestion of cell walls. Glusulase is often referred to as helicase (not to be confused with DNA helicase). ***See*** *β-Glucuronidase*

Selected entries from ***Methods in Enzymology*** [**vol**, page(s)]:
Other molecular properties: drug metabolite isolation, for, **52**, 332, 334; preparation of mitochondria, for, (*Neurospora*, for, **10**, 145; yeast, for, **10**, 135); source, **10**, 145; **12A**, 473f; **69**, 78; spheroplast preparation and, **20**, 466; **31**, 572; use in protoplast preparation, **23**, 203; use of in preparation of spores of *Saccharomyces cerevisiae*, **17A**, 69; yeast cell lysis and, **12A**, 539; yeast spheroplast, **56**, 19, 20, 129 (preparation, **22**, 121, 122)

GLUTACONATE CoA-TRANSFERASE

This transferase [EC 2.8.3.12] catalyzes the reaction of acetyl-CoA with (E)-glutaconate to produce acetate and glutaconyl-1-CoA.

(E)-glutaconate   (E)-glutaconoyl-1-CoA

Other substrates include glutarate, (R)-2-hydroxyglutarate, propenoate, and propanoate; however, note that the geometric isomer (Z)-glutaconate is not a substrate. A thioester intermediate is formed between a carboxyl group of the protein and coenzyme A.[1]

[1] W. Buckel, U. Dorn & R. Semmler (1981) *EJB* **118**, 315.

GLUTACONYL-CoA DECARBOXYLASE

This decarboxylase [EC 4.1.1.70] catalyzes the conversion of glutaconyl-CoA to crotonoyl-CoA and carbon dioxide.

The enzyme from *Acidaminococcus fermentans* is a biotinyl-protein, requires sodium ions, and acts as a sodium pump.[1] There is a retention of configuration during the decarboxylation.[2]

crotonoyl-1-CoA

The enzyme first reacts with the substrate to produce crotonoyl-CoA and the carboxybiotin form of the protein. Bicarbonate release appears to be coupled with the transport of sodium ions.

[1] B. Beatrix, K. Bendrat, S. Rospert & W. Buckel (1990) *Arch. Microbiol.* **154**, 362.

[2] W. Buckel (1986) *EJB* **156**, 259.

Selected entries from *Methods in Enzymology* [vol, page(s)]:
General discussion: 125, 547
Assay: *Acidaminococcus fermentans*, 125, 550, 553
Other molecular properties: *Acidaminococcus fermentans* (assay, 125, 550, 553; properties, 125, 556; purification, 125, 552; sodium transport activity, reconstitution into phospholipid vesicles, 125, 555)

GLUTAMATE ACETYLTRANSFERASE

This transferase [EC 2.3.1.35], also known as ornithine acetyltransferase and ornithine transacetylase, catalyzes the reversible reaction of N^2-acetyl-L-ornithine with L-glutamate to produce L-ornithine and *N*-acetyl-L-glutamate.[1–4]

N-acetyl-L-glutamate N^2-acetyl-L-ornithine

This enzyme, which has a ping pong Bi Bi kinetic mechanism, also exhibits a low hydrolytic activity (about 1% of that of the transferase activity) of N^2-acetyl-L-ornithine, producing acetate and L-ornithine. Glutamate acetyltransferase is not identical with *N*-acetylglutamate synthase (*i.e.*, amino-acid acetyltransferase, EC 2.3.1.1). The *Methanococcus jannaschii* enzyme is specific for the N^2-acetyl-L-ornithine donor whereas the *Thermotoga neapolitana* and *Bacillus stearothermophilus* enzyme can use acetyl-CoA as well.[4] ***See also*** *Amino-Acid Acetyltransferase*

[1] M. Staub & G. Dénes (1966) *BBA* **128**, 82.

[2] J. C. Jain & P. D. Shargool (1984) *Anal. Biochem.* **138**, 25.

[3] C. J. Morris & J. F. Thompson (1975) *Plant Physiol.* **55**, 960.

[4] F. Marc, P. Weigel, C. Legrain, Y. Almeras, M. Santrot, N. Glansdorff & V. Sakanyan (2000) *EJB* **267**, 5217.

Selected entries from *Methods in Enzymology* [vol, page(s)]:
General discussion: 17A, 273

GLUTAMATE ADENYLYLTRANSFERASE

This adenylyltransferase, more commonly referred to as L-glutamic acid activating enzyme or glutamate activating enzyme, catalyzes the reaction of L-glutamate with ATP to produce γ-L-glutamyladenylate (γ-L-Glu-AMP) and pyrophosphate (or, diphosphate).[1,2]

L-γ-glutamyl-AMP

This component enzyme in the membranous polyglutamyl synthetase complex of *Bacillus licheniformis* and related microorganisms participates in the biosynthesis of the poly(γ-D-glutamyl) capsule. The adenylylated product presumably remains bound to the complex and is acted upon by succeeding members of the reaction pathway.

[1] M. Kunioka (1997) *Appl. Microbiol. Biot.* **47**, 469.

[2] J. M. Gardner & F. A. Troy (1979) *JBC* **254**, 6262.

Selected entries from *Methods in Enzymology* [vol, page(s)]:
General discussion: 113, 146, 147
Assay: membrane-bound, 113, 149
Other molecular properties: membrane-bound (assays, 113, 149; product characterization, 113, 156; properties, 113, 165; purification, 113, 150, 168)

D-GLUTAMATE(D-ASPARTATE) OXIDASE

This FAD-dependent oxidoreductase [EC 1.4.3.15] catalyzes the reaction of D-glutamate with water and dioxygen to produce α-ketoglutarate (or, 2-oxoglutarate), ammonia, and hydrogen peroxide. D-Aspartate is oxidized at the same rate, producing oxaloacetate. Other amino acids are not substrates; hence, this enzyme is distinct from D-amino acid oxidase [EC 1.4.3.3].[1] ***See also*** *D-Glutamate Oxidase; D-Aspartate Oxidase; D-Amino-Acid Oxidase*

[1] S. Mitzushima (1957) *J. Gen. Appl. Microbiol.* **3**, 233.

GLUTAMATE CARBOXYPEPTIDASE

This zinc-dependent carboxypeptidase [EC 3.4.17.11], also known as carboxypeptidases G, G₁, and G₂, catalyzes the release of C-terminal glutamate residues from a wide range of *N*-acyl groups, including peptidyl, aminoacyl, benzoyl, benzyloxycarbonyl, folyl, and pteroyl groups. The G_1 and G_2 variants are very similar in their activities: however,

G_2 has higher K_m values. Carboxypeptidase G_1 also has some activity toward pteroyl compounds containing C-terminal aspartate.[1-3]

[1]P. Goldman & C. C. Levy (1967) *PNAS* **58**, 1299.

[2]K. K. Kalghatgi & J. R. Bertino (1981) in *Enzymes as Drugs* (J. S. Holcenberg & J. Roberts, eds.), Wiley, New York, p. 77.

[3]R. F. Sherwood, R. G. Melton, S. M. Alwan & P. Hughes (1985) *EJB* **148**, 447.

Selected entries from *Methods in Enzymology* [vol, page(s)]:
Other molecular properties: *Acinetobacter*, inhibition, **80**, 787, 788; properties, **248**, 223; *Pseudomonas*, **248**, 223

GLUTAMATE CARBOXYPEPTIDASE II

This metallocarboxypeptidase [EC 3.4.17.21], originally called *N*-acetylated-α-linked acidic dipeptidase, catalyzes the release of an unsubstituted, C-terminal glutamyl residue, typically from *N*-acetylaspartylglutamate ($K_m = 140$ nM), aspartylglutamate, glutamylglutamate, and γ-glutamylglutamate. The enzyme will act successively on folate-γ-pentaglutamate and on methotrexate-γ-triglutamate.[1-3]

[1]B. S. Sluster, G. Tsai, G. Yoo & J. T. Coyle (1992) *J. Comp. Neurol.* **315**, 217.

[2]V. Serval, L. Barbeito, A. Pittaluga, A. Cheramy, S. Lavielle & J. Glowinski (1990) *J. Neurochem.* **55**, 39.

[3]R. E. Carter, A. R. Feldman & J. T. Coyle (1996) *PNAS* **93**, 749.

D-GLUTAMATE CYCLASE

This enzyme [EC 4.2.1.48] catalyzes the conversion of D-glutamate to 5-oxo-D-proline and water.

D-glutamate   5-oxo-D-proline

Other substrates include various α-, β-, and γ-substituted derivatives of D-glutamate. Interestingly, D-glutamine is not a substrate.[1-3]

[1]M. Orlowsky & A. Meister (1971) *The Enzymes*, 3rd ed., **4**, 123.

[2]J. C. Unkeless & P. Goldman (1971) *JBC* **246**, 2354.

[3]A. Meister, M. W. Bukenberger & M. Strassburger (1963) *Biochem. Z.* **338**, 217.

Selected entries from *Methods in Enzymology* [vol, page(s)]:
General discussion: 17A, 860; 113, 59
Assay: 17A, 860
Other molecular properties: purification from (mouse kidney, **17A**, 862; mouse liver, **17A**, 862); purification of L-glutamate, in, **113**, 60; role, **113**, 62; specificity, **113**, 63

GLUTAMATE DECARBOXYLASE

This pyridoxal-phosphate-dependent enzyme [EC 4.1.1.15] catalyzes the conversion of L-glutamate to 4-aminobutanoate (or, γ-aminobutyrate) and carbon dioxide.[1-4] The mammalian brain enzyme also acts on L-cysteate (producing taurine), 3-sulfino-L-alanine (producing hypotaurine), and L-aspartate (generating β-alanine). The enzyme exhibits an ordered Uni Bi kinetic mechanism. The substrate binds to the enzyme and forms an aldimine with pyridoxal 5′-phosphate. Decarboxylation produces a quinoid intermediate which undergoes a second transamination to regenerate the native enzyme and release 4-aminobutanoate.

[1]M. A. Ator & P. R. Ortez de Montellano (1990) *The Enzymes*, 3rd ed., **19**, 213.

[2]D. J. Creighton & N. S. R. K. Murthy (1990) *The Enzymes*, 3rd ed., **19**, 323.

[3]A. E. Martell (1982) *AE* **53**, 163.

[4]E. A. Boeker & E. E. Snell (1972) *The Enzymes*, 3rd ed., **6**, 217.

Selected entries from *Methods in Enzymology* [vol, page(s)]:
General discussion: 17A, 857; 46, 445; 113, 3, 11
Assay: 32, 779; bacterial, 113, 11; rat brain, 113, 3
Other molecular properties: aminotransferases and, **5**, 687, 690; assay of L-glutamate, in, **3**, 465, 1048; **17B**, 631; bacterial (properties, **113**, 12; purification, **3**, 465; **113**, 12; pyridoxal phosphate-binding site, isolation, **106**, 130); chromatography on DEAE-cellulose, **113**, 5; cultured cells, **32**, 772, 773; entrapment with photo-cross-linkable resin prepolymers, **135**, 239; *Escherichia coli*, of, **17A**, 858; **27**, 383, 384, 394, 396, 399, 400, 406, 427; **113**, 12; labeling, **46**, 139; localization, **113**, 8; neurobiology, in, **32**, 765, 768, 770; prepolymer-entrapped, applications, **135**, 245; purification from (*Clostridium welchii*, **3**, 465; *Escherichia coli*, **17A**, 858; **113**, 12; rat brain, **113**, 4); rat brain, **113**, 7; separation of radiolabeled D- and L-glutamates, in, **113**, 61; solvent isotope effect, **87**, 590; transaminases and, **5**, 687, 690

GLUTAMATE DEHYDROGENASES

The several enzymes catalyzing oxidative deamination of glutamate are classified on the basis of their redox coenzyme. Glutamate dehydrogenase [EC 1.4.1.2], often abbreviated GDH, catalyzes the reversible reaction of L-glutamate with NAD^+ and water to produce α-ketoglutarate (or, 2-oxoglutarate), ammonia, and NADH with the formation and subsequent hydrolysis of α-iminoglutarate intermediate. L-Serine also undergoes deamination. The bovine liver mitochondrial enzyme is the most thoroughly investigated enzyme of this type.[1-5] Glutamate dehydrogenase ($NADP^+$) [EC 1.4.1.4] utilizes $NADP^+$ as the oxidizing coenzyme.[6-8] The *Phormidium laminiosum* enzyme is reportedly random in the reverse direction.[8] At least two forms of the $NADP^+$-specific dehydrogenase exist: a homotetrameric form as well as a homohexamer. This latter form, often found in plants and microorganisms, has an important role in ammonia assimilation. Glutamate dehydrogenase ($NAD(P)^+$) [EC 1.4.1.3] can utilize either NAD^+ or $NADP^+$ as the coenzyme substrate.[9,10]

[1]N. M. W. Brunhuber & J. S. Blanchard (1994) *Crit. Rev. Biochem. Mol. Biol.* **29**, 415.

[2]R. F. Colman (1991) *A Study of Enzymes* **2**, 173.

[3]R. F. Colman (1990) *The Enzymes*, 3rd ed., **19**, 283.

[4]E. Silverstein & G. Sulebele (1973) *Biochemistry* **12**, 2164 and (1974) **13**, 1815.

[5]H. F. Fisher (1973) *AE* **39**, 369.

[6]J. W. Coulton & M. Kapoor (1973) *Can. J. Microbiol.* **19**, 439.

[7]K. V. Sarada, N. A. Rao & T. A. Venkitasubramanian (1980) *BBA* **615**, 299.

[8]M. Martinez-Bilbao, A. Martinez, I. Urkijo, M. J. Llama & J. L. Serra (1988) *J. Bacteriol.* **170**, 4897.

[9]G. di Prisco & F. Garofano (1974) *BBRC* **58**, 683.

[10]A. H. Electricwala & F. M. Dickinson (1979) *BJ* **177**, 449.

Selected entries from **Methods in Enzymology** [vol, page(s)]:
General discussion: 2, 220; 17A, 839; 34, 126; 46, 21, 83, 87, 90, 242, 245; 113, 16

Assay: 4, 378; bovine liver, 2, 220; 113, 25; Fast Analyzer, by, 31, 816

Other molecular properties: abortive complex, 63, 420; 113, 21; affinity labeling, 87, 475; affinity partitioning with dye ligands, 228, 126, 128; alanine deamination, 64, 8; alanine dehydrogenase activity and, 5, 674, 676; 87, 405, 653; allosteric effectors, 44, 512, 513; 113, 20; aminotransferases and, 5, 686; ammonia assay, 17A, 955; 55, 209; 63, 34; assays for (D-amino acid oxidase, 63, 34; L-amino acid oxidase, 63, 34; γ-aminobutyrate, absence in, 5, 778; ammonia, 17A, 955; 55, 209; 63, 34; asparaginase, 63, 34; FGAM synthetase, 51, 196 [nonaqueous tissue fractions, in, 56, 204]; glutamate, 13, 471; 4-hydroxy-2-ketoglutarate, 17B, 275; α-ketoglutarate, 13, 455; 55, 206 [absense, 5, 778]; NAD⁺, 6, 792, 796; 18B, 3 [fluorometric recycling assay for NAD⁺, in, 18B, 28, 29]; NADP(H), 6, 792, 793; 13, 485 [enzymatic cycling of NADP⁺ and NADPH, in, 18B, 11]); bovine liver, of, 2, 220; 27, 732 (affinity precipitation, 104, 367; amidophosphoribosyltransferase assay, in, 113, 266; assay, 113, 25; hydroxylamine cleavage, 47, 140, 142; properties, 2, 224; 113, 18; purification, 2, 221; 113, 26); brain mitochondria, 55, 58; chemical cleavage at aspartyl-prolyl bond, 47, 145; conjugate (assay, 44, 508; coenzyme binding, 44, 508); denaturation studies, 44, 510; differential extraction of tissue, in, 1, 33; dissociation equilibria, dilution method, 61, 66; elimination from mitochondria, 56, 421; enzyme electrode, in, 44, 877; enzymic oxidation of NADPH, 44, 865; equilibrium constant, 2, 225; 113, 16; fluorescence and, 26, 503; glutamine:fructose-6-phosphate transamidase and, 5, 417; Hill plot, 64, 182; histidine modification, 47, 433, 441; histochemistry of, 6, 892; hysteresis, 64, 218; immobilization, 56, 483, 484 (azide method, by, 44, 907, 908; collagen, on, 44, 907, 908; porous glass beads, on, 44, 506); inhibition, 63, 399, 433, 507; 87, 392; 113, 24; interaction, gel chromatographic studies, 61, 142; intermediate, 87, 404; isolated nuclei and, 12A, 432; isomerization, 259, 214; isothermal enthalpy change interpretation, 259, 214; isotope exchange, 64, 8 (fingerprint kinetic patterns for modifier action, 87, 652); α-ketobutyrate, and, 87, 360; α-ketoglutarate assay, 13, 455; 55, 206; α-ketoglutarate dehydrogenase assay and, 1, 716; α-ketovalerate, and, 87, 360; kinetics, 87, 357, 360, 361, 364, 652; liposomes, 56, 428, 429; liver compartments, in, 37, 288; localization, 56, 205; lung lamellar bodies, in, 31, 423; lysines, active-site, 87, 404; marker enzyme, as, 56, 210, 217, 218, 222; 31, 760; mechanism, 63, 51, 157; 87, 357, 360, 361, 364, 652; 113, 20; mercuri-estradiol reaction, 36, 389; mitoplasts, in, 31, 323; modification of, 26, 589; modifiers, 87, 653; 113, 20; multienzymatic NAD(P)H recycling, in, 136, 71, 77; NAD⁺-specific, cyanylation, 47, 131; negative staining, 32, 29; *Neurospora crassa*, from (arginine blocking, 47, 160, 161; preparation, 22, 81; staphylococcal protease hydrolysis, 47, 190, 191); nitrogen fixation mutants, 69, 51; nucleoli and, 12A, 457; oligomerization, high-pressure effects, 259, 363; pH effects, 87, 392, 404; 6-phosphogluconate dehydrogenase and, 6, 800; 5-phosphoribosylpyrophosphate amidotransferase and, 6, 57, 61; preparation from *Neurospora crassa*, 22, 81; properties, 2, 224; 113, 18; protein association, 113, 19; purification from (bovine liver, 2, 221; 113,

26; chicken liver, 17A, 849; dogfish liver, 17A, 845; frog liver, 17A, 840; yeast, 17A, 852); pyridine nucleotide assay, 55, 267; reduction of NADP⁺ analogues, 44, 874; self-assembly, 48, 309; staining on gels, 22, 594; starch gel electrophoresis of, 6, 968; stereochemistry, 87, 116, 117, 405; structure, 44, 507; small-angle X-ray scattering studies, 61, 232; synergism, 87, 652; synthesis in *Chlamydomonas* cultures, 23, 73; tadpole liver, in, 17A, 843; transaminases and, 5, 686; transition-state and multisubstrate analogues, 249, 304

GLUTAMATE FORMIMIDOYLTRANSFERASE

This pyridoxal-phosphate-dependent transferase [EC 2.1.2.5], also known as glutamate formyltransferase and glutamate formiminotransferase, catalyzes the reversible reaction of 5-formimidoyltetrahydrofolate with L-glutamate to produce tetrahydrofolate and *N*-formimidoyl-L-glutamate.[1-5]

L-glutamate   *N*-formimino-L-glutamate

The enzyme will also catalyze the transfer of the formyl moiety from 5-formyltetrahydrofolate to L-glutamate. In eukaryotes, this transferase often occurs as a bifunctional enzyme with formiminotetrahydrofolate cyclodeaminase activity [EC 4.3.1.4], and substrate channeling may facilitate transfer of reactants between the separate active sites.[2]

***See also** Formiminotetrahydrofolate Cyclodeaminase*

[1]V. Schirch (1998) *CBC* 1, 211.
[2]J. Paquin, C. M. Baugh & R. E. MacKenzie (1985) *JBC* **260**, 14925.
[3]R. Beaudet & R. Mackenzie (1975) *BBA* **410**, 252.
[4]J. I. Rader & F. M. Huennekens (1973) *The Enzymes*, 3rd ed., **9**, 197.
[5]H. Tabor & L. Wyngarden (1959) *JBC* **234**, 1830.

Selected entries from **Methods in Enzymology** [vol, page(s)]:
General discussion: 5, 784, 785, 790
Assay: 5, 785; formyltransferase activity, 5, 790
Other molecular properties: properties, 5, 788, 793; purification, hog liver, 5, 786, 792; transformylase and, 5, 793

GLUTAMATE 1-KINASE

α-L-glutamyl phosphate

This phosphotransferase [EC 2.7.2.13] catalyzes the reaction of ATP with L-glutamate to produce ADP and α-L-glutamyl phosphate.[1]

[1]W.-Y. Wang, S. P. Gough & C. G. Kannangara (1981) *Carlsberg Res. Commun.* **46**, 243.

GLUTAMATE 5-KINASE

This phosphotransferase [EC 2.7.2.11], also referred to as γ-glutamyl kinase, catalyzes the reaction of ATP with L-glutamate to produce ADP and L-glutamate 5-phosphate (i.e., γ-L-glutamyl phosphate).[1–3] The product rapidly undergoes conversion to 5-oxo-L-proline and orthophosphate. cis-1-Amino-1,3-dicarboxycyclohexane is an alternative substrate for the enzyme (the phosphorylated product cannot cyclize to generate the corresponding 5-oxoproline analogue).[3]

Mothbean 1-pyrroline-5-carboxylate synthetase is a bifunctional enzyme that possesses both γ-glutamyl kinase and glutamate-γ-semialdehyde dehydrogenase activities.[4]

[1] A. Baich (1969) BBA 192, 462.
[2] R. V. Krishna & T. Leisinger (1979) BJ 181, 215.
[3] A. P. Seddon, K. Y. Zhao & A. Meister (1989) JBC 264, 11326.
[4] C. A. Hu, A. J. Delauney & D. P. Verma (1992) PNAS 89, 9354.

GLUTAMATE:METHYLAMINE LIGASE

N⁵-methyl-L-glutamine

This ligase [EC 6.3.4.12], also referred to as γ-glutamyl-methylamide synthetase, catalyzes the reaction of ATP with L-glutamate and methylamine to produce ADP, orthophosphate, and N^5-methyl-L-glutamine.[1,2] **See also** Glutamine Synthetase; Theanine Synthetase

[1] H.-F. Kung & C. Wagner (1969) JBC 244, 4136.
[2] M. E. Levitch (1977) BBRC 76, 609.

D-GLUTAMATE OXIDASE

This FAD-dependent oxidoreductase [EC 1.4.3.7] catalyzes the reaction of D-glutamate with water and dioxygen to produce α-ketoglutarate (or, 2-oxoglutarate), ammonia, and hydrogen peroxide. The oxidase from some sources will also act on D-aspartate.[1,2] **See also** D-Glutamate (D-Aspartate) Oxidase; D-Amino Acid Oxidase; D-Aspartate Oxidase

[1] K. Urich (1968) Z. Naturforsch. 23B, 1508.
[2] E. Rocca & F. Ghiretti (1958) ABB 77, 336.

Selected entries from **Methods in Enzymology** [vol, page(s)]:
General discussion: 52, 21; 56, 474

L-GLUTAMATE OXIDASE

This FAD-dependent oxidoreductase [EC 1.4.3.11] catalyzes the reaction of L-glutamate with water and dioxygen to produce α-ketoglutarate (or, 2-oxoglutarate), ammonia, and hydrogen peroxide.[1–3] The *Streptomyces endus* enzyme is very specific for L-glutamate.[1] **See also** L-Amino Acid Oxidase; L-Aspartate Oxidase

[1] A. Böhmer, A. Müller, M. Passarge, P. Liebs, H. Honeck & H.-G. Müller (1989) EJB 182, 327.
[2] T. Kamei, K. Asano, H. Kondo, M. Matsuzaki & S. Nakamura (1983) Chem. Pharm. Bull. 31, 3609.
[3] H. Kusakabe, Y. Midorikawa, T. Fujishima, A. Kuninaka & H. Yoshino (1983) Agric. Biol. Chem. 47, 1323.

Selected entries from **Methods in Enzymology** [vol, page(s)]:
General discussion: 52, 19

GLUTAMATE RACEMASE

This racemase [EC 5.1.1.3] catalyzes the interconversion of L-glutamate and D-glutamate.[1–4] The *Lactobacillus fermenti* and *L. brevis* enzymes are cofactor-independent and utilize two cysteinyl residues as acid/base catalysts during the interconversion of the glutamate enantiomers. The thiolate anion from one of the residues abstracts the α-proton, and the other cysteinyl thiol delivers a proton to the opposite face of the carbanion intermediate.[1] If the reaction is carried out in the presence of D_2O, deuterium is incorporated into the α-position.

[1] M. E. Tanner & G. L. Kenyon (1998) CBC 2, 7.
[2] N. Nakajima, K. Tanizawa, H. Tanaka & K. Soda (1986) Agric. Biol. Chem. 50, 2823.
[3] E. Adams (1976) AE 44, 69.
[4] E. Adams (1972) The Enzymes, 3rd ed., 6, 479.

Selected entries from **Methods in Enzymology** [vol, page(s)]:
General discussion: 2, 215; 17A, 873
Assay: 2, 215; 17A, 873

GLUTAMATE-1-SEMIALDEHYDE 2,1-AMINOMUTASE

This pyridoxal-phosphate-dependent mutase [EC 5.4.3.8], also known as glutamate-1-semialdehyde aminotransferase, catalyzes the interconversion of (S)-4-amino-5-oxopentanoate (or, L-glutamate 1-semialdehyde) and 5-aminolevulinate (also known as δ-aminolevulinate).[1–4]

L-glutamate α-semialdehyde   δ-aminolevulinate

This enzyme, which is absent mammals, is a potential target for selective herbicides and antibacterial agents. The catalytic mechanism is analogous to that of other aminotransferases, except that the reaction begins and ends with the coenzyme in the pyridoxamine form. The initial reaction is with the oxo-group of glutamate 1-semialdehyde. The pyridoxaldimine is formed by the net 1,3-prototropic shift from C4′ of the coenzyme to the substrate. The intermediate of the overall reaction, 4,5-diaminovalerate, can dissociate from this form of the enzyme to some extent. Under most reaction conditions, 4,5-diaminovalerate reacts rapidly with the modified coenzyme (present as an internal aldimine with a lysyl residue) to produce 5-aminolevulinate.

[1] W.-Y. Wang, D.-D. Huang, D. Stachon, S. P. Gough & C. G. Kannangara (1984) *Plant Physiol.* **74**, 569.
[2] M. A. Smith, C. G. Kannangara, B. Grimm & D. von Wettstein (1991) *EJB* **202**, 749.
[3] C. E. Pugh, J. L. Harwood & R. A. John (1992) *JBC* **267**, 1584.
[4] C. G. Kannangara, R. V. Andersen, B. Pontoppidan, R. Willows & D. von Wettstein (1994) *Ciba Found. Symp.* **180**, 3.

GLUTAMATE-5-SEMIALDEHYDE DEHYDROGENASE

This oxidoreductase [EC 1.2.1.41], also known as γ-glutamylphosphate reductase and glutamate-γ-semialdehyde dehydrogenase, catalyzes the reversible reaction of L-glutamate 5-semialdehyde with orthophosphate and NADP$^+$ to produce L-γ-glutamyl phosphate and NADPH.[1–5] The semialdehyde substrate is in equilibrium with Δ^1-pyrroline-5-carboxylate. The product cyclizes to generate 5-oxo-L-proline and orthophosphate. This protein reportedly forms a complex with glutamate 5-kinase in the proline biosynthetic pathway in a number of organism such as *Escherichia coli* (the enzyme acting as a reductase in the reverse reaction), the complex referred to as Δ^1-pyrroline-5-carboxylate synthase. The purified *E. coli* enzyme has a rapid equilibrium random Bi Bi kinetic mechanism.[2]

[1] J. J. Kramer, R. C. Gooding & M. E. Jones (1988) *Anal. Biochem.* **168**, 380.
[2] D. J. Hayzer & T. Leisinger (1983) *BBA* **742**, 391.
[3] E. Adams & L. Frank (1980) *Ann. Rev. Biochem.* **49**, 1005.
[4] R. V. Krishna, P. Beilstein & T. Leisinger (1979) *BJ* **181**, 223.
[5] A. Baich (1971) *BBA* **244**, 129.

Selected entries from *Methods in Enzymology* [vol, page(s)]:
General discussion: 113, 113
Assay: 113, 117

GLUTAMATE SYNTHASES

There are at least three distinct glutamate synthases, each utilizing a different reducing agent.[1–5]

Glutamate synthase (NADPH) [EC 1.4.1.13] is an iron-sulfur flavoprotein (containing both FAD and FMN) that catalyzes the reaction of L-glutamine with α-ketoglutarate (or, 2-oxoglutarate) and NADPH to produce NADP$^+$ and two L-glutamate molecules. Ammonia can act as the nitrogen donor substrate instead of L-glutamine, albeit weaker. The enzyme initially catalyzes the hydrolysis of L-glutamine to produce L-glutamate and nascent ammonia which then reacts with α-ketoglutarate to form an intermediate imine that is reduced by NADPH.

Glutamate synthase (NADH) [EC 1.4.1.14], which uses FMN, catalyzes the reaction of L-glutamine with α-ketoglutarate (or, 2-oxoglutarate) and NADH to produce NAD$^+$ and two L-glutamate molecules.

Glutamate synthase (ferredoxin) [EC 1.4.7.1], also known as ferredoxin-dependent glutamate synthase, is an iron-sulfur flavoprotein that catalyzes the reaction of L-glutamine with α-ketoglutarate (or, 2-oxoglutarate) and two molecules of reduced ferredoxin to produce two oxidized ferredoxin molecules and two L-glutamate molecules.[6,7] Methyl viologen can also be used as the electron donor.

[1] H. Zalkin (1993) *AE* **66**, 203.
[2] R. F. Colman (1990) *The Enzymes*, 3rd ed., **19**, 283.
[3] J. M. Buchanan (1973) *AE* **39**, 91.
[4] K. Matsuoka & K. Kimura (1986) *J. Biochem.* **99**, 1087.
[5] M. J. Boland (1979) *EJB* **99**, 531.
[6] M. Hirasawa, J. M. Boyer, K. A. Gray, D. J. Davis & D. B. Knaff (1986) *BBA* **851**, 23.
[7] M. Hirasawa & G. Tamura (1984) *J. Biochem.* **95**, 983.

Selected entries from *Methods in Enzymology* [vol, page(s)]:
General discussion: 34, 135, 136; 46, 420; 113, 327
Assay: NADH-dependent, 113, 333; NADPH-dependent, 113, 328
Other molecular properties: NADH-dependent (*Saccharomyces cerevisiae* [assay, 113, 333; catalytic properties, 113, 337; properties, 113, 336; purification, 113, 334]; stereospecificity, 87, 117); NADPH-dependent activity (*Escherichia coli* [assay, 113, 328; properties, 113, 331; purification, 113, 328]; *Klebsiella aerogenes*, 113, 331 [assay, 113, 328; catalytic behavior, 113, 331; inhibition, 113, 331; properties, 113, 331; purification, 113, 328]; stereospecificity, 87, 117); nitrogen fixation mutants, 69, 51, 52

GLUTAMINASE

This enzyme [EC 3.5.1.2], also known as L-glutamine amidohydrolase, catalyzes the hydrolysis of L-glutamine to produce L-glutamate and ammonia.[1–6] The mammalian mitochondrial enzyme is activated by orthophosphate and other phosphorylated compounds. *See also* γ-*Glutamyl Transpeptidase;* D-*Glutaminase; Glutaminase/Asparaginase*

[1] W. G. Haser, R. A. Shaphiro & N. P. Curthoys (1985) *BJ* **229**, 399.

[2]M. S. Ardawi & E. A. Newsholme (1984) *BJ* **217**, 289.
[3]H. G. Windmueller (1982) *AE* **53**, 201.
[4]S. Prusiner, J. N. Davis & E. R. Stadtman (1976) *JBC* **251**, 3447.
[5]S. C. Hartman (1971) *The Enzymes*, 3rd ed., **4**, 79.
[6]E. Robers (1960) *The Enzymes*, 2nd ed., **4**, 285.

Selected entries from *Methods in Enzymology* [**vol**, page(s)]:
General discussion: 2, 380; 11, 400; **17A**, 941; **46**, 35, 420; 113, 241; porcine brain, 113, 242; porcine kidney, 113, 242; rat kidney, 113, 242
Assay: 2, 380; 113, 242, 257
Other molecular properties: *Acinetobacter*, **27**, 349, 386, 387, 400, 401; 113, 257, 262, 263; butyl alcohol method, purification with, 1, 47, 50; *Clostridium perfringens*, from (preparation, 11, 84; properties, 11, 86); crude δ-aminovaleramidase, in, **17B**, 164; *Escherichia coli* (properties, 2, 381; purification, 2, 380, 382; **17A**, 942); glutamate decarboxylase and, 3, 465; glutaminase A, **46**, 424; glutamine:fructose-6-phosphate transamidase and, 5, 417; immune response, effect on, **44**, 705; kidney enzyme, purification, 1, 47, 50; phosphorylation potential determination, **55**, 237; porcine brain (assays, 113, 242; properties, 113, 253; purification, 113, 249); porcine kidney (assays, 113, 242; properties, 113, 253; purification, 113, 246); properties, 113, 253, 261; *Pseudomonas* 7A enzyme, 113, 262, 263; rat kidney (assays, 113, 242; properties, 113, 253; purification, 113, 251); renal mitchondria, **55**, 12

D-GLUTAMINASE

This enzyme [EC 3.5.1.35] catalyzes the hydrolysis of D-glutamine to produce D-glutamate and ammonia.[1]

[1]A. Domnas & E. C. Cantino (1965) *Phytochemistry* **4**, 273.

GLUTAMINASE/ASPARAGINASE

This enzyme [EC 3.5.1.38], also referred to as glutamin-(asparagin-)ase, catalyzes the hydrolysis of both L-glutamine and L-asparagine to produce ammonia and L-glutamate or L-aspartate, respectively.[1,2] The D-enantiomers are also hydrolyzed, albeit more slowly.

See also Glutaminase; Asparaginase

[1]E. Ortlund, M. W. Lacount, K. Lewinski & L. Lebioda (2000) *Biochemistry* **39**, 1199.
[2]J. Steckel, J. Roberts, F. S. Philips & T. C. Chou (1983) *Biochem. Pharmacol.* **32**, 971.

Selected entries from *Methods in Enzymology* [**vol**, page(s)]:
General discussion: 27, 349, 386, 387, 400, 401;113, 257
Assay: 113, 257
Other molecular properties: molecular weight, partial specific volume, 61, 56; properties, 113, 261; purification, *Acinetobacter glutaminsificans*, 113, 259

GLUTAMINE *N*-ACYLTRANSFERASE

This transferase [EC 2.3.1.68] catalyzes the reaction of an acyl-CoA with L-glutamine to produce coenzyme A and an *N*-acyl-L-glutamine.[1] Acyl donors include phenylacetyl-CoA and indole-3-acetyl-CoA, but not benzoyl-CoA.

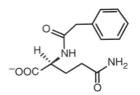

N-phenylacetyl-L-glutamine

This transferase is distinct from glycine *N*-acyltransferase [EC 2.3.1.13] and glycine *N*-benzoyltransferase [EC 2.3.1.71]. *See also* Glutamine Phenylacetyltransferase

[1]L. T. Webster, U. A. Siddiqui, S. V. Lucas, J. M. Strong & J. J. Mieyal (1976) *JBC* **251**, 3352.

Selected entries from *Methods in Enzymology* [**vol**, page(s)]:
General discussion: 77, 301
Assay: 77, 302

GLUTAMINE:FRUCTOSE-6-PHOSPHATE AMINOTRANSFERASE (ISOMERIZING)

This aminotransferase [EC 2.6.1.16; previously classified as EC 5.3.1.19], also known as hexosephosphate aminotransferase, D-fructose-6-phosphate amidotransferase, glucosamine-6-phosphate isomerase (glutamine-forming), and glucosamine-6-phosphate synthase, catalyzes the reaction of L-glutamine with D-fructose 6-phosphate to produce L-glutamate and D-glucosamine 6-phosphate.[1–3]

The initial step in the reaction mechanism involves the ring opening of fructose 6-phosphate, utilizing a histidyl residue, followed by the formation of a ketimine between D-fructose 6-phosphate and a ε-amino group of a lysyl residue of the protein. Enolization step is attended by a hydrogen transfer from C1 to C2.[3]

[1]M. Calcagno, J. A. Levy, E. Arrambide & E. Mizraji (1971) *Enzymologia* **41**, 175.
[2]S. Ghosh, H. J. Blumenthal, E. Davidson & S. Roseman (1960) *JBC* **235**, 1265.
[3]A. Teplyakov, G. Obmolova, M. A. Badet-Denisot & B. Badet (1999) *Protein Sci.* **8**, 596.

Selected entries from *Methods in Enzymology* [**vol**, page(s)]:
General discussion: 5, 414; **46**, 420; 113, 278
Assay: 5, 414;113, 278
Other molecular properties: *Escherichia coli* B (properties, 5, 417; purification, 5, 415); rat liver (properties, 113, 281; purification, 113, 279)

GLUTAMINE:*SCYLLO*-INOSOSE AMINOTRANSFERASE

This pyridoxal-phosphate-dependent aminotransferase [EC 2.6.1.50], also referred to as glutamine:keto-*scyllo*-inositol aminotransferase, catalyzes the reaction of

L-glutamine with 2,4,6/3,5-pentahydroxycyclohexanone (a *meso* compound also called *scyllo*-inosose) to produce α-ketoglutaramate (*i.e.*, 2-oxoglutaramate) and 1-amino-1-deoxy-*scyllo*-inositol (also a *meso* compound).[1,2] The enzyme participates in the biosynthesis of streptidine.

scyllo-inosose   1-amino-1-deoxy-*scyllo*-inositol

[1] J. B. Walker & M. S. Walker (1969) *Biochemistry* **8**, 763.
[2] L. A. Lucher, Y. M. Chen & J. B. Walker (1989) *Antimicrob. Agents Chemother.* **33**, 452.

Selected entries from *Methods in Enzymology* [vol, page(s)]:
General discussion: **43**, 439, 462
Assay: **43**, 440 (L-[14C]glutamine, **43**, 443; *myo*-inositol 2-dehydrogenase, **43**, 443)
Other molecular properties: biological distribution, **43**, 442; ion-exchange chromatography, **43**, 441; properties, **43**, 442, 443; reaction scheme, **43**, 439; specificity, **43**, 442, 443; stability, **43**, 442

GLUTAMINE N-PHENYLACETYLTRANSFERASE

This acyltransferase [EC 2.3.1.14] catalyzes the reaction of phenylacetyl-CoA with L-glutamine to produce α-N-phenylacetyl-L-glutamine and coenzyme A. Bovine phenylacetyltransferase reportedly uses glycine as the acceptor substrate. Other donor substrates include indoleacetyl-CoA. The Enzyme Commission also lists a glutamine acyltransferase [EC 2.3.1.68].[1] *See also Acyl-CoA:6-Aminopenicillanate Acyltransferase; Glutamine Acyltransferase*

[1] K. Moldave & A. Meister (1957) *JBC* **229**, 463.

Selected entries from *Methods in Enzymology* [vol, page(s)]:
General discussion: **77** , 301
Assay: **77**, 302; CoA determination, **77**, 303; thiolester utilization, **77**, 303)
Other molecular properties: indoleacetyl-CoA transferase activity, **77**, 302; inhibitors, **77**, 307, 308; kinetic constants, **77**, 307; mechanism, **77**, 308; molecular weight, **77**, 307; pH optima, **77**, 308; properties, **77**, 307, 308; purification, **77**, 306, 307; size, **77**, 307; specificity, **77**, 307; stability, **77**, 303, 308; stimulator, **77**, 307, 308

GLUTAMINE:PHENYLPYRUVATE AMINOTRANSFERASE

This pyridoxal-phosphate-dependent aminotransferase [EC 2.6.1.64], also known as glutamine transaminase K, catalyzes the reversible reaction of L-glutamine with phenylpyruvate to produce α-ketoglutaramate (*i.e.*, 2-oxoglutaramate) and L-phenylalanine.[1–4] At neutral or acid pH, greater than 97% of α-ketoglutaramate exists in a cyclic (or lactam) configuration that is unreactive in the reverse transamination reaction. Other amino donor substrates include L-methionine, L-histidine, and L-tyrosine. The rat kidney enzyme and kynurinine:pyruvate aminotransferase share >90% sequence identity, and both exhibit cysteine S-conjugate β-lyase activity.[4] *See also Glutamine:Pyruvate Aminotransferase*

[1] A. J. L. Cooper & A. Meister (1984) *Prog. Clin. Biol. Res.* **144B**, 3.
[2] A. J. L. Cooper & A. Meister (1974) *JBC* **249**, 2554.
[3] F. Van Leuven (1975) *EJB* **58**, 153 and (1976) **65**, 271.
[4] D. G. Abraham & A. J. L. Cooper (1996) *ABB* **335**, 311.

Selected entries from *Methods in Enzymology* [vol, page(s)]:
General discussion: **17A**, 951; **113**, 344
Assay: L-phenylalanine:α-keto-γ-methiolbutyrate assay, **113**, 346; rat kidney, **113**, 345
Other molecular properties: L-phenylalanine:α-keto-γ-methiolbutyrate, action on, **113**, 345; pyridoxal phosphate and, **17A**, 954; rat kidney (assay, **17A**, 951; **113**, 345; purification, **17A**, 952; **113**, 346, 347; substrate specificity, **113**, 349)

GLUTAMINE:PYRUVATE AMINOTRANSFERASE

This pyridoxal-phosphate-dependent enzyme [EC 2.6.1.15], also known as glutamine:pyruvate aminotransferase, glutamine:oxo-acid transaminase, and glutamine transaminase L, catalyzes the reaction of L-glutamine with pyruvate to produce α-ketoglutaramate (*i.e.*, 2-oxoglutaramate) and L-alanine.[1–4] At neutral or acid pH, greater than 97% of α-ketoglutaramate exists in a cyclic (or lactam) configuration that is unreactive in the reverse transamination reaction. Other donor substrates include L-methionine. Glyoxylate can be utilized as an acceptor substrate, generating glycine. *See also Glutamine:Phenylpyruvate Aminotransferase*

[1] A. J. L. Cooper & A. Meister (1984) *Prog. Clin. Biol. Res.* **144B**, 3
[2] A. J. L. Cooper & A. Meister (1972) *Biochemistry* **11**, 661.
[3] A. J. L. Cooper & A. Meister (1973) *JBC* **248**, 8489.
[4] A. E. Braunstein (1973) *The Enzymes*, 3rd ed., **9**, 379.

Selected entries from *Methods in Enzymology* [vol, page(s)]:
General discussion: **17A**, 1016; **113**, 338
Assay: rat liver, **113**, 338
Other molecular properties: rat liver, **2**, 382 (assays, **113**, 338; isozymic forms, occurrence, **113**, 343; properties, **113**, 342; purification, **17A**, 1019; **113**, 340, 342)

GLUTAMINE SYNTHETASE

This enzyme [EC 6.3.1.2] catalyzes the reaction of L-glutamate with MgATP and ammonia to produce L-glutamine, MgADP, and orthophosphate.[1–8] Other substrates include β-glutamate ($^-$OOCCH$_2$CH(NH$_3^+$)CH$_2$COO$^-$) and 4-methylene-L-glutamate, although there is a distinct 4-methyleneglutamine synthetase [EC 6.3.1.7]. Both stereoisomers of β-glutamine are produced, albeit with unequal rates. Hydroxylamine can replace ammonia as the nitrogen donor to form L-γ-glutamyl-hydroxamate.

Enzyme catalysis requires the simultaneous presence of all three substrates within the active site (the enzyme has a steady-state random Ter Ter kinetic mechanism[8]), and the reaction scheme is a stepwise process involving initial formation of a tightly bound acyl-phosphate intermediate which subsequently undergoes nucleophilic attack by ammonia. The participation of a γ-glutamyl-phosphate intermediate is supported by experiments on the mechanism of methionine sulfoximine inhibition, the formation of 5-oxo-L-proline in the absence of ammonia, borohydride trapping of the acyl-phosphate, positional isotope exchanges, and the enzyme's atomic structure.[2,5,7]

The enzyme from most Gram-negative bacteria is subject to an elaborate cascade system that covalently interconverts the inactive adenylylated form (dependent on the nonphysiologic metal ion Mn^{2+}) and the deadenylylated form (active with Mg^{2+}).[8]

[1] D. Eisenberg, H. S. Gill, G. M. Pfluegl & S. H. Rotstein (2000) *BBA* **1477**, 122.
[2] D. L. Purich (1998) *AE* **72**, 9.
[3] J. J. Villafranca & T. Nowak (1992) *The Enzymes*, 3rd ed., **20**, 63.
[4] S. G. Rhee, P. B. Chock & E. R. Stadtman (1989) *AE* **62**, 37.
[5] I. A. Rose (1979) *AE* **50**, 361.
[6] R. D. Allison, J. A. Todhunter & D. L. Purich (1977) *JBC* **252**, 6046.
[7] A. Meister (1974) *The Enzymes*, 3rd ed., **10**, 699.
[8] E. R. Stadtman & A. Ginsburg (1974) *The Enzymes*, 3rd ed., **10**, 755.

Selected entries from *Methods in Enzymology* [vol, page(s)]:
General discussion: 2, 337; **17A**, 900; **27**, 620; **46**, 24, 25, 66; **113**, 185, 199, 213

Assay: 2, 337; **63**, 6; **113**, 186; porcine heart muscle, **113**, 199; rat liver, **113**, 187; rat muscle, **113**, 199

Other molecular properties: adenylylated, deadenylylation of, **182**, 799; adenylylation, **17A**, 919; **113**, 213; **182**, 795 (detection methods, **107**, 187); adenylyltransferase, **17A**, 922, 933; bacterial membrane, in, **22**, 113; bicyclic cascade, **182**, 801; binding of (ADP, **17A**, 910; ATP, **17A**, 909); cascade control, **64**, 298, 310, 315, 322, 324 (cascade, identification in *Escherichia coli*, **182**, 793); chromium-nucleotides, and, **87**, 194; conformational states, in reaction sequence, **249**, 325; cooperativity, negative, **87**, 650 (false, **63**, 6, 7); crystallization, **22**, 262; cumulative feedback inhibition, **182**, 794; dead-end complexes, **87**, 649; deadenylylating enzyme, **17A**, 919, 936; electron paramagnetic resonance, **87**, 195; *Escherichia coli* (catalytic cycle and topographic studies, **113**, 213, 239; purification, **113**, 221; regulation [closed bicyclic cascade, **113**, 213; **182**, 801; feedback inhibition, **113**, 235; **182**, 794; overview, **107**, 185; oxidative inactivation, **113**, 237; steady-state mechanisms of regulation, **182**, 803]); effect of Mg^{++}/Mn^{++} on, **17A**, 920; false cooperativity, **63**, 6; feedback inhibitors, **17A**, 922; **182**, 794;

formation, regulation by UTase and P$_{II}$ protein, **182**, 807; glutamine assay, **55**, 209; γ-glutamyl transferase activity and, **2**, 264, 267, 272, 341; **87**, 522, 523; inhibition, **63**, 401, 405, 406; **87**, 650; **113**, 205 (inhibition by methionine sulfoximine, **252**, 28); intermediate, **87**, 15; isotope exchange, **64**, 5, 7, 9, 23, 33; **87**, 247, 648; **249**, 464 (fingerprint kinetic patterns for modifier action, **87**, 648); isotope trapping, **64**, 47, 48; Job plot, **87**, 522; kinetic mechanism, equilibrium isotope exchange investigation, **64**, 5, 7, 9, 23, 33; **249**, 464; large-scale isolation, **22**, 537; Leigh theory, and, **87**, 194; ligand binding sites (interaction distances, **249**, 561; intersubunit, **249**, 559); mapping, active-site, **87**, 195; mechanism, **63**, 51, 157; **64**, 5, 7, 9, 23, 33; **249**, 464 (rat liver, **113**, 192); metal-catalyzed oxidation, **258**, 392; metal studies, **87**, 194, 544; methionine sulfoximine, **87**, 16, 195; **252**, 28; modifier effects, **87**, 650; multisubstrate analog inhibitor, **249**, 292; nitrogen fixation mutants, **69**, 51; oxidation and turnover, **258**, 393; oxygen exchanges, 87, 247; partition analysis, **249**, 323; pH effect, **87**, 649; α-phenyl *N-tert*-butylnitrone, effect of, **234**, 526; (phosphonoacetylamido)-L-alanine, **87**, 15; phosphorylation potential determination, **55**, 237; physiological significance of muscle enzyme, **113**, 211; porcine muscle (assay, **113**, 199; inhibitors, **113**, 205; kinetic properties, **113**, 203; physiological role, **113**, 211; purification, **113**, 200); preparation of ^{14}C-adenylyl glutamine synthetase, **17A**, 927; porcine heart muscle, **113**, 199 (purification, **113**, 200); positional isotope exchange, **249**, 399, 453; properties (porcine heart, **113**, 203; rat liver, **113**, 190; rat muscle, **113**, 203; sheep brain, **2**, 341); purification from (*Escherichia coli*, **17A**, 915; **113**, 201; porcine heart, **113**, 200; rat liver, **113**, 188; rat muscle, **113**, 201; sheep brain, **2**, 339; **17A**, 903); rat liver (mechanism of action, **113**, 192; properties, **113**, 190; purification, **113**, 188); rat muscle (inhibitors, **113**, 205; kinetic properties, **113**, 203; physiological role, **113**, 211; purification, **113**, 201); reaction catalyzed, **249**, 398; reaction mechanism, **63**, 51, 157; **249**, 398; species separation, **107**, 192; steady-state mechanisms of regulation, **182**, 803; substrate of ATP:glutamine synthetase adenylyltransferase, **17A**, 922; substrate saturation, **63**, 12, 13; subunit structure, **17A**, 905, 918; suppression by interferon, **79**, 343; synergism, **87**, 649; transition state and multisubstrate analogues, **249**, 308

[GLUTAMINE-SYNTHETASE] ADENYLYLTRANSFERASE

This transferase [EC 2.7.7.42], also known as [glutamate:ammonia ligase] adenylyltransferase, plays a key role in controlling bacterial nitrogen metabolism in Gram-negative microorganisms by controlling the state of adenylylation of glutamine synthetase.[1–5]

O-adenylyltyrosyl residue

The enzyme catalyzes the reaction of ATP with glutamine synthetase to produce pyrophosphate (or, diphosphate) and the modified glutamine synthetase, adenylylated at a specific tyrosyl residue. The enzyme will also catalyze the deadenylylation reaction (*i.e.*, phosphorolysis of adenyl-lated glutamine synthetase to produce ADP and the deadenylylated subunit of the synthetase) at a site separate from the adenylylation site (adenylyl-[glutamine synthetase] hydrolase, EC 3.1.4.15). Regulation of the two activities is maintained by the uridylyl removing enzyme (also referred to as protein P_{II} uridylyltransferase [EC 2.7.7.59]). *See also Adenylyl-[Glutamine Synthetase] Hydrolase*

[1] S. G. Rhee, P. B. Chock & E. R. Stadtman (1989) *AE* **62**, 37.
[2] S. G. Rhee, R. Park & M. Wittenberger (1978) *Anal. Biochem.* **88**, 174.
[3] S. P. Adler, D. L. Purich & E. R. Stadtman (1975) *JBC* **250**, 6264.
[4] E. R. Stadtman & A. Ginsberg (1974) *The Enzymes*, 3rd ed., **10**, 755.
[5] E. R. Stadtman (1973) *The Enzymes*, 3rd ed., **8**, 1.

Selected entries from *Methods in Enzymology* [vol, page(s)]:
General discussion: 17A, 919, 922, 933; 34, 135; **107**, 197; **182**, 793
Other molecular properties: activation, **113**, 227; adenylylation, **17A**, 919; **113**, 213; **182**, 795, 799; (detection methods, **107**, 187); adenylyltransferase, **17A**, 922, 933; bicyclic cascade, **182**, 801; cascade control, **64**, 298, 310, 315, 322, 324 (cascade, identification in *Escherichia coli*, **182**, 793); identity with Holzer's inactivation enzyme, **182**, 798; purification from *Escherichia coli*, **17A**, 924, 934, 938; role in adenylylation and deadenylylation of glutamine synthetase, **182**, 797

GLUTAMINYL-PEPTIDE CYCLOTRANSFERASE

This cyclotransferase [EC 2.3.2.5], also referred to as glutaminyl-tRNA cyclotransferase and glutaminyl cyclase, catalyzes the conversion of an L-glutaminyl-peptide to a 5-oxo-L-prolyl-peptide and ammonia.[1-3]

glutaminyl
residue

5-oxoprolyl
residue

The enzyme participates in the formation of active thyrotropin-releasing hormone (5-oxo-L-prolyl-L-histidyl-L-prolinamide) and other biologically active peptides containing N-terminal pyroglutamyl residues. The enzyme from papaya will also act on L-glutaminyl-tRNA (producing 5-oxo-L-prolyl-tRNA and ammonia).

The enzyme promotes an intramolecular nucleophilic attack by the amino nitrogen on the carbonyl group of the amide. A proton is transferred from the α-amino group to the nitrogen of the amide group, undoubtedly aided by an acidic group of the cyclotransferase, and ammonia is then released.[4] Multiple forms of this enzyme have recently been reported, suggesting that the substrate specificity may be influenced by the nature of the adjacent aminoacyl residues.[5]

[1] W. H. Busby, G. E. Quackenbush, J. Humm, W. W. Youngblood & J. S. Kizer (1987) *JBC* **262**, 8532.
[2] W. H. Fischer & J. Spiess (1987) *PNAS* **84**, 3628.
[3] M. Messer & M. Ottesen (1964) *BBA* **92**, 409.
[4] M. Y. Gololobov, I. Song, W. Wang & R. C. Bateman, Jr., (1994) *ABB* **309**, 300.
[5] P. A. Sykes, S. J. Watson, J. S. Temple & R. C. Bateman, Jr., (1999) *FEBS Lett.* **455**, 159.

Selected entries from *Methods in Enzymology* [vol, page(s)]:
General discussion: 168, 358
Assay: 168, 358
Other molecular properties: incubations, **168**, 360; papaya, from, **11**, 408; radioimmunoassay, **168**, 358; substrate preparation, **168**, 363

GLUTAMINYL-tRNA SYNTHETASE

This enzyme [EC 6.1.1.18], also known as glutamine:tRNA ligase, catalyzes the reaction of ATP with L-glutamine and tRNA^{Gln} to produce L-glutaminyl-tRNA^{Gln}, AMP, and pyrophosphate (or, diphosphate).[1-6] The reaction mechanism proceeds via the formation of an L-glutaminyl-adenylate intermediate. Interestingly, the binding of tRNA^{Gln} is required for activation of L-glutamine by ATP. L-Glutamine is selected from L-glutamate by the enzyme's recognition of both hydrogen atoms of the nitrogen of the substrate's side chain: a tyrosyl residue and a water molecule are required for this recognition.[2] *See also Aminoacyl-tRNA Synthetases*

[1] E. A. First (1998) *CBC* **1**, 573.
[2] V. L. Rath, L. F. Silvian, B. Beijer, B. S. Sproat & T. A. Steitz (1998) *Structure* **6**, 439.
[3] W. Freist, D. H. Gauss, M. Ibba & D. Söll (1997) *Biol. Chem.* **378**, 1103.
[4] J. J. Perona, M. A. Rould & T. A. Steitz (1993) *Biochemistry* **32**, 8758.
[5] M. J. Rogers, I. Weygand-Durasevic, E. Schwob, J. M. Sherman, K. C. Rogers, T. Adachi, H. Inokuchi & D. Söll (1993) *Biochimie* **75**, 1083.
[6] J. M. Ravel, S.-F. Wang, C. Heinemeyer & W. Shive (1965) *JBC* **240**, 432.

Selected entries from *Methods in Enzymology* [vol, page(s)]:
General discussion: 113, 55
Assay: 113, 56
Other molecular properties: hydroxylamine and, **29**, 613; properties, **113**, 58; purification, **29**, 591; **113**, 57; subcellular distribution, **59**, 233, 234

GLUTAMYL AMINOPEPTIDASE

This zinc-dependent aminopeptidase [EC 3.4.11.7], also known as aminopeptidase A, aspartate aminopeptidase, and angiotensinase A, catalyzes the release of N-terminal glutamyl residues (and, to a lesser extent, aspartyl residues) from various peptides.[1-3] This enzyme, which converts angiotensin II to angiotensin III, is activated by calcium ions. The enzyme also catalyzes the hydrolysis of synthetic substrates, such as N-(α-L-glutamyl)-β-naphthylamide and N-(α-L-glutamyl)-p-nitroaniline.

[1]J. K. McDonald & A. J. Barrett (1986) in *Mammalian Proteases*, vol. **2** (J. K. McDonald & A. J. Barrett, eds.), Academic Press, New York, p. 23.
[2]A. Taylor (1996) *FASEB J.* **7**, 206.
[3]J. Wang & M. D. Cooper (1996) in *Zinc Metalloproteinases in Health and Disease* (N. M. Hooper, ed.), Taylor & Francis, London, p. 131.

Selected entries from **Methods in Enzymology** [**vol**, page(s)]:
Assay: fluorimetric assay, 248, 601
Other molecular properties: *Escherichia coli*, 248, 213, 219; properties, 248, 185, 189; rabbit intestinal brush-border membrane (mode of integration, relation to biosynthesis, 96, 417; structure, relation to mode of integration, 96, 407); structure, 248, 186

γ-GLUTAMYLCYCLOTRANSFERASE

This cyclotransferase [EC 2.3.2.4] catalyzes the conversion of a (5-L-glutamyl)-L-amino acid to 5-oxo-L-proline and an L-amino acid.[1–4] The best substrates reported for the rat liver enzyme are the L-γ-glutamyl derivatives of L-α-aminobutyrate, L-glutamine, L-alanine, and L-methionine (note that all of these amino acids are also good acceptor substrates for γ-glutamyl transpeptidase). Other substrates include the γ-glutamyl derivatives of L-glutamate, L-aspartate, and glycine. Interestingly, di-L-γ-glutamyl derivatives of several amino acids are also good substrates.

See also *γ-Glutamylamine Cyclotransferase*

[1]O. W. Griffith & A. Meister (1977) *PNAS* **74**, 3330.
[2]P. Van Der Werf & A. Meister (1975) *AE* **43**, 519.
[3]M. Orlowski & A. Meister (1971) *The Enzymes*, 3rd ed., **4**, 123.
[4]M. J. York, M. J. Crossley, S. J. Hyslop, M. L. Fisher & P. W. Kuchel (1989) *EJB* **184**, 97.

Selected entries from **Methods in Enzymology** [**vol**, page(s)]:
General discussion: 17A, 863; 19, 789; 113, 438
Assay: 19, 790; rat kidney, 113, 439
Other molecular properties: amino acid analysis, 19, 796; assay of γ-glutamylcysteine synthetase, in, 17B, 495; different forms, 19, 796; distribution, 19, 796; function, 19, 789; inhibitor, 63, 399; 113, 445; murine kidney, inhibition *in vivo*, 113, 580; pH dependence, 19, 795; 113, 443; physical and chemical properties, 19, 795; 113, 443; preparation, 19, 793, 795; 113, 438; properties, 19, 794; 113, 443; purification, 17A, 865; 87, 467; 113, 438; rat kidney (properties, 113, 443; purification, 113, 439, 441); specificity, 19, 794; 113, 444; stability, 19, 795

γ-GLUTAMYLCYSTEINE SYNTHETASE

This enzyme [EC 6.3.2.2], also known as glutamate: cysteine ligase, catalyzes the reaction of ATP with L-glutamate and L-cysteine to produce ADP, orthophosphate, and γ-L-glutamyl-L-cysteine.[1–6]

L-γ-glutamyl-L-cysteine

L-Aminohexanoate can act as a substrate in place of L-glutamate. Certain thiol-containing molecules, including glutathione and cystamine, will inhibit the mammalian enzyme.[1,4] The synthetase has been proposed to catalyze the reaction via the formation of an enzyme-bound γ-glutamyl phosphate intermediate.

[1]O. W. Griffith & R. T. Mulcahy (1999) *AE* **73**, 209.
[2]C.-S. Huang, W. R. Moore & A. Meister (1988) *PNAS* **85**, 2464.
[3]W. Moore, H. L. Wiener & A. Meister (1987) *JBC* **262**, 16771.
[4]G. F. Seelig & A. Meister (1984) *JBC* **259**, 3534.
[5]V. P. Schandle & F. B. Rudolph (1981) *JBC* **256**, 7590.
[6]A. Meister (1974) *The Enzymes*, 3rd ed., **10**, 671.

Selected entries from **Methods in Enzymology** [**vol**, page(s)]:
General discussion: 17B, 483, 495; 46, 420; 113, 379, 390
Assay: coupled assay, 17B, 484, 495; 63, 32; 113, 380; hog liver, 17B, 484; rat kidney, 17B, 495; 113, 380; 252, 68
Other molecular properties: defects in humans, 252, 91; *Escherichia coli*, characteristics, 252, 27; feedback inhibition, 113, 385; 252, 26; hog liver (properties, 17B, 487; purification, 17B, 486); inhibition, 63, 399; 77, 61, 62; 113, 385; 252, 26 (buthionine sulfoximine, 113, 388; 252, 10, 12, 29, 320; methionine sulfoximine, 113, 388; 252, 28; ophthalmic acid, 113, 385; 252, 27); isotope exchange, 64, 9; 249, 465; mechanism, 63, 51, 158; 64, 9; 113, 389; 249, 465; properties, 17B, 487, 498; 113, 383; purification, 17B, 486, 497; 113, 381, 391; rat erythrocyte, purification, 113, 391; rat kidney (assays, 113, 380; inhibition, 113, 385; 252, 27; mechanism of action, 113, 389; properties, 17B, 498; 113, 383; purification, 17B, 497; 113, 381 [purification of recombinant protein, 252, 27]; subunit structure, 252, 26); reaction catalyzed, 17B, 483, 495; 113, 379, 390; 252, 67; transition-state and multisubstrate analogues, 249, 308

γ-D-GLUTAMYL-(L)-*meso*-DIAMINOPIMELATE PEPTIDASE I

This zinc-dependent metallopeptidase [EC 3.4.19.11], also called γ-D-glutamyl-L-diamino acid endopeptidase I and endopeptidase I, catalyzes the hydrolysis of a peptide bond in peptidoglycans of chemotype A_{1g}. The enzyme will hydrolyze the γ-D-glutamyl bonds to (L)-*meso*-diaminopimelate in L-Ala-γ-D-Glu-∇-(L)-*meso*-diaminopimelate-(L)-D-Ala or in L-Ala-γ-D-Glu-∇-(L)-*meso*-diaminopimelate, where ∇ indicates the cleavage site.[1,2] No activity is seen if the diaminopimelate is replaced with lysine. It is required that the ω-amino and ω-carboxyl groups of the (L)-*meso*-diaminopimelate group are unsubstituted. Similarly, the C-terminal carboxyl group on the D-alanine residue in position P2′ should be unsubstituted. The enzyme has been isolated from *Bacillus sphaericus* and *Bacillus subtilis*. *meso*-Diaminopimelate is a competitive inhibitor (K_i is 20 μM).

[1]F. Arminjon, M. Guinand, G. Michel, J. Coyette & J.-M. Ghuysen (1976) *Biochimie* **58**, 1167.
[2]C. Valentin, M.-J. Vacheron, C. Martinez, M. Guinand & G. Michel (1983) *Biochimie* **65**, 239.

GLUTAMYL ENDOPEPTIDASE I

This endopeptidase [EC 3.4.21.19], also known as staphylococcal serine proteinase, V8 proteinase, and endoproteinase Glu-C, catalyzes the hydrolysis of Asp–Xaa and Glu–Xaa peptide bonds.[1] Under appropriate solution conditions, the specificity of bond cleavage can be restricted to Glu–Xaa. Peptide bonds involving residues with bulky hydrophobic side-chains are hydrolyzed more slowly. There are at least two variants of bacterial glutamyl endopeptidase I: a staphylococcal group and a *Bacillus* group.[1] Several glutamyl endopeptidases are activated by calcium ions. The glutamyl endopeptidases may also catalyze transacylation reactions. *See also Glutamyl Endopeptidase II; Aureolysin*

[1] V. Rolland-Fulcrand & K. Breddam (1993) *Biocatalysis* **7**, 75.

Selected entries from *Methods in Enzymology* [vol, page(s)]:
General discussion: 244, 114
Assay: 45, 471, 472; **244**, 119
Other molecular properties: activators, **45**, 473; amino acid composition, **45**, 473, 474; *Bacillus licheniformis*, purification, **244**, 117; cleavage site, **47**, 189; **244**, 115, 121; contaminant, **47**, 190; distribution, **45**, 474; esterase activity, **45**, 475; inhibitors, **244**, 120; ion dependence, **244**, 120; mass spectroscopy, protein digestion, **193**, 368; pH dependence, **244**, 119; properties, **45**, 473; purification, **45**, 469; reaction conditions, **47**, 189, 190; sequence homology, **244**, 123; specificity, **45**, 473; **47**, 190; **59**, 600; stability, **45**, 473; **47**, 191; *Staphylococcus aureus* strain V8 (biological role, **244**, 116; experimental digestion buffers, **244**, 121; mass spectroscopy, **193**, 368; purification, **244**, 115; sequence, **244**, 122; tubulin, proteolysis of, **134**, 186; unit properties, **235**, 566); substrate preparation, **244**, 118

GLUTAMYL ENDOPEPTIDASE II

This serine endopeptidase [EC 3.4.21.82], a glutamic acid-specific protease, hydrolyzes Glu–Xaa bonds preferentially in substrates containing Pro or Leu at P2 and Phe at P3.[1] Glu–Pro and Asp–Pro bonds are only slowly hydrolyzed. The enzyme's pH *versus* rate profile differentiates it from glutamyl endopeptidase I.

[1] J. J. Birktoft & K. Breddam (1994) *MIE* **244**, 114.

γ-GLUTAMYLGLUTAMATE CARBOXYPEPTIDASE

This enzyme activity, previously listed as EC 3.4.12.13 but now a deleted Enzyme Commission entry, is thought to be the consequence of γ-glutamyl hydrolase [EC 3.4.19.9] or γ-glutamyl transpeptidase [EC 2.3.2.2].

α-GLUTAMYLGLUTAMATE DIPEPTIDASE

This dipeptidase [EC 3.4.13.7], also known as Glu-Glu dipeptidase, catalyzes the hydrolysis of α-L-glutamyl-L-glutamate to produce two molecules of L-glutamate.

The enzyme will not act on γ-L-glutamyl-L-glutamate. Whether the enzyme acts on other α-glutamyl dipeptides remains to be established.[1]

[1] A. G. Pratt, E. J. Crawford & M. Friedkin (1968) *JBC* **243**, 6367.

γ-GLUTAMYLHISTAMINE SYNTHETASE

γ-L-glutamylhistamine

This enzyme [EC 6.3.2.18] catalyzes the reaction of ATP with L-glutamate and histamine to produce N^{α}-γ-L-glutamylhistamine and unspecified phosphoanhydride bond-scission products of ATP (*i.e.*, either ADP plus phosphate or AMP plus pyrophosphate).[1]

[1] C. Stein & D. Weinreich (1982) *J. Neurochem.* **38**, 204.

γ-GLUTAMYL HYDROLASE

This cysteine isopeptidase [EC 3.4.19.9; formerly EC 3.4.22.12], also known as γ-Glu-X carboxypeptidase, conjugase, folate conjugase, pteroyl-poly-γ-glutamate hydrolase, and lysosomal γ-glutamyl carboxypeptidase, catalyzes the hydrolysis of γ-glutamyl isopeptide bonds in a number of substrates.[1] Substrates include pteroyl poly-γ-glutamate (*i.e.*, folyl poly-γ-glutamate), 4-amino-10-methylpteroyl poly-γ-glutamate (*i.e.*, methotrexate poly-γ-glutamate), *p*-aminobenzoylpoly-γ-glutamate, antifolylyl poly-γ-glutamates, and poly-γ-glutamate. The enzyme will not act on α-linked polyglutamates. *See also Folylpolyglutamate Hydrolase*

[1] J. J. McGuire & J. K. Coward (1984) in *Folates and Pterins* (R. L. Blakley & S. J. Benkovic, eds.), Wiley, New York, p. 135.

Selected entries from *Methods in Enzymology* [vol, page(s)]:
General discussion: 2, 629; **66**, 655, 670
Assay: 2, 629; **66**, 672, 673 (microbiological, folic acid, **66**, 675)
Other molecular properties: chick pancreatic enzyme in releasing bound folate, use of, **18B**, 644, 645; mode of action, **66**, 674, 675; properties, **66**, 674; purification from human plasma, **66**, 673, 674

D-GLUTAMYLTRANSFERASE

This enzyme [EC 2.3.2.1], also referred to as D-glutamyl transpeptidase, catalyzes the reaction of either L- or D-glutamine with a D-glutamyl-peptide (*i.e.*, a peptide with a D-glutamyl residue at the N-terminus) to produce ammonia and a γ-L-(or D-) glutamyl-D-glutamyl-peptide.[1]

[1] W. J. Williams, J. Litwin & C. B. Thorne (1955) *JBC* **212**, 427.

γ-GLUTAMYL TRANSPEPTIDASE

This enzyme [EC 2.3.2.2], also known as γ-glutamyl-transferase, catalyzes the reaction of a γ-glutamyl substrate with an acceptor substrate to produce the γ-glutamylated acceptor derivative and the de-γ-glutamylated substrate.[1–4,6,7] γ-Glutamyl substrates include glutathione, γ-glutamylamino acid, glutamine, glutathione disulfide, leukotriene C[4], and γ-glutamyl-p-nitroanilide, of which the latter is frequently used to continuously monitor enzyme activity.

When the donor substrate is glutathione and the acceptor substrate is water, the enzyme catalyzes a glutathionase activity that produces L-glutamate and L-cysteinylglycine. When L-glutamine is the initial substrate and water is the acceptor, the enzyme exhibits a glutaminase activity that produces L-glutamate and ammonia.

When the acceptor substrate is an amino acid (particularly L-cystine, L-methionine, L-glutamine, or L-alanine), the product is the corresponding isopeptide derivative (i.e., γ-glutamyl-L-cystine, γ-glutamyl-L-methionine, γ-glutamyl-L-glutamine, and γ-glutamyl-L-alanine, respectively).[2,3] Dipeptides, certain tripeptides, and even glutathione can also serve as acceptor substrates (of the dipeptides studied, L-methionylglycine was the best acceptor substrate identified). Elevated levels of amino acids can result in substrate inhibition of all the activities (for example, the K_i value for glycine is 3.2 mM for the rat kidney enzyme).[2] The mammalian enzyme catalyzes a modified ping pong mechanism with branched reaction pathways, akin to that of glucose-6-phosphatase, the reaction proceeding through a purported γ-glutamyl-enzyme intermediate.[2] The plant and yeast enzymes have a different range of specificities than that of the mammalian enzyme mentioned above.

The degree of partitioning between the glutathionase activity and the γ-glutamyl transfer activity can be calculated from the complete enzyme rate equation containing terms for all acceptors and γ-glutamyl donors as well as terms for substrate inhibition.[2] This value agrees quite well with that obtained under experimental conditions in which all acceptor concentrations approximate physiological values.[6] Slightly more than 60% of the γ-glutamyl donor is utilized in transpeptidation reactions; the remainder is hydrolyzed.[6]

In addition to the enzyme's significant role in glutathione catabolism, generation of γ-glutamyl amino acids, as well as the production of mercapturates, γ-glutamyl transpeptidase plays a crucial role in the processing of leukotrienes.[2,3] The initial rate kinetics of the conversion of leukotriene C[4] (an S-substituted glutathione) to leukotriene D[4] (an S-substituted L-cysteinylglycine) and L-glutamate was investigated and the Michaelis constant for the leukotriene substrate was remarkably close to that of glutathione.[4] This value was later verified by other laboratories.[5] In addition, the action of γ-glutamyl transpeptidase was accelerated in the presence of amino acids and the protein also catalyzed the glutathione-dependent γ-glutamylation of leukotriene D[4] to regenerate leukotriene C[4].[4] Leukotriene E[4] (an S-substituted L-cysteine) was also observed to serve as an acceptor substrate,[4] generating a new leukotriene, originally called γ-glutamyl-leukotriene E[4] (i.e., an S-substituted L-γ-glutamyl-L-cysteine) and later named by others leukotriene F[4]. **See also** Transglutaminase; Glutamine Synthetase

[1] N. Taniguchi & Y. Ikeda (1998) AE **72**, 239.
[2] R. D. Allison (1985) MIE **113**, 419.
[3] S. S. Tate & A. Meister (1985) MIE **113**, 400.
[4] M. E. Anderson, R. D. Allison & A. Meister (1982) PNAS **79**, 1088.
[5] L. Örning & S. Hammarström (1982) BBRC **106**, 1304.
[6] R. D. Allison & A. Meister (1981) JBC **256**, 2988.
[7] S. C. Hartman (1971) The Enzymes, 3rd ed., **4**, 79.

Selected entries from **Methods in Enzymology** [vol, page(s)]:
General discussion: 2, 263, 267; **17A**, 877, 883, 894; 19, 782; 77, 237; 113, 400, 419
Assay: 2, 263, 267; 19, 782; 77, 238, 240, 241; 113, 402, 404; **252**, 67; dry reagent chemistries, with, **137**, 415; Fast Analyzer, by, **31**, 816; free sulfhydryl, 77, 242; hydrolytic, 77, 242; 113, 404; transpeptidase activity, 77, 241; 113, 403
Other molecular properties: acivicin inhibition, **252**, 10, 18, 21; activators and inhibitors, 19, 787; 113, 416, 435; affinity chromatography, 77, 245; antibody preparation, 113, 407; association with parotid plasmalemma and secretion granule membranes, 98, 85; autotranspeptidation, 77, 237; biosynthesis, 113, 412; bromelain digestion, 77, 243; brush borders, in, **31**, 122; clinical significance, **17A**, 889; comparison of bean and kidney enzymes, **17A**, 900; distribution, 19, 788; Ellman's reagent, 77, 242; endogenous metabolism, in, 77, 240; function, 19, 782; glutamine synthetase and γ-glutamyl transferase activity, 2, 264, 267, 272, 341; **17A**, 901, 920; 87, 522, 523; glutathione depletion, 77, 57; inhibition in vivo, 113, 581; inhibitors, 77, 249; 113, 416, 435; **252**, 10, 18, 21 (irreversible, 77, 249, 250; 113, 418; reversible, 77, 250, 251; 113, 416, 435; transition state analogues, 77, 250); iodoacetamide and, **25**, 437; kidney cells, in, 77, 144; kinetics, 77, 238, 87, 321; 113, 414, 419, 421; leukotrienes, and, 113, 421, 428; mercapturic acid formation, 77, 239, microvillus membrane, in, **31**, 130; molecular weight, 77, 246; 113, 409, 410; papain digestion, 77, 243, 244; 113, 405; partitioning, 113, 427; pH dependence, 19, 787; physical and chemical properties, 19, 787; 113, 408, 419; plant, 2, 263; preparation, 19, 784; processing, 113, 412; properties, 2, 266, 269, 272; 19, 785; 77, 246; purification from (Agaricus bisporus, **17A**, 879; green beans, **17A**, 896; ovine brain, 2, 270; pea, 2, 264; pigeon liver, 2, 271; porcine kidney,

17A, 885; 86, 43; *Proteus vulgaris*, 2, 269, 271; rat kidney, 77, 243; 113, 405); rat kidney (antibody preparation, 113, 407; biosynthesis and processing, 113, 412; inhibition and regulation, 113, 416, 435; mechanism, 113, 415, 419, 421, 436; properties, 113, 408, 414; purification, 77, 243; 113, 405, 406; specificity and partitioning, 113, 427); reactions catalyzed, 113, 400, 419, 422; 252, 66; role, 77, 239; sialic acid content, 77, 246, 247; size, 77, 246; solubilization, 1, 42, 50; 77, 243; 113, 405, 406; specificity, 19, 785; 113, 415, 425, 426, 427, 429; 252, 66; stability, 19, 787; structural features, 113, 408; subunits, 77, 247; 113, 409; tissue localization, 77, 238; transhydrogenase, as, 77, 238; transition state and multisubstrate analogues, 249, 304; transport function, 77, 239

GLUTAMYL-tRNA^Gln AMIDOTRANSFERASE

This transferase, also referred to as glutaminyl-tRNA synthetase (glutamine-hydrolyzing), catalyzes the ATP-dependent reaction of L-glutamyl-tRNA^Gln with L-glutamine to produce L-glutaminyl-tRNA^Gln, ADP, orthophosphate, and L-glutamate.[1–4] Note that glutaminyl-tRNA synthetase [EC 6.1.1.18] is absent in Gram-positive eubacteria, archaebacteria, and organelles. The amidotransferase from *Deinococcus* also converts L-aspartyl-tRNA^Asn to L-asparaginyl-tRNA^Asn.[1] ***See also*** *Glutaminyl-tRNA Synthetase*

[1] A. W. Curnow, D. L. Tumbula, J. T. Pelaschier, B. Min & D. Söll (1998) *PNAS* 95, 12838.
[2] K. C. Rogers & D. Söll (1995) *J. Mol. Evol.* 40, 476.
[3] H. Zalkin (1993) *AE* 66, 203.
[4] J. M. Buchanan (1973) *AE* 39, 91.

Selected entries from ***Methods in Enzymology*** [vol, page(s)]:
General discussion: 113, 303
Assay: *Bacillus subtilis*, 113, 303

GLUTAMYL-tRNA SYNTHETASE

This enzyme [EC 6.1.1.17], also known as glutamate:tRNA ligase, catalyzes the reaction of L-glutamate with tRNA^Glu and ATP to produce L-glutamyl-tRNA^Glu, AMP, and pyrophosphate (or, diphosphate).[1,2] In a number of Gram-positive bacteria and in some archaebacteria and organelles, this synthetase also catalyzes the ATP-dependent reaction of L-glutamate and tRNA^Gln, which subsequently undergoes an amido transfer reaction. ***See also*** *Aminoacyl-tRNA Synthetases*

[1] W. Freist, D. H. Gauss, D. Söll & J. Lapointe (1997) *Biol. Chem.* 378, 1313.
[2] P. Schimmel (1987) *Ann. Rev. Biochem.* 56, 125.

Selected entries from ***Methods in Enzymology*** [vol, page(s)]:
General discussion: 113, 42, 50
Assay: *Bacillus subtilis*, 113, 51; *Escherichia coli*, 113, 44
Other molecular properties: *Bacillus subtilis* (properties, 113, 53; purification, 113, 51); *Escherichia coli* (properties, 113, 47; purification, 113, 43, 47; regulatory factor, and copurification, 113, 49); subcellular distribution, 59, 233, 234

GLUTARATE SEMIALDEHYDE DEHYDROGENASE

This oxidoreductase [EC 1.2.1.20] catalyzes the reaction of glutarate semialdehyde ($^-OOC(CH_2)_3CHO$) with NAD^+ and water to produce glutarate and NADH.[1,2] Other substrates include adipate semialdehyde and α-ketoglutarate semialdehyde.

[1] Y. F. Chang & E. Adams (1977) *JBC* 252, 7979 and 7987.
[2] A. Ichihara & E. A. Ichihara (1961) *J. Biochem.* 49, 154.

Selected entries from ***Methods in Enzymology*** [vol, page(s)]:
General discussion: 17B, 166
Assay: 17B, 167 (acrylamide gels, on, 17B, 169)

GLUTARYL-CoA DEHYDROGENASE

This FAD-dependent oxidoreductase [EC 1.3.99.7] catalyzes the reaction of glutaryl-CoA with an acceptor substrate to produce crotonoyl-CoA, carbon dioxide, and the reduced acceptor.[1–4] The electron-transfer flavoprotein can serve as the acceptor substrate. The reaction mechanism proceeds via the formation of an enzyme-bound glutaconyl-CoA intermediate.

[1] T. M. Dwyer, K. S. Rao, S. I. Goodman & F. E. Frerman (2000) *Biochemistry* 39, 11488.
[2] R. A. Schaller, A. A. Mohsen, J. Vockley & C. Thorpe (1997) *Biochemistry* 36, 7761.
[3] C. M. Byron, M. T. Stankovich & M. Husain (1990) *Biochemistry* 29, 3691.
[4] M. Husain & D. J. Steenkamp (1985) *J. Bacteriol.* 163, 709.

GLUTARYL-CoA SYNTHETASE

This enzyme [EC 6.2.1.6], also referred to as glutarate:CoA ligase, catalyzes the reaction of ATP with glutarate and coenzyme A to produce ADP, orthophosphate, and glutaryl-CoA ($^-OOC(CH_2)_3COSCoA$).[1] Other nucleotide substrates include GTP and ITP.

[1] G. K. K. Menon, D. L. Friedman & J. R. Stern (1960) *BBA* 44, 375.

GLUTATHIONE *S*-ALKYLTRANSFERASE

This enzyme activity, formerly classified as EC 2.5.1.12, is now listed under glutathione transferase [EC 2.5.1.18].

GLUTATHIONE *S*-ARALKYLTRANSFERASE

This enzyme activity, formerly classified as EC 2.5.1.14, is now listed under glutathione transferase [EC 2.5.1.18].

GLUTATHIONE *S*-ARYLTRANSFERASE

This enzyme activity, formerly classified as EC 2.5.1.13, is now listed under glutathione transferase [EC 2.5.1.18].

GLUTATHIONE:CoA-GLUTATHIONE TRANSHYDROGENASE

This oxidoreductase [EC 1.8.4.3] catalyzes the reaction of coenzyme A with glutathione disulfide (G-SS-G) to produce CoA-glutathione (*i.e.*, the mixed disulfide between coenzyme A and glutathione) and glutathione (*i.e.*, γ-L-glutamyl-L-cysteinylglycine).[1,2]

[1]S. H. Chang & D. R. Wilken (1966) *JBC* **241**, 4251.
[2]R. E. Dyar & D. R. Wilken (1972) *ABB* **153**, 619.

GLUTATHIONE:CYSTINE TRANSHYDROGENASE

This transhydrogenase [EC 1.8.4.4] catalyzes the reversible reaction of two molecules of glutathione (*i.e.*, L-γ-glutamyl-L-cysteinylglycine) with cystine to produce glutathione disulfide and two molecules of cysteine. Other substrates include both D- and L-cystine, L-cystinyldiglycine, β-hydroxyethyl disulfide, diacetyl-L-cystine, L-cystine diamide, homocystine, glutathione disulfide, and dimethyl disulfide.[1,2] *See also* *Glutathione:Homocystine Transhydrogenase; Protein-Disulfide Reductase (Glutathione); Glutathione:CoA-Glutathione Transhydrogenase; Enzyme-Thiol Transhydrogenase (Glutathione Disulfide)*

[1]S. Nagai & S. Black (1968) *JBC* **243**, 1942.
[2]B. States & S. Segal (1969) *BJ* **113**, 443.

Selected entries from *Methods in Enzymology* [vol, page(s)]:
General discussion: 17B, 510; 77, 281, 423; 113, 520
Assay: 17B, 511; 77, 281, 282; 113, 521
Other molecular properties: disulfide reduction, 77, 285; human placenta (assay, 113, 521; properties, 113, 523; purification, 113, 522); kinetic parameters, 77, 284; mechanism, 77, 284; pH optima, 77, 284; properties, 17B, 514; 77, 284, 285; purification, 17B, 511; 77, 282; size, 77, 284; stability, 77, 284

GLUTATHIONE DEHYDROGENASE (ASCORBATE)

This oxidoreductase [EC 1.8.5.1], also known as glutathione:dehydroascorbate oxidoreductase and dehydroascorbate reductase, catalyzes the reaction of two molecules of glutathione (*i.e.*, L-γ-glutamyl-L-cysteinylglycine) with dehydroascorbate to produce glutathione disulfide and ascorbate. Three mammalian glutathione dependent dehydroascorbate reductases have been identified: glutaredoxin, protein disulfide isomerase, and a 32 kDa enzyme in rat liver and human erythrocytes.

In the proposed reaction mechanism, a thiohemiketal intermediate is initially formed between the enzyme and dehydroascorbate, followed either by displacement of ascorbate by glutathione (to form the mixed disulfide between the protein and glutathione) or by the action of a cysteinyl residue to displace the ascorbate (to form an intramolecular disulfide bond). The resulting internal disulfide would then be reduced by glutathione to regenerate the protein's dithiol and produce glutathione disulfide.[1–6] *See also* *Glutaredoxin; Monodehydroascorbate Reductase*

[1]M. P. Washburn & W. W. Wells (1999) *Biochemistry* **38**, 268.
[2]W. W. Wells & D. P. Xu (1994) *J. Bioenerg. Biomembr.* **26**, 369.
[3]S. Dipierro & G. Borraccino (1991) *Phytochemistry* **30**, 427.
[4]S. Shigeoka, R. Yasumoto, T. Onishi, Y. Nakano & S. Kitaoka (1987) *J. Gen. Microbiol.* **133**, 227.
[5]R. Bigley, M. Riddle, D. Layman & L. Stankova (1981) *BBA* **659**, 15.
[6]C. H. Foyer & B. Halliwell (1977) *Phytochemistry* **16**, 1347.

Selected entries from *Methods in Enzymology* [vol, page(s)]:
General discussion: 2, 847; 122, 10; 252, 30; 279, 30
Assay: 2, 847; 279, 31; incubation, 252, 32; principle, 252, 32; reagent preparation, 252, 32; spinach leaves, 122, 10
Other molecular properties: distribution, 2, 847; 252, 31; glutathione dependence, 252, 31; kinetic parameters, 252, 36; properties, 2, 850; 122, 12; 279, 33; purification, 2, 849; 122, 11; rat liver, 279, 30; reaction mechanism, 252, 37; site-directed mutagenesis, 252, 38; species distribution, 2, 847; 252, 31; spinach leaves (properties, 122, 12; purification, 122, 11)

GLUTATHIONE γ-GLUTAMYLCYSTEINYLTRANSFERASE

This transferase [EC 2.3.2.15], also known as phytochelatin synthase, catalyzes the reaction of glutathione with [γ-GluCys]$_n$-Gly (*i.e.*, a phytochelatin) to produce glycine and [γ-GluCys]$_{n+1}$-Gly.[1,2] Phytochelatins, which consist of repeating γ-glutamylcysteinyl units, ending in a C-terminal glycine, serine, or β-alanine residue, play a major role in heavy metal tolerance in plants and fungi through chelation.

[1]E. Grill, S. Löffler, E.-L. Winnacker & M. H. Zenk (1989) *PNAS* **86**, 6838.
[2]O. K. Vatamaniuk, S. Mari, Y. P. Lu & P. A. Rea (1999) *PNAS* **96**, 7110.

GLUTATHIONE:HOMOCYSTINE TRANSHYDROGENASE

This transhydrogenase [EC 1.8.4.1] catalyzes the reversible reaction of two molecules of glutathione with homocystine to produce glutathione disulfide and two molecules of homocysteine.[1] The reactions catalyzed by this and other enzymes resemble those catalyzed by glutathione transferase [EC 2.5.1.18]. *See also* *Glutathione:Cystine Transhydrogenase; Protein-Disulfide Reductase (Glutathione); Glutathione:CoA-Glutathione Transhydrogenase; Enzyme-Thiol Transhydrogenase (Glutathione Disulfide); Glutaredoxin*

[1]E. Racker (1955) *JBC* **217**, 867.

Selected entries from *Methods in Enzymology* [vol, page(s)]:
General discussion: 6, 449

Assay: 6, 449
Other molecular properties: properties, **6**, 450; purification, beef liver, **6**, 450

GLUTATHIONE OXIDASE

This FAD-dependent oxidoreductase [EC 1.8.3.3] catalyzes the reaction of two molecules of glutathione with dioxygen to produce glutathione disulfide and hydrogen peroxide.[1–5] Alternative substrates, albeit weaker, include L-cysteine, *N*-acetyl-L-cysteine, L-cysteine methyl ester, cysteamine, thiophenol, 2-mercaptoethanol, and dithiothreitol.

[1]M. C. Ostrowski & W. S. Kistler (1980) *Biochemistry* **19**, 2639.
[2]H. Kusakabe, Y. Midorikawa, A. Kuninaka & H. Yoshino (1983) *Agric. Biol. Chem.* **47**, 1385.
[3]H. Kusakaba, A. Kuninaka & H. Yoshino (1982) *Agric. Biol. Chem.* **46**, 2057.
[4]L. H. Lash & D. P. Jones (1982) *BJ* **203**, 371.
[5]K. Ormstad, P. Moldéus & S. Orrenius (1979) *BBRC* **89**, 497.

Selected entries from *Methods in Enzymology* [vol, page(s)]:
General discussion: 77, 238

GLUTATHIONE PEROXIDASE

This selenium-dependent oxidoreductase [EC 1.11.1.9] catalyzes the reaction of two molecules of glutathione with hydrogen peroxide to produce glutathione disulfide (G-SS-G) and two water molecules.[1–6] Hydrogen peroxide can be replaced by steroid and lipid hydroperoxides, albeit not as effectively. This enzyme is distinct from phospholipid-hydroperoxide glutathione peroxidase [EC 1.11.1.12]).

[1]A. B. Fisher, C. Dodia, Y. Manevich, J.-W. Chen & S. I. Feinstein (1999) *JBC* **274**, 21326.
[2]R. C. Fahey & A. R. Sundquist (1991) *AE* **64**, 1.
[3]L. Flohé (1989) in *Glutathione, Chemical, Biochemical, and Medical Aspects,* Part A (D. Dolphin, R. Poulson & O. Avramovic, eds.) p. 643, Wiley, New York.
[4]K. T. Douglas (1987) *AE* **59**, 103.
[5]L. Flohé (1982) in *Free Radicals in Biology* (W. A. Pryor, ed.) **5**, 223, Academic Press, New York.
[6]T. C. Stadtman (1979) *AE* **48**, 1.

Selected entries from *Methods in Enzymology* [vol, page(s)]:
General discussion: 52, 506; **77**, 16, 20, 230, 325; **105**, 114; **107**, 593; **113**, 490; **252**, 38
Assay: 52, 506; **77**, 325; **107**, 594; continuous, **105**, 117; fixed-time, **105**, 115, 117; polarographic, **77**, 326; thiol appearance, **77**, 326
Other molecular properties: activity in ischemia-reperfusion, **233**, 604; affinity chromatography, **77**, 329; aggregates and charge forms, **52**, 510; biological roles, **252**, 51; crystal structure, **77**, 332; definition of units, **77**, 327, 328; dendrogram of superfamily, **252**, 42; erythrocytes, from, **53**, 372; forms, **113**, 491; hydroperoxide (decomposition, **252**, 51; measurement, in, **52**, 509, 510); inhibitors, **77**, 331; isoelectric point, **77**, 331; isolation, **52**, 506, 507; kinetic constants, **77**, 331, 332; kinetics, **52**, 513; **252**, 44; localization, **113**, 492; mammalian (catalyzed reactions, **113**, 490; functional role, **113**, 494; major types, **113**, 491; natural occurrence, **113**, 492; subcellular localization, **113**, 492); marker for chromosome detection in somatic cell hybrids, as, **151**, 184; mimicry by

ebselen, **234**, 476; molecular evolution, **252**, 40; molecular weight, **52**, 512; **77**, 331; occurrence, **113**, 492; oxidized forms (differentiation, **107**, 597; preparation, **107**, 597); pH optima, **77**, 331; protection of biomolecules, **233**, 523; properties, **77**, 330; purification, **77**, 328 (ovine blood, **107**, 593; rat erythrocytes, **107**, 611; rat liver, **107**, 610); rate-limiting step, **252**, 45; reaction mechanism, **252**, 46; reactions catalyzed, **113**, 490; reduction of H_2O_2, **233**, 602; role in generation of thiyl radical metabolites of thiols, **186**, 323; selenium, **77**, 325; **252**, 39, 46, 209; selenocysteine identification, **107**, 589; **243**, 79; selenocysteine TGA codon, **252**, 39, 43; selenopeptide preparation, **107**, 617; self-rotation function, **115**, 69; specificity, **52**, 513; **77**, 330; stability, **77**, 330; subcellular distribution, **52**, 513; substrates, **77**, 18; three-dimensional structure, **252**, 48; transition state and multisubstrate analogues, **249**, 304; types, **252**, 40

GLUTATHIONE REDUCTASE (NADPH)

This FAD-dependent oxidoreductase [EC 1.6.4.2] catalyzes the reaction of NADPH with glutathione disulfide to produce NADP$^+$ and two molecules of glutathione.[1–6] In some sources (*e.g.*, the cyanobacterium *Anabaena*[7]), the reductase can also utilize NADH as the reductant.

The human erythrocyte and *Escherichia coli* enzyme are dimers with a redox-active disulfide and a FAD prosthetic group in each subunit. The enzyme has a ping pong kinetic mechanism. In the reaction mechanism, NADPH initially binds resulting in the FAD-mediated reduction of the protein's internal disulfide (a transient linkage between one of the cysteinyl residues and the C4a position of the flavin has been proposed). Glutathione disulfide subsequently binds resulting in the formation of a transient mixed disulfide intermediate between a cysteinyl residue and glutathione (and the release of the first molecule of glutathione). Nucleophilic attack by the other cysteinyl residue regenerates the internal disulfide and releases the second molecule of glutathione.

[1]R. C. Fahey & A. R. Sundquist (1991) *AE* **64**, 1.
[2]P. A. Karplus & G. E. Schulz (1989) *J. Mol. Biol.* **210**, 163.
[3]R. H. Schirmer, R. L. Krauth-Siegel & G. E. Schulz (1989) *Coenzymes Cofactors* (Pt. A) **3**, 553.
[4]K. K. Wong, M. A. Vanoni & J. S. Blanchard (1988) *Biochemistry* **27**, 7091.
[5]K. T. Douglas (1987) *AE* **59**, 103.
[6]C. H. Williams, Jr. (1976) *The Enzymes*, 3rd ed., **13**, 89.
[7]U. H. Danielson, F. Jiang, L. O. Hansson & B. Mannervik (1999) *Biochemistry* **38**, 9254.

Selected entries from *Methods in Enzymology* [vol, page(s)]:
General discussion: 2, 719, 722; **17B**, 500, 503; **34**, 4, 246; **77**, 236, 282, 327, 373; **113**, 484
Assay: 113, 485; bovine liver, **2**, 722; **113**, 485; *Escherichia coli*, **17B**, 504; plant, **2**, 719; rat liver, **113**, 485; yeast, **2**, 722; **17B**, 500
Other molecular properties: active site structure, **251**, 177; activity in ischemia-reperfusion, **233**, 602, 604; adriamycin inhibition, **251**, 183; assay of thiol-disulfide transhydrogenase, in, **17B**, 511; BCNU inhibition (affecting factors, **251**, 182; apoenzyme, **251**, 181; assays, **251**, 178; preparation of glutathione reductase-depleted erythrocytes, **251**, 180; protection by glutathione disulfide, **251**, 182); binding, to agarose-bound

nucleotides, **66**, 196; borohydride reduction, **18B**, 475; bovine liver (properties, **2**, 725; **113**, 490; purification, **2**, 724; **113**, 486); calf liver (purification, **113**, 486); catalytic forms, **251**, 177; commercial availability, **251**, 188; coupled enzyme assays, **251**, 86; cysteine inhibition, **251**, 264; defects in humans, **252**, 91; deficiency and oxidative stress, **251**, 183; 5-dehydroshikimate reductase assay and, **2**, 301; depleted erythrocytes, preparation, **251**, 180; *Escherichia coli* (mutants, **252**, 87; properties, **17B**, 508; purification, **17B**, 504; site-directed mutagenesis studies, **249**, 107); flavoprotein classification, **53**, 397; glutathione depletion, **77**, 56, 57; glutathione disulfide reaction, **77**, 381, 382; glutathione homocystine transhydrogenase and, **6**, 449; glutathione transport assay in mitochondria, **252**, 20; Hill reaction and, **2**, 721; **4**, 352; hydrogenase and, **2**, 732; intersubunit ligand binding sites, **249**, 559; lack of sulfite complexing with yeast enzyme, **18B**, 473; lactate dehydrogenase preparation, presence in, **9**, 446; marker for chromosome detection in somatic cell hybrids, as, **151**, 184; methionine reduction and, **5**, 995; NADP$^+$ (assay, **3**, 894, 897; regenerating system, **13**, 593); NADPH cofactor, **251**, 13, 86, 177; oxidation effect on activity, **251**, 1; plant, **2**, 719; preparation, **3**, 895; preparation of deuterated NADPH, in, **54**, 232; properties, **2**, 721, 724, 725; **17B**, 501, 508; **113**, 490; purification from (bovine liver, **2**, 724; **113**, 486; calf liver, **113**, 486; *Escherichia coli*, **17B**, 504; plant, **2**, 720; rat liver, **113**, 487; yeast, **2**, 723; **17B**, 501, 502); pyridine nucleotide assay, **55**, 264, 265, 267; rat liver (properties, **113**, 490; purification, **113**, 487); reaction catalyzed, **251**, 5, 13; seed protein reduction, **252**, 230, 238; stereospecificity, **87**, 118; transhydrogenase assay, **56**, 113; yeast (assay for glutathione and glutathione disulfide in biological samples, **113**, 550; properties, **2**, 724; **17B**, 501; purification, **2**, 723; **17B**, 501, 502)

GLUTATHIONE SYNTHETASE

This enzyme [EC 6.3.2.3], also known as glutathione synthase and γ-glutamylcysteine:glycine ligase, catalyzes the reaction of ATP with γ-L-glutamyl-L-cysteine with glycine to produce ADP, orthophosphate, and glutathione.[1–4] The kinetic mechanism is reportedly random Ter Ter[3] via the formation of an enzyme-bound acyl-phosphate intermediate. The enzyme also catalyzes the ATP-dependent formation of ophthalmic acid (*i.e.*, γ-L-glutamyl-L-α-aminobutyrylglycine) from γ-L-glutamyl-L-α-aminobutyrate and glycine.

[1] O. W. Griffith (1999) *Free Radic. Biol. Med.* **27**, 922.
[2] R. C. Fahey & A. R. Sundquist (1991) *AE* **64**, 1.
[3] A. Wendel & H. Heinle (1975) *Hoppe-Seyler's Z. Physiol. Chem.* **356**, 33.
[4] A. Meister (1974) *The Enzymes*, 3rd ed., **10**, 671.

Selected entries from *Methods in Enzymology* [vol, page(s)]:
General discussion: **2**, 342; **17B**, 488; **113**, 393
Assay: **2**, 343; rat kidney, **113**, 393; yeast, **17B**, 488
Other molecular properties: bakers' yeast, **17B**, 490; brewers' yeast, **2**, 345; defects in humans, **252**, 91; *Escherichia coli* mutants (B strains, **252**, 87; K strains [cell growth, **252**, 83; *gshA*, **252**, 85; *gshB*, **252**, 84]); gene transfer, **251**, 7; mechanism, **17B**, 493; **63**, 51, 157; properties, **17B**, 492; purification from (bakers' yeast, **17B**, 490; pigeon liver, **2**, 343, 344; rat kidney, **113**, 395; yeast, **2**, 345; **17B**, 490); rat kidney (properties, **113**, 398; purification, **113**, 395); substrates, **113**, 398; **251**, 4, 7; yeast mutants, **252**, 90

GLUTATHIONE THIOLESTERASE

This thiolesterase [EC 3.1.2.7] catalyzes the hydrolysis of an *S*-acylglutathione to produce glutathione

(*i.e.*, γ-L-glutamyl-L-cysteinylglycine) and a carboxylic acid anion.[1,2]

[1] W. W. Kielley & L. B. Bradley (1954) *JBC* **206**, 327.
[2] L. Uotila (1979) *BBA* **580**, 277.

Selected entries from *Methods in Enzymology* [vol, page(s)]:
Other molecular properties: acyl dehydrogenases, presence in preparations of, **5**, 555; crotonyl-CoA assay, in, **1**, 561; thio esters of glutathione, use to eliminate, **1**, 561

GLUTATHIONE *S*-TRANSFERASES

This set of enzymes [EC 2.5.1.18], also known as glutathione *S*-alkyltransferases, glutathione *S*-aryltransferases, *S*-(hydroxyalkyl)glutathione lyases, and glutathione *S*-aralkyltransferases, catalyzes the reaction of an R–X with glutathione (*i.e.*, γ-L-glutamyl-L-cysteinylglycine) to produce HX and an *S*-substituted (with R) glutathione (thus, R–S–glutathione).[1–7] R may be an aliphatic, aromatic, or heterocyclic group whereas X may be a sulfate, nitrate, or halide. In addition, some members of this set of enzymes, which typically have a broad specificity, will also catalyze the *S*-addition of glutathione to aliphatic epoxides and arene oxides. Others in this classification will catalyze the reduction of polyol nitrate by glutathione to polyol and nitrite as well as catalyze certain isomerization reactions and disulfide interchanges.

Leukotriene C$_4$ synthase, which catalyzes the reaction of leukotriene A$_4$ with glutathione to produce leukotriene C$_4$, is distinct from other members of the glutathione transferase family due to its lack of amino acid homology, limited substrate specificity, steady-state kinetics, inability to conjugate glutathione to xenobiotics, differential susceptibility to inhibitors, and failure to be immunorecognized by specific microsomal glutathione transferase antibodies.

See also *Leukotriene C$_4$ Synthase*

[1] B. Mannervik (1996) *Biochem. Soc. Trans.* **24**, 878.
[2] R. N. Armstrong (1994) *AE* **69**, 1.
[3] M. C. Wilce & M. W. Parker (1994) *BBA* **1205**, 1.
[4] R. C. Fahey & A. R. Sundquist (1991) *AE* **64**, 1.
[5] K. T. Douglas (1987) *AE* **59**, 103.
[6] B. Mannervik (1985) *AE* **57**, 357.
[7] W. B. Jakoby & T. A. Fjellstedt (1972) *The Enzymes*, 3rd ed., **7**, 199.

Selected entries from *Methods in Enzymology* [vol, page(s)]:
General discussion: **77**, 218, 231, 235, 398; **113**, 495, 499, 504, 507; **252**, 53; **327**, 429
Assay: **77**, 219, 231, 232, 398; **113**, 500, 504, 508; **252**, 65; Δ^5-androstene-3,17-dione, with, **77**, 400, 404; benzyl thiocyanate, with, **77**, 402, 404; 1-chloro-2,4-dinitrobenzene, with, **77**, 219, 400, 404; cyanide formation, **77**, 402, 403; 1,2-dichloro-4-nitrobenzene, with, **77**, 400, 404; 1,2-epoxy-3-(*p*-nitrophenoxy)propane, with, **77**, 400, 404; erythrityl tetranitrate, with, **77**, 402, 404; ethacrynic acid, with, **77**, 400, 404; ethyl thiocyanate, with, **77**, 402, 404; glutathione organic nitrate reductase, **77**, 401; human liver, **113**, 500; 1-menaphthyl sulfate, with,

77, 400, 404; methyl iodide, with, **77**, 400, 401, 404; nitrate formation, **77**, 401, 402; nitrate reductase, with, **77**, 401; *p*-nitrobenzyl chloride, with, **77**, 400, 404; nitroglycerin, with, **77**, 402, 404; *p*-nitrophenethyl bromide, with, **77**, 400; *p*-nitrophenyl acetate, with, **77**, 403; *p*-nitrophenyl trimethylacetate, with, **77**, 403; 2-nitropropane, with, **77**, 404; organic nitrate esters, with, **77**, 401, 402; *trans*-4-phenyl-3-buten-2-one, with, **77**, 400, 404; rat liver, **113**, 504; rat testis, **113**, 508; spectrophotometric, **77**, 399, 400; 2,3,4,6-tetrachloronitrobenzene, with, **77**, 404; thioether formation, **77**, 399; thiolysis, **77**, 403; titrimetric, **77**, 400, 401

Other molecular properties: Abl SH3, biotinylation, **256**, 141; ADP-ribosylation factor, preparation, **237**, 53; affinity chromatography, **77**, 233, 235 (conjugates, **252**, 54); Cdc42p, expression and purification, **256**, 287; chlorodinitrobenzene substrate, **251**, 240; *c*-Jun kinase, **255**, 350, 352; classes, **266**, 236; C3 transferase (cleavage from glutathione beads by thrombin, **256**, 178; expression in *Escherichia coli*, **256**, 175; precipitation on glutathione beads, **256**, 177; purification from bacterial cell lysates, **256**, 176); cyanide formation, **77**, 230; cytosolic and microsomal, and leukotriene C$_4$ synthase, distinction, **187**, 311; *dbl*, antisera, **256**, 352; deamidation, **77**, 231; definition of units, **77**, 219, 232; differentiation, **77**, 403; disulfide interchange, **77**, 230; drug detoxification, **251**, 6, 13, 244; fusions, **326**, 254 (fusion protein with SH3 ligands, **255**, 370, 499; fusion vector pGEX-2T, construction, **256**, 4; purification, **326**, 265, 301; protease cleavage, **326**, 268); GCN4 recombinant fusion protein, production and isolation, **218**, 527; GDI fusion protein, **250**, 131; gene families, **252**, 53; genetic control, **52**, 232; G25K, recovery, **256**, 6; glutathione assay, **251**, 214, 240; glutathione peroxidase, **77**, 230; glutathione transferase ρ (human erythrocytes, purification, **77**, 228, 229, 230; properties, **77**, 234); glutathione transferase A (purification, **77**, 220, 221); glutathione transferase AA, **113**, 497 (purification, **77**, 220, 221); glutathione transferase B, **113**, 497 (purification, **77**, 220, 221); glutathione transferase C, **113**, 497 (purification, **77**, 220, 221); glutathione transferase E, **113**, 497 (purification, **77**, 221); glutathione transferase M, **113**, 497 (purification, **77**, 220); G protein α subunit expression vector, in, **237**, 160; human, **77**, 235, 404, 405 (liver, **113**, 499 [assay, **113**, 500; properties, **113**, 503; purification, **77**, 224; **113**, 501]; placenta, **77**, 231; purification, **77**, 232 [kidney P1-1 enzyme, **252**, 60; liver enzymes A1-1, A1-2, A2-2, and M1, **113**, 497 {equipment, **252**, 55; hydroxylapatite chromatography, **252**, 57; ion-exchange chromatography, **252**, 58; reagents, **252**, 55; tissue extraction, **252**, 56}; skeletal muscle M2-2 enzyme, **252**, 61; testis M3-3 enzyme, **252**, 60; theta class enzyme expressed in *Escherichia coli* {affinity chromatography, **252**, 65; equipment, **252**, 63; purity analysis, **252**, 65; reagents, **252**, 64}]); intestinal epithelium, in, **77**, 156; isoelectric focusing, **77**, 224, 226; isoelectric point, **77**, 234; isomerization, **77**, 230; Jun protein, and, *in vitro* phosphorylation, **254**, 558; kinetic constants, **77**, 235; ligands, **77**, 229, 230; mammalian (catalyzed reactions, **113**, 498; overview, **113**, 495); mechanism, **77**, 218; mitogen-activated protein kinase, and, **255**, 288; mitogen-activated protein kinase kinase, and, **255**, 289; mitogen-activated protein kinase kinase kinase, and, **255**, 291; molecular weight, **77**, 230, 234; monochlorobimane substrate efficiency, **251**, 211; nitrate ester reduction, **77**, 230; organic peroxides, with, **77**, 332; properties, **77**, 229, 234, 235; protein kinase C, and, purification, **252**, 157, 161, 166; purification, **77**, 219, 236; Rab-GDI, and, purification, **257**, 75; Rab GTPase, and (GST cleavage, **257**, 116; production, **257**, 115); Rabphilin-3A, and, purification, **257**, 296; Rac, and, recovery, **256**, 6; Raf, and, **255**, 324, 328; Ras, and, in yeast (affinity purification, **250**, 74; expression, **250**, 73; prenylated products, HPLC, **250**, 75; proteolytic digestion, **250**, 75); rat, **77**, 235, 404, 405 (liver, **113**, 504 [assay, **113**, 504; properties, **113**, 507; purification, **77**, 219; **113**, 505]; nomenclature, **113**, 497; testis, **113**, 507 [assay, **113**, 508; properties, **113**, 510; purification, **113**, 508]; Rho, and, recovery, **256**, 6; rhodopsin kinase fusion protein, **250**, 151; Rho-GDI, and (purification, **256**, 44; recombinant, purification from *Escherichia coli*, **256**, 101); sequence similarity searching, **266**, 232, 236, 478, 480; SH2 domain, and, in assay of SH2-binding reactions, **256**, 128; size, **77**, 234; specific

activities, **77**, 404, 405; specificity, **77**, 230, 234, 235; substrates, **77**, 229, 230; subunit structure, **252**, 53; thioether formation, **77**, 230; thiolysis, **77**, 230; X-ray crystallography, pig lung enzyme (active site structure, **251**, 252; crystallization, **251**, 246; phase determination of space group of crystals, **251**, 249; protein purification, **251**, 244; structure determination, **251**, 244, 249, 253)

GLUTATHIONYLSPERMIDINE AMIDASE

This enzyme [EC 3.5.1.78], also known as glutathionyl-spermidine amidohydrolase (spermidine-forming) and γ-L-glutamyl-L-cysteinylglycine:spermidine amidase), catalyzes the hydrolysis of glutathionylspermidine to produce glutathione and spermidine. The *Escherichia coli* enzyme is bifunctional and possesses glutathionylspermidine synthetase [EC 6.3.1.8] activity, resulting in a net hydrolysis of ATP.[1,2] *See also* Glutathionylspermidine Synthetase

[1] C. H. Lin, S. Chen, D. S. Kwon, J. K. Coward & C. T. Walsh (1997) *Chem. Biol.* **4**, 859.

[2] D. S. Kwon, C.-H. Lin, S. Chen, J. K. Coward, C. T. Walsh & J. M. Bollinger, Jr. (1997) *JBC* **272**, 2429.

GLUTATHIONYLSPERMIDINE SYNTHETASE

This magnesium-dependent ligase [EC 6.3.1.8], also known as glutathione:spermidine ligase (ADP-forming) and γ-L-glutamyl-L-cysteinylglycine:spermidine ligase (ADP forming), catalyzes the reaction of glutathione with spermidine and ATP to produce N^1-glutathionylspermidine, ADP, and orthophosphate. The enzyme catalyzes the penultimate step in the biosynthesis of trypanothione in trypanosomatids.[1–3]

The *Escherichia coli* enzyme is bifunctional and possesses glutathionylspermidine amidase [EC 3.5.1.78] activity, resulting in a net hydrolysis of ATP. Interestingly, these two activities can be specifically targeted by potent, selective slow-binding inhibitors that induce time-dependent activity loss.[1,2] These studies support earlier work on the presence of a glutathionyl acyl-enzyme intermediate in the amidase reaction mechanism and interdomain communication between the two activities.

[1] C. H. Lin, S. Chen, D. S. Kwon, J. K. Coward & C. T. Walsh (1997) *Chem. Biol.* **4**, 859.

[2] D. S. Kwon, C.-H. Lin, S. Chen, J. K. Coward, C. T. Walsh & J. M. Bollinger, Jr. (1997) *JBC* **272**, 2429.

[3] E. Tetaud, F. Manai, M. P. Barrett, K. Nadeau, C. T. Walsh & A. H. Fairlamb (1998) *JBC* **273**, 19383.

Selected entries from *Methods in Enzymology* [**vol**, page(s)]:
General discussion: 17B, 815
Assay: 17B, 815

GLYCERALDEHYDE-3-PHOSPHATE DEHYDROGENASES

There are no fewer than four distinct oxidoreductases catalyzing the oxidation of glyceraldehyde 3-phosphate.[1–17]

Glyceraldehyde-3-phosphate dehydrogenase (phosphorylating) [EC 1.2.1.12], also known as triose-phosphate dehydrogenase, catalyzes the reversible reaction of D-glyceraldehyde 3-phosphate with orthophosphate and NAD$^+$ to produce 3-phospho-D-glyceroyl phosphate (or, 1,3-bisphosphoglycerate) and NADH.[1–6] The muscle enzyme, which exhibits negative cooperativity, has an ordered Ter Bi kinetic mechanism at pH 8.6, the rate-limiting step being the release of NADH (when 2-deoxyglyceraldehyde 3-phosphate is the substrate, the rate-limiting step is phosphorylation).[5] The enzyme utilizes an active-site thiol group to form a thiohemiacetal intermediate (with an *S* configuration at C1) which, upon hydride transfer, is converted to a covalently bound thiolester. Orthophosphate or arsenate attacks the thioester to produce the final product. Based on his work with this enzyme, the metabolist Ephraim Racker was among the first to propose that enzyme catalysis can take advantage of the transient formation of covalently bound intermediates. Others, including Otto Warburg himself, initially surmised *incorrectly* that formation of covalently bound intermediates would unnecessarily increase the number of bond-making/breaking steps and would decrease catalytic efficiency. D-Glyceraldehyde and several other aldehydes will also act as weaker substrates. In the presence of arsenate, D-glyceraldehyde 3-phosphate is converted to 3-phosphoglycerate. A number of thiols can replace orthophosphate, thereby releasing the corresponding thiolester.

Glyceraldehyde-3-phosphate dehydrogenase (NADP$^+$) (nonphosphorylating) [EC 1.2.1.9], also known as triose-phosphate dehydrogenase, catalyzes the reaction of D-glyceraldehyde 3-phosphate with NADP$^+$ and water to produce 3-phospho-D-glycerate and NADPH.[7–10] The spinach leaf enzyme appears to be hysteretic with a random Bi Bi kinetic mechanism.[7]

Glyceraldehyde-3-phosphate dehydrogenase (NADP$^+$) (phosphorylating) [EC 1.2.1.13], also known as triose-phosphate dehydrogenase (NADP$^+$), catalyzes the reversible reaction of D-glyceraldehyde 3-phosphate with orthophosphate and NADP$^+$ to produce 3-phospho-D-glyceroyl phosphate and NADPH.

Glyceraldehyde-3-phosphate dehydrogenase (NAD(P)$^+$) (phosphorylating) [EC 1.2.1.59], also known as NAD(P)$^+$-dependent glyceraldehyde-3-phosphate dehydrogenase and triose-phosphate dehydrogenase (NAD(P)$^+$), catalyzes the reaction of D-glyceraldehyde-3-phosphate with orthophosphate and NAD(P)$^+$ to produce 3-phospho-D-glyceroyl phosphate and NAD(P)H.[17] The coenzymes NAD$^+$ and NADP$^+$ can be used with comparable efficiency. This distinguishes EC 1.2.1.59 from EC 1.2.1.12 and EC 1.2.1.13 which are NAD$^+$- and NADP$^+$-dependent, respectively.

[1] M. W. W. Adams & A. Kletzin (1996) *Adv. Protein Chem.* **48**, 101.
[2] R. Jaenicke, H. Schurig, N. Beaucamp & R. Ostendorp (1996) *Adv. Protein Chem.* **48**, 181.
[3] H. F. Fisher (1988) *AE* **61**, 1.
[4] J. I. Harris & M. Waters (1976) *The Enzymes*, 3rd ed., **13**, 1.
[5] B. A. Orsi & W. W. Cleland (1972) *Biochemistry* **11**, 102.
[6] S. F. Velick & C. Furfine (1963) *The Enzymes*, 2nd ed., **7**, 243.
[7] A. A. Iglesias & M. Losada (1988) *ABB* **260**, 830.
[8] A. A. Iglesias, A. Serrano, M. G. Guerrro & M. Losada (1987) *BBA* **925**, 1.
[9] G. J. Kelly & M. Gibbs (1973) *Plant Physiol.* **52**, 111.
[10] J.-L. Jacob & J. D'Auzac (1972) *Eur. J. Biochem.* **31**, 255.
[11] J. Udvardy, A. Balogh & G. L. Farkas (1982) *Arch. Microbiol.* **133**, 2.
[12] M. J. O'Brien, J. S. Easterby & R. Powls (1976) *BBA* **449**, 209.
[13] R. Cerff (1978) *Phytochemistry* **17**, 2061.
[14] J. G. Zeikus, G. Fuchs, W. Kenealy & R. K. Thauer (1977) *J. Bacteriol.* **132**, 604.
[15] I. Ziegler, A. Marewa & E. Schoepe (1976) *Phytochemistry* **15**, 1627.
[16] F. E. Grissom & J. S. Kahn (1975) *ABB* **171**, 444.
[17] F. Valverde, M. Losada & A. Serrano (1997) *J. Bacteriol.* **179**, 4513.

Selected entries from *Methods in Enzymology* [vol, page(s)]:
General discussion: 1, 401, 407, 411; 9, 210; 27, 437; 34, 34, 43, 123, 148, 165, 237, 246; 41, 264, 273; 46, 18, 20, 21, 75, 87; 68, 408; 89, 301, 305, 310, 316, 319, 326, 335; 90, 479, 490, 498
Assay: 1, 401, 407, 412; 9, 210, 212; *Bacillus stearothermophilus*, from, 41, 268, 269; honey bee, 41, 274; human tissue, 9, 210; 89, 302; pea leaves, 1, 412; pea seed, 89, 320; porcine liver, 89, 310; 90, 499; porcine muscle, 90, 499; rabbit muscle, 1, 401; 41, 264, 265; 89, 306; spinach leaf, 89, 316; *Thermus thermophilus*, 89, 335; yeast, 1, 407; 89, 326
Other molecular properties: active site, 4, 269; acyl derivative of, 1, 404; adduct with 2-bromoacetamido-4-nitrophenol, difference spectra in the presence of NAD$^+$, 11, 870; affinity labeling, 47, 424, 425, 427; 87, 491; aldehyde dehydrogenase and, 1, 517; aldolase and, 1, 315; 5, 316; apoenzyme, 27, 205; assay of (2-keto-3-deoxy-6-phosphogalactonate aldolase, 9, 525; tryptophan synthase, in, 17A, 406); ATP assay, use in, 13, 491; [γ-^{32}P]ATP synthesis, 65, 516, 518, 682; *Bacillus stearothermophilus*, from, 41, 268 (amino acids, 41, 272; assay, 41, 268, 269; hydroxylamine cleavage, 47, 140; inhibition, 41, 273; properties, 41, 271; purification, 41, 269); bifunctional imidoester and, 25, 648; binding (agarose-bound nucleotides, to, 66, 196; microcalorimetric studies, 61, 302); 1,3-bisphosphoglycerate assay and, 3, 219; bisphosphoglycerate mutase assay and, 1, 425; bovine muscle, simultaneous purification with aldolase, 3-phosphoglycerate kinase, and phosphoglycerate mutase, 90, 509; carboxymethylation of, 25, 430; cGMP assay, 38, 106, 107; chiral acetate, synthesis of, 87, 141; chiral phosphate, 87, 258; chiral thiophosphate, analysis of, 87, 311, 313, 314;

conformational change, **64**, 174, 190; cooperativity, **64**, 170, 174, 190, 191, 219; cryoenzymology, **63**, 338; cyanylation, **47**, 131; derived expression cassettes (constitutive synthesis of heterologous proteins, for, **185**, 341; plasmid construction, **185**, 344); dissociation, **48**, 180; eicosapeptide containing methionine and CM-cysteine from, **11**, 394; effector changes, X-ray scattering studies, **61**, 226, 227; erythrocyte ghost sidedness assay, in, **31**, 176, 178, 179; erythrose 4-phosphate and, **6**, 484; *Escherichia coli*, from, purification, **42**, 444; gluconeogenic catalytic activity, **37**, 281, 288; glyceraldehyde-3-phosphate dehydrogenase (NADP+) (synthesis in *Chlamydomonas* cultures, **23**, 73); glyceraldehyde-3-phosphate dehydrogenase (NADP+) (phosphorylating) **1**, 411 (assay [chicken muscle, **90**, 490; pea, **1**, 412]; chicken muscle [assay, **90**, 490; purification, **90**, 492]; pea, **1**, 411 [properties, **1**, 414; purification, **1**, 413]; plant tissue, **1**, 411); GTP assay, **55**, 208; half-site reactivity, **64**, 183; hexokinase and, **5**, 232, 234; honey bee, **41**, 273 (assay, **41**, 274; chromatography, **41**, 275; inhibition, **41**, 277; molecular weight, **41**, 277; properties, **41**, 276; purification, **41**, 275, 276); human heart muscle, from (crystallization, **9**, 211; properties, **9**, 211; purification, **9**, 210; stability, **9**, 211); human tissue (assay, **89**, 302; properties, **89**, 304; purification, **89**, 303); hydrogen peroxide oxidation, **25**, 398; 4-hydroxynonenal adducts, **233**, 376, 380; hysteresis, **64**, 219; immunodiffusion and, **6**, 851; inhibition, **63**, 399; inhibition by H_2O_2, assay, **163**, 330; inorganic phosphate assay, **55**, 212; intermediate, **87**, 6; iodination of, **25**, 440; isolated nuclei and, **12A**, 445; isomorphous crystal, **9**, 215; ligand binding studies, **48**, 276; lobster tail muscle, **9**, 212 (crystallization, **9**, 213; molecular weight, **9**, 215; properties, **9**, 214; purification, **9**, 213; stability, **9**, 214); marker for chromosome detection in somatic cell hybrids, as, **151**, 184; mechanism, **63**, 51 (NAD+-specific, **63**, 52); modification (2-bromoacetamido-4-nitrophenol and, **26**, 583; 4-bromoacetamido-2-nitrophenol and, **26**, 584; [^{32}P]NAD+-dependent, **269**, 402; nitric oxide-dependent, **269**, 406; nitrophenolic organomercurials, **26**, 585); NAD+ binding (kinetics, **54**, 82; thermodynamics, **61**, 457, 458); NAD-X and, **6**, 353; negative cooperativity, **64**, 170, 174; nitric oxide effect, **269**, 406; nucleoside diphosphate glucose pyrophosphorylase assay and, **8**, 249; oxygen labeled compounds and, **4**, 913, 914; partial volume, **26**, 109, 110, 118; pea seed (assay, **89**, 320; properties, **89**, 324; purification, **89**, 321); peptide from, sequential degradation of, **11**, 472; peptide separation, **47**, 216, 218, 219; phosphoglycerate kinase assay and, **1**, 415; phosphohexokinase assay and, **1**, 307; photooxidation of, **25**, 405, 407; plasma membrane, in, **31**, 89; porcine (liver [assay, **89**, 310; **90**, 499; properties, **89**, 314; **90**, 508; purification, **89**, 312; simultaneous purification with 3-phosphoglycerate kinase and phosphoglycerate mutase, **90**, 499]; muscle [assay, **90**, 499; properties, **90**, 508; simultaneous purification with 3-phosphoglycerate kinase and phosphoglycerate mutase, **90**, 499]); preparation of [γ-^{18}O]ATP, **64**, 67; promoter-based expression vector for constitutive gene expression in yeast, **194**, 391; pyruvate kinase and, **5**, 368; rabbit muscle, of, **1**, 401; **27**, 385; **41**, 264 (assay, **1**, 401; **41**, 264, 265; **89**, 306; properties, **1**, 404; **41**, 267; **89**, 308; purification, **1**, 403; **41**, 265; **89**, 306; tRNA sequence analysis, in, **59**, 61); Raman spectroscopy, **246**, 401; reaction with 2-bromoacetamido-4-nitrophenol, **11**, 868; ribulokinase and, **5**, 298; ribulose bisphosphate carboxylase and, **5**, 266; ribulose 5-phosphate 3-epimerase, assessment of labeling by, **9**, 607; sedimentation coefficient, **61**, 108; sedimentation equilibrium, **48**, 175; separation from other enzymes, **3**, 294; sequential cooperativity, **64**, 170, 174; site-specific reagent, **47**, 483; spinach leaf (assay, **89**, 316; properties, **89**, 319; purification, **89**, 317); staining on gels, **22**, 594; starch gel electrophoresis of, **6**, 969; stereospecificity, **87**, 113, 258; succinylation of, **25**, 533, 539; synthesis in *Chlamydomonas* cultures, **23**, 73; thermostabilization, **137**, 623; *Thermus thermophilus* (assay, **89**, 335; properties, **89**, 340; purification, **89**, 336); transition state and multisubstrate analogues, **249**, 304; triose phosphate assay and, **3**, 294; use in catalytic assay for NAD+, **18B**, 3; xylulose-5-phosphate 3-epimerase and, **5**, 280; yeast, **1**, 407; **27**, 110, 204 (assay, **1**, 407; **89**, 326; properties, **1**, 410; **89**, 333; purification, **1**, 408; **89**, 330); Yeast Protein No. 2 and, **1**, 272

GLYCERATE DEHYDROGENASE

This oxidoreductase [EC 1.1.1.29], also known as hydroxypyruvate reductase and hydroxypyruvate dehydrogenase, catalyzes the reaction of (*R*)-glycerate (*i.e.*, D-glycerate) with NAD+ to produce hydroxypyruvate and NADH.[1–3] Hydroxypyruvate reductase [EC 1.1.1.81] catalyzes the same reaction but can use either NAD+ or NADP+ as the coenzyme substrate. ***See also*** *Hydroxypyruvate Reductase; 2-Hydroxy-3-ketopropionate Reductase*

[1]L. Kleczkowski & G. E. Edwards (1989) *Plant Physiol.* **91**, 278.
[2]Y. Izumi, T. Yoshida, H. Kanzaki, S.-i. Toki, S. S. Miyazaki & H. Yamada (1990) *EJB* **190**, 279.
[3]J. W. Thorner & H. Paulus (1973) *The Enzymes*, 3rd ed., **8**, 487.

Selected entries from ***Methods in Enzymology*** [vol, page(s)]:
General discussion: 9, 225;**188**, 361
Assay: 9, 225; hog spinal cord, from, **41**, 289, 290
Other molecular properties: hog spinal cord, from, **41**, 289 (assay, **41**, 289, 290; chromatography, **41**, 291, 292; properties, **41**, 292, 293; purification, **41**, 290); plants, of, **9**, 246; spinach leaves, **9**, 225 (activators, **9**, 228; inhibitors, **9**, 228; properties, **9**, 227 [kinetic, **9**, 228]; purification, **9**, 226; specificity, **9**, 227; stability, **9**, 227); stereospecificity, **87**, 105

GLYCERATE KINASE

This phosphotransferase [EC 2.7.1.31] catalyzes the reaction of ATP with (*R*)-glycerate (*i.e.*, D-glycerate) to produce ADP and 3-phospho-(*R*)-glycerate.[1–3] The rat liver[2] and spinach leaf[1] enzymes have random Bi Bi kinetic mechanisms.

A glycerate kinase from the serine-producing methylotroph *Hyphomicrobium methylovorum* GM2 produces 2-phosphoglycerate.[4]

[1]L. A. Kleczkowski, D. D. Randall & W. L. Zahler (1985) *ABB* **236**, 185.
[2]H. Katayama, Y. Kitagawa & E. Sugimoto (1980) *J. Biochem.* **88**, 765.
[3]J. W. Thorner & H. Paulus (1973) *The Enzymes*, 3rd ed., **8**, 487.
[4]T. Yoshida, K. Fukuta, T. Mitsunaga, H. Yamada & Y. Izumi (1992) *EJB* **210**, 849.

Selected entries from ***Methods in Enzymology*** [vol, page(s)]:
General discussion: 5, 352; **42**, 124; **188**, 361
Assay: 5, 352; *Escherichia coli*, from, **42**, 124; methylotrophs, assay in crude extracts of, **188**, 362
Other molecular properties: *Escherichia coli*, **42**, 124 (assay, **42**, 124; chromatography, **42**, 125; properties, **42**, 127; purification, **42**, 124); yeast, **5**, 352 (assay, **5**, 352; properties, **5**, 354; purification, **5**, 353)

GLYCEROL-1,2-CYCLIC-PHOSPHATE 2-PHOSPHODIESTERASE

glycerol-1,2-cyclic phosphate

glycerol 1-phosphate

This phosphodiesterase [EC 3.1.4.42] catalyzes the hydrolysis of either stereoisomer of glycerol 1,2-cyclic-phosphate to produce glycerol 1-phosphate.[1] Other substrates include 3′,5′-cyclic AMP and 2′,3′-cyclic AMP.

[1] N. Clarke & R. M. C. Dawson (1978) *BJ* **173**, 579.

GLYCEROL DEHYDRATASE

This cobalamin-dependent enzyme [EC 4.2.1.30] catalyzes the conversion of glycerol to 3-hydroxypropanal and water.[1–4] This and related B_{12}-dependent eliminases are characterized by the presence of a hydroxyl group on the substrate carbon atom from which a hydrogen atom is abstracted stereospecifically. The mechanism involves the formation of a radical anion intermediate, akin to that formed in the ribonucleotide reductase reaction to abstract the proton.

[1] T. Toraya (2000) *Cell. Mol. Life Sci.* **57**, 106.
[2] B. T. Golding & W. Buckel (1998) *CBC* **3**, 239.
[3] R. Daniel, T. A. Bobik & G. Gottschalk (1998) *FEMS Microbiol. Rev.* **22**, 553.
[4] R. H. Abeles (1971) *The Enzymes*, 3rd ed., **5**, 481.

Selected entries from *Methods in Enzymology* [vol, page(s)]:
General discussion: 18C, 32
Assay: 63, 34; *Aerobacter aerogenes*, from, 42, 316, 317
Other molecular properties: *Aerobacter aerogenes*, from, 42, 315 (activators and inhibitors, 42, 321; assay, 42, 316, 317; molecular weight, 42, 320; properties, 42, 320; purification, 42, 317); crude preparation, 18C, 33; inactivation by corrinoids, 18C, 32; use in determination of cobamide derivatives, 18C, 32

GLYCEROL DEHYDROGENASES

There are at least four oxidoreductases that utilize glycerol as a substrate.

Glycerol dehydrogenase [EC 1.1.1.6], also known as NAD^+-linked glycerol dehydrogenase, catalyzes the reversible reaction of glycerol with NAD^+ to produce dihydroxyacetone (*i.e.*, glycerone) and NADH.[1,2] Other substrates include 1,2-propanediol and 1,2-ethanediol. The enzyme has an ordered Bi Bi kinetic mechanism.[1]

Glycerol dehydrogenase (NADP⁺) [EC 1.1.1.72], also referred to as glycerol 1-dehydrogenase (NADP⁺), catalyzes the reversible reaction of glycerol with $NADP^+$ to produce D-glyceraldehyde and NADPH.[3–5]

Glycerol 2-dehydrogenase (NADP⁺) [EC 1.1.1.156], also referred to as dihydroxyacetone reductase, catalyzes the reversible reaction of glycerol with $NADP^+$ to produce dihydroxyacetone and NADPH.[6–8]

Glycerol dehydrogenase (acceptor) is a pyrroloquinoline quinone-dependent oxidoreductase [EC 1.1.99.22], first identified in *Gluconobacter* sp., that catalyzes the reaction of glycerol with an acceptor substrate to produce dihydroxyacetone and the reduced acceptor.[9] Other alcohol substrates include D-sorbitol, D-arabinitol, *meso*-erythritol, adonitol (*i.e.*, ribitol), galactitol, D-mannitol, and propylene glycol. 2,6-Dichlorophenolindophenol/phenazine methosulfate can act as the acceptor; however, NAD^+ does not act as an acceptor substrate (a observation that assists in distinguishing EC 1.1.99.22 from glycerol dehydrogenase [EC 1.1.1.6]).

[1] B. N. Leichus & J. S. Blanchard (1994) *Biochemistry* **33**, 14642.
[2] J. H. Marshall, J. W. May & J. Sloan (1985) *J. Gen. Microbiol.* **131**, 1581.
[3] M. Viswanath-Reddy, J. E. Pyle & E. B. Howe (1978) *J. Gen. Microbiol.* **107**, 289.
[4] A. W. Korman, R. O. Hurst & T. G. Flynn (1972) *BBA* **258**, 40.
[5] R. Schuurink, R. Busink, D. H. A. Hondmann, C. F. B. Witteveen & J. Visser (1990) *J. Gen. Microbiol.* **136**, 1043.
[6] J. H. Marshall, Y.-C. Kong, J. Sloan & J. W. May (1989) *J. Gen. Microbiol.* **135**, 697.
[7] A. Van Laere (1985) *FEMS Microbiol. Lett.* **30**, 377.
[8] A. Ben-Amotz & M. Avron (1973) *FEBS Lett.* **29**, 153.
[9] M. Ameyama, E. Shinagawa, K. Matsushita & O. Adachi (1985) *Agric. Biol. Chem.* **49**, 1001.

Selected entries from *Methods in Enzymology* [vol, page(s)]:
General discussion: NAD⁺-specific, 1, 397; NADP⁺-specific, 1-dehydrogenase, 89, 237
Assay: NAD⁺-specific, 1, 397; NADP⁺-specific, 1-dehydrogenase, 89, 238
Other molecular properties: *Aerobacter aerogenes* (properties, 1, 400; purification, 1, 398; stereospecificity, 87, 103); large-scale isolation, 22, 546; liver, in, 1, 320; NADP⁺-specific, 1-dehydrogenase (rabbit muscle [assay, 89, 238; properties, 89, 241; purification, 89, 239]; stereospecificity, 87, 109); NADP⁺-specific, 2-dehydrogenase (stereospecificity, 87, 111); stereospecificity, 87, 103, 109, 111

GLYCEROL KINASE

This phosphotransferase [EC 2.7.1.30], also known as glycerokinase and ATP:glycerol 3-phosphotransferase, catalyzes the reaction of ATP with glycerol to produce ADP and glycerol 3-phosphate.[1–5] Both dihydroxyacetone and L-glyceraldehyde can serve as substrates. The nucleoside triphosphate can be substituted by UTP and, in the case of the yeast enzyme, ITP and GTP. The *Candida mycoderma* enzyme has a random Bi Bi kinetic mechanism in which a high degree of synergistic binding between the substrates makes the mechanism appear ordered with glycerol adding first. A nucleophilic in-line transfer mechanism has been proposed based on the crystal structure.[1]

[1] C. Mao, Z. Ozer, M. Zhou & F. M. Uckun (1999) *BBRC* **259**, 640.
[2] W. B. Knight & W. W. Cleland (1989) *Biochemistry* **28**, 5728.
[3] J. D. Esko & C. R. H. Raetz (1983) *The Enzymes*, 3rd ed., **16**, 207.
[4] C. A. Janson & W. W. Cleland (1974) *JBC* **249**, 2562.
[5] J. W. Thorner & H. Paulus (1973) *The Enzymes*, 3rd ed., **8**, 487.

GLYCEROL-1-PHOSPHATASE

This phosphatase [EC 3.1.3.21] catalyzes the hydrolysis of glycerol 1-phosphate to produce glycerol and orthophosphate.[1-4] The *Dunaliella* enzyme acts more rapidly on *sn*-glycerol 1-phosphate (also known as L-glycerol 1-phosphate and D-glycerol 3-phosphate) than on the 3-phosphate (also known as D-glycerol 1-phosphate and L-glycerol 3-phosphate). The enzyme from yeast also acts on propane-1,2-diol 1-phosphate.

[1] D. Belmans & A. Van Laere (1988) *Arch. Microbiol.* **150**, 109.
[2] A. D. Brown, R. M. Lilley & T. Marengo (1982) *Z. Naturforsch.* **37C**, 1115.
[3] I. Sussman & M. Avron (1981) *BBA* **661**, 199.
[4] G. Schmidt (1961) *The Enzymes*, 2nd ed., **5**, 37.

GLYCEROL-2-PHOSPHATASE

This phosphatase [EC 3.1.3.19], also referred to as 2-phosphoglycerol phosphatase, catalyzes the hydrolysis of glycerol 2-phosphate to produce glycerol and orthophosphate.

GLYCEROL-3-PHOSPHATE O-ACYLTRANSFERASE

This transferase [EC 2.3.1.15] catalyzes the reaction of an acyl-CoA with *sn*-glycerol 3-phosphate to produce coenzyme A and a 1-acyl-*sn*-glycerol 3-phosphate.[1-7]

The acyl-CoA derivatives contain an acyl group with a chain length of at least ten carbon atoms (the *Escherichia*

coli enzyme, which exhibits cooperativity[4], has a preference of palmitoyl-CoA over oleoyl-CoA; the mammalian mitochondrial enzyme also prefers saturated fatty acyl-CoA substrates). In addition, the acyl-CoA can be substituted by an acyl-[acyl-carrier protein] derivative (note that, in plants, the chloroplast acyltransferase utilizes acyl-[acyl-carrier protein] substrates whereas the mitochondrial and cytoplasmic enzymes use acyl-CoA derivatives). A histidyl residue has been proposed to act as a general base to deprotonate the hydroxyl group of the acyl acceptor.[2]

The yeast enzyme (but not the rat enzyme) may be identical to glycerone-phosphate acyltransferase [EC 2.3.1.42].

[1] L. K. Dircks & H. S. Sul (1997) *BBA* **1348**, 17.
[2] N. Murata & Y. Tasaka (1997) *BBA* **1348**, 10.
[3] W. O. Wilkison & R. M. Bell (1997) *BBA* **1348**, 3.
[4] M. A. Scheideler & R. M. Bell (1991) *JBC* **266**, 14321.
[5] R. M. Bell & R. A. Coleman (1983) *The Enzymes*, 3rd ed., **16**, 87.
[6] J. D. Esko & C. R. H. Raetz (1983) *The Enzymes*, 3rd ed., **16**, 207.
[7] R. A. Pieringer (1983) *The Enzymes*, 3rd ed., **16**, 255.

GLYCEROL-3-PHOSPHATE CYTIDYLYLTRANSFERASE

This transferase [EC 2.7.7.39], also known as CDP-glycerol pyrophosphorylase and CDP-glycerol diphosphorylase, catalyzes the reaction of CTP with *sn*-glycerol 3-phosphate to produce pyrophosphate (or, diphosphate) and CDP-glycerol.[1,2] Other nucleotide substrates include dCTP.

[1] Y. S. Park, P. Gee, S. Sanker, E. J. Schurter, E. R. Zuiderweg & C. Kent (1997) *JBC* **272**, 15161.

[2] Y. S. Park, T. D. Sweitzer, J. E. Dixon & C. Kent (1993) *JBC* **268**, 16648.

Selected entries from *Methods in Enzymology* [vol, page(s)]:
General discussion: 8, 244
Assay: 8, 244
Other molecular properties: distribution, 8, 248; properties, 8, 247; purification, *Lactobacillus arabinosus*, 8, 246

GLYCEROL-3-PHOSPHATE DEHYDROGENASE

Glycerol-3-phosphate dehydrogenase (NAD$^+$) [EC 1.1.1.8] catalyzes the reversible reaction of *sn*-glycerol 3-phosphate with NAD$^+$ to produce glycerone phosphate (*i.e.*, dihydroxyacetone phosphate) and NADH.[1–6] 1,2-Propanediol phosphate and glycerone sulfate can also act as alternative substrates, having a weaker affinity. The *Chlamydomonas reinhardtii* enzyme[2] has a random Bi Bi kinetic mechanism with two abortive complexes; however, the salt-tolerant yeast *Debaryomyces hansenii* enzyme reportedly is ordered Bi Bi.[1] *Note*: Early literature citations often incorrectly designated this enzyme as glycerol-1-phosphate dehydrogenase.

Glycerol-3-phosphate dehydrogenase (NAD(P)$^+$) [EC 1.1.1.94], also known as NAD(P)H-dependent dihydroxyacetone-phosphate reductase, catalyzes the reversible reaction of *sn*-glycerol 3-phosphate with NAD(P)$^+$ to produce glycerone phosphate (*i.e.*, dihydroxyacetone phosphate) and NAD(P)H.[7] **See also** *Glycerol-1-Phosphate Dehydrogenase (NAD(P)$^+$)*

Glycerol-3-phosphate dehydrogenase [EC 1.1.99.5] is a flavoprotein that catalyzes the reaction of *sn*-glycerol 3-phosphate with an acceptor substrate to produce glycerone phosphate (*i.e.*, dihydroxyacetone phosphate) and the reduced acceptor.[4] The electron acceptor can be ubiquinone, phenazine methosulfate, ferricyanide, or 2,6-dichlorophenolindophenol. The porcine brain mitochondrial enzyme has a ping pong Bi Bi kinetic mechanism.[8]

[1] A. Nilsson & L. Adler (1990) *BBA* **1034**, 180.
[2] G. Klock & K. Kreuzberg (1989) *BBA* **991**, 347.
[3] J. D. Esko & C. R. H. Raetz (1983) *The Enzymes*, 3rd ed., **16**, 207.
[4] Y. Hatefi & D. L. Stiggall (1976) *The Enzymes*, 3rd ed., **13**, 175.
[5] P. Bentley & F. M. Dickinson (1974) *BJ* **143**, 11.
[6] T. Baranowski (1963) *The Enzymes*, 2nd ed., **7**, 85.
[7] M. Kito & L. I. Pizer (1969) *JBC* **244**, 3316.

Selected entries from *Methods in Enzymology* [vol, page(s)]:
General discussion: 1, 391; 5, 432; 34, 4; 41, 240, 254; 44, 337, 495, 496; 53, 405; 89, 296

Assay: 1, 391; 5, 432; 41, 37; beef liver, from, 41, 260; chicken breast muscle, of, 41, 245, 246; *Drosophila melanogaster*, 89, 297; *Escherichia coli*, from, 41, 249, 250; honey bee and other insects, from, 41, 241; pig brain mitochondria, from (flavin-dependent), 5, 432; 41, 254, 255

Other molecular properties: aldolase assay and, 1, 320; anaerobic and aerobic, mutant, 72, 695; assay for glycerol in acyl glycerols, 14, 627, 629; bacterial membrane, 31, 649; beef liver, from, 41, 259 (assay, 41, 260; chromatography, 41, 261, 262; properties, 41, 262; purification, 41, 260); brown adipose tissue mitochondria, 55, 75; cell fractions, associated with, 40, 86; chicken breast muscle, of, 41, 245 (assay, 41, 245, 246; chromatography, 41, 247; properties, 41, 248, 249; purification, 41, 246); chromatography on hydroxyapatite columns, 5, 569; deoxyribose phosphate aldolase assay and, 1, 384; dihydroxyacetone-3-thiophosphate, and, 87, 310; *Drosophila melanogaster* (assay, 89, 297; properties, 89, 301; purification, 89, 297); *Escherichia coli*, from, 41, 249 (assay, 41, 249, 250; chromatography, 41, 251, 252; distribution, 41, 254; inhibition, 41, 254; properties, 41, 253, 254; purification, 41, 250); *Escherichia coli* auxotrophs, in, 32, 864; extraction from mitochondria, 22, 205; extraction methods, effect of, 53, 409; flavin determination, 53, 429; fructose-1,6-bisphosphate aldolase, assay of, 9, 481, 487, 492; L-fuculose-1-phosphate aldolase assay, 9, 538; glyceraldehyde-3-phosphate dehydrogenase, presence in preparations of, 9, 213; glycerokinase and, 5, 355, 476; glycerophosphorylcholine diesterase assay and, 1, 668; honey bee and other insects, from, 41, 240 (amino acids, 41, 245; assay, 41, 241; inhibition, 41, 244; molecular weight, 41, 244, 245; properties, 41, 243; purification, 41, 241); induction by Clofibrate and di(2-ethylhexyl)phthalate, 72, 507; isolated nuclei and, 12A, 445; localization, 56, 230 (gels by simultaneous capture technique, in, 104, 434; on starch block, 6, 962); mechanism, 249, 199; microbodies, in, 52, 496, 501; nitrate reductase, and, anaerobic electron transfer, 56, 386; parenchyma cells, in, 32, 706; particulate, 2, 559; phosphofructokinase, assay of, 9, 425, 430, 437; phosphoketolase, assay of, 9, 515; pig brain mitochondria (flavin-dependent), from, 5, 432; 41, 254 (assay, 5, 432; 41, 254, 255; inhibitors, 41, 258; properties, 5, 436; 41, 258, 259; purification, 5, 434; 41, 255); product inhibition studies, two substrates:two products reactions, 249, 199, 202; 1,2-propanediol phosphate dehydrogenase preparations, presence in, 9, 336 (distinct from, 9, 337); pyridine nucleotide and, 1, 395; 4, 292; pyrophosphate assay, 63, 34; rabbit liver, from, cytosolic (purification, 74, 225; radioimmunoassay, 74, 225 [antibody preparation, 74, 226; application to tissue extracts, 74, 228; chromatography columns, 74, 225; enzyme inhibition studies, 74, 227; immune reaction, 74, 227; immunogen iodination, 74, 227; immunogen preparation, 74, 225; solutions, 74, 225; standard competitive binding curves, 74, 227, 228]; total liver activity, 74, 229, 230); rabbit muscle, 1, 391 (properties, 1, 396; purification, 1, 393); rat interferon production, in, 78, 173; ribulose bisphosphate carboxylase and, 5, 266; ribulose 5-phosphate 3-epimerase, 9, 605; sources, 5, 438; stain for, 224, 108; staining on gels, 22, 594; suppression by interferon, 79, 343; synthesis, inhibition by rat interferon, 78, 176; transaldolases assay, 9, 499; transketolase assay and, 1, 371, 375; triose phosphate assay and, 3, 295; triose phosphate dismutation and, 3, 293; triose phosphate isomerase assay and, 1, 387; xylulose-5-phosphate 3-epimerase and, 5, 280; xylulose-5-phosphate phosphoketolase and, 5, 262, 263

GLYCEROL-1-PHOSPHATE DEHYDROGENASE (NAD(P)$^+$)

This oxidoreductase [EC 1.1.1.261] catalyzes the reaction of *sn*-glycerol-1-phosphate with NAD(P)$^+$ to produce glycerone phosphate (*i.e.*, dihydroxyacetone phosphate), NAD(P)H, and a proton.[1] This enzyme is stereospecifically distinct from glycerol-3-phosphate dehydrogenase (NAD(P)$^+$) [EC 1.1.1.94].

[1]Y. Koga, T. Kyuragi, M. Nishihara & N. Sone (1998) *J. Mol. Evol.* **46**, 54.

GLYCEROL-3-PHOSPHATE 1-DEHYDROGENASE (NADP$^+$)

This oxidoreductase [EC 1.1.1.177] catalyzes the reaction of *sn*-glycerol 3-phosphate with NADP$^+$ to produce D-glyceraldehyde 3-phosphate and NADPH.[1] In addition to oxidation at C1, stereochemical inversion takes place at C2.

[1]E. P. Glushankov, Y. E. Epifanova & A. I. Kolotilova (1976) *Biokhimiya* **41**, 1788.

sn-GLYCEROL-3-PHOSPHATE 1-GALACTOSYLTRANSFERASE

This transferase [EC 2.4.1.96], also known as isofloridoside-phosphate synthase, catalyzes the reaction of UDP-D-galactose with *sn*-glycerol 3-phosphate to produce UDP and α-D-galactosyl-(1,1′)-*sn*-glycerol 3-phosphate.[1] The product is hydrolyzed by a phosphatase to isofloridoside. **See also** *sn-Glycerol-3-Phosphate 2-α-Galactosyltransferase*

[1]K.-S. Thomson (1983) *BBA* **759**, 154.

sn-GLYCEROL-3-PHOSPHATE 2-α-GALACTOSYLTRANSFERASE

This transferase [EC 2.4.1.137], also referred to as floridoside-phosphate synthase, catalyzes the reaction of UDP-D-galactose with *sn*-glycerol 3-phosphate to produce UDP and 2-(α-D-galactosyl)-*sn*-glycerol 3-phosphate.[1] The product is hydrolyzed by a phosphatase to generate floridoside. **See also** *sn-Glycerol-3-Phosphate 1-Galactosyltransferase*

[1]J. Meng & L. M. Srivastava (1991) *Phytochemistry* **30**, 1763.

GLYCEROL-3-PHOSPHATE:GLUCOSE PHOSPHOTRANSFERASE

This enzyme [EC 2.7.1.142] catalyzes the reaction of *sn*-glycerol 3-phosphate with D-glucose to produce glycerol and D-glucose 6-phosphate. This enzyme is an important step in the anaerobic metabolism of sugars in trypanosomes. Other sugar acceptor substrates include D-mannose and D-fructose.[1]

[1]J. K. Kiaira & R. M. Njogu (1989) *Int. J. Biochem.* **21**, 839.

GLYCEROL-3-PHOSPHATE OXIDASE

This FAD-dependent oxidoreductase [EC 1.1.3.21] catalyzes the reaction of *sn*-glycerol 3-phosphate with

dioxygen to produce glycerone phosphate (*i.e.*, dihydroxyacetone phosphate) and hydrogen peroxide.[1,2]

[1]T. W. Esders & C. A. Michrina (1979) *JBC* **254**, 2710.
[2]A. Claiborne (1986) *JBC* **261**, 14398.

GLYCEROL-3-PHOSPHATE-TRANSPORTING ATPase

This so-called ATPase [EC 3.6.3.20] catalyzes the ATP-dependent (ADP- and orthophosphate-producing) transport of *sn*-glycerol-3-phosphate$_{out}$ to produce *sn*-glycerol-3-phosphate$_{in}$.[1-3] This ABC-type (ATP-binding cassette-type) ATPase has two similar ATP-binding domains and does not undergo phosphorylation during the transport process.

The idea that a transporter is an enzyme is in keeping with a new definition of enzyme catalysis as the facilitated making/breaking of chemical bonds, not just covalent bonds.[4] This idea builds on Pauling's assertion that any long-lived, chemically distinct interaction (in this case, the persistent location of a solute with respect to the faces of a membrane) can be regarded as a chemical bond. Note also that the equilibrium constant ($K_{eq} = [sn\text{-glycerol-3-P}_{in}][ADP][P_i]/[sn\text{-glycerol-3-P}_{out}][ATP]$) does not conform to that expected for an ATPase (*i.e.*, $K_{eq} = [ADP][P_i]/[ATP]$). Thus, although the overall reaction yields ADP and orthophosphate, the enzyme is misclassified as a hydrolase, and instead should be regarded as an energase-type reaction. Energases facilitate affinity-modulated reactions by coupling the $\Delta G_{ATP\text{-hydrolysis}}$ to a force-generating or work-producing step.[4] In this case, P-O-P bond-scission supplies the energy to drive *sn*-glycerol-3-P transport. **See ATPase; Energase**

[1]M. H. Saier, Jr. (1998) *Adv. Microb. Physiol.* **40**, 81.
[2]J. K. Griffiths & C. E. Sansom (1998) *The Transporter Factsbook*, Academic Press, San Diego.
[3]H. Bahl, G. Burchhardt & A. Wienecke (1991) *FEMS Microbiol. Lett.* **65**, 83.
[4]D. L. Purich (2001) *TiBS* **26**, 417.

GLYCERONE KINASE

This enzyme [EC 2.7.1.29], also known as triokinase, acetol kinase, and dihydroxyacetone kinase, catalyzes the reaction of ATP with glycerone (*i.e.*, dihydroxyacetone) to produce ADP and glycerone phosphate (*i.e.*, dihydroxyacetone phosphate). The enzyme can also use acetol (*i.e.*, monohydroxyacetone) as a substrate.

Selected entries from **Methods in Enzymology** [vol, page(s)]:
General discussion: 9, 473; 188, 445, 451
Assay: 9, 475; 188, 451

Other molecular properties: kidney, in, **9**, 476; liver, in, **9**, 476; properties, **9**, 476; **188**, 453; purification from *Candida methylica*, **188**, 452

Selected entries from *Methods in Enzymology* [vol, page(s)]:
General discussion: **1**, 668
Assay: **1**, 668

GLYCERONE-PHOSPHATE O-ACYLTRANSFERASE

$$O = \begin{array}{c} CH_2OH \\ \rule{0pt}{1.2em} \\ CH_2OPO_3H^- \end{array}$$

dihydroxyacetone phosphate

$$O = \begin{array}{c} CH_2OCO(CH_2)_{14}CH_3 \\ \rule{0pt}{1.2em} \\ CH_2OPO_3H^- \end{array}$$

palmitoyldihydroxyacetone phosphate

This transferase [EC 2.3.1.42], also referred to as dihydroxyacetone phosphate acyltransferase, catalyzes the reaction of an acyl-CoA with glycerone phosphate (*i.e.*, dihydroxyacetone phosphate) to produce coenzyme A and an acylglycerone phosphate (*i.e.*, a 1-acyldihydroxyacetone phosphate).[1–3] Acyl-CoA substrates include palmitoyl-CoA (the best acyl-group donor), stearoyl-CoA, and oleoyl-CoA.

[1] A. K. Hajra (1997) *BBA* **1348**, 2.
[2] R. M. Bell & R. A. Coleman (1983) *The Enzymes*, 3rd ed., **16**, 87.
[3] R. M. Bell (1980) *Annu. Rev. Biochem.* **49**, 459.

Selected entries from *Methods in Enzymology* [vol, page(s)]:
General discussion: **209**, 92
Assay: **209**, 93

GLYCEROPHOSPHOCHOLINE CHOLINEPHOSPHODIESTERASE

This phosphodiesterase [EC 3.1.4.38] catalyzes the hydrolysis of *sn*-glycero-3-phosphocholine to produce glycerol and choline phosphate.[1–3] *sn*-3-Glycerophosphoethanolamine is not a substrate.

[1] J. Florin-Christensen & M. Florin-Christensen (1999) *Biochem. Mol. Biol. Int.* **47**, 283.
[2] J. Yuan & J. N. Kanfer (1994) *Neurochem. Res.* **19**, 43.
[3] R. M. Abra & P. J. Quinn (1976) *BBA* **431**, 631.

GLYCEROPHOSPHOCHOLINE PHOSPHODIESTERASE

This phosphodiesterase [EC 3.1.4.2] catalyzes the hydrolysis of *sn*-glycero-3-phosphocholine to produce choline and *sn*-glycerol 3-phosphate.[1–4] Other substrates include *sn*-glycero-3-phosphoethanolamine, glycerophosphoinositol, and glycerolphosphoserine.

[1] T. J. Larson & A. van Loo-Bhattacharya (1988) *ABB* **260**, 577.
[2] K. A. Lloyd-Davies, R. H. Michell & R. Coleman (1972) *BJ* **127**, 357.
[3] R. M. Abra & P. J. Quinn (1976) *BBA* **431**, 631.
[4] O. Hayaishi & A. Kornberg (1954) *JBC* **206**, 647.

GLYCEROPHOSPHODIESTER PHOSPHODIESTERASE

This phosphodiesterase [EC 3.1.4.46], also called glycerophosphoryl diester phosphodiesterase, catalyzes the hydrolysis of a glycerophosphodiester to produce an alcohol and *sn*-glycerol 3-phosphate.[1] Substrates include glycerophosphocholine, glycerophosphoethanolamine, glycerophosphoglycerol, bis(glycerophosphoglycerol), and cyclic *sn*-2,3-phosphoglycerol.

[1] T. J. Larson, M. Ehrmann & W. Boos (1983) *JBC* **258**, 5428.

GLYCEROPHOSPHOINOSITOL GLYCEROPHOSPHODIESTERASE

This phosphodiesterase [EC 3.1.4.44] catalyzes the hydrolysis of *sn*-glycero-3-phospho-1-inositol to produce inositol and *sn*-glycerol 3-phosphate.[1,2]

[1] W. Kusser & F. Fiedler (1984) *FEBS Lett.* **166**, 301.
[2] R. M. C. Dawson, N. Hemington, D. E. Richards & R. F. Irvine (1979) *Biochem. J.* **182**, 39.

GLYCEROPHOSPHOINOSITOL INOSITOLPHOSPHODIESTERASE

This phosphodiesterase [EC 3.1.4.43] catalyzes the hydrolysis of *sn*-glycero-3-phospho-1-inositol to produce glycerol and inositol 1-phosphate.[1]

[1] R. M. C. Dawson & N. Hemington (1977) *BJ* **162**, 241.

GLYCEROPHOSPHOLIPID ACYLTRANSFERASE (CoA-DEPENDENT)

This coenzyme A-dependent acyltransferase [EC 2.3.1.148] catalyzes the reaction of a 1-organyl-2-acyl-*sn*-glycero-3-phosphocholine with a 1-organyl-2-lyso-*sn*-glycero-3-phosphoethanolamine to produce a 1-organyl-2-acyl-*sn*-glycero-3-phosphoethanolamine and a 1-organyl-2-lyso-*sn*-glycero-3-phosphocholine, where the moiety located at position *sn*-1 on the substrates and products can be an alkyl, acyl, or alk-1-enyl group.[1] This enzyme is distinct from glycerophospholipid arachidonoyltransferase (CoA-independent) [EC 2.3.1.147] in that coenzyme A is required and that there is not a preference for the transfer of polyunsaturated acyl groups.

[1] M. Robinson, M. L. Blank & F. Snyder (1985) *JBC* **260**, 7889.

GLYCEROPHOSPHOLIPID ARACHIDONOYLTRANSFERASE (CoA-INDEPENDENT)

This acyltransferase [EC 2.3.1.147] catalyzes the reaction of a 1-organyl-2-arachidonoyl-*sn*-glycero-3-phosphocholine with a 1-organyl-2-lyso-*sn*-glycero-3-phosphoethanolamine to produce a 1-organyl-2-arachidonyl-*sn*-glycero-3-phosphoethanolamine and a 1-organyl-2-lyso-*sn*-glycero-3-phosphocholine, where the moiety located at position *sn*-1 on the substrates and products can be an alkyl, acyl, or alk-1-enyl group.[1] The acyl group located at the *sn*-2 position of the choline-containing substrate can be another polyenoic fatty acyl group. This enzyme is distinct from glycerophospholipid acyltransferase (CoA-dependent) [EC 2.3.1.148] in that coenzyme A is not required and that there is a preference for the transfer of polyunsaturated acyl groups.

[1] M. Robinson, M. L. Blank & F. Snyder (1985) *JBC* **260**, 7889.

GLYCERYL-ETHER MONOOXYGENASE

This glutathione- and phospholipid-dependent oxidoreductase [EC 1.14.16.5], also known as glyceryl-ether cleaving enzyme and glyceryl etherase, catalyzes the reaction of a 1-alkyl-*sn*-glycerol (*e.g.*, 1-*O*-hexadecyl-*sn*-glycerol) with tetrahydropteridine and dioxygen to produce a 1-(1′-hydroxyalkyl)-*sn*-glycerol, dihydropteridine, and water.[1–4] The hydroxyalkyl product spontaneously breaks down to generate a fatty aldehyde and glycerol. The active ether-lipid substrates are in micelle form or mixed micelle form.[2]

[1] H. Taguchi & W. L. Armarego (1998) *Med. Res. Rev.* **18**, 43.
[2] H. Taguchi, B. Kosar-Hashemi, B. Paal, N. Yang & W. L. Armarego (1994) *Biol. Chem. Hoppe Seyler* **375**, 329.
[3] S. Kaufman, R. J. Pollock, G. K. Summer, A. K. Das & A. K. Hajira (1990) *BBA* **1040**, 19.
[4] T. Ishibashi & Y. Imai (1983) *EJB* **132**, 23.

Selected entries from *Methods in Enzymology* [vol, page(s)]:
Assay: 141, 395

GLYCINE ACETYLTRANSFERASE

This pyridoxal-phosphate-dependent enzyme [EC 2.3.1.29], also known as glycine *C*-acetyltransferase, aminoacetone synthase, and 2-amino-3-ketobutyrate:coenzyme A ligase, catalyzes the reversible reaction of acetyl-CoA with glycine to produce coenzyme A and 2-amino-3-ketobutanoate (*i.e.*, 2-amino-3-oxobutanoate) which undergoes irreversible decarboxylation to form aminoacetone.[1] *n*-Propionyl-CoA can also act as the donor

substrate. The bovine liver enzyme has an ordered Bi Uni kinetic mechanism.[2] Three low-barrier hydrogen bonds may serve to anchor pyridoxal 5′-phosphate.[3]

[1] J. J. Mukherjee & E. E. Dekker (1987) *JBC* **262**, 14441.
[2] B. Fubara, F. Eckenrode, T. Tressel & L. Davis (1986) *JBC* **261**, 12189.
[3] H. Tong & L. Davis (1995) *Biochemistry* **34**, 3362.

Selected entries from *Methods in Enzymology* [vol, page(s)]:
General discussion: 17B, 585
Assay: 17B, 586

GLYCINE *N*-ACYLTRANSFERASES

This acyltransferase [EC 2.3.1.13] catalyzes the reaction of an acyl-CoA derivative with glycine to produce coenzyme A and an *N*-acylglycine.[1,2] The acyl-CoA derivative can be one of a number of aliphatic and aromatic acids. However, neither phenylacetyl-CoA nor indole-3-acetyl-CoA can act as substrates. *See also* Amino Acid Acetyltransferase; Glycine *N*-Choloyltransferase; Glycine N-Benzoyltransferase

[1] L. T. Webster, U. A. Siddiqui, S. V. Lucas, J. M. Strong & J. J. Mieyal (1976) *JBC* **254**, 3352.
[2] M. O. James & J. R. Bend (1978) *BJ* **172**, 285.

Selected entries from *Methods in Enzymology* [vol, page(s)]:
General discussion: 2, 346;**77**, 301
Assay: 2, 347, 349; **77**, 302
Other molecular properties: glycine N-benzoyltransferase, **77**, 301 (assay, **77**, 302; CoA determination, **77**, 303; thioester utilization, **77**, 303; inhibitors, **77**, 307, 308; kinetic constants, **77**, 307; mechanism, **77**, 308; molecular weight, **77**, 307; pH optima, **77**, 308; properties, **77**, 307, 308; purification, **77**, 304; size, **77**, 307; specificity, **77**, 307; stability, **77**, 303, 308; stimulators, **77**, 307, 308); intestinal epithelium, in, **77**, 157; purification, liver, **2**, 348

GLYCINE AMIDINOTRANSFERASE

This transferase [EC 2.1.4.1], also known as L-arginine:glycine amidinotransferase and transaminidase, catalyzes the reaction of L-arginine with glycine to produce L-ornithine and guanidinoacetate.[1,2] Canavanine can also serve as the donor substrate. In vertebrates, this enzyme, which has a ping pong Bi Bi kinetic mechanism,[3] catalyzes the rate-determining step in creatine biosynthesis. A histidyl residue is proposed to act as a general acid/general base catalyst and a cysteinyl residue acts as a nucleophile (thus, a transient amidino-enzyme intermediate is formed).[4] The enzyme adopts open and closed conformations during the reaction cycle with large conformational changes accompanying the binding of L-arginine.[5]

[1] J. B. Walker (1979) *AE* **50**, 177.
[2] J. B. Walker (1973) *The Enzymes*, 3rd ed., **9**, 497.
[3] G. Ronca, V. Vigi & E. Grazi (1966) *JBC* **241**, 2589.
[4] E. Fritsche, A. Humm & R. Huber (1997) *EJB* **247**, 483.
[5] E. Fritsche, A. Humm & R. Huber (1999) *JBC* **274**, 3026.

GLYCINE AMINOTRANSFERASE

This pyridoxal-phosphate-dependent transferase [EC 2.6.1.4] catalyzes the reaction of glycine with α-ketoglutarate (or, 2-oxoglutarate) to produce glyoxylate and L-glutamate.[1–5] **See also** *Glycine:Oxaloacetate Aminotransferase*

[1] A. E. Braunstein (1973) *The Enzymes*, 3rd ed., **9**, 379.
[2] H. I. Nakada (1964) *JBC* **239**, 468.
[3] J. S. Thompson & K. E. Richardson (1966) *ABB* **117**, 599.
[4] D. W. Rehfeld & N. E. Tolbert (1972) *JBC* **247**, 4803.
[5] N. E. Tolbert (1981) *Ann. Rev. Biochem.* **50**, 133.

GLYCINE *N*-BENZOYLTRANSFERASE

This transferase [EC 2.3.1.71] catalyzes the reaction of benzoyl-CoA with glycine to produce coenzyme A and *N*-benzoylglycine (*i.e.*, hippurate).[1] This enzyme is not identical with glycine *N*-acyltransferase [EC 2.3.1.13] or glutamine *N*-acyltransferase [EC 2.3.1.68]. The bovine liver enzyme has an ordered Bi Bi kinetic mechanism. Other acceptor substrates include glutamine and asparagine. **See also** *Amino Acid Acetyltransferase; Glycine N-Choloyltransferase*

[1] D. L. Nandi, S. V. Lucas & L. T. Webster (1979) *JBC* **251**, 7230.

GLYCINE *N*-CHOLOYLTRANSFERASE

This transferase [EC 2.3.1.65], also referred to as glycine/taurine *N*-acyltransferase and bile acid:amino acid transferase, catalyzes the reaction of choloyl-CoA with glycine to produce coenzyme A and glycocholate. Taurine can act as the acceptor substrate and several CoA derivatives of bile acids are effective as the donor substrate: for example, deoxycholoyl-CoA, lithocholoyl-CoA, chenodeoxycholoyl-CoA, and ursodeoxycholoyl-CoA. The enzyme has a ping pong Bi Bi kinetic mechanism with a covalent enzyme-cholate intermediate.[1]

[1] B. Czuba & D. A. Vessey (1980) *JBC* **255**, 5296 and (1986) **261**, 6260.

GLYCINE DEHYDROGENASE

This oxidoreductase [EC 1.4.1.10] catalyzes the reaction of glycine with water and NAD^+ to produce glyoxylate, ammonia, and NADH.[1] The equilibrium constant greatly favors the reverse reaction.

[1] D. S. Goldman & M. J. Wagner (1962) *BBA* **65**, 297.

GLYCINE DEHYDROGENASE (CYTOCHROME)

This oxidoreductase [EC 1.4.2.1], also referred to as glycine:cytochrome *c* reductase, catalyzes the reaction of glycine with water and two molecules of ferricytochrome *c* to produce glyoxylate, ammonia, and two molecules of ferrocytochrome *c*.[1]

[1] H. K. Sanders, G. E. Becker & A. Nason (1972) *JBC* **247**, 2015.

GLYCINE DEHYDROGENASE (DECARBOXYLATING)

This pyridoxal-phosphate-dependent enzyme [EC 1.4.4.2], also known as glycine decarboxylase and the glycine cleavage system P-protein (or, protein P1), catalyzes the reversible reaction of glycine with a lipoylprotein to produce *S*-aminomethyldihydrolipoylprotein and carbon dioxide.[1–4] Free lipoamide can also act as the acceptor substrate. Along with aminomethyltransferase [EC 2.1.2.10], this enzyme activity is a component of glycine synthase.

The reaction mechanism, which follows a ping pong Bi Bi kinetic scheme[2], utilizes a Schiff base formed by the reaction of glycine with the enzyme-bound pyridoxal 5′-phosphate. The amino group and the α-carbon of the glycine are then transferred to the lipoamide cofactor of the second enzyme of the complex (the H-protein) with the concomitant loss of the carboxyl group as carbon dioxide. The enzyme will also catalyze an exchange reaction between glycine and carbon dioxide. **See also** *Glycine Synthase*

[1] K. Fujiwara & Y. Motokawa (1983) *JBC* **258**, 8156.
[2] K. Hiraga & G. Kikuchi (1980) *JBC* **255**, 11664 and 11671.
[3] E. A. Boeker & E. E. Snell (1972) *The Enzymes*, 3rd ed., **6**, 217.

GLYCINE FORMIMIDOYLTRANSFERASE

This transferase [EC 2.1.2.4], also known as glycine formiminotransferase, catalyzes the reversible reaction of 5-formimidoyltetrahydrofolate with glycine to produce tetrahydrofolate and *N*-formimidoylglycine.[1–4]

[1]J. I. Rader & F. M. Huennekens (1973) *The Enzymes*, 3rd ed., **9**, 197.
[2]J. Rabinowitz & W. E. Pricer, Jr. (1957) *Fed. Proc.* **16**, 236.
[3]R. D. Sagers, J. V. Beck, W. Gruber & I. C. Gunsalus (1956) *JACS* **78**, 694.
[4]J. C. Rabinowitz & W. C. Pricer, Jr. (1956) *JACS* **78**, 5702.

Selected entries from *Methods in Enzymology* [vol, page(s)]:
General discussion: 6, 96
Assay: 6, 96

GLYCINE METHYLTRANSFERASE

This methyltransferase [EC 2.1.1.20] catalyzes the reaction of *S*-adenosyl-L-methionine with glycine to produce *S*-adenosyl-L-homocysteine and sarcosine (*i.e.*, *N*-methylglycine).[1-4] The rat liver enzyme, which has an ordered Bi Bi kinetic mechanism (*S*-adenosyl-L-methionine binding first), exhibits positive cooperativity with respect to *S*-adenosyl-L-methionine.[3] Inhibitors include acetate and 5-methyl-tetrahydrofolate.

[1]F. Takusawa, M. Fujioka, A. Spies & R. L. Schowen (1998) *CBC* I, I.
[2]H. Ogawa, T. Gomi, F. Takusawa & M. Fujioka (1998) *Int. J. Biochem. Cell Biol.* **30**, 13.
[3]H. Ogawa & M. Fujioka (1982) *JBC* **257**, 3447.
[4]J. E. Heady & S. J. Kerr (1973) *JBC* **248**, 69.

GLYCINE:OXALOACETATE AMINOTRANSFERASE

This pyridoxal-phosphate-dependent aminotransferase [EC 2.6.1.35] catalyzes the reversible reaction of glycine with oxaloacetate to produce glyoxylate and L-aspartate.[1,2] *See also Glycine Aminotransferase*

[1]A. E. Braunstein (1973) *The Enzymes*, 3rd ed., **9**, 379.
[2]R. G. Gibbs & J. G. Morris (1966) *BJ* **99**, 27P.

Selected entries from *Methods in Enzymology* [vol, page(s)]:
General discussion: 17A, 982
Other molecular properties: purification from *Micrococcus denitrificans*, **17A**, 984

GLYCINE OXIDASE

This FAD-dependent oxidoreductase reportedly catalyzes the reaction of glycine with water and dioxygen to produce glyoxylate, ammonia, and hydrogen peroxide. Other substrates include sarcosine. It should be noted that both L- and D-amino acid oxidases also exhibit this activity and glycine dehydrogenase catalyzes the NAD$^+$-dependent conversion of glycine to glyoxylate and ammonia: however, the porcine kidney enzyme is reportedly specific for glycine and sarcosine. *See L-Amino Acid Oxidase; Glycine Dehydrogenase*

Selected entries from *Methods in Enzymology* [vol, page(s)]:
General discussion: 2, 225
Assay: 2, 225

GLYCINE REDUCTASE

This oxidoreductase catalyzes the reaction of glycine with orthophosphate to produce acetyl phosphate and ammonium ion.[1-3] The oxidized form of the enzyme is subsequently reduced to its dithiol form in a NADH-linked step.

Glycine reductase is a complex of three components: protein A contains a selenocysteinyl residue, and protein B has a pyruvoyl residue. In the presence of protein B, glycine reacts with protein A to form a (carboxymethyl)selenocysteinyl intermediate. The acetyl group is subsequently transferred to protein C, and acetylated protein C then reacts with orthophosphate to generate acetyl phosphate.

[1]T. C. Stadtman (1996) *Ann. Rev. Biochem.* **65**, 83.
[2]R. A. Arkowitz & R. H. Abeles (1991) *Biochemistry* **230**, 4090.
[3]T. C. Stadtman (1979) *AE* **48**, 1.

Selected entries from *Methods in Enzymology* [vol, page(s)]:
General discussion: 17A, 959
Assay: 53, 373
Other molecular properties: activity, **53**, 373, 374, 382; *Clostridium sticklandii*, **53**, 372 (purification, **17A**, 963; **53**, 376); selenium-dependent, **53**, 373; stereochemistry, **87**, 153

GLYCINE SYNTHASE

This protein complex, also known as glycine cleavage complex, consists of four protein components.[1-3] The first protein in this complex, glycine dehydrogenase (decarboxylating) [EC 1.4.4.2] and also known as P-protein or protein P1, catalyzes the pyridoxal-phosphate-dependent reaction of glycine with a lipoylprotein to produce *S*-aminomethyldihydrolipoylprotein and carbon dioxide. The second member of the complex, H-protein or protein P2, carries the lipoamide group. The third component of the complex is aminomethyltransferase [EC 2.1.2.10], also known as glycine-cleavage system T-protein or protein P4, which catalyzes the transfer of the α-carbon from the lipoamide of H protein to tetrahydrofolate (producing (6*R*)-5,10-methylenetetrahydrofolate), with the α-amino group from the original glycine being released as ammonia. The fourth component of the complex, the L-protein or protein P3, is a flavoenzyme that utilizes NAD$^+$ to oxidize the reduced lipoate. *See also Glycine Dehydrogenase (Decarboxylating); Aminomethyltransferase*

[1]V. Schirch (1998) *CBC* 1, 211.
[2]J. I. Rader & F. M. Huennekens (1973) *The Enzymes*, 3rd ed., **9**, 197.
[3]E. A. Boeker & E. E. Snell (1972) *The Enzymes*, 3rd ed., **6**, 217.

GLYCOGEN DEBRANCHING ENZYME

This enzyme catalyzes the debranching of glycogen in a two-step process. The mammalian muscle enzyme possesses both amylo-1,6-glucosidase [EC 3.2.1.33] and 4-α-glucanotransferase [EC 2.4.1.25] activities.

The action of glycogen phosphorylase results in the formation of a limit dextrin containing branch points with a maltotetraosyl group at the 4- and 6-positions of a branch point. The first step catalyzed by the debranching enzyme is to transfer a maltotriosyl group from the maltotetraosyl group located on the 6-position of the branch glucose residue. This trisaccharide is then transferred (with retention of anomeric configuration) to the 4-position of the non-reducing end of the other chain (via a glycosyl-enzyme intermediate). The remaining glucosyl residue at the 6-position of the branch point is then hydrolyzed by the glucosidase activity to produce D-glucose (with inverted anomeric configuration) and a linear α1,4-glucan that can act as a substrate of glycogen phosphorylase.[1-3] **See** *Amylo-1,6-Glucosidase; 4-α-Glucanotransferase*

[1] B. K. Gillard & T. E. Nelson (1977) *Biochemistry* **16**, 3978.
[2] G. Davies, M. L. Sinnott & S. G. Withers (1998) *CBC* **1**, 119.
[3] E. Y. C. Lee & W. J. Whelan (1971) *The Enzymes*, 3rd ed., **5**, 191.

Selected entries from *Methods in Enzymology* [vol, page(s)]:
General discussion: **1**, 211; **8**, 515; **90**, 479
Assay: **1**, 211; **8**, 516; biopsy material, in, **8**, 527; hexokinase and, **3**, 110
Other molecular properties: α-amylase, and, starch degradation in aqueous two-phase system, **137**, 661; contaminants of, **3**, 51, 52; glycogen and, **8**, 516; polysaccharide analysis and, **3**, 51; properties, **1**, 214; **8**, 522; purification, **1**, 212; **8**, 519, 527; **90**, 479

GLYCOGENIN GLUCOSYLTRANSFERASE

This transferase [EC 2.4.1.186] catalyzes the reaction of UDP-D-glucose with glycogenin to produce UDP and glucosylglycogenin. The glycogenin subunit of glycogen synthase [EC 2.4.1.11] catalyzes this reaction. This reaction is actually autocatalytic, and glycogenin transfers D-glucose residues from UDP-D-glucose to itself, forming an α-1,4-glycan of about ten residues attached to Tyr194. The product forms a primer for glycogen via glycogen synthase.[1,2]

[1] J. Pitcher, C. Smythe & P. Cohen (1988) *EJB* **176**, 391.
[2] W. J. Whelan (1998) *Protein Sci.* **7**, 2038.

GLYCOGEN PHOSPHORYLASE

This enzyme [EC 2.4.1.1] catalyzes the reaction of [(1,4)-α-D-glucosyl]$_n$ with orthophosphate to produce [(1,4)-α-D-glucosyl]$_{n-1}$ and α-D-glucose 1-phosphate. (The EC number designates the general classification for phosphorylase, and the recommended name should be qualified with the name of the natural substance: thus, maltodextrin phosphorylase, starch phosphorylase, and glycogen phosphorylase are all classified as EC 2.4.1.1.) **See** *Phosphorylase*

GLYCOGEN SYNTHASE

This enzyme [EC 2.4.1.11], also known as glycogen (starch) synthase and UDP-glucose: glycogen glucosyltransferase, catalyzes the reaction of UDP-D-glucose with [(1,4)-α-D-glucosyl]$_n$ to produce UDP and [(1,4)-α-D-glucosyl]$_{n+1}$.[1-7] The name of the enzyme varies with the source and the nature of product: thus, the plant UDP-D-glucose-utilizing enzyme is starch synthase. Glycogen synthase from animal tissues is a complex of a catalytic subunit and the protein glycogenin, which is used as a primer in glycogen synthesis via the action of glycogenin glucosyltransferase [EC 2.4.1.186]. A similar enzyme utilizes ADP-D-glucose:starch (bacterial glycogen) synthase [EC 2.4.1.21]. **See also** *Starch (Bacterial Glycogen) Synthase*

[1] L. F. Leloir & C. E. Cardini (1962) *The Enzymes*, 2nd ed., **6**, 317.
[2] N. B. Madsen (1991) *A Study of Enzymes* **2**, 139.
[3] J. Larner (1990) *AE* **63**, 173.
[4] R. F. Colman (1990) *The Enzymes*, 3rd ed., **19**, 283.
[5] P. Cohen (1986) *The Enzymes*, 3rd ed., **17**, 461.
[6] P. J. Roach (1986) *The Enzymes*, 3rd ed., **17**, 499.
[7] W. Stalmans & H. G. Hers (1973) *The Enzymes*, 3rd ed., **9**, 309.

Selected entries from *Methods in Enzymology* [vol, page(s)]:
General discussion: **5**, 145; **8**, 374; **28**, 530; **34**, 131, 132, 164, 338
Assay: **5**, 145; **8**, 375; **28**, 539, 540; **63**, 32; glycogen synthase D, **28**, 531; glycogen synthase I, **28**, 531; swine kidney, from, **42**, 375
Other molecular properties: control studies, **37**, 293; distribution, **5**, 147; glycogen synthase D (assay, **28**, 531; properties, **28**, 537; purification, **28**, 533); glycogen synthase I (assay, **28**, 531; purification, **28**, 534); insulin effects, **39**, 99; isolation from *Escherichia coli* B, **28**, 540; liver, of, **5**, 145; **8**, 374, 382; **9**, 13 (interconversion of D and I forms, **8**, 383; properties, **5**, 147; **8**, 383; purification, **5**, 146; **8**, 382); multicascade control, **64**, 325; muscle, of, **8**, 376 (preparation of [D form, **8**, 379; I form, **8**, 379]; properties, **8**, 381; purification, **8**, 376); peptide substrates, in analysis of phosphorylase kinase specificity, **99**, 275; properties, **28**, 543; protein kinase, **38**, 328, 351; staining on gels, **22**, 599; rat liver, **5**, 145 (properties, **5**, 147; purification, **5**, 146); swine kidney, from, **42**, 375 (assay, **42**, 375; molecular weight, **42**, 380; properties, **42**, 380; purification, **42**, 377); testosterone activation, **39**, 98

[GLYCOGEN-SYNTHASE D]-PHOSPHATASE

This enzyme [EC 3.1.3.42] catalyzes the hydrolysis of [glycogen-synthase D] (*i.e.*, the phosphorylated form of glycogen synthase [at a seryl residue]) to produce [glycogen-synthase I] (*i.e.*, the dephosphorylated form: EC 2.4.1.11) and orthophosphate.[1-4] Glycogen synthase D is allosterically dependent on the presence of D-glucose

6-phosphate (hence, the designation "D"). The I-form of glycogen synthase is independent of the cooperative effects of D-glucose 6-phosphate.

Other phosphorylated substrates include the phophorylated forms of acetyl-CoA carboxylase, phosphorylase kinase, 6-phosphofructo-2-kinase/fructose-2,6-bisphosphatase, pyruvate kinase, 6-phosphofructo-1-kinase, 3-hydroxy-3-methylglutaryl-CoA reductase, 3-hydroxy-3-methyl-glutaryl-CoA reductase kinase, and the myosin light chain.

[1] L. M. Ballou & E. H. Fischer (1986) *The Enzymes*, 3rd ed., **17**, 311.
[2] S. Tsuiki, K. Kikuchi, S. Tamura, A. Hiraga & R. Shineha (1985) *Adv. Protein Phosphatases* **1**, 193.
[3] M. D. Pato (1985) *Adv. Protein Phosphatases* **1**, 367.
[4] H. C. Li (1982) *Curr. Top. Cell. Regul.* **21**, 129.

Selected entries from **Methods in Enzymology** [**vol**. page(s)]:
General discussion: 107, 102;**159**, 416, 437

GLYCOLALDEHYDE DEHYDROGENASE

This oxidoreductase [EC 1.2.1.21] catalyzes the reversible reaction of glycolaldehyde with NAD^+ and water to produce glycolate and NADH.[1] Other substrates include acetaldehyde, propanal, benzylaldehyde, and glyceraldehyde. This activity has been observed with *Escherichia coli* lactaldehyde dehydrogenase [EC 1.2.1.22].[2]

[1] H. Morita, Y. Tani, K. Ogata & H. Yamada (1978) *Agric. Biol. Chem.* **42**, 2077.
[2] E. Caballero, L. Baldoma, J. Ros, A. Boronat & J. Aguilar (1983) *JBC* **258**, 7788.

GLYCOLATE DEHYDROGENASE

This oxidoreductase [EC 1.1.99.14], also referred to as glycolate oxidoreductase, catalyzes the reaction of glycolate with an acceptor substrate to produce glyoxylate and the reduced acceptor.[1,2] Other substrates include (R)-lactate (*i.e.*, D-lactate). Examples of acceptor substrates include 2,6-dichloroindophenol, phenazine methosulfate, and cytochrome *c*.

[1] J. M. Lord (1971) *BBA* **267**, 227.
[2] A. Yokota & S. Kitaoka (1987) *Agric. Biol. Chem.* **51**, 665.

GLYCOLATE OXIDASE

This oxidoreductase, previously classified as EC 1.1.3.1, is now regarded as an activity of (S)-2-hydroxyacid oxidase [EC 1.1.3.15].

L-GLYCOL DEHYDROGENASE

This oxidoreductase [EC 1.1.1.185] catalyzes the reversible reaction of an L-glycol (for example, 2,3-pentanediol) with

$NAD(P)^+$ to produce a 2-hydroxycarbonyl compound and $NAD(P)H$.[1,2] The 2-hydroxycarbonyl compound formed can be further oxidized to a vicinal dicarbonyl compound. Substrates for the reverse reaction include 2,3-pentanedione, glyoxal, methylglyoxal, glyceraldehyde, and diacetyl. The hen muscle enzyme has an ordered Bi Bi kinetic mechanism.[2]

[1] A. Bernardo, J. Burgos & R. Martin (1981) *BBA* **659**, 189.
[2] J. G. Prieto, R. M. Sarmiento & J. Burgos (1983) *ABB* **224**, 372.

Selected entries from **Methods in Enzymology** [**vol**. page(s)]:
General discussion: 89, 523
Assay: hen muscle, **89**, 523
Other molecular properties: hen muscle (assay, **89**, 523; properties, **89**, 525; purification, **89**, 524)

GLYCOLIPID 2-α-MANNOSYLTRANSFERASE

This transferase [EC 2.4.1.131] catalyzes the transfer of an α-D-mannosyl residue from GDP-D-mannose into a lipid-linked oligosaccharide, forming an α-1,2-D-mannosyl-D-mannose linkage.[1] The two 1,2-linked mannosyl residues observed in mammalian lipid-linked oligosaccharides containing the structure $Glc_3Man_9GlcNAc_2$ are produced by this activity.

[1] J. S. Schutzbach, J. D. Springfield & J. W. Jensen (1980) *JBC* **255**, 4170.

Selected entries from **Methods in Enzymology** [**vol**, page(s)]:
Other molecular properties: enzymatic mannosylation, **247**, 120; guanosine 5'-diphosphomannose regeneration, **247**, 120; substrate specificity, **247**, 119

GLYCOLIPID 3-α-MANNOSYLTRANSFERASE

This transferase [EC 2.4.1.132], also referred to as mannosyltransferase II, catalyzes the transfer of an α-D-mannosyl residue from GDP-D-mannose into a lipid-linked oligosaccharide, producing an α-1,3-D-mannosyl-D-mannose linkage.[1,2] The 1,3-linked mannosyl residue observed in mammalian lipid-linked oligosaccharides containing the structure $Glc_3Man_9GlcNAc_2$ are produced by this activity.

[1] J. W. Jensen & J. S. Schutzbach (1981) *JBC* **256**, 12899.
[2] J. W. Jensen & J. S. Schutzbach (1984) *Biochemistry* **23**, 1115.

GLYCOPEPTIDE α-N-ACETYLGALACTOSAMINIDASE

This enzyme [EC 3.2.1.97], also known as endo-α-N-acetylgalactosaminidase and O-glycosidase, catalyzes the hydrolysis of terminal D-galactosyl-N-acetyl-α-D-galactosaminidic residues from a variety of glycopeptides and glycoproteins (*O*-linked via L-seryl or L-threonyl residues).[1,2] **See also** α-*N*-Acetylgalactosaminidase

[1]J. Umemoto, V. P. Bhavanandan & E. A. Davidson (1977) *JBC* **252**, 8609.
[2]Y. Endo & A. Kobata (1976) *J. Biochem.* **80**, 1.

Selected entries from **Methods in Enzymology** [**vol**, page(s)]:
General discussion: 50, 560
Assay: *Diplococcus pneumoniae*, 50, 561

GLYCOPEPTIDE *N*-GLYCOSIDASE

This plant enzyme, previously listed as EC 3.2.2.18, is now listed under peptide-N^4-(*N*-acetyl-β-glucosaminyl) asparagine amidase [EC 3.5.1.52].

GLYCOPROTEIN-*N*-ACETYLGALACTOS-AMINE 3-β-GALACTOSYLTRANSFERASE

This transglycosylase [EC 2.4.1.122] catalyzes the reaction of UDP-D-galactose with a glycoprotein *N*-acetyl-D-galactosamine (*i.e.*, a glycoprotein containing an *O*-seryl- or *O*-threonyl-linked, non-reducing, *N*-acetyl-D-galactosamine residue) to produce a glycoprotein D-galactosyl-(1,3)-*N*-acetyl-D-galactosamine and UDP.[1–3] Excellent acceptor substrates include large mucin glycopeptides possessing polypeptide-linked *N*-acetylgalactosyl residues.

[1]J. Mendicino, S. Sivakami, M. Davila & E. V. Chandrasekaran (1982) *JBC* **257**, 3987.
[2]K. Furukawa & S. Roth (1985) *BJ* **227**, 573.
[3]M. Granovsky, T. Bielfeldt, S. Peters, H. Paulsen, M. Meldal, J. Brockhausen & I. Brockhausen (1994) *EJB* **221**, 1039.

Selected entries from **Methods in Enzymology** [**vol**, page(s)]:
General discussion: 230, 302

GLYCOPROTEIN ENDO-α-1,2-MANNOSIDASE

This glycosidase [EC 3.2.1.130], involved in the synthesis of glycoproteins, catalyzes the hydrolysis of the terminal α-glucosyl-(1,3)-mannosyl unit from Glc-Man_9-$(GlcNAc)_2$ oligosaccharide component of the glycoprotein produced in the Golgi membrane. This endo-α-D-mannosidase provides a processing route alternative to the sequential actions of α-glucosidase II and α-mannosidase I.[1]

[1]W. A. Lubas & R. G. Spiro (1987) *JBC* **262**, 3775.

GLYCOPROTEIN *O*-FATTY-ACYLTRANSFERASE

This acyltransferase [EC 2.3.1.142] catalyzes the reaction of palmitoyl-CoA with a mucus glycoprotein to produce an *O*-palmitoylglycoprotein and Coenzyme A.[1,2]

[1]C. Kasinathan, E. Grzelinska, K. Okazaki, B. L. Slomiany & A. Slomiany (1990) *JBC* **265**, 5139.
[2]M. F. Schmidt & G. R. Burns (1989) *Biochem. Soc. Trans.* **17**, 859.

GLYCOPROTEIN-FUCOSYLGALACTOSIDE α-*N*-ACETYLGALACTOSAMINYL-TRANSFERASE

This transferase [EC 2.4.1.40], also called fucosylglycoprotein α-*N*-acetylgalactosaminyltransferase and histo-blood group A transferase, catalyzes the reaction of UDP-*N*-acetyl-D-galactosamine with a glycoprotein-α-L-fucosyl-(1,2)-D-galactose (*i.e.*, a glycoprotein containing α-L-fucosyl-(1,2)-D-galactose) to produce UDP and glycoprotein-*N*-acetyl-α-D-galactosaminyl-(1,3)-[α-L-fucosyl-(1,2)]-D-galactose, thus transferring the *N*-acetylgalactosamine group from UDP-*N*-acetyl-galactosamine to H-active structures to form A determinants.

The enzyme can use a number of 2-fucosylgalactosides as acceptor substrates: for example, α-L-fucosyl-1,2-D-galactose, 2′-fucosyllactose, and porcine submaxillary asialomucin of A-negative blood-type.

Selected entries from **Methods in Enzymology** [**vol**, page(s)]:
General discussion: 28, 511; 230, 302
Assay: 28, 511
Other molecular properties: blood type A structural determinant and, 28, 511; isolation from human milk, 28, 513; porcine submaxillary gland (assay, 83, 488; properties, 83, 493; purification, 83, 489); properties, 28, 514; purification, 28, 513

GLYCOPROTEIN-FUCOSYLGALACTOSIDE α-GALACTOSYLTRANSFERASE

This transferase [EC 2.4.1.37] (also referred to as fucosylglycoprotein 3-α-galactosyltransferase, fucosylgalactose 3-α-galactosyltransferase, histo-blood group B transferase, and blood-group substance B-dependent galactosyltransferase) catalyzes the reaction of UDP-D-galactose with a glycoprotein-α-L-fucosyl-(1,2)-D-galactose (*i.e.*, a glycoprotein containing α-L-fucosyl-(1,2)-D-galactose) to produce UDP and a glycoprotein-α-D-galactosyl-(1,3)-[α-L-fucosyl-(1,2)]-D-galactose.[1–3]

[1]L. R. Carne & W. M. Watkins (1977) *BBRC* **77**, 700.
[2]A. Betteridge & W. M. Watkins (1983) *EJB* **132**, 29.
[3]V. P. Kamath, N. O. Seto, C. A. Compston, O. Hindsgaul & M. M. Palcic (1999) *Glycoconj. J.* **16**, 599.

Selected entries from **Methods in Enzymology** [**vol**, page(s)]:
General discussion: 230, 302

GLYCOPROTEIN 6-α-L-FUCOSYLTRANSFERASE

This transferase [EC 2.4.1.68], also called GDP-fucose: glycoprotein fucosyltransferase, catalyzes the reaction of GDP-L-fucose with N^4-(N-acetyl-β-D-glucosaminyl-1,2-α-D-mannosyl-1,3-(R$_1$-α-1,6)-β-D-mannosyl-β-N-acetyl-1,4-D-glucosaminyl-1,4-N-acetyl-D-glucosaminyl)asparagine to produce GDP and N^4-(N-acetyl-β-D-glucosaminyl-1,2-α-D-mannosyl-1,3-(R$_1$-α-1,6)-β-D-mannosyl-1,4-β-N-acetyl-D-glucosaminyl-1,4-(α-L-fucosyl-1,6)-N-acetyl-D-glucosaminyl)asparagine. Asialo-agalactotransferrin glycopeptide is a substrate.[1] Although the biantennary oligosaccharide is an excellent substrate, the human platelet enzyme has been reported to be more active with a specific triantennary oligosaccharide.[2]

[1] J. A. Voynow, T. F. Scanlin & M. C. Glick (1988) Anal. Biochem. **168**, 367.
[2] J. Kaminska, M. C. Glick & J. Koscielak (1998) Glycoconj. J. **15**, 783.

Selected entries from *Methods in Enzymology* [vol, page(s)]:
General discussion: 98, 131; 230, 304
Assay: 98, 131

GLYCOPROTEIN N-PALMITOYLTRANSFERASE

This acyltransferase [EC 2.3.1.96] catalyzes the reaction of palmitoyl-CoA with a glycoprotein to produce coenzyme A and an N-palmitoylglycoprotein. The enzyme, activated by 1,4-dithiothreitol, acts on a mucus glycoprotein.[1-3]

[1] B. L. Slomiany, Y. H. Liau, S. R. Carter, J. Zielenski & A. Slomiany (1968) Arch. Oral Biol. **31**, 463.
[2] Y. H. Liau, B. L. Slomiany, A. Slomiany, A. Piasek, D. Palmer & W. S. Rosenthal (1985) Digestion **32**, 57.
[3] B. L. Slomiany, Y. H. Liau, K. Mizuta & A. Slomiany (1987) Biochem. Pharmacol. **36**, 3273.

GLYCOPROTEIN PHOSPHOLIPASE D

This phospholipase [EC 3.1.4.50], also known as glycophosphatidylinositol-phospholipase D and glycosylphosphatidylinositol-phospholipase D, catalyzes the hydrolysis of a glycoprotein phosphatidylinositol to produce a phosphatidate and a glycoprotein inositol.[1-3] Hence, it acts on the glycosylphosphatidylinositol membrane anchor of cell-surface proteins in animal tissues, thus releasing these proteins from the membrane. It is not identical with variant-surface glycoprotein phospholipase C [EC 3.1.4.47]. Lipid A, phosphatidate, and lysophosphatidate are potent inhibitors of this enzyme.[4,5]

[1] A. S. Malik & M. G. Low (1986) BJ **240**, 519.
[2] S. Stieger, S. Diem, A. Jakob & U. Brodbeck (1991) EJB **197**, 67.
[3] M. G. Low & A. R. S. Prasad (1988) PNAS **85**, 980.

[4] M. G. Low and K.-S. Huang (1993) JBC **268**, 8480.
[5] M. G. Low & P. Stutz (1999) ABB **371**, 332

Selected entries from *Methods in Enzymology* [vol, page(s)]:
General discussion: 197, 567; 250, 630
Assay: 197, 567
Other molecular properties: alkaline phosphatase substrate, 197, 568; calcium dependence, 250, 636; cDNA analysis of protein structure, 250, 633; detergent effects, 250, 637; discovery, 250, 631; glycophosphatidylinositol treatment, 230, 430; identification of glycophosphatidylinositol-anchored proteins, 250, 604 (product analysis, 250, 639; reaction conditions, 250, 637; sources of enzyme, 250, 638; substrate concentration, 250, 638); phosphorylation, 250, 634; properties, 197, 574; purification, 197, 572 (bovine plasma enzyme, 250, 633; human serum enzyme, 250, 632); substrate specificity, 250, 635; surface glycoprotein substrate, 197, 570

GLYCOSAMINOGLYCAN GALACTOSYLTRANSFERASE

This transferase [EC 2.4.1.74], which participates in the biosynthesis of galactose-containing glycosaminoglycans of *Dictyostelium discoideum*, catalyzes the reaction of UDP-D-galactose with a glycosaminoglycan to produce UDP and a D-galactosylglycosaminoglycan.[1,2] Substrates include desialylated mucin and acidic mucopolysaccharides.

[1] M. Sussman & M. J. Osborn (1964) PNAS **52**, 81.
[2] G. N. Andersson & L. C. Eriksson (1979) BBA **570**, 239.

GLYCOSPHINGOLIPID DEACYLASE

This enzyme [EC 3.5.1.69], also known as glycosphingolipid ceramide deacylase, catalyzes the hydrolysis of gangliosides and neutral glycosphingolipids (for example, ganglioside G$_{M2}$), releasing fatty acid anions to form the lyso-derivatives.[1] However, sphingolipids such as ceramide are not substrates. The enzyme is distinct from acylsphingosine deacylase [EC 3.5.1.23]. Labeled glycosphingolipids can be synthesized by the reverse reaction.[2,3]

[1] Y. Hirabayashi, M. Kimura, M. Matsumoto, K. Yamamoto, S. Kadowaki & T. Tochikura (1988) J. Biochem. **103**, 1.
[2] M. Ito, T. Kurita & K. Kita (1995) JBC **270**, 24370.
[3] T. Nakagawa, M. Tani, K. Kita & M. Ito (1999) J. Biochem. **126**, 604.

Selected entries from *Methods in Enzymology* [vol, page(s)]:
General discussion: 311, 297

GLYCOSULFATASE

This enzyme [EC 3.1.6.3], also referred to as glucosulfatase, catalyzes the hydrolysis of D-glucose 6-sulfate to produce D-glucose and sulfate.[1-3] Other substrates include D-galactose 6-sulfate, N-acetyl-D-glucosamine 6-sulfate, D-mannose 6-sulfate, L-fucose 2-, 3-, and 4-sulfates, neoarabiose 4-sulfate, neocarratetraose 4-sulfate, and adenosine 5'-sulfate.

[1] H. Hatanaka, Y. Ogawa & F. Egami (1976) *J. Biochem.* **79**, 27.
[2] H. Hatanaka, Y. Ogawa & F. Egami (1976) *BJ* **159**, 445.
[3] M. W. McLean & F. B. Williamson (1979) *EJB* **101**, 497.

Selected entries from *Methods in Enzymology* [**vol**, page(s)]:
General discussion: 2, 324, 330; 8, 670
Assay: 2, 324, 330; 8, 671

GLYCOSYLCERAMIDASE

This enzyme [EC 3.2.1.62], also known as phlorizin hydrolase and phloretin-glucosidase, catalyzes the hydrolysis of a glycosyl-*N*-acylsphingosine to produce a sugar and an *N*-acylsphingosine.[1-5] This hydrolase, which also catalyzes the hydrolysis of phlorizin to generate phloretin and D-glucose, exhibits a specificity that is even broader than glucosylceramidase [EC 3.2.1.45] or galactosylceramidase [EC 3.2.1.46]. The membrane-bound intestinal enzyme also possesses lactase [EC 3.2.1.108] activity (and is often known as lactase-phlorizin hydrolase). *See also Galactosylceramidase; Glucosylceramidase*

[1] H. Skovbjerg, H. Sjöström & O. Noren (1981) *EJB* **114**, 653.
[2] T. Kobayashi & K. Suzuki (1981) *JBC* **256**, 7768.
[3] J. Cousineau & J. R. Green (1980) *BBA* **615**, 147.
[4] S. Ramaswamy & A. N. Radhakrishnan (1975) *BBA* **403**, 446.
[5] E. Birkenmeier & D. H. Alpers (1974) *BBA* **350**, 100.

Selected entries from *Methods in Enzymology* [**vol**, page(s)]:
General discussion: 8, 591, 595; 311, 130, 276
Assay: 8, 595
Other molecular properties: intestine, from rat, 8, 595 (properties, 8, 596; purification, 8, 595); microvillous membrane, in, 31, 130

GLYCYL ENDOPEPTIDASE

This cysteine endopeptidase [EC 3.4.22.25], also known as papaya peptidase B, papaya proteinase IV, and chymopapain M, catalyzes the hydrolysis of peptide bonds in proteins with a preferential cleavage at Gly–Xaa.[1,2] There is also a preference for hydrophobic aminoacyl residues in the P2 and P3 positions. The enzyme also has an esterase activity. Chromogenic substrates, such as Boc-Ala-Ala-Gly-∇-NHPhNO$_2$ (where ∇ indicates the cleavage site), are typically used in activity assays. (Note: Powdered glycyl endopeptidase is a potent allergen and care should be exercised in its handling.)

[1] D. J. Buttle (1994) *MIE* **244**, 539.
[2] B. P. O'Hara, A. M. Hemmings, D. J. Buttle & L. H. Pearl (1995) *Biochemistry* **34**, 13190.

Selected entries from *Methods in Enzymology* [**vol**, page(s)]:
General discussion: 244, 539
Assay: activity assays (Boc-Ala-Ala-Gly-NHMec, 244, 542; Boc-Ala-Ala-Gly-NHPhNO$_2$, 244, 541)

Other molecular properties: abundance in papaya latex, 244, 549; active site titration, 244, 543; allergenic properties, 244, 549; cleavage site specificity, 244, 550, 552; discovery, 244, 540; extinction coefficient, 244, 549; homology with papain, 244, 549, 555; immunoassay, 244, 545, 548; inhibition rate constants, 244, 553; inhibitors, 244, 551, 553; purification, 244, 547; size, 244, 549; storage, 244, 548; substrate specificity, 244, 550, 555

GLYCYLPEPTIDE N-TETRADECANOYLTRANSFERASE

This acyltransferase [EC 2.3.1.97], also known as peptide *N*-myristoyltransferase and peptide *N*-tetradecanoyltransferase, catalyzes the reaction of tetradecanoyl-CoA with a glycyl-peptide (*i.e.*, a pepide with an N-terminal glycyl residue) to produce coenzyme A and an *N*-tetradecanoylglycyl-peptide.[1-4] The yeast enzyme is very specific for tetradecanoyl-CoA and for an N-terminal glycyl residue in oligopeptides containing a seryl residue in the fifth aminoacyl position. The *Saccharomyces cerevisiae* transferase has an ordered Bi Bi mechanism and is reported to have a required cysteinyl and asparaginyl residue at the active site.[5] The enzyme from mammalian heart transfers acyl groups to a number of acceptor proteins (*e.g.*, the catalytic subunit of protein kinase A and the G protein α subunit).[4]

[1] D. A. Towler, J. I. Gordon, S. P. Adams & L. Glaser (1988) *Ann. Rev. Biochem.* **57**, 69.
[2] R. S. Bhatnagar, K. Futterer, G. Waksman & J. I. Gordon (1999) *BBA* **1441**, 162.
[3] R. V. Raju & R. K. Sharma (1999) *Methods Mol. Biol.* **116**, 193.
[4] L. Zhang, E. Jackson-Machelski & J. I. Gordon (1996) *JBC* **271**, 33131.

Selected entries from *Methods in Enzymology* [**vol**, page(s)]:
General discussion: 250, 394, 405, 467
Assay: denaturing gel analysis, 250, 418, 429; Edman degradation, 250, 419, 429; gel fluorography, 250, 418; HPLC, 250, 410, 413, 415; mass spectrometry, 250, 419, 429; myristate analogues (complementation of auxotrophy and temperature-sensitive growth arrest in yeast mutant, 250, 429; *Escherichia coli*-based assay, 250, 428; inhibition of HIV replication, 250, 431)
Other molecular properties: cell-free lysate systems for *in vitro* acylation, 250, 462; chemotherapy targeting, 250, 406, 495; coexpression with G-protein α-subunits (in *Escherichia coli*, 237, 257; in *Saccharomyces cerevisiae*, 237, 149); expression factor, 237, 55; expression of yeast enzyme in *Escherichia coli* (coexpression with substrate proteins [ADP-ribosylation factor, 250, 395, 399, 402, 405, 418; evaluation of myristoylation efficiency, 250, 416; plasmids, 250, 412, 415; strains, 250, 416]; purification of product, 250, 410, 412, 481); isothermal titration calorimetry (contribution of 3'-phosphate group of CoA to binding, 250, 483; cooperativity of ligand binding, 250, 467, 484; purity of protein, 250, 477, 480; thermodynamic parameters of ligand binding, 250, 481, 483; variation of functional groups on peptide ligands, 250, 484); protein substrate sequence recognition, 250, 409; reaction mechanism, 250, 409, 480, 484; sequence homology between species, 250, 407; substrate specificity (myristate analogues, 250, 407, 421, 480, 483; proteins, 250, 400, 406, 409, 437, 480; sequence recognition, 250, 437); yeast, expression vector construction, 237, 55; yeast mutants, 250, 429

GLYCYL-tRNA SYNTHETASE

This enzyme [EC 6.1.1.14], also known as glycine:tRNA ligase, catalyzes the reaction of glycine with tRNA^{Gly} and ATP to produce glycyl-tRNA^{Gly}, AMP, and pyrophosphate (or, diphosphate).[1-4] The enzyme has an additional activity: it catalyzes the synthesis of dinucleoside polyphosphates, which may participate in the regulation of cell functions. Glycyl-tRNA synthetase is one of the few aminoacyl-tRNA synthetases which exhibit different oligomeric structures in different organisms.[2] The reaction mechanism proceeds via the formation of an enzyme-bound glycyl-adenylate intermediate, using a negatively charged pit for recognition of glycine.[1] *See also Aminoacyl-tRNA Synthetases*

[1] J. G. Arnez, A. C. Dock-Bregeon & D. Moras (1999) *J. Mol. Biol.* **286**, 1449.
[2] W. Freist, D. T. Logan & D. H. Gauss (1996) *Biol. Chem. Hoppe-Seyler* **377**, 343.
[3] D. L. Ostrem & P. Berg (1974) *Biochemistry* **13**, 1338.
[4] B. Niyomporn, J. L. Dahl & J. L. Strominger (1968) *JBC* **243**, 773.

Selected entries from *Methods in Enzymology* [vol, page(s)]:
General discussion: **17A**, 966

GLYCYRRHIZINATE β-GLUCURONIDASE

This glycosidase [EC 3.2.1.128] catalyzes the hydrolysis of glycyrrhizinate (a triterpenoid glycoside from the roots of *Glycyrrhiza* sp.) to produce 1,2-β-D-glucuronosyl-D-glucuronate and glycyrrhetinate. The enzyme from *Aspergillus niger* is very specific.[1]

[1] T. Muro, T. Kuramoto, K. Imoto & S. Okada (1986) *Agric. Biol. Chem.* **50**, 687.

GLYOXYLATE DEHYDROGENASE (ACYLATING)

This oxidoreductase [EC 1.2.1.17] catalyzes the reversible reaction of glyoxylate with coenzyme A and NADP$^+$ to produce oxalyl-CoA ($^-$OOCCOSCoA), NADPH, and a proton.[1] A cytochrome$m$ c-dependent dehydrogenase has been reported in the fungus *Tyromyces palustris*.[2]

[1] J. R. Quayle & G. A. Taylor (1961) *BJ* **78**, 611.
[2] T. Tokimatsu, Y. Nagai, T. Hattori & M. Shimada (1998) *FEBS Lett.* **437**, 117.

Selected entries from *Methods in Enzymology* [vol, page(s)]:
General discussion: **9**, 342
Assay: **9**, 343

GLYOXYLATE OXIDASE

This oxidoreductase [EC 1.2.3.5] catalyzes the reaction of glyoxylate with water and dioxygen to produce oxalate and

hydrogen peroxide.[1] A cytochrome c-dependent enzyme has been reported in the fungus *Tyromyces palustris*.[2]

[1] T. Kasai, I. Suzuki & T. Asai (1962) *Koso Kogaku Shimpojiumii* **17**, 77.
[2] T. Tokimatsu, Y. Nagai, T. Hattori & M. Shimada (1998) *FEBS Lett.* **437**, 117.

Selected entries from *Methods in Enzymology* [vol, page(s)]:
General discussion: **52**, 18, 19; **56**, 474

GLYOXYLATE REDUCTASES

Glyoxylate reductase [EC 1.1.1.26] catalyzes the reaction of glyoxylate with NADH to produce glycolate and NAD$^+$, with A-side stereospecificity.[1,2] The enzyme will also catalyze the NADH-dependent interconversion of hydroxypyruvate to D-glycerate.

Glyoxylate reductase (NADPH) [EC 1.1.1.79] catalyzes the reaction of glyoxylate with NADPH to produce glycolate and NADP$^+$, with A-side stereospecificity (as well as the hydroxypyruvate to D-glycerate conversion).[1,3-5] The enzyme can use NADH as a substrate, although not as effectively as NADPH. The glyoxylate reductase from the green algae, *Chlamydomonas reinhardtii*, does not act on hydroxypyruvate.[3]

D-Glycerate dehydrogenase [EC 1.1.1.29] also exhibits a glyoxylate reductase activity.[6]

[1] L. A. Kleczkowski (1994) *Ann. Rev. Plant Physiol. Plant Mol. Biol.* **45**, 339.
[2] I. Zelitch (1955) *JBC* **216**, 553 and (1953) **201**, 719.
[3] D. W. Husic & N. E. Tolbert (1987) *ABB* **252**, 396.
[4] L. A. Kleczkowski, D. D. Randall & D. G. Blevins (1986) *BJ* **239**, 653.
[5] A. Yokota, S. Haga & S. Kitaoka (1985) *BJ* **227**, 211.
[6] C. F. Giafi & G. Rumsby (1998) *Ann. Clin. Biochem.* **35**, 104.

Selected entries from *Methods in Enzymology* [vol, page(s)]:
General discussion: **1**, 528; **41**, 343
Assay: **1**, 532; **31**, 745; *Pseudomonas*, **41**, 343
Other molecular properties: D-glycerate dehydrogenase, comparison to, **9**, 228; marker enzyme, as, **31**, 744, 745; microbodies, in, **52**, 496, 499; *Pseudomonas*, from, **41**, 343 (assay, **41**, 343; chromatography, **41**, 345; properties, **41**, 347, 348; purification, **41**, 344); stereospecificity, **87**, 104, 109; tobacco leaves (properties, **1**, 535; purification, **1**, 533)

GLY-X CARBOXYPEPTIDASE

This metallocarboxypeptidase [EC 3.4.17.4], also called glycine carboxypeptidase, Gly–Xaa carboxypeptidase, and carboxypeptidase S, catalyzes the hydrolysis of a peptidyl-glycine to produce a peptide and glycine. The enzyme will also catalyze the release of a C-terminal amino acid from a peptide in which glycine is the penultimate aminoacyl residue (that is, in the P1 position). The enzyme is readily assayed with synthetic substrates such as Z-Gly-∇-Leu, where ∇ indicates the cleavage site.[1]

[1]D. H. Wolf & U. Weiser (1977) *EJB* **73**, 553.
Selected entries from *Methods in Enzymology* [vol, page(s)]:
Other molecular properties: Saccharomyces, **248**, 223; yeast (biochemical analysis, **185**, 385; enzymatic assay, **194**, 449; genetic analysis, **185**, 377; well test, **194**, 439)

GMP REDUCTASE

This oxidoreductase [EC 1.6.6.8], also referred to as guanosine 5′-monophosphate oxidoreductase, catalyzes the reaction of NADPH with guanosine 5′-phosphate (*i.e.*, GMP) to produce NADP$^+$, inosine 5′-phosphate (*i.e.*, IMP), and ammonia.[1,2] The human enzyme has an ordered kinetic mechanism.[1]

[1]T. Spector, T. E. Jones & R. L. Miller (1979) *JBC* **254**, 2308.
[2]A. Spadaro, A. Giacomello & C. Salerno (1986) *Adv. Exp. Med. Biol.* **195**, 321.

Selected entries from *Methods in Enzymology* [vol, page(s)]:
General discussion: 6, 114
Assay: 6, 114

GMP SYNTHETASES

There are two enzymes catalyzing the ATP-dependent synthesis of GMP from XMP.

GMP synthetase [EC 6.3.4.1], also known as GMP synthase, xanthosine-5′-phosphate:ammonia ligase, xanthosine-5′-monophosphate aminase, and XMP aminase, catalyzes the reaction of ATP with xanthosine 5′-monophosphate (*i.e.*, XMP) and ammonia to produce AMP, pyrophosphate (or, diphosphate), and guanosine 5′-monophosphate (*i.e.*, GMP).[1–3] The *Escherichia coli* enzyme reportedly has an ordered Ter Ter kinetic mechanism.[4] Initial-rate and positional isotope exchange experiments support the formation of an adenyl-XMP intermediate.[4]

GMP synthetase (glutamine-hydrolyzing) [EC 6.3.5.2], also known as GMP synthase (glutamine-hydrolyzing), catalyzes the reaction of ATP with xanthosine 5′-monophosphate (*i.e.*, XMP), L-glutamine, and water to produce AMP, pyrophosphate (or diphosphate), GMP, and L-glutamate. An essential cysteinyl residue is needed for the glutamine-dependent activity, generating a transient γ-glutamyl thioester intermediate.[5] Human GMP synthetase exhibits sigmoidal kinetics with respect to the substrate XMP. This positive cooperativity is not due to ligand-induced oligomerization, because the enzyme remains monomeric, even in the presence of substrates.[6]

[1]K. Hirai, Y. Matsuda & H. Nakagawa (1987) *J. Biochem.* **102**, 893.
[2]T. Spector, T. E. Jones, T. A. Krenitsky & R. J. Harvey (1976)*BBA* **452**, 597.

[3]T. Spector & L. M. Beacham III (1975) *JBC* **250**, 3101.
[4]W. von der Saal, C. S. Crysler & J. J. Villafranca (1985) *Biochemistry* **24**, 5343.
[5]J. Nakamura, K. Straub, J. Wu & L. Lou (1995) *JBC* **270**, 23450.
[6]J. Nakamura & L. Lou (1995) *JBC* **270**, 7347.

Selected entries from *Methods in Enzymology* [vol, page(s)]:
General discussion: 6, 111; 46, 420; 51, 213, 219; 113, 273
Assay: 6, 111; 51, 213, 214, 219; 63, 32; 113, 273
Other molecular properties: activity, **51**, 213; alternate names, **51**, 219; amino donor, **6**, 107; **51**, 213, 217; derepression of, **22**, 89; diffusion coefficient, **51**, 218; Ehrlich ascites cells, from, **51**, 219; extinction coefficient, **51**, 217; inhibitors, **51**, 217, 223, 224; kinetic properties, **51**, 217, 223, 224; molecular weight, **51**, 218, 223; partial specific volume, **51**, 218; pH optimum, **51**, 223; **113**, 277; positional isotope exchange studies, **249**, 424; properties, **6**, 113; **51**, 216, 223; **113**, 277; purification, **6**, 112; **51**, 214, 221; **113**, 274; purity, **51**, 216, 217; sedimentation coefficient, **51**, 218; storage, **51**, 223; substrate specificity, **51**, 224; subunit structure, **51**, 218; thiol group, active-site, **87**, 83

GONYAULAX LUCIFERASE

Gonyaulax luciferin

Gonyaulax oxoluciferin

This luciferase of the photosynthetic marine dinoflagellate *Gonyaulax polyedra* catalyzes the reaction of *Gonyaulax* luciferin (an open-chain tetrapyrrole) with dioxygen to produce the oxidized luciferin (oxoluciferin), water, and blue-green light.

Selected entries from *Methods in Enzymology* [vol, page(s)]:
General discussion: 133, 307; 305, 249, 258
Assay: 133, 310, 312; 305, 253, 262
Other molecular properties: gene structure, **305**, 250; glutathione S-transferase fusion protein, **305**, 253, 254; pH regulation, **305**, 249, 260; properties, **133**, 316; purification, **133**, 313; **305**, 264

G-PROTEINS

These GTP-binding/hydrolyzing proteins are involved in affinity-modulated changes in protein structure that play central roles in metabolic regulation.[1–3] The regulatory effect is maintained as long as bound GTP remains unhydrolyzed, and the system represents a timing mechanism.

The cellular plasma membrane contains highly specialized systems for integrating, amplifying, and transducing metabolic signals from extracellular hormones, neurotransmitters, other regulatory molecules, and/or physical stimulation.[1,2] A major mechanism for processing this information involves a family of guanine nucleotide-binding regulatory proteins (or G-proteins). Conformational alteration of the G-protein allows exchange of tightly bound guanosine diphosphate (an inactive ligand) for guanosine triphosphate (the activating ligand). The guanosine triphosphate-bound G-protein in turn interacts with intracellular effector molecules, such as adenylyl cyclase, and controls their functions. Vale[3] has discussed how the mechanisms of molecular switches, latches, and amplifiers conform to common mechanistic themes of G proteins and molecular motors.

Although the overall reaction yields GDP and orthophosphate, the enzyme is misclassified as a hydrolase, and instead should be regarded as an energase-type reaction. Energases facilitate affinity-modulated reactions by coupling the $\Delta G_{\text{GTP–hydrolysis}}$ to a force-generating or work-producing step. In this case, P-O-P bond-scission supplies the energy to drive protein-protein interactions needed in signal transduction.[4] The overall enzyme-catalyzed reaction is: $\text{State}_1 + \text{GTP} = \text{State}_1 + \text{GDP} + \text{P}_i$, where two affinity-modulated conformational states are indicated.[4]

[1] M. Rodbell (1992) *Curr. Top. Cell Regul.* **32**, 1.
[2] J. R. Hepler & A. G. Gilman (1992) *TiBS* **17**, 383.
[3] R. D. Vale (1996) *J. Cell Biol.* **135**, 291.
[4] D. L. Purich (2001) *TiBS* **26**, 417.

Selected entries from *Methods in Enzymology* [vol, page(s)]:
Purification and Characterization of G Proteins: application of GTP-agarose matrices, **195**, 171; purification of G proteins, **195**, 177; preparation of guanine nucleotide-free G proteins, **195**, 188; purification of recombinant G$_{sa}$, **195**, 192; purification of recombinant G$_{ia}$ and G$_{oa}$ proteins from *Escherichia coli* , **195**, 202; synthetic peptide antisera with determined specificity for G protein α or β subunits, **195**, 215; quantitation and purification of ADP-ribosylation factor, **195**, 233; soluble guanine nucleotide-dependent ADP-ribosylation factors in activation of adenylyl cyclase by cholera toxin, **195**, 243
Labeling and Quantitating of G Proteins: ADP-ribosylation of G proteins with pertussis toxin, **195**, 257; cholera toxin-catalyzed [^{32}P]ADP-ribosylation of proteins, **195**, 267; photoaffinity labeling of GTP-binding proteins, **195**, 280; identification of receptor-activated G proteins with photoreactive GTP analog, [α-^{32}P]GTP azidoanilide, **195**, 286; quantitative immunoblotting of G-protein subunits, **195**, 302; assay of G-protein $\beta\gamma$-subunit complex by catalytic support of ADP-ribosylation of G$_{o}\alpha$, **195**, 315; tryptophan fluorescence of G proteins: analysis of GTP binding and hydrolysis, **195**, 321
Guanine Nucleotide Exchange and Hydrolysis: measurement of intrinsic nucleotide exchange and GTP hydrolysis rates, **256**, 67; guanine nucleotide exchange catalyzed by *dbl* oncogene product, **256**, 77; stimulation of nucleotide exchange on ras- and rho-related proteins by

small GTP binding protein GDP dissociation stimulator, **256**, 85; interaction of Ect2 and Dbl with Rho-related GTPases, **256**, 90; solubilization of Cdc42Hs from membranes by Rho-GDP dissociation inhibitor, **256**, 98; purification and GTPase-activating protein activity of *Baculovirus* expressed p190, **256**, 105; GTPase-activated protein activity of $n(\alpha_1)$-chimaerin and effect of lipids, **256**, 114; characterization of breakpoint cluster region kinase and SH2 binding activities, **256**, 125; identification of GTPase-activating proteins by nitrocellulose overlay assay, **256**, 130; identification of 3BP-1 in cDNA expression library by SH3 domain screening, **256**, 140; purification and properties of Rab3A, **257**, 57; purification and properties of bovine Rab-GDP dissociation inhibitor, **257**, 70; expression and purification of recombinant His^6-tagged guanine nucleotide dissociation inhibitor and formation of Rab1 complex, **257**, 80; purification of GDP dissociation stimulator dss4 from recombinant bacteria, **257**, 84; expression, purification, and functional assay of Mss4, **257**, 93; expression, purification, and assay Sec12p: A Sar1p-specific GDP dissociation stimulator, **257**, 98; oligonucleotide mutagenesis of Rab GTPases, **257**, 107; high expression cloning, purification, and assay of Ypt-GTPase activating proteins, **257**, 118; preparation of recombinant ADP-ribosylation factor, **257**, 128; catalysis of guanine nucleotide exchange of Ran by RCC1 and stimulation of hydrolysis of Ran-bound GTP by Ran-GAP, **257**, 135; purification of Sec23P-Sec24p complex, **257**, 14
Heterotrimeric G Proteins (Gα Subunits): measurement of receptor-stimulated guanosine 5′-O -(γ-thio)triphosphate binding by G proteins, **236**, 1; receptor-stimulated hydrolysis of guanosine 5′-triphosphate in membrane preparations, **236**, 13; regulation of G-protein activation by mastoparans and other cationic peptides, **236**, 26; guanosine 5′-O-(γ-thio)triphosphate binding assay for solubilized G proteins, **236**, 38; activation of cholera toxin by ADP-ribosylation factors: 20-kDa guanine nucleotide-binding proteins, **236**, 44; pertussis toxin-catalyzed ADP-ribosylation of G Proteins, **236**, 63; synthesis and use of radioactive photoactivable NAD$^+$ derivatives as probes for G-protein structure, **236**, 70; photoaffinity guanosine 5′-triphosphate analogs as a tool for the study of GTP-binding proteins, **236**, 100; preparation of activated α subunits of G$_s$ and G$_{is}$: from erythrocyte to activated subunit, **236**, 110; purification and separation of closely related members of *Pertussis* toxin-substrate G proteins, **236**, 131; purification of transducin, **236**, 139; expression of G-protein α subunits in *Escherichia coli*, **236**, 146; synthesis and applications of affinity matrix containing immobilized $\beta\gamma$ subunits of G proteins, **236**, 164; purification of activated and heterotrimeric forms of Gq proteins, **236**, 174; purification of phospholipase C-activating G protein, G$_{11}$, from turkey erythrocytes, **236**, 182; purification of recombinant G9α, G$_{11}\alpha$, and G$_{16}\alpha$ from Sf9 Cells, **236**, 191; expression and purification of G-protein α subunits using *Baculovirus* expression system, **236**, 212; analysis of G-protein α and $\beta\gamma$ subunits by in vitro translation, **236**, 226; assays for studying functional properties of in vitro translated G$_5\alpha$ subunit, **236**, 239; myristoylation of G-protein α subunits, **236**, 254; specificity and functional applications of anti-peptide antisera which identify G-protein α subunits, **236**, 268; identification of receptor-activated G proteins: selective immunoprecipitation of photolabeled G-protein α subunits, **236**, 283; identification of mutant forms of G-protein α subunits in human neoplasia by polymerase chain reaction-based techniques, **236**, 295; detection of mutations and polymorphisms of G$_s\alpha$ subunit gene by denaturing gradient gel electrophoresis, **236**, 308; construction of mutant and chimeric G-protein α subunits, **236**, 321; design of degenerate oligonucleotide primers for cloning of G-protein α subunits, **236**, 327; microinjection of antisense oligonucleotides to assess G-protein subunit function, **236**, 345; inactivation of G-protein genes: double knockout in cell lines, **236**, 356; targeted inactivation of the G$_{i2}\alpha$ gene with replacement and insertion vectors: analysis in a 96-well plate format, **236**, 366; G-protein assays in *Dictyostelium*, **236**, 387; fluorescence assays for G-protein interactions, **236**, 409; specific peptide probes for G-protein interactions with receptors, **236**, 423; vaccinia virus systems for expression of Gα genes in S49 cells, **236**, 436

Heterotrimeric G Proteins (Gβγ Subunits): purification of Tβγ subunit of transducin, **236**, 449; adenylyl cyclase assay for βγ subunits of G proteins, **236**, 451; synthesis and use of biotinylated βγ complexes prepared from bovine brain G proteins, **236**, 457; design of oligonucleotide probes for molecular cloning of β and γ subunits, **236**, 471; characterization of antibodies for various G-protein β and γ subunits, **236**, 482; preparation, characterization, and use of antibodies with specificity for G-protein γ subunits, **236**, 498; isoprenylation of γ subunits and G-protein effectors, **236**, 509

Cell Expression and Analysis: analysis of *ras* protein expression in mammalian cells, **255**, 195; vaccinia virus expression of p21 *ras* asn-17, cell expression and analysis; inducible expresssion of *ras* N17 dominant inhibitory protein, **255**, 230; prenylation and palmitoylation analysis, **255**, 237; immune complex kinase assays for mitogen-activated protein kinase and MEK, **255**, 245; cell-free assay system for *ras*-dependent mek activation, **255**, 257; preparation and use of semi-intact mammalian cells for analysis of signal transduction, **255**, 265; mitogen-activated protein kinase activation after scrape loading of P21*ras*, **255**, 273; assay and expression of mitogen-activated protein kinase, MAP kinase kinase, and raf, **255**, 279; assay of MEK kinase, **255**, 290; activation of raf-1 by *ras* in intact cells, **255**, 301; *ras-raf* complexes: analyses of complexes formed *in vivo*, **255**, 310; *ras-raf* complexes *in vitro*, **255**, 323; *ras-raf* interaction: two-hybrid analysis, **255**, 331; methods for analyzing c-jun kinase, **255**, 342; use of tyrosine-phosphorylated proteins to screen bacterial expression libraries for SH2 domains, **255**, 360; detection of SH3 binding proteins in total cell lysates with glutathione S-transferase SH3 fusion proteins: SH3 blot assay, **255**, 369; inhibition of *ras* function *in vitro* and *in vivo* using inhibitors of farnesyl-protein transferase, **255**, 378

G-Proteins in Cell Growth and Transformation: analysis of guanine nucleotides associated with protooncogene *ras*, **238**, 255; measuring activation of kinases in mitogen-activated protein kinase regulatory network, **238**, 259; transcriptional activation analysis of oncogene function, **238**, 271; biological assays for cellular transformation, **238**, 277; rat embryo fibroblast complementation assay with *ras* genes, **255**, 389; biological assays for *ras* transformation, **255**, 395; *ras*-mediated transcription activation: analysis by transient cotransfection assays, **255**, 412; oocyte microinjection assay for evaluation of *ras*-induced signaling pathways, **255**, 426; mammalian cell microinjection assays, **255**, 436; detection of point mutations in ras in tumor cell lines by densturant gradient gel electrophoresis, **255**, 442; diagnostic detection of mutant *ras* genes in minor cell populations, **255**, 452; *ras* in yeast: complementation assays for test of function, **255**, 465; yeast adenylyl cyclase assays, **255**, 468; purification of human neutrophil nadph oxidase cytochrome *b*$_{588}$ and association with Rap1A, **255**, 476; use of yeast two-hybrid system to evaluate ras interactions with neurofibromin-GTPase-activating proteins, **255**, 488; screening phage-displayed randon peptide libraries for SH3 ligands, **255**, 498

Biological Activity & Cell Expression: serum induction of RhoG expression, **256**, 151; microinjection of epitope-tagged rho family cDNAs and analysis by immunolabeling, cell expression, **256**,162; purification and assay of recombinant C$_3$ transferase, cell expression, **256**, 174; *in vitro* ADP-ribosylation of Rho by bacterial ADP-ribosyltransferases, cell expression, **256**, 184; preparation of native and recombinant *Clostridium botulinum* C$_3$ ADP-ribosyltransferase and identification of rho proteins by ADP-ribosylation, cell expression, **256**, 196; *in vitro* binding assay for interactions of rho and rac with GTPase-activating proteins and effectors, cell expression, **256**, 207; purification and assay of kinases that interact with Rac/Cdc42, cell expression, **256**, 215; yeast two-hybrid system to detect protein-protein interactions with Rho GTPases, **256**, 228; assay for Rho-dependent phosphoinositide 3-kinase activity in platelet cytosol, **256**, 241; neutrophil phospholipase D: inhibition by Rho-GDP dissociation inhibitor and stimulation by small GTPase GDP dissocation stimulator, **256**, 246; measurement of Rac translocation from cytosol to membranes in activated neutrophils, **256**, 256; reconstitution of cell-free NADPH oxidase activity by purified components, **256**, 268; genetic and

biochemical analysis of Cdc42p function in *Saccharomyces cerevisiae* and *Schizosaccharomyces pombe*, **256**, 281; lymphocyte aggregation assay and inhibition by *Clostridium botulinum* C$_3$ ADP-ribosyltransferase, **256**, 290; inhibition of p21 Rho in intact cells by C$_3$ *Diphtheria* toxin chimera proteins, **256**, 297; growth factor-induced actin reorganization in Swiss 3T3 cells, **256**, 306; microinjection of Rho and Rac into quiescent Swiss 3T3 Cells, **256**, 313; inhibition of lymphocyte-mediated cytotoxicity by *Clostridium botulinum* C$_3$ transferase, **256**, 320; neutrophil chemotaxis assay and inhibition by C$_3$ ADP-ribosyltransferase, **256**, 327; cell motility assay and inhibition by Rho-GDP dissociation inhibitor, **256**, 336; cell transformation by *dbl* oncogene, **256**, 347; inhibition of Rac function using antisense oligonucleotides, **256**, 358; expression of Rab GTPase using recombinant vaccinia viruses, **257**, 155; transient expression of small GTPases to study protein transport along secretory pathway *in vivo* using recombinant T7 vaccinia virus system, **257**, 165; preparation of recombinant vaccinia virus for expression of small GTPases, **257**, 174; using oligonucleotides for cloning Rab proteins by PCR, **257**, 189; use of two-hybrid system to identify Rab binding proteins, **257**, 200; tightly regulated and inducible expression of dominant interfering dynamin mutant in stably transformed HeLa cells, **257**, 209; investigation by transient transfection of the effects on regulated exocytosis of Rab3A, **257**, 221; expression, purification, and assays of Gdi1p from recombinant *Escherichia coli*, **257**, 232; use of Rab-GDP dissociation inhibitor for solubilization and delivery of Rab proteins to biological membranes in streptolysin O-permeabilized cells, **257**, 221; reconstitution of Rab9 endosomal targeting and nucleotide exchange using purified Rab9-GDP dissociation inhibitor complexes and endosome-enriched membrane fraction, **257**, 221; localization of GTPases by indirect immunofluorescence and immunoelectron microscopy, **257**, 221; analysis of Ran/Tc4 function in nuclear protein import, **257**, 221; purification and properties of Rabphilin-3A, **257**, 221; use of antisense oligonucleotides to study Rab function *in vivo*, **257**, 221; stimulation of phospholipase D by ADP-ribosylation factor, **257**, 221; expression of Rab proteins during mouse embryonic development, **257**, 221

GRAMICIDIN S SYNTHETASE

This multienzyme complex catalyzes the overall synthesis of the cyclic decapeptide gramicidin S. The first subunit of the complex is the ATP-dependent phenylalanine racemase (also called gramicidin S synthetase 1 or GS1). The overall reaction has been proposed to act via aminoacyl-adenylate intermediates with the growing peptide chain anchored via a thioester linkage. **See *Phenylalanine Racemase***

Selected entries from ***Methods in Enzymology*** [vol, page(s)]:
General discussion: 43, 567
Assay: 43, 568
Other molecular properties: affinity chromatography, **43**, 575; amino acid-dependent ATP-[^{14}C]AMP exchange, **43**, 569, 570; amino acid-dependent ATP-^{32}PP$_i$ exchange, **43**, 569, 576; aminosulfate precipitation, **43**, 573; *Bacillus brevis* preparation, **43**, 571, 572; crude extract preparation, **43**, 573; DEAE Sephadex A-50 chromatography, **43**, 573, 574; heavy enzymes (physical properties, **43**, 577; specificity, **43**, 577); incubation mixture, **43**, 568, 569; inhibitors, **43**, 578; kinetic constants, **43**, 578; light enzyme, properties, **43**, 577, 578; molecular weight, **43**, 577; 4'-phosphopantetheine, **43**, 577; properties, **43**, 576; purification, **43**, 571; purity, **43**, 576; racemization of phenylalanine, **43**, 571; reaction scheme, **43**, 568; separation of light and heavy enzymes, **43**, 575; Sephadex G-200 chromatography, **43**, 573, 574; specificity, **43**, 577; stability, **43**, 577; streptomycin sulfate precipitation, **43**, 573; thio ester-bound amino acids, **43**, 570; thiotemplate mechanism, **43**, 568

GRANULOVIRUS CATHEPSIN

This cysteine proteinase, also known as granulovirus protease and CpGV cathepsin, was first identified from the *Cydia pomonella* granulovirus.[1]

[1] W. Kang, M. Tristem, S. Maeda, N. E. Crook & D. R. O'Reilly (1998) *J. Gen. Virol.* **79**, 2283.

GRANZYME A

This serine endopeptidase [EC 3.4.21.78], also known as cytotoxic T-lymphocyte proteinase 1 and T cell-specific proteinase 1, catalyzes the hydrolysis of peptides bonds in proteins (such as fibronectin, type IV collagen, proteoglycans, and nucleolin). The enzyme exhibits a preferential cleavage of Arg–Xaa > Lys–Xaa. Synthetic substrates such as Gly-Pro-Arg-∇-NHPhNO$_2$ are often used to assay granzyme A, which is found in cytolytic granules from cytotoxic T lymphocytes, natural killer cells, and large granular lymphocytes.[1,2]

[1] H. G. Simon & M. D. Kramer (1994) *Meth. Enzymol.* **244**, 69.
[2] M. V. Sitkovsky & P. Henkart, eds. (1993) *Cytotoxic Cells: Recognition, Effector Function, Generation and Methods*, Birkhäuser, Boston.

Selected entries from ***Methods in Enzymology*** [vol, page(s)]:
General discussion: 244, 69
Other molecular properties: biological role (B cell growth regulation, **244**, 77; control of viral infection, **244**, 77; extravasation of T lymphocytes, **244**, 76; natural killer cell-mediated cytolysis, **244**, 76); cleavage site specificity, **244**, 68, 72; expression in immunocytes, **244**, 68, 74, 77; family, **244**, 80; gene organization, **244**, 71; human (purification, **244**, 70, 79; size, **244**, 70; thioester substrates, **248**, 13); inhibitors, **244**, 73; localization in cytoplasmic granules, **244**, 74; murine (purification, **244**, 69, 78; size, **244**, 70; thioester substrates, **248**, 13); pH optimum, **244**, 73; processing, **244**, 71, 80; sequence homology between species, **244**, 71; substrate specificity, **244**, 72

GRANZYME B

This serine endopeptidase [EC 3.4.21.79], also known as cytotoxic T-lymphocyte proteinase 2 and fragmentin 1 (in rat: also called granzyme B [rodent]), catalyzes the hydrolysis of peptide bonds in proteins, exhibiting a preferential pattern of Asp–Xaa ≫ Asn–Xaa > Met–Xaa ≡ Ser–Xaa. Note that granzyme B is one of the few mammalian serine endopeptidases that prefer an aspartyl residue at the P1 position.

Synthetic substrates such as Suc-Ala-Ala-Pro-Asp-∇-NHPhNO$_2$ and Boc-Ala-Ala-Asp-∇-SBzl are frequently used to assay the enzyme (note that granzyme B has a thioesterase activity as well).[1,2]

[1] A. Caputo, M. N. James, J. C. Powers, D. Hudig & R. C. Bleackley (1994) *Nature Struct. Biol.* **1**, 364.

[2] M. V. Sitkovsky & P. Henkart, eds. (1993) *Cytotoxic Cells: Recognition, Effector Function, Generation and Methods*, Birkhäuser, Boston.

Selected entries from ***Methods in Enzymology*** [vol, page(s)]:
General discussion: 244, 80; **322**, 125
Assay: Boc-Ala-Ala-Asp-SBzl hydrolase assay, **244**, 81; glutamyl-2-naphthylamidase assay, **244**, 81
Other molecular properties: biological role, **244**, 80, 116; cellular distribution, **244**, 80; cleavage site specificity, **244**, 116; family, **244**, 80; inhibitors, **244**, 87; ion dependence, **244**, 87; lymphocyte substrate, **248**, 17; pH optimum, **244**, 87; processing, **244**, 80; purification, **244**, 82, 86; stability, **244**, 87; substrate specificity, **244**, 86; tertiary structure, **244**, 86

*Gsu*I RESTRICTION ENDONUCLEASE

This type II restriction endonuclease (subtype s) [EC 3.4.21.4], isolated from *Gluconobacter suboxydans* H-15T, catalyzes the cleavage of double-stranded DNA at the following sites: 5′ . . . CTGGAGNNNNNNNNNNNNNNN ∇ . . . 3′ and 3′ . . . GACCTCNNNNNNNNNNNNNNN ∇NN . . . 5′, where N refers to any base.[1]

[1] A. A. Janulaitis, J. Bitinaite & B. Jaskeleviciene (1983) *FEBS Lett.* **151**, 243.

GTP CYCLOHYDROLASE I

This enzyme [EC 3.5.4.16], also referred to as dihydroneopterin triphosphate synthetase, catalyzes the reaction of GTP with two water molecules to produce formate and 2-amino-4-hydroxy-6-(*erythro*-1,2,3-trihydroxypropyl) dihydropteridine triphosphate.[1–6] The reaction involves hydrolysis of two C–N bonds and isomerization of the pentose unit. The recyclization step may be nonenzymatic. This enzyme catalyzes the first reaction in the biosynthesis of tetrahydrobiopterin: it is the rate-limiting step in many mammals, but not in humans.[4]

[1] B. Thony, G. Auerbach & N. Blau (2000) *Biochem. J.* **347**, 1.
[2] M. Sawada, T. Horikoshi, M. Masada, M. Akino, T. Sugimoto, S. Matsuura & T. Nagatsu (1986) *Anal. Biochem.* **154**, 361.
[3] N. Blau & A. Niederwieser (1986) *BBA* **880**, 26.
[4] N. Blau & A. Niederwieser (1985) *J. Clin. Chem. Clin. Biochem.* **23**, 169.
[5] C. A. Nichol, G. K. Smith & D. S. Duch (1985) *Ann. Rev. Biochem.* **54**, 729.
[6] G. M. Brown & J. M. Williamson (1982) *AE* **53**, 345.

Selected entries from ***Methods in Enzymology*** [vol, page(s)]:
General discussion: 18B, 761; **34**, 4; **281**, 53
Assay: 18B, 762, 763
Other molecular properties: properties, 18B, 764, 765; purification from *Escherichia coli*, **18B**, 763, 764

GTP CYCLOHYDROLASE II

This enzyme [EC 3.5.4.25] catalyzes the reaction of GTP with three water molecules to produce formate, 2,5-diamino-6-hydroxy-4-(5-phosphoribosylamino)pyrimidine, and pyrophosphate (or, diphosphate). In this reaction, two C–N bonds are hydrolyzed, releasing formate, with the simultaneous hydrolysis of the terminal pyrophosphate.[1-3] **See also Riboflavin Synthase**

[1] N. Blau & A. Niederwieser (1985) *J. Clin. Chem. Clin. Biochem.* **23**, 169.
[2] G. M. Brown & J. M. Williamson (1982) *AE* **53**, 345.
[3] F. Foor & G. M. Brown (1975) *JBC* **250**, 3545.

Selected entries from **Methods in Enzymology [vol, page(s)]:**
General discussion: 66, 303; 280, 382
Assay: 66, 303, 304
Other molecular properties: guanosine
triphosphate-8-formylhydrolase, **43**, 515 (activators, **43**, 517;
ammonium sulfate fractionation, **43**, 517; assay, **43**, 516; cell-free
extracts, **43**, 516; DE-52 cellulose column chromatography, **43**, 517;
dissociation, **43**, 519; inhibitors, **43**, 517; Michaelis constant, **43**, 519;
molecular weight, **43**, 519; purification, **43**, 516; Sephadex G-200
column chromatography, **43**, 517; specific activity, **43**, 516; specificity,
43, 517; *Streptomyces rimosus* preparation, **43**, 516; unit definition, **43**,
516); properties, **66**, 306, 307; purification, from *Escherichia coli*,
66, 304

GTP PYROPHOSPHOKINASE

This enzyme [EC 2.7.6.5], known variously as GTP diphosphokinase, ppGpp synthetase I, stringent factor, stringent factor synthase, and guanosine $3',5'$-polyphosphate synthase, catalyzes the reaction of ATP with GTP to produce AMP and guanosine $3'$-diphosphate $5'$-triphosphate (that is, pppGpp or Magic Spot II). Other acceptor substrates include GDP, thus producing ppGpp (guanosine $3'$-diphosphate $5'$-diphosphate or Magic Spot I).[1-3]

[1] J. Justesen, T. Lund, F. S. Pedersen & N. O. Kjeldgaard (1986) *Biochimie* **68**, 715.
[2] S. Fehr & D. Richter (1981) *J. Bacteriol.* **145**, 68.
[3] J. Sy & H. Akers (1976) *Biochemistry* **15**, 4399.

GUANIDINOACETASE

This manganese-dependent enzyme [EC 3.5.3.2], also referred to as glycocyaminase, catalyzes the hydrolysis of guanidinoacetate to produce glycine and urea.[1,2]

[1] T. Yorifuji, N. Komaki, K. Oketani & E. Entani (1979) *Agric. Biol. Chem.* **43**, 55.
[2] J. Roche & G. Lacombe (1950) *BBA* **6**, 210.

GUANIDINOACETATE KINASE

This phosphotransferase [EC 2.7.3.1], also referred to as glycocyamine kinase, catalyzes the reaction of ATP with guanidoacetate to produce ADP and phosphoguanido-acetate.[1,2]

[1] Y. Shirokane, M.-O. Nakajima & K. Mizusawa (1991) *Agric. Biol. Chem.* **55**, 2235.
[2] L.-A. Pradel, R. Kassab, C. Conlay & N. van Thoai (1968) *BBA* **154**, 305.

GUANIDINOACETATE N-METHYLTRANSFERASE

This methyltransferase [EC 2.1.1.2] catalyzes the reaction of *S*-adenosyl-L-methionine with guanidoacetate to produce *S*-adenosyl-L-homocysteine and creatine.[1,2]

[1] Y. S. Im, P. K. Chiang & G. L. Cantoni (1979) *JBC* **254**, 11047.
[2] H. Ogawa, Y. Ishiguro & M. Fujioka (1983) *ABB* **226**, 265.

Selected entries from **Methods in Enzymology [vol, page(s)]:**
General discussion: 2, 260
Assay: 2, 260

1D-1-GUANIDINO-3-AMINO-1,3-DIDEOXY-SCYLLO-INOSITOL AMINOTRANSFERASE

This pyridoxal-phosphate-dependent aminotransferase [EC 2.6.1.56] catalyzes the reversible reaction of 1D-1-guanidino-3-amino-1,3-dideoxy-*scyllo*- inositol with pyruvate to produce 1D-1-guanidino-1-deoxy-3-dehydro-*scyllo*-inositol and L-alanine.[1] The reverse reaction is a step in streptidine biosynthesis (in this reverse reaction, other amino donor substrates can be used, including L-glutamate and L-glutamine).

[1] J. B. Walker & M. S. Walker (1969) *Biochemistry* **8**, 763.

Selected entries from **Methods in Enzymology [vol, page(s)]:**
General discussion: 43, 462
Assay: 43, 462

γ-GUANIDINOBUTYRALDEHYDE DEHYDROGENASE

This oxidoreductase [EC 1.2.1.54] catalyzes the reaction of 4-guanidinobutanal with NAD^+ and water to produce 4-guanidinobutanoate and NADH.[1-4] **See also Aminobutyraldehyde Dehydrogenase**

[1] A. Jann, H. Matsumoto & D. Haas (1988) *J. Gen. Microbiol.* **134**, 1043.
[2] T. Yorifuji, K. Koike, T. Sakurai & K. Yokoyama (1986) *Agric. Biol. Chem.* **50**, 2009.
[3] H. Matsuda & Y. Suzuki (1984) *Plant Physiol.* **76**, 654.
[4] A. S. Vanderbilt, N. S. Gaby & V. W. Rodwell (1975) *JBC* **250**, 5322.

GUANIDINOBUTYRASE

This manganese-dependent enzyme [EC 3.5.3.7] catalyzes the hydrolysis of 4-guanidinobutanoate to produce 4-aminobutanoate and urea.[1,2] Other substrates include 5-guanidinopentanoate and 6-guanidinohexanoate.

[1] C.-S. Chou & V. W. Rodwell (1972) *JBC* **247**, 4486.
[2] T. Yorifuji, M. Kato, T. Kobayashi, S. Ozaki & S. Ueno (1980) *Agric. Biol. Chem.* **44**, 1127.

1-GUANIDINO-1-DEOXY-SCYLLO-INOSITOL-4-PHOSPHATASE

This enzyme [EC 3.1.3.40], which participates in streptomycin and streptidine biosynthesis, catalyzes the hydrolysis of 1-guanidino-1-deoxy-*scyllo*-inositol 4-phosphate to produce 1-guanidino-1-deoxy-*scyllo*-inositol and orthophosphate.[1] Other substrates include 2-guanidino-2-deoxy-*neo*-inositol-5-phosphate and 1D-guanidino-5-amino-1,5-dideoxy-*scyllo*-inositol-4-phosphate.

[1] M. S. Walker & J. B. Walker (1971) *JBC* **246**, 7034.

Selected entries from *Methods in Enzymology* [vol, page(s)]:
General discussion: 43, 459
Assay: 43, 459

5-GUANIDINO-2-OXOPENTANOATE DECARBOXYLASE

This thiamin-pyrophosphate-dependent decarboxylase [EC 4.1.1.75], also called 2-oxo-5-guanidinopentanoate decarboxylase and 2-oxo-5-guanidinovalerate α-ketoarginine decarboxylase, catalyzes the conversion of 5-guanidino-2-oxo-pentanoate to 4-guanidinobutanoate ($^+H_2N=C(NH_2)NHCH_2CH_2CH_2COO^-$) and carbon dioxide.

GUANIDINOPROPIONASE

This manganese-dependent enzyme [EC 3.5.3.17] catalyzes the hydrolysis of 3-guanidinopropanoate to produce β-alanine and urea. Alternative substrates include taurocyamine and 4-guanidinobutanoate.[1,2]

[1] T. Yorifuji, I. Sugai, H. Matsumoto & A. Tabuchi (1982) *Agric. Biol. Chem.* **46**, 1361.
[2] T. Yorifuji & I. Sugai (1978) *Agric. Biol. Chem.* **42**, 1789.

GUANINE DEAMINASE

This enzyme [EC 3.5.4.3], also known as guanine aminohydrolase and guanase, catalyzes the hydrolysis of guanine to produce xanthine and ammonia.[1]

[1] C. L. Zielke & C. H. Suelter (1971) *The Enzymes*, 3rd ed., **4**, 47.

Selected entries from *Methods in Enzymology* [vol, page(s)]:
General discussion: 2, 480; 34, 523; 51, 512
Assay: 2, 480; 34, 526; 51, 512, 513
Other molecular properties: activity, 51, 512; guanine assay and, 2, 458, 481; guanine removal and, 2, 449; guanosine-5′-phosphate pyrophosphorylase and, 6, 146; inhibitors, 51, 516; kinetic properties, 51, 515, 516; molecular weight, 51, 517; pH optimum, 51, 515, 516; properties, 2, 482; pterin deaminase and, 6, 361, 363; purification, 2, 481; 51, 513; purine nucleoside phosphorylase and, 12A, 118; purity, 51, 515; rabbit liver, from, 51, 512; reaction mechanism, 51, 517; stability, 51, 516, 517; staining on gels, 22, 601; transition-state and multisubstrate analogues, 249, 307

GUANINE-TRANSPORTING ATPase

This so-called ATPase [EC 3.6.3.37] catalyzes the ATP-dependent (ADP- and orthophosphate-producing) transport of guanine$_{out}$ to produce guanine$_{in}$.[1–3] This ABC-type

(ATP-binding cassette-type) ATPase does not undergo phosphorylation during the transport process. The eukaryotic enzyme, which has a single ATP-binding site, also transports L-tryptophan.

The idea that a transporter is an enzyme is in keeping with a new definition of enzyme catalysis as the facilitated making/breaking of chemical bonds, not just covalent bonds.[4] This idea builds on Pauling's assertion that any long-lived, chemically distinct interaction (in this case, the persistent location of a solute with respect to the faces of a membrane) can be regarded as a chemical bond. Note also that the equilibrium constant (K_{eq} = [guanine$_{in}$][ADP][P$_i$]/[guanine$_{out}$][ATP]) does not conform to that expected for an ATPase (*i.e.*, K_{eq} = [ADP][P$_i$]/[ATP]). Thus, although the overall reaction yields ADP and orthophosphate, the enzyme is misclassified as a hydrolase, and instead should be regarded as an energase-type reaction. Energases facilitate affinity-modulated reactions by coupling the $\Delta G_{ATP\text{-hydrolysis}}$ to a force-generating or work-producing step.[4] In this case, P-O-P bond scission supplies the energy to drive guanine transport. *See ATPase; Energase*

[1] T. D. Dreesen, D. H. Johnson & S. Henikoff (1988) *Mol. Cell Biol.* **8**, 5206.
[2] R. G. Tearle, J. M. Belote, M. McKeown, B. S. Baker & A. J. Howells (1989) *Genetics* **122**, 595.
[3] J. K. Griffiths & C. E. Sansom (1998) *The Transporter Factsbook*, Academic Press, San Diego.
[4] D. L. Purich (2001) *TiBS* **26**, 417.

GUANOSINE-3′,5′-BIS(DIPHOSPHATE) 3′-PYROPHOSPHOHYDROLASE

This enzyme [EC 3.1.7.2] (also referred to as guanosine-3′,5′-bis(diphosphate) 3′-diphosphohydrolase, pentaphosphate guanosine-3′-pyrophosphohydrolase, and guanosine-3′,5′-bis(diphosphate) 3′-pyrophosphatase) catalyzes the hydrolysis of guanosine 3′,5′-bis(diphosphate) (or, ppGpp; Magic Spot I) to produce guanosine 5′-diphosphate (GDP) and pyrophosphate (or, diphosphate).[1,2]

[1] G. An, J. Justesen, R. J. Watson & J. D. Friesen (1979) *J. Bacteriol.* **137**, 1100.
[2] E. A. Heinemeyer & D. Richter (1978) *Biochemistry* **17**, 5368.

GUANOSINE DEAMINASE

This enzyme [EC 3.5.4.15], also known as guanosine aminohydrolase, catalyzes the hydrolysis of guanosine to produce xanthosine and ammonia.[1]

[1] C. L. Zielke & C. H. Suelter (1971) *The Enzymes*, 3rd ed., **4**, 47.

GUANOSINE-DIPHOSPHATASE

This enzyme [EC 3.6.1.42], also abbreviated as GDPase, catalyzes the hydrolysis of GDP to produce GMP and orthophosphate. UDP, but not other nucleoside diphosphates (NDPs) or triphosphates (NTPs), can act as an alternative substrate.[1] The *Saccharomyces cerevisiae* enzyme requires calcium ions.[2] **See also** *Nucleoside Diphosphatase*

[1] P. Raychaudhuri, S. Ghosh & U. Maitra (1985) *JBC* **260**, 8306.
[2] K. Yanagisawa, D. Resnick, C. Abeijon, P. W. Robbins & C. B. Hirschberg (1990) *JBC* **265**, 19351.

GUANOSINE PHOSPHORYLASE

This enzyme [EC 2.4.2.15] catalyzes the reaction of guanosine with orthophosphate to produce guanine and D-ribose 1-phosphate.[1] Other substrates include deoxyguanosine.

[1] E. W. Yamada (1961) *JBC* **236**, 3043.

GUANOSINE-5′-TRIPHOSPHATE, 3′-DIPHOSPHATE PYROPHOSPHATASE

This enzyme [EC 3.6.1.40] (also called guanosine-5′-triphosphate,3′-diphosphate diphosphatase, guanosine pentaphosphate phosphohydrolase, and pppGpp-5′-phosphohydrolase) catalyzes the hydrolysis of guanosine 5′-triphosphate,3′-diphosphate (*i.e.*, pppGpp; Magic Spot II) to produce guanosine 5′-diphosphate, 3′-diphosphate (*i.e.*, ppGpp; Magic Spot I) and orthophosphate. Alternative substrates include other guanosine 5′-triphosphate derivatives with at least one unsubstituted phosphate group on the 3′-position (for example, pppGp is a substrate). However, GTP, ATP, ppGpp, or pppApp are not substrates.[1]

[1] A. Hara & J. Sy (1983) *JBC* **258**, 1678.

GUANOSINE-TRIPHOSPHATE GUANYLYLTRANSFERASE

This enzyme [EC 2.7.7.45] catalyzes the reaction of two molecules of GTP (*i.e.*, guanosine 5′-triphosphate) to produce pyrophosphate (or, diphosphate) and P^1,P^4-bis(5′-guanosyl)tetraphosphate. Other substrates, albeit weaker, include GDP (producing P^1,P^3-bis(5′-guanosyl)triphosphate). The brine shrimp enzyme forms a covalent enzyme-guanylate intermediate,[2] and has two binding sites. The "donor" site is highly specific, preferring such guanine-containing substrates as GTP, dGTP, and guanosine 5′-tetraphosphate. The "acceptor" site displays a broader specificity, using GDP, ADP, XTP, and ITP.[1] Thus, a wide range of homodinucleotides and heterodinucleotides can be generated.

[1] J. J. Liu & A. G. McLennan (1994) *JBC* **269**, 11787.
[2] J. L. Cartwright & A. G. McLennan (1999) *ABB* **361**, 101.

GUANYLATE CYCLASE

This enzyme [EC 4.6.1.2], also known as guanylyl cyclase and guanyl cyclase, catalyzes the conversion of GTP to 3′,5′-cyclic GMP and pyrophosphate (or, diphosphate).[1–8] Other substrates include ITP and dGTP. Two distinct types of guanylate cyclase have been described: a soluble form and a membrane-bound form, the former being heme-dependent. The heterodimeric soluble guanylate cyclase is stimulated by nitric oxide.

[1] K. A. Lucas, G. M. Pitari, S. Kazerounian, I. Ruiz-Stewart, J. Park, S. Schulz, K. P. Chepenik & S. A. Waldman (2000) *Pharmacol. Rev.* **52**, 375.
[2] J. W. Denninger & M. A. Marletta (1999) *BBA* **1411**, 334.
[3] S. Schulz & S. A. Waldman (1999) *Vitam. Horm.* **57**, 123.
[4] D. Koesling & A. Friebe (1999) *Rev. Physiol. Biochem. Pharmacol.* **135**, 41.
[5] D. C. Foster, B. J. Wedel, S. W. Robinson & D. L. Garbers (1999) *Rev. Physiol. Biochem. Pharmacol.* **135**, 1.
[6] R. A. Johnson & J. D. Corbin, eds. (1991) *Meth. Enzymol.*, vol. **195**, Academic Press, Orlando.
[7] F. Murad, W. P. Arnold, C. K. Mittal & J. M. Braughler (1979) *Adv. Cyclic Nucleotide Res.* **11**, 175.
[8] N. D. Goldberg & M. K. Haddox (1977) *Ann. Rev. Biochem.* **46**, 823.

Selected entries from **Methods in Enzymology** [vol, page(s)]:
General discussion: 38, 192, 199; 81, 517; 195, 345, 355, 363, 373, 377, 384, 391, 397, 404, 414, 423, 436, 447, 461, 466; 315, 673, 689, 708, 718, 730, 742; 316, 558
Assay: 195, 345, 365, 378, 385, 394, 400, 406, 467 (with [³H]GTP, 195, 349; with [α-³²P]GTP, 195, 345; [α-³²P]-labeled nucleotide preparation for, 195, 29); bovine lung, 38, 192; general principles (determination of residual substrate, 38, 124, 125; divalent cations, 38, 120; enzyme concentration, 38, 117; labeled nucleoside triphosphates, 38, 118, 119; linearity of cyclic nucleotide production, 38, 122; nucleoside triphosphate concentration and volume, 38, 115; protection of cyclase activity, 38, 121, 122; purification, proof of identity and determination of product, 38, 123, 124; reduction of product degradation, 38, 120, 121; substrate degradation, 38, 119, 120; termination of cyclase reaction, 38, 122, 123); human platelets, 38, 199, 200; mammalian leukocyte (nonradiometric assay, 132, 426; radiometric assay, 132, 425); retinal, 81, 519, 525; sea urchin sperm, 38, 196
Other molecular properties: activation, 269, 142 (carbon monoxide, 268, 474; nitric oxide, 268, 474); activity with different phosphorylation states, 195, 465; ANF receptors, and, copurification from adrenal cortex, 195, 404; bovine lung (assay, 38, 192; properties, 38, 195; purification, 38, 192); calcium-regulated, protozoan (characterization, 195, 471; *Paramecium tetraurelia*, from, 195, 466; physiological regulation, 195, 472; solubilization, 195, 470; subcellular distribution, 195, 470; *Tetrahymena*, from, 195, 466); dephosphorylation, 195, 465; dephosphorylation effects, 195, 376; gel electrophoresis, apparatus for, 195, 355; human platelets (apparent activation with time, 38, 201, 202; assay, 38, 199, 200; distribution among blood cells, 38, 200; divalent cations, 38, 201; homogenate preparation, 38, 200; subcellular distribution, 38, 201); mammalian leukocyte (characterization of formed product, 132, 427; nonradiometric assay, 132, 426; radiometric assay, 132, 425); mechanism, 63, 51, 158; membrane (activation by detergents, 195, 363; molecular cloning, 195, 414; purification, 195, 373; solubilization by detergents, 195, 363); membrane vesicles, in, ³²P incorporation, 195, 464; nitric oxide binding to heme cofactor, 268, 13, 17, 23; 269, 150; nitrosyl complex detection by electron spin resonance, 268, 169, 173; particulate (ANP receptors, and, copurification from

pulmonary membranes, **195**, 397; atrial natriuretic peptide target enzyme, as, **195**, 456; kinetics, **195**, 402; purification from sea urchin, **195**, 415; radiation-inactivation analysis, **195**, 423; stimulation by atrial natriuretic peptide [cGMP production induced by, evaluation, **195**, 447; irreversibility, **195**, 460]); particulate and soluble, **269**, 139; plasma membrane, mammalian, cDNA cloning, **195**, 419 (isolation of GC-A clones, **195**, 419; isolation of GC-B clones, **195**, 421); preparation from rat lung, **195**, 364; properties, **195**, 376; retinal (assays, **81**, 519, 525; localization, **81**, 524); rod outer segments, **38**, 154, 155; sea urchin sperm (assay, **38**, 196; dephosphorylation, **195**, 461; phosphorylation, **195**, 461; ^{32}P incorporation, **195**, 463; properties, **38**, 198, 199); soluble (characteristics, **195**, 395; electrophoretic characterization, **195**, 389; heme-containing, purification from bovine lung, **195**, 377; immunoaffinity purification, **195**, 391 [from rat lung, **195**, 393]; preparation from bovine lung, **195**, 384; properties, **195**, 390; stimulation by sodium nitroprusside, **195**, 390); stimulation by GTPγS, **237**, 406

GUANYLATE KINASE

This phosphotransferase [EC 2.7.4.8], also referred to as GMP kinase, catalyzes the reversible reaction of ATP with GMP to produce ADP and GDP. dGMP can substitute for GMP and dATP can substitute for ATP.[1-5]

[1]Y. Li, Y. Zhang & H. Yan (1996) *JBC* **271**, 28038.
[2]S. W. Hall & H. Kühn (1986) *EJB* **161**, 551.
[3]E. P. Anderson (1973) *The Enzymes*, 3rd ed., **9**, 49.
[4]H. Shimono & Y. Sugino (1971) *EJB* **19**, 256.
[5]T. J. Griffith & C. W. Helleiner (1965) *BBA* **108**, 114.

Selected entries from **Methods in Enzymology** [**vol**, page(s)]:
General discussion: **51**, 473, 483
Assay: **51**, 474, 483, 484
Other molecular properties: activity, **57**, 86; affinity for calcium phosphate gel, **51**, 584; cation requirements, **51**, 481, 489; cGMP assay, **38**, 73, 78, 80, 85, 89, 106, 109; contaminant, in GMP synthetase purification, **51**, 222, 223; deoxynucleoside monophosphate kinase and, **6**, 174; determination of (cyclic GMP phosphodiesterase, **57**, 96; guanosine nucleotides, **57**, 86, 91); *Escherichia coli*, from, **51**, 473; GMP assay, **63**, 32; hog brain, from, **51**, 485; human erythrocytes, from, **51**, 483; hypoxanthine:guanine phosphoribosyltransferase assay, **63**, 32; isoelectric variants, **51**, 488; kinetic properties, **51**, 488, 489; molecular weight, **51**, 482, 488; pH optimum, **51**, 482; purification, **51**, 477, 485; purity, **51**, 482, 488; rat liver, from, **51**, 485; reaction mechanism, **51**, 490; stability, **51**, 480, 481, 487, 488; substrate specificity, **51**, 473, 474, 481, 488, 489; **57**, 93

L-GULONATE 3-DEHYDROGENASE

This oxidoreductase [EC 1.1.1.45], also referred to as L-3-aldonate dehydrogenase and L-β-hydroxyacid dehydrogenase, catalyzes the reaction of L-gulonate with NAD[+] to produce 3-dehydro-L-gulonate and NADH.[1,2] Other substrates include other L-3-hydroxyacids (for example, L-lyxonate and L-idonate).

[1]J. W. Hawes, E. T. Harper, D. W. Crabb & R. A. Harris (1997) *Adv. Exp. Med. Biol.* **414**, 395.
[2]V. J. Cannistraro, L. I. Borack & T. Chase (1979) *BBA* **569**, 1.

Selected entries from **Methods in Enzymology** [**vol**, page(s)]:
General discussion: **18A**, 29
Assay: **18A**, 33

L-GULONOLACTONE OXIDASE

This FAD-dependent oxidoreductase [EC 1.1.3.8], also known as L-gulono-γ-lactone oxidase, catalyzes the reaction of L-gulono-1,4-lactone with dioxygen to produce L-*xylo*-hexulonolactone and hydrogen peroxide.[1-4] The product spontaneously isomerizes to L-ascorbate. Phenazine methosulfate can act as the electron acceptor in place of dioxygen. In such cases, the enzyme has been referred to as a dehydrogenase; however, the dehydrogenase term usually refers to reverse reaction of glucuronolactone reductase [EC 1.1.1.20].

[1]M. Nishikimi & K. Yagi (1996) *Subcell. Biochem.* **25**, 17.
[2]M. Nishikimi, B. M. Tolbert & S. Udenfriend (1976) *ABB* **175**, 427.
[3]K. Kiuchi, M. Nishikimi & K. Yagi (1982) *Biochemistry* **21**, 5076.
[4]H. S. Bleeg & F. Christensen (1982) *EJB* **127**, 391.

Selected entries from **Methods in Enzymology** [**vol**, page(s)]:
General discussion: **6**, 339; **52**, 18; **62**, 24; **279**, 24
Assay: **6**, 340; **18A**, 30, 31; **62**, 24
Other molecular properties: expression, **279**, 24; flavin linkage, **53**, 450; properties, **6**, 340; **18A**, 31, 32; **62**, 29, 30; purification from (goat liver, **62**, 27, 28; rat liver, **62**, 25); source, **18A**, 29; staining of enzyme activity on gels, **62**, 28

H

*Hae*II RESTRICTION ENDONUCLEASE

This type II restriction endonuclease [EC 3.1.21.4], isolated from *Haemophilus aegyptius*, catalyzes the cleavage of both strands of DNA at 5′ . . . RGCGC∇Y . . . 3′, where R refers to either A or G and Y refers to either T or C.[1–3] It does not catalyze single strand cleavage. Cleavage is inhibited by the presence of 5-methylcytosine at the second cytosine within the recognition sequence.

[1] R. J. Roberts, J. B. Breitmeyer, N. F. Tabachnik & P. A. Myers (1975) *J. Mol. Biol.* **91**, 121.
[2] C.-P. D. Tu, R. Roychoudhury & R. Wu (1976) *BBRC* **72**, 355.
[3] A. Piekarowicz, R. Yuan & D. C. Stein (1989) *NAR* **17**, 10132.

Selected entries from *Methods in Enzymology* [**vol**, page(s)]:
General discussion: **65**, 23, 34, 35, 342, 363, 799
Other molecular properties: cleavage sites, on SV40 DNA, **65**, 713; purification, **65**, 94; recognition sequence, **65**, 6, 56, 57; recognition site, determination, **65**, 632; sticky end sequence, **65**, 342

*Hae*III RESTRICTION ENDONUCLEASE

This type II restriction enzyme [EC 3.1.21.4], isolated from *Haemophilus aegyptius*, catalyzes the cleavage of both strands of DNA at 5′ . . . GG∇CC . . . 3′.[1] It also catalyzes single strand cleavage.[2] The sequence adjacent to the recognition site strongly influences the rate of cleavage: AGGCCT > TGGCCA > GGGCCC ~ CGGCCG.[3] In addition, the site GGCm⁵C is cleaved but the site GGm⁵CC is resistant to cleavage.

[1] J. H. Middleton, M. H. Edgell & C. A. Hutchison, III (1972) *J. Virol.* **10**, 42.
[2] K. Horiuchi & N. D. Zinder (1975) *PNAS* **72**, 2555.
[3] H. Wolfes, A. Fliess & A. Pingoud (1985) *EJB* **150**, 105.

Selected entries from *Methods in Enzymology* [**vol**, page(s)]:
General discussion: **65**, 23, 51, 84, 166, 337, 349, 351, 363, 441, 442, 482, 483, 591, 622, 791, 799, 875
Other molecular properties: construction, of plasmid pLJ3, **65**, 237, 238; isoschizomers, **65**, 109, 126; purification, **65**, 94; reaction conditions, **65**, 130; recognition sequence, **65**, 6; restriction, of pLJ3 plasmid, **65**, 242

HALOACETATE DEHALOGENASE

This enzyme [EC 3.8.1.3] catalyzes the hydrolysis of a monohaloacetate ($X–CH_2COO^-$) to produce glycolate and the corresponding halide anion (X^-). Substrates include monofluoroacetate, monochloroacetate, monobromoacetate, and monoiodoacetate. The likely mechanism involves attack by an active-site carboxylate anion on the α-carbon of the substrate, leading to the formation of a covalently bound ester intermediate, the hydrolysis of which is probably facilitated by an active-site histidyl residue.[1]

[1] J. Q. Liu, T. Kurihara, S. Ichiyama, M. Miyagi, S. Tsunasawa, H. Kawasaki, K. Soda & N. Esaki (1998) *JBC* **273**, 30897.

2-HALOACID DEHALOGENASE

This enzyme [EC 3.8.1.2] catalyzes the hydrolysis of a short-chain (*S*)-2-haloacid anion to produce an (*R*)-2-hydroxyacid anion and the halide anion. Substrates include 2-chloropropionate and chloroacetate. The catalytic reaction involves a covalent enzyme-ester intermediate by a nucleophilic carboxylate ion of an aspartyl residue.[1,2]

[1] I. S. Ridder, H. J. Rozeboom, K. H. Kalk & B. W. Dijkstra (1999) *JBC* **274**, 30672.
[2] J. Q. Liu, T. Kurihara, M. Miyagi, N. Esaki & K. Soda (1995) *JBC* **270**, 18309.

HALOALKANE DEHALOGENASE

This dehalogenase [EC 3.8.1.5], also known as 1-chlorohexane halidohydrolase, catalyzes the hydrolysis of a short-chain 1-haloalkane to produce a primary alcohol, a halide anion, and a hydrogen ion: for example, 1,2-dichloroethane is converted to 2-chloroethanol and a chloride ion.[1–4] The enzyme has a broad specificity. Substrates include 1-chlorohexane, bromoethane, 1,2-dibromoethane, 1-iodopropane, and 4-chlorobutan-1-ol.

The *Xanthobacter* enzyme employs an aspartyl residue to displace a halide from the substrate, generating a covalent ester intermediate.[2] A histidyl residue then acts as a general base to abstract a proton from an active-site water and the incipient hydroxide ion attacks the intermediate.[2,3] **See also** Alkylhalidase

[1] K. H. Verschueren, F. Seljee, H. J. Rozeboom, K. H. Kalk & B. W. Dijkstra (1993) *Nature* **363**, 693.
[2] C. Kennes, F. Pries, G. H. Krooshof, E. Bokma, J. Kingma & D. B. Janssen (1995) *EJB* **228**, 403.
[3] J. Newman, T. S. Peat, R. Richard, L. Kan, P. E. Swanson, J. A. Affholter, I. H. Holmes, J. F. Schindler, C. J. Unkefer & T. C. Terwilliger (1999) *Biochemistry* **38**, 16105.
[4] S. D. Copley (1998) *Curr. Opin. Chem. Biol.* **2**, 613.

HALOLYSIN

This *Halobacter* subtilisin-like endopeptidase is inactive below 0.75 M NaCl but becomes fully active at 5 M salt.[1–3]

See also Archaean Serine Proteases

[1] V. M. Stepanov, G. N. Rudenskaya, L. P. Revina, Y. B. Gryaznova, E. N. Lysogorskaya, I. Y. Filippova & I. I. Ivanova (1992) *BJ* **285**, 281.
[2] J. Kim & J. S. Dordick (1997) *Biotech. Bioengin.* **55**, 471.
[3] M. Kamekura, Y. Seno & M. Dyall-Smith (1996) *BBA* **1294**, 159.

HALYSTASE

This serine proteinase, obtained from the venom of the viper *Agkistrodon halys blomhoffii* (also called *Gloydius blomhoffii*), cleaves fibrinogen (the B β chain at Arg42) and kininogen (releasing bradykinin).[1]

[1] T. Matsui, Y. Sakurai, Y. Fujimura, I. Hayashi, S. Oh-ishi, M. Suzuki, J. Hamako, Y. Yamamoto, J. Yamazaki, M. Kinoshita & K. Titani (1998) *EJB* **252**, 569.

HAMAMELOSE KINASE

This phosphotransferase [EC 2.7.1.102] catalyzes the reaction of ATP with D-hamamelose to produce ADP and D-hamamelose 2'-phosphate.[1] D-Hamamelitol is also a substrate.

[1] E. Beck, J. Wieczorek & W. Reinecke (1980) *EJB* **107**, 485.

*Hap*II RESTRICTION ENDONUCLEASE

This type II restriction enzyme [EC 3.1.21.4], isolated from *Haemophilus aphrophilus*, catalyzes the cleavage of both strands of DNA at 5′ . . . C∇CGG . . . 3′.[1,2] It can also act on single-stranded DNA.[3] If the internal cytosine is methylated, this endonuclease will barely cleave that site.

[1] H. Sugisaki & M. Takanami (1973) *Nature New Biol.* **246**, 138.
[2] M. Takanami (1974) *Methods Mol. Biol.* **7**, 113.
[3] K. Nishigaki, Y. Kaneko, H. Wakuda, Y. Husimi & T. Tanaka (1985) *NAR* **13**, 5747.

Selected entries from **Methods in Enzymology** [vol, page(s)]:
General discussion: 65, 23, 24, 799
Other molecular properties: recognition sequence, **65**, 7

HEMAGGLUTININ/PROTEASE

This metalloproteinase was first identified in *Vibrio cholerae* as the causative agent for the agglutination of erythrocytes (it has been also called mucinase, cholera lectin, and soluble hemagglutinin). It was later discovered that the protein posseses proteolytic activity. Similar proteins have been found in *Vibrio anguillarum* and *Helicobacter pylori*.

Full expression of proteolytic activity requires calcium ion. The protease reportedly cleaves a Gly–Phe bond in hemolysin and Ser–Ser and Ser–Met bonds in subunit A of bacterial heat-labile enterotoxin.[1,2] The enzyme also acts on fibronectin, ovomucin, and lactoferrin.

[1] K. Nagamune, K. Yamamoto, A. Naka, J. Matsuyama, T. Miwatani & T. Honda (1996) *Infect. Immun.* **64**, 4655.
[2] B. A. Booth, M. Boesman-Finkelstein & R. A. Finkelstein (1984) *Infect. Immun.* **45**, 558.

HEME OXYGENASE (DECYCLIZING)

This oxidoreductase [EC 1.14.99.3] catalyzes the reaction of heme (or, iron–protoporphyrin IX) with dioxygen and three electron donor molecules (AH_2) to produce biliverdin, carbon monoxide, Fe^{3+}, three molecules of the oxidized donor (A), and three water molecules.[1–4] This enzyme uses NAD(P)H as a cofactor and requires the presence of NADPH:ferrihemoprotein reductase [EC 1.6.2.4]. Other substrates include synthetic hemins, heme b, heme c, iron-mesoporphyrin IX, iron-deuteroheme IX, iron-coproheme I, and cobalt-heme.

Inducible and non-inducible heme oxygenases have been identified. The first step in the reaction mechanism is heme α-meso-hydroxylation.

[1] P. R. Montellano (2000) *Curr. Opin. Chem. Biol.* **4**, 221.
[2] H. B. Dunford (1998) *CBC* **3**, 195.
[3] S. Yamamoto & Y. Ishimura (1991) *A Study of Enzymes* **2**, 315.
[4] M. D. Maines (1988) *FASEB J.* **2**, 2557.

Selected entries from **Methods in Enzymology** [vol, page(s)]:
General discussion: 52, 367
Assay: 52, 368; **63**, 36; **300**, 322; bilirubin formation, **268**, 478; protein, **268**, 479
Other molecular properties: distribution, **52**, 367, 372; generation of cyclic GMP, **268**, 473; heme-binding site, **268**, 475; heme oxygenase 1 (agents inducing, **234**, 224; mRNA, accumulation [assay selection, **234**, 225; Northern analysis, **234**, 229; oxidative stress-induced increase, **234**, 233; transient enhancement, **234**, 224]); homogeneity, **52**, 370; inhibition by zinc-protoporphyrin, **268**, 475; inhibitors, **52**, 371; isoforms (immunodetection [antisera preparation, **268**, 481; controls, **268**, 485; endogenous peroxidase inhibition, **268**, 486; immunostaining, **268**, 483; NADPH diaphorase staining, **268**, 481; tissue preparation, **268**, 482]; selective induction, **268**, 476; types, **268**, 474, 476; Western blotting, **268**, 486); kinetics, **52**, 371; microsomal

preparations, **268**, 478; molecular weight, **52**, 370; reactivity, **52**, 367, 368; specificity, **52**, 371; stability, **52**, 370, 371

HEME *o* SYNTHASE

This enzyme catalyzes the reaction of protoheme IX with farnesyl diphosphate to produce heme *o* and diphosphate (or, pyrophosphate).[1] Heme *o* is the prosthetic group in cytochrome *o* oxidase.

[1]K. Saiki, T. Mogi, H. Hori, M. Tsubaki & Y. Anraku (1993) *JBC* **268**, 26041.

HEME-TRANSPORTING ATPase

This so-called ATPase [EC 3.6.3.41] catalyzes the ATP-dependent (ADP- and orthophosphate-producing) transport of $heme_{in}$ to produce $heme_{out}$.[1,2] It has been identified in Gram-negative bacteria. As is typical of ABC-type (ATP-binding cassette-type) ATPase possessing two similar ATP-binding domains, the enzyme does not undergo phosphorylation during the transport process. *See also ATPase, Iron-Chelated-Transporting*

The idea that a transporter is an enzyme is in keeping with a new definition of enzyme catalysis as the facilitated making/breaking of chemical bonds, not just covalent bonds.[3] This idea builds on Pauling's assertion that any long-lived, chemically distinct interaction (in this case, the persistent location of a solute with respect to the faces of a membrane) can be regarded as a chemical bond. Note also that the equilibrium constant ($K_{eq} = [heme_{out}][ADP][P_i]/[heme_{in}][ATP]$) does not conform to that expected for an ATPase (*i.e.*, $K_{eq} = [ADP][P_i]/[ATP]$). Thus, although the overall reaction yields ADP and orthophosphate, the enzyme is misclassified as a hydrolase, and instead should be regarded as an energase-type reaction. Energases facilitate affinity-modulated reactions by coupling the $\Delta G_{ATP-hydrolysis}$ to a force-generating or work-producing step.[3] In this case, P-O-P bond scission supplies the energy to drive heme transport. *See ATPase; Energase*

[1]W. Jekabsons & W. Schuster (1995) *Mol. Gen. Genet.* **246**, 166.
[2]T. M. Ramseier, H. V. Winteler & H. Hennecke (1991) *JBC* **266**, 7793.
[3]D. L. Purich (2001) *TiBS* **26**, 417.

HEPARAN-α-GLUCOSAMINIDE ACETYLTRANSFERASE

This acetyltransferase [EC 2.3.1.78], also known as acetyl-CoA:α-glucosaminide *N*-acetyltransferase, catalyzes the reaction of acetyl-CoA with heparan α-D-glucosaminide to produce heparan *N*-acetyl-α-D-glucosaminide and Coenzyme A. The enzyme acetylates glucosamine moieties within heparan sulfate and heparin from which the sulfate has been removed.[1–3]

This acetyltransferase is distinct from glucosamine acetyltransferase [EC 2.3.1.3] and glucosamine-phosphate acetyltransferase [EC 2.3.1.4].

[1]K. J. Bame & L. H. Rome (1985) *JBC* **260**, 11293.
[2]K. J. Bame & L. H. Rome (1986) *JBC* **261**, 10127.
[3]P. J. Meikle, A. M. Whittle & J. J. Hopwood (1995) *BJ* **308**, 327.
Selected entries from ***Methods in Enzymology*** [vol, page(s)]:
General discussion: 138, 607
Assay: 138, 608
Other molecular properties: diagnosis of genetic mucopolysaccharide storage disorders, in, **83**, 561; properties, **138**, 610; purification from rat liver, **138**, 609; radioactive substrates, preparation, **83**, 566

HEPARIN-GLUCOSAMINE 3-O-SULFOTRANSFERASE

This sulfotransferase [EC 2.8.2.23], also known as glucosaminyl 3-*O*-sulfotransferase, catalyzes the reaction of 3′-phosphoadenylylsulfate with heparin-glucosamine. Enzyme action at a specific glucosaminyl residue in the antithrombin-binding site produces heparin-glucosamine 3-*O*-sulfate and adenosine 3′,5′-bisphosphate. An alternative substrate is a pentasaccharide derived from the glucosamyl portion of heparin-substrate (*i.e.*, $GlcNSO_3(6-OSO_3)$-GlcA-$GlcNSO_3(6-OSO_3)$-IdoA(2-OSO_3)-$GlcNSO_3(6-OSO_3)$, where GlcA and IdoA represent D-glucuronate and L-iduronate, respectively); the central sugar unit of this pentasaccharide becomes 3-sulfated.[1] Studies with selectively desulfated heparin indicate that both IdoA 2-*O*-sulfate and GlcN 6-*O*-sulfate groups contribute to the inhibition of the 3-*O*-sulfotransferase.[2]

[1]M. Kusche, G. Bäckström, J. Riesenfeld, M. Petitou, J. Choay & U. Lindahl (1988) *JBC* **263**, 15474.
[2]N. Razi & U. Lindahl (1995) *JBC* **270**, 11267.

HEPARIN LYASE

This lyase [EC 4.2.2.7], also known as heparin eliminase and heparinase, catalyzes the eliminative cleavage of polysaccharides containing 1,4-linked glucuronate or iduronate residues and 1,4-α-linked 2-sulfoamino-2-deoxy-6-sulfo-D-glucose residues to give oligosaccharides with terminal 4-deoxy-α-D-gluc-4-enuronosyl groups at their nonreducing ends.[1,2] The enzyme often requires calcium ion for maximal activity. **See also** Platelet Heparitinase

[1] R. J. Linhardt, P. M. Galliher & C. L. Cooney (1986) Appl. Biochem. Biotechnol. **12**, 135.
[2] V. C. Yang, R. J. Linhard, H. Bernstein, C. L. Cooney & R. Langer (1985) JBC **260**, 1849.

Selected entries from **Methods in Enzymology** [vol, page(s)]:
General discussion: 28, 902
Assay: 28, 902; **137**, 516
Other molecular properties: activity, **245**, 221, 235; cleavage of heparin and heparan sulfate, **245**, 235; heparan sulfate degradation with, **245**, 228; heparinase III (activity, **245**, 221, 235; assay, proteoglycan degradation method, **248**, 50; specificity, **230**, 414); immobilization, **137**, 523; platelet heparitinase, **169**, 342 (assay, **169**, 343; properties, **169**, 350; purification, **169**, 346); preparation, **28**, 905; production, **137**, 518; properties, **28**, 910; purification, **137**, 521; reactor (construction, **137**, 525; testing in vivo, **137**, 525); specificity, **28**, 910; **230**, 414

HEPARITIN-SULFATE LYASE

This lyase [EC 4.2.2.8], also known as heparin-sulfate eliminase and heparin sulfamidase, catalyzes elimination (with C–O bond scission) of sulfate, with preference for linkages between N-acetyl-D-glucosamine and uronate, to produce a polysaccharide containing the corresponding unsaturated sugar.[1–3]

[1] P. Hovingh & A. Linker (1970) JBC **245**, 6170.
[2] H. B. Nader, C. P. Dietrich, V. Buonassi & P. Colburn (1987) PNAS **84**, 3565.
[3] R. Prinz, U. Klein, P. R. Sudhakaran, W. Sinn, K. Ullrich & K. von Figura (1980) BBA **630**, 402.

Selected entries from **Methods in Enzymology** [vol, page(s)]:
Assay: 50, 447, 448
Other molecular properties: distribution, **50**, 447; multiple sulfatase deficiency, **50**, 453, 454; Sanfilippo A syndrome and, **50**, 448, 449; specificity, **50**, 448

HEPARITIN SULFOTRANSFERASE

This sulfotransferase [EC 2.8.2.12] catalyzes the reaction of 3′-phosphoadenylylsulfate with heparitin to produce adenosine 3′,5′-bisphosphate and N-sulfoheparitin.[1–4]

[1] A. Orellana, C. B. Hirschberg, Z. Wei, S. J. Swiedler & M. Ishihara (1994) JBC **269**, 2270.
[2] E. Brandan & C. B. Hirschberg (1988) JBC **263**, 2417.
[3] D. Göhler, R. Niemann & E. Buddecke (1984) EJB **138**, 301.
[4] L. Jansson, M. Höök, A. Wasteson & U. Lindahl (1975) BJ **149**, 49.

HEPAROSAN-N-SULFATE-GLUCURONATE 5-EPIMERASE

This epimerase [EC 5.1.3.17] catalyzes the interconversion of heparosan-N-sulfate D-glucuronate and heparosan-N-sulfate L-iduronate.[1–3] Heparosan-O-sulfate D-glucuronate is also a substrate.

The enzyme, which is distinct from chondroitin-glucuronate 5-epimerase [EC 5.1.3.19], acts on D-glucuronosyl residues adjacent to sulfated D-glucosamine units in the heparin precursor. The catalyzed reaction involves reversible abstraction/readdition of the C5 proton at target hexuronate residues, and proceeds through a carbanion intermediate, such that inversion of configuration at C5 depends on the stereochemistry of proton readdition and are effected by two bases (probably lysyl residues):[4] note that, in the presence of D_2O, isotope incorporation is seen at C5. The fully N-sulfated derivative was found to be the best substrate in terms of its K_m value, which was significantly lower than that of its partially N-acetylated counterpart.[5]

[1] A. Malmström (1984) JBC **259**, 161.
[2] A. Malmström & L. Aberg (1982) BJ **201**, 489.
[3] A. Malmström, L. Roden, D. S. Feingold, I. Jacobsson, G. Bäckström & U. Lindahl (1980) JBC **255**, 3878.
[4] A. Hagner-Mcwhirter, U. Lindahl & J. P. Li (2000) BJ **347**, 69.
[5] A. Hagner-McWhirter, H. H. Hannesson, P. Campbell, J. Westley, L. Roden, U. Lindahl & J. P. Li (2000) Glycobiology **10**, 159.

HEPATITIS A VIRUS PICORNAIN 3C

This cysteine endopeptidase, also called picornain 3C, participates in proteolytic processing of polyprotein formed by hepatitis A virus. Because the proteinase domain can fold and produce active enzyme, even while in the context of the polyprotein, the enzyme exhibits autoproteolysis. The enzyme requires a glutaminyl residue in the P1 position, and there is a strong preference for substrates possessing a hydrophobic aminoacyl residue at P4. The P2 position usually contains a seryl or a threonyl residue.[1,2]

[1] D. A. Jewell, W. Swietnicki, B. M. Dunn & B. A. Malcolm (1992) Biochemistry **31**, 7862.
[2] B. A. Malcolm (1995) Protein Sci. **4**, 1439.

HEPATITIS C VIRUS ENDOPEPTIDASE 2

This endopeptidase, also called NS2-3 proteinase and Cpro-1, catalyzes the cleavage between nonstructural proteins 2 and 3 of the hepatitis C virus polyprotein (at Leu1026–Ala1027).[1]

[1] C. M. Rice (1996) in *Fields Virology* (B. N. Fields, D. M. Knipe & P. M. Howley, eds.), Raven, New York, p. 931.

HEPATITIS C VIRUS POLYPROTEIN PEPTIDASE

This serine endopeptidase [EC 3.4.21.98] (officially referred to as hepacivirin and also called NS3 serine proteinase, Cpro-2, and NS3-4A serine proteinase complex) participates in the processing of the hepatitis C virus polyprotein (of which the endopeptidase is a part). Four cleavage sites have been identified. A cysteinyl or a threonyl residue is located at position P1 while a seryl or alanyl residue is at P1'. An acidic residue (*i.e.*, Asp or Glu) is at position P6.[1–3]

[1] J. F. Bazan & R. J. Fletterick (1990) *Semin. Virol.* 1, 311.
[2] M. Houghton (1996) in *Fields Virology* (B. N. Fields, D. M. Knipe & P. M. Howley, eds.), Raven Press, New York, p. 1035.
[3] C. M. Rice (1996) in *Fields Virology* (B. N. Fields, D. M. Knipe & P. M. Howley, eds.), Raven Press, New York, p. 931.

HEPOXILIN-EPOXIDE HYDROLASE

This highly specific enzyme [EC 3.3.2.7] catalyzes the hydrolysis of hepoxilin A_3 (or, (5Z,9E,14Z)-(8ξ,11R,12S)-11,12-epoxy-8-hydroxyeicosa-5,9,14-trienoate) to produce trioxilin A_3 (or, (5Z,9E,14Z)-(8ξ,11ξ,12S)-8,11,12-trihydroxyeicosa-5,9,14-trienoate).[1] This enzyme is distinct from leukotriene A_4 hydrolase [EC 3.3.2.6].[2]

[1] C. R. Pace-Asciak (1988) *BBRC* 151, 493.
[2] C. R. Pace-Asciak & W. S. Lee (1989) *JBC* 264, 9310.

HEPSIN

This membrane-associated serine endopeptidase was identified from a human liver cDNA library and possesses a trypsin-like activity. The enzyme will catalyze the hydrolysis of such synthetic substrates as Bz-Leu-Ser-Arg-∇-NHPhNO2, Bz-Ile-Glu-Phe-Ser-Arg-∇-NHPhNO2, and Bz-Phe-Val-Arg-∇-NHPhNO2 (note: ∇ indicates the cleavage site).[1]

[1] K. Kurachi, A. Torres-Rosado & A. Tsuji (1994) *MIE* 244, 100.

Selected entries from *Methods in Enzymology* [vol, page(s)]:
General discussion: 244, 100
Other molecular properties: activation, 244, 103; biological role, 244, 109, 113; cell cycle-dependent expression, 244, 108; disulfide bridges, 244, 103; glycosylation, 244, 104; half-life, 244, 105, 108, 110; sequence, 244, 101; size, 244, 104; subcellular distribution, 244, 104; substrate specificity, 244, 108; subunits, 244, 103; tissue distribution, 244, 107; topology at cell surfaces, 244, 105

HERPESVIRUS ASSEMBLIN

This serine endopeptidase [EC 3.4.21.97], also known as HSV-1 protease, participates in the proteolytic processing of protein precursors in the herpes simplex virus. Herpesvirus assemblin is essential for viral replication. Two known cleavage sites involve Ala–Ser bonds. A common motif for cleavage sites of assemblins from the herpes group is Val/Leu-Xaa-Ala-∇-Ser where Xaa is a polar aminoacyl residue.[1] *See also Human Herpesvirus Type 6 Assemblin; Cytomegalovirus Assemblin*

[1] B. Roizman & A. Sears (1995) in *Fields Virology*, 3rd. ed. (B. N. Fields, D. M. Knipe & P. M. Howly, eds.), Lipponcott-Raven, Philadelphia, p. 2231.

Selected entries from *Methods in Enzymology* [vol, page(s)]:
Assay: 244, 411
Other molecular properties: conserved domains, 244, 403; family, 244, 403; immunoprecipitation, 244, 410; plasmid (construction, 244, 405; transfection, 244, 407); processing, 244, 400, 403, 408; substrate, 244, 400; transient transfection assay, 244, 404, 410; Western blot analysis, 244, 404, 408

HETEROGLYCAN α-MANNOSYLTRANSFERASE

This transferase [EC 2.4.1.48] catalyzes the reaction of GDP-D-mannose with a heteroglycan (containing D-mannose, D-galactose, and D-xylose) to produce GDP and a 1,2(or 1,3)-α-D-mannosylheteroglycan.[1,2] Both 1,2- and 1,3-mannosyl linkages are formed. The enzyme also uses D-mannose as an acceptor substrate to generate O-α-D-mannosyl-1,2-D-mannose.

[1] H. Ankel, E. Ankel & J. S. Schutzbach (1970) *JBC* 245, 3945.
[2] L. Lehle & W. Tanner (1974) *BBA* 350, 225.

Selected entries from
General discussion: 28, 553; 185, 469
Assay: 28, 554
Other molecular properties: capsular polysaccharide biosynthesis and, 28, 613; *Cryptococcus laurentii*, 28, 555; preparation, 28, 555; product identification, 28, 557, 614; properties, 28, 556, 614, 623; solubilized, preparation, 185, 468

HETEROTRIMERIC G-PROTEIN GTPases

Heterotrimeric G-protein GTPases [EC 3.6.1.46] are a family of GTPases in which GTP and GDP alternate in binding.[1–3] This large class of proteins are important components of receptor-mediated activation and inhibition of adenylate cyclase, the phosphatidylinositol pathway, and K^+ and Ca^{2+} channels. They are trimeric in structure: at least sixteen α subunits (the α subunit has the GTP binding site), five β subunits, and twelve γ subunits are known. G_s stimulates adenylate cyclase activity whereas G_i is inhibitory. G_p stimulates phospholipase C and G_{olf} participates in odor perception.

Although the overall reaction yields GDP and orthophosphate, the enzyme is misclassified as a hydrolase because P-O-P bond scission supplies the energy to drive a conformational change in the target protein.[4] The overall enzyme-catalyzed reaction should be regarded to be of the energase type (*e.g.*, State$_1$ + GTP = State$_1$ + GDP + P$_{i'}$, where two affinity-modulated conformational states are indicated.)[4] **See G Proteins; Energases**

[1] N. Vitale, J. Moss & M. Vaughan (1998) *JBC* **273**, 2553.
[2] E. J. Neer (1995) *Cell* **80**, 249.
[3] S. R. Sprang (1997) *Ann. Rev. Biochem.* **66**, 639.
[4] D. L. Purich (2001) *TiBS* **26**, 417.

Selected entries from **Methods in Enzymology** [vol, page(s)]:
General discussion: vols. **237** and **238**

HEXADECANAL DEHYDROGENASE (ACYLATING)

This oxidoreductase [EC 1.2.1.42], also known as fatty acyl-CoA reductase, catalyzes the reversible reaction of hexadecanoyl-CoA with NADH to produce hexadecanal, coenzyme A, and NAD$^+$.[1,2] Under most conditions, the reduction reaction is favored over the reverse reaction. Octadecanal can also be formed from octadecanoyl-CoA.

Enzymes catalyzing analogous reactions typically form a thiolacetal adduct that then undergoes oxidation to an enzyme-bound thiolester intermediate. Thiol group involvement in catalysis probably accounts for the extreme sensitivity of these enzymes toward mercuric ion and *p*-hydroxymercuribenzoate.

[1] R. C. Johnson & J. R. Gilbertson (1972) *JBC* **247**, 6991.
[2] D. Riendeau & E. Meighen (1985) *Experientia* **41**, 70.

Selected entries from **Methods in Enzymology** [vol, page(s)]:
Assay: 71, 264, 269
Other molecular properties: *Brassica oleracea*, from, **71**, 269; *Euglena gracilis*, from, **71**, 267; inhibitors and activators, **71**, 266; properties, **71**, 266, 268; specificity, **71**, 266, 268, 272; uropygial glands, from, **71**, 264

HEXADECANOL DEHYDROGENASE

This oxidoreductase [EC 1.1.1.164] catalyzes the reversible reaction of 1-hexadecanol with NAD$^+$ to produce hexadecanal and NADH. The rat liver enzyme will act on long-chain primary alcohols with eight to sixteen carbon atoms.[1] The *Euglena gracilus* Z enzyme also catalyzes

NAD$^+$-dependent conversion of long-chain aldehydes to long-chain fatty acid anions.[2]

[1] W. Stoffel, D. LeKim & G. Heyn (1970) *Hoppe-Seyler's Z. Physiol. Chem.* **351**, 875.
[2] P. E. Kolattukudy (1970) *Biochemistry* **9**, 1095.

Selected entries from **Methods in Enzymology** [vol, page(s)]:
Assay: NAD$^+$-dependent, membrane-bound, **188**, 17
Other molecular properties: NAD$^+$-dependent, membrane-bound (assay, **188**, 17; properties, **188**, 18)

2-HEXADECENAL REDUCTASE

This oxidoreductase [EC 1.3.1.27], also known as 2-alkenal reductase, catalyzes the reversible reaction of 2-*trans*-hexadecenal with NADPH (A-side[1]) to produce hexadecanal and NADP$^+$.[1–3] The enzyme acts on long chain 2-*trans*- and 2-*cis*-alkenals: the optimal substrate has between fourteen and sixteen carbon atoms. The enzyme will also act on polyunsaturated substrates, including 2-*trans*,4-*trans*-hexadienal (or, sorbaldehyde), which is converted to 4-*trans*-hexenal by the *Mucor griseo-cyanus* protein.[3]

[1] W. Stoffel & W. Därr (1974) *Hoppe-Seyler's Z. Physiol. Chem.* **355**, 54 and (1975) **356**, 385.
[2] W. Stoffel, E. Bauer & J. Stahl (1974) *Hoppe-Seyler's Z. Physiol. Chem.* **355**, 61.
[3] N. Miyamaura, H. Matsui, S. Tahara, J. Mizutani & S. Chiba (1984) *Agric. Biol. Chem.* **48**, 185.

HEXAPRENYLDIHYDROXYBENZOATE METHYLTRANSFERASE

This methyltransferase [EC 2.1.1.114], also known as 3,4-dihydroxy-5-hexaprenylbenzoate methyltransferase, catalyzes the reaction of *S*-adenosyl-L-methionine with 3-hexaprenyl-4,5-dihydroxybenzoate to produce *S*-adenosyl-L-homocysteine and 3-hexaprenyl-4-hydroxy-5-methoxybenzoate.[1]

[1] C. F. Clarke, W. Williams & J. H. Teruya (1991) *JBC* **266**, 16636.

trans-HEXAPRENYL*trans*TRANSFERASE

This transferase [EC 2.5.1.30], also known as all-*trans*-heptaprenyl-diphosphate synthase and heptaprenyl pyrophosphate synthetase, catalyzes the reaction of all-*trans*-hexaprenyl diphosphate (*i.e.*, *trans,trans*-farnesyl diphosphate) with isopentenyl diphosphate (*i.e.*, isopentenyl pyrophosphate) to produce pyrophosphate (*i.e.*, diphosphate) and all-*trans*-heptaprenyl diphosphate.[1,2] Other acceptor substrates include all-*trans*-geranylgeranyl diphosphate and all-*trans*-prenyl diphosphates of intermediate size; however, dimethylallyl diphosphate is not a substrate.

[1] I. Takahashi, K. Ogura & S. Seto (1980) *JBC* **255**, 4539.
[2] H. Fujii, T. Koyama & K. Ogura (1983) *FEBS Lett.* **161**, 257.

Selected entries from *Methods in Enzymology* [vol, page(s)]:
General discussion: **110**, 199
Assay: *Bacillus subtilis*, **110**, 199
Other molecular properties: *Bacillus subtilis* (component separation, **110**, 202; properties, **110**, 203; purification, **110**, 200)

HEXOKINASE

This prototypical sugar phosphotransferase [EC 2.7.1.1] catalyzes the ATP-dependent phosphorylation of D-glucose as well as many other aldo- and keto-hexoses (with the notable exception of D-galactose).[1-4] $MgATP^{2-}$ is the active phosphoryl donor, and the overall reaction is highly favorable (*i.e.*, K_{eq} = [ADP][D-glucose 6-phosphate]/[D-glucose][ATP] = 490 at pH 6.5; K_{eq} = 1500 at pH 7). The kinetics of the yeast enzyme are best described by the random sequential binding of the hexose and $MgATP^{2-}$ to form an enzyme · hexose · $MgATP^{2-}$ ternary complex. This was first demonstrated by the use of competitive inhibitors and isotope exchange studies at equilibrium. The hexokinase reaction mechanism is best described as a single in-line or S_N2 reaction involving facilitated attack of the 6-hydroxy group at the terminal phosphoryl of $MgATP^{2-}$.

As mentioned above, many aldo- and ketohexoses (including either anomer of D-glucose, D-mannose, D-fructose, and 2-deoxy-D-glucose) are phosphoryl-acceptor substrates. The stereochemical basis for hexose binding (as well as for the exclusion of D-galactose) stems from enzyme interactions at three subsites on each hexose.[1] Several pentoses, especially lyxose, bind in place of the hexose substrate and nonphysiologically stimulate yeast hexokinase's ATPase activity by a factor of 250,000. *See also Glucokinase; Energase*

[1] D. L. Purich, F. B. Rudolph & H. J. Fromm (1973) *AE* **39**, 249.
[2] G. Lowe, P. M. Cullis, R. L. Jarvest, B. V. L. Potter & B. S. Sproat (1981) *Phil. Trans. R. Soc. London B* **293**, 75.
[3] J. E. Wilson (1995) *Rev. Physiol. Biochem. Pharmacol.* **126**, 65.
[4] R. J. Middleton (1990) *Biochem. Soc. Trans.* **18**, 180.

Selected entries from *Methods in Enzymology* [vol, page(s)]:
General discussion: **1**, 269, 277; **5**, 226; **9**, 371; **34**, 246, 252, 328; **46**, 287; **90**, 3, 11, 16, 21
Assay: **1**, 269, 277; **3**, 873; **5**, 226; **9**, 371, 376; **63**, 37, 40; *Ascaris suum* muscle, **90**, 21; cardiac mitochondrial, **125**, 18; *Drosophila melanogaster*, **90**, 16; Fast Analyzer, by, **31**, 816; rabbit erythrocyte, **90**, 3; rat brain, **42**, 20, 21; **97**, 471; type Ia and Ib, rabbit reticulocyte, **90**, 3; type II, rat skeletal muscle, **90**, 11; yeast, **1**, 269; **9**, 376, **42**, 8
Other molecular properties: acetate kinase, and, in glucose 6-phosphate production, **136**, 52; acetyl adenylate and, **5**, 466; activation, **63**, 258, 264, 292; **87**, 654; active transport studies, **44**, 921; activity, **57**, 86 (ATP analogs, on, **44**, 876); adenosine phosphosulfate assay and, **6**, 770; adenylate kinase assay, **2**, 598; **6**, 223, 230; **38**, 162; **44**, 892; **51**, 468; **63**, 32; affinity labeling, **87**, 491; alternative substrate kinetics, **87**, 9; aluminum ion, and, **87**, 654; amylomaltase assay and, **5**, 141, 142; anomeric specificity, **63**, 378, 379; *Ascaris suum* muscle (assay, **90**, 21; properties, **90**, 24; purification, **90**, 22); aspartate residue, **87**, 403; assay of (adenylate kinase, **38**, 162; **44**, 892; **51**, 468; **63**, 32; AMP, **55**, 209; ATP, **2**, 497; **13**, 488, 490, 491; **63**, 32, 36; creatine kinase, **17A**, 995; **63**, 32; glucose, **3**, 109, 110; GTP, **55**, 208; inorganic phosphate, **55**, 212; oxidative phosphorylation, **55**, 723, 724, 747; succinyl-CoA, **55**, 210); associated with cell fractions, **40**, 86; ATPase activity, **87**, 5; ATP complex, partition analysis, **249**, 319; ATP, corrected dissociation constant, **87**, 391; ATP regeneration, in, **44**, 698; bakers' yeast, from, **1**, 269; **5**, 226; **9**, 376 (molecular weight, **9**, 381; multiple forms, **9**, 381; properties, **5**, 234; **9**, 380; purification, **5**, 228; **9**, 377); brain, bovine, purification, **1**, 281; brain, ox, from (properties, **9**, 374; purification, **9**, 372); calorimetry, reactions catalyzed, **61**, 263, 268; cell permeabilization test with, **137**, 640; chiral acetate, synthesis of, **87**, 141; chiral phosphate, **87**, 210; chromium-nucleotides, **87**, 161, 169; citrullinase and, **2**, 377; cobalt-nucleotides, **87**, 160, 161, 166, 181, 182; coentrapment, **44**, 475; competitive inhibition, **63**, 292, 476, 480; conformational change, **64**, 225, 226; conjugate, activity, **44**, 200, 271, 342; cooperativity, **64**, 218, 221; creatine kinase assay, **17A**, 995; **63**, 32; crystallization, **5**, 230, 234; cysteinyl residue, **87**, 491; determinations of guanosine nucleotides, **57**, 86, 89, 93; dialysis of, **10**, 195; *Drosophila melanogaster* (assay, **90**, 16; properties, **90**, 21; purification, **90**, 17); enhancement, rate, **87**, 281; epimerization, **87**, 179; equilibrium binding, linkage between substrate and water, **259**, 55; exchanges, **87**, 5, 654; fatty acid oxidation and, **1**, 548; galactokinase preparation free of, **9**, 410; β-galactosidase assay and, **5**, 212; GDPmannose synthesis and, **8**, 146; D-glucosamine microassay, in, **41**, 5; glucosamine 6-phosphate acetylase, assay of, **9**, 705; glucosamine 6-phosphate and, **3**, 158; **5**, 422; D-glucosamine phosphorylation, in, **9**, 38; glucose complex (catalytic competence, **249**, 316; dissociation rate, **249**, 316); glucose or fructose assay and, **3**, 52, 107, 109; glucose-6-phosphate dehydrogenase limitation in assay of, **63**, 37, 40; glucose-6-phosphate dehydrogenase preparation, need to be free of in, **9**, 388; half-site reactivity, **64**, 184; heart muscle, calf, purification, **1**, 282; hexokinase 2, affinity partitioning with dye ligands, **228**, 135; hexokinase B, yeast, self-association, **48**, 117; hexokinase C, yeast, self-association, **48**, 117; hormonal control, **37**, 293; hysteresis, **64**, 197, 215, 216, 218, 221, 222; immobilization (inert protein, on, **44**, 908; microencapsulation, by, **44**, 214; multistep enzyme system, in, **44**, 454); immobilized, glucose 6-phosphate synthesis with, **136**, 279; inhibition, **63**, 5, 348; **87**, 170, 178, 491 (determination of metal ion-nucleotide complex dissociation constants, in, **249**, 182); initial rate kinetics, **87**, 9; intestinal cells, in, **32**, 673; isolated nuclei and, **12A**, 445; isomerization, **64**, 197; isotope exchange, **64**, 4, 7, 8, 25, 29, 37, 38; **249**, 467; kinetic mechanism, equilibrium isotope exchange study, **249**, 467; kinetics, **87**, 9; kinetics, in multistep enzyme system, **44**, 432, 463; labeled ADP and, **4**, 854; liver, rat, purification, **1**, 283; mammalian brain, product inhibition, **249**, 562; D-mannose microassay, in, **41**, 3; mannose 6-phosphate and, **3**, 154; mechanism, **63**, 51; mnemonic mechanism, **64**, 221; multiple forms of, **5**, 235; myokinase and, **6**, 223, 230; negative cooperativity, **64**, 216, 221, 222; NMR, ^{31}P, **87**, 181, 182; nucleoside triphosphates and, **3**, 874; nucleotide site, **64**, 180; number of excess waters, determination by enzyme kinetics, **259**, 57, 423; oxidative phosphorylation and, **2**, 611; **6**, 266, 273, 277, 278, 285, 286; osmotic agents and conformational change, **259**, 55; osmotic pressure and substrate affinity, **259**, 58, 61; parenchyma cells, in, **32**, 706; Pasteur effect, **55**, 291; pH effects, **87**, 399, 403, 654; phosphofructokinase, in assay of, **9**, 437f; 6-phosphogluconate dehydrogenase and, **6**, 800; phosphoribosylpyrophosphate synthetase kinetics studies, in, **51**, 6; phosphorothioates, and, **87**, 200, 204, 224, 231, 300, 309; plasma membrane, **31**, 90; P/O ratio, **55**, 226; product inhibition, **63**, 433, 434, 436; **249**, 562; properties, **1**, 274, 284; **5**, 234; **9**, 374, 380; propionyl-CoA carboxylase and, **5**, 575; protein phosphokinase and, **6**,

222; purification, **1**, 270, 281, 282, 283; **5**, 228; **6**, 278; **9**, 372, 377; **10**, 31; rabbit erythrocyte (assay, **90**, 3; properties, **90**, 7; purification, **90**, 4); rat brain, of, **42**, 20 (activity assay, **97**, 471; amino acids, **42**, 24; assay, **42**, 20, 21; binding assay, **97**, 470; binding protein isolation from outer mitochondrial membrane, **97**, 472; chromatography, **42**, 22; molecular weight, **42**, 23; properties, **42**, 23; purification, **42**, 21); scrubbing by, **87**, 178; Sigma type III, **9**, 38; site-specific reagent, **47**, 482; skeletal muscle, purification, **1**, 283; slow-binding inhibitor, **63**, 450; sorbose 1-phosphate and, **3**, 179; specificity, **63**, 378, 379; **87**, 160, 161, 176, 200, 204, 210, 224, 231; spot test for, **9**, 376; staining on gels, **22**, 598; steady state assumption, **63**, 476; stereochemistry, **87**, 160, 161, 200, 204, 210, 224, 231, 300; studies of glucose-marked liposomes, in, **32**, 512; substrate inhibition (alternative, **63**, 145, 494; partial induced, **63**, 510); substrate specificity, **57**, 93; transition state and multisubstrate analogues, **249**, 305; tritium-labeled glucose 6-phosphate and, **6**, 877; type Ia and Ib, rabbit reticulocyte (assay, **90**, 3; properties, **90**, 10; purification, **90**, 8); type II, rat skeletal muscle (assay, **90**, 11; properties, **90**, 15; purification, **90**, 12); UDPglucose:fructose transglucosylase assay and, **8**, 341; wheat germ (cooperativity and slow transitions in, **249**, 545; glucose rate dependence, **249**, 329); yeast, of, **1**, 269; **5**, 226; **9**, 15, 376; **27**, 464; **42**, 6 (ADPglucose synthesis and, **28**, 279; assay, **42**, 8; chromatography, **42**, 11; coupling to sulfonated support, **104**, 61; GDPmannose synthesis and, **28**, 281, 283; isoenzymes A, B, and C, **42**, 7, 16; **48**, 117; isotope exchange study, kinetic patterns for modifier action, **87**, 654; molecular weight, **42**, 18; partition analysis, **249**, 323; preparation of labeled oligonucleotides, **59**, 75; properties, **1**, 274; **5**, 234; **9**, 380; **42**, 16; purification, **1**, 270; **5**, 228; **9**, 377; **42**, 9 [Mg^{2+}-promoted binding to immobilized dyes, by, **104**, 110]; reaction sequence, **249**, 316; reversible inhibitors as mechanistic probes, **249**, 135; thymidine diphosphate glucose-4T synthesis and, **28**, 295; tRNA labeling procedure, in, **59**, 71; tRNA sequence analysis, in, **59**, 61; solution, stability, **59**, 61)

HEXOSE OXIDASE

This copper-dependent oxidoreductase [EC 1.1.3.5] catalyzes the reaction of β-D-glucose with dioxygen to produce D-glucono-1,5-lactone and hydrogen peroxide.[1–3] Other substrates include D-galactose, D-glucose 6-phosphate, 2-deoxy-D-glucose, D-mannose, D-xylose, maltose, lactose, and cellobiose. The copper-containing enzyme from red seaweed (*Chondrus crispus*) is a flavoprotein.[4]

[1] J. D. Sullivan, Jr. & M. Ikawa (1973) *BBA* **309**, 11.
[2] R. C. Bean, G. G. Porter & B. M. Steinberg (1961) *JBC* **236**, 1235.
[3] R. C. Bean & W. Z. Hassid (1956) *JBC* **218**, 425.
[4] B. W. Groen, S. De Vries & J. A. Duine (1997) *EJB* **244**, 858.

Selected entries from *Methods in Enzymology* [vol, page(s)]:
General discussion: 52, 10; 89, 145
Assay: *Chondrus crispus*, **89**, 145

HEXOSE-1-PHOSPHATE GUANYLYLTRANSFERASE

This transferase [EC 2.7.7.29], also known as GDP-hexose pyrophosphorylase, catalyzes the reversible reaction of GTP with an α-D-hexose 1-phosphate to produce a GDP-D-hexose and pyrophosphate (or, diphosphate).[1] Substrates include α-D-glucose 1-phosphate and α-D-mannose 1-phosphate. Nucleotide substrates include ITP, UTP, and ATP.

[1] R. G. Hansen, H. Verachtert, P. Rodriguez & S. T. Bass (1966) *MIE* **8**, 269.

Selected entries from *Methods in Enzymology* [vol, page(s)]:
General discussion: 8, 269
Assay: 8, 269

H+-EXPORTING ATPase

This so-called P-type ATPase [EC 3.6.3.6; previously classified as EC 3.6.1.35], which undergoes covalent phosphorylation during the transport cycle, catalyzes the ATP-dependent (ADP- and orthophosphate-producing) transport of H^+_{in} to produce H^+_{out}.[1–8] This classification represents a group of membranal enzymes isolated from a number of protozoa, fungi, and plants that are closely related to the H^+/K^+-transporting ATPases [EC 3.6.3.10]. It participates in the generation of an electrochemical potential gradient of protons across the plasma membrane. *See also* ATP Synthase, H+-Transporting; ATPase, H+/K+ -Exchanging

The idea that a transporter is an enzyme is in keeping with a new definition of enzyme catalysis as the facilitated making/breaking of chemical bonds, not just covalent bonds.[9] This idea builds on Pauling's assertion that any long-lived, chemically distinct interaction (in this case, the persistent location of a solute with respect to the faces of a membrane) can be regarded as a chemical bond. Note also that the equilibrium constant ($K_{eq} = [H^+_{out}][ADP][P_i]/[H^+_{in}][ATP]$) does not conform to that expected for an ATPase (*i.e.*, $K_{eq} = [ADP][P_i]/[ATP]$). Thus, although the overall reaction yields ADP and orthophosphate, the enzyme is misclassified as a hydrolase, and instead should be regarded as an energase-type reaction. Energases facilitate affinity-modulated reactions by coupling the $\Delta G_{ATP-hydrolysis}$ to a force-generating or work-producing step.[9] In this case, P-O-P bond scission supplies the energy to drive proton export. *See ATPase; Energase*

[1] R. Serrano & F. Portillo (1990) *BBA* **1018**, 195.
[2] R. Serrano (1988) *BBA* **947**, 1.
[3] R. Serrano, M. C. Kielland-Brandt & G. R. Fink (1986) *Nature* **319**, 689.
[4] A. Goffeau & C. W. Slayman (1981) *BBA* **639**, 197.
[5] J. B. Dame & G. A. Scarborough (1980) *Biochemistry* **19**, 2931.
[6] A. Amory, F. Foury & A. Goffeau (1980) *JBC* **255**, 9353.
[7] J. Ahlers (1984) *Can. J. Biochem. Cell Biol.* **62**, 998.
[8] G. A. Scarborough (2000) *Cell. Mol. Life Sci.* **57**, 871.
[9] D. L. Purich (2001) *TiBS* **26**, 417.

Selected entries from *Methods in Enzymology* [vol, page(s)]:
General discussion: 157, 513, 528, 533, 544, 562, 574, 579
Assay: tonoplast-associated, 148, 110
Other molecular properties: proton translocation in *Neurospora crassa*, assay, **174**, 668, 674; tonoplast-associated (assay, **148**, 110; detection

and physiological role, **148**, 91; properties, **148**, 97, 112; purification, **148**, 96, 110; solubilization, **148**, 94; transport activity, **148**, 111); vacuolar membrane-associated (assay, **157**, 548; polypeptide composition, **157**, 558; properties, **157**, 549, 555; purification, **157**, 552)

3-HEXULOSE-6-PHOSPHATE SYNTHASE

This aldolase catalyzes the reaction of D-ribulose 5-phosphate with formaldehyde to produce D-*arabino*-3-ketohexulose 6-phosphate.

Selected entries from *Methods in Enzymology* [**vol**, page(s)]:
General discussion: **90**, 314, 319; **188**, 391, 397
Assay: *Acetobacter methanolicus* MB58, **188**, 401; *Bacillus* C1, **188**, 392; *Methylococcus capsulatus*, **90**, 314; *Methylomonas* M15, **90**, 319; *Mycobacterium gastri* MB19, **188**, 397
Other molecular properties: *Acetobacter methanolicus* MB58 (properties, **188**, 403; purification, **188**, 402); *Bacillus* C1 (properties, **188**, 395; purification, **188**, 393); *Methylococcus capsulatus* (properties, **90**, 317; purification, **90**, 316); *Methylomonas* M15 (production, **90**, 320; properties, **90**, 322; purification, **90**, 321); *Mycobacterium gastri* MB19 (properties, **188**, 400; purification, **188**, 399)

*Hga*I RESTRICTION ENDONUCLEASE

This type II restriction enzyme (subtype s) [EC 3.1.21.4], isolated from *Haemophilus gallinarum* (note that this microorganism has been reclassified as *Pasteurella volantium*), catalyzes the cleavage of DNA at 5′... GACGCNNNNN∇NNNNN... 3′ and 3′... CTGCGNNN NNNNNNN∇... 5′, where N refers to any base.[1-3] This endonuclease will also act on single-stranded DNA slowly. Note that very few restriction endonuclease produce extensions of more than four bases: this enzyme generates extensions of five bases.

[1]M. Takanami (1974) *Methods Mol. Biol.* **7**, 113.
[2]N. L. Brown & M. Smith (1977) *PNAS* **74**, 3213.
[3]H. Sugisaki (1978) *Gene* **3**, 17.

Selected entries from *Methods in Enzymology* [**vol**, page(s)]:
General discussion: **65**, 24, 392
Other molecular properties: recognition sequence, **65**, 7, 14

*Hgi*AI RESTRICTION ENDONUCLEASE

This type II restriction enzyme [EC 3.4.21.4], isolated from *Haemophilus giganteus* HP1023, catalyzes the cleavage of both strands of DNA at 5′... GWGCW∇C... 3′, where W refers to either A or T.[1,2]

[1]N. L. Brown, M. McClelland & P. R. Whitehead (1980) *Gene* **9**, 49.
[2]N. L. Brown (1980) *Biochem. Soc. Trans.* **8**, 398.

*Hha*I RESTRICTION ENDONUCLEASE

This type II restriction enzyme [EC 3.1.21.4], isolated from *Haemophilus haemolyticus*, catalyzes the cleavage of both strands of DNA at 5′... GCG∇C... 3′.[1] It will act on

single-stranded DNA at about one-half the rate observed with double-stranded DNA.[2] Note that *Hin*P1I has the same recognition sequence but generates a 5′ extension whereas this endonuclease produces a 3′ extension. *Hha*I also exhibits star activity.

[1]R. J. Roberts, P. A. Myers, A. Morrison & K. Murray (1976) *J. Mol. Biol.* **103**, 199.
[2]K. Nishigaki, Y. Kaneko, H. Wakuda, Y. Husimi & T. Tanaka (1985) *NAR* **13**, 5747.

Selected entries from *Methods in Enzymology* [**vol**, page(s)]:
General discussion: **65**, 23, 33, 53, 54, 76, 83, 175, 342, 348, 363
Other molecular properties: recognition sequence, **65**, 7; sticky end sequence, **65**, 342

*Hha*II RESTRICTION ENDONUCLEASE

This type II restriction enzyme [EC 3.1.21.4], isolated from *Haemophilus haemolyticus*, catalyzes the cleavage of both strands of DNA at 5′... G∇ANTC... 3′, where N refers to any base.[1,2]

[1]M. B. Mann, R. N. Rao & H. O. Smith (1978) *Gene* **3**, 97.
[2]S. Kelly, R. Kaddurah-Daouk & H. O. Smith (1985) *JBC* **260**, 15339.

HIGH-MANNOSE-OLIGOSACCHARIDE β-1,4-N-ACETYL-GLUCOSAMINYL-TRANSFERASE

This transferase [EC 2.4.1.197] catalyzes the transfer of an *N*-acetyl-D-glucosamine residue from UDP-*N*-acetyl-D-glucosamine to the 4-position of a mannosyl residue linked α-1,6 to the core mannose of the high-mannose oligosaccharides produced by the slime mold *Dictyostelium discoideum*.[1] The activity of the intersecting mannose residue as acceptor is dependent on two other mannosyl residues attached by α-1,3 and α-1,6 linkages. The best substrate reported is Manα(1→2)Manα(1→2) Manα(1→3)[Manα(1→2)Manα(1→3)[Manα(1→2)Manα (1→6)]Manα(1→6)]Manβ(1→4)GlcNAc.

A related mouse enzyme catalyzes the transfer of an *N*-acetyl-D-glucosamine residue from UDP-*N*-acetyl-D-glucosamine to [GlcA-GlcNAc]$_n$-GlcA-aMan, where aMan is 2,5-anhydro-D-mannose, and GlcA is glucuronic acid.[2]

[1]D. J. Sharkey & R. Kornfeld (1989) *JBC* **264**, 10411.
[2]K. Lidholt & U. Lindahl (1992) *BJ* **287**, 21.

*Hinc*II RESTRICTION ENDONUCLEASE

This type II restriction enzyme [EC 3.1.21.4], isolated from *Haemophilus influenzae* R$_c$, catalyzes the cleavage of both strands of DNA at 5′... GTY∇RAC... 3′, producing blunt ends (where Y refers to a pyrimidine base and R refers to a purine base).[1] Flanking bases are also of importance in reaction rates.[2] Overdigestion by *Hinc*II can produce star activity.

[1] A. Landy, E. Ruedisueli, L. Robinson, C. Foeller & W. Ross (1974) *Biochemistry* **13**, 2134.
[2] A. D. B. Malcolm & J. R. Moffat (1981) *BBA* **655**, 128.

Selected entries from *Methods in Enzymology* [vol, page(s)]:
General discussion: 65, 21, 46, 50, 51, 108, 363, 450, 455, 457, 459, 848, 851
Other molecular properties: digestion, 65, 456; isoschizomer, 65, 108; recognition sequence, 65, 7, 49

*Hin*I RESTRICTION ENDONUCLEASE

This type I restriction enzyme [EC 3.1.21.3], isolated from *Haemophilus influenzae* R$_d$ (exo-mutant), catalyzes the cleavage of both strands of DNA and has a recognition sequence of 5′...CAC...3′.[1,2] The sites of cleavage are random.

[1] P. H. Roy & H. O. Smith (1973) *J. Mol. Biol.* **81**, 445.
[2] R. Gromkova, J. LaPorte & S. H. Goodgal (1973) *Abstr. Gen. Meet. Am. Soc. Microbiol.* **73**, 72.

Selected entries from *Methods in Enzymology* [vol, page(s)]:
Other molecular properties: recognition sequence, 65, 7

*Hin*II RESTRICTION ENDONUCLEASE

This type II restriction enzyme [EC 3.1.21.4], isolated from *Haemophilus influenzae* R$_d$ (exo-mutant), catalyzes the cleavage of both strands of DNA at 5′...GTY∇RAC...3′, producing blunt ends (where Y refers to a pyrimidine base and R refers to a purine base).[1,2] This endonuclease will not act on single-stranded DNA.

[1] H. O. Smith & K. W. Wilcox (1970) *J. Mol. Biol.* **51**, 379.
[2] T. J. Kelly, Jr. & H. O. Smith (1970) *J. Mol. Biol.* **51**, 393.

Selected entries from *Methods in Enzymology* [vol, page(s)]:
General discussion: 65, 21, 23, 33, 49, 54, 55, 104, 166, 299, 301, 314, 339, 445, 622, 632, 706, 799, 812, 869, 876; **208**, 433
Assay: 65, 105
Other molecular properties: divalent cation requirements, 65, 108; ionic requirements, 65, 108; isoschizomer, 65, 108; pH optimum, 65, 108; properties, 65, 108; purification, 65, 105; recognition sequence, 65, 7, 55; stability, 65, 108; substrate specificity, 65, 108

*Hin*III RESTRICTION ENDONUCLEASE

This type II restriction enzyme [EC 3.1.21.4], isolated from *Haemophilus influenzae* R$_d$ (exo-mutant), catalyzes the cleavage of both strands of DNA at 5′...A∇AGCTT...3′.[1] This endonuclease will not act on single-stranded DNA and it can exhibit star activity.[2]

[1] R. Old, K. Murray & G. Roizes (1975) *J. Mol. Biol.* **92**, 331.
[2] M. Nasri & D. Thomas (1986) *NAR* **14**, 811.

Selected entries from *Methods in Enzymology* [vol, page(s)]:
General discussion: 65, 21, 23, 34, 35, 49, 104, 166, 175, 314, 324, 342, 343, 361, 363, 385, 410, 415, 418, 432, 433, 435, 440, 445, 450, 455, 457, 483, 493, 622, 706, 757, 789, 793, 812, 813, 883
Assay: 65, 105
Other molecular properties: cleavage pattern, on PM2 DNA, 65, 279; digestion, 65, 456; divalent cation requirements, 65, 108; fragment end structure produced, 65, 511; ionic requirements, 65, 108; optimal temperature, 65, 416; pH optimum, 65, 108; properties, 65, 108; purification, 65, 105; recognition sequence, 65, 7, 55, 511; restriction sites, on SV40 DNA, 65, 418, 419; stability, 65, 108; substrate specificity, 65, 108

*Hin*fI RESTRICTION ENDONUCLEASE

This type II restriction enzyme [EC 3.1.21.4], isolated from *Haemophilus influenzae* R$_f$, catalyzes the cleavage of both strands of DNA at 5′...G∇ANTC...3′, where N refers to any base.[1] This endonuclease will act on single-stranded DNA[2,3] and it can exhibit star activity.

[1] G. N. Godson & R. J. Roberts (1976) *Virology* **73**, 561.
[2] B. Hofer, G. Ruhe, A. Koch & H. Koster (1982) *Nucl. Acids Res.* **10**, 2763.
[3] K. Nishigaki, Y. Kaneko, H. Wakuda, Y. Husimi & T. Tanaka (1985) *NAR* **13**, 5747.

Selected entries from *Methods in Enzymology* [vol, page(s)]:
General discussion: 65, 342, 553, 575, 576, 799
Other molecular properties: fragment end structure produced, 65, 511; recognition sequence, 65, 7, 403, 511; sticky end sequence, 65, 342

*Hin*P1I RESTRICTION ENDONUCLEASE

This type II restriction endonuclease [EC 3.1.21.4], isolated from *Haemophilus influenzae* P$_1$, catalyzes the cleavage of both strands of DNA at 5′...G∇CGC...3′.[1] Note that *Hin*P1I produces a 5′ extension that can be effectively ligated into *Acc*I sites of cloning vectors.

[1] S. Shen, Q. Li, P. Yan, B. Zhou, S. Ye, Y. Lu & D. Wang (1980) *Sci. Sin.* **23**, 1435.

HIPPURATE HYDROLASE

This enzyme [EC 3.5.1.32], also known as benzoylglycine amidohydrolase and hippuricase, catalyzes the hydrolysis of hippurate to produce benzoate and glycine.[1,2] Other substrates include other *N*-benzoylamino acids (for example, *N*-benzoyl-L-alanine and *N*-benzoyl-L-aminobutyrate). *See also* Aminoacylase

[1] M. Röhr (1968) *Monatsh. Chem.* **99**, 2255 and 2278.
[2] E. Miyagawa, Y. Yano, T. Hamakado, Y. Kido, K. Nishimoto & Y. Motoki (1985) *Agric. Biol. Chem.* **49**, 2881.

HISTAMINE *N*-METHYLTRANSFERASE

This methyltransferase [EC 2.1.1.8] catalyzes the reaction of *S*-adenosyl-L-methionine with histamine to produce *S*-adenosyl-L-homocysteine and N^τ-methylhistamine.

Inversion of methyl group configuration during transfer by the guinea pig brain transferase was demonstrated using chiral [^{1}H,^{2}H,^{3}H-methyl]-*S*-adenosyl-L-methionine.[1-5] Initial rate and product inhibition studies on human skin[3] and bovine brain[4] enzymes indicate an ordered Bi Bi steady-state kinetic mechanism, where *S*-adenosyl-L-methionine is the leading substrate and *N*-methylhistamine is the first to dissociate.[3]

[1] R. M. Weinshilboum, D. M. Otterness & C. L. Szumlanski (1999) *Ann. Rev. Pharmacol. Toxicol.* **39**, 19.
[2] Y. Asano, R. W. Woodard, D. R. Houck & H. G. Floss (1984) *ABB* **231**, 253.
[3] D. M. Francis, M. F. Thompson & M. W. Greaves (1980) *BJ* **187**, 819.
[4] W. L. Gitomer & K. F. Tipton (1986) *BJ* **233**, 669.
[5] A. Thithapandha & V. H. Cohn (1978) *Biochem. Pharmacol.* **27**, 263.

Selected entries from ***Methods in Enzymology*** [vol, page(s)]:
General discussion: **17B**, 766
Assay: **17B**, 766

HISTIDINE *N*-ACETYLTRANSFERASE

This acetyltransferase [EC 2.3.1.33] catalyzes the reaction of acetyl-CoA with L-histidine to produce coenzyme A and *N*-acetyl-L-histidine.[1] D-Histidine can also be acetylated, albeit not as efficiently.

[1] M. H. Baslow (1966) *Brain Res.* **3**, 210.

HISTIDINE AMINOTRANSFERASE

This pyridoxal-phosphate-dependent aminotransferase [EC 2.6.1.38] catalyzes the reversible reaction of L-histidine with α-ketoglutarate (or, 2-oxoglutarate) to produce imidazol-5-yl-pyruvate and L-glutamate.[1,2] The *Pseudomonas testosteroni* enzyme has a ping pong Bi Bi kinetic mechanism.[2]

[1] J. G. Coote & H. Hassall (1969) *BJ* **111**, 237.
[2] A. J. Hacking & H. Hassall (1975) *BJ* **147**, 327.

HISTIDINE AMMONIA-LYASE

This lyase [EC 4.3.1.3], also known as histidinase, histidase, and histidine α-deaminase, catalyzes the conversion of L-histidine to urocanate and ammonia.[1-3] The enzyme engages a dehydroalanyl residue, itself generated from a serine residue in the lyase, to form a transient β-aminoalanyl intermediate.

[1] V. E. Anderson (1998) *CRC* **2**, 115.
[2] J. Retey (1996) *Naturwissenschaften* **83**, 439.
[3] K. R. Hanson & E. A. Havir (1972) *The Enzymes*, 3rd ed., **7**, 75.

Selected entries from ***Methods in Enzymology*** [vol, page(s)]:
General discussion: **2**, 228; **17B**, 47, 63, 69, 891
Assay: **2**, 228; **17B**, 47, 63, 70, 891; stratum corneum, in, **17B**, 891; **22**, 152

Other molecular properties: dehydroalanine in, **17B**, 73; L-histidine, assay of, **17B**, 895; human skin, in, **17B**, 891; hydrogen peroxide oxidation, **25**, 398; induction in microorganisms, **22**, 93; integrated rate equation, **63**, 161–162, 180; progress curve analysis, **63**, 161–162, 180; properties, **2**, 230; **17B**, 49, 68, 72; purification from (*Bacillus subtilis*, **17B**, 48; human stratum corneum, **17B**, 892; *Pseudomonas*, **2**, 229; **17B**, 64, 70); urocanate (preparation and, **3**, 626; production, **44**, 745)

HISTIDINE DECARBOXYLASE

This decarboxylase [EC 4.1.1.22] catalyzes the conversion of L-histidine to histamine and carbon dioxide.[1-4] The enzyme requires either pyridoxal phosphate or a pyruvoyl residue as a cofactor. The pyruvoyl-dependent enzyme forms a Schiff base that acts as an electron sink to facilitate loss of the α-carboxylate as carbon dioxide.[4]

[1] M. L. Hackert & A. E. Pegg (1998) *CBC* **2**, 201.
[2] M. H. O'Leary (1992) *The Enzymes*, 3rd ed., **20**, 235.
[3] D. J. Creighton & N. S. R. K. Murthy (1990) *The Enzymes*, 3rd ed., **19**, 323.
[4] E. A. Boeker & E. E. Snell (1972) *The Enzymes*, 3rd ed., **6**, 217.

Selected entries from ***Methods in Enzymology*** [vol, page(s)]:
General discussion: **2**, 187; **17B**, 663, 667; **122**, 128, 139
Assay: **17B**, 663, 673; assay by phosphorylated cellulose column chromatography, **94**, 45; pyridoxal phosphate-dependent, from *Morganella* AM-15, **122**, 139; pyruvoyl-dependent, from *Lactobacillus* 30a, **17B**, 663; **122**, 128
Other molecular properties: neurobiological studies, in, **32**, 769; occurrence, **17B**, 667; properties, **17B**, 666, 675; purification from (*Clostridium welchii*, **2**, 187; **3**, 464; *Lactobacillus* 30a, **17B**, 664; **122**, 130; mammalian tissues, **17B**, 668); pyridoxal phosphate-dependent, from *Morganella* AM-15 (assay, **122**, 139; properties, **122**, 142; purification, **122**, 140); pyruvoyl-dependent, from *Lactobacillus* 30a (purification, **17B**, 664; **122**, 130; structure and properties, **17B**, 666; **122**, 132); transaminase assay and, **2**, 171

HISTIDINOL DEHYDROGENASE

This oxidoreductase [EC 1.1.1.23] catalyzes the reaction of L-histidinol with water and two molecules of NAD$^+$ to produce L-histidine and two molecules of NADH.

The *Neurospora crassa* enzyme is multifunctional, and possesses phosphoribosyl-AMP cyclohydrolase [EC 3.5.4.19] and phosphoribosyl-ATP pyrophosphatase [EC 3.6.1.31] activities. The *Salmonella typhimurium* and cabbage enzyme have a Bi Uni Uni Bi ping pong kinetic mechanism.[1,2] L-Histidinal is an intermediate in the overall reaction. Kinetic isotope effect investigations indicate that the second hydride transfer from the aldehyde intermediate is slow relative to other steps in the reaction.[3]

[1] E. Bürger & H. Gorisch (1981) *EJB* **116**, 137.
[2] A. Kheirolomoom, J. Mano, A. Nagai, A. Ogawa, G. Iwasaki & D. Ohta (1994) *ABB* **312**, 493.
[3] C. Grubmeyer & H. Teng (1999) *Biochemistry* **38**, 7355.

Selected entries from ***Methods in Enzymology*** [vol, page(s)]:
General discussion: **17B**, 36, 43

Assay: 17B, 36
Other molecular properties: histidinol, assay of, **3**, 633f; **63**, 33; properties, **17B**, 40, 43; purification, *Salmonella typhimurium*, **17B**, 38; staining on gels, **22**, 594; stereospecificity, **87**, 104; synchronous cultures and, **21**, 469; yeast, from, antibody binding, **50**, 55, 56

HISTIDINOL PHOSPHATASE

This enzyme [EC 3.1.3.15] catalyzes the hydrolysis of L-histidinol phosphate to generate L-histidinol and orthophosphate.[1] The *Salmonella typhimurium* enzyme is bifunctional and possesses imidazoylglycerolphosphate dehydratase activity.

[1] L. L. Houston & M. E. Graham (1974) *ABB* **162**, 513.

Selected entries from *Methods in Enzymology* [vol, page(s)]:
General discussion: 17B, 29
Assay: 17B, 30; **63**, 33
Other molecular properties: *Salmonella typhimurium*, **17B**, 29 (properties, **17B**, 40; purification, **17B**, 32)

HISTIDINOL-PHOSPHATE AMINOTRANSFERASE

This pyridoxal-phosphate-dependent transferase [EC 2.6.1.9], also known as imidazolylacetolphosphate aminotransferase and imidazole acetolphosphate transaminase, catalyzes the reversible reaction of L-histidinol phosphate with α-ketoglutarate (or, 2-oxoglutarate) to produce 3-(imidazol-4-yl)-2-oxopropyl phosphate and L-glutamate.[1–4] The *Salmonella*[1] and *Bacillus*[2] enzymes have ping pong Bi Bi kinetic mechanisms.

[1] L. C. Hsu, M. Okamoto & E. E. Snell (1989) *Biochimie* **71**, 477.
[2] D. A. Weigent & E. W. Nester (1976) *JBC* **251**, 6974.
[3] A. E. Braunstein (1973) *The Enzymes*, 3rd ed., **9**, 379.
[4] B. N. Ames & B. L. Horecker (1956) *JBC* **220**, 113.

Selected entries from *Methods in Enzymology* [vol, page(s)]:
General discussion: 17B, 33, 40, 42
Assay: 17B, 33
Other molecular properties: properties, **17B**, 40, 42; purification, *Salmonella typhimurium*, **17B**, 33

HISTIDYL-tRNA SYNTHETASE

This enzyme [EC 6.1.1.21], also known as histidine: tRNA ligase, catalyzes the reaction of L-histidine with tRNA^{His} and ATP to produce L-histidyl-tRNA^{His}, AMP, and pyrophosphate (or, diphosphate).[1–3] The first step in the catalytic reaction is the in-line attack of the histidyl carboxylate at the α-phosphoryl of ATP to form the stable, enzyme-bound histidyl-AMP intermediate. *See also Aminoacyl-tRNA Synthetases*

[1] W. Freist, J. F. Verhey, A. Ruhlmann, D. H. Gauss & J. G. Arnez (1999) *Biol. Chem.* **380**, 623.

[2] A. Åberg, A. Yaremchuk, M. Tukalo, B. Rasmussen & S. Cusack (1997) *Biochemistry* **36**, 3084.
[3] P. Schimmel (1987) *Ann. Rev. Biochem.* **56**, 125.

HISTOLYSAIN

This cysteine endopeptidase [EC 3.4.22.35], also known as histolysin, catalyzes the hydrolysis of peptide bonds in proteins, exhibiting a preference for Arg-Arg-∇-Xaa bonds (*i.e.*, arginyl residues in the P1 and P2 positions). Several proteins of the extracellular matrix, including basement membrane collagen, are acted upon by histolysain. Other protozoan cysteine endopeptidases have been named histolysain.[1] *See also Amoebapain*

[1] H. Scholze & W. Schulte (1994) *Meth. Enzymol.* **244**, 512.

Selected entries from *Methods in Enzymology* [vol, page(s)]:
General discussion: 244, 512
Assay: *Entamoeba histolytica*, **244**, 514
Other molecular properties: *Entamoeba histolytica* (cellular localization, **244**, 519; effectors, **244**, 517; gene analysis, **244**, 520; substrate specificity, **244**, 518)

HISTONE ACETYLTRANSFERASE

This acetyltransferase [EC 2.3.1.48] catalyzes the reaction of acetyl-CoA with a histone to produce an acetyl-histone (at the ε-amino group of a lysyl residue) and Coenzyme A. This classification represents a group of enzymes with differing specificities toward histone acceptors. Certain of these enzymes also acetylate spermine, whereas others act only on spermidine.[1–3]

[1] D. E. Sterner & S. L. Berger (2000) *Microbiol. Mol. Biol. Rev.* **64**, 435.
[2] L. Magnaghi-Jaulin, S. Ait-Si-Ali & A. Harel-Bellan (2000) *Prog. Cell Cycle Res.* **4**, 41.
[3] M. H. Kuo & C. D. Allis (1998) *BioEssays* **20**, 615.

Selected entries from *Methods in Enzymology* [vol, page(s)]:
General discussion: 94, 325; **304**, 675, 696
Assay: 107, 231; bovine liver, **94**, 325
Other molecular properties: bovine liver (assay, **94**, 325; properties, **94**, 328; purification, **94**, 326); transition state and multisubstrate analogues, **249**, 304

HISTONE-ARGININE *N*-METHYLTRANSFERASE

This methyltransferase [EC 2.1.1.125], also known as histone protein methylase, protein-arginine and *N*-methyltransferase, catalyzes the reaction of *S*-adenosyl-L-methionine with an arginyl residue in a histone to produce *S*-adenosyl-L-homocysteine and an N^{ω}-methyl-arginyl residue in that histone.[1] Both the N^{ω}-monomethyl- and the N^{ω},N^{ω}-dimethylarginyl residues are formed; however, the N^{ω},N^{ω}-dimethylarginyl derivative is not.

[1]N. Rawal, R. Rajpurohit, W. K. Paik & S. Kim (1994) *BJ* **300**, 483.
Selected entries from *Methods in Enzymology* [vol, page(s)]:
Assay: 106, 269

HISTONE DEACETYLASE

This enzyme catalyzes the hydrolysis of acetylated histone to produce the deacetylated protein and acetate. *See also Protein-Acetyllysine Deacetylase*

Selected entries from *Methods in Enzymology* [vol, page(s)]:
General discussion: 107, 236; 304, 715
Assay: 304, 720

HISTONE-LYSINE *N*-METHYLTRANSFERASE

This methyltransferase [EC 2.1.1.43], also known as protein methylase III, catalyzes the reaction of S-adenosyl-L-methionine with an L-lysyl residue in a histone to produce S-adenosyl-L-homocysteine and a histone with an N^6-methyl-L-lysyl residue.[1-3] *See also [Cytochrome c]-Lysine N-Methyltransferase; Calmodulin-Lysine N-Methyltransferase; [Ribulose-Bisphosphate-Carboxylase]-Lysine N-Methyltransferase*

[1]M. T. Tuck, J. Z. Farooqui & W. K. Paik (1985) *JBC* **260**, 7114.
[2]M. Venkatesan & I. R. McManus (1979) *Biochemistry* **18**, 5365.
[3]W. K. Paik & S. Kim (1970) *JBC* **245**, 6010.

Selected entries from *Methods in Enzymology* [vol, page(s)]:
Assay: *Neurospora crassa*, 106, 275

H$^+$/K$^+$-EXCHANGING ATPase

This class of so-called ATPases [EC 3.6.3.10; previously classified as EC 3.6.1.36] (also known as the hydrogen/potassium-exchanging ATPase, H$^+$/K$^+$-transporting ATPase, the potassium-transporting ATPase, proton pump, and the gastric H$^+$/K$^+$ ATPase) catalyzes the ATP-dependent (ADP- and P$_i$-forming) countertransport of H$^+_{in}$ and K$^+_{out}$ to produce H$^+_{out}$, and K$^+_{in}$.[1-14] This enzyme is a P-type ATPase that undergoes covalent phosphorylation during the transport cycle. The gastric mucosal enzyme has been the best characterized. Other nucleotide substrates include dATP. The enzyme will reportedly also catalyze a proton exchange coupled with a Tl$^+$, Rb$^+$, Cs$^+$, or NH$_4^+$ transport (with varying degrees of efficiency).

The idea that a transporter is an enzyme is in keeping with a new definition of enzyme catalysis as the facilitated making/breaking of chemical bonds, not just covalent bonds.[15] This idea builds on Pauling's assertion that any long-lived, chemically distinct interaction (in this case, the persistent location of a solute with respect to the faces of a membrane) can be regarded as a chemical bond. Note also that the equilibrium constant (K_{eq} = [K$^+_{in}$][H$^+_{out}$][ADP][P$_i$]/[K$^+_{out}$][H$^+_{in}$][ATP]) does not conform to that expected for an ATPase (*i.e.*, K_{eq} = [ADP][P$_i$]/[ATP]). Thus, although the overall reaction yields ADP and orthophosphate, the enzyme is misclassified as a hydrolase, and instead should be regarded as an energase-type reaction. Energases facilitate affinity-modulated reactions by coupling the $\Delta G_{ATP\text{-}hydrolysis}$ to a force-generating or work-producing step.[15] In this case, P-O-P bond scission supplies the energy to drive hydrogen ion and potassium ion exchange. *See ATPase; Energases*

[1]F. Jaisser & A. T. Beggah (1999) *Amer. J. Physiol.* **276**, F812.
[2]R. B. Silver & M. Soleimani (1999) *Amer. J. Physiol.* **276**, F799.
[3]P. Meneton, F. Lesage & J. Barhanin (1999) *Semin. Nephrol.* **19**, 438.
[4]D. Melle-Milovanovic, N. Lambrecht, G. Sachs & J. M. Shin (1998) *Acta Physiol. Scand. Suppl.* **643**, 147.
[5]J. M. Shin, M. Besancon, K. Bamberg & G. Sachs (1997) *Ann. N. Y. Acad. Sci.* **834**, 65.
[6]M. Besanèon, J. M. Shin, F. Mercier, K. Munson, E. Rabon, S. Hersey & G. Sachs (1992) *Acta Physiol. Scand.* **146**, 77.
[7]G. Sachs, M. Besanèon, J. M. Shin, F. Mercier, K. Munson & S. Hersey (1992) *J. Bioenerg. Biomembr.* **24**, 301.
[8]E. C. Rabon & M. A. Reuben (1990) *Ann. Rev. Physiol.* **52**, 321.
[9]S. J. Hersey, A. Perez, S. Matheravidathu & G. Sachs (1989) *Am. J. Physiol.* **257**, G539.
[10]Q. Al-Awqati (1986) *Ann. Rev. Cell Biol.* **2**, 179.
[11]C. Tanford (1983) *Ann. Rev. Biochem.* **52**, 379.
[12]J. J. H. H. M. De Pont & S. L. Bonting (1981) *New Compr. Biochem.* **2**, 209.
[13]G. Sachs, T. Berglindh, E. Rabon, H. B. Stewart, M. L. Barcellona, B. Wallmark & G. Saccomani (1980) *Ann. N. Y. Acad. Sci.* **341**, 312.
[14]G. Sachs, R. H. Collier, R. L. Shoemaker & B. I. Hirschowitz (1968) *BBA* **162**, 210.
[15]D. L. Purich (2001) *TiBS* **26**, 417.

Selected entries from *Methods in Enzymology* [vol, page(s)]:
General discussion: 157, 649
Other molecular properties: blockers, effects on H$^+$ transport and ATPase, 191, 734; interactions with reversible K$^+$-site antagonists, 191, 737; membranes with (gastric sources, 191, 734; isolation from gastric oxyntic cell, 192, 151; pH gradient formation, assay, 192, 161); preparation from (porcine stomach, 157, 649; rabbit stomach, 157, 651; rat stomach, 157, 653)

HOLO-[ACYL-CARRIER PROTEIN] SYNTHASE

This transferase [EC 2.7.8.7], also known as holo-[fatty-acid synthase] synthase, catalyzes the reaction of coenzyme A with the apo-[acyl-carrier protein] to generate adenosine 3′,5′-bisphosphate and the holo-[acyl-carrier protein]. The enzyme catalyzes the transfer of 4′-phosphopantheine to a seryl residue of the acyl-carrier protein.[1-4]

[1]J. Elovson & P. R. Vagelos (1968) *JBC* **243**, 3603.
[2]D. J. Prescott & P. R. Vagelos (1972) *Adv. Enzymol.* **36**, 269.
[3]S. A. Elhussein, J. A. Miernyk & J. B. Ohlrogge (1988) *BJ* **252**, 39.

[4]R. S. Flugel, Y. Hwangbo, R. H. Lambalot, J. E. Cronan, Jr. & C. T. Walsh (2000) *JBC* **275**, 959.

Selected entries from *Methods in Enzymology* [vol, page(s)]:
General discussion: 35, 95; **252**, 39; **279**, 254
Assay: 35, 95, 99, 100; **62**, 256; **279**, 257
Other molecular properties: activators, **35**, 100; apo-peptide acceptors, **35**, 100, 101; immunochemical determination, **62**, 260; inhibitors, **35**, 100, 101; interconversion, **62**, 249 (assay, for conversion of apofatty to holofatty acid synthetase, **62**, 257 [for conversion of holofatty to apofatty acid synthetase, **62**, 259, 260]; separation, of apo and holo forms, of pigeon liver enzyme, **62**, 251 [of rat liver enzyme, **62**, 251]); mutant, **72**, 694; preparation, from rat or pigeon liver, **62**, 251, 254; properties, **62**, 255; **279**, 262; stability, **35**, 100

HOLOCYTOCHROME-c SYNTHASE

This enzyme [EC 4.4.1.17], also called cytochrome c heme-lyase, catalyzes the reversible reaction of apocytochrome c with heme to produce holocytochrome c. The reaction involves the formation of two thioether links between the heme and two cysteinyl residues of the protein. The enzyme catalyzes the covalent attachment of the heme group to apocytochrome c during mitochondrial import.[1]

[1]R. Lill, R. A. Stuart, M. E. Drygas, F. E. Nargang & W. Neupert (1992) *EMBO J.* **11**, 449.

HOMOACONITATE HYDRATASE

This hydratase [EC 4.2.1.36], also known as homoaconitase, catalyzes the reversible conversion of 1-hydroxybutane 1,2,4-tricarboxylate (*i.e.*, homoisocitrate) to but-1-ene 1,2,4-tricarboxylate (*i.e.*, *cis*-homoaconitate) and water.[1] The enzyme participates in the biosynthesis of L-lysine in fungi and euglenids.

[1]M. Strassmann & L. N. Ceci (1966) *JBC* **241**, 5401.

Selected entries from *Methods in Enzymology* [vol, page(s)]:
General discussion: 17B, 114
Assay: 17B, 114

HOMOCITRATE SYNTHASE

This enzyme [EC 4.1.3.21], also known as homocitrate condensing enzyme, catalyzes the reaction of acetyl-CoA with water and α-ketoglutarate (or, 2-oxoglutarate) to produce 2-hydroxybutane 1,2,4-tricarboxylate (or, homocitrate) and coenzyme A.[1–3] This enzyme catalyzes the first step in L-lysine biosynthesis in fungi and euglenids and L-lysine is typically a potent feedback inhibitor.

[1]H. Schmidt, R. Bode, M. Lindner & D. Birnbaum (1985) *J. Basic Microbiol.* **25**, 675.
[2]P. S. Masurekar & A. L. Demain (1974) *Appl. Microbiol.* **28**, 265.
[3]M. J. P. Higgins, J. A. Kornblatt & H. Rudney (1972) *The Enzymes*, 3rd ed., **7**, 407.

Selected entries from *Methods in Enzymology* [vol, page(s)]:
General discussion: 17B, 113

HOMOCYSTEINE DESULFHYDRASE

This pyridoxal-phosphate-dependent enzyme [EC 4.4.1.2] catalyzes the hydrolysis of L-homocysteine to produce ammonia, H_2S, and α-ketobutanoate (or, 2-oxo-butanoate).[1,2]

[1]K.-W. Thong & G. H. Coombs (1987) *Exp. Parasitol.* **63**, 143.
[2]R. E. Kallio (1951) *JBC* **192**, 371.

Selected entries from *Methods in Enzymology* [vol, page(s)]:
General discussion: 2, 318

HOMOCYSTEINE METHYLTRANSFERASE

This transferase [EC 2.1.1.10] catalyzes the reaction of S-adenosyl-L-methionine with L-homocysteine to produce S-adenosyl-L-homocysteine and L-methionine.[1,2] With the bacterial enzyme, S-methyl-L-methionine is a better substrate than S-adenosyl-L-methionine.[1] The *Saccharomyces cerevisiae* enzyme has an ordered Bi Bi kinetic mechanism.[2] *See also N^5-Methyltetrahydrofolate:Homocysteine Methyltransferase*

[1]S. K. Shapiro & D. A. Yphantis (1959) *BBA* **36**, 241.
[2]S. K. Shapiro (1958) *BBA* **29**, 405.

Selected entries from *Methods in Enzymology* [vol, page(s)]:
General discussion: 17B, 400
Assay: 17B, 401
Other molecular properties: chiral methyl groups, and, **87**, 149; properties, **17B**, 405; purification from *Saccharomyces cerevisiae*, **17B**, 402

HOMOGENTISATE 1,2-DIOXYGENASE

This ferrous-dependent oxidoreductase [EC 1.13.11.5], also known as homogentisicase and homogentisate oxygenase, catalyzes the reaction of homogentisate with dioxygen to produce 4-maleylacetoacetate.[1–3] A deficiency of this enzyme is associated with alkaptonuria.

[1]O. Hayaishi, M. Nozaki & M. T. Abbott (1975) *The Enzymes*, 3rd ed., **12**, 119.
[2]M. Nozaki (1974) in *Mol. Mech. Oxygen Activ.* (O. Hayaishi, ed.) p. 135.
[3]K. Adachi, Y. Iwayama, H. Tanioka & Y. Takeda (1966) *BBA* **118**, 88.

Selected entries from *Methods in Enzymology* [vol, page(s)]:
General discussion: 2, 292; **17A**, 638; **52**, 11
Assay: 2, 292

HOMOGLUTATHIONE SYNTHETASE

This enzyme [EC 6.3.2.23] catalyzes the reaction of ATP with γ-L-glutamyl-L-cysteine and β-alanine to produce ADP, orthophosphate, and γ-L-glutamyl-L-cysteinyl-β-alanine (*i.e.*, homoglutathione).[1,2] The enzyme is distinct from glutathione synthetase [EC 6.3.2.3]. Homoglutathione is the predominant nonprotein thiol in soybean (*Glycine max*), bean (*Phaseolus vulgaris*), and mungbean (*Vigna radiata*) nodules.[3]

[1]P. K. Macnicol (1987) *Plant Sci.* **53**, 229.
[2]S. Klapheck, H. Zopes, H. G. Levels & L. Bergmann (1988) *Physiol. Plant.* **74**, 733.
[3]M. A. Matamoros, J. F. Moran, I. Iturbe-Ormaetxe, M. C. Rubio & M. Becana (1999) *Plant Physiol.* **121**, 879.

HOMOISOCITRATE DEHYDROGENASE

This oxidoreductase [EC 1.1.1.155] catalyzes the reaction of (−)-1-hydroxy-1,2,4-butane tricarboxylate (*i.e.*, homoisocitrate) with NAD^+ to produce α-ketoadipate (*i.e.*, 2-oxoadipate), carbon dioxide, and NADH.[1] The enzyme will also catalyze the keto-enol tautomerization of tritiated α-ketoadipate.

[1]B. Rowley & A. F. Tucci (1970) *ABB* **141**, 499.

Selected entries from *Methods in Enzymology* [vol, page(s)]:
General discussion: 17B, 118
Assay: 17B, 118
Other molecular properties: properties, 17B, 119; purification from bakers' yeast, 17B, 119

HOMOSERINE O-ACETYLTRANSFERASE

This transferase [EC 2.3.1.31] catalyzes the reaction of acetyl-CoA with L-homoserine to produce coenzyme A and *O*-acetyl-L-homoserine.[1–3] The enzyme utilizes a ping pong Bi Bi kinetic mechanism in which the acetyl group is transferred to an enzyme nucleophile before the transfer to homoserine to form *O*-acetyl-L-homoserine. Rapid quench experiments confirm the occurrence of an acetyl-enzyme covalent intermediate.[3]

[1]A. Wyman & H. Paulus (1975) *JBC* **250**, 3897.
[2]S. Nagai & M. Flavin (1967) *JBC* **242**, 3884.
[3]T. L. Born, M. Franklin & J. S. Blanchard (2000) *Biochemistry* **39**, 8556.

Selected entries from *Methods in Enzymology* [vol, page(s)]:
General discussion: 17B, 442
Assay: 17B, 442
Other molecular properties: properties, 17B, 445; purification from *Neurospora*, 17B, 444

HOMOSERINE DEHYDRATASE

This enzyme activity, previously classified as EC 4.2.1.15, is now regarded as an alternative activity of cystathionine γ-lyase [EC 4.4.1.1].

HOMOSERINE DEHYDROGENASE

This oxidoreductase [EC 1.1.1.3] catalyzes the reaction of L-homoserine with $NAD(P)^+$ to produce L-aspartate 4-semialdehyde and NAD(P)H. NAD^+ is more effective in yeast, whereas $NADP^+$ is the preferred coenzyme with the *Neurospora* enzyme.[1–3] The *Escherichia coli* enzyme

is bifunctional, also possessing an aspartate kinase activity.
See also Aspartate Kinase

[1]B. DeLaBarre, P. R. Thompson, G. D. Wright & A. M. Berghuis (2000) *Nat. Struct. Biol.* **7**, 238.
[2]I. Saint-Gerons, C. Parsot, M. M. Zakin, O. Bârzy & G. N. Cohen (1988) *Crit. Rev. Biochem.* **23**, S1.
[3]P. Truffa-Bachi (1973) *The Enzymes*, 3rd ed., **8**, 509.

Selected entries from *Methods in Enzymology* [vol, page(s)]:
General discussion: 5, 824; 17A, 703
Other molecular properties: aspartate β-semialdehyde assay and, **6**, 624; *Escherichia coli*, equilibrium isotope exchange studies (kinetic mechanism, **249**, 468; modifier action, **249**, 471); induction in microorganisms, **22**, 93; ligand binding sites, **249**, 560; purification from *Rhodospirillum rubrum*, **17A**, 704; staining in gels, **22**, 594; stereospecificity, **87**, 103

HOMOSERINE KINASE

This phosphotransferase [EC 2.7.1.39] catalyzes the reaction of ATP with L-homoserine to produce ADP and *O*-phospho-L-homoserine.[1–3] The enzyme has a random Bi Bi kinetic mechanism. Some protein kinases also possess a homoserine kinase activity.

[1]X. Huo & R. E. Viola (1996) *Biochemistry* **35**, 16180.
[2]S. L. Shames & F. C. Wedler (1984) *ABB* **235**, 359.
[3]R. Miyajima & I. Shiio (1972) *J. Biochem.* **71**, 219.

Selected entries from *Methods in Enzymology* [vol, page(s)]:
Assay: 5, 954
Other molecular properties: *O*-phosphohomoserine preparation, **5**, 953; purification, **5**, 954; threonine synthase, with **5**, 952

HOMOSERINE O-SUCCINYLTRANSFERASE

This transferase [EC 2.3.1.46], also known as homoserine *O*-transsuccinylase, catalyzes the reaction of succinyl-CoA with L-homoserine to produce coenzyme A and *O*-succinyl-L-homoserine.[1–5] The *Escherichia coli* enzyme utilizes a ping pong Bi Bi kinetic mechanism in which the succinyl group of succinyl-CoA is initially transferred to an enzyme nucleophile before its transfer to L-homoserine.[5] Inhibition and chemical modification studies indicate that a cysteinyl residue is succinylated.

[1]I. Saint-Gerons, C. Parsot, M. M. Zakin, O. Bârzy & G. N. Cohen (1988) *Crit. Rev. Biochem.* **23**, S1.
[2]R. J. Rowbury & D. D. Woods (1964) *J. Gen. Microbiol.* **36**, 341.
[3]F. Röhl, J. Rabenhorst & H. Zähner (1987) *Arch. Microbiol.* **147**, 315.
[4]R. Mares, M. L. Urbanowski & G. V. Stauffer (1992) *J. Bacteriol.* **174**, 390.
[5]T. L. Born & J. S. Blanchard (1999) *Biochemistry* **38**, 14416.

HOMOSPERMIDINE SYNTHASE

This NAD^+-dependent enzyme [EC 2.5.1.44] catalyzes the reaction of two molecules of putrescine to produce *sym*-homospermidine and ammonia.[1,2] The enzyme can also catalyze the reaction of spermidine with putrescine producing *sym*-homospermidine and propane-1,3-diamine (the presence of both activities distinguishes this enzyme from homospermidine synthase (spermidine-specific) [EC 2.5.1.45]). *See also Homospermidine Synthase (Spermidine-Specific)*

[1] S. Yamamoto, S. Nagata & K. Kusaba (1993) *J. Biochem.* **114**, 45.
[2] D. Ober, D. Tholl, W. Martin & T. Hartmann (1996) *J. Gen. Appl. Microbiol.* **42**, 411.

HOMOSPERMIDINE SYNTHASE (SPERMIDINE-SPECIFIC)

This NAD$^+$-dependent enzyme [EC 2.5.1.45] catalyzes the reaction of spermidine with putrescine to produce *sym*-homospermidine (*i.e.*, *N*-(4-aminobutyl)butane-1,4-diamine) and propane-1,3-diamine.[1,2] This enzyme is more specific than homospermidine synthase [EC 2.5.1.44] since it cannot use putrescine as the donor substrate for the 4-aminobutyl group.

The first step in the reaction mechanism is the reaction of NAD$^+$ with spermidine to produce NADH and dehydrospermidine. The 4-aminobutylidene group is then transferred from this intermediate to putrescine, forming an imine which is subsequently reduced by NADH to generate the final product. This synthase is a component in the biosynthetic pathway of toxic pyrrolizidine alkaloids from ragwort. *See also Homospermidine Synthase; Deoxyhypusine Synthase*

[1] F. Böttcher, D. Ober & T. Hartmann (1994) *Can. J. Chem.* **72**, 80.
[2] D. Ober, R. Harms & T. Hartmann (2000) *Phytochemistry* **55**, 311.

HORRILYSIN

This zinc-dependent metalloendopeptidase [EC 3.4.24.47], originally called hemorrhagic proteinase IV and *Crotalus horridus* metalloendopeptidase, is isolated from the venom of the timber rattlesnake. The enzyme catalyzes peptide-bond hydrolysis in hide powder azure, type I collagen, the Aα and Bβ chains of fibrinogen, and glomerular basement membrane. A single peptide bond is hydrolyzed in each of the B and A chains of oxidized insulin: at Ala14–Leu15 of the B chain and at Ser12–Leu13 of the A chain. Three peptide bonds are hydrolyzed in bee venom melittin: Ile2–Gly3, Pro14–Ala15, and Ser18–Trp19.[1,2]

[1] D. J. Civello, J. B. Moran & C. R. Geren (1983) *Biochemistry* **22**, 755.
[2] J. B. Bjarnason & J. W. Fox (1995) *MIE* **248**, 345.

HORSERADISH PEROXIDASE

This oxidoreductase [EC 1.11.1.7] catalyzes the reaction of hydrogen peroxide with a donor substrate to produce two water molecules and the oxidized donor.[1-5] There are three enzyme types: acidic enzymes (horseradish peroxidase A), neutral enzymes (horseradish peroxidase C), and basic enzymes (horseradish peroxidase E). Dunford[1] provides a detailed description of these heme-dependent enzymes. Horseradish peroxidase also exhibits an iodoperoxidase activity. With certain substrates (for example, isobutyraldehyde), light is emitted in the peroxidase-catalyzed reaction.[6] *See also Peroxidase*

[1] H. B. Dunford (1998) *CBC* **3**, 195.
[2] O. Ryan, M. R. Smyth & C. Fágáin (1994) *Essays Biochem.* **28**, 129.
[3] M. A. Ator & P. R. Ortez de Montellano (1990) *The Enzymes*, 3rd ed., **19**, 213.
[4] S. Shahangian & L. P. Hager (1982) *JBC* **257**, 11529.
[5] S. Aibara, H. Yamashita, E. Mori, M. Kato & Y. Morita (1982) *J. Biochem.* **92**, 531.
[6] W. D. Hewson & L. P. Hager (1979) in *The Porphyrins* (D. Dolphin, ed.) **7**, 295, Academic Press, New York.

Selected entries from *Methods in Enzymology* [vol, page(s)]:
General discussion: **2**, 801
Assay: **2**, 803; activity assay, colorimetric, **70**, 431
Other molecular properties: antigen localization, in, **257**, 268; assay of (H_2O_2 production, **236**, 131 [by mitochondria, **105**, 434; by phagocytes, **132**, 395]; superoxide dismutase, **53**, 387); avidin-biotin complex, in immunoassay of HIV core proteins, **184**, 556; avidin complex, in detection of (biotinylated proteins, **184**, 435; leukocyte surface proteins, **184**, 431); binding properties, **52**, 516; biological function, **246**, 252; biotinylation, **184**, 150; carbon monoxide binding, IR studies, **54**, 310; **226**, 267, 270; catalyzed chemiluminescent reactions (enhancement of light emission [emission spectra, **133**, 346; reaction characteristics, **133**, 335; reactions, **133**, 332]; enhancers [concentration effects, **133**, 344; identification, **133**, 336; pH effects, **133**, 343; specificity, **133**, 341; synergistic action, **133**, 340]; mechanism, **133**, 332); CD studies, **54**, 269, 271, 272, 283, 284; compound II, oxidation state determination, **2**, 802; **54**, 243; conformational composition, **54**, 269, 271, 272; conjugate preparation, **70**, 431; conjugates, applications of enhanced assays, **133**, 351; conjugation to antibody, **73**, 150, 659, 660; coupled assays with luminol, **57**, 403; crude, chromatography, **52**, 518; electrochemical studies, direct, **227**, 516; electroimmunoassay, **73**, 348; endosomal marker (assay, **182**, 217; labeling of endosomal compartment, **182**, 216); enzyme channeling immunoassays with, **136**, 98; enzyme immunoassay for leukocyte interferon, **79**, 595; fluid-phase marker, as (endocytotic uptake, **146**, 413; overview, **146**, 405); glucose oxidase, and, cellobiase assay with, **160**, 110; glucose oxidase electrode, description, **137**, 37; heme group substitution, **246**, 249, 252; hole burning spectroscopy, **246**, 234, 257; hydrogen peroxide assay, in, **52**, 347, 348; immobilization, **137**, 173, 179; immunohistochemical reactions catalyzed by, **245**, 322; inhibitor, **52**, 349; lectin conjugation, **247**, 254; lectin labeling, **242**, 200, 269; luminescent assays, **133**, 331; magnetic circular dichroism, **49**, 170; marker for endocytosis and membrane recycling, as, **98**, 209; MCD studies, **54**, 296; mechanism, **2**, 802; Mössbauer spectra, **54**, 355; partition coefficients, measured and calculated, comparison, **228**, 194; peroxynitrite-dependent tyrosine nitration, **269**, 210; photochemical properties, **76**, 590; pinocytic markers in leukocytes, as, **108**, 339; properties, **2**, 808; purification, **2**, 803; quantification by partition affinity ligand assay, **92**, 503; radical generation, **234**, 284; radioiodination, **70**, 212; resonance Raman studies, **226**, 362; specificity, **2**, 807; subcellular localization, **52**, 251; sulfhydryl oxidase assay with, **143**, 122; two-dimensional NMR spectra, ^{1}H signal assignment, **239**, 501

*Hpa*I RESTRICTION ENDONUCLEASE

This type II restriction enzyme [EC 3.4.21.4], isolated from *Haemophilus parainfluenzae*, catalyzes the cleavage of both strands of DNA at 5′...GTT∇AAC...3′, producing blunt ends.[1-3] It will not act on single-stranded DNA. Under conditions of high enzyme concentration, glycerol concentration in excess of 5%, or at pH values above 8.0, *Hpa*I will exhibit star activity.
See *Star Activity*

[1]P. A. Sharp, B. Sugden & J. Sambrook (1973) *Biochemistry* **12**, 3055.
[2]D. E. Garfin & H. M. Goodman (1974) *BBRC* **59**, 108.
[3]J. L. Hines & K. L. Agarwal (1979) *Fed. Proc.* **38**, 294.
Selected entries from *Methods in Enzymology* [vol, page(s)]:
General discussion: 65, 34, 35, 153, 175, 363, 432, 445, 780, 812, 813, 824, 825
Assay: 65, 153
Other molecular properties: cation requirement, 65, 163; effect, of salt concentration, 65, 163; homogeneity, 65, 162; molecular weight, 65, 161, 162; optimal temperature, 65, 163; pH optimum, 65, 163; properties, 65, 162; purification, 65, 156; recognition sequence, 65, 8, 153; stability, 65, 162; substrate specificity, 65, 162; sulfhydryl requirement, 65, 163; use in linkage studies, 76, 816

*Hpa*II RESTRICTION ENDONUCLEASE

This type II restriction enzyme [EC 3.4.21.4], isolated from *Haemophilus parainfluenzae*, catalyzes the cleavage of both strands of DNA at 5′...C∇CGG...3′.[1-4] This endonuclease is inhibited by KCl concentrations in excess of 50 mM. Cleavage is blocked by methylation of either cytosine residue.

[1]P. A. Sharp, B. Sugden & J. Sambrook (1973) *Biochemistry* **12**, 3055.
[2]D. E. Garfin & H. M. Goodman (1974) *BBRC* **59**, 108.
[3]J. L. Hines & K. L. Agarwal (1979) *Fed. Proc.* **38**, 294.
[4]J. L. Hines, T. R. Chauncey & K. L. Agarwal (1980) *Meth. Enzymol.* **65**, 153.
Selected entries from *Methods in Enzymology* [vol, page(s)]:
General discussion: 65, 23, 34, 35, 153, 363, 385, 418, 485, 721, 741, 823
Assay: 65, 153
Other molecular properties: cation requirement, 65, 163; cleavage sites on (PM2 DNA, 65, 279; SV40 DNA, 65, 418, 419, 713); effect, of salt concentration, 65, 163; fragment end structure produced, 65, 511; properties, 65, 162; purification, 65, 159; recognition sequence, 65, 8, 153, 511; stability, 65, 162; substrate specificity, 65, 162; sulfhydryl requirement, 65, 163

*Hph*I RESTRICTION ENDONUCLEASE

This type II restriction enzyme [EC 3.4.21.4] (subtype s), isolated from *Haemophilus parahaemolyticus*, catalyzes the cleavage of both strands of DNA at 5′...GGTGANNNNNNNN∇...3′ and 3′...CCACTNN NNNNN∇N...5′, where N refers to any base.[1-3] Hydrolysis is blocked by overlapping *dam* methylation. There are reports that cleavage may occur at N_9/N_8 rather than N_8/N_7, depending on the sequence between the recognition and cleavage sites.[4,5] The restriction endonuclease also exhibits star activity.

[1]D. Kleid, Z. Humayun, A. Jeffrey & M. Ptashne (1976) *PNAS* **73**, 293.
[2]D. G. Kleid (1980) *MIE* **65**, 163.
[3]J. S. Daniels & K. S. Gates (1994) *ACS Abstracts* **208**, 66.
[4]C. Kang & C.-W. Wu (1987) *NAR* **15**, 2279.
[5]S.-H. Cho & C. Kang (1990) *Mol. Cells* **1**, 81.
Selected entries from *Methods in Enzymology* [vol, page(s)]:
General discussion: 65, 23, 163, 394
Other molecular properties: *Hph* buffer, 65, 164; purification, 65, 164; recognition sequence, 65, 8, 14, 163, 164, 166

*Hpy*99I RESTRICTION ENDONUCLEASE

This type II restriction endonuclease [EC 3.4.21.4], isolated from *Helicobacter pylori* J99, catalyzes the cleavage of both strands of DNA at 5′...CGWCG∇...3′, where W refers to A or T.[1]

[1]Q. Xu, R. D. Morgan, R. J. Roberts & M. J. Blaser (2000) *PNAS* **97**, 9671.

*Hpy*188I RESTRICTION ENDONUCLEASE

This type II restriction endonuclease [EC 3.4.21.4], isolated from *Helicobacter pylori* J188, catalyzes the cleavage of both strands of DNA at 5′...TCN∇GA...3′, where N refers to any base.[1,2] Cloning and sequence analysis reveals evidence of horizontal transfer to the *Helicobacter pylori* genome.[1]

[1]Q. Xu, S. Stickel, R. J. Roberts, M. J. Blaser & R. D. Morgan (2000) *JBC* **275**, 17086.
[2]Q. Xu, R. D. Morgan, R. J. Roberts & M. J. Blaser (2000) *PNAS* **97**, 9671.

*Hpy*CH4III RESTRICTION ENDONUCLEASE

This type II restriction endonuclease [EC 3.4.21.4], isolated from *Helicobacter pylori* CH4, catalyzes the cleavage of both strands of DNA at 5′...ACN∇GT...3′, where N refers to any base.[1]

[1]Q. Xu, R. D. Morgan, R. J. Roberts & M. J. Blaser (2000) *PNAS* **97**, 9671.

*Hpy*CH4IV RESTRICTION ENDONUCLEASE

This type II restriction endonuclease [EC 3.4.21.4], isolated from *Helicobacter pylori* CH4, catalyzes the cleavage of both strands of DNA at 5′...A∇CGT...3′.[1]

[1]Q. Xu, R. D. Morgan, R. J. Roberts & M. J. Blaser (2000) *PNAS* **97**, 9671.

*Hpy*CH4V RESTRICTION ENDONUCLEASE

This type II restriction endonuclease [EC 3.4.21.4], isolated from *Helicobacter pylori* CH4, catalyzes the cleavage of both strands of DNA at 5′...TG∇CA...3′, producing blunt-ended fragments.[1]

[1]Q. Xu, R. D. Morgan, R. J. Roberts & M. J. Blaser (2000) *PNAS* **97**, 9671.

HslVU PROTEASE

This heat-shock-associated serine endopeptidase catalyzes the hydrolysis of peptide bonds in proteins with the concomitant hydrolysis of ATP by the HslU subunit. In the presence of ATP, the endopeptidase will rapidly hydrolyze Z-Gly-Gly-Leu-∇-NHMec. ATP stimulates proteolysis, and the presence of certain proteins also stimulates the ATPase activity.[1,2] Although the overall reaction yields ADP and orthophosphate, the enzyme is not an ATPase or hydrolase, because P-O-P bond scission supplies the Gibbs free energy required to drive substrate binding interactions.[3] The overall enzyme-catalyzed reaction should be regarded to be of the energase type (*e.g.*, $State_1 + ATP = State_1 + ADP + P_i$, where two affinity-modulated conformational states are indicated).[4] Energases transduce chemical energy into mechanical work.

[1]M. Rohrwild, O. Coux, H. C. Huang, R. P. Moerschell, S. J. Yoo, J. H. Seol, C. H. Chung & A. L. Goldberg (1996) *PNAS* **93**, 5808.
[2]S. J. Yoo, J. H. Seol, D. H. Shin, M. Rohrwild, M. S. Kang, K. Tanaka, A. L. Goldberg & C. H. Chung (1996) *JBC* **271**, 14035.
[3]D. L. Purich (2001) *TiBS* **26**, 417.

*Hsu*I RESTRICTION ENDONUCLEASE

This type II restriction enzyme [EC 3.4.21.4], isolated from *Haemophilus suis*, catalyzes the cleavage of both strands of DNA at 5′...A∇AGCTT...3′.

Selected entries from *Methods in Enzymology* [**vol**, page(s)]:
Other molecular properties: recognition sequence, **65**, 8

HUMAN HERPESVIRUS TYPE 6 ASSEMBLIN

This serine endopeptidase [EC 3.4.21.97], also known as HHV-6 proteinase, catalyzes the hydrolysis of peptide bonds in a large polyprotein, of which it is a part. Peptide bonds are cleaved at a release site (R-site) and at a maturation site (M-site). These two sites form the basis of two synthetic substrates, Suc-Arg-Arg-Tyr-Ile-Lys-Ala-∇-Ser-Glu-Pro-Pro-Val-NH₂ and Suc-Arg-Arg-Ile-Leu-Asn-Ala-∇-Ser-Leu-Ala-Pro-Glu-NH₂, respectively.[1,2] **See also** *Herpesvirus Assemblin; Cytomegalovirus Assemblin*

[1]C. Lopez & R. W. Honess (1990) in *Virology* (B. N. Fields & D. M. Knipe, eds.), Raven Press, New York, p. 2055.
[2]N. Tigue, P. J. Matharu, N. A. Roberts, J. S. Mills, J. Kay & R. Jupp (1996) *J. Virol.* **70**, 4136.

HUMAN IMMUNODEFICIENCY VIRUS I RETROPEPSIN

This aspartic endopeptidase [EC 3.4.23.16], also known as HIV-1 retropepsin and HIV-1 protease, catalyzes peptide-bond hydrolysis of the viral polyprotein. Two types of cleavage sites have been observed for this dimeric enzyme. One site involves cleavage between an aromatic aminoacyl residue (that is, Tyr or Phe in position P1) and a prolyl residue (in position P1′). This cleavage site also has an asparaginyl residue at P2 and an isoleucyl residue at P2′. The second group of cleavage sites involves two hydrophobic aminoacyl residues (for example, Leu–Ala, Met–Met, Asn–Phe, Phe–Leu, Phe–Tyr, and Leu–Phe). In addition to the viral polyprotein, HIV-1 retropepsin has also been shown to catalyze the hydrolysis of peptide bonds in other proteins: for example, nuclear factor(NF)-κB and microtubule associated protein 2. **See also** *Human Immunodeficiency Virus 2 Retropepsin*

Selected entries from *Methods in Enzymology* [**vol**, page(s)]:
General discussion: 241, 70, 254, 279
Assay: 241, 128, 211, 229, 295; peptidolytic assays (chromogenic substrate technique, **241**, 54, 71; colorimetric, **241**, 52; continuous, **241**, 54, 71; fluorogenic substrate technique, **241**, 71, 73, 79; HPLC-based, **241**, 48, 70, 127; radiometric, **241**, 50; SDS-PAGE analysis, by, **241**, 47; thin-layer electrophoresis, by, **241**, 50); fluorimetric, **248**, 27; resonance energy transfer, using, **246**, 15, 18, 302, 322
Other molecular properties: acetylpepstatin complex, **241**, 170; activation, **241**, 134, 280; active site (formation, spectral probe for, **241**, 117; hydrophobicity, **241**, 158; pK_a values, mutagenesis studies, **241**, 214; protonation state, **241**, 192, 815; structure, **241**, 159, 183, 214, 276, 322; titration, **248**, 100; water molecule bound to, exploitation by inhibitors, **241**, 160, 176, 324, 349); activity (bacterial cells, in, **241**, 66; *Escherichia coli* cells, in, **241**, 12; eukaryotic cells, in, **241**, 62; factors affecting, **241**, 106; heterologous engineered substrates, on, **241**, 14; mutants, **241**, 184; protein concentration dependence, **241**, 111); amino acid sequence, comparison with eukaryotic aspartic proteases, **241**, 216; autoprocessing (analysis *in vitro*, **241**, 236; lysates from different cell types, in, **241**, 237); bacterial expression system (autoprocessing in, **241**, 33; host strain selection, **241**, 41); catalytic mechanism, **241**, 223, 297; cell culture assay, **241**, 100; chemical mechanism, **241**, 127–156; chloramphenicol acetyltransferase fusion protein, **241**, 33; cleavage assay, *in vitro*-translated precursor proteins as substrates for, **241**, 252; cleavage sites (amino acid sequence analysis, **241**, 291; peptide cleavage analysis, **241**, 264; subsite preferences, **241**, 291); coding region, **241**, 4; conformational change and catalysis, **259**, 681; crystal structure, **241**, 214 (inhibitor design, in, **241**, 321); cytotoxicity, **241**, 16; dihydrofolate reductase fusion protein, **241**, 33; dimeric, stability, **241**, 105; dissociation constant, **241**, 105, 109, 111, 231; eukaryotic aspartic proteases, and, comparison, **241**, 195; expression (cell-free expression, **241**, 229; *Escherichia coli*, in, **241**, 4, 33; yeast, in, **241**, 10); flap region, **241**, 165, 183, 256, 274, 323; fluorogenic substrates, **248**, 24 (design, **241**, 53, 75; synthesis, **241**, 76, 83); formation, **241**, 158; free energy perturbation analysis, **241**, 373; β-galactosidase fusion protein, **241**, 33; gene, synthetic, **241**, 205; genetic assays, **241**, 56; hinge region, **259**, 680; HIV-2 protease, and, specificity comparisons, **241**, 290, 299; H₂¹⁸O isotope partitioning studies, **241**, 136, 150; homodimeric structure, **241**, 9; inactivation, mechanisms, **241**, 111; inhibitor complexes (binding strength, **241**, 194; crystal structure, **241**, 165, 184,

330, 349; pepstatin, **241**, 170; structure-based design of inhibitors, **241**, 174; substrate peptide nonhydrolyzable analogues, **241**, 165; symmetrical inhibitors, **241**, 171; transition-state mimics, **241**, 167); inhibitors (1(S)-amino-2(R)-hydroxyindan P2′ peptide surrogate, **241**, 320; antiviral effect, evaluation, **241**, 102; clinical applications, **241**, 282; competitive inhibitors, evaluation, **241**, 110; complex with HIV-1 [binding strength, **241**, 194; crystal structure, **241**, 165, 184, 330, 349; X-ray structures, **241**, 157, 164]; C_2-symmetric inhibitor [binding to enzyme, **241**, 349; core unit design, **241**, 335; core unit synthesis, **241**, 341; identification via rational screening, **241**, 352; structure-activity relationships, **241**, 343]; design [strategies, **241**, 160; structure-based, **241**, 157, 174, 295; symmetry-based peptidomimetic inhibitors, **241**, 334; X-ray crystal structure used in, **241**, 321]; dissociative [evaluation, **241**, 110; subunits as, **241**, 125]; hydrogen-bonding interactions [backbone carbonyl oxygen and amide nitrogen of P2 and P2′ residues, **241**, 268; categories, **241**, 275; central atoms, at, **241**, 344; C_2-symmetric inhibitors, **241**, 349; role in substrate fit, **241**, 256, 274, 278]; hydrophobic interactions, **241**, 350; hydroxyethylamine, structure-activity relationships, **241**, 346; hydroxyethylene-containing, **241**, 319; hydroxyethylene isosteres of substrate peptides, **241**, 167, 314; identification based on rational screening, **241**, 318, 352; JG-365 [binding in active site, **241**, 316; structure, **241**, 162, 185, 315]; L-365, 505, **241**, 319; L-682, 679, **241**, 319; L-684, 434, **241**, 322; L-685, 434, **241**, 325, 329; L-687, 430, **241**, 325; L-687, 908, **241**, 319; L-689, 502, **241**, 323, 331, 389; L-704, 486, **241**, 330; L-731, 723, **241**, 389; L-735, 524, **241**, 334; lactam as lead for, **241**, 326; MVT-101 [complex with HIV-1 proteinase {molecular dynamics simulation, **241**, 193; properties, **241**, 165}; structure, **241**, 162, 185]; nonhydrolyzable analogs of substrate peptides, **241**, 162, 164, 312; P2′ peptide surrogates, **241**, 320; peptidomimetic, **241**, 162, 164, 311, 334; pseudosymmetric [binding, **241**, 349; complex with HIV-1 protease, properties, **241**, 171; design, **241**, 335; structures, **241**, 163]; relative binding free energies, **241**, 194; Ro 31-8959 [analogs, **241**, 316; oral bioavailability, **241**, 331; P2 asparagine groups, **241**, 328; structure, **241**, 315, 390; studies of retroviral protease drug resistance, in, **241**, 389]; SB-204,144 [interaction with enzyme, **241**, 172; structure, **241**, 163]; SC-52151, **241**, 316; scissile bond analogs, **241**, 167, 185; soluble C_{60} fullerene as, **241**, 176; structure, **241**, 157, 160, 162, 184; structure-based, **241**, 351; substrate-based, **241**, 312, 334; symmetry-based [complex with HIV protease, properties, **241**, 171; peptidomimetic, **241**, 334; structure, **241**, 163; structure-activity relationships, **241**, 343]; tight-binding, design, **241**, 311; transition-state analogues, **241**, 162, 167; transition-state isosteres, **241**, 312; U-75875, **241**, 163, 170, 318; U-81749, **241**, 318; U85548e [HIV-1 protease complex, molecular dynamics simulation, **241**, 192; structure, **241**, 162, 185]; U85964E, structure, **241**, 357; UCSF8 [HIV-1 protease complex, **241**, 174; structure, **241**, 163, 174]); insoluble, purification, recombinant DNA technique, **241**, 97; isolation, DNA recombinant techniques, **241**, 92; JG-365 complex, relative binding free energy, **241**, 194; kinetic assay (initial velocity studies, **241**, 130; pH rate studies, **241**, 134, 138); kinetic isotope effects (analysis, **241**, 148; chemical mechanism, in, **241**, 153; primary ^{15}N, **241**, 142, 147; secondary β-deuterium, **241**, 142, 145; solvent, **241**, 142, 148); kinetic parameters, **241**, 219; knots, **259**, 680; β-lactamase fusion protein (autoprocessing for export from cell, **241**, 16; cytotoxicity of expressed enzyme, **241**, 23; DNA sequences at fusion junctions, **241**, 18, 20; expression in Escherichia coli, **241**, 20, 33; localization in bacterial cells, **241**, 25; plasmid construction for, **241**, 17; processing, **241**, 21; solubility of expressed enzyme, **241**, 23; Western blot analysis in Escherichia coli, **241**, 21, 25); mechanism, **241**, 283; molecular dynamics simulation, **241**, 371, 376; molecular modeling, **241**, 388; multifunctional expression vector, construction, **241**, 239; mutants (analysis of enzyme activity, in, **241**, 184; expression in cell culture, **241**, 62; Gln-88 [molecular dynamics simulation, **241**, 189; role in enzymatic activity, **241**, 184]; identification, **241**, 56; preparation, structural modeling approach, **241**, 388); MVT-101 complex, molecular dynamics simulation, **241**, 193; nonequilibrium isotope exchange analysis, **241**,

136; oligopeptide substrate cleavage (FPLC/HPLC analysis, **241**, 257, 260; ultraviolet spectrophotometric analysis, **241**, 257, 261); peptide substrates, **241**, 130; peptidolytic assays (chromogenic substrate technique, **241**, 54, 71; colorimetric, **241**, 52; continuous, **241**, 54, 71; fluorogenic substrate technique, **241**, 71, 73, 79; HPLC-based, **241**, 48, 70, 127; radiometric, **241**, 50; SDS-PAGE analysis, by, **241**, 47; thin-layer electrophoresis, by, **241**, 50); pK_a values, **241**, 219; plasmid pET-11 expression vector, **241**, 28; precursor (expression in Escherichia coli, **241**, 6; processing during infection, **241**, 58; processing mutants, expression in cell culture, **241**, 62); product inhibition studies, **241**, 132; receptor docking studies, **241**, 356, 362; recombinant, preparation, **241**, 205; reverse peptidolytic reaction, **241**, 136; sedimentation equilibrium analysis, **241**, 123; soluble, recombinant DNA purification technique, **241**, 95; stability, **241**, 111; structure, **259**, 679 (comparison with aspartic proteases, **241**, 196, 255; X-ray crystallography, from, **241**, 178); structure-function analysis, molecular dynamics simulation, **241**, 178; substrates (binding, **241**, 256; inhibitors based on, **241**, 312; recognition, **241**, 48; recombinant, in vitro expression, **241**, 233; specificity, **241**, 279, 283 [active site properties related to, **241**, 158; analysis of published data, **241**, 264; mutational analysis, **241**, 276; nonviral protein substrates for, **241**, 287; role of β-hairpin loops, **241**, 165, 183, 256; viral polyprotein cleavage assay, **241**, 283]); subunit exchange, **241**, 124; tethered dimer, expression in Escherichia coli, **241**, 9; total peptide synthesis, **241**, 11; transcription-translation system in vitro, **241**, 229; transition state and multisubstrate analogues, **249**, 307; U85548e complex, molecular dynamics simulation, **241**, 192; unfolding, protein concentration dependence, **241**, 122

HUMAN IMMUNODEFICIENCY VIRUS 2 RETROPEPSIN

This aspartic endopeptidase, also known as HIV-2 retropepsin and HIV-2 protease, catalyzes peptide-bond hydrolysis in the viral polyprotein. The specificities for the cleavage sites are similar to those of HIV-1 retropepsin, although there are variations in cleavage rates. For example, Val-Ser-Gln-Asn-Tyr-∇-Val-Ile-Val is hydrolyzed thirty-five-fold faster by HIV-1 retropepsin. In addition, the synthetic inhibitor Boc-Pro-Phe-NMeHis-Leu[CH(OH)CH$_2$]Val-Ile-Amp has a K_i value of 10 nM with HIV-1 retropepsin and a value greater than 1 μM with HIV-2 retropepsin.[1,2] **See also** Human Immunodeficiency Virus 1 Retropepsin

[1] J. Tözsér, I. Bláha, T. D. Copeland, E. M. Wondrak & S. Oroszlan (1991) FEBS Lett. **281**, 77.
[2] A. G. Tomasselli, J. O. Hui, T. K. Sawyer, D. J. Staples, C. Bannow, I. M. Reardon, W. J. Howe, D. L. Camp, C. S. Craik & R. L. Heinrikson (1990) JBC **265**, 14675.

Selected entries from **Methods in Enzymology** [vol, page(s)]:
Other molecular properties: autoprocessing, in bacterial expression system, **241**, 33; dimeric, demonstration, **241**, 105; eukaryotic aspartic proteases, and, sequence comparison, **241**, 216; expression in Escherichia coli, **241**, 11, 33; HIV-1 protease, and, specificity comparisons, **241**, 290, 299; mechanism, **241**, 283; molecular modeling, **241**, 388; mutation, structural modeling approach, **241**, 388; oligomer equilibria, measurement, **241**, 105; peptidolytic assays, continuous, chromogenic substrate technique, **241**, 54; sedimentation equilibrium analysis, **241**, 123; stability during purification, **241**, 111; substrate specificity, **241**, 279, 283 (analysis of published data, **241**, 264; nonviral protein substrates for, **241**, 287)

HUMAN T-CELL LEUKEMIA VIRUS TYPE I RETROPEPSIN

This aspartic endopeptidase, also known as HTLV-1 retropepsin and HTLV-1 protease, catalyzes autoprocessing and processing of the viral polyprotein. There are two types of cleavage sites, aromatic–Pro and hydrophobic–hydrophobic, and there appears to be a slight preference for a valyl residue at the P2′ position.[1,2]

[1]S. Daenke, H. J. Scramm & C. R. M. Bangham (1994) *J. Gen. Virol.* **75**, 2233.
[2]B. G. M. Luukkonen, W. Tan, E. M. Fenyo & S. Schwartz (1995) *J. Gen. Virol.* **76**, 2169.

HURAIN

This endopeptidase, previously classified as EC 3.4.99.9, is currently listed as a variant of cucumisin [EC 3.4.21.25].[1]

[1]W. G. Jaffé (1943) *JBC* **149**, 1.

Selected entries from *Methods in Enzymology* [vol, page(s)]:
General discussion: 2, 57, 64

HYALURONATE LYASE

This enzyme [EC 4.2.2.1], also known as hyaluronidase, catalyzes the conversion of hyaluronate to *n* molecules of 3-(4-deoxy-β-D-gluc-4-enuronosyl)-*N*-acetyl-D-glucosamine. The enzyme is also capable of using chondroitins as a substrate.[1,2] *See also Hyaluronoglucosaminidase; Hyaluronoglucuronidase; Hyaluronidases*

[1]K. Ponnuraj & M. J. Jedrzejas (2000) *J. Mol. Biol.* **299**, 885.
[2]K. Meyer & M. M. Rapport (1952) *Adv. Enzymol.* **13**, 199.

Selected entries from *Methods in Enzymology* [vol, page(s)]:
General discussion: 1, 166; 8, 650, 654
Assay: 1, 169; 172; 8, 650; 235, 606

HYALURONOGLUCOSAMINIDASE

This enzyme [EC 3.2.1.35], also known as hyaluronidase and hyaluronate 4-glycanohydrolase, catalyzes the random hydrolytic scission of 1,4-linkages between *N*-acetyl-β-D-glucosamine and D-glucuronate residues in hyaluronate.[1-4] This hydrolytic protein will also catalyze the hydrolysis of 1,4-β-D-glycosidic linkages between *N*-acetylgalactosamine or *N*-acetylgalactosamine sulfate and glucuronate in chondroitin, chondroitin 4- and 6-sulfates, and dermatan (the activity previously designated as chondroitinase [EC 3.2.1.34] is now included as a part of hyaluronoglucosaminidase activity). *See also Hyaluronate Lyase; Hyaluronoglucuronidase; Hyaluronidases*

[1]G. Kreil (1995) *Protein Sci.* **4**, 1666.
[2]L. Roden, P. Campbell, J. R. Fraser, T. C. Laurent, H. Pertoft & J. N. Thompson (1989) *Ciba Found Symp.* **143**, 60 and discussion, 76, 281.

[3]K. Meyer, P. Hoffman & A. Linker (1960) *The Enzymes*, 2nd ed., **4**, 447.
[4]K. Meyer & M. M. Rapport (1952) *AE* **13**, 199.

Selected entries from *Methods in Enzymology* [vol, page(s)]:
General discussion: 1, 166; 235, 606
Assay: 8, 661; 235, 606
Other molecular properties: activity, 235, 606; animals, 8, 654 (assay, 8, 661; distribution, 8, 660; mode of action, 8, 659; preparation, 8, 661; properties, 8, 660); bacteria, of, preparation, 8, 653; mucopolysaccharides and, 3, 21, 22

HYALURONOGLUCURONIDASE

This enzyme [EC 3.2.1.36], also known as hyaluronidase and hyaluronate 3-glycanohydrolase, catalyzes the random hydrolysis of 1,3-linkages between β-D-glucuronate and *N*-acetyl-D-glucosamine residues in hyaluronate.[1]
See also Hyaluronate Lyase; Hyaluronoglucosaminidase; Hyaluronidases

[1]K. Meyer, P. Hoffman & A. Linker (1960) *The Enzymes*, 2nd ed., **4**, 447.

Selected entries from *Methods in Enzymology* [vol, page(s)]:
General discussion: 1, 166; 235, 606
Assay: 8, 661; 235, 606
Other molecular properties: activity, 235, 606; animals, 8, 654 (distribution, 8, 660; mode of action, 8, 659; preparation, 8, 661; properties, 8, 660); bacteria, of, preparation, 8, 653; mucopolysaccharides and, 3, 21, 22

HYBRID NUCLEASE

These nucleases, formerly classified as EC 3.1.4.34, are now regarded to be activities of venom exonuclease [EC 3.1.15.1], spleen exonuclease [EC 3.1.16.1], *Aspergillus* nuclease S_1 [EC 3.1.30.1], *Serratia marcescens* nuclease [EC 3.1.30.2], and micrococcal nuclease [EC 3.1.31.1].

HYDROGENASE

This oxidoreductase [EC 1.18.99.1], also known as hydrogenlyase, catalyzes the reversible reaction of H_2 with two oxidized ferredoxin molecules to produce two H^+ and two molecules of reduced ferredoxin.[1-4] These hydrogenases are iron-sulfur and/or nickel-requiring enzymes that use molecular hydrogen to reduce a variety of substances. The H_2 molecule undergoes heterolytic cleavage to produce a bound hydride ion. *See also Hydrogen Dehydrogenase; Cytochrome c_3 Hydrogenase*

[1]E. Garcin, Y. Montet, A. Volbeda, C. Hatchikian, M. Frey & F. C. Fontecilla-Camps (1998) *Biochem. Soc. Trans.* **26**, 396.
[2]M. A. Halcrow (1998) *CBC* **3**, 359.
[3]R. J. Maier (1996) *Adv. Protein Chem.* **48**, 35.
[4]G. Palmer (1975) *The Enzymes*, 3rd ed., **12**, 1.

Selected entries from *Methods in Enzymology* [vol, page(s)]:
General discussion: 2, 729, 861; 24, 423; 53, 286
Assay: 2, 729, 861; 53, 290; cyanobacterial, 167, 506; *Desulfovibrio gigas* (amperometric method, 243, 51; dithionite-reduced methyl viologen, with, 243, 52; electrochemically reduced methyl viologen, with, 243, 53;

H$_2$ evolution assay, **243**, 63; ^{2}H$_2$ or ^{3}H$_2$ exchange assay, **243**, 63; hydrogen electrode method, **243**, 51; hydrogen uptake, **243**, 63; manometric, **243**, 57; mass spectrometric method, **243**, 54; radioassays, **243**, 57; spectrophotometric method, **243**, 56); **243**, 75

Other molecular properties: activity, **53**, 287, 296; **243**, 68; amino acid composition, **53**, 294; amino acid reductase and, **2**, 218; anaerobic column chromatography of, **22**, 322; bacterial sources, **53**, 287, 297; clostridial, iron-sulfur center, **69**, 780; *Clostridium pasteurianum*, from, **2**, 729; **53**, 286 (electrochemical studies, **227**, 521; properties, 2, 731, 869; purification, **2**, 730, 869); composition, **53**, 292, 314; cyanobacterial (assay, **167**, 506; induction, **167**, 463; isolation, **167**, 502; types and distribution, **167**, 501); *Desulfovibrio*, **243**, 69; *Desulfovibrio gigas*, from, **53**, 627; **243**, 50, 60, 424; effect of pH, **53**, 296; entrapment in alginate with chloroplasts, **135**, 453; EPR sampling, **243**, 84; inhibitors, **53**, 296, 314; iron content, **53**, 290; iron hydrogenase (*Clostridium pasteurianum*, spectroscopic studies, **243**, 541; *Desulfovibrio*, **243**, 69; *Desulfovibrio vulgaris* [extraction, **243**, 45; homology with nickel-iron-selenium hydrogenase, **243**, 78; redox properties, **243**, 251; redox titration, **243**, 87; Western blot analysis, **243**, 80]; *Desulfovibrio vulgaris* (Hildenborough) [Mössbauer spectroscopy, **243**, 540; properties, **243**, 539]; *Thermotoga maritima*, spectroscopic properties, **243**, 542); iron-sulfur centers (characterization, **53**, 292; EPR characteristics, **53**, 265, 266, 271, 292, 295; reduction, **69**, 241, 242, 245, 246); kinetic properties, **53**, 294, 295; midpoint reduction potential, **243**, 83; molecular weight, **53**, 294, 313; multiple isoelectric forms, **53**, 313; nickel-iron hydrogenase (activity, **243**, 43; *Desulfovibrio*, **243**, 69; *Desulfovibrio fructosovorans*, redox state, EPR spectra and, **243**, 66; *Desulfovibrio gigas* [acid-labile sulfide, **243**, 50; activation, **243**, 60; active state, **243**, 60; activity states, **243**, 60; catalytic properties, **243**, 58; crystallization, **243**, 47; deuterium exchange reaction, mass spectrometric analysis, **243**, 54; electron acceptors, **243**, 56; electron paramagnetic resonance spectroscopy, **243**, 63; extraction, **243**, 44; forms, **243**, 48; hydrogen evolution reaction, mass spectrometric analysis, **243**, 55; hydrogen production and uptake, **243**, 43; iron-sulfur clusters, **243**, 50 {activation and oxidation/reduction states, **243**, 60}; isotope substitution methods, **243**, 48; metal content, **243**, 49; molecular weight, **243**, 49; optical absorption spectrum, **243**, 50; properties, **243**, 49; purification, **243**, 44; radioassays, **243**, 57; ready state, **243**, 60, 62; redox potential, **243**, 63–68; redox state, EPR spectra and, **243**, 66; redox titration, **243**, 87; reduction of electron acceptors with H$_2$, spectrophotometric analysis, **243**, 56; siting, **243**, 82; subunits, **243**, 49; unit of enzyme activity, **243**, 58; unready state, **243**, 60, 62]; electron donors and acceptors, **243**, 43; electron paramagnetic resonance studies, **243**, 44; forms, **243**, 44; iron-sulfur clusters, **243**, 43; nickel center, **243**, 43; *Thiocapsa roseopersicina*, **243**, 81); nickel-iron-selenium hydrogenase (*Desulfomicrobium baculatum* [activity, **243**, 77; cellular localization, **243**, 76; electron paramagnetic resonance studies, **243**, 77; extended X-ray absorption fine structure, **243**, 79; iron-sulfur clusters, **243**, 81; Mössbauer studies, **243**, 81; nickel(II) site, magnetic properties, **243**, 81; purification, **243**, 71; ^{77}Se-labeled, electron paramagnetic resonance studies, **243**, 79; selenium coordination to nickel site, **243**, 78; selenocysteine ligand to nickel in, **243**, 78; subunit genes, **243**, 77; subunits, **243**, 77]; *Desulfovibrio* [biochemical characterization, **243**, 75; categories, **243**, 69; homogeneity, measurement, **243**, 74; protein concentration, measurement, **243**, 73; purification, **243**, 71]; *Desulfovibrio baculatus* [activity, **243**, 77; D$_2$/H$^+$ exchange reactions, **243**, 77]; *Desulfovibrio salexigens* [properties, **243**, 76; purification, **243**, 71]; *Desulfovibrio vulgaris*, **243**, 82; ligand interaction, EPR monitoring, **243**, 89; redox potentiometry, **243**, 83); oxidation-reduction potentiometry, **243**, 83; oxidation-reduction titration, **243**, 84; oxygen-stable, **53**, 296 (assay, **53**, 297; properties, **53**, 313, 314; purification, **53**, 310); pH optimum, **53**, 314; potentiometric EPR titration, **243**, 83; preparation, **5**, 603; purification, **53**, 288; pyruvate synthase and, **13**, 174; reconstituted system, for hydrogen production, **69**, 626; redox potentiometry (buffers, **243**, 89; oxidants for, **243**, 88; redox titration cell, **243**, 84 [calibration, **243**, 87]); reduction of cytochrome c$_{553}$, **243**, 118;

sedimentation coefficient, **53**, 313; spectral properties, **53**, 292, 313, 314; substrate specificity, **53**, 314; succinate dehydrogenase and, **5**, 602; use in ferredoxin assay, **23**, 442; *Wolinella succinogenes*, **243**, 367

HYDROGENASE MATURATION ENDOPEPTIDASE

This endopeptidase, also known as HycI endopeptidase, catalyzes peptide-bond hydrolysis in the processing of bacterial [NiFe]-dependent hydrogenases. After the nickel has been chelated to the precursor protein, a single site is hydrolyzed in which the P1 position is occupied by either an arginyl residue (for example, in *Escherichia coli*) or a histidyl residue (for example, in *Azotobacter vinelandii* and *Desulfovibrio gigas*). The P1$'$ position contains a hydrophobic residue (*i.e.*, Met, Ile, Val, or Ala).[1,2]

[1] R. Rossmann, T. Maier, F. Lottspeich & A. Böck (1995) *EJB* **227**, 545.
[2] T. Maier & A. Böck (1996) in *Mechanisms of Metallocenter Assembly* (R. P. Hausinger, G. L. Eichhorn & L. G. Marzilli, eds.), VCH Publ., New York, p. 173.

HYDROGEN DEHYDROGENASE

This flavoprotein [EC 1.12.1.2], referred to occasionally as hydrogenase, catalyzes the reaction of H$_2$ with NAD$^+$ to produce H$^+$ and NADH.[1–5] This oxidoreductase also requires iron and nickel ions. Kinetic studies conducted anaerobically indicate that the enzyme operates by means of a ping pong Bi Bi mechanism and that substrate inhibition by H$_2$ occurs at millimolar concentrations.[5] The H$_2$ molecule undergoes heterolytic cleavage to produce a bound hydride ion, presumably at the nickel site. ***See also*** *Hydrogenase; Cytochrome c$_3$ Hydrogenase*

[1] M. A. Halcrow (1998) *CBC* **3**, 359.
[2] K. Francis, P. Patel, J. C. Wendt & K. T. Shanmugam (1990) *J. Bacteriol.* **172**, 5750.
[3] K. Schneider & H. G. Schlegel (1976) *BBA* **452**, 66.
[4] J. Pfitzner, H. A. B. Linke & H. G. Schlegel (1970) *Arch. Mikrobiol.* **71**, 67.
[5] R. G. Keefe, M. J. Axley & A. L. Harabin (1995) *ABB* **317**, 449.

Selected entries from ***Methods in Enzymology*** [vol, page(s)]: **General discussion: 137**, 232

HYDROGEN:QUINONE OXIDOREDUCTASE

This oxidoreductase [EC 1.12.99.3], also known as hydrogen:menaquinone oxidoreductase, catalyzes the reaction of molecular hydrogen with menaquinone to produce reduced menaquinone.[1] Additional substrates include other water-soluble quinones (such as, 2,3-dimethylnaphthoquinone) and viologen dyes (such as the benzyl and methyl viologens).

[1] F. Dross, V. Geisler, R. Lenger, F. Theis, T. Krafft, F. Fahrenholz, E. Kojro, A. Duchêne, D. Tripier, K. Juvenal & A. Kröger (1992) *EJB* **206**, 93.

HYDROGEN SULFIDE S-ACETYLTRANSFERASE

This transferase [EC 2.3.1.10] catalyzes the reversible reaction of acetyl-CoA with hydrogen sulfide to produce coenzyme A and thioacetate.[1]

[1]R. O. Brady & E. R. Stadtman (1954) JBC **211**, 621.

HYDROGEN SULFIDE:FERRIC ION OXIDOREDUCTASE

This glutathione-dependent enzyme, formerly called sulfur:ferric ion oxidoreductase, catalyzes the reaction of hydrogen sulfide with Fe^{3+} to produce sulfite and Fe^{2+}.[1]

[1]I. Suzuki (1994) MIE **243**, 455.

HYDROGEN SULFITE REDUCTASE

This oxidoreductase [EC 1.8.99.3], also known as bisulfite reductase, dissimilatory sulfite reductase, desulfoviridin, desulforubidin, and desulfofuscidin, catalyzes the reaction of trithionate $((O_3S\text{-}S\text{-}SO_3)^{2-})$ with an acceptor substrate, a hydroxide anion, and two water molecules to produce three HSO_3^{3-} and the reduced acceptor AH_2.[1,2] The acceptor substrate can be methyl viologen, benzyl viologen, and cytochrome c_3.

This classification represents a group of sirohemoproteins having iron-sulfur centers.

[1]G. Fauque, A. R. Lino, M. Czechowski, L. Kang, D. V. DerVartanian, J. J. G. Moura, J. LeGall & I. Moura (1990) BBA **1040**, 112.
[2]I. Moura, J. LeGall, A. R. Lino, H. D. Peck, G. Fauque, A. V. Xavier, D. V. DerVartanian, J. J. G. Moura & B. H. Huynh (1988) JACS **110**, 1075.

Selected entries from **Methods in Enzymology** [**vol**, page(s)]:
General discussion: 243, 270, 276
Assay: 243, 273, 277
Other molecular properties: desulfofuscidin (absorption maxima, **243**, 284; absorption spectra, **243**, 283; amino acid composition, **243**, 288; assay, **243**, 277; bisulfite reductase activity, temperature effects, **243**, 290; catalytic properties, **243**, 289; cellular localization, **243**, 292; composition, **243**, 286; distribution, **243**, 276; electron paramagnetic resonance spectra, **243**, 287; heme chromophore, absorption spectrum, **243**, 285; molecular weight, 243, 282; N-terminal amino acid sequence, **243**, 288; product ambiguity, **243**, 292; properties, **243**, 270, 282; purification from [Thermodesulfobacterium commune, **243**, 278; Thermodesulfobacterium mobile, **243**, 281]; purity, **243**, 282; reaction catalyzed by, **243**, 276, 292; siroheme, **243**, 285, 291, 295; specific activity, **243**, 278; subunit structure, **243**, 282; thermostability, **243**, 291; unit of enzyme activity, **243**, 278); desulforubidin (activity, **243**, 271; assay, **243**, 273; electron paramagnetic resonance studies, **243**, 274; light absorption spectra, **243**, 273; Mössbauer spectroscopy, **243**, 274; properties, **243**, 271, 294; purification from [Desulfovibrio baculatus DSM 1743, **243**, 272; Desulfovibrio desulfuricans Norway, **243**, 272]; spectroscopic studies, **243**, 274); desulfoviridin (activity, **243**, 536; distribution, **243**, 536; Mössbauer spectroscopy, **243**, 536; properties, **243**, 294, 536)

HYDROPEROXIDE DEHYDRATASE

This heme-thiolate-dependent (or, cytochrome P450-dependent) dehydratase [EC 4.2.1.92], also called hydroperoxide isomerase and allene oxide synthase, catalyzes the conversion of (9Z,11E,14Z)-(13S)-hydroperoxyoctadeca-(9,11,14)-trienoate to (9Z)-(13S)-12,13-epoxyoctadeca-9,11-dienoate and water.[1] Other unsaturated fatty-acid hydroperoxides can serve as substrates, forming the corresponding allene oxides. The enzyme reportedly requires conjugated diene hydroperoxides for catalysis.[1]

[1]C. Schneider & P. Schreier (1998) Lipids **33**, 191.

Selected entries from **Methods in Enzymology** [**vol**, page(s)]:
General discussion: 272, 250
Assay: high-performance liquid chromatography, **272**, 254; spectrophotometric, **272**, 253; substrate synthesis (materials, **272**, 251; purification, **272**, 253; quantitation, **272**, 253; soybean lipoxygenase reaction, **272**, 252)
Other molecular properties: carbon monoxide affinity, **272**, 242; concentration in monocot microsome preparations, **272**, 241; purification from flaxseed (ammonium sulfate precipitation, **272**, 256; anion-exchange chromatography, **272**, 257, 259; chromatofocusing, **272**, 259; detergent removal, **272**, 259; hydrophobic interaction chromatography, **272**, 257; solubilization, **272**, 256; tissue extraction, **272**, 256); substrates, **272**, 250

HYDROPEROXIDE ISOMERASE

This isomerase activity, originally classified as EC 5.3.99.1 and now a deleted EC entry, results from the combined action of hydroperoxide dehydratase [EC 4.2.1.92] and allene-oxide cyclase [EC 5.3.99.6].

N-HYDROXY-2-ACETAMIDOFLUORENE REDUCTASE

This oxidoreductase [EC 1.6.6.13] catalyzes the reaction of NAD(P)H with N-hydroxy-2-acetamidofluorene to produce NAD(P)$^+$ and 2-acetamidofluorene.[1,2] N-Hydroxy-4-acetamidobiphenyl is an alternative substrate.

[1]S. Kitamura & K. Tatsumi (1985) BBRC **133**, 67.
[2]H. R. Gutman & R. R. Erickson (1969) JBC **244**, 1729.

D-2-HYDROXY-ACID DEHYDROGENASES

This FAD- and zinc-dependent oxidoreductase [EC 1.1.99.6] catalyzes the reaction of (R)-lactate (that is, D-lactate) with an acceptor to produce pyruvate and the reduced acceptor.[1–4] Other substrates include a number of (R)-2-hydroxy acids as substrates (for example, D-glycerate, D-malate, and meso-tartrate). Acceptor substrates include ferricyanide, 2,6-dichlorophenolindophenol, phenazine methosulfate, cytochrome c, ubiquinone, and dioxygen.

D-2-Hydroxyisovalerate dehydrogenase, also known as 2-ketoisovalerate reductase, catalyzes the reversible reaction of D-2-hydroxyisovalerate with NADP⁺ to produce 2-ketoisovalerate and NADPH. **See also** D-*Lactate Dehydrogenase; Glycerate Dehydrogenase; D-3-Phosphoglycerate Dehydrogenase*

[1] A. R. Clarke & T. R. Dafforn (1998) *CBC* **3**, 1.
[2] Y. Hatefi & D. L. Stiggall (1976) *The Enzymes*, 3rd ed., **13**, 175.
[3] P. K. Tubbs & G. D. Greville (1959) *BBA* **34**, 290.
[4] J. A. Alvarez, J. L. Gelpi, K. Johnsen, N. Bernard, J. Delcour, A. R. Clarke, J. J. Holbrook & A. Cortes (1997) *EJB* **244**, 203.

Selected entries from *Methods in Enzymology* [**vol**, page(s)]:
General discussion: 9, 327; 41, 323; **324**, 296
Assay: 9, 327; 41, 323; **324**, 296
Other molecular properties: animal tissue, from, assay, **41**, 323; flavoprotein classification, **53**, 397; D-2-hydroxyisovalerate dehydrogenase, **324**, 293 (assay, **324**, 296; properties, **324**, 300; purification, **324**, 297); rabbit kidney, from, **41**, 325 (inhibitors, **41**, 328; preparation and purification, **41**, 325; properties, **41**, 327); sources, **9**, 332; yeast, from (kinetic properties, **9**, 331; molecular weight, **9**, 331; properties, **9**, 330; purification, **9**, 328).

(R)-3-HYDROXY-ACID ESTER DEHYDROGENASE

This oxidoreductase [EC 1.2.1.55], isolated from *Saccharomyces cerevisiae*, catalyzes the reversible reaction of ethyl (*R*)-3-hydroxyhexanoate with NADP⁺ to produce ethyl 3-oxohexanoate and NADPH.[1] The enzyme also acts on ethyl (*R*)-3-hydroxybutanoate and some other (*R*)-3-hydroxy-acid esters. This enzyme is a component of yeast fatty acid synthase multireaction complex [EC 2.3.1.86].
See also (S)-*3-Hydroxyacid Ester Dehydrogenase*

[1] J. Heidlas, K.-H. Engel & R. Tressl (1988) *EJB* **172**, 633.

(S)-3-HYDROXY-ACID ESTER DEHYDROGENASE

This oxidoreductase [EC 1.2.1.56], isolated from *Saccharomyces cerevisiae*, catalyzes the reversible reaction of ethyl (*S*)-3-hydroxyhexanoate with NADP⁺ to produce ethyl 3-oxohexanoate and NADPH. The enzyme will also act on ethyl (*S*)-3-hydroxybutyrate, (*S*)-4-hydroxypentanoate, and (*S*)-5-hydroxyhexanoate.[1] **See also** (R)-*3-Hydroxyacid Ester Dehydrogenase*

[1] J. Heidlas, K.-H. Engel & R. Tressl (1988) *EJB* **172**, 633.

HYDROXY-ACID:KETOACID TRANSHYDROGENASE

This oxidoreductase [EC 1.1.99.24], also known as hydroxyacid:oxoacid transhydrogenase, catalyzes the reaction of (*S*)-3-hydroxybutanoate with α-ketoglutarate (or, 2-oxoglutarate) to produce acetoacetate and (*R*)-2-hydroxyglutarate.[1] Other hydroxy-acid substrates include 4-hydroxybutanoate (producing succinate semialdehyde) and (*R*)-2-hydroxyglutarate. Other oxo-acid substrates include butanoate semialdehyde (or, 4-oxobutanoate).

[1] E. E. Kaufman, T. Nelson, H. M. Fales & D. M. Levin (1988) *JBC* **263**, 16872.

(S)-2-HYDROXY-ACID OXIDASE

This FMN-dependent oxidoreductase [EC 1.1.3.15] (also known as glycolate oxidase, hydroxy-acid oxidase A, and hydroxy-acid oxidase B) catalyzes the reaction of an (*S*)-2-hydroxy acid with dioxygen to produce a 2-keto-acid anion (or, 2-oxo-acid anion) and hydrogen peroxide.[1]

This oxidase exists as two major isoenzymes. The A-form preferentially oxidizes short-chain aliphatic hydroxy acids (for example, L-2-hydroxybutyrate, L-lactate, L-2-hydroxypropanoate, and glycolate) whereas the B-form preferentially oxidizes long-chain and aromatic hydroxy acids (for example, L-2-hydroxyisohexanoate, L-2-hydroxy-*n*-hexanoate, L-2-hydroxy-*n*-pentanoate, and L-2-hydroxyisopentanoate). The rat isoenzyme B also has an L-amino-acid oxidase activity.

The enzyme has a ping pong Bi Bi kinetic mechanism in which the first half-reaction proceeds via a carbanionic intermediate. Dioxygen binds in the second half-reaction and possibly forms a flavin C4a-hydroperoxide intermediate.[2]

[1] N. E. Tolbert (1981) *Ann. Rev. Biochem.* **50**, 133.
[2] B. A. Palfey & V. Massey (1998) *CBC* **3**, 83.

Selected entries from *Methods in Enzymology* [**vol**, page(s)]:
General discussion: 1, 528; 34, 302; 52, 18
Assay: 9, 339
Other molecular properties: borohydride reduction, **18B**, 475; glycolate degradation and, **4**, 609; glyoxysomes, in, **31**, 565, 569; **72**, 788; inactivation, **53**, 439, 443, 446, 447; marker enzyme, as, **23**, 679; **31**, 735, 742; mechanism, **87**, 86; microbodies, in, **52**, 496, 503; peroxisomes, in, **31**, 364, 367, 501; photosynthesis and, **4**, 894; pig liver, from, **41**, 337 (assay, **41**, 338; chromatography, **41**, 339, 340; inhibitors and activators, **41**, 342; properties, **41**, 340; purification, **41**, 338); purification over flavin-cellulose column, **18B**, 551, 552; spinach mitochondria, of, **10**, 134; structural similarities to other enzyme, **87**, 85; sulfite complex, **18B**, 473; tobacco, in, **23**, 183; wheat leaves, from (inhibitors, **9**, 342; pH optimum, **9**, 342; properties, **9**, 341; prosthetic group, **9**, 341; purification, **9**, 340)

3-HYDROXYACYL-CoA DEHYDROGENASE

This oxidoreductase [EC 1.1.1.35], also known as β-hydroxyacyl-CoA dehydrogenase and β-keto-reductase, catalyzes the reversible reaction of an (*S*)-3-hydroxyacyl-CoA (*i.e.*, L-3-hydroxyacyl-CoA) with NAD$^+$ to produce a 3-oxoacyl-CoA (*i.e.*, β-ketoacyl-CoA) and NADH.[1] Substrates include 3-hydroxybutyryl-CoA, 3-hydroxyhexanoyl-CoA, 3-hydroxyoctanoyl-CoA, and 3-hydroxylauryl-CoA. The enzyme will also utilize *S*-3-hydroxyacyl-*N*-acylthioethanolamine and *S*-3-hydroxyacylhydrolipoate as substrates. The dehydrogenase from several sources can also utilize NADP$^+$ as the coenzyme, albeit as a weaker substrate. In addition, there is a broad specificity with respect to the acyl chain length (note that there is a long-chain-length 3-hydroxyacyl-CoA dehydogenase [EC 1.1.1.211]). In general, the short-chain dehydrogenase from most sources is optimally active with six-carbon acyl-CoA substrates whereas the long-chain enzymes prefers twelve- to sixteen-carbon derivatives.

See also *3-Hydroxybutyryl-CoA Dehydrogenase; Long-Chain-3-Hydroxyacyl-CoA Dehydrogenase*

[1] J. J. Barycki, L. K. O'Brien, J. M. Bratt, R. Zhang, R. Sanishvili, A. W. Strauss & L. J. Banaszak (1999) *Biochemistry* **38**, 5786.

Selected entries from ***Methods in Enzymology*** [vol, page(s)]:
General discussion: 1, 566; **35**, 122; **71**, 378, 403
Assay: 1, 566; **35**, 123; peroxisomal, assay, **148**, 523
Other molecular properties: assay for (coenzyme A, **13**, 540; coenzyme A derivatives, **13**, 547; crotonase, **35**, 137 [αβ-unsaturated acyl-CoA derivatives, of, **35**, 140]; α-ketoacyl-[acyl-carrier-protein] synthase, **14**, 57); bromooctanoate, effect of, **72**, 568; crotonyl-CoA assay and, **1**, 561; glyoxysomes, in, **31**, 569; **72**, 788; 3-hydroxybutyryl-CoA racemase and, **5**, 557; induction by clofibrate and di(2-ethylhexyl)phthalate, **72**, 507; inhibitors, **35**, 127, 128; lactyl-CoA dehydrase preparation, present in, **9**, 683; mouse kidney, in, **39**, 458; mutant, **72**, 695; properties, **1**, 571; purification, sheep liver, **1**, 568; stability, **35**, 127; stereospecificity, **87**, 105

HYDROXYACYLGLUTATHIONE HYDROLASE

This enzyme [EC 3.1.2.6], also known as glyoxalase II, catalyzes the hydrolysis of an *S*-(2-hydroxyacyl)glutathione (for example, *S*-lactoylglutathione) to produce glutathione and a 2-hydroxy acid anion.[1–5] The enzyme will also act on *S*-acetoacetylglutathione, albeit more slowly. It has been proposed that the mechanism involves the direct nucleophilic attack of an active-site histidyl residue on the thiol ester substrate to form an acyl-imidazole intermediate that is then rapidly hydrolyzed.[1]

[1] D. L. Van der Jagt (1993) *Biochem. Soc. Trans.* **21**, 522.
[2] V. Talesa, L. Uotila, M. Koivusalo, G. Principato, E. Giovannini & G. Rosi (1988) *BBA* **955**, 103.
[3] G. B. Principato, G. Rosi, V. Talesa, E. Giovannini & L. Uotila (1987) *BBA* **911**, 349.

[4] L. Uotila (1979) *BBA* **580**, 277.
[5] L. Uotila (1973) *Biochemistry* **12**, 3938 and 3944.

Selected entries from ***Methods in Enzymology*** [vol, page(s)]:
General discussion: 77, 424; **90**, 547
Assay: murine liver, **90**, 547

2-HYDROXYACYLSPHINGOSINE 1-β-GALACTOSYLTRANSFERASE

This highly specific, endoplasmic reticulum-associated transferase [EC 2.4.1.45], also known as UDP-galactose-ceramide galactosyltransferase and cerebroside synthase, catalyzes the reaction of UDP-D-galactose with a 2-(2-hydroxyacyl)sphingosine to produce UDP and a 1-(β-D-galactosyl)-2-(2-hydroxyacyl)sphingosine.[1,2]

[1] S. Basu, A. M. Schultz, M. Basu & S. Roseman (1971) *JBC* **246**, 4272.
[2] H. Sprong, B. Kruithof, R. Leijendekker, J. W. Slot, G. van Meer & P. van der Sluijs (1998) *JBC* **273**, 25880.

Selected entries from ***Methods in Enzymology*** [vol, page(s)]:
General discussion: 71, 521; **72**, 384
Assay: 72, 384

3α-HYDROXY-5β-ANDROSTAN-17-ONE 3α-DEHYDROGENASE

This oxidoreductase [EC 1.1.1.152], also known as etiocholanolone 3α-dehydrogenase, catalyzes the reaction of 3α-hydroxy-5β-androstan-17-one (also called etiocholanolone) with NAD$^+$ to produce 5β-androstane-3,17-dione and NADH.[1]

[1] C. R. Roe & N. O. Kaplan (1969) *Biochemistry* **8**, 5093.

3-HYDROXYANTHRANILATE 3,4-DIOXYGENASE

This iron-dependent oxidoreductase [EC 1.13.11.6] catalyzes the reaction of 3-hydroxyanthranilate with dioxygen to yield 2-amino-3-carboxymuconate semialdehyde.[1–5] This product spontaneously rearranges to quinolinate. Other substrates include 4-methyl-3-hydroxyanthranilate and 4-ethyl-3-hydroxyanthranilate.

[1] K. Saito & M. P. Heyes (1996) *Adv. Exp. Med. Biol.* **398**, 485.
[2] A. M. Cesura, R. Alberati-Giani, R. Buchli, C. Broger, C. Kohler, F. Vilbois, H. W. Lahm, M. P. Heitz & P. Malherbe (1996) *Adv. Exp. Med. Biol.* **398**, 477.
[3] A. C. Foster, R. J. White & R. Schwarcz (1986) *J. Neurochem.* **47**, 23.
[4] E. Okuno, C. Köhler & R. Schwarcz (1987) *J. Neurochem.* **49**, 771.
[5] O. Hayaishi, M. Nozaki & M. T. Abbott (1975) *The Enzymes*, 3rd ed., **12**, 119.

Selected entries from ***Methods in Enzymology*** [vol, page(s)]:
General discussion: 17A, 467; **18B**, 165, 175; **52**, 11
Other molecular properties: purification from (beef liver, **17A**, 468; rat liver, **17A**, 472)

3-HYDROXYANTHRANILATE 4-C-METHYLTRANSFERASE

This methyltransferase [EC 2.1.1.97] catalyzes the reaction of S-adenosyl-L-methionine with 3-hydroxyanthranilate to produce S-adenosyl-L-homocysteine and 3-hydroxy-4-methylanthranilate. The enzyme participates in actinomycin biosynthesis in *Streptomyces antibioticus*.[1–3]

[1] F. Fawaz & J. H. Jones (1988) JBC **263**, 4602.
[2] G. H. Jones (1987) J. Bacteriol. **169**, 5575.
[3] G. H. Jones (1993) JBC **268**, 6831.

3-HYDROXYANTHRANILATE OXIDASE

This iron-dependent oxidoreductase [EC 1.10.3.5] catalyzes the reaction of 3-hydroxyanthranilate with dioxygen to produce 6-imino-5-oxocyclohexa-1,3-dienecarboxylate and hydrogen peroxide.[1–3] 4-Halo-3-hydroxyanthranilate derivatives are potent competitive inhibitors. An initial C2-peroxy intermediate which generates an iminodicarbonyl has been proposed for the reaction mechanism.[1]

[1] J. V. Schloss & M. S. Hixon (1998) CBC **2**, 43.
[2] J. S. Cook & C. I. Pogson (1983) BJ **214**, 511.
[3] C. J. Parli, P. Krieter & B. Schmidt (1980) ABB **203**, 161.

Selected entries from *Methods in Enzymology* [vol, page(s)]:
General discussion: **56**, 475

HYDROXYANTHRAQUINONE GLUCOSYLTRANSFERASE

This transferase [EC 2.4.1.181] catalyzes the reaction of UDP-D-glucose with a hydroxyanthraquinone to produce UDP and a glucosyloxyanthraquinone. Acceptor substrates include emodin, anthrapurprin, quinizarin, 2,6-dihydroanthraquinone, and 1,8-dihydroanthraquinone.[1]

[1] H. E. Khouri & R. K. Ibrahim (1987) Phytochemistry **26**, 2531.

N-HYDROXYARYLAMINE O-ACETYLTRANSFERASE

This transferase [EC 2.3.1.118], also called arylhydroxamate N,O-acetyltransferase and arylamine N-acetyltransferase, catalyzes the reaction of acetyl-CoA with an N-hydroxyarylamine to produce free coenzyme A and an N-acetoxyarylamine.[1]

[1] M. Watanabe, T. Sofuni & T. Nohmi (1992) JBC **267**, 8429.

3-HYDROXYASPARTATE ALDOLASE

This enzyme [EC 4.1.3.14] reversibly converts *erythro*-3-hydroxy-L-aspartate to glycine and glyoxylate.[1] Other substrates include allothreonine.

[1] R. G. Gibbs & J. G. Morris (1964) BBA **85**, 501.

Selected entries from *Methods in Enzymology* [vol, page(s)]:
General discussion: **17A**, 981

erythro-3-HYDROXYASPARTATE DEHYDRATASE

This pyridoxal-phosphate-dependent enzyme [EC 4.2.1.38] catalyzes the reaction of *erythro*-3-hydroxy-L-aspartate with water to produce oxaloacetate, ammonia, and water.[1,2]

[1] L. Davis & D. E. Metzler (1972) The Enzymes, 3rd ed., **7**, 33.
[2] R. G. Gibbs & J. G. Morris (1965) BJ **97**, 547.

Selected entries from
General discussion: **17A**, 987

4-HYDROXYBENZALDEHYDE DEHYDROGENASE

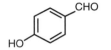
4-hydroxybenzaldehyde

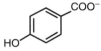
4-hydroxybenzoate

This oxidoreductase [EC 1.2.1.64], also known as *p*-hydroxybenzaldehyde dehydrogenase, catalyzes the reaction of 4-hydroxybenzaldehyde with NAD+ and water to produce 4-hydroxybenzoate, NADH, and a proton.[1,2]

[1] I. D. Bossert, G. Whited, D. T. Gibson & L. Y. Young (1989) J. Bacteriol. **171**, 2956.
[2] G. M. Whited & D. T. Gibson (1991) J. Bacteriol. **173**, 3017.

2'-HYDROXYBENZALPYRUVATE ALDOLASE

This aldolase, which participates in the degradation of many naphthalenesulfonates, catalyzes the conversion of 2'-hydroxybenzalpyruvate to salicylaldehyde and pyruvate.

2'-hydroxybenzalpyruvate

salicylaldehyde

The enzyme is a class I aldolase and the reaction proceeds via the formation of a Schiff base with the ε-amino group of a lysyl residue (the enzyme is inactivated in the presence of the substrate with NaBH4).[1] Base-assisted addition of water to the carbon–carbon double bond is followed by *syn* addition of a proton to the enamine intermediate.

[1]A. E. Kuhm, H. Knackmuss & A. Stolz (1993) *JBC* **268**, 9484.

4-HYDROXYBENZOATE DECARBOXYLASE

This monovalent-cation-dependent enzyme [EC 4.1.1.61] catalyzes the conversion of 4-hydroxybenzoate to phenol and carbon dioxide.[1,2]

4-hydroxybenzoate

A manganese-dependent enzyme, referred to as phenol carboxylase, has been isolated from anaerobes growing on phenol. It is distinct from 4-hydroxybenzoate decarboxylase.

[1]J. V. Schloss & M. S. Hixon (1998) *CBC* **2**, 43.
[2]D. J. Grant & J. C. Patel (1969) *Antonie Leeuwenhoek* **35**, 325.

4-HYDROXYBENZOATE 4-*O*-β-D-GLUCOSYLTRANSFERASE

This transferase [EC 2.4.1.194] catalyzes the reaction of UDP-D-glucose with 4-hydroxybenzoate to produce UDP and 4-(β-D-glucosyloxy)benzoate.[1]

[1]A. Bechthold, U. Berger & L. Heide (1991) *ABB* **288**, 39.

3-HYDROXYBENZOATE 2-MONOOXYGENASE

3-hydroxybenzoate   2,3-dihydroxybenzoate

This oxidoreductase [EC 1.14.99.23], also known as 3-hydroxybenzoate 2-hydroxylase, catalyzes the reaction of 3-hydroxybenzoate with dioxygen and an electron donor substrate (AH_2) to produce 2,3-dihydroxybenzoate, the oxidized donor (A), and a water molecule.[1,2]

[1]G. O. Daumy & A. S. McColl (1982) *J. Bacteriol.* **149**, 384.
[2]G. O. Daumy, A. S. McColl & G. C. Andrews (1980) *J. Bacteriol.* **141**, 293

3-HYDROXYBENZOATE 4-MONOOXYGENASE

This FAD-dependent enzyme [EC 1.14.13.23], also called 3-hydroxybenzoate 4-hydroxylase, catalyzes the reaction

of 3-hydroxybenzoate with NADPH and dioxygen to produce 3,4-dihydroxybenzoate (*i.e.*, protocatechuate), NADP$^+$, and a water molecule.[1–3]

3-hydroxybenzoate   3,4-dihydroxybenzoate

The enzyme also utilizes 3-hydroxybenzoates derivatized in the 2, 5, and 6 positions. Yomo *et al.*[4] described an enzymatic method for measuring the absolute oxygen concentration in aqueous solutions, using 4-hydroxybenzoate 3-monooxygenase and glucose oxidase.

[1]V. Massey & P. Hemmerich (1975) *The Enzymes*, 3rd ed., **12**, 191.
[2]J. M. Michalover & D. W. Ribbons (1973) *BBRC* **55**, 888.
[3]R. P. Kumar, P. V. S. Rao & C. S. Vaidyanathan (1973) *Indian J. Biochem. Biophys.* **10**, 184.
[4]T. Yomo, I. Urabe & H. Okada (1989) *Anal. Biochem.* **179**, 124.

Selected entries from **Methods in Enzymology** [**vol**, page(s)]:
General discussion: 52, 16; **188**, 138

3-HYDROXYBENZOATE 6-MONOOXYGENASE

This FAD-dependent oxidoreductase [EC 1.14.13.24], also known as 3-hydroxybenzoate 6-hydroxylase, catalyzes the reaction of 3-hydroxybenzoate with NADH and dioxygen to produce 2,5-dihydroxybenzoate (*i.e.*, gentisate), NAD$^+$, and water.[1,2]

3-hydroxybenzoate   gentisate

NADPH can also be used as a substrate, although not as effectively as NADH. The enzyme also utilizes 3-hydroxybenzoates derivatized in the 2, 4, and 5 positions. The *Pseudomonas cepacia* enzyme has a random Bi Uni Uni Bi ping pong kinetic mechanism[1]: *m*-hydroxybenzoate and NADH first bind to the monooxygenase in a random sequence, the flavoenzyme is reduced by NADH, and NAD$^+$ is released. Dioxygen subsequently binds to and reacts with the reduced flavoenzyme· *m*-hydroxybenzoate complex. Upon formation and release of water and gentisate, the oxidized holoenzyme is regenerated.

[1]Y. M. Yu, L. H. Wang & S. C. Tu (1987) *Biochemistry* **26**, 1105.
[2]V. Massey & P. Hemmerich (1975) *The Enzymes*, 3rd ed., **12**, 191.

4-HYDROXYBENZOATE 1-MONOOXYGENASE

This FAD-dependent oxidoreductase [EC 1.14.13.64], also known as 4-hydroxybenzoate 1-hydroxylase, catalyzes the reaction of 4-hydroxybenzoate with NAD(P)H and dioxygen to produce hydroquinone, NAD(P)$^+$, water, and carbon dioxide. The enzyme from *Candida parapsilosis* is specific for 4-hydroxybenzoate derivatives and prefers NADH to NADPH as the electron donor.[1,2]

[1] W. J. H. van Berkel, M. H. M. Eppink, W. J. Middelhoven, J. Vervoort & I. M. C. M. Rietjens (1994) *FEMS Microbiol. Lett.* **121**, 207.

[2] M. H. M. Eppink, S. A. Boeren, J. Vervoort & W. J. H. van Berkel (1997) *J. Bacteriol.* **179**, 6680.

4-HYDROXYBENZOATE 3-MONOOXYGENASE

This FAD-dependent oxidoreductase [EC 1.14.13.2], also known as *p*-hydroxybenzoate hydroxylase, catalyzes the reaction of 4-hydroxybenzoate with NADPH and dioxygen to produce protocatechuate (*i.e.*, 3,4-dihydroxybenzoate), NADP$^+$, and water.[1–6] Other substrates include *p*-mercaptobenzoate.

The monooxygenase has a random Bi Uni Uni Bi ping pong mechanism. The reduced coenzyme and the aromatic substrate first bind randomly to the protein: however, there is an absolute requirement that the hydroxybenzoate bind before the rapid reduction of the flavin by NADPH (binding of the aromatic substrate greatly stimulates reduction). NADP$^+$ is then released and dioxygen is activated by forming the C4a-hydroperoxyflavin intermediate (possibly via a superoxide-semiquinone complex). The distal peroxy oxygen is then incorporated into the aromatic substrate. This oxygen transfer also produces the C4a-hydroxyflavin intermediate which regenerates the flavin upon the release of water.

The enzyme from *Pseudomonas* is very specific for the coenzyme substrate whereas 4-hydroxybenzoate 3-monooxygenase (NAD(P)H) [EC 1.14.13.33] can utilize NADH and NADPH equally well. The enzyme isolated from *Corynebacterium cyclohexanicum* is highly specific for 4-hydroxybenzoate.

[1] L. Ridder, B. A. Palfey, J. Vervoort & I. M. Rietjens (2000) *FEBS Lett.* **478**, 197.

[2] G. R. Moran, B. Entsch, B. A. Palfey & D. P. Ballou (1999) *Biochemistry* **38**, 6292.

[3] B. A. Palfey & V. Massey (1998) *CBC* **3**, 83.

[4] S. Yamamoto & Y. Ishimura (1991) *A Study of Enzymes* **2**, 315.

[5] B. Entsch, D. P. Ballou & V. Massey (1976) *JBC* **251**, 2550.

[6] V. Massey & P. Hemmerich (1975) *The Enzymes*, 3rd ed., **12**, 191.

Selected entries from *Methods in Enzymology* [vol, page(s)]:
General discussion: 52, 17; 53, 543; 188, 138
Assay: 53, 545; 188, 139
Other molecular properties: activity, 53, 544; crystallization, 53, 547; flavin content, 53, 548; genetics, 188, 147; inhibitors, 53, 551; molecular weight, 53, 548; oxy form, spectrum, 52, 36; oxygenated-flavin intermediates, 53, 550, 551; properties, 188, 142; proton resonances, of aromatic amino acid residues, 66, 409; pseudomonads, from, 53, 544; *Pseudomonas fluorescens*, from, 53, 544; purification, 53, 545 (*Pseudomonas aeruginosa*, 188, 139); reaction mechanism, 52, 39; 53, 551, 552; reconstitution, 53, 549; reduction by NADPH, 53, 549, 550; reversible dissociation, 53, 433; stability, 53, 547, 548; stereospecificity, 87, 120; substrate specificity, 53, 548, 549

4-HYDROXYBENZOATE 3-MONOOXYGENASE (NAD(P)H)

This FAD-dependent oxidoreductase [EC 1.14.13.33], also known as 4-hydroxybenzoate 3-hydroxylase, catalyzes the reaction of 4-hydroxybenzoate with NAD(P)H and dioxygen to produce 3,4-dihydroxybenzoate (*i.e.*, protocatechuate), NAD(P)$^+$, and water.[1]

[1] T. Fujii & T. Kaneda (1985) *EJB* **147**, 97.

4-HYDROXYBENZOATE NONAPRENYLTRANSFERASE

This transferase [EC 2.5.1.39], also known as *p*-hydroxybenzoate polyprenyltransferase, catalyzes the reaction of solanesyl diphosphate (or, nonaprenyl diphosphate) with 4-hydroxybenzoate to produce pyrophosphate (or, diphosphate) and nonaprenyl-4-hydroxybenzoate.[1,2]

polyprenyl diphosphate; $n = 4$–9

3-polyprenyl-4-hydroxybenzoate; $n = 4$–9

This enzyme participates in the biosynthesis of ubiquinone.[1] Alternative donor substrates include decaprenyl diphosphate, octaprenyl diphosphate, hexaprenyl diphosphate, and pentaprenyl diphosphate. While nonaprenyl diphosphate is the physiological substrate for the mammalian enzyme, shorter prenyl diphosphate analogues act as the donor substrate for the yeast enzyme.

[1] A. Kalen, E. L. Appelkvist, T. Chojnacki & G. Dallner (1990) *JBC* **265**, 1158.
[2] T. Nishino & H. Rudney (1977) *Biochemistry* **16**, 605.

Selected entries from *Methods in Enzymology* [vol, page(s)]:
General discussion: 110, 327
Assay: 110, 327

2-HYDROXY-1,4-BENZOQUINONE REDUCTASE

This FMN-dependent oxidoreductase [EC 1.6.5.7] catalyzes the reaction of hydroxybenzoquinone with NADH and a proton to produce 1,2,4-trihydroxybenzene and NAD^+.[1]

[1] O. Zaborina, D. L. Daubaras, A. Zago, L. Xun, K. Saido, T. Klem, D. Nikolic & A. M. Chakrabarty (1998) *J. Bacteriol.* **180**, 4667.

4-HYDROXYBENZOYL-CoA REDUCTASE

This oxidoreductase [EC 1.3.99.20] catalyzes the reaction of 4-hydroxybenzoyl-CoA with an acceptor substrate to produce benzoyl-CoA and the reduced acceptor.[1–3]

[1] R. Glockler, A. Tschech & G. Fuchs (1989) *FEBS Lett.* **251**, 237.
[2] K. Breese & G. Fuchs (1998) *EJB* **251**, 916.
[3] R. Brackmann & G. Fuchs (1993) *EJB* **213**, 563.

4-HYDROXYBENZOYL-CoA SYNTHETASE

This ligase [EC 6.2.1.27], also known as 4-hydroxybenzoate:CoA ligase, catalyzes the reaction of ATP with 4-hydroxybenzoate and coenzyme A to produce AMP, pyrophosphate (or, diphosphate), and 4-hydroxybenzoyl-CoA. 4-Aminobenzoate is a weaker alternative substrate with the *Pseudomonas* enzyme.[1]

[1] T. Biegert, U. Altenschmidt, C. Eckerskorn & G. Fuchs (1993) *EJB* **213**, 555.

4-HYDROXYBENZOYL-CoA THIOESTERASE

This thioesterase [EC 3.1.2.23], also called 4-hydroxybenzoyl-CoA hydrolase, catalyzes the hydrolysis of 4-hydroxybenzoyl-CoA to produce 4-hydroxybenzoate and free coenzyme A.[1,2] The enzyme is a component of the bacterial 2,4-dichlorobenzoate degradation pathway.

[1] K.-H. Chang, P.-H. Liang, W. Beck, J. D. Scholten & D. Dunaway-Mariano (1992) *Biochemistry* **31**, 560.
[2] M. M. Benning, G. Wesenberg, R. Liu, K. L. Taylor, D. Dunaway-Mariano & H. M. Holden (1998) *JBC* **273**, 33572.

3-HYDROXYBENZYL-ALCOHOL DEHYDROGENASE

This oxidoreductase [EC 1.1.1.97], also known as *m*-hydroxybenzyl alcohol dehydrogenase and 3-hydroxy-benzaldehyde reductase, catalyzes the reversible reaction of 3-hydroxybenzyl alcohol with $NADP^+$ to produce 3-hydroxybenzaldehyde and NADPH.[1–3] Other substrates include 3-methoxybenzyl alcohol, 4-hydroxybenzyl alcohol, and benzyl alcohol.

[1] P. I. Forrester & G. M. Gaucher (1972) *Biochemistry* **11**, 1108.
[2] R. E. Scott, K. S. Lam & G. M. Gaucher (1986) *Can. J. Microbiol.* **32**, 167.
[3] D. J. Hopper & P. D. Kemp (1980) *J. Bacteriol.* **142**, 21.

Selected entries from *Methods in Enzymology* [vol, page(s)]:
General discussion: 43, 540
Assay: 43, 544; gentisyl-alcohol dehydrogenase activity, 43, 544
Other molecular properties: calculation, 43, 544; cell-free extract, 43, 543; gentisyl alcohol dehydrogenase assay, 43, 544; *m*-[1-^{14}C]hydroxybenzaldehyde assay, 43, 544; inhibitors, 43, 546; kinetic properties, 43, 546; Michaelis-Menten plots, 43, 546; molecular weight, 43, 546; *Penicillium urticae* preparation, 43, 542; pH optimum, 43, 546; Polyclar AT treatment, 43, 543, 545; polyketides, 43, 541; properties, 43, 545; purification, 43, 544, 545; reaction scheme, 43, 541; specificity, 43, 547; unit definition, 43, 544

2-HYDROXYBIPHENYL 3-MONOOXYGENASE

This oxidoreductase [EC 1.14.13.44], first isolated from a strain of *Pseudomonas*, catalyzes the reaction of 2-hydroxybiphenyl with NADH and dioxygen to produce 2,3-dihydroxybiphenyl, NAD^+, and water. The enzyme also converts 2,2′-dihydroxybiphenyl into 2,2′,3-trihydroxybiphenyl.[1]

[1] H.-P. E. Kohler, D. Kohler & D. D. Focht (1988) *Appl. Environ. Microbiol.* **54**, 2683.

3-HYDROXYBUTYRATE DEHYDROGENASE

This oxidoreductase [EC 1.1.1.30] catalyzes the reversible reaction of (*R*)-3-hydroxybutanoate with NAD^+ to produce acetoacetate and NADH.[1] Other 3-hydroxymonocarboxylates can act as substrates as well (however, the enantiomer is not a substrate).

[1] A. R. Clarke & T. R. Dafforn (1998) *CBC* **3**, 1.

Selected entries from *Methods in Enzymology* [vol, page(s)]:
General discussion: 14, 222, 227; 34, 238
Assay: 14, 222, 227
Other molecular properties: acetoacetyl lipoate and, 5, 654; assays for (acetoacetate, 13, 478, 480; D-3-hydroxybutyrate, 13, 476, 478); hepatoma mitochondria, 55, 82; histochemistry, 6, 892; D-3-hydroxybutyrate apodehydrogenase (adsorption chromatography, 32, 389, 390; assay, 32, 378, 379; glass-bead chromatography, 32, 384; lecithin requirement, 32, 374; precipitation, 32, 384; properties, 32, 391; protein assays, 32, 378; release from mitochondria, 32, 381; specific activity, 32, 383); 3-hydroxypropionate dehydrogenase

preparation, present in, **5**, 460; isolation, **4**, 303; liver compartments, in, **37**, 288; *Pseudomonas lemoignei*, from, **14**, 227 (activators and inhibitors, **14**, 230; properties, **14**, 230; purification, **14**, 228; specificity, **14**, 230); *Rhodopseudomonas spheroides*, from, **14**, 222 (inhibitors, **14**, 226; properties, **14**, 225; purification, **14**, 223 [by affinity chromatography on immobilized dyes, **104**, 111]; specificity, **14**, 226); stereospecificity, **54**, 229; **87**, 105

4-HYDROXYBUTYRATE DEHYDROGENASE

This oxidoreductase [EC 1.1.1.61], also known as succinate semialdehyde reductase, catalyzes the reversible reaction of 4-hydroxybutanoate with NAD^+ to produce succinate semialdehyde and NADH.[1,2] An NADPH-dependent succinate semialdehyde reductase has been purified from rat brain and found to have an ordered Bi Bi kinetic mechanism.[3]

[1]C. Matthies, F. Mayer & B. Schink (1989) *Arch. Microbiol.* **151**, 498.
[2]M. W. Nirenberg & W. B. Jakoby (1960) *JBC* **235**, 954.
[3]S. W. Cho, M. S. Song, G. Y. Kim, W. D. Kang, E. Y. Choi & S. Y. Choi (1993) *EJB* **211**, 757.

Selected entries from *Methods in Enzymology* [vol, page(s)]:
General discussion: 5, 778
Assay: 5, 778

HYDROXYBUTYRATE-DIMER HYDROLASE

This enzyme [EC 3.1.1.22] catalyzes the hydrolysis of (*R*)-3-((*R*)-3-hydroxybutanoyloxy)butanoate to produce two molecules of (*R*)-3-hydroxybutanoate (*i.e.*, D-β-hydroxybutyrate).[1-3] Oligomeric esters of 3-hydroxybutyrate are also substrates.

[1]Y. Shirakura, T. Fukui, T. Tanio, K. Nakayama, R. Matsuno & K. Tomita (1983) *BBA* **748**, 331.
[2]Y. Tanaka, T. Saito, T. Fukui, T. Tanio & K. Tomita (1981) *EJB* **118**, 177.
[3]F. P. Delafield, K. E. Cooksey & M. Doudoroff (1965) *JBC* **240**, 4023.

3-HYDROXYBUTYRYL-CoA DEHYDRATASE

This enzyme [EC 4.2.1.55], also known as crotonase, catalyzes the reversible conversion of (3*R*)-3-hydroxybutanoyl-CoA to crotonoyl-CoA and water. Other substrates include crotonoyl-*S*-pantetheine and crotonoyl-*S*-acyl-carrier protein.[1] *See also Enoyl-CoA Hydratase*

[1]G. J. Moskowitz & J. M. Merrick (1969) *Biochemistry* **8**, 2748.

4-HYDROXYBUTYRYL-CoA DEHYDRATASE

This enzyme catalyzes the conversion of 4-hydroxybutanoyl-CoA to crotonoyl-CoA and water.[1] The *Clostridium aminobutyricum* enzyme contains FAD and an iron-sulfur cluster. A transient 4-hydroxy-but-2-enoyl-CoA intermediate has been proposed to form in the reaction mechanism. *See also Enoyl-CoA Hydratase*

[1]U. Scherf & W. Buckel (1993) *EJB* **215**, 421.

3-HYDROXYBUTYRYL-CoA DEHYDROGENASE

This oxidoreductase [EC 1.1.1.157], also known as β-hydroxybutyryl-CoA dehydrogenase, catalyzes the reversible reaction of (*S*)-3-hydroxybutanoyl-CoA with $NADP^+$ to produce 3-acetoacetyl-CoA and NADPH.[1] *See also 3-Hydroxyacyl-CoA Dehydrogenase*

[1]V. K. Madan, P. Hillmer & G. Gottschalk (1973) *EJB* **32**, 51.

3-HYDROXYBUTYRYL-CoA EPIMERASE

This epimerase [EC 5.1.2.3], also called 3-hydroxyacyl-CoA epimerase and often incorrectly referred to as β-hydroxybutyryl-CoA racemase, catalyzes the interconversion of (*S*)-3-hydroxybutanoyl-CoA and (*R*)-3-hydroxybutanoyl-CoA.[1,2] The enzyme, which can use other 3-hydroxyacyl-CoA analogues, acts solely by a dehydration/hydration mechanism proceeding by way of a 2-*trans*-enoyl-CoA intermediate.

[1]T. E. Smeland, J. Li, D. Cuebas & H. Schulz (1992) *Prog. Clin. Biol. Res.* **375**, 85.
[2]A. Pramanik & H. Schulz (1983) *BBA* **750**, 41.

Selected entries from *Methods in Enzymology* [vol, page(s)]:
General discussion: 5, 557; **71**, 403, 404
Assay: 5, 558
Other molecular properties: crotonase and, **5**, 557; distribution, **5**, 559; β-oxidation of fatty acids, in, **14**, 99; properties, **5**, 559

(2R)-HYDROXYCARBOXYLATE:VIOLOGEN OXIDOREDUCTASE

This oxidoreductase catalyzes the reaction of an α-keto acid with two protons and either benzyl viologen or methyl viologen to produce the (2*R*)-hydroxyacid and the oxidized viologen.

Selected entries from *Methods in Enzymology* [vol, page(s)]:
Assay: 136, 312
Other molecular properties: 2-oxo-4-methylpentanoate reduction, **136**, 316; purification from *Proteus vulgaris*, **136**, 311; substrate specificity and kinetic data, **136**, 306

4-HYDROXYCATECHOL 1,2-DIOXYGENASE

This oxidoreductase [EC 1.13.11.44], also known as 6-chlorohydroxyquinol 1,2-dioxygenase and hydroxyquinol 1,2-dioxygenase, catalyzes the reaction of benzene-1,2,4-triol (or, 4-hydroxycatechol) with dioxygen to produce maleylacetate.[1] This reaction is an example of an *ortho*-cleavage. 6-Chlorobenzene-1,2,4-triol (*i.e.*, 6-chlorohydroxyquinol) is an alternative substrate, producing 2-chloromaleylacetate.[1] Interestingly, however, catechol, hydroquinone, and pyrogallol are not substrates.

[1] O. Zaborina, M. Latus, J. Eberspacher, L. A. Golovleva & F. Lingens (1995) *J. Bacteriol.* **177**, 229.

3α-HYDROXYCHOLANATE DEHYDROGENASE

This oxidoreductase [EC 1.1.1.52] catalyzes the reaction of 3α-hydroxy-5β-cholanate (*i.e.*, lithocholate) with NAD⁺ to produce 3-oxo-5β-cholanate and NADH.[1] Other substrates include 3α-hydroxysteroids with an acidic side chain, such as cholate, deoxycholate, dehydrocholate (3,7,12-trioxocholan-24-oate), and 3α-hydroxy-bisnorcholanate.

[1] O. Hayaishi, Y. Sato, W. B. Jakoby & E. F. Stohlman (1955) *ABB* **56**, 554.

27-HYDROXYCHOLESTEROL 7α-MONOOXYGENASE

This heme-thiolate-dependent oxidoreductase [EC 1.14.13.60], also known as 27-hydroxycholesterol 7α-hydroxylase, catalyzes the reaction of 27-hydroxycholesterol with NADPH and dioxygen to produce 7α,27-dihydroxycholesterol, NADP⁺, and water.[1] The mammalian liver enzyme has no activity toward cholesterol, thereby distinguishing it from cholesterol 7α-monooxygenase [EC 1.14.13.17].

[1] K. O. Martin, K. Budai & N. B. Javitt (1993) *J. Lipid Res.* **34**, 581.

6-*endo*-HYDROXYCINEOLE DEHYDROGENASE

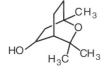

6-*endo*-hydroxycineole   6-oxocineole

This enzyme [EC 1.1.1.241], first isolated from *Rhodococcus* sp., catalyzes the reaction of 6-*endo*-hydroxycineole with NAD⁺ to produce 6-oxocineole and NADH.[1]

[1] D. R. Williams, P. W. Trudgill & D. G. Taylor (1989) *J. Gen. Microbiol.* **135**, 1957.

HYDROXYCINNAMATE 4β-GLUCOSYLTRANSFERASE

This transferase [EC 2.4.1.126] catalyzes the reaction of UDP-D-glucose with *trans*-4-hydroxycinnamate to produce UDP and 4-*O*-β-D-glucosyl-4-hydroxycinnamate.[1] Other acceptor substrates include 4-coumarate, ferulate, caffeate, and sinapate, producing a mixture of the glucoside and the glucose ester. **See also** *Sinapate 1-Glucosyltransferase*

[1] A. Fleuriet & J. J. Macheix (1980) *Z. Naturforsch.* **35c**, 967.

HYDROXYCYCLOHEXANECARBOXYLATE DEHYDROGENASE

This oxidoreductase [EC 1.1.1.166] catalyzes the reversible reaction of (1S,3R,4S)-3,4-dihydroxycyclohexane-1-carboxylate with NAD⁺ to produce (1S,4S)-4-hydroxy-3-oxocyclohexane-1-carboxylate and NADH.[1] The enzyme acts on hydroxycyclohexanecarboxylates having an equatorial carboxyl group at the C1 position, an axial hydroxyl group at C3, and an equatorial hydroxyl or carbonyl group at C4. (–)-Quinate is converted to (–)-3-dehydroquinate, and (–)-shikimate is converted to (–)-3-dehydroshikimate.

[1] G. C. Whiting & R. Coggins (1974) *BJ* **141**, 35.

4-HYDROXYCYCLOHEXANE-CARBOXYLATE DEHYDROGENASE

This oxidoreductase [EC 1.1.1.226] catalyzes the reaction of 4-hydroxycyclohexanecarboxylate with NAD⁺ to produce 4-oxocyclohexanecarboxylate and NADH (B-side).

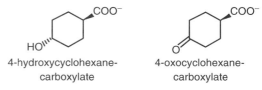

4-hydroxycyclohexane-carboxylate   4-oxocyclohexane-carboxylate

The enzyme isolated from *Corynebacterium cyclohexanicum* did not utilize the *cis* isomer as a substrate. In the reverse reaction, neither the 2-oxo nor the 3-oxo analogues were acted upon by the enzyme.[1]

[1] H. Obata, M. Uebayasi & T. Kaneda (1998) *EJB* **174**, 451.

3-HYDROXYCYCLOHEXANONE DEHYDROGENASE

This oxidoreductase [EC 1.1.99.26] catalyzes the reaction of 3-hydroxycyclohexanone with an acceptor substrate to produce cyclohexane-1,3-dione and the reduced acceptor.[1] The acceptor substrate can be 2,6-dichloroindophenol or methylene blue.

[1] W. Dangel, A. Tschech & G. Fuchs (1989) *Arch. Microbiol.* **152**, 273.

2-HYDROXYCYCLOHEXANONE 2-MONOOXYGENASE

This oxidoreductase [EC 1.14.13.66; previously classified as EC 1.14.12.6] catalyzes the reaction of 2-hydroxy-cyclohexan-1-one with NADPH and dioxygen to produce NADP$^+$, water, and 6-hydroxyhexan-6-olide, the latter decomposing spontaneously to form adipate semialdehyde (*i.e.*, 6-oxohexanoate).[1]

[1] J. F. Davey & P. W. Trudgill (1977) *EJB* **74**, 115.

(1-HYDROXYCYCLOHEXAN-1-YL)ACETYL-CoA LYASE

This enzyme [EC 4.1.3.35] catalyzes the conversion of (1-hydroxycyclohexan-1-yl)acetyl-CoA to acetyl-CoA and cyclohexanone.[1]

[1] H. J. Ougham & P. W. Trudgill (1982) *J. Bacteriol.* **150**, 1172.

2′-HYDROXYDAIDZEIN REDUCTASE

This oxidoreductase [EC 1.3.1.51] catalyzes the reaction of 2′-hydroxydaidzein with NADPH to produce 2′-hydroxydihydrodaidzein and NADP$^+$.[1]

[1] D. Fischer, C. Ebenau-Jehle & H. Grisebach (1990) *ABB* **276**, 390.

ω-HYDROXYDECANOATE DEHYDROGENASE

This oxidoreductase [EC 1.1.1.66] catalyzes the reaction of 10-hydroxydecanoate with NAD$^+$ to produce 10-oxodecanoate and NADH.[1] Other substrates include 9-hydroxynonanoate and 11-hydroxyundecanoate. The NADP$^+$-dependent potato tuber enzyme, also called ω-hydroxy-fatty-acid:NADP$^+$ oxidoreductase, with broader specificity catalyzes the reaction of an ω-hydroxy-fatty acid with NADP$^+$ to produce an ω-oxo-fatty acid and NADPH.

[1] M. A. Mitz & R. L. Heinrikson (1961) *BBA* **46**, 45.

Selected entries from **Methods in Enzymology** [vol, page(s)]:
Other molecular properties: ω-hydroxy-fatty-acid:NADP$^+$ oxidoreductase (assay, **71**, 413; wound-healing potato tuber disks, from, **71**, 411 [inhibitors, **71**, 419; properties, **71**, 417; specificity, **71**, 417, 419; stability, **71**, 417])

3-HYDROXYDECANOYL-[ACYL-CARRIER PROTEIN] DEHYDRATASE

This enzyme [EC 4.2.1.60], also known as decanoyl-thiolester dehydrase, catalyzes the conversion of (3*R*)-3-hydroxydecanoyl-[acyl-carrier protein] to water and *trans*-2-decenoyl-[acyl-carrier protein], which is then converted

to *cis*-3-decenoyl-[acyl-carrier protein].[1] The best substrate appears to be the ten-carbon 3-hydroxyacyl-ACP derivative. The ACP portion of the substrate can be substituted by *N*-acetylcysteamine or pantetheine. The acyl portion of the substrate is very specific with respect to chain length and stereochemistry. The relative amounts of product formed reportedly are dependent on the phosphate buffer concentration.

[1] K. Bloch (1971) *The Enzymes*, 3rd ed., **5**, 441.

Selected entries from **Methods in Enzymology** [vol, page(s)]:
General discussion: **14**, 73; **46**, 158
Assay: optical, **14**, 74; radioactive, **14**, 75
Other molecular properties: deficiency of, in *Escherichia coli* mutants, **32**, 859; *Escherichia coli*, from, **71**, 127 (inhibition by 3-decynoyl-*N*-acetylcysteamine, **14**, 80; **71**, 127; properties, **14**, 79; purification, **14**, 76); inactivator, **53**, 438; inhibition by 3-decynoyl *N*-acetylcysteamine, **14**, 80; **71**, 127; large-scale preparation, **22**, 536; mutant, **72**, 694; product ratios, **14**, 80; substrate specificity, **14**, 80

β-HYDROXYDECANOYL-β-HYDROXYDECANOATE L-RHAMNOSYLTRANSFERASE

This transferase catalyzes the reaction of dTDP-L-rhamnose with β-hydroxydecanoyl-β-hydroxydecanoate to produce α-L-rhamnospyranosyl-β-hydroxydecanoyl-β-hydroxydecanoate and dTDP.[1]

[1] M. M. Burger, L. Glaser & R. M. Burton (1963) *JBC* **238**, 2595.

Selected entries from **Methods in Enzymology** [vol, page(s)]:
General discussion: **8**, 441
Assay: **8**, 441

HYDROXYDECHLOROATRAZINE ETHYLAMINOHYDROLASE

This enzyme [EC 3.5.99.3], also known as hydroxy-atrazine ethylaminohydrolase and often abbreviated AtzB, catalyzes the hydrolysis of 4-(ethylamino)-2-hydroxy-6-(isopropylamino)-1,3,5-triazine to produce *N*-isopropylammelide (or, 2,4-dihydroxy-6-(isopropyl-amino)-1,3,5-triazine) and ethylamine.[1]

4-(ethylamino)-2-hydroxy-6-(isopropylamino)-1,3,5-triazine

N-isopropylammelide

The enzyme participates in the bacterial degradation of the herbicide atrazine, 2-chloro-4-(ethylamino)-6-(isopropylamino)-1,3,5-triazine.

[1]K. L. Boundy-Mills, M. L. de Souza, R. T. Mandelbaum, L. P. Wackett & M. J. Sadowsky (1997) *Appl. Environ. Microbiol.* **63**, 916.

12-HYDROXYDIHYDROCHELIRUBINE 12-O-METHYLTRANSFERASE

This methyltransferase [EC 2.1.1.120] catalyzes the reaction of *S*-adenosyl-L-methionine with 12-hydroxydihydrochelirubine to produce *S*-adenosyl-L-homocysteine and dihydromacarpine.[1]

[1]L. Kammerer, W. De-Eknamkul & M. H. Zenk (1994) *Phytochemistry* **36**, 1409.

10-HYDROXYDIHYDROSANGUINARINE 10-O-METHYLTRANSFERASE

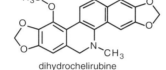

10-hydroxydihydrosanguinarine   dihydrochelirubine

This methyltransferase [EC 2.1.1.119], which participates in the plant biosynthesis of certain benzophenanthridine alkaloids, catalyzes the reaction of *S*-adenosyl-L-methionine with 10-hydroxydihydrosanguinarine to produce *S*-adenosyl-L-homocysteine and dihydrochelirubine.

13-HYDROXYDOCOSANOATE 13-β-GLUCOSYLTRANSFERASE

This transferase [EC 2.4.1.158], also known as 13-glucosyloxydocosanoate 2′-β-glucosyltransferase and UDP-glucose:13-hydroxydocosanoate glucosyltransferase, catalyzes the reaction of UDP-D-glucose with 13-hydroxydocosanoate to produce UDP and 13-β-D-glucosyloxydocosanoate.[1]

[1]T. B. Breithaupt & R. J. Light (1982) *JBC* **257**, 9622.

15-HYDROXYEICOSATETRAENOATE DEHYDROGENASE

This dehydrogenase [EC 1.1.1.232], also known as 15-hydroxyicosatetraenoate dehydrogenase, catalyzes the reaction of (15*S*)-15-hydroxy-5,8,11-*cis*-13-*trans*-eicosatetraenoate with NAD(P)+ to produce 15-oxo-5,8,11-*cis*-13-*trans*-eicosatetraenoate and NAD(P)H. The enzyme utilizes other hydroxylated eicosanoids as substrates.[1]

[1]D. E. Sok, J. B. Kang & H. D. Shin (1988) *BBRC* **156**, 524.

HYDROXYETHYLTHIAZOLE KINASE

This phosphotransferase [EC 2.7.1.50] catalyzes the reaction of ATP with 4-methyl-5-(2-hydroxyethyl)thiazole to produce ADP and 4-methyl-5-(2-phosphoethyl)-thiazole.[1,2] This protein plays a salvage role in the thiamin biosynthetic pathway.

The *Saccharomyces cerevisiae* enzyme is bifunctional, also exhibiting a thiamin-phosphate pyrophosphorylase activity. The crystal structure of the *Bacillus subtilis* enzyme suggests that phosphate transfer occurs by an in-line mechanism.[3]

[1]G. W. Camiener & G. M. Brown (1960) *JBC* **235**, 2411.
[2]Y. Kawasaki (1993) *J. Bacteriol.* **175**, 5153.
[3]N. Campobasso, I. I. Mathews, T. P. Begley & S. E. Ealick (2000) *Biochemistry* **39**, 7868.

(R)-2-HYDROXY-FATTY-ACID DEHYDROGENASE

This oxidoreductase [EC 1.1.1.98], also known as D-2-hydroxy fatty acid dehydrogenase, catalyzes the reaction of (*R*)-2-hydroxystearate with NAD+ to produce 2-ketostearate (or, 2-oxostearate) and NADH.[1]

[1]G. M. Levis (1970) *BBRC* **38**, 470.

(S)-2-HYDROXY-FATTY-ACID DEHYDROGENASE

This oxidoreductase [EC 1.1.1.99], also known as L-2-hydroxy fatty acid dehydrogenase, catalyzes the reaction of (*S*)-2-hydroxystearate with NAD+ to produce 2-ketostearate (or, 2-oxostearate) and NADH.[1]

[1]G. M. Levis (1970) *BBRC* **38**, 470.

5-HYDROXYFURANOCOUMARIN 5-O-METHYLTRANSFERASE

This methyltransferase [EC 2.1.1.69] catalyzes the reaction of *S*-adenosyl-L-methionine with 5-hydroxyfuranocoumarin (*i.e.*, bergaptol) to produce *S*-adenosyl-L-homocysteine and 5-methoxyfuranocoumarin (*i.e.*, bergapten).[1–3]

bergaptol   bergapten

The enzyme methylates the 5-hydroxyl of some hydroxy- and methylcoumarins, such as 5-hydroxyxanthotoxin, but it has little activity on non-coumarin phenols.

[1] H. J. Thompson, S. K. Sharma & S. A. Brown (1978) ABB 188, 272.
[2] S. K. Sharma, J. M. Garett & S. A. Brown (1979) Z. Naturforsch. 34, 387.
[3] S. K. Sharma & S. A. Brown (1979) Can. J. Biochem. 57, 986.

8-HYDROXYFURANOCOUMARIN 8-O-METHYLTRANSFERASE

This methyltransferase [EC 2.1.1.70] catalyzes the reaction of S-adenosyl-L-methionine with 8-hydroxy-furanocoumarin (i.e., xanthotoxol) to produce S-adenosyl-L-homocysteine and 8-methoxyfuranocoumarin (i.e., xanthotoxin).[1-3]

xanthotoxol   xanthotoxin

The enzyme catalyzes the methylation of the 8-hydroxyl group of a number of hydroxy- and methylcoumarins, such as, 8-hydroxybergapten and 7,8-dihydroxycoumarin, but has little activity on non-coumarin phenols.

[1] H. J. Thompson, S. K. Sharma & S. A. Brown (1978) ABB 188, 272.
[2] S. K. Sharma, J. M. Garett & S. A. Brown (1979) Z. Naturforsch. 34, 387.
[3] S. K. Sharma & S. A. Brown (1979) Can. J. Biochem. 57, 986.

4-HYDROXYGLUTAMATE AMINOTRANSFERASE

This aminotransferase [EC 2.6.1.23] catalyzes the reversible reaction of 4-hydroxy-L-glutamate with α-ketoglutarate (or, 2-oxoglutarate) to produce 4-hydroxy-2-ketoglutarate (or, 4-hydroxy-2-oxoglutarate) and L-glutamate.[1,2] Other amino acceptor substrates include oxaloacetate. In a number of organisms, this enzyme may be identical with aspartate aminotransferase [EC 2.6.1.1].

[1] A. Goldstone & E. Adams (1962) JBC 237, 3476.
[2] K. Kuratomi, K. Fukunaga & Y. Kobayashi (1963) BBA 78, 629.

HYDROXYGLUTAMATE DECARBOXYLASE

This pyridoxal-phosphate-dependent enzyme [EC 4.1.1.16] catalyzes the conversion of 3-hydroxy-L-glutamate to 4-amino-3-hydroxybutanoate and carbon dioxide.[1,2] The enzyme also reportedly exhibits glutamate decarboxylase activity.

[1] A. D. Homola & E. E. Dekker (1967) Biochemistry 6, 2626.
[2] W. W. Umbreit & P. Heneage (1953) JBC 201, 15.

2-HYDROXYGLUTARATE DEHYDROGENASE

This oxidoreductase [EC 1.1.99.2] catalyzes the reaction of (S)-2-hydroxyglutarate (or, L-2-hydroxyglutarate) with an acceptor substrate to produce α-ketoglutarate (or, 2-oxoglutarate) and the reduced acceptor.[1-3] Acceptor substrates include pyocyanine.

NAD(P)$^+$-specific 2-hydroxyglutarate dehydrogenases should be classified under EC 1.1.1.x.

[1] T. Suzuki, T. Uozomi & T. Beppu (1985) Agric. Biol. Chem. 49, 2939.
[2] S. Otawara, T. Oshima, N. Esaki & K. Soda (1984) Agric. Biol. Chem. 48, 1713.
[3] H. Weil-Malherbe (1937) BJ 31, 2080.

2-HYDROXYGLUTARATE SYNTHASE

This enzyme [EC 4.1.3.9] catalyzes the reaction of propanoyl-CoA with water and glyoxylate to produce 2-hydroxyglutarate and coenzyme A.[1,2]

[1] W. S. Wegener, H. C. Reeves & S. J. Ajl (1968) ABB 123, 62.
[2] H. C. Reeves & S. J. Ajl (1962) J. Bacteriol. 84, 186.

Selected entries from **Methods in Enzymology [vol, page(s)]:**
**General discussion: 13, 362
Assay: 13, 362**

(R)-2-HYDROXYGLUTARYL-CoA DEHYDRATASE

This enzyme catalyzes the conversion of (R)-2-hydroxyglutaryl-CoA to glutaconyl-CoA and water. The Acidaminococcus fermentans enzyme contains both FAD and an iron-sulfur cluster.[1] The reaction probably proceeds by a one-electron reduction to produce a ketyl anion radical intermediate which eliminates OH$^-$ to generate an enoxy radical. Deprotonation would produce a ketyl radical that is then oxidized by a second electron.[1] **See also** Enoyl-CoA Hydratase; 4-Hydroxybutyryl-CoA Dehydratase

[1] U. Müller & W. Buckel (1995) EJB 230, 698.

3α-HYDROXYGLYCYRRHETINATE DEHYDROGENASE

This oxidoreductase [EC 1.1.1.230] catalyzes the reaction of 3α-hydroxyglycyrrhetinate with NADP$^+$ to produce 3-oxoglycyrrhetinate and NADPH.

This enzyme, which is distinct from 3α-hydroxysteroid dehydrogenase (B-specific) [EC 1.1.1.50][1,2], is very specific for 3α-hydroxy derivatives.

3α-hydroxyglycyrrhetinate   3-oxoglycyrrhetinate

[1] T. Akao, T. Akao, M. Hattori, T. Namba & K. Kobashi (1988) *J. Biochem.* **103**, 504.

[2] T. Akao, T. Akao & K. Kobashi (1990) *BBA* **1042**, 241.

6-HYDROXYHEXANOATE DEHYDROGENASE

This oxidoreductase [EC 1.1.1.258], which participates in the degradation of cyclohexanol in *Acinetobacter*, catalyzes the reaction of 6-hydroxyhexanoate with NAD^+ to produce 6-oxohexanoate (or, adipate semialdehyde), NADH, and a proton.[1,2]

[1] N. A. Donoghue & P. W. Trudgill (1975) *EJB* **60**, 1.

[2] L. I. Hecker, Y. Tondeur & J. G. Farrelly (1984) *Chem. Biol. Interact.* **49**, 235.

6β-HYDROXYHYOSCYAMINE EPOXIDASE

This ascorbate- and iron-dependent oxidoreductase [EC 1.14.11.14], also known as hydroxyhyoscyamine dioxygenase, catalyzes the reaction of (6*S*)-hydroxyhyoscyamine with α-ketoglutarate (or, 2-oxoglutarate) and dioxygen to produce scopolamine, succinate, and carbon dioxide.[1]

[1] T. Hashimoto, J. Kohno & Y. Yamada (1989) *Phytochemistry* **28**, 1077.

3-HYDROXYISOBUTYRATE DEHYDROGENASE

This oxidoreductase [EC 1.1.1.31] catalyzes the reaction of 3-hydroxy-2-methylpropanoate with NAD^+ to produce 2-methyl-3-oxopropanoate (*i.e.*, methylmalonyl semialdehyde) and NADH. The *S*-stereoisomer is a significantly better substrate than its enantiomer. The rabbit liver enzyme has an ordered Bi Bi kinetic mechanism and is strongly product inhibited by NADH.[1,2]

[1] P. M. Rougraff, R. Paxton, M. J. Kuntz, D. W. Crabb & R. A. Harris (1988) *JBC* **263**, 327.

[2] J. W. Hawes, D. W. Crabb, R. M. Chan, P. M. Rougraff & R. A. Harris (1995) *Biochemistry* **34**, 4231.

Selected entries from *Methods in Enzymology* [vol, page(s)]:
General discussion: 5, 453; 324, 218
Assay: 5, 453

Other molecular properties: 3-hydroxypropionate dehydrogenase preparation, presence in, 5, 460; properties, 5, 456; purification, pig kidney, 5, 454

3-HYDROXYISOBUTYRYL-CoA HYDROLASE

This enzyme [EC 3.1.2.4] catalyzes the hydrolysis of 3-hydroxy-2-methylpropanoyl-CoA (*i.e.*, 3-hydroxyiso-butyryl-CoA) to produce coenzyme A and 3-hydroxy-2-methylpropanoate (*i.e.*, 3-hydroxyisobutyrate).[1] Other substrates include 3-hydroxypropanoyl-CoA. This enzyme participates in valine catabolism. The high liver concentration of this enzyme is thought to protect cells against the toxic effects of methacrylyl-CoA, which is in equilibrium with (*S*)-3-hydroxyisobutyryl-CoA.[2]

[1] G. Rendina & M. J. Coon (1957) *JBC* **225**, 523.

[2] Y. Shimomura, T. Murakami, N. Fujitsuka, N. Nakai, Y. Sato, S. Sugiyama, N. Shimomura, J. Irwin, J. W. Hawes & R. A. Harris (1994) *JBC* **269**, 14248.

Selected entries from *Methods in Enzymology* [vol, page(s)]:
General discussion: 324, 229

L-2-HYDROXYISOCAPROATE DEHYDROGENASE

L-2-Hydroxyisocaproate dehydrogenase, a tetrameric oxidoreductase, catalyzes the reversible reaction of L-2-hydroxyisocaproate with NAD^+ to produce 2-ketoiso-caproate (*i.e.*, 2-oxoisocaproate) and NADH. The enzyme has a broad specificity and can act on a wide range of α-hydroxy acids branched at the C4 position.[1]

[1] I. K. Feil, H. P. Lerch & D. Schomburg (1994) *EJB* **223**, 857.

2′-HYDROXYISOFLAVONE REDUCTASE

This oxidoreductase [EC 1.3.1.45] catalyzes the reaction of 2′-hydroxyformononetin, a 2′-hydroxyisoflavone, with NADPH to produce the isoflavonone known as vestitone and $NADP^+$.[1-3] Other substrates include 2′-hydroxypseudo-baptigenin (generating 7-hydroxy-3′,4′-methylenedioxy-isoflavone) and 2′,7-dihydroxy-4′,5′-methylenedioxyiso-flavone (producing sophorol). The reductase is A-side specific with respect to the pyridine coenzyme.

[1] K. Tiemann, W. Hinderer & W. Barz (1987) *FEBS Lett.* **213**, 324.

[2] D. Schlieper, K. Tiemann & W. Barz (1990) *Phytochemistry* **29**, 1519.

[3] C. L. Preisig, J. N. Bell, Y. Sun, G. Hrazdina, D. E. Matthews & H. D. Van Etten (1990) *Plant Physiol.* **94**, 1444.

3-HYDROXY-3-ISOHEXENYLGLUTARYL-CoA LYASE

This lyase [EC 4.1.3.26] catalyzes the conversion of 3-hydroxy-3-(4-methylpent-3-en-1-yl)glutaryl-CoA (*i.e.*, 3-hydroxy-3-isohexenylglutaryl-CoA) to 7-methyl-3-ketooctanoyl-CoA (*i.e.*, 7-methyl-3-oxooctanoyl-CoA) and acetate.[1] Other substrates include the hydroxy derivative of farnesoyl-CoA. The product has not only undergone an elimination reaction, but instead appears to have also lost its carbon–carbon double bond.

[1] W. Seubert & E. Fass (1964) *Biochem. Z.* **341**, 23.

4-HYDROXY-α-KETOBUTYRATE ALDOLASE

This aldolase activity, previously classified as EC 4.1.2.1, is now listed under 4-hydroxy-2-ketoglutarate aldolase [EC 4.1.3.16].

4-HYDROXY-2-KETOGLUTARATE ALDOLASE

This aldolase [EC 4.1.3.16], also known as 4-hydroxy-2-oxoglutarate aldolase, 2-oxo-4-hydroxyglutarate aldolase, and 2-keto-3-deoxyhexarate aldolase, catalyzes the reversible conversion of either stereoisomer of 4-hydroxy-2-ketoglutarate (or, 4-hydroxy-2-oxoglutarate) to pyruvate and glyoxylate.[1–5] This enzyme also catalyzes the aldol cleavage of 2-keto-4-hydroxybutyrate (or, 4-hydroxy-2-oxobutyrate) to pyruvate and formaldehyde. The enzyme also catalyzes β-decarboxylation of oxaloacetate.

$$
\begin{array}{c}
COO^- \\
\Vert O \\
H \!-\!\!\!-\! H \\
HO \!-\!\!\!-\! H \\
COO^-
\end{array}
$$

4-hydroxy-2-keto-D-glutarate

Incubation of the enzyme with either pyruvate or glyoxylate in the presence of $NaBH_4$ results in loss of aldolase activity.[1] The *Escherichia coli* enzyme, which has a "lysine-type" Schiff-base mechanism, has a catalytically essential glutamyl residue that may function as the amphoteric proton donor/acceptor in the reaction mechanism.[2]

In many organisms, this enzyme is identical to 6-phospho-2-dehydro-3-deoxygluconate aldolase [EC 4.1.2.14]. *See 6-Phospho-2-Dehydro-3-Deoxygluconate Aldolase*

[1] E. E. Dekker & R. P. Kitson (1992) *JBC* **267**, 10507.
[2] C. J. Vlahos & E. E. Dekker (1990) *JBC* **265**, 20384.
[3] M. Anderson, J. M. Scholtz & S. M. Schuster (1985) *ABB* **236**, 82.
[4] J. M. Scholtz & S. M. Schuster (1984) *Bioorg. Chem.* **12**, 229.
[5] W. A. Wood (1972) *The Enzymes*, 3rd ed., **7**, 281.

Selected entries from *Methods in Enzymology* [vol, page(s)]:
General discussion: 17B, 280; 42, 280, 285
Assay: 17B, 280; bovine liver, 42, 280, 281; *Escherichia coli*, 42, 286
Other molecular properties: bovine liver, from, 42, 280 (assay, 42, 280, 281; chromatography, 42, 282; equilibrium constant, 42, 285; inhibition, 42, 284; molecular weight, 42, 284; properties, 42, 284, 285; purification, 42, 281); *Escherichia coli*, from, 42, 285 (assay, 42, 286; chromatography, 42, 288; molecular weight, 42, 289; properties, 42, 289, 290; purification, 42, 286); rat liver (molecular weight, 17B, 283; properties, 17B, 283; purification, 17B, 281); sources, other, 17B, 284; stereochemistry, 87, 147

2-HYDROXY-6-KETO-6-PHENYLHEXA-2,4-DIENOATE REDUCTASE

This oxidoreductase [EC 1.3.1.40], also known as 2-hydroxy-6-oxo-6-phenylhexa-2,4-dienoate reductase, catalyzes the reaction of 2-hydroxy-6-keto-6-phenylhexa-2,4-dienoate (or, 2-hydroxy-6-oxo-6-phenylhexa-2,4-dienoate) with NADPH to produce 2,6-diketo-6-phenylhexanoate (or, 2,6-dioxo-6-phenylhexanoate) and $NADP^+$.[1]

[1] T. Omori, H. Ishigooka & Y. Minoda (1988) *Agric. Biol. Chem.* **52**, 503.

4-HYDROXY-2-KETOPIMELATE ALDOLASE

This enzyme catalyzes the conversion of 4-hydroxy-2-pimelate (*i.e.*, 2-oxo-4-hydroxyheptanedioate) to pyruvate and succinate semialdehyde.

Selected entries from *Methods in Enzymology* [vol, page(s)]:
General discussion: 90, 277
Assay: bacterial, 90, 277

HYDROXYLAMINE OXIDASE

This heme-dependent oxidoreductase [EC 1.7.3.4] catalyzes the four-electron reaction of hydroxylamine with dioxygen (or water) to produce nitrite and water.[1–4] The oxidase contains seven *c*-type hemes and one P460-type heme per 63 kDa subunit. This enzyme is one of the most complex hemoproteins known.[1] The P460 heme has very unusual spectroscopic properties, and after reduction, the Soret band, is at 463 nm. *See also Ammonia Monooxygenase*

[1] R. C. Prince & G. N. George (1997) *Nat. Struct. Biol.* **4**, 247.
[2] R. C. Prince & A. B. Hooper (1987) *Biochemistry* **26**, 970.
[3] K. K. Andersson, T. A. Kent, J. D. Lipscomb, A. B. Hooper & E. Münck (1984) *JBC* **259**, 6833.
[4] A. B. Hooper, M. Tran & C. Balny (1984) *EJB* **141**, 565.

Selected entries from *Methods in Enzymology* [vol, page(s)]:
General discussion: 52, 21; 56, 475

HYDROXYLAMINE REDUCTASE

This flavoprotein [EC 1.7.99.1] catalyzes the reaction of hydroxylamine with a donor substrate to produce ammonia, water, and the oxidized donor.[1,2] Reduced pyocyanine,

methylene blue, and flavins can act as the donor substrate.
See also Hydroxylamine Reductase (NADH)

[1] C. F. Cresswell, R. H. Hageman, E. J. Hewitt & D. P. Hucklesby (1965) *BJ* **94**, 40.
[2] G. C. Walker & D. J. D. Nicholas (1961) *BBA* **49**, 361.

HYDROXYLAMINE REDUCTASE (NADH)

This oxidoreductase [EC 1.6.6.11] catalyzes the reaction of hydroxylamine with NADH to produce ammonia, water, and NAD^+.[1-5] Hydroxamates, including *N*-methylhydroxylamine, *N,N*-dimethylhydroxylamine, and *N*-phenylhydroxylamine, are substrates.

[1] R. Wang & D. J. D. Nicholas (1986) *Phytochemistry* **25**, 2463.
[2] M. L. C. Bernheim (1972) *Enzymologia* **43**, 167.
[3] M. L. C. Bernheim (1969) *Arch. Biochem. Biophys.* **134**, 408.
[4] G. G. Roussos & A. Nason (1960) *JBC* **235**, 2997.
[5] M. Zucker & A. Nason (1955) *JBC* **213**, 463.

Selected entries from *Methods in Enzymology* [vol, page(s)]:
General discussion: **2**, 416
Assay: **2**, 416
Other molecular properties: adaptive nature of, **2**, 411, 417; *Neurospora crassa* (properties, **2**, 418; purification, **2**, 417); nitroaryl reductase preparation, present in, **2**, 410, 411; stereospecificity, **87**, 119

13-HYDROXYLUPININE O-TIGLOYLTRANSFERASE

This acyltransferase [EC 2.3.1.93], which participates in the biosynthesis of lupinine alkaloids, catalyzes the reaction of (*E*)-2-methylcrotonoyl-CoA (or, tigloyl-CoA) with 13-hydroxylupinine to produce coenzyme A and 13-(2-methylcrotonoyl)oxylupinine.[1] Alternative acyl-CoA substrates include benzoyl-CoA, pentanoyl-CoA, 3-methylbutanoyl-CoA, and butanoyl-CoA.

[1] M. Wink & T. Hartmann (1982) *Planta* **156**, 560.

N^6-HYDROXYLYSINE O-ACETYLTRANSFERASE

This acetyltransferase [EC 2.3.1.102] catalyzes the reaction of acetyl-CoA with N^6-hydroxy-L-lysine to produce N^6-acetyl-N^6-hydroxy-L-lysine and coenzyme A. The enzyme participates in aerobactin biosynthesis from L-lysine in certain microorganisms.[1,2]

[1] M. Coy, B. H. Paw, A. Bindereif & J. B. Neilands (1986) *Biochemistry* **25**, 2485.
[2] V. DeLorenzo, A. Bindereif, B. H. Paw & J. B. Neilands (1986) *J. Bacteriol.* **165**, 570.

HYDROXYLYSINE KINASE

This phosphotransferase [EC 2.7.1.81] catalyzes the reaction of GTP with 5-hydroxy-L-lysine to produce GDP and 5-phosphonooxy-L-lysine.[1,2] Acceptor substrates include 5-hydroxy-L-lysine and its 5-epimer, but no activity was observed with the D-stereoisomers.

[1] R. A. Hiles & L. M. Henderson (1972) *JBC* **247**, 646.
[2] A. Y. Chang (1977) *Enzyme* **22**, 230.

HYDROXYMALONATE DEHYDROGENASE

This oxidoreductase [EC 1.1.1.167], also known as tartronate dehydrogenase, catalyzes the reaction of hydroxymalonate (or, tartronate) with NAD^+ to produce ketomalonate (or, oxomalonate or mesoxalate) and NADH.[1,2]

[1] N. I. Zhukova (1975) *Biol. Nauki (Mosc.)* **18**, 113.
[2] Y. B. Philippovich, S. M. Klumova & N. I. Zhukova (1974) *Dokl. Akad. Nauk S.S.S.R.* **217**, 241.

4-HYDROXYMANDELATE OXIDASE

This FAD- and manganese-dependent oxidoreductase [EC 1.1.3.19], first isolated from *Pseudomonas convexa*, catalyzes the reaction of (*S*)-2-hydroxy-2-(4-hydroxyphenyl)acetate (or, L-4-hydroxymandelate) with dioxygen to produce 4-hydroxybenzaldehyde, carbon dioxide, and hydrogen peroxide.[1,2]

[1] S. G. Bhat & C. S. Vaidyanathan (1976) *EJB* **68**, 323.
[2] S. G. Bhat & C. S. Vaidyanathan (1976) *J. Bacteriol.* **127**, 1108.

HYDROXYMANDELONITRILE GLUCOSYLTRANSFERASE

This transferase [EC 2.4.1.178], also known as cyanohydrin glucosyltransferase, catalyzes the reaction of UDP-D-glucose with 4-hydroxymandelonitrile to produce UDP and taxiphyllin. The catalyzed reaction is the last step in the biosynthesis of cyanogenic glucosides. 3,4-Dihydroxymandelonitrile is an alternative substrate.[1]

[1] W. Hösel & O. Schiel (1984) *ABB* **229**, 177.

HYDROXYMANDELONITRILE LYASE

This enzyme [EC 4.1.2.11], also known as hydroxymandelonitrile lyase or hydroxynitrile lyase, catalyzes the conversion of (*S*)-4-hydroxymandelonitrile to cyanide and 4-hydroxybenzaldehyde.[1,2] Aliphatic hydroxynitriles do not serve as substrates for this enzyme, unlike that of mandelonitrile lyase [EC 4.1.2.10] and (*S*)-hydroxynitrilase [EC 4.1.2.39]. Vanillin cyanohydrin is also a substrate.[3]

[1] C. Bové & E. E. Conn (1961) *JBC* **236**, 207.
[2] M. K. Seely, R. S. Criddle & E. E. Conn (1966) *JBC* **241**, 4457.
[3] H. Wajant & F. Effenberger (1996) *Biol. Chem.* **377**, 611.

Selected entries from **Methods in Enzymology** [**vol**, page(s)]:
General discussion: 17B, 239
Assay: 17B, 239

6-HYDROXYMELLEIN O-METHYLTRANSFERASE

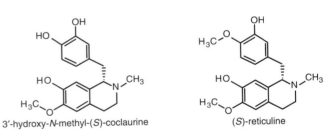

6-hydroxymellein   6-methoxymellein

This methyltransferase [EC 2.1.1.108] catalyzes the reaction of S-adenosyl-L-methionine with 6-hydroxymellein to produce S-adenosyl-L-homocysteine and 6-methoxymellein, a phytoalexin in carrots.[1,2] 3,4-Dehydro-6-hydroxymellein is an alternative substrate.

[1] F. Kurosaki & A. Nishi (1988) FEBS Lett. 227, 183.
[2] F. Kurosaki, Y. Kizawa & A. Nishi (1989) Phytochemistry 28, 1843.

2-(HYDROXYMETHYL)-3-(ACETAMIDOMETHYLENE)-SUCCINATE HYDROLASE

This enzyme [EC 3.5.1.66] catalyzes the hydrolysis of 2-(hydroxymethyl)-3-(acetamidomethylene)succinate with two water molecules to produce acetate, 2-(hydroxymethyl)-4-oxobutanoate, ammonia, and carbon dioxide.[1]

[1] M. S. Huynh & E. E. Snell (1985) JBC 260, 2379.

4-(HYDROXYMETHYL)BENZENE-SULFONATE DEHYDROGENASE

This oxidoreductase [EC 1.1.1.257] catalyzes the reaction of 4-(hydroxymethyl)benzenesulfonate with NAD^+ to produce 4-formylbenzenesulfonate, NADH, and a proton.[1]

[1] F. Junker, E. Saller, H. R. Schläfli Oppenberg, P. M. Kroneck, T. Leisinger & A. M. Cook (1996) Microbiology 142, 2419.

3-HYDROXY-2-METHYLBUTYRYL-CoA DEHYDROGENASE

This oxidoreductase [EC 1.1.1.178], also known as 2-methyl-3-hydroxybutyryl-CoA dehydrogenase, catalyzes the reaction of (2S,3S)-3-hydroxy-2-methylbutanoyl-CoA with NAD^+ to produce 2-methylacetoacetyl-CoA and NADH.[1] Other substrates include (2S,3S)-2-hydroxy-3-methylpentanoyl-CoA and 3-hydroxybutanoyl-CoA.

[1] R. S. Conrad, L. K. Massey & J. R. Sokatch (1974) J. Bacteriol. 4, 103.

3-HYDROXYMETHYLCEPHEM CARBAMOYLTRANSFERASE

This ATP-dependent transferase [EC 2.1.3.7] catalyzes the reaction of carbamoyl phosphate with a 3-hydroxymethylceph-3-em-4-carboxylate to produce orthophosphate and a 3-carbamoyloxymethylcephem.[1] Substrates include a wide variety of 3-hydroxymethylcephems, a subclass of the cephalosporin antibiotics; these include (3-hydroxymethyl)deacetylcephalosporin C, decarbamoylcefuroxime, and 7β-acylamino-3-hydroxymethylcephem.

[1] S. J. Brewer, P. M. Taylor & M. K. Turner (1980) BJ 185, 555.

3β-HYDROXY-4α-METHYLCHOLESTENE-CARBOXYLATE 3-DEHYDROGENASE (DECARBOXYLATING)

This oxidoreductase [EC 1.1.1.170], also known as sterol 4α-carboxylic decarboxylase, catalyzes the reaction of 3β-hydroxy-4α-methyl-5α-cholest-7-ene-4β-carboxylate with NAD^+ to produce 4α-methyl-5α-cholest-7-en-3-one, carbon dioxide, and NADH.[1] Other substrates include 3β-hydroxy-5α-cholest-7-ene-4α-carboxylate.

[1] A. D. Rahimtula & J. L. Gaylor (1972) JBC 247, 9.

3′-HYDROXY-N-METHYL-(S)-COCLAURINE 4′-O-METHYLTRANSFERASE

3′-hydroxy-N-methyl-(S)-coclaurine   (S)-reticuline

This methyltransferase [EC 2.1.1.116], a component of plant isoquinoline biosynthesis, catalyzes the reaction of S-adenosyl-L-methionine with 3′-hydroxy-N-methyl-(S)-coclaurine to produce S-adenosyl-L-homocysteine and (S)-reticuline. Alternative substrates include (R,S)-laudanosoline, (S)-3′-hydroxycoclaurine, and (R,S)-7-O-methylnoraudanosoline.[1]

[1] T. Frenzel & M. H. Zenk (1990) Phytochemistry 29, 3505.

HYDROXYMETHYLGLUTARYL-CoA HYDROLASE

This enzyme [EC 3.1.2.5] catalyzes the hydrolysis of (S)-3-hydroxy-3-methylglutaryl-CoA to produce coenzyme A and 3-hydroxy-3-methylglutarate.[1,2] *See also* *Hydroxymethylglutaryl-CoA Lyase*

[1] A. B. Sipat & J. R. Sabine (1981) *BBA* **666**, 181.
[2] E. E. Dekker, M. J. Schlesinger & M. Coon (1958) *JBC* **233**, 434.

3-HYDROXY-3-METHYLGLUTARYL-CoA LYASE

This lyase [EC 4.1.3.4], also known as β-hydroxy-β-methylglutaryl-CoA cleavage enzyme, catalyzes the conversion of (*S*)-3-hydroxy-3-methylglutaryl-CoA to acetyl-CoA and acetoacetate in a Claisen-type reaction.[1–4] The enzyme also catalyzes the enolization of acetyl-CoA.

[1] D. J. Creighton & N. S. R. K. Murthy (1990) *The Enzymes*, 3rd ed., **19**, 323.
[2] P. R. Kramer & H. M. Miziorko (1983) *Biochemistry* **22**, 2353.
[3] P. R. Kramer & H. M. Miziorko (1980) *JBC* **255**, 11023.
[4] M. J. P. Higgins, J. A. Kornblatt & H. Rudney (1972) *The Enzymes*, 3rd ed., **7**, 407.

Selected entries from *Methods in Enzymology* [vol, page(s)]:
General discussion: 5, 900; 17A, 823; 71, 498, 504; **166**, 219; **324**, 139, 150
Assay: 5, 900
Other molecular properties: interference in assay of HMG-CoA synthase, **35**, 157, 162; β-methylcrotonyl carboxylase, assay for, **5**, 897; properties, **5**, 902; purification from (bovine liver, **17A**, 825; pig heart, **5**, 901); stereochemistry, **87**, 147

3-HYDROXY-3-METHYLGLUTARYL-CoA REDUCTASE

β-Hydroxy-β-methylglutaryl-CoA reductase (NADPH) [EC 1.1.1.34] catalyzes the reaction of (*S*)-3-hydroxy-3-methylglutaryl-CoA with two molecules of NADPH to produce (*R*)-mevalonate, coenzyme A, and two $NADP^+$ molecules.[1–5] The enzyme is inactivated by [3-hydroxy-3-methylglutaryl-CoA reductase (NADPH)] kinase [EC 2.7.1.109] and reactivated by [hydroxymethylglutaryl-CoA reductase (NADPH)]-phosphatase [EC 3.1.3.47]. The mammalian reductase catalyzes the rate-limiting step in sterol and ubiquinone biosynthesis and is a therapeutic target for lowering plasma cholesterol.

β-Hydroxy-β-methylglutaryl-CoA reductase (NADH) [EC 1.1.1.88], first isolated from *Mycobacterium* sp. (S4) and *Pseudomonas mevalonii*, catalyzes the reversible reaction of 3-hydroxy-3-methylglutaryl-CoA with two molecules of NADH to produce (*S*)-mevalonate, coenzyme A, and two NAD^+ molecules.

[1] K. Frimpong & V. W. Rodwell (1994) *JBC* **269**, 11478.
[2] D. M. Gibson & R. A. Parker (1987) *The Enzymes*, 3rd ed., **18**, 179.
[3] T.-Y. Chang (1983) *The Enzymes*, 3rd ed., **16**, 491.
[4] V. W. Rodwell, D. J. McNamara & D. J. Shapiro (1973) *AE* **38**, 373.
[5] T. Kawachi & H. Rudney (1970) *Biochemistry* **9**, 1700.

Selected entries from *Methods in Enzymology* [vol, page(s)]:
General discussion: 71, 480; 324, 259
Assay: 71, 488, 498, 499 (dose response, 71, 507; mononuclear cells, in, 71, 501); *Hevea brasiliensis* latex, in, 110, 44; human fibroblast, assay, 129, 556; *Manduca sexta*, 110, 52; murine macrophage, 129, 556; NADPH-dependent, assay in cultured cells, 98, 255; pea seedlings, microsomal and plastid, 110, 30, 36; *Pseudomonas*, from, 71, 481; rat liver, from, 71, 463
Other molecular properties: activation by endogenous phosphatase, 71, 499; cross-reactivity among species, 74, 328; diurnal variation of activity, 71, 500; gene expression, regulation, 128, 59; *Hevea brasiliensis* latex, in (assay, 110, 44; characteristics, 110, 46; collection, 110, 41; isolation, 110, 43; measurement, 110, 45); human fibroblast, assay, 129, 556; immunodiffusion studies, 74, 327; immunotitration, 74, 320 (antibody production, 74, 325; assay, mevalonate formation, 74, 330; procedure, 74, 329; soluble extract, preparation, 74, 328); K_m for HMG-CoA, 71, 507, 508; kinetics, 87, 357; *Manduca sexta* (assay, 110, 52; product isolation, 110, 57; properties, 110, 54; purification, 110, 53); mechanism, 87, 357; murine macrophage, assay, 129, 556; pea seedlings (compartmentation, 110, 27; isolation variables, 110, 28; microsomal and plastid [assays, 110, 30, 36; interrelations, 110, 39; isolation, 110, 34, 38; properties, 110, 35, 39]); phosphatase inhibitors, 71, 499; photoinduction of the NADPH-dependent enzyme, 214, 275; *Pseudomonas*, from, 71, 480 (assay, 71, 481; physical and kinetic properties, 71, 483; stoichiometry, 71, 484); purification, 74, 320 (ammonium sulfate precipitation, 74, 322; chromatographic purification, 74, 322; enzyme solubilization, 74, 321; homogeneity, 74, 325; microsome isolation, 74, 320; resin variability, 74, 322); rat liver, from, 71, 462 (assays, 71, 463; methods, 71, 465, 467, 472, 475, 477 [comparison, 71, 478]; orientation, 71, 471; properties, 71, 468; solubilization, 71, 466); regulation by disulfide bond oxidation, 251, 13, 26; regulation *in vitro*, analysis with oxysterols, 110, 11; reversible phosphorylation, 71, 486 (HMG-CoA reductase assay, 71, 488; HMG-CoA reductase kinase assay, 71, 489; HMG-CoA reductase kinase kinase assay, 71, 490; HMG-CoA reductase phosphatase assay, 71, 492; phosphorylase phosphatase assay, 71, 491; reductase kinase a, 71, 495; reductase kinase b, 71, 496; reductase kinase phosphatase assay, 71, 493); specific activity and recovery, 71, 500; stereospecificity, 87, 105; subcellular distribution, 71, 499

[HYDROXYMETHYLGLUTARYL-CoA REDUCTASE (NADPH)] KINASE

This phosphotransferase [EC 2.7.1.109], also known simply as reductase kinase, catalyzes the reaction of ATP with the enzyme 3-hydroxy-3-methylglutaryl-CoA reductase (NADPH), at a seryl residue, producing ADP and the phosphorylated form of the reductase.[1–4] This phosphorylation inactivates the reductase. The enzyme, which is activated by AMP, also exhibits a histone kinase activity.

[1] J. Weekes, K. L. Ball, F. B. Caudwell & D. G. Hardie (1993) *FEBS Lett.* **334**, 335.
[2] Z. H. Beg, J. A. Stonik & H. B. Brewer (1987) *JBC* **262**, 13228.
[3] H. J. Harwood, K. G. Brandt & V. W. Rodwell (1984) *JBC* **259**, 2810.
[4] Z. H. Beg, J. A. Stonik & H. B. Brewer, Jr. (1978) *PNAS* **75**, 3678.

Selected entries from *Methods in Enzymology* [vol, page(s)]:
General discussion: 200, 362
Assay: 71, 489, 495, 496; **200**, 364
Other molecular properties: identity with AMP-activated protein kinase, **200**, 362; properties and function, **107**, 95; **200**, 368; purification, **200**, 367; purity and stability, **200**, 368

{[HYDROXYMETHYLGLUTARYL-CoA REDUCTASE (NADPH)] KINASE} PHOSPHATASE

This enzyme catalyzes the hydrolysis of the phosphorylated form of [hydroxymethylglutaryl-CoA reductase (NADPH)] kinase to produce the dephosphorylated protein and orthophosphate.

Selected entries from *Methods in Enzymology* [vol, page(s)]:
Assay: 71, 493

[HYDROXYMETHYLGLUTARYL-CoA REDUCTASE (NADPH)] PHOSPHATASE

This phosphatase [EC 3.1.3.47], also known simply as reductase phosphatase, catalyzes the hydrolysis of hydroxymethylglutaryl-CoA reductase (NADPH) (containing a phosphorylated seryl residue) to produce orthophosphate and the dephosphorylated hydroxymethylglutaryl-CoA reductase (NADPH).[1,2] This reaction activates the reductase [EC 1.1.1.34].

[1] M. Sitges, G. Gil & F. G. Hegardt (1984) *J. Lipid Res.* 25, 497.
[2] G. Gil, M. Sitges & F. G. Hegardt (1981) *BBA* 663, 211.

Selected entries from *Methods in Enzymology* [vol, page(s)]:
Assay: 71, 492

3-HYDROXY-3-METHYLGLUTARYL-CoA SYNTHASE

This enzyme [EC 4.1.3.5] catalyzes the reaction of acetyl-CoA with acetoacetyl-CoA and water to generate (*S*)-3-hydroxy-3-methylglutaryl-CoA and coenzyme A.[1-4] The reaction mechanism for this enzyme can be written as three steps: (a) transfer of an acetyl group to an active-site sulfhydryl group to form Enz–S–COCH$_3$; (b) base-catalyzed proton abstraction from the methyl carbon, followed by nucleophilic attack on C3 of acetoacetyl-CoA to generate Enz–S–COCH$_2$C(OH)(CH$_3$)CH$_2$COSCoA; and (c) hydrolysis of the enzyme's thiolester linkage to yield the overall reaction product.

[1] F. G. Hegardt (1999) *BJ* 338, 569.
[2] D. J. Creighton & N. S. R. K. Murthy (1990) *The Enzymes*, 3rd ed., 19, 323.
[3] K. D. Clinkenbeard, T. Sugiyama, W. D. Reed & M. D. Lane (1975) *JBC* 250, 3124.
[4] M. J. P. Higgins, J. A. Kornblatt & H. Rudney (1972) *The Enzymes*, 3rd ed., 7, 407.

Selected entries from *Methods in Enzymology* [vol, page(s)]:
General discussion: 35, 155, 160, 173; 110, 19
Assay: 35, 155, 161, 162, 174; 110, 20
Other molecular properties: chicken liver (assays, 110, 20; properties, 110, 24; purification, 110, 22); cytosolic, from chicken liver, 35, 160; mitochondrial, from chicken liver, 35, 155; stereochemistry, 87, 147, 157; yeast, 35, 173

4-HYDROXY-4-METHYL-2-KETOGLUTARATE ALDOLASE

This aldolase [EC 4.1.3.17], also known as 4-hydroxy-4-methyl-2-oxoglutarate aldolase and 4-hydroxy-4-methyl-2-oxoglutarate pyruvate-lyase, catalyzes the conversion of 4-hydroxy-4-methyl-2-ketoglutarate (*i.e.*, 4-hydroxy-4-methyl-2-oxoglutarate) to two molecules of pyruvate.[1-4] 4-Hydroxy-4-methyl-2-ketoadipate and 4-carboxy-4-hydroxy-2-ketohexadioate can serve as substrates as well.

[1] B. F. Tack, P. J. Chapman & S. Dagley (1972) *JBC* 247, 6444.
[2] C. Schuld Ritter, P. J. Chapman & S. Dagley (1973) *J. Bacteriol.* 113, 1064.
[3] W. A. Wood (1972) *The Enzymes*, 3rd ed., 7, 281.
[4] L. M. Shannon & A. Marcus (1962) *JBC* 237, 3342.

Selected entries from *Methods in Enzymology* [vol, page(s)]:
General discussion: 90, 272
Assay: *Pseudomonas putida*, 90, 273
Other molecular properties: *Pseudomonas putida* (assay, 90, 273; properties, 90, 276; purification, 90, 274)

3-HYDROXY-2-METHYL-PYRIDINECARBOXYLATE DIOXYGENASE

This FAD-dependent oxidoreductase [EC 1.14.12.4], also known as methylhydroxypyridinecarboxylate oxidase and methylhydroxypyridine carboxylate dioxygenase, catalyzes the reaction of 3-hydroxy-2-methylpyridine-5-carboxylate with NAD(P)H and dioxygen to produce 2-(acetamidomethylene)succinate and NAD(P)$^+$.[1-3]

3-hydroxy-2-methyl-pyridine-5-carboxylate

2-(acetamidomethylene)-succinate

One atom of oxygen from dioxygen and one atom of oxygen from water is incorporated into the product: thus, the enzyme catalyzes both the monooxygenation and the hydrolytic ring opening of the pyridine ring to yield the final product.[4] Dioxygen first reacts with the enzyme to form a flavin hydroperoxide intermediate which is proposed to be converted to the hydroxyflavin. The pyridine substrate is then hydroxylated at the 2-position and the resulting intermediate is primed for hydrolysis of the ring.

[1] G. M. Kishore & E. E. Snell (1981) *JBC* 256, 4228 and 4234.
[2] M. S. Flashner & V. Massey (1974) in *Mol. Mech. Oxygen Activ.* (O. Hayaishi, ed.), p. 245.
[3] L. G. Sparrow, P. P. K. Ho, T. K. Sundaram, D. Zach, E. J. Nyns & E. E. Snell (1969) *JBC* 244, 2590.

[4]P. Chaiyen, P. Brissette, D. P. Ballou & V. Massey (1997) *Biochemistry* **36**, 8060.

Selected entries from *Methods in Enzymology* [vol, page(s)]:
General discussion: 18A, 654
Assay: 18A, 654; **52**, 16

3-HYDROXY-2-METHYLPYRIDINE-4,5-DICARBOXYLATE 4-DECARBOXYLASE

3-hydroxy-2-methyl-
pyridine-4,5-dicarboxylate

3-hydroxy-2-methyl-
pyridine-5-carboxylate

This decarboxylase [EC 4.1.1.51] catalyzes the conversion of 3-hydroxy-2-methylpyridine-4,5-dicarboxylate to 3-hydroxy-2-methylpyridine-5-carboxylate and carbon dioxide.[1]

[1]E. E. Snell, A. A. Smucker, E. Ringelmann & F. Lynen (1964) *Biochem. Z.* **341**, 109.

Selected entries from *Methods in Enzymology* [vol, page(s)]:
General discussion: 18A, 652
Assay: 18A, 652

5-HYDROXYMETHYLPYRIMIDINE KINASE

This phosphotransferase [EC 2.7.1.49] catalyzes the reaction of ATP with 4-amino-2-methyl-5-hydroxymethyl-pyrimidine to produce 4-amino-2-methyl-5-phospho-methylpyrimidine and ADP.[1-3] Other donor substrates include CTP, UTP, and GTP. The *Escherichia coli* enzyme will also act on pyridoxine, pyridoxal, and pyridoxamine; in fact, the *E. coli* enzyme may be identical to pyridoxal kinase.[3]

[1]E. P. Anderson (1973) *The Enzymes*, 3rd ed., **9**, 49.
[2]I. M. Lewis & G. M. Brown (1961) *JBC* **236**, 2768.
[3]T. Mizote & H. Nakayama (1989) *BBA* **991**, 109.

4-HYDROXYMUCONATE-SEMIALDEHYDE DEHYDROGENASE

This oxidoreductase [EC 1.2.1.61], which participates in the degradation pathway for 4-nitrophenol, catalyzes the reaction of 4-hydroxymuconate semialdehyde with NAD^+ and water to produce maleylacetate, NADH, and a proton.[1]

[1]J. C. Spain & D. T. Gibson (1991) *Appl. Environ. Microbiol.* **57**, 812.

2-HYDROXYMUCONATE-SEMIALDEHYDE HYDROLASE

This enzyme [EC 3.7.1.9] catalyzes the hydrolysis of 2-hydroxymuconate semialdehyde to produce formate and 2-oxopent-4-enoate (or, 2-ketopent-4-enoate).[1]

[1]E. Diaz & K. N. Timmis (1995) *JBC* **270**, 6403.

1-HYDROXY-2-NAPHTHOATE 1,2-DIOXYGENASE

This iron-dependent oxidoreductase [EC 1.13.11.38], also known as 1-hydroxy-2-naphthoate-degrading enzyme and 1-hydroxy-2-naphthoic acid dioxygenase, catalyzes the reaction of 1-hydroxy-2-naphthoate and dioxygen to produce (3*E*)-4-(2-carboxyphenyl)-2-oxobut-3-enoate.[1,2]

[1]H. Kiyohara & K. Nagao (1978) *J. Gen. Microbiol.* **105**, 69.
[2]E. A. Barnsley (1983) *J. Bacteriol.* **154**, 113.

6-HYDROXYNICOTINATE REDUCTASE

This oxidoreductase [EC 1.3.7.1], also known as 6-oxotetrahydronicotinate dehydrogenase, catalyzes the reversible reaction of 6-hydroxynicotinate with reduced ferredoxin to produce 1,4,5,6-tetrahydro-6-oxonicotinate and oxidized ferredoxin.[1]

[1]J. S. Holcenberg & L. Tsai (1969) *JBC* **244**, 1204.

Selected entries from *Methods in Enzymology* [vol, page(s)]:
General discussion: 18B, 239
Assay: 18B, 239

(R)-6-HYDROXYNICOTINE OXIDASE

This FAD-dependent oxidoreductase [EC 1.5.3.6], also known as D-6-hydroxynicotine oxidase, catalyzes the reaction of (*R*)-6-hydroxynicotine with water and dioxygen to produce 1-(6-hydroxypyrid-3-yl)-4-(methylamino)butan-1-one and hydrogen peroxide.[1-4]

(*R*)-6-hydroxynicotine

1-(6-hydroxypyrid-3-yl)-4-
(methylamino)butan-1-one

The enzyme isolated from *Arthrobacter oxidans* is stereospecific for the (*R*)-stereoisomer and contains a covalently bound FAD.

[1]K. Decker (1976) *Trends Biochem. Sci.* **1**, 184.
[2]M. Brühmüller, H. Möhler & K. Decker (1972) *EJB* **29**, 143.

[3]K. Decker, V. D. Dai, H. Möhler & M. Brühmüller (1972) Z. Naturforsch. **27B**, 1072.
[4]K. Decker & H. Bleeg (1965) BBA **105**, 313.

Selected entries from **Methods in Enzymology** [**vol**, page(s)]:
General discussion: **52**, 18; **56**, 475; **280**, 413

(S)-6-HYDROXYNICOTINE OXIDASE

(S)-6-hydroxynicotine

1-(6-hydroxypyrid-3-yl)-4-(methylamino)butan-1-one

This FAD-dependent oxidoreductase [EC 1.5.3.5], also known as L-6-hydroxynicotine oxidase, catalyzes the reaction of (S)-6-hydroxynicotine with water and dioxygen to produce 1-(6-hydroxypyrid-3-yl)-4-(methylamino)-butan-1-one and hydrogen peroxide.[1-4] **See also** *(R)-6-Hydroxynicotine Oxidase*

[1]K. Decker, V. D. Dai, H. Möhler & M. Brühmüller (1972) Z. Naturforsch. **27B**, 1072.
[2]G. Palmer & V. Massey (1968) in *Biological Oxidations* (T. P. Singer, ed.), p. 263.
[3]K. Decker & V. D. Dai (1967) EJB **3**, 132.
[4]K. Decker & H. Bleeg (1965) BBA **105**, 313.

Selected entries from **Methods in Enzymology** [**vol**, page(s)]:
General discussion: **52**, 18; **56**, 475

HYDROXYNITRILASE

This enzyme [EC 4.1.2.39], also called hydroxynitrile lyase and oxynitrilase, catalyzes the conversion of 2-hydroxyisobutyronitrile to cyanide and acetone.

3-HYDROXYOCTANOYL-[ACYL-CARRIER PROTEIN] DEHYDRATASE

This dehydratase [EC 4.2.1.59] catalyzes the conversion of (3R)-3-hydroxyoctanoyl-[acyl-carrier protein] to 2-octenoyl-[acyl-carrier-protein] and water.[1] The enzyme is specific for 3-hydroxyacyl-[acyl-carrier protein] derivatives of six to twelve carbon atoms in length (hence, (3R)-3-hydroxyhexanoyl-[acyl-carrier protein], (3R)-3-hydroxyheptanoyl-[acyl-carrier protein], (3R)-3-hydroxynonanoyl-[acyl-carrier protein], (3R)-3-hydroxydecanoyl-[acyl-carrier protein], *etc.*). **See also** *3-Hydroxypalmitoyl-[Acyl-Carrier Protein] Dehydratase*

[1]M. Mizugaki, A. C. Swindell & S. J. Wakil (1968) BBRC **33**, 520.

HYDROXYOXOBUTYRATE ALDOLASE

This aldolase activity, previously classified as EC 4.1.2.1, is now listed under 4-hydroxy-2-ketoglutarate aldolase [EC 4.1.3.16].

2-HYDROXY-3-OXOPROPIONATE REDUCTASE

This oxidoreductase [EC 1.1.1.60], also known as tartronate semialdehyde reductase and 2-hydroxy-3-ketopropionate reductase, catalyzes the reversible reaction of 2-hydroxy-3-oxopropanoate (or, tartronate semialdehyde) with NAD(P)H to produce (R)-glycerate and NAD(P)$^+$.[1-3] Other substrates include malonate semialdehyde and mesoxalate semialdehyde. **See also** *Glycerate Dehydrogenase; Hydroxypyruvate Reductase*

[1]C. van der Drift & F. E. de Windt (1983) *Antonie Leeuwenhoek* **49**, 167.
[2]L. Kohn (1968) JBC **243**, 4426.
[3]A. M. Gotto & H. L. Kornberg (1961) BJ **81**, 273.

Selected entries from **Methods in Enzymology** [**vol**, page(s)]:
General discussion: **9**, 240
Assay: **9**, 240
Other molecular properties: crystallization, **22**, 250; *Pseudomonas ovalis*, **9**, 240 (activators, **9**, 246; inhibitors, **9**, 246; kinetic properties, **9**, 246; pH optimum, **9**, 246; properties, **9**, 245; purification, **9**, 243; specificity of, **9**, 245; stability, **9**, 246); stereospecificity, **87**, 109

3-HYDROXYPALMITOYL-[ACYL-CARRIER PROTEIN] DEHYDRATASE

This enzyme [EC 4.2.1.61], also known as β-hydroxyacyl thioester dehydrase or β-hydroxyacyl-[acyl-carrier protein] dehydratase, is the fatty-acid synthase component that catalyzes the conversion of (3R)-3-hydroxypalmitoyl-[acyl-carrier protein] to 2-hexadecenoyl-[acyl-carrier protein] and water.[1-4] This enzyme displays specificity toward 3-hydroxyacyl-[acyl-carrier protein] derivatives (with chain lengths from twelve to sixteen carbon atoms), with highest activity on the palmitoyl derivative. **See also** *Fatty Acid Synthase*

[1]M. A. Ator & P. R. Ortez de Montellano (1990) *The Enzymes*, 3rd ed., **19**, 213.
[2]S. J. Wakil & J. K. Stoops (1983) *The Enzymes*, 3rd ed., **16**, 3.
[3]K. Bloch (1971) *The Enzymes*, 3rd ed., **5**, 441.
[4]M. Mizugaki, A. C. Swindell & S. J. Wakil (1968) BBRC **33**, 520.

5-HYDROXYPENTANOATE CoA-TRANSFERASE

This acyltransferase [EC 2.8.3.14], also known as 5-hydroxyvalerate CoA-transferase, catalyzes the reversible reaction of acetyl-CoA with 5-hydroxypentanoate to form 5-hydroxypentanoyl-CoA and acetate.[1] The *Clostridium aminovalericum* enzyme exhibits ping-pong initial rate kinetics, suggesting thiolester formation between a specific carboxyl group of the enzyme and coenzyme A during catalysis.

[1] U. Eikmanns & W. Buckel (1990) *Biol. Chem. Hoppe-Seyler* **371**, 1077.

4-HYDROXYPHENYLACETALDEHYDE DEHYDROGENASE

This oxidoreductase [EC 1.2.1.53] catalyzes the reaction of 4-hydroxyphenylacetaldehyde with NAD^+ and water to produce 4-hydroxyphenylacetate and NADH.[1]

[1] C. M. Cuskey, V. Peccoraro & R. H. Olsen (1987) *J. Bacteriol.* **169**, 2398.

4-HYDROXYPHENYLACETALDEHYDE-OXIME ISOMERASE

(*E*) isomer   (*Z*) isomer

This thiol-dependent isomerase [EC 5.2.1.11] catalyzes the interconversion of (*E*)-4-hydroxyphenylacetaldehyde oxime and (*Z*)-4-hydroxyphenylacetaldehyde oxime.[1]

[1] B. A. Halkier, C. E. Olsen & B. L. Moller (1989) *JBC* **264**, 19487.

4-HYDROXYPHENYLACETALDEHYDE OXIME MONOOXYGENASE

This oxidoreductase [EC 1.14.13.68] (also known as 4-hydroxybenzeneacetaldehyde oxime monooxygenase, cytochrome P450II-dependent monooxygenase, NADPH-cytochrome P450 reductase, and CYP71E1), catalyzes the reaction of 4-hydroxyphenylacetaldehyde oxime with NADPH, H^+, and dioxygen to produce 4-hydroxymandelonitrile, $NADP^+$, and two water molecules.[1,2]

[1] I. J. MacFarlane, E. M. Lees & E. E. Conn (1975) *JBC* **250**, 4708.
[2] M. Shimada & E. E. Conn (1977) *ABB* **180**, 199.

3-HYDROXYPHENYLACETATE 6-MONOOXYGENASE

This FAD-dependent oxidoreductase [EC 1.14.13.63], also known as 3-hydroxyphenylacetate 6-hydroxylase, catalyzes the reaction of 3-hydroxyphenylacetate with NAD(P)H and dioxygen to produce homogentisate, $NAD(P)^+$, and water.[1]

[1] W. J. H. Van Berkel & W. J. J. Van den Tweel (1991) *EJB* **201**, 585.

4-HYDROXYPHENYLACETATE 1-MONOOXYGENASE

This FAD-dependent oxidoreductase [EC 1.14.13.18], also known as 4-hydroxyphenylacetate 1-hydroxylase, catalyzes the reaction of 4-hydroxyphenylacetate with NAD(P)H and dioxygen to produce homogentisate, $NAD(P)^+$ and water.[1] Other substrates include 4-hydroxyhydratropate (producing 2-methylhomogentisate) and 4-hydroxyphenoxyacetate (generating hydroquinone and glycolate).

[1] W. A. Hareland, R. L. Crawford, P. J. Chapman & S. Dagley (1975) *J. Bacteriol.* **121**, 272.

Selected entries from *Methods in Enzymology* [vol, page(s)]:
General discussion: 52, 17

4-HYDROXYPHENYLACETATE 3-MONOOXYGENASE

This FAD-dependent oxidoreductase [EC 1.14.13.3], also known as *p*-hydroxyphenylacetate 3-hydroxylase, catalyzes the reaction of 4-hydroxyphenylacetate with NADH and dioxygen to produce 3,4-dihydroxyphenylacetate, NAD^+, and water.[1-3] Other substrates include 3-hydroxyphenylacetate and 4-fluorophenylacetate.

[1] B. A. Palfey & V. Massey (1998) *CBC* **3**, 83.
[2] U. Arunachalam, V. Massey & S. M. Miller (1994) *JBC* **269**, 150.
[3] S. G. Raju, A. V. Kamath & C. S. Vaidyanathan (1988) *BBRC* **154**, 537.

HYDROXYPHENYLACETONITRILE 2-MONOOXYGENASE

This heme-thiolate-dependent oxidoreductase (that is, a cytochrome P450 system) [EC 1.14.13.42], also known as 4-hydroxyphenylacetonitrile hydroxylase, catalyzes the reaction of 4-hydroxyphenylacetonitrile with NADPH and dioxygen to produce 4-hydroxymandelonitrile, $NADP^+$, and water.[1,2]

[1] W. Hösel & O. Schiel (1984) *ABB* **229**, 177.
[2] B. A. Halkier & B. L. Moller (1990) *JBC* **265**, 21114.

2-(2-HYDROXYPHENYL)-BENZENESULFINATE HYDROLASE

This enzyme [EC 3.1.2.24], also known as gene *dsz* B-encoded hydrolase, catalyzes the hydrolysis of 2-(2-hydroxyphenyl)benzenesulfinate to produce 2-hydroxybiphenyl and sulfite.[1]

[1]C. Oldfield, O. Pogrebinsky, J. Simmonds, E. S. Ölson & C. F. Kulpa (1997) *Microbiology* **143**, 2961.

D-4-HYDROXYPHENYLGLYCINE AMINOTRANSFERASE

This pyridoxal-phosphate-dependent aminotransferase [EC 2.6.1.72] catalyzes the reaction of D-4-hydroxyphenyl-glycine with α-ketoglutarate (or, 2-oxoglutarate) to produce 4-hydroxyphenylglyoxylate and L-glutamate.[1]

[1]W. J. J. van den Tweel, M. N. Widjojoatmodjo & J. A. M. de Bont (1988) *Arch. Microbiol.* **150**, 471.

(R)-4-HYDROXYPHENYLLACTATE DEHYDROGENASE

This manganese-dependent dehydrogenase [EC 1.1.1.222], also known as (R)-aromatic lactate dehydrogenase, catalyzes the reversible reaction of (R)-3-(4-hydroxy-phenyl)lactate with NAD$^+$ or NADP$^+$) to produce 3-(4-hydroxyphenyl)pyruvate and NADH or NADPH). Alternative substrates include (R)-3-phenyllactate, (R)-3-(indole-3-yl)lactate, and (R)-lactate.[1,2]

[1]R. Bode, A. Lippoldt & D. Birnbaum (1986) *Biochem. Physiol. Pflanz.* **181**, 189.
[2]S. Leelayoova, D. Marbury, P. M. Rainey, N. E. MacKenzie & J. E. Hall (1992) *J. Protozool.* **39**, 350.

4-HYDROXYPHENYLPYRUVATE DIOXYGENASE

This iron-dependent oxidoreductase [EC 1.13.11.27], also known as *p*-hydroxyphenylpyruvate oxidase, catalyzes the reaction of 4-hydroxyphenylpyruvate with dioxygen to produce homogentisate and carbon dioxide.[1-4]

[1]M. Rundgren (1983) *EJB* **133**, 657.
[2]P. A. Roche, T. J. Moorehead & G. A. Hamilton (1982) *ABB* **216**, 62.
[3]B. Lindblad, G. Lindstedt, S. Lindstedt & M. Rundgren (1977) *JBC* **252**, 5073.
[4]O. Hayaishi, M. Nozaki & M. T. Abbott (1975) *The Enzymes*, 3rd ed., **12**, 119.

Selected entries from *Methods in Enzymology* [vol, page(s)]:
General discussion: 52, 12, 38; **142**, 132, 139, 143, 148; **324**, 342
Assay: 142, 133, 139, 143, 148
Other molecular properties: properties, **142**, 136, **142**, 147, 153; purification from (chicken liver, **142**, 150; human liver, **142**, 139; porcine liver, **142**, 135; *Pseudomonas* sp., **142**, 145)

4-HYDROXYPHENYLPYRUVATE OXIDASE

This oxidoreductase [EC 1.2.3.13] catalyzes the reaction of 4-hydroxyphenylpyruvate and 1/2 dioxygen to produce 4-hydroxyphenylacetate and carbon dioxide.[1] This enzyme participates in the pathway for the degradation of L-tyrosine in *Arthrobacter*.

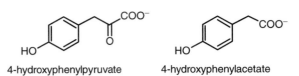

4-hydroxyphenylpyruvate   4-hydroxyphenylacetate

4-Hydroxyphenylpyruvate dioxygenase [EC 1.13.11.27] is occasionally incorrectly referred to by this name.

[1]E. R. Blakley (1977) *Can. J. Microbiol.* **23**, 1128.

HYDROXYPHENYLPYRUVATE REDUCTASE

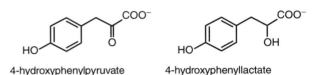

4-hydroxyphenylpyruvate   4-hydroxyphenyllactate

This oxidoreductase [EC 1.1.1.237], first isolated from the common house plant *Coleus blumei*, catalyzes the reaction of 3-(4-hydroxyphenyl)pyruvate with NADH to produce 3-(4-hydroxyphenyl)lactate and NAD$^+$. 3-(3,4-Dihydroxyphenyl)lactate will also act as an alternative substrate in the reverse direction.[1,2]

[1]M. Petersen & A. W. Alfermann (1988) *Z. Naturforsch.* **43c**, 501.
[2]E. Häusler, M. Petersen & A. W. Alfermann (1991) *Z. Naturforsch.* **46c**, 371.

p-HYDROXYPHENYLPYRUVATE *enol-keto* TAUTOMERASE

This tautomerase catalyzes the conversion of the keto forms of *para*- and *meta*-hydroxyphenylpyruvate to the corresponding enol forms.

4-hydroxyphenyl-(*keto*)-pyruvate   4-hydroxyphenyl-(*enol*)-pyruvate

This enzyme also acts on phenylpyruvate and certain dihy-droxyphenylpyruvates. Rosengren *et al.*[1] recently reported that the enzyme catalyzing the conversion of D-dopachrome to 5,6-dihydroxyindole is a phenylpyruvate tautomerase.
See also *Phenylpyruvate Tautomerase*

[1]E. Rosengren, S. Thelin, P. Aman, C. Hansson, L. Jacobsson & H. Rorsman (1997) *Melanoma Res.* **7**, 517.

Selected entries from *Methods in Enzymology* [vol, page(s)]:
General discussion: 2, 289
Assay: 2, 289; crude tissue preparations, in, **2**, 290

HYDROXYPHYTANATE OXIDASE

This oxidoreductase [EC 1.1.3.27] catalyzes the reaction of L-2-hydroxyphytanate with dioxygen to produce α-ketophytanate (or, 2-oxophytanate) and hydrogen peroxide.[1,2]

[1]J. P. Draye, F. van Hoof, E. de Hoffmann & J. Vamecq (1987) *EJB* **167**, 573.
[2]J. Vamecq & J. P. Draye (1988) *Biomed. Environ. Mass Spectrom.* **15**, 345.

3-HYDROXYPIMELOYL-CoA DEHYDROGENASE

This oxidoreductase [EC 1.1.1.259] catalyzes the reaction of 3-hydroxypimeloyl-CoA with NAD⁺ to produce 3-oxopimeloyl-CoA (*i.e.*, 3-ketopimeloyl-CoA), NADH, and a proton.[1]

[1]C. S. Harwood & J. Gibson (1997) *J. Bacteriol.* **179**, 301.

17α-HYDROXYPROGESTERONE ALDOLASE

This aldolase [EC 4.1.2.30] catalyzes the conversion of 17α-hydroxyprogesterone to 4-androstene-3,17-dione and acetaldehyde.[1] The rate-limiting step is product release.[2]

[1]E. Nowotny, R. D. Sananez, G. Nattero, C. Yantorno & M. G. Faillaci (1974) *Hoppe-Seyler's Z. Physiol. Chem.* **355**, 716.
[2]H. Tagashira, S. Kominami & S. Takemori (1995) *Biochemistry* **34**, 10939.

16α-HYDROXYPROGESTERONE DEHYDRATASE

This dehydratase [EC 4.2.1.98], also called hydroxyprogesterone dehydroxylase, catalyzes the conversion of 16α-hydroxyprogesterone to 16,17-didehydroprogesterone and water. 16α-Hydroxypregnenolone is an alternative substrate.[1]

[1]T. L. Glass & R. S. Lamppa (1985) *BBA* **837**, 103.

trans-L-3-HYDROXYPROLINE DEHYDRATASE

trans-3-hydroxy-L-proline   Δ¹-pyrroline-2-carboxylate

This enzyme [EC 4.2.1.77] reportedly catalyzes the conversion of *trans*-L-3-hydroxyproline to Δ¹-pyrroline-2-carboxylate and a water molecule. 2,3-Dehydroproline is a presumptive intermediate.[1]

[1]S. G. Ramaswamy (1983) *Fed. Proc.* **42**, 2232.

4-HYDROXYPROLINE-2-EPIMERASE

This epimerase [EC 5.1.1.8] catalyzes the interconversion of *trans*-4-hydroxy-L-proline to *cis*-4-hydroxy-D-proline.[1–4]

trans-4-hydroxy-L-proline   cis-4-hydroxy-D-proline

The enzyme also interconverts *trans*-4-hydroxy-D-proline and *cis*-4-hydroxy-L-proline. In the presence of D₂O, solvent protons are incorporated at the α-position. In addition, an overshoot of the equilibrium value is observed in the change in optical rotation. The enzyme utilizes two cysteinyl residues in the epimerization reaction.[1]

[1]M. E. Tanner & G. L. Kenyon (1998) *CBC* **2**, 7.
[2]E. Adams (1976) *Adv. Enzymol.* **44**, 69.
[3]R. Kuttan & A. N. Radhakrishnan (1973) *Adv. Enzymol.* **37**, 273.
[4]E. Adams (1972) *The Enzymes*, 3rd ed., **6**, 479.
Selected entries from *Methods in Enzymology* [vol, page(s)]:
General discussion: 17B, 286
Assay: 17B, 286
Other molecular properties: properties, 17B, 289; purification from *Pseudomonas putida*, 17B, 288

3-HYDROXYPROPIONATE DEHYDROGENASE

This oxidoreductase [EC 1.1.1.59] catalyzes the reversible reaction of 3-hydroxypropanoate with NAD⁺ to produce 3-oxopropanoate (or, malonate semialdehyde) and NADH.[1,2]

[1]S. Miyakoshi, H. Uchiyama, T. Someya, T. Satoh & T. Tabuchi (1987) *Agric. Biol. Chem.* **51**, 2381.
[2]H. Den, W. G. Robinson & M. J. Coon (1959) *JBC* **234**, 1671.

Selected entries from *Methods in Enzymology* [vol, page(s)]:
General discussion: 5, 457
Assay: 5, 457

9-HYDROXYPROSTAGLANDIN DEHYDROGENASE

This oxidoreductase catalyzes the reaction of NAD⁺ with 13-keto-13,14-dihydroprostaglandin F₂α to produce NADH and 15-keto-13,14-dihydroprostaglandin E₂. Note that 3α-hydroxysteroid dehydrogenase (B-specific) [EC 1.1.1.50] also exhibits this activity. *See also 3α-Hydroxysteroid Dehydrogenase (B-Specific)*

Selected entries from *Methods in Enzymology* [vol, page(s)]:
Assay: 63, 36; rat kidney, **86**, 114

15-HYDROXYPROSTAGLANDIN DEHYDROGENASES

Several different 15-hydroxyprostaglandin dehydrogenases have been identified and classified.

15-Hydroxyprostaglandin dehydrogenase (NAD$^+$) [EC 1.1.1.141], a significant factor in prostaglandin inactivation, catalyzes the reaction of prostaglandin E$_1$ (or, (13E)-(15S)-11α,15-dihydroxy-9-oxoprost-13-enoate) with NAD$^+$ to produce 15-ketoprostaglandin E$_1$ (or, (13E)-11α-hydroxy-9,15-dioxoprost-13-enoate) and NADH.[1,2]

prostaglandin E$_1$ 15-ketoprostaglandin E$_1$

Other substrates include prostaglandin E$_2$, prostaglandin F$_{2\alpha}$, and prostaglandin B$_1$: but prostaglandin D$_2$ is not a substrate. The enzyme is distinct from that of 15-hydroxyprostaglandin-D dehydrogenase (NADP$^+$) [EC 1.1.1.196] and 15-hydroxyprostaglandin dehydrogenase (NADP$^+$) [EC 1.1.1.197]. The kinetic mechanism of the enzyme is ordered Bi Bi.[3,4] A threonyl residue is needed for appropriate binding of the coenzyme[5] and cysteinyl, seryl, tyrosyl, and lysyl residues are reportedly needed for catalytic activity.[6-8]

15-Hydroxyprostaglandin-D dehydrogenase (NADP$^+$) [EC 1.1.1.196], also known as prostaglandin-D 15-dehydrogenase (NADP$^+$), is specific for prostaglandin D and catalyzes the reaction of prostaglandin D$_2$ (or, (5Z,13E)-(15S)-9α,15-dihydroxy-11-oxoprosta-5,13-dienoate) with NADP$^+$ to produce 15-ketoprostaglandin D$_2$ (or, (5Z,13E)-9α-hydroxy-11,15-dioxoprosta-5,13-dienoate) and NADPH.[9,10]

prostaglandin D$_2$ 15-ketoprostaglandin D$_2$

The enzyme is distinct from 15-hydroxyprostaglandin dehydrogenase (NAD$^+$) [EC 1.1.1.141] and 15-hydroxyprostaglandin dehydrogenase (NADP$^+$) [EC 1.1.1.197]. Myristate and arachidonate are competitive inhibitors.[11]

15-Hydroxyprostaglandin dehydrogenase (NADP$^+$) [EC 1.1.1.197] catalyzes the reaction of prostaglandin E$_1$ (or, (13E)-(15S)-11α,15-dihydroxy-9-oxoprost-13-enoate) with NADP$^+$ to produce (13E)-11α-hydroxy-9,15-dioxoprost-13-enoate and NADPH.[1] Substrates include prostaglandin E$_2$, prostaglandin F$_{2\alpha}$, and B$_1$: prostaglandin D$_2$ is not a substrate. Hence, the enzyme is distinct from that of 15-hydroxyprostaglandin dehydrogenase (NAD$^+$) [EC 1.1.1.141] and 15-hydroxyprostaglandin-D dehydrogenase (NADP$^+$) [EC 1.1.1.196]. However, it may be identical with prostaglandin E$_2$ 9-reductase [EC 1.1.1.189].[12]

15-Hydroxyprostaglandin-I dehydrogenase (NADP$^+$) [EC 1.1.1.231] is specific for prostaglandin I$_2$ (prostacyclin). It catalyzes the reaction of (5Z,13E)-(15S)-6,9α-epoxy-11α,15-dihydroxyprosta-5,13-dienoate (or, prostaglandin I$_2$) with NADP$^+$ to produce (5Z,13E)-6,9α-epoxy-11α-hydroxy-15-oxoprosta-5,13-dienoate (or, 15-ketoprostaglandin I$_2$) and NADPH.[13]

prostaglandin I$_2$ 15-ketoprostaglandin I$_2$

[1] S.-C. Lee & L. Levine (1975) *JBC* **250**, 548.
[2] S. S. Braithwaite & J. Jarabak (1975) *JBC* **250**, 2315.
[3] M. F. Ruckrich, A. Wendel, W. Schlegel, R. Jackisch & A. Jung (1975) *Hoppe-Seyler's Z. Physiol. Chem.* **356**, 799.
[4] D. T. Kung-Chao & H. H. Tai (1980) *BBA* **614**, 14.
[5] H. Zhou & H. H. Tai (1999) *BBRC* **257**, 414.
[6] C. M. Ensor & H. H. Tai (1996) *ABB* **333**, 117.
[7] C. M. Ensor & H. H. Tai (1996) *BBRC* **220**, 330.
[8] C. M. Ensor & H. H. Tai (1994) *BBA* **1208**, 151.
[9] K. Watanabe, T. Shimizu, S. Iguchi, H. Wakatsuka, M. Hayashi & O. Hayaishi (1980) *JBC* **255**, 1779.
[10] H. Tokumoto, K. Watanabe, D. Fukushima, T. Shimizu & O. Hayaishi (1982) *JBC* **257**, 13576.
[11] H. Osama, S. Narumiya, O. Hayaishi, H. Iinuma, T. Takeuchi & H. Umezawa (1983) *BBA* **752**, 251.
[12] D. G. Chang, M. Sun & H. H. Tai (1981) *BBRC* **99**, 745.
[13] J. M. Korff & J. Jarabak (1982) *JBC* **257**, 2177.

Selected entries from *Methods in Enzymology* [vol, page(s)]:
General discussion: 14, 215; **86**, 126, 131, 147, 152

Assay: NAD$^+$-dependent (human placental, **86**, 126; microassay in tissue samples, **86**, 135; porcine lung, **14**, 215; (15*S*)-[15-^{3}H]prostaglandin E$_2$, with, **86**, 131); porcine kidney, **86**, 143; prostaglandin D$_2$-specific (porcine brain, **86**, 148); prostaglandin I$_2$-specific (rabbit kidney, **86**, 152)

Other molecular properties: NAD$^+$-dependent (human placental [assay, **86**, 126; porcine lung, **14**, 215; properties, **14**, 218; **86**, 129; purification, **86**, 126]; stereospecificity, **87**, 111); porcine kidney (assay, **86**, 143; properties, **86**, 145; purification, **86**, 144); porcine lung, **14**, 215 (properties, **14**, 218; purification, **14**, 216; specificity, **14**, 218); prostaglandin D$_2$-specific (porcine brain [assay, **86**, 148; properties, **86**, 150; purification, **86**, 149; reaction sequence, **86**, 151]); prostaglandin I$_2$-specific (rabbit kidney [assay, **86**, 152; properties, **86**, 155; purification, **86**, 153])

2-HYDROXYPYRIDINE 5-MONOOXYGENASE

This oxidoreductase [EC 1.14.99.26] catalyzes the reaction of 2-hydroxypyridine with reduced acceptor AH$_2$ (including NADH) and dioxygen to produce 2,5-dihydroxypyridine, oxidized acceptor A (including NAD$^+$), and water.[1-3] 2,5-Dihydroxypyridine is also a substrate, but the enzyme does not act on 3-hydroxypyridine, 4-hydroxypyridine, or 2,6-dihydroxypyridine.

[1]M. L. Sharma, S. M. Kaul & O. P. Shukla (1984) *Biol. Mem.* **9**, 43.
[2]O. P. Shukla & S. M. Kaul (1986) *Can. J. Microbiol.* **32**, 330.
[3]C. Houghton & R. B. Cain (1972) *BJ* **130**, 879.

Selected entries from *Methods in Enzymology* [vol, page(s)]:
General discussion: 52, 20

3-HYDROXYPYRIDINE 6-MONOOXYGENASE

This oxidoreductase catalyzes the reaction of 3-hydroxypyridine with reduced acceptor substrate AH$_2$ and dioxygen to produce 2,5-dihydroxypyridine, an oxidized acceptor A, and water.[1]

[1]C. Houghton & R. B. Cain (1972) *BJ* **130**, 879.

Selected entries from *Methods in Enzymology* [vol, page(s)]:
General discussion: 52, 20

4-HYDROXYPYRIDINE 3-HYDROXYLASE

This FAD-dependent oxidoreductase catalyzes the reaction of 4-hydroxypyridine with a reduced acceptor substrate AH$_2$ and dioxygen to produce 3,4-dihydroxypyridine, an oxidized acceptor A, and water.[1,2]

[1]C. Houghton & R. B. Cain (1972) *BJ* **130**, 879.
[2]G. K. Watson, C. Houghton & R. B. Cain (1974) *BJ* **140**, 277.

Selected entries from *Methods in Enzymology* [vol, page(s)]:
General discussion: 52, 17, 20

HYDROXYPYRUVATE DECARBOXYLASE

This decarboxylase [EC 4.1.1.40] catalyzes the conversion of hydroxypyruvate to glycolaldehyde and carbon dioxide.[1] The enzyme also catalyzes the decarboxylation of bromopyruvate to form chloroacetaldehyde and carbon dioxide.

[1]J. L. Hedrick & H. J. Sallach (1964) *Arch. Biochem. Biophys.* **105**, 261.

HYDROXYPYRUVATE ISOMERASE

This isomerase [EC 5.3.1.22] catalyzes the interconversion of hydroxypyruvate and 2-hydroxy-3-oxopropanoate (*i.e.*, tartronate semialdehyde or hydroxymalonate semialdehyde).[1-3] The equilibrium constant slightly favors the semialdehyde ([tartronate semialdehyde]/[hydroxypyruvate] = 2.5 at 30°C and pH 7.1).[1]

[1]F. E. De Windt & C. Van der Drift (1980) *BBA* **613**, 556.
[2]W. Braun & H. Kaltwasser (1979) *Arch. Microbiol.* **121**, 129.
[3]G. Krakow, S. S. Barkulis & J. A. Hayashi (1961) *J. Bacteriol.* **81**, 509.

HYDROXYPYRUVATE REDUCTASE

This oxidoreductase [EC 1.1.1.81], also known as D-glycerate dehydrogenase, catalyzes the reversible reaction of hydroxypyruvate with NAD(P)H to produce D-glycerate and NAD(P)$^+$.[1]

hydroxypyruvate   D-glycerate

Other substrates, albeit weaker, include oxaloacetate, glyoxylate, acetoin, and diacetyl. The NAD$^+$-specific enzyme is D-glycerate dehydrogenase [EC 1.1.1.29]. **See also** *Glycerate Dehydrogenase; 2-Hydroxy-3-ketopropionate Reductase*

[1]L. A. Kleczkowski (1994) *Ann. Rev. Plant Physiol. Plant Mol. Biol.* **45**, 339.

Selected entries from *Methods in Enzymology* [vol, page(s)]:
General discussion: 9, 221, 229; **41**, 289; **89**, 341; **188**, 373
Assay: 9, 221, 229; **89**, 341; **188**, 373
Other molecular properties: calf liver, of, **9**, 221 (activators, **9**, 225; inhibitors, **9**, 225; properties, **9**, 224 [kinetics, **9**, 224]; specificity, **9**, 224); glyoxysomes, in, **31**, 569; marker enzyme, as, **31**, 735, 741; peroxisomes, marker for, **23**, 680; microbodies, in, **52**, 496, 499; properties, **188**, 377; *Pseudomonas acidovorans* (assay, **9**, 229; **89**, 341; properties, **9**, 232; **89**, 344; purification, **9**, 229; **89**, 342); purification from *Methylobacterium extorquens* AMI, **188**, 374; stereospecificity, **87**, 110

8-HYDROXYQUERCETIN 8-*O*-METHYLTRANSFERASE

This methyltransferase [EC 2.1.1.88], also known as flavonol 8-*O*-methyltransferase, catalyzes the reaction of *S*-adenosyl-L-methionine with 3,3′,4′,5,7,8-hexahydroxyflavone (*i.e.*, 8-hydroxyquercetin) to produce *S*-adenosyl-L-homocysteine and 3,3′,4′,5,7-pentahydroxy-8-methoxyflavone (*i.e.*, 8-methoxyquercetin). 8-Hydroxykaempferol is an alternative substrate, generating 8-methoxykaempferol; however, the corresponding 8-hydroxyflavonol glycosides are not substrates. The kinetic mechanism is either ordered Bi Bi or mono-Iso Theorell-Chance Bi Bi.[1]

[1] M. Jay, V. De Luca & R. K. Ibrahim (1985) *EJB* **153**, 321.

HYDROXYQUINOL 1,2-DIOXYGENASE

This iron-dependent oxidoreductase [EC 1.13.11.37] catalyzes the reaction of benzene-1,2,4-triol (*i.e.*, hydroxyquinol) with dioxygen to produce 3-hydroxy-*cis,cis*-muconate (which isomerizes to 2-maleylacetate [*i.e.*, *cis*-hex-2-enedioate]).[1–3] Catechol and pyrogallol react at less than one percent of the rate of benzene-1,2,4-triol.

[1] J. A. Buswell, K.-E. Eriksson, J. K. Gupta, S. G. Hamp & I. Nordh (1982) *Arch. Microbiol.* **131**, 366.
[2] I. S.-Y. Sze & S. Dagley (1984) *J. Bacteriol.* **159**, 353.
[3] D. L. Daubaras, K. Saido & A. M. Chakrabarty (1996) *Appl. Environ. Microbiol.* **62**, 4276.

2-HYDROXYQUINOLINE 5,6-DIOXYGENASE

This oxidoreductase [EC 1.14.12.16] (also known as 2-oxo-1,2-dihydroquinoline 5,6-dioxygenase, quinolin-2-ol 5,6-dioxygenase, and quinolin-2(1*H*)-one 5,6-dioxygenase) catalyzes the reaction of quinolin-2-ol with NADH and dioxygen to produce 2,5,6-trihydroxy-5,6-dihydroquinoline and NAD+.[1] Alternative substrates include 3-methylquinolin-2-ol, quinolin-8-ol, and quinolin-2,8-diol.

[1] S. Schach, B. Tshisuaka, S. Fetzner & F. Lingens (1995) *EJB* **232**, 536.

2-HYDROXYQUINOLINE 8-MONOOXYGENASE

There are two distinct enzyme activities bearing this name. The first iron-dependent oxidoreductase [EC 1.14.13.61], also known as 2-oxo-1,2-dihydroquinoline 8-monooxygenase, catalyzes the reaction of quinolin-2-ol (existing largely as the quinolin-2(1*H*)-one tautomer) with NADH and dioxygen to produce quinoline-2,8-diol (*i.e.*, 8-hydroxy-2-oxo-1,2-dihydroquinoline), NAD+, and water.[1–3]

A second oxidoreductase [EC 1.14.13.65] catalyzing the same reaction, but uses 3-methyl-quinolin-2-ol as an alternative substrate.

[1] B. Rosche, B. Tshisuaka, S. Fetzner & F. Lingens (1995) *JBC* **270**, 17836.
[2] S. Schach, B. Tshisuaka, S. Fetzner & F. Lingens (1995) *EJB* **232**, 536.
[3] B. Rosche, B. Tshisuaka, B. Hauer, F. Lingens & S. Fetzner (1997) *J. Bacteriol.* **179**, 3549.

4-HYDROXYQUINOLINE 3-MONOOXYGENASE

This oxidoreductase [EC 1.14.13.62], also known as quinolin-4(1*H*)-one 3-monooxygenase and 1-*H*-4-oxoquinoline 3-monooxygenase, catalyzes the reaction of quinolin-4-ol with NADH and dioxygen to produce quinoline-3,4-diol, NAD+, and water.

5α-HYDROXYSTEROID DEHYDRATASE

ergosta-7,22-diene-3β,5α-diol   ergosterol

This enzyme [EC 4.2.1.62] catalyzes the conversion of 5α-ergosta-7,22-diene-3β-diol to ergosterol and water.[1]

[1] R. W. Topham & J. L. Gaylor (1970) *JBC* **245**, 2319.

3α-HYDROXYSTEROID DEHYDROGENASE (A-SPECIFIC)

androsterone   5α-androstane-3,17-dione

This oxidoreductase [EC 1.1.1.213] catalyzes the reaction of androsterone with NAD(P)+ to produce 5α-androstane-3,17-dione and NAD(P)H. A number of other 3α-hydroxysteroids will serve as alternative substrates. The enzyme is specific for the A-face of NAD+ or NADP+. The B-specific enzyme is classified under EC 1.1.1.50. Much of the early literature on these two dehydrogenases does not distinguish between the two different stereochemistries.

The hamster liver[1] and rat liver[2] enzymes are reported to have ordered Bi Bi kinetic mechanisms. ***See 3α-Hydroxysteroid Dehydrogenase (B-Specific)***

[1]H. Sawada, A. Hara, M. Ohmura, T. Nakayama & Y. Deyashiki (1991)
J. Biochem. **109**, 770.
[2]L. J. Askonas, J. W. Ricigliano & T. M. Penning (1991) *BJ* **278**, 835.

Selected entries from *Methods in Enzymology* [**vol**, page(s)]:
Other molecular properties: staining on gels, **22**, 594; stereospecificity,
87, 107

3α-HYDROXYSTEROID DEHYDROGENASE (B-SPECIFIC)

This oxidoreductase [EC 1.1.1.50], also known as hydroxyprostaglandin dehydrogenase, catalyzes the reversible reaction of androsterone with NAD(P)$^+$ to produce 5α-androstane-3,17-dione and NAD(P)H.[1–4] Other 3α-hydroxysteroids can act as substrates as can 9-, 11- and 15-hydroxyprostaglandins. The stereochemistry is B-specific with respect to the pyridine coenzymes, but much of the earlier literature did not distinguish these dehydrogenases on the basis of stereochemistry. Other substrates include 5α-dihydrotestosterone, 5β-dihydrotestosterone, 3α-hydroxy-5α-pregnan-20-one, glycolithocholate, cholate, and etiocholan-3α-ol-17-one. **See also** 3α-*Hydroxysteroid Dehydrogenase (A-Specific); 9-Hydroxyprostaglandin Dehydrogenase*

[1]T. M. Penning (1999) *J. Steroid Biochem. Mol. Biol.* **69**, 211.
[2]L. J. Askonas, J. W. Ricigliano & T. M. Penning (1991) *BJ* **278**, 835.
[3]A. Hara, Y. Inoue, M. Nakagawa, F. Naganeo & H. Sawada (1988)
J. Biochem. **103**, 1027.
[4]G. H. Jacobi, R. J. Moore & J. D. Wilson (1977) *J. Steroid Biochem.* **8**, 719.

Selected entries from *Methods in Enzymology* [**vol**, page(s)]:
General discussion: **5**, 512, 520; **15**, 651; **34**, 558, 564
Assay: **5**, 513, 520; **15**, 282, 557, 651
Other molecular properties: bile acids, assay of, **15**, 255, 281, 283;
5β-cholestane-3α,7α,12α-triol, formation of, **15**, 557; chromatography of, **15**, 653; cortisone 5β-reductase and, **5**, 501, 502; hepatic, **5**, 520; **15**, 651 (assay, **15**, 557, 651; properties, **15**, 654; purification, **15**, 560, 652; purification steps, **15**, 561, 654; pyridine nucleotide requirement, **15**, 655; relative rates of reaction with different steroids, **15**, 560; steroid specificity, **15**, 656); multienzymatic NADH recycling, in, **136**, 73; properties, **15**, 654; *Pseudomonas testosteroni*, from, **15**, 255, 280 (hydrophobic interaction chromatographic behavior, prediction, **228**, 289; properties, **5**, 515; purification, **5**, 514; **15**, 281, 282); rat liver, of, **5**, 520; **15**, 651 (properties, **15**, 521, 654; purification, **5**, 521, **15**, 560, 652); staining on gels, **22**, 594; stereospecificity, **87**, 106

3β-HYDROXY-Δ^5-STEROID DEHYDROGENASE

This oxidoreductase [EC 1.1.1.145], also known as 3β-hydroxy-5-ene steroid dehydrogenase and progesterone reductase, catalyzes the reversible reaction of a 3β-hydroxy-Δ^5-steroid with NAD$^+$ to produce a 3-oxo-Δ^5-steroid and NADH.[1–6] Substrates include 3β-hydroxyandrost-5-en-17-one (*i.e.*, prasterone, producing androst-4-ene-3,17-dione), 4-androstene-3β,17β-diol, and

3β-hydroxypregn-5-en-20-one (*i.e.*, pregnenolone, producing progesterone).

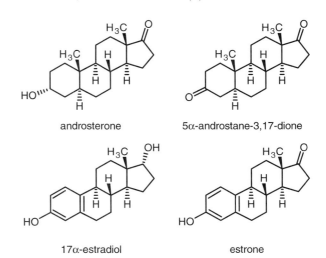

pregnenolone   progesterone

The mammalian enzyme also has a Δ^5/Δ^4 isomerase activity. It has been proposed that the reduced coenzyme product, NADH, induces a conformational change in the protein resulting in the activation of the isomerase activity.[1,2] **See also** 3β*(or 17β)-Hydroxysteroid Dehydrogenase*

[1]J. L. Thomas, C. Frieden, W. E. Nash & R. C. Strickler (1995) *JBC* **270**, 21003.
[2]J. I. Mason, D. Naville, B. W. Evans & J. L. Thomas (1998) *Endocr. Res.* **24**, 549.
[3]A. Hiwatashi, I. Hamamoto & Y. Ichikawa (1985) *J. Biochem.* **98**, 1519.
[4]T. Rabe, K. Brandstetter, J. Kellermann & B. Runnebaum (1982)
J. Steroid Biochem. **17**, 427.
[5]D. G. Armstrong (1982) *J. Steroid Biochem.* **17**, 225.
[6]M. Vincent, J. Gallay, C. de Paillerets, A. Alfsen & J. F. Biellmann (1978)
BBA **525**, 1.

Selected entries from *Methods in Enzymology* [**vol**, page(s)]:
Other molecular properties: staining on gels, **22**, 594; transition-state and multisubstrate analogues, **249**, 304

3(OR 17)α-HYDROXYSTEROID DEHYDROGENASE

This oxidoreductase [EC 1.1.1.209] catalyzes the reversible reaction of androsterone with NAD(P)$^+$ to produce 5α-androstane-3,17-dione and NAD(P)H.[1]

androsterone   5α-androstane-3,17-dione

17α-estradiol   estrone

The enzyme will also act on the 3α-hydroxyl group of androgens of the 5α-androstane series, as well as the 17α-hydroxyl group of both androgenic and estrogenic substrates, albeit more weakly. Other substrates include 17α-estradiol (producing estrone) and epitestosterone (producing androst-4-ene-3,17-dione). **See also** *3(or 17)β-Hydroxysteroid Dehydrogenase*

[1]P. C. Lau, D. S. Layne & D. G. Williamson (1982) *JBC* **257**, 9444 and 9450.

Selected entries from *Methods in Enzymology* [vol, page(s)]:
General discussion: 15, 719
Assay: 15, 719
Other molecular properties: properties, **15**, 720; purification, **15**, 720; staining on gels, **22**, 594

3(OR 17)β-HYDROXYSTEROID DEHYDROGENASE

This oxidoreductase [EC 1.1.1.51] catalyzes the reaction of testosterone with NAD(P)$^+$ to produce androst-4-ene-3,17-dione and NAD(P)H (greater activity is observed with NADP$^+$ than with NAD$^+$). The enzyme can also use other 3β- or 17β-hydroxysteroids as substrates. Evidence has been presented that the 3β and 17β activities occur at the same site, the reactions occurring alternately.[1] The enzyme is reported to have an ordered Bi Bi kinetic mechanism.[2,3] Other substrates include 3β-hydroxy-5β-pregnane-20-one and 3β-hydroxy-5α-androst-16-ene.

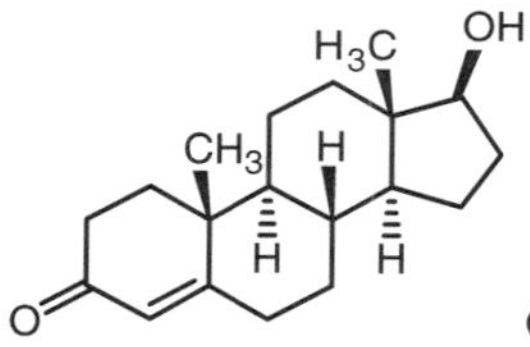

testosterone

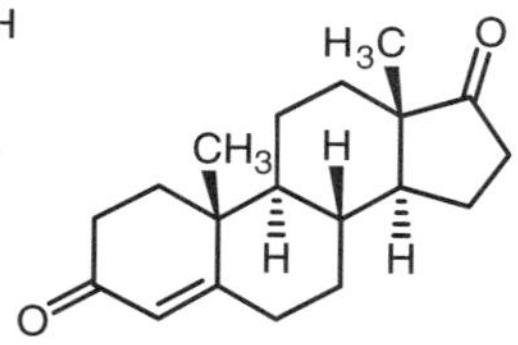

androst-4-ene-3,17-dione

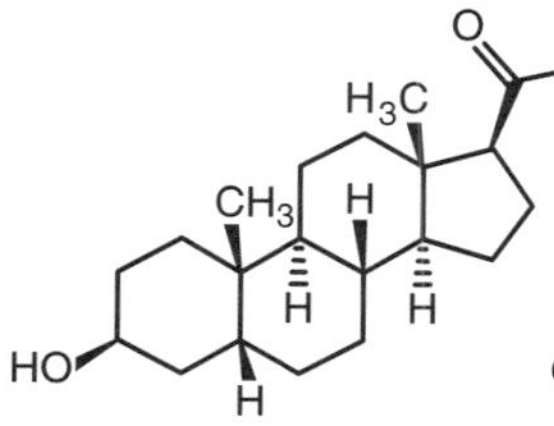

3β-hydroxy-5β-pregnan-20-one

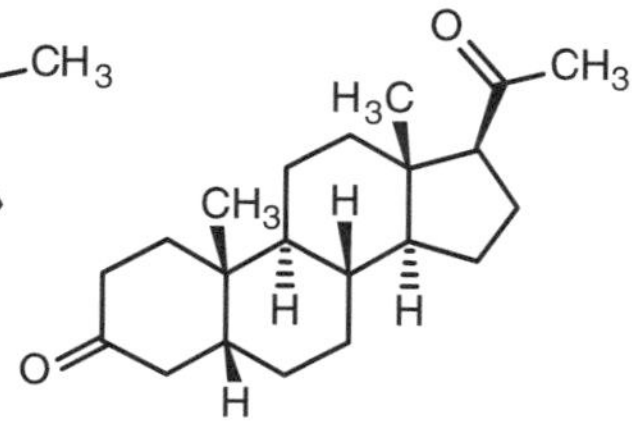

5β-pregnane-3,20-dione

This oxidoreductase is distinct from 3(or 17)α-hydroxysteroid dehydrogenase [EC 1.1.1.209], which acts on the α-hydroxysteroid derivatives. Regrettably, the earlier literature often does not distinguish between dehydrogenases that acted on both the 3 and 17 positions from those that acted only on one position (for example,

3β-hydroxy-Δ^5-steroid dehydrogenase, testosterone 17β-dehydrogenase, and estradiol 17β-dehydrogenase). **See also** *3α(or 17β)-Hydroxysteroid Dehydrogenase; 3β-Hydroxy-Δ^5-Steroid Dehydrogenase; Estradiol 17β-Dehydrogenase; Testosterone 17β-Dehydrogenase*

[1]P. Minard, M. D. Legoy & D. Thomas (1985) *FEBS Lett.* **188**, 85.
[2]M. A. Levy, D. A. Holt, M. Brandt & B. W. Metcalf (1987) *Biochemistry* **26**, 2270.
[3]H. Sawada, A. Hara, T. Nakayama & M. Hayashibara (1982) *J. Biochem.* **92**, 185.

Selected entries from *Methods in Enzymology* [vol, page(s)]:
General discussion: 5, 513; 34, 558, 564
Other molecular properties: bile acids, assay of, **15**, 255, 281; histochemistry, **6**, 893; 3β-hydroxysteroids, assay of, **15**, 281; identification of isolated Leydig cells, in, **109**, 285; luteinization, in, **39**, 215; *Pseudomonas testosteroni*, hydrophobic interaction chromatographic behavior, prediction, **228**, 289; staining on gels, **22**, 594; stereospecificity, **87**, 108

3α(17β)-HYDROXYSTEROID DEHYDROGENASE (NAD$^+$)

This dehydrogenase [EC 1.1.1.239], first isolated from hamster liver,[1] catalyzes the reaction of testosterone with NAD$^+$ to produce androst-4-ene-3,17-dione and NADH. NADP$^+$ was reported to provide only 15% of the activity observed with NAD$^+$.

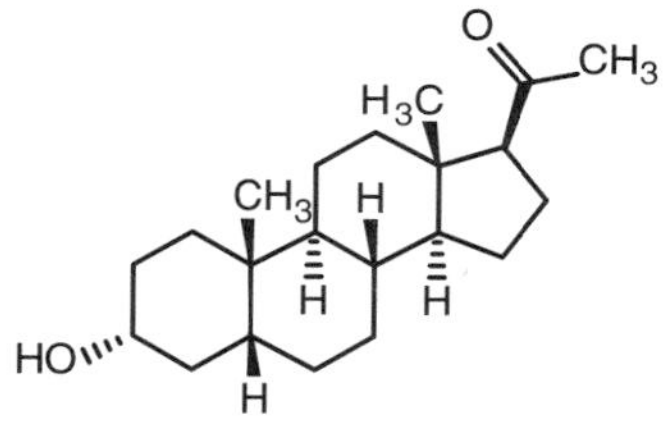

5β-pregnan-3α-ol-20-one

The enzyme also acts on other 17β-hydroxysteroids, on the 3α-hydroxy group of pregnanes and bile acids, and on benzene dihydrodiol. Interestingly, the dehydrogenase is reported to be a random-on, ordered-off Bi Bi kinetic mechanism.[2] Other substrates include 5β-pregnane-3α-ol-20-one and 5β-androstane-3α,17β-diol. This dehydrogenase is distinct from 3α-hydroxysteroid dehydrogenase [EC 1.1.1.50] nor to 3α-hydroxysteroid dehydrogenase (A-specific) [EC 1.1.1.213]. However, 3(or 17)β-hydroxysteroid dehydrogenase [EC 1.1.1.51] catalyzes a similar reaction. In addition, the bovine liver cytosolic carbonyl reductase has been reported to have this activity as well.[3] **See also** *3β-Hydroxy-Δ^5-Steroid Dehydrogenase; Estradiol 17β-Dehydrogenase; Testosterone 17β-Dehydrogenase; 3(or 17β)-Hydroxysteroid Dehydrogenase*

[1] M. Ohmura, A. Hara, M. Nakagawa & H. Sawada (1990) *BJ* **266**, 583.

[2] H. Sawada, A. Hara, M. Ohmura, T. Nakayama & Y. Deyashiki (1991) *J. Biochem. (Tokyo)* **109**, 770.

[3] T. Terada, N. Niwase, I. Koyama, M. Imamura, K. Shinagawa, H. Toya & T. Mizoguchi (1993) *Int. J. Biochem.* **25**, 1233.

Selected entries from *Methods in Enzymology* [vol, page(s)]:
Other molecular properties: staining on gels, **22**, 594

3α(OR, 20β)-HYDROXYSTEROID DEHYDROGENASE

This oxidoreductase [EC 1.1.1.53], also known as cortisone reductase and (*R*)-20-hydroxysteroid dehydrogenase, catalyzes the reversible reaction of androstane-3α,17β-diol with NAD$^+$ to produce 17β-hydroxyandrostan-3-one (*i.e.*, stanolone) and NADH.[1–5]

androstane-3α,17β-diol

stanolone

4-pregnene-17α,20β,21-triol-3,11-dione

cortisone

Under physiological conditions, the reverse reaction occurs frequently. Substrates include the 3α-hydroxyl group or 20β-hydroxyl group of pregnane and androstane steroids: for example, 4-pregnene-17α,20β,21-triol-3,11-dione (*i.e.*, Reichstein's substance U) is also converted reversibly to cortisone. Other substrates include progesterone, 17α-hydroxyprogesterone, pregnenolone, deoxycorticosterone, and 17,20-dihydroxy-4-pregnen-3-one-21-oate. The crystal structure of the prokaryotic enzyme identifies conserved tyrosyl and lysyl residues at the active site. It has been suggested that these residues, together with a seryl residue, promote electrophilic attack on the C20 carbonyl oxygen atom (in the reverse direction). This attack enables the carbon atom to accept a hydride from the reduced coenzyme. The enzyme has a random Bi Bi kinetic mechanism with a partial preference for a certain order of substrate binding. ***See also*** *3α-Hydroxysteroid Dehydrogenase; 3(or 17)α-Hydroxysteroid Dehydrogenase*

[1] D. Ghosh, Z. Wawrzak, C. M. Weeks, W. L. Duax & M. Erman (1994) *Structure* **2**, 629.

[2] B. Tyrakowska, R. M. D. Verhaert, R. Hilhorst & C. Veeger (1990) *EJB* **187**, 81.

[3] S. Nakajin, S. Ohno & M. Shinoda (1988) *J. Biochem.* **104**, 565.

[4] F. Sweet & B. R. Samant (1980) *Biochemistry* **19**, 978.

[5] C. A. F. Edwards & J. C. Orr (1978) *Biochemistry* **17**, 4370.

Selected entries from *Methods in Enzymology* [vol, page(s)]:
General discussion: **5**, 516; **135**, 475; **136**, 150
Assay: **5**, 517; **136**, 221
Other molecular properties: active site studies, **36**, 390 (modified amino acid, **36**, 400); affinity labeling, **36**, 387; **87**, 478; crystallization, **5**, 519; entrapped in reverse micelles (assay, **136**, 221; intrinsic rate parameters, **136**, 223; preparation, **136**, 221); immobilization, **136**, 152; immobilized subunits (preparation, **135**, 476; properties, **135**, 478); isotope exchange, **64**, 8; oxidation at C20, use in synthetic, **15**, 53; progesterone, assay for, **15**, 190; properties, **5**, 520; purification from (*Streptomyces hydrogenans*, **5**, 517, 518); staining on gels, **22**, 594; stereospecificity, **87**, 108; steroid transformation in two-phase systems, **136**, 154

3β(OR 20α)-HYDROXYSTEROID DEHYDROGENASE

This oxidoreductase [EC 1.1.1.210], also called progesterone reductase and 3-β-HSD, catalyzes the reversible reaction of 5α-androstane-3β,17β-diol with NADP$^+$ to produce 17β-hydroxy-5α-androstan-3-one and NADPH.

5α-androstane-3β,17β-diol

17β-hydroxy-5α-androstan-3-one

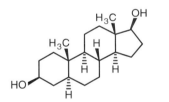

4-pregnen-20α-ol-3-one

progesterone

20α-Hydroxysteroids can act as alternative substrates. In the reverse reaction, progesterone is a substrate, producing 4-pregnen-20α-ol-3-one. The 3β and the 20α activities take place at the same active site.[1,2]

[1] M. A. Sharaf & F. Sweet (1982) *Biochemistry* **21**, 4615.

[2] Q. X. Chen, L. O. Rosik, C. D. Nancarrow & F. Sweet (1989) *Biochemistry* **28**, 8856.

6α-HYDROXYSTEROID DEHYDROGENASE

6α-hydroxyestrone   6-ketoestrone

This oxidoreductase catalyzes the reaction of 6α-hydroxyestrone with NAD(P)$^+$ to produce 6-ketoestrone (*i.e.*, 6-oxoestrone), NAD(P)H, and a proton.

Selected entries from **Methods in Enzymology** [**vol**, page(s)]:
General discussion: 15, 716
Assay: 15, 716
Other molecular properties: properties, 15, 717; staining on gels, 22, 594

6β-HYDROXYSTEROID DEHYDROGENASE

6β-hydroxyestrone   6-ketoestrone

This oxidoreductase catalyzes the reaction of 6β-hydroxyestrone with NAD(P)$^+$ to produce 6-ketoestrone (or, 6-oxoestrone), NAD(P)H, and a proton.

Selected entries from **Methods in Enzymology** [**vol**, page(s)]:
General discussion: 15, 716
Assay: 15, 716
Other molecular properties: properties, 15, 717; staining on gels, 22, 594

7α-HYDROXYSTEROID DEHYDROGENASE

This oxidoreductase [EC 1.1.1.159] catalyzes the reaction of 3α,7α,12α-trihydroxy-5β-cholanate (*i.e.*, cholate) with NAD$^+$ to produce 3α,12α-dihydroxy-7-oxo-5β-cholanate and NADH.[1–5]

cholate   3α,12α-dihydroxy-7-oxo-5β-cholanate

The 7α-hydroxyl group of bile acids and alcohols, both in their free and conjugated forms, is oxidized: for example, glycocholate, taurocholate, chenodeoxycholate, *etc.* The *Bacteroides fragilis* and *Clostridium* enzymes can also utilize NADP$^+$ as the coenzyme substrate.

A microbial NADP$^+$-dependent enzyme has been reported to have an ordered Bi Bi kinetic mechanism.[2] **See also** *7β-Hydroxysteroid Dehydrogenase (NADP$^+$)*

[1] T. Tanabe, N. Tanaka, K. Uchikawa, T. Kabashima, K. Ito, T. Nonaka, Y. Mitsui, M. Tsuru & T. Yoshimoto (1998) *J. Biochem.* **124**, 634.
[2] C. V. Franklund, P. de Prada & P. B. Hylemon (1990) *JBC* **265**, 9842.
[3] I. A. MacDonald & P. D. Roach (1981) *BBA* **665**, 262.
[4] I. A. MacDonald, C. N. Williams, D. E. Mahoney & W. M. Christie (1975) *BBA* **384**, 12.
[5] I. A. MacDonald, C. N. Williams & D. E. Mahony (1973) *BBA* **309**, 243.

Selected entries from **Methods in Enzymology** [**vol**, page(s)]:
General discussion: 15, 716
Assay: 15, 716
Other molecular properties: properties, 15, 717; staining on gels, 22, 594

11β-HYDROXYSTEROID DEHYDROGENASE

This oxidoreductase [EC 1.1.1.146] catalyzes the reaction of an 11β-hydroxysteroid with NADP$^+$ to produce an 11-oxosteroid (*i.e.*, 11-ketosteroid) and NADPH.[1–7]

cortisol   cortisone

Substrates include corticosterone (producing 11-dehydrocorticosterone), cortisol (producing cortisone), and 11β-hydroxyprogesterone (generating 11-oxoprogesterone). The type I isoform acts physiologically in both directions and has a prominent role in biosynthesis of the active glucocorticoids cortisol or corticosterone. The type II isozyme acts primarily in generating inactive 11-keto metabolites.[1] The liver dehydrogenase has an ordered Bi Bi kinetic mechanism.[5]

[1] Z. Krozowski, K. X. Li, K. Koyama, R. E. Smith, V. R. Obeyesekere, A. Stein-Oakley, H. Sasano, C. Coulter, T. Cole & K. E. Sheppard (1999) *J. Steroid Biochem. Mol. Biol.* **69**, 391.
[2] Z. Krozowski (1999) *Mol. Cell Endocrinol.* **151**, 121.
[3] P. M. Stewart & Z. S. Krozowski (1999) *Vitam. Horm.* **57**, 249.
[4] R. Benediktsson & C. R. W. Edwards (1996) *Essays in Biochem.* **31**, 23.
[5] C. Monder, V. Lakshmi & Y. Miroff (1991) *BBA* **1115**, 23.
[6] V. Lakshmi & C. Monder (1988) *Endocrinology* **123**, 2390.

[7]I. E. Bush, S. A. Hunter & R. A. Meigs (1968) *BJ* **107**, 239.

Selected entries from *Methods in Enzymology* [vol, page(s)]:
General discussion: 5, 522; 15, 716
Assay: 5, 522; 15, 716
Other molecular properties: preparation, 5, 523; properties, 15, 717; staining on gels, 22, 594

12α-HYDROXYSTEROID DEHYDROGENASE

This oxidoreductase [EC 1.1.1.176] catalyzes the reaction of 3α,7α,12α-trihydroxy-5β-cholanate (*i.e.*, cholate) with NADP[+] to produce 3α,7α-dihydroxy-12-oxo-5β-cholanate and NADPH.[1–3]

cholate 3α,7α-dihydroxy-12-oxo-5β-cholanate

Substrates include the bile acids, both in their free and conjugated form, as well as bile alcohols.

[1]M. Braun, H. Lünsdorf & A. F. Bückmann (1991) *EJB* **196**, 439.
[2]I. A. MacDonald, J. F. Jellett, D. E. Mahony & L. V. Holdeman (1979) *Appl. Environ. Microbiol.* **37**, 992.
[3]J. N. Harris & P. B. Hylemon (1978) *BBA* **528**, 148.

Selected entries from *Methods in Enzymology* [vol, page(s)]:
Other molecular properties: immobilization, 136, 152; staining on gels, 22, 594; steroid transformation in two-phase systems, 136, 154

12β-HYDROXYSTEROID DEHYDROGENASE

This oxidoreductase [EC 1.1.1.238], first studied in *Clostridium* sp.[1], catalyzes the reaction of 3α,7α,12β-trihydroxy-5β-cholanate with NADP[+] to produce 3α,7α-dihydroxy-12-oxo-5β-cholanate and NADPH.

3α,7β,12β-trihydroxy-5β-cholanate 3α,7β-dihydroxy-12-oxo-5β-cholanate

The enzyme acts on a number of bile acids, both in their free and conjugated forms.

[1]R. Edenharder & A. Pfützner (1988) *BBA* **962**, 362.

Selected entries from *Methods in Enzymology* [vol, page(s)]:
Other molecular properties: staining on gels, 22, 594

15α-HYDROXYSTEROID DEHYDROGENASE

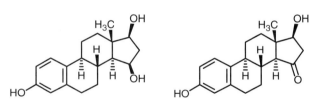

15α-hydroxy-17β-estradiol 15-keto-17β-estradiol

This oxidoreductase catalyzes the reaction of 15α-hydroxy-17β-estradiol with NAD(P)[+] to produce 15-keto-17β-estradiol (*i.e.*, 15-oxo-17β-estradiol), NAD(P)H, and a proton.

Selected entries from *Methods in Enzymology* [vol, page(s)]:
General discussion: 15, 716
Assay: 15, 716
Other molecular properties: properties, 15, 717; staining on gels, 22, 594

15β-HYDROXYSTEROID DEHYDROGENASE

15β-hydroxy-17β-estradiol 15-keto-17β-estradiol

This oxidoreductase catalyzes the reaction of 15β-hydroxy-17β-estradiol with NAD(P)[+] to produce 15-keto-17β-estradiol (*i.e.*, 15-oxo-17β-estradiol), NAD(P)H, and a proton.

Selected entries from *Methods in Enzymology* [vol, page(s)]:
General discussion: 15, 716
Assay: 15, 716
Other molecular properties: properties, 15, 717; staining on gels, 22, 594

16α-HYDROXYSTEROID DEHYDROGENASE

This oxidoreductase [EC 1.1.1.147], also known as 16α-hydroxysteroid oxidoreductase, estriol 16α-dehydrogenase, and 16α-hydroxyestrogen oxidoreductase, catalyzes the reversible reaction of a 16α-hydroxysteroid with NAD(P)[+] to produce a 16-oxosteroid and NAD(P)H.[1,2]

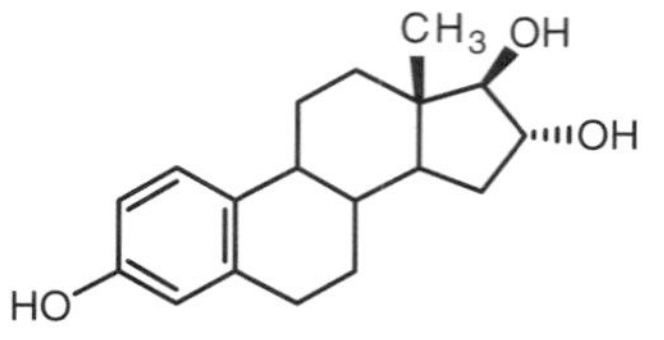

estriol

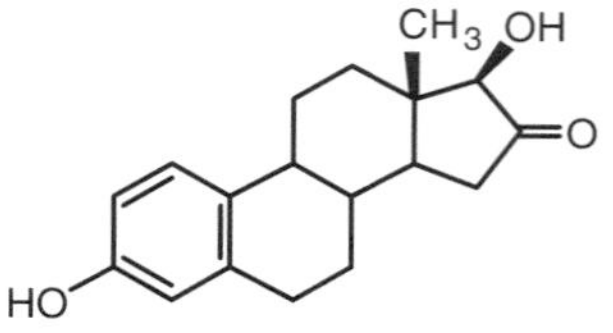

16-keto-17β-estradiol

Substrates include estriol (producing 16-keto-17β-estradiol) and 16α,17β-dihydroxyandrost-4-en-3-one (producing 16-ketotestosterone).

[1] R. A. Meigs & K. J. Ryan (1966) *JBC* **241**, 4011.
[2] P. Preumont & M. Smuk (1975) *Acta Endocrinol.* **78**, 760.

Selected entries from *Methods in Enzymology* [vol, page(s)]:
General discussion: 15, 718, 753
Assay: 15, 753
Other molecular properties: distribution, **15**, 718; properties, **15**, 755; purification, **15**, 754; staining on gels, **22**, 594; substrate specificity, **15**, 755

16β-HYDROXYSTEROID DEHYDROGENASE

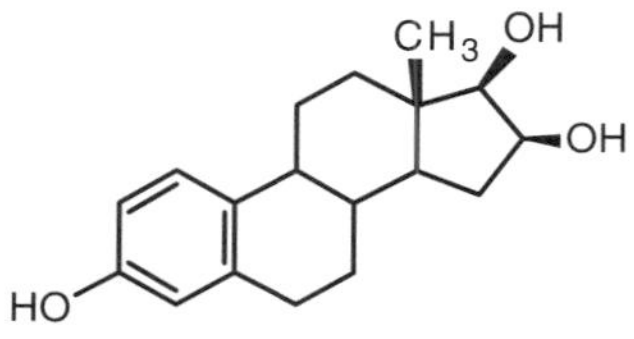

16-epiestriol

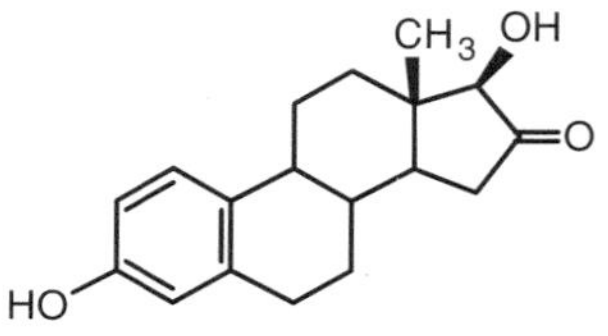

16-keto-17β-estradiol

This oxidoreductase, also known as 16β-hydroxyestrogen oxidoreductase, catalyzes the reversible reaction of 16-epiestriol with NAD(P)$^+$ to produce 16-keto-17β-estradiol and NAD(P)H.

Selected entries from *Methods in Enzymology* [vol, page(s)]:
General discussion: 15, 716, 718
Assay: 15, 716, 718
Other molecular properties: distribution, **15**, 719; properties, **15**, 717, 719; purification, **15**, 719; staining on gels, **22**, 594

21-HYDROXYSTEROID DEHYDROGENASE

21-Hydroxysteroid dehydrogenase (NAD$^+$) [EC 1.1.1.150] catalyzes the reversible reaction of pregnan-21-ol with NAD$^+$ to produce pregnan-21-al and NADH.[1-3]

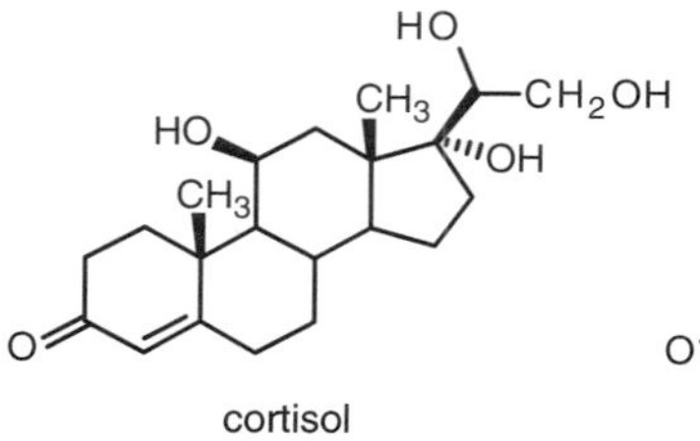

cortisol

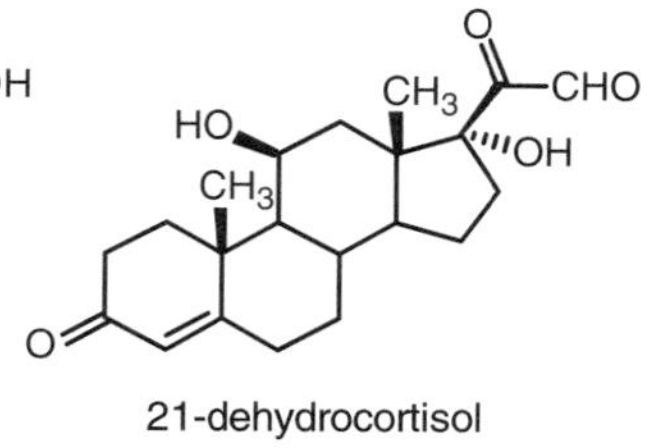

21-dehydrocortisol

21-Hydroxysteroid dehydrogenase (NADP$^+$) [EC 1.1.1.151] catalyzes the reaction of pregnan-21-ol with NADP$^+$ to produce pregnan-21-al and NADPH.[3]

With both dehydrogenases, other 21-hydroxycorticosteroids can also serve as substrates, including cortisol which is converted to 21-dehydrocortisol.

[1] J. C. Orr & C. Monder (1975) *JBC* **250**, 7547.
[2] J. C. Orr & C. Monder (1975) *J. Steroid Biochem.* **6**, 297.
[3] C. Monder & A. White (1965) *JBC* **240**, 71.

Selected entries from *Methods in Enzymology* [vol, page(s)]:
General discussion: 15, 667
Assay: 15, 667
Other molecular properties: NAD$^+$-dependent, of lamb liver (properties, **15**, 670; purification, **15**, 668; purification steps, **15**, 670); NADH-dependent, of bovine adrenal (properties, **15**, 675; purification procedure, **15**, 673; purification steps, **15**, 674); NADPH-dependent, of lamb liver (properties, **15**, 672; purification procedure, **15**, 671; purification steps, **15**, 672); staining on gels, **22**, 594; stereospecificity, **87**, 111

7β-HYDROXYSTEROID DEHYDROGENASE (NADP$^+$)

This oxidoreductase [EC 1.1.1.201], also known as NADP$^+$-dependent 7β-hydroxysteroid dehydrogenase, catalyzes the reversible reaction of a 7β-hydroxysteroid with NADP$^+$ to produce a 7-ketosteroid (*i.e.*, 7-oxosteroid) and NADPH.[1-3]

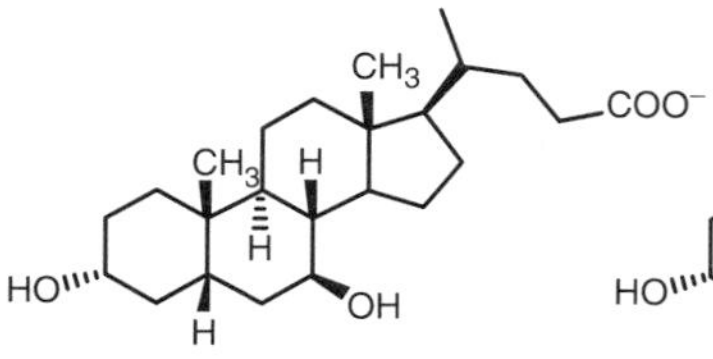

ursodeoxycholate

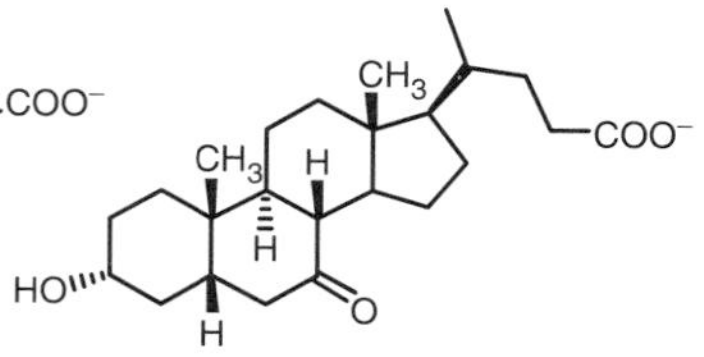

7-ketolithocholate

Substrates include bile acids such as ursodeoxycholate and bile acid conjugates that contain a 7β-hydroxyl group.

[1] I. A. Macdonald & P. D. Roach (1981) *BBA* **665**, 262.
[2] J. D. Sutherland & C. N. Williams (1985) *J. Lipid Res.* **26**, 344.
[3] R. Edenharder, A. Pfützner & R. Hammann (1989) *BBA* **1004**, 230.

Selected entries from *Methods in Enzymology* [vol, page(s)]:
General discussion: 15, 716
Assay: 15, 716
Other molecular properties: properties, **15**, 717; staining on gels, **22**, 594

16-HYDROXYSTEROID EPIMERASE

This epimerase [EC 5.1.99.2] catalyzes the interconversion of a 16α-hydroxysteroid and a 16β-hydroxysteroid. 16-Hydroxyestrone is also a substrate.[1]

[1] K. Dahm, M. Lindlau & H. Breuer (1968) *BBA* **159**, 377.

16α-hydroxyestrone   16β-hydroxyestrone

4α-HYDROXYTETRAHYDROBIOPTERIN DEHYDRATASE

This dehydratase [EC 4.2.1.96], also called tetrahydro-biopterin dehydratase and pterin-4α-carbinolamine dehydratase, catalyzes the conversion of (6R)-6-(L-*erythro*-1,2-dihydroxypropyl)-5,6,7,8-tetrahydro-4α-hydroxypterin to 3(6R)-6-(L-erythro-1,2-dihydroxypropyl)-7,8-dihydro-6*H*-pterin and water. Other 4α-hydroxytetrahydropterins can act as alternative substrates.

The protein is a bifunctional enzyme involved in regeneration of tetrahydrobiopterin and as a cofactor for the transcription factor HNF-1α. Two histidyl residues act as general acid catalysts for the stereospecific elimination of the hydroxyl group. A third histidyl residue serves in substrate binding and as a component of base catalysis.[1]

[1] I. Rebrin, B. Thony, S. W. Bailey & J. E. Ayling (1998) *Biochemistry* **37**, 11246.

4-HYDROXYTHREONINE-4-PHOSPHATE DEHYDROGENASE

This oxidoreductase [EC 1.1.1.262], also known as NAD^+-dependent threonine 4-phosphate dehydrogenase, L-threonine 4-phosphate dehydrogenase, and 4-(phospho-hydroxy)-L-threonine dehydrogenase, catalyzes the reaction of 4-(phosphonooxy)threonine with NAD^+ to produce 2-amino-3-oxo-4-phosphonooxybutyrate, NADH, and a proton.[1]

[1] D. E. Cane, Y. Hsiung, J. A. Cornish, J. K. Robinson & I. D. Spenser (1998) *JACS* **120**, 1936.

5-HYDROXYVALERYL-CoA DEHYDRATASE

This enzyme catalyzes the reversible conversion of 5-hydroxyvaleryl-CoA to 4-pentenoyl-CoA and water. The color of *Clostridium aminovalericum* enzyme is green, owing to the presence of a flavin charge-transfer complex.[1]

See also *Enoyl-CoA Hydratase; 4-Hydroxybutyryl-CoA Dehydratase*

[1] U. Eikmanns & W. Buckel (1991) *EJB* **197**, 661.

25-HYDROXYVITAMIN D₃ 1α-MONOOXYGENASE

This ferredoxin-dependent oxidoreductase [EC 1.14.13.13], also known as 25-hydroxycholecalciferol-1-hydroxylase, 25-hydroxycholecalciferol 1-monooxygenase, and officially as calcidiol-1-monooxygenase, catalyzes the reaction of calcidiol (*i.e.*, 25-hydroxycholecalciferol) with NADPH and dioxygen to produce calcitriol (*i.e.*, 1α,25-dihydroxycholecalciferol), $NADP^+$, and water. A member of the cytochrome P450 superfamily, this enzyme undergoes negative feedback regulation through complexation with vitamin D receptor.[1]

[1] K. Takeyama, S. Kitanaka, T. Sato, M. Kobori, J. Yanagisawa & S. Kato (1997) *Science* **277**, 1827.

Selected entries from *Methods in Enzymology* [**vol**, page(s)]:
General discussion: 52, 15; 67, 445; 282, 200, 213
Assay: 67, 433, 442, 445; 123, 161; 282, 200
Other molecular properties: induction in chicks, **67**, 530; isolation, of mammalian kidney (properties, **123**, 160); microsomes, **67**, 411, 442; properties, **67**, 428, 446, 531; **123**, 160; **282**, 201; reactions, **67**, 532, 533

25-HYDROXYVITAMIN D₃ 24-MONOOXYGENASE

This heme-thiolate- and ferredoxin-dependent oxidoreductase (also known as 25-hydroxycholecalciferol-24-hydroxylase, 25-hydroxycholecalciferol 24-monooxygenase, and calcidiol 24-monooxygenase) catalyzes the oxidation of calcidiol (*i.e.*, 25-hydroxycholecalciferol) with NADPH and dioxygen to produce 24,25-dihydroxycholecalciferol, $NADP^+$, and water.

Selected entries from *Methods in Enzymology* [**vol**, page(s)]:
General discussion: 67, 445; 282, 200, 213
Assay: 67, 433, 442, 445; 282, 200
Other molecular properties: induction in chicks, **67**, 530, 531; isolation, of microsomes, **67**, 411, 442; properties, **67**, 428, 446; **282**, 201; reactions, **67**, 538

HYGROMYCIN B KINASE

This phosphotransferase [EC 2.7.1.119], also known as hygromycin B phosphotransferase and APH(7″), catalyzes the reaction of ATP with hygromycin B to produce ADP and 7″-*O*-phosphohygromycin B.[1] Alternative substrates include 1-*N*-methylhygromycin B and destomycin, but not hydromycin B2.

[1] M. Zalacain, J. M. Pardo & A. Jiménez (1987) *EJB* **162**, 419.

Selected entries from *Methods in Enzymology* [**vol**, page(s)]:
Other molecular properties: gene, **237**, 358; selectable marker for eukaryotic expression system, as, **245**, 303

HYICOLYSIN

This zinc-dependent metalloproteinase (also called *Staphylococcus* neutral protease) and neutral metalloendopeptidase, exhibits a broad activity, but prefers to cleave substrates at Thr–Val peptide bonds.[1,2]

[1] S. Ayora & F. Götz (1994) *Mol. Gen. Genet.* **242**, 421.
[2] S. Ayora, P.-E. Lindgren & F. Götz (1994) *J. Bacteriol.* **176**, 3218.

HYOSCYAMINE (6S)-DIOXYGENASE

This ascorbate- and iron-dependent oxidoreductase [EC 1.14.11.11], also known as hyoscyamine (6S)-hydroxylase and hyoscyamine 6β-hydroxylase, catalyzes the reaction of L-hyoscyamine with α-ketoglutarate (*i.e.*, 2-oxoglutarate) and dioxygen to produce (6S)-hydroxyhyoscyamine, succinate, and carbon dioxide.[1,2]

This enzyme catalyzes the first reaction in the biosynthetic pathway from hyoscyamine to scopolamine in several solanaceous plants. Alternative substrates include isobutyltropine, apoatropine, L-norhyoscyamine, and noratropine-*N*-acetic acid.

[1] T. Hashimoto & Y. Yamada (1987) *EJB* **164**, 277.
[2] T. Hashimoto, A. Hayashi, Y. Amano, J. Kohno, H. Iwanari, S. Usuda & Y. Yamada (1991) *JBC* **266**, 4648.

HYPODERMIN C

This serine endopeptidase [EC 3.4.21.49], also known as *Hypoderma* collagenase, catalyzes peptide-bond hydrolysis in proteins, including native collagen, and prefers to act on Xaa–Ala bonds.[1,2] Little hydrolytic activity can be measured using low-molecular-weight synthetic peptide substrates.

[1] A. Lecroisey, A.-M. Gilles, A. De Wolf & B. Keil (1987) *JBC* **262**, 7546.
[2] A. Lecroisey & B. Keil (1985) *EJB* **152**, 123.

Selected entries from *Methods in Enzymology* [vol, page(s)]:
General discussion: **19**, 32
Other molecular properties: crystals, salt effect, **276**, 44; MARK II, **276**, 281; preparation and properties, **19**, 107

HYPONITRITE REDUCTASE

This metallo-oxidoreductase [EC 1.6.6.6], isolated from *Neurospora crassa*, catalyzes the reaction of two molecules of NADH with hyponitrite ($N_2O_2^{2-}$) to produce two molecules of NAD^+ and two molecules of hydroxylamine.[1]

[1] A. Medina & D. J. D. Nicholas (1957) *Nature* **179**, 533.

HYPOTAURINE AMINOTRANSFERASE

This aminotransferase activity catalyzes the reaction of hypotaurine with α-ketoglutarate (*i.e.*, 2-oxoglutarate) to produce sulfinoacetaldehyde and L-glutamate. Sulfinoacetaldehyde decomposes readily to SO_2 and acetaldehyde: hence, this aminotransferase reaction is essentially irreversible.

Selected entries from *Methods in Enzymology* [vol, page(s)]:
General discussion: **143**, 183
Assay: 143, 184

HYPOTAURINE DEHYDROGENASE

This heme- and molybdenum-dependent oxidoreductase [EC 1.8.1.3] catalyzes the reaction of hypotaurine with water and NAD^+ to produce taurine and NADH.[1]

[1] K. Sumizu (1962) *BBA* **63**, 210.

HYPOTAUROCYAMINE KINASE

This phosphotransferase [EC 2.7.3.6], first isolated from *Phascolosoma vulgare*, catalyzes the reaction of ATP with hypotaurocyamine to produce ADP and N^{ω}-phosphohypotaurocyamine.[1] Taurocyamine is also a substrate.

[1] N. van Thoai, Y. Robin & L.-A. Pradel (1963) *BBA* **73**, 437.

Selected entries from *Methods in Enzymology* [vol, page(s)]:
General discussion: **17A**, 1002

HYPOXANTHINE:GUANINE PHOSPHORIBOSYLTRANSFERASE

This enzyme [EC 2.4.2.8], also known as hypoxanthine phosphoribosyltransferase, IMP pyrophosphorylase, transphosphoribosidase, and guanine phosphoribosyltransferase, catalyzes the purine salvage reaction of hypoxanthine with 5-phospho-α-D-ribose 1-diphosphate to produce IMP and pyrophosphate (or, diphosphate).[1–7]

hypoxanthine   IMP

The enzyme also utilizes guanine and 6-mercaptopurine as substrates, activities that are important in guanine salvage by neurons and prodrug activation in cancer chemotherapy, respectively. Other alternative substrates include xanthine, 6-thioguanine, and 6-azahypoxanthine. The enzyme, which has an ordered Bi Bi kinetic mechanism,[2,3] has been proposed to have an S_N1-type reaction with the formation of an unstable ribooxocarbenium ion intermediate: however, an associative S_N2-type scheme has not been ruled out.[1]

[1] S. P. Craig III & A. E. Eakin (2000) *JBC* **275**, 20231.

[2] L. Yuan, S. P. Craig III, J. H. McKerrow & C. C. Wang (1992) *Biochemistry* **31**, 806.

[3] Y. Xu, J. Eads, J. C. Sacchettini & C. Grubmeyer (1997) *Biochemistry* **36**, 3700.

[4] C. Montero & P. Lorente (1991) *BJ* **275**, 327.

[5] C. M. Schimandle, L. A. Mole & I. W. Sherman (1987) *Mol. Biochem. Parasitol.* **23**, 39.

[6] M. Nagy & A.-M. Ribet (1977) *EJB* **77**, 77.

[7] W. N. Kelley & J. B. Wyngaarden (1974) *Adv. Enzymol.* **41**, 1.

Selected entries from *Methods in Enzymology* [vol, page(s)]:
General discussion: 6, 144; 34, 731; 51, 543, 549
Assay: 6, 144; 51, 543, 544, 550; 63, 32

Other molecular properties: activity, 51, 543, 549; adenosine-5′-phosphate pyrophosphorylase and, 6, 143; amino acid composition, 51, 549; cation requirement, 51, 557; cellular localization, 51, 557; Chinese hamster brain, from, 51, 543; deficiency in cell hybridization, 32, 576, 577; 39, 123, 124; enteric bacteria, from, 51, 549; gene (expression, inhibition by antisense RNA, 151, 519; retrovirus-mediated transfection, 149, 13); human erythrocytes, from, 51, 543; hybridoma production, for, 70, 34, 35, 65, 136; 73, 3; hybrid selection, 58, 352; hydrogen bonds, low barrier, 308, 304; inhibitors, 51, 557, 558; isozymes, 51, 548; kinetic properties, 6, 148; 51, 548, 557; molecular weight, 51, 548; mRNA, detection by microinjection and complementation of mutant cells, 151, 377; pH optimum, 51, 548, 557; phosphoribosylpyrophosphate synthetase assay, 51, 5, 13; physiological function, 51, 558; properties, 6, 148; purification, 6, 146; 51, 545, 553, 558; purity, 51, 549, 555; reaction stoichiometry, 51, 555; regulation, 51, 550; sedimentation coefficient, 51, 548; selectable marker for gene targeting experiments, as, 245, 391; stability, 51, 548, 555; substrate specificity, 6, 148; 51, 548, 549, 550, 557; subunit structure, 51, 548

I

D-IDITOL 2-DEHYDROGENASE

This oxidoreductase [EC 1.1.1.15], also known as D-sorbitol dehydrogenase, catalyzes the reaction of D-iditol with NAD^+ to produce D-sorbose and NADH.[1] The enzyme also catalyzes the NAD^+-dependent conversion of xylitol to L-xylulose and L-glucitol to L-fructose. **See also** *D-Sorbitol Dehydrogenase (Acceptor)*

[1] D. R. D. Shaw (1956) *BJ* **64**, 394.

L-IDONATE 2-DEHYDROGENASE

This oxidoreductase [EC 1.1.1.128], also known as 5-ketogluconate 2-reductase, catalyzes the reaction of 5-dehydro-D-gluconate with NADPH to produce L-idonate and $NADP^+$.[1,2]

[1] T. Chiyonobu, E. Shinagawa, O. Adachi & M. Ameyama (1975) *Agric. Biol. Chem.* **39**, 2425.
[2] Y. Takagi (1962) *Agric. Biol. Chem.* **26**, 719.

L-IDONATE 5-DEHYDROGENASE

This oxidoreductase [EC 1.1.1.264] catalyzes the reaction of L-idonate with $NAD(P)^+$ to produce 5-dehydrogluconate (or, 5-ketogluconate), NAD(P)H, and a proton.[1]

L-idonate   5-ketogluconate

Note that the *Escherichia coli* enzyme will not act on D-gluconate which is acted on by gluconate 5-dehydrogenase [EC 1.1.1.69].

[1] C. Bausch, N. Peekhaus, C. Utz, T. Blais, E. Murray, T. Lowary & T. Conway (1998) *J. Bacteriol.* **180**, 3704.

IDURONATE 2-SULFATASE

This sulfatase [EC 3.1.6.13], also known as iduronate-2-sulfate sulfatase and chondroitinsulfatase, catalyzes the hydrolysis of the 2-sulfate groups on the L-iduronate 2-sulfate units of dermatan sulfate, heparan sulfate, and heparin.[1–3] The enzyme can be assayed with the smaller, artificial substrate O-(α-L-idopyranosyluronate 2-sulfate)-$(1 \rightarrow 4)$-2,5-anhydro-D-[^{3}H]mannitol 6-sulfate.

[1] H. Kresse & J. Glössl (1987) *Adv. Enzymol.* **60**, 217.
[2] P. Di Natale & L. Ronsisvalle (1981) *BBA* **661**, 106.
[3] T. Yutaka, A. L. Fluharty, R. L. Stevens & H. Kihara (1982) *J. Biochem.* **91**, 433.

Selected entries from **Methods in Enzymology** [**vol**, page(s)]:
General discussion: **83**, 573
Assay: **50**, 444, 445; **83**, 573
Other molecular properties: distribution, **50**, 444; human plasma (assay, **83**, 573; properties, **83**, 576; purification, **83**, 574); Hunter syndrome, **50**, 446, 447; α-L-iduronidase, **50**, 444, 447; multiple sulfatase deficiency, **50**, 453, 454; radioactive substrates, **50**, 150; specific activity, **50**, 445

L-IDURONIDASE

This enzyme [EC 3.2.1.76] catalyzes the hydrolysis of α-L-iduronosidic linkages in desulfated dermatan.[1–4] Small substrates such as phenyl-α-L-iduronide, anhydromannitol iduronide, and 4-methylumbelliferyl α-L-iduronide are also hydrolyzed.

[1] H. Kresse & J. Glössl (1987) *Adv. Enzymol.* **60**, 217.
[2] L. H. Rome, A. J. Garvin & E. F. Neufeld (1978) *ABB* **189**, 344.
[3] L. J. Shapiro, C. W. Hall, I. G. Leder & F. Neufeld (1976) *ABB* **172**, 156.
[4] B. Weissmann & R. Santiago (1972) *BBRC* **46**, 1430.

Selected entries from **Methods in Enzymology** [**vol**, page(s)]:
General discussion: **83**, 578
Assay: **50**, 148; colorimetric, **50**, 442, 443; human kidney, **83**, 578; radioactive, **50**, 443, 444
Other molecular properties: human (kidney [assay, **83**, 578; properties, **83**, 581; purification, **83**, 579]; urine, isozymes (properties, **83**, 586; separation, **83**, 585; uptake and binding assays, **83**, 583); mucopolysaccharidosis I, **50**, 443; radioactive substrates, **50**, 150; synthetic substrates, **50**, 141

IMIDAZOLEACETATE 4-MONOOXYGENASE

This FAD-dependent oxidoreductase [EC 1.14.13.5], also known as imidazoleacetate hydroxylase, catalyzes the

4-imidazole acetate   5-hydroxy-4-imidazole acetate

reaction of 4-imidazole acetate with NADH and dioxygen to produce 5-hydroxy-4-imidazole acetate, NAD^+, and water.[1-3]

[1]S. Yamamoto & Y. Ishimura (1991) *A Study of Enzymes* **2**, 315.
[2]T. Watanabe, H. Kambe, I. Imamura, Y. Taguchi, T. Tamura & H. Wada (1983) *Anal. Biochem.* **130**, 321.
[3]M. S. Flashner & V. Massey (1974) in *Mol. Mech. Oxygen Activ.* (O. Hayaishi, ed.), p. 245.

Selected entries from *Methods in Enzymology* [vol, page(s)]:
General discussion: **17B**, 773; **52**, 17
Assay: **17B**, 773
Other molecular properties: properties, **17B**, 777; purification from *Pseudomonas*, **17B**, 774

IMIDAZOLE N-ACETYLTRANSFERASE

This acetyltransferase [EC 2.3.1.2], also known as imidazole acetylase, catalyzes the reaction of acetyl-CoA with imidazole to produce N-acetylimidazole and coenzyme A. Propanoyl-CoA can also act as the acyl donor substrate.[1]

[1]S. C. Kinsky (1960) *JBC* **235**, 94.

IMIDAZOLEGLYCEROL-PHOSPHATE DEHYDRATASE

This enzyme [EC 4.2.1.19] catalyzes the conversion of D-*erythro*-1-(imidazol-4-yl)glycerol 3-phosphate to 3-(imidazol-4-yl)-2-oxopropyl phosphate and water.[1-6] The reaction mechanism probably proceeds via a diazafulvene intermediate.[1]

[1]K. Gohda, Y. Kimura, I. Mori, D. Ohta & T. Kikuchi (1998) *BBA* **1385**, 107.
[2]K. Struhl & R. W. Davis (1977) *PNAS* **74**, 5255.
[3]R. D. Glaser & L. L. Houston (1974) *Biochemistry* **13**, 5145.
[4]D. R. Brady & L. L. Houston (1973) *JBC* **248**, 2588.
[5]R. G. Martin, M. A. Berberich, B. N. Ames, W. W. Davis, R. F. Goldberger & J. D. Yourno (1971) *Meth. Enzymol.* **17B**, 3.
[6]B. N. Ames (1957) *JBC* **228**, 131.

Selected entries from *Methods in Enzymology* [vol, page(s)]:
General discussion: **17B**, 29
Assay: **17B**, 29
Other molecular properties: reporter, **326**, 107; *Salmonella typhimurium*, **17B**, 29, 40 (properties, **17B**, 40; purification, **17B**, 32)

IMIDAZOLONEPROPIONASE

This enzyme [EC 3.5.2.7], also known as imidazolone-5-propionate hydrolase, catalyzes the hydrolysis of 4-imidazolone-5-propanoate to generate N-formimino-L-glutamate.[1-3]

[1]G. R. Smith, Y. S. Halpern & B. Magasanik (1971) *JBC* **246**, 3320.

[2]S. H. Snyder, O. L. Silva & M. W. Kies (1961) *JBC* **236**, 2996.
[3]D. R. Rao & D. M. Greenberg (1961) *JBC* **236**, 1758.

Selected entries from *Methods in Enzymology* [vol, page(s)]:
General discussion: **17B**, 55, 92
Assay: **17B**, 55, 92

3-(IMIDAZOL-5-YL)LACTATE DEHYDROGENASE

This oxidoreductase [EC 1.1.1.111] catalyzes the reaction of (*S*)-3-(imidazol-5-yl)lactate with $NAD(P)^+$ to produce 3-(imidazol-5-yl)pyruvate and NAD(P)H.[1]

[1]J. G. Coote & H. Hasall (1969) *BJ* **111**, 237.

IMMUNOGLOBULIN A₁-SPECIFIC METALLOENDOPEPTIDASE

This metalloendopeptidase [EC 3.4.24.13], also known as immunoglobulin A₁ proteinase and IgA protease, has been identified in a number of microorganisms often found in the human oral cavity and in the upper respiratory tract. The enzyme catalyzes the hydrolysis of peptide bonds in the hinge region of the heavy chain of human immunoglobulin A₁ (IgA₁) at Pro227–Thr228. Human IgA₂ is not a substrate and chimpanzee and gorilla IgA₁ are weak substrates.[1-3] **See also** *Immunoglobulin A₁-Specific Serine-Endopeptidase*

[1]M. H. Mulks & R. J. Shober (1994) *MIE* **235**, 543.
[2]A. J. Plaut & A. Wright (1995) *MIE* **248**, 634.
[3]M. Kilian, J. Reinholdt, H. Lomholt, K. Poulsen & E. V. G. Frandsen (1996) *APMIS* **104**, 321.

Selected entries from *Methods in Enzymology* [vol, page(s)]:
General discussion: **248**, 634
Assay: **248**, 647
Other molecular properties: bacteria producing, **248**, 634; biological activity, **248**, 641; *Capnocytophaga*, **248**, 634 (antibody inhibition, **248**, 641; chemical inhibition, **248**, 640; cleavage specificity, **248**, 635; inhibition, **248**, 634); classification, **248**, 634; cleavage of IgA₁, **248**, 635; *Haemophilus influenzae*, chemical inhibition, **248**, 640; inhibitors (antibodies, **248**, 640; chemical, **248**, 640); properties, **248**, 207, 210, 227, 634; secretion, **248**, 636; stability, **248**, 638; storage, **248**, 638; *Streptococcus*, **248**, 207, 210, 634 (amino acid sequence, **248**, 639; antibody inhibition, **248**, 640; biological effects, **248**, 641; catalytic mechanism, **248**, 638; chemical inhibition, **248**, 640; cleavage specificity, **248**, 635; purification, **248**, 637; structure, **248**, 638); substrate specificity, **248**, 636; synthesis, **248**, 636; types, **248**, 634

IMMUNOGLOBULIN A₁-SPECIFIC SERINE ENDOPEPTIDASE

This serine endopeptidase [EC 3.4.21.72], also called IgA protease and immunoglobulin A₁ protease, catalyzes the hydrolysis of peptide bonds in the hinge region of human immunoglobulin A₁ (IgA₁). The bacterial proteases act at Pro–Xaa bonds in this region of the substrate.

There are at least two types of IgA$_1$-specific serine endo-peptidase. The type 1 protease from *Haemophilus influenzae* will act at Pro231–Ser232 (in the sequence Pro227-Thr-Pro-Ser-Pro-Ser-Thr-Pro-Pro-Thr-Pro-Ser-Pro239 in the hinge region of the heavy chain). The type 1 enzymes from *Neisseria gonorrhoeae* and *Neisseria meningitidis* cleave at Pro237–Ser238. The type 2 enzymes from all three microorganisms act at Pro235–Thr236. The respective seryl residues are *O*-glycosylated. The only natural substrates that have been identified are human, chimpanzee, and gorilla IgA$_1$. By contrast, IgA$_2$ is not a substrate.[1–3]

See also Immunoglobulin A$_1$-Specific Metalloendopeptidase

[1]A. G. Plaut (1983) *Ann. Rev. Microbiol.* **37**, 603.
[2]M. H. Mulks (1985) in *Bacterial Enzymes and Virulence* (I. A. Holder, ed.), CRC Press, Boca Raton, p. 81.
[3]M. Kilian & J. Reinholdt (1986) in *Medical Microbiology*, vol. **5** (C. S. F. Easmon & J. Jeljaszewics, eds.), Academic Press, New York, p. 173.

Selected entries from **Methods in Enzymology** [vol, page(s)]:
General discussion: 244, 137
Assay: 116, 55; **244**, 143; qualitative, **235**, 548; quantitative (immunoassay procedure, **235**, 550; SDS-PAGE procedure, **235**, 546); substrate preparation, **235**, 545
Other molecular properties: active site residues, **244**, 145; activity, **235**, 543; bacterial (cleavage of IgA, **116**, 60; properties and assays, **116**, 55; purification, **116**, 58); catalytic mechanism, **244**, 146; epidemiological investigation, **244**, 150; genes, **244**, 139, 145; *Haemophilus influenzae*, chemical inhibition, **248**, 640; IgA$_1$ substrate purification, **244**, 143; infections from bacterial synthesis, **244**, 137; inhibitors (antibodies, **244**, 148; synthetic peptides, **244**, 147); isolation, **235**, 551; isotypes, **244**, 142; *Neisseria gonorrhoeae*, chemical inhibition, **248**, 640; neisserial, production and isolation, **165**, 117; peptide bonds cleaved by, **235**, 543; processing, **244**, 34, 141; production, **235**, 551; proline cleavage specificity, **244**, 141; properties, **235**, 553; purification, **244**, 144; role in infectious process, **244**, 150; secretion pathways, **244**, 140; sequence homology, **235**, 543; **244**, 145; storage, **244**, 145; structure, **244**, 145; substrates, **235**, 554; substrate specificity, **244**, 141

IMP CYCLOHYDROLASE

This enzyme [EC 3.5.4.10], also known as inosinicase and IMP synthetase, catalyzes the conversion of 5-formamido-1-(5-phosphoribosyl)imidazole-4-carboxamide to produce IMP and water.[1] In many species, this enzyme is bifunctional, also exhibiting a 5-aminoimidazole-4-carboxamide ribonucleotide formyltransferase activity. Channeling of the intermediate between the two active sites has been proposed.[2]

[1]R. Geiger & H. Guglielmi (1975) *Hoppe-Seyler's Z. Physiol. Chem.* **356**, 819.
[2]E. Szabados & R. I. Christopherson (1998) *Int. J. Biochem. Cell Biol.* **30**, 933.

Selected entries from **Methods in Enzymology** [vol, page(s)]:
General discussion: 6, 53, 94
Assay: 6, 94
Other molecular properties: absence in *Escherichia coli* mutant, **51**, 214; 5-amino-4-imidazolecarboxamide ribotide transformylase and, **6**, 92, 93; 5-formamido-4-imidazolecarboxamide ribotide and, **6**, 702; 5-formylamino-4-imidazolecarboxamide ribotide and, **6**, 89; glycinamide ribotide transformylase and, **6**, 65; properties, **6**, 95; purification, **6**, 92, 95

IMP DEHYDROGENASE

This oxidoreductase [EC 1.1.1.205], also known as inosine-5'-monophosphate dehydrogenase and inosinate dehydrogenase, catalyzes the reaction of IMP with NAD$^+$ and water to produce xanthosine 5'-phosphate and NADH.[1–4]

[1]J. A. Digits & L. Hedstrom (1999) *Biochemistry* **38**, 2295.
[2]G. D. Markham, C. L. Bock & C. Schalk-Hihi (1999) *Biochemistry* **38**, 4433.
[3]A. Hampton, L. W. Brox & M. Bayer (1969) *Biochemistry* **8**, 2303.
[4]E. Heyde & J. F. Morrison (1976) *BBA* **429**, 661.

Selected entries from **Methods in Enzymology** [vol, page(s)]:
General discussion: 6, 107; **46**, 301, 302
Assay: 6, 107
Other molecular properties: *Aerobacter aerogenes*, **6**, 107 (properties, **6**, 110; purification, **6**, 108); derepression, **22**, 89; *Escherichia coli*, purification by affinity chromatography on immobilized dyes, **104**, 109; kinetics, **87**, 359; stereospecificity, **87**, 114

INDANOL DEHYDROGENASE

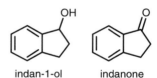

indan-1-ol   indanone

This oxidoreductase [EC 1.1.1.112] catalyzes the reaction of indan-1-ol with NAD(P)$^+$ to produce indanone and NAD(P)H.[1–4] The enzyme will also use 3(20)-α-hydroxysteroids as alternative substrates.[4]

[1]R. E. Billings, H. R. Sullivan & R. E. McMaron (1971) *JBC* **246**, 3512.
[2]A. P. Kulkarni, B. H. Strohm & W. H. Houser (1985) *Xenobiotica* **15**, 513.
[3]A. Hara, K. Kariya, M. Nakamura, T. Nakayama & H. Sawada (1986) *ABB* **249**, 225.
[4]A. Hara, M. Nakagawa, H. Taniguchi & H. Sawada (1989) *J. Biochem.* **106**, 900.

INDOLE-3-ACETALDEHYDE OXIDASE

This heme-dependent oxidoreductase [EC 1.2.3.7] catalyzes the reaction of two molecules of indole-3-acetaldehyde with dioxygen to produce two molecules of indole-3-acetate and two water molecules.[1–3] Other substrates include indole-3-aldehyde and acetaldehyde, albeit more slowly.

[1]P. J. Bower, H. M. Browns & W. K. Purves (1978) *Plant Physiol.* **61**, 107.
[2]S. Miyata, Y. Suzuki, S. Kamisaka & Y. Masuda (1981) *Physiol. Plant.* **51**, 402.
[3]R. Rajagopal (1971) *Physiol. Plant.* **24**, 272.

INDOLE-3-ACETALDEHYDE REDUCTASE (NADH)

This oxidoreductase [EC 1.1.1.190] catalyzes the reaction of indole-3-acetaldehyde with NADH to produce indole-3-ethanol and NAD+.[1,2]

[1] H. M. Brown & W. K. Purves (1976) JBC 251, 907.
[2] H. M. Brown & W. K. Purves (1980) Plant Physiol. 65, 107.

INDOLE-3-ACETALDEHYDE REDUCTASE (NADPH)

This oxidoreductase [EC 1.1.1.191] catalyzes the reaction of indole-3-acetaldehyde with NADPH to produce indole-3-ethanol and NADP+.[1–4]

[1] J. Ludwig-Müller & W. Hilgenberg (1990) Physiol. Plant. 80, 541.
[2] J. Ludwig-Müller, P. Schramm & W. Hilgenberg (1990) Physiol. Plant. 80, 472.
[3] H. M. Brown & W. K. Purves (1976) JBC 251, 907.
[4] H. M. Brown & W. K. Purves (1980) Plant Physiol. 65, 107.

INDOLEACETALDOXIME DEHYDRATASE

This enzyme [EC 4.2.1.29] catalyzes the conversion of 3-indoleacetaldoxime to 3-indoleacetonitrile and water.[1]

[1] P. S. Shulka & S. Mahadevan (1970) ABB 137, 166.

INDOLE-3-ACETATE β-GLUCOSYLTRANSFERASE

This transferase [EC 2.4.1.121] catalyzes the reaction of UDP-D-glucose with indole-3-acetate to produce UDP and indole-3-acetyl-β-1-D-glucose.[1–3] Naphthalene-1-acetate is another glucosyl acceptor.

[1] L. Michalczuk & R. S. Bandurski (1982) BJ 207, 273.
[2] A. J. Leznicki & R. S. Bandurski (1988) Plant Physiol. 88, 1474 and 1481.
[3] S. Kowalczyk & R. S. Bandurski (1991) BJ 279, 509.

INDOLEACETYLGLUCOSE:INOSITOL O-ACYLTRANSFERASE

This transferase [EC 2.3.1.72] catalyzes the reaction of indole-3-acetyl-β-1-D-glucose with myo-inositol to produce D-glucose and indole-3-acetyl-myo-inositol.[1–3]

[1] J. M. Kesy & R. S. Bandurski (1990) Plant Physiol. 94, 1598.
[2] L. Michalczuk & R. S. Bandurski (1982) BJ 207, 273.
[3] L. Michalczuk & R. S. Bandurski (1980) BBRC 93, 588.

N-(INDOLE-3-ACETYL)-L-LYSINE SYNTHETASE

This ligase [EC 6.3.2.20], also known as indoleacetate:lysine ligase, catalyzes the reaction of ATP with indole-3-acetate and L-lysine to produce ADP, orthophosphate, and N^6-[(indole-3-yl)acetyl]-L-lysine.[1]

[1] O. Hutzinger & T. Kosuga (1968) Biochemistry 7, 601.

INDOLEAMINE-PYRROLE 2,3-DIOXYGENASE

This heme-dependent oxidoreductase [EC 1.13.11.42], also known as indoleamine 2,3-dioxygenase, catalyzes the reaction of L-tryptophan with dioxygen to produce N-formylkynurenine.[1–4] Alternative substrates include many substituted and unsubstituted indoleamines, including melatonin, D- and L-tryptophan, tryptamine, serotonin and D- and L-5-hydroxytryptophan.[1] Many publications fail to make a clear distinction between this enzyme and the more specific tryptophan 2,3-dioxygenase [EC 1.13.11.11], and in some organisms, they may be the same enzyme. **See also** Tryptophan 2,3-Dioxygenase

[1] S. Yamamoto & O. Hayaishi (1967) JBC 242, 5260.
[2] F. Hirata, O. Hayaishi, T. Tokuyama & S. Senoh (1974) JBC 249, 1311.
[3] R. Yoshida & O. Hayaishi (1978) PNAS 75, 3998.
[4] Y. Kudo & C. A. R. Boyd (2000) BBA 1500, 119.

Selected entries from **Methods in Enzymology** [vol, page(s)]:
General discussion: 52, 13, 16, 20
Assay: 105, 63; 142, 192
Other molecular properties: catalyzed reaction, 142, 189; intestinal epithelium, in, 77, 156; physiological significance, 105, 67; properties, 105, 62; 142, 194; purification from rabbit intestine, 105, 63; 142, 193; reaction mechanism, role of superoxide anion, 105, 65

INDOLE 2,3-DIOXYGENASE

This oxidoreductase [EC 1.13.11.17], also known as indole oxidase, catalyzes the reaction of indole with dioxygen to produce 2-formylaminobenzaldehyde.[1–3] The jasmin enzyme is a flavoprotein that contains copper and produces anthranilate as the final product. The Tecoma enzyme uses three atoms of oxygen per molecule of indole, generating anthranil (3,4-benzisoxazole). A second nonflavin-dependent enzyme from Tecoma utilizes four atoms of oxygen, and anthranilate is the final product.[2]

[1] C. S. Pundir, G. K. Garg & V. S. Rathore (1984) Phytochemistry 23, 2423.
[2] S. P. Kunapuli & C. S. Vaidyanathan (1983) Plant Physiol. 71, 19.
[3] T. Ohnishi, F. Hirata & O. Hayaishi (1977) JBC 252, 4643.

Selected entries from **Methods in Enzymology** [vol, page(s)]:
General discussion: 52, 9 (skatole 2,3-dioxygenase, 52, 11)

INDOLE ETHANOL OXIDASE

This oxidoreductase, also known as tryptophol oxidase, catalyzes the reaction of indole-3-ethanol (or, tryptophol)

with dioxygen to produce indole-3-acetaldehyde and hydrogen peroxide.[1,2]

[1] J. S. Zhu & G. K. Scott (1995) *Biochem. Mol. Biol. Int.* **35**, 423.
[2] L. E. Vickery & W. K. Purves (1972) *Plant Physiol.* **49**, 716.

Selected entries from *Methods in Enzymology* [vol, page(s)]:
General discussion: 52, 21

INDOLEGLYCEROLPHOSPHATE ALDOLASE

This aldolase, previously classified as EC 4.1.2.8, is now regarded as a side reaction of tryptophan synthase [EC 4.2.1.20].

INDOLE-3-GLYCEROL-PHOSPHATE SYNTHASE

This enzyme [EC 4.1.1.48] catalyzes the conversion of 1-(2-carboxyphenylamino)-1-deoxy-D-ribulose 5-phosphate to 1-(indol-3-yl)glycerol 3-phosphate, carbon dioxide, and water.[1-3] In some organisms, this enzyme is part of a multifunctional protein together with one or more enzymes leading to the biosynthesis of tryptophan.

[1] B. Darimont, C. Stehlin, H. Szadkowski & K. Kirschner (1998) *Protein Sci.* **7**, 1221.
[2] C. N. Hankins, M. Largen & S. E. Mills (1975) *Anal. Biochem.* **69**, 510.
[3] T. E. Creighton & C. Yanofsky (1968) *JBC* **241**, 4616.

Selected entries from *Methods in Enzymology* [vol, page(s)]:
General discussion: 5, 799; **142**, 386
Assay: 5, 799; **17A**, 370
Other molecular properties: anthranilic deoxyribulotide analysis and, **6**, 594; phosphoribosylanthranilate isomerase and (assays, **142**, 389; properties, **142**, 394; purification from *Escherichia coli*, **5**, 800; **142**, 392; substrate preparation, **142**, 387); properties, **5**, 800; purification from (*Escherichia coli*, **5**, 800; **17A**, 372; **142**, 392; *Neurospora crassa*, **17A**, 393)

INDOLE-3-LACTATE DEHYDROGENASE

This oxidoreductase [EC 1.1.1.110] catalyzes the reversible reaction of indole-3-lactate with NAD^+ to produce indole-3-pyruvate and NADH.[1] Other substrates include phenyllactate and *p*-hydroxyphenyllactate.

[1] M. Jean & R. D. DeMoss (1968) *Can. J. Microbiol.* **14**, 429.

INDOLEPYRUVATE DECARBOXYLASE

This thiamin pyrophosphate-dependent decarboxylase [EC 4.1.1.74], also known as indole-3-pyruvate decarboxylase, catalyzes the conversion of 3-(indol-3-yl)pyruvate to 2-(indol-3-yl)acetaldehyde and carbon dioxide.[1] This enzyme is more specific than pyruvate decarboxylase [EC 4.1.1.1].

[1] J. Koga, T. Adachi & H. Hidaka (1992) *JBC* **267**, 15823.

INDOLE-3-PYRUVATE C-METHYLTRANSFERASE

This methyltransferase [EC 2.1.1.47], also known as indolepyruvate methyltransferase, catalyzes the reaction of *S*-adenosyl-L-methionine with indole-3-pyruvate to produce *S*-adenosyl-L-homocysteine and (*S*)-β-methylindole-3-pyruvate.[1]

[1] M. K. Speedie, U. Hornemann & H. G. Floss (1975) *JBC* **250**, 7819.

Selected entries from *Methods in Enzymology* [vol, page(s)]:
General discussion: 43, 498
Assay: 43, 499
Other molecular properties: DEAE-Sephadex chromatography, **43**, 501; inhibitors, **43**, 512; kinetic properties, **43**, 502; molecular weight, **43**, 502; pH effect, **43**, 501, 502; purification, **43**, 500, 501; reaction scheme, **43**, 498; specificity, **43**, 502; *Streptomyces griseus* preparation, **43**, 500

INDOLYLACETYLINOSITOL ARABINOSYLTRANSFERASE

This transferase [EC 2.4.2.34], also known as arabinosylindolylacetylinositol synthase, catalyzes the reaction of UDP-L-arabinose with indol-3-ylacetyl-*myo*-inositol to produce indol-3-ylacetyl-*myo*-inositol L-arabinoside and UDP.[1] Corn kernels are a ready source of this glycosyltransferase.

[1] L. J. Curcuera & R. S. Bandurski (1982) *Plant Physiol.* **70**, 1664.

INDOLYLACETYL-*myo*-INOSITOL GALACTOSYLTRANSFERASE

This transferase [EC 2.4.1.156], also known as indol-3-ylacetyl-*myo*-inositol galactoside synthase, catalyzes the reaction of UDP-D-galactose with indol-3-ylacetyl-*myo*-inositol to produce 5-*O*-(indol-3-ylacetyl-*myo*-inositol)-D-galactoside and UDP.[1]

[1] L. J. Corcuera, L. Michalczuk & R. S. Bandurski (1982) *BJ* **207**, 283.

INFECTIOUS PANCREATIC NECROSIS VIRUS ENDOPEPTIDASE

This endopeptidase participates in the processing of the polyprotein of the barnavirus, infectious pancreatic necrosis virus.[1]

[1] D. S. Manning, C. L. Mason & J. C. Leong (1990) *Virology* **179**, 9.

INITIATION FACTORS

Initiation factors (IF in prokaryotes and eIF in eukaryotes) are treated as a class of protein-synthesizing GTPases [EC 3.6.1.48] that assist in the ribosome-dependent translation of mRNA.[1] In prokaryotes, there are at least three principal initiation factors: IF-1, IF-2 (existing in two forms, IF-2a and IF-2b), and IF-3. IF-2b displays so-called GTPase activity, protects formylmethionyl-tRNA$_f^{Met}$, and promotes its binding to the 70S ribosomal complex. IF-3 binds to the 30S ribosomal subunit and enhances the availability of the 30S subunit.

Although the overall reaction yields ADP and orthophosphate, the enzyme is misclassified as a ATPase or hydrolase, because P-O-P bond scission supplies the Gibbs free energy required to drive protein folding.[2] The overall enzyme-catalyzed reaction should be regarded to be of the energase type (e.g., State$_1$ + ATP = State$_1$ + ADP + P$_i$, where two affinity-modulated conformational states are indicated.)[2] Energases transduce chemical energy into mechanical work.

Many initiation factors have been characterized from eukaryotes: e.g., eIF-1, eIF-2, eIF-3, eIF-4A, etc. The GTP binding factor is eIF-2. *See Energases; G Proteins*

[1] T. V. Kurzchalia, U. A. Bommer, G. T. Babkina & G. G. Karpova (1984) *FEBS Lett.* **175**, 313.
[2] D. L. Purich (2001) *TiBS* **26**, 417.

Selected entries from ***Methods in Enzymology*** [vol, page(s)]:
Assay: 20, 273; 30, 7, 32, 157, 198; 60, 17, 94, 101; crude, 20, 236, 252; factor f1, 20, 237; factor f2, 20, 238; ribosome-dependent GTPase (definition of unit and specific activity, 20, 295; procedure, 20, 294, 324; reagents, 20, 294)

Other molecular properties: activity in polysomes, 59, 368 (assay, 59, 367, 368); affinity labeling, 46, 182, 713; binding site, on ribosomes, 61, 250; crude (crude fraction, preparation, 59, 363, 839, 840; isolation, 20, 239, 251, 272; phase partition and, 20, 275; preparation, 20, 406; 30, 80, 373; proteins in, 30, 381); eukaryotic, see also specific initiation factors below (Krebs ascites cells, 60, 87 [assay, 60, 93, 94; isolation, 60, 89; purification, 60, 94]; pig liver, 60, 240; rabbit reticulocyte [assay {amino acid incorporation, 60, 126; globin synthesis, 60, 126, 127, 170, 202, 203; methionylpuromycin synthesis, 60, 406, 407; polyphenylalanine synthesis, 60, 115}; complexes, 60, 32, 33, 71, 72, 75, 82, 122, 123; distribution in lysate, 60, 80; electrophoretic purity, 60, 28, 159, 204; functional role, 60, 32, 33, 63, 72, 77, 108, 136, 147, 161, 166, 334, 335; identification by gel electrophoresis, 60, 175, 521; molecular weight, 60, 28, 29, 63, 159, 166; nomenclature, 60, 63, 88, 108, 109, 137, 138; phosphorylation, 60, 505; pool size, 60, 78; protein content, 60, 31 {purification, 60, 19, 67, 101, 124, 125, 128, 129, 137, 146, 167, 257, 258, 283, 599; radiolabeled, 60, 34, 35, 68, 69, 75, 122, 123, 208}; radioiodination, 70, 237; wheat germ, 60, 181]); function, 20, 261; gel electrophoresis, 30, 5; globin synthesis and, 30, 120; initial dipeptide synthesis and, 30, 118; initiation factor 1 (assay of, 30, 7, 32; crystallization, 30, 38; eukaryotic, 30, 197 [assays, 30, 198; properties, 30, 205; purification, 30, 201; ribosonal subunit preparation, 30, 199]; function of, 30, 32; properties, 30, 38; purification, 30, 8, 34; radioactive labeling, 30, 24); initiation factor 2 (assay of activity, 30, 5, 27; phosphorylation of, 30, 29; purification, 30, 11, 26; resolution of two proteins, 30, 13, 27); initiation factor 3 (assay of, 30, 7, 45, 54 [materials, 30, 55; procedures, 30, 55]; cistron specificity, measurement of, 30, 55; complete separation from interference factors, 30, 61; definition of, 30, 54; fractionation and isolation of

interference factors, 30, 56; fractions, messenger specificity of, 30, 57, 66; functions, 30, 45; properties, 30, 53; purification, 30, 15, 49; radioactive, 30, 24, 39 [assay, 30, 40; properties, 30, 44; purification, 30, 41]; resolution of two proteins, 30, 51); initiation factor I (function, 30, 154; routine assay of, 30, 157; stability of, 30, 160; ternary complex with [^{35}S]methionyl-tRNA and GTP, 30, 160); initiation factor II (assay, 30, 157, 162; function, 30, 154, 163); initiation factor A (purification, 20, 277); initiation factor B (assay, 20, 275; purification, 20, 277; synthetic oligonucleotides and, 20, 271); initiation factor C (assay of, 20, 275; mRNA binding and, 20, 265; purification, 20, 276); initiation factor eIF-1 (assay [amino acid incorporation, 60, 116; globin synthesis, 60, 17, 18, 31, 94, 101]; functional role, 60, 565; molecular weight, 60, 100, 565; purification, 60, 20, 21, 28, 29, 91, 92, 95, 100, 132, 156, 562, 563); initiation factor eIF-2, 79, 142, 171 (affinity for guanosine nucleotides, 60, 583, 584, 589; assay, 60, 37, 46; 79, 284; binding to mRNA, 60, 381, 395, 396, 398; complex with cofactor Co-EIF-1, 60, 57; dephosphorylation, 60, 527; electrophoretic purity, 60, 46, 189, 267; functional role, 60, 36, 37, 46, 49, 58, 59, 60, 74, 181, 182, 265, 266, 380, 381; inactivation by heme-regulated translational inhibitor, 60, 481; mediated translation in expression of cDNA genes in mammalian cells, 185, 504; molecular weight, 60, 45, 48, 245, 395, 396; phosphorylation, 60, 267, 274, 275, 468, 470, 476, 490, 492, 493, 525, 532; 79, 285 [initiation factor kinase, eIF-2α dsRNA-dependent, from rabbit reticulocyte lysates {assays, 99, 349; properties, 99, 357; purification, 99, 351}; protein kinase, 79, 142, 147, 169, 194, 293]; pool size, 60, 78; preparation [Artemia salina cysts, from, 60, 310, 311; gel electrophoresis, by, 60, 267, 399, 400, 470; heparin-Sepharose chromatography, including, 60, 131, 132; ribonucleic acid-cellulose chromatography, including, 60, 247; pig liver, from, 60, 240; rabbit reticulocyte lysate, from, 60, 20, 21, 24, 25, 39, 47, 48, 107, 154, 406, 468, 581; wheat germ, from, 60, 187, 199]; reduced, in interferon-treated cells, 79, 293; separation and purification, 60, 39, 40, 42, 43; small subunit [molecular weight, 79, 285; phosphorylation, 79, 285]; subunits, dephosphorylation, 60, 532, 533; ternary complex [binding to 40A ribosomal subunit, 60, 275, 461, 563, 586; dissociation by magnesium ion, 60, 562, 570; formation, 60, 460, 461, 565, 585, 586]); initiation factor eIF-3 (assay by [globin synthesis, 60, 17, 18, 31, 93, 94, 121, 122; methionylpuromycin synthesis, 60, 68]; functional role, 60, 408, 409, 533; nomenclature, 60, 88; preparation from [Krebs ascites cells, 60, 91, 92, 95, 97, 98; rabbit reticulocytes, 60, 20, 107, 127, 149, 152; wheat germ, 60, 198, 199]; properties, 60, 88; subunits [acetylation, 60, 539; dephosphorylation, 60, 533, 534; phosphorylation, 60, 526, 527]); initiation factor eIF-5 (assay [catalysis of complex formation, 60, 38; GTP hydrolysis, of, 60, 108, 110, 117, 118; methionylpuromycin synthesis, 60, 18, 19, 31, 38, 68, 170]; functional role, 60, 36, 38, 50, 51, 60, 74, 574; molecular weight, 60, 52; purification, 60, 20, 21, 24, 39, 131, 132, 156); initiation factor eIF-2A (assay [aminoacyl-tRNA binding, 60, 116, 182, 183; methionylpuromycin synthesis, 60, 119, 120; phenylalanylpuromycin synthesis, 60, 121]; functional role, 60, 181, 182; molecular weight, 60, 189, 191; separation and purification, 60, 107, 164, 165, 187); initiation factor eIF-4A (assay by globin synthesis, 60, 17, 18, 31, 93, 121, 122, 170; purification, 60, 20, 21, 26, 27, 94, 107, 132, 156); initiation factor eIF-4B (assay by globin synthesis, 60, 17, 18, 31, 121, 122; preparation, 60, 20, 127, 149, 152); initiation factor eIF-4C (assay, 60, 18, 19, 31, 68, 94, 101, 170; molecular weight, 60, 100; preparation, 60, 20, 21, 27, 91, 92, 95, 98, 132, 156, 164); initiation factor eIF-4D (assay by methionylpuromycin synthesis, 60, 18, 19, 31, 68, 69; functional role, 60, 164; hypusine detection, 106, 348; preparation, 60, 20, 21, 27, 28, 132); initiation factor f1 (properties, 20, 243; purification, 20, 240 [alternative method, 20, 246]; radioactive, preparation of, 20, 245); initiation factor f2 (properties, 20, 244; purification, 20, 243 [alternative method, 20, 247]); initiation factor FI (properties, 20, 258; purification, 20, 256); initiation factor FII (properties, 20, 258; purification, 20, 256); initiation factor FIII (properties, 20, 259; purification, 20, 257); initiation factor IF-1 (assay, 60, 5, 6, 209; catalysis of exchange of ribosomal subunits, **60**, 212; functional role, **60**, 20, 204, 205; interaction with ribosomal particles, **60**, 13, 15; isotopically labeled, **60**, 13; radiolabeled [preparation, **59**, 790, 791; ribosomal binding activity, **59**, 792]; separation and purification, **60**, 6, 206, 207; stimulation of formylmethionyl-tRNA binding to 30S ribosomal subunits, **60**, 209 [of recycling of initiation factor IF-2, **60**, 211, 212]; yield and purity, **60**, 8); initiation factor IF-2 (assay, **60**, 5, 226, 227; binding of formylmethionyl-tRNA, **60**, 14, 15, 215, 217; molecular weight, **60**, 229, 245; purification from bovine mitochondria [extract preparation, **264**, 254; high-performance liquid chromatography, **264**, 250, 255; mitochondria isolation, **264**, 253; yield, **264**, 261]; radiolabeled, **60**, 208 [preparation, **59**, 790, 791]; recycling, **60**, 211, 212; ribosomal initiation complex formation assay, **264**, 251; separation and purification, **60**, 6, 7, 206, 207, 225, 339; yield and purity, **60**, 8, 206, 229); initiation factor IF-3 (assay, **60**, 5, 6, 230; binding sites in MS-2 RNA, **60**, 343; content of *Escherichia coli* cells, **60**, 237, 238; *Escherichia coli*, reductive alkylation, from **47**, 475; functions, **60**, 13, 230, 437; interaction with rRNA, physical studies, **164**, 238; photoaffinity labeling, for, **59**, 798; radiolabeling, **59**, 784, 790, 791, 794, 795; **60**, 208; recycling, **60**, 221; separation and purification, **60**, 6, 206, 207, 234, 338, 339, 344, 345; ternary complex with 30S ribosomal subunits and MS-2 RNA, **60**, 443; yield and purity, **60**, 8, 206); initiation factor kinases (eIF-2α dsRNA-dependent, from rabbit reticulocyte lysates [assays, **99**, 349; properties, **99**, 357; purification, **99**, 351]); initiation factor M₁ (liver [purification, **30**, 176, 180; separation, **30**, 174, 183]; reticulocyte, purification of, **30**, 131; ribosomal subunit reassociation and, **30**, 191; separation of, **30**, 130); initiation factor M₂ (liver, separation of, **30**, 186); initiation factor M₂A (GTP, and, **30**, 114; purification, **30**, 133, 177, 180); initiation factor M₂(A+B) (elution of, **30**, 130; separation of, **30**, 174); initiation factor M₂B (purification, **30**, 135, 178, 180); initiation factor M₃ (elution of, **30**, 130; purification, **30**, 136, 179, 180 [buffers, **30**, 137; discussion, **30**, 141; procedure, **30**, 137; reagents, **30**, 137]; separation, **30**, 174); materials used, **30**, 3; methionyl-puromycin synthesis and, **30**, 113, 119; methionyl-tRNA binding to natural templates and, **30**, 116; muscle, preparation of, **30**, 673; nomenclature, **30**, 39; polyphenylalanine synthesis and, **30**, 110; preparation of, **30**, 656; purified, characteristics of, **30**, 151; rabbit reticulocyte (assay and biological function, **30**, 160; assay methods, **30**, 157; buffers and reagents, **30**, 155; purification procedures, **30**, 158; results and discussion, **30**, 164); reagents for purification, **20**, 235, 248, 271; removal of, **30**, 5; ribosomal subunit reassociation, **30**, 186 (assay, **30**, 189; characteristics of reaction, **30**, 194; preparation of subunits, **30**, 188; reagents, **30**, 188; remarks, **30**, 196); wheat germ (assays, **118**, 114, 135; functional properties, **118**, 127; physical properties, **118**, 124; purification, **118**, 110, 116, 130)

INNER MEMBRANE PROTEASE

This serine endopeptidase, often abbreviated IMP, catalyzes the processing of proteins targeted for the inner mitochondrial membrane. Two of these enzymes have been identified in *Saccharomyces cerevisiae*, IMP1 and IMP2. Each processes its own subset of mitochondrial precursors: IMP1 processes cytochrome *c* oxidase subunit II and cytochrome b_1 while IMP2 processes cytochrome c_1. IMP1 appears to have a preference for an asparaginyl residue in the P1 position.[1-3]

[1] E. Pratje, K. Esser & G. Michaelis (1994) in *Signal Peptidases* (G. von Heijne, ed.), R. G. Landes, Austin, Texas, p. 105.
[2] J. Nunnari, T. D. Fox & P. Walter (1993) *Science* **262**, 1997.
[3] A. Schneider, W. Oppliger & P. Jenö (1994) *JBC* **269**, 8635.

INORGANIC PYROPHOSPHATASE

Inorganic pyrophosphatase [EC 3.6.1.1] plays a central role in phosphorus metabolism by catalyzing the hydrolysis of the phosphoanhydride bond of inorganic pyrophosphate (or, diphosphate, often abbreviated PP_i) to produce two molecules of orthophosphate.[1-6] The actual substrate is the magnesium-pyrophosphate complex.

The enzyme's specificity varies with the nature of the metal ion: for example, yeast pyrophosphatase is very specific for pyrophosphate in the presence of Mg^{2+} but will catalyze the hydrolysis of ATP and tripolyphosphate in the presence of Zn^{2+} or Mn^{2+}. The yeast enzyme's mechanism proceeds *via* a pyrophosphoryl- and a phosphoryl-enzyme intermediates.

This hydrolytic cleavage reaction provides an additional thermodynamic impetus for driving pyrophosphate-forming biosynthetic reactions.

The vacuolar H^+-translocating pyrophosphatases are members of a new class of energized ion translocases. This electrogenic proton pump is found in most land plants, but only in some alga, protozoa, bacteria, and archaebacteria.[7] Although the overall reaction yields 2 molecules of orthophosphate, this vacuolar H^+-translocating pyrophosphatases is misclassified as a hydrolase, because P-O-P bond scission supplies the energy to drive proton translocation.[8] The overall enzyme-catalyzed reaction should be regarded to be of the energase type (*e.g.*, $[H]_{out} + PP_i = [H]_{in} + 2P_i$). Energases transduce the favorable Gibbs free energy of covalent bond scission into mechanical work, in this case the work associated in creating or maintaining an ion gradient.[8]

[1] A. A. Baykov, B. S. Cooperman, A. Goldman & R. Lahti (1999) *Prog. Mol. Subcell. Biol.* **23**, 127.
[2] P. Heikinheimo, J. Lehtonen, A. Baytov, R. Lahti, B. S. Cooperman & A. Goldman (1996) *Structure* **4**, 1491.
[3] B. S. Cooperman, A. A. Baykov & R. Lahti (1992) *Trends Biochem. Sci.* **17**, 262.
[4] J. Josse & S. C. K. Wong (1971) *The Enzymes*, 3rd ed., **4**, 499.
[5] L. G. Butler (1971) *The Enzymes*, 3rd ed., **4**, 529.
[6] M. Kunitz & P. W. Robbins (1961) *The Enzymes*, 2nd ed., **5**, 169.
[7] M. Maeshima (2000) *BBA* **1465**, 37.
[8] D. L. Purich (2001) *TiBS* **26**, 417.

Selected entries from **Methods in Enzymology** [vol, page(s)]:
General discussion: 2, 570; 87, 526.
Assay: 2, 570; 4, 373; 32, 397; 87, 526; *Rhodospirillum rubrum*, membrane-bound, **126**, 541; tonoplast-associated, **148**, 114.
Other molecular properties: aceto-coenzyme A kinase, not present in preparation of, 5, 464; active-site, 87, 539; adenosine diphosphate sulfurylase, elimination in assay of, 5, 977; adenosine-5'-phosphate

pyrophosphorylase and, **6**, 143; adenylate ribosidase and, **6**, 121; alumina gel, elution from, **9**, 9; amino acid activation and, **5**, 705, 706, 719, 722; amino acid incorporation enzymes and, **5**, 735; application in DNA sequencing, **216**, 339; bioluminescence, effect on, **57**, 29, 33; bovine heart mitochondria, membrane-bound, **126**, 449; immobilized, UDPglucose synthesis with, **136**, 279; magnesium and, **2**, 669; mechanism, **87**, 526; ^{18}O-labeled, **64**, 61, 65; partition coefficients, **64**, 82; preparative acrylamide gel electrophoresis of, **22**, 431; purification, **87**, 526; reaction pathway, **64**, 76; *Rhodospirillum rubrum*, membrane-bound , **126**, 541; specificity, **87**, 528; staining on gels, **22**, 601; structure, **87**, 527; triphosphatase and, **2**, 581; yeast (active site studies, **87**, 539; assay, **87**, 526; binding sites, **87**, 529; catalysis of P$_i$:PP$_i$ equilibration, **87**, 531; catalytic activities and substrate specificity, **87**, 528; mechanism, **87**, 545; properties, **2**, 575; purification, **2**, 571, 575; **87**, 526; structure, **87**, 527)

scyllo-INOSAMINE KINASE

This phosphotransferase [EC 2.7.1.65] catalyzes the reaction of ATP with 1-amino-1-deoxy-*scyllo*-inositol (that is, *scyllo*-inosamine) to produce ADP and 1-amino-1-deoxy-*scyllo*-inositol 4-phosphate (that is, *scyllo*-inosamine 4-phosphate).[1] Other substrates include streptamine, 2-deoxystreptamine, and 1D-1-guanidino-3-amino-1,3-dideoxy-*scyllo*-inositol.

[1] J. B. Walker & M. S. Walker (1967) *Biochemistry* **6**, 3821.

Selected entries from *Methods in Enzymology* [vol, page(s)]:
General discussion: 43, 444
Assay: method I, **43**, 445 (extract preparation, **43**, 446; mycelia growth, **43**, 446; L-ornithine in incubation mixture, **43**, 445; paper chromatography, **43**, 446, 447); method II, **43**, 447 (L-arginine:inosamine-phosphate amidinotransferase, **43**, 447); radiochemical method, **43**, 451
Other molecular properties: aminodeoxy-*scyllo*-inositol, **43**, 448; L-arginine:inosamine-phosphate amidinotransferase assay, **43**, 447; [γ-^{32}P]ATP assay, **43**, 451; biological distribution, **43**, 450; 2-deoxystreptamine from kanamycin, **43**, 449, 450; equations, **43**, 444; inhibitors, **43**, 451; monoamidinated streptamines, **43**, 449; paper chromatography, **43**, 446, 447; properties, **43**, 450, 451; purification, **43**, 450; specificity, **43**, 450; streptamine and streptidine from dihydro-streptomycin, **43**, 448, 449; *Streptomyces bikiniensis*, from, **17A**, 1014

INOSAMINE-PHOSPHATE AMIDINOTRANSFERASE

This transferase [EC 2.1.4.2] catalyzes the reaction of L-arginine with 1-amino-1-deoxy-*scyllo*-inositol 4-phosphate (that is, *scyllo*-inosamine 4-phosphate) to produce L-ornithine and 1-guanidino-1-deoxy-*scyllo*-inositol 4-phosphate.[1-3] Other substrates include 1D-1-guanidino-3-amino-1,3-dideoxy-*scyllo*-inositol 6-phosphate, streptamine phosphate, and 2-deoxystreptamine phosphate. Canavanine can substitute for L-arginine.

[1] J. B. Walker (1974) *JBC* **249**, 2397.
[2] J. B. Walker (1973) *The Enzymes*, 3rd ed., **9**, 497.
[3] M. S. Walker & J. B. Walker (1966) *JBC* **241**, 1262.

Selected entries from *Methods in Enzymology* [vol, page(s)]:
General discussion: 43, 451

Other molecular properties: assay (canavanine, **43**, 454, 455; L-[*guanidino*-^{14}C]arginine, **43**, 452; hydroxylamine, **43**, 455, 456); ATP:inosamine phosphotransferase assay, **43**, 447; biological distribution, **43**, 457; canavanine:ammonium hydroxide transamination, **43**, 458; chemically phosphorylated inosamine derivative, **43**, 453, 454; inhibitors, **43**, 458; natural acceptors preparation, **43**, 453; properties, **43**, 457, 458; purification, **43**, 453, 456, 457; specific activity unit, **43**, 456; specificity, **43**, 457; unit definition, **43**, 456

INOSINATE NUCLEOSIDASE

This nucleosidase [EC 3.2.2.12], first isolated from *Aspergillus oryzae*, catalyzes the hydrolysis of IMP to produce hypoxanthine and D-ribose 5-phosphate.[1]

[1] A. Kuninaka (1957) *Koso Kagaku Shinpojiumu* **12**, 65.

INOSINE KINASE

This phosphotransferase [EC 2.7.1.73], also known as guanosine–inosine kinase, catalyzes the reaction of ATP with inosine to generate IMP and ADP.[1-3] Guanosine is also phosphorylated to produce GMP.

[1] A. Combes, J. Lefleuriel & F. Le Floc'h (1989) *Plant Physiol. Biochem.* **27**, 729.
[2] E. P. Anderson (1973) *The Enzymes*, 3rd ed., **9**, 49.
[3] K. J. Pierre & G. A. LePage (1968) *Proc. Soc. Exp. Biol. Med.* **127**, 432.

INOSINE NUCLEOSIDASE

This nucleosidase [EC 3.2.2.2], also called inosinase, catalyzes the hydrolysis of inosine to generate hypoxanthine and D-ribose.[1-6] Other substrates include xanthosine, nebularine, and 8-azainosine.

[1] G. Davies, M. L. Sinnott & S. G. Withers (1998) *CBC* **1**, 119.
[2] A. Guranowski (1982) *Plant Physiol.* **70**, 344.
[3] F. Le Floc'h & J. Lafleuriel (1981) *Phytochemistry* **20**, 2127.
[4] M. Yoshino, T. Tsukada & K. Tsushima (1978) *Arch. Microbiol.* **119**, 59.
[5] A. L. Koch (1956) *JBC* **223**, 535.
[6] B. A. Horenstein & V. L. Schramm (1993) *Biochemistry* **32**, 7089.

Selected entries from *Methods in Enzymology* [vol, page(s)]:
Other molecular properties: inhibitor design, **308**, 200; neural network analysis, **308**, 416; Raman spectroscopy, difference, **308**, 200; transition state, **308**, 199, 418, 422

INOSITOL-1,3-BISPHOSPHATE 3-PHOSPHATASE

This magnesium-independent phosphatase [EC 3.1.3.65] catalyzes the hydrolysis of D-*myo*-inositol 1,3-bisphosphate to produce D-*myo*-inositol 1-monophosphate and orthophosphate.[1,2] This enzyme is distinct from inositol-1,4-bisphosphate 1-phosphatase [EC 3.1.3.57], inositol-1,4,5-trisphosphate 1-phosphatase [EC 3.1.3.61], and inositol-3,4-bisphosphate 4-phosphatase [EC 3.1.3.66].

[1] S. Howell, R. J. Barnaby, T. Rowe, C. I. Ragan & N. S. Gee (1989) *EJB* **183**, 169.

[2]K. K. Caldwell, D. L. Lips, V. S. Bansal & P. W. Majerus (1991) *JBC* **266**, 18378.

INOSITOL-1,4-BISPHOSPHATE 1-PHOSPHATASE

This phosphatase [EC 3.1.3.57], also called inositol polyphosphate 1-phosphatase, catalyzes the hydrolysis of D-*myo*-inositol 1,4-bisphosphate to produce D-*myo*-inositol 4-phosphate and orthophosphate.[1–4] An alternative substrate is D-*myo*-inositol 1,3,4-trisphosphate (forming D-*myo*-inositol 3,4-bisphosphate). The enzyme does not act on inositol 1-phosphate, inositol 1,4,5-trisphosphate, or inositol 1,3,4,5-tetrakisphosphate. In addition, there is also a sigmoidal dependence upon magnesium ion, suggesting that Mg^{2+} binds cooperatively.[1] Lithium ion acts as an uncompetitive inhibitor.[3,4] D-*myo*-Inositol 4-phosphate is a noncompetitive inhibitor and orthophosphate is competitive.[4]

Inositol-1,4,5-trisphosphate 1-phosphatase [EC 3.1.3.61] catalyzes the hydrolysis of D-*myo*-inositol 1,4,5-trisphosphate to produce D-*myo*-inositol 4,5-bisphosphate and orthophosphate.

[1]R. C. Inhorn & P. W. Majerus (1987) *JBC* **262**, 15946.
[2]C. A. Hansen, T. Inubushi, M. T. Williamson & J. R. Williamson (1989) *BBA* **1001**, 134.
[3]P. C. Inhorn & P. W. Majerus (1988) *JBC* **263**, 14559.
[4]A. Delvaux, J. E. Dumont & C. Erneux (1988-1989) *Second Messengers Phosphoproteins* **12**(5–6), 281.

INOSITOL-3,4-BISPHOSPHATE 4-PHOSPHATASE

This phosphatase [EC 3.1.3.66], also known as inositol polyphosphate 4-phosphatase, catalyzes the hydrolysis of D-*myo*-inositol 3,4-bisphosphate to produce D-*myo*-inositol 3-monophosphate and orthophosphate.[1] This enzyme is distinct from inositol-1,4-bisphosphate 1-phosphatase [EC 3.1.3.57], inositol-1,4,5-trisphosphate 1-phosphatase [EC 3.1.3.61], and inositol-1,3-bisphosphate 3-phosphatase [EC 3.1.3.65].

[1]S. Howell, R. J. Barnaby, T. Rowe, C. I. Ragan & N. S. Gee (1989) *EJB* **183**, 169.

myo-INOSITOL 2-DEHYDROGENASE

This oxidoreductase [EC 1.1.1.18] catalyzes the reaction of *myo*-inositol with NAD^+ to produce 2,4,6/3,5-pentahydroxycyclohexanone (or, *scyllo*-inosose) and NADH.[1,2] The enzyme has an ordered Bi Bi kinetic mechanism.[2]

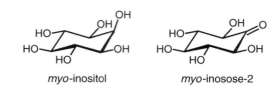

myo-inositol   *myo*-inosose-2

Other substrates include α-D-glucose (producing D-glucono-1,5-lactone) and D-xylose (generating D-xylono-1,4-lactone). ***See also*** *Inosose Reductase*

[1]R. Ramaley, Y. Fujita & E. Freese (1979) *JBC* **254**, 7684.
[2]M. Vidal-Leiria & N. Van Uden (1973) *BBA* **293**, 295.

Selected entries from ***Methods in Enzymology*** [vol, page(s)]:
General discussion: 5, 326; 43, 433, 443
Assay: 5, 326; **43**, 434; aminodeoxy-*scyllo*-inositol, **43**, 434; extract preparations, **43**, 436; high-voltage paper electrophoresis, **43**, 436, 437; *myo*-inositol (labeled, **43**, 434; nonlabeled, **43**, 437, 438); mycelia growth, **43**, 435; reagents, **43**, 434, 438
Other molecular properties: biological distribution, **43**, 438; properties, **5**, 328; **43**, 438; purification, **5**, 326; **43**, 439; reaction scheme, **43**, 433; specificity, **5**, 328; **43**, 439; stereospecificity, **87**, 103

INOSITOL EPIMERASE

This NADPH-dependent epimerase catalyzes the conversion of *myo*-inositol to either *scyllo*-inositol or *neo*-inositol.

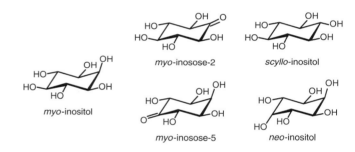

myo-inosose-2   *scyllo*-inositol

myo-inositol

myo-inosose-5   *neo*-inositol

The first step is an $NADP^+$-dependent oxidation to either *myo*-inosose-2 or *myo*-inosose-5 (the production of inosose-2 is the $NADP^+$ version of *myo*-inositol 2-dehydrogenase [EC 1.1.1.18]). The second step involves NADPH-dependent reduction of the carbonyl group generating the *scyllo*- and *neo*-epimers, respectively. ***See also*** *Inosose Reductase; myo-Inositol 2-Dehydrogenase*

Selected entries from ***Methods in Enzymology*** [vol, page(s)]:
General discussion: 89, 593
Assay: 89, 594

INOSITOL 1-α-GALACTOSYLTRANSFERASE

This transferase [EC 2.4.1.123], also known as galactinol synthase, catalyzes the reaction of UDP-D-galactose with *myo*-inositol to produce UDP and 1-*O*-α-D-galactosyl-D-*myo*-inositol.[1,2]

[1] D. M. Pharr, H. N. Sox, R. D. Locy & S. C. Huber (1981) *Plant Sci. Lett.* **23**, 25.

[2] N. Sprenger & F. Keller (2000) *Plant J.* **21**, 249.

myo-INOSITOL 1-KINASE

This phosphotransferase [EC 2.7.1.64] catalyzes the reaction of ATP with *myo*-inositol to produce ADP and 1L-*myo*-inositol 1-phosphate.[1–3]

[1] M. W. Loewus, K. Sasaki, A. L. Leavitt, L. Munsell, W. R. Sherman & F. A. Loewus (1982) *Plant Physiol.* **70**, 1661.

[2] P. D. English, M. Dietz & P. Albersheim (1966) *Science* **151**, 198.

[3] M. Dietz & P. Albersheim (1965) *BBRC* **19**, 598.

myo-INOSITOL 1-*O*-METHYLTRANSFERASE

This methyltransferase [EC 2.1.1.39] catalyzes the reaction of *S*-adenosyl-L-methionine with *myo*-inositol to produce *S*-adenosyl-L-homocysteine and 1-methyl-*myo*-inositol (also known as L-bornesitol).[1,2] L-*chiro*-Inositol is also a substrate.

[1] H. Hofmann & O. Hoffmann-Ostenhof (1969) *Hoppe-Seyler's Z. Physiol. Chem.* **350**, 1465.

[2] F. Koller & O. Hoffmann-Ostenhof (1976) *Hoppe-Seyler's Z. Physiol. Chem.* **357**, 1465.

myo-INOSITOL 3-*O*-METHYLTRANSFERASE

This methyltransferase [EC 2.1.1.40] catalyzes the reaction of *S*-adenosyl-L-methionine with *myo*-inositol to produce *S*-adenosyl-L-homocysteine and 3-methyl-*myo*-inositol (also known as D-bornesitol).[1,2] Other substrates include L-*chiro*-inositol and D-*chiro*-inositol.

[1] I. Wagner, H. Hofmann & O. Hoffmann-Ostenhof (1969) *Hoppe-Seyler's Z. Physiol. Chem.* **350**, 1460.

[2] F. Koller & O. Hoffmann-Ostenhof (1976) *Hoppe-Seyler's Z. Physiol. Chem.* **357**, 1465.

myo-INOSITOL 4-*O*-METHYLTRANSFERASE

This methyltransferase [EC 2.1.1.129], first isolated from *Mesembryanthemum crystallinum*, catalyzes the reaction of *S*-adenosyl-L-methionine with *myo*-inositol to produce *S*-adenosyl-L-homocysteine and 4-methyl-*myo*-inositol.[1]

[1] D. M. Vernon & H. J. Bohnert (1992) *EMBO J.* **11**, 2077.

myo-INOSITOL 6-*O*-METHYLTRANSFERASE

This methyltransferase [EC 2.1.1.134] catalyzes the reaction of *S*-adenosyl-L-methionine with *myo*-inositol to produce *S*-adenosyl-L-homocysteine and 1D-4-*O*-methyl-*myo*-inositol.[1]

[1] W. Wanek & A. Richter (1995) *Planta* **197**, 427.

myo-INOSITOL OXYGENASE

This iron-dependent oxidoreductase [EC 1.13.99.1] catalyzes the reaction of *myo*-inositol with dioxygen to produce D-glucuronate and water.[1–3] There are strong suggestions that L-*myo*-inosose-1 is a reaction intermediate.[1]

[1] N. I. Naber, J. S. Swan & G. A. Hamilton (1986) *Biochemistry* **25**, 7201.

[2] C. C. Reddy, P. A. Pierzchala & G. A. Hamilton (1981) *JBC* **256**, 8519.

[3] F. C. Charalampous (1959) *JBC* **234**, 220.

Selected entries from **Methods in Enzymology** [vol, page(s)]:
General discussion: 5, 329; **52**, 12
Assay: 5, 329

1L-*myo*-INOSITOL-1(OR 4)-PHOSPHATASE

This enzyme [EC 3.1.3.25], also known as *myo*-inositol-1(or 4)-monophosphatase and *myo*-inositol-1-phosphatase, catalyzes the hydrolysis of *myo*-inositol 1-monophosphate to produce *myo*-inositol and orthophosphate.[1–4] Both enantiomers of *myo*-inositol 1-phosphate and *myo*-inositol 4-phosphate can act as substrates. However, the enzyme does not hydrolyze inositol bisphosphates, trisphosphates, or tetrakisphosphates. This phosphatase has a two-metal-catalyzed mechanism in which an activated water molecule serves as the nucleophile. Lithium, an effective therapeutic drug for manic depression, is an uncompetitive inhibitor.

[1] J. R. Atack (1996) *Brain Res. Brain Res. Rev.* **22**, 183.

[2] J. R. Atack, H. B. Broughton & S. J. Pollack (1995) *FEBS Lett.* **361**, 1.

[3] P. W. Majerus (1992) *Ann. Rev. Biochem.* **61**, 225.

[4] J. K. Shute, R. Baker, D. C. Billington & D. Gani (1988) *J. Chem. Soc. Chem. Commun.* **1988**, 422 and 626.

Selected entries from **Methods in Enzymology** [vol, page(s)]:
General discussion: 9, 698; 141, 127
Assay: 9, 699

myo-INOSITOL-1-PHOSPHATE SYNTHASE

This NAD$^+$-dependent enzyme [EC 5.5.1.4] (also known as D-glucose 6-phosphate cycloaldolase, glucose 6-phosphate cyclase, and glucocycloaldolase) catalyzes the conversion of D-glucose 6-phosphate to 1L-*myo*-inositol 1-phosphate.[1–5]

[1] A. L. Majumder, M. D. Johnson & S. A. Henry (1997) *BBA* **1348**, 245.

[2] D. J. Creighton & N. S. R. K. Murthy (1990) *The Enzymes*, 3rd ed., **19**, 323.

[3] M. W. Loewus, D. L. Bedgar & F. A. Loewus (1984) *JBC* **259**, 7644.

[4] J. E. G. Barnett & D. L. Corina (1968) *BJ* **108**, 125.

[5] J. E. G. Barnett, A. Rasheed & D. L. Corina (1973) *BJ* **131**, 21.

Selected entries from **Methods in Enzymology** [vol, page(s)]:
General discussion: 9, 698; 90, 309; 141, 127
Assay: bovine testis, 90, 309; yeast, 9, 698
Other molecular properties: bovine testis (assay, 90, 309; properties, **90**, 312; purification, 90, 310); rat testis, purification, **141**, 131; stereospecificity, 87, 122; yeast, from (properties, 9, 703; purification, 9, 700)

INOSITOLPHOSPHORYL CERAMIDE SYNTHETASE

This enzyme catalyzes the reaction of phosphatidylinositol with a ceramide to produce a diacylglycerol and inositolphosphoryl ceramide.

Selected entries from *Methods in Enzymology* [vol, page(s)]:
General discussion: 311, 123
Assay: 311, 124

1D-*myo*-INOSITOL-TETRAKISPHOSPHATE 1-KINASE

This phosphotransferase [EC 2.7.1.134], also known as 1D-*myo*-inositol-tetraphosphate 1-kinase, catalyzes the reaction of ATP with 1D-*myo*-inositol 3,4,5,6-tetrakisphosphate to produce ADP and 1D-*myo*-inositol 1,3,4,5,6-pentakisphosphate.[1-3] Inositol 1,3,4-trisphosphate inhibits this 1-kinase.

[1] A. Craxton, C. Erneux & S. B. Shears (1994) *JBC* **269**, 4337.
[2] L. R. Stephens, P. T. Hawkins, A. J. Morris & C. P. Downes (1988) *BJ* **249**, 283.
[3] Z. Tan, K. S. Bruzik & S. R. Shears (1997) *JBC* **272**, 2285.

1D-*myo*-INOSITOL-TETRAKISPHOSPHATE 5-KINASE

This phosphotransferase [EC 2.7.1.140] catalyzes the reaction of ATP with 1D-*myo*-inositol 1,3,4,6-tetrakisphosphate to produce 1D-*myo*-inositol 1,3,4,5,6-pentakisphosphate and ADP.[1]

[1] S. B. Shears (1989) *JBC* **264**, 19879.

INOSITOL-1,3,4,5-TETRAKISPHOSPHATE 3-PHOSPHATASE

This phosphatase [EC 3.1.3.62] catalyzes the hydrolysis of D-*myo*-inositol 1,3,4,5-tetrakisphosphate to produce D-*myo*-inositol 1,4,5-trisphosphate and orthophosphate.[1,2] Alternative substrates include inositol 1,3,4,5,6-pentakisphosphate and inositol hexakisphosphate: the kinetic properties of this enzyme indicate that the pentakisphosphate and hexakisphosphate are likely to be the preferred substrates *in vivo*.[2]

[1] P. J. Cullen, R. F. Irvine, B. K. Drobak & A. P. Dawson (1989) *BJ* **259**, 931.
[2] K. Nogimori, P. J. Hughes, M. C. Glennon, M. E. Hodgson, J. W. Putney & S. B. Shears (1991) *JBC* **266**, 16499.

1D-*myo*-INOSITOL-TRISPHOSPHATE 3-KINASE

This calcium-dependent phosphotransferase [EC 2.7.1.127] catalyzes the reaction of ATP with 1D-*myo*-inositol 1,4,5-trisphosphate to produce ADP and 1D-*myo*-inositol 1,3,4,5-tetrakisphosphate.[1-3] There are at least two isozymes of the 3-kinase which are differentially regulated by calcium/calmodulin and via phosphorylation by protein kinase C or the cyclic AMP-dependent protein kinase.[3]

[1] C. A. Hansen, S. Mah & J. R. Williamson (1986) *JBC* **261**, 8100.
[2] R. F. Irvine, A. J. Letcher, J. P. Heslop & M. J. Berridge (1986) *Nature* **320**, 631.
[3] P. J. Woodring & J. C. Garrison (1997) *JBC* **272**, 30447.

1D-*myo*-INOSITOL-TRISPHOSPHATE 5-KINASE

This phosphotransferase [EC 2.7.1.139] catalyzes the reaction of ATP with 1D-*myo*-inositol 1,3,4-trisphosphate to produce ADP and 1D-*myo*-inositol 1,3,4,5-tetrakisphosphate.[1,2] The rat liver enzyme is likely to be identical to 1D-*myo*-inositol-trisphosphate 6-kinase [EC 2.7.1.133].[2] Inositol 3,4,5,6-tetrakisphosphate is an inhibitor of this 5-kinase.[3]

[1] S. B. Shears (1989) *JBC* **264**, 19879.
[2] M. Abdullah, P. J. Hughes, A. Craxton, R. Gigg, T. Desai, J. F. Marecek, G. D. Prestwich & S. B. Shears (1992) *JBC* **267**, 22340.
[3] P. J. Hughes, A. R. Hughes, J. W. Putney & S. B. Shears (1989) *JBC* **264**, 19871.

1D-*myo*-INOSITOL-TRISPHOSPHATE 6-KINASE

This phosphotransferase [EC 2.7.1.133] catalyzes the reaction of ATP with 1D-*myo*-inositol 1,3,4-trisphosphate to produce ADP and 1D-*myo*-inositol 1,3,4,6-tetrakisphosphate.[1-3] The rat liver enzyme is likely to be identical to rat liver 1D-*myo*-inositol-trisphosphate 5-kinase [EC 2.7.1.139].[2] The highly purified enzyme exhibits both the 5- and 6-kinase activities in a ratio of 1:5. Inhibitors of the enzyme include inositol-3,4,5,6-tetrakisphosphate, inositol-1,3,4,5-tetrakisphosphate, and inositol-1,3,4,6-tetrakisphosphate.

[1] C. A. Hansen, S. vom Dahl, B. Huddell & J. R. Williamson (1988) *FEBS Lett.* **236**, 53.
[2] M. Abdullah, P. J. Hughes, A. Craxton, R. Gigg, T. Desai, J. F. Marecek, G. D. Prestwich & S. B. Shears (1992) *JBC* **267**, 22340.
[3] S. B. Shears (1989) *JBC* **264**, 19879.

INOSITOL-1,4,5-TRISPHOSPHATE 1-PHOSPHATASE

This phosphatase [EC 3.1.3.61] catalyzes the hydrolysis of D-*myo*-inositol 1,4,5-trisphosphate to produce D-*myo*-inositol 4,5-bisphosphate and orthophosphate.[1] *See also Inositol-1,4,5-Trisphosphate 5-Phosphatase*

[1] M. M. Van Lookeren Campagne, C. Erneux, R. Van Eijk & P. J. Van Haastert (1988) *BJ* **254**, 343.

INOSITOL-1,4,5-TRISPHOSPHATE 5-PHOSPHATASE

This phosphatase [EC 3.1.3.56] (also known as inositol trisphosphate phosphomonoesterase, inositol polyphosphate 5-phosphatase, and 5PTase) catalyzes the hydrolysis of D-*myo*-inositol 1,4,5-trisphosphate to produce D-*myo*-inositol 1,4-bisphosphate and orthophosphate.[1–4] This is a key enzyme in the polyphosphoinositide signal transduction pathway. The type I variant of this enzyme (but not the type II enzyme) will also catalyze the hydrolysis of inositol 1,3,4,5-tetrakisphosphate to produce inositol 1,3,4-trisphosphate and orthophosphate.[3] However, neither variant will act on inositol 1,4-bisphosphate or inositol 1,3,4-trisphosphate. Interestingly, L-*myo*-inositol 1,4,5-trisphosphate is a good substrate.

[1] M. M. Van Lookeren Campagne, C. Erneux, R. Van Eijk & P. J. Van Haastert (1988) *BJ* **254**, 343.
[2] C. P. Downes, M. C. Mussat & R. H. Michell (1982) *BJ* **203**, 169.
[3] C. A. Hansen, R. A. Johanson, M. T. Williamson & J. R. Williamson (1987) *JBC* **262**, 17319.
[4] D. Communi & C. Erneux (1996) *BJ* **320**, 181.

myo-INOSOSE-2 DEHYDRATASE

This enzyme [EC 4.2.1.44] catalyzes the conversion of 2,4,6/3,5-pentahydroxycyclohexanone (also known as *myo*-inosose-2 or 2-keto-*myo*-inositol) to 3,5/4-trihydroxycyclohexa-1,2-dione and water.[1] The enzyme requires a metal cofactor such as cobalt or manganese ion.

[1] T. Berman & B. Magasanik (1966) *JBC* **241**, 800.

INOSOSE REDUCTASE

This NADPH-dependent oxidoreductase is a component of inositol epimerase and catalyzes the reaction of *myo*-inosose-2 with NADPH to produce *scyllo*-inositol and NADP$^+$. The cockroach enzyme uses NADH. **See** *Inositol Epimerase; myo-Inositol 2-Dehydrogenase*

Selected entries from *Methods in Enzymology* [vol, page(s)]:
Assay: 89, 594

INSULYSIN

This zinc-dependent metalloproteinase [EC 3.4.24.56] (also called insulinase, insulin-degrading enzyme, and insulin protease) catalyzes peptide-bond hydrolysis in a number of polypeptides, most notably insulin (insulysin is the primary protease for the degradation of insulin in cell lysates; its role *in vivo* isn't as clear).[1,2] Insulysin will act on intact insulin (K_m of about 0.1 μM), the insulin B chain (at Glu13–Ala14 and Tyr16–Leu17), oxidatively damaged hemoglobin, β-endorphin, β-amyloid peptide, glucagon, atrial natriuretic factor, transforming growth factor α, and dynorphin A$_{1-13}$ (at Arg6–Arg7).

[1] A. B. Becker, L. Ding & R. A. Roth (1996) in *Diabetes Mellitus: A Fundamental and Clinical Text* (D. LeRoith, J. Olefsky & S. Taylor, eds.), Lippincott-Raven, Philadelphia, p. 242.
[2] A. B. Becker & R. A. Roth (1995) *Meth. Enzymol.* **248**, 693.

Selected entries from *Methods in Enzymology* [vol, page(s)]:
General discussion: 248, 693
Assay: insulin degradation, 248, 698; plate binding method, 248, 699
Other molecular properties: active site, 248, 697; activity, 248, 693; amino acid sequence, 248, 695, 697, 715; atrial natriuretic factor degradation, 248, 694; cDNA, 248, 695; cellular localization, 248, 697, 702; divalent cation requirement, 248, 696; *Drosophila*, 248, 695 (hydrolysis of synthetic substrate QF27, 248, 689); *Eimeria bovis*, 248, 695; evolution, 248, 695; expression in eukaryotic cells, 248, 698; forms, 248, 698; gene knock-out studies, 248, 698, 702; glucagon degradation, 248, 694; hemoglobin degradation, 248, 694; human, homologues, 248, 695; HXXEH sequence, 248, 696; inactivation, 37, 213; insulin degradation, 248, 693 (assay, 248, 698; plate binding method, 248, 699); nomenclature, 248, 693; overexpression studies, 248, 698, 702; peroxisomal targeting sequence, 248, 697; phylogenesis, 248, 216; physiological functions, 248, 693, 698, 702; pitrilysin and, similarity, 248, 695; properties, 248, 212, 684; purification, 248, 695; rat, 248, 695; sequence analysis, 248, 695; tissue distribution, 248, 694; transforming growth factor α degradation, 248, 694; translation, initiation site, 248, 698; zinc content, 248, 696

INULINASE

This enzyme [EC 3.2.1.7], also known as 2,1-β-D-fructan fructanohydrolase and inulase, catalyzes the endohydrolysis of 2,1-β-D-fructosidic linkages in inulin (hence, it is an endoinulinase; note that exoinulinases also exist).[1] β-Fructofuranosidase [EC 3.2.1.26] has an exoinulinase activity (from the nonreducing end). **See also** *Inulin Fructotransferase (Depolymerizing); Inulin Fructotransferase (Depolymerizing, Difructofuranose-1,2':2',1-Dianhydride-forming); Levanase; Fructan β-Fructosidase; 2,6-β-Fructan 6-Levanbiohydrolase*

[1] E. J. Vandamme & D. G. Derycke (1983) *Adv. Appl. Microbiol.* **29**, 139.

Selected entries from *Methods in Enzymology* [vol, page(s)]:
General discussion: 8, 625
Assay: 8, 625

INULIN FRUCTOTRANSFERASE (DEPOLYMERIZING)

This transferase [EC 2.4.1.93], also known as inulase II and inulinase II, catalyzes the transfer of a terminal D-fructosyl-D-fructofuranosyl group to the terminal 3-position, forming a cyclic anhydride (that is, a cyclic 1,2':2,3'-dianhydride) and leaving a residual di-, tri-, or oligosaccharide.[1–3] **See also** *Inulin Fructotransferase (Depolymerizing, Difructofuranose-1,2′:2′,1-Dianhydride-Forming)*

[1] A. Yokota, K. Enomoto & F. Tomita (1991) *J. Ferment. Bioeng.* **72**, 262.
[2] T. Uchiyama, S. Niwa & K. Tanaka (1973) *BBA* **315**, 412.
[3] T. Uchiyama (1975) *BBA* **397**, 153.

INULIN FRUCTOTRANSFERASE (DEPOLYMERIZING, DIFRUCTOFURANOSE-1,2′:2′,1-DIANHYDRIDE-FORMING)

This transferase [EC 2.4.1.200], also known as inulin fructotransferase (DFA-I-producing), catalyzes the successive removal of terminal D-fructosyl-D fructofuranosyl groups from inulin as the cyclic 1,2′:2,1-dianhydride, leaving a residual tetra- or pentasaccharide.[1] The *Arthrobacter globiformis* gene has been sequenced.[2] **See also** *Inulin Fructotransferase (Depolymerizing)*

[1] K. Seki, K. Haraguchi, M. Kishimoto, S. Koboyashi & K. Kainuma (1989) *Agric. Biol. Chem.* **53**, 2089.
[2] K. Haraguchi, K. Seki, M. Kishimoto, T. Nagata, T. Kasumi, K. Kainuma & S. Kobayashi (1995) *Biosci. Biotechnol. Biochem.* **59**, 1809.

INULOSUCRASE

This enzyme [EC 2.4.1.9], also known as sucrose 1-fructosyltransferase, catalyzes the reaction of sucrose with $[(2,1)\text{-}\beta\text{-D-fructosyl}]_n$ to produce D-glucose and $[(2,1)\text{-}\beta\text{-D-fructosyl}]_{(n+1)}$.[1,2] This enzyme also participates in the conversion of sucrose into inulin and D-glucose. Other fructosyl-containing sugars can serve as substrates as well instead of sucrose.

[1] G. Mooser (1992) *The Enzymes*, 3rd ed., **20**, 187.
[2] J. Edelmann & J. S. D. Bacon (1951) *BJ* **49**, 529.

IODIDE PEROXIDASE

This heme-dependent oxidoreductase [EC 1.11.1.8] (also known as iodotyrosine deiodase, iodinase, iodoperoxidase and thyroid peroxidase) catalyzes the reaction of three iodide anions with hydrogen peroxide and two protons to produce triiodide and two water molecules.[1–3] This reactive product can now participate in the iodination of organic substrates via the formation of active species such as HOI. Hence, this enzyme is frequently used in the iodination of tyrosine, thyroglobulin, and tyrosyl residues in many proteins. Horseradish peroxidase has an iodoperoxidase activity.

Note: Iodide reacts with hydrogen peroxide non-enzymatically to form triiodide and water and the triiodide product will react with excess hydrogen peroxide to produce dioxygen, three iodide anions, and two protons. **See also** *Chloroperoxidase; Lactoperoxidase; Peroxidase; Bromoperoxidase; Horseradish Peroxidase*

[1] A. Taurog (1999) *Biochimie* **81**, 557.
[2] M. Mahajani, I. Haldar & A. G. Datta (1973) *EJB* **37**, 541.
[3] M. L. Coval & A. Taurog (1967) *JBC* **242**, 5510.

Selected entries from **Methods in Enzymology** [**vol**, page(s)]:
General discussion: **17A**, 653, 658; **107**, 445
Assay: **107**, 489
Other molecular properties: properties, **107**, 496; purification from bovine thyroid, **107**, 492

IODOPHENOL *O*-METHYLTRANSFERASE

This methyltransferase [EC 2.1.1.26] catalyzes the reaction of *S*-adenosyl-L-methionine with 2-iodophenol to produce *S*-adenosyl-L-homocysteine and 2-iodophenol methyl ether.[1] Other substrates include 3,5-diiodo-4-hydroxybenzoate, 3,5,3′,5′-tetraiodothyroacetate, 3,5,3′-triiodo-L-thyronine, thyroxine, and 3,5,3′-triiodothyroacetate.

[1] K. Tonita, M. C. Cha & H. A. Lardy (1964) *JBC* **239**, 1202.

IRON-CHELATE-TRANSPORTING ATPase

This so-called ATPase [EC 3.6.3.34], also known as ferric-chelate-transporting ATPase, catalyzes the ATP-dependent (ADP- and orthophosphate-producing) transport of [iron chelate]$_{out}$ to produce [iron chelate]$_{in}$.[1–3] This ABC-type (ATP-binding cassette-type) ATPase has two similar ATP-binding domains and does not undergo phosphorylation during the transport process. The *Escherichia coli* enzyme will catalyze the import of ferric enterobactin, Fe-dicitrate, ferric hydroxamate, and other ferrisiderophores. **See also** *ATPase, Mn²⁺-Transporting; ATPase, Heme-Transporting; ATPase, Fe³⁺-Transporting*

The idea that a transporter is an enzyme is in keeping with a new definition of enzyme catalysis as the facilitated making/breaking of chemical bonds, not just covalent bonds.[4] This idea builds on Pauling's assertion that any long-lived, chemically distinct interaction (in this case, the persistent location of a solute with respect to the faces of a membrane) can be regarded as a chemical bond. Note also that the equilibrium constant ($K_{eq} = [\text{iron-chelate}_{in}][\text{ADP}][\text{P}_i]/[\text{iron-chelate}_{out}][\text{ATP}]$) does not conform to that expected for an ATPase (*i.e.*, $K_{eq} = [\text{ADP}][\text{P}_i]/[\text{ATP}]$). Thus, although the overall reaction yields ADP and orthophosphate, the enzyme is misclassified as a hydrolase, and instead should be regarded as an energase-type reaction. Energases facilitate affinity-modulated reactions by coupling the $\Delta G_{\text{ATP-hydrolysis}}$ to a force-generating or work-producing step.[4] In this case, P-O-P bond scission supplies the energy to drive iron-chelate transport. **See ATPase; Energase**

[1] C. M. Shea & M. A. McIntosh (1991) *Mol. Microbiol.* **5**, 1415.
[2] W. Koster & B. Böhm (1992) *Mol. Gen. Genet.* **232**, 399.
[3] A. Mademidis & W. Koster (1998) *Mol. Gen. Genet.* **258**, 156.
[4] D. L. Purich (2001) *TiBS* **26**, 417.

IRON:CYTOCHROME-c REDUCTASE

This oxidoreductase [EC 1.9.99.1] catalyzes the reaction of ferrocytochrome *c* with an Fe^{3+} ion to produce ferricytochrome *c* and the Fe^{2+} ion.[1]

[1] M.-G. Yates & A. Nason (1966) *JBC* **241**, 4872.

ISOAMYLASE

This enzyme [EC 3.2.1.68], also called debranching enzyme, catalyzes the hydrolysis of α-(1,6)-D-glucosidic branch linkages in glycogen, amylopectin, and their β-limit dextrins.[1–4] This particular enzyme is distinguished from α-dextrin endo-(1,6)-α-glucosidase [EC 3.2.1.41] by the inability of isoamylase to act on pullulan, and by its limited action on α-limit dextrins. Nevertheless, the action on glycogen is complete in contrast to the limited action by α-dextrin enzyme. Isoamylase will hydrolyze only 1,6-linkages at a branch point (note that maltose is the smallest sugar that can be released from a branch point).

[1] G. Mooser (1992) *The Enzymes*, 3rd ed., **20**, 187.
[2] I. S. Pretorius, M. G. Lambrechts & J. Marmur (1991) *Crit. Rev. Biochem. Mol. Biol.* **26**, 53.
[3] M. Vihinen & P. Mäntsälä (1989) *Crit. Rev. Biochem. Mol. Biol.* **24**, 329.
[4] E. Y. C. Lee & W. J. Whelan (1971) *The Enzymes*, 3rd ed., **5**, 191.

ISOBUTYRALDOXIME O-METHYLTRANSFERASE

This methyltransferase [EC 2.1.1.91], also known as aldoxime methyltransferase, catalyzes the reaction of *S*-adenosyl-L-methionine with 2-methylpropanal oxime (*i.e.*, isobutyraldoxime: $(CH_3)_2CHCH{=}N{-}OH$) to produce *S*-adenosyl-L-homocysteine and 2-methylpropanal *O*-methyloxime (*i.e.*, isobutyraldoxime methyl ether: $(CH_3)_2CHCH{=}N{-}OCH_3$). Other aldoxime substrates of the *Pseudomonas* enzyme include 2-methylbutyraldoxime, 3-methylbutyraldoxime, *n*-propionaldoxime,

n-butyraldoxime, *n*-pentanaldoxime, and methacrylaldoxime.[1] Ketoximes were not substrates.

[1] D. B. Harper & J. T. Kennedy (1985) *BJ* **226**, 147.

ISOBUTYRYL-CoA MUTASE

This vitamin B_{12}-dependent enzyme [EC 5.4.99.13] catalyzes the interconversion of 2-methylpropanoyl-CoA (or, isobutyryl-CoA) and butanoyl-CoA.[1,2] The coenzyme produces a 5′-deoxyadenosyl radical which abstracts a hydrogen from a methyl group of the substrate to produce a substrate radical. This radical fragments into propene and the carbonyl-CoA radical; recombination generates the butanoyl-CoA radical which receives a hydrogen from 5′-deoxyadenosine.

[1] B. T. Golding & W. Buckel (1998) *CBC* **3**, 239.
[2] G. Brendelberger, J. Retey, D. M. Ashworth, K. Reynolds, F. Willenbrock & J. A. Robinson (1988) *Angew. Chem.* **100**, 1122.

ISOCHORISMATASE

This enzyme [EC 3.3.2.1], also known as 2,3-dihydro-2,3-dihydroxybenzoate synthase, catalyzes the hydrolysis of isochorismate to produce 2,3-dihydro-2,3-dihydroxybenzoate and pyruvate.[1] Chorismate is also a substrate.[2,3]

[1] I. G. Young & F. Gibson (1969) *BBA* **177**, 401.
[2] F. Rusnak, J. Liu, N. Quinn, G. A. Berchtold & C. T. Walsh (1990) *Biochemistry* **29**, 1425.
[3] A. M. Gehring, K. A. Bradley & C. T. Walsh (1997) *Biochemistry* **36**, 8495.

ISOCHORISMATE MUTASE

This enzyme [EC 5.4.99.6], also known as isochorismate synthase and isochorismate hydroxymutase, catalyzes the interconversion of chorismate and isochorismate.[1–3]

chorismate   isochorismate

This reaction is the first step in the biosynthesis of the powerful iron-chelating agent enterobactin. A different isozyme participates in the biosynthesis of menaquinone (vitamin K_2).

[1] J. Liu, N. Quinn, G. A. Berchtold & C. T. Walsh (1990) *Biochemistry* **29**, 1417.
[2] R. Bentley (1990) *Crit. Rev. Biochem. Mol. Biol.* **25**, 307.
[3] L. G. Young & F. Gibson (1969) *BBA* **177**, 401.

ISOCITRATE DEHYDROGENASES

There are two isocitrate dehydrogenases, based on the redox coenzyme and the ability to react with added oxalosuccinate.[1–7]

Isocitrate dehydrogenase (NAD$^+$) [EC 1.1.1.41], also known as β-ketoglutaric:isocitric carboxylase, catalyzes the reaction of isocitrate (or, (1R,2S)-1-hydroxypropane 1,2,3-tricarboxylate) with NAD$^+$ to produce α-ketoglutarate (or, 2-oxoglutarate), carbon dioxide, and NADH. The enzyme will not decarboxylate oxalosuccinate.

Isocitrate dehydrogenase (NADP$^+$) [EC 1.1.1.42], also known as oxalosuccinate decarboxylase, catalyzes the reaction of isocitrate with NADP$^+$ to produce α-ketoglutarate, carbon dioxide, and NADPH. Oxalosuccinate is an enzyme-bound intermediate and the enzyme is able to decarboxylate added oxalosuccinate.

[1] A. R. Clarke & T. R. Dafforn (1998) *CBC* **3**, 1.
[2] J. V. Schloss & M. S. Hixon (1998) *CBC* **2**, 43.
[3] R. F. Colman (1990) *The Enzymes*, 3rd ed., **19**, 283.
[4] D. J. Creighton & N. S. R. K. Murthy (1990) *The Enzymes*, 3rd ed., **19**, 323.
[5] V. J. C. Willson & K. F. Tipton (1980) *EJB* **109**, 411 and (1981) **113**, 477.
[6] K. Dalziel (1975) *The Enzymes*, 3rd ed., **11**, 1.
[7] G. W. E. Plaut (1963) *The Enzymes*, 2nd ed., **7**, 105.

Selected entries from *Methods in Enzymology* [**vol**, page(s)]:
General discussion: 1, 699, 705, 707, 710; 5, 645; 34, 164, 246
Assay: 10, 457; Fast Analyzer, by, 31, 816; NADP$^+$-specific, 1, 699, 705; 5, 645; NAD$^+$-specific, 1, 707, 710
Other molecular properties: activation, 63, 293; allosterism, 64, 162; assays for (citrate, with aconitase 13, 446; isocitrate, 13, 453, 605; radioactive citrate and isocitrate, 13, 528); associated with cell fractions, 40, 86; beef heart, from, affinity labeling, 47, 428; binding, to agarose-bound nucleotides, 66, 196; carboxymethylation, 25, 431; cell permeabilization test with, 137, 641; citrate assay, 55, 211; cyanylation, 47, 131; deuterium transfer by, 13, 588; electron acceptors and, 4, 330; enzyme contaminants of, 3, 430; equilibrium perturbation, 64, 123; *Euglena* cultures, in, 23, 77 (hydropyrimidine dehydrogenase and, 12A, 54); extraction from mitochondria, 10, 4; heme oxygenase activity, in, 52, 368; hydroxycitrate substrate for, 13, 618; inhibition, 55, 206 (by KCN, 10, 5); isocitrate assay and, 6, 876; isolated nuclei and, 12A, 445; isotope effect, 63, 111; isotope exchange, 64, 8; labeled α-ketoglutarate synthesis with, 13, 592; ligand binding sites, interaction distances, 249, 561; liver compartments, in, 37, 288; malic enzyme, similarity to, 1, 699, 703; marker enzyme for peroxisomes, as, 23, 682; marker for chromosome detection in somatic cell hybrids, as, 151, 186; membrane fraction with, 10, 441, 457; muscle mitochondria, in, 10, 89; NADPH-generating system, in, 52, 274, 359, 362, 378, 380, 401, 414; NADP$^+$-specific, 1, 699, 705; 13, 30 (brain mitochondria, 55, 58; isotope exchange, 64, 8; isotope exchanges, inhibition, 87, 657; kinetic isotope effect in tritiated water, 13, 593; microbodies, in, 52, 496, 499; NAD$^+$ kinase, assay of, 2, 652; NADPH preparation, 2, 411; NADP$^+$, assay of, 3, 892, 897; NAD(P)$^+$ transhydrogenase, assay of, 2, 681; 6, 434; preparation of deuterated or tritiated α-ketoglutarate and succinate, 13, 592, 595, 600; propanediol dehydrogenase, not present in preparation of, 9, 336; properties, 1, 703, 706; 5, 650; purification [pig heart, 1, 702; 5, 646; yeast, 1, 705]; stability, 13, 33; stereospecificity, 87, 106; substrate inhibition, 13, 33; toluene-treated mitochondria, 56, 547; yeast [purification, 1, 705]); NADP$^+$, use in assay for, 3, 892, 897; 18B, 149, 156, 157; NAD$^+$-specific (beef heart enzyme, 1, 710; 13, 34 [activation by ADP, 13, 39; hydroxylapatite columns, behavior on, 13, 36; inhibition by ATP, 13, 40 {mercurials, by, and protection by isocitrate and Mn^{++}, 13, 40; by NADH, 13, 40}; molecular weight, 13, 39; nucleotide specificity of activators and inhibitors, 13, 40; potentiation of NADH inhibition by NADPH, 13, 40; properties, 1, 713; purification, 1, 711; stability, 13, 38; stereochemistry of hydrogen transfer, 13, 41; substrate specificity, 13, 41]; brain mitochondria, 55, 58; intact mitochondria, in, Ca^{2+}-sensitive properties, assays, 174, 107; isotope effect, 87, 397; *Neurospora crassa* enzyme, 13, 42 [activation by AMP, citrate, and isocitrate, 13, 46; growth of organism, 13, 43; inhibition by glutamate and α-ketoglutarate, 13, 46; loss of regulatory site at pH 6.5, 13, 46; molecular weight, 13, 46; specificity of activators and inhibitors, 13, 46]; pea mitochondria enzyme, 13, 47 [citrate as activator, 13, 50; divalent metal ion requirement, 13, 51; inhibition by {monovalent anions, 13, 51; NADH, 13, 51}; sigmoid kinetics, 13, 50, 51; stability of enzyme, 13, 50]; pH effect, 87, 397; properties, 1, 709; purification [yeast, 1, 708]; stereospecificity, 87, 106; sticky substrates, 87, 396; toluene-treated mitochondria, 56, 547; yeast, 1, 707); occurrence of, 3, 893; oxalosuccinate carboxylase activity of, 1, 700; oxalosuccinate reductase (NADPH) 5, 650 (assay, 5, 645; metals and, 5, 651); 6-phosphogluconate dehydrogenase and, 6, 800; purification by elution with substrate, 22, 385; pyridine nucleotide assay, 55, 264, 265, 267; pyridine nucleotide transhydrogenase and, 2, 681; 6, 434; reduction of NADP$^+$ analogues, 44, 874; sodium laurate hydroxylation, in, 52, 320; specific radioactivity determination, 13, 529; stain for, 224, 108; staining on gels, 22, 595; starch gel electrophoresis of, 6, 969; substrate inhibition, 63, 504; succinate preparation with, 13, 595, 600; transhydrogenase assay and, 2, 682; yeast, 1, 705, 707

[ISOCITRATE DEHYDROGENASE (NADP$^+$)] KINASE

This phosphotransferase [EC 2.7.1.116], also known as isocitrate dehydrogenase kinase and isocitrate dehydrogenase kinase/phosphatase, catalyzes the reaction of ATP with isocitrate dehydrogenase [EC 1.1.1.42] to produce ADP and phosphorylated isocitrate dehydrogenase, the inactive form of the enzyme.[1,2] The *Escherichia coli* enzyme is bifunctional and catalyzes a phosphatase activity in the presence of a divalent metal ion and ADP or ATP.[3]

[1] J. Y. Wang & D. E. Koshland (1982) *ABB* **218**, 59.
[2] G. A. Nimmo, A. C. Borthwick, W. H. Holms, H. G. Nimmo (1984) *EJB* **141**, 401.
[3] G. A. Nimmo & H. G. Nimmo (1984) *EJB* **141**, 409.

ISOCITRATE O-DIHYDROXYCINNAMOYLTRANSFERASE

This acyltransferase [EC 2.3.1.126] catalyzes the reaction of caffeoyl-CoA with isocitrate to produce coenzyme A and 2-caffeoylisocitrate.[1] Alternative acyl-CoA substrates include feruloyl-CoA and 4-coumaroyl-CoA.

[1] D. Strack, P. Leicht, M. Bokern, V. Wray & L. Grotjahn (1987) *Phytochemistry* **26**, 2919.

ISOCITRATE EPIMERASE

This epimerase [EC 5.1.2.6] catalyzes the interconversion of (1R,2S)-1-hydroxypropane-1,2,3-tricarboxylate and (1S,2S)-1-hydroxypropane-1,2,3-tricarboxylate.[1]

[1]S. Hoshiko, Y. Kunimoto, K. Arima & T. Beppu (1982) *Agric. Biol. Chem.* **46**, 143.

ISOCITRATE LYASE

This lyase [EC 4.1.3.1] (also known as isocitrase, isocitritase, and isocitratase) catalyzes the reversible conversion of isocitrate (specifically, (1R,2S)-1-hydroxypropane 1,2,3-tricarboxylate; actually, the magnesium ion complex is the true substrate) to succinate and glyoxylate.[1–6] In addition to the metal ion–substrate complex, Mg^{2+} acts as a nonessential activator as well. The enzyme has an ordered Uni Bi kinetic mechanism.[2]

[1]J. V. Schloss & M. S. Hixon (1998) *CBC* **2**, 43.
[2]E. Perdiguero, D. de Arriaga, F. Busto & J. Soler (1995) *Biochemistry* **34**, 6059.
[3]D. J. Creighton & N. S. R. K. Murthy (1990) *The Enzymes*, 3rd ed., **19**, 323.
[4]P. Vanni, E. Giachetti, G. Pinzauti & B. A. McFadden (1990) *Comp. Biochem. Physiol. B* **95**, 431.
[5]J. V. Schloss & W. W. Cleland (1982) *Biochemistry* **21**, 4420.
[6]L. B. Spector (1972) *The Enzymes*, 3rd ed., **7**, 357.

Selected entries from *Methods in Enzymology* [vol, page(s)]:
General discussion: 5, 628; 13, 163
Assay: 5, 628; 13, 164
Other molecular properties: anaplerotic functions, 13, 163; colorimetric assay, glyoxylate, 13, 164; glyoxysomes, in, 31, 565, 568, 743; 72, 788; inhibitors, 13, 170; 63, 399, 400; labeled succinate configuration, 13, 601; labeling, 46, 139; marker enzyme, as, 31, 735, 743; microbodies, in, 52, 496; properties, 5, 632; 13, 168; purification (*Pseudomonas aeruginosa*, 5, 629; *Pseudomonas indigofera*, 13, 166); sources, 5, 622; substrate inhibition, 13, 170; synthesis of deuterated succinate with, 13, 599; transition-state and multisubstrate analogues, 249, 307

ISOFLAVONE-7-O-β-GLUCOSIDE 6″-O-MALONYLTRANSFERASE

This acyltransferase [EC 2.3.1.115], also called flavone/flavonol 7-O-β-D-glucoside malonyltransferase, catalyzes the reaction of malonyl-CoA with biochanin A 7-O-glucoside to produce coenzyme A and biochanin A 7-O-glucoside-6″-O-malonylester.[1–3] Formononetin 7-O-glucoside can also act as an acceptor substrate, as can other 7-O-glucosides of isoflavones, flavones, and flavonols.

[1]U. Matern, C. Feser & D. Hammer (1983) *ABB* **226**, 206.
[2]U. Matern, J. R. Potts & K. Hahlbrock (1981) *ABB* **208**, 233.
[3]J. Koester, R. Bussmann & W. Barz (1984) *ABB* **234**, 513.

ISOFLAVONE 7-O-GLUCOSYLTRANSFERASE

This transferase [EC 2.4.1.170] catalyzes the reaction of UDP-D-glucose with an isoflavone to produce UDP and an isoflavone 7-O-β-D-glucoside.[1] Substrates include biochanin A, formononetin and, more slowly, the 4′-hydroxy isoflavones genistein and daidzein. The enzyme will not act on isoflavonones, flavones, flavonones, flavonols, or coumarins.

[1]J. Köster & W. Barz (1981) *ABB* **212**, 98.

ISOFLAVONE 2′-HYDROXYLASE

This heme-thiolate-dependent oxidoreductase [EC 1.14.13.53], which participates in the biosynthesis of the pterocarpin phytoalexins medicarpin and maackiain, catalyzes the reaction of formononetin with NADPH and dioxygen to produce 2′-hydroxyformononetin, $NADP^+$, and water.[1,2] Substrates are isoflavones with a 4′-methoxy group in the B ring.

[1]W. Hinderer, U. Flentje & W. Barz (1987) *FEBS Lett.* **214**, 101.
[2]S. Daniel, K. Tiemann, U. Wittkampf, W. Bless, W. Hinderer & W. Barz (1990) *Planta* **182**, 270.

ISOFLAVONE 3′-HYDROXYLASE

This heme-thiolate-dependent oxidoreductase [EC 1.14.13.52] catalyzes the reaction of formononetin with NADPH and dioxygen to produce calycosin, $NADP^+$, and water.[1,2] Biochanin A and other isoflavones with a 4′-methoxy group are alternative substrates (biochanin A produces pratensein).

[1]W. Hinderer, U. Flentje & W. Barz (1987) *FEBS Lett.* **214**, 101.
[2]S. Daniel, K. Tiemann, U. Wittkampf, W. Bless, W. Hinderer & W. Barz (1990) *Planta* **182**, 270.

ISOFLAVONE 4′-O-METHYLTRANSFERASE

This methyltransferase [EC 2.1.1.46] catalyzes the reaction of S-adenosyl-L-methionine with isoflavone to produce S-adenosyl-L-homocysteine and 4′-O-methylisoflavone.[1,2] Acceptor substrates include daidzein (or, 7,4′-dihydroxyisoflavone) and genistein. The enzyme has an ordered Bi Bi kinetic mechanism.[1] An isoflavone O-methyltransferase from alfalfa acts on isoflavones with a free 7-hydroxyl group, but can also methylate the 5-hydroxyl group of genistein.[3]

[1]H. Wengenmayer, J. Ebel & H. Grisebach (1974) *EJB* **50**, 135.
[2]R. Edwards & R. A. Dixon (1991) *Phytochemistry* **30**, 2597.
[3]X. Z. He & R. A. Dixon (1996) *ABB* **336**, 121.

ISOHEXENYLGLUTACONYL-CoA HYDRATASE

This hydratase [EC 4.2.1.57] catalyzes the reaction of 3-(4-methylpent-3-en-1-yl)pent-2-enedioyl-CoA with water to produce 3-hydroxy-3-(4-methylpent-3-en-1-yl)glutaryl-CoA.[1] Other substrates include dimethyl-acryloyl-CoA and farnesoyl-CoA.

[1] W. Seubert & E. Fass (1964) *Biochem. Z.* **341**, 23.

ISOLEUCYL-tRNA SYNTHETASE

This enzyme [EC 6.1.1.5], also known as isoleucine:tRNA ligase and isoleucine translase, catalyzes the reaction of L-isoleucine with tRNA^{Ile} and ATP to produce L-isoleucyl-tRNA^{Ile}, AMP, and pyrophosphate (or, diphosphate).[1-3] This class I aminoacyl-tRNA synthetase, which has an editing mechanism, catalyzes the reaction via the formation of an enzyme-bound isoleucyl-adenylate intermediate.

[1] W. Freist (1988) *Angew. Chem. Int. Ed. Engl.* **27**, 773.
[2] P. Schimmel (1987) *Ann. Rev. Biochem.* **56**, 125.
[3] D. Söll & P. R. Schimmel (1974) *The Enzymes*, 3rd ed., **10**, 489.

Selected entries from ***Methods in Enzymology*** [**vol**, page(s)]:
General discussion: 34, 4, 170; 46, 90, 152, 172, 173, 177, 179, 180
Other molecular properties: active site, **47**, 66; affinity labeling, **87**, 474; isolation, **5**, 711; properties, **5**, 715; isotope exchange, **64**, 9; microcalorimetric studies, on binding, **61**, 304; protein–tRNA interaction studies, in, **59**, 342; purification, **5**, 729; **29**, 570, 592; **59**, 262, 266; separation, **29**, 562; subcellular distribution, **59**, 233, 234; tritium labeling experiments, in, **59**, 342; velocity constants, **29**, 640

ISOMALTULOSE SYNTHASE

This enzyme [EC 5.4.99.11; originally it was classified under EC 5.4.99.10], also called sucrose glucosylmutase, catalyzes the interconversion of sucrose and 6-*O*-α-D-glucopyranosyl-D-fructofuranose (or, isomaltulose).[1]

[1] P. S. Cheetham (1984) *BJ* **220**, 213.

ISONOCARDICIN SYNTHASE

This enzyme [EC 2.5.1.38], which participates in the biosynthetic pathway for the β-lactam antibiotic nocardicin A, catalyzes the reaction of *S*-adenosyl-L-methionine with nocardicin E to produce 5′-methylthioadenosine and isonocardicin A.[1]

[1] B. A. Wilson, S. Bantia, G. M. Salituro, A. McE. Reeve & C. A. Townsend (1988) *JACS* **110**, 8238.

ISOORIENTIN 3′-*O*-METHYLTRANSFERASE

This methyltransferase [EC 2.1.1.78], which participates in flavone biosynthesis, catalyzes the reaction of *S*-adenosyl-L-methionine with isoorientin to produce *S*-adenosyl-L-homocysteine and isoscoparin.[1] Isoorientin 2″-*O*-rhamnoside is an alternative substrate.

[1] J. van Brederode, R. Kamps-Heinsbroek & O. Mastenbroek (1982) *Z. Pflanzenphysiol.* **106**, 43.

ISOPENICILLIN N SYNTHASE

This iron-dependent oxidoreductase catalyzes the reaction of δ-(L-α-amino-δ-adipoyl)-L-cysteinyl-D-valine with dioxygen to produce isopenicillin N and two water molecules, resulting in a double ring closure. The peptide substrate initially binds to the enzyme in which the cysteine sulfur is ligated to Fe^{2+}. Subsequently, dioxygen binds to generate a ferric-superoxide adduct. Electron rearrangements results in the release of water and the formation of a transient $Fe^{4+}=O$ species. The β-lactam ring then forms followed by a radical intermediate that produces the final penam ring.[1,2]

[1] J. E. Baldwin & E. P. Abraham (1988) *Nat. Prod. Rep.* **5**, 129.
[2] J. E. Baldwin & M. Bradley (1990) *Chem. Rev.* **90**, 1079.

ISOPENTENYL-PYROPHOSPHATE Δ-ISOMERASE

This isomerase [EC 5.3.3.2] catalyzes the interconversion of isopentenyl pyrophosphate (or, Δ^3-isopentenyl diphosphate) and dimethylallyl pyrophosphate (also known as dimethylallyl diphosphate and Δ^2-isopentenyl diphosphate).[1-5] Other substrates include 3-ethylbut-3-enyl pyrophosphate and *trans*-3-methylpent-3-enyl pyrophosphate. The reaction proceeds via a carbocation intermediate.[3]

[1] R. A. Gibbs (1998) *CBC* **1**, 31.
[2] A. C. Ramos-Valdivia, R. van der Heijden & R. Verpoorte (1997) *Nat. Prod. Rep.* **14**, 591.
[3] J. E. Reardon & R. H. Abeles (1986) *Biochemistry* **25**, 5609.
[4] P. W. Holloway (1972) *The Enzymes*, 3rd ed., **6**, 565.
[5] B. W. Agranoff, H. Eggerer, U. Henning & F. Lynen (1960) *JBC* **235**, 326.

Selected entries from ***Methods in Enzymology*** [**vol**, page(s)]:
General discussion: 5, 489; 15, 425; 110, 92, 209
Assay: 5, 493; 15, 428, 429, 430, 456; chicken liver, 110, 93; *Claviceps purpurea*, 110, 93; tomato fruit chromoplasts, 110, 212
Other molecular properties: analysis of reaction products, 15, 431; assay coupled with geranyl transferase, 15, 429; characterization, 214, 355; chicken liver (assay, 110, 93; properties, 110, 98; purification, 110, 94); *Claviceps purpurea* (assay, 110, 93; properties, 110, 98; purification, 110, 96); distribution in ammonium sulfate fractions, 15, 429; preparation, 5, 498; 15, 434; properties, 15, 437; purification, 15, 430, 435; separation from *Capsicum* phytoene synthase complex, 214, 353; stereochemistry, 87, 150; tomato fruit (*Lycopersicon esculentum*) chromoplasts (assay, 110, 212; extraction and partial purification, 110, 217; properties, 110, 219); transition state and multisubstrate analogues, 249, 308

ISOPIPERITENOL DEHYDROGENASE

This dehydrogenase [EC 1.1.1.223] catalyzes the reaction of (−)-*trans*-isopiperitenol with NAD$^+$ to produce (−)-isopiperitenone and NADH. Alternative substrates include (+)-*trans*-isopiperitenol, (+)-*trans*-piperitenol, and (+)-*trans*-pulegol.[1]

(–)-*trans*-isopiperitenol   (–)-isopiperitenone

[1]R. B. Kjonaas, K. V. Venkatachalam & R. Croteau (1985) *ABB* **238**, 49.

ISOPIPERITENONE Δ-ISOMERASE

isopiperitenone   piperitenone

This isomerase [EC 5.3.3.11] catalyzes the interconversion of isopiperitenone and piperitenone, involving an intramolecular 1,3-hydrogen transfer.[1]

[1]R. B. Kjonaas, K. V. Venkatachalam & R. Croteau (1985) *ABB* **238**, 49.

ISOPRENYLATED PROTEIN ENDOPEPTIDASE

This cysteine endopeptidase, also known as prenyl-protein-specific protease and AAX endopeptidase, is reportedly associated with intracellular microsomal membranes and participates in the processing of prenylated Ras proteins. Those proteins having a C-terminal motif of CAAX can be acted upon by this endopeptidase resulting in the removal of the AAX amino acids (where A represents any aliphatic aminoacyl residue and C is a cysteinyl residue that has been either farnesylated or geranylgeranylated, depending on the nature of the terminal amino acid X; for example, if X is leucine, the protein will have been geranylgeranylated).[1,2] The unprenylated protein is not acted upon by the endopeptidase. The human prenyl-protein-specific endopeptidase has recently been cloned.[3]

See also *Ste24 Protease; Ras and a-Factor Converting Protein*

[1]S. Clarke (1992) *Ann. Rev. Biochem.* **61**, 355.
[2]Y.-T. Ma, A. Chaudhuri & R. R. Rando (1992) *Biochemistry* **31**, 11772.
[3]J. C. Otto, E. Kim, S. G. Young & P. J. Casey (1999) *JBC* **274**, 8379.

Selected entries from *Methods in Enzymology* [vol, page(s)]:
General discussion: 250, 189, 206
Assay: 244, 635; avidin-based assay, **250**, 204; assays, interference from other proteases, **250**, 247; coupled assay with yeast carboxylmethyltransferase, **250**, 262; direct assay (controls, **250**, 244;

reaction conditions, **250**, 244; substrate, **250**, 243, 245); HPLC-based assays **250**, 203 (amino-terminal reaction product identification, **250**, 245; carboxyl-terminal reaction product identification, **250**, 247); indirect coupled assay (controls, **250**, 241; inhibitor analysis, **250**, 242; reaction conditions, **250**, 241; substrate, **250**, 240); linearity, **250**, 205; microsomal membranes, in, **255**, 64; pH, **250**, 204
Other molecular properties: cleavage site specificity, **244**, 633; enzyme sources, **244**, 636; inhibitors, **244**, 634 (assay, **250**, 215; inhibition constants, **250**, 208; structures, **250**, 208; synthesis, **250**, 209); product radiolabeling *in vivo*, **250**, 249; sequence recognition, **250**, 207; species distribution, **250**, 190, 235; specific activity, **250**, 204, 207, 235, 237, 251, 264, 266; stereospecificity, **244**, 639; substrate specificity, **244**, 633, 638; synthesis of ECB-C(S-farnesyl)-VI-[^{3}H]S substrate, **250**, 191, 197; synthesis of peptide substrates, **244**, 637

ISOPROPANOL DEHYDROGENASE (NADP⁺)

This oxidoreductase [EC 1.1.1.80] catalyzes the reaction of propan-2-ol (*i.e.*, isopropanol) with NADP⁺ to produce acetone and NADPH.[1–3] Alternative substrates include other short-chain secondary alcohols. Primary alcohols are slowly acted upon.

[1]I. Ploc & L. Starka (1979) *J. Chromatogr.* **172**, 374.
[2]K. Hoshino (1960) *Nippon Nogei Kagaku Kaishi* **34**, 606.
[3]L. S. al-Kassim & C. S. Tsai (1990) *Biochem. Cell Biol.* **68**, 907.

N-ISOPROPYLAMMELIDE ISOPROPYLAMINOHYDROLASE

N-isopropylammelide   cyanuric acid   isopropylamine

This enzyme [EC 3.5.99.4], often abbreviated AtzC, catalyzes the hydrolysis of *N*-isopropylammelide (*i.e.*, 2,4-dihydroxy-6-(isopropylamino)-1,3,5-triazine) to produce cyanuric acid and isopropylamine.[1]

[1]M. J. Sadowsky, Z. Tong, M. de Souza & L. P. Wackett (1998) *J. Bacteriol.* **180**, 152.

3-ISOPROPYLMALATE DEHYDRATASE

This enzyme [EC 4.2.1.33], also known as isopropylmalate hydro-lyase, catalyzes the reversible conversion of 3-isopropylmalate to 2-isopropylmaleate and water.[1,2] The enzyme will also catalyze the reaction of 2-isopropylmaleate with water to produce 3-hydroxy-4-methyl-3-carboxypentanoate.

[1]R. Bigelis & H. E. Umbarger (1976) *JBC* **251**, 3545.
[2]S. R. Gross, R. O. Burns & H. E. Umbarger (1963) *Biochemistry* **2**, 1046.

Selected entries from *Methods in Enzymology* [vol, page(s)]:
General discussion: 17A, 782; 166, 423
Assay: 166, 424

3-ISOPROPYLMALATE DEHYDROGENASE

This oxidoreductase [EC 1.1.1.85] catalyzes the reaction of 3-carboxy-2-hydroxy-4-methylpentanoate and NAD^+ to produce 3-carboxy-4-methyl-2-ketopentanoate (*i.e.*, 3-carboxy-4-methyl-2-oxopentanoate) and NADH. The product then decarboxylates, generating 4-methyl-2-ketopentanoate (*i.e.*, 4-methyl-2-oxopentanoate).[1,2]

[1]K. Imada, K. Inagaki, H. Matsunami, H. Kawaguchi, H. Tanaka, N. Tanaka & K. Namba (1998) *Structure* 6, 971.
[2]A. M. Dean & L. Dvorak (1995) *Protein Sci.* 4, 2156.

Selected entries from *Methods in Enzymology* [vol, page(s)]:
General discussion: 17A, 793; 166, 225, 429; 324, 301
Assay: 166, 225, 430

2-ISOPROPYLMALATE SYNTHASE

This potassium-dependent enzyme [EC 4.1.3.12] catalyzes the reaction of acetyl-CoA with 3-methyl-2-ketobutanoate (*i.e.*, 3-methyl-2-oxobutanoate or α-ketoisovalerate) and water to produce 3-carboxy-3-hydroxy-4-methylpentanoate (*i.e.*, 2-isopropylmalate) and coenzyme A.[1,2] The spinach chloroplast enzyme is strongly feedback-inhibited by L-leucine.[1]

[1]P. Hagelstein & G. Schultz (1993) *Biol. Chem. Hoppe-Seyler* 374, 1105.
[2]M. J. P. Higgins, J. A. Kornblatt & H. Rudney (1972) *The Enzymes*, 3rd ed., 7, 407.

Selected entries from *Methods in Enzymology* [vol, page(s)]:
General discussion: 17A, 771; 34, 4; 166, 414
Assay: 166, 415

ISOPULLULANASE

This enzyme [EC 3.2.1.57] catalyzes the hydrolysis of pullulan to produce isopanose (6-α-maltosylglucose).[1–3] Panose (4-α-isomaltosylglucose) is a substrate and is hydrolyzed to isomaltose and D-glucose. The enzyme acts very weakly on starch. **See also** α-Dextrin Endo-1,6-α-Glucosidase; Pullulanase; Neopullulanase

[1]M. Vihinen & P. Mäntsälä (1989) *Crit. Rev. Biochem. Mol. Biol.* 24, 329.
[2]G. Okada, T. Takayanagi, S. Miyahara & T. Sawai (1988) *Agric. Biol. Chem.* 52, 829.
[3]Y. Sakano, M. Higuchi & T. Kobayashi (1972) *Arch. Biochem. Biophys.* 153, 180.

ISOQUINOLINE 1-OXIDOREDUCTASE

This iron- and molybdenum-dependent oxidoreductase [EC 1.3.99.16] catalyzes the reaction of isoquinoline with an acceptor substrate and water to produce isoquinolin-1(2*H*)-one and the reduced acceptor.[1,2] Electron acceptor substrates include 1,2-benzoquinone, cytochrome *c*, ferricyanide, iodonitrotetrazolium chloride, nitroblue tetrazolium, Meldola blue, and phenazine methosulfate.

[1]M. Lehmann, B. Tshisuaka, S. Fetzner & F. Lingens (1995) *JBC* 270, 14420.

ISOVALERYL-CoA DEHYDROGENASE

This FAD-dependent oxidoreductase [EC 1.3.99.10] catalyzes the reaction of 3-methylbutanoyl-CoA (*i.e.*, isovaleryl-CoA) with an electron-transferring flavoprotein to produce 3-methylbut-2-enoyl-CoA and the reduced electron-transferring flavoprotein.[1–4] Together with the electron-transferring flavoprotein and the electron-transferring flavoprotein dehydrogenase [EC 1.5.5.1], this enzyme forms a multienzyme system that can reduce ubiquinone and other acceptors. Isovaleryl-CoA dehydrogenase is not identical with butyryl-CoA dehydrogenase, acyl-CoA dehydrogenase, or 2-methylacyl-CoA dehydrogenase. Other substrates include *n*-valeryl-CoA (*i.e.*, pentanoyl-CoA).

[1]G. Finocchiaro, M. Ito & K. Tanaka (1987) *JBC* 262, 7982.
[2]Y. Ikeda & K. Tanaka (1983) *JBC* 258, 1077.
[3]Y. Ikeda, C. Dabrowski & K. Tanaka (1983) *JBC* 258, 1066.
[4]W. J. Rhead, C. Hall & K. Tanaka (1981) *JBC* 256, 1616.

Selected entries from *Methods in Enzymology* [vol, page(s)]:
General discussion: 166, 155, 374
Assay: 166, 375
Other molecular properties: inhibition by hypoglycin, 72, 611, 616; mutant, in isovaleric acidemia fibroblasts, assay and properties, 166, 155; properties, 166, 384; purification from rat liver, 166, 377

ISOVITEXIN β-GLUCOSYLTRANSFERASE

This transferase [EC 2.4.1.106], isolated from *Silene alba*, catalyzes the reaction of UDP-D-glucose with isovitexin to produce UDP and isovitexin 2″-*O*-β-D-glucoside.[1,2]

[1]R. Heinsbroek, J. van Brederode, G. van Nigtevecht, J. Maas, J. Kamsteeg, E. Besson & J. Chopin (1980) *Phytochemistry* 19, 1935.
[2]J. van Brederode, J. Chopin, J. Kamsteeg, G. van Nigtevecht & R. Heinsbroek (1979) *Phytochemistry* 18, 655.

ITACONATE OXIDASE

This oxidoreductase, first isolated from *Aspergillus terreus*, reportedly catalyzes the reaction of itaconate ($H_2C{=}C(COO^-)CH_2COO^-$) with dioxygen to produce itatartarate ($HOCH_2C(OH)(COO^-)CH_2COO^-$).[1,2]

[1]J. Arpai (1958) *Nature* 182, 661.
[2]J. Arpai (1959) *J. Bacteriol.* 78, 153.

Selected entries from *Methods in Enzymology* [vol, page(s)]:
Other molecular properties: preparation and properties, 5, 596

ITACONYL-CoA HYDRATASE

This enzyme [EC 4.2.1.56] catalyzes the reaction of itaconyl-CoA ($H_2C{=}C(COO^-)CH_2COSCoA$) with water to produce citramalyl-CoA ($H_3CC(OH)(COO^-)CH_2COS$-CoA).[1]

[1] R. A. Cooper & H. L. Kornberg (1964) *BJ* **91**, 82.

Selected entries from *Methods in Enzymology* [**vol**, page(s)]:
General discussion: 13, 317

J

JARARHAGIN

This zinc-dependent fibrinolytic metalloproteinase [EC 3.4.24.73], also known as HF₂-proteinase (hemolytic factor-2) and jararafibrase 1 (JF 1), is isolated from the venom of the jararaca snake (*Bothrops jararaca*; the Brazilian pit viper). It catalyzes the hydrolysis of peptide bonds in the α chain of fibrinogen and in von Willebrand factor. The insulin B chain is hydrolyzed at His10-Leu11, Ala14-Leu15, Leu15-Tyr16, and Phe24-Phe25.[1]

[1] M. J. Paine, H. P. Desmond, R. D. Theakston & J. M. Crampton (1992) *J. Biol. Chem.* **267**, 22869.

Selected entries from *Methods in Enzymology* [**vol**, page(s)]:
Other molecular properties: properties, 248, 192, 200, 351, 362; structure, **248**

JUGLONE 3-MONOOXYGENASE

This oxidoreductase [EC 1.14.99.27] catalyzes the reaction of 5-hydroxy-1,4-naphthoquinone (*i.e.*, juglone) with a reduced acceptor substrate (AH₂) to produce 3,5-dihydroxy-1,4-naphthoquinone, the oxidized acceptor (A), and water.[1]

juglone   3,5-dihydroxy-1,4-naphthoquinone

Other substrates include 1,4-naphthoquinone, naphthazarin, and 2-chloro-1,4-naphthoquinone.

[1] H. Rettenmaier & F. Lingens (1985) *Biol. Chem. Hoppe-Seyler* **366**, 637.

JUVENILE-HORMONE ESTERASE

This esterase [EC 3.1.1.59], often abbreviated JH esterase, catalyzes the hydrolysis of methyl (2*E*,6*E*)-(10*R*,11*S*)-10,11-epoxy-3,7,11-trimethyltrideca-2,6-dienoate to produce methanol and (2*E*,6*E*)-(10*R*,11*S*)-10,11-epoxy-3,7,11-trimethyltrideca-2,6-dienoate.

C-17 JH

(2*E*,6*E*)-(10*R*,11*S*)-10,11-epoxy-3,7,11-trimethyltrideca-2,6-dienoate

While the enzyme demethylates the insect juvenile hormones, JHI and JHIII, it does not hydrolyze the analogous ethyl or isopropyl esters.[1,2] The K_m value for JHIII and the *Drosophila melanogaster* enzyme is 89 nM.[3]

[1] C. A. D. de Kort & N. A. Granger (1981) *Ann. Rev. Entomol.* **26**, 1.
[2] T. Mitsui, L. M. Riddiford & G. Bellamy (1979) *Insect Biochem.* **9**, 637.
[3] P. M. Campbell, J. G. Oakeshott & M. J. Healy (1998) *Insect Biochem. Mol. Biol.* **28**, 501.

Selected entries from *Methods in Enzymology* [**vol**, page(s)]:
General discussion: 111, 487
Assay: partition assay, 111, 490; TLC assay, 111, 493

K

KALLIKREIN, PLASMA

This serine endopeptidase [EC 3.4.21.34], also known as kininogenin and serum kallikrein, catalyzes peptide-bond hydrolysis with a preference for Arg–Xaa and Lys–Xaa bonds (in small molecule substrates, Arg > Lys in the P1 position). Kallikrein will catalyze the release of bradykinin from kininogen. The enzyme also has an esterase activity (for example, it catalyzes the hydrolysis of Tos-Arg-OMe). Plasma kallikrein will also act on coagulation factor XII, pro-u-plasminogen activator, prorenin, plasminogen, and complement component C1 and will convert coagulation factor XIIa to factor XIIf.[1,2]

Plasma kallikrein is generated from a proenzyme called plasma prokallikrein or plasma prekallikrein. Coagulation factor XIIa mediates this conversion (factor XIIa is augmented by H-kininogen) and the bond being hydrolyzed is Arg371–Ile372. Kallikrein also autocatalyzes a cleavage at Lys140–Ala141 to produce β-kallikrein.[3] *See also Kallikrein, Tissue*

[1] W. E. Hathaway, L. P. Belhasen & H. S. Hathaway (1965) *Blood* **26**, 521.
[2] Y. T. Wachtfogel, R. A. DeLa Cadena & R. W. Colman (1993) *Thromb. Res.* **72**, 1.
[3] D. W. Chung, K. Fujikawa, B. A. McMullen & E. W. Davie (1986) *Biochemistry* **25**, 2410.

Selected entries from *Methods in Enzymology* [vol, page(s)]:
General discussion: 80, 385
Assay: 45, 304; 80, 359; 163, 87; 263, 162; 267, 30; bovine, 80, 359; human, 45, 304; 80, 357; peptide *p*-nitroanilides, with, 80, 354, 357, 358
Other molecular properties: activation of Factor XII, 80, 208, 209; 163, 66; activation of plasminogen, kinetic analysis, 80, 413; affinity chromatography, 34, 5, 432; affinity labeling, 80, 830, 838; 87, 484; bovine (assay, 80, 359; chain structure, 80, 172; inhibition, 80, 172); C1 inhibitor, purification and properties, 163, 179; compared to glandular, 80, 467; fragmentation of high molecular weight kininogen, 80, 185; β hairpin loop mutations, accommodation, 202, 74; human (assay, 80, 357; chain structure, 80, 172); immobilization for inhibitor-binding studies, 267, 38; inhibition, 45, 314; 46, 206; 80, 51, 299, 467, 763 (by sea anemone SAI 5-II, 80, 819); kininogen as substrate, 80, 194;

kinin-releasing enzyme, 80, 174, 175; lipoprotein-associated coagulation inhibitor, phage display library screening, 267, 44, 49; molecular weight, 45, 311; pH optimum, 45, 311; prekallikrein, from, activation, 45, 315; properties, 45, 309; 163, 91, 93 (physical, 45, 311); protein C, and, 80, 325; proteins binding, isolation of cDNA clones, 217, 334; purification, 45, 306; 163, 93 (alcohol fractionation, 45, 307; chromatography, 45, 308); reaction with α2-antiplasmin, 80, 404; regulator of HMW kininogen cofactor activity, 80, 198; role, 45, 303; stability, 45, 309; structure, 163, 91; substrates (natural, 45, 312; synthetic, 45, 313); thioester substrates, 248, 13; trapped by α2-macroglobulin, 80, 747, 748; zymogen, 80, 157

KALLIKREIN, TISSUE

This serine endopeptidase [EC 3.4.21.35], also called glandular kallikrein and kininogenin, catalyzes the hydrolysis of peptide bonds.[1–4] There is a preference for cleavage of Arg–Xaa bonds in small molecule substrates (*e.g.*, arginyl residues are preferred in position P1). There is also a preference for substrates containing large, bulky hydrophobic residues in P2. Human tissue kallikrein will catalyze the release of kallidin (lysyl-bradykinin) from kininogens. Rat and mouse tissue kallikreins will release bradykinin. The enzyme also has an esterase activity and can be assayed with Tos-Arg-OMe. It is also assayed with Bz-Pro-Phe-Arg-∇-NHPhNO2, DL-Val-Leu-Arg-∇-NHPhNO2, Pro-Phe-Arg-∇-NHMec, and Val-Leu-Arg-∇-NHMec.[1,2]

Human kallikrein hK2 is very similar in sequence to semenogelase [EC 3.4.21.77], which cleaves substrates C-terminal of single or double arginyl residues.[3,4] *See also Kallikrein, Plasma*

[1] H. Kato, E. Nakawishi, K. Enjyoi, I. Hayashi, S. Oh-Ish & S. Iwanaga (1987) *J. Biochem.* **102**, 1389.
[2] K. Hosoi, S. Tsunasawa, K. Kurihara, H. Aoyama, T. Ueha, T. Murai & F. Sakiyama (1994) *J. Biochem.* **115**, 137.
[3] D. Deperthes, P. Chapdelaine, R. R. Tremblay, C. Brunet, J. Berton, J. Hebert, C. Lazure & J. Y. Dube (1995) *BBA* **1245**, 311.
[4] J. Lovgren, K. Airas & H. Lilja (1999) *EJB* **262**, 781.

Selected entries from *Methods in Enzymology* [vol, page(s)]:
General discussion: 19, 681
Assay: 19, 682; 80, 467; 163, 104, 129; ^{125}I-aprotinin, with, 163, 160; kallikrein A (pig pancreatic, 45, 300 [spectrophotometric, 45, 303; Technician autoanalyser, with, 45, 300]); (kallikrein B, 45, 300); peptide *p*-nitroanilides, with, 80, 354, 357, 358; porcine, 80, 493 (pancreatic, 80, 357, 359, 497); salivary, 80, 357; urinary, 80, 357, 359, 467, 480, 488, 489, 491

Other molecular properties: activators and inhibitors, **19**, 698; active center, **19**, 699; distribution, **19**, 699; **80**, 467; immunohistochemical localization, **163**, 143; inhibition, **80**, 466, 763 (by sea anemone SAI 5-II, **80**, 819); α-kallikrein (porcine glandular, **80**, 493, 494); β-kallikrein (porcine pancreatic, **80**, 493); γ-kallikrein (porcine pancreatic, **80**, 493, 494); kallikrein A (porcine pancreatic [assay, **45**, 300 {spectrophotometric, **45**, 303}; carbohydrate composition, **80**, 520; kinetics, **45**, 298, 299; molecular weight, **45**, 297; properties, **45**, 293, 294, 299; purification, **45**, 289; purity, **45**, 294; role, **45**, 289; specific activity, **45**, 293; stability, **45**, 293; structure, **45**, 294]; porcine submandibular, molecular weight, **80**, 516; porcine urinary, molecular weight, **80**, 516); β-kallikrein A (porcine pancreatic [carbohydrate composition, **80**, 516; isolation, **80**, 506, 507; molecular weight, **80**, 517]); kallikrein B (porcine pancreatic [assay, **45**, 300; composition, **45**, 294 {amino acid, **45**, 295, 296, 299}; carbohydrate, **45**, 295; kinetics, **45**, 298, 299; molecular weight, **45**, 297; properties, **45**, 293, 294, 299; purification, **45**, 289; purity, **45**, 294; role, **45**, 289; specific activity, **45**, 293; stability, **45**, 293; structure, **45**, 294]; porcine submandibular, molecular weight, **80**, 516; urinary, molecular weight, **80**, 516); β-kallikrein B (porcine pancreatic [amino acid sequence, **80**, 523, 524; carbohydrate composition, **80**, 516; catalytic properties, **80**, 526, 527; isolation, **80**, 506, 507; molecular weight, **80**, 517]); α-kallikressin B′ (porcine pancreatic [amino acid composition, **80**, 522; catalytic properties, **80**, 526, 527; molecular weight, **80**, 518, 519; purification, **80**, 500; purity, **80**, 506]); β-kallikrein C (porcine pancreatic, **80**, 516 [molecular weight, **80**, 517]); β-kallikrein III (porcine pancreatic, **80**, 507 [carbohydrate composition, **80**, 516, 517; molecular weight, **80**, 517]); kinetics, **19**, 698; latent form (assay, **163**, 96; purification from human urine, **163**, 99); physical properties, **19**, 697; properties, **163**, 138; proteins binding, isolation of cDNA clones, **217**, 334; porcine, **80**, 493 (amino acid composition, **80**, 521; assay, **80**, 493; homology to crab collagenase, **80**, 729; inhibitors, **80**, 530, 531; isoelectric points, **80**, 521; microheterogeneity, **80**, 521; molecular heterogeneity, **80**, 515, 516; molecular weights, **80**, 516; nomenclature, **80**, 493; pancreatic [amino acid composition, **80**, 484, 485; assay, **80**, 357, 359, 497; fragmentation of HMW kininogen, **80**, 190; hydrolysis of amino acid esters, **80**, 524; inhibition, **80**, 801; N-terminal amino acid sequence, **80**, 486; radioiodination, **80**, 498; structure, **80**, 521]; properties, **80**, 515 [optical, **80**, 521]; submandibular [amino acid composition, **80**, 522; catalytic properties, **80**, 526, 527; purification, **80**, 507]; unit of activity, **80**, 495; urinary [amino acid composition, **80**, 522; catalytic properties, **80**, 526, 527; purification, **80**, 510; purity, **80**, 515; stability, **80**, 515]); purification, **19**, 688 (human urine, **163**, 112, 162; rat submandibular gland, **163**, 150; rat urine, **163**, 135); purity and stability criteria, **163**, 114; specificity, **19**, 698; stability, **19**, 697; thioester substrates, **248**, 13; urinary, **80**, 466

KANAMYCIN 6′-N-ACETYLTRANSFERASE

This acetyltransferase [EC 2.3.1.82; formerly EC 2.3.1.55], also known as aminoglycoside $N^{6'}$-acetyltransferase and 6′-aminoglycoside-N-acetyltransferase, catalyzes the reaction of acetyl-CoA with kanamycin to produce coenzyme A and $N^{6'}$-acetylkanamycin.[1–3] Substrates include kanamycin A, kanamycin B, neomycin B, neomycin C, gentamicin C1a, gentamicin C2, hybrimycin A1-3, hybrimycin B1-3, nebramycin factor 4, sisomicin, and tobramycin (the 6-amino group of the purpurosamine ring is acetylated). The enzyme catalyzes a rapid equilibrium random Bi Bi kinetic mechanism.[3]

[1]F. Le Goffic & A. Martel (1974) Biochimie **56**, 893.

[2]J. F. Meyer & B. Wiedemann (1985) J. Antimicrob. Chemother. **15**, 271.

[3]A. Martel, M. Masson, N. Moreau & F. Le Goffic (1983) EJB **133**, 515.

Selected entries from **Methods in Enzymology** [**vol**, page(s)]:
General discussion: 43, 611, 616, 623
Assay: 43, 612, 615
Other molecular properties: cofactors, **43**, 617; crude extract preparation, **43**, 620; DEAE-cellulose chromatography, **43**, 622, 623; enzymic lysis by (lysozyme Escherichia coli, **43**, 621; lysostaphin, **43**, 621, 622); Escherichia coli, from (activity, **43**, 623; buffer, **43**, 623; kinetic data, **43**, 623; pH optimum, **43**, 623; stability, **43**, 623); French pressure cell, **43**, 621; molecular weight, **43**, 623; osmotic shock, **43**, 620; polymyxin B, **43**, 622; properties, **43**, 623; purification, **43**, 622, 623; Pseudomonas aeruginosa, from, **43**, 624; purification, **43**, 622, 623; sonication, **43**, 620; streptomycete, growth medium preparation, **43**, 619

KANAMYCIN KINASE

This phosphotransferase [EC 2.7.1.95], also known as aminoglycoside 3′-phosphotransferase and neomycin-kanamycin phosphotransferase, catalyzes the reaction of ATP with kanamycin to produce ADP and kanamycin 3′-phosphate. Other substrates include neomycin, paromomycin, neamine, paromamine, vistamycin, and gentamicin A. The Pseudomonas aeruginosa enzyme also acts on butirosin. The enterococci and staphylococci enzymes are reported to have a Theorell-Chance Bi Bi kinetic mechanism.[1,2]

[1]G. A. McKay & G. D. Wright (1995) JBC **270**, 24686.
[2]G. A. McKay & G. D. Wright (1996) Biochemistry **35**, 8680.

Selected entries from **Methods in Enzymology** [**vol**, page(s)]:
General discussion: 43, 611
Assay: 43, 616, 618; plant transformants, in, **153**, 303
Other molecular properties: Escherichia coli, from, **43**, 626; expression in rice, assay, **216**, 436; neomycin phosphotransferase I, **43**, 616; neomycin phosphotransferase II, **43**, 616; Pseudomonas aeruginosa, from, **43**, 626; selectable marker for eukaryotic expression system, as, **245**, 303; Staphylococcus aureus, from, **43**, 627

KasI RESTRICTION ENDONUCLEASE

This type II restriction endonuclease [EC 3.1.21.4], isolated from Kluyvera ascorbata, catalyzes the hydrolysis of both strands of DNA at 5′…G∇GCGCC…3′, producing a four-base 5′ extension.

KATANIN

This microtubule-severing ATPase [EC 3.6.4.3] catalyzes ATP-dependent microtubule (MT) severing into tubulin dimers.[1,2] Although the overall reaction yields ADP and orthophosphate, the enzyme is misclassified as a hydrolase because P-O-P bond scission supplies the energy to drive microtubule scission. The overall enzyme-catalyzed reaction should be regarded to be of the energase type (*e.g.*, $MT_n + ATP = MT_m + MT_{n-m} + ADP + P_i$, where two severed microtubules contain m and (n-m) tubulin dimers.)[3] **See Energases**

[1]F. J. McNally & R. D. Vale (1993) *Cell* **75**, 419.
[2]J. J. Hartman, J. Mahr, K. McNally, K. Okawa, A. Iwamatsu, S. Thomas, S. Cheesman, J. Heuser, R. D. Vale & F. J. McNally (1998) *Cell* **93**, 277.
[3]D. L. Purich (2001) *TiBS* **26**, 417.

Selected entries from *Methods in Enzymology* [vol, page(s)]:
General discussion: 298, 206

KAURENE SYNTHASE

This enzyme, also known as *ent*-kaurene synthase, catalyzes the conversion of *trans*-geranylgeranyl pyrophosphate to (−)-kaurene, a tetracyclic diterpene, and pyrophosphate (or, diphosphate).[1-4]

geranylgeranyl diphosphate   copalyl diphosphate

(−)-kaurene

The reaction proceeds through the formation of a transient carbocation (formed by protonation of a double bond) which leads to formation of the cyclic intermediate, copalyl pyrophosphate. A second cyclization occurs with concomitant elimination and alkyl migration, generating (−)-kaurene.

[1]R. A. Gibbs (1998) *CBC* **1**, 31.
[2]K. A. Wickham & C. A. West (1992) *ABB* **293**, 320.
[3]C. A. West (1981) in *Biosynthesis of Isoprenoid Compounds* (J. W. Ported & S. L. Spurgeon, eds.), p. 373, Wiley, New York.
[4]P. D. Simcox, D. T. Dennis & C. A. West (1975) *BBRC* **66**, 166.

Selected entries from *Methods in Enzymology* [vol, page(s)]:
General discussion: 15, 481
Assay: 15, 481

KERATAN-SULFATE ENDO-1,4-β-GALACTOSIDASE

This galactosidase [EC 3.2.1.103], also known as keratanase and endo-β-galactosidase, catalyzes the endohydrolysis of 1,4-β-D-galactosidic linkages in keratan sulfate.[1,2] The 1,4-β-D-galactosyl linkages that are hydrolyzed are located adjacent to a 1,3-β-D-*N*-acetylglucosaminyl residue. The enzyme will only act on blood group substances after 1,2-linked fucosyl residues have been removed. **See** *Blood-Group-Substance Endo-1,4-β-Galactosidase*.

[1]D. S. P. Q. Horton & Y. M. Michelacci (1986) *EJB* **161**, 139.
[2]M. N. Fukuda (1981) *JBC* **256**, 3900.

Selected entries from *Methods in Enzymology* [vol, page(s)]:
General discussion: 83, 610, 619
Assay: 179, 496; *Escherichia freundii*, **83**, 611; proteoglycan degradation method, **248**, 50
Other molecular properties: *Escherichia freundii* (application for oligosaccharide structural analysis, **83**, 615; assay, **83**, 611; glycosphingolipid analysis, for, **230**, 385; properties, **83**, 614; purification, **83**, 612; specificity, **230**, 414; *Flavobacterium keratolyticus* (properties, **83**, 622; purification, **83**, 620); modification of erythrocytes, **236**, 226; releasing Gal(α1 → 3)Gal (assay, **179**, 496; properties, **179**, 499; purification from *Clostridium perfringens*, **179**, 498); specificity, **230**, 414

KERATAN SULFOTRANSFERASE

This sulfotransferase [EC 2.8.2.21] catalyzes the reaction of 3′-phosphoadenylylsulfate with keratan to produce adenosine 3′,5′-bisphosphate and keratan 6′-sulfate.[1] The sulfation takes place at the 6-position of galactosyl and *N*-acetylglucosaminyl residues in keratan, a proteoglycan. The enzyme is distinct from chondroitin 4-sulfotransferase [EC 2.8.2.5], choline sulfotransferase [EC 2.8.2.6] and chondroitin 6-sulfotransferase [EC 2.8.2.17].

[1]E. R. Rüter & H. Kresse (1984) *JBC* **259**, 11771.

3-KETOACID CoA-TRANSFERASE

This transferase [EC 2.8.3.5], also known as 3-oxo-acid CoA-transferase and succinyl-CoA:3-ketoacid-CoA transferase, catalyzes the reversible reaction of succinyl-CoA with a 3-keto acid (or, 3-oxo acid) to produce succinate and a 3-keto-acyl-CoA.[1-4] The 3-keto acid substrate can be acetoacetate, 3-oxopropanoate, 3-oxopentanoate, 3-oxo-4-methylpentanoate, or 3-oxohexanoate. Malonyl-CoA can act instead of succinyl-CoA. The transferase catalyzes a ping pong Bi Bi kinetic mechanism.[1] The reaction proceeds via a covalent enzyme thiol ester intermediate, E-CoA, and the binding energy from noncovalent interactions with coenzyme A increases k_{cat}/K_m by approximately 10^{10} times.[3]

[1]L. B. Hersh & W. P. Jencks (1967) *JBC* **242**, 3481.
[2]J. A. Sharp & M. R. Edwards (1983) *BJ* **213**, 179.
[3]A. Whitty, C. A. Fierke & W. P. Jencks (1995) *Biochemistry* **34**, 11678.
[4]W. P. Jencks (1973) *The Enzymes*, 3rd ed., **9**, 483.

Selected entries from *Methods in Enzymology* [vol, page(s)]:
General discussion: 1, 573, 574; **13**, 75
Assay: 1, 574; **13**, 75
Other molecular properties: isotope exchange, **64**, 10; mechanism, **13**, 81; **63**, 51; molecular weight, partial specific volume, **61**, 56; properties,

I, 580; **13**, 80; purification, **I**, 577; **13**, 77; substrate specificity, **13**, 80; transition-state and multisubstrate analogues, **249**, 305

3-KETOACYL-[ACYL-CARRIER PROTEIN] REDUCTASE

This oxidoreductase [EC 1.1.1.100], also known as 3-oxo-acyl-[acyl-carrier protein] reductase, catalyzes the reaction of a 3-ketoacyl-[acyl-carrier protein] (also called 3-oxo-acyl-[acyl-carrier protein]) with NADPH to produce a (3R)-3-hydroxyacyl-[acyl-carrier protein] and $NADP^+$.[1-3] This is the first reductase step in fatty acid biosynthesis. Although 3-oxoacyl-CoA derivatives will serve as substrates, there is a marked preference for acyl-carrier protein derivatives. *See also* Fatty Acid Synthase

[1] P. S. Sheldon, R. G. O. Kekwick, C. G. Smith, C. Sidebottom & A. R. Slabas (1992) BBA **1130**, 151.
[2] T. Shimakata & P. K. Stumpf (1982) Arch. Biochem. Biophys. **218**, 77.
[3] S. J. Wakil & J. K. Stoops (1983) The Enzymes, 3rd ed., **16**, 3.

Selected entries from **Methods in Enzymology** [vol, page(s)]:
General discussion: 14, 60
Assay: 14, 60
Other molecular properties: Escherichia coli (properties, **14**, 62; purification, **14**, 61); stereospecificity, **87**, 110

3-KETOACYL-[ACYL-CARRIER PROTEIN] REDUCTASE (NADH)

This oxidoreductase [EC 1.1.1.212], also known as 3-oxo-acyl-[acyl-carrier protein] reductase (NADH), forms a part of the fatty acid synthase complex in many plants and catalyzes the reaction of 3-ketoacyl-[acyl-carrier protein] with NADH to produce (3R)-3-hydroxyacyl-[acyl-carrier protein] and NAD^+. The enzyme can be separated from 3-ketoacyl-[acyl-carrier protein] reductase [EC 1.1.1.100], which utilizes NADPH. The NADH-dependent enzyme has been isolated from the avocado.[1]

[1] I. Caughey & R. G. Kekwick (1982) EJB **123**, 553.

3-KETOACYL-[ACYL-CARRIER PROTEIN] SYNTHASE

This enzyme [EC 2.3.1.41], also known as 3-oxoacyl-[acyl-carrier protein] synthase, catalyzes the reaction of an acyl-[acyl-carrier protein] with the neighboring malonyl-[acyl-carrier protein] to produce a 3-ketoacyl-[acyl-carrier protein], carbon dioxide, and a free, neighboring acyl-carrier protein.[1-5]

This enzyme functions as a component of fatty acid biosynthesis and is responsible for chain elongation.

Note that decarboxylation of malonyl-[acyl-carrier protein] generates an enol form of the thiol ester. The ensuing condensation reaction is of the Claisen type. The overall reaction mechanism utilizes an acyl-enzyme intermediate.[3]
See also *Fatty Acid Synthase*

[1] J. L. Garwin, A. L. Klages & J. E. Cronan (1980) JBC **255**, 11949.
[2] T. Shimakata & P. L. Stumpf (1983) ABB **220**, 39.
[3] G. D'Agnolo, I. S. Rosenfeld & P. R. Vagelos (1975) JBC **250**, 5283.
[4] P. R. Vagelos (1973) The Enzymes, 3rd ed., **8**, 155.
[5] S. J. Wakil & J. K. Stoops (1983) The Enzymes, 3rd ed., **16**, 3.

Selected entries from **Methods in Enzymology** [vol, page(s)]:
General discussion: 14, 57
Assay: 14, 57
Other molecular properties: assay for acyl carrier protein, **14**, 44; cerulenin inhibition, **71**, 117; **72**, 520, 529; *Escherichia coli*, from (inhibition by cerulenin, **72**, 520, 529; properties, **14**, 59; purification, **14**, 59)

3-KETOADIPATE CoA-TRANSFERASE

This transferase [EC 2.8.3.6], also known as 3-oxoadipate CoA-transferase and β-ketoadipate:succinyl-CoA transferase, catalyzes the reversible reaction of succinyl-CoA with 3-ketoadipate (also called 3-oxoadipate) to produce succinate and 3-ketoadipyl-CoA.[1-3] The enzyme reportedly will not act on malonate, fumarate, oxalate, or acetate. ***See also*** *3-Ketoacid CoA-Transferase; Propionate CoA-Transferase*

[1] M. Katagiri & O. Hayaishi (1957) JBC **226**, 439.
[2] W.-K. Yeh & L. N. Ornston (1981) JBC **256**, 1565.
[3] W.-K. Yeh & L. N. Ornston (1984) Arch. Microbiol. **138**, 102.

Selected entries from **Methods in Enzymology** [vol, page(s)]:
General discussion: 35, 235, 299
Other molecular properties: acetate assay, in, **35**, 299; specificity, **35**, 242; stability, **35**, 241; succinyl-CoA assay, in, **35**, 235, 236, 299

3-KETOADIPATE ENOL-LACTONASE

This enzyme [EC 3.1.1.24], also known as 3-oxoadipate enol-lactonase, carboxymethylbutenolide lactonase, and β-ketoadipate enol-lactone hydrolase, catalyzes the hydrolysis of 3-ketoadipate enol-lactone to produce 3-ketoadipate (also called 3-oxoadipate).[1,2] The substrate is the product of 4-carboxymuconolactone decarboxylase [EC 4.1.1.44].

[1] L. N. Ornston (1966) JBC **241**, 3787.
[2] W. K. Yeh & L. N. Ornston (1984) Arch. Microbiol. **138**, 102.

Selected entries from **Methods in Enzymology** [vol, page(s)]:
General discussion: 2, 273, 282; **17A**, 529, 546
Assay: 2, 282

2-KETOADIPATE REDUCTASE

This oxidoreductase [EC 1.1.1.172], also known as 2-oxo-adipate reductase, catalyzes the reaction of 2-ketoadipate (also called 2-oxoadipate) with NADH to produce 2-hydroxyadipate and NAD^+.[1,2]

[1] T. Suda, J. C. Robinson & T. A. Fjellstedt (1976) *ABB* **176**, 610.
[2] T. Suda, J. C. Robinson & T. A. Fjellstedt (1977) *BBRC* **77**, 586.

2-KETOALDEHYDE DEHYDROGENASE

This oxidoreductase [EC 1.2.1.23], also known as 2-oxoaldehyde dehydrogenase (NAD⁺), and methylglyoxal dehydrogenase, catalyzes the reaction of a 2-ketoaldehyde (or, 2-oxoaldehyde) with NAD^+ and water to produce a 2-keto acid (or, 2-oxo acid) and NADH.[1-5] The enzyme is distinct from 2-ketoaldehyde dehydrogenase (NADP⁺) [EC 1.2.1.49]. Aldehyde substrates include methylglyoxal and 2-keto-3-deoxyglucose. **See also** *2-Ketoaldehyde Dehydrogenase (NADP⁺)*

[1] M. Ray & S. Ray (1982) *JBC* **257**, 10571.
[2] J. Dunkerton & S. P. James (1975) *BJ* **149**, 609.
[3] A. J. Willetts & J. M. Turner (1970) *BBA* **222**, 668.
[4] E. Jellum (1968) *BBA* **165**, 357.
[5] C. Monder (1967) *J. Biol. Chem.* **242**, 4003.

Selected entries from **Methods in Enzymology** [vol, page(s)]:
General discussion: 89, 513

2-KETOALDEHYDE DEHYDROGENASE (NADP⁺)

This oxidoreductase [EC 1.2.1.49], also known as 2-oxoaldehyde dehydrogenase (NADP⁺), α-ketoaldehyde dehydrogenase (NADP⁺), and methylglyoxal dehydrogenase, catalyzes the reaction of an α-ketoaldehyde (or, 2-oxoaldehyde) with NADP⁺ and water to produce an α-keto acid anion (or, 2-oxo acid anion) and NADPH.[1-7] Methylglyoxal and NADP⁺ are converted to pyruvate and NADPH. The parsley enzyme will also act on 3-deoxyglucosone, phenylglyoxal, glucosone, and DL-glyceraldehyde.[8] The enzyme is distinct from 2-ketoaldehyde dehydrogenase (NAD⁺) [EC 1.2.1.23]. The sheep and pig liver enzymes can use either NAD^+ or $NADP^+$ as the coenzyme substrate.[4-7]

[1] S. Ray & M. Ray (1982) *JBC* **257**, 10566.
[2] M. Ray & S. Ray (1982) *JBC* **257**, 10571.
[3] D. L. Vander Jagt & L. M. Davidson (1977) *BBA* **484**, 260.
[4] J. Dunkerton & S. P. James (1976) *BJ* **153**, 503.
[5] J. Dunkerton & S. P. James (1975) *BJ* **149**, 609.
[6] E. Jellum (1968) *BBA* **165**, 357.
[7] C. Monder (1967) *JBC* **242**, 4003.
[8] L. Q. Liang, F. Hayase, T. Nishimura & H. Kato (1990) *Agri. Biol. Chem.* **54**, 319.

Selected entries from **Methods in Enzymology** [vol, page(s)]:
General discussion: 89, 513
Assay: 89, 514

2-KETOBUTYRATE SYNTHASE

propionyl-CoA   2-ketobutyrate

This enzyme [EC 1.2.7.2], also known as 2-oxobutyrate synthase, catalyzes the reversible reaction of propanoyl-CoA with bicarbonate and reduced ferredoxin to produce 2-ketobutanoate (or, 2-oxobutanoate), coenzyme A, and oxidized ferredoxin.[1-3]

[1] R. S. Bush & F. D. Sauer (1976) *BJ* **157**, 325.
[2] B. B. Buchanan (1971) *The Enzymes*, 3rd ed., **6**, 193.
[3] B. B. Buchanan (1969) *JBC* **244**, 4218.

2-KETO-3-DEOXY-L-FUCONATE DEHYDROGENASE

This enzyme catalyzes the conversion of 2-keto-3-deoxy-L-fuconate to two molecules of L-lactate.

Selected entries from **Methods in Enzymology** [vol, page(s)]:
General discussion: 89, 219
Assay: 89, 219
Other molecular properties: porcine liver (assays, **89**, 219; properties, **89**, 224; purification, **89**, 222)

α-KETOGLUTARATE DEHYDROGENASE COMPLEX

This multienzyme system catalyzes the overall reaction of α-ketoglutarate with NAD^+ and coenzyme A to produce succinyl-CoA, NADH, and carbon dioxide.[1-6]

α-ketoglutarate   succinyl-CoA

The complex, which uses five cofactors (NAD⁺, coenzyme A, thiamin pyrophosphate, lipoamide, and FAD), contains three distinct enzyme activities. α-Ketoglutarate dehydrogenase (lipoamide) [EC 1.2.4.2], often referred to as E_1, is thiamin-pyrophosphate-dependent and catalyzes the reaction of α-ketoglutarate with lipoamide to produce *S*-succinyldihydrolipoamide and carbon dioxide. Dihydrolipoamide *S*-succinyltransferase [EC 2.3.1.61], often referred to as E_2, catalyzes the reaction of *S*-succinyldihydrolipoamide with coenzyme A to produce dihydrolipoamide and succinyl-CoA. Dihydrolipoamide dehydrogenase [EC 1.8.1.4], known as E_3, is an FAD-dependent oxidoreductase that catalyzes the reaction of dihydrolipoamide with NAD^+ to produce lipoamide and NADH.

[1] K. F. Sheu & J. P. Blass (1999) *Ann. N. Y. Acad. Sci.* **893**, 61.
[2] R. N. Perlman (1991) *Biochemistry* **30**, 8501.
[3] M. Hamada & H. Tanenaka (1991) *A Study of Enzymes* **2**, 227.
[4] S. J. Yeaman (1986) *Trends Biochem. Sci.* **11**, 293.
[5] L. J. Reed & D. J. Cox (1970) *The Enzymes*, 3rd ed., **1**, 213.
[6] R. L. Schowen (1998) *CBC* **2**, 217.

Selected entries from *Methods in Enzymology* [**vol**, page(s)]:
General discussion: 1, 714; 9, 252; 13, 52, 55
Assay: 1, 714; 9, 252; 13, 52, 56
Other molecular properties: Amytal and, 4, 282; assay for (acetyl-coenzyme A, 3, 916; 13, 497, 543; acyl-coenzyme A, 13, 544, 550; (-)-carnitine, 14, 616; coenzyme A, 13, 497, 536, 539, 541, 548; radioactive α-ketoglutarate, 13, 497, 528; succinate thiokinase, 13, 65); bacterial particles and, 5, 55; brain mitochondria, 55, 58; carbon dioxide and, 2, 841; characteristics of enzyme complex, 13, 55, 56, 57; coenzyme A assay, 55, 209; 56, 371; cofactors, 13, 55; component of enzyme complex, 1, 718; 13, 52, 56; 251, 437; isoelectric precipitation, 13, 60; lipoyl dehydrogenase, in preparation of, 9, 274; lipoyl domain, in *Bacillus stearothermophilus* (lipoylation [recognition site, 251, 446; role in catalysis, 251, 436, 445]; purification, 251, 438; reductive acetylation, assay, 251, 439; size, 251, 444); mitochondria, Ca^{2+}-sensitive properties, in intact, 174, 104; properties, 1, 718; 13, 54, 61; purification (calcium phosphate gel cellulose, on, 22, 342; *Escherichia coli*, 9, 252; 13, 58; pig heart, from, 1, 716; 13, 53); pyruvate oxidase and, 1, 718; reaction sequence, 13, 56; succinyl coenzyme A deacylase assay and, 1, 603; toluene-treated mitochondria, 56, 547

α-KETOGLUTARATE DEHYDROGENASE (LIPOAMIDE)

This thiamin-pyrophosphate-dependent oxidoreductase [EC 1.2.4.2], also known as oxoglutarate dehydrogenase (lipoamide) and oxoglutarate decarboxylase, catalyzes the reaction of α-ketoglutarate (or, 2-oxoglutarate) with lipoamide to produce *S*-succinyldihydrolipoamide and carbon dioxide.[1-4] This protein is the E_1 component of the multienzyme α-ketoglutarate dehydrogenase complex. Lipoamide refers to the lipoic acid covalently linked to the ε-amino group of a lysyl residue of dihydrolipoamide succinyltransferase [EC 2.3.1.61], another component of the complex.

lipoamide   *S*-succinyldihydrolipoamide

The thiazolium form of the coenzyme undergoes general base-catalyzed deprotonation. This intermediate reacts with α-ketoglutarate to form an α-hydroxyglutarylthiamin pyrophosphate intermediate. Subsequently, decarboxylation generates a carboanionic center at the α-carbon of the

adduct, which is stabilized by the thiazolium nucleus. The electrons are delocalized and a reactive enamine intermediate can form. Two possibilities have been suggested for the reaction of the enamine intermediate with lipoamide. (1) The enamine acts as a nucleophile (the α-carbon attacking the sulfur) resulting in disulfide bond fission and C–S bond formation; proton reorganization results in elimination of thiamin pyrophosphate and formation of *S*-succinyldihydrolipoamide. (2) An electron transfer occurs between the enamine and lipoamide to form the succinylthiamin pyrophosphate intermediate and a deprotonated dihydrolipoamide. The final product is now formed via a simple acetyl transfer reaction. **See also** α-*Ketoglutarate Dehydrogenase Complex*

[1] K. F. Sheu & J. P. Blass (1999) *Ann. N. Y. Acad. Sci.* **893**, 61.
[2] S. E. Severin, L. S. Khailova & V. S. Gomazkova (1986) *Adv. Enzyme Regul.* **25**, 347.
[3] S. J. Yeaman (1986) *Trends Biochem. Sci.* **11**, 293.
[4] L. J. Reed & D. J. Cox (1970) *The Enzymes*, 3rd ed., **1**, 213.

Selected entries from *Methods in Enzymology* [**vol**, page(s)]:
General discussion: 13, 52, 55

α-KETOGLUTARATE DEHYDROGENASE (NADP⁺)

This oxidoreductase [EC 1.2.1.52], also known as oxoglutarate dehydrogenase ($NADP^+$), catalyzes the reaction of α-ketoglutarate (or, 2-oxoglutarate) with coenzyme A and $NADP^+$ to produce succinyl-CoA, carbon dioxide, and NADPH.[1-3] The *Euglena gracilis* enzyme[3] can also use NAD^+ as the coenzyme substrate, albeit at 20% of the rate of $NADP^+$.

[1] C. Lochmeyer & G. Fuchs (1990) *Arch. Microbiol.* **153**, 226.
[2] K. Ziegler, K. Braun, A. Böckler & G. Fuchs (1987) *Arch. Microbiol.* **149**, 62.
[3] H. Inui, K. Miyatake, Y. Nakano & S. Kitaoka (1984) *J. Biochem.* **96**, 931.

α-KETOGLUTARATE SYNTHASE

This enzyme [EC 1.2.7.3], also known as 2-oxoglutarate synthase and α-ketoglutarate-ferredoxin oxidoreductase, catalyzes the reversible reaction of succinyl-CoA with carbon dioxide and reduced ferredoxin to produce α-ketoglutarate (or, 2-oxoglutarate), coenzyme A, and oxidized ferredoxin.[1-4] The enzyme contains an iron-sulfur complex and the reaction proceeds through a radical intermediate.[1]

[1] L. Kerscher & D. Oesterhelt (1981) *EJB* **116**, 587 and 595.
[2] M. J. Allison, I. M. Robinson & A. L. Baetz (1979) *J. Bacteriol.* **140**, 980.
[3] U. Gehring & D. I. Arnon (1972) *JBC* **247**, 6963.
[4] B. B. Buchanan (1971) *The Enzymes*, 3rd ed., **6**, 193.

Selected entries from *Methods in Enzymology* [**vol**, page(s)]:
General discussion: 13, 177
Assay: 13, 178

Other molecular properties: photoreduction of ferredoxin, **13**, 181; preparation of chloroplasts, **13**, 181; properties, **13**, 179; purification from (*Chlorobium thiosulfatophilum*, **13**, 179; *Rhodospirillum rubrum*, **13**, 179); reductive carboxylic acid cycle, in, **13**, 179; requirement for thiamin pyrophosphate, **13**, 179; sources for ferredoxin, **13**, 179, 180

KETOHEXOKINASE

This phosphotransferase [EC 2.7.1.3], also known as hepatic fructokinase and fructo-1-kinase, catalyzes the reaction of ATP with D-fructose to produce ADP and D-fructose 1-phosphate. Other substrates include D-sorbose, D-tagatose, D-psicose, D-xylulose, D-agatose, and 5-dehydro-D-fructose. The beef liver enzyme is specific for the β-furanose anomer.[1] The enzyme has a random Bi Bi kinetic mechanism.[2] *See also* Fructokinase

[1]F. M. Raushel & W. W. Cleland (1973) *JBC* **248**, 8174.
[2]F. M. Raushel & W. W. Cleland (1977) *Biochemistry* **16**, 2169.

Selected entries from **Methods in Enzymology** [vol, page(s)]:
General discussion: 1, 286; **41**, 61
Assay: 1, 286; **63**, 32
Other molecular properties: chromium-ATP, **87**, 179; estimation, in crude tissue preparations, **41**, 61; isotope exchange, **64**, 9; isotope trapping, **64**, 52; myokinase, not present in preparations of, **6**, 230; partition analysis, **249**, 323; pH effects, **87**, 403; product inhibition, **63**, 433; properties, 1, 289; purification, beef liver, 1, 288; specificity, **63**, 373

α-KETOISOCAPROATE DEHYDROGENASE

This enzyme, formerly classified as EC 1.2.4.3 and now a deleted entry, is now included with branched-chain α-keto acid dehydrogenase [EC 1.2.4.4].

2-KETOISOVALERATE DEHYDROGENASE (ACYLATING)

This oxidoreductase [EC 1.2.1.25], also known as 2-oxoisovalerate dehydrogenase (acylating), catalyzes the reaction of 3-methyl-2-oxobutanoate (α-ketoisovalerate) with coenzyme A and NAD$^+$ to produce 2-methylpropanoyl-CoA, carbon dioxide, and NADH.[1] Other substrates include (S)-3-methyl-2-ketopentanoate and 4-methyl-2-ketopentanoate. In many organisms, this enzyme is identical to 2-ketoisovalerate dehydrogenase (lipoamide) [EC 1.2.4.4].

[1]Y. Namba, K. Yoshizawa, A. Ejima, T. Hayashi & T. Kaneda (1969) *JBC* **244**, 4437.

2-KETOISOVALERATE DEHYDROGENASE (LIPOAMIDE)

This thiamin-pyrophosphate-dependent oxidoreductase [EC 1.2.4.4] (also known as 3-methyl-2-oxobutanoate dehydrogenase (lipoamide), 3-methyl-2-ketobutanoate dehydrogenase (lipoamide), 2-oxoisovalerate dehydrogenase, branched-chain α-keto acid dehydrogenase, and branched-chain α-keto acid decarboxylase) catalyzes the reaction of 3-methyl-2-oxobutanoate (that is, α-ketoisovalerate) with lipoamide to produce S-(2-methylpropanoyl)dihydrolipoamide and carbon dioxide. It is the E$_1$ component of the branched-chain α-keto acid dehydrogenase complex that participates in the degradation of L-valine, L-isoleucine, and L-leucine. Other substrates include (S)-3-methyl-2-ketopentanoate (or, (S)-3-methyl-2-ketovalerate) and 4-methyl-2-ketopentanoate (or, α-ketoisocaproate). The reaction has been proposed to proceed via a reactive enamine intermediate.[1–3] *See Branched-Chain α-Keto Acid Dehydrogenase Complex*

[1]P. J. Randle, P. A. Patston & J. Espinal (1987) *The Enzymes*, 3rd ed., **18**, 97.
[2]L. J. Reed, Z. Damuni & M. L. Merryfield (1985) *Curr. Top. Cell. Regul.* **27**, 41.
[3]A. E. Harper, R. H. Miller & K. P. Block (1984) *Ann. Rev. Nutr.* **4**, 409.

Selected entries from **Methods in Enzymology** [vol, page(s)]:
General discussion: **166**, 114, 146, 189, 298, 303, 309, 313, 342, 350

KETOL-ACID REDUCTOISOMERASE

This oxidoreductase [EC 1.1.1.86], also known as dihydroxyisovalerate dehydrogenase (isomerizing), α-keto-β-hydroxylacyl reductoisomerase, and acetohydroxy acid isomeroreductase, catalyzes the reaction of (S)-2-hydroxy-2-methyl-3-oxobutanoate (or, (S)-2-hydroxy-2-methyl-3-ketobutanoate) with NADPH to produce (R)-2,3-dihydroxy-3-methylbutanoate and NADP$^+$. This enzyme also catalyzes the NADPH-dependent conversion of 2-aceto-2-hydroxybutyrate (or, 2-hydroxy-2-ethyl-3-oxobutanoate) to (R)-2,3-dihydroxyacid-3-methylvalerate (or, 2,3-dihydroxy-3-methylpentanoate). The *Escherichia coli* enzyme has an ordered Bi Bi kinetic mechanism.[1] Results of pH and kinetic isotope effect experiments suggest that a base-catalyzed proton shuttle mechanism for the alkyl migration reaction is followed by an acid-assisted ketone reduction by the coenzyme.[2]

[1]S. K. Chunduru, G. T. Mrachko & K. C. Calvo (1989) *Biochemistry* **28**, 486.
[2]G. T. Mrachko, S. K. Chunduru & K. C. Calvo (1992) *ABB* **294**, 446.

Selected entries from **Methods in Enzymology** [vol, page(s)]:
General discussion: **17A**, 745, 751, 765
Assay: **17A**, 765

3-KETOLAURATE DECARBOXYLASE

This decarboxylase [EC 4.1.1.56], also known as 3-oxolaurate decarboxylase, β-ketolaurate decarboxylase, and β-ketoacyl decarboxylase, catalyzes the conversion of 3-oxododecanoate (or, 3-ketolaurate) to 2-undecanone and carbon dioxide.[1,2] Other substrates include 3-ketocaprate, 3-ketomyristate, and 3-ketopalmitate.

[1]W. Franke, A. Platzeck & G. Eichhorn (1961) *Arch. Microbiol.* **40**, 73.
[2]D. H. Hwang, Y. J. Lee & J. E. Kinsella (1976) *Int. J. Biochem.* **7**, 165.

KETOPANTOALDOLASE

This aldolase [EC 4.1.2.12] catalyzes the reaction of 3-methyl-2-oxobutanoate (also known as 3-methyl-2-keto-butanoate) with formaldehyde to produce 2-dehydro-pantoate.[1]

[1]E. N. McIntosh, M. Purko & W. A. Wood (1957) *JBC* **228**, 499.

5-KETOPENT-3-ENE-1,2,5-TRICARBOXY-LATE DECARBOXYLASE

This decarboxylase [EC 4.1.1.68], also known as 5-oxo-pent-3-ene-1,2,5-tricarboxylate decarboxylase, catalyzes the conversion of 5-oxopent-3-ene-1,2,5-tricarboxylate (or, 5-ketopent-3-ene-1,2,5-tricarboxylate) to produce 2-oxo-hept-3-enedioate (*i.e.*, 2-ketohept-3-enedioate) and carbon dioxide.[1] The *Escherichia coli* enzyme also exhibits a 2-hydroxyhepta-2,4-diene-1,7-dioate isomerase activity.[2]

[1]A. Garrido-Pertierra & R. A. Cooper (1981) *EJB* **117**, 581.
[2]D. I. Roper & R. A. Cooper (1993) *EJB* **217**, 575.

2-KETOPENT-4-ENOATE HYDRATASE

This hydratase [EC 4.2.1.80], also known as 2-oxopent-4-enoate hydratase, catalyzes the reversible reaction of 2-ketopent-4-enoate (or, 2-oxopent-4-enoate) with water to produce 4-hydroxy-2-ketopentanoate (or, 4-hydroxy-2-oxopentanoate).[1,2] Another substrate is *cis*-2-ketohex-4-enoate.

[1]D. A. Kunz, D. W. Ribbons & P. J. Chapman (1981) *J. Bacteriol.* **148**, 72.
[2]P. Marcotte & C. Walsh (1978) *Biochemistry* **17**, 5620.

2-KETOPHYTODIENOATE REDUCTASE

This oxidoreductase [EC 1.3.1.42], also known as 12-oxophytodienoate reductase, catalyzes the reaction of (15*Z*)-12-ketophyto-10,15-dienoate with NADPH to produce 8-[(1*R*,2*R*)-3-keto-2-{(*Z*)-pent-2-enyl}cyclopentyl] octanoate and NADP+.[1–3] The enzyme participates in the conversion of linolenate into jasmonate in corn.

[1]B. A. Vick & D. C. Zimmerman (1986) *Plant Physiol.* **80**, 202.
[2]B. A. Vick & D. C. Zimmerman (1984) *Plant Physiol.* **75**, 458.
[3]F. Schaller & E. W. Weiler (1997) *EJB* **245**, 294.

15-KETOPROSTAGLANDIN 13-REDUCTASE

This oxidoreductase [EC 1.3.1.48], also known as 15-ketoprostaglandin Δ^{13}-reductase and 15-oxoprostaglandin

13-reductase, catalyzes the reaction of (5*Z*)-(15*S*)-11α-hydroxy-9,15-diketoprosta-13-enoate with NAD(P)H to produce (5*Z*)-(15*S*)-11α-hydroxy-9,15-diketoprostanoate and NAD(P)+.[1–6] Other substrates include 15-ketoprosta-glandin B$_1$, 15-ketoprostaglandin E$_2$, 15-ketoprostaglandin F$_{2\alpha}$, 15-ketoprostaglandin F$_{1\alpha}$, 15-ketoprostaglandin A$_1$, and 6,15-diketoprostaglandin F$_{1\alpha}$. The bovine lung enzyme can use either NADH or NADPH whereas the human placenta enzyme is specific for NADH. There are five isozymes in rat liver.[4] Glutathione disulfide inhibits the rabbit kidney enzyme.[5]

[1]C. Westbrook & J. Jarabak (1975) *BBRC* **66**, 541.
[2]H. S. Hansen (1979) *BBA* **574**, 136.
[3]C. Westbrook & J. Jarabak (1978) *Arch. Biochem. Biophys.* **185**, 429.
[4]S. Kitamura, H. Katsura & K. Tatsumi (1996) *Prostaglandins* **52**, 35.
[5]S. Sakuma, Y. Fujimoto, H. Nishida, T. Sumiya, I. Yamamoto & T. Fujita (1992) *Prostaglandins* **43**, 435.
[6]C. R. Pace-Asciak & W. L. Smith (1983) *The Enzymes*, 3rd ed., **16**, 543.

Selected entries from *Methods in Enzymology* [vol, page(s)]:
General discussion: **86**, 156, 163
Assay: bovine lung, **86**, 157; human placental, **86**, 164
Other molecular properties: bovine lung (assay, **86**, 157; properties, **86**, 160; purification, **86**, 159); human placental (assay, **86**, 164; properties, **86**, 166; purification, **86**, 164)

3-KETOSTEROID-Δ^1-DEHYDROGENASE

This oxidoreductase [EC 1.3.99.4], also known as 3-oxo-steroid 1-dehydrogenase (the name recommended by the Enzyme Commission) and Δ^{1-2}-dehydrogenase, cata-lyzes the reaction of a 3-ketosteroid (or a 3-oxosteroid) with an acceptor substrate to produce a 3-keto-Δ^1-steroid and the reduced acceptor.[1] Steroid substrates include Δ^4-androstene-3,17-dione, testosterone, 11α-hydroxytestosterone, Δ^4-pregnene-3,20-dione, cortisone, corticosterone, and 17α-hydroxyprogesterone. Essential tyrosyl[2] and histidyl[3] residues have been identified.

[1]E. Itagaki, T. Wakabayashi & T. Hatta (1990) *BBA* **1038**, 60.
[2]C. Fujii, S. Morii, M. Kadode, S. Sawamoto, M. Iwami & E. Itagaki (1999) *J. Biochem.* **126**, 662.
[3]H. Matsushita & E. Itagaki (1992) *J. Biochem.* **111**, 594.

Selected entries from *Methods in Enzymology* [vol, page(s)]:
General discussion: **5**, 533; **44**, 184, 471
Assay: **5**, 533; **15**, 660
Other molecular properties: activity assay, **44**, 186; cofactor system, **44**, 188; properties, **5**, 538; purification, **5**, 536

3-KETO-5α-STEROID Δ^4-DEHYDROGENASE

This oxidoreductase [EC 1.3.99.5], also known as 3-oxo-5α-steroid 4-dehydrogenase (the preferred Enzyme Com-mission name) and steroid 5α-reductase, catalyzes the reaction of a 3-keto-5α-steroid (or, 3-oxo-5α-steroid)

with an acceptor substrate to produce a 3-keto-Δ^4-steroid and the reduced acceptor.[1-3] The enzyme irreversibly converts testosterone to the biologically inactive 5α-dihydrotestosterone.

[1]Y. Farkash, H. Soreq & J. Orly (1988) *PNAS* **85**, 5824.
[2]H. R. Levy & P. Talalay (1959) *JBC* **234**, 2014.
[3]V. C. Aries, P. Goddard & M. J. Hill (1971) *BBA* **248**, 482.

Selected entries from *Methods in Enzymology* [vol, page(s)]:
General discussion: 5, 533; **36**, 466
Assay: 5, 533; **36**, 466
Other molecular properties: preparation, **5**, 536; **36**, 468; properties, **5**, 538; **36**, 470

3-KETO-5β-STEROID Δ^4-DEHYDROGENASE

This oxidoreductase [EC 1.3.99.6], also known as 3-oxo-5β-steroid 4-dehydrogenase, Δ^4-3-ketosteroid 5β-reductase, and Δ^4-5β-steroid dehydrogenase, catalyzes the reaction of a 3-keto-5β-steroid (or, 3-oxo-5β-steroid) with an acceptor substrate to produce a 3-keto-Δ^4-steroid and the reduced acceptor.[1-5] Steroid substrates include 5β-androstane-3,17-dione, Δ^1-5β-androstene-3,17-dione, 5β-pregnane-3,20-dione, 21-hydroxy-5β-pregnane-3,20-dione, and 17β-hydroxy-5β-androstan-3-one.

[1]S. J. Davidson & P. Talalay (1966) *JBC* **241**, 906.
[2]N. A. Stokes & P. B. Hylemon (1985) *BBA* **836**, 255.
[3]V. C. Aries, P. Goddard & M. J. Hill (1971) *BBA* **248**, 482.
[4]Y. J. Abul-Hajj (1978) *JBC* **253**, 2356.
[5]A. Okuda & K. Okuda (1984) *JBC* **259**, 7519.

Selected entries from *Methods in Enzymology* [vol, page(s)]:
General discussion: 15, 656
Assay: 15, 657

KETOSTEROID MONOOXYGENASE

This FAD-dependent oxidoreductase [EC 1.14.13.54], also known as steroid-ketone monooxygenase, catalyzes the reaction of a ketosteroid with NADPH and dioxygen to produce a steroid ester or lactone, NADP$^+$, and water.[1,2] There are actually three types of activities catalyzed by these monooxygenases (analogous to Baeyer-Villiger reactions or rearrangements[3-5]). The first is the oxidative esterification of a number of derivatives of pregn-4-ene-3,20-dione (progesterone) to produce the corresponding 17α-hydroxysteroid 17-acetate ester (for example, the enzyme catalyzes the reaction of progesterone with NADPH and dioxygen to produce testosterone acetate, NADP$^+$, and water). The second activity is the oxidative lactonization of a number of derivatives of androst-4-ene-3,17-dione (androstenedione) to produce the corresponding 13,17-secoandrosteno-17,13α-lactone

(for example, the enzyme catalyzes the reaction of androstenedione with NADPH and dioxygen to produce 3-oxo-13,17-secoandrost-4-eno-17,13α-lactone (testololactone), NADP$^+$, and water). The third activity is the oxidative cleavage of the 17β side chain of 17α-hydroxypregn-4-ene-3,20-dione (that is, 17α-hydroxyprogesterone) to produce androst-4-ene-3,17-dione (or, androstenedione) and acetate: *i.e.*, the enzyme catalyzes the reaction of 17α-hydroxyprogesterone with NADPH and dioxygen to produce androstenedione, acetate, NADP$^+$, and water.

Note that the first activity is also exhibited by progesterone monooxygenase [EC 1.14.99.4] whereas the other two activities are also catalyzed by androst-4-ene-3,17-dione monooxygenase [EC 1.14.99.12]. In some organisms or tissues, a single enzyme may be responsible for all the activities.

[1]S. Morii, S. Sawamoto, Y. Yamauchi, M. Miyamoto, M. Iwami & E. Itagaki (1999) *J. Biochem.* **126**, 624.
[2]E. Itagaki (1986) *J. Biochem.* **99**, 815.
[3]A. Baeyer & V. Villiger (1899) *Ber.* **32**, 3625.
[4]A. Baeyer & V. Villiger (1900) *Ber.* **33**, 858.
[5]B. Plesnicar (1978) in *Oxidation in Organic Chemistry* (W. S. Trahanovsky, ed.), pt. C, Academic Press, New York, p. 254.

KETOTETROSE-PHOSPHATE ALDOLASE

This aldolase [EC 4.1.2.2], also known as phosphoketotetrose aldolase and erythrulose-1-phosphate synthetase, catalyzes the reversible conversion of erythrulose 1-phosphate to dihydroxyacetone phosphate (or, glycerone phosphate) and formaldehyde.[1,2]

[1]F. C. Charalampous & G. C. Mueller (1953) *JBC* **201**, 161.
[2]E. J. Wawszkiewicz (1968) *Biochemistry* **7**, 683.

Selected entries from *Methods in Enzymology* [vol, page(s)]:
General discussion: 5, 283
Assay: 5, 283

3-KETOVALIDOXYLAMINE C-N-LYASE

This calcium-dependent enzyme [EC 4.3.3.1] catalyzes the conversion of 4-nitrophenyl-3-ketovalidamine to 4-nitroaniline and 5-D-(5/6)-5-*C*-(hydroxymethyl)-2,6-dihydroxycyclohex-2-en-1-one.[1,2] The enzyme also catalyzes the release of 4-nitrophenol from 4-nitrophenyl-α-D-3-dehydroglucoside.

[1]M. Takeuchi, N. Asano, Y. Kameda & K. Matsui (1985) *J. Biochem.* **98**, 1631.
[2]M. Takeuchi, N. Asano, Y. Kameda & K. Matsui (1986) *J. Biochem.* **99**, 1571.

KEXIN

This calcium-ion-activated enzyme [EC 3.4.21.61], also called protease YscF and Kex2 protease, catalyzes the hydrolysis of peptide bonds at LysArg–Xaa and ArgArg–Xaa to process yeast α-factor pheromone and killer toxin precursors. Kexin is the KEX2 gene product (for $\underline{k}$iller $\underline{ex}$pression since it was associated with the secretion of the killer toxin activity of *Saccharomyces cerevisiae*). The P1 position is very selective for arginyl residues. When a lysyl residue is at P1, the corresponding k_{cat}/K_m value is decreased by 10^2- to 10^4-fold. If citrulline is at P1, activity drops by more than 10^5. There is also a preference for basic aminoacyl residues at the P2 position.[1,2]

[1] C. Brenner, A. Bevan & R. S. Fuller (1994) *Meth. Enzymol.* **244**, 152.
[2] R. S. Fuller, R. E. Sterne & J. Thorner (1988) *Ann. Rev. Physiol.* **50**, 345.

Selected entries from ***Methods in Enzymology*** [**vol**, page(s)]:
General discussion: 244, 152
Assay: 244, 157; halo assay, **244**, 162; quantitative mating assay, **244**, 162
Other molecular properties: cleavage site specificity, **244**, 152, 155; C-terminal retention signal, **244**, 153 (processing, **244**, 155; truncation, **244**, 154); family, **244**, 152, 155; halo assay, **244**, 162; homology with furin, **244**, 168; inhibitors, **244**, 159; ion dependence, **244**, 167; physiological substrates, **244**, 152; prepro-α-factor (assay of proteolysis, **244**, 162; processing by Kex2, **244**, 161; sequence, **244**, 161; size, **244**, 160; structure, **244**, 161); quantitative mating assay, **244**, 162; site-directed mutagenesis (cassette mutagenesis, **244**, 163; detection of mutations, **244**, 165; gene manipulation, **244**, 162; one-step method, **244**, 163; oxyanion hole, **244**, 164; P2 lysyl residue, **244**, 163); soluble enzyme (active sites, titration, **244**, 158; assay, **244**, 157; genetic engineering, **244**, 155; pH dependence, **244**, 159; preparation, yeast culture for, **244**, 156; quantitation by amino acid analysis, **244**, 157; stability assay, **244**, 159); subcellular localization, **244**, 153; substrate specificity, **244**, 154

KIEVITONE HYDRATASE

kievitone   kievitone hydrate

This hydratase [EC 4.2.1.95] catalyzes the reaction of kievitone, an isoflavonoid phytoalexin, with water to produce kievitone hydrate, which is less toxic.[1]

[1] C. S. Turbek, D. A. Smith & C. L. Schardl (1992) *FEMS Microbiol. Lett.* **94**, 187.

KINESIN AND KINESIN ATPase

Kinesin is a molecular motor associated with microtubules that utilizes an ATPase activity to drive vectorial transport along a eukaryote microtubule.[1-3] Plus-end-directed kinesin ATPase [EC 3.6.4.4] catalyzes the hydrolysis of ATP (or GTP) and moves along microtubules toward the plus end (in contast to dynein). It participates in organelle movement, mitosis, and meiosis. Minus-end-directed kinesin ATPase [EC 3.6.4.5] also catalyzes the hydrolysis of ATP but directs movement along the microtubule to the minus end. It is structurally very similar to microtubule-severing ATPase [EC 3.6.4.3].

Although the overall reaction yields ADP and orthophosphate, the enzyme is misclassified as a hydrolase because P-O-P bond scission supplies the energy to drive motility. The overall enzyme-catalyzed reaction should be regarded to be of the energase type (*e.g.*, $Position_1 + ATP = Position_1 + ADP + P_i$), where two successive kinesin binding sites on the microtubule lattice are indicated.[4]

[1] S. Khan & M. Sheetz (1997) *Ann. Rev. Biochem.* **66**, 785.
[2] J. R. McIntosh & M. E. Porter (1989) *JBC* **264**, 6001.
[3] E. P. Sablin, R. B. Case, S. C. Dai, C. L. Hart, A. Ruby, R. D. Vale & R. J. Fletterick (1998) *Nature* **395**, 813.
[4] D. L. Purich (2001) *TiBS* **26**, 417.

Selected entries from ***Methods in Enzymology*** [**vol**, page(s)]:
General discussion: 298, 133, 154
Other molecular properties: enrichment by microtubule affinity, **196**, 162; extraction from bovine brain, **196**, 177; isolation with batch ion exchange, **196**, 175; microtubule affinity purification, **196**, 178; microtubule-kinesin, ATP binding pathway, **249**, 52, 56; microtubule-stimulated ATPase activity, assay, **196**, 157; purification from bovine brain, **196**, 157

*Kpn*I RESTRICTION ENDONUCLEASE

This type II restriction enzyme [EC 3.1.21.4], isolated from *Klebsiella pneumoniae*, catalyzes the hydrolysis of double-stranded DNA at 5′ ... GGTAC∇C ... 3′.[1] Note that *Kpn*I produces fragments with a four-base 3′ extension.

[1] J. Tomassini, R. Roychoudhury, R. Wu & R. J. Roberts (1978) *NAR* **5**, 4055.

K⁺-TRANSPORTING ATPase

This so-called P-type ATPase [EC 3.6.3.12], also known as K⁺-translocating Kdp-ATPase and multisubunit K⁺-transport ATPase, catalyzes the ATP-dependent (ADP- and orthophosphate-producing) transport of K_{out}^+ to produce

K_{in}^{+}.[1,2] This enzyme undergoes covalent phosphorylation during the transport cycle. With the *Escherichia coli* enzyme, the phosphoenzyme intermediate was tentatively identified as an acylphosphate on the basis of its alkali lability and its sensitivity to hydroxylamine.[1] *See also ATPase, H^{+}/K^{+}-Exchanging*

The idea that a transporter is an enzyme is in keeping with a new definition of enzyme catalysis as the facilitated making/breaking of chemical bonds, not just covalent bonds.[3] This idea builds on Pauling's assertion that any long-lived, chemically distinct interaction (in this case, the persistent location of a solute with respect to the faces of a membrane) can be regarded as a chemical bond. Note also that the equilibrium constant ($K_{eq} = [K_{in}^{+}][ADP][P_i]/[K_{out}^{+}][ATP]$) does not conform to that expected for an ATPase (*i.e.*, $K_{eq} = [ADP][P_i]/[ATP]$). Thus, although the overall reaction yields ADP and orthophosphate, the enzyme is misclassified as a hydrolase, and instead should be regarded as an energase-type reaction. Energases facilitate affinity-modulated reactions by coupling the $\Delta G_{ATP-hydrolysis}$ to a force-generating or work-producing step.[3] In this case, P-O-P bond scission supplies the energy to drive K^{+} transport. *See ATPase; Energases*

[1] A. Siebers & K. Altendorf (1989) *JBC* **264**, 5831.
[2] M. Gassel, A. Siebers, W. Epstein & K. Altendorf (1998) *BBA* **1415**, 77.
[3] D. L. Purich (2001) *TiBS* **26**, 417.

Selected entries from ***Methods in Enzymology*** [vol, page(s)]:
General discussion: 157, 655; **294**, 445
Other molecular properties: *Escherichia coli*, from (genetic analysis, **157**, 655; phosphorylation, **157**, 677; properties, **157**, 676; solubilization and purification, **157**, 671); plant plasma membrane marker, as, assay, **148**, 543

KYNURENATE-7,8-DIHYDRODIOL DEHYDROGENASE

7,8-dehydro-7,8-dihydroxykynurenate   7,8-dihydroxykynurenate

This oxidoreductase [EC 1.3.1.18] catalyzes the reaction of 7,8-dihydro-7,8-dihydroxykynurenate with NAD^{+} to produce 7,8-dihydroxykynurenate and NADH.[1]

[1] H. Taniuchi & O. Hayaishi (1963) *JBC* **238**, 283.

KYNURENATE 7,8-HYDROXYLASE

kynurenate   7,8-dehydro-7,8-dihydroxykynurenate

This oxidoreductase [EC 1.14.99.2] catalyzes the reaction of kynurenate with dioxygen and a electron donor substrate (AH$_2$) to produce 7,8-dihydro-7,8-dihydroxykynurenate and the oxidized donor (A). Both NADH and NADPH can act as AH$_2$.[1,2]

[1] H. Taniuchi & O. Hayaishi (1963) *JBC* **238**, 283.
[2] D. T. Gibson (1971) *Meth. Microbiology* **6A**, 463.

Selected entries from ***Methods in Enzymology*** [vol, page(s)]:
General discussion: 52, 11

KYNURENINASE

This pyridoxal-phosphate-dependent enzyme [EC 3.7.1.3], also known as L-kynurenine hydrolase, catalyzes the hydrolysis of L-kynurenine to produce anthranilate and L-alanine.[1–6]

L-kynurenine   anthranilate

Other substrates include 3'-hydroxykynurenine and other (3-arylcarbonyl)alanines. (2S,3S)-*erythro*-β-Methyl-L-kynurenine is a slow substrate for the *Pseudomonas fluorescens* enzyme whereas (2S,3R)-*threo*-β-methyl-L-kynurenine is significantly poorer.[1] Studies have indicated that the rate-limiting step in the reaction occurs subsequent to the release of anthranilate, is general base catalyzed, and involves transfer of only a single proton. The rate-limiting step has been suggested to be the C4' deprotonation of the pyruvate pyridoxamine 5'-phosphate ketimine intermediate.[2,3]

[1] L. V. Cyr, M. G. Newton & R. S. Phillips (1999) *Bioorg. Med. Chem.* **7**, 1497.
[2] S. V. Koushik, J. A. Moore, 3rd, B. Sundararaju & R. S. Phillips (1998) *Biochemistry* **37**, 1376.
[3] R. S. Phillips, B. Sundararaju & S. V. Koushik (1998) *Biochemistry* **37**, 8783.
[4] A. J. L. Cooper (1998) *AE* **72**, 199.
[5] A. E. Martell (1982) *AE* **53**, 163.
[6] K. Soda & K. Tanizawa (1979) *AE* **49**, 1.

KYNURENINE AMINOTRANSFERASE

This pyridoxal-phosphate-dependent aminotransferase [EC 2.6.1.7], also known as kynurenine:oxoglutarate aminotransferase and kynurenine transaminase, catalyzes the irreversible reaction of L-kynurenine with α-ketoglutarate (or, 2-oxoglutarate) to produce 4-(2-aminophenyl)-2,4-dioxobutanoate (which then undergoes intramolecular dehydration to form kynurenate, a neuroprotectant) and L-glutamate.[1,2]

L-kynurenine   o-aminobenzoylpyruvate   kynurenate

L-3-Hydroxykynurenine, L-phenylalanine, L-tyrosine, and 3,5-diiodo-L-tyrosine can act as an alternative substrates.

See also *Kynurenine:Glyoxylate Aminotransferase*

[1] A. J. L. Cooper (1998) *Adv. Enzymol.* **72**, 199.
[2] A. E. Braunstein (1973) *The Enzymes*, 3rd ed., **9**, 379.

KYNURENINE:GLYOXYLATE AMINOTRANSFERASE

This pyridoxal-phosphate-dependent aminotransferase [EC 2.6.1.63] catalyzes the reaction of L-kynurenine with glyoxylate to produce 4-(2-aminophenyl)-2,4-dioxobutanoate, which then undergoes intramolecular dehydration to form kynurenate and glycine.[1] The enzyme will also act on L-phenylalanine, L-histidine and L-tyrosine. ***See also*** *Kynurenine Aminotransferase*

[1] I. Harada, T. Noguchi & R. Kido (1978) *Hoppe-Seyler's Z. Physiol. Chem.* **359**, 481.

KYNURENINE 3-MONOOXYGENASE

This FAD-dependent oxidoreductase [EC 1.14.13.9], also known as kynurenine 3-hydroxylase, catalyzes the reaction of L-kynurenine with NADPH and dioxygen to produce 3-hydroxy-L-kynurenine, NADP+, and water.[1–7]

L-kynurenine   3-hydroxy-L-kynurenine

Both *o*-hydroxybenzoyl-DL-alanine and *o*-nitrobenzoyl-DL-alanine are weak substrates. A random kinetic mechanism has been proposed for the human enzyme.[6]

[1] H. Okamoto & O. Hayaishi (1967) *BBRC* **29**, 394.
[2] Y. Saito, O. Hayaishi & S. Rothenberg (1957) *JBC* **229**, 921.
[3] Y. Nisimoto, F. Takeuchi & Y. Shibata (1977) *J. Biochem.* **81**, 1413.
[4] H.-H. Schott & H. Staudinger (1971) *Hoppe-Seyler's Z. Physiol. Chem.* **352**, 1654.
[5] E. Stratakis (1981) *J. Comp. Physiol.* **141**, 451.
[6] J. Breton, N. Avanzi, S. Magagnin, N. Covini, G. Magistrelli, L. Cozzi & A. Isacchi (2000) *EJB* **267**, 1092.
[7] V. Massey & P. Hemmerich (1975) *The Enzymes*, 3rd ed., **12**, 191.

L

LACCASE

This copper-dependent enzyme [EC 1.10.3.2] catalyzes the overall reaction of four molecules of benzenediol with dioxygen to produce four molecules of benzosemiquinone and two water molecules.[1-4] This classification number represents a group of enzymes with low specificity for the diol substrate. Substrates include 1,4-benzenediol, pyrocatechol (or, 1,2-dihydroxybenzene), p-aminophenol, DOPA, p-cresol, and urushiols. The latter are pyrocatechol metabolites substituted at the 3-position with a fifteen or seventeen carbon, saturated or unsaturated, chain. The potent inhibitor fluoride binds at the copper center. Laccase has a two-site ping pong Bi Bi kinetic mechanism. *See also Catechol Oxidase; Monophenol Monooxygenase*

[1]A. Messerschmidt (1998) *CBC* **3**, 401.
[2]A. Messerschmidt & R. Huber (1990) *EJB* **187**, 341.
[3]A. M. Mayer (1987) *Phytochemistry* **26**, 11.
[4]B. G. Malmström, L.-E. Andréasson & B. Reinhammer (1975) *The Enzymes*, 3rd ed., **12**, 507.

Selected entries from *Methods in Enzymology* [vol, page(s)]:
General discussion: 52, 10
Other molecular properties: complex, reaction mechanism, **52**, 39; derivatives, dioxygen reactivity, **226**, 31; immobilization, by microencapsulation, **44**, 214; metal-hybrid, preparation, **158**, 92; monophenolase and, **2**, 825; oxy form, spectrum, **52**, 36; prosthetic group, **52**, 4; Raman frequencies and assignments, **49**, 144; reaction mechanism, **52**, 39; resonance Raman spectrum, **49**, 142; trinuclear copper site (binding of azide, **226**, 17; magnetic CD spectroscopy, **226**, 9)

LACTALDEHYDE DEHYDROGENASE

This oxidoreductase [EC 1.2.1.22] catalyzes the reaction of (S)-lactaldehyde (CH$_3$CH(OH)CHO) with NAD$^+$ and water to sproduce (S)-lactate (L-lactate) and NADH.[1-4] The enzyme also catalyzes the NAD$^+$-dependent conversion of L-glyceraldehyde to L-glycerate as well as glycolaldehyde to glycolate. (R)-Lactaldehyde is a substrate, but reacts more slowly than its enantiomer.

[1]Y. Inoue, K. Watanabe, M. Shimosaka, T. Saikusa, Y. Fukuda, K. Murata & A. Kimura (1985) *EJB* **153**, 243.
[2]L. Baldoma & J. Aguilar (1987) *JBC* **262**, 13991.
[3]S. Sridhara & T. T. Wu (1969) *JBC* **244**, 5233.
[4]A. J. Willetts & J. M. Turner (1970) *BBA* **222**, 234.

D-LACTALDEHYDE DEHYDROGENASE

This oxidoreductase [EC 1.1.1.78], also known as methylglyoxal reductase, catalyzes the reaction of (R)-lactaldehyde (D-lactaldehyde) with NAD$^+$ to produce methylglyoxal (CH$_3$COHCHO) and NADH.[1-3] NADP$^+$ also can serve as the redox coenzyme.

[1]S.-M. Ting, O. N. Miller & O. Z. Sellinger (1965) *BBA* **97**, 407.
[2]M. Ray & S. Ray (1984) *BBA* **802**, 119.
[3]Y. Inoue, H. Rhee, K. Watanabe, K. Murata & A. Kimura (1988) *EJB* **171**, 213.

LACTALDEHYDE REDUCTASE (NADH)

This oxidoreductase [EC 1.1.1.77], also known as propanediol oxidoreductase, catalyzes the reversible reaction of (R)- or (S)-lactaldehyde with NADH to produce the corresponding (R)- or (S)-propane-1,2-diol and NAD$^+$.[1,2] *See also Lactaldehyde Reductase (NADPH); 1,3-Propanediol Dehydrogenase*

[1]S.-M. Ting, O. Z. Sellinger & O.-N. Miller (1964) *BBA* **89**, 217.
[2]A. Boronat & J. Aguilar (1979) *J. Bacteriol.* **140**, 320.

LACTALDEHYDE REDUCTASE (NADPH)

This oxidoreductase [EC 1.1.1.55] catalyzes the reaction of L-lactaldehyde with NADPH to produce propane-1,2-diol and NADP$^+$.[1] In some organisms, this enzyme may be identical to alcohol dehydrogenase (NADP$^+$) [EC 1.1.1.2].

The Enzyme Commission lists EC 1.1.1.55 as acting on L-lactaldehyde (as well as utilizing NADPH and not NADH), and this activity was reported with respect to the pig kidney enzyme many years ago.[2] However, the stereoisomer had been mischaracterized, and the NADPH-dependent enzyme was discovered not to act on the L-lactaldehyde.[3] *See also Lactaldehyde Reductase (NADH); 1,3-Propanediol Dehydrogenase*

[1]P. J. Weimer (1984) *Appl. Environ. Microbiol.* **47**, 263.
[2]N. R. Gupta & W. G. Robinson (1960) *JBC* **235**, 1609.
[3]W. G. Robinson (1966) *MIE* **9**, 332

Selected entries from *Methods in Enzymology* [vol, page(s)]:
General discussion: 9, 332
Assay: 9, 332
Other molecular properties: pig kidney, from (inhibitors, **9**, 336; properties, **9**, 336; purification, **9**, 333; specificity, **9**, 336); stereospecificity, **87**, 112

β-LACTAMASES

These enzymes [EC 3.5.2.6], including penicillinase and cephalosporinase, act with varying specificity in their catalysis of β-lactam hydrolysis, producing a substituted β-amino acid.[1–4] Several act more efficiently on penicillins, whereas others prefer cephalosporin-type substrates. Class B lactamases possess one or two zinc ions that stabilize the dianionic tetrahedral intermediate. Classes A, C, and D β-lactamases employ active-site seryl residues as the catalytic nucleophile. *See also specific β-Lactamase*

[1]Z. Wang, W. Fast, A. M. Valentine & S. J. Benkovic (1999) *Curr. Opin. Chem. Biol.* **3**, 614.
[2]M. A. Ator & P. R. Ortez de Montellano (1990) *The Enzymes*, 3rd ed., **19**, 213.
[3]N. Citri (1971) *The Enzymes*, 3rd ed., **4**, 23.
[4]M. R. Pollock (1960) *The Enzymes*, 2nd ed., **4**, 269.

Selected entries from *Methods in Enzymology* [vol, page(s)]:
General discussion: 2, 120; **34**, 5; **43**, 69, 86, 640, 652, 664, 672, 678; **46**, 531; γ-penicillinase, **43**, 640
Assay: 2, 120; **46**, 533 (acidimetric methods, **43**, 77, 83, 84; alkimetric methods, **43**, 77; biological, **43**, 81, 82; of DD-carboxypeptidase-transpeptidase, **45**, 613; gene, **68**, 5; hydroxylamine, **43**, 81, 85; indicator method, **43**, 78, 79; iodometric method, **43**, 74, 83, 84; macroiodometric determination, **43**, 74; macroiodometric method of Perret, **43**, 83; manometric measurement of CO_2, **43**, 79, 80; microbiological, **43**, 81; microiodometric determination, **43**, 76, 77; pH stat titration method, **43**, 79)
Other molecular properties: Affinity labeling, **46**, 533; **87**, 472; *Bacillus cereus*, from, **2**, 122; **43**, 640; *Bacillus licheniformis*, from, **43**, 653 (kinetic properties, **43**, 658, 659; Michaelis constants, **43**, 659; modification, **43**, 662; molecular weight, **43**, 662; physical properties, **43**, 662; plasma membrane bound, purification, **43**, 663; purification, **43**, 653; secretion, **43**, 663; specific activity, **43**, 659; stability, **43**, 658); enzyme electrode, in, **44**, 592, 597; *Escherichia coli*, from, **43**, 672 (β-lactamase IIIa, **43**, 672 [centrifugation steps, **43**, 675; DEAE-cellulose column chromatography, **43**, 676; EDTA in sucrose purification, **43**, 673; G-75 Sephadex column, **43**, 676; properties, **43**, 676, 677; purification, **43**, 673; R factor-mediated, **43**, 673; specificity, **43**, 676, 677; stability, **43**, 676; ultrasonic treatment, **43**, 675]); fluorescence for testing antibiotics resistance to, **43**, 208; immobilization, on glass beads, **44**, 670; kinetics, **46**, 535; β-lactamase la, **43**, 673; β-lactamase II, metallointermediates, low-temperature stopped-flow spectroscopy, **226**, 563; β-lactamase-less mutant, **43**, 89; measurement of bacterial outer membrane permeability, in, **125**, 271; molecular variants, detection of, **43**, 94; neutralization analysis, **43**, 90; penicillin acylase preparations, in, **43**, 700; pH effect, **46**, 534, 535; physiological efficiency, **43**, 660; reaction scheme, **43**, 70; R factor-mediated, **43**, 678; R_{TEM}-mediated, **43**, 678; specific anti-β-lactamase sera for quantitative study, **43**, 86; staining on gels, **22**, 601; *Staphylococcus aureus*, from, **43**, 664; *Streptomyces albus*, from, **43**, 687 (kinetic properties, **43**, 689, 690; K_m and V_{max} values, **43**, 689, 690;

metal ion requirements, **43**, 696; pH optimum, **43**, 696; physical properties, **43**, 696, 697; properties, **43**, 696, 698; purification, **43**, 692; specific activity, **43**, 693; specificity, **43**, 697; stability, **43**, 696; unit definition, **43**, 689); substrates, **44**, 615; synthesis, **46**, 533 (synthesis in *Escherichia coli* system, **217**, 133); transition-state and multisubstrate analogues, **249**, 307

LACTASE

β-Galactosidase [EC 3.2.1.23], also called lactase, catalyzes the hydrolysis of terminal, nonreducing β-D-galactose residues in β-D-galactosides. (*See β-Galactosidase*) Lactase [EC 3.2.1.108] catalyzes the hydrolysis of lactose to produce D-glucose and D-galactose.[1–5] The enzyme from intestinal mucosa is isolated as a complex which also catalyzes the reaction of glycosylceramidase [EC 3.2.1.62]. *See also Glycosylceramidase*

[1]S. Schlegel-Haueter, P. Hore, K. R. Kerry & G. Semenza (1972) *BBA* **258**, 506.
[2]S. Ramswamy & A. N. Radhakrishnan (1975) *BBA* **403**, 446.
[3]H. Skovbjerg, H. Sjöström & O. Noren (1981) *EJB* **114**, 653.
[4]M. W. Bauer, S. B. Halio & R. M. Kelly (1996) *Adv. Protein Chem.* **48**, 271.
[5]E. H. Van Beers, H. A. Büller, R. J. Grand, A. W. C. Einerhand & J. Dekker (1995) *Crit. Rev. Biochem. Mol. Biol.* 30, **197**.

Selected entries from *Methods in Enzymology* [vol, page(s)]:
General discussion: 1, 231; **8**, 584
Assay: 63, 33

LACTATE ALDOLASE

This enzyme [EC 4.1.2.36], isolated from rat liver and also known as lactate synthase, catalyzes the reaction of formate with acetaldehyde to produce (*S*)-lactate (or, L-lactate).[1]

[1]M. F. Gulyi & N. V. Silonova (1987) *Ukr. Biokhim. Zh.* **59**, 29.

D-LACTATE DEHYDROGENASE

This NAD$^+$-dependent oxidoreductase (EC 1.1.1.28) catalyzes the reversible reaction of (*R*)-lactate with NAD$^+$ to produce pyruvate, NADH, and a proton. Hydride transfer proceeds to or from the A-side of the pyridine ring.[1–3] The kinetic reaction mechanism is ordered Bi Bi.[2,3]

[1]E. L. Garvie (1980) *Microbiol. Rev.* **44**, 106.
[2]K. B. Storey (1977) *Comp. Biochem. Physiol.* **56B**, 181.
[3]S. Kochhar, V. S. Lamzin, A. Razeto, M. Delley, H. Hottinger & J. E. Germond (2000) *EJB* **267**, 1633.

Selected entries from *Methods in Enzymology* [vol, page(s)]:
General discussion: 5, 426; **31**, 742; **41**, 293, 299, 313; **53**, 519
Assay: 5, 426; **53**, 519, 520 (by Fast Analyzer, **31**, 816); *Butyribacterium rettgeri*, **41**, 299; fungi, **41**, 294, 295; horseshoe crab, **41**, 313; *Peptostreptococcus elsdenii*, **41**, 309, 310
Other molecular properties: activity, **53**, 519; bacterial membrane, in, **31**, 649; biological role, **53**, 519; *Butyribacterium rettgeri*, from, **41**, 299 (activators and inhibitors, **41**, 303; assay, **41**, 299; properties, **41**, 302, 303; purification, **41**, 300); cytosol marker, as, **31**, 406, 408, 410, 740; *Escherichia coli*, from, **53**, 519; flavin content, **53**, 524; fungi, from, **41**,

293 (assay, **41**, 294, 295; chromatography, **41**, 296; inhibitors and regulation, **41**, 298; molecular weight, **41**, 298; properties, **41**, 297, 298; purification, **41**, 295); horseshoe crab, from, **41**, 313 (assay, **41**, 313; chromatography, **41**, 314; inhibitors, **41**, 316; molecular weight and subunit structure, **41**, 316; properties, **41**, 316, 318; purification, **41**, 313); immunochemical studies, **53**, 526, 527; inactivation, **53**, 439, 440; inhibitors, **53**, 525; kinetic properties, **53**, 524, 525; D-lactate determination, in, **89**, 29; *Lactobacillus plantarum*, **5**, 426; *Leuconostoc mesenteroides*, of, **9**, 326; molecular weight, **53**, 524; *Peptostreptococcus elsdenii*, from, **41**, 309 (activators and cofactors, **41**, 312; assay, **41**, 309, 310; chromatography, **41**, 311; properties, **41**, 312; purification, **41**, 310); pH optimum, **53**, 524; purification, **5**, 427; **53**, 520; purity, **53**, 524; reconstitution studies, **53**, 525, 526; spectral properties, **53**, 524; stereospecificity, **87**, 104; substrate specificity, **53**, 524, 525

L-LACTATE DEHYDROGENASE

This oxidoreductase [EC 1.1.1.27] catalyzes the reversible reaction of (*S*)-lactate, the enantiomer formed in mammalian anaerobic glycolysis, with NAD$^+$ to produce pyruvate, NADH, and a proton.[1–8] Other substrates include such (*S*)-2-hydroxymonocarboxylates as 2-hydroxybutyrate, 2-hydroxypentanoate, glycolate, and 2-hydroxyglutarate. NADP$^+$ is a much poorer substrate than NAD$^+$ with the mammalian, but not the bacterial, enzyme. The kinetic reaction mechanism is ordered Bi Bi, with the formation of Enz·NAD$^+$·pyruvate and Enz·NADH·lactate abortive complexes.

[1]R. Jaenicke, H. Schurig, N. Beaucamp & R. Ostendorp (1996) *Adv. Protein Chem.* **48**, 181.
[2]N. J. Oppenheimer & A. L. Handlon (1992) *The Enzymes*, 3rd ed., **20**, 453.
[3]M. A. Ator & P. R. Ortez de Montellano (1990) *The Enzymes*, 3rd ed., **19**, 213.
[4]H. F. Fisher (1988) *Adv. Enzymol.* **61**, 1.
[5]J. J. Holbrook, A. Liljas, S. J. Steindel & M. J. Rossman (1975) *The Enzymes*, 3rd ed., **11**, 191.
[6]K. Dalziel (1975) *The Enzymes*, 3rd ed., **11**, 1.
[7]M. G. Rossman, A. Liljas, C.-I. Brändén & L. J. Banaszak (1975) *The Enzymes*, 3rd ed., **11**, 61.
[8]J. Everse & N. O. Kaplan (1973) *Adv. Enzymol.* **37**, 61.

Selected entries from *Methods in Enzymology* [vol, page(s)]:
General discussion: **I**, 441, 449; **5**, 426; **9**, 278, 288; **27**, 20, 69, 76, 262; **34**, 165, 237, 239, 246, 249, 252, 491, 595, 598, 599; **46**, 20, 21, 145, 162; **89**, 345, 351, 359, 362
Assay: **I**, 441, 449; **4**, 377; **5**, 426; **9**, 278, 289, 296; **58**, 169, 170; *Bacillus subtilis*, **41**, 304; *Homarus americanus*, **89**, 359; murine isozymes, **89**, 352; sweet potato root, **89**, 345; *Streptococcus cremoris*, **89**, 362
Other molecular properties: abortive complex formation, **63**, 420, 421; affinity chromatography of NAD$^+$, **66**, 43, 44; affinity labeling, **47**, 424; **87**, 489; affinity partitioning, **228**, 138; analysis by continuous-flow method, **22**, 7, 8; assessment of leakage during neutrophil electropermeabilization, in, **256**, 330; binding, to agarose-bound nucleotides, **66**, 196; bovine, indefinite, self-association, **48**, 112; bovine heart, **I**, 450; **9**, 286; coupling to 1,1′-carbonyldiimidazole-activated support, **135**, 115; cultured cells and, **5**, 120; cytosol marker, assay, **182**, 220; dead-end competitive inhibition, **63**, 481; determination of isoenzyme levels, **41**, 47; diagnostic enzyme, as, **31**, 24; electrophoresis, **5**, 47; **228**, 142; enzyme electrode, in, **44**, 591, 877; gluconeogenic catalytic activity, **37**, 281, 288; H$_4$-lactate dehydrogenase, **34**, 600; immobilization, **44**, 44; inactivation, **53**, 439; inhibitor, **63**, 398, 399, 411; isoenzymes, **34**, 595, 599, 605; **58**, 169, 170; isolated nuclei and, **12A**,

445; isotope effects, **87**, 620, 631, 635; isotope exchange, **64**, 8, 19, 20; kinetics (inhibition, of, **44**, 428; multistep enzyme system, in, **44**, 467, 468); L-lactate assay and determination, **41**, 42; **89**, 29; marker enzyme, as, **56**, 210, 211, 217, 218, 222; mechanism, **63**, 51; microcalorimetric studies, of binding, **61**, 304; M$_4$-lactate dehydrogenase, **34**, 600; NAD$^+$ assay and, **6**, 792, 796; NADH oxidation catalyzed by, enzyme progress curves, nonlinear regression analysis, **249**, 75; polarography, **56**, 478; progress curve analysis, **63**, 181; properties, **I**, 443, 452; proton-exchange experiments, **176**, 343; purification, **I**, 441, 450; **9**, 279, 280, 281; reaction equilibrium in cultured cells, **264**, 457; reduction of NAD$^+$-analogues, **44**, 873, 874; sedimentation coefficient, **61**, 108; stereospecificity, **87**, 104; substrate inhibition, **63**, 421; temperature dependence, **63**, 246; transition state, **308**, 189; transition-state and multisubstrate analogues, **249**, 304; s-triazine binding, **44**, 52; tumors, clinical assesment of, **9**, 569; use in fluorometric recycling assay for NAD$^+$, **18B**, 28, 29

D-LACTATE DEHYDROGENASE (CYTOCHROME)

This FAD-dependent enzyme [EC 1.1.2.4], also known as D-lactate cytochrome *c* reductase, catalyzes the reaction of (*R*)-lactate with two molecules of ferricytochrome *c* to produce pyruvate and two molecules of ferrocytochrome *c*.[1–3] Other substrates include DL-2-hydroxybutyrate and DL-2-hydroxypentanoate.

[1]S. T. Olson & V. Massey (1979) *Biochemistry* **18**, 4714.
[2]Y. Hatefi & D. L. Stiggall (1976) *The Enzymes*, 3rd ed., **13**, 175.
[3]A. P. Nygaard (1963) *The Enzymes*, 2nd ed., **7**, 557.

Selected entries from *Methods in Enzymology* [vol, page(s)]:
General discussion: **9**, 302; **41**, 309; **243**, 21
Assay: **9**, 302; *Peptostreptococcus elsdenii*, **41**, 309
Other molecular properties: bakers' yeast, from (intact mitochondria, in, **9**, 312; molecular weight, **9**, 310; properties, **9**, 310; kinetic, **9**, 311; purification, **9**, 306; turnover number, **9**, 312); flavoprotein classification, **53**, 397; light sensitivity, **9**, 308; *Peptostreptococcus elsdenii*, from, **41**, 309 (activators and cofactors, **41**, 312; assay, **41**, 309, 310; chromatography, **41**, 311; properties, **41**, 312; purification, **41**, 310)

L-LACTATE DEHYDROGENASE (CYTOCHROME)

This FMN/protoheme IX-dependent oxidoreductase [EC 1.1.2.3], also known as flavocytochrome b_2, catalyzes the reaction of (*S*)-lactate with two molecules of the monoelectronic acceptor ferricytochrome *c* to produce pyruvate and two ferrocytochrome *c* molecules.[1–6] In the presence of only L-lactate and a keto acid, the enzyme catalyzes a flavin-dependent transhydrogenation, and the substrate α-hydrogen is transferred not only to the solvent but also in part to the keto acid.[2]

The enzyme will also catalyze bromopyruvate reduction, yielding pyruvate in a reductive elimination that proceeds by way of a carbanion intermediate.[4]

[1]B. A. Palfey & V. Massey (1998) *CBC* **3**, 83.
[2]K. S. Rao & F. Lederer (1998) *Protein Sci.* **7**, 1531.

[3] F. Lederer (1991) in *Flavins and Flavoproteins 1990*, Proceedings of the Tenth International Symposium, Como, Italy, July 15–20 (B. Curti, S. Ronchi & G. Zanetti, eds.), p. 773, Walter de Gruyter, New York.
[4] P. Urban & F. Lederer (1984) *EJB* **144**, 345.
[5] Y. Hatefi & D. L. Stiggall (1976) *The Enzymes*, 3rd ed., **13**, 175.
[6] A. P. Nygaard (1963) *The Enzymes*, 2nd ed., **7**, 557.

Selected entries from *Methods in Enzymology* [vol, page(s)]:
General discussion: 1, 444; 9, 314; 53, 238; 243, 21
Assay: 1, 444; 9, 314; 53, 238, 239, 240
Other molecular properties: bakers' yeast, 9, 321; cleaved crystalline form, 53, 240; conformational composition, 54, 269, 272; cytochrome b_2 and, 1, 448; 2, 746; extinction coefficient, 4, 280; 53, 240; flavoprotein and, 4, 288; *Hansenula anomala*, from, 53, 238, 249; inactivation of, 9, 321; 53, 439; intact noncrystalline form, 53, 244; kinetic properties, 53, 244; lactate dehydrogenase and, 2, 746; leader sequence in yeast, 264, 397; magnetic circular dichroism, 49, 170; 54, 295; molar activity, 53, 244; molecular weight, 53, 244; pH optimum, 53, 244; photosensitivity, 53, 254, 255; reaction with cytochrome c, 53, 156; reduction by ferredoxin-NADP$^+$ reductase, 23, 446; *Saccharomyces cerevisiae*, from, 53, 238, 240; spectral properties, 53, 246; subunit structure, 53, 244; whole yeast cells, assay, 1, 444; 9, 315; yeast, 1, 444; 2, 744; 53, 238, 439

D-LACTATE DEHYDROGENASE (CYTOCHROME c-553)

This oxidoreductase [EC 1.1.2.5], isolated from *Desulfovibrio vulgaris*, catalyzes the reaction of (R)-lactate with two molecules of ferricytochrome c-553 to produce pyruvate and two ferrocytochrome c-553 molecules.[1] DL-2-Hydroxybutyrate is an alternative substrate.

[1] M. Ogata, K. Arihara & T. Yagi (1981) *J. Biochem.* **89**, 1423.

D-LACTATE DEHYDROGENASE (NAD$^+$-INDEPENDENT)

This FMN-dependent oxidoreductase, first isolated from *Lactobacillus arabinosis*, catalyzes the reaction of (R)-lactate with an oxidized acceptor substrate to produce pyruvate and the reduced acceptor. NAD$^+$ will not act as a cosubstrate but 2,6-dichlorophenolindophenol will.

Selected entries from *Methods in Enzymology* [vol, page(s)]:
General discussion: 9, 321
Assay: 9, 321

L-LACTATE DEHYDROGENASE (NAD$^+$-INDEPENDENT)

This FMN-dependent oxidoreductase, first isolated from *Lactobacillus arabinosis*, catalyzes the reaction of (S)-lactate with an oxidized acceptor substrate to produce pyruvate and the reduced acceptor. NAD$^+$ will not act as a cosubstrate, but 2,6-dichlorophenolindophenol will.

Selected entries from *Methods in Enzymology* [vol, page(s)]:
General discussion: 9, 321
Assay: 9, 321

Other molecular properties: electron acceptors, 9, 326; optimum pH, 9, 326; properties, 9, 326; prosthetic group, 9, 326; purification, 9, 322; specificity, 9, 326; stability, 9, 326

LACTATE:MALATE TRANSHYDROGENASE

This transhydrogenase [EC 1.1.99.7], also known as malate:lactate transhydrogenase, catalyzes the reversible reaction of L-lactate with oxaloacetate to produce pyruvate and L-malate.[1–3] Thus, there is a hydrogen transfer from three-carbon or four-carbon carboxylic acids, converting (S)-2-hydroxy acids to 2-keto acids. The enzyme contains tightly bound NAD$^+$ or NADH in its active center: the removal of this nucleotide will result in enzyme denaturation.

[1] S. H. G. Allen (1966) *JBC* **241**, 5266.
[2] S. H. G. Allen & J. R. Patil (1972) *JBC* **247**, 909.
[3] S. H. G. Allen (1973) *EJB* **35**, 338.

Selected entries from *Methods in Enzymology* [vol, page(s)]:
General discussion: 13, 262; 89, 367
Assay: 13, 262; *Veillonella alcalescens*, 89, 368
Other molecular properties: mechanism, 13, 267; properties, 13, 267; 89, 372; prosthetic group requirements, 13, 267; purification, (*Micrococcus lactilyticus*, 13, 265; *Veillonella alcalescens*, 89, 370); substrate specificity, 13, 267; *Veillonella alcalescens* (assays, 89, 368; properties, 89, 372; purification, 89, 370)

LACTATE 2-MONOOXYGENASE

This FMN-dependent enzyme [EC 1.13.12.4], also known as lactate oxidative decarboxylase, lactate oxidase, and lactate oxygenase, catalyzes the reaction of L-lactate with dioxygen to produce acetate, carbon dioxide, and water.[1–8] The *Mycobacterium smegmatis* enzyme will also act on glycolate, producing formate, carbon dioxide, and water in a scheme similar to that observed with L-lactate; however, in this case, approximately 20% of the reactants are observed to generate glyoxylate and hydrogen peroxide as well.[2] Enantiomeric flavin-N^5-glycolyl adducts are generated via carbanion intermediates in the reaction mechanism.

L-Lactate first binds to the oxidized enzyme to produce a reduced flavin · pyruvate complex. Pyruvate dissociates extremely slowly from this complex. Dioxygen then binds to generate the oxidized flavin · pyruvate · H$_2$O$_2$ complex. Pyruvate and hydrogen peroxide then react, in a partially rate-determining step, to generate water, carbon dioxide, and acetate which subsequently dissociate.[2]

Although this enzyme is widely known as lactate oxidase, another lactate oxidase catalyzes the reaction of

L-lactate with dioxygen to produce pyruvate and hydrogen peroxide.[1,2]

[1]K. Yorita, H. Misaki, B. A. Palfey & V. Massey (2000) *Proc. Natl. Acad. Sci. U.S.A.* **97**, 2480.
[2]B. A. Palfey & V. Massey (1998) *CBC* **3**, 83.
[3]M. A. Ator & P. R. Ortez de Montellano (1990) *The Enzymes*, 3rd ed., **19**, 213.
[4]S. Ghisla & V. Massey (1980) *J. Biol. Chem.* **255**, 5688.
[5]V. Massey, S. Ghisla & K. Kieschke (1980) *J. Biol. Chem.* **255**, 2796.
[6]V. Massey & P. Hemmerich (1975) *The Enzymes*, 3rd ed., **12**, 191.
[7]M. S. Flashner & V. Massey (1974) in *Mol. Mech. Oxygen Activ.* (O. Hayaishi, ed.), p. 245.
[8]S. Yamamoto & Y. Ishimura (1991) *A Study of Enzymes* **2**, 315.

Selected entries from *Methods in Enzymology* [vol, page(s)]:
General discussion: 41, 329
Assay: 41, 329
Other molecular properties: apoenzyme, assay of FMN, **53**, 422; decarboxylating (borohydride reduction, **18B**, 475; sulfite complex, **18B**, 473); inactivation, **53**, 439 (protection, **53**, 447); *Mycobacterium phlei*, from, **41**, 329 (assay, **41**, 329, 330; molecular weight, **41**, 332; preparation, **41**, 330; properties, **41**, 332, 333; purification, **41**, 330); transition-state and multisubstrate analogues, **249**, 304

LACTATE RACEMASE

This enzyme [EC 5.1.2.1], also known as lacticoracemase and hydroxyacid racemase, catalyzes the interconversion of L-lactate and D-lactate.[1–3]

[1]M. E. Tanner & G. L. Kenyon (1998) *CBC* **2**, 7.
[2]E. Adams (1976) *Adv. Enzymol.* **44**, 69.
[3]L. Glaser (1972) *The Enzymes*, 3rd ed., **6**, 355.

Selected entries from *Methods in Enzymology* [vol, page(s)]:
General discussion: 5, 426, 430
Assay: 5, 430

D-LACTATE-2-SULFATASE

This sulfatase [EC 3.1.6.17] catalyzes the hydrolysis of (*R*)-2-*O*-sulfolactate (that is, D-lactate 2-sulfate) to produce (*R*)-lactate and sulfate.[1] L-Lactate 2-sulfate is not a substrate. In $H_2^{18}O$, inorganic sulfate is exclusively labeled.

[1]A. M. Crescenzi, K. S. Dodgson & G. F. White (1984) *BJ* **223**, 487.

LACTO-*N*-BIOSIDASE

This enzyme [EC 3.2.1.140], also known as oligosaccharide lacto-*N*-biosylhydrolase, catalyzes the hydrolysis of lacto-*N*-tetraose (or, β-D-Gal-(1→3)-β-D-GlcNAc-(1→3)-β-D-Gal-(1→4)-D-Glc) to produce lacto-*N*-biose (or, β-D-Gal-(1→3)-D-GlcNAc and lactose. The enzyme from *Streptomyces* specifically hydrolyzes the terminal lacto-*N*-biosyl residue (β-D-Gal-(1→3)-D-GlcNAc) from the non-reducing end of oligosaccharides with the structure β-D-Gal-(1→3)-β-D-GlcNAc-(1→3)-β-D-Gal-(1→R), where R is the point of attachment.[1] For example,

lacto-*N*-hexaose (β-D-Gal-(1→3)-β-D-GlcNAc-(1→3)-β-D-Gal-(1→3)-β-D-GlcNAc-(1→3)-β-D-Gal-(1→4)-D-Glc) is hydrolyzed to lacto-*N*-tetraose plus lacto-*N*-biose, which is subsequently converted to lactose. In this way, the enzyme specifically hydrolyzes lacto-*N*-tetraose and the *N*-acetyllactosamine type of triantennary sugar chain with the type 1 chain, but does not hydrolyze type 2 chain oligosaccharides or the type 1 chain oligosaccharides with fucose or sialic acid including lacto-*N*-fucopentaose I and II and α-2,3-sialyllacto-*N*-tetraose.

[1]M. Sano, K. Hayakawa & I. Kato (1993) *JBC* **268**, 18560.

LACTOCEPIN

This serine endopeptidase [EC 3.4.21.96], also known as extracellular lactococcal proteinase and lactococcal cell envelope-associated proteinase, catalyzes peptide-bond hydrolysis with a very broad specificity. There are three types of lactocepins (types I, III, and I/III; type II was later discovered to be an artifact of temperature and pH). Both lactocepins I and III hydrolyze β- and κ-caseins, albeit with different specificities. Slight differences in their action may be partly responsible for imparting distinct flavor qualities to certain cheeses. Large hydrophobic aminoacyl residues are preferred in positions P1 and P4. A prolyl residue is preferred in the P2 position. Lactocepin I readily catalyzes the hydrolysis of MeOSuc-Arg-Pro-Tyr-∇-NHPhNO₂ (Note: ∇ indicates the cleavage site), whereas lactocepin III requires the presence of 10% (w/v) NaCl. Conversely, lactocepin III will act on Suc-Ala-Glu-Pro-Phe-∇-NHPhNO₂ while lactocepin I has negligible activity.[1–6]

[1]T. D. Thomas & G. G. Pritchard (1987) *FEMS Microbiol. Rev.* **46**, 245.
[2]J. Kok (1990) *FEMS Microbiol. Rev.* **87**, 15.
[3]G. G. Pritchard & T. Coolbear (1993) *FEMS Microbiol. Rev.* **12**, 179.
[4]S. Visser (1993) *J. Dairy Sci.* **76**, 329.
[5]J. Kok & W. M. de Vos (1994) in *Genetics and Biotechnology of Lactic Acid Bacteria* (M. J. Gasson & W. M. de Vos, eds.), Blackie and Professional, Glasgow, p. 169.
[6]W. Bockelmann (1995) *Int. Dairy J.* **5**, 977.

1,4-LACTONASE

This calcium-dependent enzyme [EC 3.1.1.25], also known as γ-lactonase, catalyzes the hydrolysis of a 1,4-lactone to produce a 4-hydroxy-acid anion.[1] All known substrates are 1,4-lactones consisting of four to eight carbon atoms; these include γ-butyrolactone, γ-valerolactone, γ-caprolactone, and γ-octalactone. Simple aliphatic esters, acetylcholine, sugar lactones, or substituted aliphatic lactones (including 3-hydroxy-4-butyrolactone) are not substrates. **See also** *specific lactonases*

[1]W. N. Fishbein & S. P. Bessman (1966) *JBC* **241**, 4835 and 4842.

LACTOPEROXIDASE

This variant of peroxidase [EC 1.11.1.7] found in milk catalyzes the reaction of a donor with hydrogen peroxide to produce the oxidized donor and two water molecules; for example, it will oxidize iodide to iodine.[1,2] The enzyme also possesses a bromoperoxidase activity. **See also** *Peroxidase; Bromoperoxidase; Iodide Peroxidase*

[1]E. I. Thomas, P. M. Bozeman & D. B. Learn (1991) in *Peroxidases in Chemistry and Biology* (J. Everse, K. E. Everse & M. B. Grisham, eds.) **1**, p. 123, CRC Press, Boca Raton.
[2]K. M. Pruitt & J. O. Tenovuo, eds. (1985) *The Lactoperoxidase System, Chemistry, and Biological Significance*, Marcel Dekker, New York.

Selected entries from *Methods in Enzymology* [vol, page(s)]:
General discussion: 2, 813; 17A, 653; 68, 437
Assay: 2, 813; chemiluminescence assay, 57, 427
Other molecular properties: antibody (conjugation, **73**, 151; iodination, **50**, 58); chemiluminescence assay, **57**, 427; determination of membrane proteins, in, **32**, 103; direct iodination of proteins, **182**, 726; generation of cysteine and glutathione thiyl radicals, **186**, 324; γ-globulin iodination and, **30**, 636; glucose oxidase, and, in ^{125}I labeling of interferons, **119**, 263; hormone iodination, **37**, 229, 232, 233, 325; hormone labeling, in, **37**, 147; induced erythrocyte cytolysis, continuous assay, **132**, 491; intracellular iodination of lysosomal membrane, in, **98**, 404 (application to membrane recycling, **98**, 414); iodination and coupling reactions, **107**, 476 (dissociation, **107**, 483; product identification, **107**, 485); iodination, **230**, 41 (acidic and basic fibroblast growth factor, of, **198**, 475; chloroplast membranes, of, by photosystem II, **69**, 504; direct iodination of proteins, **182**, 726; transforming growth factor β1, of, **198**, 305); lactoperoxidase-catalyzed radioiodination, **70**, 214; **73**, 448, 449, 672 (factors affecting labeling, **70**, 215; glucose oxidase procedure, **70**, 330; procedure, **74**, 355, 455, 534; product isolation, **70**, 216; protein localization, **70**, 218 [procedure, **70**, 215, 216 {for membranes, **70**, 219, 220, 258}; recommendations, **70**, 219, 220]); lectin label, as, **32**, 616; nitrogen content, **2**, 799; properties, **2**, 816; purification from bovine milk, **2**, 813; **17A**, 655; radiolabeling of, in (lymphocyte receptors, **150**, 408; murine interferon, **119**, 322; neuropeptides, **124**, 24); tyrosine modification, for, **53**, 137, 148, 149; uptake and determination in internalized plasma membranes, **98**, 546

LACTOSE DEHYDROGENASE

This flavin-dependent oxidoreductase catalyzes the reaction of lactose with a reducing agent to produce lactobionic-δ-lactone and the oxidized agent. **See also** *Gluconolactonase; L-Arabinonolactonase*

Selected entries from *Methods in Enzymology* [vol, page(s)]:
General discussion: 9, 73
Assay: 9, 73

LACTOSE SYNTHASE

This enzyme [EC 2.4.1.22] catalyzes the reaction of UDP-D-galactose with D-glucose to generate UDP and lactose.[1–4] In many tissues, lactose synthase is a complex of two proteins, designated A and B. In the absence of the α-lactalbumin (component B), the enzyme (component A) catalyzes the transfer of D-galactose from UDP-D-galactose to *N*-acetyl-D-glucosamine, both in the free form of the sugar or as glycoproteins and glycolipids. In the presence of α-lactalbumin, D-glucose is the preferred acceptor substrate. Alternative substrates for D-glucose include D-xylose, maltose, and 2-deoxy-D-glucose.

Whether or not α-lactalbumin is present, the enzyme's kinetic mechanism is random Bi Bi. No positional isotope exchanges are observed and there is a large secondary kinetic isotope effect, suggesting that the anomeric carbon of the galactosyl moiety has substantial sp^2 character in the transition state.[2–4]

[1]N. J. Kuhn, M. Stankiewicz & S. Ward (1992) *Biochem. Soc. Trans.* **20**, 714.
[2]R. L. Hill & K. Brew (1975) *AE* **43**, 411.
[3]K. E. Ebner (1973) *The Enzymes*, 3rd ed., **9**, 363.
[4]K. Brew (1970) *Essays in Biochem.* **6**, 93.

Selected entries from *Methods in Enzymology* [vol, page(s)]:
General discussion: 8, 346; 28, 500; 34, 4, 359, 360
Assay: 8, 346; 28, 501, 502
Other molecular properties: activation, **87**, 359; affinity labeling, **87**, 473; isolation from bovine milk, **8**, 347; **28**, 503; kinetics, **87**, 359; mechanism, **63**, 51; physiological significance, **28**, 500; plasma membrane glycoconjugate labeling, in, **98**, 416; properties, **8**, 350; **28**, 506; purification from bovine milk, **8**, 347; **28**, 503; specificity, **28**, 638; spectrophotometric assay, **28**, 502

LACTOSYLCERAMIDE 1,3-*N*-ACETYL-β-D-GLUCOSAMINYLTRANSFERASE

This transferase [EC 2.4.1.206] catalyzes the reaction of UDP-*N*-acetyl-D-glucosamine with D-galactosyl-1,4-β-D-glucosylceramide (that is, lactosylceramide) to produce UDP and *N*-acetyl-D-glucosaminyl-1,3-β-D-galactosyl-1,4-β-D-glucosylceramide.[1–3]

[1]D. K. Chou & F. B. Jungalwala (1993) *JBC* **268**, 21727.
[2]J. Gottfries, A. K. Percy, J.-E. Mansson, P. Fredman, C. J. Wikstrand, H. S. Friedman, D. D. Bigner & L. Svennerholm (1991) *BBA* **1081**, 253.
[3]E. H. Holmes, S.-i. Hakomori & G. K. Ostrander (1987) *JBC* **262**, 15649.

Selected entries from *Methods in Enzymology* [vol, page(s)]:
General discussion: 138, 597

LACTOSYLCERAMIDE β-1,3-GALACTOSYLTRANSFERASE

This transferase [EC 2.4.1.179] catalyzes the reaction of UDP-D-galactose with D-galactosyl-1,4-β-D-glucosyl-R, where R is an oligosaccharide or glycolipid, to produce D-galactosyl-1,3-β-D-galactosyl-1,4-β-D-glucosyl-R and UDP.[1-3] Lactose can act as a poor acceptor substrate. This enzyme participates in the elongation of oligosaccharide chains, especially in glycolipids.

[1]P. Bailly, F. Piller & J. P. Cartron (1988) *EJB* **173**, 417.
[2]M. Messer & K. R. Nicholas (1991) *BBA* **1077**, 79.
[3]E. H. Holmes (1989) *ABB* **270**, 630.

LACTOSYLCERAMIDE α-2,3-SIALYLTRANSFERASE

This sialyltransferase [EC 2.4.99.9], also known as CMP-*N*-acetylneuraminate:lactosylceramide α-2,3-sialyltransferase and G$_{M3}$ synthase, catalyzes the reaction of CMP-*N*-acetylneuraminate with lactosylceramide (or, β-D-galactosyl-1,4-β-D-glucosylceramide) to produce ganglioside G$_{M3}$ (or, α-*N*-acetylneuraminyl-2,3-β-D-galactosyl-1,4-β-D-glucosylceramide) and CMP.[1,2] Lactose is not a substrate. *See also Sialyltransferases; α-2,3-Sialyltransferases*

[1]P. H. Fishman, R. M. Bradley & R. C. Henneberry (1976) *ABB* **172**, 618.
[2]L. J. Melkerson-Watson & C. C. Sweeley (1991) *JBC* **266**, 4448.

Selected entries from *Methods in Enzymology* [vol, page(s)]:
General discussion: 138, 601

LACTOSYLCERAMIDE α-2,6-*N*-SIALYLTRANSFERASE

This sialyltransferase [EC 2.4.99.11] catalyzes the reaction of CMP-*N*-acetylneuraminate with a β-D-galactosyl-1,4-β-D-glucosylceramide to produce CMP and an α-*N*-acetylneuraminyl-2,6-β-D-galactosyl-1,4-β-D-glucosylceramide.[1] *See also Sialyltransferases; α-2,6-Sialyltransferases*

[1]I. Albarracin, F. E. Lassaga & R. Caputto (1988) *BJ* **254**, 559.

LACTOSYLCERAMIDE SYNTHASE

This transferase, also known as glucosylceramide galactosyltransferase and Gal-T1, catalyzes the reaction of UDP-D-galactose with D-glucosylceramide to produce lactosylceramide (or, β-D-galactosyl-1,4-β-D-glucosylceramide) and UDP.

Selected entries from *Methods in Enzymology* [vol, page(s)]:
General discussion: 311, 73
Assay: 311, 75
Other molecular properties: properties, 311, 77; role, 311, 78

LACTOYL-CoA DEHYDRATASE

This enzyme [EC 4.2.1.54] catalyzes the reversible conversion of lactoyl-CoA to acryloyl-CoA and water.[1,2]

The *Clostridium propionicum* enzyme contains iron-sulfur centers and requires ATP.[3-5]

[1]R. L. Baldwin, W. A. Wood & R. S. Emery (1965) *BBA* **97**, 202.
[2]P. R. Vagelos, J. M. Earl & E. R. Stadtman (1959) *JBC* **234**, 2272.
[3]R. D. Kuchta & R. H. Abeles (1985) *JBC* **260**, 13181.
[4]R. D. Kuchta, G. R. Hanson, B. Holmquist & R. H. Abeles (1986) *Biochemistry* **25**, 7301.
[5]A. E. Hofmeister & W. Buckel (1992) *EJB* **206**, 547.

Selected entries from *Methods in Enzymology* [vol, page(s)]:
General discussion: 9, 683
Assay: 9, 683

LACTOYLGLUTATHIONE LYASE

This lyase [EC 4.4.1.5], also known as methylglyoxalase, aldoketomutase, glyoxalase I, and ketone-aldehyde mutase, catalyzes the reversible conversion of (*R*)-*S*-lactoylglutathione to glutathione and methylglyoxal (CH$_3$COCHO).[1-6] In the reverse reaction, the substrate is not methylglyoxal; rather, the thiohemiacetal is formed upon reaction of methylglyoxal with glutathione. The enzyme catalyzes the isomerization of this thiohemiacetal to *S*-D-lactoylglutathione. A *cis*-enediolate intermediate is formed along the reaction pathway.[1] *See also Hydroxyacylglutathione Hydrolase*

[1]A. D. Cameron, M. Ridderström, B. Olin, M. J. Kavarana, D. J. Creighton & B. Mannervik (1999) *Biochemistry* **38**, 13480.
[2]A. Kimura & Y. Inoue (1993) *Biochem. Soc. Trans.* **21**, 518.
[3]B. Mannervik & M. Ridderstrom (1993) *Biochem. Soc. Trans.* **21**, 515.
[4]D. L. Van der Jagt (1989) *Coenzymes Cofactors (Glutathione, Chem. Biochem. Med. Aspects, Pt. A)* **3**, 597.
[5]B. Mannervik (1987) *Biochem. Soc. Trans.* **15**, 717.
[6]K. T. Douglas (1987) *AE* **59**, 103.

Selected entries from *Methods in Enzymology* [vol, page(s)]:
General discussion: 1, 454; 77, 236, 297, 424; 90, 535, 542
Assay: 1, 454; 77, 297, 298; human erythrocyte, 90, 536; murine liver, 90, 542
Other molecular properties: affinity chromatography, 77, 299; definition of units, 77, 298; exchanges, proton, 87, 87; formaldehyde dehydrogenase, in preparation of, 9, 359; glutathione assay, 2, 343, 719; 3, 605; 77, 380; human erythrocyte (assay, 90, 536; properties, 90, 540; purification, 90, 536); intermediate, 87, 84, 87; kinetics, 77, 300, 301 (models, 87, 380); marker for chromosome detection in somatic cell hybrids, as, 151, 185; methylglyoxal assay and, 3, 296; murine liver (assay, 90, 542; properties, 90, 545; purification, 90, 543); properties, 1, 457; 77, 300, 301; purification, 77, 298 (bakers' yeast, 1, 455); size, 77, 301; substrates, 77, 301; transition-state and multisubstrate analogues, 249, 308; triose-phosphate isomerase and, 3, 294

LAMINARIBIOSE PHOSPHORYLASE

This phosphorylase [EC 2.4.1.31] catalyzes the reversible reaction of laminaribiose (or, 3-β-D-glucosylglucose) with orthophosphate to produce D-glucose and α-D-glucose 1-phosphate. In the presence of arsenate, laminaribiose is completely converted into D-glucose. Other substrates include 1,3-β-D-oligoglucans (for example, laminaritriose, laminaritetraose, and laminaripentaose).

This phosphorylase, which has an ordered Bi Bi kinetic mechanism,[1] has a different specificity spectrum than 1,3-β-oligoglucan phosphorylase [EC 2.4.1.30] and laminarin phosphorylase [EC 2.4.1.97].

[1]M. Kitaoka, T. Sasaki & H. Taniguchi (1993) ABB **304**, 508.

Selected entries from *Methods in Enzymology* [vol, page(s)]:
General discussion: 28, 953
Assay: 28, 954

LANOSTEROL SYNTHASE

This enzyme [EC 5.4.99.7], also known as 2,3-epoxy-squalene:lanosterol cyclase, oxidosqualene:lanosterol cyclase, 2,3-oxidosqualene cyclase, and squalene-2,3-oxide:lanosterol cyclase, catalyzes the conversion of (*S*)-2,3-epoxysqualene to lanosterol.

(3*S*)-2,3-oxidosqualene   lanosterol

Note that lanosterol synthase and squalene monooxygenase are often known collectively as 2,3-oxidosqualene cyclase and squalene oxidocyclase.[1-3] Enzyme catalysis proceeds by a carbocationic cyclization reaction, with subsequent rearrangements. The substrate possesses only one chiral center; the fact that the reaction product has seven chiral centers indicates the high degree of regiospecificity and stereospecificity. Cyclization results in the formation of a protosterol cation intermediate which then undergoes two 1,2-methyl shifts; elimination of a proton ensues, with two concomitant 1,2-hydride shifts.

[1]L. Cattel & M. Ceruti (1998) Crit. Rev. Biochem. Mol. Biol. **33**, 353.
[2]I. Abe, M. Rohmer & G. D. Prestwich (1993) Chem. Rev. **93**, 2189.
[3]S. Yamamoto, K. Lin & K. Bloch (1969) PNAS **63**, 110.

Selected entries from *Methods in Enzymology* [vol, page(s)]:
General discussion: 15, 497; squalene monooxygenase, with, 5, 490
Assay: 15, 497; squalene monooxygenase, with, 5, 493
Other molecular properties: hog liver, from, purification steps, 15, 495; incubations, 15, 498; pea seedlings, from, purification steps, 15, 500; preparation 5, 498 (hog liver, from, 15, 495; pea seedling, from, 15, 498); properties, 15, 496; solubilization, 15, 495, 499

LATHOSTEROL OXIDASE

This oxidoreductase [EC 1.3.3.2], also known as lathosterol 5-desaturase and Δ^7-sterol Δ^5-dehydrogenase, catalyzes the reaction of lathosterol (or, 5α-cholest-7-en-3β-ol) with dioxygen to produce 7-dehydrocholesterol (or, cholesta-5,7-dien-3β-ol) and hydrogen peroxide.[1-4] The reaction proceeds with a initial burst, followed by the steady-state phase.[5]

[1]T. Ishibashi & K. Bloch (1981) JBC **256**, 12962.
[2]M. E. Dempsey, K. E. McCoy & H. N. Barker (1981) JBC **256**, 1867.
[3]S. Kawata, J. M. Traskos & J. L. Gaylor (1985) JBC **260**, 6609.
[4]G. F. Grinstead & J. L. Gaylor (1982) JBC **257**, 13937.
[5]H. Nishino, J. Nakaya, S. Nishi, T. Kurosawa & T. Ishibashi (1997) ABB **339**, 298.

Selected entries from *Methods in Enzymology* [vol, page(s)]:
General discussion: 15, 501; 56, 474
Assay: isotopic, 15, 503; spectrophotometric, 15, 505
Other molecular properties: inhibitors, 15, 513; preparation, 15, 509; properties, 15, 512; purification steps, 15, 511

LATIA-LUCIFERIN MONOOXYGENASE (DEMETHYLATING)

This flavin-dependent oxidoreductase [EC 1.14.99.21], also known as *Latia* luciferase, catalyzes the reaction of *Latia* luciferin with an electron donor substrate (AH$_2$) and two molecules of dioxygen to produce the oxidized *Latia* luciferin, carbon dioxide, formate, the oxidized donor (A), water, and light.[1,2]

Latia luciferin

Latia luciferin is (*E*)-2-methyl-4-(2,6,6-trimethyl-1-cyclo-hex-1-yl)-1-buten-1-ol formate. The overall reaction may be the result of two enzyme activities (an oxygenase and a monooxygenase) acting in sequence.

[1]O. Shimomura, F. H. Johnson & Y. Kohama (1972) PNAS **69**, 2086.
[2]O. Shimomura & F. H. Johnson (1968) Biochemistry **7**, 1734.

*Lca*I RESTRICTION ENDONUCLEASE

This type II restriction enzyme [EC 3.1.21.4], isolated from *Lactobacillus casei*, catalyzes the cleavage of both strands of DNA at 5′...AT∇CGAT...3′.

LEGHEMOGLOBIN REDUCTASE

This oxidoreductase [EC 1.6.2.6] catalyzes the reaction of NAD(P)H with two molecules of ferrileghemoglobin to produce NAD(P)$^+$ and two ferroleghemoglobin molecules.[1,2] NADH is the more effective of the two coenzyme substrates.

[1] L. L. Saari & R. V. Klucas (1984) *ABB* **231**, 102.
[2] M. Becana & R. V. Klucas (1990) *PNAS* **87**, 7295.

LEGUMAIN

This cysteine endopeptidase [EC 3.4.22.34] (also known as bean endopeptidase, asparaginyl endopeptidase, vicilin peptidohydrolase, and phaseolin, catalyzes the hydrolysis of peptide bonds in proteins such as azocasein. The endopeptidase exhibits preferential cleavage at Asn–Xaa, where Xaa can be any amino acid.[1] Legumain will not act on the peptide bond when the asparaginyl residue is the N-terminal residue or if the asparaginyl residue is *N*-glycosylated.

[1] S.-I. Ishii (1994) *MIE* **244**, 604.

LEISHMANOLYSIN

This zinc- and calcium-dependent endopeptidase [EC 3.4.24.36], also known as promastigote surface endopeptidase (PSP) and glycoprotein gp63, is a glycosylated, membrane-bound enzyme that has an extended substrate-binding site. It catalyzes the hydrolysis of Gly8–Ser9 and Tyr16–Leu17 in the insulin B chain and Tyr10–Ser11, Asp15–Ser16, and Trp25–Leu26 in glucagon. The enzyme is autoactivated by cleaving and activating its own proenzyme at Val100–Val101.[1]

[1] J. Bouvier, P. Schneider & R. Etges (1995) *MIE* **248**, 614.

Selected entries from *Methods in Enzymology* [vol, page(s)]:
General discussion: **248**, 614.
Assay: **248**, 624 (peptide substrate, **248**, 625 [TLC analysis, **248**, 630]; in solution, **248**, 625)
Other molecular properties: amastigote, **248**, 617, 632; *Crithidia*, **248**, 616, 618, 632; expression, regulation, **248**, 618, 632; fluorogenic substrate for, **248**, 617, 627; genes, **248**, 618, 632; *Herpetomonas*, **248**, 616, 630; immunological applications, **248**, 618; inhibition by synthetic peptide, **248**, 617; *Leishmania* (amastigote homologue, **248**, 617; structure, **248**, 616); peptide bond preference, **248**, 633 (assay, **248**, 627); pH optimum, **248**, 632; physiological functions, **248**, 617, 632; promastigote, **248**, 616, 632 (isotopic labeling, **248**, 623); properties, **248**, 207, 617, 618; proteolytic activity, on peptides, **248**, 627; purification, **248**, 620; structure, **248**, 616, 633; substrate specificity, **248**, 617, 632; vaccine, as, **248**, 618; zymography, **248**, 624

LEUCINE *N*-ACETYLTRANSFERASE

This acetyltransferase [EC 2.3.1.66] catalyzes the reaction of acetyl-CoA with L-leucine to produce coenzyme A and *N*-acetyl-L-leucine.[1]

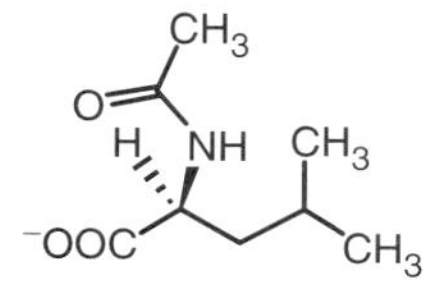

N-acetyl-L-leucine

Other acceptor substrates include L-arginine, L-valine, L-phenylalanine, L-methionine, and peptides containing N-terminal L-leucyl residues (particularly, L-leucyl-L-leucine). Propionyl-CoA can serve as the acyl donor, albeit weakly.

[1] K. Suzukake, H. Hayashi, M. Hori & H. Umezawa H. (1980) *J. Antibiot.* **33**, 857.

LEUCINE 2,3-AMINOMUTASE

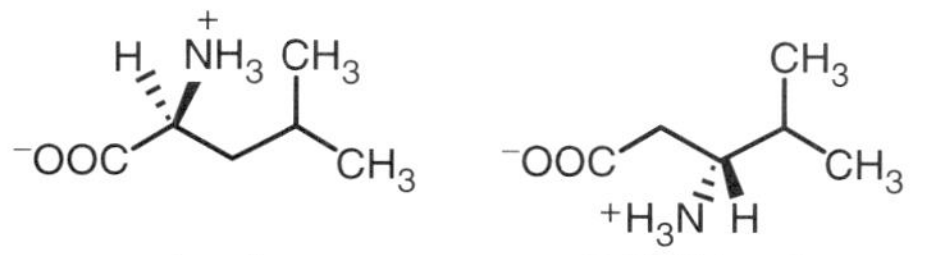

L-leucine   (3R)-β-leucine

This cobalamin-dependent mutase [EC 5.4.3.7] catalyzes the interconversion of (2S)-α-leucine and (3R)-β-leucine, presumably via radical intermediates.[1–3]

[1] O. W. Griffith (1986) *Ann. Rev. Biochem.* **55**, 855.
[2] J. M. Poston (1986) *Adv. Enzymol.* **58**, 173.
[3] B. M. Babior & J. S. Krouwer (1979) *Crit. Rev. Biochem.* **6**, 35.

Selected entries from *Methods in Enzymology* [vol, page(s)]:
General discussion: **166**, 130
Assay: **166**, 130
Other molecular properties: clinical associations, **166**, 135; occurrence and distribution, **166**, 134; preparation, **166**, 133

LEUCINE AMINOTRANSFERASE

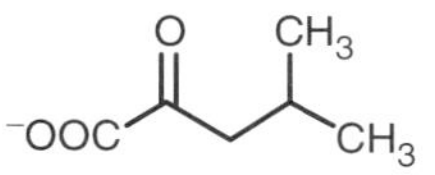

4-methyl-2-ketopentanoate

This pyridoxal-phosphate-dependent enzyme [EC 2.6.1.6] catalyzes the reversible reaction of L-leucine with α-ketoglutarate (or, 2-oxoglutarate) to produce 4-methyl-2-ketopentanoate (*i.e.*, 4-methyl-2-oxopentanoate) and L-glutamate.[1,2] *See also* Branched-Chain Amino Acid Aminotransferase

[1] A. E. Braunstein (1973) *The Enzymes*, 3rd ed., **9**, 379.
[2] K. Aki, K. Ogawa & A. Ichihara (1968) *BBA* **159**, 276.

Selected entries from **Methods in Enzymology** [vol, page(s)]:
General discussion: 17A, 802, 814; 113, 71; 166, 269
Assay: 113, 72
Other molecular properties: purification from rat liver, 17A, 815

LEUCINE DEHYDROGENASE

This oxidoreductase [EC 1.4.1.9] catalyzes the reversible reaction of L-leucine with NAD^+ and water to produce 4-methyl-2-ketopentanoate (or, 4-methyl-2-oxopentanoate), ammonia, and NADH (B-side specific).[1-4] Other substrates include L-isoleucine, L-valine, L-norvaline, and L-norleucine. The *Bacillus sphaericus* enzyme is octameric and has an ordered Bi Ter kinetic mechanism, the coenzyme binding first.[4]

[1] A. R. Clarke & T. R. Dafforn (1998) CBC 3, 1.
[2] P. J. Baker, A. P. Turnbull, S. E. Sedelnikova, T. J. Stillman & D. W. Rice (1995) Structure 3, 693.
[3] N. M. W. Brunhuber & J. S. Blanchard (1994) Crit. Rev. Biochem. Mol. Biol. 29, 415.
[4] T. Ohshima, H. Misono & K. Soda (1978) JBC 253, 5719.

Selected entries from **Methods in Enzymology** [vol, page(s)]:
General discussion: 17A, 799; 166, 282; 228, 608
Assay: 166, 283; 228, 611
Other molecular properties: assay of branched-chain amino and keto acids, in, 166, 3; Bacillus cereus (affinity partitioning and extraction, 228, 613; assay, 228, 611; extraction in PEG-salt systems, 228, 612; isolation, 228, 608; partitioning behavior and extraction, 228, 611; partitioning experiments, 228, 610; production, 228, 608; purification, 17A, 800; 228, 608 [methods, 228, 609, 614; yield, 228, 617]); catalyzed reaction, 228, 608; production, 166, 284; properties, 166, 286; purification from (Bacillus cereus, 17A, 800; Bacillus subtilis, 166, 284); stereospecificity, 87, 117

LEUCOLYSIN

This zinc-dependent metalloendopeptidase [EC 3.4.24.6] was the first venom proteinase purified to a high degree of homogeneity.[1-3] The enzyme has both peptidase and esterase activity. It acts on both Bz-L-Arg-OEt and L-Leu-NHPhNO₂. In the insulin B chain, leucolysin catalyzes the hydrolysis of Phe1–Val2, His5–Leu6, His10–Leu11, Ala14–Leu15, Gly20–Glu21, Gly23–Phe24, and Phe24–Phe25[1] (however, a different hydrolysis pattern was more recently reported[3]: Leu6–Cys7, Leu11–Val12, Leu15–Tyr16, Phe24–Phe25, and Phe25–Tyr26).

[1] A. M. Spiekerman, K. K. Fredericks, F. W. Wagner & J. M. Prescott (1973) BBA 293, 464.
[2] J. B. Bjarnason & J. W. Fox (1995) MIE 248, 345.
[3] F. W. Wagner (1998) in Handbook of Proteolytic Enzymes (A. J. Barrett, N. D. Rawlings & J. F. Woessner, eds.), p. 1291, Academic Press, San Diego.

Selected entries from **Methods in Enzymology** [vol, page(s)]:
General discussion: 45, 397
Other molecular properties: cleavage specificity on insulin B chain, 248, 350, 361; properties, 248, 350, 360

LEUCYL AMINOPEPTIDASE

This zinc-dependent metalloaminopeptidase [EC 3.4.11.1] (also known as LAP, cytosol aminopeptidase, leucine aminopeptidase, and peptidase S) catalyzes the release of an N-terminal amino acid from a peptide or protein with a preference for N-terminal leucyl residues. Linderstrøm-Lang originally called the enzyme dipeptidase II since it hydrolyzed leucylglycine twentyfold faster than glycylglycine.[1] Interestingly, leucyl aminopeptidase was one of the first seven proteins crystallized in Sumner's laboratory, although its identity was not known for fifty years. It was known as the football protein since the crystals resembled an American football (that is, a prolate ellipsoid). Leucyl aminopeptidase is the first dizinc enzyme that has had its crystal structure determined.[2]

As mentioned above, all N-terminal amino acids are released but there is a preference for leucyl residues in the P1 position (that is, Leu-Xaa). However, peptides containing a prolyl residue in the P1' position are not hydrolyzed. Amino acid amides and esters are also hydrolyzed.[3-7] **See also Vacuolar Aminopeptidase I**

[1] K. Linderstrøm-Lang (1929) Z. Physiol. Chem. 182, 151.
[2] S. K. Burley, P. R. David, A. Taylor & W. N. Lipscomb (1990) PNAS 87, 6878.
[3] S. R. Himmelhoch (1969) ABB 134, 597.
[4] R. J. Delange & E. L. Smith (1971) The Enzymes, 3rd ed., 3, 81.
[5] H. Hanson & M. Frohne (1976) MIE 45, 504.
[6] H. Kim & W. N. Lipscomb (1994) AE 68, 153.
[7] A. Taylor (1993) FASEB J. 7, 290.

Selected entries from **Methods in Enzymology** [vol, page(s)]:
General discussion: 2, 88; 11, 426; 19, 508, 534, 741; 45, 504; 46, 28, 431; 47, 542, 577
Assay: 2, 88; 19, 508, 748; 45, 518
Other molecular properties: action, 45, 504; activation with magnesium ion, 11, 658; activators and inhibitors, 2, 93; 19, 754; activity, 45, 513; adjunct in phenylisothiocyanate degradation, as, 11, 436; amino acid sequence determination and, 6, 840; bacterial homologues, 248, 219; check for endopeptidase activity in, 11, 659; clinical aspects, 17B, 875; complete hydrolysis with, 11, 433; contaminating activities present in crude preparations, 11, 657; esterase activity, 45, 517; extent of hydrolysis with, 11, 427; glycopeptide digestion and, 8, 28; isolation, 45, 505; kinetics, 45, 515; limited proteolysis, in, 11, 656; macrocrystals, 45, 505; molecular weight, 19, 755; peptide sequence determination, 25, 250 (enzyme assays, 25, 257; enzyme purification, 25, 254; enzyme specificity, 25, 258; peptide hydrolysis, 25, 258; sequence studies, 25, 260); pH optimum, 2, 93; 19, 755; plant homologues, 248, 219; polysaccharide release and, 28, 79; porcine kidney specificity, from, 59, 600; preparation, 45, 506 (suitable for sequence studies, 11, 427; suitable for modification of proteins, 11, 655); procedure for modification of proteins with, 11, 659; properties, 2, 91; 11, 432; 19, 513; 25, 254; 45, 507; 248, 213, 219; purification, 2, 89; 19, 509, 751; purity, 19, 755; 45, 507; rate of hydrolysis of amino acid amides, 25, 255; sequence studies with, 11, 434; serum, in assay of, 17B, 877;

specificity, **2**, 91; **11**, 426; **19**, 754; **45**, 515; **59**, 600; stability, **19**, 755; **45**, 509; staining on gels, **22**, 585, 592; structure, **45**, 511; substrate, fluorogenic, **248**, 23; swine kidney, from, **19**, 508; thioester substrate, **248**, 15; transition-state and multisubstrate analogues, **249**, 306; use in sequence elucidation, **11**, 426 (procedure for, **11**, 432); *Vibrio*, **248**, 223, 225; yeast, **248**, 222

LEUCYL AMINOPEPTIDASE (BACTERIAL)

This metalloaminopeptidase [EC 3.4.11.10], also known as aminopeptidase I (*Escherichia coli*) and peptidase A (*Salmonella typhimurium* and *E. coli*), catalyzes the release of N-terminal aminoacyl residues from protein/peptide substrates having leucyl or methionyl residues in the P1 position. The *S. typhimurium* enzyme will not act on the peptide bond when the adjacent P1′ position is occupied by a prolyl residue. Unlike the mammalian enzyme, the bacterial enzyme is manganese-dependent and is inhibited by zinc ions.[1-3] **See also** *Vibrio Aminopeptidase*

[1] N. D. Rawlings & A. J. Barrett (1995) *Meth. Enzymol.* **248**, 183.
[2] T. Gonzales & J. Robert-Baudouy (1996) *FEMS Microbiol. Rev.* **18**, 319.
[3] C. G. Miller (1996) in *Escherichia coli and Salmonella Cellular and Molecular Biology* (F. C. Neidhardt, R. Curtiss III, J. L. Ingraham, E. C. C. Lin, K. Brooks Low, Jr., B. Magasanik, W. S. Reznikoff, M. Riley, M. Schaechter & H. E. Umbarger, eds.), ASM Press, Washington, p. 938.

Selected entries from *Methods in Enzymology* [vol, page(s)]:
General discussion: **45**, 530; **248**, 183

LEUCYL ENDOPEPTIDASE

This serine endopeptidase [EC 3.4.21.57], also known as plant Leu-proteinase and leucine-specific serine proteinase, catalyzes peptide-bond hydrolysis in protein/peptide substrates at Leu–Xaa bonds. Acylation is the rate-determining step. The enzyme can be readily assayed with synthetic substrates such as Bz-Leu-∇-NHPhNO₂ and Z-Leu-∇-OPhNO₂.[1,2] (Note: ∇ indicates the cleavage site.) Leucyl endopeptidase also possesses an esterase activity.

[1] P. Aducci, P. Ascenzi, M. Pierini & A. Ballio (1986) *Plant Physiol.* **81**, 812.
[2] P. Aducci, P. Ascenzi & A. Ballio (1986) *Plant Physiol.* **82**, 591.

LEUCYLTRANSFERASE

This monovalent cation-requiring transferase [EC 2.3.2.6], also known as leucyl/phenylalanyl-tRNA-protein transferase, catalyzes the reaction of L-leucyl-tRNA with a protein to produce tRNA and the L-leucyl-protein (*i.e.*, a protein with a new N-terminal residue).[1-3] Other donor substrates include L-phenylalanyl-tRNA and L-methionyl-tRNA. Acceptor substrates are peptides and proteins containing an N-terminal arginyl, lysyl, or histidyl residue.

[1] R. L. Soffer (1973) *JBC* **248**, 8424.
[2] R. C. Scarpulla, C. E. Deutch & R. L. Soffer (1976) *BBRC* **71**, 584.
[3] I. E. Ichetovkin, G. Abramochkin & T. E. Shrader (1997) *JBC* **272**, 33009.

Selected entries from *Methods in Enzymology* [vol, page(s)]:
General discussion: **106**, 198
Assay: **106**, 198

LEUCYL-tRNA SYNTHETASE

This enzyme [EC 6.1.1.4], also called leucine:tRNA ligase and leucine-activating enzyme, catalyzes the reaction of L-leucine with tRNA^Leu and ATP to produce L-leucyl-tRNA^Leu, AMP, and pyrophosphate (or, diphosphate).[1]

[1] E. A. First (1998) *CBC* **1**, 573.

Selected entries from *Methods in Enzymology* [vol, page(s)]:
General discussion: **5**, 711; **34**, 170; **166**, 260
Assay: **166**, 262
Other molecular properties: bacterial membrane, in, **22**, 113; mechanism, **63**, 52; preparation from CHO cells, **166**, 263; properties, **166**, 263; purification, **5**, 711, 729; **29**, 591; **59**, 262; subcellular distribution, **59**, 233, 234

LEUKOTRIENE A₄ HYDROLASE

This zinc-containing enzyme [EC 3.3.2.6] catalyzes the hydrolysis of leukotriene A₄ (or, (7E,9E,11Z,14Z)-(5S,6S)-5,6-epoxyeicosa-7,9,11,14-tetraenoate) to form leukotriene B₄ (or, (6Z,8E,10E,14Z)-(5S,12R)-5,12-dihydroxyeicosa-6,8,10,14-tetraenoate).[1-4] Other substrates include LTA₅ and LTA₃. (Note: This enzyme activity is distinct from epoxide hydrolase [EC 3.3.2.3].)

Interestingly, this enzyme also exhibits protease activity that can be conveniently assayed with *p*-nitroanilide and β-naphthylamide derivatives of L-alanine, L-arginine, L-leucine, and L-proline.[1,2]

[1] J. Z. Haeggström (1990) *Meth. Enzymol.* **187**, 324.
[2] J. Z. Haeggström, A. Wetterholm, J. F. Medina & B. Samuelsson (1993) *J. Lipid Mediat.* **6**, 1.
[3] J. Z. Haeggström (1997) in *SRS-A to Leukotrienes* (S. Holgate & S.-E. Dahlén, eds.), Blackwell Science, Oxford, p. 85.
[4] O. Rådmark & J. Z. Haeggström (1990) *Adv. Prostaglandin Thrombox. Leukot. Res.* **20**, 35.

Selected entries from *Methods in Enzymology* [vol, page(s)]:
General discussion: **187**, 286, 324
Assay: **187**, 289
Other molecular properties: cDNA cloning, **187**, 486; kinetics, **187**, 291, 332; properties, **187**, 290; **248**, 185, 189; purification from (guinea pig liver cytosol, **187**, 328; human lung, **187**, 286); structure, **248**, 186; subcellular distribution in hepatic tissue, **187**, 327; substrate specificity, **187**, 327

LEUKOTRIENE B₄ 20-MONOOXYGENASE

This heme-thiolate-dependent oxidoreductase [EC 1.14.13.30], known variously as leukotriene-B₄ ω-hydroxylase (LTB₄ ω-hydroxylase) and leukotriene-B₄ 20-hydroxylase (LTB₄ 20-hydroxylase), catalyzes the reaction of leukotriene B₄ (*i.e.*, (6Z,8E,10E,14Z)-(5S,12R)-5,12-dihydroxyeicosa-6,8,10,14-tetraenoate) with NADPH and dioxygen to produce 20-hydroxyleukotriene B₄ (or, (6Z,8E,10E,14Z)-(5S,12R)-5,12,20-trihydroxy-eicosa-6,8,10,14-tetraenoate), NADP⁺, and water.[1,2]

A number of variants of this monooxygenase have also been reported that further catalyze the NADPH-dependent oxidation to form the 20-aldehyde derivative. This cytochrome P450-dependent enzyme is inhibited by metyrapone and α-naphthoflavone[3] as well as by terminal acetylenic analogues of fatty acids (such as 17-octadecynoic acid).[4] The monooxygenase is highly specific; the triene bond configuration and the chirality of the hydroxyl groups is crucial for recognition by the enzyme.[5] The enzyme, which is not identical with leukotriene-E₄ 20-monooxygenase [EC 1.14.13.34], has also been reported to catalyze the 20-hydroxylation of prostaglandin A₁.[6] This activity is found in most of the members of the CYP4F subfamily of P450 proteins. The mouse enzyme cyp4f14, which has an amino acid sequence 95% similar to the rat enzyme (CYP4F1) will also catalyze the ω-hydroxylation of 6-*trans*-leukotriene B₄, lipoxin A₄, prostaglandin A₁, and 8-hydroxyeicosatetraenoate.[7]

[1] W. S. Powell (1984) *J. Biol. Chem.* **259**, 3082.
[2] R. J. Soberman, T. W. Harper, R. C. Murphy & K. F. Austen (1985) *PNAS* **82**, 2292.
[3] M. C. Romano, R. D. Eckardt, P. E. Bender, T. B. Leonard, K. M. Straub & J. F. Newton (1987) *JBC* **262**, 1590.
[4] S. Shak, N. O. Reich, I. M. Goldstein & P. R. Ortiz de Montellano (1985) *JBC* **260**, 13023.
[5] R. J. Soberman, R. T. Okita, B. Fitzsimmons, J. Rokach, B. Spur & K. F. Austen (1987) *JBC* **262**, 12421.
[6] H. Sumimoto, E. Kusunose, Y. Gotoh, M. Kusunose & S. Minakami (1988) *J. Biochem.* **108**, 215.
[7] Y. Kikuta, H. Kasyu, E. Kusunose & M. Kusunose (2000) *ABB* **383**, 225.

Selected entries from *Methods in Enzymology* [vol, page(s)]:
General discussion: 163, 349
Assay: 163, 350

LEUKOTRIENE C₄ SYNTHASE

This enzyme [EC 2.5.1.37] catalyzes the reaction of leukotriene A₄ (*i.e.*, 5(S),6(S)-oxido-7,9-*trans*-11,14-*cis*-eicosatetraenoate) with glutathione (*i.e.*, L-γ-glutamyl-L-cysteinylglycine) to produce leukotriene C₄ (or,

5(S)-hydroxy-6(R)-glutathionyl-7,9-*trans*-11,14-*cis*-eicosatetraenoate) and water.[1–4] The synthase is not identical to glutathione transferase [EC 2.5.1.18], even though that enzyme also produces an *S*-glutathione adduct. The kinetic mechanism of leukotriene C₄ synthase is rapid equilibrium random Bi Uni, with strong substrate inhibition by leukotriene A₄.[3]

This enzyme is distinct from other members of the glutathione transferase family due to its lack of amino acid homology, limited substrate specificity, steady-state kinetics, inability to conjugate glutathione to xenobiotics, differential susceptibility to inhibitors, and failure to be immunorecognized by specific microsomal glutathione transferase antibodies.

[1] M. K. Bach, J. R. Brashler & D. R. Morton (1984) *ABB* **230**, 455.
[2] T. Yoshimoto, R. J. Soberman, B. Spur & K. F. Austen (1988) *J. Clin. Invest.* **81**, 866.
[3] N. Gupta, M. J. Gresser & A. W. Ford-Hutchinson (1998) *BBA* **1391**, 157.
[4] B. K. Lam, J. F. Penrose, K. Xu, M. H. Baldasaro & K. F. Austen (1997) *JBC* **272**, 13923.

Selected entries from *Methods in Enzymology* [vol, page(s)]:
General discussion: 187, 306, 335, 353
Assay: 163, 354; 187, 336; murine mastocytoma cell, 187, 307
Other molecular properties: murine mastocytoma cell (assay, 187, 307; characterization, 187, 306; preparation, 187, 310); properties, 163, 356; 187, 337; purification from (guinea pig lung, 187, 335; rat basophilic leukemia cells, 163, 355)

LEUKOTRIENE-E₄ 20-MONOOXYGENASE

This oxidoreductase [EC 1.14.13.34], also known as leukotriene-E₄ ω-hydroxylase, catalyzes the reaction of leukotriene E₄ (or, 7E,9E,11Z,14Z)-(5S,6R)-6-(cystein-*S*-yl)-5-hydroxyeicosa-7,9,11,14-tetraenoate) with NADPH and dioxygen to produce 20-hydroxyleukotriene E₄, NADP⁺, and water.[1] *N*-Acetylleukotriene E₄ is an alternative substrate. The enzyme is not identical with leukotriene-B₄ 20-monooxygenase [EC 1.14.13.30].

[1] L. Örning (1987) *EJB* **170**, 77.

LEVANASE

This enzyme [EC 3.2.1.65], also known as 2,6-β-D-fructan fructanohydrolase and levan-splitting enzyme, catalyzes the random hydrolytic scission of 2,6-β-D-fructofuranosidic linkages in 2,6-β-D-fructans (levans) containing more than three fructosyl units. The levanases from several sources have been reported to have both endo- and exo-β-fructofuranosidase activities while the enzymes from other sources exhibit specificity for only one type of hydrolytic action.

A levanase has been isolated from *Pseudomonas* that acts as an exoglycosidase and produces levanbiose as the product[1] whereas another levanase produced levanoctaose.[2] *See also 2,6-β-Fructan 6-Levanbiohydrolase; Levansucrase; Inulinase; Fructan β-Fructosidase*

[1]E. J. Kang, S. O. Lee, J. D. Lee & T. H. Lee (1999) *Biotechnol. Appl. Biochem.* **29**, 263.
[2]S. K. Kang, S. O. Lee, Y. S. Lim, K. L. Jang & T. H. Lee (1998) *Biotechnol. Appl. Biochem.* **27**, 159.

Selected entries from *Methods in Enzymology* [vol, page(s)]:
General discussion: 8, 621
Assay: 8, 621

LEVANSUCRASE

This enzyme [EC 2.4.1.10], also known as sucrose 6-fructosyltransferase, catalyzes the reaction of sucrose with $[(2,6)-\beta-D-fructosyl]_n$ to produce D-glucose and $[(2,6)-\beta-D-fructosyl]_{(n+1)}$.[1-5] A few other fructosyl-containing sugars can act as substrates.

The enzyme has a modified ping pong mechanism (similar to that observed with glucose-6-phosphatase) with a transient fructosyl-enzyme intermediate.[4,5] This intermediate is stable at low pH in the *Bacillus subtilis* enzyme: a fructosylated aspartyl residue has been identified.

Most levansucrases studied have a transfructosylation activity in which the fructosyl residue of sucrose is transferred to a variety of acceptors: for example, water (thus, catalyzing sucrose hydrolysis), D-glucose (hence, an exchange reaction), fructan (a polymerase activity), and sucrose (oligofructoside synthesis). In a number of sources of the enzyme, the hydrolytic activity is suppressed completely in 60% acetonitrile.[1]

[1]G. Davies, M. L. Sinnott & S. G. Withers (1998) *CBC* **1**, 119.
[2]G. Mooser (1992) *The Enzymes*, 3rd ed., **20**, 187.
[3]A. S. Jarnagin & E. Ferrari (1992) *Biotechnology* **22**, 189.
[4]R. Chambert, G. Treboul & R. Dedonder (1974) *EJB* **41**, 285.
[5]L. Hernandez, J. Arrieta, C. Menendez, R. Vazquez, A. Coego, V. Suarez, G. Selman, M. F. Petit-Glatron & R. Chambert (1995) *BJ* **309**, 113.

Selected entries from *Methods in Enzymology* [vol, page(s)]:
General discussion: 1, 186; 8, 500
Assay: 1, 186; 8, 500

LICHENINASE

This enzyme [EC 3.2.1.73; previously classified as EC 3.2.1.5], also known as lichenase, β-glucanase, endo-β-1,3-1,4-glucanase, 1,3-1,4-β-D-glucan 4-glucanohydrolase, and mixed linkage β-glucanase, catalyzes the hydrolysis of 1,4-β-D-glycosidic linkages in β-D-glucans containing 1,3- and 1,4-bonds: thus, it acts on lichenin and cereal β-D-glucans, but not on β-D-glucans containing only 1,3- or 1,4-linkages.[1,2]

Glutamyl residues have been identified that act both as the catalytic nucleophile and the general acid-base. The *Bacillus licheniformis* enzyme, which has four sub-sites, forms a catalytic glucosyl-enzyme intermediate.[2]

[1]J. D. Erfle, R. M. Teather, P. J. Wood & J. E. Irvin (1988) *BJ* **255**, 833.
[2]C. Malet & A. Planas (1997) *Biochemistry* **36**, 13838.

Selected entries from *Methods in Enzymology* [vol, page(s)]:
General discussion: 8, 613; 160, 572
Assay: 8, 613; 160, 573 (with dye-labeled CM-cellulose, 160, 78)
Other molecular properties: cellobiase, presence in preparation of, 1, 178; preparation, 8, 614; properties, 8, 615; 160, 574; purification from *Bacillus subtilis*, 160, 573

LICODIONE 2′-O-METHYLTRANSFERASE

This methyltransferase [EC 2.1.1.65], first isolated from *Glycyrrhiza echinata* (licorice), catalyzes the reaction of *S*-adenosyl-L-methionine with licodione (*i.e.*, 1-(2,4-dihydroxyphenyl)-3-(4-hydroxyphenyl)-1,3-propanedione) to produce *S*-adenosyl-L-homocysteine and 2′-O-methyllicodione.[1-3] Other substrates, albeit weaker, include isoliquiritigenin, 2″-hydroxylicodione, and dihydroisoliquiritigenin.

[1]M. Ichimura, T. Furuno, T. Takahashi, R. A. Dixon & S. Ayabe (1997) *Phytochemistry* **44**, 991.
[2]S. Ayabe, A. Udagawa, K. Iida, T. Yoshikawa & T. Furuya (1987) *Plant Cell Rep.* **6**, 16.
[3]S. Ayabe, T. Yoshikawa, M. Kobayashi & T. Furuya (1980) *Phytochemistry* **19**, 2331.

LIGNAN PEROXIDASE

This heme-dependent enzyme [EC 1.11.1.14], also known as diarylpropane peroxidase, diarylpropane oxygenase, and ligninase I, catalyzes the reaction of 1,2-bis(3,4-dimethoxyphenyl)propane-1,3-diol with hydrogen peroxide to produce veratraldehyde (or, 1-(3,4-dimethylphenyl)ethane-1,2-diol) and four water molecules.[1-5]

The enzyme facilitates oxidative C–C bond cleavage in a number of model compounds and also oxidizes benzyl alcohols to aldehydes or ketones. It participates in the breakdown of lignin by white rot Basidiomycetes. Other substrates include 3,4-dimethoxybenzyl alcohol (*i.e.*, veratryl alcohol, producing 3,4-dimethoxybenzaldehyde), 1-(3′,4′-diethoxyphenyl)-1,3-dihydroxy-2-(4″-methoxyphenyl)propane (that is, diarylpropane, a lignin-model compound, producing 1-(4'-methoxyphenyl)-1,2-dihydroxyethane and 3,4-diethoxybenzaldehyde), 1-(4'-ethoxy-3'-methoxyphenyl)propane (producing 1-(4'-ethoxy-3'-methoxyphenyl)1-hydroxypropane), and 1-(4'-ethoxy-3'-methoxyphenyl)-1,2-propene (producing 1-(4'-ethoxy-3'-methoxyphenyl)-1,2-dihydroxypropane).

[1] H. B. Dunford (1998) CBC 3, 195.
[2] F. P. Guengerich (1990) Crit. Rev. Biochem. Mol. Biol. 25, 97.
[3] A. Paszczynski, V. B. Huynh & R. Crawford (1986) ABB 244, 750.
[4] V. Renganathan, K. Miki & M. H. Gold (1985) ABB 241, 304.
[5] L. A. Andersson, V. Renganathan, A. A. Chiu, T. M. Loehr & M. H. Gold (1985) JBC 260, 6080.

Selected entries from **Methods in Enzymology** [vol, page(s)]:
General discussion: 188, 154
Assay: 137, 693; 161, 242; 188, 160
Other molecular properties: cDNA clones, identification with synthetic oligonucleotide probes, 161, 228; chromatofocusing, 161, 269; 2D NMR spectra, proton signal assignment, 239, 501; production, 188, 164 (by Phanerochaete chrysosporium, 137, 690, 695); properties, 161, 246; 188, 168; purification from Phanerochaete chrysosporium, 161, 243; 188, 166; storage, 161, 246

LIGNOSTILBENE αβ-DIOXYGENASE

This iron-dependent oxidoreductase [EC 1.13.11.43] catalyzes the reaction of 1,2-bis(4-hydroxy-3-methoxyphenyl)ethylene with dioxygen to produce two molecules of vanillin (or, 4-hydroxy-3-methoxybenzaldehyde) by oxidative cleavage of an interphenyl double bond.[1,2] The enzyme is responsible for the degradation of a diarylpropane-type structure in lignin.

[1] S. Kamoda, N. Habu, M. Samejima & T. Yoshimoto (1989) Agric. Biol. Chem. 53, 2757.
[2] S. Kamoda & M. Samejima (1991) Agric. Biol. Chem. 55, 1411.

LIMIT DEXTRINASE

This enzyme [EC 3.2.1.142], also known as R-enzyme and amylopectin-1,6-glucosidase, catalyzes the hydrolysis of (1→6)-α-D-glucosidic linkages in α- and β-limit dextrins of amylopectin and glycogen.[1,2] The enzyme will also act on amylopectin and pullulan. The plant enzyme exhibits little or no action on glycogen. The action on amylopectin is incomplete; however, action on α-limit dextrins is complete. Maltose is the smallest sugar that can release from an α(1→6)-linkage.

The term "limit dextrinase" has also been used for oligo-1,6-glucosidase [EC 3.2.1.10] and α-dextrin endo-1,6-α-glucosidase [EC 3.2.1.41].

[1] R. W Gordon, D. J. Manners & J. R. Stark (1975) Carbohydr. Res. 42, 125.
[2] D. J. Manners (1997) J. Appl. Glycosci. 44, 83.

LIMONENE-1,2-EPOXIDE HYDROLASE

This enzyme [EC 3.3.2.8], also called limonene oxide hydrolase, catalyzes the reaction of limonene-1,2-epoxide with water to produce limonene-1,2-diol.[1] Other substrates include 1-methylcyclohexene oxide, indene oxide, and cyclohexene oxide.

[1] M. J. van der Werf, K. M. Overkamp & J. A. M. de Bont (1998) J. Bacteriol. 180, 5052.

(−)-LIMONENE 3-MONOOXYGENASE

This heme-thiolate-dependent oxidoreductase [EC 1.14.13.47], also known as (−)-limonene 3-hydroxylase, catalyzes the reaction of (−)-limonene with NADPH and dioxygen to produce (−)-trans-isopiperitenol, NADP+, and water.[1-3] NADH can also serve as the coenzyme, albeit less effectively.

This cytochrome P450-dependent system also can utilize other, albeit weaker, substrates such as (+)-limonene (producing (+)-trans-isopiperitenol), (−)-p-menth-1-ene (i.e., (−)-8,9-dihydrolimonene, producing (−)-trans-piperitol), and (+)-p-menth-1-ene (i.e., (+)-8,9-dihydro-limonene, producing (+)-trans-piperitol).

[1] F. Karp, C. A. Mihaliak, J. L. Harris & R. Croteau (1990) ABB 276, 219.
[2] R. Croteau, F. Karp, K. C. Wagschal, M. Satterwhite, D. C. Hyatt & C. B. Skotland (1991) Plant Physiol. 96, 744.
[3] J. Gershenzon, D. McCaskill, J. I. M. Rajaonarivony, C. Mihaliak, F. Karp & R. Croteau (1992) Anal. Biochem. 200, 130.

(−)-LIMONENE 6-MONOOXYGENASE

This heme-thiolate-dependent oxidoreductase [EC 1.14.13.48], also known as (−)-limonene 6-hydroxylase, catalyzes the reaction of (−)-limonene with NADPH and dioxygen to produce (−)-trans-carveol, NADP+, and water.[1-3] NADH can also serve as the coenzyme, albeit less effectively.

This cytochrome P450-dependent system also can utilize other substrates, including (+)-limonene (producing (+)-cis-carveol), (−)-p-menth-1-ene (i.e., (−)-8,9-dihydro-limonene, producing (−)-trans-carvotanacetol), and (+)-p-menth-1-ene (i.e., (+)-8,9-dihydrolimonene, producing (+)-cis-carvotanacetol).

[1] F. Karp, C. A. Mihaliak, J. L. Harris & R. Croteau (1990) ABB 276, 219.
[2] R. Croteau, F. Karp, K. C. Wagschal, M. Satterwhite, D. C. Hyatt & C. B. Skotland (1991) Plant Physiol. 96, 744.
[3] J. Gershenzon, D. McCaskill, J. I. M. Rajaonarivony, C. Mihaliak, F. Karp & R. Croteau (1992) Anal. Biochem. 200, 130.

(−)-LIMONENE 7-MONOOXYGENASE

This heme-thiolate-dependent oxidoreductase [EC 1.14.13.49], also known as (−)-limonene 7-hydroxylase, catalyzes the reaction of (−)-limonene with NADPH and dioxygen to produce (−)-perillyl alcohol, $NADP^+$, and water.[1] NADH can also serve as the coenzyme, albeit less effectively. (+)-Limonene will also be hydroxylated by this cytochrome P450-dependent system.

[1] F. Karp, C. A. Mihaliak, J. L. Harris & R. Croteau (1990) *ABB* **276**, 219.

(−)-(4S)-LIMONENE SYNTHASE

This enzyme [EC 4.2.3.16, previously classified as EC 4.1.99.10] (also known as monoterpene synthase, limonene cyclase, and monoterpene cyclase) is dependent on the presence of divalent cations such as Mn^{2+} and catalyzes the conversion of geranyl diphosphate to limonene and pyrophosphate (or, diphosphate).[1]

geranyl diphosphate   limonene

Stereochemical investigations with deuterated substrates indicate that the acyclic double bond in limonene is formed via proton elimination from the *cis*-methyl group.

[1] J. Bohlmann, C. L. Steele & R. Croteau (1997) *JBC* **272**, 21784.

Selected entries from *Methods in Enzymology* [vol, page(s)]:
Assay: 110, 407
Other molecular properties: *Citris limonum* (assay, **110**, 407; product identification, **110**, 409; properties, **110**, 415; purification, **110**, 413; radioactive substrates, synthesis, **110**, 410)

LIMONIN-D-RING-LACTONASE

This lactonase [EC 3.1.1.36] catalyzes the hydrolysis of limonoate D-ring-lactone to produce the triterpenoid limonoate.[1–3]

[1] S. Hasegawa, M. N. Patel & R. C. Snyder (1982) *J. Agric. Food Chem.* **30**, 509.
[2] S. Hasegawa (1976) *J. Agric. Food Chem.* **24**, 24.
[3] V. P. Maier, S. Hasegawa & E. Hera (1969) *Phytochemistry* **8**, 405.

LIMONOID GLUCOSYLTRANSFERASE

This transferase [EC 2.4.1.210] catalyzes the reaction of UDP-D-glucose with limonin to produce D-glucosyl-limonin and UDP.[1]

[1] M. K. Sim & B. C. Lim (1997) *Phytochemistry* **46**, 33.

LIMULUS CLOTTING ENZYME

This serine endopeptidase [EC 3.4.21.86], a part of the coagulation cascade in the horseshoe crab (or, *Limulus*) hemolymph, is activated from the proenzyme by active *Limulus* coagulation factor B (which hydrolyzes the Arg98–Ile99 bond in the proclotting enzyme) or by active *Limulus* coagulation factor G.

Elevated concentrations of trypsin will also activate the clotting enzyme *in vitro*. The activated enzyme will catalyze the hydrolysis of the Arg18–Thr19 and Arg47–Gly48 peptide bonds in coagulogen to produce coagulin and fragments.[1,2] *Limulus* clotting enzyme will also convert bovine prothrombin to α-thrombin, and it hydrolyzes the synthetic substrates Tos-Ile-Glu-Gly-Arg-∇-NHPhNO₂ and Boc-Leu-Gly-Arg-NHPhNO₂, where ∇ indicates the cleavage site.

[1] T. Muta & S. Iwanaga (1996) *Curr. Opin. Immunol.* **8**, 41.
[2] T. Nakamura, T. Morita & S. Iwanaga (1985) *J. Biochem.* **97**, 1561.

LIMULUS COAGULATION FACTOR B

This serine endopeptidase [EC 3.4.21.85], also known as *Limulus* clotting factor B, is a part of the coagulation cascade in the horseshoe crab (or, *Limulus*) hemolymph. The zymogen form of factor B is converted to the active form (or, *Limulus* coagulation factor B̄) by the hydrolysis of Arg103–Ser104 and Ile124–Ile125 bonds by *Limulus* coagulation factor C̄. The active form of factor B catalyzes the hydrolysis of the (Thr94Thr95Thr96Thr97)-Arg98–Ile99 peptide bond in the *Limulus* preclotting enzyme to produce the *Limulus* clotting enzyme [EC 3.4.21.86]. The active factor will also act on the synthetic substrates Boc-Met-Thr-Arg-∇-NHMec, Bz-Thr-Thr-Arg-NHMec, and Bz-Ser-Thr-Arg-∇-NHMec.[1,2] (Note: ∇ indicates the cleavage site.)

[1] T. Muta & S. Iwanaga (1996) *Curr. Opin. Immunol.* **8**, 41.
[2] T. Nakamura, T. Horiuchi, T. Morita & S. Iwanaga (1986) *J. Biochem.* **99**, 847.

LIMULUS COAGULATION FACTOR C

This serine endopeptidase [EC 3.4.21.84], also called *Limulus* clotting factor C, catalyzes peptide-bond hydrolysis and is isolated from the horseshoe crab (or, *Limulus*). Purified factor C consists of two forms: a single-chain form (123 kDa) and a heterodimeric form (an H chain of 80 kDa and an L chain of 43 kDa). In the presence of the lipopolysaccharide found in the outer membranes of Gram-negative bacteria, both forms of factor C undergo autoproteolysis (at a Phe–Ile bond) to form activated factor C̄, which consists of three chains: H (80 kDa), A (7.9 kDa), and B (34 kDa). This activated form of the factor catalyzes the hydrolysis of Arg103–Ser104 and Ile124–Ile125 bonds in *Limulus* clotting factor B to form activated factor B (*i.e.*, *Limulus* coagulation factor B̄). Synthetic substrates such as Boc-Val-Pro-Arg-∇-NHPhNO$_2$ and Boc-Val-Pro-Arg-∇-NHMec can be used to assay the enzyme. (Note: ∇ indicates the cleavage site.) The activated form of the factor is inhibited by antithrombin III but not by α_2-plasmin inhibitor.[1–3]

[1]S. Iwanaga (1993) *Curr. Opin. Immunol.* **5**, 74.
[2]T. Muta & S. Iwanaga (1996) *Curr. Opin. Immunol.* **8**, 41.
[3]T. Nakamura, T. Morita & S. Iwanaga (1986) *EJB* **154**, 511.

Selected entries from ***Methods in Enzymology*** [vol, page(s)]:
Assay: 223, 354

LIMULUS COAGULATION FACTOR G

This serine endopeptidase is the β-1,3-glucan-sensitive component of the coagulation cascade in the horseshoe crab (or, *Limulus*) hemolymph. The factor is activated by certain β-1,3-glucans (such as curdlan or paramylon), producing *Limulus* coagulation factor Ḡ. This activated form of factor G activates *Limulus* clotting enzyme [EC 3.4.21.86], probably by hydrolysis of the Arg98–Ile99 bond. The enzyme also acts on synthetic substrates such as Boc-Glu(OBzl)-Gly-Arg-∇-NHMec, Boc-Ser(OBzl)-Ala-Arg-∇-NHMec, and Boc-Met-Thr-Arg-∇-NHMec.[1,2] (Note: ∇ indicates the cleavage site.)

[1]T. Muta & S. Iwanaga (1996) *Curr. Opin. Immunol.* **8**, 41.
[2]T. Muta, N. Seki, Y. Takaki, R. Hashimoto, T. Oda, A. Iwanaga, T. Tokunaga & S. Iwanaga (1995) *JBC* **270**, 892.

LINALOOL 8-MONOOXYGENASE

This cytochrome P450 (or, heme-thiolate) oxidoreductase [EC 1.14.99.28] catalyzes the reaction of linalool (or, 3,7-dimethylocta-1,6-dien-3-ol) with reduced acceptor AH$_2$ and dioxygen to produce 8-hydroxylinalool, (or, (*E*)-3,7-dimethylocta-1,6-diene-3,8-diol), oxidized acceptor A, and water.[1–3] Both stereoisomers of linalool can serve as the substrate.

[1]P. K. Bhattacharyya, T. B. Samanta, A. H. J. Ullah & I. C. Gunsalus (1984) *Proc. Indian Acad. Sci. Chem. Sci.* **93**, 1289.
[2]A. H. J. Ullah, R. I. Murray, P. K. Bhattacharyya, G. C. Wagner & I. C. Gunsalus (1990) *JBC* **265**, 1345.
[3]J. D. Ropp, I. C. Gunsalus & S. G. Sligar (1993) *J. Bacteriol.* **175**, 6028.

LINAMARIN SYNTHASE

This enzyme [EC 2.4.1.63], first isolated from flax (*Linum usitatissimum*), catalyzes the reaction of UDP-D-glucose with 2-hydroxy-2-methylpropanenitrile (that is, acetone cyanohydrin; CH$_3$C(OH)(CN)CH$_3$) to produce UDP and linamarin.[1] Other acceptor substrates include the cyanohydrins of butanone (*i.e.*, 2-hydroxy-2-methylbutanenitrile) and pentan-3-one (*i.e.*, 2-hydroxy-2-ethylbutanenitrile).

[1]K. Hahlbrock & E. E. Conn (1970) *JBC* **245**, 917.

LINOLEATE DIOL SYNTHASE

This oxidoreductase [EC 1.13.11.44], also known as linoleate (8*R*)-dioxygenase, catalyzes the reaction of linoleate with dioxygen to produce (9*Z*,12*Z*)-(7*S*,8*S*)-dihydroxyoctadeca-9,12-dienoate.[1–4] Other substrates include oleate and linolenate. Stearate, elaidate, γ-linolenate, arachidonate, and eicosapentaenate are not substrates. Note that lipoxygenase [EC 1.13.11.12] will also catalyze the reaction of dioxygen with linoleate; however, in that case, a 9,11-dienoate hydroperoxy product is formed. The *Gaeumannomyces graminis* enzyme converts linoleate sequentially to 8*R*-hydroperoxylinoleate through an 8-dioxygenase activity by insertion of dioxygen and to 7*S*,8*S*-dihydroxylinoleate through a hydroperoxide isomerase activity by intramolecular oxygen transfer. Studies suggest the presence of ferryl intermediates and a protein radical during the reaction pathway.[4]

[1]I. D. Brodowsky, M. Hamberg & E. H. Oliw (1992) *JBC* **267**, 14738.
[2]M. Hamberg, L.-Y. Zhang, I. D. Brodowsky & E. H. Oliw (1994) *ABB* **309**, 77.
[3]C. Su & E. H. Oliw (1996) *JBC* **271**, 14112.
[4]C. Su, M. Sahlin & E. H. Oliw (1998) *JBC* **273**, 20744.

LINOLEATE ISOMERASE

This isomerase [EC 5.2.1.5] catalyzes the interconversion of 9-*cis*,12-*cis*-octadecadienoate (*i.e.*, linoleate) and 9-*cis*,11-*trans*-octadecadienoate. Other substrates include 9-*cis*,12-*cis*,15-*cis*-octadecatrienoate (*i.e.*, α-linolenate), 6-*cis*,9-*cis*,12-*cis*-octadecatrienoate (*i.e.*, γ-linolenate), and 9-*cis*,12-*cis*-heptadecadienoate.[1–3]

[1]S. Seltzer (1972) *The Enzymes*, 3rd ed., **6**, 381.
[2]P. Kemp & R. M. Dawson (1968) *BJ* **109**, 477.
[3]C. R. Kepler & S. B. Tove (1967) *JBC* **242**, 5686.

Selected entries from ***Methods in Enzymology*** [vol, page(s)]:
General discussion: 14, 105
Assay: 14, 105

LINOLEATE 11-LIPOXYGENASE

This oxidoreductase [EC 1.13.11.45], also known as linoleate dioxygenase and manganese lipoxygenase, catalyzes the reaction of linoleate (or, (9Z,12Z)-octadeca-9,12-dienoate) with dioxygen to produce (9Z,12Z)-(11S)-11-hydroperoxyoctadeca-9,12-dienoate.[1,2] The product is converted slowly into (9Z,11E)-(13R)-13-hydroperoxyoctadeca-9,11-dienoate. Other substrates include α-linolenate (i.e., (9Z,12Z,15Z)-octadeca-9,12,15-trienoate) and, more weakly, γ-linolenate (i.e., (6Z,9Z,12Z)-octadeca-6,9,12-trienoate). Arachidonate and oleate are not substrates.

[1] M. Hamberg, C. Su & E. H. Oliw (1998) JBC 273, 13080.
[2] C. Su & E. H. Oliw (1998) JBC 273, 13072.

LINOLEOYL-CoA DESATURASE

This iron-dependent oxidoreductase [EC 1.14.19.3; previously classified as EC 1.14.99.25], also known as Δ^6-desaturase, catalyzes the reaction of linoleoyl-CoA with AH_2 and dioxygen to produce γ-linolenoyl-CoA, A, and two water molecules (where AH_2 is the reduced acceptor substrate and A is the oxidized acceptor).[1–3] The rat liver enzyme is an enzyme system involving cytochrome b_5 and cytochrome b_5 reductase [EC 1.6.2.2].

[1] T. Okayasu, M. Nagao, T. Ishibashi & Y. Imai (1981) ABB 206, 21.
[2] R. Jeffcoat, A. P. Dunton & A. T. James (1978) BBA 528, 28.
[3] J. A. Browse & C. R. Slack (1981) FEBS Lett. 131, 111.

LIPID-A-DISACCHARIDE SYNTHASE

This transferase [EC 2.4.1.182] catalyzes the reaction of UDP-2,3-bis(3-hydroxytetradecanoyl)-D-glucosamine with 2,3-bis(3-hydroxytetradecanoyl)-β-D-glucosaminyl 1-phosphate to produce UDP and 2,3-bis(3-hydroxytetradecanoyl)-D-glucosaminyl-1,6-β-D-2,3-bis(3-hydroxytetradecanoyl)-β-D-glucosaminyl 1-phosphate. It participates with acyl-[acyl-carrier-protein]:UDP-N-acetylglucosamine O-acyltransferase [EC 2.3.1.129] and tetraacyldisaccharide 4′-kinase [EC 2.7.1.130] in the biosynthesis of the phosphorylated glycolipid, lipid A, in Escherichia coli and other Gram-negative bacteria.[1] Unlike most enzymes of glycerophospholipid synthesis, the lipid-A-disaccharide synthase does not require the presence of a detergent for catalytic activity.[2]

[1] B. L. Ray, G. Painter & C. R. Raetz (1984) JBC 259, 4852.
[2] K. Radika & C. R. Raetz (1988) JBC 263, 14859.

Selected entries from Methods in Enzymology [vol, page(s)]:
General discussion: 209, 455

Assay: Escherichia coli, 209, 455
Other molecular properties: Escherichia coli (assay, 209, 455; purification, 209, 458; properties, 209, 460)

LIPOAMIDASE

This enzyme catalyzes the hydrolysis of the lipoamide group in certain α-keto acid dehydrogenase complexes (e.g., pyruvate dehydrogenase complex and α-ketoglutarate dehydrogenase complex) producing lipoate and the apoprotein. The substrate is the E_2 component of these complexes.

Selected entries from Methods in Enzymology [vol, page(s)]:
General discussion: 18A, 292; 34, 292; 279, 202
Assay: 18A, 292; 279, 202
Other molecular properties: properties, 18A, 297, 298; 279, 202; purification, 18A, 295; 279, 202

LIPOPOLYSACCHARIDE N-ACETYL-GLUCOSAMINYLTRANSFERASE

This enzyme [EC 2.4.1.56] catalyzes the reaction of UDP-N-acetyl-D-glucosamine with a lipopolysaccharide to produce UDP and an N-acetyl-D-glucosaminyl-lipopolysaccharide;[1] hence, this enzyme transfers N-acetylglucosaminyl residues to a D-galactosyl residue in the partially completed lipopolysaccharide core.

[1] M. J. Osborn & L. D'Ari (1964) BBRC 16, 568.

Selected entries from Methods in Enzymology [vol, page(s)]:
General discussion: 8, 460
Assay: 8, 460

LIPOPOLYSACCHARIDE N-ACETYL-MANNOSAMINOURONOSYL-TRANSFERASE

This transferase [EC 2.4.1.180] catalyzes the reaction of UDP-N-acetylmannosaminouronate with a lipopolysaccharide to produce UDP and an N-acetyl-β-D-mannosaminouronosyl-1,4-lipopolysaccharide. The enzyme participates in the biosynthesis of the enterobacterial common antigen in Escherichia coli.[1]

[1] K. Barr, S. Ward, U. Meier-Dieter, H. Mayer & P. D. Rick (1988) J. Bacteriol. 170, 228.

LIPOPOLYSACCHARIDE DEACYLATING ENZYME

This hydrolase, also called acyloxyacyl hydrolase, catalyzes the removal of fatty acyl groups from acyloxyacyl positions in lipopolysaccharides in a number of microorganisms.

LIPOPOLYSACCHARIDE GALACTOSYLTRANSFERASE

This enzyme [EC 2.4.1.44] catalyzes the reaction of UDP-D-galactose with a lipopolysaccharide to produce a D-galactosyl-lipopolysaccharide and UDP by acting on D-glucosyl residues in the partially completed core of a lipopolysaccharide.[1–3]

[1]A. Endo & L. Rothfield (1969) *Biochemistry* **8**, 3500.
[2]E. Müller, A. Hinckley & L. Rothfield (1972) *JBC* **247**, 2614.
[3]I. R. Beacham & D. F. Silbert (1973) *JBC* **248**, 5310.

LIPOPOLYSACCHARIDE GLUCOSYLTRANSFERASE I

Lipopolysaccharide glucosyltransferase I [EC 2.4.1.58] catalyzes the reaction of UDP-D-glucose with a lipopolysaccharide to produce a D-glucosyl-lipopolysaccharide and UDP by transferring glucosyl residues to the backbone portion of a lipopolysaccharide.[1,2]

[1]L. Rothfield, M. J. Osborn & B. L. Horecker (1964) *JBC* **239**, 2788.
[2]E. Müller, A. Hinckley & L. Rothfield (1972) *JBC* **247**, 2614.

LIPOPOLYSACCHARIDE GLUCOSYLTRANSFERASE II

Lipopolysaccharide glucosyltransferase II [EC 2.4.1.73] catalyzes the reaction of UDP-D-glucose with a lipopolysaccharide to produce UDP and a D-glucosyl-lipopolysaccharide by transfer of the glucosyl residue to the D-galactosyl-D-glucosyl side chains in the partially completed core of a lipopolysaccharide.[1]

[1]R. D. Edstrom & E. C. Heath (1967) *JBC* **242**, 3581.

LIPOPOLYSACCHARIDE-TRANSPORTING ATPase

This so-called ATPase [EC 3.6.3.39] found in Gram-negative bacteria catalyzes the ATP-dependent (ADP- and orthophosphate-producing) transport of a lipopolysaccharide$_{in}$ (or lipooligosaccharide$_{in}$) to produce lipopolysaccharide$_{out}$.[1–3] This ABC-type (ATP-binding cassette-type) ATPase has two similar ATP-binding domains and does not undergo phosphorylation during the transport process.

The idea that a transporter is an enzyme is in keeping with a new definition of enzyme catalysis as the facilitated making/breaking of chemical bonds, not just covalent bonds.[4] This idea builds on Pauling's assertion that any long-lived, chemically distinct interaction (in this case, the persistent location of a solute with respect to the faces of a membrane) can be regarded as a chemical bond. Note also that the equilibrium constant, defined by K_{eq} = [lipopolysaccharide$_{out}$][ADP][P$_i$]/ [lipopolysaccharide$_{in}$][ATP], does not conform to that expected for an ATPase (*i.e.*, K_{eq} = [ADP][P$_i$]/[ATP]). Thus, although the overall reaction yields ADP and orthophosphate, the enzyme is misclassified as a hydrolase, and instead should be regarded as an energase-type reaction. Energases facilitate affinity-modulated reactions by coupling the $\Delta G_{ATP\text{-}hydrolysis}$ to a force-generating or work-producing step.[4] In this case, P-O-P bond scission supplies the energy to drive lipopolysaccharide transport. **See** *ATPase; Energase*

[1]M. J. Fath & R. Kolter (1993) *Microbiol. Rev.* **57**, 995.
[2]M. Fernandez-Lopez, W. D'Haeze, P. Mergaert, C. Verplancke, J. C. Prome, M. Van Montagu & M. Holstens (1996) *Mol. Microbiol.* **20**, 993.
[3]I. T. Paulsen, A. M. Beness & M. H. Saier, Jr. (1997) *Microbiology* **143**, 2685.
[4]D. L. Purich (2001) *TiBS* **26**, 417.

LIPOPROTEIN LIPASE

This enzyme [EC 3.1.1.34] (also called clearing factor lipase, diglyceride lipase, and diacylglycerol lipase) catalyzes the hydrolysis of a triacylglycerol to produce a diacylglycerol and a fatty acid anion.[1–6] This lipase hydrolyzes triacylglycerols in chylomicrons and in very low-density lipoproteins as well as acts on diacylglycerols. **See also** *Triacylglycerol Lipase*

[1]R. Potenz, J.-Y. Lo, E. Zsigmond, L. C. Smith & L. Chan (1996) *MIE* **263**, 319.
[2]H. Okuda (1991) *A Study of Enzymes* **2**, 579.

[3]A. Bensadoun (1991) *Ann. Rev. Nutr.* **11**, 217.

[4]T. Olivecrona & G. Bengtsson (1984) in *Lipases* (B. Borgström & H. L. Brockman, eds.), 205.

[5]R. L. Jackson (1983) *The Enzymes*, 3rd ed., **16**, 141.

[6]P. Desnuelle (1972) *The Enzymes*, 3rd ed., **7**, 575.

Selected entries from *Methods in Enzymology* [**vol**, page(s)]:
General discussion: **5**, 542; **34**, 4; **35**, 181; **129**, 691; **263**, 64, 319; **286**, 102
Assay: **5**, 543; **72**, 331; **197**, 317; bovine milk, **129**, 692; human plasma, **129**, 692; human platelet, **86**, 12, 653
Other molecular properties: activation (serum, by, **72**, 335; sodium dodecyl sulfate, by, **72**, 326); activator apolipoproteins, **72**, 327; activity, **72**, 325 (assay, **72**, 331; postheparin plasma, in, **72**, 325); analysis by radiation inactivation, **197**, 284; antibodies, **263**, 325; apolipoprotein C-II, **263**, 188, 207; bovine milk (assay, **129**, 692; assay for very-low-density lipoprotein metabolism in whole plasma, in, **129**, 704; properties, **129**, 703; purification, **129**, 698); catalyzed reactions (kinetics, **129**, 757; substrates for, **129**, 746); diacylglycerol hydrolase activity (hydrolysis of [dioleoylglycerol, assay, **141**, 310; palmitoylacetylglycerol, assay, **141**, 311]; preparation from various sources, **141**, 303); discovery, **128**, 4; distribution, **5**, 545; function, **5**, 542; human plasma (assay, **129**, 692; properties, **129**, 703; purification, **129**, 698); human platelet (assay, **86**, 12, 653; properties, **86**, 15); human postheparin plasma, isolation and characterization, **197**, 339; mechanism, analytical techniques, **129**, 738; metabolic function, **128**, 17; milk (incubation conditions, **197**, 353; purification from bovine, **197**, 347); muscle mitochondria, **55**, 13; properties, **5**, 545; purification, **5**, 544; **263**, 320; radioiodination, **70**, 232; quantitation (chromatography, **263**, 324; immunochemical, **263**, 327); role in arachidonate liberation from phospholipids, **191**, 677; structural domains, **129**, 741; surface density, **64**, 377; transition-state and multisubstrate analogues, **249**, 305; very-low-density lipoprotein, **263**, 319

LIPOXYGENASE

This iron-dependent oxidoreductase [EC 1.13.11.12] (also known as lipoxidase, carotene oxidase, and lipoperoxidase) catalyzes the reaction of linoleate with dioxygen to produce $(9Z,11E)$-$(13S)$-13-hydroperoxyoctadeca-9,11-dienoate.[1-6] This enzyme can also oxidize other methylene-interrupted polyunsaturated fatty acids. The plant 13-lipoxygenase corresponds to the mammalian 15-lipoxygenase (both acting at the $\omega - 6$ position). ***See also specific lipoxygenase***

[1]A. R. Brash (1999) *JBC* **274**, 23679.

[2]B. G. Fox (1998) *CBC* **3**, 261.

[3]M. A. Ator & P. R. Ortez de Montellano (1990) *The Enzymes*, 3rd ed., **19**, 213.

[4]T. Schewe, S. M. Rapoport & H. Kühn (1986) *Adv. Enzymol.* **58**, 191.

[5]H. Kühn, T. Schewe & S. M. Rapoport (1986) *Adv. Enzymol.* **58**, 273.

[6]O. Hayaishi, M. Nozaki & M.T. Abbott (1975) *The Enzymes*, 3rd ed., **12**, 119.

Selected entries from *Methods in Enzymology* [**vol**, page(s)]:
General discussion: **5**, 539; **34**, 4; **52**, 11
Assay: **5**, 539; chemiluminescence method, **105**, 129; rabbit reticulocytes, from, **71**, 430
Other molecular properties: brain, inactivation by microwave irradiation, **86**, 635; eicosanoid generation, effects of flavonoids and coumarins, **234**, 443–454; electronic absorption spectroscopy, **226**, 37; enzymatic activity, **31**, 522, 523; generated products of arachidonate, HPLC, **187**, 98; induced lipid peroxidation, **269**, 389; inhibition, **31**, 524; nitric oxide effect, **269**, 101; oxidation of tocopherols, in, **186**, 203; oxygen radical generation, **234**, 423; phospholipase A$_2$, and, oxidative effects on LDL molecular species, **234**, 517; plants, in, overcoming problems, **31**, 520; products, HPLC, **187**, 570; rabbit reticulocytes, from, **71**, 430 (assays, **71**, 430; immunological characterization, **71**, 439; inhibitors, **71**, 441; properties, **71**, 437; specificity, **71**, 440; stability, **71**, 439); reaction products (HPLC, **86**, 518; TLC, **86**, 32); reaction with eicosanoids, chiral analysis, **187**, 194; soybeans, from, **71**, 441 (iron content, **71**, 448; lipoxygenase-1, **71**, 443, 445; lipoxygenase-2, **71**, 444, 446; lipoxygenase-3, **71**, 444, 446; nitrosyl complex, detection by electron spin resonance, **268**, 174; properties, **5**, 541; **71**, 447; purification, **5**, 541; substrate specificity, **71**, 450)

LISTERIA METALLOPROTEINASE MPL

This zinc-dependent metalloendopeptidase catalyzes peptide-bond hydrolysis of proteins such as casein. The enzyme is also present in other pathogenic *Listeria* species and may have a role in virulence.[1,2]

[1]E. Domann, M. Leimeister-Wächter, W. Goebel & T. Chakraborty (1991) *Infect. Immun.* **59**, 65.

[2]J. Mengaud, C. Geoffroy & P. Cossart (1991) *Infect. Immun.* **59**, 1043.

LOGANATE *O*-METHYLTRANSFERASE

loganate   loganin

This methyltransferase [EC 2.1.1.50] catalyzes the reaction of *S*-adenosyl-L-methionine with loganate to produce *S*-adenosyl-L-homocysteine and loganin.[1] Secologanate is also a substrate.

[1]K. M. Madyastha, R. Guarnaccia, C. Baxter & C. L. Coscia (1973) *JBC* **248**, 2497.

LOMBRICINE KINASE

This phosphotransferase [EC 2.7.3.5] catalyzes the reaction of ATP with lombricine (or, guanidinoethylphospho-*O*-serine) to produce ADP and *N*-phospholombricine.[1-4] Both the D- and L-lombricine are substrates. Other substrates include L-thalassemine (*i.e.*, guanidinoethylphospho-*O*-(α-*N*,*N*-dimethyl)serine), taurocyamine, and guanidinoethylphosphate; however, the specificity of the enzyme varies from species to species.

[1]E. Der Terrossian, L.-A. Pradel, R. Kassab & G. Desvages (1974) *EJB* **45**, 243.

[2]N. van Thoai, Y. Robin & Y. Guillou (1972) *Biochemistry* **11**, 3890.

[3]R. Kassab, L.-A. Pradel & N. van Thoai (1965) *BBA* **99**, 397.
[4]T. J. Gaffney, H. Rosenberg & A. H. Ennor (1964) *BJ* **90**, 170.

Selected entries from *Methods in Enzymology* [vol, page(s)]:
General discussion: 34, 546; 17A, 1002

LONG-CHAIN ACYL-CoA DEHYDROGENASE

This FAD-dependent oxidoreductase [EC 1.3.99.13] catalyzes the reaction of a long-chain acyl-CoA (such as palmitoyl-CoA) with the electron-transferring flavoprotein (ETF) to produce a long-chain 2,3-dehydroacyl-CoA and the reduced ETF.[1-4] The preferred acyl-CoA chain length is six to twenty-two carbon atoms. When combined with ETF and the electron-transferring-flavoprotein dehydrogenase, the complex will reduce ubiquinone and other acceptors. Binding of the acyl-CoA substrate results in production of the reduced form of the enzyme which is reoxidized by the electron-transferring flavoprotein. The enzyme is not identical with butyryl-CoA dehydrogenase [EC 1.3.99.2], acyl-CoA dehydrogenase [EC 1.3.99.3], isovaleryl-CoA dehydrogenase [EC 1.3.99.10], or 2-methylacyl-CoA dehydrogenase [EC 1.3.99.12]. *See also Acyl-CoA Dehydrogenases; Butyryl-CoA Dehydrogenase*

[1]Y. Matsubara, Y. Indo, E. Naito, H. Ozasa, R. Glassberg, J. Vockley, Y. Ikeda, J. Kraus & K. Tanaka (1989) *JBC* **264**, 16321.
[2]H. Suzuki, J. Yamada, T. Watanabe & T. Suga (1989) *BBA* **990**, 25.
[3]V. Dommes & W.-H. Kunau (1984) *JBC* **259**, 1789.
[4]Y. Ikeda, K. Okamura-Ikeda & K. Tanaka (1985) *JBC* **260**, 1311.

Selected entries from *Methods in Enzymology* [vol, page(s)]:
General discussion: 5, 546; 53, 402; 71, 375
Assay: 5, 546

N-(LONG-CHAIN-ACYL)-ETHANOLAMINE DEACYLASE

This deacylase [EC 3.5.1.60], also known as *N*-acylethanolamine amidohydrolase, catalyzes the hydrolysis of an *N*-(long-chain-acyl)ethanolamine to produce a long-chain carboxylic anion and ethanolamine. *N*-Acylsphingosine and *N,O*-diacylethanolamine are not substrates.[1]

[1]P. C. Schmid, M. L. Zuzarte-Augustin & H. H. Schmid (1985) *JBC* **260**, 14145.

LONG-CHAIN-ALCOHOL DEHYDROGENASE

This oxidoreductase [EC 1.1.1.192], also known as fatty alcohol oxidoreductase, catalyzes the reaction of a long-chain alcohol with water and two molecules of NAD+ to produce a long-chain carboxylate anion and two molecules of NADH.[1,2] Substrates include *n*-hexadecanol. All sources studied act on C_8-C_{16} primary alcohols. *See also Alcohol Dehydrogenase*

[1]T.-C. Lee (1979) *JBC* **254**, 2892.
[2]T. Yamada, H. Nawa, S. Kawamoto, A. Tanaka & S. Fukui (1980) *Arch. Microbiol.* **128**, 145.

Selected entries from *Methods in Enzymology* [vol, page(s)]:
General discussion: 188, 171

LONG-CHAIN-ALCOHOL O-FATTY-ACYLTRANSFERASE

This acyltransferase [EC 2.3.1.75], also known as wax synthase, catalyzes the reaction of a long-chain acyl-CoA with a long-chain alcohol to produce an wax-like ester.[1,2] The saturated or unsaturated donor-substrate usually contains eighteen to twenty carbons, such as stearyl-CoA, arachidyl-CoA, and *cis*-11-eicosenoyl-CoA. The best acceptor substrate reported is *cis*-11-eicosen-1-ol, but other acceptors include 1-decanol, 1-dodecanol, 1-tetradecanol, 1-hexadecanol, 1-octadecanol, *cis*-9-octadecenol, 1-eicosanol, and *cis*-13-eicosen-1-ol.

[1]X.-Y. Wu, R. A. Moreau & P. K. Stumpf (1981) *Lipids* **16**, 897.
[2]W. S. Garver, J. D. Kemp & G. D. Kuehn (1992) *Anal. Biochem.* **207**, 335.

LONG-CHAIN ALCOHOL OXIDASE

This oxidoreductase [EC 1.1.3.20] catalyzes the reaction of dioxygen with two molecules of a long-chain primary alcohol to produce two molecules of the long-chain aldehyde and two water molecules.[1] Substrates include 1-dodecanol (that is, lauryl alcohol) and 1-tetradecanol (*i.e.*, myristyl alcohol). The more effective substrates contain between six and fourteen carbon atoms, and for enzymes from most sources, activity decreases with longer chains. The *Tanacetum vulgare* enzyme exhibits enhanced activity if the substrate has a Δ^2 double bond.

[1]F. M. Dickinson & C. Wadforth (1992) *BJ* **282**, 325.

Selected entries from *Methods in Enzymology* [vol, page(s)]:
General discussion: 71, 804, 805, 809
Assay: 71, 809, 810

LONG-CHAIN ALDEHYDE DEHYDROGENASE

This oxidoreductase [EC 1.2.1.48] catalyzes the reaction of a long-chain aldehyde with NAD+ to produce a long-chain carboxylate anion and NADH.[1] Substrates include hexanal, octanal, nonanal, decanal, dodecanal, tetradecanal, and hexadecanal.

[1] M. Vedadi, R. Szittner, L. Smillie & E. Meighen (1995) *Biochemistry* **34**, 16725.

Selected entries from *Methods in Enzymology* [vol, page(s)]:
General discussion: 71, 804, 805, 811; **188**, 176
Assay: 71, 811, 812
Other molecular properties: properties, 71, 812

LONG-CHAIN-ENOYL-CoA HYDRATASE

This enzyme [EC 4.2.1.74] catalyzes the reaction of *trans*-2-enoyl-CoA with water to produce (3*S*)-3-hydroxyacyl-CoA.[1-3] The best substrate is oct-2-enoyl-CoA. Alternative substrates include other 2-enoyl-CoA derivatives containing between eight and sixteen carbon atoms in the acyl chain. Crotonoyl-CoA, a substrate of enoyl-CoA hydratase [EC 4.2.1.17], is not a substrate of this enzyme.

See also *Enoyl-CoA Hydratase*

[1] F. C. Fong & H. Schulz (1977) *JBC* **252**, 542.
[2] H. Schulz (1974) *JBC* **249**, 2704.
[3] F. R. Beadle, C. C. Gallen, R. S. Conway & R. M. Waterson (1979) *JBC* **254**, 4387.

Selected entries from *Methods in Enzymology* [vol, page(s)]:
General discussion: 71, 390

LONG-CHAIN-FATTY-ACID:LUCIFERIN-COMPONENT LIGASE

This ligase component of the bacterial luciferase system [EC 6.2.1.19], also called acyl-protein synthetase, catalyzes the reaction of ATP with an acid and a particular protein component to produce acyl-protein thiolester, AMP, and pyrophosphate (or, diphosphate). The reaction proceeds through an acyl-adenylate intermediate. The combined action of this ligase and long-chain-fatty-acyl-CoA reductase [EC 1.2.1.50] produces the substrate of light-emitting alkanal monooxygenase (FMN-linked) [EC 1.14.14.3].[1]

[1] A. Rodriguez, L. Wall, S. Raptis, C. G. Zarkadas & E. Meighen (1988) *BBA* **964**, 266.

Selected entries from *Methods in Enzymology* [vol, page(s)]:
Assay: *Photobacterium phosphoreum*, 133, 174

LONG-CHAIN-FATTY-ACYL-CoA REDUCTASE

This oxidoreductase component of the bacterial luciferase system [EC 1.2.1.50], also known as acyl-CoA reductase, catalyzes the reaction of a long-chain acyl-CoA (*e.g.*, tetradecanoyl-CoA) with NADPH to produce a long-chain aldehyde (*e.g.*, tetradecanal), coenzyme A, and NADP$^+$.[1-4] Together with long-chain-fatty-acid:luciferin-component ligase [EC 6.2.1.19], this reductase forms a fatty acid reductase system which produces the substrate of light-emitting bacterial alkanal monooxygenase (FMN-linked) [EC 1.14.14.3].

[1] D. Riendeau & E. Meighen (1981) *Can. J. Biochem.* **59**, 440.
[2] D. Riendeau, A. Rodriguez & E. Meighen (1982) *JBC* **257**, 6908.
[3] A. Rodriguez, D. Riendeau & E. Meighen (1983) *JBC* **258**, 5233.
[4] L. Wall & E. Meighen (1989) *Biochem. Cell Biol.* **67**, 163.

Selected entries from *Methods in Enzymology* [vol, page(s)]:
General discussion: 133, 172
Assay: *Photobacterium phosphoreum*, 133, 173
Other molecular properties: *Photobacterium phosphoreum* (assays, 133, 173; properties, 133, 177; purification, 133, 175)

LONG-CHAIN FATTY ACYL-CoA SYNTHETASE

This enzyme [EC 6.2.1.3], known variously as long-chain-fatty-acid:CoA ligase, acyl-activating enzyme, acyl-CoA synthetase, fatty acid thiokinase (long-chain), and lignoceroyl-CoA synthetase, catalyzes the reaction of ATP with a long-chain carboxylic acid anion and coenzyme A to produce a fatty acyl-CoA, AMP, and pyrophosphate (or, diphosphate).[1-5] Substrates include a wide range of long-chain saturated and unsaturated fatty acid anions; however, the enzymes from different tissues exhibit variations in specificity. The mammalian liver enzyme acts on acid anions from C_6 (*n*-caproate) to C_{20} (arachidate), whereas the brain synthetase shows high activity up to C_{24} (lignoceroate). ***See also*** *Acyl-CoA Synthetase (GDP-Forming); Butyryl-CoA Synthetase*

[1] H. Tomoda, K. Igarashi & S. Omura (1987) *BBA* **921**, 595.
[2] N. Morisaki, T. Kanzaki, Y. Saito & S. Yoshida (1986) *BBA* **875**, 311.
[3] P. T. Normann, M. S. Thomassen, E. N. Christiansen & T. Flatmark (1981) *BBA* **664**, 416.
[4] D. P. Philipp & P. Parsons (1979) *JBC* **254**, 10776.
[5] J. C. Londesborough & L. T. Webster (1974) *The Enzymes*, 3rd ed., **10**, 469.

Selected entries from *Methods in Enzymology* [vol, page(s)]:
General discussion: 5, 467; 14, 91; 35, 117; 71, 44, 325, 334
Assay: 5, 467; 14, 622; cardiac mitochondrial, 125, 18; GTP-specific, 14, 91 (nitroprusside assay, 14, 92); myristate analogs as substrate, with (enzyme-coupled assay, 250, 423, 425; product purification by HPLC, 250, 422); peroxisomal, 148, 521
Other molecular properties: activators, 35, 122; arachidonate:CoA ligase, and, separation, 187, 240; assay for carnitine, 14, 613, 614; brown adipose tissue mitochondria, 55, 76; cardiac mitochondrial, assay, 125, 18; coenzyme A assay, use in, 13, 537, 540; esterase, not present in preparation of, 71, 242; expression in *Escherichia coli*, 250, 426; fat cells, in, 31, 67, 68; GTP-specific, 14, 91; lipid content, 35, 121; localization, 55, 9; location of, 10, 462; mutant, 72, 695; rat liver, from, 71, 334 (amino terminus, 71, 341; assay methods, 71, 334; effect of dimethyl sulfoxide, 71, 338;

kinetic properties, **71**, 340; microsomes, from, **71**, 337; mitochondrial fraction, from, **71**, 338; molecular weight, **71**, 339; quantitation of fatty acid or CoA, **71**, 340; specificity, **71**, 340; stoichiometry, **71**, 340); sequence alignment between species, **250**, 424; stability, **35**, 120, 121; substrate specificity, **250**, 426; synthesis reactions (fatty acyl-CoA analogs, **250**, 462; tritiated myristoyl-CoA, **250**, 410); types I and II (*Candida lipolytica*, from, **71**, 325; properties [of type I, **71**, 331 {molecular weight, **71**, 331; specificity and kinetic properties, **71**, 332; stoichiometry, **71**, 331; use in quantitation of fatty acid or CoA, **71**, 332}; of type II, **71**, 332]

LONG-CHAIN-FATTY-ACYL-GLUTAMATE DEACYLASE

This enzyme [EC 3.5.1.55], also called long-chain aminoacylase, catalyzes the hydrolysis of an *N*-(long-chain-fatty-acyl)-L-glutamate to produce a fatty acid anion and L-glutamate. The enzyme will not act on acyl derivatives of other amino acids, and the optimum chain length of the fatty acyl residue is between twelve and sixteen carbon atoms.[1,2]

[1]H. Fukuda, S. Iwade & A. Kimura (1982) *J. Biochem.* **91**, 1731.
[2]Y. Shintani, H. Fukuda, N. Okamoto, K. Murata & A. Kimura (1984) *J. Biochem.* **96**, 637.

LONG-CHAIN-3-HYDROXYACYL-CoA DEHYDROGENASE

This oxidoreductase [EC 1.1.1.211], also known as β-hydroxyacyl-CoA dehydrogenase (long-chain), catalyzes the reaction of (*S*)-3-hydroxyacyl-CoA with NAD$^+$ to produce 3-oxoacyl-CoA (*i.e.*, 3-ketoacyl-CoA) and NADH. The best acyl-CoA substrates are those containing between ten and sixteen carbons in the chain.[1,2] In addition to the dehydrogenase activity, this enzyme is also reported to exhibit long-chain enoyl-CoA hydratase and long-chain 3-ketothiolase activities.[2,3] **See also** *3-Hydroxyacyl-CoA Dehydrogenase*

[1]M. El-Fakhri & B. Middleton (1982) *BBA* **713**, 270.
[2]K. Carpenter, R. J. Pollitt & B. Middleton (1992) *BBRC* **183**, 443.
[3]K. Carpenter, R. J. Pollitt & B. Middleton (1993) *Biochem. Soc. Trans.* **21**, 35S.

[LOW-DENSITY LIPOPROTEIN RECEPTOR] KINASE

This phosphotransferase [EC 2.7.1.131], also known as LDL receptor kinase, catalyzes the reaction of ATP with a seryl residue in the low-density lipoprotein receptor (Ser833) to produce ADP and [low-density lipoprotein receptor] *O*-phospho-L-serine.[1,2] This phosphotransferase is inhibited by polylysine and heparin.

[1]A. Kishimoto, M. S. Brown, C. A. Slaughter & J. L. Goldstein (1987) *JBC* **262**, 1344.

[2]A. Kishimoto, J. L. Goldstein & M. S. Brown (1987) *JBC* **262**, 9367.

L-PEPTIDASE

This cysteine endopeptidase [EC 3.4.22.46], also called foot-and-mouth disease virus L-peptidase and leader proteinase, acts on the host initiation factor eIF-4G, a subunit of the cap-binding protein complex required for the translation of cap-dependent mRNAs. The initial cleavage occurs at Gly479–Arg480, followed by scission at Lys318–Arg319.[1,2]

[1]R. Kirchweger, E. Ziegler, B. J. Lamphear, D. Waters, H. D. Liebig, W. Sommergruber, F. Sobrino, C. Hohenadl, D. Blaas, R. E. Rhoads & T. Skern (1994) *J. Virol.* **68**, 5677.
[2]M. E. Piccone, M. Zellner, T. F. Kumosinski, P. W. Mason & M. J. Grubman (1995) *J. Virol.* **69**, 4950.

*Lsp*I RESTRICTION ENDONUCLEASE

This type II restriction enzyme [EC 3.1.21.4], isolated from *Lactobacillus* species, catalyzes the hydrolysis of both strands of DNA at 5′ . . . TT∇CGAA . . . 3′.

LUMAZINE SYNTHASE

This enzyme catalyzes the condensation of 5-amino-6-ribitylamino-2,4(1*H*,3*H*)-pyrimidinedione with 3,4-dihydroxy-2-butanone 4-phosphate to form 6,7-dimethyl-8-ribityllumazine and orthophosphate.[1–4] Subsequent dismutation by riboflavin synthase converts lumazine to riboflavin and a pyrimidine, such that every second pyrimidine molecule must be processed twice by lumazine synthase.[3]

[1]G. Neuberger & A. Bacher (1986) *BBRC* **139**, 1111.
[2]R. Volk & A. Bacher (1988) *JACS* **110**, 3651.
[3]K. Kis, R. Volk & A. Bacher (1995) *Biochemistry* **34**, 2883.
[4]S. Mörtl, M. Fischer, G. Richter, J. Tack, S. Weinkauf & A. Bacher (1996) *JBC* **271**, 33201.

Selected entries from **Methods in Enzymology** [**vol**, page(s)]:
General discussion: 280, 389

LUPANINE HYDROXYLASE

lupanine   17-hydroxylupanine

This oxidoreductase catalyzes the reaction of lupanine (or, 2-oxosparteine) with dioxygen and an acceptor substrate to produce 17-hydroxylupanine, water, and the reduced acceptor. The acceptor can be ferricyanide or 2,6-dichlorophenolindophenol.

Selected entries from **Methods in Enzymology** [vol, page(s)]:
General discussion: **52**, 21; **280**, 98

LUTEOLIN-7-O-DIGLUCURONIDE 4′-O-GLUCURONOSYLTRANSFERASE

This transferase [EC 2.4.1.191] catalyzes the reaction of UDP-D-glucuronate with luteolin 7-O-β-D-glucuronide to produce UDP and luteolin 7-O-[β-D-glucuronosyl-(1→2)-β-D-glucuronide]-4′-O-β-D-glucuronide.[1,2]

[1]M. Schulz & G. Weissenböck (1988) *Phytochemistry* **27**, 1261.
[2]S. Anhalt & G. Weissenböck (1992) *Planta* **187**, 83.

LUTEOLIN-7-O-GLUCURONIDE 7-O-GLUCURONOSYLTRANSFERASE

This transferase [EC 2.4.1.190] catalyzes the reaction of UDP-D-glucuronate with luteolin 7-O-glucuronide to produce UDP and luteolin 7-O-β-D-diglucuronide.[1,2]

[1]M. Schulz & G. Weissenböck (1988) *Phytochemistry* **27**, 1261.
[2]S. Anhalt & G. Weissenböck (1992) *Planta* **187**, 83.

LUTEOLIN 7-O-GLUCURONOSYL-TRANSFERASE

This transferase [EC 2.4.1.189] catalyzes the reaction of UDP-D-glucuronate with luteolin to produce UDP and luteolin 7-O-β-D-glucuronide.[1]

[1]M. Schulz & G. Weissenböck (1988) *Phytochemistry* **27**, 1261.

LUTEOLIN O-METHYLTRANSFERASE

This methyltransferase [EC 2.1.1.42], also known as O-dihydric phenol methyltransferase, catalyzes the reaction of S-adenosyl-L-methionine with 5,7,3′,4′-tetrahydroxyflavone (that is, luteolin) to produce S-adenosyl-L-homocysteine and 5,7,4′-trihydroxy-3′-methoxyflavone (*i.e.*, chrysoeriol).[1–3] Other substrates include luteolin-7-O-β-D-glucoside, caffeate, eriodictyol, and quercetin.

[1]J. Ebel, K. Hahlbrock & H. Grisebach (1972) *BBA* **269**, 313.
[2]J. E. Poulton, K. Hahlbrock & H. Grisebach (1977) *ABB* **180**, 543.
[3]J. Ebel & K. Hahlbrock (1977) *EJB* **75**, 201.

*Lwe*I RESTRICTION ENDONUCLEASE

This type II restriction enzyme (subtype s) [EC 3.1.21.4], isolated from *Listeria welshimeri* RFL 131, catalyzes the hydrolysis of both strands of DNA at 5′...GCATCNNNNN∇NNNN...3′ and 3′...

CGTAGNNNNNNNNNN∇...5′, where N refers to any base.

LYSINE N-ACETYLTRANSFERASE

L-lysine   N⁶-acetyl-L-lysine

This acetyltransferase [EC 2.3.1.32], also known as lysine N⁶-acetyltransferase, catalyzes the reaction of acetyl phosphate with L-lysine to produce orthophosphate and N⁶-acetyl-L-lysine.[1] L-Ornithine is also acetylated.

[1]W. K. Paik & S. Kim (1964) *ABB* **108**, 221.

LYSINE-2,3-AMINOMUTASE

This aminomutase [EC 5.4.3.2] catalyzes the interconversion of L-lysine and (3S)-3,6-diaminohexanoate (or, L-β-lysine).[1–6]

L-lysine   (3S)-β-lysine

The enzyme is stimulated by S-adenosyl-L-methionine and pyridoxal 5′-phosphate. In the transfer of hydrogen between C2 and C3, there is no exchange with solvent protons. A radical is generated in the reaction mechanism from a reversible reaction between the iron-sulfur monocation complex and S-adenosyl-L-methionine at the active site to form [4Fe-4S]²⁺, the 5′-deoxyadenosyl radical, and L-methionine as intermediates. The 5′-deoxyadenosyl radical is postulated to abstract a hydrogen atom from C3 of L-lysine, which is bound to pyridoxal 5′-phosphate as an aldimine.[1]

[1]W. Wu, S. Booker, K. W. Lieder, V. Bandarian, G. H. Reed & P. A. Frey (2000) *Biochemistry* **39**, 9561.
[2]P. A. Frey (1997) *Curr. Opin. Chem. Biol.* **1**, 347.
[3]P. A. Frey & G. H. Reed (1993) *AE* **66**, 1.
[4]E. J. Brush & J. W. Kozarich (1992) *The Enzymes*, 3rd ed., **20**, 317.
[5]T. C. Stadtman (1973) *AE* **38**, 413.
[6]T. P. Chirpich, V. Zappia, R. N. Costilow & H. A. Barker (1970) *JBC* **245**, 1778.

Selected entries from **Methods in Enzymology** [vol, page(s)]:
General discussion: **17B**, 215; **258**, 362
Assay: **17B**, 215
Other molecular properties: cofactors, **258**, 362; electron paramagnetic resonance (anisotropy, **258**, 370; Fourier transform,

258, 374; isotopic substitution, **258**, 367, 377; liquid phase measurements, **258**, 366; π-radical properties, **258**, 370; rapid-mix/ freeze-quench trapping of intermediates, **258**, 365; resolution enhancement, **258**, 373; signal characterization, **258**, 367; slow substrate trapping of intermediates, **258**, 365; spectral properties, **258**, 363, 367; spectral simulation, **258**, 376; structure determination, **258**, 379); inactivation by oxygen, **258**, 367; properties, **17B**, 220; purification from *Clostridium* SB₄, **17B**, 218; reaction catalyzed, **258**, 362; reaction mechanism, **258**, 363; spectrum, **17B**, 221

β-LYSINE 5,6-AMINOMUTASE

This cobalamin-dependent aminomutase [EC 5.4.3.3], also known as β-lysine mutase, catalyzes the interconversion of (3*S*)-3,6-diaminohexanoate (*i.e.*, L-β-lysine) and (3*S*,5*S*)-3,5-diaminohexanoate.[1–4]

(3*S*)-β-lysine   (3*S*,5*S*)-3,5-diaminohexanoate

Pyridoxal 5'-phosphate is required as an additional cofactor. The substrate initially forms a Schiff base between the pyridoxal 5'-phosphate and the ε-amino group. A cobalamin-mediated radical intermediate is generated, rearrangement occurs, and the product is released with the concomitant regeneration of the pyridoxal form of the coenzyme.

[1] B. T. Golding & W. Buckel (1998) *CBC* **3**, 239.
[2] P. A. Frey & G. H. Reed (1993) *AE* **66**, 1.
[3] T. C. Stadtman (1973) *AE* **38**, 413.
[4] T. C. Stadtman (1972) *The Enzymes*, 3rd ed., **6**, 539.

Selected entries from **Methods in Enzymology** [**vol**, page(s)]:
General discussion: **17B**, 206
Assay: **17B**, 206

D-LYSINE 5,6-AMINOMUTASE

This cobalamin- and pyridoxal-phosphate-dependent aminomutase [EC 5.4.3.4], also called D-α-lysine mutase, catalyzes the interconversion of D-lysine and 2,5-diaminohexanoate.[1–5] The reaction proceeds via the formation of both Schiff base and radical intermediates. **See also** D-*Ornithine 4,5-Aminomutase*

[1] C. H. Chang & P. A. Frey (2000) *JBC* **275**, 106.
[2] P. A. Frey & G. H. Reed (1993) *AE* **66**, 1.
[3] T. C. Stadtman (1973) *AE* **38**, 413.
[4] T. C. Stadtman (1972) *The Enzymes*, 3rd ed., **6**, 539.
[5] C. G. D. Morley & T. C. Stadtman (1970) *Biochemistry* **9**, 4890 and (1972) **11**, 600.

Selected entries from **Methods in Enzymology** [**vol**, page(s)]:
General discussion: **17B**, 211
Assay: **17B**, 211

Other molecular properties: α-(5,6-dimethylbenzimidazolyl)-Co-5'-deoxyadenosylcobamin and, **17B**, 214; preparation from *Clostridium sticklandii*, **17B**, 212; properties, **17B**, 214

LYSINE 6-AMINOTRANSFERASE

This aminotransferase [EC 2.6.1.36], also called lysine ε-aminotransferase, catalyzes the pyridoxal-phosphate-dependent reaction of L-lysine with α-ketoglutarate (or, 2-oxoglutarate) to produce 2-aminoadipate 6-semi-aldehyde and L-glutamate.[1] The aldehyde product, also known as allysine, then undergoes dehydration to form 1-piperideine 6-carboxylate. L-Ornithine can also be converted to glutamate γ-semialdehyde. **See also** *Lysine:Pyruvate 6-Aminotransferase*

[1] A. E. Braunstein (1973) *The Enzymes*, 3rd ed., **9**, 379.

Selected entries from **Methods in Enzymology** [**vol**, page(s)]:
General discussion: **17B**, 222; **113**, 96
Assay: **17B**, 222
Other molecular properties: extinction coefficients, **17B**, 227; *Flavobacterium lutescens* (proton abstraction, stereochemistry, **113**, 100; purification, **113**, 96; subunit structure, **113**, 98); properties, **17B**, 226; purification from *Achromobacter liquidum*, **17B**, 224; pyridoxal phosphate and, **17B**, 227

LYSINE(ARGININE) CARBOXYPEPTIDASE

This zinc-dependent metallocarboxypeptidase [EC 3.4.17.3], also known as carboxypeptidase N, lysine carboxypeptidase, arginine carboxypeptidase, kininase I, catalyzes the release of a C-terminal lysine or arginine from a peptide or protein. Rates of hydrolysis are greater for lysine-containing peptides than for arginine-containing peptides. The enzymes prefer substrates with alanyl residues in the P1 position. The chromogenic dipeptides furoacryloyl-Ala-Lys and dansyl-Ala-Arg can be used to assay the enzyme.[1–3]

[1] T. H. Plummer, Jr. & E. G. Erdös (1981) *MIE* **80**, 442.
[2] R. A. Skidgel (1995) *MIE* **248**, 653.
[3] R. A. Skidgel (1996) in *Zinc Metalloproteases in Health and Disease* (N. M. Hooper, ed.), Taylor & Francis, London, p. 241.

Selected entries from **Methods in Enzymology** [**vol**, page(s)]:
General discussion: **80**, 442, 788; **248**, 653
Assay: **163**, 186; **248**, 653; human plasma, **80**, 443, 444
Other molecular properties: activation by metal ion, **248**, 661; activity, effect of trypsin, **80**, 449; activity units, calculation, **248**, 654; cDNA sequence, **248**, 661; concentration, **248**, 659; discovery, **248**, 653, 663; distribution, **163**, 190; human plasma, **80**, 442 (amino acid composition, **80**, 448; assay, **80**, 443, 444; carbohydrate composition, **80**, 448, 449; concentration, **80**, 447; decrease, **80**, 449; deficiency, **80**, 449; increase, in pregnancy, **80**, 449; isoelectric forms, **80**, 449; kinetics of activation, **80**, 449; molecular weight, **80**, 448; properties, **80**, 448, 449; purification, **80**, 444; **163**, 188; quantitation, **80**, 447; stability, **80**, 447, 448); inhibition, **80**, 449; **163**, 191; **248**, 661; kinetics, **248**, 661; leucine-rich repeat region, **248**, 661; molecular weight, **248**, 660; pH optimum, **248**, 661; physiological functions,

248, 653; properties, **163**, 189; **248**, 213, 216, 217, 660 (enzymatic, **248**, 660); purification, **248**, 654; quantification, **163**, 189; storage, **248**, 659; substrates, **248**, 661; subunits (isolation, **248**, 659; separation, **248**, 659); transition-state and multisubstrate analogues, **249**, 306; zinc content, **248**, 661

LYSINE CARBAMOYLTRANSFERASE

This transferase [EC 2.1.3.8], also known as lysine transcarbamylase, catalyzes the reaction of carbamoyl phosphate with L-lysine to produce orthophosphate and L-homocitrulline.[1] The enzyme is not identical with ornithine carbamoyltransferase [EC 2.1.3.3].

[1] F. A. Hommes, A. G. Eller, D. F. Scott & A. L. Carter (1983) *Enzyme* **29**, 271.

LYSINE DECARBOXYLASE

This pyridoxal-phosphate-dependent enzyme [EC 4.1.1.18] catalyzes the conversion of L-lysine to cadaverine and carbon dioxide.[1] 5-Hydroxy-L-lysine and the L-lysine isostere *S*-aminoethyl-L-cysteine are also decarboxylated.

[1] E. A. Boeker & E. E. Snell (1972) *The Enzymes*, 3rd ed., **6**, 217.

Selected entries from *Methods in Enzymology* [vol, page(s)]:
General discussion: 2, 188; **17B**, 677; **94**, 180
Assay: *Bacterium cadaveris*, **17B**, 677; *Escherichia coli*, **94**, 181
Other molecular properties: assay of L-lysine, 3, 465, 1048; **17B**, 631; **44**, 277, 278; *Bacterium cadaveris*, in, 3, 464; **17B**, 631, 677 (properties, **17B**, 681; purification, 3, 464; **17B**, 678); conjugate, activity, **44**, 271; diaminopimelate decarboxylase, presence in assay of, **5**, 866; diaminopimelate racemase and, **5**, 861; *Escherichia coli* (assay, **94**, 181; deficient mutants, mass screening, **94**, 89; properties, **94**, 183; purification, 2, 188; **94**, 182); transaminase assay and, 2, 171

LYSINE 6-DEHYDROGENASE

This oxidoreductase [EC 1.4.1.18] catalyzes the reaction of L-lysine with NAD^+ to produce allysine ($^-OOCCH(NH_3^+)(CH_2)_3CHO$) and NADH.[1-3] The allysine product spontaneously converts to 1,6-didehydropiperidine-2-carboxylate. *See also Protein-Lysine 6-Oxidase*

[1] H. Hashimoto, H. Misono, S. Nagata & S. Nagasaki (1989) *J. Biochem.* **106**, 76.
[2] H. Misono, H. Hashimoto, H. Uehigashi, S. Nagata & S. Nagasaki (1989) *J. Biochem.* **105**, 1002.
[3] H. Misono, H. Uehigashi, E. Morimoto & S. Nagasaki (1985) *Agric. Biol. Chem.* **49**, 2253.

LYSINE DEHYDROGENASE (α-DEAMINATING)

This oxidoreductase [EC 1.4.1.15] catalyzes the reaction of L-lysine with NAD^+ to produce 1,2-didehydropiperidine 2-carboxylate, ammonia, and NADH.[1,2]

[1] N. M. W. Brunhuber & J. S. Blanchard (1994) *Crit. Rev. Biochem. Mol. Biol.* **29**, 415.
[2] W. Buergi, R. Richterich & J. P. Colombo (1966) *Nature* **211**, 854.

L-LYSINE-LACTAMASE

This enzyme [EC 3.5.2.11], also known as L-α-aminocaprolactam hydrolase and L-lysinamidase, catalyzes the hydrolysis of L-lysine 1,6-lactam to produce L-lysine.[1] Other substrates include L-lysinamide.

[1] T. Fukumura, G. Talbot, H. Misono, Y. Teramura, K. Kato & K. Soda (1978) *FEBS Lett.* **89**, 298.

ε-N-L-LYSINE METHYLTRANSFERASE

This methyltransferase, also known as *S*-adenosylmethionine:ε-*N*-L-lysine methyltransferase, catalyzes the reaction of *S*-adenosyl-L-methionine with L-lysine to produce *S*-adenosyl-L-homocysteine and N^ε-methyl-L-lysine. Di- and trimethyl products are also produced, possibly by the same enzyme. This reaction is similar to the corresponding reaction in the pathway for carnitine formation in mammals, except that the *Neurospora crassa* enzyme acts on free L-lysine, and not on a lysyl residue in a protein.[1]

[1] H. P. Broquist (1986) *MIE* **123**, 303.

Selected entries from *Methods in Enzymology* [vol, page(s)]:
General discussion: 123, 303
Assay: 123, 304

LYSINE 2-MONOOXYGENASE

This FAD-dependent oxidoreductase [EC 1.13.12.2] catalyzes the reaction of L-lysine with dioxygen to produce 5-aminopentanamide ($H_2NCO(CH_2)_4NH_3^+$), carbon dioxide, and water.[1-4] Other diamino acid substrates include 7,8-diaminoheptanoate, δ-hydroxylysine, 5-methyllysine, and L-ornithine.

[1] S. Yamamoto & Y. Ishimura (1991) *A Study of Enzymes* **2**, 315.
[2] V. Massey & P. Hemmerich (1975) *The Enzymes*, 3rd ed., **12**, 191.
[3] M. S. Flashner & V. Massey (1974) in *Mol. Mech. Oxygen Activ.* (O. Hayaishi, ed.) p. 245.
[4] M. I. S. Flashner & V. Massey (1974) *JBC* **249**, 2579.

Selected entries from *Methods in Enzymology* [vol, page(s)]:
General discussion: 17B, 154; **52**, 16
Assay: 17B, 154

L-LYSINE 6-MONOOXYGENASE

This FAD-dependent oxidoreductase [EC 1.13.12.10], also known as lysine N^6-hydroxylase, catalyzes the reaction of L-lysine with dioxygen to produce N^6-hydroxy-L-lysine and water.[1] L-Ornithine is an alternative substrate.

L-Lysine 6-monooxygenase (NADPH) [EC 1.14.13.59] catalyzes the same hydroxylation reaction but requires NADPH as a cosubstrate.[2] **See also** *Protein-Lysine 6-Oxidase*

[1]V. de Lorenzo, A. Bindereif, B. H. Paw & J. B. Neilands (1986) *J. Bacteriol.* **165**, 570.
[2]H.-J. Plattner, P. Pfefferle, A. Romaguera, S. Waschütza & H. Diekmann (1989) *Biol. Met.* **2**, 1.

L-LYSINE 6-MONOOXYGENASE (NADPH)

This FAD-dependent oxidoreductase [EC 1.14.13.59], also known as lysine N^6-hydroxylase, catalyzes the reaction of L-lysine with NADPH and dioxygen to produce N^6-hydroxy-L-lysine, NADP$^+$, and water.[1] The enzyme from *Escherichia coli* (strain EN 222) is highly specific for L-lysine and will not utilize L-ornithine or L-homolysine. L-Lysine 6-monooxygenase [EC 1.13.12.10] catalyzes the same reaction without NADPH. **See also** *Protein-Lysine 6-Oxidase*

[1]A. M. Thariath, K. L. Fatum, M. A. Valvano & T. Viswanatha (1993) *BBA* **1203**, 27.

L-LYSINE OXIDASE

This FAD-dependent oxidoreductase [EC 1.4.3.14] catalyzes the reaction of L-lysine and dioxygen and water to produce 6-amino-2-ketohexanoate (or, 6-amino-2-oxohexanoate), ammonia, and hydrogen peroxide.[1] Other substrates include L-ornithine, L-phenylalanine, L-tyrosine, L-arginine, and L-histidine. **See also** *Lysine 2-Monooxygenase; Protein-Lysine 6-Oxidase*

[1]H. Kusakabe, K. Kodama, A. Kuninaka, H. Yoshino, H. Misono & K. Soda (1980) *JBC* **255**, 976.

LYSINE:PYRUVATE 6-AMINOTRANSFERASE

This aminotransferase [EC 2.6.1.71] catalyzes the reaction of L-lysine with pyruvate to produce 2-aminoadipate 6-semialdehyde and L-alanine.[1,2] **See also** *Lysine 6-Aminotransferase*

[1]H. Schmidt, R, Bode & D. Birnbaum (1987) *J. Basic Microbiol.* **27**, 595.
[2]H. Schmidt, R. Bode & D. Birnbaum (1988) *FEMS Microbiol. Lett.* **49**, 203.

LYSINE RACEMASE

This racemase [EC 5.1.1.5] catalyzes the interconversion of L-lysine and D-lysine.[1,2]

[1]F. P. Guengerich & H. P. Broquist (1973) *Biochemistry* **12**, 4270.

[2]Y.-F. Chang & E. Adams (1974) *J. Bacteriol.* **117**, 753.
Selected entries from *Methods in Enzymology* [vol, page(s)]:
Other molecular properties: *Clostridium sticklandii*, in, **17B**, 201, 205

LYSOLECITHIN ACYLMUTASE

This mutase [EC 5.4.1.1], also known as lysolecithin migratase, catalyzes the interconversion of a 2-lysolecithin to a 3-lysolecithin.[1]

[1]M. Uziel & D. J. Hanahan (1957) *JBC* **226**, 789.

LYSOPHOSPHATIDATE PHOSPHATASE

This enzyme catalyzes the hydrolysis of a lysophosphatidylglycerophosphate to produce a monoacylglycerol and orthophosphate. **See also** *Phosphatidate Phosphatase*

Selected entries from *Methods in Enzymology* [vol, page(s)]:
General discussion: 209, 228
Assay: 209, 228

LYSOPHOSPHOLIPASE

This phospholipase [EC 3.1.1.5] (also known as lecithinase B, lysolecithinase, and phospholipase B) catalyzes the hydrolysis of 2-lysophosphatidylcholine to produce glycerophosphocholine and a fatty acid anion.[1-3]

Care should be exercised whenever using the term "lysophospholipid." While often referring to a 1-acyl-*sn*-glycero-3-phospho derivative (such as 1-palmitoyl-*sn*-glycero-3-phosphocholine, a 2-lysophospholipid), this term may also refer to the 2-acyl-*sn*-glycero-3-phospho compound (*e.g.*, 2-palmitoyl-*sn*-glycero-3-phosphocholine). Both lysophospholipids are hydrolyzed by this enzyme. The *Penicillum notatum* enzyme reportedly will also act on the diacyl lipid (for example, phosphatidylcholine).

[1]A. Wang & E. A. Dennis (1999) *BBA* **1439**, 1.
[2]E. A. Dennis (1983) *The Enzymes*, 3rd ed., **16**, 307.
[3]J. D. Esko & C. R. H. Raetz (1983) *The Enzymes*, 3rd ed., **16**, 207.

Selected entries from *Methods in Enzymology* [vol, page(s)]:
General discussion: 71, 513; 197, 446
Assay: 163, 34; 197, 320; *Penicillium notatum*, 197, 448
Other molecular properties: enzymatic activity, 31, 522; hydrolysis of lecithin monolayers, 14, 643, 646; identity with Charcot-Leyden crystal protein, 163, 41; immunochemical assay, 163, 41; inhibition, 31, 524; localization in eosinophils and basophils, 163, 43; lysophospholipase L$_1$, (assay, 197, 438; properties, 197, 445; purification, 197, 444); lysophospholipase L$_2$ (assay, 197, 438; primary structure, 197, 442; properties, 197, 442; purification, 197, 440); lysophospholipase-transacylase, 71, 513 (assay, rat lung, from, 71, 513;

rabbit myocardium, purification and assay, **197**, 475; rat lung, from, **71**, 513 [*O*-acyl cleavage mechanism, **71**, 520; acyl transfer reaction specificity, **71**, 520; assays, **71**, 513; covalent acylenzyme intermediate, **71**, 520; dual activities, **71**, 518; properties, **71**, 518; stability, **71**, 519]); *Penicillium notatum* (assay, **197**, 448; characterization, **197**, 450; inhibition, **197**, 455; phospholipase B activity, assay, **197**, 447; purification, **197**, 449; stimulation, **197**, 455; substrate preparation, **197**, 454; substrate specificity, **197**, 453); phospholipase assay and, **I**, 661; plant extracts, in, **31**, 527; properties, **163**, 39; purification from human eosinophils, **163**, 36; rabbit myocardium, purification and assay, **197**, 475; reaction products, chromatography, **197**, 164; thioester substrate, **248**, 16; type I (bovine liver [assay, **197**, 468; properties, **197**, 473; purification, **197**, 470]; from P388D₁ cells [assay, **197**, 457; properties, **197**, 465; purification, **197**, 459]); type II (bovine liver [assay, **197**, 468; properties, **197**, 473; purification, **197**, 472]; from P388D₁ cells [assay, **197**, 457; properties, **197**, 465; purification, **197**, 461])

D-LYSOPINE DEHYDROGENASE

This oxidoreductase [EC 1.5.1.16], also known as D-lysopine synthase, catalyzes the reversible reaction of N^2-(D-1-carboxyethyl)-L-lysine (*i.e.*, D-lysopine) with NADP⁺ and water to produce L-lysine, pyruvate, and NADPH. In the reverse reaction, a number of L-amino acids are substrates, producing the corresponding N^2-(D-1-carboxyethyl) metabolites (*i.e.*, opines).[1]

[1]J. Thompson & S. P. F. Miller (1991) *Adv. Enzymol.* **64**, 317.

LYSOSOMAL PRO-XAA CARBOXYPEPTIDASE

This serine carboxypeptidase [EC 3.4.16.2], also known as proline carboxypeptidase, angiotensinase C, lysosomal carboxypeptidase C, lysosomal Pro-X carboxypeptidase, and prolylcarboxypeptidase [PCP], catalyzes the release of a C-terminal amino acid from a protein or polypeptide. Enzyme activity requires that the penultimate prolyl residue Pro–Xaa, where Xaa is an aromatic or aliphatic amino acid. The enzyme also acts on angiotensins II and III and can be conveniently assayed through the use of the synthetic substrates, such as Z-Pro-Ala, Z-Pr-Val, Z-Pro-Leu.[1–3]

[1]C. E. Odya & E. G. Erdös (1981) *MIE* **80**, 460.
[2]K. Breddam (1986) *Carlsberg Res. Commun.* **51**, 83.
[3]F. Tan, P. W. Morris, R. A. Skidgel & E. G. Erdös (1993) *JBC* **268**, 16631.

Selected entries from *Methods in Enzymology* [**vol**, page(s)]:
General discussion: 80, 460
Other molecular properties: active site residues, **244**, 45; human, **80**, 460 (activity units, **80**, 462; assay, **80**, 460; distribution, **80**, 466; kidney [inhibition, **80**, 465, 466; molecular weight, **80**, 465; properties, **80**, 465, 466; purification, **80**, 462; stability, **80**, 465]); substrate specificity, **80**, 461

LYSOZYME

This enzyme [EC 3.2.1.17], also called muramidase and endoacetylmuramidase, catalyzes the hydrolysis of the 1,4-β-linkages between *N*-acetyl-D-glucosamine and *N*-acetylmuramic acid in peptidoglycan heteropolymers of prokaryotic cell walls.[1–5] Some chitinases [EC 3.2.1.14] also exhibit this activity. Hen egg white lysozyme was the first enzyme to have its X-ray crystal structure determined.[4]

The lysozyme of bacteriophage T4 is distinct from the hen egg white enzyme in that it is an inverting glycosidase: the α-anomer is released.[1]

[1]G. Davies, M. L. Sinnott & S. G. Withers (1998) *CBC* **I**, 119.
[2]G. Mooser (1992) *The Enzymes*, 3rd ed., **20**, 187.
[3]H. A. McKenzie & F. H. White, Jr. (1991) *Adv. Protein Chem.* **41**, 174.
[4]C. C. F. Blake, D. F. Koenig, G. A. Mair, A. C. T. North, D. C. Phillips & V. R. Sarma (1965) *Nature* **206**, 757.
[5]P. Jolles (1960) *The Enzymes*, 2nd ed., **4**, 431.

Selected entries from *Methods in Enzymology* [**vol**, page(s)]:
General discussion: 5, 137; 8, 685; 27, 197, 200, 206, 392, 428, 473, 514, 539, 549, 561, 564, 565, 572, 582, 675, 695, 700, 705, 729, 743, 747, 748, 816, 821, 831, 832, 835, 882, 907; 34, 4, 5, 639, 705; 46, 403; 154, 511
Assay: 5, 137; 202, 344; Fast Analyzer, by, 31, 816
Other molecular properties: *N*-acetylglucosamine interaction, **64**, 275; acid-induced difference spectra, **11**, 764; active site mapping, **227**, 56; active sites, determination by selective oxidative free radical probes, **105**, 496; acylation, **25**, 547 (comparison of reagents, **25**, 552; techniques, **47**, 149); adsorption of, **5**, 30; adenylate cyclase extraction, **38**, 163, 164; affinity labeling, **46**, 410; **87**, 473; algal lysis, use in, **31**, 682; alkaline phosphatase release and, **12B**, 215 (purification, in, **9**, 640); amatoxin conjugates (biological properties, **112**, 233; preparation by mixed anhydride coupling, **112**, 231); amidinated, **25**, 591, 595; ancestral (molecular biology, **224**, 577; purification, **224**, 582; thermostability, **224**, 584); antibody interaction, measurement, **224**, 507; antigenic sites, **70**, 46, 47; aromatic sulfonyl chlorides and, **25**, 645; assay for assessment of phagocytic function, **132**, 120, 160; azobenzene-2-sulfenyl bromide and, **25**, 490; bacterial cell disruption, in, **8**, 690, 694; **51**, 287; bacterial DNA and, **12A**, 544, 546; **12B**, 136; bacterial extraction by, **1**, 161; **2**, 421, 786; bacterial lysis, for, **52**, 153; bacteriophage, *in vitro* synthesis, **20**, 531, 532, 533, 539, 541; bacteriophage T4 (assembly details, **177**, 226; crystals, **276**, 6; 1D and 2D isotope-directed NOE measurements, **177**, 69; disulfide bond formation, assay, **202**, 343; expression system, **177**, 46; first domain structure, solution, **177**, 239; folding and stability, genetic analysis, **131**, 265; forbidden-echo spectra, **177**, 66; intermediate accessible volumes, analysis, **177**, 229; isotopic enrichment, determination, **177**, 61; multiple disulfide bonds, with [activity, **202**, 352; mutant construction, **202**, 350; thermal stability, **202**, 352]; mutated [expression and purification, **202**, 342; oxidation and reduction, **202**, 344; thermal stability, **202**, 344]; ¹⁵N-labeled, NMR, **177**, 65; production by bacterial hosts bearing pHSe plasmids, **177**, 72; selective ¹⁵N labeling, **177**, 60; single disulfide bond, with [assay, **202**, 347; bond formation, assessment, **202**, 345; mutant construction, **202**, 345; redox properties of disulfide bond, **202**, 349; thermal stability, **202**, 349]; site-directed mutagenesis, **202**, 340; structure and thermal stability,

synthesis (assay of, **30**, 663; initiation factor assay and, **20**, 275); synthetic, **34**, 644, 645; temperature-induced difference spectra, **11**, 767; thiol-disulfide interchange in, thiol pK_a values, **143**, 134; transition-state and multisubstrate analogues, **249**, 306; translational diffusion coefficient, **48**, 420; translational motion, measurement by pulse-gradient spin-echo NMR, **176**, 439; tryptophan content, **49**, 161, 162; ultraviolet-endonuclease preparation and, **21**, 246; ultraviolet-exonuclease preparation and, **21**, 251; use in bacterial lysis, **17B**, 52, 60; use in cell breakage, **22**, 491

LYSYL AMINOPEPTIDASE (BACTERIA)

This metalloaminopeptidase, commonly known as peptidase N or PepN, catalyzes the release of the N-terminal aminoacyl residue in proteins and peptides. The enzyme exhibits a broad specificity for peptides. The *Lactococcus lactis* enzyme acts on di- and tripeptides as well, showing a preference for N-terminal arginyl residues whereas a lysyl residue is preferred over an arginyl residue with the *Lactobacillus helveticus* enzyme.[1,2]

Some databanks list these bacterial enzymes under EC 3.4.11.2; however, this is inappropriate, because they are clearly distinct from membrane alanyl aminopeptidases.

[1]B. Poolman, E. R. S. Kunji, A. Hagting, V. Juillard & W. N. Konings (1995) *J. Appl. Bacteriol. Symp. Suppl.* **24**, 65S.
[2]G. W. Niven, S. A. Holder & P. Strøman (1995) *Appl. Microbiol. Biotechnol.* **44**, 100.

Selected entries from **Methods in Enzymology** [vol, page(s)]:
Other molecular properties: *Lactococcus* (properties, **248**, 185; structure, **248**, 186)

LYSYL ENDOPEPTIDASE

This serine endopeptidase [EC 3.4.21.50], also called *Achromobacter* proteinase I (API), endoproteinase Lys-C, and lysyl bond specific proteinase, catalyzes peptide-bond hydrolysis in proteins with a preference for Lys–Xaa bonds, including Lys–Pro bonds. The enzyme also has an esterase and a transpeptidation activity.[1,2]

[1]F. Sakiyama & T. Masaki (1994) *Meth. Enzymol.* **244**, 126.
[2]S. Norioka, S. Ohata, T. Ohara, S.-I. Lim & P. Sakiyama (1994) *JBC* **269**, 17025.

Selected entries from **Methods in Enzymology** [vol, page(s)]:
General discussion: 244, 126
Assay: 244, 126
Other molecular properties: activation, **244**, 134; active site residues, **244**, 34, 133, 136; assay methods, **244**, 126; catalytic rate, **244**, 129; cleavage site specificity, **244**, 126; disulfide bridges, **244**, 133; family relationships, **244**, 34; gene, **244**, 134; human insulin semisynthesis, **244**, 132; immobilized (assay, **136**, 163; preparation, **136**, 165; preparation of, in [human insulin, **136**, 168; porcine dealanine insulin, **136**, 167]); inhibitors, **244**, 129; mass spectroscopy, protein digestion, **193**, 363; pH optimum, **244**, 129; processing, **244**, 34, 134; protein digestion for MS analysis,

193, 363; purification, **244**, 127, 135; sequence, **244**, 133; solvent inactivation, **244**, 131; stability, **244**, 132; tertiary structure, **244**, 136

LYSYLTRANSFERASE

This transferase [EC 2.3.2.3], also known as lysyl–tRNA:phosphatidylglycerol transferase, catalyzes the reaction of L-lysyl-tRNA with phosphatidylglycerol to produce tRNA and 3-phosphatidyl-1'-(3'-O-L-lysyl)-glycerol.[1,2]

[1]R. L. Soffer (1974) *AE* **40**, 91.
[2]W. J. Lennarz, P. P. M. Bonsen & L. L. M. Van Deenen (1967) *Biochemistry* **6**, 2307.

LYSYL-tRNA SYNTHETASE

This enzyme [EC 6.1.1.6], also known as lysine:tRNA ligase and lysine translase, catalyzes the reaction of L-lysine with ATP and tRNA^{Lys} to produce L-lysyl-tRNA^{Lys}, AMP, and pyrophosphate (or, diphosphate).[1,2] L-Lysine is first activated by forming a lysyl-adenylate intermediate before acyl transfer to the cognate tRNA.

[1]W. Freist & D. H. Gauss (1995) *Biol. Chem. Hoppe-Seyler* **376**, 451.
[2]Y. Nakamura & K. Ito (1993) *Mol. Microbiol.* **10**, 225.

Selected entries from **Methods in Enzymology** [vol, page(s)]:
General discussion: 34, 170
Other molecular properties: purification, **5**, 729, 730; **29**, 591; subcellular distribution, **59**, 233, 234; tRNA interaction, X-ray studies, **61**, 227

α-LYTIC ENDOPEPTIDASE

This serine endopeptidase [EC 3.4.21.12], also called α-lytic protease (α-LP), catalyzes peptide-bond hydrolysis in proteins. It was first isolated from *Lysobacter enzymogenes* (originally misclassified as *Myxobacter* 495 and *Sorangium* sp.). Specificity studies indicate that there is a preference for small hydrophobic aminoacyl residues in the P1 position (*e.g.*, alanyl or valyl). There is also a preference for a prolyl residue in position P2 and an alanyl residue in P1'. Synthetic substrates such as Ac-Ala-Pro-Ala-∇-NHPhNO$_2$ and MeO-Suc-Ala-Ala-Pro-Ala-∇-NHMec can be used to assay the enzyme.[1,2] (Note: ∇ indicates the cleavage site.)

His36 participates in extracting the proton from Ser143 and in donating a proton to the leaving group. This same histidyl residue is hydrogen bonded to that seryl residue, polarizing the hydroxyl group and increasing its nucleophilicity.

[1] H. Kaplan & D. R. Whitaker (1969) *Can. J. Biochem.* **47**, 305.
[2] K. Morihara (1974) *AE* **41**, 179.

Selected entries from *Methods in Enzymology* [vol, page(s)]:
General discussion: 19, 599
Assay: 19, 600
Other molecular properties: catalytic serine codon, **244**, 33; family members, **244**, 27, 32; *Myxobacterium*, **19**, 599; processing, **244**, 32; sequence homology, **244**, 32; solvent isotope effect, **87**, 592, 596; *Sorangium* α-lytic protease, **19**, 140, 599 (activation and inactivation, **19**, 609; amino acid sequence and composition, **19**, 606, 610; assay methods for, **19**, 600; crystallographic properties, **19**, 609; distribution, **19**, 612; kinetic properties, **19**, 612; physical properties, **19**, 608; production and purification, **19**, 601; purity, **19**, 608; specificity, **19**, 609; stability, **19**, 608)

β-LYTIC METALLOENDOPEPTIDASE

This zinc-dependent metalloendopeptidase [EC 3.4.24.32], also known as achromopeptidase component and β-lytic endopeptidase, catalyzes the cleavage of *N*-acetylmuramoyl-∇-Ala and Gly-∇-(ε-amino)Lys bonds in bacterial cell walls, ∇ indicates the cleavage site. The enzyme also catalyzes hydrolysis of casein and the Gly23–Phe24 and, more slowly, the Val18–Cys19-(SO_3^-) of the oxidized insulin B chain. The synthetic substrates FA-Gly-∇-Leu-NH$_2$ and Z-Gly-∇-Phe-NH$_2$ are also hydrolyzed.[1] *See also Myxobacter AL-1 Protease; Staphylolysin*

[1] E. Kessler (1995) *MIE* **248**, 740.

Selected entries from *Methods in Enzymology* [vol, page(s)]:
General discussion: 248, 740

Assay: 248, 741, 742
Other molecular properties: *Achromobacter*, **248**, 224, 740 (amino acid sequence, **248**, 750; molecular weight, **248**, 751; properties, **248**, 747, 751; purification, **248**, 743; synthesis, **248**, 751); activity, **248**, 740; *Aeromonas hydrophila*, **248**, 224, 740 (amino acid sequence, **248**, 750; bacteriolytic activity, **248**, 755; cleavage specificity, **248**, 755; inhibitors, **248**, 756; kinetic constants, **248**, 756; molecular weight, **248**, 755; physicochemical properties, **248**, 755; properties, **248**, 747, 755; purification, **248**, 746; stability, **248**, 756); amino acid sequence, **248**, 749; classification, **248**, 756); *Lysobacter enzymogenes*, **248**, 740 (action on isolated cell walls, **248**, 749; amino acid sequence, **248**, 749; bacteriolytic activity, **248**, 749 [assay, **248**, 742]; elastolytic activity, **248**, 751; inhibitors, **248**, 751; properties, **248**, 747; proteolytic activity, **248**, 751; purification, **248**, 743; spectrophotometric assay, **248**, 742); occurrence, **248**, 740; properties, **248**, 223, 747; purification, **248**, 743; *Sorangium* β-lytic protease, AL-1 protease compared to, **19**, 599

D-LYXOSE KETOL-ISOMERASE

This isomerase [EC 5.3.1.15], also known as lyxose ketol-isomerase, catalyzes the interconversion of D-lyxose and D-xylulose.[1,2] D-Mannose is also converted to D-fructose.

[1] E. A. Noltmann (1972) *The Enzymes*, 3rd ed., **6**, 271.
[2] R. L. Anderson & D. P. Allison (1965) *JBC* **240**, 2367.

Selected entries from *Methods in Enzymology* [vol, page(s)]:
General discussion: 9, 593
Assay: 9, 593
Other molecular properties: *Aerobacter aerogenes*, from (activators, **9**, 596; molecular weight, **9**, 596; pH optimum, **9**, 596; properties, **9**, 596; purification, **9**, 594; specificity, **9**, 596)

MACERASE

The proprietory name for the pectinase from *Rhizopus* sp. used for isolating intact plant cells and for preparing protoplasts.[1] **See also** *Polygalacturonase*

[1] I. Takebe (1968) *Plant Cell Physiol.* **9**, 115.

MACROCIN *O*-METHYLTRANSFERASE

This methyltransferase [EC 2.1.1.101] catalyzes the reaction of *S*-adenosyl-L-methionine with macrocin to produce *S*-adenosyl-L-homocysteine and tylosin. The enzyme also catalyzes lactenosin methylation to generate desmyocin. The *Streptomyces fradiae* enzyme has an ordered Bi Bi kinetic mechanism.[1]

[1] N. J. Bauer, A. J. Kreuzman, J. E. Dotzlaf & W. K. Yeh (1988) *JBC* **263**, 15619.

MACROLIDE 2′-KINASE

This phosphotransferase [EC 2.7.1.136] catalyzes the ATP-dependent phosphorylation of oleandomycin, yielding ADP and oleandomycin 2′-*O*-phosphate. Other substrates include erythromycin A, tylosin, midekamycin, leucomycin A3, and spiramycin.[1-3]

[1] P. F. Wiley, L. Baczynskyj, L. A. Dolak, J. I. Cialdella & V. P. Marshall (1987) *J. Antibiot.* **40**, 195.
[2] K. O'Hara, T. Kanda & M. Kono (1988) *J. Antibiot.* **41**, 823.
[3] K. O'Hara, T. Kanda, K. Ohmiya, T. Ebisu & M. Kono (1989) *Antimicrob. Agents Chemother.* **33**, 1354.

MAGAININASE

This metallopeptidase of *Xenopus laevis* catalyzes the hydrolysis of a single Xaa–Lys peptide bond in magainin peptide precursors. The lysyl residue must be precisely positioned relative to the hydrophobic face of the α helix (of at least twelve residues in length).[1]

[1] N. M. Resnick, W. L. Maloy, H. R. Guy & M. Zasloff (1991) *Cell* **66**, 541.

MAGNESIUM-PROTOPORPHYRIN *O*-METHYLTRANSFERASE

This transferase [EC 2.1.1.11] catalyzes the reaction of *S*-adenosyl-L-methionine with magnesium protoporphyrin (or the corresponding zinc and calcium ion species) to produce *S*-adenosyl-L-homocysteine and magnesium (or zinc or calcium) protoporphyrin monomethyl ester.[1,2] Exchange between unlabeled *S*-adenosyl-L-methionine and labeled *S*-adenosyl-L-homocysteine is consistent with a ping pong-type mechanism for the wheat enzyme.[3]

[1] K. D. Gibson, A. Neuberger & G. H. Tait (1963) *Biochem. J.* **88**, 325.
[2] S. Granick & S. I. Beale (1978) *AE* **46**, 33.
[3] W. C. Yee, S. J. Eglsaer & W. R. Richards (1989) *BBRC* **162**, 483.

Selected entries from ***Methods in Enzymology*** [vol, page(s)]:
General discussion: 17A, 222

MAGNOLYSIN

This metalloendopeptidase [EC 3.4.24.62], known variously as pro-oxytocin convertase, pro-oxytocin/neurophysin-converting enzyme, and neurosecretory granule protease, catalyzes the hydrolytic processing of oxytocin and neurophysin at the Arg–Ala bond in the Gly-Lys-Arg-Ala-Val motif. The enzyme prefers basic L-aminoacyl doublets in both positions P1 and P2, and Lys-Arg-∇-Xaa appears to be the most effective substrate.[1,2]

[1] M. Rholam, N. Brakch, D. Germain, D. Thomas, C. Fahy, H. Boussetta, G. Boileau & P. Cohen (1995) *EJB* **227**, 707.
[2] N. Brakch, M. Rholam, H. Boussetta & P. Cohen (1993) *Biochemistry* **32**, 4925.

MALATE DEHYDROGENASE

Several different enzymes catalyze the oxidation of malate to oxaloacetate, and others facilitate decarboxylation to generate pyruvate.

Malate dehydrogenase [EC 1.1.1.37], also called malic dehydrogenase, catalyzes the reversible reaction of (*S*)-malate (or, L-malate) with NAD^+ to produce oxaloacetate and NADH.[1-4] Other 2-hydroxydicarboxylic acids, such as 2-hydroxymalonate, can act as substrates as well. This oxidoreductase has an ordered Bi Bi kinetic mechanism[3,4] with the coenzyme binding first (partially ordered at pH 9 with the porcine heart enzyme[4]).

Malate dehydrogenase (NADP$^+$) [EC 1.1.1.82] catalyzes the reaction of (S)-malate with NADP$^+$ to produce oxaloacetate and NADPH.[5–7] The plant enzyme is activated by light.

Malate dehydrogenase (acceptor) [EC 1.1.99.16] catalyzes the FAD-dependent reaction of (S)-malate with an acceptor substrate to produce oxaloacetate and the reduced acceptor.[8,9] 2,6-Dichloroindophenol can serve as the electron acceptor.

Malate dehydrogenase (oxaloacetate-decarboxylating) [EC 1.1.1.38], also referred to as malic enzyme and pyruvic-malic carboxylase, catalyzes the reaction of (S)-malate with NAD$^+$ to produce pyruvate, carbon dioxide, and NADH.[10–12] The enzyme will also decarboxylate added oxaloacetate.

Malate dehydrogenase (decarboxylating) [EC 1.1.1.39], also known as the malic enzyme, catalyzes the reaction of (S)-malate with NAD$^+$ to produce pyruvate, carbon dioxide, and NADH. This oxidoreductase will not decarboxylate added oxaloacetate.[13–15] The cauliflower enzyme has a random kinetic mechanism.[13]

Malate dehydrogenase (oxaloacetate-decarboxylating, NADP$^+$) [EC 1.1.1.40], also referred to as malic enzyme and pyruvic-malic carboxylase, catalyzes the reaction of (S)-malate with NADP$^+$ to produce pyruvate, carbon dioxide, and NADPH.[16] The enzyme will also decarboxylate added oxaloacetate.

D-Malate dehydrogenase (decarboxylating) [EC 1.1.1.83] catalyzes the reaction of (R)-malate (or, D-malate) with NAD$^+$ to produce pyruvate, carbon dioxide, and NADH.[17–19] The Rhodopseudomonas sphaeroides enzyme also exhibits an NAD$^+$-dependent L-tartrate dehydrogenase [EC 1.1.1.93] activity, producing oxaloglycolate and NADH.[18]

[1] A. R. Clarke & T. R. Dafforn (1998) CBC 3, 1.
[2] J. V. Schloss & M. S. Hixon (1998) CBC 2, 43.
[3] K. E. Crow, T. J. Braggins & M. J. Hardman (1983) ABB 225, 621.
[4] E. Silverstein & G. Sulebele (1969) Biochemistry 8, 2543.
[5] K. Fickenscher & R. Scheibe (1988) ABB 260, 771.
[6] T. Kagawa & P. L. Bruno (1988) ABB 260, 674.
[7] N. Ferte, J.-P. Jaquot & J.-C. Meunier (1986) EJB 154, 587.
[8] T. L. P. Reddy, P. S. Murthy & T. A. Venkitasubramanian (1975) BBA 376, 210.
[9] K. Imai & A. F. Brodie (1973) JBC 248, 7487.
[10] M. Iwakura, M. Tokushige, H. Katsuki & S. Muramatsu (1978) J. Biochem. 83, 1387.
[11] M. Yamaguchi, M. Tokushige, K. Takeo & H. Katsuki (1974) J. Biochem. 76, 1259.
[12] M. Schütz & F. Radler (1973) Arch. Mikrobiol. 91, 183.
[13] P. F. Canellas & R. T. Wedding (1984) ABB 229, 414.
[14] R. T. Wedding & M. K. Black (1983) Plant Physiol. 72, 1021.
[15] S. D. Grover, P. F. Canellas & R. T. Wedding (1981) ABB 209, 396.
[16] G. E. Edwards & C. S. Andreo (1992) Phytochemistry 31, 1845.
[17] M. Lähdesmäki & P. Mäntsälä (1980) BBA 613, 266.
[18] F. Gifforn & A. Kuhn (1983) J. Bacteriol. 155, 281.
[19] W. Knichel & F. Radler (1982) EJB 123, 547.

Selected entries from *Methods in Enzymology* [vol, page(s)]:

General discussion: 1, 735; 13, 99, 106, 116, 123, 129, 135, 141, 145, 148; 31, 642, 768; 34, 246, 252; 46, 21, 66; malate dehydrogenase (decarboxylating), 1, 739, 748; 13, 230

Assay: 1, 735; 4, 378, 379; 10, 457; 13, 99, 106, 116, 123, 129, 135, 141, 145, 148; 44, 458, 474; 57, 181, 210; malate dehydrogenase (acceptor), 13, 129; malate dehydrogenase (decarboxylating), 1, 739, 748; 13, 230; malate dehydrogenase (NADP$^+$), 252, 243

Other molecular properties: acetyl coenzyme A assay, 5, 466; 13, 501, 545; activation, 13, 105; 87, 654; activity, 57, 181; affinity partitioning with dye ligands, 228, 126, 128; *Ascaris suum*, 9, 15 (conformational states, in reaction sequence, 249, 325; isotope trapping, 64, 51, 52; partition analysis, 249, 324); assays using malate dehydrogenase for (acetate, 35, 299, 302; acetylcarnitine, 13, 509; acetylcarnitine, 14, 617; acetyl-CoA, 5, 466; 13, 501, 545; acetyl-CoA synthetase, 13, 375; L-asparaginase, 44, 690; aspartate, 13, 473; binding, to agarose-bound nucleotides, 66, 196; carboxyltransferase, 35, 32; carnitine acetyltransferase, 13, 388; citrate, 13, 450; 55, 211; citrate synthase, 13, 3, 7, 12, 20, 365; coenzyme A, 13, 540; 55, 210; fumarate, 13, 463, 13, 465; 13, 529; α-ketoglutarate, 13, 458; 55, 206; long-chain fatty acyl-CoA, 35, 273; malate, 13, 466, 529; 57, 187; lactate:malate transhydrogenase, 1, 748; 13, 262; methylmalonyl-coenzyme A mutase, 13, 208; methylmalonyl-coenzyme A racemase, 13, 195; oxaloacetate, 13, 468; 55, 207; 57, 187; 63, 33; oxaloacetate transcarboxylase, 13, 215; phosphoenolpyruvate carboxykinase, 13, 270; phosphoenolpyruvate carboxylase, 13, 277, 283, 288, 292; phosphoenolpyruvate carboxykinase, 63, 33; phosphoenolpyruvate carboxytransphosphorylase, 13, 297; pyruvate carboxylase, 13, 235, 250, 251, 258); bacterial particles and, 5, 55; brain mitochondria, 55, 58; carboxymethylation of, 25, 431; channelling, 308, 142; chiral methyl group analysis, 87, 137, 142; citrate-cleavage enzyme and, 5, 641, 644; citrate synthase, and, 308, 142; comparison, from different sources, 13, 115, 147; component of multienzyme complex, 71, 60; condensing enzyme assay and, 1, 687; conformational change, 87, 404; conformational states, in reaction sequence, 249, 325; crotonyl coenzyme A assay and, 1, 562; crystallization, 22, 250; enzyme electrode, in, 44, 595; equilibrium perturbation, 64, 104, 105, 122, 123; *Euglena* cultures, in, 23, 78; ferricyanide and, 4, 332; fluorescence, 26, 503; fluorocitrate synthesis, 13, 655; glutamate:aspartate transaminase and, 5, 677; glyoxysomes, in, 31, 568, 569; 72, 788; half-site reactivity, 64, 184; heart, in, 3, 430; histochemistry, 6, 892; hormone receptor studies, 37, 212; immobilization (in multistep enzyme system, 44, 181, 182, 457, 472; on nylon tube, 44, 120); isocitrate dehydrogenase preparations, malic enzyme in, 3, 430; isolated nuclei, 12A, 445; isotope effect, 87, 397, 404, 634; isotope exchange, 64, 8, 23, 25, 26 (kinetic patterns for modifier action, 87, 653); isotope trapping, 64, 51, 52; kinetics, in multistep enzyme system, 44, 434, 466; lactate dehydrogenase contamination of malic enzyme, 13, 235; labeling, 46, 139; malate dehydrogenase (acceptor) (anion activation, 13, 133; assay, 13, 129, 135; electron transport inhibitors, 13, 133; *Mycobacterium phlei*

ETP, **55**, 184, 185, 186); malate dehydrogenase (decarboxylating), **1**, 739, 748; **13**, 230 (coenzyme binding, **13**, 235; contamination with lactate dehydrogenase, **13**, 235; distribution, **1**, 747; equilibrium perturbation, **64**, 104, 105, 122, 123; properties, **1**, 746, 752; **13**, 234; purification [*Lactobacillus arabinosis*, **1**, 749; pigeon liver, **1**, 742; **13**, 231; wheat germ, **1**, 744]; stereospecificity, **87**, 106; Theorell-Chance mechanism, **87**, 355); malate dehydrogenase (FAD-linked) (anion activation, **13**, 133; assay, **13**, 129, 135; electron transport inhibitors, **13**, 133); malate dehydrogenase (NADP$^+$) (activation by thioredoxin, **252**, 240 [dithiothreitol system, **252**, 244; NADP$^+$ inhibition, **252**, 249; photochemical activation, **252**, 244]; assay, **252**, 243; C405A/C417A mutant protein [activation kinetics, **252**, 252; purification, **252**, 251; site-directed mutagenesis, **252**, 249; storage, **252**, 252; thioredoxin reductase assay, **252**, 250]; cloning by PCR, **252**, 245; purification, C$_4$ plant enzyme [affinity chromatography, **252**, 242; ammonium sulfate fractionation, **252**, 242; extraction from leaves, **252**, 241; hydrophobic interaction chromatography, **252**, 243; storage, **252**, 243]; purification from *Escherichia coli* expression system [affinity chromatography, **252**, 248; ammonium sulfate fractionation, **252**, 248; cell growth, **252**, 247; N-terminal sequencing, **252**, 248]; stereospecificity, **87**, 110); malate dehydrogenase (oxaloacetate-decarboxylating) (NADP$^+$) (marker for chromosome detection in somatic cell hybrids, as, **151**, 187; stereospecificity, **87**, 106); malate:vitamin K reductase, *Mycobacterium phlei* ETP, **55**, 184, 185, 186; marker enzyme for peroxisomes, as, **23**, 681; marker for chromosome detection in somatic cell hybrids, as, **151**, 187; matrix, **55**, 101, 104 (assay, **55**, 102); membrane fraction bound, **10**, 441, 457; microbodies, in, **52**, 496, 499; mitochondrial subfractions, **55**, 94, 99; mitoplasts, in, **31**, 321, 322; multienzymatic NADH recycling, in, **136**, 70, 74; NAD-coimmobilized (assay, **136**, 28; L-malate production with, **136**, 29; preparation, **136**, 26); oxaloacetate assay, **5**, 618; **63**, 33; oxaloacetate decarboxylase activity, **1**, 741; **87**, 404; parenchyma cells, in, **32**, 706; partition analysis, **249**, 324; pea epicotyls, **13**, 148; pH effects, **87**, 361, 394, 653; phosphoenolpyruvate carboxykinase, assay of, **63**, 33; 6-phosphogluconate dehydrogenase, malic enzyme in preparation of, **6**, 800; preparation from (*Acetobacter xylinum*, **13**, 130; *Bacillus subtilis*, **13**, 142; beef heart, **13**, 100, 123; beef kidney, **13**, 118, 120; chicken heart, **13**, 107; *Escherichia coli*, **13**, 145; *Micrococcus denitrificans*, **17A**, 983; NADP$^+$, activation, thioredoxin *m* assay, **69**, 385, 386, 388; pea epicotyl, **13**, 148; pig heart, **1**, 726; *Pseudomonas ovalis*, **13**, 137; tuna heart, **13**, 113); properties, **1**, 738; **13**, 104, 115, 122, 128, 132, 140, 144, 147, 150; *Pseudomonas ovalis*, **13**, 135; rapid equilibrium, **87**, 361; reduction of NAD$^+$ analogues, **44**, 873, 874; rhein, **55**, 459; *Rhodopseudomonas spheroides*, purification, **104**, 111; sedimentation coefficient, **61**, 108; solvent perturbation, **87**, 404; stain for, **224**, 109; staining on gels, **22**, 595; starch gel electrophoresis, **6**, 967, 968; steroid hydroxylation, **55**, 10; stereochemistry, **87**, 105, 106, 110, 148; sticky substrates, **87**, 396, 404; substrate channelling, **308**, 142; substrate specificity, **13**, 104, 112, 144; subunits, **13**, 145; succinate dehydrogenase and, **5**, 599, 613; synthesis of (deuterated citrate, **13**, 586; deuterated malate, **13**, 582); Theorell-Chance mechanism, **87**, 355; toluene-treated mitochondria, **56**, 547; transition-state and multisubstrate analogues, **249**, 304; tryptophan content, **13**, 106, 112, 116

MALATE OXIDASE

This FAD-dependent oxidoreductase [EC 1.1.3.3], also referred to as malic oxidase, catalyzes the reaction of (S)-malate (or, L-malate) with dioxygen to produce oxaloacetate and hydrogen peroxide.[1]

[1]S. Narindrasorasak, A. H. Goldie & B. D. Sanwal (1979) *JBC* **254**, 1540.

Selected entries from ***Methods in Enzymology*** [**vol**, page(s)]:
General discussion: **13**, 135; **52**, 21; **56**, 474
Assay: **13**, 135

MALATE SYNTHASE

This enzyme [EC 4.1.3.2], also known as malate condensing enzyme and glyoxylate transacetylase, catalyzes the reaction of acetyl-CoA with glyoxylate and water to produce (S)-malate and coenzyme A.[1–5]

[1]B. R. Howard, J. A. Endrizzi & S. J. Remington (2000) *Biochemistry* **39**, 3156.
[2]R. Kluger (1992) *The Enzymes*, 3rd ed., **20**, 271.
[3]D. J. Creighton & N. S. R. K. Murthy (1990) *The Enzymes*, 3rd ed., **19**, 323.
[4]H. Durchschlag, G. Biedermann & H. Eggerer (1981) *EJB* **114**, 255.
[5]M. J. P. Higgins, J. A. Kornblatt & H. Rudney (1972) *The Enzymes*, 3rd ed., **7**, 407.

Selected entries from ***Methods in Enzymology*** [**vol**, page(s)]:
General discussion: **5**, 633; **13**, 362, 365
Assay: **5**, 633; **87**, 133; polarographic, **13**, 365; radioactive, **13**, 362
Other molecular properties: chiral methyl groups, and, **87**, 129, 147, 159; glyoxysomes, in, **31**, 565, 569, 743; **72**, 788; kinetic isotope effects, **87**, 130; marker enzyme, as, **31**, 735, 743; microbodies, in, **52**, 496; properties, **5**, 636; purification, **87**, 134 (bakers' yeast, **5**, 634); small-angle X-ray scattering studies, **61**, 235; stability, **87**, 136; stereochemistry, **87**, 147, 156, 157

MALEATE HYDRATASE

This enzyme [EC 4.2.1.31] catalyzes the reaction of maleate (*i.e.*, *cis*-butenedioate) with water to produce (R)-malate.[1,2]

[1]J. S. Britten, H. Morell & J. V. Taggart (1969) *BBA* **185**, 220.
[2]J.-L. Dreyer (1985) *EJB* **150**, 145.

Selected entries from ***Methods in Enzymology*** [**vol**, page(s)]:
Other molecular properties: malate formation, **13**, 584

MALEATE ISOMERASE

This enzyme [EC 5.2.1.1] catalyzes the interconversion of maleate and fumarate.[1–3]

[1]S. Seltzer (1972) *The Enzymes*, 3rd ed., **6**, 381.
[2]Y. Takamura, T. Takamura, M. Soejima & T. Uemura (1969) *Agric. Biol. Chem.* **33**, 718.
[3]W. Scher & W. B. Jacoby (1969) *JBC* **244**, 1878.

Selected entries from ***Methods in Enzymology*** [**vol**, page(s)]:
Other molecular properties: staining on gels, **22**, 585, 593

MALEIMIDE HYDROLASE

This enzyme [EC 3.5.2.16], also called imidase and cyclic imide hydrolase, catalyzes the reversible reaction of maleimide with water to produce maleamate.[1] Other substrates include succinimide, glutarimide, and sulfur-containing cyclic imides such as rhodanine and 2,4-thiazolidinedione. Dihydrouracil and hydantoin are poor substrates.

[1]J. Ogawa, C. L. Soong, M. Honda & S. Shimizu (1997) *EJB* **243**, 322.

MALEYLACETATE REDUCTASE

This oxidoreductase [EC 1.3.1.32], also referred to as maleolylacetate reductase, catalyzes the reaction of 2-maleylacetate ($^-$OOCCH$_2$COCH=CHCOO$^-$) with NAD(P)H to produce β-ketoadipate (or, 3-oxoadipate) and NAD(P)$^+$.[1] 2-Chloromaleylacetate and other intermediates formed from the degradation of chloroaromatic compounds are acted upon by this reductase.

[1] A. B. Gaal & H. Y. Neujahr (1980) *Biochem. J.* **185**, 783.

MALEYLACETOACETATE ISOMERASE

This enzyme [EC 5.2.1.2] catalyzes the interconversion of 4-maleylacetoacetate and 4-fumarylacetoacetate.[1–4] Maleylpyruvate can also act as a substrate. The enzyme from some sources requires glutathione.

[1] J. V. Schloss & M. S. Hixon (1998) *CBC* **2**, 43.
[2] K. T. Douglas (1987) *AE* **59**, 103.
[3] W. S. Morrison, G. Wong & S. Seltzer (1976) *Biochemistry* **15**, 4229.
[4] S. Seltzer (1972) *The Enzymes*, 3rd ed., **6**, 381.

Selected entries from *Methods in Enzymology* [vol, page(s)]:
General discussion: 2, 295
Assay: 2, 295

MALEYLPYRUVATE ISOMERASE

This enzyme [EC 5.2.1.4] catalyzes the interconversion of 3-maleylpyruvate and 3-fumarylpyruvate.[1–3] Maleylacetoacetate and fumarylacetoacetate are also interconverted. Glutathione is a required in many organisms.

[1] K. T. Douglas (1987) *AE* **59**, 103.
[2] S. R. Hagedorn, G. Bradley & P. J. Chapman (1985) *J. Bacteriol.* **163**, 640.
[3] S. Seltzer (1972) *The Enzymes*, 3rd ed., **6**, 381.

MALONATE CoA-TRANSFERASE

This transferase [EC 2.8.3.3] catalyzes the reversible reaction of acetyl-CoA with malonate to produce acetate and malonyl-CoA. The *Pseudomonas ovalis* enzyme also exhibits a malonyl-CoA decarboxylase [EC 4.1.1.9] activity.[1]

[1] Y. Takamura & Y. Kitayama (1981) *Biochem. Int.* **3**, 483.

MALONATE-SEMIALDEHYDE DEHYDRATASE

This dehydratase [EC 4.2.1.27] catalyzes the reversible conversion of 3-oxopropanoate (*i.e.*, malonate semialdehyde, or $^-$OOCCH$_2$CHO) to propynoate ($^-$OOCC≡CH) and water.[1]

[1] E. W. Yamada & W. B. Jakoby (1959) *JBC* **234**, 941.

MALONATE-SEMIALDEHYDE DEHYDROGENASE

This oxidoreductase [EC 1.2.1.15] catalyzes the reaction of 3-oxopropanoate (or, malonate semialdehyde) with NAD(P)$^+$ and water to produce malonate and NAD(P)H.[1]

[1] K. Nakamura & F. Bernheim (1961) *BBA* **50**, 147.

MALONATE-SEMIALDEHYDE DEHYDROGENASE (ACETYLATING)

This oxidoreductase [EC 1.2.1.18] catalyzes the reaction of 3-oxopropanoate (or, malonate semialdehyde) with coenzyme A and NAD(P)$^+$ to produce acetyl-CoA, carbon dioxide, and NAD(P)H.[1–5] Other substrates include 2-methyl-3-oxopropanoate, which generates propionyl-CoA.[1]

[1] G. W. Goodwin, P. M. Rougraff, E. J. Davis & R. A. Harris (1989) *JBC* **264**, 14965.
[2] W. B. Jakoby (1963) *The Enzymes*, 2nd ed., **7**, 203.
[3] O. Hayaishi, Y. Nishizuka, M. Tatibana, M. Takeshita & S. Kuno (1961) *JBC* **236**, 781.
[4] E. W. Yamada & W. B. Jakoby (1960) *JBC* **235**, 589.
[5] O. W. Griffith (1986) *Ann. Rev. Biochem.* **55**, 855.

MALONYL-CoA CARBOXYLTRANSFERASE

This enzyme, previously classified as EC 2.1.3.4, is now a deleted Enzyme Commission entry.

MALONYL-CoA DECARBOXYLASE

This decarboxylase [EC 4.1.1.9], also known as malonyl-CoA carboxy-lyase, catalyzes the conversion of malonyl-CoA to acetyl-CoA and carbon dioxide.[1–4] The *Pseudomonas ovalis* enzyme also exhibits a malonate CoA-transferase [EC 2.8.3.3] activity.[2]

[1] A. R. Hunaiti & P. E. Kolattukudy (1984) *ABB* **229**, 426.
[2] Y. S. Kim & P. E. Kolattukudy (1978) *ABB* **190**, 234.
[3] A. H. Koeppen, E. J. Mitzen & A. A. Abdel (1974) *Biochemistry* **13**, 3589.
[4] C. Landriscina, G. V. Gnoni & E. Quagliariello (1971) *EJB* **19**, 573.

Selected entries from *Methods in Enzymology* [vol, page(s)]:
General discussion: 35, 75, 81; **71**, 150
Assay: spectrophotometric, **71**, 151
Other molecular properties: contaminant of fatty acid synthase, **71**, 106; duck, **71**, 155; fatty acid synthase, presence in preparation of, **5**, 449; geese, **71**, 152, 155; interference with assay for fatty acid synthase, **35**, 75; *Mycobacterium tuberculosis*, **71**, 150, 160 (inhibitors, **71**, 162; properties, **71**, 162; substrate specificity, **71**, 162); rat liver mitochondria, **71**, 150, 157 (extraction of enzyme, **71**, 158; inhibitors, **71**, 160; preparation of mitochondria, **71**, 157; subcellular localization, **71**, 160; substrate specificity, **71**, 160); rat mammary gland, **71**, 150; water fowl, **71**, 150, 152, 155 (amino acid composition, **71**, 155; inhibitors, **71**, 155; kinetic properties, **71**, 155; stability, **71**, 154; subcellular localization, **71**, 157; substrate specificity, **71**, 155; subunit structure, **71**, 156)

MALTASE

This enzyme [EC 3.2.1.20] catalyzes the hydrolysis of terminal, nonreducing 1,4-linked D-glucosyl residues, resulting in the release of D-glucose. The enzyme also catalyzes the hydrolysis of maltose to produce two molecules of D-glucose.[1–3]

While this EC entry indicates a set of enzymes favoring exohydrolysis of 1,4-α-glucosidic linkages, some variants hydrolyze oligosaccharides more rapidly than polysaccharides. The intestinal enzyme exhibits a glucan 1,4-α-glucosidase [EC 3.2.1.3] activity. **See** α-*Glucosidase*

[1] E. H. Van Beers, H. A. Büller, R. J. Grand, A. W. C. Einerhand & J. Dekker (1995) *Crit. Rev. Biochem. Mol. Biol.* **30**, 197.
[2] G. Mooser (1992) *The Enzymes*, 3rd ed., **20**, 187.
[3] M. Vihinen & P. Mäntsälä (1989) *Crit. Rev. Biochem. Mol. Biol.* **24**, 329.

MALTODEXTRIN PHOSPHORYLASE

This variant of phosphorylase [EC 2.4.1.1] catalyzes the reversible reaction of maltodextrin with orthophosphate to produce α-D-glucose 1-phosphate and a maltodextrin containing one less glucosyl unit.[1–4] Short substrates such as maltoheptaose are also hydrolyzed, but the minimum length of the substrate is five glucosyl units. The enzyme catalyzes a random Bi Bi kinetic mechanism.

Escherichia coli maltodextrin phosphorylase differs from mammalian glycogen phosphorylase by not being regulated by phosphorylation or allosteric effectors; nevertheless, it is catalytically quite similar and utilizes pyridoxal 5′-phosphate as a cofactor. **See also** *Glycogen Phosphorylase; Phosphorylase; Starch Phosphorylase*

[1] R. Schinzel & B. Nidetzky (1999) *FEMS Microbiol. Lett.* **171**, 73.
[2] F. Bartl, D. Palm, R. Schinzel & G. Zundel (1999) *Eur. Biophys. J.* **28**, 200.
[3] D. Linder, G. Kurz, H. Bender & K. Wallenfels (1976) *EJB* **70**, 291.
[4] S. Tanabe, M. Kobayashi & K. Matsuda (1988) *Agric. Biol. Chem.* **52**, 757.

MALTOSE O-ACETYLTRANSFERASE

This acetyltransferase [EC 2.3.1.79] catalyzes the reaction of acetyl-CoA with maltose to produce coenzyme A and acetylmaltose. The enzyme is not identical with galactoside O-acetyltransferase [EC 2.3.1.18].[1,2] Other acceptor substrates include D-glucose, D-mannose, and D-galactose.

[1] S. Freundlieb & W. Boos (1982) *Ann. Microbiol.* **133A**, 181.
[2] B. Brand & W. Boos (1991) *JBC* **266**, 14113.

MALTOSE α-D-GLUCOSYLTRANSFERASE

This transferase [EC 5.4.99.16], also known as trehalose synthase and maltose glucosylmutase, catalyzes the interconversion of maltose and α,α-trehalose.[1] It was first isolated from *Pseudomonas putida* and *Pimelobacter* sp.

[1] T. Nishimoto, M. Nakano, S. Ikegami, H. Chaen, S. Fukuda, T. Sugimoto, M. Kurimoto & Y. Tsujisaka (1995) *Biosci. Biotech. Biochem.* **59**, 2189.

MALTOSE 3-GLYCOSYLTRANSFERASE

This transferase, previously classified as EC 2.4.1.6, is now a deleted Enzyme Commission entry.

MALTOSE-6′-PHOSPHATE GLUCOSIDASE

This glucosidase [EC 3.2.1.122], also known as phospho-α-glucosidase, catalyzes the hydrolysis of maltose 6′-phosphate to produce D-glucose 6-phosphate and D-glucose. Other substrates include α′, α-trehalose 6-phosphate, sucrose 6-phosphate, and p-nitrophenyl-α-D-glucopyranoside 6-phosphate.[1]

[1] J. Thompson, C. R. Gentry-Weeks, N. Y. Nguyen, J. E. Folk & S. A. Robrish (1995) *J. Bacteriol.* **177**, 2505.

MALTOSE PHOSPHORYLASE

This phosphorylase [EC 2.4.1.8] catalyzes the reversible reaction of maltose with orthophosphate to produce D-glucose and β-D-glucose 1-phosphate.[1–4] In the presence of arsenate, maltose is converted to two molecules of D-glucose.

[1] J. J. Mieyal & R. H. Abeles (1972) *The Enzymes*, 3rd ed., **7**, 515.
[2] M. Doudoroff (1961) *The Enzymes*, 2nd ed., **5**, 229.
[3] C. Fitting & M. Doudoroff (1952) *JBC* **199**, 153.
[4] M. Cohn (1961) *The Enzymes*, 2nd ed., **5**, 179.

Selected entries from *Methods in Enzymology* [vol, page(s)]:
General discussion: 1, 229; **230**, 304
Assay: 1, 229

MALTOSE SYNTHASE

This enzyme [EC 2.4.1.139], first isolated from spinach (*Spinacia oleracea*), catalyzes the reaction of two molecules of α-D-glucose 1-phosphate to produce maltose and two molecules of orthophosphate.[1]

[1] N. Schilling (1982) *Planta* **154**, 87.

MALTOSE-TRANSPORTING ATPase

This so-called ATPase [EC 3.6.3.19] catalyzes the ATP-dependent (ADP- and orthophosphate-producing) transport of maltose$_{out}$ to produce maltose$_{in}$.[1,2] The bacterial system will also catalyze the import of maltose oligosaccharides. This ABC-type (ATP-binding cassette-type) ATPase has two similar ATP-binding domains and does not undergo phosphorylation during the transport process.

The idea that a transporter is an enzyme is in keeping with a new definition of enzyme catalysis as the facilitated making/breaking of chemical bonds, not just covalent bonds.[3] This idea builds on Pauling's assertion that any long-lived, chemically distinct interaction (in this case, the persistent location of a solute with respect to the faces of a membrane) can be regarded as a chemical bond. Note also that the equilibrium constant ($K_{eq} = $ [maltose$_{in}$][ADP][P$_i$]/[maltose$_{out}$][ATP]) does not conform to that expected for an ATPase (i.e., $K_{eq} = $ [ADP][P$_i$]/[ATP]). Thus, although the overall reaction yields ADP and orthophosphate, the enzyme is misclassified as a hydrolase, and instead should be regarded as an energase-type reaction. Energases facilitate affinity-modulated reactions by coupling the $\Delta G_{ATP\text{-hydrolysis}}$ to a force-generating or work-producing step.[3] In this case, P-O-P bond scission supplies the energy to drive maltose import. **See ATPase; Energase**

[1]C. F. Higgins (1992) Ann. Rev. Cell Biol. **8**, 67.
[2]E. Dassa & S. Muir (1993) Mol. Microbiol. **7**, 29.
[3]D. L. Purich (2001) TiBS **26**, 417.

MALYL-CoA LYASE

This enzyme [EC 4.1.3.24] catalyzes the reversible conversion of (3S)-3-carboxy-3-hydroxypropanoyl-CoA (or, malyl-CoA) to acetyl-CoA and glyoxylate.[1,2]

[1]L. B. Hersh (1973) JBC **248**, 7295 and (1974) **249**, 5208.
[2]A. J. Hacking & J. R. Quayle (1974) Biochem. J. **139**, 399.

Selected entries from **Methods in Enzymology** [vol, page(s)]:
General discussion: 188, 361, 379, 386
Assay: 188, 379; in crude extracts of methylotrophs, 188, 364

MALYL-CoA SYNTHETASE

This enzyme [EC 6.2.1.9], also known as malate:CoA ligase and malate thiokinase, catalyzes the reversible reaction of ATP with malate and coenzyme A to produce ADP, orthophosphate and malyl-CoA.[1-4] Succinate can also act as the acceptor substrate. The enzyme is thought to operate by means of a random kinetic mechanism.[1]

[1]K. K. Surendranathan & L. B. Hersch (1983) JBC **258**, 3794.
[2]L. B. Hersh & K. K. Surendranathan (1982) JBC **257**, 11633.
[3]L. B. Hersh & M. Peet (1981) JBC **256**, 1732.
[4]M. Elwell & L. B. Hersh (1979) JBC **254**, 2434.

MANDELAMIDE AMIDASE

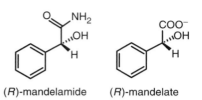

(R)-mandelamide   (R)-mandelate

This amidase [EC 3.5.1.86] catalyzes the hydrolysis of (R)-mandelamide to produce (R)-mandelate and ammonia.[1]

[1]K. Yamamoto, K. Oishi, I. Fujimatsu & K. Komatsu (1991) Appl. Environ. Microbiol. **57**, 3028.

MANDELATE DEHYDROGENASE

This oxidoreductase, first isolated from *Pseudomonas fluorescens*, reportedly catalyzes the reaction of (S)-mandelate (or, L-mandelate) with 0.5 dioxygen to produce benzoylformate and water.[1]

(S)-mandelate   benzylformate

An NAD$^+$-dependent enzyme that acts on the D-stereoisomer has been reported in a Gram-positive organism[2] and in *Rhodotorula graminis*[3] (which also has an NAD$^+$-independent L-mandelate dehydrogenase[4]). Both NAD(P)$^+$-independent D- and L-specific enzymes have been isolated from *Acinetobacter calcoaceticus*.[5] The D-mandelate enzyme is very similar to D-lactate dehydrogenase.

[1]R. Y. Stanier, C. F. Gunsalus & I. C. Gunsalus (1953) J. Bacteriol. **66**, 543.
[2]C. Dorner & B. Schink (1991) Arch. Microbiol. **156**, 302.
[3]D. P. Baker & C. A. Fewson (1989) J. Gen. Microbiol. **135**, 2035.
[4]D. R. Durham (1984) J. Bacteriol. **160**, 778.
[5]N. Allison, M. J. O'Donnell & C. A. Fewson (1985) Biochem. J. **231**, 407.

Selected entries from **Methods in Enzymology** [vol, page(s)]:
General discussion: 2, 273, 277
Assay: 2, 277
Other molecular properties: mandelate racemase, in preparation of, 2, 276; preparation from *Pseudomonas putida*, 2, 278; **17A**, 672, 676

MANDELATE 4-MONOOXYGENASE

(S)-mandelate   (S)-4-hydroxymandelate

This iron-dependent enzyme [EC 1.14.16.6], also known as L-mandelate 4-hydroxylase, catalyzes the reaction of (S)-2-hydroxy-2-phenylacetate (or, L-mandelate) with tetrahydropteridine and dioxygen to produce (S)-4-hydroxymandelate, dihydropteridine, and water.[1–3]

[1] S. G. Bhat, M. Ramanarayanan & C. S. Vaidyanathan (1973) BBRC **52**, 834.
[2] S. G. Bhat & C. S. Vaidyanathan (1976) ABB **176**, 314.
[3] S. G. Bhat & C. S. Vaidyanathan (1976) J. Bacteriol. **127**, 1108.

MANDELATE RACEMASE

This racemase [EC 5.1.2.2] catalyzes the interconversion of the stereoisomers of mandelate: that is, the interconversion of (S)-mandelate (i.e., L-mandelate) and (R)-mandelate (or, D-mandelate).[1–7] The enzyme will also act on the stereoisomers of p-hydroxymandelate. There is an absolute requirement for a divalent metal ion (Mg^{2+} is clearly preferred). In the presence of D_2O, deuterium is incorporated at the α-position. The C–H bond breaking step is partially rate-determining.

(R)-mandelate   (S)-mandelate

The *Pseudomonas putida* enzyme has a lysyl residue positioned to act as the (S)-base catalyst (the pK_a value of the ε-amino group of this active-site lysyl residue is about 6) and a histidyl residue (with an unperturbed pK_a) positioned to function as the (R)-specific base catalyst.[4] An enolic intermediate is generated upon proton abstraction. While proton abstraction adjacent to a carboxyl group is kinetically and thermodynamically unfavorable, this racemase exhibits a remarkable rate enhancement of about 1.7×10^{15}.[3]

[1] M. E. Tanner & G. L. Kenyon (1998) CBC **2**, 7.
[2] J. V. Schloss & M. S. Hixon (1998) CBC **2**, 43.
[3] S. L. Bearne & R. Wolfenden (1997) Biochemistry **36**, 1646.
[4] S. L. Schafer, W. C. Barrett, A. T. Kallarakal, B. Mitra, J. W. Kozarich, J. A. Gerlt, J. G. Clifton, G. A. Petsko & G. L. Kenyon (1996) Biochemistry **35**, 5662.
[5] G. L. Kenyon & G. D. Hegeman (1979) AE **50**, 325.
[6] E. Adams (1976) AE **44**, 69.
[7] L. Glaser (1972) The Enzymes, 3rd ed., **6**, 355.

Selected entries from **Methods in Enzymology** [vol, page(s)]:
General discussion: 2, 273, 276; **17A**, 671; **46**, 541
Assay: 2, 276
Other molecular properties: effect of metal ion, **46**, 546, 547; mandelate, effect of growth on, **2**, 274; properties, **2**, 277; purification from *Pseudomonas putida*, **17A**, 671; residue identification, **46**, 547, 548; stoichiometry of binding, **46**, 545, 546

MANDELONITRILE LYASE

This enzyme [EC 4.1.2.10], also known as hydroxynitrile lyase and oxynitrilase, catalyzes the reversible conversion of mandelonitrile to cyanide and benzaldehyde.[1]

mandelonitrile   benzaldehyde

The enzyme from the bitter almond (*Prunus amygdalus*) utilizes the *R*-stereoisomer, whereas the *Sorghum bicolor* enzyme uses the (*S*)-stereoisomer. In many organisms, this lyase is a flavoprotein. In the reverse reaction, the enzyme has an ordered Bi Uni kinetic mechanism, benzaldehyde binding before cyanide. The availability of both (*R*)- and (*S*)-specific enzymes has facilitated asymmetric synthesis of a variety of organic molecules.[2,3] **See also** *Hydroxynitrilase*

[1] B. A. Palfey & V. Massey (1998) CBC **3**, 83.
[2] H. Griengl, H. Schwab & M. Fechter (2000) Trends Biotechnol. **18**, 252.
[3] D. V. Johnson, A. A. Zabelinskaja-Mackova & H. Griengl (2000) Curr. Opin. Chem. Biol. **4**, 103.

Selected entries from **Methods in Enzymology** [vol, page(s)]:
Other molecular properties: activities, **228**, 584; borohydride reduction, **18B**, 475; catalyzed reaction, **228**, 584; enantiomers, purification (almonds, from, **228**, 584, 586; *Sorghum bicolor*, from, **228**, 584, 587); purification from plants, **228**, 584; sulfite complex, **18B**, 473

MANGANESE PEROXIDASE

This heme-dependent oxidoreductase [EC 1.11.1.13], also known as Mn-dependent peroxidase, catalyzes the reaction of two ions of free, uncomplexed Mn(II) with hydrogen peroxide and two protons to produce two Mn(III) ions and two water molecules.[1–4] The enzyme participates in the degradation of lignins (the Mn(III) product diffuses from the active site to oxidize lignan: note that diarylpropane peroxidase [EC 1.11.1.14] will also degrade lignan). Hydrogen peroxide oxidizes the ferric form of the enzyme to an Fe(IV)-porphyrin-π-cation state which acts on two ions of Mn^{2+} via two one-electron steps. The enzyme reportedly has a ping-pong mechanism rather than an ordered scheme.[1] **See also** *Peroxidase*

[1] H. Wariishi, K. Valli & M. H. Gold (1992) JBC **267**, 23688.
[2] H. Wariishi, L. Akileswaran & M. H. Gold (1988) Biochemistry **27**, 5365.
[3] M. H. Gold, H. L. Youngs & M. D. Gelpke (2000) Met. Ions Biol. Syst. **37**, 559.
[4] M. D. Sollewijn Gelpke, P. Moenne-Loccoz & M. H. Gold (1999) Biochemistry **38**, 11482.

Selected entries from **Methods in Enzymology** [vol, page(s)]:

General discussion: 161, 258, 264
Assay: 161, 259, 266

MANNAN ENDO-1,4-β-MANNOSIDASE

This endoglycosidase [EC 3.2.1.78], also known as β-mannanase and endo-1,4-mannanase, catalyzes the random hydrolytic scission of 1,4-β-D-mannosidic linkages in mannans, galactomannans, glucomannans, and galactoglucomannans.[1-3] **See also** β-Mannosidase

[1] B. V. McCleary (1978) Phytochemistry 17, 651 and (1979) 18, 757.
[2] T. Akino, N. Nakamura & K. Horikoshi (1988) Agric. Biol. Chem. 52, 773.
[3] P. Halmer, J. D. Bewley & T. A. Thorpe (1975) Nature 258, 716.

MANNAN ENDO-1,6-β-MANNOSIDASE

This endoglycosidase [EC 3.2.1.101], also known as endo-1,6-β-mannanase, catalyzes the random hydrolytic scission of 1,6-β-D-mannosidic linkages in unbranched 1,6-mannans. A mannan endo-1,6-α-mannosidase has also been reported.[1] **See also** β-Mannosidase

[1] T. Nakajima, S. K. Maitra & C. E. Ballou (1976) JBC 251, 174.

Selected entries from **Methods in Enzymology** [**vol**, page(s)]:
Other molecular properties: purification from Bacillus circulans, **185**, 461

MANNAN EXO-1,2-1,6-α-MANNOSIDASE

This enzyme [EC 3.2.1.137], first isolated from Cellulomonas sp., catalyzes the hydrolytic scission of 1,2-α-D- and 1,6-α-D-linkages in yeast mannan, releasing D-mannose.[1] Mannose residues linked α-D-1,3- are also released, albeit quite slowly.

[1] K. Takegawa, S. Miki, T. Jikibara & S. Iwahara (1989) BBA **991**, 431.

MANNAN 1,4-β-MANNOBIOSIDASE

This enzyme [EC 3.2.1.100], also known as exo-1,4-β-mannobiohydrolase, catalyzes the hydrolytic scission of 1,4-β-D-mannosidic linkages in 1,4-β-D-mannans, so as to remove successive mannobiose residues from the non-reducing ends of the polymer.[1,2] **See also** β-Mannosidase; Mannan Endo-1,4-β-Mannosidase

[1] T. Araki & M. Kitamikado (1982) J. Biochem. **91**, 1181.
[2] B. V. McCleary (1983) Phytochemistry **22**, 649.

MANNAN 1,2-(1,3)-α-MANNOSIDASE

This enzyme [EC 3.2.1.77], also known as exo-1,2-1,3-α-mannosidase, catalyzes the hydrolytic scission of 1,2- and 1,3-linkages in yeast mannans, releasing D-mannose.[1,2] Ultimately, the final product is a 1,6-α-D-mannan backbone. This enzyme will also act on those α-1,6-linkages, albeit slowly.

[1] J. F. Preston, E. Lapis & J. E. Gander (1973) Arch. Microbiol. **88**, 71.
[2] G. H. Jones & C. E. Ballou (1969) JBC **244**, 1043 and 1052.

MANNITOL 1-DEHYDROGENASE

This oxidoreductase [EC 1.1.1.255], also known as NAD$^+$-dependent mannitol dehydrogenase, catalyzes the reaction of D-mannitol with NAD$^+$ to produce D-mannose, NADH, and a proton.[1-3] The Apium graveolens (celery) enzyme will act on other alditols: however, there is a specificity requirement of 2R stereochemistry.

[1] J. M. H. Stoop & D. M. Pharr (1992) ABB **298**, 612.
[2] J. M. H. Stoop, J. D. Williamson, M. A. Conkling & D. M. Pharr (1995) Plant Physiol. **108**, 1219.
[3] J. M. H. Stoop, W. S. Chilton & D. M. Pharr (1996) Phytochemistry **43**, 1145.

MANNITOL 2-DEHYDROGENASE

This oxidoreductase [EC 1.1.1.67], also known as mannitol dehydrogenase, catalyzes the reversible reaction of D-mannitol with NAD$^+$ to produce D-fructose and NADH. Other substrates include sorbitol and D-arabinitol. The Pseudomonas fluorescens enzyme has an ordered Bi Bi kinetic mechanism at pH 8.2 in which the release of NADH from the E·NADH binary complex is the rate-limiting step.[1] At pH 7.0, the rate of hydride transfer is partially rate-limiting.

[1] M. Slatner, B. Nidetzky & K. D. Kulbe (1999) Biochemistry **38**, 10489.

Selected entries from **Methods in Enzymology** [**vol**, page(s)]:
General discussion: 9, 143; 41, 138
Assay: 9, 143; 41, 138
Other molecular properties: crystallization, 9, 145; Lactobacillus brevis, from, 9, 143 (coenzyme, 9, 145; crystallization, 9, 145; pH optima, 9, 146; properties, 9, 146; purification, 9, 143; specificity, 9, 145; stability, 9, 146); Leuconostoc mesenteroides, from, 41, 138 (chromatography, 41, 140; molecular weight, 41, 141; properties, 41, 141, 142; purification, 41, 139; sources, 41, 139); properties, 9, 146; purification (Lactobacillus brevis, 9, 143); sorbitol induction, 5, 325; stereospecificity, 87, 109

MANNITOL 2-DEHYDROGENASE (CYTOCHROME)

This oxidoreductase [EC 1.1.2.2], also known as polyol dehydrogenase, catalyzes the reaction of D-mannitol with ferricytochrome c to produce D-fructose and ferrocytochrome c.[1] Efficient substrates are polyols (or alcohol sugars) having a D-lyxo configuration. Other substrates include D-sorbitol and D-arabinitol.

[1] A. C. Arcus & N. L. Edson (1956) Biochem. J. **64**, 385.

Selected entries from **Methods in Enzymology** [**vol**, page(s)]:

General discussion: 9, 147
Assay: manometric, **9**, 147; spectrophotometric, **9**, 148

MANNITOL 2-DEHYDROGENASE (NADP⁺)

This oxidoreductase [EC 1.1.1.138], also known as mannitol dehydrogenase (NADP⁺) and NADP⁺:*lyxo*-dehydrogenase, catalyzes the reaction of D-mannitol with NADP⁺ to produce D-fructose and NADPH.[1–3] Other substrates include D-arabinitol and D-sorbitol.

[1] N. Morton, A. G. Dickerson & J. B. W. Hammond (1985) *J. Gen. Microbiol.* **131**, 2885.
[2] J. Boutelje, K. Hult & S. Gatenbeck (1983) *Eur. J. Appl. Microbiol. Biotechnol.* **17**, 7.
[3] W. G. Niehaus, Jr. & R. P. Dilts, Jr. (1982) *J. Bacteriol.* **151**, 243.

Selected entries from **Methods in Enzymology** [vol, page(s)]:
General discussion: 9, 177
Assay: 9, 171
Other molecular properties: 2-ketogluconate reductase preparation, presence in, **9**, 198; 5-ketogluconate reductase preparation, absence in, **9**, 201; pH optimum, **9**, 177; purification, *Gluconobacter oxydans*, **9**, 171; specificity, **9**, 176, 177; stability, **9**, 174

MANNITOL KINASE

This phosphotransferase, formerly classified as EC 2.7.1.57, is now a deleted Enzyme Commission entry.

MANNITOL-1-PHOSPHATASE

This enzyme [EC 3.1.3.22] catalyzes the hydrolysis of D-mannitol 1-phosphate to produce D-mannitol and orthophosphate.[1,2]

[1] H. Yamada, K. Okamoto, K. Kodama, F. Noguchi & S. Tanaka (1961) *J. Biochem.* **49**, 404.
[2] D. H. Jennings (1984) *Adv. Microb. Physiol.* **25**, 149.

MANNITOL-1-PHOSPHATE 5-DEHYDROGENASE

This oxidoreductase [EC 1.1.1.17], also known as mannitol-1-phosphate dehydrogenase and fructose-6-phosphate reductase, catalyzes the reversible reaction of D-mannitol 1-phosphate with NAD⁺ to produce D-fructose 6-phosphate and NADH.[1–3]

[1] T. Chase, Jr. (1986) *Biochem. J.* **239**, 435.
[2] W. Teschner, M. C. Serre & J. R. Garel (1990) *Biochimie* **72**, 33.
[3] R. C. Kiser & W. G. Niehaus, Jr. (1981) *ABB* **211**, 613.

Selected entries from **Methods in Enzymology** [vol, page(s)]:
General discussion: 1, 346; **9**, 150, 155; **27**, 80
Assay: 1, 346; **9**, 150, 155
Other molecular properties: *Aerobacter aerogenes*, **9**, 150 (pH optima, **9**, 153; properties, **9**, 152; purification, **9**, 151; sedimentation constant, **9**, 153; stimulation, **9**, 152; specificity, **9**, 153); *Bacillus subtilis*, **9**, 155 (pH optima, **9**, 158; properties, **9**, 158; purification, **9**, 156; specificity, **9**, 158); *Escherichia coli* (inhibition, **1**, 348; pH optimum, **1**, 348; properties, **1**, 348; purification, **1**, 347; reversibility, **1**, 348; specificity, **1**, 348);

fructose assay and, **3**, 107; induced cooperativity in, **202**, 714; stereospecificity, **87**, 103

MANNOKINASE

This phosphotransferase [EC 2.7.1.7] catalyzes the reaction of ATP with D-mannose to produce ADP and D-mannose 6-phosphate.[1–3] Other acceptor substrates include D-fructose and D-glucose. The *Streptomyces* enzyme has a random Bi Bi kinetic mechanism with two dead-end complexes.[2]

[1] B. Sabater, J. Sebastian & C. Asensio (1972) *BBA* **284**, 406.
[2] C. Coulombel, M.-J. Foglietti & F. Percheron (1982) *BBA* **706**, 117.
[3] J. Sebastian & C. Asensio (1972) *ABB* **151**, 227.

Selected entries from **Methods in Enzymology** [vol, page(s)]:
General discussion: 42, 39
Assay: 42, 39; coupled enzyme assay, **63**, 32, 33
Other molecular properties: *Leuconostoc mesenteroides*, from, **42**, 39 (chromatography, **42**, 42; molecular weight, **42**, 43; properties, **42**, 42, 43; purification, **42**, 40)

MANNONATE DEHYDRATASE

This enzyme [EC 4.2.1.8], also known as D-mannonate hydro-lyase and altronate hydrolase, catalyzes the conversion of D-mannonate to 2-dehydro-3-deoxy-D-gluconate (or, 2-keto-3-deoxy-D-gluconate) and water.[1,2] This dehydratase requires the presence of a loosely bound Fe^{2+}.[3]

[1] J. M. Robert-Baudouy & F. R. Stoeber (1973) *BBA* **309**, 473.
[2] W. A. Wood (1971) *The Enzymes*, 3rd ed., **5**, 573.
[3] J. L. Dreyer (1987) *Eur. J. Biochem.* **166**, 623.

Selected entries from **Methods in Enzymology** [vol, page(s)]:
General discussion: 5, 203; **90**, 288
Assay: 5, 203; **90**, 288
Other molecular properties: *Escherichia coli* (properties, **5**, 205; **90**, 291; purification, **5**, 203; **90**, 289)

MANNONATE DEHYDROGENASE (NAD(P)⁺)

This oxidoreductase, formerly classified as EC 1.1.1.180, is now a deleted entry, and the activity is listed under mannuronate reductase [EC 1.1.1.131].

MANNOSE ISOMERASE

This isomerase [EC 5.3.1.7], also known as D-mannose ketol-isomerase, catalyzes the interconversion of D-mannose and D-fructose.[1–4] Other substrates include D-lyxose (producing D-xylulose) and D-rhamnose (producing D-rhamnulose).

[1] E. A. Noltmann (1972) *The Enzymes*, 3rd ed., **6**, 271.
[2] A. Hey-Ferguson & A. D. Elbein (1970) *J. Bacteriol.* **101**, 777.
[3] Y. Takasaki, S. Takano & O. Tanabe (1964) *Agric. Biol. Chem.* **28**, 605.
[4] N. J. Palleroni & M. Doudoroff (1955) *JBC* **218**, 535.

MANNOSE-1-PHOSPHATE GUANYLYLTRANSFERASE

This enzyme [EC 2.7.7.13], also known as GTP:mannose-1-phosphate guanylyltransferase, catalyzes the reversible reaction of GTP with α-D-mannose 1-phosphate to produce GDP-D-mannose and pyrophosphate (or, diphosphate).[1–3] The bacterial enzyme can also use ITP and dGTP as the phosphoryl donors. Glucose 1-phosphate can act as the acceptor substrate with the pig enzyme but not with the mycobacterial protein.

Many bacterial enzymes are multifunctional, also exhibiting phosphomannose isomerase and phosphoglucoisomerase activities; however, the mycobacterial enzyme is not multifunctional.[4]

[1] T. Szumilo, R. R. Drake, J. L. York & A. D. Elbein (1993) *JBC* 268, 17943.
[2] D. Shinabarger, A. Berry, T. B. May, R. Rothmel, A. Fialho & A. M. Chakrabarty (1991) *JBC* 266, 2080.
[3] J. W. Smoot & G. S. Serif (1985) *EJB* 148, 83.
[4] B. Ning & A. B. Elbein (1999) *ABB* 362, 339.

MANNOSE-1-PHOSPHATE GUANYLYLTRANSFERASE (GDP)

This transferase [EC 2.7.7.22], also known as GDP-mannose pyrophosphorylase and GDP-mannose phosphorylase, catalyzes the reaction of GDP with D-mannose 1-phosphate to produce orthophosphate and GDP-D-mannose.[1]

[1] H. Carminatti & E. Cabib (1961) *BBA* 53, 417.

MANNOSE-6-PHOSPHATE ISOMERASE

This zinc-dependent isomerase [EC 5.3.1.8], also known as phosphomannose isomerase, phosphohexoisomerase, and phosphohexomutase, catalyzes the interconversion of D-mannose 6-phosphate and D-fructose 6-phosphate.[1–4] The β-anomer of D-mannose 6-phosphate is the better substrate of the two diastereoisomers.

[1] K. U. Yüksel & R. W. Gracy (1991) *A Study of Enzymes* 2, 457.
[2] D. J. Creighton & N. S. R. K. Murthy (1990) *The Enzymes*, 3rd ed., 19, 323.
[3] I. A. Rose (1975) *AE* 43, 491.
[4] E. A. Noltmann (1972) *The Enzymes*, 3rd ed., 6, 271.

MANNOSE-6-PHOSPHATE 6-REDUCTASE

This oxidoreductase [EC 1.1.1.224] catalyzes the reaction of D-mannose 6-phosphate with NADPH to produce D-mannitol 1-phosphate and $NADP^+$.[1–3] This enzyme participates in the biosynthesis of mannitol in celery (*Apium graveolens*) leaves.

[1] M. E. Rumpho, G. D. Edwards & W. H. Loescher (1983) *Plant Physiol.* 73, 869.
[2] W. H. Loescher, R. H. Tyson, J. D. Everard, J. Redgewell & R. L. Bieleski (1992) *Plant Physiol.* 98, 1396.
[3] J. D. Everard, C. Cantini, R. Grumet, J. Plummer & W. H. Loescher (1997) *Plant Physiol.* 113, 1427.

α-MANNOSIDASE

This hydrolase [EC 3.2.1.24], also known as exo-α-mannosidase, catalyzes the hydrolytic scission of terminal, nonreducing α-D-mannose residues in α-D-mannosides (this includes glycopeptides and glycoproteins containing α-mannosides).[1–7] In addition, α-D-lyxosides and heptopyranosides (with the same configuration at C2, C3, and C4 as D-mannose) will serve as substrates as well. α-Mannosidases have been reported to be specific for the type of linkage (*e.g.*, α1,2 and α1,3) or specific for the aglycon portion of the substrate. *See also specific enzyme*

[1] M. W. Bauer, S. B. Halio & R. M. Kelly (1996) *Adv. Protein Chem.* 48, 271.
[2] S. C. Fry (1995) *Ann. Rev. Plant Physiol. Plant Mol. Biol.* 46, 497.
[3] P. F. Daniel, B. Winchester & C. D. Warren (1994) *Glycobiology* 4, 551.
[4] E. Conzelmann & K. Sandhoff (1987) *AE* 60, 89.
[5] P. M. Dey & E. del Campillo (1984) *AE* 56, 141.
[6] K. A. Presper & E. C. Heath (1983) *The Enzymes*, 3rd ed., 16, 449.
[7] H. M. Flowers & N. Sharon (1979) *AE* 48, 29.

91; Golgi apparatus marker, assay, **182**, 219; hen oviduct (properties, **28**, 780; purification, **28**, 777); identification, mannosidase inhibitors in, **230**, 324; inhibitors, **8**, 582; **230**, 324 (in glycoprotein function studies, **230**, 327; structure, **230**, 324); jack bean, **28**, 704 (glycophosphatidylinositol glycan digestion, **230**, 441; incubation conditions, **230**, 284; purification, **230**, 283, 286; specificity, **230**, 284); lysosomal α-D-mannosidase (assay, **179**, 500; properties, **179**, 503; purification from rat epididymis, **179**, 501); lysosomes, **8**, 513; 1,2-α-mannosidase (analysis with inhibitors, **230**, 325; *Aspergillus niger* [properties, **28**, 746, 748; purification, **28**, 745]; *Aspergillus saitoi* [incubation conditions, **230**, 284; purification, **230**, 283; specificity, **230**, 284, 290, 298]; assay, **28**, 744; **83**, 423; glycoprotein processing pathway, in, **230**, 320; inhibitor, **230**, 320; purification, glycosidase inhibitors as affinity ligands in, **230**, 323; rat liver [assay, **83**, 423; properties, **83**, 426; purification, **83**, 424]); mannosidosis, **50**, 452; marker for chromosome detection in somatic cell hybrids, as, **151**, 188; mucolipidoses II and III, **50**, 455; *p*-nitrophenyl α-mannoside, **8**, 581; nonspecific, purification from *Oerskovia* sp., **185**, 464; *Phaseolus vulgaris* (purification, **28**, 721; substrate specificity, **28**, 727); pig kidney (properties, **28**, 794; purification, **8**, 581; **28**, 793); properties, **8**, 580; **28**, 709; **160**, 623; rat liver, **50**, 489 (assay, **83**, 423, 427; Golgi, **50**, 501; lysosomes, **50**, 495; properties, **83**, 426; purification, **83**, 424); reaction specificity, **247**, 73; *Rhodococcus erythropolis*, **160**, 621; staining on gels, **22**, 591; sweet almond emulsin (properties, **28**, 700; purification, **28**, 699); type I (assay, **179**, 464; properties, **179**, 467; purification from mung bean microsomes, **179**, 465); types I and II (assay, **138**, 780; properties, **138**, 784; purification from *Aspergillus*, **138**, 782); yeast vacuole, assay, **194**, 657

β-MANNOSIDASE

β-Mannosidase [EC 3.2.1.25], also known as mannanase, mannase and exo-β-D-mannosidase, catalyzes the hydrolysis of terminal, nonreducing β-D-mannose residues in β-D-mannosides.[1-3] Small alternative substrates include 4-methylumbelliferyl β-D-mannopyranoside, *p*-nitrophenyl-β-D-mannopyranoside, naphthyl-β-D-mannopyranoside, β-D-mannotriitol, and β-D-mannotetraitol. The enzyme also acts on mannose-containing glycopeptides and glycoproteins. *See also Mannan Endo-1,4-β-Mannosidase; Mannan Endo-1,6-β-Mannosidase*

[1]M. W. Bauer, S. B. Halio & R. M. Kelly (1996) *Adv. Protein Chem.* **48**, 271.

[2]Y. Iwasaki, A. Tsuji, K. Omura & Y. Suzuki (1989) *J. Biochem.* **106**, 331.

[3]P. M. Dey & E. Del Campillo (1984) *AE* **56**, 141.

Selected entries from *Methods in Enzymology* [vol, page(s)]:
General discussion: **160**, 589, 614; **330**, 224, 238
Assay: **28**, 769; 777; **160**, 76, 590, 599, 611, 616
Other molecular properties: carbohydrate structure, **28**, 19, 557; hen oviduct (properties, **28**, 780; purification, **28**, 777); hyperthermophilic, **330**, 224, 238; inhibition by aldonolactones, **8**, 573; β-D-mannopyranoside substrates, synthesis, **160**, 515; preparation of manno- and glucomannooligosaccharides, in, **160**, 518; properties, **160**, 593, 617 (*Aspergillus* enzyme, **160**, 608; *Bacillus* enzyme, **160**, 604; guar seed enzyme, **160**, 602; *Helix* enzyme, **160**, 603; *Irpex* enzyme, **160**, 609; lucerne seed enzyme, **160**, 601; *Streptomyces* enzyme, **160**, 614); purification from (*Aspergillus niger*, **160**, 604; *Bacillus circulans*, **185**, 461; *Bacillus subtilis*, **160**, 603; guar seeds, **160**, 591, 601; *Helix pomatia*, **160**, 602, 616; *Irpex lacteus*, **160**, 609; lucerne seeds, **160**, 600; *Streptomyces*, **160**, 613); snails (properties, **28**, 771; purification, **28**, 769); staining in gels, **160**, 616; substrate preparation, synthesis, **160**, 515

MANNOSYL-GLYCOPROTEIN β-1,4-N-ACETYLGLUCOSAMINYL-TRANSFERASE

This transferase [EC 2.4.1.201], also known as *N*-glycosyl-oligosaccharide-glycoprotein and *N*-acetylglucosaminyltransferase V, catalyzes the reaction of UDP-*N*-acetyl-D-glucosamine with *N*-acetyl-β-D-glucosaminyl-1,6-β-D-(*N*-acetyl-D-glucosaminyl-1,2)-β-D-mannosyl-R to produce UDP and *N*-acetyl-β-D-glucosaminyl-1,6-β-D-(*N*-acetyl-D-glucosaminyl-1,2-β)-(*N*-acetyl-D-glucosaminyl-1,4-β)-D-mannosyl-R, where R represents the remainder of the *N*-oligosaccharide core of asparagine-linked glycopeptides. This reaction is required in the biosynthesis of penta-antennary oligosaccharides of hen ovomucoid.[1] *See also α-1,3-Mannosyl-Glycoprotein β-1,2-N-Acetylglucosaminyltransferase; α-1,6-Mannosyl-Glycoprotein β-1,2-N-Acetylglucosaminyltransferase; β-1,4-Mannosyl-Glycoprotein β-1,4-N-Acetylglucosaminyltransferase; α-1,3-Mannosyl-Glycoprotein β-1,4-N-Acetylglucosaminyltransferase; β-1,3-Galactosyl-O-Glycosyl-Glycoprotein β-1,3-N-Acetylglucosaminyltransferase*

[1]I. Brockhausen, E. Hull, O. Hindsgaul, H. Schachter, R. N. Shah, S. W. Michnick & J. P. Carver (1989) *JBC* **264**, 11211.

Selected entries from *Methods in Enzymology* [vol, page(s)]:
General discussion: **179**, 351
Assay: **179**, 380
Other molecular properties: properties, **179**, 381

α-1,3-MANNOSYL-GLYCOPROTEIN β-1,2-N-ACETYLGLUCOSAMINYL-TRANSFERASE

This transferase [EC 2.4.1.101], also known as *N*-glycosyl-oligosaccharide-glycoprotein *N*-acetylglucosaminyltransferase I and UDP-*N*-acetylglucosamine:α-D-mannoside β1,2-*N*-acetylglucosaminyltransferase I, catalyzes the reaction of UDP-*N*-acetyl-D-glucosamine with an α-D-mannosyl-1,3-(R_1)-β-D-mannosyl-R_2 to produce *N*-acetyl-β-D-glucosaminyl-1,2-α-D-mannosyl-1,3-(R_1)-β-D-mannosyl-R_2 and UDP. The acceptor substrate is the core oligosaccharide of glycoprotein *N*-oligosaccharides containing 2-*N*-acetyl-D-glucosamine, three to five D-mannose residues, and with or without one L-fucosyl residue. Thus, the enzyme transfers an *N*-acetylglucosamine in β-1,2-linkage to a mannosyl-1,3-β-mannosyl-terminus. Acceptor substrates include dehexoso-orosomucoid, ovalbumin, β-galactosidase- and β-*N*-acetylhexosamidase-treated asialo-fetuin and asialotransferrin, desialodegalacto-dehexosamine-orosomucoid, and free and protein-matrix-bound glycans.

The kinetic mechanism for the rabbit liver enzyme is essentially ordered Bi Bi.[1] *See also α-1,6-Mannosyl-Glycoprotein β-1,2-N-Acetylglucosaminyltransferase; α-1,4-Mannosyl-Glycoprotein β-1,4-N-Acetylglucosaminyltransferase; and α-1,3-Mannosyl-Glycoprotein β-1,4-N-Acetylglucosaminyltransferase*

[1]Y. Nishikawa, W. Pegg, H. Paulsen & H. Schachter (1988) *JBC* **263**, 8270.

Selected entries from *Methods in Enzymology* [vol, page(s)]:
General discussion: **83**, 508; **98**, 110; **230**, 302
Assay: **83**, 508; **98**, 111; **179**, 363

Other molecular properties: acceptor preparation, **98**, 105; bovine colostrum, **98**, 110 (assay, **98**, 111; properties, **98**, 113; purification, **98**, 112; specificity, **98**, 115); N- and O-glycan synthesis, in (HPLC assay, **179**, 351; substrate preparation, **179**, 360); Golgi apparatus, in, **22**, 142; **31**, 191; porcine liver, **98**, 113; rabbit liver, **98**, 113; **230**, 315 (assay, **83**, 508, **179**, 363; properties, **83**, 513, **179**, 367; purification, **83**, 510, **179**, 364); reaction in Golgi apparatus, **22**, 142; substrate preparation, **98**, 102; synthesis of GlcNAcβ1→2Manα1→3[Manα1→6]Manβ-OR, in, **230**, 314

α-1,3-MANNOSYL-GLYCOPROTEIN β-1,4-N-ACETYLGLUCOSAMINYL-TRANSFERASE

This transferase [EC 2.4.1.145], also known as N-glycosyl-oligosaccharide-glycoprotein N-acetylglucosaminyltransferase IV and UDP-N-acetylglucosamine:α-D-mannoside β-1,2-N-acetylglucosaminyltransferase IV, catalyzes the reaction of UDP-N-acetyl-D-glucosamine with (N-acetyl-β-D-glucosaminyl-1,2)-α-D-mannosyl-1,3-(β-N-acetyl-D-glucosaminyl-1,2-α-D-mannosyl-1,6)-β-D-mannosyl-R to produce UDP and N-acetyl-β-D-glucosaminyl-1,4-(N-acetyl-D-glucosaminyl-1,2)-α-D-mannosyl-1,3-(β-N-acetyl-D-glucosaminyl-1,2-α-d-mannosyl-1,6)-β-d-mannosyl-R (where R represents the remainder of the N-oligosaccharide core, with or without a fucosyl residue). This enzyme initiates the formation of the third antenna or branch in the oligosaccharide portion of a glycoprotein.[1] *See also α-1,3-Mannosyl-Glycoprotein β-1,2-N-Acetylglucosaminyltransferase; α-1,4-Mannosyl-Glycoprotein β-1,4-N-Acetylglucosaminyltransferase; α-1,6-Mannosyl-Glycoprotein β-1,2-N-Acetylglucosaminyltransferase; α-1,3(6)-Mannosyl-Glycoprotein β-1,6-N-Acetylglucosaminyltransferase; and Mannosyl-Glycoprotein β-1,4-N-Acetylglucosaminyltransferase*

[1]P. A. Gleeson & H. Schachter (1983) *JBC* **258**, 6162.

Selected entries from *Methods in Enzymology* [vol, page(s)]:
General discussion: **98**, 120; **179**, 405; **230**, 302
Assay: **98**, 120; **179**, 377, 405

Other molecular properties: acceptor preparation, **98**, 105; chicken oviduct, properties, **98**, 120; **179**, 377; distribution in rat tissues, **179**, 406; properties, **179**, 405; substrate preparation, **98**, 102

α-1,3(6)-MANNOSYL-GLYCOPROTEIN β-1,6-N-ACETYLGLUCOSAMINYL-TRANSFERASE

This transferase [EC 2.4.1.155], also called α-mannoside β-1,6-N-acetylglucosaminyltransferase and N-acetylglu-cosaminyltransferase V, catalyzes the reaction of UDP-N-acetyl-D-glucosamine with N-acetyl-β-D-glucosaminyl-1,2-α-D-mannosyl-1,3(6)-(N-acetyl-β-D-glucosaminyl-1,2-α-D-mannosyl-1,6(3))-β-D-mannosyl-1,4-N-acetyl-β-D-glucosaminyl-R to produce UDP and N-acetyl-β-D-glucosaminyl-1,2-(N-acetyl-β-D-glucosaminyl-1,6)-1,2-α-D-mannosyl-1,3(6)-(N-acetyl-β-D-glucosaminyl-1,2-α-D-mannosyl-1,6(3))-β-D-mannosyl-1,4-N-acetyl-β-D-glucos-aminyl-R.[1] Acceptor substrates are branched mannose glycopeptides with non-reducing N-acetylglucosamine terminal residues: the enzyme is most active with triantennary and biantennary sugar systems. There is inversion of configuration at the anomeric center. *See also α-1,3-Mannosyl-Glycoprotein β-1,2-N-Acetylglucosaminyltransferase; α-1,6-Mannosyl-Glycoprotein β-1,2-N-Acetylglucosaminyltransferase; α-1,3-Mannosyl-Glycoprotein β-1,4-N-Acetylglucosaminyltrans-ferase; α-1,4-Mannosyl-Glycoprotein β-1,4-N-Acetylglucos-aminyltransferase; and Mannosyl-Glycoprotein β-1,4-N-Acetyl-glucosaminyltransferase.*

[1]N. Zhang, K. C. Peng, L. Chen, D. Puett & M. Pierce (1997) *JBC* **272**, 4225

Selected entries from *Methods in Enzymology* [vol, page(s)]:
General discussion: **179**, 351, 397; **230**, 302
Assay: **179**, 378, 406; activity assay, **247**, 223; enzyme-linked immunosorbent assay, **247**, 221; radiochemical, **247**, 220

Other molecular properties: acceptor specificity, **247**, 215; affinity chromatography (enzyme purification, **247**, 225; ligands, **247**, 216; preparation of supports, **247**, 219); distribution in rat tissues, **179**, 407; properties, **179**, 379, 407; reaction catalyzed, **247**, 215

β-1,4-MANNOSYL-GLYCOPROTEIN β-1,4-N-ACETYLGLUCOSAMINYL-TRANSFERASE

This transferase [EC 2.4.1.144], also known as N-glycosyl-oligosaccharide-glycoprotein N-acetylglucosaminyltrans-ferase III and UDP-N-acetylglucosamine:α-D-mannoside β1,2-N-acetylglucosaminyltransferase III, catalyzes the reaction of UDP-N-acetyl-D-glucosamine with an N-acetyl-β-D-glucosaminyl-1,2-α-D-mannosyl-1,3-(N-acetyl-β-D-glucosaminyl-1,2-α-D-mannosyl-1,6)-β-D-mannosyl-1,4-N-acetyl-β-D-glucosaminyl-R to produce UDP and N-acetyl-β-D-glucosaminyl-1,2-α-D-mannosyl-1,3-(N-acetyl-β-D-glucosaminyl-1,2-α-D-mannosyl-1,6)-(N-acetyl-β-D-glucosaminyl-1,4)-β-D-mannosyl-1,4-N-acetyl-β-D-glu-cosaminyl-R (where R represents the remainder of the

N-oligosaccharide core, with or without a fucosyl residue): thus, the enzyme incorporates an N-acetylglucosamine residue into branching regions of glycoproteins. **See also** *α-1,3-Mannosyl-Glycoprotein β-1,2-N-Acetylglucosaminyltransferase; α-1,6-Mannosyl-Glycoprotein β-1,2-N-Acetylglucosaminyltransferase; α-1,3-Mannosyl-Glycoprotein β-1,4-N-Acetylglucosaminyltransferase; α-1,3(6)-Mannosyl-Glycoprotein β-1,6-N-Acetylglucosaminyltransferase; and Mannosyl-Glycoprotein β-1,4-N-Acetylglucosaminyltransferase.*

Selected entries from **Methods in Enzymology** [**vol**, page(s)]:
General discussion: 98, 117; **179**, 400; **230**, 302
Assay: 98, 117; **179**, 374, 400
Other molecular properties: acceptor preparation, 98, 105; chicken oviduct, properties, 98, 119; **179**, 375; distribution in rat tissues, **179**, 404; properties, **179**, 402; substrate preparation, 98, 102

α-1,6-MANNOSYL-GLYCOPROTEIN β-1,2-N-ACETYLGLUCOSAMINYL-TRANSFERASE

This transferase [EC 2.4.1.143], also known as N-glycosyl-oligosaccharide-glycoprotein N-acetylglucosaminyltransferase II and UDP-N-acetylglucosamine:α-D-mannoside β1,2-N-acetylglucosaminyltransferase II, catalyzes the reaction of UDP-N-acetyl-D-glucosamine with an α-D-mannosyl-1,6-(N-acetyl-β-D-glucosaminyl-1,2-α-D-mannosyl-1,3)-β-D-mannosyl-R to produce UDP and N-acetyl-β-D-glucosaminyl-1,2-α-D-mannosyl-1,6-(N-acetyl-β-D-glucosaminyl-1,2-α-D-mannosyl-1,3)-β-D-mannosyl-R (where R represents the remainder of the N-oligosaccharide core, with or without a fucosyl residue).[1] Thus, this enzyme catalyzes the initiation of the formation of the second antenna or branch of the oligosaccharide portion of glycoproteins. It has largely an ordered Bi Bi kinetic mechanism. **See also** *α-1,3-Mannosyl-Glycoprotein β-1,2-N-Acetylglucosaminyltransferase; α-1,4-Mannosyl-Glycoprotein β-1,4-N-Acetylglucosaminyltransferase; α-1,3-Mannosyl-Glycoprotein β-1,4-N-Acetylglucosaminyltransferase; α-1,3(6)-Mannosyl-Glycoprotein β-1,6-N-Acetylglucosaminyltransferase; and Mannosyl-Glycoprotein β-1,4-N-Acetylglucosaminyltransferase.*

[1] B. Bendiak & H. Schachter (1987) JBC **262**, 5784.

Selected entries from **Methods in Enzymology** [**vol**, page(s)]:
General discussion: 98, 116; **179**, 368, 472; **230**, 302
Assay: 98, 116; **179**, 368, 472; **230**, 316
Other molecular properties: bovine colostrum (properties, 98, 116; purification, 98, 116); N- and O-glycan synthesis, in (HPLC assay, **179**, 351; substrate preparation, **179**, 360); mung bean microsomal (assay, **179**, 472; properties, **179**, 474; purification, **179**, 473); rat liver (assay, **179**, 368; properties, **179**, 373; purification, **179**, 369); substrate preparation, 98, 102

MANNOSYL-GLYCOPROTEIN ENDO-β-N-ACETYLGLUCOSAMINIDASES

These hydrolases [EC 3.2.1.96], also called endo-β-N-acetylglucosaminidase and di-N-acetylchitobiosyl β-N-acetylglucosaminidases, catalyze the endohydrolysis of the di-N-acetylchitobiosyl unit in high-mannose glycopeptides and glycoproteins containing the -[Man(N-acetylglucosamine)$_2$]Asn- structure.[1,2] One N-acetyl-D-glucosamine residue remains attached to the protein; the rest of the oligosaccharide is released intact. Different variants of the enzyme activity are often specific for the branching in the oligosaccharide portion of the structure. Glycoprotein substrates include ribonuclease B, ovalbumin, yeast carboxypeptidase and invertase, bromelain, ovomucoid, fetuin, and transferrin. **See also** *β-N-Acetylhexosaminidase; Chitinase; β-N-Acetylglucosaminidase; Endo-β-N-Acetylglucosaminidases*

[1] E. Conzelmann & K. Sandhoff (1987) AE **60**, 89.
[2] H. M. Flowers & N. Sharon (1979) AE **48**, 29.

Selected entries from **Methods in Enzymology** [**vol**, page(s)]:
Assay: 50, 557, 568, 575; **83**, 604; **138**, 771; **179**, 506
Other molecular properties: applications, **179**, 515; bacterial cell wall degradation and, 8, 690; *Clostridium perfringens* enzyme C$_I$ (assay, 50, 568, 569; properties, 50, 571; purification, 50, 569; reaction, 50, 567); *Clostridium perfringens* enzyme C$_{II}$ (assay, 50, 568; properties, 50, 571; purification, 50, 569; reaction, 50, 567); *Diplococcus pneumoniae* enzyme (assay, 50, 556; properties, 50, 557; purification, 50, 556, 557); endo-β-N-acetylglucosaminidase D (*Diplococcus pneumoniae*, from [assay, 50, 556; hydrolysis of rhodopsin glycoprotein T1, 81, 217; properties, 50, 557; purification, 50, 556, 557]; reconstitution of ER-Golgi transport in vitro, in, 221, 229; sensitivity of ER-cis-Golgi transport, acquisition during reconstitution, 219, 112); endo F (activation of latent TGF-β1, 198, 333; assay, 138, 771; 230, 46 [synthetic fluorescent substrates for, 230, 55]; oligosaccharide release from glycoprotein, in, 230, 142; properties, 138, 775; 230, 45; purification, 138, 772; 179, 506; specificity, 230, 44; structural determinants, 230, 45); endo F$_1$ (assay, 230, 46; deglycosylation efficiency, 230, 53; properties, 230, 52; purification, 230, 48; purity, 230, 51; specificity, 230, 45, 52, 56); endo F$_2$ (assay, 230, 46; deglycosylation efficiency, 230, 53; properties, 230, 53; purification, 230, 48; specificity, 230, 45, 53); endo F$_3$ (assay, 230, 46; deglycosylation efficiency, 230, 53; properties, 230, 53; purification, 230, 48; specificity, 230, 45, 53, 56); endo H (analysis of yeast glycoproteins, in, 194, 694; assay for transport-coupled oligosaccharide processing in vitro, in, 98, 42; identification of occupied N-glycosylation sites, in, 230, 343; oligosaccharide release from glycoproteins by, 230, 142, 218; postincubation of VSV-infected cells with, 219, 123; properties, 138, 769; purification from *Streptomyces plicatus*, 138, 765; specificity, 230, 44, 52; VSV-G postincubation with, 257, 171); properties on glycoprotein substrates, 179, 511; purification from *Flavobacterium meningosepticum*, 179, 506; *Streptomyces plicatus* enzyme (assay, 50, 575, 576; properties, 50, 578; purification, 50, 576); substrate specificities, 230, 200; type D, hydrolysis of rhodopsin glycoprotein T1, 81, 217; type F (assay, 138, 771; properties, 138, 775; purification from *Flavobacterium*, 138, 772; 179, 506); type H (properties, 138, 769; purification from *Streptomyces plicatus*, 138, 765); type L, *Streptomyces plicatus* (assay, 83, 604; properties, 83, 607; purification, 83, 606)

MANNOSYL-OLIGOSACCHARIDE GLUCOSIDASE

This glucosidase [EC 3.2.1.106], also known as processing A-glucosidase I and trimming glucosidase I, catalyzes the exohydrolysis of the non-reducing terminal glucose residue in the mannosyl-oligosaccharide $Glc_3Man_9GlcNAc_2$.[1-3] The enzyme also acts on the corresponding glycolipids and glycopeptides.

[1] E. Bause, R. Erkens, J. Schweden & L. Jaenicke (1986) FEBS Lett. **206**, 208.
[2] R. D. Kilker, B. Saunier, J. S. Tkacz & A. Herscovics (1981) JBC **256**, 5299.
[3] H. Hettkamp, G. Legler & E. Bause (1984) EJB **142**, 85.

Selected entries from *Methods in Enzymology* [vol, page(s)]:
General discussion: 83, 429; 179, 456; 230, 325
Assay: 83, 419, 429; 179, 456
Other molecular properties: glycoprotein processing, in, **230**, 317; inhibitors (preparation of oligosaccharide substrates, in, **230**, 317; role in glycoprotein function, **230**, 325); properties, **83**, 431; **179**, 459; purification from (hen oviduct, **83**, 430; mung bean microsomes, **179**, 457); purification, glycosidase inhibitors as affinity ligands in, **230**, 323; substrate preparation, **179**, 455

MANNOSYL-OLIGOSACCHARIDE GLUCOSIDASE II

This glucosidase, also known as glucosidase II, catalyzes the sequential exohydrolysis of the non-reducing terminal glucose residues in the mannosyl-oligosaccharide $Glc_2Man_9GlcNAc_2$.[1-3] Substrates therefore include $GlcMan_9GlcNAc_2$. The enzyme also acts on the corresponding glycolipids and glycopeptides.

Selected entries from *Methods in Enzymology* [vol, page(s)]:
General discussion: 83, 419; 179, 461
Assay: 83, 419; 179, 461
Other molecular properties: glycoprotein processing, in, **230**, 317; inhibitors (preparation of oligosaccharide substrates, in, **230**, 317); properties, **83**, 421; **179**, 463; purification from (bovine liver, **83**, 420; mung bean microsomes, **179**, 462); role in glycoprotein function, **230**, 325

MANNOSYL-OLIGOSACCHARIDE 1,2-α-MANNOSIDASE

This glycosidase [EC 3.2.1.113], also known as mannosidase 1A and mannosidase 1B, catalyzes the hydrolysis of the terminal 1,2-linked α-D-mannose residues in the oligo-mannose oligosaccharide $Man_9(GlcNAc)_2$.[1-3] The *Saccharomyces cerevisiae* enzyme catalyzes the release of β-D-mannose, thereby demonstrating that the enzyme is of the inverting type.[2] **See also** *Mannan 1,2-(1,3)-α-Mannosidase*

[1] I. Tabas & S. Kornfeld (1979) JBC **254**, 11655.
[2] F. Lipari, B. J. Gour-Salin & A. Herscovics (1995) BBRC **209**, 322

[3] T. Yoshida, K. Maeda, M. Kobayashi & E. Ichishima (1994) BJ **303**, 97

Selected entries from *Methods in Enzymology* [vol, page(s)]:
General discussion: 138, 779; 179, 446, 452, 464
Assay: 179, 446, 468
Other molecular properties: identification, mannosidase inhibitors in, **230**, 324; inhibitors, **230**, 324 (in glycoprotein function studies, **230**, 327; structure, **230**, 324); properties, **179**, 449; purification from *Aspergillus phoenicis*, **185**, 463; purification from rat liver Golgi membranes, **179**, 447

MANNOSYL-OLIGOSACCHARIDE 1,3-1,6-α-MANNOSIDASE

This glycosidase [EC 3.2.1.114], also known as mannosidase II, catalyzes the hydrolytic scission of the terminal 1,3- and 1,6-linked α-D-mannose residues in the mannosyl-oligosaccharide $Man_5GlcNAc_3$. This mannosidase is potently inhibited by swainsonine and less strongly by 1,4-dideoxy-1,4-imino-D-mannitol.[1]

[1] G. P. Kaushal, T. Szumilo, I. Pastuszak & A. D. Elbein (1990) Biochemistry **29**, 2168.

Selected entries from *Methods in Enzymology* [vol, page(s)]:
General discussion: 138, 779; 179, 452
Assay: 179, 468; acting on yeast mannoproteins, general assay, 185, 461
Other molecular properties: analysis with inhibitors, **230**, 325; glycoprotein processing pathway, in, **230**, 320; inhibitors, **230**, 320; properties, **179**, 470; purification from mung bean microsomes, **179**, 469

β-MANNOSYLPHOSPHODECAPRENOL: MANNOOLIGOSACCHARIDE 6-MANNOSYLTRANSFERASE

This transferase [EC 2.4.1.199] catalyzes the reaction of β-D-mannosylphosphodecaprenol with a 1,6-α-D-mannosyloligosaccharide to produce decaprenol phosphate and a 1,6-α-D-mannosyl-1,6-α-D-mannosyloligosaccharide.[1]

[1] K. Yokoyama & C. E. Ballou (1989) JBC **264**, 21621.

MANNOTETRAOSE 2-α-N-ACETYLGLUCOSAMINYLTRANSFERASE

This transferase [EC 2.4.1.138], also known as α-N-acetylglucosaminyltransferase, catalyzes the reaction of UDP-N-acetyl-D-glucosamine with 1,3-α-D-mannosyl-1,2-α-D-mannosyl-1,2-α-D-mannosyl-D-mannose to produce UDP and 1,3-α-D-mannosyl-1,2-(N-acetyl-α-D-glucosaminyl-α-D-mannosyl)-1,2-α-D-mannosyl-D-mannose.[1] Acceptor substrates include mannoproteins containing one or more mannotetraose portions in the oligosaccharide chain.

[1] R. H. Douglas & C. E. Ballou (1982) Biochemistry **21**, 1561.

MANNURONAN C-5 EPIMERASE

This calcium-dependent epimerase catalyzes the conversion of D-mannuronate residues in polymannuronate to L-glucuronate residues.

mannuronan

subunit of alginate

This is the main step in the synthesis of alginate, a linear polysaccharide in brown algae and certain bacteria that consists of D-mannuronate and L-glucuronate residues.[1,2] If the reaction is carried out in the presence of deuterium oxide, the isotope is incorporated at C5. **See also** Heparosan-N-Sulfate-Glucuronate 5-Epimerase

[1] G. Skjåk-Bræk & B. Larsen (1985) Carbohyd. Res. **139**, 273.
[2] H. Ertesvåg, B. Doseth, B. Larsen, G. Skjåk-Bræk & S. Valla (1994) J. Bacteriol. **176**, 2846.

MANNURONATE REDUCTASE

This oxidoreductase [EC 1.1.1.131], also known as mannonate dehydrogenase, catalyzes the reversible reaction of D-mannuronate with NAD(P)H to produce D-mannonate and NAD(P)$^+$.[1]

[1] J. J. Farmer & R. G. Eagon (1969) J. Bacteriol. **97**, 97.

Selected entries from **Methods in Enzymology** [vol, page(s)]:
General discussion: 5, 197
Assay: 5, 197
Other molecular properties: D-altronate dehydrogenase preparation, not present in, **5**, 195; properties, **5**, 199; purification, Escherichia coli, **5**, 198; uronate isomerase assay, in, **5**, 190, 192

MATRILYSIN

This zinc-dependent metalloendopeptidase [EC 3.4.24.23] also known as matrin, uterine metalloendopeptidase, matrix metalloproteinase 7 (MMP-7), putative metalloproteinase-1 (PUMP-1), and punctuated metalloproteinase (lacking a C-terminal hemopexin domain), catalyzes peptide-bond hydrolysis in many proteins that are a part of the extracellular matrix: for example, gelatins I, III,

IV, V, fibronectin, collagen IV, elastin, laminin, aggrecan, fibulin 1 and 2, and vitronectin. It also hydrolyzes the Ala14–Leu15 and Tyr16–Leu17 bonds in the oxidized insulin B chain. The enzyme exhibits a preference for large hydrophobic aminoacyl residues in the P1′ position. A prolyl residue is preferred at P3. The best synthetic substrate is Ac-Pro-Leu-Glu-∇-Leu-Arg-Ala-NH$_2$, where ∇ indicates the cleavage site.[1–3]

[1] J. F. Woessner, Jr. (1995) Meth. Enzymol. **248**, 485.
[2] C. L. Wilson & L. M. Matrisian (1996) Int. J. Biochem. Cell Biol. **28**, 123.
[3] H. Nagase & G. B. Fields (1996) Biopolymers **40**, 399.

Selected entries from **Methods in Enzymology** [vol, page(s)]:
General discussion: 248, 485.
Assay: Azocoll, with, **248**, 485; Mca-peptide, with, **248**, 488; proteoglycan degradation method, **248**, 52; transferrin, with, **248**, 487
Other molecular properties: activation, **248**, 493; active-site titration, **248**, 100; discovery, **248**, 485; human (activity, **248**, 490; purification, **248**, 492; recombinant, **248**, 493; substrate specificity, **248**, 494); inhibition, **248**, 495; isoelectric point, **248**, 493; molecular weight, **248**, 485, 493; nomenclature, **248**, 485; pH profile, **248**, 493; polypeptide chain structure, **248**, 204; properties, **248**, 192, 203, 511 (enzymatic, **248**, 494; physical, **248**, 493); purification (after expression of recombinant DNA, **248**, 493; from cell culture media, **248**, 492); rat uterus, **248**, 485 (activity, **248**, 490; properties, **248**, 491; purification, **248**, 490; purity, **248**, 491; substrate specificity, **248**, 494; stability, **248**, 493; substrate specificity, **248**, 494); tissue extracts, measurement, in, **248**, 523; zymography, **248**, 489

MATRILYSIN-2

This metalloproteinase, also called endometase and matrix metalloproteinase 26, has been shown to act on type IV collagen, fibronectin, fibrinogen, type I gelatin, and α$_1$-proteinase inhibitor as well as the substrates of matrixins and tumor necrosis factor-α converting enzyme.[1,2]

[1] J. A. Uría & C. López-Otín (2000) Cancer Res. **60**, 4745.
[2] H. I. Park, J. Ni, F. E. Gerkema, D. Liu, V. E. Belozerov & Q. X. A. Sang (2000) JBC **275**, 20540.

MATRIPTASE

This serine endopeptidase, also called epithin and membrane-type serine protease 1, catalyzes peptide-bond hydrolysis with trypsin-like activity, preferring arginyl or lysyl residues in the P1 position.[1] Small aminoacyl residues (e.g., Ala or Gly) are preferred in the P2 position. The enzyme is able to act on proteins in the extracellular matrix. In addition, matriptase can activate hepatocyte growth factor/scattering factor, can activate c-Met tyrosine phosphorylation in A549 human lung carcinoma cells, and can activate urokinase plasminogen activator.[2]

[1] C. Y. Lin, J. Anders, M. Johnson, Q. X. A. Sang & R. B. Dickson (1999) JBC **274**, 18231.
[2] S. L. Lee, R. B. Dickson & C. Y. Lin (2000) JBC **275**, 36720.

MATRIX METALLOPROTEINASE 19

This metalloproteinase, often abbreviated MMP-19, is proposed to act on proteins associated with the basement membrane. The catalytic domain of MMP-19 has been shown to act on type IV collagen, laminin, nidogen, the large tenascin-C isoform, fibronectin, and type I gelatin *in vitro*.[2]

[1] A. M. Pendás, V. Knäuper, X. S. Puente, E. Llano, M. G. Mattei, S. Apte, G. Murphy & C. López-Otín (1997) JBC 272, 4281.
[2] J. O. Stracke, M. Hutton, M. Stewart, A. M. Pendás, B. Smith, C. López-Otín, G. Murphy & V. Knäuper (2000) JBC 275, 14809.

MATRIX METALLOPROTEINASE 23

This metalloproteinase, often abbreviated MMP-23 (it had earlier been known as MMP-21 and MMP-22) and also called femalysin, is predominantly expressed in ovary, testis, and prostate.[1]

[1] G. Velasco, A. M. Pendás, A. Fueyo, V. Knäuper, G. Murphy & C. López-Otín (1999) JBC 274, 4570.

MATRIX METALLOPROTEINASE 27

This metalloproteinase, often abbreviated MMP-27 and previously called MMP-22, acts on peptide bonds in casein and gelatin.[1]

[1] M. Z. Yang & M. Kurkinen (1998) JBC 273, 17893.

MATRIX METALLOPROTEINASE 28

This metalloproteinase, abbreviated MMP-28, has a broad range of expression in carcinomas as well as normal adult and fetal tissues.[1]

[1] G. N. Marchenko & A. Y. Strongin (2001) Gene 265, 87.

*Mbo*I RESTRICTION ENDONUCLEASE

This type II restriction enzyme [EC 3.1.21.4], isolated from *Moraxella bovis*, catalyzes the hydrolysis of both strands of DNA at 5′ ... ∇GATC ... 3′ (it also catalyzes single strand cleavage).[1] *Mbo*I is blocked by *dam* methylation. The 5′-GATC extension produced by this endonuclease can be ligated to cleaved fragments generated by a number of other endonucleases (for example, *Bam*HI, *Bcl*I, and *Dpn*II).

[1] R. E. Gelinas, P. A. Myers & R. J. Roberts (1977) J. Mol. Biol. 114, 169.

Selected entries from **Methods in Enzymology** [vol, page(s)]:
General discussion: 65, 82, 348, 350, 351, 485
Other molecular properties: isoschizomers, 65, 137; purification, 65, 94; recognition sequence, 65, 8, 14

*Mbo*II RESTRICTION ENDONUCLEASE

This type II restriction enzyme [EC 3.1.21.4], isolated from *Moraxella bovis*, catalyzes the hydrolysis of both strands of DNA at 5′ ... GAAGANNNNNNNN∇ ... 3′ and 3′ ... CTTCTNNNNNNN∇N ... 5′ (where N refers to any base), producing a single-base 3′-extension.[1–4] This monomeric endonuclease has star activity and can also catalyze the hydrolysis of single-stranded DNA. It is blocked by overlapping *dam* methylation. **See** *"Star" Activity*

[1] R. E. Gelinas, P. A. Myers & R. J. Roberts (1977) J. Mol. Biol. 114, 169.
[2] N. L. Brown, C. A. Hutchison, III, & M. Smith (1980) J. Mol. Biol. 140, 143.
[3] M. McClelland, M. Nelson & C. R. Cantor (1985) NAR 13, 7171.
[4] M. Sektas, T. Kaczorowski & A. J. Podhajska (1992) NAR 20, 433.

Selected entries from **Methods in Enzymology** [vol, page(s)]:
General discussion: 65, 82, 348, 350, 351, 485
Other molecular properties: recognition sequence, 65, 8

M.*Bst*I TYPE II METHYLASE

This methylating enzyme, also known as *Bst*I methyltransferase, is associated with the *Bst*I restriction endonuclease type II system. This enzyme recognizes the sequence 5′ ... GGATCC ... 3′ and catalyzes the methylation of the adenine residue, using S-adenosyl-L-methionine.[1] **See also** *Site-Specific DNA-Methyltransferase (Adenine-Specific)*

[1] W. P. Levy & N. E. Welker (1981) Biochemistry 20, 1120.

Selected entries from **Methods in Enzymology** [vol, page(s)]:
General discussion: 65, 170

M.*Bsu*RI METHYLTRANSFERASE

This site-specific type II DNA-methyltransferase (cytosine-specific) [EC 2.1.1.73], obtained from *Bacillus subtilis* strain R, recognizes the DNA sequence 5′ ... GGCC ... 3′ and catalyzes the methylation of C at position 3, using S-adenosyl-L-methionine, producing a 5-methylcytosine residue and S-adenosyl-L-homocysteine.[1] **See also** *Site-Specific DNA-Methyltransferase (Cytosine-Specific)*

[1] R. Lauster, T. A. Trautner & M. Noyer-Weidner (1989) J. Mol. Biol. 206, 305.

M.*Bbv*SI ORPHAN METHYLASE

This site-specific methyltransferase (cytosine-specific) [EC 2.1.1.73], first isolated from *Bacillus brevis* S and formerly known as *Bbv*SI, catalyzes the transfer of a methyl group from S-adenosyl-L-methionine to the cytosine in the number-two base position in the DNA sequence 5′ ... GCWGC ... 3′, where W refers to either A or T, to produce S-adenosyl-L-homocysteine and the sequence containing 5-methylcytosine.[1] *See also Site-Specific DNA-Methyltransferase (Cytosine-Specific)*

[1] B. F. Vanyushin & A. P. Dobritsa (1975) *BBA* **407**, 61.

Selected entries from *Methods in Enzymology* [vol, page(s)]:
Other molecular properties: recognition sequence, **65**, 3

M.*Eco*RI METHYLTRANSFERASE

This site-specific DNA-adenine N^6-methyltransferase [EC 2.1.1.72], obtained from *Escherichia coli* RY13, catalyzes the reaction of *S*-adenosyl-L-methionine with the third base in the double-stranded DNA sequence 5′ ... GAATCC ... 3′ at the N^6-position, producing the methylated DNA and *S*-adenosyl-L-homocysteine.[1–5]

adenine residue $\qquad$ N^6-methyladenine residue

The enzyme, also called *Eco*RI methyltransferase and *Eco*RI methylase, has a random Bi Bi kinetic mechanism.[1] Fluorescence and kinetic investigations demonstrate that the enzyme catalyzes the "flipping-out" of the base prior to methylation.[3] There is less than a 4 ms delay between enzyme binding and target base flipping.[4] Investigations also suggest that M.*Eco*RI proceeds to do a one-dimensional search along the DNA polymer for the recognition sequence.[5] Upon location of the target base, the DNA is distorted and methylation occurs, possibly assisted with general base catalysis. *See also Site-Specific DNA-Methyltransferase (Adenine-Specific)*

[1] N. O. Reich & N. Mashhoon (1993) *JBC* **268**, 9191.
[2] A. L. Pogolotti, A. Ono, R. Subramaniam & D. V. Santi (1988) *JBC* **263**, 7461.
[3] B. W. Allan, J. M. Beechem, W. M. Lindstrom & N. O. Reich (1998) *JBC* **273**, 2368.
[4] B. W. Allan, N. O. Reich & J. M. Beechem (1999) *Biochemistry* **38**, 5308.
[5] M. A. Surby & N. O. Reich (1996) *Biochemistry* **35**, 2201 and 2209.

Selected entries from *Methods in Enzymology* [vol, page(s)]:
Other molecular properties: *Eco*RI, use of, **68**, 264; partial digestion of DNA, **225**, 627

MELILOTATE 3-MONOOXYGENASE

This FAD-dependent oxidoreductase [EC 1.14.13.4], also known as melilotate 3-hydroxylase and 2-hydroxyphenylpropionate hydroxylase, catalyzes the reaction of 3-(2-hydroxyphenyl)propanoate (that is, melilotate) with NADH and dioxygen to produce 3-(2,3-dihydroxyphenyl)propanoate, NAD^+, and water.[1–7]

melilotate $\qquad$ 3-(2,3-dihydroxy-phenyl)propionate

Other substrates include 2-hydroxycinnamate. The enzyme has an ordered Ter Bi kinetic mechanism, melilotate binding first, followed by NADH and finally O_2. The dissociation of NAD^+ is partially rate-limiting. The overall reaction proceeds via the formation of an oxygenated flavin intermediate (flavin-C-4a-hydroperoxide).

[1] B. A. Palfey & V. Massey (1998) *CBC* **3**, 83.
[2] L. M. Schopfer & V. Massey (1980) *JBC* **255**, 5355.
[3] L. M. Schopfer & V. Massey (1979) *JBC* **254**, 10634.
[4] V. Massey & P. Hemmerich (1975) *The Enzymes*, 3rd ed., **12**, 191.
[5] S. Strickland, L. M. Schopfer & V. Massey (1975) *Biochemistry* **14**, 2230.
[6] M. S. Flashner & V. Massey (1974) in *Mol. Mech. Oxygen Archiv.* (O. Hayaishi, ed.) p. 245.
[7] S. Strickland & V. Massey (1973) *JBC* **248**, 2953.

Selected entries from *Methods in Enzymology* [vol, page(s)]:
General discussion: 52, 16
Assay: 53, 552, 553
Other molecular properties: activity, **53**, 552, 556; *Arthrobacter*, **53**, 552; effectors, **53**, 557; flavin content, **53**, 556; kinetic properties, **53**, 552; optical properties, **53**, 556; oxy form, spectrum, **52**, 36; oxygenated-flavin intermediates, **53**, 557, 558; *Pseudomonas*, **53**, 552; purification, **53**, 553; purity, **53**, 556; reconstitution, **53**, 555; reduction, **53**, 557; staining on gels, **22**, 595; stereospecificity, **54**, 229; **87**, 120; substrate specificity, **53**, 556, 557; subunits, **53**, 556; uncoupling, **53**, 557

MEMAPSINS

Memapsin 1 (also called BACE2 and β-site APP-cleaving enzyme 2) and memapsin 2 (also called BACE1 and β-site APP-cleaving enzyme) are aspartic endopeptidases that catalyze the hydrolysis of a peptide bond in the amyloid-β precursor protein.[1–3]

[1] X. L. Lin, C. Koelsch, S. L. Wu, D. Downs, A. Dashti & J. Tang (2000) *PNAS* **97**, 1456.
[2] M. Haniu, P. Denis, Y. Young, E. A. Mendiaz, J. Fuller, J. O. Hui, B. D. Bennett, S. Kahn, S. Ross, T. Burgess, V. Katta, G. Rogers, R. Vassar & M. Citron (2000) *JBC* **275**, 21099.
[3] L. Hong, G. Koelsch, X. L. Lin, S. L. Wu, S. Terzyan, A. K. Ghosh, X. C. Zhang & J. Tang (2000) *Science* **290**, 150.

MEMBRANE ALANYL AMINOPEPTIDASE

This zinc-dependent metalloaminopeptidase [EC 3.4.11.2] (also known as microsomal aminopeptidase, aminopeptidase M, aminopeptidase N, particle-bound aminopeptidase, amino-oligopeptidase, membrane aminopeptidase I, and peptidase E) catalyzes the release of an N-terminal amino acid from a peptide, amide, or arylamide. There is a preference for N-terminal alanyl residues but most other amino acids are released as well.

The order of preference appears to be Ala, Phe, Tyr, Leu, Arg, Thr, Trp, Lys, Ser, Asp, His, and Val (note, however, that leucinamide is a poor substrate). N-terminal prolyl, α-glutamyl, and γ-glutamyl residues are weak substrates. If the N-terminal aminoacyl residue is hydrophobic and bulky (for example, Leu, Tyr, or Trp) and the adjacent residue is prolyl, the Xaa-Pro dipeptide can be released.

The enzyme also catalyzes a dipeptidase activity; for example, cysteinylglycine is readily hydrolyzed.[1-3] **See also** *Cysteinylglycine Dipeptidase*

[1] J. Wang & M. D. Cooper (1996) in *Zinc Metalloproteinases in Health and Disease* (N. M. Hooper, ed.), Taylor & Francis, London, p. 131.
[2] S. S. Tate (1985) *Meth. Enzymol.* **113**, 471.
[3] J. K. McDonald & A. J. Barrett (1986) *Mammalian Proteases: A Glossary and Bibliography*, vol. **2**, Academic Press, New York, p. 59.

Selected entries from *Methods in Enzymology* [**vol**, page(s)]:
General discussion: 19, 514; 47, 542, 577; 113, 474
Assay: 113, 474
Other molecular properties: activity, 248, 270; activity assay, 47, 77; amino acid analysis of flavin peptides, for, 53, 456; complete hydrolysis, 47, 41, 43; *Escherichia coli* (properties, 248, 190; structure, 248, 186); esterified carboxyl groups and, 25, 613; immobilization, 47, 44; inhibitors (analgesic responses induced by, 248, 272; antinociceptive effects, 248, 272; behavioral effects, 248, 279; clinical applications, 248, 282; dependence, 248, 276; development, 248, 270; properties, 248, 264, 270; side effects, 248, 276; tolerance, 248, 276); properties, 25, 254; rate of hydrolysis of amino acid derivatives, 25, 256; rat kidney (assay, 113, 474; properties, 113, 480; purification, 113, 479; substrate specificity, 113, 482)

MEMBRANE DIPEPTIDASE

This membrane-bound, zinc-dependent dipeptidase [EC 3.4.13.19], also known as microsomal dipeptidase, renal dipeptidase, and dehydropeptidase I, catalyzes the hydrolysis of dipeptides. The enzyme has a broad specificity, but hydrophobic dipeptides, including Pro–Xaa, are cleaved preferentially. Interestingly, the C-terminal amino acid can be either the D- or L-stereoisomer (for example, glycyl-D-phenylalanine is readily hydrolyzed). This hydrolase also acts on dehydropeptides such as glycyldehydrophenylalanine. The enzyme is inhibited by cilastatin (IC$_{50}$ = 0.1 mM).[1,2] **See also** *Aminoacylase*

[1] B. J. Campbell (1970) *MIE* **19**, 722.
[2] S. Keynan, N. M. Hooper & A. J. Turner (1996) in *Zinc Metalloproteases in Health and Disease* (N. H. Hooper, ed.), Taylor & Francis, London, p. 285.

Selected entries from *Methods in Enzymology* [**vol**, page(s)]:
General discussion: 2, 107, 108, 109; 19, 722
Assay: 2, 109
Other molecular properties: activators, 19, 728; amino acid composition, 19, 729; distribution, 19, 729; exocystine desulfhydrase, present in preparation of, 2, 319; inhibitors, 19, 728; kinetic properties, 19, 728; 248, 222; physical properties and purity, 19, 727; properties, 2, 113; purification, 2, 112; 19, 724; rat kidney (assays, 113, 472; properties, 113, 480; purification, 113, 475; substrate specificity, 113, 482); specificity, 19, 728; stability, 19, 727

MEMBRANE-OLIGOSACCHARIDE GLYCEROPHOSPHOTRANSFERASE

This transferase [EC 2.7.8.21], also known as periplasmic phosphoglycerotransferase and phosphoglycerol cyclase, catalyzes the transfer of a glycerophospho group from one membrane-derived oligosaccharide to another.[1] β-Glucose residues in simple glucosides, such as gentiobiose, also act as acceptor substrates. If the acceptor concentration is low, free cyclic 1,2-phosphoglycerol is formed.

[1] D. E. Goldberg, M. K. Rumley & E. P. Kennedy (1981) *PNAS* **78**, 5513.

MEMBRANE PRO-Xaa CARBOXYPEPTIDASE

This membrane-bound glycoprotein [EC 3.4.17.16] (also known as carboxypeptidase P, microsomal carboxypeptidase, and membrane Pro-X carboxypeptidase) catalyzes the release of a C-terminal residue adjacent to a prolyl residue (that is, Pro–Xaa). C-terminal proline, 4-hydroxyproline, and leucine are not released. Other aminoacyl residues can be tolerated at the P1 position besides a prolyl residue: for example, Z-Ala-Tyr, Z-Gly-Tyr, Z-Ala-Phe, and Z-Gly-Phe are all hydrolyzed.

The enzyme also exhibits dipeptidase activity, catalyzing the hydrolysis of prolylalanine and prolylphenylalanine. Its pH optimum is 7.75 and the enzyme is not inhibited by diisopropylfluorophosphate: thus, it is readily distinguishable from serine-type carboxypeptidases and from lysosomal Pro-Xaa carboxypeptidase [EC 3.4.16.2].[1,2] **See also** *Carboxypeptidase P*

[1] P. Dehm & A. Nordwig (1970) *EJB* **17**, 372.
[2] S. Hedeager-Sørensen & A. J. Kenny (1985) *BJ* **229**, 251.

Selected entries from *Methods in Enzymology* [**vol**, page(s)]:
Other molecular properties: bovine kidney, 80, 788

MEMBRANE-TYPE MATRIX METALLOPROTEINASES

Six membrane-type matrix metalloproteinases (MT1-MMP, MT2-MMP, MT3-MMP, MT4-MMP, MT5-MMP, and MT6-MMP) have been identified in mammals. They correspond to matrix metalloproteins MMP-14 through MMP-17, MMP-24, and MMP-25, respectively (the molecular masses of the activated forms of the first four enzymes are 53.7 kDa, 61.2 kDa, 55.4 kDa, and 57.8 kDa). All catalyze peptide bond hydrolysis in matrix proteins. MT1-MMP will activate progelatinase A by cleaving the Asn66–Leu67 bond. MT1-MMP will also degrade fibronectin, vitronectin, elastin, gelatin, interstitial collagens, and casein. MT3-MMP will activate progelatinase A.[1-3]

[1] H. Sato, T. Takino, Y. Okada, J. Cao, A. Shinagawa, E. Yamamoto & M. Seiki (1994) *Nature* **370**, 61.

[2] H. Will, S. J. Atkinson, G. S. Butler, B. Smith & G. Murphy (1996) *JBC* **271**, 17119.

[3] T. Kinoshita, H. Sato, T. Takino, T. Itoh, T. Akizawa & M. Seiki (1996) *Cancer Res.* **56**, 2535.

(−)-MENTHOL DEHYDROGENASE

This oxidoreductase [EC 1.1.1.207], also known as monoterpenoid dehydrogenase, catalyzes the reaction of (−)-menthone with NADPH to produce (−)-menthol and NADP⁺.[1,2]

(−)-menthone   (−)-menthol

Other substrates include other cyclohexanones and cyclohexenones. The enzyme is not identical with (+)-neomenthol dehydrogenase [EC 1.1.1.208], which also catalyzes the NADPH-dependent reaction of (−)-menthone.

[1] R. Kjonaas, C. Martinkus-Taylor & R. Croteau (1982) *Plant Physiol.* **69**, 1013.

[2] R. Croteau & J. N. Winters (1982) *Plant Physiol.* **69**, 975.

(−)-MENTHOL MONOOXYGENASE

This oxidoreductase [EC 1.14.13.46] catalyzes the reaction of (−)-menthol with NADPH and dioxygen to produce *p*-menthane-3,8-diol, NADP⁺, and water.[1]

[1] K. M. Madyastha & V. Srivatsan (1988) *Drug Metab. Disp.* **16**, 765.

MEPRIN A

This membrane-bound, zinc-dependent metalloendopeptidase [EC 3.4.24.18], also called endopeptidase-2 and PABA-peptide hydrolase, catalyzes peptide-bond hydrolysis in proteins and peptides.[1-3] There appears to be a preference for hydrophobic aminoacyl residues at the P1 position (*i.e.*, Xaa in Xaa–Xbb), and uncharged residues in P1′. The oxidized insulin B chain is hydrolyzed at His5–Leu6, Leu6–Cys7, Gly8–Ser9, His10–Leu11, Ala14–Leu15, Leu15–Tyr16, Leu17–Val18, Gly20–Glu21, Phe24–Phe25, and Phe25–Tyr26. Protein substrates include parathroid hormone, bradykinin, the angiotensins, substance P, neurotensin, type IV collagen, α-melanocyte stimulating hormone, *etc.* The enzyme is not inhibited by thiorphan or phosphoramidon, both of which inhibit neprilysin [EC 3.4.24.11]. Captopril, which inhibits angiotensin I-converting enzyme [EC 3.4.15.1], does not inhibit meprin A. Meprin's oligomeric nature helps in distinguishing this enzyme from other proteases as does its responses (or, lack of responce) to certain inhibitors. *See also Meprin B; Meprin, Human*

[1] J. S. Bond & R. J. Beynon (1995) *Protein Sci.* **4**, 1247.

[2] R. L. Wolz & J. S. Bond (1995) *MIE* **248**, 325.

[3] P. Marchand & J. S. Bond (1996) in *Intracellular Protein Catabolism* (K. Suzuki & J. S. Bond, eds.), Plenum, New York, p. 13.

Selected entries from *Methods in Enzymology* [vol, page(s)]:
General discussion: 248, 325
Assay: with azocasein, **248**, 329; with fluorogenic peptides, **248**, 332; with nitrobradykinin, **248**, 331
Other molecular properties: activity, **248**, 326, 337; domain structure, deduced from cDNAs, **248**, 335; endopeptidase activity, effects of trypsin treatment, **248**, 337; hydrolysis of biologically active peptides, **248**, 339; inhibitors, **248**, 341; isoelectric point, **248**, 334; kinetic characteristics, **248**, 338; metal requirements, **248**, 334; molecular properties, **248**, 332; molecular weight, **248**, 332; primary sequence, **248**, 335; purification, **248**, 326; substrates, **248**, 329; substrate specificity, comparison with astacin, **248**, 341; subunits, **248**, 326 (carbohydrate, **248**, 333)

MEPRIN B

This membrane-bound, zinc-dependent metalloendopeptidase [EC 3.4.24.63], also called meprin β subunit, catalyzes peptide-bond hydrolysis in proteins and peptides. Meprin B has none of the meprin α subunits found in meprin A. Meprin B will hydrolyze the oxidized insulin B chain at His5–Leu6, Leu6–Cys7, Ala14–Leu15, and Cys19–Gly20 (note that the cleavage pattern is different from that of meprin A). In addition, meprin B will not act on small peptides of ten or fewer aminoacyl residues (for example, bradykinin) which can be hydrolyzed by meprin A. Meprin B is inhibited by EDTA and 1,10-phenanthroline, but not by phosphoramidon, captopril, or thiorphan.[1-3] *See also Meprin A; Meprin, Human*

[1] J. S. Bond & R. J. Beynon (1995) *Protein Sci.* **4**, 1247.

[2]R. L. Wolz & J. S. Bond (1995) *MIE* **248**, 325.
[3]P. Marchand & J. S. Bond (1996) in *Zinc Metalloproteinases in Health and Disease* (N. M. Hooper, ed.), Taylor & Francis, London, p. 23.

Selected entries from *Methods in Enzymology* [vol, page(s)]:
General discussion: 248, 325
Assay: 248, 329; azocasein assay, 248, 329
Other molecular properties: activity, 248, 326; domain structure, deduced from cDNAs, 248, 335; endopeptidase activity, effects of trypsin treatment, 248, 337; inhibitors, 248, 341; kinetic characteristics, 248, 338; metal requirements, 248, 334; molecular weight, 248, 332; primary sequence, 248, 335; purification, 248, 328; substrates, 248, 329; subunits, 248, 326 (carbohydrate, 248, 333)

MEPRIN, HUMAN

First observed for its hydrolytic action on Bz-Tyr-*p*-aminobenzoic acid (hence the alternative name PABA-peptide hydrolase), this metalloproteinase catalyzes peptide-bond hydrolysis in a number of proteins, including angiotensin I, angiotensin II, parathyroid hormone, the B chain of insulin, bradykinin, and substance P. There is an apparent preference for an aromatic aminoacyl residue in position P1.[1,2]

[1]J. S. Bond & R. J. Beynon (1995) *Protein Sci.* **4**, 1247.
[2]E. Sterchi, E. Dumermuth, J. Eldering & J. Grünberg (1994) in *Mammalian Brush Border Membrane Proteins II* (M. J. Lentze, H. Y. Naim & R. Grand, eds.), Georg Thième, Stuttgart, p. 141.

Selected entries from *Methods in Enzymology* [vol, page(s)]:
Other molecular properties: properties, 248, 194

3-MERCAPTOPYRUVATE SULFURTRANSFERASE

This zinc-dependent enzyme [EC 2.8.1.2] catalyzes the reaction of 3-mercaptopyruvate to produce pyruvate and elemental sulfur (S^0); in the presence of cyanide, elemental sulfur then forms thiocyanate (^-SCN).[1–4]

HS—CH₂—C(=O)—COO⁻ 3-mercaptopyruvate
H₃C—C(=O)—COO⁻ pyruvate

Other substrates include sulfite, sulfinates, mercapto-ethanol, and mercaptopyruvate. The kinetic mechanism appears to be rapid equilibrium ordered when 3-mercaptopyruvate is the sulfur donor substrate[1]; with mercapto-ethanol, however, the enzyme exhibits a random Bi Bi kinetic mechanism.[2] As noted above, sulfur is not directly discharged from the enzyme as thiocyanate; rather, elemental S^0 is released, and thiocyanate is subsequently formed nonenzymatically at a rate determined by the cyanide concentration.[1,3]

[1]R. Jarabak & J. Westley (1980) *Biochemistry* **19**, 900.

[2]R. Jarabak (1981) *MIE* **77**, 291.
[3]J. V. Schloss & M. S. Hixon (1998) *CBC* **2**, 93.
[4]A. J. L. Cooper (1983) *Ann. Rev. Biochem.* **52**, 187.

Selected entries from *Methods in Enzymology* [vol, page(s)]:
General discussion: 5, 987, 990; 77, 291
Assay: 5, 987, 990; 77, 292, 293 (with cyanide, 5, 987; 77, 293; with 2-mercaptoethanol, 77, 292; pyruvate formation, 77, 292; thiocyanate formation, 77, 293)
Other molecular properties: copper content, 77, 296; definition of units, 77, 293; kinetic constants, 77, 296; mechanism, 77, 291, 296, 297; properties, 5, 989, 991; 77, 296, 297; purification, 5, 988, 991; 77, 293; size, 77, 296

MERCURY(II) REDUCTASE

This FAD-dependent enzyme [EC 1.16.1.1], also known as mercuric reductase, catalyzes the reaction of Hg(II) with NADPH to produce elemental Hg^0, $NADP^+$, and a proton.[1–5] This reaction protects thiol-containing metabolites/proteins against Hg^{2+} mercaptide formation. The complex series of electron transfer steps begins with NADPH and forms intermediates, first involving FAD, followed by transfer to the enzyme's redox-active disulfides; the latter reaction yields four thiol groups from which two electrons are transferred to Hg^{2+}. A flavin-C4a-thiol adduct is formed transiently in this scheme.[2]

[1]A. R. Clarke & T. R. Dafforn (1998) *CBC* **3**, 1.
[2]B. A. Palfey & V. Massey (1998) *CBC* **3**, 83.
[3]M. J. Moore, M. D. Distefano, L. D. Zydowsky, R. T. Cummings & C. T. Walsh (1990) *Acc. Chem. Res.* **23**, 301.
[4]L. Sahlman, A.-M. Lambeir, S. Lindskog & H. B. Dunford (1984) *JBC* **259**, 12403.
[5]S. J. Rinderle, J. E. Booth & J. W. Williams (1983) *Biochemistry* **22**, 869.

MESENTERICOPEPTIDASE

This serine endopeptidase, isolated from *Bacillus pumilus* (formerly *Bacillus mesentericus*), resembles subtilisin and is currently classified under the same EC number [EC 3.4.21.62].[1] The enzyme has an alkaline pH optimum.

[1]I. Svendsen, N. Genov & K. Idakieva (1986) *FEBS Lett.* **196**, 228.

METALLOCARBOXYPEPTIDASE D

This zinc-dependent metallocarboxypeptidase [EC 3.4.17.22], also known as carboxypeptidase D, catalyzes the release of C-terminal arginine and lysine residues from polypeptides. This enzyme prefers alanyl residues in the penultimate position (that is, position P1). The large enzyme (180 kDa) is stimulated by millimolar levels of Co^{2+} and has a pH optimum of 5.5 to 6.5.[1]

[1]L. Song & L. D. Fricker (1995) *JBC* **270**, 25007.

METHANE MONOOXYGENASE

This oxidoreductase [EC 1.14.13.25], which has a broad specificity, catalyzes the reaction of methane with NAD(P)H and dioxygen to produce methanol, NAD(P)$^+$, and water.[1-8] Many alkanes can be hydroxylated, and alkenes are converted to the corresponding epoxides. Carbon monoxide is oxidized to carbon dioxide; ammonia is oxidized to hydroxylamine; and some aromatic compounds and cyclic alkanes can also be hydroxylated, albeit not as effectively.

Whether a radical intermediate is formed is still debated. There are three distinct proteins in the soluble methylococcal enzyme: an FAD-containing protein that has a [2Fe:2S] cluster (electrons are transferred to the flavin from NADH via the iron-sulfur cluster), a protein that serves as a coupling factor between the electron transfers and the hydroxylation of methane, and a protein with a hydroxo-iron center that binds dioxygen and methane.

[1] L. Westerheide, M. Pascaly & B. Krebs (2000) *Curr. Opin. Chem. Biol.* **4**, 235.
[2] A. R. Clarke & T. R. Dafforn (1998) *CBC* **3**, 1.
[3] B. G. Fox (1998) *CBC* **3**, 261.
[4] A. Messerschmidt (1998) *CBC* **3**, 401.
[5] J. B. Howard & D. C. Rees (1991) *Adv. Protein Chem.* **42**, 199.
[6] M. J. Rataj, J. E. Kauth & M. I. Donnelly (1991) *JBC* **266**, 18684.
[7] J. Green & H. Dalton (1989) *JBC* **264**, 17698.
[8] Y. Jin & J. D. Lipscomb (1999) *Biochemistry* **38**, 6178.

Selected entries from *Methods in Enzymology* [vol, page(s)]:
General discussion: 188, 181, 191
Assay: 188, 181, 183, 191
Other molecular properties: *Methylosinus trichosporium* OB3b (assay, **188**, 191; component protein concentrations, optimization for assays, **188**, 192; electron nuclear double resonance spectroscopy, **246**, 581; properties, **188**, 200; purification, **188**, 197); reductase NADH oxidoreductase, component of methane monooxygenase, assay, **188**, 195; soluble, from *Methylococcus capsulatus* Bath (assay, **188**, 181; component proteins, assay and purification, **188**, 183; properties, **188**, 186; purification, **188**, 183)

METHANE OXIDASE

This oxidoreductase reportedly catalyzes the reaction of methane with NADH and two molecules of dioxygen to produce formate, NAD$^+$, and two molecules of water.[1]

[1] D. W. Ribbons (1975) *J. Bacteriol.* **122**, 1351.

Selected entries from *Methods in Enzymology* [vol, page(s)]:
General discussion: 52, 20

METHANETHIOL OXIDASE

This oxidoreductase [EC 1.8.3.4], also known as methylmercaptan oxidase, catalyzes the reaction of methanethiol (CH_3SH) with dioxygen and water to produce formaldehyde, hydrogen sulfide, and hydrogen peroxide.[1-3] Ethanethiol is converted to acetaldehyde.

[1] G. M. H. Suylen, P. J. Large, J. P. van Dijken & J. G. Kuenen (1987) *J. Gen. Microbiol.* **133**, 2989.
[2] W. D. Gould & T. Kanagawa (1992) *J. Gen. Microbiol.* **138**, 217.
[3] N. A. Smith & D. P. Kelly (1988) *J. Gen. Microbiol.* **134**, 3031.

METHANOL DEHYDROGENASE

This pyrroloquinoline-quinone-dependent enzyme [EC 1.1.99.8], also known as alcohol dehydrogenase (acceptor), catalyzes the reaction of a primary alcohol with an acceptor substrate to produce an aldehyde and the reduced acceptor (the reduced protein is olive-green in color due to the cofactor).[1-7] A wide range of primary alcohols (*e.g.*, methanol and ethanol) can serve as substrates. The acceptor substrate can be cytochrome *c* or artificial electron acceptors such as phenazine ethosulfate (ammonia or methylamine are required as activators when the enzyme is assayed with artificial acceptors). The enzyme has a ping pong Bi Bi mechanism[7] in which there are two single-electron transfer steps following the release of methanol, resulting in the reoxidation of pyrroloquinoline-quinone (PQQ).

A PQQ-independent methanol dehydrogenase [EC 1.1.1.244] catalyzes the reaction of methanol with NAD$^+$ to produce formaldehyde and NADH.[8-10] This enzyme will also use ethanol, 1-propanol, and 1-butanol as alternative substrates.

[1] C. Anthony & M. Ghosh (1998) *Prog. Biophys. Mol. Biol.* **69**, 1.
[2] C. Anthony (1998) *CBC* **3**, 155.
[3] C. Anthony (1996) *Biochem. J.* **320**, 697.
[4] J. P. Klinman & D. Mu (1994) *Ann. Rev. Biochem.* **63**, 299.
[5] M. A. Ator & P. R. Ortez de Montellano (1990) *The Enzymes*, 3rd ed., **19**, 213.
[6] J. A. Duine, J. Frank, & J. A. Jongejan (1987) *AE* **59**, 169.
[7] J. A. Duine & J. Frank (1980) *Biochem. J.* **187**, 213.
[8] N. Arfman, E. M. Watling, W. Clement, R. J. Van Oosterwijk, G. E. De Vries, W. Harder, M. M. Attwood & L. Dijkhuizen (1989) *Arch. Microbiol.* **152**, 280.
[9] J. Vonck, N. Arfman, G. E. De Vries, J. Van Beeumen, E. F. J. Van Bruggen & L. Dijkhuizen (1991) *JBC* **266**, 3949.
[10] N. Arfman, J. Van Beeumen, G. E. De Vries, W. Harder & L. Dijkhuizen (1991) *JBC* **266**, 3955.

Selected entries from *Methods in Enzymology* [vol, page(s)]:
General discussion: 18B, 808; 89, 450; 188, 33, 202, 210, 216; 249, 373; PQQ-independent, 188, 223
Assay: 18B, 808, 809; 188, 203, 210, 212, 217, 223
Other molecular properties: *Bacillus* C1 (assay, **188**, 223; properties, **188**, 225; purification, **188**, 224); calcium role, **258**, 213; cloning from bacteria, mutant complementation, **258**, 217; electron acceptors, **258**, 195; genes, bacterial (expression systems, **258**, 221; mutant construction, **258**, 224); *Hyphomicrobium* X (assay, **188**, 203; properties, **188**, 206; purification, **188**, 204); *Methylobacterium extorquens* AM1 (assay, **188**, 210; properties, **188**, 215; purification, **188**, 213); modifier protein (assay, **188**, 217; properties, **188**, 221; purification from

The Enzyme Reference

557

METHANOL:5-HYDROXY-BENZIMIDAZOLYLCOBAMIDE Co-METHYLTRANSFERASE

This methyltransferase [EC 2.1.1.90] catalyzes the reaction of methanol with 5-hydroxybenzimidazolylcobamide to produce water and Co-methyl-Co-5-hydroxybenzimidazolylcob(I)amide.[1–3] The enzyme from *Methanosarcina barkeri* contains three to four molecules of bound 5-hydroxybenzimidazolylcobamide which act as the methyl acceptor. The methyltransferase is inactivated by dioxygen and other oxidizing agents, and reactivated by catalytic amounts of ATP (or GTP or CTP) and hydrogen with ferredoxin and a hydrogenase.[1,3] The enzyme has also been identified in *Eubacterium limosum*, *Methanobacterium bryantii*, and *Methanobacterium thermoautotrophicum*.

[1]P. van der Meijden, B. W. te Brommelstroet, C. M. Poirot, C. van der Drift & G. D. Vogels (1984) *J. Bacteriol.* **160**, 629.
[2]P. van der Meijden, H. J. Heythuysen, A. Pouwels, F. Houwen, C. van der Drift & G. D. Vogels (1983) *Arch. Microbiol.* **134**, 238.
[3]P. van der Meijden, C. van der Lest, C. van der Drift & G. D. Vogels (1984) *BBRC* **118**, 760.

METHANOL OXIDASE

This FAD-dependent oxidoreductase [EC 1.1.3.31] catalyzes the reaction of methanol with dioxygen to produce formaldehyde and hydrogen peroxide. In some organisms, this enzyme may be identical to alcohol oxidase [EC 1.1.3.13]. The enzyme in yeast peroxisomes contains a modified FAD. **See also** *Alcohol Oxidase*

Selected entries from *Methods in Enzymology* [**vol**, page(s)]:
General discussion: **161**, 322; **188**, 420
Assay: **161**, 323; **188**, 421
Other molecular properties: fungal production, **161**, 323; immunocytochemical detection, **188**, 417; localization in *Hansenula polymorpha*, **188**, 414; microbodies, in, **52**, 496; properties, **161**, 325; **188**, 424; purification from *Hansenula polymorpha* CBS 4732, **188**, 422; purification from *Phanerochaete chrysosporium*, **161**, 324

5,10-METHENYLTETRAHYDROFOLATE CYCLOHYDROLASE

This enzyme [EC 3.5.4.9] catalyzes the reaction of 5,10-methenyltetrahydrofolate with water to produce 10-formyltetrahydrofolate.[1–4] In many eukaryotes (for example, human, pig, and chicken) this cyclohydrolase is trifunctional, also exhibiting methyltetrahydrofolate dehydrogenase (NADP$^+$) [EC 1.5.1.5] and formate tetrahydrofolate ligase [EC 6.3.4.3] activities. In certain prokaryotes, the cyclohydrolase is bifunctional, possessing either an additional methyltetrahydrofolate dehydrogenase [EC 1.5.1.5] or formiminotetrahydrofolate cyclodeaminase [EC 4.3.1.4] activity.

[1]V. Schirch (1998) *CBC* **I**, 211.
[2]J. N. Pelletier & R. E. MacKenzie (1995) *Biochemistry* **34**, 12673.
[3]S. W. Ragsdale (1991) *Crit. Rev. Biochem. Mol. Biol.* **26**, 261.
[4]J. I. Rader & F. M. Huennekens (1973) *The Enzymes*, 3rd ed., **9**, 197.

Selected entries from *Methods in Enzymology* [**vol**, page(s)]:
General discussion: **6**, 386; **18B**, 789; **66**, 599; **122**, 385; **281**, 146, 171
Assay: **6**, 386; **18B**, 790; **66**, 603, 609; **122**, 386
Other molecular properties: dihydrofolate reductase, not present in preparation of, **6**, 368; formiminoglycine formiminotransferase (presence in assay of, **6**, 97; presence in purification of, **6**, 98); formiminotetrahydrofolate cyclodeaminase, presence in assay of, **5**, 789; **6**, 381; N^{10}-formyltetrahydrofolate dehydrogenase, in assay of, **6**, 373; glycinamide ribotide transformylase, not present in preparation of, **6**, 68; present in impure methylenetetrahydrofolate dehydrogenase, **18B**, 605; properties, **6**, 387; **18B**, 792; **66**, 606, 615; **122**, 390; purification from (*Clostridium formicoaceticum*, **122**, 388; *Clostridium thermoaceticum*, **66**, 605; beef liver, **18B**, 790; pig liver, **66**, 611; rabbit liver, **5**, 785f; **6**, 387)

5,10-METHENYLTETRAHYDROFOLATE SYNTHETASE

This enzyme [EC 6.3.3.2], also known as 5-formyltetrahydrofolate cyclo-ligase, catalyzes the reaction of ATP with 5-formyltetrahydrofolate and H_3O^+ to produce ADP, orthophosphate, two water molecules, and 5,10-methenyltetrahydrofolate.[1–6]

The rabbit liver enzyme has a random Bi Ter kinetic mechanism.[4] A reaction scheme in which a phosphorylated substrate intermediate forms has been proposed.[1]

[1]V. Schirch (1998) *CBC* **I**, 211.
[2]R. Bertrand, R. E. MacKenzie & J. Jolivot (1987) *BBA* **911**, 154.
[3]F. M. Huennekens, G. B. Henderson, K. S. Vitols & C. E. Grimshaw (1984) *Adv. Enzyme Regul.* **27**, 3.
[4]S. Hopkins & V. Schirch (1984) *JBC* **259**, 5618.
[5]C. E. Grimshaw, G. B. Henderson, G. G. Soppe, G. Hansen, E. J. Mathur & F. M. Huennekens (1984) *JBC* **259**, 2728.
[6]J. I. Rader & F. M. Huennekens (1973) *The Enzymes*, 3rd ed., **9**, 197.

Selected entries from *Methods in Enzymology* [**vol**, page(s)]:
General discussion: **6**, 383; **18B**, 786; **281**, 146, 162
Assay: **6**, 383; **18B**, 786, 787
Other molecular properties: properties, **6**, 385; **18B**, 789; purification from liver, **6**, 384; **18B**, 787

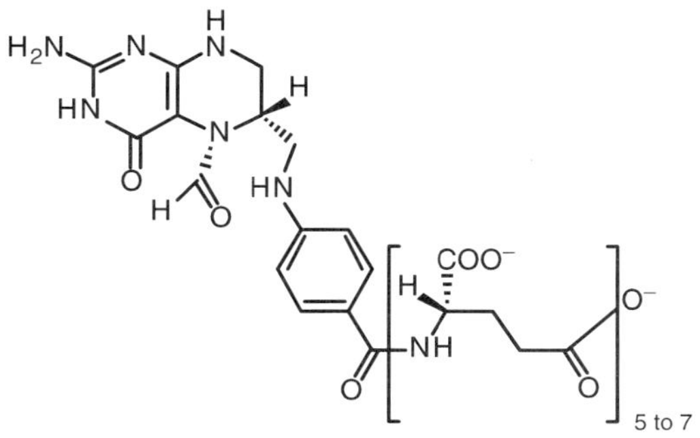

5-formyltetrahydrofolate

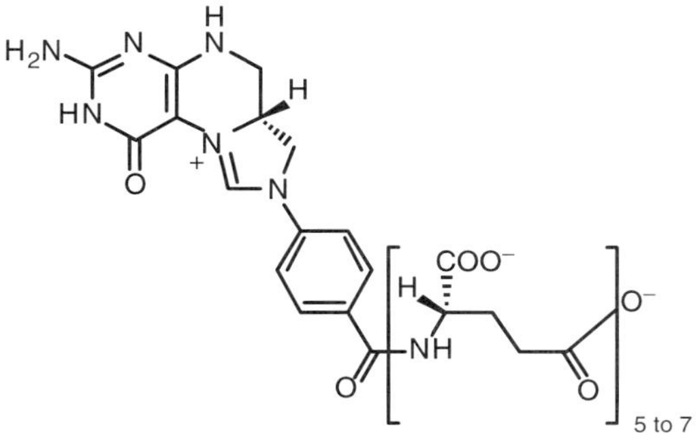

5,10-methenyltetrahydrofolate

METHENYLTETRAHYDRO-METHANOPTERIN CYCLOHYDROLASE

This enzyme [EC 3.5.4.27] catalyzes the reversible hydrolysis of 5,10-methenyl-5,6,7,8-tetrahydromethanopterin to produce N^5-formyl-5,6,7,8-tetrahydromethanopterin (the equilibrium constant actually favors the formation of the 5,10-methenyl derivative). The cyclohydrolase participates in the formation of methane from carbon dioxide in *Methanobacterium thermoautotrophicum*. The nonenzymatic hydrolysis product is N^{10}-formyl-5,6,7,8-tetrahydromethanopterin.[1-4]

[1] M. I. Donnelly, J. C. Escalante-Semerena, K. L. Rinehart & R. S. Wolfe (1985) *ABB* **242**, 430.
[2] A. A. DiMarco, M. I. Donnelly & R. S. Wolfe (1986) *J. Bacteriol.* **168**, 1372.
[3] B. W. te Brommelstroet, C. M. Hensgens, W. J. Geerts, J. T. Keltjens, C. van der Drift & G. D. Vogels (1990) *J. Bacteriol.* **172**, 564.
[4] J. T. Keltjens, A. J. Brugman, J. M. Kesseleer, B. W. te Brommelstroet, C. van der Drift & G. D. Vogels (1992) *Biofactors* **3**, 249.

N^5,N^{10}-METHENYLTETRAHYDRO-METHANOPTERIN HYDROGENASE

This oxidoreductase [EC 1.12.99.4], also known as H$_2$-forming N^5,N^{10}-methylenetetrahydromethanopterin dehydrogenase, catalyzes the reaction of N^5,N^{10}-methylenetetrahydromethanopterin with a proton to produce N^5,N^{10}-methenyltetrahydromethanopterin and

dihydrogen.[1-3] The enzyme does not catalyze the reduction of artificial dyes with H$_2$; nor it does catalyze an H$_2 \leftrightarrow$ H$^+$ exchange reaction in the absence of the methanopterin derivative.

[1] K. Ma, C. Zirngibl, R. K. Thauer, D. Linder & K. O. Stetter (1991) *Arch. Microbiol.* **156**, 43.
[2] C. Zirngibl, R. Hedderich & R. K. Thauer (1990) *FEBS Lett.* **261**, 112.
[3] A. Klein, V. M. Fernandez & R. K. Thauer (1995) *FEBS Lett.* **368**, 203.

METHIONINE ADENOSYLTRANSFERASE

This enzyme [EC 2.5.1.6], also known as *S*-adenosyl-methionine synthetase, catalyzes the reaction of ATP with L-methionine and water to produce *S*-adenosyl-L-methionine (AdoMet or SAM), orthophosphate, and pyrophosphate (or, diphosphate).[1-4]

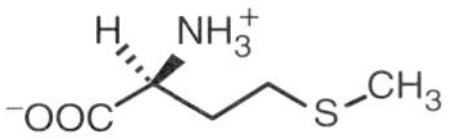

L-methionine 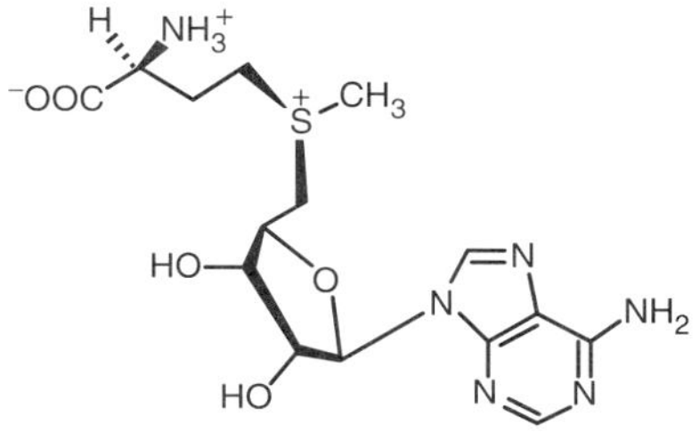   *S*-adenosyl-L-methionine

The biologically active $S(+)$ form readily undergoes nonenzymatic epimerization, even in the solid state, and care must be exercised whenever using SAM in investigations.

[1] J. M. Mato, L. Alvarez, P. Ortiz & M. A. Pajares (1997) *Pharmacol. Ther.* **73**, 265.
[2] M. Kotb & N. M. Kredich (1985) *JBC* **260**, 3923.
[3] C. W. Tabor & H. Tabor (1984) *AE* **56**, 251.
[4] S. H. Mudd (1973) *The Enzymes*, 3rd ed., **8**, 121.

Selected entries from *Methods in Enzymology* [vol, page(s)]:
General discussion: 2, 254; **17B**, 393; **94**, 219, 223
Assay: 2, 254; **17B**, 394; **63**, 36; **94**, 219, 223
Other molecular properties: *S*-adenosylethionine, synthesis of, **6**, 573; *S*-adenosylmethionine, synthesis of, **6**, 573; **17B**, 393; *Escherichia coli* (assay, **94**, 219; deficient mutants, screening, **94**, 91; gene cloning, **94**, 117; properties, **94**, 221; purification, **17B**, 395; **94**, 220); isozymes, rat liver (assay, **94**, 223; properties, **94**, 226; purification, **94**, 224); rabbit liver (properties, **2**, 256; purification, **2**, 255); transition-state and multisubstrate analogues, **249**, 305; yeast, purification, **6**, 570

METHIONINE DECARBOXYLASE

This decarboxylase [EC 4.1.1.57] catalyzes the conversion of L-methionine to carbon dioxide and 3-methyl-thiopropanamine.[1-4] The *Streptomyces*[1,4] and fern[1-3] enzymes are pyridoxal-phosphate-dependent.

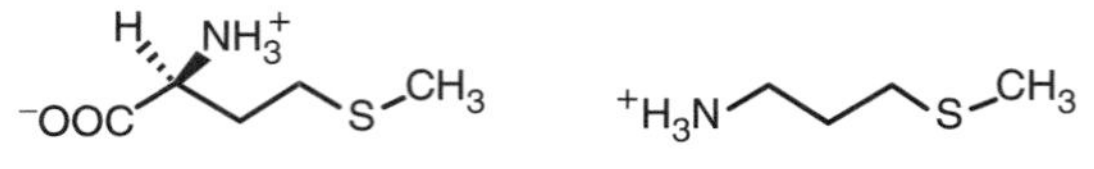

L-methionine   3-methylthiopropanamine

The fern enzyme will also decarboxylate L-valine and L-leucine. With deuterated L-methionine labeled at the C2 position, the enzyme is found to retain configuration at C2. Isotope effect investigations indicate that the solvent-sensitive transition state occurs before the first irreversible step, the release of carbon dioxide.

[1] D. E. Stevenson, M. Akhtar & D. Gani (1990) *Biochemistry* **29**, 7660.
[2] M. Akhtar, D. E. Stevenson & D. Gani (1990) *Biochemistry* **29**, 7648.
[3] D. E. Stevenson, M. Akhtar & D. Gani (1990) *Biochemistry* **29**, 7631.
[4] H. Hagino & K. Nakayama (1968) *Agric. Biol. Chem.* **32**, 727.

METHIONINE:GLYOXYLATE AMINOTRANSFERASE

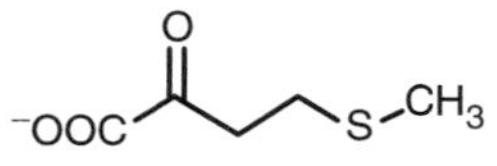

4-methylthio-2-oxobutanoate

This aminotransferase [EC 2.6.1.73] catalyzes the reversible reaction of L-methionine with glyoxylate to produce 4-methylthio-2-oxobutanoate and glycine.[1,2]

[1] J. R. Glover, C. C. S. Chapple, S. Rothwell, I. Tober & B. E. Ellis (1988) *Phytochemistry* **27**, 1345.
[2] C. C. S. Chapple, J. R. Glover & B. E. Ellis (1990) *Plant Physiol.* **94**, 1887.

L-METHIONINE γ-LYASE

This pyridoxal-phosphate-dependent enzyme [EC 4.4.1.11], also known as L-methioninase, catalyzes the conversion of L-methionine to methanethiol (CH_3SH), ammonia, and α-ketobutyrate (or, 2-oxobutanoate).[1–3]

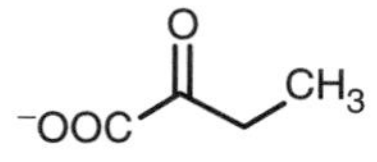

2-oxobutanoate

The enzyme also catalyzes a rapid exchange of the α- and β-hydrogens of L-methionine with deuterium from the solvent.[2]

[1] M. A. Ator & P. R. Ortez de Montellano (1990) *The Enzymes*, 3rd ed., **19**, 213.
[2] N. Esaki, T. Nakayama, S. Sawada, H. Tanaka & K. Soda (1985) *Biochemistry* **24**, 3857.
[3] T. Nakayama, N. Esaki, H. Tanaka & K. Soda (1988) *Biochemistry* **27**, 1587.

Selected entries from *Methods in Enzymology* [vol, page(s)]:
General discussion: 143, 459
Assay: 143, 460

Other molecular properties: bacteria, in, **77**, 256; properties, **143**, 462; purification from (*Aeromonas* sp., **143**, 462; *Pseudomonas putida*, 143, 460)

METHIONINE S-METHYLTRANSFERASE

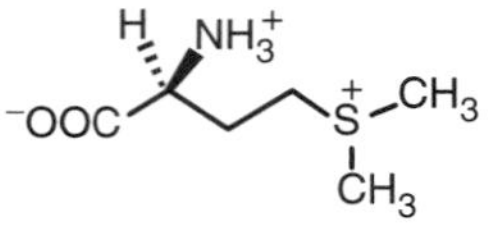

S-methyl-L-methionine

This methyltransferase [EC 2.1.1.12] catalyzes the reaction of S-adenosyl-L-methionine with L-methionine to produce S-adenosyl-L-homocysteine and S-methyl-L-methionine.[1]

[1] F. James, K. D. Nolte & A. D. Hanson (1995) *JBC* **270**, 22344.

METHIONINE S-OXIDE REDUCTASE

This enzyme [EC 1.8.4.5], also known as methionine sulfoxide reductase, catalyzes the reversible reaction of L-methionine S-oxide with reduced thioredoxin to produce L-methionine and oxidized thioredoxin.

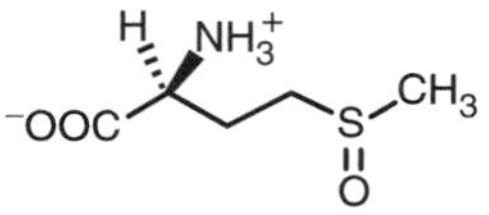

L-methionine S-oxide

Dithiothreitol can substitute for reduced thioredoxin.[1,2] In addition, other methyl sulfoxides can replace methionine sulfoxide.

[1] S. Ejiri, H. Weissbach & N. Brot (1980) *Anal. Biochem.* **102**, 393.
[2] S. Black, E. M. Harte, B. Hudson & L. Wartolofsky (1960) *JBC* **235**, 2910.

Selected entries from *Methods in Enzymology* [vol, page(s)]:
General discussion: 5, 992; 300, 239
Assay: 5, 992; 251, 463
Other molecular properties: amino acid sequence, **251**, 463; distribution and physiological function, **107**, 359; encoding gene (cloning, **251**, 464; glutathione S-transferase fusion protein, **251**, 466; overexpression, **251**, 465; polymerase chain reaction, **251**, 466; sequencing, **251**, 465); *Escherichia coli* (properties and characteristics, **107**, 356; purification, **107**, 355; **251**, 463, 468); human leukocyte, purification, **107**, 358; purification from *Escherichia coli* enzyme, **251**, 463, 468; species distribution, **251**, 463; substrate specificity, **251**, 462; yeast (preparation, **5**, 993; properties, **5**, 996)

D-METHIONINE:PYRUVATE AMINOTRANSFERASE

This pyridoxal-phosphate-dependent aminotransferase [EC 2.6.1.41], also known as D-methionine aminotransferase, catalyzes the reaction of D-methionine with pyruvate to produce 4-methylthio-2-oxobutanoate and L-alanine.[1,2]

4-methylthio-2-oxobutanoate

Oxaloacetate or α-ketoglutarate can replace pyruvate as the acceptor substrate, albeit not as effectively.

[1] J. I. Durham, P. W. Morgan, J. M. Prescott & C. M. Lyman (1973) *Phytochemistry* **12**, 2123.
[2] L. W. Mapson, J. F. March & D. A. Wardale (1969) *Biochem. J.* **115**, 653.

METHIONINE RACEMASE

L-methionine   D-methionine

This pyridoxal-phosphate-dependent racemase [EC 5.1.1.2] catalyzes the interconversion of L-methionine and D-methionine.[1]

[1] R. E. Kallio & A. D. Larson (1955) in *A Symposium on Amino Acid Metabolism* (W. D. McElroy & H. B. Glass, eds.) p. 616, Johns Hopkins Press, Baltimore.

[METHIONINE SYNTHASE]-COBALAMIN METHYLTRANSFERASE (COB(II)ALAMIN REDUCING)

This flavoprotein [EC 2.1.1.135], also known as methionine synthase cob(II)alamin reductase (methylating) and methionine synthase reductase, catalyzes the reaction of [methionine synthase]-cob(II)alamin with NADPH and S-adenosyl-L-methionine to produce [methionine synthase]-methylcob(I)alamin, S-adenosyl-L-homocysteine, and NADP$^+$.[1] Electrons are probably transferred from NADPH to FAD to FMN in the course of the reaction.

[1] D. Leclerc, A. Wilson, R. Dumas, C. Gafuik, D. Song, D. Watkins, H. H. Q. Heng, J. M. Rommens, S. W. Scherer, D. S. Rosenblatt & R. A. Gravel (1998) *PNAS* **95**, 3059.

METHIONYL AMINOPEPTIDASE

There are two classes of this metalloaminopeptidase [EC 3.4.11.18], also known as methionine aminopeptidase and peptidase M. Both catalyze the release of N-terminal amino acids, preferentially methionine residues, from peptides, nascent proteins, and arylamides.[1,2] Methionyl aminopeptidase type I (MetAP I), found in prokaryotes and eukaryotes but not in archaebacteria, has a preference for aminoacyl residues in the adjacent position (that is, P1′)

having a low radius of gyration (for example, glycyl, alanyl, seryl, threonyl, prolyl, valyl, or cysteinyl residues). MetAP I is inhibited by EDTA and is stimulated by cobalt ions.

Methionyl aminopeptidase type II (MetAP II) has the same specificity requirements as MetAP I. The type II enzymes are distinguished by the presence of a long insert in the catalytic domain. They have also been observed in some archaebacteria.

[1] S. M. Arfin & R. A. Bradshaw (1995) in *The Encyclopedia of Molecular Biology: Fundamentals and Applications* (J. Kendrew, ed.), VCH Verlag, Weinheim, p. 346.
[2] R. A. Bradshaw & S. M. Arfin (1996) in *The Aminopeptidases*, vol. **6** (A. Taylor, ed.), R. G. Landes, Georgetown, p. 91.
Selected entries from *Methods in Enzymology* [vol, page(s)]:
Other molecular properties: *Bacillus subtilis*, **248**, 221; *Escherichia coli*, **248**, 220; homologues, **248**, 221; isolation from prokaryotes, **185**, 407; properties, **248**, 213; *Salmonella typhimurium*, **248**, 221; yeast, **248**, 220

METHIONYL-tRNA FORMYLTRANSFERASE

This enzyme [EC 2.1.2.9] catalyzes the reaction of 10-formyltetrahydrofolate with L-methionyl-tRNA and water to produce tetrahydrofolate and N-formylmethionyl-tRNA.[1–6] Asparaginyl, histidyl, and aspartyl residues have been identified to be essential for catalytic activity.[1]

[1] D. T. Newton & D. Mangroo (1999) *BJ* **339**, 63.
[2] T. Meinnel, Y. Mechulam & S. Blanquet (1993) *Biochimie* **75**, 1061.
[3] P. Crosti, A. Gambini, G. Lucchini & R. Bianchetti (1977) *BBA* **477**, 356.
[4] C. E. Samuel & J. C. Rabinowitz (1974) *J. Bacteriol.* **118**, 21.
[5] J. I. Rader & F. M. Huennekens (1973) *The Enzymes*, 3rd ed., **9**, 197.
[6] H. W. Dickerman, E. Steers, B. G. Redfield & H. Weissbach (1967) *JBC* **242**, 1522.
Selected entries from *Methods in Enzymology* [vol, page(s)]:
General discussion: 12B, 681; 20, 182; 106, 141
Assay: 12B, 682; 20, 183; 106, 142
Other molecular properties: absence, in *Escherichia coli* mutant, **51**, 214; anions and, **12B**, 687; *Escherichia coli* (ligand binding stoichiometries, **106**, 148; properties, **12B**, 686; **106**, 146; purification, **12B**, 684; **20**, 36, 185, 250; **106**, 144; specificity, **106**, 152); inhibitors, **20**, 189; preparation, **20**, 185; **29**, 536; reversal, **20**, 183, 191

METHIONYL-tRNA SYNTHETASE

This enzyme [EC 6.1.1.10], also known as methionine:tRNA ligase, catalyzes the reaction of L-methionine with tRNA^{Met} and ATP to produce L-methionyl-tRNA^{Met}, AMP, and pyrophosphate (or, diphosphate).[1–5] Both L-homocysteine and L-norleucine have been reported to act as alternative substrates. Surprisingly, this is the only known aminoacyl-tRNA synthetase that does not require the presence of Mg^{2+}. *See also Aminoacyl-tRNA Synthetases*

[1] E. A. First (1998) *CBC* 1, 573.

[2] B. Senger & F. Fasiolo (1996) *Biochimie* **78**, 597.
[3] J. B. Armstrong & J. A. Fairfield (1975) *Can. J. Microbiol.* **21**, 754.
[4] F. Lawrence (1973) *EJB* **40**, 493.
[5] F. Lawrence, S. Blanquet, M. Poiret, M. Robert-Gero & J. P. Waller (1973) *EJB* **36**, 234.

Selected entries from *Methods in Enzymology* [**vol**, page(s)]:
General discussion: 5, 708; 34, 506
Other molecular properties: binding constants, **29**, 640; bound reactants and products, equilibrium constant determination, **177**, 372; equilibrium constant, **177**, 372; hydroxamate formation rate, **29**, 613; maleylated, analysis of, **25**, 197; properties, **5**, 715; purification, **5**, 714; **20**, 36, 250; **29**, 591; **34**, 511, 512; sequence analysis, **47**, 331; small-angle X-ray scattering studies, **61**, 241; spacer, **34**, 508; stereochemistry, **87**, 206; subcellular distribution, **59**, 233, 234

4-METHOXYBENZOATE MONOOXYGENASE (*O*-DEMETHYLATING)

This oxidoreductase [EC 1.14.99.15] catalyzes the reaction of 4-methoxybenzoate (or, *p*-anisate) with dioxygen and an electron donor (AH_2) to produce 4-hydroxybenzoate, formaldehyde, the oxidized donor (A), and water.[1,2] The electron donor is typically NADH. Other aromatic substrates include 4-ethoxybenzoate, *N*-methyl-4-amino-benzoate, 3-nitro-4-methoxybenzoate, and toluate.

4-methoxybenzoate   4-hydroxybenzoate

The *Pseudomonas putida* enzyme is a two-component system, consisting of a reductase (NADH:putidamonooxin reductase, an iron-sulfur flavoprotein) and putidamonooxin (similar to ferredoxin), containing a Rieske center and mononuclear iron. The fungal enzyme acts best on vera-trate (3,4-dimethoxybenzoate). The reaction involves the hydroxylation of the methyl group of 4-methoxybenzoate resulting in the formation of an unstable hemiacetal. This hemiacetal breaks down into formaldehyde and 4-hydroxybenzoate. ***See also*** *Putidamonooxin*

With certain alternative substrates, the enzyme also has a dioxygenase activity. For example, it catalyzes the reaction of dioxygen with 4-vinylbenzoate to produce 4-(1′,2′-dihydroxyethyl)benzoate.

[1] B. G. Fox (1998) *CBC* **3**, 261.
[2] V. Ullrich & W. Duppel (1975) *The Enzymes*, 3rd ed., **12**, 253.

Selected entries from *Methods in Enzymology* [**vol**, page(s)]:
General discussion: 52, 12; 161, 281

Assay: 161, 287
Other molecular properties: components (assays, **161**, 287; interactions, **161**, 291; properties, **161**, 288; purification from *Pseudomonas putida*, **161**, 284; substrate specificity and reaction mechanisms, **161**, 292); importance in bacterial metabolism, **161**, 281

16-METHOXY-2,3-DIHYDRO-3-HYDROXY-TABERSONINE *N*-METHYLTRANSFERASE

This methyltransferase [EC 2.1.1.99] catalyzes the reaction of *S*-adenosyl-L-methionine with 16-methoxy-2,3-dihydro-3-hydroxytabersonine to produce *S*-adenosyl-L-homocysteine and deacetoxyvindoline (or, N^1-methyl-16-methoxy-2,3-dihydro-3-hydroxytabersonine).[1,2] This enzyme participates in the biosynthesis of vindoline from tabersonine in the rosy periwinkle.

[1] V. De Luca & A. J. Cutler (1987) *Plant Physiol.* **85**, 1099.
[2] V. De Luca, J. Balsevich, R. T. Tyler, U. Eilert, B. D. Panchuk & W. G. W. Kurz (1986) *J. Plant Physiol.* **125**, 147.

2-METHYLACYL-CoA DEHYDROGENASE

This FAD-dependent oxidoreductase [EC 1.3.99.12], also known as branched-chain acyl-CoA dehydrogenase, catalyzes the reaction of 2-methylbutanoyl-CoA with an acceptor substrate to produce 2-methylbut-2-enoyl-CoA and the reduced acceptor.[1-3] Other substrates include 2-methylpropanoyl-CoA and 2-methylvaleryl-CoA. Acceptor substrates include the electron-transfer flavoprotein and phenazine methosulfate. An essential glu-tamyl residue has been identified at the active site.[4] The enzyme is not identical with butyryl-CoA dehydrogenase [EC 1.3.99.2], acyl-CoA dehydrogenase [EC 1.3.99.3], isovaleryl-CoA dehydrogenase [EC 1.3.99.10], or long-chain acyl-CoA dehydrogenase [EC 1.3.99.13]. There are isozymes that are specific for short-, medium-, long-, and very long-branched-chain acyl-CoA substrates.

[1] R. Komuniecki, S. Fekete & J. Thissen-Parra (1985) *JBC* **260**, 4770.
[2] Y. Ikeda & K. Tanaka (1983) *JBC* **258**, 9477.
[3] Y. Ikeda, C. Dabrowski & K. Tanaka (1983) *JBC* **258**, 1066.
[4] B. Binzak, J. Willard & J. Vockley (1998) *BBA* **1382**, 137.

Selected entries from *Methods in Enzymology* [**vol**, page(s)]:
General discussion: 166, 360; 324, 241
Assay: 166, 361

α-METHYLACYL-CoA RACEMASE (OR, EPIMERASE)

This epimerase [EC 5.1.99.4] catalyzes the interconversion of (2*S*)-2-methylacyl-CoA and (2*R*)-2-methylacyl-CoA. The term "racemase" used by the Enzyme Commission is incorrect, because this enzyme does not catalyze the interconversion of two enantiomers; rather, the substrate and product are diastereoisomers.

α-Methyl-branched acyl-CoA derivatives with chain lengths of more than ten carbons are substrates. However, free acids are not substrates. Alternative substrates include some aromatic CoA derivatives (*e.g.*, ibuprofen-CoA) and bile acid-CoA intermediates (for example, trihydroxycoprostanoyl-CoA). The enzyme also catalyzes a rapid exchange of the hydrogen atom in the 2-position of the fatty acyl-CoA derivative against a proton from water. The mammalian enzyme is responsible for the conversion of pristanoyl-CoA and C_{27}-bile acyl-CoAs to their (*S*)-stereoisomers, which are the only stereoisomers that can be degraded via the β-oxidation pathway in the peroxisome.[1,2]

[1]W. Schmitz, R. Fingerhut & E. Conzelmann (1994) *EJB* **222**, 313.
[2]W. Schmitz, C. Albers, R. Fingerhut & E. Conzelmann (1995) *EJB* **231**, 815.

1-METHYLADENOSINE NUCLEOSIDASE

This nucleosidase [EC 3.2.2.13] catalyzes the hydrolysis of 1-methyladenosine to produce 1-methyladenine and D-ribose.[1,2]

[1]H. L. A. Tarr (1973) *J. Fish. Res. Board Can.* **30**, 1861.
[2]H. Shirai & H. Kanatani (1972) *Exp. Cell Res.* **75**, 79.

N-METHYL-L-ALANINE DEHYDROGENASE

This oxidoreductase [EC 1.4.1.17] catalyzes the reversible reaction of *N*-methyl-L-alanine with NADP$^+$ and water to produce pyruvate, methylamine, and NADPH.[1,2]

[1]N. M. W. Brunhuber & J. S. Blanchard (1994) *Crit. Rev. Biochem. Mol. Biol.* **29**, 415.
[2]M. C. M. Lin & C. Wagner (1975) *JBC* **250**, 3746.

METHYLAMINE:GLUTAMATE METHYLTRANSFERASE

This FMN-dependent transferase [EC 2.1.1.21], also known as *N*-methylglutamate synthase, catalyzes the reversible reaction of methylamine and L-glutamate to produce ammonia and *N*-methyl-L-glutamate.[1-3]

N-methyl-L-glutamate

The enzyme has a ping pong Bi Bi kinetic mechanism with an α-ketoglutarate enzyme-bound intermediate.

[1]B. A. Palfey & V. Massey (1998) *CBC* **3**, 83.
[2]R. J. Pollock & L. B. Hersh (1973) *JBC* **248**, 6724.
[3]M. Schuman Jorns & L. B. Hersh (1975) *JBC* **250**, 3620.
Selected entries from *Methods in Enzymology* [vol, page(s)]:
General discussion: 17A, 868; 113, 36
Assay: 113, 36
Other molecular properties: properties, 113, 39; purification, *Pseudomonas*, 17A, 871; 113, 38

N-METHYL-L-AMINO-ACID OXIDASE

This flavin-dependent oxidoreductase [EC 1.5.3.2] catalyzes the reaction of an *N*-methyl-L-amino acid with water and dioxygen to produce an L-amino acid, formaldehyde, and hydrogen peroxide.[1] Substrates include N^α-methyl-L-tryptophan, *O,N*-dimethyl-DL-tyrosine, and *N*-methyl-DL-phenylalanine.

[1]M. Moritani, T.-C. Tung, S. Fujii, H. Mito, N. Izumiya, K. Kenmochi & R. Hirohata (1954) *JBC* **209**, 485.

Selected entries from *Methods in Enzymology* [vol, page(s)]:
General discussion: 52, 18; 56, 475

METHYLARSONATE REDUCTASE

This oxidoreductase [EC 1.97.1.7] catalyzes the reaction of methylarsonate with two molecules of glutathione (that is, γ-L-glutamyl-L-cysteinylglycine) to produce methylarsonite ($CH_3AsO_2H^-$) and glutathione disulfide.[1]

[1]R. A. Zakharyan & H. V. Aposhian (1999) *Chem. Res. Toxicol.* **12**, 1278.

METHYLARSONITE METHYLTRANSFERASE

This methyltransferase [EC 2.1.1.138] catalyzes the reaction of *S*-adenosyl-L-methionine with methylarsonite to produce *S*-adenosyl-L-homocysteine and dimethylarsinate (that is, cacodylate), which is less toxic than arsenate.[1,2]

[1]R. A. Zakharyan & H. V. Aposhian (1999) *Chem. Res. Toxicol.* **121**, 1278.
[2]R. A. Zakharyan, F. Ayala-Fierro, W. R. Cullen, D. M. Carter & H. V. Aposhian (1999) *Toxicol. Appl. Pharmacol.* **158**, 9.

β-METHYLASPARTASE

This cobalamin-dependent enzyme [EC 4.3.1.2], also known as β-methylaspartate ammonia-lyase, catalyzes the conversion of L-*threo*-3-methylaspartate to mesaconate (or, 2-methylfumarate) and ammonia.[1-6] Other 3-alkyl-aspartates are also substrates. The enzyme reportedly also has an aspartate ammonia-lyase activity as well.

This methylaspartase exhibits a random Uni Bi kinetic mechanism and catalyzes the exchange of the C3 hydrogen of the substrate with solvent hydrogen.[1] Kinetic isotope effect investigations indicate that a thiolate acts as the base for deprotonation at C3.[5] Interestingly, the mechanism changes from concerted to a carbocationic amino-enzyme elimination scheme upon changing the stereochemistry at the C3 position of the substrate.[6] The enzyme is not stereo-specific in its action at position 3; both the 3S and 3R substrates are converted to the same product. The elimination reaction from (3S)-3-methyl-L-aspartate is *anti*, and from (3R)-3-methyl-L-aspartate, it is *syn*.

(3S)-3-methyl-L-aspartate

(3R)-3-methyl-L-aspartate

(E)-methylfumarate

[1] N. P. Botting & D. Gani (1992) *Biochemistry* **31**, 1509.
[2] N. P. Botting, M. Akhtar, M. A. Cohen & D. Gani (1988) *Biochemistry* **27**, 2953.
[3] K. R. Hanson & E. A. Havir (1972) *The Enzymes*, 3rd ed., **7**, 75.
[4] M. F. Winkler & V. R. Williams (1967) *BBA* **146**, 287.
[5] J. R. Pollard, S. Richardson, M. Akhtar, P. Lasry, T. Neal, N. P. Botting & D. Gani (1999) *Bioorg. Med. Chem.* **7**, 949.
[6] D. Gani, C. H. Archer, N. P. Botting & J. R. Pollard (1999) *Bioorg. Med. Chem.* **7**, 977.

Selected entries from *Methods in Enzymology* [vol, page(s)]:
General discussion: 5, 827; 13, 347
Assay: 5, 827; 13, 347
Other molecular properties: assay for glutamate mutase, 13, 319, 325, 327; effect of divalent metal ions, 13, 353; inhibition by N-ethylmaleimide, 13, 353; mechanism, 13, 353; photooxidation, 25, 407; properties, 5, 831; 13, 351; purification, *Clostridium tetanomorphum*, 5, 828; 13, 348; subunit structure, 13, 352

β-METHYLASPARTATE MUTASE

This cobalamin-dependent enzyme [EC 5.4.99.1], also known as glutamate mutase or methylaspartate-glutamate mutase, catalyzes the interconversion of L-*threo*-3-methyl-aspartate (*i.e.*, (2S,3S)-3-methylaspartate) and L-glutamate (*i.e.*, (S)-glutamate).[1–5]

This rearrangement proceeds via the formation of free radical intermediates that are generated by homolysis of adenosylcobalamin. The generation of adenosyl radical and proton abstraction are either strongly coupled or are concerted events. Abstraction of a hydrogen from the methyl group generates the β-methylaspartate radical which produces a glycyl radical and acrylate.[1,6] Recombination produces the glutamate radical which receives a hydrogen from 5'-deoxyadenosine.

[1] B. T. Golding & W. Buckel (1998) *CBC* **3**, 239.
[2] R. L. Switzer (1982) in *B12* (D. Dolphin, ed.) **2**, 289.
[3] B. M. Babior & J. S. Krouwer (1979) *Crit. Rev. Biochem.* **6**, 35.
[4] H. A. Barker (1972) *The Enzymes*, 3rd ed., **6**, 509.
[5] H. A. Barker, V. Rooze & A. A. Iodice (1964) *JBC* **239**, 3260.
[6] H.-W. Chih & E. N. G. Marsh (1999) *Biochemistry* **38**, 13684.

Selected entries from *Methods in Enzymology* [vol, page(s)]:
General discussion: 13, 319; 113, 121
Assay: 13, 319, 321, 325; 113, 121
Other molecular properties: *Clostridium tetanomorphum* (components S and E [interactions, 113, 131; properties, 13, 325, 328; 113, 126, 130; purification, 13, 321, 326; 113, 124, 128]); coenzyme specificity, 13, 329; 113, 132; component E, 13, 321; 113, 128; component S, 13, 325; 113, 124; inhibitor, 63, 399; mechanism, 13, 214; properties, 13, 325, 328, 329; 113, 131; substrate specificity, 13, 329; 113, 132

METHYLATED-DNA:[PROTEIN]-CYSTEINE S-METHYLTRANSFERASE

This methyltransferase [EC 2.1.1.63], also known as 6-O-methylguanine-DNA methyltransferase and O-6-methylguanine-DNA-alkyltransferase, catalyzes the reaction of DNA, containing a 6-O-methylguanine residue, with an L-cysteinyl residue in a protein to produce the DNA that now contains a demethylated guanine and a protein containing an S-methyl-L-cysteinyl residue.[1–5]

[1] Z. Yu, J. Chen, B. N. Ford, M. E. Brackley & B. W. Glickman (1999) *Environ. Mol. Mutagen.* **33**, 3.
[2] R. D. Wood (1996) *Ann. Rev Biochem.* **65**, 135.
[3] A. E. Pegg, M. E. Dolan & R. C. Moschel (1995) *Prog. Nucleic Acid Res. Mol. Biol.* **51**, 167.
[4] A. E. Pegg & T. L. Byers (1992) *FASEB J.* **6**, 2302.
[5] M. Olsson & T. Lindahl (1980) *JBC* **255**, 10569.

2-METHYL-BRANCHED-CHAIN-ENOYL-CoA REDUCTASE

This FAD-dependent oxidoreductase [EC 1.3.1.52] catalyzes the reaction of 2-methylcrotonoyl-CoA with NADH to produce 2-methylbutanoyl-CoA and NAD^+. Catalysis only occurs in the presence of the electron-transferring flavoprotein.[1]

[1] R. Komuniecki, J. McCrury, J. Thissen & N. Rubin (1989) *BBA* **975**, 127.

3-METHYLBUTANAL REDUCTASE

This oxidoreductase [EC 1.1.1.265], also known as iso-amylaldehyde reductase, isopentanal reductase, isovaleral reductase, and isovaleraldehyde reductase, catalyzes the reaction of 3-methylbutanal with NADPH and a proton to produce 3-methylbutanol and $NADP^+$.[1,2] The *Saccharomyces cerevisiae* enzyme will act on a number of aldehydes, particularly aromatic aldehydes, including 3-pyridine carboxaldehyde.

[1]L. Ven Nedervelde, V. Verlinden, D. Philipp & A. Debourg (1997) *Proc. 26th Congr.-Eur. Brew. Conv.*, 447.

[2]M. F. M. van Iersel, M. H. M. Eppink, W. J. H. van Berkel, F. M. Rombouts & T. Abee (1997) *Appl. Environ. Microbiol.* **63**, 4079.

2-METHYLCITRATE DEHYDRATASE

This dehydratase [EC 4.2.1.79] catalyzes the conversion of 2-hydroxybutane-1,2,3-tricarboxylate (that is, 2-methylcitrate) to (Z)-but-2-ene-1,2,3-tricarboxylate and water.[1,2] This enzyme is distinct from that of citrate dehydratase [EC 4.2.1.4].

[1]H. Aoki & T. Tabuchi (1981) *Agric. Biol. Chem.* **45**, 2831.

[2]T. Tabuchi, H. Aoki, H. Uchiyama & T. Nakahara (1981) *Agric. Biol. Chem.* **45**, 2823.

2-METHYLCITRATE SYNTHASE

This enzyme [EC 4.1.3.31] catalyzes the reaction of propanoyl-CoA with water and oxaloacetate to produce 2-methylcitrate and coenzyme A.[1] Other substrates include acetyl-CoA (thus, generating citrate), butanoyl-CoA, and pentanoyl-CoA. The enzyme is distinct from that of citrate synthase [EC 4.1.3.7].

[1]H. Uchiyama & T. Tabuchi (1976) *Agric. Biol. Chem.* **40**, 1411.

METHYL-S-COENZYME-M METHYLREDUCTASE

This nickel-dependent oxidoreductase, isolated from methanogens, catalyzes the reaction of methyl-*S*-coenzyme M with 7-mercaptoheptanoylthreonine phosphate (HTP-SH) to produce methane and the mixed disulfide of coenzyme M and 7-mercaptoheptanoylthreonine phosphate (HTP-S-S-CoM). The mixed disulfide is then reduced by a disulfide reductase. The enzyme has also been demonstrated *in vitro* to catalyze the reduction of alkyl halides to produce both alkanes and alkenes.

Most methanogenic bacteria express two different types of methyl-*S*-coenzyme-M methylreductase, which appear to be expressed at different points in the organisms' growth cycle. The catalytically essential nickel is part of a dodecahydroporphyrin prosthetic group known as factor 430 (F_{430}), so-named since the free Ni^{2+}-containing cofactor has a λ_{max} of 430 nm (the λ_{max} shifts to 420 nm when inserted into the enzyme). Much remains unclear about the course of the reaction. It has been suggested that HTP-SH forms a nickel-sulfur bridge with the cofactor and generates the Ni^{1+} form of the cofactor and the corresponding sulfur radical form of the substrate. Methyl-*S*-coenzyme M now

binds and a disulfide radical is produced. Reaction with the Ni^{1+} cofactor results in the formation of the mixed disulfide and a methylated Ni^{2+} cofactor. Methane is released upon binding of a new HTP-SH.[1,2]

[1]B. Juan (1990) *Helv. Chim. Acta* **73**, 2209.

[2]A. Berkessel (1991) *Bioorg. Chem.* **19**, 101.

β-METHYLCROTONYL-CoA CARBOXYLASE

This biotin-dependent enzyme [EC 6.4.1.4] catalyzes the reaction of MgATP with 3-methylcrotonyl-CoA and bicarbonate to produce MgADP, orthophosphate, and 3-methylglutaconyl-CoA.[1–5] The enzyme will also convert crotonyl-CoA to glutaconyl-CoA and 3-ethylcrotonyl-CoA to 3-ethylglutaconyl-CoA. The carboxylase from most organisms is composed of two nonidentical subunits, one of which contains the biotin. The enzyme has a ping pong mechanism in which MgATP and bicarbonate bind in sequence to first form the carboxybiotin intermediate.[5] This intermediate then reacts with 3-methylcrotonyl-CoA to generate the final product.

[1]F. Lynen (1979) *Crit. Rev. Biochem.* **7**, 103.

[2]A. W. Alberts & P. R. Vagelos (1972) *The Enzymes*, 3rd ed., **6**, 37.

[3]J. Moss & M. D. Lane (1971) *AE* **35**, 321.

[4]F. Lynen, J. Knappe, E. Lorch, G. Jüttig, E. Ringelmann & J.-P. Lachance (1961) *Biochem. Z.* **335**, 123.

[5]T. A. Diez, E. S. Wurtele & B. J. Nikolau (1994) *ABB* **310**, 64.

Selected entries from *Methods in Enzymology* [vol, page(s)]:
General discussion: 5, 896; 71, 781, 791, 800; 324, 280
Assay: 5, 897; 71, 792, 800
Other molecular properties: *Achromobacter*, 71, 781 (apoenzyme, 71, 789; avidin, 71, 788; biotin-containing subunit, 71, 787 [activity, 71, 789]; biotin-free carboxylation, 71, 781, 788 [subunit, 71, 787]; cultivation of *Achromobacter* IV S, 71, 783; exchange reactions, 71, 788; holoenzyme, 71, 789; reaction mechanism, 71, 782; reconstitution of active enzyme from subunits, 71, 790; regulation of degree of dissociation of enzyme complex, 71, 790); bovine kidney, 71, 800 (assay, 71, 800; biotin content, 71, 803; properties, 71, 803); chicken liver (properties, 5, 900; purification, 5, 898); β-hydroxy-β-methylglutaryl coenzyme A cleavage enzyme, not present in preparation of, 5, 902; *Pseudomonas citronellis*, 71, 791 (assays, 71, 792; biotin-dependent CO_2-fixation reaction, 71, 792; growth of cells, 71, 794; properties, 71, 796; specificity, 71, 799); sources, 5, 896

METHYLCYSTEINE SYNTHASE

The enzyme, previously classified as EC 4.2.1.23 but now a deleted Enzyme Commission entry, is a side reaction of cystathionine β-synthase [EC 4.2.1.22].

5-METHYLDEOXYCYTIDINE-5'-PHOSPHATE KINASE

This phosphotransferase [EC 2.7.4.19] catalyzes the reaction of ATP with 5-methyldeoxycytidine 5'-phosphate (that is, 5-methyl-dCMP; m^5-dCMP) to produce ADP and 5-methyldeoxycytidine diphosphate (5-methyl-dCDP; m^5-dCDP).[1]

[1] R. Y.-H. Wang, L.-H. Huang & M. Ehrlich (1982) *BBA* **696**, 31.

METHYLENEDIUREA DEAMINASE

This enzyme [EC 3.5.3.21], also called methylenediurease, catalyzes the hydrolysis of methylenediurea (NH$_2$CONHCH$_2$NHCONH$_2$) with two molecules of water to produce *N*-(hydroxymethyl)urea, carbon dioxide, and two ammonia molecules.

The reaction proceeds through an *N*-(carboxyaminomethyl) urea intermediate and then to an aminated methylurea (*N*-(aminomethyl)urea), which spontaneously hydrolyzes to *N*-(hydroxymethyl)urea. The enzyme from *Ochrobactrum anthropi* also hydrolyzes dimethylenetriurea, trimethylenetetraurea, ureidoglycolate, and allantoate.[1]

[1] T. Jahns, R. Schepp & H. Kaltwasser (1997) *Can. J. Microbiol.* **43**, 1111.

METHYLENE-FATTY-ACYL-PHOSPHOLIPID SYNTHASE

This methyltransferase [EC 2.1.1.16] (also called unsaturated-phospholipid methyltransferase and cyclopropane synthetase) catalyzes the reaction of *S*-adenosyl-L-methionine with a phospholipid olefinic fatty acid to produce *S*-adenosyl-L-homocysteine and a phospholipid methylene fatty acid.[1]

The enzyme reportedly transfers a methyl group to the 10-position of a Δ^9-olefinic acyl chain in phosphatidylglycerol and phosphatidylinositol. Phosphatidylethanolamine is a weak substrate (thus, has a different specificity compared to cyclopropane-fatty-acyl-phospholipid synthase [EC 2.1.1.79]). Subsequent proton transfer produces a 10-methylene group. **See also** *Cyclopropane-Fatty-Acyl-Phospholipid Synthase*

[1] Y. Akamatsu & J. H. Law (1970) *JBC* **245**, 701.

4-METHYLENEGLUTAMINASE

This enzyme [EC 3.5.1.67], also known as 4-methyleneglutamine deamidase, catalyzes the hydrolysis of 4-methylene-L-glutamine to produce 4-methylene-L-glutamate and ammonia.[1–3]

[1] S. A. Ibrahim, P. J. Lea & L. Fowden (1984) *Phytochemistry* **23**, 1545.
[2] H. C. Winter & E. E. Dekker (1991) *Plant Physiol.* **95**, 206.
[3] G. K. Powell & E. E. Dekker (1983) *JBC* **258**, 8677.

METHYLENEGLUTAMINE SYNTHETASE

This ligase [EC 6.3.1.7], also called 4-methyleneglutamate: ammonia ligase, catalyzes the reaction of ATP with 4-methylene-L-glutamate and ammonia to produce AMP, pyrophosphate (or, diphosphate), and 4-methylene-L-glutamine.[1,2] L-Glutamine can serve as the nitrogen-donor substrate, albeit more weakly (this reaction is also catalyzed by glutamine synthetase).

[1] H. C. Winter, T. Z. Su & E. E. Dekker (1983) *BBRC* **111**, 484.
[2] H. C. Winter & E. E. Dekker (1986) *JBC* **261**, 11189.

2-METHYLENEGLUTARATE MUTASE

This cobalamin-dependent enzyme [EC 5.4.99.4], first isolated from *Clostridium barkeri*, catalyzes the interconversion of 2-methyleneglutarate and 2-methylene-3-methylsuccinate (that is, (*R*)-3-methylitaconate).[1–5]

2-methyleneglutarate   (*R*)-3-methylitaconate

This enzyme utilizes a radical mechanism and contains two different Co(II) species: adenosylcobalamin and the oxygen-stable cob(II)alamin.[6] Light inactivates the enzyme.[7] The 5'-deoxyadenosine radical abstracts the *proR* hydrogen from the C4 position of the substrate to produce the substrate radical. This radical generates acrylate and the 2-acrylate radical. Recombination produces the 2-methylene-3-methylsuccinate radical which receives a hydrogen from 5'-deoxyadenosine. An alternative mechanism involving a cyclopropylcarbinyl radical has also been proposed.[1]

[1] B. T. Golding & W. Buckel (1998) *CBC* **3**, 239.
[2] B. M. Babior & J. S. Krouwer (1979) *Crit. Rev. Biochem.* **6**, 35.
[3] H. A. Barker (1972) *The Enzymes*, 3rd ed., **6**, 509.
[4] H.-F. Kung & L. Tsai (1971) *JBC* **246**, 6436.
[5] H.-F. Kung & T. C. Stadtman (1971) *JBC* **246**, 3378.
[6] O. Zelder & W. Buckel (1993) *Biol. Chem. Hoppe-Seyler* **374**, 85.
[7] C. Michel & W. Buckel (1991) *FEBS Lett.* **281**, 108.

3-METHYLENEOXINDOLE REDUCTASE

This oxidoreductase [EC 1.3.1.17] catalyzes the reaction of 3-methyleneoxindole with NADPH to produce 3-methyloxindole and NADP$^+$.[1–3]

[1] P. S. Basu & V. Tuli (1972) *Plant Physiol.* **50**, 503.
[2] H. S. Moyed & V. Williamson (1967) *JBC* **242**, 1075.
[3] H. S. Moyed & V. Williamson (1967) *Plant Physiol.* **42**, 510.

5,10-METHYLENETETRAHYDROFOLATE DEHYDROGENASE

Two oxidoreductases catalyze the conversion of 5,10-methylenetetrahydrofolate to 5,10-methenyltetrahydrofolate.

5,10-Methylenetetrahydrofolate dehydrogenase (NADP$^+$) [EC 1.5.1.5] catalyzes the reversible reaction of 5,10-methylenetetrahydrofolate with NADP$^+$ to produce 5,10-methenyltetrahydrofolate and NADPH.[1–6] This oxidoreductase has an ordered Bi Bi kinetic mechamism.[5] The pig, mouse, rabbit, sheep, and human enzymes have all been reported to be trifunctional, also exhibiting methenyltetrahydrofolate cyclohydrolase and formyltetrahydrofolate synthetase activities. The *Clostridium thermoaceticum* enzyme is bifunctional, also having a methenyltetrahydrofolate cyclohydrolase activity.

5,10-Methylenetetrahydrofolate dehydrogenase (NAD$^+$) [EC 1.5.1.15] catalyzes the reversible reaction of 5,10-methylenetetrahydrofolate with NAD$^+$ to produce 5,10-methenyltetrahydrofolate and NADH. The *Saccharomyces cerevisiae* and *Acetobacterium woodii* enzymes reportedly are monofunctional proteins whereas the mammalian enzyme has a methenyltetrahydrofolate cyclohydrolase activity as well (this is in addition to the presence of the trifunctional enzyme discussed in the previous paragraph). The *Acetobacterium woodii* enzyme has a ping pong Bi Bi kinetic mechanism[7] whereas the ascites tumor cells has an equilibrium ordered Bi Bi scheme.[8]

[1] V. Schirch (1998) *CBC* 1, 211.
[2] S. W. Ragsdale (1991) *Crit. Rev. Biochem. Mol. Biol.* 26, 261.
[3] R. E. MacKenzie, N. Mejia & X.-M. Yang (1988) *Adv. Enzyme Regul.* 27, 31.
[4] S. J. Benkovic (1980) *Ann. Rev. Biochem.* 49, 227.
[5] L. Cohen & R. E. MacKenzie (1978) *BBA* 522, 311.
[6] R. J. Harvey & I. K. Dev (1975) *Adv. Enzyme Regul.* 13, 99.
[7] S. W. Ragsdale & L. G. Ljungdahl (1984) *JBC* 259, 3499.
[8] E. M. Rios-Orlandi & R. E. MacKenzie (1988) *JBC* 263, 4662.

Selected entries from **Methods in Enzymology** [vol, page(s)]:
General discussion: 6, 368; 18B, 605; 66, 599, 609, 616; 281, 146, 171, 178
Assay: 6, 369; 66, 601, 602, 609, 616
Other molecular properties: "active formaldehyde" assay, 6, 809; assay of (N^5,N^{10}-methylenetetrahydrofolate and tetrahydrofolate, 18B, 605; chicken liver, in 5-methyl-L-tetrahydrofolate assay (NADP$^+$ enzyme), 122, 382; α-methylserine transhydroxymethylase, 17B, 341; serine transhydroxymethylase, 17B, 335); dihydrofolate reductase and, 6, 368; N^{10}-formyltetrahydrofolate deacylase and, 6, 373, 375; properties, 6, 372; 66, 607, 615, 624; purification (Bakers' yeast, 18B, 606; chicken liver, 6, 370; *Clostridium cylindrosporum*, 18B, 606, 607; *Clostridium formicoaceticum*, 66, 604, 605; pig liver, 66, 611; sheep liver, 66, 618; yeast, 66, 621); pyridine nucleotides and, 6, 368; serine hydroxymethylase and, 5, 838, 840

5,10-METHYLENETETRAHYDROFOLATE REDUCTASE

Two oxidoreductases catalyze the conversion of 5,10-methylenetetrahydrofolate to 5-methyltetrahydrofolate.

The FAD-dependent 5,10-methylenetetrahydrofolate reductase (NADPH) [EC 1.5.1.20] catalyzes the reaction of 5,10-methylenetetrahydrofolate with NADPH to produce 5-methyltetrahydrofolate and NADP$^+$.[1–4] The pig liver oxidoreductase has a ping pong Bi Bi kintic mechanism.[4] The reduction reaction proceeds via the formation of a N^5-iminium cation intermediate. Electrons are transferred to the iminium cation from the pteridine ring resulting in the generation of a quinonoid intermediate which now accepts electrons from the reduced flavin to produce the final product.

5,10-Methylenetetrahydrofolate reductase (FADH₂) [EC 1.7.99.5], also a FAD-dependent oxidoreductase, catalyzes the reaction of 5,10-methylenetetrahydrofolate with a reduced donor substrate to produce 5-methyltetrahydrofolate and the oxidized acceptor.[5] The donor substrate can be free FADH₂. This enzyme was originally classified as EC 1.1.1.68, based on the mistaken idea that the enzyme only uses NADH or NADPH as the donor substrate.

[1] V. Schirch (1998) *CBC* 1, 211.
[2] R. G. Matthews, C. Sheppard & C. Goulding (1998) *Eur. J. Pediatr.* 157, Suppl 2, S54.
[3] S. W. Ragsdale (1991) *Crit. Rev. Biochem. Mol. Biol.* 26, 261.
[4] M. A. Vanoni, D. P. Ballou & R. G. Matthews (1983) *JBC* 258, 11510.
[5] J. E. Clark & L. G. Ljungdahl (1984) *JBC* 259, 10845.

Selected entries from **Methods in Enzymology** [vol, page(s)]:
General discussion: 17B, 371; 122, 372, 392
Assay: 17B, 372; 122, 373; FADH₂ enzyme, 122, 393; 5-methyl-L-tetrahydrofolate assay, 122, 382
Other molecular properties: FADH₂ enzyme (*Clostridium formicoaceticum* [assay, 122, 393; properties, 122, 397; purification, 122, 395]); NADPH enzyme (bovine liver, in 5-methyl-L-tetrahydrofolate assay, 122, 382; porcine liver [assays, 122, 373; properties, 122, 377; purification, 122, 376]); purification from pig liver, 17B, 372; stereospecificity, 87, 117

5,10-METHYLENETETRAHYDROFOLATE: tRNA (URACIL-5-)-METHYLTRANSFERASE (FADH₂ OXIDIZING)

This enzyme [EC 2.1.1.74], also known as folate-dependent ribothymidyl synthase, catalyzes the reaction of 5,10-methylenetetrahydrofolate with tRNA (containing UΨC) and FADH₂ to produce tetrahydrofolate, tRNA (containing TΨC), and FAD.[1–4]

[1] D. Söll & L. K. Kline (1982) *The Enzymes*, 3rd ed., 15, 557.

[2]A. S. Delk, D. P. Nagle & J. C. Rabinowitz (1980) JBC 255, 4387.
[3]A. S. Delk, D. P. Nagle & J. C. Rabinowitz (1979) BBRC 86, 244.
[4]A. S. Delk, J. M. Romeo, D. P. Nagle & J. C. Rabinowitz (1976) JBC 251, 7649.

METHYLENETETRAHYDRO-METHANOPTERIN DEHYDROGENASE

This oxidoreductase [EC 1.5.99.9] catalyzes the reversible reaction of 5,10-methylenetetrahydromethanopterin with coenzyme F420 to produce 5,10-methenyltetrahydromethanopterin and reduced coenzyme F420.[1–4] This enzyme participates in the biosynthesis of methane from carbon dioxide in *Methanobacterium thermoautotrophicum*, *Methanosarcina thermophila*, and *Archaeoglobus fulgidus*.

[1]P. L. Hartzell, G. Zvilius, J. C. Escalante-Semerena & M. I. Donelly (1985) BBRC 133, 884.
[2]B. Mukhopadhyay & L. Daniels (1988) Can. J. Microbiol. 35, 499.
[3]P. E. Jablonski, A. A. DiMarco, T. A. Bobik, M. C. Cabell & J. G. Ferry (1990) J. Bacteriol. 172, 1271.
[4]D. Moeller-Zinkhan, G. Boerner & R. K. Thauer (1989) Arch. Microbiol. 152, 362.

3-METHYLGLUTACONYL-CoA HYDRATASE

This enzyme [EC 4.2.1.18] catalyzes the conversion of (S)-3-hydroxy-3-methylglutaryl-CoA to *trans*-3-methylglutaconyl-CoA and water.[1,2]

[1]B. Messner, H. Eggerer, J. W. Cornforth & R. Mallaby (1975) EJB 53, 255.
[2]H. Hilz, J. Knappe, E. Ringelmann & F. Lynen (1958) Biochem. Z. 329, 476.

Selected entries from *Methods in Enzymology* [vol, page(s)]:
General discussion: 166, 214
Assay: 166, 214
Other molecular properties: *Achromobacter* enzyme, binds to calcium phosphate gel, 71, 784; properties, 166, 217

METHYLGLUTAMATE DEHYDROGENASE

This oxidoreductase [EC 1.5.99.5] catalyzes the reaction of *N*-methyl-L-glutamate with water and an acceptor substrate to produce L-glutamate, formaldehyde, and the reduced acceptor. Acceptor substrates include 2,6-dichloroindophenol, ferricyanide, and cytochrome *c*.[1–5] Other *N*-substituted substrates include *N*-methylaspartate, *N*-methylalanine, sarcosine, *N*-methylvaline, *N*-methylphenylalanine, and *N*-methylisoleucine.

[1]C. A. Boulton, G. W. Haywood & P. J. Large (1980) J. Gen. Microbiol. 117, 293.
[2]C. W. Bamforth & P. J. Large (1977) Biochem. J. 161, 357.
[3]C. W. Bamforth & P. J. Large (1975) Biochem. Soc. Trans. 3, 1066.
[4]L. B. Hersh, M. J. Stark, S. Worthen & M. K. Fiero (1972) ABB 150, 219.
[5]L. B. Hersh, J. A. Peterson & A. A. Thompson (1971) ABB 145, 115.

METHYLGLYOXAL SYNTHASE

This enzyme [EC 4.2.3.3; previously classified as EC 4.2.99.11] catalyzes the conversion of dihydroxyacetone phosphate (or, glycerone phosphate) to methylglyoxal and orthophosphate. D-Glyceraldehyde 3-phosphate is not a substrate. The C3 proton (*pro-S*) of dihydroxyacetone phosphate is stereospecifically removed (an aspartyl residue acts as the catalytic base) resulting in the formation of a transient ene-diol(ate)-enzyme intermediate which collapses to form a 2-hydroxy 2-propenal enol intermediate and orthophosphate.[1]

[1]D. Saadat & D. H. Harrison (2000) Biochemistry 39, 2950.

Selected entries from *Methods in Enzymology* [vol, page(s)]:
Assay: 41, 502, 503
Other molecular properties: distribution, 41, 508; enol product, 87, 90; *Escherichia coli*, 41, 504 (preparation, 41, 504; properties, 41, 506, 508; purification, 41, 507); inhibition, 41, 508; intermediate, 87, 97; molecular weights, 41, 508; *Pseudomonas saccharophila*, 41, 503 (preparation, 41, 503, 504; purification, 41, 505); stereochemistry, 87, 153

METHYLGUANIDINASE

This enzyme [EC 3.5.3.16] catalyzes the hydrolysis of methylguanidine ($^+H_2N=C(NH_2)NHCH_3$) to produce methylamine and urea.[1] Other substrates include other alkylguanidines (for example, ethylguanidine, *n*-propylguanidine, and agmatine).

[1]M. Nakajima, Y. Shirokane & K. Mizusawa (1980) FEBS Lett. 110, 43.

N-METHYLHYDANTOINASE (ATP-HYDROLYZING)

This enzyme [EC 3.5.2.14], also called *N*-methylhydantoin amidohydrolase, catalyzes the reaction of ATP with *N*-methylimidazolidine-2,4-dione and two water molecules to produce ADP, orthophosphate, and *N*-carbamoylsarcosine.[1,2]

[1]J. M. Kim, S. Shimizu & H. Yamada (1987) BBRC 142, 1006.
[2]J. Ogawa, J. M. Kim, W. Nirdnoy, Y. Amano, H. Yamada & S. Shimizu (1995) EJB 229, 284.

N-METHYLHYDRAZINE DEMETHYLASE

This heme-thiolate-dependent oxidoreductase (i.e., a cytochrome P450) catalyzes the demethylation of methylhydrazine using dioxygen and NADPH.[1] Other substrates include 1,1-dimethylhydrazine and 1,2-dimethylhydrazine. Free radicals have been suggested to be generated as intermediates.

[1]J. A. Wittkop, R. A. Prough & D. J. Reed (1969) ABB 134, 308.

Selected entries from *Methods in Enzymology* [vol, page(s)]:
General discussion: **52**, 15

4-METHYL-3-HYDROXYANTHRANILATE ADENYLYLTRANSFERASE

This ligase, also called actinomycin synthetase I and 4-methyl-3-hydroxyanthranilate:AMP ligase, catalyzes the reaction of 4-methyl-3-hydroxyanthranilate with ATP to produce pyrophosphate and the adenylate of the 4-methyl-3-hydroxyanthranilate.[1] This activated product is a substrate of actinomycin synthetase II. *See* Actinomycin Synthetase II

[1] U. Keller & W. Schlumbohm (1992) *JBC* **267**, 11745.

2-METHYLISOCITRATE DEHYDRATASE

This enzyme [EC 4.2.1.99], also known as (2S,3R)-3-hydroxybutane-1,2,3-tricarboxylate hydro-lyase, catalyzes the conversion of (2S,3R)-3-hydroxybutane-1,2,3-tricarboxylate (that is, *threo*-D$_S$-2-methylisocitrate) to (Z)-but-2-ene-1,2,3-tricarboxylate (or, 2-methyl-*cis*-aconitate) and water. This dehydratase, first isolated from the fungus *Yarrowia lipolytica*, does not use isocitrate as an alternative substrate.[1]

[1] H. Aoki, H. Uchiyama, H. Umetsu & T. Tabuchi (1995) *Biosci. Biotech. Biochem.* **59**, 1825.

METHYLISOCITRATE LYASE

This lyase [EC 4.1.3.30] catalyzes the conversion of (2S,3R)-3-hydroxybutane-1,2,3-tricarboxylate (that is, methylisocitrate) to pyruvate and succinate.[1,2] Other substrates include *threo*-D-2-methylisocitrate (but not *threo*-D-isocitrate, *threo*-DL-isocitrate, or *erythro*-L-isocitrate).

[1] H. Uchiyama, M. Ando, Y. Toyonaka & T. Tabuchi (1982) *EJB* **125**, 523.
[2] T. Tabuchi & T. Satoh (1977) *Agric. Biol. Chem.* **41**, 169.

METHYLITACONATE Δ-ISOMERASE

This enzyme [EC 5.3.3.6] catalyzes the interconversion of methylitaconate and dimethylmaleate.[1–3] The reaction has been proposed to proceed via a carbonium ion or a radical mechanism.

[1] H. F. Kung & L. Tsai (1971) *JBC* **246**, 6436.
[2] H. K. Kung & T. C. Stadtman (1971) *JBC* **246**, 3378.
[3] G. Hartrampf & W. Buckel (1986) *EJB* **156**, 301.

[3-METHYL-2-KETOBUTANOATE DEHYDROGENASE (LIPOAMIDE)] KINASE

This mitochondrial phosphotransferase [EC 2.7.1.115], also known as [3-methyl-2-oxobutanoate dehydrogenase (lipoamide)] kinase and branched-chain α-ketoacid dehydrogenase kinase, catalyzes the reaction of ATP with 3-methyl-2-oxobutanoate dehydrogenase (lipoamide) to produce ADP and [3-methyl-2-oxobutanoate dehydrogenase (lipoamide)] phosphate (that is, the phosphorylated form of the dehydrogenase complex [EC 1.2.4.4], thus inactivating the complex).[1–4] The Ser293 of the E1 α-subunit of the complex is phosphorylated.

[1] L. J. Reed, Z. Damuni & M. L. Merryfield (1985) *Curr. Top. Cell Regul.* **27**, 41.
[2] Y. Shimomura, M. J. Kuntz, M. Suzuki, T. Ozawa & R. A. Harris (1988) *ABB* **266**, 210.
[3] H. Y. Lee, T. B. Hall, S. M. Kee, H. Y. Tung & L. J. Reed (1991) *Biofactors* **3**, 109.
[4] R. A. Harris, J. W. Hawes, K. M. Popov, Y. Zhao, Y. Shimomura, J. Sato, J. Jaskiewicz & T. D. Hurley (1997) *Adv. Enzyme Regul.* **37**, 271.

Selected entries from *Methods in Enzymology* [vol, page(s)]:
General discussion: **166**, 114, 166, 313; **324**, 48, 162, 491, 498
Assay: **166**, 166, 314; **324**, 48; potential sources of error, **166**, 167
Other molecular properties: complex (properties, **166**, 319; purification from [rabbit heart, **166**, 319; rabbit liver, **166**, 316]); inhibition by α-chloroisocaproate, **166**, 114; properties and function, **107**, 94

[3-METHYL-2-KETOBUTANOATE DEHYDROGENASE (LIPOAMIDE)]-PHOSPHATASE

This mitochondrial phosphatase [EC 3.1.3.52], also known as [3-methyl-2-oxobutanoate dehydrogenase (lipoamide)]-phosphatase and branched-chain keto-acid dehydrogenase phosphatase, catalyzes the hydrolysis of the phosphorylated form of 3-methyl-2-ketobutanoate dehydrogenase (lipoamide) to produce 3-methyl-2-oxobutanoate dehydrogenase (lipoamide) and orthophosphate.[1,2]

[1] H. R. Fatania, P. A. Patston & P. J. Randle (1983) *FEBS Lett.* **158**, 234.
[2] Z. Damuni & L. J. Reed (1987) *JBC* **262**, 5129.

Selected entries from *Methods in Enzymology* [vol, page(s)]:
General discussion: **166**, 321
Assay: **166**, 321
Other molecular properties: inhibitor protein (assay, **166**, 322; properties, **166**, 327; purification, **166**, 326); properties, **166**, 325; purification from bovine kidney, **166**, 323

3-METHYL-2-KETOBUTANOATE HYDROXYMETHYLTRANSFERASE

This transferase [EC 2.1.2.11] (also known as α-ketoisovalerate hydroxymethyltransferase, ketopantoate hydroxymethyltransferase, 3-methyl-2-oxobutanoate hydroxymethyltransferase, and dehydropantoate hydroxymethyltransferase) catalyzes the reaction of 5,10-methylenetetrahydrofolate with 3-methyl-2-ketobutanoate (or, 3-methyl-2-oxobutanoate) to produce tetrahydrofolate and 2-dehydropantoate.[1–3]

[1]G. M. Brown & J. M. Williamson (1982) *Adv. Biochem.* **53**, 345.
[2]J. H. Teller, S. G. Powers & E. E. Snell (1976) *JBC* **251**, 3780.
[3]S. G. Powers & E. E. Snell (1976) *JBC* **251**, 3786.
Selected entries from *Methods in Enzymology* [vol, page(s)]:
General discussion: **62**, 204
Assay: **62**, 205, 206
Other molecular properties: properties, **62**, 208, 209; purification from *Escherichia coli*, **62**, 206

N-METHYL-2-KETOGLUTARAMATE HYDROLASE

This enzyme [3.5.1.36], also known as 5-hydroxy-*N*-methylpyroglutamate synthase, catalyzes the hydrolysis of *N*-methyl-2-oxoglutaramate (or, *N*-methyl-2-ketoglutaramate) to form 2-ketoglutarate and methylamine. This chemical reaction has the interesting property that the products can combine nonenzymatically to form the cyclic product 2-hydroxy-*N*-methyl-5-oxoproline.[1]

[1]L. B. Hersh (1970) *JBC* **245**, 3526.

N[6]-METHYL-L-LYSINE OXIDASE

This oxidoreductase [EC 1.5.3.4] catalyzes the reaction of *N*[6]-methyl-L-lysine with water and dioxygen to produce L-lysine, formaldehyde, and hydrogen peroxide.[1,2] Other substrates include 2-keto-*N*-methyl-6-aminohexanoate and *N*[6]-ethyl-L-lysine.

[1]W. K. Paik & S. Kim (1974) *ABB* **165**, 369.
[2]S. Kim, L. Benoiton & W. K. Paik (1964) *JBC* **239**, 3790.

Selected entries from *Methods in Enzymology* [vol, page(s)]:
General discussion: **52**, 21; **56**, 475
Assay: **106**, 283
Other molecular properties: rat kidney (assay, **106**, 283; properties, **106**, 286; purification, **106**, 285)

(*R*)-2-METHYLMALATE DEHYDRATASE

citraconate   (*R*)-citramalate

This iron-dependent enzyme [EC 4.2.1.35], also known as citraconase and citraconate hydratase, catalyzes the reversible reaction of 2-methylmaleate (or, citraconate) with water to produce (*R*)-2-methylmalate.[1] **See also** *(S)-2-Methylmalate Dehydratase*

[1]S. S. Subramanian & M. R. R. Rao (1968) *JBC* **243**, 2367.

Selected entries from *Methods in Enzymology* [vol, page(s)]:
General discussion: **13**, 314, 318

(*S*)-2-METHYLMALATE DEHYDRATASE

mesaconate   (*S*)-citramalate

This enzyme [EC 4.2.1.34], also known as mesaconase and mesaconate hydratase, catalyzes the reversible reaction of 2-methylfumarate (that is, mesaconate) and water to produce (*S*)-2-methylmalate.[1,2] The enzyme will also catalyze the hydration of fumarate to (*S*)-malate. **See also** *(R)-2-Methylmalate Dehydratase*

[1]S. Suzuki, T. Osumi & K. Katsuki (1977) *J. Biochem.* **81**, 1917.
[2]C. C. Wang & H. A. Barker (1969) *JBC* **244**, 2516 and 2527.

Selected entries from *Methods in Enzymology* [vol, page(s)]:
General discussion: **13**, 314, 318, 331
Assay: **13**, 331, 338
Other molecular properties: component I, **13**, 332; component II, **13**, 338 (activators, **13**, 342, 344); interaction of components, **13**, 331, 342; properties, **13**, 336, 341, 342 (component I, **13**, 336; component II, **13**, 341); purification, *Clostridium tetanomorphum*, **13**, 333, 338 (component I, **13**, 333; component II, **13**, 338)

METHYLMALONATE-SEMIALDEHYDE DEHYDROGENASE (ACYLATING)

This oxidoreductase [EC 1.2.1.27] catalyzes the reaction of 2-methyl-3-ketopropanoate (that is, methylmalonate semialdehyde or 2-methyl-3-oxopropanoate) with coenzyme A and NAD[+] to produce propanoyl-CoA, carbon dioxide, and NADH.[1,2] The enzyme will also convert propanal to propanoyl-CoA and malonate semialdehyde to acetyl-CoA. This enzyme will also catalyze the hydrolysis of *p*-nitrophenyl acetate.[3]

[1]G. W. Goodwin, P. M. Rougraff, E. J. Davis & R. A. Harris (1989) *JBC* **264**, 14965.
[2]D. Bannerjee, L. E. Sanders & J. R. Sokatch (1970) *JBC* **245**, 1828.
[3]K. M. Popov, N. Y. Kedishvili & R. A. Harris (1992) *BBA* **1119**, 69.

Selected entries from *Methods in Enzymology* [vol, page(s)]:
General discussion: **166**, 389; **324**, 207
Assay: **166**, 390
Other molecular properties: properties, **166**, 393; purification from *Pseudomonas aeruginosa*, **166**, 391

METHYLMALONYL-CoA CARBOXYLTRANSFERASE

This biotin-dependent transferase [EC 2.1.3.1], also known as oxaloacetate transcarboxylase and transcarboxylase, catalyzes the reversible reaction of (*S*)-methylmalonyl-CoA with pyruvate to produce propanoyl-CoA and oxaloacetate.[1,2] The enzyme, which also contains tightly bound Zn^{2+} and Co^{2+}, has a two-site ping pong Bi Bi kinetic mechanism. (*S*)-Methylmalonyl-CoA first reacts with the enzyme-bound biotin to produce propionyl-CoA and the carboxy form of the coenzyme. Pyruvate then binds and is carboxylated to form oxaloacetate and regenerate the biotin form of the protein. Isotope effect studies indicate that the carboxylation of pyruvate proceeds by a stepwise mechanism involving the intermediate formation of the substrate carbanion.[1]

[1] S. J. O'Keefe & J. R. Knowles (1986) *Biochemistry* **25**, 6077.
[2] H. G. Wood (1972) *The Enzymes*, 3rd ed., **6**, 83.

Selected entries from *Methods in Enzymology* [vol, page(s)]:
General discussion: 13, 215; 46, 21
Assay: 13, 215
Other molecular properties: alternative product, **87**, 94; assay for (methylmalonyl-coenzyme A mutase, **13**, 208; methylmalonyl-coenzyme A epimerase, **13**, 195); biotin content, **13**, 225; bound metal, **13**, 226; conformational study, **49**, 354, 356; dianions, effect of, **13**, 217; inhibitors, **13**, 227; intermediate, **87**, 94; mechanism, **13**, 228; **63**, 51; product inhibition, **63**, 435; properties, **13**, 224; purification, **13**, 217; specificity, **13**, 227; stability of carboxylated biotin-enzyme, **13**, 229; stereochemistry, **87**, 94, 148; transition-state and multisubstrate analogues, **249**, 304

METHYLMALONYL-CoA DECARBOXYLASE

This enzyme [EC 4.1.1.41], also known as propionyl-CoA carboxylase, catalyzes the conversion of (*S*)-2-methyl-3-oxopropanoyl-CoA (that is, methylmalonyl-CoA) to propanoyl-CoA and carbon dioxide.[1–4] Malonyl-CoA can also be decarboxylated to generate acetyl-CoA.

This enzyme is often confused with propionyl-CoA carboxylase [EC 6.4.1.3]. This decarboxylase initially binds methylmalonyl-CoA to produce propionyl-CoA and the carboxybiotin form of the enzyme; release of bicarbonate is coupled with the transport of sodium ions. *See also Propionyl-CoA Carboxylase*

[1] J. N. Earnhardt & D. N. Silverman (1998) *CBC* **1**, 495.
[2] A. Hoffmann & P. Dimroth (1987) *FEBS Lett.* **220**, 121.
[3] W. Hilpert & P. Dimroth (1984) *EJB* **138**, 579.
[4] W. Hilpert & P. Dimroth (1983) *EJB* **132**, 579.

Selected entries from *Methods in Enzymology* [vol, page(s)]:
General discussion: 125, 540
Assay: 125, 541
Other molecular properties: isomerase and, **5**, 587; *Veillonella alcalescens* (assays, **125**, 541; properties, **125**, 543; purification, **125**, 542; sodium transport activity [properties, **125**, 546; reconstitution in proteoliposome, **125**, 544])

METHYLMALONYL-CoA EPIMERASE

This epimerase [EC 5.1.99.1] catalyzes the interconversion of (*R*)-2-methyl-3-oxopropanoyl-CoA (or, (*R*)-methyl-malonyl-CoA) and (*S*)-2-methyl-3-oxopropanoyl-CoA (or, (*S*)-methylmalonyl-CoA).[1–6]

This enzyme is frequently incorrectly referred to as methylmalonyl-CoA racemase even though the substrate and product are diastereoisomers, not enantiomers (the chiral centers in the coenzyme portion of the molecules do not undergo inversion).

Exchange and isotope effect studies of the *Propionibacterium shermanii* enzyme suggest that an active-site base acts to remove the proton from the substrate and a second residue provides the new proton. Isotopic exchange between enzyme-bound intermediates and solvent protons is slow.[1,3,4]

[1] M. E. Tanner & G. L. Kenyon (1998) *CBC* **2**, 7.
[2] S. P. Stabler, P. D. Marcell & R. H. Allen (1985) *ABB* **241**, 252.
[3] J. Q. Fuller & P. F. Leadlay (1983) *Biochem. J.* **213**, 643.
[4] P. F. Leadlay & J. Q. Fuller (1983) *Biochem. J.* **213**, 635.
[5] E. Adams (1976) *AE* **44**, 69.
[6] L. Glaser (1972) *The Enzymes*, 3rd ed., **6**, 355.

Selected entries from *Methods in Enzymology* [vol, page(s)]:
General discussion: 13, 190, 194; 166, 400
Assay: 13, 190, 194; 166, 401
Other molecular properties: assay for methylmalonyl-CoA mutase, **13**, 199, 208; mechanism, **13**, 193, 197; properties, **13**, 193, 197; **166**, 406; purification (*Propionibacterium shermanii*, **13**, 196; rat liver, **166**, 403; sheep liver, **13**, 192)

(S)-METHYLMALONYL-CoA HYDROLASE

This enzyme [EC 3.1.2.17], also known as D-methylmalonyl-CoA hydrolase and methylmalonyl-CoA deacylase, catalyzes the hydrolysis of (*S*)-methylmalonyl-CoA to produce methylmalonate and free coenzyme A.[1] The (*R*)-isomer (or L-isomer) is not a substrate. Many acyl-CoA derivatives, including succinyl-CoA, are competitive inhibitors.

[1] R. J. Kovachy, S. D. Copley & R. H. Allen (1983) *JBC* **258**, 11415.

Selected entries from *Methods in Enzymology* [vol, page(s)]:
General discussion: 166, 393
Assay: 166, 394
Other molecular properties: properties, **166**, 399; propionyl-CoA carboxylase, not present in preparation of, **5**, 575, 579; purification from rat liver, **166**, 396

METHYLMALONYL-CoA MUTASE

This cobalamin-dependent enzyme [EC 5.4.99.2] catalyzes the interconversion of (*R*)-2-methyl-3-oxopropanoyl-CoA (or, (*R*)-methylmalonyl-CoA) to succinyl-CoA.[1–3]

The reaction proceeds via radical intermediates generated by substrate-induced homolysis of the coenzyme carbon-cobalt bond, forming cob(II)alamin and the 5'-deoxyadenosyl radical. This radical abstracts a hydrogen from the methyl group of the substrate to produce the substrate radical which generates acrylate and the carbonyl-CoA radical. Recombination produces succinyl-CoA radical which receives a hydrogen from 5'-deoxyadenosine.

[1] N. H. Thoma, P. R. Evans & P. F. Leadlay (2000) *Biochemistry* **39**, 9213.
[2] S. Chowdhury & R. Banerjee (2000) *Biochemistry* **39**, 7998.
[3] H. A. Barker (1972) *The Enzymes*, 3rd ed., **6**, 509.

Selected entries from ***Methods in Enzymology*** [vol, page(s)]:
General discussion: 5, 581;13, 198, 207; **166**, 407
Assay: 5, 581;13, 198, 207; radioisotopic, **13**, 198; **166**, 407
Other molecular properties: assay for methylmalonyl-CoA epimerase, **13**, 190, 195; coenzyme content, **13**, 206, 212; coenzyme protective effect on apoenzyme, **13**, 207; mechanism, 5, 586; **13**, 213; properties, 5, 585; **13**, 206, 212; **166**, 413; purification from (human placenta, **166**, 410; *Propionibacterium shermanii*, **13**, 209; sheep kidney, 5, 583; sheep liver, **13**, 199); resolution of apoenzyme and holoenzyme, **13**, 204; stereospecificity, 5, 585; synthesis of stereospecifically deuterated succinate, **13**, 600

6-O-METHYLNORLAUDANOSOLINE 5'-O-METHYLTRANSFERASE

This methyltransferase [EC 2.1.1.121] catalyzes the reaction of *S*-adenosyl-L-methionine with 6-*O*-methylnorlaudanosoline to produce *S*-adenosyl-L-homocysteine and nororientaline, a precursor in the biosynthesis of the alkaloid papaverine.[1]

[1] M. Rueffer, N. Nagakura & M. H. Zenk (1983) *Planta Medica* **49**, 196.

METHYL-ONN-AZOXYMETHANOL GLUCOSYLTRANSFERASE

This transferase [EC 2.4.1.171], also known as cycasin synthase, catalyzes the reaction of UDP-D-glucose with methylazoxymethanol ($CH_3-N^+(-O^-)=N-CH_2OH$) to produce UDP and cycasin, a toxic substance in the leaves of Japanese cycad.[1]

[1] K. Tadera, F. Yagi, M. Arima & A. Kobayashi (1985) *Agric. Biol. Chem.* **49**, 2827.

4-METHYLOXALOACETATE ESTERASE

This esterase [EC 3.1.1.44] catalyzes the hydrolysis of oxaloacetate 4-methyl ester to produce oxaloacetate and methanol.[1]

[1] M. I. Donnelly & S. Dagley (1980) *J. Bacteriol.* **142**, 916.

METHYLPHENYLTETRAHYDROPYRIDINE N-MONOOXYGENASE

This FAD-dependent oxidoreductase [EC 1.13.12.11] catalyzes the reaction of 1-methyl-4-phenyl-1,2,3,6-tetrahydropyridine with dioxygen to produce 1-methyl-4-phenyl-1,2,3,6-tetrahydropyridine *N*-oxide and methanol.[1,2]

[1] K. Chiba, E. Kubota, T. Miyakawa, Y. Kato & T. Ishizaki (1988) *J. Pharmacol. Exp. Ther.* **246**, 1108.
[2] I. Brockhausen, E. Hull, O. Hindsgaul, H. Schachter, R. N. Shah, S. W. Michnick & J. P. Carver (1989) *JBC* **264**, 11211.

N-METHYLPHOSPHOETHANOLAMINE CYTIDYLYLTRANSFERASE

This transferase [EC 2.7.7.57] catalyzes the reaction of CTP with *N*-methylethanolamine phosphate to produce pyrophosphate (or, diphosphate) and CDP-*N*-methylethanolamine.[1] *N*,*N*-Dimethylethanolamine phosphate is an alternative substrate.

[1] A. H. Datko & S. H. Mudd (1988) *Plant Physiol.* **88**, 1338.

METHYLPHOSPHOTHIOGLYCERATE PHOSPHATASE

This phosphatase [EC 3.1.3.14] catalyzes the hydrolysis of *S*-methyl-3-phospho-1-thio-D-glycerate to produce *S*-methyl-1-thio-D-glycerate and orthophosphate.[1]

[1] S. Black & N. G. Wright (1956) *JBC* **221**, 171.

METHYLQUERCETAGETIN 6-O-METHYLTRANSFERASE

This methyltransferase [EC 2.1.1.84], also known as flavonol 6-*O*-methyltransferase, catalyzes the reaction of *S*-adenosyl-L-methionine with 3',4',5,6-tetrahydroxy-3,7-dimethoxyflavone to produce *S*-adenosyl-L-homocysteine and 3',4',5-trihydroxy-3,6,7-trimethoxyflavone.[1,2] 3,7,3'-Trimethylquercetagetin, 3,7-dimethylquercetagetin, and 8-hydroxykaempferol are alternative substrates.

[1] V. De Luca & R. K. Ibrahim (1985) *ABB* **238**, 596.
[2] V. De Luca & R. K. Ibrahim (1985) *ABB* **238**, 606.

3-METHYLQUERCETIN 7-O-METHYLTRANSFERASE

This methyltransferase [EC 2.1.1.82], also known as flavonol 7-*O*-methyltransferase, catalyzes the reaction of *S*-adenosyl-L-methionine with 3',4',5,7-tetrahydroxy-3-methoxyflavone (that is, 3-methylquercetin) to produce *S*-adenosyl-L-homocysteine and 3',4',5-trihydroxy-3,7-dimethoxyflavone (that is, 3,7-dimethylquercetin).[1,2]

This enzyme participates with quercetin 3-O-methyltransferase [EC 2.1.1.76] and 3,7-dimethylquercetin 4'-O-methyltransferase [EC 2.1.1.83] in the methylation of quercetin to 3,7,4'-trimethylquercetin in the golden saxifrage (*Chrysosplenium americanum*). Flavones, dihydroflavonols, or their glucosides are not substrates. This methyltransferase is reported to have an ordered Bi Bi kinetic mechanism.[1]

[1] H. E. Khouri, V. De Luca & R. K. Ibrahim (1988) *ABB* **265**, 1.
[2] V. De Luca & R. K. Ibrahim (1985) *ABB* **238**, 596.

6-METHYLSALICYLATE DECARBOXYLASE

6-methylsalicylate   *m*-cresol

This decarboxylase [EC 4.1.1.52] catalyzes the conversion of 6-methylsalicylate (or, 2,6-cresotate) to 3-cresol and carbon dioxide.[1]

[1] R. J. Light (1969) *BBA* **191**, 430.
Selected entries from *Methods in Enzymology* [vol, page(s)]:
General discussion: 43, 530
Assay: fluorescence, 43, 531; radioactive, 43, 531, 532
Other molecular properties: ammonium sulfate fractionation, 43, 536, 537; DEAE-Sephadex A-50 column chromatography, 43, 437; homogeneity, 43, 538; hydroxyapatite column chromatography, 43, 437, 438; inhibitors, 43, 540; isotope effects, 43, 539; kinetic properties, 43, 539; 6-methylsalicylate (isolation, 43, 533; preparation, 43, 535; chemical synthesis, 43, 532, 533); molecular weight, 43, 539; *Penicillium patulum* preparation, 42, 534; properties, 43, 538; purification, 43, 536; Sephadex G-100 gel filtration, 43, 437; specificity, 43, 539; stability, 43, 538; ultracentrifugation, 43, 536; unit definition, 43, 532

6-METHYLSALICYLATE SYNTHASE

This multienzyme complex, similar to the fatty acid synthase complex, catalyzes the overall reaction of acetyl-CoA with three molecules of malonyl-CoA, NADPH, and a proton to produce 6-methylsalicylate, three molecules of carbon dioxide, four molecules of coenzyme A, NADP$^+$, and water.

Selected entries from *Methods in Enzymology* [vol, page(s)]:
General discussion: 43, 520
Assay: 43, 521; fluorometric, 43, 521
Other molecular properties: ammonium sulfate fractionation, 43, 525; cell breakage, 43, 524; DEAE-cellulose chromatography, 43, 526, 527; homogeneity, 43, 528; hydroxyapatite chromatography, 43, 525, 526; inhibitors, 43, 528; kinetic properties, 43, 528; molecular weight, 43, 528; *Penicillium patulum*, preparation, 43, 523; polyethylene glycol 1500 fractionation, 43, 524, 525; polyethylene glycol 6000 precipitation, 43, 524; properties, 43, 528; purification, 43, 524; reaction scheme, 43, 529, 530; Sepharose 6B, gel filtration, 43, 526; shake culture, 43,

524, 534; specificity, 43, 529; stability, 43, 528; unit definition, 43, 523

METHYLSTEROL MONOOXYGENASE

This oxidoreductase [EC 1.14.99.16], also known as methylsterol hydroxylase, catalyzes the reaction of 4,4-dimethyl-5α-cholest-7-en-3β-ol with dioxygen and an electron donor substrate (AH$_2$) to produce 4α-methyl-4β-hydroxymethyl-5α-cholest-7-en-3β-ol, the oxidized donor (A), and water.[1,2]

[1] J. L. Gaylor & H. S. Mason (1968) *JBC* **243**, 4966.
[2] W. L. Miller, M. E. Kalafer, J. L. Gaylor & C. V. Delwiche (1967) *Biochemistry* **6**, 2673.

N^5-METHYLTETRAHYDROFOLATE: CORRINOID/IRON–SULFUR PROTEIN METHYLTRANSFERASE

This methyltransferase, first isolated from *Clostridium thermoaceticum*, catalyzes the reaction of N^5-methyltetrahydrofolate with the cobalt center of a corrinoid/iron–sulfur protein producing methylcob(III)amide and tetrahydrofolate.

This enzyme catalyzes the first step in the Wood-Ljungdahl pathway. The Co(I) of the corrinoid/iron–sulfur protein attacks the methyl group in a rate-limiting S$_N$2 reaction to form the products.[1] The kinetic mechanism is random Bi Bi. The corrinoid/iron–sulfur protein is a heterodimeric protein in which the larger subunit contains a [4Fe-4S] cluster which participates in reductive activation of the cobalt center while the smaller subunit contains the cobalt center, 5'-methoxybenzimidazolylcobamide.

[1] J. Seravalli, S. Zhao & S. W. Ragsdale (1999) *Biochemistry* **38**, 5728.

N^5-METHYLTETRAHYDROFOLATE: HOMOCYSTEINE METHYLTRANSFERASE

This cobalamin-dependent enzyme [EC 2.1.1.13], also known as methionine synthase and tetrahydropteroylglutamate methyltransferase, catalyzes the reaction of 5-methyltetrahydrofolate with L-homocysteine to produce tetrahydrofolate and L-methionine.[1–5] The enzyme utilizes mono- or triglutamylated derivatives of tetrathydrofolate. The bacterial enzyme is reported to require *S*-adenosyl-L-methionine and FADH$_2$.

The enzyme has a ping pong Bi Bi kinetic mechanism in which L-homocysteine reacts with the methylcob(III)alamin form of the enzyme to produce L-methionine and the cob(I)alamin form. The second half-reaction has the cob(I)alamin reacting with 5-methyltetrahydrofolate to regenerate the original form of the enzyme and release tetrahydrofolate. Each methylation half-reaction exhibits inversion of configuration, suggesting that both are S_N2-like reactions. **See also** Tetrahydropteroyltriglutamate Methyltransferase; Homocysteine Methyltransferase

[1] F. Takusagawa, M. Fujioka, A. Spies & R. L. Schowen (1998) CBC 1, 1.
[2] V. Schirch (1998) CBC 1, 211.
[3] B. T. Golding & W. Buckel (1998) CBC 3, 239.
[4] M. L. Ludwig & R. G. Matthews (1997) Ann. Rev. Biochem. 66, 269.
[5] R. T. Taylor & H. Weissbach (1973) The Enzymes, 3rd ed., 9, 121.

Selected entries from **Methods in Enzymology** [vol, page(s)]:
General discussion: 17B, 371, 379, 388; 281, 189, 196
Assay: 17B, 373, 379, 388
Other molecular properties: S-adenosylmethionine and, 17B, 373, 378; assay of N^5-methyltetrahydrofolate, 18B, 611; properties, 17B, 377, 384, 392; purification from (Escherichia coli, 12B, 685; 17B, 380; 18B, 612, 613; liver, pig, 17B, 375; yeast, 17B, 390); spectrum, 17B, 384; vitamin B_{12}, 17B, 378, 386

METHYLTETRAHYDROPROTOBERBERINE 14-MONOOXYGENASE

This heme-thiolate-dependent oxidoreductase [EC 1.14.13.37], also known as methyltetrahydroprotoberberine 14-hydroxylase and (S)-cis-N-methyltetrahydroprotoberberine-14-hydroxylase, catalyzes the reaction of (S)-N-methylcanadine with NADPH and dioxygen to produce allocryptopine, $NADP^+$, and water.[1] Alternative substrates include (S)-cis-N-methylstylopine, (S)-cis-N-methyltetrahydrothalifendine, (S)-cis-N-methyltetrahydropalmitine, and (S)-cis-N-methylcorydalmine.

[1] M. Rueffer & M. H. Zenk (1987) Tetrahedron Lett. 28, 5307.

5-METHYLTETRAHYDROPTEROYL-TRIGLUTAMATE:HOMOCYSTEINE S-METHYLTRANSFERASE

This orthophosphate-dependent transferase [EC 2.1.1.14], also known as methionine synthase and tetrahydropteroyltriglutamate methyltransferase, catalyzes the reaction of 5-methyltetrahydropteroyltri-L-glutamate with L-homocysteine to produce tetrahydropteroyltri-L-glutamate and L-methionine.[1–4] **See also** N^5- Methyltetrahydrofolate:Homocysteine Methyltransferase

[1] J. T. Madison (1990) Methods Plant Biochem. 3, 361.
[2] I. Saint-Gerons, C. Parsot, M. M. Zakin, O. Bârzy & G. N. Cohen (1988) Crit. Rev. Biochem. 23, 51.
[3] R. T. Taylor & H. Weissbach (1973) The Enzymes, 3rd ed., 9, 121.
[4] C. D. Whitefield, E. J. Steers & H. Weissbach (1970) JBC 245, 390.

Selected entries from **Methods in Enzymology** [vol, page(s)]:
General discussion: 17B, 388
Assay: 17B, 388

METHYLTHIOADENOSINE NUCLEOSIDASE

This enzyme [EC 3.2.2.16], also known as methylthioadenosine nucleosidase, catalyzes the hydrolysis of methylthioadenosine to produce adenine and 5-methylthio-D-ribose.[1,2] Although S-adenosylhomocysteine is not a substrate, methylthioinosine is hydrolyzed. **See also** S-Adenosylhomocysteine Nucleosidase

[1] F. Schlenk (1983) AE 54, 195.
[2] A. B. Guranowski, P. K. Chiang & G. L. Cantoni (1981) EJB 114, 293.

Selected entries from **Methods in Enzymology** [vol, page(s)]:
General discussion: 17B, 410; 94, 365
Assay: 17B, 411; 94, 365

5'-METHYLTHIOADENOSINE PHOSPHORYLASE

This phosphorylase [EC 2.4.2.28] catalyzes the reaction of 5'-methylthioadenosine with orthophosphate to produce adenine and 5-methylthio-D-ribose 1-phosphate.[1–5] Other substrates include 5'-deoxyadenosine, 5'-n-butylthioadenosine, and 5'-isobutylthioadenosine. The enzyme has an ordered Bi Bi kinetic mechanism.

[1] G. Cacciapuoti, M. Porcelli, C. Bertoldo & V. Zappia (1992) Life Chem. Rep. 10, 75.
[2] V. Zappia, F. Della Ragione, G. Pontoni, V. Gragnaniello & M. Carteni-Farina (1988) Adv. Exp. Med. Biol. 250, 165.
[3] F. Schlenk (1983) AE 54, 195.
[4] L. Shugart, L. Mahoney & B. Chastain (1981) Int. J. Biochem. 13, 559.
[5] D. L. Garbers (1978) BBA 523, 82.

Selected entries from **Methods in Enzymology** [vol, page(s)]:
General discussion: 94, 356
Assay: 94, 356
Other molecular properties: Caldariella acidophila (assay, 94, 356; properties, 94, 359; purification, 94, 357)

5-METHYLTHIORIBOSE KINASE

This phosphotransferase [EC 2.7.1.100] catalyzes the reaction of ATP with S^5-methyl-5-thio-D-ribose, yielding ADP and S^5-methyl-5-thio-D-ribose 1-phosphate.[1–3] CTP can replace ATP, albeit not as effectively. Other acceptor substrates include 5-ethylthioribose, 5-propylthioribose, 5-isobutylthioribose, and other 5-alkylthioribose analogues.

[1] A. Guranowski (1983) Plant Physiol. 71, 932.
[2] A. J. Ferro, A. Barrett & S. K. Shapiro (1978) JBC 253, 6021.
[3] F. Schlenk (1983) AE 54, 195.

Selected entries from **Methods in Enzymology** [vol, page(s)]:
General discussion: 94, 361
Assay: 94, 362

Other molecular properties: *Enterobacter aerogenes* (assay, **94**, 362; properties, **94**, 364; purification, **94**, 363)

5-METHYLTHIORIBOSE-1-PHOSPHATE ISOMERASE

This isomerase [EC 5.3.1.23] catalyzes the conversion of 5-methylthio-5-deoxy-D-ribose 1-phosphate to 5-methyl-thio-5-deoxy-D-ribulose 1-phosphate.[1,2]

[1]P. C. Trackman & R. Abeles (1983) *JBC* **258**, 6717.
[2]E. S. Furfine & R. H. Abeles (1988) *JBC* **263**, 9598.

METHYLUMBELLIFERYL-ACETATE DEACETYLASE

This deacetylase [EC 3.1.1.56], also called esterase D and sialic acid-specific *O*-acetylesterase, catalyzes the hydrolysis of 4-methylumbelliferyl acetate to produce 4-methylumbelliferone and acetate.[1,2] Alternative substrates include other short-chain acyl esters of 4-methyl-umbelliferone; however, naphthyl, indoxyl, and thio-choline esters are not substrates.

The ability of this enzyme to hydrolyze 4-nitrophenyl acetate as well as 4-nitrophenyl butyrate affords a convenient continuous spectrophotometric assay.

[1]Y. Okada & K. Wakabayashi (1988) *ABB* **263**, 130.
[2]K. Matsuo, K. Kobayashi, K. Hagiwara & T. Kajii (1985) *EJB* **153**, 217.

METRIDIN

This serine-dependent endopeptidase [EC 3.4.21.3], also called *Metridium* proteinase A and sea anemone protease A, is isolated from the sea anemone (*Metridium senile*) and catalyzes peptide bond hydrolysis (also ester bond hydrolysis) with an action resembling α-chymotrypsin (that is, at Tyr–Xaa, Phe–Xaa, and Leu–Xaa). However, there is considerably less preference for tryptophanyl residues in the P1 position (as measured by k_{cat}/K_m data). Glucagon is hydrolyzed at Phe6–Thr7, Tyr10–Ser11, Tyr13–Leu14, Phe22–Val23, and Leu26–Met27. The Trp25–Leu26 bond was not hydrolyzed.[1]

[1]D. Gibson & G. H. Dixon (1969) *Nature* **222**, 753.

Selected entries from *Methods in Enzymology* [**vol**, page(s)]:
General discussion: 80, 682
Other molecular properties: purification and properties, **19**, 107

MET-Xaa DIPEPTIDASE

This dipeptidase [EC 3.4.13.12] (also known as Met-X dipeptidase, methionyl dipeptidase, and dipeptidase M) catalyzes the hydrolysis of Met-Xaa dipeptides. It does not act on tripeptides or larger structures. N-blocked derivatives are also not substrates. Although manganese ions have been shown to stabilize the activity, cobalt ions seem to have a greater effect with respect to certain dipeptide substrates. Some non-methionyl-containing dipeptides are also substrates: for example, alanylalanine, leucylalanine, and serylalanine. Leucylleucine is a very poor substrate. The *Escherichia coli* enzyme is stabilized with 2-mercaptoethanol (thus, distinguishing Met-Xaa dipeptidase from cytosolic nonspecific dipeptidase [EC 3.4.13.18]).[1,2]

[1]J. L. Brown (1973) *JBC* **248**, 409.
[2]S. Simmonds, K. S. Szeto & C. G. Fletterick (1976) *Biochemistry* **15**, 261.

METYRAPONE REDUCTASE

This oxidoreductase catalyzes the reaction of 2-methyl-1,2-di-3-pyridyl-1-propanone (or, metyrapone) with NADPH to produce 2-methyl-1,2-di-3-pyridyl-1-propanol and NADP$^+$. The enzyme may be identical to carbonyl reductase.

Selected entries from *Methods in Enzymology* [**vol**, page(s)]:
Other molecular properties: stereospecificity, **87**, 112

MEVALDATE REDUCTASE

Mevaldate   Mevalonate

Two enzymes catalyze the reduction of mevaldate. Mevaldate reductase [EC 1.1.1.32] catalyzes the reaction of mevaldate with NADH to produce (*R*)-mevalonate and NAD$^+$.[1,2] Mevaldate reductase (NADPH) [EC 1.1.1.33] catalyzes the reaction of mevaldate with NADPH to produce (*R*)-mevalonate and NADP$^+$.[3] Other substrates include pyridine-3-carboxaldehyde and 4-cyano-benzaldehyde. There are reports that this enzyme may be identical with alcohol dehydrogenase (NADP$^+$) [EC 1.1.1.2].

[1]M. J. Schlesinger & M. J. Coon (1961) *JBC* **236**, 2421.
[2]H.-L. Nigan & G. Popjak (1975) *Bioorg. Chem.* **4**, 166.
[3]A. S. Beedle, H. H. Rees & T. W. Goodwin (1974) *BJ* **139**, 205.

Selected entries from *Methods in Enzymology* [**vol**, page(s)]:
General discussion: 15, 393, 394
Assay: 15, 394
Other molecular properties: pig liver (preparation of 5-labeled mevalonate, **15**, 398; purification, **15**, 395); rat liver (purification, **15**, 397); stereospecificity, **87**, 105

MEVALONATE KINASE

This phosphotransferase [EC 2.7.1.36] catalyzes the reaction of ATP with (R)-mevalonate to produce ADP and (R)-5-phosphomevalonate.[1–4] The nucleotide substrate can also be provided with CTP, GTP, ITP, dATP, XTP, or UTP. The *Catharanthus roseus* enzyme has an ordered Bi Bi kinetic mechanism.[1]

[1] A. E. Schulte, R. van der Heijden & R. Verpoorte (2000) *ABB* **378**, 287.
[2] R. A. Dixon, P. M. Dey & C. J. Lamb (1983) *AE* **55**, 1.
[3] G. J. Schroepfer, Jr. (1981) *Ann. Rev. Biochem.* **50**, 585.
[4] E. D. Beytía & J. W. Porter (1976) *Ann. Rev. Biochem.* **45**, 113.

Selected entries from *Methods in Enzymology* [vol, page(s)]:
General discussion: 5, 489; 15, 402; 110, 71
Assay: 5, 490; 15, 402; 110, 60, 72
Other molecular properties: mevalonate assay, 6, 505; 15, 445n; phosphomevalonate kinase and, 5, 496; phosphomevalonate preparation, 6, 506; 15, 411; plant tissues, 13, 559; porcine liver (assays, 15, 402; 110, 60, 72; inhibition, 110, 77; properties, 15, 410; 110, 76; purification, 15, 407; 110, 60, 74; radioenzymatic assay for plasma mevalonate, in, 110, 63; reaction mechanism, 110, 77); properties, 15, 410; 110, 76; purification, 5, 494; 15, 407, 444; 110, 60, 74; specificity, 15, 400, 410

MEVALONATE-5-PYROPHOSPHATE DECARBOXYLASE

This decarboxylase [EC 4.1.1.33] (also known as diphosphomevalonate decarboxylase, pyrophosphomevalonate decarboxylase and mevalonate diphosphate decarboxylase) catalyzes the ATP-dependent reaction with (R)-5-pyrophosphomevalonate (or, (R)-5-diphosphomevalonate) to produce isopentenyl pyrophosphate, ADP, orthophosphate, and carbon dioxide.[1–4]

mevalonate 5-pyrophosphate   isopentenyl pyrophosphate

The transition state of the decarboxylation step has considerable carbocationic character, consistent with the potent inhibition by *N*-methyl-*N*-carboxymethyl-2-pyrophosphoethanolamine, a probable transition-state analogue having a positively charged nitrogen atom in place of the carbocation.[1] The enzyme displays an ordered sequential mechanism with mevalonate 5-pyrophosphate as the leading substrate, the reaction proceeding via *anti*-elimination following the phosphorylation by ATP of the β-hydroxyl group.[2]

[1] S. Dhe-Paganon, J. Magrath & R. H. Abeles (1994) *Biochemistry* **33**, 13355.
[2] A. M. Jabalquinto & E. Cardemil (1989) *BBA* **996**, 257.
[3] R. A. Gibbs (1998) *CBC* **1**, 31.
[4] R. A. Dixon, P. M. Dey & C. J. Lamb (1983) *AE* **55**, 1.

Selected entries from *Methods in Enzymology* [vol, page(s)]:
General discussion: 5, 489, 490; 15, 419; 110, 86

Assay: chromatographic assay, 5, 490; radiometric assay, 15, 420; 110, 86; spectrophotometric assay, 5, 490
Other molecular properties: chicken liver (assays, 110, 86; properties, 110, 91; purification, 110, 89); properties, 15, 425; 110, 91; purification, 15, 423; 110, 89; yeast enzyme, purification of, 5, 494

MEXICANAIN

This protease, also called mexicain and previously classified as EC 3.4.99.14, is currently classified as a variant of papain [EC 3.4.22.2].

*Mfe*I RESTRICTION ENDONUCLEASE

This type II restriction endonuclease [EC 3.1.21.4], isolated from *Mycoplasma fermentas*, catalyzes the hydrolysis of both strands of DNA at 5′ ... C∇AATTG ... 3′.[1,2] Note that this enzyme generates *Eco*RI-compatible cohesive ends.

[1] N. F. Halden, J. B. Wolf & W. J. Leonard (1989) *NAR* **17**, 3491.
[2] J. Maniloff, K. Dybvig & T. L. Sladek (1992) in *Mycoplasmas: Molecular Biology and Pathogenesis* (J. Maniloff, ed.). p. 325, Amer. Soc. Microbiol., Rochester, NY.

MG^{2+}-DEPENDENT (OR MG^{2+}-ACTIVATED) ATPase

This enzyme activity, orginally classified as EC 3.6.1.4 but now a deleted Enzyme Commission entry, is listed under ATPase [EC 3.6.1.3]

MG^{2+}-DEPENDENT (PHOSPHOLIPID-FLIPPING) ATPase

This so-called P-type ATPase [EC 3.6.3.1], also known as flippase, catalyzes the ATP-dependent (ADP- and orthophosphate-producing) transport of phospholipids from one lipid bilayer face to the other.[1–3] The enzyme undergoes covalent phosphorylation during the transport cycle.

The idea that a transporter is an enzyme is in keeping with a new definition of enzyme catalysis as the facilitated making/breaking of chemical bonds, not just covalent bonds.[4] This idea builds on Pauling's assertion that any long-lived, chemically distinct interaction (in this case, the persistent location of a solute with respect to the faces of a membrane) can be regarded as a chemical bond. Note also that the equilibrium constant, defined by $K_{eq} = $ [phospholipid$_{State-2}$][ADP][P_i]/[phospholipid$_{State-1}$][ATP], does not conform to that expected for an ATPase (*i.e.*, $K_{eq} = $ [ADP][P_i]/[ATP]). Thus, although the overall reaction yields ADP and orthophosphate, the enzyme is misclassified as a hydrolase, and instead should be regarded as an energase-type reaction. Energases facilitate affinity-modulated reactions by coupling the $\Delta G_{\text{ATP-hydrolysis}}$ to a force-generating or work-producing step.[4] In this case, P-O-P bond scission supplies the energy to drive phospholipid transport. **See** ATPase; Energase

[1]M. B. Morris, M. E. Auland, Y. H. Xu & B. D. Roufogalis (1993) *Biochem. Mol. Biol. Int.* **31**, 823.

[2]W. P. Vermeulen, J. J. Briede & B. Rolofsen (1996) *Mol. Membr. Biol.* **13**, 95.

[3]H. Suzuki, M. Kamakura, M. Morii & N. Takeguchi (1997) *JBC* **272**, 10429.

[4]D. L. Purich (2001) *TiBS* **26**, 417.

Mg^{2+}-IMPORTING ATPase

This so-called P-type ATPase [EC 3.6.3.2], which undergoes covalent phosphorylation during the transport cycle, catalyzes the ATP-dependent (ADP- and orthophosphate-producing) transport of Mg$^{2+}_{\text{out}}$ to produce Mg$^{2+}_{\text{in}}$.[1,2] This enzyme, which occurs in both Gram-positive and Gram-negative bacteria, has been identified by three types: CorA, MgtA, and MgtB. CorA will also catalyze the import of Co^{2+} and Ni^{2+}.

The idea that a transporter is an enzyme is in keeping with a new definition of enzyme catalysis as the facilitated making/breaking of chemical bonds, not just covalent bonds.[3] This idea builds on Pauling's assertion that any long-lived, chemically distinct interaction (in this case, the persistent location of a solute with respect to the faces of a membrane) can be regarded as a chemical bond. Note also that the equilibrium constant $(K_{\text{eq}} = [\text{Mg}^{2+}_{\text{in}}][\text{ADP}][\text{P}_i]/[\text{Mg}^{2+}_{\text{out}}][\text{ATP}])$ does not conform to that expected for an ATPase (*i.e.*, $K_{\text{eq}} = [\text{ADP}][\text{P}_i]/[\text{ATP}]$). Thus, although the overall reaction yields ADP and orthophosphate, the enzyme is misclassified as a hydrolase, and instead should be regarded as an energase-type reaction. Energases facilitate affinity-modulated reactions by coupling the $\Delta G_{\text{ATP-hydrolysis}}$ to a force-generating or work-producing step.[3] In this case, P-O-P bond scission supplies the energy to drive Mg^{2+} transport. **See** ATPase; Energase

[1]T. Tao, M. D. Snavely, S. G. Farr & M. E. Maguire (1995) *J. Bacteriol.* **177**, 2654.

[2]R. L. Smith, M. A. Szegedy, L. M. Kucharski, C. Walker, R. M. Wiet, A. Redpath, M. T. Kaczmarek & M. E. Maguire (1998) *JBC* **273**, 28663.

[3]D. L. Purich (2001) *TiBS* **26**, 417.

m^{7}G(5′)pppN PYROPHOSPHATASE

This pyrophosphatase [EC 3.6.1.30], also known as m^{7}G(5′)pppN diphosphatase and decapase, catalyzes the hydrolysis of 7-methylguanosine 5′-triphospho-5′-polynucleotide (*i.e.*, a polynucleotide [typically mRNA] possessing a 7-methylguanosine "cap" at the 5′ end) to produce 7-methylguanosine 5′-phosphate and a polynucleotide that is said to be "decapped."[1–3]

[1]D. L. Nuss, Y. Furiuchi & G. Koch (1975) *Cell* **6**, 21.

[2]D. L. Nuss & Y. Furuichi (1977) *JBC* **252**, 2815.

[3]D. L. Nuss, R. E. Altschuler & A. J. Peterson (1982) *JBC* **257**, 6224.

M.HaeIII METHYLTRANSFERASE

This site-specific methyltransferase [EC 2.1.1.73], isolated from *Haemophilus aegyptus*, catalyzes the reaction of *S*-adenosyl-L-methionine with the first cytosine in the sequence 5′ ... GGCC... 3′ in double-stranded DNA to produce *S*-adenosyl-L-homocysteine and the sequence containing a 5-methylcytosine residue.[1–3] **See also** Site-Specific DNA-Methyltransferase (Cytosine-Specific)

[1]R. J. Roberts & X. Cheng (1998) *Ann. Rev. Biochem.* **67**, 181.

[2]K. M. Reinisch, L. Chen, G. L. Verdine & W. N. Lipscomb (1995) *Cell* **82**, 143.

[3]M. B. Mann & H. O. Smith (1977) *NAR* **4**, 4211.

M.HhaI METHYLTRANSFERASE

This site-specific methyltransferase [EC 2.1.1.73], isolated from *Haemophilus haemolyticus*, catalyzes the reaction of *S*-adenosyl-L-methionine with the first cytosine in the sequence 5′ ... GCGC... 3′ in double-stranded DNA to produce *S*-adenosyl-L-homocysteine and the sequence containing a 5-methylcytosine residue.[1–3] The transferase has an ordered Bi Bi kinetic mechanism in which *S*-adenosyl-L-methionine is the first substrate to bind and *S*-adenosyl-L-homocysteine is the last product to dissociate.[1] **See also** Site-Specific DNA-Methyltransferase (Cytosine-Specific)

[1]J. C. Wu & D. V. Santi (1987) *JBC* **262**, 4778.

[2]R. J. Roberts & X. Cheng (1998) *Ann. Rev. Biochem.* **67**, 181.

[3]M. O'Gara, S. Klimasauskas, R. J. Roberts & X. Cheng (1996) *J. Mol. Biol.* **261**, 634.

M.HindIV METHYLTRANSFERASE

This site-specific methyltransferase [EC 2.1.1.72] is an orphan methylase (subtype α) isolated from *Haemophilus influenzae* R$_d$ (exo-mutant). The recognition sequence is 5′ ... GAT... 3′ and the transferase, which has had its gene sequenced (as has the entire genome), catalyzes the *S*-adenosyl-L-methionine-dependent methylation of the adenosine (producing 6-methyladenosine).[1] *See also Site-Specific DNA-Methyltransferase (Adenine-Specific)*

[1] P. H. Roy & H. O. Smith (1973) *J. Mol. Biol.* **81**, 427 and 445.

MICROCOCCAL NUCLEASE

This enzyme [EC 3.1.31.1], also known as staphylococcal nuclease and micrococcal endonuclease, catalyzes the endonucleolytic cleavage of double-stranded or single-stranded polynucleotides (DNA and RNA) to nucleoside 3′-phosphate and 3′-phosphooligonucleotide end products.[1-5] Similar enzymes are *Chlamydomonas* nuclease, spleen exonuclease, and spleen endonuclease.

[1] C. B. Anfinsen, P. Cuatrecasas & H. Taniuchi (1971) *The Enzymes*, 3rd ed., **4**, 177.

[2] F. A. Cotton & E. E. Hazen (1971) *The Enzymes*, 3rd ed., **4**, 153.

[3] J. A. Gerlt (1992) *The Enzymes*, 3rd ed., **20**, 95.

[4] A. S. Mildvan & D. C. Fry (1987) *AE* **59**, 241.

[5] N. H. Williams (1998) *CBC* **1**, 543.

Selected entries from *Methods in Enzymology* [vol, page(s)]:
General discussion: **12A**, 257; **12B**, 207; **27**, 729, 732, 832; **34**, 492, 495, 496, 592, 632; **46**, 358; **68**, 214
Other molecular properties: affinity labeled peptides, **34**, 187; affinity labeling, **87**, 472, 477; agarose-linked, **34**, 186; assembly of chromatin with oocyte extracts, in, **170**, 611; analysis of RNA-protecting factor interactions, in, **180**, 465; azo linkage, **34**, 106; bifunctional aryl halides and, **25**, 640; ^{13}C chemical shift, **239**, 369; chemical shift, in structural analysis, **239**, 405; circular dichroic spectrum, **49**, 161; cleavage of tryptophanyl peptide bond, **25**, 422; determination of activity, **12A**, 258; deuterated (enzymatic activity, **26**, 616; nuclear magnetic resonance spectra, **26**, 617, 630); digestion of nuclei, **170**, 329; digest mix, **68**, 136; DNA-free crude extracts and, **21**, 202; DNA sequence analysis and, **29**, 241, 242, 243; electron density maps, interpretation, **115**, 194, 199; encoding gene, placement under control of strong inducible promoter, **177**, 76; fingerprinting procedure and, **12A**, 369; folding and stability, genetic analysis, **131**, 266; footprinting (chromatin [chicken, **274**, 254; yeast, **274**, 208]; reconstituted dinucleosomes, **274**, 269); fragment complementing systems (folding kinetics and intermediates, **131**, 205; fragment exchange in, **131**, 207; preparation, **131**, 193); heteronuclear relaxation studies, **239**, 564; H^N-H^α J couplings, quantitative J correlation, **239**, 102; histidine nuclear magnetic resonance spectra, **26**, 645; gene cloning, **125**, 146, 148; inactivation of histone-specific RNA processing, **181**, 82; incorporation of isotopically labeled amino acids, **177**, 77; inhibitor, **63**, 396, 398; labeled DNA, digestion and analysis, **65**, 64; mechanism, **87**, 197; microorganism, **12A**, 258; modification of tryptophan residues, **25**, 421; near-UV circular dichroism, **61**, 377; nitration of, **25**, 518, 519; MOPC-315, **46**, 155; NOE cross-peaks, residue-type assignment, **177**, 284; nuclease-T (nuclease-T-(6-48), synthetic, **34**, 632; nuclease-T-(49-149), native, **34**, 632; nuclease-T-agarose, **34**, 635; semisynthetic, **34**, 632; thermolysin hydrolysis, **47**, 188); occurrence, **12A**, 257; oligonucleotide identification and, **12A**, 282, 289; polydeoxythymidylate preparation and, **21**, 313; preparation of chromatin, in, **170**, 127; properties, **12A**, 260; purification, **12A**, 259; **34**, 50 (affinity absorption, **25**, 229; affinity chromatography, **22**, 353, 355, 360, 368, 371, 373, 377); purification by affinity chromatography, **22**, 353, 355, 360, 368, 371, 373, 377; reagents for affinity labeling, **46**, 358; reverse transcriptase and, **29**, 160; RNA polymerase polynucleotide complex and, **21**, 526; ribosubstituted deoxyribonucleic acid fragments, **29**, 304 (both phosphatase and phosphodiesterase treatment required, **29**, 320; desalting, **29**, 313; digestion, **29**, 310, 321; elution from thin-layer plates, **29**, 312; marker

mix preparation, **29**, 308; other applications, **29**, 321; phosphatase-sensitive ^{32}P determination, **29**, 315; resolution of spleen phosphodiesterase and monoesterase, **29**, 317; specific examples, **29**, 307; spleen phosphodiesterase treatment, **29**, 316); selectively deuterated, **26**, 608; site-directed mutants, nonadditivity (antagonistic) effects, **249**, 118; staphylococcal nuclease A (concanavalin A mutant S28G, **240**, 631, 642; heat capacity of denaturation, **240**, 641; pressure effect on structure, **259**, 411; unfolding analysis, **240**, 628, 631; unfolding monitoring, **259**, 505); sulfenyl halides and, **25**, 493; synthetic, **34**, 632; 3′-terminal dinucleotide analysis and, **29**, 290; 3′-terminal identification and, **29**, 279; 5′-terminus sequence analysis and, **29**, 283; thermolysin hydrolysis, **47**, 188; trifluoroacetylation, **11**, 322; yeast mRNA titration, in, **194**, 539

MICROCOCCUS LUTEUS EXONUCLEASE

This exonuclease is currently classified as a variant of exodeoxyribonuclease VII [EC 3.1.11.6].

Selected entries from *Methods in Enzymology* [vol, page(s)]:
General discussion: **21**, 249
Assay: **21**, 250

MICROCOCCUS LUTEUS UV ENDONUCLEASE

This endonuclease is now classified as a variant of DNA-(apurinic or apyrimidinic site) lyase [EC 4.2.99.18].

Selected entries from *Methods in Enzymology* [vol, page(s)]:
General discussion: **21**, 244
Assay: **21**, 244

MIMOSINASE

This plant enzyme [EC 3.5.1.61], also known as mimosine amidohydrolase, catalyzes the hydrolysis of L-mimosine (or, (S)-2-amino-3-(3-hydroxy-4-oxo-4H-pyridin-1-yl)propanoate) to form 3-hydroxy-4H-pyrid-4-one and L-serine.[1]

[1] B. Tangendjaja, J. B. Lowry & R. B. H. Wills (1986) *J. Sci. Food Agric.* **37**, 523.

L-MIMOSINE SYNTHASE

This plant enzyme [EC 4.2.99.15] catalyzes the reaction of O^3-acetyl-L-serine with 3,4-dihydroxypyridine to produce 3-(3,4-dihydroxypyridin-1-yl)-L-alanine (or, L-mimosine) and acetate.[1,2] Care should be exercized with the acetylated substrate: *O*-acetylserine undergoes a rapid *O*-to-*N* shift above pH 8. L-Mimosine synthase is not identical with β-pyrazolylalanine synthase (acetylserine) [EC 4.2.99.14].

See also Uracilylalanine Synthase; β-Pyrazolylalanine Synthase (Acetylserine)

[1] I. Murakoshi, H. Kuramoto & J. Haginiwa (1972) *Phytochemistry* **11**, 177.

[2] I. Murakoshi, F. Ikegami, Y. Hinuma & Y. Hanma (1984) *Phytochemistry* **23**, 1905.

MITOCHONDRIAL INTERMEDIATE PEPTIDASE

This zinc-dependent endopeptidase [EC 3.4.24.59], often abbreviated MIP, has been isolated from human (HMIP), rat (RMIP), *Schizophyllum commune* (SMIP), and *Saccharomyces cerevisiae* (YMIP). The enzyme catalyzes the release of an N-terminal octapeptide at the second stage of processing of some proteins imported into the mitochondrion.[1]

[1]G. Isaya & F. Kalousek (1995) *MIE* **248**, 556.

Selected entries from *Methods in Enzymology* [vol, page(s)]:
General discussion: **248**, 556
Assay: **248**, 560 (principles, **248**, 560; substrate, **248**, 560)
Other molecular properties: activity, **248**, 556; amino acid sequence, **248**, 208, 556, 566; evolution, **248**, 552; homologs, **248**, 551, 566; homology with oligopeptidase M, **248**, 531; inhibition, **248**, 560 (by synthetic octapeptides, **248**, 562); isolation, **248**, 556; molecular characterization, **248**, 556, 564; precursor, mitochondrial import and processing, **248**, 565; properties, **248**, 207, 215, 556; rat, homologs, **248**, 591; rat liver (assay, **248**, 560; cDNA, isolation, **248**, 564; inhibitors, **248**, 560; metal requirements, **248**, 560; pH optimum, **248**, 560; properties, **248**, 560; purification, **248**, 557; sequence analysis, **248**, 566); structurally related enzymes, **248**, 576; structure, **248**, 556, 560; substrate specificity, **248**, 560

MITOCHONDRIAL PROCESSING PEPTIDASE

This zinc-dependent metalloendopeptidase [EC 3.4.24.64], often abbreviated MPP and called processing enhancing peptidase, catalyzes the release of N-terminal transit peptides from precursor proteins imported into the mitochondrion. The dimeric enzyme exhibits a preference for an arginyl residue at the P2 position and, less frequently, at the P3 position. The transit signal is often characterized by a high content of basic aminoacyl residues that often form α-helical structure.[1,2]

[1]H. P. Braun & U. K. Schmitz (1995) *Trends Biochem. Sci.* **20**, 171.
[2]M. Brunner & W. Neupert (1995) *MIE* **248**, 717.

Selected entries from *Methods in Enzymology* [vol, page(s)]:
General discussion: **248**, 214, 717
Assay: **248**, 723 (conditions, variation, **248**, 724; processing of [^{35}S]methionine-labeled precursor proteins, by, **248**, 723; processing of pb2(167)Δ19-DHFR, by, **248**, 724)
Other molecular properties: activity, **248**, 556, 717; α subunit, **248**, 717 (properties, **248**, 213; sequence identity to insulin-degrading enzymes, **248**, 696; structure, **248**, 715); β subunit, **248**, 717 (properties, **248**, 213; sequence identity to insulin-degrading enzymes, **248**, 696); homologs, **248**, 215; *Neurospora crassa*, **248**, 717 (activity, **248**, 723; a subunit [expression in *Escherichia coli*, **248**, 718; purification, **248**, 718; recombinant expression system, **248**, 718]; assay, **248**, 723 [conditions, variation, **248**, 724; processing of [^{35}S]methionine-labeled precursor proteins, by, **248**, 723; processing of pb2(167)Δ19-DHFR, by, **248**, 724]; β subunit, purification, **248**, 720; chimeric preprotein pb2(167)Δ19-DHFR [expression in *Escherichia coli*, **248**, 722; purification, **248**, 722]; competitive inhibition of processing by presequence-derived peptides, **248**, 725; kinetics, **248**, 725, 727; preprotein substrates and, interaction, **248**, 725; purification, **248**, 718);

potato, **248**, 718; properties, **248**, 684; rat, **248**, 718; structure, **248**, 717; yeast, **248**, 718

MITOCHONDRIAL PROTEIN-TRANSPORTING ATPase

This set of so-called ATPases [EC 3.6.3.51] catalyzes the ATP-dependent (ADP- and orthophosphate-producing) transport of preproteins into the mitochondria using the Tim protein complex.[1,2] Tim is the protein transport machinery of the inner mitochondrial membrane that contains three essential Tim proteins: Tim17 and Tim23 are thought to build a preprotein translocation channel while Tim44, a peripheral protein, interacts transiently with the matrix heat-shock protein Hsp70 to form an ATP-driven import motor. Examples of preproteins transported include the precursors of cytochrome b_2 and $F_1\beta$, the F_1-ATPase β subunit.

The idea that a transporter is an enzyme is in keeping with a new definition of enzyme catalysis as the facilitated making/breaking of chemical bonds, not just covalent bonds.[3] This idea builds on Pauling's assertion that any long-lived, chemically distinct interaction (in this case, the persistent location of a solute with respect to the faces of a membrane) can be regarded as a chemical bond. Note also that the equilibrium constant ($K_{eq} = [\text{protein}_{out}][\text{ADP}][P_i]/[\text{protein}_{in}][\text{ATP}]$) does not conform to that expected for an ATPase (*i.e.*, $K_{eq} = [\text{ADP}][P_i]/[\text{ATP}]$). Thus, although the overall reaction yields ADP and orthophosphate, the enzyme is misclassified as a hydrolase, and instead should be regarded as an energase-type reaction. Energases facilitate affinity-modulated reactions by coupling the $\Delta G_{\text{ATP-hydrolysis}}$ to a force-generating or work-producing step.[3] In this case, P-O-P bond scission supplies the energy to drive protein transport into mitochondria. **See** *ATPase; Energase*

[1]U. Bömer, M. Meijer, A. C. Maarse, A. Hönlinger, P. J. Dekker, N. Pfanner & J. Rassow (1997) *EMBO J.* **16**, 2205.
[2]W. Voos, H. Martin, T. Krimmer & N. Pfanner (1999) *BBA* **1422**, 235.
[3]D. L. Purich (2001) *TiBS* **26**, 417.

MITOGEN-ACTIVATED PROTEIN KINASE

As integral elements of the signal transduction pathway connecting the binding of a protein growth factor on the extracellular surface to specific gene expression, these protein kinases, often known as MAP kinases, migrate from cytosol to nucleus and phosphorylate various transcription factors, including Jun/AP-1, Fos, and Myc.

The kinases are turned on by a number of proliferation- and/or differentiation-inducing signals. **See also** *S6 Kinase*

Selected entries from ***Methods in Enzymology*** [vol, page(s)]:
General discussion: 200, 344; **255**, 273, 279, 290
Assay: anion-exchange chromatography, **238**, 260; electrophoretic mobility shift assay, **255**, 227, 283; gel renaturable activity assay, **255**, 275, 283; immune complex kinase assay (antisera, **255**, 247; cell extract preparation, **255**, 249; p42^mapk mutant, expression and purification, **255**, 248; sensitivity, **255**, 256); immunoblotting, **255**, 283, 432; immunoprecipitates, **255**, 276, 282; myelin basic protein as substrate, **255**, 229, 272, 275, 280; tissue extracts, **255**, 281
Other molecular properties: activation, **238**, 258; functions, **238**, 258; glutathione *S*-transferase fusion protein, purification, **255**, 288; phosphorylation and activation, **255**, 227, 245, 277, 290, 432; recombinant, preparation (cell growth, **238**, 266; expression plasmid, **238**, 265; histidine tagging, **238**, 267; induction, **238**, 266; MEK-1 assay substrate, **238**, 262; purification, **238**, 237); solubility, **238**, 265; Western blot analysis, **255**, 254, 256, 277

MITOGEN-ACTIVATED PROTEIN KINASE KINASE

One of the downstream elements of the signal transduction pathway, mitogen-activated protein kinase kinase (MAPKK or MEK) is phosphorylated by a Ser/Thr protein kinase, known as Raf (or, MAP kinase kinase kinase), and phosphorylated MAPKK then catalyzes phosphoryl transfer to MAP kinase.

Selected entries from ***Methods in Enzymology*** [vol, page(s)]:
General discussion: **255**, 279
Assay: **255**, 272, 279, 284
Other molecular properties: autophosphorylation, **255**, 309; immune complex kinase assay (antisera, **255**, 247; cell extract preparation, **255**, 249; filter assay, **255**, 252; gel analysis, **255**, 251; p42^mapk mutant, expression and purification, **255**, 248; sensitivity, **255**, 256); MAPK-glutathione *S*-transferase fusion protein, purification, **255**, 289; MEK-1 (activity assay, **238**, 262; coupled activity assay, **238**, 263; mitogen-activated protein kinase specificity, **238**, 259; phosphorylation, **238**, 260; recombinant, preparation, **238**, 237, 265; solubility, **238**, 265); phosphorylation and activation, **255**, 290; purification of recombinant protein, **255**, 259; structure, **255**, 245; Western blot analysis, **255**, 254, 256

MITOGEN-ACTIVATED PROTEIN KINASE KINASE KINASE

This phosphotransferase, abbreviated as MAPKKK and often known as Raf, catalyzes the ATP-dependent phosphorylation of mitogen-activated protein kinase kinase (MAPKK). There are at least three different Raf proteins. Raf-1 has the MAPKK kinase activity and mediates signals associated with cell growth, transformation, and differentiation.

Selected entries from ***Methods in Enzymology*** [vol, page(s)]:
Assay: **238**, 270 (autophosphorylation, **255**, 296; gel autoradiography, **255**, 287, 295; ion-exchange chromatography, 255, 297, 300)
Other molecular properties: activation by growth factors, **255**, 291, 296; domains, **255**, 291; immunoprecipitation, **255**, 294; isoforms, **255**, 291; MEK-1 activation, **238**, 260; purification, **238**, 270; purification of amino-terminal GST fusion protein (affinity chromatography, **255**, 293; cell growth, **255**, 292; plasmid construction, **255**, 291)

MIXED-FUNCTION OXIDASE

Mixed-function oxidases is another term for the class of oxidoreductases known as monooxygenases and hydroxylases.

*Mlu*I RESTRICTION ENDONUCLEASE

This type II restriction endonuclease [EC 3.1.21.4], isolated from *Micrococcus luteus*, catalyzes the hydrolysis of both strands of DNA at $5' \ldots A\nabla CGCGT \ldots 3'$.[1] The enzyme also exhibits star activity. **See** *"Star" Activity*

[1] H. Sugisaki & S. Kanazawa (1981) *Gene* **16**, 73.

*Mly*I RESTRICTION ENDONUCLEASE

This type II restriction endonuclease (subtype s) [EC 3.1.21.4], isolated from *Micrococcus lylae*, catalyzes the hydrolysis of both strands of DNA at $5' \ldots GAGTCNNNNN\nabla \ldots 3'$ and $3' \ldots CTCAGNNNN N\nabla \ldots 5'$, where N refers to any base. Note that blunt-ended fragments are generated.[1]

[1] P. Brinkley, D. S. Bautista & F. L. Graham (1991) *Gene* **100**, 267.

*Mnl*I RESTRICTION ENDONUCLEASE

This type II restriction enzyme (subtype s) [EC 3.1.21.4], isolated from *Moraxella nonliquefaciens*, catalyzes the hydrolysis of both strands of DNA. The recognition sequence and site of cleavage is at $5' \ldots CCTCNNNNNNN\nabla \ldots 3'$ and $3' \ldots GGAGNNN NNN\nabla N \ldots 5'$, where N refers to any base.[1]

[1] P. Brinkley, D. S. Bautista & F. L. Graham (1991) *Gene* **100**, 267.
Selected entries from ***Methods in Enzymology*** [vol, page(s)]:
Other molecular properties: purification, **65**, 94; recognition sequence, **65**, 8; use in hemoglobin S gene analysis, **76**, 817

*Mno*I RESTRICTION ENDONUCLEASE

This type II restriction enzyme [EC 3.1.21.4], isolated from *Moraxella nonliquefaciens*, catalyzes the hydrolysis of both strands of DNA at $5' \ldots C\nabla CGG \ldots 3'$.[1]

[1] B. R. Baumstark, R. J. Roberts & U. L. RajBhandary (1979) *JBC* **254**, 8943.

Mn^{2+}-TRANSPORTING ATPase

This so-called ATPase [EC 3.6.3.35] catalyzes the ATP-dependent (ADP- and orthophosphate-producing) transport of Mn^{2+}_{out} to produce Mn^{2+}_{in}.[1,2] The bacterial system will also catalyze the import of Zn^{2+} and iron chelates. This ABC-type (ATP-binding cassette-type) ATPase has two similar ATP-binding domains and does not undergo phosphorylation during the transport process.

The idea that a transporter is an enzyme is in keeping with a new definition of enzyme catalysis as the facilitated making/breaking of chemical bonds, not just covalent bonds.[1] This idea builds on Pauling's assertion that any long-lived, chemically distinct interaction (in this case, the persistent location of a solute with respect to the faces of a membrane) can be regarded as a chemical bond. Note also that the equilibrium constant ($K_{eq} = [Mn^{2+}_{in}][ADP][P_i]/[Mn^{2+}_{out}][ATP]$) does not conform to that expected for an ATPase (*i.e.*, $K_{eq} = [ADP][P_i]/[ATP]$). Thus, although the overall reaction yields ADP and orthophosphate, the enzyme is misclassified as a hydrolase, and instead should be regarded as an energase-type reaction. Energases facilitate affinity-modulated reactions by coupling the $\Delta G_{ATP\text{-}hydrolysis}$ to a force-generating or work-producing step.[3] In this case, P-O-P bond scission supplies the energy to drive Mn^{2+} transport. **See** *ATPase; Energase*

[1]R. Novak, J. S. Braun, E. Charpentier & E. Tuomanen (1998) *Mol. Microbiol.* **29**, 1285.

[2]P. E. Kolenbrander, R. N. Andersen, R. A. Baker & H. F. Jenkinson (1998) *J. Bacteriol.* **180**, 290.

[1]D. L. Purich (2001) *TiBS* **26**, 417.

MOLYBDATE-TRANSPORTING ATPase

This so-called ATPase [EC 3.6.3.29] catalyzes the ATP-dependent (ADP- and orthophosphate-producing) transport of molybdate$_{out}$ to produce molybdate$_{in}$.[1,2] The molybdate (MoO_4^{2-}) transport systems in *Escherichia coli*, *Haemophilus influenzae*, *Azotobacter vinelandii*, and *Rhodobacter capsulatus* are very similar, with the lowest K_m value reported to be about 50 nM.[1] These ABC-type (ATP-binding cassette-type) ATPases have two similar ATP-binding domains and do not undergo phosphorylation during the transport process.

The idea that a transporter is an enzyme is in keeping with a new definition of enzyme catalysis as the facilitated making/breaking of chemical bonds, not just covalent bonds.[3] This idea builds on Pauling's assertion that any long-lived, chemically distinct interaction (in this case, the persistent location of a solute with respect to the faces of a membrane) can be regarded as a chemical bond. Note also that the equilibrium constant ($K_{eq} = [molybdate_{in}][ADP][P_i]/[molybdate_{out}][ATP]$) does not conform to that expected for an ATPase (*i.e.*, $K_{eq} = [ADP][P_i]/[ATP]$). Thus, although the overall reaction yields ADP and orthophosphate, the enzyme is misclassified as a hydrolase, and instead should be regarded as an energase-type reaction. Energases facilitate affinity-modulated reactions by coupling the $\Delta G_{ATP\text{-}hydrolysis}$ to a force-generating or work-step.[3] In this case, P-O-P bond scission supplies the energy to drive molybdate transport. **See** *ATPase; Energase*

[1]A. M. Grunden & K. T. Shanmugam (1997) *Arch. Mikrobiol.* **168**, 345.

[2]J. K. Griffiths & C. E. Sansom (1998) *The Transporter Factsbook*, Academic Press, San Diego.

[3]D. L. Purich (2001) *TiBS* **26**, 417.

MOLONEY MURINE LEUKEMIA VIRUS RETROPEPSIN

This aspartic endopeptidase, also called Moloney MLV retropepsin and Moloney MLV protease, participates in polyprotein processing events. The enzyme catalyzes the hydrolysis of the Tyr–Pro bond in the synthetic peptide Val-Ser-Gln-Asn-Tyr-Pro-Ile-Val-Gln. This retropepsin has a greater preference for large hydrophobic aminoacyl residues in positions P4 and P2 than does HIV retropepsin.[1,2] **See also** *Retropepsins*

[1]S. Oroszlan & R. B. Luftig (1990) *Curr. Top. Microbiol. Immunol.* **157**, 153.

[2]L. Menéndez-Arias, I. T. Weber, J. Soss, R. W. Harrison, D. Gotte & S. Oroszlan (1994) *JBC* **269**, 16795.

Selected entries from *Methods in Enzymology* [vol, page(s)]:
Other molecular properties: cleavage site sequence, **241**, 297; eukaryotic aspartic proteases, amino acid sequence comparison, and, **241**, 216; Gag-pol polyprotein synthesis, **241**, 227; identification, **241**, 91; purification from virus particles, **241**, 94; transcription-translation system *in vitro*, **241**, 228

MONODEHYDROASCORBATE REDUCTASE (NADH)

This oxidoreductase [EC 1.6.5.4] catalyzes the reaction of NADH with two molecules of monodehydroascorbate to produce NAD^+ and two molecules of ascorbate.[1-6] A monodehydroascorbate radical reacts with the fully reduced form of reductase with a diffusion-controlled rate.[1] This activity is also associated with thioltransferases (glutaredoxins) and protein disulfide isomerases in animal cells. These enzymes catalyze the glutathione-dependent, two-electron regeneration of ascorbate.[7]

[1]K. Kobayashi, S. Tagawa, S. Sano & K. Asada (1995) *JBC* **270**, 27551.

[2]S. Shigeoka, R. Yasumoto, T. Onishi, Y. Nakano & S. Kitaoka (1987) *J. Gen. Microbiol.* **133**, 227.

[3]G. Borraccino, S. Dipierro & O. Arrigoni (1986) *Planta* **167**, 521.

[4]M. A. Hossain & K. Asada (1985) *JBC* **260**, 12920.

[5]G. Oehler, W. Weis & H. Staudinger (1972) *Hoppe-Seyler's Z. Physiol. Chem.* **353**, 495.

[6]H.-U. Schulze, H.-H. Schott & H. Staudinger (1972) *Hoppe-Seyler's Z. Physiol. Chem.* **353**, 1931.

[7]W. W. Wells & D. P. Xu (1994) *J. Bioenerg. Biomembr.* **26**, 369.

Selected entries from *Methods in Enzymology* [vol, page(s)]:
General discussion: **279**, 30

MONOMETHYL-SULFATASE

This sulfatase [EC 3.1.6.16] catalyzes the hydrolysis of monomethyl sulfate to produce methanol and sulfate. Monoethyl sulfate, monoisopropyl sulfate, and monododecyl sulfate are not substrates.[1]

[1]O. Ghisalba & M. Kuenzi (1983) *Experientia* **39**, 1257.

MONOPHENOL MONOOXYGENASE

This copper-dependent oxidoreductase [EC 1.14.18.1] (also known as tyrosinase, phenolase, monophenol oxidase, and cresolase) catalyzes the reaction of L-tyrosine with L-DOPA and dioxygen to produce L-DOPA (carbon skeleton from L-tyrosine), dopaquinone (carbon skeleton from L-DOPA$_{substrate}$), and water,[1–5] likely via a radical mechanism. Dioxygen binds at the binuclear copper site in a side-on (μ-η^2:η^2) position. The phenolic substrate then coordinates directly with one of the copper ions and oxygen transfer occurs at the *ortho* position.

This classification actually includes those copper proteins that also exhibit catechol oxidase [EC 1.10.3.1] activity when only 1,2-benzenediols are available as substrates. Thus, the enzyme will also catalyze pyrocatechol (1,2-dihydroxybenzene) formation from phenol, 4-methyl-*o*-benzoquinone from 4-methylcatechol, as well as a large number of other derivatives from monophenol analogues.

See also Catechol Oxidase; Tyrosinase

[1]A. Messerschmidt (1998) *CBC* **3**, 401.

[2]J. V. Schloss & M. S. Hixon (1998) *CBC* **2**, 43.

[3]C. W. van Gelder, W. H. Flurkey & H. J. Wichers (1997) *Phytochemistry* **45**, 1309.

[4]A. Sanchez-Ferrer, J. N. Rodriguez-Lopez, F. Garcia-Canovas & F. Garcia-Carmona (1995) *BBA* **1247**, 1.

[5]A. M. Mayer (1987) *Phytochemistry* **26**, 11.

Selected entries from *Methods in Enzymology* [vol, page(s)]:
General discussion: 2, 817, 827; **17A**, 615, 620, 626, 628; **34**, 6; **44**, 886; **52**, 9; **142**, 154, 165
Assay: 2, 817, 827; **17A**, 615; **31**, 534; **58**, 568; **142**, 160, 165; **233**, 499
Other molecular properties: action on enzymes, **4**, 265; generation of VP-16 phenoxyl radicals, **234**, 632; histochemical method for, **2**, 828; **4**, 383; inhibitors, **13**, 558, 562; **31**, 538; interactions with VP-16 phenoxyl radicals, **234**, 640; invertase, inactivation of, **1**, 257; isozymes, **17A**, 620, 623, 631; luminescence, **226**, 535; mammalian, **2**, 827; melanocytes, **58**, 564; *Neurospora*, S^β-(2-histidyl)cysteine properties and assay, **106**, 355; oxidation of plant phenols, **13**, 556; plants, **2**, 817, 830; plant tissues, from, **13**, 562; **31**, 528; polyphenol oxidase activity, **2**, 817 (assay, **31**, 534; inhibitor, **31**, 538; peroxisomes, **52**, 503); preparation, **233**, 496; properties, **142**, 164, 168; purification from (hamster melanoma, **17A**, 622; murine melanoma, **142**, 156; mushroom, **9**, 14; **17A**, 628; *Neurospora crassa*, **17A**, 616; **142**, 166); purification on modified cellulose, **22**, 347; staining on gels, **22**, 595; studies, **31**, 544; superoxide role, **233**, 501; synthesis of, in melanin formation, **31**, 389, 394; toxicity, **58**, 567

MONOPRENYL ISOFLAVONE EPOXIDASE

This FAD-dependent oxidoreductase [EC 1.14.99.34], also known as monoprenyl isoflavone monooxygenase, catalyzes the reaction of 7-*O*-methylluteone with NADPH, H$^+$, and dioxygen to produce dihydrofurano derivatives, NADP$^+$, and water.[1] 7-*O*-Methylluteone is 3-(2,4-dihydroxyphenyl)-5-hydroxy-7-methoxy-6-(3-methyl-2-butenyl)-4*H*-1-benzopyran-4-one.

Note that the oxygen atom in the epoxidation product is derived from dioxygen, not water. The enzyme product is slowly and nonenzymically converted into the corresponding dihydrofurano derivative.

[1]M. Tanaka & S. Tahara (1997) *Phytochemistry* **46**, 433.

MONOSACCHARIDE-TRANSPORTING ATPase

This so-called ATPase [EC 3.6.3.17] catalyzes the ATP-dependent (ADP- and orthophosphate-producing) transport of monosaccharide$_{out}$ to produce monosaccharide$_{in}$.[1–5] The bacterial system catalyzes the import of D-ribose, D-xylose, D-arabinose, D-galactose, and methyl-D-galactoside. This ABC-type (ATP-binding cassette-type) ATPase has two similar ATP-binding domains and does not undergo phosphorylation during the transport process.

That this transporter is an enzyme is in keeping with a new definition of enzyme catalysis as the facilitated making/breaking of chemical bonds, not just covalent bonds.[6] This idea builds on Pauling's assertion that any long-lived, chemically distinct interaction (in this case, the persistent location of a solute with respect to the faces of a membrane) can be regarded as a chemical bond. Note also that the equilibrium constant, defined by $K_{eq} =$ [monosaccharide$_{in}$][ADP][P$_i$]/[monosaccharide$_{out}$][ATP], does not conform to that expected for an ATPase (*i.e.*, $K_{eq} =$ [ADP][P$_i$]/[ATP]). Thus, although the overall reaction yields ADP and orthophosphate, the enzyme is misclassified as a hydrolase, and instead should be regarded as an energase-type reaction. Energases facilitate affinity-modulated reactions by coupling the $\Delta G_{\text{ATP-hydrolysis}}$ to a force-generating or work-producing step.[6] In this case, P-O-P bond scission supplies the energy to drive monosaccharide transport. **See ATPase; Energase**

[1] C. F. Higgins (1992) *Ann. Rev. Cell Biol.* **8**, 67.
[2] G. Kuan, E. Dassa, N. Saurin, M. Hofnung & M. H. Saier, Jr. (1995) *Res. Microbiol.* **146**, 271.
[3] J. M. Kemner, S. Liang & E. W. Nester (1997) *J. Bacteriol.* **179**, 2452.
[4] S. Song & C. Park (1998) *FEMS Microbiol. Lett.* **163**, 255.
[5] J. K. Griffiths & C. E. Sansom (1998) *The Transporter Factsbook*, Academic Press, San Diego.
[6] D. L. Purich (2001) *TiBS* **26**, 417.

MONOSIALOGANGLIOSIDE SIALYLTRANSFERASE

This sialyltransferase [EC 2.4.99.2] catalyzes the reaction of CMP-*N*-acetylneuraminate with D-galactosyl-*N*-acetyl-D-galactosaminyl-(*N*-acetylneuraminyl)-D-galactosyl-D-glucosylceramide (that is, ganglioside G$_{M1}$) to produce CMP and *N*-acetylneuraminyl-D-galactosyl-*N*-acetyl-D-galactosaminyl-(*N*-acetylneuraminyl)-D-galactosyl-D-glucosylceramide (ganglioside D$_{1a}$).[1–3] The porcine enzyme reportedly has a random Bi Bi kinetic mechanism.[1] This pig protein may be the same as β-galactosamide α-2,3-sialyltransferase [EC 2.4.99.4]. **See also** *β-Galactosamide α-2,3-Sialyltransferase; Sialyltransferases; α-2,3-Sialyltransferases*

[1] J. I. Rearick, J. E. Sadler, J. C. Paulson & R. L. Hill (1979) *JBC* **254**, 4444.
[2] T.-J. Gu, X.-B. Gu, T. Ariga & R. K. Yu (1990) *FEBS Lett.* **275**, 83.
[3] M. Trinchera, B. Pirovano, R. Ghidoni (1990) *JBC* **265**, 18242.

MONOTERPENOL *O*-ACETYLTRANSFERASE

This transferase [EC 2.3.1.69], also known as menthol transacetylase, catalyzes the reaction of acetyl-CoA with a monoterpenol to produce coenzyme A and a monoterpenol acetate ester.[1,2] Substrates include borneol, cyclohexanol, *n*-decanol, (−)-menthol, isomenthol, and (+)-neomenthol.

[1] C. Martinkus & R. Croteau (1981) *Plant Physiol.* **68**, 99.
[2] R. Croteau & C. L. Hooper (1978) *Plant Physiol.* **61**, 737.

MONOTERPENOL β-GLUCOSYLTRANSFERASE

This transferase [EC 2.4.1.127], first isolated from Black Mitcham peppermint (*Mentha piperita*), catalyzes the reaction of UDP-D-glucose with (−)-menthol to produce

UDP and (−)-menthyl *O*-β-D-glucoside.[1] Other acceptor substrates include (+)-neomenthol.

[1] C. Martinkus & R. Croteau (1981) *Plant Physiol.* **68**, 99.

MONOTERPENYL-PYROPHOSPHATASE

This enzyme [EC 3.1.7.3] (also known as monoterpenyl-diphosphatase, bornyl pyrophosphate hydrolase, and bornyl diphosphate hydrolase) catalyzes the hydrolysis of a monoterpenyl diphosphate to produce a monoterpenol and pyrophosphate (or, diphosphate).[1] This classification represents a group of enzymes with varying specificity for the monoterpenol moiety. One variant (for example, from sage [*Salvia officinalis*]) has highest activity on sterically hindered compounds such as (+)-bornyl diphosphate whereas another has significant activity on the diphosphates of primary allylic alcohols such as geraniol.

[1] R. Croteau & F. Karp (1979) *ABB* **198**, 523.

MORPHINE 6-DEHYDROGENASE

This oxidoreductase [EC 1.1.1.218], also known as naloxone reductase, catalyzes the reaction of morphine with NAD(P)$^+$ to produce morphinone and NAD(P)H (the *Pseudomonas putida* enzyme uses only NADP$^+$ as the coenzyme substrate).

morphine   morphinone

Alternative substrates include codeine, nalorphine, normorphine, and ethylmorphine. The enzyme also reduces naloxone to the 6-α-hydroxy analogue in the reverse direction. The guinea pig enzyme is activated by 2-mercaptoethanol.[1]

[1] S. Yamano, E. Kageura, T. Ishida & S. Toki (1985) *JBC* **260**, 5259.

Selected entries from *Methods in Enzymology* [vol, page(s)]:
Other molecular properties: stereochemistry, **87**, 113

MORPHOL DIOXYGENASE

This iron-dependent oxidoreductase reportedly catalyzes the reaction of 3,4-dihydroxyphenanthrene (that is, morphol) with dioxygen to produce *cis*-4-(1-hydroxynaphth-2-yl)-2-oxobut-3-enoate.[1]

3,4-dihydroxyphenanthrene

cis-4-(1-hydroxynaphth-2-yl)-2-oxobut-3-enoate

[1]W. C. Evans, H. N. Fernley & E. Griffiths (1965) *Biochem. J.* **95**, 819.

Selected entries from *Methods in Enzymology* [vol, page(s)]:
General discussion: 52, 11

mRNA (ADENOSINE-$O^{2'}$-)-METHYLTRANSFERASE

This methyltransferase, previously classified as EC 2.1.1.58, is now regarded as an activity of mRNA (nucleoside-$O^{2'}$-)-methyltransferase [EC 2.1.1.57].

mRNA (GUANINE-N^{7}-)-METHYLTRANSFERASE

This enzyme [EC 2.1.1.56], also known as cap methyltransferase, catalyzes the addition of the N^{7}-methyl group to the cap of mRNA: thus, S-adenosyl-L-methionine reacts with G(5′)pppR-RNA to produce S-adenosyl-L-homocysteine and m^{7}G(5′)pppR-RNA (where the R group is typically a guanosine or adenosine residue).[1–5]

[1]P. Narayan & F. M. Rottman (1992) *AE* **65**, 255.
[2]S. P. Wang & S. Shuman (1997) *JBC* **272**, 14683.
[3]K. Mizumoto & Y. Kaziro (1987) *Prog. Nucleic Acid Res. Mol. Biol.* **34**, 1.
[4]C. Locht, J.-L. Beaudart & J. Delcour (1983) *EJB* **134**, 117.
[5]N. Saha, B. Schwer & S. Shuman (1999) *JBC* **274**, 16553.

Selected entries from *Methods in Enzymology* [vol, page(s)]:
Assay: activity of vaccinia virus mRNA capping enzyme, **181**, 175

mRNA GUANYLYLTRANSFERASE

This nucleotidyltransferase [EC 2.7.7.50], also known as mRNA capping enzyme and GTP:RNA guanylyltransferase, catalyzes the reaction of GTP with (5′)pp-Pur-mRNA (that is, a mRNA containing a purine residue at the 5′-end) to produce G(5′)ppp-Pur-mRNA (that is, a mRNA containing a guanosine residue linked 5′ via three phosphoryl groups to the 5′-purine) and pyrophosphate (or, diphosphate).[1–10] Synthetic poly(A) and poly(G) are substrates that give rise to the products G(5′)pppA_{n} and G(5′)pppG_{n}, respectively.

The reaction mechanism proceeds through a guanyl-enzyme intermediate (to an ε-amino group of a lysyl residue in the rat liver enzyme).

[1]M. Bisaillon & G. Lemay (1997) *Virology* **236**, 1.
[2]S. Shuman (1995) *Prog. Nucleic Acid Res. Mol. Biol.* **50**, 101.
[3]Y. Shibagaki, E. Gilboa, N. Itho, H. Yamada, S. Nagata & K. Mizumoto (1992) *JBC* **267**, 9521.
[4]D. D. Dunigan & M. Zaitlin (1990) *JBC* **265**, 7779.
[5]N. Itho, H. Yamada, Y. Kaziro, & K. Mizumoto (1987) *JBC* **262**, 1989.
[6]N. Itoh, K. Mizumoto & Y. Kaziro (1984) *JBC* **259**, 13923 and 13930.
[7]S. Shuman & J. Hurwitz (1982) *The Enzymes*, 3rd ed., **15**, 245.
[8]S. Venkatesan & B. Moss (1980) *JBC* **255**, 2835.
[9]Y. Groner & H. Aviv (1978) *Biochemistry* **17**, 977.
[10]G. Monroy, E. Spencer & J. Hurwitz (1978) *JBC* **253**, 4481 and 4490.

mRNA (2′-O-METHYLADENOSINE-N^{6}-)-METHYLTRANSFERASE

This methyltransferase [EC 2.1.1.62] catalyzes the reaction of S-adenosyl-L-methionine with m^{7}G(5′)pppAm (that is, an mRNA containing an N^{7}-guanosine cap linked 5′ via three phosphoryl groups to a $O^{2'}$-methyladenosine residue) to produce S-adenosyl-L-homocysteine and m^{7}G(5′)pppm^{6}Am (that is, an mRNA containing an N^{7}-guanosine cap linked 5′ via three phosphoryl groups to an $N^{6},O^{2'}$-dimethyladenosine residue).[1,2]

[1]F. M. Rottman, J. A. Bokar, P. Narayan, M. E. Shambaugh & R. Ludwiczak (1994) *Biochimie* **76**, 1109.
[2]J. M. Keith, M. J. Ensinger & B. Moss (1978) *JBC* **253**, 5033.

mRNA (NUCLEOSIDE-2′-O-)-METHYLTRANSFERASE

This methyltransferase [EC 2.1.1.57], also known as mRNA (adenosine-2′-O-)-methyltransferase, catalyzes the reaction of S-adenosyl-L-methionine with m^{7}G(5′)pppR-RNA (that is, an mRNA containing an N^{7}-methylguanosine cap linked 5′ via three phosphoryl groups to R, either a guanosine or an adenosine residue) to produce S-adenosyl-L-homocysteine and m^{7}G(5′)pppR^{m}-RNA.[1–3]

Homopolyribonucleotides containing the 7-methylguanosine cap are also methylated with poly(A) and poly(I) being the best acceptor substrates. The vaccinia virus enzyme has a random Bi Bi kinetic mechanism.[1] Two distinct 2′-O-methyltransferases have been identified in HeLa cells (additional methylation producing m^{7}G(5′)pppR^{m}pR^{m}-RNA).[2]

[1]E. Barbosa & B. Moss (1978) *JBC* **253**, 7692 and 7698.
[2]S. R. Langberg & B. Moss (1981) *JBC* **256**, 10054.
[3]K. Mizumoto & Y. Kaziro (1987) *Prog. Nucleic Acid Res. Mol. Biol.* **34**, 1.

M.Sau3AI METHYLTRANSFERASE

This site-specific type II DNA-methyltransferase (cytosine-specific) [EC 2.1.1.73], obtained from *Staphylococcus aureus* 3A, recognizes and acts on DNA at

5′ ... GATC ... 3′ at position 4 using *S*-adenosyl-L-methionine to produce DNA with a 5-methylcytosine residue and *S*-adenosyl-L-homocysteine.[1] **See also** Site-Specific DNA-Methyltransferase (Cytosine-Specific)

[1] A. Y. Lebenka & Y. A. Rackus (1989) *Biokhimiia* **54**, 1009.

Selected entries from **Methods in Enzymology** [**vol**, page(s)]:
Other molecular properties: nucleosome mapping, **274**, 224, 226

M.*Sau*3AI METHYLTRANSFERASE

This site-specific type II DNA-methyltransferase (cytosine-specific) [EC 2.1.1.73], obtained from *Staphylococcus aureus* 3A, recognizes and acts on DNA at 5′ ... GATC ... 3′ at position 4 using *S*-adenosyl-L-methionine to produce DNA with a 5-methylcytosine residue and *S*-adenosyl-L-homocysteine.[1] **See also** Site-Specific DNA-Methyltransferase (Cytosine-Specific)

[1] A. Y. Lebenka & Y. A. Rackus (1989) *Biokhimiia* **54**, 1009.

Selected entries from **Methods in Enzymology** [**vol**, page(s)]:
Other molecular properties: nucleosome mapping, **274**, 224, 226

*Msc*I RESTRICTION ENDONUCLEASE

This type II restriction enzyme [EC 3.1.21.4], isolated from *Micrococcus* species (NEB 502), catalyzes the hydrolysis of both strands of DNA at 5′ ... TGG∇CCA ... 3′.[1] The enzyme is blocked by overlapping *dcm* methylation.

[1] C. Polisson (1989) *NAR* **17**, 5858.

*Mse*I RESTRICTION ENDONUCLEASE

This type II restriction enzyme [EC 3.1.21.4], isolated from *Micrococcus* species (NEB 446), catalyzes the hydrolysis of both strands of DNA at 5′ ... T∇TAA ... 3′. Note that the recognition site is free of G and C bases.[1]

[1] R. D. Morgan (1988) *NAR* **16**, 3104.

*Msl*I RESTRICTION ENDONUCLEASE

This type II restriction enzyme [EC 3.1.21.4], isolated from *Moraxella osloensis* (NEB 722), catalyzes the hydrolysis of both strands of DNA at 5′ ... CAYNN∇NNRTG ... 3′, where Y refers to C or T, N refers to any base, and R refers to A or G. Note that blunt-ended fragments are generated.

*Msp*AII RESTRICTION ENDONUCLEASE

This type II restriction enzyme [EC 3.1.21.4], isolated from *Moraxella* species A1, catalyzes the hydrolysis of both strands of DNA at 5′ ... CMG∇CKG ... 3′, where M refers to A or C and K refers to G or T. Note that blunt-ended fragments are generated.

*Msp*I RESTRICTION ENDONUCLEASE

This type II restriction enzyme [EC 3.1.21.4], isolated from *Moraxella* species, catalyzes the hydrolysis of both strands of DNA at 5′ ... C∇CGG ... 3′. The endonuclease can also catalyze cleavage of both methylated and unmethylated sites (that is, when the internal cytosine is methylated; if the external cytosine is modified, cleavage will not occur)[1,2] as well as act on single strands.[3]

[1] C. Waalwijk & R. A. Flavell (1978) *NAR* **5**, 3231.
[2] M. Busslinger, E. deBoer, S. Wright, F. G. Grosveld & R. A. Flavell (1983) *NAR* **11**, 3559.
[3] O. J. Yoo & K. L. Agarwal (1980) *JBC* **255**, 10559.

M.*Sss*I METHYLTRANSFERASE

This site-specific type II DNA-methyltransferase (cytosine-specific) [EC 2.1.1.73] acts on DNA at 5′ ... GC ... 3′ at position 2 using *S*-adenosyl-L-methionine to produce DNA with a 5-methylcytosine residue and *S*-adenosyl-L-homocysteine.[1,2] **See also** Site-Specific DNA-Methyltransferase (Cytosine-Specific)

[1] P. Renbaum & A. Razin (1992) *FEBS Lett.* **313**, 243.
[2] P. Renbaum, D. Abrahamove, A. Fainsod, G. G. Wilson, S. Rottem & A. Razin (1990) *NAR* **18**, 1145.

Selected entries from **Methods in Enzymology** [**vol**, page(s)]:
General discussion: **304**, 431

*Mst*I RESTRICTION ENDONUCLEASE

This type II restriction enzyme [EC 3.1.21.4], isolated from *Microcoleus* species, catalyzes the hydrolysis of both strands of DNA at 5′ ... TGC∇GCA ... 3′, producing blunt-ended fragments.[1]

[1] T. R. Gingeras, J. P. Milazzo & R. J. Roberts (1978) *NAR* **5**, 4105.

M.*Taq*I METHYLTRANSFERASE

This site-specific DNA-adenine N^6-methyltransferase (subtype γ) [EC 2.1.1.72], obtained from *Thermus aquaticus* YTI, catalyzes the reaction of *S*-adenosyl-L-methionine with the adenyl residue in the double-stranded DNA sequence 5′ ... TCGA ... 3′ at the N^6-position, producing the methylated DNA and *S*-adenosyl-L-homocysteine.[1–7]

The enzyme has a random Bi Bi kinetic mechanism.[1] Fluorescence investigations demonstrate that M.*Taq*I catalyzes the "flipping-out" of the base prior to methylation.[5,6] A phenylalanyl residue has been shown to play an important role in stabilizing the extrahelical adenine and a tyrosyl residue has been suggested to position the target base in an optimal position.[7] **See also** *Site-Specific DNA-Methyltransferase (Adenine-Specific)*

[1] N. O. Reich & N. Mashhoon (1991) *Biochemistry* **30**, 2933.
[2] M. McClelland (1981) *NAR* **9**, 6795.
[3] G. Schluckebier, J. Labahn, J. Granzin & W. Saenger (1995) *J. Biomol. Struct. Dyn.* **12**, A206.
[4] X. Cheng (1995) *Ann. Rev. Biophys. Biomol. Struct.* **24**, 293.
[5] B. Holz, H. Pues, J. Wolcke & E. Weinhold (1997) *FASEB J.* **11**, A1151.
[6] R. J. Roberts & X. Cheng (1998) *Ann. Rev. Biochem.* **67**, 181.
[7] H. Pues, N. Bleimling, B. Holz, J. Wolcke & E. Weinhold (1999) *Biochemistry* **38**, 1426.

Selected entries from **Methods in Enzymology** [**vol**, page(s)]:
General discussion: 216, 244

MUCINAMINYLSERINE MUCINAMINIDASE

This enzyme [EC 3.2.1.110] catalyzes the hydrolysis of D-galactosyl-3-(N-acetyl-β-D-galactosaminyl)-L-serine to produce D-galactosyl-3-N-acetyl-β-D-galactosamine and L-serine.[1–3] The enzyme will also catalyze the hydrolysis of D-galactosyl-3-(N-acetyl-β-D-galactosaminyl)-L-threonine.

[1] Y. Endo & A. Kobata (1976) *J. Biochem.* **80**, 1.
[2] J. Umemoto, K. L. Matta, J. J. Barlow & V. P. Bhavanandan (1978) *Anal. Biochem.* **91**, 186.
[3] J. Umemoto, V. P. Bhavanandan & E. A. Davidson (1977) *JBC* **252**, 8609.

MUCONATE CYCLOISOMERASE

This manganese-dependent cycloisomerase [EC 5.5.1.1], also known as muconate lactonizing enzyme and *cis,cis*-muconate lactonizing enzyme I, catalyzes the interconversion of 2,5-dihydro-5-oxofuran-2-acetate (or, muconolactone) and *cis,cis*-hexadienedioate (or, *cis,cis*-muconate).[1–5] In the reverse reaction, 3-methyl-*cis,cis*-hexadienedioate can serve as a substrate as can *cis,trans*-hexadienedioate, albeit weakly.

muconolactone   *cis,cis*-muconate

The enzyme differs from chloromuconate cycloisomerase [EC 5.5.1.7] and dichloromuconate cycloisomerase [EC 5.5.1.11]. The laconizing enzyme utilizes acid/base catalysis in which there is a catalyzed proton removal from the lactone and protonation of *cis,cis*-muconate, respectively.

This is consistent with the stereospecific incorporation of solvent deuterium into the *pro-5R* position of muconolactone. In the reverse direction, the departure of the carboxylic oxygen atom and proton from the C4 and C5 positions follows a *syn* route.

[1] J. B. Powlowsi, J. Ingebrand & S. Dagley (1985) *J. Bacteriol.* **164**, 1136.
[2] K. L. Ngai, L. N. Ornston & R. G. Kallen (1983) *Biochemistry* **22**, 5223.
[3] K. L. Ngai & R. G. Kallen (1983) *Biochemistry* **22**, 5231.
[4] A. Gaal & H. Y. Neujahr (1980) *Biochem. J.* **191**, 37.
[5] W. R. Sistrom & R. Y. Stanier (1954) *JBC* **210**, 821.

Selected entries from **Methods in Enzymology** [**vol**, page(s)]:
General discussion: 2, 273, 282; **17A**, 529, 533; **188**, 126
Assay: 2, 282; **188**, 127
Other molecular properties: *cis,cis*-muconate and, **2**, 273, 282; properties, **2**, 284; **188**, 129; purification from *Pseudomonas putida*, **2**, 283; **17A**, 531, 534; **188**, 128

MUCONOLACTONE Δ-ISOMERASE

This isomerase [EC 5.3.3.4] catalyzes the interconversion of 2,5-dihydro-5-oxofuran-2-acetate (or, muconolactone) and 3,4-dihydro-5-oxofuran-2-acetate.[1] The *Pseudomonas putida* enzyme is comprised of ten 96-residue subunits.[2]

[1] R. B. Meagher, K.-L. Ngai & L. N. Ornston (1990) *Meth. Enzymol.* **188**, 130.
[2] S. K. Katti, B. A. Katz & H. W. Wyckoff (1989) *J. Mol. Biol.* **205**, 557.

Selected entries from **Methods in Enzymology** [**vol**, page(s)]:
General discussion: 17A, 529, 530, 536; **188**, 130
Assay: 188, 131
Other molecular properties: properties, **188**, 132; purification from *Pseudomonas putida*, **17A**, 531, 537; **188**, 132

MUCORPEPSIN

This aspartic endopeptidase [EC 3.4.23.23], a major milk-clotting enzyme used in industry, catalyzes peptide-bond hydrolysis in proteins (for example, casein and denatured hemoglobin). The enzyme also acts on the synthetic substrate Leu-Ser-Phe(NO₂)-∇-Nle-Ala-Leu-Ome, where ∇ indicates the cleavage site. Mucorpepsin prefers substrates with hydrophobic aminoacyl residues at positions P1 and P1'. It cleaves the Phe105–Met106 bond in κ-casein.[1,2]

See also *Chymosin*

[1] K. Arima, J. Yu & S. Iwasaki (1970) *Meth. Enzymol.* **19**, 446.
[2] T. Beppu, Y.-N. Park, J. Aikawa, M. Nishiyama & S. Horinouchi (1995) in *Aspartic Proteinase: Structure, Function, Biology and Medical Implications* (T. Takahashi, ed.), Plenum, New York, p. 559.

Selected entries from **Methods in Enzymology** [**vol**, page(s)]:
General discussion: 19, 189, 373, 446, 459
Assay: 19, 448
Other molecular properties: active site, **19**, 457; amino acid composition, **19**, 456; cheese-making experiments with, **19**, 449; clotting activity, **19**, 454; physical properties, **19**, 454, 457, 459; potential use in cheese-making, **19**, 375; purification and crystallization, **19**, 450, 459; specificity, **19**, 458; stability, **19**, 451

MUCROLYSIN

This zinc-dependent metalloendopeptidase [EC 3.4.24.54], also called *Trimeresurus* metalloendopeptidase A and mucrotoxin A, is isolated from the venom of the Chinese habu snake (*Trimeresurus mucrosquamatus*). It has hemorrhagic and fibrinolytic activity and will hydrolyze the oxidized insulin B chain at Ser9–His10, His10–Leu11, Ala14–Leu15, Leu15–Tyr16, and Tyr16–Leu17. It will also act on the Aα and Bβ chain of cattle fibrinogen.

Mucrolysin also exhibits a dipeptidase activity, catalyzing the hydrolysis of glycylglycine, glycylproline, leucyltyrosine, tyrosylleucine, and histidylleucine.[1,2] **See also** *Trimerelysin I; Trimerelysin II*

[1] M. Kishida, T. Nikai, N. Mori, S. Kohmura & H. Sugihara (1985) *Toxicon* **23**, 637.
[2] J. B. Bjarnason & J. W. Fox (1995) *Meth. Enzymol.* **248**, 345.

Selected entries from **Methods in Enzymology** [vol, page(s)]:
General discussion: **248**, 345

MUNG BEAN NUCLEASE

This zinc-dependent nuclease, currently regarded as a variant of *Aspergillus* nuclease S₁ [EC 3.1.30.1], catalyzes the endonucleolytic cleavage of polynucleotides (preferentially single-stranded) to yield nucleoside-5′-monophosphates and 5′-phosphooligonucleotide products. Mung bean nuclease has proven to be extremely useful for removing 3′ and 5′ extensions from DNA or RNA termini.[1–3]

[1] D. Kowalski, W. D. Kroeker & M. Laskowski, Sr. (1976) *Biochemistry* **15**, 4457
[2] T. F. McCutchan, J. L. Hansen, J. B. Dame & J. A. Mullins (1984) *Science* **225**, 626.
[3] I. R. Lehman (1981) *The Enzymes*, 3rd ed., **14**, 193.

Selected entries from **Methods in Enzymology** [vol, page(s)]:
Other molecular properties: hairpin loop cleavage in cDNA, **152**, 334; RNA structure probing, specificity and properties, for, **180**, 197; structural probing of DNA *in vitro*, **212**, 299, 301

MURAMOYLPENTAPEPTIDE CARBOXYPEPTIDASE

This carboxypeptidase [originally classified as EC 3.4.12.6 and now reclassified as EC 3.4.17.8] catalyzes the hydrolysis of UDP-N-acetylmuramoyl-L-alanyl-D-γ-glutamyl-6-carboxy-L-lysyl-D-alanyl-D-alanine to produce D-alanine and UDP-N-acetylmuramoyl-L-alanyl-D-γ-glutamyl-6-carboxy-L-lysyl-D-alanine. The enzyme will not catalyze the release of D-alanine from the product.

A zinc-dependent D-Ala-D-Ala carboxypeptidase [EC 3.4.17.14] has been isolated from *Streptomyces albus*.

See also *Zinc* D-Ala-D-Ala *Carboxypeptidase; Serine-Type* D-Ala-D-Ala *Carboxypeptidase*

Selected entries from **Methods in Enzymology** [vol, page(s)]:
General discussion: **8**, 487; **17A**, 182
Other molecular properties: properties, **8**, 489; **248**, 213

MURAMOYLTETRAPEPTIDE CARBOXYPEPTIDASE

This carboxypeptidase [EC 3.4.17.13], also called murein tetrapeptide LD-carboxypeptidase and carboxypeptidase IIW, catalyzes the hydrolysis of a peptide bond between L-aminoacyl residues and D-aminoacyl residues in precursors of murein.[1–3] These lipid-linked precursors are in the tetrapeptide sequence found in the murein sacculus: for example, the L-lysyl-D-alanine bond in N-acetyl-D-glucosaminyl-N-acetylmuramoyl-L-alanyl-D-glutamyl-6-carboxyl-L-lysyl-D-alanine. It will also act on UDP-N-acetylmuramoyl-L-Ala-D-Glu-m-A₂pm-∇-D-Ala, on undecaprenylpyrophosphatidyl-N-acetylmuramoyl-L-Ala-D-Glu-m-A₂pm-∇-D-Ala, and the -L-Ala-D-Ala-m-A₂pm-∇-D-Ala- sequence in the murein sacculus, where m-A₂pm represents a *meso*-diaminopimelyl residue and ∇ indicates the cleavage site.

[1] J.-M. Ghuysen & R. Hakenbeck (1994) in *New Comprehensive Biochemistry* (A. Neuberger & L. L. M. van Deenen, eds.), vol. **27**, Elsevier, Amsterdam.
[2] H. J. Rogers, H. R. Perkins & J. B. Ward (1980) *Microbial Cell Walls and Membranes*, Chapman & Hall, London.
[3] J. I. Leguina, J. C. Quintela & M. A. de Pedro (1994) *FEBS Lett.* **339**, 249.

MUTALYSINS

Mutalysins are zinc-dependent metalloendopeptidases isolated from the venom of the bushmaster snake (*Lachesis muta muta*). Two enzymes with hemorrhagic activity have been identified in the venom and are also known as *Lachesis* hemorrhagic factors I and II (LHF-I and LHF-II) with molecular masses of 100 kDa and 22.5 kDa, respectively. Mutalysin I, which has potent hemorrhagic activity, catalyzes the hydrolysis of the Ala14–Leu15 bond in the oxidized insulin B chain at a relatively low rate. In contrast, mutalysin II, which has a low hemorrhagic effect, will cleave the Ala14–Leu15 bond rapidly, followed by Phe24–Phe25, His10–Leu11, and His5–Leu6. Both enzymes will act on the Aα chain of fibrinogen faster than the Bβ chain.[1–3]

[1] J. B. Bjarnason & J. W. Fox (1995) *Meth. Enzymol.* **248**, 345.
[2] E. F. Sanchez, A. Magalhaes & C. R. Diniz (1987) *Toxicon* **25**, 611.
[3] E. F. Sanchez, A. Magalhaes, F. R. Mandelbaum & C. R. Diniz (1991) *BBA* **1074**, 347.

Selected entries from *Methods in Enzymology* [**vol**, page(s)]:
General discussion: 248, 345

*Mwo*I RESTRICTION ENDONUCLEASE

This type II restriction enzyme [EC 3.1.21.4], isolated from *Methanobacterium wolfeii*, catalyzes the hydrolysis of both strands of DNA at 5′ . . . GCNNNNN∇NNGC . . . 3′, where N refers to any base.[1,2]

[1] K. D. Lunnen, R. D. Morgan, C. J. Timan, J. A. Krzycki, J. N. Reeve & G. G. Wilson (1989) *Gene* **77**, 11.
[2] K. R. Sowers (1995) in *Archaea: Methanogens: A Laboratory Manual* (K. R. Sowers & H. J. Schreier, eds.). p. 505, Cold Spring Harbor Lab. Press, Plainview, NY.

MYCOCEROSATE SYNTHASE

This acyltransferase [EC 2.3.1.111] catalyzes the reaction of an acyl-CoA with n molecules of methylmalonyl-CoA and $2n$ molecules of NADPH to produce a multimethyl-branched acyl-CoA, n molecules of coenzyme A, n molecules of carbon dioxide, and $2n$ molecules of NADP$^+$.[1,2] This enzyme both elongates fatty acyl-CoA derivatives (typically starting with six to twenty carbon atoms) and incorporates methyl branches by using methylmalonyl-CoA instead of malonyl-CoA.

[1] D. L. Rainwater & P. E. Kolattukudy (1985) *JBC* **260**, 616.
[2] M. Mathur & P. E. Kolattukudy (1992) *JBC* **267**, 19388.

MYCODEXTRANASE

This enzyme [EC 3.2.1.61] catalyzes the endohydrolysis of 1,4-α-D-glucosidic linkages in α-D-glucans containing both 1,3- and 1,4-linkages (for example, nigeran, which contains alternating α-1,3- and α-1,4-linkages). The ultimate products are nigerose (or, 3-*O*-α-D-glucopyranosyl-D-glucose) and 4α-D-nigerosylglucose.[1–4] The enzyme will not act on α-D-glucans containing only 1,3- or 1,4-linkages.

[1] K. K. Tung, A. Rosenthal & J. H. Nordin (1971) *JBC* **246**, 6722.
[2] A. L. Rosenthal & J. H. Nordin (1975) *JBC* **256**, 5295.
[3] R. H. Marchessault, J.-F. Revol, F. Bobbitt & J. H. Nordin (1980) *Biopolymers* **19**, 1069.
[4] K. Okazaki, T. Abe, K. Saruwatari, F. Kato, K. Maruyama & K. Tagawa (1992) *Biosci. Biotechnol. Biochem.* **56**, 1401.

MYCOLYSIN

This zinc-dependent metalloendopeptidase [EC 3.4.24.31], also known as pronase component (since it was first isolated from a commercial preparation known as pronase) and *Streptomyces griseus* neutral proteinase, catalyzes peptide-bond hydrolysis with a preference for hydrophobic aminoacyl residues in position P1′.[1,2] In the oxidized insulin B chain, there are nine bonds broken: Phe1–Val2, His5–Leu6,

His10–Leu11, Ala14–Leu15, Tyr16–Leu17, Leu17–Val18, Gly23–Phe24, Phe24–Phe25, and Phe25–Tyr26.

[1] H. Tsuyuki, K. Kajiwara, A. Fujita, T. Kumazaki & S. Ishii (1991) *J. Biochem.* **110**, 339.
[2] K. Kajiwara, T. Kumazaki, E. Sato, Y. Kanaoka & S. Ishii (1991) *J. Biochem.* **110**, 345.

Selected entries from *Methods in Enzymology* [**vol**, page(s)]:
Other molecular properties: inhibition, **248**, 187; production, **248**, 187; properties, **248**, 185, 187; structure, **248**, 186; synthesis, **248**, 187

MYELIN-ASSOCIATED METALLOPROTEINASE

This metalloproteinase, isolated from brain myelin preparations, is a glycoprotein that catalyzes limited cleavage of myelin basic protein. It has also been known as MADM (for mammalian disintegrin metalloprotease) and ADAM 10 (for a disintegrin and metalloprotease). This metalloproteinase will hydrolyze a number of bonds in the myelin basic protein, one of which is prominent: Pro73–Gln74.[1]

[1] L. Howard & P. Glynn (1995) *Meth. Enzymol.* **248**, 388.

Selected entries from *Methods in Enzymology* [**vol**, page(s)]:
General discussion: 248, 388
Assay: 248, 389, 393
Other molecular properties: bovine (activity, **248**, 388; CG4 monoclonal antibody to, **248**, 389; inhibition, **248**, 388; isolation, **248**, 388; pH profile, **248**, 389; properties, **248**, 192, 200, 389); bovine kidney, isolation, **248**, 392; species distribution, **248**, 389

[MYELIN BASIC PROTEIN]-ARGININE N-METHYLTRANSFERASE

This methyltransferase [EC 2.1.1.126] (also known as myelin basic protein methylase, protein-arginine N-methyltransferase, and protein methylase I; the last alternative name should not be used as there are several proteins known as methylase I) catalyzes the reaction of S-adenosyl-L-methionine with an arginyl residue in the myelin basic protein to produce S-adenosyl-L-homocysteine and [myelin basic protein]-N-methylarginine.[1]

[1] S. K. Ghosh, W. K. Paik & S. Kim (1988) *JBC* **263**, 19024.

Selected entries from *Methods in Enzymology* [**vol**, page(s)]:
Assay: 32, 329; **106**, 269
Other molecular properties: wheat germ (assay, **106**, 269; properties, **106**, 272; purification, **106**, 271)

MYELIN-PROTEOLIPID O-PALMITOYLTRANSFERASE

This acyltransferase [EC 2.3.1.100], also known as myelin PLP acyltransferase and acyl-protein synthase, catalyzes the reaction of palmitoyl-CoA with a myelin proteolipid to produce coenzyme A and a [myelin proteolipid]

O-palmitoylprotein. Alternative acyl-CoA substrates for the rat brain enzyme include stearoyl-CoA, oleoyl-CoA, and myristoyl-CoA.[1]

[1]O. A. Bizzozero, J. F. McGarry & M. B. Lees (1987) *JBC* **262**, 2138.

MYELOBLASTIN

This serine endopeptidase [EC 3.4.21.76] (also called proteinase 3, and leukocyte proteinase 3) catalyzes peptide-bond hydrolysis in proteins as well as acts as an esterase. The activity is similar to that of elastase and prefers to act on substrates with small aliphatic aminoacyl residues in both the P1 and P1′ positions (that is, Ala, Val, Ser, Met). Myeloblastin is inhibited by α_1-proteinase inhibitor and α_2-macroglobulin but not by secretory leukoprotease inhibitor.[1]

[1]J. R. Hoidal, N. V. Rao & B. Gray (1995) *Meth. Enzymol.* **244**, 61.

Selected entries from *Methods in Enzymology* [vol, page(s)]:
General discussion: 244, 61
Assay: 244, 62, 63
Other molecular properties: abundance, 244, 61; active site titration, 244, 63; biological role, 244, 62; cation exchange chromatography, 244, 65; dye-ligand affinity chromatography, 244, 64; gene, 244, 67; HPLC/FPLC chromatography, 244, 65; human leukocyte elastase, comparison, and, 244, 63, 66; immunoaffinity chromatography, 244, 66; inhibitors, 244, 66; isoforms, 244, 66; kinetic constants, 248, 10, 12; peptide thioester substrate, 248, 10; purification, 244, 64; size, 244, 61, 66; source, 248, 7; spectrophotometric assay, 244, 63; substrate specificity, 244, 62

MYELOPEROXIDASE

This heme-dependent oxidoreductase [EC 1.11.1.7], found in polymorphonuclear leukocytes, catalyzes the reaction of a donor substrate with hydrogen peroxide to produce an oxidized donor and two water molecules.[1–6] The enzyme originally was called verdoperoxidase, because of its green color. *See also Peroxidase; Chloroperoxidase; Bromoperoxidase*

[1]H. B. Dunford (1998) *CBC* **3**, 195.
[2]J. K. Hurst (1991) in *Peroxidases in Chemistry and Biology* (J. Everse, K. E. Everse & M. B. Grisham, eds.) I, p. 37, CRC Press, Boca Raton.
[3]S. J. Klebanoff (1991) in *Peroxidases in Chemistry and Biology* (J. Everse, K. E. Everse & M. B. Grisham, eds.) I, p. 1, CRC Press, Boca Raton.
[4]E. Cadenas (1989) *Ann. Rev. Biochem.* **58**, 79.
[5]J. K. Hurst & W. C. Barrette, Jr. (1989) *Crit. Rev. Biochem. Mol. Biol.* **24**, 271.
[6]A. Naqui, B. Chance & E. Cadenas (1986) *Ann. Rev. Biochem.* **55**, 137.

Selected entries from *Methods in Enzymology* [vol, page(s)]:
General discussion: 2, 794
Assay: 2, 794; 31, 347, 348; 132, 274, 370; 233, 495
Other molecular properties: activation of xenobiotics, 186, 579; activity, 233, 502, 540, 639; antimicrobial activity, measurement, 105, 399; assay as measure of leukocyte infiltration in Arthus reaction, 162, 480; assay of leukocyte infiltration of postischemic intestinal mucosa, in, 186, 729; bromination of proteins, in, 107, 441, 444; cellular

distribution, 233, 502; chlorination activity (pH range, 233, 505; substrates, 233, 505); chlorination assays (applications, 233, 504; ascorbate method, 233, 506, 510; comparison, 233, 509; general considerations, 233, 504; hydrogen peroxide electrode method, 233, 504, 506; monochlorodimedon method, 233, 506, 510; taurine chloramine method, 233, 506, 508; tetramethylbenzidine assay, 233, 506, 511); deficiency, detection with cellular chemiluminescence, 133, 496; effects of aminosalicylates, 234, 571; human leukocyte (assays, 132, 274, 370; hemi-enzyme production, 132, 374; isolation, 132, 371; physicochemical properties, 132, 377; purification, 132, 372); inactivation by hydrogen peroxide, 233, 505; isolated, hypochlorous acid production by, 233, 508; luminescence assay, 233, 495; modification of proteins and lipids, 300, 88; neutrophil granule-associated, assay, 162, 547; -oxidase, generation of cysteine thiyl radicals, 186, 325; phagocytic cell chemiluminescence, 57, 462, 488, 489; PMN granules, in, 31, 346, 352; preparation, 233, 495; preparation of chloramines, in, 132, 575; properties, 2, 798; purification, 2, 797; 233, 504; reaction catalyzed by, 233, 502; reaction mechanism, 233, 502; staining on gels, 22, 595; tyrosine nitration, myeloperoxidase-catalyzed peroxynitrite-dependent, 269, 210

MYOSIN AND MYOSIN ATPase

Myosin, which has an inherent ATPase activity (other nucleoside triphosphates hydrolyzed include GTP, UTP, ITP, and CTP), is a major component of the contractile apparatus of both muscle and nonmuscle cells.[1–5] In striated muscle, it is the major protein of the thick filaments. Myosin consists of two identical heavy chains and two pairs of light chains.

The ATPase activity of myosin [EC 3.6.4.1; previously classified as EC 3.6.1.32] is directly responsible for muscle contraction. In the absence of actin filaments, myosin is a feeble ATPase with a k_{cat} of only $0.05\ \text{s}^{-1}$, because product release is much slower than the rapid release of a proton in the P-O-P bond-cleavage step forming ADP and orthophosphate from bound ATP. Interaction with actin filaments, however, strongly stimulates the ATPase activity, increasing k_{cat} by a factor of about 200. This acceleration occurs as a consequence of the expedited release of orthophosphate and ADP.

[1]I. Rayment (1996) *JBC* **271**, 15850.
[2]K. A. Johnson (1992) *The Enzymes*, 3rd ed., **20**, 1.
[3]S. A. Endow & M. A. Titus (1992) *Ann. Rev. Cell Biol.* **8**, 29.
[4]J. R. Sellers & R. S. Adelstein (1987) *The Enzymes*, 3rd ed., **18**, 381.
[5]W. W. Kielley (1961) *The Enzymes*, 2nd ed., **5**, 159.

Selected entries from *Methods in Enzymology* [vol, page(s)]:
General discussion: 2, 582; 27, 217, 309, 333; 34, 490; 44, 882; 46, 318; 64, 76, 82, 83, 247, 248; 85, 357; 298, 3
Assay: 2, 582
Other molecular properties: *Acanthamoeba castellanii* (assays, 85, 357; properties, 85, 360, 363; purification, 85, 359, 361); actin depolymerization, effect on, 215, 74; actin interaction, 64, 82; active-site trapping with (thiol crosslinking reagents, 85, 93, 114; vanadate, 85, 116, 122); aggregation (assays, 85, 29; by dialysis, 85, 23; by dilution, 85, 24; ionic strength effects, 85, 26; pH effects, 85, 25; protein concentration

effects, **85**, 29; time effects, **85**, 27); antibody binding, **50**, 55 (competitive, **134**, 443; dissociation constant, determination, **134**, 437; localization, **134**, 440, 444); ATPase activity of, **2**, 582; **46**, 86, 260, 287, 318; **64**, 82; **85**, 123 (activation energy, **63**, 243; biotinylation, **184**, 412; butanol and, **1**, 45; cryoenzymology, **63**, 38; inactivation [energy of activation, **63**, 256; entropy of activation, **63**, 256]; inhibition by [antibodies, **134**, 446; vanadate, **134**, 485]; inhibitor, **63**, 398; kinetic analysis [chemical quench-flow methods, **134**, 692; stopped-flow methods, **134**, 678]; kinetic constants, **63**, 244; myosin S$_1$ ATPase catalyzed ^{18}O exchange, detection by NMR, **176**, 352; platelet, **215**, 81; protection, **1**, 43; pseudouridine diphosphate and, **12B**, 519, 520; reaction with arylazido-β-alanine ATP, **46**, 273; site [biotinylation, **184**, 412; visualization, **184**, 409]; steady-state rate measurements, **85**, 698; temperature, solvent, and pressure effects, **63**, 239; transient kinetic measurements, **85**, 701); bipolar filament assembly, antibody inhibition, **134**, 444; cell-free synthesis (amino acid incorporation system, **30**, 670; identification of myosin, **30**, 673; solutions, **30**, 669; special procedures, **30**, 670); cGMP assay, **38**, 106, 109; circular dichroism, **12B**, 328; coated beads, movement, antibody inhibition, **134**, 452; contraction of cytoplasmic extracts, antibody inhibition, **134**, 449; crude (actin removal, **85**, 134; phosphofructokinase removal, **85**, 134; preparation, **85**, 133; separation from C, H, and X proteins, **85**, 135); cyanylation, **47**, 131, 132; divalent cation-binding sites, detection with Mn^{2+} probe, **85**, 619; EPR, **85**, 594; *Escherichia coli* expressing, lysis, **196**, 378; exchanges, **87**, 304; expression in *Escherichia coli*, **196**, 368 (growth conditions for, **196**, 377; vector-related variability, **196**, 370); filament hybrids, copolymerization, **85**, 41; flow birefringence, **26**, 311; fluorescently labeled, microinjection, **196**, 503; gel electrophoresis, **85**, 127; heart cells, in, **32**, 741, 744, 745; heart mitochondria, **55**, 47; heavy chain (MLC1 binding site on, mapping, **196**, 384; SDS complex, properties, **48**, 6); hydrolysis, enzyme surface, **87**, 535; interaction with (actin, EPR, **85**, 613; substrates, optical activity measurements, **85**, 679); K$^+$-EDTA, related activity of platelet myosin, **215**, 82; labeling with tetramethylrhodamine iodoacetamide, **196**, 497; light chains (alkali, labeling, **85**, 92; essential exchanges, **85**, 78; fractionation, **85**, 76; MLC1, binding site on heavy chain, mapping, **196**, 384; mixed, isolation, **85**, 74; phosphorylation, assay, **215**, 83; phosphorylation sites, phosphopeptide mapping, **215**, 85; purification, **85**, 84 [recombinant light chain produced in *Escherichia coli*, **153**, 480]; regulatory, purification, **85**, 82; storage, **85**, 78); measurement, **32**, 744; methylated lysines and 3-methylhistidine (biosynthesis, analytical techniques, **106**, 290, 293; determination, **106**, 289; functional and research role, **106**, 292; peptides with, isolation strategy, **106**, 295; tissue-specific occurrence, **106**, 287); Mg-, actin-activated, related activity of platelet myosin, **215**, 81; minifilaments, aggregation, **85**, 42; molecular weight calibration standard, as, **237**, 92; monomeric, **27**, 323; movement in algal cells, assay (with coated beads, **134**, 531; video tracking device for, **134**, 538); myosin I (antibodies, reaction with brush border extract, **196**, 8; assay, **196**, 3, 15; purification from [*Acanthamoeba castellanii*, **196**, 12; intestinal brush border, **196**, 3]); myosin II (purification, **196**, 77); L-myosin and "crystalline" myosin, **2**, 583; negative staining for electron microscopy, **85**, 129; nonfilamentous, aggregation, **85**, 45; nonmuscle (concentration, **85**, 352; oxygen exchanges, **87**, 304; phosphate release, **87**, 304; phosphorothioates, and, **87**, 200, 207, 210, 261, 316; phosphorylation effect on thick filament stabilization, **85**, 364; purification, **85**, 331, 364 (from macrophages, **162**, 259); storage, **85**, 354); partial volume, **26**, 109; phosphorus-31 magnetic resonance, **64**, 392; photochemical analysis, **196**, 459; photocleavage mediated by vanadate (active site, at, **196**, 444, 447; simultaneous action at V1 and V2 sites, **196**, 448; V2 site, at, **196**, 444, 448); photomodification at Ser-180, **196**, 446; photooxidation of myosin A, **25**, 407; platelet (ATPase activity, assay, **215**, 81; characterization, **215**, 78; dephosphorylation, **215**, 82; phosphorylation, **215**, 82; preparation, **215**, 79; purification, **215**, 78); P-light chain phosphorylation, quantitation, **102**, 65, 75; polyalanyl myosin (preparation, **25**, 558; solubility, **25**, 554); polymer, **27**, 313, 318, 319, 327, 330; potassium ion-EDTA, related activity of platelet myosin, **215**, 82; preparation, **196**,

400; properties, **2**, 586; **85**, 68; purification from (platelets, **85**, 355; **215**, 78; skeletal muscle, **2**, 583; **85**, 56; **98**, 371; smooth muscle, **85**, 292); rabbit fast skeletal muscle, expression in *Escherichia coli*, **217**, 8; rabbit muscle, viscosity, **48**, 54; rate, alteration of, **87**, 210; reaction pathway, **64**, 76; recombinant subfragments, expression, **196**, 332; removal from sarcoplasmic reticulum, **32**, 292; reversible association, **27**, 312, 330; rod aggregates, formation and assay, **85**, 48; rod fragments, expressed in *Escherichia coli*, purification, **196**, 385; saturation-transfer studies, **49**, 492; **85**, 612; scattering light, **27**, 314; self association, **27**, 328; S1 fragments, expressed in *Escherichia coli*, purification, **196**, 383; simplest model, **27**, 312; solution, **27**, 313; spin labeling, **85**, 607; stereochemistry, **87**, 207, 261, 265, 300, 316; storage, **196**, 400; subfragment 1 (crystallization, **276**, 178; decoration of actin filaments, **236**, 480; gel filtration, **196**, 402); subfragments (equilibrium binding to [F-actin, **85**, 710; regulated actin, **85**, 715]; interactions with substrates, optical activity measurements, **85**, 679; properties, **85**, 68; purification from skeletal muscle, **85**, 59; **98**, 373; steady-state binding to [F-actin, **85**, 717; regulated actin, **85**, 722]); subunits, coexpression to map MLC-1 binding site on MHC, **196**, 384; sulfhydryl groups (intramolecular crosslinking, **85**, 95; modification and labeling, **85**, 84); surfaces coated with, actin sliding over, assay, **196**, 399; tetramethylrhodamine-labeled (ATPase activity, **196**, 501; characterization, **196**, 497; preparation, **196**, 497; self-assembly, **196**, 501); thiophosphate, chiral, **87**, 265, 300; triphosphatase in, **2**, 580; vanadate-mediated photocleavage, **196**, 442

[MYOSIN HEAVY-CHAIN] KINASE

This phosphotransferase [EC 2.7.1.129] catalyzes the reaction of ATP with the myosin heavy-chain to produce ADP and [myosin heavy-chain] phosphate (that is, the phosphorylated heavy chain).[1-7]

[1] H. Maruta & E. D. Korn (1977) *JBC* **252**, 8329.
[2] J. A. Hammer, J. P. Albanesi & E. D. Korn (1983) *JBC* **258**, 10168.
[3] G. P. Cote & U. Bukiejko (1987) *JBC* **262**, 1065.
[4] Q. G. Medley, W. L. Bagshaw, T. Truong & G. P. Cote (1992) *BBA* **1175**, 7.
[5] S. Ravid & J. A. Spudich (1989) *JBC* **264**, 15144.
[6] H. Brzeska & E. D. Korn (1996) *JBC* **271**, 16983.
[7] J. L. Tan, S. Ravid & J. A. Spudich (1992) *Ann. Rev. Biochem.* **61**, 721.

Selected entries from *Methods in Enzymology* [vol, page(s)]:
General discussion: **139**, 105; **196**, 12, 23
Assay: **139**, 106; **196**, 15, 25
Other molecular properties: calmodulin-dependent (assays, **139**, 106; isolation from chicken intestine, **139**, 110; properties, **139**, 113); myosin I heavy-chain kinase (assay, **196**, 15; purification from *Acanthamoeba castellanii*, **196**, 12); myosin II heavy-chain kinase A (assay, **196**, 25; phosphorylated (purification, **196**, 31; stability, **196**, 33)); purification from *Dictyostelium discoideum*, **196**, 23; regulation, **196**, 34; unphosphorylated (assay, **196**, 31; purification, **196**, 26; stability, **196**, 30)

[MYOSIN LIGHT-CHAIN] KINASE

This calcium- and calmodulin-dependent phosphotransferase [EC 2.7.1.117], also known as myosin kinase and smooth-muscle-myosin-light-chain kinase, catalyzes the reaction of ATP with the myosin light chain to produce ADP and [myosin light-chain] phosphate (that is, the phosphorylated form of the myosin light chain).[1,2] Smooth muscle myosin light chain is the best substrate, but light chain from other myosins can also be phosphorylated. Positive cooperative kinetics are observed for the skeletal muscle myosin light chain.

[1] J. R. Sellers & R. S. Adelstein (1987) *The Enzymes*, 3rd ed., **18**, 381.

[2] J. T. Stull, M. H. Nunnally & C. H. Michnaff (1986) *The Enzymes*, 3rd ed., **17**, 113.

Selected entries from *Methods in Enzymology* [vol, page(s)]:
General discussion: **85**, 298; **196**, 34
Assay: **85**, 307; **99**, 279; **196**, 43
Other molecular properties: affinity chromatography, **200**, 186; autophosphorylation, role in enzyme regulation, **139**, 696; Ca^{2+}-independent, preparation, **196**, 42; calmodulin-binding domain (assay, **139**, 115; corresponding peptides [effects on calmodulin-dependent enzymes, **139**, 125; extraction from gizzard smooth muscle, **196**, 37; interactions with calmodulin, **139**, 124; preparation from (bovine arterial smooth muscle, **196**, 41; brain, **196**, 44; enzyme, **139**, 119; human platelets, **196**, 44; porcine myometrium, **196**, 42; rabbit skeletal muscle, **196**, 46; rat pancreas, **196**, 44; turkey gizzard smooth muscle, **196**, 36); synthesis, **139**, 120; structure-function studies, **139**, 123]); inhibitors (effects on transmembrane Ca^{2+} signaling, **139**, 576); nucleotide-binding site, affinity labeling, **139**, 188; properties and function, **107**, 90; **196**, 34; smooth muscle (assays, **85**, 307; **99**, 279; properties, **85**, 305; **99**, 286; purification, **85**, 298; **99**, 281; **196**, 34; regulation, **99**, 287; substrate specificity, **99**, 286)

[MYOSIN LIGHT-CHAIN]-PHOSPHATASE

This phosphatase [EC 3.1.3.53], also known as myosin light-chain kinase phosphatase and myosin phosphatase, catalyzes the hydrolysis of phosphorylated myosin light-chain (or, [myosin light-chain] phosphate) to produce the myosin light chain and orthophosphate.[1–5] The phosphatase consists of three subunits. The holoenzyme catalyzes the dephosphorylation of myosin light chains and myosin-light-chain kinase, but not of myosin. The catalytic subunit acts on all three substrates. Other substrates include heavy meromyosin, phosphorylase *a*, and phosphorylase kinase.

cGMP-dependent protein kinase Iα interacts with the myosin-binding subunit of myosin phosphatase. Uncoupling of this interaction prevents cGMP-dependent dephosphorylation of the myosin light chain.[4]

[1] M. D. Pato & R. S. Adelstein (1983) *JBC* **258**, 7047.

[2] D. Alessi, L. K. MacDougall, M. M. Sola, M. Ikebe & P. Cohen (1992) *EJB* **210**, 1023.

[3] H. Shimizu, M. Ito, M. Miyahara, K. Ichikawa, S. Okubo, T. Konishi, M. Naka, T. Tanaka, K. Hirano & D. J. Hartshorne (1994) *JBC* **269**, 30407.

[4] H. K. Surks, N. Mochizuki, Y. Kasai, S. P. Georgescu, K. M. Tang, M. Ito, T. M. Lincoln & M. E. Mendelsohn (1999) *Science* **286**, 1583.

[5] J. L. Tan, S. Ravid & J. A. Spudich (1992) *Ann. Rev. Biochem.* **61**, 721.

Selected entries from *Methods in Enzymology* [vol, page(s)]:
General discussion: **159**, 446
Assay: **159**, 447
Other molecular properties: properties, **159**, 447; purification from turkey gizzard, **159**, 448

MYRCENE SYNTHASE

This enzyme [EC 4.2.3.15; previously classified as EC 4.1.99.9], also known as monoterpene synthase, catalyzes the conversion of geranyl diphosphate to produce myrcene and pyrophosphate (or, diphosphate).[1]

[1] J. Bohlmann, C. L. Steele & R. Croteau (1997) *JBC* **272**, 21784.

MYRISTOYL-CoA 11-(*E*) DESATURASE

This oxidoreductase [EC 1.14.99.31], which participates in the synthesis of sex pheromones, catalyzes the reaction of myristoyl-CoA with NAD(P)H, H^+, and dioxygen to produce (*E*)-11-tetradecenoyl-CoA, $NADP^+$, and two water molecules.[1]

myristoyl-CoA

(*E*)-11-tetradecenoyl-CoA

This desaturase forms the (*E*)-geometric isomer (that is, *trans*) by removing the hydrogens from the *pro*-(*R*) positions at C11 and the *pro*-(*S*) position at C12. Myristoyl-CoA 11-(*Z*) desaturase [EC 1.14.99.32] removes the *pro*-(*R*) hydrogens from both C11 and C12.

[1] I. Navaro, I. Font, G. Fabrias & F. Camps (1997) *JACS* **119**, 11335.

MYRISTOYL-CoA 11-(*Z*) DESATURASE

This oxidoreductase [EC 1.14.99.32] catalyzes the reaction of myristoyl-CoA with NAD(P)H, H^+, and dioxygen to produce (*Z*)-11-tetradecenoyl-CoA, $NAD(P)^+$, and two water molecules.[1]

(*Z*)-11-tetradecenoyl-CoA

This desaturase forms the (*Z*)-geometric isomer (*i.e.*, *cis*) by removing the hydrogens from the *pro*-(*R*) positions at C11 and C12. Myristoyl-CoA 11-(*E*) desaturase [EC 1.14.99.31] removes the *pro*-(*R*) hydrogen from C11 and the *pro*-(*S*) hydrogen from C12.

[1] I. Navaro, I. Font, G. Fabrias & F. Camps (1997) *JACS* **119**, 11335.

MYROSULFATASE

This enzyme, previously classified as EC 3.1.6.5, is now a deleted Enzyme Commission entry.

Selected entries from *Methods in Enzymology* [vol, page(s)]:
General discussion: 2, 324

MYTILIDASE

This endopeptidase, previously classified as EC 3.4.99.34 and currently not classified by the Enzyme Commission, catalyzes the hydrolysis of peptide bonds in proteins (for example, hemoglobin). It has a very low pH_{opt} (1.6-2.6).[1,2]

[1] I. F. Dumitru, D. Iordächescu & S. Niculescu (1978) *Comp. Biochem. Physiol.* **59B**, 81.
[2] D. Iordächescu, I. F. Dumitru & S. Niculescu (1978) *Comp. Biochem. Physiol.* **61B**, 119.

MYXOBACTER AL-1 PROTEASE

This zinc-dependent metalloendopeptidase was previously classified as EC 3.4.99.29 and has an activity very similar to β-lytic metalloendopeptidase [EC 3.4.24.32].[1,2] *See also β-Lytic Metalloendopeptidase*

[1] E. Kessler (1995) *Meth. Enzymol.* **248**, 740.
[2] R. L. Jackson & G. R. Matsueda (1970) *Meth. Enzymol.* **19**, 591.

Selected entries from *Methods in Enzymology* [vol, page(s)]:
General discussion: 19, 591; **248**, 740
Assay: 19, 591
Other molecular properties: amino acid composition, **19**, 598; distribution, **19**, 599; inactivators, **19**, 597; pH optimum, **19**, 597; purification, **19**, 593; purity and physical properties, **19**, 595; specificity, **19**, 598; stability, **19**, 597

N

NAD⁺ ADP-RIBOSYLTRANSFERASE

This transferase [EC 2.4.2.30], also known as poly-(adenosine diphosphate ribose) polymerase, ADP-ribosyltransferase (polymerizing), and poly(ADP-ribose) synthetase, catalyzes the reaction of NAD^+ with an [ADP-D-ribosyl]$_n$–acceptor to produce nicotinamide and the [ADP-D-ribosyl]$_{n+1}$–acceptor.[1–4] The ADP-D-ribosyl group of NAD^+ is transferred to an acceptor carboxyl group on a histone or on the enzyme itself (at glutamyl residues). Additional ADPR groups are then transferred to the 2′-position of the terminal adenosine moiety, ultimately generating a polymer with an average chain length of 20–30 units. Protein acceptor substrates include chromatin, DNA ligase, DNA-polymerase, histone H1, histones H2A and H2B, other DNA-binding proteins, chromatin, terminal deoxynucleotidyltransferase, topoisomerases I and II, and human neutrophil GTP-binding protein G22K.

[1] M. E. Smulson, C. M. Simbulan-Rosenthal, A. H. Boulares, A. Yakovlev, B. Stoica, S. Iyer, R. Luo, B. Haddad, Z. Q. Wang, T. Pang, M. Jung, A. Dritschilo & D. S. Rosenthal (2000) Adv. Enzyme Regul. 40, 183.

[2] M. K. Jacobson & E. L. Jacobson (1999) TiBS 24, 415.

[3] H. Ushiro, Y. Yokoyama & Y. Shizuta (1987) JBC 262, 2352.

[4] N. J. Oppenheimer & A. L. Handlon (1992) The Enzymes, 3rd ed., 20, 453.

Selected entries from **Methods in Enzymology** [vol, page(s)]:
General discussion: 18B, 223; 235, 617
Assay: 18B, 224, 225, 230, 231; 106, 501
Other molecular properties: activation in oxidant-exposed cells, assay, 163, 337; preparation of nuclear fraction, 18B, 224, 225, 231, 232; product, 18B, 223; properties, 18B, 232, 233; 106, 504; purification from bovine thymus and rat liver, 106, 502

NADase

This multifunctional enzyme [EC 3.2.2.5] possessing three different catalytic activities (NADase, ADP-ribosyl cyclase, and cADPR hydrolase/glycohydrolase) participates in the metabolism of cADPR. This recently discovered cyclic nucleotide consists of a ribose molecule linked through its C5 hydroxyl to the β-phosphate of ADP and through its C1 hydroxy to the N1 position of the adenine ring.

The NADase, also known as NAD^+ nucleosidase, DPNase, DPN hydrolase, and NAD^+ glycohydrolase, catalyzes the hydrolysis of NAD^+ to produce nicotinamide and ADP-D-ribose.[1–4] This enzyme forms an ADPR-enzyme intermediate during the hydrolase reaction, a property that Zatman et al.[1] first exploited to synthesize [^{14}C]NAD^+ from [^{14}C]nicotinamide and NAD^+. NADases are ecto-enzymes with their active sites facing the outer cell surface and having no access to intracellular NAD^+.[2] NADase has also proven to be a valuable plasma membrane marker protein. **See** NAD(P)⁺ Nucleosidase

The bovine spleen enzyme also catalyzes the transformation of NAD^+ into cADPR at a rate that is less than two percent of the NADase rate. This cyclase activity is studied more directly by using nicotinamide guanine dinucleotide (NGD^+) as a substrate, because the rate of production of the corresponding guanine-containing metabolite cGDPR is much higher. An enzyme-stabilized oxocarbenium intermediate, generated from $NA(G)D^+$ during the rate-limiting step, undergoes an intramolecular cyclization reaction involving attack by nitrogen N1 of adenine ring and nitrogen N7 of guanine.[2] The formation of cADPR (minor pathway) and cGDPR compete with the hydrolysis of the intermediate which forms A(G)DPR or with methanolysis which forms methyl A(G)DPR.

This multifunctional enzyme also catalyzes cADPR hydrolysis to produce ADP-D-ribose; however, because cGDPR is not readily converted into GDP-D-ribose, NGD^+ accumulates from the cyclase activity.

[1] L. J. Zatman, N. O. Kaplan & S. P. Colowick (1953) JBC 200, 197.

[2] H. M. Muller-Steffner, A. Augustin & F. Schuber (1996) JBC 271, 23967.

[3] G. Davies, M. L. Sinnott & S. G. Withers (1998) CBC 1, 119.

[4] N. J. Oppenheimer & A. L. Handlon (1992) The Enzymes, 3rd ed., 20, 453.

Selected entries from **Methods in Enzymology** [vol, page(s)]:
General discussion: 2, 660, 664; 66, 137, 144, 151; 122, 173
Assay: 2, 660, 664; 66, 144, 153; 122, 169, 173

Other molecular properties: ADPribosyl cyclase, 280, 331; analogue preparation and, 3, 902; animal tissue, 2, 653, 660, 683; 3, 902; assay of ADP-ribosylation, in, 235, 646; bovine seminal plasma (assay, 66, 144; properties, 66, 148; purification, 66, 147, 148); bovine spleen (purification, 2, 661); *Bungarus fasciatus* snake venom (assays, 122, 169, 173; immobilization and transglycosidation, 122, 178; properties, 122, 176; purification, 122, 175); cholera toxin ADP-ribosylation in presence of, 165, 241; diphtheria toxin activity as, assay, 165, 224; G proteins in presence of, ADP-ribosylation, 195, 274; isotope exchange, 64, 36; lactate dehydrogenase, 87, 620; large-scale isolation, 22, 537; mitochondrial preparations, in, 10, 111; *Neurospora*, 2, 664; 3, 875; 27, 77 (properties, 2, 665; purification, 2, 664); nicotinamide and, 2, 662, 670, 683; nicotinamide mononucleotide assay and, 3, 899; pertussis toxin-associated, 237, 63; porcine brain (purification, 2, 662); potato, 2, 655; properties, 2, 662; protein inhibitor, 66, 137 (assay, 66, 139; properties, 66, 142; purification, 66, 141, 152); pyridine nucleotide assay and, 3, 891, 896; rat liver nuclear envelope (assay, 66, 153; properties, 66, 153, 154; purification, 66, 151, 152); staining on gels, 22, 600

NAD$^+$-DEPENDENT HISTONE/PROTEIN DEACETYLASE

This newly discovered zinc-dependent transferase, which appears to be a critical factor in gene silencing, catalyzes the reaction of ε-N-acetyllysine residues on histones with NAD$^+$ to produce the corresponding deacetylated histone, nicotinamide, and 1-O-acetyl-ADP-ribose.[1,2] The modified amino acid ε-N-acetyl-lysine is also deacetylated in an NAD$^+$-dependent manner. The enzyme also promotes an NAD$^+$ ↔ nicotinamide exchange reaction that requires ε-N-acetyl-lysine or ε-N-acetylhistone. The nonhydrolyzable analogue carba-NAD$^+$ is a competitive inhibitor of the enzyme.

The enzyme appears to operate by a unique two-step catalytic mechanism: (1) the glycosidic bond between nicotinamide and ADPR in NAD$^+$ is cleaved at one subsite (with an ε-N-acetyllysine residue serving as an essential activator) to generate nicotinamide and an enzyme-bound ADPribose oxocarbenium ion intermediate; (2) amide hydrolysis at a second subsite generates enzyme-bound acetate which then attacks the oxocarbenium intermediate to generate the novel acylal product.

[1] K. G. Tanner, J. Landry, R. Sternglanz, and J. M. Denu (2000) *PNAS* 97, 14178.
[2] J. Min, J. Landry, R. Sternglanz & R. M. Xu (2001) *Cell* 105, 269.

NAD$^+$:[DINITROGEN-REDUCTASE] ADP-D-RIBOSYLTRANSFERASE

This transferase [EC 2.4.2.37], also known as NAD$^+$: azoferredoxin (ADP-ribose)transferase, catalyzes the reaction of NAD$^+$ with dinitrogen reductase (also known as nitrogenase) to produce nicotinamide and

ADP-D-ribosyl-[dinitrogen reductase]. Dinitrogen reductase is modified at an arginyl residue of one of the subunits. ADP-ribosyl-[dinitrogen reductase] hydrolase [EC 3.2.2.24], controls the level of activity of nitrogenase [EC 1.18.6.1].[1] MgADP is an activator of the ADPribosyltransferase.[2]

[1] R. G. Lowery & P. W. Ludden (1988) *JBC* 263, 16714.
[2] R. G. Lowery & P. W. Ludden (1989) *Biochemistry* 28, 4956.

NAD$^+$:DIPHTHAMIDE ADP-RIBOSYLTRANSFERASE

This transferase [EC 2.4.2.36] catalyzes the reaction of NAD$^+$ with a peptide diphthamide to produce nicotinamide and a peptide N-(ADP-D-ribosyl)diphthamide. Diphtheria toxin and other related bacterial toxins inhibit translation by catalyzing the ADPribosylation of the diphthamide residue in elongation factor 2.[1-3] *See also NAD(P)$^+$:Arginine ADP-Ribosyltransferase*

[1] Y. Sanai, K. Morihara, H. Tsuzuki, J. Y. Homma & I. Kato (1980) *FEBS Lett.* 120, 131.
[2] H. Lee & W. J. Iglewski (1984) *PNAS* 81, 2703.
[3] M. S. Weiss, S. R. Blanke, R. J. Collier & D. Eisenberg (1995) *Biochemistry* 34, 773.

Selected entries from *Methods in Enzymology* [vol, page(s)]:
Assay: ADP-ribosyltransferase activity, 165, 222; 235, 617, 631; NAD$^+$ glycohydrolase activity, 165, 224; toxicity assays, 165, 219
Other molecular properties: A chain (coupling to antibodies, 112, 218; influenza virosome-mediated delivery into BHK-21 cells, 220, 328; isolation, 112, 212; microinjection by erythrocyte ghost-cell fusion, 221, 313; radiolabeling, 112, 213); active site residues, photoaffinity labeling, 235, 633; ADP-ribosylation of elongation factor 2, 106, 379, 392; antibody conjugates (biological evaluation, 112, 221; cytotoxicity, 112, 222; physicochemical characterization, 112, 220; preparation, 112, 213; purification and storage, 112, 219); based idiotype-specific immunotoxins for B-cell neoplasms, preparation, 178, 358; catalysis of ADP ribosylation, 60, 677, 678, 706, 710; chain elongation factor T-II assay and, 20, 332; characteristics, 178, 407; *Corynebacterium diphtheriae*, from (ADP-ribosyltransferase activity, 235, 617; assay, 235, 618; purification, 165, 69); coupling to antibodies with (chlorambucil-N-hydroxysuccinimidyl ester, 112, 214; N-hydroxysuccinimidyl ester of iodoacetic acid, 112, 217; N-succinimidyl-3-(2-pyridyldithio)propionate, 112, 215); cytopathogenic effects, 235, 682; elongation phase and, 30, 287, 289; fragment A (gonadotropin-coupled [characterization, 165, 208; preparation, 165, 204]; growth factor-conjugated [cell variant isolation with, 147, 386; preparation, 147, 384]; purification, 147, 382); fragment B (binding to receptor, 256, 298; fusion protein with C3 exoenzyme [cytopathic effect on Vero cells, 256, 303; expression from *Escherichia coli*, 256, 300; genetic construction, 256, 299; immunoblotting, 256, 303; purification, 256, 301, 302; SDS-PAGE, 256, 303]); handling, safety considerations, 112, 224; inhibition of host protein synthesis, 60, 780; 235, 647; mechanism of action, 235, 680; mutant (derived immunotoxins [ADP-ribosylation assay, 178, 414; cytotoxicity assay, 178, 415; inactivation of protein synthesis, kinetics, 178, 418; synthesis and purification, 178, 411]; preparation, 178, 407); myeloperoxidase and, 3, 801; preparation, 20, 331; properties, 165, 74; protein synthesis and, 30, 266, 282; purification, 112, 209; 256, 301 (*Corynebacterium diphtheriae*, from, 165, 69); radioiodination, 70, 238;

radiolabeling, 112, 213; resistant strains, 58, 314; toxicity assays, 165, 219

NADH DEHYDROGENASE AND NADH DEHYDROGENASE (UBIQUINONE)

This flavin- and iron-sulfur-dependent oxidoreductase [EC 1.6.99.3; previously classified as EC 1.6.2.1], also known as cytochrome c reductase and type I dehydrogenase, catalyzes the reaction of NADH with an acceptor substrate to produce NAD^+ and the reduced acceptor.[1-3] Acceptor substrates include ubiquinone-1, 2,6-dichlorophenol indophenol, benzoquinone, and naphthoquinone. This activity is present in a mitochondrial complex as NADH dehydrogenase (ubiquinone) [EC 1.6.5.3]. Cytochrome c also acts as the acceptor substrate.

NADH dehydrogenase (ubiquinone) [EC 1.6.5.3], also known as ubiquinone reductase, type I dehydrogenase, and complex I, catalyzes the reaction of NADH with ubiquinone to produce NAD^+ and ubiquinol, with hydride transfer occurring from the B-side.[4-7] This FMN- and iron-sulfur-dependent multiprotein complex is associated with the inner mitochondrial membrane. Partial degradation produces NADH dehydrogenase [EC 1.6.99.3]. Under normal reaction conditions, two protons are also translocated from the cytoplasm or from the intramitochondrial or stromal compartment. *See also* other *NADH Dehydrogenases; Cytochrome b_5 Reductase*

[1] T. Ohshima, M. Ohshima & G. Drews (1984) *Z. Naturforsch.* **39c**, 68.
[2] S. Kitajima, Y. Yasukochi & S. Minakami (1981) *ABB* **210**, 330.
[3] L. I. Hochstein & B. P. Dalton (1973) *BBA* **302**, 216.
[4] U. Brandt (1997) *BBA* **1318**, 79.
[5] H. Weiss, T. Friedrich, G. Hofhaus & D. Preis (1991) *EJB* **197**, 563.
[6] Y. Hatefi & D. L. Stiggall (1976) *The Enzymes*, 3rd ed., **13**, 175.
[7] G. Palmer (1975) *The Enzymes*, 3rd ed., **12**, 1.

Selected entries from *Methods in Enzymology* [vol, page(s)]:
General discussion: 46, 260, 284, 285; 53, 11, 15, 58, 397, 405, 406, 407, 408, 413, 414, 415, 416, 418; 56, 577; 126, 360; 260, 3, 14
Assay: 10, 226, 235, 275, 276, 278, 294, 297, 427; 18B, 68, 69; 31, 20; 32, 397, 398, 400; 52, 244; 53, 9, 10, 13, 14, 20, 21, 156, 406; 56, 102, 585, 586; 260, 23; spectrophotometric, 264, 491; ubiquinol-cytochrome-*c* oxidoreductase complex, assay, 264, 492
Other molecular properties: activation by phospholipid, 10, 426; activators, 53, 20; activity, 53, 11, 17, 18; 56, 587; antisera for, 10, 698; assay for ubiquinone, 18C, 153, 154; assembly intermediates, 260, 9, 11; bacterial, 3, 693; blue-native gel electrophoresis, 260, 196, 199; bovine heart mitochondria enzyme (assay, 260, 23; isolation, 260, 15 [ammonium sulfate fractionation, 260, 18, 21; mitochondria preparation, 260, 17; subcomplex isolation, 260, 22]; structure, 260, 16; subcomplex activity, 260, 16, 25; subunits [extraction with organic solvents, 260, 25; gel electrophoresis, 260, 26; post-translational modification, 260, 33; transfer to membranes, 260, 26]; water-soluble extract, preparation, 260, 33); chaotropic ion solubilization, 31, 778; chromatographic separation, 32, 401; components, 54, 145; 56, 580, 581; composition,

53, 12, 17; deficiency (polarographic assay, 264, 480; skin fibroblast culture testing, 264, 455); diagnostic enzyme, as, 31, 20, 24, 33, 36, 37, 148; diaphorase activity of, 3, 708, 711; effect of DABS or PMPS labeling, 56, 621; electron transfer flavoprotein, of, 10, 308; endoplasmic reticulum, in, 31, 165, 185; energy conservation site, 53, 1, 413; 30° enzyme preparation (assay method, 10, 276; kinetics, 10, 289; molecular weight, 10, 288; preparation, 10, 283; properties, 10, 288; purification, 10, 287; spectra, 10, 288); EPR signal characteristics, 54, 138, 140, 141, 143, 144, 146, 148; EPR spectrum, effect of alcohols, 54, 150; extraction, 22, 229 (from mitochondria, 22, 205, 212, 215); fatty acids on, effect of, 10, 416; ferricyanide reductase activity, 53, 13, 14; 260, 8; FMN content of, 10, 229; freezing effect on activity, 264, 490; gel electrophoresis of, 10, 679; gene organization, 260, 3; heart and, 3, 688; 5, 65; high molecular weight form, 53, 414; immunoprecipitation of human enzyme, 260, 202 (antibody incubation, 260, 208; antisera preparation, 260, 206; assembly mutants, analysis, 260, 204; gel electrophoresis, 260, 208; immunoadsorbent preparation, 260, 207; mitochondria lysis, 260, 208; protein radiolabeling, 260, 205); inhibitors, 10, 225, 231, 239; 53, 8, 9, 13, 20; 56, 585, 586; 260, 3, 9, 22, 25; inhibitor sensitivity, 10, 32; intestinal mucosa, in, 32, 667, 671; iron-sulfur centers, 53, 12, 16; iron-sulfur center, EPR spectra, 54, 146, 147; iron-sulfur protein, 53, 15; isolation of, 6, 422; Keilin-Hartree preparation, 55, 126, 127; low molecular weight form, 53, 413, 415; Mackler type preparation (assay method, 10, 278; molecular weight, 10, 289; preparation, 10, 289; properties, 10, 290; specificity, 10, 290; stability, 10, 291); marker enzyme, as, 31, 262, 746; membrane marker, as, 32, 393, 402; microbial supernatant and, 6, 291; microbodies, in, 52, 496; microsomal fractions, in, 52, 88; mitochondrial, resolution into polypeptides, 126, 362; mitochondrial, two-dimensional crystals (electron microscopy, 126, 345; image analysis, 126, 348; preparation, 126, 345); mixed-function oxidase substrate, as, 31, 233; molecular weight, 53, 18, 19; NADPH enzyme and, 3, 710; *Neurospora crassa* enzyme (isolation [complex without 15-kDa subunit, 260, 12; large membrane arm intermediate, 260, 14; peripheral arm, 260, 13; wild type, 260, 12]; NADH:2,6-dichloro-phenolindophenol, 18B, 69 (assay, 18B, 69; distribution, 18B, 69; purification from pig heart, 18B, 70, 71); Nitro-BT and, 6, 892; nuclear membrane, in, 31, 290; particulate preparations, 3, 710; pH effects, 53, 20; phosphorylation associated with, 10, 31; polypeptides, 56, 587, 588, 589; preparation, 10, 300; 53, 3, 4, 6, 7, 11, 12; 55, 716; properties, 10, 230, 237; 53, 7, 12, 13; purification, 10, 227, 236, 292, 300; 53, 3, 5, 15 (complex I–III, from, 53, 11, 12; complex III, of, 53, 86; spectra, 10, 293; stability, 10, 294); purity, 53, 18; reaction with (arylazido-β-alanine NAD$^+$, 46, 284, 285; arylazido-β-alanine NADH, 46, 284, 285); redox groups, 260, 3; reductant (of azurin, 53, 661; of cytochrome *c*-551, 53, 661); reversible dissociation, 53, 431, 433, 436; rotenone binding, 55, 455; rotenone inhibition, 264, 477, 492, 500; rotenone-insensitive, outer membrane, 55, 101; rotenone-sensitive, inner membrane, 53, 413; 55, 101; soluble, properties, 53, 413; specific activity, 10, 227, 236, 229, 276; spectral properties, 10, 239; 53, 19; spectrophotometric assay, 264, 491; stability, 10, 238, 302; 53, 17; stereospecificity, 10, 741; 87, 119; submitochondrial particles, 55, 109; substrate specificity, 53, 12, 13, 17; subunit structure, 260, 3, 15, 19, 27, 199; tartronate semialdehyde reductase, assay for, 9, 242, 244; thiourea type preparation 10, 278; toluene-treated mitochondria, 56, 547; transhydrogenase, not in preparation of, 10, 321; ubiquinol-cytochrome-*c* oxidoreductase complex, assay, 264, 492; ubiquinone, 18C, 141, 142; 53, 576; yeast enzyme preparation (assay, 10, 294; flavin of, 10, 295; properties, 10, 295; purification, 10, 295)

NADH DEHYDROGENASE (QUINONE)

This oxidoreductase [EC 1.6.99.5] catalyzes the reaction of NADH with an acceptor substrate to produce NAD^+ and the reduced acceptor. The acceptor substrate can be menaquinone, 1,2-naphthoquinone, and benzaquinone.[1,2] This NADH dehydrogenase is inhibited by AMP and 2,4-dinitrophenol but not by dicoumarol or folate derivatives.

[1]A. K. Koli, C. Yearby, W. Scott & K. O. Donaldson (1969) JBC **244**, 621.
[2]P. Pupillo & L. de Luca (1982) Dev. Plant Biol. **7**, 321.

NADH KINASE

This phosphotransferase [EC 2.7.1.86; previously also classified under EC 2.7.1.96], activated by acetate, catalyzes the reaction of ATP with NADH to produce ADP and NADPH. Other phosphoryl donor substrates include CTP, ITP, UTP, and GTP.[1–3] **See also** NAD$^+$ Kinase

[1]M. M. Griffiths & C. Bernofsky (1972) JBC **247**, 1473.
[2]Y. Iwahashi & T. Nakamura (1989) J. Biochem. **105**, 916 and 922.
[3]Y. Iwahashi, A. Hitoshio, N. Tajima & T. Nakamura (1989) J. Biochem. **105**, 588.

NADH PEROXIDASE

This FAD-dependent enzyme [EC 1.11.1.1] catalyzes the reaction of NADH with hydrogen peroxide to produce NAD$^+$ and two water molecules.[1–4] Alternative substrates for hydrogen peroxide include several quinones, such as 1,4-naphthoquinone. The enzyme will also catalyze the reaction of hydrogen peroxide with ferricyanide.

In addition to the FAD, the *Enterococcus faecalis* enzyme contains a cysteine-sulfinic acid redox center which undergoes reversible two-electron reduction and oxidation by hydrogen peroxide (the sulfinic acid form is the resting state of the enzyme).[5,6] The sulfinic acid is first activated with stoichiometric amounts of NADH to generate the active cysteinyl residue via the flavin in a priming reaction. The thiol form of the enzyme, containing oxidized FAD, now binds NADH. The thiol reacts with hydrogen peroxide, reforming the sulfenic acid. The bound NADH reduces the FAD, producing NAD$^+$ and enzyme-bound FADH$_2$. This reduced flavin now reacts with the sulfinic acid to regenerate the thiol and a new catalytic round can commence.

[1]B. A. Palfey & V. Massey (1998) CBC **3**, 83.
[2]C. Cox, P. Camus, J. Buret & J. Duvivier (1982) Anal. Biochem. **119**, 185.
[3]M. I. Dolin (1977) BBRC **78**, 393.
[4]M. I. Dolin (1957) JBC **225**, 557.
[5]E. J. Crane, 3rd, J. Vervoort & A. Claiborne (1997) Biochemistry **36**, 8611.
[6]E. J. Crane, 3rd, D. Parsonage, L. B. Poole & A. Claiborne (1995) Biochemistry **34**, 14114.

NAD$^+$ KINASE

This phosphotransferase [EC 2.7.1.23] catalyzes the reaction of ATP with NAD$^+$ to produce ADP and NADP$^+$.[1–4] This reaction has the potential for altering the intracellular [NADP$^+$]/[NADPH] redox potential by converting NAD$^+$ to NADP$^+$. The kinetic mechanism for this kinase is reportedly random Bi Bi for the pigeon liver and *C. utilis* enzymes and ping pong Bi Bi for the *A. vinelandii* kinase. **See also** NADH Kinase

[1]G. Magni, A. Amici, M. Emanuelli, N. Raffaelli & S. Ruggieri (1999) AE **73**, 135.
[2]E. T. McGuiness & J. R. Butler (1985) Int. J. Biochem. **17**, 1.
[3]E. P. Anderson (1973) The Enzymes, 3rd ed., **9**, 49.
[4]D. K. Apps (1968) EJB **5**, 444.

Selected entries from **Methods in Enzymology** [vol, page(s)]:
General discussion: 2, 652; 18B, 20, 149, 156; 66, 101
Assay: 18B, 43, 44, 149
Other molecular properties: Azotobacter vinelandii, 18B, 151 (properties, 18B, 152; purification, 18B, 151, 152); calf liver, 18B, 161, 162; immobilization by radiation polymerization, 135, 148; isocitrate dehydrogenase assay, 18B, 149, 156, 157; mammalian tissues, 18B, 43 (assay, 18B, 43, 44; properties and activities, 18B, 44, 45; tissue extract, 18B, 43); NADP$^+$ preparation and, 3, 882; preparation, from Azotobacter vinelandii, 66, 87; properties, 66, 104; purification, from sea urchin eggs, 66, 101; rabbit liver, 18B, 154 (properties, 18B, 155, 156; purification, 18B, 154, 155); rat liver, 18B, 157 (properties, 18B, 159; purification, 18B, 157)

NAD(P)$^+$:ARGININE ADP-RIBOSYLTRANSFERASE

This ADP-ribosyltransferase [EC 2.4.2.31] catalyzes the reaction of NAD$^+$ (or NADP$^+$) with L-arginine to produce nicotinamide and N^2-(ADP-D-ribosyl)-L-arginine.[1–3] Other acceptor substrates include arginyl residues in proteins (*e.g.*, actin, adenylate cyclase, and RNA polymerase), agmatine, arginine methyl ester, guanidine, and guanidinobutyrate. This activity will also result in the activation of adenylate cyclase [EC 4.6.1.1]. The enzyme has a random Bi Bi kinetic mechanism.[3] **See also** NAD$^+$ ADP-Ribosyltransferase; NAD$^+$:Diphthamide ADP-Ribosyltransferase; Cholera Toxin

[1]J. C. Osborne, Jr., S. J. Stanley & J. Moss (1985) Biochemistry **24**, 5235.
[2]N. J. Oppenheimer & A. L. Handlon (1992) The Enzymes, 3rd ed., **20**, 453.
[3]J. Moss & M. Vaughan (1988) AE **61**, 303.

Selected entries from **Methods in Enzymology** [vol, page(s)]:
General discussion: 106, 403, 418, 430
Assay: 195, 255

NADPH:CYTOCHROME C_2 REDUCTASE

This FAD-dependent oxidoreductase [EC 1.6.2.5] catalyzes the reaction of NADPH with two molecules of ferricytochrome c_2 to produce $NADP^+$ and two molecules of ferrocytochrome c_2.[1,2]

[1] D. J. Sabo & J. A. Orlando (1968) *JBC* **243**, 3742.
[2] L. Norrheim, H. Sorensen, K. Gautvik, J. Bremer & O. Spydevold (1990) *BBA* **1051**, 319.

NADPH:CYTOCHROME *f* REDUCTASE

This oxidoreductase system catalyzes the reaction of NADPH with ferricytochrome *f* to produce $NADP^+$ and ferrocytochrome *f*. Other acceptor substrates include ferricyanide and a number of indophenols.

Selected entries from *Methods in Enzymology* [vol, page(s)]:
General discussion: 23, 447

NADPH DEHYDROGENASE

This oxidoreductase [EC 1.6.99.1], also known as Old Yellow enzyme and NADPH diaphorase, catalyzes the reaction of NADPH with an acceptor substrate to produce $NADP^+$ and the reduced acceptor.[1-6] The yeast enzyme uses FMN as a cofactor whereas the plant enzyme uses FAD. Electron acceptor substrates include dioxygen, cytochrome *c*, ferricyanide, methylene blue, benzoquinone, cyclohexa-2-en-1-one, and menadione. Interestingly, α-NADPH is actually a better substrate than the natural β-isomer.

Although the Old Yellow enzyme was the first flavoprotein to be purified, its physiologic role remains uncertain, because the natural substrates are unknown. The enzyme is typically assayed as an NADPH oxidase, which operates by a ping pong Bi Bi kinetic mechanism. Upon binding of NADPH, a charge-transfer complex is formed between NADPH and the flavin. The *pro-(R)* hydride of NADPH is then transferred to the *si*-face of the flavin.[5,6]

[1] B. A. Palfey & V. Massey (1998) *CBC* **3**, 83.
[2] P. A. Karplus, K. M. Fox & V. Massey (1995) *FASEB J.* **9**, 1518.
[3] L. M. Schopfer & V. Massey (1991) *A Study of Enzymes* **2**, 247.
[4] T. Honma & Y. Ogura (1977) *BBA* **484**, 9.
[5] H. J. Bright & D. J. T. Porter (1975) *The Enzymes*, 3rd ed., **12**, 421.
[6] Å. Åkeson, A. Ehrenberg & H. Theorell (1963) *The Enzymes*, 2nd ed., **7**, 477.

Selected entries from *Methods in Enzymology* [vol, page(s)]:
General discussion: 2, 712; 6, 430; 52, 19, 45, 46
Assay: 2, 712; 6, 430; 52, 90, 243 (by dual-wavelength stopped-flow spectroscopy, 52, 221)
Other molecular properties: activity, 52, 95; *Ah* locus, 52, 232; biphenyl hydroxylation, 52, 399; borohydride reduction, 18B, 475; chloroplasts, 6, 430; chromatographic purification, 52, 93, 94; contaminant, of cytochrome P-450$_{LM}$ forms, 52, 117; cytochrome P-450 assay, in, 52, 110, 205; detergent-solubilized, properties, 52, 241, 245, 246; dissociation kinetics, 53, 430; DT-diaphorase activity, 186, 297; flavin content, 52, 90, 94, 95; histochemistry, 6, 893; holoenzyme

reconstitution, 53, 436; inhibitors, 52, 96; microsomal fractions, in, 52, 88, 201; molecular weight, 52, 90, 95; NAD(P)$^+$ transhydrogenase and, 6, 439; optimal purification method for active preparation, 52, 203, 204; properties, 2, 715; 6, 433; purification, 2, 713; 6, 431; 10, 364; 52, 89, 90, 92; requirement by mitochondrial P-450, 37, 304; reversible dissociation, 53, 429, 433; stability, 52, 95; staining of nitric oxide synthase, 268, 500; nitro blue tetrazolium staining, 268, 491, 493, 501, 513; stereospecificity, 87, 119; sulfite complexing, lack of, 18B, 473

NADPH DEHYDROGENASE (FLAVIN)

This oxidoreductase [EC 1.6.8.2] catalyzes the reaction of NADPH with riboflavin to produce $NADP^+$ and reduced riboflavin.[1] The enzyme from *Entamoeba histolytica* reduces galactoflavin, FMN, FAD, and methylene blue. NADH can be used as the reducing agent, albeit at one-twentieth of the rate of NADPH.

[1] H.-S. Lo & R. E. Reeves (1980) *Mol. Biochem. Parasitol.* **2**, 23.

NAD(P)H DEHYDROGENASE (FMN)

This oxidoreductase [EC 1.6.8.1], also known as FMN reductase, catalyzes the reaction of NADPH (or NADH) with FMN to produce $NAD(P)^+$ and $FMNH_2$.[1-3] The enzyme from luminescent bacteria (*e.g.*, *Photobacterium fischeri*) also reduces riboflavin and FAD, albeit more slowly. Some organisms have distinct NADPH-specific and NADH-specific enzymes.

The *Vibrio harveyi* enzyme exhibits a ping pong Bi Bi kinetic pattern when NADPH oxidation is assayed spectrophotometrically; however, in a luciferase-coupled assay, the kinetics yield decidedly convergent-line data with K_m values 26-times tighter for FMN and 1000-times tighter for NADPH.[3] These data suggest the possibility of metabolic channeling.

[1] B. Nefsky & M. DeLuca (1982) *ABB* **216**, 10.
[2] E. Jablonski & M. DeLuca (1978) *Biochemistry* **17**, 672.
[3] G. A. Michaliszyn, S. S. Wing & E. A. Meighen (1977) *JBC* **252**, 7495.

Selected entries from *Methods in Enzymology* [vol, page(s)]:
Assay: 57, 205
Other molecular properties: immobilization, 57, 204, 205; luciferase-coimmobilized (bioluminescent assays with, 136, 88; preparation, 136, 85); purification, 57, 203; stereospecificity, 87, 122

NADPH DEHYDROGENASE (QUINONE)

This FAD-dependent oxidoreductase [EC 1.6.99.6] catalyzes the reaction of NADPH with an acceptor substrate to produce $NADP^+$ and the reduced acceptor.[1,2] Acceptor substrates include menaquinone, menadione, and 1,4-naphthoquinone. The enzyme is inhibited by dicoumarol and folate derivatives, but 2,4-dinitrophenol is without effect.

[1] A. K. Koli, C. Yearby, W. Scott & K. O. Donaldson (1969) JBC **244**, 621.
[2] P. Pupillo & L. de Luca (1982) Dev. Plant Biol. **7**, 321.

NAD(P)H DEHYDROGENASE (QUINONE)

This FAD-dependent oxidoreductase [EC 1.6.99.2], known variously as NAD(P)H:quinone reductase, DT diaphorase,[1] quinone reductase, azoreductase, phylloquinone reductase, and menadione reductase, catalyzes the reaction of NAD(P)H with an acceptor substrate to produce $NAD(P)^+$ and the reduced acceptor.[2–5] Acceptor substrates include vitamin K, menadione, methylene blue, several quinones, ferricyanide, and 2,6-dichlorophenolindophenol. The enzyme has an important protective role during oxidative stress and in vitamin K metabolism. The enzyme from all sources studied exhibits ping pong kinetics, with the reduced pyridine coenzyme binding first.[4,5] Evidence has been presented that the reaction proceeds by a two-electron transfer.[2]

[1] The D and T in this unusual name relates to the enzyme's action on NADH and NADPH, previously known as DPNH and TPNH, respectively.
[2] B. A. Palfey & V. Massey (1998) CBC **3**, 83.
[3] E. Cadenas, P. Hochstein & L. Ernster (1992) AE **65**, 97.
[4] H. Petitdemange, R. Marczak, G. Raval & R. Gay (1980) Can. J. Microbiol. **26**, 324.
[5] M. Hayashi, K. Hasegawa, Y. Oguni & T. Unemoto (1990) BBA **1035**, 230.

Selected entries from **Methods in Enzymology** [vol, page(s)]:
General discussion: 2, 707; 10, 315; 161, 271; 186, 287
Assay: 10, 297, 314; 161, 271; 186, 292
Other molecular properties: acceptors for, 10, 313; Ah locus, 52, 232; biological significance, 186, 297; distribution of, 10, 310; DT-diaphorase activity, 186, 297; electron acceptor, 2, 728; halophile membranes, 56, 403; Mycobacterium, 10, 161; NAD(P)$^+$ transhydrogenase and, 6, 439; occurrence and intracellular distribution, 186, 288; properties, 10, 312; 161, 272; 186, 293; purification, 10, 301, 310; 186, 290; relation to DT diaphorase, 10, 315; stability, 10, 312; soluble coupling enzyme and, 6, 283; stereospecificity of, 10, 741

NADPH:FERRIHEMOPROTEIN REDUCTASE

This FAD- and FMN-dependent oxidoreductase [EC 1.6.2.4], also known as NADPH:cytochrome P450 reductase and ferrihemoprotein P450 reductase, catalyzes the reaction of NADPH with two molecules of a ferricytochrome to produce $NADP^+$ and two molecules of a ferrocytochrome.[1–6] This reductase acts on a number of heme-thiolate-dependent monooxygenases (or, cytochromes P450). It also catalyzes the reduction of cytochrome b_5 and cytochrome c. Non-cytochrome electron-acceptor substrates include ferricyanide, 2,6-dichlorophenolindophenol, menadione, vitamin K_3,

and neotetrazolium chloride. **See also** NADPH:Cytochrome c_2 Reductase

[1] H. Kojima, K. Takahashi, F. Sakane & J. Koyama (1987) J. Biochem. **102**, 1083.
[2] I. Benveniste, B. Gabriac & F. Durst (1986) BJ **235**, 365.
[3] N. Takahashi, T. Saito, Y. Goda & K. Tomita (1986) J. Biochem. **99**, 513.
[4] M. S. Johnson & S. A. Kuby (1991) in A Study of Enzymes (S. A. Kuby, ed.), vol. 2, p. 285, CRC Press, Boca Raton.
[5] B. A. Palfey & V. Massey (1998) CBC **3**, 83.
[6] C. H. Williams, Jr. (1976) The Enzymes, 3rd ed., **13**, 89.

Selected entries from **Methods in Enzymology** [vol, page(s)]:
General discussion: 2, 699, 704; 10, 565; 34, 302; 52, 45
Assay: 31, 743; 52, 243, 362; 62, 397, 398, 400; 110, 375; 187, 376; yeast endoplasmic reticulum, 194, 658
Other molecular properties: antibody against, 52, 241, 242; coexpression with CYP3A4 in baculovirus-insect cell system, 272, 88, 91; detergent-solubilized, properties, 52, 241, 245, 246; effect of extraction methods in properties, 53, 408, 409; electron transfer, 272, 368; extraction from microsomes, 22, 205, 218; flavin determination, 53, 429; FMN assay, 3, 958; 10, 497; hemoglobin-catalyzed reactions with, 231, 573 (hemoglobin concentration and, 231, 579; hemoglobin native structure and, 231, 579; inhibition by carbon monoxide, 231, 579; NADPH concentration and, 231, 579; optimal conditions, 231, 578; oxygen concentration and, 231, 579; pH dependence, 231, 580; reaction mixtures, 231, 578; reductase concentration and, 231, 580; temperature dependence, 231, 579); hepatocyte endoplasmic reticulum, biosynthesis, 96, 536; homology with nitric oxide synthase, 268, 312, 334, 420; immunological screening of microsomal and mitochondrial extracts, 52, 248; inhibition, by anti-NADPH-cytochrome c reductase, 52, 246, 247; intestinal epithelium, in, 77, 156; kinetics, 10, 573; lipid peroxidation, 52, 303; liver, in, 31, 97; marker enzyme, as, 31, 743; membrane marker, as, 32, 393; microsomal, 10, 566; 22, 205, 218; 52, 88, 99; 55, 101, 103; mixed-function oxidase substrate, as, 31, 233; modification by symmetrical disulfide radical, 251, 100; nuclear membrane, in, 31, 290; outer membranes, 55, 103, 104; partial purification, 52, 364; plasma membrane, in, 31, 89, 91, 100, 225; preparation, 10, 569; 231, 577; preparative gel-density gradient electrophoresis of, 22, 436; properties, 10, 572; protease-solubilized, properties, 52, 241, 245, 246; rat liver (assay, 110, 375; purification, 110, 376); removal from microsomal protein, 52, 362; rough microsomal subfractions, in, 52, 75, 76; synthesis, free and membrane-bound polysomes and, 30, 326; unit of specific activity, 10, 567; yeast endoplasmic reticulum, assay, 194, 658

NADPH OXIDASE

This enzyme system, also known as respiratory burst oxidase, is found in neutrophils, monocytes, and macrophages shortly after phagocytosis. The system catalyzes the overall reaction of NADPH with two molecules of dioxygen to produce $NADP^+$, two superoxide ions, and a proton.[1–6]

The term has also been used to refer to other systems catalyzing the reaction of NADPH with a proton and 0.5 dioxygen to produce $NADP^+$ and water. **See also** NAD(P)H Dehydrogenase

[1] C. Lamb & R. A. Dixon (1997) Ann. Rev. Plant Physiol. Plant Mol. Biol. **48**, 251.
[2] A. Abo & A. W. Segal (1995) MIE **256**, 268.
[3] S. J. Chanock, J. E. Benna, R. M. Smith & B. M. Babior (1994) JBC **269**, 24519.

[4]B. M. Babior (1992) *AE* **65**, 49.
[5]B. M. Shapiro & P. B. Hopkins (1991) *AE* **64**, 291.
[6]J. K. Hurst & W. C. Barrette, Jr. (1989) *Crit. Rev. Biochem. Mol. Biol.* **24**, 271.

Selected entries from *Methods in Enzymology* [vol, page(s)]:
General discussion: 105, 364; 255, 746; 256, 268
Assay: chemiluminescence, 233, 226; cytochrome *c* reduction method, 233, 223; fluorescent probes for, 233, 227; oxygen consumption method, 233, 228; Rac2 protein, for, 256, 24
Other molecular properties: activators, 233, 222; activity, 233, 222, 540; amylomaltase assay, not present in, 5, 142; cell-free activity (assay, 256, 22, 276; reconstitution by purified components, 256, 268; superoxide production, assay, 256, 276); cellular distribution, 233, 222; fructose-1,6-bisphosphatase, presence in assay of, 5, 273; guinea pig neutrophil (assay, 132, 366; properties, 132, 368; purification, 132, 367); hexokinase, not present in assay of, 9, 371; human neutrophil (characterization, 132, 359; preparation, 132, 356; resolution into components, 132, 358); 2-keto-6-phosphogluconate reductase, in assay of, 5, 295; microsomal fractions, in, 52, 89; NADPH dehydrogenase, and, 2, 712; NADPH oxidase–cytochrome b-558, 255, 746; oxygen radical generation, 234, 423; particles in human neutrophils, in (assay, 105, 364; isolation, 105, 363); phagocytes, constituents, in, 256, 15; plasma membrane and, 31, 90; 233, 222; properties, 233, 222; regulation by Rac proteins, 256, 358; role in phagocyte respiratory burst, 57, 464, 488; stereochemistry, 87, 122

NADPH PEROXIDASE

This peroxidase [EC 1.11.1.2] catalyzes the reaction of NADPH with hydrogen peroxide to produce NADP$^+$ and two water molecules.[1,2] *See also NADH Peroxidase*

[1]E. E. Conn, L. M. Kraemer, P.-N. Liu & B. Vennesland (1952) *JBC* **194**, 143.
[2]J. Coves, M. Eschenbrenner & M. Fontecave (1991) *BBRC* **178**, 54.

NADPH:QUINONE REDUCTASE

This zinc-dependent oxidoreductase [EC 1.6.5.5], also known as quinone oxidoreductase and ζ-crystallin, catalyzes the reaction of NADPH and a proton with a quinone to produce NADP$^+$ and a semiquinone in a one-electron transfer. Orthoquinones are highly effective substrates. Dicoumarol (K_i = 13 μM) and nitrofurantoin (K_i = 14 μM) are competitive inhibitors with respect to the quinone substrate. The enzyme is abundant in the lens of the mammalian eye; in fact, the guinea pigs comprises 10% of the total lens protein.[1,2] The enzyme is thought to be distinct from NAD(P)H dehydrogenase (quinone) [EC 1.6.99.2]. *See also NAD(P)H Dehydrogenase (Quinone)*

[1]P. V. Rao & J. S. Zigler, Jr., (1991) *ABB* **284**, 181.
[2]P. V. Rao, C. M. Krishna & J. S. Zigler, Jr., (1992) *JBC* **267**, 96.

NAD(P)$^+$ NUCLEOSIDASE

This pyridine nucleosidase [EC 3.2.2.6] catalyzes the hydrolysis of NAD(P)$^+$ to produce nicotinamide and ADP-ribose(P).[1,2] The enzyme also catalyzes the transfer of ADP-ribose(P) residues to acceptor substrates. *See also NADase*

[1]I. H. Mather & M. Knight (1972) *BJ* **129**, 141.
[2]M. Bröcker, J. Schindelmeiser & H. Pape (1979) *FEMS Microbiol. Lett.* **6**, 245.

Selected entries from *Methods in Enzymology* [vol, page(s)]:
General discussion: 18B, 132
Assay: 18B, 132; conditions, 15, 636
Other molecular properties: beef spleen, 18B, 132; glucose-6-phosphate dehydrogenase, should not be in preparations of, 9, 371; inhibition by nicotinamide, 15, 636; pig brain, 18B, 135; purification from calf spleen, 66, 106; use (formation of NAADP, 66, 105; separation of α and β NADP$^+$, 66, 89, 90)

NAD(P)$^+$ TRANSHYDROGENASE

This FAD-dependent oxidoreductase [EC 1.6.1.1], also known as NAD(P)$^+$ transhydrogenase (B-specific), pyridine nucleotide transhydrogenase, and nicotinamide nucleotide transhydrogenase, catalyzes the reversible hydride transfer from NADPH to NAD$^+$ to produce NADP$^+$ and NADH via a ping pong Bi Bi kinetic mechanism.[1–6] This enzyme exhibits B-type stereospecificity in hydride transfer to and from the *si*-face of the pyridine ring in both NAD$^+$ and NADP$^+$ (hence the name: "BB-transhydrogenase"). Deamino analogues of the above coenzymes can also serve as substrates. *See also NAD(P)$^+$ Transhydrogenase (AB-Specific)*

[1]A. R. Clarke & T. R. Dafforn (1998) *CBC* **3**, 1.
[2]F. Widmer & N. O. Kaplan (1976) *Biochemistry* **15**, 4693.
[3]J. Rydström, J. B. Hoek & L. Ernster (1976) *The Enzymes*, 3rd ed., **13**, 51.
[4]H. W. J. van den Broek & C. Veeger (1971) *EJB* **24**, 72.
[5]A. E. Chung (1970) *J. Bacteriol.* **102**, 438.
[6]P. T. Cohen & N. O. Kaplan (1970) *JBC* **245**, 2825 and 4666.

Selected entries from *Methods in Enzymology* [vol, page(s)]:
General discussion: 2, 681; 6, 434; 10, 317
Assay: 2, 681; 6, 434; 10, 317, 741; 55, 276, 788, 790, 811; 126, 354; principle, 55, 261; procedure, 55, 263
Other molecular properties: activity, 15, 651, 652; ATPase inhibitor, 55, 407, 412, 413; ATP-dependent, assay, in mutants, 56, 113, 114; brown adipose tissue mitochondria, 55, 77; DCCD, 55, 499; dehydrogenase localization, 6, 962; diaphorase and, 2, 710, 711; digitonin solubilization of, 22, 225; effect of oligomycin on, 10, 545; energy-dependent, assay, 55, 789, 790; enzyme preparations, 55, 263; equilibrium constant, 2, 311; extraction of, 1, 38; 22, 229; hydroxysteroid dehydrogenase and, 5, 522, 525; *Leuconostoc mesenteroides* and, 1, 331; localization, 55, 261; mitochondrial, rhein, 55, 459; properties, 2, 685, 687; 6, 437; 55, 282, 283; purification (beef heart, 2, 686; procedure, 55, 277; properties, 55, 282, 283; *Pseudomonas fluorescens*, 2, 683; reagents, 55, 276; spinach, 6, 435); quinate dehydrogenase, presence in preparation of, 2, 310; reconstitution (beef heart mitochondria, from, [assay, 55, 811; 126, 354; comments, 55, 815, 816; enzyme preparation, 2, 686; 55, 811; 126, 354; procedure, 55, 812]; *Escherichia coli*, from, 55, 787, 788); shikimate reduction and, 2, 308; stereospecificity, 4, 841; 87, 118; steroid hydroxylation, 55, 10; submitochondrial particles and, 5, 277; substrate analogues, 55, 262, 270; venturicidin, 55, 506

NAD(P)$^+$ TRANSHYDROGENASE (AB-SPECIFIC)

This oxidoreductase [EC 1.6.1.2], also known as pyridine nucleotide transhydrogenase and transhydrogenase, catalyzes the reversible reaction of NADPH with NAD$^+$ to produce NADP$^+$ and NADH.[1–7] Interestingly, the enzyme from heart mitochondria is A-specific (that is, *pro-R*) with respect to NAD$^+$ and B-specific (that is, *pro-S*) with respect to NADP$^+$. This observation distinguishes this enzyme from NAD(P)$^+$ transhydrogenase [EC 1.6.1.1].

These transhydrogenases link proton pumping to hydride transfer between NADP$^+$ and NAD$^+$ across cytosolic or mitochondrial membranes: note that proton pumping is linked directly to the NAD$^+$ to NADH conversion, does not utilize another energy source, and is governed by the relative concentrations of the pyridine nucleotides on either side of the membrane. In addition, note that transhydrogenation can proceed in the absence of a proton-motive force, although the latter accelerates the reaction and affects the overall equilibrium constant. The kinetic mechanism of the bovine heart mitochondrial enzyme is random Bi Bi and hydride transfer between coenzymes is direct, involving no stable intermediates.[7] *See also NAD(P)$^+$ Transhydrogenase*

[1]J. B. Jackson, T. M. Lever, J. Rydström, B. Persson & E. Carlenor (1991) *Biochem. Soc. Trans.* **19**, 573.
[2]J. Rydström, J. B. Hoek & L. Ernster (1976) *The Enzymes*, 3rd ed., **13**, 51.
[3]J. B. Hoek & J. Rydström (1988) *BJ* **254**, 1.
[4]R. R. Fisher & S. R. Earle (1982) in *The Pyridine Nucleotide Coenzymes* (J. Everse, B. Andersson & K.-S. You, eds.), Academic Press, New York, p. 279.
[5]J. Rydström, B. Persson & E. Carlenor (1987) in *Pyridine Nucleotide Coenzymes, Chemical, Biochemical and Medical Aspects* (D. Dolphin, R. Poulson & O. Avramovic, eds.) vol. **2B**, Wiley, New York, p. 433.
[6]J. Rydström (1981) in *Mitochondria and Microsomes* (C. P. Lee & G. Dallner, eds.), Addison-Wesley, Reading, Massachusetts, p. 317.
[7]K. Enander & J. Rydstrom (1982) *JBC* **257**, 14760.

Selected entries from *Methods in Enzymology* [**vol**, page(s)]:
General discussion: **55**, 261, 275, 811; **126**, 353
Assay: **10**, 317
Other molecular properties: bovine heart (antibody for, **10**, 322; assay, **10**, 317; **126**, 354; properties, **10**, 321; purification, **10**, 318; **126**, 354; reconstitution, **126**, 357; stereospecificity, **10**, 321; types, **10**, 289, 321); stereospecificity, **87**, 118

NAD$^+$ PYROPHOSPHATASE

This enzyme [EC 3.6.1.22], also known as NAD$^+$ diphosphatase, NADP$^+$ pyrophosphatase, and NADH pyrophosphatase, catalyzes NAD$^+$ hydrolysis to produce AMP and nicotinamide mononucleotide (NMN).[1,2] Other substrates include NADP$^+$ as well as 3-acetylpyridine and the thionicotinamide analogues of NAD$^+$ and NADP$^+$.

Note: The enzyme's name can lead to confusion with NAD$^+$ pyrophosphorylase [EC 2.7.7.1], which is more widely known as nicotinamide-nucleotide adenylyltransferase.

[1]R. Wagner, F. Feth & K. G. Wagner (1986) *Planta* **167**, 226 and **168**, 408.
[2]Y. Nakajima, N. Fukunaga, S. Sasaki & S. Usami (1973) *BBA* **293**, 242.

Selected entries from *Methods in Enzymology* [**vol**, page(s)]:
Other molecular properties: cleavage of NAD$^+$, **66**, 65; coupling, to Sepharose, **66**, 65; hydrolysis, of NAD$^+$ to NMN, **66**, 63, 120; isolation, from potatoes, **66**, 63; luciferase assay and, **3**, 873

NAD$^+$ SYNTHETASES

NAD$^+$ synthetase (ammonia-utilizing) [EC 6.3.1.5] catalyzes the reaction of ATP with deamido-NAD$^+$ and ammonia (or L-glutamine) to produce NAD$^+$, AMP, (L-glutamate), and pyrophosphate.[1–5] With the *Escherichia coli* enzyme, the Michaelis constant for ammonia is 65 μM *versus* 16 mM for L-glutamine).

NAD$^+$ synthetase (glutamine-utilizing) [EC 6.3.5.1] catalyzes the reaction of ATP with deamido-NAD$^+$, L-glutamine, and water to produce NAD$^+$, L-glutamate, AMP, and pyrophosphate (or, diphosphate).[1–6] Ammonia can also act as the nitrogen donor; for the *Saccharomyces cerevisiae* enzyme, the Michaelis constant is 6.4 and 5 mM for ammonia and L-glutamine, respectively.

[1]G. Magni, A. Amici, M. Emanuelli, N. Raffaelli & S. Ruggieri (1999) *AE* **73**, 135.
[2]H. Zalkin (1993) *AE* **66**, 203.
[3]K. T. Hughes, M. O. Baldomero & J. R. Roth (1988) *J. Bacteriol.* **170**, 2113.
[4]J. M. Buchanan (1973) *AE* **39**, 91.
[5]R. L. Spencer & J. Preiss (1967) *JBC* **242**, 385.
[6]H. Zalkin & J. L. Smith (1998) *AE* **72**, 87.

Selected entries from *Methods in Enzymology* [**vol**, page(s)]:
General discussion: **6**, 345; **18B**, 20; **113**, 297; **280**, 211
Assay: **6**, 350; **18B**, 42; **46**, 420; **113**, 297; **280**, 211
Other molecular properties: *Escherichia coli* (assay, **113**, 297; properties, **113**, 302; purification, **113**, 298); properties and activities in mammalian tissues, **18B**, 42, 43; *Saccharomyces cerevisiae* (assay, **6**, 350; **113**, 297; properties, **6**, 352; **113**, 302; purification, **6**, 351; **113**, 300); tissue extract, **18B**, 42

*Nae*I RESTRICTION ENDONUCLEASE

This type II site-specific endonuclease [EC 3.1.21.4], isolated from *Nocardia aerocolonigenes*, catalyzes the hydrolysis of both strands of DNA at 5′...GCC∇GGC...3′, producing blunt-ended fragments. *Nae*I requires simultaneous interaction with two copies of the recognition sequence before cleavage occurs, product release being the rate-determining step.[1] In addition, the endonuclease exhibits marked site preferences with the flanking bases influencing the activity.[2]

[1]C. C. Yang, B. K. Baxter & M. D. Topal (1994) *Biochemistry* **33**, 14918.
[2]C. C. Yang & M. D. Topal (1992) *Biochemistry* **31**, 9657.

[3]M. De Luca, L. C. Dunlop, R. K. Andrews, J. F. Flannery, Jr., R. Ettling, D. A. Cumming, G. M. Veldman & M. C. Berndt (1995) *JBC* **270**, 26734.

Na$^+$-EXPORTING ATPase

This so-called ATPase [EC 3.6.3.7] catalyzes the ATP-dependent (ADP- and orthophosphate-producing) transport of Na$^+_{in}$ to produce Na$^+_{out}$ in Gram-positive bacteria and in yeast.[1-3] The enzyme is considered to be a P-type ATPase that undergoes covalent phosphorylation during the transport cycle. **See also** *ATPase, Na$^+$/K$^+$-Exchanging*

The idea that a transporter is an enzyme is in keeping with a new definition of enzyme catalysis as the facilitated making/breaking of chemical bonds, not just covalent bonds.[4] This idea builds on Pauling's assertion that any long-lived, chemically distinct interaction (in this case, the persistent location of a solute with respect to the faces of a membrane) can be regarded as a chemical bond. Note also that the equilibrium constant (K_{eq} = [Na$^+_{out}$] [ADP][P$_i$]/[Na$^+_{in}$][ATP]) does not conform to that expected for an ATPase (*i.e.*, K_{eq} = [ADP][P$_i$]/[ATP]). Thus, although the overall reaction yields ADP and orthophosphate, the enzyme is misclassified as a hydrolase, and instead should be regarded as an energase-type reaction. Energases facilitate affinity-modulated reactions by coupling the $\Delta G_{ATP-hydrolysis}$ to a force-generating or work-producing step.[4] In this case, P-O-P bond scission supplies the energy to drive sodium transport. **See** *ATPase; Energase*

[1]J. Wieland, A. M. Nitsche, J. Strayle, H. Steiner & H. K. Rudolph (1995) *EMBO J.* **14**, 3870.
[2]P. Catty, A. de Kerchove d'Exaerde & A. Goffeau (1997) *FEBS Lett.* **409**, 325.
[3]J. Cheng, A. A. Guffanti & T. A. Krulwich (1997) *Mol. Microbiol.* **23**, 1107.
[4]D. L. Purich (2001) *TiBS* **26**, 417.

NAJALYSINS

These metalloendopeptidases are isolated from the venom of members of the *Naja* genus of snakes (that is, cobras). The enzyme obtained from *Naja nigricollis* (spitting cobra), also called proteinase F1, catalyzes peptide-bond hydrolysis in fibrinogen (in the Aα chain).[1,2] The enzyme isolated from *Naja mocambique mocambique* catalyzes Tyr10–Asp11 hydrolysis in the P-selectin glycoprotein ligand-1. This venom proteinase, also called mocarhagin, also cleaves the platelet glycoprotein GPIb-V-IX complex at Glu282–Asp283.[3]

[1]H. J. Evans (1984) *BBA* **802**, 49.
[2]H. J. Evans & A. J. Barrett (1988) in *Hemostasis and Animal Venoms* (H. Pirkle & F. S. Markland, eds.), Marcel Dekker, New York, p. 213.

Na$^+$/K$^+$-EXCHANGING ATPase

This so-called ATPase activity [EC 3.6.3.9; previously classified as EC 3.6.1.37] (also known as the sodium/potassium-transporting ATPase and Na$^+$/K$^+$-transporting ATPase) catalyzes the ATP-dependent (ADP- and orthophosphate-producing) countertransport of Na$^+_{in}$, and K$^+_{out}$ to produce Na$^+_{out}$, and K$^+_{in}$.[1-13] Ouabain is a specific inhibitor. This protein complex, ubiquitous in animal cells, is a P-type ATPase that undergoes covalent phosphorylation during the transport cycle.[3] It catalyzes the efflux of three Na$^+$ and influx of two K$^+$ per ATP hydrolyzed and participates in the generation of the plasma membrane electrical potential.

The idea that a transporter is an enzyme is in keeping with a new definition of enzyme catalysis as the facilitated making/breaking of chemical bonds, not just covalent bonds.[14] This idea builds on Pauling's assertion that any long-lived, chemically distinct interaction (in this case, the persistent location of a solute with respect to the faces of a membrane) can be regarded as a chemical bond. Note also that the equilibrium constant (K_{eq} = [Na$^+_{out}$][K$^+_{in}$][ADP][P$_i$]/[Na$^+_{in}$][K$^+_{out}$][ATP]) does not conform to that expected for an ATPase (*i.e.*, K_{eq} = [ADP][P$_i$]/[ATP]). Thus, although the overall reaction yields ADP and orthophosphate, the enzyme is misclassified as a hydrolase, and instead should be regarded as an energase-type reaction. Energases facilitate affinity-modulated reactions by coupling the $\Delta G_{ATP-hydrolysis}$ to a force-generating or work-producing step.[14] In this case, P-O-P bond scission supplies the energy to drive ion transport. **See** *ATPase; Energase*

Nerve stimulation results in a net influx of sodium ions, and normal conditions are restored by the outward transport of sodium ions against an electrochemical gradient. While several earlier workers had identified ATPases in the sheath of giant squid axons, it was Skou[1,2,13] who first connected the Na$^+$/K$^+$-ATPase with the ion flux of neurons. (This discovery culminated in his sharing the Nobel Prize in 1997 with Boyer and Walker who were cited for their studies on ATP synthase.) Kanazawa *et al.*[4] measured the rates of formation and decomposition of the phosphorylated intermediate in the reaction catalyzed by Na$^+$/K$^+$ ATPase. **See also** *ATPase, Na$^+$-Exporting*

[1] J. C. Skou (1957) *BBA* **23**, 394.
[2] J. C. Skou (1960) *BBA* **42**, 6.
[3] R. L. Post, A. K. Sen & A. S. Rosenthal (1965) *JBC* **240**, 1437.
[4] T. Kanazawa, M. Saito & Y. Tonomura (1970) *J. Biochem. (Tokyo)* **67**, 693.
[5] J. D. Robinson & M. S. Flashner (1979) *BBA* **549**, 145.
[6] F. S. Stekhoven & S. L. Bonting (1981) *Physiol. Rev.* **61**, 1.
[7] F. M. A. H. S. Stekhoven & S. L. Bonting (1981) *New Compr. Biochem.* **2**, 159.
[8] L. C. Cantley (1981) *Curr. Top. Bioenerg.* **11**, 201.
[9] M. C. Trachtenberg, D. J. Packey & T. Sweeney (1981) *Curr. Top. Cell. Regul.* **19**, 159.
[10] B. M. Anner (1985) *BBA* **832**, 335.
[11] K. J. Sweadner (1989) *BBA* **988**, 185.
[12] J. G. Norby (1989) *Biochem. Soc. Trans.* **17**, 806.
[13] J. C. Skou (1990) *FEBS Lett.* **268**, 314.
[14] D. L. Purich (2001) *TiBS* **26**, 417.

Selected entries from *Methods in Enzymology* [vol, page(s)]:
General discussion: **10**, 15, 762; **74**, 469; **156**, 1, 29, 43, 46, 48, 80, 379, 417

Assay: **32**, 285; **156**, 36, 51, 105, 109; **191**, 543; avian salt gland, **96**, 638; coupled assay, **156**, 116; erythrocyte inside-out vesicles, in (enzyme assays, **156**, 177; **173**, 372; transport assays, **156**, 175); isolated cells, in, **173**, 676; muscle, in, **173**, 695; parietal cells, in, **192**, 78; reconstituted vesicles, in, **156**, 135; sarcolemmal vesicles, in, **157**, 32

Other molecular properties: acidic pH gel electrophoresis, **156**, 124; active cation pumping, induction by electric field, **157**, 240; active site, **156**, 14 (studies, **36**, 489); active site phosphorylation by inorganic phosphate, assay, **173**, 688; amiloride and analogues, characterization with, **191**, 744; amino acid sequence, **156**, 15; antiserum preparation, **156**, 95; ATP and ADP binding (centrifugation assay, **156**, 200; rate of dialysis assay, **156**, 193); ATP binding during ATP hydrolysis, assay, **157**, 238; avian salt gland (assay, **96**, 638; cytochemical localization with [antibodies, **96**, 646; ouabain, **96**, 643; various substrates, **96**, 640]); 8-azido-ATP, **56**, 646; back-door phosphorylation, **156**, 121 (assay, **156**, 121; data analysis, **156**, 122); bioassay of arachidonic acid metabolites, in, **187**, 378; Ca^{2+}-activated ATPase compared, **32**, 305, 306; cardiac glycoside binding, assay, **156**, 201; **173**, 678; cation binding during ATP hydrolysis, assay, **157**, 233; $C_{12}E_8$-solubilized, incorporation into liposomes (determination of sidedness, **156**, 160; procedure, **156**, 157); circular dichroism studies, **32**, 230; conformational states (eosin-based fluorescence assay, **156**, 278; equilibrium fluorescence assay, **156**, 273; stopped-flow fluorescence assay, **156**, 275); conformational transition, **156**, 4; coupling of chemical reactions and transport, **156**, 17; crystallization, **156**, 16; cultured heart cells, in (characterization, **173**, 639; effects of ouabain and low K^+ levels, **173**, 651; inhibition, correlation with glycoside receptor occupation, **173**, 649; number of molecules, determination, **173**, 643; relationship to positive inotropic action of glycosides, **173**, 655); detergent effect, **32**, 289; detergent-dispersed, preparation, **156**, 335; digitalis binding site (affinity labels, **156**, 323; covalent labeling, **156**, 331; mapping on a subunit, **156**, 332); description, **156**, 7; distribution of, **10**, 762; effects of Na^+ and K^+, **156**, 2; electrolyte transport studies, **192**, 381; endosome marker, as, determination, **109**, 267; erythrocyte inside-out vesicles, in (enzyme assays, **156**, 177; **173**, 372; stoichiometry and coupling, **173**, 377; transport assays, **156**, 175); ESR studies, **156**, 371; front-door phosphorylation, **156**, 121; gel electrophoresis, **32**, 288, 289; identification as Na^+,K^+-pump, **156**, 2; immunoassays, **156**, 120; inactivation and neuropathy, **252**, 150; inactivation by detergents, protective agents, **156**, 76; incorporated into sealed vesicles, density gradient purification, **156**, 139; inhibitors, **156**, 9; isolated cells, in, assay, **173**, 676; isolation, **32**, 277 (incubation, **32**, 281, 282; tissue preparation, **32**, 280, 281); isozyme α(+), preparation from mammalian axolemma, **156**, 65; kinetic properties, **10**, 767 (effects of inhibitors and modifiers, **156**, 248; general considerations, **156**, 237; monovalent cation interactions, **156**, 244; substrate interactions, **156**, 239); labeling with

fluorescein 5′-isothiocyanate, **156**, 272; liberated phosphate, measurement, **156**, 107; lipid requirement for activity, **156**, 12; membrane-bound (crystallization in two dimensions, **156**, 80; electron microscopic analysis, **156**, 83, 430; image analysis of two-dimensional crystals, **156**, 85; preparation, **156**, 81; proteolysis for structural and conformational analyses, **156**, 291; subunit interactions, detection with cross-linking reagents, **156**, 345); methionine-containing peptide purification, **91**, 374; microcalorimetric studies of binding, **61**, 303; microscopic localization (autoradiographic method for ouabain-binding sites, **156**, 428; cytochemical method, **156**, 417; immunocytochemical method, **156**, 429); microsomal fractions, in, **52**, 88; molecular activity, **32**, 286; molecular weight, **156**, 13; monoclonal antibodies (applications, **156**, 407; isotype identification, **156**, 98; preparation, **156**, 96, 98, 394; screening procedures, **156**, 399); muscle, in, **173**, 695; muscle fibers, in, stoichiometric ratio, **173**, 707; NMR studies (nuclear relaxation with paramagnetic probes, **156**, 355; transferred nuclear Overhauser enhancement measurements, **156**, 367); ouabain binding (assay in intact cells, **156**, 219, 223; calcium effects, **156**, 223; capacity, **32**, 287, 288; equilibrium binding conditions, **156**, 219; equilibrium binding constant, **156**, 219, 221; potassium effects, **156**, 222; specificity, **156**, 215); parietal cells, in, assay, **192**, 78; phosphatase activities, **156**, 7 (K^+-dependent, assay, **156**, 109; Na^+/K^+-activated, assay, **156**, 110); phosphoenzyme intermediate, **64**, 76; phosphorylation effect, **252**, 149, 151; photoaffinity labeling (applications, **156**, 319; procedures, **156**, 317); physiological function, **156**, 1; ^{32}P labeling of, **10**, 773; potassium, sodium, and rubidium binding (centrifugation assay, **156**, 230; characteristics, **156**, 235; criteria for specific binding, **156**, 233); preparative problems (inactivation by detergent, **156**, 31; latency in plasma membrane preparations, **156**, 30; purity and integrity criteria, **156**, 31); properties, **32**, 279; **36**, 438, 439; **156**, 60; purification, **10**, 765; **32**, 282 (brine shrimp, from, **156**, 48; duck salt gland, from, **156**, 46; mammalian kidney, from, **156**, 29, 88; overview, **156**, 10, 16; shark rectal gland, from, **156**, 43); purity criteria (molar activity, **156**, 112; specific activity, **156**, 110; subunit purity, **156**, 114); quaternary structure determination, **156**, 333; rate measurement with radioactive tracers, **173**, 700; rat parotid gland, activity in plasmalemma and secretion granule membranes, **98**, 85; reaction with ATP, **156**, 5; reconstitution by (cholate dialysis, **156**, 127; freeze-thaw sonication, **156**, 141); release of occluded ions, measurement technique, **156**, 281; solubilization, **156**, 72; solubilized, purification and analysis, **156**, 77; stability after purification, **156**, 43; steroid effects, **36**, 434; α subunit (folding through membrane and functional domain identification, **156**, 300; gene cloning with antibody probes, **156**, 379; orientation of proteolytic fragments, **156**, 298); subunits (isolation, **156**, 90; peptide preparation from, **156**, 91; preparation for immunization, **156**, 95); sulfhydryl group modification (assays, **156**, 309; procedures, **156**, 307); thioinosine triphosphate derivatives as affinity labels, **156**, 310; transport assays, **156**, 120, 136; transport characteristics in erythrocytes, **173**, 80; turnover rate, measurement, **173**, 712; turtle colon basolateral membrane, in, analysis, **192**, 740; zonal centrifugation, **32**, 283, 284, 289

NAPHTHALENE 1,2-DIOXYGENASE

This oxidoreductase [EC 1.14.12.12] catalyzes the reaction of naphthalene with NADH and dioxygen to produce (1R,2S)-1,2-dihydronaphthalene-1,2-diol and NAD^+.[1-5] This protein complex contains an iron-sulfur FAD-dependent reductase, an iron-sulfur oxygenase, and ferredoxin.

1,2-Dihydronaphthalene can also be oxidized, producing both the 1,2-dihydro- and 1,2,3,4-tetrahydrodiol products. Other substrates include indole and indane.

naphthalene   (1*R*,2*S*)-1,2-dihydronaphthalene-1,2-diol

[1] A. M. Jeffrey, H. J. Yeh, D. M. Jerina, T. R. Patel, J. F. Davey & D. T. Gibson (1975) *Biochemistry* **14**, 575.

[2] B. E. Haigler & D. T. Gibson (1990) *J. Bacteriol.* **172**, 457 and 465.

[3] R. E. Parales, J. V. Parales & D. T. Gibson (1999) *J. Bacteriol.* **181**, 1831.

[4] R. E. Parales, K. Lee, S. M. Resnick, H. Jiang, D. J. Lessner & D. T. Gibson (2000) *J. Bacteriol.* **182**, 1641.

[5] B. G. Fox (1998) *CBC* **3**, 261.

Selected entries from *Methods in Enzymology* [vol, page(s)]:
General discussion: **188**, 46

NAPHTHOATE SYNTHASE

This highly specific enzyme [EC 4.1.3.36], a component of the menaquinone biosynthetic pathway, catalyzes the conversion of *O*-succinylbenzoyl-CoA to 1,4-dihydroxy-2-naphthoate and coenzyme A.[1] The second step in the ring closure reaction is rate-limiting.[2]

[1] R. Meganathan & R. Bentley (1979) *J. Bacteriol.* **140**, 92.

[2] U. Igbavboa & E. Leistner (1990) *EJB* **192**, 441.

NARDILYSIN

This metalloendopeptidase[1] [EC 3.4.24.61], also known as N-arginine dibasic convertase and NRD convertase, catalyzes peptide bond hydrolysis in polypeptides. The enzyme has a preference for the Xaa–Arg bond in the motif Xaa-Arg-Lys. The enzyme also acts at Arg–Arg, Lys–Arg, and Arg–Lys. Nardilysin cleaves somatostatin-28, dynorphin A, dynorphin B, preproneurotensin, and α-neoendorphin. Interestingly, this metalloenzyme is inhibited by bestatin, amastin, and ethylmaleimide.[2–4]

[1] The unusual name is an acronym describing enzyme's specificity for the N-terminal side of an arginyl residue in dibasic sites.

[2] A. R. Pierotti, A. Prat, V. Chesneau, F. Gaudoux, A.-M. Leseney, T. Foulon & P. Cohen (1994) *PNAS* **91**, 6078.

[3] V. Chesneau, A. R. Pierotti, N. Barré, C. Créminon, C. Tougard & P. Cohen (1994) *JBC* **269**, 2056.

[4] V. Chesneau, A. Prat, D. Segretain, V. Hospital, A. Dupaix, T. Foulon, B. Jégou & P. Cohen (1996) *J. Cell Sci.* **109**, 2737.

Selected entries from *Methods in Enzymology* [vol, page(s)]:
General discussion: **248**, 703
Assay: 248, 704
Other molecular properties: activity, **248**, 703; aminopeptidase B activity and, **248**, 703; cDNA cloning, **248**, 713; cDNA sequencing, **248**, 713; cleavage specificity, **248**, 703; discovery, **248**, 703; distribution, **248**, 715; inhibitor profile, **248**, 708; isoelectric point, **248**, 714; pH profile, **248**, 708; pitrilysin family and, **248**, 716; properties, **248**, 214, 684, 708; purification, **248**, 706; reactivation by cations, **248**, 708; structure, **248**, 711; substrate specificity, **248**, 710

NARINGENIN-CHALCONE SYNTHASE

This enzyme [EC 2.3.1.74], also known as chalcone synthase, flavonone synthase, and 6′-deoxychalcone synthase, catalyzes the reaction of three molecules of malonyl-CoA with 4-coumaroyl-CoA to produce four molecules of coenzyme A, naringenin-chalcone, and three molecules of carbon dioxide.[1–4] 4,2′,4′,6′-Tetrahydroxychalcone is generated and spontaneously converted to the flavanone naringenin. In the presence of NADH and a reductase, 6′-deoxychalcone is produced. The enzyme from certain sources also catalyzes the reaction of caffeoyl-CoA with malonyl-CoA to produce eriodictyol, coenzyme A, and carbon dioxide.

[1] C. R. Martin (1993) *Int. Rev. Cytol.* **147**, 233.

[2] G. Hrazdina, E. Lifson & N. F. Weeden (1986) *ABB* **247**, 414.

[3] W. Knogge, E. Schmelzer & G. Weissenbröck (1986) *ABB* **250**, 364.

[4] R. A. Dixon, P. M. Dey & C. J. Lamb (1983) *AE* **55**, 1.

NARINGENIN 3-DIOXYGENASE

This iron- and ascorbate-dependent oxidoreductase [EC 1.14.11.9], also known as flavanone 3-dioxygenase and flavanone 3β-hydroxylase, catalyzes the reaction of (2*S*)-naringenin with α-ketoglutarate (or, 2-oxoglutarate) and dioxygen to produce (2*R*,3*R*)-dihydrokaempferol, succinate, and carbon dioxide.[1–3]

naringenin   dihydrokaempferol

Other substrates include (2*S*)-eriodictyol, producing (2*R*,3*R*)-dihydroquercitin.

[1] A. G. Prescott & P. John (1996) *Ann. Rev. Plant Physiol. Plant Mol. Biol.* **47**, 245.

[2] L. Britsch (1990) *ABB* **282**, 152.

[3] L. Britsch & H. Grisebach (1986) *EJB* **156**, 569.

*Nar*I RESTRICTION ENDONUCLEASE

This type II site-specific endonuclease [EC 3.1.21.4], isolated from *Nocardia argentinensis*, catalyzes the hydrolysis of both strands of DNA at 5′ . . . GG∇CGCC . . . 3′. *Nar*I requires simultaneous interaction with two copies of the recognition sequence before cleavage occurs. In addition, the endonuclease exhibits marked site preferences.

Na$^+$-TRANSPORTING TWO-SECTOR ATPase

This so-called ATPase [EC 3.6.3.15], also known as vacuolar-type Na$^+$-ATPase, Na$^+$-translocating ATPase, Na$^+$-translocating F$_1$F$_o$-ATPase, and vacuolar-type Na$^+$-translocating ATPase, is a multisubunit non-phosphorylated enzyme that catalyzes the ATP-dependent (ADP- and orthophosphate-producing) transport of sodium cations.[1–3] The complex found in certain bacteria (for example, *Enterococcus hirae* and *Acetobacterium woodii*) is similar to H$^+$-transporting two-sector ATPase (better known as ATP synthase, H$^+$-transporting) [EC 3.6.3.14] where the sodium ion has taken the place of the proton.

The idea that a transporter is an enzyme is in keeping with a new definition of enzyme catalysis as the facilitated making/breaking of chemical bonds, not just covalent bonds.[4] This idea builds on Pauling's assertion that any long-lived, chemically distinct interaction (in this case, the persistent location of a solute with respect to the faces of a membrane) can be regarded as a chemical bond. Note also that the equilibrium constant ($K_{eq} = $ [Na$^+_{out}$][ADP][P$_i$]/[Na$^+_{in}$][ATP]) does not conform to that expected for an ATPase (*i.e.*, $K_{eq} = $ [ADP][P$_i$]/[ATP]). Thus, although the overall reaction yields ADP and orthophosphate, the enzyme is misclassified as a hydrolase, and instead should be regarded as an energase-type reaction. Energases facilitate affinity-modulated reactions by coupling the $\Delta G_{ATP\text{-hydrolysis}}$ to a force-generating or work-producing step.[4] In this case, P-O-P bond scission supplies the energy to drive sodium ion transport. *See ATPase; Energase*

[1] M. Solioz & K. Davies (1994) *JBC* **269**, 9453.
[2] K. Takase, S. Kakinuma, I. Yamato, K. Konishi, K. Igarashi & Y. Kanikuma (1994) *JBC* **269**, 11037.
[3] S. Rahlfs & V. Müller (1997) *FEBS Lett.* **404**, 269.
[4] D. L. Purich (2001) *TiBS* **26**, 417.

*Nci*I RESTRICTION ENDONUCLEASE

This type II site-specific endonuclease [EC 3.1.21.4], isolated from *Neisseria cinerea*, catalyzes the hydrolysis of both strands of DNA at 5′...CC∇SGG...3′, where S refers to G or C.[1] The DNA fragments have a single base 5′ extension which is more difficult to ligate than blunt-ended fragments. The enzyme also exhibits star activity. Cleavage is not blocked by CpG methylation, although the reaction rates are affected.[2] *See "Star" Activity*

[1] R. Watson, M. Zuker, S. M. Martin & L. P. Visentin (1980) *FEBS Lett.* **118**, 47.
[2] R. R. Meehan, E. Ulrich & A. P. Bird (1993) *NAR* **21**, 5517.

*Nco*I RESTRICTION ENDONUCLEASE

This type II site-specific endonuclease [EC 3.1.21.4], isolated from *Nocardia corallina*, catalyzes the hydrolysis of both strands of DNA at 5′...C∇CATGG...3′.[1] The enzyme also exhibits star activity. *See "Star" Activity*

[1] R. J. Watson, I. Schildkraut, B.-Q. Qiang, S. M. Martin & L. P. Visentin (1982) *FEBS Lett.* **150**, 114.

*Nde*I RESTRICTION ENDONUCLEASE

This type II site-specific endonuclease [EC 3.1.21.4], isolated from *Neisseria denitrificans*, catalyzes the hydrolysis of both strands of DNA at 5′...CA∇TATG...3′. The enzyme also exhibits star activity. *See "Star" Activity*

NEOLACTOTETRAOSYLCERAMIDE α-2,3-SIALYLTRANSFERASE

This sialyltransferase [EC 2.4.99.10] catalyzes the reaction of CMP-*N*-acetylneuraminate with a β-D-galactosyl-1,4-*N*-acetyl-β-D-glucosaminyl-1,3-β-D-galactosyl-1,4-β-D-glucosylceramide (also known as neolactotetraosylceramide or ganglioside LC4) to produce CMP and an α-*N*-acetylneuraminyl-2,3-β-D-galactosyl-1,4-*N*-acetyl-β-D-glucosaminyl-1,3-β-D-galactosyl-1,4-β-D-glucosylceramide (or, ganglioside LM1). The enzyme will also catalyze the synthesis of ganglioside G$_{M3}$ from ganglioside G$_{M1}$.[1–4]

[1] M. Basu, S. Basu, A. Stoffyn & P. Stoffyn (1982) *JBC* **257**, 12765.
[2] H. Higashi, M. Basu & S. Basu (1985) *JBC* **260**, 824.
[3] S. K. Basu, R. L. Whisler & A. J. Yates (1986) *Biochemistry* **25**, 2577.
[4] S. Dasgupta, J. L. Chien & E. L. Hogan (1986) *BBA* **876**, 363.

Selected entries from *Methods in Enzymology* [**vol**, page(s)]:
General discussion: 8, 365
Assay: 8, 366

(+)-NEOMENTHOL DEHYDROGENASE

This oxidoreductase [EC 1.1.1.208], also known as monoterpenoid dehydrogenase, catalyzes the reaction of (−)-menthone with NADPH to produce (+)-neomenthol and NADP$^+$.[1,2]

Other substrates include (+)-isomenthone and other cyclohexanones. This enzyme is distinct from that of (−)-menthol dehydrogenase [EC 1.1.1.207], even though the substrates are identical.

(−)-menthone   (+)-neomenthol

[1] R. Kjonaas, C. Martinkus-Taylor & R. Croteau (1982) *Plant Physiol.* **69**, 1013.

[2] R. Croteau & J. N. Winters (1982) *Plant Physiol.* **69**, 975.

NEOPULLULANASE

This glycosidase [EC 3.2.1.135] catalyzes the hydrolysis of pullulan, a linear glucan of maltotriosyl units linked through $\alpha(1 \to 6)$-glucosidic linkages, to produce panose (6-α-D-glucosylmaltose). Reportedly, the enzyme functions through three steps: (1) the enzyme acts only on $\alpha(1 \to 4)$-glucosidic linkages on the nonreducing side of $\alpha(1 \to 6)$ linkages of pullulan and produces panose and several intermediate oligomers of panose; (2) the enzyme hydrolyzes either $\alpha(1 \to 4)$ or $\alpha(1 \to 6)$ linkages in 6(2)-*O*-α-(6(3)-*O*-α-glucosylmaltotriosyl)-maltose to produce panose plus maltose; and (3) the enzyme hydrolyzes $\alpha(1 \to 4)$ linkage of 6(3)-*O*-α-glucosyl-maltotriose to yield glucose and another panose.[1]

A single active site on the enzyme participates in the dual activity toward $\alpha(1 \to 4)$- and $\alpha(1 \to 6)$-glucosidic linkages.[2] This same site also catalyzes transglycosylation reactions: to form $\alpha(1 \to 4)$-glucosidic linkages and to form $\alpha(1 \to 6)$-glucosidic linkages.[3] ***See also*** *α-Dextrin Endo-1,6-α-Glucosidase (formerly Pullulanase); Isopullulanase*

[1] T. Imanaka & T. Kuriki (1989) *J. Bacteriol.* **171**, 369.

[2] T. Kuriki, H. Takata, S. Okada & T. Imanaka (1991) *J. Bacteriol.* **173**, 6147.

[3] H. Takata, T. Kuriki, S. Okada, Y. Takesada, M. Iizuka, N. Minamiura & T. Imanaka (1992) *JBC* **267**, 18447.

NEPENTHESIN

This aspartic endopeptidase [EC 3.4.23.12], also called *Nepenthes* aspartic proteinase, catalyzes peptide-bond hydrolysis in a various proteins, including casein, fibrin, and ovalbumin.[1–4] Nepenthesin has a broad specificity but prefers to cleave at Xaa–Asp and Asp–Xaa bonds. This endopeptidase also acts at Lys–Arg, Ala–Xaa, Xaa–Ala, and Leu–Xaa, and cleaves the following bonds in oxidized insulin B chain: Leu6–Cys7, Glu13–Ala14, Leu15–Tyr16, Tyr16–Leu17, and Phe24–Phe25.

[1] F. E. Lloyd (1942) *The Carnivorous Plants*, Ronald Press, New York.

[2] S. Amagase, S. Nakayama & A. Tsugita (1969) *J. Biochem.* **66**, 431.

[3] S. Amagase (1972) *J. Biochem.* **72**, 73.

[4] J. F. Woessner (1998) in *Handbook of Proteolytic Enzymes* (A. J. Barrett, N. D. Rawlings & J. F. Woessner, eds.), Academic Press, New York, p. 846.

NEPRILYSIN

This zinc-dependent endopeptidase [EC 3.4.24.11], known variously as membrane metalloendopeptidase, neutral endopeptidase [NEP], kidney brush-border neutral proteinase, and enkephalinase, catalyzes the hydrolysis of peptide bonds on polypeptides consisting of forty or fewer aminoacyl residues. The principal substrates of this membrane-bound glycoprotein appear to be substance P and related peptides, the enkephalins, and members of the atrial natriuretic family.[1,2] This enzyme represents 4% of the total protein in renal brush border membranes.

The best substrates have a leucyl residue in the P1′ position. The smallest peptide substrate identified so far is *N*-formylmethionylleucylphenylalanine which is cleaved at Met–Leu. Thiorphan is a potent inhibitor ($K_i = 2.3$ nM) as is phosphoramidon.

[1] E. G. Erdös & R. A. Skidgel (1989) *FASEB J.* **3**, 145.

[2] B. P. Roques, F. Noble, V. Daugé, M.-C. Fourié-Zaluski & A. Beaumont (1993) *Pharmacol. Rev.* **45**, 87.

Selected entries from ***Methods in Enzymology*** [vol, page(s)]:
General discussion: 248, 253, 263
Assay: fluorimetric, 248, 257, 602; quenched fluorescent substrate method, 248, 258; radiolabeled enkephalin substrate method, 248, 255; synthetic chromogenic substrate method, 248, 257

Other molecular properties: active site, substrate binding, 248, 264; activity, 248, 253, 263, 282, 595; amino acid sequence, 248, 264; cellular localization, 248, 254; detergent-solubilized, purification, 248, 260; discovery, 248, 253; distribution, 248, 188; family M13, 248, 183–186, 188, 227, 592; functional residues, identification, 248, 262; homologs, 248, 590; inactivation of atrial natriuretic peptide, inhibition, 248, 281; inhibitors (analgesic responses induced by, 248, 272; antinociceptive effects, 248, 272; behavioral effects, 248, 279; clinical applications, 248, 282; dependence, 248, 276; development, 248, 266; mechanism of action, 248, 266; opioid effects, mechanism of action, 248, 266; properties, 248, 264, 264, 270; side effects, 248, 276; tolerance, 248, 276); *Lactococcus*, 248, 580; mechanism of action, 248, 264; papain-solubilized, purification, 248, 259; properties, 248, 188, 253; purification, 248, 259; structure, 248, 186, 188, 253, 264; substrate (fluorigenic, 248, 34; physiological, 248, 272); substrate specificity, 248, 254, 264, 588; tissue distribution, 248, 254, 282; transition-state and multisubstrate analogues, 249, 307

NEUROLYSIN

This endopeptidase [EC 3.4.24.16], also known as neurotensin endopeptidase, oligopeptidase M, and endopeptidase 24.16, catalyzes the hydrolysis of peptide bonds in oligopeptides. All known peptide substrates contain seventeen or fewer aminoacyl residues. Interestingly, dynorphin A$_{1-17}$, which is an inhibitor of thimet oligopeptidase [EC 3.4.24.15], is cleaved at Pro10–Tyr11 by neurolysin.[1,2]

[1] A. J. Barrett, M. A. Brown, P. M. Dando, C, G. Knight, N. McKie, N. D. Rawlings & A. Serizawa (1995) *MIE* **248**, 529.
[2] F. Checler, H. Barelli, P. Dauch, V. Dive, B. Vincent & J. P. Vincent (1995) *MIE* **248**, 593.

Selected entries from **Methods in Enzymology** [vol, page(s)]:
General discussion: 248, 529, 593
Assay: 248, 535, 594, 603 (fluorimetric, 248, 607 [in whole tissue homogenates, 248, 613; HPLC, by, 248, 603, 607; substrate {Mcc-Pro-Leu-Gly-Pro-D-Lys-Dnp (QFS) as, 248, 606; neurotensin as, 248, 603}])
Other molecular properties: activity, 248, 593; amino acid composition, 248, 542; amino acid sequence, 248, 543; cerebral regionalization, 248, 611; discovery, 248, 531; distribution, 248, 554; dynorphin, interactions, and, 248, 549; evolution, 248, 551; homologues, 248, 551; homology with mitochondrial intermediate peptidase, 248, 531; inhibitors, 248, 548, 609; intracellular localization, 248, 531, 554; ontogeny, 248, 613; physiological functions, 248, 593; properties, 248, 206, 531, 532; purification, 248, 594 (rat brain synaptic membranes, from, 248, 595; rat liver mitochondria, from, 248, 539; whole tissue homogenates, from, 248, 600); structure, 248, 542; substrate specificity, 248, 536, 547; thiol activation, 248, 546; tissue distribution, 248, 593, 613

*Ngo*MIV RESTRICTION ENDONUCLEASE

This type II site-specific endonuclease [EC 3.1.21.4], isolated from *Neisseria gonorrhoeae* MS11, catalyzes the hydrolysis of both strands of DNA at 5′...G∇CCGGC...3′. The enzyme also exhibits star activity. This enzyme was previously called *Ngo*MI restriction endonuclease. *Ngo*MIV is the first tetrameric endonuclease to have the enzyme–DNA complex crystal structure determined. Two primary dimers are arranged back to back with two oligonucleotides bound in clefts on opposite sides of the tetramer.[1] **See** *"Star" Activity*

[1] M. Deibert, S. Grazulis, G. Sasnauskas, V. Siksnys & R. Huber (2000) *Nat. Struct. Biol.* **7**, 792.

NICKEL-TRANSPORTING ATPase

This so-called ATPase [EC 3.6.3.24] catalyzes the ATP-dependent (ADP- and orthophosphate-producing) transport of Ni$^{2+}_{out}$ to produce Ni$^{2+}_{in}$.[1-3] The nickel-transporting enzyme is considered to be a ABC-type (ATP-binding cassette-type) ATPase that has two similar ATP-binding domains and does not undergo phosphorylation during the transport process.

The idea that a transporter is an enzyme is in keeping with a new definition of enzyme catalysis as the facilitated making/breaking of chemical bonds, not just covalent bonds.[4] This idea builds on Pauling's assertion that any long-lived, chemically distinct interaction (in this case, the persistent location of a solute with respect to the faces of a membrane) can be regarded as a chemical bond. Note also that the equilibrium constant (K_{eq} = [Ni$^{2+}_{in}$] [ADP][P$_i$]/[Ni$^{2+}_{out}$][ATP]) does not conform to that expected for an ATPase (*i.e.*, K_{eq} = [ADP][P$_i$]/[ATP]). Thus, although the overall reaction yields ADP and orthophosphate, the enzyme is misclassified as a hydrolase, and instead should be regarded as an energase-type reaction. Energases facilitate affinity-modulated reactions by coupling the $\Delta G_{ATP-hydrolysis}$ to a force-generating or work-producing step.[4] In this case, P-O-P bond scission supplies the energy to drive Ni^{2+} transport. **See** *ATPase; Energase*

[1] J. K. Hendricks & H. L. Mobley (1997) *J. Bacteriol.* **179**, 5892.
[2] M. H. Saier, Jr. (1998) *Adv. Microb. Physiol.* **40**, 81.
[3] J. K. Griffiths & C. E. Sansom (1998) *The Transporter Factsbook*, Academic Press, San Diego.
[4] D. L. Purich (2001) *TiBS* **26**, 417.

*Nhe*I RESTRICTION ENDONUCLEASE

This type II site-specific endonuclease [EC 3.1.21.4], isolated from *Neisseria mucosa heidelbergensis*, catalyzes the hydrolysis of both strands of DNA at 5′...G∇CTAGC...3′. The enzyme generates DNA fragments that have a 5′ CTAG extension that is easily ligated to fragments generated by *Avr*II, *Spe*I, or *Xba*I. The enzyme also exhibits star activity. **See** *"Star" Activity*

NICOTIANAMINE SYNTHASE

nicotianamine

This enzyme [EC 2.5.1.43] catalyzes the reaction of three molecules of *S*-adenosyl-L-methionine to produce three molecules of 5′-*S*-methyl-5′-thioadenosine and nicotianamine.[1]

[1] K. Higuchi, K. Kanazawa, N.-K. Nishizawa, M. Chino & S. Mori (1994) *Plant and Soil* **165**, 173.

NICOTINAMIDASE

This enzyme [EC 3.5.1.19] catalyzes the hydrolysis of nicotinamide to produce nicotinate and ammonia.[1,2]

[1]G. Magni, A. Amici, M. Emanuelli, N. Raffaelli & S. Ruggieri (1999) AE **73**, 135.

[2]L. M. Henderson (1983) Ann. Rev. Nutr. **3**, 289.

Selected entries from *Methods in Enzymology* [vol, page(s)]:
General discussion: **18B**, 33, 180, 185; **66**, 132
Assay: **18B**, 33, 34, 180, 184, 185; **66**, 132, 133
Other molecular properties: assay of nicotinamide, **66**, 3, 4; *Flavobacterium peregrinum*, preparation, **66**, 133; hydrolysis of nicotinamide to nicotinate, **66**, 120; mammalian tissues (assays, **18B**, 33, 34, 185; properties and activities, **18B**, 34, 35; **66**, 136; tissue extract, **18B**, 33); rabbit liver, **18B**, 189 (properties, **18B**, 191, 192; purification, **18B**, 189); *Saccharomyces cerevisiae*, preparation, **66**, 3, 4; *Torula cremoris*, **18B**, 180 (activators and inhibitors, **18B**, 184, 185; assay, **18B**, 180, 181; properties, **18B**, 183, 184; purification, **18B**, 181)

NICOTINAMIDE *N*-METHYLTRANSFERASE

This methyltransferase [EC 2.1.1.1] catalyzes the reaction of *S*-adenosyl-L-methionine with nicotinamide to produce *S*-adenosyl-L-homocysteine and 1-methylnicotinamide.[1–3] Other acceptor substrates include thionicotinamide, 3-acetylpyridine, quinoline, isoquinoline, and 4-methylnicotinamide, with the latter affording a sensitive fluorometric assay.[4]

[1]T. A. Alston & R. H. Abeles (1988) ABB **260**, 601.

[2]G. L. Cantoni (1951) JBC **189**, 203.

[3]J. Rini, C. Szumlanski, R. Guerciolini & R. M. Weinshilboum (1990) Clin. Chim. Acta **186**, 359.

[4]A. Sano, N. Takimoto, & S. Takitani (1989) Chem. Pharm. Bull. (Tokyo) **37**, 3330.

Selected entries from *Methods in Enzymology* [vol, page(s)]:
General discussion: **2**, 257
Assay: **2**, 257

NICOTINAMIDE MONONUCLEOTIDE ADENYLYLTRANSFERASE

This transferase [EC 2.7.7.1], also known as NAD+ pyrophosphorylase, catalyzes the reaction of ATP with nicotinamide ribonucleotide to produce NAD+ and pyrophosphate.[1–4] Nicotinate nucleotide and 3-acetyl-pyridine-nicotinamide ribonucleotide are also substrates. The human placental enzyme has an ordered Bi Bi kinetic mechanism.[2] The enzyme reaction proceeds with inversion of configuration, such that the adenylyl group is transferred directly from ATP by means of an in-line mechanism.[3]

Nicotinate-nucleotide adenylyltransferase [EC 2.7.7.18] uses nicotinate ribonucleotide as the pyridine substrate, thereby producing deamido-NAD+ and pyrophosphate.

[1]G. Magni, N. Raffaelli, M. Emanuelli, A. Amici, P. Natalini & S. Ruggieri (1998) AE **73**, 240.

[2]M. Emanuelli, F. Carnevali, F. Saccucci, F. Pierella, A. Amici, & G. Magni (2001) JBC **276**, 406.

[3]G. Lowe & G. Tansley (1983) EJB **132**, 117.

[4]M. R. Atkinson, J. F. Jackson & R. K. Morton (1961) BJ **80**, 318.

Selected entries from *Methods in Enzymology* [vol, page(s)]:
General discussion: **2**, 670; **18B**, 20, 127; **280**, 211, 241, 248
Assay: **2**, 670; **18B**, 40, 41, 131, 132
Other molecular properties: *Lactobacillus fructosus* (assay, **18B**, 131, 132; properties, **18B**, 132; purification, **18B**, 132; use in synthesis of radioactive NAD, **18B**, 62, 63, 65, 66); localization, **12A**, 425; *Methanococcus jannaschii* (assay, **331**, 294; isolation, **331**, 294); nuclei, in, **31**, 253, 259, 261, 262; nucleoli, in, **10**, 11; **12A**, 457; rat liver (assay, **2**, 670; **18B**, 40, 41; properties and activities, **2**, 672; **18B**, 41; purification, **2**, 671); *Sulfolobus solfataricus* (assay, *331*, 283; isolation, *331*, 285); tissue extract, **18B**, 40

NICOTINAMIDE-NUCLEOTIDE AMIDASE

This enzyme [EC 3.5.1.42], also known as NMN deamidase, catalyzes the hydrolysis of β-nicotinamide D-ribonucleotide to produce β-nicotinate D-ribonucleotide and ammonia.[1–3] Other substrates include β-nicotinamide D-ribonucleoside, albeit more slowly.

[1]G. Magni, A. Amici, M. Emanuelli, N. Raffaelli & S. Ruggieri (1999) AE **73**, 135.

[2]T. Imai (1973) J. Biochem. **73**, 139.

[3]M. Kuwahara, Y. Ishida & M. Okatani (1983) J. Ferment. Technol. **61**, 61.

Selected entries from *Methods in Enzymology* [vol, page(s)]:
General discussion: **18B**, 192

NICOTINAMIDE PHOSPHORIBOSYLTRANSFERASE

This transferase [EC 2.4.2.12], also known as NMN pyrophosphorylase, catalyzes the reaction of nicotinamide with 5-phospho-α-D-ribose 1-diphosphate to produce nicotinamide D-ribonucleotide and pyrophosphate.[1–5]

[1]M. Rocchigiani, V. Micheli, J. A. Duley & H. A. Simmonds (1992) Anal. Biochem. **205**, 334.

[2]W. D. L. Musick (1981) Crit. Rev. Biochem. **11**, 1.

[3]L.-F. H. Lin, S. J. Lan, A. H. Richardson & L. V. M. Henderson (1972) JBC **247**, 8016.

[4]L. S. Dietrich & O. Muniz (1972) Biochemistry **11**, 1691.

[5]J. Preiss & P. Handler (1957) JBC **225**, 759.

Selected entries from *Methods in Enzymology* [vol, page(s)]:
General discussion: **18B**, 20, 127, 144; **280**, 211
Assay: **18B**, 35, 36, 127, 128
Other molecular properties: *Lactobacillus fructosus* (assay, **18B**, 127, 128; properties, **18B**, 129; purification, **18B**, 128; use in synthesis of radioactive NAD, **18B**, 65, 66); rat tissue, **18B**, 35 (assay, **18B**, 35, 36; properties and activities, **18B**, 36, 37; tissue extract, **18B**, 35)

NICOTINATE DEHYDROGENASE

This iron- and flavin-dependent oxidoreductase [EC 1.5.1.13], also known as nicotinate hydroxylase, catalyzes the reaction of nicotinate with NADP+ and water to produce 6-hydroxynicotinate and NADPH.[1–4] Other substrates include NADPH. The *Clostridium barkeri* enzyme contains a molybdenum-selenium center.[1]

[1]V. N. Gladyshev, S. V. Khangulov & T. C. Stadtman (1996) *Biochemistry* **35**, 212.
[2]T. C. Stadtman (1980) *Ann. Rev. Biochem.* **49**, 93.
[3]J. S. Holcenberg & E. R. Stadtman (1969) *JBC* **244**, 1194.
[4]M. Nagel & J. R. Andreesen (1990) *Arch. Microbiol.* **154**, 605.

Selected entries from *Methods in Enzymology* [**vol**, page(s)]:
General discussion: 18B, 233

NICOTINATE GLUCOSYLTRANSFERASE

This transferase [EC 2.4.1.196] catalyzes the reversible reaction of UDP-D-glucose with nicotinate to produce UDP and *N*-glucosylnicotinate. The reverse reaction releases nicotinate for use in NAD^+ and $NADP^+$ biosynthesis.[1,2]

[1]B. Upmeier, J. E. Thomzik & W. Barz (1988) *Z. Naturforsch.* **43c**, 835.
[2]S. Köster, B. Upmeier, D. Komossa & W. Barz (1989) *Z. Naturforsch.* **44c**, 623.

NICOTINATE *N*-METHYLTRANSFERASE

This methyltransferase [EC 2.1.1.7] catalyzes the reaction of *S*-adenosyl-L-methionine with nicotinate (or, pyridine-3-carboxylate), the only known methyl group acceptor for this enzyme to produce *S*-adenosyl-L-homocysteine and 1-methylnicotinate (or, trigonelline).[1,2] Nicotinamide is not a substrate.[2]

[1]J. G. Joshi & P. Handler (1960) *JBC* **235**, 2981.
[2]B. Upmeier, W. Gross, S. Koster & W. Barz (1988) *ABB* **262**, 445.

NICOTINATE-NUCLEOTIDE ADENYLYLTRANSFERASE

This enzyme [EC 2.7.7.18], also known as nicotinate-mononucleotide adenylyltransferase and deamido-NAD^+ pyrophosphorylase, catalyzes the reaction of ATP with nicotinate ribonucleotide to produce deamido-NAD^+ and pyrophosphate.[1,2] Nicotinamide and 3-acetylpyridine ribonucleotide are also substrates.

In some organisms, this adenylyltransferase and nicotinamide-mononucleotide adenylyltransferase [EC 2.7.7.1] may be activities of the same enzyme.

[1]J. Imsande (1961) *JBC* **236**, 1494.
[2]W. Dahmen, B. Webb & J. Preiss (1967) *ABB* **120**, 440.

Selected entries from *Methods in Enzymology* [**vol**, page(s)]:
General discussion: 6, 349; **18B**, 20, 38; **280**, 211.
Assay: 6, 349; **18B**, 39

NICOTINATE-NUCLEOTIDE:DIMETHYL-BENZIMIDAZOLE PHOSPHORIBOSYLTRANSFERASE

This transferase [EC 2.4.2.21], also known as CobT, catalyzes the reaction of β-nicotinate D-ribonucleotide with dimethylbenzimidazole to produce nicotinate and N^1-(5-phospho-α-D-ribosyl)-5,6-dimethylbenzimidazole.[1-4] This enzyme catalyzes a key step in the biosynthesis of the cobalamins. Benzimidazole, 5,6-dichlorobenzimidazole, and 5(6)-nitrobenzimidazole also serve as acceptor substrates. The *Clostridium sticklandii* enzyme acts on adenine to produce 7-α-D-ribosyladenine 5′-phosphate.

[1]H. C. Friedmann (1965) *JBC* **240**, 413.
[2]J. A. Fyfe & H. C. Friedmann (1969) *JBC* **244**, 1659.
[3]H. C. Friedmann & J. A. Fyfe (1969) *JBC* **244**, 1667.
[4]B. Cameron, F. Blanche, M.-C. Rouyez, D. Bisch, A. Famechon, M. Couder, L. Cauchois, D. Thibaut, L. Debussche & J. Crouzet (1991) *J. Bacteriol.* **173**, 6066.

Selected entries from *Methods in Enzymology* [**vol**, page(s)]:
General discussion: 18B, 197
Assay: 18B, 197

NICOTINATE-NUCLEOTIDE PYROPHOS-PHORYLASE (CARBOXYLATING)

This enzyme [EC 2.4.2.19], also known as nicotinate-nucleotide diphosphorylase (carboxylating) and quinolinate phosphoribosyltransferase (decarboxylating), catalyzes the reversible reaction of nicotinate D-ribonucleotide with pyrophosphate (or, diphosphate) and carbon dioxide to produce pyridine-2,3-dicarboxylate (or, quinolinate) and 5-phospho-α-D-ribose 1-diphosphate. The enzyme has an ordered Bi Bi kinetic mechanism.[1-4]

[1]G. Magni, A. Amici, M. Emanuelli, N. Raffaelli & S. Ruggieri (1999) *AE* **73**, 135.
[2]R. Bhatia & K. C. Calvo (1996) *ABB* **325**, 270.
[3]L. M. Henderson (1983) *Ann. Rev. Nutr.* **3**, 289.
[4]W. D. L. Musick (1981) *Crit. Rev. Biochem.* **11**, 1.

Selected entries from *Methods in Enzymology* [**vol**, page(s)]:
General Discussion: 17A, 491, 501; **18B**, 138; **66**, 96
Assays: *Astasia longa*, **18B**, 45; mammalian tissue, **18B**, 45
Other molecular properties: *Astasia longa* (properties and activities, **18B**, 46; tissue extract, **18B**, 45); crystallization, **22**, 250; mammalian tissue, **18B**, 45, 46, 165 (properties, **18B**, 46; **66**, 99, 100, 101; tissue extract, **18B**, 45); purification from (beef liver, **17A**, 496; hog liver, **66**, 96; *Pseudomonas*, **17A**, 502)

NICOTINATE PHOSPHORIBOSYLTRANSFERASE

This transferase [EC 2.4.2.11] (also known as nicotinate mononucleotide glycohydrolase, nicotinate mononucleotide pyrophosphorylase, and niacin ribonucleotidase) catalyzes the reaction of nicotinate with 5-phospho-α-D-ribose 1-diphosphate to produce nicotinate D-ribonucleotide and pyrophosphate (or, diphosphate).[1-8] The yeast (*Saccharomyces cerevisiae*) enzyme has an ordered-on (PRPP followed by nicotinate), random-off Bi Bi kinetic mechanism. The enzyme also possesses an energase-type

ATPase activity, proceeding by means of a phospho-histidine intermediate, that alters the kinetics of the main reaction.[7,8] With the ATPase activity, the yeast enzyme has an ordered-on Uni Uni Bi Ter (random-off Q-R) ping pong kinetic mechanism.[1,7] Phosphorylation of the enzyme lowers the K_m values for the substrates and minimizes the contribution from dead-end complexes.

[1] L. S. Hanna, S. L. Hess & D. L. Sloan (1983) *JBC* **258**, 9745.
[2] W. D. L. Musick (1981) *Crit. Rev. Biochem.* **11**, 1.
[3] A. Kosaka, H. O. Spivey & R. K. Gholson (1977) *ABB* **179**, 334.
[4] Z. N. Gaut & H. M. Solomon (1971) *Biochem. Pharmacol.* **20**, 2903.
[5] J. Imsande (1961) *JBC* **236**, 1494.
[6] J. Imsande & P. Handler (1961) *JBC* **236**, 525.
[7] J. W. Gross, M. Rajavel & C. Grubmeyer (1998) *Biochemistry* **37**, 4189.
[8] J. Gross, M. Rajavel, E. Segura & C. Grubmeyer (1996) *Biochemistry* **35**, 3917.

Selected entries from *Methods in Enzymology* [vol, page(s)]:
General discussion: 6, 346; **18B**, 20, 123; **280**, 211; **308**, 28
Assay: 6, 346; **18B**, 37, 38, 123, 124, 138, 139, 144, 145; **308**, 39
Other molecular properties: active site, **308**, 40; *Astasia longa*, **18B**, 138 (assay, **18B**, 138, 139; growth of organism, **18B**, 138; properties, **18B**, 139, 140; sonic lysate, **18B**, 138, 139); ATP dependence, **308**, 37, 43; bakers' yeast, **18B**, 36, 123 (assay, **18B**, 123, 124; properties, **18B**, 125; purification, **18B**, 124, 125); beef liver (properties, 6, 348; purification, 6, 346); energy coupling, **308**, 28; erythrocytes, **18B**, 56; mammalian tissues in general, **18B**, 37 (assay, **18B**, 37, 38; properties and activities, **18B**, 38, 39; tissue extract, **18B**, 37; pH effects, **308**, 40; rat liver, **18B**, 144 (assay, **18B**, 144, 145; properties, **18B**, 147; purification, **18B**, 145); side reactions, **308**, 43

NICOTINE DEHYDROGENASE

This FAD- or FMN-dependent oxidoreductase [EC 1.5.99.4] catalyzes the reaction of nicotine with water and an acceptor substrate to produce (*S*)-6-hydroxynicotine and the reduced acceptor.[1] Acceptor substrates include 2,6-dichlorophenolindophenol, brilliant cresyl blue, menadione, vitamin K_5, methylene blue, and 5-hydroxy-1,4-naphthoquinone.

[1] L. L. Hochstein & B. P. Dalton (1967) *BBA* **139**, 56.

NICOTINE 2-HYDROXYLASE

This FMN-dependent oxidoreductase catalyzes the reaction of nicotine with dioxygen to produce 6-hydroxynicotine.[1] Other substrates include nicotine-2-oxide, myosamine, and anagasine.

[1] L. I. Hochstein & B. P. Dalton (1967) *BBA* **139**, 56.

Selected entries from *Methods in Enzymology* [vol, page(s)]:
General discussion: 52, 17

NITRATE-ESTER REDUCTASE

This activity, formerly classified as EC 1.8.6.1, is now a deleted Enzyme Commission entry, and the activity is included with glutathione transferase [EC 2.5.1.18].

NITRATE REDUCTASES

There are at least five distinct nitrate reductases, all catalyzing the conversion of nitrate (NO_3^-) to nitrite (NO_2^-).[1-8]

Nitrate reductase (NADH) [EC 1.6.6.1], also known as assimilatory nitrate reductase, catalyzes the reaction of NADH with nitrate to produce NAD^+, nitrite, and water. This enzyme uses FAD or FMN, a cytochrome, an iron-sulfur cluster (not present in higher plants), and molybdenum as cofactors. The *Chlorella fusca* enzyme has a random Bi Bi kinetic mechanism.[9] In higher plants, the mechanism is ping pong. The cytochrome present is a type-*b* cytochrome. In the course of the reaction, electron transfer from the flavin to the cytochrome heme does not occur until the NAD^+ product has dissociated from the flavin site and that is likely to be the rate-limiting step of catalysis.[10] The actual site of nitrate reduction is at the molybdenum cofactor, in which Mo(VI) → Mo(IV).

Nitrate reductase (NAD(P)H) [EC 1.6.6.2], also known as assimilatory nitrate reductase, catalyzes the reaction of NAD(P)H with nitrate to produce $NAD(P)^+$, nitrite, and water.[5,11] This enzyme uses FAD or FMN, an iron-sulfur cluster (not present in higher plants), a cytochrome, and molybdenum as cofactors. The *Ankistrodesmus braunii* enzyme reportedly has a ping pong iso Bi Bi kinetic mechanism.[12]

Nitrate reductase (NADPH) [EC 1.6.6.3] catalyzes the reaction of NADPH with nitrate to produce $NADP^+$, nitrite, and water. This enzyme uses FAD, a cytochrome, and molybdenum as cofactors. The *Aspergillus nidulans* enzyme has a random Bi Bi kinetic mechanism.[13] The *Penicillium chrysogenum* enzyme is suggested to have a two-site random mechanism in which FAD and NADPH bind randomly at site 1 and FAD and nitrate binding randomly at site 2.[14]

Nitrate reductase (cytochrome) [EC 1.9.6.1] catalyzes the reaction of nitrate with ferrocytochrome to produce nitrite and ferricytochrome.[15-17] In many organisms, this nitrate reductase is probably identical to nitrate reductase (acceptor) [EC 1.7.99.4]. Substitute substrates for ferrocytochrome include reduced ferrodoxin, quinols, reduced methylene blue, and reduced benzyl viologen.

Nitrate reductase (acceptor) [EC 1.7.99.4], also known as respiratory nitrate reductase, catalyzes the reaction of nitrate with a reduced acceptor to produce nitrite and the oxidized acceptor. Acceptor substrates include ferrocyanide, reduced ferredoxin, reduced benzyl viologen, reduced methyl viologen, quinols, and reduced methylene blue. The *Pseudomonas* enzyme is a cytochrome-dependent system, but the enzyme from *Micrococcus halodenitrificans* is an iron protein containing molybdenum. The *Escherichia coli* enzyme exhibits a Theorell–Chance Bi Bi kinetic mechanism if reduced viologen dyes are used as the electron donor and a two-site mechanism with quinols as the cosubstrate.[18] As mentioned in the previous paragraph, EC 1.7.99.4 and EC 1.9.6.1 are probably identical in many organisms.

[1] A. R. Clarke & T. R. Dafforn (1998) *CBC* **3**, 1.

[2] P. E. Baugh, D. Collison, C. D. Garner & J. A. Joule (1998) *CBC* **3**, 377.

[3] H. C. Huppe & D. H. Turpin (1994) *Ann. Rev. Plant Physiol. Plant Mol. Biol.* **45**, 577.

[4] L. P. Solomonson & M. J. Barber (1990) *Ann. Rev. Plant Physiol. Plant Mol. Biol.* **41**, 225.

[5] M. G. Guerrero, J. M. Vega & M. Losada (1981) *Ann. Rev. Plant Physiol.* **32**, 169.

[6] R. C. Bray (1980) *AE* **51**, 107.

[7] G. Palmer (1975) *The Enzymes*, 3rd ed., **12**, 1.

[8] A. Nason (1963) *The Enzymes*, 2nd ed., **7**, 587.

[9] W. D. Howard & L. P. Solomonson (1981) *JBC* **256**, 12725.

[10] K. Ratnam, N. Shiraishi, W. H. Campbell & R. Hille (1997) *JBC* **272**, 2122.

[11] C. R. Hipkin, A. H. Ali & A. Cannons (1986) *J. Gen. Microbiol.* **132**, 1997.

[12] M. A. De la Rosa (1983) *Mol. Cell. Biochem.* **50**, 65.

[13] D. W. McDonald & A. Coddington (1974) *EJB* **46**, 169.

[14] F. Renosto, D. M. Ornitz, D. Peterson & I. H. Segel (1981) *JBC* **256**, 8616.

[15] J.-P. Rosso, P. Forget & F. Pichinoty (1973) *BBA* **321**, 443.

[16] M. B. Kemp, B. A. Haddock & P. B. Garland (1975) *BJ* **148**, 329.

[17] C. H. MacGregor, C. A. Schnaitman, D. E. Normasell & M. G. Hodgins (1974) *JBC* **249**, 5321.

[18] F. F. Morpeth & D. H. Boxer (1985) *Biochemistry* **24**, 40.

Selected entries from **Methods in Enzymology** [**vol**, page(s)]:
General discussion: 2, 411; 23, 491
Assay: 2, 411; *in vitro*, 23, 495; 53, 349, 350; 69, 272, 273; *in vivo*, 69, 274, 275
Other molecular properties: adaptive, 2, 415; antibody against (preparation, **53**, 352; subunit studies, **53**, 352); apoprotein reconstitution, in assay of molybdenum cofactor, **122**, 403; assay methods *in vitro*, **23**, 495; assay of glutathione S-transferase, **77**, 401; characteristics (cellular location, **69**, 280; flavin requirement, **69**, 278; induction, **69**, 280; inhibitors, **69**, 279, 280; involvement, of metals, **69**, 278; kinetics, **69**, 279; molecular weight, **69**, 277; nitrite formation for nitric oxide assay, **268**, 142, 147, 150, 242, 246 (*Escherichia coli* enzyme, **268**, 145, 148; *Pseudomonas oleovorans* enzyme, **268**, 149); phosphate requirement, **69**, 278; pyridine nucleotide specificity, **69**, 277, 278; sequence of electron transfer, **69**, 278, 279; stability, **69**, 280; structure, **69**, 279; substrate affinity, **69**, 277; subunits, **69**, 279; sulfhydryl, **69**, 278); deficient cell lines, in selection of somatic hybrids, **221**, 388; *Escherichia coli*, from, **53**, 347; immobilization, by microencapsulation, **44**, 214; isoelectric point, **53**, 643; mechanism, **2**, 415; metal content, **53**, 352, 355; *Micrococcus halodenitrificans*, from, **53**, 643; molecular weight, **53**, 350, 351, 643; mutant, activation, **69**, 790; *Neurospora crassa*, preparation, **2**, 413; **22**, 81, 88; *Paracoccus denitrificans*, from, **53**, 642; pH optimum, **53**, 643; preparation of, **23**, 491 (extraction, **69**, 270, 271; material, **69**, 270; special protectants, **69**, 271, 272); properties, **23**, 500; *Pseudomonas perfectomarinus*, from, **53**, 642; purification, **23**, 499; **53**, 348, 349, 355, 642, 643; **69**, 275; radiolabeling, **53**, 354; respiratory-deficient cells, **56**, 175, 176; sedimentation coefficient, **53**, 351; soybean, **1**, 63; **22**, 414; staining on gels, **22**, 595; stereospecificity, **87**, 119; subunit composition, **53**, 351, 352, 355; synthesis, **53**, 355; turnover number, **2**, 413.

NITRATE-TRANSPORTING ATPase

This so-called ATPase [EC 3.6.3.26] catalyzes the ATP-dependent (ADP- and orthophosphate-producing) transport of nitrate$_{out}$ to produce nitrate$_{in}$.[1–3] In addition to catalyzing the import of nitrate anions (NO_3^-), the system from many organisms will also import nitrite (NO_2^-) and cyanate (OCN^-) anions. This nitrate-transporting system is an ABC-type (ATP-binding cassette-type) ATPase that has two similar ATP-binding domains and does not undergo phosphorylation during the transport process.

The idea that a transporter is an enzyme is in keeping with a new definition of enzyme catalysis as the facilitated making/breaking of chemical bonds, not just covalent bonds.[4] This idea builds on Pauling's assertion that any long-lived, chemically distinct interaction (in this case, the persistent location of a solute with respect to the faces of a membrane) can be regarded as a chemical bond. Note also that the equilibrium constant (K_{eq} = [nitrate$_{in}$][ADP][P_i]/[nitrate$_{out}$][ATP]) does not conform to that expected for an ATPase (*i.e.*, K_{eq} = [ADP][P_i]/[ATP]). Thus, although the overall reaction yields ADP and orthophosphate, the enzyme is misclassified as a hydrolase, and instead should be regarded as an energase-type reaction. Energases facilitate affinity-modulated reactions by coupling the $\Delta G_{\text{ATP-hydrolysis}}$ to a force-generating or work-producing step.[4] In this case, P-O-P bond scission supplies the energy to drive nitrate transport. **See ATPase; Energase**

[1] T. Omata (1995) *Plant Cell Physiol.* **36**, 207.

[2] M. H. Saier, Jr. (1998) *Adv. Microb. Physiol.* **40**, 81.

[3] J. K. Griffiths & C. E. Sansom (1998) *The Transporter Factsbook*, Academic Press, San Diego.

[4] D. L. Purich (2001) *TiBS* **26**, 417.

NITRIC OXIDE DIOXYGENASE

This heme- and FAD-dependent oxidoreductase [EC 1.14.12.17] catalyzes the reaction of two molecules of nitric oxide (·NO) with NAD(P)H, H$^+$, and two molecules of dioxygen to produce $NAD(P)^+$, two protons, and two nitrate ions ($NO_3{}^-$).[1,2] FAD participates by carrying the electrons from NAD(P)H to the heme.

[1] P. R. Gardner, G. Costantino & A. L. Salzman (1998) *JBC* **273**, 26528.
[2] P. R. Gardner, A. M. Gardner, L. A. Martin & A. L. Salzman (1998) *PNAS* **95**, 10378.

NITRIC-OXIDE REDUCTASE

This oxidoreductase [EC 1.7.99.7; previously classified as 1.7.99.2] catalyzes the reaction of two molecules of nitric oxide ($\cdot$NO) with a reduced acceptor substrate to produce nitrous oxide (N_2O), the oxidized acceptor, and water.[1-4] One reduced acceptor substrate is reduced phenazine methosulfate. During the course of the reaction, the first intermediate formed is a ferric–NO complex. The reaction is irreversible; nitrous oxide cannot be reconverted to nitric oxide.

The enzyme is a heterodimer of cytochromes *b* and *c*. Cytochrome c_{550} may be the physiological acceptor substrate. Excess nitric oxide inhibits the reductase, apparently by binding to the oxidized form of the enzyme.[5,6]

[1] M. A. Grant & L. I. Hochstein (1984) *Arch. Microbiol.* **137**, 79.
[2] B. Heiss, K. Frunzke & W. G. Zumft (1989) *J. Bacteriol.* **171**, 3288.
[3] M. Dermastia, T. Turk & T. C. Hollocher (1991) *JBC* **266**, 10899.
[4] G. J. Carr & S. J. Ferguson (1990) *BJ* **269**, 423.
[5] P. Girsch & S. de Vries (1997) *BBA* **1318**, 202.
[6] M. Koutny & I. Kucera (1999) *BBRC* **262**, 562.

Selected entries from *Methods in Enzymology* [vol, page(s)]:
General discussion: 2, 420
Assay: 53, 645

NITRIC-OXIDE SYNTHASE

This heme-dependent enzyme [EC 1.14.13.39] (also known as NO synthase, endothelium-derived relaxation factor-forming enzyme, and endothelium-derived relaxing factor synthase), which has a tightly bound tetrahydrobiopterin as well as FAD and FMN, catalyzes the reaction of L-arginine with 1.5 NADPH, 0.5 H$^+$, and two molecules of dioxygen to produce L-citrulline, nitric oxide ($\cdot$NO), 1.5 NADP$^+$, and two water molecules.[1-5] Unlike the enzyme induced in lung or liver by endotoxin, the brain enzyme requires calcium ions. The inducible murine macrophage enzyme contains tightly bound calmodulin.

L-arginine   L-citrulline

N^{ω}-Hydroxy-L-arginine is an intermediate in the reaction, formed from L-arginine, dioxygen, NADPH, and a proton: a heme-Fe^{II}-O$_2$ intermediate is converted to the active Fe^{V}=O. In the second half of the overall reaction, nitric oxide synthase acts as a dioxygenase. Heme-Fe^{II}-O$_2$ reacts with N^{ω}-hydroxy-L-arginine to produce the final products, possibly via the formation of four-membered dioxetane-like ring.[1]

[1] H. B. Dunford (1998) *CBC* **3**, 195.
[2] L. Packer, ed. (1996) *MIE*, vols. **268** and **269**.
[3] M. Marletta (1993) *JBC* **268**, 12231.
[4] S. Kaufman (1993) *Ann. Rev. Nutr.* **13**, 261.
[5] D. J. Stuehr & O. W. Griffith (1992) *AE* **65**, 287.

Selected entries from *Methods in Enzymology* [vol, page(s)]:
Assay: 233, 250, 265; **301**, 70, 92, 504; automated flow injection technique with Griess reagent, **268**, 155; cell culture with Griess reagent, **268**, 410; citrulline radioassay, **233**, 252, 255; **268**, 341, 395, 414, 434; **269**, 132; cofactor dependence, **268**, 458; Griess reaction microtiter plate assay, **268**, 331; hemoglobin method, **233**, 252; myoglobin microtiter plate assay, **268**, 163; NADPH consumption, **268**, 165, 333; nitrite/nitrate method, **233**, 252, 256; oral neutrophil enzyme, **268**, 506; oxyhemoglobin spectrophotometric assay, **268**, 332, 457; reduction of electron acceptors, **268**, 333

Other molecular properties: activity in disease states, **268**, 376, 404, 409; baculovirus expression system (enzyme extraction, **268**, 455, 457; levels of expression, **268**, 450; plasmid construction, **268**, 421; Sf9 cell infection, growth, and harvesting, **268**, 421, 435, 454); binding domains (arginine, **268**, 316, 319; calmodulin, **268**, 313, 319, 325, 429, 431; flavin adenine dinucleotide, **268**, 320; flavin mononucleotide, **268**, 312, 320; heme-binding site, **268**, 315, 467; modularity, **268**, 467; radioligand binding studies, **268**, 322; syntrophin-homology domain, **268**, 314, 432; tetrahydrobiopterin, **268**, 314, 316, 319, 468); biopterin content, determination, **233**, 262; chemical reaction, **268**, 83, 312, 324, 334, 349, 365, 460; chondrocyte, **269**, 76; cloning, **268**, 428, 450, 454; cofactors (calcium dependence, **268**, 458, 461; dependence of isoforms, **268**, 458, 460; heme role in reaction, **268**, 462; prosthetic groups, **268**, 325; quantitation, **268**, 358; requirements, **233**, 251, 259; tetrahydrobiopterin role in reaction, **268**, 314, 334); crude enzyme preparation from rat brain, **268**, 155; cytochrome P450 reductase homology, **268**, 312, 334, 420, 467; distribution, **233**, 250, 258; effect on prostaglandin production, **269**, 13; endothelial isoform (activation, **268**, 428; acylation [fatty acid analysis, **268**, 447; hydroxylamine cleavage, **268**, 447; metabolic labeling with tritiated fatty acids, **268**, 445; myristoylation, **268**, 444; palmitoylation, **268**, 444]; complementary DNA, isolation, **268**, 450, 454; membrane association, **268**, 325, 334; phosphorylation [effect on activity, **268**, 437; immunoprecipitation of labeled enzyme, **268**, 440; metabolic labeling with orthophosphate, **268**, 439; phosphoamino acid analysis, **268**, 441; phosphopeptide mapping, **268**, 438, 441]; purification, bovine enzyme [2',5'-ADP Sepharose chromatography, **268**, 339; endothelial cell culture and extraction, **268**, 337; gel filtration, **268**, 339]; purification, human placenta enzyme, **268**, 343); endothelial, expression in heterologous systems, **269**, 55; enzymatic activity, **269**, 412; evolution, **268**, 311; flavin content, determination, **233**, 262; gene expression (modulation by antisense oligonucleotides, **269**, 422; targeted disruption, **269**, 420); gene structure, **268**, 311; N-hydroxy-L-arginine intermediate (nitric oxide reactivity, **268**, 374; oxidation, **268**, 372; reduction, **268**, 374; synthesis, **268**, 367); immunolocalization of isoforms, **269**, 410 (antibodies [commercial availability, **268**, 502; epitope targeting, **268**, 351; generation of specific antibodies, **268**, 351, 502]; endothelial enzyme immunolocalization, **268**, 356, 515; immunofluorescence, **268**, 512;

NITRILE HYDRATASE

There are a several enzyme-catalyzed reactions classified under nitrile hydratase or nitrilase.[1–7]

Nitrilase [EC 3.5.5.1], also known as nitrile aminohydrolase and nitrile hydratase, catalyzes the hydrolysis of a nitrile to produce a carboxylate anion and ammonia.[1–5] The enzyme acts on a wide range of aromatic nitriles (for example, (indole-3-yl)-acetonitrile and benzonitrile) as well as some aliphatic nitriles. The reaction proceeds via the formation of a thiolester at a catalytically essential cysteinyl residue.[1]

Another nitrile hydratase [EC 4.2.1.84], a metallohydratase also called nitrilase, catalyzes the reaction of a short-chain aliphatic nitrile with water to produce an aliphatic amide (*e.g.*, acetonitrile is converted to acetamide). Both cobalt- and iron-containing nitrile hydratases have been reported. Cysteine-sulfenic acid residues have been identified in some native nitrile hydratases where it functions in both iron coordination and nitric oxide binding.[6] Iron-dependent nitrile hydratases have an octahedral coordinated ferric site in which one of the coordination sites contains a solvent-exchangeable hydroxide ligand. This metal-bound hydroxide attacks the nitrile group of the substrate generating a transient amide tautomer coordinated to the iron. This intermediate either rearranges to the product and dissociates or is attacked by solvent water while still bound to the metal ion.[1] *See also* Nitrilase; Aliphatic Nitrilase; Ricinine Nitrilase; Cyanoalanine Nitrilase; Arylacetonitrilase; Bromoxynil Nitrilase

[1] B. G. Fox (1998) *CBC* **3**, 261.
[2] T. Rausch & W. Hilgenberg (1980) *Phytochemistry* **19**, 747.
[3] D. B. Harper (1977) *BJ* **167**, 685 and **165**, 309.
[4] A. Arnaud, P. Galzy & J. C. Jallageas (1977) *Agric. Biol. Chem.* **41**, 2183.
[5] K. V. Thimann & S. Mahadevan (1964) *ABB* **105**, 133.
[6] A. Claiborne, J. I. Yeh, T. C. Mallett, J. Luba, E. J. Crane, 3rd, V. Charrier & D. Parsonage (1999) *Biochemistry* **38**, 15407.
[7] J. A. Duine & J. A. Jongejan (1989) *Ann. Rev. Biochem.* **58**, 403.

Selected entries from *Methods in Enzymology* [vol, page(s)]:
General discussion: 136, 523
Assay: 136, 524
Other molecular properties: electron nuclear double resonance spectroscopy (nitrogen spectra, **246**, 584, 586; proton exchange, **246**, 587; proton spectra, **246**, 586); iron ligation, **246**, 584, 588; nitrilase (acrylamide production with, **136**, 530; assay, **136**, 524; properties, **136**, 527)

NITRITE REDUCTASE

There are at least four enzyme-catalyzed reactions classified under nitrite reductase.[1–5]

Nitrite reductase (NAD(P)H) [EC 1.6.6.4] catalyzes the reaction of nitrite (NO_2^-) with three molecules of NAD(P)H to yield ammonia, two molecules of water, and three molecules of NAD(P)$^+$. Cofactors include FAD, non-heme iron, and siroheme.[6–10] The initial hydride transfer is to FAD, the electrons then being transferred to the iron-sulfur cluster followed by the siroheme, and then to nitrite.

Nitrite reductase (cytochrome) [EC 1.7.2.1] is a copper-dependent system that catalyzes the reaction of nitrite with two molecules of ferrocytochrome c to produce nitric oxide, water, and two molecules of ferricytochrome c.[11,12] Cytochrome c_{552} or cytochrome c_{553} from *Pseudomonas denitrificans* acts as the cytochrome substrate.

Ferredoxin:nitrite reductase [EC 1.7.7.1] is a heme- and iron-dependent enzyme that catalyzes the reaction of nitrite with three molecules of reduced ferredoxin to produce ammonia and three molecules of oxidized ferredoxin.[13,14] Other reducing agents include flavodoxin, methyl viologen, and benzyl viologen. The spinach enzyme has a random Bi Bi kinetic mechanism.[13]

Nitrite reductase (acceptor) [EC 1.7.99.3] is a copper- and FAD-dependent enzyme that catalyzes the reaction of two ions of nitrite with a reduced donor substrate to produce two molecules of nitric oxide, two water molecules, and the oxidized acceptor substrate.[15–18] Donor substrates can be reduced pyocyanine, flavins (*e.g.*, FMNH$_2$ and FADH$_2$), and reduced methylene blue. The *Alcaligenes xylosoxidans* enzyme has an ordered kinetic mechanism in which nitrite binds to the oxidized type 2 copper centers before electron transfer from the reduced type 1 center occurs. It has been suggested that particular residues (for example, histidyl) are utilized in the transfer of electrons between the copper sites.[19] *See also Cytochrome Oxidase, Pseudomonas*

[1] A. R. Clarke & T. R. Dafforn (1998) *CBC* **3**, 1.
[2] A. Messerschmidt (1998) *CBC* **3**, 401.
[3] E. T. Adman (1991) *Adv. Protein Chem.* **42**, 145.
[4] T. E. Meyer & M. D. Kamen (1982) *Adv. Protein Chem.* **35**, 105.
[5] Y. Hatefi & D. L. Stiggall (1976) *The Enzymes*, 3rd ed., **13**, 175.
[6] R. H. Jackson, J. A. Cole & A. Cornish-Bowden (1982) *BJ* **203**, 505.
[7] R. H. Jackson, J. A. Cole & A. Cornish-Bowden (1981) *BJ* **199**, 171.
[8] K. J. Coleman, A. Cornish-Bowden & J. A. Cole (1978) *BJ* **175**, 483.
[9] K. N. Prodouz & R. H. Garrett (1981) *JBC* **256**, 9711.
[10] R. A. Lazzarini & D. E. Atkinson (1961) *JBC* **236**, 3330.
[11] W. P. Michalski & D. J. D. Nicholas (1985) *BBA* **828**, 130.
[12] Y. Lam & D. J. D. Nicholas (1969) *BBA* **180**, 459.
[13] B. Mikami & S. Ida (1989) *J. Biochem.* **105**, 47.
[14] M. G. Guerrero, J. M. Vega & M. Losada (1981) *Ann. Rev. Plant Physiol.* **32**, 169.
[15] T. Kakutani, H. Watanabe, K. Arima & T. Beppu (1981) *J. Biochem.* **89**, 453.
[16] G. C. Walker & D. J. D. Nicholas (1961) *BBA* **49**, 350.
[17] H. Iwasaki & T. Matsubara (1972) *J. Biochem.* **71**, 645.
[18] H. Iwasaki, S. Noji & S. Shidara (1975) *J. Biochem.* **78**, 355.
[19] R. W. Strange, L. M. Murphy, F. E. Dodd, Z. H. Abraham, R. R. Eady, B. E. Smith & S. S. Hasnain (1999) *J. Mol. Biol.* **287**, 1001.

Selected entries from **Methods in Enzymology** [**vol**, page(s)]:
General discussion: 23, 487, 491; 69, 255; 243, 303
Assay: methods *in vitro*, 23, 487; 53, 643, 649
Other molecular properties: assay method for, 23, 487; cyanide and, 2, 412; dissimilatory, cytochrome cd_1 type, 243, 533; hexaheme nitrite reductase (assay, *Desulfovibrio desulfuricans*, 243, 304; *Desulfovibrio desulfuricans* (absorption spectra, 243, 307; amino acid composition, 243, 310; assay, 243, 304; catalytic activities, 243, 313; electrochemical properties, 243, 318; electron donors, 243, 312; electron paramagnetic resonance studies, 243, 314; heme and iron content, 243, 309; heme prosthetic group, 243, 308; molecular weight, 243, 310; Mössbauer spectroscopy, 243, 317; pH optimum, 243, 312; properties, 243, 307; purification, 243, 305; substrate specificity, 243, 312; temperature effects, 243, 312; topography, 243, 311); distribution, 243, 303; *Escherichia coli* K-12, 243, 303, 311, 313; *Vibrio fischeri*, 243, 303, 311; *Wolinella succinogenes*, 243, 303, 311, 313); hydroxylamine reductase and, 2, 418; inhibitor, 53, 646; nitrate reductase and, 2, 412; nitroaryl reductase and, 2, 410, 411; properties, 23, 490; *Paracoccus denitrificans*, from, 53, 643; purification, 23, 488; 53, 643; stereospecificity, 87, 119

p-NITROACETOPHENONE REDUCTASE

This oxidoreductase catalyzes the reaction of *p*-nitroacetophenone with NADPH to produce 2-hydroxyethyl-*p*-nitrobenzene and NADP$^+$. In many organisms, this activity is catalyzed by carbonyl reductase [EC 1.1.1.184]. *See also Carbonyl Reductase*

β-NITROACRYLATE REDUCTASE

This oxidoreductase [EC 1.3.1.16], first isolated from *Penicillium atrovenetum*, catalyzes the reaction of β-nitro-acrylate ($O_2NCH=CHCOO^-$) with NADPH to produce 3-nitropropanoate ($O_2NCH_2CH_2COO^-$) and NADP$^+$.[1]

[1] P. D. Shaw (1967) *Biochemistry* **6**, 2253.

NITROARYL NITROREDUCTASE

This enzyme catalyzes the NADH-dependent reduction of nitro-containing aromatic compounds (for example, 1,4-dinitrobenzene, 4-nitrophenol, 2,4-dinitrotoluene, and 4-nitrobenzoate) to produce the amino-containing product (*e.g.*, 4-nitroaniline and 4-aminophenol). The *Enterobacter cloacae* enzyme catalyzes a ping pong Bi Bi kinetic mechanism,[1] as does the nitroreductase from *Escherichia coli*.[2,3]

[1] R. L. Koder & A. F. Miller (1998) *BBA* **1387**, 395.
[2] S. Zenno, H. Koike, M. Tanokura & K. Saigo (1996) *J. Biochem.* **120**, 736.

[3]S. Zenno, H. Koike, A. N. Kumar, R. Jayaraman, M. Tanokura & K. Saigo (1996) *J. Bacteriol.* **178**, 4508.

Selected entries from *Methods in Enzymology* [vol, page(s)]:
General discussion: 2, 406
Assay: 2, 406

p-NITROBENZALDEHYDE REDUCTASE

This oxidoreductase catalyzes the reaction of 4-nitro-benzaldehyde with NADPH to produce 4-nitrobenzyl alcohol and NADP$^+$.[1] The rat and rabbit liver enzymes display an A-side stereospecificity, whereas the rabbit liver reductase with a low p*I* value has B-side stereospecificity. In a number of organisms, this enzyme activity may be due to carbonyl reductase, aldehyde reductase, or one of the alcohol dehydrogenases. *See Alcohol Dehydrogenase (NADP$^+$)*

[1]A. R. Clarke & T. R. Dafforn (1998) *CBC* **3**, 1.

NITROETHANE OXIDASE

This oxidoreductase [EC 1.7.3.1], sometimes known as nitroethane reductase and nitroethanase, catalyzes the reaction of nitroethane with water and dioxygen to produce acetaldehyde, nitrite, and hydrogen peroxide.[1] Other substrates include 1-nitropropane, 2-nitropropane, and 3-nitropropionate.

[1]H. N. Little (1951) *JBC* **193**, 347.

Selected entries from *Methods in Enzymology* [vol, page(s)]:
General discussion: 2, 400
Assay: 2, 400

NITROGENASE

There are at least three different types of iron-sulfur-dependent nitrogenases: molybdenum-dependent, vanadium-dependent, and molybdenum/vanadium-independent. The Enzyme Commission currently assigns two classification numbers to the nitrogenases.[1–9]

Nitrogenase (ferredoxin) [EC 1.18.6.1], an iron-sulfur protein, catalyzes the overall reaction of eight molecules of reduced ferredoxin molecules with eight protons, dinitrogen, and sixteen ATP molecules to produce eight oxidized ferredoxin molecules, two ammonia molecules, sixteen ADP molecules, and sixteen ions of orthophosphate. This nitrogenase can be either molybdenum- or vanadium-dependent. Other substrates include acetylene, propyne, cyanide, N$_2$O, and azide. Acetaldehyde can act as an acceptor substrate; in the absence of any acceptor substrate, protons will generate H$_2$.

The iron-sulfur- and molybdenum-dependent nitrogenase (flavodoxin) [EC 1.19.6.1] catalyzes the overall reaction of eight reduced flavodoxin molecules with eight protons, dinitrogen, and sixteen molecules of ATP to produce eight oxidized flavodoxin molecules, two ammonia molecules, sixteen molecules of ADP, and sixteen orthophosphate ions.

Molybdenum-dependent nitrogenases can also reduce acetylene to ethene (vanadium nitrogenases can reduce HC≡CH further to ethane), azide to ammonia and dinitrogen, cyanide to methane and ammonia, and nitrous oxide to dinitrogen and water. While hydrazine has been suggested to be an intermediate in the reaction catalyzed by vanadium nitrogenases, this does not appear to be the case for the molybdenum enzymes.

The molybdenum nitrogenase of *Klebsiella pneumoniae* catalyzes the reaction via an electron transfer sequence from the Fe-protein subunit to the MoFe-protein subunit. This later subunit undergoes reduction by three electrons and three protons prior to the binding of N$_2$.

[1]D. C. Rees & J. B. Howard (1999) *J. Mol. Biol.* **293**, 343.
[2]G. J. Leigh (1998) *CBC* **3**, 349.
[3]P. E. Baugh, D. Collison, C. D. Garner & J. A. Joule (1998) *CBC* **3**, 377.
[4]L. E. Mortenson, L. C. Seefeldt, T. V. Morgan & J. T. Bolin (1993) *AE* **67**, 299.
[5]J. T. Bolin, N. Campobasso, S. W. Muchmore, T. V. Morgan & L. E. Mortenson (1993) in *Molybdenum Enzymes, Cofactors & Model Systems* (E. I. Stiefel, D. Coucouvanis & W. E. Newton, eds.) p. 186, Amer. Chem. Soc., Washington.
[6]J. B. Howard & D. C. Rees (1991) *Adv. Protein Chem.* **42**, 199.
[7]D. W. Emerich, R. V. Hageman & R. H. Burris (1981) *AE* **52**, 1.
[8]R. C. Bray (1980) *AE* **51**, 107.
[9]G. Palmer (1975) *The Enzymes*, 3rd ed., **12**, 1.

Selected entries from *Methods in Enzymology* [vol, page(s)]:
General discussion: 24, 446, 456, 480; **69**, 753; **118**, 511
Assay: 24, 456, 476, 481; **53**, 329; **69**, 754, 755; **118**, 554; **167**, 460, 475
Other molecular properties: acetylene reduction, **69**, 744; *Anabaena cylindrica*, from, **53**, 328; *Azotobacter chroococcum*, from, **53**, 319, 326; **69**, 761, 762; *Azotobacter vinelandii*, from, **53**, 319, 324; **69**, 759; *Bacillus polymyxa*, from, **53**, 315; *Chromatium vinosum*, from, **69**, 762, 763; *Clostridium pasteurianum*, from, **53**, 319, 321, 322; **69**, 756, 757; component (criteria of purity, **69**, 763; nomenclature, **69**, 754); component 1, **53**, 314 (purification and crystallization, **69**, 793); component 2, **53**, 314; cyanobacterial (assays, **167**, 460, 475; electron donation to, assay, **167**, 499; induction, **167**, 463); electron paramagnetic resonance, **24**, 493; **49**, 514; **53**, 265, 272; **54**, 123, 131; FeMo cofactor, isolation, **69**, 793; Fe proteins, properties, **69**, 773; freezing effects, **31**, 532; genetic locus, **69**, 47; heterocysts, **69**, 812; hydrogen evolution, **69**, 731; induction, in liquid cultures of rhizobia, **69**, 752, 753; iron-molybdenum cofactor, **246**, 648; iron-sulfur center, **69**, 779, 780; *Klebsiella pneumoniae*, from, **53**, 319, 323, 324; **69**, 758, 759 (assays, **118**, 514; properties, **118**, 518; purification, **118**, 516); low-temperature EPR studies, **54**, 123, 131; measurement, of protein, metal and sulfide content, **69**, 755, 756; Mo-Fe protein, **53**, 317–327 (ligand exchange, **69**, 788, 789; properties, **69**, 763); molybdoferredoxin (crystallization, **53**, 325; purification, **53**, 317, 320, 322, 327; specific

activity, **53**, 319); Mössbauer spectroscopy, **54**, 364, 365, 371; mutant, activation, **69**, 792; nodular preparation, **24**, 470; properties, **24**, 467, 489; protein components, **53**, 314 (specific activities, **53**, 319); purification, **24**, 446, 460, 484; *Rhodospirillum rubrum*, from, **53**, 318; soybean, from, **53**, 328; *Spirillum lipoferum*, from, **53**, 328; temperature, **69**, 749; visual and UV spectra, **24**, 492; X-ray absorption spectroscopy, **54**, 332, 338; **246**, 648

2-NITROPHENOL 2-MONOOXYGENASE

This oxidoreductase [EC 1.14.13.31], also known as nitrophenol oxygenase, catalyzes the reaction of 2-nitrophenol with NADPH and dioxygen to produce catechol, nitrite, NADP$^+$, and water. The *Pseudomonas putida* enzyme metabolizes nitro-aromatic compounds.[1,2]

[1]J. Zeyer, H. P. Kocher & K. N. Timmis (1986) *Appl. Environ. Microbiol.* **52**, 334.
[2]J. Zeyer & H. P. Kocher (1988) *J. Bacteriol.* **170**, 1789.

4-NITROPHENOL 2-MONOOXYGENASE

This FAD-dependent oxidoreductase [EC 1.14.13.29], also known as 4-nitrophenol-2-hydroxylase and 4-nitrophenol hydroxylase, catalyzes the reaction of 4-nitrophenol with NADH and dioxygen to produce 4-nitrocatechol, NAD$^+$, and water.[1,2] NADPH is a slightly less effective substrate.

[1]D. Mitra & C. S. Vaidyanathan (1984) *Biochem. Int.* **8**, 609.
[2]D. R. Koop (1986) *Mol. Pharmacol.* **29**, 399.

4-NITROPHENYLPHOSPHATASE

This enzyme [EC 3.1.3.41] catalyzes the hydrolysis of 4-nitrophenyl phosphate to produce 4-nitrophenol and orthophosphate.[1–3] This hydrolytic enzyme is partially distinguished by the organophosphates that are not substrates: for example, phenyl phosphate, 4-nitrophenyl sulfate, acetyl phosphate, phosphotyrosine, and glycerol phosphate.

Other enzymes readily hydrolyze 4-nitrophenyl phosphate, including alkaline phosphatase [EC 3.1.3.1], acid phosphatase [EC 3.1.3.2], nucleotidase [EC 3.1.3.31], [acetyl-CoA carboxylase]-phosphatase [EC 3.1.3.44], 2′,3′-cyclic-nucleotide 2′-phosphodiesterase [EC 3.1.4.16], the potassium-dependent ATPase, and the calcium-dependent ATPase [EC 3.6.3.8].

[1]A. K. Verma & J. T. Penniston (1984) *Biochemistry* **23**, 5010.
[2]S. Kashiwamata, S. Goto, R. K. Semba & F. N. Suzuki (1979) *JBC* **254**, 4577.
[3]J. Attias & J. L. Bonnet (1972) *BBA* **268**, 422.

2-NITROPROPANE DIOXYGENASE

This iron- and FAD-dependent oxidoreductase [EC 1.13.11.32], also known as nitroalkane oxidase, catalyzes

the reaction of dioxygen with two molecules of 2-nitropropane to produce two nitrite ions and two molecules of acetone.[1–4] Other substrates include nitroethane (producing acetaldehyde), 1-nitropropane (generating propanal), 2-nitro-1-propanol, 2-nitro-1-butanol, and 3-nitro-2-pentanol.

Superoxide is a essential catalytic intermediate in this reaction.[5] The *Hansenula mrakii* and *Fusarium oxysporum* enzymes are FAD-dependent whereas the *Neurospora crassa* protein utilizes FMN. The kinetic mechanism of the *F. oxysporum* enzyme, which catalyzes the reaction of a nitroalkane with dioxygen and water to produce the corresponding aldehyde or ketone, nitrite, and hydrogen peroxide, is iso bi ter ping pong[6] and is activated by imidazole.

[1]T. Kido, K. Soda, T. Suzuki & K. Asada (1976) *JBC* **251**, 6994.
[2]T. Kido, K. Soda & K. Asada (1978) *JBC* **253**, 226.
[3]T. Kido, K. Hashizume & K. Soda (1978) *J. Bacteriol.* **133**, 53.
[4]T. Kido, K. Tanizawa, M. Ishida, K. Inagaki & K. Soda (1984) *Agric. Biol. Chem.* **48**, 1361.
[5]T. Kido & K. Soda (1984) *ABB* **234**, 468.
[6]G. Gadda & P. F. Fitzpatrick (2000) *Biochemistry* **39**, 1400.

3-*aci*-NITROPROPANOATE OXIDASE

This FMN-dependent oxidoreductase [EC 1.7.3.5] catalyzes the reaction of 3-*aci*-nitropropanoate ($^-$OOCCH$_2$CH=NO$_2$) with dioxygen to produce 3-oxopropanoate (or, malonate semialdehyde), nitrite, and hydrogen peroxide.[1] The primary products of the enzymatic reaction are probably the nitropropanoate free radical and superoxide.

[1]D. J. T. Porter & H. J. Bright (1987) *JBC* **262**, 14428.

NITROQUINOLINE-*N*-OXIDE REDUCTASE

This oxidoreductase [EC 1.6.6.10] catalyzes the reaction of 4-nitroquinoline-*N*-oxide with two molecules of NAD(P)H to produce 4-hydroxyaminoquinoline-*N*-oxide and two molecules of NAD(P)$^+$. An epidemiologic correlation has been established between this enzyme activity and the frequency of chemical carcinogenesis.[1]

[1]D. R. Booth (1990) *Cell Tissue Kinet.* **23**, 331.

NITROUS-OXIDE REDUCTASE

This copper-dependent oxidoreductase [EC 1.7.99.6] catalyzes the reaction of nitrous oxide (N$_2$O) with a reduced donor substrate to produce dinitrogen, water, and the oxidized donor, a two-electron reduction.[1–4] Donor substrates include reduced viologens or methylene blue.

The *Pseudomonas* enzyme is a homodimer with four copper atoms per monomer. Two copper ions form a Cu_A center while the other two form a Cu_Z center.

[1] C. L. Coyle, W. G. Zumft, P. M. H. Kroneck, H. Körner & W. Jakob (1985) *EJB* **153**, 459.
[2] W. P. Michalski, D. H. Hein & D. J. D. Nicholas (1986) *BBA* **872**, 50.
[3] S. Ferretti, J. G. Grossmann, S. S. Hasnain, R. R. Eady & B. E. Smith (1999) *EJB* **259**, 651.
[4] E. T. Adman (1991) *Adv. Protein Chem.* **42**, 145.

*Nla*III RESTRICTION ENDONUCLEASE

This type II site-specific endonuclease [EC 3.1.21.4], isolated from *Neisseria lactamica*, catalyzes the hydrolysis of both strands of DNA at 5′ . . . CATG▽ . . . 3′.[1] The products have a 3′ CATG extension that can be ligated to *Sph*I-cleaved DNA. *Nla*III is inhibited by NaCl and KCl but not $(NH_4)_2SO_4$.

[1] B.-Q. Qiang & I. Schildkraut (1986) *NAR* **14**, 1991.

*Nla*IV RESTRICTION ENDONUCLEASE

This type II site-specific endonuclease [EC 3.1.21.4], isolated from *Neisseria lactamica*, catalyzes the hydrolysis of both strands of DNA at 5′ . . . GGN▽NCC . . . 3′, where N refers to any base.[1] This enzyme, which produces blunt-ended fragments, is inhibited by NaCl and KCl.

[1] B.-Q. Qiang & I. Schildkraut (1986) *NAR* **14**, 1991.

NMN NUCLEOSIDASE

This enzyme [EC 3.2.2.14], also known as NMNase and nicotinamide mononucleotide hydrolase, catalyzes the hydrolysis of nicotinamide D-ribonucleotide to produce nicotinamide and D-ribose 5-phosphate.[1,2] Guanine nucleotides such as pppGpp and GTP have been shown to activate the *Azotobacter vinelandii* enzyme.[1]

[1] T. Imai (1979) *J. Biochem.* **85**, 887 and (1987) **101**, 153 and 163.
[2] R. Wagner, F. Feth & K. G. Wagner (1986) *Planta* **167**, 226 and **168**, 408.

NOCARDICIN-A EPIMERASE

This epimerase [EC 5.1.1.14] catalyzes the interconversion of isonocardicin A and nocardicin A, a monocyclic β-lactam antibiotic obtained from the fermentation broth of a strain of actinomycetes.[1]

[1] B. A. Wilson, S. Bantia, G. M. Salituro, A. M. Reeve & C. A. Townsend (1988) *JACS* **10**, 8238.

NODAVIRUS ENDOPEPTIDASE

This aspartic endopeptidase [EC 3.4.23.44], first studied in the black beetle virus (and called Black Beetle virus endopeptidase), is unusual in that each active site is utilized only once. Hence, it is not a true catalyst and thus not a true enzyme. The cleavage site in nodaviruses[1,2] is Asn363–Ala364 whereas in the *Nudaurelia capensis* ω virus (a tetravirus), which utilizes a glutamyl residue instead of an aspartyl, the site is at Asn–Phe.[3]

[1] M. V. Hosur, T. Schmidt, R. C. Tucker, J. E. Johnson, T. M. Gallagher, B. H. Selling & R. R. Rueckert (1987) *Proteins: Struct. Funct. Genet.* **2**, 167.
[2] A. Zlotnick, V. S. Reddy, R. Dasgupta, A. Schneemann, W. J. Ray, R. R. Rueckert & J. E. Johnson (1994) *JBC* **269**, 13680.
[3] D. K. Agrawal & J. E. Johnson (1995) *Virology* **207**, 89.

NON-CHAPERONIN MOLECULAR CHAPERONE ATPases

This enzyme classification [EC 3.6.4.10], also known as molecular chaperone Hsc70 ATPases, represents a highly diverse set of enzymes that perform many functions that are similar to those of chaperonins, with the concomitant hydrolysis of ATP to ADP and orthophosphate.[1–3] Included in this classification are a number of heat-shock cognate proteins. Other members of this set of ATPases are those proteins active in clathrin uncoating and in the oligomerization of actin. Examples include the mammalian 70-kilodalton heat-shock cognate protein (Hsc70 or Hsp70), the molecular chaperone BiP, the ClpX heat-shock protein of *Escherichia coli*, and calnexin.

Although the overall reaction yields ADP and orthophosphate, the enzyme is misclassified as a hydrolase, because P-O-P bond scission supplies the Gibbs free energy required to drive protein folding.[4] The overall enzyme-catalyzed reaction should be regarded to be of the energase type (*e.g.*, State₁ + ATP = State₁ + ADP + Pᵢ, where two affinity-modulated conformational states are indicated).[4] Energases transduce chemical energy into mechanical work.

[1] S. Sadis & L. E. Hightower (1992) *Biochemistry* **31**, 9406.
[2] S. Blond-Elguindi, A. M. Fourie, J. F. Sambrook & M. J. Gething (1993) *JBC* **268**, 12730.
[3] A. Wawrzynow, D. Wojtkowiak, J. Marszalek, B. Banecki, M. Jonsen, B. Graves, C. Georgopoulos & M. Zylicz (1995) *EMBO J.* **14**, 1867.
[4] D. L. Purich (2001) *TiBS* **26**, 417.

NONPOLAR-AMINO-ACID-TRANSPORTING ATPase

This set of so-called ATPases [EC 3.6.3.22] catalyzes the ATP-dependent (ADP- and orthophosphate-producing) transport of a [nonpolar amino acid]_out to produce a [nonpolar amino acid]_in.[1–3] The set of bacterial enzymes catalyzes the import of L-leucine, L-valine, and L-isoleucine. These ABC-type (ATP-binding cassette-type) ATPases have two similar ATP-binding domains and do not undergo phosphorylation during the transport process.

The idea that a transporter is an enzyme is in keeping with a new definition of enzyme catalysis as the facilitated making/breaking of chemical bonds, not just covalent bonds.[4] This idea builds on Pauling's assertion that any long-lived, chemically distinct interaction (in this case, the persistent location of a solute with respect to the faces of a membrane) can be regarded as a chemical bond. Note also that the equilibrium constant (K_{eq} = [amino acid$_{in}$][ADP][P$_i$]/[amino acid$_{out}$][ATP]) does not conform to that expected for an ATPase (*i.e.*, K_{eq} = [ADP][P$_i$]/[ATP]). Thus, although the overall reaction yields ADP and orthophosphate, the enzyme is misclassified as a hydrolase, and instead should be regarded as an energase-type reaction. Energases facilitate affinity-modulated reactions by coupling the $\Delta G_{\text{ATP-hydrolysis}}$ to a force-generating or work-producing step.[4] In this case, P-O-P bond scission supplies the energy to drive nonpolar amino acid transport. **See ATPase; Energase**

[1] G. Kuan, E. Dassa, N. Saurin, M. Hofnung & M. H. Saier, Jr. (1995) *Res. Microbiol.* **146**, 271.
[2] M. H. Saier, Jr. (1998) *Adv. Microb. Physiol.* **40**, 81.
[3] J. K. Griffiths & C. E. Sansom (1998) *The Transporter Factsbook*, Academic Press, San Diego.
[4] D. L. Purich (2001) *TiBS* **26**, 417.

NON-STEREOSPECIFIC DIPEPTIDASE

This dipeptidase [EC 3.4.13.17], also known as D-(or L-)aminoacyl-dipeptidase and peptidyl-D-amino acid hydrolase, catalyzes the hydrolysis of dipeptides containing either D- or L-amino acids or both. At pH 8, the enzyme catalyzes the hydrolysis of L-alanyl-D-alanine and L-alanyl-L-alanine at nearly equal rates.[1]

[1] A. D'Aniello & L. Strazzullo (1984) *JBC* **259**, 4237.

NOPALINE DEHYDROGENASE

This oxidoreductase [EC 1.5.1.19], also known as D-nopaline synthase, catalyzes the reversible reaction of L-arginine with NADPH and α-ketoglutarate (or, 2-oxoglutarate) to produce D-nopaline (N^2-(D-1,3-dicarboxypropyl)-L-arginine), NADP$^+$, and water.[1,2] The enzyme also converts L-ornithine to D-ornaline (N^2-(D-1,3-dicarboxypropyl)-L-ornithine) in an analogous manner.

The dehydrogenase has a rapid equilibrium random Ter Bi kinetic mechanism.[1]

[1] J. D. Kemp, D. W. Sutton & E. Hack (1979) *Biochemistry* **18**, 3755.
[2] J. Thompson & S. P. F. Miller (1991) *AE* **64**, 317.

Selected entries from **Methods in Enzymology** [**vol**, page(s)]:
Assay: 153, 311

(R,S)-NORCOCLAURINE 6-O-METHYLTRANSFERASE

This methyltransferase [EC 2.1.1.128] catalyzes the reaction of S-adenosyl-L-methionine with (R,S)-norcoclaurine (that is, 6,7-dihydroxy-1-[(4-hydroxyphenyl)methyl]-1,2,3,4-tetrahydroisoquinoline) to produce S-adenosyl-L-homocysteine and (R,S)-coclaurine.[1,2]

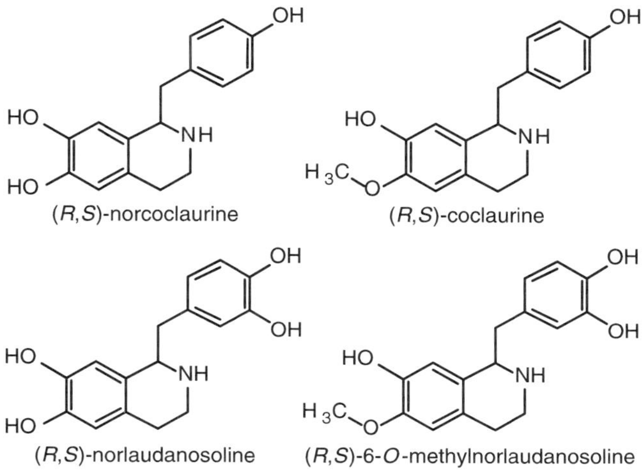

(R,S)-norcoclaurine

(R,S)-coclaurine

(R,S)-norlaudanosoline

(R,S)-6-O-methylnorlaudanosoline

(R,S)-Norlaudanosoline is an alternative substrate, producing 6-O-methylnorlaudanosoline. Interestingly, the methyltransferase from *Coptis japonica* has a ping pong Bi Bi kinetic mechanism.[2]

[1] M. Rueffer, N. Nagakura & M. H. Zenk (1983) *Planta Medica* **49**, 131.
[2] F. Sato, T. Tsujita, Y. Katagiri, S. Yoshida & Y. Yamada (1994) *EJB* **225**, 125.

(S)-NORCOCLAURINE SYNTHASE

This enzyme [EC 4.2.1.78], also known as (S)-norlaudanosoline synthase, catalyzes the reaction of 4-(2-aminoethyl)benzene-1,2-diol with 4-hydroxyphenylacetaldehyde to produce (S)-norcoclaurine and water.[1,2] The enzyme also catalyzes the reaction of 4-(2-aminoethyl)benzene-1,2-diol with (3,4-dihydroxyphenyl)acetaldehyde to generate (S)-norlaudanosoline. Four isozymes have been identified in *Eschscholtzia tenuifolia*.[2]

[1] M. Rüffer, H. El-Shagi, N. Nagakura & M. H. Zenk (1981) *FEBS Lett.* **129**, 5.
[2] H.-M. Schumacher, M. Rüffer, N. Nagakura & M. H. Zenk (1983) *J. Med. Plant Res.* **48**, 212.

sym-NORSPERMIDINE SYNTHASE

This enzyme [EC 2.5.1.23] catalyzes the reaction of *S*-adenosylmethioninamine and propane-1,3-diamine to produce 5′-methylthioadenosine and bis(3-aminopropyl)-amine (also referred to as *sym*-norspermidine; $^+H_3NCH_2CH_2CH_2NHCH_2CH_2CH_2NH_3{}^+$).[1] The enzyme is distinct from spermidine synthase [EC 2.5.1.16] and spermine synthase [EC 2.5.1.22].

[1]A. Aleksijevic, J. Grove & F. Schuber (1979) *BBA* **565**, 199.

*Not*I RESTRICTION ENDONUCLEASE

This type II site-specific endonuclease [EC 3.1.21.4], isolated from *Nocardia otitidis-caviarum*, catalyzes the hydrolysis of both strands of DNA at 5′ . . . GC∇GGCCG C . . . 3′.[1]

[1]L. A. Sznyter & J. E. Brooks (1988) *Heredity* **61**, 308.

Selected entries from ***Methods in Enzymology*** [vol, page(s)]: **General discussion: 155**, 15

NOVOZYM

This is a proprietary name for a mixture of enzymes from *Trichoderma harzianum* useful in lysing cell walls and in the preparation of protoplasts from yeast and fungi. Novozym contains cellulase, chitinase, xylanase, proteinases, and laminarinase.

*Nru*I RESTRICTION ENDONUCLEASE

This type II site-specific endonuclease [EC 3.1.21.4], isolated from *Nocardia rubra*, catalyzes the hydrolysis of both strands of DNA at 5′ . . . TCG∇CGA . . . 3′, producing blunt-ended fragments. The enzyme is blocked by overlapping *dam* methylation.

*Nsi*I RESTRICTION ENDONUCLEASE

This type II site-specific endonuclease [EC 3.1.21.4], isolated from *Neisseria sicca*, catalyzes the hydrolysis of both strands of DNA at 5′ . . . ATGCA∇T . . . 3′.

*Nsp*I RESTRICTION ENDONUCLEASE

This type II site-specific endonuclease [EC 3.1.21.4], isolated from *Nostoc* species C, catalyzes the hydrolysis of both strands of DNA at 5′ . . . RCATG∇Y . . . 3′, where R refers to A or G and Y refers to T or C.[1]

[1]J. Reaston, M. G. C. Duyvesteyn & A. de Waard (1982) *Gene* **20**, 103.

NUATIGENIN 3-β-GLUCOSYLTRANSFERASE

This transferase [EC 2.4.1.192] catalyzes the reaction of UDP-D-glucose with nuatigenin (or, (20*S*,22*S*,25*S*)-22,25-epoxyfurost-5-ene-3β,26-diol) to produce nuatigenin 3β-D-monoglucoside (or, (20*S*,22*S*,25*S*)-22,25-epoxyfurost-5-ene-3β,26-diol 3-*O*-β-D-glucoside) and UDP.[1,2]

This enzyme participates in the biosynthesis of plant saponins. Other substrates include isonuatigenin, chlorogenin, solanidine, and pregn-5-en-3β-ol-20-one (or, pregnenolone). The enzyme is not identical with sterol glucosyltransferase [EC 2.4.1.173] or sarsapogenin 3-β-glucosyltransferase [EC 2.4.1.193].

[1]M. Kalinowska & Z. A. Wojciechowski (1986) *Phytochemistry* **25**, 2525.
[2]M. Kalinowska & Z. A. Wojciechowski (1988) *Plant Sci.* **55**, 239.

NUCLEOPLASMIN ATPase

This enzyme [EC 3.6.4.11] catalyzes the ATP-dependent (ADP- and orthophosphate-forming) assembly of nucleosome cores as well as the decompaction/decondensation of sperm chromatin and other chromatin-reorganizing processes.[1,2]

Although the overall reaction yields ADP and orthophosphate, the enzyme is misclassified as a hydrolase, because P-O-P bond scission supplies the Gibbs free energy required to drive protein folding. The overall enzyme-catalyzed reaction should be regarded to be of the energase type (*e.g.*, $State_1 + GTP = State_1 + GDP + P_i$, where two affinity-modulated conformational states are indicated). Energases transduce chemical energy into mechanical work.[3]

[1]T. Ito, J. K. Tyler, M. Bulger, R. Kobayashi & J. T. Kadonaga (1996) *JBC* **271**, 25041.
[2]R. A. Laskey, A. D. Mills, A. Philpott, G. H. Leno, S. M. Dilworth & C. Dingwall (1993) *Phil. Trans. R. Soc. London B Biol. Sci.* **339**, 263.
[3]D. L. Purich (2001) *TiBS* **26**, 417.

NUCLEOSIDE DEOXYRIBOSYLTRANSFERASE

This transferase [EC 2.4.2.6], also known as nucleoside:purine(pyrimidine) deoxy-D-ribosyltransferase, catalyzes the reaction of 2-deoxy-D-ribosyl-base$_1$ with base$_2$ to produce 2-deoxy-D-ribosyl-base$_2$ and base$_1$ in which base$_1$ and base$_2$ are purines and/or pyrimidines.[1,2] The enzyme has a ping pong Bi Bi kinetic mechanism in which a covalent deoxyribosyl-enzyme intermediate is formed at an active-site glutamyl residue (thus, a 2-deoxyribofuranosylated carboxylate). An aspartyl residue functions as a general acid in the catalytic mechanism.[3] In the absence of a base[2], the *Lactobacillus leishmanii* enzyme slowly hydrolyzes nucleosides.

Two distinct transferases have been identified in *Lactobacillus leichmannii*, one of which is specific for purines as the base[1] and base[2] entities.[4]

[1] C. Danzin & R. Cardinaud (1974) *EJB* **48**, 255.
[2] C. Danzin & R. Cardinaud (1976) *EJB* **62**, 365.
[3] S. A. Short, S. R. Armstrong, S. E. Ealick & D. J. T. Porter (1996) *JBC* **271**, 4978.
[4] J. Becker & M. Brendel (1996) *Biol. Chem. Hoppe-Seyler* **377**, 357.

Selected entries from *Methods in Enzymology* [vol, page(s)]:
General discussion: 2, 464; 51, 446
Assay: 2, 464; 51, 446
Other molecular properties: activity, 51, 446; distribution, 51, 452; isozymes, 51, 452; kinetic properties, 51, 452; *Lactobacillus helveticus*, from, 2, 467; 51, 446; molecular weight, 51, 450; pH optimum, 51, 452; stability, 51, 450; substrate specificity, 2, 468; 51, 450, 452

NUCLEOSIDE DIPHOSPHATASE

This enzyme [EC 3.6.1.6], also known as inosine diphosphatase and abbreviated as NDPase, catalyzes the hydrolysis of a nucleoside diphosphate to produce a nucleoside monophosphate and orthophosphate.[1-4] Nucleoside diphosphate substrates include IDP, GDP, UDP, as well as D-ribose 5-diphosphate and thiamin diphosphate. **See also** *Guanosine Diphosphatase*

[1] S. Sano, Y. Matsuda & H. Nakagawa (1988) *EJB* **171**, 231.
[2] S. Sano, Y. Matsuda, S. Miyamoto & H. Nakagawa (1984) *BBRC* **118**, 292.
[3] K. O'Toole (1982) *Enzyme* **28**, 362.
[4] I. Ohkubo, T. Ishibashi, N. Taniguchi & A. Makita (1980) *EJB* **112**, 111.

Selected entries from *Methods in Enzymology* [vol, page(s)]:
General discussion: 6, 231
Assay: 6, 231; 135, 155
Other molecular properties: chromatographic separation, 32, 401, 402, 404; complex dissociation, 64, 161, 162; coupling enzyme, as a, 63, 32; endoplasmic reticulum, of, 10, 15; Golgi apparatus, in, 10, 15; 22, 144; membrane marker, as, 32, 393, 403; microsomal fractions, in, 52, 83, 88; myokinase, not present in preparation of, 6, 230; plant Golgi apparatus, in, 22, 148; preparation, 6, 232; 135, 155; properties, 6, 235; Sepharose-coupled (continuous reaction with, 135, 160; preparation, 135, 157; properties, 135, 157)

NUCLEOSIDE-DIPHOSPHATE KINASE

This phosphotransferase [EC 2.7.4.6], also known as nucleoside diphosphokinase and nucleoside 5'-diphosphate phosphotransferase, catalyzes the reversible reaction of ATP with a nucleoside diphosphate (NDP) to produce ADP and a nucleoside triphosphate (NTP).[1-5] ATP can be substituted by a number of nucleoside triphosphate and deoxynucleoside triphosphate compounds: for example, GTP, CTP, UTP, ITP, dTTP, dCTP, and dGTP. Acceptor substrates include ADP, GDP, CDP, UDP, IDP, dCDP, dADP, dGDP, and dTDP. The 3'-hydroxyl group is important for catalysis.[1] It has been suggested that the broad specificity may be due to a mixture of isozymes, each having a narrower specificity.[4] The kinase has a ping pong Bi Bi kinetic mechanism[5] with a phospho-enzyme intermediate (at His122 in the *Dictyostelium* enzyme: at the δ-N).

NM23 proteins, which bind DNA and regulate a diverse array of cellular events including growth and development, have a nucleoside-diphosphate kinase activity.

[1] P. Gonin, Y. Xu, L. Milon, S. Dabernat, M. Morr, R. Kumar, M. L. Lacombe, J. Janin & I. Lascu (1999) *Biochemistry* **38**, 7265.
[2] A. M. Chakrabarty (1998) *Mol. Microbiol.* **28**, 875.
[3] N. B. Ray & C. K. Mathews (1992) *Curr. Top. Cell. Reg.* **33**, 343.
[4] R. E. Parks, Jr., & R. P. Agarwal (1973) *The Enzymes*, 3rd ed., **8**, 307.
[5] E. Garces & W. W. Cleland (1969) *Biochemistry* **8**, 633.

Selected entries from *Methods in Enzymology* [vol, page(s)]:
General discussion: 6, 163; 10, 769; 51, 371, 376; retinal, 316, 87
Assay: 6, 163; 51, 371, 376; 55, 287; 57, 30; microtubule-associated, 85, 431; NDP:NTP phosphotransferase (radioisotopic, 69, 347, 348; spectrophotometric, 69, 347)
Other molecular properties: acetate kinase and, 87, 18; activity, 51, 371, 376, 386; 57, 86; activity by ADP-ATP exchange enzyme, 10, 542; adenosine phosphosulfate kinase, contaminant in preparation of, 5, 975; adenosine 5'-O-(1-thiodiphosphate), and, 87, 200, 205, 300; adenosine triphosphate-phosphate exchange and, 6, 323; affinity for calcium phosphate gel, 51, 584; amino acid incorporation enzymes and, 5, 734, 735; cation requirements, 51, 385; CDP kinase (CDP reductase, presence in preparation of, 12A, 163); contaminant, in luciferase preparation, 57, 35, 82, 62; determination of guanosine nucleotides, 57, 86, 92; diffraction, 276, 282; exchanges, 64, 8, 34; N^{10}-formyltetrahydrofolate synthase and, 6, 379; human erythrocytes, from, 51, 376; inosine diphosphatase and, 6, 233; interference by, in ATP assay, 57, 29; isoelectric variation, 51, 378, 379; kinetic properties, 51, 386; mechanism, 51, 375, 384; 64, 8, 34; 87, 3, 6, 8, 9, 300; microtubule-associated, assay, 85, 431; molecular weight, 51, 375, 385; nucleotide kinase (nucleoside triphosphate preparation and, 29, 104, 112; preparation, 29, 111; terminal, identification of, 29, 280); pH effects, 6, 166; 51, 386; phosphoenolpyruvate carboxykinase, and, 1, 759; properties, 6, 165; purification, 6, 164; 51, 373; 60, 581; reaction mechanism, 51, 375, 384; 87, 3, 6, 8, 9, 300; retinal, 316, 87; *Salmonella typhimurium*, from, 51, 371; stability, 51, 375; stereochemistry, 87, 200, 205, 231, 300; substrate specificity, 6, 165; 51, 375, 384; synergism quotient, 87, 8; temperature dependence, 51, 385; thiol group activity, 51, 385, 386

NUCLEOSIDE OXIDASE

This FAD-dependent oxidoreductase [EC 1.1.3.28] catalyzes the reaction of two molecules of inosine with two molecules of dioxygen to produce two molecules of 9-ribouronosylhypoxanthine and four molecules of water.[1,2] Other purine nucleoside substrates include xanthosine, guanosine, adenosine, uridine, deoxyguanosine, deoxyadenosine, and deoxyinosine. Pyrimidine nucleoside substrates include cytidine, deoxycytidine, and thymidine. Ribose and nucleotides are not substrates. With inosine as the model substrate, 5'-dehydroinosine is an intermediate in the reaction and reacts with the second molecule of dioxygen to generate the final product.

This enzyme is distinguished from nucleoside oxidase [EC 1.1.3.39] in that hydrogen peroxide is not a product.

[1]Y. Isono, T. Sudo & M. Hoshino (1989) Agric. Biol. Chem. 53, 1663 and 1671.
[2]Y. Isono & M. Hoshino (1989) Agric. Biol. Chem. 53, 2197.

NUCLEOSIDE OXIDASE (HYDROGEN-PEROXIDE-FORMING)

This heme- and FAD-containing oxidoreductase [EC 1.1.3.39] catalyzes the reaction of adenosine with two molecules of dioxygen to produce 9-ribouronosyladenine and two molecules of hydrogen peroxide.[1] Other substrates include other purine and pyrimidine nucleosides, 2'-deoxynucleosides, and arabinosides. Ribose and nucleotides are not substrates. Production of hydrogen peroxide distinguishes this enzyme from nucleoside oxidase [EC 1.1.3.28].

In the reaction scheme, the nucleoside first reacts with one molecule of dioxygen to generate a 5'-dehydronucleoside intermediate and the first molecule of hydrogen peroxide. This intermediate then reacts with the second dioxygen molecule to produce the final product and the second molecule of hydrogen peroxide.

[1]S. Koga, J. Ogawa, L. Y. Cheng, Y. M. Choi, H. Yamada & S. Shimizu (1997) Appl. Environ. Microbiol. 63, 4282.

NUCLEOSIDE-PHOSPHATE KINASE

This phosphotransferase [EC 2.7.4.4], also known as NMP-kinase, catalyzes the reaction of ATP with a nucleoside phosphate (or, nucleoside monophosphate, NMP) to produce ADP and a nucleoside diphosphate (NDP). NMP acceptor substrates include AMP and GMP.[1,2] See also Adenylate Kinase; Guanylate Kinase; Uridylate Kinase; Cytidylate Kinase

[1]H. Yan & M. D. Tsai (1999) AE 73, 103.
[2]D. M. Gibson, P. Ayengar & D. R. Sanadi (1956) BBA 21, 86.

Selected entries from Methods in Enzymology [vol, page(s)]:
Other molecular properties: adenosine triphosphate-phosphate exchange system, not present in, 6, 323; cytidine 5'-diphosphate regeneration system, 247, 114; intermediate, 87, 3; nucleotide kinase (nucleoside triphosphate preparation and, 29, 104, 112; preparation, 29, 111; terminal, identification of, 29, 280); occurrence, 2, 603

NUCLEOSIDE PHOSPHOACYLHYDROLASE

This enzyme [EC 3.6.1.24] catalyzes the hydrolysis of mixed phospho-anhydride bonds: for example, substrates include ribonucleoside 5'-nitrophenylphosphates (producing p-nitrophenol and the nucleoside monophosphate) and acetyladenylate (producing acetate and AMP).[1]

[1]P. F. Spahr & R. F. Gesteland (1970) EJB 12, 270.

NUCLEOSIDE PHOSPHOTRANSFERASE

This transferase [EC 2.7.1.77] catalyzes the reaction of a nucleotide (that is, NMP) with a 2'-deoxynucleoside to produce a nucleoside and a 2'-deoxynucleoside 5'-monophosphate (dNMP).[1-3] The nucleotide substrate can be substituted with phenyl phosphate (for example, phenyl phosphate will react with uridine to produce phenol and UMP) and nucleoside 3'-phosphates, albeit not as effectively. Acceptor substrates include adenosine, deoxyadenosine, uridine, cytidine, ribothymidine, thymidine, guanosine, and deoxyguanosine (in many organisms, there is a decided preference for 2'-deoxynucleoside acceptor substrates).

The barley (Hordeum vulgare) enzyme has a branched ping pong Bi Bi kinetic mechanism involving a phospho-enzyme intermediate,[3] a kinetic scheme similar to that observed with glucose-6-phosphatase and γ-glutamyl transpeptidase. This intermediate can transfer its phosphoryl group either to a nucleoside or to water, the latter accounting for its intrinsic hydrolase activity.

[1]P. A. Frey (1992) The Enzymes, 3rd ed., 20, 141.
[2]P. A. Frey (1989) AE 62, 119.
[3]D. C. Prasher, M. C. Carr, D. H. Ives, T. C. Tsai & P. A. Frey (1982) JBC 257, 4931.

Selected entries from Methods in Enzymology [vol, page(s)]:
General discussion: 51, 387
Assay: 51, 387, 388
Other molecular properties: activity, 51, 387, 390, 391; amino acid analysis, 51, 390; carrot, from, 51, 387; cations and anions, effect of, 51, 393; hydrolase activity, characteristics, 51, 392; isoelectric point, 51, 390; kinetic properties, 51, 391; molecular weight, 51, 390; pH optimum, 51, 392; properties, 51, 390; purification, 51, 388; purity, 51, 390; stability, 51, 394; stereochemistry, 87, 206, 231, 300; substrate specificity, 51, 391, 392

NUCLEOSIDE RIBOSYLTRANSFERASE

This transferase [EC 2.4.2.5], also known as nucleoside transribosidase, catalyzes the reversible reaction of

a nucleoside (that is, a D-ribosyl-base$_1$) with a base$_2$ to produce a new nucleoside (*i.e.*, D-ribosyl-base$_2$) and base$_1$ where base$_1$ and base$_2$ represent various purines and pyrimidines.[1,2] Nucleoside substrates include adenosine, inosine, xanthosine, guanosine, and uridine. Acceptor substrates include adenine, hypoxanthine, guanine, xanthine, and uracil. *See also Nucleoside Deoxyribosyltransferase*

[1]A. L. Koch (1956) *JBC* **223**, 535.
[2]A. Kamimura, K. Mitsugi & S. Okumura (1973) *Agric. Biol. Chem.* **37**, 2063.

NUCLEOSIDE TRIPHOSPHATASE

This general class of so-called phosphohydrolases [EC 3.6.1.15] cleaves the terminal phosphoryl from a variety of nucleoside 5′-triphosphates.[1–5] Because the "hydrolase" activity is almost invariably linked to some other noncovalent substrate-like and product-like state, these reactions are not ATPases! Rather, they represent the "energases", a newly recognized class of enzymes that transduce chemical energy into mechanical work. While the overall reaction yields ADP and orthophosphate, P-O-P bond scission supplies the Gibbs free energy required to drive protein folding, ion and solute transport, cytoskeletal filament assembly, proteasome reactions, as well as other affinity-modulated reactions involving the broad class of G-proteins.[6] The overall enzyme-catalyzed reaction should be regarded to be of the energase type (*e.g.*, State$_1$ + ATP (or GTP) = State$_1$ + ADP (or GDP) + P$_i$, where two affinity-modulated conformational states are indicated.) *See also F$_o$F$_1$ ATP Synthase; ATPase; ATPase, Calcium-Dependent; Myosin ATPase; Actin ATPase; G-Protein GTPases; Energases*

[1]H. C. Schröder, M. Rottmann, M. Bachmann & W. E. G. Müller (1986) *JBC* **261**, 663.
[2]D. R. McCarty & B. Selman (1986) *ABB* **248**, 523.
[3]F. Harper, F. Lamy & R. Calvert (1978) *Can. J. Biochem.* **56**, 565.
[4]S. Matsushita & J. D. Raake (1968) *BBA* **166**, 707.
[5]M. Lewis & S. Weisman (1965) *ABB* **109**, 490.
[6]D. L. Purich (2001) *TiBS* **26**, 417.

NUCLEOSIDE TRIPHOSPHATE: ADENYLATE KINASE

This phosphotransferase [EC 2.7.4.10], also known as GTP: AMP phosphotransferase, catalyzes the reaction of GTP (or rather MgGTP) with AMP to produce GDP (or, MgGDP) and ADP.[1] Many nucleoside triphosphates besides GTP can act as donor substrates: for example, ITP, UTP, CTP, and dGTP (interestingly, ATP is a poor donor substrate). The enzyme is fairly specific for AMP as the acceptor substrate, although dAMP will also be phosphorylated.

[1]G. J. Albrecht (1970) *Biochemistry* **9**, 2462.

NUCLEOSIDE TRIPHOSPHATE: HEXOSE-1-PHOSPHATE NUCLEOTIDYLTRANSFERASE

This transferase [EC 2.7.7.28], also known as NDP-hexose pyrophosphorylase, catalyzes the reversible reaction of a nucleoside triphosphate (NTP) with a hexose 1-phosphate to produce a NDP-hexose and pyrophosphate.[1,2] In the reverse reaction the NDP-hexose can be, in decreasing order of activity, guanosine, inosine, and adenosine diphosphate hexoses in which the sugar is either D-glucose or D-mannose.

[1]K. A. Presper & E. C. Heath (1983) *The Enzymes*, 3rd ed., **16**, 449.
[2]H. Verachtert, P. Rodriguez, S. T. Bass & R. G. Hansen (1966) *JBC* **241**, 2007.

NUCLEOSIDE TRIPHOSPHATE PYROPHOSPHATASE

This enzyme [EC 3.6.1.19], also known as nucleoside-triphosphate diphosphatase, catalyzes the hydrolysis of a nucleoside 5′-triphosphate (including ITP, XTP, GTP, dGTP. dITP, UTP, and dUTP) to produce the corresponding nucleoside monophosphate and pyrophosphate.[1–3] The enzyme may be identical to nucleotide pyrophosphatase [EC 3.6.1.9]. *See also ATP Pyrophosphatase; Apyrase*

[1]C. J. Chern, A. B. McDonald & A. J. Morris (1969) *JBC* **244**, 5489.
[2]J. K. Wang & A. J Morris (1974) *ABB* **161**, 118.
[3]S. A. Fuller & A. J. Morris (1981) *Biochem. Genet.* **19**, 955.
Selected entries from *Methods in Enzymology* [**vol**, page(s)]:
General discussion: 51, 275
Assay: 51, 275, 276

NUCLEOTIDASE

This hydrolase [EC 3.1.3.31] (also known as nucleotide phosphatase, nucleotide phosphohydrolase, and acid nucleotidase) catalyzes the hydrolysis of a nucleotide (or, nucleoside monophosphate, NMP) to produce a nucleoside and orthophosphate.[1–3] The enzyme exhibits a wide specificity for 2′-, 3′-, and 5′-nucleotides. Other substrates include glycerol phosphate and 4-nitrophenyl phosphate, the latter chromogenic substrate affording a convenient visible spectrophotometric assay. Examples of nucleotide substrates include 5′-AMP, 3′-AMP, 2′-AMP, 3′-UMP, 5′-dUMP, and 5′-dTMP.

[1]H. Tjershaugen (1979) *Acta Chem. Scand.* **B33**, 384.
[2]P. Fritzson (1978) *Adv. Enzym. Regul.* **16**, 43.
[3]J. Delaunay, S. Fischer, J.-P. Piau, M. Tortolero & G. Schapira (1978) *BBA* **527**, 425.
Selected entries from *Methods in Enzymology* [**vol**, page(s)]:
General discussion: 51, 271

Assay: 51, 272
Other molecular properties: acid nucleotidase, 51, 271 (activators, 51, 274; assay, 51, 272; inhibitors, 51, 274; kinetic properties, 51, 274; molecular weight, 51, 275; pH optimum, 51, 274; purification, 51, 273, 274; purity, 51, 274; rat liver lysosomes, from, 51, 271; substrate specificity, 51, 274; temperature optimum, 51, 274); adenylate cyclase systems, in, 31, 111; assay for assessment of phagocytic function, 132, 121, 162; comparative properties, 51, 290

3'-NUCLEOTIDASE

This enzyme [EC 3.1.3.6] catalyzes the hydrolysis of a 3'-ribonucleotide to produce a ribonucleoside and orthophosphate.[1–4] The enzyme exhibits a broad substrate specificity toward 3'-nucleotides, including 3'-AMP, 3'-IMP, 3'-GMP, 3'-UMP, 3'-CMP. *See also Aspergillus Nuclease S₁; Penicillium Nuclease P₁*

[1]T. T. Nguyen, M. M. Palcic & D. Hadziyev (1987) *J. Chromatogr.* **391**, 257.
[2]K. Endo, Y. Umeyama, J. Nakajima & H. Kawai (1980) *Agric. Biol. Chem.* **44**, 1545.
[3]G. I. Drummond & M. Yamamoto (1971) *The Enzymes*, 3rd ed., **4**, 337.
[4]L. Shuster & N. O. Kaplan (1953) *JBC* **201**, 535.

Selected entries from *Methods in Enzymology* [vol, page(s)]:
General discussion: 2, 551
Assay: 2, 551
Other molecular properties: action, 3, 742; ADP, in assay of, 63, 32; adsorbents (alumina and calcium phosphate gel) and, 1, 95, 96; 3'-CMP, action on, 3, 755; 5'-CMP, lack of action on, 3, 755; coenzyme A, action on, 3, 930; coupling enzyme, as a, 63, 32; di(dinitrophenyl) phosphate and, 2, 524; 2',5'-diphosphoadenosine, lack of action on, 3, 906; 3',5'-diphosphoadenosine, action on, 3, 906; heat lability of, 2, 554; rye grass, 3'-phosphoadenosine-5'-phosphosulfate, action on, 6, 772; plasma membrane and, 31, 90; properties, 2, 553; purification, rye grass, 2, 552; sulfhydryl inhibition, 3, 930; terminal 3'-phosphomonoester assay of oligonucleotides, and, 3, 765

5'-NUCLEOTIDASE

This enzyme [EC 3.1.3.5] catalyzes the hydrolysis of a 5'-ribonucleotide to produce a ribonucleoside and orthophosphate.[1–3] The enzyme exhibits a broad substrate specificity toward 5'-nucleotides, including 5'-AMP, 5'-UMP, 5'-CMP, 5'-GMP, 5'-IMP, 5'-XMP, 5'-dAMP. 5'-dGMP, 5'-dCMP, 5'-dIMP, 5'-dUMP, and 5'-dTMP. Nevertheless, substrate specifity can vary among different isozymes.

[1]H. Zimmermann (1992) *BJ* **285**, 345.
[2]G. I. Drummond & M. Yamamoto (1971) *The Enzymes*, 3rd ed., **4**, 337.
[3]L. A. Heppel (1961) *The Enzymes*, 2nd ed., **5**, 49.

Selected entries from *Methods in Enzymology* [vol, page(s)]:
General discussion: 2, 546; 32, 368
Assay: 2, 547, 549; 31, 91, 92; 62, 124, 368, 369
Other molecular properties: aceto-coenzyme A kinase, not present in preparation of, 5, 464; activity and distribution, 32, 124; adenosine 5'-phosphosulfate, action on, 6, 772; agarose chromatography,

32, 372, 373; assay for HL60 cell differentiation, in, 190, 129; bacterial membranes, in, 22, 113; borate inhibition, 12A, 101; brush borders, in, 31, 122; cGMP assay, 38, 88; chiral thiophosphate, preparation, 87, 257, 265, 300; cyclic nucleotide phosphodiesterase histo- and cytochemistry, in, 159, 478; 2-deoxy-D-ribose 5-phosphate, action on, 3, 187; diagnostic enzyme, as, 31, 19, 24, 35, 93, 262, 729; diesterase and, 2, 561; end group determination and, 3, 766; endosome marker, determination, as, 109, 267; extraction from rat liver microsomes, 22, 206; FAD pyrophosphatase, and, 280, 424; fat-cell plasma membrane, in, 31, 65, 67; Golgi apparatus, 22, 140, 141; HeLa cells, in, 22, 127, 128; histochemical method for, 4, 386; laminin binding properties, 144, 506; liver, 31, 97; marker enzyme, as, 32, 124, 393 (membrane flow during pinocytosis, for, 98, 395); microsomal fractions, in, 52, 88 (microsomal peroxide production system, in, 52, 344; oxidation procedure, 52, 92 [in reductase assay, 52, 91]); nucleoside diphosphates and, 3, 742; oligonucleotide identification and, 12A, 283; osmotic shock and, 12B, 844; perchloric acid, 55, 201; phagolysosomes, in, 31, 344, 345; phosphodiesterase I, removal of in preparation of, 6, 239; phosphodiesterase, assay, 2, 561; 38, 133, 205, 210, 218, 224, 225, 240, 249, 250, 263, 264; plasma membranes, in, 22, 127, 135, 141; 31, 88, 91, 100, 101, 148; 32, 91; potato, 2, 550; properties, 2, 549, 550; 32, 126, 373, 374; purification (potato, 2, 550; seminal plasma, 2, 547; snake venom, 2, 549; sphingomyelin complex, as, 32, 368); purine nucleotides and, 3, 756; pyrimidine nucleotides and, 3, 755; release from membranes by phospholipase C, 71, 739, 745; snake venom enzyme used in hydrolysis of ribonucleotides of nicotinate and nicotinamide, 18B, 60; specificity of, 3, 762; staining on gels, 22, 601; stereochemistry, 87, 207, 258, 265, 300, 316; viscosity barrier centrifugation, 55, 133, 134

NUCLEOTIDE PYROPHOSPHATASE

This enzyme [EC 3.6.1.9], also known as nucleotide diphosphatase, catalyzes the hydrolysis of a dinucleotide to produce two mononucleotide molecules. Substrates include NAD^+, NADH, $NADP^+$, and FAD.[1–4] The enzyme will also act on certain mononucleotides such as ATP, ADP, thiamin pyrophosphate, and coenzyme A. *See also Nucleoside-Triphosphate Pyrophosphatase; Inorganic Pyrophosphatase*

[1]K. B. Jacobson & N. O. Kaplan (1957) *JBC* **226**, 427.
[2]K. Decker & E. Bischoff (1972) *FEBS Lett.* **21**, 95.
[3]P. S. Bachorik & L. S. Dietrich (1972) *JBC* **247**, 5071.
[4]R. K. Haroz, J. S. Twu & R. K. Bretthauer (1972) *JBC* **247**, 1452.

Selected entries from *Methods in Enzymology* [vol, page(s)]:
General discussion: 2, 655
Assay: 2, 655
Other molecular properties: cleavage of capped mRNA pyrophosphate linkage, 180, 171; flavin adenine pyrophosphorylase, presence in preparation of, 2, 675; inhibition by ATP, 2, 673; NADH, action on by snake venom enzyme, 2, 654; 5-phosphoribosylpyrophosphate and, 6, 477; phosphorothioates, and, 87, 222; potato enzyme used in hydrolysis of NAD, 18B, 53; preparation, potato, 2, 656; properties, 2, 659; purification of flavin peptides, 53, 455

NUCLEOTIDE PYROPHOSPHOKINASE

This enzyme [EC 2.7.6.4], also known as nucleotide diphosphokinase, catalyzes the reaction of ATP with a nucleoside 5′-phosphate (NMP; for example, AMP and GMP) to produce AMP and a 5′-phosphonucleoside 3′-diphosphate.[1,2] In addition to monophosphate derivatives, nucleoside 5′-diphosphates (for example, GDP, IDP, and ADP) and nucleoside 5′-triphosphates (e.g., GTP, ITP, CTP, UTP, and ATP) are also substrates.

[1] T. Nishino & S. Murao (1975) *Agric. Biol. Chem.* **39**, 1827.
[2] T. Oki, A. Yoshimoto, S. Sato & A. Takamatsu (1975) *BBA* **410**, 262.

OCTADECANAL DECARBONYLASE

This enzyme [EC 4.1.99.5] catalyzes the conversion of *n*-octadecanal to *n*-heptadecane and carbon monoxide.[1]

[1] T. M. Cheesbrough & P. E. Kolattukudy (1988) *JBC* **263**, 2738.

OCTANOL DEHYDROGENASE

This oxidoreductase [EC 1.1.1.73] catalyzes the reaction of 1-octanol with NAD^+ to produce 1-octanal and NADH.[1,2] Other substrates include 1-heptanol, 1-hexanol, 1-pentanol, and 1-butanol.

[1] F. Sieber, D. J. Fox & H. Ursprung (1972) *FEBS Lett.* **26**, 274.
[2] B. Roche & E. Azoulay (1969) *EJB* **8**, 426.

Selected entries from *Methods in Enzymology* [**vol**, page(s)]:
General discussion: **188**, 171

trans-OCTAPRENYL*trans*TRANSFERASE

This enzyme [EC 2.5.1.11], also known as all-*trans*-nonaprenyl-diphosphate synthase and solanesyl-diphosphate synthase, catalyzes the reaction of all-*trans*-octaprenyl diphosphate with isopentenyl diphosphate to produce pyrophosphate (or, diphosphate) and all-*trans*-nonaprenyl diphosphate.[1-3]

octaprenyl diphosphate

nonaprenyl diphosphate

Other donor substrates include geranyl diphosphate and all-*trans*-prenyl diphosphates of intermediate size. The concentration of isopeptenyl diphosphate affects the chain length distribution of the products formed by this enzyme.[3]

[1] S.-i. Ohnuma, T. Koyama & K. Ogura (1991) *JBC* **266**, 23706.
[2] T. Gotoh, T. Koyama & K. Ogura (1992) *J. Biochem.* **112**, 20.
[3] S. Ohnuma, T. Koyama & K. Ogura (1992) *J. Biochem.* **112**, 743.

Selected entries from *Methods in Enzymology* [**vol**, page(s)]:
General discussion: 110, 206
Assay: 110, 206
Other molecular properties: *Micrococcus luteus* (assay, **110**, 206; properties, **110**, 208; purification, **110**, 207)

OCTOPAMINE DEHYDRATASE

This dehydratase [EC 4.2.1.87] catalyzes the conversion of 1-(4-hydroxyphenyl)-2-aminoethanol (or, octopamine) to (4-hydroxyphenyl)acetaldehyde, ammonia, and water.

octopamine   (4-hydroxyphenyl)acetaldehyde

The enzyme-catalyzed reaction is believed to be dehydration to an enamine which spontaneously hydrolyzes to an aldehyde and ammonia.[1]

[1] S. M. Cuskey, V. Peccoraro & R. H. Olsen (1987) *J. Bacteriol.* **169**, 2398.

D-OCTOPINE DEHYDROGENASE

This oxidoreductase [EC 1.5.1.11], also known as D-octopine synthase, catalyzes the reversible reaction of N^2-(D-1-carboxyethyl)-L-arginine (that is, octopine) with NAD^+ and water to produce L-arginine, pyruvate, and NADH.[1-6] In the reverse direction, the enzyme acts on L-ornithine, L-lysine, L-homoarginine, L-histidine, L-canavaline, and L-citrulline. The sea mollusc *Concholepas concholepas* enzyme has a random Bi Ter kinetic mechanism[1] and the enzyme isolated from the giant scallop (*Pecten maximus*) has an ordered bi ter mechanism.[2] A cysteinyl residue has been identified as essential for catalytic activity.[3]

[1] N. Carvajal & E. Kessi (1988) *BBA* **953**, 14.
[2] M. O. Doublet & A. Olomucki (1975) *EJB* **59**, 175.

[3]S. Sheikh & S. S. Katiyar (1993) *BBA* **1202**, 251.

[4]J. Thompson & J. A. Donkersloot (1992) *Ann. Rev. Biochem.* **61**, 517.

[5]J. Thompson & S. P. F. Miller (1991) *AE* **64**, 317.

[6]H. F. Fisher (1988) *AE* **61**, 1.

Selected entries from *Methods in Enzymology* [vol, page(s)]:
Other molecular properties: assay in monocot transformants, **153**, 311; cooperativity and slow transitions in, **249**, 545; product inhibition studies, **249**, 209; stereospecificity, **54**, 231; **87**, 118

*Okr*AI RESTRICTION ENDONUCLEASE

This type II restriction enzyme [EC 3.1.21.4], isolated from *Oceanospirillum kriegii*, catalyzes the hydrolysis of DNA and acts on both strands at 5′ . . . G∇GATCC . . . 3′.

OLEATE HYDRATASE

This enzyme [EC 4.2.1.53] catalyzes the reversible conversion of (*R*)-10-hydroxystearate to oleate and water as well as the conversion of palmitoleate to oleate and water.[1,2]

[1]A. Kisic, Y. Miura & G. J. Schroepfer (1971) *Lipids* **6**, 541.

[2]W. G. Niehaus, A. Kisic, A. Torkelson, D. J. Bednarczyk & G. J. Schroepfer (1970) *JBC* **245**, 3790.

OLEOYL-[ACYL-CARRIER PROTEIN] HYDROLASE

This enzyme [EC 3.1.2.14], also known as *S*-acyl fatty acid synthase thioesterase and acyl-[acyl-carrier protein] hydrolase, catalyzes the hydrolysis of oleoyl-[acyl-carrier protein] to produce the free acyl-carrier protein and oleate.[1,2] Fatty acids linked to the acyl-carrier protein as thiolesters can contain a 12- to 18-carbon chain length that is either saturated or unsaturated. The corresponding oleoyl derivative is hydrolyzed much more rapidly than other acyl derivatives. *See also Palmitoyl-CoA Hydrolase*

[1]J. B. Ohlrogge, W. E. Shine & P. K. Stumpf (1978) *ABB* **189**, 382.

[2]I. Löhden & M. Frentzen (1988) *Planta* **176**, 506.

Selected entries from *Methods in Enzymology* [vol, page(s)]:
General discussion: 35, 101
Assay: 71, 180
Other molecular properties: fatty acid synthase of uropygial gland, from, **71**, 230 (amino acid composition, **71**, 233; effect of bovine serum albumin, **71**, 233; immunological comparison, **71**, 233; inhibitors, **71**, 233; molecular weight, **71**, 233; substrate specificity, **71**, 233; trypsin treatment, **71**, 231); interference in butyryl-CoA dehydrogenase assay, **71**, 360; long-chain, from rat, **71**, 181 (activators and inhibitors, **71**, 187; comparison of liver and mammary gland enzymes, **71**, 188; immunochemical studies, **71**, 186; location of thioesterase I domains, **71**, 186; properties, **71**, 186; separation of intact and nicked polypeptides, **71**, 185; specificity, **71**, 187; trypsinization of fatty acid synthase, **71**, 183); medium-chain (lactating goat mammary gland, from, **71**, 200, 219 [effect of mammary microsomal fractions from different species, **71**, 224; pattern of fatty acids synthesized, **71**, 221; properties, **71**, 219; specificity, **71**, 229]; lactating rabbit mammary gland, from, **71**, 200, 204 [chain termination assay, **71**, 201; concentration, **71**, 212; 5,5′-dithiobis(2-nitrobenzoic acid) assay, **71**, 200; interaction with fatty acid synthase, **71**, 214; preparation of substrates, **71**, 203; properties,

71, 209; purification of fatty acid synthase, **71**, 204; reactivation assay, **71**, 202; specificity, **71**, 209]; lactating rat mammary gland, from, **71**, 188 [anti-rat fatty acid synthase immunoglobulins, **71**, 191; assay of thioesterase II, **71**, 192; inhibitors, **71**, 199; properties, **71**, 198; specificity, **71**, 198; trypsinized fatty acid synthase core, **71**, 190]); safflower, from, **71**, 178 (assay, **71**, 179; properties, **71**, 180; specificity, **71**, 180)

OLIGODEOXYRIBONUCLEATE EXONUCLEASE

This activity, originally classified as EC 3.1.4.29, is now a deleted Enzyme Commission entry.

OLIGOGALACTURONIDE LYASE

This lyase [EC 4.2.2.6] catalyzes the conversion of 4-(4-deoxy-β-D-gluc-4-enuronosyl)-D-galacturonate to two molecules of 5-dehydro-4-deoxy-D-glucuronate.[1,2] It also catalyzes the removal of unsaturated terminal residues from oligosaccharides of D-galacturonate.

[1]F. Moran, S. Nasuno & M. P. Starr (1968) *ABB* **125**, 734.

[2]C. Hatanaka & J. Ozawa (1970) *Agric. Biol. Chem.* **34**, 1618.

1,3-β-OLIGOGLUCAN PHOSPHORYLASE

This enzyme [EC 2.4.1.30], also known as β-1,3-oligoglucan:orthophosphate glucosyltransferase II and first isolated from *Euglena gracilis*, catalyzes the reversible reaction of [(1,3)-β-D-glucosyl]$_n$ (that is, a polymer of glucose linked β1 → 3) with orthophosphate to produce [(1,3)-β-D-glucosyl]$_{n-1}$ and α-D-glucose 1-phosphate.[1] The enzyme, which does not act on the seaweed β-1,3-glucan laminarin, differs in specificity from laminaribiose phosphorylase [EC 2.4.1.31] and 1,3-β-glucan phosphorylase [EC 2.4.1.97].

[1]L. R. Marechal (1967) *BBA* **146**, 417 and 431.

Selected entries from *Methods in Enzymology* [vol, page(s)]:
General discussion: 28, 957
Assay: 28, 957
Other molecular properties: isolation from *Euglena gracilis*, **28**, 957; specificity of, **28**, 958

OLIGO-1,6-GLUCOSIDASE

This enzyme [EC 3.2.1.10], also known as sucrase-isomaltase, *O*-glycosyl glycosidase, isomaltase, and limit dextrinase, catalyzes the hydrolysis of 1,6-α-D-glucosidic linkages in isomaltose and dextrins generated by α-amylase action on starch and glycogen. This glucosidase catalyzes the release of an unsubstituted glucose residue linked α1,6 to a main polysaccharide chain.[1–5] Isomaltulose (or, palatinose), isomaltotriose, and panose are also substrates. This enzyme will not act on glycogen.

Oligo-1,6-glucosidase is distinguished from amylo-α-1,6-glucosidase [EC 3.2.1.33], both by its preference for short-chain substrates and by not requiring the 6-glucosylated residue to be at a branch point. The intestinal mucosal enzyme exhibits sucrose α-glucosidase [EC 3.2.1.48] activity. The term "limit dextrinase" should be avoided to prevent confusion with α-dextrin endo-1,6-α-glucosidase [EC 3.2.1.41] and limit dextrinase [EC 3.2.1.142]. *See also Sucrose α-Glucosidase*

[1] G. Semenza (1991) *Indian J. Biochem. Biophys.* **28**, 331.
[2] A. R. Plant, S. Parratt, R. M. Daniel & H. W. Morgan (1988) *BJ* **255**, 865.
[3] Y. Suzuki & Y. Tomura (1986) *EJB* **158**, 77.
[4] H. Wacker, R. Aggeler, N. Kretchmer, B. O'Neill, Y. Takesue & G. Semenza (1984) *JBC* **259**, 4878.
[5] Y. Suzuki, R. Aoki & H. Hayashi (1982) *BBA* **704**, 476.

Selected entries from **Methods in Enzymology** [vol, page(s)]:
General discussion: 8, 562; 46, 377, 379
Assay: 8, 562
Other molecular properties: affinity labeling, **87**, 473; brush borders, in, **31**, 129, 130; properties, **8**, 565; purification from (intestine, **8**, 585; yeast, **8**, 563); sucrase complex, small-intestinal brush border membrane, **46**, 377, 379 (anchoring segment, orientation of N-terminal amino acid, **96**, 392; assembly, **96**, 399; biosynthesis [in vitro, **96**, 400; mechanism, **96**, 405]; lipid bilayer-embedded subunit, identification, **96**, 388); transition-state and multisubstrate analogues, **249**, 306

OLIGONUCLEOTIDASE

This enzyme [EC 3.1.13.3] catalyzes the exonucleolytic cleavage of oligonucleotides to yield nucleoside 5'-phosphates as well as hydrolyzing NAD^+ to NMN and AMP.[1-3] The enzyme prefers to act on short oligoribonucleotides, attacking first at the 3' end.

[1] S. K. Niyogi & A. K. Datta (1975) *JBC* **250**, 7307.
[2] A. K. Datta & S. N. Niyogi (1975) *JBC* **250**, 7313.
[3] M. Futai & D. Mizuno (1967) *JBC* **242**, 5301.

OLIGOPEPTIDASE A

This zinc-dependent metalloendopeptidase [EC 3.4.24.70], also known as OpdA, catalyzes peptide bond hydrolysis in N-blocked oligopeptides containing at least four aminoacyl residues.[1] If there are five or more residues present, the N-terminal amino group can be unblocked: Ac-(Ala)$_4$ and (Ala)$_5$ are substrates but (Ala)$_4$ is not. Aminoacyl residues distant from the scissile bond are also important: Z-Gly-Pro-Gly-∇-Gly-Pro-Ala is cleaved, but not Z-(Gly)$_5$. (Note: ∇ indicates the cleavage site.)

[1] C. A. Conlin & C. G. Miller (1995) *MIE* **248**, 567.

Selected entries from **Methods in Enzymology** [vol, page(s)]:
General discussion: 248, 567
Assay: nonhydrin method, **248**, 568; spectrophotometric aminopeptidase-coupled method, **248**, 569
Other molecular properties: activation by metal ions, **248**, 575; activity, **248**, 567; amino acid sequence, **248**, 208; detection on gels,

248, 571; dipeptidyl carboxypeptidase and, relationship, **248**, 575; discovery, **248**, 567; evolution, **248**, 552; heat-shock protein, as, **248**, 577; homologues, **248**, 551, 591; inhibition, **248**, 574; physiological functions, **248**, 555, 577; properties, **248**, 207; purification, **248**, 572; structurally related enzymes, **248**, 575; substrate specificity, **248**, 573

OLIGOPEPTIDASE B

This serine endopeptidase [EC 3.4.21.83], also known as protease II, catalyzes peptide-bond hydrolysis at -Arg–Xaa- and -Lys–Xaa- , even when Xaa (that is, the P1' position) is a prolyl residue.[1-3] The descriptor "B" refers to a preference for *basic* aminoacyl residues at the P1 position. The enzyme has an esterase activity as well and will catalyze the hydrolysis of Bz-Arg-OEt and Tos-Arg-OMe. It will also cleave Bz-Arg-NHPhNO$_2$, Bz-Lys-NHNap, and Bz-Arg-NHMec. It will not act on substrates larger than 6 kDa.[1,2] The oligopeptidase has been isolated from *Escherichia coli*, *Rhodococcus erythropolis*, soybean seed, and *Trypanosoma brucei*.

Another enzyme bearing the same name (endo-oligopeptidase B) is now listed as prolyl oligopeptidase [EC 3.4.21.26].

[1] D. Tsuru & T. Yoshimoto (1994) *MIE* **244**, 201.
[2] N. D. Rawlings & A. J. Barrett (1994) *MIE* **244**, 19.
[3] B. Burleigh & N. W. Andrews (1995) *JBC* **270**, 5172.

Selected entries from **Methods in Enzymology** [vol, page(s)]:
General discussion: 80, 681; 244, 201
Assay: fluorometric, **244**, 202; spectrophotometric, **244**, 201
Other molecular properties: *Escherichia coli* (active site labeling, **244**, 208; active site residues, **244**, 214; encoding gene [cloning, **244**, 203; expression, **244**, 203; locus, **244**, 214; screening, **244**, 202; sequence, **244**, 206, 210; structure, **244**, 206, 208]; family, **244**, 209, 211; homology with other proteases, **244**, 209, 211, 212; inhibitors, **244**, 205; pH optimum, **244**, 205; purification, **244**, 203, 205; sequence, **244**, 206, 210; size, **244**, 205, 208; substrate specificity, **244**, 205, 207, 213); inhibitors, **244**, 201; species distribution, **244**, 201; substrate specificity, **244**, 201, 213

OLIGOPEPTIDASE E

This cysteine endopeptidase, also called PepE, catalyzes peptide-bond hydrolysis on oligopeptides; it will not act on casein or β-casomorphin. Met-enkephalin is hydrolyzed at Gly3–Phe4 and, to a lesser extent, at Gly2–Gly3. Bradykinin is cleaved at Gly4–Phe5.[1,2]

[1] E. R. S. Kunji, I. Mierau, A. Hagting, B. Poolman & W. N. Konings (1996) *Antoine van Leeuwenhoek* **70**, 187.
[2] K. M. Fenster, K. L. Parkin & J. L. Steele (1997) *J. Bacteriol.* **179**, 2529.

OLIGOPEPTIDASE F

This metalloendopeptidase, also known as PepF and LEP1, catalyzes peptide-bond hydrolysis in peptides containing between seven and twenty-three aminoacyl residues. Extended incubation of longer oligopeptides with oligopeptidase F will also result in cleavage; nevertheless, proteins and short peptides are not substrates. The cleavage site specificity appears to be fairly broad. The best substrate reported is bradykinin, with hydrolysis occurring at Phe5–Ser6.[1,2]

[1]V. Monnet (1995) *MIE* **248**, 579.
[2]R. Baankreis, S. van Schalkwijk, A. C. Alting & F. A. Exterkate (1995) *Appl. Microbiol. Biotechnol.* **44**, 386.

Selected entries from *Methods in Enzymology* [vol, page(s)]:
General discussion: **248**, 579
Assay: *Lactococcus*, **248**, 581
Other molecular properties: evolution, **248**, 552; homologues, **248**, 551; *Lactococcus* (amino acid composition, **248**, 587; assay, **248**, 581; gene, cloning, **248**, 591; homologs, **248**, 591; properties, **248**, 207, 580; purification, **248**, 582, 585; substrate specificity, **248**, 588)

OLIGOPEPTIDASE O

This metallopeptidase, also called PepO, LEPII, oligoendopeptidase, and neutral thermolysin-like oligoendopeptidase (NOP), catalyzes peptide bond hydrolysis in various neuropeptides as well as peptide fragments of β- and κ-casein. The peptidase does not act on intact α-, β-, or κ-casein; nor will it hydrolyzes di-, tri- and tetra-peptides. The substrate size can range from five to thirty aminoacyl residues and the enzyme prefers a hydrophobic residue in the P1′ position.[1] *See also* Aspergillopepsin O

[1]V. Monnet (1995) *MIE* **248**, 579.

Selected entries from *Methods in Enzymology* [vol, page(s)]:
General discussion: **248**, 579
Assay: *Lactococcus*, **248**, 581
Other molecular properties: *Lactococcus*, **248**, 580 (amino acid composition, **248**, 587; assay, **248**, 581; biological function, 248, 184; gene, cloning, **248**, 591; isoelectric point, **248**, 587; physiological functions, **248**, 592; properties, **248**, 185, 189; purification, **248**, 582; structure, **248**, 186; substrate specificity, **248**, 588)

OLIGOPEPTIDE-TRANSPORTING ATPase

This so-called ATPase [EC 3.6.3.23] catalyzes the ATP-dependent (ADP- and orthophosphate-producing) transport of oligopeptide$_{out}$ to produce the oligopeptide$_{in}$.[1–3] The bacterial enzyme, also known as oligopeptide permease, will also catalyze the import of dipeptides. This ABC-type (ATP-binding cassette-type) ATPase that has two similar ATP-binding domains and does not undergo phosphorylation during the transport process. *See also* ATPase, Peptide-Transporting

The idea that a transporter is an enzyme is in keeping with a new definition of enzyme catalysis as the facilitated making/breaking of chemical bonds, not just covalent bonds.[4] This idea builds on Pauling's assertion that any long-lived, chemically distinct interaction (in this case, the persistent location of a solute with respect to the faces of a membrane) can be regarded as a chemical bond. Note also that the equilibrium constant (K_{eq} = [oligopeptide$_{in}$][ADP][P$_i$]/[oligopeptide$_{out}$][ATP]) does not conform to that expected for an ATPase (*i.e.*, K_{eq} = [ADP][P$_i$]/[ATP]). Thus, although the overall reaction yields ADP and orthophosphate, the enzyme is misclassified as a hydrolase, and instead should be regarded as an energase-type reaction. Energases facilitate affinity-modulated reactions by coupling the $\Delta G_{ATP\text{-}hydrolysis}$ to a force-generating or work-producing step.[4] In this case, P-O-P bond scission supplies the energy to drive oligopeptide transport. *See ATPase; Energase*

[1]S. R. Pearce, M. L. Mimmack, M. P. Gallagher, U. Gileadi, S. C. Hyde & C. F. Higgins (1992) *Mol. Microbiol.* **6**, 47.
[2]M. H. Saier, Jr. (1998) *Adv. Microb. Physiol.* **40**, 81.
[3]J. K. Griffiths & C. E. Sansom (1998) *The Transporter Factsbook*, Academic Press, San Diego.
[4]D. L. Purich (2001) *TiBS* **26**, 417.

OLIGOSACCHARIDE-DIPHOSPHO-DOLICHOL PYROPHOSPHATASE

This enzyme [EC 3.6.1.44], also called oligosaccharide-diphosphodolichol diphosphatase, catalyzes the hydrolysis of an oligosaccharide-diphosphodolichol to produce an oligosaccharide phosphate and dolichyl phosphate. The oligosaccharide portion of the substrate is often oligomannosides containing a chitobiose phosphate at the reducing end.[1]

[1]M. Belard, R. Cacan & A. Verbert (1988) *BJ* **255**, 235.

OLIGOSACCHARIDE 4-α-D-GLUCOSYLTRANSFERASE

This glucosyltransferase [EC 2.4.1.161], also known as amylase III, catalyzes the transfer of the non-reducing terminal α-D-glucosyl residue from a 1,4-α-D-glucan to the 4-position of an α-D-glucan.[1] Substrates include amylose, amylopectin, glycogen, and maltooligosaccharides. Maltose is not a substrate. No detectable free glucose is formed in the course of the reaction.

[1]P. Nebinger (1986) *Biol. Chem. Hoppe-Seyler* **367**, 161 and 169.

OLIGOSACCHARIDE-TRANSPORTING ATPase

This so-called ATPase [EC 3.6.3.18] catalyzes the ATP-dependent (ADP- and orthophosphate-producing)

transport of an oligosaccharide$_{out}$ to produce an oligosaccharide$_{in}$.[1-4] Raffinose and disaccharides, such as lactose and melibiose, are also imported. This ABC-type (ATP-binding cassette-type) ATPase has two similar ATP-binding domains and does not undergo phosphorylation during the transport process. **See also ATPase, Maltose-Transporting**

The idea that a transporter is an enzyme is in keeping with a new definition of enzyme catalysis as the facilitated making/breaking of chemical bonds, not just covalent bonds.[5] This idea builds on Pauling's assertion that any long-lived, chemically distinct interaction (in this case, the persistent location of a solute with respect to the faces of a membrane) can be regarded as a chemical bond. Note also that the equilibrium constant, defined by K_{eq} = [oligosaccharide$_{in}$][ADP][P$_i$]/[oligosaccharide$_{out}$][ATP], does not conform to that expected for an ATPase (*i.e.*, K_{eq} = [ADP][P$_i$]/[ATP]). Thus, although the overall reaction yields ADP and orthophosphate, the enzyme is misclassified as a hydrolase, and instead should be regarded as an energase-type reaction. Energases facilitate affinity-modulated reactions by coupling the $\Delta G_{ATP\text{-hydrolysis}}$ to a force-generating or work-producing step.[5] In this case, P-O-P bond scission supplies the energy to drive oligosaccharide transport. **See ATPase; Energase**

[1] C. F. Higgins (1992) *Ann. Rev. Cell Biol.* **8**, 67.
[2] G. Kuan, E. Dassa, N. Saurin, M. Hofnung & M. H. Saier, Jr. (1995) *Res. Microbiol.* **146**, 271.
[3] M. H. Saier, Jr. (1998) *Adv. Microb. Physiol.* **40**, 81.
[4] S. G. Williams, J. A. Greenwood & C. W. Jones (1992) *Mol. Microbiol.* **6**, 1755.
[5] D. L. Purich (2001) *TiBS* **26**, 417.

OLIGOXYLOGLUCAN β-GLYCOSIDASE

This highly specific glycosidase [EC 3.2.1.120], also known as isoprimeverose-producing oligoxyloglucan hydrolase, catalyzes the hydrolysis of 1,4-β-D-glucosidic links in oligoxyloglucans so as to remove successive isoprimeverose (that is, α-xylo-1,6-β-D-glucosyl-) residues from the non-reducing ends of the chains. It was first isolated from *Aspergillus oryzae*.[1]

[1] Y. Kato, J. Matsushita, T. Kubodera & K. Matsuda (1985) *J. Biochem.* **97**, 801.

OMPTIN

This serine endopeptidase [EC 3.4.21.87] (also called protease VII, protease a, OmpT, and outer membrane protein 3B) catalyzes peptide bond hydrolysis. Its name is derived

from its location, an <u>o</u>uter <u>m</u>embrane <u>prote</u>in of many Gram-negative bacteria. Omptin catalyzes the cleavage of Xaa–Yaa in which both Xaa and Yaa are arginyl or lysyl residues. In addition, Arg–Met and Arg–Val bonds have been reported to be hydrolyzed. It will also act on plasminogen to generate plasmin.[1] Other proteins acted upon by omptin include T7 RNA polymerase, creatine kinase, human interferon γ, and tryptophan synthetase. A homologue of omptin (called Pla [for <u>pl</u>asminogen <u>a</u>ctivator activity]) is a major virulence factor in *Yersinia pestis*, the microorganism responsible for bubonic plague.[2]

[1] W. F. Mangel, D. L. Toledo, M. T. Brown, K. Worzalla, M. Lee & J. J. Dunn (1994) *MIE* **244**, 384.
[2] O. A. Sodeinde, Y. V. B. K. Subrahmanyam, K. Stark, T. Quan, Y. Bao & J. D. Goguen (1992) *Science* **258**, 1004.

Selected entries from *Methods in Enzymology* [vol, page(s)]:
General discussion: 244, 384
Assay: fluorogenic substrates, **244**, 387; inhibition assay, **244**, 387; optimization, **244**, 396; plasminogen activator activity, **244**, 386, 393, 396; sensitivity, **244**, 388; T7 RNA polymerase cleavage, **244**, 387
Other molecular properties: autolytic cleavage, **244**, 395; cleavage site specificity, **244**, 397; cloning (expression vector, **244**, 391; polymerase chain reaction amplification, **244**, 390; primers, **244**, 391; restriction site introduction, **244**, 390); family (biological role, **244**, 59; inhibitors, **244**, 58; sequence homology, **244**, 59); gene, **244**, 384; homology with protein E, **244**, 384; inhibitors, **244**, 397; kinetic parameters, **244**, 396; pI, **244**, 386; processing, **244**, 384; purification (boiling step, **244**, 395; cell growth, **244**, 392; cell lysis, **244**, 393; extraction from membrane, **244**, 393; polyacrylamide gel electrophoresis, **244**, 395); size, **244**, 386; substrate specificity, **244**, 397

OPHELINE KINASE

This phosphotransferase [EC 2.7.3.7] catalyzes the reaction of ATP with opheline (or, guanidinoethyl methyl phosphate) to produce phosphoopheline (or, N'-phosphoguanidinoethyl methyl phosphate) and ADP.[1] Other phosphoryl acceptor substrates include taurocyamine, lombricine, and 2-guanidinoethyl phosphate.

[1] N. van Thoai, F. di Jeso, Y. Robin & E. der Terrossian (1966) *BBA* **113**, 542.

Selected entries from *Methods in Enzymology* [vol, page(s)]:
General discussion: 17A, 1002

OPINE DEHYDROGENASE

This oxidoreductase [EC 1.5.1.28] catalyzes the reversible reaction of (2S)-2-{[1-(R)-carboxyethyl]amino}pentanoate with NAD$^+$ and water to produce L-2-aminopentanoate, pyruvate, and NADH. The *Arthrobacter* enzyme will also act on secondary amine dicarboxylates such as N-(1-carboxyethyl)methionine and N-(1-carboxyethyl)phenylalanine.[1,2] An imine is generated which then converts to the amino acid and pyruvate. Interestingly, in the reverse direction, the enzyme will use neutral amino acids as an amino donor (examples include L-2-aminopentanoate, L-2-aminobutyrate, L-2-aminohexanoate, L-3-chloroalanine, L-O-acetylserine, L-methionine, L-isoleucine, L-valine, L-phenylalanine, L-leucine, and L-alanine. Amino acceptor substrates (once again, in the reverse reaction) include oxaloacetate, glyoxylate, and α-ketobutyrate (or, 2-oxobutyrate) as well as pyruvate.

[1]Y. Asano, K. Yamaguchi & K. Kondo (1989) J. Bacteriol. **171**, 4466.
[2]T. Dairi & Y. Asano (1995) Appl. Environ. Microbiol. **61**, 3169.

OPSIN KINASE

Opsin kinase was originally described as a phosphotransferase activity that catalyzed the reaction of ATP with photo-bleached rhodopsin to produce ADP and phosphorylated rhodopsin. It was reported not to act on unbleached rhodopsin, histones, or phosvitin and was originally given the classification number EC 2.7.1.97: however, that Enzyme Commission entry is now deleted. The activity is now listed as rhodopsin kinase [EC 2.7.1.125]. **See** Rhodopsin Kinase

ORCINOL 2-MONOOXYGENASE

This FAD-dependent oxidoreductase [EC 1.14.13.6], also known as orcinol hydroxylase, catalyzes the reaction of orcinol (or, 3,5-dihydroxytoluene) with NADH and dioxygen to produce 2,3,5-trihydroxytoluene, NAD^+, and water.[1-4]

orcinol   2,3,5-trihydroxytoluene

Alternative substrates include 3-trifluoromethylphenol, 3-chlorophenol, and 3-ethylphenol.

[1]S. R. Sahasrabudhe, D. Lala & V. V. Modi (1986) Can. J. Microbiol. **32**, 535.
[2]Y. Ohta & D. W. Ribbons (1976) EJB **61**, 259.
[3]Y. Ohta, I. J. Higgins & D. W. Ribbons (1975) JBC **250**, 3814.
[4]V. Massey & P. Hemmerich (1975) The Enzymes, 3rd ed., **12**, 191.

Selected entries from **Methods in Enzymology** [vol, pages(s)]: **General discussion: 52**, 17

ORNITHINE 4,5-AMINOMUTASE

This activity, orginally classified as EC 5.4.3.1 and now a deleted entry, results from the combined action of ornithine racemase [EC 5.1.1.12] and D-ornithine 4,5-aminomutase [EC 5.4.3.5].

D-ORNITHINE 4,5-AMINOMUTASE

This pyridoxal-phosphate- and cobalamin-dependent aminomutase [EC 5.4.3.5] catalyzes the interconversion of D-ornithine and D-threo-2,4-diaminopentanoate: a 1,2-amino shift.[1-5]

D-ornithine   D-threo-2,4-diaminopentanoate

Enzyme activity also requires the presence of dithiothreitol. The reaction mechanism first involves the formation of a Schiff base followed by cobalamin-mediated generation of a radical intermediate. **See also** D-Lysine 5,6-Aminomutase

[1]R. Somack & R. N. Costilow (1973) Biochemistry **12**, 2597.
[2]B. T. Golding & W. Buckel (1998) CBC **3**, 239.
[3]P. A. Frey & G. H. Reed (1993) AE **66**, 1.
[4]B. M. Babior & J. S. Krouwer (1979) Crit. Rev. Biochem. **6**, 35.
[5]T. C. Stadtman (1972) The Enzymes, 3rd ed., **6**, 539.

L-ORNITHINE AMINOTRANSFERASE

This pyridoxal-phosphate-dependent aminotransferase [EC 2.6.1.13], also known as ornithine:oxo-acid aminotransferase, glutamate:ornithine transaminase, and ornithine: keto acid aminotransferase, catalyzes the reversible reaction of L-ornithine with an α-keto acid (or 2-oxo acid) to produce L-glutamate 5-semialdehyde and an L-amino acid.[1-4]

L-ornithine   L-glutamate γ-semialdehyde

The best α-keto acid acceptor substrate is α-ketoglutarate (or, 2-oxoglutarate), which upon transamination yields L-glutamate. Other acceptor substrates include pyruvate and oxaloacetate.

The enzyme, which displays ping pong bi bi kinetic mechanism, is structurally similar to aspartate aminotransferase except in their N-terminal regions. **See also** Ornithine: α-Ketoglutarate 2-Aminotransferase; Ornithine(Lysine) Aminotransferase; N-Acetylornithine Aminotransferase

[1]J. M. Burcham, C. S. Giometti, S. L. Tollaksen & C. Peraino (1988) ABB **262**, 501.
[2]T. Ohura, E. Kominami, K. Tada & N. Katunuma (1982) J. Biochem. **92**, 1785.
[3]H. J. Strecker (1965) JBC **240**, 1225.
[4]A. E. Braunstein (1973) The Enzymes, 3rd ed., **9**, 379.

Selected entries from **Methods in Enzymology** [vol, page(s)]:
General discussion: 17A, 281; 113, 76
Assay: 113, 77

ORNITHINE *N*-BENZOYLTRANSFERASE

L-ornithurate

This acyltransferase [EC 2.3.1.127], first isolated from Japanese quail (*Coturnix coturnix japonica*), catalyzes the reaction of two molecules of benzoyl-CoA with L-ornithine to produce ornithuric acid (or, N^2,N^5-dibenzoyl-L-ornithine) and two molecules of coenzyme A.[1]

[1] M. A. Seymour, P. Millburn & G. H. Tait (1987) *Biochem. Soc. Trans.* 15, 1108.

ORNITHINE CARBAMOYLTRANSFERASE

This transferase [EC 2.1.3.3], also known as ornithine transcarbamylase and citrulline phosphorylase, catalyzes the reaction of carbamoyl phosphate with L-ornithine to produce orthophosphate and L-citrulline.[1–11]

L-citrulline

In mammals, this enzyme is a component of the urea cycle. L-Lysine can act as an alternative acceptor substrate, producing L-homocitrulline. The carrot,[2] *Salmonella*,[3] and *Mycobacterium*[1] enzymes have ordered bi bi kinetic mechanisms[1] whereas the *Escherichia coli*[4] enzyme reportedly has a Theorell–Chance bi bi mechanism. In the human enzyme, a cysteinyl residue interacts with the δ-amino group of L-ornithine, and an aspartyl residue facilitates attack by the amino group on carbamoyl phosphate.[8]

The plant enzyme appears to be multifunctional and catalyzes the reactions of putrescine carbamoyltransferase [EC 2.1.3.6], carbamate kinase [EC 2.7.2.2], and agmatine deiminase [EC 3.5.3.12], thereby acting as a putrescine synthase.

[13]N-Labeled and ^{11}C-labeled L-citrulline can be conveniently synthesized by utilizing immobilized ornithine carbamoyltransferase.[9]

[1] S. Ahmad, R. K. Bhatnagar & T. A. Venkitasubramanian (1986) *Biochem. Cell Biol.* 64, 1349.
[2] S. R. Baker & R. J. Yon (1983) *Phytochemistry* 22, 2171.
[3] A. T. Abdelal, E. H. Kennedy & O. Nainan (1977) *J. Bacteriol.* 129, 1387.
[4] C. Legrain & V. Stalon (1976) *EJB* 63, 289.
[5] X. F. Xiong & P. M. Anderson (1989) *ABB* 270, 198.
[6] W. E. Kurtin, S. H. Bishop & A. Himoe (1971) *BBRC* 45, 551.
[7] V. Carunchio, A. M. Girelli & A. Messina (1999) *Biomed. Chromatogr.* 13, 65.
[8] D. Shi, H. Morizono, Y. Ha, M. Aoyagi, M. Tuchman & N. M. Allewell (1998) *JBC* 273, 34247.
[9] A. S. Gelbard, D. S. Kaseman, K. C. Rosenspire & A. Meister (1985) *Int. J. Nucl. Med. Biol.* 12, 235.
[10] H. J. Vogel & R. H. Vogel (1974) *AE* 40, 65.
[11] S. Ratner (1973) *AE* 39, 1.

Selected entries from **Methods in Enzymology** [vol, page(s)]:
General discussion: 2, 350; 5, 910; 17A, 286, 295
Assay: 2, 350; 5, 910
Other molecular properties: acid fractionation steps in purification, 5, 906; allosteric, generation by site-directed mutagenesis *in vitro*, 202, 717, 722; altered permeability and, 12B, 842; aspartate carbamoyltransferase, in preparation of, 5, 915; assay of (carbamate kinase, 17A, 229; carbamoyl-P synthetase, 5, 917, 923; 17A, 244; 51, 105, 112); carbamoyl-P synthetase, assay of, 5, 917, 923; 17A, 244 (preparation of, in, 5, 909); carbamoylphosphokinase assay and, 5, 907, 910; citrulline decomposition and arsenolysis due to, 5, 925; clinical aspects, 17B, 885; protoplasts and, 5, 133; hepatic mitochondrial (intracellular transport, 97, 402; posttranslational uptake and processing *in vitro*, 97, 405; synthesis, 97, 396); induced cooperativity in, 202, 713; properties, 5, 912, 924; purification from (*Mycoplasma*, 17A, 296; *Streptococcus faecalis*, 5, 911; 17A, 232, 287; vertebrates, 2, 353; 5, 924); serum, assay of, in, 17B, 886; synthesis in *Chlamydomonas* cultures, 23, 73; transition-state and multisubstrate inhibitors, 249, 304; vertebrates, in, 5, 924

ORNITHINE CYCLODEAMINASE

This NAD$^+$-dependent enzyme [EC 4.3.1.12], also known as ornithine cyclase, catalyzes the conversion of L-ornithine to L-proline and ammonia. The α-amino group of L-ornithine is eliminated in this reaction.[1–4] NAD$^+$, which stabilizes the enzyme and is absolutely required for catalytic activity, is transiently reduced, resulting in the formation of a 2-imino-5-aminopentanoate intermediate. Subsequent cyclization forms Δ^1-pyrroline 2-carboxylate, which reoxidizes the coenzyme and generates L-proline.[3,4]

[1] R. N. Costilow & L. Laycock (1971) *JBC* 246, 6655.
[2] W. L. Muth & R. N. Costilow (1974) *JBC* 249, 7457.
[3] N. Sans, U. Schindler & J. Schröder (1988) *EJB* 173, 123.
[4] A. R. Clarke & T. R. Dafforn (1998) *CBC* 3, 1.

ORNITHINE DECARBOXYLASE

This pyridoxal-phosphate-dependent decarboxylase [EC 4.1.1.17], which catalyzes a key step in the synthesis of putrescine and the polyamines, converts L-ornithine to putrescine and carbon dioxide.[1-6]

putrescine

Depending on its source, the enzyme often catalytically converts L-lysine to cadaverine.

[1] A. Katz & C. Kahana (1988) JBC **263**, 7604.
[2] T. Kitani & H. Fujisawa (1988) J. Biochem. **103**, 547.
[3] J. J. DiGangi, M. Seyfzadeh & R. H. Davis (1987) JBC **262**, 7889.
[4] B. M. Guirard & E. E. Snell (1980) JBC **255**, 5960.
[5] R. A. John (1998) CBC **2**, 173.
[6] E. A. Boeker & E. E. Snell (1972) The Enzymes, 3rd ed., **6**, 217.

Selected entries from **Methods in Enzymology** [vol, page(s)]:
General discussion: 2, 189; 94, 125, 135, 140, 154, 158, 162, 185, 193
Assay: 2, 189; 79, 345; **103**, 592; Escherichia coli (biosynthetic and biodegradative forms), **94**, 125; germinated barley seeds, **94**, 162; murine kidney, **94**, 159; Physarum polycephalum, **94**, 141; rat liver, **94**, 154; retinoid modulation of phorbol ester effects in skin, **190**, 348; Saccharomyces cerevisiae, **94**, 135
Other molecular properties: antizyme inhibitor, rat liver (assay, **94**, 189; properties, **94**, 191; purification, **94**, 190); antizymes, **94**, 193; citrullinase and, **2**, 378; Clostridium septicum, **3**, 464; deficient mutants (Chinese hamster ovary cell, selection, **94**, 108; Escherichia coli, mass screening, **94**, 88; Saccharomyces cerevisiae, mass screening, **94**, 104); Escherichia coli, **94**, 120; hormone induction of biorhythms, **36**, 480; induction, **79**, 345 mechanisms of stimulation, **103**, 601; modifying protein, Physarum polycephalum (assay, **94**, 145; characteristics, **94**, 146; preparation, **94**, 145); murine kidney (assay, **94**, 159; autoradiographic localization, **94**, 169; immunolocalization, **94**, 166; labeling with α-[5-14C]difluoromethylornithine, **94**, 206; properties, **94**, 161); L-ornithine, assay for, **3**, 1048; Physarum polycephalum (assay, **94**, 141; phosphorylation by polyamine-dependent protein kinase, **99**, 366; properties, **94**, 144); purification, **3**, 464; **87**, 467; **94**, 127, 136, 141, 145, 156; rat liver (assay, **94**, 154; labeling with α-[5-14C]difluoromethylornithine, **94**, 206; properties, **94**, 158); rat ovary and placenta, immunocytochemical localization, **94**, 166; regulatory properties of polyamine-dependent protein kinase, **107**, 154; resolution of, **2**, 189; Saccharomyces cerevisiae (assay, **94**, 135; properties, **94**, 139; purification, **94**, 136); simultaneous assay with histidine decarboxylase, **94**, 45; transaminase assay and, **2**, 171

ORNITHINE(LYSINE) AMINOTRANSFERASE

This aminotransferase [EC 2.6.1.68], first isolated from *Trichomonas vaginalis*, catalyzes the reaction of L-ornithine with α-ketoglutarate (or, 2-oxoglutarate) to produce 3,4-dihydro-2*H*-pyrrole-2-carboxylate and L-glutamate. The pyrrole product probably forms nonenzymatically.

3,4-dihydro-2*H*-pyrrole-2-carboxylate

L-Lysine can serve as an alternative substrate to produce 2,3,4,5-tetrahydropyridine-2-carboxylate.[1] **See also**

Ornithine Aminotransferase; N-Acetylornithine Aminotransferase

[1] P. N. Lowe & A. F. Rowe (1986) Mol. Biochem. Parasitol. **21**, 65.

ORNITHINE RACEMASE

L-ornithine   D-ornithine

This enzyme [EC 5.1.1.12] catalyzes the interconversion of L-ornithine and D-ornithine.[1]

[1] R. Somack & R. N. Costilow (1973) Biochemistry **12**, 2597.

OROTATE PHOSPHORIBOSYL-TRANSFERASE

This magnesium-dependent transferase [EC 2.4.2.10], also known as orotidylate phosphorylase, orotidine-5'-phosphate pyrophosphorylase, and OMP synthase, catalyzes the reversible reaction of orotate with 5-phospho-α-D-ribose 1-diphosphate to produce orotidine 5'-phosphate and pyrophosphate (or, diphosphate).[1-8]

orotate   orotidylate

The mammalian protein also exhibits an orotidine-5'-phosphate decarboxylase [EC 4.1.1.23] activity. Initial rate, product inhibition, and equilibrium exchange studies show that the *Salmonella typhimurium* enzyme has a random Bi Bi kinetic mechanism; phosphoribosyl transfer (for an S_N2-like scheme) is followed by a slow conformational change linked to product release.[1,3] The true pyrimidine substrate is the lactim tautomer (*i.e.*, the 2-hydroxy form).[7] The yeast enzyme exhibits a ping pong Bi Bi mechanism.[6]

[1] G. P. Wang, C. Lundegaard, K. F. Jensen & C. Grubmeyer (1999) Biochemistry **38**, 275.
[2] M. B. Bhatia & C. Grubmeyer (1993) ABB **303**, 321.
[3] M. B. Bhatia, A. Vinitsky & C. Grubmeyer (1990) Biochemistry **29**, 10480.
[4] M. Shimosaka, Y. Fukuda, K. Murata & A. Kimura (1985) J. Biochem. **98**, 1689.
[5] M. E. Jones (1980) Ann. Rev. Biochem. **49**, 253.
[6] J. Victor, L. B. Greenberg & D. L. Sloan (1979) JBC **254**, 2647.

[7]G. Davies, M. L. Sinnott & S. G. Withers (1998) *CBC* **1**, 119.

[8]W. D. L. Musick (1981) *Crit. Rev. Biochem.* **11**, 1.

Selected entries from *Methods in Enzymology* [**vol**, page(s)]:
General discussion: 6, 148; 51, 69, 135, 155
Assay: 6, 148; 51, 69, 70, 135, 136, 141, 143, 156, 157
Other molecular properties: activators, 51, 74, 152; animal sources, 51, 72, 73; assay of phosphoribosyl pyrophosphate, in, 18B, 31; equilibrium constant, 6, 151; 51, 164; inhibitors, 51, 74, 153, 163, 166; kinetic properties, 51, 74, 140, 152; molecular weight, 51, 73; orotate phosphoribosyltransferase:orotidine-5'-phosphate decarboxylase complex, 51, 135 (assay, 51, 135, 136, 156; dissociation, 51, 141; Ehrlich ascites cells, from, 51, 155; erythrocyte, from, 51, 143; kinetic properties, 6, 151; 51, 140, 141, 166, 167; molecular weight, 51, 137, 138, 154, 164; purification, 51, 137, 138, 146, 160; purity, 51, 167; *Serratia marcescens*, 51, 135; stability, 51, 167); pH optimum, 6, 151; 51, 73, 152; purification, 6, 149; 51, 70; purity, 51, 73; pyrophosphorolysis, 51, 154; stability, 51, 73, 153, 154; substrate specificity, 6, 151; 51, 152; yeast, from, 6, 149; 51, 69

OROTATE REDUCTASE (NADH)

This FMN- and FAD-dependent oxidoreductase [EC 1.3.1.14], also known as dihydroorotate dehydrogenase and L-5,6-dihydroorotate:NAD$^+$ oxidoreductase, catalyzes the reversible reaction of NADH with orotate to produce NAD$^+$ and (*S*)-dihydroorotate.[1–3] The reverse reaction proceeds by anti-elimination.

orotate   (*S*)-dihydroorotate

The bacteria *Lactococcus lactis* and *Enterococcus faecalis* are the only known organisms containing two functional enzymes that can catalyze the conversion of orotate to dihydroorotate. Both isozymes from *Enterococcus faecalis* have ping pong Bi Bi kinetic mechanisms.[4,5] The *Clostridium oroticum* enzyme (class IB), which has a ping pong scheme, reportedly has a concerted mechanism in which the C5-*proS* proton transfer and the C6-hydride transfer occur in a single, partially rate-limiting step.[6] *See also Dihydroorotate Dehydrogenase*

[1]H. C. Friedmann & B. Vennesland (1958) *JBC* **233**, 1398.

[2]I. Lieberman & A. Kornberg (1953) *BBA* **12**, 223.

[3]P. Blattmann & J. Retey (1972) *EJB* **30**, 130.

[4]J. Marcinkeviciene, L. M. Tinney, K. H. Wang, M. J. Rogers & R. A. Copeland (1999) *Biochemistry* **38**, 13129.

[5]J. Marcinkeviciene, W. Jiang, G. Locke, L. M. Kopcho, M. J. Rogers & R. A. Copeland (2000) *ABB* **377**, 178.

[6]A. Argyrou, M. W. Washabaugh & C. M. Pickart (2000) *Biochemistry* **39**, 10373.

Selected entries from *Methods in Enzymology* [**vol**, page(s)]:
General discussion: 2, 493; 6, 181, 197
Assay: 2, 493; 6, 197

OROTATE REDUCTASE (NADPH)

This flavoenzyme [EC 1.3.1.15] catalyzes a reversible NADPH-dependent conversion of orotate to yield (*S*)-dihydroorotate and NADP$^+$.[1,2]

[1]W. H. Taylor, M. L. Taylor & D. F. Eames (1966) *J. Bacteriol.* **91**, 2251.

[2]S. Udaka & B. Vennesland (1962) *JBC* **237**, 2018.

OROTIDINE-5'-PHOSPHATE DECARBOXYLASE

This enzyme [EC 4.1.1.23] (also known as orotidylic decarboxylase, OMP decarboxylase, uridine 5'-monophosphate synthase, and UMP synthase) catalyzes the decarboxylation of orotidine 5'-phosphate to produce UMP and carbon dioxide.[1–4]

OMP   UMP

The enzyme's catalytic proficiency is among the highest of any known enzyme.[5] Based on ^{15}N kinetic isotope effect studies, the transition state is apt to be a carbanion intermediate stabilized by electrostatic interaction with Lys93. The driving force for the reaction appears to be ground-state destabilization resulting from charge repulsion between the carboxyl of the substrate and Asp91.[6]

The mammalian enzyme is a bifunctional protein also exhibiting orotate phosphoribosyltransferase [EC 2.4.2.10] activity. *See also Orotate Phosphoribosyltransferase*

[1]I. Lieberman, A. Kornberg & E. S. Simms (1955) *JBC* **215**, 403.

[2]J. A. Fyfe, R. L. Miller & T. A. Krenitsky (1973) *JBC* **248**, 3801.

[3]W. T. Shoaf & M. E. Jones (1973) *Biochemistry* **12**, 4039.

[4]G. K. Brown, R. M. Fox & W. J. O'Sullivan (1975) *JBC* **250**, 7352.

[5]A. Radzicka & R. Wolfenden (1995) *Science* **267**, 90.

[6]M. A. Rishavy & W. W. Cleland (2000) *Biochemistry* **39**, 4569.

Selected entries from *Methods in Enzymology* [**vol**, page(s)]:
General discussion: 51, 74
Assay: 51, 74, 75, 136, 145
Other molecular properties: acceleration, 308, 10, 358; assay of (phosphoribosyl pyrophosphate, 18B, 31); free energy profile, 308, 10; inhibitors, 51, 79, 153, 163, 166; kinetics, 51, 78, 79, 140; molecular weight, 51, 78; orotidine-5'-phosphate pyrophosphorylase and, 6, 148, 150; orotate phosphoribosyltransferase assay, in, 51, 69; pH optimum, 51, 78, 152; properties, 51, 78, 79; purification, 51, 75; purity, 51, 78;

stability, **51**, 78, 153, 154; storage, **51**, 70; synchronous cultures and, **21**, 469; transition-state and multisubstrate analogues, **249**, 307; yeast (anionic intermediate, analogues, **249**, 289; from, **51**, 74)

ORSELLINATE DECARBOXYLASE

o-orsellinate   orcinol

This enzyme [EC 4.1.1.58] catalyzes the decarboxylation of o-orsellinate (or, 2,4-dihydroxy-6-methylbenzoate) to form orcinol and carbon dioxide.[1,2]

[1]G. Petterson (1965) *Acta Chem. Scand.* **19**, 2013.
[2]K. Mosbach & J. Schultz (1971) *EJB* **22**, 485.

Selected entries from *Methods in Enzymology* [vol, page(s)]:
Assay: 44, 348

ORSELLINATE-DEPSIDE HYDROLASE

This enzyme [EC 3.1.1.40], also known as lecanorate hydrolase, catalyzes the hydrolysis of orsellinate depside to produce two molecules of orsellinate.[1]

orsellinate depside   o-orsellinate

Other substrates include only those compounds structurally similar to 2,4-dihydroxy-6-methylbenzoate and having a free hydroxyl *ortho* to the depside linkage (*e.g.*, gyrophorate and lecanorate).

[1]J. Schultz & K. Mosbach (1971) *EJB* **22**, 153.

ORYZIN

This serine endopeptidase [EC 3.4.21.63] (also known as *Aspergillus* alkaline proteinase, elastinolytic proteinase, and aspergillopeptidase B) catalyzes protein peptide-bond hydrolysis with a broad specificity. The fungal enzyme has a substrate spectrum resembling that of subtilisin. It is easily assayed with Suc-Ala-Ala-Pro-Leu-∇-NHPhNO$_2$, where ∇ indicates the cleavage site. Oryzin also has an esterase activity; reactivity with Bz-Arg-OEt exceeds Ac-Tyr-OEt. This extracellular, alkaline enzyme

(pH$_{opt}$ = 8.0–9.0) has been identified in five species of *Aspergillus*. The enzyme is potently inhibited (K_i = 7 nM) by potato chymotrypsin inhibitor 1.[1,2]

[1]M. V. Ramesh, T. Sirakova & P. E. Kolattukudy (1994) *Infect. Immun.* **62**, 79.
[2]U. Reichard, S. Buttner, H. Eiffert, F. Staib & R. Ruchel (1990) *J. Med. Microbiol.* **33**, 243.

Selected entries from *Methods in Enzymology* [vol, page(s)]:
Other molecular properties: amino acid composition, **45**, 423; *Aspergillus* spp., from, **19**, 581; definition, **19**, 376; purification, **19**, 586

OXALATE CoA-TRANSFERASE

This transferase [EC 2.8.3.2], also known as succinyl-β-ketoacyl-CoA transferase, catalyzes the reversible reaction of succinyl-CoA with oxalate to produce succinate and oxalyl-CoA.[1]

[1]J. R. Quayle, D. B. Keech & G. A. Taylor (1961) *BJ* **78**, 225.

OXALATE DECARBOXYLASE

This decarboxylase [EC 4.1.1.2] catalyzes the conversion of oxalate to formate and carbon dioxide.[1–3]

[1]W. B. Jakoby, E. Ohmura & O. Hayashi (1956) *JBC* **222**, 435.
[2]E. B. Lillehoj & F. G. Smith (1965) *ABB* **109**, 216.
[3]E. Emiliani & B. Riera (1968) *BBA* **167**, 414.

Selected entries from *Methods in Enzymology* [vol, page(s)]:
General discussion: 5, 637
Assay: 5, 638

OXALATE OXIDASE

This flavoenzyme [EC 1.2.3.4] catalyzes the reaction of oxalate with dioxygen to produce hydrogen peroxide and two molecules of carbon dioxide.[1–3] The barley (*Hordeum vulgare*) enzyme requires manganese ion.[4]

[1]H. Koyama (1988) *Agric. Biol. Chem.* **52**, 743.
[2]C. S. Pundir & R. Nath (1984) *Phytochemistry* **9**, 1871.
[3]J. Chiriboga (1966) *ABB* **116**, 516.
[4]L. Requena & S. Bornemann (1999) *BJ* **343**, 185.

Selected entries from *Methods in Enzymology* [vol, page(s)]:
General discussion: 52, 19; **56**, 474

4-OXALMESACONATE HYDRATASE

This hydratase [EC 4.2.1.83], also known as 4-carboxy-2-oxohexenedioate hydratase, catalyzes the reversible reaction of (*E*)-4-oxobut-1-ene-1,2,4-tricarboxylate with water to produce 2-hydroxy-4-oxobutane-1,2,4-tricarboxylate.[1]

[1]K. Maruyama (1985) *BBRC* **128**, 271.

OXALOACETASE

This enzyme [EC 3.7.1.1], also known as oxalacetate hydrolase, catalyzes the hydrolysis of oxaloacetate to produce oxalate and acetate.[1–4]

[1]H. Lenz, P. Wunderwald & H. Eggerer (1976) *EJB* **65**, 225.
[2]D. R. Houck & E. Inamine (1987) *ABB* **259**, 58.
[3]O. Hayaishi, H. Shimazono, M. Katagiri & Y. Saito (1956) *JACS* **78**, 5126.
[4]D. J. Creighton & N. S. R. K. Murthy (1990) *The Enzymes*, 3rd ed., **19**, 323.

OXALOACETATE DECARBOXYLASE

This decarboxylase [EC 4.1.1.3], also known as oxalate β-decarboxylase, catalyzes the conversion of oxaloacetate to pyruvate and carbon dioxide.[1-5] While the *Klebsiella aerogenes* is biotin-dependent, several biotin-independent oxaloacetate decarboxylases are known. Initial-rate studies indicate a ping pong mechanism.[1] Other substrates in the biotin-dependent reaction are glutaconyl-CoA and methylmalonyl-CoA. This oxaloacetate decarboxylase is a class II decarboxylase. The first half-reaction involves the reaction of oxaloacetate with enzyme-bound biotin to form pyruvate and enzyme-bound carboxybiotin. The second half-reaction involves the transport of sodium ions and the breakdown of carboxybiotin: ^-O_2C–biotin–$E + H^+ + 2Na^+_{in} \leftrightarrow$ biotin–$E + HCO_3^- + 2Na^+_{out}$.

The enzyme from some animal sources and some anaerobic prokaryotes require divalent cations such as manganese ions for maximal activity. When associated with phospholipid vesicles, the enzyme acts as a sodium pump.

[1]P. Dimroth & A. Thomer (1986) *EJB* **156**, 157.
[2]N. E. Labrou & Y. D. Clonis (1999) *ABB* **365**, 17.
[3]J. N. Earnhardt & D. N. Silverman (1998) *CBC* **1**, 495.
[4]M. H. O'Leary (1992) *The Enzymes*, 3rd ed., **20**, 235.
[5]D. J. Creighton & N. S. R. K. Murthy (1990) *The Enzymes*, 3rd ed., **19**, 323.

Selected entries from *Methods in Enzymology* [vol, page(s)]:
General discussion: **1**, 753; **46**, 23; **125**, 530
Assay: **1**, 753
Other molecular properties: carboxyltransferase partial reaction, oxaloacetate decarboxylase-associated (assay, **125**, 535; isolation, **125**, 536); citrate assay, **13**, 517; immobilization and assay, **136**, 489; *Klebsiella aerogenes* (assays, **125**, 531; carboxyltransferase subunit [assay, **125**, 535; isolation, **125**, 536]; purification, **125**, 532; properties, **125**, 534, 539; reconstitution in proteoliposomes, **125**, 538; sodium transport activity, assays, **125**, 536, 538); "malic" enzyme and, **1**, 741, 749; malic enzyme, pH effect, **87**, 404; *Micrococcus lysodeikticus* (now *Micrococcus luteus*), **1**, 753 (properties, **1**, 756; purification, **1**, 754); pyruvate kinase and, **87**, 142; specific radioactivity determination, use in, **13**, 529; stereochemistry, **87**, 148; transition-state and multisubstrate analogues, **249**, 307

OXALOACETATE TAUTOMERASE

This tautomerase [EC 5.3.2.2], also known as oxalacetate *keto-enol* isomerase and oxaloacetate *keto-enol* tautomerase, catalyzes the interconversion of *keto*-oxaloacetate and *enol*-oxaloacetate.[1-6]

oxaloacetate   *enol*-oxaloacetate

Note that inactive bovine aconitase and *Escherichia coli* fumarase A also have an oxaloacetate tautomerase activity.[4,5]

[1]J. D. Johnson, D. J. Creighton & M. R. Lambert (1986) *JBC* **261**, 4535.
[2]J. C. Wesenberg, A. Chaudhari & R. G. Annett (1976) *Can. J. Biochem.* **54**, 233.
[3]R. G. Annett & G. W. Kosicki (1969) *JBC* **244**, 2059.
[4]Y. O. Belikova, A. B. Kotlyar & A. D. Vinogradov (1989) *FEBS Lett.* **246**, 17.
[5]D. H. Flint (1993) *Biochemistry* **32**, 799.
[6]J. V. Schloss & M. S. Hixon (1998) *CBC* **2**, 43.

4-OXALOCROTONATE DECARBOXYLASE

This decarboxylase [EC 4.1.1.77] catalyzes the conversion of 4-oxalocrotonate (or, (*Z*)-5-oxohex-2-enedioate) to 2-oxopent-4-enoate and carbon dioxide.[1]

(*Z*)-5-oxohex-2-enedioate   2-oxopent-4-enoate

The enzyme participates in the *meta*-cleavage pathway for the degradation of phenols, cresols, and catechols. Recent studies suggest that the true product of this decarboxylase is the enol form, 2-hydroxy-2,4-pentadienoate, which is converted to 2-oxopent-4-enoate by the action of vinylpyruvate hydratase.[2] The decarboxylation step is nearly rate-limiting.[2]

[1]V. Shingler, J. Powlowski & U. Marklund (1992) *J. Bacter.* **174**, 711.
[2]T. M. Stanley, W. H. Johnson, Jr., E. A. Burks, C. P. Whitman, C. C. Hwang & P. F. Cook (2000) *Biochemistry* **39**, 718.

OXALOGLYCOLATE REDUCTASE (DECARBOXYLATING)

This oxidoreductase [EC 1.1.1.92] catalyzes the reaction of 2-hydroxy-3-oxosuccinate (or, oxaloglycolate) with NAD(P)H to produce D-glycerate, NAD(P)$^+$, and carbon dioxide.[1-3]

oxaloglycolate   D-glycerate

Other substrates include hydroxypyruvate (producing D-glycerate) and glyoxylate (generating glycolate).

[1]K. H. do Nascimento & D. D. Davies (1975) *BJ* **149**, 553.
[2]L. D. Kohn & W. B. Jakoby (1968) *JBC* **243**, 2486.
[3]J. V. Schloss & M. S. Hixon (1998) *CBC* **2**, 43.

OXALOMALATE LYASE

This enzyme [EC 4.1.3.13] catalyzes the reaction of oxaloacetate with glyoxylate to produce 3-oxalomalate.[1] There have been no reports of the reverse reaction.

[1]Y. Sekizawa, M. Maragoudakis, T. E. King & V. H. Cheldelin (1966) *Biochemistry* **5**, 2392.

OXALOSUCCINATE REDUCTASE

The reaction of oxalosuccinate with NADPH to produce isocitrate and $NADP^+$ is an alternative activity of isocitrate dehydrogenase ($NADP^+$) [EC 1.1.1.42] and isocitrate dehydrogenase (NAD^+) [EC 1.1.1.41].

OXALYL-CoA DECARBOXYLASE

This thiamin-pyrophosphate-dependent decarboxylase [EC 4.1.1.8] catalyzes the conversion of oxalyl-CoA to formyl-CoA and carbon dioxide.[1,2]

[1]J. R. Quayle (1963) *BJ* **89**, 492.
[2]A. L. Baetz & M. J. Allison (1989) *J. Bacteriol.* **171**, 2605.

Selected entries from *Methods in Enzymology* [**vol**, page(s)]:
General discussion: 13, 369
Assay: 13, 369
Other molecular properties: activation by divalent metal ions, 13, 371; dependence on thiamin pyrophosphate, 13, 371; mechanism, 13, 369; properties, 13, 371; purification, 13, 370; species distribution, 13, 369

OXALYL-CoA SYNTHETASE

This enzyme [EC 6.2.1.8], also known as oxalate:CoA ligase, catalyzes the reaction of ATP, oxalate, and coenzyme A to form oxalyl-CoA, AMP, and pyrophosphate.[1,2]

[1]J. Giovanelli (1966) *BBA* **118**, 124.
[2]R. N. Adsule & G. K. Barat (1977) *Experientia* **33**, 416.

OXAMATE CARBAMOYLTRANSFERASE

This transferase [EC 2.1.3.5], also known as oxamate transcarbamylase, catalyzes the reversible reaction of carbamoyl phosphate with oxamate ($^-OOCCONH_2$) to produce orthophosphate and oxalureate ($^-OOCCONH CONH_2$).[1,2]

[1]R. Bojanowski, E. Gaudy, R. C. Valentint & R. S. Wolfe (1964) *J. Bacteriol.* **87**, 75.
[2]C. V. Wauven, J.-P. Simon, P. Slos & V. Stalon (1986) *Arch. Microbiol.* **145**, 386.

Selected entries from *Methods in Enzymology* [**vol**, page(s)]:
General discussion: 5, 904

*Oxa*NI RESTRICTION ENDONUCLEASE

This type II restriction enzyme [EC 3.1.21.4], isolated from *Oerskovia xanthineolytica* N, catalyzes the hydrolysis of DNA. The recognition sequence and the site of action on both strands is at 5′ . . . CCVTNAGG . . . 3′ where N represents A, G, C, or T.

OXIMINOTRANSFERASE

This enzyme [EC 2.6.3.1], also known as transoximinase, oximase, and pyruvate:acetone oximinotransferase, catalyzes the reaction of pyruvate oxime ($HON=C(CH_3)COO^-$) with acetone to produce pyruvate and acetone oxime ($HON=C(CH_3)_2$).[1–3] Other acceptor substrates include acetaldehyde and α-ketoglutarate. D-Glucose oxime can act as the donor substrate.

[1]K. Yamafuji, K. Y. Omura & M. Miura (1953) *Enzymologia* **16**, 75.
[2]K. Yamafuji & M. Eto (1953) *Enzymologia* **16**, 247.
[3]K. Yamafuji, M. Shimamura & H. Omura (1956) *Enzymologia* **17**, 359.

OXISURAN REDUCTASE

This oxidoreductase catalyzes the reaction of oxisuran (or, 2-[(methylsulfinyl)-acetyl]pyridine) with NADPH to produce oxisuranol (α-(methylsulfinyl)methyl-2-pyridinemethanol) and $NADP^+$.[1,2]

[1]N. R. Bachur & R. L. Felsted (1976) *Drug Metab. Dispos.* **4**, 239.
[2]R. L. Felsted, D. R. Richter, D. M. Jones & N. R. Bachur (1980) *Biochem. Pharmacol.* **29**, 1503.

6-OXOCINEOLE DEHYDROGENASE

This oxidoreductase [EC 1.14.13.51] catalyzes the reaction of 6-oxocineole with NADPH and dioxygen to produce 1,6,6-trimethyl-2,7-dioxabicyclo[3.2.2]nonan-3-one, $NADP^+$, and water.[1] The product undergoes nonenzymatic cleavage and subsequent ring closure to form the lactone 4,5-dihydro-5,5-dimethyl-4-(3-oxobutyl)furan-2(3*H*)-one.

[1]D. R. Williams, P. W. Trudgill & D. G. Taylor (1989) *J. Gen. Microbiol.* **135**, 1957.

8-OXOCOFORMYCIN REDUCTASE

This oxidoreductase [EC 1.1.1.235], also called 8-ketocoformycin reductase and first isolated from *Streptomyces antibioticus*, catalyzes the reaction of 8-oxocoformycin with NADPH (B-side specific) to produce coformycin and $NADP^+$.[1] 8-Oxodeoxy-coformycin will also act as an alternative substrate, producing the antibiotic deoxycoformycin.

[1] J. C. Hanvey, E. S. Hawkins, D. C. Baker & R. J. Suhadolnik (1988) *Biochemistry* **27**, 5790.

6-OXOHEXANOATE DEHYDROGENASE

This oxidoreductase [EC 1.2.1.63], also known as adipate semialdehyde dehydrogenase, catalyzes the reaction of 6-oxohexanoate (or, adipate semialdehyde) with $NADP^+$ and water to produce adipate, NADPH, and a proton.[1,2] In *Acinetobacter*, this enzyme participates in the degradation of cyclohexanol.

[1] J. F. Davey & P. W. Trudgill (1977) *EJB* **74**, 115.
[2] N. A. Donoghue & P. W. Trudgill (1975) *EJB* **60**, 1.

5-OXOPROLINASE (ATP-HYDROLYZING)

5-oxo-L-proline   L-glutamate

This enzyme [EC 3.5.2.9], also known as pyroglutamase (ATP-hydrolyzing) and 5-oxo-L-prolinase, catalyzes the reaction of ATP with 5-oxo-L-proline and two water molecules to produce ADP, orthophosphate, and L-glutamate.[1–4]

[1] A. P. Seddon, L. Li & A. Meister (1984) *JBC* **259**, 8091.
[2] J. M. Williamson & A. Meister (1982) *JBC* **257**, 9161 and 12039.
[3] O. W. Griffith & A. Meister (1982) *JBC* **257**, 4392.
[4] P. Van Der Werf & A. Meister (1975) *AE* **43**, 519.

Selected entries from *Methods in Enzymology* [**vol**, page(s)]:
General discussion: 113, 445, 451, 468
Assay: *Pseudomonas putida*, 113, 452; rat kidney, 113, 447
Other molecular properties: inhibition *in vivo*, 113, 468, 580; inhibitor, 63, 399; partition analysis, 249, 324; *Pseudomonas putida* (components β^5 and F^8 [properties, 113, 456; purification, 113, 454; separation, 113, 457; stability, 113, 455]); rat kidney (mechanism, 113, 451; properties, 113, 449; purification, 113, 447; substrate specificity, 113, 450)

4-OXOPROLINE REDUCTASE

4-oxo-L-proline   4-hydroxy-L-proline

This oxidoreductase [EC 1.1.1.104], also known as hydroxyproline oxidase, catalyzes the reaction of 4-oxo-L-proline with NADH to produce 4-hydroxy-L-proline and NAD^+.[1]

[1] T. E. Smith & C. Mitoma (1962) *JBC* **237**, 1177.

P

PacI RESTRICTION ENDONUCLEASE

This type II site-specific endonuclease [EC 3.1.21.4], isolated from *Pseudomonas alcaligenes*, catalyzes the hydrolysis of both strands of DNA at 5′ ... TAAT∇TTAA ... 3′: note that the octanucleotide recognition site contains neither G nor C.[1]

[1]D. T. Bilcock, L. E. Daniels, A. J. Bath & S. E. Halford (1999) *JBC* **274**, 36379.

PaeR7I RESTRICTION ENDONUCLEASE

This type II site-specific endonuclease [EC 3.1.21.4], isolated from *Pseudomonas aeruginosa* PA0303, catalyzes the hydrolysis of both strands of DNA at 5′ ... C∇TCGAG ... 3′. Surprisingly, the sequence CTCTCGAG is resistant to cleavage by *PaeR7I*.

PALMITOYL-CoA HYDROLASE

This enzyme [EC 3.1.2.2], also known as long-chain fatty-acyl-CoA hydrolase, catalyzes the hydrolysis of palmitoyl-CoA to produce free coenzyme A and palmitate. Other long-chain coenzyme A thioesters can serve as alternative substrates (for the rat brain enzyme, C_8- through C_{18}-acyl-CoA analogues are all hydrolyzed[1]). The *Escherichia coli* enzyme also exhibits an arylesterase activity.[2]

[1]J. Yamada, T. Furihata, H. Tamura, T. Watanabe & T. Suga (1996) *ABB* **326**, 106.
[2]Y. L. Lee, J. C. Chen & J. F. Shaw (1997) *BBRC* **231**, 452.

Selected entries from ***Methods in Enzymology*** [vol, page(s)]:
General discussion: 35, 102, 106, 107
Assay: 35, 102, 103, 107; *Mycobacterium smegmatis*, 71, 242; rat liver mitochondria, 71, 234
Other molecular properties: effect on elongation activity of fatty acid synthase, 35, 89, 90; fatty acid synthesis and, 5, 450; inhibitors, 35, 106, 109; interference in assay of palmitoyl-CoA synthetase, 35, 118; *Mycobacterium smegmatis*, from, 71, 242 (inhibitors, 71, 246; properties, 71, 245; purity, 71, 245; specificity, 71, 246; stabilization, 71, 245); rat liver mitochondria, from, 71, 234 (properties, 71, 241; stability, 71, 239; substrate specificity, 71, 241, 242); stability, 35, 105, 109; 71, 239; substrate specificity, 35, 102, 106, 109; 71, 241, 242

PALMITOYL-PROTEIN HYDROLASE

This enzyme [EC 3.1.2.22], also known as palmitoyl-protein thioesterase, catalyzes the hydrolysis of a palmitoyl-protein to produce palmitate and the depalmitated protein. Other long-chain fatty acylated proteins (acylated on a cysteinyl residue) can serve as substrates as can palmitoylcysteine and palmitoyl-CoA. Didemnin B inhibits the human lysosomal enzyme uncompetitively.[1]

[1]L. Meng, N. Sin & C. M. Crews (1998) *Biochemistry* **37**, 10488.

Selected entries from ***Methods in Enzymology*** [vol, page(s)]:
General discussion: 250, 336
Assay: using H-Ras as substrate (detergent addition, 250, 340; HPLC analysis, 250, 340; lipid extraction, 250, 339; substrate preparation [expression in Sf9 cells, 250, 337; labeling with tritiated palmitic acid, 250, 337; purification, 250, 338; solubilization, 250, 340])

PalI RESTRICTION ENDONUCLEASE

This type II site-specific deoxyribonuclease [EC 3.1.21.4], isolated from *Providencia alcalifaciens*, catalyzes the hydrolysis of both strands of DNA at 5′ ... GG∇CC ... 3′, producing blunted end fragments.[1,2]

[1]K. Baksi & G. W. Rushizky (1979) *Anal. Biochem.* **99**, 207.
[2]G. W. Rushizky (1981) *Gene Amplif. Anal.* **1**, 239.

Selected entries from ***Methods in Enzymology*** [vol, page(s)]:
Other molecular properties: isoschizomers, 65, 109, 126; recognition sequence, 65, 9

PANTETHEINASE

This enzyme, also known as pantetheine hydrolase, catalyzes the hydrolysis of D-pantetheine to produce D-pantothenate and cysteamine.[1-3]

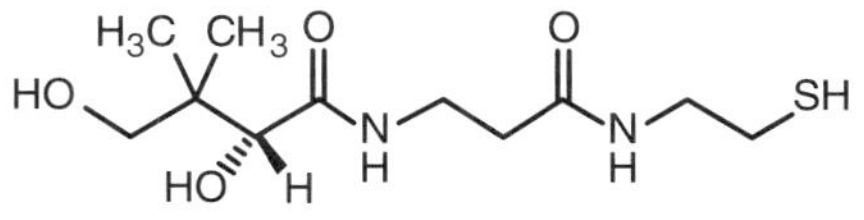

D-pantetheine

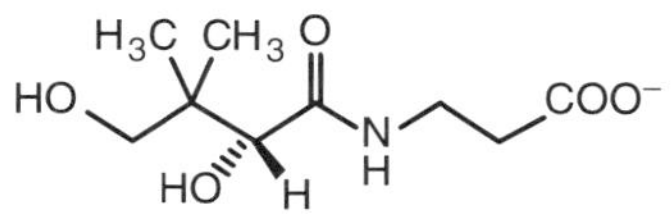

D-pantothenate

The enzyme is inhibited by reagents that inactivate thiol groups.

[1] G. Pitari, G. Maurizi, P. Ascenzi, G. Ricci & S. Dupre (1994) *EJB* **226**, 81.
[2] G. Maurizi, G. Pitari & S. Dupre (1995) *J. Protein Chem.* **14**, 373.
[3] G. Pitari, G. Antonini, R. Mancini & S. Dupre (1996) *BBA* **1298**, 31.

Selected entries from *Methods in Enzymology* [vol, page(s)]:
General discussion: 62, 262; 122, 36

PANTETHEINE KINASE

This phosphotransferase [EC 2.7.1.34] catalyzes the reaction of ATP with pantetheine to produce ADP and pantetheine 4'-phosphate.[1]

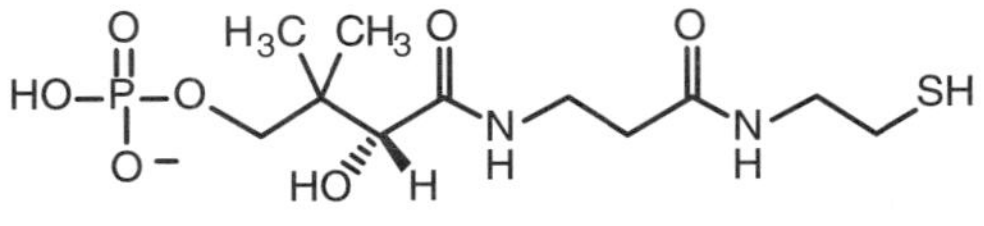

4'-phospho-D-pantetheine

[1] G. D. Novelli (1953) *Fed. Proc.* **12**, 675.

Selected entries from *Methods in Enzymology* [vol, page(s)]:
General discussion: 2, 633
Assay: 2, 633
Other molecular properties: properties, 2, 635; purification, 2, 634

PANTETHEINE-PHOSPHATE ADENYLYLTRANSFERASE

This adenylyltransferase [EC 2.7.7.3], also known as phosphopantetheine adenylyltransferase and dephospho-CoA pyrophosphorylase, catalyzes the reaction of ATP with pantetheine 4'-phosphate to produce pyrophosphate (or, diphosphate) and dephospho-coenzyme A.[1-4]

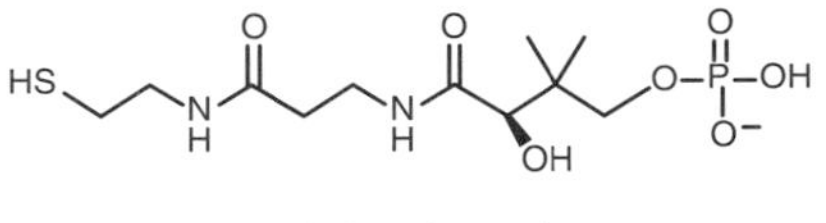

D-pantetheine 4'-phosphate

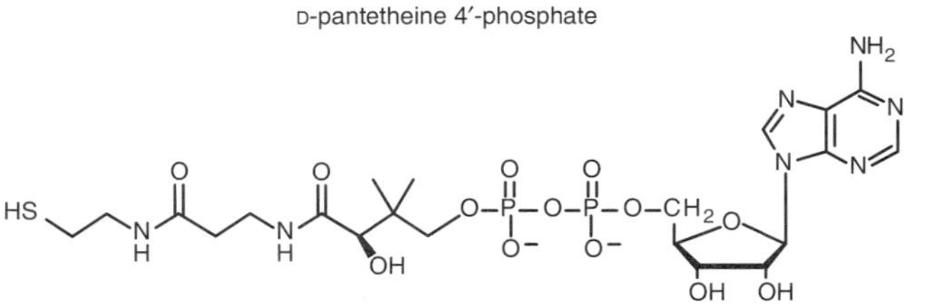

dephospho-coenzyme A

The pig liver enzyme reportedly is a bifunctional enzyme also catalyzing a dephospho-CoA kinase activity [EC 2.7.1.24].[4]

[1] S. Skrede & O. Halvorsen (1983) *EJB* **131**, 57.
[2] D. P. Martin & D. G. Drueckhammer (1993) *BBRC* **192**, 1155.
[3] M. B. Hoagland & G. D. Novelli (1954) *JBC* **207**, 767.
[4] D. M. Worrall & P. K. Tubbs (1983) *BJ* **215**, 153.

Selected entries from *Methods in Enzymology* [vol, page(s)]:
General discussion: 2, 667; 18A, 358
Assay: 2, 667; 18A, 359, 360
Other molecular properties: bifunctional nature, 18A, 358; properties, 2, 669; 18A, 361; purification, 2, 668; 18A, 360, 361; role in CoA biosynthesis, 62, 237

PANTOATE 4-DEHYDROGENASE

This oxidoreductase [EC 1.1.1.106], also known as panthothenase, catalyzes the reversible reaction of (R)-pantoate with NAD+ to produce (R)-4-dehydropantoate and NADH and a proton.[1-3]

[1] C. T. Goodhue & E. E. Snell (1966) *Biochemistry* **5**, 403.
[2] P. Mäntsälä (1978) *BBA* **526**, 25.
[3] T. Myöhänen & P. Mäntsälä (1980) *BBA* **614**, 266.

(R)-PANTOLACTONE DEHYDROGENASE (FLAVIN)

This FMN-dependent oxidoreductase [EC 1.1.99.27] (also known as 2-dehydropantolactone reductase (flavin), 2-dehydropantoyl-lactone reductase (flavin), and (R)-pantoyllactone dehydrogenase (flavin)) catalyzes the reaction of (R)-pantolactone with an acceptor substrate to produce 2-dehydropantolactone and the reduced acceptor. The acceptor substrate can be fulfilled by phenazine methosulfate and nitrotetrazolium blue. The membrane-bound enzyme from *Nocardia* is induced by 1,2-propanediol.[1]

[1] M. Kataoka, S. Shimizu & H. Yamada (1992) *EJB* **204**, 799.

PANTOTHENASE

This enzyme [EC 3.5.1.22] catalyzes the hydrolysis of pantothenate to produce pantoate and β-alanine.[1,2] A strain of *Pseudomonas fluorescens* exhibited a second pantothenase activity, producing β-alanine and pantoyl lactone.[3] (Note: The term panthothenase refers not to this enzyme, but pantoate 4-dehydrogenase instead.)

[1] V. Nurmikko, E. Salo, H. Hakola, U. Mähinen & E. E. Snell (1966) *Biochemistry* **5**, 399.
[2] R. K. Airas (1974) *BJ* **157**, 415 and (1988) **250**, 447.

Selected entries from *Methods in Enzymology* [vol, page(s)]:
General discussion: 62, 201, 267; 122, 33
Assay: 62, 268, 269; 122, 33

PANTOTHENATE KINASE

This phosphotransferase [EC 2.7.1.33] catalyzes the reaction of ATP with D-pantothenate to produce ADP and D-4'-phosphopantothenate.[1-4] Kinetic investigations with the *Escherichia coli* enzyme indicated an ordered Bi Bi kinetic mechanism with positive cooperativity.[4]

[1]G. M. Brown (1959) *JBC* **234**, 370.
[2]O. Halvorsen & S. Skrede (1982) *EJB* **124**, 211.
[3]M. N. Fisher, J. D. Robishaw & J. R. Neely (1985) *JBC* **260**, 15745.
[4]W. J. Song & S. Jackowski (1994) *JBC* **269**, 27051.
Selected entries from *Methods in Enzymology* [vol, page(s)]:
Other molecular properties: induction by clofibrate and di(2-ethyl-hexyl)phthalate, **72**, 507; phosphorylation of pantothenate, **62**, 236

PANTOTHENATE SYNTHETASE

This enzyme [EC 6.3.2.1], also known as pantoate: β-alanine ligase and pantoate activating enzyme, catalyzes the reaction of ATP with (*R*)-pantoate and β-alanine to produce AMP, pyrophosphate, and (*R*)-pantothenate.[1–3]

[1]W. K. Maas (1952) *JBC* **198**, 23.
[2]K. Miyatake, Y. Nakano & S. Kitaoka (1973) *ABC* **37**, 1205.
[3]G. M. Brown & J. M. Williamson (1982) *AE* **53**, 345.
Selected entries from *Methods in Enzymology* [vol, page(s)]:
General discussion: **2**, 619; **62**, 215
Assay: **2**, 619; **62**, 215, 216
Other molecular properties: properties, **2**, 621; **62**, 218, 219; purification, from *Escherichia coli*, **2**, 619; **62**, 216

PANTOTHENOYLCYSTEINE DECARBOXYLASE

This decarboxylase [EC 4.1.1.30] catalyzes the conversion of *N*-((*R*)-pantothenoyl)-L-cysteine to pantotheine and carbon dioxide.[1–3] Both the bacterial and mammalian enzymes are pyruvoyl-dependent proteins, that form Schiff base intermediates. *See Phosphopantothenoylcysteine Decarboxylase*

[1]M. B. Hoagland & G. D. Novelli (1954) *JBC* **207**, 767.
[2]G. M. Brown (1957) *JBC* **226**, 651.

PAPAIN

This cysteine endopeptidase [EC 3.4.22.2], isolated from the latex of the papaya (*Carica papaya*), catalyzes the hydrolysis of peptide bonds in proteins and exhibits broad specificity.[1–9] This enzyme prefers to bind to substrates bearing a large hydrophobic side chain at the P2 position and will not accept a valyl residue in the P1 position. Papain also exhibits an esterase and an amidase activity. Under favorable circumstances, papain catalyzes peptide-bond formation.

The widely held view is that catalytically active papain contains the Cys25 thiolate ion paired to the imidazolium ion of His159 and that the reaction proceeds by way of a thiol ester intermediate. Acylation is generally rate-determining for amide and anilide substrates, whereas deacylation is rate-limiting with ester and thionoester substrates.[1,3,4]

The tetrahedral intermediate generated by the action of a cysteinyl residue on the peptide bond is stabilized by an oxyanion hole. A histidyl imidazolium group acts as a general acid to assist both the departure of amide product and formation of the thiol ester. The thiol ester is attacked by water (with general base catalysis of an imidazole group) to release the final product and the thiolate anion of the cysteinyl residue which reprotonates.[1,4]

[1]K. Brocklehurst, A. B. Watts, M. Patel, C. Verma & E. W. Thomas (1998) *CBC* **1**, 381.
[2]E. Shaw (1990) *AE* **63**, 271.
[3]K. Brocklehurst, F. Willenbrock & E. Salih (1988) *New Compr. Biochem.*, **16**, 39.
[4]K. Brocklehurst, F. Willenbrock & E. Salih (1987) in *Hydrolytic Enzymes* (A. Neuberger & K. Brocklehurst, eds.), Elsevier, Amsterdam, p. 39.
[5]J. S. Fruton (1982) *AE* **53**, 239.
[6]A. N. Glazer & E. L. Smith (1971) *The Enzymes*, 3rd ed., **3**, 501.
[7]J. Drenth, J. N. Jansonius, R. Koekoek & B. G. Wolters (1971) *The Enzymes*, 3rd ed., **3**, 485.
[8]E. L. Smith & J. R. Kimmel (1960) *The Enzymes*, 2nd ed., **4**, 133.
[9]J. R. Kimmel & E. L. Smith (1957) *AE* **19**, 267.
Selected entries from *Methods in Enzymology* [vol, page(s)]:
General discussion: **2**, 56, 59; **19**, 226; **34**, 5, 105, 532; **46**, 22, 155, 206, 221; **80**, 552; **244**, 486, 539
Assay: **19**, 226; amidase activity, **19**, 228; esterase activity, **19**, 227; **80**, 825; method comparison, **19**, 229; *p*-nitrophenol *N*-benzyloxycarbonyl glycinate, with, **45**, 6; proteoglycan degradation method, **248**, 51; proteolytic activity, **19**, 226; thiol assay (accessibility of thiol groups in proteins, **251**, 235; degassing of solutions, **251**, 232, 236; effect of substrate concentration, **251**, 236; papain preparation, **251**, 231; rate of thiol reaction, **251**, 234; reaction conditions, **251**, 232; role of exogenous cystamine, **251**, 230, 233; sensitivity, **251**, 229, 233, 237)
Other molecular properties: activation, **19**, 235; active site, **19**, 240, 243 (size of, **19**, 241; residues, **87**, 460, 480, 491; titration, **80**, 543, 544 [with disulfides, **248**, 90; with epoxypeptides, **248**, 92]); active site thiol group, pK, **251**, 229, 367; affinity labeling, **47**, 427, 428; **87**, 480, 491; amino acid sequence, **19**, 238; antipain, inhibited, **45**, 683; chromogenic substrates, **251**, 230, 236; cleavage of peptides and proteins, in, **11**, 227; covalent chromatography, **34**, 538; cryoenzymology, **63**, 338, 347; cyanylation, **47**, 131; immobilization, **47**, 44; immobilized, peptide synthesis with, **136**, 183; immobilized system, oscillatory phenomena, **135**, 557; inactivation, **19**, 235; **80**, 821 (α$_1$-proteinase inhibitor, **80**, 763); inhibitor from egg white, **19**, 891; kinetic properties, **19**, 242; **44**, 425, 437; **63**, 238; leupeptin inhibition, **45**, 679; ligand interactions, evaluation, **202**, 522; mercuripapain, **19**, 232, 956; **27**, 278; (rate of thiol exchange, **251**, 230, 234); nitrogen isotope effect, **64**, 95; *p*-purification, **2**, 59; **19**, 230; **87**, 467; radioiodination, **70**, 212, 244; reaction with *O*-methylisourea, **11**, 586; residues, active-site, **87**, 460, 480, 491; stability, **19**, 234; structure and conformation, **19**, 239; substrate, fluorigenic, **248**, 34; sulfhydryl group, protection, **44**, 274; thiol assay (accessibility of thiol groups in proteins, **251**, 235); transition state analogues, **249**, 307; water-insoluble derivatives of, **19**, 956

*Pci*I RESTRICTION ENDONUCLEASE

This type II site-specific endonuclease [EC 3.1.21.4], isolated from *Planococcus citreus* SE-F45, catalyzes the hydrolysis of both strands of DNA at 5′ ... A∇CAT GT ... 3′.

PECTATE LYASE

This calcium ion-dependent enzyme [EC 4.2.2.2], also known as pectate transeliminase and endopectate lyase, catalyzes the eliminative cleavage of pectate to produce oligosaccharides with 4-deoxy-α-D-gluc-4-enuronosyl groups at the non-reducing ends.[1–3] Other polygalacturonides can serve as substrates: however, the enzyme will not act on pectin. Pectate lyase C has a unique structural motif of parallel β strands coiled into a large helix; within the core, the amino acid residues form linear stacks and a novel asparagine ladder. Since the pectate lyases exhibit sequence similarities with many other proteins, the parallel β helix motif may be a common structural element.[3]

[1] M. Sato & A. Kaji (1977) ABB 41, 2193.
[2] R. E. McCarthy, S. F. Kotarsky & A. A. Salyers (1985) J. Bacteriol. 161, 493.
[3] M. D. Yoder, N. T. Keen & F. Jurnak (1993) Science 260, 1503.

Selected entries from Methods in Enzymology [vol, page(s)]:
General discussion: 1, 162
Assay: 1, 163; 161, 334, 381
Other molecular properties: analysis with activity stain, 161, 330; biological function, 161, 384; digestion of polygalacturonate, 179, 567; induction in microorganisms, 22, 94; properties, 1, 165; 161, 383; purification, 1, 164 (Erwinia aroideae, 161, 382); secretion in Erwinia chrysanthemi, analysis by TnphoA mutagenesis, 235, 441

PECTINESTERASE

This enzyme [EC 3.1.1.11] (also known as pectin methylesterase, pectin demethoxylase, and pectin methoxylase) catalyzes the reaction of pectin with n molecules of water to produce n molecules of methanol and pectate.[1–3] The Aspergillus niger enzyme preferentially catalyzes the hydrolysis of methyl esters located on the internal galacturonate residues of the substrate, followed by hydrolysis of the methyl esters toward the reducing end: however, the methyl ester of the galacturonate moiety at the non-reducing end is not hydrolyzed.[1]

[1] H. C. Kester, J. A. Benen, J. Visser, M. E. Warren, R. Orlando, C. Bergmann, D. Magaud, D. Anker & A. Doutheau (2000) BJ 346, 469.
[2] H. Deuel & E. Stutz (1958) AE 20, 341.
[3] H. Lineweaver & E. F. Jansen (1951) AE 11, 267.

Selected entries from Methods in Enzymology [vol, page(s)]:
General discussion: 1, 159; 161, 355
Assay: 1, 159; 161, 355, 366
Other molecular properties: chromatography with Spheron ion exchangers, 161, 388; properties, 1, 161; 161, 359, 369; purification from (Botrytis cinerea, 161, 367; Phytophthora infestans, 161, 356; tomato pulp, 1, 160); substrates, 3, 28

PECTIN LYASE

This enzyme [EC 4.2.2.10], also known as pectin transeliminase, catalyzes the eliminative cleavage of pectin to produce oligosaccharides with terminal 4-deoxy-6-methyl-α-D-galact-4-enuronosyl groups.[1,2] The enzyme will not utilize de-esterified pectin as a substrate. The three-dimensional structure of Aspergillus niger pectin lyase B shows the presence of a large right-handed cylinder (a parallel β helix) similar to that seen in pectate lyase [EC 4.2.2.2], but the substrate binding site of pectin lyase is less hydrophilic than pectate lyase.[3]

[1] A. F. Schlemmer, C. F. Ware & N. T. Keen (1987) J. Bacteriol. 169, 4493.
[2] J. Y. Lim, Y. Fujio & S. Ueda (1983) J. Appl. Biochem. 5, 91.
[3] J. Vitali, B. Schick, H. C. Kester, J. Visser & F. Jurnak (1998) Plant Physiol. 116, 69.

Selected entries from Methods in Enzymology [vol, page(s)]:
General discussion: 8, 628
Assay: 8, 628; 161, 351
Other molecular properties: applications, 161, 354; chromatography with Spheron ion exchangers, 161, 388; mode of action, 8, 630; nomenclature and terminology, 161, 350; properties, 8, 630; 161, 354; purification (Aspergillus fonsecaeus, 8, 629; Phoma medicaginis, 161, 352)

PENICILLIN AMIDASE

This amidase [EC 3.5.1.11], also known as penicillin acylase, catalyzes the hydrolysis of penicillin to produce a fatty acid anion and 6-aminopenicillanate.[1–3] Penicillin G is also a substrate in a reaction producing phenylacetate and 6-aminopenicillanate. The enzyme has an essential seryl residue (the N-terminal residue of the B chain of the heterodimer) at the active site; however, there are no histidyl residues (or other bases) near the seryl hydroxyl group, as is seen in the serine endopeptidases. The nearest base to the hydroxyl of this serine is its own α-amino group.[1]

[1] H. J. Duggleby, S. P. Tolley, C. P. Hill, E. J. Dodson, G. Dodson & P. C. Moody (1995) Nature 373, 264.
[2] F. Valle, P. Balbas, E. Merino & F. Bolivar (1991) TiBS 16, 36.
[3] C. W. Wharton (1998) CBC 1, 345.

Selected entries from Methods in Enzymology [vol, page(s)]:
General discussion: 43, 206, 485, 698
Assay: 43, 699 (biochromatographic, 43, 703; buffer, 43, 700; chromatographic assay, 43, 700; hydroxylamine assay, 43, 669, 700, 701, 702; β-lactamase as contaminant, 43, 700, 701); Escherichia coli, 43, 699; fungal, 43, 699
Other molecular properties: Achromobacter, from, 43, 719; activity assay, 44, 237, 761; Alcaligenes faecalis, from, 43, 718; Bacillus megaterium, from, 43, 711; Bacillus subtilis preparation, 43, 705; benzylpenicillin deacylation in aqueous two-phase system, 137, 665; Erwinia aroideae, from, 43, 718; Escherichia coli, from, 43, 705; fiber entrapment, industrial application, 44, 241; Penicillium chrysogenum acylase, 43, 722, 723 [6-aminopenicillanic acid, 43, 722; culture, 43, 722, 723; extraction, 43, 723; specific activity, 43, 723]; Penicillium fusarium acylase, 43, 723 [activators, 43, 725; chemical properties, 43, 725; culture, Fusarium semitectum, 43, 724; extraction, 43, 724; fractionation, 43, 724; inhibitors, 43, 725; kinetic properties, 43, 726; molecular weight, 43, 725; phenoxymethylpenicillin hydrolysis rate, 43, 726; pH optimum, 43, 725, 726; properties, 43, 725; stability, 43, 725]; specificity, 43, 726; immobilization, 44, 194; Micrococcus roseus, from, 43, 719; Nocardia, from, 43, 720; pH optimum, 43, 700; Pleurotus ostreatus, from, 43, 721;

Proteus rettgeri, from, **43**, 718; *Pseudomonas melanogenum*, from, **43**, 719; *Streptomyces lavendulae*, from, **43**, 720; substrate, **43**, 703

purification, **45**, 436; purity, **45**, 439; side-chain modeling, **202**, 185; specificity, **45**, 450; stability, **45**, 443; tertiary structure, **45**, 445

PENICILLIUM NUCLEASE P$_I$

This endonuclease, obtained from *Penicillium citrinum* and currently classified as a variant of *Aspergillus* nuclease S$_1$ [EC 3.1.30.1], catalyzes the hydrolysis of single-stranded RNA or DNA to generate nucleoside 5′-monophosphates and 5′-phosphooligonucleotide end products.[1,2] The reaction requires three zinc ions and results in stereochemical inversion at the phosphorus reaction center. **See also Aspergillus Nuclease S$_I$**

[1] N. H. Williams (1998) *CBC* **1**, 543.
[2] J. E. Coleman (1992) *ARB* **61**, 897.

Selected entries from *Methods in Enzymology* [vol, page(s)]:
Other molecular properties: analysis, of nucleoside 3′,5′-diphosphates, **65**, 676; commercial source, **65**, 668; determination of RNA branch trinucleotide core composition, in, **181**, 188; hydrolysis of DNA, **234**, 20; oligonucleotide digestion, in, **59**, 79; *Penicillium citrinum*, in tRNA sequence analysis, from, **59**, 61; post-Fenton reaction digestion of DNA, **234**, 54; preparation of α-^{32}P-labeled nucleotides, in, **195**, 34; solution, stability, **59**, 61

PENICILLOPEPSIN

This aspartic endopeptidase [EC 3.4.23.20] catalyzes the hydrolysis of peptide bonds in proteins, acting preferentially when hydrophobic side-chain residues are in the P1 and P1′ positions.[1-3] The rate-determining step is likely to involve a conformational change induced by hydrogen-bond formation upon occupancy of subsites S2′ and S3 as well as protonic reorganizations resulting from this conformational change.

Two penicillopepsins have been isolated from *Penicillium janthinellum*. Variants have also been obtained from *P. roqueforti*, *P. duponti*, and *P. camemberti*.

[1] T. D. Meek (1998) *CBC* **1**, 327.
[2] B. Allen, M. Blum, A. Cunningham, G.-C. Tu & T. Hofmann (1990) *JBC* **265**, 5060.
[3] M. N. G. James, A. R. Sielecki & T. Hofmann (1985) in *Aspartic Proteinases and Their Inhibitors* (V. Kostka, ed.), Walter de Gruyter, Berlin, p. 163.

Selected entries from *Methods in Enzymology* [vol, page(s)]:
General discussion: 45, 434
Assay: 19, 376; **45**, 441
Other molecular properties: activity, **45**, 451; amino acid composition, **19**, 382; **45**, 445 (N-terminal sequence, **80**, 578; sequence, **45**, 445; **241**, 216); bovine serum albumin, **45**, 443 (trypsinogen, **45**, 441); characteristics, **45**, 434; β hairpin loop mutations, accommodation, **202**, 77, 81; inhibitors, **45**, 450; pH optimum, **45**, 452; *Penicillium janthinellum* (amino acid composition, **19**, 382; enzymatic properties, **19**, 382; molecular properties, **19**, 381; nomenclature, **19**, 373, 375; N-terminal sequence, **80**, 578; pepsin compared to, 19, 375; purification, **19**, 379; stability, **19**, 381; trypsinogen activation by, **19**, 374, 376); pH optimum, **45**, 452; properties, **45**, 443; **248**, 108, 111 (molecular, **45**, 444);

PENTACHLOROPHENOL MONOOXYGENASE

This oxidoreductase [EC 1.14.13.50], also known as pentachlorophenol hydroxylase (PCP hydroxylase) and pentachlorophenol dehalogenase, catalyzes the reaction of pentachlorophenol (PCP) with NADPH, a proton, and dioxygen to produce tetrachlorohydroquinone, NADP$^+$, and chloride ion.[1] In ^{18}O-labeled water, the product becomes labeled; however, if labeled dioxygen is used, no label is incorporated into the product.[2] The enzyme also catalyzes the release of iodide from triiodophenol.[3]

[1] T. Schenk, R. Muller, F. Morsberger, M. K. Otto & F. Lingens (1989) *J. Bacteriol.* **171**, 5487.
[2] T. Schenk, R. Muller & F. Lingens (1990) *J. Bacteriol.* **172**, 7272.
[3] L. Xun & C. S. Orser (1991) *J. Bacteriol.* **173**, 4447.

PENTALENENE SYNTHASE

This magnesium-dependent enzyme [EC 4.2.3.7; previously classified as EC 4.6.1.5] catalyzes the conversion of 2-*trans*,6-*trans*-farnesyl diphosphate to pentalenene and pyrophosphate.

$^{2-}HO_3PO_3P\text{—}O$

farnesyl diphosphate   humulene   pentalenene

The synthase, which forms humulene as an intermediate, participates in the biosynthesis of pentalenolactone and related antibiotics.[1-3] Neither product is a strong inhibitor of the *Streptomyces* enzyme, but significant inhibition occurs when both are present.[2]

[1] D. E. Cane, J. S. Oliver, P. H. M. Harrison, C. Abell, B. R. Hubbard, C. T. Kane & R. Lattman (1990) *JACS* **112**, 4513.
[2] D. E. Cane & C. Pargellis (1987) *ABB* **254**, 421.
[3] R. A. Gibbs (1998) *CBC* **1**, 31.

PENTANAMIDASE

This enzyme [EC 3.5.1.50], also known as valeramidase, catalyzes the hydrolysis of pentanamide ($CH_3(CH_2)_3$-$CONH_2$) to produce pentanoate ($CH_3(CH_2)_3COO^-$) and ammonia.[1] Other short-chain aliphatic amides are hydrolyzed, albeit slowly. Pentanamidase is distinct from formamidase [EC 3.5.1.49].

[1] F. G. Friedrich & G. Mitrenga (1981) *J. Gen. Microbiol.* **125**, 367.

trans-PENTAPRENYL*trans*TRANSFERASE

This transferase [EC 2.5.1.33], also known as *trans*-penta-prenyltransferase and all-*trans*-hexaprenyl-diphosphate synthase, catalyzes the reaction of all-*trans*-pentaprenyl diphosphate with isopentenyl pyrophosphate (or, isopentenyl diphosphate) to produce pyrophosphate and all-*trans*-hexaprenyl diphosphate.[1–3] The *pro-R* hydrogen at C2 of isopentenyl pyrophosphate is eliminated. This enzyme participates in coenzyme Q biosynthesis.

[1] H. Fujii, T. Koyama & K. Ogura (1982) *JBC* **257**, 14610.
[2] I. Yoshida, T. Koyama & K. Ogura (1989) *BBRC* **160**, 448.
[3] R. A. Gibbs (1998) *CBC* **1**, 31.

Selected entries from *Methods in Enzymology* [vol, page(s)]:
General discussion: 110, 192
Assay: *Micrococcus luteus* B-P 26, **110**, 193
Other molecular properties: *Micrococcus luteus* B-P 26 (assay, **110**, 193; properties, **110**, 197; purification, **110**, 194; resolution into essential components, **110**, 194)

PENTOXIFYLLINE REDUCTASE

This oxidoreductase, also known as 3,7-dimethyl-1-(5-oxohexyl)xanthine reductase, catalyzes the reaction of 3,7-dimethyl-1-(5-oxohexyl)xanthine (also known as pentoxifylline) with NADPH (stereospecifically to and from the B-side) and a proton to produce 3,7-dimethyl-1-(5-hydroxyhexyl)xanthine and NADP$^+$.

pentoxifylline   3,7-dimethyl-1-(5-hydroxyhexyl)xanthine

PEPSIN A

This protypical aspartic endopeptidase [EC 3.4.23.1], also simply known as pepsin, catalyzes peptide-bond hydrolysis in peptides and proteins.[1–5] The principal acid protease of the stomach in vertebrates, pepsin prefers hydrophobic (especially aromatic) aminoacyl residues in the P1 and P1′ positions (*e.g.*, Phe and Leu).[1–5] The pH range for peptide hydrolysis is between 1 and 6; pH-rate studies indicate active-site pK_a values of 1.6 and 5.0.[4,5] A hydrogen-bonded water molecule near aspartyl residues Asp32 and Asp215 probably acts as a nucleophile on the peptide carbonyl group, with Asp215 donating a proton to the nitrogen atom during hydrolysis.

[1] J. S. Fruton (1971) *The Enzymes*, 3rd ed., **3**, 119.
[2] J. S. Fruton (1976) *AE* **44**, 1.
[3] J. Tang, ed. (1976) *Acid Proteases, Structure, Function, and Biology, Adv. Exp. Med. Biol.*, vol. **95**, Plenum, New York.

[4] T. D. Meek (1998) *CBC* **1**, 327.
[5] J. S. Fruton (1982) *AE* **53**, 239.

Selected entries from *Methods in Enzymology* [vol, page(s)]:
General discussion: 2, 3; 19, 316, 337, 347, 358, 364, 406; 27, 473; 34, 5; 45, 452; 248, 105, 107
Assay: 2, 3; 19, 316; 241, 213; (in presence of gastricsin, 19, 407; using hemoglobin, 19, 406); chicken, from, 19, 347, 350; cow, from, 19, 337; proteoglycan degradation method, 248, 51
Other molecular properties: acetylation, 25, 496, 502; activation energy, 63, 243; activators and inactivators, 19, 333, 334; active halogen compounds and, 25, 607; active site, structure, 241, 214; active-site titration, 248, 100; canine, 45, 458; carbodiimide binding, 44, 77; carboxyl groups, 25, 598; cow, from, 19, 337 (preparation, 19, 344; properties, 19, 345); cyanate and, 25, 584; diethylpyrocarbonate treatment, 47, 433; digestion of proteins, 40, 35; 2-hydroxy-5-nitrobenzyl bromide and, 25, 493; immobilization, 44, 141; 47, 42; inactivators, 19, 334; kinetic parameters, 63, 244; 241, 222; kinin-releasing enzyme, 80, 174; optical rotatory dispersion of, 6, 938; oxidation with NBS, effect on tryptophan in, 11, 517; pencillopepsin compared to, 19, 375; pepsin D, 19, 336 (amino composition, 19, 333; assay, 19, 316, 317; chicken, from, 19, 362, 363; definition, 19, 316; dogfish, from, 19, 371; purification, 19, 327; purity and physical properties, 19, 329; specificity, 19, 335); pH profile (for proteolytic activity, 44, 79; for stability, 44, 79, 80); pK_a values, 241, 222; porcine, 19, 316 (homology to cathepsin D, 80, 578; protein engineering, 241, 197; structural analysis with thermolysin, 47, 184); preparation of monovalent antibody fragments, 74, 248, 528, 541; processing enzymes, 248, 136; properties, 19, 329; protein phosphokinase and, 6, 222; purification, 19, 326; purity and physical properties, 19, 329; retropepsin, 248, 105, 108, 112; specificity, 11, 224; 19, 334; specificity of catalysis, 248, 111; spectrokinetic probe, 61, 323; stability, 19, 329; structure, 248, 105, 108; tight-binding inhibition, 63, 462, 466; transition-state and multisubstrate inhibitors, 249, 307; use in selective cleavage of peptides, 11, 423

PEPSIN B

This aspartic endopeptidase [EC 3.4.23.2], originally known as parapepsin I (and found, so far, only in pig), catalyzes peptide-bond hydrolysis with a more restricted activity than pepsin A, exhibiting only 4% of the pepsin A activity on denatured hemoglobin.[1,2] The enzyme degrades gelatin more readily than pepsin A; in fact, its activity on gelatin may have first been recognized in the classical work of Northrop.[3]

[1] A. P. Ryle (1970) *MIE* **19**, 316.
[2] P. K. Nielsen & B. Foltmann (1995) *ABB* **322**, 417.
[3] J. H. Northrop (1932) *J. Gen. Physiol.* **15**, 29.

Selected entries from *Methods in Enzymology* [vol, page(s)]:
General discussion: 19, 316
Assay: 19, 316, 319

PEPSIN F

This aspartic proteinase is expressed in the neonatal stomach and placental yolk sac.[1,2] The enzyme exhibits broad specificity, but prefers peptide bonds containing a phenyl-alanyl or leucyl residue.

¹T. Kageyama, M. Ichinose, S. Tsukada-Kato, M. Omata, Y. Narita, A. Moriyama & S. Yonezawa (2000) *BBRC* **267**, 806.
²T. Kageyama, K. Tanabe & O. Koiwai (1990) *JBC* **265**, 17031.

PEPTIDASE T

This metallopeptidase catalyzes the release of the N-terminal amino acid of various tripeptides. While this enzyme is relatively nonspecific in its action, Pro-Gly-Gly is not a substrate.[1]

¹K. L. Strauch, T. H. Carter & C. G. Miller (1983) *J. Bacteriol.* **156**, 743.

Selected entries from *Methods in Enzymology* [vol, page(s)]:
Other molecular properties: *Salmonella*, **248**, 223

PEPTIDE-N⁴-(N-ACETYL-β-GLUCOSAMINYL)ASPARAGINE AMIDASE

This amidase [EC 3.5.1.52] (also known as glycopeptide *N*-glycosidase, *N*-glycosidase F, glycopeptidase, *N*-oligosaccharide glycopeptidase, and *N*-glycanase) catalyzes the hydrolysis of an N^4-(N-acetyl-β-D-glucosaminyl)asparaginyl residue in a glycopeptide (in which the *N*-acetyl-D-glucosamine moiety may be further glycosylated), to produce *N*-acetyl-β-D-glucosaminylamine (or the corresponding glycosylated compound) and the peptide containing an aspartyl residue.[1–3] The glycopeptide substrate must have more than two aminoacyl residues: N^4-(N-acetyl-β-glucosaminyl)asparagine is not a substrate. This single amino acid-containing substrate is hydrolyzed by N^4-(β-N-acetylglucosaminyl)-L-asparaginase [EC 3.5.1.26].

The active site of peptide-N^4-(N-acetyl-β-D-glucosaminyl)asparagine amidase F has three essential acidic residues: Asp60, Glu206, and Glu118. The aspartyl residue appears to have a catalytic role, which makes this amidase mechanistically different from L-asparaginase and glycosylasparaginase, both of which utilize a threonyl residue.[3]

¹M. Makino, T. Kojima, T. Ohgushi & I. Yamashina (1968) *J. Biochem.* **63**, 186.
²N. Takahashi & H. Nishibe (1978) *J. Biochem.* **84**, 1467.
³P. Kuhn, C. Guan, T. Cui, A. L. Tarentino, T. H. Plummer, Jr. & P. Van Roey (1995) *JBC* **270**, 29493.

Selected entries from
General discussion: **138**, 770
Assay: **138**, 771; **179**, 508; **230**, 46
Other molecular properties: applications, **179**, 515; deglycosylation, **230**, 292; deglycosylation efficiency, **230**, 53; deglycosylation of polysialylated neural cell adhesion molecule, **138**, 636; digestion reaction, **247**, 48; identification of occupied *N*-glycosylation sites, in, **230**, 343; oligosaccharide release from glycoproteins with, **230**, 142, 253, 218; properties, **138**, 775; **230**, 51; purification, **230**, 45; purity, **230**, 51; specificity, **230**, 44, 52; **247**, 45;

structural determinants, **230**, 45; treatment of *O*-GlcNAc on proteins, **230**, 449

PEPTIDE α-N-ACETYLTRANSFERASE

This transferase [EC 2.3.1.88], also known as β-endorphin acetyltransferase, catalyzes the reaction of acetyl-CoA with a peptide to produce coenzyme A and an *N*-acetylpeptide. The enzyme will act on N-terminal alanyl, seryl, methionyl, and glutamyl residues in a number of peptides and proteins including β-endorphin, corticotropins, and melanotropin. The hen oviduct enzyme requires a substrate containing at least ten aminoacyl residues[1] and is associated with a 7S RNA subunit.[2] *See also Tubulin N-Acetyltransferase; Ribosomal-Protein-Alanine N-Acetyltransferase*

¹K. Kamitani, K. Narita & F. Sakiyama (1989) *JBC* **264**, 13188.
²K. Kamitani & F. Sakiyama (1989) *JBC* **264**, 13194.

Selected entries from *Methods in Enzymology* [vol, page(s)]:
Assay: **106**, 194; **60**, 535
Other molecular properties: catalytic effects, **60**, 537, 539, 540; isolation, **60**, 536; rat, distribution in various organs, **106**, 176; rat pituitary (activity in anterior and posterior lobes, **106**, 177; kinetic constants, **106**, 178; properties, **106**, 173; subcellular localization, **106**, 175); varieties, **60**, 537, 539; wheat germ, **106**, 194.

PEPTIDE ANTIBIOTIC LACTONASE

This enzyme catalyzes the hydrolysis of a cyclic peptide lactone to produce a linear peptide containing a carboxyl group at one terminus and an alcohol at the other. Substrates include the antibiotics echinomycin, etamycin, stendomycin, staphylomycin, thiostrepton, and vernamycin Bα.

Selected entries from *Methods in Enzymology* [vol, page(s)]:
General discussion: **43**, 767
Assay: **43**, 770
Other molecular properties: agar diffusion bioassay, **43**, 770, 771; calcium phosphate gel chromatography, **43**, 771; DEAE-cellulose column chromatography, **43**, 771, 772; inhibitors, **43**, 772; kinetic properties, **43**, 772; pH optimum, **43**, 772; properties, **43**, 772; purification, **43**, 771; specific activity, **43**, 772, 773; specificity, **43**, 772

PEPTIDE-ASPARTATE β-DIOXYGENASE

This iron-dependent oxidoreductase [EC 1.14.11.16], also known as aspartate β-hydroxylase and aspartyl/asparaginyl β-hydroxylase, catalyzes the reaction of a peptide aspartyl residue with α-ketoglutarate (or, 2-oxoglutarate) and dioxygen, thereby producing a 3-hydroxy-L-aspartyl residue (*erythro-*) within the peptide, succinate, and carbon dioxide.[1–3] Certain asparaginyl residues in specific peptide substrates also serve as acceptor substrates. Peptide/protein substrates include some vitamin K-dependent coagulation factors and synthetic peptides based on the structure of the first epidermal growth factor domain of human coagulation factor IX or X. An active-site histidyl residue has been identified.[4]

[1] R. S. Gronke, W. J. VanDusen, V. M. Garsky, J. W. Jacobs, M. K. Sardana, A. M. Stern & P. A. Friedman (1989) *PNAS* **86**, 3609.

[2] R. S. Gronke, D. J. Welsch, W. J. VanDusen, V. M. Garsky, M. K. Sardana, A. M. Stern & P. A. Friedman (1990) *JBC* **265**, 8558.

[3] Q. P. Wang, W. J. VanDusen, C. J. Petroski, V. M. Garsky, A. M. Stern & P. A. Friedman (1991) *JBC* **266**, 14004.

[4] S. Jia, K. McGinnis, W. J. VanDusen, C. J. Burke, A. Kuo, P. R. Griffin, M. K. Sardana, K. O. Elliston, A. M. Stern & P. A. Friedman (1994) *PNAS* **91**, 7227.

PEPTIDE-TRANSPORTING ATPase

This family of so-called ATPases [EC 3.6.3.43] catalyzes the ATP-dependent (ADP- and orthophosphate-producing) transport of peptide$_{in}$ to produce peptide$_{out}$.[1–3] Examples of exported peptides include α-hemolysin, cyclolysin, colicin V, and siderophores from Gram-negative bacteria, and bacteriocin, subtilin, competence factor, and pediocin from Gram-positive bacteria. ABC-type (ATP-binding cassette-type) ATPases have two similar ATP-binding domains and do not undergo phosphorylation during transport. *See also* ATPase, Oligopeptide-Transporting

The idea that a transporter is an enzyme is in keeping with a new definition of enzyme catalysis as the facilitated making/breaking of chemical bonds, not just covalent bonds.[4] This idea builds on Pauling's assertion that any long-lived, chemically distinct interaction (in this case, the persistent location of a solute with respect to the faces of a membrane) can be regarded as a chemical bond. Note also that the equilibrium constant ($K_{eq} =$ [peptide$_{out}$][ADP][P$_i$]/[peptide$_{in}$][ATP]) does not conform to that expected for an ATPase (*i.e.*, $K_{eq} =$ [ADP][P$_i$]/[ATP]). Thus, although the overall reaction yields ADP and orthophosphate, the enzyme is misclassified as a hydrolase, and instead should be regarded as an energase-type reaction. Energases facilitate affinity-modulated reactions by coupling the $\Delta G_{ATP\text{-}hydrolysis}$ to a force-generating or work-producing step.[4] In this case, P-O-P bond scission supplies the energy to drive peptide transport. *See* ATPase; Energase

[1] C. Klein & K. D. Entian (1994) *Appl. Environ. Microbiol.* **60**, 2793.

[2] F. Momburg, J. Roelse, J. C. Howard, G. W. Butcher, G. J. Hammerling & J. J. Neefjes (1994) *Nature* **367**, 648.

[3] R. Binet, S. Létoffé, J. M. Ghigo, P. Delepaire & C. Wanderman (1997) *Gene* **192**, 7.

[4] D. L. Purich (2001) *TiBS* **26**, 417.

PEPTIDE-TRYPTOPHAN 2,3-DIOXYGENASE

This oxidoreductase [EC 1.13.11.26], also known as pyrrolooxygenase and tryptophan pyrrolooxygenase, catalyzes the reaction of dioxygen with a tryptophanyl residue in a peptide, thereby producing a formylkynureninyl residue in the peptide.[1–4] The dioxygenase can also use L- or D-tryptophan as a substrate, thereby exhibiting a tryptophan 2,3-dioxygenase [EC 1.13.11.11] activity.

[1] B. Frydman, R. B. Frydman & M. L. Tomaro (1973) *Mol. Cell. Biochem.* **2**, 121.

[2] R. B. Frydman, M. L. Tomaro & B. Frydman (1972) *BBA* **284**, 63.

[3] B. Camoretti-Mercado & R. B. Frydman (1986) *EJB* **156**, 317.

[4] A. R. Sburlati & R. B. Frydman (1983) *Plant Physiol.* **71**, 822.

PEPTIDOGLYCAN β-N-ACETYLMURAMIDASE

This enzyme [EC 3.2.1.92], also known as exo-β-N-acetylmuramidase, catalyzes the hydrolysis of terminal, non-reducing N-acetylmuramate residues in peptidoglycans.[1,2] The synthetic substrate 4-methylumbelliferyl-2-acetamido-3-O-(D-1-carboxyethyl)-2-deoxy-β-d-glucose is converted to 4-methylumbelliferone and 2-acetamido-3-O-(D-1-carboxyethyl)-2-deoxy-β-D-glucose (or, N-acetyl-muramate).

[1] L. A. Del Rio & R. C. W. Berkeley (1976) *EJB* **65**, 3.

[2] L. A. Del Rio, R. C. W. Berkeley, S. J. Brewer & S. E. Roberts (1973) *FEBS Lett.* **37**, 7.

PEPTIDOGLYCAN GLYCOSYLTRANSFERASE

This transferase [EC 2.4.1.129], a component in peptidoglycan biosynthesis, catalyzes the reaction of *n* molecules of bactoprenyldiphospho-N-acetylmuramoyl-(N-acetyl-D-glucosaminyl)-L-alanyl-D-glutamyl-*meso*-diaminopimelyl-D-alanyl-D-alanine to produce a polymer containing *n* linked disaccharide subunits.[1–3] The enzyme exhibits transpeptidase activities.

[1] A. Taku, M. Stuckey & D. P. Fan (1982) *JBC* **257**, 5018.

[2] W. Park & M. Matsuhashi (1984) *J. Bacteriol.* **157**, 538.

[3] J. Nakagawa, S. Tamaki, S. Tomioka & M. Matsuhashi (1984) *JBC* **259**, 13937.

Selected entries from *Methods in Enzymology* [**vol**, page(s)]:
General discussion: 28, 687
Assay: 28, 687
Other molecular properties: isolation from *Bacillus megaterium*, **28**, 689; peptidoglycan synthesis and, **28**, 687; properties, **28**, 690

PEPTIDYLAMIDOGLYCOLATE LYASE

This lyase [EC 4.3.2.5], also known as α-hydroxyglycine amidating dealkylase, catalyzes the conversion of a peptidylamidoglycolate (the product of peptidylglycine monooxygenase [EC 1.14.17.3][1]) into a peptidyl amide and glyoxylate. The combined action of these two enzymes generates a peptidyl amide by removing the elements of CH_2COO^- from a C-terminal glycine. While the rat and frog enzymes are bifunctional, the *Drosophila* enzymes are separate polypeptides.[2]

[1] A. G. Katopodis, D. Ping & S. W. May (1990) *Biochemistry* **29**, 6115.
[2] A. S. Kolhekar, M. S. Roberts, N. Jiang, R. C. Johnson, R. E. Mains, B. A. Eipper & P. H. Taghert (1997) *J. Neurosci.* **17**, 1363.

PEPTIDYL-Asp METALLOENDOPEPTIDASE

This metalloendopeptidase [EC 3.4.24.33], also known as endoproteinase Asp-N and X-Asp metalloendopeptidase, is isolated from *Pseudomonas fragi* and catalyzes the hydrolysis of Xaa–Asp or Xaa–Cya bonds (where Cya refers to cysteic acid). Cleavage can occur at Xaa–Glu peptide bonds under certain conditions; however, Xaa–Asp bonds are kinetically preferred.[1]

[1] M.-L. Hagmann, U. Geuss, S. Fischer & G.-B. Kresse (1995) *MIE* **248**, 782.

Selected entries from *Methods in Enzymology* [vol, page(s)]:
General discussion: 248, 782
Assay: 248, 782
Other molecular properties: *Pseudomonas fragi* (activators, **248**, 786; amino acid sequence, **248**, 785; applications, **248**, 787; assay, **248**, 782; distribution, **248**, 787; inhibitors, **248**, 786; molecular weight, **248**, 785; pH dependence, **248**, 785; properties, **248**, 784; purification, **248**, 783; purity, **248**, 784; specificity, **248**, 786; stability, **248**, 784)

PEPTIDYL-DIPEPTIDASE A

This zinc-dependent dipeptidase [EC 3.4.15.1] (also known as dipeptidyl carboxypeptidase I, peptidyl dipeptidase I, angiotensin I-converting enzyme (ACE), kininase II, peptidase P, and carboxycathepsin) catalyzes the release of C-terminal dipeptides from oligopeptides. Enzymatic hydrolysis of oligopeptide-Xaa-Xbb produces an oligopeptide (with two fewer aminoacyl residues) and Xaa-Xbb, as long as Xaa is not a prolyl residue, and Xbb is neither aspartate nor glutamate. Peptidyl-dipeptidase A converts angiotensin I to angiotensin II and histidylleucine by cleaving the Phe–His bond. Other substrates include bradykinin (Arg-Pro-Pro-Gly-Phe-Ser-Pro-^-Phe-Arg), neurotensin, [Met[5]]enkephalin, β-neoendorphin, dynorphins, and the insulin B chain.[1–4]

Another variant of peptidyl-dipeptidase A has been found in invertebrates, first identified in the heads of houseflies (*Musca domestica*) and then in fruit flies (*Drosophila melanogaster*), the leech (*Theromyzon tessulatum*),

Haematobia irritans, and *Boophilus microplus*. Compared to the mammalian enzyme, it has a lower carbohydrate content, is not bound to a membrane, and will not cross-react with antibodies to the pig enzyme.[5] **See also** Peptidyl Dipeptidase Dcp

[1] M. R. W. Ehlers & J. F. Riordan (1989) *Biochemistry* **28**, 5311.
[2] M. R. W. Ehlers & J. F. Riordan (1990) in *Hypertension Pathophysiology: Diagnosis and Management* (J. H. Laragh & B. M. Benner, eds.), Raven Press, New York, p. 1217.
[3] E. G. Erdös & R. A. Skidgel (1987) *Lab. Invest.* **56**, 345.
[4] N. M. Hooper (1991) *Int. J. Biochem.* **23**, 576.
[5] R. E. Isaac, D. Coates, T. A. Williams & L. Schoofs (1998) in *Arthropod Endocrinology: Perspectives and Recent Advances* (G. M. Coast & S. G. Webster, eds.), Cambridge Univ. Press, Cambridge, p. 357.

Selected entries from *Methods in Enzymlogy* [vol, page(s)]:
General discussion: 80, 788
Assay: 163, 196; **248**, 286; fluorimetric assay, **248**, 602; human, **80**, 450
Other molecular properties: active site, **248**, 285; activity, **248**, 283; anchorage to plasma membrane, molecular mechanism, **248**, 298; angiotensin I-converting enzyme, as, **248**, 283; anion activation, **248**, 286; binding site determination, **202**, 153; biological function, **248**, 190; chelator inhibition, mechanism, **248**, 233; competitive inhibitor interaction, **248**, 295; fluorimetric assay, **248**, 602; inhibitors, **248**, 263; **249**, 293; intracellular localization, **248**, 284, 302; isoforms, **248**, 284; monobromobimane as substrate, **251**, 147; polypeptide sequencing, **91**, 516; properties, **163**, 203; **248**, 190, 284; purification, **248**, 287; structurally related enzymes, **248**, 576; structure, **248**, 287; structure-function relationships, **248**, 291; substrates, **248**, 283; substrate specificity, **248**, 285; testis-specific, **248**, 190; transition-state and multisubstrate inhibitors, **249**, 306; zinc dissociated from, half-life, **248**, 236; zinc ion in, **248**, 285; zinc replacement with other metals, **248**, 240

PEPTIDYL-DIPEPTIDASE B

This zinc-dependent, membrane-bound peptidase [EC 3.4.15.4] (also known as dipeptidyl carboxyhydrolase, atrial peptide convertase, and atrial di-(tri-)peptidyl carboxylase) catalyzes the release of a C-terminal dipeptide from an oligopeptide, analogous to the activity observed with peptidyl-dipeptidase A [EC 3.4.15.1]. However, peptidyl-dipeptidase B will also act on atriopeptin II to produce atriopeptin I and phenylalanylarginine.[1,2]

[1] R. B. Harris & I. B. Wilson (1985) *Peptides* **6**, 393.
[2] D. F. Soler & R. B. Harris (1989) *Peptides* **10**, 63.

PEPTIDYL-DIPEPTIDASE Dcp

This zinc-dependent metallopeptidase [EC 3.4.15.5], formerly known as dipeptidyl carboxypeptidase (Dcp) and peptidyl dipeptidase II, catalyzes the release of unblocked, C-terminal dipeptides from oligopeptides, with a broad specificity: thus, it catalyzes the hydrolysis of oligopeptidyl-Xaa-Xbb-Xcc to produce oligopeptidyl-Xaa and Xbb-Xcc provided that Xbb is not a prolyl residue or both Xaa and Xbb are not glycyl residues. Note, however, that tripeptides are not substrates.[1,2]

[1]A. Yaron (1976) *MIE* **45**, 599.
[2]C. A. Conlin & C. G. Miller (1995) *MIE* **248**, 567.

Selected entries from *Methods in Enzymology* [vol, page(s)]:
General discussion: 45, 599; 248, 567
Assay: 248, 567; benzyloxycarbonyltetraalanine, 45, 600; ninhydrin method, 45, 599; 248, 568; potentiometric method, 45, 602
Other molecular properties: activation by metal ions, 248, 574; detection after gel electrophoresis, 248, 571; *Escherichia coli* (activity, 248, 567; amino acid sequence, 248, 208; assay, 248, 567 [ninhydrin method, 248, 568]; detection, 248, 567; discovery, 248, 567; preparation, 91, 514; substrate, 248, 567); eubacterial, 248, 206; gene; homologs, 248, 591; immunodiffusion and immunoelectrophoresis, 45, 607; inhibition, 248, 574; kinetics, 45, 608; metal ion requirement, 45, 607; molecular weight, 45, 607; oligopeptidase A and, relationship, 248, 575; physiological functions, 248, 577; polypeptide sequencing, 91, 516; properties, 45, 607; purification, 45, 603, 604; 248, 572; purity, 45, 607; stability, 45, 607; structurally related enzymes, 248, 575; substrate, fluorogenic, 248, 24; substrate specificity, 45, 608, 609; 248, 573

PEPTIDYL-DIPEPTIDASE (*STREPTOMYCES*)

This metallopeptidase, first isolated from *Streptomyces* sp., catalyzes the release of prolylproline from the C-terminus of proline-containing peptides (for example, Boc-Pro-Pro-Pro-Pro and Leu-Pro-Pro-Pro-Pro-Pro). The C-terminal amino acid (that is, the P2′ position) in the substrate can be 4-hydroxyproline or 3,4-dehydroproline (but not D-proline or pipecolate).[1,2]

[1]S. Miyoshi, G. Nomura, M. Suzuki, F. Fukui, H. Tanaka & M. Maruyama (1992) *J. Biochem.* **112**, 253.
[2]S. Maruyama, S. Miyoshi, G. Nomura, M. Suzuki, H. Tanaka & H. Maeda (1993) *BBA* **1162**, 72.

PEPTIDYL-GLUTAMINASE

This enzyme [EC 3.5.1.43], also known as peptidoglutaminase I, catalyzes the hydrolysis of an α-*N*-peptidyl-L-glutamine (that is, a C-terminal glutamine of a peptide) to produce an α-*N*-peptidyl-L-glutamate and ammonia.[1] The enzyme is specific for the γ-amide of glutamine (substitution has to be present on the α-amino group). Small substrates include glycyl-L-glutamine, *N*-acetyl-L-glutamine, and L-leucylglycyl-L-glutamine. *See also Protein-Glutamine Glutaminase*

[1]M. Kikuchi & K. Sakaguchi (1973) *ABC* **37**, 827 and 1813.

Selected entries from *Methods in Enzymology* [vol, page(s)]:
General discussion: 45, 485

PEPTIDYL-GLYCINAMIDASE

This hydrolase [EC 3.4.19.2], also known as carboxyamidase, carboxyamidopeptidase, and peptidyl carboxyamidase, catalyzes the hydrolysis of a peptidylglycinamide to produce a peptide and glycinamide. Naturally occurring substrates include vasopressin and oxytocin. Peptides containing other C-terminal amino acid amides can also serve as substrate: for example, methioninamide can be released from substance P. In addition, the enzyme will catalyze the hydrolysis of aminoacylamides (such as phenylalaninamide) and act as an esterase (for example, with Bz-Arg-OEt and Ac-Tyr-OEt).[1,2]

[1]W. H. Simmons & R. Walter (1980) *Biochemistry* **19**, 39.
[2]W. H. Simmons & R. Walter (1981) in *Neurohypophyseal Peptide Hormones and Other Biologically Active Peptides* (D. H. Schlesinger, ed.), Elsevier, New York, p. 151.

PEPTIDYLGLYCINE MONOOXYGENASE

This copper-dependent oxidoreductase [EC 1.14.17.3], also known as peptidyl α-amidating enzyme and peptidylglycine 2-hydroxylase, catalyzes the reaction of a peptidylglycine (that is, a peptide with a C-terminal glycine residue) with ascorbate and dioxygen to produce peptidyl(2-hydroxyglycine), dehydroascorbate, and water.[1,2] The best peptide substrates contain a neutral aminoacyl residue in the penultimate position. This enzyme, which is involved in the final step of α-melanotropin biosynthesis (and of related peptides), acts with peptidylamidoglycolate lyase [EC 4.3.2.5] to convert a glycine-extended propeptide into the corresponding α-hydroxyglycine derivative and then into the amidated peptide plus glyoxylate. In several organisms, the two activities are part of a bifunctional protein.

[1]A. F. Bradbury, M. D. Finnie & D. G. Smyth (1982) *Nature* **298**, 686.
[2]K. Suzuki, H. Shimoi, T. Iwasaki, T. Kawahara, Y. Matsuura & Y. Nishikawa (1990) *EMBO J.* **9**, 4259.

Selected entries from *Methods in Enzymology* [vol, page(s)]:
General discussion: 279, 35
Assay: 168, 355; 279, 36

PEPTIDYL-Lys METALLOENDOPEPTIDASE

This zinc-dependent metalloendopeptidase [EC 3.4.24.20], also known as *Armillaria mellea* neutral proteinase, catalyzes the hydrolysis of Xaa–Lys peptide bonds in proteins, even in cases where Xaa is a prolyl residue. However, C-terminal lysine residues are not released.[1–3]

[1]R. A. Shipolini, G. L. Callewwart, R. C. Cottrell & C. A. Vernon (1974) *EJB* **48**, 465.
[2]S. Doonan, H. J. Doonan, R. Hanford, C. A. Vernon, J. M. Walker, L. P. da S. Airoldi, F. Bossa, D. Barra, M. Carloni, P. Fasella & F. Riva (1975) *BJ* **149**, 497.
[3]T. Nonaka, N. Dohmae, Y. Tsumuraya, Y. Hashimoto & K. Takio (1997) *JBC* **272**, 30032.

PEPTIDYLPROLYL ISOMERASE

This isomerase [EC 5.2.1.8], also known as peptidylprolyl *cis-trans* isomerase (PPIase) and cyclophilin, catalyzes the *cis*-to-*trans* interconversion of proline imidic peptide bonds in oligopeptides: thus, the interconversion of peptidylproline ($\omega = 180°$) and peptidylproline ($\omega = 0°$).[1–5] Kinetic β-deuterium isotope effects suggest a covalent mechanism.[1] The presence of the isomerase assists in the refolding of denatured proteins[2] such as ribonuclease A, cytochrome *c*, and types III and IV collagen (*cis-trans* isomerization is the rate-determining step in the folding of type IV collagen[3]). Prolyl isomerases act in concert with other folding enzymes, chaperones, and protein disulfide isomerases.

[1] G. Fischer, E. Berger & H. Bang (1989) *FEBS Lett.* **250**, 267.
[2] K. Lang, F. X. Schmid & G. Fischer (1987) *Nature* **329**, 268.
[3] J. M. Davis, B. A. Boswell & H. P. Bachinger (1989) *JBC* **264**, 8956.
[4] R. L. Stein (1993) *Adv. Protein Chem.* **44**, 1.
[5] F. X. Schmid, L. M. Mayr, M. Mücke & E. R. Schönbrunner (1993) *Adv. Protein Chem.* **44**, 25.

Selected entries from *Methods in Enzymology* [vol, page(s)]:
General discussion: 290, 84

PEPTIDYLTRANSFERASE

This ribosomal enzyme [EC 2.3.2.12] catalyzes the acyl transfer from a peptidyl-tRNA' to an aminoacyl-tRNA to produce tRNA' and a peptidylaminoacyl-tRNA.[1–4]

[1] H. F. Noller (1993) *J. Bacteriol.* **175**, 5297.
[2] B. S. Cooperman, T. Wooten, D. P. Romero & R. R. Traut (1995) *Biochem. Cell Biol.* **73**, 1087.
[3] H. F. Noller, V. Hoffarth & L. Zimniak (1992) *Science* **256**, 1416.
[4] R. L. Soffer (1974) *AE* **40**, 91.

Selected entries from *Methods in Enzymology* [vol, page(s)]:
General discussion: 5, 731; 12B, 704; 19, 797; 20, 472, 481, 490; 30, 489; 46, 184, 627, 638, 676, 703
Assay: 12B, 704; 20, 490; *Escherichia coli*, 19, 797; fragment reaction assay, 59, 779, 780; reconstituted ribosomal subunits and, 30, 560
Other molecular properties: acetylleucyl oligonucleotide preparation and, 20, 474; assay for ribosome dissociation factor, 60, 296, 297; deuterated ribosomal 50S subunit, in, assay, 164, 146; distribution, 20, 481; *Escherichia coli*, 19, 797 (activators and inhibitors, 19, 803; distribution, 19, 803; physical properties and purity, 19, 802; properties, 19, 802; purification, 19, 801; specificity, 19, 802; stability, 19, 802); ester and polyester formation (activity assays, 30, 491; polyuridylate-directed synthesis, 30, 494; puromycin analogs, 30, 489; R17 ribonucleic acid-directed synthesis, 30, 495); inhibitor binding, 20, 481; liver ribosomal subunits, 20, 438; location, 30, 283; monovalent cations and, 30, 409; photoaffinity labeling in ribosomes, 164, 361; properties, 12B, 706; purification, 12B, 705; sulfhydryl reagents and, 30, 462

PERILLYL-ALCOHOL DEHYDROGENASE

This oxidoreductase [EC 1.1.1.144] catalyzes the reversible reaction of perillyl alcohol with NAD^+ to produce perillyl aldehyde (or, perillaldehyde) and NADH and a proton.[1,2] Other alcohol substrates include 8-hydroxyphellandrol, cumic alcohol, *p*-ethyl benzyl alcohol, and *p*-methyl benzyl alcohol.

perillyl alcohol   perillaldehyde

Preferred substrates are primary alcohols in which the hydroxy group is allylic to an endocyclic double bond and a six-membered ring, either aromatic or nonaromatic.

[1] N. R. Ballal, P. K. Bhattacharyya & P. N. Rangachari (1966) *BBRC* **23**, 473.
[2] N. R. Ballal, P. K. Bhattacharyya & P. N. Rangachari (1968) *Indian J. Biochem.* **5**, 1.

PEROXIDASE

This heme-dependent enzyme [EC 1.11.1.7] catalyzes the reaction of a donor substrate with hydrogen peroxide to produce the oxidized donor and two water molecules.[1–6] There are several variants of peroxidase: for example, myeloperoxidase (formerly known as verdoperoxidase because of its green color) occurs in polymorphonuclear cells and can oxidize halogen ions to the free halogen. Lactoperoxidase is found in milk and can oxidize I^- to iodine as well. **See also** specific peroxidase

[1] I. S. Isaac & J. H. Dawson (1999) *Essays Biochem.* **34**, 51.
[2] H. B. Dunford (1998) *CBC* **3**, 195.
[3] J. Everse, K. E. Everse & M. B. Grisham, eds., (1991) *Peroxidases Chem. Biol.*, CRC Press, Boca Raton.
[4] M. A. Ator & P. R. Ortez de Montellano (1990) *The Enzymes*, 3rd ed., **19**, 213.
[5] W. D. Hewson & L. P. Hager (1979) *The Porphyrins* **7**, 295.
[6] K. G. Paul (1963) *The Enzymes*, 2nd ed., **8**, 227.

Selected entries from *Methods in Enzymology* [vol, page(s)]:
General discussion: 2, 764, 791, 794, 801, 813; 4, 306, 311; 17A, 648, 653, 658; 27, 729, 730; 34, 336, 338; 77, 16
Assay: 2, 769, 794, 801, 803, 813; 4, 293; assay with luminol, 57, 405; assay reagent, 44, 460, 474, 476, 632; coupled assay, 63, 34; microperoxidase, chemiluminescence, 57, 426; plant, 228, 668
Other molecular properties: assay of hydrogen peroxide, in, 17B, 690, 741; assay for (β-galactosidase, 14, 157; singlet oxygen, 319, 59); azide and, 25, 400; catalase and, 2, 793; catalyst, in luminol chemiluminescence, 57, 449; electrophoresis, 5, 43; enzyme electrode, in, 44, 589; hydrogen peroxide assay, in, 52, 347, 348; immobilization, 137, 173, 179; inhibitor, 52, 349; luminescent assays, 133, 331; marker for endocytosis and membrane recycling, as, 98, 209; monoclonal antibodies, assay, 121, 220; mutarotase assay and, 5, 220; oxidation of (plant phenols, 13, 556; tocopherols, 186, 201); protein oxidation and, 4, 264; reaction mechanism, 2, 770, 802; stepwise amplified immunoenzymatic staining, 121, 855; substance P, and, bispecific monoclonal antibodies, assay,

121, 217; temperature study, **63**, 240; stain for detection, **44**, 713; thiyl radicals of cysteine and glutathione, generation of, **186**, 319; tyrosine nitration catalyzed by horseradish peroxidase, **269**, 210

PEROXISOME-ASSEMBLY ATPases

This energase-type enzyme [EC 3.6.4.7] assists in the ATP-dependent assembly of peroxisome components into the peroxisome organelle.[1-3] Although the overall reaction yields ADP and orthophosphate, the enzyme is currently misclassified as a hydrolase, because P-O-P bond scission supplies the energy to drive import and assembly.[4] **See also Vesicle-Fusing ATPases; Energases**

[1]Y. J. Lee & R. B. Wickner (1992) *Yeast* **8**, 787.
[2]T. Tsukamoto, S. Miura, T. Nakai, S. Yokota, N. Shimozawa, Y. Suzuki, T. Orii, Y. Fujiki, F. Sakai, A. Bogaki, H. Yasumo & T. Osumi (1995) *Nat. Genet.* **11**, 395.
[3]T. Yahraus, N. Braverman, G. Dodt, J. E. Kalish, J. C. Morrell, H. W. Moser, D. Valle & S. J. Gould (1996) *EMBO J.* **15**, 2914.
[4]D. L. Purich (2001) *TiBS* **26**, 417.

PERTUSSIS TOXIN

This enterotoxin catalyzes the NAD$^+$-dependent ADP-ribosylation of the α-subunit of the G protein G$_i$.[1-3]

[1]N. J. Oppenheimer & A. L. Handlon (1992) *The Enzymes*, 3rd ed., **20**, 453.
[2]J. Moss & M. Vaughan (1988) *AE* **61**, 303.
[3]C.-Y. Lai (1986) *AE* **58**, 99.

Selected entries from *Methods in Enzymology* [vol, page(s)]:
General discussion: 195, 267; 235, 617; 237, 63
Assay: 106, 416; 109, 560; 235, 629, 684; bioassay, 165, 125
Other molecular properties: A component, 237, 63; activation, 237, 64; activation for *in vitro* ADP-ribosylation, 195, 261; active site residues, photoaffinity labeling, 235, 633; ADP-ribosylation of (G proteins, of, 195, 257; membrane N$_s$ and N$_i$ components, 109, 566); ADP-ribosyltransferase activity, 235, 617, 628, 632; 237, 63; (comparison with cholera toxin activity, 165, 242; effect on adenylate cyclase system, 106, 411; G protein substrates, 237, 132; 238, 371; G protein α subunits [α$_i$, α$_o$, and α$_t$ subfamilies, 237, 25; βγ-dependent processes, 237, 236; G$_q$ class, 238, 237; specific cysteine residues, 237, 71]; preparation for, 237, 117; transducin by (N)-[^{125}I]AIPP, 237, 77, 93); cytopathogenic effects, 235, 682, 684; effect on neutrophil chemotaxis, 236, 67; gel electrophoretic analysis, 165, 125; G protein sensitivity, 238, 144; labeling of G proteins, 237, 63; mechanism of action, 235, 680; probe for (adenylate cyclase function, as, 109, 563; G protein role in phagocyte chemotactic activation, 162, 276; polyphosphoinositide breakdown, 141, 266); purification, 109, 561; 165, 122; structure, 235, 628; 238, 371; substrate G proteins, 237, 133; substrates, 237, 64

*Pfl*FI RESTRICTION ENDONUCLEASE

This type II site-specific endonuclease [EC 3.1.21.4], isolated from *Pseudomonas fluorescens* F, catalyzes the hydrolysis of both strands of DNA at 5′ . . . GACN∇NNG TC . . . 3′, where N refers to any base.

*Pfl*MI RESTRICTION ENDONUCLEASE

This type II site-specific endonuclease [EC 3.1.21.4], isolated from *Pseudomonas fluorescens*, catalyzes the hydrolysis of both strands of DNA at 5′ . . . CCANNNN∇NTG G . . . 3′, where N refers to any base. The enzyme is blocked by overlapping *dcm* methylation.[1]

[1]R. A. Sturm & P. Yaciuk (1989) *NAR* **17**, 3615.

PHASEOLLIDIN HYDRATASE

This hydratase [EC 4.2.1.97] catalyzes the reaction of phaseollidin, an isoflavonoid phytoalexin, with water to produce phaseollidin hydrate.[1] The enzyme is distinct from kievitone hydratase [EC 4.2.1.95].[1]

[1]C. S. Turbek, D. A. Smith & C. L. Schardl (1992) *FEMS Microbiol. Lett.* **94**, 187.

PHENOL β-GLUCOSYLTRANSFERASE

This transferase [EC 2.4.1.35], also known as UDP-glucosyltransferase, catalyzes the reaction of UDP-D-glucose with a phenol to produce UDP and an aryl β-D-glucoside.[1-3] Phenol, 2-aminophenol, 4-nitrophenol, as well as benzyl alcohol and 2-phenylethanol are substrates, but 1-naphthol is preferred in enzyme assays.

[1]G. J. Dutton (1966) *ABB* **116**, 399.
[2]U. Keil & P. Schreier (1989) *Phytochemistry* **28**, 2281.
[3]M. V. Vaisanen, P. I. Mackenzie & O. O. Hanninen (1983) *EJB* **130**, 141.

PHENOL *O*-METHYLTRANSFERASE

This methyltransferase [EC 2.1.1.25] catalyzes the reaction of *S*-adenosyl-L-methionine with phenol to produce *S*-adenosyl-L-homocysteine and anisole.[1,2] Other substituted phenols are also methyl group acceptor substrates.

[1]J. Axelrod & J. Daly (1968) *BBA* **159**, 472.
[2]P. A. Pazmino & R. M. Weinshilboum (1978) *Clin. Chim. Acta* **89**, 317.

PHENOL 2-MONOOXYGENASE

This FAD-dependent oxidoreductase [EC 1.14.13.7], also known as phenol hydroxylase and phenol *o*-hydroxylase, catalyzes the reaction of phenol with NADPH, a proton, and dioxygen to produce catechol, NADP$^+$, and water.[1-9]

phenol   catechol

Other substrates include resorcinol, *o*-cresol, and *o*-chlorophenol (as well as *m*- and *p*-). The enzyme reportedly catalyzes an ordered Bi Uni Uni Bi ping pong kinetic mechanism with substrate inhibition. Monovalent anions are uncompetitive inhibitors, apparently by stabilization of a proton on the N5 position of the flavin.[3] The structure of the monooxygenase is consistent with a hydroxyl transfer mechanism via a flavin-C4a-hydroxide intermediate.[5] *See also* Resorcinol 6-Hydroxylase; Monophenol Monooxygenase

[1] H. Y. Neujahr & A. Gaal (1973) Eur. J. Biochem. **35**, 386.
[2] H. Y. Neujahr & K. G. Kjellen (1978) JBC **253**, 8835.
[3] K. Detmer & V. Massey (1984) JBC **259**, 11265.
[4] B. A. Palfey & V. Massey (1998) CBC **3**, 83.
[5] B. G. Fox (1998) CBC **3**, 261.
[6] J. V. Schloss & M. S. Hixon (1998) CBC **2**, 43.
[7] J. H. Law, P. E. Dunn & K. J. Kramer (1977) AE **45**, 389.
[8] V. Massey & P. Hemmerich (1975) The Enzymes, 3rd ed., **12**, 191.
[9] V. Ullrich & W. Duppel (1975) The Enzymes, 3rd ed., **12**, 253.

PHENOXAZINONE SYNTHASE

This copper-dependent enzyme catalyzes the conversion of 3-hydroxyanthranilate to cinnabarinate.

3-hydroxyanthranilate   cinnabarinate

Other phenoxazinones can be synthesized from other 3-hydroxyanthranilate derivatives. This enzyme catalyzes a crucial step in the biosynthesis of the actinomycins.

Selected entries from *Methods in Enzymology* [**vol**, page(s)]:
General discussion: 17A, 549, 554
Assay: 17A, 550, 554, 557
Other molecular properties: properties, 17A, 556, 559; purification from (rat liver, 17A, 557; Streptomyces antibioticus, 17A, 550; Tecoma stans, 17A, 555)

PHENYLACETALDEHYDE DEHYDROGENASE

This oxidoreductase [EC 1.2.1.39] catalyzes the reaction of phenylacetaldehyde with NAD^+ and water to produce phenylacetate and NADH and a proton.[1]

[1] A. Ferrandez, M. A. Prieto, J. L. Garcia & E. Diaz (1997) FEBS Lett. **406**, 23.

Selected entries from *Methods in Enzymology* [**vol**, page(s)]:
General discussion: 17A, 593

PHENYLACETYL-CoA HYDROLASE

This enzyme, isolated from *Penicillium chrysogenum*, catalyzes the hydrolysis of phenylacetyl-CoA to produce phenylacetate and coenzyme A.

Selected entries from *Methods in Enzymology* [**vol**, page(s)]:
General discussion: 43, 482
Assay: 43, 483
Other molecular properties: ammonium sulfate precipitation, 43, 485; buffer, 43, 487; DEAE-cellulose column, 43, 485; molecular weight, 43, 486; Penicillium chrysogenum preparation, 43, 484; phenylacetate, activation, 43, 432, 487; pH optimum, 43, 486; properties, 43, 485, 486; purification, 43, 484, 485; substrate, 43, 476, 486; thiol effect on enzyme activity, 43, 486

PHENYLACETYL-CoA SYNTHETASE

This ligase [EC 6.2.1.30], also known as phenyl-acetate:CoA ligase, catalyzes the reaction of ATP with phenylacetate and coenzyme A to produce AMP, pyrophosphate, and phenylacetyl-CoA. Weaker alternative substrates include acetate, propanoate, and butanoate. This enzyme catalyzes the first step in the catabolism of phenylacetate in *Pseudomonas putida* and other organisms.[1]

Another phenylacetate:CoA ligase (or, phenylacetyl-CoA synthetase), designated EC 6.2.1.21, reportedly catalyzes the conversion of ATP, phenylacetate and coenzyme A to phenylacetyl-CoA plus apparently undefined products of ATP breakdown.

[1] H. Martinez-Blanco, A. Reglero, L. B. Rodriguez-Aparicio & J. M. Luengo (1990) JBC **265**, 7084.

Selected entries from *Methods in Enzymology* [**vol**, page(s)]:
General discussion: 43, 476
Assay: 43, 447, 478

PHENYLALANINE *N*-ACETYLTRANSFERASE

This acetyltransferase [EC 2.3.1.53], first isolated from *Escherichia coli*, catalyzes the reaction of acetyl-CoA with L-phenylalanine to produce coenzyme A and *N*-acetyl-L-phenylalanine.[1]

phenylalanine   *N*-acetylphenylalanine

Other acetyl acceptor substrates, albeit weaker, include L-histidine and L-alanine. The enzyme will not utilize D-phenylalanine.

[1] R. V. Krishna, P. R. Krishnaswamy, & D. R. Rao (1971) BJ **124**, 905.

PHENYLALANINE ADENYLYLTRANSFERASE

This transferase [EC 2.7.7.54] catalyzes the reaction of ATP with L-phenylalanine to produce pyrophosphate and *N*-adenylyl-L-phenylalanine in the biosynthetic pathway leading to the alkaloid cyclopeptin in *Penicillium cyclopium*.[1,2]

[1]W. Lerbs & M. Luckner (1985) *J. Basic Microbiol.* **25**, 387.
[2]M. Gerlach, N. Schwelle, W. Lerbs & M. Luckner (1985) *Phytochemistry* **24**, 1935.

PHENYLALANINE AMMONIA-LYASE

This lyase [EC 4.3.1.5] catalyzes the conversion of L-phenylalanine to *trans*-cinnamate and ammonia.[1–4] The enzyme from certain sources may also act on L-tyrosine, producing *trans-p*-coumarate. The ammonia-lyase from higher plants exhibits cooperativity. A dehydroalanyl residue, derived from a seryl residue, is thought to be an essential for catalysis.[1–4]

phenylalanine   *trans*-cinnamate

A rival mechanism predicts the formation of an enzyme-intermediate complex via a Friedel-Crafts-type reaction between the dehydroalanyl residue and the *ortho*-carbon to facilitate abstraction of the (3*S*)-β-proton.[5,6]

[1]V. E. Anderson (1998) *CBC* **2**, 115.
[2]R. A. Dixon, P. M. Dey & C. J. Lamb (1983) *AE* **55**, 1.
[3]E. L. Camm & G. H. N. Towers (1973) *Phytochemistry* **12**, 961.
[4]K. R. Hanson & E. A. Havir (1972) *The Enzymes*, 3rd ed., **7**, 75.
[5]A. Lewandowicz, J. Jemielity, M. Kanska, J. Zon & P. Paneth (1999) *ABB* **370**, 216.
[6]A. Gloge, B. Langer, L. Poppe & J. Retey (1998) *ABB* **359**, 1.

Selected entries from *Methods in Enzymology* [vol, page(s)]:
General discussion: 17A, 575, 581; 46, 67; 142, 242, 248
Assay: 142, 244, 249
Other molecular properties: dehydroalanine and, 17A, 580; glyoxysomes, in, 52, 500; overview, 142, 242; properties, 142, 252; purification from (potato tubers, 17A, 576; *Rhodotorula glutinis*, 142, 246; soybean, 142, 249; *Ustilago hordei*, 17A, 582)

PHENYLALANINE DECARBOXYLASE

This pyridoxal-phosphate-dependent decarboxylase [EC 4.1.1.53] catalyzes the conversion of L-phenylalanine to phenethylamine and carbon dioxide.[1,2]

phenylalanine   phenethylamine

Other substrates include L-tyrosine and other aromatic amino acids. *See also* Aromatic-L-Amino Acid Decarboxylase

[1]J.-C. David, W. Dairman & S. Udenfried (1974) *ABB* **160**, 561.
[2]C. J. Haley & A. E. Harper (1978) *ABB* **189**, 524.

Selected entries from *Methods in Enzymology* [vol, page(s)]:
General discussion: 17A, 585; 44, 271

PHENYLALANINE DEHYDROGENASE

This oxidoreductase [EC 1.4.1.20] catalyzes the reversible reaction of L-phenylalanine with water and NAD$^+$ to produce phenylpyruvate, ammonia, and NADH and a proton.[1–4]

L-phenylalanine   phenylpyruvate

The *Bacillus badius*[1] and *Sporosarcina ureae*[2] enzymes are highly specific for L-phenylalanine, whereas the *B. sphaericus*[2] also acts on L-tyrosine. The *B. sphaericus*[2] has B-side (*pro-S*) specificity with respect to NADH. The kinetic mechanism for the *Rhodococcus maris* enzyme is ordered with abortive complexes present.[3]

[1]Y. Asano, A. Nakazawa, K. Endo, Y. Hibino, M. Ohmori, N. Numao & K. Kondo (1987) *EJB* **168**, 153.
[2]Y. Asano, A. Nakazawa & K. Endo (1987) *JBC* **262**, 10346.
[3]H. Misono, J. Yonezawa, S. Nagata & S. Nagasaki (1989) *J. Bacteriol.* **171**, 30.
[4]T. Ohshima, H. Takada, T. Yoshimura, N. Esaki & K. Soda (1991) *J. Bacteriol.* **173**, 3943.

PHENYLALANINE (HISTIDINE) AMINOTRANSFERASE

This pyridoxal-phosphate-dependent aminotransferase [EC 2.6.1.58], also known as phenylalanine aminotransferase, catalyzes the reversible reaction of L-phenylalanine with pyruvate to produce phenylpyruvate and L-alanine.[1] Other substrates include L-histidine and L-tyrosine. In the reverse reaction, L-methionine, L-serine, and L-glutamine can replace L-alanine. *See also* Aromatic Amino Acid Aminotransferase; Aromatic Amino Acid:Glyoxylate Aminotransferase

[1]Y. Minatogawa, T. Noguchi & R. Kido (1977) *Hoppe-Seyler's Z. Physiol. Chem.* **358**, 59.

Selected entries from *Methods in Enzymology* [vol, page(s)]:
General discussion: 5, 695; 17A, 586

PHENYLALANINE 2-MONOOXYGENASE

This FAD-dependent oxidoreductase [EC 1.13.12.9], also known as L-phenylalanine oxidase (deaminating and decarboxylating), catalyzes the reaction of L-phenylalanine with dioxygen to produce 2-phenylacetamide, carbon dioxide, and water.[1,2] Alternative substrates include L-tyrosine, DL-*o*-tyrosine, DL-*m*-tyrosine, *p*-fluoro-DL-phenylalanine, and β-2-thienyl-DL-alanine.[2] The enzyme reportedly also catalyzes a reaction similar to that of L-amino acid oxidase [EC 1.4.3.2]. Alternative substrates for this second activity include L-methionine and L-norleucine.[2]

[1] H. Koyama (1982) *J. Biochem.* **92**, 1235.
[2] H. Koyama (1984) *J. Biochem.* **96**, 421.

PHENYLALANINE 4-MONOOXYGENASE

This iron-dependent enzyme [EC 1.14.16.1], known as phenylalaninase and more commonly known as phenylalanine 4-hydroxylase, catalyzes the reaction of L-phenylalanine with tetrahydrobiopterin and dioxygen to produce L-tyrosine, dihydrobiopterin, and water.[1–7] This enzyme is responsible for the disposition of approximately three-fourths of L-phenylalanine, which is a nutritionally essential amino acid in mammals. The *Chromobacterium violaceum* enzyme has a random BC Ter Bi kinetic mechanism in which dioxygen is the first substrate to bind.[7] No kinetic isotope effect is observed with deuterated phenylalanine, suggesting that the rate-determining step is formation of the hydroxylating intermediate (possibly a 4a-peroxytetrahydropterin derivative or an intermediate containing a peroxy bridge between the pterin and Fe^{2+}).[3] The formation of L-tyrosine proceeds with an NIH shift due to the initial formation of an arene oxide.

[1] P. F. Fitzpatrick (2000) *AE* **74**, 235.
[2] P. F. Fitzpatrick (1998) *CBC* **3**, 181.
[3] S. Kaufman (1993) *AE* **67**, 77.
[4] S. Kaufman (1987) *The Enzymes*, 3rd ed., **18**, 217.
[5] V. Massey & P. Hemmerich (1975) *The Enzymes*, 3rd ed., **12**, 191.
[6] V. Ullrich & W. Duppel (1975) *The Enzymes*, 3rd ed., **12**, 253.
[7] S. O. Pember, K. A. Johnson, J. J. Villafranca & S. J. Benkovic (1989) *Biochemistry* **28**, 2124.

Selected entries from *Methods in Enzymology* [vol, page(s)]:
General discussion: 5, 809; 17A, 597, 603; 52, 12; 53, 278; 142, 3, 17, 27, 35, 44, 50
Assay: 5, 809; 53, 278; 142, 5, 18, 28, 35, 50
Other molecular properties: activity, 53, 278; alternative substrate study, 63, 498; cascade control, 64, 325; cofactor of, 5, 811; 142, 15; inhibitors, 53, 285; iron, 53, 286; isozymes, 53, 286; kinetic properties, 53, 285; molecular weight, 53, 286; purification from (bovine liver, 142, 39; *Chromobacterium violaceum*, 142, 46, 52; human liver, 142, 29; *Pseudomonas*, 17A, 599; (rat liver, 5, 811, 814; 17A, 605, 53, 280; 142, 8, 19); rat liver, 5, 809; 53, 278; 142, 8, 19; rat liver to assay reduced L-erythrobiopterin, 18B, 614); substrate specificity, 53, 283, 284; tetrahydropterin specificity, 53, 284, 285; 142, 15

PHENYLALANINE RACEMASE

This racemase [EC 5.1.1.11], also known officially as phenylalanine racemase (ATP-hydrolyzing), catalyzes the reaction of L-phenylalanine with ATP to produce D-phenylalanine, AMP, and pyrophosphate.[1,2] (This enzyme is essential for the synthesis of gramicidin S, the cyclic decapeptide antibiotic [structure: *cyclo*(-Val-Orn-Leu-D-Phe-Pro-Val-Orn-Leu-D-Phe-Pro-)], hence the alternative name gramicidin S synthetase 1; TycA has a similar role in the biosynthesis of tyrocidine).

L-phenylalanine   D-phenylalanine

In this unusual mechanism, an aminoacyl adenylate intermediate is formed, and the acyl group is then transferred to the thiol group of an enzyme-bound pantotheine moiety. The well-known lability of α-C–H bonds of thiol esters leads to racemization, thereby permitting reversible inversion of stereochemistry at the α-carbon. Subsequent hydrolysis of the thiol-ester yields D-phenylalanine and regenerates the active enzyme. ***See also** Gramicidin S Synthetase*

D-Phenylalanine is also a substrate for this enzyme. Interestingly, incubation of the enzyme with either enantiomer will not result in a racemic mixture: a mixture of approximately 4:1 in favor of the D-enantiomer is obtained (the actual ratio being dependent on the reaction conditions).[1]

[1] M. E. Tanner & G. L. Kenyon (1998) *CBC* **2**, 7.
[2] E. Adams (1972) *The Enzymes*, 3rd ed., **6**, 479.

Selected entries from *Methods in Enzymology* [vol, page(s)]:
General discussion: 43, 567, 571
Assay: 43, 568
Other molecular properties: affinity chromatography, 43, 575; amino acid-dependent ATP-[14C]AMP exchange, 43, 569, 570; amino acid-dependent ATP-32PPi exchange, 43, 569, 576; aminosulfate precipitation, 43, 573; *Bacillus brevis* preparation, 43, 571, 572; crude extract preparation, 43, 573; incubation mixture, 43, 568, 569; inhibitors, 43, 578; kinetic constants, 43, 578; light enzyme, properties, 43, 577, 578; molecular weight, 43, 577; 4'-phosphopantetheine, 43, 577; properties, 43, 576; racemization of phenylalanine, 43, 571; reaction scheme, 43, 568; separation of light and heavy enzymes, 43, 575; Sephadex G-200 chromatography, 43, 573, 574; specificity, 43, 577; stability, 43, 577; thiolester-bound amino acids, 43, 570; thiotemplate mechanism, 43, 568

PHENYLALANYL-tRNA SYNTHETASE

This ligase [EC 6.1.1.20], also known as phenylalanine:tRNA ligase, catalyzes the reaction of ATP with L-phenylalanine and tRNA^{Phe} to produce AMP, pyrophosphate, and L-phenylalanyl-tRNA^{Phe}. This ligase is the only class II aminoacyl-tRNA synthetase that preferentially aminoacylates the 2'-hydroxyl group of the terminal ribose of the tRNA rather than the 3'-hydroxyl. The enzyme also catalyzes the formation of P^1,P^4-di(adenosine-5')-tetraphosphate (ApppA). An editing mechanism is present to increase the fidelity of tRNA aminoacylation. L-Tyrosine can act as a weak alternative substrate; however, the noncognate L-tyrosyl-AMP is released and subsequently hydrolyzed.

Fasiolo and Fersht[1,2] used rapid-mix/quench and steady-state kinetic experiments to show that yeast phenylalanyl-tRNA synthetase catalyzes phenylalanyl-tRNA^{Phe} formation by means of an enzyme·phenylalanyl-acyladenylate (enzyme·F-AMP) intermediate. Their results were most consistent with the formation of F-tRNA^{Phe} by an initial rapid formation of an aminoacyladenylate complex followed by the slow, rate-determining, transfer of the phenylalanyl moiety to tRNA^{Phe}. Later studies from the same laboratory examined the transfer of amino acid to tRNA by bacterial phenylalanyl-tRNA synthetase mutants with replacements of Ala294 in the α-subunit to modify amino acid specificity. While steady-state analysis of tRNA charging gave little evidence of a difference between wild-type and altered enzymes, pre-steady-state kinetics revealed higher rates of tRNA^{Phe} charging by both the A294S enzyme·F-AMP and the A294G enzyme·*p*-chloro-F-AMP. The authors suggested that the decreased energy required to form the transition state of amino acid transfer in these mutants may well be related to a weaker binding of the amino acid in the aminoacyl adenylate complex. They suggest that a compromise is apt to exist between amino acid activation and tRNA charging, because slowing down the first step increases the rate of the second step, perhaps as a consequence of decreased stability of the enzyme·amino acid-AMP complex. ***See also*** *Aminoacyl-tRNA Synthetases*

[1] F. Fasiolo & A. R. Fersht (1978) *EJB* **85**, 85.
[2] M. Ibba, C. M. Johnson, H. Hennecke & A. R. Fersht (1995) *FEBS Lett.* **358**, 293.

Selected entries from ***Methods in Enzymology*** [**vol**, page(s)]:
General discussion: 20, 220; 34, 5, 167, 513; 46, 85; 59, 178; 60, 618
Assay: aminoacylation, 59, 183
Other molecular properties: hydroxamate formation rate, 29, 613; misactivation studies, 59, 291; preparation, 12B, 698; preparation of aminoacylated tRNA, 59, 126, 127; purification, 34, 169, 170; 59, 262, 266; subcellular distribution, 59, 233, 234; tRNA interaction, X-ray studies, 61, 231, 232

PHENYLETHANOLAMINE *N*-METHYLTRANSFERASE

This methyltransferase [EC 2.1.1.28], also known as noradrenaline *N*-methyltransferase and norepinephrine

N-methyltransferase, catalyzes the reaction of S-adenosyl-L-methionine with phenylethanolamine to produce S-adenosyl-L-homocysteine and *N*-methylphenylethanolamine.[1]

phenylethanolamine   *N*-methylphenylethanolamine

Other phenylethanolamines can act as methyl acceptors: for example, converting norepinephrine into epinephrine, normetanephrine into metanephrine, norfenefine into neosynephrine, and octopamine into synephrine. The enzyme has an essential active-site cysteinyl residue.

[1] R. J. Connett & N. Kirshner (1970) *JBC* **245**, 329.
Selected entries from ***Methods in Enzymology*** [**vol**, page(s)]:
General discussion: 17B, 761; 74, 374; 142, 655
Assay: 17B, 761; 142, 656
Other molecular properties: catecholamine radioimmunoassay with, 142, 559; octopamine, assay of, 63, 36; preparation from (bovine adrenal gland, 17B, 762; 142, 554; rabbit adrenal gland, 142, 657); properties, 17B, 763; 142, 658

PHENYLGLYOXYLATE DEHYDROGENASE (ACYLATING)

This thiamin pyrophosphate- and FAD-dependent oxidoreductase [EC 1.2.1.58] catalyzes the reaction of phenylglyoxylate with NAD$^+$ and coenzyme A to produce benzoyl-CoA, carbon dioxide, and NADH and a proton.[1] The enzyme from the denitrifying bacterium *Azoarcus evansii* is fairly specific for phenylglyoxylate: α-ketoisovalerate (or, 2-oxoisovalerate) is oxidized at 15% of the rate for phenylglyoxylate.

[1] W. Hirsch, H. Schagger & G. Fuchs (1998) *EJB* **251**, 907.

PHENYLPYRUVATE DECARBOXYLASE

This thiamin-pyrophosphate-dependent decarboxylase [EC 4.1.1.43] catalyzes the conversion of phenylpyruvate to phenylacetaldehyde and carbon dioxide.[1]

phenylpyruvate   phenylacetaldehyde

Indole-3-pyruvate can also act as a substrate to produce indole-3-ethanal.

[1] T. Asakawa, H. Wada & T. Yamano (1968) *BBA* **170**, 375.

Selected entries from ***Methods in Enzymology*** [**vol**, page(s)]:
General discussion: 17A, 589

PHENYLPYRUVATE TAUTOMERASE

This tautomerase [EC 5.3.2.1] interconverts *keto*-phenylpyruvate and *enol*-phenylpyruvate. Because the enzyme also acts on other arylpyruvates, it may be identical to *p*-hydroxyphenylpyruvate *keto-enol* tautomerase.[1–3] The beef kidney tautomerase catalyzes stereospecific exchange of one of the enantiotopic methylene protons with bulk water, thereby providing a convenient method for the preparation of chirally labeled phenylpyruvates.[1] *See also p-Hydroxyphenylpyruvate Keto-Enol Tautomerase*

keto-phenylpyruvate   enol-phenylpyruvate

Macrophage migration inhibitory factor (MIF), an immunoregulatory protein, exhibits a phenylpyruvate tautomerase (PPT) activity. The catalytic mechanism of this enzyme is of interest in determining whether there is a relationship between the PPT activity and the role of MIF in various immune and inflammatory processes.[2]

[1] J. Retey, K. Bartl, E. Ripp & W. E. Hull (1977) *Eur. J. Biochem.* **72**, 251.
[2] W. H. Johnson, Jr., R. M. Czerwinski, S. L. Stamps & C. P. Whitman (1999) *Biochemistry* **38**, 16024.
[3] J. V. Schloss & M. S. Hixon (1998) *CBC* **2**, 43.

PHENYLSERINE ALDOLASE

This pyridoxal-phosphate-dependent enzyme [EC 4.1.2.26] catalyzes the conversion of L-*threo*-3-phenylserine to glycine and benzaldehyde.[1]

L-threo-3-phenylserine   benzaldehyde

The D-stereoisomer is not a substrate.

[1] F. H. Bruns & L. Fiedler (1958) *Nature* **181**, 1533.

PHLORETIN HYDROLASE

This enzyme [EC 3.7.1.4] catalyzes the hydrolysis of phloretin to produce phloretate and phloroglucinol.[1–3]

phloretin   phloroglucinol   phloretate

Other C-acylated phenols similar to phloretin can act as substrates: for example, 3-methylphloracetophenone and 2,4,4-trihydroxydihydrochalcone. The monkey enzyme will also act on phlorizin, the glucoside of phloretin.

[1] S. Ramaswamy & A. N. Radhakrishnan (1975) *BBA* **403**, 446.
[2] T. Minamikawa, N. P. Jaysankar, B. A. Bohm, I. E. P. Taylor & G. H. N. Towers (1970) *BJ* **116**, 889.
[3] A. K. Chatterjee & L. N. Gibbins (1969) *J. Bacteriol.* **100**, 594.

PHLOROGLUCINOL REDUCTASE

This oxidoreductase [EC 1.3.1.57], which participates in the bacterial degradation of gallate, catalyzes the reaction of phloroglucinol with NADPH and a proton to produce dihydrophloroglucinol and $NADP^+$.[1]

[1] J. D. Haddock & J. G. Ferry (1989) *JBC* **264**, 4423.

PHLOROISOVALEROPHENONE SYNTHASE

This enzyme [EC 2.3.1.156], also known as valerophenone synthase and 3-methyl-1-(trihydroxyphenyl)butan-1-one synthase, catalyzes the reaction of isovaleryl-CoA with three molecules of malonyl-CoA to produce four molecules of coenzyme A, three molecules of carbon dioxide, and 3-methyl-1-(2,4,6-trihydroxyphenyl)butan-1-one (that is, phloroisovalerophenone). The enzyme is closely related to chalcone synthase [EC 2.3.1.74].[1,2]

[1] S. Y. Fung, K. W. M. Zuurbier, N. B. Paniego, J. J. C. Scheffer & R. Verpoorte (1997) *Proc. 26th Congr. Eur. Brew. Conv.*, 215.
[2] K. W. M. Zuurbier, J. Leser, T. Berger, A. J. P. Hofte, G. Schroder, R. Verpoorte & J. Schroder (1998) *Phytochemistry* **49**, 1945.

PHOLAS LUCIFERASE

This copper-dependent oxidoreductase, isolated from the mollusk *Pholas dactylus*, catalyzes the reaction of pholasin (a glycoprotein containing a tightly bound organic molecule) with dioxygen to produce oxypholasin and light.[1]

[1] J. P. Henry, C. Monny & A. M. Michelson (1975) *Biochemistry* **14**, 3458.

Selected entries from *Methods in Enzymology* [vol, page(s)]:
General discussion: 57, 385
Assay: 57, 385, 386
Other molecular properties: catalase, assay for, **57**, 404; complex formation with luciferin, **57**, 393; copper content, **57**, 388; electrochemical oxidation, **57**, 401; glucose, assay for, **57**, 403; glucose oxidase, assay for, **57**, 403; hydrogen peroxide, assay for, **57**, 403; kinetics, **57**, 394; luciferase-mediated assay of antibody binding sites, **57**, 402, 403; molecular weight, **57**, 388, 391; peroxidase activity, **57**, 397, 398; pH optimum, **57**, 391, 393; physical properties, **57**, 388, 391; properties, **57**, 388; protein–protein interactions, **57**, 393; purification, **57**, 387, 388

PHORBOL-DIESTER HYDROLASE

This enzyme [EC 3.1.1.51], also known as diacylphor-bate 12-hydrolase, catalyzes the hydrolysis of phorbol 12,13-dibutanoate to produce phorbol 13-butanoate and butanoate.[1] Because 12-*O*-tetradecanoylphorbol-13-ace-tate is an active tumor promoter, this hydrolase inactivates this carcinogenic agent.

[1]M. Shoyab, T. C. Warren & G. L. Torado (1981) *JBC* **256**, 12529.

PHOSPHATE ACETYLTRANSFERASE

This acyltransferase [EC 2.3.1.8], also known as phos-photransacetylase, catalyzes the reversible reaction of acetyl-CoA with orthophosphate to produce coenzyme A and acetylphosphate. Other short-chain acyl-CoA meta-bolites and short-chain acyl phosphates are substrates, albeit at reduced rates. The kinetic mechanism of phos-photransacetylase is of the rapid equilibrium random Bi Bi type,[1,2] without formation of any acyl-enzyme intermediate.[3] Because acetylphosphate can be chemi-cally synthesized by reaction of orthophosphate with acetic anhydride or ketene, this enzymatic reaction affords a convenient and cost-effective route to the synthesis of radiolabeled acyl-CoA derivatives in high yield. The enzyme may also be employed as an auxilliary enzyme to maintain the acetyl-CoA concentration in various acetyl-CoA-dependent processes.

[1]R. A. Pelroy & H. R. Whiteley (1972) *J. Bacteriol.* **111**, 47.
[2]K. Kreuzberg, H. Umlauf & H. P. Blaschkowski (1985) *BBA* **842**, 22.
[3]J. Henkin & R. H. Abeles (1976) *Biochemistry* **15**, 3472.

Selected entries from *Methods in Enzymology* [vol, page(s)]:
General discussion: 1, 596; 13, 381; **34**, 267; **243**, 94
Assay: 1, 596; 13, 381; **18A**, 316
Other molecular properties: acetyl coenzyme A, assay of, **3**, 935; **13**, 497, 499, 546, 548; **122**, 43; acetyl coenzyme A deacylase, assay of, 1, 606; acetyl coenzyme A regenerating system, **13**, 584, 588; **17A**, 788; acetyl phosphate and acetyl-CoA, in determination of, **122**, 44; acylphosphatase, not present in preparation of, **6**, 327; aldehyde dehydrogenase, assay of, 1, 518; amines, in acetylation of, 1, 608; amino acid, in acetylation of, 1, 616, 618; amino acid acetylase, in assay of, 1, 616; ammonium ion requirement, 13, 385; butyrate oxidation and, 1, 551; chiral methyl group analysis, in, **87**, 130; choline acetylase, assay of, 1, 622; citrate synthase, assay of, 1, 685; **13**, 3; *Clostridium kluyveri*, from, 1, 596; **13**, 381; coenzyme A assay and, **3**, 915; **13**, 539, 540, 543; **18A**, 314; **55**, 210; coenzyme A transphorase, assay of, 1, 599; condensing enzyme assay and, 1, 685; dephosphocoenzyme A kinase, assay of, **2**, 649; glucosamine, in acetylation of, 1, 613; heat inactivation of, 1, 492, 602; imidazole catalyzing same reaction of, **6**, 765; inhibition by coenzyme A analogues, **18A**, 325; lipoate transacetylase, assay of, **5**, 652, 654, 655; **9**, 262; properties, 1, 598; **13**, 384; purification from (*Clostridium kluyveri*, 1, 597; **13**, 383; *Escherichia coli* B, **18A**, 315); pyruvate dehydrogenase complex, assay of, **9**, 248; pyruvate oxidase, assay of, 1, 486, 487; pyruvate phosphoroclastic system, in, **243**, 96, 99; pyruvate synthase, assay of, **13**, 173; sodium ion inhibition, 1, 607; synthesis of acetyl-coenzyme A, in, **17B**, 788

PHOSPHATE BUTANOYLTRANSFERASE

This acyltransferase [EC 2.3.1.19], also known as phos-photransbutyrylase, catalyzes the reversible reaction of butanoyl-CoA with orthophosphate to produce coenzyme A and butanoyl phosphate.[1,2] This enzymatic activ-ity is distinguished from phosphate acetyltransferase on the basis of its use of isovaleryl-CoA and valeryl-CoA as alternative substrates.[2] Otherwise, the mecha-nisms of phosphotransacetylase and phosphotransbuty-rylase are probably quite similar. Acetyl-CoA and propionyl-CoA also act as substrates, albeit at a much reduced rate.

[1]R. C. Valentine & R. S. Wolfe (1960) *JBC* **235**, 1948.
[2]D. P. Wiesenborn, F. B. Rudolph & E. T. Papoutsakis (1989) *AEM* **55**, 317.

PHOSPHATE-TRANSPORTING ATPase

This so-called ATPase [EC 3.6.3.27], catalyzes the ATP-dependent (ADP- and orthophosphate-producing) trans-port of phosphate$_{out}$ to produce phosphate$_{in}$.[1–3] ABC-type (ATP-binding cassette-type) ATPases have two similar ATP-binding domains and do not undergo phosphorylation during the transport process.

The idea that a transporter is an enzyme is in keep-ing with a new definition of enzyme catalysis as the facilitated making/breaking of chemical bonds, not just covalent bonds.[4] This idea builds on Pauling's assertion that any long-lived, chemically distinct interaction (in this case, the persistent location of a solute with respect to the faces of a membrane) can be regarded as a chemi-cal bond. Note also that the equilibrium constant ($K_{eq} =$ [phosphate$_{in}$][ADP][P$_i$]/[phosphate$_{out}$][ATP]) does not conform to that expected for an ATPase (*i.e.*, $K_{eq} =$ [ADP][P$_i$]/[ATP]). Thus, although the overall reaction yields ADP and orthophosphate, the enzyme is misclas-sified as a hydrolase, and instead should be regarded as an energase-type reaction. Energases facilitate affinity-modulated reactions by coupling the $\Delta G_{ATP-hydrolysis}$ to a force-generating or work-producing step.[4] In this case, P-O-P bond scission supplies the energy to drive phosphate transport. *See* ATPase; Energase

[1]D. C. Webb, H. Rosenberg & G. B. Cox (1992) *JBC* **267**, 24661.
[2]M. Braibant, P. LeFevre, L. de Wit, J. Ooms, P. Peirs, K. Huygen, R. Wattiez & J. Content (1996) *FEBS Lett.* **394**, 206.
[3]J. K. Griffiths & C. E. Sansom (1998) *The Transporter Factsbook*, Academic Press, San Diego.
[4]D. L. Purich (2001) *TiBS* **26**, 417.

PHOSPHATIDATE CYTIDYLYLTRANSFERASE

This transferase [EC 2.7.7.41], also known as CDP-diacylglycerol synthase and CTP:phosphatidate cytidylyltransferase, catalyzes the reaction of CTP with phosphatidate to produce pyrophosphate and CDP-diacylglycerol.[1-7] The *Bacillus subtilis* enzyme has a ping pong Bi Bi kinetic mechanism[1] whereas the *Escherichia coli*[2,3] and *Saccharomyces cerevisiae*[4] enzymes are not.

[1]J.-L. Gaillard, B. Lubochinsky & D. Rigomier (1983) *BBA* **753**, 372.
[2]C. P. Sparrow & C. R. H. Raetz (1985) *JBC* **260**, 12084.
[3]K. E. Langley & E. P. Kennedy (1978) *J. Bacteriol.* **136**, 85.
[4]M. J. Kelley & G. M. Carman (1987) *JBC* **262**, 14563.
[5]C. Kent (1995) *ARB* **64**, 315.
[6]J. D. Esko & C. R. H. Raetz (1983) *The Enzymes*, 3rd ed., **16**, 207.
[7]R. A. Pieringer (1983) *The Enzymes*, 3rd ed., **16**, 255.

Selected entries from *Methods in Enzymology* [vol, page(s)]:
General discussion: 209, 237, 242
Assay: *Escherichia coli*, **209**, 237; plant endoplasmic reticulum, **148**, 591; rat liver, **109**, 474; yeast, **209**, 244
Other molecular properties: *Escherichia coli* (assay, **209**, 237; handling, **209**, 241; properties, **209**, 242; purification, **209**, 237; solubilization, **209**, 239; storage, **209**, 241); mutant, **72**, 694; plant endoplasmic reticulum, in, (assay, **148**, 591; properties, **148**, 592); yeast (assay, **209**, 244; properties, **209**, 247; purification, **209**, 244; synthetic and analytical uses, **209**, 247)

PHOSPHATIDATE PHOSPHATASE

This enzyme [EC 3.1.3.4] catalyzes the hydrolysis of a 3-*sn*-phosphatidate to generate a 1,2-diacyl-*sn*-glycerol and orthophosphate.[1-4] *See also Lysophosphatidate Phosphatase*

[1]D. N. Brindley & D. W. Waggoner (1998) *JBC* **273**, 24281.
[2]R. M. Bell & R. A. Coleman (1983) *The Enzymes*, 3rd ed., **16**, 87.
[3]J. D. Esko & C. R. H. Raetz (1983) *The Enzymes*, 3rd ed., **16**, 207.
[4]R. A. Pieringer (1983) *The Enzymes*, 3rd ed., **16**, 255.

Selected entries from *Methods in Enzymology* [vol, page(s)]:
General discussion: 5, 482; 14, 185; 197, 548, 553; 209, 219; 312, 373
Assay: 5, 482; 14, 185; 197, 553 (in rat liver, rationale, 197, 555); yeast, 197, 548; 209, 220, 228
Other molecular properties: activators and inhibitors, 14, 187; 35, 506; distribution, 14, 187; mitochondrial, yeast (properties, 209, 223; purification, 209, 220; radiochemical assay, 209, 220; substrate preparation, 209, 220); properties, 5, 483; 14, 186; 197, 551; 209, 223, 229; purification (chicken liver, 5, 483; pig brain, 14, 186; yeast, 197, 549); specificity, 5, 483; 14, 186; yeast (assays, 197, 548; properties, 197, 551; purification, 197, 549; substrates, preparation, 197, 548)

PHOSPHATIDYLCHOLINE DESATURASE

This enzyme [EC 1.3.1.35] (also known as oleate desaturase, oleoyl-CoA desaturase and linoleate synthase) catalyzes the reaction of 1-acyl-2-oleoyl-*sn*-glycero-3-phosphocholine with NAD^+ to produce 1-acyl-2-linoleoyl-*sn*-glycero-3-phosphocholine and NADH and a proton.[1,2]

[1]C. R. Slack, P. G. Roughan & J. Browse (1979) *BJ* **179**, 649.
[2]E. L. Pugh & M. Kates (1975) *BBA* **380**, 442.

PHOSPHATIDYLCHOLINE:DOLICHOL *O*-ACYLTRANSFERASE

This acyltransferase [EC 2.3.1.83] catalyzes the reaction of 3-*sn*-phosphatidylcholine with dolichol (dolichol-16 to dolichol-22) to produce 1-acyl-*sn*-glycero-3-phosphocholine and an acyldolichol.[1]

[1]R. W. Keenan & M. E. Kruczek (1976) *Biochemistry* **15**, 1586.

PHOSPHATIDYLCHOLINE 12-MONOOXYGENASE

This oxidoreductase [EC 1.14.13.26], also known as ricinoleic acid synthase and oleate Δ^{12}-hydroxylase, catalyzes the reaction of a 1-acyl-2-oleoyl-*sn*-glycero-3-phosphocholine with NADH, a proton, and dioxygen to produce a 1-acyl-2-[(*S*)-12-hydroxyoleoyl]-*sn*-glycero-3-phosphocholine, NAD^+, and water.[1,2] The enzyme from *Lesquerella fendleri* also exhibits desaturase activity.[3]

[1]R. A. Moreau & P. K. Stumpf (1981) *Plant Physiol.* **67**, 672.
[2]J. T. Lin, T. A. McKeon, M. Goodrich-Tanrikulu & A. E. Stafford (1996) *Lipids* **31**, 571.
[3]P. Broun, S. Boddupalli & C. Somerville (1998) *Plant J.* **13**, 201.

PHOSPHATIDYLCHOLINE:RETINOL *O*-ACYLTRANSFERASE

This acyltransferase [EC 2.3.1.135], also known as lecithin:retinol acyltransferase, catalyzes the reaction of phosphatidylcholine with retinol-[cellular-retinol-binding-protein] to produce 2-acylglycerophosphocholine and retinyl-ester-[cellular-retinol-binding-protein].[1-3] The *sn*-1-acyl moiety of phosphatidylcholine is transferred; nevertheless, the fatty acyl group in the *sn*-2-position is important in substrate recognition.[1] The enzyme, which also acts on unbound all-*trans*-retinol, catalyzes a ping pong Bi Bi kinetic mechanism with an acyl-enzyme intermediate.[3]

[1]J. C. Saari & D. L. Bredberg (1989) *JBC* **264**, 8636.
[2]P. N. MacDonald & D. E. Ong (1988) *JBC* **263**, 12478.
[3]Y. Q. Shi, I. Hubacek & R. R. Rando (1993) *Biochemistry* **32**, 1257.

Selected entries from *Methods in Enzymology* [vol, page(s)]:
General discussion: 190, 156; 316, 400
Assay: 189, 450; 190, 156

PHOSPHATIDYLCHOLINE:STEROL ACYLTRANSFERASE

This acyltransferase [EC 2.3.1.43], more commonly known as lecithin:cholesterol acyltransferase (abbreviated as

LCAT) as well as phospholipid:cholesterol acyltransferase, catalyzes the reaction of phosphatidylcholine with a sterol (such as cholesterol) to produce a sterol ester and 1-acylglycerophosphocholine. The enzyme transfers palmitoyl, oleoyl, or linoleoyl groups. The bacterial enzyme has phospholipase A_2 [EC 3.1.1.4] and lysophospholipase [EC 3.1.1.5] activity.

Selected entries from *Methods in Enzymology* [vol, page(s)]:
General discussion: 15, 543; 34, 747; 71, 753; 72, 375; 111, 274; 129, 763, 783, 790
Assay: 15, 544; 72, 375; apolipoprotein A-I, 72, 380, 382; cholesteryl ester transfer protein, 72, 380; gas-liquid chromatography, 72, 376; human plasma, 111, 268; 129, 773; incorporation of labeled fatty acids into cholesterol esters, 72, 378; specificity, 72, 379; thin-layer chromatography, 72, 377
Other molecular properties: activation by synthetic peptide analogues of apolipoproteins, 128, 644; association with high-density lipoproteins, 71, 759; human plasma, from, 15, 543; 71, 753; inhibitors, 15, 547; 71, 766; isolation, 197, 428; metabolic function, 128, 17; partial purification, 15, 545; 72, 381 (concentration by adsorption spectrum, 72, 383; HDL 3-Sepharose affinity column, 72, 382); phospholipase activity (assay, 197, 429; enzymatic properties, 197, 431); plasma, 15, 543; properties, 15, 547; purification, 15, 545; 72, 381; stereospecificity at phosphorus, 197, 258

PHOSPHATIDYLETHANOLAMINE METHYLTRANSFERASE

This methyltransferase [EC 2.1.1.17] catalyzes the reaction of *S*-adenosyl-L-methionine with phosphatidylethanolamine to produce *S*-adenosyl-L-homocysteine and phosphatidyl-*N*-methylethanolamine.[1–5] The rat liver enzyme has an ordered Bi Bi kinetic mechanism[3] whereas the human erythrocyte enzyme has a random Bi Bi scheme.[4] *See also Phosphatidyl-N-Methylethanolamine Methyltransferase*

[1] L. B. Tijburg, M. J. Geelen & L. M. van Golde LM. (1989) *BBA* 1004, 1.
[2] N. D. Ridgway & D. E. Vance (1987) *JBC* 262, 17231.
[3] N. D. Ridgway & D. E. Vance (1988) *JBC* 263, 16864.
[4] R. C. Reitz, D. J. Mead, R. A. Bjur, A. H. Greenhouse & W. H. Welch, Jr. (1989) *JBC* 264, 8097.
[5] J. D. Esko & C. R. H. Raetz (1983) *The Enzymes*, 3rd ed., 16, 207.

Selected entries from *Methods in Enzymology* [vol, page(s)]:
General discussion: 14, 125; 71, 581; 209, 366
Assay: 14, 125; castor bean endosperm, 71, 608; hepatic, rat, 209, 366; plant endoplasmic reticulum, 148, 589
Other molecular properties: bovine adrenal medulla, substrates and products, from, 71, 585; castor bean endosperm, from, 71, 608 (assay, 71, 608; detergent effect, 71, 609; inhibitors, 71, 609; intracellular distribution, 71, 609; products, 71, 609; properties, 71, 608); chromatographic separation of products, 14, 126; comparison of sources, 14, 128; endoplasmic reticulum of liver, from, 71, 581; hepatic, rat (assay, 209, 366; properties, 209, 372; purification, 209, 368); inhibitors, 14, 128; methyl group transfer to (phosphatidyl-*N*,*N*-dimethylethanolamine, 71, 582; phosphatidylethanolamine, 71, 582; phosphatidyl-*N*-monomethylethanolamine, 71, 582); plant endoplasmic reticulum, in (assay, 148, 589; properties, 148, 590)

PHOSPHATIDYLETHANOLAMINE:L-SERINE PHOSPHATIDYLTRANSFERASE

This enzyme catalyzes the reaction of phosphatidylethanolamine with L-serine to produce phosphatidyl-L-serine and ethanolamine.[1,2] Note that phosphatidylserine decarboxylase has been reported to catalyze this exchange reaction.[1] L-Serine can be replaced with ethanolamine, choline, monomethylethanolamine, and dimethylethanolamine.

***See also** Phosphatidylserine Decarboxylase*

[1] A. M. Cook, E. Low & M. Ishijimi (1972) *Nature (New Biol.)* 239, 150.
[2] J. D. Esko & C. R. H. Raetz (1983) *The Enzymes*, 3rd ed., 16, 207.

Selected entries from *Methods in Enzymology* [vol, page(s)]:
General discussion: 209, 341
Assay: 71, 594; castor bean endosperm, 71, 610; plant endoplasmic reticulum, 148, 590
Other molecular properties: brain, 71, 588; castor bean endosperm, from, 71, 609 (assay, 71, 610; intracellular distribution, 71, 611; metal ion specificity, 71, 611; products, 71, 611; properties, 71, 610); plant endoplasmic reticulum, in (assay, 148, 590; properties, 148, 591); properties, 71, 596, 610; 148, 591; solubilization, 71, 589

PHOSPHATIDYLGLYCEROL:MEMBRANE-OLIGOSACCHARIDE GLYCEROPHOSPHOTRANSFERASE

This transferase [EC 2.7.8.20], also known as phosphoglycerol transferase, catalyzes the reaction of a phosphatidylglycerol with a membrane-derived-oligosaccharide D-glucose to produce 1,2-diacyl-*sn*-glycerol and a membrane-derived-oligosaccharide 6-(glycerophospho)-D-glucose.[1,2] The actual acceptor substrates are 1,2-β- and 1,6-β-linked glucose residues in membrane polysaccharides and in synthetic glucosides (for example, *p*-hydroxyphenyl-β-D-glucoside). The enzyme requires Mg^{2+} or another divalent cation.

[1] B. J. Jackson & E. P. Kennedy (1983) *JBC* 258, 2394.
[2] J. P. Bohin & E. P. Kennedy (1984) *JBC* 259, 8388.

PHOSPHATIDYLGLYCEROL:PROLIPO-PROTEIN DIACYLGLYCERYLTRANSFERASE

This transferase catalyzes the reaction of a phosphatidylglycerol with a cysteinyl residue near the N-terminus of a prolipoprotein to produce a diacylglyceryl-prolipoprotein and glycerophosphate. The protein product of this reaction is the substrate for signal peptidase II [EC 3.4.23.36].

Selected entries from *Methods in Enzymology* [vol, page(s)]:
Assay: denaturing gel electrophoresis, 250, 684, 686; reaction conditions, 250, 687; substrate (Braun's prolipoprotein, 250, 684, 686; peptide, 250, 688; storage, 250, 686; synthesis, 250, 686)
Other molecular properties: crude enzyme preparation, 250, 687; detergent sensitivity, 250, 688; lethal mutations, 250, 697; sequence from *Salmonella typhimurium*, 250, 690; substrate specificity, 250, 688

PHOSPHATIDYLGLYCEROPHOSPHATASE

This enzyme [EC 3.1.3.27], also known as phosphatidyl-glycerol phosphate phosphohydrolase, catalyzes the reaction of phosphatidylglycerophosphate with water to form phosphatidylglycerol and orthophosphate.[1-4] The presence of unsaturated fatty acids stimulates the rat heart enzyme.[2]

[1] B. D. Cain, T. J. Donohue, W. D. Sheperd & S. Kaplan (1984) *JBC* **259**, 942.
[2] S. G. Cao & G. M. Hatch (1994) *Lipids* **29**, 475.
[3] J. D. Esko & C. R. H. Raetz (1983) *The Enzymes*, 3rd ed., **16**, 207.
[4] R. A. Pieringer (1983) *The Enzymes*, 3rd ed., **16**, 255.

Selected entries from *Methods in Enzymology* [vol, page(s)]:
General discussion: **209**, 224
Assay: *Escherichia coli*, **209**, 227; plant endoplasmic reticulum, **148**, 593
Other molecular properties: castor bean endosperm, from, **71**, 611; *Escherichia coli* (assay, **209**, 227; substrate preparation, **209**, 225); localization, **209**, 229; plant endoplasmic reticulum, in (assay, **148**, 593; properties, **148**, 593); properties, **209**, 228

PHOSPHATIDYLINOSITOL N-ACETYLGLUCOSAMINYLTRANSFERASE

This transferase [EC 2.4.1.198] catalyzes the reaction of UDP-*N*-acetyl-D-glucosamine with a phosphatidylinositol to produce UDP and *N*-acetyl-D-glucosaminylphosphatidylinositol.[1]

[1] T. L. Doering, W. J. Masterson, P. T. Englund & G. W. Hart (1989) *JBC* **264**, 11168.

PHOSPHATIDYLINOSITOL-BISPHOSPHATASE

This enzyme [EC 3.1.3.36], also known as triphosphoinositide phosphatase, catalyzes the hydrolysis of phosphatidyl-*myo*-inositol 4,5-bisphosphate to produce phosphatidylinositol 4-phosphate and orthophosphate.[1-4] The enzyme will also act on *myo*-inositol 1,4,5-trisphosphate (producing *myo*-inositol 1,4-bisphosphate) and *myo*-inositol 1,3,4,5-tetrakisphosphate (producing *myo*-inositol 1,3,4-trisphosphate).

[1] P. D. Roach & F. B. S. C. Palmer (1981) *BBA* **661**, 323.
[2] F. B. S. C. Palmer (1981) *Can. J. Biochem.* **59**, 469.
[3] T. M. Connolly, V. S. Bansal, T. E. Bross, R. F. Irvine & P. W. Majerus (1987) *JBC* **262**, 2146.
[4] R. S. Rana, M. C. Sekar, L. E. Hokin & M. J. MacDonald (1986) *JBC* **261**, 5237.

1-PHOSPHATIDYLINOSITOL-4,5-BISPHOSPHATE PHOSPHODIESTERASE

This enzyme [EC 3.1.4.11], also known as triphosphoinositide phosphodiesterase and phosphoinositide-specific phospholipase C, catalyzes the hydrolysis of 1-phosphatidyl-1D-*myo*-inositol 4,5-bisphosphate to produce D-*myo*-inositol 1,4,5-trisphosphate and 1,2-diacylglycerol.[1-3] Other substrates include phosphatidylinositol 4-phosphate, producing D-*myo*-inositol 1,4-bisphosphate and 1,2-diacylglycerol. There are at least six forms of this phosphodiesterase, all participating in various signal-transducing pathways. **See also Phospholipase C; 1-Phosphatidylinositol Phosphodiesterase**

[1] Y. Banno & Y. Nozawa (1987) *BJ* **248**, 95.
[2] P. Wang, S. Toyoshima & T. Osawa (1986) *J. Biochem.* **100**, 1015.
[3] R. A. Akhtar & A. A. Abdel-Latif (1978) *BBA* **527**, 159.

Selected entries from *Methods in Enzymology* [vol, page(s)]:
General discussion: **141**, 92; **187**, 226
Assay: **71**, 741, 742, 743; **109**, 476; **141**, 108; **187**, 226; **197**, 497, 504, 527
Other molecular properties: *Bacillus cereus*, from, **71**, 731 (acetylcholinesterase-releasing activity, **71**, 733; alkaline phosphatase-releasing activity, **71**, 732; assay, **71**, 731; **197**, 497; effect, of metal ions, **71**, 737; osmotic fragility, of, **71**, 741; phosphorus-release assay, **71**, 732; properties, **71**, 736; **197**, 499; purification, **197**, 494; purity, **71**, 735; stability, **71**, 737); bovine cerebral (assays, **197**, 504; purification, **197**, 506); comparison of metal ion activation, **71**, 745; effect on erythrocytes and erythrocyte ghosts, **71**, 745; effect on plasminogen activator receptor amphiphilicity, **223**, 218; glycophosphatidylinositol treatment, **230**, 428; human platelet (assay, **86**, 8; **197**, 519; properties, **197**, 525; purification, **197**, 520); hydroxylapatite chromatography, **238**, 204; immunoblot monitoring of enzyme release, **238**, 161; ion-exchange chromatography, **238**, 203, 206; rat cerebral cortical membrane, activation, **197**, 183; rat liver, assay, in, **109**, 476; release of enzymes by, from membranes, **71**, 745; specificity, **71**, 745; stereospecificity at phosphorus, **197**, 258; treatment of whole U937 cells, **223**, 217

PHOSPHATIDYLINOSITOL DEACYLASE

This deacylase [EC 3.1.1.52], also known as phosphatidylinositol phospholipase A$_2$, catalyzes the hydrolysis of 1-phosphatidyl-1D-*myo*-inositol to produce 1-acyl-glycerophosphoinositol and a fatty acid anion.[1]

[1] N. C. C. Gray & K. P. Strickland (1981) *Can. J. Biochem.* **60**, 108.

PHOSPHATIDYLINOSITOL:*myo*-INOSITOL PHOSPHATIDYLTRANSFERASE

This enzyme, also known as phosphatidylinositol exchange enzyme, catalyzes the reaction of phosphatidyl-*myo*-inositol with labeled *myo*-inositol to produce labeled phosphatidyl-*myo*-inositol and *myo*-inositol.

Selected entries from *Methods in Enzymology* [vol, page(s)]:
Assay: castor bean endosperm, **71**, 606; Mn^{2+}-activated, in rat liver, **109**, 475; plant endoplasmic reticulum, **148**, 595
Other molecular properties: castor bean endosperm, from, **71**, 606 (assay, **71**, 606; detergent effects, **71**, 607; intracellular distribution, **71**, 607; metal ion specificity, **71**, 607; nucleotide activation, **71**, 607; phospholipid effects, **71**, 607; properties, **71**, 607); plant endoplasmic reticulum, in (assay, **148**, 595; properties, **148**, 596)

1-PHOSPHATIDYLINOSITOL 3-KINASE

This phosphotransferase [EC 2.7.1.137] (also known as phosphatidylinositol 3-kinase, PI$_3$-kinase, and PtdIns-3-kinase) catalyzes the reaction of ATP with 1-phosphatidyl-1D-*myo*-inositol to produce ADP and 1-phosphatidyl-1D-*myo*-inositol 3-phosphate.[1–3]

[1]S. J. Morgan, A. D. Smith & P. J. Parker (1990) *EJB* **191**, 761.
[2]C. L. Carpenter, B. C. Duckworth, K. R. Auger, B. Cohen, B. S. Schaffhausen & L. C. Cantley (1990) *JBC* **265**, 19704.
[3]F. Shibasaki, Y. Homma & T. Takenawa (1991) *JBC* **266**, 8108.

Selected entries from *Methods in Enzymology* [**vol**, page(s)]:
General discussion: **256**, 241
Assay: **201**, 79
Other molecular properties: plate preparation for, **201**, 77; immunoprecipitation from (insulin-stimulated CHO cells, **201**, 78; PDGF-stimulated smooth muscle cells, **198**, 83); insulin-stimulated, determination in phosphotyrosine antibody immunoprecipitates, **201**, 76; Rho-dependent, assay in platelet cytosol, **256**, 241

1-PHOSPHATIDYLINOSITOL 4-KINASE

This phosphotransferase [EC 2.7.1.67] catalyzes the reaction of ATP with 1-phosphatidyl-1D-*myo*-inositol to produce ADP and 1-phosphatidyl-1D-*myo*-inositol 4-phosphate.[1–3] The *Saccharomyces cerevisiae* enzyme has an ordered Bi Bi kinetic mechanism.[1,2]

[1]C. J. Belunis, M. Bae-Lee, M. J. Kelley & G. M. Carman (1988) *JBC* **263**, 18897.
[2]J. T. Nickels, Jr., R. J. Buxeda & G. M. Carman (1992) *JBC* **267**, 16297.
[3]R. A. Anderson, I. V. Boronenkov, S. D. Doughman, J. Kunz & J. C. Loijens (1999) *JBC* **274**, 9907.

Selected entries from *Methods in Enzymology* [**vol**, page(s)]:
General discussion: **141**, 210; **209**, 183, 202
Assay: **109**, 476; **141**, 111, 212
Other molecular properties: cerebral membrane, bovine, fluorescent assay and purification, **209**, 204; membrane-associated, assay, **141**, 108, 111; properties, **141**, 215; purification from rabbit reticulocytes, **141**, 213; yeast (product identification, **209**, 188; properties, **209**, 188; purification, **209**, 185; radiochemical assay, **209**, 183; synthetic uses, **209**, 189)

PHOSPHATIDYL-*myo*-INOSITOL α-MANNOSYLTRANSFERASE

This transferase [EC 2.4.1.57] catalyzes the transfer of one or more α-D-mannose units from GDP-D-mannose to positions 2, 6, and others in 1-phosphatidyl-*myo*-inositol.[1]

[1]P. Brennan & C. E. Ballou (1968) *BBRC* **30**, 69.

PHOSPHATIDYLINOSITOL-3-PHOSPHATASE

This phosphatase [EC 3.1.3.64], often abbreviated PtdIns 3-phosphatase, catalyzes the hydrolysis of phosphatidylinositol 3-phosphate to produce phosphatidylinositol and orthophosphate.[1,2] Inositol 1,3-bisphosphate is also a substrate, producing inositol 1-phosphate.

[1]D. L. Lips & P. W. Majerus (1989) *JBC* **264**, 19911.
[2]K. K. Caldwell, D. L. Lips, V. S. Bansal & P. W. Majerus (1991) *JBC* **266**, 18378.

1-PHOSPHATIDYLINOSITOL-4-PHOSPHATE 5-KINASE

This phosphotransferase [EC 2.7.1.68], also known as diphosphoinositide kinase, catalyzes the reaction of ATP with 1-phosphatidyl-1D-*myo*-inositol 4-monophosphate to produce ADP and 1-phosphatidyl-1D-*myo*-inositol 4,5-bisphosphate.[1,2]

[1]C. L. Carpenter & L. C. Cantley (1996) *Curr. Opin. Cell Biol.* **8**, 153.
[2]J. C. Loijens, I. V. Boronenkov, G. J. Parker & R. A. Anderson (1996) *Adv. Enzyme Regul.* **36**, 115.

Selected entries from *Methods in Enzymology* [**vol**, page(s)]:
General discussion: **209**, 189, 202
Assay: **209**, 191, 207
Other molecular properties: antibodies, preparation, **209**, 192; cerebral membrane, bovine (properties, **209**, 210; purification, **209**, 208; radiochemical assay, **209**, 207); cytosolic (properties, **209**, 195; purification, **209**, 192; effect of modulators, **209**, 198; kinetic properties, **209**, 199); membrane-associated, assay, **141**, 108, 111; membrane-bound (type I, purification, **209**, 195; type II, purification and properties, **209**, 195; types I and II, comparison, **209**, 197)

1-PHOSPHATIDYLINOSITOL PHOSPHODIESTERASE

This phosphodiesterase [EC 3.1.4.10], also known as monophosphatidylinositol phosphodiesterase and phosphatidylinositol phospholipase C, catalyzes the conversion of 1-phosphatidyl-1D-*myo*-inositol to D-*myo*-inositol 1,2-cyclic phosphate and 1,2-diacylglycerol.[1–4]

phosphatidylinositol   inositol 1,2-cyclic phosphate   inositol 1-phosphate

The animal phosphodiesterase, but not the bacterial enzyme, hydrolyzes the cyclic phosphate, forming inositol 1-phosphate as the final product. The eukaryotic enzyme also requires Ca^{2+}. **See also** *Phospholipase C; 1-Phosphatidylinositol-4,5-Bisphosphate Phosphodiesterase*

[1]W. Sies & E. G. Lapetina (1983) *BBA* **752**, 329.
[2]A. A. Abdel-Latif, B. Luke & J. P. Smith (1980) *BBA* **614**, 425.
[3]R. Sundler, A. W. Alberts & P. R. Vagelos (1978) *JBC* **253**, 4175.
[4]R. J. Hondal, Z. Zhao, A. V. Kravchuk, H. Liao, S. R. Riddle, X. Yue, K. S. Bruzik & M. D. Tsai (1998) *Biochemistry* **37**, 4568.

Other molecular properties: amino acid sequence, **209**, 354; *Escherichia coli*, from, **71**, 571 (covalently bound pyruvic acid, **71**, 576; effect of detergents, **71**, 575; gene dosage effect, **71**, 574; inhibitors, **71**, 575; membrane-bound enzyme, **71**, 572, 575; mutant variants, **71**, 576; overproducing strains, **71**, 574; properties, **209**, 352; purification, **209**, 350; specificity, **71**, 575; thermal inactivation, **71**, 576; Triton-extracted enzyme, **71**, 575); hepatic, rat (properties, **209**, 365; standard assay, **209**, 362; substrate preparation, **209**, 360); mixed micelle, **64**, 355; mutant, **72**, 694

PHOSPHATIDYLINOSITOL-3,4,5-TRISPHOSPHATE 3-PHOSPHATASE

This magnesium-dependent phosphatase [EC 3.1.3.67], often abbreviated PtdIns(3,4,5)P$_3$ phosphatase, catalyzes the hydrolysis of phosphatidylinositol-3,4,5-trisphosphate to produce phosphatidylinositol-4,5-bisphosphate and orthophosphate.[1]

[1]Y. Kabuyama, N. Nakatsu, Y. Homma & Y. Fukui (1996) *EJB* **238**, 350.

PHOSPHATIDYL-N-METHYLETHANOL-AMINE N-METHYLTRANSFERASE

This methyltransferase [EC 2.1.1.71] catalyzes the reaction of *S*-adenosyl-L-methionine with phosphatidyl-*N*-methylethanolamine to produce *S*-adenosyl-L-homocysteine and phosphatidyl-*N*,*N*-dimethylethanolamine.[1-5] The enzyme will also act on the product, thereby producing phosphatidylcholine. **See also** *Phosphatidylethanolamine Methyltransferase*

[1]P. M. Gaynor & G. M. Carman (1990) *BBA* **1045**, 156.
[2]P. K. Dudeja, E. S. Foster & T. A. Brasitus (1986) *BBA* **875**, 493.
[3]C. Prasad & R. M. Edwards (1981) *JBC* **256**, 13000.
[4]W. J. Schneider & D. E. Vance (1979) *JBC* **254**, 3886.
[5]F. Hirata, O. H. Viveros, E. J. Diliberto & J. Axelrod (1978) *PNAS* **75**, 1718.

Selected entries from *Methods in Enzymology* [vol, page(s)]:
General discussion: **14**, 125

PHOSPHATIDYLSERINE DECARBOXYLASE

This decarboxylase [EC 4.1.1.65] catalyzes the conversion of phosphatidyl-L-serine to phosphatidylethanolamine and carbon dioxide.[1-5] An enzyme-bound pyruvoyl group reacts with the amino group of phosphatidylserine to form the Schiff base which subsequently undergoes decarboxylation; hydrolysis ensues and release of phosphatidylethanolamine regenerates the pyruvoyl group.

[1]T. Suda & M. Matsuda (1974) *BBA* **369**, 331.
[2]T. G. Warner & E. A. Dennis (1975) *JBC* **250**, 8004.
[3]M. L. Hackert & A. E. Pegg (1998) *CBC* **2**, 201.
[4]J. D. Esko & C. R. H. Raetz (1983) *The Enzymes*, 3rd ed., **16**, 207.
[5]R. A. Pieringer (1983) *The Enzymes*, 3rd ed., **16**, 255.

Selected entries from *Methods in Enzymology* [vol, page(s)]:
General discussion: **71**, 571; **209**, 348, 360; **280**, 81
Assay: **209**, 348, 362

PHOSPHATIDYLSERINE TRANSFER PROTEINS

These proteins, also known as phosphatidyl exchange proteins and phospholipid exchange enzymes, reportedly facilitate the incorporation of phosphatidylserine into membranes lacking phosphatidylserine.

Selected entries from *Methods in Enzymology* [vol, page(s)]:
Assay: **209**, 516
Other molecular properties: isolation, **209**, 519; properties, **209**, 522

PHOSPHINOTHRICIN ACETYLTRANSFERASE

This transferase catalyzes the reaction of phosphinothricin (also known as glufosinate; structural formula: CH$_3$P(O$_2^-$) CH$_2$CH$_2$CH(NH$_3^+$)COO$^-$) with acetyl-CoA to produce *N*-acetylphosphinothricin.[1] Phosphinothricin, first isolated from *Streptomyces viridichromogenes* and *Streptomyces hygroscopicus*, is an active ingredient in a number of herbicides and is a potent inhibitor of glutamine synthetase.

[1]J. Botterman, V. Gossele, C. Thoen & M. Lauwereys (1991) *Gene* **102**, 33.

Selected entries from *Methods in Enzymology* [vol, page(s)]:
General discussion: **216**, 417

PHOSPHOACETYLGLUCOSAMINE MUTASE

This mutase [EC 5.4.2.3], also known as acetylglucosamine phosphomutase and *N*-acetylglucosamine-phosphate mutase, uses *N*-acetyl-D-glucosamine 1,6-bisphosphate as a cofactor and catalyzes the interconversion of *N*-acetyl-D-glucosamine 1-phosphate and *N*-acetyl-D-glucosamine 6-phosphate.[1-3] The enzyme also interconverts glucose 1-phosphate and glucose 6-phosphate as well as mannose 1-phosphate and mannose 6-phosphate.

[1]P.-W. Cheng & D. M. Carlson (1979) *JBC* **254**, 8353.
[2]C. P. Selitrennikoff & D. R. Sonneborn (1976) *BBA* **451**, 408.
[3]A. Fernandez-Sorensen & D. M. Carlson (1971) *JBC* **246**, 3485.

Selected entries from *Methods in Enzymology* [vol, page(s)]:
General discussion: **8**, 175, 179
Assay: *Neurospora crassa*, **8**, 175; pig submaxillary gland, **8**, 179
Other molecular properties: *Neurospora crassa*, of, **8**, 175 (distribution, **8**, 178; properties, **8**, 177; purification, **8**, 176); pig submaxillary gland, of, **8**, 179 (distribution, **8**, 182; properties, **8**, 182; purification, **8**, 180)

PHOSPHO-*N*-ACETYLMURAMOYL-PENTAPEPTIDE-TRANSFERASE

This enzyme [EC 2.7.8.13], also known as UDP-MurNAc-pentapeptide phosphotransferase, catalyzes the reversible transfer of a phosphoryl group in the reaction of UDP-*N*-acetylmuramoyl-L-Ala-D-Glu-L-Lys-D-Ala-D-Ala with undecaprenyl phosphate to form *N*-acetylmuramoyl-L-Ala-D-Glu-L-Lys-D-Ala-D-Ala-diphosphoundecaprenol and UMP.[1]

[1]W. A. Weppner & F. C. Neuhaus (1977) *JBC* **252**, 2296.

PHOSPHOADENYLYLSULFATASE

This manganese-dependent sulfatase [EC 3.6.2.2], also known as 3-phosphoadenylyl sulfatase and PAPS sulfatase, catalyzes the hydrolysis of 3′-phosphoadenylylsulfate to produce adenosine 3′,5′-bisphosphate and sulfate.[1–3]

[1]H. B. Denner, A. M. Stokes, F. A. Rose & K. S. Dodgson (1973) *BBA* **315**, 394.
[2]A. S. Balasubramanian & B. K. Bachhawat (1962) *BBA* **59**, 389.
[3]A. B. Roy (1971) *The Enzymes*, 3rd ed., **5**, 1.

Selected entries from *Methods in Enzymology* [vol, page(s)]:
General discussion: 17B, 546

3′-PHOSPHOADENYLYLSULFATE 3′-PHOSPHATASE

This enzyme, formerly classified EC 3.1.3.30, has been deleted from the Enzyme Commission classification.

PHOSPHOADENYLYL-SULFATE REDUCTASE (THIOREDOXIN)

This oxidoreductase [EC 1.8.4.8; previously classified as 1.8.99.4] (also known as PAPS reductase [thioredoxin-dependent], PAPS reductase, 3′-phosphoadenylylsulfate reductase, and phosphoadenosine-phosphosulfate reductase) catalyzes the reaction of 3′-phosphoadenylyl sulfate with reduced thioredoxin to produce adenosine 3′,5′-bisphosphate, sulfite, and oxidized thioredoxin.[1,2]

The *Escherichia coli* enzyme exhibits a special ping pong mechanism with adenosine 3′-phosphate 5′-phosphosulfate reacting with the reduced enzyme isomer in a Theorell-Chance-type mechanism.[1]

The reductase can be coupled with thioredoxin reductase [EC 1.6.4.5], resulting in the NADPH-dependent reduction of PAPS.

[1]U. Berendt, T. Haverkamp, A. Prior & J. D. Schwenn (1995) *EJB* **233**, 347.

[2]J. D. Schwenn, F. A. Krone & K. Husmann (1988) *Arch. Microbiol.* **150**, 313.
Selected entries from *Methods in Enzymology* [vol, page(s)]:
General discussion: 17B, 546
Assay: NADPH, with, 17B, 547

PHOSPHOAMIDASE

This hydrolase [EC 3.9.1.1] catalyzes the hydrolysis of *N*-phosphocreatine, *N*-phosphoarginine, and other phosphoamides to yield orthophosphate and the corresponding amine-containing metabolite.[1–3] The microsomal rat kidney enzyme was shown to be identical to alkaline phosphatase.[4]

[1]M. F. Singer & J. S. Fruton (1957) *JBC* **229**, 111.
[2]S. M. Dudkin, R. K. Ledneva, Z. A. Shabarova & M. A. Prokofiev (1971) *FEBS Lett.* **16**, 48.
[3]R. Parvin & R. A. Smith (1969) *Biochemistry* **8**, 1748.
[4]M. Nishino, S. Tsujimura, M. Kuba & A. Kumon (1994) *ABB* **312**, 101.
Selected entries from *Methods in Enzymology* [vol, page(s)]:
Assay: histochemical, 4, 387
Other molecular properties: histochemical method for, 4, 387; phosphoramidate:hexose phosphotransferase, in preparations of, 9, 406; rennin, phosphoamidase activity of, 2, 77; specificity and acid phosphatase, 4, 388

6-PHOSPHO-2-DEHYDRO-3-DEOXYGALACTONATE ALDOLASE

This aldolase [EC 4.1.2.21], also known as 2-dehydro-3-deoxyphosphogalactonate aldolase and 6-phospho-2-keto-3-deoxygalactonate aldolase, catalyzes the reversible conversion of 2-dehydro-3-deoxy-D-galactonate 6-phosphate to pyruvate and D-glyceraldehyde 3-phosphate.[1,2]

[1]W. A. Wood (1972) *The Enzymes*, 3rd ed., **7**, 281.
[2]D. J. Creighton & N. S. R. K. Murthy (1990) *The Enzymes*, 3rd ed., **19**, 323.
Selected entries from *Methods in Enzymology* [vol, page(s)]:
General discussion: 9, 524
Assay: *Pseudomonas saccharophila*, 9, 524; 90, 264
Other molecular properties: *Pseudomonas saccharophila* (assay, 9, 524; 90, 264; properties, 9, 527; purification, 9, 525; 90, 265; specificity, 9, 528); stereochemistry, 87, 147

6-PHOSPHO-2-DEHYDRO-3-DEOXYGLUCONATE ALDOLASE

This aldolase [EC 4.1.2.14] (also known as 2-dehydro-3-deoxyphosphogluconate aldolase, phospho-2-keto-3-deoxygluconate aldolase, and 2-keto-3-deoxy-6-phosphogluconate aldolase) catalyzes the reversible conversion of 2-dehydro-3-deoxy-D-gluconate 6-phosphate to pyruvate and D-glyceraldehyde 3-phosphate.[1–4] The enzyme, which has an ordered Uni Bi kinetic mechanism (pyruvate is the last product released), also catalyzes the exchange of methyl hydrogens of pyruvate with solvent protons (due to the enolization of pyruvate). In many organisms (*e.g.*, *Escherichia coli*), this enzyme is identical with 4-hydroxy-2-ketoglutarate aldolase [EC 4.1.3.16]. The *Azotobacter vinelandii* enzyme also has an 8-phospho-2-dehydro-3-deoxyoctanate aldolase [EC 4.1.2.16] activity.[2]

See also *6-Phospho-5-Dehydro-2-Deoxygluconate Aldolase*

[1]W. A. Wood (1972) *The Enzymes*, 3rd ed., **7**, 281.
[2]T. S. Taha & T. L. Deits (1994) *BBRC* **200**, 459.
[3]K. N. Allen (1998) *CBC* **2**, 135.
[4]D. J. Creighton & N. S. R. K. Murthy (1990) *The Enzymes*, 3rd ed., **19**, 323.

Selected entries from *Methods in Enzymology* [vol, page(s)]:
General discussion: 9, 520; 42, 258; 46, 132
Assay: *Pseudomonas fluorescens*, 9, 520; *Pseudomonas putida*, 42, 259, 260
Other molecular properties: affinity labeling, 87, 474; anomeric specificity, 63, 378; enol intermediates, and, 87, 86; ketodeoxygluconate kinase (assay of, 5, 205; preparation free of, 5, 208); 6-phospho-2-dehydro-3-deoxygluconate preparation, in, 9, 51; 6-phosphogluconate dehydrase, assay of, 9, 653; *Pseudomonas fluorescens*, from (crystallization, 9, 523; molecular weight, 9, 524; properties, 9, 523; purification, 5, 206; 9, 521); *Pseudomonas putida*, from, 42, 258 (amino acids, 42, 262; assay, 42, 259, 260; molecular weight, 42, 262; properties, 42, 262; purification, 42, 260); sodium pyruvate synthesis in TOH or D$_2$O, in, 41, 106; stereochemistry, 87, 147

6-PHOSPHO-5-DEHYDRO-2-DEOXYGLUCONATE ALDOLASE

This aldolase [EC 4.1.2.29], also known as 5-dehydro-2-deoxyphosphogluconate aldolase and 6-phospho-5-keto-2-deoxygluconate aldolase, catalyzes the conversion of 5-dehydro-2-deoxy-D-gluconate 6-phosphate to glycerone phosphate (or, dihydroxyacetone phosphate) and malonate semialdehyde ($^-$OOCCH$_2$CHO).[1]

[1]W. A. Anderson & B. Magasanik (1971) *JBC* **246**, 5662.

7-PHOSPHO-2-DEHYDRO-3-DEOXYHEPTONATE ALDOLASE

This aldolase [EC 4.1.2.15] (also known as 2-dehydro-3-deoxyphosphoheptonate aldolase, 3-deoxy-D-*arabino*-heptulosonate-7-phosphate synthase, and 7-phospho-2-keto-3-deoxyheptonate aldolase) catalyzes the reaction of phosphoenolpyruvate with D-erythrose 4-phosphate and water to produce 2-dehydro-3-deoxy-D-*arabino*-heptonate 7-phosphate and orthophosphate.[1–4]

This enzyme catalyzes the first step in the shikimate acid pathway. The tryptophan-sensitive enzyme from *Escherichia coli* exhibits sigmoidal kinetics,[3] but the *Neurospora crassa* enzyme reportedly has a rapid-equilibrium ordered Bi Bi kinetic mechanism.[4]

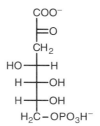

2-dehydro-3-deoxy-D-*arabino*-
heptonate 7-phosphate

[1]P. R. Srinivasan & D. B. Sprinson (1954) *JBC* **234**, 716.
[2]H. Görisch & F. Lingens (1971) *BBA* **242**, 617 and 630.
[3]J. P. Akowski & R. Bauerle (1997) *Biochemistry* **36**, 15817.
[4]G. A. Nimmo & J. R. Coggins (1981) *BJ* **199**, 657.

Selected entries from *Methods in Enzymology* [vol, page(s)]:
General discussion: 5, 394; 17A, 349, 392; 34, 394, 397
Assay: 5, 395; 17A, 349, 392
Other molecular properties: active-site labeling, 46, 139; dehydroquinate synthetase and, 5, 401; phenylalanine-sensitive, 34, 397; 8-phospho-2-dehydro-3-deoxyoctonate aldolase and, 8, 217; properties, 5, 397; purification by affinity chromatography, 22, 357, 359; purification from (*Escherichia coli*, 5, 396; *Neurospora crassa*, 17A, 393); significance, 5, 394; tyrosine-sensitive, 34, 397

8-PHOSPHO-2-DEHYDRO-3-DEOXYOCTONATE ALDOLASE

This aldolase [EC 4.1.2.16] (also known as 2-dehydro-3-deoxyphosphooctonate aldolase, phospho-2-keto-3-deoxyoctonate aldolase, and 3-deoxy-D-*manno*-octulosonate 8-phosphate synthase) catalyzes the reaction of phosphoenolpyruvate with D-arabinose 5-phosphate and water to produce 2-dehydro-3-deoxy-D-octonate 8-phosphate and orthophosphate.[1–4] This enzyme catalyzes a key step in the biosynthesis of lipopolysaccharides in Gram-negative bacteria. Both C3 protons are retained throughout the reaction. When [^{18}O]enol-labeled phosphoenolpyruvate is used as a substrate, all of the label is recovered in orthophosphate. Neither a [^{32}P]phosphate ↔ phosphoenolpyruvate exchange nor a scrambling of bridge ^{18}O to non-bridging positions in [^{18}O]phosphoenolpyruvate is observed in the presence or absence of D-arabinose 5-phosphate.[4] An acyclic intermediate has been proposed.[4]

[1]D. H. Levin & E. Racker (1959) *JBC* **234**, 2532.
[2]P. Ray (1980) *J. Bacteriol.* **141**, 635.
[3]P. D. Rick & D. A. Young (1982) *J. Bacteriol.* **150**, 447.
[4]L. Hedstrom & R. Abeles (1988) *BBRC* **157**, 816.

Selected entries from *Methods in Enzymology* [vol, page(s)]:
General discussion: 8, 216; 83, 525
Assay: 8, 217; 83, 526
Other molecular properties: *Escherichia coli* (properties, 83, 530; purification, 83, 528); 7-phospho-2-dehydro-3-deoxyheptonate aldolase

and, **8**, 217; *Pseudomonas aeruginosa* (properties, **8**, 220; purification, **8**, 218)

PHOSPHODIESTERASE I

This enzyme [EC 3.1.4.1], also known as 5′-exonuclease and 5′-nucleotide phosphodiesterase, catalyzes the hydrolytic removal of 5′-nucleotides successively from the 3′-hydroxy termini (that is, a nonphosphorylated 3′ end; an oligonucleotide with a 3′-phosphate terminus inhibits hydrolysis) of 3′-hydroxy-terminated oligonucleotides; hence, oligonucleotides will get converted to 5′-nucleotides.[1–4] A covalent phosphorylated intermediate has been identified[3] and a group with an apparent pK_a value of 6.85 controls the rate-determining step.[4] The phosphorylated residue has been reported to be threonine. *See also Phosphodiesterase; Venom Exonuclease; Spleen Exonuclease; 5′-Nucleotidase*

[1]C. Kanno, Y. Ohmura & H. Yanagisawa (1989) *ABC* **53**, 607.
[2]M. Landt & L. G. Butler (1978) *Biochemistry* **17**, 4130.
[3]O. A. Moe, Jr. & L. G. Butler (1983) *JBC* **258**, 6941.
[4]J. S. Culp, H. J. Blytt, M. Hermodson & L. G. Butler (1985) *JBC* **260**, 8320.

Selected entries from *Methods in Enzymology* [vol, page(s)]:
General discussion: 6, 237; **136**, 517
Assay: 6, 238; **31**, 91, 93
Other molecular properties: classification, 6, 236; digestion of tRNA, in, 59, 59, 60, 79, 81, 83, 87, 89, 94, 103, 122, 126, 127, 132, 140, 175, 187; gel electrophoresis, 32, 83, 90; kidney, 6, 238 (purification, 6, 242; tissue distribution, 6, 249); liver, in, 31, 97; marker enzyme, as, 31, 93; plasma membrane, in, 31, 91, 100, 101; ribonuclease I preparation free of, 6, 252; storage, as solution, 59, 60; substrates, 6, 237; uses of, 6, 237; venom, 6, 245 (points of attack, 6, 249; properties, 6, 240; purification, 6, 238)

PHOSPHOENOLPYRUVATE CARBOXYKINASE

There are three distinct phosphoenolpyruvate carboxykinases, differing primarily with respect the origin of the transferred phosphoryl group.[1–11]

Phosphoenolpyruvate carboxykinase (GTP) [EC 4.1.1.32] (also known as phosphoenolpyruvate carboxylase, PEP carboxykinase, and PEPCK) catalyzes the reaction of GTP (or ITP) with oxaloacetate to produce GDP (or IDP), phosphoenolpyruvate, and carbon dioxide.[1–3] This reaction is the rate-determining step of gluconeogenesis in eukaryotes. The enzyme also has an oxaloacetate decarboxylase activity. Based on X-ray crystallographic analysis the nucleotide binding site of PEPCK is structurally unique and may represent a new motif for protein-nucleotide interactions.[3] Matte *et al.*[1] suggest that it is unclear whether the conversion of oxaloacetate to phosphoenolpyruvate is a two-step process

(involving oxaloacetate decarboxylation to form an enolate anion intermediate, followed by phosphorylation) or whether it occurs as a single concerted reaction.

Phosphoenolpyruvate carboxykinase (ATP) [EC 4.1.1.49] (also known as phosphoenolpyruvate carboxylase, PEP carboxykinase, and PEPCK) catalyzes the reaction of ATP with oxaloacetate to produce ADP, phosphoenolpyruvate, and carbon dioxide.[1,2] The *Escherichia coli* enzyme has a random kinetic mechanism.[4,5] While the GTP-dependent carboxykinases are found in most eukaryotes, the ATP-dependent enzymes are found in bacteria, yeast, plants, and in trypanomastid parasites. Most of these enzymes are multimeric (except for ATP-dependent PEPCK of bacteria) whereas the GTP-dependent class are monomeric.

Phosphoenolpyruvate carboxykinase (pyrophosphate) [EC 4.1.1.38] (also known as phosphoenolpyruvate carboxykinase (diphosphate), phosphoenolpyruvate carboxylase, and phosphoenolpyruvate carboxytransphosphorylase) catalyzes the reversible reaction of pyrophosphate with oxaloacetate to produce orthophosphate, phosphoenolpyruvate, and carbon dioxide.[2] The enzyme also catalyzes the reaction of phosphoenolpyruvate with orthophosphate to produce pyruvate and pyrophosphate (in the absence of bicarbonate or carbon dioxide).

[1]A. Matte, L. W. Tari, H. Goldie & L. T. J. Delbaere (1997) *JBC* **272**, 8105.
[2]M. F. Utter & H. M. Kolenbrander (1972) *The Enzymes*, 3rd ed., **6**, 117.
[3]A. Matte, H. Goldie, R. M. Sweet & L. T. J. Delbaere (1996) *JMB* **256**, 126.
[4]A. Krebs & W. A. Bridger (1980) *Can. J. Biochem.* **58**, 309.
[5]D. R. Arnelle & M. H. O'Leary (1992) *Biochemistry* **31**, 4363.
[6]H. G. Wood, W. E. O'Brien & G. Michaels (1977) *AE* **45**, 85.
[7]J. V. Schloss & M. S. Hixon (1998) *CBC* **2**, 43.
[8]R. W. Hanson & Y. M. Patel (1994) *AE* **69**, 203.
[9]J. J. Villafranca & T. Nowak (1992) *The Enzymes*, 3rd ed., **20**, 63.
[10]S. J. Pilkis, M. R. El-Maghrabi & T. H. Claus (1988) *ARB* **57**, 755.
[11]H. G. Hers & L. Hue (1983) *ARB* **52**, 617.

Selected entries from *Methods in Enzymology* [vol, page(s)]:
General discussion: ATP-dependent, 5, 619; GTP-dependent, 1, 758; 13, 270; pyrophosphate-dependent, 13, 297
Assay: 63, 33, 36; ATP-dependent, 5, 619 (in presence of phosphoenolpyruvate carboxylase, 5, 620); 14C-bicarbonate fixation, 13, 271; 14C exchange, 13, 273; coupling assay, 63, 33, 36; GTP-dependent, 1, 758; 13, 270; pyrophosphate-dependent, 13, 297; spectrophotometry (carboxylating, 13, 271; decarboxylating, 13, 272); radioiodination, 70, 232
Other molecular properties: cooperativity, 64, 218; GTP-dependent (positional isotope exchange studies, 249, 424; properties, 1, 762; 13, 275; purification [chicken liver, 1, 760; pig liver, 13, 273]); hysteresis, 64, 218; oxaloacetate, determination of, 13, 529; 63, 33; phosphoenolpyruvate, determination of, 63, 36; production increase in

Escherichia coli upon addition of pyruvate as a carbon source, 22, 88; pyrophosphate-dependent, 13, 297 (effect of thiols, 13, 298, 307; inhibition by [buffers, 13, 307; malate, 13, 307]; mechanism, 13, 308; metal ion requirements, inhibitors, and activators, 13, 306; properties, 13, 305; purification, *Propionibacterium shermanii*, 13, 299); specific radioactivity determination of oxaloacetate with, 13, 529

PHOSPHOENOLPYRUVATE CARBOXYLASE

This carboxylase [EC 4.1.1.31], also known as PEP carboxylase, catalyzes the reaction of water with phosphoenolpyruvate and carbon dioxide to produce orthophosphate and oxaloacetate.[1–4] Bicarbonate, and not carbon dioxide, binds to the enzyme. The *Zea mays* enzyme has a random Bi Bi kinetic mechanism with synergistic binding of the substrates.[1] Kinetic isotope effect and alternative substrate investigations suggest that carboxylation of phosphoenolpyruvate occurs via attack of the enolate of pyruvate on carbon dioxide rather than directly on the carboxyphosphate that was formed after the binding of bicarbonate. The bridging oxygen is in a vibrationally stiffer environment in the transition state for the association reaction.[2,3] Occasionally, the enzyme has been incorrectly termed phosphoenolpyruvate carboxykinase. **See also** *Phosphoenolpyruvate Carboxykinase*

[1] J. W. Janc, M. H. O'Leary & W. W. Cleland (1992) *Biochemistry* 31, 6421.
[2] J. W. Janc, J. L. Urbauer, M. H. O'Leary & W. W. Cleland (1992) *Biochemistry* 31, 6432.
[3] E. Gawlita, W. S. Caldwell, M. H. O'Leary, P. Paneth & V. E. Anderson (1995) *Biochemistry* 34, 2577.
[4] M. F. Utter & H. M. Kolenbrander (1972) *The Enzymes*, 3rd ed., 6, 117.

Selected entries from **Methods in Enzymology** [vol, page(s)]:
General discussion: 5, 617; 13, 277, 283, 288, 292; 188, 361
Assay: 5, 617; 13, 277, 283, 288, 292; phosphoenolpyruvate carboxykinase, in the presence of, 5, 620
Other molecular properties: acetyl-CoA-independent, assay in crude extracts of methylotrophs, 188, 363; component of multienzyme complex, 71, 60; enol product, 87, 90; *Escherichia coli* K12, from, 13, 288; *Escherichia coli* strain W, from, 13, 288 (activation by acetyl-CoA and other acyl-CoA derivatives, 13, 292; inhibition by L-aspartate, 13, 288, 292; properties, 13, 291; purification, 13, 289; requirement for divalent metal ions, 13, 292); gluconeogenic catalytic activity, 37, 281; isotope effects, 64, 96; mesophyll cells, 69, 57; *Pseudomonas* AM1, from, 13, 292 (properties, 13, 296; purification, 13, 295; substrate specificity, 13, 296); *Salmonella typhimurium*, strain LT 2, from, 13, 283 (activation by [acetyl-CoA, fructose 1,6-bisphosphate, and nucleotides, 13, 288; organic solvents, 13, 287]; binding of polycations, 13, 287, 288; inhibition by aspartate, 13, 288; properties, 13, 287; purification, 13, 284); species distribution, 13, 277; spinach (purification, 5, 618); synthesis in *Chlamydomonas* cultures, enzyme, 23, 73

PHOSPHOENOLPYRUVATE:FRUCTOSE PHOSPHOTRANSFERASE

This enzyme, formerly classified as EC 2.7.1.98, is now a deleted Enzyme Commission entry.

PHOSPHOENOLPYRUVATE:GLYCERONE PHOSPHOTRANSFERASE

This phosphotransferase [EC 2.7.1.121], also known as phosphoenolpyruvate:dihydroxyacetone phosphotransferase, catalyzes the reaction of phosphoenolpyruvate with glycerone (or, dihydroxyacetone) to produce pyruvate and glycerone phosphate (or, dihydroxyacetone phosphate).[1]

[1] R. Z. Jin & E. C. Lin (1984) *J. Gen. Microbiol.* 130, 83.

PHOSPHOENOLPYRUVATE MUTASE

This enzyme [EC 5.4.2.9], also known as phosphoenolpyruvate phosphomutase, catalyzes the interconversion of phosphoenolpyruvate and 3-phosphonopyruvate.[1–6] The *Tetrahymena pyriformis* enzyme requires a divalent cation (*e.g.*, Mg^{2+}) which has to bind prior to the binding of the pyruvate derivative (hence, ordered binding).[1] The retention of phosphorus stereochemistry combined with strong inhibition displayed by the pyruvyl enolate analogue oxalate, have been cited as suggestive of a phosphoenzyme-pyruvyl enolate intermediate.[2] However, evidence against such an intermediate has been presented.[3] Site-directed mutagenesis studies suggest the participation of an aspartyl residue in nucleophilic catalysis.[4]

[1] E. D. Bowman, M. S. McQueney, J. D. Scholten & D. Dunaway-Mariano (1990) *Biochemistry* 29, 7059.
[2] H. M. Seidel & J. R. Knowles (1994) *Biochemistry* 33, 5641.
[3] J. Kim & D. Dunaway-Mariano (1996) *Biochemistry* 35, 4628.
[4] Y. Jia, Z. Lu, K. Huang, O. Herzberg & D. Dunaway-Mariano (1999) *Biochemistry* 38, 14165.
[5] J. V. Schloss & M. S. Hixon (1998) *CBC* 2, 43.
[6] A. C. Hengge (1998) *CBC* 1, 517.

PHOSPHOENOLPYRUVATE PHOSPHATASE

This phosphatase [EC 3.1.3.60], first isolated from mung bean (*Vigna radiata*) and black mustard (*Brassica nigra*), catalyzes the hydrolysis of phosphoenolpyruvate to produce pyruvate and orthophosphate.[1–3] Other monophosphates are alternative substrates, albeit at slower rates. ATP and ADP are both slowly hydrolyzed in the presence of the enzyme.

[1] O. P. Malhotra & A. M. Kayastha (1989) *Plant Sci.* 65, 161.
[2] S. M. G. Duff, D. D. Lefebvre & W. C. Plaxton (1989) *Plant Physiol.* 90, 734.
[3] S. M. Duff, D. D. Lefebvre & W. C. Plaxton (1991) *ABB* 286, 226.

PHOSPHOENOLPYRUVATE:PROTEIN PHOSPHOTRANSFERASE

This phosphotransferase [EC 2.7.3.9], also known as enzyme I of the phosphoenolpyruvate-dependent glucose phosphotransferase system, catalyzes the reaction of phosphoenolpyruvate with an L-histidyl residue in a low-molecular-weight (9.11 kDa in *Escherichia coli*) heat-stable protein (HPr) to produce pyruvate and HPr with an N^{π}-phospho-L-histidyl residue. NMR studies have identified the binding surface on the transferase for HPr.[1–5] An essential active-site histidyl residue is needed for catalysis. Pyruvate, phosphate dikinase is a homologue of this phosphotransferase. Another component of this phosphotransferase system is protein-N^{π}-phosphohistidine:sugar phosphotransferase [EC 2.7.1.69].

Because the energy of P–O bond scission in phosphoenolpyruvate drives the hexose transport, the reaction may be regarded as an energase-type reaction (hexose$_{out}$ + 2-phosphoenolpyruvate = hexose 6-phosphate$_{inside}$ + pyruvate, where the transport is indicated by the subscripts).[6]

See *Protein-N$^{\pi}$-Phosphohistidine:Sugar Phosphotransferase*

[1] F. Chauvin, L. Brand & S. Roseman (1996) *Res. Microbiol* **147**, 471.
[2] W. Hengstenberg, B. Reiche, R. Eisermann, R. Fischer, U. Kessler, A. Tarrach, W. M. De Vos, H. R. Kalbitzer & S. Glaser (1989) *FEMS Microbiol. Rev.* **5**, 35.
[3] G. T. Robillard (1982) *Mol. Cell Biochem.* **46**, 3.
[4] P. W. Postma & S. Roseman (1976) *BBA* **457**, 213.
[5] D. S. Garrett, Y. J. Seok, A. Peterkofsky, G. M. Clore & A. M. Gronenborn (1997) *Biochemistry* **36**, 4393.
[6] D. L. Purich (2001) *TiBS* **26**, 417.

Selected entries from **Methods in Enzymology** [vol, page(s)]:
General discussion: 9, 396; 90, 431
Assay: 9, 396

PHOSPHOETHANOLAMINE N-METHYLTRANSFERASE

This methyltransferase [EC 2.1.1.103] catalyzes the reaction of *S*-adenosyl-L-methionine with ethanolamine phosphate to produce *S*-adenosyl-L-homocysteine and *N*-methylethanolamine phosphate.[1,2]

[1] C. Andriamampandry, R. Massarelli & J. N. Kanfer (1992) *BJ* **288**, 267.
[2] A. H. Datko & H. S. Mudd (1988) *Plant Physiol.* **88**, 1338.

1-PHOSPHOFRUCTOKINASE

This phosphotransferase [EC 2.7.1.56], also known as fructose-1-phosphate kinase, catalyzes the reaction of ATP with D-fructose 1-phosphate to produce ADP and D-fructose 1,6-bisphosphate. Other nucleotide phosphoryl donors include ITP, GTP, and UTP. One should not confuse this enzyme with phosphofructo-1-kinase (or, 6-phosphofructokinase). **See also** *6-Phosphofructokinase*

Selected entries from **Methods in Enzymology** [vol, page(s)]:
General discussion: 42, 63; 90, 82
Assay: *Aerobacter aerogenes*, 42, 63; *Clostridium pasteurianum*, 90, 82

Other molecular properties: *Aerobacter aerogenes*, from, 42, 63 (chromatography, 42, 65; molecular weight, 42, 66; properties, 42, 66; purification, 42, 64); *Clostridium pasteurianum* (properties, 90, 85; purification, 90, 84)

6-PHOSPHOFRUCTOKINASE

This phosphotransferase [EC 2.7.1.11] (also known as phosphohexokinase, phosphofructokinase I, and phosphofructo-1-kinase), which exhibits cooperativity, catalyzes the reaction of ATP with D-fructose 6-phosphate to produce ADP and D-fructose 1,6-bisphosphate.[1–4] The rabbit muscle enzyme, which has a random Bi Bi kinetic mechanism,[1] is specific for the β-anomer of D-fructose 6-phosphate. Other phosphoryl donors include UTP, CTP, and ITP. Phosphoryl acceptor substrates include D-tagatose 6-phosphate and sedoheptulose 7-phosphate. The enzyme is not identical to either 6-phosphofructo-2-kinase [EC 2.7.1.105] or 1-phosphofructokinase [EC 2.7.1.56]. **See also** *1-Phosphofructokinase*

[1] R. L. Hanson, F. B. Rudolph & H. A. Lardy (1973) *JBC* **248**, 7852.
[2] S. J. Pilkis, T. H. Claus, P. D. Kountz & M. R. El-Maghrabi (1987) *The Enzymes*, 3rd ed., **18**, 3.
[3] E. Van Schaffingen (1987) *AE* **59**, 315.
[4] K. Uyeda (1979) *AE* **48**, 193.

Selected entries from **Methods in Enzymology** [vol, page(s)]:
General discussion: 1, 306; 10, 28; 9, 425, 430, 436; 34, 5, 532; 42, 67, 71, 78, 86, 91, 99, 111; 46, 90; 90, 35, 39, 44, 49, 60, 70, 77, 82
Assay: 1, 306; 9, 425, 430, 436; 228, 145; *Ascaris* muscle, 90, 44; *Bacillus licheniformis*, 90, 71; *Clostridium pasteurianum*, 42, 86, 87; Ehrlich ascites tumor, 90, 35; *Escherichia coli*, 42, 91 (types 1 and 2, 90, 60); heart muscle, 9, 430; human erythrocytes, 42, 110, 111; mammalian tissues, 42, 99, 105; oyster adductor muscle, 90, 39; porcine liver and/or kidney, 42, 99; rabbit liver, 42, 67; rabbit muscle, 1, 306; 9, 425; 42, 71, 72; *Streptococcus lactis*, 90, 77; swine kidney, 42, 382; yeast, 9, 436; 42, 78, 79; 90, 49
Other molecular properties: abortive complexes, 63, 433; activation, 64, 220, 221; affinity labeling, 87, 474; allosteric constant, 240, 807; anomeric specificity, 63, 373, 374, 376; *Ascaris* muscle (assay, 90, 44; properties, 90, 47; purification, 90, 45); ATP binding sites, 228, 272; *Bacillus licheniformis* (assay, 90, 71; properties, 90, 73; purification, 90, 72); *Bacteroides symbiosis*, from, 42, 63; bakers' yeast, of (inhibitors, 9, 440; properties, 9, 440; purification, 9, 438; specificity, 9, 440); bifurcation analysis, 240, 807; binding site, 46, 76; burst analysis, 64, 197; calmodulin, interaction with (binding assays, 139, 749; enzyme activity assays, 139, 754; enzyme polymerization studies, 139, 760); cAMP binding, 37, 286; cascade control, 64, 325; chromium nucleotides, and, 87, 176; *Clostridium pasteurianum*, from, 42, 86 (assay, 42, 86, 87; chromatography, 42, 88; molecular weight, 42, 89; properties, 42, 89, 91; purification, 42, 87); dimer-tetramer equilibrium, 64, 220; equilibrium shift, 64, 220; *Escherichia coli*, from, 42, 91 (assay, 42, 91; chromatography, 42, 94, 95; electrophoresis, 42, 96; fluorescent derivatives, 42, 97; molecular weight, 42, 97; properties, 42, 84, 96; purification, 42, 93; types 1 and 2 [assay, 90, 60; immunological relationship to other bacterial phosphofructokinases, 90, 68; properties, 90, 64, 67; purification, 90, 63, 65]); fractional precipitation with poly(ethylene glycol) 6000, 228, 149; fructose bisphosphate sensitivity, 240, 808; gel filtration, 228, 153; human erythrocytes, from, 42, 110; hysteresis, 64, 195, 217; immobilized system (global behavior, numerical prediction, 135, 564; oscillatory phenomena, 135, 560); inactivation, 64,

221; isotope exchange, **87**, 655; Pasteur effect, **55**, 291, 293; phosphate, chiral, **87**, 210; physiological damping, **64**, 217, 220, 221; product inhibition and abortive complexes, **63**, 433; properties, **74**, 232, 235; rabbit liver, from, **42**, 67; rabbit muscle, from, **1**, 306; **9**, 425; **42**, 71; rate properties, initial, **63**, 156; rat liver, substrate-directed regulation of cAMP-dependent phosphorylation, **99**, 215; site-specific reagent, **47**, 483; staining on gels, **22**, 599; *Streptococcus lactis* (assay, **90**, 77; properties, **90**, 80; purification, **90**, 78); substrate analogues, **63**, 373, 374; substrate inhibition, allosteric, **63**, 501 (steady-state random mechanism, **63**, 51; **249**, 540); yeast, from, **9**, 436; **42**, 78

6-PHOSPHOFRUCTO-2-KINASE/FRUCTOSE-2,6-BISPHOSPHATASE

6-Phosphofructo-2-kinase [EC 2.7.1.105], also known as phosphofructokinase 2, catalyzes the reaction of ATP with D-fructose 6-phosphate to produce ADP and D-fructose 2,6-bisphosphate.[1–3] This enzymatic activity should not be confused with 6-phosphofructokinase [EC 2.7.1.11]. This phosphotransferase also possesses a fructose-2,6-bisphosphatase [EC 3.1.3.46] activity (also known as fructose-2,6-bisphosphate 2-phosphatase), catalyzing the hydrolysis of D-fructose 2,6-bisphosphate to produce D-fructose 6-phosphate and orthophosphate via an ordered Uni Bi kinetic mechanism. Again, this bisphosphatase activity should not be confused with fructose-2,6-bisphosphate 6-phosphatase [EC 3.1.3.54].

[1] H. Mizuguchi, P. F. Cook, C. A. Hasemann & K. Uyeda (1997) *Biochemistry* **36**, 8775.
[2] S. J. Pilkis, T. H. Claus, P. D. Kountz & M. R. El-Maghrabi (1987) *The Enzymes*, 3rd ed., **18**, 3.
[3] E. Van Schaftingen (1987) *AE* **59**, 315.

6-PHOSPHO-β-GALACTOSIDASE

This galactosidase [EC 3.2.1.85], also known as β-D-phosphogalactoside galactohydrolase, catalyzes the hydrolysis of a 6-phospho-β-D-galactoside to produce 6-phospho-D-galactose and an alcohol.[1–3] Lactose 6-phosphate is hydrolyzed by this enzyme, as is the artificial substrate *o*-nitrophenyl-β-D-galactopyranoside-6-phosphate. Most enzymes investigated so far indicate the formation of a covalent glycosyl-enzyme intermediate in the reaction mechanism, and an active-site glutamyl residue has been suggested to function as a nucleophile in the *Staphylococcus aureus* enzyme.[3]

[1] B. G. Hall (1979) *J. Bacteriol.* **138**, 691.
[2] W. Hengstenberg, W. K. Penberthy & M. L. Morse (1970) *EJB* **14**, 27.
[3] P. Staedtler, S. Hoenig, R. Frank, S. G. Withers & W. Hengstenberg (1995) *EJB* **232**, 658.

Selected entries from *Methods in Enzymology* [vol, page(s)]:
General discussion: 42, 491
Assay: 42, 491

Other molecular properties: amino acids, **42**, 494; chromatography, **42**, 492, 493; distribution, **42**, 493; molecular weight, **42**, 494; properties, **42**, 494; purification, **42**, 491

PHOSPHOGALACTOSYLTRANSFERASE

This enzyme catalyzes the reaction of UDP-D-galactose with a capsular polysaccharide, resulting in the incorporation of a D-galactosyl residue. The enzyme also catalyzes the reaction of UDP-D-galactose with undecaprenyl phosphate to produce UMP and α-D-galactosyl-diphosphoundecaprenol (an undecaprenyl-phosphate galactose phosphotransferase activity).

Selected entries from *Methods in Enzymology* [vol, page(s)]:
Other molecular properties: biosynthesis of capsular polysaccharide and, **28**, 610; isolation from *Aerobacter aerogenes*, **28**, 602; product identification, **28**, 613; properties, **28**, 610; reaction, **28**, 612

PHOSPHOGLUCOKINASE

This phosphotransferase [EC 2.7.1.10], also known as glucose-phosphate kinase, catalyzes the reaction of ATP with D-glucose 1-phosphate to produce ADP and D-glucose 1,6-bisphosphate.[1,2] Rabbit muscle 6-phosphofructokinase [EC 2.7.1.11] also catalyzes this reaction. While the two enzymes may be identical in some organisms, a distinct phosphoglucokinase is found in porcine tissues.[3]

[1] A. C. Paladini, R. Caputto, L. F. Leloir, R. E. Trucco & C. E. Cardini (1949) *Arch. Biochem.* **23**, 55.
[2] P. Eyer, H. W. Hofer, E. Krystek & D. Pette (1971) *EJB* **20**, 153.
[3] F. Climent, M. Carreras & J. Carreras (1985) *Comp. Biochem. Physiol. B* **81**, 737.

Selected entries from *Methods in Enzymology* [vol, page(s)]:
General discussion: 1, 354
Assay: 1, 355
Other molecular properties: properties, 1, 356; purification, rabbit muscle, 1, 356

PHOSPHOGLUCOMUTASE

This mutase [EC 5.4.2.2], also known as glucose phosphomutase, catalyzes the interconversion of α-D-glucose 1-phosphate and α-D-glucose 6-phosphate and requires α-D-glucose 1,6-bisphosphate as a cofactor.[1–5] This latter is formed by transferring a phosphoryl group from the enzyme's active-site seryl residue to the substrate; the bisphosphate intermediate then reorients and transfers its other phosphoryl group to the seryl residue. Other substrates include mannose 1-phosphate, galactose 1-phosphate, glucosamine 1-phosphate, and *N*-acetylglucosamine 1-phosphate. The enzyme also interconverts α-D-ribose 1-phosphate and α-D-ribose 5-phosphate. *See also Phosphopentomutase; Phosphoglucosamine Mutase*

[1] W. J. Ray & E. J. Peck (1972) *The Enzymes*, 3rd ed., **6**, 407.
[2] V. A. Najjar (1962) *The Enzymes*, 2nd ed., **6**, 161.
[3] S. Shankar, R. W. Ye, D. Schlictman & A. M. Chakrabarty (1995) *AE* **70**, 221.
[4] J.-B. Dai, Y. Liu, W. J. Ray, Jr. & M. Konno (1992) *JBC* **267**, 6322.
[5] I. A. Rose (1975) *AE* **43**, 491.

Selected entries from *Methods in Enzymology* [vol, page(s)]:
General discussion: 1, 294; 89, 599; 90, 479
Assay: 1, 294; yeast, 89, 599
Other molecular properties: cadmium dependence, **240**, 442; catalysis steps, **308**, 196; chromatography, DEAE-cellulose, **5**, 24; diagnostic enzyme, as, **31**, 24; galactose 1-phosphate, assay of, **5**, 175; galactose-1-phosphate uridylyltransferase, and, **87**, 20, 31 (assay of, **9**, 708; **63**, 33; contamination of, **5**, 184); glucosamine 6-phosphate and, **3**, 162; glucose 1,6-bisphosphate assay and, **3**, 143; glucose 1-phosphate assay and, **2**, 675, 676; **3**, 968; **5**, 176, 179; **63**, 33; inhibitor, **63**, 400; inorganic phosphate assay, **55**, 212; intermediate, **87**, 271, 272; isotope exchange, **64**, 10; marker for chromosome detection in somatic cell hybrids, as, **151**, 190; NMR exchange studies, **240**, 439; phosphomannomutase and, **8**, 185; phosphoribomutase activity, **1**, 361; ^{31}P NMR studies, **87**, 271; **107**, 72; purification, **1**, 296; rabbit muscle, **1**, 296; rate constants determined by induced transport, **249**, 237; ribose phosphate formation and, **2**, 503; stain for, **224**, 110; staining on gels, **22**, 584, 588, 599; substrate inhibition, **87**, 25; transition-state and multisubstrate inhibitors, **249**, 308; yeast, **89**, 599

β-PHOSPHOGLUCOMUTASE

This mutase [EC 5.4.2.6] catalyzes the interconversion of β-D-glucose 1-phosphate and β-D-glucose 6-phosphate.[1,2]

[1] E. Belocopitow & L. R. Maréchal (1974) *EJB* **46**, 631.
[2] R. Ben-Zvi & M. Schramm (1961) *JBC* **236**, 2186.

PHOSPHOGLUCOMUTASE (GLUCOSE-COFACTOR)

This mutase [EC 5.4.2.5] catalyzes the interconversion of α-D-glucose 1-phosphate and α-D-glucose 6-phosphate.[1] In contrast to phosphoglucomutase [EC 5.4.2.2], this enzyme is activated by D-glucose, which probably acts as a phosphoryl acceptor from the substrate α-D-glucose 1-phosphate and is itself converted to the product.

[1] A. Fujimoto, P. Ingram & R. A. Smith (1965) *BBA* **96**, 91.

6-PHOSPHOGLUCONATE DEHYDRATASE

This dehydratase [EC 4.2.1.12], also known as phosphogluconate dehydratase, catalyzes the conversion of 6-phospho-D-gluconate to 2-dehydro-3-deoxy-6-phospho-D-gluconate and water.[1,2]

[1] M. Rodriguez, A. G. Wedd & R. K. Scopes (1996) *Biochem. Mol. Biol. Int.* **38**, 783.
[2] W. A. Wood (1971) *The Enzymes*, 3rd ed., **5**, 573.

Selected entries from *Methods in Enzymology* [vol, page(s)]:
General discussion: 9, 653; 89, 98
Assay: 9, 653
Other molecular properties: enol product, **87**, 90; *Pseudomonas fluorescens*, from (properties, **9**, 656; purification, **9**, 654); *Pseudomonas putida* (2-keto-3-deoxygluconate 6-phosphate synthesis, **89**, 98; purification, **89**, 99)

6-PHOSPHOGLUCONATE 2-DEHYDROGENASE

This oxidoreductase [EC 1.1.1.43], also known as 2-keto-6-phosphogluconate reductase, catalyzes the reversible reaction of 6-phospho-D-gluconate with NAD(P)$^+$ to produce 6-phospho-2-dehydro-D-gluconate and NAD(P)H and a proton.[1]

[1] E. W. Frampton & W. A. Wood (1961) *JBC* **236**, 2571.

Selected entries from *Methods in Enzymology* [vol, page(s)]:
General discussion: 5, 295
Assay: 5, 295
Other molecular properties: 2-ketogluconokinase (assay of, **5**, 291; preparation free of, **5**, 294); *Pseudomonas fluorescens* (properties, **5**, 297; purification, **5**, 296)

PHOSPHOGLUCONATE DEHYDROGENASE (DECARBOXYLATING)

This oxidoreductase [EC 1.1.1.44], also known as 6-phosphogluconate dehydrogenase and 6-phosphogluconate carboxylase, catalyzes the reaction of 6-phospho-D-gluconate with NADP$^+$ to produce D-ribulose 5-phosphate, carbon dioxide and NADPH and a proton.[1–5] Isotope investigations indicate that the oxidative decarboxylation of 6-phosphoglutaconate to the 1,2-enediol of ribulose 5-phosphate proceeds in a stepwise mechanism with hydride transfer preceding decarboxylation.[2]

The enzyme will also catalyze the reduction of other oxidants: for example, D-ribulose 5-phosphate reacts with NADPH and tetranitromethane to produce 1,2-dioxopentitol 5-phosphate, nitroformate, and NADPH.

[1] A. R. Rendina, J. D. Hermes & W. W. Cleland (1984) *Biochemistry* **23**, 6257.
[2] C. C. Hwang, A. J. Berdis, W. E. Karsten, W. W. Cleland & P. F. Cook (1998) *Biochemistry* **37**, 12596.
[3] M. H. O'Leary (1992) *The Enzymes*, 3rd ed., **20**, 235.
[4] D. J. Creighton & N. S. R. K. Murthy (1990) *The Enzymes*, 3rd ed., **19**, 323.
[5] K. Dalziel (1975) *The Enzymes*, 3rd ed., **11**, 1.

Selected entries from *Methods in Enzymology* [vol, page(s)]:
General discussion: 1, 323, 328; 9, 137, 141; 41, 214, 220, 227, 232, 237; 89, 282; 188, 335, 339
Assay: 1, 323, 332; 4, 379; 9, 137; 41, 215, 220, 227, 232, 237, 238; 89, 278, 283, 291; *Arthrobacter globiformis*, **188**, 343; *Methylobacillus flagellatum*, **188**, 337; clinical, **9**, 141
Other molecular properties: *Arthrobacter globiformis*, **188**, 343; *Bacillus stearothermophilus*, **89**, 283; *Candida utilis*, **9**, 137; **41**, 237; clinical significance, **9**, 141; enol intermediate, **87**, 88; human erythrocytes, from, **41**, 220; isolated nuclei and, **12A**, 445; *Leuconostoc mesenteroides*, **1**, 328; NADP$^+$ assay and, **6**, 792, 794; NADP$^+$ and NADPH, cycling of, **18B**, 12; *Neurospora crassa*, **41**, 227; partially random mechanism, **63**, 157; 6-phosphogluconate assay and, **3**, 173; photooxidation, **25**, 407; *Pseudomonas fluorescens*, **89**, 278; D-ribulose 5-phosphate preparation, in,

3, 190; **9**, 45; sheep liver, from, **41**, 214; stain for, **224**, 110; staining on gels, **22**, 595; staining solution in starch gel electrophoresis, **6**, 968; stereospecificity, **87**, 106; *Streptococcus faecalis*, from, **41**, 232; use in enzymatic cycling of NADP[+] and NADPH, **18B**, 12

6-PHOSPHOGLUCONOLACTONASE

This enzyme [EC 3.1.1.31], also known as 6-phospho-glucono-δ-lactone hydrolase, catalyzes the hydrolysis of 6-phospho-D-glucono-1,5-lactone to produce 6-phospho-D-gluconate.

Selected entries from *Methods in Enzymology* [vol, page(s)]:
General discussion: 18A, 52
Assay: 18A, 52

PHOSPHOGLUCOSAMINE MUTASE

This mutase [EC 5.4.2.10], which participates in bacterial cell-wall peptidoglycan and lipopolysaccharide synthesis, interconverts D-glucosamine 1-phosphate and D-glucosamine 6-phosphate and requires D-glucosamine 1,6-bisphosphate as a cofactor.[1] The *Escherichia coli* enzyme is activated by phosphorylation and can be autophosphorylated *in vitro* by D-glucosamine 1,6-bisphosphate, D-glucose 1,6-bisphosphate, or ATP.[2] The enzyme's kinetic mechanism is ping pong Bi Bi.[1] *See also Phosphoglucomutase*

[1] L. Jolly, P. Ferrari, D. Blanot, J. van Heijenoort, F. Fassy & D. Mengin-Lecreulx (1999) *EJB* **262**, 202.

[2] L. Jolly, F. Pompeo, J. van Heijenoort, F. Fassy & D. Mengin-Lecreulx (2000) *J. Bacteriol.* **182**, 1280.

6-PHOSPHO-β-GLUCOSIDASE

This glucosidase [EC 3.2.1.86], also known as phosphocellobiase, catalyzes the hydrolysis of 6-phospho-β-D-glucosyl-(1,4)-D-glucose (or, 6-phosphocellobiose) to produce D-glucose 6-phosphate and D-glucose.[1-4] Alternative substrates include other phospho-β-D-glucosides, but not their unphosphorylated forms. Interestingly, the *Escherichia coli* 6-phospho-β-D-glucosidase, CelF, requires the presence of NAD[+] which does not appear to be a reactant.[3]

[1] G. Wilson & C. F. Fox (1974) *JBC* **249**, 5586.

[2] R. E. Palmer & R. L. Anderson (1972) *JBC* **247**, 3420.

[3] J. Thompson, S. B. Ruvinov, D. I. Freedberg & B. G. Hall (1999) *J. Bacteriol.* **181**, 7339.

[4] M. W. Bauer, S. B. Halio & R. M. Kelly (1996) *Adv. Protein Chem.* **48**, 271.

Selected entries from *Methods in Enzymology* [vol, page(s)]:
General discussion: 42, 494

3-PHOSPHOGLYCERATE DEHYDROGENASE

This oxidoreductase [EC 1.1.1.95], also known simply as phosphoglycerate dehydrogenase, catalyzes the reversible reaction of 3-phosphoglycerate with NAD[+] to produce 3-phosphohydroxypyruvate and NADH (stereospecifically from the A-side) and a proton.[1-3] The *Escherichia coli* enzyme is a V_{max}-type allosteric enzyme subject to feedback inhibition by L-serine.[1,2]

[1] G. A. Grant, D. J. Schuller & L. J. Banaszak (1996) *Protein Sci.* **5**, 34.

[2] D. J. Schuller, G. A. Grant & L. J. Banaszak (1995) *Nat. Struct. Biol.* **2**, 69.

[3] A. R. Clarke & T. R. Dafforn (1998) *CBC* **3**, 1.

Selected entries from *Methods in Enzymology* [vol, page(s)]:
General discussion: 9, 216; **17B**, 325; **41**, 278, 282, 285
Assay: 9, 216; *Escherichia coli*, **17B**, 325; hog spinal cord, **41**, 282; *Pisum sativum* seedlings, **41**, 278, 279; wheat germ, **41**, 285, 286
Other molecular properties: chicken liver, from, **9**, 220; *Escherichia coli*, **17B**, 325; hog spinal cord, from, **41**, 282; *Pisum sativum* seedlings, from, **41**, 278; stereospecificity, **87**, 110; wheat germ, from, **41**, 285

PHOSPHOGLYCERATE KINASE

This glycolytic/gluconeogenic pathway enzyme [EC 2.7.2.3] catalyzes the reversible reaction of ATP with 3-phospho-D-glycerate to produce ADP and 3-phospho-D-glyceroyl phosphate.[1-4] The enzyme has a random Bi Bi kinetic mechanism.[3]

[1] C. C. Blake & D. W. Rice (1981) *Philos. Trans. R. Soc. Lond. B Biol. Sci.* **293**, 93.

[2] R. K. Scopes (1973) *The Enzymes*, 3rd ed., **8**, 335.

[3] C. A. Janson & W. W. Cleland (1974) *JBC* **249**, 2567.

[4] P. A. Frey (1992) *The Enzymes*, 3rd ed., **20**, 141.

Selected entries from *Methods in Enzymology* [vol, page(s)]:
General discussion: 1, 415; **34**, 246; **42**, 127, 134, 139, 144; **90**, 103, 110, 115, 121, 126
Assay: *Bacillus stearothermophilus*, **90**, 126; bakers' yeast, from, **42**, 134; brewer's yeast, **1**, 415; bovine liver, **90**, 115; coupled enzyme assay, **63**, 33; *Escherichia coli*, **42**, 139; human erythrocytes, **42**, 144, 145; murine muscle and testis, isozymes and genetic variants from, **90**, 121; nonaqueous tissue fractions, in, **56**, 204; porcine liver and muscle, **90**, 499; skeletal muscle, **42**, 127; spinach, **90**, 110; vertebrate tissue, **90**, 103; yeast, **90**, 110, 115
Other molecular properties: activation, **63**, 258; activity on ATP analogues, **44**, 876; affinity partitioning with dye ligands, **228**, 135; ATP synthesis and, **4**, 914; [[γ-^{18}O]ATP synthesis, **64**, 67; [γ-^{32}P]ATP synthesis, **65**, 516, 518, 682]; *Bacillus stearothermophilus*, **90**, 126; Bakers' yeast, **42**, 134; 1,3-bisphospho-D-glycerate, dissociation, **87**, 5; brewer's yeast, **1**, 415; bound reactants and products, equilibrium constant determination, **177**, 371; bovine liver, **90**, 115; chiral ATP thiophosphate synthesis, **87**, 265, 300, 313; *Escherichia coli*, **42**, 139; exchange kinetics, **87**, 4; fluorometric assay for ATP, **13**, 491; GTP assay, **55**, 208; human erythrocyte, **42**, 144; inorganic phosphate assay, **55**, 212; intermediate debunked, **87**, 4; phosphorothioates, and, **87**, 258, 313; porcine liver and muscle, **90**, 499; properties, **90**, 508; skeletal muscle, from, **42**, 127; spinach, **90**, 110; stereochemistry, **87**, 258, 265, 300, 311, 313; yeast, **90**, 110, 115

PHOSPHOGLYCERATE KINASE (GTP)

This phosphotransferase [EC 2.7.2.10] catalyzes the reversible reaction of GTP with 3-phospho-D-glycerate to produce GDP and 3-phospho-D-glyceroyl phosphate.[1]

[1] R. E. Reeves & D. J. South (1974) *BBRC* **58**, 1053.

PHOSPHOGLYCERATE MUTASE

This mutase [EC 5.4.2.1], also known as phosphoglycerate phosphomutase and phosphoglyceromutase, catalyzes the reversible reaction of 2-phospho-D-glycerate with 2,3-bisphosphoglycerate to produce 3-phospho-D-glycerate and 2,3-bisphosphoglycerate (hence, catalyzing the net interconversion of 2-phospho-D-glycerate and 3-phospho-D-glycerate).[1–5] The enzymes from mammals and from yeast are phosphorylated at a histidyl residue by (2R)-2,3-bisphosphoglycerate, which also functions as an intermediate (thus, 2,3-bisphosphoglycerate-dependent mutases). Dissociation of the bisphosphate intermediate from the rabbit muscle enzyme is much slower than the overall isomerization. The cofactor-independent castor plant enzyme also has essential histidyl residues that participate in the catalytic reaction.[4] A 2,3-bisphosphoglycerate-independent mutase has been isolated from *Bacillus stearothermophilus* that reportedly utilizes a phosphorylated seryl residue.[5] The enzyme will also exhibit a bisphosphoglycerate mutase activity [EC 5.4.2.4], albeit slowly.

[1] S. I. Winn, H. C. Watson, R. N. Harkins & L. A. Fothergill (1981) *Philos. Trans. R. Soc. Lond. B Biol. Sci.* **293**, 121.
[2] Z. B. Rose (1980) *AE* **51**, 211.
[3] W. J. Ray, Jr. & E. J. Peck, Jr. (1972) *The Enzymes*, 3rd ed., **6**, 407.
[4] Y. Huang & D. T. Dennis (1995) *EJB* **229**, 395.
[5] M. J. Jedrzejas, M. Chander, P. Setlow & G. Krishnasamy (2000) *EMBO J.* **19**, 1419.

Selected entries from *Methods in Enzymology* [vol, page(s)]:
General discussion: 1, 423; 5, 236; 42, 139, 429, 435; 87, 42; 90, 479, 490, 498
Assay: 5, 236; chicken muscle, 90, 490; *Escherichia coli*, 42, 139; human erythrocytes, 42, 450; rabbit muscle, 1, 423; porcine liver and muscle, 42, 436; 90, 499; wheat germ, 42, 429
Other molecular properties: activity determination, 47, 497; affinity labeling, 47, 423, 496, 497; 87, 473; affinity partitioning with dye ligands, 228, 135; amino acids, 42, 445; baker's yeast, 5, 241; 2,3-bisphosphoglycerate assay and, 6, 484; chicken muscle, 90, 490; distribution, 5, 242; *Escherichia coli*, from, 42, 139; glycerate-2,3-bisphosphatase and, 5, 245, 247; human erythrocytes, from, 42, 450; intermediates, 87, 42; kinetics, 87, 43; liver, from, molecular weight, 42, 445; molecular weight, 42, 446; phosphorothioates, and, 87, 280; pig heart, purification, from, 42, 438, 445; pig kidney, 42, 443; porcine liver and muscle, 90, 499; rabbit muscle, 42, 446, 447; rice germ, 42, 435; sheep muscle, 42, 446; site-specific reagent, 47, 483; stability, 5, 242; stereochemistry, 87, 300; types, 5, 236; wheat germ, 42, 429; yeast, 42, 435

PHOSPHOGLYCERATE PHOSPHATASE

This enzyme [EC 3.1.3.20] catalyzes the hydrolysis of D-glycerate 2-phosphate to produce D-glycerate and orthophosphate. Other substrates include D-glycerate 2,3-bisphosphate (producing D-glycerate 3-phosphate and orthophosphate), phosphoglycolate, and *O*-phospho-L-serine.[1]

[1] H. J. Fallon & W. L. Byrne (1965) *BBA* **105**, 43.
[2] D. P. Kestler, B. C. Mayne, T. B. Ray, L. D. Goldstein, R. H. Brown & C. C. Black (1975) *BBRC* **66**, 1439.

3-PHOSPHOGLYCERATE PHOSPHATASE

This phosphatase [EC 3.1.3.38] catalyzes the hydrolysis of D-glycerate 3-phosphate to produce D-glycerate and orthophosphate.[1,2] The enzyme has a wide specificity, but 3-phosphoglycerate is the most effective substrate: other substrates include phosphoenolpyruvate and *p*-nitrophenylphosphate.

[1] R. M. Mulligan & N. E. Tolbert (1980) *Plant Physiol.* **66**, 1169.
[2] D. D. Randall & N. E. Tolbert (1971) *JBC* **246**, 5510.

Selected entries from *Methods in Enzymology* [vol, page(s)]:
General discussion: 42, 405
Assay: 42, 405, 406

PHOSPHOGLYCEROL GERANYLGERANYLTRANSFERASE

This transferase [EC 2.5.1.41], also known as geranylgeranylglyceryl-phosphate synthase, catalyzes the reaction of geranylgeranyl diphosphate (or, geranylgeranyl pyrophosphate) with (*S*)-*sn*-glyceryl phosphate to produce diphosphate (or, pyrophosphate) and *sn*-3-*O*-(geranylgeranyl)glyceryl 1-phosphate.[1,2] Farnesyl diphosphate, farnesylgeranyl diphosphate, and phytyl diphosphate are also substrates, albeit less effective than geranylgeranyl diphosphate.

[1] D.-L. Zhang, L. Daniels & C. D. Poulter (1990) *JACS* **112**, 1264.
[2] A. Chen, D. Zhang & C. D. Poulter (1993) *JBC* **268**, 21701.

3-PHOSPHOGLYCEROYL-PHOSPHATE: POLYPHOSPHATE PHOSPHOTRANSFERASE

This phosphotransferase [EC 2.7.4.17] catalyzes the reaction of 3-phospho-D-glyceroyl phosphate with polyphosphate (that is, [phosphate]$_n$) to produce 3-phospho-D-glycerate and [phosphate]$_{n+1}$.[1]

[1] I. S. Kulaev, M. A. Bobyk, N. N. Nikolaev, N. S. Sergeev & S. O. Uryson (1971) *Biokhimiya* **36**, 943.

PHOSPHOGLYCOLATE PHOSPHATASE

This phosphatase [EC 3.1.3.18] catalyzes the hydrolysis of 2-phosphoglycolate to produce glycolate and orthophosphate.[1–3] Alternative substrates include ethylphosphate, L-phospholactate, *n*-propyl-phosphate, and isopropyl-phosphate. A phosphoenzyme intermediate has been detected.[4]

[1] D. H. Huh, A. V. Kursell & H. D. Husic (1989) *Phytochemistry* **28**, 719.

[2]C. Verin-Vergeau, P. Baldy & G. Cavalie (1980) *Phytochemistry* **19**, 763.
[3]J. T. Christeller & N. E. Tolbert (1978) *JBC* **253**, 1780.
[4]S. N. Seal & Z. B. Rose (1987) *JBC* **262**, 13496.

Selected entries from *Methods in Enzymology* [vol, page(s)]:
General discussion: 9, 646
Assay: 9, 646; 31, 735

PHOSPHOKETOLASE

This thiamin-pyrophosphate-dependent enzyme [EC 4.1.2.9] (also known as D-xylulose-5-phosphate, D-glyceraldehyde-3-phosphate-lyase, and xylulose-5-phosphate phosphoketolase) catalyzes the reaction of D-xylulose 5-phosphate with orthophosphate to produce acetyl phosphate, D-glyceraldehyde 3-phosphate, and water. With arsenate in place of orthophosphate, the reaction products are D-glyceraldehyde 3-phosphate and acetate. The enzyme also catalyzes the reaction of D-fructose 6-phosphate with orthophosphate to produce acetyl phosphate and D-erythrose 4-phosphate. The stereochemical course of the reaction was investigated using double-labeled D-fructose 6-phosphate. Retention of configuration in the methyl group of the acetyl phosphate product was observed with the *Bifidium globosum* enzyme.[1]

[1]I. Merkler & J. Retey (1981) *EJB* **120**, 593.

Selected entries from *Methods in Enzymology* [vol, page(s)]:
General discussion: 5, 261; 9, 515
Assay: 5, 261; 9, 515, 516
Other molecular properties: L-arabinose isomerase free of, **9**, 600; distribution, **5**, 266; *Leuconostoc mesenteroides*, from (properties, **9**, 519; purification, **9**, 517; **41**, 413, 414); properties, **5**, 265; **9**, 519; purification, **5**, 263; **9**, 517; **41**, 413, 414; ribulose-5-phosphate 3-epimerase, in assay of, **9**, 605, 607; ribulose-5-phosphate 4-epimerase, in assay of, **5**, 253; xylulose-5-phosphate 3-epimerase, purification byproduct, **5**, 250

PHOSPHOLIPASE A$_1$

This calcium-dependent enzyme [EC 3.1.1.32] catalyzes phosphatidylcholine hydrolysis to produce a 2-acylglycerophosphocholine and a fatty acid anion (*i.e.*, the removal of the fatty acyl residue at C1 of the glycerol moiety).[1–4] Other substrates include phosphatidylethanolamine, phosphatidylglycerol, phosphatidylinositol, phosphatidyl-L-serine, and cardiolipin. Phospholipase A$_1$ is less specific than phospholipase A$_2$ [EC 3.1.1.4].

[1]K. Y. Hostetler, M. F. Gardner & J. R. Giordano (1986) *Biochemistry* **25**, 6456.
[2]G. Nalbone, K. Y. Hostetler, J. Leonardi, M. Trotz & H. Lafont (1986) *BBA* **877**, 88.
[3]J. D. Esko & C. R. H. Raetz (1983) *The Enzymes*, 3rd ed., **16**, 207.
[4]E. A. Dennis (1983) *The Enzymes*, 3rd ed., **16**, 307.

Selected entries from *Methods in Enzymology* [vol, page(s)]:
General discussion: 1, 660; 14, 167; 71, 674

Assay: 14, 167; **197**, 319; detergent-resistant, from *Escherichia coli* membranes, **197**, 310; rat liver lysosomes, from, **71**, 678; rat liver plasma membrane, from, **71**, 684
Other molecular properties: bacterial membrane, in, **31**, 649; detergent effects, **197**, 103; detergent-resistant, from *Escherichia coli* membranes (assay, **197**, 310; properties, **197**, 313; purification, **197**, 311); dihydroorotate dehydrogenase solubilization, for, **51**, 66; *Escherichia coli* membrane, from, **32**, 91; lysosomal, rat kidney, **197**, 329; rat brain, **14**, 169; rat liver lysosomes, from, **71**, 674, 678; rat liver plasma membrane, from, **71**, 674, 684; reaction products, chromatography, **197**, 159; snake venom, ubiquinone reductase activity, **53**, 13; specific for the α′ position of lecithin, **14**, 167; thiolester substrate, **248**, 16

PHOSPHOLIPASE A$_2$

This calcium-dependent enzyme [EC 3.1.1.4] (also known as phosphatidylcholine 2-acylhydrolase, lecithinase A, phosphatidase, and phosphatidolipase) catalyzes the hydrolysis of a phosphatidylcholine to produce a 1-acyl-glycerophosphocholine and a fatty acid anion (*i.e.*, removal of the fatty acyl residue located at the C2 position of the glycerol moiety).[1–6]

phosphatidylcholine   1-acylglycerophosphocholine

Other substrates include phosphatidylethanolamine, phosphatidyl-L-serine, choline plasmalogen, and phosphatides. The snake venom enzymes have a broad specificity as does the mammalian pancreatic phospholipase. The membrane-bound phospholipase has an important role in eicosanoid biosynthesis and is stimulated by a number of agents. Note that the phospholipase reaction is typically assayed at lipid-water interfaces and the experimental approaches in the study of enzyme kinetics are somewhat different from that of homogeneous-phase enzyme reactions. *See also Phospholipase A; Phospholipase*

[1]E. A. Dennis (1983) *The Enzymes*, 3rd ed., **16**, 307.
[2]D. J. Hanahan (1971) *The Enzymes*, 3rd ed., **5**, 71.
[3]C. C. Leslie (1997) *JBC* **272**, 16709.
[4]M. K. Jain, M. H. Gelb, J. Rogers & O. G. Berg (1995) *MIE* **249**, 567.
[5]D. L. Scott & P. B. Sigler (1994) *Adv. Protein Chem.* **45**, 53.
[6]J. D. Esko & C. R. H. Raetz (1983) *The Enzymes*, 3rd ed., **16**, 207.

Selected entries from *Methods in Enzymology* [vol, page(s)]:
General discussion: 1, 660; 14, 170, 178; 32, 147; 64, 340; 71, 690, 698, 703; 72, 411, 415; 187, 216; 249, 567; 286, 168
Assay: 14, 170, 179; 72, 372; acidimetric assay, **197**, 11; anthroyloxyundecanoylphosphatidylcholine, **72**, 374; assay substrates (phospholipids, **197**, 3; selection, **197**, 30); autoclaved *Escherichia coli*, with, sensitivity, **197**, 27; bacterial, **165**, 299; bee venom, from, conductimetric assay, **71**, 699; bioluminescence assays, **72**, 338, 341; cobra venom, **71**, 703; **197**, 359; cytosolic, **197**, 401; electron spin

resonance assay, **197**, 21; fluorometric assays, **197**, 19; fluorescent phosphatidylcholine, **72**, 373; fluorophore-labeled lipid substrates, with, **197**, 90; hepatic mitochondrial, rat (assay, **197**, 366; human platelet, assay, **86**, 653; intrinsic activity, scooting assay, **197**, 123; macrophage, sn-2-arachidonate-specific, **187**, 216; monolayer assay, **197**, 14; plasmalogen-specific, coupled enzyme assays, **197**, 79; polarographic assay, **197**, 14; presence of inhibitors, in, **197**, 21; radiometric assays, **197**, 12; rheumatoid synovial fluid, human, **197**, 373; spectrophotometric assays, **197**, 14; spectrophotometric assay with acyloxynitrobenzoic acid substrates, **197**, 75; splenic, **197**, 390; strategies and methods, **197**, 3; thio-based assay, **197**, 65; titrametric assay, **197**, 11; trinitrophenyl-ω-aminolauroylphosphatidylcholine, **72**, 372

Other molecular properties: acting on mixed substrate/inhibitor monomolecular films, inhibition, **197**, 59; activation of, **1**, 299; activation on lipid bilayers, **197**, 249; adsorption, **64**, 347, 348; aggregation, **64**, 387, 388; alkyl ether substrates, preparation, **197**, 134; analysis by radiation inactivation, **197**, 287; anchoring site, **64**, 390; antibodies, preparation and characterization, **197**, 270; applications in photoreceptor membrane studies, **81**, 320; assay of succinate dehydrogenase, **53**, 468; ATPase and, **2**, 590; bacterial, **1**, 50, 299; beef tissue, from, **14**, 176; bee venom, from, **71**, 698; **72**, 340; bilayer, **64**, 355; binding to (aggregated lipid, thermodynamics, **197**, 241; calcium, thermodynamics, **197**, 237); bioluminescence assays, **72**, 338, 341; Ca²⁺ dependency, relationship to phosphoinositide cascade, **191**, 682; catalysis of vesicle hydrolysis in scooting mode, **197**, 112; catalytic site calcium ion, **249**, 573, 575; cell localization, **197**, 269; chemical modification, **64**, 387, 390; cobra venom (Naja naja naja), from, **71**, 703 (assay, **71**, 703; **197**, 359; effect on erythrocytes, **32**, 137, 138; inhibitors, **71**, 709; kinetic parameters, **71**, 709; properties, **71**, 707; **197**, 362; protein determination, **71**, 705; purification, **197**, 360; specificity, **71**, 708; stability and storage, **71**, 708); competitive inhibition, **64**, 347, 354; conformational change, **64**, 348, 387; cross-linking, **64**, 387; Crotalus adamanteus, **14**, 178; cytosolic, **197**, 401; detection methods, **197**, 9; detergent effects, **197**, 103; diethylpyrocarbonate treatment, **47**, 433; diffusion coefficient, **64**, 351; dihydroorotate dehydrogenase solubilization, for, **51**, 63; dimerization, **64**, 387; dimethylsuberimidate modification, **64**, 387; dissection, **197**, 201; distribution, **14**, 179; **249**, 573; disulfide bridges, reduction and alkylation, **197**, 209; dual-phospholipid model, **64**, 388; effects of short-chain phosphatidylcholine mixed micelles, **197**, 111; hepatic mitochondrial, **197**, 366; human platelet, assay, **86**, 653; human tissue, from, **14**, 172; hydrolytic reaction catalyzed by, **249**, 569; hydrolytic action in monolayer phase transition region, **197**, 52; i-face, **249**, 573; impurities in biological samples, detection, **197**, 123; inhibition by calcium-binding protein CAB-27, **139**, 78; inhibitory lipocortins, analytical methods, **163**, 23; interface activation, **64**, 343, 344; interface recognition site, **64**, 390; interfacial activation, detection, **197**, 100; interfacial catalysis, **64**, 343, 370 (activation, kinetic basis, **249**, 610; chemical step [irreversibility, **249**, 595; rate-limiting step for maximal turnover, as, **249**, 598]; competitive inhibitors, kinetic characterization, **249**, 604; competitive substrate specificity, **249**, 601; components, properties, **249**, 570; discrete enzyme forms in (E* or E*L), **249**, 588 [spectroscopic signatures of, **249**, 593]; divalent cations in, **249**, 599; effect of covalent modifiers, **249**, 603; enzyme binding to interface, **249**, 592; equilibrium constants, **249**, 587 [by protection method, **249**, 588]; equilibrium parameters, **249**, 587; forward commitment to, **249**, 598; integrated Michaelis-Menten equation applied to, **249**, 595; interfacial Michaelis constant K_m^*, **249**, 596 [determination, **249**, 597]; kinetic basis for, **249**, 567 [interpretation, **249**, 578]; kinetic parameters, **249**, 594; kinetic problems at monolayer interfaces, **249**, 610; monomeric enzyme activity in, **249**, 600; neutral diluents [for measurement of equilibrium constants, **249**, 588; properties, **249**, 575]; nonspecific activators, **249**, 607; nonspecific inhibitors, **249**, 606; phospholipid-detergent mixed micelles, **249**, 578, 580, 607; phospholipid vesicles, on, **249**, 578, 586;

products, properties, **249**, 572; rate constants, deconvolution, **249**, 594; short-chain phospholipids dispersed as monomers, **249**, 610; structure-function relationships in, site-directed mutagenesis studies, **249**, 605; substrates, properties, **249**, 575; zwitterionic vesicle hydrolysis, **249**, 605); intrinsic activity, scooting assay, **197**, 123; kinetics (analysis with short-chain phosphatidylcholine, **197**, 100; analysis with zero-order troughs, **197**, 52; monolayer-based studies, **197**, 49; NMR studies, **197**, 31); kinetics, knots, **259**, 691; lag period, product effect, **64**, 375; ligand interactions, thermodynamics, **197**, 234; lipolytic activity, lag periods, **197**, 57; micelle interaction, **64**, 355, 360, 361, 363; mixed monolayer substrates, **197**, 58; monoclonal antibodies (application, **197**, 228; preparation, **197**, 223); monolayer activity, induction, **197**, 51; Naja naja, from, **32**, 137; **71**, 703; penetration, **64**, 390; plasmalogen-specific, coupled enzyme assays, **197**, 79; platelet, rat, **197**, 229; pressure effect, **64**, 352, 353, 366, 367, 373, 374; presteady state, **64**, 350; products, NMR analysis, **197**, 31; protein–protein interactions, thermodynamics, **197**, 247; purity considerations, **197**, 206; radiolabeled Escherichia coli as substrate, **197**, 24; reaction products, chromatography, **197**, 159; residues 1-52, analysis, **197**, 210; short-chain phospholipids, on, assays, **197**, 95; single-site model, **64**, 354; site-directed mutants, acid-base catalysis, **249**, 114; snake venom, **71**, 690; spectrophotometric assay with acyloxynitrobenzoic acid substrates, **197**, 75; spleen enzyme, **197**, 390; stereospecificity at phosphorus, **197**, 258; structure, **249**, 572; substrate binding site, **249**, 573; super-substrate binding site, **64**, 389; thio-based assay, **197**, 65; thiolester substrate, **248**, 16; transition-state analogues, **249**, 305, 574; ultraviolet difference spectra, **64**, 390; unit of activity, **32**, 149; vinyl ether substrates, preparation, **197**, 134; zymogen form, **32**, 147

PHOSPHOLIPASE B

The term phospholipase B refers to lysophospholipase [EC 3.1.1.5]. In the past, this name also referred to enzyme preparations containing both phospholipases A_1 and A_2. More recently, a phospholipase B from *Penicillium notatum* was found to catalyze the hydrolysis of phosphatidylcholine to produce glycerophosphocholine and two molecules of fatty acid anions.[1] The enzyme will also act on 1-acyllysophosphatidylcholine and 2-acyllysophosphatidylcholine to produce a fatty acid anion and glycerophosphocholine. **See** *Lysophospholipase*

[1] K. Saito, J. Sugatani & T. Okumura (1991) MIE **197**, 446.

PHOSPHOLIPASE C

This zinc-dependent enzyme [EC 3.1.4.3] (also known as lipophosphodiesterase I and lecithinase C) catalyzes the hydrolysis of a phosphatidylcholine to produce a 1,2-diacylglycerol and choline phosphate (thus, it is a phosphodiesterase catalyzing the hydrolysis of the P–O bond between the glycerol moiety and the phosphoryl group).[1–3] The bacterial enzyme also acts on sphingomyelin and phosphatidylinositol; however, the enzyme from seminal plasma does not act on phosphatidylinositol. *Clostridium welchii* α-toxin and *Clostridium oedematiens* β- and γ-toxins are phospholipase C proteins.

phosphatidylcholine

1,2-diacylglycerol   phosphocholine

Related enzymes include 1-phosphatidylinositol phosphodiesterase [EC 3.1.4.10] and 1-phosphatidylinositol-4,5-bisphosphate phosphodiesterase [EC 3.1.4.11]. *See also Variant-Surface-Glycoprotein Phospholipase C; 1-Phosphatidylinositol-4,5-Bisphosphate Phosphodiesterase*

[1] E. A. Dennis (1983) *The Enzymes*, 3rd ed., **16**, 307.
[2] D. J. Hanahan (1971) *The Enzymes*, 3rd ed., **5**, 71.
[3] J. D. Esko & C. R. H. Raetz (1983) *The Enzymes*, 3rd ed., **16**, 207.

Selected entries from *Methods in Enzymology* [vol, page(s)]:
General discussion: 1, 670; 14, 188; 32, 154; 71, 710, 725; 187, 226
Assay: 1, 670; 14, 189; 72, 347; 237, 176, 194; acid production, 14, 189, 192; antithromboplastic effect, 14, 189, 191; assay in presence of other lipid hydrolases, 197, 125; bioluminescence, 72, 338, 341, 348; clostridial (hemolytic assay, 165, 294; titrimetric assay, 165, 297; turbidimetric assay, 165, 296); *Dictyostelium discoideum*, 238, 208, 218; egg yolk turbidimetry, 14, 189, 190; hemolytic properties, 14, 189; human platelet, 86, 653; lethality, 14, 189; phosphatase coupled, 72, 347, 349; phosphatidylinositide-specific, 71, 742, 743; 109, 476; 141, 108; 197, 497, 504, 527; phosphorus release, 14, 189, 193; radioactive, 72, 347, 350; spectrophotometric, 72, 348
Other molecular properties: activation, 269, 157; activity, role of arachidonate and metabolites, 191, 683; applications in photoreceptor membrane studies, 81, 320; assay in presence of other lipid hydrolases, 197, 125; *Bacillus cereus*, from, 14, 189, 190, 191, 193, 194; 32, 131, 139, 161; 71, 725; 72, 340; catalyzed inositol trisphosphate production by neutrophil membranes, assay, 187, 232; cellular distribution, 238, 154; cloning, 238, 131, 195; *Clostridium perfringens*, from, 14, 189, 190, 191, 196; 71, 710 (activators and inhibitors, 71, 721; affinity chromatography, 71, 711, 712; assay, 71, 723; biological activities, 71, 722; calcium ion requirement, 71, 721; cationic substrate requirement, 14, 190, 197; detergent requirement, 71, 721; effect on neutrophil chemotaxis, 236, 76; hemolytic activity, 71, 722; isoelectric focusing, 71, 718; lethal toxicity, 71, 722; necrotizing and hemolytic action, 71, 710; pathogenesis of gas gangrene, 71, 722; platelet-aggregating activity, 71, 722; properties, 71, 715; purification, 165, 91; specificity, 71, 720; stability, 71, 720; use in preparing isolated fat cells, 35, 556; vascular permeability-increasing activity, 71, 722); *Clostridium welchii*, from, 32, 139, 159; cytosolic, purification, 197, 523; *Dictyostelium discoideum* (assay, 238, 208; calcium effects, 238, 213; gel electrophoresis, 197, 294; glycophosphatidylinositol, with [*See Glycophosphatidylinositol-Phospholipase C*]); glycophosphatidylinositol-specific, 230, 420;

G protein-activating, purification, 237, 182; G protein regulatory effects, assays (applications, 238, 133, 137; calcium effects, 238, 139; data interpretation, 238, 134, 140; detergent interference, 238, 138; Dowex chromatography, 238, 133; enzyme concentration, 238, 138; reaction stopping, 238, 132; substrates [endogenous, 238, 132; exogenous, 238, 135]; temperature, 238, 138); Gα subunit purification, in, 237, 193; GTPγ binding assay, in, 237, 35; GTPγS regulation, 237, 406; high-performance liquid chromatography, 238, 230; human platelet (assay, 86, 653; preparation, 86, 8); hydrolysis (lecithin in monolayers, 14, 633, 646; lipid products, isolation and TLC, 197, 126; phosphoinositides in suspension, assay, 187, 230; phospholipids, 187, 206; water-soluble products [assay, 197, 191; isolation and HPLC, 197, 130]); immunoprecipitated, phosphatase treatment, 197, 297; immunoprecipitation, 197, 293, 295; intact cell, substrates and products, analysis, 197, 149; interfacial catalysis, 249, 613; molecular weight, 71, 745; rat cerebral cortical membrane, activation, 197, 183; rat liver, assay, in, 109, 476; release of enzymes by, from membranes, 71, 745; specificity, 71, 745; stereospecificity at phosphorus, 197, 258; *Staphylococcus aureus*, from, 71, 741; stereospecificity at phosphorus, 197, 258; stimulation by Gβγ subunits, 237, 452; substrates, 32, 159, 160; substrate specificity, 238, 131, 159; temperature sensitivity, 238, 138; treatment of microsomes, 71, 539; Triton X-100, 64, 361

PHOSPHOLIPASE D

This enzyme [EC 3.1.4.4] (also known as lipophosphodiesterase II, lecithinase D, and choline phosphatase) catalyzes the hydrolysis of a phosphatidylcholine to produce choline and a phosphatidate. This phosphodiesterase therefore catalyzes the hydrolysis of the bond between phosphorus of the glycerophospholipid portion of the substrate and the nitrogenous base.[1-5] Other substrates include other phosphatidyl esters: for example, phosphatidyl-L-serine, lysophosphatidylcholine, phosphatidylethanolamine, and cardiolipin. Sphingomyelin can also be hydrolyzed. *See also Glycoprotein Phospholipase D*

[1] E. A. Dennis (1983) *The Enzymes*, 3rd ed., **16**, 307.
[2] M. Heller (1978) *Adv. Lipid Res.* **16**, 267.
[3] J. H. Exton (1997) *JBC* **272**, 15579.
[4] W. D. Singer, H. A. Brown & P. C. Sternweis (1997) *ARB* **66**, 475.
[5] G. A. M. Cross (1990) *Ann. Rev. Cell Biol.* **6**, 1.

Selected entries from *Methods in Enzymology* [vol, page(s)]:
General discussion: 1, 672; 14, 197, 208; 35, 226; 71, 746; 256, 246; 257, 313
Assay: 1, 672; 14, 197, 208; 35, 227, 228; 256, 249; assay in presence of other lipid hydrolases, 197, 125; bacterial assay, 165, 301; exogenous substrate, with, 257, 314; glycosylphosphatidylinositol-specific, 197, 567; plastid preparations, in, 14, 198; reconstitution, 238, 167; soluble preparations, in, 14, 199; substrate labeling, 238, 167
Other molecular properties: activators and inhibitors, 14, 203; 35, 230, 231; activity, 257, 313; activity from different sources, 14, 201; applications in photoreceptor membrane studies, 81, 320; assay in presence of other lipid hydrolases, 197, 125; distribution, 14, 203; enzymatic activity, 31, 521; exchange reaction, 72, 428, 430; glycosylphosphatidylinositol-specific (alkaline phosphatase substrate, 197, 568; assays, 197, 567; properties, 197, 574; purification, 197, 572); GTP-activated, stimulation by smgGDS, 256, 255; GTP-binding protein requirement, 238, 167; GTPγS-stimulated, inhibition by Rho-GDI, 256, 254; hydrolysis (lecithin in monolayers, 14, 633, 645; lipid products, isolation and TLC, 197, 126; water-soluble products

[assay, **197**, 191; isolation and HPLC, **197**, 132]); isolation from white cabbage, **72**, 634; lag period, **64**, 367, 375; monolayer action, **64**, 367; pH optimum and K_m, **35**, 232; phosphatidic preparation, in, **32**, 142, 143; phospholipid spin labeling, in, **32**, 192, 193; polyol dehydrogenase, lack of solubilization, **9**, 179; preparation (HL-60 cells, from, **257**, 320; rat cerebral membrane, from, **257**, 322); preparation of 1-O-1'-alkenyl-2-oleoyl-sn-glycero-3-phosphoethanolamine, **197**, 139; product effect, **64**, 367, 375; properties, **14**, 202, 210; purification, **14**, 200, 210; **35**, 228; rat, purification, solubilization, and assay, **197**, 575; rat brain, from, **71**, 746 (activators and inhibitors, **71**, 750; assay, **71**, 746; properties, **71**, 750; specificity, **71**, 750); reaction products, chromatography, **197**, 163; Savoy cabbage, from, **14**, 208; specificity, **14**, 202, 203; **35**, 231; stability, **35**, 232; stereospecificity at phosphorus, **197**, 258; stimulation by ARF, **257**, 317; transphosphatidylase activity, **14**, 203; **35**, 232 (mechanism, **14**, 203)

PHOSPHOLIPID HYDROPEROXIDE GLUTATHIONE PEROXIDASE

This selenium-dependent oxidoreductase [EC 1.11.1.12] catalyzes the reaction of two molecules of glutathione with a lipid hydroperoxide to produce glutathione disulfide, a lipid, and two molecules of water.[1–3] Acceptor substrates include linoleate hydroperoxide, all phospholipid hydroperoxides, fatty acid anion hydroperoxides, tert-butyl hydroperoxide, and cholesterol hydroperoxides. This peroxidase has a ping pong mechanism.

[1] F. Ursini, M. Maiorini & C. Gregolin (1985) BBA **839**, 62.
[2] M. Maiorino, A. Roveri, C. Gregolin & F. Ursini (1986) ABB **251**, 600.
[3] E. Panfili, G. Sandri & L. Ernster (1991) FEBS Lett. **290**, 35.

Selected entries from **Methods in Enzymology** [vol, page(s)]:
General discussion: **186**, 448
Assay: spectrophotometric, **186**, 453; **233**, 204; tissues, in, **186**, 455
Other molecular properties: activity, **233**, 202; antibodies, production, **233**, 206; characteristics, **186**, 448; ELISA, **233**, 206; membrane-bound, properties, **233**, 203; properties, **233**, 202; purification from porcine heart, **186**, 455; reactivity with hydroperoxides, **233**, 202; sample preparation, **233**, 204; soluble and membrane fractions of rat tissues, **233**, 210, 212; spectrophotometric assay, **186**, 453; **233**, 204; Western blotting, **233**, 210

PHOSPHOLIPID TRANSFER PROTEINS

These proteins, also known as phosphatidyl exchange proteins and phospholipid exchange enzymes, reportedly facilitate the incorporation of phosphatidyl lipids into membranes lacking phosphatidyl lipids.[1] **See specific protein; for example, Phosphatidylcholine Transfer Protein; Phosphatidylserine Transfer Protein**

[1] A. Tall (1995) ARB **64**, 235.

Selected entries from **Methods in Enzymology** [vol, page(s)]:
Assay: higher plant, **209**, 523; yeast, **209**, 515
Other molecular properties: higher plant (assay, **209**, 523; properties, **209**, 528; purification, **209**, 525); yeast (assay, **209**, 515; purification, **209**, 517)

PHOSPHOMANNAN MANNOSEPHOSPHOTRANSFERASE

This transferase [EC 2.7.8.9], first isolated from Hansenula holstii, catalyzes the reaction of GDP-D-mannose with (phosphomannan)$_n$ to produce GMP and (phosphomannan)$_{n+1}$.[1]

[1] R. K. Bretthauer, L. P. Kozak & W. E. Irwin (1969) BBRC **37**, 820.

PHOSPHOMANNOMUTASE

This mutase [EC 5.4.2.8], also known as phosphomannose mutase, catalyzes the interconversion of D-mannose 1-phosphate and D-mannose 6-phosphate.[1–3] D-Mannose 1,6-bisphosphate is a required cofactor, but D-glucose 1,6-bisphosphate is also effective. The enzyme will also catalyze the interconversion of D-glucose 1-phosphate and D-glucose 6-phosphate.[1] Two isozymes have been identified in mammals, both proceeding through a phospho-enzyme intermediate.[2]

[1] S. K. Guha & Z. B. Rose (1985) ABB **243**, 168.
[2] M. Pirard, Y. Achouri, J. F. Collet, E. Schollen, G. Matthijs & E. Van Schaftingen (1999) BJ **339**, 201.
[3] S. Shankar, R. W. Ye, D. Schlictman & A. M. Chakrabarty (1995) AE **70**, 221.

Selected entries from **Methods in Enzymology** [vol, page(s)]:
General discussion: **8**, 183
Assay: **8**, 183

PHOSPHOMETHYLPYRIMIDINE KINASE

This phosphotransferase [EC 2.7.4.7], also known as hydroxymethylpyrimidine phosphokinase, catalyzes the reaction of ATP with 4-amino-2-methyl-5-phosphomethyl-pyrimidine to produce ADP and 4-amino-2-methyl-5-diphospho-methylpyrimidine.[1]

[1] L. M. Lewin & G. M. Brown (1961) JBC **236**, 2768.

PHOSPHOMEVALONATE KINASE

This phosphotransferase [EC 2.7.4.2] catalyzes the reaction of ATP with (R)-5-phosphomevalonate to produce ADP and (R)-5-diphosphomevalonate.[1–5] The pig liver enzyme has an ordered Bi Bi kinetic mechanism. Stereospecificity studies using phosphorothioate analogues of ATP indicate that the metal is chelated with the β-phosphate of the nucleotide in the active site and that the chelate structure of metal-ATP substrate is the δ,β,γ-bidentate complex.[4]

[1] D. N. Skilleter & R. G. O. Kekwick (1971) BJ **124**, 407.
[2] S. Bazaes, E. Beytia, A. M. Jabalquinto, F. Solis de Ovando, I. Gomez & J. Eyzaguirre (1980) Biochemistry **19**, 2300.
[3] S. Bazaes, E. Beytia, A. M. Jabalquinto, F. Solis de Ovando & I. Gomez (1980) Biochemistry **19**, 2305.

[4] C. S. Lee & W. J. O'Sullivan (1985) *JBC* **260**, 13909.
[5] E. D. Beytía & J. W. Porter (1976) *ARB* **45**, 113.

Selected entries from *Methods in Enzymology* [vol, page(s)]:
General discussion: 5, 489, 490; 15, 413; 110, 78
Assay: 5, 490; 15, 414; 110, 78
Other molecular properties: plant tissues, from, 13, 559; porcine liver (assays, 110, 78; properties, 110, 84; purification, 110, 82); properties, 15, 416; purification, 5, 494; 15, 415; pyrophosphomevalonate, preparation of, 6, 507; pyrophosphomevalonate decarboxylase not present in preparation of, 5, 497

PHOSPHONATE-TRANSPORTING ATPase

This so-called ATPase [EC 3.6.3.28], catalyzes the ATP-dependent (ADP- and orthophosphate-producing) transport of a phosphonate$_{out}$ to produce a phosphonate$_{in}$.[1-3] ABC-type (ATP-binding cassette-type) ATPases have two similar ATP-binding domains and do not undergo phosphorylation during the transport process.

The idea that a transporter is an enzyme is in keeping with a new definition of enzyme catalysis as the facilitated making/breaking of chemical bonds, not just covalent bonds.[4] This idea builds on Pauling's assertion that any long-lived, chemically distinct interaction (in this case, the persistent location of a solute with respect to the faces of a membrane) can be regarded as a chemical bond. Note also that the equilibrium constant (K_{eq} = [phosphonate$_{in}$][ADP][P$_i$]/[phosphonate$_{out}$][ATP]) does not conform to that expected for an ATPase (*i.e.*, K_{eq} = [ADP][P$_i$]/[ATP]). Thus, although the overall reaction yields ADP and orthophosphate, the enzyme is misclassified as a hydrolase, and instead should be regarded as an energase-type reaction. Energases facilitate affinity-modulated reactions by coupling the $\Delta G_{ATP\text{-}hydrolysis}$ to a force-generating or work-producing step.[4] In this case, P-O-P bond scission supplies the energy to drive phosphonate transport. *See ATPase; Energase*

[1] B. L. Wanner & W. W. Metcalf (1992) *FEMS Microbiol. Lett.* **79**, 133.
[2] M. H. Saier, Jr. (1998) *Adv. Microb. Physiol.* **40**, 81.
[3] J. K. Griffiths & C. E. Sansom (1998) *The Transporter Factsbook*, Academic Press, San Diego.
[4] D. L. Purich (2001) *TiBS* **26**, 417.

PHOSPHONOACETALDEHYDE HYDROLASE

This enzyme [EC 3.11.1.1], also known as phosphonatase, catalyzes the hydrolysis of phosphonoacetaldehyde to produce acetaldehyde and orthophosphate.[1-4]

phosphonoacetaldehyde   acetaldehyde

Use of labeled sodium borohydride has provided evidence the enzyme proceeds via the formation of a Schiff base with an active-site lysyl residue, followed by phosphoryl transfer to an aspartyl residue and generation of an enamine that is subsequently converted to an imine. Hydrolysis of the imine and acyl phosphate then ensues.[2-4]

[1] D. B. Olsen, T. W. Hepburn, M. Moos, P. S. Mariano & D. Dunaway-Mariano (1988) *Biochemistry* **27**, 2229.
[2] J. M. La Nauze, J. R. Coggins & H. B. F. Dixon (1977) *BJ* **165**, 409.
[3] A. S. Baker, M. J. Ciocci, W. W. Metcalf, J. Kim, P. C. Babbitt, B. L. Wanner, B. M. Martin & D. Dunaway-Mariano (1998) *Biochemistry* **37**, 9305.
[4] K. N. Allen (1998) *CBC* **2**, 135.

Selected entries from *Methods in Enzymology* [vol, page(s)]:
Other molecular properties: transition-state and multisubstrate analogues, 249, 307

PHOSPHONOACETATE HYDROLASE

This zinc-dependent enzyme [EC 3.11.1.2], first isolated from *Pseudomonas fluorescens*, catalyzes the hydrolysis of phosphonoacetate (^-O_3P–CH_2COO^-) to produce acetate and orthophosphate.[1]

[1] J. W. McGrath, G. B. Wisdom, G. McMullan, M. J. Larkin & J. P. Quinn (1995) *EJB* **234**, 225.

PHOSPHOPANTOTHENOYLCYSTEINE DECARBOXYLASE

This pyruvoyl-dependent decarboxylase [EC 4.1.1.36] catalyzes the conversion of *N*-((*R*)-4-phosphopantothenoyl)-L-cysteine to pantotheine 4′-phosphate and carbon dioxide.[1-3] The substrate binds and forms a Schiff base with the enzyme pyruvoyl group, and decarboxylation occurs with retention of configuration, as indicated by experiments with substrates that are deuterated at position C2 in the cysteine moiety.[3] *See also Pantothenoylcysteine Decarboxylase*

[1] H. Yang & R. H. Abeles (1987) *Biochemistry* **26**, 4076.
[2] R. Scandurra, L. Politi, L. Santoro & V. Consalvi (1987) *FEBS Lett.* **212**, 79.
[3] D. J. Aberhart, P. K. Ghoshal, J.-A. Cotting & D. J. Russell (1985) *Biochemistry* **24**, 7178.

Selected entries from *Methods in Enzymology* [vol, page(s)]:
General discussion: 18A, 354; 62, 245
Assay: 18A, 354, 355; 62, 245
Other molecular properties: properties, 62, 248; purification from horse liver, 62, 247, 248; rat liver, from, 18A, 354 (properties, 18A, 356; purification, 18A, 355, 356); synthesis of 4′-phosphopantothenoyl-L-[*U*-^{14}C]cysteine, 62, 246

PHOSPHOPANTOTHENOYLCYSTEINE SYNTHETASE

This enzyme [EC 6.3.2.5], also known as phospho-pantothenate:cysteine ligase, catalyzes the reaction of CTP with (R)-4′-phosphopantothenate and L-cysteine to produce (R)-4′-phosphopantothenoyl-L-cysteine and yet-to-be-identified products of CTP.[1,2] L-Cysteine can be replaced by β-mercaptoethylamine and α-methylcysteine.

[1] Y. Abiko (1967) *J. Biochem.* **61**, 290.
[2] G. W. E. Plaut, C. M. Smith & W. L. Alworth (1974) *ARB* **43**, 899.

Selected entries from *Methods in Enzymology* [vol, page(s)]:
General discussion: 18A, 350; 62, 236
Assay: rat liver, 18A, 350, 351
Other molecular properties: rat liver, from, 18A, 350 (properties, 18A, 353, 354; purification, 18A, 351, 353); role, in CoA biosynthesis, 62, 236

PHOSPHOPENTOMUTASE

This mutase [EC 5.4.2.7], also known as phospho-ribomutase and phosphodeoxyribomutase, catalyzes the interconversion of D-ribose 1-phosphate and D-ribose 5-phosphate.[1-3] Other substrates include 2-deoxy-D-ribose 1-phosphate (into 2-deoxy-D-ribose 5-phosphate). The enzyme also requires a bisphosphorylated cofactor, such as D-ribose 1,5-bisphosphate, 2-deoxy-D-ribose 1,5-bisphosphate, and α-D-glucose 1,6-bisphosphate. *See also Phosphoglucomutase*

[1] K. Hammer-Jespersen & A. Munch-Petersen (1970) *EJB* **17**, 397.
[2] M. G. Tozzi, R. Catalani, P. L. Ipata & U. Mura (1982) *Anal. Biochem.* **123**, 265.
[3] W. J. Ray, Jr. & E. J. Peck, Jr. (1972) *The Enzymes*, 3rd ed., **6**, 407.

Selected entries from *Methods in Enzymology* [vol, page(s)]:
General discussion: 1, 361
Assay: 1, 361
Other molecular properties: action of, 3, 186, 187; 2-deoxy-D-ribose-5-phosphate, preparation of, 3, 186; properties, 1, 362; purification, muscle, 1, 362; purine nucleoside phosphorylase, not present in preparations of beef liver, 12A, 115; purine nucleoside phosphorylase, present in preparations of fish muscle, 12A, 118; regulation, 51, 438, 517

PHOSPHOPOLYPRENOL GLUCOSYLTRANSFERASE

This transferase [EC 2.4.1.78] catalyzes the reaction of UDP-D-glucose with a polyprenyl phosphate to produce UDP and a polyprenylphosphate-D-glucose.[1] The best polyprenol phosphate substrate is ficaprenyl phosphate, a mixture of C50, C55, and C66 *cis*- and *trans*-polyprenol phosphates.

[1] W. Jankowski, T. Mankowski & T. Chojnacki (1974) *BBA* **337**, 153.

PHOSPHORAMIDATE:HEXOSE PHOSPHOTRANSFERASE

This enzyme [EC 2.7.1.62] catalyzes the reaction of phosphoramidate ($^{-}O_3PNH_3^+$) with a hexose to produce ammonia and hexose 1-phosphate, proceeding via a phos-phoenzyme intermediate.[1] In some organisms, this enzyme may be identical to glucose-6-phosphatase [EC 3.1.3.9]. Hexose substrates include D-glucose, 2-deoxy-D-glucose, D-mannose, and D-fructose.

[1] J. R. Stevens-Clark, K. A. Conklin, A. Fujimoto & R. A. Smith (1968) *JBC* **243**, 4474.

Selected entries from *Methods in Enzymology* [vol, page(s)]:
General discussion: 9, 403
Assay: 9, 403

PHOSPHORIBOKINASE

This phosphotransferase [EC 2.7.1.18] catalyzes the reaction of ATP with D-ribose 5-phosphate to produce ADP and D-ribose 1,5-bisphosphate.[1]

[1] H. M. Kalckar (1953) *BBA* **12**, 250.

Selected entries from *Methods in Enzymology* [vol, page(s)]:
General discussion: 42, 120
Assay: *Pseudomonas saccharophila*, 42, 120, 121
Other molecular properties: interference, in phosphoribosylpyrophosphate synthetase assay, 51, 3; *Pseudomonas saccharophila*, from, 42, 120 (assay, 42, 120, 121; inhibition, 42, 123; properties, 42, 123; purification, 42, 121)

5-PHOSPHORIBOSYLAMINE SYNTHETASE

This enzyme [EC 6.3.4.7], also known as ribose-5-phosphate:ammonia ligase and ribose-5-phosphate amino-transferase, catalyzes the reaction of ATP with D-ribose 5-phosphate and ammonia to produce ADP, orthophos-phate, and 5-phosphoribosylamine.[1,2]

[1] G. H. Reem (1968) *JBC* **243**, 5695.
[2] E. K. Holmes, G. L. King, A. Leyva & S. C. Singer (1976) *PNAS* **73**, 2458.

PHOSPHORIBOSYLAMINOIMIDAZOLE-CARBOXAMIDE FORMYLTRANSFERASE

This formyltransferase [EC 2.1.2.3], also known as 5-amino-4-imidazolecarboxamide ribotide transformy-lase (or, AICAR transformylase), catalyzes the reaction of 10-formyltetrahydrofolate with 5′-phosphoribosyl-5-amino-4-imidazolecarboxamide (also known as 5-amino-1-(5-phospho-D-ribosyl)imidazole-4-carboxamide) to pro-duce tetrahydrofolate and 5′-phosphoribosyl-5-formamido-4-imidazolecarboxamide (also known as 5-formamido-1-(5-phospho-D-ribosyl)imidazole-4-carboxamide).[1-6] The chicken liver enzyme reportedly has an ordered Bi Bi kinetic mechanism.[4]

In many organisms, this is a bifunctional protein, also exhibiting IMP cyclohydrolase [EC 3.5.4.10] activity.

[1]J. E. Baggott & C. L. Krumdieck (1979) *Biochemistry* **18**, 1036.
[2]J. E. Baggott, W. H. Vaughn & B. B. Hudson (1986) *BJ* **236**, 193.
[3]K. Iwai, Y. Fujisawa & N. Suzuki (1972) *ABC* **36**, 398.
[4]W. T. Mueller & S. J. Benkovic (1981) *Biochemistry* **20**, 337.
[5]V. Schirch (1998) *CBC* **1**, 211.
[6]J. I. Rader & F. M. Huennekens (1973) *The Enzymes*, 3rd ed., **9**, 197.

Selected entries from *Methods in Enzymology* [vol, page(s)]:
General discussion: 2, 504; 6, 89
Assay: 6, 89

PHOSPHORIBOSYLAMINOIMIDAZOLE CARBOXYLASE

This enzyme [EC 4.1.1.21], also known as 5-aminoimidazole ribotide carboxylase (AIR carboxylase), catalyzes the reversible reaction of 1-(5-phosphoribosyl)-5-amino-imidazole with carbon dioxide to produce 1-(5-phosphoribosyl)-5-amino-4-imidazolecarboxylate (also known as 5-amino-1-(5-phospho-D-ribosyl)imidazole).[1,2] Biotin is not a cofactor of this carboxylase. In some organisms, this enzyme is a component of the multifunctional protein ade2 that also contains a phosphoribosylaminoimidazole-succinocarboximide synthetase [EC 6.3.2.6] activity.

[1]C. A. H. Patey & G. Shaw (1973) *BJ* **135**, 543.
[2]L. N. Lukens & J. M. Buchanan (1959) *JBC* **234**, 1799.

Selected entries from *Methods in Enzymology* [vol, page(s)]:
General discussion: 6, 79
Assay: 6, 79

PHOSPHORIBOSYLAMINOIMIDAZOLE-SUCCINOCARBOXAMIDE SYNTHETASE

This enzyme [EC 6.3.2.6], also known as *N*-(5-amino-1-ribosyl-4-imidazolylcarbonyl) L-aspartic acid 5'-phosphate synthetase and succino-AICAR synthetase, catalyzes the reaction of ATP with 1-(5-phosphoribosyl)-4-carboxy-5-aminoimidazole and L-aspartate to produce ADP, orthophosphate, and 1-(5-phosphoribosyl)-4-(*N*-succino-carboxamide)-5-aminoimidazole (also known as (*S*)-2-[5-amino-1-(5-phospho-D-ribosyl)imidazole-4-carboxamido]-succinate).[1,2]

In some organisms, this enzyme is a component of the multifunctional protein ade2 that also contains a phosphoribosylaminoimidazole carboxylase [EC 4.1.1.21] activity.

[1]C. A. H. Patey & G. Shaw (1973) *BJ* **135**, 543.
[2]L. N. Lukens & J. M. Buchanan (1959) *JBC* **234**, 1799.

Selected entries from *Methods in Enzymology* [vol, page(s)]:
General discussion: 6, 82; 51, 186
Assay: 6, 82; 51, 187, 188
Other molecular properties: activity, 51, 186; bacterial sources, 51, 190; chicken liver, from, 51, 190; properties, 6, 84; 51, 193; purification, chicken liver, 6, 83; 51, 190; stability, 51, 192

PHOSPHORIBOSYL-AMP CYCLOHYDROLASE

This enzyme [EC 3.5.4.19] catalyzes the hydrolysis of 5-phosphoribosyl-AMP to produce 5-(5-phospho-D-ribosylaminoformimino)-1-(5-phosphoribosyl)imidazole-4-carboxamide.[1,2] The enzyme from *Neurospora crassa* also catalyzes the reactions of histidinol dehydrogenase [EC 1.1.1.23] and phosphoribosyl-ATP pyrophosphatase [EC 3.6.1.31]. The *Methanococcus vannielii* enzyme has a zinc ion that is essential for catalytic activity.[3]

[1]A. C. Minson & E. H. Creaser (1969) *BJ* **114**, 49.
[2]J. K. Keesey, R. Bigelis & G. R. Fink (1979) *JBC* **254**, 7427.
[3]R. L. D'Ordine, T. J. Klem & V. J. Davisson (1999) *Biochemistry* **38**, 1537.

Selected entries from *Methods in Enzymology* [vol, page(s)]:
General discussion: 2, 504; 17B, 24
Assay: 17B, 24

PHOSPHORIBOSYLANTHRANILATE ISOMERASE

This isomerase [EC 5.3.1.24], also known as *N*-(5'-phosphoribosyl)anthranilate isomerase (PRAI), catalyzes the conversion of *N*-(5-phospho-β-D-ribosyl)-anthranilate to 1-(2-carboxyphenylamino)-1-deoxy-D-ribulose 5-phosphate.[1–4] The reaction is an example of a reversible Amadori rearrangement.[2]

In some organisms, this enzyme is part of a multifunctional protein together with one or more components of the system for biosynthesis of tryptophan (anthranilate phosphoribosyltransferase [EC 2.4.2.18], indole-3-glycerol-phosphate synthase [EC 4.1.1.48], anthranilate synthase [EC 4.1.3.27], and tryptophan synthase [EC 4.2.1.20]). The enzyme from *Thermotoga maritima* is extremely stable and active as a homodimer.[3] The *Escherichia coli* enzyme is bifunctional and monomeric, also catalyzing the indole-3-glycerol-phosphate synthase reaction.[4]

[1]T. E. Creighton & C. Yanofsky (1970) *MIE* **17A**, 365.
[2]U. Hommel, M. Eberhard & K. Kirschner (1995) *Biochemistry* **34**, 5429.
[3]R. Sterner, G. R. Kleemann, H. Szadkowski, A. Lustig, M. Hennig & K. Kirschner (1996) *Protein Sci.* **5**, 2000.
[4]J. P. Priestle, M. G. Grutter, J. L. White, M. G. Vincent, M. Kania, E. Wilson, T. S. Jardetzky, K. Kirschner & J. N. Jansonius (1987) *PNAS* **84**, 5690.

PHOSPHORIBOSYL-ATP PYROPHOSPHATASE

This enzyme [EC 3.6.1.31], also known as phosphoribosyl-ATP diphosphatase, catalyzes the hydrolysis of 1-(5-phosphoribosyl)-ATP to produce 1-(5-phosphoribosyl)-AMP and pyrophosphate.[1,2] The *Neurospora crassa* and *Saccharomyces cerevisiae* enzymes also catalyze the reactions of histidinol dehydrogenase [EC 1.1.1.23] and phosphoribosyl-AMP cyclohydrolase [EC 3.5.4.19]. There are two separate domains for the phosphoribosyl-ATP pyrophosphatase and phosphoribosyl-AMP cyclohydrolase activities in the *Arabidopsis* enzyme.[3]

[1] D. W. E. Smith & B. N. Ames (1965) *JBC* 240, 3056.
[2] J. K. Keesey, R. Bigelis & G. R. Fink (1979) *JBC* 254, 7427.
[3] K. Fujimori & D. Ohta (1998) *Plant Physiol.* 118, 275.

N-(5'-PHOSPHO-D-RIBOSYLFORMIMINO)-5-AMINO-1-(5''-PHOSPHORIBOSYL)-4-IMIDAZOLE CARBOXAMIDE ISOMERASE

This isomerase [EC 5.3.1.16] catalyzes the interconversion of N-(5'-phospho-D-ribosylformimino)-5-amino-1-(5''-phosphoribosyl)-4-imidazolecarboxamide and N-(5'-phospho-D-1'-ribulosylformimino)-5-amino-1-(5''-phosphoribosyl)-4-imidazolecarboxamide: an Amadori rearrangement.[1,2]

[1] M. N. Margolies & R. F. Goldberger (1967) *JBC* 242, 256.
[2] M. N. Margolies & R. F. Goldberger (1966) *JBC* 241, 3262.

PHOSPHORIBOSYLFORMYLGLYCINAMIDINE CYCLO-LIGASE

This ligase [EC 6.3.3.1], also known as phosphoribosylaminoimidazole synthetase and 5'-aminoimidazole ribonucleotide synthetase (AIR synthetase), catalyzes the reaction of ATP with 5'-phosphoribosylformylglycinamidine (or, 2-(formamido)-N^1-(5-phospho-D-ribosyl)-acetamidine) to produce ADP, orthophosphate, and 5'-phosphoribosyl-5-aminoimidazole (also referred to as 5-amino-1-(5-phospho-D-ribosyl)imidazole).[1,2]

In some organisms (*e.g.*, chicken liver), this activity is a part of a trifunctional protein exhibiting phosphoribosylglycinamide synthetase [EC 6.3.4.13] and phosphoribosylglycinamide formyltransferase [EC 2.1.2.2] activities.[3]

[1] J. L. Schrimsher, F. J. Schendel & J. Stubbe (1986) *Biochemistry* 25, 4356.
[2] J. L. Schrimsher, F. J. Schendel, J. Stubbe & J. M. Smith (1986) *Biochemistry* 25, 4366.
[3] C. S. Daubner, J. L. Schrimsher, F. J. Schendel, M. Young, S. Henikoff, D. Patterson, J. Stubbe & S. J. Benkovic (1985) *Biochemistry* 24, 7059.

PHOSPHORIBOSYLFORMYL-GLYCINAMIDINE SYNTHETASE

This enzyme [EC 6.3.5.3], also known as FGAM synthetase and formylglycinamide ribotide amidotransferase (FGAR amidotransferase), catalyzes the reaction of ATP with 5'-phosphoribosylformylglycinamide, L-glutamine, and water to produce ADP, orthophosphate, 5'-phosphoribosylformylglycinamidine (or, 2-(formamido)-N^1-(5-phospho-D-ribosyl)acetamidine), and L-glutamate.[1-3] The amidotransferase, which has a minor glutaminase activity, has an ordered kinetic mechanism[1]; however, the enzyme from ascites tumor cells is reportedly ping pong.[2] Binding of L-glutamine to the enzyme results in the formation of a transient γ-glutamyl thioester intermediate.

[1] H.-C. Li & J. M. Buchanan (1971) *JBC* 246, 4720.
[2] S. Y. Chu & J. F. Henderson (1972) *Can. J. Biochem.* 50, 484.
[3] J. M. Buchanan (1973) *AE* 39, 91.

PHOSPHORIBOSYLGLYCINAMIDE FORMYLTRANSFERASE

This formyltransferase [EC 2.1.2.2] (also known as 5'-phosphoribosylglycinamide transformylase · glycinamide ribonucleotide transformylase, and GAR transformylase) catalyzes the reaction of 10-formyltetrahydrofolate with 5′-phosphoribosylglycinamide to produce tetrahydrofolate and 5′-phosphoribosyl-*N*-formylglycinamide. The transferase has an ordered Bi Bi kinetic mechanism.[1–6] Site-directed mutagenesis, pH-dependency, and isotope-effect investigations of the *Escherichia coli* enzyme show that a histidyl residue acts in a salt bridge with an aspartyl residue as a general acid catalyst with a pK_a of 9.7. This same aspartyl residue also participates in the preparation of the active site geometry for catalysis. The rate-limiting step of catalysis appears to be those steps involving the formation of tetrahedral intermediates.[4] The oxyanion tetrahedral intermediate formed by the nucleophilic attack of amine on the formyl group is stabilized by hydrogen bonds to a histidyl residue and to a water molecule hydrogen bonded to an aspartyl residue.

In some organisms, the enzyme is trifunctional, also exhibiting phosphoribosylformylglycinamidine cyclo-ligase [EC 6.3.3.1] and phosphoribosylglycinamide synthetase [EC 6.3.4.13] activities.

[1] J. Inglese, D. L. Johnson, A. Shiau, J. M. Smith & S. J. Benkovic (1990) *Biochemistry* **29**, 1436.
[2] C. A. Caperelli (1985) *Biochemistry* **24**, 1316.
[3] C. A. Caperelli (1989) *JBC* **264**, 5053.
[4] J. H. Shim & S. J. Benkovic (1999) *Biochemistry* **38**, 10024.
[5] V. Schirch (1998) *CBC* **1**, 211.
[6] J. I. Rader & F. M. Huennekens (1973) *The Enzymes*, 3rd ed., **9**, 197.

Selected entries from *Methods in Enzymology* [vol, page(s)]:
General discussion: 2, 504; 6, 65
Assay: 6, 65
Other molecular properties: properties, 6, 68; purification, 6, 66; transition-state and multisubstrate analogues, 249, 304

PHOSPHORIBOSYLGLYCINAMIDE SYNTHETASE

This enzyme [EC 6.3.4.13; formerly classified as EC 6.3.1.3], also known as phosphoribosylamine:glycine ligase and glycinamide ribonucleotide synthetase (GAR synthetase), catalyzes the reaction of ATP with 5-phosphoribosylamine and glycine to produce ADP, orthophosphate, and 5′-phosphoribosylglycinamide.[1–3] The *Escherichia coli* enzyme reportedly has an ordered Ter Ter kinetic mechanism.[4]

The chicken,[5] *Escherichia coli*,[6] hamster,[7] mouse,[8] and human[9] enzymes are trifunctional, also exhibiting phosphoribosylformylglycinamidine cyclo-ligase [EC 6.3.3.1]

and phosphoribosylglycinamide formyltransferase [EC 2.1.2.2] activities.

[1] N. R. Gandhi & C. A. Westby (1972) *Enzymologia* **42**, 185.
[2] P. B. Rowe, E. McCairns, G. Madsen, D. Sauer & H. Elliott (1978) *JBC* **253**, 7711.
[3] Y. S. Cheng, M. Murray, F. Schendel, J. Otvos, J. Wehrli & J. Stubbe (1987) *Adv. Enzyme Regul.* **26**, 319.
[4] Y. S. Cheng, Y. Shen, J. Rudolph, M. Stern, J. Stubbe, K. A. Flannigan & J. M. Smith (1990) *Biochemistry* **29**, 218.
[5] C. S. Daubner, J. L. Schrimsher, F. J. Schendel, M. Young, S. Henikoff, D. Patterson, J. Stubbe & S. J. Benkovic (1985) *Biochemistry* **24**, 7059.
[6] J. Aimi, H. Qiu, J. Williams, H. Zalkin & J. E. Dixon (1990) *NAR* **18**, 6665.
[7] F. H. Chang, T. S. Barnes, D. Schild, A. Gnirke, J. Bleskan & D. Patterson (1991) *Somat. Cell Mol. Genet.* **17**, 411.
[8] J. L. Kan, M. Jannatipour, S. M. Taylor & R. G. Moran (1993) *Gene* **137**, 195.
[9] M. T. Poch, W. Qin & C. A. Caperelli (1998) *Protein Expr. Purif.* **12**, 17.

Selected entries from *Methods in Enzymology* [vol, page(s)]:
General discussion: 2, 504; 6, 61; 51, 179
Assay: 51, 179
Other molecular properties: *Aerobacter aerogenes*, from, 51, 179; amidophosphoribosyltransferase assay, 51, 172; inhibitors, 51, 185; kinetic properties, 51, 185; molecular weight, 51, 185; pH optimum, 51, 185; pigeon liver, from, 51, 180; properties, 51, 185; purification, 51, 182

1-(5-PHOSPHORIBOSYL)IMIDAZOLE-4-ACETATE SYNTHETASE

This enzyme [EC 6.3.4.8], also called imidazole-acetate:phosphoribosyldiphosphate ligase, catalyzes the reaction of ATP with imidazole-4-acetate and 5-phosphoribosyl-pyrophosphate to produce ADP, orthophosphate, 1-(5-phosphoribosyl)imidazole-4-acetate, and pyrophosphate.[1]

[1] G. M. Crowley (1964) *JBC* **239**, 2593.

Selected entries from *Methods in Enzymology* [vol, page(s)]:
General discussion: 17B, 770
Assay: 17B, 770

PHOSPHORIBULOKINASE

This phosphotransferase [EC 2.7.1.19], also known as phosphopentokinase, catalyzes the reaction of ATP with D-ribulose 5-phosphate to produce ADP and D-ribulose 1,5-bisphosphate.[1–3] The enzymes from prokaryotes and eukaryotes are structurally different and are regulated by different mechanisms.[1]

[1] H. M. Miziorko (2000) *AE* **74**, 95.
[2] J. A. Runquist, D. H. Harrison & H. M. Miziorko (1999) *Biochemistry* **38**, 13999.
[3] A. T. H. Abdelal & H. G. Schlegel (1974) *BJ* **139**, 481.

Selected entries from *Methods in Enzymology* [vol, page(s)]:
General discussion: 5, 258
Assay: *Chromatium*, 42, 115, 116; spinach leaves, 5, 258
Other molecular properties: adenylic acid ribosidase and, 6, 123; *Chromatium*, from, 42, 115 (assay, 42, 115, 116; chromatography, 42,

118; electrophoresis, **42**, 119; inhibition, **42**, 119; molecular weight, **42**, 119; properties, **42**, 119; purification, **42**, 116); interference, in phosphoribosylpyrophosphate synthetase assay, **51**, 3; ribulose diphosphate carboxylase in assay of, **5**, 270; ribulose diphosphate formation and, **3**, 193; spinach leaves, **5**, 260

PHOSPHORYLASE

This glucan phosphorylase [EC 2.4.1.1] catalyzes the reversible reaction of $[(1,4)\text{-}\alpha\text{-D-glucosyl}]_n$ with orthophosphate to produce $[(1,4)\text{-}\alpha\text{-D-glucosyl}]_{n-1}$ and α-D-glucose 1-phosphate with overall retention of anomeric configuration.[1-3] The enzyme will not act on residues situated within four glucosyl units of a branch point, such that exhaustive phosphorylase action on glycogen produces a limit dextrin. Positional isotope exchange experiments suggest that the enzyme may act via a glycosyl-enzyme intermediate; however, a mechanism involving a stabilized oxocarbenium ion intermediate has also been proposed. Pyridoxal 5′-phosphate, held firmly in an imine linkage to a specific ε-amino side-chain in the enzyme, stabilizes the structure and its phosphate group may function in the acid-base reactions that promote orthophosphate attack on the polysaccharide substrate.[3] (Note: The EC number for phosphorylases is a general classification number, and the common name must indicate the natural polysaccharide substrate, (*e.g.*, maltodextrin phosphorylase, starch phosphorylase, and glycogen phosphorylase). *See also Starch Phosphorylase; Maltodextrin Phosphorylase; Phosphorylase a; Phosphorylase b; Sugar Nucleotide Phosphorylase; etc.*

[1] L. N. Johnson & D. Barford (1990) *JBC* **265**, 2409.
[2] N. D. Madsen (1986) *The Enzymes*, 3rd ed., **17**, 365.
[3] D. J. Graves & J. H. Wang (1972) *The Enzymes*, 3rd ed., **7**, 435.

Selected entries from *Methods in Enzymology* [vol, page(s)]:
General discussion: I, 192, 200, 215; **27**, 389; **28**, 960, 963; **34**, 131; **44**, 501, 887; **230**, 304
Assay: I, 192, 200, 215; **28**, 960, 964; biopsy material, in, **8**, 526; chicken muscle, **90**, 490
Other molecular properties: activity unit, II, 672; aggregates, small-angle X-ray scattering studies, **61**, 239, 240; allosteric control, **64**, 162; **249**, 554; alternative substrate, **64**, 327; analysis of bound and free pyridoxal 5-phosphate in, **5**, 682; II, 673; bacterial, I, 185; bicyclic cascade, **64**, 310, 325; cAMP derivatives, **38**, 399, 400; carboxymethylation of, **25**, 430; chicken muscle (assay, **90**, 490; isolation, **90**, 492); conformational change, **64**, 223; control studies, **37**, 293; conversion of *b* into *a*, **28**, 963; cyanate and, **25**, 584; debranching enzyme effect, **64**, 24; 459; end-labeled substrate, **64**, 255; fat cells, in, **31**, 68; histochemical method for, **4**, 388; hysteresis, **64**, 194, 197; "inactive," **5**, 377 (cAMP inhibitor assay, **38**, 279); isotope exchange, **64**, 9, 24; kinetic constant determination, **64**, 215; ligand binding sites, interaction distances, **249**, 561; liver, I, 215; **3**, 40, 137; noninvasive indicator of cytoplasmic calcium, as, **141**, 18; oxygen-labeled phosphates and, **4**, 913, 914; ^{31}P NMR studies, **107**, 75; potato, I, 62, 194; **4**, 504; **8**, 550; properties, **28**, 962; pyridoxal 5-phosphate in, II, 671; rabbit muscle, of, **27**, 110; rabbit muscle, hydroxylamine cleavage, from, **47**,

140; reductive alkylation, **47**, 474, 475; reduction with sodium borohydride, **5**, 682; II, 671 (procedure at high salt concentration, II, 674 [at low pH, II, 673; at pH 6.5, II, 674]); staining on gels, **22**, 598; starch phosphorylase, I, 192; **8**, 550; swine kidney, from, **42**, 375, 389 (assay, **42**, 389; molecular weight, **42**, 393; properties, **42**, 393, 394; purification, **42**, 390); transition-state and multisubstrate analogues, **249**, 304; yeast, I, 199 (isolation from bakers' yeast, **28**, 961; purification, I, 199)

PHOSPHORYLASE *a*

Phosphorylase *a* is the phosphorylated and more active form of glycogen phosphorylase produced from phosphorylase *b* by the action of phosphorylase kinase. *See also Phosphorylase; Phosphorylase b; Starch Phosphorylase*

Selected entries from *Methods in Enzymology* [vol, page(s)]:
General discussion: I, 202, 205; **8**, 526, 537
Other molecular properties: AMP assay, **55**, 209; branching enzyme and, **8**, 395; conversion to phosphorylase *b*, **5**, 377; **8**, 541, 546; dephosphophosphorylase kinase and, **5**, 389; formation of, **5**, 369, 373, 376; gel electrophoresis, **32**, 101; inorganic phosphate assay, **55**, 212; isolation from rabbit muscle, **9**, 14; isotope exchange, **64**, 9; **87**, 654; molecular weight, **53**, 350, 351; perfused muscle, in, **39**, 73; phosphorylase *b* and, **8**, 540; phosphorylase *b* kinase and, **8**, 543; phosphorylase phosphatase and, **8**, 546; preparation, **107**, 108; properties of isozymes, I, 205; **8**, 542; pyrophosphate assay, in, **87**, 527; radioiodination, **70**, 212; ^{35}S-labeled (preparation from phosphorylase *b*, **159**, 348; substrate for protein phosphatases, as, **159**, 352)

PHOSPHORYLASE *b*

Phosphorylase *b* is unphosphorylated and less active form of glycogen phosphorylase produced from phosphorylase *a* by the action of phosphorylase phosphatase. AMP activates this enzyme. *See also Phosphorylase; Phosphorylase a; Starch Phosphorylase*

Selected entries from *Methods in Enzymology* [vol, page(s)]:
General discussion: I, 202, 205; **5**, 369; **8**, 537; **34**, 127, 128; **46**, 241, 293, 316
Assay: **5**, 369; **8**, 538
Other molecular properties: activation, **46**, 293; activity of, **5**, 337; apophosphorylase *b* (preparation from phosphorylase *b*, II, 675; properties, II, 677); comparison of properties with sodium borohydride-reduced phosphorylase *b*, II, 674, 675; conversion to phosphorylase *a*, **8**, 540; crystallization, **5**, 370; dephosphophosphorylase kinase and, **5**, 389; isotope exchange, **64**, 9; isozymes, **8**, 537; mechanism, **63**, 51; molecular weight standard, **237**, 92, 115; muscle, protein phosphokinase and, **6**, 222; phosphorylase *b* kinase and, **8**, 543; phosphorylase phosphatase and, **8**, 546; preparation of apophosphorylase *b* by use of deforming agents, II, 675; properties of isozymes, I, 205; **5**, 371; **8**, 542; purification, I, 204; **5**, 369; **8**, 539; rabbit muscle, from, reductive alkylation, **47**, 474, 475; sedimentation coefficient, **61**, 108; site-specific reagent, **47**, 481, 484

PHOSPHORYLASE KINASE

This phosphotransferase [EC 2.7.1.38], also known as dephosphophosphorylase kinase, catalyzes the reaction of four molecules of ATP with two molecules of the dimeric phosphorylase *b* to produce four molecules of ADP and the tetrameric phosphorylase a.[1-4] No other known protein kinase catalyzes this reaction. The enzyme reportedly has a random Bi Bi kinetic mechanism.[5]

[1] E. G. Krebs (1986) *The Enzymes*, 3rd ed., **17**, 3.
[2] C. A. Picket-Gies & D. A. Walsh (1986) *The Enzymes*, 3rd ed., **17**, 395.
[3] G. M. Carlson, P. J. Bechtel & D. J. Graves (1979) *AE* **50**, 41.
[4] D. A. Walsh & E. G. Krebs (1973) *The Enzymes*, 3rd ed., **8**, 555.
[5] Y. J. Farrar & G. M. Carlson (1991) *Biochemistry* **30**, 10274.

Selected entries from *Methods in Enzymology* [vol, page(s)]:
General discussion: **5**, 373, 384; **8**, 543; **99**, 243, 250, 259, 268; **102**, 219
Assay: **5**, 373, 384; **8**, 543; bovine heart, **99**, 252; rabbit skeletal muscle, **8**, 543; **99**, 244
Other molecular properties: activation, **8**, 540, 545; adenosine 3',5'-phosphate and, **5**, 386; autophosphorylation, role in enzyme regulation, **139**, 690; bovine heart, **99**, 252; calmodulin-bound, quantitation, **102**, 220; calmodulin subunits, interactions, and, **102**, 223; inhibition, **63**, 481, 482; ligand binding sites, **249**, 560; perfused muscle, in, **39**, 73; phosphorylase *a*, production of, **5**, 369, 376; preparation, **107**, 109; properties and function, **5**, 390; **8**, 545; **107**, 89; protein kinase inhibitor, **38**, 350, 351; purification, **5**, 374, 379, 387, 388; **8**, 544; **99**, 246, 269; rabbit, **28**, 963, 967; rabbit skeletal muscle (properties, **5**, 375; **8**, 545; **99**, 248; purification, **5**, 374; **8**, 544; **99**, 246, 269); specificity analysis with peptide substrates, **99**, 268; stabilization, **276**, 136; γ subunit (HPLC-isolated, renaturation, **200**, 438; reactivated, isolation from calmodulin, **200**, 440); subunits, HPLC, **200**, 437; subunit dissociation, **99**, 262 (assays, **99**, 260; physicochemical characterization, **99**, 261; preparation of [active αγδ complex, **99**, 263; active γδ complex, **99**, 265]); thiophosphorylation of phosphorylase *b*, in, **159**, 348; troponin C interactions with, **102**, 225

PHOSPHORYLASE PHOSPHATASE

This enzyme [EC 3.1.3.17] catalyzes the hydrolysis of the tetrameric phosphorylase *a* to produce two molecules of dimeric phosphorylase *b* and four molecules of orthophosphate.[1-4] The enzyme also dephosphorylates a number of other phosphoproteins, including phosphorylated forms of phosphorylase kinase, histone H1, 3-hydroxy-3-methylglutaryl-CoA reductase, troponin I, myosin light chain, glycogen synthase, *etc*.

[1] L. M. Ballou & E. H. Fischer (1986) *The Enzymes*, 3rd ed., **17**, 311.
[2] N. D. Madsen (1986) *The Enzymes*, 3rd ed., **17**, 365.
[3] W. Merlevede & J. Di Salvo, eds. (1985, 1986) *Adv. Protein Phosphatases*, vols. **1**, **2**, and **3**.
[4] E. Y. Lee, S. R. Silberman, M. K. Ganapathi, S. Petrovic & H. Paris (1980) *Adv. Cyclic Nucleotide Res.* **13**, 95.

Selected entries from *Methods in Enzymology* [vol, page(s)]:
General discussion: **1**, 206; **5**, 377; **8**, 546; **107**, 102; **159**, 368, 377, 390, 427
Assay: **1**, 206; **5**, 377; **8**, 546; **159**, 379
Other molecular properties: activation, **64**, 339; AMP, inhibition, **64**, 339; alternative substrate effect, **64**, 336; cascade control, **64**, 299; characterization, **159**, 386; dog liver (properties, **5**, 383; purification, **5**, 379); glucose 1-phosphate, inhibition, **64**, 339; glycogen, activation, **64**, 339; glycogen phosphorylase and, **28**, 963, 967; magnesium ion effect, **64**, 339; phosphorylase *a*, *b*, and, **5**, 377; **8**, 541; phosphorylase kinase and, **5**, 376; properties, **5**, 383; **8**, 549; purification, **1**, 208; **5**, 379; **8**, 548 (rabbit muscle, from, **1**, 208; **159**, 381)

PHOSPHOSERINE AMINOTRANSFERASE

This pyridoxal-phosphate-dependent aminotransferase [EC 2.6.1.52] catalyzes the reversible reaction of *O*-phospho-L-serine with α-ketoglutarate (or, 2-oxoglutarate) to produce 3-phosphonooxypyruvate and L-glutamate.[1-3] The enzyme has a ping pong Bi Bi kinetic mechanism.[2]

[1] H. Hirsch & D. M. Greenberg (1967) *JBC* **242**, 2283.
[2] M.-J. Basurko, M. Marche, M. Darriet & A. Cassaigne (1989) *Biochem. Soc. Trans.* **17**, 787.
[3] A. E. Braunstein (1973) *The Enzymes*, 3rd ed., **9**, 379.

Selected entries from *Methods in Enzymology* [vol, page(s)]:
General discussion: **17B**, 331
Assay: **17B**, 331

PHOSPHOSERINE PHOSPHATASE

This enzyme [EC 3.1.3.3] catalyzes the hydrolysis of phosphoserine (both stereoisomers are substrates) to produce serine and orthophosphate.[1] Phosphoserine phosphatase has a branched-chain ping pong kinetic mechanism similar to that of glucose-6-phosphatase.[2] The reaction proceeds via the formation of an acylphosphate with an aspartyl residue.[3]

[1] W. L. Byrne (1961) *The Enzymes*, 2nd ed., **5**, 73.
[2] R. C. Nordlie (1982) *MIE* **87**, 319.
[3] J. F. Collet, V. Stroobant & E. Van Schaftingen (1999) *JBC* **274**, 33985.

Selected entries from *Methods in Enzymology* [vol, page(s)]:
General discussion: **6**, 215
Assay: **6**, 215
Other molecular properties: branched mechanism, **87**, 321, 322; isotope exchange, **64**, 9; properties, **6**, 217; purification, **6**, 216

3-PHOSPHOSHIKIMATE 1-CARBOXYVINYLTRANSFERASE

This transferase [EC 2.5.1.19], also known as 5-enolpyruvylshikimate-3-phosphate synthase (EPSP synthase) and 3-enolpyruvoylshikimate-5-phosphate synthase, catalyzes the reaction of phosphoenolpyruvate with 3-phosphoshikimate to produce orthophosphate and 5-*O*-(1-carboxyvinyl)-3-phosphoshikimate.[1-7] The enzyme has a random Bi Bi kinetic mechanism with substrate synergism.[1,2]

This enzyme also catalyzes the exchange of solvent protons with the C3 position of phosphoenolpyruvate. The enzyme appears to form a tetrahedral intermediate produced by the addition of the hydroxyl group of 3-phosphoshikimate to the double bond of phosphoenolpyruvate (note that the C2 carbon of phosphoenolpyruvate has been converted to sp^3 hybridization in this intermediate). A recent report[5] offers an alternative mechanism in which an enzyme-enolpyruvate intermediate followed by a enzyme-ketal intermediate are present in the mechanism: however, this alternative mechanism has been challenged.[6,7]

[1]K. J. Gruys, M. C. Walker & J. A. Sikorski (1992) *Biochemistry* **31**, 5534.
[2]K. J. Gruys, M. R. Marzabadi, P. D. Pansegrau & J. A. Sikorski (1993) *ABB* **304**, 345.
[3]Q. K. Huynh (1987) *ABB* **258**, 233.
[4]H. C. Steinrücken & N. Amrhein (1984) *EJB* **143**, 341 and 351.
[5]D. R. Studelska, L. M. McDowell, M. P. Espe, C. A. Klug & J. Schaefer (1997) *Biochemistry* **36**, 15555.
[6]D. L. Jakeman, D. J. Mitchell, W. A. Shuttleworth & J. N. Evans (1998) *Biochemistry* **37**, 12012.
[7]J. Lewis, K. A. Johnson & K. S. Anderson (1999) *Biochemistry* **38**, 7372.

Selected entries from ***Methods in Enzymology*** **[vol**, page(s)]:
General discussion: 142, 342, 348
Assay: 142, 343, 349
Other molecular properties: intermediates, detection, **249**, 57; properties, **142**, 346, 352; purification from (*Escherichia coli*, **142**, 344; *N. crassa*, **17A**, 393; pea, **142**, 350, 353); quench-flow experiments, quenching agent, **249**, 42; transition-state and multisubstrate analogues, **249**, 305

α,α-PHOSPHOTREHALASE

This enzyme [EC 3.2.1.93], also known as trehalose-6-phosphate hydrolase, catalyzes the hydrolysis of α,α-trehalose 6-phosphate to produce D-glucose and D-glucose 6-phosphate.[1,2] Unlike the *Escherichia coli* enzyme,[3] the *Bacillus subtilis* hydrolase[4] will also hydrolyze trehalose.[3]

[1]S. O. Salminen & J. G. Streeter (1986) *Plant Physiol.* **81**, 538.
[2]A. Bhumiratana, R. L. Anderson & R. N. Costilow (1974) *J. Bacteriol.* **119**, 484.
[3]M. Rimmele & W. Boos (1994) *J. Bacteriol.* **176**, 5654.
[4]S. Gotsche & M. K. Dahl (1995) *J. Bacteriol.* **177**, 2721.

PHOTINUS-LUCIFERIN 4-MONOOXYGENASE (ATP-HYDROLYZING)

This oxidoreductase [EC 1.13.12.7], also known as firefly luciferase and *Photinus pyralis* luciferase, catalyzes the reaction of *Photinus* luciferin with dioxygen and ATP to produce the oxidized *Photinus* luciferin, carbon dioxide, water, AMP, pyrophosphate, and light.[1–4]

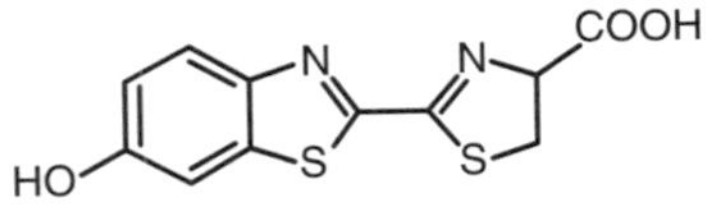

Photinus luciferin

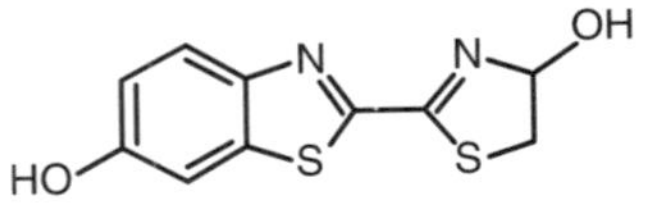

Photinus oxyluciferin

Photinus luciferin is (*S*)-4,5-dihydro-2-(6-hydroxy-2-benzothiazoloyl)-4-thiazolecarboxylate. The reaction proceeds via the initial formation of an acid anhydride between the carboxylic group and AMP, with the concomitant release of pyrophosphate: (1) E + ATP + D-luciferin → E(luciferyl-adenylate) + pyrophosphate. This is followed by: (2) E(luciferyl-adenylate) + O_2 → E(oxyluciferin* · AMP) + carbon dioxide; (3) E(oxyluciferin* · AMP) → E(oxyluciferin · AMP) + photon; and (4) E(oxyluciferin · AMP) → E + oxyluciferin + AMP.

[1]M. DeLuca (1976) *AE* **44**, 37.
[2]S. Rajgopal & M. A. Vijayalakshmi (1984) *Enzyme Microb. Technol.* **6**, 482.
[3]F. R. Leach (1981) *J. Appl. Biochem.* **3**, 473.
[4]L. J. Kricka (1988) *Anal. Biochem.* **175**, 14.

Selected entries from ***Methods in Enzymology*** **[vol**, page(s)]:
General discussion: 2, 851, 854; **6**, 445; **46**, 151, 287, 537; **57**, 3; **305**, 180, 188, 346
Assay: 2, 851; **6**, 445; **57**, 29, 76, 77; **217**, 636; **255**, 421
Other molecular properties: acetone powder, preparation, **57**, 6, 7; activity, **57**, 3, 37, 58, 74; aequorin fusion protein, **305**, 482, 494, 498; age, **56**, 537; **133**, 58; albumin, effect of, **133**, 57, 58; application in preimplantation development studies, **225**, 412; arsenate, effect of, **57**, 13; assay of (adenylate energy charge, **57**, 73; AMP, **55**, 209; ATP, **3**, 871; **55**, 539; **57**, 36; **133**, 3, 15, 36; **305**, 346, 356, 360, 364 [calibration, **305**, 354]; ATPase, **56**, 530; **57**, 50; ATP/phosphocreatine, **305**, 364; bacterial content in fluids, **57**, 65; biomass, **57**, 73; **305**, 356; calmodulin, **57**, 107; cAMP, **57**, 94; cGMP, **57**, 94; creatine kinase isoenzymes, **57**, 56; **305**, 366; cyclic nucleotide phosphodiesterases, **57**, 94; FMN, **57**, 215; GDP, **57**, 85; glycerol, **305**, 365; GMP, **57**, 85; GTP, **57**, 85; guanosine nucleotides, **57**, 85; integrated light-flux method, **57**, 77, 90; ligand binding, **57**, 113; malate, **57**, 187; malate dehydrogenase, **57**, 181; modulator protein, **57**, 107; myristic acid, **57**, 189; NAD^+, **133**, 237; NADH, **57**, 215; NADPH, **57**, 215; oxaloacetate, **57**, 187; oxidative phosphorylation, **57**, 41; **305**, 368; peak height method, **57**, 77, 89, 90; photophosphorylation, **57**, 50; **305**, 368; protease, **57**, 200; protein-binding reactions, **57**, 113; promoter/enhancer activity in murine ova, in, **225**, 419; pyrophosphate, **305**, 367; urea, **305**, 366); bidirectional vector construction, **273**, 320; buffers, **133**, 56, 64; cDNA (construction, **133**, 5; expression in *Escherichia coli*, **133**, 10, 11; screening, **133**, 8); characterization of 2′ P-ADP, **66**, 112, 119; cloning, **133**, 3; coimmobilized with other enzymes, bioluminescent assays with, **136**, 82; commercial reagents (applications, **133**, 62, 68; development, **133**, 51; evaluation, **133**, 53; suppliers, **133**, 53, 65); commercial sources, **57**, 37; **133**, 52; continuous-flow bioluminescent assay for NAD^+, in, **133**, 237; continuous measurement of ATPase, **56**, 530, 531 (materials and instrumentation, **56**, 531; precautions, **56**, 541; qualitative aspects, **56**, 533; quantitation, **56**, 534; rapid kinetics, **56**, 540, 541; rate of change of ATP concentration, **56**, 537); crude extract preparation, **57**, 6; crystalline (preparation, **6**, 447; **57**, 8; stability, **6**, 448; **57**, 11); effect on light production by α-bungarotoxin-luciferin conjugates, **133**, 45; enzyme classification, **57**, 126; expression in plant cells (assay, **216**, 406; influencing factors, **216**, 411); extinction coefficient, **57**, 11; half-life, alteration, **216**, 414; immobilized, **133**, 248; inhibitors, **2**, 856; **57**, 14, 70; integrated light-flux method, **57**, 77, 90; kinetics, **57**, 11; **305**, 350; light reaction, kinetics, **57**, 11; luciferase-luciferin (choice of concentrations, **56**, 541, 543; sources, **56**, 531); luciferase-luciferin enzyme preparations (commercial, sensitivity comparison, **57**, 554, 555, 559; endogenous light, decay pattern, **57**, 552); luciferase reagent, preparation, **57**, 55, 76, 88; luciferyl adenylate hydrolase, luciferase and, **6**, 448; metal ions in, **133**, 57; molecular weight, **57**, 11; mRNA transcript, analysis, **216**, 412; oxy form, spectrum, **52**, 36; pH optimum,

2, 856; **57**, 15; **133**, 56; properties, **2**, 856; **6**, 448; purification, **2**, 854; **6**, 446; **57**, 3, 28, 30, 51, 61, 62, 82; **305**, 180; reaction catalyzed, **2**, 851; **6**, 445; **56**, 531; **57**, 3; **133**, 3; reaction kinetics, **57**, 11; **305**, 350; reagents, **305**, 349, 369; reporter gene (assay in mammalian cells, **216**, 386; **305**, 559, 571; multiple, application, **216**, 412; sensitivity, **273**, 324; transient expression analysis in plants, in, **216**, 397); salt concentration, effect of, **57**, 14, 15; specificity, **2**, 856; **44**, 583; **57**, 14, 50, 82; **133**, 60; stability, **57**, 6, 7, 13; **255**, 421; studies, in, of, (ATP synthesis in mitochondria, **57**, 39; protein-ligand binding, **57**, 113; rapid kinetics of ATP synthesis, **57**, 45); substrate specificity, **2**, 856; **44**, 583; **57**, 14, 50, 82; **133**, 60; targeting to cellular components, **216**, 413

PHOTOSYNTHETIC REACTION CENTER

The photosynthetic reaction center is a transmembrane protein complex that is responsible for photosynthesis in bacteria. Also associated with this complex are the bacteriochlorophylls, bacteriopheophytins, ubiquinone, cytochromes, and menaquinone.[1-3] **See also** *Photosystem II*

[1] J. Deisenhofer & J. R. Norris, eds. (1993) *The Photosynthetic Reaction Center*, vols. I & II, Academic Press, Orlando.
[2] J. B. Howard & D. C. Rees (1991) *Adv. Protein Chem.* **42**, 199.
[3] D. C. Rees & D. Farrelly (1990) *The Enzymes*, 3rd ed., **19**, 37.

Selected entries from *Methods in Enzymology* [vol, page(s)]:
General discussion: 23, 688, 696; **297**, 57, 310
Other molecular properties: absorption bands, assignment, **246**, 225; charge transfer band, **246**, 159, 164; components (algae, **246**, 238; photosystem II, **246**, 225, 650; purple bacteria, **246**, 222); dimerization of primary donor, **246**, 224, 226, 244; electron paramagnetic resonance, **246**, 533; electron–phonon coupling, **246**, 244; electron transfer (mechanism, **246**, 244; rate, **246**, 164, 223, 225); far-infrared spectroscopy, **246**, 166; fluorescence spectroscopy, **246**, 6, 164; hole burning spectroscopy, **246**, 244; low-temperature spectroscopy, **246**, 164; photoinduction during spectroscopy, **246**, 750; *Rhodopseudomonas*, structure, **239**, 516; time-resolved resonance Raman spectroscopy, **246**, 500; transient absorption spectroscopy, **246**, 222; X-ray absorption spectroscopy, **246**, 650, 670

PHOTOSYSTEM I

This transmembrane, macromolecular complex allows plants to harvest solar photic energy as a means for driving the photooxidation of plastocyanin, a copper-containing thylakoid membrane protein that undergoes electron transfer reactions.[1-5] There are at least ten subunits of photosystem I in the chloroplasts of the higher plants, green algae, and cyanobacteria. The primary reaction center of photosystem I is P700. Also present are chlorophyll a_0, PsaA and PsaB, the phylloquinone A_1, and a number of iron-sulfur proteins. **See also** *Photosystem II*

[1] C. Tommos & G. T. Babcock (1998) *Acc. Chem. Res.* **31**, 18.
[2] J. E. Penner-Hahn (1998) *CBC* **3**, 439.
[3] J. Barber, ed. (1992) *The Photosystems: Structure, Function & Molecular Biology*, Elsevier, New York.
[4] L. Bogorad & I. K. Vasil (eds.) (1991) *The Photosynthetic Apparatus*, Academic Press, Orlando.
[5] A. San Pietro, ed. (1971) *MIE* vol. **23**, Academic Press, New York.

Selected entries from *Methods in Enzymology* [vol, page(s)]:
General discussion: 6, 308; **297**, 18, 95, 124
Other molecular properties: acceptors of electrons, **69**, 415; chains and equilibria, **24**, 236; chlorophyll-protein complexes, isolation (from barley chlorina $f2$ mutant, **118**, 348; pea, from, **118**, 342); cyclic electron flow, DBMIB, **69**, 689, 690; dichlorophenolindophenol, **69**, 418; donor system, **69**, 425; DBMIB, **69**, 687; electron donors, reaction type IV, **24**, 160; electron transport inhibition, **24**, 361; ferricyanide, **69**, 417, 418; fluorescence lifetimes, **54**, 25; glutaraldehyde-treated chloroplasts, **69**, 620, 621; inhibitors, of acceptor site, **69**, 676; iron-sulfur proteins, associated, EPR analysis in cyanobacteria, **167**, 432; kinetics, general principles, **24**, 230; laser exciting, **24**, 35; modulated electrode action spectrum, **24**, 134; net charge, electron transfer, **69**, 219; particles, plastoquinones, **69**, 378; photophosphorylation, **69**, 648 (assay for ATP formation, **69**, 655; reaction conditions, **69**, 654, 655); photoreductant, **24**, 221; photoreduction, **69**, 245, 246 (DBMIB, **69**, 689); pools, general consideration, **24**, 224; pool size, **24**, 226, 227; proteoliposomes, **55**, 760, 761, 768; reaction center complex (individual subunits [antibody preparation, **118**, 358; isolation, **118**, 358; quantitative estimation, **118**, 360; structure and function, **118**, 357]; preparation from *Synechococcus*, **167**, 264; properties, **118**, 357; purification, **118**, 354; **126**, 286; reconstituted in liposomes [preparation, **126**, 288; photophosphorylation in, **126**, 290; proton translocation, assays, **126**, 288]; synthesis during development, analytical techniques, **118**, 367); reaction systems, for noncyclic electron transport, requirements, **69**, 425; separation by digitonin method of subchloroplast fragment isolation, **23**, 268; studies on aldehyde-fixed chloroplasts, **23**, 248; xanthosome vesicles (isolation from *Ectocarpus siliculosus*, **214**, 405; properties, **214**, 410)

PHOTOSYSTEM II

Photosystem II functions with photosystem I in the photosynthetic process occurring in the thylakoid membranes of higher plants, green algae, and cyanobacteria. Components of the multiprotein photosystem II include the proteins D_1 and D_2, the pheophytin acceptor, plastoquinone acceptors Q_a and Q_b, chlorophyll a molecules, and cytochrome b_{559}. **See also** *Photosystem I*

Selected entries from *Methods in Enzymology* [vol, page(s)]:
General discussion: 258, 303; **297**, 293, 320
Assay: 23, 279
Other molecular properties: acceptors of electrons, **69**, 419; assembly and function, role of thylakoid polypeptide phosphorylation and turnover, **97**, 554; calcium effects, analysis in cyanobacteria, **167**, 280; chains and equilibria, **24**, 236; chemiosmotic coupling, **69**, 433; chlorophyll-protein complexes, isolation from (barley chlorina $f2$ mutant, **118**, 348; pea, **118**, 342; spinach, **118**, 351); chloroplast membrane conformational change, **69**, 507; donors of electrons, **69**, 429; EF particles, **69**, 537; electron acceptors, **69**, 414; electron donors, **24**, 163; electron flow, **69**, 349, 350 (cyclic, DBMIB, **69**, 689); electron transfer, **258**, 304; electron transport inhibition, **24**, 359; glutaraldehyde-treated chloroplasts, **69**, 618; Hill reaction, **24**, 159; isotopic labeling of tyrosine in cyanobacteria, **258**, 307; kinetics, general principles, **24**, 230; modulated electrode action spectrum, **24**, 134; oxygen-evolving preparations (isolation from [*Phormidium laminosum*, **167**, 273; *Synechococcus*, **167**, 267]; properties, **167**, 276); particles, plastoquinones, **69**, 378, 379; photophosphorylation (assay, **69**, 580; noncyclic, **69**, 580; preparation of KCN/Hg and KCN/Hg/NH$_2$OH inhibited chloroplasts, **69**, 577; properties, **69**, 582); photoreductant, **24**, 221; photoreductions, DBMIB, **69**, 687; phycobilisome complexes, preparation from (*Porphyridium*, **167**, 286; *Synechococcus*, **167**, 289); pools, general consideration, **24**, 224; pool size, **24**, 226, 229;

purification from cyanobacteria (anion-exchange chromatography, **258**, 309; thylakoid membrane preparation, **258**, 308); reaction center complexes, preparation from *Synechococcus*, **167**, 266; reaction center mutations, construction in *Synechocystis* 6803, **167**, 766; separation by digitonin method of subchloroplast fragment isolation, **23**, 268; studies on aldehyde-fixed chloroplasts, **23**, 248; subunit structure, **258**, 304; tyrosyl radicals (D site, **258**, 304; electron paramagnetic resonance [apparatus, **258**, 312; D°, **258**, 312; deuteration effects, **258**, 314, 316; M⁺, **258**, 306, 315; Z°, **258**, 312]; Fourier transform infrared spectroscopy, **258**, 316; M⁺ site, **258**, 305, 319; redox kinetics, **258**, 305; Z site, **258**, 304)

PHTHALATE 4,5-*cis*-DIHYDRODIOL DEHYDROGENASE

This oxidoreductase [EC 1.3.1.64] catalyzes the reaction of *cis*-4,5-dihydroxycyclohexa-1(6),2-diene-1,2-dicarboxylate with NAD⁺ to produce 4,5-dihydroxyphthalate, NADH, and a proton.[1]

[1] C. J. Batie, E. LaHaie & D. P. Ballou (1987) *JBC* **262**, 1510.

PHTHALATE 4,5-DIOXYGENASE

This oxidoreductase [EC 1.14.12.7], also known as phthalate dioxygenase (PDO), catalyzes the reaction of phthalate with NADH and dioxygen to produce *cis*-4,5-dihydroxycyclohexa-1(6),2-diene-1,2-dicarboxylate and NAD⁺.[1–5]

phthalate   *cis*-4,5-dihydroxycyclohexa-1,6-diene-1,2-dicarboxylate

A Rieske-type [2Fe-2S] cluster acts as the electron-transferring cofactor, and a mononuclear iron site, chelated to two histidyl residues, is the putative site of substrate oxygenation. A reductase, which contains FMN and a plant-type [2Fe-2S] ferredoxin domain, transfers electrons from NADH to the Rieske center.[3] *See also Benzene 1,2-Dioxygenase*

[1] C. J. Batie, E. LaHaie & D. P. Ballou (1987) *JBC* **262**, 1510.
[2] R. J. Gurbiel, C. J. Batie, M. Sivaraja, A. E. True, J. A. Fee, B. M. Hoffman & D. P. Ballou (1989) *Biochemistry* **28**, 4861.
[3] E. D. Coulter, N. Moon, C. J. Batie, W. R. Dunham & D. P. Ballou (1999) *Biochemistry* **38**, 11062.
[4] B. A. Palfey & V. Massey (1998) *CBC* **3**, 83.
[5] B. G. Fox (1998) *CBC* **3**, 261.

Selected entries from *Methods in Enzymology* [vol, page(s)]:
General discussion: 188, 61
Assay: 188, 62

PHTHALATE DIOXYGENASE REDUCTASE

The phthalate dioxygenase system catalyzes the reaction of phthalate with dioxygen and NADH to produce

cis-4,5-dihydroxycyclohexa-1(6),2-diene-1,2-dicarboxylate and NAD⁺. The FMN-, [2Fe-2S]-, and ferredoxin-dependent reductase participates in the transfer of electrons from NADH to the dioxygenase. Interestingly, the rate of electron transfer from the charge transfer complex of NAD⁺ and FMNH₂ and the oxidized iron-sulfur cluster is relatively slow. It has been suggested that release of NAD⁺ would increase the rate of transfer.[1,2] *See Phthalate 4,5-Dioxygenase*

[1] B. G. Fox (1998) *CBC* **3**, 261.
[2] B. A. Palfey & V. Massey (1998) *CBC* **3**, 83.

Selected entries from *Methods in Enzymology* [vol, page(s)]:
Assay: 188, 62

PHTHALYL AMIDASE

This enzyme [EC 3.5.1.79], also known as *O*-phthalyl amidase, catalyzes the hydrolysis of a phthalylamide (that is, a 2-carboxybenzoylamide; for example, 4-(2′-carboxy-*N*-benzoyl)amino-2-carboxynitrobenzene) to produce phthalate and a substituted amine.[1,2]

[1] T. D. Black, B. S. Briggs, R. Evans, W. L. Muth, S. Vangala & M. J. Zmijewski (1996) *Biotech. Lett.* **18**, 875.
[2] B. S. Briggs, A. J. Kreuzman, C. Whitesitt, W.-K. Yeh & M. Zmijewski (1996) *J. Mol. Cat. B: Enzym.* **2**, 53.

PHYLLOQUINONE MONOOXYGENASE (2,3-EPOXIDIZING)

This oxidoreductase [EC 1.14.99.20], also known as phylloquinone epoxidase, catalyzes the reaction of phylloquinone (vitamin K) with dioxygen and an electron donor substrate (AH₂) to produce 2,3-epoxyphylloquinone (vitamin K 2,3-oxide), water, and the oxidized electron donor (A).

vitamin K₁   vitamin K₁ 2,3-oxide

While NADH can serve as the donor substrate, it is not required if hydroquinone is the substrate.[1–3]

[1] A. K. Willingham & J. T. Matschiner (1974) *BJ* **140**, 435.
[2] J. W. Suttie, L. O. Geweke, S. L. Martin & A. K. Willingham (1980) *FEBS Lett.* **109**, 267.
[3] J. J. McTigue & J. W. Suttie (1986) *FEBS Lett.* **200**, 71.

PHYSAROPEPSIN

This aspartic endopeptidase [EC 3.4.23.27], also known as *Physarum* aspartic proteinase, catalyzes peptide-bond hydrolysis in proteins and has a very acidic pH optimum (about 1.7). The oxidized insulin B chain is hydrolyzed at Gly8–Ser9 rapidly, followed by Leu11–Val12, Cya19–Gly20, Gly20–Glu21, and Phe24–Phe25. A Pepstatin A has no appreciable inhibitory effect on physaropepsin.[1]

[1] K. Murakami-Murofushi, T. Takahashi, Y. Minowa, S. Iino, T. Takeuchi, H. Kitagaki-Ogawa, H. Murofushi & K. Takahashi (1990) *JBC* **265**, 19898.

PHYSARUM POLYCEPHALUM RIBONUCLEASE

This slime mold ribonuclease [EC 3.1.26.1] catalyzes the endonucleolytic cleavage of RNA to 5′-phospho-oligonucleotides (5′-phosphomonoesters).[1] Similar enzymes are pig liver nuclease, HeLa cell ribonuclease, *Escherichia coli* ribonuclease, and bovine adrenal cortex ribonuclease. The endoribonuclease PhyM from *Physarum polycephalum* cleaves almost exclusively at Up-N and Ap-N.[2]

[1] M. Hiramaru, T. Uchida & F. Egami (1969) *J. Biochem.* **65**, 701.
[2] H. Donis-Keller (1980) *NAR* **8**, 3133.

PHYTANOYL-CoA DIOXYGENASE

This Fe^{2+}- and ascorbate-dependent oxidoreductase [EC 1.14.11.18], also known as phytanoyl-CoA hydroxylase, catalyzes the reaction of phytanoyl-CoA (or, 3,7,11,15-tetramethylhexadecanoyl-CoA) with α-ketoglutarate (or, 2-oxoglutarate) and dioxygen to produce 2-hydroxy-phytanoyl-CoA, succinate, and carbon dioxide.[1–3]

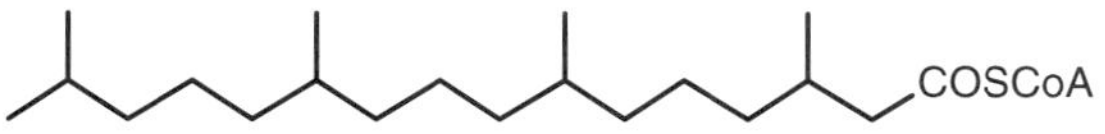

phytanoyl-CoA

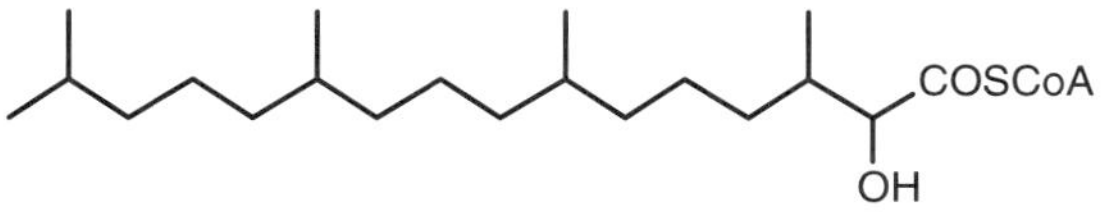

2-hydroxyphytanoyl-CoA

The enzyme participates in the α-oxidation pathway of the peroxisome.

[1] G. A. Jansen, S. J. Mihalik, P. A. Watkins, C. Jakobs, H. W. Moser & R. J. A. Wanders (1998) *Clin. Chim. Acta* **271**, 203.
[2] G. A. Jansen, S. J. Mihalik, P. A. Watkins, H. W. Moser, C. Jakobs, S. Denis & R. J. A. Wanders (1996) *BBRC* **229**, 205.
[3] S. J. Mihalik, A. M. Rainville & P. A. Watkins (1995) *EJB* **232**, 545.

PHYTANOYL-CoA SYNTHETASE

This ligase [EC 6.2.1.24], also known as phytanate:CoA ligase, catalyzes the reaction of ATP with phytanate and coenzyme A to produce AMP, pyrophosphate, and phytanoyl-CoA.

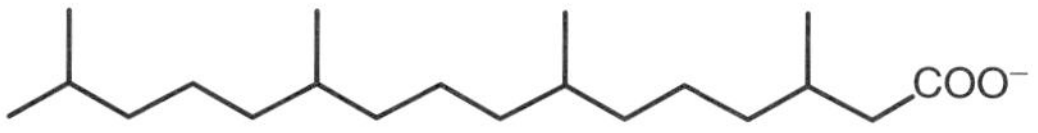

phytanate

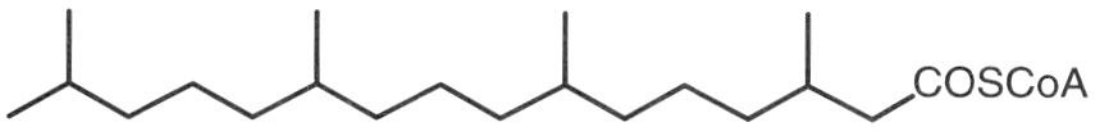

phytanoyl-CoA

The enzyme is not identical with acyl-[acyl-carrier protein] synthetase [EC 6.2.1.20]. It requires magnesium ions for activity.[1]

[1] F. N. Muralidharan & V. B. Muralidharan (1986) *Biochem. Int.* **13**, 123.

3-PHYTASE

This enzyme [EC 3.1.3.8], also known as phytate 3-phosphatase and *myo*-inositol-hexaphosphate 3-phosphohydrolase, catalyzes the hydrolysis of *myo*-inositol hexakisphosphate to produce D-*myo*-inositol 1,2,4,5,6-pentakisphosphate (or, 1L-*myo*-inositol 1,2,3,4,5-pentakisphosphate) and orthophosphate.[1]

[1] D. J. Cosgrove (1969) *Ann. N.Y. Acad. Sci.* **165**, 677.

6-PHYTASE

This enzyme [EC 3.1.3.26], also known as phytate 6-phosphatase and *myo*-inositol-hexaphosphate 6-phosphohydrolase, catalyzes the hydrolysis of *myo*-inositol hexakisphosphate to produce 1-*myo*-inositol 1,2,3,4,5-pentakisphosphate and orthophosphate.[1,2] Product release is the rate-limiting step in the reaction catalyzed by the *Aspergillus niger* enzyme.[1]

[1] A. Tomschy, M. Wyss, D. Kostrewa, K. Vogel, M. Tessier, S. Hofer, H. Burgin, A. Kronenberger, R. Remy, A. P. van Loon & L. Pasamontes (2000) *FEBS Lett.* **472**, 169.
[2] J. Dvorakova (1998) *Folia Microbiol. (Praha)* **43**, 323.

PHYTEPSIN

This aspartic endopeptidase [EC 3.4.23.40], also known as barley aspartic proteinase (it was first isolated from barley seeds, *Hordeum vulgare*), catalyzes peptide-bond hydrolysis in proteins. It prefers to cleave between two hydrophobic aminoacyl residues; however, it also acts on Phe–Asp and Asp–Asp bonds in 2S albumin from plant seeds. The oxidized insulin B chain is cleaved at Leu11–Val12, Ala14–Leu15, Leu15–Tyr16, Phe24–Phe25, and Phe25–Tyr26. In glucagon, hydrolysis is at Asp9–Tyr10, Tyr13–Leu14, Leu14–Asp15, Ser16–Arg17,

Arg18–Ala19, Phe22–Val23, Trp25–Leu26, and Leu26–Met27. Phytepsin possibly has a role in the hydrolysis of storage proteins in germinating grains and in protein processing.[1]

[1]J. Kervinen, K. Törmäkangas, P. Runeberg-Roos, K. Guruprasad, T. Blundell & T. H. Teeri (1995) in *Aspartic Proteinases: Structure, Function, Biology, and Biomedical Implications* (K. Takahashi, ed.), Plenum Press, New York, p. 241.

PHYTOENE SYNTHASE COMPLEX

Phytoene synthase complex [EC 2.5.1.32] (also known as geranylgeranyl-diphosphate geranylgeranyltransferase and prephytoene-diphosphate synthase) catalyzes the reaction of two molecules of geranylgeranyl diphosphate to produce pyrophosphate and prephytoene diphosphate.[1–4]

geranylgeranyl diphosphate

prephytoene diphosphate

cis-phytoene

Prephytoene diphosphate is then converted to phytoene and pyrophosphate.

In higher plants, this enzyme is found in a complex with isopentenyl-diphosphate isomerase and geranylgeranyl-diphosphate synthase. Phytoene synthase is similar to squalene synthase; however, there is an absolute requirement for Mn^{2+} and NADPH is not utilized to form a dihydro derivative of phytoene. The reaction mechanism has been proposed to include rearrangement and, in the conversion of prephytoene diphosphate to the product, the H_S proton on the cyclopropane ring is removed. In many organisms the product formed is *cis*-phytoene. Nevertheless, some variants of this synthase have been reported that produce the *trans*-isomer.

[1]D. E. Gregonis & H. C. Rilling (1974) *Biochemistry* **13**, 1538.
[2]O. Dogbo, A. Laferriere, A. D'Harlingue & B. Camara (1988) *PNAS* **85**, 7054.
[3]R. A. Gibbs (1998) *CCBC* **1**, 31.

[4]J. Chappel (1995) *Ann. Rev. Plant Physiol. Plant Mol. Biol.* **46**, 521.

Selected entries from *Methods in Enzymology* [vol, page(s)]:
General discussion: 110, 209; **214**, 352
Other molecular properties: *Capsicum* complex, component enzymes (biogenesis, **214**, 352, 364; immunochemistry, **214**, 359; immunology, **214**, 352; separation, **214**, 353); preparation from tomato, **15**, 458; properties, **15**, 459; purification steps, **15**, 459; putative, CrtE protein, sequence analysis, **214**, 304; tomato fruit chromoplasts (extraction and partial purification, **110**, 216; properties, **110**, 218)

PHYTOGLYCOGEN SYNTHASE

This enzyme, distinct from starch synthase, catalyzes the incorporation of D-glucose from ADP-D-glucose into a phytoglycogen polymer. Phytoglycogen is a highly branched glucan found in fungi, corn, maize, and rice. It is distinct from amylopectin. **See also** *Glycogen Synthase; Starch Synthase; Starch (Bacterial Glycogen) Synthase*

Selected entries from *Methods in Enzymology* [vol, page(s)]:
General discussion: 8, 391
Assay: 8, 391

PICORNAIN 2A

This cysteine endopeptidase [EC 3.4.22.29], also known as picornavirus endopeptidase 2A, poliovirus protease 2A, and rhinovirus protease 2A, catalyzes peptide-bond hydrolysis in the processing of a number of virus polyproteins. The rhinovirus and enterovirus 2A picornains have a preference for a glycyl residue at the P1′ position, and an isoleucyl, leucyl, or valyl residue at the P4 position. The P1 position specificity is somewhat broad: Tyr, Thr, Ala, Phe, and Val have all been observed to be acceptable. The enzymes from human rhinovirus and from coxsackievirus B4 will also catalyze cleavage of the eukaryotic initiation factor eIF-4G at Arg486–Gly487.[1–3]

[1]T. Skern, W. Sommergruber, H. Auer, P. Volkmann, M. Zorn, H.-D. Liebig, F. Fessl, D. Blass & E. Kuechler (1991) *Virology* **181**, 46.
[2]W. Sommergruber, H. Ahorn, A. Zöphel, I. Maurer-Fogy, F. Fessl, G. Schnorrenberg, H.-D. Liebig, D. Blass, E. Kuechler & T. Skern (1992) *JBC* **267**, 22639.
[3]W. Sommergruber, H. Ahorn, H. Klump, I. Seipelt, A. Zoephel, F. Fessl, C. Krystek, D. Blass, E. Kuechler, H.-D. Liebig & T. Skern (1994) *Virology* **198**, 741.

Selected entries from *Methods in Enzymology* [vol, page(s)]:
General discussion: 244, 583
Other molecular properties: assay substrates, **244**, 587; encephalomyelitis virus, **248**, 118, 120; protease activity, **244**, 583; purification of recombinant protein (cell lysis, **244**, 593; expression, **244**, 592; gel filtration, **244**, 593; Mono Q chromatography, **244**, 593)

PICORNAIN 3C

This cysteine endopeptidase [EC 3.4.22.28], also known as picornavirus endopeptidase 3C, poliovirus protease 3C, rhinovirus protease 3C, and foot-and-mouth protease 3C, catalyzes peptide-bond hydrolysis at specific sites in viral polyproteins. The poliovirus picornain 3C cleaves exclusively at eight Gln–Gly sites (out of a total of thirteen Gln–Gly pairs), whereas the encephalomyocarditis virus enzyme acts at Gln–Gly, Glu–Ser, and Glu–Ala. Picornain 3C from foot-and-mouth disease virus cleaves at Gln–Leu and Gln–Ile. In such enteroviruses as poliovirus, coxsackievirus, and human rhinoviruses, the enzyme prefers a prolyl residue at the P2 position.[1–3] *See also Hepatitis A Virus Picornain 3C*

[1] B. Dunn & J. Kay (1990) *BBA* **1048**, 1.
[2] R. W. Carrell & A. M. Lesk (1994) *Nature* **334**, 528.
[3] M. Allaire & M. N. G. James (1994) *Nature Struct. Biol.* **1**, 505.

Selected entries from *Methods in Enzymology* [vol, page(s)]:
General discussion: 244, 583

PIMELOYL-CoA DEHYDROGENASE

This oxidoreductase [EC 1.3.1.62], which participates in the bacterial degradation of benzoate, catalyzes the reaction of pimeloyl-CoA with NAD^+ to produce 6-carboxyhex-2-enoyl-CoA, NADH, and a proton.[1]

[1] C. Gallus & B. Schink (1994) *Microbiology* **140**, 409.

α-PINENE-OXIDE DECYCLASE

This enzyme [EC 5.5.1.10], also known as α-pinene oxide lyase, catalyzes the conversion of α-pinene oxide to (*Z*)-2-methyl-5-isopropylhexa-2,5-dienal ((CH$_3$)$_2$CHC(=CH$_2$)-CH$_2$CH=C(CH$_3$)CHO).

α-pinene oxide

α-Pinene oxide is an intermediate in the degradation of α-pinene by *Nocardia* sp. and some *Pseudomonas* strains.[1] Both rings of pinene are cleaved in the reaction, and decyclization is likely to be initiated by protonation of the epoxide oxygen.[1]

[1] E. T. Griffiths, P. C. Harries, R. Jeffcoat & P. W. Trudgill (1987) *J. Bacteriol.* **169**, 4980.

PINENE SYNTHASE

This enzyme [EC 4.2.3.14; previously classified as EC 4.1.99.8], also known as β-geraniolene synthase, (−)-(1*S*,5*S*)-pinene synthase, and monoterpene cyclase, catalyzes the conversion of geranyl diphosphate to pinene (actually, a mixture of α- and β-pinene) and pyrophosphate.[1,2]

OPO$_3$PO$_3$H^{2-}

geranyl diphosphate

(−)-α-pinene   (−)-β-pinene   (−)-limonene   β-myrcene   (−)-camphene

Pinene synthase II from sage produces (−)-α-pinene, (−)-β-pinene, and minor amounts of (−)-limonene, myrcene, and (−)-camphene. Isotope effect studies demonstrated that all the products were synthesized by the same enzyme and not by a mixture of hard-to-separate proteins. (+)-α-Pinene is synthesized via pinene synthase I.

[1] J. Bohlmann, C. L. Steele & R. Croteau (1997) *JBC* **272**, 21784.
[2] K. C. Wagschal, H. J. Pyun, R. M. Coates & R. Croteau (1994) *ABB* **308**, 477.

Selected entries from *Methods in Enzymology* [vol, page(s)]:
Assay: 110, 407
Other molecular properties: *Citris limonum* (assay, 110, 407; product identification, 110, 409; properties, 110, 415; purification, 110, 413; radioactive substrates, synthesis, 110, 410)

D-PINITOL DEHYDROGENASE

This oxidoreductase [EC 1.1.1.142] catalyzes the reaction of 5D-5-*O*-methyl-*chiro*-inositol (or, D-pinitol) with $NADP^+$ to produce 5D-5-*O*-methyl-2,3,5/4,6-pentahydroxycyclohexanone and NADPH (stereospecifically to and from the A-side) and a proton.[1]

[1] H. Ruis & O. Hoffmann-Osterhof (1969) *EJB* **7**, 442.

PINOSYLVIN SYNTHASE

This acyltransferase [EC 2.3.1.146], also known as stilbene synthase, catalyzes the reaction of three molecules of malonyl-CoA with cinnamoyl-CoA to produce four molecules of coenzyme A, three molecules of carbon dioxide, and pinosylvin (that is, 3,5-dihydroxystilbene).[1] The enzyme, first isolated from Scotch pine (*Pinus sylvestris*), is not identical with naringenin-chalcone synthase [EC 2.3.1.74] or trihydroxystilbene synthase [EC 2.3.1.95].

[1] A. Schoeppner & H. Kindl (1979) *FEBS Lett.* **108**, 349.

L-PIPECOLATE DEHYDROGENASE

This oxidoreductase [EC 1.5.99.3] catalyzes the reaction of L-pipecolate with an acceptor substrate to produce 2,3,4,5-tetrahydropyridine-2-carboxylate and the reduced acceptor (2,6-dichlorophenolindophenol can act as the acceptor).[1] The product then reacts with water to gerate 2-aminoadipate 6-semialdehyde (or, 2-amino-6-oxohexanoate). *See also* L-Pipecolate Oxidase

[1]M. L. Baginsky & V. W. Rodwell (1967) J. Bacteriol. **94**, 1034.
Selected entries from *Methods in Enzymology* [vol, page(s)]:
General discussion: 17B, 183, 188
Assay: 17B, 183

L-PIPECOLATE OXIDASE

This oxidoreductase [EC 1.5.3.7] catalyzes the reaction of L-pipecolate with dioxygen to produce piperidine-6-carboxylate (or, 2,3,4,5-tetrahydropyridine-2-carboxylate) and hydrogen peroxide.[1-3] The pyridine product then reacts with water to form 2-aminoadipate 6-semialdehyde (or, 2-amino-6-oxohexanoate). *See also* L-Pipecolate Dehydrogenase

[1]V. V. Rao & Y.-F. Chang (1990) BBA **1038**, 295.
[2]R. J. A. Wanders, G. J. Romeyn, R. B. H. Schutgens & J. M. Tager (1989) BBRC **164**, 550.
[3]S. J. Mihalik & W. J. Rhead (1989) JBC **264**, 2509.
Selected entries from *Methods in Enzymology* [vol, page(s)]:
General discussion: 52, 18

Δ1-PIPERIDEINE-2-CARBOXYLATE REDUCTASE

This oxidoreductase [EC 1.5.1.21] catalyzes the reaction of Δ1-piperideine-2-carboxylate with NADPH to produce L-pipecolate and NADP+.[1] *See also* L-Pipecolate Dehydrogenase; 1,2-Didehydropipecolate Reductase

[1]C. W. Payton & Y.-F. Chang (1982) J. Bacteriol. **149**, 864.
Selected entries from *Methods in Enzymology* [vol, page(s)]:
General discussion: 5, 878
Assay: 5, 878
Other molecular properties: distribution, 5, 879, 882; properties, 5, 881; purification, 5, 879

PIPERIDINE N-PIPEROYLTRANSFERASE

This acyltransferase [EC 2.3.1.145] catalyzes the reaction of (E,E)-piperoyl-CoA with piperidine to produce coenzyme A and N-[(E,E)-piperoyl]-piperidine.[1] Alternative acceptor substrates include pyrrolidine and 3-pyrroline.

[1]J. G. Geisler & G. G. Gross (1990) Phytochemistry **29**, 489.

PITRILYSIN

This zinc-dependent metalloendopeptidase [EC 3.4.24.55], also known as protease III and protease Pi (since it was a periplasmic activity from *Escherichia coli*), catalyzes peptide-bond hydrolysis in proteins. This enzyme preferentially degrades β-galactosidase fragments that have masses of less than 7 kDa, it hydrolyzes the oxidized insulin B chain at Tyr16–Leu17 (note that pitrilysin is the main insulin-degrading activity of *E. coli*), and it cleaves pig vasoactive intestinal fragment (YTRLRKQ MAVKKYLNSILN) at Leu–Arg, Arg–Lys, Leu–Asn, and Ser–Ile. The smallest known substrate is substance P, the undecapeptide hormone that promotes muscle contraction and reduces blood pressure.

Selected entries from *Methods in Enzymology* [vol, page(s)]:
General discussion: 80, 682, 693; 248, 684, 693
Assay: 248, 684 (fluorimetric, 248, 685; insulin degradation method, 248, 684; quenched fluorescence method, 248, 685; substrate, 248, 684)
Other molecular properties: active site, 248, 696; activity, 248, 684 (requirements for, 248, 688); amino acid sequence, 248, 214; N-arginine dibasic convertase as family member, 248, 716; cellular localization, 248, 687; discovery, 248, 684; expression in *Escherichia coli*, 248, 686; gene, 248, 688; hydrolysis of synthetic substrate QF27, 248, 688; inhibition, 248, 688, 691; insulysin and, similarity, 248, 695; lack of activity on proteins, 248, 691; metal dependence, 248, 692; mitochondrial processing peptidase subfamily, 248, 212, 215; peptide cleavage, 248, 688 (kinetics, 248, 689); pH optimum, 248, 688; phylogenesis, 248, 216; pitrilysin subfamily, 248, 212; properties, 248, 211; purification, 248, 686; signal peptides, 248, 227; site-directed mutagenesis, 248, 696; structure, 248, 688, 715; substrate specificity, 248, 688; zinc binding, 248, 688; zinc-binding residue, 248, 696; zinc content, 248, 688

PLASMALOGEN SYNTHASE

This enzyme [EC 2.3.1.25] catalyzes the reaction of an acyl-CoA with a 1-O-alk-1-enylglycero-3-phosphocholine to produce coenzyme A and a plasmenylcholine.[1] Although the enzyme usually acts on linoleoyl-CoA, arachidonoyl-CoA, and oleoyl-CoA, the guinea pig heart enzyme appears to be specific for linoleoyl-CoA.[2,3]

[1]P. C. Choy, M. Skrzypczak, D. Lee & F. T. Jay (1997) BBA **1348**, 124.
[2]G. Arthur, L. L. Page, C. L. Zaborniak & P. C. Choy (1987) BJ **242**, 171.
[3]K. Waku & W. E. M. Lands (1968) JBC **243**, 2654.
Selected entries from *Methods in Enzymology* [vol, page(s)]:
General discussion: 209, 86

PLASMANYLETHANOLAMINE DESATURASE

This magnesium-dependent oxidoreductase [EC 1.14.99.19], also known as alkylacylglycerophosphoethanolamine desaturase, catalyzes the reaction of an O-1-alkyl-2-acyl-sn-glycero-3-phosphoethanolamine with an electron donor (AH2) and dioxygen to produce an O-1-alk-1-enyl-2-acyl-sn-glycero-3-phosphoethanolamine, the oxidized donor (A), and two water molecules.[1–3] The enzyme, which may involve the participation of cytochrome b_5,[2] can use either NADPH or NADH as the electron donor. There also appears to be a requirement for the presence of ATP.[2]

[1] F. Paltauf & A. Holasek (1973) *JBC* **248**, 1609.
[2] R. L. Wykle, M. L. Blank, B. Malone & F. Snyder (1972) *JBC* **247**, 5442.
[3] F. C. Lee, R. L. Wykle, M. L. Blank & F. Snyder (1973) *BBRC* **55**, 574.

Selected entries from *Methods in Enzymology* [vol, page(s)]:
General discussion: 52, 13; **209**, 390
Assay: 209, 391

PLASMEPSIN I

This aspartic endopeptidase [EC 3.4.23.38], also known as aspartic hemoglobinase I and PfAPG, was first obtained from the human malaria pathogen *Plasmodium falciparum* (it is found in the digestive vacuole). It catalyzes peptide-bond hydrolysis in human native hemoglobin as well as in acid-denatured globin. Initial cleavage in native hemoglobin is in the α chain at Phe33–Leu34. There appears to be a preference for a phenylalanyl residue in the P1 position.[1–3]

[1] D. E. Goldberg, A. F. G. Slater, R. Beavis, B. Chait, A. Cerami & G. B. Henderson (1991) *J. Exp. Med.* **173**, 961.
[2] I. Y. Glutzman, S. E. Francis, A. Oksman, C. Smith, K. Duffin & D. E. Goldberg (1994) *J. Clin. Invest.* **93**, 1602.
[3] P. L. Olliaro & D. E. Goldberg (1995) *Parasitol. Today* **11**, 294.

PLASMEPSIN II

This aspartic endopeptidase [EC 3.4.23.39], also known as aspartic hemoglobinase II and PfAPD, catalyzes peptide-bond hydrolysis in native human hemoglobin and in acid-denatured globin at three or four sites. There is a preference for hydrophobic aminoacyl residues in positions P1, P2, and P1'. The Phe33–Leu34 bond cleaved in the α chain by plasmepsin I is also cleaved by plasmepsin II.[1,2]

[1] I. Y. Glutzman, S. E. Francis, A. Oksman, C. Smith, K. Duffin & D. E. Goldberg (1994) *J. Clin. Invest.* **93**, 1602.
[2] P. L. Olliaro & D. E. Goldberg (1995) *Parasitol. Today* **11**, 294.

PLASMIN

Plasmin [EC 3.4.21.7], also known as fibrinase and fibrinolysin, is a serine endopeptidase catalyzing the hydrolysis of peptide bonds with a preference for Lys–Xaa > Arg–Xaa bonds (there is a greater selectivity than displayed by trypsin). A wide variety of proteins are acted upon by plasmin. Its physiological substrates are fibrin and fibrinogen but it has also been shown to act on extracellular matrix proteins (*e.g.*, fibronectin, laminin, and type IV collagen),

casein, kininogen, proinsulin, *etc.*[1–6] The reaction proceeds via an acyl-enzyme intermediate. Esters and amides of arginine and lysine are also cleaved.

Two peptide bonds are hydrolyzed in plasminogen to generate plasmin. In human plasminogen, the first is at Arg561–Val562, catalyzed by one of the plasminogen activators. The second cleavage is at Lys77–Lys78, catalyzed by the plasmin that is formed upon activation.

[1] F. J. Castellino (1981) *Chem. Rev.* **81**, 431.
[2] F. J. Castellino (1983) *Bioscience* **33**, 647.
[3] F. J. Castellino (1984) *Semin. Thromb. Haemast.* **10**, 18.
[4] F. J. Castellino (1995) in *Molecular Basis of Thrombosis and Hemostasis* (K. A. High & H. R. Roberts, eds.), Marcel Dekker, New York, p. 495.
[5] D. Collen & H. R. Lijnen (1986) *CRC Crit Rev. Oncol. Hematol.* **4**, 249.
[6] F. B. Ablondi & J. J. Hagen (1960) *The Enzymes*, 2nd ed., **4**, 176.

Selected entries from *Methods in Enzymology* [vol, page(s)]:
General discussion: 19, 184; **34**, 5; **46**, 206; **80**, 365, 379
Assay: 19, 184; **45**, 257; **267**, 29; assay, with peptide chromogenic and fluorogenic substrates, **80**, 345, 352, 353, 358
Other molecular properties: activation (complement zymogens, of, **80**, 365; Factor XII, of, **80**, 208; latent TGF-β1, of, **198**, 332; plasminogen, from, **45**, 272; rabbit plasminogen, of, **45**, 282); activators and inhibitors, **2**, 165; active, determination of concentration, **80**, 370; active-center serine residue, **45**, 270; active-site titrations, **80**, 419, 420 (with fluorescein diester, **80**, 417); affinity labeling, **80**, 837; amidase and esterase kinetic parameters, **80**, 382, 383; amino acid composition, **19**, 193, 194; **45**, 267, 269 (SCM-light-chain derivative, of, **19**, 52, 195); amino terminal sequences, **45**, 271; α₂-antiplasmin, complex with, **80**, 404; antithrombin-heparin cofactor neutralization, **45**, 668; B chain, streptokinase assay, **80**, 354; bovine, assay with fluorogenic substrates, **80**, 358, 359; crystallized (biological activity, assay, **112**, 121; preparation, **112**, 120; release studies, **112**, 122); chains, amino acid composition, **45**, 267, 269; clot dissolution and, **2**, 140; formation of, **2**, 165; histidine residues, **45**, 270; human, **80**, 379 (assay, **80**, 345, 352, 353, 358; binding to staphylokinase, stoichiometry, **223**, 163; conversion to plasminogen, **80**, 369, 370; Glu¹-plasmin, **80**, 379 [amidase and esterase kinetic parameters, **80**, 383; human, preparation, **80**, 380; physical properties, **80**, 376]; homology to crab collagenase, **80**, 729; hydrodynamic properties, **45**, 265, 266; isoelectric focusing, **45**, 264; N-terminal sequence, **80**, 631; preparation, **45**, 262; **223**, 160; reference standard, **80**, 371); immobilization for inhibitor-binding studies, **267**, 38; immobilization on (aldehyde-Sephadex, **137**, 556; soluble dextran, **137**, 557); immunochemical properties, **19**, 193; inactivation of C1 inhibitor, **80**, 365; inhibition, **19**, 197; **45**, 263, 826; **80**, 51, 395, 763 (by leupeptin, **45**, 679); inhibitor (α-plasmin inhibitor [human {functional assay, **223**, 188; functions, **223**, 186; immunochemical assay, **223**, 187; non-plasminogen-binding form, **223**, 187; purification, **223**, 190; recombinant expression, **223**, 193}; interactions, **87**, 76; recombinant, purification and analysis, **223**, 196]; reaction with Lys⁷⁷- and Val⁴⁴²-plasmin, **80**, 387); kinetic properties, **19**, 199; kinin-releasing enzyme, **80**, 174, 190; lipoprotein-associated coagulation inhibitor, phage display library screening, **267**, 44, 49; Lys⁷⁷-plasmin, **80**, 379 (amidase and esterase kinetic parameters, **80**, 383; formation, **80**, 377, 378; human, preparation, **80**, 380; physical properties, **80**, 376; reaction with α₂-plasmin inhibitor, **80**, 387; recombinant, preparation, **80**, 381, 382); Lys⁷⁷-plasmin-streptokinase (amidase and esterase kinetic parameters, **80**, 383; assay, **80**, 354; kinetic parameters of plasminogen activation, **80**, 384; recombinant, preparation, **80**, 381, 382); nomenclature forms, **45**, 258; origin as zymogen, **19**, 33; **45**, 272;

porcine, inhibition, **80**, 801; preparation, **19**, 190 (from plasminogen, **34**, 429, 430); purity and physical properties, **19**, 192; **45**, 256; rabbit (amino acid composition, **45**, 285; sequence, **45**, 283; inhibitors, **45**, 285; kinetics properties, **45**, 285; molecular weight, **45**, 284; preparation, **45**, 278; properties, **45**, 284, 285; purity, **45**, 284; storage, **45**, 284); radioiodination, **70**, 212; radiolabeling, **45**, 263; recombinant, **80**, 379; regulator of HMW kininogen cofactor activity, **80**, 198; removal from column, in factor I isolation, **223**, 29; Sephadex-immobilized, thrombolytic activity, **137**, 562; serine protease, as, **19**, 32; specificity, **19**, 198; **45**, 257; stability, **19**, 192; streptokinase-activated, **1**, 165; **45**, 269; substitute for Factor D in alternate pathway, **80**, 365; terminal amino acid residues, **19**, 195; titration with p-nitrophenyl p'-guanidinobenzoate HCl, **19**, 20; trapped by α_2-macroglobulin, **80**, 747, 748; Val^{442}-plasmin, **80**, 379 (amidase and esterase kinetic parameters, **80**, 383; human, preparation, **80**, 380; reaction with α_2-plasmin inhibitor, **80**, 387); Val^{561}-plasmin, **80**, 379 (amidase and esterase kinetic parameters, **80**, 383; human, preparation, **80**, 380, 381)

PLASMINOGEN ACTIVATOR

This serine endopeptidase [EC 3.4.21.73], officially known as u-plasminogen activator (and also known as uPA, urokinase, urinary plasminogen activator, and cellular plasminogen activator), catalyzes peptide-bond hydrolysis at Arg560–Val561 in plasminogen to form plasmin. The K_m value for plasminogen is significantly lower (180-times) when uPA is associated with the uPA receptor on the cell surface. The enzyme can be assayed with synthetic peptide substrates that mimic the naturally occurring sequence (Cys-Pro-Gly-Arg-Val-Gly-Gly-Cys) or with small substrates such as Glp-Gly-Arg-NHPhNO$_2$ and Glt-Gly-Arg-NHMec.[1-4] *See also t-Plasminogen Activator*

[1] K. Danø, P. A. Andreasen, J. Grøndahl, P. Kristensen, L. S. Nielsen & L. Skriver (1985) *Adv. Cancer Res.* **44**, 139.
[2] V. Ellis, N. Behrendt & K. Danø (1991) *JBC* **266**, 12752.
[3] K. Danø, N. Behrendt, N. Brünner, V. Ellis, M. Ploug & C. Pyke (1994) *Fibrinolysis* **8** (suppl. 1), 189.
[4] V. Ellis, M. Ploug, T. Plesner & K. Danø (1995) in *Fibrinolysis in Disease: Molecular and Hemovascular Aspects of Fibrinolysis* (P. Glas-Greenwalt, ed.), CRC Press, Boca Raton, p. 30.

Selected entries from *Methods in Enzymology* [vol, page(s)]:
General discussion: 19, 665, 821, 834, 838; **34**, 451
Assay: 19, 665; **34**, 453; **45**, 239, 240; **163**, 294; assay using FDE to monitor kinetics, **80**, 420; assay, with peptide chromogenic and fluorogenic substrates, **80**, 345, 352, 353, 358; salivary, from vampire bat, **223**, 234; streptokinase-plasminogen and SK-plasmin complexes, assay based on, **223**, 150
Other molecular properties: activation, **80**, 377; amidase kinetic parameters, **80**, 382, 383; assay, with peptide chromogenic and fluorogenic substrates, **80**, 345, 352, 353, 358; distribution, **19**, 672; **45**, 243; human, active-site titration, **80**, 423; immobilization on acrylamide-acrylic acid copolymer, **137**, 558; inhibitors, **19**, 671; **45**, 242 (assay, **163**, 298, 307; purification, **163**, 301; type I, inactivation of recombinant bat species II salivary PA, **223**, 248); isoelectric focusing, **45**, 249; plasmin-streptokinase activator complex, isolation and properties of, **19**, 197; preparation, of activator-free plasmin, **45**, 278, 279; properties, **45**, 242; purification, **19**, 667; **45**, 241; **163**, 301; radioiodination, **70**, 212; reaction with α_2-antiplasmin, **80**, 404; specific activity, **45**, 247; specificity of, **19**, 671; **45**, 242; thiolester substrates, **248**, 13

PLASTOQUINOL:PLASTOCYANIN REDUCTASE

This iron-dependent oxidoreductase [EC 1.10.99.1], also known as cytochrome b_6/f complex, catalyzes the reaction of plastoquinol-1 with two molecules of oxidized plastocyanin to produce plastoquinone and two molecules of reduced plastocyanin.[1-3] Other substrates include plastoquinol-9 and ubiquinol. Cytochrome c_{552} can replace plastocyanin, albeit more slowly.

[1] G. Hauska, E. Hurt, N. Gabellini & W. Lockau (1983) *BBA* **726**, 97.
[2] R. D. Clark & G. Hind (1983) *JBC* **258**, 10348.
[3] M. Krinner, G. Hauska, E. Hurt & W. Lockau (1982) *BBA* **681**, 110.

Selected entries from *Methods in Enzymology* [vol, page(s)]:
General discussion: 126, 271; **167**, 341, 779
Assay: 126, 279, 283; **167**, 348
Other molecular properties: chloroplast (assays, **126**, 283); cyanobacterial (assays, **126**, 279; isolation, **126**, 279); genes, isolation and characterization from *Nostoc*, **167**, 779; isolation, **126**, 277, 279; properties, **167**, 344; purification from cyanobacteria, **167**, 342

*Ple*I RESTRICTION ENDONUCLEASE

This type II site-specific endonuclease (subtype s) [EC 3.1.21.4], isolated from *Pseudomonas lemoignei*, catalyzes the hydrolysis of both strands of DNA at 5′ ... GAGTCNNNN∇ ... 3′ and 3′ ... CTCAGNNNNN∇ ... 5′, where N refers to any base.

*Pme*I RESTRICTION ENDONUCLEASE

This type II site-specific endonuclease [EC 3.1.21.4], isolated from *Pseudomonas mendocina*, catalyzes the hydrolysis of both strands of DNA at 5′ ... GTTT∇AAAC ... 3′, producing blunt-ended fragments. The enzyme can exhibit star activity. *See "Star" Activity*

*Pml*I RESTRICTION ENDONUCLEASE

This type II site-specific endonuclease [EC 3.1.21.4], isolated from *Pseudomonas maltophilia*, catalyzes the hydrolysis of both strands of DNA at 5′ ... CAC∇GTG ... 3′, producing blunt-ended fragments.

POLAR-AMINO-ACID-TRANSPORTING ATPase

This set of so-called ATPases [EC 3.6.3.21] catalyzes the ATP-dependent (ADP- and orthophosphate-producing) transport of a [polar amino acid]$_{out}$ to produce a [polar amino acid]$_{in}$.[1-4] Bacterial proteins have been identified that will catalyze the import of L-histidine (hence, histidine permease), L-arginine, L-lysine, L-glutamate, L-glutamine,

L-aspartate, L-ornithine, octopine (*i.e.*, N^2-(D-1-carboxy-ethyl)-L-arginine), and nopaline (*i.e.*, N^2-(1,3-dicarboxy-propyl)-L-arginine). ABC-type (ATP-binding cassette-type) ATPases have two similar ATP-binding domains and do not undergo phosphorylation during the transport process.

The idea that a transporter is an enzyme is in keeping with a new definition of enzyme catalysis as the facilitated making/breaking of chemical bonds, not just covalent bonds.[5] This idea builds on Pauling's assertion that any long-lived, chemically distinct interaction (in this case, the persistent location of a solute with respect to the faces of a membrane) can be regarded as a chemical bond. Note also that the equilibrium constant (K_{eq} = [amino acid$_{in}$][ADP][P$_i$]/[amino acid$_{out}$][ATP]) does not conform to that expected for an ATPase (*i.e.*, K_{eq} = [ADP][P$_i$]/[ATP]). Thus, although the overall reaction yields ADP and orthophosphate, the enzyme is misclassified as a hydrolase, and instead should be regarded as an energase-type reaction. Energases facilitate affinity-modulated reactions by coupling the $\Delta G_{ATP\text{-}hydrolysis}$ to a force-generating or work-producing step.[5] In this case, P-O-P bond scission supplies the energy to drive amino acid transport. *See* ATPase; Energase

[1]G. Kuan, E. Dassa, N. Saurin, M. Hofnung & M. H. Saier, Jr. (1995) *Res. Microbiol.* **146**, 271.
[2]M. H. Saier, Jr. (1998) *Adv. Microb. Physiol.* **40**, 81.
[3]K. Nikaido, P. Q. Liu & G. Ferro-Luzzi Ames (1997) *JBC* **272**, 27745.
[4]D. L. Walshaw, S. Lowthorpe, A. East & P. S. Poole (1997) *FEBS Lett.* **414**, 397.
[5]D. L. Purich (2001) *TiBS* **26**, 417.

POLY(ADP-RIBOSE) GLYCOHYDROLASE

This glycohydrolase [EC 3.2.1.143] catalyzes the hydrolysis of glycosidic bonds (at $1''$–$2'$) on protein–poly(ADP-D-ribosyl)$_n$ to produce protein–ADP-D-ribose and ($n - 1$) molecules of ADP-D-ribose.[1–4]

[1]K. Ueda & O. Hayaishi (1985) *ARB* **54**, 73.
[2]O. Hayaishi & K. Ueda (1977) *ARB* **46**, 95.
[3]M. Miwa & T. Sugimura (1971) *JBC* **246**, 6362.
[4]W. Lin, J. C. Ame, N. Aboul-Ela, E. L. Jacobson & M. K. Jacobson (1997) *JBC* **272**, 11895.
Selected entries from *Methods in Enzymology* [vol, page(s)]:
Assay: 106, 505
Other molecular properties: analysis of poly(ADP-ribosyl)ated proteins, in, 106, 508

POLYAMINE OXIDASE

This iron- and FAD-dependent oxidoreductase [EC 1.5.3.11] catalyzes the reaction of N^1-acetylspermine with dioxygen and water to produce N^1-acetylspermidine, 3-aminopropanal, and hydrogen peroxide.[1–3] Other substrates include N^1-acetylspermidine, N^1,N^{12}-diacetylspermine (producing N^1-acetylspermidine and 3-acetamidopropanal), and spermine (producing spermidine and 3-aminopropanal).

[1]E. Höltta (1977) *Biochemistry* **16**, 91.
[2]D. M. L. Morgan (1987) *Essays in Biochem.* **23**, 82.
[3]C. W. Tabor & H. Tabor (1984) *ARB* **53**, 749.
Selected entries from *Methods in Enzymology* [vol, page(s)]:
General discussion: 94, 306
Assay: oat seedling, 94, 312; rat liver, 94, 306
Other molecular properties: microbodies, in, 52, 496; oat seedling (assay, 94, 312; properties, 94, 313; purification, 94, 312); rat liver (assay, 94, 306; properties, 94, 309; purification, 94, 307)

POLYAMINE-TRANSPORTING ATPase

This so-called ATPase [EC 3.6.3.31] catalyzes the ATP-dependent (ADP- and orthophosphate-producing) transport of polyamine$_{out}$ to produce polyamine$_{in}$.[1] Polyamines imported include putrescine and spermidine. ABC-type (ATP-binding cassette-type) ATPases have two similar ATP-binding domains and do not undergo phosphorylation during the transport process.

The idea that a transporter is an enzyme is in keeping with a new definition of enzyme catalysis as the facilitated making/breaking of chemical bonds, not just covalent bonds.[2] This idea builds on Pauling's assertion that any long-lived, chemically distinct interaction (in this case, the persistent location of a solute with respect to the faces of a membrane) can be regarded as a chemical bond. Note also that the equilibrium constant (K_{eq} = [polyamine$_{in}$][ADP][P$_i$]/[polyamine$_{out}$][ATP]) does not conform to that expected for an ATPase (*i.e.*, K_{eq} = [ADP][P$_i$]/[ATP]). Thus, although the overall reaction yields ADP and orthophosphate, the enzyme is misclassified as a hydrolase, and instead should be regarded as an energase-type reaction. Energases facilitate affinity-modulated reactions by coupling the $\Delta G_{ATP\text{-}hydrolysis}$ to a force-generating or work-producing step.[2] In this case, P-O-P bond scission supplies the energy to drive polyamine transport. *See* ATPase; Energase

[1]K. Kashiwagi, S. Miyamoto, E. Nukui, H. Kobayashi & K. Igarashi (1993) *JBC* **268**, 19358.
[2]D. L. Purich (2001) *TiBS* **26**, 417.

POLY(A)-SPECIFIC RIBONUCLEASE

This ribonuclease [EC 3.1.13.4] catalyzes the exonucleolytic cleavage of poly(A), releasing 5′-AMP. The substrate can be either single- or double-stranded.[1–3] The poly(A)

polymerase from *Vigna unguiculata* seedlings is a bifunctional enzyme responsible for both polymerizing and hydrolyzing activities.[4]

[1]H. C. Schröder, R. K. Zahn, K. Dose & W. E. G. Müller (1980) JBC 255, 4535.
[2]H. C. Schröder, M. Bachmann, R. Messer & W. E. G. Müller (1985) Prog. Mol. Subcell. Biol. 9, 53.
[3]M. Wormington (1993) Curr. Opin. Cell Biol. 5, 950.
[4]Y. Tarui & T. Minamikawa (1989) EJB 186, 591.

POLYGALACTURONASE

This enzyme [EC 3.2.1.15], also known as pectin depolymerase and pectinase, catalyzes the random hydrolysis of 1,4-α-D-galactosiduronic linkages in pectate and other galacturonans.[1–5] The polymeric substrate must have at least three galacturonate residues at the non-reducing end. Digestion by polygalacturonase results in the ultimate formation of di-, tri-, tetra-, penta-, and hexagalacturonates. The enzyme from *Botrytis cinerea* will also generate monogalacturonate. *See also Galacturonide 1,4-α-Galacturonidase*

[1]C. Lang & H. Dornenburg (2000) Appl. Microbiol. Biotechnol. 53, 366.
[2]R. A. Prade, D. Zhan, P. Ayoubi & A. J. Mort (1999) Biotechnol. Genet. Eng. Rev. 16, 361.
[3]K. A. Hadfield & A. B. Bennett (1998) Plant Physiol. 117, 337.
[4]H. Deuel & E. Stutz (1958) AE 20, 341.
[5]H. Lineweaver & E. J. Jansen (1951) AE 11, 267.

Selected entries from *Methods in Enzymology* [vol, page(s)]:
General discussion: 1, 158; 8, 636; 161, 329, 361, 366
Assay: 8, 636; 161, 333, 337, 361, 371
Other molecular properties: analysis with activity stain, 161, 330; chromatography with Spheron ion exchangers, 161, 388; digestion of polygalacturonate, in, 179, 567; isolation (chloroplast, 69, 123; plant cell, 69, 59); isozymes, chromatography with Mono S cation exchanger, 161, 396; mode of action, 8, 636, 639; preparation, 8, 638; properties, 8, 639; 161, 364, 372; protoplasts, plant cell, 69, 78; purification from (*Botrytis cinerea*, 161, 372; *Colletotrichum lindemuthianum*, 118, 10; *Corticium rolfsii*, 161, 363); substrates, 3, 28; use in protoplast preparation, 23, 200, 202

POLYGALACTURONATE 4-α-GALACTURONOSYLTRANSFERASE

This transferase [EC 2.4.1.43] catalyzes the reaction of UDP-D-galacturonate with [(1,4)-α-D-galacturonosyl]$_n$ to produce [(1,4)-α-D-galacturonosyl]$_{n+1}$ and UDP.[1]

[1]C. L. Villemez, A. L. Swanson & W. Z. Hassid (1966) ABB 116, 446.

POLYGLUCURONIDE LYASE

This activity [formerly classified as EC 4.2.99.5] is now a deleted Enzyme Commission entry.

POLY(GLYCEROL-PHOSPHATE) α-GLUCOSYLTRANSFERASE

This glucosyltransferase [EC 2.4.1.52], also known as UDP-glucose:polyglycerolteichoic acid glucosyltransferase, catalyzes the reaction of UDP-D-glucose with poly(glycerol phosphate) to produce α-D-glucosylpoly(glycerol phosphate) and UDP.[1]

[1]L. Glaser & M. M. Burger (1964) JBC 239, 3187.

Selected entries from *Methods in Enzymology* [vol, page(s)]:
General discussion: 8, 436
Assay: 8, 436

POLY(α-L-GULURONATE) LYASE

This lyase [EC 4.2.2.11], also known as alginase II, catalyzes the eliminative cleavage of polysaccharides containing a terminal α-L-guluronate group, producing oligosaccharides with 4-deoxy-α-L-*erythro*-hex-4-enuronosyl groups at their non-reducing ends.[1,2]

[1]J. Boyd & J. R. Turvey (1977) Carbohydr. Res. 57, 163.
[2]I. W. Davidson, I. W. Sutherland & C. J. Lawson (1976) BJ 159, 707.

POLYMANNURONATE HYDROLASE

This glycosidase [EC 3.2.1.121], first isolated from *Pseudomonas aeruginosa*, catalyzes the endohydrolysis of D-mannuronide linkages of polymannuronates, but, will not act on alginic acid, a copolymer of polymannuronate.[1]
See also Alginate Lyase

[1]W. M. Dunne & F. L. Buckmire (1985) AEM 50, 562.

Selected entries from *Methods in Enzymology* [vol, page(s)]:
General discussion: 34, 3; 235, 304

POLYNUCLEOTIDE ADENYLYLTRANSFERASE

This adenylyltransferase [EC 2.7.7.19] (also known as NTP polymerase, RNA adenylating enzyme, and poly(A) polymerase) catalyzes the reaction of *n* molecules of ATP with (nucleotide)$_m$ to produce *n* molecules of pyrophosphate and (nucleotide)$_{m+n}$.[1–3] The polynucleotide undergoes a template-independent extension of the 3′-end of a strand one nucleotide at a time.

The adenylyltransferase cannot initiate the synthesis of a polynucleotide chain *de novo*. A primer polynucleotide (either RNA or DNA depending on the enzyme source or even an oligo(A) with a free 3′-hydroxyl group) is required.

[1]T. Wittmann & E. Wahle (1997) BBA 1350, 293.
[2]M. Edmonds (1982) The Enzymes, 3rd ed., 15, 217.
[3]R. J. Mans & T. J. Walter (1971) BBA 247, 113.

Selected entries from *Methods in Enzymology* [vol, page(s)]:
General discussion: 29, 70
Assay: 181, 167; calf thymus, 29, 72; Vaccinia virus (assay, [cap-specific (nucleoside-2′-O-)-methyltransferase, 275, 225; polymerase, 275, 224])
Other molecular properties: antibodies, effect on polyadenylation *in vitro*, 181, 167; calf thymus (assay, 29, 72; homogeneous preparation,

29, 75; properties, **29**, 78; purification, **29**, 73); properties, **181**, 164; purification, **181**, 162; Vaccinia virus (adenylyltransferase activity, **275**, 219; assay, [cap-specific (nucleoside-2′-*O*-)-methyltransferase, **275**, 225; polymerase, **275**, 224]; metal ion dependence, **275**, 219; pH optimum, **275**, 218; processivity, **275**, 218; purification [virion extract preparations, **275**, 221; VP39 overexpressed in *Escherichia coli*, **275**, 223; VP39 overexpressed in HeLa cells, **275**, 223; VP55 overexpressed in HeLa cells, **275**, 222]; RNA 3′ end labeling reactions, **275**, 226; substrate specificity, **275**, 219; subunits, **275**, 217)

POLYNUCLEOTIDE-5′-HYDROXYL-KINASE

This phosphotransferase [EC 2.7.1.78], also known as DNA kinase, catalyzes the reaction of ATP with 5′-dephospho-DNA (either single-stranded or double-stranded) to produce ADP and 5′-phospho-DNA.[1–3] 5′-Dephospho-RNA can also be phosphorylated (there are eukaryotic enzymes that prefer DNA and others that prefer RNA). In addition, 5′-hydroxyoligonucleotides and 3′-mononucleotides are substrates as well. Nucleosides are not phosphorylated.

The human and bacteriophage T4 enzymes also exhibit 3′-phosphatase activity, whereas kinases acting preferentially on RNA substrates lack phosphatase activity.

[1] F. Karimi-Busheri, G. Daly, P. Robins, B. Canas, D. J. Pappin, J. Sgouros, G. G. Miller, H. Fakhrai, E. M. Davis, M. M. Le Beau & M. Weinfeld (1999) *JBC* **274**, 24187.
[2] C. C. Richardson (1981) *The Enzymes*, 3rd ed., **14**, 299.
[3] K. Kleppe & J. R. Lillehaug (1979) *AE* **48**, 245.

Selected entries from *Methods in Enzymology* [vol, page(s)]:
General discussion: 12B, 207; **34**, 469
Assay: 12B, 207
Other molecular properties: deoxyribonuclease contamination in, **21**, 334; deoxyribonucleic acid labeling and, **29**, 282, 283, 284; deoxyribonucleic acid ligase substrate and, **21**, 334; deoxyribonucleic acid molecular weight and, **12B**, 416; dodecanucleotide labeling and, **29**, 250, 251; 5′-end labeling of (DNA, **212**, 162; RNA, **180**, 194; **275**, 374); exchange assay, of Class II restriction endonucleases, **65**, 28 (procedure, **65**, 31, 32); exchange reaction (DNA fragment size, **65**, 34; nucleotide composition, **65**, 35); labeling of oligonucleotides, **208**, 450; labeling of RNA digests for fingerprinting, **180**, 135; optimal conditions, **65**, 31, 32; phosphorylation of (adaptors, **152**, 348; linkers, **152**, 344); polydeoxythymidylate and, **21**, 313; preparation, **21**, 323 (of α-³²P-labeled nucleotides, **195**, 34 [reaction monitoring, **195**, 38]); properties, **12B**, 210; purification procedure, **12B**, 208; specificity, **65**, 33; T4-polynucleotide kinase, **68**, 94, 134 (assay, of restriction endonucleases, **65**, 28; commercial source, **65**, 681; end labeling of [DNA, **65**, 75, 267, 481, 507, 521, 557, 584, 848; plasmid pLJ3, **65**, 242; RNA, **65**, 669, 683, 726]; messenger RNA phosphorylation, **79**, 615; preparation, **65**, 66 [[5′-³²P]nucleoside 3′,5′-diphosphate, **65**, 67; synthesis of (2′-5′)-oligoadenylic acid analogue, in, **119**, 525; tritium derivative method, **65**, 647, 648]; tRNA labeling procedure, from, **59**, 58, 61, 71, 75, 102, 103)

POLYNUCLEOTIDE 3′-PHOSPHATASE

This phosphatase [EC 3.1.3.32], also known as 2′(3′)-polynucleotidase, catalyzes the hydrolysis of a 3′-phosphopolynucleotide to produce a polynucleotide with a 3′-hydroxyl group and orthophosphate.[1] The enzyme also catalyzes nucleoside 2′-, 3′-, and 5′-mono-phosphate hydrolysis. In a number of sources, the enzyme also has a polynucleotide 5′-hydroxyl-kinase activity.

[1] Y. Habraken & W. G. Verly (1988) *EJB* **171**, 59.

Selected entries from *Methods in Enzymology* [vol, page(s)]:
Assay: 65, 690, 691

POLYNUCLEOTIDE 5′-PHOSPHATASE

This phosphatase [EC 3.1.3.33], also known as 5′-poly-nucleotidase and polynucleotide 5′-triphosphatase, catalyzes the hydrolysis of a 5′-phosphopolynucleotide (*i.e.*, a polynucleotide with a 5′-phosphoryl terminus) to produce a polynucleotide with a 5′-hydroxyl group as well as orthophosphate.[1] The enzyme does not act on nucleoside monophosphates.

[1] A. Becker & J. Hurwitz (1967) *JBC* **242**, 936.

POLYOL DEHYDROGENASE (NADP⁺)

This activity, previously classified under EC 1.1.1.139, is now regarded as an activity of aldehyde reductase [EC 1.1.1.21].

POLYPEPTIDE N-ACETYLGALACTOS-AMINYLTRANSFERASE

This calcium and manganese-dependent transferase [EC 2.4.1.41], also known as protein:UDP acetylgalac-tosaminyltransferase, catalyzes the reaction of UDP-*N*-acetyl-D-galactosamine with a polypeptide (or, protein) to produce an *N*-acetyl-D-galactosaminyl-polypeptide (or, -protein) and UDP. The site modified is the hydroxyl group on either a threonyl or a seryl residue. Substrates include the polypeptide core of submaxillary mucin, κ-casein, apofetuin, apoantifreeze glycoproteins, asialomucin, and myelin basic protein.

This enzyme catalyzes the first step in the formation of the saccharide portion of many *O*-linked glycoproteins. Interestingly, a member of this enzyme family has recently been characterized and found to require the presence of another *N*-acetylglutamate on the protein, suggesting that *O*-glycosylation of multisite substrates may proceed in a hierarchical manner.[1]

[1] K. G. Ten Hagen, D. Tetaert, F. K. Hagen, C. Richet, T. M. Beres, J. Gagnon, M. M. Balys, B. VanWuyckhuyse, G. S. Bedi, P. Degand & L. A. Tabak (1999) *JBC* **274**, 27867.

POLYPHOSPHATE:GLUCOSE PHOSPHOTRANSFERASE

This enzyme [EC 2.7.1.63], also known as polyphosphate glucokinase, catalyzes the reaction of (phosphate)$_n$ (that is, polyphosphate) with D-glucose to produce (phosphate)$_{n-1}$ and D-glucose 6-phosphate.[1–4] The reaction requires the presence of a neutral salt (such as potassium chloride) for maximum activity. D-Glucosamine can also serve as a substrate. The *Mycobacterium tuberculosis*[2] and *Propionibacterium shermanii*[3] enzymes have ordered Bi Bi kinetic mechanisms. Depending on the source, enzyme catalysis can be processive.[4]

[1] N. F. Phillips, P. C. Hsieh & T. H. Kowalczyk (1999) *Prog. Mol. Subcell. Biol.* **23**, 101.
[2] P. C. Hsieh, T. H. Kowalczyk & N. F. Phillips (1996) *Biochemistry* **35**, 9772.
[3] T. H. Kowalczyk, P. J. Horn, W. H. Pan & N. F. Phillips (1996) *Biochemistry* **35**, 6777.
[4] C. A. Pepin & H. G. Wood (1986) *JBC* **261**, 4476.

POLYPHOSPHATE KINASE

This phosphoprotein [EC 2.7.4.1] catalyzes the reaction of ATP with (phosphate)$_n$ to produce ADP and (phosphate)$_{n+1}$.[1–5] Best polyphosphate substrates have more than 130 phosphoanhydride bonds; no activity is observed for substrates with four or fewer phosphoanhydride bonds. The mechanism of elongation of the polyphosphate is strictly processive.[4,5]

A number of microorganisms utilize the reverse reaction to regenerate ATP. The enzyme also catalyzes a general nucleoside diphosphate kinase activity; for example, the reaction of GDP with (phosphate)$_n$ produces GTP and (phosphate)$_{n-1}$. Furthermore, guanosine 5′-tetraphosphate can be synthesized in the following reaction: GDP + (phosphate)$_n$ = ppppG and (phosphate)$_{n-2}$.[2] The initial rate kinetics for ppppG synthesis fits a rapid equilibrium random Bi Bi kinetic mechanism.[2]

[1] T. Shiba, K. Tsutsumi, K. Ishige & T. Noguchi (2000) *Biochemistry (Mosc).* **65**, 315.
[2] C. M. Tzeng & A. Kornberg (2000) *JBC* **275**, 3977.
[3] H. G. Wood & J. E. Clark (1988) *ARB* **57**, 235.
[4] N. A. Robinson, J. E. Clark & H. G. Wood (1987) *JBC* **262**, 5216.
[5] N. A. Robinson & H. G. Wood (1986) *JBC* **261**, 4481.

POLY(RIBITOL-PHOSPHATE) *N*-ACETYL-GLUCOSAMINYLTRANSFERASE

This transferase [EC 2.4.1.70], a key participant in the biosynthesis of teichoic acids, catalyzes the reaction of UDP-*N*-acetyl-D-glucosamine with poly(ribitol phosphate) to produce *N*-acetyl-D-glucosaminylpoly(ribitol phosphate) and UDP.

POLY(RIBITOL-PHOSPHATE) β-GLUCOSYLTRANSFERASE

This transferase [EC 2.4.1.53], first isolated from *Bacillus subtilis*, catalyzes the reaction of UDP-D-glucose with poly(ribitol phosphate) to produce β-D-glucosylpoly(ribitol phosphate) and UDP.[1]

[1] T. Chin, M. M. Burger & L. Glaser (1966) *ABB* **116**, 358.

POLYRIBONUCLEOTIDE NUCLEOTIDYLTRANSFERASE

This transferase [EC 2.7.7.8], more widely known as polynucleotide phosphorylase, catalyzes the reversible reaction of a poly(ribonucleotide)$_{n+1}$ (that is, RNA with $n+1$ bases) with orthophosphate to produce poly-(ribonucleotide)$_n$ and a nucleoside diphosphate (NDP).[1–7] The NDP products formed (or, donor substrates in the reverse reaction) can be ADP, IDP, GDP, UDP, and CDP.

This enzyme participates in the degradation of mRNA. The *Escherichia coli* enzyme exhibits a preferential degradation of polyadenylated and polyuridinylated RNAs.[1] The reverse reaction can also be used to synthesize homopolymers and copolymers ribonucleotides.[3] In this reverse reaction, the highly processive enzyme incorporates a nucleoside diphosphate specifically into the 5′ end of the polymer.[2]

[1] I. Lisitsky & G. Schuster (1999) *EJB* **261**, 468.
[2] M. Sulewski, S. P. Marchese-Ragona, K. A. Johnson & S. J. Benkovic (1989) *Biochemistry* **28**, 5855.
[3] U. Z. Littauer & H. Soreq (1982) *The Enzymes*, 3rd ed., **15**, 517.
[4] T. Godefroy-Colburn & M. Grunenberg-Manago (1972) *The Enzymes*, 3rd ed., **7**, 533.
[5] J. Y. Chou & M. F. Singer (1971) *JBC* **246**, 7486, 7497, and 7505.
[6] S. Ochoa & S. Mii (1961) *JBC* **236**, 3303.
[7] M. Grunberg-Manago, P. J. Ortiz & S. Ochoa (1956) *BBA* **20**, 269.

Other molecular properties: *Azotobacter vinelandii* (properties, **6**, 10; purification, **6**, 5); copolymers of guanylic acid, **12B**, 522, 529; *Escherichia coli* B, **12B**, 513; **65**, 690 (properties, **12B**, 518; purification, **12B**, 515; **65**, 691; synthesis, of oligodeoxyribonucleotides, of defined sequence, **65**, 694, 700); exchange reactions, **87**, 248; hydroxylapatite chromatography in RNA polymerase preparation, separation by, **12B**, 569; labeling nucleoside diphosphates by, **5**, 735; manganese ion inhibition, **12B**, 819; methylated nucleotides as substrates, **12A**, 150; NMR, ^{31}P, **87**, 248; phosphate buffer, inhibition by, **12B**, 556; oligonucleotide synthesis and, **30**, 295, 307; polyguanylic acid, synthesis of, **6**, 715; **12B**, 522; polypseudouridylic acid, synthesis of, **12B**, 519; polyribonucleotides, synthesis, **6**, 713; preparation, **20**, 107; preparation of polynucleotides, **6**, 713; **78**, 327; primer-dependent, in oligonucleotide sequence analysis, **59**, 88; properties (purity, **6**, 10; **12B**, 518; specificity, **6**, 11; **12B**, 519; stability, **12B**, 518); purification procedure, **6**, 5; **12B**, 515; RNA polymerase, in preparation of, **12B**, 556; sigma factor assay and, **21**, 502; staining on gels, **22**, 599; stereochemistry, **87**, 203, 212; substrates and nonsubstrates of, **6**, 11, 716; tRNA interaction, **20**, 106 (procedure, **20**, 110; reagents, **20**, 107)

POLYSACCHARIDE O-METHYLTRANSFERASE

This methyltransferase [EC 2.1.1.18] catalyzes the reaction of *S*-adenosyl-L-methionine with a 1,4-α-D-gluco-oligosaccharide to produce *S*-adenosyl-L-homocysteine and an oligosaccharide containing 6-methyl-D-glucose units.[1] The best oligosaccharide substrates contain between seven and ten glucosyl units.

[1]J. A. Ferguson & C. E. Ballou (1970) *JBC* **245**, 4213.

POLYSIALIC-ACID O-ACETYLTRANSFERASE

This acetyltransferase [EC 2.3.1.136] catalyzes the reaction of acetyl-CoA with an α-2,8-linked polymer of sialic acid to produce coenzyme A and a polysialic acid acetylated at *O*-7 or *O*-9. The polymeric substrate must have at least fourteen sialosyl residues. This enzyme catalyzes the modification of capsular polysaccharides in some strains of *Escherichia coli*.[1]

[1]H. H. Higa & A. Varki (1988) *JBC* **263**, 8872.

POLYVINYL-ALCOHOL DEHYDROGENASE (ACCEPTOR)

This pyrroloquinoline-quinone-dependent oxidoreductase [EC 1.1.99.23], often abbreviated PVA dehydrogenase, catalyzes the reaction of polyvinyl alcohol with an acceptor substrate to produce the oxidized polyvinyl alcohol and the reduced acceptor.[1,2] Acceptor substrates include cytochrome *c*, phenazine methosulfate, phenazine ethosulfate, 2,6-dichlorophenolindophenol, and 2,6-dichloroindophenol. Other alcohol substrates, albeit much weaker, include 2-hexanol and some other secondary alcohols. ***See also*** *Alcohol Dehydrogenase (Acceptor); Alkan-1-ol Dehydrogenase (Acceptor)*

[1]M. Shimao, S. Onishi, N. Kato & C. Sakazawa (1989) *AEM* **55**, 275.
[2]M. Shimao, K. Ninomiya, O. Kuno, N. Kato & C. Sakazawa (1986) *AEM* **51**, 268.

POLYVINYL-ALCOHOL OXIDASE

This oxidoreductase [EC 1.1.3.30], often abbreviated PVA oxidase, catalyzes the reaction of polyvinyl alcohol with dioxygen to produce the oxidized polyvinyl alcohol and hydrogen peroxide.[1,2]

[1]M. Shimao, Y. Nishimura, N. Kato & C. Sakazawa (1985) *AEM* **49**, 8.
[2]M. Shimao, S. Onishi, N. Kato & C. Sakazawa (1989) *AEM* **55**, 275.

PORPHOBILINOGEN DEAMINASE

This dipyrromethane-dependent deaminase [EC 4.3.1.8], also known as hydroxymethylbilane synthase (the preferred Enzyme Commission name) and pre-uroporphyrinogen synthase, catalyzes the reaction of four molecules of porphobilinogen with water to produce hydroxymethylbilane and four ammonia molecules.[1–4] In the presence of uroporphyrinogen-III synthase [EC 4.2.1.75], often known as uroporphyrinogen-III cosynthase, the product is cyclized to form uroporphyrinogen III. There are five forms of the deaminase in rat and human.

The dipyrromethane cofactor is covalently linked to a cysteinyl residue. This cofactor serves as a primer for the assembly of the tetrapyrrole product.[2] Elongation of the pyrrole chain appears to occur by either a translocation of the growing oligopyrrole chain within the active site cleft or by protein conformational changes that bring the appropriate catalytic and/or dipyrromethane-containing domains into play.[2]

[1]P. M. Shoolingin-Jordan (1998) *Biochem. Soc. Trans.* **26**, 326.
[2]G. V. Louie, P. D. Brownlie, R. Lambert, J. B. Cooper, T. L. Blundell, S. P. Wood, V. N. Malashkevich, A. Hadener, M. J. Warren & P. M. Shoolingin-Jordan (1996) *Proteins* **25**, 48.
[3]P. M. Shoolingin-Jordan (1995) *J. Bioenerg. Biomembr.* **27**, 181.
[4]S. Granick & S. I. Beale (1978) *AE* **46**, 33.

Selected entries from ***Methods in Enzymology*** [**vol**, page(s)]:
General discussion: 5, 885; 281, 317
Assay: 5, 885
Other molecular properties: control sequence in nitric oxide synthase assay, **269**, 416; mammalian (coupled-enzyme assay for porphobilinogen synthase, in, **123**, 343; purification, **123**, 340); properties, **5**, 891; purification, **5**, 889

PORPHOBILINOGEN SYNTHASE

This zinc-dependent enzyme [EC 4.2.1.24], also known as δ-aminolevulinate dehydratase, catalyzes the reaction of two molecules of 5-aminolevulinate to produce porphobilinogen and two water molecules.[1–6] The plant enzyme and the protein from certain microorganisms have a Zn^{2+} and Mg^{2+} or a solely Mg^{2+} metal ion requirement. The first 5-aminolevulinate that binds to the enzyme is the molecular entity that supplies the propionate-containing portion of the product[3] (the first 5-aminolevulinate forms a Schiff base intermediate with the ε-amino group of a lysyl residue: Lys252 in the bovine enzyme and Lys247 in the *Escherichia coli* protein). The kinetic mechanism is ordered Bi Uni, even though the two substrates are identical. Binding of the second 5-aminolevulinate molecule is metal ion-dependent. The catalyzed reaction is thought to proceed via a di-Schiff base intermediate.

[1]P. M. Shoolingin-Jordan (1998) *Biochem. Soc. Trans.* **26**, 326.
[2]K. N. Allen (1998) *CBC* **2**, 135.
[3]P. M. Jordan & P. N. B. Gibbs (1985) *BJ* **227**, 1015.
[4]B. L. Vallee & A. Galdes (1984) *AE* **56**, 283.
[5]S. Granick & S. I. Beale (1978) *AE* **46**, 33.
[6]D. Shemin (1972) *The Enzymes*, 3rd ed., **7**, 323.

Selected entries from *Methods in Enzymology* [vol, page(s)]:
General discussion: 5, 883; 17A, 205; 44, 844; 123, 339, 427
Assay: 5, 883; 123, 429
Other molecular properties: bovine liver (activation, 123, 429; assay, 123, 428; properties, 123, 433; purification, 17A, 217; 123, 429); genetic variations, mouse liver, 17A, 213; immobilization, on Sepharose 4B, 44, 846, 847; mammalian, coupled-enzyme assay, 123, 339; mouse liver, from, 17A, 212; properties, 44, 844; rat Harderian gland, from, 17A, 220; *Rhodopseudomonas spheroides*, from, 5, 883; 17A, 206 (properties, 5, 884; purification, 5, 884); *Saccharomyces cerevisiae*, from, 17A, 219; synthesis of porphobilinogen, in, 17A, 221; uroporphyrinogen decarboxylase preparation, present in, 5, 895

POTYVIRUS HELPER-COMPONENT PROTEINASE

This cysteine endopeptidase [EC 3.4.22.45], also known as helper-component proteinase (or HC-Pro), participates in polyprotein processing (cleaving at Gly–Gly near the C-terminus) in such potyviruses as tobacco etch virus (TEV) and tobacco vein mottling virus.[1–3]

[1]J. L. Reichmann, S. Lain & J. A. Garcia (1992) *J. Gen. Virol.* **73**, 1.
[2]W. G. Dougherty & B. L. Semler (1993) *Microbiol. Rev.* **57**, 781.
[3]H.-G. Kräusslich & E. Wimmer (1988) *ARB* **57**, 701.

POTYVIRUS NIα PROTEASE

This cysteine endopeptidase [EC 3.4.22.44], also known as potyvirus nuclear inclusion protein α and nuclear-inclusion-α endopeptidase (NIα), catalyzes peptide-bond hydrolysis in the processing of the viral polyprotein. The cleavage motif for the enzyme from tobacco etch virus is Glu-Xaa-Xaa-Tyr-Xaa-Gln-^-(Ser/Gly),

for the tobacco vein mottling virus enzyme it is Val-(Arg,Lys)-Phe-Gln-^-(Ser/Gly), and for the plum pox potyvirus enzyme it is Val-Xaa-His-Gln-^-Yaa where Yaa is a small aminoacyl residue (such as Gly, Ala, Ser, or Thr). Note that the P1 position is occupied by a glutaminyl residue in each case.[1,2]

[1]J. A. Lindbo & W. G. Dougherty (1994) in *Encyclopedia of Virology*, vol. 3 (R. G. Webster & A. Granoff, eds.) Academic Press, New York, p. 1148.
[2]D. D. Skukla, C. W. Ward & A. A. Brunt (1994) *The Potyviridae*, CAB International, Wallingford, CT, p. 80.

*Ppu*MI RESTRICTION ENDONUCLEASE

This type II site-specific endonuclease [EC 3.1.21.4], isolated from *Pseudomonas putida* M, catalyzes the hydrolysis of both strands of DNA at $5' \ldots RG\nabla GWCCY \ldots 3'$, where R refers to G or A, W refers to A or T, and Y refers to T or C. This enzyme is blocked by overlapping *dcm* methylation.

PRECORRIN-3B C17-METHYLTRANSFERASE

This methyltransferase [EC 2.1.1.131], also known as precorrin-3 methyltransferase and precorrin-3 methylase, catalyzes the reaction of *S*-adenosyl-L-methionine with precorrin-3B to produce *S*-adenosyl-L-homocysteine and precorrin 4.[1]

[1]L. Debussche, D. Thibaut, B. Cameron, J. Crouzet & F. Blanche (1993) *J. Bacteriol.* **175**, 7430.

PRECORRIN-2 C20-METHYLTRANSFERASE

This methyltransferase [EC 2.1.1.130], also known as *S*-adenosyl-L-methionine:precorrin-2 methyltransferase, catalyzes the reaction of *S*-adenosyl-L-methionine with precorrin-2 to produce *S*-adenosyl-L-homocysteine and precorrin-3A.[1]

[1]J. Crouzet, B. Cameron, L. Cauchois, S. Rigault, M.-C. Rouyez, F. Blanche, D. Thibaut & L. Debussche (1990) *J. Bacteriol.* **172**, 5980.

PRECORRIN-4 C11-METHYLTRANSFERASE

This methyltransferase [EC 2.1.1.133], also known as precorrin-3 methylase, catalyzes the reaction of *S*-adenosyl-L-methionine with precorrin-4 to produce *S*-adenosyl-L-homocysteine and precorrin 5: the C11 position on precorrin-4 is methylated.[1,2]

[1]J. R. Roth, J. G. Lawrence, M. Rubenfield, S. Kieffer-Higgins & G. M. Church (1993) *J. Bacteriol.* **175**, 3303.
[2]L. Debussche, D. Thibaut, B. Cameron, J. Crouzet & F. J. Blanche (1993) *J. Bacteriol.* **175**, 7430.

PRECORRIN-8X METHYLMUTASE

This enzyme [EC 5.4.1.2] catalyzes the conversion of precorrin-8X to hydrogenobyrinate. It was first isolated from *Pseudomonas denitrificans*.[1]

precorrin 8X   hydrogenobyrinate

The enzyme is also known as precorrin isomerase and hydrogenobyrinate-binding protein.

[1] D. Thibaut, M. Couder, A. Famechon, L. Debussche, F. Blanche, B. Cameron & J. Crouzet (1992) *J. Bacteriol.* **174**, 1043.

PRECORRIN-6X REDUCTASE

This oxidoreductase [EC 1.3.1.54] catalyzes the reaction of precorrin-6X with NADPH to produce precorrin-6Y and NADP+.[1]

[1] F. Blanche, D. Thibaut, A. Famechon, L. Debussche, B. Cameron & J. Crouzet (1992) *J. Bacteriol.* **174**, 1036.

PRECORRIN-6Y C5,15-METHYL-TRANSFERASE (DECARBOXYLATING)

This methyltransferase [EC 2.1.1.132], also known as precorrin-6 methyltransferase and precorrin-6Y methylase, reportedly catalyzes the reaction of two molecules of S-adenosyl-L-methionine with precorrin-6Y to produce two molecules of S-adenosyl-L-homocysteine, precorrin-8X, and carbon dioxide (a C-12 decarboxylation).[1] The enzyme has been isolated from *Pseudomonas denitrificans*.

[1] F. Blanche, A. Famechon, D. Thibaut, L. Debussche, B. Cameron & J. Crouzet (1992) *J. Bacteriol.* **174**, 1050.

PRENYLCYSTEINE LYASE

This enzyme [EC 4.4.1.18] catalyzes the hydrolysis of a prenyl-L-cysteine to produce a prenol and L-cysteine.[1] Substrates include farnesyl-L-cysteine and geranylgeranyl-L-cysteine. Smaller S-prenylated cysteinyl-containing peptides are not substrates.

[1] L. Zhang, W. R. Tschantz & P. J. Casey (1997) *JBC* **272**, 23354.

PRENYL-PYROPHOSPHATASE

This enzyme [EC 3.1.7.1], also known as prenyl-diphosphatase, catalyzes the hydrolysis of a prenyl-pyrophosphate (or prenyl-diphosphate) to produce a prenol and pyrophosphate.[1,2] The best substrate for the rat enzyme is farnesyl pyrophosphate, but isopentenyl pyrophosphate is hydrolyzed more slowly.

[1] S. C. Tsai & J. L. Gaylor (1966) *JBC* **241**, 4043.
[2] L. M. Pérez, G. Taucher & O. Cori (1980) *Phytochemistry* **19**, 183.

PREPHENATE AMINOTRANSFERASE

This pyridoxal-phosphate-dependent aminotransferase catalyzes the reversible reaction of prephenate with L-glutamate to produce L-arogenate and α-ketoglutarate (or 2-oxoglutarate).[1]

prephenate   L-arogenate

L-Aspartate is another amino group donor in this reaction.

[1] R. Bentley (1990) *Crit. Rev. Biochem. Mol. Biol.* **25**, 307.

Selected entries from *Methods in Enzymology* [**vol**, page(s)]:
General discussion: 142, 479
Assay: 142, 480

PREPHENATE DEHYDRATASE

This dehydratase [EC 4.2.1.51] catalyzes the conversion of prephenate to phenylpyruvate, water, and carbon dioxide.[1,2]

prephenate   phenylpyruvate

The enterobacterial enzyme is bifunctional and contains distinct chorismate mutase [EC 5.4.99.5] and prephenate dehydratase domains. The *Escherichia coli* enzyme, designated P-protein, also possesses a domain for feedback regulation by L-phenylalanine.

[1] S. Zhang, D. B. Wilson & B. Ganem (2000) *Biochemistry* **39**, 4722.
[2] R. Bentley (1990) *Crit. Rev. Biochem. Mol. Biol.* **25**, 307.

Selected entries from *Methods in Enzymology* [vol, page(s)]:
General discussion: 17A, 564; 142, 432, 440, 507
Assay: 142, 432, 507
Other molecular properties: chorismate mutase, with (assays, 142, 432; properties, 142, 435; purification from *Escherichia coli*, 142, 433); inhibition, 63, 180; monofunctional (assay, 142, 507; properties, 142, 511; purification from *Bacillus subtilis*, 142, 509); product inhibition, progress curve analysis, 63, 180; 249, 72; purification from *Aerobacter aerogenes*, 17A, 571

PREPHENATE DEHYDROGENASE

This oxidoreductase [EC 1.3.1.12] catalyzes the reaction of prephenate with NAD^+ to produce 4-hydroxyphenyl-pyruvate, carbon dioxide, and NADH and a proton.[1,2]

prephenate   4-hydroxyphenylpyruvate

In enteric bacteria, this enzyme also possesses chorismate mutase activity and converts chorismate into prephenate. The prephenate dehydrogenase activity of bacterial bifunctional enzyme chorismate mutase-prephenate dehydrogenase catalyzes hydride transfer to the B side of NAD^+. [13]C primary kinetic isotope effect studies[1] suggest that the decarboxylation mechanism is concerted.

Prephenate dehydrogenase [EC 1.3.1.13] catalyzes the reaction of prephenate with $NADP^+$ to produce 4-hydroxy-phenylpyruvate, carbon dioxide, and NADPH.

[1] J. D. Hermes, P. A. Tipton, M. A. Fisher, M. H. O'Leary, J. F. Morrison & W. W. Cleland (1984) *Biochemistry* 23, 6263.
[2] M. H. O'Leary (1992) *The Enzymes*, 3rd ed., 20, 235.
Selected entries from *Methods in Enzymology* [vol, page(s)]:
General discussion: 17A, 562, 564; 142, 440, 503
Assay: 17A, 562; 142, 442, 503
Other molecular properties: chorismate mutase, with (assays, 142, 442; properties, 142, 448; purification from *Escherichia coli*, 142, 444; substrate preparation, 142, 440); monofunctional (assay, 142, 503; properties, 142, 507; purification from *Alcaligenes eutrophus*, 142, 504); purification from *Aerobacter aerogenes*, 17A, 567

PREPILIN PEPTIDASE

This aspartic peptidase [EC 3.4.23.43], also known as type IV prepilin peptidase and type IV prepilin leader peptidase, participates in the processing of type IV pili (fimbriae) produced by many Gram-negative pathogens. Pili consist of repeating subunits that are first synthesized

as precursors (type IV prepilin). The endopeptidase catalyzes the release of the short (5-8 aminoacyl residues) leader peptide. The consensus cleavage site is at -Gly-^-(Phe/Met)-Thr-Leu-(Ile/Leu)-Glu-. The glycyl residue at the P1 position is absolutely required.[1-3] Following hydrolysis, the enzyme methylates the new N-terminus using *S*-adenosyl-L-methionine, producing the methylated protein and *S*-adenosyl-L-homocysteine.

[1] A. P. Pugsley (1993) *Microbiol. Rev.* 57, 50.
[2] M. Hobbs & J. S. Mattick (1993) *Mol. Microbiol.* 10, 233.
[3] C. F. Lapointe & R. K. Taylor (2000) *JBC* 275, 1502.
Selected entries from *Methods in Enzymology* [vol, page(s)]:
General discussion: 235, 527; 244, 484
Assay: 244, 485

PRESENILIN 1

This aspartic endopeptidase, also known as γ-secretase, regulates the processing of β-amyloid precursor protein C-terminal fragments and the generation of amyloid β-protein in the endoplasmic reticulum and Golgi body.[1] A deficiency of presenilin 1 is associated with a decrease in the production of amyloid β-protein.

[1] W. M. Xia, J. M. Zhang, B. L. Ostaszewski, W. T. Kimberly, P. Seubert, E. H. Koo, J. Shen & D. J. Selkoe (1998) *Biochemistry* 37, 16465.

PRESENILIN 2

This aspartic endopeptidase, also known as γ-secretase (putative), assists presenilin 1 in the generation of amyloid β-protein.[1]

[1] W. T. Kimberly, W. Xia, T. Rahmati, M. S. Wolfe & D. J. Selkoe (2000) *JBC* 275, 3173.

PROCOLLAGEN C-ENDOPEPTIDASE

This metalloendopeptidase [EC 3.4.24.19], also known as procollagen C-proteinase (PCP) and carboxy-terminal procollagen peptidase, catalyzes the cleavage of the C-terminal propeptide at Ala–Asp in type I and II procollagens (on the α1(I), α2(I), and α1(II) chains) and at Arg–Asp in chick type III collagen (on the α1(III) chain; in human type III collagen, cleavage is at Gly–Asp). It will also act on the respective procollagen processing intermediates containing the C-terminal propeptides but not the N-propeptides. The enzyme is specific for an aspartyl residue at position P1'. Interestingly, it will also act on pro-lysyl oxidase (at Gly–Asp) and on the prolaminin γ2 chain (at Gly–Asp). The pH_{opt} is between 8.0 and 8.5 and the enzyme requires calcium ions for activity.[1]

[1] K. E. Kadler & R. B. Watson (1995) *MIE* 248, 771.

Selected entries from *Methods in Enzymology* [vol, page(s)]:
General discussion: 248, 771
Assay: 82, 305; 248, 776
Other molecular properties: activity, 248, 771; catalyzed reactions, 248, 772; chicken (properties, 248, 774; purification, 248, 777; storage, 248, 779); discovery, 248, 773; forms, 248, 774; murine fibroblast (properties, 248, 773; purification, 248, 779; storage, 248, 781); nomenclature, 248, 775; pH optimum, 248, 775; properties, 248, 775; purification techniques, 82, 314

PROCOLLAGEN III N-ENDOPEPTIDASE

This metalloendopeptidase catalyzes the cleavage of the N-terminal propeptide from native procollagen III (at a Pro–Gln bond), but not from procollagens I, IV, or from denatured procollagen III. Enzyme activity requires the presence of calcium ions.[1,2]

[1] B. V. Nusgens, Y. Goebels, H. Shinkai & C. Lapière (1980) *BJ* 191, 699.
[2] R. Halila & L. Peltonen (1984) *Biochemistry* 23, 1251.

PROCOLLAGEN N-ENDOPEPTIDASE

This zinc-dependent metalloendopeptidase [EC 3.4.24.14], also known as procollagen N-proteinase, catalyzes the cleavage of the N-propeptide of collagen chain $\alpha 1$(I) in type I procollagen and the $\alpha 1$(II) chain in type II procollagen, both at Pro–Gln. It will also catalyze hydrolysis of an Ala–Gln peptide bond in the $\alpha 2$(I) chain in type I procollagen. Interestingly, no cleavages are observed with heat denatured procollagens. Procollagen III and collagen XIV are not substrates. Calcium ions are required for activity.[1,2]

[1] L. Tuderman, K. I. Kivirikko & D. J. Prockop (1978) *Biochemistry* 17, 2948.
[2] A. Colige, S.-W. Li, A. L. Sieron, B. V. Nusgens, D. J. Prockop & C. M. Lapière (1997) *PNAS* 94, 2374.

Selected entries from *Methods in Enzymology* [vol, page(s)]:
Assay: 82, 305; type I/II, 248, 763; type III, 248, 764
Other molecular properties: activity, 248, 756, 760; catalyzed reactions, 248, 757; discovery, 248, 758; disease, in, 248, 761; inhibition, 248, 760; nomenclature, 248, 763; physiological functions, 248, 756, 761; procollagen substrate, 248, 758; properties, 248, 758; purification techniques, 82, 312; 248, 756; type I/II, 248, 756 (assay, 248, 763; purification, 248, 765 [from bovine, 248, 768]; storage, 248, 768); type III, 248, 757, 760 (assay, 248, 764; purification, 248, 770; storage, 248, 771); types, 248, 758

PROCOLLAGEN GALACTOSYLTRANSFERASE

This transferase [EC 2.4.1.50], also known as hydroxylysine galactosyltransferase, catalyzes the reaction of UDP-D-galactose with a 5-hydroxy-L-lysyl residue in procollagen to produce UDP and procollagen with a 5-(galactosyloxy)-L-lysyl residue.[1] *See also Procollagen Glucosyltransferase*

[1] R. Myllylä, L. Risteli & K. I. Kivirikko (1975) *EJB* 52, 401.

Selected entries from *Methods in Enzymology* [vol, page(s)]:
General discussion: 28, 632; 82, 245, 284
Assay: 28, 634; 82, 287, 291
Other molecular properties: isolation from rat kidney, 28, 635; plasma membrane, in, 22, 127, 128; properties, 28, 635; 82, 301; purification, 82, 299; sources, 28, 636

PROCOLLAGEN GLUCOSYLTRANSFERASE

This glucosyltransferase [EC 2.4.1.66], also known as galactosylhydroxylysine-glucosyltransferase, catalyzes the reaction of UDP-D-glucose with a 5-(D-galactosyloxy)-L-lysyl residue in procollagen to produce a 1,2-D-glucosyl-5-D-(galactosyloxy)-L-lysyl residue in procollagen plus UDP. The enzyme works in conjunction with procollagen galactosyltransferase [EC 2.4.1.50] to produce glucosylgalactosyl-hydroxylysyl units of collagens, basement membranes, and serum glycoproteins. The glucosyltransferase does not appear to act on the intact triple helix. The enzyme has an ordered Bi Bi kinetic mechanism.[1] Evidence for a random Bi Bi mechanism with a dead-end complex has also been presented.[2] *See also Procollagen Galactosyltransferase*

[1] K. I. Kivirikko & R. Myllyla (1982) *MIE* 82, 245.
[2] D. F. Smith, C. Wu & G. A. Jamieson (1977) *BBA* 483, 263.

Selected entries from *Methods in Enzymology* [vol, page(s)]:
General discussion: 28, 625; 40, 353; 82, 245, 284
Assay: 28, 627; 82, 287, 291
Other molecular properties: isolation from rat kidney, 28, 625; plasma membrane, in, 22, 127; properties, 82, 297; purification, 82, 294; sources, 28, 631

PROCOLLAGEN-LYSINE 5-DIOXYGENASE

This iron- and ascorbate-dependent oxidoreductase [EC 1.14.11.4] (also known as procollagen-lysine:α-ketoglutarate 5-dioxygenase, procollagen-lysine:2-oxoglutarate 5-dioxygenase, and lysine hydroxylase) catalyzes the reaction of an L-lysyl residue in procollagen with α-ketoglutarate (or, 2-oxoglutarate) and dioxygen to produce procollagen containing a 5-hydroxy-L-lysyl residue, succinate, and carbon dioxide.[1–6] The lysyl residue is always within the sequence –Xaa–Lys–Gly–. The enzyme has an ordered Ter Ter kinetic mechanism.[1]

[1] U. Puistola, T. M. Turpeenniemi-Hujanen, R. Myllylä & K. I. Kivirikko (1980) *BBA* 611, 40 and 51.
[2] K. I. Kivirikko & T. Pihlajaniemi (1998) *AE* 72, 325.
[3] B. G. Fox (1998) *CBC* 3, 261.
[4] S. Englard & S. Seifter (1986) *Ann. Rev. Nutr.* 6, 365.
[5] O. Hayaishi, M. Nozaki & M. T. Abbott (1975) *The Enzymes*, 3rd ed., 12, 119.
[6] V. Ullrich & W. Duppel (1975) *The Enzymes*, 3rd ed., 12, 253.

Selected entries from *Methods in Enzymology* [vol, page(s)]:
General discussion: 52, 12; 82, 245

PROCOLLAGEN-PROLINE: α-KETOGLUTARATE 3-DIOXYGENASE

This iron- and ascorbate-dependent oxidoreductase [EC 1.14.11.7], also known as procollagen-proline 3-dioxygenase and prolyl 3-hydroxylase, catalyzes the reaction of an L-prolyl residue in procollagen with α-ketoglutarate (or, 2-oxoglutarate) and dioxygen to produce a *trans*-3-hydroxy-L-prolyl residue in procollagen, succinate, and carbon dioxide.[1–3] Specificity studies indicate that the prolyl residue must be adjacent to a 4-hydroxyprolyl-glycyl sequence: that is, –Pro–4Hyp–Gly–. The rate of collagen synthesis appears to affect the extent of proline 3-hydroxylation in newly synthesized collagen.[1]

[1] K. Majamaa, R. Myllylä, K. Alitalo & A. Vaheri (1982) *BJ* **206**, 499.
[2] H. Mori, T. Shibasaki, K. Yano & A. Ozaki (1997) *J. Bacteriol.* **179**, 5677.
[3] K. I. Kivirikko & T. Pihlajaniemi (1998) *AE* **72**, 325.

Selected entries from *Methods in Enzymology* [**vol**, page(s)]:
General discussion: **82**, 245
Assay: **82**, 248

PROCOLLAGEN-PROLINE: α-KETOGLUTARATE 4-DIOXYGENASE

This iron- and ascorbate-dependent oxidoreductase [EC 1.14.11.2], also known as procollagen-proline:2-oxo-glutarate-4-dioxygenase and prolyl hydroxylase, catalyzes the reaction of an L-prolyl residue in procollagen (at –Xaa–Pro–Gly– or –Xaa–Pro–Ala–) with α-ketoglutarate and dioxygen to produce a *trans*-4-hydroxy-L-prolyl residue, succinate, and carbon dioxide.[1–7] One of the subunits of the dioxygenase is the multifunctional protein disulfide isomerase, but the isomerase is not needed for hydroxylation of prolyl residues.

[1] K. I. Kivirikko & T. Pihlajaniemi (1998) *AE* **72**, 325.
[2] B. G. Fox (1998) *CBC* **3**, 261.
[3] J. V. Schloss & M. S. Hixon (1998) *CBC* **2**, 43.
[4] O. Hayaishi, M. Nozaki & M. T. Abbott (1975) *The Enzymes*, 3rd ed., **12**, 119.
[5] V. Ullrich & W. Duppel (1975) *The Enzymes*, 3rd ed., **12**, 253.
[6] G. J. Cardinale & S. Udenfriend (1974) *AE* **41**, 245.
[7] R. Kuttan & A. N. Radhakrishnan (1973) *AE* **37**, 273.

Selected entries from *Methods in Enzymology* [**vol**, page(s)]:
General discussion: **17B**, 306; **52**, 12; **82**, 245; **107**, 361; **144**, 96
Assay: **82**, 248, 262
Other molecular properties: antibody preparation, **144**, 101; catalytic properties, **82**, 280; cDNA clones, preparation, **144**, 103; collagen (activity with carboxymethylated *Ascaris* cuticle collagen, **17B**, 316; assay of, **17B**, 307; decarboxylation of α-ketoglutarate and, **17B**, 306; hydroxylation of proline polypeptides, **17B**, 307, 314; purification from

skin, rat, **17B**, 306, 312); properties, **82**, 267; **144**, 103; purification techniques, **82**, 266; **144**, 97; β subunit-related protein (molecular properties, **82**, 273; purification techniques, **82**, 271); subunits, isolation, **144**, 100

PROGESTERONE MONOOXYGENASE

This oxidoreductase [EC 1.14.99.4], also known as progesterone hydroxylase, catalyzes the reaction of progesterone with dioxygen and an electron donor (AH₂; this donor can be NADPH, which is probably the physiological substrate) to produce testosterone acetate, water, and the oxidized donor (A; that is, $NADP^+$).

progesterone   testosterone acetate

The enzyme has a wide specificity, including substrates such as pregna-4,16-diene-3,20-dione, 17α-hydroxypregn-4-ene-3,20-dione, deoxycorticosterone, and 16α,17α-oxidopregn-4-ene-3,20-dione.

A single enzyme from *Cylindrocarpon radicicola*, known as ketosteroid monooxygenase [EC 1.14.13.54], catalyzes both this reaction and that of androst-4-ene-3,17-dione monooxygenase [EC 1.14.99.12].[1,2]

[1] M. A. Rahim & C. J. Sih (1966) *JBC* **241**, 3615.
[2] E. Itagaki (1986) *J. Biochem.* **99**, 815 and 825.

PROGESTERONE 11α-MONOOXYGENASE

This oxidoreductase [EC 1.14.99.14], also known as progesterone 11α-hydroxylase, catalyzes the reaction of progesterone with dioxygen and an electron donor (AH₂: such as NADPH) to produce 11α-hydroxyprogesterone, water, and the oxidized donor (A: such as $NADP^+$).[1,2]

progesterone   11α-hydroxyprogesterone

In a number of organisms (for example, *Aspergillus ochraceus* and *Rhizopus nigricans*), this is a cytochrome P450-dependent system.

[1]D. Ghosh & T. B. Samanta (1981) *J. Steroid Biochem.* **14**, 1063.

[2]C. R. Jayanthi, P. Madyastha & K. M. Madyastha (1982) *BBRC* **106**, 1262.

Selected entries from *Methods in Enzymology* [vol, page(s)]:
General discussion: 52, 20

and, **17B**, 253; cytochrome *c* and, **17B**, 252, 253; preparation from [kidney, beef, **17B**, 268; liver, rat, **17B**, 252]; Δ^1-pyrroline-3-hydroxy-5-carboxylate preparation, **17B**, 268; requirement for electron acceptor, **17B**, 253; ubiquinone and, **17B**, 253)

PROGESTERONE 5α-REDUCTASE

This oxidoreductase [EC 1.3.1.30], also known as steroid 5α-reductase, catalyzes the reaction of progesterone with NADPH to produce 5α-pregnane-3,20-dione and NADP$^+$.[1–3]

progesterone   5α-pregnane-3,20-dione

Other substrates include testosterone and 20α-hydroxy-4-pregnen-3-one. The enzyme has an ordered Bi Bi kinetic mechanism.[2,3]

[1]P. J. Bertics & H. J. Karavolas (1984) *J. Steroid Biochem.* **21**, 305.

[2]J. S. Campbell & H. J. Karavolas (1989) *J. Steroid Biochem.* **32**, 283.

[3]J. S. Campbell, P. J. Bertics & H. J. Karavolas (1986) *J. Steroid Biochem.* **24**, 801.

PROLINE DEHYDROGENASE

This FAD-dependent oxidoreductase [EC 1.5.99.8], also known as proline oxidase, catalyzes the reaction of L-proline with water and an acceptor substrate (*e.g.*, 2,6-dichlorophenolindophenol, phenazine methosulfate, ferricyanide, menadione, and cytochrome *c*) to produce (*S*)-1-pyrroline-5-carboxylate and the reduced acceptor.[1–4]

L-proline   (*S*)-1-pyrroline-5-carboxylate

4-Methylene-L-proline is a potent mechanism-based inhibitor of the rat liver enzyme.[1] In many organisms, the enzyme will also act on *trans*-4-hydroxy-L-proline to produce Δ^1-pyrroline-3-hydroxy-5-carboxylate. *See also Allohydroxy-D-Proline Oxidase*

[1]D. Tritsch, H. Mawlawi & J. F. Biellmann (1993) *BBA* **1202**, 77.

[2]S. B. Graham, J. T. Stephenson & J. M. Wood (1984) *JBC* **259**, 2656.

[3]L. Meile & T. Leisinger (1982) *Eur. J. Biochem.* **129**, 67.

[4]R. C. Scarpulla & R. L. Soffer (1978) *JBC* **253**, 5997.

Selected entries from *Methods in Enzymology* [vol, page(s)]:
Other molecular properties: hydroxyproline oxidase, rat kidney, **17B**, 267, 268; proline oxidase, **17B**, 251 (assay, **17B**, 251; coenzyme Q$_{10}$

PROLINE RACEMASE

This enzyme [EC 5.1.1.4] catalyzes the interconversion of L-proline and D-proline.[1–5] The bacterial enzyme is a homodimer with residues from both subunits comprising a single active site. Catalysis occurs by a two-base mechanism: proton abstraction alternates from the two faces of the proline ring, and the transition state is an enzyme-proline carbanion sandwiched between the two catalytic thiols.

[1]T. C. Stadtman & P. Elliott (1957) *JBC* **228**, 983.

[2]G. Rudnick & R. H. Abeles (1975) *Biochemistry* **14**, 4515.

[3]L. M. Fisher, J. G. Belasco, T. W. Bruice, W. J. Albery & J. R. Knowles (1986) *Biochemistry* **25**, 2543.

[4]M. E. Tanner & G. L. Kenyon (1998) *CBC* **2**, 7.

[5]E. Adams (1972) *The Enzymes*, 3rd ed., **6**, 479.

Selected entries from *Methods in Enzymology* [vol, page(s)]:
General discussion: 5, 875; **46**, 23
Assay: 5, 875
Other molecular properties: *Clostridium* (catalysis, proton transfer in, **249**, 501; rate constants determined by induced transport, **249**, 237); free energy profile, **308**, 11; ligand-independent recycling, **249**, 328; proline reductase and, **5**, 870, 872; properties, **5**, 877; purification, **5**, 877; transition-state and multisubstrate analogues, **249**, 308

D-PROLINE REDUCTASE

This activity, originally classified as EC 1.4.1.6, is now included with D-proline reductase (dithiol) [EC 1.4.4.1]. *See D-Proline Reductase (Dithiol)*

D-PROLINE REDUCTASE (DITHIOL)

This pyruvoyl-dependent oxidoreductase [EC 1.4.4.1] catalyzes the reaction of D-proline with dihydrolipoate to produce 5-aminopentanoate and lipoate.[1–4]

D-proline   5-aminopentanoate

Other dithiols can function as reducing agents. The *Clostridium sticklandii* protein is a selenoenzyme.[1] In the course of the reaction, ^{18}O atoms from the carboxyl group of D-proline are not lost during conversion to product and it has been suggested that reduction of proline proceeds without the formation of an acyl enzyme intermediate.[2] The substrate forms a Schiff base with the pyruvoyl group.[4]

[1]U. C. Kabisch, A. Grantzdorffer, A. Schierhorn, K. P. Rucknagel, J. R. Andreesen & A. Pich (1999) *JBC* **274**, 8445.

[2]R. A. Arkowitz, S. Dhe-Paganon & R. H. Abeles (1994) *ABB* **311**, 457.

[3]P. D. van Poelje & E. E. Snell (1990) *ARB* **59**, 29.

[4]P. A. Recsei & E. E. Snell (1984) *ARB* **53**, 357.

Selected entries from *Methods in Enzymology* [vol, page(s)]:
General discussion: 5, 870; 17B, 317
Assay: 5, 870; 17B, 317
Other molecular properties: proline racemase assay and, 5, 876; properties, 5, 874; 17B, 320; purification from *Clostridium sticklandii*, 5, 872; 17B, 318; pyruvate in, 17B, 321

PROLYL AMINOPEPTIDASE

This serine-dependent aminopeptidase [EC 3.4.11.5] (also known as proline aminopeptidase, proline iminopeptidase (PIP), Pro-Xaa aminopeptidase, and cytosol aminopeptidase V) catalyzes the release of an N-terminal proline from a peptide.[1–5] The enzyme isolated from some sources will also catalyze the release of an N-terminal hydroxyproline. Manganese ion, originally described as an activator of the *Escherichia coli* enzyme, inhibits most prolyl aminopeptidases. The best substrates are di- and tripeptides (although there are reports that the bacterial enzyme will act on polyproline). The presence of this enzyme in mammals has been controversial.[4] The mammalian enzyme may be identical to leucyl aminopeptidase [EC 3.4.11.1].

[1]A. Kitazono, K. Ito & T. Yoshimoto (1994) *J. Biochem.* **116**, 943.

[2]K. Mäkinen (1969) *Acta Chem. Scand.* **23**, 1409.

[3]K. Ninomiya, K. Kawatani, S. Tanaka, S. Kawata & S. Makisumi (1982) *J. Biochem.* **92**, 413.

[4]E. Heymann & K. Peter (1993) *Biol. Chem. Hoppe-Seyler* **374**, 1033.

[5]R. J. DeLange & E. L. Smith (1971) *The Enzymes*, 3rd ed., **3**, 81.

PROLYL OLIGOPEPTIDASE

This serine endopeptidase [EC 3.4.21.26], also known as prolyl endopeptidase and post-proline cleaving enzyme) catalyzes the hydrolysis of Pro–Xaa bonds in oligopeptides.[1,2] Naturally occurring substrates include bradykinin, substance P, neurotensin, angiotensin II, vasopressin, oxytocin, dynorphin, gonadoliberin, thyrotropin-releasing hormone, luteinizing hormone-releasing hormone, and α-melanocyte-stimulating hormone. Large, denatured proteins (for example, denatured casein or collagen) are not substrates.

Prolyl oligopeptidase, which has been discovered to be widely represented in vertebrates, plants, and microorganisms, can be assayed with Z-Gly-Pro-NHNap, Z-Gly-Pro-NHPhNO$_2$, and Z-Gly-Pro-NHMec.

[1]E. Shaw (1990) *AE* **63**, 271.

[2]J. S. Bond & P. E. Butler (1987) *ARB* **56**, 333.

Selected entries from *Methods in Enzymology* [vol, page(s)]:
General discussion: 244, 188; 330, 445
Assay: 168, 369, 370; fluorescence detection, 244, 189; fluorimetric, 248, 601; principle, 244, 188; LHRH-degrading, assay, 103, 546; substrates, 244, 189
Other molecular properties: active site residues, 244, 41, 200; *cis-trans* isomeric specificity, 244, 195; cleavage site specificity, 244, 196; concentration, calculation, 244, 190; deuterium isotope effects, 244, 199; family members, 244, 42; hyperthermophilic, 330, 445; inhibitors, 244, 197; interference with glutaminylpeptide cyclase assay, 168, 363; ionic strength effects, 244, 194; LHRH-degrading, assay, 103, 546; mechanism of action, 244, 198; organic solvent effects, 244, 194; pH dependence, 244, 198; processing, 244, 43; purification, 244, 191; species distribution, 244, 40, 42, 193; structure, 244, 200; substrate specificity, 244, 195; thiol dependence, 244, 43, 198

PROLYL-tRNA SYNTHETASE

This class II aminoacyl-tRNA synthetase [EC 6.1.1.15], also known as proline:tRNA ligase and proline translase, catalyzes the reaction of ATP with L-proline and tRNA^{Pro} to produce L-prolyl-tRNA^{Pro}, AMP, and pyrophosphate.[1,2] L-Azetidine-2-carboxylate can also serve as a substrate.

The genome sequences of certain archaea do not contain a recognizable gene for cysteinyl-tRNA synthetase. Studies have indicated that archaeal prolyl-tRNA synthetase can synthesize both cysteinyl-tRNA^{Cys} and prolyl-tRNA^{Pro}.[3,4] In several eukaryotes, a glutamyl-tRNA synthetase activity and aprolyl-tRNA synthetase activity are two different domains of a single polypeptide chain.[5] *See also Aminoacyl-tRNA Synthetases*

[1]T. S. Papas & A. H. Mehler (1971) *JBC* **246**, 5924.

[2]S. J. Norton (1964) *ABB* **106**, 147.

[3]C. Stathopoulos, T. Li, R. Longman, U. C. Vothknecht, H. D. Becker, M. Ibba & D. Soll (2000) *Science* **287**, 479.

[4]M. Yarus (2000) *Science* **287**, 440.

[5]S. M. Ting, P. Bogner & J. D. Dignam (1992) *JBC* **267**, 17701.

Selected entries from *Methods in Enzymology* [vol, page(s)]:
General discussion: 34, 170
Other molecular properties: large-scale isolation, 22, 535, 541; mechanism, 29, 629; subcellular distribution, 59, 233, 234

PRONASE

Pronase, a commercial mixture of extracellular enzymes, has proven its value in the proteolytic fragmentation of proteins. *See Streptogrisin A; Streptogrisin B; Mycolysin; Streptomyces griseus Aminopeptidase*

Selected entries from *Methods in Enzymology* [vol, page(s)]:
General discussion: 19, 651; 44, 390; 46, 426
Assay: 19, 651; aminopeptidase activity, 19, 655; carboxypeptidase activity, 19, 657; esterase activity, 19, 653; proteinase activity, 19, 651
Other molecular properties: alkaline phosphatase and, 12B, 218; aminopeptidase hydrolysis and, 25, 259; basic protein removal and, 12B, 656; chromatography on immobilized leupeptin derivative, 80, 847, 848; cleavage of peptides and proteins, in, 11, 230 (procedure for, 11, 231);

coupling to 1,1'-carbonyldiimidazole-activated support, **135**, 115; detection, zymographic technique, **235**, 578; digestion of proteins, **40**, 35 (concentration and incubation period, **40**, 37); dissociation of gastric mucosa into individual cells, in, **171**, 446; immobilization, **47**, 44; immobilized, in analysis of protein structural changes, **135**, 598; infectious ribonucleic acid and, **12B**, 551; inhibition, assay, **45**, 861, 870; iodinated proteins and, **25**, 443; kinetics, of substrate diffusion, **44**, 437; lens proteins and, **30**, 688; modification of erythrocytes, **236**, 226; mouse mRNA preparation, in, **79**, 94; mRNA extraction and, **30**, 651; nucleic acid preparation and, **12B**, 109, 110, 114; oxidation with NBS, **11**, 521; polysaccharide release and, **28**, 78; preparation of (adenovirus 2 DNA, **65**, 773; chloroplast DNA, **65**, 786); pronase B in adenovirus DNA preparation, **79**, 313; properties, **19**, 663; purification, **19**, 658; recipient cell purification and, **21**, 160, 161; ribonuclease and, **12B**, 433; ribonucleoprotein and, **30**, 348; ribosome digestion by, **30**, 581; source, **32**, 707; specificity, **11**, 230; glycopeptide preparation, role in, **28**, 14); tissue culture, **58**, 125, 126, 129; trypsin and, **19**, 613; use in affinity chromatography, **22**, 349; zymography in nondissociating gels with copolymerized substrates, **235**, 588

PROPANEDIOL DEHYDRATASE

This cobalamin-dependent dehydratase [EC 4.2.1.28], also known as diol dehydratase, catalyzes the conversion of propane-1,2-diol ($CH_3CH(OH)CH_2OH$; both stereoisomers are substrates) to propanal and water.[1-5] The enzyme will also catalyze the dehydration of ethylene glycol to acetaldehyde and glycerol to 3-hydroxypropanal. A potassium-ion-dependent mechanism has been proposed in which K^+ assists in the 1,2-shift of the hydroxyl group in the substrate-derived radical intermediate.[4] The coenzyme-mediated abstraction of the *proR* hydrogen of the C1 position of the substrate produces the substrate radical. Loss of the proton from the hydroxyl group yields the anion radical. Elimination of the neighboring hydroxyl is followed by hydration of the aldehyde and transfer of a hydrogen from the coenzyme. This produces propane-1,1-diol. Loss of the *proS* hydroxyl generates the final product.

[1] T. Toraya & S. Fukui (1982) in B_{12} (D. Dolphin, ed.) **2**, p. 233, Wiley, New York.
[2] T. Toraya, T. Shirakashi, T. Kosuga & S. Fukui (1976) BBRC **69**, 475.
[3] R. H. Abeles (1971) The Enzymes, 3rd ed., **5**, 481.
[4] T. Toraya, K. Yoshizawa, M. Eda & T. Yamabe (1999) J. Biochem. **126**, 650.
[5] B. T. Golding & W. Buckel (1998) CBC **3**, 239.

Selected entries from *Methods in Enzymology* [vol, page(s)]:
General discussion: 9, 686; 188, 26
Assay: 9, 686
Other molecular properties: Aerobacter aerogenes (properties, **9**, 689; purification, **9**, 687); inhibition, **63**, 399; role of B_{12} coenzyme, **13**, 214; stereochemistry, **87**, 153

PROPANEDIOL 2-DEHYDROGENASE

This enzyme reportedly catalyzes the reaction of propane-1,2-diol with NAD^+ to produce hydroxyacetone and

NADH (with A-side stereospecificity) and a proton. **See also** *Propanediol-Phosphate Dehydrogenase*

1,3-PROPANEDIOL DEHYDROGENASE

This oxidoreductase [EC 1.1.1.202], also known as 3-hydroxypropionaldehyde reductase and 1,3-propanediol oxidoreductase, catalyzes the reversible reaction of propane-1,3-diol with NAD^+ to produce 3-hydroxypropanal and NADH and a proton.[1,2]

[1] E. A. Johnson & E. C. C. Lin (1987) J. Bacteriol. **169**, 2050.
[2] T. L. Talarico, L. T. Axelsson, J. Novotny, M. Fiuzat & W. J. Dobrogosz (1990) AEM **56**, 943.

PROPANEDIOL-PHOSPHATE DEHYDROGENASE

This oxidoreductase [EC 1.1.1.7] catalyzes the reversible reaction of propane-1,2-diol 1-phosphate with NAD^+ to produce hydroxyacetone phosphate and NADH and a proton.[1]

[1] O. Z. Sellinger & O. N. Miller (1959) JBC **234**, 1641.

Selected entries from *Methods in Enzymology* [vol, page(s)]:
General discussion: 9, 336
Assay: 9, 337

PROPANE OXYGENASE

This oxidoreductase catalyzes the reaction of propane with dioxygen and NADH to produce propanol, water, and NAD^+. It also catalyzes the NADH- and dioxygen-dependent conversion of propene to 1,2-epoxypropane.[1]

[1] W. Ashraf, A. Mihdhir & J. C. Murrell (1994) FEMS Microbiol. Lett. **122**, 1.

Selected entries from *Methods in Enzymology* [vol, page(s)]:
General discussion: 188, 26
Assay: 188, 28

PROPIOIN SYNTHASE

This aldolase [EC 4.1.2.35], isolated from bakers' yeast and also known as 4-hydroxy-3-hexanone aldolase, catalyzes the reaction of two molecules of propanal to generate 4-hydroxy-3-hexanone (or, propioin).[1]

[1] S. Morimoto, K. Azuma, T. Oshima & M. Sakamoto (1988) J. Ferment. Technol. **66**, 7.

PROPIONATE CoA-TRANSFERASE

This transferase [EC 2.8.3.1], also known as CoA-transferase, catalyzes the reversible reaction of acetyl-CoA with propanoate to produce acetate and propanoyl-CoA.[1,2] Other acceptor substrates include butanoate and lactate.

[1] E. R. Stadtman (1952) Fed. Proc. **11**, 291.

[2]G. Schweiger & W. Buckel (1984) *FEBS Lett.* **171**, 79.

Selected entries from *Methods in Enzymology* [vol, page(s)]:
General discussion: 1, 599
Assay: 1, 599
Other molecular properties: amino acid acetylation and, 1, 618; assay of (methylmalonyl-CoA mutase, **13**, 207; methylmalonyl-CoA epimerase, **13**, 194); butyrate oxidation and, 1, 551; properties, 1, 602; purification, *Clostridium kluyveri*, 1, 600; thiolester assay and, **3**, 935

PROPIONYL-CoA CARBOXYLASE

This biotin-dependent carboxylase [EC 6.4.1.3], often abbreviated PCCase, catalyzes the reaction of ATP with bicarbonate and propanoyl-CoA to produce ADP, orthophosphate, and (S)-methylmalonyl-CoA.[1–4] The enzyme also catalyzes the carboxylation of butanoyl-CoA (forming (S)-ethylmalonyl-CoA), isobutyryl-CoA (forming dimethylmalonyl-CoA), and crotonyl-CoA (forming glutaconyl-CoA). The first two substrates, ATP and bicarbonate, bind to the enzyme to generate the carboxybiotin-enzyme intermediate, and ADP and orthophosphate are released before the acyl-CoA binds. The α-carbon of the acyl-CoA appears to lose a proton to form a catalytically active carbanion intermediate.[4]

[1]F. Kalousek, M. D. Darigo & L. E. Rosenberg (1980) *JBC* **255**, 60.
[2]A. W. Alberts & P. R. Vagelos (1972) *The Enzymes*, 3rd ed., **6**, 37.
[3]J. Moss & M. D. Lane (1971) *AE* **35**, 1.
[4]J. Stubbe, S. Fish & R. H. Abeles (1980) *JBC* **255**, 236.

Selected entries from *Methods in Enzymology* [vol, page(s)]:
General discussion: 5, 570, 576; **13**, 181
Assay: 5, 570, 576; **13**, 181
Other molecular properties: assay for (methylmalonyl-CoA mutase, **13**, 199; methylmalonyl-CoA racemase, **13**, 190); crystallization, **13**, 185; intermediate, **87**, 88; kinetics, **87**, 355; mechanism, **13**, 189; mitochondrial, 5, 576; properties, 5, 575, 580; **13**, 188; purification (bovine liver, 5, 578; pig heart, 5, 572; **13**, 184)

PROPIONYL-CoA SYNTHETASE

This enzyme [EC 6.2.1.17], also known as propionate:CoA ligase, catalyzes the reaction of ATP with propanoate and coenzyme A to produce AMP, diphosphate, and propanoyl-CoA. Other substrates include propenoate, acetate, and butyrate. *See also Acetyl-CoA Synthetase*

Selected entries from *Methods in Enzymology* [vol, page(s)]:
General discussion: 188, 26
Assay: 14, 625; 188, 31

PROPIONYL-CoA C^2-TRIMETHYL-TRIDECANOYLTRANSFERASE

This transferase [EC 2.3.1.154], also known as 3-oxo-pristanoyl-CoA hydrolase, 3-oxopristanoyl-CoA thiolase, oxopristanoyl-CoA thiolase, and peroxisomal 3-oxoacyl coenzyme A thiolase, catalyzes the reversible reaction of 4,8,12-trimethyltridecanoyl-CoA with propanoyl-CoA to produce 3-oxopristanoyl-CoA and coenzyme A.[1,2]

4,8,12-trimethyltridecanoyl-CoA

3-oxopristanoyl-CoA

Sterols bind tightly to this enzyme, which is sometimes called sterol carrier protein. The acyltransferase also participates in the peroxisomal β-oxidation of branched chain fatty acids. Because this enzyme displays little activity toward 3-oxoacyl-CoA thioesters of 2-methyl-branched chain fatty acids, it is also distinct from acetyl-CoA *C*-myristoyltransferase [EC 2.3.1.155] and acetyl-CoA *C*-acyltransferase [EC 2.3.1.16].

[1]U. Seedorf, P. Brysch, T. Engel, K. Schrage & G. Assmann (1994) *JBC* **269**, 21277.
[2]R. J. A. Wanders, S. Denis, F. Wouters, K. W. A. Wirtz & U. Seedorf (1997) *BBRC* **236**, 565.

PROPROTEIN CONVERTASES

Proprotein convertase 1 [EC 3.4.21.93], also known as PC1, prohormone convertase I, neuroendocrine convertase 1 (or NEC 1), is a serine endopeptidase catalyzing the release of protein hormones, neuropeptides, and renin from their respective precursors. PC1 catalyzes peptide-bond cleavages in proteins with sequences: (Lys/Arg)-Arg-^-Xaa or (Arg/Lys)-(Xaa)$_n$-Arg-^-Xaa, where n is 2, 4, or 6. The enzyme requires calcium ion, and its proenzyme undergoes autocatalytic activation. Substrates include proopiomelanocortin, prorenin, proenkephalin, prodynorphin, prosomatostatin, prothyrotropin-releasing hormone, and proinsulin. The active form is stored in dense-core secretory vesicles. Unlike PC2, this enzyme does not cleave proluteinizing-hormone-releasing hormone.[1–3]

Proprotein convertase 2 [EC 3.4.21.94] (also known as PC2, prohormone convertase II, and neuroendocrine convertase 2 (NEC 2)) catalyzes the release of protein hormones and neuropeptides from their precursors. As with PC1, this serine endopeptidase preferentially cleaves peptide bonds in sequences: (Lys/Arg)-Arg-^-Xaa or (Arg/Lys)-(Xaa)$_n$-Arg-^-Xaa, where n is 2, 4, or 6;

however, PC2 does not act on prorenin or prosomato-statin. PC2 also requires calcium ion, and its proenzyme undergoes autocatalytic activation. Substrates include pro-opiomelanocortin, proenkephalin, prodynorphin, prothyrotropin-releasing hormone, and proinsulin. The specificity of PC2 appears to be broader than that of PC1. PC1 and PC2 often act together.[1,2,4]

Proprotein convertase 3 was a term often used to describe PC1.[5] Although protein PC4 has a specificity almost identical to PC1, it does not act on proenkephalin or proopiomelanocortin.[6–10] PC5 is also a serine endopeptidase that requires calcium ions, a neutral pH, and prefers substrates containing the sequence (Arg/Lys)-(Xaa)$_n$-Arg-$^\wedge$-Xaa where n is 2, 4, or 6. It acts on the sequence (Val-Gln-Arg-Glu-Lys-Arg-$^\wedge$-Ala-Val-Gly-Leu) between gp120 and gp41 within the human immunodeficiency virus-1 gp160 precursor. The enzyme has been mistakenly called PC6 and PC5/6 (PC6 proved to be identical to proprotein convertase 5).[11–13] PC7 (also known as LPC, PC8, and SPC7) is a serine endopeptidase with a neutral pH optimum, a calcium-ion dependency, and a preference for (Lys/Arg)-Arg-$^\wedge$-Xaa or (Arg/Lys)-(Xaa)$_n$-Arg-$^\wedge$-Xaa where n is 2, 4, or 6. With some substrates, the specificity resembles that of kexin and furin.[14–17] *See also* specific protein; Furin; Kex2 Protease; Kexin

[1] N. G. Seidah (1995) in *Intramolecular Chaperones and Protein Folding* (U. Shinde & M. Inouye, eds.), R. G. Landes, Austin, Texas, p. 181.
[2] N. G. Seidah & M. Chrétien (1994) *MIE* **244**, 175.
[3] D. F. Steiner, Y. Rouille, Q. Gong, S. Martin, R. Carroll & S. J. Chan (1996) *Diabetes Metab.* **22**, 94.
[4] Y. Rouille, S. J. Duguay, K. Lund, M. Furuta, Q. M. Gong, G. Lipkind, A. A. Oliva, S. J. Chan & D. F. Steiner (1995) *Front. Neuroendocrinol.* **16**, 322.
[5] S. P. Smeekens, A. S. Avruch, J. LaMendola, S. J. Chan & D. F. Steiner (1991) *PNAS* **88**, 340.
[6] K. Nakayama, W. S. Kim, S. Torii, M. Hosaka, T. Nakagawa, I. Ikemizu, T. Baba & K. Murakami (1992) *JBC* **267**, 5897.
[7] N. G. Seidah, R. Day, J. Hamelin, A. Gaspar, M. W. Collard & M. Chrétien (1992) *Mol. Endocrinol.* **6**, 1559.
[8] N. G. Seidah, R. Day & M. Chrétien (1994) *Biochimie* **76**, 197.
[9] M. Mbikay, M.-L. Raffin-Samson, H. Tadros, F. Sirois, N. G. Seidah & M. Chrétien (1994) in *Functions of Somatic Cells in the Testis* (A. Barthe, ed.), Springer-Verlag, New York, p. 388.
[10] R. E. Mains, C. A. Berard, J. B. Denault, A. Zhao, R. C. Johnson & R. Leduc (1997) *BJ* **321**, 587.
[11] J. Lusson, D. Vieau, J. Hamelin, R. Day, M. Chrétien & N. G. Seidah (1993) *PNAS* **90**, 6691.
[12] T. Nakagawa, M. Hosaka, S. Torii, T. Watanabe, K. Murakami & K. Nakayama (1993) *J. Biochem.* **113**, 132.
[13] N. G. Seidah, R. Day & M. Chrétien (1993) *Biochem. Soc. Trans.* **21**, 685.
[14] N. G. Seidah, J. Hamelin, M. Mamarbachi, W. Dong, H. Tadros, M. Mbikay, M. Chrétien & R. Day (1996) *PNAS* **93**, 3388.
[15] A. Bruzzaniti, K. Goodge, P. Jay, S. A. Taviaux, M. H. C. Lam, P. Berta, T. J. Martin, J. M. Moseley & M. T. Gillespie (1996) *BJ* **314**, 727.
[16] J. Meerabux, M. L. Yaspo, A. J. Roebroek, W. J. Van de Ven, T. A. Lister & B. D. Young (1996) *Cancer Res.* **56**, 448.
[17] D. B. Constam, M. Calfon & E. J. Robertson (1996) *J. Cell Biol.* **134**, 181.

Selected entries from **Methods in Enzymology** [vol, page(s)]:
General discussion: 244, 175
Other molecular properties: active site residues, 244, 176; biological roles, 244, 183, 187; calcium dependence, 244, 181; cleavage site specificity, 244, 180; comparative structural elements, 244, 177; detection, oligonucleotides for (consensus sequence, 244, 179; screening, 244, 176, 179); domains, 244, 180; evolution, 244, 187; family, 244, 177; genes (differential splicing, 244, 187; loci, 244, 185; promoter, 244, 186; structure, 244, 186); N-glycosylation, 244, 180; processing, 244, 178; substrates, 244, 183, 187; tissue distribution, 244, 183; tyrosine sulfation, 244, 180

3-PROPYLMALATE SYNTHASE

This enzyme [EC 4.1.3.11] catalyzes the reaction of pentanoyl-CoA with water and glyoxylate to produce 3-propylmalate ($^-$OOCCH(OH)CH(CH$_2$CH$_2$CH$_3$)COO$^-$) and coenzyme A.[1,2]

[1] K. Imai, H. C. Reeves & S. J. Ajl (1963) *JBC* **238**, 3193.
[2] W. S. Wegener, P. Furmanski & S. J. Ajl (1967) *BBA* **144**, 34.

Selected entries from **Methods in Enzymology** [vol, page(s)]:
General discussion: 13, 362
Assay: 13, 362

PROSTAGLANDIN-A$_1$ Δ-ISOMERASE

prostaglandin A$_1$ prostaglandin C$_1$

This isomerase [EC 5.3.3.9], also known as prostaglandin A isomerase, catalyzes the interconversion of prostaglandin A$_1$ (or, (13E)-(15S)-15-hydroxy-9-oxoprosta-10,13-dienoate) and prostaglandin C$_1$ (or, (13E)-(15S)-15-hydroxy-9-oxoprosta-11,13-dienoate).[1,2]

[1] R. L. Jones, S. Cammock & E. W. Horton (1972) *BBA* **280**, 588.
[2] P. P. K. Ho, R. D. Towner & H. R. Sullivan (1974) *Prep. Biochem.* **4**, 257.

PROSTAGLANDIN-D SYNTHASE

This glutathione-dependent enzyme [EC 5.3.99.2], also known as prostaglandin-H$_2$ D-isomerase, prostaglandin D$_2$ synthase (PGD$_2$ synthase), and PGH-PGD isomerase, catalyzes the conversion of prostaglandin H$_2$ (or, (5Z,13E)-(15S)-9α,11α-epidioxy-15-hydroxyprosta-5,13-dienoate) to prostaglandin D_2 (or, (5Z,13E)-(15S)-9α,15-dihydroxy-11-oxoprosta-5,13-dienoate).[1–6]

prostaglandin H_2

prostaglandin D_2

arachidonate

prostaglandin G_2

prostaglandin H_2

This enzyme also converts prostaglandin G_2 to 15-hydroperoxyprostaglandin D_2. A unique cleft, which distinguishes this enzyme from other glutathione S-transferases, is thought to be responsible for the specific isomerization from PGH_2 to PGD_2.[3] The enzyme apparently has two functions: first as a PGD_2-producing enzyme, and second as a high-affinity lipophilic ligand-binding protein for retinoids, thyroids, and bile pigments.[4,5]

[1] T. Shimizu, S. Yamamoto & O. Hayaishi (1979) JBC 254, 5222.
[2] D. H. Nugteren & E. Hazelhof (1973) BBA 326, 448.
[3] Y. Kanaoka, H. Ago, E. Inagaki, T. Nanayama, M. Miyano, R. Kikuno, Y. Fujii, N. Eguchi, H. Toh, Y. Urade & O. Hayaishi (1997) Cell 90, 1085.
[4] Y. Urade & O. Hayaishi (2000) Vitam. Horm. 58, 89.
[5] Y. Urade, K. Watanabe & O. Hayaishi (1995) J. Lipid Mediat. Cell Signal. 12, 257.
[6] C. R. Pace-Asciak & W. L. Smith (1983) The Enzymes, 3rd ed., 16, 543.

Selected entries from **Methods in Enzymology** [vol, page(s)]:
General discussion: 86, 73, 77
Assay: 86, 73, 78

PROSTAGLANDIN ENDOPEROXIDE SYNTHASE

This heme-dependent oxidoreductase [EC 1.14.99.1], also known as prostaglandin synthase and prostaglandin G/H synthase (PG synthase), catalyzes the reaction of arachidonate with a reduced donor (AH_2; typically, two molecules of glutathione) and two molecules of dioxygen to produce prostaglandin H_2, the oxidized donor (A; e.g., glutathione disulfide), and water.[1–15]

The synthase first produces the hydroperoxide, prostaglandin G_2, by the sequential reaction of two dioxygens with a free radical of arachidonate. The radical of prostaglandin G_2 is converted to prostaglandin G_2 by hydrogen atom abstraction. The peroxidase activity of the synthase generates prostaglandin H_2 from this intermediate. Constitutive and inducible isozymes are known as isozymes 1 and 2, respectively.

[1] H.-H. Tai, C. L. Tai & C. S. Hollander (1976) BJ 154, 257.
[2] Y. Friedman, M. Lang & G. Burke (1975) BBA 397, 331.
[3] P. Wlodawer & B. Samuelsson (1973) JBC 248, 5673.
[4] H. B. Dunford (1998) CBC 3, 195.
[5] W. L. Smith, R. M. Garavito & D. L. DeWitt (1996) JBC 271, 33157.
[6] D. Picot, P. J. Loll & M. Garavito (1994) Nature 367, 243.
[7] A.-L. Tsai, L. C. Hsu, R. J. Kulmacz, G. Palmer & W. L. Smith (1994) JBC 269, 5085.
[8] S. Yamamoto & Y. Ishimura (1991) in A Study of Enzymes (S. A. Kuby, ed.), vol. 2, p. 315, CRC Press, Boca Raton.
[9] F. P. Guengerich (1990) Crit. Rev. Biochem. Mol. Biol. 25, 97.
[10] E. Cadenas (1989) ARB 58, 79.
[11] J. A. Stubbe (1989) ARB 58, 257.
[12] P. Needleman, J. Turk, B. A. Jakschik, A. R. Morrison & J. B. Lefkowith (1986) ARB 55, 69.
[13] A. Naqui, B. Chance & E. Cadenas (1986) ARB 55, 137.
[14] C. R. Pace-Asciak & W. L. Smith (1983) The Enzymes, 3rd ed., 16, 543.
[15] O. Hayaishi, M. Nozaki & M. T. Abbott (1975) The Enzymes, 3rd ed., 12, 119.

Selected entries from **Methods in Enzymology** [vol, page(s)]:
General discussion: 52, 19; **73**, 277, 278; **186**, 283; **319**, 67
Assay: apoenzyme, **187**, 479, 480; bovine seminal vesicle, **86**, 55; human platelet, **86**, 653; ovine seminal vesicle, **86**, 61
Other molecular properties: apoenzyme (assay, **187**, 479, 480; preparation and proteolysis, **187**, 479); bovine seminal vesicle (assay, **86**, 55; properties, **86**, 59; purification, **86**, 58); cyclooxygenase activity, activation, in assay of hydroperoxides, **186**, 435; eicosanoid generation, effects of flavonoids and coumarins, **234**, 443; fatty acid, heme requirement, **231**, 562; generation of (glutathione thiyl radicals, **186**, 324; oxygen radicals, assays, **105**, 412); heme requirement, **231**, 562; hemoglobin activity, **231**, 562; human platelet, quantification with [acetyl-^{3}H]aspirin, **86**, 397; initiation assay for hydroperoxides, **186**, 431; intermediates, **319**, 67; murine (cloning, **187**, 469; primary structure, **187**, 477); nitric oxide, interaction, **269**, 12; nitrosyl complex, detection by electron spin resonance, **268**, 169, 173; ovine (cloning, **187**, 469; primary structure, **187**, 477); ovine seminal vesicle (apoenzyme [assay, **86**, 69; preparation from holoenzyme, **86**, 71]; assay, **86**, 61; immunocytofluorescence, **86**, 218; immunoelectron microscopy, **86**, 220; immunoprecipitation, **86**, 216; immunoradiometric assay, **86**, 236; isolation, **186**, 432; monoclonal antibodies [preparation, **86**, 230; purification, **86**, 235]; properties, **86**, 65; purification, **86**, 63; rabbit antisera [applications, **86**, 216; IgG isolation, **86**, 215; preparation, **86**, 214]; radioimmunoassay, **86**, 222); oxidation of reducing cofactors, **186**, 286; oxygenation of arachidonate, **186**, 285; proteolytic fragments, peptide mapping, **187**, 484; rat brain, inactivation by microwave irradiation, **86**, 635; reaction products, HPLC, **86**, 518; trypsin-cleaved, reversed-phase HPLC, **187**, 483; trypsin digestion, **187**, 482

PROSTAGLANDIN-E$_2$ 9-REDUCTASE

This oxidoreductase [EC 1.1.1.189], also known as 9-ketoprostaglandin E$_2$ reductase and prostaglandin-E$_2$ 9-oxoreductase, catalyzes the reaction of prostaglandin E$_2$ (or, (5Z,13E)-(15S)-11,15-dihydroxy-9-oxoprosta-5,13-dienoate) with NADPH and a proton to produce prostaglandin F$_{2\alpha}$ (or, (5Z,13E)-(15S)-9,11,15-trihydroxy-prosta-5,13-dienoate) and NADP$^+$.[1-8]

prostaglandin E$_2$ prostaglandin F$_{2\alpha}$

A number of other 9-oxo- and 15-oxoprostaglandin derivatives can also be reduced to the corresponding hydroxy compound (for example, prostaglandin A$_1$ and 15-ketoprostaglandin E$_2$). This enzyme may be identical with 15-hydroxyprostaglandin dehydrogenase (NADP$^+$) [EC 1.1.1.197]. In many organisms, EC 1.1.1.189 is identical with carbonyl reductase [EC 1.1.1.184].[6,7] *See also Carbonyl Reductase (NADPH); 15-Hydroxyprostaglandin Dehydrogenases*

[1]S. Krüger & W. Schlegel (1986) *EJB* **157**, 481.
[2]W. Schlegel, S. Krüger & K. Korte (1984) *FEBS Lett.* **171**, 141.
[3]D. G.-B. Chang, M. Sun & H.-H. Tai (1981) *BBRC* **99**, 745.
[4]L. P. Thuy & M. P. Carpenter (1978) *BBRC* **81**, 322.
[5]L. Kaplan, S.-C. Lee & L. Levine (1975) *ABB* **167**, 287.
[6]A. Schieber, R. W. Frank & S. Ghisla (1992) *EJB* **206**, 491.
[7]A. Schieber & S. Ghisla (1992) *Eicosanoids* **5** Suppl, S37.
[8]C. R. Pace-Asciak & W. L. Smith (1983) *The Enzymes*, 3rd ed., **16**, 543.

Selected entries from *Methods in Enzymology* [vol, page(s)]:
General discussion: 86, 113, 142
Assay: 86, 143
Other molecular properties: porcine kidney (assay, **86**, 143; properties, **86**, 145; purification, **86**, 144)

PROSTAGLANDIN-E SYNTHASE

This glutathione-dependent enzyme [EC 5.3.99.3] (also known as prostaglandin-H$_2$ E-isomerase (PGH-PGE isomerase), PGE isomerase, and endoperoxide isomerase) catalyzes the conversion of prostaglandin H$_2$ (or, (5Z,13E)-(15S)-9α,11α-epidioxy-15-hydroxyprosta-5,13-dienoate) to prostaglandin E$_2$ (or, (5Z,13E)-(15S)-11α,15-dihydroxy-9-oxoprosta-5,13-dienoate).[1-5]

The same enzyme also converts prostaglandin G$_2$ to 15-hydroperoxyprostaglandin E$_2$. Glutathione-independent prostaglandin-E synthases have also been isolated.[4]

prostaglandin H$_2$ prostaglandin E$_2$

[1]Y. Tanaka, S. L. Ward & W. L. Smith (1987) *JBC* **262**, 1374.
[2]N. Ogino, T. Miyamoto, S. Yamamoto & O. Hayaishi (1977) *JBC* **252**, 890.
[3]Y. Urade, K. Watanabe & O. Hayaishi (1995) *J. Lipid Mediat. Cell Signal.* **12**, 257.
[4]K. Watanabe, K. Kurihara, Y. Tokunaga & O. Hayaishi (1997) *BBRC* **235**, 148.
[5]C. R. Pace-Asciak & W. L. Smith (1983) *The Enzymes*, 3rd ed., **16**, 543.

Selected entries from *Methods in Enzymology* [vol, page(s)]:
General discussion: 86, 84
Assay: 86, 85

PROSTAGLANDIN-F SYNTHASE

This oxidoreductase [EC 1.1.1.188], also known as prostaglandin-D$_2$ 11-reductase, prostaglandin-D$_2$ 11-ketoreductase (PGD$_2$ 11-ketoreductase), and prostaglandin 11-ketoreductase, catalyzes the reaction of prostaglandin D$_2$ (or, (5Z,13E)-(15S)-9α,15-dihydroxy-11-oxoprosta-5,13-dienoate) with NADPH and a proton to produce prostaglandin F$_{2\alpha}$ (or, (5Z,13E)-(15S)-9α,11α,15-trihydroxyprosta-5,13-dienoate) and NADP$^+$.[1-4]

prostaglandin H$_2$ prostaglandin D$_2$

prostaglandin F$_{2\alpha}$

The same enzyme reduces prostaglandin H$_2$, to produce prostaglandin F$_{2\alpha}$, but prostaglandin D$_2$ is not an intermediate. The enzyme has a ping pong Bi Bi kinetic mechanism.[5]

[1]Y. Urade, K. Watanabe & O. Hayaishi (1995) *J. Lipid Mediat. Cell Signal.* **12**, 257.

[2]K. Watanabe, R. Yoshida, T. Shimizu & O. Hayaishi (1985) *JBC* **260**, 7035.
[3]L.-Y. Chen, K. Watanabe & O. Hayaishi (1992) *ABB* **296**, 17.
[4]P. Y.-K. Wong (1981) *BBA* **659**, 169.
[5]O. A. Barski & K. Watanabe (1993) *FEBS Lett.* **320**, 107.

Selected entries from *Methods in Enzymology* [vol, page(s)]:
General discussion: 86, 117

PROSTAGLANDIN-I SYNTHASE

This heme-thiolate-dependent or cytochrome P450 enzyme [EC 5.3.99.4], also known as prostacyclin synthase and PGI$_2$ synthase, catalyzes the conversion of prostaglandin H$_2$ (or, (5Z,13E)-(15S)-9α,11α-epidioxy-15-hydroxyprosta-5,13-dienoate) to prostaglandin I$_2$ (or, prostacyclin or (5Z,13E)-(15S)-6,9α-epoxy-11α,15-dihydroxyprosta-5,13-dienoate)[1–5]

prostaglandin H$_2$ prostaglandin I$_2$

The enzyme appears to convert its substrate to a radical intermediate, followed by ionic rearrangement to produce the final product.[4]

[1]T. Tanabe & V. Ullrich (1995) *J. Lipid Mediat. Cell Signal.* **12**, 243.
[2]D. L. DeWitt & W. L. Smith (1983) *JBC* **258**, 3285.
[3]W. L. Smith, D. L. DeWitt & J. S. Day (1983) *Adv. Prost. Throm. Leuk. Res.* **11**, 87.
[4]M. Hecker & V. Ullrich (1989) *JBC* **264**, 141.
[5]C. R. Pace-Asciak & W. L. Smith (1983) *The Enzymes*, 3rd ed., **16**, 543.

Selected entries from *Methods in Enzymology* [vol, page(s)]:
General discussion: 86, 91, 99, 240
Assay: bovine aortic, immunoradiometric assay, 86, 243; porcine aortic, 86, 94
Other molecular properties: bovine aortic (immunoradiometric assay, 86, 243; monoclonal antibodies, 86, 242; detection in lymphocytes, 187, 583; porcine aortic (assays, 86, 94 [with [5,6-^{3}H]prostaglandin H$_2$, 86, 99]; purification, 86, 93)

PROTAMINE KINASE

This cAMP-dependent phosphotransferase [EC 2.7.1.70], also known as histone kinase, catalyzes the reaction of ATP with many seryl residues in a protamine (and histones) to produce ADP and phosphorylated protamine.[1–3] The enzyme reportedly has a ping pong Bi Bi kinetic mechanism, proceeding through a phosphoenzyme intermediate.[2,3] *See also cdc2 Kinase*

[1]M. Nakashima, Y. Takai, H. Yamamura & Y. Nishizuka (1975) *BBA* **397**, 117.

[2]A. G. Gabibov, S. N. Kochetkov, L. P. Sashchenko, I. V. Smirnov & E. S. Severin (1981) *EJB* **115**, 297.
[3]A. G. Gabibov, S. N. Kochetkov, L. P. Sashchenko, I. V. Smirnov & E. S. Severin (1983) *EJB* **135**, 491.

Selected entries from *Methods in Enzymology* [vol, page(s)]:
General discussion: 34, 261

PROTEASOME AND PROTEASOME ATPases

Proteasomes are large multi-subunit complexes that catalyze the hydrolysis of peptide bonds in a wide variety of proteins. Often, many of the proteins degraded by proteasomes are tagged with ubiquitin. The proteasome ATPase activity [EC 3.6.4.8] is associated with channel gating and polypeptide unfolding before proteolysis occurs.[1,2] Thus, although the overall reaction yields ADP and orthophosphate, this energase-type enzyme is misclassified as a hydrolase, because P-O-P bond scission supplies the energy to drive peptide release.[3] *See also Vesicle-Fusing ATPases; Energases; specific Proteasome; for example, Archaean Proteasome; Bacterial Proteasome; Eukaryotic 20S Proteasome; Eukaryotic 26S Proteasome*

[1]A. J. Rivett, G. G. Mason, R. Z. Murray & J. Reidlinger (1997) *Mol. Biol. Rep.* **24**, 99.
[2]G. G. Mason, R. Z. Murray, D. Pappin & A. J. Rivett (1998) *FEBS Lett.* **430**, 269.
[3]D. L. Purich (2001) *TiBS* **26**, 417.

PROTEIN N-ACETYLGLUCOSAMINYL-TRANSFERASE

This transferase [EC 2.4.1.94], often abbreviated N-GlcNAc transferase, catalyzes the reaction of UDP-N-acetyl-D-glucosamine with an asparaginyl residue in a protein (at a sequence of –Asn–Xaa–(Ser/Thr)–) to produce UDP and a protein containing a 4-N-(N-acetyl-D-glucosaminyl)asparaginyl residue.[1–3] Protein substrates include pancreatic ribonuclease A.

[1]H. Arakawa & S. Mookerjea (1984) *EJB* **140**, 297.
[2]Z. Khalkhali, R. D. Marshall, F. Reuvers, C. Habets-Willems & P. Boer (1976) *BJ* **160**, 37.
[3]Z. Khalkhali & R. D. Marshall (1975) *BJ* **146**, 299.

Selected entries from *Methods in Enzymology* [vol, page(s)]:
Other molecular properties: purification, 230, 445

PROTEIN-ARGININE DEIMINASE

This deiminase [EC 3.5.3.15] catalyzes the hydrolysis of an L-arginyl residue within a protein to produce ammonia and a protein containing an L-citrullinyl residue. Protein substrates include trichohyalin, keratin, myelin basic protein, histones, glycogen phosphorylase *b*, and polyarginine. Low-molecular-weight substrates include

N-acyl-L-arginine (for example, α-*N*-benzoyl-L-arginine, producing ammonia and α-*N*-benzoyl-L-citrulline) and L-arginine esters, albeit more slowly.[1,2] Deimination renders myelin basic protein more susceptible to degradation by cathepsin D.[3] Deiminated glycogen phosphorylase *b* exhibits positive cooperativity for FMN binding and shows less capability to form tetramers in the presence of AMP.[4]

[1]M. Fujisaki & K. Sugawara (1981) *J. Biochem.* **89**, 257.
[2]J. Kubilus & H. P. Baden (1983) *BBA* **745**, 285.
[3]L. B. Pritzker, S. Joshi, J. J. Gowan, G. Harauz & M. A. Moscarello (2000) *Biochemistry* **39**, 5374.
[4]T. B. Eronina, N. B. Livanova, N. A. Chebotareva, B. I. Kurganov, S. Luo & D. J. Graves (1996) *Biochimie* **78**, 253.

Selected entries from *Methods in Enzymology* [vol, page(s)]:
General discussion: 107, 624

PROTEINASE K

This serine endopeptidase [EC 3.4.21.64], also known as endopeptidase K, is an alkaline proteinase isolated from the culture medium of the mold *Tritirachium album* Limber. The enzyme catalyzes peptide-bond hydrolysis in keratin (hence, the descriptor "K") and other proteins. Proteinase K will also catalyze hydrolysis of peptide amides and esters. The specificity of proteinase K is rather broad, resembling that of subtilisin. Studies with the oxidized insulin B chain indicate a preference for aromatic and hydrophobic aminoacyl residues in the P1 position. It will not hydrolyze a peptide bond if a prolyl residue is in the P1′ position.[1–3]

Full enzymatic activity requires the presence of two calcium ions. The enzyme undergoes autolysis when stored in low concentrations (less than 10 μg/mL). It is significantly more stable at concentrations above 1.0 mg/mL.

Salivary gland proteinase K, also known as proteinase B, is another serine endopeptidase, but not a member of the same clan as the *Tritirachium* enzyme. The salivary gland enzyme from rat and mouse submandibular glands and from cattle erythrocyte membranes catalyzes the release of T-kinin (Ile-Ser-bradykinin) from T-kininogen (hence, the descriptor K). It has also been known as T-kininogenase, antigen γ, and kallikrein k10. The enzyme exhibits a preference for Arg–Xaa bonds in small peptides. Either polar or nonpolar aminoacyl residues can be accommodated at the P2 position.[4]

[1]P. J. Sweeney & J. M. Walker (1993) *Meth. Mol. Biol.* **16**, 305.
[2]E. Kraus, H. H. Kiltz & U. F. Femfert (1976) *Z. Physiol. Chem.* **357**, 233.
[3]D. Brömme, K. Peters, S. Fink & S. Fittkau (1986) *ABB* **244**, 439.
[4]J. R. Chagas, I. Y. Hirata, M. A. Juliano, W. Xiong, C. Wang, J. Chao, L. Juliano & E. S. Prado (1992) *Biochemistry* **31**, 4969.

Selected entries from *Methods in Enzymology* [vol, page(s)]:
Other molecular properties: commercial source, **65**, 719; extraction of radiolabeled RNA, in, **59**, 836; mouse mRNA preparation, in, **79**, 94; nerve growth factor mRNA purification, in, **198**, 53; PK buffer, **65**, 720; preparation of (cellular RNA, **65**, 582; herpes simplex virus DNA, **79**, 313; ^{32}P-labeled polyoma virus DNA, **65**, 722; polyoma virus-induced cellular RNA, **65**, 727; retrovirus RNA, **65**, 687); protein–RNA interactions, in studies of, **59**, 576; rapid DNA isolation buffer, **65**, 411; removal, of protein contaminants of DNA fragment preparations, **65**, 124; ribonuclease inactivation, for, **31**, 720; synthesis, cell-free, DNA directed, **65**, 802; *Tritirachium album*, specificity, from, **59**, 600; tubulin, proteolysis of, **134**, 186; viral RNA preparation, in, **79**, 317

PROTEIN DISULFIDE-ISOMERASE

This isomerase [EC 5.3.4.1], also known as S-S rearrangase and disulfide bond isomerase (dsbA), catalyzes intrachain and interchain disulfide bond rearrangement within proteins, yielding proteins in their native conformational states.[1–8] In eukaryotes, protein disulfide isomerase (PDI) facilitates proper folding and disulfide bond formation in nascent proteins within the endoplasmic reticulum. Secreted PDI associates with the cell's outer surface. The catalytic activity of the isomerase is due to two cysteinyl residues. The rate accelerations observed are small relative to other enzymes: for example, the acceleration of ribonuclease A isomerization is only fifty-fold.

Escherichia coli dsbA is a strongly oxidizing thiol reagent with one catalytic disulfide bridge and an intrinsic redox potential of −0.089 V.[5] This enzyme mediates disulfide bond formation in newly secreted proteins. Addition of thiol reagents to purified dsbA reduces its disulfide bond, resulting in active disulfide isomerase upon removal of the thiol reagent. DsbA catalyzes the conversion of stable misfolded proteins to the correctly folded conformation under physiological conditions. This conversion is the result of breaking and re-forming disulfide bonds. The uncatalyzed rate is undetectable. The K_m for misfolded insulin-like growth factor-I (IGF-I) is 43 μM and the k_{cat} is 0.2 min^{-1}.[6] The oxidized form of dsbA stimulates the oxidative folding of completely reduced IGF-I at pH 7.0. Thus, dsbA has two functions that depend on its redox state. The reduced form of the protein is a disulfide isomerase, whereas the oxidized protein facilitates disulfide bond formation in reduced substrates under physiological conditions.[6]

The oxidizing properties of dsbA are thought to result from a tense conformation of its oxidized state, which is converted to the relaxed, reduced state upon oxidation of thiols by dsbA.[5] Couprie *et al.*[7] synthesized a stable cysteine-homoalanine thioether bond in place of the labile disulfide bond normally formed between dsbA and polypeptides during catalysis. They then used NMR spectroscopy to detect the progressive increase in stability along the dsbA catalytic pathway, a finding that strongly supports a thermodynamically driven mechanism.

[1]R. C. Fahey & A. R. Sundquist (1991) AE **64**, 1.
[2]W. W. Wells, Y. Yang & T. L. Deits (1993) AE **66**, 149.
[3]N. J. Bulleid (1993) Adv. Prot. Chem. **44**, 125.
[4]H. F. Gilbert (1998) CBC **1**, 609.
[5]M. Wunderlich, R. Jaenicke & R. Glockshuber (1993) J. Mol. Biol. **233**, 559.
[6]J. C. Joly & J. R. Swartz (1994) Biochemistry **33**, 4231.
[7]J. Couprie, F. Vinci, C. Dugave, E. Quemeneur & M. Moutiez (2000) Biochemistry **39**, 6732.
[8]K. I. Kivirikko & T. Pihlajaniemi (1998) AE **72**, 325.

Selected entries from **Methods in Enzymology** [vol, page(s)]:
General discussion: 144, 96; 251, 397; 290, 26, 50
Assay: 107, 282; 144, 112; calculation of activity, 251, 405; reaction conditions, 251, 405; ribonuclease substrate (characterization, 251, 404; preparation, 251, 403); substrates, 251, 402
Other molecular properties: bovine liver (pI, 251, 402; pK value of reactive thiol, 251, 402; purification, 107, 288; 251, 399; 252, 36; size, 251, 401); dehydroascorbate reductase activity, 252, 31, 36; disulfide bond formation and, 253, 434, 438; oxidation potential, 251, 17, 401; properties, 107, 291; 144, 114; purification, 87, 468 (bovine liver, 107, 288; 251, 399; chick embryo, 144, 113); reaction catalyst, as, 251, 397, 402; redox potential, 252, 201; specificity of action, 107, 292; subcellular location, 107, 287; tissue distribution, 107, 286

PROTEIN-DISULFIDE REDUCTASE (GLUTATHIONE)

This oxidoreductase [EC 1.8.4.2], also known as glutathione:insulin transhydrogenase and insulin reductase, catalyzes the reaction of two molecules of glutathione with a disulfide bond in a protein to produce glutathione disulfide and a protein containing two new thiol groups (i.e., two cysteinyl residues).[1–6] Protein substrates include insulin and proinsulin; certain peptides are also substrates (e.g., oxytoxin and vasopressin). The enzyme has a random kinetic mechanism. It is distinct from protein-disulfide isomerase [EC 5.3.4.1]. **See also** Enzyme-Thiol Transhydrogenase (Glutathione Disulfide); Protein-Disulfide Reductase (NAD(P)H); Protein-Mixed Disulfide Reductase

[1]P. T. Varandani (1989) Coenzymes Cofactors, **3** pt. A, 753.
[2]P. T. Varandani (1978) Dev. Biochem. **1**, 29.
[3]M. L. Chandler & P. T. Varandani (1975) Biochemistry **14**, 2107.
[4]P. T. Varandani (1972) BBA **286**, 126.
[5]M. L. Chandler & P. T. Varandani (1972) BBA **286**, 136.
[6]P. D. Spolter & J. M. Vogel (1968) BBA **167**, 525.

Selected entries from **Methods in Enzymology** [vol, page(s)]:
General discussion: 17B, 515; 113, 541
Assay: 17B, 515
Other molecular properties: activity and specific enzymes, 233, 405; physiological role, 113, 543; properties, 17B, 519; 113, 543; protein-disulfide isomerase, separation from, 87, 468; purification (bovine liver, 17B, 516; 87, 468)

PROTEIN-DISULFIDE REDUCTASE (NAD(P)H)

This oxidoreductase [EC 1.6.4.4] catalyzes the reaction of NAD(P)H with a disulfide bond (that is, a cystinyl residue) in a protein to produce NAD(P)$^+$ and a protein containing two new thiol groups (i.e., two cysteinyl residues).[1] Care should be exercised in investigations of this enzyme that glutathione reductase activity is not present. **See also** Protein-Disulfide Reductase (Glutathione); Protein-Mixed Disulfide Reductase

[1]M. D. Hatch & J. F. Turner (1960) BJ **76**, 556.

PROTEIN FARNESYLTRANSFERASE

This zinc- and magnesium-dependent transferase catalyzes the reaction of farnesyl diphosphate with a cysteinyl residue located near the C-terminus of a protein to produce pyrophosphate and a protein with a farnesyl moiety linked to the sulfur of the cysteine.[1–4] The C-terminus of the protein is –Cys-Aaa-Aaa-Xaa where Aaa represents aliphatic amino acyl residues and the C-terminal Xaa can be any amino acid (Ras is an example of such a protein). If Xaa is Met, Ser, Cys, Ala, or Gln, the cysteinyl residue is farnesylated. If Xaa is a leucyl residue, the sulfur is geranylgeranylated. Cis isomers of farnesyl diphosphate inhibit farnesylation of Ras.[5]

farnesyl diphosphate

farnesylated protein

Although the mammalian enzyme has a random Bi Bi kinetic mechanism, isotope trapping studies suggest that there is a preference in the binding order: farnesyl diphosphate binding first, followed by the peptide substrate. Product release is rate limiting. Using deutero-labeled farnesyl diphosphate in the C1 position, investigations demonstrate that the reaction proceeds with inversion of configuration at C1.[6] **See** Protein Geranylgeranyltransferase I; Protein Geranylgeranyltransferase II

[1]R. A. Gibbs (1998) CBC **1**, 31.
[2]P. J. Casey & M. G. Seabra (1996) JBC **271**, 5289.
[3]F. L. Zhang & P. J. Casey (1996) ARB **65**, 241.
[4]S. Clarke (1992) ARB **61**, 355.

[5]N. P. Das & C. M. Allen (1991) *BBRC* **181**, 729.
[6]Y.-Q. Mu, C. A. Omer & R. A. Gibbs (1996) *JACS* **118**, 1817.

Selected entries from *Methods in Enzymology* [vol, page(s)]:

Assay: continuous fluorescence assay (dansylation of substrate, **250**, 31; effect of organic solvents, **250**, 41; farnesylation of substrate, **250**, 33; farnesyl diphosphate stock solution, **250**, 34; fluorescence monitoring, **250**, 31; inhibition by detergents, **250**, 38; inner filter effect checking, **250**, 38; K_m of substrate, **250**, 42; maximal velocity, **250**, 43; quantitation of assay, **250**, 36; quartz cuvette handling, **250**, 35; reaction conditions, **250**, 36; standard curve, **250**, 35; substrate inhibition analysis, **250**, 41; substrate stock solution, **250**, 34); gel fluorography, **250**, 59; *in vivo* yeast assay (cell cycle arrest in *GPA1* mutant, **250**, 46; growth media, **250**, 45; heat-shock sensitivity in *RAS2*[Val-19] mutant, **250**, 48; inhibitor, **250**, 46; yeast strains, **250**, 46); Ras protein filter assay (buffers, **250**, 14; *Escherichia coli* extract, **250**, 62; reaction conditions, **250**, 14, 58; solubility of product, **250**, 31; substrate preparation, **250**, 15; yeast extract, **250**, 58)

Other molecular properties: baculovirus expression system (disadvantages, **250**, 11; transfection, **250**, 16; vectors, **250**, 16); chemotherapy targeting, **250**, 20, 44, 52, 111; crosslinking of H-Ras to β subunit (inhibition by substrate, **250**, 29; reaction conditions, **250**, 29; zinc requirement, **250**, 30); expression in *Escherichia coli* (plasmid, **250**, 3, 7; ribosomal binding site incorporation, **250**, 3, 7, 11, 31; yeast enzyme, **250**, 32, 53, 56, 61); farnesyl pyrophosphate and (binding assay, **250**, 28; complex [affinity, **250**, 24; formation, **250**, 23; isolation by gel filtration, **250**, 23; stability, **250**, 24; stoichiometry of binding, **250**, 24]; exchange rate, **250**, 27; transfer to acceptors [p21[H-ras], **250**, 25, 28; peptides, **250**, 26]); geranylgeranyl pyrophosphate affinity, **250**, 25; immunoblotting of yeast recombinant protein, **250**, 66; inhibitor (assays [cell assays, **255**, 42, 380, 382; *in vitro*, **255**, 40, 381; Ras-processing assay, **255**, 383; SDS-PAGE, **255**, 380; soft agar assay, **255**, 384; yeast assay, **255**, 84, 90]; cancer therapy, **255**, 379, 386; design, **255**, 379; selectivity, **255**, 380); inhibitor screening, **250**, 44, 50; isoprenoid product (chemical cleavage, **250**, 60; identification by HPLC, **250**, 60); metal dependence, **255**, 38; metal requirement, **250**, 13, 25; mutant, generation in yeast (characterization, **255**, 89; gene library construction, **255**, 87; inhibitor resistance, **255**, 90; selection for altered substrate specificity, **255**, 88, 91); purification, from bovine brain (anion-exchange chromatography, **250**, 17; extraction, **250**, 17; phenyl-Sepharose chromatography, **250**, 17; yield, **250**, 17); purification, recombinant protein, **255**, 41 (human enzyme from *Escherichia coli*, **250**, 4 [cell growth, **250**, 8; cell lysis, **250**, 9; Mono Q chromatography, **250**, 10; yield, **250**, 11; YL1/2 chromatography, **250**, 9]; from Sf9 cells, **255**, 41 [anion-exchange chromatography, **250**, 20; cell harvesting, **250**, 19; phenyl-Superose chromatography, **250**, 20; yield, **250**, 20]; yeast enzyme from *Escherichia coli* [cell growth, **250**, 32, 63; chromatography, **250**, 33, 64; extraction, **250**, 33, 64; yield, **250**, 65]); reaction mechanism, **255**, 39; sequence recognition, **250**, 12, 43, 79, 158; site-directed mutagenesis (effect on substrate specificity, **250**, 66; primers, **250**, 65); structure, **255**, 38, 83; substrate recognition, **257**, 30; substrate specificity, **255**, 38, 48; subunits (encoding genes in yeast, **250**, 45, 52; structure, **250**, 12, 22, 45, 79); synthesis of prenylated peptides, **250**, 161

PROTEIN GERANYLGERANYL-TRANSFERASE I

This magnesium- and zinc-dependent transferase catalyzes the reaction of geranylgeranyl diphosphate with a cysteinyl residue located near the C-terminus of a protein to produce pyrophosphate and a protein with a geranylgeranyl moiety linked to the sulfur of the cysteine.[1–4] The C-terminus of the protein is: protein–Cys-Aaa-Aaa-Xaa,

where Aaa represents aliphatic amino acyl residues. When the C-terminal Xaa residue is Met, Ser, Cys, Ala, or Gln, the cysteinyl residue is farnesylated. When Xaa is a leucyl residue, the sulfur is geranylgeranylated. The kinetic mechanism is ordered Bi Bi in which geranylgeranyl diphosphate binds to the free enzyme. *See Protein Farnesyltransferase; Protein Geranylgeranyltransferase II*

[1]R. A. Gibbs (1998) *CBC* **1**, 31.
[2]P. J. Casey & M. G. Seabra (1996) *JBC* **271**, 5289.
[3]F. L. Zhang & P. J. Casey (1996) *ARB* **65**, 241.
[4]S. Clarke (1992) *ARB* **61**, 355.

Selected entries from *Methods in Enzymology* [vol, page(s)]:
General discussion: **257**, 21, 30
Assay: gel fluorography, **250**, 59; Ras protein filter assay (buffers, **250**, 14; conditions, **250**, 14, 58; *Escherichia coli* extract, **250**, 62; substrate preparation, **250**, 15; yeast extract, **250**, 58)

Other molecular properties: baculovirus expression system (disadvantages, **250**, 11; transfection, **250**, 16; vectors, **250**, 16); expression in *Escherichia coli* (plasmid, **250**, 3, 7; ribosomal binding site incorporation, **250**, 3, 7, 11; yeast enzyme, **250**, 53, 56, 61); farnesyl pyrophosphate affinity, **250**, 25; geranylgeranyl pyrophosphate and (complex, **250**, 25; exchange rate, **250**, 27; transfer to acceptors, **250**, 27); isoprenoid product (chemical cleavage, **250**, 60; identification by HPLC, **250**, 60); metal requirement, **250**, 13, 25; purification, from bovine brain (anion-exchange chromatography, **250**, 17; extraction, **250**, 17; Gγ affinity chromatography [column preparation, **250**, 15; elution, **250**, 18]; phenyl-Sepharose chromatography, **250**, 17; yield, **250**, 17); purification, recombinant protein (human enzyme from *Escherichia coli*, **250**, 4 [cell growth, **250**, 8; cell lysis, **250**, 9; Mono Q chromatography, **250**, 10; yield, **250**, 11; YL1/2 chromatography, **250**, 9]; from Sf9 cells [anion-exchange chromatography, **250**, 20; cell harvesting, **250**, 19; phenyl-Superose chromatography, **250**, 20]; yeast enzyme from *Escherichia coli* [cell growth, **250**, 63; chromatography, **250**, 64; extraction, **250**, 64; yield, **250**, 65]); Ras protein modification, **255**, 48; sequence recognition, **250**, 3, 12, 22, 52, 79, 107, 158, 439; structure, **255**, 83; substrate recognition, **257**, 30; substrate specificity, **255**, 48, 60, 83, 380; subunits (encoding genes in yeast, **250**, 45, 52; structure, **250**, 12, 22, 79); synthesis of prenylated peptides, **250**, 161

PROTEIN GERANYLGERANYL-TRANSFERASE II

This transferase, also known as Rab geranylgeranyltransferase, catalyzes the reaction of geranylgeranyl diphosphate with a cysteinyl residue located near the C-terminus of a protein to produce pyrophosphate and a protein with a geranylgeranyl moiety linked to the sulfur of the cysteine.

geranylgeranyl diphosphate

geranylgeranylated protein

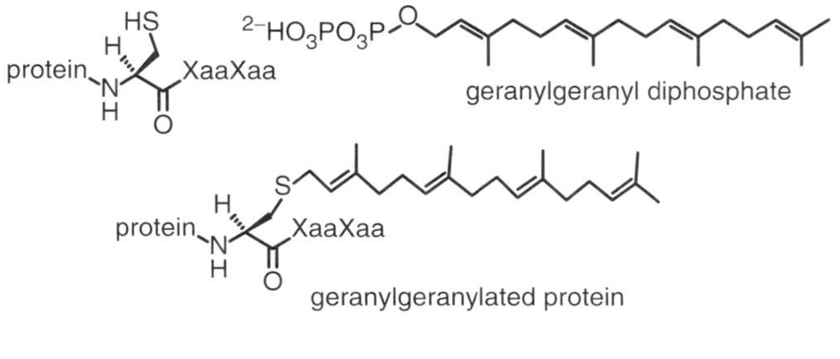

Neither geranyl diphosphate nor farnesyl diphosphate can serve as the prenyl donor substrate. The C-terminus of the protein has the sequence –Cys–Cys, –Cys–Xaa–Cys, or –Cys–Cys–Xaa–Xaa. Both cysteinyl residues can be geranylgeranylated. This transferase is responsible for the prenylation of Rab G-proteins involved in vesicle transport and fusion (the transferase requires the presence of the Rab escort protein). *See Protein Farnesyltransferase; Protein Geranylgeranyltransferase I*

Selected entries from **Methods in Enzymology** [vol, page(s)]:
General discussion: 257, 21, 30
Assay: activity in yeast lysates, assay, 257, 22; *in vitro*, 257, 39
Other molecular properties: α and β subunits, baculovirus encoding, construction, 257, 33; genes encoding subunits in yeast, 250, 52; modification of prenylating enzymes, 250, 13; mutation in choroidemia, 250, 108; Rab geranylgeranyltransferase (assay, *in vitro*, 257, 39; α and β subunits, baculovirus encoding, construction, 257, 33; recombinant [production in Sf9 cells, 257, 32, 34; purification from Sf9 cytosol, 257, 35]; role in geranylgeranylation, 257, 31); Rab protein modification, 255, 61; recombinant (production in Sf9 cells, 257, 32, 34; purification from Sf9 cytosol, 257, 35); reconstitution *in vitro* (Bet2p and Bet4p coexpression in *Escherichia coli*, 257, 26; expression vector preparation, 257, 24; recombinant Mrs6p expression in *Escherichia coli*, 257, 27; recombinant subunits, with, 257, 27); role in geranylgeranylation, 257, 31; sequence recognition, 250, 52, 107, 158; structure, 255, 83; substrate specificity, 255, 48, 61, 83, 380; yeast, structure, 257, 22

PROTEIN-GLUCOSYLGALACTOSYL-HYDROXYLYSINE GLUCOSIDASE

This enzyme [EC 3.2.1.107] catalyzes the hydrolysis of an α-D-glucosyl-1,2-β-D-galactosyl-L-hydroxylysyl residue in a protein to produce D-glucose and a protein containing a β-D-galactosyl-L-hydroxylysyl residue.[1–3] Substrate specificity studies indicate that the L-hydroxylysyl residue must have a free, positively charged ε-amino group. Low-molecular-weight substrates include free and N^α-acylated 2-O-α-D-glucopyranosyl-O-β-D-galactopyranosyl-hydroxylysine.

[1] H. Hamazaki & K. Hotta (1980) *EJB* 111, 587.
[2] H. Hamazaki & K. Hotta (1979) *JBC* 254, 9682.
[3] M. Sternberg & R. G. Spiro (1979) *JBC* 254, 10329.

PROTEIN-GLUTAMATE METHYLESTERASE

This esterase [EC 3.1.1.61], also known as chemotaxis-specific methylesterase, catalyzes the hydrolysis of a protein L-glutamate O^4-methyl ester (that is, of a methyl ester glutamyl residue in a protein) to produce a protein L-glutamate (that is, a protein with a normal glutamyl residue) and methanol.[1,2] This enzyme acts on the product of protein-glutamate O-methyltransferase [EC 2.1.1.80].

The active site of the esterase is reported to contain a crucial seryl residue.[3] Interestingly, changing a particular glycyl residue to a valyl will result in complete loss of activity.[4]

[1] M. R. Kehry, T. G. Doak & F. W. Dahlquist (1984) *JBC* 259, 11828.
[2] C. Gagnon, D. Harbour & R. Camato (1984) *JBC* 259, 10212.
[3] J. K. Krueger, J. Stock & C. E. Schutt (1992) *Biochim. Biophys. Acta* 1119, 322.
[4] S. A. Simms, E. W. Cornman, J. Mottonen & J. Stock (1987) *JBC* 262, 29.

Selected entries from **Methods in Enzymology** [vol, page(s)]:
General discussion: 106, 321
Assay: 106, 322

PROTEIN-GLUTAMATE O-METHYLTRANSFERASE

This methyltransferase [EC 2.1.1.80; previously classified as EC 2.1.1.24], also known as chemotaxis protein O-methyltransferase, catalyzes the reaction of S-adenosyl-L-methionine with a glutamyl residue in a protein to produce S-adenosyl-L-homocysteine and a protein containing a glutamyl methyl ester residue.[1–5] The *Salmonella typhimurium* enzyme has a random Bi Bi kinetic mechanism.[1]

[1] S. A. Simms & K. Subbaramaiah (1991) *JBC* 266, 12741.
[2] S. A. Simms, A. M. Stock & J. B. Stock (1987) *JBC* 262, 8537.
[3] W. R. Springer & D. E. Koshland (1977) *PNAS* 74, 533.
[4] A. Burgess-Cassler, A. H. Ullah & G. W. Ordal (1982) *JBC* 257, 8412.
[5] A. H. Ullah & G. W. Ordal (1981) *BJ* 199, 795.

Selected entries from **Methods in Enzymology** [vol, page(s)]:
General discussion: 106, 295, 310
Assay: 102, 164; 106, 296, 312
Other molecular properties: differences, 106, 318; effects of attractants, 106, 318; *Escherichia coli* and *Salmonella typhimurium* chemotaxis-related (parital purification, 106, 318); product identification, 106, 316; properties, 106, 297, 307, 310; purification from brain and erythrocytes, 102, 165; 106, 301

PROTEIN-GLUTAMINE γ-GLUTAMYLTRANSFERASE

This calcium-dependent transferase [EC 2.3.2.13], also known as transglutaminase and fibrinoligase, catalyzes the reaction of an alkylamine with the γ-carboxyamide group of a glutaminyl residue of a protein to produce ammonia and a protein containing an N^5-alkylglutaminyl residue.[1–4] The alkylamine substrate can be the ε-amino groups of lysyl residues in proteins or peptides, thus generating intra- and intermolecular N^6-(5-glutamyl)lysyl crosslinks. The enzyme has a branched reaction pathway (often referred to as a modified ping pong Bi Bi mechanism), where the acyl-enzyme intermediate undergoes hydrolysis slowly. *See also Coagulation Factor XIII$_a$*

[1] J. E. Folk & S. I. Chung (1973) *AE* 38, 109.

[2]J. E. Folk (1983) *AE* **54**, 1.
[3]J. E. Folk (1980) *ARB* **49**, 517.
[4]J. E. Folk & J. S. Finlayson (1977) *Adv. Protein Chem.* **31**, 1.

Selected entries from *Methods in Enzymology* [vol, page(s)]:
General discussion: 5, 833; 17A, 889; 46, 151; 80, 333, 334; 87, 36; 113, 358; 223, 378; 309, 172; 322, 433
Assay: 322, 433; guinea pig, 5, 833; 113, 372; 113, 359; horseshoe crab, 223, 380; human coagulation factor XIII$_a$, 113, 364; human epidermis, 113, 370; type-I, 190, 43
Other molecular properties: branched reaction pathway, 87, 322; fibrinogen clotting, 45, 35, 36; fibrin stabilizing factor system, 19, 770 (action of, 19, 770; assay of, 19, 772 [amine incorporation, by, 19, 776; clot stability, by, 19, 774]; conversion of zymogen to fibrinoligase enzyme, 19, 782; properties, 19, 780; purification procedure, 19, 778); filter paper method for titration, 45, 186; fluorescent labeling, 48, 358; function, 45, 177; glycosyl-asparagine incorporation into glutamine residues, 138, 413; guinea pig hair follicle (properties, 113, 374; purification, 113, 373); guinea pig liver (properties, 5, 837; 113, 363; purification, 5, 836; 113, 361); horseshoe crab (amino acid sequence, 223, 386; assay, 223, 380; cDNA, nucleotide sequence, 223, 386; properties, 223, 383; purification, 223, 380); human coagulation factor XIII$_a$ (isolation, 113, 366; properties, 113, 369); human epidermis (assay, 113, 370; properties, 113, 372; purification, 113, 371); lobster muscle transpeptidase, 19, 765 (assay, 19, 765 [by clotting, 19, 765; by monodansylcadaverine incorporation into α-casein, 19, 766]; ester hydrolysis by, 19, 770; isoelectric point, 19, 769; molecular weight, 19, 769; purification procedure, 19, 767); mechanism, 63, 51; 87, 36, 322; occurrence, 5, 837; plasma transglutaminase, thrombin and, 19, 181; purification from guinea pig liver, 17A, 890; radioiodination, 70, 232; substrates, synthetic, 45, 179; trimethylacetyl complex (bond nature and acyl group position, 87, 41; formation and isolation, 87, 37); type I (assay, 190, 43; down-regulation by retinoids, 190, 42)

PROTEIN-HISTIDINE KINASE

There are two types of kinases catalyzing the phosphorylation of histidyl residues in proteins.[1–6]

Protein-histidine *pros*-kinase [EC 2.7.3.11] catalyzes the reaction of ATP with a histidyl residue in a particular protein to produce ADP and protein N^π-phospho-L-histidine. His75 of histone H4 is phosphorylated in this manner.[1,2] The enzyme operates by an ordered Bi Bi kinetic mechanism.[3]

Protein-histidine *tele*-kinase [EC 2.7.3.12] catalyzes the reaction of ATP with a histidyl residue in a particular protein to produce ADP and protein N^τ-phospho-L-histidine. His18 of histone H4 is phosphorylated in this manner.[2]

See also Protein Kinase

[1]V. D. Huebner & H. R. Matthews (1985) *JBC* **260**, 16106.
[2]J. M. Fujikati, G. Fung, E. Y. Oh & R. A. Smith (1981) *Biochemistry* **20**, 3658.
[3]J. M. Huang, Y. F. Wei, Y. H. Kim, L. Osterberg & H. R. Matthews (1991) *JBC* **266**, 9023.
[4]T. W. Grebe & J. B. Stock (1999) *Adv. Microb. Physiol.* **41**, 139.
[5]W. F. Loomis, G. Shaulsky & N. Wang (1997) *J. Cell Sci.* **110**, 1141.
[6]L. A. Alex & M. I. Simon (1994) *Trends Genet.* **10**, 133.

Selected entries from *Methods in Enzymology* [vol, page(s)]:
Other molecular properties: purification, 200, 388; yeast, from, 200, 404

PROTEIN-HISTIDINE N-METHYLTRANSFERASE

This methyltransferase [EC 2.1.1.85], also known as protein methylase IV, catalyzes the reaction of *S*-adenosyl-L-methionine with an L-histidyl residue in a protein (that is, protein-L-histidine) to produce *S*-adenosyl-L-homocysteine and an N^τ-methyl-L-histidyl residue in the protein.[1,2] His73 of actin is an excellent methyl acceptor substrate. A synthetic peptide corresponding to residues 69 to 77 of actin is also a substrate (Tyr-Pro-Ile-Glu-His-Gly-Ile-Ile-Thr).

[1]C. Vijayasarathy & B. S. N. Rao (1987) *BBA* **923**, 156.
[2]M. Raghavan, U. Lindberg, C. Schutt (1992) *EJB* **210**, 311.

PROTEIN-L-ISOASPARTATE(D-ASPARTATE) O-METHYLTRANSFERASE

This methyltransferase [EC 2.1.1.77], variously known as protein L-isoaspartyl methyltransferase, protein D-aspartate methyltransferase, protein β-aspartate *O*-methyltransferase, and L-isoaspartyl protein carboxyl methyltransferase, catalyzes the reaction of *S*-adenosyl-L-methionine with an L-β-aspartyl residue in a protein to produce *S*-adenosyl-L-homocysteine and a protein containing an L-β-aspartyl methyl ester residue.[1–3]

D-Aspartyl (but not L-aspartyl) residues in proteins also act as methyl acceptors. The enzyme has a role in repair or degradation of proteins that contain atypical L-isoaspartyl or D-aspartyl residues due to damage and/or age. Using a synthetic substrate, Trp-Ala-Gly-Gly-isoAsp-Ala-Ser-Gly-Glu, the bovine brain enzyme has a random Bi Bi kinetic mechanism.[1] Interestingly, two atoms of ^{18}O from labeled water are incorporated into the product using an L-isoaspartyl-containing peptide. This suggests that more than a single cycle of methylation and demethylation occurs, via a hydrolyzable succinimide intermediate, for repair of the peptide.[2]

[1]B. A. Johnson & D. W. Aswad (1993) *Neurochem. Res.* **18**, 87.
[2]J. A. Lindquist & P. N. McFadden (1994) *J. Protein Chem.* **13**, 553.
[3]J. D. Lowenson & S. Clarke (1992) *JBC* **267**, 5985.

PROTEIN-S ISOPRENYLCYSTEINE O-METHYLTRANSFERASE

This methyltransferase [EC 2.1.1.100], also known as isoprenylcysteine carboxylmethyltransferase, catalyzes the reaction of *S*-adenosyl-L-methionine with the carboxyl group of a C-terminal *S*-farnesyl-L-cysteine residue in a protein to produce *S*-adenosyl-L-homocysteine and the methyl ester of that C-terminal *S*-farnesyl-L-cysteine.[1–4] C-Terminal *S*-geranylgeranylcysteine residues and C-terminal *S*-geranylcysteine residues are also methylated, albeit less effectively.

Many signaling proteins, such as the small monomeric G proteins and the γ subunits of heterotrimeric G proteins, undergo this form of posttranslational modification. This transferase catalyzes an ordered Bi Bi kinetic mechanism.[2] The rat kidney enzyme is a metalloprotein.[3]

[1]M. R. Philips, M. H. Pillinger, R. Staud, C. Volker, M. G. Rosenfeld, G. Weissmann & J. B. Stock (1993) *Science* **259**, 977.
[2]Y. Q. Shi & R. R. Rando (1992) *JBC* **267**, 9547.
[3]R. R. Desrosiers, Q. T. Nguyen & R. Beliveau (1999) *BBRC* **261**, 790.
[4]D. Boivin, W. Lin & R. Beliveau (1997) *Biochem. Cell Biol.* **75**, 63.

Selected entries from *Methods in Enzymology* [vol, page(s)]:
General discussion: 250, 216, 226, 251; 255, 65; 256, 49
Assay: 250, 232 (AFC as substrate, 250, 258; G25K methylation, 250, 259; microsomal membranes, 255, 64; turnover rate, 255, 68)
Other molecular properties: effects of non-binding farnesylcysteine analogs (neutrophils, 250, 234; platelets, 250, 234); eukaryotes, in, types, 255, 66; human neutrophil membranes, in (activity, 256, 53; detergent effects, 256, 53; detergent-extracted [partial purification, 256, 61; reconstitution in liposomes, 256, 59]; endogenous Ras-related protein substrates, 256, 53; kinetics, 256, 57; localization, 256, 58; prenylcysteine analogs as substrates, 256, 55; recombinant Ras-related protein substrates, 256, 54); inhibitors, 250, 218, 228 (assay, 250, 232; specificity, 255, 76; synthesis, 250, 228); mechanism, 250, 227; pH optima, 250, 217; 255, 73; product labeling with tritiated methionine, 250, 332; role in G protein signal transduction, 250, 221; species distribution, 250, 217, 253; stereospecificity, 250, 227; structure, 255, 74; subcellular distribution, 250, 218; substrate hydrophobicity effect on affinity, 250, 220; substrate specificity, 250, 216, 227, 251; 255, 67, 72, 76; tissue distribution, 255, 73; yeast (coupled assay for determination of protease activity, 250, 262; expression of TrpE fusion protein in *Escherichia coli* [expression vector, 250, 254; extract preparation, 250, 257; induction, 250, 256]; gene, 250, 253; structure, 250, 253; synthesis of methylated derivatives *in vitro*, 250, 258)

PROTEIN KINASE A (OR, PKA)

Formerly known as cAMP-dependent or cAMP-stimulated protein kinase, this kinase catalyzes the reaction of a serine/threonine hydroxyl group with ATP to produce an *O*-phosphoserine/*O*-phosphothreonine residue in various protein substrates.[1–7]

Type I PKA is a complex, containing two regulatory subunits and two catalytic subunits (*i.e.*, an R_2C_2 quaternary structure), which dissociates and becomes active when 3′,5′-cyclic-AMP binds to the regulatory subunit. The free energy of hydrolysis of the cyclic nucleotide activator is large ($\Delta G \approx -13$ kcal/mol [or, -54 kJ/mol]), allowing the 3′,5′-cAMP to be converted virtually irreversibly to AMP by a specific phosphodiesterase. Originally discovered by Nobelists Edwin Krebs and Edward Fischer, PKA type I is the prototype for over two thousand members of the protein kinase superfamily. Because dilution also results in dissociation and activation, type I PKA is often known as a cAMP-stimulated rather than cAMP-dependent protein kinase.

Type II PKA contains a similar catalytic subunit, but its tightly bound regulatory subunit directs the kinase to various membrane-associated anchoring proteins. ***See also** Protein Kinase*

[1]S. S. Taylor, E. Radzio-Andzelm, Madhusudan, X. Cheng, L. Ten Eyck & N. Narayana (1999) *Pharmacol. Ther.* **82**, 133.
[2]D. Bossemeyer, R. A. Engh, V. Kinzel, H. Ponstingl & R. Huber (1993) *EMBO J.* **12**, 849.
[3]J. Larner (1990) *AE* **63**, 173.
[4]S. I. Walaas & P. Greengard (1987) *The Enzymes*, 3rd ed., **18**, 285.
[5]S. J. Beebe & J. D. Corbin (1986) *The Enzymes*, 3rd ed., **17**, 43.
[6]G. M. Carlson, P. J. Bechtel & D. J. Graves (1979) *AE* **50**, 41.
[7]D. A. Walsh & E. G. Krebs (1973) *The Enzymes*, 3rd ed., **8**, 555.

Selected entries from *Methods in Enzymology* [vol, page(s)]:
Assay: 38, 287, 308, 323, 324; 90, 201; 99, 3, 52, 163, 134; 191, 650; 201, 341; bovine rod outer segment, 81, 497; dot-blot assay, 200, 85; grasshopper, 99, 72; liver cytosol, 38, 323, 324; luciferase, with, 305, 410; microtubule protein-associated, 85, 431; rabbit reticulocytes, 38, 316, 317; renaturation and assay after electrophoresis, 200, 417; swine kidney, from, 42, 394; synthetic peptide substrates, with, 99, 134; yeast cAMP cascade system mutants, assay, 159, 39
Other molecular properties:
Protein Kinase A, Type I: activation in assays of cyclic nucleotides and analogues, 159, 78; activation in intact cells, 38, 358, 359 (hormonal regulation, 38, 363; materials, 38, 359; methodology for [adipose tissue, 38, 359; other tissues, 38, 365]); activation, kinetic assay, 99, 163; affinity chromatography, 200, 178, 182; affinity labeling, 87, 474; alternative substrate effect, 64, 327; amino acid sequence alignment, 200, 56; assays (with low-molecular-weight substrates, 200, 113; with phosphocellulose-binding peptides, 200, 115; phosphocellulose paper, 200, 161; polyacrylamide gel electrophoresis, 200, 162; trichloroacetic acid precipitation, 200, 161); 8-azido-cAMP, 56, 646; binding cooperativity and selectivity, analysis with analogues, 99, 168; blotted (activity assessment, 200, 430; renaturation, 200, 431); cAMP assay, 31, 106 (activation [method, 38, 66; modifications, 38, 72, 73; sensitivity, reproducibility and validation, 38, 70]); cAMP inhibitor, 38, 281; carotenoid droplet protein p57 labeling, 214, 53; catalytic subunit (affinity chromatography, 200, 185; anti-peptide antibodies, generation and use, 200, 463; cDNA cloning, 159, 311; critical sites, identification, 200, 472; database, 200, 38 [in PK classification, 200, 59]; fluoresceinated, in regulatory subunit cytochemistry, 159, 255; free, direct cytochemical localization, 159, 236; hexagonal crystals, 200, 514; monoclinic crystals, 200, 511; orthorhombic crystals, 200, 516; preparation, 38, 306, 326, 387; porcine kidney [purification from {cytosol, 90, 203; particulate fraction, 90, 205}; properties, 90, 206]; small-angle neutron scattering in solution, 200, 517; topology, 200, 471); catalyzed reaction, 107, 81; cellular compartmentalization, immunogold electron microscopy, 159, 225; CHO cell, physiological role, genetic analysis, 99, 197; classification, 38, 290, 291; 107, 86; 200, 3 (catalytic domain database, with, 200, 59; criteria, 38, 291 [generality, 38, 293]);

cobalt nucleotides, **87**, 176, 178; conserved residues (identification, **200**, 56; regions with, selection for targeting by oligonucleotide probes, **200**, 526); contaminants, as, **60**, 477, 478; cooperativity, **64**, 180; cross-reactivity with type 2 protein kinase, **74**, 318; C_α subunit (expression in *Escherichia coli*, **200**, 581; mammalian, purification from yeast, **200**, 617; recombinant, kinetic parameters and posttranslational modifications, **200**, 594); cyclase preparations, **38**, 143; cyclic nucleotide exchange, kinetic assay, **99**, 163; deactivation (equilibrium assay, **99**, 164; kinetic assay, **99**, 163); dissociated catalytic subunit, fluoresceinated PK inhibitor as marker for, **201**, 316; dual messenger system, as, effect on cAMP-dependent PK inhibitor, **201**, 311; encoding clones, identification with oligonucleotide probes, **200**, 525; erythrocyte, cAMP derivatives, **38**, 298; follicle-stimulating hormone activation, **39**, 253; holoenzyme purification with RI subunit, **200**, 591; hormonal activation, detection in gastric parietal cells, **191**, 652; identification (activators and inhibitors, with, **107**, 133; addition of purified kinases, by, **107**, 134; in genetic systems, in, **200**, 427; tissue distribution analysis, by, **107**, 134); identification of pseudosubstrate region, **201**, 294; inhibition by (calphostin C, **201**, 346; staurosporine, **201**, 344; UCN-01, **201**, 344); inhibitor (analysis of cAMP-mediated cellular processes, **201**, 304; applications, **201**, 314; cAMP assay, for, **31**, 105; **38**, 49, 52, 53; effective concentration, relationship to K_i for cAMP-dependent protein kinase, **201**, 312; effects of [dual protein kinase messenger system, **201**, 311; multiple mechanisms of cAMP action, **201**, 312]; fluoresceinated [in localization of cAMP-dependent protein kinase free catalytic subunit, **159**, 247; preparation, **159**, 238]; heat-stable cAMP-dependent, rabbit skeletal muscle [assay, **99**, 77, 81; physiological role, **99**, 91; properties, **99**, 90; purification, **99**, 78, 84]; heat-stable cAMP-dependent [rabbit tissues, levels, in, and regulation, **99**, 92]; peptide chain initiation, of, **79**, 142, 147; preparation, **38**, 52; protein inhibitor, **38**, 296, 350, 351, 360, 361 [assay, **38**, 351; interaction with enzyme, **38**, 356, 357; interactions with substrates and activators, **38**, 357, 358; properties, **38**, 355, 356; purification, **38**, 353]; sources, **201**, 313; specificity, **201**, 311); isozymes, hormonal activation, analysis, **159**, 97, 105; isozyme identification, **191**, 651; K_i, relationship to cAMP-dependent protein kinase inhibitor concentration, **201**, 312; ligand binding sites, **249**, 560; liver cytosol (assay, **38**, 323, 324; mechanism of cAMP action, **38**, 323; properties, **38**, 327, 328; purification, **38**, 324); macropurification, **200**, 159, 167; mammalian (C_α subunit, purification from yeast, **200**, 617; functional expression in yeast, **200**, 605); micropurification, **200**, 159, 162; microtubule protein-associated, assay, **85**, 431; modified residues, identification, **200**, 504; monocyclic cascade control, **64**, 299; natural substrates, identification, **107**, 134; NMR, **87**, 178; nomenclature, **60**, 507; **107**, 86 (phosphorylation [histone, **60**, 524; initiation factors, **60**, 507, 525, 532, 533]); partition analysis, **249**, 324; pathway, analysis by lipopolyamine-based gene transfer, **217**, 613; peptide-based affinity labeling, **200**, 500; peptide substrates, preparation and assay, **107**, 138; phosphate removal from proteins, in, **99**, 14; phospholamban kinase, Ca^{2+}-ATPase-phospholamban complex, purification, **102**, 269; phosphorylation, **79**, 141, 142 (phosphorylation reaction, evaluation, **107**, 99); phosphorylation of (basic FGF *in vitro*, **198**, 140; CRE-binding protein, **269**, 161; human interferon γ, **119**, 296; large synthetic polypeptides reacting with ATP, **200**, 114; protein phosphatase I_G, **201**, 420); phosphorylation sites (consensus specificity motifs, **200**, 75; sequences, **200**, 63); porcine kidney (assay, **90**, 201; catalytic subunit purification from [cytosol, **90**, 203; particulate fraction, **90**, 205]; properties, **90**, 206); properties, **74**, 299, 301, 310; protein blot, on, renaturation, **200**, 423; protein inhibitor, **38**, 296, 350, 351, 360, 361 (assay, **38**, 351; interaction with enzyme, **38**, 356, 357 [with substrates and activators, **38**, 357, 358]; properties, **38**, 355, 356; purification, **38**, 353); protein modulation, **99**, 206; protein substrates, properties, **107**, 83, 145; purification, **74**, 300 (porcine skeletal muscle, **198**, 140; synthetic polymers, with, **200**, 111); purified, phosphorylation of EGF receptor *in vitro*, **198**, 235; rabbit reticulocyte, **38**, 297, 315, 316, 322; **79**, 148 (assay, **38**, 316, 317; crude preparation, **38**, 317, 318; purification of [kinase I, **38**, 318, 319; kinase II, **38**, 319]); rabbit skeletal

muscle, **38**, 299 (catalytic subunit, **38**, 306; holoenzyme, **38**, 301); radiolabeling of protein substrates (effect on protein activity, **262**, 436; reaction conditions, **262**, 436; specific activity, **262**, 436; stoichiometry of phosphate incorporation, **262**, 436); rat tissues, activity ratio determination, from, **99**, 227; recognition sequence (Michaelis constant for phosphate transfer, **262**, 434; protein incorporation for ^{32}P end-labeling, **262**, 434, 437); reconstituted, **74**, 300; regulation, **107**, 83 (cAMP, **60**, 484, 496; regulatory subunit [inhibition, **38**, 296; preparation, **38**, 327]); regulatory subunit I (antibody production, **74**, 301; bovine heart, regulatory subunit purification, **99**, 55; comparison with catabolite gene activator protein, **159**, 284; direct cytochemical localization, **159**, 255; expression and mutagenesis in *Escherichia coli*, **159**, 325; immunoreactivity, effect of cAMP, **74**, 307; intact cultured cells, regulatory subunit metabolism, analytical techniques, in, **99**, 233; preparation by nondenaturing method, **159**, 208; purification, **74**, 300; rabbit skeletal muscle, regulatory subunit purification, **99**, 55; radioimmunoassay, **74**, 299 [antisera characterization, **74**, 303; application, **74**, 300, 307; assay requirements, **74**, 303; data analysis, **74**, 303; precision, **74**, 304; procedure, **74**, 303; second antibody stage, **74**, 303; sensitivity, **74**, 304, 305; standard curve, **74**, 304, 305; tissue extract preparation, **74**, 302]; radioiodination, **74**, 301; specificity, species and tissue, **74**, 304); related mutations in Y1 adrenal tumor cells, characterization, **109**, 355; removal from ribosomes, **30**, 601; renaturation and assay after electrophoresis, **200**, 417; ribosomal proteins and, **30**, 539, 563, 564; role in cellular regulation, molecular genetic analysis, **159**, 299; storage, **200**, 169; structure and function, carbodiimides as probes, **200**, 487; subunits, **38**, 299 (HPLC, **200**, 436; nucleotide photoaffinity labeling, **200**, 477; renaturation, **200**, 436); swine kidney, from, **42**, 375, 394 (assay, **42**, 394; molecular weight, **42**, 397; properties, **42**, 397; purification, **42**, 396, 397); synergistic responses to site-selective cAMP analogues in intact cells, **159**, 118; synthetic peptide inhibitors, **159**, 173; synthetic substrates, preparation and assay, **107**, 136; thiophosphorylation of histone, in, **159**, 350; utilization of nucleotide triphosphates, **60**, 287, 503, 508

Protein Kinase A, Type II: activation in assays of cyclic nucleotides and analogues, **159**, 78; activation in intact cells, **38**, 358, 359 (hormonal regulation, **38**, 363; materials, **38**, 359; methodology for [adipose tissue, **38**, 359; other tissues, **38**, 365]); activity assay, **99**, 52; affinity chromatography, **200**, 178, 182; affinity labeling, **87**, 474; amino acid sequence alignment, **200**, 56; antibody production, **74**, 311; assays (with low-molecular-weight substrates, **200**, 113; with phosphocellulose-binding peptides, **200**, 115; phosphocellulose paper, **200**, 161; polyacrylamide gel electrophoresis, **200**, 162; trichloroacetic acid precipitation, **200**, 161); ATP-binding site, affinity labeling, **99**, 141; 8-azido-cAMP, **56**, 646; binding cooperativity and selectivity, analysis with analogues, **99**, 168; blotted (activity assessment, **200**, 430; renaturation, **200**, 431); bovine brain (immunoprecipitation, **99**, 190, 191; purification, **99**, 189; radioimmunoassay, **99**, 191); bovine heart (autophosphorylation, **99**, 177, 179, 183, 185; dephosphorylation, **99**, 183; immunoprecipitation, **99**, 191; purification, **99**, 176); bovine rod outer segment (assay, **81**, 497; properties, **81**, 501; purification, **81**, 499); cAMP-binding site, affinity labeling, **99**, 147; cAMP-free and cAMP-bound holoenzyme, purification, **159**, 202; cAMP inhibitor, **38**, 281; carotenoid droplet protein p57 labeling, **214**, 53; catalytic subunit (affinity chromatography, **200**, 185; anti-peptide antibodies, generation and use, **200**, 463; bound substrates, NMR, **99**, 98; cDNA cloning, **159**, 311; critical sites, identification, **200**, 472; crystallization, **200**, 508; cubic crystals, **200**, 512; database, **200**, 38 (in PK classification, **200**, 59); fluoresceinated, in regulatory subunit cytochemistry, **159**, 255; free, direct cytochemical localization, **159**, 236; hexagonal crystals, **200**, 514; interaction with regulatory subunit, **99**, 115; magnetic resonance spectroscopy, **99**, 94; monoclinic crystals, **200**, 511; orthorhombic crystals, **200**, 516; preparation, **38**, 314, 315; purification, **99**, 53; small-angle neutron scattering in solution, **200**, 517; substrate exchange kinetics, **99**, 113; topology, **200**, 471); catalyzed reaction, **107**, 81; cellular compartmentalization, immunogold electron microscopy, **159**,

225; CHO cell, physiological role, genetic analysis, **99**, 197; classification, **38**, 290, 291; **107**, 86; **200**, 3 (catalytic domain database, with, **200**, 59; criteria, **38**, 291 [generality, **38**, 293]); cobalt nucleotides, **87**, 176, 178; competitive displacement by regulatory subunit, **74**, 316, 317; conserved residues (identification, **200**, 56; regions with, selection for targeting by oligonucleotide probes, **200**, 526); contaminants, as, **60**, 477, 478; cooperativity, **64**, 180; cross-reactivity among tissues, **74**, 317 (between species, **74**, 319; with type I protein kinase, **74**, 318); cyclase preparations, **38**, 143; dissociated catalytic subunit, fluoresceinated PK inhibitor as marker for, **201**, 316; dual messenger system, as, effect on cAMP-dependent PK inhibitor, **201**, 311; encoding clones, identification with oligonucleotide probes, **200**, 525; erythrocyte, cAMP derivatives, **38**, 298; follicle-stimulating hormone activation, **39**, 253; hormonal activation, detection in gastric parietal cells, **191**, 652; identification (activators and inhibitors, with, **107**, 133; addition of purified kinases, by, **107**, 134; in genetic systems, in, **200**, 427; tissue distribution analysis, by, **107**, 134); identification of pseudosubstrate region, **201**, 294; inhibition by (calphostin C, **201**, 346; cAMP antagonist (*Rp*)-cAMPS, **159**, 163; staurosporine, **201**, 344; UCN-01, **201**, 344); inhibitor (analysis of cAMP-mediated cellular processes, **201**, 304; applications, **201**, 314; effective concentration, relationship to K_i for cAMP-dependent protein kinase, **201**, 312; effects of [dual protein kinase messenger system, **201**, 311; heat-stable cAMP-dependent [effect on enzyme autophosphorylation, **99**, 179; rabbit tissues, levels, in, and regulation, **99**, 92]; multiple mechanisms of cAMP action, **201**, 312; peptide chain initiation, of, **79**, 142, 147; preparation, **38**, 52; sources, **201**, 313; specificity, **201**, 311); isozymes, hormonal activation, analysis, **159**, 97, 105; isozyme identification, **191**, 651; K_i, relationship to cAMP-dependent protein kinase inhibitor concentration, **201**, 312; ligand binding sites, **249**, 560; macropurification, **200**, 159, 167; mammalian (C_α subunit, purification from yeast, **200**, 617; functional expression in yeast, **200**, 605); micropurification, **200**, 159, 162; microtubule protein-associated, assay, **85**, 431; modified residues, identification, **200**, 504; monocyclic cascade control, **64**, 299; natural substrates, identification, **107**, 134; NMR, **87**, 178; nomenclature, **60**, 507; **107**, 86 (phosphorylation [histone, **60**, 524; initiation factors, **60**, 507, 525, 532, 533]); partition analysis, **249**, 324; pathway, analysis by lipopolyamine-based gene transfer, **217**, 613; peptide-based affinity labeling, **200**, 500; peptide substrates, preparation and assay, **107**, 138; phosphate removal from proteins, in, **99**, 14; phospholamban kinase, Ca^{2+}-ATPase-phospholamban complex, purification, **102**, 269; phosphorylation, **79**, 141, 142 (phosphorylation reaction, evaluation, **107**, 99); phosphorylation of (basic FGF *in vitro*, **198**, 140; CRE-binding protein, **269**, 161; cytochrome P450 *in vitro*, **206**, 309; human interferon γ, **119**, 296; protein phosphatase I_G, **201**, 420; large synthetic polypeptides reacting with ATP, **200**, 114); phosphorylation sites (consensus specificity motifs, **200**, 75; sequences, **200**, 63); preparation, **38**, 66, 67; properties and function, **107**, 86; properties, **38**, 313; protein blot, on, renaturation, **200**, 423; protein modulation, **99**, 206; protein substrates, properties, **107**, 83, 145; purification, **38**, 310 (*Drosophila*, **159**, 215; synthetic polymers, with, **200**, 111); purified, phosphorylation of EGF receptor *in vitro*, **198**, 235; rabbit reticulocyte, **38**, 297, 315, 316, 322; **79**, 148 (assay, **38**, 316, 317; crude preparation, **38**, 317, 318; purification of [kinase I, **38**, 318, 319; kinase II, **38**, 319]); radioimmunoassay, **74**, 310 (antiserum dilution, **74**, 314, 315; applications, **74**, 316; data analysis, **74**, 314, 316; materials, **74**, 313; principle, **74**, 311; procedure, **74**, 314; radioiodination procedure, **74**, 312; second antibody stage, **74**, 314; solution storage, **74**, 313; standard curve, **74**, 314); radiolabeling of protein substrates (effect on protein activity, **262**, 436; reaction conditions, **262**, 436; specific activity, **262**, 436; stoichiometry of phosphate incorporation, **262**, 436); rat tissues, activity ratio determination, from, **99**, 227; recognition sequence (Michaelis constant for phosphate transfer, **262**, 434; protein incorporation for ^{32}P end-labeling, **262**, 434, 437); regulatory subunit, **107**, 83 (antibody production, **74**, 312; binding proteins, identification, **159**, 183; bovine brain [antiserum preparation, **99**, 190; purification, **99**,

189]; bovine heart, autophosphorylation, **99**, 178; cAMP, **60**, 484, 496; cDNA cloning, **159**, 318; comparison with cAMP-binding protein from *Trypanosoma*, **159**, 294; comparison with catabolite gene activator protein, **159**, 284; competitive displacement of holoenzyme, **74**, 316, 317; competitive displacement by regulatory subunit, **74**, 316, 317; direct cytochemical localization, **159**, 255; interaction with catalytic subunit, **99**, 115; NMR, **99**, 116; phosphorylation in intact cells, **159**, 139; photoaffinity labeling, **99**, 154; preparation by nondenaturing method, **159**, 208; radioimmunoassay, **74**, 310 [data analysis, **74**, 314, 316; principle, **74**, 311; procedure, **74**, 314; second antibody stage, **74**, 314; sensitivity, **74**, 317; standard curve, **74**, 314]; radioiodination procedure, **74**, 312); relative affinities of tissue enzymes, **74**, 317; renaturation and assay after electrophoresis, **200**, 417; role in cellular regulation, molecular genetic analysis, **159**, 299; storage, **200**, 169; structure and function, carbodiimides as probes, **200**, 487; subclass identification, **99**, 192; substrate motif screening in combinatorial peptide library (phosphorylation of peptide library, **267**, 214; radiolabeled beads [detection by autoradiography, **267**, 214; isolation, **267**, 214]); subunits (HPLC, **200**, 436; nucleotide photoaffinity labeling, **200**, 477; renaturation, **200**, 436); synergistic responses to site-selective cAMP analogues in intact cells, **159**, 118; synthetic substrates, preparation and assay, **107**, 136; utilization of nucleotide triphosphates, **60**, 287, 503, 508

Other: amphibian oocytes, analytical microinjection techniques, in, **99**, 219; grasshopper (assay, **99**, 72; properties, **99**, 76; purification, **99**, 73; stability, **99**, 75); prokaryotes, properties and purification, **200**, 223; purification from *Drosophila*, **159**, 215; rabbit reticulocyte, **38**, 297, 315, 316, 322; **79**, 148 (assay, **38**, 316, 317; crude preparation, **38**, 317, 318; purification of [kinase I, **38**, 318, 319; kinase II, **38**, 319]); yeast cAMP cascade system mutants, assay, **159**, 39

PROTEIN KINASE C (PKC)

This phosphotransferase, regulated by calcium ions and 1,2-diacylglycerol, catalyzes the ATP-dependent phosphorylation of seryl or threonyl residues in various proteins as a consequence of signal tranduction mechanisms.[1-4] ***See also*** *Protein Kinase A; Ca^{2+}/Calmodulin-Dependent Protein Kinase*

[1] U. Kikkawa & Y. Nishizuka (1986) *The Enzymes*, 3rd ed., **17**, 167.
[2] J. Larner (1990) *AE* **63**, 173.
[3] T. Hunter & B. M. Sefton, eds. (1991) *MIE*, vols. **200** and **201**, Academic Press, San Diego.
[4] J. F. Kuo, ed. (1994) *Protein Kinase C*, Oxford Univ. Press, New York.

Selected entries from ***Methods in Enzymology*** [**vol**, page(s)]:
Assay: **38**, 287; **99**, 52, 290; **124**, 350; **141**, 403, 426; **168**, 289, 348; **169**, 437; **200**, 243; **201**, 341; **252**, 137
Other molecular properties: activation, **252**, 124, 132, 150, 156 (role of arachidonate metabolism, **191**, 684); affinity chromatography, **200**, 184; atomic absorption spectroscopy, **252**, 131, 159; binding to [^{3}H]phorbol dibutyrate, **201**, 342; biology mediated by, evaluation, **201**, 325; Ca^{2+}-dependent, purification and properties, **99**, 243 (properties and function, **107**, 89); Ca^{2+}-independent, purification and properties, **99**, 308; concentration, determination, **252**, 127, 159; depleted pituitary cells, characterization, **168**, 291; effects of G_{M3} and derivatives in A431 cells, **179**, 536; extended X-ray absorption fine structure (data processing, **252**, 128; spectra, **252**, 129); extraction from pituitary cells, **168**, 289; family classification, **252**, 154; inhibition by (calphostin C, **201**, 346; sphingosine, **201**, 316 [override by diacylglycerol and phorbol ester, **201**, 323; structure-function relationship, **201**, 323]; staurosporine, **201**, 344; UCN-01, **201**, 344); inhibitors, **252**, 132 (analysis of bacterial invasion, for, **236**, 471, 475; effects on transmembrane Ca^{2+} signaling, **139**, 577); intracellular signaling, and matrix assembly, in, **245**, 528; isotypes α, β_1, β_2, γ, and ε, expression

and purification, **200**, 670; isozymes (activation by phospholipids, sensitivity, **200**, 250; antibodies specific to, preparation, characterization, and use, **200**, 454; assay, **200**, 243; cerebral, purification from [bovine, **200**, 234; rat, **200**, 244]; expressed in transfected COS-7 cells [assay, **200**, 231; separation, **200**, 230]; expression, **200**, 228; immunochemical identification in cells, **200**, 462; inactivation by phospholipids, sensitivity, **200**, 250; protein substrate specificities, comparison, **200**, 252; stimulation by diacylglycerol and phorbol 12,13-dibutyrate, **200**, 251; tryptic proteolysis, sensitivity, **200**, 248); membrane complex (hormone-induced stabilization, assay, **141**, 405; phorbol ester-induced stabilization, assay, **141**, 400); mixed micelle assay, **124**, 353; neuroendocrine tissue, in, activation by diacylglycerols, **124**, 60; nucleic acid aptamers, *in vitro* selection, **267**, 338, 357; peripheral nerve activity, **252**, 148; phosphorylation activity in intact cells, analysis by (cell treatment with enzyme modulators, **141**, 414; enzyme assay in crude tissue extracts, **141**, 421; production of enzyme-deficient cells, **141**, 419; use of phosphoprotein marker for enzyme activation, **141**, 417); phosphorylation of basic FGF *in vitro*, **198**, 140; platelet (assay, **169**, 437; partial purification and properties, **169**, 439); pseudosubstrate region identification, **201**, 301; purification (bovine brain, from, **198**, 140; calpain-free enzyme, **252**, 134; glutathione *S*-transferase fusion protein [affinity chromatography, **252**, 158; cell growth optimization, **252**, 157, 161, 166; plasmid construction, **252**, 157]; oxidatively modified enzyme, **252**, 138; protein kinase Cγ from Sf9 cells, **252**, 157); purified, phosphorylation of EGF receptor *in vitro*, **198**, 235; rat brain (activated form, properties, **102**, 284; assay, **99**, 290; **124**, 350; **168**, 348; preparation for Ca²⁺-protease I assay, **102**, 282; properties and function, **99**, 294; **107**, 91; purification, **99**, 292; **124**, 351; **141**, 426; subspecies resolution, **168**, 347); redistribution in leukocytes in response to chemoattractants, assay, **162**, 282; refolded, phorbol ester binding, effect of zinc, **256**, 119; regulatory domain (conserved sequence, **252**, 124, 133, 154; oxidation, **252**, 133; phorbol ester binding assay, **252**, 137, 158; zinc binding, **252**, 126, 156); role in (arachidonate release from phospholipids, **191**, 682; stimulus-response coupling in pituitary, **168**, 287); sphingosine effects, **312**, 361; storage, **252**, 135; structure-function analysis with diacylglycerol analogs, **141**, 313; subspecies-specific anti-peptide antibodies, preparation, **200**, 447; substrate phosphorylation, effect of sphingosine, **201**, 322; thiol modification (catalytic domain, **252**, 133, 137, 143; chelerythrine, **252**, 143; enzyme preparation, **252**, 135; *N*-ethylmaleimide, **252**, 144, 146; hydrogen peroxide, **252**, 139; inhibitors, **252**, 136, 141; metaperiodate, **252**, 140; nitric oxide, **252**, 145; removal of excess reagent, **252**, 136; sanguinarine, **252**, 143); X-ray fluorescence (data processing, **252**, 125; sample preparation, **252**, 125; spectra, **252**, 126); zinc (role, **252**, 131; stoichiometry, **252**, 160, 166)

PROTEIN KINASE, Ca²⁺/CALMODULIN-DEPENDENT

This enzyme [EC 2.7.1.123], also known as calcium/calmodulin-dependent protein kinase type II and microtubule-associated protein 2 kinase, catalyzes the reaction of ATP with a protein (at a seryl or threonyl residue) to produce ADP and an *O*-phosphoprotein.[1,2] The enzyme requires calcium ions and calmodulin. Proteins that can serve as substrates include vimentin, synapsin, myelin basic protein, glycogen synthase, the myosin light-chains, microtubule-associated protein 2, and the microtubule-associated tau protein. This enzyme is distinct from myosin light-chain kinase [EC 2.7.1.117], caldesmon kinase [EC 2.7.1.120],

and [tau protein] kinase [EC 2.7.1.135]. **See also** *Protein Kinase; Protein Kinase C*

[1] S. I. Walaas & P. Greengard (1987) *The Enzymes*, 3rd ed., **18**, 285.
[2] J. T. Stull, M. H. Nunnally & C. H. Michnaff (1986) *The Enzymes*, 3rd ed., **17**, 113.

Selected entries from **Methods in Enzymology** [vol, page(s)]:
Other molecular properties: autophosphorylation, role in enzyme regulation, **139**, 690, 701; Ca²⁺/CaM kinase II, inhibitory effects of KN-62, molecular mechanisms, **201**, 331; calmodulin-dependent (autophosphorylation, role in enzyme regulation, **139**, 690, 701; detection in GH₃ cell nuclear matrix, **139**, 665; identification of pseudosubstrate region, **201**, 297); purification from rat brain, **139**, 709

PROTEIN KINASE, cGMP-DEPENDENT

This general class of ATP-dependent protein kinases (abbreviated cGPK) catalyzes 3′,5′-cyclic-GMP-stimulated phosphorylation of protein substrates.[1-4] cGPK's are single-chain kinases containing two copies of the cyclic nucleotide-binding domain in their N-terminal regions. The nucleotide specificity of cAMP-protein kinases and cGMP-protein kinases can be traced to the conserved region of β-barrel 7: a threonyl residue is invariant in cGPK, and an alanyl residue is invariant in most cAPK's. One preferred substrate is vasodilator-stimulated phosphoprotein (VASP), an adapter in actin-based motility mechanisms. VASP's Ser239 appears to be the preferred cGMP-dependent protein kinase phosphorylation site in intact cells and platelets. **See also** *Protein Kinase*

[1] P. Ruth (1999) *Pharmacol. Ther.* **82**, 355.
[2] A. M. Edelman, D. K. Blumenthal & E. G. Krebs (1987) *ARB* **56**, 567.
[3] S. J. Beebe & J. D. Corbin (1986) *The Enzymes*, 3rd ed., **17**, 43.
[4] D. A. Flockhart & J. D. Corbin (1982) *Crit. Rev. Biochem.* **12**, 133.

Selected entries from **Methods in Enzymology** [vol, page(s)]:
General discussion: **38**, 329, 330
Assay: **99**, 63; **200**, 333
Other molecular properties: activation in intact tissues, **159**, 150; assay, **200**, 333; bovine lung (assay, **99**, 63; purification, **99**, 64); characterization, **38**, 335; cross-reactivity with other protein kinases, **74**, 304, 305; molecular weight, **74**, 301; peptide substrates, preparation and assay, **107**, 139; phospholamban, substrate for cGMP-dependent protein kinase, **269**, 159; preparation, **38**, 91; properties and function, **107**, 88; purification, **38**, 330; **74**, 301; standard assay, **38**, 330; subunit, **38**, 344; type Iα, type Iβ, and type Iβ proteolyzed monomeric enzyme, purification, **200**, 332

PROTEIN-LYSINE 6-OXIDASE

This copper-dependent oxidoreductase [EC 1.4.3.13], also known as lysyl oxidase, catalyzes the reaction of water and dioxygen with a L-lysyl residue in a peptide or protein (*i.e.*, peptidyl-L-lysyl-peptide) to produce ammonia, hydrogen peroxide, and a peptide or protein containing an allysyl residue (*i.e.*, an α-aminoadipic-δ-semialdehyde residue).[1-5] The enzyme also acts on 5-hydroxylysyl residues.

Protein substrates include collagen and elastin. The enzyme has been proposed to have a ping pong kinetic mechanism which utilizes an unusual cofactor, lysine tyrosylquinone (in which a modified tyrosyl residue is crosslinked to a lysyl residue).[2]

[1] L. I. Smith-Mungo & H. M. Kagan (1998) *Matrix Biol.* **16**, 387.
[2] S. X. Wang, M. Mure, K. F. Medzihradszky, A. L. Burlingame, D. E. Brown, D. M. Dooley, A. J. Smith, H. M. Kagan & J. P. Klinman (1996) *Science* **273**, 1078.
[3] J. P. Klinman (1996) *JBC* **271**, 27189.
[4] M. A. Ator & P. R. Ortez de Montellano (1990) *The Enzymes*, 3rd ed., **19**, 213.
[5] G. J. Cardinale & S. Udenfriend (1974) *AE* **41**, 245.

Selected entries from *Methods in Enzymology* [vol, page(s)]:
General discussion: 40, 321; 52, 10; 258, 122; 280, 98
Assay: bovine aortic, 82, 639; chicken embryo, 82, 315
Other molecular properties: active site peptide (cyanogen bromide cleavage, 258, 128; gel electrophoresis, 258, 128; labeling, 258, 127; proteolytic digestion, 258, 130; sequencing, 258, 128); activity of enzymes, 40, 347; biological role, 258, 123; bovine aortic, 82, 639; carbonyl reagent sensitivity, 258, 122; chicken embryo, 82, 315; cofactor characterization, 258, 124, 132; ethylenediamine inhibition, 258, 127; oxidative half reaction, 258, 124; purification from bovine aorta, 258, 125; reductive half reaction, 258, 124

PROTEIN-METHIONINE-S-OXIDE REDUCTASE

This oxidoreductase [EC 1.8.4.6], also known as protein-methionine sulfoxide reductase and peptide-methionine sulfoxide reductase, catalyzes the reaction of a L-methionine S-oxide residue in a protein or peptide with reduced thioredoxin (dithiothreitol will also act as a substrate; however, 2-mercaptoethanol will not) to produce oxidized thioredoxin and an L-methionyl residue in the protein or peptide.[1,2]

[1] N. Brot, L. Weissbach, J. Werth & H. Weissbach (1981) *PNAS* **78**, 2155.
[2] H. Fliss, G. Vasanthakumar, E. Schiffmann, H. Weissbach & N. Brot (1982) *BBRC* **109**, 194.

Selected entries from *Methods in Enzymology* [vol, page(s)]:
General discussion: 107, 352; 251, 462

PROTEIN-N^{π}-PHOSPHOHISTIDINE:SUGAR PHOSPHOTRANSFERASE

This phosphotransferase [EC 2.7.1.69], also known as enzyme II of the phosphoenolpyruvate:glycose phosphotransferase system, catalyzes the reaction of a sugar (typically, a hexose) with an *N*-phosphohistidyl residue in a protein to produce a sugar phosphate and the dephosphorylated histidyl residue in the protein.[1] This classification includes a group of related enzymes, some with different specificities for the sugar substrate. The small, heat-stable protein substrate is known as HPr, and transfer of the phosphoryl group from HPr proceeds via a phosphoenzyme intermediate. The bacterial enzyme also catalyzes the transmembrane translocation of the sugar as it is phosphorylated. The phosphoryl acceptors include aldohexoses (for example, D-glucose) and their glycosides (*e.g.*, methyl-α-D-glucopyranoside), deoxy sugars (*e.g.*, 2-deoxy-D-glucose), sugar amines (*e.g.*, *N*-acetylglucosamine), and alditols (*e.g.*, mannitol). All of these phosphorylations are at the O6 position. Other substrates include fructose and sorbose (at the O1 position), glycerol, and disaccharides.

Because the energy of P–O bond scission in phosphoenolpyruvate drives the hexose transport, the reaction may be regarded as an energase-type reaction (hexose$_{out}$ + 2-phosphoenolpyruvate = hexose-6-phosphate$_{in}$ + pyruvate, where the transport is indicated by the subscripts).[2]
See *Phosphoenolpyruvate:Glycose Phosphotransferase System; Phosphoenolpyruvate:Protein Phosphotransferase*

[1] H. Hüdig & W. Hengstenberg (1980) *FEBS Lett.* **114**, 103.
[2] D. L. Purich (2001) *TiBS* **26**, 417.

Selected entries from *Methods in Enzymology* [vol, page(s)]:
General discussion: 9, 396; 90, 423
Assay: 9, 396; 90, 421, 423
Other molecular properties: *Escherichia coli* (properties, 9, 402; purification, 9, 399); *Salmonella typhimurium* (properties, 90, 444; purification, 90, 439); *Staphylococcus aureus* (properties, 90, 455; purification, 90, 453)

PROTEIN-SECRETING ATPase

This set of so-called ATPases [EC 3.6.3.50] catalyze the ATP-dependent (ADP- and orthophosphate-producing) secretion of protein.[1-4] This classification of enzymes consists of a number of families: for example, Sec or Type II systems (present in bacteria, archaea, and eukarya) are ATPases of the general secretory pathway (for example, the secretion of pectinase and cellulase by *Erwinia carotovora* and lipase and phospholipase C of *Pseudomonas aeruginosa*); Type III consists of ATPases of the virulence-related secretory pathway (hence, the factors secreted subvert normal host cell functions via mechanisms that are of benefit to the invading microorganism; for example, invasin of *Shigella flexneri*); and Type IV systems participate in the conjugal DNA-protein transfer pathway. These latter two families are present in bacteria where they form components of a multi-subunit complex.

The idea that a transporter is an enzyme is in keeping with a new definition of enzyme catalysis as the facilitated making/breaking of chemical bonds, not just covalent bonds.[5] This idea builds on Pauling's assertion that any long-lived, chemically distinct interaction (in this case, the persistent location of a solute with respect to the faces of a membrane) can be regarded as a chemical bond. Note also that the equilibrium constant ($K_{eq} = [\text{protein}_{out}][ADP][P_i]/[\text{protein}_{in}][ATP]$) does not conform to that expected for an ATPase (*i.e.*, $K_{eq} = [ADP][P_i]/[ATP]$). Thus, although the overall reaction yields ADP and orthophosphate, the enzyme is misclassified as a hydrolase, and instead should be regarded as an energase-type reaction. Energases facilitate affinity-modulated reactions by coupling the $\Delta G_{\text{ATP-hydrolysis}}$ to a force-generating or work-producing step.[5] In this case, P-O-P bond scission supplies the energy to drive protein transport. **See** *ATPase; Energase*

[1]J. Mecsas & E. J. Strauss (1996) *Emerg. Infect. Diseases* **2**, 270.
[2]J. D. Thomas, P. J. Reeves & G. P. Salmond (1997) *Microbiology* **143**, 713.
[3]B. Baker, P. Zambryski, B. Staskawicz & S. P. Dinesh-Kumar (1997) *Science* **276**, 726.
[4]A. Martinez, P. Ostrovsky & D. N. Nunn (1998) *Mol. Microbiol.* **28**, 1235.
[5]D. L. Purich (2001) *TiBS* **26**, 417.

PROTEIN-SERINE EPIMERASE

This epimerase [EC 5.1.1.16], also incorrectly called protein-serine racemase, catalyzes the conversion of an L-seryl residue in a protein (or peptide) to a D-seryl residue. The enzyme from the venom of the funnel web spider (*Agelenopsis aperta*) specifically interconverts the configuration of Ser46 of ω-agatoxin-KT; Ser28 remains unchanged. The enzyme is capable of isomerizing L-seryl, D-seryl, L-cysteinyl, L-*O*-methylseryl, and L-alanyl residues in the middle of peptide chains provided that they share a common Leu-Xaa-Phe-Ala recognition site.[1,2] A two-base mechanism has been suggested in which abstraction of a proton from one face is concomitant with delivery from the opposite face by the conjugate acid of the second enzymic base.[1]

[1]Y. Shikata, T. Watanabe, T. Teramoto, A. Inoue, Y. Kawakami, Y. Nishizawa, K. Katayama & M. Kuwada (1995) *JBC* **270**, 16719.
[2]S. D. Heck, W. S. Faraci, P. R. Kelbaugh, N. A. Saccomano, P. F. Thadeio & R. A. Volkmann (1996) *PNAS* **93**, 4036.

PROTEIN-SYNTHESIZING GTPases

Protein-synthesizing GTPases [EC 3.6.1.48] assist in the translation of mRNA. This category includes initiation factors, elongation factors, and peptide-release or termination factors in both prokaryotic and eukaryotic protein synthesis.[1–5] In the prokaryotic initiation factor complex, it is IF-2b that displays the GTPase activity. In eukaryotes, it is eIF-2. In the elongation phase, the GTP-hydrolyzing proteins are the EF-Tu polypeptide of the prokaryotic transfer factor, the eukaryotic elongation factor EF-1a, the prokaryotic EF-G, the eukaryotic EF-2, and the signal recognition particle that play a role in endoplasmic reticulum protein synthesis. EF-Tu and EF-1a catalyze binding of aminoacyl-tRNA to the ribosomal A-site, while EF-G and EF-2 catalyze the translocation of peptidyl-tRNA from the A-site to the P-site. GTPase activity is also involved in polypeptide release from the ribosome with the aid of the prokaryotic and eukaryotic releasing factors, pRF and eRF, respectively.

These reactions are often misclassified as GTPases, but the reaction is of the energase type: Conformational State$_1$ + GTP = Conformational State$_1$ + GDP + P$_i$, where two affinity-modulated states are indicated.[6] Energases tranduce the Gibbs free energy of P-O-P bond scission into mechanical work that modulates the strength of protein–protein interactions. **See** *specific enzyme; G Proteins; Elongation Factors; Releasing Factors; Initiation Factors*

[1]I. M. Krab & A. Parmeggiani (1998) *BBA* **1443**, 1.
[2]D. V. Freistroffer, M Y. Pavlov, J. MacDougall, R. H. Buckingham & M. Ehrenberg (1997) *EMBO J.* **16**, 4126.
[3]M. V. Rodnina, A. Savelsberg, V. I. Katunin & W. Wintermeyer (1997) *Nature* **385**, 37.
[4]T. V. Kurzchalia, U. A. Bommer, G. T. Babkina & G. G. Karpova (1984) *FEBS Lett.* **175**, 313.
[5]L. L. Kisselev & L. Y. Frolova (1995) *Biochem. Cell Biol.* **73**, 1079.
[6]D. L. Purich (2001) *TiBS* **26**, 417.

PROTEIN-TYROSINE KINASE

This phosphotransferase [EC 2.7.1.112], also known as tyrosylprotein kinase or simply tyrosine kinase, catalyzes the reaction of ATP with a tyrosyl residue in a protein to produce a protein containing an *O*-phosphotyrosyl residue plus ADP. The protein-tyrosine kinases participate in cell signaling, growth, and oncogenesis.[1–6]

[1]G. Hardie & S. Hanks (1995) *The Protein Kinase Facts Book, Protein-Tyrosine Kinases*, Academic Press, New York.
[2]S. A. Courtneidge (1994) *Protein Kinases* (J. R. Woodgett, ed.) p. 212, IRL Press, Oxford.
[3]T. Hunter & B. M. Sefton, eds. (1991) *MIE*, vols. **200** and **201**, Academic Press, San Diego.
[4]J. Larner (1990) *AE* **63**, 173.
[5]T. Hunter & J. A. Cooper (1986) *The Enzymes*, 3rd ed., **17**, 191.
[6]M. F. White & C. R. Kahn (1986) *The Enzymes*, 3rd ed., **17**, 247.

PROTEIN-TYROSINE-PHOSPHATASE

This phosphatase [EC 3.1.3.48], also known as phosphotyrosine phosphatase, catalyzes the hydrolysis of a phosphorylated tyrosyl residue to produce a protein lacking the phosphorylated tyrosine plus orthophosphate.[1-5] These phosphorylated protein substrates are the products of protein-tyrosine kinase [EC 2.7.1.112]. Phosphoprotein substrates include: α-casein, epidermal growth factor, insulin-like growth factor 1, insulin receptor, membrane protein 3, and myelin basic protein. *See also* *Serine/Threonine Protein Phosphatase*

[1] A. C. Hengge (1998) *CBC* 1, 517.
[2] Z. Y. Zhang & J. E. Dixon (1994) *AE* 68, 1.
[3] M. J. Gresser, A. S. Tracey & P. J. Stankiewicz (1987) *Adv. Protein Phosphatases* (W. Merlevede & J. Di Salvo, eds.) 4, 35.
[4] N. K. Tonks, C. D. Diltz & E. H. Fischer (1987) *Adv. Protein Phosphatases* (W. Merlevede & J. Di Salvo, eds.) 4, 431.
[5] L. M. Ballou & E. H. Fischer (1986) *The Enzymes*, 3rd ed., 17, 311.

PROTEIN-TYROSINE SULFOTRANSFERASE

This sulfotransferase [EC 2.8.2.20], also known as tyrosylprotein sulfotransferase, catalyzes the reaction of 3'-phosphoadenylylsulfate with a tyrosyl residue in a protein to produce adenosine 3',5'-bisphosphate and protein tyrosine-O-sulfate.[1-3] This enzyme participates in post-translational processing of a variety of secretory and membrane proteins as well as of biologically active peptides. Specificity studies indicate that acidic aminoacyl residues near the tyrosyl site promote sulfation by increasing the affinity of enzyme-substrate binding and have little effect on catalytic rate.[4]

[1] S. William, P. Ramaprasad & C. Kasinathan (1997) *ABB* 338, 90.
[2] C. Niehrs, M. Kraft, R. W. H. Lee & W. B. Huttner (1990) *JBC* 265, 8525.
[3] R. W. Lee & W. B. Huttner (1983) *J. Biol. Chem.* 258, 11326.
[4] W. H. Lin, K. Larsen, G. L. Hortin & J. A. Roth (1992) *JBC* 267, 2876.

PROTEIN XYLOSYLTRANSFERASE

This transferase [EC 2.4.2.26] catalyzes the transfer of a D-xylosyl group from UDP-D-xylose to a seryl residue in a protein.[1,2] The enzyme, which has an ordered Bi Bi kinetic mechanism,[2] participates in the biosynthesis of the linkage region of proteochondroitin sulfate. Silk fibroin from *Bombyx mori* is also an acceptor substrate. Rat kidney glycogenin also has a xylosyltransferase activity.[3]

[1] J. D. Sandy (1979) *BJ* 177, 569.
[2] A. E. Kearns, S. C. Campbell, J. Westley & N. B. Schwartz (1991) *Biochemistry* 30, 7477.
[3] L. Roden, S. Ananth, P. Campbell, S. Manzella & E. Meezan (1994) *JBC* 269, 11509.

PROTOAPHIN-AGLUCONE DEHYDRATASE (CYCLIZING)

This dehydratase [EC 4.2.1.73] catalyzes the conversion of protoaphin aglucone to xanthoaphin and water.[1] Xanthophin is then converted nonenzymically to erythroaphin, an aphid pigment. The enzyme participates in pigment transformations in the wooly aphid (*Eriosoma lanigerum*).

[1] D. W. Cameron, W. H. Sawyer & V. M. Trikojus (1977) *Aust. J. Biol. Sci.* 30, 173.

PROTOCATECHUATE DECARBOXYLASE

This decarboxylase [EC 4.1.1.63], also known as 3,4-dihydroxybenzoate decarboxylase, catalyzes the reversible conversion of 3,4-dihydroxybenzoate (protocatechuate) to catechol (1,2-benzenediol) and carbon dioxide.[1]

3,4-dihydroxybenzoate   catechol

Studies indicate that the *Clostridium hydroxybenzoicum* enzyme does not require thiamin pyrophosphate, pyridoxal 5′-phosphate, or a pyruvoyl group.[2]

[1]D. J. W. Grant & J. C. Patel (1969) *Antonie Leeuwenhoek* **35**, 325.
[2]Z. He & J. Wiegel (1996) *J. Bacteriol.* **178**, 3539.

PROTOCATECHUATE 3,4-DIOXYGENASE

This iron-dependent intradiol-cleaving catecholic dioxygenase [EC 1.13.11.3], also known as protocatechuate oxygenase, catalyzes the reaction of 3,4-dihydroxybenzoate with dioxygen to produce 3-carboxy-*cis,cis*-muconate (3-carboxy-*cis,cis*-2,4-hexadienedioate).[1–10]

3,4-dihydroxybenzoate   3-carboxy-*cis,cis*-muconate

The enzyme's burgundy color is attributed to ligand-to-metal charge transfer from tyrosinate ligands to Fe(III). The relatively broad spectrum occurs because both Tyr108 and Tyr147 contribute discrete charge-transfer bands at about 435 and 525 nm.[1] Crystallographic studies indicate that the substrate binds at a trigonal pyrimidal iron site that contains a coordinated hydroxide. The formation of the Fe^{3+} chelate activates the substrate for attack by dioxygen. Dioxygen attack is also promoted by ketonization of the 3-position of the substrate.[8]

[1]R. W. Frazee, A. M. Orville, K. B. Dolbeare, H. Yu, D. H. Ohlendorf & J. D. Lipscomb (1998) *Biochemistry* **37**, 2131.
[2]J. D. Lipscomb & A. M. Orville (1992) in *Metal Ions in Biological Systems* (H. Sigel & A. Sigel, eds.) p. 243, Marcel Dekker, New York.
[3]L. Que, Jr. (1989) in *Iron Carriers and Iron Proteins* (T. M. Loehr, ed.) p. 467, VCH, New York.
[4]L. Que, Jr. (1993) in *Bioinorganic Catalysis* (J. Reedijk, ed.) p. 347, Marcel Dekker, New York.

[5]L. Que, Jr. & R. Y. N. Ho (1996) *Chem. Rev.* **96**, 2607.
[6]L. Que, Jr. (1983) *Adv. Inorg. Biochem.* **5**, 167.
[7]H. Fujisawa & O. Hayaishi (1968) *JBC* **243**, 2673.
[8]B. G. Fox (1998) *CBC* **3**, 261.
[9]J. B. Howard & D. C. Rees (1991) *Adv. Protein Chem.* **42**, 199.
[10]O. Hayaishi, M. Nozaki & M. T. Abbott (1975) *The Enzymes*, 3rd ed., **12**, 119.

Selected entries from *Methods in Enzymology* [**vol**, page(s)]:
General discussion: 2, 273, 284; **17A**, 526; **52**, 11; **188**, 82
Assay: 2, 285; **188**, 83
Other molecular properties: iron content, **17A**, 529; Mössbauer studies, **54**, 379; oxy form, spectrum of, **52**, 36; properties, **2**, 286; **188**, 87; purification from (*Brevibacterium fuscum*, **188**, 85; *Pseudomonas*, **2**, 285; **17A**, 527); transition-state and multisubstrate analogues, **249**, 304

PROTOCATECHUATE 4,5-DIOXYGENASE

This iron-dependent oxidoreductase [EC 1.13.11.8], also known as protocatechuate 4,5-oxygenase, catalyzes the reaction of protocatechuate with dioxygen to produce 4-carboxy-2-hydroxymuconate semialdehyde.[1–6]

3,4-dihydroxybenzoate   4-carboxy-2-hydroxy-*cis,cis*-muconate 6-semialdehyde

[1]D. M. Arciero & J. D. Lipscomb (1986) *JBC* **261**, 2170.
[2]L. Que, Jr. (1983) *Adv. Inorg. Biochem.* **5**, 167.
[3]R. Zabinski, E. Münck, P. N. Champion & J. M. Wood (1972) *Biochemistry* **11**, 3212.
[4]B. G. Fox (1998) *CBC* **3**, 261.
[5]J. V. Schloss & M. S. Hixon (1998) *CBC* **2**, 43.
[6]O. Hayaishi, M. Nozaki & M. T. Abbott (1975) *The Enzymes*, 3rd ed., **12**, 119.

Selected entries from *Methods in Enzymology* [**vol**, page(s)]:
General discussion: 52, 11; **188**, 89
Assay: 188, 89
Other molecular properties: properties, **188**, 92; purification from *Pseudomonas testosteroni*, **188**, 90

PROTOCHLOROPHYLLIDE REDUCTASE

This oxidoreductase [EC 1.3.1.33], also known as NADPH:protochlorophyllide oxidoreductase, catalyzes the reaction of protochlorophyllide with NADPH to produce chlorophyllide *a* (which has the (7*S*,8*S*)-configuration in the D ring) and $NADP^+$.[1–3]

The substrate's D-ring undergoes a two-step, light-dependent *trans*-reduction: the initial single-quantum event generates an unstable free-radical intermediate that spontaneously converts to chlorophyllide *a*.[1]

protochlorophyllide   chlorophyllide a

[1] N. Lebedev & M. P. Timko (1999) *PNAS* **96**, 9954.
[2] T. P. Begley & H. Young (1989) *JACS* **111**, 3095.
[3] W. T. Griffiths (1978) *BJ* **174**, 681.

PROTOPINE 6-MONOOXYGENASE

This heme-thiolate-dependent oxidoreductase [EC 1.14.13.55], also known as protopine 6-hydroxylase, catalyzes the reaction of protopine with NADPH and dioxygen to produce 6-hydroxyprotopine, $NADP^+$, and water.[1] The cytochrome P450-dependent system participates in benzophenanthridine alkaloid biosynthesis in higher plants.

[1] T. Tanahashi & M. H. Zenk (1990) *Phytochemistry* **29**, 1113.

PROTOPORPHYRINOGEN OXIDASE

This FAD-dependent oxidoreductase [EC 1.3.3.4], also known as protoporphyrinogen IX oxidase and protoporphyrinogenase, catalyzes the reaction of protoporphyrinogen IX with dioxygen to produce protoporphyrin IX and hydrogen peroxide.[1,2]

protoporphyrinogen IX   protoporphyrin IX

The enzyme will also act on mesoporphyrinogen IX (producing mesoporphyrin IX) and hematoporphyrinogen, albeit not as effectively. The mammalian liver enzyme also exhibits substrate inhibition at millimolar levels of protoporphyrinogen IX. A similar enzyme is found in *Desulfovibrio gigas*; however, that enzyme does not use dioxygen as the electron acceptor. Although 2,6-dichlorophenolindophenol can act as the acceptor substrate, NAD^+, $NADP^+$, FMN, and FAD cannot.[3]

[1] J. M. Camadro, N. G. Abraham & R. D. Levere (1985) *ABB* **242**, 206.

[2] S. Granick & S. I. Beale (1978) *AE* **46**, 33.
[3] D. J. Klemm & L. L. Barton (1987) *J. Bacteriol.* **169**, 5209.
Selected entries from *Methods in Enzymology* [**vol**, page(s)]:
General discussion: 281, 340

Pro-Xaa DIPEPTIDASE

This dipeptidase [EC 3.4.13.8], known variously as Pro-X dipeptidase, prolyl dipeptidase, iminodipeptidase, prolinase, and L-prolylglycine dipeptidase, catalyzes the hydrolysis of Pro–Xaa dipeptides and hydroxyprolyl-Xaa dipeptides. These substrates are also hydrolyzed by cytosol nonspecific dipeptidase [EC 3.4.13.18] and early studies were also contaminated with prolyl aminopeptidase [EC 3.4.11.5]. For this reason, it has been suggested that this EC number be discontinued.[1] However, others have disagreed. The prolinase from *Lactobacillus rhamnosus*[2] may not be the cytosol nonspecific dipeptidase.

[1] J. F. Woessner (1998) in *Handbook of Proteolytic Enzymes* (A. J. Barrett, N. D. Rawlings & J. F. Woessner, eds.), Academic Press, New York, p. 1517.
[2] P. Varmanen, T. Rantanen, A. Palva & S. Tynkkynen (1998) *AEM* **64**, 1831.
Selected entries from *Methods in Enzymology* [**vol**, page(s)]:
General discussion: 2, 97; **19**, 748
Assay: 2, 97; **19**, 756

PRUNASIN β-GLUCOSIDASE

This highly specific glycosidase [EC 3.2.1.118] catalyzes the hydrolysis of (*R*)-prunasin to produce mandelonitrile and D-glucose.[1–4]

(*R*)-prunasin

An epimer of (*R*)-prunasin, (*S*)-sambunigrin, is a weak substrate, but amygdalin, linamarin, or gentiobiose are not substrates. The enzyme can be assayed by using 2-nitrophenyl-β-D-glucopyranoside as a chromogenic substrate. *See also β-Glucosidase*

[1] G. Kuroki, P. A. Lizotte & J. E. Poulton (1983) *Z. Naturforsch.* **39c**, 232.
[2] E. Swain, C. P. Li & J. E. Poulton (1992) *Plant Physiol.* **100**, 291.
[3] C. P. Li, E. Swain & J. E. Poulton (1992) *Plant Physiol.* **100**, 282.
[4] G. Kuroki & J. E. Poulton (1987) *ABB* **255**, 19.

PSEUDOLYSIN

This zinc-dependent metalloendopeptidase [EC 3.4.24.26], also known as *Pseudomonas* elastase and *Pseudomonas* neutral metalloproteinase, catalyzes peptide-bond hydrolysis in many proteins, including casein, collagen types III and IV, coagulation factor XII, complement components elastin, fibronectin, hemoglobin, immunoglobulins G and A, and α_1-proteinase inhibitor.[1-3] Pseudolysin prefers hydrophobic or aromatic aminoacyl residues at position P1' (Phe > Leu > Tyr > Val > Ile). Alanyl residues are preferred at the P1 and P2' positions. Synthetic substrates such as FA-Gly-^-Leu-NH$_2$, FA-Gly-^-Leu-Ala-NH$_2$, or Abz-Ala-Gly-^-Leu-Ala-Nba (where Abz refers to the *o*-aminobenzoyl group and Nba refers to 4-nitrobenzylamide) can be used to assay the enzyme. Calcium ions are required for enzyme stability.

Pseudolysin is a major virulence factor of *P. aeruginosa*, resulting in tissue destruction and damage of cell functions. It also interferes with the host's defense mechanisms.

[1] K. Morihara (1995) *MIE* **248**, 242.
[2] K. Morihara & J. Y. Homma (1985) in *Bacterial Enzymes and Virulence* (I. A. Holder, ed.), CRC Press, Boca Raton, p. 41.
[3] D. R. Galloway (1991) *Mol. Microbiol.* **5**, 2315.
Selected entries from *Methods in Enzymology* [**vol**, page(s)]:
General discussion: 248, 242
Assay: 248, 242; casein-digestion method, 248, 242; elastin-Congo Red method, 235, 557; elastin-digestion method, 248, 242; elastin-nutrient agar method, 235, 561; fluorogenic substrate method, 235, 559; hydrolysis of synthetic peptides, by, 248, 242, 245; principle, 248, 242
Other molecular properties: activities, 235, 554; amino acid sequence, 248, 411; effect on neutrophil chemotaxis, 236, 77; inhibition, 248, 406; isolation, 248, 245; pathogenesis, in, 235, 555; pathogenic activity, 248, 251; production, 248, 245; properties, 248, 185, 187, 247 (activity, with synthetic substrates, 248, 248; enzymatic, 248, 247; inhibition, 248, 248; pH optimum, 248, 247; substrate specificity, 248, 247); proteinase activity, 248, 406; purification, 235, 555; 248, 246; related enzymes, 248, 242; strains producing, pulmonary persistence, 235, 134, 381; structure, 248, 186, 249 (thermolysin structure, and, comparison, 248, 249); substrate specificity, 248, 187, 247; trapped by α_2-macroglobulin, 80, 748

PSEUDOMONAPEPSIN

This serine endopeptidase [EC 3.4.21.100; formerly EC 3.4.23.37], known also as *Pseudomonas* sp. pepstatin-insensitive carboxyl proteinase, catalyzes peptide-bond hydrolysis in proteins. With casein as a substrate, the pH optimum is 3.0. Oxidized insulin B is cleaved at Glu13–Ala14, Leu18–Tyr19, and Phe25–Tyr26. Angiotensin I is hydrolyzed at Tyr4–Ile5. Synthetic substrates include Lys-Pro-Ile-Glu-Phe-^-Phe(NO$_2$)-Arg-Leu and Lys-Pro-Ala-Leu-Phe-^-Phe(NO$_2$)-Arg-Leu. While pseudomonapepsin is not inhibited by pepstatin, diazoacetylnorleucine methyl ester, or 1,2-epoxy-3(*p*-nitrophenoxy)propane (EPNP), the enzyme is inhibited by tyrostatin ($K_i = 2.5$ nM).[1]

[1] K. Oda, S. Takahashi, T. Shin & S. Murao (1995) in *Aspartic Proteinases: Structure, Function, Biology, and Biomedical Implications* (K. Takahashi, ed.), Plenum Press, New York, p. 529.

PSEUDOMONAS CYTOCHROME OXIDASE

This heme-dependent oxidoreductase [EC 1.9.3.2], also known as cytochrome *cd*, cytochrome *cd*$_1$, and nitrite reductase, catalyzes the reaction of four molecules of ferrocytochrome c_2 with dioxygen to produce four molecules of ferricytochrome c_2 and two water molecules.[1-5] Other acceptor substrates include hydroxylamine and nitrite, with the latter leading to the production of nitric oxide and hydroxide anion. The natural substrate is *Pseudomonas* ferricytochrome c_{551}.

[1] R. K. Poole (1983) *BBA* **726**, 205.
[2] R. Timkovich, R. Dhesi, K. J. Martinkus, M. K. Robinson & T. M. Rea (1982) *ABB* **215**, 47.
[3] D. Barber, S. R. Parr & C. Greenwood (1976) *BJ* **157**, 431.
[4] J. Singh (1973) *BBA* **333**, 28.
[5] J. C. Gudat, J. Singh & D. C. Wharton (1973) *BBA* **292**, 376.
Selected entries from *Methods in Enzymology* [**vol**, page(s)]:
Other molecular properties: *Thiobacillus denitrificans*, Mössbauer spectroscopy, 243, 533

PSEUDOURIDINE KINASE

This phosphotransferase [EC 2.7.1.83] catalyzes the reaction of ATP with pseudouridine (Ψ) to produce ADP and pseudouridine 5'-phosphate (that is, pseudouridylate).[1,2]

[1] L. R. Solomon & T. R. Breitmann (1971) *BBRC* **44**, 299.
[2] E. P. Anderson (1973) *The Enzymes*, 3rd ed., **9**, 49.

PSEUDOURIDYLATE SYNTHASE

This enzyme [EC 4.2.1.70], also known as pseudouridine synthase and uracil hydrolyase, catalyzes the reaction of uracil with D-ribose 5-phosphate to produce pseudouridine 5'-phosphate (that is, pseudouridylate) and water.[1-4]

uracil   pseudouridine 5'-phosphate

There are pseudouridine synthases that catalyze the isomerization of specific uridines present in a variety of RNAs to pseudouridine. *Escherichia coli* pseudouridine synthase I, which catalyzes two or three conversions in *E. coli* tRNA,

is a dimeric protein containing two RNA-binding clefts.[3]
See *tRNA Pseudouridine Synthase I*

[1] T. Suziki & R. M. Hochster (1966) *Can. J. Biochem.* **44**, 259.
[2] R. L. Heinrikson & E. Goldwasser (1964) *JBC* **239**, 1177.
[3] P. G. Foster, L. Huang, D. V. Santi & R. M. Stroud (2000) *Nat. Struct. Biol.* **7**, 23.
[4] L. K. Kline & D. Söll (1982) *The Enzymes*, 3rd ed., **15**, 567.

PSYCHOSINE SULFOTRANSFERASE

This transferase [EC 2.8.2.13] catalyzes the reaction of 3'-phosphoadenylylsulfate with galactosylsphingosine (or, psychosine) to produce adenosine 3',5'-bisphosphate and psychosine sulfate.[1]

[1] J.-L. Nussbaum & P. Mandel (1972) *J. Neurochem.* **19**, 1789.

PTERIDINE OXIDASE

This oxidoreductase [EC 1.17.3.1], first isolated from castor bean (*Ricinus communis*) seeds, catalyzes the reaction of 2-amino-4-hydroxypteridine with dioxygen to produce 2-amino-4,7-dihydroxypteridine and another product.[1]

2-amino-4-hydroxypteridine

2-amino-4,7-dihydroxypteridine

Hypoxanthine is not an alternative substrate (however, both hypoxanthine and xanthine are inhibitors). The enzyme is distinct from xanthine oxidase [EC 1.1.3.22].

[1] Y.-N. Hong (1980) *Plant Sci. Lett.* **18**, 169.

PTERIDINE REDUCTASE

This enzyme [EC 1.1.1.253] reversibly catalyzes the reaction of biopterin with two molecules of NADPH (stereospecifically to and from the B-side) to generate 5,6,7,8-tetrahydrobiopterin and NADP[+] molecules.[1,2] The *Leishmania* enzyme can also act on folate and a wide variety of unconjugated pterins; moreover, 7,8-dihydropterins and 7,8-dihydrofolate can also be converted to the corresponding tetrahydro forms. This reductase will not act on the quinonoid form of dihydrobiopterin, as observed with both dihydrofolate reductase [EC 1.5.1.3] and dihydropteridine reductase [EC 1.6.99.7].

[1] A. R. Bello, B. Nare, D. Freedman, L. Hardy & S. M. Beverley (1994) *PNAS* **91**, 11442.
[2] B. Nare, L. W. Hardy & S. M. Beverley (1997) *JBC* **272**, 13883.

PTERIN DEAMINASE

This deaminase [EC 3.5.4.11] catalyzes the hydrolysis of 2-amino-4-hydroxypteridine to produce 2,4-dihydroxypteridine and ammonia.[1-3] The rat liver enzyme is specific for pterin, isoxanthopterin, and tetrahydropterin.[3]

[1] B. Wurster, F. Bek & U. Butz (1981) *J. Bacteriol.* **148**, 183.
[2] H. Rembold & F. Simmersbach (1969) *BBA* **184**, 589.
[3] B. Levenberg & O. Hayaishi (1959) *JBC* **234**, 955.

Selected entries from *Methods in Enzymology* [vol, page(s)]:
General discussion: 6, 359; **66**, 652
Assay: 6, 359

PTEROCARPIN SYNTHASE

This enzyme [EC 1.1.1.246], first[1,2] obtained from chickpea (*Cicer arietinum*), catalyzes the reaction of vestitone with NADPH to produce medicarpin and NADP[+].

vestitone   medicarpin

This is the last step in the pathway in the biosynthesis of the pterocarpin phytoalexin medicarpin. In alfalfa, pterocarpin synthase activity results from the action of two separate enzymes.[3]

[1] W. Bless & W. Barz (1988) *FEBS Lett.* **235**, 47.
[2] S. Daniel, K. Tiemann, U. Wittkampf, W. Bless, W. Hinderer & W. Barz (1990) *Planta* **182**, 270.
[3] I. Guo, R. A. Dixon & N. L. Paiva (1994) *JBC* **269**, 22372.

PULLULANASE

This glucosidase [EC 3.2.1.41] (also known as α-dextrin endo-1,6-α-glucosidase, pullulan 6-glucanohydrolase, limit dextrinase (a misnomer), debranching enzyme, and amylopectin 6-glucanohydrolase) catalyzes the hydrolysis of α(1,6)-glucosidic linkages in pullulan and amylopectin.[1-3] This hydrolase is the starch debranching enzyme. Pullulan is an extracellular glucan produced by certain fungi containing both α(1 → 6) and α(1 → 4) linkages.
See also *Limit Dextrinase; α-Dextrin Endo-1,6-α-Glucosidase; Debranching Enzyme*

[1]G. Mooser (1992) *The Enzymes*, 3rd ed., **20**, 187.
[2]S. Kusano, N. Nagahata, S. Takahashi, D. Fujimoto & Y. Sakano (1988) *Agric. Biol. Chem.* **52**, 2293.
[3]E. Y. C. Lee & W. J. Whelan (1972) *The Enzymes*, 3rd ed., **5**, 191.

Selected entries from *Methods in Enzymology* [vol, page(s)]:
General discussion: 8, 555
Assay: 8, 555
Other molecular properties: immobilized, 44, 470; preparation, 8, 556; properties, 8, 558; substrate, 8, 358; 64, 255

PURINE IMIDAZOLE-RING CYCLASE

This enzyme [EC 4.3.2.4] catalyzes the conversion of 4,6-diamino-5-formamidopyrimidine in DNA to an adenine in that DNA and the release of water. The reaction results in imidazole-ring reclosure in purines damaged by gamma-rays. The enzyme also acts on 2,6-diamino-5-formamido-3,4-dihydro-4-oxopyrimidine residues in DNA, producing a guanine residue in the polynucleotide.[1]

[1]C. J. Chetsanga & C. Grigorian (1985) *PNAS* **82**, 633.

PURINE NUCLEOSIDASE

This enzyme [EC 3.2.2.1], also known as inosine-uridine preferring nucleoside hydrolase, catalyzes the hydrolysis of an *N*-D-ribosylpurine to produce a purine and D-ribose.[1,2]

adenosine   D-ribose   adenine

Substrates include adenosine, guanosine, inosine, xanthosine, AMP, NAD$^+$, and NADP$^+$. The purine-specific enzyme from *Trypanosoma brucei brucei* does not act on AMP and displays a rapid equilibrium random Uni Bi kinetic mechanism.[3]

[1]C. A. Atkins, P. J. Storer & B. J. Shelp (1989) *J. Plant Physiol.* **134**, 447.
[2]M. Kuwahara & T. Fujii (1978) *Can. J. Biochem.* **56**, 345.
[3]D. W. Parkin (1996) *JBC* **271**, 21713.

Selected entries from *Methods in Enzymology* [vol, page(s)]:
General discussion: 2, 456, 462
Assay: 2, 462

PURINE-NUCLEOSIDE PHOSPHORYLASE

This enzyme [EC 2.4.2.1], also known as inosine phosphorylase, catalyzes the reversible reaction of a purine nucleoside with orthophosphate to produce a purine and α-D-ribose 1-phosphate.[1–11] The enzyme also exhibits the ribosyltransferase reactions of nucleoside ribosyltransferase [EC 2.4.2.5]. Purine nucleoside substrates include inosine, deoxyinosine, guanosine, deoxyguanosine, and xanthosine. In the presence of arsenate, the enzyme will produce a purine and D-ribose.[9] Electrostatic potential surfaces of the transition states indicate that the arsenate anion is more effective in neutralizing the oxocarbenium ion than is water.[6] Substrate trapping experiments show that there is no detectable forward commitment to catalysis for inosine hydrolysis, whereas bound inosine is significantly more likely to form hypoxanthine and α-D-ribose 1-phosphate than to dissociate when the enzyme-inosine complex is exposed to saturating concentrations of orthophosphate.[6] The enzyme has a random Bi Bi kinetic mechanism in chicken liver[1] and bovine lens (nonlinear kinetics),[2] a Theorell-Chance mechanism with the human erythrocyte[3] and bovine brain[4] enzymes, and an ordered mechanism from human erythrocytes (albeit random in the reverse reaction).[5]

inosine   α-D-ribose 1-phosphate   hypoxanthine

The transition-state structure is characterized by an elevated pK_a at N7 of the purine ring, making that position favorable for protonation or hydrogen-bond interaction with the enzyme and favoring the oxocarbenium ion formation in the ribosyl ring.[6] The orthophosphate substrate reportedly also functions as a component of a catalytic triad (with glutamyl and histidyl residues).[7] Studies with transition-state inhibitors that bind to only one-third of the sites of the human erythrocyte and calf spleen trimeric proteins suggest a sequential mechanism similar to that of F_1 ATPase.[8] *See also Thymidine Phosphorylase*

[1]K. Murakami & K. Tsushima (1975) *BBA* **384**, 390.
[2]D. Barsacchi, M. Cappiello, M. G. Tozzi, A. Del Corso, M. Peccatori, M. Camici, P. L. Ipata & U. Mura (1992) *BBA* **1160**, 163.
[3]A. S. Lewis & B. A. Lowy (1979) *JBC* **254**, 9927.
[4]A. S. Lewis & M. D. Glantz (1976) *Biochemistry* **15**, 4451.
[5]J. D. Stoeckler, R. P. Agarwal, K. C. Agarwal & R. E. Parks (1978) *MIE* **51**, 530.
[6]P. C. Kline & V. L. Schramm (1995) *Biochemistry* **34**, 1153.
[7]M. D. Erion, J. D. Stoeckler, W. C. Guida, R. L. Walter & S. E. Ealick (1997) *Biochemistry* **36**, 11735.
[8]R. W. Miles, P. C. Tyler, R. H. Furneaux, C. K. Bagdassarian & V. L. Schramm (1998) *Biochemistry* **37**, 8615.

[9]P. C. Kline & V. L. Schramm (1993) *Biochemistry* **32**, 13212.
[10]R. E. Parks & R. P. Agarwal (1972) *The Enzymes*, 3rd ed., **7**, 483.
[11]G. Davies, M. L. Sinnott & S. G. Withers (1998) *CBC* **I**, 119.

Selected entries from *Methods in Enzymology* [vol, page(s)]:
General discussion: 2, 448; 12A, 113; 18B, 204; 51, 517, 524, 530, 538
Assay: 2, 449; 51, 518, 525, 526, 531, 532, 538
Other molecular properties: active site, 51, 529; activity, 51, 517, 524, 530, 531, 542; affinity for calcium phosphate gel, 51, 584; amino acid composition, 51, 536, 538; arsenolysis, 308, 341; assays during phosphorolysis of pyridine nucleosides (radioactive nicotinate or nicotinamide formed, 18B, 205, 206; radioactive phosphate converted to ribose 1-phosphate, 18B, 206, 207); Chinese hamster, from, 51, 538; commitment factors, 308, 341; crystallization, 51, 534, 535; deoxyribose 1-phosphate and, 3, 183; deoxyribose 5-phosphate and, 3, 186; exchange studies, 87, 249; extinction coefficient, 51, 536; frictional ratio, 51, 536; gene therapy, 326, 145; human erythrocytes, from, 51, 530; inhibitors, 51, 530, 536, 537; 308, 349, 353, 394; isoelectric point, 51, 529, 543; isotope effects, 308, 341; isozymes, 51, 518, 536, 537; kinetic properties, 51, 537, 543; marker for chromosome detection in somatic cell hybrids, as, 151, 188; molecular weight, 51, 520, 530, 536, 543; NMR, 87, 249; occurrence, 12A, 113; partial specific volume, 51, 536; pH optimum, 51, 521, 529; photooxidation, 51, 530; preparation and properties of, 2, 450; 12A, 114; properties, 18B, 209, 210; purification, 18B, 207; 51, 519, 527, 532, 541, 542; purity, 51, 529; pyrimidine requirement and, 6, 154; rabbit liver, from, 51, 524; rat liver, from, 18B, 207; 51, 521; reaction mechanism, 51, 537; regulation, 51, 438, 517; removal, with calcium phosphate gel, 51, 503, 504; ribose 1-phosphate and, 3, 181; *Salmonella typhimurium*, from, 51, 517; secondary structure from CD data, 51, 536; sedimentation coefficients, 51, 536, 543; sources, 51, 524, 525; stability, 51, 521, 530, 536; stain for, 224, 110; Stokes radius, 51, 536; substrate specificity, 51, 521, 529, 537; subunit structure, 51, 520, 530, 543; transition state, 308, 341; transition-state and multisubstrate analogues, 249, 305; translational diffusion coefficient, 51, 536

PUTIDAREDOXIN REDUCTASE

This oxidoreductase, which is associated with a cytochrome P450 system, is a flavoprotein that catalyzes the NADH-dependent reduction of putidaredoxin.

Selected entries from *Methods in Enzymology* [vol, page(s)]:
General discussion: 52, 173
Other molecular properties: cytochrome P-450$_{cam}$ activity assay, in, 52, 156; expression in *Escherichia coli*, 206, 38; fluorometric measurement of flavin content, 52, 176; gel electrophoresis, 52, 175; isolation and purification, 52, 176 (flow chart, 52, 173; procedure, 52, 176; reagents, 52, 175); optical properties, 52, 184, 185; partial purification, 52, 154, 155; product inhibition studies, 249, 210; purity criteria, 52, 184; stability, 52, 183, 184

PUTRESCINE CARBAMOYLTRANSFERASE

This transferase [EC 2.1.3.6] catalyzes the reaction of carbamoyl phosphate with putrescine ($^+H_3NCH_2CH_2CH_2CH_2NH_3^+$) to produce orthophosphate and N-carbamoylputrescine ($^+H_3NCH_2CH_2CH_2CH_2NHCONH_2$).[1,2] The plant enzyme also catalyzes the reactions of ornithine carbamoyltransferase [EC 2.1.3.3], carbamate kinase [EC 2.7.2.2], and agmatine deiminase [EC 3.5.3.12], thus acting as putrescine synthase.

This enzyme also carbamoylates ornithine, 1,3-diaminopropane, 1,6-diaminohexane, spermine, spermidine, and cadaverine, albeit with lower efficiency. Experiments on substrate protection against thermal inactivation, product inhibition, and dead-end inhibition provide evidence for a random addition of the substrates, with a preferred pathway in which carbamoylphosphate is the leading substrate.[1]

[1]B. Wargnies, N. Lauwers & V. Stalon (1979) *EJB* **101**, 143.
[2]C. W. Tabor & H. Tabor (1984) *ARB* **53**, 749.

Selected entries from *Methods in Enzymology* [vol, page(s)]:
General discussion: 34, 339
Assay: 94, 340
Other molecular properties: *Streptococcus faecalis* (assay, 94, 340; properties, 94, 342; purification, 94, 340)

PUTRESCINE N-HYDROXY-CINNAMOYLTRANSFERASE

This acyltransferase [EC 2.3.1.138], first isolated from American tobacco (*Nicotinia tabacum*), catalyzes the reaction of caffeoyl-CoA with putrescine to produce coenzyme A and N-caffeoylputrescine.[1–4]

N-caffeoylputrescine

Alternative acyl-CoA substrates include feruloyl-CoA, cinnamoyl-CoA, and sinapoyl-CoA.

[1]J. Negrel (1989) *Phytochemistry* **28**, 477.
[2]B. Meurer-Grimes, J. Berlin & D. Strack (1989) *Plant Physiol.* **89**, 488.
[3]J. Negrel, M. Paynot & F. Javelle (1992) *Plant Physiol.* **98**, 1264.
[4]J. Negrel, F. Javelle & M. Paynot (1991) *Phytochemistry* **30**, 1089.

PUTRESCINE N-METHYLTRANSFERASE

This methyltransferase [EC 2.1.1.53] catalyzes the reaction of S-adenosyl-L-methionine with putrescine to produce S-adenosyl-L-homocysteine and N-methylputrescine.[1,2] N-Methylputrescine displays about one-twelfth the activity of putrescine.[2]

[1]F. Feth, H.-A. Arfmann, V. Wray & K. G. Wagner (1985) *Phytochemistry* **24**, 921.
[2]S. Mizusaki, Y. Tanabe, M. Noguchi & E. Tamaki (1971) *Plant Cell Physiol.* **12**, 633.

PUTRESCINE OXIDASE

This FAD-dependent oxidoreductase [EC 1.4.3.10] catalyzes the reaction of putrescine with dioxygen and water to produce 4-aminobutanal, ammonia, and hydrogen peroxide; the aldehyde-containing product then condenses non-enzymatically to generate 1-pyrroline.[1–3]

1-pyrroline

Other substrates include cadaverine, spermidine, 1,3-diaminopropane, and *N*-acetylputrescine. **See also** Diamine Oxidase; Amine Oxidase (Copper-Containing)

[1]M. Okada, S. Kawashima & K. Imahori (1979) *J. Biochem.* **86**, 97.
[2]M. Okada, S. Kawashima & K. Imahori (1980) *J. Biochem.* **88**, 481.
[3]W. A. Gahl & H. C. Pitot (1982) *BJ* **202**, 603.

Selected entries from **Methods in Enzymology** [vol, page(s)]:
General discussion: 17B, 726; 52, 18; 94, 301
Assay: assay, with *Diplocardia* bioluminescence, 57, 381; *Micrococcus rubens*, 17B, 726; 94, 301
Other molecular properties: flavin adenine dinucleotide and, 17B, 729; *Micrococcus rubens* (properties, 17B, 729; purification, 17B, 727 [by affinity chromatography, 94, 302]); spectrum, 17B, 729

PYCNOPOROPEPSIN

This aspartic endopeptidase [EC 3.4.23.30], also known as proteinase Ia and Ib, catalyzes peptide-bond hydrolysis in proteins. Pycnoporopepsin Ia and Ib cleave the oxidized and native insulin B chain at Ala14–Leu15, Tyr16–Leu17, and Phe24–Phe25. The native A chain is not acted upon. Hydrolysis of angiotensin II is at Tyr4–Ile5. The enzyme exhibits a preference for hydrophobic aminoacyl residues (*e.g.*, Leu or Ile) at the P1′ position and a hydrophobic residue (*e.g.*, Tyr or Val) at the P2′ position. This enzyme has a role in fungal digestion of wood and is similar to aspergillopepsin I, but with a different specificity.[1,2]

[1]E. Ichishima, H. Kumagai & K. Tomoda (1980) *Curr. Microbiol.* **3**, 333.
[2]E. Ichishima, M. Emi, E. Majima, Y. Mayumi, H. Kumagai, K. Hayashi & K. Tomoda (1982) *BBA* **700**, 247.

Selected entries from **Methods in Enzymology** [vol, page(s)]:
General discussion: 19, 373

PYRANOSE OXIDASE

This FAD-dependent oxidoreductase [EC 1.1.3.10], also known as glucose 2-oxidase, catalyzes the reaction of D-glucose (both anomers are substrates) with dioxygen to produce 2-dehydro-D-glucose (that is, 2-keto-D-glucose) and hydrogen peroxide.[1–5]

β-D-glucose   2-keto-β-D-glucose

Other substrates include D-xylose, D-glucono-1,5-lactone, and L-sorbose, the latter undergoing oxidation at C5 rather than C2 (producing 5-keto-D-fructose). Interestingly, a ping pong Bi Bi kinetic mechanism was reported for the *Phanerochaete chrysosporium* enzyme.[1]

[1]M. J. Artolozaga, E. Kubatova, J. Volc & H. M. Kalisz (1997) *Appl. Microbiol. Biotechnol.* **47**, 508.
[2]Y. Izumi, Y. Furuya & H. Yamada (1990) *ABC* **54**, 799 and 1393.
[3]K. E. Eriksson, J. Pettersson, J. Volc & V. Musilik (1986) *Appl. Microbiol. Biotechnol.* **23**, 257.
[4]T. Taguchi, K. Ohwaki & J. Okuda (1985) *J. Appl. Biochem.* **7**, 289.
[5]Y. Machida & T. Nakanishi (1984) *ABC* **48**, 2463.

Selected entries from **Methods in Enzymology** [vol, page(s)]:
General discussion: 41, 170; 52, 21; 161, 316
Assay: 41, 170; 137, 601; 161, 317
Other molecular properties: amidination, 137, 602; amino groups, assay, 137, 603; crosslinked (characterization, 137, 609; gel electrophoresis, 137, 604; preparation, 137, 602); glucosone effects, 137, 608, 611; *Polyporus obtusus*, from, 41, 170 (assay, 41, 170; production, 41, 171; properties, 41, 172; purification, 41, 172, 173); properties, 161, 319; purification from *Phanerochaete chrysosporium*, 161, 318; thermodenaturation, 137, 605, 615

PYRAZOLYLALANINE SYNTHASE

This pyridoxal-phosphate-dependent enzyme [EC 4.2.1.50] catalyzes the reaction of L-serine with pyrazole to produce 3-(pyrazol-1-yl)-L-alanine and water.[1,2]

pyrazole   3-(pyrazol-1-yl)-L-alanine

O-Acetyl-L-serine is also reported to be a substrate (but see β-pyrazolylalanine synthase [acetylserine]).

[1]I. Murakoshi, F. Ikegami, Y. Hinuma & Y. Hanma (1984) *Phytochemistry* **23**, 973.
[2]P. M. Dunnill & L. Fowden (1963) *J. Exp. Bot.* **14**, 237.

β-PYRAZOLYLALANINE SYNTHASE (ACETYLSERINE)

This enzyme [EC 4.2.99.14], often abbreviated BPA synthase, catalyzes the reaction of O^3-acetyl-L-serine with pyrazole to produce 3-(pyrazol-1-yl)-L-alanine and acetate.

pyrazole   3-(pyrazol-1-yl)-L-alanine

The synthase, which is not identical to L-mimosine synthase [EC 4.2.99.15], is highly specific for acetylserine and pyrazole.[1,2] In some sources, this enzyme may be identical to pyrazolylalanine synthase [EC 4.2.1.50]. The acetylated substrate rapidly undergoes a *O*-acetyl to *N*-acetyl shift above pH 8. **See also** L-*Mimosine Synthase; Uracilylalanine Synthase*

[1] I. Murakoshi, H. Kuramoto & J. Haginiwa (1972) *Phytochemistry* **11**, 177.
[2] I. Murakoshi, F. Ikegami, Y. Hinuma & Y. Hanma (1984) *Phytochemistry* **23**, 973.

PYRIDINE *N*-METHYLTRANSFERASE

pyridine   *N*-methylpyridinium cation

This methyltransferase [EC 2.1.1.87] catalyzes the reaction of *S*-adenosyl-L-methionine with pyridine to produce *S*-adenosyl-L-homocysteine and *N*-methylpyridinium.[1]

[1] L. A. Damani, M. S. Shaker, P. A. Crooks, C. S. Godin & C. Nwosu (1986) *Xenobiotica* **16**, 645.

PYRIDOXAL 4-DEHYDROGENASE

pyridoxal   4-pyridoxolactone

This oxidoreductase [EC 1.1.1.107] catalyzes the reaction of pyridoxal with NAD^+ to produce 4-pyridoxolactone and NADH and a proton.[1]

[1] R. W. Burg & E. E. Snell (1969) *JBC* **244**, 2585.

Selected entries from *Methods in Enzymology* [vol, page(s)]:
General discussion: 18A, 648; **122**, 115
Assay: 18A, 648, 649; human liver, **122**, 115
Other molecular properties: properties, 18A, 650, 651; purification from *Pseudomonas* sp., 18A, 649, 650

PYRIDOXAL KINASE

This phosphotransferase [2.7.1.35] catalyzes the reaction of ATP and pyridoxal to form ADP and pyridoxal 5′-phosphate.[1-3]

Other phosphoryl acceptors are pyridoxine and pyridoxamine. The brain enzyme[1] exhibits a random Bi Bi kinetic mechanism whereas the liver enzyme[2] is ordered.

pyridoxal   pyridoxal 5-phosphate

[1] J. E. Churchich & C. Wu (1981) *JBC* **256**, 780.
[2] M. Tagaya, K. Yamano & T. Fukui (1989) *Biochemistry* **28**, 4670.
[3] S. L. Ink & L. M. Henderson (1984) *Ann. Rev. Nutr.* **4**, 455.

Selected entries from *Methods in Enzymology* [vol, page(s)]:
General discussion: 2, 646; 18A, 611, 619; **34**, 5
Assay: 2, 646; 18A, 611, 619; human liver, **122**, 111
Other molecular properties: properties, 18A, 617, 625; purification from (brewer's yeast, 2, 648; *Escherichia coli*, 18A, 622; *Lactobacillus casei* and *Streptococcus faecalis*, 18A, 614, 615; liver and brain, 18A, 615; rat liver, **122**, 100); rat liver (purification, **122**, 100; synthesis of [^{14}C]pyridoxal 5-phosphate, in, **122**, 101)

PYRIDOXAL OXIDASE

pyridoxal   4-pyridoxate

This molybdopterin-dependent enzyme [EC 1.2.3.8] catalyzes the reaction of pyridoxal with water and dioxygen to produce 4-pyridoxate and hydrogen peroxide.[1,2]

[1] C. K. Warner & V. Finnerty (1981) *Mol. Gen. Genet.* **184**, 92.
[2] E. W. Hanly (1980) *Mol. Gen. Genet.* **180**, 455.

Selected entries from *Methods in Enzymology* [vol, page(s)]:
Assay: human liver, **122**, 115

PYRIDOXAL 5′-PHOSPHATE HYDROLASE

This enzyme activity, also known as pyridoxal 5′-phosphatase, refers to the hydrolysis of pyridoxal 5′-phosphate, producing pyridoxal and orthophosphate. The activity is probably due to alkaline phosphatase. **See also** *Alkaline Phosphatase*

Selected entries from *Methods in Enzymology* [vol, page(s)]:
General discussion: 62, 574
Assay: 62, 575; human liver, **122**, 114
Other molecular properties: preparation, of enzyme, from rat liver, 62, 579; properties, 62, 581, 582

PYRIDOXAMINE:OXALOACETATE AMINOTRANSFERASE

This aminotransferase [EC 2.6.1.31] catalyzes the reversible reaction of pyridoxamine with oxaloacetate to produce pyridoxal and L-aspartate.[1,2]

pyridoxamine   pyridoxal

[1]H. Wada & E. E. Snell (1962) *JBC* **237**, 127.
[2]H. L. C. Wu & M. Mason (1964) *JBC* **239**, 1492.

PYRIDOXAMINE-PHOSPHATE AMINOTRANSFERASE

This aminotransferase [EC 2.6.1.54] catalyzes the reversible reaction of pyridoxamine 5-phosphate with α-ketoglutarate (or, 2-oxoglutarate) to produce pyridoxal 5'-phosphate and D-glutamate.[1] Pyridoxamine exhibits about one-fourth of the activity observed with pyridoxamine 5-phosphate.

[1]Y. Tani, M. Ukita & K. Ogata (1972) *ABC* **36**, 181.

PYRIDOXAMINE 5'-PHOSPHATE HYDROLASE

This enzyme activity, also known as pyridoxamine 5'-phosphatase, refers to the hydrolysis of pyridoxamine 5'-phosphate, producing pyridoxamine and orthophosphate. The activity is probably due to alkaline phosphatase.

See also *Alkaline Phosphatase*

Selected entries from *Methods in Enzymology* [**vol**, page(s)]:
General discussion: **62**, 574
Assay: **62**, 575
Other molecular properties: preparation, of enzyme, from rat liver, **62**, 579; properties, **62**, 581, 582

PYRIDOXAMINE-5-PHOSPHATE OXIDASE

This FMN-dependent oxidoreductase [EC 1.4.3.5], also known as pyridoxine(pyridoxamine)-5-phosphate oxidase, catalyzes the reaction of pyridoxamine 5-phosphate with water and dioxygen to produce pyridoxal 5'-phosphate, ammonia, and hydrogen peroxide.[1–3]

pyridoxamine 5-phosphate   pyridoxal 5-phosphate

Other substrates include pyridoxine 5-phosphate and pyridoxine. The brain enzyme is activated by 3-hydroxyanthranilate and 3-hydroxykynurenine, both tryptophan metabolites.

[1]J.-D. Choi, M. Bowers-Komro, M. D. Davis, D. E. Edmondson & D. B. McCormick (1983) *JBC* **258**, 840.
[2]M. N. Kazarinoff & D. B. McCormick (1975) *JBC* **250**, 3436.
[3]H. Wada & E. E. Snell (1961) *JBC* **236**, 2089.
Selected entries from *Methods in Enzymology* [**vol**, page(s)]:
General discussion: **6**, 331; **18A**, 626; **34**, 302; **52**, 18; **53**, 403; **56**, 474; **62**, 568; **122**, 110, 116
Assay: **6**, 331; **18A**, 627; **62**, 568, 569; human liver, **122**, 113; rabbit liver, **18A**, 627; **122**, 117
Other molecular properties: properties, **6**, 332; **18A**, 629; **62**, 572; purification, from rabbit liver, **6**, 332; **18A**, 627; **62**, 569; **122**, 118

PYRIDOXAMINE:PYRUVATE AMINOTRANSFERASE

pyridoxamine   pyridoxal

This pyridoxal-phosphate-dependent aminotransferase [EC 2.6.1.30] catalyzes the reversible reaction of pyridoxamine with pyruvate to produce pyridoxal and L-alanine.[1–4]

[1]P. J. Gilmer & J. F. Kirsch (1977) *Biochemistry* **16**, 5246.
[2]J. E. Ayling & E. E. Snell (1968) *Biochemistry* **7**, 1626.
[3]H. Wada & E. E. Snell (1962) *JBC* **237**, 133.
[4]A. E. Braunstein (1973) *The Enzymes*, 3rd ed., **9**, 379.
Selected entries from *Methods in Enzymology* [**vol**, page(s)]:
General discussion: **18A**, 643; **46**, 21
Assay: **18A**, 643
Other molecular properties: multisubstrate analogue inhibitor, **249**, 292, 305; properties, **18A**, 645; purification from *Pseudomonas* sp., **18A**, 644

5-PYRIDOXATE DIOXYGENASE

This flavin-dependent oxidoreductase [EC 1.14.12.5] catalyzes the reaction of 3-hydroxy-4-hydroxymethyl-2-methylpyridine-5-carboxylate with NADPH and dioxygen to produce 2-(acetamidomethylene)-3-(hydroxymethyl)-succinate and $NADP^+$.[1]

[1]M. J. K. Nelson & E. E. Snell (1986) *JBC* **261**, 15115.
Selected entries from *Methods in Enzymology* [**vol**, page(s)]:
General discussion: **52**, 16

PYRIDOXINE 4-DEHYDROGENASE

This oxidoreductase [EC 1.1.1.65], also known as pyridoxal reductase, catalyzes the reversible reaction of pyridoxine with $NADP^+$ to produce pyridoxal and NADPH and a proton.[1]

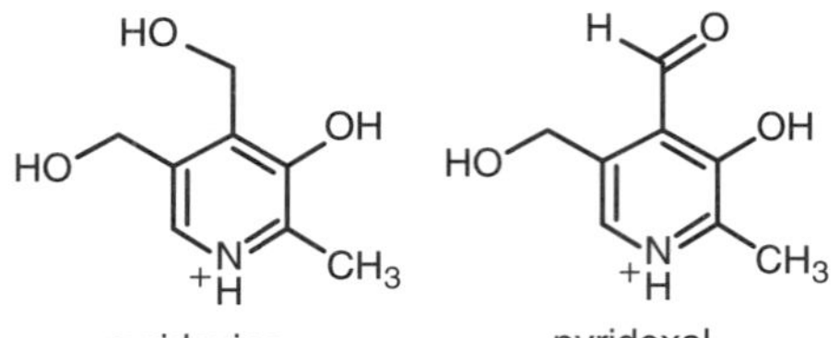

pyridoxine   pyridoxal

Pyridoxine 5-phosphate is also be oxidized to pyridoxal 5′-phosphate. The equilibrium lies far in the direction of pyridoxine formation.[2] **See also** *Pyridoxine 4-Oxidase*

[1]H. Holzer & S. Schneider (1961) *BBA* **48**, 71.
[2]B. M. Guirard & E. E. Snell (1988) *Biofactors* **1**, 187.

PYRIDOXINE 5-DEHYDROGENASE

This FAD-dependent oxidoreductase [EC 1.1.99.9], also known as pyridoxol-5-dehydrogenase, catalyzes the reaction of pyridoxine with an acceptor substrate to produce isopyridoxal and the reduced acceptor.[1,2]

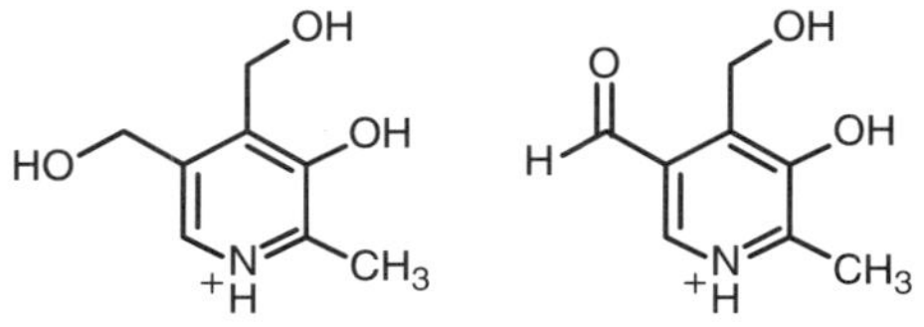

pyridoxine   isopyridoxal

Acceptor substrates can be 2,6-dichlorophenolindophenol, dioxygen, or quinones.

[1]Y.-J. Jong, M. J. K. Nelson & E. E. Snell (1986) *JBC* **261**, 15102.
[2]T. K. Sundaram & E. E. Snell (1969) *JBC* **244**, 2577.

PYRIDOXINE 5′-O-β-D-GLUCOSYLTRANSFERASE

This transferase [EC 2.4.1.160], also known as UDP-glucose-pyridoxine glucosyltransferase, catalyzes the reaction of UDP-D-glucose with pyridoxine to produce UDP and 5′-O-β-D-glucosylpyridoxine.

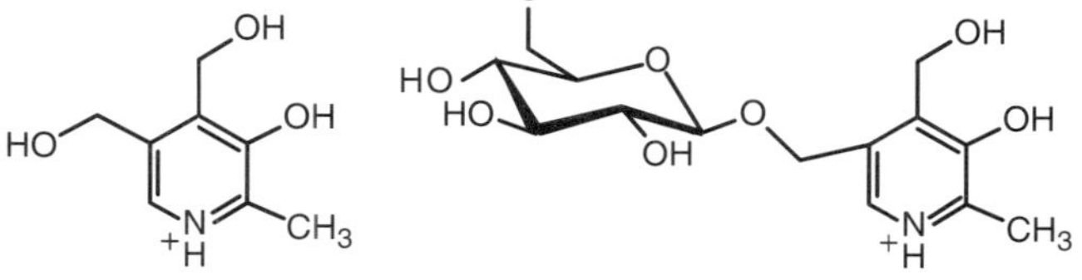

pyridoxine   5′-O-β-D-glucosylpyridoxine

Other substrates include 4′-deoxypyridoxine and pyridoxamine. Substrate specificity studies demonstrated the importance of substituents at the C4 position on the reaction rate.[1]

[1]K. Tadera, F. Yagi & A. Kobayashi (1982) *J. Nutr. Sci. Vitaminol.* **28**, 359.

PYRIDOXINE 4-OXIDASE

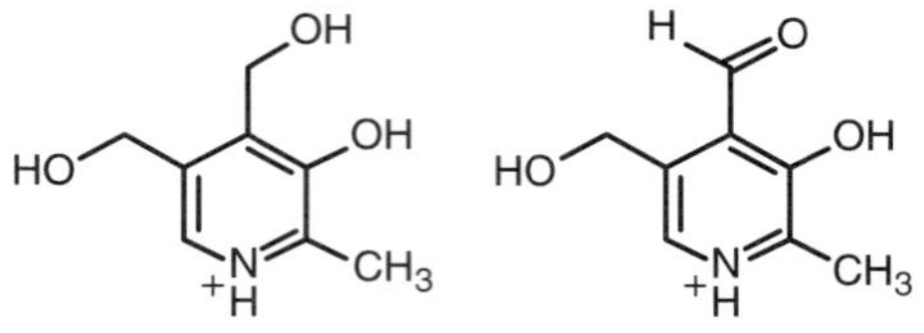

pyridoxine   pyridoxal

This FAD-dependent oxidoreductase [EC 1.1.3.12] catalyzes the reaction of pyridoxine with dioxygen to produce pyridoxal and hydrogen peroxide.[1] 2,6-Dichloroindophenol can replace dioxygen.

[1]T. K. Sundaram & E. E. Snell (1969) *JBC* **244**, 2577.

Selected entries from **Methods in Enzymology** [vol, page(s)]:
General discussion: 18A, 646; **52**, 18; **56**, 474
Assay: 18A, 646
Other molecular properties: properties, 18A, 647; purification, 18A, 647

4-PYRIDOXOLACTONASE

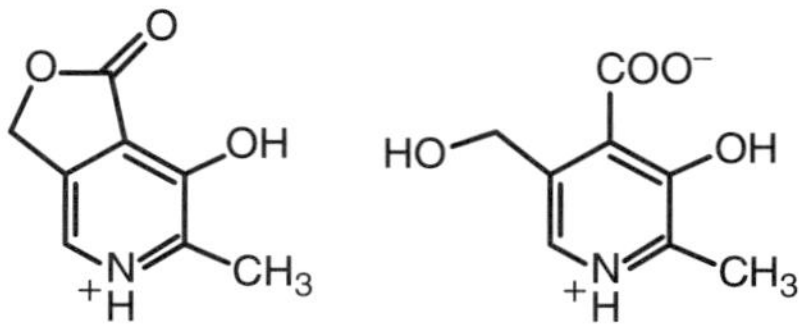

4-pyridoxolactone   4-pyridoxate

This enzyme [EC 3.1.1.27] catalyzes the hydrolysis of 4-pyridoxolactone to produce 4-pyridoxate.[1,2]

[1]Y.-J. Jong & E. E. Snell (1986) *JBC* **261**, 15112.
[2]R. W. Burg & E. E. Snell (1969) *JBC* **244**, 2585.

Selected entries from **Methods in Enzymology** [vol, page(s)]:
General discussion: 18A, 651
Assay: 18A, 651

PYRIMIDINE-DEOXYNUCLEOSIDE 1′-DIOXYGENASE

This ascorbate and iron-dependent oxidoreductase [EC 1.14.11.10], also known as pyrimidine-deoxynucleoside, 2-oxoglutarate 1′-dioxygenase, catalyzes the reaction of 2-deoxyuridine with α-ketoglutarate (or, 2-oxoglutarate) and dioxygen to produce uracil, dioxyribonolactone, succinate, and carbon dioxide.[1] **See also** *Pyrimidine-Deoxynucleoside 2′-Dioxygenase*

[1]J. Stubbe (1985) *JBC* **260**, 9972.

PYRIMIDINE-DEOXYNUCLEOSIDE 2′-DIOXYGENASE

This iron- and ascorbate-dependent enzyme [EC 1.14.11.3], known variously as pyrimidine-deoxynucleoside 2-oxoglutarate 2′-dioxygenase, thymidine 2′-hydroxylase, thymidine 2-oxoglutarate dioxygenase, and deoxyuridine 2′-hydroxylase, catalyzes the reaction of 2-deoxyuridine with α-ketoglutarate (or, 2-oxoglutarate) and dioxygen to produce uridine, succinate, and carbon dioxide.[1] The enzyme also acts on thymidine.

[1] O. Hayaishi, M. Nozaki & M. T. Abbott (1975) *The Enzymes*, 3rd ed., **12**, 119.

PYRIMIDINE NUCLEOSIDE PHOSPHORYLASE

This enzyme [EC 2.4.2.2] catalyzes the reaction of a pyrimidine nucleoside with orthophosphate to produce a pyrimidine and α-D-ribose 1-phosphate.[1] Nucleoside substrates include uridine, thymidine, and 5-bromouridine. In some organisms, this enzyme may be identical to uridine phosphorylase [EC 2.4.2.3]. In many lower organisms, such as *Bacillus stearothermophilus*, the enzyme will accept thymidine and uridine as substrates, whereas in mammalian organisms there are more specific proteins.

See also *Thymidine Phosphorylase; Uridine Phosphorylase*

[1] P. P. Saunders, B. A. Wilson & G. F. Saunders (1969) *JBC* **244**, 3691.

Selected entries from **Methods in Enzymology** [vol, page(s)]:
General discussion: 51, 432
Assay: 51, 432
Other molecular properties: activity, **51**, 432, 437; *Bacillus stearothermophilus*, from, **51**, 437; cytosine nucleoside deaminase preparation, presence in, **2**, 480; *Haemophilus influenzae*, from, **51**, 432; kinetic properties, **51**, 436, 437; pH effects, **51**, 435; purification, **51**, 433; purine nucleoside phosphorylase and, **2**, 448; stability, **51**, 435; substrate specificity, **51**, 435, 436; use of, **12A**, 118

PYRIMIDINE-5′-NUCLEOTIDE NUCLEOSIDASE

This enzyme [EC 3.2.2.10] catalyzes the hydrolysis of a pyrimidine 5′-nucleotide to produce a pyrimidine and D-ribose 5-phosphate.[1,2] Substrates include UMP, CMP, dUMP, dTMP, and dCMP.

[1] L. Bankel, G. Lindstedt & S. Lindstedt (1976) *FEBS Lett.* **71**, 147.
[2] A. Imada (1967) *J. Gen. Appl. Microbiol.* **13**, 267.

PYRIMIDODIAZEPINE SYNTHASE

This oxidoreductase [EC 1.5.4.1] catalyzes the reaction of 6-pyruvoyltetrahydropterin with two molecules of glutathione to produce the acetyldihydro derivative

of pyrimidodiazepine and glutathione disulfide. This enzyme participates in the biosynthesis of the eye pigment drosopterin in the fruit fly (*Drosophila melanogaster*).[1]

[1] G. J. Wiederrecht & G. M. Brown (1984) *JBC* **259**, 14121.

PYRITHIAMIN DEAMINASE

This deaminase [EC 3.5.4.20] catalyzes the hydrolysis of 1-[(4-amino-2-methylpyrimid-5-yl)methyl]-3-(2-hydroxyethyl)-2-methylpyridinium bromide (or, pyrithiamin) to produce 1-[(4-hydroxy-2-methylpyrimid-5-yl)methyl]-3-(2-hydroxyethyl)-2-methylpyridinium bromide and ammonia.[1,2]

[1] A. K. Sinha & G. C. Chatterjee (1968) *Enzymologia* **35**, 298.
[2] A. K. Sinha & G. C. Chatterjee (1968) *BJ* **107**, 165.

o-PYROCATECHUATE DECARBOXYLASE

This decarboxylase [EC 4.1.1.46], also known as 2,3-dihydroxybenzoate decarboxylase, catalyzes the conversion of 2,3-dihydroxybenzoate (or, *o*-pyrocatechuate) to catechol and carbon dioxide.[1–3]

o-pyrocatechuate   catechol

Other substrates include 2,3,6-trihydroxybenzoate and 2,3,5-trihydroxybenzoate.

[1] A. V. Kamath, N. A. Rao & C. S. Vaidyanathan (1989) *BBRC* **165**, 20.
[2] J. J. Anderson & S. Dagley (1981) *J. Bacteriol.* **146**, 291.
[3] P. V. Subba Rao, K. Moore & G. H. N. Towers (1967) *ABB* **122**, 466.

Selected entries from **Methods in Enzymology** [vol, page(s)]:
General discussion: 17A, 514
Other molecular properties: purification from *Aspergillus niger*, **17A**, 515

PYROGALLOL HYDROXYLTRANSFERASE

This molybdenum-dependent oxidoreductase [EC 1.97.1.2], also known as transhydroxylase, catalyzes the reversible reaction of 1,2,3,5-tetrahydroxybenzene with 1,2,3-trihydroxybenzene (or, pyrogallol) to produce 1,3,5-trihydroxybenzene (or, phloroglucinol) and 1,2,3,5-tetrahydroxybenzene.[1–3] The enzyme will also catalyze the reaction of 1,2,4-trihydroxybenzene (or, hydroxyhydroquinone) with pyrogallol to produce resorcinol (or, 1,3-dihydroxybenzene) and 1,2,3,5-tetrahydroxybenzene. Many other similar reactions are possible. The process involves intermolecular transfer of a hydroxyl moiety from the cosubstrate 1,2,3,5-tetrahydroxybenzene to the substrate pyrogallol, thus producing phloroglucinol while regenerating the cosubstrate.

[1] A. Brune & B. Schink (1990) J. Bacteriol. **172**, 1070.
[2] A. Brune & B. Schink (1992) Arch. Microbiol. **157**, 417.
[3] A. Brune, S. Schnell & B. Schink (1992) AEM **58**, 1861.

PYROGALLOL 1,2-OXYGENASE

1,2,3-trihydroxybenzene   (Z)-5-oxohex-2-enedioate

This iron-dependent oxidoreductase [EC 1.13.11.35] catalyzes the reaction of 1,2,3-trihydroxybenzene (or, pyrogallol) with dioxygen to produce (Z)-5-oxohex-2-enedioate.[1]

[1] E. E. Groseclose & D. W. Ribbons (1981) J. Bacteriol. **146**, 460.

PYROGLUTAMYL-PEPTIDASE I

This cysteine peptidase [EC 3.4.19.3] (also known as 5-oxoprolyl-peptidase, pyrrolidone-carboxylate peptidase, and pyroglutamyl aminopeptidase) catalyzes the hydrolysis of a 5-oxoprolyl-peptide to produce 5-oxoproline and a peptide: thus, it removes 5-oxoproline from various penultimate aminoacyl residues of peptides.[1–3] Note that, if an L-prolyl residue is the penultimate residue, the reaction will not be catalyzed (except for the case of the *Klebsiella cloacae* enzyme which can hydrolyze 5-oxoprolyl-L-proline). Both the bacterial and mammalian enzymes exhibit a broad specificity toward 5-oxoprolyl-peptides. There appears to be a strict specificity for the L-5-oxoprolyl residue as D-5-oxoprolyl-peptides are not substrates. Originally the enzyme was classified as an aminopeptidase under EC 3.4.11.8; but the substrate does not contain a free amino group and the enzyme was reclassified as an ω-peptidase.[1]

[1] A. Awadé, P. Cleuziat, T. Gonzalès & J. Robert-Baudouy (1994) Protein Struct. Funct. Genet. **20**, 34.
[2] E. Shaw (1990) AE **63**, 271.
[3] R. J. DeLange & E. L. Smith (1971) The Enzymes, 3rd ed., **3**, 81.

Selected entries from *Methods in Enzymology* [**vol**, page(s)]:
General discussion: **19**, 555; **330**, 394
Assay: cerebral, types I and II, simultaneous assay, **168**, 367; fluorimetric assay, **248**, 601; serum, assay, **168**, 368
Other molecular properties: action of, **19**, 555; amino acids in, **19**, 566; bacterial (preparation, **25**, 235; properties, **25**, 240); catalytic residues, **244**, 483; distribution and function of, **19**, 568; hyperthermophilic, **330**, 394; inhibitors, **244**, 483; interference

with glutaminylpeptide cyclase assay, **168**, 363; occurrence, **25**, 233; PCA-peptides and PCA-proteins, **19**, 563; physicochemical properties, **19**, 565; preparation, **19**, 560; specificity, **19**, 566; stability, **19**, 565; transition-state and multisubstrate analogues, **249**, 306

PYROGLUTAMYL-PEPTIDASE II

This metallopeptidase [EC 3.4.19.6], also known as thyroliberinase and pyroglutamyl aminopeptidase II, catalyzes the release of the N-terminal 5-oxoproline group from 5-oxoprolylhistidyl-Xaa tripeptides and 5-oxoprolylhistidyl-Xaa-glycyl tetrapeptides; hence, it will act on thyrotropin-releasing hormone (also known as thyroliberin and 5-oxoprolylhistidylprolylamide).[1,2] Other biologically active 5-oxoprolyl-containing peptides such as neurotensin, gastrin, bombesin, or 5-oxoprolylglutamylprolinamide are not hydrolyzed. Interestingly, the decapeptide amide luteinizing hormone-releasing hormone, which contains a 5-oxoprolyl-His bond, is not hydrolyzed, but inhibits the hydrolysis of thyrotropin-releasing hormone.

Pyroglutamyl-peptidase II is a membrane-bound protein that can readily be distinguished from the smaller, soluble pyroglutamyl-peptidase I that has a much broader specificity. A variant has been identified in pig serum that is produced from the membrane enzyme and is unrelated to pyroglutamyl-peptidase I. **See also** *Pyroglutamyl-Peptidase I*

[1] G. O'Cuinn, B. O'Connor & M. Elmore (1990) J. Neurochem. **54**(1), 1.
[2] R. O'Leary & B. O'Connor (1995) J. Neurochem. **65**, 953.

Selected entries from *Methods in Enzymology* [**vol**, page(s)]:
Assay: cerebral, types I and II, simultaneous assay, **168**, 367; fluorimetric assay, **248**, 601; serum, assay, **168**, 368

2-PYRONE-4,6-DICARBOXYLATE LACTONASE

2-pyrone-4,6-dicarboxylate   4-carboxy-2-hydroxyhexa-2,4-dienedioate

This highly specific lactonase [EC 3.1.1.57], first isolated from *Pseudomonas ochraceae*, catalyzes the hydrolysis of 2-pyrone-4,6-dicarboxylate to produce 4-carboxy-2-hydroxyhexa-2,4-dienedioate, which then isomerizes to 4-oxalmesaconate.[1]

[1] K. Maruyama (1983) J. Biochem. **93**, 557.

PYROPHOSPHATE:FRUCTOSE-6-PHOSPHATE 1-PHOSPHOTRANSFERASE

This enzyme [EC 2.7.1.90], also known as diphosphate: fructose-6-phosphate 1-phosphotransferase, 6-phosphofructokinase (pyrophosphate), and pyrophosphate-dependent 6-phosphofructose-1-kinase, catalyzes the reversible reaction of pyrophosphate with D-fructose 6-phosphate to produce orthophosphate and D-fructose 1,6-bisphosphate.[1-6]

α-D-fructose 6-phosphate   α-D-fructose 1,6-bisphosphate

The kinetic mechanism of this phosphotransferase is random Bi Bi, based on inhibitor studies[1] and equilibrium isotope exchange measurements.[2] The *Giardia lamblia* enzyme, however, has an ordered Bi Bi kinetic mechanism.[3]

[1]B. L. Bertagnolli & P. F. Cook (1984) *Biochemistry* **23**, 4101.
[2]Y. K. Cho, T. O. Matsunaga, G. L. Kenyon, B. L. Bertagnolli & P. F. Cook (1988) *Biochemistry* **27**, 3320.
[3]N. F. Phillips & Z. Li (1995) *Mol. Biochem. Parasitol.* **73**, 43.
[4]H. C. Huppe & D. H. Turpin (1994) *Ann. Rev. Plant Physiol. Plant Mol. Biol.* **45**, 577.
[5]E. Van Schaffingen (1987) *AE* **59**, 315.
[6]H. G. Wood, W. E. O'Brien & G. Michaels (1977) *AE* **45**, 85.

Selected entries from *Methods in Enzymology* [vol, page(s)]:
General discussion: 90, 91, 97
Assay: *Entamoeba histolytica*, **90**, 98; mung bean, **90**, 91
Other molecular properties: *Entamoeba histolytica* (assay, **90**, 98; properties, **90**, 101; purification, **90**, 99); mung bean (assays, **90**, 91; properties, **90**, 96; purification, **90**, 94); pyrophosphate, used in assay of, **63**, 34

PYROPHOSPHATE:GLYCEROL PHOSPHOTRANSFERASE

This phosphotransferase [EC 2.7.1.79], also known as diphosphate:glycerol phosphotransferase, catalyzes the reaction of glycerol with pyrophosphate to produce glycerol 1-phosphate and orthophosphate.[1] Depending on the enzyme source, this enzyme is often identical with glucose 6-phosphatase [EC 3.1.3.9].

[1]M. R. Stetten (1970) *BBA* **208**, 394.

PYROPHOSPHATE:PROTEIN PHOSPHOTRANSFERASE

This transphosphorylase [EC 2.7.1.104], also known as diphosphate:protein phosphotransferase, catalyzes the pyrophosphate-dependent phosphorylation of endoplasmic reticulum-membrane protein(s) to produce ER-membrane protein-phosphate and orthophosphate.[1]

[1]K. S. Lam & C. B. Kasper (1980) *PNAS* **77**, 1927.

PYROPHOSPHATE:SERINE PHOSPHOTRANSFERASE

This bacterial enzyme [EC 2.7.1.80], also known as diphosphate:serine phosphotransferase, catalyzes the reaction of L-serine with pyrophosphate to produce *O*-phospho-L-serine and orthophosphate.[1] This enzyme is especially useful in assays of L-serine in biological fluids.[2]

[1]L. M. Cagen & H. C. Friedmann (1972) *JBC* **247**, 3382.
[2]R. D. Hurst, K. Lund & R. W. Guynn (1981) *Anal. Biochem.* **117**, 339.

1-PYRROLINE-5-CARBOXYLATE DEHYDROGENASE

This widely distributed oxidoreductase [EC 1.5.1.12] (also known as 1-pyrroline dehydrogenase, Δ1-pyrroline-5-carboxylate dehydrogenase, pyrroline-5-carboxylic acid dehydrogenase, and L-pyrroline-5-carboxylate:NAD^+ oxidoreductase) catalyzes the NAD^+-dependent oxidation of 1-pyrroline-5-carboxylate to produce L-glutamate and NADH and a proton.[1-5]

Δ1-pyrroline 5-carboxylate

Another substrate is Δ1-pyrroline-3-hydroxy-5-carboxylate, producing 4-hydroxyglutamate. If the reverse reaction occurs, it occurs at a rate 1/15,000 of the rate for glutamate formation.[4]

[1]K. Isobe, T. Matsuzawa & K. Soda (1987) *ABC* **51**, 1947.
[2]A. Herzfeld, V. A. Mezl & W. E. Knox (1977) *BJ* **166**, 95.
[3]D. W. Lundgren & M. Ogur (1973) *BBA* **297**, 246.
[4]W. C. Small & M. E. Jones (1990) *JBC* **265**, 18668.
[5]E. Adams & L. Frank (1980) *ARB* **49**, 1005.

Selected entries from *Methods in Enzymology* [vol, page(s)]:
General discussion: 17B, 262
Assay: 17B, 262

PYRROLINE-2-CARBOXYLATE REDUCTASE

This oxidoreductase [EC 1.5.1.1] catalyzes the reaction of 1-pyrroline-2-carboxylate with NAD(P)H and a proton to produce L-proline and $NAD(P)^+$.[1,2]

1-pyrroline 2-carboxylate   L-proline

This reductase will also catalyze the NAD(P)H-dependent reduction of 1,2-didehydropiperidine-2-carboxylate to L-pipecolate.

[1]G. Garweg, D. von Rehren & U. Hintze (1980) *J. Neurochem.* **35**, 616.
[2]A. Meister, A. N. Radhakrishnan & S. D. Buckley (1957) *JBC* **229**, 789.
Selected entries from ***Methods in Enzymology*** [vol, page(s)]:
General discussion: 5, 878
Assay: 5, 878

PYRROLINE-5-CARBOXYLATE REDUCTASE

This oxidoreductase [EC 1.5.1.2] catalyzes the reaction of 1-pyrroline-5-carboxylate with NAD(P)H and a proton to produce L-proline and NAD(P)$^+$.[1]

1-pyrroline 5-carboxylate   L-proline

The enzyme will also catalyze the NAD(P)H-dependent reduction of 1-pyrroline-3-hydroxy-5-carboxylate to L-4-hydroxyproline.

[1]E. Adams & L. Frank (1980) *Ann. Rev. Biochem.* **49**, 1005.

Selected entries from ***Methods in Enzymology*** [vol, page(s)]:
General discussion: 5, 959; 17B, 258
Assay: 5, 960; 17B, 258
Other molecular properties: distribution, 5, 959; properties, 5, 963; 17B, 261; purification from (calf liver, 17B, 259; *Neurospora*, 5, 962; rat liver, 5, 961)

1-PYRROLINE-4-HYDROXY-2-CARBOXYLATE DEAMINASE

1-pyrroline-4-hydroxy-2-carboxylate

This deaminase [EC 3.5.4.22] catalyzes the hydrolysis of 1-pyrroline-4-hydroxy-2-carboxylate to produce 2,5-dioxopentanoate (or, α-ketoglutarate semialdehyde) and ammonia.[1]

[1]R. M. M. Singh & E. Adams (1965) *JBC* **240**, 4344 and 4352.

Selected entries from ***Methods in Enzymology*** [vol, page(s)]:
General discussion: 17B, 296
Assay: 17B, 296

PYRUVATE CARBOXYLASE

This biotin-dependent enzyme [EC 6.4.1.1] catalyzes the reaction of ATP with pyruvate and bicarbonate to produce ADP, orthophosphate, and oxaloacetate.[1–6] The yeast enzyme requires zinc, whereas the mammalian enzyme requires both manganese and acetyl-CoA. The overall reaction involves two partial reactions at spatially distinct subsites within the active site, such that covalently bound biotin acts as a mobile carboxyl-group carrier. In the first partial reaction, biotin is carboxylated using ATP and bicarbonate (forming carboxyphosphate), and the second partial reaction entails the transfer of the carboxyl group from carboxybiotin to pyruvate.

[1]S. Jitrapakdee & J. C. Wallace (1999) *BJ* **340**, 1.
[2]J. C. Wallace, S. Jitrapakdee & A. Chapman-Smith (1998) *Int. J. Biochem. Cell Biol.* **30**, 1.
[3]F. Lynen (1979) *Crit. Rev. Biochem.* **7**, 103.
[4]M. F. Utter, R. E. Barden & B. L. Taylor (1975) *AE* **42**, 1.
[5]M. C. Scrutton & M. R. Young (1972) *The Enzymes*, 3rd ed., **6**, 1.
[6]J. Moss & M. D. Lane (1971) *AE* **35**, 321.

Selected entries from ***Methods in Enzymology*** [vol, page(s)]:
General discussion: 13, 235, 250, 258
Assay: 13, 235, 250, 258
Other molecular properties: activation, 13, 257; 63, 270; carboxybiotin-enzyme intermediate, 13, 246; chiral methyl groups, and, 87, 137, 148; comparison of enzyme from different sources, 13, 246, 248, 250, 257, 261; competition with pyruvate dehydrogenase for pyruvate entry into Krebs cycle, 177, 423; decarboxylation reaction, 13, 246; distribution, 13, 248; exchange reactions, 13, 246; 87, 5; gluconeogenic catalytic activity, 37, 281, 282; hysteresis, 64, 218; inhibitors, 13, 248, 257; intermediate, 87, 88; isotope effect, 87, 137; kinetics, 13, 247, 257, 262; labeling, 46, 139; mechanism, 13, 236; 63, 52; product inhibition, 63, 435; properties, 13, 245, 257, 261; purification from (bakers' yeast, 13, 252; chicken liver, 13, 237, 239; *Pseudomonas*, 13, 260); specificity, 13, 245, 247; stability, 13, 244; stereochemistry, 87, 148; storage properties, 13, 244; thermodynamics, 13, 247; transition-state and multisubstrate analogues, 249, 308

PYRUVATE DECARBOXYLASE

This thiamin-pyrophosphate-dependent decarboxylase [EC 4.1.1.1], also known as α-carboxylase and α-keto acid carboxylase, catalyzes the conversion of an α-keto acid anion (or, 2-oxo acid anion) to produce an aldehyde and carbon dioxide. Substrates include pyruvate (generating acetaldehyde) and hydroxypyruvate (producing glycoaldehyde).[1–5] The enzyme also catalyzes acyloin formation. The *Saccharomyces cerevisiae* enzyme is a hysteretically regulated enzyme (in which pyruvate binds at a regulatory site and a catalytic site) whereas the bacterial enzyme (for example, *Zymomonas mobilis*) is not.[3]

Note that the pyruvate dehydrogenase (lipoamide) [EC 1.2.4.1] component of the pyruvate dehydrogenase complex has also been known as pyruvate decarboxylase.

[1]F. Jordan (1999) *FEBS Lett.* **457**, 298.
[2]J. M. Candy & R. G. Duggleby (1998) *BBA* **1385**, 323.
[3]R. L. Schowen (1998) *CBC* **2**, 217.
[4]J. V. Schloss & M. S. Hixon (1998) *CBC* **2**, 43.

[5]B. Vennesland (1951) *The Enzymes*, 1st ed., **2**, 183.

Selected entries from *Methods in Enzymology* [vol, page(s)]:
General discussion: **1**, 460, 465; **9**, 258; **18A**, 109; **46**, 50, 51; **90**, 528; **279**, 131
Assay: **1**, 460, 465; **9**, 258; **18A**, 112; sweet potato root, **90**, 528
Other molecular properties: apocarboxylase, **1**, 464 (thiamin pyrophosphate assay and, **2**, 637, 638); carbon dioxide and, **2**, 840; *Escherichia coli*, from (cofactors, **9**, 261; multiple forms, **9**, 261; physical constants, **9**, 261; properties, **9**, 261; resolution from pyruvate dehydrogenase complex, **9**, 259; specificity, **9**, 261); plant, **1**, 460; **2**, 192, 193; primary product, **2**, 840; properties, **1**, 463, 465; **9**, 261; pyruvate degradation and, **4**, 605; reactions, **243**, 97; specificity, **3**, 414; sweet potato root (assay, **90**, 528; properties, **90**, 531; purification, **90**, 529); urea lability, **9**, 256; wheat germ, **1**, 465; yeast, from, **18A**, 109 (apoenzyme, **18A**, 112, 116; properties, **1**, 463; preparation, **1**, 461; **18A**, 110; use in binding studies on thiamin derivatives, **18A**, 117, 238)

PYRUVATE DEHYDROGENASE COMPLEX

This multienzyme system catalyzes the overall reaction of pyruvate with NAD^+ and coenzyme A to produce acetyl-CoA, NADH, and carbon dioxide.[1–4] The complex uses five cofactors: NAD^+, coenzyme A, thiamin pyrophosphate, lipoamide, and FAD. Pyruvate dehydrogenase (lipoamide) [EC 1.2.4.1] catalyzes the thiamin-pyrophosphate-dependent reaction of pyruvate with lipoamide to produce *S*-acetyldihydrolipoamide and carbon dioxide. Dihydrolipoamide *S*-acetyltransferase [EC 2.3.1.12] catalyzes the reaction of *S*-acetyldihydrolipoamide with coenzyme A to produce dihydrolipoamide and acetyl-CoA. Dihydrolipoamide dehydrogenase [EC 1.8.1.4] is an FAD-dependent oxidoreductase that catalyzes the reaction of dihydrolipoamide with NAD^+ to produce lipoamide and NADH.

The PDH complex is highly regulated by [pyruvate dehydrogenase (lipoamide)] kinase [EC 2.7.1.99] and [pyruvate dehydrogenase (lipoamide)] phosphatase [EC 3.1.3.43].

[1]J. V. Schloss & M. S. Hixon (1998) *CBC* **2**, 43.
[2]L. J. Reed & S. J. Yeaman (1987) *The Enzymes*, 3rd ed., **18**, 77.
[3]G. Palmer (1975) *The Enzymes*, 3rd ed., **12**, 1.
[4]L. J. Reed & D. J. Cox (1970) *The Enzymes*, 3rd ed., **1**, 213.

Selected entries from *Methods in Enzymology* [vol, page(s)]:
General discussion: **1**, 486, 490; **9**, 247, 265; **166**, 330; **251**, 436; **279**, 131
Assay: **1**, 486, 490; **9**, 248; *Bacillus*, **89**, 401; bovine heart and kidney, **89**, 376; broccoli mitochondrial, **89**, 408; *Escherichia coli*, **89**, 392; *Hansenula miso*, **89**, 420; *Neurospora crassa*, **89**, 387; pigeon breast muscle, **89**, 414
Other molecular properties: *Bacillus* (assay, **89**, 401; properties, **89**, 405; purification, **89**, 402); bacterial particles and, **5**, 55; bovine (heart [assay, **89**, 376; properties, **89**, 386; purification, **89**, 383]; kidney [assay, **89**, 376; properties, **89**, 382; purification, **89**, 377]); brain mitochondria, **55**, 58; branched-chain 2-keto acid dehydrogenase complex (assay, **166**, 331; properties, **166**, 337; purification from *Bacillus subtilis*, **166**, 332); cascade control, **64**, 298, 306, 324, 325; competition with pyruvate carboxylase for pyruvate entry into Krebs cycle, **177**, 423; components, **251**, 437; control studies, **37**, 293, 294; deficiency, skin fibroblast culture

testing, **264**, 455; energy transfer studies, **48**, 378; *Escherichia coli*, from, **9**, 247; **27**, 620 (assay, **89**, 392; behavior on ethanol-Sepharose 2B, **89**, 398; crystallization, **9**, 270; dimensions, **9**, 253; molecular weight, **9**, 252, 271; physical constants, **9**, 252; properties, **9**, 252; purification, **9**, 249; **89**, 393; specificity, **9**, 252); *Hansenula miso* (assay, **89**, 420; properties, **89**, 423; purification, **89**, 421); induction by pyruvate, **13**, 59; kinetics, **87**, 354; molecular weight, **61**, 218; mutant, **72**, 695 (acetate auxotroph, **72**, 695); *Neurospora crassa* (assay, **89**, 387; properties, **89**, 390; purification, **89**, 389); nonpolar intermediates, analogues, **249**, 291; pigeon breast muscle (assay, **89**, 414; properties, **89**, 418; purification, **89**, 416); product inhibition, **63**, 435; purification on calcium phosphate gel-cellulose, **22**, 342; reactions, **243**, 97; separation from α-ketoglutarate dehydrogenase, **13**, 61; sources, **9**, 247; subunits, size, **61**, 223; white adipose tissue mitochondria, **55**, 16

Pyruvate Dehydrogenase (Lipoamide)
Assay: *Bacillus*, **89**, 401; bovine heart and kidney, **89**, 376; broccoli mitochondrial, **89**, 408; *Escherichia coli*, **89**, 392; *Hansenula miso*, **89**, 420; *Neurospora crassa*, **89**, 387; pigeon breast muscle, **89**, 414
Other molecular properties: *Bacillus* (assay, **89**, 401; properties, **89**, 405; purification, **89**, 402); bovine (heart [assay, **89**, 376; properties, **89**, 386; purification, **89**, 383]; kidney [assay, **89**, 376; properties, **89**, 382; purification, **89**, 377]); broccoli mitochondrial (assay, **89**, 408; properties, **89**, 412; purification, **89**, 411); cauliflower mitochondrial (assay, **89**, 408; properties, **89**, 410; purification, **89**, 409); *Escherichia coli* (assay, **89**, 392; behavior on ethanol-Sepharose 2B, **89**, 398; purification, **89**, 393); *Hansenula miso* (assay, **89**, 420; properties, **89**, 423; purification, **89**, 421); *Neurospora crassa* (assay, **89**, 387; properties, **89**, 390; purification, **89**, 389); pigeon breast muscle (assay, **89**, 414; properties, **89**, 418; purification, **89**, 416)

Dihydrolipoamide Dehydrogenase
Assay: **9**, 253
Other molecular properties: *Escherichia coli*, from, **9**, 253 (inhibitors, **9**, 257; physical constants, **9**, 257; properties, **9**, 257; resolution from pyruvate dehydrogenase complex, **9**, 254; specificity, **9**, 257)

Dihydrolipoamide Acetyltransferase
Assay: **9**, 262
Other molecular properties: *Escherichia coli*, from, **9**, 262 (multiple forms of, **9**, 265; physical constants, **9**, 264; properties, **9**, 264; prosthetic group, **9**, 264; purification, **9**, 263; specificity, **9**, 264; stability, **9**, 264)

[Pyruvate Dehydrogenase (Lipoamide)] Kinase
Assay: bovine kidney, **90**, 196; **99**, 332
Other molecular properties: bovine kidney (assay, **90**, 196; **99**, 332; properties, **90**, 199; **99**, 335; purification, **90**, 197; **99**, 333); properties and function, **107**, 94

[Pyruvate Dehydrogenase (Lipoamide)]-Phosphatase
Assay: bovine heart, **90**, 402
Other molecular properties: bovine heart (assay, **90**, 402; properties, **90**, 406; purification, **90**, 404)

PYRUVATE DEHYDROGENASE (CYTOCHROME)

This thiamin-pyrophosphate-dependent oxidoreductase [EC 1.2.2.2], first isolated in *Escherichia coli*, catalyzes the reaction of pyruvate with membrane-bound ferricytochrome b_1 and water to produce carbon dioxide, acetate, and ferrocytochrome b_1.[1–3]

[1]R. L. Houghton & B. E. P. Swoboda (1973) *FEBS Lett.* **30**, 277.
[2]C. C. Cunningham & L. P. Hager (1971) *JBC* **246**, 1575 and 1583.
[3]F. R. Williams & L. P. Hager (1966) *ABB* **116**, 168.

Selected entries from *Methods in Enzymology* [vol, page(s)]:
General discussion: **9**, 265
Assay: **9**, 265

Other molecular properties: crystallization, **9**, 265; properties, **9**, 271; purification, **9**, 267; turnover number, **9**, 272

PYRUVATE DEHYDROGENASE (LIPOAMIDE)

This thiamin-pyrophosphate-dependent oxidoreductase [EC 1.2.4.1], also known as pyruvate decarboxylase and pyruvate dehydrogenase, catalyzes the reaction of pyruvate with lipoamide to produce S-acetyldihydrolipoamide and carbon dioxide.[1-7] This enzyme is component E_1 of the multienzyme pyruvate dehydrogenase complex. Lipoamide is a lipoyl group linked to the ε-amino group of a lysyl residue of dihydrolipoamide acetyltransferase, component E_2 of the pyruvate dehydrogenase complex.

lipoamide S-acetyldihydrolipoamide

The thiazolium form of the coenzyme undergoes general-base catalyzed deprotonation. This intermediate reacts with pyruvate to form an α-lactylthiamin pyrophosphate intermediate. Subsequently, decarboxylation generates a carboanionic center at the α-carbon of the adduct, which is stabilized by the thiazolium nucleus. The electrons are delocalized and a reactive enamine intermediate can form. Two possibilities have been suggested for the reaction of the enamine intermediate with lipoamide. (1) The enamine acts as a nucleophile (the α-carbon attacking the sulfur) resulting in disulfide bond fission and C–S bond formation; proton reorganization results in elimination of thiamin pyrophosphate and formation of S-acetyldihydrolipoamide. (2) An electron transfer occurs between the enamine and lipoamide to form the acetyl-thiamin pyrophosphate intermediate and a deprotonated dihydrolipoamide. The final product is now formed via a simple acetyl transfer reaction.[8,9]

[1] A. de Kok, A. F. Hengeveld, A. Martin & A. H. Westphal (1998) BBA **1385**, 353.
[2] L. J. Reed & S. J. Yeaman (1987) The Enzymes, 3rd ed., **18**, 77.
[3] S. J. Yeaman (1986) Trends Biochem. Sci. **11**, 293.
[4] S. E. Severin, L. S. Khailova & V. S. Gomazkova (1986) Adv. Enzyme Regul. **25**, 347.
[5] L. J. Reed, Z. Damuni & M. L. Merryfield (1985) Curr. Top. Cell. Regul. **27**, 41.
[6] P. J. Randle (1978) Trends Biochem. Sci. **3**, 217.
[7] L. J. Reed & D. J. Cox (1970) The Enzymes, 3rd ed., **1**, 213.
[8] D. S. Fluorney & P. A. Frey (1986) Biochemistry **25**, 6036.
[9] Y.-S. Yang & P. A. Frey (1989) ABB **268**, 465.

Selected entries from **Methods in Enzymology** [vol, page(s)]:
General discussion: 9, 247; 89, 376, 386, 391, 399, 408, 414, 420
Assay: Bacillus, **89**, 401; bovine heart and kidney, **89**, 376; broccoli mitochondrial, **89**, 408; Escherichia coli, **89**, 392; Hansenula miso, **89**, 420; Neurospora crassa, **89**, 387; pigeon breast muscle, **89**, 414
Other molecular properties: Bacillus (assay, **89**, 401; properties, **89**, 405; purification, **89**, 402); bovine (heart [assay, **89**, 376; properties, **89**, 386; purification, **89**, 383]; kidney [assay, **89**, 376; properties, **89**, 382; purification, **89**, 377]); broccoli mitochondrial (assay, **89**, 408; properties, **89**, 412; purification, **89**, 411); cauliflower mitochondrial (assay, **89**, 408; properties, **89**, 410; purification, **89**, 409); Escherichia coli (assay, **89**, 392; behavior on ethanol-Sepharose 2B, **89**, 398; purification, **89**, 393); Hansenula miso (assay, **89**, 420; properties, **89**, 423; purification, **89**, 421); Neurospora crassa (assay, **89**, 387; properties, **89**, 390; purification, **89**, 389); pigeon breast muscle (assay, **89**, 414; properties, **89**, 418; purification, **89**, 416)

[PYRUVATE DEHYDROGENASE (LIPOAMIDE)] KINASE

This phosphotransferase [EC 2.7.199], associated with the eukaryotic pyruvate dehydrogenase complex, catalyzes the reaction of ATP with certain seryl residues in pyruvate dehydrogenase (lipoamide) [EC 1.2.4.1], the E_1 component of the pyruvate dehydrogenase complex, to produce ADP and [pyruvate dehydrogenase (lipoamide) containing phosphorylated seryl residues.[1-4] Phosphorylation results in inactivation of the eukaryotic complex. The enzyme reportedly has an essential histidyl residue critical for catalysis.[1]

[1] B. P. Mooney, N. R. David, J. J. Thelen, J. A. Miernyk & D. D. Randall (2000) BBRC **267**, 500.
[2] T. E. Pratt & T. E. Roche (1979) JBC **254**, 7191.
[3] T. E. Roche & L. J. Reed (1974) BBRC **59**, 1341.
[4] F. Hucho, D. D. Randall, T. E. Roche, M. W. Burgett, J. W. Pelley & L. J. Reed (1972) ABB **151**, 328.

Selected entries from **Methods in Enzymology** [vol, page(s)]:
General discussion: 90, 195; 99, 331
Assay: bovine kidney, 90, 196; 99, 332
Other molecular properties: bovine kidney (assay, 90, 196; 99, 332; properties, **90**, 199; **99**, 335; purification, 90, 197; 99, 333); properties and function, **107**, 94

[PYRUVATE DEHYDROGENASE (LIPOAMIDE)]-PHOSPHATASE

This enzyme [EC 3.1.3.43], associated with the pyruvate dehydrogenase complex, catalyzes the hydrolysis of certain O-phospho-L-seryl residues in pyruvate dehydrogenase (lipoamide) [EC 1.2.4.1], the E_1 component of the pyruvate dehydrogenase complex, to produce orthophosphate and pyruvate dehydrogenase (lipoamide) containing dephosphorylated seryl residues.[1,2] Dephosphorylation reactivates the pyruvate dehydrogenase complex.

[1] L. J. Reed & S. J. Yeaman (1987) The Enzymes, 3rd ed., **18**, 77.
[2] L. J. Reed, Z. Damuni & M. L. Merryfield (1985) Curr. Top. Cell. Regul. **27**, 41.

PYRUVATE DEHYDROGENASE (NADP⁺)

This iron-sulfur-, FAD-, and thiamin-pyrophosphate-dependent oxidoreductase [EC 1.2.1.51] catalyzes the reaction of pyruvate with coenzyme A and NADP⁺ to produce acetyl-CoA, carbon dioxide, and NADPH.[1-6] The *Euglena gracilis* enzyme can also use methyl viologen or benzyl viologen as the acceptor, albeit more slowly. The *Euglena* enzyme is reported to have a two-site ping-pong mechanism having substrate inhibition.[3] The enzyme is inhibited by dioxygen, apparently affecting the iron-sulfur domain of the protein complex.[1]

[1]H. Inui, R. Yamaji, H. Saidoh, Y. Nakano & S. Kitaoka (1991) *ABB* **286**, 270.
[2]H. Inui, K. Miyatake, Y. Nakano & S. Kitaoka (1990) *ABB* **280**, 292.
[3]H. Inui, K. Miyatake, Y. Nakano & S. Kitaoka (1989) *ABB* **274**, 434.
[4]H. Inui, K. Ono, K. Miyatake, Y. Nakano & S. Kitaoka (1987) *JBC* **262**, 9130.
[5]E. Brostedt & S. Nordlund (1991) *BJ* **279**, 155.
[6]J. G. Zeikus, G. Fuchs, W. Kenealy & R. K. Thauer (1977) *J. Bacteriol.* **132**, 604.

PYRUVATE KINASE

This phosphotransferase [EC 2.7.1.40], also known as phosphoenolpyruvate kinase, catalyzes the reaction of phosphoenolpyruvate with MgADP to produce MgATP and pyruvate.[1-3] Potassium or ammonium ions are required for enzymatic activity, the latter yielding a maximum velocity that is 72% of that observed with potassium ion. Phosphoryl acceptors from phosphoenolpyruvate can also be UDP, GDP, CDP, IDP, and dADP. The enzyme will also catalyze the phosphorylation of hydroxylamine and fluoride in the presence of bicarbonate, magnesium ion, and potassium ion. The equilibrium constant written in the direction of ATP synthesis is approximately 3000 in the presence of excess magnesium ion.

The kinetic mechanism of rabbit muscle pyruvate kinase utilizes the random addition of MgADP and phosphoenolpyruvate (PEP) to form the catalytically active ternary complex. A β-phosphoryl oxygen of MgADP carries out nucleophilic attack on the phosphorus atom of PEP, displacing enolpyruvate which undergoes tautomerization and drives ATP synthesis. In the reverse direction, loss of a methyl proton occurs during conversion of pyruvate to phosphoenolpyruvate. Enolization of pyruvate precedes phosphoryl transfer, based on exchange studies with tritiated substrates.

[1]J. J. Villafranca & T. Nowak (1992) *The Enzymes*, 3rd ed., **20**, 63.
[2]A. S. Mildvan (1979) *AE* **49**, 103.
[3]F. J. Kayne (1973) *The Enzymes*, 3rd ed., **8**, 353.

multisubstrate analogues, **249**, 305; yeast, from, **42**, 176 (assay, **42**, 176; electrophoresis, **42**, 181; molecular weight, **42**, 181; properties, **42**, 181, 182; purification, **42**, 176)

[PYRUVATE KINASE]-PHOSPHATASE

This phosphatase [EC 3.1.3.49] catalyzes the hydrolysis of *O*-phospho-L-seryl residue(s) in pyruvate kinase (which had been phosphorylated by a protein kinase) to produce orthophosphate and pyruvate kinase containing dephosphorylated L-seryl residue(s), thereby restoring the catalytic activity of pyruvate kinase.[1-3] The muscle isozyme is not subject to phosphorylation/dephosphorylation control.

[1]M.-F. Jett, L. Hue & H.-G. Hers (1981) *FEBS Lett.* **132**, 183.
[2]L. Lopez-Alarcon, M. Mojena, L. Monge & J. E. Feliu (1986) *BBRC* **134**, 292.
[3]T. S. Ingebritsen, A. A. Stewart & P. Cohen (1983) *EJB* **132**, 297.

PYRUVATE, ORTHOPHOSPHATE DIKINASE

This phosphotransferase [EC 2.7.9.1], also known as pyruvate, phosphate dikinase, catalyzes the reaction of ATP with pyruvate and orthophosphate to produce AMP, phosphoenolpyruvate, and pyrophosphate.[1-6] The kinetic mechanism, depending on the source, is either a three-site Hexa Uni ping pong kinetic mechanism, which proceeds via pyrophosphoryl-enzyme and phosphoryl-enzyme intermediates (at a histidyl residue)[3] or a two-site Bi Bi Uni Uni ping pong scheme[1] (also with pyrophosphoryl- and phosphoryl-enzyme intermediates) in which the phosphohistidyl domain swivels between the two sites.[2] The phosphorylated dikinase also catalyzes the enolization of pyruvate. *See also Pyruvate, Water Dikinase*

[1]S. H. Thrall & D. Dunaway-Mariano (1994) *Biochemistry* **33**, 1103.
[2]O. Herzberg, C. C. Chen, G. Kapadia, M. McGuire, L. J. Carroll, S. J. Noh & D. Dunaway-Mariano (1996) *PNAS* **93**, 2652.
[3]N. H. Goss & H. G. Wood (1982) *MIE* **87**, 51.
[4]R. A. Cooper & H. L. Kornberg (1974) *The Enzymes*, 3rd ed., **10**, 631.
[5]P. A. Frey (1992) *The Enzymes*, 3rd ed., **20**, 141.
[6]H. G. Wood, W. E. O'Brien & G. Michaels (1977) *AE* **45**, 85.

Selected entries from *Methods in Enzymology* [vol, page(s)]:
General discussion: **42**, 187, 192, 199, 212; **87**, 51; **308**, 149
Assay: **87**, 56; *Acetobacter xylinum*, **42**, 192; *Bacteroides symbiosus*, **42**, 187, 188, 199, 200; leaves, **42**, 212; *Propionibacterium shermanii*, **42**, 199, 200
Other molecular properties: *Acetobacter xylinum*, from, **42**, 192 (assay, **42**, 192; chromatography, **42**, 196; distribution, **42**, 199; inhibitors, **42**, 198; properties, **42**, 197; purification, **42**, 196, 197); ATP/AMP site, covalent labeling, **87**, 61; *Bacteroides symbiosus*, from, **42**, 187, 199 (activators and inhibitors, **42**, 191; assay, **42**, 187, 188, 199, 200; chromatography, **42**, 205; inhibitors, **42**, 208; mechanism, **42**, 208; molecular weight, **42**, 191, 207; properties, **42**, 191, 206; protein determination, **42**, 206; purification, **42**, 188, 200); energetics, **308**, 149; enzyme-phosphoryl intermediate, **308**, 149 (isolation, **87**, 61; properties and functions, **87**, 61); enzyme-pyrophosphoryl intermediate, **87**, 6, 54; **308**, 149 (formation, **87**, 57; isolation, **87**, 57; properties and functions, **87**, 59); mechanism, **63**, 52; **87**, 51; **308**, 151, 162; phosphoenolpyruvate/pyruvate site, covalent labeling, **87**, 65;

phosphorylhistidine-containing peptide, sequence, **87**, 63; positional isotope exchange studies, **249**, 424; preparation from *Bacillus symbiosus*, **87**, 55; properties, **87**, 56; *Propionibacterium shermanii*, from, **42**, 209 (assay, **42**, 199, 200; chromatography, **42**, 210, 211; inhibitors, **42**, 211; mechanism, **42**, 212; molecular weight, **42**, 211; properties, **42**, 211, 212; protein determination, **42**, 210; purification, **42**, 209); pyruvate enolization, monovalent anion requirement, **87**, 64

PYRUVATE OXIDASE

This thiamin-pyrophosphate- and FAD-dependent oxidoreductase [EC 1.2.3.3] catalyzes the reaction of pyruvate with orthophosphate, dioxygen, and water to produce acetyl phosphate, carbon dioxide, and hydrogen peroxide.[1-4] FAD binds in a two-step mechanism and facilitates the formation of the hydroxyethylthiamin pyrophosphate intermediate, but not the transfer of electrons from this molecular entity to dioxygen.[3]

[1]B. Sedewitz, K. H. Schleifer & F. Götz (1984) *J. Bacteriol.* **160**, 273 and 462.
[2]J. Carlsson & U. Kujala (1985) *FEMS Microbiol. Lett.* **25**, 53.
[3]K. Tittmann, D. Proske, M. Spinka, S. Ghisla, R. Rudolph, G. Hubner & G. Kern (1998) *JBC* **273**, 12929.
[4]R. L. Schowen (1998) *CBC* **2**, 217.

Selected entries from *Methods in Enzymology* [vol, page(s)]:
General discussion: **1**, 482; **52**, 19; **279**, 131
Assay: **1**, 483

PYRUVATE OXIDASE (CoA-ACETYLATING)

This FAD-dependent oxidoreductase [EC 1.2.3.6] catalyzes the reaction of pyruvate with coenzyme A and dioxygen to produce acetyl-CoA, carbon dioxide, and hydrogen peroxide.[1] In some species, this enzyme is identical with pyruvate synthase [EC 1.2.7.1].

[1]T. Takeuchi, E. C. Weinbach & L. S. Diamond (1975) *BBRC* **65**, 591.

Selected entries from *Methods in Enzymology* [vol, page(s)]:
General discussion: **126**, 87

PYRUVATE SYNTHASE

This thiamin pyrophosphate-dependent enzyme [EC 1.2.7.1], also known as pyruvate:ferredoxin oxidoreductase and pyruvate oxidoreductase, catalyzes the reaction of pyruvate with coenzyme A and oxidized ferredoxin to produce acetyl-CoA, carbon dioxide, and reduced ferredoxin.[1-4] A unique feature of the catalytic mechanism is the formation of a stable radical intermediate.[3]

[1]G. Eden & G. Fuchs (1983) *Arch. Microbiol.* **135**, 68.
[2]B. B. Buchanan (1971) *The Enzymes*, 3rd ed., **6**, 193.
[3]M. H. Charon, A. Volbeda, E. Chabriere, L. Pieulle & J. C. Fontecilla-Camps (1999) *Curr. Opin. Struct. Biol.* **9**, 663.
[4]E. J. Brush & J. W. Kozarich (1992) *The Enzymes*, 3rd ed., **20**, 317.

Selected entries from *Methods in Enzymology* [vol, page(s)]:
General discussion: **13**, 172; **41**, 334; **167**, 496; **243**, 94
Assay: **13**, 172; **41**, 334

Other molecular properties: catalyzed reactions, role of thiyl radicals, **186**, 328; *Clostridium acidi-urici*, from, **41**, 334 (assay, **41**, 334; properties, **41**, 336, 337; purification, **41**, 334); electron donor to nitrogenase, assay in cyanobacteria, as, **167**, 500; flavodoxin, **69**, 392, 393; occurrence, **13**, 177; properties, **13**, 176; purification (*Chlorobium thiosulfatophilum*, **13**, 174; *Rhodospirillum rubrum*, **13**, 175); pyruvate phosphoroclastic system, in, **243**, 96; reductive carboxylic acid cycle, in, **13**, 172; requirement for (ferredoxin, **13**, 175; thiamin pyrophosphate, **13**, 176); thiyl radicals (generation, **251**, 108; role of, **186**, 328); transition-state and multisubstrate analogues, **249**, 304

PYRUVATE, WATER DIKINASE

This manganese-dependent phosphotransferase [EC 2.7.9.2], also known as phosphoenolpyruvate synthetase, catalyzes the reversible reaction of ATP with pyruvate and water to produce AMP, phosphoenolpyruvate, and orthophosphate.[1–4] A phosphohistidine intermediate forms during catalysis.[1] *See also* Pyruvate, Orthophosphate Dikinase

[1] S. Narindrasorasak & W. A. Bridger (1977) *JBC* **252**, 3121.
[2] R. A. Cooper & H. L. Kornberg (1974) *The Enzymes*, 3rd ed., **10**, 631.
[3] K. M. Berman & M. Cohn (1970) *JBC* **245**, 5309 and 5319.
[4] P. A. Frey (1992) *The Enzymes*, 3rd ed., **20**, 141.

Selected entries from *Methods in Enzymology* [vol, page(s)]:
General discussion: **13**, 309; **249**, 305
Assay: orthophosphate formation, pyruvate-dependent, **13**, 310; pyruvate formation, lactate dehydrogenase coupled, **13**, 311; pyruvate removal, ATP-dependent, **13**, 309

6-PYRUVOYLTETRAHYDROPTERIN 2′-REDUCTASE

This oxidoreductase [EC 1.1.1.220] catalyzes the reaction of 6-pyruvoyl-tetrahydropterin with NADPH and a proton to produce 6-lactoyl-5,6,7,8-tetrahydropterin and NADP⁺. The reduction is at the C2′-oxo group of the substrate; it is inactive toward the C1′-oxo group.[1] This reductase

is not identical with sepiapterin reductase [EC 1.1.1.153]. When sepiapterin reductase activity is limiting, a large proportion of tetrahydrobiopterin synthesis proceeds through this 6-lactoyl intermediate.[1] The human enzyme is identical to aldose reductase.[2]

[1] S. Milstien & S. Kaufman (1989) *JBC* **264**, 8066.
[2] P. Steinerstauch, B. Wermuth, W. Leimbacher & H. C. Curtius (1989) *BBRC* **164**, 1130.

6-PYRUVOYLTETRAHYDROPTERIN SYNTHASE

This zinc- and magnesium ion-dependent enzyme [EC 4.2.3.12; previously classified as EC 4.6.1.10], often abbreviated PTPS, catalyzes the conversion of 6-(L-*erythro*-1,2-dihydroxypropyl 3-triphosphate)-7,8-dihydropterin (also called dihydroneopterin triphosphate) to 6-(1,2-dioxopropyl)-5,6,7,8-tetrahydropterin (or, 6-pyruvoyltetrahydrobiopterin) and triphosphate. This enzyme is a component in the tetrahydrobiopterin synthetic pathway. Interestingly, the reaction results in the incorporation of solvent protons into positions C6 and C3′ of the enzyme product. The Zn(II) ion activates the protons of the substrate, stabilizes the intermediates, and disfavors the breaking of the C1′-C2′ bond in the pyruvoyl side chain.[1] The human enzyme is phosphorylated at Ser19 by cGMP-dependent protein kinase type II.[2] This covalent modification is required for maximal activity.

[1] T. Ploom, B. Thony, J. Yim, S. Lee, H. Nar, W. Leimbacher, J. Richardson, R. Huber & G. Auerbach (1999) *J. Mol. Biol.* **286**, 851.
[2] T. Scherer-Oppliger, W. Leimbacher, N. Blau & B. Thony (1999) *JBC* **274**, 31341.

Selected entries from *Methods in Enzymology* [vol, page(s)]:
General discussion: **281**, 53, 123

Q

Qβ RNA REPLICASE

This RNA-directed RNA polymerase [EC 2.7.7.48] plays an essential role in bacteriophage Qβ replication, which is usually carried out in *Escherichia coli*.[1,2] **See** *RNA-Directed RNA Polymerase*

[1]T. Blumenthal (1982) *The Enzymes*, 3rd ed., **15**, 267.
[2]T. Blumenthal & G. G. Carmichael (1979) *ARB* **48**, 525.

Selected entries from **Methods in Enzymology** [vol, page(s)]:
General discussion: **12B**, 530, 540; **60**, 628
Assay: **12B**, 531, 541; polyguanylate polymerase activity, **60**, 629, 630
Other molecular properties: binding of mRNA, **60**, 14, 15, 337; elongation factors and, **30**, 219; functional role elongation, **60**, 426, 427, 436; initiation, **60**, 417, 418, 426, 430, 431, 436, 446, 447, 456; isolation from *Escherichia coli* ribosomes, **60**, 448, 456; phage replication, **275**, 512; properties, **12B**, 538; **60**, 634, 635; purification, **60**, 632; subunits, **60**, 418, 628, 637; template inhibitor screening, **275**, 519

QUATERNARY-AMINE-TRANSPORTING ATPase

This so-called ATPase [EC 3.6.3.32], also known as betaine-transporting ATPase, catalyzes the ATP-dependent (ADP- and orthophosphate-producing) transport of a [quaternary amine]$_{out}$ to produce a [quaternary amine]$_{in}$.[1] The bacterial enzyme (for example, from *Bacillus subtilis*) catalyzes the import of glycine betaine, an osmoprotectant. ABC-type (ATP-binding cassette-type) ATPases have two similar ATP-binding domains and do not undergo phosphorylation during the transport process.

The idea that a transporter is an enzyme is in keeping with a new definition of enzyme catalysis as the facilitated making/breaking of chemical bonds, not just covalent bonds.[2] This idea builds on Pauling's assertion that any long-lived, chemically distinct interaction (in this case, the persistent location of a solute with respect to the faces of a membrane) can be regarded as a chemical bond. Note also that the equilibrium constant

(K_{eq} = [Amine$_{in}$][ADP][P$_i$]/[Amine$_{out}$][ATP]) does not conform to that expected for an ATPase (*i.e.*, K_{eq} = [ADP][P$_i$]/[ATP]). Thus, although the overall reaction yields ADP and orthophosphate, the enzyme is misclassified as a hydrolase, and instead should be regarded as an energase-type reaction. Energases facilitate affinity-modulated reactions by coupling the $\Delta G_{ATP\text{-hydrolysis}}$ to a force-generating or work-producing step.[2] In this case, P-O-P bond scission supplies the energy to drive quaternary amine transport. **See** *ATPase; Energase*

[1]B. Kempf, J. Gade & E. Bremer (1997) *J. Bacteriol.* **179**, 6213.
[2]D. L. Purich (2001) *TiBS* **26**, 417.

QUERCETIN-3,3′-BISSULFATE 7-SULFOTRANSFERASE

This sulfotransferase [EC 2.8.2.28] catalyzes the reaction of 3′-phosphoadenylylsulfate with quercetin 3,3′-bissulfate to produce adenosine 3′,5′-bisphosphate and quercetin 3,3′,7-trissulfate.[1–3]

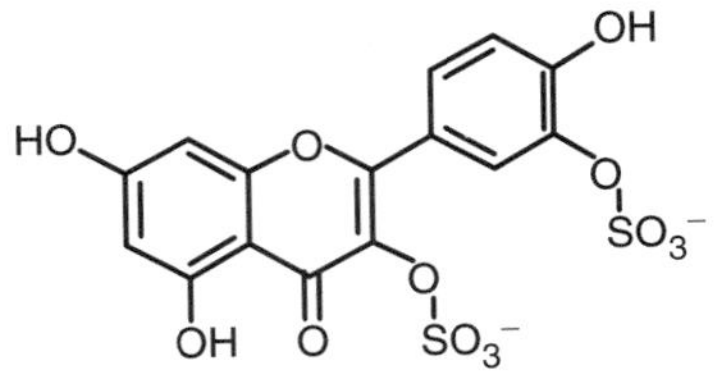

quercetin 3,3′-bissulfate

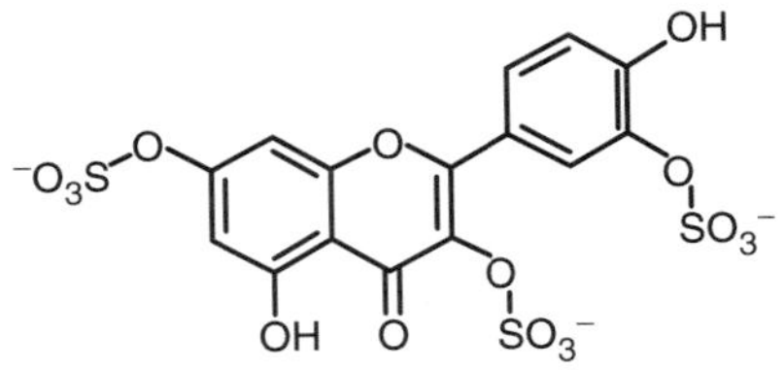

quercetin 3,3′,7-trissulfate

Quercetin 3,4′-bissulfate can also act as an alternative substrate in this reaction.

[1]L. Varin & R. K. Ibrahim (1991) *Plant Physiol.* **95**, 1254.
[2]L. Varin (1988) *Bull. Liaison-Groupe Polyphenols* **14**, 249.
[3]L. Varin, V. DeLuca, R. K. Ibrahim & N. Brisson (1992) *PNAS* **89**, 1286.

QUERCITIN 2,3-DIOXYGENASE

This copper-dependent oxidoreductase [EC 1.13.11.24] catalyzes the reaction of quercetin, a flavonol, with dioxygen to produce 2-protocatechuoylphloroglucinol carboxylate and carbon monoxide.[1–3]

quercetin

2-protocatechuoylphloro-
glucinol carboxylate

The *Aspergillus flavus* enzyme will catalyze the conversion of a number of 3-hydroxyflavones to the corresponding depsides.

[1] T. Oka, F. J. Simpson, J. J. Child & S. C. Mills (1971) *Can. J. Microbiol.* **17**, 111.
[2] A. Messerschmidt (1998) *CBC* **3**, 401.
[3] O. Hayaishi, M. Nozaki & M. T. Abbott (1975) *The Enzymes*, 3rd ed., **12**, 119.

Selected entries from *Methods in Enzymology* [**vol**, page(s)]:
General Discussion: 52, 9

QUERCETIN 3-O-METHYLTRANSFERASE

This methyltransferase [EC 2.1.1.76], also known as flavonol 3-O-methyltransferase and quercetin O^3-methyltransferase, catalyzes the reaction of S-adenosyl-L-methionine with 3,5,7,3′,4′-pentahydroxyflavone (that is, quercetin) to produce S-adenosyl-L-homocysteine and 3-methoxy-5,7,3′,4′-tetrahydroxyflavone.[1,2]

3-methoxy-5,7,3′,4′-tetrahydroxyflavone

Rhamnetin is a poor alternative substrate. Related enzymes catalyze 3-O-methylation of other flavonols such as galangin and kaempferol.

[1] V. De Luca, G. Brunet, H. Khouri & R. Ibrahim (1982) *Z. Naturforsch.* **37c**, 134.
[2] R. K. Ibrahim & V. De Luca (1982) *Naturwissenschaften* **69**, 41.

QUERCETIN-3-SULFATE 3′-SULFOTRANSFERASE

This sulfotransferase [EC 2.8.2.26] catalyzes the reaction of 3′-phosphoadenylylsulfate with quercetin 3-sulfate to produce adenosine 3′,5′-bisphosphate and quercetin 3,3′-bissulfate.[1,2]

quercetin 3-sulfate

quercetin 3,3′-bissulfate

Tamarixetin 3-sulfate and patuletin 3-sulfate are alternative substrates.

[1] L. Varin & R. K. Ibrahim (1989) *Plant Physiol.* **90**, 977.
[2] L. Varin (1988) *Bull. Liaison-Groupe Polyphenols* **14**, 248.

QUERCETIN-3-SULFATE 4′-SULFOTRANSFERASE

This sulfotransferase [EC 2.8.2.27] catalyzes the reaction of 3′-phosphoadenylylsulfate with quercetin 3-sulfate to produce adenosine 3′,5′-bisphosphate and quercetin 3,4′-bissulfate.[1–3]

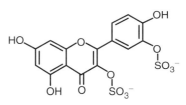

quercetin 3,4′-bis-sulfate

Isorhamnetin 3-sulfate and kaempferol 3-sulfate are alternative substrates.

[1]L. Varin & R. K. Ibrahim (1989) *Plant Physiol.* **90**, 977.
[2]L. Varin (1988) *Bull. Liaison-Groupe Polyphenols* **14**, 248.
[3]L. Varin, V. DeLuca, R. K. Ibrahim & N. Brisson (1992) *PNAS* **89**, 1286.

QUERCITRINASE

This enzyme [EC 3.2.1.66] catalyzes the hydrolysis of quercitrin (that is, quercetin 3-rhamnoside) to produce L-rhamnose and quercetin.[1]

[1]D. W. S. Westlake (1963) *Can. J. Microbiol.* **9**, 211.

Selected entries from ***Methods in Enzymology*** [**vol**, page(s)]:
General Discussion: 52, 38

QUESTIN MONOOXYGENASE

This oxidoreductase [EC 1.14.13.43], also known as questin oxygenase, catalyzes the reaction of questin with NADPH and dioxygen to produce sulochrin, $NADP^+$, and water.[1]

[1]I. Fujii, Y. Ebizuka & U. Sankawa (1988) *J. Biochem.* **103**, 878.

QUEUINE tRNA-RIBOSYLTRANSFERASE

This zinc-dependent transferase [EC 2.4.2.29], also known as tRNA-guanine transglycosylase and guanine insertion enzyme, catalyzes the reaction of tRNA guanine residue (located in the first position of the anticodon of certain undermodified tRNAs such as tRNA^{Asn} and tRNA^{Tyr}) with queuine to produce tRNA queuine and guanine.[1-3]

queuine

The enzyme will also catalyze the exchange of precursors of queuine and of guanine itself for the guanine located in the first position of certain tRNA anticodons. The *Escherichia coli* enzyme tolerates a wide variety of substituents at the 5-position.[4]

[1]N. Okada, S. Noguchi, H. Kasai, N. Shindo-Okada, T. Ohgi, T. Goto & S. Nishimura (1979) *JBC* **254**, 3067.
[2]N. K. Howes & W. R. Farkas (1978) *JBC* **253**, 9082.
[3]N. Shindo-Okada, N. Okada, T. Ohgi, T. Goto & S. Nishimura (1980) *Biochemistry* **19**, 395.
[4]G. C. Hoops, L. B. Townsend & G. A. Garcia (1995) *Biochemistry* **34**, 15381.

QUINALDATE 4-OXIDOREDUCTASE

This oxidoreductase [EC 1.3.99.18], also known as quinaldic acid 4-oxidoreductase, catalyzes the reaction of quinaldate with an acceptor substrate and water to produce kynurenate and the reduced acceptor.[1] The *Pseudomonas* enzyme will also act on quinoline-8-carboxylate, whereas the *Serratia marcescens* utilizes nicotinate as an alternative substrate. Electron acceptor substrates include 1,4-benzoquinone, cytochrome *c*, menadione, 1,2-naphthoquinone, nitroblue tetrazolium, phenazine methosulfate, thionine, and 2,4,6-trinitrobenzenesulfonate.

[1]S. Fetzner & F. Lingens (1993) *Biol. Chem. Hoppe-Seyler* **374**, 363.

QUINATE 5-DEHYDROGENASE

This oxidoreductase [EC 1.1.1.24] catalyzes the reaction of quinate with NAD^+ to produce 5-dehydroquinate and NADH.[1,2]

[1]J. L. Barea & N. H. Giles (1978) *BBA* **524**, 1.
[2]R. B. Cain (1972) *BJ* **127**, 15P.

Selected entries from ***Methods in Enzymology*** [**vol**, page(s)]:
General Discussion: 2, 307
Assay: 2, 307

QUINATE DEHYDROGENASE (PYRROLOQUINOLINE-QUINONE)

This PQQ-dependent oxidoreductase [EC 1.1.99.25], also known as $NAD(P)^+$-independent quinate dehydrogenase, catalyzes the reaction of quinate with pyrroloquinoline-quinone to produce 5-dehydroquinate and the reduced pyrroloquinoline-quinone.[1]

[1]M. A. G. van Kleef & J. A. Duine (1988) *Arch. Microbiol.* **150**, 32.

QUINATE *O*-HYDROXYCINNAMOYL-TRANSFERASE

This acyltransferase [EC 2.3.1.99], also known as hydroxycinnamoyl Coenzyme A:quinate transferase, catalyzes the reaction of feruloyl-CoA with quinate to produce Coenzyme A and *O*-feruloylquinate.[1-4] Other acyl-CoA substrates include caffeoyl-CoA and 4-coumaroyl-CoA.

[1]M. J. C. Rhodes & L. S. C. Wooltorton (1976) *Phytochemistry* **15**, 947.
[2]D. Strack, H. Keller & G. Weissenböck (1987) *J. Plant Physiol.* **131**, 61.
[3]L. Lotfy, A. Fleuriet & J.-J. Macheix (1992) *Phytochemistry* **31**, 767.
[4]D. Strack, W. Gross, V. Wray & L. Grotjahn (1987) *Plant Physiol.* **83**, 475.

QUININE 3-MONOOXYGENASE

This oxidoreductase [EC 1.14.13.67], also known as quinine 3-hydroxylase, catalyzes the reaction of quinine with

NADPH, H^+, and dioxygen to produce 3-hydroxyquinine, $NADP^+$, and water.[1-3]

quinine   3-hydroxyquinine

[1] H. Zhang, P. F. Coville, R. J. Walker, J. O. Miners, D. J. Birkett & S. Wanwimolruk (1997) *Brit. J. Clin. Pharmacol.* **43**, 245.
[2] X.-J. Zhao, T. Kawashiro & T. Ishizaki (1998) *Drug Metab. Dispos.* **26**, 188.
[3] X.-J. Zhao, H. Yokoyama, K. Chiba, S. Wanwimolruk & T. Ishizaki (1996) *J. Pharmacol. Exp. Ther.* **279**, 1327.

QUINOHEMOPROTEIN ALCOHOL DEHYDROGENASE

This pyrroloquinoline-quinone-dependent (PQQ-dependent) oxidoreductase catalyzes the reaction of a primary or secondary alcohol with an acceptor substrate (including certain redox-active dyes) to produce the corresponding aldehyde and the reduced acceptor.[1] *See also* Quinoprotein Alcohol Dehydrogenase

[1] J. A. Duine, J. Frank & J. A. Jongejan (1987) *AE* **59**, 169.

Selected entries from **Methods in Enzymology** [vol, page(s)]:
Assay: 188, 36

QUINOLINE-4-CARBOXYLATE 2-OXIDOREDUCTASE

This molybdenum-dependent oxidoreductase [EC 1.3.99.19], also known as quinoline-4-carboxylic acid 2-oxidoreductase and quinaldic acid 4-oxidoreductase, catalyzes the reaction of quinoline-4-carboxylate with an acceptor substrate and water to produce quinolin-2(1H)-one and the reduced acceptor.[1,2] The enzyme will also act on 4-methylquinoline and quinoline. Electron acceptor substrates include 4-chloroquinoline, 1,4-benzoquinone, 2,6-dichlorophenolindophenol, iodonitrotetrazolium chloride, menadione, and thionine.

[1] M. Sauter, B. Tshisuaka, S. Fetzner & F. Lingens (1993) *Biol. Chem. Hoppe-Seyler* **374**, 1037.
[2] S. Fetzner & F. Lingens (1993) *Biol. Chem. Hoppe-Seyler* **374**, 363.

QUINOLINE 2-OXIDOREDUCTASE

This FAD-, molybdenum-, and iron-dependent oxidoreductase [EC 1.3.99.17] catalyzes the reaction of quinoline with an acceptor substrate and water to produce isoquinolin-1(2H)-one and the reduced acceptor.[1] Alternative substrates include quinolin-2-ol, quinolin-7-ol, quinolin-8-ol, 3-, 4- and 8-methylquinolines, and 8-chloroquinoline. Iodonitrotetrazolium chloride is an effective electron acceptor.[2]

[1] B. Tshisuaka, R. Kappl, J. Huttermann & F. Lingens (1993) *Biochemistry* **32**, 12928.
[2] S. Schach, B. Tshisuaka, S. Fetzner & F. Lingens (1995) *EJB* **232**, 536.

QUINOPROTEIN ALCOHOL DEHYDROGENASE

This pyrroloquinoline-quinone- (or, PQQ-) dependent oxidoreductase catalyzes the reaction of a primary or secondary alcohol with an acceptor substrate to produce the corresponding aldehyde and the reduced acceptor.[1] In contrast to quinohemoprotein alcohol dehydrogenase, the enzyme is activated by ammonia or primary amines. **See also** *Quinohemoprotein Alcohol Dehydrogenase*

[1] J. A. Duine, J. Frank & J. A. Jongejan (1987) *AE* **59**, 169.

Selected entries from **Methods in Enzymology** [vol, page(s)]:
Assay: 188, 33

R

RAFFINOSE:RAFFINOSE α-GALACTOSYLTRANSFERASE

This transferase [EC 2.4.1.166] catalyzes the reaction of two molecules of raffinose to produce 1^F-α-D-galactosylraffinose (or, lychnose) and sucrose.[1]

[1] H. Hopf, G. Gruber, A. Zinn & O. Kandler (1984) *Planta* **162**, 283.

RAN GTPase

This small nuclear GTP-regulatory protein regulates the cell cycle and mRNA transport. The reaction is currently misclassified as a GTPase [EC 3.6.1.47], but the reaction is of the energase type: $State_1 + GTP = State_1 + GDP + P_i$, where two affinity-modulated conformational states are indicated.[1] **See G Proteins**

[1] D. L. Purich (2001) *TiBS* **26**, 417.

RAS/a-FACTOR CONVERTING ENZYME

This membrane-bound metalloendopeptidase, often abbreviated RACE, catalyzes the removal of the C-terminal tripeptide from either isoprenylated Ras or yeast **a**-factor hormone.[1] **See Isoprenylated Protein Endopeptidase; Ste24 Protease**

[1] V. L. Boyartchuk, M. N. Ashby & J. Rine (1997) *Science* **275**, 1796

Selected entries from ***Methods in Enzymology*** [**vol**, page(s)]:
Assay: assays, interference from other proteases, **250**, 247; direct assay (controls, **250**, 244; reaction conditions, **250**, 244; substrate, **250**, 243, 245); HPLC-based assays (amino-terminal reaction product identification, **250**, 245; carboxyl-terminal reaction product identification, **250**, 247); indirect coupled assay (controls, **250**, 241; inhibitor analysis, **250**, 242; reaction conditions, **250**, 241; substrate, **250**, 240)

RAS GTPase

This 21 kDa host-cell GTP/GDP-dependent regulatory protein, possessing intrinsic GTPase activity, is related to the gene products of the retrovirus and sarcoma virus oncogenes. Ras binds either GTP or GDP with high affinity, but only the GTP-bound form actively participates in signal transduction.[1] The intrinsic GTPase activity serves as a timer that controls the activation of Ras-associated signal transduction processes. Replacement of Ras-bound GDP by GTP is a very slow exchange process that can be accelerated by guanine nucleotide exchange factors (GEFs).[2] Activated *ras* oncogenes are common in human cancer, are associated with point mutations corresponding to residues 12 and 13 in the gene product, and are responsible for arresting the GTPase activity.

The reaction is currently misclassified as a GTPase [EC 3.6.1.47], but the reaction is of the energase type: $State_1 + GTP = State_1 + GDP + P_i$, where two affinity-modulated conformational states are indicated.[3] **See also G Proteins**

[1] W. E. Balch, C. J. Der & A. Hall, eds. (1995) *MIE*, **255-257**, Academic Press, San Diego.
[2] M. Geyer & A. Wittinghofer (1997) *Curr. Opin. Struct. Biol.* **7**, 786.
[3] D. L. Purich (2001) *TiBS* **26**, 417.

Selected entries from ***Methods in Enzymology*** [**vol**, page(s)]:
Assay: activated charcoal, **255**, 164; filter assay, **255**, 164; immunoprecipitated Ras, **255**, 112; solvent partitioning, **255**, 163; thin-layer chromatography, **255**, 165

RAUCAFFRICINE β-GLUCOSIDASE

This highly specific glycosidase [EC 3.2.1.125] catalyzes the hydrolysis of raucaffricine to produce vomilenine and D-glucose.[1,2] Alternative substrates include 17-*O*-deacetyl-1,2-dihydroraucaffricine, 1,2-dihydroraucaffricine, 1,2-dihydro-1-methylraucaffricine, and 1,2,19,20-tetrahydroraucaffricine.

[1] H. Schübel & J. Stöckigt (1986) *Helv. Chim. Acta* **69**, 538.
[2] H. Warzecha, P. Obitz & J. Stöckigt (1999) *Phytochemistry* **50**, 1099.

REDOXYENDONUCLEASES

These enzymes recognize and remove oxidatively modified DNA bases.[1] The best known example of these proteins is probably *Escherichia coli* deoxyribonuclease III. **See specific enzyme**

[1] A. Sancar & G. B. Sancar (1988) *ARB* **57**, 29.

RELEASING FACTORS

These termination factors facilitate the release of the newly formed polypeptide chain from the ribosome during protein biosynthesis. The Enzyme Commission currently classifies releasing factors possessing GTPase activities as protein-synthesizing GTPases [EC 3.6.1.48]. This reaction is currently misclassified as a GTPase [EC 3.6.1.49], but the reaction is of the energase type: $State_1 + GTP = State_1 + GDP + P_i$, where two affinity-modulated conformational states are indicated.[1] *See* G Proteins

[1] D. L. Purich (2001) *TiBS* **26**, 417.

RENILLA-LUCIFERIN 2-MONOOXYGENASE

This oxidoreductase [EC 1.13.12.5], widely known as aequorin, catalyzes the reaction of the *Renilla* luciferin (or, coelenterazine) with dioxygen to produce oxidized *Renilla* luciferin (or, coelenteramide), carbon dioxide, and a photon of energy corresponding to 480 nm.[1,2] Aequorin, first isolated from the jellyfish *Aequorea* sp., was one of the earliest recognized Ca^{2+}-regulated photoprotein.

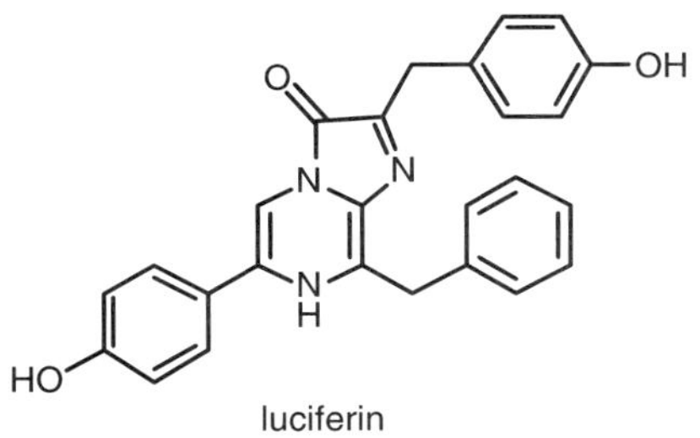

luciferin
(coelenterazine)

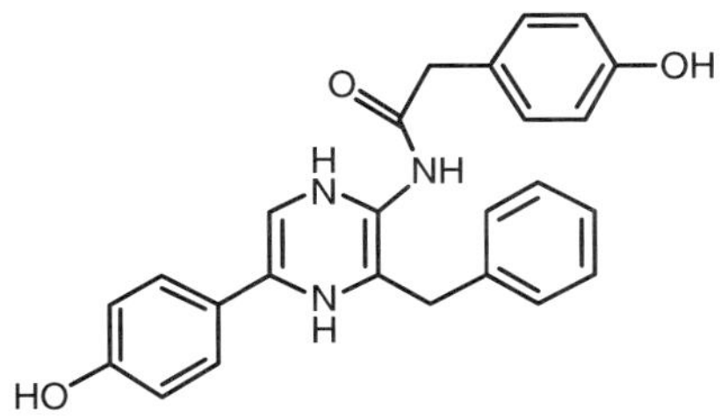

oxyluciferin
(coelenteramide)

Renilla luciferin is structurally related to 8-benzyl-2-(4-hydroxybenzyl)-6-(4-hydroxyphenyl)imidazo[1,2a] pyrazin-3(7*H*)-one, containing a low-molecular-weight substituent in place of the benzyl group. This monooxygenase is unusually rich in aromatic amino acids: tryptophan, phenylalanine, and tyrosine comprise 31 of its 312 amino acid residues.[2]

[1] R. C. Hart, J. C. Matthews, K. Hori & M. J. Cormier (1979) *Biochem.* **11**, 2202.
[2] H. Charbonneau & M. J. Cormier (1979) *JBC* **254**, 769.

RENILLA-LUCIFERIN SULFOTRANSFERASE

This sulfotransferase [EC 2.8.2.10], also known as luciferin sulfotransferase, catalyzes the reaction of 3'-phosphoadenylylsulfate with the *Renilla* luciferin to produce adenosine 3',5'-bisphosphate and the *Renilla* luciferyl sulfate.[1]

[1] M. J. Cormier, K. Hori & Y. D. Karkhanis (1970) *Biochem.* **9**, 1184.

RENIN

This aspartic endopeptidase [EC 3.4.23.15], also known as angiotensin-forming enzyme, catalyzes the hydrolysis of the Leu10–Val11 bond in human angiotensinogen to generate angiotensin I.[1–3] The minimal substrate contains at least eight aminoacyl residues (PFHLLVYS). Human renin hydrolyzes the Leu10–Leu11 bond in pig, horse, and sheep angiotensinogen. The pH 5.5–7.5 optimum of renin is unusual for aspartic proteinases which are typically active at significantly lower pH.

[1] T. Inagami (1989) *J. Hypertens.* **7** (suppl. 2), S3.
[2] A. Fukamizu & K. Murakami (1995) *TEM* **6**, 279.
[3] T. D. Meek (1998) *CBC* **1**, 327.

Other molecular properties: active site, 241, 255; angiotensin I radioimmunoassay, 80, 436; human (isoelectric point, 80, 440; kinetics of activation, 80, 439; molecular weight, 80, 440; optimum pH, 80, 440; 241, 214; properties, 80, 440; purification, 80, 433; stability, 80, 435); inhibition, 80, 441, 442; Michaelis constants, 19, 704; mouse submaxillary, 80, 427 (active site, 80, 442; direct radioimmunoassay, 80, 436, 438; isoelectric point, 80, 440; molecular weight, 80, 440; optimum pH, 80, 440; properties, 80, 440; purification, 80, 429); pH optima, 19, 705; 241, 214; physiological function, 80, 427; plasma, activation, 80, 429; stability, 80, 439; substrate, 2, 124, 135; 46, 235 (fluorigenic, 248, 34); substrate specificity, 80, 439; 241, 279; transition-state and multisubstrate analogues, 249, 307

γ-RENIN

This serine endopeptidase [EC 3.4.21.54], obtained from the submandibular gland of male mice, catalyzes peptide-bond hydrolysis preferentially at Leu–Leu linkages. γ-Renin is not active on natural angiotensinogen but readily hydrolyzes Bz-Arg-*p*-nitroanilide.[1] **See also** *Renin*

[1] M. Poe, J. K. Wu, J. R. Florance, J. A. Rodkey, C. D. Bennett & K. Hoogsteen (1983) *JBC* **258**, 2209.

REPRESSOR LexA

This serine endopeptidase [EC 3.4.21.88] catalyzes the hydrolysis of the Ala84–Gly85 peptide bond in the repressor LexA. In alkaline pH conditions, the LexA protein undergoes autocatalytic cleavage (a process known as autodigestion). The same bond is hydrolyzed in the RecA-mediated reaction, the RecA protein termed a co-protease having an indirect role in the cleavage (RecA had been previously classified as EC 3.4.99.37). While the cleavage reaction can be either intramolecular or intermolecular, the 4-mM K_m value for the intermolecular path is obviously nonphysiologic. LexA-mediated cleavage is a step in the so-called SOS response to DNA damage.[1–4]

[1] J. W. Little & D. W. Mount (1982) *Cell* **29**, 11.
[2] J. W. Little, B. Kim, K. L. Roland, M. H. Smith, L.-L. Lin & S. N. Slilaty (1994) *MIE* **244**, 266.
[3] J. W. Little (1993) *J. Bacteriol.* **175**, 4943.
[4] E. C. Friedberg, G. C. Walker & W. Siede (1995) *DNA Repair and Mutagenesis*, Amer. Soc. Microbiol. Press, Washington, D.C.

RETICULINE OXIDASE

This FAD-dependent oxidoreductase [EC 1.5.3.9], also known as berberine-bridge-forming enzyme and tetrahydroprotoberberine synthase, catalyzes the reaction of (*S*)-reticuline with dioxygen to produce (*S*)-scoulerine and hydrogen peroxide (H_2O_2).[1–3] Studies on a series of (*S*)-reticuline derivatives suggest that ring closure proceeds in two steps: formation of the methylene iminium ion and subsequent ring closure via an ionic mechanism.[4]

[1] P. Steffens, N. Nagakura & M. H. Zenk (1985) *Phytochemistry* **24**, 2577.
[2] T. Frenzel, J. M. Beale, M. Kobayashi, M. H. Zenk & H. G. Floss (1988) *JACS* **110**, 7878.
[3] M. Amann, G. Wanner & M. H. Zenk (1986) *Planta* **167**, 310.
[4] T. M. Kutchan & H. Dittrich (1995) *JBC* **270**, 24475.

RETINAL DEHYDROGENASE

This FAD-dependent oxidoreductase [EC 1.2.1.36] catalyzes the reaction of retinal with NAD⁺ and water to produce retinoate and NADH.[1,2] Both the all-*trans*- and 13-*cis*-forms of retinal are substrates. Rat retinal dehydrogenase can utilize retinal bound to retinol-binding protein as a substrate.[2]

[1] D. J. Moffa, F. J. Lotspeich & R. F. Krause (1970) *JBC* **245**, 439.
[2] K. C. Posch, R. D. Burns & J. L. Napoli (1992) *JBC* **267**, 19676.

RETINAL ISOMERASE

This isomerase [EC 5.2.1.3], also known as retinene isomerase and retinoid isomerase, catalyzes the interconversion of all-*trans*-retinal and 11-*cis*-retinal.[1–3] The enzyme also interconverts all-*trans*-retinol and 11-*cis*-retinol. **See also** *Retinol Isomerase*

[1] R. Hubbard (1956) *J. Gen. Physiol.* **39**, 935.
[2] P. S. Bernstein, W. C. Law & R. R. Rando (1987) *JBC* **262**, 16848.
[3] J. Heller (1972) *The Enzymes*, 3rd ed., **6**, 573.
Selected entries from *Methods in Enzymology* [vol, page(s)]:
General discussion: **189**, 494, 503
Assay: **189**, 498, 505; eye, **189**, 494; retinal pigment epithelium, **189**, 503

RETINAL OXIDASE

This oxidoreductase [EC 1.2.3.11], also known as retinene oxidase, catalyzes the reaction of retinal (all-*trans* and 9-*cis*) with dioxygen to produce retinoate and hydrogen peroxide.[1–4]

[1] G. S. Kishore & R. K. Boutwell (1980) *BBRC* **94**, 1381.
[2] P. V. Bhat, L. Poissant, P. Falardeau & L. Lacroix (1988) *Biochem. Cell Biol.* **66**, 735.
[3] J. Hupert, S. Mobarhan, T. J. Layden, V. M. Papa & D. J. Lucchesi (1991) *Biochem. Cell Biol.* **69**, 509.
[4] P. V. Bhat, L. Poissant & A. Lacroix (1988) *BBA* **967**, 211.

RETINOL DEHYDROGENASE

This oxidoreductase [EC 1.1.1.105] catalyzes the reaction of retinol with NAD⁺ to produce retinal and NADH.[1–4] Kinetic properties and the inhibitory action of 13-*cis*-retinoate on recombinant human and mouse 9-*cis*-retinol dehydrogenase and bovine 11-*cis*-retinol dehydrogenase indicate that these activities are associated with the same enzyme.[1]

[1] M. V. Gamble, N. L. Mata, A. T. Tsin, J. R. Mertz & W. S. Blaner (2000) *BBA* **1476**, 3.

[2]W. S. Blaner & J. E. Churchich (1980) *BBRC* **94**, 820.
[3]C. Nicotra & M. A. Livrea (1982) *JBC* **257**, 11836.
[4]S. Ishiguro, Y. Suzuki, M. Tamai & K. Mizuno (1991) *JBC* **266**, 15520.

Selected entries from *Methods in Enzymology* [vol, page(s)]:
General discussion: 189, 436, 470, 520
Other molecular properties: differentiation from cytosolic alcohol dehydrogenases, 189, 478; NAD$^+$-dependent, detection in hepatic microsomes and characterization, 189, 520

RETINOL *O*-FATTY-ACYLTRANSFERASE

This acyltransferase [EC 2.3.1.76], also known as acyl-CoA:retinol *O*-acyltransferase, catalyzes the reaction of an acyl-CoA derivative with retinol to generate Coenzyme A and the retinyl ester.[1–4] Palmitoyl-CoA and other long-chain fatty-acyl-CoA metabolites are effective substrates.

[1]A. C. Ross (1982) *JBC* **257**, 2453.
[2]H. Mueller & K. R. Norum (1986) *Br. J. Nutr.* **55**, 37.
[3]L. R. Chaudhary & E. C. Nelson (1987) *BBA* **917**, 24.
[4]R. Blomhoff, M. H. Green & K. R. Norum (1992) *Ann. Rev. Nutr.* **12**, 37.

Selected entries from *Methods in Enzymology* [vol, page(s)]:
General discussion: 189, 442, 446; 190, 156
Assay: 189, 442, 446; bovine retinal pigment epithelial microsomes, in, assay, 190, 156

RETINOL 4-HYDROXYLASE

This oxidoreductase catalyzes the reaction of all-*trans*-retinol with NADPH and dioxygen to produce 4-hydroxy-retinol, NADP$^+$, and water.

Selected entries from *Methods in Enzymology* [vol, page(s)]:
General discussion: 189, 483
Assay: 189, 485

RETINOL ISOMERASE

This isomerase [EC 5.2.1.7] interconverts all-*trans*-retinol and 11-*cis*-retinol. This reaction can take place in the dark and reverses the effect of the *cis-trans* isomerization of retinal isomerase [EC 5.2.1.3]. Stereochemical inversion at C15 accompanies enzyme-catalyzed isomerization.[1] *See also Retinal Isomerase*

[1]W. C. Law & R. R. Rando (1988) *Biochem.* **27**, 4147.

Selected entries from *Methods in Enzymology* [vol, page(s)]:
General discussion: 189, 494, 503
Assay: 189, 498, 505; eye, 189, 494; retinal pigment epithelium, 189, 503

RETINYL-PALMITATE ESTERASE

This esterase [EC 3.1.1.21] catalyzes the hydrolysis of retinyl palmitate to produce retinol and palmitate. Substrates include all-*trans*-retinyl palmitate, 13-*cis*-retinyl palmitate, 9-*cis*-retinyl palmitate, 9,13-*cis*-retinyl palmitate, and 11-*cis*-retinyl palmitate as well as retinyl esters of other fatty acids.[1–3] Three different esterases, each

with its own distinct pH optimum, have been identified in bovine retinal pigment epithelium membranes.[4]

[1]S. Mahadevan, N. I. Ayyoub & O. A. Roels (1966) *JBC* **241**, 57.
[2]D. A. Cooper & J. A. Olson (1988) *ABB* **260**, 705.
[3]W. S. Blaner, S. R. Das, P. Gouras & M. T. Flood (1987) *JBC* **262**, 53.
[4]M. C. Gueli, C. M. Nicotra, A. M. Pintaudi, A. Paganini, L. Pandolfo, G. De Leo & M. A. Di Bella (1991) *ABB* **288**, 572.

RETINYL-PALMITATE ESTERASE

Two distinct esterases hydrolyze retinyl-palmitate ester. The 11-*cis*-retinyl-palmitate esterase [EC 3.1.1.63], which is activated by such bile salts as cholate, taurocholate, or glycocholate, catalyzes the hydrolysis of 11-*cis*-retinyl palmitate to produce 11-*cis*-retinol and palmitate.[1–3] A second esterase [EC 3.1.1.64], which requires detergents for activity, catalyzes the hydrolysis of all-*trans*-retinyl palmitate to produce all-*trans*-retinol and palmitate.[1,2]

[1]W. S. Blaner, S. R. Das, P. Gouras & M. T. Flood (1987) *JBC* **262**, 53.
[2]N. L. Mata, A. T. Tsin & J. P. Chambers (1992) *JBC* **267**, 9794.
[3]W. S. Blaner, J. H. Prystowsky, J. E. Smith & D. S. Goodman (1984) *BBA* **794**, 419.

Selected entries from *Methods in Enzymology* [vol, page(s)]:
Assay: bile salt-independent, membrane-associated, 189, 460; hepatic, 189, 490

RETROPEPSINS

These aspartic endoproteinases [EC 3.4.23.16] process the formation of individual protein units upon endoproteolysis of polyproteins biosynthesized by positive-strand RNA viruses.

RETROTRANSPOSON PROTEASES

These proteases, encoded by long terminal repeat retrotransposons, are functionally similar to retroviral proteases.

REVERSE TRANSCRIPTASE

RNA-directed DNA polymerases [EC 2.7.7.49], commonly known as reverse transcriptases and also known as DNA nucleotidyltransferase (RNA-directed) and revertase, catalyze the RNA-template-directed extension of the 3′-end of a DNA strand by one deoxynucleotide. These enzymes catalyze the reaction of n molecules of deoxynucleoside triphosphates to produce n molecules of pyrophosphate (or, diphosphate) and a DNA polymer containing n bases.[1–7] Reverse transcriptases cannot initiate *de novo* chain synthesis, because the presence of an RNA or DNA primer is required. *See DNA Polymerase; Avian Myeloblastosis Virus Reverse Transcriptase; Hepatitis B Virus Reverse Transcriptase; Human Immunodeficiency Virus*

Type I Reverse Transcriptase; Moloney Murine Leukemia Virus Reverse Transcriptase; Ribonuclease H; Ribonuclease H*

[1]C. M. Joyce & T. A. Steitz (1994) *ARB* **63**, 777.
[2]A. M. Skalka & S. P. Goff, eds. (1993) *Reverse Transcriptase*, Cold Spring Harbor Press, Cold Spring Harbor.
[3]J. A. Peliska & S. J. Benkovic (1992) *Science* **258**, 1112.
[4]I. M. Verma (1981) *The Enzymes*, 3rd ed., **14**, 87.
[5]H. M. Temin & S. Mizutani (1974) *The Enzymes*, 3rd ed., **10**, 211.
[6]D. Baltimore (1970) *Nature* **226**, 1209.
[7]H. M. Temin & S. Mizutani (1970) *Nature* **226**, 1211.

Selected entries from **Methods in Enzymology** [vol, page(s)]:
General discussion: 29, 119, 125, 130, 143, 150, 173; **34**, 5, 472, 473, 728, 729; **65**, 562, 568, 576, 579; **79**, 603
Assay: 79, 426, 427, 455, 456 [**See also DNA Polymerase**]; avian myeloblastosis virus, 29, 125, 144, 161; DNA polymerase assay in bacterial colonies *in situ*, 262, 356; DNA strand transfer, 275, 301, 303; DNA synthesis on homopolymers (autoradiography detection, 262, 348; bacterial extracts, 262, 351; plate replication, 262, 352; reaction conditions, 262, 349, 351; sensitivity, 262, 350; stock solution storage, 262, 351; substrates, 262, 141, 143, 348; tissue culture, 262, 348); DNA synthesis *in virion* (detergent concentration, 262, 354; reaction conditions, 262, 354; sensitivity, 262, 353; transfer RNA primer tagging, 262, 355; virus preparation, 262, 353); polymerase/ribonuclease H coupling, 275, 292, 295; processivity (ribonuclease H activity, 262, 279; transcriptase activity, 262, 143); pyrophosphate exchange, 275, 283, 285, 287; pyrophosphorolysis, 275, 283, 285; ribonuclease H, 275, 288, 307, 309, 336, 338
Other molecular properties: avian myeloblastosis virus reverse transcriptase (fidelity, 275, 536; inhibition assays [acyclic nucleoside phosphonates, 275, 473, 486; chain-terminating nucleoside analog 5′-triphosphates, 275, 478; oligonucleotide ligands, 275, 509]); hepatitis B virus reverse transcriptase (chain-terminating nucleoside analog 5′-triphosphate IC$_{50}$ values, determination, 275, 423; duck enzyme, 275, 335; human enzyme, expression in baculovirus-insect cell system [priming reaction, 275, 203; protein purification, 275, 204; vector, 275, 203]; priming reaction, 275, 195, 198, 201, 207); human immunodeficiency virus type I reverse transcriptase (active sites, distance of separation, 275, 379; catalytic activities, 275, 123, 261, 297, 376, 441, 472; drug resistance and viral mutation [catalytic efficiency of mutants, 275, 502; clinical impact, 275, 558, 562, 593, 595, 599; inhibitors [acyclic nucleoside phosphonates, 275, 473, 486; chain-terminating nucleoside analogue 5′-triphosphates, 275, 409, 415, 423, 478; nonnucleosides [α-anilinophenylacetamide compounds, 275, 465; arylsulfonylindoles, 275, 446; benzophenone compounds, 275, 469; benzothiadiazines, 275, 469; binding sites, 275, 443, 466, 468, 470, 476; bis(heteroaryl) piperazines, 275, 449, 452; dipyridodiazepinones, 275, 446; nucleoside-like inhibitors, 275, 458, 475; phenylethylthiazolylthiourea compounds, 275, 468; pyridinones, 275, 454; quinazolinones, 275, 460; tetrahydroimidazo[4,5,1-j,k]-[1,4]benzodiazepin-2(1H)one derivatives, 275, 452, 475; thiazoloisoindolones, 275, 466, 468]; kinetic mechanism, 275, 377; molecular dynamics simulation, 241, 381; presteady-state rate constants, 275, 378; proteolytic processing, 275, 123, 132; product inhibition studies, 249, 189; recombinant expression systems, 275, 124; recombinant protein purification, 262, 131; stabilization, 276, 136; subunit overexpression in *Escherichia coli*, 262, 131, 172; subunits, 275, 123, 261; template lesion by-pass, 262, 255; template secondary structure, effects on activity, 275, 380

L-RHAMNONATE DEHYDRATASE

This dehydratase [EC 4.2.1.90] catalyzes the conversion of L-rhamnonate to 2-dehydro-3-deoxy-L-rhamnonate and water.[1,2]

[1]L. U. Rigo, L. R. Marechal, M. M. Vieira & L. A. Veiga (1985) *Can. J. Microbiol.* **31**, 817.
[2]A. L. Twerdochlip, F. O. Pedrosa, S. Funayama & L. U. Rigo (1994) *Can. J. Microbiol.* **40**, 896.

L-RHAMNONO-1,4-LACTONASE

This lactonase [EC 3.1.1.65] catalyzes the hydrolysis of L-rhamnono-1,4-lactone to produce L-rhamnonate.[1,2]

[1]L. U. Rigo, L. R. Marechal, M. M. Vieira & L. A. Veiga (1985) *Can. J. Microbiol.* **31**, 817.
[2]A. L. Twerdochlib, F. O. Pedrosa, S. Funayama & L. U. Rigo (1994) *Can. J. Microbiol.* **40**, 896.

L-RHAMNOSE 1-DEHYDROGENASE

This oxidoreductase [EC 1.1.1.173] catalyzes the reversible reaction of L-rhamnofuranose with NAD$^+$ to produce L-rhamno-1,4-lactone and NADH.[1,2]

[1]L. U. Rigo, M. Nakano, L. A. Veiga & D. S. Feingold (1976) *BBA* **445**, 286.
[2]F. Pittner & P. L. Turecek (1987) *Appl. Biochem. Biotechnol.* **16**, 15.

L-RHAMNOSE ISOMERASE

This isomerase [EC 5.3.1.14], also known as L-rhamnulose isomerase, catalyzes the interconversion of L-rhamnose and L-rhamnulose.[1–3]

[1]Y. Takagi & H. Sawada (1964) *BBA* **92**, 10.
[2]K. Izumori, M. Mitchell & A. D. Elbein (1976) *J. Bacteriol.* **126**, 553.
[3]E. A. Noltmann (1972) *The Enzymes*, 3rd ed., **6**, 271.

Selected entries from **Methods in Enzymology** [vol, page(s)]:
General discussion: 9, 579
Assay: 9, 579
Other molecular properties: *Lactobacillus plantarum*, from (cofactors of, 9, 581; pH optimum, 9, 582; properties, 9, 581; purification, 9, 580; specificity, 9, 582); rhamnulokinase, in preparation of, 9, 466; rhamnulose-1-phosphate aldolase, assay of, 9, 545; sources, 9, 582

α-L-RHAMNOSIDASE

This enzyme [EC 3.2.1.40] catalyzes the hydrolysis of terminal non-reducing α-L-rhamnosyl residues in α-L-rhamnosides.[1–3] Substrates include *p*-nitrophenyl α-L-rhamnoside, rutinose (that is, 6-*O*-α-L-rhamnosyl-D-glucopyranose), methyl 3-*O*-α-L-rhamnopyranosyl-α-D-xylopyranoside, and methyl 3-*O*-α-L-rhamnopyranosyl-α-D-mannopyranoside. Note: Narginase is a complex of β-D-glucosidase and α-L-rhamnosidase.

[1]P. M. Dey & E. del Campillo (1984) *AE* **56**, 141
[2]Y. Kurosawa, K. Ikeda & F. Egami (1973) *J. Biochem.* **73**, 31.
[3]F. Miake, T. Satho, H. Takesue, F. Yanagida, N. Kashige & K. Watanabe (2000) *Arch. Microbiol.* **173**, 65.

L-RHAMNOSYL-β-HYDROXYDECANOYL-β-HYDROXYDECANOATE L-RHAMNOSYLTRANSFERASE

This transferase catalyzes the reaction of dTDP-L-rhamnose with L-rhamnosyl-β-hydroxydecanoyl-β-hydroxydecanoate to produce α-L-rhamnopyranosyl-1 → 2-α-L-rhamnopyranosyl-β-hydroxydecanoyl-β-hydroxydecanoate and dTDP.[1]

[1] M. M. Burger, L. Glaser & R. M. Burton (1963) JBC **238**, 2595.

Selected entries from *Methods in Enzymology* [vol, page(s)]:
General discussion: 8, 441
Assay: 8, 441

RHAMNULOKINASE

This phosphotransferase [EC 2.7.1.5] catalyzes the reaction of ATP with L-rhamnulose to produce ADP and L-rhamnulose 1-phosphate.[1]

[1] W.-D. Fessner, J. Badia, O. Eyrisch, A. Schneider & G. Sinerius (1992) *Tetrahedron Lett.* **33**, 5231.

Selected entries from *Methods in Enzymology* [vol, page(s)]:
General discussion: 9, 464
Assay: 9, 464

RHAMNULOSE-1-PHOSPHATE ALDOLASE

This zinc-dependent aldolase [EC 4.1.2.19] catalyzes the conversion of L-rhamnulose 1-phosphate to dihydroxyacetone phosphate and (S)-lactaldehyde.[1-4] The ene-diolate intermediate can react with dioxygen, thereby conferring a minor oxygenase activity akin to that of D-ribulose 1,5-bisphosphate carboxylase.[2]

[1] D. S. Feingold & P. A. Hoffee (1972) *The Enzymes*, 3rd ed., **7**, 303.
[2] M. Hixon, G. Sinerius, A. Schneider, C. Walter, W. D. Fessner & J. V. Schloss (1996) *FEBS Lett.* **392**, 281.
[3] J. V. Schloss & M. S. Hixon (1998) *CBC* **2**, 43.
[4] D. J. Creighton & N. S. R. K. Murthy (1990) *The Enzymes*, 3rd ed., **19**, 323.

Selected entries from *Methods in Enzymology* [vol, page(s)]:
General discussion: 9, 542; **42**, 264
Assay: 9, 542; **42**, 264
Other molecular properties: activators and inhibitors, **42**, 268, 269; enediol substrate, possible, **87**, 97; *Escherichia coli*, from, **42**, 265 (pH optimum, **9**, 544; properties, **9**, 544; purification, **9**, 543); molecular weight, **42**, 269

RHIZOPUSPEPSIN

This aspartic endopeptidase [EC 3.4.23.21], also known as *Rhizopus* aspartic proteinase, catalyzes peptide-bond hydrolysis in proteins with a broad specificity similar to that of pepsin A.[1-3] The enzyme exhibits a preference for aromatic or bulky aminoacyl residues at positions P1 and P1′. Oxidized insulin B chain is cleaved at Phe1–Val2, His5–Leu6, His10–Leu11, Leu11–Val12, Val12–Glu13,

Glu13–Ala14, Ala14–Leu15, Leu15–Tyr16, Tyr16–Leu17, Gly20–Glu21, Gly23–Phe24, Phe24–Phe25, and Phe25–Tyr26.

[1] B. M. Dunn (1998) in *Handbook of Proteolytic Enzymes* (A. J. Barrett, N. D. Rawlings & J. F. Woessner, eds.), Academic Press, New York, p. 887.
[2] T. D. Meek (1998) *CBC* **1**, 327.
[3] D. R. Davies (1990) *Ann. Rev. Biophys. Biophys. Chem.* **19**, 189.

Selected entries from *Methods in Enzymology* [vol, page(s)]:
General discussion: 19, 373
Assay: 19, 391; **241**, 213
Other molecular properties: kinetic parameters, **241**, 222; molecular properties, **19**, 395; pepstatin binding, thermodynamics, **259**, 654, 670; pKa values, **241**, 222; purification, **19**, 391; recombinant (purification, **241**, 210; synthesis, **241**, 210); trypsinogen-activation by, **19**, 374

RHODANESE

This sulfurtransferase [EC 2.8.1.1], known officially as thiosulfate sulfurtransferase, catalyzes the reaction of thiosulfate ($S\text{-}SO_3^{2-}$) with cyanide to produce sulfite and thiocyanate (SCN^-). The enzyme has a ping pong Bi Bi kinetic mechanism in which a covalent sulfur-enzyme intermediate forms.[1-4]

[1] S. Oi (1975) *J. Biochem.* **78**, 825.
[2] K. Alexander & M. Volini (1987) *JBC* **262**, 6595.
[3] B. A. Aird, R. L. Heinrikson & J. Westley (1987) *JBC* **262**, 17327.
[4] A. J. L. Cooper (1983) *ARB* **52**, 187.

Selected entries from *Methods in Enzymology* [vol, page(s)]:
General discussion: 2, 334; **77**, 285; **143**, 235; **243**, 501
Assay: 2, 334; **77**, 285, 286; continuous spectrophotometric method, **243**, 505; discontinuous determination of thiocyanate product, by, **243**, 505; fluorogenic substrates, **143**, 239
Other molecular properties: affinity chromatography, **77**, 289; cooperativity, **64**, 218; definition of units, **77**, 286; hysteresis, **64**, 218; inhibitors, **2**, 337; **77**, 290; kinetic constants, **77**, 290; mercaptopyruvate transsulfurase and, **5**, 989; oxidation with NBS, **11**, 521; phototrophic sulfur metabolism, in, **243**, 401; properties, **2**, 336; **77**, 289; **243**, 505; purification, **2**, 335; **77**, 286; size, **77**, 289; specificity, **2**, 336; **77**, 290

RHODOPSIN KINASE

This phosphotransferase [EC 2.7.1.125] catalyzes the reaction of ATP with rhodopsin (bleached or activated) to produce phosphorylated rhodopsin and ADP.[1-5] While up to seven phosphoryls are transferred to the carboxyl terminal region of freshly bleached rhodopsin, only Ser358 within photoactivated octopus rhodopsin is phosphorylated by the octopus rhodopsin kinase.[4] Initial phosphoryl group incorporation is slower than the formation of more highly phosphorylated species. Rhodopsin kinase reportedly undergoes autophosphorylation.[6,7] *See also* [β-Adrenergic Receptor] Kinase

[1] M. Weller, N. Virmaux & P. Mandel (1975) *PNAS* **72**, 381.
[2] H. Shichi & R. L. Somers (1978) *JBC* **253**, 7040.
[3] K. Palczewski, J. H. McDowell & P. A. Hargrave (1988) *JBC* **263**, 14067.

[4]H. Ohguro, N. Yoshida, H. Shindou, J. W. Crabb, K. Palczewski & M. Tsuda (1998) *Photochem. Photobiol.* **68**, 824.

[5]G. Adamus, A. Arendt, P. A. Hargrave, T. Heyduk & K. Palczewski (1993) *ABB* **304**, 443.

[6]R. H. Lee, B. M. Brown & R. N. Lolley (1982) *Biochemistry* **21**, 3303.

[7]D. J. Kelleher & G. L. Johnson (1990) *JBC* **265**, 2632.

Selected entries from *Methods in Enzymology* [**vol**, page(s)]:
General discussion: 99, 362
Assay: 81, 498; 99, 362
Other molecular properties: antibody preparation, 250, 151; bovine rod outer segment (properties, 81, 506; 99, 365; purification, 81, 499; 99, 364); expression systems, 250, 150; mutagenesis by PCR, 250, 151; polyisoprenoid analysis by HPLC, 250, 153; translocation, 250, 149

RHODOTORULAPEPSIN

This aspartic endopeptidase [EC 3.4.23.26], also known as *Rhodotorula* aspartic proteinase, catalyzes peptide-bond hydrolysis with a specificity similar to that of pepsin A. The enzyme will act on the synthetic substrates Z-Lys-∇-Ala-Ala-Ala and Z-Lys-∇-Leu-Ala-Ala, but not Z-Lys-Ala-Ala, where ∇ indicates the cleavage site.[1,2] *See also Paecilomyces varioti Aspartic Proteinase*

[1]K. Morihara & T. Oka (1973) *ABB* **157**, 561.

[2]E. Majima, K. Oda, S. Murao & E. Ichishima (1988) *Agric. Biol. Chem.* **52**, 787.

RIBITOL 2-DEHYDROGENASE

This oxidoreductase [EC 1.1.1.56] catalyzes the reaction of ribitol with NAD^+ to produce D-ribulose and NADH. (Note: ribitol is a *meso* compound.) This was the first enzyme reaction examined by product inhibition methods for the expressed purpose of deducing the order of substrate addition and product release.[1] In either direction of this reversible reaction, the leading substrate is the coenzyme.

[1]H. J. Fromm & D. R. Nelson (1962) *JBC* **237**, 215.

Selected entries from *Methods in Enzymology* [**vol**, page(s)]:
General discussion: 9, 180
Assay: 9, 180
Other molecular properties: abortive complex, 63, 419, 420, 424; 64, 40; *Aerobacter aerogenes*, of, 9, 180 (activators, 9, 183; K_m values, 9, 184; mechanism, 63, 51; molecular weight, 9, 184; pH optimum, 9, 183; product inhibition, 63, 411; properties, 9, 183; purification, 9, 182; D-ribulose preparation, 9, 39; specificity, 9, 183; stability, 9, 183; stereospecificity, 87, 109); D-arabinose isomerase, assay of, 9, 583; L-arabinose isomerase, absent in, 9, 600

RIBITOL-5-PHOSPHATE CYTIDYLYLTRANSFERASE

This transferase [EC 2.7.7.40], also known as CDP-ribitol pyrophosphorylase and CDP-ribitol diphosphorylase, catalyzes the reaction of CTP with D-ribitol 5-phosphate to produce CDP-ribitol and pyrophosphate.[1,2]

[1]D. R. D. Shaw (1962) *BJ* **82**, 297.

[2]S.-C. Cheah, H. Hussey & J. Baddiley (1981) *EJB* **118**, 497.

Selected entries from *Methods in Enzymology* [**vol**, page(s)]:
General discussion: 8, 244, 248
Assay: 8, 244

RIBITOL-5-PHOSPHATE 2-DEHYDROGENASE

This oxidoreductase [EC 1.1.1.137] catalyzes the reaction of D-ribitol 5-phosphate with $NAD(P)^+$ to produce D-ribulose 5-phosphate and $NAD(P)H$.[1,2]

[1]L. Glaser (1963) *BBA* **67**, 252.

[2]S. Z. Hausman & J. London (1987) *J. Bacteriol.* **169**, 1651.

Selected entries from *Methods in Enzymology* [**vol**, page(s)]:
General discussion: 8, 240
Assay: 8, 240

RIBOFLAVINASE

This enzyme [EC 3.5.99.1] catalyzes the hydrolysis of riboflavin to produce ribitol and lumichrome.[1,2]

[1]T. Yanagita & J. W. Foster (1956) *JBC* **221**, 593.

[2]C. S. Yang & D. B. McCormick (1967) *BBA* **132**, 511.

Selected entries from *Methods in Enzymology* [**vol**, page(s)]:
General discussion: 18B, 571
Assay: 18B, 572

RIBOFLAVIN KINASE

This phosphotransferase [EC 2.7.1.26], also known as flavokinase, catalyzes the reaction of ATP with riboflavin to produce ADP and FMN (flavin mononucleotide).[1–3] Rat liver riboflavin kinase has an ordered Bi Bi kinetic mechanism.[1] The *Corynebacterium ammoniagenes* enzyme is bifunctional, also exhibiting an FMN adenylyltransferase activity. Interestingly, the FMN product of the riboflavin kinase activity has to come off the enzyme surface and then rebind prior to the adenylyltransferase activity.[3]

[1]Y. Yamada, A. H. Merrill, Jr., & D. B. McCormick (1990) *ABB* **278**, 125.

[2]E. P. Anderson (1973) *The Enzymes*, 3rd ed., **9**, 49.

[3]I. Efimov, V. Kuusk, X. Zhang & W. S. McIntire (1998) *Biochem.* **37**, 9716.

Selected entries from *Methods in Enzymology* [**vol**, page(s)]:
General discussion: 2, 640; 18B, 544, 553; 66, 323
Assay: 2, 640; 18B, 544; 66, 323; coupled enzyme assay, 63, 36; radiosubstrate HPLC assay, 122, 238
Other molecular properties: FAD synthesis and, 2, 675; properties, 2, 644; 18B, 546; 66, 326, 327; purification (beer yeast, 2, 642; *Megasphaera elsdenii*, from, 66, 325, 326; modified cellulose, on, 22, 347; rat liver, from [classic fractionation, 18B, 545, 546; flavin-cellulose derivatives, 18B, 550])

RIBOFLAVIN PHOSPHOTRANSFERASE

This phosphotransferase [EC 2.7.1.42] catalyzes the reaction of D-glucose 1-phosphate with riboflavin to produce D-glucose and FMN.[1]

[1]H. Katagiri, H. Yamada & K. Imai (1959) *J. Biochem.* **46**, 1119.

RIBOFLAVIN SYNTHASE

This flavin-dependent enzyme [EC 2.5.1.9] catalyzes the conversion of two molecules of 6,7-dimethyl-8-(1-D-ribityl)lumazine to riboflavin and 4-(1-D-ribitylamino)-5-amino-2,6-dihydroxypyrimidine.[1,2] *See GTP Cyclohydrolase II*

[1] G. M. Brown & J. M. Williamson (1982) *Adv. Biochem.* **53**, 345.
[2] G. W. E. Plaut, C. M. Smith & W. L. Alworth (1974) *ARB* **43**, 899.

Selected entries from *Methods in Enzymology* [vol, page(s)]:
General discussion: 18B, 515, 539; 66, 307; 122, 192; 280, 389
Assay: 18B, 521, 539, 540; heavy, from *Bacillus subtilis*. 122, 193
Other molecular properties: chromatography, 18B, 524; purification, 122, 194; subunit isolation, 122, 196; properties from (spinach, 18B, 542, 543; yeast, 18B, 533); purification from (*Eremothecium ashbyii*, 66, 310; spinach, 18B, 540; yeast, 18B, 527); riboflavin synthesis, 66, 314; transition-state and multisubstrate analogues, 249, 305

RIBOKINASE

This phosphotransferase [EC 2.7.1.15], also known as pentokinase, catalyzes the reaction of ATP with D-ribose to produce ADP and D-ribose 5-phosphate.[1-3] Other substrates include 2-deoxy-D-ribose and 2-deoxy-D-ribitol. Conformational changes accompany the binding of the sugar substrate.[4]

[1] B. W. Agranoff & R. O. Brady (1956) *JBC* **219**, 221.
[2] A. Ginsburg (1959) *JBC* **234**, 481.
[3] S. D. Schimmel, P. Hoffee & B. L. Horecker (1974) *ABB* **164**, 560.
[4] J. A. Sigrell, A. D. Cameron & S. L. Mowbray (1999) *J. Mol. Biol.* **290**, 1009.

Selected entries from *Methods in Enzymology* [vol, page(s)]:
General discussion: 1, 357; 5, 302
Assay: 1, 357; 5, 302
Other molecular properties: deoxyribokinase and, 5, 305, 309; interference, in phosphoribosylpyrophosphate synthetase assay, 51, 3; properties, 1, 360; 5, 305; purification from (*Aerobacter aerogenes*, 3, 189; calf liver, 5, 303; *Lactobacillus plantarum*, 5, 305; yeast, 1, 358); ribose 5-phosphate preparation and, 3, 189, 190

RIBONUCLEASE

This class of RNA phosphodiester bond-cleaving hydrolases includes both exoribonucleases, which catalyze the removal of mononucleotides from either the 5′ or 3′ end of the polymer, as well as endoribonucleases, which catalyze hydrolysis with the RNA chain. *See Ribonuclease, Pancreatic; also specific enzyme*

Selected entries from *Methods in Enzymology* [vol, page(s)]:
Assay: filter paper disk assay for, 12B, 171; liver nuclei, 6, 250; stopped-time coupled enzyme assay, 63, 36

RIBONUCLEASE III

This enzyme [EC 3.1.26.3], also known as ribonuclease O, ribonuclease D, and RNase III, catalyzes the endonucleolytic cleavage of RNA to 5′-phosphomonoester, recognizing double-stranded RNA followed by single-stranded scissions.[1-3] The enzyme acts on multimeric tRNA precursors at the spacer region; hence, it participates in the processing of precursor tRNA as well as of rRNA, hnRNA, and early T7-mRNA. It will also catalyzes the hydrolysis of double-stranded DNA.

[1] J. J. Dunn (1982) *The Enzymes*, 3rd ed., 15, 485.
[2] R. Kole & S. Altman (1982) *The Enzymes*, 3rd ed., 15, 469.
[3] R. A. Srivastava & N. Srivastava (1996) *Indian J. Biochem. Biophys.* 33, 253.

Selected entries from *Methods in Enzymology* [vol, page(s)]:
General discussion: 59, 824
Assay: 59, 825, 826; 79, 143; 181, 422
Other molecular properties: cleavage of 30 S pre-rRNA, in, 59, 832, 833, 837; cloned processing site, analysis, 181, 196; digestion of affinity-labeled RNA, in, 59, 808; properties, 181, 425; purification, 59, 826 (*Escherichia coli*, from, 181, 192, 424); storage, 59, 829, 833; test substrates, 181, 190

RIBONUCLEASE IV

This enzyme [EC 3.1.26.6], also known as endoribonuclease IV and poly(A)-specific ribonuclease, catalyzes the endonucleolytic cleavage of poly(A) to fragments terminated by 3′-hydroxyl and 5′-phosphate groups.[1-3] The products are oligonucleotides with an average chain length of ten bases. The enzyme will also act on the poly(A) segments on mRNA, leaving a stretch of five adenine nucleotides on the mRNA.

[1] W. E. G. Müller (1976) *EJB* 70, 241.
[2] W. E. G. Müller, G. Seibert, R. Steffen & R. K. Zahn (1976) *EJB* 70, 249.
[3] V. Shen & D. Schlessinger (1982) *The Enzymes*, 3rd ed., 15, 501.

RIBONUCLEASE V

This enzyme [EC 3.1.27.8], also known as endoribonuclease V and originally isolated from calf thymus, catalyzes the hydrolysis of poly(A), forming oligoribonucleotides and ultimately generating 3′-AMP.[1] This enzyme also acts on certain AU-rich sequences[2] and on poly(U), the latter reaction yielding 3′-UMP.

[1] H. C. Schroeder, K. Dose, R. K. Zahn & W. E. G. Mueller (1980) *JBC* 255, 5108.
[2] C. Jochum, R. Voth, S. Rossol, K. H. Meyer zum Buschenfelde, G. Hess, H. Will, H. C. Schroder, R. Steffen & W. E. Muller (1990) *J. Virol.* 64, 1956.

RIBONUCLEASE IX

This endoribonuclease [EC 3.1.26.10] catalyzes the endonucleolytic cleavage of poly(U) or poly(C) to fragments terminated by 3′-hydroxyl and 5′-phosphate groups. Poly(C) has the higher affinity of the two polynucleotides. The enzyme will not act on poly(A) or poly(G).[1]

[1] D. C. Sideris & E. G. Fragoulis (1987) *EJB* 164, 309.

RIBONUCLEASE ALPHA

This enzyme [EC 3.1.26.2], also known as ribonuclease α, catalyzes the endonucleolytic cleavage of RNA to 5′-phosphomonoesters and is specific for 2′-O-methylated RNA.[1]

[1]J. Norton & J. S. Roth (1967) *JBC* **242**, 2029.

RIBONUCLEASE, *ENTEROBACTER*

This ribonuclease [EC 3.1.27.6] catalyzes the endonucleolytic cleavage of RNA to produce nucleoside 3′-phosphates and 3′-phosphooligonucleotides with 2′,3′-cyclic phosphate intermediates.[1,2] The enzyme has a preference for cleavage at CpA. Homopolymers of A, U, or G are not hydrolyzed. **See also** *Ribonuclease*

[1]T. P. Karpetsky, K. K. Shriver & C. C. Levy (1980) *JBC* **255**, 2713.
[2]J. J. Frank, I. A. Hawk & C. C. Levy (1976) *BBA* **432**, 369.

RIBONUCLEASE F

This enzyme [EC 3.1.27.7], also called ribonuclease L, catalyzes the endonucleolytic cleavage of an RNA precursor, leaving 5′-hydroxyl and 3′-phosphate groups.[1,2,] The preferential site of cleavage is between a cytosine and an adenine. (2′-5′)-Oligoadenylate and (3′-5′)-oligoadenylate activate the enzyme.

[1]M. Gurevitz, N. Watson & D. Apirion (1982) *EJB* **124**, 553.
[2]N. H. Williams (1998) *CBC* **1**, 543.

Selected entries from *Methods in Enzymology* [**vol**, page(s)]:
General discussion: **79**, 139, 147, 149
Assay: **119**, 491
Other molecular properties: activation by (2′-5′)-oligoadenylate, **79**, 138, 147, 149, 150, 185, 250; binding (2′-5′)-oligoadenylate, **79**, 138; crosslinking of (2′-5′)(A)$_3$[^{32}P]pCp to, **119**, 497; discrimination between RNAs, **79**, 140, 141; latent ribonuclease, **79**, 138; localization of activity, **79**, 141; molecular weight, **79**, 140; presence in absence of interferon, **79**, 137, 147; purification from murine EAT cells, **119**, 493

RIBONUCLEASE H

This Mg^{2+}- or Mn^{2+}-dependent enzyme [EC 3.1.26.4], also known as endoribonuclease H and abbreviated RNase H, catalyzes the endonucleolytic cleavage of RNA to 5′-phosphomonoesters. The oligonucleotides produced are typically between two and nine bases in length.[1,2] **See also** *Ribonuclease*

[1]R. C. Haberkern & G. L. Cantoni (1973) *Biochem.* **12**, 2389.
[2]N. H. Williams (1998) *CBC* **1**, 543.

Selected entries from *Methods in Enzymology* [**vol**, page(s)]:
Assay: **59**, 826; acid precipitation, **275**, 289; activity gel assay, **275**, 336, 338, 343; degradation product detection, **275**, 289; filter assay, **275**, 289; gel electrophoresis, **275**, 290; polymerase coupling assays (presteady-state analysis, **275**, 295; trapping assays, **275**, 293); radioactive assay, **262**, 583; substrate preparation, **275**, 289

RIBONUCLEASE M5

This hydrolase [EC 3.1.26.8], also known as 5S ribosomal maturation nuclease, catalyzes the endonucleolytic cleavage of RNA, removing 21 and 42 nucleotides, respectively, from the 5′- and 3′-termini of a 5S-rRNA precursor.[1,2]

[1]M. L. Sogin, B. Pace & N. R. Pace (1977) *JBC* **252**, 1350.
[2]B. Pace, D. A. Stahl & N. R. Pace (1984) *JBC* **259**, 11454.

Selected entries from *Methods in Enzymology* [**vol**, page(s)]:
General discussion: **181**, 366
Assay: **181**, 371

RIBONUCLEASE P

This divalent cation-dependent (optimally Mg^{2+} or Mn^{2+}) ribozyme [EC 3.1.26.5] catalyzes the endonucleolytic cleavage of RNA, removing 5′-extra nucleotide from tRNA precursor.[1–5] An essential step in tRNA processing, ribonuclease P generates the 5′-termini of mature tRNA molecules. A similar activity is observed with ribonuclease P3 (RNase P3). The catalytic activity of ribonuclease P resides in RNA, and the protein component exhibits no catalytic activity.

Ribonuclease P was the first such nonprotein-catalyzed reaction and is an example of a catalytic RNA, also known as a ribozyme. Ribonuclease P is unlike other ribozymes in that the catalyzed reaction is a direct hydrolysis reaction, and not a transphosphorylation reaction. **See also** *Ribozymes*

[1]J. A. Grasby (1998) *CBC* **1**, 563.
[2]S. Altman (1989) *AE* **62**, 1.
[3]T. R. Cech & B. L. Bass (1986) *ARB* **55**, 599.
[4]S. Altman, M. F. Baer, M. Bartkiewicz, H. Gold, C. Guerrier-Takada, L. A. Kirsebom, N. Lumelsky & K. Peck (1989) *Gene* **82**, 63.
[5]R. Kole & S. Altman (1982) *The Enzymes*, 3rd ed., **15**, 469.

Selected entries from *Methods in Enzymology* [**vol**, page(s)]:
General discussion: **264**, 86
Assay: cleavage activity, assays, **181**, 580
Other molecular properties: *Bacillus subtilis* (pre-tRNA binary complex formation, assay, **249**, 28, 30; kinetic mechanism, determination from transient kinetics, **249**, 28; RNA component, pre-tRNA hydrolysis catalyzed by, transient kinetics, **249**, 28); catalytic mechanism, transient kinetics experiments, **249**, 28; reaction catalyzed, **249**, 27; RNA component, pre-tRNA hydrolysis catalyzed by, time course of, **249**, 28; RNA, design by phylogenetic comparison, **203**, 500; *Saccharomyces cerevisiae* mitochondria (activity assay, **264**, 91, 93; components, **264**, 86; gel electrophoresis, **264**, 95; protein-to-RNA ratio, **264**, 89; purification **264**, 93)

RIBONUCLEASE P4

This phosphodiesterase [EC 3.1.26.7], first identified in *Escherichia coli*, catalyzes the endonucleolytic cleavage of RNA, thereby removing 3′-extranucleotides from tRNA precursors.[1]

[1]T. Sekiya, R. Contreras, T. Takeya & H. G. Khorana (1979) *JBC* **254**, 5802.

RIBONUCLEASE, PANCREATIC

This phosphodiesterase [EC 3.1.27.5], often abbreviated RNase and known as ribonuclease I or ribonuclease A, catalyzes the endonucleolytic cleavage of RNA to produce nucleoside 3′-phosphates and 3′-phosphooligonucleotides ending in Cp or Up with 2′,3′-cyclic phosphate intermediates. Hydrolysis occurs primarily between a pyrimidine nucleoside 3-phosphate residue and the neighboring ribose group of the adjoining residue.[1–8]

RNA hydrolysis by pancreatic ribonuclease occurs by general acid/base catalysis. During the enzyme-catalyzed reaction, nucleoside-2′,3′-cyclic phosphodiesters accumulate, indicating a two-step reaction: first, facilitated attack of the 2′-hydroxyl of RNA forms the cyclic intermediate (with stereochemical inversion at phosphorus); second, cyclic phosphodiester hydrolysis yields the 3′-monoester (again with inversion at phosphorus).

[1]N. H. Williams (1998) *CBC* **1**, 543.
[2]S. A. Benner & R. K. Allemann (1989) *Trends Biochem Sci.* **14**, 396.
[3]P. Blackburn & S. Moore (1982) *The Enzymes*, 3rd ed., **15**, 317.
[4]V. Shen & D. Schlessinger (1982)*The Enzymes*, 3rd ed., **15**, 501.
[5]F. M. Richards & H. W. Wyckoff (1971) *The Enzymes*, 3rd ed., **4**, 647.
[6]C. B. Anfinsen & F. H. White (1961) *The Enzymes*, 2nd. ed., **5**, 95.
[7]H. G. Khorana (1961) *The Enzymes*, 2nd ed., **5**, 79.
[8]R. Breslow (1991) *Acc. Chem. Res.* **24**, 317.

Selected entries from *Methods in Enzymology* [vol, page(s)]:
General discussion: **1**, 119; **2**, 427
Assay: **2**, 427; coupled enzyme assay, **63**, 36; liver nuclei, **6**, 250
Other molecular properties: activators, **2**, 433; affinity chromatography, **34**, 5, 119, 120; affinity labeling, **34**, 64, 185; **46**, 362; **64**, 185; **87**, 470, 471, 473, 485, 491, 494; anion binding site (electrostatic effects on proton uptake and release, **130**, 433; pH-dependent development, **130**, 431); bovine pancreatic, **2**, 427 (activity, modification, **47**, 160; disulfide cleavage, **47**, 112; internally labeled RNA, **180**, 134; dinitrophenylation (kinetics, **11**, 553; preparation, **11**, 555); folding intermediates, methanol stabilization, **240**, 625; isoelectric point, determination by cross-partitioning, **228**, 228, 232; kinetics, **2**, 433; mechanism, **87**, 197, 211, 223, 233, 494; stereochemistry, **87**, 212; thermal transitions, analysis with immobilized carboxypeptidase, **135**, 597; transesterification, **87**, 199, 212; transition-state and multisubstrate analogues, **249**, 306

RIBONUCLEASE (POLY(U)-SPECIFIC)

This riboendonuclease [EC 3.1.26.9] catalyzes the endonucleolytic cleavage of poly(U) to fragments terminated by 3′-hydroxyl and 5′-phosphate groups, producing oligonucleotides with a chain length of six to twelve nucleotides.[1]

[1]M. Bachmann, F. Trautmann, R. Messer, R. K. Zahn, K. H. Meyer zum Büschenfelde & W. E. Müller (1983) *EJB* **136**, 447.

RIBONUCLEASE T₁

This heat/acid-stable enzyme [EC 3.1.27.3], also known as guanyloribonuclease and *Aspergillus oryzae* ribonuclease, catalyzes the two-stage endonucleolytic cleavage of RNA to nucleoside 3′-phosphates and 3′-phosphooligonucleotides. The first step involves the cleavage of internucleotide bonds between 3′-guanyl groups and the 5′-hydroxyl groups of the adjacent nucleotides. This results in the intermediary formation of guanosine 2′,3′-cyclic phosphate. The second step is the hydrolysis of the cyclic phosphate to produce 3′-guanylate. Thus, the final products are 3′-phosphomononucleotides and 3′-phosphooligonucleotides ending in Gp. The enzyme from many organisms will also catalyze the hydrolysis of guanosine 2′,3′-cyclic phosphate, producing guanosine-3′-phosphate. The reaction mechanism is very similar to pancreatic ribonuclease. His92 protonates the leaving group.[1–3] *See also Ribonuclease, Pancreatic; Ribonuclease T₂; Ribonuclease*

[1]T. Uchida & F. Egami (1971) *The Enzymes*, 3rd ed., **4**, 205.
[2]K. Takahashi & S. Moore (1982) *The Enzymes*, 3rd ed., **15**, 435.
[3]J. Sevcik, R. G. Sanishvili, A. G. Pavlovsky & K. M. Polyakov (1990) *Trends Biochem. Sci.* **15**, 158.

Selected entries from *Methods in Enzymology* [vol, page(s)]:
General discussion: **46**, 177, 178, 671
Assay: **65**, 681

RIBONUCLEASE T₂

This enzyme [EC 3.1.27.1], also known as ribonuclease II, catalyzes the two-stage endonucleolytic cleavage of RNA to nucleoside 3′-phosphates and 3′-phosphooligonucleotides.[1–3] *See also Ribonuclease; Ribonuclease T₁*

[1]T. Uchida & F. Egami (1971) *The Enzymes*, 3rd ed., **4**, 205.
[2]V. Shen & D. Schlessinger (1982) *The Enzymes*, 3rd ed., **15**, 501.
[3]P. M. Kaiser, L. Bonacker, H. Witzel & A. Holy (1975) *Hoppe-Seyler's Z. Physiol. Chem.* **356**, 143.

Selected entries from *Methods in Enzymology* [vol, page(s)]:
General discussion: **12B**, 224; **46**, 175; **65**, 672; **181**, 188

RIBONUCLEASE U₂

This enzyme [EC 3.1.27.4], first isolated from the smut fungus *Ustilago sphaerogena*, catalyzes the two-stage endonucleolytic cleavage of RNA to nucleoside 3'-phosphates and 3'-phosphooligonucleotides ending in Ap or Gp with 2',3'-cyclic phosphate intermediates.[1-4] **See also** *Ribonuclease*

[1]T. Uchida & F. Egami (1971) *The Enzymes*, 3rd ed., **4**, 205.
[2]T. Yasuda & Y. Inoue (1982) *Biochem.* **21**, 364.
[3]T. Uchida & Y. Shibata (1981) *J. Biochem.* **90**, 463.
[4]S. Minato & A. Hirai (1979) *J. Biochem.* **85**, 327.

RIBONUCLEASE, YEAST

This enzyme [EC 3.1.14.1] catalyzes the exonucleolytic cleavage of RNA to produce nucleoside 3'-phosphates.[1] A similar enzyme is ribonuclease U_4 of *Ustilago sphaerogena*.[2] **See also** *Ribonuclease*

[1]J. K. Shetty, R. C. Weaver & J. E. Kinsella (1980) *Biochem. J.* **189**, 363.
[2]A. Blank & C. A. Dekker (1972) *Biochemistry* **11**, 3962.

RIBONUCLEOSIDE DIPHOSPHATE REDUCTASE

This iron- and ATP-dependent oxidoreductase [EC 1.17.4.1], also known as ribonucleotide reductase, catalyzes the reaction of a ribonucleoside diphosphate (ADP, CDP, UDP, and GDP) with reduced thioredoxin to produce a 2'-deoxyribonucleoside diphosphate, oxidized thioredoxin, and water.[1-8] There are at least three different classes of these reductases.[6] These reductases differ in their substrate and cofactor requirements. Although their detailed mechanisms of action are different, each reductase class promotes substrate reactivity by forming a radical intermediate. The protein radical abstracts the hydrogen atom from the 3'-C position on the substrate. The 2'-hydroxyl group undergoes protonation, water is released, and a cation radical intermediate is generated. This intermediate is reduced by a dithiol to produce the 2'-deoxyribonucleotide 3'-radical. This intemediate is subsequently reduced and the protein radical is regenerated.

The *Escherichia coli* reductase contains a binuclear iron center which functions in the generation and maintenance of the radical. Adenosylcobalamin-dependent reductases use the homolysis of the carbon-cobalt bond to generate cob(II)alamin and a 5'-deoxyadenosyl radical, which proceeds to generate the protein radical. Anaerobically grown *Escherichia coli* contains an [Fe4:S4] cluster and requires adenosylmethionine in the production of the protein radical. **See also** *Ribonucleoside Triphosphate Reductase*

[1]B. T. Golding & W. Buckel (1998) *CBC* **3**, 239.
[2]B. G. Fox (1998) *CBC* **3**, 261.
[3]J. E. Penner-Hahn (1998) *CBC* **3**, 439.
[4]P. Reichard (1993) *Science* **260**, 1773.
[5]M. Fontecave, P. Nordlund, H. Eklund & P. Reichard (1992) *AE* **65**, 147.
[6]E. J. Brush & J. W. Kozarich (1992) *The Enzymes*, 3rd ed., **20**, 317.
[7]J. Stubbe (1990) *AE* **63**, 349.
[8]M. A. Ator & P. R. Ortez de Montellano (1990) *The Enzymes*, 3rd ed., **19**, 213.

Selected entries from *Methods in Enzymology* [**vol**, page(s)]:
General discussion: 12A, 155; **34**, 254; **46**, 321; **51**, 227, 237; **258**, 278
Assay: 12A, 157; **51**, 227, 238, 239

Other molecular properties: active site, **51**, 237; allosteric regulation, **51**, 235, 236; assay of glutaredoxin, **113**, 526; **252**, 287; B1 catalytic properties, **51**, 235; effector binding, **51**, 236; *Escherichia coli*, **51**, 227; (amino acid sequence, **243**, 219; assay for glutaredoxins, **113**, 526; diferric-tyrosyl radical cofactor assembly [glycyl radical evidence, **258**, 361; mechanism, **258**, 302; rapid freeze-quench electron paramagnetic resonance, **258**, 286, 303; rapid freeze-quench Mössbauer spectroscopy, **258**, 294, 301, 303; reducing equivalents, **258**, 278; stopped-flow spectroscopy, **258**, 280, 293; structure, **258**, 278]; diiron site, **243**, 223; spectroscopic studies, **243**, 223); kinetic properties, enzyme concentration, **51**, 243, 244; redox properties, **51**, 236; specificity sites, **51**, 236; storage, **51**, 241, 242; substrate binding, **51**, 236; substrate specificity, **51**, 235, 236, 245

RIBONUCLEOSIDE TRIPHOSPHATE REDUCTASE

This ATP-dependent oxidoreductase [EC 1.17.4.2], also known as ribonucleotide reductase, catalyzes the reaction of a ribonucleoside triphosphate (CTP, GTP, ATP, and UTP) with reduced thioredoxin to produce a 2'-deoxyribonucleoside triphosphate, oxidized thioredoxin, and water.[1-3] The enzyme from *Lactobacillus leichmannii* also requires dNTPs as allosteric effectors; hence, the 2'-deoxyribonucleoside triphosphates are both products and effectors of the same system. The reductase utilizes adenosylcobalamin as a cofactor and catalyzes the exchange of tritium from [5'-^{3}H]adenosylcobalamin with solvent in addition to producing 2'-deoxyribonucleoside triphosphates. The role of the adenosylcobalamin is to generate a thiyl radical which, in turn, abstracts the 3'-hydrogen atom of the substrate.[3] The generation of a radical anion results in the elimination of the 2' hydroxyl group. **See also** *Ribonucleoside Diphosphate Reductase*

[1]J. Stubbe (1990) *AE* **63**, 349.
[2]A. Holmgren (1981) *Curr. Top. Cell. Regul.* **19**, 47.
[3]S. S. Licht, S. Booker & J. Stubbe (1999) *Biochemistry* **38**, 1221.

Selected entries from *Methods in Enzymology* [**vol**, page(s)]:
General discussion: 34, 254; **51**, 246
Assay: 51, 247
Other molecular properties: absorption coefficient, **51**, 250; activators, **51**, 258; adenosylcobalamin requirement, **51**, 246, 247, 258; kinetic properties, **51**, 258; *Lactobacillus leichmannii*, from, **51**, 246; ligand binding, **51**, 259; mammalian, **34**, 258; molecular weight, **51**, 257; purification, **34**, 253; **51**, 250; purity, **51**, 257; reaction mechanism, **51**, 259; reducing substrates, **51**, 258; regulation, **51**, 257, 258; stability, **51**, 257; substrate specificity, **51**, 257, 258

RIBOSE 1-DEHYDROGENASE (NADP⁺)

This oxidoreductase [EC 1.1.1.115], also known as NADP⁺:pentose dehydrogenase, catalyzes the reaction of D-ribose with NADP⁺ and water to produce D-ribonate and NADPH.[1] Other substrates, although weaker, include D-xylose, L-arabinose, D-arabinose, digitoxose, and 2-deoxy-D-ribose.

[1]H. W. Schiwara, W. Domschke & G. F. Domagk (1968) *Hoppe-Seyler's Z. Physiol. Chem.* **349**, 1575.

RIBOSE ISOMERASE

This isomerase [EC 5.3.1.20] catalyzes the interconversion of D-ribose and D-ribulose. The enzyme will also act on D-lyxose and L-rhamnose.[1,2]

[1]K. Izumori, A. W. Rees & A. D. Elbein (1975) *JBC* **250**, 8085.
[2]K. Izumori, M. Mitchell & A. D. Elbein (1976) *J. Bacteriol.* **126**, 553.

Selected entries from *Methods in Enzymology* [vol, page(s)]:
General discussion: 89, 547
Assay: *Mycobacterium smegmatis*, **89**, 547

RIBOSE-5-PHOSPHATE ADENYLYLTRANSFERASE

This transferase [EC 2.7.7.35], also known as ADP-ribose phosphorylase, catalyzes the reaction of orthophosphate with ADP-D-ribose to produce ADP and D-ribose 5-phosphate.[1,2]

[1]W. R. Evans & A. S. Pietro (1966) *ABB* **113**, 236.
[2]A. I. Stern & M. Avron (1966) *BBA* **118**, 577.

Selected entries from *Methods in Enzymology* [vol, page(s)]:
General discussion: 23, 566
Assay: 23, 566
Other molecular properties: *Euglena gracilis*, from, purification, **23**, 566

RIBOSE-5-PHOSPHATE ISOMERASE

This isomerase [EC 5.3.1.6], also known as ribose 5-phosphate epimerase, phosphopentose isomerase, and D-ribose-5-phosphate ketol-isomerase, catalyzes the interconversion of D-ribose 5-phosphate and D-ribulose 5-phosphate.[1–3] (Although the Enzyme Commission recommended the name ribose-5-phosphate epimerase, the substrate and product are not epimers.) Other substrates include D-ribose 5-diphosphate and D-ribose 5-triphosphate.

[1]M. E. Kiely, A. L. Stuart & T. Wood (1973) *BBA* **293**, 534.
[2]M. K. Essenberg & R. A. Cooper (1975) *EJB* **55**, 323.
[3]E. A. Noltmann (1972) *The Enzymes*, 3rd ed., **6**, 271.

Selected entries from *Methods in Enzymology* [vol, page(s)]:
General discussion: 1, 363; 41, 424, 427; 89, 571
Assay: 1, 363; 41, 424, 427; 89, 572

Other molecular properties: bacterial (properties, 89, 576; purification, 89, 574; regulation and inhibition, 89, 577); *Candida utilis*, from, 41, 427 (inhibitors, 41, 429; properties, 41, 429; purification, 41, 428, 429); phosphoribulokinase and, 5, 258, 261; properties, 1, 364; 41, 426, 429; purification, 1, 364; 5, 248, 258; skeletal muscle, from, 41, 424 (chromatography, 41, 425; inhibitors, 41, 426; properties, 41, 426; purification, 41, 425, 426); source, 8, 241; transition-state and multisubstrate analogues, 249, 308; xylulokinase and, 5, 211; xylulose 5-phosphate preparation, 3, 193

RIBOSE-5-PHOSPHATE PYROPHOSPHOKINASE

This enzyme [EC 2.7.6.1] (also known as ribose-phosphate diphosphokinase, phosphoribosyl pyrophosphate synthetase [PRPP synthetase], and phosphoribosyl diphosphate synthetase) catalyzes the reaction of ATP (or, dATP) with D-ribose 5-phosphate to produce AMP (or, dAMP) and 5-phospho-α-D-ribose 1-diphosphate (or, 5-phospho-α-D-ribose 1-pyrophosphate, PRPP).[1–6] The enzyme exhibits an ordered Bi Bi kinetic mechanism.[2]

[1]R. L. Switzer (1974) *The Enzymes*, 3rd ed., **10**, 607.
[2]I. H. Fox & W. N. Kelley (1972) *JBC* **247**, 2126.
[3]J. J. Villafranca & F. M. Raushel (1980) *Ann. Rev. Biophys. Bioengin.* **9**, 363.
[4]M. A. Becker, K. O. Raivio & J. E. Seegmiller (1979) *AE* **49**, 281.
[5]A. S. Mildvan (1979) *AE* **49**, 103.
[6]W. N. Keley & J. B. Wyngaarden (1974) *AE* **41**, 1.

Selected entries from *Methods in Enzymology* [vol, page(s)]:
General discussion: 2, 501, 504; 6, 158; 51, 3, 12
Assay: 6, 158; 51, 3, 12
Other molecular properties: activators, 51, 11, 16; activity, 51, 3, 12; anthranilic deoxyribulotide formation and, 5, 797; imidazole glycerolphosphate synthesis and, 6, 581; inhibitors, 51, 11, 17; isotope exchange, 64, 26; kinetics, 51, 10, 11, 17; preparation, 6, 474; 17B, 9; properties, 6, 161; 51, 9, 16, 17; states of aggregation, 51, 11; storage, 51, 9, 10, 16; substrate specificity, 51, 10, 16

RIBOSOMAL-PROTEIN-ALANINE N-ACETYLTRANSFERASE

This acetyltransferase [EC 2.3.1.128] catalyzes the reaction of acetyl-CoA with the N-terminal alanyl residue of specific ribosomal proteins (for example, *Escherichia coli* ribosomal proteins S18 and S5) to produce Coenzyme A and an *N*-acetyl-L-alanyl residue in the protein.[1] **See also** *Peptide α-N-Acetyltransferase*

[1]A. Yoshikawa, S. Isono, A. Sheback & K. Isono (1987) *Mol. Gen. Genet.* **209**, 481.

RIBOSYLHOMOCYSTEINASE

This enzyme [EC 3.2.1.148; formerly EC 3.3.1.3] catalyzes the hydrolysis of *S*-ribosyl-L-homocysteine to produce D-ribose and L-homocysteine.[1–3]

[1]J. A. Duerre & C. H. Miller (1966) *J. Bacteriol.* **91**, 1210.
[2]C. H. Miller & J. A. Duerre (1968) *JBC* **243**, 92.
[3]R. D. Walker & J. A. Duerre (1975) *Can. J. Biochem.* **53**, 312.

RIBOSYLNICOTINAMIDE KINASE

This phosphotransferase [EC 2.7.1.22] catalyzes the reaction of ATP with N-ribosylnicotinamide to produce ADP and nicotinamide ribonucleotide (NMN).[1]

[1]J. W. Rowen & A. Kornberg (1951) JBC 193, 497.

Selected entries from **Methods in Enzymology** [vol, page(s)]:
General discussion: 18B, 141
Assay: 18B, 141

RIBOSYLPYRIMIDINE NUCLEOSIDASE

This enzyme [EC 3.2.2.8] catalyzes the hydrolysis of an N-D-ribosylpyrimidine to produce a pyrimidine and D-ribose.[1-3] Substrates include cytidine (producing cytosine and D-ribose) and uridine (producing uracil and D-ribose). Purine D-ribonucleosides are also hydrolyzed, albeit more slowly.

[1]M. Terada, M. Tatibana & O. Hayaishi (1967) JBC 242, 5578.
[2]G. W. Koszalka & T. A. Krenitsky (1979) JBC 254, 8185.
[3]T. A. Krenitsky, G. W. Koszalka, J. V. Tuttle, D. L. Adamczyk, G. B. Elion & J. J. Marr (1980) Adv. Exp. Med. Biol. 122, 51.

Selected entries from **Methods in Enzymology** [vol, page(s)]:
General discussion: 2, 456
Assay: 2, 456

D-RIBULOKINASE

This phosphotransferase [EC 2.7.1.47] catalyzes the reaction of ATP with D-ribulose to produce ADP and D-ribulose 5-phosphate.[1,2] The enzyme displays a steady-state random Bi Bi kinetic mechanism.[2]

[1]H. J. Fromm (1959) JBC 234, 3097.
[2]M. M. Stayton & H. J. Fromm (1979) JBC 254, 3765.

Selected entries from **Methods in Enzymology** [vol, page(s)]:
General discussion: 9, 446
Assay: 9, 446

L-RIBULOKINASE

This phosphotransferase [EC 2.7.1.16] catalyzes the reaction of ATP with L-ribulose to produce ADP and L-ribulose 5-phosphate.[1-3] Other substrates include D-ribulose, ribitol, and L-arabinitol.

[1]F. J. Simpson & W. A. Wood (1956) JACS 78, 5452.
[2]D. P. Burman & B. L. Horecker (1958) JBC 231, 1039.
[3]N. Lee & I. Bendet (1967) JBC 242, 2043.

Selected entries from **Methods in Enzymology** [vol, page(s)]:
General discussion: 5, 298; 9, 449
Assay: 5, 298; 9, 449
Other molecular properties: Aerobacter aerogenes (properties, 5, 301; purification, 5, 299); L-arabinose isomerase (assay for, 9, 597; preparation free of, 9, 600); Escherichia coli, from (crystallization, 9, 452; gene, 9, 15; isolation of mutants, 9, 15; kinetic constants, 9, 453; molecular weight, 9, 453; properties, 9, 453; purification, 9, 50, 450); glutathione stimulation, 5, 294, 301; induction, 9, 49; Lactobacillus,

plantarum, purification, 5, 254; L-ribulose-5-phosphate preparation, 5, 253, 255; 9, 48, 50; L-ribulose 5-phosphate 4-epimeraseless mutant of Escherichia coli, from, (growth of, 9, 49; induction, 9, 49; preparation, 9, 50; L-ribulose 5-phosphate preparation, 9, 48); sources, 9, 454

[RIBULOSE-BISPHOSPHATE-CARBOXYLASE]-LYSINE N-METHYLTRANSFERASE

This methyltransferase [EC 2.1.1.127], also known as rubisco methyltransferase and ribulose-bisphosphate-carboxylase/oxygenase N-methyltransferase, catalyzes the reaction of S-adenosyl-L-methionine with a lysyl residue in ribulose-1,5-bisphosphate carboxylase to produce S-adenosyl-L-homocysteine and an N^6-methyl-L-lysyl residue in the carboxylase.[1]

[1]P. Wang, M. Royer & R. L. Houtz (1995) Prot. Expr. Purif. 6, 528.

RIBULOSE 1,5-BISPHOSPHATE CARBOXYLASE/OXYGENASE

Probably the most abundant naturally occurring catalyst, this enzyme [EC 4.1.1.39], also known as ribulose-bisphosphate carboxylase, "rubisco," and ribulose 1,5-diphosphate carboxylase, catalyzes the reaction of D-ribulose 1,5-bisphosphate with carbon dioxide to produce two molecules of 3-phospho-D-glycerate. The enzyme can also utilize dioxygen instead of carbon dioxide, producing 3-phospho-D-glycerate and 2-phosphoglycolate.[1-6] A number of reaction mechanisms have been proposed for this enzyme; for example, the generation of a carbanion in the carboxylation reaction or a carbonate mechanism in which CO_2 attacks a metal-stabilized enediol to form a carbamate which undergoes a rearrangement reaction.

[1]J. V. Schloss & M. S. Hixon (1998) CBC 2, 43.
[2]F. C. Hartman & M. R. Harpel (1993) AE 67, 1.
[3]M. H. O'Leary (1992) The Enzymes, 3rd ed., 20, 235.
[4]T. Akazawa, T. Takabe & H. Kobayashi (1984) Trends Biochem. Sci. 9, 380.
[5]H. M. Miziorko & G. H. Lorimer (1983) ARB 52, 507.
[6]M. I. Siegel, M. Wishnick & M. D. Lane (1972) The Enzymes, 3rd ed., 6, 169.

Selected entries from **Methods in Enzymology** [vol, page(s)]:
General discussion: 5, 266; 23, 570; 31, 768; 42, 457, 461, 472, 481; 46, 23, 142, 152, 388, 389; 69, 335, 336; 90, 522
Assay: 5, 266; 23, 570; 42, 457, 458, 462, 463, 472, 482; 69, 331, 332 (procedure, 69, 333, 334; reagents, 69, 332, 333); 89, 47; 90, 522; 118, 427
Other molecular properties: affinity labeling, 87, 473; inhibitor, 5, 270; 63, 401; intermediate, 87, 84; isotope effects, 64, 95, 96; isotope trapping, 64, 58; labeling lysyl residues, 46, 395; leaves, from, 5, 266; 42, 481 (extraction, 42, 481; purification, 42, 482); oxygen, 69, 573; oxygenase (assay, 42, 484; properties, 42, 487; purification, 42, 486; reactions catalyzed, 69, 326 [reagents, 69, 334, 335]); physicochemical properties, 69, 330, 331 (polarographic analysis, 69, 335); pressure effect on dissociation, 259, 413; purification from spinach leaves, 5, 267; 23, 570; reactions catalyzed, 69, 326 (reagents, 69, 334, 335); removal,

from thylakoids, **69**, 535; *Rhodospirillum rubrum*, from, **42**, 457, 468 (assay, **90**, 522; **118**, 427; chromatography, **42**, 459, 460, 469, 470; expression, analytical techniques, **118**, 425, 427; gene cloning, **118**, 420; inhibitors, **42**, 461; small-angle X-ray scattering, **61**, 226); spinach enzyme, **5**, 266; **42**, 472; *Synechococcus*, large and small subunits, **118**, 413; synthesis in *Chlamydomonas* cultures, **23**, 73; tobacco leaf, purification and crystallization, **90**, 520; tomato leaf, purification, **90**, 521; transition-state and multisubstrate analogues, **249**, 307

D-RIBULOSE-5-PHOSPHATE 3-EPIMERASE

This epimerase [EC 5.1.3.1] (also known as phosphoribulose epimerase, erythrose-4-phosphate epimerase, phosphoketopentoepimerase, pentose-5-phosphate 3-epimerase, and xylulose-5-phosphate 3-epimerase) catalyzes the interconversion of D-ribulose 5-phosphate and D-xylulose 5-phosphate.[1-6]

[1] A. Karmali, A. F. Drake & N. Spencer (1983) *BJ* **211**, 617.
[2] J. Kopp, S. Kopriva, K. H. Suss & G. E. Schulz (1999) *J. Mol. Biol.* **287**, 761.
[3] Y. R. Chen, F. W. Larimer, E. H. Serpersu & F. C. Hartman (1999) *JBC* **274**, 2132.
[4] M. E. Tanner & G. L. Kenyon (1998) *CBC* **2**, 7.
[5] E. Adams (1976) *AE* **44**, 69.
[6] L. Glaser (1972) *The Enzymes*, 3rd ed., **6**, 355.

Selected entries from *Methods in Enzymology* [vol, page(s)]:
General discussion: 5, 247, 280; 9, 605
Assay: 5, 248, 250, 280; 9, 605; 41, 37, 63
Other molecular properties: action of, **3**, 193; L-arabinose isomerase free of, **9**, 600; detection, **5**, 247; ribulokinase preparation, not in, **5**, 301; ribulose-5-phosphate, in assay of, **5**, 282; **41**, 37, 40; properties, **5**, 252, 282; **9**, 607; purification, **5**, 249, 251, 264, 281; **9**, 606; D-xylulokinase, presence in preparation of, **5**, 211; xylulose-5-phosphate phosphoketolase and, **5**, 262, 264; xylulose-5-phosphate (assay, **41**, 37, 40; preparation, **3**, 193; **5**, 262; **9**, 42); yeast, **9**, 608

L-RIBULOSE-5-PHOSPHATE 4-EPIMERASE

This epimerase [EC 5.1.3.4], also known as D-xylulose-5-phosphate 4-epimerase, catalyzes the interconversion of L-ribulose 5-phosphate and D-xylulose 5-phosphate.[1-9] Isotope effect investigations indicate the reaction proceeds by first catalyzing an aldol cleavage to the enediolate of dihydroxyacetone (metal-bound) and glycolaldehyde phosphate. This step is followed by rotation of the aldehyde group and condensation to the epimer at the C-4 position.[4-7] The substrate binds to the enzyme at a site containing Mn^{2+} or Zn^{2+}.

[1] D. P. Burma & B. L. Horecker (1958) *JBC* **231**, 1053.
[2] M. J. Wolin, F. J. Simpson & W. A. Wood (1958) *JBC* **232**, 559.
[3] N. Lee, J. W. Patrick & M. Masson (1968) *JBC* **243**, 4700.
[4] A. E. Johnson & M. E. Tanner (1998) *Biochemistry* **37**, 5746.
[5] L. V. Lee, M. V. Vu & W. W. Cleland (2000) *Biochemistry* **39**, 4808.
[6] L. V. Lee, R. R. Poyner, M. V. Vu & W. W. Cleland (2000) *Biochemistry* **39**, 4821.
[7] M. E. Tanner & G. L. Kenyon (1998) *CBC* **2**, 7.
[8] E. Adams (1976) *AE* **44**, 69.
[9] L. Glaser (1972) *The Enzymes*, 3rd ed., **6**, 355.

Selected entries from *Methods in Enzymology* [vol, page(s)]:
General discussion: 5, 253; 41, 412, 419
Assay: 5, 253; 41, 412, 419, 420

RICININE NITRILASE

This enzyme [EC 3.5.5.2] catalyzes the hydrolysis of ricinine to produce 3-carboxy-4-methoxy-*N*-methyl-2-pyridone and ammonia.[1,2]

ricinine   3-carboxy-4-methoxy-*N*-methyl-2-pyridone

[1] W. G. Robinson & R. H. Hook (1964) *JBC* **239**, 4257.
[2] R. H. Hook & W. G. Robinson (1964) *JBC* **239**, 4263.

Selected entries from *Methods in Enzymology* [vol, page(s)]:
General discussion: 17B, 244
Assay: 17B, 245

RIFAMYCIN-B-OXIDASE

This oxidoreductase [EC 1.10.3.6] catalyzes the reaction of rifamycin B with dioxygen to produce rifamycin O and hydrogen peroxide.[1-3] Rifamycin O will the undergo spontaneous hydrolysis to rifamycin S under neutral aqueous conditions.[1] Alternative substrates include benzene-1,4-diol and other phosphoquinols. The enzyme is not identical with catechol oxidase [EC 1.10.3.1], laccase [EC 1.10.3.2], *o*-aminophenol oxidase [EC 1.10.3.4], or 3-hydroxyanthranilate oxidase [EC 1.10.3.5].

[1] M. H. Han, B.-L. Seong, H.-J. Son & T.-I. Mheen (1983) *FEBS Lett.* **151**, 36.
[2] B. L. Seong, H. J. Son, T. I. Mheen, Y. H. Park & M. H. Han (1985) *J. Ferment. Technol.* **63**, 515.
[3] G. M. Lee & C. Y. Choi (1984) *Biotechnol. Lett.* **6**, 143.

RNA-DIRECTED RNA POLYMERASE

This polymerase [EC 2.7.7.48], also known as RNA nucleotidyltransferase (RNA-directed) and ribonucleic acid synthetase, catalyzes the RNA-template-directed extension of the 3'-end of an RNA strand by one nucleotide at a time; thus, the enzyme catalyzes the reaction of *n* molecules of nucleoside triphosphate to produce *n* molecules of pyrophosphate and an RNA polymer with *n* bases.[1] The enzyme can initiate the synthesis of an RNA chain *de novo*, as can DNA-directed RNA polymerase [EC 2.7.7.6]. *See also Qβ RNA Replicase*

[1] T. Blumenthal (1982) *The Enzymes*, 3rd ed., **15**, 267.

Selected entries from *Methods in Enzymology* [vol, page(s)]:
General discussion: **12B**, 540, 572; **27**, 318, 333; **60**, 628
Assay: **12B**, 541, 572; **275**, 257
Other molecular properties: biologically active viral RNA synthesis (assay for infectious RNA, **12B**, 550; preparation of enzyme, **12B**, 542; preparation of virus and RNA, **12B**, 549; properties, **12B**, 551); Coronavirus, **275**, 83; expression systems, **275**, 89; purification of recombinant protein from *Escherichia coli* [cell growth, **275**, 38; GTP-agarose chromatography, **275**, 38, 44; phosphocellulose chromatography, **275**, 38; reagents, **275**, 37]; reconstitution systems for RNA replication (advantages, **275**, 36; primer requirements, **275**, 44, 48; templates, **275**, 48); properties, **12B**, 575; purification, **12B**, 573

RNA LIGASE (ATP)

This ligase [EC 6.5.1.3], also known as polyribonucleotide synthetase (ATP) and ribonucleic ligase, catalyzes the reaction of ATP with a ribonucleotide containing m bases and a ribonucleotide containing n bases to produce AMP, pyrophosphate and a ribonucleotide with $(m + n)$ bases.[1-4] This enzyme also converts linear RNA into a circular form by transferring the 5'-phosphate to the 3'-hydroxyl terminus.

[1] L. Pick, H. Furneaux & J. Hurwitz (1986) *JBC* **261**, 6694.
[2] K. K. Perkins, H. Furneaux & J. Hurwitz (1985) *PNAS* **82**, 684.
[3] P. A. Frey (1992) *The Enzymes*, 3rd ed., **20**, 141.
[4] O. C. Uhlenbeck & R. I. Gumport (1982) *The Enzymes*, 3rd ed., **15**, 31.

Selected entries from *Methods in Enzymology* [vol, page(s)]:
Other molecular properties: AMP complex, assay and characterization, **181**, 497; bacteriophage T4 (assay, **100**, 40; DNA joining reactions, **100**, 44; purification, **100**, 41; RNA 3'-end labeling with, **152**, 103; RNA joining reactions, **100**, 52); 3'-end labeling of RNA, **180**, 196; **275**, 375; *Leishmania tarentolae* RNA editing, assay in, **264**, 114; radioactive (2'-5')-oligoadenylate derivative synthesis, **79**, 217; stereochemistry, **87**, 212; T4 (activity, **65**, 65; RNA labeling, **65**, 65; source and purification, **65**, 66; **202**, 324); T4, radiolabeling of transfer RNA, **260**, 318; wheat germ, **181**, 488

RNA-3'-PHOSPHATE CYCLASE

This enzyme [E.C. 6.5.1.4], also known as RNA 3'-terminal phosphate cyclase and RNA cyclase, catalyzes the Mg^{2+}-dependent reaction of ATP with an RNA 3'-terminal-phosphate to produce AMP, pyrophosphate, and an RNA terminal-2',3'-cyclic-phosphate[1]. Although ATP(γ-S) is a substrate, dATP is not.

The enzyme employs a ping-pong mechanism via the initial formation of an adenylyl-enzyme intermediate. Interestingly, the enzyme does not catalyze the hydrolysis of ATP to AMP and PP$_i$ in the absence of RNA; nor does it catalyze exchange reactions that are typical for ping-pong enzymes.[2-4]

[1] W. Filipowicz, M. Konarska, H. J. Gross & A. Shatkin (1983) *NAR* **11**, 1405.
[2] W. Filipowicz, K. Strugala, M. Konarska & A. J. Shatkin (1985) *PNAS* **82**, 1316.
[3] D. Reinberg, J. Arenas & J. Hurwicz (1985) *JBC* **260**, 6088.
[4] O. Vicente & W. Filipowicz (1988) *EJB* **176**, 431.

Selected entries from *Methods in Enzymology* [vol, page(s)]:
General discussion: **181**, 499
Assay: **181**, 500

RNA POLYMERASE I

This polymerase is a eukaryotic DNA-directed RNA polymerase [EC 2.7.7.6] that is found in the nucleolus and is insensitive to α-amanitin. *See DNA-Directed RNA Polymerase*

Selected entries from *Methods in Enzymology* [vol, page(s)]:
Assay: 273, 236
Other molecular properties: activation by transcriptional activators, **273**, 25; comparison to prokaryotic holoenzyme, **273**, 28; gene specificity, **273**, 14, 165, 170; promoter binding (promoter structural elements, **273**, 21; recognition proteins, **273**, 21, 28, 171, 233); purification from mouse **273**, 236

RNA POLYMERASE II

This α-amanitin-sensitive, eukaryotic DNA-directed RNA polymerase [EC 2.7.7.6] catalyzes the synthesis of precursors of mRNAs within the nucleus. *See DNA-Directed RNA Polymerase*

Selected entries from *Methods in Enzymology* [vol, page(s)]:
Assay: distinguishing initiation from elongation effects (3'-extended template assays, **274**, 423; promoter-specific initiation, **274**, 427); elongation rate determination, **274**, 429; elongation stimulation by factors, **274**, 421, 439; mapping of 3' ends, **274**, 430; pausing, **274**, 431; promoter clearance, **274**, 428; readthrough, **274**, 432; RNA cleavage by factors, **274**, 433, 436; template preparation, **274**, 60; termination, **274**, 431
Other molecular properties: abortive initiation assay, **273**, 106; activation by transcriptional activators, **273**, 23; C-terminal domain phosphorylation (assay [casein kinase II assay, **273**, 185; direct transfer assay, **273**, 188; mobility shift assay, **273**, 188, 190; yeast holoenzyme, **273**, 176]; kinase identification, **273**, 187); closed complex formation (holoenzyme binding, **273**, 18; minimal promoter sequence, **273**, 15; recognition proteins, **273**, 15, 28, 101, 110, 168; Sarkosyl effects, **273**, 102); comparison to prokaryotic holoenzyme, **273**, 28; epitope tagging, **194**, 517; fractionation with Polymin P, **194**, 517; gene specificity, **273**, 14, 165, 167; immunoaffinity purification (wheat germ enzyme [crude extract preparation, **274**, 521; immunoaffinity chromatography, **274**, 522; Polymin P precipitation, **274**, 521; polyol-responsive monoclonal antibody column preparation, **274**, 514, 520, 525]; yeast enzyme and associated factors, **274**, 523); open complex formation (assay, **273**, 104; nucleotide triphosphate requirement, **273**, 101, 103; permanganate probing, **273**, 103, 106); plant, *in vitro* transcription (efficiency, **273**, 273; primer extension assay of transcripts, **273**, 272; reaction conditions, **273**, 271; template preparation, **273**, 270); preinitiation complex (assembly, **273**, 16, 110; gel filtration, **273**, 116; purification using immobilized DNA template [DNA immobilization, **273**, 113; immunoblot analysis, **273**, 114; principle, **273**, 112; transcriptionally competent complex formation on DNA, **273**, 114; troubleshooting, **273**, 116]); promoter detection by GRAIL, **266**, 272, 277; protein affinity chromatography with initiation factors, **274**, 120; purification from HeLa cells (core enzyme purification, **274**, 98; holoenzyme purification,

274, 87); reconstitution, 274, 69; regulation of transcription elongation, 274, 419; subunits, 274, 97, 520; yeast enzyme, 273, 175, 183.

RNA POLYMERASE III

This eukaryotic DNA-directed RNA polymerase [EC 2.7.7.6] catalyzes the synthesis of precursors of 5S rRNA, tRNAs, and other small RNAs within the nucleus. *See DNA-Directed RNA Polymerase*

Selected entries from *Methods in Enzymology* [vol, page(s)]:
Assay: gel-retardation assay, 273, 252; nonspecific transcription assay, 273, 251; reconstituted specific transcription assay, 273, 250; single-round transcription assay, 273, 251
Other molecular properties: activation by transcriptional activators, 273, 24; comparison to prokaryotic holoenzyme, 273, 28; gene specificity, 273, 14, 18, 165, 249; plant, *in vitro* transcription, 273, 277; promoter binding (preinitiation complex formation, 273, 19; promoter elements [class 1, 273, 19; class 2, 273, 20; class 3, 273, 20]; recognition proteins, 273, 19, 28, 167, 249); purification from yeast 273, 255; subunit structure, 273, 257; transcription from chromatin and cloned gene templates *in vitro*, analysis, in, 170, 347

RNA POLYMERASE, *BACILLUS SUBTILIS*

This DNA-dependent RNA polymerase is probably similar to other prokaryotic polymerases that transcribe all classes of DNA to form the corresponding RNA transcript. *See DNA-Directed RNA Polymerase*

Selected entries from *Methods in Enzymology* [vol, page(s)]:
Assay: polydeoxyadenylic-thymidylic acid template, 273, 159; runoff transcription assay, 273, 159
Other molecular properties: $E\sigma^E$ purification, 273, 155; holoenzyme forms (growth phase, 273, 150; sporulating phase, 273, 151); polyacrylamide gel electrophoresis, 273, 160

RNA POLYMERASE 3D^{POL}, POLIOVIRUS

This polymerase catalyzes RNA synthesis from an RNA template. *See RNA-Directed RNA Polymerase*

Selected entries from *Methods in Enzymology* [vol, page(s)]:
Assay: 275, 39, 44

RNA POLYMERASE, *ESCHERICHIA COLI*

This DNA-dependent RNA polymerase is responsible for transcribing all classes of DNA to form the corresponding RNA transcript. *See DNA-Directed RNA Polymerase*

Selected entries from *Methods in Enzymology* [vol, page(s)]:
Other molecular properties: elongation arrest suppressors, (GreA, 274, 315, 318, 320; GreB, 274, 318, 320, 323; N antiterminator, 274, 364, 375, 386, 388, 390); elongation rate, 274, 334; epitope mapping of subunits α subunit, 274, 507; β′ subunit, 274, 505; *Escherichia coli* Eσ^{70}, 185, 40; fluorescence spectroscopy (kinetic assay of RNA polymerases, 274, 475; labeling of subunits, 274, 476; nucleotide probes [binding to *Escherichia coli* RNA polymerase, 274, 463, 469, 474; resonance energy transfer studies with rifampicin as acceptor, 274, 465; synthesis, 274, 458, 472, 477]; polarization studies of protein binding, 274, 500; tryptophan [intrinsic protein fluorescence, 274, 456; time-resolved emission in transcription factors, 274, 457]); guanosine 5′-diphosphate

3′-diphosphate binding and regulatory effects, 274, 471; immunoprinting of N complex, 274, 367; pausing assays, 274, 349; phage-induced ADP-ribosylation, 106, 418; promoter (binding, 273, 5; consensus sequences, 273, 30; representations of specificity, 273, 31; sequence alignment, 273, 36); proteolytic digestion, 208, 242; purification of reconstituted enzyme, 273, 128; subunits, 274, 403, 476, 503.

RNA POLYMERASE, HEPATITIS C VIRUS

This polymerase catalyzes RNA synthesis from an RNA template. *See RNA-Directed RNA Polymerase*

Selected entries from *Methods in Enzymology* [vol, page(s)]:
General discussion: 275, 58
Assay: 275, 60

RNA POLYMERASE, N4

This heterodimeric RNA polymerase II of coliphage N4 is a DNA-dependent RNA polymerase that transcribes DNA into the corresponding RNA transcript. *See DNA-Directed RNA Polymerase*

Selected entries from *Methods in Enzymology* [vol, page(s)]:
Other molecular properties: early promoters (mapping by primer extension analysis, 274, 14; plasmid preparation, 274, 12; structural probing on supercoiled templates, 274, 14); purification, 274, 11; subunits, 274, 9; transcription assay *in vitro*, 274, 12

RNA POLYMERASE, *SACCHAROMYCES CEREVISIAE* MITOCHONDRIA

This DNA-dependent RNA polymerase is responsible for transcribing DNA into its corresponding RNA transcript. *See DNA-Directed RNA Polymerase*

Selected entries from *Methods in Enzymology* [vol, page(s)]:
Other molecular properties: core polymerase, nonselective transcription assay, 264, 58; dilution, 264, 63; holoenzyme, selective transcription assay, 264, 59; promoter consensus sequence, 264, 62; recombinant core polymerase purification (cell growth, 264, 60; DEAE chromatography, 264, 60; extract preparation, 264, 60; Mono Q chromatography, 264, 61; phenyl-Superose chromatography, 264, 61; phosphocellulose chromatography, 264, 60; Superose 6 chromatography, 264, 61; yield, 264, 59); recombinant specificity factor purification (DEAE chromatography, 264, 62; extract preparation, 264, 62; induction, 264, 61; Mono S chromatography, 264, 61; phosphocellulose chromatography, 264, 62; yield, 264, 61); storage, 264, 63; subunits and purification, 264, 57; transcription *in vitro*, 264, 65; transcription template, 264, 62

[RNA-POLYMERASE]-SUBUNIT KINASE

This phosphotransferase [EC 2.7.1.141], also known as CTD kinase (for C-terminal domain kinase), catalyzes the reaction of ATP with DNA-directed RNA polymerase to produce ADP and phospho-[DNA-directed RNA polymerase] which is the phosphorylated form of the largest subunit of eukaryotic DNA-directed RNA polymerase [EC 2.7.7.6].[1-4]

[1] J. M. Lee & A. L. Greenleaf (1989) *PNAS* **86**, 3624.
[2] A. Stevens & M. K. Maupin (1989) *BBRC* **159**, 508.
[3] N. Feuerstein (1991) *JBC* **266**, 16200.
[4] J. M. Payne & M. E. Dahmus (1993) *JBC* **268**, 80.

Selected entries from **Methods in Enzymology** [**vol**, page(s)]:
General discussion: 200, 301
Other molecular properties: E1 isozyme, purification, **200**, 308; E2 isozyme, purification, **200**, 310; enzymatic properties, **200**, 319; filter-binding assay, **200**, 304; p34^{cdc2} component, detection, **200**, 315; purification from (murine ascites tumor cells, **200**, 306; yeast, **200**, 322); sodium dodecyl sulfate gel assay, **200**, 305

RNA POLYMERASE, T4-MODIFIED

This polymerase is responsible for producing phage-specific mRNA transcripts after injection of double-stranded bacteriophage T4 DNA. **See DNA-Directed RNA Polymerase**

Selected entries from **Methods in Enzymology** [**vol**, page(s)]:
Other molecular properties: ADP-ribosylation, **274**, 44; AsiA binding to σ^{70}, **274**, 44, 50; modification effect on activity, **274**, 55; purification from infected cells (DNA affinity chromatography, **274**, 54; Polymin P extraction, **274**, 54); transcription assay, **274**, 45, 55

RNA (RIBOSE-$O^{2'}$-)-METHYLTRANSFERASE

This methyltransferase, also known as *Vaccinia* protein 39 (or VP 39), catalyzes the reaction of *S*-adenosyl-L-methionine with the 2'-hydroxyl group of the ribose of the first transcribed nucleotide in host-cell RNA to produce *S*-adenosyl-L-homocysteine and the 2'-methoxy derivative.[1] **See also rRNA (Adenosine-$O^{2'}$)-Methyltransferase**

[1] A. E. Hoden, P. D. Gershon, X. Shi & F. A. Quiocho (1996) *Cell* **85**, 247.

RNA URIDYLYLTRANSFERASE

This transferase [EC 2.7.7.52], also known as terminal uridylyltransferase (or TUT), catalyzes the reaction of UTP with RNA to produce pyrophosphate and an RNA polymer with one more base.[1-3] The enzyme requires an oligoribonucleotide or polyribonucleotide with a free terminal 3'-OH as a primer.

[1] N. Bakalara, A. M. Simpson & L. Simpson (1989) *JBC* **264**, 18679.
[2] N. C. Andrews & D. Baltimore (1986) *PNAS* **83**, 221.
[3] P. Zabel, L. Dorssers, K. Wernars & A. van Kammen (1981) *NAR* **9**, 2433.

Selected entries from **Methods in Enzymology** [**vol**, page(s)]:
Assay: 264, 114
Other molecular properties: detection of T-complexes, **264**, 119

ROSMARINATE SYNTHASE

This acyltransferase [EC 2.3.1.140] catalyzes the reaction of caffeoyl-CoA with 3-(3,4-dihydroxyphenyl)lactate to produce Coenzyme A and rosmarinate.[1-3]

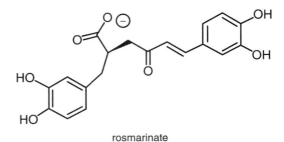

rosmarinate

4-Coumaroyl-CoA can substitute for caffeoyl-CoA and 4-hydroxyphenyllactate can serve as the acceptor substrate.

[1] M. Petersen & A. W. Alfermann (1988) *Z. Naturforsch.* **43c**, 501.
[2] M. Petersen, E. Häusler, B. Karwatzki & J. Meinhard (1993) *Planta* **189**, 10.
[3] M. S. Petersen (1991) *Phytochemistry* **30**, 2877.

ROUS SARCOMA VIRUS RETROPEPSIN

This viral protease, originally called p15 protein, acts at eight cleavage sites in the Gag and Gag-Pol polyproteins. The enzyme prefers substrates with bulky hydrophobic aminoacyl side chains at positions P1, P1', and P3'.[1-3]

[1] A. M. Skalka (1989) *Cell* **56**, 911.
[2] B. Grinde, C. Cameron, J. Leis, I. Weber, A. Wlodawer, H. Burstein & A. Skalka (1992) *JBC* **267**, 9491.
[3] C. Cameron, B. Grinde, P. Jacques, J. Jentoft & J. Leis (1993) *JBC* **268**, 11711.

Selected entries from **Methods in Enzymology** [**vol**, page(s)]:
Other molecular properties: active site, **241**, 276; amino acid sequence, comparison with eukaryotic aspartic proteases, **241**, 216; cleavage site sequence, **241**, 297; crystal structure, **202**, 728; efficiency, **241**, 229; *Escherichia coli* expression system, **241**, 11; structure, **241**, 157, 254; substrate based on, in modeling of HIV-1 protease, **202**, 727; substrate specificity, mutational analysis, **241**, 276; subunit exchange, **241**, 124

rRNA (ADENINE-N^6-)-METHYLTRANSFERASE

This methyltransferase [EC 2.1.1.48] catalyzes the reaction of *S*-adenosyl-L-methionine with an adenine residue in an rRNA at specific sites to produce *S*-adenosyl-L-homocysteine and an rRNA containing an N^6-methyladenine residue.[1-3] The enzyme, which will also dimethylate appropriately substituted adenine residues, has a random Bi Bi kinetic mechanism.

[1] J. E. Sipe, W. M. Anderson, C. N. Remy & S. H. Love (1972) *J. Bact.* **110**, 81.
[2] R. Skinner, E. Cundliffe & F. J. Schmidt (1983) *JBC* **258**, 12702.
[3] C. Denoya & D. Dubnau (1989) *JBC* **264**, 2615.

rRNA (ADENOSINE-$O^{2'}$-)-METHYLTRANSFERASE

This methyltransferase [EC 2.1.1.66], also known as RNA-pentose methylase and thiostrepton-resistance methylase, catalyzes the reaction of *S*-adenosyl-L-methionine with a specific adenosine residue in an rRNA to produce *S*-adenosyl-L-homocysteine and an rRNA containing a single residue of 2′-*O*-methyladenosine (*e.g.*, adenosine-1067 in *Escherichia coli* 23S rRNA).[1,2] Methylation is involved in thiostrepton resistance. ***See also*** *RNA (Ribose-O²′-)-Methyltransferase*

[1]J. Thompson & E. Cundliffe (1981) *J. Gen. Microbiol.* **124**, 291.
[2]J. Thompson, F. Schmidt & E. Cundliffe (1982) *JBC* **257**, 7915.

rRNA ENDONUCLEASE

This endoribonuclease [EC 3.1.27.10], also known as α-sarcin, catalyzes the hydrolysis of the phosphodiester linkage between guanosine and adenosine residues at one specific position in the 28S rRNA from rat ribosomes.[1,2]

[1]J. Lacadena, A. Martinez del Pozo, A. Martinez-Ruiz, J. M. Perez-Canadillas, M. Bruix, J. M. Mancheno, M. Onaderra & J. G. Gavilanes (1999) *Proteins* **37**, 474.
[2]Y. Endo & K. Tsurugi (1988) *JBC* **263**, 8735.

rRNA *N*-GLYCOSYLASE

This glycosylase [EC 3.2.2.22], also called rRNA *N*-glycosidase, catalyzes the endohydrolysis of the *N*-glycosidic bond at one specific adenosine residue on the 28S rRNA. Naked rRNA is attacked more slowly than intact ribosomal RNA. *Escherichia coli* rRNA is cleaved at a corresponding position (at A-2600; but only on naked rRNA). The ricin A-chain from castor beans (at A-4324 of rat 28S rRNA)[1] and related toxins exhibit this activity.

[1]Y. Endo & K. Tsurugi (1988) *JBC* **263**, 8735.

rRNA (GUANINE-*N*¹-)-METHYLTRANSFERASE

This methyltransferase [EC 2.1.1.51] catalyzes the reaction of *S*-adenosyl-L-methionine with a guanine residue in an rRNA to produce *S*-adenosyl-L-homocysteine and an rRNA containing an *N*¹-methylguanine residue.[1]

[1]L. A. Isaksson (1973) *BBA* **312**, 122.

rRNA (GUANINE-*N*²-)-METHYLTRANSFERASE

This methyltransferase [EC 2.1.1.52] catalyzes the reaction of *S*-adenosyl-L-methionine with a guanine residue in an rRNA to produce *S*-adenosyl-L-homocysteine and an rRNA containing an *N*²-methylguanine residue.[1]

[1]L. A. Isaksson (1973) *BBA* **312**, 122.

*Rsa*I RESTRICTION ENDONUCLEASE

This type II restriction enzyme [EC 3.1.21.4], isolated from *Rhodopseudomonas sphaeroides*, catalyzes the hydrolysis of both strands of DNA at 5′ . . . GT∇AC . . . 3′, producing blunt-ended fragments.[1] (The symbol ∇ indicates the cleavage site.) In addition to the recognition site, at least two flanking bases must be present.[2]

[1]S. P. Lynn, L. K. Cohen, S. Kaplan & J. F. Gardner (1980) *J. Bacteriol.* **142**, 380.
[2]K. Majumder (1989) *Biophys. J.* **55**, 48a.

*Rsh*I RESTRICTION ENDONUCLEASE

This type II restriction enzyme [EC 3.1.21.4], isolated from *Rhodopseudomonas sphaeroides* strain 2.4.1, catalyzes the hydrolysis of both strands of DNA at 5′ . . . CGAT∇CG . . . 3′.[1] (The symbol ∇ indicates the cleavage site.)

[1]S. P. Lynn, L. K. Cohen, J. F. Gardner & S. Kaplan (1979) *J. Bacteriol.* **138**, 505.

Selected entries from ***Methods in Enzymology*** [**vol**, page(s)]: **Other molecular properties:** recognition sequence, 65, 9

*Rsr*II RESTRICTION ENDONUCLEASE

This type II restriction enzyme [EC 3.1.21.4], isolated from *Rhodopseudomonas sphaeroides*, catalyzes the hydrolysis of both strands of DNA at 5′ . . . CG∇GWCCG . . . 3′, where W refers to A or T.[1] (The symbol ∇ indicates the cleavage site.)

[1]C. D. O'Connor, E. Metcalf, C. J. Wrighton, T. J. R. Harris & J. R. Saunders (1984) *NAR* **12**, 6701.

Selected entries from ***Methods in Enzymology*** [**vol**, page(s)]: **General discussion: 155**, 11

RUBBER *cis*-POLYPRENYL*cis*TRANSFERASE

This transferase [EC 2.5.1.20], also known as rubber allyltransferase and rubber transferase, catalyzes the reaction of poly-*cis*-polyprenyl diphosphate (or rubber particles) with isopentenyl diphosphate to produce pyrophosphate and a poly-*cis*-polyprenyl diphosphate longer by one five-carbon isoprenyl unit.[1–4] The rubber substrate must contain a terminal allylic diphosphate group. The reaction is stereospecific: only the *cis*-polyisoprene is generated from isopentenyl diphosphate.

[1]A. I. McMullen & G. P. McSweeney (1966) *BJ* **101**, 42.
[2]C. R. Benedict, S. Madhavan, G. A. Greenblatt, K. V. Venkatachalam & M. A. Foster (1989) *Plant Physiol.* **92**, 816.
[3]R. A. Gibbs (1998) *CBC* **1**, 31.
[4]E. D. Beytía & J. W. Porter (1976) *ARB* **45**, 113.

Selected entries from *Methods in Enzymology* [**vol**, page(s)]:
General discussion: 15, 476
Assay: 15, 477

RUBERLYSIN

This zinc-dependent metalloendopeptidase [EC 3.4.24.48], also called *Crotalus ruber* metalloendopeptidase II and hemorrhagic toxin II (HT-2), is a hemorrhagic toxin with fibrinogenolytic activity from the venom of the red rattlesnake (*Crotalus ruber ruber*).[1–3] Ruberlysin catalyzes peptide-bond hydrolysis of the oxidized insulin B chain at His10–Leu11, Ala14–Leu15, Tyr16–Leu17, and Gly23–Phe24. It also acts on His6–Pro7 and Pro7–Phe8 of angiotensin I and Gly3–Phe4 of [Met[5]]enkephalin. Luteinizing-hormone-releasing hormone is cleaved at Trp3–Ser4 and Gly6–Leu7, and the enzyme hydrolyzes the Aα and Bβ chains of fibrinogen.

[1]N. Marsh (1994) *Thromb. Haemost.* **71**, 793.
[2]N. Mori, T. Nikai, H. Sugihara & A. T. Tu (1987) *ABB* **253**, 108.
[3]H. Takeya, A. Onikura, T. Nikai, H. Sugihara & S. Iwanaga (1990) *J. Biochem.* **108**, 711.

RUBREDOXIN:NAD(P)⁺ REDUCTASE

This oxidoreductase [EC 1.18.1.4] catalyzes the reaction of oxidized rubredoxin with NAD(P)H to produce reduced rubredoxin and NAD(P)$^+$.[1] This enzyme may be identical to rubredoxin:NAD$^+$ reductase [EC 1.18.1.1].
See *Rubredoxin:NAD$^+$ Reductase*

[1]H. Petitdemange, H. Blusson & R. Gay (1981) *Anal. Biochem.* **116**, 564.

RUBREDOXIN:NAD⁺ REDUCTASE

This flavin-dependent oxidoreductase [EC 1.18.1.1], also known as rubredoxin reductase, catalyzes the reaction of oxidized rubredoxin with NADH to produce reduced rubredoxin and NAD$^+$.[1–4]

[1]J. Le Gall (1968) *Ann. Inst. Pasteur* **114**, 109.
[2]R. Claus, O. Asperger & H.-P. Kleber (1979) *Z. Allg. Mikrobiol.* **19**, 695.
[3]H. J. Lee, J. Basran & N. S. Scrutton (1998) *Biochem.* **37**, 15513.
[4]G. Palmer (1975) *The Enzymes*, 3rd ed., **12**, 1.

Selected entries from *Methods in Enzymology* [**vol**, page(s)]:
Other molecular properties: assay of rubredoxin, 53, 617, 618; component of *Pseudomonas oleovorans* hydrocarbon monooxygenase

system, isolation, 53, 356; 188, 6; *Desulfovibrio gigas*, 53, 619, 631; 243, 205; purification, 53, 619; strain specificity, 53, 618, 633

RUSSELL'S VIPER VENOM FACTOR V ACTIVATOR

This serine endopeptidase [EC 3.4.21.95] (also known as snake venom factor V activator, RVV-V component, factor V-activating proteinase) catalyzes the full activation of pure human clotting factor V by a single cleavage at the Trp-Tyr-Leu-Arg1545-∇-Ser1546-Asn-Asn-Gly bond, where ∇ indicates the cleavage site.[1,2] Bovine factor V is also cleaved. No other proteins of the clotting system are attacked. The enzyme also exhibits an esterase activity with Bz-Arg-OEt and Tos-Arg-OMe. Amidase activity is observed with Phe-pipecolyl-Arg-NHPhNO$_2$.

[1]F. Tokunaga, K. Nagasawa, S. Tamura, T. Miyata, S. Iwanagawa & W. Kisiel (1988) *JBC* **263**, 17471.
[2]W. Kisiel & W. M. Canfield (1981) *MIE* **80**, 275.

RUSSELLYSIN

This zinc-dependent and calcium-dependent metalloendopeptidase [EC 3.4.24.58], also known as Russell's viper venom factor X activator (RVV-X) and Russell's venom coagulation factor X-activating enzyme, catalyzes the activation of several components of the blood clotting system, including coagulation factor X (catalyzing the hydrolysis of the Arg51–Ile52 bond), coagulation factor IX, and protein C, also by cleavage of Arg–Xaa bonds. No action on the oxidized insulin B chain is observed.[1,2]

[1]B. C. Furie & B. Furie (1976) *MIE* **45**, 191.
[2]G. W. Amphlett, R. Byrne & F. J. Castellino (1982) *Biochemistry* **21**, 125.

Selected entries from *Methods in Enzymology* [**vol**, page(s)]:
General discussion: 19, 719; 34, 592; 45, 191
Assay: 19, 719; 45, 192; continuous spectrophotometric, 45, 194; one stage, 45, 192; two stage, 45, 193
Other molecular properties: activation of protein C, 80, 320, 325, 327; activator factor X, bovine, 45, 90, 100; activators and inhibitors, 19, 722; characteristics, 45, 202; hemorrhagic activity, 248, 353, 356; kinetics, 45, 203; properties, 19, 721; 45, 202; 248, 194 (metal binding, 45, 205; physical, 45, 204); purification, 19, 720; 45, 196, 200, 201; specificity, 45, 202; structure, 248, 364

S

SABINENE-HYDRATE SYNTHASE

This enzyme [EC 4.2.3.11; previously listed as EC 4.6.1.9] catalyzes the conversion of geranyl diphosphate to sabinene hydrate (both *cis-* and *trans-* isomers) and pyrophosphate.

sabinene hydrate

The reaction proceeds by the initial ionization and isomerization of the substrate (a suprafacial process) to produce the (3R)-linalyl diphosphate intermediate, which undergoes subsequent cyclization to form a monocyclic (4R)-α-terpinyl cation. A 1,2-hydride shift follows and a second cyclization with water capture of the resulting cation generates the final product(s).[1]

[1]T. W. Hallahan & R. Croteau (1989) *ABB* **269**, 313.

(+)-SABINOL DEHYDROGENASE

This plant dehydrogenase [EC 1.1.1.228], which participates in the biosynthesis of (+)-3-thujone, catalyzes the reaction of (+)-*cis*-sabinol with NAD^+ to produce (+)-sabinone and NADH.[1]

(+)-*cis*-sabinol   (+)-sabinone

NADP$^+$ can also served as the hydride-transfer coenzyme substrate, albeit more weakly than NAD^+.

[1]S. S. Dehal & R. Croteau (1987) *ABB* **258**, 287.

SACCHAROLYSIN

This zinc-dependent metalloendopeptidase [EC 3.4.24.37; previously classified as a cysteine peptidase under EC 3.4.22.22], also known as proteinase yscD (referring to yeast *Saccharomyces cerevisiae*), oligopeptidase yscD, and yeast cysteine proteinase D, catalyzes peptide-bond hydrolysis in proteins.[1] Specificity studies indicate a preference for hydrophobic side-chain residues at position P1 and P3 as well as a prolyl residue at P2′. Bradykinin is cleaved at Phe5–Ser6, and in this case, both the P3 and P2′ postions are occupied by prolyl residues. Synthetic substrates include Bz-Pro-∇-Phe-Arg-NHPhNO$_2$, Ac-Ala-∇-Ala-Pro-Met-NHPhNO$_2$, MeOSuc-Ala-∇-Ala-Pro-Met-NHPhNO$_2$, Ac-Ala-∇-Ala-Pro-Phe-NHPhNO$_2$, and Pz-Pro-Leu-∇-Gly-Pro-D-Arg, where ∇ indicates the cleavage site.

[1]M. Büchler, U. Tisljar & D. H. Wolf (1994) *EJB* **219**, 627.

Selected entries from *Methods in Enzymology* [**vol**, page(s)]:
Molecular properties: amino acid sequence, **248**, 208; evolution, **248**, 552; homologues, **248**, 551, 591; properties, **248**, 206

SACCHAROPEPSIN

This abundant yeast aspartic endopeptidase [EC 3.4.23.25], also known as yeast endopeptidase A and proteinase yscA (for yeast *Saccharomyces cerevisiae*), catalyzes peptide-bond hydrolysis with a broad specificity similar to pepsin A [EC 3.4.23.1] and cathepsin D [EC 3.4.23.5].[1,2] The enzyme prefers a phenylalanyl, leucyl, or glutamyl residue in the P1 position and phenylalanyl, isoleucyl, leucyl, or alanyl residue in the P1′ position.

[1]T. Dreyer (1989) *Carlsberg Res. Commun.* **54**, 85.
[2]H. B. van den Hazel, J. R. Winther & M. C. Kielland-Brandt (1996) *Yeast* **12**, 1.

Selected entries from *Methods in Enzymology* [**vol**, page(s)]:
General discussion: **194**, 428
Assay: enzymatic assays, **185**, 380; **194**, 444

Other molecular properties: Bakers' yeast, from, **19**, 372; genetic analysis in yeast, **185**, 379; overlay test, **194**, 442

SACCHAROPINE DEHYDROGENASE

There are four saccharopine dehydrogenases that differ with respect to coenzyme (NAD^+ or $NADP^+$) and product (lysine or glutamate) formed.[1-10]

saccharopine

Saccharopine dehydrogenase (NAD^+, L-glutamate forming) [EC 1.5.1.9] catalyzes the reversible reaction of saccharopine (or, N^6-(L-1,3-dicarboxypropyl)-L-lysine) with NAD^+ and water to produce L-glutamate, 2-aminoadipate 6-semialdehyde, and NADH.[1,2]

Saccharopine dehydrogenase (NAD^+, L-lysine forming) [EC 1.5.1.7], also known as lysine:2-oxoglutarate reductase and L-lysine:α-ketoglutarate reductase, catalyzes the reversible reaction of saccharopine with NAD^+ and water to produce L-lysine, α-ketoglutarate, and NADH.[3]

Saccharopine dehydrogenase ($NADP^+$, L-glutamate forming) [EC 1.5.1.10] catalyzes the reversible reaction of L-glutamate with 2-aminoadipate 6-semialdehyde and NADPH to produce saccharopine, $NADP^+$, and water.[4-6]

Saccharopine dehydrogenase ($NADP^+$, L-lysine forming) [EC 1.5.1.8] catalyzes the reversible reaction of L-lysine with α-ketoglutarate and NADPH to produce saccharopine, $NADP^+$, and water.[7-9]

[1]T. A. Fjellstedt & J. C. Robinson (1975) ABB **171**, 191.
[2]J. Hutzler & J. Dancis (1970) BBA **206**, 205.
[3]M. Fujioka (1984) ABB **230**, 553.
[4]E. E. Jones & H. P. Broquist (1966) JBC **241**, 3430.
[5]D. R. Storts & J. K. Bhattacharjee (1987) J. Bacteriol. **169**, 416.
[6]H. Schmidt, R. Bode & D. Birnbaum (1989) Antonie Leeuwenhoek **56**, 337.
[7]P. Arruda, L. Sodek & W. J. Da Silva (1982) Plant Physiol. **69**, 988.
[8]C. Noda & A. Ichihara (1978) BBA **525**, 307.
[9]J. Hutzler & J. Dancis (1975) BBA **377**, 42.
[10]J. Thompson & S. P. F. Miller (1991) AE **64**, 317.

Selected entries from **Methods in Enzymology** [vol, page(s)]:
General discussion: NAD^+-dependent, lysine-forming, **17B**, 124; $NADP^+$-dependent, L-glutamate-forming, **17B**, 121
Assay: NAD^+-dependent, lysine-forming, **17B**, 125; $NADP^+$-dependent, L-glutamate-forming, **17B**, 121
Other molecular properties: mechanism, **63**, 157; properties (NAD^+-dependent and lysine-forming enzyme, **17B**, 127; $NADP^+$-dependent, L-glutamate-forming, **17B**, 123); purification from Bakers' yeast (NAD^+-dependent and lysine-forming enzyme, **17B**, 126;

$NADP^+$-dependent, L-glutamate-forming, **17B**, 122); stereospecificity, NAD^+-dependent and lysine-forming enzyme, **87**, 117; synchronous cultures and, **21**, 469

*Sac*I RESTRICTION ENDONUCLEASE

This type II restriction endonuclease [EC 3.1.21.4] is obtained from *Streptomyces achromogenes* and acts on both strands of DNA at $5'\ldots$GAGCT∇C$\ldots 3'$.[1,2] (The symbol ∇ indicates the cleavage site.) The enzyme also exhibits star activity. It is inhibited by salt concentrations above 10 mM.

[1]M. Nasri & D. Thomas (1986) Biomed. Biochim. Acta **45**, 997.
[2]R. Vazquez, J. Brito & M. Guerra (1996) Biotechnol. Apl. **13**, 21.

Selected entries from **Methods in Enzymology** [vol, page(s)]:
General discussion: 65, 173
Other molecular properties: recognition sequence, **65**, 10

*Sac*II RESTRICTION ENDONUCLEASE

This type II restriction endonuclease [EC 3.1.21.4] is obtained from *Streptomyces achromogenes* and acts on both strands of DNA at $5'\ldots$CCGC∇GG$\ldots 3'$. (The symbol ∇ indicates the cleavage site.) It is inhibited at salt concentrations above 10 mM.

Selected entries from **Methods in Enzymology** [vol, page(s)]:
Other molecular properties: recognition sequence, **65**, 10

SALICYL-ALCOHOL GLUCOSYLTRANSFERASE

This glucosyltransferase [EC 2.4.1.172] catalyzes the reaction of UDP-D-glucose with salicyl alcohol (or, 2-hydroxybenzenemethanol) to produce UDP and salicin.[1,2]

salicyl alcohol   salicin

[1]H. Mizukami, T. Terao & H. Ohashi (1985) Planta Med. **1985**, 104.
[2]T. Terao, H. Ohashi & H. Mizukami (1984) Plant Sci. Lett. **33**, 47.

SALICYLALDEHYDE DEHYDROGENASE

This oxidoreductase [EC 1.2.1.65] catalyzes the reaction of salicylaldehyde with NAD^+ and water to produce salicylate, NADH, and a proton.[1]

salicylaldehyde   salicylate

In some microorganisms, this enzyme participates in naphthalene degradation.

[1] R. Eaton & P. J. Chapman (1992) *J. Bacteriol.* **174**, 7542.

SALICYLATE 1-MONOOXYGENASE

This FAD-dependent oxidoreductase [EC 1.14.13.1], also known as salicylate hydroxylase, catalyzes the reaction of salicylate with NADH and dioxygen to produce catechol, NAD^+, water, and carbon dioxide.[1–7] Alternative substrates include 2,3-dihydroxybenzoate, 2,4-dihydroxybenzoate, 2,5-dihydroxybenzoate, 2,6-dihydroxybenzoate, *p*-aminosalicylate, and 1-hydroxy-2-naphthoate.

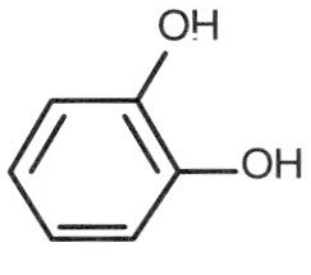

catechol

The enzyme will also deiodinate and oxidize *o*-iodophenol to catechol. *o*-Benzoquinone is formed as a nascent product.[3]

[1] K. Suzuki, S. Takemori & M. Katagiri (1969) *BBA* **191**, 77.
[2] S. Takemori, H. Yasuda, K. Mihara, K. Suzuki & M. Katagiri (1969) *BBA* **191**, 58.
[3] K. Suzuki, T. Gomi & E. Itagaki (1991) *J. Biochem.* **109**, 791.
[4] S. Takemori, K. Hon-Nami, F. Kawahara & M. Katagiri (1974) *BBA* **342**, 137.
[5] S. Takemori, M. Nakamura, K. Suzuki, M. Katagiri & T. Nakamura (1972) *BBA* **284**, 382.
[6] B. A. Palfey & V. Massey (1998) *CBC* **3**, 83.
[7] V. Massey & P. Hemmerich (1975) *The Enzymes*, 3rd ed., **12**, 191.

Selected entries from *Methods in Enzymology* [vol, page(s)]:
General discussion: 52, 17; 53, 527; 188, 138
Assay: 53, 529, 530
Other molecular properties: activity, 53, 527, 528, 535; altered form from mutated strain, 53, 542, 543; external monooxygenase, as, 53, 527; flavin determination, 53, 429, 528, 534; fluorescence and, 26, 503; inhibition by salts, 53, 541; kinetic properties, 53, 535; molecular weight, 53, 528, 535; oxy form, spectrum, 52, 36; oxygenated flavin intermediate, 53, 540, 541; partial specific volume, 53, 535; *Pseudomonas*, from, 53, 527; pseudosubstrates, 53, 528, 537; purification, 53, 532; purity, 53, 532; reaction mechanism, 53, 540, 541; reduced enzyme-substrate complex, reaction with oxygen, 53, 540, 541; reduction studies, 53, 541, 542; sedimentation coefficient, 53, 535; stereospecificity, 87, 120; substrate specificity, 53, 535, 536, 538, 539; subunits, 53, 528, 535; uncoupling effect, 53, 536, 549

*Sal*PI RESTRICTION ENDONUCLEASE

This type II restriction endonuclease [EC 3.1.21.4] is obtained from *Streptomyces albus* and acts on both strands of DNA at 5′...CTGCA∇G...3′. (The symbol ∇ indicates the cleavage site.) The enzyme has the same specificity as *Pst*I restriction endonuclease.[1,2]

[1] J. A. Carter, K. F. Chater, C. J. Bruton & N. L. Brown (1980) *NAR* **8**, 4943.
[2] K. F. Chater (1977) *NAR* **4**, 1989.

Selected entries from *Methods in Enzymology* [vol, page(s)]:
Other molecular properties: recognition sequence, 65, 10

*Sal*I RESTRICTION ENDONUCLEASE

This type II restriction endonuclease [EC 3.1.21.4] is obtained from *Streptomyces albus* G and acts on both strands of DNA at 5′...G∇TCGAC...3′.[1,2] (The symbol ∇ indicates the cleavage site.) It also exhibits star activity. The site GTm^5CGAC is resistant to cleavage.
See *"Star" Activity*

[1] J. R. Arrand, P. A. Myers & R. J. Roberts (1978) *J. Mol. Biol.* **118**, 127.
[2] B. Hinsch & M.-R. Kula (1981) *NAR* **9**, 3159.

Selected entries from *Methods in Enzymology* [vol, page(s)]:
General discussion: 65, 363, 789, 793; **208**, 433
Other molecular properties: fragment end structure produced, **65**, 511; recognition sequence, **65**, 10, 511

SALUTARIDINE REDUCTASE (NADPH)

This enzyme [EC 1.1.1.248] catalyzes the reversible reaction of salutaridine with NADPH (B-side) to produce salutaridinol and $NADP^+$.

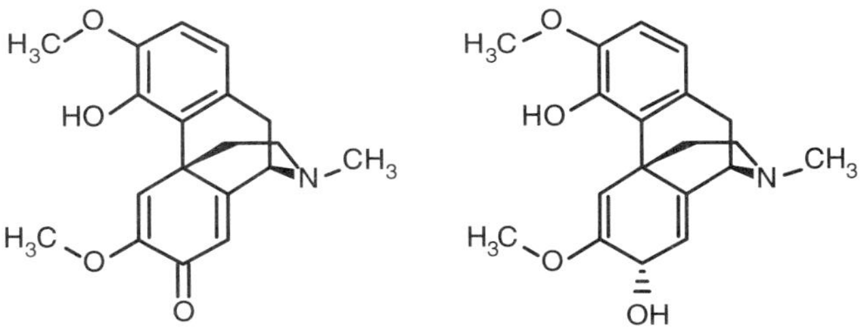

salutaridine   salutaridinol

Salutaridinol is a precursor of the morphinan alkaloids in the poppy plant, *Papaver somniferum*.[1]

[1] R. Gerardy & M. H. Zenk (1993) *Phytochem.* **34**, 125.

SALUTARIDINE SYNTHASE

This oxidoreductase [EC 1.1.3.35], also known as (*R*)-reticuline oxidase (C–C phenol-coupling), catalyzes the reaction of (*R*)-reticuline with NADPH and dioxygen to produce salutaridine, $NADP^+$, and two water molecules.

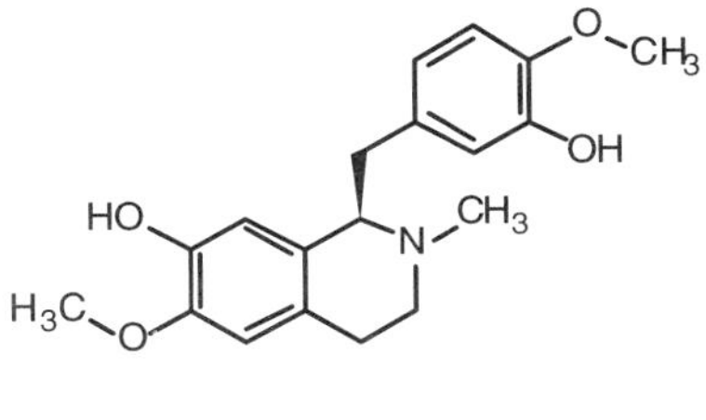

(*R*)-reticuline

Heme-thiolate is a cofactor. Note that there is no incorporation of oxygen in the morphinan alkaloid product.

SALUTARIDINOL 7-O-ACETYLTRANSFERASE

This acetyltransferase [EC 2.3.1.150], isolated from the opium poppy, catalyzes the reaction of acetyl-CoA with salutaridinol to produce Coenzyme A and 7-O-acetylsalutaridinol.

7-O-acetylsalutaridinol   thebaine

At elevated pH values the product spontaneously generates the morphine precursor thebaine.[1]

[1]R. Lenz & M. H. Zenk (1995) JBC **270**, 31091.

SanDI RESTRICTION ENDONUCLEASE

This type II restriction endonuclease [EC 3.1.21.4] is obtained from *Streptomyces* species and acts on both strands of DNA at 5′...GG∇GWCCC...3′, where W refers to either A or T, and the symbol ∇ indicates the cleavage site.[1] This endonuclease is useful for megabase mapping and vector constructions.

[1]T. G. Simcox, L. Fabian, K. Kretz, V. Hedden & M. E. C. Simcox (1995) *Gene* **155**, 129.

α-SANTONIN 1,2-REDUCTASE

This oxidoreductase [EC 1.3.1.47] catalyzes the reaction of α-santonin with NAD(P)H to produce 1,2-dihydrosantonin and NAD(P)$^+$.[1]

α-santonin   1,2-dihydro-α-santonin

Enzyme-catalyzed reduction occurs via a transient zwitterionic intermediate.

[1]U. Naik & S. Mavuinkurve (1986) *Can. J. Microbiol.* **33**, 658.

SapI RESTRICTION ENDONUCLEASE

This type II restriction endonuclease [EC 3.1.21.4] is obtained from a *Saccharopolyspora* species (NEB 597) and acts on DNA at 5′...GCTCTTCN∇...3′ and 3′...CGAGAAGNNNN∇...5′, where N refers to any base. (The symbol ∇ indicates the cleavage site.)

SARCOSINE DEHYDROGENASE

This flavin-dependent oxidoreductase [EC 1.5.99.1] catalyzes the reaction of sarcosine ($^-$OOCCH$_2$NHCH$_3$) with an acceptor substrate and water to produce glycine, formaldehyde, and the reduced acceptor.[1,2] In addition to a covalently bound FAD, the enzyme binds a folate derivative tightly. In the presence of tetrahydrofolate, N^5,N^{10}-methylenetetrahydrofolate is released rather than formaldehyde. Electron acceptor substrates can include 2,6-dichlorophenolindophenol, phenazine methosulfate, methylene blue, meldola blue, nile blue, and potassium ferricyanide.

[1]V. Schirch (1998) *CBC* **1**, 211.
[2]M. H. Stipanuk (1986) *Ann. Rev. Nutr.* **6**, 179.

Selected entries from **Methods in Enzymology** [**vol**, page(s)]:
General discussion: 5, 738; 10, 302; **17A**, 976; 53, 402; rat liver, **122**, 255
Assay: 5, 738
Other molecular properties: assay for electron-transferring flavoprotein, **14**, 111, 115; distribution, 5, 741; flavin linkage, **53**, 450; properties, 5, 741; inhibition, **46**, 163; purification from (monkey liver, **17A**, 979; rat liver, 5, 739; **17A**, 977; **122**, 257); rat liver (properties, **122**, 259; purification, **17A**, 977; **122**, 257); reaction with electron transfer flavoprotein, **10**, 309; reduced, substrate for electron-transferring flavoprotein, **14**, 118

SARCOSINE OXIDASE

This FAD-dependent oxidoreductase [EC 1.5.3.1] catalyzes the reaction of sarcosine ($^-$OOCCH$_2$NHCH$_3$) with dioxygen and water to produce glycine, formaldehyde, and hydrogen peroxide.[1-4] The *Corynebacterium* enzyme reportedly has a ping pong Bi Bi kinetic mechanism[1] and contains a covalently bound FMN that reacts with molecular oxygen.[2] Noncovalently bound FAD reacts with sarcosine and NAD$^+$. In the presence of tetrahydrofolate, N^5,N^{10}-methylenetetrahydrofolate is released rather than formaldehyde.

Sarcosine oxidase also oxidizes L-proline and L-pipecolate slowly. The enzyme's absorption spectrum during the steady-state offers no evidence of semiquinoid flavin intermediates. Stopped-flow kinetic data indicate that 70% of the reaction flux occurs *via* the fully reduced enzyme— the remainder proceeding through a side pathway without forming the fully reduced enzyme.[3]

[1] S. Hayashi, M. Suzuki & S. Nakamura (1983) *BBA* **742**, 630.
[2] M. S. Jorns (1985) *Biochem.* **24**, 3189.
[3] Y. Kawamura-Konishi & H. Suzuki (1987) *BBA* **915**, 346.
[4] V. Schirch (1998) *CBC* **1**, 211.

Selected entries from *Methods in Enzymology* [vol, page(s)]:
General discussion: 52, 18; 56, 474
Other molecular properties: reconstituted system (components, **14**, 118; inhibitors, **14**, 118)

SARSAPOGENIN 3-β-GLUCOSYLTRANSFERASE

This transferase [EC 2.4.1.193] catalyzes the reaction of UDP-D-glucose with sarsapogenin (or (25*S*)-5β-spirostan-3β-ol) to produce UDP and (25*S*)-5β-spirostan-3β-ol 3-*O*-β-D-glucoside.[1] The enzyme participates in the biosynthesis of plant saponins and is specific to 5β-spirostanols. Another substrate is (25*R*)-5β-spirostan-3β-ol (*i.e.*, smilagenin).

This transferase is not identical with sterol glucosyltransferase [EC 2.4.1.173] or nuatigenin 3-β-glucosyltransferase [EC 2.4.1.192].

[1] C. Paczkowski & Z. A. Wojciechowski (1988) *Phytochemistry* **27**, 2743.

*Sau*3AI RESTRICTION ENDONUCLEASE

This type II restriction endonuclease [EC 3.1.21.4] is obtained from *Streptomyces aureus* 3A and acts on both strands of DNA at 5′ ... ∇GATC... 3′.[1,2] *Sau*3AI is an isoschizomer of *Mbo*I restriction endonuclease; however, *Sau*3AI is not blocked by *dam* methylation.

[1] J. S. Sussenbach, C. H. Monfoort, R. Schiphof & E. E. Stobberingh (1976) *NAR* **3**, 3193.
[2] E. E. Stobberingh, R. Schiphof & J. S. Sussenbach (1977) *J. Bacteriol.* **131**, 645.

Selected entries from *Methods in Enzymology* [vol, page(s)]:
Other molecular properties: recognition sequence, **65**, 9

*Sau*I RESTRICTION ENDONUCLEASE

This type II restriction endonuclease [EC 3.1.21.4] is obtained from *Streptomyces aureofaciens* and acts on both strands of DNA at 5′ ... CC∇TNAGG... 3′ where N represents any base.[1] (The symbol ∇ indicates the cleavage site.)

[1] J. Timko, A. H. Horwitz, J. Zelinka & G. Wilcox (1981) *J. Bacteriol.* **145**, 873.

Selected entries from *Methods in Enzymology* [vol, page(s)]:
Other molecular properties: fragment end structure produced, **65**, 511; recognition sequence, **65**, 511

*Sau*96I RESTRICTION ENDONUCLEASE

This type II restriction endonuclease [EC 3.1.21.4] is obtained from *Streptomyces aureus* PS96 and acts on both strands of DNA at 5′ ... G∇GNCC... 3′ where N represents any base.[1] (The symbol ∇ indicates the cleavage site.) The reaction can be blocked by overlapping *dam* methylation.

[1] J. S. Sussenbach, P. H. Steenbergh, J. A. Rost, W. J. van Leeuwen & J. D. A. van Embden (1978) *NAR* **5**, 1153.

Selected entries from *Methods in Enzymology* [vol, page(s)]:
Other molecular properties: recognition sequence, **65**, 10

*Sbf*I RESTRICTION ENDONUCLEASE

This type II restriction endonuclease [EC 3.1.21.4] is obtained from *Streptomyces* species Bf-61 and acts on both strands of DNA at 5′ ... CCTGCA∇GG... 3′. (The symbol ∇ indicates the cleavage site.)

*Sca*I RESTRICTION ENDONUCLEASE

This type II restriction endonuclease [EC 3.1.21.4] is obtained from *Streptomyces caespitosus* and acts on both strands of DNA at 5′ ... AGT∇ACT... 3′, producing blunt ends.[1,2] (The symbol ∇ indicates the cleavage site.) High enzyme concentrations may result in star activity.
See *"Star" Activity*

[1] H. Takahashi, H. Kojima & H. Saito (1985) *Biochem. J.* **231**, 229.
[2] K. Kita, N. Hiraoka, F. Kimizuka, A. Obayashi, H. Kojima, H. Takahashi & H. Saito (1985) *NAR* **13**, 7015.

SCOPOLETIN GLUCOSYLTRANSFERASE

This glucosyltransferase [EC 2.4.1.128] catalyzes the reaction of UDP-D-glucose with the plant natural product scopoletin to produce UDP and scopolin.[1]

scopoletin   scopolin

[1] F. Hino, M. Okazaki & Y. Miura (1982) *Plant Physiol.* **69**, 810.

(S)-SCOULERINE 9-*O*-METHYLTRANSFERASE

This plant methyltransferase [EC 2.1.1.117], which participates in the biosynthesis of protoberberine alkaloids, catalyzes the reaction of *S*-adenosyl-L-methionine with (*S*)-scoulerine to produce *S*-adenosyl-L-homocysteine and (*S*)-tetrahydrocolumbamine.[1,2]

(S)-scoulerine   (S)-tetrahydrocolumbamine

[1] S. Muemmler, M. Rueffer, N. Nagakura & M. H. Zenk (1985) *Plant Cell Rep.* **4**, 36.
[2] T. Hashimoto & Y. Yamada (1994) *Ann. Rev. Plant Physiol. Plant Mol. Biol.* **45**, 257.

*Scr*FI RESTRICTION ENDONUCLEASE

This type II restriction endonuclease [EC 3.1.21.4] is obtained from *Streptococcus cremoris* F and acts on both strands of DNA at 5′ ... CC∇NGG ... 3′ where N refers to any base.[1] (The symbol ∇ indicates the cleavage site.) Note that the resulting DNA fragments have a single-base 5′ extension that is more difficult to ligate than blunt-ended fragments. The enzyme is blocked by overlapping *dcm* methylation.

[1] G. F. Fitzgerald, C. Daly, L. R. Brown & T. R. Gingeras (1982) *NAR* **10**, 8171.

SCYTALIDOPEPSIN A

This pepstatin-insensitive aspartic endopeptidase [EC 3.4.23.31], also known as *Scytalidium* aspartic proteinase A, catalyzes peptide-bond hydrolysis in proteins with a specificity that is similar, but not identical, to pepsin A [EC 3.4.23.1].[1,2]

[1] K. Oda & S. Murao (1991) in *Structure and Function of the Aspartic Proteinases* (B. M. Dunn, ed.), Plenum, New York, p. 185.
[2] K. Oda, S. Takahashi, T. Shin & S. Murao (1995) in *Aspartic Proteinases: Structure, Function, Biology, and Biomedical Implications* (K. Takahashi, ed.), Plenum, New York, p. 529.

SCYTALIDOPEPSIN B

This aspartic endopeptidase [EC 3.4.23.32], also called *Scytalidium* aspartic proteinase B and *Scytalidium* pepstatin-insensitive carboxyl proteinase B, catalyzes peptide-bond hydrolysis in proteins with a broad specificity.[1] The enzyme cleaves the oxidized insulin B chain at Glu13–Ala14, Ala14–Leu15, Phe24–Phe25, and Tyr26–Thr27 (scytalidopepsin B appears to be the only known aspartic proteinase that will hydrolyze the Tyr26–Thr27 bond). The enzyme will also hydrolyze the His6–Pro7 bond in angiotensin I. Scytalidopepsin B is insensitive to pepstatin and is inhibited by 1,2-epoxy-3-(*p*-nitrophenoxy)propane (modifying Glu53 which replaces one of the aspartyl residues at the active site; thus, with Asp98, scytalidopepsin B is both an aspartic and a glutamic endopeptidase).

[1] K. Oda, S. Takahashi, T. Shin & S. Murao (1995) in *Aspartic Proteinases: Structure, Function, Biology, and Biomedical Implications* (K. Takahashi, ed.), Plenum, New York, p. 529.

SCYTALONE DEHYDRATASE

This dehydratase [EC 4.2.1.94] catalyzes the conversion of scytalone to 1,3,8-trihydroxynaphthalene and water.

scytalone   1,3,8-trihydroxynaphthalene

Together with tetrahydroxynaphthalene reductase [EC 1.1.1.252], the dehydratase participates in melanin biosynthesis in certain pathogenic fungi. Two tyrosyl residues assist in protonation of the substrate's carbonyl group through a water molecule, and kinetic isotope effect data suggest that the steps after dehydration are rate-contributing.[1,2]

[1] G. S. Basarab, J. J. Steffens, Z. Wawrzak, R. S. Schwartz, T. Lundqvist & D. B. Jordan (1999) *Biochemistry* **38**, 6012.
[2] V. E. Anderson (1998) *CBC* **2**, 115.

SECONDARY-ALCOHOL OXIDASE

This iron-dependent oxidoreductase [EC 1.1.3.18] catalyzes the reaction of many polyvinyl alcohols (220-to-1500 molecular weight range) with dioxygen to produce ketone products and hydrogen peroxide.[1,2]

[1] M. Shimao, T. Tsuda, M. Takahashi, N. Kato & C. Sakazawa (1983) *FEMS Microbiol. Lett.* **20**, 429.
[2] T. Suzuki (1978) *Agric. Biol. Chem.* **42**, 1187.

SEDOHEPTULOKINASE

This phosphotransferase [EC 2.7.1.14], also known as heptulokinase, catalyzes the ATP-dependent phosphorylation of sedoheptulose to form ADP and sedoheptulose 7-phosphate.[1] Mouse liver sedoheptulokinase phosphorylates at the C1-position.[2]

[1] M. Ebata, R. Sato & T. Bak (1955) *J. Biochem.* **42**, 715.
[2] H. Iwai, A. Kiyomoto & M. Takeshita (1968) *BBA* **176**, 302.

SEDOHEPTULOSE-1,7-BISPHOSPHATASE

This enzyme [EC 3.1.3.37] catalyzes the hydrolysis of sedoheptulose 1,7-bisphosphate to produce sedoheptulose 7-phosphate and orthophosphate.

[3] N. Esaki, N. Karai, T. Nakamura, H. Tanaka & K. Soda (1985) *ABB* **238**, 418.

[4] H. Mihara, T. Kurihara, T. Yoshimura, K. Soda & N. Esaki (1997) *JBC* **272**, 22417.

[5] G. M. Lacourciere & T. C. Stadtman (1998) *JBC* **273**, 30921.

[6] H. Mihara, T. Kurihara, T. Watanabe, T. Yoshimura & N. Esaki (2000) *JBC* **275**, 6195.

[7] R. Daher & F. Van Lente (1992) *J. Trace Elem. Electrolytes Health Dis.* **6**, 189.

[8] T. C. Stadtman (1990) *ARB* **59**, 111.

Selected entries from *Methods in Enzymology* [vol, page(s)]:
General discussion: 143, 415, 493
Assay: 143, 415
Other molecular properties: properties, 143, 418, 494; purification from (*Citrobacter freundii*, 143, 493; porcine liver, 143, 416); selenocysteine degradation, 107, 581

SELF-PROCESSING N-TERMINAL NUCLEOPHILE AMIDOHYDROLASES

This category of enzymes, also known as N-terminal-nucleophile hydrolases (Ntn-hydrolases) and Ntn-amido-hydrolases, engages peptide-bond cleavage to generate the catalytically active form of the enzyme. Members of this group undergo autoproteolysis of the nascent chain to generate a new N-terminal cysteinyl, threonyl, or seryl residue required for catalytic activity.[11,22] There are three subsets of these enzymes:

(a) The first subset includes those proenzymes that are processed autocatalytically to generate two or more polypeptide chains, where the new *N*-terminus of one chain becomes the active-site nucleophile for the active enzyme. Examples include glycosylasparaginase, plant asparaginase, penicillin acylase, cephalosporin acylase, and acyl-CoA:isopenicillin

(b) The second subset involves proenzymes that simply undergo autocatalytic release of an *N*-terminal peptide fragment, thus exposing the new N-terminal nucleophile. Examples include the proteasome and some forms of glutamine:PRPP amidotransferases (such as amidophosphoribosyltransferase, EC 2.4.2.14).

(c) The last subset includes those proenzymes whose essential nucleophilic aminoacyl residue is the second residue in the proenzyme sequence, adjacent to the initiation *N*-terminal methionyl residue. Upon action of methionine aminopeptidase, this aminoacyl residue becomes exposed.

[1] J. A. Brannigan, G. Dodson, H. J. Duggleby, P. C. E. Moody, J. L. Smith, D. R. Tomchick & A. G. Murzin (1995) *Nature* **378**, 416.

[2] N. N. Aronson, Jr., (1996) *Glycobiology* **6**, 669.

SELENIDE, WATER DIKINASE

This magnesium-dependent enzyme [EC 2.7.9.3], also known as selenophosphate synthetase and selenium donor protein, catalyzes the reaction of ATP with selenide and water to produce AMP, selenophosphate, and orthophosphate.[1-4] Positional isotope exchange data indicate that the reaction proceeds via a phosphoryl-enzyme intermediate.[2,3] In the absence of selenide, ATP is converted quantitatively to AMP and two molecules of orthophosphate in a slow partial reaction.[3]

[1] Z. Veres, I. Y. Kim, T. D. Scholz & T. C. Stadtman (1994) *JBC* **269**, 10597.
[2] G. M. Lacourciere (1999) *Biofactors* **10**, 237.
[3] H. Walker, J. A. Ferretti & T. C. Stadtman (1998) *PNAS* **95**, 2180.
[4] T. C. Stadtman (1996) *ARB* **65**, 83.

Selected entries from *Methods in Enzymology* [vol, page(s)]:
Assay: 252, 313
Other molecular properties: gene, 252, 309; metal dependence, 252, 315; purification of recombinant *Escherichia coli* enzyme, 252, 312; stability, 252, 315; substrate specificity, 252, 315

SELENOCYSTEINE LYASE

This pyridoxal-phosphate-dependent enzyme [EC 4.4.1.16], also called selenocysteine reductase, catalyzes the reaction of L-selenocysteine with a reduced acceptor to produce hydrogen selenide, L-alanine, and the oxidized acceptor.[1-8] The reduced acceptor can be dithiothreitol (DTT) or 2-mercaptoethanol. During the course of the reaction, the enzyme produces elemental selenium which is then nonenzymatically reduced to H_2Se by dithiothreitol or another reducing agent.[3] Note that cysteine desulfurase (or, cysteine sulfinate desulfinase) of *Azotobacter vinelandii* acts indiscriminately on both cysteine and selenocysteine.[4] The lyase reportedly functions as well as a selenium delivery protein to selenophosphate synthetase in selenoprotein biosynthesis.[5,6]

L-Cysteine is a competitive inhibitor for the pig liver and *Citrobacter freundii* enzymes[1,2] but a noncompetitive inhibitor of the human enzyme.[7] β-Chloro-L-alanine, which reacts to form ammonia, pyruvate, and chloride ion, irreversibly inactivates the *Citrobacter freundii* enzyme.[2]

[1] N. Esaki, T. Nakamura, H. Tanaka & K. Soda (1982) *JBC* **257**, 4386.
[2] P. Chocat, N. Esaki, K. Tanizawa, K. Nakamura, H. Tanaka & K. Soda (1985) *J. Bact.* **163**, 669.

SEMENOGELASE

This serine endopeptidase [EC 3.4.21.77], also known as prostate-specific antigen [PSA], γ-seminoprotein, seminin, and P-30 antigen, catalyzes chymotrypsin-like peptide-bond hydrolysis. Specificity studies indicate that the P1 position is either a bulky, hydrophobic aminoacyl residue (such as Tyr, Phe, Leu, Val) or a basic residue (His, Lys, Arg). Semenogelase forms a complex with α_1-proteinase inhibitor.[1,2] **See also Glutamate Carboxypeptidase II**

[1] H. Lilja (1993) *World J. Urol.* **11**, 188.
[2] J. Malm & H. Lilja (1995) *Scand. J. Clin. Lab. Invest.* **221**, 15.

SENESCENCE CYSTEINE PROTEINASES

These plant cysteine endopeptidases, abbreviated SenCP, are isolated from senescing petals and ovaries as well as yellowing leaves. They catalyze peptide-bond hydrolysis in a number of proteins (for example, D-ribulose 1,5-bisphosphate carboxylase/oxidase). Most of these proteases are monomeric and have an acidic pH optimum (although some have a neutral optima).[1] In many plants, SebCp is regulated by ethylene.

[1] A. Granell, M. Cercós & J. Carbonell (1998) in *Handbook of Proteolytic Enzymes* (A. J. Barrett, N. D. Rawlings & J. F. Woessner, eds.), Academic Press, New York, p. 578.

SEPIAPTERIN DEAMINASE

This deaminase [EC 3.5.4.24] catalyzes the reaction of sepiapterin with water to produce xanthopterin-B2 and ammonia.[1–3] Isosepiapterin is also an alternative substrate.

[1] M. Tsusue & T. Mazda (1977) *Experientia* **33**, 854.
[2] M. Tsusue (1971) *J. Biochem.* **69**, 781.
[3] M. Tsusue (1967) *Experientia* **23**, 116.

SEPIAPTERIN REDUCTASE

This oxidoreductase [EC 1.1.1.153] catalyzes the reaction of sepiapterin with NADPH to produce 7,8-dihydrobiopterin and $NADP^+$.[1,2]

[1] E. R. Werner, H. Wachter & G. Werner-Felmayer (1997) *MIE* **281**, 53.
[2] C. A. Nichol, G. K. Smith & D. S. Duch (1985) *ARB* **54**, 729.

Selected entries from *Methods in Enzymology* [vol, page(s)]:
Other molecular properties: conversion of sepiapterin to dihydrobiopterin, **18B**, 617

SEQUOYITOL DEHYDROGENASE

This oxidoreductase [EC 1.1.1.143] catalyzes the reaction of 5-O-methyl-*myo*-inositol with NAD^+ to produce 5D-5-O-methyl-2,3,5/4,6-pentahydroxycyclohexanone and NADH with B-specificity with respect to the coenzyme.[1]

[1] H. Ruis & O. Hoffmann-Ostenhof (1969) *EJB* **7**, 442.

SERINE O-ACETYLTRANSFERASE

This acetyltransferase [EC 2.3.1.30] catalyzes the reaction of acetyl-CoA with L-serine to produce Coenzyme A and O-acetyl-L-serine.[1,2] The *Salmonella typhimurium* enzyme has a ping pong Bi Bi kinetic mechansm.[3] The *Escherichia coli* enzyme exists as a dimer of trimers.[4] Two types of plant serine O-acetyltransferase have been identified; only one is allosterically inhibited by L-cysteine.

[1] N. M. Kredich & G. M. Tomkins (1966) *JBC* **241**, 4955.
[2] I. K. Smith & J. F. Thompson (1971) *BBA* **227**, 288.
[3] L. S. Leu & P. F. Cook (1994) *Biochem.* **33**, 2667.
[4] V. J. Hindson, P. C. Moody, A. J. Rowe & W. V. Shaw (2000) *JBC* **275**, 461.

Selected entries from *Methods in Enzymology* [vol, page(s)]:
General discussion: 17B, 459, 464
Assay: 17B, 464

SERINE CARBOXYPEPTIDASE

This is a large class of carboxypeptidases, all of which have an essential seryl residue at the active site. **See Carboxypeptidase C; Carboxypeptidase D; Carboxypeptidase Y**

D-SERINE DEHYDRATASE

This pyridoxal-phosphate-dependent enzyme [EC 4.2.1.14], also known as D-hydroxyamino acid dehydratase, catalyzes the reaction of D-serine with water to produce pyruvate, ammonia, and water.[1–6] D-Threonine is an alternative substrate, producing α-ketobutyrate. Isotope effect investigations disclosed that the rate-determining step in the catalytic pathway varied with the substrate structure and pH. D-Serine, in addition to being the best substrate, was processed faster than the other substrates through steps in the catalytic pathway which were not rate-controlling.[4]

[1] D. Dupourque, W. A. Newton & E. E. Snell (1966) *JBC* **241**, 1233.
[2] S. Kikuchi & M. Ishimoto (1978) *J. Biochem.* **84**, 1133.
[3] W. Dowhan, Jr., & E. E. Snell (1970) *JBC* **245**, 4618.
[4] C. S. Federiuk, R. Bayer & J. A. Shafer (1983) *JBC* **258**, 5379.
[5] A. E. Braunstein & E. V. Goryachenkova (1984) *AE* **56**, 1.
[6] L. Davis & D. E. Metzler (1972) *The Enzymes*, 3rd ed., **7**, 33.

Selected entries from *Methods in Enzymology* [vol, page(s)]:
General discussion: 2, 322; 17B, 356
Assay: 2, 322; 17B, 356
Other molecular properties: D-glucosaminate dehydratase, similarity to, **9**, 660; properties, **2**, 323; **17B**, 360; purification from *Escherichia coli*, **2**, 322; **17B**, 357; stereochemistry, **87**, 149

L-SERINE DEHYDRATASE

This pyridoxal-phosphate-dependent enzyme [EC 4.2.1.13], also known as serine deaminase and L-hydroxyamino acid dehydratase, catalyzes the reaction of L-serine with water to produce pyruvate, ammonia, and water.[1-6] The enzyme isolated from a number of sources also has an L-threonine dehydratase activity, producing α-ketobutyrate.

[1] P. N. Porter, M. S. Grishaver & O. W. Jones (1974) BBA **364**, 128.
[2] J. Hoshino, D. Simon & H. Kröger (1972) EJB **27**, 388.
[3] M. E. Farias, A. M. Strasser de Saad, A. A. Pesce de Ruiz Holgado & G. Oliver (1985) J. Gen. Appl. Microbiol. **31**, 563.
[4] E. B. Newman & V. Kapoor (1980) Can. J. Biochem. **58**, 1292.
[5] A. E. Braunstein & E. V. Goryachenkova (1984) AE **56**, 1.
[6] L. Davis & D. E. Metzler (1972) The Enzymes, 3rd ed., **7**, 33.

Selected entries from **Methods in Enzymology** [vol, page(s)]:
General discussion: **2**, 319; **5**, 942; **17B**, 346, 351
Assay: **2**, 319; **5**, 942; **17B**, 346, 351
Other molecular properties: D-glucosaminate dehydratase, similarity to, **9**, 660; properties, **2**, 322; **5**, 946; **17B**, 349, 355; purification from (Clostridium acidiurici, **17B**, 353; Neurospora crassa, **2**, 320; rat liver, **5**, 944; **17B**, 347); separation from cystathionine synthase, **17B**, 450, 458

SERINE DEHYDROGENASE

This oxidoreductase [EC 1.4.1.7] catalyzes the reaction of L-serine with water and NAD^+ to produce 3-hydroxypyruvate, ammonia, and NADH.[1,2]

[1] V. L. Kretovich & K. M. Stepanowich (1966) Izv. Akad. Nauk S.S.S.R. Ser. Biol. **2**, 295.
[2] N. M. W. Brunhuber & J. S. Blanchard (1994) Crit. Rev. Biochem. Mol. Biol. **29**, 415.

SERINE:ETHANOLAMINE BASE-EXCHANGE ENZYME

This transferase catalyzes the reversible reaction of phosphatidylserine with ethanolamine to produce phosphatidylethanolamine and serine. Other substrates include phosphatidylcholine, phosphatidylmonomethylethanolamine, and phosphatidyldimethylethanolamine.

Selected entries from **Methods in Enzymology** [vol, page(s)]:
General discussion: **71**, 588; **209**, 341
Assay: **71**, 594; **209**, 345
Other molecular properties: brain, from, **71**, 592; cerebral, rat (assay, **209**, 345; properties, **209**, 346; separation and solubilization, **209**, 342); properties, **71**, 596; **209**, 346; purification, **71**, 596; **209**, 342

SERINE-ETHANOLAMINEPHOSPHATE PHOSPHODIESTERASE

This phosphodiesterase [EC 3.1.4.13] catalyzes the hydrolysis of serine-phosphoethanolamine to produce serine and ethanolamine phosphate.[1,2] The enzyme will act only on those phosphodiesters having ethanolamine as a component, such as L-threonine ethanolamine phosphate, bisethanolamine phosphate, and L-lombricine.

[1] D. D. Hagerman, H. Rosenberg, A. H. Ennor, P. Schiff & S. Inoue (1965) JBC **240**, 1108.
[2] J. M. Chalovich & M. Barany (1980) ABB **199**, 615.

SERINE:GLYOXYLATE AMINOTRANSFERASE

This pyridoxal-phosphate-dependent aminotransferase [EC 2.6.1.45] catalyzes the reaction of L-serine with glyoxylate to produce 3-hydroxypyruvate and glycine.[1] Alternative substrates for glyoxylate include α-ketomalonate and pyruvate. The enzyme from rye seedlings has a ping pong Bi Bi kinetic mechanism.[2]

[1] I. K. Smith (1973) BBA **321**, 156.
[2] A. Paszkowski (1991) Acta Biochim. Pol. **38**, 437.

Selected entries from **Methods in Enzymology** [vol, page(s)]:
General discussion: **188**, 361
Assay: assay in crude extracts of methylotrophs, **188**, 362

SERINE HYDROXYMETHYLTRANSFERASE

This pyridoxal-phosphate-dependent transferase [EC 2.1.2.2], also known as serine hydroxymethylase, serine transhydroxymethylase, and by the EC-recommended name glycine hydroxymethyltransferase, catalyzes the reversible reaction of 5,10-methylenetetrahydrofolate with glycine and water to produce tetrahydrofolate and L-serine by means of an aldol condensation reaction.[1-9] The enzyme catalyzes the exchange of H_S proton on glycine with the solvent; the pro-(2R) proton will also exchange at a slower rate. The enzyme also catalyzes the reaction of glycine with acetaldehyde to form L-threonine. In the case of 4-trimethylammoniobutanal, the product is 3-hydroxy-N^6,N^6,N^6-trimethyl-L-lysine.

[1] R. J. Ulevitch & R. G. Kallen (1977) Biochemistry **16**, 5342, 5350, and 5355.
[2] M. Akhtar & H. A. El-Obeid (1972) BBA **791**, 791.
[3] K. S. Ramesh & N. A. Rao (1980) BJ **187**, 623.
[4] M. K. Mitchell, P. H. S. Reynolds & D. G. Blevins (1986) Plant Physiol. **81**, 553.
[5] V. Schirch (1998) CBC **1**, 211.
[6] R. G. Matthews & J. T. Drummond (1990) Chem. Rev. **90**, 1275.
[7] L. Schirch (1982) AE **53**, 83.
[8] S. J. Benkovic (1980) ARB **49**, 227.
[9] J. I. Rader & F. M. Huennekens (1973) The Enzymes, 3rd ed., **9**, 197.

Selected entries from **Methods in Enzymology** [vol, page(s)]:
General discussion: **5**, 838, 931; **17B**, 335; **188**, 365; **281**, 146
Assay: **5**, 838; **17B**, 335; **188**, 366
Other molecular properties: detection in serine cycle methylotrophs, **188**, 372; dihydrofolate reductase, free of in preparation of, **6**, 368; Escherichia coli, site-directed mutants, acid-base catalysis, **249**, 117; N^{10}-formyltetrahydrofolate deacylase, in assay of, **6**, 373; glyoxylate-activated, purification, **188**, 368; glyoxylate-insensitive, purification, **188**, 369; inhibition, **32**, 576; properties, **5**, 843; **17B**, 339;

188, 370; purification of liver enzyme from (chicken, **5**, 840; rabbit, **17B**, 336)

SERINE *C*-PALMITOYLTRANSFERASE

This pyridoxal-phosphate-dependent transferase [EC 2.3.1.50] catalyzes the reaction of palmitoyl-CoA with L-serine to produce Coenzyme A, 3-dehydro-D-sphinganine ($HOCH_2CH(NH_3^+)CO(CH_2)_{14}CH_3$), and carbon dioxide.[1,2]

[1] W. M. Holleran, M. L. Williams, W. N. Gao & P. M. Elias (1990) *J. Lipid Res.* **31**, 1655.
[2] R. D. Williams, E. Wang & A. H. Merrill (1984) *ABB* **228**, 282.

Selected entries from *Methods in Enzymology* [vol, page(s)]:
General discussion: 209, 427; **311**, 3, 130
Assay: 209, 428

SERINE-PHOSPHOETHANOLAMINE SYNTHASE

This enzyme [EC 2.7.8.4] catalyzes the reaction of CDP-ethanolamine with L-serine to produce CMP and L-serine-phosphoethanolamine.[1,2]

[1] A. K. Allen & H. Rosenberg (1968) *BBA* **151**, 504.
[2] H. Rosenberg & A. H. Ennor (1966) *BBA* **115**, 23.

SERINE:PYRUVATE AMINOTRANSFERASE

This pyridoxal-phosphate-dependent aminotransferase [EC 2.6.1.51] catalyzes the reversible reaction of L-serine with pyruvate to produce 3-hydroxypyruvate and L-alanine.[1–3] The rabbit, rat, human, and dog liver enzymes are identical with alanine:glyoxylate aminotransferase [EC 2.6.1.44].

[1] D. W. Rehfeld & N. E. Tolbert (1972) *JBC* **247**, 4803.
[2] T. Noguchi & S. Fujiwara (1988) *JBC* **263**, 182.
[3] T. Oda, H. Miyajima, Y. Suzuki, T. Ito, S. Yokota, M. Hoshino & A. Ichiyama (1989) *J. Biochem.* **106**, 460.

Selected entries from *Methods in Enzymology* [vol, page(s)]:
Other molecular properties: marker enzyme for peroxisomes, **23**, 682

SERINE RACEMASE

This pyridoxal-phosphate-dependent enzyme catalyzes the interconversion of L-serine and D-serine. Other substrates include alanine.[1] *See also Protein-Serine Epimerase; Alanine Racemase*

[1] M.-L. Svensson & S. Gatenbeck (1981) *Arch. Microbiol.* **129**, 213.

SERINE-SULFATE AMMONIA-LYASE

This enzyme [EC 4.3.1.10] catalyzes the reaction of L-serine *O*-sulfate with water to produce pyruvate, ammonia, and sulfate.[1–3]

[1] J. H. Thomas & N. Tudball (1967) *BJ* **105**, 467.

[2] N. Tudball, P. Thomas & R. Bailey-Wood (1971) *BJ* **121**, 747.
[3] N. Tudball & P. Thomas (1972) *BJ* **128**, 41.

Selected entries from *Methods in Enzymology* [vol, page(s)]:
General discussion: 17B, 361
Assay: 17B, 362
Other molecular properties: properties, **17B**, 366; purification from rat liver, **17B**, 364

SERINE/THREONINE-SPECIFIC PROTEIN PHOSPHATASE

This phosphatase [EC 3.1.3.16], also known as phosphoprotein phosphatase and protein phosphatase-1, catalyzes the hydrolysis of a phosphoprotein (in which the phosphoryl group is esterified to a seryl or threonyl hydroxyl group) to produce a protein and orthophosphate.[1–5] Interestingly, the spleen enzyme reportedly acts on phenolic phosphates and phosphamides as well (that is, protein-tyrosine phosphatase [EC 3.1.3.48] and phosphoamidase activities [EC 3.9.1.1]). *See also Protein-Tyrosine-Phosphatase; specific protein phosphatase [for example, Phosphorylase Phosphatase]*

[1] E. Y. Lee, L. Zhang, S. Zhao, Q. Wei, J. Zhang, Z. Q. Qi & E. R. Belmonte (1999) *Front. Biosci.* **4**, D270.
[2] A. C. Hengge (1998) *CBC* **1**, 517.
[3] E. Waelkens, P. A. Goris & W. Merlevede (1987) *Adv. Enzyme Regul.* **26**, 241.
[4] S. I. Walaas & P. Greengard (1987) *The Enzymes*, 3rd ed., **18**, 285.
[5] L. M. Ballou & E. H. Fischer (1986) *The Enzymes*, 3rd ed., **17**, 311.

Selected entries from *Methods in Enzymology* [vol, page(s)]:
General discussion: 6, 211; **79**, 142, 147; **107**, 102; **139**, 79; **159**, 356, 390, 409
Assay: 60, 526; calmodulin-dependent, **102**, 231; cell extracts, in, **201**, 389; Mg^{2+}-dependent, **159**, 417, 438; porcine kidney, **90**, 408; PPI in cell extracts, **201**, 392; PPI$_G$, **201**, 420; PP2A in cell extracts, **201**, 392; PP2C in cell extracts, **201**, 392; serine/threonine-specific, assay, **159**, 341; turkey gizzard smooth muscle, **85**, 314; types 1 and 2, **107**, 110; type 1, **159**, 392; type 2A, **159**, 392; type 2B, calcium-dependent, **159**, 409; type 2C, Mg^{2+}-dependent, **159**, 417; types C-I and C-II, **159**, 359; tyrosine-specific, **159**, 340; yeast cAMP cascade system mutants, in, **159**, 41
Other molecular properties: action on enzymes, **4**, 265; activity, **60**, 530; alternative substrate method, **64**, 327; analysis with calmodulin derivatives, **159**, 605; assay in cell extracts, **201**, 389; cascade control, **64**, 299; classification, **107**, 103; **201**, 397; functional role, **60**, 522; gene family, amino acid sequence, **201**, 405; identification, **107**, 118; inhibition by (double-stranded RNA, **79**, 142; okadaic acid in intact cells, **201**, 469; poly(I):poly(C), **79**, 179); Mg^{2+}-dependent (assays, **159**, 438; properties, **159**, 444; purification from rat liver, **159**, 440; related enzymes, **159**, 445); molecular weights, **60**, 529; phosphoprotein phosphatase 2Ac (assay, **262**, 544; purification from bovine heart, **262**, 544; role in SV40 DNA replication, **262**, 543); phosphoprotein substrates, preparation, **107**, 108; plasma membrane, **31**, 90; porcine kidney (assay, **90**, 408; properties, **90**, 412; purification, **90**, 410); PPI (assay in cell extracts, **201**, 392; cDNA probes, in screening PP isoforms and homologs, **201**, 403; cloning, **201**, 398; recombinant baculovirus-expressed, reactivation, **201**, 408); PPI$_G$ (assay, **201**, 420; binding to glycogen, assay, **201**, 417; C and G subunits, dissociation, **201**, 419; G subunit, immunoblotting, **201**, 426; immunoprecipitation, **201**, 425; phosphopeptide mapping, **201**, 421; phosphorylation, **201**, 420, 421; purification, **201**, 416); PP2A (assay in cell extracts, **201**, 392; cloning, **201**, 398); PP2B, cloning

procedures, **201**, 402; PP2C (assay in cell extracts, **201**, 392; cloning procedures, **201**, 402); prokaryotic, properties and purification, **200**, 227; protein phosphokinase and, **6**, 218; purification, **60**, 528; protein-phosphotyrosine-phosphatase inhibitor, vanadate as, **201**, 477; types 1 and 2 (assays, **107**, 110; isolation, **107**, 120; protein modulators, purification, **107**, 124; purification, **107**, 123; regulators, assay, **107**, 116; substrate specificity, **107**, 104; tissue distribution, **107**, 121); type 1 (assay, **159**, 392; properties, **159**, 406; purification from rabbit muscle, **159**, 395, 404); type 2A (assay, **159**, 392; properties, **159**, 406; purification from rabbit muscle, **159**, 395, 400); type 2B, calcium-dependent (assay, **159**, 409; properties, **159**, 414; purification from rabbit muscle, **159**, 411); type 2C, Mg^{2+}-dependent (assay, **159**, 417; purification from [rabbit liver, **159**, 423; rabbit muscle, **159**, 418]); types C-I and C-II (assays, **159**, 359; properties, **159**, 358; purification from rabbit muscle, **159**, 360); tyrosine-specific, assay, **159**, 340; yeast cAMP cascade system mutants, in, assay, **159**, 41

SERINE-TYPE D-Ala-D-Ala CARBOXYPEPTIDASE

This carboxypeptidase [EC 3.4.16.4], also known as D-ala-nyl-D-alanine carboxypeptidase and DD-peptidase, catalyzes the hydrolysis of R-D-alanyl-D-alanine to produce molecules of D-alanine and R-D-alanine. It also has a D-alanyl-D-alanine dipeptidase activity. Other substrates include $N^{\alpha},N^{\varepsilon}$-diacetyl-Lys-D-Ala-D-Ala and $N^{\alpha},N^{\varepsilon}$-diacetyl-Lys-D-Ala-D-lactate. The enzyme also catalyzes the transpeptidation of peptidyl-alanyl moieties that are N-acetyl-substituents of D-alanine. The enzyme mechanism proceeds through the classical acyl enzyme pathway of serine proteases. *See also Penicillin-Binding Protein 5; Zinc-Type D-Ala-D-Ala Carboxypeptidase*

Selected entries from *Methods in Enzymology* [vol, page(s)]:
General discussion: 8, 487; 28, 692; 34, 398, 405; 43, 690
Assay: 8, 487; 28, 692; 43, 690; 45, 611 (for D,D-carboxylase activity, 45, 611; for β-lactamase, 45, 613)
Other molecular properties: affinity labeling, 87, 472; D-alanine carboxypeptidase I and II, 17A, 182 (purification from *Escherichia coli*, 17A, 182); alanine, chemical estimation of free, 43, 690, 691; D-alanine, enzymic estimation of, 43, 691; cell wall biosynthesis and, 28, 692; crude extract preparation, 43, 692; enzymatic properties, 28, 696; fluorodinitrobenzene assay, 43, 690; incubation conditions, 43, 690; isolation from *Bacillus subtilis*, 28, 693; β-lactam antibiotics and, 28, 692, 696; penicillin G, and, 87, 472; properties, 8, 489; radioactive [^{14}C]Ac$_2$-L-Lys-D-Ala dipeptide estimation, 43, 691, 692; *Streptomyces* preparation, 43, 692; 45, 610, 614, 619 (electrophoresis, polyacrylamide gel, 45, 620; hydrolysis reaction, 45, 631; β-lactam antibiotics, by, inhibition, 45, 635; molecular weight, 45, 619, 620; properties, 45, 619, purification, 45, 615, 617, 618); *Streptomyces* K15 protein (enzyme production, 244, 252; purification, 244, 255); *Streptomyces* R61 protein (enzyme production, 244, 252; isoforms, 244, 257; purification, 244, 256); substrates, 43, 690; substrate specificity, 244, 249, 266; unit definition, 43, 690

SEROTONIN SULFOKINASE

This enzyme catalyzes the reaction of serotonin with 3′-phosphoadenylylsulfate to produce serotonin O-sulfate and adenosine 3′,5′-bisphosphate. The enzyme is probably

identical to aryl sulfotransferase [EC 2.8.2.1]. *See Aryl Sulfotransferase; Tyrosine-Ester Sulfotransferase*

Selected entries from *Methods in Enzymology* [vol, page(s)]:
General discussion: 17B, 825
Assay: 17B, 825
Other molecular properties: properties, 17B, 828; purification from rabbit liver, 17B, 827

SERRALYSIN

This zinc-dependent metalloendopeptidase [EC 3.4.24.40], first isolated from *Serratia* sp. E-15, catalyzes peptide-bond hydrolysis with a broad specificity, preferring small to medium-size hydrophobic aminoacyl residues (for example, glycyl or alanyl residues) in the P1′ position. Hydrophobic residues are also favored in the P2 and P2′ positions.[1,2]

[1] H. Maeda & K. Morihara (1995) *MIE* **248**, 395.
[2] W. L. Mock (1998) *CBC* **1**, 425.

Selected entries from *Methods in Enzymology* [vol, page(s)]:
General discussion: 248, 395
Assay: caseinolysis method, 248, 397; fluorescence polarization method, 248, 398; fluorimetric, with fluorogenic peptide substrates, 248, 399

SERRATIA MARCESCENS NUCLEASE

This endonuclease [EC 3.1.30.2], also known as "endonu-clease (*Serratia marcescens*)", catalyzes the endonucleo-lytic cleavage of double- and single-stranded RNA and DNA to form the corresponding 5′-phosphomononucleotide and 5′-phosphooligonucleotide end-products.[1–3]

[1] A. Stevens & R. J. Hilmoe (1960) *JBC* **235**, 3016 and 3023.
[2] M. Pietrzak, H. Cudny & M. Maluszynski (1980) *BBA* **614**, 102.
[3] G. Meiss, F. U. Gast & A. M. Pingoud (1999) *J. Mol. Biol.* **288**, 377.

L-SERYL-tRNA^{SEC} SELENIUM TRANSFERASE

This pyridoxal-phosphate-dependent transferase [EC 2.9.1.1] (also known as selenocysteine synthase [SeC synthase], L-selenocysteinyl-tRNA^{Sel} synthase, L-seleno-cysteinyl-tRNA^{Sec} synthase, cysteinyl-tRNA^{Sel}-selenium transferase, and cysteinyl-tRNA^{Sec}-selenium transferase) catalyzes the reaction of L-seryl-tRNA^{Sec} with selenophos-phate to produce L-selenocysteinyl-tRNA^{Sec}, water, and orthophosphate.[1] This enzyme is responsible for the inser-tion of selenocysteine into selenoproteins.

[1] K. Forchhammer & A. Bock (1991) *JBC* **266**, 6324.

SERYL-tRNA SYNTHETASE

This class II aminoacyl synthetase [EC 6.1.1.11], also known as serine:tRNA ligase and serine translase, catalyzes the reaction of ATP with L-serine and tRNA^{Ser} to produce

AMP, pyrophosphate, and L-seryl-tRNA^{Ser}.[1-4] Seryl-tRNA synthetase (and other aminoacyl-tRNA synthetases) also catalyzes the synthesis of P^1,P^4-di(adenosine-5') tetraphosphate (or Ap$_4$A), a cellular signal of stress conditions. In this reaction mechanism, ATP reacts with the enzyme-bound L-seryl-adenylate intermediate (*i.e.*, nucleophilic attack by the γ-phosphate of ATP to generate a pentacoordinated transition state) to form L-serine and Ap$_4$A). *See also Aminoacyl-tRNA Synthetases*

[1] J. R. Katze & W. Konigsberg (1970) *JBC* **245**, 923.
[2] E. A. Boeker & G. L. Cantoni (1973) *Biochemistry* **12**, 2384.
[3] E. A. Boeker, A. P. Hays & G. L. Cantoni (1973) *Biochemistry* **12**, 2378.
[4] H. Jakubowski & J. Pawelkiewicz (1975) *EJB* **52**, 301.

Selected entries from *Methods in Enzymology* [vol, page(s)]:
General discussion: 34, 167, 170
Other molecular properties: aminoacylation with, 267, 390, 404; preparation, 12B, 697; subcellular distribution, 59, 233, 234

SESQUITERPENE CYCLASE

This category represents a class of enzymes that catalyze the cyclization of sesquiterpenes. *See specific enzyme*

*Sfi*I RESTRICTION ENDONUCLEASE

This type II restriction endonuclease [EC 3.1.21.4] is obtained from *Streptomyces fimbriatus* and cleaves both strands of DNA at 5' ... GGCCNNNN∇NGGCC ... 3' where N refers to any base.[1-3] (The symbol ∇ indicates the cleavage site.) The endonuclease will convert a substrate with two recognition sites directly into the product cleaved at both sites, without liberating DNA cut at a single site: *i.e.*, two copies of the recognition site are required for the DNA cleavage reaction (*Sfi*I is a tetramer that interacts with two sites before cleaving DNA).[4,5]

[1] B.-Q. Qiang & I. Schildkraut (1984) *NAR* **12**, 4507.
[2] L. M. Wentzell, M. Oram & S. E. Halford (1994) *Biochem. Soc. Trans.* **22**, 302S.
[3] T. F. Nobbs, L. M. Wentzell, M. D. Szczelkum & S. E. Halford (1997) *The NEB Transcript* **8**, 10.
[4] D. T. Bilcock, L. E. Daniels, A. J. Bath & S. E. Halford (1999) *JBC* **274**, 36379.
[5] N. A. Gormley, A. J. Bath & S. E. Halford (2000) *JBC* **275**, 6928.

Selected entries from *Methods in Enzymology* [vol, page(s)]:
General discussion: 155, 15

*Sfo*I RESTRICTION ENDONUCLEASE

This type II restriction endonuclease [EC 3.1.21.4] is obtained from *Serratia fonticola* and acts on both strands of DNA at 5' ... GGC∇GCC ... 3', producing blunt-ended fragments. (The symbol ∇ indicates the cleavage site.)

*Sgf*I RESTRICTION ENDONUCLEASE

This type II restriction endonuclease [EC 3.1.21.4] is obtained from *Streptomyces griseoruber* and acts on both strands of DNA at 5' ... GCGAT∇CGC ... 3'.[1] (The symbol ∇ indicates the cleavage site.) This enzyme also exhibits star activity. *See "Star" Activity*

[1] J. R. Kappelman, M. Brady, K. Knoche, E. Murray, T. Schoenfeld, R. Williams & N. Vesselinova (1995) *Gene* **160**, 55.

*Sgr*AI RESTRICTION ENDONUCLEASE

This type II restriction endonuclease [EC 3.1.21.4] is obtained from *Streptomyces griseus* and acts on both strands of DNA at 5' ... CR∇CCGGYG ... 3', where R refers to A or G and Y refers to T or C.[1] (The symbol ∇ indicates the cleavage site.) This enzyme also exhibits star activity. *See "Star" Activity*

[1] N. Tautz, K. Kaluza, B. Frey, M. Jarsch, G. G. Schmitz & C. Kessler (1990) *NAR* **18**, 3087.

SHIKIMATE 5-DEHYDROGENASE

This oxidoreductase [EC 1.1.1.25], also known as 5-dehydroshikimate reductase, catalyzes the reversible reaction of shikimate with NADP$^+$ to produce 5-dehydroshikimate and NADPH.[1-4]

shikimate   5-dehydroshikimate

[1] E. J. Lourenco & V. A. Neves (1984) *Phytochemistry* **23**, 497.
[2] T. Koshiba (1978) *BBA* **522**, 10.
[3] G. W. Sanderson (1966) *BJ* **98**, 248.
[4] D. Balinsky & D. D. Davies (1961) *BJ* **80**, 292.

Selected entries from *Methods in Enzymology* [vol, page(s)]:
General discussion: 2, 301; 17A, 354, 387; 142, 315
Assay: 2, 301; 17A, 391; 142, 316
Other molecular properties: distribution, 2, 304; isotope exchange, 64, 8; properties, 2, 304; 142, 318; purification from (*Escherichia coli*, 2, 303; 142, 317; *Neurospora crassa*, 17A, 393; peas, 17A, 355); stereospecificity, 87, 104; tea leaves, from, 13, 559

SHIKIMATE *O*-HYDROXYCINNAMOYLTRANSFERASE

This transferase [EC 2.3.1.133] catalyzes the reaction of 4-coumaroyl-CoA with shikimate to produce Coenzyme A and 4-coumaroylshikimate.[1-4] Alternative acyl-CoA substrates include caffeoyl-CoA, feruloyl-CoA, and sinapoyl-CoA. The enzyme will also act on quinate, albeit not as efficiently as shikimate.

[1] B. Ulbrich & M. H. Zenk (1980) *Phytochemistry* **19**, 1625.
[2] D. Strack, H. Keller & G. Weissenböck (1987) *J. Plant Physiol.* **131**, 61.
[3] D. Strack, W. Gross, J. Heilemann, H. Keller & S. Ohm (1987) *Z. Naturforsch.* **43c**, 32.
[4] L. Lotfy, A. Fleuriet & J.-J. Macheix (1992) *Phytochemistry* **31**, 767.

SHIKIMATE KINASE

This phosphotransferase [EC 2.7.1.71] catalyzes the reaction of ATP with shikimate to produce ADP and shikimate 3-phosphate.[1–4]

[1] L. Huang, A. L. Montoya & E. W. Nester (1975) *JBC* **250**, 7675.
[2] J. R. Bowen & T. Kosuge (1979) *Plant Physiol.* **64**, 382.
[3] R. C. DeFeyter & J. Pittard (1986) *J. Bacteriol.* **165**, 331.
[4] R. Bentley (1990) *Crit. Rev. Biochem. Mol. Biol.* **25**, 307.

Selected entries from *Methods in Enzymology* [vol, page(s)]:
General discussion: 17A, 387; 142, 355
Assay: 142, 355
Other molecular properties: enzyme 1 (assay, 142, 355; properties, 142, 359; purification from *Escherichia coli*, 142, 357); enzyme 2 (assay, 142, 355; partial purification from *Escherichia coli*, 142, 359); *Neurospora crassa* purification, 17A, 393

SIALATE O-ACETYLESTERASE

This esterase [EC 3.1.1.53] catalyzes the hydrolysis of *N*-acetyl-*O*-acetylneuraminate to produce *N*-acetylneuraminate and acetate.[1,2] The hydrolase acts on free and glycosidically bound *N*-acetyl- or *N*-glycoloylneuraminate, mainly at 4-*O*- and 9-*O*-positions.

[1] R. Schauer, G. Reuter & S. Stoll (1988) *Biochimie* **70**, 1511.
[2] R. Schauer, G. Reuter, Stoll, F. Posadas del Rio, G. Herrler & H.-D. Klenk (1988) *Biol. Chem. Hoppe-Seyler* **369**, 1121.

Selected entries from *Methods in Enzymology* [vol, page(s)]:
General discussion: 138, 611
Assay: 179, 411; 230, 189
Other molecular properties: isolation, 138, 622; properties, 138, 623; 179, 413; purification from rat liver, 179, 412; sialate removal from glycoconjugates, 230, 194; sources, 230, 194

O-SIALOGLYCOPROTEIN ENDOPEPTIDASE

This metalloendopeptidase [EC 3.4.24.57], abbreviated glycoprotease, catalyzes the hydrolysis of peptide bonds in *O*-sialoglycoproteins (specifically, proteins that are rich in sialoglycans *O*-linked to threonyl or seryl residues). The enzyme does not act on unglycosylated proteins, proteins that have been desialylated, or proteins that are *N*-glycosylated.[1]

[1] A. Mellors & R. Y. C. Lo (1995) *MIE* **248**, 728.

Selected entries from *Methods in Enzymology* [vol, page(s)]:
General discussion: 248, 728
Assay: 248, 731 (κ-casein hydrolysis method, 248, 731; glycophorin A hydrolysis method, 248, 731)

Other molecular properties: *Pasteurella* (*Yersinia*), 248, 223, 728; amino acid sequence, 248, 728; cleavage of *O*-sialoglycoproteins from surface of live cells, 248, 736; cleavage specificity, 248, 728; gene, 248, 728; glycophorin A hydrolysis, 248, 735; homologues, 248, 224; inhibition, 248, 738; isolation, 248, 730; metal ion cofactor, 248, 738; occurrence, 248, 728; pathogenic role, 248, 738; pH optimum, 248, 739; positional specificity, 248, 735; properties, 248, 738; recombinant, 248, 739; stability, 248, 738; substrate specificity, 248, 729

SIGNAL PEPTIDASE I

This serine endopeptidase [EC 3.4.21.89] (also known as bacterial leader peptidase I, SPase I, leader peptidase, and phage-procoat-leader peptidase) catalyzes the hydrolysis of N-terminal leader sequences from secreted and periplasmic proteins precursor. The *Escherichia coli* enzyme is assayed with Phe-Ser-Ala-Ser-Ala-Leu-Ala-∇-Lys-Ile, where ∇ indicates the cleavage site. There is a preference for small aminoacyl residues in the P1 position and uncharged or aliphatic residues at P3. The P1 and P3 positions appear to be critical for most signal peptidases. Interestingly, the enzyme is unaffected by agents that typically inhibit serine, cysteine, aspartic and metalloproteinases.[1–4] *See also* *Signal Peptidase (Eukaryote); Signal Peptidase SipS; Signal Peptidase II*

[1] W. R. Tschantz & R. E. Dalbey (1994) *MIE* **244**, 285.
[2] P. B. Wolfe, C. Zwizinski & W. Wickner (1983) *MIE* **97**, 40.
[3] R. E. Dalbey & G. von Heijne (1992) *Trends Biochem. Sci.* **17**, 474.
[4] G. von Heijne, ed. (1994) *Signal Peptidases*, R. G. Landes, Austin, Texas.

Selected entries from *Methods in Enzymology* [vol, page(s)]:
General discussion: 97, 40; 244, 285
Assay: 96, 788; *Escherichia coli*, 97, 41
Other molecular properties: active site residues, 244, 50, 53; canine pancreatic, 96, 786; catalytic residues, 244, 299; cleavage site specificity, 244, 50; *Escherichia coli*, 97, 41; half-life of mutant enzyme, 244, 296; inhibitors, 244, 293; membrane topology, 244, 291; mitochondrial, properties, 248, 215; peptide assay, 244, 286, 290; processing assays, 244, 296; pulse-chase stability assay, 244, 299; purification, 244, 287; stability, 244, 289; substrates, 244, 285; substrate specificity, 244, 291

SIGNAL PEPTIDASE II

This aspartic endopeptidase [EC 3.4.23.36] (also known as Spase II, premurein-leader peptidase, prolipoprotein-signal peptidase, leader peptidase II, and lipoprotein-specific signal peptidase) catalyzes the cleavage of N-terminal leader sequences from membrane prolipoproteins (note: signal peptides only of bacterial lipoproteins). The scissile bond in the majority of the bacterial lipoproteins is . . . -Xaa-Xbb-Xcc-∇-Cys-. . . , (∇ = cleavage site) in which Xaa is a hydrophobic residue (preferably leucyl), Xbb is usually a seryl or an alanyl residue, Xcc is often a glycyl, seryl, or an alanyl residue, and the cysteinyl residue in the P1' position is alkylated on sulfur with a diacylglyceryl group.[1,2] **See also** *Leader Peptidase II*

[1] K. Sankaran & H. C. Wu (1994) in *Signal Peptidases* (G. von Heijne, ed.), R. G. Landes Co., Austin, Texas, p. 17.
[2] K. Sankaran & H. C. Wu (1995) *MIE* **248**, 169.

Selected entries from **Methods in Enzymology** [vol, page(s)]:
General discussion: 248, 109, 115, 116, 169;
Assay: 248, 170 (substrate, preparation, 248, 171)
Other molecular properties: amino acid sequence, 248, 170, 177; biological function, 248, 169; defective *Escherichia coli* strains, 250, 687; hydropathy profile, 248, 177; inhibition, 248, 107, 116, 176, 180; kinetics, 248, 175; membrane topology, 248, 176, 180; processing of bacterial lipoprotein, 250, 683; properties, 248, 109, 116, 174, 179; purification, 248, 172; solubilization, 248, 172; structure-function relationship, 248, 177; substrate specificity, 248, 175, 180

SIGNAL PEPTIDASE (EUKARYOTE)

This serine endopeptidase, often called microsomal signal peptidase, catalyzes the cleavage of precursor proteins targeted for the endoplasmic reticulum. The natural substrate for this peptidase is a precursor protein that is in the process of being synthesized by a ribosome attached to the endoplasmic reticulum (ER) and which is being actively transported into the ER lumen.[1–4]

[1] R. E. Dalbey & G. von Heijne (1992) *Trends Biochem. Sci.* **17**, 474.
[2] G. von Heijne, ed. (1994) *Signal Peptidases*, R. G. Landes, Austin, Texas.
[3] R. E. Dalbey, M. O. Lively, S. Bron & J. M. van Dijl (1997) *Protein Sci.* **6**, 1129.
[4] M. O. Lively, A. L. Newsome & M. Nusier (1994) *MIE* **244**, 301.

Selected entries from **Methods in Enzymology** [vol, page(s)]:
General discussion: 244, 301
Assay: 96, 784; autoradiography quantitation, 244, 303; cell-free translation system, 244, 303, 305; linearity, 244, 305; post-translational assay, 244, 304; substrates, 244, 304 (preplacental lactogen preparation, 244, 304; preprolactin, 244, 304)
Other molecular properties: active site residues, 244, 50; biological role, 244, 301; chicken, purification, 244, 302; cleavage-site specificity, 244, 50; endoplasmic reticulum localization, 244, 301; mechanism, 244, 313; sequence homologies, 244, 53, 312; sources of pure enzyme, 244, 301; specificity, 244, 303; subunit structure, 244, 53

SIGNAL PEPTIDASE SipS

This variant of signal peptidase I [EC 3.4.21.89], with the descriptor SipS referring to <u>s</u>ignal <u>p</u>eptidase of *Bacillus <u>s</u>ubtilis*, catalyzes peptide-bond hydrolysis in secretory precursor proteins.[1,2]

[1] J. M. van Dijl, A. de Jong, J. Vehmaanperä, G. Venema & S. Bron (1992) *EMBO J.* **11**, 2819.
[2] A. Bolhuis, A. Sorokin, V. Azevedo, S. D. Ehrlich, P. G. Braun, A. de Jong, G. Venema, S. Bron & J. M. van Dijl (1996) *Mol. Microbiol.* **22**, 605.

SIGNAL-RECOGNITION-PARTICLE GTPase

This GTP-hydrolyzing enzyme is associated with the signal-recognition particle, a protein- and RNA-containing structure involved in endoplasmic-reticulum-associated protein synthesis.[1–3] The reaction is currently misclassified as a GTPase [EC 3.6.1.49], but the reaction is of the energase type: $State_1 + GTP = State_1 + GDP + P_i$, where two affinity-modulated conformational states are indicated.[4]

See *G Proteins*

[1] T. Connolly & R. Gilmore (1989) *Cell* **57**, 599.
[2] J. D. Miller, H. Wilhelm, L. Gierasch, R. Gilmore & P. Walter (1993) *Nature* **366**, 351.
[3] D. M. Freymann, R. J. Keenan, R. M. Stroud & P. Walter (1997) *Nature* **385**, 361.
[4] D. L. Purich (2001) *TiBS* **26**, 417.

SINAPATE 1-GLUCOSYLTRANSFERASE

This glucosyltransferase [EC 2.4.1.120] catalyzes the reaction of UDP-D-glucose with sinapate to produce UDP and 1-sinapoyl-D-glucose.[1,2] Substrates include other hydroxycinnamates, such as 4-coumarate, ferulate, and caffeate.

[1] D. Strack (1979) *Z. Naturforsch.* **35c**, 204.
[2] G. Nurmann & D. Strack (1981) *Z. Pflanzenphysiol.* **102**, 11.

SINAPINE ESTERASE

This esterase [EC 3.1.1.49] catalyzes the hydrolysis of sinapoylcholine (that is, sinapine) to produce sinapate and choline.[1–3] Other substrates include: sinapoylglucose, feruloylcholine, *p*-coumaroylcholine, cinnamoylcholine, and caffeoylcholine.

[1] G Nurmann & D. Strack (1979) *Z. Naturforsch.* **34c**, 715.
[2] D. Strack, G. Nurmann & G. Sachs (1980) *Z. Naturforsch.* **35c**, 963.
[3] L. M. Larsen, T. H. Nielsen, A. Plöger & H. Sorensen (1988) *J. Chromatogr.* **450**, 121.

SINAPOYLGLUCOSE:CHOLINE O-SINAPOYLTRANSFERASE

This acyltransferase [EC 2.3.1.91], also known as sinapine synthase, catalyzes the reaction of 1-O-sinapoyl-β-D-glucoside with choline to produce D-glucose and O-sinapoylcholine (or sinapine).[1–4]

[1] W. Gräwe & D. Strack (1986) *Z. Naturforsch.* **41c**, 28.
[2] D. Strack, W. Knogge & B. Dahlbender (1982) *Z. Naturforsch.* **38c**, 21.
[3] J. Regenbrecht & D. Strack (1985) *Phytochemistry* **24**, 407.
[4] T. Vogt, R. Aebersold & B. Ellis (1993) *ABB* **300**, 622.

SINAPOYLGLUCOSE:MALATE O-SINAPOYLTRANSFERASE

This transferase [EC 2.3.1.92] catalyzes the reaction of 1-O-sinapoyl-β-D-glucoside with (*S*)-malate (that is, L-malate) to produce D-glucose and sinapoyl-(*S*)-malate.[1–4] Other glucoside substrates are 1-O-feruloyl-β-D-glucoside and 1-O-caffeoyl-β-D-glucoside.

[1]H.-P. Mock, T. Vogt & D. Strack (1992) *Z. Naturforsch.* **47c**, 680.
[2]D. Strack, B. E. Ellis, W. Gräwe & J. Heilemann (1990) *Planta* **180**, 217.
[3]N. Tkotz & D. Strack (1980) *Z. Naturforsch.* **35c**, 835.
[4]W. Gräwe, P. Bachhuber, H.-P. Mock & D. Strack (1992) *Planta* **187**, 236.

SINAPOYLGLUCOSE:SINAPOYLGLUCOSE O-SINAPOYLTRANSFERASE

This transferase [EC 2.3.1.103] catalyzes the reaction of two molecules of 1-O-sinapoyl-β-D-glucoside to produce D-glucose and 1,2-bis-O-sinapoyl-β-D-glucoside.[1] Alternative substrates are 1-(p-coumaroyl)-β-D-glucoside and 1-feruloyl-β-D-glucoside.

[1]B. Dahlbender & D. Strack (1986) *Phytochemistry* **25**, 1043.

SINDBIS VIRUS nsP2 ENDOPEPTIDASE

This cysteine endopeptidase, also known as nsP2 proteinase, participates in the release of *Sindbis* nonstructural proteins from the initial viral polyprotein. The cleavage sites are represented by the motif (Ala,Ile)-Gly-(Ala,Cys,Gly)-∇-(Ala,Tyr), where the P2 glycyl residue is essential for cleavage and ∇ indicates the cleavage site.[1,2]

[1]E. G. Strauss, R. J. de Groot, R. Levinson & J. H. Strauss (1992) *Virology* **191**, 932.
[2]R. J. de Groot, W. R. Hardy, Y. Shirako & J. H. Strauss (1990) *EMBO J.* **9**, 2631.

SINIGRIN SULFOHYDROLASE

This hydrolase activity, originally listed by the Enzyme Commission as EC 3.1.6.5, is now a deleted entry.

SITE-SPECIFIC DNA METHYLTRANSFERASE (ADENINE-SPECIFIC)

These methyltransferases [EC 2.1.1.72], also known as N^6-adenine-specific DNA methylase, modification methylase, and restriction-modification system, catalyze the reaction of S-adenosyl-L-methionine with an adenine residue in DNA to produce S-adenosyl-L-homocysteine and a 6-methylaminopurine residue in DNA.[1,2] *See specific enzyme*

[1]C. Kessler & V. Manta (1990) *Gene* **92**, 1.
[2]R. Yuan (1981) *ARB* **50**, 285.

SITE-SPECIFIC DNA METHYLTRANSFERASE (CYTOSINE-SPECIFIC)

These methyltransferases [EC 2.1.1.73], also known as C-5 cytosine-specific DNA methylase and modification methylase, catalyzes the reaction of S-adenosyl-L-methionine with a cytosine residue in DNA to produce S-adenosyl-L-homocysteine and a 5-methylcytosine residue in the DNA.[1,2] *See specific enzyme*

[1]C. Kessler & V. Manta (1990) *Gene* **92**, 1.
[2]R. Yuan (1981) *ARB* **50**, 285.

SITE-SPECIFIC DNA-METHYLTRANSFERASE (CYTOSINE-N⁴-SPECIFIC)

This enzyme designation [EC 2.1.1.113] (also known as N^4-cytosine-specific DNA methylase, modification methylase, and restriction-modification system) represents a large group of enzymes that form, with the restriction endonucleases (type I site-specific deoxyribonuclease [EC 3.1.21.3], type II site-specific deoxyribonuclease [EC 3.1.21.4], and type III site-specific deoxyribonuclease [EC 3.1.21.5]), restriction-modification systems. The variants of EC 2.1.1.113 catalyze the reaction of S-adenosyl-L-methionine with a specific cytosine residue in DNA to produce S-adenosyl-L-homocysteine and an N^4-methylcytosine residue in the DNA.[1,2] *See specific enzyme*

[1]R. J. Roberts & D. Macelis (1992) *Nucleic Acids Res.* **20** (Suppl.), 2167.
[2]R. Yuan (1981) *ARB* **50**, 285.

S6 KINASE

This ATP-dependent protein kinase catalyzes phosphorylation of the 40S ribosomal protein S6, at a number of seryl residues.

Selected entries from *Methods in Enzymology* [vol, page(s)]:
General discussion: **200**, 252, 268
Assay: mitogen-activated, **200**, 271; S6 kinase II, **200**, 253
Other molecular properties: mitogen-activated, **200**, 181; S6 kinase II, **200**, 253; transition-state and multisubstrate inhibitors, **249**, 305

*Sla*I RESTRICTION ENDONUCLEASE

This type II restriction endonuclease [EC 3.1.21.4] is obtained from *Streptomyces lavendulae* and acts on both strands of DNA at 5′...C∇TCGAG...3′.[1] (The symbol ∇ indicates the cleavage site.)

[1]H. Takahashi, M. Shimizu, H. Saito, Y. Ikeda & H. Sugisaki (1979) *Gene* **5**, 9.

Selected entries from *Methods in Enzymology* [vol, page(s)]:
Other molecular properties: recognition sequence, **65**, 10

SMALL MONOMERIC GTPases

These low-molecular-weight, monomeric enzymes comprise a family of apppproximately fifty enzymes that are distantly related to the structure of the α-subunit of heterotrimeric G-protein GTPase [EC 3.6.1.46].[1–4]

Small molecule GTPases participate in cell-growth regulation (Ras),[3] membrane vesicle traffic and uncoating (Rab and ARF subfamilies), nuclear protein import (Ran subfamily), and organization of the cytoskeleton (Rho and Rac subfamilies).[2]

This reaction is currently misclassified as a GTPase [EC 3.6.1.47], but the reaction is of the energase type: $State_1 + GTP = State_1 + GDP + P_i$, where two affinity-modulated conformational states are indicated. **See** *specific protein; Energase; G Proteins*

[1] H. R. Bourne, D. A. Sanders & F. McCormick (1991) *Nature* **349**, 117.
[2] A. Hall (1994) *Ann. Rev. Cell Biol.* **10**, 31.
[3] M. Geyer & A. Wittinghofer (1997) *Curr. Opin. Struct. Biol.* **7**, 786.
[4] N. Vitale, J. Moss & M. Vaughan (1998) *JBC* **273**, 2553.
[5] D. L. Purich (2001) *TiBS* **26**, 417.

*Sma*I RESTRICTION ENDONUCLEASE

This type II restriction endonuclease [EC 3.1.21.4] is obtained from *Streptomyces marcescens*. It acts on both strands of DNA at 5′...CCC∇GGG...3′ to give blunt ends. (The symbol ∇ indicates the cleavage site.) The endonuclease is an isoschizomer of *Xma*I, which does not produce blunt-ended fragments. Interestingly, the two isoschizomers bend DNA in opposite orientations.[1] The rate-limiting step is the very slow dissociation of the enzyme-product complex.[2] *Sma*I also exhibits star activity. **See** *"Star" Activity*

[1] B. E. Withers & J. C. Dunbar (1993) *NAR* **21**, 2571.
[2] B. Hinsch & M.-R. Kula (1981) *NAR* **9**, 3159.

Selected entries from **Methods in Enzymology** [vol, page(s)]:
General discussion: **65**, 21, 363, 433, 434, 485, 757, 758, 778, 780, 783, 784, 789, 793
Other molecular properties: cleavage sites, on adenovirus 2 DNA, **65**, 769; recognition sequence, **65**, 9

*Sml*I RESTRICTION ENDONUCLEASE

This type II restriction endonuclease [EC 3.1.21.4] is obtained from *Stenotrophomonas maltophilia* and acts on both strands of DNA at 5′...C∇TYRAG...3′ where Y refers to C or T and R refers to G or A.[1] (The symbol ∇ indicates the cleavage site.)

[1] R. J. Roberts & D. Macelis (1997) *NAR* **25**, 248.

*Sna*BI RESTRICTION ENDONUCLEASE

This type II restriction endonuclease [EC 3.1.21.4] is obtained from *Sphaerotilus natans* and acts on both strands of DNA at 5′...TAC∇GTA...3′ to produce blunt-ended fragments.[1] (The symbol ∇ indicates the cleavage site.)

The enzyme can exhibit star activity. *Sna*BI is s sensitive to methylation of cytosine within its recognition sequence.[1] **See** *"Star" Activity*

[1] Y. Yang & Q. Li (1990) *NAR* **18**, 3083.

SNAPALYSIN

This metalloendopeptidase [EC 3.4.24.77], whose name is derived from *Streptomyces* <u>n</u>eutral <u>p</u>roteinase <u>A</u> and is also called SnpA and Prt, catalyzes peptide-bond hydrolysis on milk proteins.[1] The enzyme prefers a tyrosyl or phenylalanyl residue in the P1′ position.

[1] M. J. Butler (1998) in *Handbook of Proteolytic Enzymes* (A. J. Barrett, N. D. Rawlings & J. F. Woessner, eds.), p. 1134, Academic Press, New York.

Selected entries from **Methods in Enzymology** [vol, page(s)]:
Other molecular properties: database code, **248**, 195; sequence conservation around Zn ligands and catalytic residues, **248**, 192

SOLANAIN

This enzyme, previously listed by the Enzyme Commission as EC 3.4.99.21, is now a deleted entry.

SORBITOL DEHYDROGENASE

This oxidoreductase [EC 1.1.1.14], also known as L-iditol 2-dehydrogenase (the name preferred by the Enzyme Commission) and polyol dehydrogenase, catalyzes the reaction of L-iditol with NAD^+ to produce L-sorbose and NADH.[1–3] The enzyme will also act on D-glucitol (that is, sorbitol, producing D-fructose) and xylitol (producing xylulose). The bovine lens and sheep liver enzymes operate by a Theorell-Chance Bi Bi kinetic mechanism whereas the rat liver enzyme reportedly has a rapid equilibrium random mechanism.[3] **See also** *Polyol Dehydrogenase; Aldehyde Reductase*

[1] J. Jeffery, L. Cummins, M. Carlquist & H. Jörnvall (1981) *EJB* **120**, 229.
[2] R. I. Lindstad, L. F. Hermansen & J. S. McKinley-McKee (1992) *Eur. J. Biochem.* **210**, 641.
[3] J. Jeffery & H. Jornvall (1988) *AE* **61**, 47.

Selected entries from **Methods in Enzymology** [vol, page(s)]:
General discussion: **1**, 348; **9**, 155, 159, 163, 170; **89**, 135
Assay: *Bacillus subtilis*, **9**, 155; *Candida utilis*, **9**, 163; *Gluconobacter*, **9**, 171; rat liver, **1**, 348; **89**, 135; rat spermatozoa, **9**, 159
Other molecular properties: *Bacillus subtilis*, from, **9**, 155 (pH optima, **9**, 159; properties, **9**, 159; purification, **9**, 158; specificity, **9**, 159); borate and, **12A**, 101; *Candida utilis*, of, **9**, 163; *Gluconobacter* (activators, **9**, 177; properties, **9**, 174; purification, **9**, 171); nomenclature and, **5**, 317; rat liver (assay, **1**, 348; **89**, 135; properties, **1**, 349; **89**, 137; purification, **1**, 349; **89**, 136); sheep liver, reversible inhibitors as mechanistic probes, **249**, 137; specificity of, Bertrand-Hudson rule, **9**, 170; spermatozoa, from, **9**, 159 (Michaelis constants, **9**, 163; properties, **9**, 162; purification, **9**, 160; specificity, **9**, 162; stability, **9**, 162); stain for, **224**, 111; stereospecificity, **87**, 103

D-SORBITOL DEHYDROGENASE (ACCEPTOR)

This FAD-dependent, membrane-bound oxidoreductase [EC 1.1.99.21] catalyzes the reaction of D-sorbitol with an acceptor to produce L-sorbose and the reduced acceptor.[1] Electron acceptor substrates include potassium ferricyanide, 2,6-dichlorophenolindophenol/phenazine methosulfate, nitro blue tetrazolium, and tetramethyl-*p*-phenylenediamine. **See also** D-Iditol Dehydrogenase

[1] E. Shinagawa, K. Matsushita, O. Adachi & M. Ameyama (1982) *Agric. Biol. Chem.* **46**, 135.

Selected entries from **Methods in Enzymology** [vol, page(s)]:
General discussion: **89**, 141
Assay: *Gluconobacter suboxydans* membrane-bound, **9**, 178; **89**, 141

SORBITOL-6-PHOSPHATASE

This enzyme [EC 3.1.3.50] catalyzes the hydrolysis of sorbitol 6-phosphate to produce sorbitol and orthophosphate.[1]

[1] C. Grant & T. A. Rees (1981) *Phytochemistry* **20**, 1505.

SORBITOL-6-PHOSPHATE 1-DEHYDROGENASE

This oxidoreductase catalyzes the reaction of sorbitol 6-phosphate with $NADP^+$ to produce glucose 6-phosphate and NADPH.[1]

[1] M. Hirai (1979) *Plant Physiol.* **63**, 715 and **67**, 221.

SORBITOL-6-PHOSPHATE 2-DEHYDROGENASE

This oxidoreductase [EC 1.1.1.140], also called ketose-phosphate reductase and glucitol-6-phosphate dehydrogenase, catalyzes the reaction of D-sorbitol 6-phosphate with NAD^+ to produce D-fructose 6-phosphate and NADH.[1-3]

[1] M. Liss, S. B. Horwitz & N. O. Kaplan (1962) *JBC* **237**, 1342.
[2] P. J. Du Toit & J. P. Kotze (1970) *BBA* **206**, 333.
[3] M. J. Novotny, J. Reizer, F. Esch & M. Saier (1984) *J. Bacteriol.* **159**, 986.

Selected entries from **Methods in Enzymology** [vol, page(s)]:
General discussion: **9**, 150
Assay: **9**, 150

SORBOSE DEHYDROGENASE

This oxidoreductase [EC 1.1.99.12] catalyzes the reaction of L-sorbose with an acceptor substrate to produce 5-dehydro-D-fructose and the reduced acceptor.[1] The electron acceptor substrate can be 2,6-dichloroindophenol.

[1] K. Sato, Y. Yamada, K. Aida & T. Uemura (1969) *J. Biochem.* **66**, 521.

SORBOSE 5-DEHYDROGENASE (NADP+)

This oxidoreductase [EC 1.1.1.123], also known as 5-keto-fructose reductase, catalyzes the reaction of 5-dehydro-D-fructose with NADPH to produce L-sorbose and $NADP^+$.[1,2] **See also** *Fructose 5-Dehydrogenase (NADP+)*

[1] S. Yagi, K. Kobayahi & T. Sonoyama (1989) *J. Ferment. Bioeng.* **67**, 212.
[2] S. Englard, G. Kaysen & G. Avigad (1970) *JBC* **245**, 1311.

Selected entries from **Methods in Enzymology** [vol, page(s)]:
General discussion: **41**, 132
Assay: yeast, **41**, 132
Other molecular properties: *Bacillus subtilis*, **41**, 137; yeast, from, **41**, 132 (assay, **41**, 132; chromatography, **41**, 134; inhibitors, **41**, 137; properties, **41**, 136; purification, **41**, 133; stereospecificity, **87**, 111)

L-SORBOSE OXIDASE

This oxidoreductase [EC 1.1.3.11] catalyzes the reaction of L-sorbose with dioxygen to produce 5-dehydro-D-fructose and hydrogen peroxide.[1] Other substrates include D-glucose (producing 2-dehydro-D-glucose) and D-galactose (producing 2-dehydro-D-galactose); however, D-fructose is not a substrate. 2,6-Dichloroindophenol can also substitute for dioxygen.

[1] Y. Yamada, K. Iizuka, K. Aida & T. Uemura (1967) *J. Biochem.* **62**, 223.

Selected entries from **Methods in Enzymology** [vol, page(s)]:
General discussion: **52**, 21

L-SORBOSE-1-PHOSPHATE REDUCTASE

This oxidoreductase catalyzes the reaction of L-sorbose 1-phosphate with NAD(P)H to produce D-glucitol 6-phosphate and $NAD(P)^+$. The enzyme also catalyzes the NAD(P)H-dependent conversion of D-fructose 1-phosphate to generate D-mannitol 6-phosphate along with $NAD(P)^+$.

Selected entries from **Methods in Enzymology** [vol, page(s)]:
General discussion: **89**, 248
Assay: *Klebsiella pneumoniae*, **89**, 248

*Spe*I RESTRICTION ENDONUCLEASE

This type II restriction endonuclease [EC 3.1.21.4] is obtained from *Sphaerotilus* species (ATCC 13923) and acts on both strands of DNA at 5′...A∇CTAGT...3′. (The symbol ∇ indicates the cleavage site.) Note also that this enzyme generates fragments with a 5′ CTAG extension which can be ligated to DNA fragments generated by the restriction endonucleases *Avr*II, *Nhe*I, and *Xba*I.

SPERMIDINE DEHYDROGENASE

This FAD/heme-dependent oxidoreductase [EC 1.5.99.6], also known as spermidine oxidase, catalyzes the reaction

of spermidine ($^+H_3N(CH_2)_4NH(CH_2)_3NH_3^+$) with water and a suitable electron acceptor substrate (including ferricyanide, 2,6-dichloroindophenol, phenazine methosulfate, or cytochrome c.) to produce 1,3-diaminopropane, 4-aminobutanal (which then cyclizes to form 1-pyrroline), and the reduced acceptor.[1-3]

[1] M. Okada, S. Kawashima & K. Imahori (1979) J. Biochem. **85**, 1235.
[2] H. Tabor & C. W. Tabor (1972) AE **36**, 203.
[3] C. W. Tabor & P. D. Kellogg (1970) JBC **245**, 5424.
Selected entries from *Methods in Enzymology* [vol, page(s)]:
General discussion: **17B**, 746; **52**, 18; **94**, 303
Assay: **17B**, 746; **94**, 303

SPERMIDINE SYNTHASE

This enzyme [EC 2.5.1.16], also known as putrescine aminopropyltransferase and aminopropyltransferase, catalyzes the reaction of S-adenosylmethioninamine with putrescine ($^+H_3N(CH_2)_4NH_3^+$) to produce 5′-methylthioadenosine and spermidine ($^+H_3N(CH_2)_4NH(CH_2)_3NH_3^+$).[1-3] While the mammalian enzyme is highly specific, the bacterial enzyme can synthesize spermine and use other acceptors. Spermidine synthase is distinct from spermine synthase [EC 2.5.1.22].

[1] H. Tabor & C. W. Tabor (1972) AE **36**, 203.
[2] V. Zappia, G. Cacciapuoti, G. Pontoni & A. Oliva (1980) JBC **255**, 7276.
[3] F. Schlenk (1983) AE **54**, 195.
Selected entries from *Methods in Enzymology* [vol, page(s)]:
General discussion: **5**, 761; **94**, 260, 265, 270, 279, 286, 294
Assay: **5**, 761; **94**, 257, 260, 265, 271, 281
Other molecular properties: bovine brain (properties, **94**, 275; purification, **94**, 272); Chinese cabbage enzyme, **94**, 283; *Escherichia coli* enzyme, **94**, 267; inhibitors, **94**, 294; properties, **5**, 764; **94**, 269, 275, 285; purification, **5**, 762; **94**, 267, 272, 274, 283; rat prostate, **94**, 275; substrate synthesis, **94**, 286; transition-state and multisubstrate inhibitors, **249**, 305

SPERMINE SYNTHASE

This enzyme [EC 2.5.1.22], also known as spermidine aminopropyltransferase, catalyzes the reaction of S-adenosylmethioninamine with spermidine ($^+H_3N(CH_2)_4NH(CH_2)_3NH_3^+$) to produce 5′-methylthioadenosine and spermine ($^+H_3N(CH_2)_3HN(CH_2)_4NH(CH_2)_3NH_3^+$).[1-3] The enzyme is not identical with spermidine synthase [EC 2.5.1.16] or *sym*-norspermidine synthase [EC 2.5.1.23].

[1] R.-L. Pajula (1983) BJ **215**, 669.
[2] A. E. Pegg, K. Shuttleworth & H. Hibasami (1981) BJ **197**, 315.
[3] F. Schlenk (1983) AE **54**, 195.
Selected entries from *Methods in Enzymology* [vol, page(s)]:
General discussion: **94**, 276, 286, 294
Assay: **94**, 257, 260, 276
Other molecular properties: bovine brain (properties, **94**, 278; purification, **94**, 277); inhibitors, **94**, 294; *Saccharomyces cerevisiae*,

deficient mutants, mass screening, **94**, 107; substrate synthesis, **94**, 286; transition-state and multisubstrate inhibitors, **249**, 305

SPERMOSIN

This trypsin-like endopeptidase [EC 3.4.21.99], localized in the sperm head, has a role in ascidian fertilization (the enzyme is released during sperm activation).[1,2] It catalyzes the hydrolysis of . . . −Arg−∇-Xaa−. . . , where ∇ indicates the cleavage site and a prolyl residue resides in the P2 position.

[1] H. Sawada, K. Iwasaki, F. Kihara-Negishi, H. Ariga & H. Yokosawa (1996) BBRC **222**, 499.
[2] H. Sawada & T. Someno (1996) Mol. Reprod. Dev. **45**, 240.

SPHINGANINE KINASE

This phosphotransferase [EC 2.7.1.91], also known as dihydrosphingosine kinase, catalyzes the reaction of ATP with sphinganine (or $HOCH_2CH(NH_3^+)CH(OH)(CH_2)_{14}CH_3$) to produce ADP and sphinganine 1-phosphate (or $^-HO_3POCH_2CH(NH_3^+)CH(OH)(CH_2)_{14}CH_3$).[1,2]

[1] W. Stoffel, E. Bauer & J. Stahl (1974) Hoppe-Seyler's Z. Physiol. Chem. **355**, 61.
[2] B. M. Buehrer & R. M. Bell (1992) JBC **267**, 3154.

SPHINGANINE-1-PHOSPHATE ALDOLASE

This pyridoxal-phosphate-dependent enzyme [EC 4.1.2.27] catalyzes the conversion of sphinganine 1-phosphate ($^-HO_3POCH_2CH(NH_3^+)CH(OH)(CH_2)_{14}CH_3$) to phosphoethanolamine and palmitaldehyde.[1,2]

[1] W. Stoffel, D. LeKim & G. Sticht (1969) Hoppe-Seyler's Z. Physiol. Chem. **350**, 1233.
[2] P. P. van Veldhoven & G. P. Mannaerts (1993) Adv. Lipid Res. **26**, 69.

SPHINGOMYELIN PHOSPHODIESTERASE

This phosphodiesterase [EC 3.1.4.12], also known as acid sphingomyelinase and neutral sphingomyelinase, catalyzes the hydrolysis of sphingomyelin to produce N-acylsphingosine and choline phosphate.[1,2]

[1] J. D. Esko & C. R. H. Raetz (1983) The Enzymes, 3rd ed., **16**, 207.
[2] R. O. Brady (1983) The Enzymes, 3rd ed., **16**, 409.
Selected entries from *Methods in Enzymology* [vol, page(s)]:
General discussion: **14**, 131, 144; **28**, 874; **71**, 735, 741; **197**, 536, 540; **311**, 149, 156, 164, 168, 176, 184; **322**, 382
Assay: **14**, 131, 144; **28**, 874; **50**, 465 (fluorescent derivatives of sphingomyelin, with, **72**, 356; preparation of substrates, **72**, 358; trinitrophenylaminolauroyl sphingomyelin, with, **72**, 354); human urine, **197**, 536, 541; **322**, 382
Other molecular properties: analysis of glycophosphatidylinositol-anchor proteins, **250**, 602; deficiency, **50**, 465; detergent effects, **14**, 147; diagnosis of Niemann-Pick disease, **138**, 736;

human urine, **197**, 536; hydrolysis of brain ganglioside, in, **14**, 138; inhibitors, **14**, 149; isolation from human liver, **28**, 877; magnesium dependent, **50**, 467, 468; Niemann-Pick disease and, **14**, 133, 148; **28**, 879; **50**, 465; **138**, 736; properties, **28**, 879; rat brain, from, **14**, 144; rat liver, from, **14**, 131; removal of ceramidase during extraction, **14**, 148; removal of phospholipase during extraction, **14**, 148; sphingomyelinase C, effect on neutrophil chemotaxis, **236**, 76; *Staphylococcus aureus*, cytolytic mechanisms, **235**, 658

SPHINGOMYELIN PHOSPHODIESTERASE D

This phosphodiesterase [EC 3.1.4.41] catalyzes the hydrolysis of sphingomyelin to ceramide phosphate and choline.[1–3] This enzyme also catalyzes the hydrolysis of 2-lysophosphatidylcholine to choline and 2-lysophosphatidate, but will not act on phosphatidylcholine.

[1] G. Kurpiewski, L. J. Forrester, J. T. Barrett & B. J. Campbell (1981) *BBA* **678**, 467.
[2] E. O. Onon (1979) *BJ* **177**, 181.
[3] A. Soucek, C. Michalec & A. Soukova (1971) *BBA* **227**, 116.

SPHINGOSINE *N*-ACYLTRANSFERASE

This acyltransferase [EC 2.3.1.24], also known as ceramide synthase, catalyzes the reaction of an acyl-CoA (for example, palmitoyl-CoA, stearoyl-CoA, or lignoceroyl-CoA) with sphingosine (or the 2-epimer) to produce Coenzyme A and an *N*-acylsphingosine.[1,2]

[1] M. Sribney (1966) *BBA* **125**, 542.
[2] K. Hirschberg, J. Rodger & A. H. Futerman (1993) *BJ* **290**, 751.

Selected entries from *Methods in Enzymology* [vol, page(s)]:
General discussion: 209, 427; **311**, 15
Assay: 209, 432; **322**, 378

*Sph*I RESTRICTION ENDONUCLEASE

This type II restriction endonuclease [EC 3.1.21.4] is obtained from *Streptomyces phaeochromogenes* and acts on both strands of DNA at 5′ . . . GCATG▽C . . . 3′.[1] Note that this enzyme produces fragments having a 3′ CATG extension which can be ligated to DNA fragments generated by *Nla*III restriction endonuclease.

[1] L. Y. Fuchs, L. Covarrubias, L. Escalante, S. Sanchez & F. Bolivar (1980) *Gene* **10**, 39.

SPLEEN EXONUCLEASE

This exonuclease [EC 3.1.16.1], also known as 3′-exonuclease and phosphodiesterase II, catalyzes the exonucleolytic cleavage in the 5′- to 3′-direction of polynucleotides (with a preference for single-stranded substrates) to yield nucleoside 3′-phosphates.[1,2]

[1] A. Bernardi & G. Bernardi (1971) *The Enzymes*, 3rd ed., **4**, 329.
[2] H. G. Khorana (1961) *The Enzymes*, 2nd ed., **5**, 79.

Selected entries from *Methods in Enzymology* [vol, page(s)]:
General discussion: 2, 565; **3**, 755, 756, 758, 764; **6**, 245; **29**, 319; 473
Assay: 2, 565; **6**, 245
Other molecular properties: calf spleen, nearest-neighbor base sequences and, **6**, 741; classification, **6**, 236; cyanoethyl tRNA nucleotides and, **20**, 165; cyclic nucleotides, action on, **3**, 766, 809, 810; DNA sequence analysis and, **29**, 242, 243, 245, 247, 248; hydrolysis of DNA, **234**, 6; labeled oligodeoxynucleotides and, **29**, 325, 331, 334, 339; labeled oligonucleotides and, **29**, 358; nucleotide sequence analysis and, **12A**, 370; oligonucleotide identification and, **12A**, 283; partial digestion, of short DNA fragments, **65**, 627; 3′-penultimate nucleotide determination and, **29**, 348; preparation, **2**, 565; **6**, 246; **29**, 343, 356, 474; properties, **2**, 565; **6**, 247; ribonuclease I preparation free of, **6**, 252; specificity, **29**, 350, 354; 3′-terminal determination and, **29**, 279, 345, 346; 5′-terminus sequence analysis and, **29**, 283, 291, 294; tissue distribution, **6**, 249; tritium derivative method, **65**, 647, 660; tRNA fragments and, materials and electrophoretic apparatus, **29**, 474 (digestion, **29**, 655; principle, **29**, 473; procedure, **29**, 474; remarks, **29**, 477)

SPUMAPEPSIN

This aspartic endopeptidase (variants have also been called human spumaretrovirus polyprotein peptidase and simian foamy virus polyprotein peptidase) participates in viral polyprotein processing.[1,2] The enzyme has been shown to act at the Asn–Thr bond in ArgAlaValAsnThrValThrGln.

[1] K. I. Pfrepper, J. Reed, H. R. Rackwitz, M. Schnölzer & R. M. Flügel (2001) *Virus Genes* **22**, 61.
[2] G. Fenyöfalvi, P. Bagossi, T. D. Copeland, S. Oroszlan, P. Boross & J. Tözser (1999) *FEBS Lett.* **462**, 397.

SQUALENE HYDROXYLASE

This enzyme activity, originally classified as EC 1.14.1.3, results from the combined action of squalene monooxygenase [EC 1.14.99.7] and lanosterol synthase [EC 5.4.99.7].

SQUALENE MONOOXYGENASE

This FAD-dependent oxidoreductase [EC 1.14.99.7], also known as squalene epoxidase, catalyzes the reaction of squalene with dioxygen and an electron donor (AH$_2$) to produce (S)-squalene-2,3-epoxide, water and the oxidized substrate (A).[1–3] In a number of organisms, the electron donor is NADPH or NADH. Squalene monooxygenase, together with lanosterol synthase [EC 5.4.99.7], was formerly known as squalene oxidocyclase.

[1] T. Ono, K. Nakazono & H. Kosaka (1982) *BBA* **709**, 84.
[2] N. S. Ryder & M.-C. Dupont (1985) *BJ* **230**, 765.
[3] N. S. Ryder & M.-C. Dupont (1984) *BBA* **794**, 466.

Selected entries from *Methods in Enzymology* [vol, page(s)]:
General discussion: 5, 493; **15**, 497; **110**, 375
Assay: rat liver, **110**, 376

SQUALENE SYNTHASE

This membrane-bound enzyme [EC 2.5.1.21] is also known as presqualene-diphosphate synthase and farnesyl-diphosphate farnesyltransferase. The latter is the preferred EC name which can be confused easily with botryococcene synthase, another well-known farnesyl-diphosphate farnesyltransferase.

farnesyl diphosphate

presqualene diphosphate

squalene

Squalene synthase catalyzes the conversion of two molecules of farnesyl diphosphate (or, farnesyl pyrophosphate) to produce first presqualene diphosphate and diphosphate (or, pyrophosphate); NADPH then binds and squalene, pyrophosphate, and NADP$^+$ are released.[1-4] Isotope labeling established the stereochemical course of the squalene synthase reaction.[5-7]

[1] K. Sasiak & H. C. Rilling (1988) *ABB* **260**, 622.
[2] I. Shechter, E. Klinger, M. L. Rucker, R. G. Engstrom, J. A. Spirito, M. A. Islam, B. R. Boettcher & D. B. Weinstein (1992) *JBC* **267**, 8628.
[3] I. Shechter & K. Bloch (1971) *JBC* **246**, 7690.
[4] R. Kluger (1992) *The Enzymes*, 3rd ed., **20**, 271.
[5] J. W. Cornforth (1971) *Chem. Soc. Rev.* **1**, 1.
[6] G. Popják & J. W. Cornforth (1966) *BJ* **101**, 553.
[7] K. A. Mookhtiar, S. S. Kalinowski, D. Zhang & C. D. Poulter (1994) *JBC* **269**, 11201.

Selected entries from *Methods in Enzymology* [vol, page(s)]:
General discussion: **15**, 359, 450; **110**, 359, 373
Assay: yeast, **110**, 363
Other molecular properties: preparation, **15**, 452; properties, **15**, 453; role in formation of 12-*cis*-dehydrosqualene, **110**, 373; stereochemistry, **87**, 122; transition-state and multisubstrate inhibitors, **249**, 305; yeast enzyme, **110**, 360

*Srf*I RESTRICTION ENDONUCLEASE

This type II restriction endonuclease [EC 3.1.21.4] is obtained from *Streptomyces* species and acts on both strands of DNA at 5′ ... GCCC▽GGGC ... 3′, producing blunt ends.[1] (The symbol ▽ indicates the cleavage site.)

[1] T. G. Simcox, S. J. Marsh, E. A. Gross, W. Lernhardt, S. Davis & M. E. C. Simcox (1991) *Gene* **109**, 121.

sRNA ADENYLYLTRANSFERASE

This enzyme activity, originally classified by the now deleted number EC 2.7.7.20, is identical with tRNA adenylyltransferase [EC 2.7.7.25].

*Ssp*I RESTRICTION ENDONUCLEASE

This type II restriction endonuclease [EC 3.1.21.4] is obtained from *Sphaerotilus* species and acts on both strands of DNA at 5′ ... AAT▽ATT ... 3′ to produce blunt-ended fragments.[1] (The symbol ▽ indicates the cleavage site.) Note the absence of G and C in the recognition sequence. If the reaction conditions consist of low ionic strength, pH greater than 8.0, high glycerol concentrations, or high enzyme concentrations, *Ssp* I may exhibit star activity. **See** *"Star" Activity*

[1] W. Warren & M. Merion (1988) *BioChromatography* **3**, 225.

Selected entries from *Methods in Enzymology* [vol, page(s)]:
General discussion: **65**, 9

*Sst*I RESTRICTION ENDONUCLEASE

This type II restriction endonuclease [EC 3.1.21.4] is obtained from *Streptomyces stanford* and acts on both strands of DNA at 5′ ... GAGCT▽C ... 3′.[1] (The symbol ▽ indicates the cleavage site.) The enzyme also exhibits star activity. **See** *"Star" Activity*

[1] S. P. Goff & A. Rambach (1978) *Gene* **13**, 347.

Selected entries from *Methods in Enzymology* [vol, page(s)]:
General discussion: **65**, 170, 721
Assay: **65**, 170, 171
Other molecular properties: isoschizomer, **65**, 173; properties, **65**, 172, 173; purification, **65**, 171, 172; recognition sequence, **65**, 11; stability, **65**, 172

*Sst*II RESTRICTION ENDONUCLEASE

This type II restriction endonuclease [EC 3.1.21.4] is obtained from *Streptomyces stanford* and acts on both strands of DNA at 5′ ... CCGC▽GG ... 3′.[1] (The symbol ▽ indicates the cleavage site.)

[1] S. P. Goff & A. Rambach (1978) *Gene* **13**, 347.

Selected entries from *Methods in Enzymology* [vol, page(s)]:
Other molecular properties: recognition sequence, **65**, 11

STAPHYLOPAIN

This cysteine endopeptidase, isolated from *Staphylococcus aureus*, catalyzes peptide-bond hydrolysis with a very broad specificity, whereas its action on small synthetic substrates is more specific.[1,2] It does not act on *p*-nitroanilide derivatives commonly used with trypsin or chymotrypsin.

[1] A. Björklind & H. Jörnvall (1974) *BBA* **370**, 542.
[2] J. Potempa, A. Dubin, G. Korzus & J. Travis (1988) *JBC* **263**, 2664.

"STAR" ACTIVITY

This term describes an undesirable condition wherein a restriction endonuclease exhibits reduced ability to recognize specific DNA sequences.[1-3] For example, while *Bam*HI normally recognizes the GGATCC sequence, this enzyme exhibits broader action, including scission of NGATCC, GPuATCC and GGNTCC at low ionic strength. Star activity is frequently observed (a) upon prolonged exposure of DNA to enzyme; (b) at high enzyme:DNA concentration ratios; (c) in the presence of organic solvents (e.g., ethanol or certain enzyme-stabilizing agents, like glycerol and dimethylsulfoxide); (d) at low ionic strength; (e) at high pH; and (f) when Mn^{2+} or Co^{2+} is used in place of the physiologic metal ion, usually Mg^{2+}.

[1] M. Nasri & D. Thomas (1986) *NAR* **14**, 811.
[2] J. George & J. G. Chirikjian (1982) *PNAS* **79**, 2432.
[3] M. Nasri & D. Thomas (1987) *NAR* **15**, 7677.

STARCH (BACTERIAL GLYCOGEN) SYNTHASE

This transferase [EC 2.4.1.21], also known as glycogen synthase, phytoglycogen synthase, and ADP-glucose-starch glucosyltransferase, catalyzes the reaction of ADP-D-glucose with [(1,4)-α-D-glucosyl]$_n$ to produce ADP and [(1,4)-α-D-glucosyl]$_{n+1}$.[1-3] The enzyme is named according to the source and the nature of the product. Different isoforms of starch synthase are present in potato, varying in affinity for ADPglucose and the glucan substrates, activation by amylopectin, response to citrate, thermosensitivity, and the processivity of glucan chain extension.[2] Glycogen synthase [EC 2.4.1.11] uses UDP-D-glucose as the glucosyl donor. *See also Glycogen Synthase*

[1] C. Pollock & J. Preiss (1980) *ABB* **204**, 578.
[2] A. Edwards, A. Borthakur, S. Bornemann, J. Venail, K. Denyer, D. Waite, D. Fulton, A. Smith & C. Martin (1999) *EJB* **266**, 724.
[3] J. Preiss (1982) *Ann. Rev. Plant Physiol.* **33**, 431.
Selected entries from *Methods in Enzymology* [vol, page(s)]:
General discussion: 8, 384, 387; 28, 545
Assay: 8, 384, 388, 391; 28, 545
Other molecular properties: *Arthrobacter* enzyme, **8**, 385; bean enzyme, **8**, 389; **9**, 14; glutinous rice grain enzyme, **9**, 14; occurrence, **8**, 389, 394; potato tuber enzyme, **8**, 389; **228**, 179; properties, **8**, 389, 393; **28**, 547; purification (beans, **8**, 389; corn enzyme, **8**, 389, 392; leaf enzyme, **8**, 389; soybean enzyme, **8**, 389); spinach enzyme, **28**, 546; staining on gels, **22**, 598

STARCH PHOSPHORYLASE

This enzyme [EC 2.4.1.1] catalyzes the reaction of starch (that is, [(1,4)-α-D-glucosyl]$_n$) with orthophosphate to produce [(1,4)-α-D-glucosyl]$_{n-1}$ and α-D-glucose 1-phosphate.[1-3] The Enzyme Commission number is a general classification number for phosphorylase in which the recommended name is qualified with the name of the natural substance: thus, maltodextrin phosphorylase, starch phosphorylase, and glycogen phosphorylase are all classified as EC 2.4.1.1. *See also Phosphorylase; Phosphorylase a; Phosphorylase b; Glycogen Phosphorylase; Maltodextrin Phosphorylase*

[1] J. Preiss (1982) *Ann. Rev. Plant Physiol.* **33**, 431.
[2] M. Cohn (1961) *The Enzymes*, 2nd ed., **5**, 179.
[3] D. H. Brown & C. F. Cori (1961) *The Enzymes*, 2nd ed., **5**, 207.
Selected entries from *Methods in Enzymology* [vol, page(s)]:
General discussion: 1, 194; 8, 550
Assay: 1, 194; 8, 550
Other molecular properties: glucose 1-phosphate, labeled, preparation of, **4**, 504; potato, **8**, 550 (assay, **1**, 194; **8**, 550; properties, **1**, 196; **8**, 553; purification, **1**, 62, 194; **8**, 551)

STEAROYL-CoA Δ^9-DESATURASE

This non-heme iron-dependent desaturase [EC 1.14.19.1; previously EC 1.14.99.5], also known as acyl-CoA desaturase, fatty acid desaturase, and Δ^9-desaturase, catalyzes the reaction of stearoyl-CoA with a electron donor substrate (AH_2) and dioxygen to produce oleoyl-CoA, water, and the oxidized factor (A).[1-3] Other fatty acyl-CoA substrates include myristoyl-CoA and palmitoyl-CoA. The rat liver enzyme system utilizes cytochrome b_5 and cytochrome b_5 reductase. Note that steareoyl-[acyl-carrier protein] Δ^9-desaturase will also act on stearoyl-CoA.

This reaction is the rate-limiting step in the synthesis of monounsaturated fatty acids. *See also Acyl-[Acyl-Carrier Protein] Desaturase*

[1] J. M. Ntambi (1999) *J. Lipid Res.* **40**, 1549.
[2] A. J. Fulco & K. Bloch (1964) *JBC* **239**, 993.
[3] J. Shanklin, E. Whittle & B. G. Fox (1994) *Biochemistry* **33**, 12787.
Selected entries from *Methods in Enzymology* [vol, page(s)]:
General discussion: 35, 253; 52, 13, 46, 47, 188; 71, 252, 258
Assay: 35, 254; 52, 188 (vesicle preparations, in, **52**, 207, 208); chicken liver, **71**, 253; non-substrate-binding protein stimulator, **71**, 259
Other molecular properties: chicken liver, from, **71**, 252 (cytochrome b_5 of, **71**, 252; dietary and hormonal induction, **71**, 257; inhibitors, **71**, 257; NADH-cytochrome b_5 reductase of, **71**, 252; properties, **71**, 255; specificity, **71**, 257; Δ^9-terminal desaturase, **71**, 252); inhibitors, **35**, 260, 261; kinetics, **52**, 211; mechanism, **35**, 261, 262; non-substrate-binding protein stimulator, **71**, 258; pH optimum and K_m values, **35**, 261; reactivity, **52**, 101; reconstitution, **52**, 206; specificity, **35**, 260; stability, **52**, 192; substrate specificity, **52**, 192, 193; Triton X-100 removal, **52**, 209, 210

D-STEREOSPECIFIC AMINOPEPTIDASE

This aminopeptidase [EC 3.4.11.19], also known as D-aminopeptidase, catalyzes the release of an N-terminal D-amino acid from a peptide, preferring N-terminal D-alanyl, D-seryl, or D-threonyl residues. The enzyme, first isolated from a strain of *Ochrobacterium anthropi*, can be assayed by hydrolysis of D-amino acid amides (such as D-alaninamide; large D-amino acid amides and L-derivatives are not substrates; glycinamide is a substrate) or synthetic substrates such as D-Ala-NHPhNO2. The enzyme also exhibits an esterase activity and is inhibited by ceratin β-lactams.[1,2]

[1] Y. Asano, Y. Kato, A. Nakazawa & K. Kondo (1992) *Biochemistry* **31**, 2316.
[2] Y. Asano, A. Nakazawa, Y. Kato & K. Kondo (1989) *JBC* **264**, 14233.

STERIGMATOCYSTIN 7-O-METHYLTRANSFERASE

This methyltransferase [EC 2.1.1.110], which participates in fungal biosynthesis of aflatoxins, catalyzes the reaction of *S*-adenosyl-L-methionine with sterigmatocystin to produce *S*-adenosyl-L-homocysteine and 7-*O*-methylsterigmatocystin.[1,2] Dihydrosterigmatocystin can also undergo methylation.

[1] K. Yabe, Y. Ando, J. Hashimoto & T. Hamasaki (1989) *Appl. Environ. Microbiol.* **55**, 2172.
[2] D. Bhatnagar, A. H. J. Ullah & T. E. Cleveland (1988) *Prep. Biochem.* **18**, 321.

STEROID N-ACETYLGLUCOS-AMINYLTRANSFERASE

This transferase [EC 2.4.1.39], also known as sucrose-glucan glucosyltransferase, catalyzes the reaction of UDP-*N*-acetyl-D-glucosamine with estradiol-17α 3-D-glucuronoside to produce UDP and 17α-(*N*-acetyl-D-glucosaminyl)estradiol 3-D-glucuronoside.[1,2] Other steroid substrates include 16-epiestriol-3-glucuronoside and 16,17-epiestriol-3-glucuronoside.

[1] D. C. Collins, H. Jirku & D. S. Layne (1968) *JBC* **243**, 2928.
[2] R. S. Labow, D. G. Williamson & D. S. Layne (1973) *Biochemistry* **12**, 1548.

STEROID Δ-ISOMERASE

This isomerase [EC 5.3.3.1], also known as Δ5-3-ketosteroid isomerase, catalyzes the conversion of a 3-keto-Δ5-steroid (or, 3-oxo-Δ5-steroid; for example, 5-androstene-3,17-dione) to a 3-keto-Δ4-steroid (or, 3-oxo-Δ4-steroid; thus, producing 4-androstene-3,17-dione).[1–9]

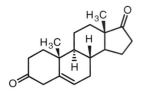

5-androstene-3,17-dione   4-androstene-3,17-dione

The reaction proceeds through an intermediate dienolate. It has been suggested that an aspartyl residue acts as a base to abstract a proton from C4.[5]

[1] F. S. Kawahara, S. Wang & P. Talalay (1962) *JBC* **237**, 1500.
[2] P. Geynet, J. Gallay & A. Alfsen (1972) *EJB* **31**, 464.
[3] A. G. Smith & C. J. W. Brooks (1977) *BJ* **167**, 121.
[4] T. M. Penning, D. F. Covey & P. Talalay (1981) *JBC* **256**, 6842.
[5] F. Henot & R. M. Pollack (2000) *Biochemistry* **39**, 3351.
[6] J. V. Schloss & M. S. Hixon (1998) *CBC* **2**, 43.
[7] M. A. Ator & P. R. Ortez de Montellano (1990) *The Enzymes*, 3rd ed., **19**, 213.
[8] D. J. Creighton & N. S. R. K. Murthy (1990) *The Enzymes*, 3rd ed., **19**, 323.
[9] P. Talalay & A. M. Benson (1972) *The Enzymes*, 3rd ed., **6**, 591.

Selected entries from *Methods in Enzymology* [vol, page(s)]:
General discussion: 5, 527; 15, 642; 34, 557; 46, 23, 89, 159, 461
Assay: 5, 327; 15, 642
Other molecular properties: affinity labeling, 46, 476; amino acid composition, 15, 649; chelating agents, 15, 648; chromatography, 15, 645; crystallization, 15, 646; extraction from *Pseudomonas testosteroni*, 5, 528; 15, 644; inhibitor, 63, 401, 404; labeling by photoexcited steroid ketones, 46, 469; mechanism of reaction, 15, 650; molecular weight, 15, 648; precipitation by cations, 15, 649; properties, 5, 531; 15, 647; *Pseudomonas testosteroni*, 5, 527; 15, 642; purification, 5, 528; 15, 643; 34, 560; relative rates of reaction of substrates, 15, 648; site-directed mutants (additivity effects, 249, 118; catalysis, Brønsted relationship, 249, 501); stereospecificity, 15, 650; transition-state and multisubstrate inhibitors, 249, 308; ultraviolet absorption spectrum, 15, 649

STEROID-LACTONASE

This enzyme [EC 3.1.1.37] catalyzes the hydrolysis of testololactone to produce testolate.[1]

[1] C. E. Holmlund & R. H. Blank (1965) *ABB* **109**, 29.

STEROID 2α-MONOOXYGENASE

A 2α-hydroxylase activity was first described from guinea pig adrenal tissue that catalyzes the conversion of cortisol to 2α-hydroxycortisol.

Selected entries from *Methods in Enzymology* [vol, page(s)]:
General discussion: 15, 637, 598; 52, 14

STEROID 2β-MONOOXYGENASE

This oxidoreductase reportedly catalyzes the hydroxylation of a steroid to produce a 2β-hydroxysteroid.[1]

[1] S. Burstein, B. R. Bhavnani & M. Gut (1965) *JBC* **240**, 2843.

Selected entries from *Methods in Enzymology* [vol, page(s)]:
General discussion: 52, 14

STEROID 6α-MONOOXYGENASE

This oxidoreductase reportedly catalyzes the hydroxylation of a steroid to produce a 6α-hydroxysteroid.[1,2]

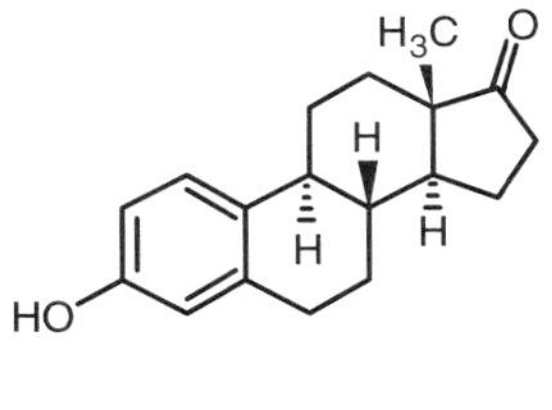

estrone

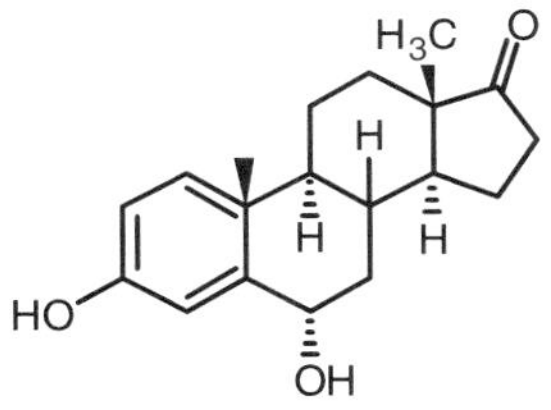

6α-hydroxyestrone

The rat liver enzyme catalyzes the reaction of estrone with NADPH and dioxygen to produce 6α-hydroxyestrone, NADP[+], and water. Other substrates include 17β-estradiol and estriol.

[1]A. J. Liston & P. Toft (1972) *BBA* **273**, 52.
[2]A. S. Meyer, M. Hayano, M. C. Lindberg, M. Gut & O. G. Rodgers (1955) *Acta Endocrinol.* **18**, 148.

Selected entries from *Methods in Enzymology* [**vol**, page(s)]:
General discussion: 15, 598, 709; **52**, 14
Assay: 15, 709
Other molecular properties: estrone activity, 15, 709; partially purified, 5, 504; 15, 709; properties, 15, 710

STEROID 6β-MONOOXYGENASE

This oxidoreductase reportedly catalyzes the hydroxylation of a steroid to produce a 6β-hydroxysteroid.[1–3]

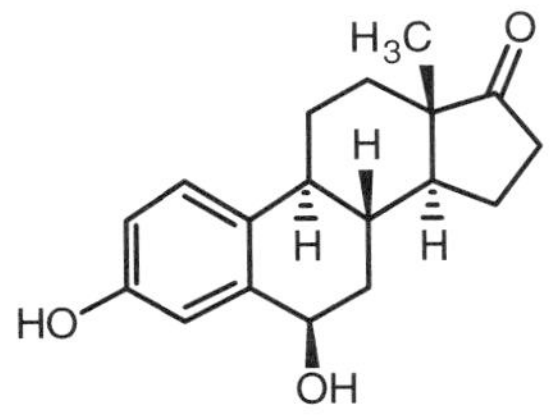

6β-hydroxyestrone

The rat liver enzyme catalyzes the reaction of estrone with NADPH and dioxygen to produce 6β-hydroxyestrone, NADP[+], and water. Other substrates include 17β-estradiol.

[1]A. J. Liston & P. Toft (1972) *BBA* **273**, 52.
[2]A. S. Meyer, M. Hayano, M. C. Lindberg, M. Gut & O. G. Rodgers (1955) *Acta Endocrinol.* **18**, 148.
[3]G. C. Mueller & G. Rumney (1957) *JACS* **79**, 1004.

Selected entries from *Methods in Enzymology* [**vol**, page(s)]:
General discussion: 5, 504; 15, 598, 710; **52**, 14
Assay: 15, 710

STEROID 7α-MONOOXYGENASE

This oxidoreductase reportedly catalyzes the hydroxylation of a steroid to produce a 7α-hydroxysteroid.[1,2]

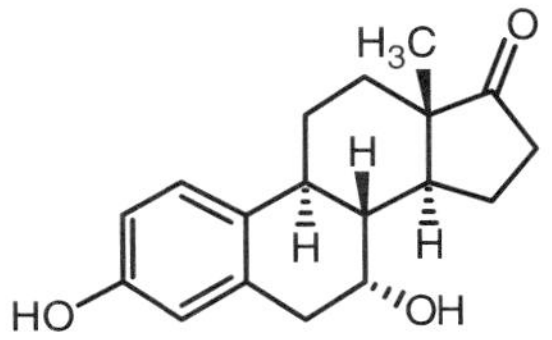

7α-hydroxyestrone

The adrenal and liver enzyme catalyzes the reaction of estrone with NADPH and dioxygen to produce 7α-hydroxyestrone, NADP[+], and water.

[1]A. J. Liston & P. Toft (1972) *BBA* **273**, 52.
[2]J. van Cantfort, J. Renson, J. Gielen & B. Robaye (1975) *EJB* **55**, 23, 33, and 41.

Selected entries from *Methods in Enzymology* [**vol**, page(s)]:
General discussion: 15, 711; **52**, 14
Assay: 15, 711
Other molecular properties: properties, 15, 711

STEROID 9α-MONOOXYGENASE

This iron-sulfer- and FMN-dependent oxidoreductase [EC 1.14.99.24], also known as steroid 9α-hydroxylase, catalyzes the reaction of pregna-4,9(11)-diene-3,20-dione with AH$_2$ and dioxygen to produce 9,11-α-epoxypregn-4-ene-3,20-dione, A, and water (where AH$_2$ is the reduced acceptor substrate and A is the oxidized acceptor).[1,2]

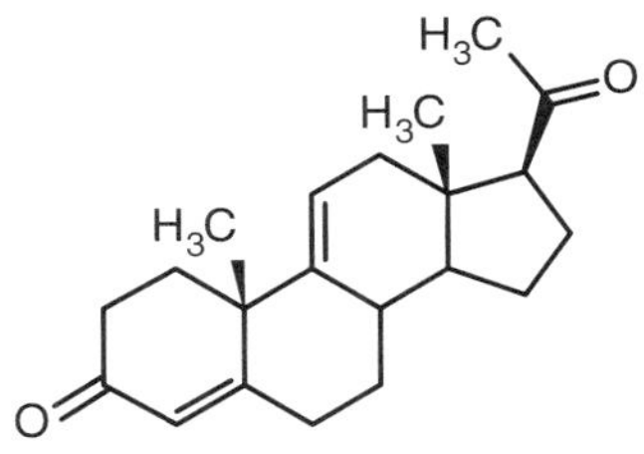

pregna-4,9(11)-diene-3,20-dione

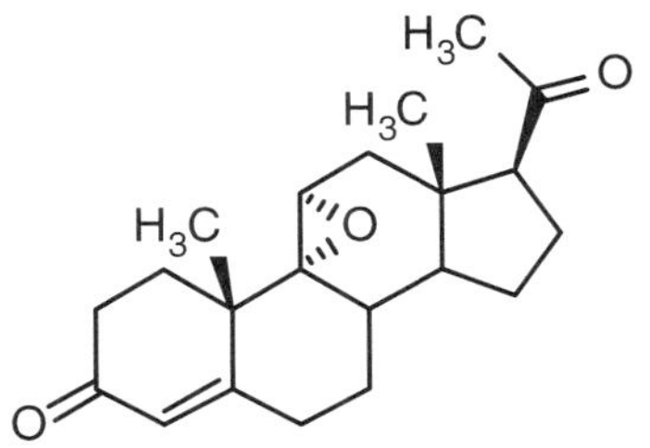

9,11α-epoxypregn-4-ene-3,20-dione

An alternative substrate is 4-androstene-3,17-dione.

[1]A. Strijewski (1982) *EJB* **128**, 125.
[2]K. Sonomoto, N. Usui, A. Tanaka & S. Fukui (1983) *Eur. J. Appl. Microbiol. Biotechnol.* **17**, 203.

STEROID 11β-MONOOXYGENASE

This heme-thiolate-dependent oxidoreductase [EC 1.14.15.4] (also referred to as steroid 11β-hydroxylase, steroid 11β/18-hydroxylase, cytochrome P450 11B1, and CYP11B1), catalyzes the reaction of a steroid with dioxygen and a reduced adrenal ferredoxin to produce an 11β-hydroxysteroid, water, and the oxidized adrenal ferredoxin.[1–3] The enzyme will also catalyze the hydroxylation of steroids at the 18-position as well as convert 18-hydroxycorticosterone into aldosterone: hence, this monooxygenase will also convert 4-pregnen-21-ol-3,20-dione to aldosterone (via corticosterone and 18-hydroxy-corticosterone).

Two 18-unsaturated progesterone derivatives, 18-vinyl-progesterone and 18-ethynylprogesterone have proved to be potent inhibitors of the bovine enzyme, suggesting that 18-vinylprogesterone inactivates by destroying the heme rather than by modifying the protein.[1]

An ox adrenal enzyme has been reported to catalyze the oxidation of estrone to 18-hydroxyestrone.

[1]C. Delorme, A. Piffeteau, F. Sobrio & A. Marquet (1997) Eur. J. Biochem. **248**, 252.
[2]M. J. Coon & D. R. Koop (1983) The Enzymes, 3rd ed., **16**, 645.
[3]V. Ullrich & W. Duppel (1975) The Enzymes, 3rd ed., **12**, 253.
[4]K. H. Loke, G. F. Marrian & E. J. D. Watson (1959) Biochem. J. **71**, 43.

Selected entries from *Methods in Enzymology* [**vol**, page(s)]:
General discussion: 5, 505; 15, 604, 712; 44, 184, 471; 52, 6, 14, 125
Assay: 15, 712; steroid 18-hydroxylase activity, 15, 616, 618, 715; 52, 126, 127
Other molecular properties: activity assay, 44, 187; 52, 124, 131; adrenodoxin antibody, effect of, 52, 244; binding reaction, 15, 611; cofactors, 15, 607; complexes, g value, 52, 254; cytochrome P450, 37, 309; immunologic properties, 52, 124, 131; inhibition, 15, 611; 52, 132; inhibitors with intact mitochondria, 15, 606; intact mitochondria, with, 15, 605; kinetics, 52, 131; low-spin species (compounds causing shift, 52, 131; preparation, 52, 130); Michaelis constant, 15, 611; molecular weight, 52, 130; partially purified, 5, 504; properties, 15, 712; reconstitution, 15, 610; resolution, 15, 608; specific activity, 15, 611; spectral properties, 52, 131; stability, 52, 131; steroid 18-hydroxylase activity, 15, 614 (assay, 15, 616, 618, 715; carbon monoxide inhibition, 15, 621; cofactors, 15, 620; cytochrome P450, 37, 309; 11β-hydroxylase activity, 15, 608; inhibitors, 15, 621; initial linear reaction rate, 15, 623; preparation, 15, 615, 618; properties, 15, 620; properties of preparations, 15, 617; purification, 15, 616; rat adrenals, by, 15, 623; relative activity in mammalian adrenals, 15, 622; species distribution, 15, 614; specific activity, 15, 623; tissue preparations, 15, 615, 618); submitochondrial particles, with, 15, 608; substrate specificity, 52, 132

STEROID 12α-MONOOXYGENASE

This oxidoreductase reportedly catalyzes the hydroxylation of a steroid to produce a 12α-hydroxysteroid.[1]

[1]G. Nicolav, B. I. Cohen, G. Salen & E. H. Mosbach (1976) Lipids **11**, 148.

Selected entries from *Methods in Enzymology* [**vol**, page(s)]:
General discussion: 52, 14

STEROID 14α-MONOOXYGENASE

A beef adrenal oxidoreductase has been reported that catalyzes the reaction of estrone with NADPH and dioxygen to produce 14α-hydroxyestrone, NADP⁺, and water.

estrone

14α-hydroxyestrone

Selected entries from *Methods in Enzymology* [**vol**, page(s)]:
General discussion: 15, 712
Assay: 15, 712
Other molecular properties: properties, 15, 713

STEROID 15α-MONOOXYGENASE

A beef adrenal oxidoreductase has been reported to catalyze the reaction of estrone with NADPH and dioxygen to produce 15α-hydroxyestrone, NADP⁺, and water. Other substrates include 17β-estradiol.

15α-hydroxyestrone

Selected entries from *Methods in Enzymology* [**vol**, page(s)]:
General discussion: 15, 713
Assay: 15, 713
Other molecular properties: properties, 15, 713

STEROID 15β-MONOOXYGENASE

This oxidoreductase reportedly catalyzes the hydroxylation of a steroid to produce a 15β-hydroxysteroid.[1,2]

[1]A. Berg, J.-Å. Gustafsson & M. Ingelman-Sundberg (1976) JBC **249**, 2831.
[2]J.-Å. Gustafsson & M. Ingelman-Sundberg (1974) JBC **249**, 1940.

Selected entries from *Methods in Enzymology* [**vol**, page(s)]:
General discussion: 52, 14

STEROID 16α-MONOOXYGENASE

This oxidoreductase reportedly catalyzes the hydroxylation of a steroid to produce a 16α-hydroxysteroid.[1] The liver enzyme also catalyzes the reaction of 17β-estradiol with

NADPH and dioxygen to produce estriol, NADP[+], and water.

17β-Estradiol

Estriol

[1]A. J. Liston & P. Toft (1972) *Biochim. Biophys. Acta* **273**, 52.

Selected entries from *Methods in Enzymology* [vol, page(s)]:
General discussion: 15, 713; **52**, 14
Assay: 15, 713
Other molecular properties: properties, 15, 714

STEROID 16β-MONOOXYGENASE

This oxidoreductase reportedly catalyzes the hydroxylation of a steroid to produce a 16β-hydroxysteroid.[1,2] A chicken liver hydroxylase catalyzes the reaction of 17β-estradiol with NADPH and dioxygen to produce 16-epiestrol, NADP[+], and water. Other substrates include estrone.

16-Epiestriol

[1]K. N. Wynne & A. G. C. Renwick (1976) *BJ* **156**, 419.
[2]M. Ingelman-Sundberg, A. Rarre & J.-ÅGustafsson (1975) *Biochemistry* **14**, 429.

Selected entries from *Methods in Enzymology* [vol, page(s)]:
General discussion: 15, 714; **52**, 14
Assay: 15, 714
Other molecular properties: properties, 15, 715

STEROID 17α-MONOOXYGENASE

This heme-thiolate or cytochrome P450 monooxygenase, [EC 1.14.99.9] also known as steroid 17α-hydroxylase/17,20 lyase, cytochrome P450 17A1 (CYP17A1), steroid 17α-hydroxylase, catalyzes the reaction of a steroid with dioxygen and an electron donor (AH2) to produce a 17α-hydroxysteroid, water, and the oxidized donor (A).[1–5] NADPH or NADH can serve as the electron donor, and pregnenolone is also a steroid substrate.

The enzyme will also exhibit a C17–C20 lyase activity on a number of steroid substrates. ***See also*** *Estradiol 17α-Dehydrogenase*

[1]S. Nakajin & P. F. Hall (1981) *JBC* **256**, 3871.
[2]D.-F. Fan, H. Oshima, B. R. Troen & P. Troen (1974) *BBA* **360**, 88.
[3]S. Nakajin, M. Shinoda, M. Haniu, J. E. Shively & P. F. Hall (1984) *JBC* **259**, 3971.
[4]T. Yamazaki, T. Ohno, T. Sakaki, M. Akiyoshi-Shibata, Y. Yabusaki, T. Imai & S. Kominami (1998) *Biochem.* **37**, 2800.
[5]M. J. Coon & D. R. Koop (1983) *The Enzymes*, 3rd ed., **16**, 645.

Selected entries from *Methods in Enzymology* [vol, page(s)]:
General discussion: 15, 636; **52**, 14

STEROID 19-MONOOXYGENASE

This oxidoreductase reportedly catalyzes the hydroxylation of a steroid to produce a 19-hydroxysteroid.[1–4]

[1]I. C. Gunsalus, T. C. Pederson & S. G. Sligar (1975) *ARB* **44**, 377.
[2]J. E. Longchampt, C. Gual, M. Ehrenstein & R. I. Dorfman (1960) *Endocrinology* **66**, 416.
[3]M. Hayano & R. I. Dorfman (1955) *ABB* **55**, 289.
[4]A. S. Meyer (1955) *Experentia* **11**, 99.

Selected entries from *Methods in Enzymology* [vol, page(s)]:
General discussion: 15, 598; **52**, 14
Other molecular properties: partially purified, 5, 504

STEROID 20-MONOOXYGENASE

This oxidoreductase reportedly catalyzes the hydroxylation of a steroid to produce a 20-hydroxysteroid.[1,2]

[1]S. Burstein, B. S. Middleditch & M. Gut (1974) *BBRC* **61**, 642.
[2]G. Constantopoulos & T. T. Tchen (1961) *JBC* **236**, 65.

Selected entries from *Methods in Enzymology* [vol, page(s)]:
General discussion: 15, 598; **52**, 14

STEROID 21-MONOOXYGENASE

This heme-thiolate-dependent (or, a cytochrome P450 system) oxidoreductase [EC 1.14.99.10], also referred to as steroid 21-hydroxylase and cytochrome P450 21A1 (CYP21A1), catalyzes the reaction of a steroid with dioxygen and an electron donor (AH2) to produce a 21-hydroxysteroid, water, and the reduced donor (A).[1–4] The electron donor can be NADPH or NADH. Steroid substrates include progesterone, 11β-hydroxyprogesterone, 17α-hydroxyprogesterone, and 11,17-dihydroxyprogesterone.

[1]K. J. Ryan & L. L. Engel (1957) *JBC* **225**, 103.
[2]A. Hiwatashi & Y. Ichikawa (1981) *BBA* **664**, 33.
[3]L. Ponticorvo, N. Greenfield, A. Wolfson, F. Chasalow & S. Lieberman (1980) *ABB* **200**, 223.
[4]M. J. Coon & D. R. Koop (1983) *The Enzymes*, 3rd ed., **16**, 645.

Selected entries from *Methods in Enzymology* [vol, page(s)]:
General discussion: 5, 510; 15, 598, 624; **52**, 14; **206**, 166, 548
Assay: 5, 510; 15, 624; **206**, 169
Other molecular properties: ascorbate inhibition, 15, 629; carbon monoxide inhibition, 15, 629; enzymatic properties, 15, 628; hemoprotein content, 15, 634; inhibitors, 15, 629; Michaelis constants, 15, 628; pH optimum, 15, 628; preparation and incubation procedures, 5, 511; 15, 626; properties, 5, 511; 15, 628; species distribution,

15, 637; specific activity, 15, 628; spectrophotometric of titration, 15, 631; storage, 15, 627; substrate specificity, 15, 628, 632

STEROID 22-MONOOXYGENASE

This oxidoreductase reportedly catalyzes the hydroxylation of a steroid to produce a 22-hydroxysteroid.[1]

[1]G. Constantopoulos & T. T. Tchen (1961) JBC 236, 65.

Selected entries from **Methods in Enzymology** [**vol**, page(s)]:
General discussion: 15, 598; 52, 14

STEROID 25-MONOOXYGENASE

This oxidoreductase reportedly catalyzes the hydroxylation of a steroid to produce a 25-hydroxysteroid.[1,2]

[1]I. Björkhem, H. Danielsson & K. Wikvall (1976) JBC 251, 3495.
[2]D. S. Frederickson (1956) BBA 22, 183.

Selected entries from **Methods in Enzymology** [**vol**, page(s)]:
General discussion: 52, 14
Other molecular properties: partially purified, 5, 504

STEROID 26-MONOOXYGENASE

This oxidoreductase reportedly catalyzes the hydroxylation of a steroid to produce a 26-hydroxysteroid.[1–4]

[1]I. Björkhem, H. Danielsson & K. Wikvall (1976) JBC 251, 3495.
[2]I. Björkhem & J.-Å Gustafsson (1973) EJB 36, 201.
[3]G. Nicolav, B. I. Cohen, G. Salen & E. H. Mosbach (1976) Lipids 11, 148.
[4]D. S. Frederickson (1956) BBA 22, 183.

Selected entries from **Methods in Enzymology** [**vol**, page(s)]:
General discussion: 52, 14
Other molecular properties: partially purified, 5, 504

STEROID SULFOTRANSFERASE

This sulfotransferase [EC 2.8.2.15], also known as steroid sulfokinase, catalyzes the reaction of 3′-phosphoadenylylsulfate with a phenolic steroid or hydroxysteroid (e.g., dehydroepiandrosterone, androst-5-ene-3β,17α-diol, estrone, androsterone, testosterone, and pregnenolone) to produce adenosine 3′,5′-bisphosphate and the corresponding steroid O-sulfate.[1,2] **See also** Aryl Sulfotransferase; Alcohol Sulfotransferase

[1]J. B. Adams & D. McDonald (1979) BBA 567, 144; 615, 275.
[2]C. N. Falany, M. E. Vasquez & J. M. Kalb (1989) BJ 260, 641.

Selected entries from **Methods in Enzymology** [**vol**, page(s)]:
General discussion: 5, 980; 143, 201
Assay: 5, 980; 143, 201; application in crude tissue preparations, 5, 982
Other molecular properties: androstenolone sulfotransferase, 15, 732; binding assay, 143, 206; preparation, 5, 982; properties, 5, 982; specificity, 5, 982

STEROID-TRANSPORTING ATPase

This so-called ATPase [EC 3.6.3.45], also known as pleiotropic-drug-resistance protein and PDR protein,

catalyzes the ATP-dependent (ADP- and orthophosphate-producing) transport of steroid$_{in}$ to produce steroid$_{out}$.[1,2] The yeast protein will also catalyze the export of a variety of xenobiotics. Substrates exported include estradiol, actinomycin D, cerulenin, and cytochalasin B. ABC-type (ATP-binding cassette-type) ATPases have two similar ATP-binding domains and do not undergo phosphorylation during transport. **See also** ATPase, Xenobiotic-Transporting

The idea that a transporter is an enzyme is in keeping with a new definition of enzyme catalysis as the facilitated making/breaking of chemical bonds, not just covalent bonds.[3] This idea builds on Pauling's assertion that any long-lived, chemically distinct interaction (in this case, the persistent location of a solute with respect to the faces of a membrane) can be regarded as a chemical bond. Note also that the equilibrium constant ($K_{eq} = $[Steroid$_{out}$][ADP][P$_i$]/[Steroid$_{in}$][ATP]) does not conform to that expected for an ATPase (i.e., $K_{eq} = $[ADP][P$_i$]/[ATP]). Thus, although the overall reaction yields ADP and orthophosphate, the enzyme is misclassified as a hydrolase, and instead should be regarded as an energase-type reaction. Energases facilitate affinity-modulated reactions by coupling the $\Delta G_{ATP\text{-hydrolysis}}$ to a force-generating or work-producing step.[3] In this case, P-O-P bond scission supplies the energy to drive steroid transport. **See** ATPase; Energase

[1]K. Nagao, Y. Taguchi, M. Arioka, H. Kadokura, A. Takatsuki, K. Yoda & M. Yamasaki (1995) J. Bacteriol. 177, 1536.
[2]Y. Mahé, Y. Lemoine & K. Kuchler (1996) JBC 271, 25167.
[3]D. L. Purich (2001) TiBS 26, 417.

Δ$^{14(15)}$-STEROL DEHYDROGENASE

This oxidoreductase catalyzes the NADP$^+$-dependent formation of 14-ene-sterols: for example, the conversion 8-cholesten-3β-ol to 8,14-cholestdien-3β-ol and the conversion of 7-cholesten-3β-ol to 7,14-cholestdien-3β-ol.[1]

[1]B. C. Wilton, I. A. Watkinson & M. Akhtar (1970) BJ 119, 673.

STEROL GLUCOSYLTRANSFERASE

This glucosyltransferase [EC 2.4.1.173] catalyzes the reaction of UDP-D-glucose with a sterol to produce an O-glucosylsterol and UDP. Steroid substrates include sitosterol, poriferasterol, epiandrosterone, pregnenolone, stigmasterol, campesterol, cholesterol, and α-spinasterol.[1–3] This enzyme is not identical with nuatigenin 3-β-glucosyltransferase [EC 2.4.1.192] or sarsapogenin 3-β-glucosyltransferase [EC 2.4.1.193].

[1]K. Murakami-Murofushi & J. Ohta (1989) *BBA* **992**, 412.
[2]T. W. Esders & R. J. Light (1972) *JBC* **247**, 7494.
[3]Z. A. Wojciechowski, J. Zimowski, J. G. Zimowski & A. Lyznik (1979) *BBA* **570**, 363.

Selected entries from *Methods in Enzymology* [vol, page(s)]:
General discussion: 28, 478
Assay: 28, 478; product characterization, 28, 481

Δ²⁴-STEROL C-METHYLTRANSFERASE

This glutathione-dependent methyltransferase [EC 2.1.1.41] catalyzes the reaction of S-adenosyl-L-methionine with 5α-cholesta-8,24-dien-3β-ol (or, zymosterol) to produce S-adenosyl-L-homocysteine and 24-methylene-5α-cholest-8-en-3β-ol (or, fecosterol).[1–5] Other substrates include cycloartenol and desmosterol. The *Saccharomyces cerevisiae* enzyme exhibits cooperative kinetics and the reaction involves a 1,2-hydride shift of H24 to C25 from the *re*-face of the original 24,25- double bond.[5]

[1]J. T. Moore & J. L. Gaylor (1969) *JBC* **244**, 6334.
[2]W. D. Nes, G. G. Janssen & A. Bergenstrahle (1991) *JBC* **266**, 15202.
[3]H.-K. Lin & H. W. Knoche (1976) *Phytochem.* **15**, 683.
[4]M. A. Ator, S. J. Schmidt, J. L. Adams & R. E. Dolle (1989) *Biochemistry* **28**, 9633.
[5]W. D. Nes, B. S. McCourt, W. X. Zhou, J. Ma, J. A. Marshall, L. A. Peek & M. Brennan (1998) *ABB* **353**, 297.

STEROL Δ²⁴-REDUCTASE

This oxidoreductase catalyzes the NADPH-dependent conversion of desmosterol (Δ⁵,²⁴-cholestadien-3β-ol) to cholesterol.

Selected entries from *Methods in Enzymology* [vol, page(s)]:
General discussion: 15, 514
Assay: 15, 515
Other molecular properties: incubation conditions, 15, 516; inhibition, 15, 513; inhibitor action, 15, 519; insects, in, 15, 522; mechanism of action, 15, 520; occurrence, 15, 521; role in sterol metabolism, 15, 521; stereospecificity, 87, 116; substrate specificity, 15, 519

STERYL-β-GLUCOSIDASE

This glucosidase [EC 3.2.1.104] catalyzes the hydrolysis of cholesteryl-β-D-glucoside to produce cholesterol and D-glucose.[1] Sitosteryl-β-D-glucoside is also a substrate.

[1]M. Kalinowska & Z. A. Wojciechowski (1978) *Phytochemistry* **17**, 1533.

STERYL SULFATASE

These sulfatases [EC 3.1.6.2] (also referred to as steroid sulfatase, steryl-sulfate sulfohydrolase, estrone sulfatase, and arylsulfatase C) catalyze the hydrolysis of 3β-hydroxyandrost-5-en-17-one 3-sulfate to produce 3β-hydroxyandrost-5-en-17-one and sulfate.[1–3] Other substrates include dehydroepiandrosterone sulfate, cortisone 21-sulfate, cholesteryl sulfate, estrone sulfate, β-estradiol-3-sulfate, and dehydroandrosterone sulfate.[1,2]

Specific steryl sulfatases have been reported and all are currently classified under EC 3.1.6.2. Note that a number of arylsulfatases [EC 1.3.6.1] also exhibit steryl sulfatase activity. Estrone-3-O-sulfamate is a potent active-site inhibitor.[3]

[1]H. Noel, L. Plante, G. Bleau, A. Chapdelaine & K. D. Roberts (1983) *J. Steroid Biochem.* **19**, 1591.
[2]A. B. Roy (1971) *The Enzymes*, 3rd ed., **5**, 1.
[3]A. Purohit, G. J. Williams, N. M. Howarth, B. V. Potter & M. J. Reed (1995) *Biochemistry* **34**, 11508.

Selected entries from *Methods in Enzymology* [vol, page(s)]:
General discussion: 2, 324, 330, 332; 15, 684; 143, 361
Assay: 15, 685; 143, 362
Other molecular properties: cholesterol sulfate sulfatase, multiple sulfatase deficiency, 50, 474; dehydroepiandrosterone sulfatase, multiple sulfatase deficiency, 50, 474; estrone sulfatase, multiple sulfatase deficiency, 50, 474; *Helix pomatia* purification, 143, 363; human placenta, of, 15, 687; lipophilic Sephadex chromatography, 35, 389, 393; multiple sulfatase deficiency, 50, 454, 474; *Patella vulgata*, of, 15, 687; placental defect, 39, 251; properties, 15, 688; 143, 365; purification, 15, 687; sources, 15, 689; substrate specificity, 15, 688

STIPITATONATE DECARBOXYLASE

This decarboxylase [EC 4.1.1.60] catalyzes the conversion of stipitatonate to stipitatate and carbon dioxide.[1]

[1]R. Bentley & C. P. Thiessen (1963) *JBC* **238**, 3811.

STIZOLOBATE SYNTHASE

This zinc-dependent oxidoreductase [EC 1.13.11.29] catalyzes the reaction of 3,4-dihydroxy-L-phenylalanine with dioxygen to produce 4-(L-alanin-3-yl)-2-hydroxy-*cis,cis*-muconate 6-semialdehyde.[1] The product undergoes ring closure and oxidation, in the presence of NAD(P)⁺ as the acceptor, to form stizolobate.

[1]K. Saito & A. Komamine (1978) *EJB* **82**, 385.

STIZOLOBINATE SYNTHASE

This zinc-dependent oxidoreductase [EC 1.13.11.30] catalyzes the reaction of 3,4-dihydroxy-L-phenylalanine with dioxygen to produce 5-(L-alanin-3-yl)-2-hydroxy-*cis,cis*-muconate 6-semialdehyde.[1] The product undergoes ring closure and oxidation, in the presence of NAD(P)⁺ as the acceptor, to form stizolobinate.

[1]K. Saito & A. Komamine (1978) *EJB* **82**, 385.

STREPTOGRISIN A, B, AND C

Streptogrisin A [EC 3.4.21.80] (also referred to as *Streptomyces griseus* protease A [SGPA], elastase-like enzyme II, alkaline protease A, and Pronase enzyme A) is one of the serine proteinase components of commercial Pronase. Streptogrisin A catalyzes peptide-bond hydrolysis with a specificity resembling that of chymotrypsin, but is not inhibited by Tos-Phe-CH$_2$Cl. While the enzyme prefers a tyrosyl or phenylalanyl residue in the P1 position, other residues are tolerated (Phe > Tyr > Leu > Val > Ala > Gly).[1-4] The enzyme has an extended binding domain consisting of at least seven subsites.

Streptogrisin B [EC 3.4.21.81] (also called *Streptomyces griseus* proteinase B [SGPB], Pronase elastase, and Pronase enzyme B), is one of the serine proteinase components of commercial Pronase. Streptogrisin B catalyzes peptide-bond hydrolysis with a specificity resembling that of chymotrypsin. The enzyme prefers a tyrosyl or phenylalanyl residue in the P1 position; however, Met, Leu, Trp, Ala, Val, and Gly are all tolerated. Increasing the substrate chain length greatly increases enzyme efficiency; there appears to be at least seven subsites present.[2,5] Streptogrisin B is unusually stable, still retaining activity in 6 M guanidinium chloride and in 8 M urea. It also has a broad pH range, with measurable activity below pH 2.5.

Streptogrisin C, also isolated from *Streptomyces griseus*, has a specificity similar to streptogrisin B and may bind chitin.[6]

Streptogrisin D is a dimeric serine endopeptidase and has a specificity similar to that of the other streptogrisins.[6]

[1] G. D. Brayer, L. T. J. Delbaere, M. N. G. James, C.-A. Bauer & R. C. Thompson (1979) *PNAS* **76**, 96.
[2] W. Lu, I. Apostol, M. A. Qasim, N. Warne, R. Wynn, W. L. Zhang, S. Anderson, Y. W. Chiang, I. Rothberg, K. Ryan & M. Laskowski, Jr., (1997) *JMB* **266**, 441.
[3] C.-A. Bauer, R. C. Thompson & E. R. Blout (1976) *Biochemistry* **15**, 1291 & 1296.
[4] Y. Narahashi (1970) *MIE* **19**, 651.
[5] K. Huang, W. Lu, S. Anderson, M. Laskowski & M. N. G. James (1995) *Protein Sci.* **4**, 1985.
[6] S. S. Sidhu, G. B. Kalmar, L. G. Willis & T. J. Borgford (1994) *JBC* **269**, 20167.

STREPTOMYCES GRISEUS AMINOPEPTIDASE

This zinc-dependent metalloaminopeptidase, isolated from *Streptomyces griseus*, is a component of commercial Pronase. The enzyme prefers hydrophobic N-terminal aminoacyl residues as well as hydrophobic residues in the adjacent P1′ position. Leu-NHPhNO$_2$ is a useful synthetic substrate for activity assays.[1]

[1] W. M. Awad, Jr., (1998) in *Handbook of Proteolytic Enzymes* (A. J. Barrett, N. D. Rawlings & J. F. Woessner, eds.), Academic Press, New York, p. 1431.

STREPTOMYCES K15 D-Ala-D-Ala TRANSPEPTIDASE

This serine-dependent enzyme catalyzes both a DD-transpeptidation reaction (in which an acyl-D-alanyl group is transferred from one peptide to another) and a DD-carboxypeptidase activity. During catalysis, the enzyme is acylated at a seryl residue. If the substrate is Ac$_2$-L-Lys-D-Ala-D-Ala, the acylated enzyme intermediate is Ac$_2$-L-Lys-D-Ala-enzyme, with the concomitant release of D-alanine. If the second substrate is instead water, then hydrolysis ensues, allowing the enzyme to exhibit DD-carboxypeptidase activity. The D-alanine generated in the first step can compete with water, thereby regenerating the peptide donor. In the presence of millimolar levels of glycylglycine, transpeptidation results in the formation of Ac$_2$-L-Lys-D-Ala-Gly-Gly. Besides glycylglycine and D-alanine, other acceptor substrates include glycyl-L-alanine, *meso*-diaminopimelate, D-aminoadipate, and glycine. The enzyme is also inactivated by β-lactams such as penicillin.[1,2] **See also** *Streptomyces R61 D-Ala-D-Ala Carboxypeptidase*

[1] J.-M. Ghuysen, P. Charlier, J. Coyette, C. Duez, E. Fonzé, C. Fraipont, C. Goffin, B. Joris & M. Nguyen-Distèche (1996) *Microb. Drug Resistance: Mech. Epidemiol. Dis.* **2**, 163.
[2] J.-M. Ghuysen (1997) *Int. J. Antimicrob. Agents* **8**, 45.

STREPTOMYCES R61 D-Ala-D-Ala CARBOXYPEPTIDASE

This membrane-bound enzyme resembles D-Ala-D-Ala carboxypeptidase [EC 3.4.16.4] and catalyzes the hydrolysis of R-D-Ala-D-Ala to produce R-D-Ala and D-alanine. **See also** *Streptomyces K15 D-Ala-D-Ala Transpeptidase*

STREPTOMYCIN 3″-ADENYLYLTRANSFERASE

This nucleotidyltransferase [EC 2.7.7.47] catalyzes the reaction of ATP with streptomycin to form 3″-adenylylstreptomycin and pyrophosphate (or, diphosphate).[1] Spectinomycin and bluensomycin are also substrates.

[1] M. Kono, K. Ohmiya, T. Kanda, N. Noguchi & K. O'Hara (1987) *FEMS Microbiol. Lett.* **40**, 223.

Selected entries from **Methods in Enzymology** [vol, page(s)]:
General discussion: **43**, 616, 618, 619, 625, 626

STREPTOMYCIN 3″-KINASE

This phosphotransferase [EC 2.7.1.87], also known as streptomycin 3″-phosphotransferase, catalyzes the reaction of ATP with streptomycin to produce ADP and streptomycin 3″-phosphate.[1,2] Other substrates include 3′-deoxydihydrostreptomycin, dihydrostreptomycin, and corresponding 6-phosphate derivatives.

[1] J. B. Walker & M. Skovarga (1973) JBC 248, 2435.
[2] P. Heinzel, O. Werbitzky, J. Distler & W. Piepersberg (1988) Arch. Microbiol. 150, 184.

Selected entries from **Methods in Enzymology** [vol, page(s)]:
General discussion: 43, 616, 632
Assay: 43, 618, 619, 633, 634
Other molecular properties: biological distribution, 43, 631; Escherichia coli, from, 43, 627; Pseudomonas aeruginosa, from, 43, 627; reaction scheme, 43, 632; specificity, 43, 634; stability, 43, 633

STREPTOMYCIN 6-KINASE

This phosphotransferase [EC 2.7.1.72], also known as streptidine kinase and streptomycin 6-phosphotransferase, catalyzes the reaction of ATP (or dATP) with streptomycin to produce ADP (or dADP) and streptomycin 6-phosphate.[1,2] Other substrates include 2-deoxystreptidine, dihydrostreptomycin, and streptidine.

[1] J. B. Walker & M. Skorvaga (1973) JBC 248, 2435.
[2] M. Sugiyama, M. Sakamoto, H. Mochizuki & O. R. Nimi (1983) J. Gen. Microbiol. 129, 1683.

Selected entries from **Methods in Enzymology** [vol, page(s)]:
General discussion: 43, 628
Assay: 43, 629, 632
Other molecular properties: biological distribution, 43, 631; properties, 43, 631, 632; reaction scheme, 43, 628; specificity, 43, 631; stability, 43, 631; Streptomyces bikiniensis, from, 17A, 1014

STREPTOMYCIN-6-PHOSPHATASE

This phosphatase [EC 3.1.3.39] catalyzes the hydrolysis of streptomycin 6-phosphate to produce streptomycin and orthophosphate.[1] Other substrates include N-amidinostreptamine phosphate, dihydrostreptomycin 3′α,6-bisphosphate, and streptidine 6-phosphate.

[1] M. S. Walker & J. B. Walker (1971) JBC 246, 7034.

Selected entries from **Methods in Enzymology** [vol, page(s)]:
General discussion: 43, 460, 465, 468, 476
Assay: alternative, 43, 470; [3′,α-³H]dihydrostreptomycin 6-phosphate assay, 43, 466, 467; streptomycin-6-phosphatase assay, 43, 467, 468

STREPTOPAIN

This cysteine endopeptidase [EC 3.4.22.10] (also known as streptopapain, streptococcal cysteine proteinase, streptococcal peptidase A, and streptococcal proteinase) catalyzes peptide-bond hydrolysis in proteins with a preference for hydrophobic aminoacyl residues at the P2, P1, and P1′ positions.[1,2] The enzyme also exhibits an esterase activity.

[1] T.-Y. Liu & S. D. Elliott (1971) The Enzymes, 3rd ed., 3, 609.
[2] K. Brocklehurst, F. Willenbroek & E. Salih (1987) in Hydrolytic Enzymes (A. Neuberger & K. Brocklehurst, eds.), p. 39, Elsevier, Amsterdam.

Selected entries from **Methods in Enzymology** [vol, page(s)]:
General discussion: 19, 252
Assay: 19, 252 (esterase activity, 19, 253; peptidase activity, 19, 252; proteolytic activity, 19, 252)
Other molecular properties: activators and inactivators, 19, 258; active site, 19, 261; affinity chromatography, 244, 641; family, 244, 469; kinetic properties, 19, 259; preparation and purification, 19, 253; processing, 244, 469; purity and physical properties, 19, 257; stability, 19, 257; structural analysis with thermolysin, 47, 184; substrate specificity, 19, 258; 244, 469; zymogen, 19, 260

STRICTOSIDINE β-GLUCOSIDASE

This glucosidase [EC 3.2.1.105] catalyzes the hydrolysis of strictosidine to produce strictosidine aglycone and D-glucose.[1]

[1] T. Hemscheidt & M. H. Zenk (1988) FEBS Lett. 110, 187.

STRICTOSIDINE SYNTHASE

This ligase [EC 4.3.3.2] catalyzes a Pictet-Spengler-type reaction, wherein imine formation induces secologenin's α-carbon to form a carbanion which attacks the indole ring of tryptamine in a stereochemically controlled manner, thereby forming 3-α(S)-strictosidine.[1]

[1] T. Hashimoto & Y. Yamada (1994) Ann. Rev. Plant Physiol. Plant Mol. Biol. 45, 257.

Selected entries from **Methods in Enzymology** [vol, page(s)]:
General discussion: 136, 342

STROMBINE DEHYDROGENASE

This oxidoreductase [EC 1.5.1.22] catalyzes the reversible reaction of N-(carboxymethyl)-D-alanine with NAD⁺ and water to produce glycine, pyruvate, and NADH.[1,2] The enzyme exhibits a weakly alanopine dehydrogenase [EC 1.5.1.17] activity, such that 2,2′-iminodipropanoate reacts with NAD⁺ and water to produce L-alanine, pyruvate, and NADH.

[1] J. Blackstock & M. C. Burdass (1987) Biochem. Soc. Trans. 15, 383.
[2] J. Thompson & S. P. F. Miller (1991) AE 64, 317.

STROMELYSIN 1

This zinc-dependent metalloendopeptidase [EC 3.4.24.17], also known as matrix metalloproteinase 3 (MMP-3), proteoglycanase, and transin, catalyzes peptide-bond hydrolysis in extracellular matrix proteins. (The name finds its origin from the fact that the metalloproteinase was originally derived from stromal cells.) The enzyme prefers substrates with hydrophobic residues at the P2, P1′, and P2′ positions, as well as a prolyl residue in the P3 position. Protein substrates include proteoglycans, aggrecan, laminin, fibronectin, large tenascin C, and collagen types III, IV, V, IX, and X (it does not act on types I or II collagen).[1] *See also* Stromelysin 2; Stromelysin 3

[1]H. Nagase (1995) *MIE* **248**, 449.

Selected entries from **Methods in Enzymology** [vol, page(s)]:
General discussion: 248, 449
Assay: proteoglycan degradation method, **248**, 52; human (fluorogenic peptide, with, **248**, 453; methods, **248**, 451)
Other molecular properties: active-site titration, **248**, 100, 460, 502; activity, **248**, 450; activation, **248**, 458 (4-aminophenylmercuric acetate, by, **248**, 458; endopeptidase, by, **248**, 459); heat method, **248**, 460; promatrix metalloproteinases, of other, **248**, 467; active site titration, **248**, 460, 502; calcium ion requirement, **248**, 462; cleavage sites on protein substrates, **248**, 464; inhibitors, **248**, 467; kinetics, **248**, 454, 466; pH optimum, **248**, 462; polypeptide chain structure, **248**, 204; procollagenase activation, **248**, 467; proenzyme, **248**, 450, 461; progelatinase B activation, **248**, 468, 473, 481; properties, **248**, 461; purification, **248**, 454; stability, **248**, 462; structure, **248**, 461; substrate specificity, **248**, 464; synthesis, **248**, 461; transferrin radioassay, **248**, 451

STROMELYSIN 2

This zinc-dependent metalloendopeptidase [EC 3.4.24.22], also referred to as matrix metalloproteinase 10 (MMP-10) and transin-2, catalyzes peptide-bond hydrolysis in extracellular matrix proteins, such as aggrecan, collagen (types III, IV, and V), fibronectin, gelatins, and proteoglycans. This enzyme acts on the proenzymes of interstitial collagenase and neutrophil collagenase. Stromelysins 1 and 2 share 75% identity in their amino acid sequences.[1]

[1]L. J. Windsor, H. Grenett, B. Birkedal-Hansen, M. K. Bodden, J. A. Engler & H. Birkedal-Hansen (1993) *JBC* **268**, 17341.

Selected entries from **Methods in Enzymology** [vol, page(s)]:
General discussion: 248, 449
Assay: proteoglycan degradation method, **248**, 52; human, **248**, 468
Other molecular properties: human, **248**, 449 (assay, **248**, 468; proenzyme, **248**, 450 [activation, **248**, 468; properties, **248**, 470; sources, **248**, 468]; purification, **248**, 468; synthesis, **248**, 450); properties, **248**, 192, 195, 511; secretion, **248**, 450

STROMELYSIN 3

This metalloendopeptidase, also called matrix metalloproteinase 11 (MMP-11), is distantly related to stromelysins

1 and 2. The mature enzyme hydrolyzes the Ala350–Met351 bond in the α_1-proteinase inhibitor, resulting in complete loss of antiproteolytic activity.[1,2] *See also* Stromelysin 1; Stromelysin 2

[1]N. Rouyer, C. Wolf, M. P. Chenard, M. C. Rao, P.Chambon, J. P. Bellocq & P. Basset (1995) *Invasion Metastasis* **14**, 269.
[2]D. Pei, G. Majmudar & S. J. Weiss (1994) *JBC* **269**, 25849.

*Stu*I RESTRICTION ENDONUCLEASE

This type II restriction endonuclease [EC 3.1.21.4] is obtained from *Streptomyces tubercidicus* and acts on both strands of DNA at 5′...AGG∇CCT...3′, generating blunt-ended fragments.[1] (The symbol ∇ indicates the cleavage site.) The enzyme is blocked by overlapping *dcm* methylation.[2]

[1]H. Shimotsu, H. Takahashi & H. Saito (1980) *Gene* **11**, 219.
[2]Y.-H. Song, T. Rueter & R. Geiger (1988) *NAR* **16**, 2718.

(S)-STYLOPINE SYNTHASE

This oxidoreductase [EC 1.1.3.32], also known as (S)-cheilanthifoline oxidase (methylenedioxy-bridge-forming), catalyzes the reaction of (S)-cheilanthifoline with NADPH and dioxygen to produce (S)-stylopine, NADP+, and two water molecules.[1]

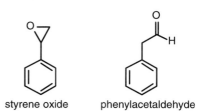

(S)-cheilanthifoline   (S)-stylopine

Note that this enzyme uses heme-thiolate (or, cytochrome P450) as a cofactor and does not incorporate oxygen into the protoberberine alkaloid product.

[1]W. Bauer & M. H. Zenk (1991) *Phytochemistry* **30**, 2953.

STYRENE-OXIDE ISOMERASE

This isomerase [EC 5.3.99.7] catalyzes the conversion of styrene oxide to phenylacetaldehyde.[1]

styrene oxide   phenylacetaldehyde

[1]S. Hartmans, J. P. Smits, M. J. van der Werf, W. F. Volkering & J. A. M. de Bont (1989) *Appl. Environ. Microbiol.* **55**, 2850.

*Sty*I RESTRICTION ENDONUCLEASE

This type II restriction endonuclease [EC 3.1.21.4] is obtained from *Salmonella typhi* 27 and acts on both strands of DNA at 5′...C∇CWWGG...3′, where W refers to A or T.[1] The symbol ∇ indicates the position of bond cleavage.

[1]K. Mise & K. Nakajima (1985) *Gene* **33**, 357.

SUBTILISIN

This enzyme [EC 3.4.21.62], a serine endopeptidase that evolved independently of chymotrypsin, was first isolated from *Bacillus subtilis*. The enzyme, which catalyzes the peptide-bond hydrolysis in proteins, displays a broad specificity, preferring large uncharged aminoacyl residues in the P1 position.[1-6] Subtilisin also catalyzes the hydrolysis of peptide amides and exhibits an esterase activity. The rate-limiting step in amide bond hydrolysis is acylation, whereas deacylation limits the rate of ester hydrolysis. Like trypsin and chymotrypsin, subtilisin possesses an oxyanion hole that stabilizes tetrahedral transition-state intermediates. The enzyme has an alkaline pH-optimum and requires a deprotonated histidyl residue ($pK_a \sim 7$) for activity. Calcium ions stabilize the enzyme, and assays are carried out in the presence of 2 mM of this divalent cation. **See also** *Thiolsubtilisin; specific types of subtilisin (e.g., Subtilisin Carlsberg)*

[1]J. S. Fruton (1982) *AE* **53**, 239.
[2]R. J. Coll, P. D. Compton & A. L. Fink (1982) *MIE* **87**, 66.
[3]J. Kraut (1971) *The Enzymes*, 3rd ed., **3**, 547.
[4]F. S. Markland, Jr., & E. L. Smith (1971)*The Enzymes*, 3rd ed., **3**, 562.
[5]L. Polgár (1987) in *Mechanisms of Protease Action* (A. Neuberger & K. Brocklehurst, eds.), CRC Press, Boca Raton.
[6]C. W. Wharton (1998) *CBC* **1**, 345.

Selected entries from **Methods in Enzymology** [vol, page(s)]:
General discussion: 19, 32, 199, 215; 34, 417; 44, 936; 46, 22, 27, 75, 206, 207
Assay: 19, 201 (esterolytic activity, 19, 201; proteolytic activity, 19, 202)
Other molecular properties: activators, 19, 210; affinity labeling, 87, 480; amino acid composition, 45, 423; boronic acid adducts, 87, 74; carboxypeptidase A and, 25, 529; cleavage of peptides and proteins, in, 11, 229; cleavage-site specificity, 244, 114; cryoenzymology, 63, 338; 87, 71; denaturation, 26, 21; intermediates, 87, 71; transition-state and multisubstrate analogues, 249, 306; use in detergents, 19, 201; use in preparation of ribonuclease S, 11, 641

SUBTILISIN BPN′

This enzyme is a subtilisin variant [EC 3.4.21.62] isolated from *Bacillus amyloliquefaciens*. **See** Subtilisin; Thiolsubtilisin

Selected entries from **Methods in Enzymology** [vol, page(s)]:
General discussion: 46, 215

Other molecular properties: active-site titration, 248, 96, 100; amino acid composition, 19, 205; esterase kinetics, 19, 213; inhibitors, 19, 210; isolation, 19, 204; physicochemical constants, 19, 207; secondary and tertiary structure, 19, 208; similarity to subtilisin Novo, 19, 200, 205; substrate, design, synthetic peptide library in, 248, 33; synonyms, 19, 200; thiolsubtilisin BPN′, 19, 215 (activity, 19, 214; affinity labeling, 87, 491; assay, 19, 216; inhibitors, 19, 220; kinetic properties, 19, 219; mechanism of action, 19, 220; purification, 19, 218; pyridyl disulfides, and, 87, 461; specificity, 19, 219; stability, 19, 220)

SUBTILISIN CARLSBERG

This enzyme is a subtilisin variant [EC 3.4.21.62], also known as subtilisin A, subtilopeptidase A, and Alcalase, from *Bacillus licheniformis*. **See also** *Subtilisin; Subtilisin BPN′; Subtilopeptidase A*

Selected entries from **Methods in Enzymology** [vol, page(s)]:
Assay: 19, 202
Other molecular properties: chemical modification, 19, 214; EMA-subtilopeptidase A derivative, preparation and properties, 19, 960; esterase activity, 19, 212 (kinetics, 19, 213); inhibitors, 19, 210; isolation, 19, 202; kinetics, of substrate diffusion, 44, 437; optical rotation, 19, 208; peptidase activity, 19, 212; physicochemical constants, 19, 207; similarity to subtilisin BPN′, 19, 208; Soman-inactivated, reactivation with 1-methyl-2-hydroxyiminomethylpyridinium methanesulfonate, 11, 701; trapped by α_2-macroglobulin, 80, 748; water-insoluble derivatives, 19, 960

SUBTILISIN NOVO

This variant of subtilisin [EC 3.4.21.62] is virtually identical to subtilisin BPN′. **See also** *Subtilisin; Subtilisin BPN′*

Selected entries from **Methods in Enzymology** [vol, page(s)]:
Other molecular properties: boronic acid adduct, 87, 74; chemical modification, 19, 214; esterase kinetics, 19, 213; hemiketal adduct, 87, 74; inhibitors, 19, 210; optical rotation, 19, 208; physicochemical constants, 19, 207; synonyms, 19, 200; trapped by α_2-macroglobulin, 80, 748

SUCCINATE:CITRAMALATE CoA-TRANSFERASE

This transferase [EC 2.8.3.7], also known as itaconate CoA-transferase and citramalate CoA-transferase, catalyzes the reversible reaction of succinyl-CoA with citramalate to produce succinate and citramalyl-CoA.[1] The enzyme will also transfer the CoA group to itaconate. In various organisms, this enzyme is indistinguishable from citramalate CoA-transferase [EC 2.8.3.11]. **See also** *Citramalate-CoA Transferase*

[1]R. A. Cooper & H. L. Kornberg (1964) *BJ* **91**, 82.

Selected entries from **Methods in Enzymology** [vol, page(s)]:
General discussion: 13, 314

SUCCINATE:CYTOCHROME c REDUCTASE

The activity, often referred to as succinate:cytochrome c reductase, is a multireaction system containing both succinate dehydrogenase (ubiquinone) (complex II; EC 1.3.5.1] and ubiquinol:cytochrome *c* reductase (complex III; EC 1.10.2.2).

Selected entries from *Methods in Enzymology* [vol, page(s)]:
General discussion: 10, 213
Assay: 10, 202, 213, 216, 222, 265, 427; 13, 84; 55, 102
Other molecular properties: acetone-extracted particles, of, 10, 265; activation energy, 10, 207; bovine heart mitochondria (iron-sulfur protein-depleted complex, purification, 126, 216; isolation, 126, 214; reconstitutively active iron-sulfur protein, purification, 126, 214); brush borders, in, 31, 121; composition, 10, 220; effect of phospholipase, 10, 416; electron transport particle, assay of, 6, 417; inhibitors, 10, 219; iron-sulfur center, redox titration, 53, 490; isolation, 6, 422; Keilin-Hartree preparation, 10, 202; 55, 126; liver, in, 31, 97; NADH-cytochrome *c* reductase and, 6, 413; pH optima, 10, 219; plasma membrane, in, 31, 90, 92, 100, 185, 186; preparation, 10, 214, 217; 53, 183; properties, 10, 215, 219; purification of complex III, 53, 87; reconstitution, 10, 222; 53, 22, 25, 32, 49, 50; resonance Raman studies, 54, 245; soluble coupling enzyme and, 6, 283; specific activity, 10, 214, 217; spectra, 10, 220; stability, 10, 220; submitochondrial particles, 55, 106; toluene-treated mitochondria, 56, 547; turnover number, 10, 219

SUCCINATE DEHYDROGENASE (UBIQUINONE) AND SUCCINATE DEHYDROGENASE

Succinate dehydrogenase (ubiquinone) [EC 1.3.5.1] is an FAD- and iron-sulfur-dependent complex (often referred to as complex II) that catalyzes the reaction of succinate with ubiquinone to produce fumarate and ubiquinol.[1–5] The complex can be converted to form succinate dehydrogenase [EC 1.3.99.1], also referred to as fumarate reductase, fumarate dehydrogenase, and fumaric hydrogenase. The form designated EC 1.3.99.1 cannot react with ubiquinone, yet retains the ability to catalyze the FAD-dependent reaction of succinate with an electron acceptor (such as phenazine methosulfate, ferricyanide, and methylene blue) to produce fumarate and reduced acceptor.

[1] K. M. Robinson & B. D. Lemire (1995) *MIE* 260, 34.
[2] Y. Hatefi & D. L. Stiggall (1976) *The Enzymes*, 3rd ed., 13, 175.
[3] G. Palmer (1975) *The Enzymes*, 3rd ed., 12, 1.
[4] T. P. Singer, E. B. Kearney & W. C. Kenney (1973) *AE* 37, 189.
[5] J. V. Schloss & M. S. Hixon (1998) *CBC* 2, 43.

Selected entries from *Methods in Enzymology* [vol, page(s)]:
General discussion: 1, 722; 5, 597; 10, 231, 322; 13, 81; 31, 742; 53, 21, 22, 25, 27, 397, 405, 407, 466; 126, 377, 399; 148, 491; 260, 34
Assay: 1, 722; 5, 597, 610; 10, 231, 322; 13, 81; 53, 26, 27, 34, 35, 406, 466; avian salt gland, 96, 638; cardiac mitochondrial, 125, 18; electron-transferring flavoprotein, in, 53, 466, 471, 474; *Escherichia coli*, 126, 380; ferricyanide, with, 10, 34; histochemical method for, 4, 381; 6, 889, 891; manometric, 2, 748; 13, 82; phenazine methosulfate, assay with, 5, 536; 13, 82; short reaction time in assay, 5, 562; spectrophotometric assay, 13, 83; 264, 493; *Wolinella succinogenes*, 126, 397; 260, 39
Other molecular properties: activators, 53, 25, 33; active sites of, 5, 600, 607; activity, 53, 21, 24, 25, 27; brain homogenates, in, 31, 464, 466; cardiac mitochondrial, assay, 125, 18; catalytic site, structure, 53, 479, 482, 483; cofactors in bacteria, 251, 476; components, 54, 145; 56, 580,

581; composition, 53, 24, 30, 476, 484, 485; cytochemical staining, 264, 512, 520; cytochrome *b* and, 1, 728; 2, 740, 744, 748; electrochemical studies, direct, 227, 521; electron-transfer components, thermodynamic parameters, 53, 490; electron-transferring flavoprotein, activity studies, 53, 410, 411, 414, 416, 418, 466, 471, 474; enzyme-inhibitor complexes, 13, 90; EPR characteristics, 53, 265, 486, 488; 54, 148; *Escherichia coli* (assays, 126, 380; auxotrophs, in, 32, 864; dimer purification, 126, 381; holoenzyme purification, 126, 385); ferricyanide system (inhibitor sensitivity, 10, 35; phosphorylation accompanying, 10, 33); inhibitors, 10, 51, 330; 13, 88, 90, 525; 53, 26, 33, 34; iron-sulfur centers, 53, 31; 69, 249 (EPR characteristics, 53, 486; 54, 138, 143; redox titration, 53, 490, 491); isolation, 9, 274; isolation of flavin peptide, 53, 554; isotope effects, 87, 129; Keilin-Hartree preparation, 55, 124; kinetic properties, 53, 31, 32, 406; membrane-bound and soluble, comparative studies, 53, 410; mitochondrial, 31, 65, 165, 166, 735; 32, 712 (properties, 148, 494; preparation, 1, 725); 10, 326; 20, 312; 53, 3, 4, 6; properties, 1, 726; 5, 608; 10, 234, 329; 13, 88; spectra, 4, 285-286; 10, 235, 330; 53, 31; spectral properties, 53, 25; stability, 10, 325; stereochemistry, 87, 129, 157; structure, 260, 34; subsites, 87, 500, 507; substrate specificity, 13, 90, 525; 53, 24, 482, 483); succinate dehydrogenase (activity studies, 53, 410, 411; purification, 53, 27, 478); tetrazolium salts and, 6, 889, 891; turnover number, 53, 32, 411, 412, 477, 478; ubiquinone, 53, 576 (extraction, 53, 576); ubiquinone Q₁ assay of, 53, 466

SUCCINATE:HYDROXYMETHYL-GLUTARATE CoA-TRANSFERASE

This transferase [EC 2.8.3.13] catalyzes the reversible reaction of succinyl-CoA with (*S*)-3-hydroxy-3-methylglutarate to produce succinate and 3-hydroxy-3-methylglutaryl-CoA.[1,2] Malonyl-CoA, adipyl-CoA, and glutaryl-CoA can also act as donor substrates, albeit more slowly. The enzyme has a rapid equilibrium random Bi Bi kinetic mechanism.[1]

[1] M. A. Francesconi, A. Donella-Deana, V. Furlanetto, L. Cavallini, P. Palatini & R. Deana (1989) *BBA* 999, 163.
[2] R. Deana (1992) *Biochem. Int.* 26, 767.

SUCCINATE-SEMIALDEHYDE DEHYDROGENASE

Two enzyme activities with this name[1–3] have been classified as: (a) [EC 1.2.1.24], which catalyzes the reaction of succinate semialdehyde with NAD^+ and water to produce succinate and $NADH^{1,2}$, and (b) [EC 1.2.1.16], which catalyzes the same reaction but uses $NAD(P)^+$.

[1] A. J. Rivett & K. F. Tipton (1981) *EJB* 117, 187.
[2] C. Cash, L. Ciesielski, M. Maitre & P. Mandel (1977) *Biochimie* 59, 257.
[3] B. Sohling & G. Gottschalk (1993) *EJB* 212, 121.

Selected entries from *Methods in Enzymology* [vol, page(s)]:
General discussion: 55, 51; NADP⁺, with, 5, 774
Assay: 5, 774
Other molecular properties: γ-aminobutyrate:glutamate transaminase, in assay of, 5, 771, 772; brain mitochondria, 55, 58; properties, 5, 775; purification, 5, 775

O-SUCCINYLBENZOATE:CoA LIGASE

This enzyme [EC 6.2.1.26], also known as *O*-succinyl-benzoyl-CoA synthetase, catalyzes the reaction of ATP with

O-succinylbenzoate and Coenzyme A to produce AMP, pyrophosphate, and *O*-succinylbenzoyl-CoA.[1–3]

[1]R. Meganathan & R. Bentley (1979) *J. Bact.* **140**, 92.
[2]O. Kwon, D. K. Bhattacharyya & R. Meganathan (1996) *J. Bact.* **178**, 6778.
[3]D. K. Bhattacharyya, O. Kwon & R. Meganathan (1997) *J. Bact.* **179**, 6061.

SUCCINYL-CoA HYDROLASE

This enzyme [EC 3.1.2.3], also referred to as succinyl-CoA acylase, catalyzes the hydrolysis of succinyl-CoA to produce Coenzyme A and succinate.[1]

[1]J. Gergely, P. Hele & C. V. Ramakrishnan (1952) *JBC* **198**, 323.

Selected entries from *Methods in Enzymology* [vol, page(s)]:
General discussion: **1**, 602
Assay: **1**, 602
Other molecular properties: α-ketoglutarate dehydrogenase assay and, **1**, 715; heart, presence in, **1**, 574; methylmalonyl-CoA isomerase preparations, presence in, **5**, 581, 585; properties, **1**, 605; purification, **1**, 604

SUCCINYL-CoA SYNTHETASE (ADP-FORMING)

This enzyme [EC 6.2.1.5], also referred to as succinate:CoA ligase (ADP-forming) and succinate thiokinase, catalyzes the reaction of ATP with succinate and Coenzyme A to produce ADP, succinyl-CoA, and orthophosphate.[1–3] Alternative acceptor substrates include itaconate and α-methylsuccinate. *See also Succinyl-CoA Synthetase (GDP-Forming)*

[1]L. P. Hager (1962) *The Enzymes*, 2nd ed., **6**, 387.
[2]J. S. Nishimura (1972) *AE* **36**, 183.
[3]W. A. Bridger (1974) *The Enzymes*, 3rd ed., **10**, 581.

Selected entries from *Methods in Enzymology* [vol, page(s)]:
General discussion: **1**, 718; **13**, 70; **44**, 882
Assay: **1**, 719; **13**, 70
Other molecular properties: affinity labeling, **47**, 427; ATPase activity, **13**, 74; fluorometric assay with, for (carnitine, **13**, 505; succinate, **13**, 458); isotope exchange, **64**, 9; **87**, 656; lipoic transsuccinylase and, **5**, 655; mechanism, **249**, 453; nucleotide specificity, **13**, 74; partial reactions, **87**, 7; phospho-histidine intermediate, **10**, 769; **13**, 74; preparation of from *Rhodopseudomonas spheroides*, **17A**, 196; properties, **1**, 721; **13**, 74; purification (*Escherichia coli*, **13**, 71; pig heart, **1**, 720); regulation, **87**, 656; side reactions, **13**, 75; substrate for α-ketoglutarate synthase, **13**, 178; substrate synergism, **64**, 38; **87**, 7, 656; succinyl phosphate intermediate, **13**, 75; specificity, **13**, 74; uv spectrum, **13**, 74

SUCCINYL-CoA SYNTHETASE (GDP-FORMING)

This enzyme [EC 6.2.1.4], also referred to as succinate:CoA ligase (GDP-forming), catalyzes the reaction of GTP with succinate and Coenzyme A to produce GDP, succinyl-CoA, and orthophosphate.[1–4] Itaconate can replace succinate, and ITP can replace GTP. *See also Succinyl-CoA Synthetase (ADP-Forming)*

[1]L. P. Hager (1962) *The Enzymes*, 2nd ed., **6**, 387.
[2]J. S. Nishimura (1972) *AE* **36**, 183.
[3]W. A. Bridger (1974) *The Enzymes*, 3rd ed., **10**, 581.
[4]P. A. Frey (1992) *The Enzymes*, 3rd ed., **20**, 141.

Selected entries from *Methods in Enzymology* [vol, page(s)]:
General discussion: **13**, 62
Assay: **13**, 62
Other molecular properties: arsenolysis reactions, **13**, 69; assay with, for succinyl-CoA, **13**, 547; cGMP assay, **38**, 73, 74, 78, 81, 82; mechanism, **13**, 69; metabolism of itaconate and mesaconate with, in mammalian liver, **13**, 315; properties, **13**, 68; purification, pig heart, **13**, 65; succinyl-CoA generating system, in, **52**, 352

SUCCINYLDIAMINOPIMELATE AMINOTRANSFERASE

This pyridoxal-phosphate-dependent aminotransferase [EC 2.6.1.17], also known as succinyldiaminopimelate transferase, catalyzes the reversible reaction of α-ketoglutarate (or, 2-oxoglutarate) with *N*-succinyl-L-2,6-diaminoheptanedioate to produce *N*-succinyl-L-2-amino-6-oxoheptanedioate and L-glutamate.[1–3]

[1]B. Peterkofsky & C. Gilvarg (1961) *JBC* **236**, 1432.
[2]T. L. Born & J. S. Blanchard (1999) *Curr. Opin. Chem. Biol.* **3**, 607.
[3]G. Scapin & J. S. Blanchard (1998) *AE* **72**, 279.

Selected entries from *Methods in Enzymology* [vol, page(s)]:
General discussion: **5**, 853
Assay: **5**, 853
Other molecular properties: distribution, **5**, 858; properties, **5**, 857; purification, **5**, 855

SUCCINYLDIAMINOPIMELATE DESUCCINYLASE

This hydrolase [EC 3.5.1.18] catalyzes the conversion of *N*-succinyl-LL-2,6-diaminoheptanedioate to succinate and LL-2,6-diaminoheptanedioate.[1–4] The enzyme is metal-ion dependent and solvent kinetic isotope effect studies yielded inverse isotope effects.[2]

[1]S. H. Kindler & C. Gilvarg (1960) *JBC* **235**, 3532.
[2]T. L. Born, R. Zheng & J. S. Blanchard (1998) *Biochemistry* **37**, 10478.
[3]Y. K. Lin, R. Myhrman, M. L. Schrag & M. H. Gelb (1988) *JBC* **263**, 1622.
[4]G. Scapin & J. S. Blanchard (1998) *AE* **72**, 279.

Selected entries from *Methods in Enzymology* [vol, page(s)]:
General discussion: **5**, 851
Assay: **5**, 851
Other molecular properties: activators, **5**, 853; Michaelis constants, **5**, 853; preparation, **5**, 852; specificity, **5**, 853; **248**, 223; transaminase assay and, **5**, 853

S-SUCCINYLGLUTATHIONE HYDROLASE

This enzyme [EC 3.1.2.13] catalyzes the hydrolysis of *S*-succinylglutathione to produce glutathione and succinate.[1,2]

[1]L. Uotila (1979) *JBC* **254**, 7024.

[2]V. Talesa, L. Uotila, M. Koivusalo, G. Principato, E. Giovannini & G. Rosi (1988) *BBA* **955**, 103.

Selected entries from *Methods in Enzymology* [vol, page(s)]:
General discussion: 77, 424

O-SUCCINYLHOMOSERINE (THIOL)-LYASE

This pyridoxal-phosphate-dependent enzyme [EC 4.2.99.9], also known as cystathionine γ-synthase, catalyzes the reaction of *O*-succinyl-L-homoserine with L-cysteine to produce cystathionine and succinate.[1–5] The enzyme can also use hydrogen sulfide and methanethiol as substitutes for L-cysteine (thus, generating L-homocysteine and L-methionine, respectively). In the absence of a thiol, the enzyme can also catalyze β,γ-elimination to form α-ketobutanoate (or, 2-oxobutanoate), succinate, and ammonia. A pyridoxamine derivative of vinylglyoxylate is a key intermediate in the γ-elimination and γ-replacement reactions.

[1]M. Flavin & C. Slaughter (1967) *BBA* **132**, 400.
[2]J. L. Wiebers & H. R. Garner (1967) *JBC* **242**, 12.
[3]M. A. Ator & P. R. Ortez de Montellano (1990) *The Enzymes*, 3rd ed., **19**, 213.
[4]D. J. Creighton & N. S. R. K. Murthy (1990) *The Enzymes*, 3rd ed., **19**, 323.
[5]A. E. Martell (1982) *AE* **53**, 163.

Selected entries from *Methods in Enzymology* [vol, page(s)]:
General discussion: 17B, 425, 450
Assay: *Salmonella typhimurium*, **17B**, 425
Other molecular properties: *O*-acetylhomoserine sulfhydrylase and, **17B**, 453; *Neurospora*, in, **17B**, 453; pyridoxal phosphate and, **17B**, 431; *Salmonella typhimurium*, from (properties, **17B**, 430; purification, **17B**, 427); spectrum, **17B**, 431, 433

SUCROSE 1[F]-FRUCTOSYLTRANSFERASE

This transferase [EC 2.4.1.99] catalyzes the reaction of two molecules of sucrose to produce D-glucose and 1[F]-β-D-fructosylsucrose (or, 1-kestose).[1–3]

[1]R. J. Henry & B. Darbyshire (1980) *Phytochemistry* **19**, 1017.
[2]N. Shiomi, H. Kido & S. Kiriyama (1985) *Phytochemistry* **24**, 695.
[3]P. M. Chevalier & R. A. Rupp (1993) *Plant Physiol.* **101**, 589.

SUCROSE 6[F]-α-GALACTOSYL-TRANSFERASE

This galactosyltransferase [EC 2.4.1.167] catalyzes the reaction of UDP-D-galactose with sucrose to produce UDP and 6[F]-α-D-galactosylsucrose (that is, planteose). The enzyme participates in the biosynthesis of the trisaccharide planteose and higher analogues in sesame seeds (*Sesamum indicum*).[1]

[1]H. Hopf, M. Spanfelner & O. Kandler (1984) *Z. Pflanzenphysiol.* **114**, 485.

SUCROSE:1,6-α-GLUCAN 3-α-GLUCOSYLTRANSFERASE

This transferase catalyzes the reaction of sucrose with a [(1,6)-α-D-glucosyl]$_n$ to produce D-fructose and a [(1,6)-α-D-glucosyl]$_n$ with a glucosyl residue at the 3-position of one of the glucosyl residues in the 1,6-α-glucan. A glycosyl-enzyme intermediate has been trapped with the *Streptococcus sobrinus* protein.[1] **See also** *Sucrose:1,6-α-Glucan 3(6)-α-Glucosyltransferase*

[1]G. Mooser, S. A. Hefta, R. J. Paxton, J. E. Snively & T. D. Lee (1991) *JBC* **266**, 8916.

SUCROSE:1,6-α-GLUCAN 3(6)-α-GLUCOSYLTRANSFERASE

This transferase [EC 2.4.1.125], also known as water-soluble glucan synthase, catalyzes the reaction of sucrose with a [(1,6)-α-D-glucosyl]$_n$ to produce D-fructose and a [(1,6)-α-D-glucosyl]$_{n+1}$.[1–4] The enzyme also catalyzes the transfer of glucosyl residues to the 3-position on D-glucosyl residues in glucans, thus generating a highly branched 1,6-α-D-glucan. **See also** *Sucrose:1,6-α-Glucan 3-α-Glucosyltransferase*

[1]A. Shimamura, H. Tsumori & H. Mukasa (1982) *BBA* **702**, 72.
[2]H. Mukasa, A. Shimamura & H. Tsumori (1982) *BBA* **719**, 81.
[3]H. Tsumori, A. Shimamura & H. Mukasa (1985) *J. Gen. Microbiol.* **131**, 3347.
[4]T. P. Binder & J. F. Robyt (1985) *Carbohydr. Res.* **140**, 9.

SUCROSE α-GLUCOSIDASE

This enzyme [EC 3.2.1.48], also known as sucrose α-glucohydrolase and sucrase, catalyzes the hydrolysis of sucrose (producing D-glucose and D-fructose) and maltose (producing two molecules of D-glucose) via an α-D-glucosidase-type action.[1–6] The enzyme isolated from intestinal mucosa is synthesized as a single polypeptide chain and will also catalyze the hydrolysis of isomaltose (an oligo-1,6-glucosidase activity similar to the action of EC 3.2.1.10) This intestinal enzyme, which is often known as sucrase-isomaltase, contains three distinct domains: an anchor peptide, a sucrase subunit, and an isomaltase subunit. Kinetic isotope effect investigations suggest that an oxocarbenium ion-like transition state is present with both activities.[5,6] **See also** *Oligo-1,6-Glucosidase*

[1]H. Sigrist, P. Ronner & G. Semenza (1975) *BBA* **406**, 433.
[2]J. Kolinska & J. Kraml (1972) *BBA* **284**, 235.
[3]R. K. Montgomery, M. A. Sybicki, A. G. Forcier & R. J. Grand (1981) *BBA* **661**, 346.
[4]K. A. Conklin, K. M. Yamashiro & G. M. Gray (1975) *JBC* **250**, 5735.
[5]E. H. Van Beers, H. A. Büller, R. J. Grand, A. W. C. Einerhand & J. Dekker (1995) *Crit. Rev. Biochem. Mol. Biol.* **30**, 197.

[6]H. Hauser & G. Semenza (1983) *Crit. Rev. Biochem.* **14**, 319.

Selected entries from *Methods in Enzymology* [vol, page(s)]:
Other molecular properties: affinity labeling, **87**, 473; immobilization, by bead polymerization, **44**, 195; isomalltase complex, **46**, 377, 379 (small-intestinal brush border membrane [assembly, **96**, 399; biosynthesis {*in vitro*, **96**, 400; mechanism, **96**, 405}; isomalltase-anchoring segment, orientation of N-terminal amino acid, **96**, 392; lipid bilayer-embedded subunit, identification, **96**, 388]); transition-state and multisubstrate analogues, **249**, 306

SUCROSE-PHOSPHATASE

This phosphatase [EC 3.1.3.24] catalyzes the hydrolysis of sucrose 6^F-phosphate to produce sucrose and orthophosphate.[1–3]

[1]D. P. Whitaker (1984) *Phytochemistry* **23**, 2429.
[2]J. S. Hawker & M. D. Hatch (1966) *BJ* **99**, 102.
[3]J. Preiss (1982) *Ann. Rev. Plant Physiol.* **33**, 431.

Selected entries from *Methods in Enzymology* [vol, page(s)]:
General discussion: 42, 341
Assay: plants, 42, 341; *Streptococcus mutans*, 90, 556
Other molecular properties: plants, from, 42, 341 (activators and inhibitors, **42**, 346; properties, **42**, 346; purification, **42**, 344); *Streptococcus mutans* (properties, **90**, 559; purification, **90**, 557)

SUCROSE PHOSPHATE SYNTHASE

This enzyme [EC 2.4.1.14], also known as UDP-glucose: fructose-phosphate glucosyltransferase and sucrose phosphate:UDP-glucosyltransferase, catalyzes the reaction of UDP-D-glucose with D-fructose 6-phosphate to produce UDP and sucrose 6-phosphate.[1–4] The enzyme exhibits either an ordered Bi Bi or Theorell-Chance kinetic mechanism.[1,2] A histidyl residue is reported to be essential for catalytic activity.[3]

[1]S. Harbron, C. Foyer & D. Walker (1981) *ABB* **212**, 237.
[2]G. L. Salerno & H. G. Pontis (1977) *ABB* **180**, 298.
[3]A. K. Sinha, U. V. Pathre & P. V. Sane (1998) *BBA* **1388**, 397.
[4]S. C. Huber & J. L. Huber (1996) *Ann. Rev. Plant Physiol. Plant Mol. Biol.* **47**, 431.

Selected entries from *Methods in Enzymology* [vol, page(s)]:
General discussion: 5, 170
Assay: 5, 170
Other molecular properties: Michaelis constants, **5**, 171; pH optimum, **5**, 171; purification from wheat germ, **5**, 170

SUCROSE PHOSPHORYLASE

This enzyme [EC 2.4.1.7], also known as sucrose glucosyltransferase, catalyzes the reaction of sucrose with orthophosphate to produce D-fructose and α-D-glucose 1-phosphate.[1–5] In the forward reaction, arsenate can replace phosphate as the substrate: however, the resulting product is unstable in aqueous solutions. Various ketoses and L-arabinose may replace D-fructose in the reverse reaction: in fact, the reverse reaction has often been used

in the laboratory for synthetic purposes. The enzyme has a ping pong Bi Bi kinetic mechanism.

Evidence for a covalent glucose-enzyme intermediate includes: (a) trapping of [^{14}C]glucosyl moiety after the enzyme is combined with radiolabeled sucrose and rapidly quenched by addition of hot methanol or by rapidly adjusting the pH to 3; and (b) observation that sodium periodate rapidly inactivates sucrose phosphorylase when the substrate is present.[2] The mechanism most likely involves dissociation of the glucosidic bond to form an incipient carbocation that is stabilized by a carboxyl group on the enzyme. Whether this stabilization occurs by means of a strained ester bond or a tight ion pair remains to be determined.

[1]G. Mooser (1992) *The Enzymes*, 3rd ed., **20**, 187.
[2]J. J. Mieyal & R. H. Abeles (1972) *The Enzymes*, 3rd ed., **7**, 515.
[3]G. Davies, M. L. Sinnott & S. G. Withers (1998) *CBC* I, 119.
[4]M. Cohn (1961) *The Enzymes*, 2nd ed., **5**, 179.
[5]M. Doudoroff (1961) *The Enzymes*, 2nd ed., **5**, 229.

Selected entries from *Methods in Enzymology* [vol, page(s)]:
General discussion: 1, 225; 28, 935; **230**, 304
Assay: *Pseudomonas sacchrophilia*, 1, 225; 28, 935
Other molecular properties: acylal intermediate, **87**, 6; *Agrobacterium tumefaciens*, from, glucose-1-phosphate-3T formation and, **28**, 288; carboxyl groups, **25**, 621; intermediate, **87**, 6, 11; isotope exchange, **64**, 9; labeled sucrose and, **4**, 502; *Leuconostoc mesenteroides* (properties, I, 229; purification, I, 227); oxygen labeled glucose phosphate and, **4**, 914; *Pseudomonas sacchrophilia* (assay, I, 225; 28, 935; properties, I, 229; 28, 941; purification, I, 226; 28, 936); side reaction, **63**, 8; staining on gels, **22**, 584, 589, 599; transglycosylase side reaction, **63**, 8; trapping of intermediate, **87**, 11

SUCROSE SYNTHASE

This enzyme [EC 2.4.1.13], also known as UDP-glucose:fructose glucosyltransferase and sucrose:UDP-glucosyltransferase, catalyzes the reversible reaction of UDP-D-glucose with D-fructose to produce UDP and sucrose.[1–3] Based on positional isotope exchange (PIX) experiments, one may conclude that either sucrose synthase does not catalyze the cleavage of the scissile carbon-oxygen bond of UDP-D-glucose in the absence of fructose or, alternatively, the β-phosphoryl group of the newly formed UDP does not rotate.[3]

[1]R. W. Wolosiuk & H. G. Pontis (1971) *FEBS Lett.* **16**, 237.
[2]D. P. Delmer (1972) *JBC* **247**, 3822.
[3]A. N. Singh, L. S. Hester & F. M. Raushel (1987) *JBC* **262**, 2554.

Selected entries from *Methods in Enzymology* [vol, page(s)]:
General discussion: 5, 167; 8, 341
Assay: 5, 167; 8, 341
Other molecular properties: inhibitors and, **5**, 169; **8**, 345; properties, **5**, 169; **8**, 344; purification (sugar beet roots, **8**, 342; wheat germ, **5**, 168)

SUGAR PHOSPHATASE

This phosphatase [EC 3.1.3.23] catalyzes the hydrolysis of a sugar phosphate to produce a sugar and orthophosphate. Substrates include aldohexose 1-phosphates, ketohexose 1-phosphates, aldohexose 6-phosphates, ketohexose 6-phosphates, both phosphate ester bonds of fructose 1,6-bisphosphate, α-D-glucuronate 1-phosphate, phosphoric esters of disaccharides (e.g., α-gentiobiose 1-phosphate and α-lactose 1-phosphate), xylitol 5-phosphate, as well as various pentose and triose phosphates. **See also** *Sugar-Terminal Phosphatase*

SUGAR-TERMINAL-PHOSPHATASE

This phosphatase [EC 3.1.3.58], also known as xylitol-5-phosphatase, catalyzes the hydrolysis of D-glucose 6-phosphate to produce D-glucose and orthophosphate.[1] Quite a few sugars and sugar alcohols phosphorylated on the terminal carbon position are substrates for the bacterial enzyme, which exhibits a preference for D-sugars in the *erythro* configuration. Substrates include D-glucose 6-phosphate, D-xylitol 5-phosphate, D-mannose 6-phosphate, 6-phospho-D-gluconate, D-fructose 6-phosphate, and D-erythrose 4-phosphate. Little or no activity is observed with phosphomonoesters located at the C1 position. **See also** *Sugar-Phosphatase*

[1] J. London, S. Z. Hausman & J. Thompson (1985) *J. Bacteriol.* **163**, 951.

SULCATONE REDUCTASE

This oxidoreductase [EC 1.1.1.260] catalyzes the reaction[1,2] of sulcatone (or, 6-methylhept-5-en-2-one) with NADH and a proton to produce sulcatol (or, 6-methylhept-5-en-2-ol) and NAD^+.

sulcatone   sulcatol

Studies on enzymes from *Clostridium pasteurianum*, *C. tyrobutyricum*, and *Lactobacillus brevis* suggest the presence of multiple isozymes.

[1] E. C. Tidswell, G. J. Salter, D. B. Kell & J. G. Morris (1997) *Enzyme Microb. Technol.* **21**, 143.
[2] E. C. Tidswell, A. N. Thompson & J. G. Morris (1991) *J. Appl. Microbiol. Biotechnol.* **35**, 317.

SULFATE ADENYLYLTRANSFERASE

This adenylyltransferase [EC 2.7.7.4], also known as sulfate adenylate transferase and ATP-sulfurylase, catalyzes the reversible reaction of ATP with sulfate (or SeO_4^{2-}) to produce pyrophosphate and adenylylsulfate (or AMP-selenate).[1-6] The rat liver and spinach leaf enzymes have a random-on (MgATP and sulfate), ordered-off (Mg-pyrophosphate before adenylylsulfate) Bi Bi kinetic mechanism.[2,3] The *Desulfovibrio* enzyme contains both cobalt and zinc ions.[4]

Depending on reaction conditions, the mass action ratio of the sulfurylase reaction can be shifted by as much as 10^5 in favor of adenylylsulfate formation.[4,6] Note that this reaction is coupled to a so-called GTPase activity, which is an energase-type reaction that modulates the enzyme's kinetic parameters.[6] Reactions of the energase type have the following stoichiometry: $State_1 + GTP = State_1 + GDP + P_i$, where two affinity-modulated conformational states are indicated.

[1] H. D. Peck (1974) *The Enzymes*, 3rd ed., **10**, 651.
[2] M. Yu, R. L. Martin, S. Jain, L. J. Chen & I. H. Segel (1989) *ABB* **269**, 156.
[3] O. Y. Gavel, S. A. Bursakov, J. J. Calvete, G. N. George, J. J. Moura & I. Moura (1998) *Biochemistry* **37**, 16225.
[4] T. S. Leyh (1993) *Crit. Rev. Biochem. Mol. Biol.* **28**, 515.
[5] D. L. Purich (2001) *TiBS* **26**, 417.
[6] T. S. Leyh (1999) *MIE* **308**, 48.

Selected entries from **Methods in Enzymology** [vol, page(s)]:
General discussion: 5, 964; 143, 329, 334; 243, 331, 400; 308, 52
Assay: 5, 968, 976; 143, 335; *Archaeoglobus fulgidus*, 243, 332; ATP formation method, 243, 408; coupled spectrophotometric method, 243, 407
Other molecular properties: activity, 57, 245; 243, 507; adenosine 5'-phosphosulfate, assay of, 6, 769; *Archaeoglobus fulgidus* (assay, 243, 332; dissimilatory sulfate reduction, in, 243, 331; properties, 243, 338; purification, 243, 334); bond breaking, 308, 56; catalyzed reaction, 143, 334; central complex, 308, 54; *Chlorobium limicola f. thiosulfatophilum* (partial purification, 243, 418; properties, 243, 419); GTPase activity, 308, 53; mechanism, 308, 54; *Penicillium chrysogenum* purification, 143, 344; phototrophic sulfur metabolism, in, 243, 401, 403

SULFATE ADENYLYLTRANSFERASE (ADP)

This adenylyltransferase [EC 2.7.7.5], also known as adenosine diphosphate sulfurylase and ADP-sulfurylase, catalyzes the reaction of orthophosphate with adenylylsulfate to produce ADP and sulfate.[1-6] In the absence of sulfate, the enzyme will catalyze an ADP ↔ orthophosphate exchange reaction. The reverse reaction is not observed and a number of investigators have suggested that the name be changed. The *Thiobacillus denitrificans* enzyme catalyzes a ping pong Bi Bi kinetic mechanism with a covalently bound AMP as a reaction intermediate.[5]

[1] H. D. Peck (1974) *The Enzymes*, 3rd ed., **10**, 651.
[2] U. Bias & H. G. Trüper (1987) *Arch. Microbiol.* **147**, 406.
[3] R. Nicholls (1977) *BJ* **165**, 149.
[4] S. Khanna & D. J. D. Nicholas (1983) *J. Gen. Microbiol.* **129**, 1365.

[5]T. Bruser, T. Selmer & C. Dahl (2000) *JBC* **275**, 1691.
[6]A. Schmidt & K. Jäger (1992) *Ann. Rev. Plant Physiol. Plant Mol. Biol.* **43**, 325.

SULFATE-TRANSPORTING ATPase

This so-called ATPase [EC 3.6.3.25] catalyzes the ATP-dependent (ADP- and orthophosphate-producing) transport of sulfate$_{out}$ to produce sulfate$_{in}$.[1,2] In addition to catalyzing the import of sulfate anions, the *Escherichia coli* protein will also catalyze the import of thiosulfate ($S_2O_3^{2-}$) anions. ABC-type (ATP-binding cassette-type) ATPases have two similar ATP-binding domains and do not undergo phosphorylation during the transport process.

The idea that a transporter is an enzyme is in keeping with a new definition of enzyme catalysis as the facilitated making/breaking of chemical bonds, not just covalent bonds.[3] This idea builds on Pauling's assertion that any long-lived, chemically distinct interaction (in this case, the persistent location of a solute with respect to the faces of a membrane) can be regarded as a chemical bond. Note also that the equilibrium constant ($K_{eq} = $ [Sulfate$_{in}$][ADP][P$_i$]/[Sulfate$_{out}$][ATP]) does not conform to that expected for an ATPase (*i.e.*, $K_{eq} = $ [ADP][P$_i$]/[ATP]). Thus, although the overall reaction yields ADP and orthophosphate, the enzyme is misclassified as a hydrolase, and instead should be regarded as an energase-type reaction. Energases facilitate affinity-modulated reactions by coupling the $\Delta G_{ATP\text{-hydrolysis}}$ to a force-generating or work-producing step.[3] In this case, P-O-P bond scission supplies the energy to drive sulfate transport. **See** *ATPase; Energase*

[1]A. Sirko, M. Zatyka, E. Sadowy & D. Hulanicka (1995) *J. Bacteriol.* **177**, 4134.
[2]M. H. Saier, Jr. (1998) *Adv. Microb. Physiol.* **40**, 81.
[3]D. L. Purich (2001) *TiBS* **26**, 417.

SULFIDE:CYTOCHROME *c* OXIDOREDUCTASE

This oxidoreductase, also known as flavocytochrome *c*, catalyzes the sulfide-dependent conversion of ferricytochrome *c* to ferrocytochrome *c*. **See also** *Sulfite Dehydrogenase*

SULFINOALANINE DECARBOXYLASE

This pyridoxal-phosphate-dependent decarboxylase [EC 4.1.1.29], also known as cysteine-sulfinate decarboxylase, catalyzes the conversion of 3-sulfino-L-alanine (or, cysteine sulfinate) to hypotaurine and carbon dioxide.[1–3] Other substrates include L-cysteate, producing taurine. In early reports, this enzyme was misnamed as sulfoalanine decarboxylase.

[1]J. G. Jacobsen, L. L. Thomas & L. H. Smith, Jr. (1964) *BBA* **85**, 103.
[2]O. W. Griffith (1983) *JBC* **258**, 1591.
[3]C. L. Weinstein & O. W. Griffith (1987) *JBC* **262**, 7254.

SULFITE DEHYDROGENASE

This oxidoreductase [EC 1.8.2.1] catalyzes the reaction of two molecules of ferricytochrome *c* with sulfite and water to produce sulfate and two molecules of ferrocytochrome *c*.[1–3] Cytochrome *c*$_{551}$ is associated with this system. The *Thiobacillus novellus* enzyme contains molybdopterin and exhibits a ping pong kinetic mechanism involving two reactive sites.[4] **See also** *Sulfite Oxidase; Sulfide:Cytochrome c Oxidoreductase*

[1]W.-P. Lu & D. P. Kelly (1984) *J. Gen. Microbiol.* **130**, 1683.
[2]A. M. Charles & I. Suzuki (1966) *BBA* **128**, 522.
[3]E. J. Kurek (1985) *Arch. Microbiol.* **143**, 277.
[4]U. Kappler, B. Bennett, J. Rethmeier, G. Schwarz, R. Deutzmann, A. G. McEwan & C. Dahl (2000) *JBC* **275**, 13202.

SULFITE OXIDASE

This heme- and molybdenum-dependent oxidoreductase [EC 1.8.3.1] catalyzes the reaction of sulfite with dioxygen and water to produce sulfate and hydrogen peroxide.[1–4]

In various organisms, cytochrome *c* suppresses reaction with dioxygen, and the enzyme exhibits its sulfite dehydrogenase activity. *See also* *Sulfite Dehydrogenase*

[1]R. C. Bray (1975) *The Enzymes*, 3rd ed., **12**, 299.
[2]C. A. Kipke, M. A. Cusanovich, G. Tolin, R. A. Sunde & J. H. Enemark (1988) *Biochemistry* **27**, 2918.
[3]P. E. Baugh, D. Collison, C. D. Garner & J. A. Joule (1998) *CBC* **3**, 377.
[4]R. C. Bray (1980) *AE* **51**, 107.

Selected entries from *Methods in Enzymology* [vol, page(s)]:
General discussion: 52, 15; 56, 475; 243, 447
Other molecular properties: sulfite assay with, **143**, 11; *Thiobacillus denitrificans*, **243**, 454; toluene-treated mitochondria, **56**, 547

SULFITE REDUCTASE

There are three types of sulfite reductase reactions that differ with respect to redox cofactor involvement. *See also* *Desulfofuscidin; Desulforubidin; Desulfoviridin; P-582*

Sulfite reductase (NADPH) [EC 1.8.1.2], a heme-, FAD-, and FMN-dependent system, catalyzes the reaction of sulfite with three molecules of NADPH to produce hydrogen sulfide, three molecules of NADP$^+$, and three molecules of water.[1–3] The enzyme will also catalyze the nonphysiologic conversion of nitrite to ammonia.

The iron-dependent sulfite reductase (ferredoxin) [EC 1.8.7.1] catalyzes the reaction of sulfite with three molecules of reduced ferredoxin to produce hydrogen sulfide, three molecules of oxidized ferredoxin, and three molecules of water.[1,4–6] The enzyme will also catalyze the ferredoxin-dependent conversion of *S*-sulfoglutathione to glutathione persulfide.

The iron-dependent sulfite reductase [EC 1.8.99.1] catalyzes the reaction of sulfite with an acceptor substrate and three water molecules to produce hydrogen sulfide and the reduced acceptor.[1,7–9] Using oxidized methyl viologen as the acceptor substrate, six molecules of reduced methyl viologen was produced per molecule of sulfide formed.

[1]Y. Hatefi & D. L. Stiggall (1976) *The Enzymes*, 3rd ed., **13**, 175.
[2]L. M. Siegel, D. C. Rueger, M. J. Barber, R. J. Krueger, N. R. Orme-Johnson & W. H. Orme-Johnson (1982) *JBC* **257**, 6343.
[3]W. Dott & H. G. Trüper (1976) *Arch. Microbiol.* **108**, 99.
[4]O. Koguchi & G. Tamura (1988) *Agric. Biol. Chem.* **52**, 373.
[5]O. Koguchi & G. Tamura (1989) *Agric. Biol. Chem.* **53**, 783.
[6]R. J. Krueger & L. M. Siegel (1982) *Biochemistry* **21**, 2892 and 2905.
[7]E. Saito & G. Tamura (1971) *Agric. Biol. Chem.* **35**, 491.
[8]H. D. Peck, S. Tedro & M. D. Kamen (1974) *PNAS* **71**, 2404.
[9]B. H. Huynh, L. Kang, D. V. DerVatanian, H. D. Peck & J. LeGall (1984) *JBC* **259**, 15373.

Selected entries from *Methods in Enzymology* [vol, page(s)]:
General discussion: **17B**, 520, 528, 539; **243**, 296, 331, 400, 404, 422, 423; *Archaeoglobus fulgidus*, **243**, 332
Assay: **17B**, 520, 528, 540; **243**, 332, 423
Other molecular properties: antisera against, cross-reactions, **243**, 340; *Archaeoglobus fulgidus* (assay, **243**, 332; dissimilatory sulfate reduction, in, **243**, 331; gene [amino-terminal sequence determination, **243**, 342; cloning, **243**, 340; internal amino acid sequence determination, **243**, 342; oligonucleotide probe, synthesis, **243**, 342]; properties, **243**, 338; purification, **243**, 334, 337); *Chromatium vinosum* D (properties, **243**, 413; purification, **243**, 412) cytochromeless strains, **56**, 122; EPR characteristics, **53**, 265; **76**, 321; dissimilatory, **243**, 276, 296; *Escherichia coli*, siroheme in, of, **52**, 437, 439; flavin determination, **53**, 429; heme-deficient mutants, **56**, 560; low-spin (*Desulfovibrio vulgaris* Hildenborough, **243**, 296; dissimilatory sulfite reductase, comparison, and, **243**, 296, 303; enzymatic activity, **243**, 300; EPR studies, **243**, 300; genetic studies, **243**, 300; mechanistic studies, **243**, 301; Mössbauer studies, **243**, 301; NMR studies, **243**, 300; properties, **17B**, 525, 535, 538, 543; **243**, 298; purification, **17B**, 523, 533, 537, 541; **243**, 297; siroheme, **243**, 296); phototrophic sulfur metabolism, in, **243**, 404; purification from (*Escherichia coli*, **17B**, 541; spinach, **17B**, 533, 537; yeast, **17B**, 523); *Pyrobaculum islandicum*, **243**, 340, 347; reaction catalyzed by, **243**, 296; reverse siroheme enzyme (phototrophic sulfur metabolism, in, **243**, 401; *Thiobacillus denitrificans* [absorption spectra, **243**, 426; assay, **243**, 423; catalytic properties, **243**, 426; iron-sulfur clusters, **243**, 426; molecular properties, **243**, 426; properties, **243**, 426; purification, **243**, 424]); reversible dissociation, **53**, 436; sulfide quinone reductase (phototrophic sulfur metabolism, in, **243**, 401); *Thermodesulfobacterium commune*, purification, **243**, 334; yeast, **17B**, 520

SULFOACETALDEHYDE LYASE

This thiamin-pyrophosphate-dependent enzyme [EC 4.4.1.12] catalyzes the reaction of sulfoacetaldehyde ($^-$O$_3$SCH$_2$CHO) with water to produce sulfite and acetate.[1,2]

[1]H. Kondo & M. Ishimoto (1975) *J. Biochem.* **78**, 317.
[2]G. Shimamoto & R. S. Berk (1980) *BBA* **632**, 121.

4-SULFOBENZOATE 3,4-DIOXYGENASE

This oxidoreductase [EC 1.14.12.8] catalyzes the reaction of 4-sulfobenzoate with NADH and dioxygen to produce 3,4-dihydroxybenzoate, sulfite, and NAD$^+$. This dioxygenase is a protein complex containing an iron-sulfur FMN-dependent reductase and an iron-sulfur oxygenase, but no independent ferredoxin.[1]

[1]H. H. Locher, T. Leisinger & A. M. Cook (1991) *BJ* **274**, 833.

N-SULFOGLUCOSAMINE-3-SULFATASE

This enzyme [EC 3.1.6.15], also known as chondroitin-sulfatase, catalyzes the hydrolysis of the 3-sulfate groups of the *N*-sulfo-D-glucosamine 3-*O*-sulfate residues of heparin.[1–3] The enzyme does not act on the 4- or 6-*O*-sulfate esters.

The *Flavobacterium heparinum* enzyme also catalyzes the hydrolysis of *N*-acetyl-D-glucosamine 3-*O*-sulfate, whereas the human enzyme acts only on the disulfated acetoamino sugar.[2] Neither acts on 3-*O*-sulfo-D-glucosamine.

[1] I. G. Leder (1980) *BBRC* **94**, 1183.
[2] J. S. Bruce, M. W. McLean, W. F. Long & F. B. Williamson (1985) *EJB* **148**, 359.
[3] H. Kresse & J. Glössl (1987) *AE* **60**, 217.

N-SULFOGLUCOSAMINE SULFOHYDROLASE

This enzyme [EC 3.10.1.1], also known as sulfoglucosamine sulfamidase and sulfamidase, catalyzes the hydrolysis of *N*-sulfo-D-glucosamine to produce D-glucosamine and sulfate.[1,2] The enzyme will also act on *N*-sulfo-D-glucosamine-6-*O*-sulfate and on *N*-sulfo-D-glucosamine residues of heparin sulfate. The human liver sulfohydrolase appears to be an exoenzyme.[2]

[1] D. S. Anson & J. Bielicki (1999) *Int. J. Biochem. Cell Biol.* **31**, 363.
[2] C. Freeman & J. J. Hopwood (1986) *BJ* **234**, 83.

SULFUR DIOXYGENASE

This iron-dependent oxidoreductase [EC 1.13.11.18] catalyzes the reaction of sulfur with dioxygen and water to produce sulfite.[1,2]

[1] I. Suzuki & M. Silver (1966) *BBA* **122**, 22.
[2] A. Kletzin (1989) *J. Bacteriol.* **171**, 1638.

Selected entries from *Methods in Enzymology* [vol, page(s)]:
General discussion: 52, 11; 243, 501
Assay: 243, 509
Other molecular properties: lithotrophs and heterotrophs, in, (assay, 243, 509; glutathione-independent, assay, **243**, 509)

SULFUR REDUCTASE

This oxidoreductase catalyzes the conversion of sulfur, which often appears in nature as a eight-membered ring or as a colloidal species, to hydrogen sulfide (H_2S), which, in addition to mobilizing sulfur into metabolic pathways, serves as a natural defense against heavy metal ion toxicity.[1] *See also Sulfur Dioxygenase; Sulfur Oxidizing Enzymes*

[1] R. J. Maier (1996) *Adv. Protein Chem.* **48**, 35.

Selected entries from *Methods in Enzymology* [vol, page(s)]:
General discussion: 243, 353, 367
Assay: 243, 354
Other molecular properties: *Desulfomicrobium baculatum* DSM 1743, **243**, 362 (purification, **243**, 361; reduction of colloidal sulfur, **243**, 364); *Desulfomicrobium baculatum* Norway 4, **243**, 354 (assay, **243**, 354; catalytic properties, **243**, 362; pH optimum, **243**, 362; physicochemical characteristics, **243**, 361; purification, **243**, 356; purity, **243**, 361; reduction of colloidal sulfur, **243**, 364 [mechanism of attack, **243**, 362];

specific activity, **243**, 356; spectral properties, **243**, 362; stability, **243**, 361; temperature effects, **243**, 362; unit of enzyme activity, **243**, 356); *Desulfovibrio*, reduction of colloidal sulfur, **243**, 364; *Desulfuromonas acetoxidans* DSM 1675, **243**, 362; flavocytochrome c, **243**, 463 (assay, **243**, 471); mesophilic sulfur-reducing eubacterial extracts, activity, in, **243**, 379; spirilloid mesophilic sulfur-reducing bacteria, from, **243**, 367; *Sulfospirillum deleyianum*, **243**, 362, 376 (activity, **243**, 378; cellular localization, **243**, 378; properties, **243**, 382; purification, **243**, 380); thiophilic sulfate reducer extracts, activity, in, **243**, 379; *Wolinella succinogenes* DSM 1740, **243**, 362

SULOCHRIN OXIDASE ((+)-BISDECHLOROGEODIN-FORMING)

This copper-containing oxidoreductase [EC 1.10.3.7] catalyzes the reaction of two molecules of sulochrin with dioxygen to produce two molecules of (+)-bisdechlorogeodin and two water molecules.[1] This enzyme also acts on several diphenols (for example, catechol) and phenylenediamines (for example, *p*-phenylenediamine, *o*-phenylenediamine, *N,N*-dimethyl-*p*-phenylenediamine, and *N,N,N',N'*-tetramethyl-*p*-phenylenediamine), albeit with lower efficiency.

[1] H. Nordlöv & S. Gatenbeck (1982) *Arch. Microbiol.* **131**, 208.

SULOCHRIN OXIDASE ((−)-BISDECHLOROGEODIN-FORMING)

This copper-containing oxidoreductase [EC 1.10.3.8], first obtained from *Oospora sulphurea-ochracea*, catalyzes the reaction of two molecules of sulochrin with dioxygen to produce two molecules of (−)-bisdechlorogeodin and two water molecules.[1] This enzyme will also act on several diphenols and phenylenediamines (for example, *N,N,N',N'*-tetramethyl-*p*-phenylenediamine), albeit with low affinity.

[1] H. Nordlöv & S. Gatenbeck (1982) *Arch. Microbiol.* **131**, 208.

SUPEROXIDE DISMUTASE

This enzyme [EC 1.15.1.1] catalyzes the reaction of two ions of superoxide ($2O_2^-$) with two H^+ to produce dioxygen and hydrogen peroxide.[1–9] There are three types of superoxide dismutases (SODs): (a) the manganese-containing, (b) the iron-containing, and (c) the copper/zinc-containing enzymes. The Cu-Zn SODs have been largely isolated from eukaryotes, and each of its two identical subunits (total molecular mass = 32 kDa) possesses one copper atom (catalytic) and one zinc atom (structural). The other eukaryotic enzyme, MnSOD, is a tetramer (22 kDa subunit mass) that protects mitochondria from oxidative damage. Its catalytic pathway appears to proceed through reactions in which the manganese ion cycles between oxidized and reduced states, Mn(III) and Mn(II), respectively.

The *Escherichia coli* superoxide dismutase, which contains pentacoordinate Fe^{3+} in its active site, binds superoxide as a transient six-coordinate complex. Inner-sphere electron transfer occurs producing dioxygen and a pentacoordinated Fe^{2+}. The second superoxide anion radical binds to this complex and a second electron transfer occurs, generating a peroxy-Fe^{3+} species. Two protons are transferred and hydrogen peroxide is released.[1]

The manganese-superoxide dismutases of human mito-chondria and *Thermus thermophilus* contain a pentacoordi-nated Mn^{3+}. The complex in *Bacillus stearothermophilus* is somewhat different in that it is a distorted tetrahedral structure. The redox cycling mechanism observed with iron-SOD appears to be similar to that found in Mn-SOD.[3]

The copper/zinc superoxide dismutases contain a cupric ion (Cu^{2+}) in each subunit. As was seen with the iron- and manganese-dependent systems, the copper ion undergoes a reduction to Cu^+ and an oxidation back to Cu^{2+}.[4-9]

[1] B. G. Fox (1998) *CBA* **3**, 261.
[2] A. Messerschmidt (1998) *CBA* **3**, 401.
[3] J. E. Penner-Hahn (1998) *CBA* **3**, 439.
[4] E. Cadenas, P. Hochstein & L. Ernster (1992) *AE* **65**, 97.
[5] E. T. Adman (1991) *Adv. Protein Chem.* **42**, 145.
[6] J. A. Tainer, V. A. Roberts, C. L. Fisher, R. A. Hallewell & E. D. Getzoff (1991) in *A Study of Enzymes* (S. A. Kuby, ed.), vol. **2**, p. 499, CRC Press, Boca Raton.
[7] I. Fridovich (1986) *AE* **58**, 61.
[8] B. G. Malmström, L.-E. Andréasson & B. Reinhammer (1975) *The Enzymes*, 3rd ed., **12**, 507.
[9] I. Fridovich (1974) *AE* **41**, 35.

Selected entries from ***Methods in Enzymology*** [vol, page(s)]:
General discussion: 51, 69; **53**, 382; **54**, 93; **56**, 475; **105**, 88, 105; **186**, 249

Assay: 53, 383 (with luminol, **57**, 404, 405); **105**, 92, 95, 101, 457; cuprozinc enzyme, comparison of assay techniques, **74**, 360; erythrocyte, **179**, 573; chemiluminescence assay, **186**, 227; **305**, 384; hematoxylin autoxidation, **186**, 220; NAD(P)H oxidation-based chemical system, **186**, 209; extracellular (in plasma, automated assay, **186**, 232); whole bacterial cells, in, **186**, 237
Other molecular properties: action mechanism, **105**, 27; activity, **53**, 382; **233**, 213; ascorbate, **69**, 426, 428, 430, 431; autoxidizable electron acceptors, **69**, 426, 428; biological role, **53**, 382; catalyzed peroxynitrite-dependent tyrosine nitration, **269**, 210; cobalt-substituted, magnetic circular dichroism, **49**, 168, 175; Cu^{2+} site, charge distribution, **243**, 586; differentiation from Mn-superoxide dismutase, **186**, 229; electroimmunoassay, **105**, 97; diffusion-controlled reactions, **202**, 489; electron paramagnetic resonance studies, **49**, 514; inhibition, pH effects, **186**, 224; iron-enzyme, (electronic absorption spectroscopy, **226**, 35; isolation from bacteria, **105**, 91); kinetic properties, **53**, 386, 387; magnetic circular dichroism, **49**, 175; measurement, of electron transport, **69**, 653; metal content, **53**, 383, 389, 393; proton NMR spectrum, **177**, 249; properties, general, **105**, 27; spectral properties, **53**, 389, 393; stability, **53**, 389, 393; stain for, **224**, 111; ultraviolet resonance Raman spectroscopy, **226**, 396

Swal RESTRICTION ENDONUCLEASE

This type II restriction endonuclease [EC 3.1.21.4] is obtained from *Staphylococcus warneri* and acts on both strands of DNA at $5' \ldots ATTT\nabla AAAT \ldots 3'$.[1] (The symbol ∇ indicates the cleavage site.) Note that the recognition site, which generates blunt-ended fragments, does not contain G or C.

[1] M. Lechner, B. Frey, F. Laue, W. Ankenbauer & G. Schmitz (1992) *Fresenius Z. Anal. Chem.* **343**, 121.

SYNEPHRINE DEHYDRATASE

This dehydratase [EC 4.2.1.88] catalyzes the conver-sion of 1-(4-hydroxyphenyl)-2-(methylamino)ethanol (or, ($\pm$)-synephrine) to a 2,3-enamine which, in turn, hydro-lyzes to (4-hydroxyphenyl)acetaldehyde and methylamine.[1]

[1] M. Veeraswamy, N. A. Devi, R. K. Kutty & P. V. Rao (1976) *BJ* **159**, 807.

T

TABERNAMONTANAIN

This serine endopeptidase from the Venezuelan shrub *Tabernamontana grandiflora*, originally classified as EC 3.4.99.23, is currently regarded as a variant of cucumisin [EC 3.4.21.25].[1]

[1]W. G. Jaffe (1943) *Rev. Brasil Biol.* **3**, 149.

TAGATOSE-1,6-BISPHOSPHATE ALDOLASE

This aldolase [EC 4.1.2.40], also known as D-tagatose-1,6-bisphosphate aldolase and tagatose-1,6-diphosphate aldolase, catalyzes the conversion of D-tagatose 1,6-bisphosphate to glycerone phosphate (or, dihydroxyacetone phosphate) and D-glyceraldehyde 3-phosphate.[1] Class I aldolases are insensitive to such metal ion chelators as EDTA, and class II aldolases are EDTA-sensitive metallo-enzymes.

[1]V. L. Crow & T. D. Thomas (1982) *J. Bact.* **151**, 600.

Selected entries from *Methods in Enzymology* [vol, page(s)]:
General discussion: 90, 228, 232
Assay: class I, **90**, 228; class II, **90**, 232

TAGATOSE 3-EPIMERASE

This epimerase catalyzes the interconversion of D-tagatose and D-sorbose by means of a 2-ene-2,3-diol intermediate.

CH₂OH   CH₂OH
D-tagatose   D-sorbose

This enzyme also acts on a number of other 2-ketohexoses and 2-ketopentoses.[1]

[1]H. Itoh, H. Okaya, A. R. Khan, S. Tajima, S. Hayakawa & K. Izumori (1994) *Biosci. Biotech. Biochem.* **58**, 2168.

TAGATOSE KINASE

This phosphotransferase [EC 2.7.1.101] catalyzes the reaction of ATP with D-tagatose to produce ADP and D-tagatose 6-phosphate.[1]

[1]T. Szumilo (1981) *BBA* **660**, 366.

TAGATOSE-6-PHOSPHATE KINASE

This phosphotransferase [EC 2.7.1.144], also called phosphotagatokinase, catalyzes the reaction of ATP with D-tagatose 6-phosphate to produce ADP and D-tagatose 1,6-bisphosphate.[1]

[1]D. L. Bissett & R. L. Anderson (1980) *JBC* **255**, 8745.

Selected entries from *Methods in Enzymology* [vol, page(s)]:
General discussion: 90, 87
Assay: 90, 87

TAGATURONATE REDUCTASE

This oxidoreductase [EC 1.1.1.58], also known as tagaturonate dehydrogenase and altronate dehydrogenase, catalyzes the reversible reaction of D-tagaturonate with NADH to produce D-altronate and NAD⁺.

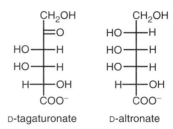

D-tagaturonate   D-altronate

The enzyme operates by a steady-state ordered Bi Bi kinetic mechanism.[1]

[1]R. C. Portalier (1972) *EJB* **30**, 211.

Selected entries from *Methods in Enzymology* [vol, page(s)]:
General discussion: 5, 195; **89**, 210
Assay: 5, 195; **89**, 210
Other molecular properties: *Escherichia coli* (properties, **89**, 213; purification, **89**, 212); mannonate dehydrogenase, with, 5, 199; properties, 5, 197; **89**, 213; purification, 5, 195; **89**, 212; uronate isomerase and, 5, 190, 192

TAKA-DIASTASE

A multienzyme preparation from *Aspergillus oryzae* containing α-amylase, adenosine deaminase, arylsulfatases, ribonucleases, and a number of peptidases.

Selected entries from *Methods in Enzymology* [vol, page(s)]:
General discussion: 19, 373
Assay: 19, 386
Other molecular properties: adenosine deaminase in, 2, 475; arylsulfatase in, 2, 328; folic acid conjugase in, 2, 630; metaphosphatase in, 2, 579; nonspecific adenosine deaminase from, 12A, 126; nuclease S_1 of, 29, 400; purification and properties of, 19, 386; ribonuclease T_1 from, 12A, 228; ribonuclease T_2 from, 12A, 239; triphosphatase in, 2, 580, 582

TANNASE

This enzyme [EC 3.1.1.20] catalyzes the hydrolysis of (a) digallate to produce two gallate molecules and (b) the hydrolysis of tannic acid to form glucose and *n* gallate molecules.

digallate   gallate

Taq POLYMERASE

This thermostable DNA-directed DNA polymerase [EC 2.7.7.7] isolated from *Thermus aquaticus* that exhibits a $5' \rightarrow 3'$ polymerase activity as well as $5'$ nuclease activity, but lacks the $3' \rightarrow 5'$ exonuclease activity of other DNA polymerases.[1,2] *See also DNA Polymerase*

[1]F. G. Pluthero (1993) *NAR* 21, 4850.
[2]Y. Kim, S. H. Eom, J. Wang, D. S. Lee, S. W. Suh & T. A. Steitz (1995) *Nature* 376, 612.

*Taq*I RESTRICTION ENDONUCLEASE

This type II restriction endonuclease [EC 3.1.21.4] is obtained from *Thermus aquaticus* and recognizes and acts on both strands of DNA at $5' \ldots T\nabla CGA \ldots 3'$, where ∇ indicates the site of bond scission.[1] It is a thermostable endonuclease. If the cytosine is methylated, *Taq*I restriction endonuclease will still catalyze double-strand cleavage.[2] The enzyme will also act on DNA-RNA hybrids. In the presence of some organic solvents, the enzyme also catalyzes single-strand and double-strand cuts at sequences termed *Taq*I star sites. All of the star recognition sequences differed from TCGA by a single base. This suggests that sequence discrimination is governed by eight hydrogen bonds formed between the endonuclease and the cognate nucleotides within the major groove.[3,4] Steady-state studies show that the enzyme follows Michaelis–Menten kinetics at 60°C. At 0°C, the enzyme is completely inactive. However, at 15°C, turnover generated nicked substrate as the major product in excess of enzyme suggesting that dissociation occurs between nicking events.[5] Both Mg^{2+} and high temperature are required to attain the correct endonuclease conformation to form the tight complex observed in the steady state.[5] Additional studies demonstrate that sequence-specific phosphate contacts are formed in the transition state, thus amplifying the apparent contribution of base contacts to transition-state stabilization.[6] **See "Star" Activity**

Taq^{α}I restriction enzyme has a two amino acid replacement at its amino terminus. This permits a higher level of expression of the endonuclease.

[1]S. Sato, C. A. Hutchison, III, & J. I. Harris (1977) *PNAS* 74, 542.
[2]R. E. Streeck (1980) *Gene* 12, 267.
[3]F. Barany (1988) *Gene* 65, 149.
[4]J. A. Zebala, J. Choi, G. L. Trainor & F. Barany (1992) *JBC* 267, 8106.
[5]J. Zebala, J. Choi & F. Barany (1992) *JBC* 267, 8097.
[6]A. N. Mayer & F. Barany (1994) *JBC* 269, 29067.

Selected entries from *Methods in Enzymology* [vol, page(s)]:
General discussion: 65, 175, 342, 450, 462

*Taq*II RESTRICTION ENDONUCLEASE

This type II restriction endonuclease [EC 3.1.21.4], subtype s, is obtained from *Thermus aquaticus* and recognizes and acts on both strands of DNA at $5' \ldots$ GACCGANNNNN NNNNNN$\nabla \ldots 3'$ and $3' \ldots$ CTGGCTNNNNNNNNNN∇ NN $\ldots 5'$ where N is any base, and ∇ indicates the site of bond scission. It also acts at $5' \ldots$ CACCCANNNNNNN NNNN$\nabla \ldots 3'$ and $3' \ldots$ GTGGGTNNNNNNNNNN∇ NN $\ldots 5'$.[1] It is a thermostable endonuclease.

[1]D. Barker, M. Hoff, A. Oliphant & R. White (1984) *NAR* 12, 5567.

Selected entries from *Methods in Enzymology* [vol, page(s)]:
General discussion: 65, 11
Other molecular properties: purification, 65, 94

TARTRATE DECARBOXYLASE

This decarboxylase [EC 4.1.1.73] catalyzes the conversion of (*R,R*)-tartrate to D-glycerate and carbon dioxide. While the *Pseudomonas* enzyme requires NAD^+, NADH formation occurs at only 1% the rate of decarboxylation. The enzyme also catalyzes the NAD^+-linked oxidative decarboxylation of D-malate.[1]

[1] S. Furuyoshi, Y. Nawa, N. Kawabata, H. Tanaka & K. Soda (1991) *J. Biochem.* **110**, 520.

L(+)-TARTRATE DEHYDRATASE

This iron-dependent dehydratase [EC 4.2.1.32] catalyzes the conversion of (R,R)-tartrate to produce oxaloacetate and water.[1] The inactive low-molecular weight form is converted into active enzyme in the presence of Fe^{2+} and thiol. Neither D(−)-tartrate nor *meso*-tartrate is a substrate. **See also** D(-)-*Tartrate Dehydratase*

[1] R. E. Hurlbert & W. B. Jakoby (1965) *JBC* **240**, 2772.

Selected entries from **Methods in Enzymology** [vol, page(s)]:
General discussion: 9, 680
Assay: 9, 680

D(−)-TARTRATE DEHYDRATASE

This iron-dependent dehydratase [EC 4.2.1.81] catalyzes the conversion of (S,S)-tartrate to oxaloacetate and water.[1] Neither L(+)-tartrate nor *meso*-tartrate is a substrate. **See also** L(+)-*Tartrate Dehydratase*

[1] H. Rode & F. Giffhorn (1982) *J. Bact.* **150**, 1061 and 1602.

TARTRATE DEHYDROGENASE

This manganese- and monovalent-cation-dependent oxidoreductase [EC 1.1.1.93] catalyzes the reaction of tartrate with NAD^+ to produce oxaloglycolate (which exists in tautomeric equilibrium with dihydroxyfumarate) and NADH.[1-4] Both *meso*-tartrate and (R,R)-tartrate (L(+)-tartrate) reportedly act as substrates. The *Rhodobacter sphaeroides* enzyme is bifunctional, acting also as a D-(+)-malate dehydrogenase (decarboxylating) [EC 1.1.1.83].[1] The *Pseudomonas putida* enzyme also produces D-glycerate and carbon dioxide from *meso*-tartrate in what is formally a decarboxylation reaction, occurring without net oxidation or reduction.[3,4] The kinetic mechanism for the first two activities is ordered Bi Bi.[3] **See also** *meso-Tartrate Dehydrogenase*

[1] F. Giffhorn & A. Kuhn (1983) *J. Bact.* **155**, 281.
[2] L. D. Kohn, P. M. Packman, R. H. Allen & W. B. Jakoby (1968) *JBC* **243**, 2479.
[3] P. A. Tipton & J. Peisach (1990) *Biochem.* **29**, 1749.
[4] P. A. Tipton (1996) *Biochem.* **35**, 3108.

Selected entries from **Methods in Enzymology** [vol, page(s)]:
General discussion: 9, 236
Assay: 9, 236

meso-TARTRATE DEHYDROGENASE

This oxidoreductase [EC 1.3.1.7] catalyzes the reaction of *meso*-tartrate with NAD^+ to produce dihydroxyfumarate

(which exists in tautomeric equilibrium with oxaloglycolate) and NADH. Both *meso*-tartrate and (R,R)-tartrate (L(+)-tartrate) reportedly act as substrates. This enzyme activity may be related to tartrate dehydrogenase [EC 1.1.1.93]. **See** *Tartrate Dehydrogenase*

Selected entries from **Methods in Enzymology** [vol, page(s)]:
General discussion: 9, 236
Assay: 9, 236

TARTRATE EPIMERASE

This epimerase [EC 5.1.2.5] catalyzes the interconversion of (R,R)-tartrate (or, L(+)-tartrate) and *meso*-tartrate.[1]

[1] S. Ranjan, K. K. Patnaik & M. M. Laloraya (1961) *Naturwissenschaften* **48**, 406.

TARTRONATE O-HYDROXYCINNAMOYL-TRANSFERASE

This acyltransferase [EC 2.3.1.106], also known as tartronate sinapoyltransferase, catalyzes the reaction of sinapoyl-CoA with hydroxymalonate (or, tartronate; $^-OOCCH(OH)COO^-$) to produce Coenzyme A and sinapoyltartronate.[1]

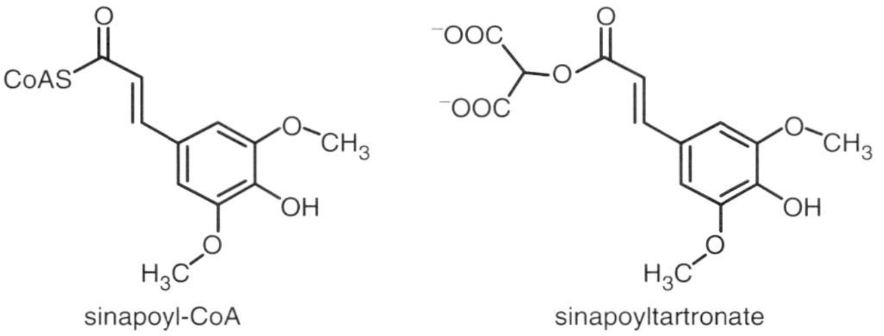

sinapoyl-CoA

sinapoyltartronate

Other acyl-CoA substrates include 4-coumaroyl-CoA, caffeoyl-CoA, and feruloyl-CoA.

[1] D. Strack, R. Ruhoff & W. Gräwe (1986) *Phytochemistry* **25**, 833.

TARTRONATE-SEMIALDEHYDE SYNTHASE

This FAD- and thiamin-pyrophosphate-dependent enzyme [EC 4.1.1.47], also known as tartronate semialdehyde carboxylase and glyoxylate carboligase, catalyzes the reaction of two molecules of glyoxylate to produce tartronate semialdehyde and carbon dioxide.[1-3] The enzyme proceeds via a 2-hydroxymethylthiamin pyrophosphate intermediate.

[1] T. H. Cromartie & C. T. Walsh (1976) *JBC* **251**, 329.
[2] N. B. Gupta & B. Vennesland (1966) *ABB* **113**, 255.
[3] B. A. Palfey & V. Massey (1998) *CBC* **3**, 83.

Selected entries from **Methods in Enzymology** [vol, page(s)]:
General discussion: 9, 693
Assay: 9, 693

*Tat*I RESTRICTION ENDONUCLEASE

This type II restriction endonuclease [EC 3.1.21.4] is obtained from *Thermus aquaticus* CBA1-331 and recognizes and acts on both strands of DNA at 5′...W∇GTACW...3′ where W refers to either A or T.[1] (The symbol ∇ indicates the cleavage site.) The enzyme also exhibits star activity. **See "Star" Activity**

[1]R. J. Roberts & D. Macelis (1996) *NAR* **24**, 223.

[TAU PROTEIN] KINASE

This phosphotransferase [EC 2.7.1.135], also known at τ protein kinase, catalyzes the reaction of ATP with tau protein, a microtubule-associated protein, to produce ADP and tau protein containing a phosphorylated seryl residue.[1–4] The kinase is activated by tubulin but is not activated by calmodulin, Ca^{2+}, or cyclic nucleotides, unlike calcium/calmodulin-dependent protein kinase [EC 2.7.1.123]. Human glycogen synthase kinase 3β is identical to human [tau protein] kinase I.[4]

[1]K. Ishiguro, Y. Ihara, T. Uchida & K. Imahori (1988) *J. Biochem.* **104**, 319.
[2]K. Ishiguro, M. Takamatsu, K. Tomizawa, A. Omori, M. Takahashi, M. Arioka, T. Uchida & K. Imahori (1992) *JBC* **267**, 10897.
[3]J. S. Song & S. D. Yang (1995) *J. Protein Chem.* **14**, 95.
[4]K. Ishiguro, A. Shiratsuchi, S. Sato, A. Omori, M. Arioka, S. Kobayashi, T. Uchida & K. Imahori (1993) *FEBS Lett.* **325**, 167.

TAURINE AMINOTRANSFERASE

This pyridoxal-phosphate-dependent aminotransferase [EC 2.6.1.55] catalyzes the reversible reaction of taurine ($^+H_3NCH_2CH_2SO_3^-$) with α-ketoglutarate to produce sulfoacetaldehyde ($O{=}CHCH_2SO_3^-$) and L-glutamate. Other substrates include DL-3-aminoisobutanoate, β-alanine, 3-aminopropanesulfonate, and 4-aminobutyrate.

Selected entries from *Methods in Enzymology* [vol, page(s)]:
General discussion: 113, 102
Assay: 113, 102

TAURINE DEHYDROGENASE

This oxidoreductase [EC 1.4.99.2] catalyzes the reaction of taurine ($^+H_3NCH_2CH_2SO_3^-$) with water and an acceptor substrate to produce sulfoacetaldehyde ($O{=}CHCH_2SO_3^-$), ammonia, and the reduced acceptor. Phenazine methosulfate can serve as the acceptor substrate.

Selected entries from *Methods in Enzymology* [vol, page(s)]:
General discussion: 143, 496
Assay: 143, 496

TAURINE DIOXYGENASE

This Fe^{2+}-dependent oxidoreductase [EC 1.14.11.17] catalyzes the reaction of taurine with dioxygen and α-ketoglutarate (or, 2-oxoglutarate) to produce sulfite, aminoacetaldehyde, succinate, and carbon dioxide.[1] Other substrates for the *Escherichia coli* enzyme, albeit weaker, include pentanesulfonate, 3-(*N*-morpholino)propanesulfonate, and 2-(1,3-dioxoisoindolin-2-yl)ethanesulfonate.

[1]E. Eichhorn, J. R. Van Der Poeg, M. A. Kertesz & T. Leisinger (1997) *JBC* **272**, 23031.

TAURINE-TRANSPORTING ATPase

This so-called ATPase [EC 3.6.3.36] catalyzes the ATP-dependent (ADP- and orthophosphate-producing) transport of taurine$_{out}$ to produce taurine$_{in}$.[1] ABC-type (ATP-binding cassette-type) ATPases have two similar ATP-binding domains and do not undergo phosphorylation during the transport process.

The idea that a transporter is an enzyme is in keeping with a new definition of enzyme catalysis as the facilitated making/breaking of chemical bonds, not just covalent bonds.[2] This idea builds on Pauling's assertion that any long-lived, chemically distinct interaction (in this case, the persistent location of a solute with respect to the faces of a membrane) can be regarded as a chemical bond. Note also that the equilibrium constant ($K_{eq} = [\text{Taurine}_{in}][\text{ADP}][P_i]/[\text{Taurine}_{out}][\text{ATP}]$) does not conform to that expected for an ATPase (*i.e.*, $K_{eq} = [\text{ADP}][P_i]/[\text{ATP}]$). Thus, although the overall reaction yields ADP and orthophosphate, the enzyme is misclassified as a hydrolase, and instead should be regarded as an energase-type reaction. Energases facilitate affinity-modulated reactions by coupling the $\Delta G_{\text{ATP-hydrolysis}}$ to a force-generating or work-producing step.[2] In this case, P-O-P bond scission supplies the energy to drive taurine transport. **See ATPase; Energase**

[1]J. R. van der Ploeg, M. A. Weiss, E. Saller, H. Nashimoto, N. Saito, M. A. Kertesz & T. Leisinger (1996) *J. Bacteriol.* **178**, 5438.
[2]D. L. Purich (2001) *TiBS* **26**, 417.

TAUROCYAMINE KINASE

This phosphotransferase [EC 2.7.3.4] catalyzes the reaction of ATP with taurocyamine ($^+H_2N{=}C(NH_2)NHCH_2CH_2SO_3^-$) to produce ADP and *N*-phosphotaurocyamine ($^+H_2N{=}C(NHPO_3^{2-})NHCH_2CH_2SO_3^-$).[1]

[1]B. Surholt (1979) *EJB* **93**, 279.

TAURODEOXYCHOLATE 7α-HYDROXYLASE

A cytochrome P450 system that catalyzes the hydroxylation of taurodeoxycholate to produce taurocholate.[1,2]

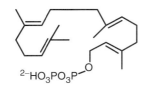

taurodeoxycholate   taurocholate

See also Cholesterol 7α-Monooxygenase

[1]D. Trülzsch, H. Greim, P. Czygan, F. Hutterer, F. Schaffner, H. Popper, D. Y. Cooper & O. Rosenthal (1973) Biochemistry 12, 76.
[2]K. Murakami & K. Okuda (1981) JBC 256, 8658.

Selected entries from **Methods in Enzymology** [vol, page(s)]:
General discussion: 52, 14

TAUROPINE DEHYDROGENASE

This oxidoreductase [EC 1.5.1.23] catalyzes the reaction of tauropine with NAD^+ and water to produce taurine, pyruvate, and NADH.[1-4] Alanine can replace taurine in the reverse reaction, albeit more slowly.

tauropine   taurine

α-Ketobutanoate and α-ketopentanoate can replace pyruvate.

[1]G. Gäde (1987) Biol. Chem. Hoppe-Seyler 368, 1519.
[2]G. Gäde (1986) EJB 160, 311.
[3]N. Kanno, M. Sato, E. Nagahisa & Y. Sato (1996) Comp. Biochem. Physiol. B Biochem. Mol. Biol. 114, 409.
[4]J. Thompson & S. P. F. Miller (1991) AE 64, 317.

TAXADIENE SYNTHASE

$^{2-}HO_3PO_3P$—O

geranylgeranyl diphosphate

taxadiene

This enzyme, also called taxadiene cyclase, catalyzes the conversion of geranylgeranyl diphosphate to taxadiene, a diterpene precursor of the anticancer drug taxol (pacitaxelin).[1]

[1]X. Lin, M. Hezari, A. E. Koepp, H. G. Floss & R. Croteau (1996) Biochemistry 35, 2968.

TAXIFOLIN 8-MONOOXYGENASE

This oxidoreductase [EC 1.14.13.19] catalyzes the reaction of taxifolin with NAD(P)H and dioxygen to produce 2,3-dihydrogossypetin, $NAD(P)^+$, and water.[1] The substrate is a flavonol and the product is a flavonone.

taxifolin

2,3-dihydrogossypetin

[1]A. M. Jeffrey, M. Knight & W. C. Evans (1972) BJ 130, 373.

TEICHOIC-ACID-TRANSPORTING ATPase

This so-called ATPase [EC 3.6.3.40] catalyzes the ATP-dependent (ADP- and orthophosphate-producing) transport of teichoic $acid_{in}$ to produce teichoic $acid_{out}$ in Gram-positive bacteria.[1-3] ABC-type (ATP-binding cassette-type) ATPases have two similar ATP-binding domains and do not undergo phosphorylation during the transport process.

The idea that a transporter is an enzyme is in keeping with a new definition of enzyme catalysis as the facilitated making/breaking of chemical bonds, not just covalent bonds.[4] This idea builds on Pauling's assertion that any long-lived, chemically distinct interaction (in this case, the persistent location of a solute with respect to the faces of a membrane) can be regarded as a chemical bond. Note also that the equilibrium constant (K_{eq} = [Teichoic-$Acid_{out}$][ADP][P_i]/[Teichoic-$Acid_{in}$][ATP]) does not conform to that expected for an

ATPase (*i.e.*, $K_{eq} = [ADP][P_i]/[ATP]$). Thus, although the overall reaction yields ADP and orthophosphate, the enzyme is misclassified as a hydrolase, and instead should be regarded as an energase-type reaction. Energases facilitate affinity-modulated reactions by coupling the $\Delta G_{ATP\text{-hydrolysis}}$ to a force-generating or work-producing step.[4] In this case, P-O-P bond scission supplies the energy to drive teichoic acid transport. *See ATPase; Energase*

[1] M. J. Fath & R. Kolter (1993) *Microbiol. Rev.* **57**, 995.
[2] V. Lazarevic & D. Karamoto (1995) *Mol. Microbiol.* **16**, 345.
[3] J. K. Griffiths & C. E. Sansom (1998) *The Transporter Factsbook*, Academic Press, San Diego.
[4] D. L. Purich (2001) *TiBS* **26**, 417.

TENEBRIO α-PROTEINASE

This serine proteinase, previously classified as EC 3.4.21.18, is now a deleted entry.

TENTOXILYSIN

This zinc-dependent metalloendopeptidase [EC 3.4.24.68], also called tetanus neurotoxin, catalyzes hydrolysis of the Gln76–Phe77 bond in synaptobrevin. Short peptide analogues of the sequence in synaptobrevin are not substrates. This enzyme is thought to be the sole cause of the symptoms observed in tetanus.[1,2]

[1] C. Montecucco, ed. (1995) *Clostridial Neurotoxins. Current Topics in Microbiology and Immunology*, vol. **195**, Springer-Verlag, Heidelberg.
[2] W. E. van Heyningen (1968) *Sci. Amer.* **218**(4), 69.

TEREPHTHALATE 1,2-*cis*-DIHYDRODIOL DEHYDROGENASE

This enzyme [EC 1.3.1.61] catalyzes the reaction of *cis*-4,5-dihydroxycyclohexa-1(6),2-diene-1,4-dicarboxylate with NAD^+ to produce 3,4-dihydroxybenzoate, carbon dioxide, NADH, and a proton.[1]

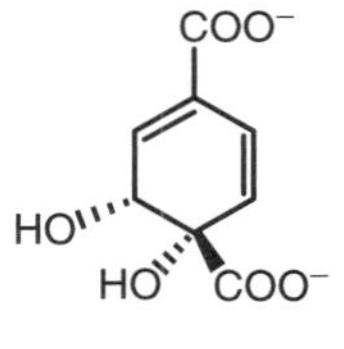

cis-4,5-dihydroxycyclohexa-1(6),2-diene-1,4-dicarboxylate

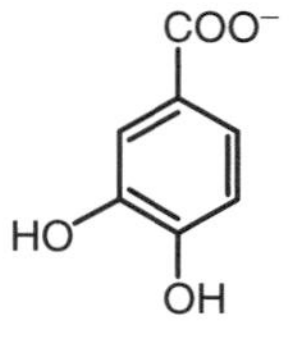

3,4-dihydroxybenzoate

[1] Y. Z. Wang, Y. Zhou & G. J. Zylstra (1995) *Environ. Health Perspect.* **103**, Suppl. 5, 9.

TEREPHTHALATE 1,2-DIOXYGENASE

This oxidoreductase [EC 1.14.12.15], also known as benzene-1,4-dicarboxylate 1,2-dioxygenase and

1,4-dicarboxybenzoate 1,2-dioxygenase, catalyzes the reaction of terephthalate (or, benzene-1,4-dicarboxylate) with NADH, H^+, and dioxygen to produce $(1R,6S)$-dihydroxycyclohexa-2,4-diene-1,4-dicarboxylate and NAD^+.[1] The *Comamonas testosteroni* enzyme contains a Rieske [2Fe-2S] center.

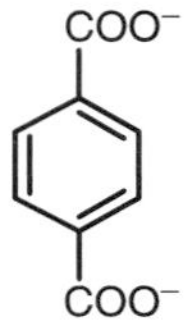

terephthalate

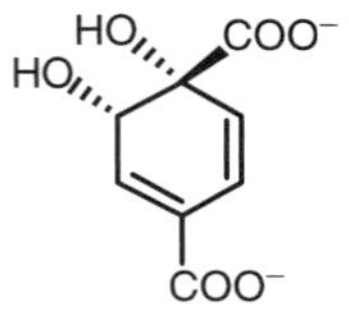

1,6-dihydroxycyclohexa-2,4-diene-1,4-dicarboxylate

[1] H. R. Schläfli, M. A. Weiss, T. Leisinger & A. M. Cook (1994) *J. Bact.* **176**, 6644.

TESTOSTERONE 17β-DEHYDROGENASE

This oxidoreductase [EC 1.1.1.63] catalyzes the reaction of testosterone with NAD^+ to produce androst-4-ene-3,17-dione and NADH. The coenzyme stereospecificity for the *Cylindrocarpon radicicola* enzyme is for the *pro-S* position (that is, the B side).[1–3]

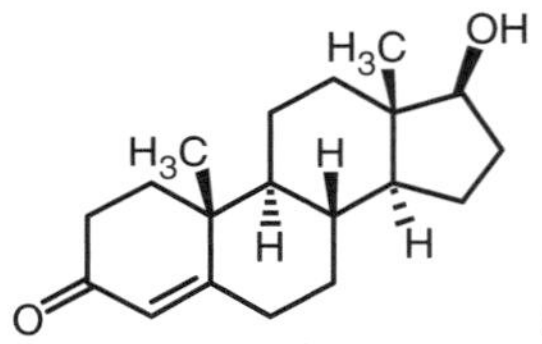

testosterone

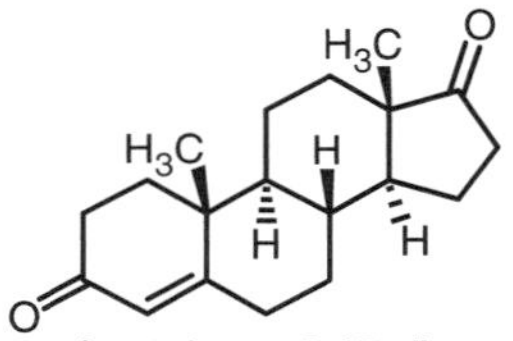

androst-4-ene-3,17-dione

See also 3α(or 17β)-Hydroxysteroid Dehydrogenase; Estradiol 17β-Dehydrogenase; 3(or 17β)-Hydroxysteroid Dehydrogenase; Testosterone 17β-Dehydrogenase (NADP+)

[1] E. Itagaki & T. Iwaya (1988) *J. Biochem.* **103**, 1039.
[2] D. W. Payne & P. Talalay (1985) *JBC* **260**, 13648.
[3] C. H. Blomquist & C. E. Kotts (1978) *Steroids* **32**, 399.

TESTOSTERONE 17β-DEHYDROGENASE (NADP+)

This oxidoreductase [EC 1.1.1.64] catalyzes the reaction of testosterone with $NADP^+$ to produce androst-4-ene-3,17-dione and NADPH. The enzyme will also convert 3-hydroxyhexobarbital to 3-oxohexobarbital. The pig testis enzyme is B side specific[1,2] whereas the mouse and guinea pig liver enzymes are A side specific.[3,4] The pig enzyme exhibits a random Bi Bi kinetic mechanism[1] whereas the mechanism for the murine liver enzyme is ordered Bi Bi.[3] *See also Testosterone 17β-Dehydrogenase*

[1] H. Inano & B.-I. Tamaoki (1986) *Steroids* **48**, 3.
[2] H. Inano & B.-I. Tamaoki (1975) *EJB* **53**, 319.
[3] A. Hara, T. Nakayama, M. Nakagawa, Y. Inoue & H. Sawada (1987) *J. Biochem.* **102**, 1585.
[4] T. Nakayama, A. Hara, K.-I. Kariya & H. Sawada (1985) *J. Biochem.* **98**, 1131.

TETRAACYLDISACCHARIDE 4'-KINASE

This phosphotransferase [EC 2.7.1.130], also known as lipid-A 4'-kinase, catalyzes the reaction of ATP with 2,3-bis(3-hydroxytetradecanoyl)-D-glucosaminyl-(β-D-1,6)-2,3-bis(3-hydroxytetradecanoyl)-D-glucosaminyl β-phosphate to produce ADP and 2,3,2',3'-tetrakis(3-hydroxytetradecanoyl)-D-glucosaminyl-1,6-β-D-glucosamine 1,4'-bisphosphate. The enzyme participates along with acyl-[acyl-carrier protein]:UDP-N-acetylglucosamine O-acyltransferase [EC 2.3.1.129] and lipid-A-disaccharide synthase [EC 2.4.1.182] in the biosynthesis of the phosphorylated glycolipid, lipid A in the outer membrane of *Escherichia coli*.[1]

[1] B. L. Ray & C. R. Raetz (1987) *JBC* **262**, 1122.

TETRACHLOROETHENE DEHALOGENASE

This cobalamin- and iron-sulfur-dependent dehalogenase, isolated from *Dehalospirillium multivorans*, catalyzes the reaction of tetrachloroethene ($Cl_2C{=}CCl_2$) with an electron donor (such as H_2) to produce (Z)-dichloroethene (*i.e.*, *cis*-dichloroethene) and two chloride anions, proceeding via trichloroethene. The reaction mechanism appears to involve the nucleophilic attack of the cofactor's cobalt ion to form either a trichlorovinylcob(III)alamin or a tetrachloroethylcobalamin intermediate.[1]

[1] A. Neumann, G. Wohlfahrt & G. Diekert (1996) *JBC* **241**, 16515.

TETRAHYDROBERBERINE OXIDASE

This oxidoreductase [EC 1.3.3.8], often abbreviated THB oxidase, catalyzes the reaction of (S)-tetrahydroberberine (or, (S)-canadine) with two molecules of dioxygen to produce berberine and two molecules of hydrogen peroxide.[1–3]

(S)-tetrahydroberberine   berberine

The *Berberis wilsonia* enzyme is a flavoprotein, while that from *Coptis japonica* is not. Other substrates include (S)-tetrahydrojatrorrhizine.

[1] M. Amann, N. Nagakura & M. H. Zenk (1984) *Tetrahedron Lett.* **25**, 953.
[2] N. Okada, A. Shinmyo, H. Okada & Y. Yamada (1988) *Phytochemistry* **27**, 979.
[3] Y. Kobayashi, M. Hara, H. Fukui & M. Tabata (1991) *Phytochemistry* **30**, 3605.

TETRAHYDROCOLUMBAMINE 2-O-METHYLTRANSFERASE

This methyltransferase [EC 2.1.1.89] catalyzes the reaction of S-adenosyl-L-methionine with 5,8,13,13a-tetrahydrocolumbamine (or, isocorypalmine) to produce S-adenosyl-L-homocysteine and tetrahydropalmatine.[1]

5,8,13,13a-tetrahydrocolumbamine   tetrahydropalmatine

[1] C. W. W. Beecher & W. J. Kelleher (1984) *Tetrahedron Lett.* **25**, 4595.

TETRAHYDRODIPICOLINATE N-ACETYLTRANSFERASE

This acetyltransferase [EC 2.3.1.89], also known as tetrahydrodipicolinate acetylase, catalyzes the reaction of acetyl-CoA with L-2,3,4,5-tetrahydrodipicolinate to produce Coenzyme A and L-2-acetamido-6-oxoheptanedioate.[1]

L-2,3,4,5-tetrahydro-dipicolinate   L-2-acetamido-6-oxoheptanedioate

The enzyme participates in L-lysine biosynthesis in *Bacillus megaterium*.

[1] S. P. Chatterjee & P. J. White (1982) *J. Gen. Microbiol.* **128**, 1073.

TETRAHYDROMETHANOPTERIN S-METHYLTRANSFERASE

This transferase [EC 2.1.1.86] catalyzes the reaction of 5-methyl-5,6,7,8-tetrahydromethanopterin with

2-mercaptoethanesulfonate ($HSCH_2CH_2SO_3^-$) to produce 5,6,7,8-tetrahydromethanopterin and 2-(methylthio)ethanesulfonate ($CH_3SCH_2CH_2SO_3^-$). The enzyme participates in the formation of methane from carbon monoxide in *Methanobacterium thermoautotrophicum* and other methanogenic microorganisms.[1–4]

[1] F. D. Sauer (1986) *BBRC* **136**, 542.
[2] P. Gartner, A. Ecker, R. Fischer, D. Linder, G. Fuchs & R. K. Thauer (1993) *EJB* **213**, 537.
[3] P. Gartner, D. S. Weiss, U. Harms & R. K. Thauer (1994) *EJB* **226**, 465.
[4] M. L. Ludwig & R. G. Matthews (1997) *ABB* **66**, 269.

(S)-TETRAHYDROPROTOBERBERINE *N*-METHYLTRANSFERASE

This methyltransferase [EC 2.1.1.122], also known as tetrahydroprotoberberine *cis-N*-methyltransferase, catalyzes the reaction of *S*-adenosyl-L-methionine with *(S)*-7,8,13,14-tetrahydroprotoberberine to produce *S*-adenosyl-L-homocysteine and *cis-N*-methyl-*(S)*-7,8,13,14-tetrahydroprotoberberine.[1]

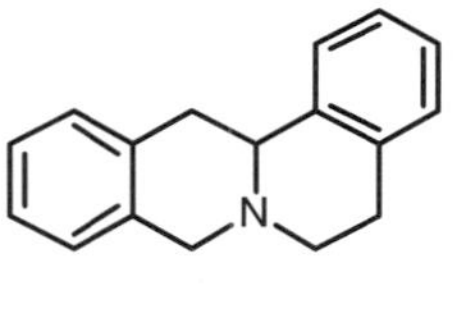

tetrahydroproto-
berberine

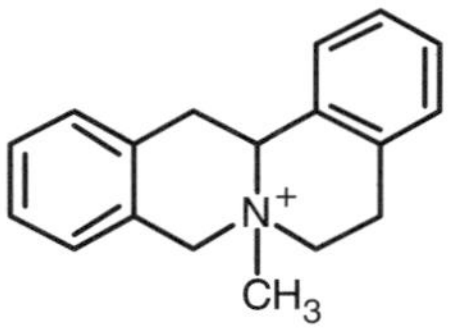

cis-N-methyl-tetrahydro-
protoberberine

[1] M. Rueffer, G. Zumstein & M. H. Zenk (1990) *Phytochemistry* **29**, 3727.

2,3,4,5-TETRAHYDROPYRIDINE-2,6-DICARBOXYLATE *N*-SUCCINYLTRANSFERASE

The Enzyme Commission describes this acyltransferase [EC 2.3.1.117] as the catalyst of the reversible reaction of succinyl-CoA with 2,3,4,5-tetrahydropyridine-2-carboxylate to produce Coenzyme A and *N*-succinyl-L-2-amino-6-oxoheptanedioate;[1,2] however, to generate the desired product, the acyl acceptor must be tetrahydrodipicolinate (or, 2,3,4,5-tetrahydropyridine-2,6-dicarboxylate).

This enzyme participates in the biosynthesis of L-lysine in bacteria, blue-green algae, and higher plants. Substrate and transition-analogue studies of the bacterial enzyme suggest that upon binding of the L-configuration of the pyridine substrate, hydration of the imine group occurs to give 2-hydroxypiperidine-2,6-dicarboxylate in which the

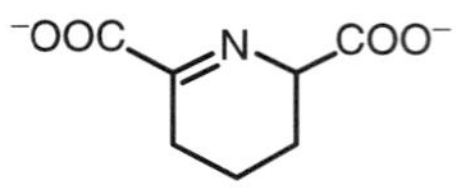

2,3,4,5-tetrahydropyridine-
2,6-dicarboxylate

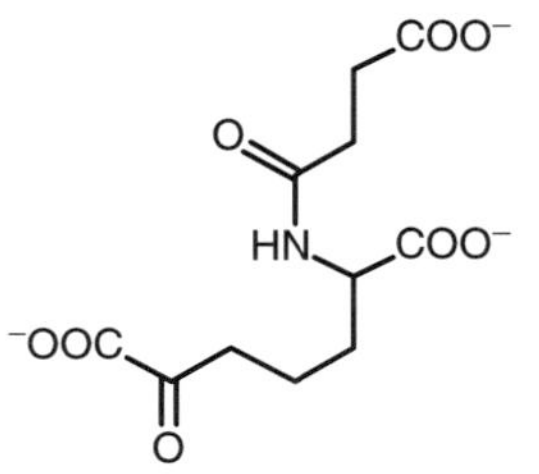

N-succinyl-L-2-amino-
6-oxoheptanedioate

two carboxyl groups are *trans*. Succinylation then occurs and the ring opens to give the acyclic product.[2]

[1] S. A. Simms, W. H. Voige & C. Gilvarg (1984) *JBC* **259**, 2734.
[2] D. A. Berges, W. E. DeWolf, G. L. Dunn, D. J. Newman, S. J. Schmidt, J. J. Taggart & C. Gilvarg (1986) *JBC* **261**, 6160.

TETRAHYDROXYNAPHTHALENE REDUCTASE

This enzyme [EC 1.1.1.252] catalyzes the reaction of 1,3,6,8-tetrahydroxynaphthalene with NADPH to produce (+)-scytalone and $NADP^+$.[1]

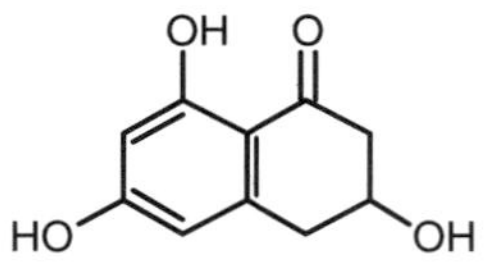

scytalone

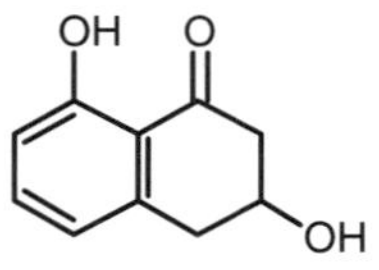

vermelone

This enzyme is a target protein for several commercially important fungicides used to prevent blast disease in rice plants. **See** *Trihydroxynaphthalene Reductase*

[1] A. Vidal-Cros, F. Viviani, G. Labesse, M. Boccara & M. Gaudry (1994) *EJB* **219**, 985.

TETRAHYDROXYPTERIDINE CYCLOISOMERASE

This NAD^+-dependent isomerase [EC 5.5.1.3] catalyzes the conversion of tetrahydroxypteridine to xanthine-8-carboxylate.[1]

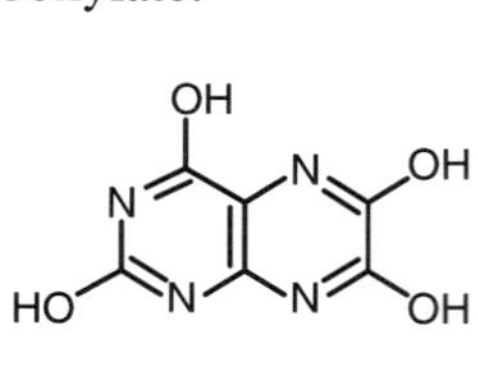

tetrahydroxypteridine

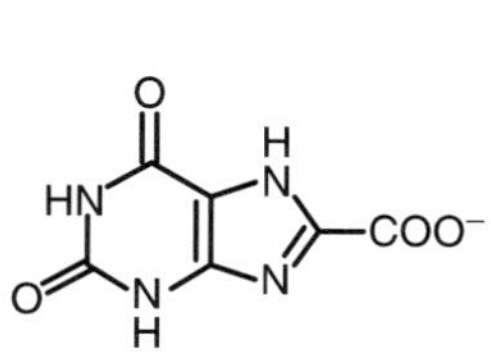

xanthine 8-carboxylate

[1] W. S. McNutt & S. P. Damle (1964) *JBC* **239**, 4272.

*Tfi*I RESTRICTION ENDONUCLEASE

This type II restriction endonuclease [EC 3.1.21.4], obtained from *Thermis filiformis*, recognizes and acts on both strands of DNA at 5′ . . . G∇AWTC . . . 3′ where W refers to A or T. (The symbol ∇ indicates the cleavage site.) The enzyme can also exhibit star activity. **See** *"Star" Activity*

*Tha*I RESTRICTION ENDONUCLEASE

This type II restriction endonuclease [EC 3.1.21.4] is obtained from *Thermoplasma acidophilum*, a thermophilic mycoplasma, and acts on both strands of DNA at 5′ . . . CG∇CG . . . 3′,[1] producing blunt ends. (The symbol ∇ indicates the cleavage site.)

[1] D. J. McConnell, D. G. Searcy & J. G. Sutcliffe (1978) *NAR* **5**, 1729.

Selected entries from **Methods in Enzymology** [vol, page(s)]:
Other molecular properties: recognition sequence, **65**, 11

THEANINE HYDROLASE

This enzyme [EC 3.5.1.65] catalyzes the hydrolysis of N^5-ethyl-L-glutamine (or, theanine) to produce L-glutamate and ethylamine.[1]

L-theanine

L-glutamate   ethylamine

Other substrates include N^5-*n*-propyl-L-glutamine, N^5-*n*-butyl-L-glutamine, N^5-isobutyl-L-glutamine, and N^5-*n*-pentyl-L-glutamine.

[1] T. Tsushida & T. Takeo (1985) *Agric. Biol. Chem.* **49**, 2913.

THEANINE SYNTHETASE

This enzyme [EC 6.3.1.6], also known as glutamate:ethylamine ligase and N^5-ethyl-L-glutamine synthetase, catalyzes the reaction of ATP with L-glutamate and ethylamine to produce ADP, orthophosphate, and N^5-ethyl-L-glutamine (also known as theanine).[1] **See also** *Glutamine Synthetase; Glutamate:Methylamine Ligase*

[1] K. Sasaoka, M. Kito & Y. Onishi (1965) *Agric. Biol. Chem.* **29**, 984.

THERMITASE

This extracellular serine protease [EC 3.4.21.66] catalyzes peptide-bond hydrolysis in collagen and elastin.[1,2] There is preference for hydrophobic aminoacyl residues (particularly Phe) at the P1 position and small residues (*e.g.*, Gly or Ala) at the P1′ position. Prolyl residues are not tolerated in the P1, P1′, or P3 positions. Thermitase has optimal temperatures of 60°C (with Ac-Ala-Ala-Ala-Ala-∇-OMe), 65°C (with Ac-Ala-Ala-Ala-∇-NHPhNO2), and 85°C (with azocasein). (Note: ∇ indicates the cleavage site.)

[1] D. Brömme, K. Peters, S. Fink & S. Fittkau (1986) *ABB* **244**, 439.
[2] R. Bott & C. Betzel, eds. (1996) *Subtilisin Enzymes, Practical Protein Engineering. Adv. Exp. Med. Biol.*, vol. **379**, Plenum, New York.

THERMOLYSIN

This extracellular zinc-dependent metalloendopeptidase [EC 3.4.24.27], first isolated from *Bacillus thermoproteolyticus*, catalyzes peptide-bond hydrolysis in proteins, preferentially at the N-terminal side of Leu, Phe, Ile, or Val.[1–5] Thermolysin is potently inhibited at pH 7 by phosphoamidon, a naturally occurring phosphoramidate ($K_i = 30$ nM).[1–5]

The amino acid composition and properties of the thermostable enzyme from *Streptomyces rectus* var. *proteolyticus* are closely similar.

[1] B. W. Matthews (1988) *Acc. Chem. Res.* **21**, 333.
[2] R. Roche & G. Voorduow (1978) *CRC Crit. Rev. Biochem.* **5**, 1.
[3] K. A. Walsh, Y. Burnstein & M. K. Pangburn (1974) *MIE* **34**, 435.
[4] D. W. Christiansen (1991) *Adv. Protein Chem.* **42**, 281.
[5] B. L. Vallee & A. Galdes (1984) *AE* **56**, 283.

Selected entries from **Methods in Enzymology** [vol, page(s)]:
General discussion: 19, 642; **34**, 435, 592
Assay: 19, 642, 561; **34**, 436 (by spectroscopy, **44**, 89); proteoglycan degradation method, **248**, 51
Other molecular properties: active-site titration, **248**, 100; acyl azide binding, **44**, 95; affinity labeling, **87**, 472; inhibition, **19**, 648; **248**, 406; kinetic properties, **19**, 650; peptide thioester substrate, **248**, 15; phosphoramidon inhibition, **45**, 693; thermostability, **47**, 181; transition-state analogue complexes, characterization, **249**, 294, 307; zinc replacement by other metals, **248**, 240

THERMOMYCOLIN

This thermostable fungal serine protease [EC 3.4.21.65], originally named thermomycolase, catalyzes peptide-bond hydrolysis. There is no distinctive preference for a particular aminoacyl residue in the P1 position; however, with small *N*-blocked *p*-nitrophenyl esters, the order of reactivity is Ala > Tyr > Phe ≫ Gly ≫ Leu ≫ Val. The enzyme's exceptional thermostability ($t_{1/2} = 110$ min at 73°C) is thought to be the consequence of calcium ion binding.[1]

[1]G. Voordouw, G. M. Gaucher & R. S. Roche (1974) *Can. J. Biochem.* **52**, 981.

Selected entries from *Methods in Enzymology* [vol, page(s)]:
General discussion: **45**, 415
Assay: **45**, 418

THERMOPSIN

This aspartic endopeptidase [EC 3.4.23.42; previously EC 3.4.99.43], catalyzes peptide-bond hydrolysis in a wide variety of proteins with an optimal pH of 2.0 and an optimal temperature of around 80°C.[1] Oxidized insulin B chain is hydrolyzed at Leu11–Val12, Leu15–Tyr16, Phe24–Phe25, and Phe26–Tyr27. The enzyme prefers bulky hydrophobic aminoacyl residues on either side of the scissile bond (positions P1 and P1′). The enzyme is inhibited by pepstatin.

[1]X. Lin & J. Tang (1995) *MIE* **248**, 156.

Selected entries from *Methods in Enzymology* [vol, page(s)]:
General discussion: **248**, 109, 115; 156
Assay: **248**, 156 (hemoglobin substrate, **248**, 156; synthetic substrate, **248**, 158)

THERMUS STRAIN Rt41A PROTEASE

This serine protease, also known as Rt41A protease and Pretaq (a commercial name), is isolated from the thermophilic eubacterium *Thermus* strain Rt41A. It catalyzes peptide-bond hydrolysis in proteins and is stabilized by calcium ions, which protects against thermal denaturation at 10 μM levels and prevents autolysis at 5 mM levels. The protease has a preference for aromatic or small aliphatic residues at the P1 position; however, the main cleavage in the oxidized insulin B chain is Leu15–Tyr16.[1,2]

[1]K. Peek, R. M. Daniel, C. Monk, L. Parker & T. Coolbear (1992) *EJB* **207**, 1035.
[2]R. M. Daniel, H. S. Toogood & P. L. Bergquist (1995) *Biotechnol. Genet. Eng. Rev.* **13**, 51.

THETIN:HOMOCYSTEINE METHYLTRANSFERASE

This methyltransferase [EC 2.1.1.3], also known as dimethylthetin:homocysteine methylpherase, catalyzes the reaction of dimethylsulfonioacetate (*i.e.*, dimethylthetin; $(CH_3)_2SH^+CH_2COO^-$) with L-homocysteine to produce *S*-methylthioglycolate and L-methionine.[1]

[1]G. A. Maw (1958) *BJ* **70**, 168.

Selected entries from *Methods in Enzymology* [vol, page(s)]:
General discussion: **5**, 743
Assay: **5**, 744

THIAMINASE

This enzyme [EC 3.5.99.2], also known as thiaminase II, catalyzes the hydrolysis of thiamin to produce 4-amino-5-hydroxymethyl-2-methylpyrimidine and 5-(2-hydroxyethyl)-4-methylthiazole.

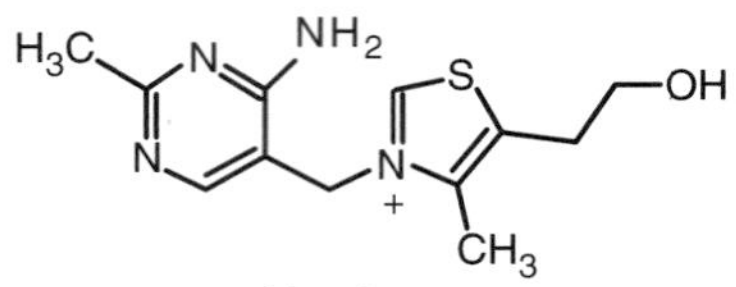

thiamin

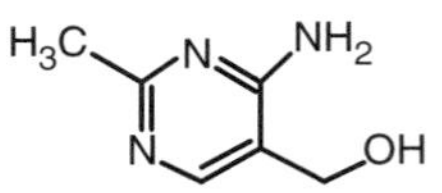

4-amino-5-hydroxymethyl-
2-methylpyrimidine

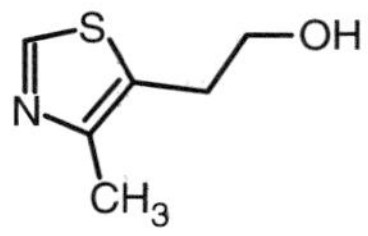

5-(2-hydroxyethyl)-
4-methylthiazole

Other substrates include pyrimidinemethylaniline (producing pyrimidinemethanol and aniline), thiothiamin (producing 2-thiothiazole and pyrimidine), and 2-northiamin (generating 2-norhydroxymethylpyrimidine and hydroxyethylthiazole). ***See also*** *Thiamin Pyridinylase*

Selected entries from *Methods in Enzymology* [vol, page(s)]:
General discussion: **2**, 622; **18A**, 234; **62**, 113
Assay: **2**, 622, 626; **18A**, 235; **62**, 113
Other molecular properties: differentiation I and II, **62**, 115; properties, **2**, 624, 627; **18A**, 237, 238; purification, **2**, 623, 626 (*Bacillus aneurinolyticus*, from, **18A**, 236, 237); sources, **18A**, 236

THIAMIN DIPHOSPHATE KINASE

This phosphotransferase [EC 2.7.4.15] catalyzes the reaction of ATP with thiamin diphosphate (or, thiamin pyrophosphate) to produce ADP and thiamin triphosphate.

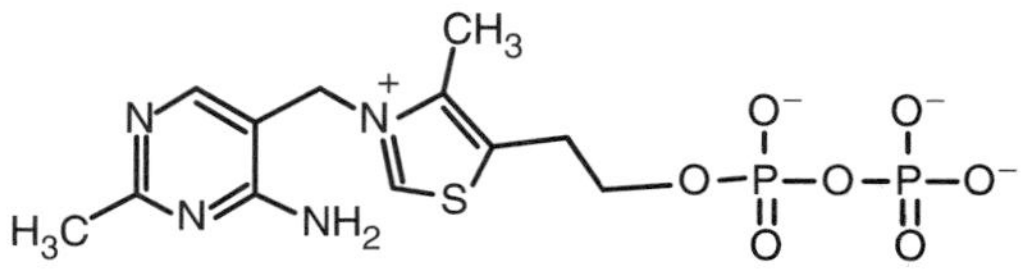

thiamin diphosphate

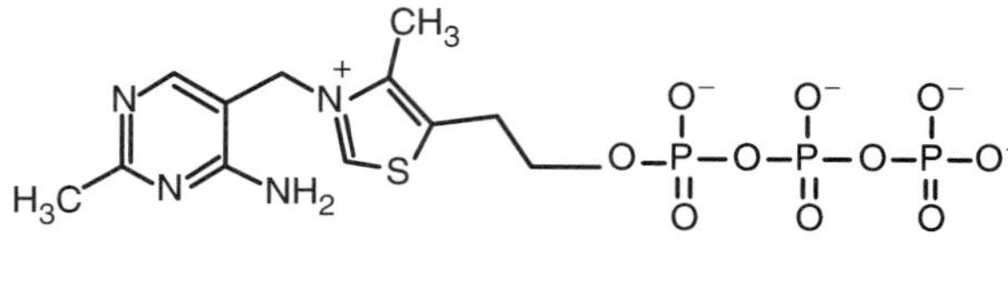

thiamin triphosphate

The bovine enzyme reportedly acts only on protein-bound thiamin diphosphate.[1]

[1]K. Nishino, Y. Itokawa, N. Nishino, K. Piros & J. R. Cooper (1983) *JBC* **258**, 11871.

Selected entries from *Methods in Enzymology* [vol, page(s)]:
General discussion: 18A, 226; 122, 24
Assay: 18A, 226; 122, 24

THIAMIN KINASE

This phosphotransferase [EC 2.7.1.89] catalyzes the reaction of ATP with thiamin to produce ADP and thiamin phosphate.[1] Care should be exercised when one comes across the name of this enzyme in the literature.

thiamin   thiamin phosphate

The term "thiamin kinase" has also been used indiscriminantly in reference to thiamin pyrophosphokinase [EC 2.7.6.2].

[1] A. Iwashima, H. Nishino & Y. Nose (1972) *BBA* **258**, 333.

THIAMIN OXIDASE

This FAD-dependent oxidoreductase [EC 1.1.3.23], also known as thiamin dehydrogenase, catalyzes the reaction of thiamin with two molecules of dioxygen to produce oxidized thiamin and two molecules of hydrogen peroxide.

thiamin   thiamin acetate

The two-step oxidation proceeds without the release of the intermediate aldehyde from the enzyme when thiamin concentrations are at or below saturating conditions.[1]

[1] C. Gomez-Moreno & D. E. Edmondson (1985) *ABB* **239**, 46.

Selected entries from *Methods in Enzymology* [vol, page(s)]:
Properties: flavin linkage, **53**, 450; flavin peptide (amino acid analysis, **53**, 457; fluorescence, **53**, 462; interactions, **53**, 465)

THIAMIN-PHOSPHATE KINASE

This phosphotransferase [EC 2.7.4.16], also known as thiamin-monophosphate kinase, catalyzes the reaction of ATP with thiamin phosphate to produce ADP and thiamin diphosphate (or, thiamin pyrophosphate).[1]

[1] H. Nishino (1972) *J. Biochem.* **72**, 1093.

THIAMIN-PHOSPHATE PYROPHOSPHORYLASE

This enzyme [EC 2.5.1.3] (also known as thiamin-phosphate diphosphorylase and thiamin-phosphate synthase; the alternative names TMP pyrophosphorylase and TMP diphosphorylase should be avoided) catalyzes the reaction of 2-methyl-4-amino-5-hydroxymethylpyrimidine diphosphate (or, 2-methyl-4-amino-5-hydroxymethyl-pyrimidine pyrophosphate) with 4-methyl-5-(2-phosphonooxyethyl)thiazole to produce diphosphate (or, pyrophosphate) and thiamin monophosphate. The *Saccharomyces cerevisiae* enzyme also has a hydroxyethylthiazole kinase activity.[1]

[1] Y. Kawasaki (1993) *J. Bact.* **175**, 5153.

Selected entries from *Methods in Enzymology* [vol, page(s)]:
General discussion: 18A, 207; 62, 69
Assay: 18A, 207, 208; 62, 69, 70
Other molecular properties: production, 22, 89; properties, 18A, 211, 212; 62, 72, 73; purification 62, 70 (yeast, from, 18A, 209)

THIAMIN PYRIDINYLASE

This enzyme [EC 2.5.1.2], also known as pyrimidine transferase and thiaminase I, catalyzes the reaction of thiamin with pyridine to produce heteropyrithiamin and 4-methyl-5-(2-hydroxyethyl)thiazole.[1]

heteropyrithiamin   4-methyl-5-(2-hydroxy-ethyl)thiazole

Other bases acting in place of pyridine include aniline, imidazole, 2-aminopyrimidine, 3-aminopyrimidine, 4-aminopyrimidine, ethanolamine, quinoline, adenine, proline, *o*-toluidine, and 2-mercaptobenzoic acid.

[1] N. Campobasso, C. A. Costello, C. Kinsland, T. P. Begley & S. E. Ealick (1998) *Biochemistry* **37**, 15981.

Selected entries from *Methods in Enzymology* [vol, page(s)]:
General discussion: 18A, 229
Assay: 18A, 229

THIAMIN PYROPHOSPHATASE

This enzyme, also known as thiamin diphosphatase, catalyzes the hydrolysis of thiamin pyrophosphate to produce thiamin monophosphate and orthophosphate.[1,2] Other enzymes catalyzing this hydrolytic activity include nucleoside-diphosphatase [EC 3.6.1.6], nucleoside-triphosphatase [EC 3.6.1.15], apyrase [EC 3.6.1.5], and nucleotide pyrophosphatase [EC 3.6.1.9].

[1] J. R. Cooper & M. M. Kini (1972) *J. Neurochem.* **19**, 1809.
[2] K. T. Shetty & Veeranna (1991) *Neurochem. Internat.* **19**, 33.

Selected entries from *Methods in Enzymology* [vol, page(s)]:
General discussion: 18A, 227
Assay: 18A, 228; 22, 138, 141

THIAMIN PYROPHOSPHOKINASE

This transferase [EC 2.7.6.2], also known as thiamin diphosphokinase (the term thiamin kinase should be avoided as that best describes EC 2.7.1.89), catalyzes the reaction of ATP with thiamin to produce AMP and thiamin diphosphate (or, thiamin pyrophosphate).[1-3] The pig brain enzyme reportedly has a partial ping pong Bi Bi kinetic mechanism.[1]

[1]J. W. Peterson, C. J. Gubler & S. A. Kuby (1975) *BBA* **397**, 377.
[2]A. I. Voskoboyev, I. M. Artsukevich & Y. M. Ostrovsky (1983) *Acta Vitaminol. Enzymol.* **5**, 105.
[3]R. L. Switzer (1974) *The Enzymes*, 3rd ed., **10**, 607.

Selected entries from *Methods in Enzymology* [vol, page(s)]:
General discussion: 2, 636; **18A**, 219, 221; **62**, 103, 105, 107
Assay: 2, 636; **18A**, 219, 221, 224; **62**, 101, 104; membrane-bound, assay, **62**, 118
Other molecular properties: affinity chromatography, **62**, 105; membrane-bound (assay, **62**, 118; source, **62**, 119); occurrence, **18A**, 225; preparation, from plants, **62**, 108; properties, 2, 640; **18A**, 221; **62**, 111; purification from (rat liver, **18A**, 220; rat brain, **62**, 106, 107; yeast, **2**, 638)

THIAMIN-TRIPHOSPHATASE

This enzyme [EC 3.6.1.28] catalyzes the hydrolysis of thiamin triphosphate to produce thiamin diphosphate and orthophosphate.[1,2]

[1]A. F. Makarchikov & I. P. Chernikevich (1992) *BBA* **1117**, 326.
[2]L. Bettendorff, C. Grandfils, P. Wins & E. Schoffeniels (1989) *J. Neurochem.* **53**, 738.

THIAZOLIDINECARBOXYLATE DEHYDROGENASE

This enzyme catalyzes the reaction of thiazolidinecarboxylate with an acceptor substrate to produce the reduced acceptor and *N*-formylcysteine ($^-$OOCCH(NHCHO)CH$_2$SH).

thiazolidinecarboxylate

Proline dehydrogenase also catalyzes this activity.

Selected entries from *Methods in Enzymology* [vol, page(s)]:
General discussion: 5, 736
Assay: 5, 736

THIMET OLIGOPEPTIDASE

This zinc-dependent oligopeptidase [EC 3.4.24.15] (also known as soluble metallo-endopeptidase, endo-oligopeptidase A, endopeptidase 24.15, and Pz-peptidase) catalyzes peptide-bond hydrolysis in oligopeptides with preferential cleavage of bonds with hydrophobic

aminoacyl residues (other than isoleucyl) at positions P1 and P3′ and a prolyl residue at P2′. Substrates are oligopeptides of six to seventeen residues. Bradykinin is hydrolyzed at Phe5–Ser6 and neurotensin is acted on at Arg8–Arg9. Thiol compounds activate at low concentrations (in fact, the term thimet is an abbreviation of <u>thi</u>o-sensitive <u>met</u>allopeptidase).[1]

[1]A. J. Barrett, M. A. Brown, P. M. Dando, C. G. Knight, N. McKie, N. D. Rawlings & A. Scrizawa (1995) *MIE* **248**, 529.

Selected entries from *Methods in Enzymology* [vol, page(s)]:
General discussion: 248, 529
Assay: 248, 532 (Bz-Gly-Ala-Ala-Phe-*p*-aminobenzoate, with, **248**, 533; chromogenic, **248**, 602; quenched fluorescence method, with QF01 and QF02, **248**, 534; specificity, **248**, 536)
Other molecular properties: activity, **248**, 545, 595, 606; distribution, **248**, 551; human erythrocyte, purification, **248**, 536; inhibitors, **248**, 548; pH optimum, **248**, 545; physiological functions, **248**, 554; properties, **248**, 206, 226, 529, 530; purification, **248**, 536, 541; rat, homologues, **248**, 591; rat, structurally related enzymes, **248**, 576; rat testis (purification, **248**, 536; recombinant, purification, **248**, 536, 538); reactivation by metal ions, **248**, 546; substrate, fluorigenic, **248**, 34; substrate specificity, **248**, 546; thiol activation, **248**, 546

THIOCYANATE HYDROLASE

This hydrolase [EC 3.5.5.8] catalyzes the reaction of thiocyanate (SCN$^-$) with two water molecules to produce carbonyl sulfide (COS), ammonia, and hydroxide ion.[1,2]

[1]Y. Katayama, Y. Narahara, Y. Inoue, F. Amano, T. Kanagawa & H. Kuraishi (1992) *JBC* **267**, 9170.
[2]Y. Katayama, Y. Matsushita, M. Kaneko, M. Kondo, T. Mizuno & H. Nyunoya (1998) *J. Bact.* **180**, 2583.

Selected entries from *Methods in Enzymology* [vol, page(s)]:
General discussion and assay: 243, 506

THIOCYANATE ISOMERASE

This isomerase [EC 5.99.1.1], also known as isothiocyanate isomerase, catalyzes the interconversion of benzyl isothiocyanate and benzyl thiocyanate.[1]

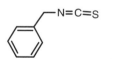

benzyl isothiocyanate   benzyl thiocyanate

[1]A. I. Virtanen (1962) *ABB* Suppl.1, 200.

THIOETHANOLAMINE S-ACETYLTRANSFERASE

This acetyltransferase [EC 2.3.1.11], also known as thioltransacetylase B, catalyzes the reaction of acetyl-CoA with thioethanolamine (or, cysteamine) to produce Coenzyme A and *S*-acetylthioethanolamine

$(CH_3COSCH_2CH_2NH_2)$.[1] 2-Mercaptoethanol also serves as a substrate, producing S-acetyl-2-hydroxy-1-thioethane.

[1] R. O. Brady & E. R. Stadtman (1954) JBC **211**, 621.

THIOETHER S-METHYLTRANSFERASE

This methyltransferase [EC 2.1.1.96] catalyzes the reaction of S-adenosyl-L-methionine with dimethyl sulfide $(H_3C-S-CH_3)$ to produce S-adenosyl-L-homocysteine and trimethylsulfonium ion $(^+S(CH_3)_3)$.[1] Other substrates include dimethyl selenide, dimethyl telluride, diethyl sulfide, 1,4-dithiane, and 2-methylthioethylamine; however, ethionine is not a substrate.

[1] N. M. Mozier, K. P. McConnell & J. L. Hoffman (1988) JBC **263**, 4527.

THIOGLUCOSIDASE

This enzyme [EC 3.2.1.147; formerly EC 3.2.3.1], variously known as myrosinase, sinigrinase, and sinigrase, catalyzes the hydrolysis of a thioglucoside to produce a thiol and glucose.[1-4] The enzyme has a wide specificity for thioglycoside substrates, including sinigrin (or, 2-propenylglucosinolate), p-nitrophenyl-β-D-glucoside, and 2-hydroxybut-3-enylglucosinolate.[1] The reaction mechanism proceeds via a glycosyl-enzyme intermediate in which the sugar ring is bound via an α-glycosidic linkage to a glutamyl residue, the catalytic nucleophile of the enzyme.[2] When glucosinolates such as sinigrin are used as substrates, one of the products (the other product is D-glucose) undergoes a thio-Lossen rearrangement to produce an alkyl isothiocyanate.

[1] M. Shikita, J. W. Fahey, T. R. Golden, W. D. Holtzclaw & P. Talalay (1999) BJ **341**, 725.
[2] W. P. Burmeister, S. Cottaz, H. Driguez, R. Iori, S. Palmieri & B. Henrissat (1997) Structure **5**, 663.
[3] G. Davies, M. L. Sinnott & S. G. Withers (1998) CBC **1**, 119.
[4] M. W. Bauer, S. B. Halio & R. M. Kelly (1996) Adv. Protein Chem. **48**, 271.

THIOHYDROXIMATE β-D-GLUCOSYLTRANSFERASE

This transferase [EC 2.4.1.195] catalyzes the reaction of UDP-D-glucose with phenylthioacetohydroximate to produce UDP and desulfoglucotropeolin. Alternative substrates include propiothiohydroximate, butyrothiohydroximate, isobutyrothiohydroximate, benzothiohydroximate, and 4-methylthiobutyrothiohydroximate.[1-3]

[1] J. C. Jain, D. W. Reed, J. W. Groot Wassink & E. W. Underhill (1989) Anal. Biochem. **178**, 137.
[2] J. C. Jain, M. R. Michayluk, J. W. D. Groot Wassink & E. W. Underhill (1989) Plant Sci. **64**, 25.

[3] J. C. Jain, J. W. D. Groot Wassink, D. W. Reed & E. W. Underhill (1990) J. Plant Physiol. **136**, 356.

THIOL S-METHYLTRANSFERASE

This methyltransferase [EC 2.1.1.9] catalyzes the reaction of S-adenosyl-L-methionine with a thiol to produce S-adenosyl-L-homocysteine and a thioether.[1] The methyl acceptor substrate can be a wide variety of alkyl, aryl (for example, 6-thiopurine and thiophenol), and heterocyclic thiols. Hydrogen sulfide and such hydroxy thiols as 2-mercaptoethanol are effective substrates. **See also** *Thiopurine S-Methyltransferase*

[1] R. A. Weisiger & W. B. Jakoby (1979) ABB **196**, 631.

Selected entries from **Methods in Enzymology** [vol, page(s)]:
General discussion: 77, 257
Assay: 77, 257
Other molecular properties: distribution, **77**, 262; inhibitors, **77**, 262; intestinal epithelium, in, **77**, 156; kinetic constants, **77**, 261; mercapturic acid formation, **77**, 262; pH optima, **77**, 260, 261; properties, **77**, 260; purification, **77**, 258; rat liver microsomes, from, **71**, 583; specificity, **77**, 261; stability, **77**, 258; substrates, **77**, 261

THIOL OXIDASE

This oxidoreductase [EC 1.8.3.2], also known as sulfhydryl oxidase, catalyzes the reaction of dioxygen with four molecules of a thiol (that is, $R'C(R)SH$) to produce two water molecules and two disulfides (that is, $R'C(R)S-S(R)CR'$). In addition to a variety of other substituents, the R group may even be =S or =O. The enzyme displays little or no specificity for R'. Substrates include cysteamine, 5,5'-dithiobis(2-nitrobenzoate), ergothioneine, glycylglycyl-L-cysteine, D-penicillamine, and thiohistidine.

Selected entries from **Methods in Enzymology** [vol, page(s)]:
General discussion: 44, 525; 143, 504, 510
Assay: 143, 119, 509, 511
Other molecular properties: enzyme reactors, characteristics, **136**, 427; immobilization, **136**, 425; properties, **143**, 514; purification from (bovine milk, **136**, 423; **143**, 508; rat skin, **143**, 512)

THIOLSUBTILISIN

This chemically altered subtilisin containing a thiol in place of the active-site serine hydroxyl group catalyzes the hydrolysis of p-nitrophenyl esters and imidazole amides, but is inactive with many other ester and amide substrates. **See also** *Subtilisin*

Selected entries from **Methods in Enzymology** [vol, page(s)]:
General discussion: 19, 215
Assay: 19, 216
Other molecular properties: activity, **19**, 214; affinity labeling, **87**, 491; inhibitors, **19**, 220; kinetic properties, **19**, 219; mechanism of action, **19**, 220; purification, **19**, 218; pyridyl disulfides, and, **87**, 461; specificity, **19**, 219; stability, **19**, 220

THIOL SULFOTRANFERASE

This transferase [EC 2.8.2.16] catalyzes the reaction of 3'-phosphoadenylylsulfate with a dithiol to produce adenosine 3',5'-bisphosphate and a dithiol sulfate.[1,2] Other substrates include dithioerythritol, glutathione, and 2,3-mercaptopropanol.

[1] A. Schmidt & U. Christen (1979) Z. Naturforsch. **34c**, 222.
[2] A. Schmidt & U. Christen (1978) Planta **140**, 239.

THIOMORPHOLINE-CARBOXYLATE DEHYDROGENASE

This oxidoreductase [EC 1.5.1.25], also known as ketimine reductase, catalyzes the reaction of thiomorpholine-3-carboxylate with NAD(P)$^+$ to produce 3,4-dehydro-1,4-thiomorpholine-3-carboxylate and NAD(P)H.[1,2] The product is the cyclic imine of the α-ketoacid corresponding to S-(2-aminoethyl)cysteine.

thiomorpholine   3,4-dehydrothiomorpholine
3-carboxylate   3-carboxylate

The kinetic mechanism of this dehydrogenase is ping pong Bi Bi. Substrates in the reverse reaction include S-aminoethylcysteine ketimine, lanthionine ketimine (producing 1,4-thiomorpholine 3,5-dicarboxylate), and cystathionine ketimine (producing cyclothionine).

[1] M. Nardini, G. Ricci, A. M. Caccuri, S. P. Solinas, L. Vesci & D. Cavallini (1988) EJB **173**, 689.
[2] M. Nardini, G. Ricci, L. Vesci, L. Pecci & D. Cavallini (1988) BBA **957**, 286.

THIOPHENE-2-CARBONYL-CoA MONOOXYGENASE

This molybdenum-dependent oxidoreductase [EC 1.14.99.35], also known as thiophene-2-carboxyl-CoA dehydrogenase and thiophene-2-carboxyl-CoA hydroxylase, catalyzes the reaction of thiophene-2-carbonyl-CoA with a reduced substrate and dioxygen to produce 5-hydroxythiophene-2-carbonyl-CoA, the oxidized substrate, and water.[1] Tetrazolium salts can act as cosubstrates.

[1] A. Bambauer, F. A. Rainey, E. Stackebrandt & J. Winter (1998) Arch. Microbiol. **169**, 293.

THIOPURINE S-METHYLTRANSFERASE

This methyltransferase [EC 2.1.1.67] catalyzes the reaction of S-adenosyl-L-methionine with a thiopurine to produce S-adenosyl-L-homocysteine and a thiopurine S-methylether.[1–3] The enzyme, which is distinct from thiol S-methyltransferase [EC 2.1.1.9], also acts more slowly on thiopyrimidines and aromatic thiols. Specific substrates include 2-amino-6-thiopurine, 2-bromothiophenol, 2,8-dihydroxy-6-thiopurine, 6-thioguanine, 6-thiopurine, 2-thiothymine, 2-thiouracil, and 6-thiourate. **See also Thiol S-Methyltransferase**

[1] C. N. Remy (1963) JBC **238**, 1078.
[2] G. Loo & J. T. Smith (1985) BBRC **126**, 1201.
[3] M. M. Ames, C. D. Selassie, L. C. Woodson, J. A. Van Loon, C. Hansch & R. M. Weinshilboum (1986) J. Med. Chem. **29**, 354.

THIOREDOXIN REDUCTASE (FERREDOXIN)

This [4Fe–4S] oxidoreductase catalyzes the reaction of reduced [2Fe–2S] ferredoxin with oxidized thioredoxin (that is, the disulfide form) to produce oxidized ferredoxin and reduced thioredoxin (that is, the dithiol form).[1,2] The catalytic mechanism involves novel sulfur-based cluster chemistry with a stable one electron intermediate being generated over the course of the reaction. The electron transfer is facilitated to the active-site disulfide resulting in covalent attachment of the electron-transfer cysteinyl residue and thiol-disulfide interchange with thioredoxin.[2]

[1] C. R. Staples, E. Ameyibor, E. Fu, L. Gardet-Salvi, A. L. Stritt-Etter, P. Schurmann, D. B. Knaff & M. K. Johnson (1996) Biochemistry **35**, 11425.
[2] C. R. Staples, E. Gaymard, A. L. Stritt-Etter, J. Telser, B. M. Hoffman, P. Schurmann, D. B. Knaff & M. K. Johnson (1998) Biochemistry **37**, 4612.

Selected entries from **Methods in Enzymology [vol, page(s)]:**
Assay: **167**, 423; **252**, 280; method, **69**, 389 (procedure, **69**, 389, 390; reagents, **69**, 389)
Other molecular properties: absorption spectra, **252**, 282; active site thiols, **252**, 275; distribution, **69**, 391; function, **69**, 382; properties, **167**, 427; purification, **69**, 390, 391 (Nostoc muscorum, from, **167**, 426; spinach, **252**, 281); stability, **69**, 391

THIOREDOXIN REDUCTASE (NADPH)

This FAD-dependent oxidoreductase [EC 1.6.4.5] catalyzes the reaction of NADPH with oxidized thioredoxin (disulfide form) to produce NADP$^+$ and reduced thioredoxin (dithiol form).[1] Mammalian cytosolic thioredoxin reductase has a redox center consisting of two cysteinyl residues adjacent to the flavin ring of FAD and another cysteinyl residue and a selenocysteinyl residue near the C terminus.[2] The selenocysteinyl residue plays a critical redox function during the catalytic cycle.[2,3]

[1] C. H. Williams (1976) The Enzymes, 3rd ed., **13**, 89.
[2] S. R. Lee, S. Bar-Noy, J. Kwon, R. L. Levine, T. C. Stadtman & S. G. Rhee (2000) PNAS **97**, 2521.
[3] S. N. Gorlatov & T. C. Stadtman (2000) Biofactors **11**, 79.

THIOSULFATE DEHYDROGENASE

This iron-dependent oxidoreductase [EC 1.8.2.2], also
known as tetrathionate synthase and thiosulfate oxidase,
catalyzes the reaction of two molecules of thiosulfate
($S-SO_3^{2-}$) with two molecules of ferricytochrome c to
produce tetrathionate ($S_4O_6^{2-}$) and two molecules of
ferrocytochrome c.[1–3]

[1] I. Suzuki (1974) *Ann. Rev. Microbiol.* 28, 85.
[2] L. B. Schook & R. S. Berk (1979) *J. Bact.* 140, 306.
[3] P. A. Trudinger (1967) *J. Bact.* 93, 550.

THIOSULFATE-FORMING ENZYME

This enzyme reportedly catalyzes the reaction of bisulfite
(HSO_3^-) with trithionate ($S_3O_6^{2-}$) to produce thiosulfate
($S-SO_3^{2-}$).[1] *See also Hydrogensulfite Reductase*

[1] H. L. Drake & J. M. Akagi (1977) *J. Bact.* 132, 132.

THIOSULFATE REDUCTASE

Two enzymes catalyze the reduction of thiosulfate, one
that is glutathione-dependent and the other glutathione-
independent.

Glutathione-dependent thiosulfate reductase [EC 2.8.1.3],
officially known as thiosulfate:thiol sulfurtransferase, catalyzes the reaction of thiosulfate ($S-SO_3^{2-}$) with two
molecules of glutathione to produce sulfite, glutathione
disulfide, and sulfide. The initial product is glutathione
hydrodisulfide, which reacts with glutathione to generate
glutathione disulfide and sulfide. L-Cysteine can also act
as acceptor substrate, thereby producing cystine. The yeast
enzyme exhibits a rapid equilibrium ordered Bi Bi kinetic
mechanism.[1]

This second thiosulfate reductase [EC 2.8.1.5], officially
known as thiosulfate:dithiol sulfurtransferase, catalyzes
the reaction of thiosulfate with dithioerythritol to produce sulfite, 4,5-*cis*-dihydroxy-1,2-dithiacyclohexane, and
sulfide.[2]

[1] T. R. Chauncey & J. Westley (1983) *JBC* 258, 15037.
[2] A. Schmidt, I. Erdle & B. Gamon (1984) *Planta* 162, 243.

L-THREONATE DEHYDROGENASE

This oxidoreductase [EC 1.1.1.129], also known as
L-threonate 3-dehydrogenase, catalyzes the reaction of
L-threonate with NAD^+ to produce 3-dehydro-L-threonate
and NADH.[1]

L-threonate   3-dehydro-L-threonate

The lactone form of threonate is not a substrate.

[1] A. J. Aspen & W. B. Jakoby (1964) *JBC* 239, 710.

D-THREONINE ALDOLASE

This pyridoxal-phosphate-dependent enzyme [EC 4.1.2.5]
catalyzes the reversible conversion of D-threonine to
glycine and acetaldehyde.[1] Other substrates include
D-allothreonine.

[1] M. Kataoka, M. Ikemi, T. Morikawa, T. Miyoshi, K. Nishi, M. Wada,
H. Yamada & S. Shimizu (1997) *EJB* 248, 385.

L-THREONINE ALDOLASE

This pyridoxal-phosphate-dependent enzyme [EC 4.1.2.5] catalyzes the reversible conversion of L-threonine to glycine and acetaldehyde. Other substrates include L-allothreonine and L-serine. Some confusion arises when serine hydroxymethyltransferase [EC 2.1.2.1] is referred to as L-threonine aldolase.

Selected entries from *Methods in Enzymology* [vol, page(s)]:
General discussion: 5, 931
Assay: 5, 931

THREONINE DEHYDRATASE

This pyridoxal-phosphate-dependent enzyme [EC 4.2.1.16], also known as threonine deaminase and L-serine dehydratase, catalyzes the reaction of L-threonine with water to produce α-ketobutyrate, ammonia, and water.[1–3]

L-threonine   α-ketobutyrate

The enzyme will also use L-serine as a substrate (thus, the activity of serine dehydratase).

[1] Y. Shizuta & O. Hayaishi (1976) *Curr. Top. Cell. Regul.* **11**, 99.
[2] H. E. Umbarger (1973) *AE* **37**, 349.
[3] L. Davis & D. E. Metzler (1972) *The Enzymes*, 3rd ed., **7**, 33.

Selected entries from *Methods in Enzymology* [vol, page(s)]:
General discussion: 2, 319; 5, 946; 17B, 555; 34, 5; 46, 64
Assay: 2, 319; 5, 947; 17B, 555, 561, 566, 571, 576
Other molecular discussion: hysteresis, **64**, 219; purification from (*Bacillus subtilis*, **17B**, 562; *Clostridium tetanomorphum*, **17B**, 572 *Escherichia coli*, **17B**, 577; *Rhodopseudomonas spheroides*, **17B**, 567; *Salmonella typhimurium*, **17B**, 557; sheep liver, **5**, 947); serine dehydratase and, **17B**, 350; staining of active enzyme on gels, **22**, 601

L-THREONINE 3-DEHYDROGENASE

This oxidoreductase [EC 1.1.1.103] catalyzes the reaction of L-threonine with NAD^+ to produce NADH and L-2-amino-3-ketobutanoate which spontaneously decarboxylates to aminoacetone ($^+H_3NCH_2COCH_3$).[1–4]

L-threonine   L-2-amino-3-
                     ketobutanoate

The enzyme displays an ordered Bi Bi kinetic mechanism.[1,2] Because carbon dioxide and aminoacetone are released prior to NADH, the kinetic mechanism can be formally described as ordered Bi Ter.[3]

[1] D. Hartshorne & D. M. Greenberg (1964) *ABB* **105**, 173.
[2] Y. Aoyama & Y. Motokawa (1981) *JBC* **256**, 12367.
[3] T. Tressel, R. Thompson, L. R. Zieske, M. I. Menendez & L. Davis (1986) *JBC* **261**, 16428.
[4] E. P. Epperly & E. E. Dekker (1991) *JBC* **266**, 6086.

Selected entries from *Methods in Enzymology* [vol, page(s)]:
General discussion: 17B, 580
Assay: 17B, 580

THREONINE RACEMASE

This enzyme [EC 5.1.1.6] catalyzes the interconversion of L-threonine and D-threonine.[1] Note that both chiral centers are reported to undergo inversion.

[1] H. Amos (1954) *JACS* **76**, 3858.

THREONINE SYNTHASE

This pyridoxal-phosphate-dependent enzyme [EC 4.2.3.1; previously classified as EC 4.2.99.2] catalyzes the reaction of *O*-phospho-L-homoserine with water to produce L-threonine and orthophosphate, where the latter is released by nonhydrolytic elimination.[1]

[1] L. Davis & D. E. Metzler (1972) *The Enzymes*, 3rd ed., **7**, 33.

Selected entries from *Methods in Enzymology* [vol, page(s)]:
General discussion: 5, 951
Assay: 5, 952

THREONYL-tRNA SYNTHETASE

This enzyme [EC 6.1.1.3], also known as threonine:tRNA ligase and threonine translase, catalyzes the reaction of ATP with L-threonine and $tRNA^{Thr}$ to produce AMP, pyrophosphate, and L-threonyl-$tRNA^{Thr}$. This class II aminoacyl-tRNA synthetase has a Bi Uni Uni Bi ping pong kinetic mechanism and is selectively inhibited by the antibiotic borrelidin.[1] ***See also*** *Aminoacyl-tRNA Synthetases*

[1] W. Freist & D. H. Gauss (1995) *Biol. Chem. Hoppe-Seyler* **376**, 213.

Selected entries from *Methods in Enzymology* [vol, page(s)]:
General discussion: 34, 170
Other molecular properties: hydroxylamine and, **29**, 613; mechanism, **29**, 629; **63**, 52; preparation, **12B**, 697; purification, **59**, 262, 263, 267; subcellular distribution, **59**, 233, 234

THROMBIN

This serine endoprotease [EC 3.4.21.5], also known as fibrinogenase and coagulation factor IIa, catalyzes peptide-bond hydrolysis and preferentially cleaves Arg–Gly bonds.[1–5] There is greater selectivity, compared to trypsin, for the P3, P2, P2′, and P3′ positions. In addition to the preference for an arginyl residue at position P1, there is also a preference for a prolyl residue at P2. Phe and Val are the best residues at P3 (interestingly, D-aminoacyl aromatic residues are even better at P3). Phe is preferred at P2′ and Arg or Lys at P3′.

Thrombin is derived from its proenzyme (or, zymogen) forms by the action of coagulation factor Xa [EC 3.4.21.6] in the presence of factor Va, calcium ions, and a phospholipid surface.

[1] R. J. Grand, A. S. Turnell & P. W. Grabham (1996) BJ 313, 353.
[2] C. W. Wharton (1998) CBC 1, 345.
[3] L. J. Berliner (1992) Thrombin, Structure & Function, Plenum Press, New York.
[4] E. W. Davie, K. Fujikawa, K. Kurachi & W. Kisiel (1979) AE 48, 277.
[5] S. Magnusson (1971) The Enzymes, 3rd ed., 3, 278.

Selected entries from *Methods in Enzymology* [vol, page(s)]:
General discussion: 2, 156; 34, 5, 445; 45, 123; 46, 128, 206, 207; 80, 365
Assay: 2, 156; 19, 145, 170; 34, 447, 448; 45, 126, 156 (clotting time, 45, 159; hirudin, 45, 677; substrates, 45, 160); 80, 297, 345 (with peptide chromogenic and fluorogenic substrates, 80, 344, 358, 359 [effects of pH and solution composition, 80, 345, 348, 349]); 267, 30
Other molecular properties: activation, measurement in vitro, 222, 522; activators and inactivators, 19, 156, 180; active-site titration, benzyl p-guanidinothiobenzoate in, 248, 14; affinity labeling, 80, 835, 838; 87, 472, 484; amidase reaction mechanism, 259, 132; chymotrypsin, compared to, 80, 294, 295; clotting mechanism and, 2, 139, 140, 152, 156; clotting time measurement, 19, 147; hirudin as inhibitor, 19, 924; 45, 166; 80, 424; 223, 317; immobilization for inhibitor-binding studies, 267, 38; kinetics, 19, 157, 181 (allosteric effectors, 259, 133, 135, 139; Michaelis-Menten parameters, 259, 132, 136, 140; steady state solution, 259, 133); physical properties of, 19, 155, 177; prothrombin assay and, 2, 141; purification, 2, 157; 19, 172; 34, 447; 45, 175 (porcine, 45, 166); 215, 168; reaction with p-nitrophenyl p′-guanidinobenzoate, 19, 171; role, 45, 156; 163, 711; synthetic substrates for, 19, 157, 182; 80, 297, 298; thioester substrates, 248, 13; α-thrombin, 80, 287, 295, 303, 319

THROMBOXANE A SYNTHASE

This heme-thiolate-dependent (cytochrome P450) enzyme [EC 5.3.99.5] catalyzes the conversion of (5Z,13E)-(15S)-9α,11α-epidioxy-15-hydroxyprosta-5,13-dienoate (prostaglandin H_2) to (5Z,13E)-(15S)-9α,11α-epoxy-15-hydroxythromba-5,13-dienoate (thromboxane A_2).[1,2]

prostaglandin H_2

thromboxane A_2

[1] H. B. Dunford (1998) CBC 3, 195.
[2] C. R. Pace-Asciak & W. L. Smith (1983) The Enzymes, 3rd ed., 16, 543.

Selected entries from *Methods in Enzymology* [vol, page(s)]:
General discussion: 86, 106
Assay: bovine platelet, 86, 107; human platelet, 86, 653
Other molecular properties: bovine platelet (properties, 86, 109; purification, 86, 108); inhibitor, 63, 401, 404 (assay methods, 86, 110)

THYLAKOIDAL PROCESSING PEPTIDASE

This serine endopeptidase, often abbreviated TPP, catalyzes the processing of proteins targeted to the thylakoid lumen. TPP generates the mature proteins. Plastocyanin and subunits of photosystem II have been identified as proteins processed by this peptidase.[1,2]

[1] C. Robinson (1994) in Signal Peptidases (G. von Heijne, ed.), R. G. Landes, Austin, Texas, p. 113.
[2] J. B. Shackleton & C. Robinson (1991) JBC 266, 12152.

THYMIDINE KINASE

This phosphotransferase [EC 2.7.1.21] catalyzes the reaction of ATP with thymidine to produce ADP and thymidine 5′-phosphate.[1] Other acceptor substrates include deoxyuridine. The deoxypyrimidine kinase complex induced by herpes simplex virus also catalyzes this reaction, as do AMP:thymidine kinase [EC 2.7.1.114], ADP:thymidine kinase [EC 2.7.1.118], and thymidylate kinase [EC 2.7.4.9].

[1] E. P. Anderson (1973) The Enzymes, 3rd ed., 9, 49.

Selected entries from *Methods in Enzymology* [vol, page(s)]:
General discussion: 34, 5; 46, 89; 51, 354, 360, 365
Assay: 12B, 173; 51, 354, 365
Other molecular properties: cation requirement, 51, 358, 370; inhibitor, 51, 371; kinetic properties, 51, 359, 360, 370; labeled nucleotide preparation, 64, 128; marker for gene targeting experiments, 237, 358, 366; molecular weight, 51, 359, 369, 370; purification, 51, 356, 362; reaction mechanism, 51, 371; regulation, 51, 354; selectable marker for eukaryotic expression system, as, 245, 303; stability, 51, 358, 369; substrate specificity, 51, 358, 370; transition-state and multisubstrate analogues, 249, 305

THYMIDINE PHOSPHORYLASE

This enzyme [EC 2.4.2.4], also known as pyrimidine phosphorylase, catalyzes the reaction of thymidine with orthophosphate to produce thymine and 2-deoxy-D-ribose 1-phosphate.[1–3] In some tissues, this enzyme also catalyzes deoxyribosyltransferase reactions akin to nucleoside deoxyribosyltransferase [EC 2.4.2.6]. The *Salmonella typhimurium* enzyme has an ordered Bi Bi kinetic mechanism with the nucleoside binding first,[1] whereas the *Escherichia coli* enzyme is either ordered Bi Bi with orthophosphate binding first[2] or random Bi Bi with a dead-end ternary complex.[3] *See also Pyrimidine Nucleoside Phosphorylase*

[1]J. G. Blank & P. A. Hoffee (1975) *ABB* **168**, 259.
[2]M. Schwartz (1978) *MIE* **51**, 442.
[3]M. H. Iltzsch, M. H. el Kouni & S. Cha (1985) *Biochemistry* **24**, 6799.

Selected entries from **Methods in Enzymology** [vol, page(s)]:
General discussion: 3, 185; **12A**, 119; **51**, 437, 442
Assay: arsenolysis of thymidine, **12A**, 119; thyminolysis of deoxyribose 1-phosphate, **12A**, 119; unit, **12A**, 120; **51**, 438, 442, 443
Other molecular properties: activity, **51**, 437, 438, 442; *Escherichia coli*, from, **12A**, 120; **51**, 442; *Haemophilus influenzae*, from, **51**, 432; kinetic properties, **51**, 441, 445; liver, **2**, 453; **3**, 185; pH optimum, **51**, 442, 445; *Salmonella typhimurium*, from, **51**, 437; stability, **51**, 442, 445; substrate specificity, **12A**, 121; **51**, 441, 444; transferase activity, **51**, 431

THYMIDINE-TRIPHOSPHATASE

This enzyme [EC 3.6.1.39] catalyzes the hydrolysis of dTTP to produce dTDP and orthophosphate.[1–3] Weaker substrates include dUTP and UTP. Bacteriophage T7 DNA helicase also exhibits a thymidine-triphosphatase activity. This ring-shaped hexameric protein catalyzes duplex DNA unwinding by using the Gibbs free energy of dTTP hydrolysis.[4]

[1]D. H. Roscoe (1969) *Virology* **38**, 520.
[2]N. Dahlmann (1984) *Hoppe-Seyler's Z. Physiol. Chem.* **365**, 1263.
[3]N. Dahlmann (1982) *Biochemistry* **21**, 6634.
[4]M. M. Hingorani, M. T. Washington, K. C. Moore & S. S. Patel (1997) *PNAS* **94**, 5012.

THYMIDYLATE 5′-PHOSPHATASE

This enzyme [EC 3.1.3.35], also known as thymidylate 5′-nucleotidase, catalyzes the hydrolysis of thymidylate to produce thymidine and orthophosphate.[1,2] Other alternative substrates include 5-methyl-dCMP, dGMP, and dUMP.

[1]H. Vasken Aposhian & G. Y. Tremblay (1966) *JBC* **241**, 5095.
[2]A. R. Price & S. M. Fogt (1973) *JBC* **248**, 1372.

Selected entries from **Methods in Enzymology** [vol, page(s)]:
General discussion: **51**, 285
Other molecular properties: activity, **51**, 285, 290; *Bacillus subtilis*, from, **51**, 285; inhibitors, **51**, 289; kinetic properties, **51**, 289; molecular weight, **51**, 290; pH optimum, **51**, 289; purification, **51**, 286; stability, **51**, 290; substrate specificity, **51**, 285, 288, 289; temperature optimum, **51**, 289

THYMIDYLATE SYNTHASE

Thymidylate synthase [EC 2.1.1.45] reductively methylates 2′-deoxyuridine 5′-monophosphate to form 2′-deoxythymidine 5′-monophosphate in the following folate-dependent reaction: dUMP reacts with N^5,N^{10}-methylenetetrahydrofolate to produce dTMP and dihydrofolate.[1–5] 5-Fluorouracil is a potent inhibitor. **See also** *Dihydrofolate Reductase-Thymidylate Synthase*

[1]C. W. Carreras & D. V. Santi (1995) *ARB* **64**, 721.
[2]D. L. Birdsall, J. Finer-Moore & R. M. Stroud (1996) *JMB* **255**, 522.

[3]V. Schirch (1998) *CBC* **1**, 211.
[4]M. A. Ator & P. R. Ortez de Montellano (1990) *The Enzymes*, 3rd ed., **19**, 213.
[5]J. I. Rader & F. M. Huennekens (1973) *The Enzymes*, 3rd ed., **9**, 197.

Selected entries from **Methods in Enzymology** [vol, page(s)]:
General discussion: 6, 124; **51**, 90; **64**, 125
Assay: 6, 124; 34, 520; 46, 164, 307; **51**, 90, 101; **66**, 721
Other molecular properties: affinity labeling, **87**, 474; dehalogenation activity, **64**, 132; 5-fluoro-2′-deoxyuridylate 5,10-methyleneTH$_4$ complex with, **46**, 309; induced substrate inhibition, **63**, 508, 509; induction in microorganisms, **22**, 92; inhibitors, **36**, 576; **51**, 96; **63**, 398, 508; intermediate, **64**, 125, 132; **87**, 84; kinetic properties, **51**, 96; *Lactobacillus casei*, from, **51**, 90 (site-directed mutants, acid-base catalysis, **249**, 114); mechanism, **64**, 126, 132, 133; pH optimum, **51**, 96; preparation, **64**, 128; purification, **51**, 92, 98 (calf thymus, **66**, 716; *Escherichia coli*, 6, 129; **66**, 715); quantitation, **34**, 522, 523; spectral properties, **51**, 96; stability, **51**, 96; stabilization by thiols, **51**, 96, 97; stereochemistry, **87**, 152; subunit structure, **51**, 96; ternary complex formation, **51**, 97; tight-binding ligand, **63**, 462

THYMINE DIOXYGENASE

This iron- and ascorbate-dependent oxidoreductase [EC 1.14.11.6], also known as thymine:2-oxoglutarate dioxygenase and thymine 7-hydroxylase, catalyzes the reaction of thymine with α-ketoglutarate and dioxygen to produce 5-hydroxymethyluracil, succinate, and carbon dioxide. The enzyme also acts on the –CH$_2$OH group of 5-hydroxymethyluracil, first forming the aldehyde 5-formyluracil and then oxidizing this intermediate to the carboxyl product 5-carboxyuracil. Each of these steps still requires dioxygen and α-ketoglutarate. Kinetic isotope effect investigations demonstrated that α-ketoglutarate is irreversibly altered before the bond-breaking occurs in thymine.[1–4] An ordered Ter Ter kinetic mechanism has been reported for the *Neurospora crassa* enzyme; however the enzyme from *Rhodotorula glutinis* has a different kinetic mechanism.[2]

[1]E. Holme (1982) *BBA* **707**, 259.
[2]B. G. Fox (1998) *CBC* **3**, 261.
[3]S. England & S. Seifter (1986) *Ann. Rev. Nutr.* **6**, 365.
[4]O. Hayaishi, M. Nozaki & M. T. Abbott (1975) *The Enzymes*, 3rd ed., **12**, 119.

Selected entries from **Methods in Enzymology** [vol, page(s)]:
General discussion: **12A**, 47; **52**, 12
Assay: **12A**, 48

THYROID HORMONE AMINOTRANSFERASE

3,5,3′-triiodothyronine   3,5,3′-triiodothyropyruvate

This pyridoxal-phosphate-dependent enzyme [EC 2.6.1.26], also known as 3,5-dinitrotyrosine aminotransferase, catalyzes the reversible reaction of L-3,5,3'-triiodothyro-nine with α-ketoglutarate to produce 3,5,3'-triiodothyro-pyruvate and L-glutamate.[1,2]

Other substrates include monoiodotyrosine, diiodotyro-sine, triiodothyronine, thyroxine, and dinitrotyrosine (note that diiodotyrosine aminotransferase [EC 2.6.1.24] does not act on dinitrotyrosine). **See also** *Diiodotyrosine Amino-transferase*

[1] R. L. Soffer, P. Hechtman & M. Savage (1973) JBC **248**, 1224.
[2] A. E. Braunstein (1973) The Enzymes, 3rd ed., **9**, 379.
Selected entries from **Methods in Enzymology** [vol, page(s)]:
General discussion: 17A, 660

THYROID PEROXIDASE

This important variant of iodoperoxidase catalyzes the iodination of tyrosyl residues in thyroglobulin.[1-3] **See** *Peroxidase; Iodide Peroxidase*

[1] R. P. Magnussen (1991) in Peroxidases in Chemistry and Biology (J. Everse, K. E. Everse & M. B. Grisham, eds.) I, p. 199, CRC Press, Boca Raton.
[2] D. R. Doerge, A. Taurog & M. L. Dorris (1994) ABB **315**, 90.
[3] A. Taurog, M. L. Dorris & D. R. Doerge (1994) ABB **315**, 82.

THYROXINE AMINOTRANSFERASE

This enzyme, formerly classified as EC 2.6.1.25, is now included under diiodotyrosine aminotransferase [EC 2.6.1.24].

THYROXINE DEIODINASE

This deiodinase [EC 3.8.1.4], also known as iodothy-ronine 5'-monodeiodinase and iodothyronine outer ring monodeiodinase, catalyzes the reaction of L-thyroxine with a reduced acceptor substrate AH_2 to produce 3,5,3'-L-triiodo-L-thyronine, iodide, the oxidized acceptor A, and H^+.

L-thyroxine   L-3,5,3'-triiodothyronine

This classification currently represents a group of enzymes that catalyzes the sequential removal of iodine atoms from thyroxine. Distinct enzymes may act on the so-called inner and outer rings of iodinated thyronine.

T_2-INDUCED DEOXYNUCLEOTIDE KINASE

This phosphotransferase [EC 2.7.4.12] catalyzes the reaction of ATP with dGMP (or dTMP) to produce ADP and dGDP (or dTDP).[1] Other substrates include 5'-hydroxymethyl-dCMP and dUMP. dATP can substitute for ATP.

[1] L. J. Bello & M. J. Bessman (1963) JBC **238**, 1777.

tissue PLASMINOGEN ACTIVATOR (tPA)

This serine endopeptidase [EC 3.4.21.68], also called tissue plasminogen activator, catalyzes peptide-bond hydrolysis of a single Arg561–Val562 bond in plasminogen to form plasmin. Fibrin activates this enzyme; in its absence, the K_m value of plasminogen is 65 μM, whereas in its presence, the K_m value is lowered to 0.16 μM.[1-3] **See also** *Plasminogen Activator*

[1] D. Collen (1980) Thromb. Haemost. **43**, 77.
[2] D. C. Rijken & D. Collen (1981) JBC **256**, 7035.
[3] M. Hoylaerts, D. C. Rijken, H. R. Lijnen & D. Collen (1982) JBC **257**, 2912.
Selected entries from **Methods in Enzymology** [vol, page(s)]:
General discussion: 19, 821
Assay: 19, 822; 163, 305; 168, 424; chromogenic assays, 223, 271
Other molecular properties: activation, measurement in vivo, 222, 524; activation of Glu-plasminogen, 223, 246; kinetic analysis, 223, 198; presence of fibrin-like stimulators, in, 223, 202; chromogenic assays, 223, 271; detection, 217, 55; plasminogen, molecular interactions, 223, 197; purification, 19, 829; 163, 300

*Tli*I RESTRICTION ENDONUCLEASE

This type II restriction endonuclease [EC 3.1.21.4], obtained from *Thermococcus litoralis*, recognizes and acts on both strands of DNA at 5'...CVTCGAG...3'. This enzyme is highly thermostable: $T_{opt} = 75°C$ and has less than 10% activity at 37°C.

TOCOPHEROL METHYLTRANSFERASE

This methyltransferase [EC 2.1.1.95] catalyzes the reaction of S-adenosyl-L-methionine with γ-tocopherol to produce S-adenosyl-L-homocysteine and α-tocopherol.[1] β- and δ-tocopherol are poor substrates.[1-3]

[1] A. d'Harlingue & B. Camara (1985) JBC **260**, 15200.
[2] S. Shigeoka, H. Ishiko, Y. Nakano & T. Mitsunaga (1992) BBA **1128**, 220.
[3] M. Michalowski & W. Janiszowska (1993) Acta Biochim. Pol. **40**, 116.

TOLUENE DIOXYGENASE

This oxidoreductase [EC 1.14.12.11], also known as toluene 1,2-dioxygenase, catalyzes the reaction of toluene, NADH, two protons, and dioxygen to produce cis-(1S,2R)-3-methylcyclohexa-3,5-diene-1,2-diol and NAD^+.[1-6] This is a protein complex containing an iron-sulfur, FAD-dependent reductase, an iron-sulfur oxygenase, and a Rieske-type ferredoxin.

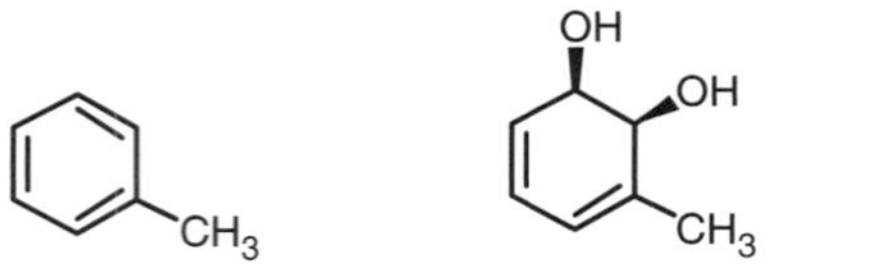

toluene

(1*S*,2*R*)-3 methylcyclohexa-3,5-diene-1,2-diol

The iron-sulfur oxygenase of the *Pseudomonas putida* F1 enzyme complex consists of two subunits, α and β. The α subunit can accept electrons from reduced ferredoxin but is catalytically inactive in the absence of the β subunit.[5]

[1]V. Subramanian, T. N. Liu, W. K. Yeh, C. M. Serdar, L. P. Wackett & D. T. Gibson (1985) *JBC* **260**, 2355.
[2]L. P. Wackett, L. D. Kwart & D. T. Gibson (1988) *Biochemistry* **27**, 1360.
[3]V. Subramanian, T. N. Liu, W. K. Yeh & D. T. Gibson (1979) *BBRC* **1**, 1131.
[4]W. K. Yeh, D. T. Gibson & T. N. Liu (1977) *BBRC* **78**, 401.
[5]H. Jiang, R. E. Parales & D. T. Gibson (1999) *Appl. Environ. Microbiol.* **65**, 315.
[6]B. G. Fox (1998) *CBC* **3**, 261.

Selected entries from *Methods in Enzymology* [vol, page(s)]:
General discussion: **188**, 39
Assay: **188**, 40
Other molecular properties: properties, **188**, 45; purification from *Pseudomonas putida* F1, **188**, 42

TOLUENE MONOOXYGENASES

At least three distinct forms of this iron-sulfur-dependent oxidoreductase exist, each specifically catalyzing the hydroxylation at the 2-, 3-, or 4-position of toluene. (The methyl group is hydroxylated by xylene monooxygenase.)

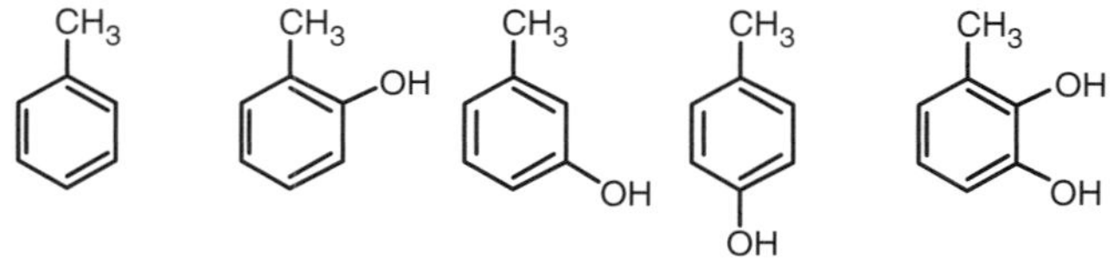

toluene   *o*-cresol   *m*-cresol   *p*-cresol   3-methylcatechol

Toluene 2-monooxygenase has an flavin-dependent iron-sulfur reductase and catalytic effector protein. Toluene 4-monooxygenase requires a Rieske-type ferredoxin and a flavin-dependent reductase. Both of these enzymes will also catalyze the oxidation of trichloroethylene. The 4-monooxygenase will also act on chloroform and 1,2-dichloroethane. In addition, an "NIH-shift" has been observed with this enzyme when utilizing [4-^{2}H]toluene as a substrate. Toluene 2-monooxygenase can catalyze two successive hydroxylation reactions, resulting in the formation of 3-methylcatechol.[1–3] ***See also*** *Xylene Monooxygenase*

[1]L. M. Newman & L. P. Wackett (1995) *Biochemistry* **34**, 14066.

[2]J. D. Pikus, J. M. Studts, C. Achim, K. E. Kauffmann, E. Munck, R. J. Steffan, K. McClay & B. G. Fox (1996) *Biochemistry* **35**, 9106.
[3]A. M. Byrne, J. J. Kukor & R. H. Olsen (1995) *Gene* **154**, 65.

TOYOCAMYCIN NITRILE HYDROLASE

This enzyme catalyzes the hydrolysis of the nitrile in toyocamycin to produce sangivamycin.

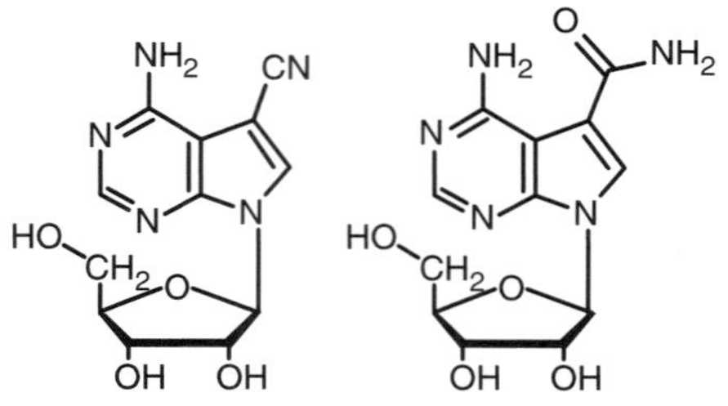

toyocamycin   sangivamycin

Selected entries from *Methods in Enzymology* [vol, page(s)]:
General discussion: **43**, 759
Assay: **43**, 759, 760
Other molecular properties: activators, **43**, 762; inhibitors, **43**, 762; metal ions effect, **43**, 762; Michaelis constant, **43**, 762; pH optimum, **43**, 762; purification, **43**, 760; specific activity, **43**, 760; specificity, **43**, 762

TRANSALDOLASE

This enzyme [EC 2.2.1.2], also known as dihydroxy-acetone transferase and glycerone transferase, catalyzes the reversible reaction of sedoheptulose 7-phosphate with D-glyceraldehyde 3-phosphate to produce D-erythrose 4-phosphate and D-fructose 6-phosphate.[1–4] Additional donor substrates that can replace sedoheptulose 7-phosphate include D-fructose 6-phosphate, D-fructose, L-sorbose 6-phosphate, 2,5-D-*threo*-diketohexose, sedoheptulose, and octulose 8-phosphate. Additional acceptor substrates include D-erythrose 4-phosphate, D-glyceraldehyde, L-glyceraldehyde 3-phosphate, hydroxypyruvic aldehyde, D-erythrose, and D-ribose 5-phosphate.

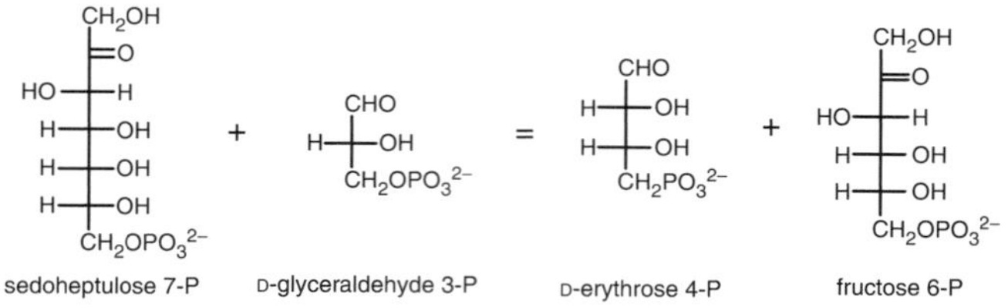

sedoheptulose 7-P   D-glyceraldehyde 3-P   D-erythrose 4-P   fructose 6-P

The enzyme displays a ping pong Bi Bi kinetics, consistent with the formation of Schiff-base intermediate (at a lysyl residue) that can be trapped by borohydride reduction.[4]

[1]K. N. Allen (1998) *CBC* **2**, 135.
[2]R. Kluger (1992) *The Enzymes*, 3rd ed., **20**, 271.
[3]O. Tsolas & B. L. Horecker (1972) *The Enzymes*, 3rd ed., **7**, 259.

[4]B. L. Horecker, S. Pontremoli, C. Ricci & T. Cheng (1961) *PNAS* **47**, 1949.

Selected entries from *Methods in Enzymology* [vol, page(s)]:
General discussion: 1, 381; **9**, 499; **46**, 50, 51; **182**, 788; **188**, 405
Assay: 1, 381; **9**, 499; **188**, 406

TRANSKETOLASE

This thiamin-pyrophosphate-dependent enzyme [EC 2.2.1.1], also known as glycolaldehyde transferase, catalyzes the reversible reaction of sedoheptulose 7-phosphate with D-glyceraldehyde 3-phosphate to produce D-ribose 5-phosphate and D-xylulose 5-phosphate.[1–6] The enzyme, which has a ping pong Bi Bi kinetic mechanism, exhibits a broad specificity for both substrates.

sedoheptulose 7-P   D-glyceraldehyde 3-P   D-ribose 5-P   D-xylulose 5-P

The enzyme also catalyzes the reaction of hydroxypyruvate with R–CHO to produce carbon dioxide and R–CH(OH)–CO–CH$_2$OH.

[1]R. L. Schowen (1998) *CBC* **2**, 217.
[2]J. V. Schloss & M. S. Hixon (1998) *CBC* **2**, 43.
[3]Y. Lindqvist & G. Schneider (1993) *Curr. Opin. Struct. Biol.* **3**, 896.
[4]Y. A. Muller, Y. Lindqvist, W. Furey, G. E. Schulz, F. Jordan & G. Schneider (1993) *Structure* **1**, 95.
[5]R. Kluger (1992) *The Enzymes*, 3rd ed., **20**, 271.
[6]C. J. Gubler (1991) in *A Study of Enzymes* (S. A. Kuby, ed.), vol. 2, p. 117, CRC Press, Boca Raton.

Selected entries from *Methods in Enzymology* [vol, page(s)]:
General discussion: 1, 371, 375; **49**, 50, 51; **89**, 43; **90**, 209, 223; **279**, 131
Assay: 1, 371, 375; **9**, 508; ferricyanide reduction, via, **89**, 43; human erythrocyte, **90**, 223; porcine liver, **90**, 220; rat liver, **90**, 217; yeast, 1, 375; **90**, 209

TRANSLOCATOR-ASSOCIATED BACTERIOCIN LEADER PEPTIDASE

This cysteine peptidase catalyzes the removal of a leader peptide from bacteriocin precursors during ATP-dependent export. All known substrates contain a Gly residue in the P2 position as well as a Gly, Ala, or Ser residue in the P1 position.[1]

[1]R. W. Jack, G. Bierbaum & H.-G. Sahl (1998) in *Handbook of Proteolytic Enzymes* (A. J. Barrett, N. D. Rawlings & J. F. Woessner, eds.), Academic Press, New York, p. 768.

TRANSPOSASE

An enzyme that catalyzes the multistep transposition process by relying on both protein–DNA and protein–protein interactions. The ends of a transposon are bound, synapsed, and cleaved by two or more transposase molecules acting in concert; then, the protein-transposon complex recognizes the target DNA and catalyzes strand transfer.

Selected entries from *Methods in Enzymology* [vol, page(s)]:
Other molecular properties: altered target specificity mutant (fused to *Ptac* [in mini-Tn*10* derivative, **204**, 163; in Tn*10* derivative, **204**, 155]; mini-Tn*10* with, **204**, 157]; loss in minitransposons, **235**, 387; Tn*10*, mutation decreasing target site specificity, **204**, 150; wild-type (fused to *Ptac*, in Tn*10* derivative, **204**, 160; mini-Tn*10* constructs with, **204**, 162)

α,α-TREHALASE

This enzyme [EC 3.2.1.28], also simply known as trehalase, catalyzes the hydrolysis of α,α-trehalose to produce two molecules of D-glucose.[1,2]

α,α-trehalose

[1]E. H. Van Beers, H. A. Büller, R. J. Grand, A. W. C. Einerhand & J. Dekker (1995) *Crit. Rev. Biochem. Mol. Biol.* **30**, 197.
[2]P. M. Dey & E. del Campillo (1984) *AE* **56**, 141.

Selected entries from *Methods in Enzymology* [vol, page(s)]:
General discussion: 8, 600; **28**, 996
Assay: 8, 600

TREHALOSE O-MYCOLYLTRANSFERASE

This acyltransferase [EC 2.3.1.122] catalyzes the reversible reaction of two molecules of α,α′-trehalose 6-mycolate to produce α,α′-trehalose and α,α′-trehalose 6,6′-bismycolate.[1]

[1]N. Sathyamoorthy & K. Takayama (1987) *JBC* **262**, 13417.

TREHALOSE PHOSPHATASE

This enzyme [EC 3.1.3.12], also known as trehalose-6-phosphate phosphatase, catalyzes the hydrolysis of trehalose 6-phosphate to produce trehalose and orthophosphate.[1,2]

[1]M. Matula, M. Mitchell & A. D. Elbein (1971) *J. Bact.* **107**, 217.
[2]S. Friedman (1971) *JBC* **246**, 4122.

Selected entries from *Methods in Enzymology* [vol, page(s)]:
General discussion: 8, 372
Assay: 8, 374; **28**, 520

α,α-TREHALOSE PHOSPHATE SYNTHASE

There are at least two distinct enzymes that catalyze the formation of trehalose 6-phosphate: one producing UDP, while the other produces GDP. α,α-Trehalose-phosphate synthase (UDP-forming) [EC 2.4.1.15], variously known as UDPglucose-glucose-phosphate glucosyltransferase, trehalose phosphate-UDP glucosyltransferase, and trehalosephosphate-UDP glucosyltransferase, catalyzes the reaction of UDP-D-glucose with D-glucose 6-phosphate to produce UDP and α,α-trehalose 6-phosphate.[1]

α,α-Trehalose-phosphate synthase (GDP-forming) [EC 2.4.1.36], also known as GDP-glucose-glucosephosphate glucosyltransferase, catalyzes the reaction of GDP-D-glucose with D-glucose 6-phosphate to produce GDP and α,α-trehalose 6-phosphate.[2]

[1]K. Lippert, E. A. Galinski & H. G. Truper (1993) *Antonie Van Leeuwenhoek* **63**, 85.
[2]A. D. Elbein (1967) *JBC* **242**, 403.

Selected entries from *Methods in Enzymology* [vol, page(s)]:
General discussion: 5, 165; 28, 515
Assay: 5, 165; 28, 515

α,α-TREHALOSE PHOSPHORYLASE

This enzyme [EC 2.4.1.64] catalyzes the reversible reaction of α,α-trehalose with orthophosphate to produce D-glucose and β-D-glucose 1-phosphate.[1,2] Initial-rate kinetics rule out a ping pong kinetic mechanism.[3] The *Schizophyllum commune* enzyme proceeds with net *retention* of anomeric configuration, yielding α-D-glucose 1-phosphate and α-D-glucose as the products.

[1]L. R. Maréchal & E. Belocopitow (1972) *JBC* **247**, 3223.
[2]Y. Kitamoto, H. Akashi, H. Tanaka & N. Mori (1988) *FEMS Microbiol. Lett.* **55**, 147.
[3]C. Eis & B. Nidetzky (1999) *BJ* **341**, 385.

TRIACETATE-LACTONASE

This enzyme [EC 3.1.1.38] catalyzes the hydrolysis of triacetate lactone to produce triacetate.[1,2]

[1]S. Kato, H. Ueda, S. Nonomura & C. Tatsumi (1968) *Nippon Nogeikagaku Kaishi* **42**, 596.
[2]Y. Eto, S. Nishioka, H. Horitsu & M. Tomoyeda (1977) *Nippon Nogeikagaku Kaishi* **51**, 655.

TRIACYLGLYCEROL LIPASE

This enzyme [EC 3.1.1.3], variously known as triacylglycerol acylhydrolase, lipase, triglyceride lipase, and tributyrase, catalyzes the hydrolysis of a triacylglycerol to produce a diacylglycerol and a fatty acid anion.[1-10] The pancreatic enzyme acts only at an ester-water interface, with the outer ester linkages preferentially hydrolyzed. Other substrates include 1,2-diacylglycerols, and *p*-nitrophenylacetate.

Activation of the pancreatic enzyme requires the presence of pancreatic colipase, which prevents surface denaturations and assists in the formation of the oxyanion hole. **See also** Lipoprotein Lipase

[1]P. Desnuelle (1972) *The Enzymes*, 3rd ed., **7**, 575.
[2]M. A. Wells & N. A. DiRenzo (1983) *The Enzymes*, 3rd ed., **16**, 113.
[3]M. Hamosh (1984) in *Lipases* (B. Borgström & H. L. Brockmann, eds.), 49.
[4]R. Verger (1984) in *Lipases* (B. Borgström & H. L. Brockmann, eds.), 84.
[5]M. Semevira & P. Desnuelle (1979) *AE* **84**, 319.
[6]D. M. Quinn & S. R. Feaster (1998) *CBC* **1**, 455.
[7]Z. S. Derewenda (1994) *Adv. Protein Chem.* **45**, 1.
[8]P. Strålfars, H. Olsson & P. Belfrage (1987) *The Enzymes*, 3rd ed., **18**, 147.
[9]R. L. Jackson (1983) *The Enzymes*, 3rd ed., **16**, 141.
[10]J. C. Khoo & D. Steinberg (1983) *The Enzymes*, 3rd ed., **16**, 183.

Selected entries from *Methods in Enzymology* [vol, page(s)]:
General discussion: 1, 627, 657; 64, 340, 370; 71, 619, 627, 675; 129, 716; 284, 3, 28, 61, 85, 107, 157, 185, 194, 220, 232, 246, 272, 285, 298, 317, 353; 286, 45, 67, 80, 126, 153, 190, 252, 263, 306, 351, 386, 443, 473, 495, 509
Assays: 1, 627; 72, 331, 338, 340, 351 (preparation of substrates, 72, 353; trinitrophenyl ω-aminolauroyl glycerides, with, 72, 351); bioluminescence assays, 72, 338, 340; 197, 332
Other molecular properties: action (bulk conditions, 64, 348, 349; critical packing density, 64, 353, 386; induction time, 64, 348, 350, 378, 379, 386, 387; presteady state condition, 64, 350, 375); analysis by radiation inactivation, 197, 280; bioassay, 57, 193, 194; *Candida cyclindracea*, from, 72, 340; cat, 71, 636; Celite-adsorbed, triglyceride interesterification with, 136, 405; chicken adipose tissue, 71, 627 (activation by phosphorylation, 71, 628 [cAMP-dependent protein kinase, 71, 629; cGMP-dependent protein kinase, 71, 630]; deactivation by dephosphorylation, 71, 628; diglyceride lipase activity, 71, 631, 633; lipase phosphatase, 71, 631; lipoprotein lipase removal, 71, 632; phosphorylation by cAMP-dependent protein kinase, 71, 635; triglyceride lipase activity, 71, 633); cholic acid activation and, 5, 474; chylomicrons and, 5, 545; determination of K_m value for water, 242, 78; entrapment with (photocrosslinkable resin prepolymers, 135, 237; urethane prepolymers, 135, 241); enzymatic activity, 31, 521, 522; enzyme solubilization by, 1, 39; esterase activity, assay, 242, 73; ester synthesis activity, assay, 242, 73; extraction of enzymes from membranes, use in, 22, 217; fat cells, in, 31, 67, 68; guinea pig (phospholipase A₁ activity, assay, 197, 316; properties, 197, 323; purification, 197, 321); hepatic lipase (activation [apolipoprotein A-II, by, 263, 14; apolipoprotein E, by, 263, 173]; antibodies, 263, 334; ribonuclease protection assay, 263, 359; sandwich immunoassay, 263, 333); hexokinase extraction and, 1, 281; hormone sensitive, 35, 181 (activation by cyclic AMP-dependent protein kinase, 35, 187; assay, 35, 182; inhibitors, 35, 189; phosphorylation, 35, 188, 189; purification, 35, 185; species differences, 35, 189); human, 71, 636; human liver, isolation and characterization, 197, 339; hydrolysis of lipids in monolayers, 14, 640; hydrolysis of riboflavin fatty acid esters, 18B, 451, 452; hydrolysis of triglycerides and cholesteryl esters in plasma, 213, 342; identification of luminous bacteria, 57, 161, 162; immobilization (microencapsulation, by, 44, 214; polyacrylamide, on, 44, 442, 447; stereoselective esterification of DL-menthol, 136, 299); inactivation thermodynamics, 63, 256; inhibitors, 1, 638; 5, 545; 31, 524; isolated nuclei and, 12A, 445; liver, from, 72, 325 (assay, 72, 331; inhibition by sodium dodecyl sulfate,

72, 326, 329; measurement of activity, **72**, 325; postheparin plasma, in, **72**, 325; porcine pancreas, **71**, 619 [assay, **71**, 621; effect of bile salts, **71**, 619]; preparation [of colipase, **71**, 620; of delipidated pancreas, **71**, 622]); localization in gels, **104**, 438; magnetized, preparation, **242**, 81; microbodies, in, **52**, 496; microcalorimetric studies, of binding, **61**, 304; microsome damage, **31**, 195; mitochondria, **31**, 591, 593; modification, effect on activity, **44**, 447; monoglyceride, **35**, 181; mouse, **71**, 636; NMR studies of membranes, in, **32**, 211; other species, from, **71**, 635; pancreatic, **31**, 47, 48; **35**, 181, 273, 317, 320, 324 (assay, **35**, 182; [long-chain fatty acyl-CoA, of, **35**, 273]; identification of lipids, **14**, 602; inactivation thermodynamics, **63**, 256); phosphorylation, **72**, 325; plasma membranes, in, **31**, 88; polyethylene glycol modification (activated PEG2, with, **242**, 74; activated PM, with, **242**, 80; activity effects, **242**, 71, 75; *Candida* lipase, **242**, 79; heat stability effects, **242**, 81; PEG succinimide, with, **242**, 79; *Pseudomonas* lipase, **242**, 74; solubility effects, **242**, 71, 80; substrate specificity effects, **242**, 77); preparation of, in (racemic esters, **136**, 134; racemic secondary alcohols, **136**, 127); prepolymer-entrapped, applications, **135**, 245; properties, **1**, 635; **22**, 218; **71**, 625; purification, **1**, 634; **10**, 554, 567; rat adipose tissue, **71**, 636 (activation, **71**, 645; cyclic AMP-dependent protein kinase, **71**, 645; phosphorylation, **71**, 645; properties, **71**, 644; specificity, **71**, 645; substrate emulsion, **71**, 637, 638); rat liver (assay, **197**, 332; purification, **197**, 331; substrate specificity, **197**, 336); *Rhizopus arrhizus*, treatment of choline and ethanolamine glycerophospholipids, **197**, 138; salt-resistant, in human liver and plasma (assays, **129**, 719, 732; characterization, **129**, 726; purification, **129**, 718); specificity, **71**, 627; staining on gels, **22**, 585, 592, 601; 12-*O*-tetradecanoylphorbol-3-acetate hydrolase activity (preparation, **141**, 303; TPA hydrolysis, assays, **141**, 308); transacylase, and, bacterial, interfacial catalysis, **249**, 613; wheat germ, **15**, 332 (α-glycerophosphate dehydrogenase, ineffective solubilizing agent for, **5**, 439)

TRIACYLGLYCEROL:STEROL *O*-ACYLTRANSFERASE

This transferase [EC 2.3.1.77] catalyzes the reaction of a triacylglycerol with a 3β-hydroxysterol to produce a diacylglycerol and a 3β-hydroxysterol ester.[1,2] While tripalmitoylglycerol is the preferred donor substrate, other triacylglycerols containing six to twenty-two carbon fatty acyl groups can also act as the donor substrate, some being weak donors. The best acceptor substrates are 3β-hydroxysterols such as cholesterol and cholestanol.

[1] M. Kalinowska & Z. A. Wojciechowski (1984) *Phytochemistry* **23**, 2485.
[2] J. Zimowski & Z. A. Wojciechowski (1981) *Phytochemistry* **20**, 1799.

TRICHODIENE SYNTHASE

farnesyl pyrophosphate

nerolidyl pyrophosphate

bisabolyl cation

trichodiene

This enzyme [EC 4.2.3.6; previously classified as EC 4.1.99.6] catalyzes the conversion of 2-*trans*,6-*trans*-farnesyl diphosphate to trichodiene and pyrophosphate.[1–4]

The reaction pathway is believed to proceed via a (3*R*)-nerolidyl pyrophosphate intermediate, thereby generating a second bisabolyl carbocation intermediate that undegoes two methyl migrations and a 1,4 hydride shift.[4]

[1] D. E. Cane, G. Yang, Q. Xue & J. H. Shim (1995) *Biochemistry* **34**, 2471.
[2] D. E. Cane & T. E. Bowser (1999) *Bioorg. Med. Chem. Lett.* **9**, 1127.
[3] D. E. Cane, H. T. Chiu, P. H. Liang & K. S. Anderson (1997) *Biochemistry* **36**, 8332.
[4] R. A. Gibbs (1998) *CBC* **1**, 31.

TRICHOPHYTON MENTAGROPHYTES KERATINASE

This metalloendopeptidase, originally classified as EC 3.4.24.10, is now a deleted EC entry. **See** *Cell-Associated Endopeptidase of Trichophyton*

TRICHOPHYTON SCHOENLENINII COLLAGENASE

This metallocollagenase, originally classified as EC 3.4.24.9, is now a deleted EC entry.

TRICORN PROTEASE

This serine protease catalyzes peptide-bond hydrolysis, acting on both trypsin substrates (*e.g.*, Bz-Val-Gly-Arg-∇-NHMec) and chymotrypsin substrates (such as Ala-Ala-Phe-∇-NHMec).[1] (Note: ∇ indicates the cleavage site.)

[1] T. Tamura & W. Baumeister (1998) in *Handbook of Proteolytic Enzymes* (A. J. Barrett, N. D. Rawlings & J. F. Woessner, eds.), Academic Press, New York, p. 465.

TRIGLUCOSYLALKYLACYLGLYCEROL SULFOTRANSFERASE

This sulfotransferase [EC 2.8.2.19] catalyzes the reaction of 3′-phosphoadenylylsulfate with α-D-glucosyl-1,6-α-D-glucosyl-1,6-α-D-glucosyl-1,3-*O*-alkyl-2-*O*-acylglycerol to produce adenosine 3′,5′-bisphosphate and 6-sulfo-α-D-glucosyl-1,6-α-D-glucosyl-1,6-α-D-glucosyl-1,3-*O*-alkyl-2-*O*-acylglycerol.[1,2]

[1] Y. H. Liau, E. Zdebska, A. Slomiany & B. L. Slomiany (1982) *JBC* **257**, 12019.
[2] B. L. Slomiany, Y. H. Liau, E. Zdebska, V. L. Murty & A. Slomiany (1983) *BBRC* **113**, 817.

3α,7α,12α-TRIHYDROXYCHOLESTAN-26-AL 26-OXIDOREDUCTASE

This oxidoreductase [EC 1.2.1.40], also known as 3α,7α,12α-trihydroxycholestan-26-al 26-dehydrogenase, catalyzes the reaction of 3α,7α,12α-trihydroxy-5β-cholestan-26-al (or, trihydroxycoprostanal) with NAD$^+$ and water to produce 3α,7α,12α-trihydroxy-5β-cholestan-26-oate and NADH.[1–3]

3α,7α,12α-trihydroxy-5β-cholestan-26-al

3α,7α,12α-trihydroxy-5β-cholestan-26-oate

In some organisms, this enzyme may be identical to aldehyde dehydrogenase (NAD$^+$) [EC 1.2.1.3].

[1] T. Masui, R. Herman & E. Staple (1966) BBA **117**, 266.
[2] K. Okuda, E. Higuchi & R. Fukuba (1973) BBA **293**, 15.
[3] K. Okuda & N. Takigawa (1968) BBRC **33**, 788.

3α,7α,12α-TRIHYDROXY-5B-CHOLESTANATE:CoA LIGASE

This enzyme [EC 6.2.1.29], also known as THCA:CoA ligase and 3α,7α,12α-trihydroxy-5β-cholestanoyl-CoA synthetase, catalyzes the reaction of ATP with Coenzyme A and 3α,7α,12α-trihydroxy-5β-cholestanate to produce AMP, pyrophosphate (or, diphosphate), and 3α,7α,12α-trihydroxy-5β-cholestanoyl-CoA.[1]

3α,7α,12α-trihydroxy-5β-cholestanoate

This enzyme is not identical to either acyl-CoA synthetase [EC 6.2.1.3] or choloyl-CoA synthetase [EC 6.2.1.7].

[1] L. Schepers, M. Casteels, K. Verheyden, G. Parmentier, S. Asselberghs, H. J. Eyssen & G. P. Mannaerts (1989) BJ **257**, 221.

TRIHYDROXYPTEROCARPAN DIMETHYLALLYLTRANSFERASE

This transferase [EC 2.5.1.36], also known as glyceollin synthase and first identified in soybean (*Glycine max*), catalyzes the reaction of dimethylallyl diphosphate (or, dimethylallyl pyrophosphate) with (6aS,11aS)-3,6a,9-trihydroxypterocarpan to produce diphosphate (or, pyrophosphate) and glyceollin.[1,2]

[1] J. Leube & H. Grisebach (1983) Z. *Naturforsch.* **38c**, 730.
[2] U. Zähringer, E. Schaller & H. Grisebach (1981) Z. *Naturforsch.* **36c**, 234.

TRIHYDROXYSTILBENE SYNTHASE

This acyltransferase [EC 2.3.1.95], also known as resveratrol synthase and stilbene synthase, reportedly catalyzes the reaction of three molecules of malonyl-CoA with 4-coumaroyl-CoA to produce four molecules of Coenzyme A, three carbon dioxide, and 3,4′,5-trihydroxystilbene (that is, resveratrol).[1] The enzyme is not identical with naringenin-chalcone synthase [EC 2.3.1.74] or pinosylvin synthase [EC 2.3.1.146]. A cysteinyl residue is essential for enzyme activity.[2]

[1] A. Schoppner & H. Kindl (1984) JBC **259**, 6806.
[2] T. Lanz, S. Tropf, F. J. Marner, J. Schroder & G. Schroder (1991) JBC **266**, 9971.

TRIMERELYSIN I

This hemorrhagic, zinc-dependent metalloendopeptidase [EC 3.4.24.52], also known as hemorrhagic proteinase HR1A, catalyzes peptide-bond hydrolysis exhibiting a preference for bulky hydrophobic residues in the P1′ position (such as Leu or Met). The best two synthetic peptide substrates reported are Abz-Thr-Glu-Lys-∇-Leu-Val-2,4-dinitroanilinoethylamide and Abz-Ser-Pro-∇-Met-Leu-2,4-dinitroanilinoethylamide. (Note: ∇ indicates the cleavage site.) The oxidized insulin B chain is hydrolyzed at His10–Leu11 and Ala14–Leu15. Other protein substrates include type IV collagen, fibronectin, laminin, and nidogen. The related HR1B proteinase has a similar specificity.[1] **See also** *Metalloendopeptidase, Habu Pit Viper; Trimerelysin II*

[1] S.-i. Kawabata & S. Iwanaga (1998) in *Handbook of Proteolytic Enzymes* (A. J. Barrett, N. D. Rawlings & J. F. Woessner, eds.), Academic Press, New York, p. 1270.

TRIMERELYSIN II

This non-hemorrhagic, zinc-dependent metalloendopeptidase [EC 3.4.24.53] from the venom of the habu snake (*Trimeresurus flavoviridis*) catalyzes peptide-bond hydrolysis with preference for bulky hydrophobic residues in the P1′ position (such as Leu or Met). The oxidized insulin B chain is hydrolyzed at Asn3–Gln4, His10–Leu11, and Ala14–Leu15. The related protein HR2a is hemorrhagic and will act on these same synthetic peptides, albeit with different k_{cat}/K_m values.[1] **See also** *Metalloendopeptidase, Habu Pit Viper; Trimerelysin I*

[1]S.-i. Kawabata & S. Iwanaga (1998) in *Handbook of Proteolytic Enzymes* (A. J. Barrett, N. D. Rawlings & J. F. Woessner, eds.), Academic Press, New York, p. 1273.

TRIMETAPHOSPHATASE

This enzyme [EC 3.6.1.2] catalyzes the hydrolysis of trimetaphosphate to produce triphosphate.[1-4] **See also** *Endopolyphosphatase; Exopolyphosphatase*

[1]L. M. Busman & M. A. Tabatabai (1985) *Soil Sci. Soc. Am. J.* **49**, 630.
[2]S. R. Kornberg (1956) *JBC* **218**, 23.
[3]S. B. Doty, C. E. Smith, A. R. Hand & C. Oliver (1977) *J. Histochem. Cytochem.* **25**, 1381.
[4]O. Meyerhof, R. Shatas & A. Kaplan (1953) *BBA* **12**, 121.

Selected entries from **Methods in Enzymology** [vol, page(s)]:
General discussion: 2, 577
Assay: 2, 577

TRIMETHYLAMINE DEHYDROGENASE

This FMN-dependent oxidoreductase [EC 1.5.99.7] catalyzes the reaction of trimethylamine with water and an acceptor substrate to produce dimethylamine, formaldehyde, and the reduced acceptor. A number of alkyl-substituted derivatives of trimethylamine can also act as donor substrates. Electron acceptor substrates include phenazine methosulfate and 2,6-dichloroindophenol.[1,2]

[1]M. H. Jang, J. Basran, N. S. Scrutton & R. Hille (1999) *JBC* **274**, 13147.
[2]J. Basran, M. H. Jang, M. J. Sutcliffe, R. Hille & N. S. Scrutton (1999) *JBC* **274**, 13155.

Selected entries from **Methods in Enzymology** [vol, page(s)]:
Assay: 188, 251
Other molecular properties: iron-sulfur center, **69**, 780 (identification, **53**, 273); other chromophores, **69**, 788; properties, **188**, 257; purification from bacterium W3A1, **188**, 253

TRIMETHYLAMINE N-OXIDE ALDOLASE

This aldolase [EC 4.1.2.32] catalyzes the reversible reaction of trimethylamine N-oxide $(CH_3)_3NO$ to produce dimethylamine $((CH_3)_2NH)$ and formaldehyde.[1-3]

[1]P. J. Large (1971) *FEBS Lett.* **18**, 297.
[2]P. A. Myers & L. J. Zatman (1971) *BJ* **121**, 10P.
[3]C. A. Boulton & P. J. Large (1979) *FEMS Microbiol. Lett.* **5**, 159.

TRIMETHYLAMINE N-OXIDE REDUCTASE

This molybdenum-, iron-, and zinc-dependent oxidoreductase [EC 1.6.6.9], also known as trimethylamine oxidase, catalyzes the reaction of NADH with trimethylamine N-oxide $((CH_3)_3NO)$ to produce NAD^+, trimethylamine $((CH_3)_3N)$, and water.[1-3]

[1]H. S. Kwan & E. L. Barrett (1983) *J. Bact.* **155**, 1147 and 1455.
[2]I. Yamamoto, N. Okubo & M. Ishimoto (1986) *JB* **99**, 1773.
[3]G. J. Clarke & F. B. Ward (1988) *J. Gen. Microbiol.* **13**, 379.

4-TRIMETHYLAMMONIOBUTYRALDEHYDE DEHYDROGENASE

This oxidoreductase [EC 1.2.1.47] catalyzes the reaction of 4-trimethylammoniobutanal with NAD^+ and water to produce 4-trimethylammoniobutanoate and NADH.[1]

[1]J. D. Hulse & L. M. Henderson (1980) *JBC* **255**, 1146.

TRIMETHYLLYSINE DIOXYGENASE

This iron- and ascorbate-dependent oxidoreductase [EC 1.14.11.8], also known as trimethyllysine,2-oxoglutarate dioxygenase and trimethyllysine,α-ketoglutarate dioxygenase, catalyzes the reaction of N^6,N^6,N^6-trimethyl-L-lysine with α-ketoglutarate and dioxygen to produce 3-hydroxy-N^6,N^6,N^6-trimethyl-L-lysine, succinate, and carbon dioxide.[1-4]

[1]J. D. Hulse, S. R. Ellis & L. M. Henderson (1978) *JBC* **253**, 1654.
[2]L. M. Henderson, P. J. Nelson & L. Henderson (1982) *Fed. Proc.* **41**, 2843.
[3]D. S. Sachan & C. L. Hoppel (1980) *BJ* **188**, 529.
[4]S. Englard & S. Seifter (1986) *Ann. Rev. Nutr.* **6**, 365.

TRIMETHYLSULFONIUM: TETRAHYDRO-FOLATE N-METHYLTRANSFERASE

This transferase [EC 2.1.1.19] catalyzes the reaction of trimethylsulfonium cation $((CH_3)_3S^+)$ with tetrahydrofolate to produce dimethylsulfide and 5-methyltetrahydrofolate.[1]

[1]C. Wagner, S. M. Lusty, H. Kung & N. L. Rogers (1967) *JBC* **242**, 1287.

TRIOKINASE

This phosphotransferase [EC 2.7.1.28], also known as triose kinase, catalyzes the reaction of ATP with D-glyceraldehyde to produce ADP and D-glyceraldehyde 3-phosphate.[1-3]

[1]H. G. Hers & T. Kusaka (1953) *BBA* **11**, 427.
[2]M. A. G. Sillero & A. Sols (1969) *EJB* **10**, 345.
[3]E. K. Frandsen & N. Grunnet (1971) *EJB* **23**, 588.

Selected entries from **Methods in Enzymology** [vol, page(s)]:
General discussion: 5, 362; **188**, 445
Assay: 5, 362; **188**, 445

TRIOSE-PHOSPHATE ISOMERASE

This enzyme [EC 5.3.1.1], also known as triosephosphate mutase (TIM) and phosphotriose isomerase, catalyzes the interconversion of D-glyceraldehyde 3-phosphate and dihydroxyacetone phosphate (IUPAC: glycerone phosphate).

D-glyceraldehyde
3-phosphate

dihydroxyacetone
phosphate

As pointed out by Rose,[1] this enzyme is chiefly responsible for the largely symmetrical conversion of the two three-carbon segments of glucose to lactate and for the nearly uniform distribution of ^{14}C from pyruvate in the glucosyl units of liver glycogen. The reaction favors dihydroxy-acetone phosphate, DHAP, with an equilibrium constant of 22. This enzyme is extraordinarily efficient, displaying a specific activity of 6700 μmol/min/mg at saturating D-glyceraldehyde 3-phosphate (G3P) concentrations.[2] A *cis*-enediolate or ene-diol intermediate is a key catalytic feature,[3] and Albery and Knowles[4] used kinetic isotope effect measurements to analyze the energetics of TIM catalysis. In fact, this enzyme and proline racemase are among the most extensively characterized enzymes in terms of the energetics of the catalytic reaction cycle.[5-7]

[1] I. A. Rose (1969) *Comprehensive Biochem.* **17**, 126.
[2] E. A. Noltmann (1972) *The Enzymes*, 3rd ed., **6**, 271.
[3] I. A. Rose (1982) *MIE* **87**, 84.
[4] W. J. Albery & J. R. Knowles (1976) *Biochemistry* **15**, 5631.
[5] J. R. Knowles (1991) *Nature* **350**, 121.
[6] R. C. Davenport, P. A. Bash, B. A. Seaton, M. Karplus, G. A. Petsko & D. Ringe (1991) *Biochemistry* **30**, 5821.
[7] J. V. Schloss & M. S. Hixon (1998) *CBC* **2**, 43.

Selected entries from *Methods in Enzymology* [vol, page(s)]:
General discussion: 1, 387; 41, 430, 438, 442, 447; 44, 337; 46, 18, 23, 140, 141, 143, 381, 382, 432; 47, 66; 89, 579
Assay: 1, 387; 25, 670; 41, 37, 430, 434, 435, 438, 442, 447; assay reagent, 44, 495, 496; 89, 579; 90, 490
Other molecular properties: affinity labeling, 47, 497, 498; 87, 88, 473; chloroacetol phosphate and, 25, 670; conjugate, activity, 44, 271; deoxyribose aldolase assay and, 1, 384; diffusion-controlled reactions, 202, 493; enolization, 87, 86; immobilization, on inert protein, 44, 909; inhibition, 63, 401; 87, 89, 92, 393; intermediate, 87, 85; mechanism, 87, 95; rate constants determined by induced transport, 249, 237; transition-state and multisubstrate analogues, 87, 89; 249, 308; triokinase and, 5, 362; yeast, from, 41, 434 (assay, 41, 434, 435; chromatography, 41, 436; inhibitors, 41, 438; properties, 41, 437, 438; purification, 41, 435)

TRIPEPTIDE AMINOPEPTIDASE

This zinc-dependent metalloaminopeptidase [EC 3.4.11.4], originally classified as an α-aminopeptide aminoacidohydrolase [EC 3.4.1.3], and also known as tripeptidase, aminotripeptidase, aminoexotripeptidase, lymphopeptidase, imidoendopeptidase, and peptidase B, catalyzes the release of the N-terminal aminoacyl residue from a tripeptide. Tripeptides with N-terminal lysyl or glutamyl residues are poor substrates. In addition, those peptides with a prolyl residue at the P1' position are not hydrolyzed. The enzyme will also not hydrolyze dipeptides or tripeptide amides.[1,2]

[1] E. L. Smith (1955) *MIE* **2**, 83.
[2] M. Harada (1998) in *Handbook of Proteolytic Enzymes* (A. J. Barrett, N. D. Rawlings & J. F. Woessner, eds.), Academic Press, New York, p. 1510.

Selected entries from *Methods in Enzymology* [vol, page(s)]:
General discussion: 2, 83; 19, 534, 748
Assay: 2, 83
Other molecular properties: dipeptidases in, 2, 87; prolinase and, 2, 97; properties, 2, 87; 19, 756; purification, 2, 84; solubility, 2, 84; specificity, 2, 87

TRIPEPTIDYL-PEPTIDASE I

This lysosomal serine peptidase [EC 3.4.14.9], also known as tripeptidyl aminopeptidase, catalyzes the release of an N-terminal tripeptide from a polypeptide or protein. Virtually any aminoacyl residue can occupy the P1 and P1' position, with the exception of prolyl residues; accordingly, angiotensin III is a substrate, but bradykinin is not.[1]

[1] J. K. McDonald (1998) in *Handbook of Proteolytic Enzymes* (A. J. Barrett, N. D. Rawlings & J. F. Woessner, eds.), Academic Press, New York, p. 539.

TRIPEPTIDYL-PEPTIDASE II

This serine peptidase [EC 3.4.14.10], also known as tripeptidyl aminopeptidase and tripeptidyl peptidase, catalyzes the release of an N-terminal tripeptide from a polypeptide. The substrate specificity is fairly broad; however, prolyl residues are not tolerated in the P1 or P1' positions. The enzyme can be assayed with Arg-Arg-Ala-∇-NHPhNO$_2$ and labeled Arg-Arg-Ala-∇-Ser-Val-Ala.[1,2] (Note: ∇ indicates the cleavage site.)

[1] R.-M. Bälöw, B. Tomkinson, U. Ragnarsson & Ö. Zetterqvist (1986) *JBC* **261**, 2409.
[2] B. Tomkinson (1994) *BJ* **304**, 517.

TRIPHOSPHATASE

This enzyme [EC 3.6.1.25] catalyzes the hydrolysis of triphosphate (also called tripolyphosphate) to produce pyrophosphate and orthophosphate.[1,2]

[1] A. M. Umnov, S. N. Egorov, S. E. Mansurova & I. S. Kulaev (1974) *Biokhimiya* **39**, 373.
[2] L. V. Trilisenko, J. Novotna, V. Erban, V. Behal, Z. Hostalek & I. S. Kulaev (1987) *Folia Microbiol.* **32**, 402.

Selected entries from *Methods in Enzymology* [vol, page(s)]:
General discussion: 2, 580
Assay: 2, 580

TRITHIONATE HYDROLASE

This enzyme [EC 3.12.1.1] catalyzes the hydrolysis of trithionate ($^-O_3S-S-SO_3{}^-$) to produce thiosulfate, sulfate, and two protons.[1,2]

[1] W.-P. Lu & D. P. Kelly (1988) *J. Gen. Microbiol.* **134**, 877.
[2] R. Meulenberg, J. T. Pronk, J. Frank, W. Hazeu, P. Bos & J. G. Kuenen (1992) *EJB* **209**, 367.

TRITHIONATE REDUCTASE

This flavodoxin-dependent and bisulfite-reductase-dependent oxidoreductase catalyzes the conversion of trithionate ($S_3O_6^{2-}$) to thiosulfate ($S_2O_3^{2-}$).

Selected entries from *Methods in Enzymology* [vol, page(s)]:
General discussion: 243, 260
Assay: 243, 265
Other molecular properties: bisulfite reductase-dependent, **243**, 260 (*Desulfotomaculum nigrificans*, **243**, 266; electron carriers, **243**, 267; inhibitors, **243**, 267; molecular weight, **243**, 266; properties, **243**, 266; purification, **243**, 265; stability, **243**, 266; storage, **243**, 266); role in trithionate pathway, **243**, 293; *Thermodesulfobacterium*, **243**, 295

tRNA (ADENINE-N^1-)-METHYLTRANSFERASE

This methyltransferase [EC 2.1.1.36] catalyzes the reaction of *S*-adenosyl-L-methionine with a tRNA to produce *S*-adenosyl-L-homocysteine and a tRNA containing an N^1-methyladenine residue.[1,2] The enzymes from different sources are specific for different adenine residues in tRNA. Adenine 58, a conserved residue in the TΨC loop of tRNAs, is methylated by the *Thermus thermophilus* enzyme.[3]

[1] D. Söll & L. K. Kline (1982) *The Enzymes*, 3rd ed., **15**, 557.
[2] S. J. Kerr & E. Borek (1973) *The Enzymes*, 3rd ed., **9**, 167.
[3] N. Yamazaki, H. Hori, K. Ozawa, S. Nakanishi, T. Ueda, I. Kumagai, K. Watanabe & K. Nishikawa (1992) *Nucl. Acids Symp. Ser.* **1992**, 141.

tRNA (ADENINE-N^6-)-METHYLTRANSFERASE

This methyltransferase [EC 2.1.1.55] catalyzes the reaction of *S*-adenosyl-L-methionine with a tRNA to produce *S*-adenosyl-L-homocysteine and a tRNA containing an N^6-methyladenine residue.[1]

[1] O. K. Sharma (1973) *BBA* **299**, 415.

tRNA ADENYLYLTRANSFERASE

This nucleotidyltransferase [EC 2.7.7.25], also known as tRNA CCA-pyrophosphorylase and CCA-adding enzyme, catalyzes the reaction of ATP with a tRNA (containing *n* bases) to produce pyrophosphate (or, diphosphate) and a tRNA with *n* + 1 bases.[1] The enzyme may be identical with tRNA cytidylyltransferase [EC 2.7.7.21]. *See also tRNA Cytidylyltransferase*

[1] M. P. Deutscher (1982) *The Enzymes*, 3rd ed., **15**, 183.

Selected entries from *Methods in Enzymology* [vol, page(s)]:
General discussion: 12B, 579; **29**, 706; **181**, 434

Assay: 12B, 579; **29**, 707; **34**, 254
Other molecular properties: liver, rat (assay, **12B**, 579; properties, **12B**, 584; purification, **12B**, 581); properties, **181**, 438; purification, **181**, 436; rabbit liver, **29**, 706 (assay method, **29**, 707; catalytic properties, **29**, 714; properties, **29**, 711); stereochemistry, **87**, 203, 212

tRNA CYTIDYLYLTRANSFERASE

This nucleotidyltransferase [EC 2.7.7.21], also known as tRNA CCA-pyrophosphorylase and tRNA CCA-diphosphorylase, catalyzes the reaction of CTP with a tRNA (containing *n* bases) to produce pyrophosphate and a tRNA containing *n* + 1 bases.[1] The enzyme may be identical with tRNA adenylyltransferase [EC 2.7.7.25]. CMP and AMP are incorporated into tRNA molecules from which all or part of 3′-terminal trinucleotide sequence-C-C-A has been removed. *See also tRNA Adenylyltransferase*

[1] M. P. Deutscher (1982) *The Enzymes*, 3rd ed., **15**, 183.

Selected entries from *Methods in Enzymology* [vol, page(s)]:
General discussion: 12B, 579; **29**, 706; **181**, 434
Assay: 12B, 579; **29**, 707; **59**, 123 (nucleotide incoporation, of, **59**, 182; tRNA hydrolysis, of, **59**, 183)
Other molecular properties: activity, **59**, 135; *Escherichia coli*, from, **59**, 61, 135; incorporation of terminal nucleoside analogues, in, **59**, 188; liver, rat (assay, **12B**, 579; properties, **12B**, 584; purification, **12B**, 581); purification, **59**, 123, 183, 262; reconstitution of tRNA, in, **59**, 127, 128, 133, 134, 140, 141; stability, **59**, 185; tRNA end-group labeling, in, **59**, 103; tRNA nucleotidyltransferase and, **29**, 716; yeast, from, **59**, 123

tRNA (CYTOSINE-5-)-METHYLTRANSFERASE

This transferase [EC 2.1.1.29] catalyzes the reaction of *S*-adenosyl-L-methionine with a tRNA to produce *S*-adenosyl-L-homocysteine and a tRNA containing 5-methylcytosine.[1–5] Good acceptor substrates include tRNA^{Val} and tRNA^{fMet}. The *Saccharomyces cerevisiae* enzyme is capable of methylating cytosine at several positions in different yeast tRNAs and pre-tRNAs containing intron.

[1] J. Hurwitz, M. Gold & M. Anders (1964) *JBC* **239**, 3474.
[2] Y. Motorin & H. Grosjean (1999) *RNA* **5**, 1105.
[3] G. R. Björk, J. U. Ericson, C. E. D. Gustafsson, T. G. Hagervall, Y. H. Jönsson & P. M. Wikström (1987) *ARB* **56**, 263.
[4] D. Söll & L. K. Kline (1982) *The Enzymes*, 3rd ed., **15**, 557.
[5] S. J. Kerr & E. Borek (1973) *The Enzymes*, 3rd ed., **9**, 167.

tRNA (GUANINE-N^1-)-METHYLTRANSFERASE

This transferase [EC 2.1.1.31] catalyzes the reaction of *S*-adenosyl-L-methionine with a tRNA to produce *S*-adenosyl-L-homocysteine and a tRNA containing an N^1-methylguanine residue.[1–4] Studies with the *Escherichia coli* enzyme and tRNA^{Leu} demonstrate that the gross tRNA structure is important for full enzyme activity.[1] The presence of an anticodon G adjacent to the G being methylated is also important for enzyme recognition.

[1] W. M. Holmes, C. Andraos-Selim, I. Roberts & S. Z. Wahab (1992) *JBC* **267**, 13440.
[2] G. R. Björk, J. U. Ericson, C. E. D. Gustafsson, T. G. Hagervall, Y. H. Jönsson & P. M. Wikström (1987) *ARB* **56**, 263.
[3] D. Söll & L. K. Kline (1982) *The Enzymes*, 3rd ed., **15**, 557.
[4] S. J. Kerr & E. Borek (1973) *The Enzymes*, 3rd ed., **9**, 167.

Selected entries from *Methods in Enzymology* [vol, page(s)]:
General discussion: 12B, 480; 29, 716
Assay: 12B, 480; 29, 716
Other molecular properties: 1-methyladenine-specific of rat tissues, purification of, 29, 721; products formed by, 12B, 488; properties (inhibitors, 29, 726; specificity, 29, 724; stability, 29, 725); purification (*Escherichia coli*, 12B, 483; rat tissues, 29, 721; yeast, 29, 717, 721); stability, 29, 720

tRNA (GUANINE-N^2-)-METHYLTRANSFERASE

This methyltransferase [EC 2.1.1.32], also known as N^2,N^2-dimethylguanine tRNA methyltransferase, catalyzes the reaction of S-adenosyl-L-methionine with a tRNA to produce S-adenosyl-L-homocysteine and a tRNA containing an N^2-methylguanine residue.[1-5]

[1] P. Izzo & R. Gantt (1977) *Biochemistry* **16**, 3576.
[2] J. Kraus & M. Staehelin (1974) *Nucleic Acids Res.* **1**, 1455.
[3] J. M. Glick, V. M. Averyhart & P. S. Leboy (1978) *BBA* **518**, 158.
[4] D. Söll & L. K. Kline (1982) *The Enzymes*, 3rd ed., **15**, 557.
[5] S. J. Kerr & E. Borek (1973) *The Enzymes*, 3rd ed., **9**, 167.

Selected entries from *Methods in Enzymology* [vol, page(s)]:
General discussion: 29, 716
Assay: 29, 716
Other molecular properties: inhibitors, 29, 726; ionic requirements, 29, 725; purification (rat tissues, 29, 721; yeast, 29, 717, 721); specificity, 29, 724; stability, 29, 725

tRNA (GUANINE-N^7-)-METHYLTRANSFERASE

This methyltransferase [EC 2.1.1.33] catalyzes the reaction of S-adenosyl-L-methionine with a tRNA to produce S-adenosyl-L-homocysteine and a tRNA containing an N^7-methylguanine residue.[1,2] The enzyme specifically methylates the unpaired guanylate residue of the extra arm.[2]

[1] A. Colonna, G. Ciliberto, R. Santamaria, F. Cimino & F. Salvatore (1983) *Mol. Cell. Biochem.* **52**, 97.
[2] D. Söll & L. K. Kline (1982) *The Enzymes*, 3rd ed., **15**, 557.

tRNA (GUANOSINE-2′-O-)-METHYLTRANSFERASE

This methyltransferase [EC 2.1.1.34] catalyzes the reaction of S-adenosyl-L-methionine with a tRNA to produce

S-adenosyl-L-homocysteine and a tRNA containing a 2′-O-methylguanosine residue.[1-4] The enzyme methylates the 2′-hydroxyl group of a guanosine present in a GG sequence in the tRNA.

[1] I. Kumagai, K. Watanabe & T. Oshima (1982) *JBC* **257**, 7388.
[2] T. Matsumoto, K. Watanabe & T. Ohta (1984) *Nucleic Acids Symp. Ser.* **15**, 131.
[3] T. Matsumoto, T. Ohta, I. Kumagai, T. Oshima, K. Murao, T. Hasegawa, H. Ishikura & K. Watanabe (1987) *JB* **101**, 1191.
[4] T. Matsumoto, K. Nishikawa, H. Hori, T. Ohta, K. Miura & K. Watanabe (1990) *JB* **107**, 331.

tRNA GUANYLYLTRANSFERASE

This guanylyltransferase catalyzes the reaction of GTP with the 5′-end of tRNA^{His} (that is 5′-pGCC-tRNA) to produce pyrophosphate and pGpGCC-tRNA. **See also** *Queuine tRNA-Ribosyltransferase*

Selected entries from *Methods in Enzymology* [vol, page(s)]:
Assay: 181, 451
Other molecular properties: partial purification from (*Drosophila*, 181, 454; yeast, 181, 453); properties, 181, 456

tRNA-INTRON ENDONUCLEASE

This endoribonuclease [EC 3.1.27.9], also called tRNA-splicing endonuclease, catalyzes the endonucleolytic cleavage of pre-tRNA, producing 5′-hydroxyl and 2′,3′-cyclic phosphate termini, and specifically removing the intron.[1-3] This is the final stage in the maturation of tRNA molecules.

[1] L. D. Thompson & C. J. Daniels (1988) *JBC* **263**, 17951.
[2] F. Miao & J. Abelson (1993) *JBC* **268**, 672.
[3] M. I. Baldi, E. Mattoccia, E. Bufardeci, S. Fabbri & G. P. Tocchini-Valentini (1992) *Science* **255**, 1404.

Selected entries from *Methods in Enzymology* [vol, page(s)]:
General discussion: 181, 471, 510
Assay: 181, 472, 512
Other molecular properties: *Saccharomyces cerevisiae* (assay, 181, 472; polypeptide components, 181, 480; purification, 181, 473); *Xenopus laevis* (assay, 181, 512; purification, 181, 515)

tRNA ISOPENTENYLTRANSFERASE

This transferase [EC 2.5.1.8], also known as isopentenyl-diphosphate:tRNA isopentenyltransferase, catalyzes the reaction of isopentenyl pyrophosphate (or, isopentenyl diphosphate) with a tRNA to produce pyrophosphate (or, diphosphate) and a tRNA containing a 6-isopentenyl-adenosine residue.[1]

[1] J. Holtz & D. Klämbt (1978) *Hoppe-Seyler's Z. Physiol. Chem.* **359**, 89.

tRNA LIGASE

This enzyme catalyzes the ligation of the two precursor halves of a tRNA (the 5′ half of the molecule and the 3′

half) with the concomitant conversion of ATP to AMP and pyrophosphate. Use of the term ligase is misleading, because an intact tRNA is not being ligated. **See also** *tRNA-Intron Endonuclease*

Selected entries from *Methods in Enzymology* [vol, page(s)]:
General discussion: 181, 463
Assay: 181, 463

tRNA (5-METHYLAMINOMETHYL-2-THIOURIDYLATE)-METHYLTRANSFERASE

This methyltransferase [EC 2.1.1.61] catalyzes the reaction of *S*-adenosyl-L-methionine with a tRNA to produce *S*-adenosyl-L-homocysteine and a tRNA containing a 5-methylaminomethyl-2-thiouridylate.[1,2] The enzyme is specific for the terminal methyl group of 5-methylaminomethyl-2-thiouridylate.

[1] Y. Taya & S. Nishimura (1973) *BBRC* 51, 1062.
[2] T. G. Hagervall, C. G. Edmonds, J. A. McCloskey & G. R. Björk (1987) *JBC* 262, 8488.

tRNA NUCLEOTIDYLTRANSFERASE

This transferase [EC 2.7.7.56], also known as phosphate-dependent exonuclease and RNase PH, catalyzes the reaction of a tRNA$_{n+1}$ with orthophosphate to produce a tRNA$_n$ and a nucleoside diphosphate.[1-3] The enzyme mediates the final exonucleolytic trimming of the 3'-terminus of tRNA precursors in *Escherichia coli* by a phosphorolysis, thereby producing a mature 3'-terminus in the tRNA. This nucleotidyltransferase is not identical with polyribonucleotide nucleotidyltransferase [EC 2.7.7.8].

[1] H. Cudny & M. P. Deutscher (1988) *JBC* 263, 1518.
[2] K. O. Kelly & M. P. Deutscher (1992) *JBC* 267, 17153.
[3] M. P. Deutscher (1982) *The Enzymes*, 3rd ed., 15, 183.

tRNA-PSEUDOURIDINE SYNTHASE I (ψSI)

This enzyme [EC 5.4.99.12], also known as tRNA-uridine isomerase and tRNA pseudouridylate synthase I, catalyzes the interconversion of tRNA uridine residues to tRNA pseudouridine residues. The uridylate residues at positions 38, 39, and 40 of nearly all tRNAs are isomerized to pseudouridine. This classification actually represents a family of enzymes that catalyze this reaction. For example, in *Saccharomyces cerevisiae*, there are three different enzymes that generate pseudouridine residues at tRNA positions 13, 32, and 55, respectively.[1] Binding of tRNA to the *Salmonella typhimurium* enzyme apparently induces dimer formation.[2]

The synthase forms a covalent (FUra)-tRNA^{phe} adduct containing 5-fluorouracil (FUra) in place of Ura to form a putative analogue of the catalytic intermediate in the normal reaction pathway.[3] The conserved aspartate of the enzyme adds to the 6-position of the target FUra to form a stable covalent adduct, which undergoes *O*-acyl hydrolytic cleavage to form the observed product. An analogous covalent complex is most likely formed in the normal enzyme reaction, thus defining the synthase's complete mechanism.

[1] T. Samuelsson & M. Olsson (1990) *JBC* 265, 8782.
[2] F. Arena, G. Ciliberto, S. Ciampi & R. Cortese (1978) *NAR* 5, 4523.
[3] X. Gu, Y. Liu & D. V. Santi (1999) *PNAS* 96, 14270.

tRNA (PURINE-2- OR -6-)-METHYLTRANSFERASE

This methyltransferase activity is now thought to be that of tRNA (guanine-N^2-)-methyltransferase [EC 2.1.1.32].

tRNA-QUEUOSINE β-MANNOSYLTRANSFERASE

This mannosyltransferase [EC 2.4.1.110] catalyzes the reaction of GDP-D-mannose with the queuosine residue in tRNA^{Asp} to produce GDP and tRNA^{Asp}-*O*-5''-β-D-mannosylqueuosine.[1] There is a concomitant inversion of anomeric configuration at C1-carbon in the sugar.

[1] N. Okada & S. Nishimura (1977) *Nucleic Acids Res.* 4, 2931.

tRNA SULFURTRANSFERASE

This transferase [EC 2.8.1.4] catalyzes the reaction of L-cysteine with a suitably activated tRNA to produce L-serine and a tRNA containing a thionucleotide residue.[1-5] This EC classification currently includes a group of enzymes that transfer sulfur to various nucleotides in a tRNA chain. With some members in this group, mercaptopyruvate acts as the sulfur donor. The *Escherichia coli* enzyme contains pyridoxal phosphate as a cofactor.[4]

[1] J. W. Abrell, E. E. Kaufman & M. N. Lipsett (1971) *JBC* 246, 294.
[2] T.-W. Wong, S. B. Weiss, G. L. Eliceiri & J. Bryant (1970) *Biochemistry* 9, 2376.
[3] C. L. Harris & E. B. Titchener (1971) *Biochemistry* 10, 4207.
[4] R. Kambampati & C. T. Lauhon (1999) *Biochemistry* 38, 16561.
[5] L. K. Kline & D. Söll (1982) *The Enzymes*, 3rd ed., 15, 567.

tRNA (URACIL-5-)-METHYLTRANSFERASE

This methyltransferase [EC 2.1.1.35] catalyzes the reaction of a tRNA with *S*-adenosyl-L-methionine to produce *S*-adenosyl-L-homocysteine and a tRNA containing thymine.[1–3] The *Escherichia coli* enzyme is reported to have a random Bi Bi kinetic mechanism.[1] The enzyme also catalyzes an *S*-adenosylmethionine-independent exchange reaction between [5-³H]uridine-labeled substrate tRNA and protons of water at a rate that is 100 times lower than the normal methylation reaction.[2]

[1] L. Shugart (1978) *Biochemistry* **17**, 1068.
[2] J. T. Kealey & D. V. Santi (1995) *Biochemistry* **34**, 2441.
[3] G. R. Björk, J. U. Ericson, C. E. D. Gustafsson, T. G. Hagervall, Y. H. Jönsson & P. M. Wikström (1987) *ARB* **56**, 263.

Selected entries from *Methods in Enzymology* [vol, page(s)]:
General discussion: 59, 190
Assay: 59, 190, 191
Other molecular properties: bacterial, **59**, 190 (assay, **59**, 190, 191; inhibitor, **59**, 202; product analysis, by fluorography, **59**, 192, 193; specificity, **59**, 202); product, **12B**, 488; purification, **12B**, 483, 484; uridine 5-oxyacetic acid methyl ester methyltransferase (assay, **59**, 191; inhibition, **59**, 203; purification, **59**, 200, 201)

tRNA-URIDINE AMINOCARBOXY-PROPYLTRANSFERASE

This transferase [EC 2.5.1.25] catalyzes the reaction of *S*-adenosyl-L-methionine with a uridine residue in a tRNA to produce 5′-methylthioadenosine and a tRNA with a 3-(3-amino-3-carboxypropyl)uridine residue.[1,2]

[1] S. Nishimura, Y. Taya, Y. Kuchino & Z. Ohashi (1974) *BBRC* **57**, 702.
[2] L. K. Kline & D. Söll (1982) *The Enzymes*, 3rd ed., **15**, 567.

TROPINE DEHYDROGENASE

This dehydrogenase [EC 1.1.1.206], also known as tropinone reductase-I, catalyzes the reversible reaction of tropine with NADP⁺ to produce tropinone and NADPH. Other tropan-3α-ols, but not the corresponding β-derivatives, will act as alternative substrates.[1] Note that tropinone reductase [EC 1.1.1.236] catalyzes the same reverse reaction, albeit pseudotropine is formed.

See Tropinone Reductase

[1] B. A. Bartholomew, M. J. Smith, M. T. Long, P. J. Darcy, P. W. Trudgill & D. J. Hopper (1995) *BJ* **307**, 603.

TROPINESTERASE

atropine   tropine   tropate

This enzyme [EC 3.1.1.10], also known as atropinesterase and atropine acylhydrolase, catalyzes the hydrolysis of atropine to produce tropine and tropate.[1–3]

Other substrates include cocaine and related tropine-containing esters.

[1] P. Moog & K. Krisch (1974) *Hoppe-Seyler's Z. Physiol. Chem.* **355**, 529.
[2] L. F. M. van Zutphen (1972) *Enzymologia* **42**, 201.
[3] A. C. M. van der Drift, W. Sluiter & F. Berends (1987) *BBA* **912**, 167.

TROPINONE REDUCTASE

Tropinone reductase I [EC 1.1.1.206] and II [EC 1.1.1.236] catalyze the reversible reaction of tropinone with NADPH (B-side specific) to produce NADP⁺ and the stereoisomeric alkamines tropine [in the case of EC 1.1.1.206] and pseudotropine [in the case of EC 1.1.1.236].[1–5] The enzyme will also use *N*-methyl-4-piperidone, tetrahydrothiopyran-4-one, and 4-piperidone as alternative substrates.[1–3] The stereospecificity of the reductase appears to be determined by the orientation of the tropinone substrate at the substrate-binding site.[4]

[1] B. Dräger, T. Hashimoto & Y. Yamada (1988) *Agric. Biol. Chem.* **52**, 2663.
[2] T. Hashimoto, K. Nakajima, G. Ongena & Y. Yamada (1992) *Plant Physiol.* **100**, 836.
[3] K. Nakajima, T. Hashimoto & Y. Yamada (1993) *PNAS* **90**, 9591.
[4] A. Yamashita, H. Kato, S. Wakatsuki, T. Tomizaki, T. Nakatsu, K. Nakajima, T. Hashimoto, Y. Yamada & J. Oda (1999) *JB* **38**, 7630.
[5] T. Hashimoto & Y. Yamada (1994) *Ann. Rev. Plant Physiol. Plant Mol. Biol.* **45**, 257.

TROPOMYOSIN KINASE

This phosphotransferase [EC 2.7.1.132] catalyzes the reaction of ATP with tropomyosin to produce ADP and [tropomyosin] *O*-phospho-L-serine (*i.e.*, tropomyosin containing a phosphorylated seryl residue: Ser283 in chicken tropomyosin). Casein is an effective alternative substrate.[1,2]

[1] I. deBelle & A. S. Mak (1987) *BBA* **925**, 17.
[2] M. H. Watson, A. K. Taneja, R. S. Hodges & A. S. Mak (1988) *Biochemistry* **27**, 4506.

TRYPANOTHIONE REDUCTASE

This FAD-dependent oxidoreductase [EC 1.6.4.8], also known as N^1,N^8-bis(glutathionyl)spermidine reductase, catalyzes the reversible reaction of NADPH with trypanothione disulfide (or N^1,N^8-bis(glutathionyl)spermidine disulfide) to produce NADP⁺ and reduced trypanothione.[1–4]

Catalysis relies on a redox-active cystinyl residue at the active site. This reductase helps to maintain an intracellular reducing environment in trypanosomatids, a group of protozoan parasites.

trypanothione disulfide

[1] M. C. Jockers-Scherubl, R. H. Schirmer & R. L. Krauth-Siegel (1989) *EJB* **180**, 267.

[2] F. X. Sullivan, S. L. Shames & C. T. Walsh (1989) *Biochemistry* **28**, 4986.

[3] R. L. Krauth-Siegel, B. Enders, G. B. Henderson, A. H. Fairlamb & R. H. Schirmer (1987) *EJB* **164**, 123.

[4] B. A. Palfey & V. Massey (1998) *CBC* **3**, 83.

Selected entries from *Methods in Enzymology* [vol, page(s)]:
Assay: 251, 292
Other molecular properties: assay of glutathionylspermidines, **251**, 290, 292; inhibition by BCNU, **251**, 182; pH optimum, **251**, 292

TRYPANOTHIONE SYNTHETASE

This ligase [EC 6.3.1.9] catalyzes the reaction of glutathione with N^1-glutathionylspermidine and ATP to produce N^1,N^8-bisglutathionylspermidine (or, trypanothione), ADP, and orthophosphate. Trypanothione is involved in maintaining intracellular thiol redox and in defense against oxidants in organisms such as Protozoa of the order Kinetoplastida.[1]

[1] G. B. Henderson, M. Yamaguchi, L. Novoa, A. H. Fairlamb & A. Cerami (1990) *Biochemistry* **29**, 3924.

TRYPSIN

This serine endopeptidase [EC 3.4.21.4] catalyzes the hydrolysis of peptide bonds at Arg–Xaa and Lys–Xaa. The enzyme also catalyzes an esterase activity.[1–10] An arginyl residue in the P1 position of the substrate is typically favored over a lysyl residue by a factor of 2–10, based on k_{cat}/K_m values of numerous substrates. With ester substrates, the degree of discrimination is less. Formation of the acyl-enzyme intermediate is usually the rate-determining step in amide hydrolysis. However, intermediate hydrolysis is rate-determining for the esterase activity. Small, synthetic substrates are often used to assay the enzyme: for example, Bz-Arg-OEt, Tos-Arg-OMe,

Bz-Arg-NHPhNO$_2$, Suc-Ala-Ala-Pro-Arg-NHPhNO$_2$, and Z-Gly-Pro-Arg-NHPhNO$_2$.

The precursor of mammalian trypsin is a preproenzyme produced by the acinar cells of the pancreas. The proenzyme (or zymogen) trypsinogen is stored in secretory granules. Once released into the gut, trypsinogen is acted upon by either enteropeptidase or trypsin at a site near the N-terminus.

[1] L. Hedstrom (1996) *Biol. Chem.* **377**, 465.

[2] J. J. Perona & C. S. Craik (1995) *Protein Sci.* **4**, 337.

[3] C. W. Wharton (1998) *CBC* **1**, 345.

[4] D. R. Corey & C. S. Craik (1992) *JACS* **114**, 1784.

[5] A. A. Kossiakoff (1987) in *Biological Macromolecules & Assemblies* (F. A. Jurnak & A. McPherson, eds.) **3**, 369, Wiley, New York.

[6] E. T. Kaiser, D. S. Scott & S. E. Rokita (1985) *ARB* **54**, 565.

[7] T. A. Steitz & R. G. Schulman (1982) *Ann. Rev. Biophys. Bioeng.* **11**, 419.

[8] J. S. Fruton (1982) *AE* **53**, 239.

[9] J. Kraut (1977) *ARB* **46**, 331.

[10] B. Keil (1971) *The Enzymes*, 3rd ed., **3**, 250.

Selected entries from *Methods in Enzymology* [vol, page(s)]:
General discussion: 2, 26; **19**, 37, 41; **34**, 5, 417, 418, 440; **46**, 96, 128, 155, 156, 197, 206, 207, 215, 229, 233
Assay: 2, 32; **11**, 232; **19**, 20, 42, 60, 61; **34**, 445; by Fast Analyzer, **31**, 816; microcalorimetric assay, **44**, 664, 665; with peptide chromogenic and fluorogenic substrates, **80**, 354, 358, 359; proteoglycan degradation method, **248**, 51
Other molecular properties: ε-*N*-acetylated, **19**, 57; acetylation of with *N*-acetylimidazole, **11**, 576; activation of zymogens, in, **19**, 33, 42, 64; active-site titration, **80**, 414, 415, 419, 420; **248**, 87 (benzyl *p*-guanidinothiobenzoate in, **248**, 14; fluorescein diester, with, **80**, 417; NPGB, with, **80**, 417); acyltrypsin, **87**, 69; adrenal cortex cell isolation, in, **32**, 674, 689, 690, 692; affinity labeling, **87**, 472, 480, 492; affinophoresis, **271**, 213; alkaline pH, at (hydrolysis, **47**, 170; stability, **47**, 172); alanylation of, **11**, 581; allotypes of, **19**, 34; amidinated, **25**, 591; amidinated proteins and, **25**, 595; amino acid analogue substrates, **47**, 430; amino acid sequences of, **19**, 56; **45**, 296; analysis of, in (membrane insertion, **96**, 136; protein interaction with organelle membranes, **96**, 143); anhydrotrypsin (affinity adsorbent for tryptic peptides, as, **91**, 381; immobilization on Sepharose, **91**, 381; preparation, **91**, 379; purification, **91**, 379); anionic (antibody production, **74**, 278; clinical significance, **74**, 273; molecular form in normal serum, **74**, 274); autodigestion, **49**, 454; autolysis, rate determination, **44**, 572, 573; azo linkage, **34**, 104, 105; bacterial, structure, **80**, 640, 641; BCF$_O$-BCF$_I$, **55**, 195; behavior on ultrafiltration, **11**, 902; bovine, **80**, 357 (action on oxidized insulin B chain, **80**, 649; activation of plasminogen, **80**, 413; active-site titration, **80**, 423; amino acid composition, **80**, 640; amino-terminal sequence, **80**, 641, 651; chromatography on immobilized leupeptin derivatives, **80**, 846, 847; pancreas, specificity, from, **59**, 601; reductive methylation, **47**, 475); carboxyl-terminal peptide recognition and, **25**, 153, 154; carboxy terminal sequence analysis, **240**, 705, 708, 710, 712; carboxymethylation of, **25**, 431; cationic (clinical significance, **74**, 273; diisopropylfluorophosphate-inactivated [standard, **74**, 277; tracer, **74**, 281]; extinction coefficient, **74**, 278; labeling with TLCK, **74**, 283; α$_2$-macroglobulin-bound [assays, in, **74**, 280, 281; pathologic serum, in, **74**, 275]; molecular form in normal serum, **74**, 273 [pathologic serum, in, **74**, 274, 276]; α$_1$-protease inhibitor-bound [assays, in, **74**, 280, 281; pathologic serum, in, **74**, 275, 276]; radioimmunoassay, **74**, 272 [antibody production, **74**, 278, 279; data analysis, **74**, 286, 289; equilibration conditions, **74**, 284; principle, **74**, 277, 280; procedure, **74**, 286; second antibody stage, **74**, 284; standards, **74**, 277; tracer **74**, 280]; specificity, **74**, 288); chromatography, **1**, 110; **19**, 51 (analytical scale,

TRYPTASE

This serine endopeptidase [EC 3.4.21.59], also known as mast cell tryptase and mast cell protease II, catalyzes peptide-bond hydrolysis at Arg–Xaa and Lys–Xaa and displays greater specificity than trypsin.[1] Protein substrates include H-kininogen (at Arg431–Asp432), complement C3, fibrinogen, vasoactive intestinal peptide, kinetensin, and prothrombin. Interestingly, tryptase is not inhibited by α₁-proteinase inhibitor, antithrombin III, soybean trypsin inhibitor, or α₂-macroglobulin. Aprotinin, the well known cattle pancreatic trypsin inhibitor, does not inhibit human lung tryptase, but this agent does inhibit dog, rat, and cattle tryptases.[2]

[1] J. S. Bond & P. E. Butler (1987) *ARB* **56**, 333.
[2] L. B. Schwartz (1994) *MIE* **244**, 88.

Selected entries from *Methods in Enzymology* [vol, page(s)]:
General discussion: **244**, 88; rat, **80**, 588
Assay: **80**, 597, 598; **163**, 327; **244**, 90
Other molecular properties: bovine (inhibitors, **244**, 100; size, **244**, 100); canine (inhibitors, **244**, 100; size, **244**, 100); cleavage site specificity, **244**, 91; heparin binding, **244**, 91; human, purification, **244**, 89; human lung (active-site titration, benzyl *p*-guanidinothiobenzoate in, **248**, 14; thioester substrates, **248**, 13); human mast cells, from, **163**, 321; mast cell (abundance, **244**, 96; expression during differentiation, **244**, 96; marker of cell activation, **244**, 88, 96); molecular weight, **80**, 601; physiological role, **80**, 604; properties, **80**, 599; **163**, 314; purification, **80**, 598; specific activity, **80**, 601; stabilization, **80**, 601; **244**, 91; structure, **80**, 601; substrate specificity, **80**, 603; **244**, 91, 93

D-TRYPTOPHAN *N*-ACETYLTRANSFERASE

This acetyltransferase [EC 2.3.1.34] catalyzes the reaction of acetyl-CoA with D-tryptophan to produce Coenzyme A and *N*-acetyl-D-tryptophan.[1] The L-stereoisomer is not a substrate.

[1] M. H. Zenk & J. Schmitt (1964) *Naturwissenschaften* **51**, 510.

TRYPTOPHANAMIDASE

This manganese-dependent amidase [EC 3.5.1.57] catalyzes the hydrolysis of L-tryptophanamide to produce L-tryptophan and ammonia.[1] Alternative substrates include L-phenylalaninamide, and L-tyrosinamide, and some tryptophan dipeptides. The enzyme exhibits a weak glutaminase activity.

[1] A. Iwayama, T. Kimura, O. Adachi & M. Ameyama (1983) *Agric. Biol. Chem.* **47**, 2483.

TRYPTOPHAN AMINOTRANSFERASE

This pyridoxal-phosphate-dependent transferase [EC 2.6.1.27] catalyzes the reaction of L-tryptophan with α-ketoglutarate to produce indole-3-pyruvate and L-glutamate.[1-3] Other substrates include 5-hydroxytryptophan and the phenyl amino acids (for example, L-phenylalanine, L-tyrosine, DL-*p*-fluorophenylalanine, and L-3,4-dihydroxyphenylalanine). *See also Tryptophan: Phenylpyruvate Aminotransferase*

[1] H. George & S. Gabay (1968) *BBA* **167**, 555.
[2] S. R. O'Neil & R. D. DeMoss (1968) *ABB* **127**, 361.
[3] A. E. Braunstein (1973) *The Enzymes*, 3rd ed., **9**, 379.

TRYPTOPHANASE

This enzyme [EC 4.1.99.1], also known as L-tryptophan indole-lyase, catalyzes the hydrolysis of L-tryptophan to generate indole, pyruvate, and ammonia.[1-4] The reaction requires the presence of pyridoxal phosphate and potassium ions. The enzyme also catalyzes tryptophan synthesis from indole and serine as well as 2,3-elimination and β-replacement reactions with certain indole-substituted tryptophan analogues of L-cysteine and L-serine.

[1] A. E. Braunstein & E. V. Goryachenkova (1984) *AE* **56**, 1.
[2] A. E. Martell (1982) *AE* **53**, 163.
[3] E. E. Snell (1975) *AE* **42**, 287.
[4] L. Davis & D. E. Metzler (1972) *The Enzymes*, 3rd ed., **7**, 33.

Selected entries from *Methods in Enzymology* [vol, page(s)]:
General discussion: **2**, 238; **17A**, 439; **46**, 445; **142**, 414
Assay: **2**, 238; **142**, 414
Other molecular properties: activation, **142**, 417; assay of tryptophan, **17A**, 407; induction in microorganisms, **22**, 94; kynureninase and, **2**, 253; properties, **2**, 241; **18A**, 509; **142**, 420; purification from *Escherichia coli*, **2**, 239; **17A**, 442; **18A**, 507, 508; **142**, 415; pyridoxal phosphate and, **17A**, 444; stereochemistry, **87**, 149; transition-state and multisubstrate analogues, **249**, 307; use in microassay for pyridoxal 5-phosphate, **18A**, 505, 611, 619

TRYPTOPHAN DECARBOXYLASE

This enzyme, originally classified as EC 4.1.1.27, is now included under aromatic L-amino acid decarboxylase [EC 4.1.1.28]. *See Aromatic L-Amino Acid Decarboxylase*

TRYPTOPHAN DEHYDROGENASE

This calcium-dependent oxidoreductase [EC 1.4.1.19] catalyzes the reversible reaction of L-tryptophan with $NAD(P)^+$ to produce (indol-3-yl)pyruvate, ammonia, and NAD(P)H. The equilibrium constant favors the reverse direction.[1-4]

[1] M. K. El Bahr, M. Kutacek & Z. Opatrny (1987) *Biochem. Physiol. Pflanz.* **182**, 213.
[2] M. Kutacek (1985) *Biol. Plant.* **27**, 145.
[3] K. Vackova, A. Mehta & M. Kutacek (1985) *Biol. Plant.* **27**, 154.
[4] N. M. W. Brunhuber & J. S. Blanchard (1994) *Crit. Rev. Biochem. Mol. Biol.* **29**, 415.

TRYPTOPHAN DIMETHYLALLYLTRANSFERASE

This transferase [EC 2.5.1.34], also known as dimethylallyltryptophan synthase, catalyzes the reaction of dimethylallyl pyrophosphate with L-tryptophan to produce pyrophosphate and 4-(3-methylbut-2-enyl)-L-tryptophan.[1-3] The *Claviceps* enzyme has a random Bi Bi kinetic mechanism.[3]

[1] S.-L. Lee, H. G. Floss & P. Heinstein (1976) *ABB* **177**, 84.
[2] M. Shibuya, H.-M. Chou, M. Fountoulakis, S. Hassam, S.-U. Kim, K. Kobayashi, H. Otsuka, E. Rogalska, J. M. Cassady & H. G. Floss (1990) *JACS* **112**, 297.
[3] W. A. Cress, L. T. Chayet & H. C. Rilling (1981) *JBC* **256**, 10917.

Selected entries from *Methods in Enzymology* [vol, page(s)]:
General discussion: 110, 335
Assay: 110, 336

TRYPTOPHAN 2′-DIOXYGENASE

This heme-dependent oxidoreductase [EC 1.13.99.3], also known as indole-3-alkane α-hydroxylase and tryptophan side-chain oxidase, catalyzes the reaction of L-tryptophan with dioxygen to produce 3-indolylglycolaldehyde, carbon dioxide, and ammonia.[1–3]

3-indoleglycolaldehyde

The enzyme will act on a number of indolyl-3-alkane derivatives, oxidizing the 3-side-chain in the 2′-position (for example, 3-methylindole or skatole, tryptamine, 5-hydroxytryptamine, melantonin, 3-indolepropionate, 3-indolelactate, 3-indolepyruvate, 3-indolethanol, 3-indoleacetaldehyde, 3-indolemethanol, and 3-indoleacetamide). Best substrates are L-tryptophan and 5-hydroxy-L-tryptophan.

[1] Y. Noda, K. Takai, T. Tokuyama, S. Narumiya, H. Ushiro & O. Hayaishi (1977) *JBC* **252**, 4413.
[2] K. Takai, Y. Sasai, H. Morimoto, H. Yamazaki, H. Yoshii & S. Inoue (1984) *JBC* **259**, 4452.
[3] S. Narumiya, K. Takai, T. Tokuyama, Y. Noda, H. Ushiro & O. Hayaishi (1979) *JBC* **254**, 7007.

Selected entries from *Methods in Enzymology* [vol, page(s)]:
General discussion: 142, 195
Assay: 142, 201

TRYPTOPHAN 2,3-DIOXYGENASE

This heme-dependent oxidoreductase [EC 1.13.11.11], also known as tryptophan pyrrolase, tryptophanase, tryptophan oxygenase, tryptamine 2,3-dioxygenase, and tryptophan peroxidase, catalyzes the reaction of L-tryptophan with dioxygen to produce L-formylkynurenine.[1–4] The cooperative enzyme displays broad specificity towards tryptamine

N-formyl-L-kynurenine

and tryptophan derivatives including D-tryptophan, 5-hydroxytryptophan, and serotonin.

Note: Peptide-tryptophan 2,3-dioxygenase [EC 1.13.11.26] also catalyzes this reaction as does indoleamine-pyrrole 2,3-dioxygenase [EC 1.13.11.42], the latter displaying a broader range of substrates. Numerous publications and internet search engines often fail to distinguish between these enzymes. *See also Indoleamine-Pyrrole 2,3-Dioxygenase*

[1] A. A.-B. Badawy (1979) *Biochem. Soc. Trans.* **7**, 575.
[2] P. Feigelson & F. O. Brady (1974) in *Mol. Mech. Oxygen Activ.* (O. Hayaishi, ed.) p. 87.
[3] H. B. Dunford (1998) *CBC* **3**, 195.
[4] O. Hayaishi, M. Nozaki & M. T. Abbott (1975) *The Enzymes*, 3rd ed., **12**, 119.
Selected entries from *Methods in Enzymology* [vol, page(s)]:
General discussion: 2, 242; **17A**, 415, 421, 429, 434; **52**, 13
Assay: 2, 242; differential assay of various forms, **17A**, 415; immunochemical assay for, **17A**, 418
Other molecular properties: activation (by ascorbic acid, **17A**, 415, 425, 433, 438; by hydrogen peroxide, **17A**, 433, 434); hydrocortisone administration *in vivo*, effect on enzyme concentration, **17A**, 422; oxy form, spectrum, **52**, 36; purification, **2**, 244 (*Pseudomonas fluorescens*, **17A**, 430; rabbit intestine, **17A**, 436; rat liver, **17A**, 419); stabilization of by α-methyl-D,L-tryptophan, **17A**, 428

D-TRYPTOPHAN N-MALONYLTRANSFERASE

This acyltransferase [EC 2.3.1.112] catalyzes the reaction of malonyl-CoA with D-tryptophan (but not L-tryptophan) to produce Coenzyme A and N^2-malonyl-D-tryptophan.[1] Interestingly, 1-aminocyclopropane-1-carboxylate is an effective alternative substrate.

[1] U. Matern, C. Feser & W. Heller (1984) *ABB* **235**, 218.

TRYPTOPHAN 2-C-METHYLTRANSFERASE

This methyltransferase [EC 2.1.1.106] catalyzes the reaction of S-adenosyl-L-methionine with L-tryptophan to produce S-adenosyl-L-homocysteine and L-2-methyltryptophan.[1] Alternative substrates, albeit weaker, include D-tryptophan and indole-3-pyruvate. Contrary to the action of most methyltransferases, methyl group transfer occurs with retention of configuration.[1]

[1] T. Frenzel, P. Zhou & H. G. Floss (1990) *ABB* **278**, 35.
Selected entries from *Methods in Enzymology* [vol, page(s)]:
General discussion: 142, 235
Assay: 142, 236

TRYPTOPHAN 2-MONOOXYGENASE

This FAD-dependent oxidoreductase [EC 1.13.12.3] catalyzes the reaction of L-tryptophan with dioxygen to produce indole-3-acetamide, carbon dioxide, and water.[1–6] The kinetic mechanism is sequential with tryptophan and ping-pong with phenylalanine and methionine.[4]

[1] T. Kosuge, M. G. Heskett & E. E. Wilson (1966) *JBC* **241**, 3738.
[2] L. Comai & T. Kosuge (1980) *J. Bact.* **143**, 950.
[3] S. W. Hutcheson & T. Kosuge (1985) *JBC* **260**, 6281.
[4] J. J. Emanuele & P. F. Fitzpatrick (1995) *Biochemistry* **34**, 3710.
[5] J. J. Emanuele & P. F. Fitzpatrick (1995) *Biochemistry* **34**, 3716.
[6] P. F. Fitzpatrick (2000) *AE* **74**, 235.

TRYPTOPHAN 5-MONOOXYGENASE

This iron-dependent oxidoreductase [EC 1.14.16.4], also known as tryptophan 5-hydroxylase, catalyzes the reaction of L-tryptophan with tetrahydropteridine and dioxygen to produce 5-hydroxy-L-tryptophan, dihydropteridine, and water.[1,2]

5-hydroxy-L-tryptophan

[1] S. Kaufman (1987) *The Enzymes*, 3rd ed., **18**, 217.
[2] S. Koizumi, Y. Matsushima, T. Nagatsu, H. Hnuma, T. Takeuchi & H. Umezawa (1984) *BBA* **789**, 111.

Selected entries from *Methods in Enzymology* [vol, page(s)]:
General discussion: 5, 819; **17A**, 449; **52**, 12; **142**, 83, 88, 93
Assay: 5, 819; **39**, 387, 388; **142**, 83, 89, 93

TRYPTOPHAN α,β-OXIDASE

This heme-dependent oxidoreductase [EC 1.4.3.17], also known as L-tryptophan α,β-dehydrogenase and tryptophan 2′,3′-oxidase, catalyzes the reaction of L-tryptophan with dioxygen to produce α,β-didehydrotryptophan and hydrogen peroxide.[1,2]

α,β-didehydrotryptophan

[1] R. Genet, C. Denoyelle & A. Menez (1994) *JBC* **269**, 18177.
[2] R. Genet, P. H. Benetti, A. Hammadi & A. Menez (1995) *JBC* **270**, 23540.

TRYPTOPHAN:PHENYLPYRUVATE AMINOTRANSFERASE

This aminotransferase [EC 2.6.1.28] catalyzes the reversible reaction of L-tryptophan with phenylpyruvate to produce indole-3-pyruvate and L-phenylalanine.[1,2] Valine, leucine, and isoleucine can replace L-phenylalanine. *See also Tryptophan Aminotransferase*

[1] N. K. Sukanya & C. S. Vaidyanathan (1964) *BJ* **92**, 594.
[2] Y. Koide, M. Honma & T. Shimomura (1980) *Agric. Biol. Chem.* **44**, 2013.

TRYPTOPHAN SYNTHASE

This pyridoxal-phosphate-dependent enzyme [EC 4.2.1.20], also known as tryptophan desmolase, catalyzes the reaction of L-serine with 1-(indol-3-yl)glycerol 3-phosphate to produce L-tryptophan, glyceraldehyde 3-phosphate, and water.[1–8] The enzyme will also catalyze (a) the conversion of serine and indole into tryptophan and water and (b) conversion of indoleglycerol phosphate into indole and glyceraldehyde phosphate. In some organisms, this enzyme is part of a multifunctional protein together with one or more components of the system for biosynthesis of tryptophan (anthranilate phosphoribosyltransferase [EC 2.4.2.18], indole-3-glycerol-phosphate synthase [EC 4.1.1.48], anthranilate synthase [EC 4.1.3.27], and phosphoribosylanthranilate isomerase [EC 5.3.1.24]).

A important aspect of the structure of the protein complex is a tunnel that channels and connects two distinct active sites. The indole intermediate is synthesized at one site and channeled to the other site containing the aldimine of aminoacrylate (originating from serine).[4,6]

[1] A. Lane & K. Kirschner (1983) *EJB* **129**, 561.
[2] W. F. Drewe, Jr., & M. F. Dunn (1985) *Biochemistry* **24**, 3977.
[3] H. S. Ro & E. W. Miles (1999) *JBC* **274**, 31189.
[4] R. A. John (1998) *CBC* **2**, 173.
[5] K. A. Johnson (1992) *The Enzymes*, 3rd ed., **20**, 1.
[6] E. W. Miles (1991) *AE* **64**, 93.
[7] E. W. Miles (1979) *AE* **49**, 127.
[8] C. Yanofsky & I. P. Crawford (1972) *The Enzymes*, 3rd ed., **7**, 1.

Selected entries from *Methods in Enzymology* [vol, page(s)]:
General discussion: 2, 233; 5, 801; **17A**, 365, 406; **46**, 31, 32; **142**, 398
Assay: 2, 233; 5, 801; **17A**, 370; **142**, 400
Other molecular properties: $\alpha_2\beta_2$ complex of, **17A**, 375, 379; purification from (*Escherichia coli*, 5, 803; **17A**, 375, 451; *Neurospora crassa*, 2, 234; **17A**, 408; **22**, 81, 83; subunits, *Escherichia coli*, 5, 803; **17A**, 375); pyridoxal phosphate and, **17A**, 378, 412; rapid scanning stopped-flow spectroscopy (indole reaction mechanism, **246**, 196; ligand-induced conformational change, **246**, 198; pyridoxal phosphate reaction mechanism, **246**, 194; reactions, 5, 801; **246**, 193; staining active enzyme on gels, **22**, 585, 593; stereochemistry, **87**, 149, 156; substrate channeling, **246**, 193; α subunit (isolation, **142**, 406; separation from β₂ subunit, **142**, 405); transition-state and multisubstrate analogues, **249**, 308

TRYPTOPHANYL AMINOPEPTIDASE

This metalloaminopeptidase [EC 3.4.11.17], also called tryptophan aminopeptidase, is reported to catalyze the preferential release of an N-terminal tryptophanyl residue from peptide.[1,2]

[1] O. Adachi (1998) in *Handbook of Proteolytic Enzymes* (A. J. Barrett, N. D. Rawlings & J. F. Woessner, eds.), Academic Press, New York, p. 1515.
[2] A. Iwayama, T. Kimura, O. Adachi & M. Ameyama (1983) *Agric. Biol. Chem.* **47**, 2483.

TRYPTOPHANYL-tRNA SYNTHETASE

This ligase [EC 6.1.1.2], also known as tryptophan:tRNA ligase and tryptophan translase, catalyzes the reaction of ATP with L-tryptophan and tRNA^{Trp} to produce AMP, pyrophosphate (or, diphosphate), and L-tryptophanyl-tRNA^{Trp}.[1-5] The reaction proceeds via the formation of a tryptophanyl adenylate intermediate.[3] **See also** *Aminoacyl-tRNA Synthetases*

[1] H. Jakubowski & J. Pawelkiewicz (1975) *EJB* **52**, 301.
[2] D. Andrews, V. Trezeguet, M. Merle, P. Graves, K. H. Muench & B. Labouesse (1988) *EJB* **146**, 201.
[3] M. Merle, V. Trezeguet, P. Graves, D. Andrews, K. H. Muench & B. Labouesse (1986) *Biochemistry* **25**, 1115.
[4] M. Merle, V. Trezeguet, J. C. Gandar & B. Labouesse (1988) *Biochemistry* **27**, 2244.
[5] P. Schimmel (1987) *ARB* **56**, 125.

Selected entries from **Methods in Enzymology** [vol, page(s)]:
General discussion: 5, 718; 59, 234
Assay: 5, 718; 59, 236
Other molecular properties: aminoacylation test of tRNA hydrolysis, in, 59, 132, 133; covalently bound tryptophan, with, 59, 246, 247; diffusion constant, 59, 244; extinction coefficient, 59, 239; inhibitors, 59, 249; kinetic properties, 59, 247; multiple forms, 59, 245; paracrystals, 59, 246; pH optimum, 59, 247; preparation from *Neurospora crassa*, 22, 81, 83; stability, 59, 244; subcellular distribution, 59, 233, 234; substrate binding, 59, 247, 248; substrate specificity, 59, 249; subunit structure, 59, 244, 245; tertiary structure, 59, 245

TseI RESTRICTION ENDONUCLEASE

This type II restriction endonuclease [EC 3.1.21.4], obtained from *Thermus* species 93170, acts on both strands of DNA at 5'...G∇CWGC...3' sequences, where ∇ indicates the cleavage site and W refers to A or T.

Tsp PROTEASE

This serine protease [EC 3.4.21.102] (also called C-terminal processing peptidase (CtpA protein), tail-specific protease (hence, Tsp), Prc (named after its gene, *prc*: for processing involving C-terminal cleavage), and protease Re) catalyzes peptide-bond hydrolysis in proteins with C-termini of nonpolar amino acids. Alanine is preferred at the C-terminus and there is a preference for an alanyl or tyrosyl residue at the penultimate position. The third position from the end is typically an alanyl or leucyl residue. Although these residues are the recognition criteria, proteolysis can occur at a peptide bond far removed from the C-terminus (and the distance form the C-terminus is variable). The cleavage site specificity is broad: Ala/Ser/Val > Ile/Leu/Lys/Arg at the P1 position and Met, Tyr, or Trp at the P1' site. Tsp is a component of the ssrA RNA protein-tagging pathway in certain microorganisms for the removal and degradation of incorrectly-synthesized proteins. Proteins "tagged" with a C-terminal sequence (for example, Leu-Ala-Ala) are targeted for degradation.[1]

The C-terminal processing peptidase from the unicellular cyanobacterium *Synechocystis* sp., catalyzes the processing of D1, one of the protein subunits of photosystem II in the photosynthetic membranes of cyanobacteria.[2] A similar activity has been identified in the thylakoid membranes of spinach and other plants.[3] The cleavage site in the D1 precursor is at Ala344–Ala345.[4,5]

[1] K. C. Keiler & R. T. Sauer (1998) in *Handbook of Proteolytic Enzymes* (A. J. Barrett, N. D. Rawlings & J. F. Woessner, eds.), Academic Press, New York, p. 460.
[2] S. V. Shestakov, P. R. Anbudurai, G. E. Stanbekova, A. Gadzhiev, L. K. Lind & H. B. Pakrasi (1994) *JBC* **269**, 19354.
[3] J. B. Marder, P. Goloubinoff & M. Edelman (1984) *JBC* **259**, 3900.
[4] M. Takahashi, T. Shiraishi & K. Asada (1988) *FEBS Lett.* **240**, 6.
[5] F. Taguchi, Y. Yamamoto & K. Satoh (1995) *JBC* **270**, 10711.

Tsp45I RESTRICTION ENDONUCLEASE

This type II restriction endonuclease [EC 3.1.21.4], obtained from *Thermus* species YS45, recognizes and acts on both strands of DNA at 5'...∇GTSAC...3' where S refers to C or G.[1] (The symbol ∇ indicates the cleavage site.) Approximately 10% activity is observed upon incubation at 37°C.

[1] N. D. H. Raven, R. A. D. Williams, K. E. Smith, C. D. Kelly & N. D. Carter (1993) *NAR* **21**, 4397.

Tsp509I RESTRICTION ENDONUCLEASE

This type II restriction endonuclease [EC 3.1.21.4], obtained from *Thermus* species ITI 346, recognizes and acts on both strands of DNA at 5'...∇AATT...3'. (The symbol ∇ indicates the cleavage site.) This is one of the relatively few restriction endonucleases that have only A and T in the recognition site. It is sensitive to methylation by *Eco*RI methylase.

TspRI RESTRICTION ENDONUCLEASE

This type II restriction endonuclease [EC 3.1.21.4], obtained from *Thermus* species strain R, recognizes and acts on both strands of DNA at 5′ . . . NNCASTGNN∇ . . . 3′ where S refers to G or C and N refers to A, T, G, or C. (The symbol ∇ indicates the cleavage site.) Note that this endonuclease produces DNA fragments which have a nine-base 3′ extension.

*Tth*IIII RESTRICTION ENDONUCLEASE

This type II restriction endonuclease [EC 3.1.21.4], obtained from *Thermus thermophilus* 111, recognizes and acts on both strands of DNA at 5′ . . . GACN∇NNGTC . . . 3′ where N refers to any base.[1] (The symbol ∇ indicates the cleavage site.) The enzyme also exhibits star activity. **See** *"Star" Activity*

[1] T. Shinomiya & S. Sato (1980) *NAR* **8**, 43.

TUBULIN GTPase

This αβ-heterodimeric cytoskeletal protein undergoes an indefinite series of head-to-tail polymerization reactions to form microtubules.[1–3] Tubulin contains two guanine nucleotide sites that can be represented as $\alpha_{N \cdot GTP}\beta_{E \cdot GTP}$. The subscript N indicates the nonexchangeable site, where is lodged within the subunit-subunit interface and cannot exchange with GTP or GDP in the medium. The subscript E indicates the exchangeable nucleotide site (located exclusively on the β subunit) which undergoes hydrolysis during tubulin polymerization. Binding of $\alpha_{N \cdot GTP}\beta_{E \cdot GTP}$ at a terminal growth site induces the hydrolysis of the underlying Tb_EGTP dimer to form penultimate Tb_EGDP. This assembly-induced GTP hydrolysis reaction accounts for the boundary-stabilizing properties of microtubules grown in the presence of GTP.[2]

Tubulin is currently misclassified as a GTPase [EC 3.6.1.51], because the reaction is of the energase type: $State_1 + GTP = State_2 + GDP + P_i$, where two affinity-modulated conformational states are indicated.[3] Energases transduce the Gibbs free energy of NTP hydrolysis into mechanical work. Under normal conditions tubulin is not a GTPase; however, microtubule assembly can become uncoupled from exchangeable-site nucleotide hydrolysis in the presence of colchicine and other assembly-inhibiting agents.[1]

[1] D. L. Purich & J. M. Angelastro (1994) *AE* **69**, 121.
[2] T. L. Karr, A. E. Podrasky & D. L. Purich (1979) *PNAS* **76**, 5475.
[3] D. L. Purich (2001) *TiBS* **26**, 417.

Selected entries from *Methods in Enzymology* [vol, page(s)]:
General discussion: 34, 623; 46, 567; 298, 218; 308, 93

Other molecular properties: assembly, energetics, 308, 93; biotin-labeled, preparation, 196, 479; caged, 291, 348; 298, 125; cycling, 196, 478; depolymerization, 308, 93; detyrosinated (assay, 196, 257; binding of MAPs, assay, 196, 260; energetics, 308, 93; functional assay, 196, 254; preparation, 106, 231, 233; properties, 106, 237; taxol-dependent purification from HeLa cells, 196, 255); disassembly, energetics, 308, 93; elongation, 308, 96, 106; fluorescent labeling (characterization, 134, 524, 525; preparation, 134, 522, 526); fluorochrome-labeled, preparation, 196, 480; GTP-binding site, photoaffinity studies, 196, 454; guanine nucleotide complexes (preparation, 85, 417; properties, 85, 423); polymerization, energetics, 308, 93; polymerization with 8-azido-GTP, 56, 646; self-association patterns in velocity sedimentation, 130, 4

TUBULIN *N*-ACETYLTRANSFERASE

This acetyltransferase [EC 2.3.1.108], also known as α-tubulin acetylase, catalyzes the reaction of acetyl-CoA with the ε-amino group of a lysyl residue in α-tubulin to produce Coenzyme A and an N^6-acetyl-L-lysyl residue in α-tubulin.[1–3]

[1] K. Greer, H. Maruta, S. W. L'Hernault & J. L. Rosenbaum (1985) *JCB* **101**, 2081.
[2] H. Maruta, K. Greer & J. L. Rosenbaum (1986) *JCB* **103**, 571.
[3] T. H. MacRae (1997) *EJB* **244**, 265.

TUBULIN:TYROSINE LIGASE

This enzyme [EC 6.3.2.25] catalyzes the ATP-dependent religation of L-tyrosine to the C-terminal L-glutamate residue in the detyrosinylated αβ-tubulin dimer, yielding ADP, orthophosphate, and the religated tubulin dimer.[1,2] L-Phenylalanine and 3,4-dihydroxy-L-phenylalanine are also substrates. This is the only known C-terminal peptide bond-forming reaction involving a protein substrate. The ligase catalyzes a random Ter Ter kinetic mechanism.[1]

[1] N. L. Deans, R. D. Allison & D. L. Purich (1992) *BJ* **286**, 243.
[2] B. J. Terry & D. L. Purich (1982) *AE* **53**, 113.

Selected entries from *Methods in Enzymology* [vol, page(s)]:
Assay: 85, 432; 106, 225; 134, 171
Other molecular properties: applications, 106, 231; properties, 106, 231; purification from porcine brain, 106, 227; 134, 173, 176

TUBULINYL-TYROSINE CARBOXYPEPTIDASE

This carboxypeptidase [EC 3.4.17.17], also known as carboxypeptidase-tubulin, soluble carboxypeptidase, and tubulin-tyrosine carboxypeptidase, catalyzes the hydrolysis of the C-terminal tyrosine residue from native tubulin (acting on . . . -Glu-Tyr of α-tubulin). The enzyme is inactive on Z-Glu-Tyr. It will also release a phenylalanine residue that has replaced the C-terminal tyrosine.[1–3] The true protein substrate appears to be polymerized tubulin (that is, microtubules) rather than the free dimer.

[1] C. E. Argarana, H. S. Barra & R. Caputto (1978) *Mol. Cell Biochem.* **19**, 17.

[2] G. G. Deanin, S. F. Preston, R. K. Hanson & M. W. Gordon (1980) *Eur. J. Biochem.* **109**, 207.

[3] H. S. Barra, C. E. Argarana & R. Caputto (1982) *J. Neurochem.* **38**, 112.

Selected entries from **Methods in Enzymology** [vol, page(s)]:
Assay: 106, 234
Other molecular properties: properties, 106, 236; purification from bovine brain, 106, 235; substrate preparation, 106, 234

TUMOR NECROSIS FACTOR α-CONVERTING ENZYME

This metalloproteinase, also known as TACE (acronym for tumor alpha-converting enzyme), catalyzes hydrolysis of the 26 kDa tumor necrosis factor α at the processing site, PLAQA-∇-VRSSS. If the P1 alanyl residue is substituted with isoleucyl, no hydrolysis occurs.[1] (The symbol ∇ indicates the cleavage site.)

[1] R. A. Black & J. D. Becherer (1998) in *Handbook of Proteolytic Enzymes* (A. J. Barrett, N. D. Rawlings & J. F. Woessner, eds.), Academic Press, New York, p. 1315.

TYPE I SITE-SPECIFIC DEOXYRIBONUCLEASE

These deoxyribonucleases [EC 3.1.21.3], also known as type I restriction enzymes, catalyze the endonucleolytic cleavage of DNA to give random double-stranded fragments with a terminal 5'-phosphate; ATP (or, dATP) is simultaneously hydrolyzed to ADP (or, dADP) and orthophosphate.[1-4] This classification represents many enzymes possessing an absolute requirement for ATP (or dATP) and *S*-adenosyl-L-methionine. Each recognizes sequence-specific short DNA sequences and cleaves at sites that are remote from these recognition sequences. Enzymes in this category also catalyze site-specific methyltransfer that is adenine-specific [EC 2.1.1.72] or cytosine-specific [EC 2.1.1.73]. **See specific restriction endonuclease**

[1] B. Endlich & S. Linn (1981) *The Enzymes*, 3rd ed., **14**, 137.

[2] J.-E. Sjöström, S. Löfdahl & L. Philipson (1978) *J. Bact.* **133**, 1144.

[3] T. A. Bickle (1982) in *Nucleases* (S. M. Linn & R. J. Roberts, eds.), p. 85.

[4] G. P. Davies, I. Martin, S. S. Sturrock, A. Cronshaw, N. E. Murray & D. T. Dryden (1999) *JMB* **290**, 565.

TYPE II SITE-SPECIFIC DEOXYRIBONUCLEASE

These magnesium-dependent hydrolases [EC 3.1.21.4], also known as type II restriction enzymes, catalyze endonucleolytic DNA cleavage, yielding double-stranded fragments having terminal 5'-phosphates. This classification represents many enzymes, each recognizing specific short DNA sequences and cleaving either within or at a short specific distance from its recognition site.[1-5] **See Restriction Endonucleases; specific enzyme**

[1] R. D. Wells, R. D. Klein & C. K. Singleton (1981) *The Enzymes*, 3rd ed., **14**, 157.

[2] P. L. Molloy & R. H. Symons (1980) *NAR* **8**, 2939.

[3] A. Kiss, G. Posfai, C. C. Keller, P. Venetianer & R. J. Roberts (1985) *NAR* **13**, 6403.

[4] G. G. Wilson (1988) *Trends Genet.* **4**, 314.

[5] P. Modrich & R. J. Roberts (1982) in *Nucleases* (S. M. Linn & R. J. Roberts, eds.), 109.

TYPE III SITE-SPECIFIC DEOXYRIBONUCLEASE

These hydrolases [EC 3.1.21.5], also known as type III restriction enzymes, catalyze the endonucleolytic cleavage of DNA to give specific double-stranded fragments with terminal 5'-phosphates.[1] Each enzyme has an absolute requirement for ATP, but does not hydrolyze the nucleotide. In addition, *S*-adenosyl-L-methionine stimulates the reaction but is not absolutely required. These enzymes recognize specific short DNA sequences and cleave a short distance away from the recognition sequence. The type III enzymes form complexes with site-specific methyltransferase (adenine-specific) [EC 2.1.1.72] and site-specific methyltransferase (cytosine-specific) [EC 2.1.1.73], enzymes having similar site specificities. **See Restriction Endonucleases; specific enzyme**

[1] T. A. Bickle (1982) in *Nucleases* (S. M. Linn & R. J. Roberts, eds.), 85.

TYRAMINE *N*-FERULOYLTRANSFERASE

This acyltransferase [EC 2.3.1.110] catalyzes the reaction of feruloyl-CoA with tyramine to produce Coenzyme A and *N*-feruloyltyramine.[1,2] Acyl donor substrates include cinnamoyl-CoA, 4-coumaroyl-CoA, and sinapoyl-CoA. Aromatic amine substrates include *N*-methyltyramine, dopamine, norepinephrine, octopamine, 3-methoxytyramine, and homotyramine.

[1] J. Negrel & C. Martin (1984) *Phytochemistry* **23**, 2797.

[2] J. Fleurence & J. Negrel (1989) *Phytochemistry* **28**, 733.

TYRAMINE *N*-METHYLTRANSFERASE

This methyltransferase [EC 2.1.1.27] catalyzes the reaction of *S*-adenosyl-L-methionine with tyramine to produce *S*-adenosyl-L-homocysteine and *N*-methyltyramine.[1] α-Methyltyramine (or, paredrine) and β-phenethanolamine are also methylated.

[1] J. D. Mann & S. H. Mudd (1963) *JBC* **238**, 381.

TYRAMINE OXIDASE

This enzyme, originally classified as EC 1.4.3.9, is now listed under amine oxidase (flavin-containing) [EC 1.4.3.4].

TYROCIDINE SYNTHETASE SYSTEM

This group of enzymes participates in the biosynthesis of the cyclic decapeptide tyrocidin, first isolated from *Bacillus brevis*. Tyrocidine synthetase I produces D-phenylalanine and is thus a phenylalanine racemase.

Selected entries from *Methods in Enzymology* [**vol**, page(s)]:
General discussion: **43**, 585
Assay: **43**, 589, 590
Other molecular properties: amino acid activating subunits, **43**, 598; ATP-PP$_i$ exchange, **43**, 588, 589, 596; *Bacillus brevis* preparation, **43**, 587, 588; enzymatic synthesis, **43**, 588, 589; molecular weight, **43**, 595; purification, **43**, 592); intermediate enzyme, **43**, 592; ornithine-dependent PP$_i$-ATP exchange, **43**, 588, 589, 596

TYROSINASE

This term has been used for both monophenol monooxygenase [EC 1.14.18.1] and catechol oxidase [EC 1.10.3.1].

TYROSINE 2,3-AMINOMUTASE

This ATP-dependent mutase [EC 5.4.3.6], also known as tyrosine α,β-mutase, catalyzes the conversion of L-tyrosine to 3-amino-3-(4-hydroxyphenyl)propanoate.[1]

[1] Z. Kurylo-Borowska & T. Abramsky (1972) *BBA* **264**, 1.

TYROSINE AMINOTRANSFERASE

This pyridoxal-phosphate-dependent aminotransferase [EC 2.6.1.5], also known as tyrosine transaminase, catalyzes the reversible reaction of L-tyrosine with α-ketoglutarate to produce 4-hydroxyphenylpyruvate and L-glutamate.[1–3] L-Phenylalanine is an alternative substrate. The mitochondrial enzyme may be identical with aspartate aminotransferase [EC 2.6.1.1]. There are three isoenzymic forms which can be interconverted with stem bromelain [EC 3.4.22.32] or fruit bromelain [EC 3.4.22.33]. The kinetic mechanism is pseudo-terreactant and ping pong (wherein the coenzyme freely dissociates from E or F in the ping pong mechanism).[1]

[1] G. Litwack & W. W. Cleland (1968) *Biochemistry* **7**, 2072.
[2] H. Hayashi, H. Wada, T. Yoshimura, N. Esaki & K. Soda (1990) *ARB* **59**, 87.
[3] A. E. Braunstein (1973) *The Enzymes*, 3rd ed., **9**, 379.

Selected entries from *Methods in Enzymology* [**vol**, page(s)]:
General discussion: **2**, 289; **17A**, 633; **34**, 295; **46**, 445
Assay: **18B**, 114, 115

Other molecular properties: diurnal rhythm, **36**, 477, 479 (hormone factors, **36**, 480); hepatocyte cell lines, in, **32**, 740; hormone induction, **32**, 733; induction by (hydrocortisone, **39**, 39; nicotinamide, **18B**, 113); kinetics, **87**, 355; properties of induction system, **18B**, 116; pseudo-terreactant kinetics, **87**, 355; purification from rat liver, **17A**, 634; rat interferon production, in, **78**, 173; staining of active enzyme on gels, **22**, 599

TYROSINE CARBOXYPEPTIDASE

This enzyme, originally classified as EC 3.4.16.3, is now included with carboxypeptidase C [EC 3.4.16.5].

TYROSINE DECARBOXYLASE

This pyridoxal-phosphate-dependent decarboxylase [EC 4.1.1.25] catalyzes the conversion of L-tyrosine to tyramine and carbon dioxide.[1–4] The bacterial enzyme also acts on 3-hydroxytyrosine and 3-hydroxyphenylalanine. Aromatic-L-amino-acid decarboxylase [EC 4.1.1.28] catalyzes this reaction as well.

[1] A. P. H. Phan, T. T. Ngo & H. M. Lenhoff (1983) *Appl. Biochem. Biotechnol.* **8**, 127.
[2] I. A. Marques & P. E. Brodelius (1988) *Plant Physiol.* **88**, 46 and 52.
[3] D. J. Creighton & N. S. R. K. Murthy (1990) *The Enzymes*, 3rd ed., **19**, 323.
[4] E. A. Boeker & E. E. Snell (1972) *The Enzymes*, 3rd ed., **6**, 217.

Selected entries from *Methods in Enzymology* [**vol**, page(s)]:
General discussion: **2**, 185, 188; **46**, 445
Other molecular properties: enzyme electrode, use in, **44**, 589; immobilization, on inert protein, **44**, 909; preparation, **2**, 188; **3**, 464; 963; properties of apoenzyme, **18A**, 512; purification of apoenzyme, **2**, 647; **18A**, 511; pyridoxal phosphate assay and, **3**, 963; **18A**, 509; resolution of, **2**, 189; transaminase assay and, **2**, 171

TYROSINE-ESTER SULFOTRANSFERASE

This sulfotransferase [EC 2.8.2.9], also known as aryl sulfotransferase IV, catalyzes the reaction of 3′-phospho-adenylylsulfate with L-tyrosine methyl ester to produce adenosine 3′,5′-bisphosphate and L-tyrosine methyl ester 4-sulfate. Phenols (including 2-naphthol, 2-chlorophenol, 3-chlorophenol, 4-chlorophenol, 3-methylphenol, and 4-nitrophenol) also act as acceptors.[1–3] This enzyme exhibits a rapid equilibrium random Bi Bi kinetic mechanism.[1,2] *See also Aryl Sulfotransferase*

[1] M. W. Duffel & W. B. Jakoby (1981) *JBC* **256**, 11123.
[2] P. Mattock, D. J. Barford, J. M. Basford & J. G. Jones (1970) *BJ* **116**, 805.
[3] A. D. Marshall, J. F. Darbyshire, P. McPhie & W. B. Jakoby (1998) *Chem. Biol. Interact.* **109**, 107.

Selected entries from *Methods in Enzymology* [**vol**, page(s)]:
General discussion: **77**, 197; serotonin sulfotransferase, **17B**, 825
Assay: **17B**, 825

TYROSINE 3-MONOOXYGENASE

This iron-dependent oxidoreductase [EC 1.14.16.2], also known as tyrosine 3-hydroxylase, catalyzes the reaction of L-tyrosine with tetrahydropteridine and dioxygen to produce 3,4-dihydroxy-L-phenylalanine, dihydropteridine, and water.[1–8] The rat enzyme has an ordered Ter Bi kinetic mechanism with the following binding order: coenzyme, dioxygen, and tyrosine.[5] L-Phenylalanine can also act as a substrate, producing both 3-hydroxy-L-phenylalanine and the minor product L-tyrosine. The reaction is attended by a NIH shift.[5] The cleavage of the oxygen-oxygen bond occurs in a separate step from amino acid hydroxylation.[4] Tyrosine 3-monooxygenase kinase [EC 2.7.1.124] catalyzes the activation of this enzyme by the seryl residue phosphorylation.

[1] S. Kaufman (1987) *The Enzymes*, 3rd ed., **18**, 217.
[2] T. J. Nelson & S. Kaufman (1987) *ABB* **257**, 69.
[3] M. Ikeda, M. Levitt & S. Udenfriend (1967) *ABB* **120**, 420.
[4] H. R. Ellis, S. C. Daubner & P. F. Fitzpatrick (2000) *Biochemistry* **39**, 4174.
[5] P. F. Fitzpatrick (1998) *CBC* **3**, 181.
[6] S. Kaufman (1995) *AE* **70**, 103.
[7] V. Massey & P. Hemmerich (1975) *The Enzymes*, 3rd ed., **12**, 191.
[8] V. Ullrich & W. Duppel (1975) *The Enzymes*, 3rd ed., **12**, 253.

Selected entries from *Methods in Enzymology* [vol, page(s)]:
General discussion: 17A, 609; **34**, 6; **52**, 12; **142**, 56, 63, 71
Assay: 32, 785; **142**, 57, 63, 72

TYROSINE *N*-MONOOXYGENASE

This heme-thiolate-dependent oxidoreductase [EC 1.14.13.41], also known as tyrosine *N*-hydroxylase and cytochrome P450$_{Tyr}$ (or, CYP79A1), catalyzes the reaction of L-tyrosine with NADPH and dioxygen to produce *N*-hydroxy-L-tyrosine, NADP$^+$, and water.[1–4]

N-hydroxytyrosine

2-nitroso-3-(*p*-hydroxy-phenyl)propionate

(*E*)-*p*-hydroxyphenyl-acetaldehyde oxime

This multifunctional enzyme catalyzes a second hydroxylation step, thereby generating *N,N*-dihydroxy-L-tyrosine which is subsequently dehydrated to 2-nitroso-3-(*p*-hydroxyphenyl)propionate. Decarboxylation then follows to generate *p*-hydroxyphenylacetaldehyde oxime (the resulting *E/Z* ratio of the oxime produced is 69:31).[4] The dehydration and decarboxylation reactions may proceed nonenzymatically. This enzyme catalyzes the committed step in the biosynthesis of the cyanogenic glucoside dhurrin.

[1] B. A. Halkier & B. L. Moller (1990) *JBC* **265**, 21114.
[2] O. Sibbesen, B. Koch, B. A. Halkier & B. L. Moller (1994) *PNAS* **91**, 9740.
[3] B. A. Halkier, O. Sibbesen, B. Koch & B. L. Moller (1995) *Drug Metabol. Drug Interact.* **12**, 285.
[4] O. Sibbesen, B. Koch, B. A. Halkier & B. L. Moller (1995) *JBC* **270**, 3506.

Selected entries from *Methods in Enzymology* [vol, page(s)]:
General discussion: 272, 268

TYROSINE 3-MONOOXYGENASE KINASE

This phosphotransferase [EC 2.7.1.124] catalyzes the reaction of ATP with tyrosine 3-monooxygenase to produce ADP and [tyrosine-3-monooxygenase] phosphate, thus activating that enzyme. The kinase copurifies with the monooxygenase.[1,2]

[1] D. Pigeon, R. Drissi-Daoudi, F. Gros & J. Thibault (1986) *Compt. Rend. Acad. Sci. Paris Ser. 3*, **302**, 435.
[2] D. Pigeon, P. Ferrara, F. Gros & J. Thibault (1987) *JBC* **262**, 6155.

TYROSINE PHENOL-LYASE

This pyridoxal-phosphate-dependent lyase [EC 4.1.99.2], also known as β-tyrosinase, catalyzes the reaction of L-tyrosine with water to produce phenol, pyruvate, and ammonia.[1–4] Other less effective substrates include D-tyrosine, *S*-methyl-L-cysteine, L-cysteine, L-serine, and D-serine. Intriguingly, this lyase also catalyzes racemization of L-alanine. Isotope effect investigations indicate that proton abstraction from the 2-position of the substrate is partially rate limiting.[1] In addition, while both the protonated and unprotonated enzyme can bind the substrate, and may be interconverted directly, only the unprotonated Michaelis complex is catalytically competent.[1] **See also** *Tyrosinase*

[1] D. M. Kiick & R. S. Phillips (1988) *Biochemistry* **27**, 7333.
[2] H. Chen, P. Gollnick & R. S. Phillips (1995) *EJB* **229**, 540.
[3] E. T. Adman (1991) *Adv. Protein Chem.* **42**, 145.
[4] L. Davis & D. E. Metzler (1972) *The Enzymes*, 3rd ed., **7**, 33.

Selected entries from *Methods in Enzymology* [vol, page(s)]:
General discussion: 17A, 642; **44**, 886; **52**, 6, 9
Assay: coupled enzyme assay, **63**, 34
Other molecular properties: coupling enzyme, use as a, **63**, 34; oxy form, spectrum, **52**, 36; prosthetic group, **52**, 4; reaction mechanism, **52**, 39; stereochemistry, **87**, 149

TYROSINE:PYRUVATE AMINOTRANSFERASE

This enzyme activity, originally classified as EC 2.6.1.20, is now a deleted EC entry.

TYROSYLARGININE SYNTHETASE

This ligase [EC 6.3.2.24], also known as tyrosine:arginine ligase and kyotorphin synthetase, catalyzes the reaction of ATP with L-tyrosine and L-arginine to produce AMP, pyrophosphate (or, diphosphate), and L-tyrosyl-L-arginine.[1]

[1] H. Ueda, Y. Yoshihara, N. Fukushima, H. Shiomi, A. Nakamura & H. Takagi (1987) JBC **262**, 8165.

TYROSYL-tRNA SYNTHETASE

This enzyme [EC 6.1.1.1], also known as tyrosine:tRNA ligase and tyrosine translase, catalyzes the reaction of ATP with L-tyrosine and tRNA^{Tyr} to produce AMP, pyrophosphate (or, diphosphate), and L-tyrosyl-tRNA^{Tyr}.[1–5] This class I aminoacyl-tRNA synthetase has a random Bi Uni Uni Bi ping pong kinetic mechanism in which either L-tyrosine or ATP can bind to the free enzyme, the tyrosyl-adenylate intermediate is produced, pyrophosphate is released, and is followed by binding of tRNA^{Tyr}. **See also** *Aminoacyl-tRNA Synthetases*

[1] V. Buonocore, M. H. Harris & S. Schlesinger (1972) JBC **247**, 4843.
[2] E. A. First & A. R. Fersht (1995) Biochemistry **34**, 5030.
[3] E. A. First (1998) CBC **1**, 573.
[4] K. A. Johnson (1992) The Enzymes, 3rd ed., **20**, 1.
[5] P. Schimmel (1987) Ann. Rev. Biochem. **56**, 125.

Selected entries from **Methods in Enzymology** [**vol**, page(s)]:
General discussion: 5, 722; **29**, 547; **34**, 170, 503; **46**, 90
Assay: 5, 722
Other molecular properties: efficiency, site-directed mutagenesis studies, **249**, 105; hydroxamate formation rate, **29**, 613; phosphocellulose capacity for, **34**, 167; properties, **5**, 725; purification, **5**, 723; **29**, 564; **59**, 262; separation, **29**, 559; stereochemistry, **87**, 206; subcellular distribution, **59**, 233, 234

UBIQUINOL:CYTOCHROME c REDUCTASE

This enzyme complex [EC 1.10.2.2], also known as cytochrome bc_1 and complex III, catalyzes the reaction of ubiquinol with two ferricytochrome c to produce ubiquinone and two ferrocytochrome c. The complex also contains cytochrome b_{562}, cytochrome b_{566}, cytochrome c_1, and a two-iron ferredoxin.[1–4]

ubiquinol (50)   ubiquinone (50)

Ferricyanide and 2,6-dichlorophenolindophenol can be used in place of cytochrome c. Depending on the organism and physiological conditions, the complex will also catalyze the movement of either two or four protons from the cytoplasmic side of the membrane to the non-cytoplasmic compartment. **See also** Cytochrome bc_1

[1] J. V. Schloss & M. S. Hixon (1998) *CBC* **2**, 43.
[2] U. Brandt & B. Trumpower (1994) *CRBMB* **29**, 165.
[3] B. L. Trumpower & R. B. Gennis (1994) *ARB* **63**, 675.
[4] G. Palmer (1975) *The Enzymes*, 3rd ed., **12**, 1.

Selected entries from *Methods in Enzymology* [**vol**, page(s)]:
General discussion: 34, 301; **53**, 35, 80, 92, 98, 113; **56**, 577; **125**, 86; **126**, 191, 201, 224, 293; **260**, 51, 63, 70, 82
Assay: 10, 239; **31**, 19; **53**, 38, 90, 91; cytochrome c reduction, **260**, 54, 68, 77; NADH dehydrogenase complex, assay, **264**, 492; processing peptidase activity, **260**, 78; proton transfer, **260**, 68, 77; *Rhodopseudomonas sphaeroides*, **126**, 283; spectrophotometric assay, **264**, 494; succinate-ubiquinone oxidoreductase complex, assay, **264**, 493; yeast mitochondrial, **126**, 178
Other molecular properties: activity, **53**, 35, 37, 90; antimycin inhibition, **264**, 477, 500; bovine heart mitochondria, **53**, 35, 80 (isolation and properties, **126**, 295; photosynthetic reaction center hybrid system, **126**; components, **56**, 580, 582, 583); cytochrome a_3 assay and, **2**, 737; cytochrome bc_1 particle or complex (assay [*Rhodopseudomonas sphaeroides*, **126**, 283; yeast mitochondrial, **126**, 178]; beef heart complex III, from, **53**, 92 [bovine heart mitochondria {iron-sulfur protein, purification, **126**, 212; protein subunits, isolation,

126, 227; purification, **126**, 183, 186, 225; reaction with *N,N'*-dicyclohexylcarbodiimide, **125**, 102; redox Bohr effects, measurement, **126**, 338; resolution, **126**, 218, 231}]; binding of cytochrome c, **53**, 104, 105; composition, **53**, 116, 117; heme centers, properties, **53**, 110, 112; inhibitors, **55**, 461, 462; iron-sulfur center, EPR signal properties, **54**, 138, 140, 141, 143, 144, 146, 147; isolation, **53**, 92; kinetic properties, **53**, 117; mitochondrial, inhibition by [antimycin, **126**, 267; 2,3-dimercaptopropanol, **126**, 270; 5,5'-dithiobis(2-nitrobenzoic acid), **126**, 271; funiculosin, **126**, 268; hydroxyquinoline *N*-oxides, **126**, 270; hydroxyquinones, **126**, 263; β-methoxyacrylate, **126**, 257]; *Neurospora crassa* [isolation, **126**, 203; structural properties, **126**, 206]; *Paracoccus denitrificans* [cytochrome b purification, **126**, 326; properties, **126**, 323]; peptide [composition, **53**, 95, 96, 106, 107; molecular weight, **53**, 117; physicochemical data, **53**, 97, 98; preparation, **10**, 222; properties, **10**, 223; purification, **53**, 223; **126**, 318 {affinity chromatography, by, **53**, 100; ammonium sulfate fractionation, by, **53**, 114, 115}]; *Rhodopseudomonas sphaeroides* [assays, **126**, 283; cytochrome b purification, **126**, 326; purification, **126**, 183, 186, 282]; small [amino acid composition, **53**, 110; definition, **53**, 99; peptides [amino acid compositions, **53**, 110; molar ratios, **53**, 111; molecular weights, **53**, 111]; purification, **53**, 105; spectral properties, **53**, 112]; spectral properties, **53**, 112, 117; spectrum, **10**, 224; ubiquinone extraction, **53**, 576; yeast, from, **53**, 114 [mitochondrial {assays, **126**, 178; properties, **126**, 178; purification, **126**, 174, 183, 185; reconstitution, **126**, 179}]); electron spin resonance of, **6**, 914; electron transport particle and, **5**, 742; **6**, 418; extraction of, **1**, 35; kinetic studies, using monospecific antibodies, **74**, 260; marker enzyme, as, **31**, 735, 743, 745; microbial supernatant and, **6**, 291; $NADP^+$ assay and, **3**, 894; plasma membranes, in, **31**, 165, 166; pyridine nucleotide and, **4**, 303; solubilization, **1**, 39; sources, **1**, 29, 34, 39, 50; supernatant fraction and, **5**, 56

UBIQUINOL OXIDASE

Ubiquinol oxidase is the term used with respect to the membrane-bound complexes such as those from *Paracoccus denitrificans* that catalyze the oxidation of ubiquinol to ubiquinone. The *P. denitrificans* complex contains cytochromes b, c_1, a, a_3, and c-552.

Selected entries from *Methods in Enzymology* [**vol**, page(s)]:
General discussion: 126, 305
Assay: 126, 106, 114, 121, 306
Other molecular properties: cytochrome o, **52**, 15 (*Acetobacter suboxydans*, from, **53**, 208; *Escherichia coli*, from, **53**, 207; extraction from bacterial membrane, **22**, 206; *Micrococcus pyogenes* var. *albus*, from, **53**, 207; photosynthetic bacteria, of, **23**, 363; prosthetic group, **52**, 5, 15; terminal oxidase complex, *Escherichia coli* [oxidase activity, effect of oxygen concentration, **126**, 109; oxidation-reduction potential, **126**, 105; polypeptide composition, **126**, 99; purification, **126**, 95; reconstitution, **126**, 109; spectral properties, **126**, 101; ubiquinol oxidase assay, **126**, 106]; *Vitreoscilla*, from, **53**, 209); cytochrome o oxidase (assay, **126**, 114, 121; containing proteoliposomes, characterization, **126**, 121; *Escherichia coli* [assays, **126**, 114, 121;

properties, **126**, 119; purification, **126**, 115; reconstitution, **126**, 118, 124]); *Paracoccus denitrificans* (assays, **126**, 306; properties, **126**, 313; purification, **126**, 307)

UBIQUITIN:CALMODULIN LIGASE

This ligase [EC 6.3.2.21], also known as ubiquityl-calmodulin synthetase, catalyzes the reaction of *n*(ATP) with calmodulin and *n*(ubiquitin) to produce *n*(AMP), *n*(pyrophosphate), and (ubiquitin)*n*-calmodulin. At least three ubiquitin molecules become covalently linked to lysyl residues in calmodulin.[1,2]

[1]H. P. Jennissen & M. Laub (1988) *Hoppe-Seyler Biol. Chem.* **369**, 1325.
[2]M. Majetschak, M. Laub, C. Klocke, J. A. Steppuhn & H. P. Jennissen (1998) *EJB* **255**, 492.

UBIQUITIN CARBOXYL-TERMINAL HYDROLASE

This hydrolase [EC 3.4.19.12; previously classified as the thiolesterase, EC 3.1.2.15] and also referred to as ubiquitin thiolesterase or ubiquitin carboxy-terminal esterase, catalyzes the hydrolysis of a ubiquitin C-terminal thiolester (*i.e.*, at the C-terminal Gly76), producing ubiquitin and a thiol.[1-4] The enzyme will also act on AMP-ubiquitin. Enzyme inactivation by sodium borohydride or hydroxylamine (when ubiquitin is present) suggests a mechanism proceeding through a ubiquitinyl-enzyme intermediate. Some variants fail to hydrolyze ubiquitin when a proline is in the P1′ position. *See also Ubiquitin Isopeptidase T*

[1]M. Hochstrasser (1996) *Cell* **84**, 813.
[2]A. Hershko & A. Ciechanover (1998) *ARB* **67**, 425.
[3]K. D. Wilkinson, M. J. Cox, A. N. Mayer & T. Frey (1986) *Biochemistry* **25**, 6644.
[4]C. M. Pickart & I. A. Rose (1986) *JBC* **261**, 10210.

UBIQUITIN ISOPEPTIDASE T

This peptidase, presently considered to be a variant of ubiquitin carboxyl-terminal hydrolase [EC 3.1.2.15] by IUBMB, is one of the deubiquitinating cysteine peptidases acting on proteins targeted for proteolysis by the 26S proteosome. Isopeptidase T catalyzes the disassembly of polyubiquitin chains by facilitating hydrolysis of Gly76–Lys48 isopeptide bonds.[1,2]

[1]K. D. Wilkinson (1995) *ARN* **15**, 161.
[2]M. Hochstrasser (1996) *Cell* **84**, 813.

UBIQUITIN:PROTEIN LIGASE

This ligase [EC 6.3.2.19], also referred to as ubiquitin-conjugating enzyme and ubiquitin-activating enzyme, catalyzes the reaction of ATP with ubiquitin and a protein lysyl residue to produce AMP, pyrophosphate, and an N^{ε}-ubiquitylated protein.[1-4] The isopeptide bond joins the C-terminal glycine of ubiquitin to the ε-amino group.

[1]A. Hershko & A. Ciechanover (1998) *ARB* **67**, 425
[2]K. D. Wilkinson (1995) *ARN* **15**, 161.
[3]A. Hershko & A. Ciechanover (1992) *ARB* **61**, 761.
[4]M. Rechsteiner (1987) *ARCB* **3**, 1.

UBIQUITIN THIOLESTERASE

This enzyme [EC 3.1.2.15], also referred to as ubiquitin carboxy-terminal esterase, catalyzes hydrolysis of a ubiquitin C-terminal thiolester to produce ubiquitin and a free thiol.[1] The thiolesterase acts on thiolesters formed between thiols such as dithiothreitol or glutathione and the C-terminal glycine residue of the polypeptide ubiquitin. In many instances, this enzyme appears to be identical to ubiquitin carboxyl-terminal hydrolase [EC 3.4.19.12].
See Ubiquitin Carboxy-Terminal Hydrolase

[1]I. A. Rose & J. V. B. Warms (1983) *Biochemistry* **22**, 4234.

UDP-*N*-ACETYLGALACTOSAMINE-4-SULFATE SULFOTRANSFERASE

This transferase [EC 2.8.2.7] catalyzes the reaction of 3′-phosphoadenylylsulfate with UDP-*N*-acetyl-D-galactosamine 4-sulfate to produce adenosine 3′,5′-bisphosphate and UDP-*N*-acetyl-D-galactosamine 4,6-bissulfate.[1,2]

[1]T. Harada, S. Shimizu, Y. Nakanishi & S. Suzuki (1967) *JBC* **242**, 2288.
[2]K. Otsu, H. Inoue, Y. Nakanishi, S. Kato, M. Tsuji & S. Suzuki (1984) *JBC* **259**, 6403.

UDP-*N*-ACETYLGLUCOSAMINE 1-CARBOXYVINYLTRANSFERASE

This transferase [EC 2.5.1.7], also referred to as enoylpyruvate transferase and UDP-*N*-acetylglucosamine enolpyruvyltransferase, catalyzes the reaction of phosphoenolpyruvate with UDP-*N*-acetyl-D-glucosamine to produce UDP-*N*-acetyl-3-*O*-(1-carboxyvinyl)-D-glucosamine and orthophosphate.[1,2] As the committed step in the biosynthesis of bacterial cell wall peptidoglycan, the reaction is feedback inhibited by the UDP-muramoyl pentapeptide. The reaction proceeds by an addition-elimination mechanism *via* a tetrahedral ketal intermediate where a cysteinyl residue acts as an active-site general acid. The stereochemistry of elimination of the tetrahedral intermediate of the reaction is *syn*.[3]

[1]K. J. Gruys & J. A. Sikorski (1998) *CBC* **1**, 273.
[2]J. V. Schloss & M. S. Hixon (1998) *CBC* **2**, 43.
[3]T. Skarzynski, D. H. Kim, W. J. Lees, C. T. Walsh & K. Duncan (1998) *Biochemistry* **37**, 2572.

UDP-*N*-ACETYLGLUCOSAMINE 6-DEHYDROGENASE

This oxidoreductase [EC 1.1.1.136] catalyzes the reaction of UDP-*N*-acetyl-D-glucosamine with water and two molecules of NAD$^+$ to produce UDP-*N*-acetyl-2-amino-2-deoxy-D-glucuronate and two molecules of NADH.[1,2]

[1] D.-F. Fan, C. E. John, J. Zalitis & D. S. Feingold (1969) *ABB* **135**, 45.
[2] T. Kawamura, N. Ichihara, S. Sugiyama, H. Yokota, N. Ishimoto & E. Ito (1985) *JB* **98**, 105.

Selected entries from *Methods in Enzymology* [vol, page(s)]:
General discussion: 28, 435
Assay: 28, 435
Other molecular properties: properties, 28, 436

UDP-*N*-ACETYLGLUCOSAMINE:DOLICHYL-PHOSPHATE *N*-ACETYLGLUCOSAMINE-PHOSPHOTRANSFERASE

This transferase [EC 2.7.8.15], also referred to as *N*-acetylglucosamine-1-phosphate transferase and GlcNAc-1-P transferase, catalyzes the reaction of UDP-*N*-acetyl-D-glucosamine with dolichyl phosphate to produce UMP and *N*-acetyl-D-glucosaminyl-diphosphodolichol.[1–4]

[1] C. B. Sharma, L. Lehle & W. Tanner (1982) *EJB* **126**, 319.
[2] C. L. Villemez & P. L. Carlo (1980) *JBC* **255**, 8174.
[3] K. Shailubhai, B. Dong-Yu, E. S. Saxena & I. K. Vijay (1988) *JBC* **263**, 15964.
[4] K. A. Presper & E. C. Heath (1983) *The Enzymes*, 3rd ed., **16**, 449.

UDP-*N*-ACETYLGLUCOSAMINE 2-EPIMERASE

This epimerase [EC 5.1.3.14] catalyzes the interconversion of UDP-*N*-acetyl-D-glucosamine and UDP-*N*-acetyl-D-mannosamine.[1–4] The mammalian enzyme also catalyzes the conversion of UDP-*N*-acetyl-D-mannosamine to UDP and *N*-acetyl-D-mannosamine. The rat liver enzyme is a bifunctional catalyst possessing *N*-acetylmannosamine kinase activity.[1]

[1] S. Hinderlich, R. Stasche, R. Zeitler & W. Reutter (1997) *JBC* **272**, 24313.
[2] M. E. Tanner & G. L. Kenyon (1998) *CBC* **2**, 7.
[3] E. Adams (1976) *AE* **44**, 69.
[4] L. Glaser (1972) *The Enzymes*, 3rd ed., **6**, 355.

Selected entries from *Methods in Enzymology* [vol, page(s)]:
General discussion: 9, 612; 83, 515
Assay: 9, 612; 83, 515
Other molecular properties: *Escherichia coli* (assay, 83, 515; properties, 83, 518; purification, 83, 516); liver, rat, (properties, 9, 615; purification, 9, 613)

UDP-*N*-ACETYLGLUCOSAMINE 4-EPIMERASE

This NAD$^+$-dependent epimerase [EC 5.1.3.7], also known as UDP-*N*-acetylgalactosamine 4-epimerase, catalyzes the interconversion of UDP-*N*-acetyl-D-glucosamine and UDP-*N*-acetyl-D-galactosamine. The reaction presumably proceeds via the UDP-*N*-acetyl-4-keto-D-glucosamine intermediate. In some organisms, UDP-glucose 4-epimerase and UDP-*N*-acetylglucosamine 4-epimerase activities are associated with a single polypeptide.[1]

[1] F. Piller, M. H. Hanlon & R. L. Hill (1983) *JBC* **258**, 10774.

Selected entries from *Methods in Enzymology* [vol, page(s)]:
General discussion: 8, 277
Assay: 8, 277
Other molecular properties: occurrence, 8, 280; properties, 8, 281; purification, 8, 280

UDP-*N*-ACETYLGLUCOSAMINE: GLYCOPROTEIN *N*-ACETYLGLUCOS-AMINYLTRANSFERASE

This transferase subclass was originally listed as EC 2.4.1.51 and included enzymes catalyzing the reaction of UDP-*N*-acetyl-D-glucosamine with a glycoprotein to produce UDP and an *N*-acetyl-D-glucosaminylglycoprotein. These enzymes are now individually reclassified.

UDP-*N*-ACETYLGLUCOSAMINE: LYSOSOMAL ENZYME *N*-ACETYLGLUCOS-AMINEPHOSPHOTRANSFERASE

This Mg^{2+}/Mn^{2+}-dependent transferase [EC 2.7.8.17], also referred to as *N*-acetylglucosaminylphosphotransferase, catalyzes the reaction of UDP-*N*-acetyl-D-glucosamine with a lysosomal enzyme-bound D-mannose to produce UMP and a lysosomal enzyme *N*-acetyl-D-glucosaminyl-phospho-D-mannose.[1] An α-linked *N*-acetylglucosamine 1-phosphate is transferred to the 6-hydroxyl group of a mannose in high mannose-containing oligosaccharides of glycoproteins. Protein substrates include cathepsin D, uteroferrin, and β-*N*-acetylhexosaminidase.

[1] M. Bao, B. J. Elmendorf, J. L. Booth, R. R. Drake & W. M. Canfield (1996) *JBC* **271**, 31446.

Selected entries from *Methods in Enzymology* [vol, page(s)]:
General discussion: 107, 163
Assay: 107, 163

UDP-*N*-ACETYLGLUCOSAMINE PYROPHOSPHORYLASE

This divalent cation-dependent transferase [EC 2.7.7.23], also known as UDP-*N*-acetylglucosamine diphosphorylase and *N*-acetylglucosamine-1-phosphate uridylyltransferase, catalyzes the reversible reaction of UTP with *N*-acetyl-α-D-glucosamine 1-phosphate to produce pyrophosphate and UDP-*N*-acetyl-D-glucosamine.[1–3] The *Escherichia coli* protein is bifunctional and exhibits glucosamine-1-phosphate acetyltransferase activity.[4] Note that the *E. coli* enzyme can produce UDP-*N*-acetyl-ᴅ-glucosamine from UTP, ᴅ-glucosamine 1-phosphate, and acetyl-CoA. Acetyl transfer precedes uridylyl transfer at separate active sites.[5]

[1] T. N. Pattabiraman & B. K. Bachhawat (1961) *BBA* **50**, 129.
[2] J. L. Strominger & M. S. Smith (1959) *JBC* **234**, 1822.
[3] K. Yamamoto, M. Moriguchi, H. Kawai & T. Tochikura (1980) *BBA* **614**, 367.
[4] D. Mengin-Lecreulx & J. van Heijenoort (1994) *J. Bact.* **176**, 5788.
[5] A. M. Gehring, W. J. Lees, D. J. Mindiola, C. T. Walsh & E. D. Brown (1996) *Biochemistry* **35**, 579.

UDP-*N*-ACETYL-ᴅ-MANNOSAMINE DEHYDROGENASE

This oxidoreductase catalyzes the reaction of UDP-*N*-acetyl-ᴅ-mannosamine with two molecules of NAD$^+$ to produce UDP-*N*-acetyl-ᴅ-mannosaminuronate and two NADH molecules. The *Escherichia coli* enzyme is reportedly quite specific for UDP-*N*-acetyl-ᴅ-mannosamine. The reaction is analogous to that of UDP-*N*-acetyl-ᴅ-glucosamine dehydrogenase [EC 1.1.1.136].

Selected entries from *Methods in Enzymology* [vol, page(s)]:
General discussion: 83, 519
Assay: 83, 519
Other molecular properties: *Escherichia coli* (properties, **83**, 521; purification, **83**, 520)

UDP-*N*-ACETYLMURAMATE DEHYDROGENASE

This FAD-dependent oxidoreductase [EC 1.1.1.158], also known as UDP-*N*-acetylenolpyruvoylglucosamine reductase, catalyzes the reversible reaction of UDP-*N*-acetylmuramate with NADP$^+$ to produce UDP-*N*-acetyl-3-*O*-(1-carboxyvinyl)-ᴅ-glucosamine and NADPH.[1–4] The reverse reaction is often the physiological observed process, and reducing agents such as sodium dithionite, sodium borohydride, and NADH can replace NADPH with diminished catalytic activity. The *Escherichia coli* enzyme exhibits ping pong Bi Bi kinetics as well as substrate inhibition.[3] In the reduction of the enolpyruvyl-UDP-*N*-acetylglucosamine, a hydride ion is transferred from the flavin N^5 position to the substrate, thereby forming an enediol intermediate.[4]

[1] A. Taku, K. G. Gunetileke & R. A. Anwar (1970) *JBC* **245**, 5012.
[2] A. Taku & R. A. Anwar (1973) *JBC* **248**, 4971.
[3] A. M. Dhalla, J. Yanchunas, Jr., H. T. Ho, P. J. Falk, J. J. Villafranca & J. G. Robertson (1995) *Biochemistry* **34**, 5390.
[4] T. E. Benson, C. T. Walsh & V. Massey (1997) *Biochemistry* **36**, 796.

UDP-*N*-ACETYLMURAMOYL-ʟ-ALANINE SYNTHETASE

This enzyme [EC 6.3.2.8], also known as UDP-*N*-acetylmuramate:alanine ligase and ʟ-alanine-adding enzyme, catalyzes the reaction of ATP with UDP-*N*-acetylmuramate and ʟ-alanine to produce UDP-*N*-acetylmuramoyl-ʟ-alanine, ADP, and orthophosphate.[1] The reaction most likely proceeds through an acyl phosphate intermediate.[2]

[1] F. Nosal, A. Masson, R. Legrand, D. Blanot, B. Schoot, J. van Heijenoort & C. Parquet (1998) *FEBS Lett.* **426**, 309.
[2] J. J. Emanuele, Jr., H. Jin, J. Yanchunas, Jr., & J. J. Villafranca (1997) *Biochemistry* **36**, 7264.

UDP-*N*-ACETYLMURAMOYL-ʟ-ALANYL-ᴅ-GLUTAMATE SYNTHETASE

This enzyme [EC 6.3.2.9], also known as UDP-*N*-acetylmuramoylalanine:ᴅ-glutamate ligase and ᴅ-glutamate-adding enzyme, catalyzes the reaction of ATP with UDP-*N*-acetylmuramoyl-ʟ-alanine and ᴅ-glutamate to produce ADP, orthophosphate, and UDP-*N*-acetylmuramoyl-ʟ-alanyl-ᴅ-glutamate.[1–3]

UDP-*N*-acetylmuramoyl-ʟ-alanine

UDP-*N*-acetylmuramoyl-ʟ-alanyl-ᴅ-glutamate

[1] C. Michaud, D. Blanot, B. Flouret & J. Van Heijenoort (1987) *EJB* **166**, 631.
[2] D. Mengin-Lecreulx, B. Flouret & J. Van Heijenoort (1982) *J. Bact.* **151**, 1109.
[3] S. G. Nathenson, J. L. Strominger & E. Ito (1964) *JBC* **239**, 1773.

UDP-*N*-ACETYLMURAMOYL-ʟ-ALANYL-ᴅ-GLUTAMYL-*meso*-2,6- DIAMINOPIMELATE SYNTHETASE

This ligase [EC 6.3.2.13], also known as UDP-*N*-acetylmuramoylalanyl-ᴅ-glutamate:2,6-diaminopimelate ligase and UDP-*N*-acetylmuramyl-tripeptide synthetase, catalyzes the reaction of ATP with UDP-*N*-acetylmuramoyl-ʟ-alanyl-ᴅ-glutamate and *meso*-2,6-diaminoheptanedioate to produce UDP-*N*-acetylmuramoyl-ʟ-alanyl-ᴅ-glutamyl-*meso*-2,6-diaminoheptanedioate, ADP, and orthophosphate.[1]

[1] C. Michaud, D. Mengin-Lecreulx, J. van Heijenoort & D. Blanot (1990) *EJB* **194**, 853.

Selected entries from *Methods in Enzymology* [vol, page(s)]:
General discussion: 8, 324; 17B, 150
Assay: 17B, 150

UDP-*N*-ACETYLMURAMOYL-L-ALANYL-D-GLUTAMYL-*meso*-2,6-DIAMINOPIMELOYL-D-ALANYL-D-ALANINE SYNTHETASE

This ligase [EC 6.3.2.15], also known as UDP-*N*-acetylmuramoyl-L-alanyl-D-glutamyl-2,6-*meso*-diaminopimelate:D-alanyl-D-alanine ligase and UDP-MurNAc-pentapeptide synthetase, catalyzes the reaction of ATP with UDP-*N*-acetylmuramoyl-L-alanyl-D-glutamyl-*meso*-2,6-diaminoheptanedioate and D-alanyl-D-alanine to produce UDP-*N*-acetylmuramoyl-L-alanyl-D-glutamyl-6-carboxy-L-lysyl-D-alanyl-D-alanine, ADP, and orthophosphate.[1]

[1] M. S. Anderson, S. S. Eveland, H. R. Onishi & D. L. Pompliano (1996) *Biochemistry* 35, 16264.

UDP-*N*-ACETYLMURAMOYL-L-ALANYL-D-GLUTAMYL-L-LYSINE SYNTHETASE

This enzyme [EC 6.3.2.7], also known as UDP-*N*-acetylmuramoyl-L-alanyl-D-glutamate:L-lysine ligase and L-lysine-adding enzyme, catalyzes the reaction of UDP-*N*-acetylmuramoyl-L-alanyl-D-glutamate with ATP and L-lysine to produce UDP-*N*-acetylmuramoyl-L-alanyl-D-glutamyl-L-lysine, ADP, and orthophosphate. The product orthophosphate reportedly is also an activator.[1]

[1] R. A. Anwar & M. Vlaovic (1986) *Biochem Cell Biol.* 64, 297.

UDP-*N*-ACETYLMURAMOYL-L-ALANYL-D-GLUTAMYL-L-LYSYL-D-ALANYL-D-ALANINE SYNTHETASE

This ligase [EC 6.3.2.10], also known as UDP-*N*-acetylmuramoyl-L-alanyl-D-glutamyl-L-lysine:D-alanyl-D-alanine ligase and UDP-MurNAc-L-Ala-D-Glu-L-Lys:D-Ala-D-Ala ligase, catalyzes the reaction of ATP with UDP-*N*-acetylmuramoyl-L-alanyl-D-glutamyl-L-lysine and D-alanyl-D-alanine to produce UDP-*N*-acetylmuramoyl-L-alanyl-D-glutamyl-L-lysyl-D-alanyl-D-alanine, ADP, and orthophosphate.[1,2]

[1] B. Oppenheim & A. Patchornik (1974) *FEBS Lett.* 48, 172.
[2] P. E. Linnett & D. J. Tipper (1974) *J. Bact.* 120, 342.

UDP-*N*-ACETYLMURAMOYLPENTAPEPTIDE-LYSINE *N*[6]-ALANYLTRANSFERASE

This transferase [EC 2.3.2.10] catalyzes the reaction of L-alanyl-tRNA[Ala] with UDP-*N*-acetylmuramoyl-L-alanyl-D-glutamyl-L-lysyl-D-alanyl-D-alanine to produce tRNA[Ala] and UDP-*N*-acetylmuramoyl-L-alanyl-D-glutamyl-*N*[6]-(L-alanyl)-L-lysyl-D-alanyl-D-alanine.[1,2] The enzyme can also utilize L-seryl-tRNA[Ser] as the donor substrate.

[1] R. Plapp & J. L. Strominger (1970) *JBC* 245, 3673.
[2] R. L. Soffer (1974) *AE* 40, 91.

UDP-4-AMINO-2-ACETAMIDO-2,4,6-TRIDEOXYGLUCOSE AMINOTRANSFERASE

This pyridoxal-phosphate-dependent transferase [EC 2.6.1.34] catalyzes the reversible reaction of UDP-2-acetamido-4-amino-2,4,6-trideoxyglucose with α-ketoglutarate to produce UDP-2-acetamido-4-dehydro-2,6-dideoxyglucose and L-glutamate.[1]

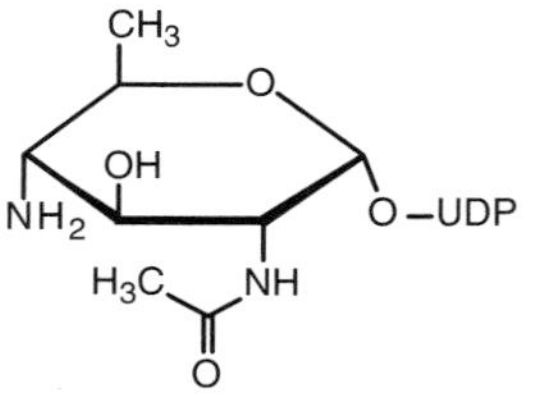

UDP-2-acetamido-4-amino-2,4,6-trideoxyglucose

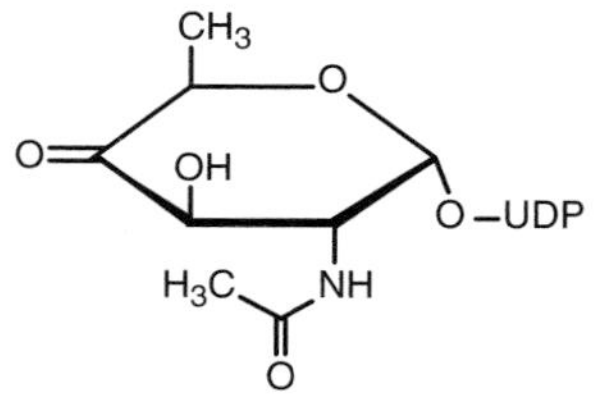

UDP-2-acetamido-4-dehydro-2,6-dideoxyglucose

[1] J. Distler, B. Kaufman & S. Roseman (1966) *ABB* 116, 466.

UDP-ARABINOSE 4-EPIMERASE

This epimerase [EC 5.1.3.5], also known as UDP-xylose 4-epimerase, catalyzes the reversible interconversion of UDP-L-arabinose and UDP-D-xylose, reportedly by means of an UDP-4-ketoxylose intermediate.[1–3]

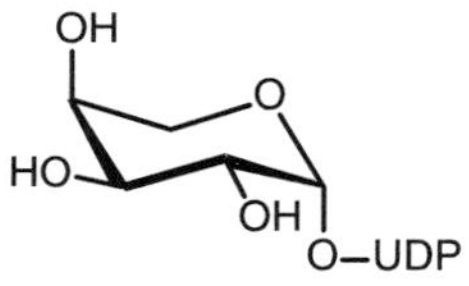

UDP-L-arabinose

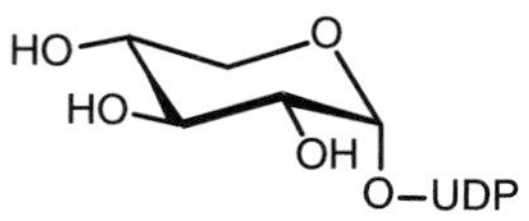

UDP-D-xylose

[1] D. S. Feingold, E. F. Neufeld & W. Z. Hassid (1960) *JBC* 235, 910.
[2] D. Fan & D. S. Feingold (1970) *Plant Physiol.* 46, 592.
[3] G. Dalessandro & D. H. Northcote (1977) *Phytochemistry* 16, 853.

Selected entries from *Methods in Enzymology* [vol, page(s)]:
General discussion: 6, 787; 28, 422
Assay: 28, 422
Other molecular properties: apiose and, 28, 445; equilibrium constant, 6, 787; isolation from wheat germ, 28, 423; properties, 28, 424; separation of substrate and product, 6, 784, 786; uridine diphosphate pentose pyrophosphorylase and, 6, 784

UDP-GALACTOPYRANOSE MUTASE

This FAD-dependent mutase [EC 5.4.99.9] catalytically interconverts UDP-D-galactopyranose and UDP-D-galacto-1,4-furanose.[1,2]

UDP-D-galactopyranose   UDP-D-galacto-1,4-furanose

[1]A. G. Trejo, T. G. J. F. Chittenden, J. G. Buchanan & J. Baddile (1970) *BJ* **117**, 637.
[2]P. M. Nassau, S. L. Martin, R. E. Brown, A. Weston, D. Monsey, M. R. McNeil & K. Duncan (1996) *J. Bact.* **178**, 1047.

UDP-GALACTOSE:*N*-ACETYL-D-GLUCOS-AMINE β1,3-GALACTOSYLTRANSFERASE

Catalysis of the reaction of UDP-D-galactose with *N*-acetyl-D-glucosamine to produce D-galactosyl-β(1,3)-*N*-acetyl-D-glucosamine and UDP represents a minor activity of glycoprotein-*N*-acetylgalactosamine 3-β-galactosyltransferase [EC 2.4.1.122].

N-acetyl-D-glucosamine

D-galactosyl-β(1,3)-
N-acetyl-D-glucosamine

Selected entries from *Methods in Enzymology* [vol, page(s)]:
Assay: **98**, 125
Other molecular properties: purifcation of porcine tracheal and rat intestinal enzymes, **98**, 126

UDP-GALACTOSE:*N*-ACETYL-D-GLUCOS-AMINE 4-β-D-GALACTOSYLTRANSFERASE

This enzyme activity, formerly classified as EC 2.4.1.98, is now included with *N*-acetyllactosamine synthase [EC 2.4.1.90].

UDP-GALACTOSE:UDP-*N*-ACETYL-GLUCOSAMINE GALACTOSE-PHOSPHOTRANSFERASE

This transferase [EC 2.7.8.18] catalyzes the reaction of UDP-D-galactose with UDP-*N*-acetyl-D-glucosamine (or *N*-acetylglucosamine end-groups in glycoproteins) to produce

UMP and UDP-*N*-acetyl-6-(D-galactose-1-phospho)-D-glucosamine (or the analogous glycoprotein product).[1]

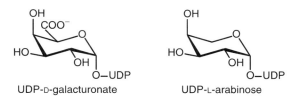

UDP-*N*-acetyl-D-glucosamine   UDP-*N*-acetyl-6(D-galactose-1-phospho)-D-glucosamine

[1]Y. Nakanishi, K. Otsu & S. Suzuki (1983) *FEBS Lett.* **151**, 15.

UDP-GALACTURONATE DECARBOXYLASE

This decarboxylase [EC 4.1.1.67] catalyzes the conversion of UDP-D-galacturonate to UDP-L-arabinose and carbon dioxide.[1]

UDP-D-galacturonate   UDP-L-arabinose

[1]D.-F. Fan & D. S. Feingold (1972) *ABB* **148**, 576.
Selected entries from *Methods in Enzymology* [vol, page(s)]:
General discussion: **28**, 438
Assay: **28**, 438

UDP-GALACTURONOSYLTRANSFERASE

This transferase [EC 2.4.1.75], also known as *p*-nitrophenol conjugating enzyme, catalyzes the reaction of UDP-D-galacturonate with an acceptor substrate to produce the acceptor β-D-galacturonide and UDP.[1,2]

[1]D. A. Vessey & D. Zakim (1973) *BBA* **315**, 43.
[2]R. H. Tukey & T. R. Tephly (1981) *ABB* **209**, 565.
Selected entries from *Methods in Enzymology* [vol, page(s)]:
General discussion: **77**, 177

UDP-GLUCOSAMINE EPIMERASE

This epimerase [EC 5.1.3.16] catalyzes the interconversion of UDP-D-glucosamine and UDP-D-galactosamine.[1] The enzyme reportedly also acts on UDP-*N*-acetyl-D-glucosamine, an activity of UDP-*N*-acetylglucosamine 4-epimerase, EC 5.1.3.7.[2]

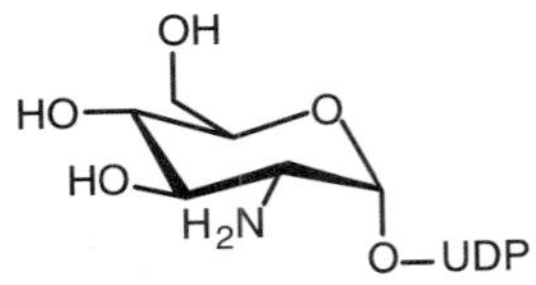

UDP-D-glucosamine 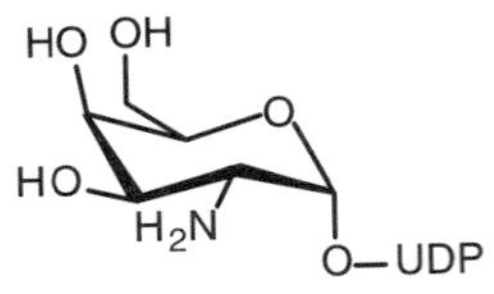   UDP-D-galactosamine

[1]J. E. Silbert & D. H. Brown (1961) *BBA* **54**, 590.
[2]F. Maley & G. F. Maley (1959) *BBA* **31**, 577.

UDP-GLUCOSE 4,6-DEHYDRATASE

This dehydratase [EC 4.2.1.76] catalyzes the conversion of UDP-D-glucose to UDP-4-dehydro-6-deoxy-D-glucose and water.[1]

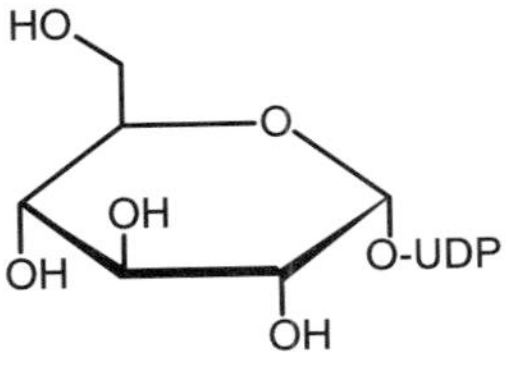

UDP-D-glucose 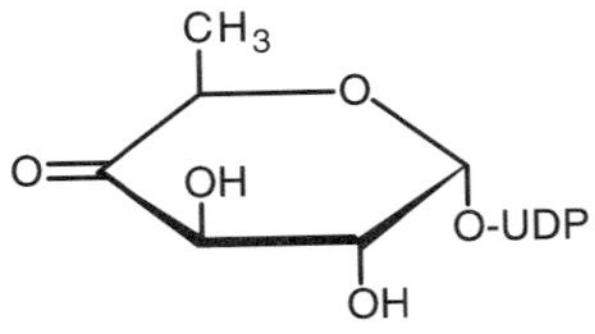   UDP-4-dehydro-6-deoxy-D-glucose

[1]J. Kamsteeg, J. Van Brederode & G. Van Nigtevecht (1978) *FEBS Lett.* **91**, 281.

UDP-GLUCOSE 6-DEHYDROGENASE

This oxidoreductase [EC 1.1.1.22] catalyzes the reaction of UDP-D-glucose with water and two molecules of NAD$^+$ to produce UDP-D-glucuronate and two molecules of NADH. Other substrates include UDP-2-deoxy-D-glucose.[1]

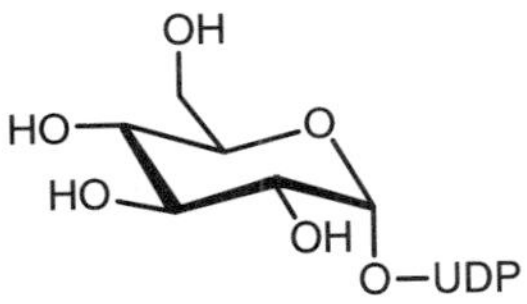

UDP-D-glucose 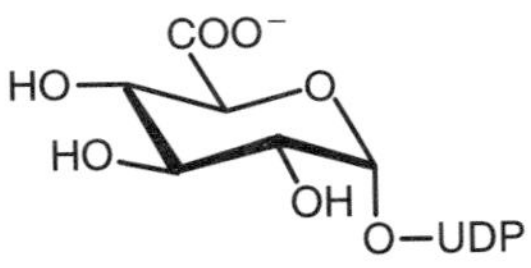   UDP-D-glucuronate

[1]R. E. Campbell, R. F. Sala, I. van de Rijn & M. E. Tanner (1997) *JBC* **272**, 3416.

Selected entries from *Methods in Enzymology* [vol, page(s)]:
General discussion: 28, 430
Assay: 28, 430
Other molecular properties: chemical and physical properties, 28, 433; galactokinase assay and, 5, 176; galactose 1-phosphate determination, 5, 176; galactose-1-phosphate uridylyltransferase assay, 87, 22; purification, 28, 432; UDP-glucose 4-epimerase assay and, 5, 185; 8, 235; UDP-glucose assay and, 2, 677

UDP-GLUCOSE 4-EPIMERASE

This NAD$^+$-dependent epimerase [EC 5.1.3.2], also known as UDP-galactose 4-epimerase and galactowaldenase,

catalyzes the interconversion of UDP-D-glucose and UDP-D-galactose.[1–4]

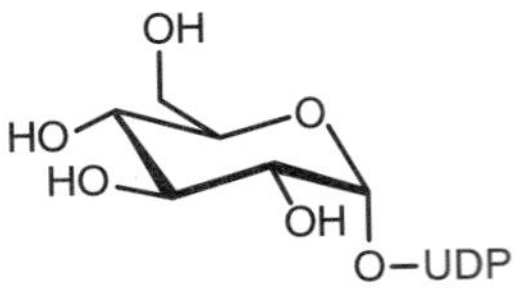

UDP-glucose 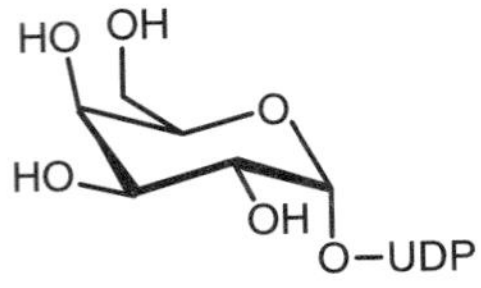   UDP-galactose

Other substrates include UDP-2-deoxy-D-glucose and UDP-6-deoxy-D-glucose. Catalysis is initiated by reduction of enzyme-bound NAD$^+$. The oxidized intermediate UDP-4-ketopyranose then undergoes a spatial reorientation, followed by the stereospecific return of the hydride ion and the release of product. Arginyl, histidyl, and cysteinyl residues have been identified as essential.[1] Certain microbial enzymes exhibit no NAD$^+$ requirement.

[1]U. Bhattacharyya, G. Dhar & A. Bhaduri (1999) *JBC* **274**, 14573.
[2]M. E. Tanner (1998) *CBC* **3**, 76.
[3]E. Adams (1976) *AE* **44**, 69.
[4]L. Glaser (1972) *The Enzymes*, 3rd ed., **6**, 355.

Selected entries from *Methods in Enzymology* [vol, page(s)]:
General discussion: 1, 293; 5, 178, 185; 8, 229, 235; 89, 584
Assay: 1, 293; 5, 178, 185; 8, 235; 89, 589
Other molecular properties: inhibitors, 63, 401; kinetics, 8, 240; large-scale isolation of, 22, 491, 537; Leloir pathway, 87, 21; properties, 1, 294; 8, 239; purification, 5, 186; 8, 236; *Saccharomyces cerevisiae* enzyme, 89, 590; sugar nucleotide regeneration system, 247, 110; sugar nucleotide synthesis, 28, 296, 298; synthetic galactosylation of oligosaccharides, 247, 110; UDPglucose assay and, 2, 677; 3, 968

UDP-GLUCOSE:GLYCOPROTEIN GLUCOSEPHOSPHOTRANSFERASE

This transferase [EC 2.7.8.19] catalyzes the reaction of UDP-D-glucose with penultimate D-mannose residues on oligomannose-type glycoproteins to produce UMP and a glycoprotein 6-(D-glucose-1-phospho)-D-mannose.[1]

[1]L. A. Koro & R. B. Marchase (1982) *Cell* **31**, 739.

UDP-GLUCOSE:HEXOSE-1-PHOSPHATE URIDYLYLTRANSFERASE

This zinc-dependent uridylyltransferase [EC 2.7.7.12], also known as hexose-1-phosphate uridylyltransferase, uridylyl removing enzyme, and galactose-1-phosphate uridylyltransferase, catalyzes the reaction of UDP-D-glucose with α-D-galactose 1-phosphate to produce α-D-glucose 1-phosphate and UDP-D-galactose. The enzyme operates by means of a ping pong Bi Bi kinetic mechanism with a uridylyl-enzyme intermediate.[1,2] *See also* other uridylyltransferases; *UTP:Hexose-1-Phosphate Uridylyltransferase*

The Enzyme Reference

[1] P. A. Frey, L.-J. Wong, K.-F. Sheu & S.-L. Yang (1982) *MIE* **87**, 20.
[2] S. Geeganage & P. A. Frey (1999) *Biochemistry* **38**, 13398.

Selected entries from *Methods in Enzymology* [vol, page(s)]:
General discussion: 5, 179; 9, 708, 713; 87, 20; 89, 584
Assay: 5, 179; 9, 708, 715; 28, 271; 87, 22; coupled assay, 63, 33;
  Saccharomyces cerevisiae, 89, 584
Other molecular properties: calf liver enzyme, 5, 180; 9, 709; human
  erythrocyte enzyme, 9, 712; *Escherichia coli* enzyme (characterization,
  87, 31; exchange reactions, 64, 10; 87, 27; properties, 5, 184;
  purification, 5, 181; 87, 30; steady-state kinetics, 87, 22;
  stereochemistry, 87, 36, 203, 224; uridylylation stoichiometry, 87, 31);
  isotope exchange, 64, 10; positional isotope exchange studies, 249, 413,
  416; *Saccharomyces cerevisiae* (properties, 89, 588; purification, 89, 586);
  sugar nucleotide regeneration, 247, 110; synthetic galactosylation of
  oligosaccharides, 247, 110; UDP-galactosamine synthesis and, 28, 271;
  UDP-galactose and, 1, 293; uridine 5'-*O*-(1-thiodiphosphate)glucose, 87,
  224, 229; uridine 5'-*O*-(1-thiotriphosphate), 87, 203, 229

UDP-GLUCOSE:β-XYLOSIDE TRANSGLUCOSYLASE

This transferase catalyzes the reaction of UDP-D-glucose
with *O*-xylosylserine to produce glucosylxylosylserine
and UDP.

Selected entries from *Methods in Enzymology* [vol, page(s)]:
General discussion: 28, 482
Assay: 28, 482

UDP-GLUCURONATE:BILIRUBIN-GLUCURONOSIDE GLUCURONOSYL-TRANSFERASE

This transferase, now listed as glucuronosyltransferase [EC
2.4.1.17] in place of its former number EC 2.4.1.77, catalyzes the reaction of UDP-D-glucuronate with bilirubin-
glucuronoside to produce UDP and bilirubin bisglucuronoside.

UDP-GLUCURONATE DECARBOXYLASE

This enzyme [EC 4.1.1.35] catalyzes the conversion of
UDP-D-glucuronate to produce UDP-D-xylose and carbon
dioxide.[1–3]

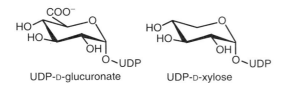

UDP-D-glucuronate   UDP-D-xylose

The enzyme from a number of sources has an absolute
requirement for NAD⁺.

[1] K. V. John, J. S. Schutzbach & H. Ankel (1977) *JBC* **252**, 8013.
[2] M. H. O'Leary (1992) *The Enzymes*, 3rd ed., **20**, 235.
[3] D. J. Creighton & N. S. R. K. Murthy (1990) *The Enzymes*, 3rd ed., **19**, 323.

Selected entries from *Methods in Enzymology* [vol, page(s)]:
General discussion: 8, 287
Assay: 8, 287, 290

Other molecular properties: *Cryptococcus laurentii* enzyme
  (preparation, 8, 290; properties, 8, 292); UDP-xylose, preparation of,
  6, 782; wheat germ enzyme, 8, 287 (properties, 8, 290; purification,
  8, 288)

UDP-GLUCURONATE:1,2-DIACYLGLY-CEROL GLUCURONOSYLTRANSFERASE

This transferase was formerly classified as EC 2.4.1.84
and described as catalyzing the reaction of UDP-D-
glucuronate with 1,2-diacylglycerol to produce UDP and
1,2-diacylglycerol 3-D-glucuronoside. The Enzyme Commission now lists this activity with glucuronosyltransferase
[EC 2.4.1.17].

UDP-GLUCURONATE 4-EPIMERASE

This epimerase [EC 5.1.3.6] catalyzes the interconversion of UDP-D-glucuronate and UDP-D-galacturonate,
reportedly by means of an UDP-4-keto-D-glucuronate
intermediate.[1,2]

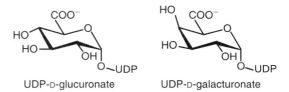

UDP-D-glucuronate   UDP-D-galacturonate

[1] M. A. Gaunt, U. S. Maitra & H. Ankel (1974) *JBC* **249**, 2366.
[2] R. Munoz, R. Lopez, M. de Frutos & E. Garcia (1999) *Mol. Microbiol.*
  31, 703.

Selected entries from *Methods in Enzymology* [vol, page(s)]:
General discussion: 8, 276; 28, 426, 428
Assay: 8, 276; 28, 426

UDP-GLUCURONATE 5'-EPIMERASE

This NAD⁺-dependent epimerase [EC 5.1.3.12] catalyzes
the interconversion of UDP-D-glucuronate and UDP-L-
iduronate.[1]

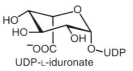

UDP-L-iduronate

[1] E. Adams (1976) *AE* **44**, 69.

Selected entries from *Methods in Enzymology* [vol, page(s)]:
General discussion: 8, 281
Assay: 8, 281

UDP-GLUCURONATE:PHENOL GLUCURONOSYLTRANSFERASE

This transferase, formerly classified as EC 2.4.1.108, is
now classified as glucuronosyltransferase [EC 2.4.1.17].

UDP-GLUCURONATE:TESTOSTERONE GLUCURONOSYLTRANSFERASE

This transferase, formerly classified as EC 2.4.1.107, is now classified as glucuronosyltransferase [EC 2.4.1.17].

UDP-SUGAR DIPHOSPHATASE

This enzyme [EC 3.6.1.45] (variously known as UDP-sugar pyrophosphatase, UDP-sugar hydrolase, nucleoside diphosphate sugar hydrolase, nucleoside diphosphate-sugar pyrophosphatase, and nucleoside diphosphate-sugar diphosphatase) catalyzes the hydrolysis of a UDP-sugar (such as UDP-D-glucose) to produce UMP and a sugar 1-phosphate.[1,2]

[1]I. R. Beacham & M. S. Wilson (1982) *ABB* **218**, 603.
[2]L. Glaser, A. Melo & R. Paul (1967) *JBC* **242**, 1944.

Selected entries from *Methods in Enzymology* [**vol**, page(s)]:
General discussion: 28, 970
Assay: 28, 977

Ulp1 ENDOPEPTIDASE

This cysteine-dependent endopeptidase is an Smt3-specific protease and participates in cell-cycle progression.[1,2] In yeast, Ulp1 has been shown to associate with nucleoporins.[1]

[1]Y. Takahashi, J. Mizoi, A. Toh-E & Y. Kikuchi (2000) *J. Biochem.* **128**, 723.
[2]S. J. Li & M. Hochstrasser (1999) *Nature* **398**, 246.

Ulp2 ENDOPEPTIDASE

This cysteine-dependent endopeptidase processes the SUMO (Smt3p) precursor and deconjugates SUMO from its substrates in *Saccharomyces cerevisiae*.[1]

[1]I. Schwienhorst, E. S. Johnson & R. J. Dohmen (2000) *Mol. Gen. Genet.* **263**, 771.

UmuD PROTEIN

This serine-dependent proteinase, also called UmuD g.p. (*Escherichia coli*), undergoes a RecA-mediated self-cleavage to produce UmuD′ which is active in *SOS* mutagenesis.[1,2] The proteinase activity may modify DNA replication machinery to allow bypass synthesis across a damaged template.

[1]J. T. Konola, A. Guzzo, J. B. Gow, G. C. Walker & K. L. Knight (1998) *J. Mol. Biol.* **276**, 405.
[2]T. S. Peat, E. G. Frank, J. P. Mcdonald, A. S. Levine, R. Woodgate & W. A. Hendrickson (1996) *Nature* **380**, 727.

UNDECAPRENOL KINASE

This phosphotransferase [EC 2.7.1.66], also known as iso-prenoid-alcohol kinase, catalyzes the reaction of ATP with undecaprenol to produce ADP and undecaprenyl phosphate.[1,2]

[1]J. R. Kalin & C. M. Allen (1979) *BBA* **574**, 112.
[2]J. R. Kalin & C. M. Allen (1980) *BBA* **619**, 76.

Selected entries from *Methods in Enzymology* [**vol**, page(s)]:
General discussion: 32, 439
Assay: 32, 439

UNDECAPRENYL-PHOSPHATE GALACTOSEPHOSPHOTRANSFERASE

This transferase [EC 2.7.8.6], also known as poly(iso-prenol)-phosphate galactosephosphotransferase, catalyzes the reaction of UDP-D-galactose with undecaprenyl phosphate to produce UMP and α-D-galactosyl-diphosphoundecaprenol.[1,2]

α-D-galactosyl-diphosphoundecaprenol

[1]A. Wright, M. Dankert, P. Pennessy & P. W. Robbins (1967) *Biochemistry* **57**, 1798.
[2]M. J. Osborn & R. Tze-Yuen (1968) *JBC* **243**, 5145.

UNDECAPRENYL-PHOSPHATE MANNOSYLTRANSFERASE

This transferase [EC 2.4.1.54], which requires phosphatidylglycerol, catalyzes the reaction of GDP-D-mannose with undecaprenyl phosphate to produce GDP and D-mannosyl-1-phosphoundecaprenol.[1-3]

[1]M. Lahav, T. H. Chiu & W. J. Lennarz (1969) *JBC* **244**, 5890.
[2]A. F. Clark & C. L. Villemez (1973) *FEBS Lett.* **32**, 84.
[3]W. T. Forsee & A. D. Elbein (1973) *JBC* **248**, 2858.

Selected entries from *Methods in Enzymology* [**vol**, page(s)]:
Assay: 28, 563

UNDECAPRENYL-PYROPHOSPHATASE

This enzyme [EC 3.6.1.27], also known as undecaprenyl-diphosphatase, catalyzes the hydrolysis of undecaprenyl pyrophosphate to produce undecaprenyl phosphate and orthophosphate.[1,2]

[1]K. J. Stone & J. L. Strominger (1971) *PNAS* **68**, 3223.
[2]R. Goldman & J. L. Strominger (1972) *JBC* **247**, 5116.

UNDECAPRENYL-PYROPHOSPHATE SYNTHASE

This transferase [EC 2.5.1.31] (variously known as di-*trans*-poly-*cis*-decaprenylcistransferase, di-*trans*-poly-*cis*-undecaprenyl-diphosphate synthase, and bactoprenyl-diphosphate synthase) catalyzes the reaction of di-*trans*-poly-*cis*-decaprenyl pyrophosphate with isopentenyl pyrophosphate to produce pyrophosphate and di-*trans*-poly-*cis*-undecaprenyl pyrophosphate.[1–5]

n = 8: di-*trans*,poly-*cis*-decaprenyl pyrophosphate
n = 9: di-*trans*,poly-*cis*-undecaprenyl pyrophosphate

Other donor substrates include *trans*,*trans*-farnesyl pyrophosphate and di-*trans*,poly-*cis*-prenyl pyrophosphates of intermediate size. The two *trans* bonds in the substrate and product are those furthest from the pyrophosphate group.

[1] J. D. Muth & C. M. Allen (1984) *ABB* **230**, 49.
[2] I. Takahashi & K. Ogura (1982) *JB* **92**, 1527.
[3] T. Koyama, I. Yoshida & K. Ogura (1988) *JB* **103**, 867.
[4] T. Baba & C. M. Allen (1978) *Biochemistry* **17**, 5598.
[5] C. M. Allen & J. D. Muth (1977) *Biochemistry* **16**, 2908.

Selected entries from *Methods in Enzymology* [vol, page(s)]:
General discussion: 110, 281
Assay: 110, 120, 282
Other molecular properties: *Lactobacillus plantarum* (assays, 110, 120, 282; photoinactivation, 110, 122; photolabile pyrophosphate analogue activity as substrates and inhibitors, 110, 121; photolysis, 110, 120; properties, 110, 294; purification, 110, 120, 289; separation and identification in mixtures of prenyltransferases, 110, 287)

UNP PEPTIDASE

This cysteine-dependent peptidase, also called Unp oncoprotein, is ubiquitin-specific and efficiently cleaves the ubiquitin-proline bond.[1,2]

[1] C. A. Gilchrist, D. A. Gray & R. T. Baker (1997) *JBC* **272**, 32280.
[2] C. A. Gilchrist & R. T. Baker (2000) *BBA* **1481**, 297.

UNSPECIFIC MONOOXYGENASE

This heme-thiolate enzyme [EC 1.14.14.1] (known variously as a cytochrome P450, microsomal monooxygenase, xenobiotic monooxygenase, aryl-4-monooxygenase, and flavoprotein-linked monooxygenase) catalyzes the reaction of R–H with a reduced flavoprotein and dioxygen to produce R–OH, the oxidized flavoprotein, and water. The acceptor substrate (R–H) includes a wide range of substances (such as xenobiotics, steroids, fatty acids, vitamins, and prostaglandins). Aryl hydrocarbon substrates include benzo[*a*]pyrene, ethoxyresuforin, biphenyl, *p*-nitroanisole, acetanilide, 2-acetylaminofluorene, 2-ethoxycoumarin, 17β-estradiol, and testosterone. The reactions include hydroxylation, epoxidation, *N*-oxidation, sulfooxidation, N-, S-, and O-dealkylations, desulfation, deamination, and reduction of azo, nitro, and N-oxide groups.[1–3]

[1] M. A. Lang & D. W. Nebert (1981) *JBC* **256**, 12058.
[2] M. A. Lang, J. E. Gielen & D. W. Nebert (1981) *JBC* **256**, 12068.
[3] A. D. Theoharides & D. Kupfer (1981) *JBC* **256**, 2168.

Selected entries from *Methods in Enzymology* [vol, page(s)]:
General discussion: 5, 816; 52, 15
Assay: 52, 234, 372, 409, 410, 413; microassay, **52**, 236; sensitivity, **52**, 236; standard curves, **52**, 238, 239
Other molecular properties: Ah[b] allele, **52**, 231; assay for analysis of interferon effects on hepatic drug metabolism, **119**, 723; cytochrome P450IA1-dependent activity, cells with, selection, **206**, 381; differences in basal and aromatic hydrocarbon-induced forms, **52**, 235; fatty acid hydroxylation, for, **52**, 319; hormone induction, **39**, 39; inducible, FACS analysis, **108**, 239; inhibitors, **52**, 410; liver microsomes, in, components, **52**, 201; subcellular localization, **52**, 238

URACIL-5-CARBOXYLATE DECARBOXYLASE

This decarboxylase [EC 4.1.1.66] catalyzes the conversion of uracil 5-carboxylate to uracil and carbon dioxide.[1,2]

uracil 5-carboxylate   uracil

[1] W. R. Griswold, V. O. Madrid, P. M. Shaffer, D. C. Tappen, C. S. G. Pugh & M. T. Abbott (1976) *J. Bact.* **125**, 1040.
[2] R. D. Palmatier, R. P. McCroskey & M. T. Abbott (1970) *JBC* **245**, 6706.

URACIL DEHYDROGENASE

This oxidoreductase [EC 1.1.99.19], also known as uracil oxidase and uracil-thymine oxidase, catalyzes the reaction of hydrated uracil with an acceptor substrate to produce barbiturate and the reduced acceptor.[1,2] Thymine is also converted to 5-methylbarbiturate. The acceptor substrate can be 2,6-dichlorophenolindophenol or methylene blue, and dioxygen will also act as an acceptor when methylene blue is present.

[1] O. Hayaishi & A. Kornberg (1952) *JBC* **197**, 717.
[2] N. P. Bharat & T. P. West (1987) *FEMS Microbiol. Lett.* **40**, 33.

Selected entries from *Methods in Enzymology* [vol, page(s)]:
General discussion: 2, 490
Assay: 2, 490

URACIL-DNA GLYCOSYLASE

This hydrolytic enzyme [EC 3.2.2.-] catalyzes the release of uracil from DNA by hydrolysis of the base-sugar glycosidic bond.[1–4] The DNA duplex undergoes a base-flipping step, thus facilitating uracil extrusion.[1,2]

[1] J. T. Stivers, K. W. Pankiewicz & K. A. Watanabe (1999) *Biochemistry* **38**, 952.
[2] A. C. Drohat, J. Jagadeesh, E. Ferguson & J. T. Stivers (1999) *Biochemistry* **38**, 11866.
[3] T. Lindahl (1982) *ARB* **51**, 61.
[4] B. K. Duncan (1981) *The Enzymes*, 3rd ed., **14**, 565.
Selected entries from *Methods in Enzymology* [vol, page(s)]:
Other molecular properties: *Escherichia coli*, **65**, 284 (mechanism, of action, **65**, 289; molecular weight, **65**, 288, 289; purification, **65**, 286, 289; reaction conditions, **65**, 289; substrate preparation, **65**, 286; substrate specificity, **65**, 289); transition-state and multisubstrate analogues, **249**, 306

URACIL PHOSPHORIBOSYLTRANSFERASE

This transferase [EC 2.4.2.9], also known as UMP pyrophosphorylase and UMP diphosphorylase, catalyzes the reversible reaction of uracil with 5-phospho-α-D-ribose 1-pyrophosphate to produce UMP and pyrophosphate.[1,2] The *Escherichia coli* enzyme is activated by GTP.[1] The *Bacillus subtilis* enzyme has also been shown to mediate transcriptional attenuation at three separate sites within the pyrimidine nucleotide biosynthetic operon.[2]

[1] K. F. Jensen & B. Mygind (1996) *EJB* **240**, 637.
[2] R. J. Turner, E. R. Bonner, G. K. Grabner & R. L. Switzer (1998) *JBC* **273**, 5932.
Selected entries from *Methods in Enzymology* [vol, page(s)]:
General discussion: **6**, 152
Assay: **6**, 152

URACILYLALANINE SYNTHASE

This enzyme [EC 4.2.99.16], also known as willardiine synthase and isowillardiine synthase, catalyzes the reaction of O^3-acetyl-L-serine with uracil to produce 3-(uracil-1-yl)-L-alanine (L-willardiine) and acetate.[1,2]

3-(uracil-1-yl)-L-alanine

The enzyme is not identical with cysteine synthase [EC 4.2.99.8]. The acetylated substrate *O*-acetylserine undergoes a rapid *O*-to-*N* acyl shift above pH 8. *See also* L-*Mimosine Synthase; β-Pyrazolylalanine Synthase (Acetylserine)*

[1] I. Murakoshi, F. Ikegami, N. Ookawa, T. Ariki & J. Haginiwa (1978) *Phytochemistry* **17**, 1571.
[2] M. A. S. Ahmmad, C. S. Maskall & E. G. Brown (1984) *Phytochemistry* **23**, 265.

URATE OXIDASE

This oxidoreductase [EC 1.7.3.3], also known as uricase, catalyzes the reaction of urate with dioxygen and water to produce 5-hydroxyisourate and hydrogen peroxide (H_2O_2).[1–3] The 5-hydroxyisourate then decomposes to form allantoin.

uric acid   allantoin

The enzyme reaction proceeds by forming urate hydroperoxide from dioxygen and the urate dianion, followed by collapse of the peroxide to form dehydrourate and subsequent hydration to form 5-hydroxyisourate.[1]

[1] K. Kahn & P. A. Tipton (1998) *Biochemistry* **37**, 11651.
[2] K. Kahn & P. A. Tipton (1997) *Biochemistry* **36**, 4731.
[3] N. E. Tolbert (1981) *ARB* **50**, 133.
Selected entries from *Methods in Enzymology* [vol, page(s)]:
General discussion: **2**, 485; **52**, 10, 13; **77**, 18
Assay: **2**, 458, 485; **31**, 820, 821
Other molecular properties: activity, effect of xanthine concentration, **44**, 918, 919; glyoxysomes, in, **31**, 565, 569; guanase assay, effect on, **2**, 481; immobilization, **44**, 905, 908; peroxisomes, in, **10**, 14; **31**, 364, 367, 370, 374; properties, **2**, 489; purification, **2**, 486; sources, **56**, 468; therapeutic use, **44**, 695, 696, 704

URATE-RIBONUCLEOTIDE PHOSPHORYLASE

This phosphorylase [EC 2.4.2.16] catalyzes the reaction of urate D-ribonucleotide with orthophosphate to produce urate and D-ribose 1-phosphate.[1]

[1] L. Laster & A. Blair (1963) *JBC* **238**, 3348.

UREA CARBOXYLASE

This biotin-dependent enzyme [EC 6.3.4.6], variously known as urease (ATP-hydrolyzing), urea carboxylase (hydrolyzing), ATP:urea amidolyase, and urea amidolyase, catalyzes the reaction of ATP with urea and carbon dioxide to produce urea-1-carboxylate ($H_2NCONH_2COO^-$), ADP, and orthophosphate.[1–4] The reaction initially proceeds by the MgATP-dependent formation of *N*-carboxybiotin, which is then attacked by urea. The yeast enzyme also exhibits allophanate hydrolase [EC 3.5.1.54] activity, which brings about the ATP-dependent hydrolysis of urea to carbon dioxide and ammonia.[1]

[1] R. A. Sumrada & T. G. Cooper (1982) *JBC* **257**, 9119.
[2] J. N. Earnhardt & D. N. Silverman (1998) *CBC* **1**, 495.

[3]T. G. Cooper (1984) AE **56**, 91.
[4]H. G. Wood & R. E. Barden (1977) ARB **46**, 385.

UREASE

This nickel$_2$-dependent enzyme [EC 3.5.1.5] catalyzes the hydrolysis of urea to ammonia and carbamate ($H_2NCO_2^-$), which then degrades nonenzymatically to form carbon dioxide and a second molecule of ammonia.[1–5] One of the nickel ions binds urea and activates it for nucleophilic attack. The nucleophile appears to be a hydroxy ligand bound to the other nickel. The rate increase of the enzyme-catalyzed reaction over uncatalyzed aqueous ureolysis is at least 10^{14}.

[1]J. B. Sumner (1926) JBC **69**, 435 and **70**, 97.
[2]M. A. Halcrow (1998) CBC **1**, 506.
[3]M. W. Lubbers, S. B. Rodriguez, N. K. Honey & R. J. Thornton (1996) Can. J. Microbiol. **42**, 132.
[4]B. Zerner (1991) Bioorg. Chem. **19**, 116.
[5]F. J. Reithel (1971) The Enzymes, 3rd ed., **4**, 1.

Selected entries from **Methods in Enzymology** [**vol**, page(s)]:
General discussion: 2, 378; **17A**, 317; **34**, 105
Assay: 2, 379; activity assay, **70**, 444; immunosorbent, **326**, 297; spectrophotometric, **44**, 104
Other molecular properties: activation energy, **63**, 243; ammonia release in isotopic studies, **4**, 484; arginolytic activity, **2**, 370; assay of urea, in, **2**, 370; **17A**, 328; carbon dioxide product, **2**, 841; enzyme electrode, in, **44**, 587, 597, 600; fiber entrapment, biomedical application, **44**, 242; fluorometric urea assay, in, **44**, 630; immobilization **44**, 167; microencapsulation, **44**, 214, 330; photoinactivation, **10**, 626; purification, **2**, 378; **17A**, 320; **44**, 389; **87**, 467; **326**, 301; urea quantitation, with urease and firefly luciferase, **305**, 366; urea synthesis with, in water-organic solvent mixtures, **136**, 234; uric acid degradation and, **4**, 636, 637

UREASE (ATP-HYDROLYZING)

This bicarbonate-dependent enzyme, originally classified by the Enzyme Commission as EC 3.5.1.45, is now deleted.
See Urea Carboxylase (Hydrolyzing); Allophanase

UREIDOGLYCOLATE DEHYDROGENASE

This oxidoreductase [EC 1.1.1.154] catalyzes the reaction of (S)-ureidoglycolate with $NAD(P)^+$ to produce oxalureate ($^-OOCCONHCONH_2$) and NAD(P)H.[1]

[1]C. Van der Drift, P. E. M. Van Helvoort & G. D. Vogels (1971) ABB **145**, 465.

UREIDOGLYCOLATE HYDROLASE

This hydrolase [EC 3.5.3.19] converts (−)-ureidoglycolate into glyoxylate, carbon dioxide, and two ammonia molecules.[1,2] This enzyme can be assayed by using lactate dehydrogenase or glyoxylate reductase as coupling enzymes.[3]

[1]X. E. Wells & E. M. Lees (1991) ABB **287**, 151.
[2]R. G. Winkler, D. G. Blevins & D. D. Randall (1988) Plant Physiol. **86**, 1084.
[3]M. Pineda, P. Piedras & J. Cardenas (1994) Anal. Biochem. **222**, 450.

UREIDOGLYCOLATE LYASE

This enzyme [EC 4.3.2.3] catalyzes the conversion of (−)-ureidoglycolate to glyoxylate and urea.[1]

[1]T. G. Cooper (1984) AE **56**, 91.

β-UREIDOPROPIONASE

This enzyme [EC 3.5.1.6], also called β-alanine synthase, catalyzes the hydrolysis of N-carbamoyl-β-alanine to produce β-alanine, carbon dioxide, and ammonia.[1,2] The rat liver enzyme also acts on β-ureidoisobutyrate.

[1]N. Tamaki, N. Mizutani, M. Kikugawa, S. Fujimoto & C. Mizota (1987) EJB **169**, 21.
[2]O. W. Griffith (1986) ARB **55**, 855.

Selected entries from **Methods in Enzymology** [**vol**, page(s)]:
General discussion: 6, 177; **324**, 399
Assay: 324, 400
Other molecular properties: properties, **324**, 409; purification, **12A**, 58; **324**, 403

UREIDOSUCCINASE

This enzyme [EC 3.5.1.7] catalyzes the hydrolysis of N-carbamoyl-L-aspartate to produce L-aspartate, carbon dioxide, and ammonia.[1]

[1]I. Lieberman & A. Kornberg (1955) JBC **212**, 909.

Selected entries from **Methods in Enzymology** [**vol**, page(s)]:
General discussion: 6, 177
Assay: 6, 177
Other molecular properties: irreversibility, **2**, 497; dihydroorotate, in metabolism of, **2**, 497; properties, **6**, 179; purification, **6**, 178

URETHANASE

This iron-containing hydrolase [EC 3.5.1.75], also known as urethane hydrolase, catalyzes the hydrolysis of urethane ($H_2NCOOCH_2CH_3$) to produce ethanol, carbon dioxide, and ammonia. Other carbamates (including n-butyl, phenyl, and benzyl groups in place of the ethyl radical) are actually hydrolyzed with greater efficiency than urethane.[1]

[1]C. J. Zhao & K. Kobashi (1994) Biol. Pharm. Bull. **17**, 773.

URIDINE KINASE

This phosphotransferase [EC 2.7.1.48], also known as uridine monophosphokinase, catalyzes the reaction of ATP with uridine to produce ADP and UMP.[1–3] Cytidine is also a phosphoryl acceptor, and GTP and ITP can substitute for ATP.

[1] A. Orengo (1969) *JBC* **244**, 2204.
[2] R. C. Payne, N. Cheng & T. W. Traut (1985) *JBC* **260**, 10242.
[3] E. P. Anderson (1973) *The Enzymes*, 3rd ed., **9**, 49.

Selected entries from *Methods in Enzymology* [vol, page(s)]:
General discussion: 6, 194; 51, 299, 308, 314
Assay: 6, 194; 51, 299, 308, 309, 314, 315
Other molecular properties: cation requirements, 51, 305, 319; *Escherichia coli* enzyme, 51, 308; kinetic properties, 51, 306, 314, 320, 321; molecular weight, 51, 314, 320; Novikoff ascites enzyme, 51, 299; pH optimum, 51, 314; purification, 6, 195; 51, 301, 311, 317; regulation, 51, 307, 314, 320; substrate specificity, 51, 305

URIDINE NUCLEOSIDASE

This enzyme [EC 3.2.2.3] catalyzes the hydrolysis of uridine (also 5-methyluridine) to produce uracil (also 5-methyluracil) and D-ribose.[1] The enzyme also acts on 5-methyluridine.

[1] G. Magni, E. Fioretti, P. L. Ipata & P. Natalini (1975) *JBC* **250**, 9.

Selected entries from *Methods in Enzymology* [vol, page(s)]:
General discussion: 2, 461; 51, 290
Assay: 2, 461; 51, 291, 292
Other molecular properties: activity, 51, 290; inhibitors, 51, 294, 295; isoelectric point, 51, 294; metal content, 51, 294; molecular weight, 51, 294; pH optimum, 2, 462; 51, 294; purification, 2, 462; 51, 292, 293; stability, 51, 293, 294; substrate specificity, 2, 462; 51, 294

URIDINE PHOSPHORYLASE

This enzyme [EC 2.4.2.3], also known as pyrimidine phosphorylase, catalyzes the reaction of uridine with orthophosphate to produce uracil and α-D-ribose 1-phosphate.[1–3] In some organisms, the enzyme may be identical to pyrimidine-nucleoside phosphorylase [EC 2.4.2.2]. **See also** *Pyrimidine Nucleoside Phosphorylase*

[1] K. R. Albe & B. E. Wright (1989) *Exp. Mycol.* **13**, 13.
[2] R. Bose & E. W. Yamada (1974) *Biochemistry* **13**, 2051.
[3] A. Vita, C. Y. Huang & G. Magni (1983) *ABB* **226**, 687.

Selected entries from *Methods in Enzymology* [vol, page(s)]:
General discussion: 6, 189; 12A, 118; 51, 423
Assay: 12A, 123; 51, 424; 58, 25, 26; reverse direction assay, 51, 425
Other molecular properties: activity, 51, 423; inhibitors, 51, 429; kinetic properties, 51, 430; molecular weight, 51, 429; pentosyl transfer, 51, 430, 431; pH optima, 51, 429; product inhibition, 63, 432, 433; properties of, 12A, 124; purification, 12A, 123; 51, 426; rat liver, from, 51, 423; reaction mechanism, 51, 431; reverse direction assay, 51, 425; sources, 51, 423; stability, 51, 428, 429; substrate specificity, 51, 429; thymidine phosphorylase and, 12A, 121; tissue distribution, 51, 431

URIDYLYL REMOVING ENZYME

This enzyme [EC 2.7.7.59] (also known as [protein-P_{II}] uridylyltransferase and P_{II} uridylyltransferase) is a bifunctional catalyst for (a) the reaction of UTP with [protein-P_{II}] to produce pyrophosphate (PP$_i$; also known as diphosphate) and uridylyl-[protein-P_{II}], and (b) the hydrolysis of uridylylated P_{II} protein to yield [protein-P_{II}] and UMP. The macromolecular substrate is a small trimeric protein component of the regulatory cascade controlling the adenylylation state of glutamine synthetase from *Escherichia coli* and several related bacilli.[1–3] While the transferase operates by an ordered Bi Bi kinetic mechanism,[4] the uridylyl-removing activity reportedly exhibits rapid equilibrium substrate binding followed by random order release of products.[4] Both enzyme activities are activated by ATP and α-ketoglutarate. Glutamine inhibits the transferase reaction, and the very same nitrogen-rich amino acid is a nonessential mixed-type activator of deuridylation; glutamine's reciprocal effects serve to minimize futile UTP hydrolysis to UMP and PP$_i$.

[1] E. Garcia & S. G. Rhee (1983) *JBC* **258**, 2246.
[2] S. H. Francis & E. G. Engleman (1978) *ABB* **191**, 590.
[3] S. P. Adler, D. L. Purich & E. R. Stadtman (1975) *JBC* **250**, 6264.
[4] P. Jiang, J. A. Peliska & A. J. Ninfa (1998) *Biochemistry* **37**, 12782.

Selected entries from *Methods in Enzymology* [vol, page(s)]:
General discussion: 107, 198; 113, 231
Assay: 107, 198; 113, 231
Other molecular properties: *Escherichia coli* (assays, 113, 231; properties, 113, 234; purification, 113, 232); regulation of glutamine synthetase formation, 182, 807

UROCANATE HYDRATASE

This NAD$^+$-dependent enzyme [EC 4.2.1.49], also known as urocanase and imidazolonepropionate hydrolase, catalyzes the reversible hydrolysis of urocanate to produce 4,5-dihydro-4-oxo-5-imidazolepropanoate.[1,2]

urocanate   imidazolonepropionate

The enzyme uses a tightly bound NAD$^+$ molecule as an electrophile rather than as a redox agent. This mechanism accounts for the urocanase-catalyzed exchange of the C5 proton of urocanate and of β-(imidazol-4-yl)propionate.[1] The actual product of the enzyme-catalyzed reaction is the enol species, which undergoes tautomerization after its release from the enzyme.

[1] J. Klepp, A. Fallert-Muller, K. Grimm, W. E. Hull & J. Retey (1990) *EJB* **192**, 669.
[2] A. R. Clarke & T. R. Dafforn (1998) *CBC* **3**, 1.

Selected entries from *Methods in Enzymology* [vol, page(s)]:
General discussion: 2, 231; 17B, 51, 73, 84
Assay: 2, 231; 17B, 51, 73, 84
Other molecular properties: induction in microorganisms, 22, 94; properties, 2, 233; 17B, 53, 78, 87; purification from (*Bacillus subtilis*, 17B, 51); liver, beef, 6, 582; 17B, 85; urocanate assay and, 6, 581, 584

URONATE DEHYDROGENASE

This oxidoreductase [EC 1.1.1.203] catalyzes the reaction of D-galacturonate with NAD$^+$ and water to produce D-galactarate and NADH. Other substrates include D-glucuronate, producing D-glucarate.[1,2]

[1]D. F. Bateman, T. Kosuge & W. W. Kilgore (1970) *ABB* **136**, 97.
[2]G. Wagner & S. Hollmann (1976) *EJB* **61**, 589.

URONOLACTONASE

This enzyme [EC 3.1.1.19] catalyzes the hydrolysis of D-glucurono-6,2-lactone, producing D-glucuronate.[1,2]

D-glucurono-6,2-lactone   D-glucuronate

[1]D. Mukherjee, N. C. Kar, N. Sasmal & G. C. Chatterjee (1968) *BJ* **106**, 627.
[2]J. Winkelman & A. L. Lehninger (1958) *JBC* **233**, 794.

UROPORPHYRIN-III C-METHYLTRANSFERASE

This methyltransferase [EC 2.1.1.107], also known as urogen III methylase, SUMT, and uroporphyrinogen III methylase, catalyzes the reaction of two molecules of *S*-adenosyl-L-methionine with uroporphyrin III to produce two molecules of *S*-adenosyl-L-homocysteine and sirohydrochlorin (it catalyzes two successive methylations at positions 2 and 7).[1-3]

uroporphyrinIII   sirohydrochlorin

[1]T. Leustek, M. Smith, M. Murillo, D. P. Singh, A. G. Smith, S. C. Woodcock, S. J. Awan & M. J. Warren (1997) *JBC* **272**, 2744.
[2]F. Blanche, L. Debussche, D. Thibaut, J. Crouzet & B. Cameron (1989) *J. Bact.* **171**, 4222.
[3]M. J. Warren, C. A. Roessner, P. J. Santander & A. I. Scott (1990) *BJ* **265**, 725.

UROPORPHYRINOGEN DECARBOXYLASE

This decarboxylase [EC 4.1.1.37], also known as uroporphyrinogen III decarboxylase and porphyrinogen

carboxy-lyase, converts uroporphyrinogen III to coproporphyrinogen and four molecules of carbon dioxide.[1,2] A number of porphyrinogens can act as substrates.

uroporphyrinogen III   coproporphyrinogen III

[1]J. Luo & C. K. Lim (1993) *BJ* **289**, 529.
[2]S. Granick & S. I. Beale (1978) *AE* **46**, 33.

Selected entries from *Methods in Enzymology* [vol, page(s)]:
General discussion: 5, 893; 123, 415; 281, 349
Assay: 5, 893; 123, 416
Other molecular properties: chicken erythrocyte enzyme, 123, 417; rabbit reticulocyte enzyme 5, 893; stereospecificity, 87, 148

UROPORPHYRINOGEN III SYNTHASE

This synthase [EC 4.2.1.75], also known as uroporphyrinogen-III cosynthase and hydroxymethylbilane hydro-lyase (cyclizing), catalyzes the conversion of hydroxy-methylbilane to uroporphyrinogen-III and water.

hydroxymethylbilane   uroporphyrinogen III

In the presence of hydroxymethylbilane synthase [EC 4.3.1.8], this system rapidly forms uroporphyrinogen III from porphobilinogen.[1,2]

[1]P. M. Shoolingin-Jordan (1998) *Biochem. Soc. Trans.* **26**, 326.
[2]S. Granick & S. I. Beale (1978) *AE* **46**, 33.

Selected entries from *Methods in Enzymology* [vol, page(s)]:
General discussion: 5, 891; 281, 327
Assay: 5, 891; 281, 327

(S)-USNATE REDUCTASE

This oxidoreductase [EC 1.1.1.199] catalyzes the reaction of (*S*)-usnate with NAD(P)H producing NAD(P)$^+$ and 2-acetyl-6-(3-acetyl-2,4,6-trihydroxy-5-methyl-phenyl)-3-hydroxy-6-methylcyclohexa-2,4-dienone.[1]

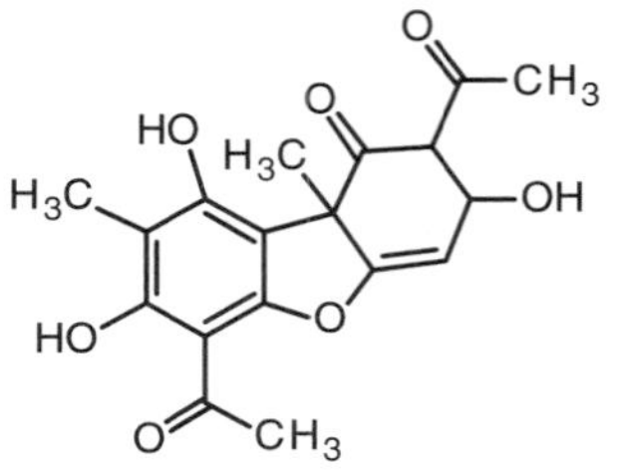

usnic acid

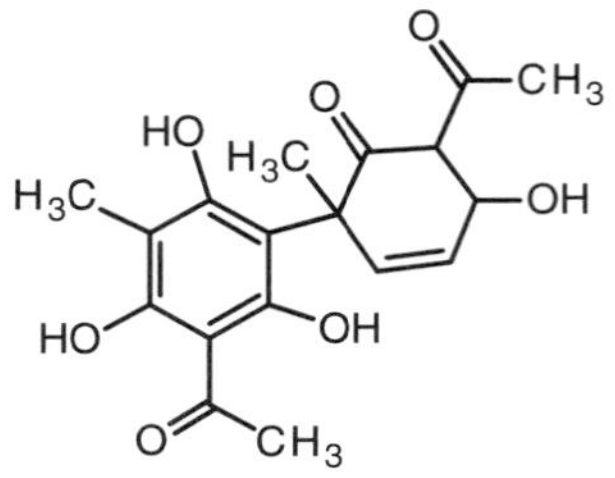

reduced usnic acid

[1] M. P. Estevéz, E. Legaz, L. Olmeda, F. J. Pérez & C. Vicente (1981) *Z. Naturforsch.* **36c**, 35.

UTP:GLUCOSE 1-PHOSPHATE URIDYLYLTRANSFERASE

This transferase [EC 2.7.7.9], also known as UDP-glucose pyrophosphorylase and glucose-1-phosphate uridylyltransferase, catalyzes the reversible reaction of UTP with α-D-glucose 1-phosphate to produce pyrophosphate, and UDP-glucose.[1] *See also UTP:Hexose-1-Phosphate Uridylyltransferase*

[1] K. K. Tsuboi, K. Fukunaga & J. C. Petricciani (1969) *JBC* **244**, 1008.

Selected entries from *Methods in Enzymology* [vol, page(s)]:
General discussion: **2**, 675; **6**, 355; **8**, 248
Assay: **2**, 675; **6**, 355; **8**, 248
Other molecular properties: activity, **57**, 86; immobilized, UDP-glucose synthesis with, **136**, 279; inhibitor, **57**, 93; localization in gels by simultaneous capture technique, **104**, 437; partition analysis, **249**, 323; positional isotope exchange studies, **249**, 413; purification, **2**, 677; **6**, 356; **8**, 250; reaction products, separation on ion exchange paper, **8**, 115; stereochemistry, **87**, 203, 212, 224; substrate specificity, **57**, 93; sugar nucleotide regeneration system, **247**, 110; synthetic galactosylation of oligosaccharides, **247**, 110; UDP-glucose determination, **2**, 676

UTP:HEXOSE-1-PHOSPHATE URIDYLYLTRANSFERASE

This transferase [EC 2.7.7.10], also known as galactose-1-phosphate uridylyltransferase, catalyzes the reversible reaction of UTP with α-D-galactose 1-phosphate to produce pyrophosphate, and UDP-D-galactose. α-D-Glucose 1-phosphate is a less effective substrate. *See also UTP:Glucose-1-Phosphate Uridylyltransferase; UDP-Glucose: Hexose-1-Phosphate Uridylyltransferase*

Selected entries from *Methods in Enzymology* [vol, page(s)]:
General discussion: **90**, 552
Assay: **63**, 33

*Uur*960I RESTRICTION ENDONUCLEASE

This type II restriction endonuclease [EC 3.1.21.4], obtained from *Ureaplasma urealyticum* 960, recognizes and cleaves both strands of DNA at 5′ . . . GC∇NGC . . . 3′, where N refers to any base.[1]

[1] B. G. Cocks & L. R. Finch (1987) *Int. J. Syst. Bacteriol.* **37**, 451.

VALINE DECARBOXYLASE

This pyridoxal-phosphate-dependent decarboxylase [EC 4.1.1.14] catalyzes the conversion of L-valine to 2-methyl-propanamine and carbon dioxide. The enzyme will also catalyze the decarboxylation of L-leucine, producing isoamylamine.[1,2]

[1]T. Hartmann (1972) *Phytochemistry* **11**, 1327.
[2]C. R. Sutton & H. K. King (1962) *ABB* **96**, 360.

VALINE DEHYDROGENASE (NADP+)

This oxidoreductase [EC 1.4.1.8] catalyzes the reaction of L-valine with water and NADP+ to produce 3-methyl-2-ketobutanoate, ammonia, and NADPH.[1–5]

[1]C. G. Hyun, S. S. Kim, K. H. Park & J. W. Suh (2000) *FEMS Microbiol. Lett.* **182**, 29.
[2]A. R. Clarke & T. R. Dafforn (1998) *CBC* **3**, 1.
[3]N. M. W. Brunhuber & J. S. Blanchard (1994) *CRBMB* **29**, 415.
[4]T. Ohshima & K. Soda (1993) *Adv. Biochem. Eng. Biotech.* **1162**, 221.
[5]I. Vancurová, A. Vancura, J. Volc, J. Neuzil, M. Flieger, G. Basarová & V. Bêhal (1993) *Arch. Microbiol.* **150**, 438.

VALINE:3-METHYL-2-KETOVALERATE AMINOTRANSFERASE

This pyridoxal-phosphate-dependent transferase [EC 2.6.1.32], also known as valine:isoleucine transaminase and valine:isoleucine aminotransferase, catalyzes the reversible reaction of L-valine with (*S*)-3-methyl-2-ketopentanoate to produce 3-methyl-2-ketobutanoate and L-isoleucine.[1]

[1]Z. S. Kagan, A. S. Dronov & V. L. Kretovich (1967) *Dokl. Akad. Nauk S.S.S.R.* **175**, 1171 and (1968) *Dokl. Akad. Nauk S.S.S.R.* **179**, 1236.

VALINE:PYRUVATE AMINOTRANSFERASE

This pyridoxal-phosphate-dependent transferase [EC 2.6.1.66], also known as transaminase C and alanine:valine transaminase, catalyzes the reversible reaction of L-valine with pyruvate to produce 3-methyl-2-ketobutanoate

and L-alanine.[1] This enzyme is distinct from branched-chain amino acid aminotransferase [EC 2.6.1.42].

[1]D. Rudman & A. Meister (1953) *JBC* **200**, 591.

VALYL-tRNA SYNTHETASE

This ligase [EC 6.1.1.9], also known as valine:tRNA ligase, catalyzes the reaction of ATP with L-valine and tRNA^{Val} to produce AMP, pyrophosphate, and L-valyl-tRNA^{Val}. The CCA sequence at the 3′-terminal of the tRNA, particularly the base portion of the terminal adenosine residue, plays an important role not only in aminoacylation efficiency with valine but also in preventing misaminoacylation by hydrolyzing misactivated threonyl-tRNA^{Val} as well as misactivated threonyladenylate (hence, an editing mechanism).[1] **See also** *Aminoacyl-tRNA Synthetases*

[1]K. Tamura, N. Nameki, T. Hasegawa, M. Shimizu & H. Himeno (1994) *JBC* **269**, 22173.

Selected entries from **Methods in Enzymology** [vol, page(s)]:
General discussion: 5, 708; 29, 547; 34, 170; 46, 90, 152
Other molecular properties: binding constants, **29**, 640; isolation of, **5**, 711; mechanism, **29**, 634; misactivation studies, in, **59**, 289; properties, **5**, 715; purification, **29**, 567; **59**, 262, 266, 267; radioiodination, **73**, 121; separation, **29**, 561; solvent isotope effects, **87**, 580; subcellular distribution, **59**, 233, 234

VANILLATE O-DEMETHYLASE

This oxidoreductase [EC 1.2.3.12] catalyzes the reaction of vanillate with NAD(P)H and dioxygen to produce protocatechuate (3,4-dihydroxybenzoate), NAD(P)+, formaldehyde, and water.[1,2]

COO⁻   COO⁻

vanillate   protocatechuate

[1]F. Brunel & J. Davison (1988) *J. Bact.* **170**, 4924.
[2]H. Priefert, J. Rabenhorst & A. Steinbuchel (1997) *J. Bact.* **179**, 2595.

Selected entries from **Methods in Enzymology** [vol, page(s)]:
General discussion: 161, 294
Assay: 161, 295

Other molecular properties: properties, 161, 300; preparation from *Pseudomonas*, 161, 297

VANILLATE HYDROXYLASE

This oxidoreductase has been reported to catalyze the reaction of vanillate with NAD(P)H and dioxygen to produce 2-methoxyhydroquinone, carbon dioxide, NAD(P)$^+$, and water.

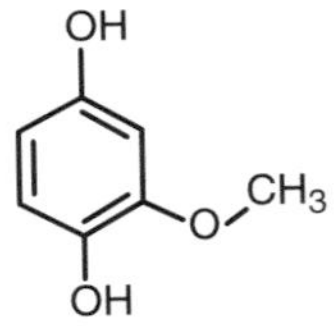

2-methoxyhydroquinone

Selected entries from **Methods in Enzymology** [vol, page(s)]:
General discussion: 161, 274
Assay: 161, 274

VANILLIN DEHYDROGENASE

This oxidoreductase [EC 1.2.1.67] catalyzes the reaction of vanillin (that is, 4-hydroxy-3-methoxybenzaldehyde) with NAD$^+$ and water to produce vanillate (*i.e.*, 4-hydroxy-3-methoxybenzoate), NADH, and a proton.[1]

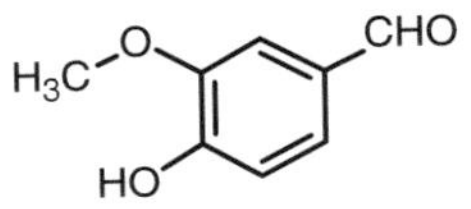

vanillin

[1] A. L. Pometto & D. L. Crawford (1983) *Appl. Environ. Microbiol.* **45**, 1582.

VANILLIN SYNTHASE

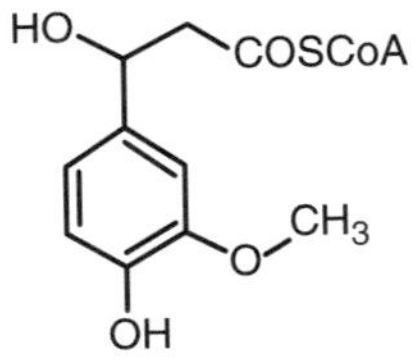

3-hydroxy-3-(4-hydroxy-3-methoxyphenyl)propionyl-CoA

This enzyme [EC 4.1.2.41] reportedly catalyzes the conversion of 3-hydroxy-3-(4-hydroxy-3-methoxyphenyl)-propionyl-CoA to vanillin and acetyl-CoA.[1]

[1] A. Nabid & M. J. Gasson (1998) *Microbiology* **144**, 1397.

VANILLYL-ALCOHOL OXIDASE

This FAD-dependent oxidase [EC 1.1.3.38], also called 4-hydroxy-2-methoxybenzyl alcohol oxidase, catalyzes the reaction of vanillyl alcohol with dioxygen to produce vanillin and hydrogen peroxide.[1]

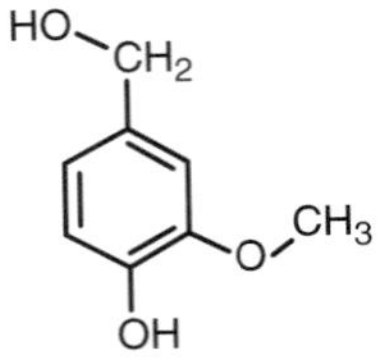

vanillyl alcohol

A number of 4-hydroxybenzyl alcohols and 4-hydroxy-benzylamines can serve as alternative substrates, being converted into the corresponding aldehydes. Interestingly, the allyl group of 4-allylphenols is also converted into R-CH=CH-CH$_2$OH. The FAD coenzyme is covalently linked *via* His422. With 4-methylphenol as an alternative substrate in a series of kinetic isotope effect studies, an ordered Bi Bi mechanism has been proposed, proceeding via a fluorescent 5-(4′-hydroxybenzyl)-FAD intermediate.[2]

[1] M. W. Fraaije, R. H. van den Heuvel, W. J. van Berkel & A. Mattevi (1999) *JBC* **274**, 35514.
[2] M. W. Fraaije, R. H. van den Heuvel, J. C. Roelofs & W. J. van Berkel (1998) *EJB* **253**, 712.

*Van*911 RESTRICTION ENDONUCLEASE

This type II restriction endonuclease [EC 3.1.21.4], obtained from *Vibrio anguillarum* RFL91, recognizes and cleaves DNA at 5′...CCANNNN∇NTGG...3′, where N refers to any base and ∇ indicates the position of bond scission.

VanX D,D-DIPEPTIDASE

This zinc-dependent dipeptidase catalyzes the hydrolysis of the bacterial cell wall intermediate D-alanyl-D-alanine. Good substrates for the dipeptidase contain D-alanyl or D-seryl residues in the P1 position and glycine, D-serine, D-alanine, D-phenylalanine, D-valine, or D-asparagine at position P1′.[1] Interestingly, the enzyme will not catalyze the hydrolysis of the ester analogue D-Ala-D-lactate, which one might suspect to be a kinetically and thermodynamically favored substrate.

[1] C. T. Walsh, S. L. Fisher. L.-S. Park, M. Prahalad & Z. Wu (1996) *Chem. Biol.* **3**, 21.

VARIANT-SURFACE-GLYCOPROTEIN PHOSPHOLIPASE C

This enzyme [EC 3.1.4.47] catalyzes the hydrolysis of a variant-surface-glycoprotein 1,2-didecanoyl-*sn*-phosphatidylinositol to produce 1,2-didecanoylglycerol and a soluble variant-surface-glycoprotein.[1,2] The enzyme is not identical with glycoprotein phospholipase D [EC 3.1.4.50].

[1] D. Hereld, J. L. Krakow, J. D. Bangs, G. W. Hart & P. T. Englund (1986) *JBC* **261**, 13813.
[2] F. Fouchier, T. Baltz & G. Rougon (1990) *BJ* **269**, 321.

Selected entries from *Methods in Enzymology* [vol, page(s)]:
Assay: reaction conditions, **250**, 645; variant surface glycoprotein as substrate (labeling with tritiated myristate, **250**, 643; purification, **250**, 644)

VARICELLA ZOSTER VIRUS ASSEMBLIN

This serine protease [EC 3.4.21.97] is released from herpes virus polyprotein by autocatalytic cleavage. The enzyme also acts at a maturation site near the C-terminal end of the protease precursor and the *Varicella zoster* virus assembly protein. Synthetic substrates, based upon the sequences of the release and maturation sites, are hydrolyzed at Ala–Ser bonds.[1,2] *See also Cytomegalovirus Assemblin; Herpesvirus Assemblin*

[1] W. Gibson, A. R. Welch & M. J. Ludford (1994) *MIE* **244**, 399.
[2] D. J. McMillan, J. Kay & J. S. Mills (1997) *J. Gen. Virol.* **78**, 2153.

Selected entries from *Methods in Enzymology* [vol, page(s)]:
General discussion: 244, 399

VENOMBIN A

This serine endopeptidase [EC 3.4.21.74], isolated from certain snake venoms, catalyzes the hydrolysis of the Arg16–Gly17 bond in the fibrinogen Aα chain, leading to the release of fibrinopeptides A, AY, and AP, and the conversion of fibrinogen to a fibrin clot. The specificity of further degradation of fibrinogen varies with species origin of the enzyme. Enzymes in this class differ from thrombin in that they catalyze the release of fibrinopeptide A but not fibrinopeptide B from fibrinogen (hence, the descriptor "A" in the name). In addition, they do not activate coagulation factor XIII or induce platelet aggregation.[1-3] *See also Batroxobin; Ancrod; Crotalase*

[1] C. Nolan, L. S. Hall & G. H. Barlow (1976) *MIE* **45**, 205.
[2] K. Stocker & G. H. Barlow (1976) *MIE* **45**, 214.
[3] F. S. Markland (1976) *MIE* **45**, 223.

VENOMBIN AB

This serine endopeptidase [EC 3.4.21.55], also known as *Bitis gabonica* serine proteinase and gabonase, catalyzes the release of both fibrinopeptides A and B from fibrinogen (hence, the descriptor "AB" in the name; venombin A [EC 3.4.21.74] releases only fibrinopeptide A). The enzyme catalyzes the cleavage of Arg16–Gly17 peptide bond in the

Aα chain of fibrinogen. At a slower rate, another Arg–Gly bond is hydrolyzed near the N-terminus of the Bβ chain, generating fibrinopeptide B. Unlike thrombin [EC 3.4.21.5], this enzyme is not inhibited by antithrombin III/heparin or hirudin.[1,2]

[1] H. Pirkle, I. Theodor, D. Miyada & G. Simmons (1986) *JBC* **261**, 8830.
[2] H. Pirkle & I. Theodor (1998) in *The Enzymology of Snake Venoms* (G. S. Bailey, ed.), Alaken, Fort Collins.

VENOM EXONUCLEASE

This enzyme [EC 3.1.15.1], also called venom phosphodiesterase, catalyzes the exonucleolytic cleavage of polynucleic acids (with a preference for a single-stranded substrate) in the 3′- to 5′-direction to yield nucleoside 5′-phosphates.[1] *See also Phosphodiesterase; Phosphodiesterase I*

[1] M. Laskowski, Sr. (1971) *The Enzymes*, 3rd ed., **4**, 313.

Selected entries from *Methods in Enzymology* [vol, page(s)]:
General discussion: 2, 561; **6**, 237; **29**, 477; **34**, 605; **237**, 74
Assay: 2, 561; **6**, 238; **65**, 279; **106**, 507
Other molecular properties: hydrolysis of (deamido NAD$^+$, **18B**, 135; NAD$^+$, **18B**, 53, 60); nucleoside end determination and, **12B**, 224, 227; nucleoside trialcohols and, **12B**, 239; preparation, **29**, 343, 478; reduced transfer ribonucleic acid and, **20**, 136; RNA, action on, **3**, 767; **29**, 477; 5′ terminal determination and, **29**, 347; tRNA hydrolysis and, **29**, 477, 479, 495

VENOM METALLOENDOPEPTIDASE

This metalloendopeptidase, isolated from the venom of the broad-banded copperhead (*Agistrodon contortrix laticinctus*), is nonhemorrhagic, has a low molecular weight, and contains three disulfide bonds.[1]

[1] H. S. Selistre-de-Araujo, E. L. De Souza, L. M. Beltramini, C. L. Ownby & D. H. F. Souza (2000) *Protein Express. Purif.* **19**, 41.

VENOM PLASMINOGEN ACTIVATOR (*TRIMERESURUS STEJNEJERI*)

This serine-dependent proteinase, isolated from the venom of the Chinese green tree viper (*Trimeresurus stejnejeri*), specifically activates plasminogen.[1,2] It is not inhibited by bovine pancreatic trypsin inhibitor.

[1] S. Braud, M. A. A. Parry, R. Maroun, C. Bon & A. Wisner (2000) *JBC* **275**, 1823.
[2] M. A. A. Parry, U. Jacob, R. Huber, A. Wisner, C. Bon & W. Bode (1998) *Structure* **6**, 1195.

VENT POLYMERASE

Vent polymerase is a proprietory name for a thermophilic DNA polymerase that possesses 3′-to-5′ proofreading exonuclease activity. The polymerase, obtained from the archaea *Thermococcus litoralis* (isolated from submarine thermal vents) and strains of *Escherichia coli* that carry the gene from *T. litoralis*, is very heat stable (90% of the activity remains after one hour incubation at 95°C).[1-3] The enzyme has proved very useful in primer extension applications, in dideoxy sequencing, and in thermal cycle sequencing. *See also DNA Polymerase*

[1] F. B. Perler, D. G. Comb, W. E. Jack, L. S. Moran, B. Qiang, R. B. Kucera, J. Benner, B. E. Slatko, D. O. Nwankwo, S. K. Hempstead, C. K. S. Carlow & H. Jannasch (1992) *PNAS* **89**, 5577.
[2] P. Mattila, J. Korpela, T. Tenkanen & K. Pitkanen (1991) *NAR* **19**, 4967.
[3] H. Kong, R. B. Kucera & W. E. Jack (1993) *JBC* **268**, 1965.

VERY LONG CHAIN FATTY ACID α-HYDROXYLASE

This oxidoreductase catalyzes the reaction of very long chain carboxylic acids (typically C_{22} to C_{24} acids) with dioxygen and NADPH to produce α-hydroxyacids (R-CHOH-COO$^-$), NADP$^+$, and water.

Selected entries from *Methods in Enzymology* [vol, page(s)]:
General discussion: 52, 310
Assay: 52, 310
Other molecular properties: activators and inhibitors, **52**, 316, 317; kinetic properties, **52**, 317; liver, in, **52**, 317; reaction mechanism, **52**, 317; specificity, **52**, 317, 318; stability, **52**, 316; subcellular localization, **52**, 316

VESICLE-FUSING ATPases

These enzymes [EC 3.6.4.6] catalyze the hydrolysis of ATP to form ADP and orthophosphate and participate in the heterotypic fusion of membrane vesicles with target membranes and the homotypic fusion of various membrane compartments.[1-3] *See also Peroxisome-Assembly ATPase*

This enzyme is currently misclassified as a ATPase [EC 3.6.1.51], because the reaction is of the energase type: Vesicle Fusion State$_1$ + ATP = Vesicle Fusion State$_2$ + ADP + P$_i$, where two affinity-modulated conformational states are indicated.[4] Energases transduce the Gibbs free energy of NTP hydrolysis into mechanical work, in this case vesicle fusion. *See Energases*

[1] F. Confalonieri & M. Duguet (1995) *BioEssays* **17**, 639.
[2] A. Imamura, S. Tamura, N. Shimoyawa, Y. Suzuki, Z. Zhang, T. Tsukamoto, T. Orii, N. Kondo, T. Osumi & Y. Fujiki (1998) *Hum. Mol. Genet.* **7**, 2089.
[3] W. Babst, B. Wendland, E. J. Estepa & S. D. Emr (1998) *EMBO J.* **17**, 2982.
[4] D. L. Purich (2001) *TiBS* **26**, 417.

VIBRIO AMINOPEPTIDASE

This zinc-dependent metalloaminopeptidase [EC 3.4.11.10], also known as *Aeromonas proteolytica* aminopeptidase, catalyzes the release of an N-terminal amino acid, preferentially leucine (although even basic residues and proline are also released). N-terminal glutamyl, aspartyl, or cysteic acid residues are not released.[1,2]

[1] J. M. Prescott & S. H. Wilkes (1976) *MIE* **44**, 530.
[2] B. Chevrier, H. D'Orchymont, C. Schalk, C. Tarnus & D. Moras (1996) *EJB* **237**, 393.

VIBRIO COLLAGENASE

This zinc-dependent collagenase isolated from *Vibrio algi-nolyticus* (formerly *Achromobacter iophagus*) was previously classified as EC 3.4.24.8 but is now classified as EC 3.4.24.3. *Vibrio* collagenases hydrolyzes collagen much more rapidly than interstitial collagenase.[1] *See Collagenase, Microbial; Clostridium Collagenases*

[1] B. Keil (1992) *Matrix Suppl.* **1**, 127.

Selected entries from *Methods in Enzymology* [vol, page(s)]:
Assay: 235, 600; 248, 207
Other molecular properties: activity, **235**, 595; HEXXH motif, **248**, 207; preparation for assay, **235**, 566; synthesis with signal peptides and propeptides, **248**, 227

VIBRIOLYSIN

This zinc-dependent metalloproteinase [EC 3.4.24.25], also known as aeromonolysin and *Aeromonas proteolytica* neutral proteinase, catalyzes the hydrolysis of peptide bonds with a preference for bulky hydrophobic aminoacyl residues in the P2 and P1' positions of the substrate.[1,2] The enzyme was first isolated from *Vibrio proteolyticus* (formerly *Aeromonas proteolytica*), but IUBMB still lists the enzyme as aeromonolysin.

[1] S. H. Wilkes & J. M. Prescott (1976) *MIE* **45**, 404.
[2] C. C. Häse & R. A. Finklestein (1993) *Microbiol. Rev.* **57**, 823.

VICIANIN β-GLUCOSIDASE

This glycosidase [EC 3.2.1.119] catalyzes the hydrolysis of (R)-vicianin to produce mandelonitrile and vicianose (that is, the disaccharide 6-O-α-L-arabinopyranosyl-D-glucose).[1,2]

vicianin   vicianose   mandelonitrile

Alternative, but weak, substrates include (*R*)-amygdalin and (*R*)-prunasin.

[1] G. Kuroki, P. A. Lizotte & J. E. Poulton (1983) *Z. Naturforsch.* **39c**, 232.
[2] P. A. Lizotte & J. E. Poulton (1988) *Plant Physiol.* **86**, 322.

VIGNAIN

This cysteine endopeptidase, also called bean proteinase A, participates in the mobilization of protein reserves in plant seed germination.[1–3]

[1] M. Asano, S. Suzuki, M. Kawai, T. Miwa & H. Shibai (1999) *J. Biochem.* **126**, 296.
[2] T. Okamoto & T. Minamikawa (1998) *J. Plant Physiol.* **152**, 675.
[3] C. Gietl, B. Wimmer, J. Adamec & F. Kalousek (1997) *Plant Physiol.* **113**, 863.

VIMELYSIN

This metalloproteinase, obtained from *Vibrio* str. T1800, is resistant to ethanol denaturation and is active on a number of proteins.[1,2] It acts on the insulin B chain at the Phe24–Phe25 and Tyr16–Leu17 bonds, and slightly at His5–Leu6, His10–Leu11, Ala14–Leu15, and Gly23–Phe24. There is a preference for a phenylalanyl residue at P1′. Surprisingly, k_{cat} has an inverse temperature relationship. The proteinase is more active at lower temperatures.

[1] S. Kunugi, A. Koyasu, M. Kitayaki, S. Takahashi & K. Oda (1996) *EJB* **241**, 368.
[2] S. Takahashi, K. Okayama, S. Kunugi & K. Oda (1996) *Biosci. Biotechnol. Biochem.* **60**, 1651.

VINYLACETYL-CoA Δ-ISOMERASE

This isomerase [EC 5.3.3.3] catalyzes the interconversion of vinylacetyl-CoA and crotonoyl-CoA (or, *trans*-but-2-enoyl-CoA).[1–7]

The enzyme also interconverts 3-methylvinylacetyl-CoA and 3-methylcrotonoyl-CoA. The 4-hydroxybutyryl-CoA dehydratases of *Clostridium aminobutyricum* and *Clostridium kluyveri* have an intrinsic vinylacetyl-CoA Δ-isomerase activity.[5,6]

[1] H. C. Rilling & M. J. Coon (1960) *JBC* **235**, 3087.
[2] R. G. Barsch & H. A. Barker (1961) *ABB* **92**, 122.
[3] H. Hashimoto, H. Günther & H. Simon (1973) *FEBS Lett.* **33**, 81.
[4] E. Schleicher & H. Simon (1976) *Hoppe-Seyler's Z. Physiol. Chem.* **357**, 535.
[5] U. Scherf & W. Buckel (1993) *EJB* **215**, 421.
[6] U. Scherf, B. Sohling, G. Gottschalk, D. Linder & W. Buckel (1994) *Arch. Microbiol.* **161**, 239.
[7] D. J. Creighton & N. S. R. K. Murthy (1990) *The Enzymes*, 3rd ed., **19**, 323.

Selected entries from *Methods in Enzymology* [vol, page(s)]:
General discussion: 71, 403

VIOMYCIN KINASE

This phosphotransferase [EC 2.7.1.103], also known as capreomycin phosphotransferase, catalyzes the reaction of ATP with the cyclic peptide antibiotic viomycin to produce ADP and *O*-phosphoviomycin.[1]

[1] R. H. Skinner & E. Cundliffe (1980) *J. Gen. Microbiol.* **120**, 95.

VITAMIN A ESTERASE

This hydrolase, originally listed as EC 3.1.1.12, is now classified as carboxylesterase [EC 3.1.1.1].

VITAMIN B₁₂-TRANSPORTING ATPase

This so-called ATPase [EC 3.6.3.33], also known as cobalamin-transporting ATPase, catalyzes the ATP-dependent (ADP- and orthophosphate-producing) transport of [vitamin B₁₂]out to produce [vitamin B₁₂]in.[1] Several derivatives of cobalamin can be imported by this system. ABC-type (ATP-binding cassette-type) ATPases have two similar ATP-binding domains and do not undergo phosphorylation during the transport process.

The idea that a transporter is an enzyme is in keeping with a new definition of enzyme catalysis as the facilitated making/breaking of chemical bonds, not just covalent bonds.[2] This idea builds on Pauling's assertion that any long-lived, chemically distinct interaction (in this case, the persistent location of a solute with respect to the faces of a membrane) can be regarded as a chemical bond. Note also that the equilibrium constant (K_{eq} = [Vitamin-B₁₂in][ADP][Pi]/[Vitamin-B₁₂out][ATP]) does not conform to that expected for an ATPase (*i.e.*, K_{eq} = [ADP][Pi]/[ATP]). Thus, although the overall reaction yields ADP and orthophosphate, the enzyme is misclassified as a hydrolase, and instead should be regarded as an energase-type reaction. Energases facilitate affinity-modulated reactions by coupling the $\Delta G_{ATP-hydrolysis}$ to a force-generating or work-producing step.[2] In this case, P-O-P bond scission supplies the energy to drive vitamin B₁₂ transport. *See ATPase; Energase*

[1] M. J. Friedrich, L. C. de Veaux & R. J. Kadner (1986) *J. Bacteriol.* **167**, 928.
[2] D. L. Purich (2001) *TiBS* **26**, 417.

VITAMIN-K-EPOXIDE REDUCTASE (WARFARIN-INSENSITIVE)

This oxidoreductase [EC 1.1.4.2] catalyzes the reaction of 2,3-epoxy-2,3-dihydro-2-methyl-3-phytyl-1,4-naphthoquinone with 1,4-dithiothreitol to produce 3-hydroxy-2-methyl-3-phytyl-2,3-dihydronaphthoquinone (also, the 2-hydroxy form), oxidized dithiothreitol, and water.[1,2] Thioredoxin is the probable physiological cofactor.

[1] E. F. Hildebrandt, P. C. Preusch, J. L. Patterson & J. W. Suttie (1984) ABB **228**, 480.
[2] R. E. Olson (1984) ARN **4**, 281.

VITAMIN-K-EPOXIDE REDUCTASE (WARFARIN-SENSITIVE)

This oxidoreductase [EC 1.1.4.1] catalyzes the reaction of 2,3-epoxy-2,3-dihydro-2-methyl-1,4-naphthoquinone (or, vitamin K₁ 2,3-epoxide) with 1,4-dithiothreitol to produce 2-methyl-3-phytyl-1,4-naphthoquinone (or, vitamin K₁), oxidized dithiothreitol, and water.[1–3] The naturally occurring reducing factor may be thioredoxin. The enzyme is strongly inhibited by warfarin, while a related epoxide reductase [EC 1.1.4.2].

[1] E. F. Hildebrandt, P. C. Preusch, J. L. Patterson & J. W. Suttie (1984) ABB **228**, 480.
[2] R. B. Silverman & D. L. Nandi (1990) J. Enzyme Inhib. **3**, 289.
[3] R. E. Olson (1984) ARN **4**, 281.

VITEXIN β-GLUCOSYLTRANSFERASE

This transferase [EC 2.4.1.105] catalyzes the reaction of UDP-glucose with vitexin (a flavonoid of *Cannabis sativa* and *Silex alba*) to produce UDP and vitexin 2″-*O*-β-D-glucoside.[1]

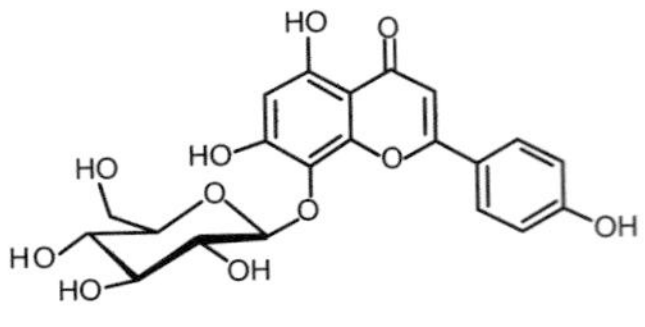

vitexin

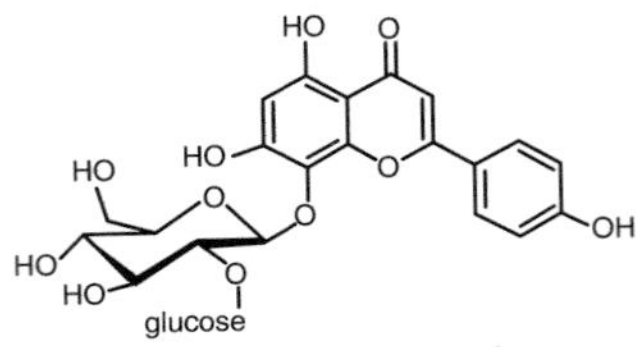

vitexin 2″-*O*-β-D-glucoside

[1] R. Heinsbroek, J. van Brederode, G. van Nigtevecht, J. Maas, J. Kamsteeg, E. Besson & J. Chopin (1980) Phytochemistry **19**, 1935.

VOMIFOLIOL 4′-DEHYDROGENASE

This dehydrogenase [EC 1.1.1.221] catalyzes the reaction of (±)-6-hydroxy-3-oxo-α-ionol with NAD⁺ to produce (±)-6-hydroxy-3-oxo-α-ionone and NADH as well as the reaction of vomifoliol with NAD⁺ to yield dehydrovomifoliol and NADH.[1]

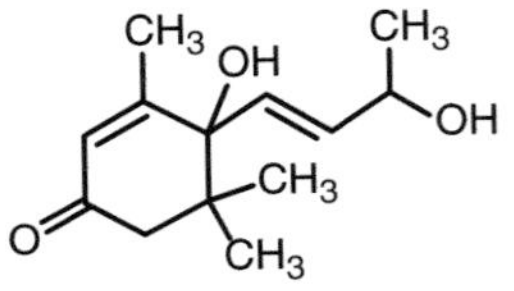

(±)-6-hydroxy-3-oxo-α-ionol

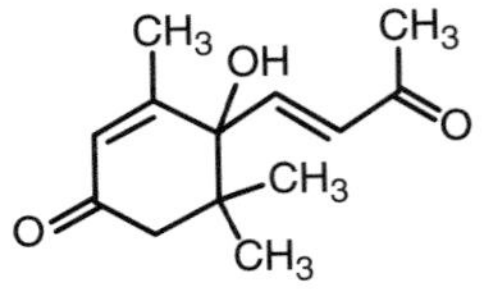

(±)-6-hydroxy-3-oxo-α-ionone

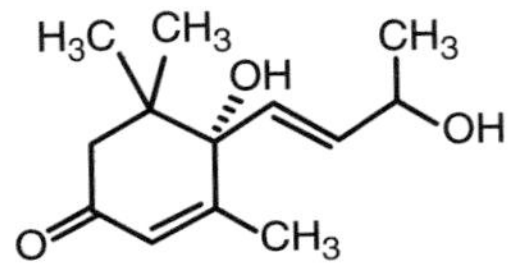

vomifoliol

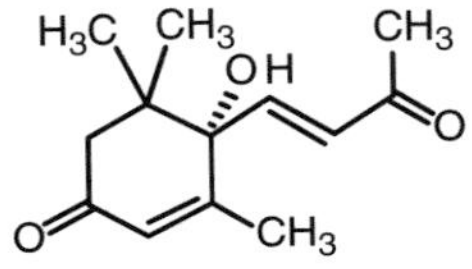

dehydrovomifoliol

[1] S. Hasegawa, S. M. Poling, V. P. Maier & R. D. Bennett (1984) Phytochemistry **23**, 2769.

WATASEMIA-LUCIFERIN 2-MONOOXYGENASE

This oxidoreductase [EC 1.13.12.8], also known as *Watasemia*-type luciferase and luciferase, catalyzes the reaction of the *Watasemia* luciferin with dioxygen to produce the oxidized *Watasemia* luciferin, and carbon dioxide, with the release of a photon.[1]

Watasemia luciferin

Watasemia oxyluciferin

Watasemia luciferin is 8-(phenylmethyl)-6-(4-sulfo-oxyphenyl)-2-[(4-sulfooxyphenyl)methyl]imidazo[1,2-*a*]-pyrazin-3(7*H*)-one.[1]

[1]S. Inoue, H. Kakoi & T. Goto (1976) *Tetrahedron. Lett.* **34**, 2971.

Selected entries from *Methods in Enzymology* [vol, page(s)]:
General discussion: 57, 344

WAX ESTER HYDROLASE

This enzyme [EC 3.1.1.50] catalyzes the hydrolysis of a wax ester to produce a long-chain alcohol and a long-chain fatty acid anion. For example, hexadecanyl palmitate $(CH_3(CH_2)_{14}COO(CH_2)_{15}CH_3)$ is hydrolyzed to hexadecanol and palmitate. The enzyme also utilizes long-chain monoacylglycerols as substrates, but not diacyl- or triacylglycerols.

Selected entries from *Methods in Enzymology* [vol, page(s)]:
General discussion: 71, 804, 805, 809
Other molecular properties: properties, **71**, 809

X

Xaa-Arg DIPEPTIDASE

This metallopeptidase [EC 3.4.13.4], also called N^{α}-(γ-aminobutyryl)lysine hydrolase, X-Arg dipeptidase, and aminoacyl-lysine dipeptidase, catalyzes the hydrolysis of dipeptides containing a dibasic amino acid (that is, arginine, lysine, or ornithine) at the C-terminus.[1] The highest activity is observed with N^{α}-(γ-aminobutyryl)arginine. If the α-amino group is blocked, the dipeptide cannot be hydrolyzed.

[1] A. Kumon, Y. Matsuoka, Y. Kakimoto, T. Nakajima & I. Sano (1970) BBA **200**, 466.

Xaa-His DIPEPTIDASE

This zinc dependent hydrolase [EC 3.4.13.3] (also known as tissue carnosinase, carnosinase, homocarnosinase, and aminoacylhistidine dipeptidase) catalyzes the hydrolysis of Xaa–His dipeptides.[1-3] There appear to be considerable differences in substrate specificity depending on the organism; however, carnosine (β-alanylhistidine) is consistently hydrolyzed. Anserine (β-alanyl-1-methylhistidine) is a poor substrate and homocarnosine (γ-aminobutyrylhistidine) is not hydrolyzed at all by the mouse kidney enzyme.[4]

See also Carnosinase; Cytosol Nonspecific Dipeptidase

[1] H. T. Hanson & E. L. Smith (1949) JBC **179**, 789.
[2] A. Rosenberg (1960) ABB **88**, 83.
[3] A. Rosenberg (1960) BBA **45**, 297.
[4] F. L. Margolis, M. Grillo, N. Grannot-Reisfeld & A. I. Farbinan (1983) BBA **744**, 237.

Selected entries from **Methods in Enzymology** [**vol**, page(s)]:
General discussion: 2, 93
Assay: 2, 93
Other molecular properties: database code, **248**, 222; lack of signal or propeptide, **248**, 223, 225, 227; properties, **2**, 96; purification, **2**, 94; source, **248**, 225

Xaa-METHYL-His DIPEPTIDASE

This dipeptidase [originally classified as EC 3.4.3.4, but reclassifed as EC 3.4.13.5], also referred to as X-methyl-His dipeptidase, aminoacylmethylhistidine dipeptidase, and anserinase, catalyzes the hydrolysis of anserine (or, β-alanyl-N^{π}-methyl-L-histidine). Carnosine, homocarnosine, glycylleucine, and closely related dipeptides are alternative substrates. The enzyme activity once assigned the name acetylhistidine deacetylase and classified as EC 3.5.1.24 is now known to be Xaa-methyl-His dipeptidase.[1]

[1] J. F. Lenney, M. H. Baslow & G. H. Sugiyama (1978) Comp. Biochem. Physiol. B Comp. Biochem. **61**, 253.

Xaa-Pro AMINOPEPTIDASE

This metalloaminopeptidase [EC 3.4.11.9] (also known as X-Pro aminopeptidase, proline aminopeptidase, aminopeptidase P, peptidase P (PepP), and aminoacylproline aminopeptidase) catalyzes the release of any N-terminal amino acid, including proline, that is linked with proline, even from a dipeptide or tripeptide (that is, catalyzing the hydrolysis of Xaa-^-Pro-...). Thus, it will catalyze the release of proline from poly-L-proline.[1-3]

[1] A. Yaron & A. Berger (1970) MIE **19**, 521.
[2] A. Yaron & F. Naider (1993) CRBMB **28**, 31.
[3] T. Yashimoto, A. T. Orawski & W. H. Simmons (1994) ABB **311**, 28.

Selected entries from **Methods in Enzymology** [**vol**, page(s)]:
Assay: 19, 521; 136, 171, 173
Other molecular properties: absorption spectrum, **19**, 528; amino acid composition, **19**, 528; Escherichia coli, **248**, 213, 220; immobilization and properties, **136**, 172; kinetic parameters, **19**, 531; metal ion requirement, **19**, 530; occurrence, **19**, 533; pH profile, **19**, 529; purification, **19**, 524; specificity, **19**, 531; stability, **19**, 527

Xaa-Pro DIPEPTIDASE

This metallodipeptidase [EC 3.4.13.9] (also called X-Pro dipeptidase (XPD), proline dipeptidase, imidodipeptidase, prolidase, peptidase D, and γ-peptidase) catalyzes the hydrolysis of Xaa-Pro dipeptides. The specificity for a proline residue at position P1' is fairly narrow. Thiazolidinecarboxylate is accepted; but hydroxyproline or sarcosine residues at P1' are poor dipeptide substrates. The specificity for the P1 position is broad; however, prolylproline is not a good substrate. The enzyme will cleave only the *trans* rotamer of the acyl-proline linkage.[1]

[1] W. L. Mock & P. C. Green (1990) JBC **265**, 19606.

Selected entries from *Methods in Enzymology* [vol, page(s)]:
General discussion: 2, 100
Assay: 2, 100
Other molecular properties: hydrolysis, 47, 41, 43;
iminodipeptidase, present in preparation of, 2, 99;
immobilization, 47, 44; kidney, 2, 98, 101, 103; properties,
2, 104; 248, 213, 220; protein hydrolysis and, 25, 259;
purification, 2, 101 (equine erythrocytes, 2, 102; swine kidney,
2, 103)

Xaa-Pro DIPEPTIDYL-PEPTIDASE

This serine peptidase [EC 3.4.14.11], also known as X-Pro dipeptidyl-peptidase, X-prolyl dipeptidyl aminopeptidase, and PepX, catalyzes the release of Xaa-Pro dipeptides from unblocked N-termini of peptides lacking prolyl residues in positions P2 and P1'. The enzyme also catalyzes the slower release of Xaa-Ala and Xaa-Gly dipeptides. While this enzymatic activity resembles that of peptidyl-peptidase IV [EC 3.4.14.5], the two enzymes have different sequences. Xaa-Pro dipeptidyl-peptidases are isolated from lactic acid bacteria.[1,2]

[1] R. J. Lloyd & G. G. Pritchard (1991) *J. Gen. Microbiol.* 137, 49.
[2] J.-F. Chich, P. Rigolet, M. Nardi, J.-C. Gripon, L. Ribadeau-Dumas & S. Brunie (1995) *Proteins* 23, 278.

Xaa-Trp AMINOPEPTIDASE

This zinc-dependent metalloproteinase [EC 3.4.11.16], also known as X-Trp aminopeptidase, aminopeptidase X-Trp, and aminopeptidase W, catalyzes the release of a variety of amino acids, especially glutamate and leucine, from the N-terminus of relatively short peptides. Substrates must contain a tryptophanyl residue (or, at least phenylalanyl or tyrosyl residue) in the P1' position. The enzyme acts on short peptides and will also catalyze the hydrolysis of certain dipeptides, including leucyltryptophan and glutamyltryptophan. N-Blocked dipeptides are not hydrolyzed. Asp-∇-Phe-NH$_2$ (where ∇ indicates the cleavage site) is also a substrate.[1,2]

[1] N. S. Gee & A. J. Kenny (1987) *Biochem. J.* 246, 97.
[2] N. M. Hooper (1993) in *Biological Barriers to Protein Delivery* (K. L. Audus & T. J. Raub, eds.), Plenum Press, New York, p. 23.

XANTHAN LYASE

This lyase [EC 4.2.2.12] catalyzes the eliminative cleavage of the terminal β-D-mannosyl-β-D-1,4-glucuronosyl linkage of the side-chain of the exopolysaccharide xanthan, leaving a 4-deoxy-α-L-*threo*-hex-4-enuronosyl group at the terminus of the side-chain. The combined action of this lyase and endoglucanase yields a series of oligosaccharides, each with a side-chain terminating in an unsaturated uronate.[1]

[1] I. W. Sutherland (1987) *J. Gen. Microbiol.* 133, 3129.

XANTHINE DEHYDROGENASE

This molybdenum-dependent and FAD-containing enzyme [EC 1.1.1.204] catalyzes the reaction of xanthine with NAD$^+$ and water to produce urate and NADH.[1-6] A variety of purines and aldehydes can act as substrates. The mammalian enzyme can be interconverted to xanthine oxidase [EC 1.1.3.22] by storage at $-20°C$, by treatment with proteolytic agents or organic solvents, or by thiol reagents such as Cu^{2+}, *N*-ethylmaleimide, or 4-hydroxymercuribenzoate. The feature distinguishing xanthine oxidase from xanthine dehydrogenase is the nature of the protein environment surrounding the FAD. **See also** *Xanthine Oxidase*

[1] B. A. Palfey & V. Massey (1998) *CBC* 3, 83.
[2] P. E. Baugh, D. Collison, C. D. Garner & J. A. Joule (1998) *CBC* 3, 377.
[3] N. E. Tolbert (1981) *ARB* 50, 133.
[4] R. C. Bray (1980) *AE* 51, 107.
[5] T. C. Stadtman (1980) *ARB* 49, 93.
[6] R. C. Bray (1975) *The Enzymes*, 3rd ed., 12, 299.
Selected entries from *Methods in Enzymology* [vol, page(s)]:
Other molecular properties: flavin determination, 53, 428; in microbodies, 52, 496, 500; oxidants and, 12A, 5; reversible dissociation, 53, 435; stereospecificity, 87, 114

XANTHINE-GUANINE PHOSPHORIBOSYLTRANSFERASE

This transferase [EC 2.4.2.22], also known as xanthine phosphoribosyltransferase and xanthosine-5'-phosphate pyrophosphorylase (XMP pyrophosphorylase), catalyzes the reaction of 5-phospho-α-D-ribose 1-diphosphate with xanthine to produce (9-D-ribosylxanthine)-5'-phosphate (or, xanthosine 5'-monophosphate, XMP) and pyrophosphate.[1-4] The enzyme will also act on guanine (producing GMP) and, in many cases, hypoxanthine (forming IMP).

[1] T. A. Krenitsky, S. M. Neil & R. L. Miller (1970) *JBC* 245, 2605.
[2] R. L. Miller, D. L. Adamczyk, J. A. Fyfe & G. B. Elion (1974) *ABB* 165, 349.
[3] J. V. Tuttle & T. A. Krenitsky (1980) *JBC* 255, 909.
[4] S. W. Liu & G. Milman (1983) *JBC* 258, 7469.

XANTHINE OXIDASE

This enzyme [EC 1.1.3.22], also known as hypoxanthine oxidase, catalyzes the reaction of xanthine with dioxygen and water to produce urate and hydrogen peroxide.[1-7] This enzyme has two iron-sulfur centers, FAD, and molybdenum. Hypoxanthine and some other purines and pterins can act as substrates. In addition, aldehydes can act as substrates as well (the enzyme is reported to exhibit an aldehyde oxidase activity); it has been suggested that the enzyme probably acts on the hydrated forms of these compounds. Under certain conditions, the product is mainly superoxide rather than hydrogen peroxide: thus, R–H reacts with two dioxygen and water to produce R–OH, two H^+, and two O_2^-. The *Micrococcus* enzyme can use ferredoxin as the acceptor substrate. The enzyme from mammalian tissues can be interconverted to xanthine dehydrogenase [EC 1.1.1.204].

The feature distinguishing xanthine oxidase and xanthine dehydrogenase is the nature of the protein environment surrounding the FAD. The redox potential of this bound coenzyme is different between the two forms of the protein. Xanthine oxidase produces only small amounts of the semiquinone. The two-electron flavin potential is too high to be an effective reductant of NAD^+. **See also** Xanthine *Dehydrogenase*

[1] B. A. Palfey & V. Massey (1998) *CBC* **3**, 83.
[2] P. E. Baugh, D. Collison, C. D. Garner & J. A. Joule (1998) *CBC* **3**, 377.
[3] C. Kisker, H. Schindelin & D. C. Rees (1997) *ARB* **66**, 233.
[4] B. G. Malmström (1982) *ARB* **51**, 21.
[5] R. C. Bray (1980) *AE* **51**, 107.
[6] R. C. Bray (1975) *The Enzymes*, 3rd ed., **12**, 299.
[7] G. Palmer (1975) *The Enzymes*, 3rd ed., **12**, 1.

Selected entries from **Methods in Enzymology** [vol, page(s)]:
General discussion: 2, 482; **12A**, 5; **18B**, 210, 590; 34, 6, 527, 528; **52**, 6, 17; **53**, 397, 401, 404; **54**, 93; **56**, 474; **136**, 254; **251**, 69
Assay: 2, 458, 482; **4**, 380; biological tissues, in, **186**, 651; biopsy, in, **22**, 151, 152; bovine small intestine, in, **12A**, 6; chemiluminescence assay, **305**, 385
Other molecular properties: assay of (caffeine, **272**, 124, 130; nucleoside deoxyribosyltransferase, **51**, 447; inosine, **63**, 32; purine nucleoside phosphorylase, **51**, 518, 531; superoxide dismutase, **53**, 384; **57**, 405; **186**, 227); cryoenzymology, **63**, 338; cyanide inhibition, **9**, 368; determination, **18B**, 590; diaphorase activity of, **2**, 711; electron acceptors and, **4**, 330; electron spin resonance, **6**, 914; **49**, 514; **243**, 31, 34; flavin determination, **53**, 428; homocysteine oxidation and, **5**, 753; hypoxanthine, removal of, **2**, 448; immobilization, **44**, 908; **136**, 256, 258, 259; inhibitors, **233**, 603; purine nucleoside phosphorylase and, **12A**, 114; radioiodination, **70**, 232; reductive titration, **54**, 124; reversible dissociation, **53**, 434; stereospecificity, **87**, 114; substrate specificity, **243**, 30

XANTHOMMATIN REDUCTASE

This oxidoreductase [EC 1.3.1.41], first isolated from fruit flies (*Drosophila melanogaster*), catalyzes the reaction of

xanthommatin   5,12-dihydroxanthommatin

xanthommatin with NADH to produce 5,12-dihydroxanthommatin and NAD^+.[1,2]

[1] P. Santoro & G. Parisi (1986) *J. Exp. Zool.* **239**, 169.
[2] P. Santoro & G. Parisi (1987) *Insect Biochem.* **17**, 635.

XANTHOMONAPEPSIN

This serine endopeptidase [EC 3.4.21.101; formerly EC 3.4.23.33], also called *Xanthomonas* aspartic proteinase and *Xanthomonas* pepstatin-insensitive carboxyl proteinase, catalyzes the hydrolysis of peptide bonds in a number of peptides and proteins. Casein is hydrolyzed with a pH_{opt} of about 2.7. Also hydrolyzed is the chromogenic substrate Lys-Pro-Ala-Leu-Phe-∇-Nph-Arg-Leu, where ∇ indicates the cleavage site. The enzyme is similar in activity to pseudomonapepsin [EC 3.4.21.100]. However, the specificity requirements for xanthomonapepsin are less strict.[1]

[1] K. Oda, S. Takahashi, T. Shin & S. Murao (1995) in *Aspartic Proteinases: Structure, Function, Biology, and Biomedical Implications* (K. Takahashi, ed.), Plenum Press, New York, p. 529.

XANTHOTOXOL O-METHYLTRANSFERASE

This methyltransferase [EC 2.1.1.93], first isolated from parsley (*Petroselinum crispum*), catalyzes the reaction of S-adenosyl-L-methionine with xanthotoxol to produce S-adenosyl-L-homocysteine and O-methylxanthotoxol (also called xanthotoxin and methoxalen).[1]

xanthotoxol   xanthotoxin

With 5-hydroxyxanthotoxol as an alternative acceptor substrate, isopimpinellin is produced.

[1] K. D. Hauffe, K. Hahlbrock & D. Scheel (1986) *Z. Naturforsch.* **41C**, 228.

XbaI RESTRICTION ENDONUCLEASE

This type II restriction endonuclease [EC 3.1.21.4] is obtained from *Xanthomonas badrii* and recognizes and acts on both strands of DNA at 5′ . . . T∇CTAGA . . . 3′.[1] (The symbol ∇ indicates the cleavage site.) The endonuclease is blocked by overlapping *dam* methylation of the recognition sequence.

[1] B. S. Zain & R. J. Roberts (1977) *J. Mol. Biol.* **115**, 249.

Selected entries from **Methods in Enzymology** [vol, page(s)]:
General Discussion: 65, 485

Other molecular properties: fragment end structure produced, **65**, 511; recognition sequence, **65**, 11, 511.

*Xcm*I RESTRICTION ENDONUCLEASE

This type II restriction endonuclease [EC 3.1.21.4] is obtained from *Xanthomonas campestris*, and recognizes and acts on both strands of DNA at 5′... CCANNNNN∇ NNNNTGG... 3′, where N represents any base.[1] (The symbol ∇ indicates the cleavage site.) Note that the DNA fragments generated have only a single base 3′ extension.

[1] P. C. Shaw & Y. K. Mok (1993) *Gene* **133**, 85.

XENOBIOTIC-TRANSPORTING ATPase

This so-called ATPase [EC 3.6.3.44], also known as multidrug-resistance protein and P-glycoprotein, catalyzes the ATP-dependent (ADP- and orthophosphate-producing) transport of a xenobiotic$_{in}$ to produce xenobiotic$_{out}$.[1–6] This protein system is an ABC-type (ATP-binding cassette-type) ATPase, characterized by the presence of two similar ATP-binding domains. Several variants of this transport system may be present in a single eukaryotic cell, each catalyzing the transport of a number of drugs and xenobiotics with distinct specificities. Certain variants are specific for glutathione adducts (*i.e.*, *S*-substituted glutathione). Other members of this class also exhibit a flippase activity (*i.e.*, catalyzing the movement of phospholipids from one lipid bilayer face to the other).[5] Phosphorylation of the enzyme has not been observed during the transport process. Examples of xenobiotics include verapamil, vincristine, cyclosporin, and leukotriene C_4.

The idea that a transporter is an enzyme is in keeping with a new definition of enzyme catalysis as the facilitated making/breaking of chemical bonds, not just covalent bonds.[7] This idea builds on Pauling's assertion that any long-lived, chemically distinct interaction (in this case, the persistent location of a solute with respect to the faces of a membrane) can be regarded as a chemical bond. Note also that the equilibrium constant ($K_{eq} = [\text{Xenobiotic}_{out}][\text{ADP}][\text{P}_i]/[\text{Xenobiotic}_{in}][\text{ATP}]$) does not conform to that expected for an ATPase (*i.e.*, $K_{eq} = [\text{ADP}][\text{P}_i]/[\text{ATP}]$). Thus, although the overall reaction yields ADP and orthophosphate, the enzyme is misclassified as a hydrolase, and instead should be regarded as an energase-type reaction. Energases facilitate affinity-modulated reactions by coupling the $\Delta G_{\text{ATP-hydrolysis}}$ to a force-generating or work-producing step.[7] In this case,

P-O-P bond scission supplies the energy to drive xenobiotic transport. *See ATPase; Energase*

[1] J. K. Griffiths & C. E. Sansom (1998) *The Transporter Factsbook*, Academic Press, San Diego.
[2] D. W. Loe, R. G. Deeley & S. P. Cole (1998) *Cancer Res.* **58**, 5130.
[3] H. W. van Veen & W. N. Konings (1998) *BBA* **1365**, 31.
[4] W. T. Bellamy (1996) *Ann. Rev. Pharmac. Toxicol.* **36**, 161.
[5] C. M. Frijters, R. Ottenhoff, M. J. Van Wijland, C. Van Nieuwkerk, A. K. Groen & R. P. Oude-Elferink (1996) *Adv. Enzym. Regul.* **36**, 351.
[6] D. Keppler, J. König & M. Buchler (1996) *Adv. Enzym Regul.* **37**, 321.
[7] D. L. Purich (2001) *TiBS* **26**, 417.

*Xho*I RESTRICTION ENDONUCLEASE

This type II restriction endonuclease [EC 3.1.21.4] is obtained from *Xanthomonas holcicola* and recognizes and acts on both strands of DNA at 5′...C∇TCGAG...3′, where the symbol ∇ indicates the cleavage site.[1] Evidence has been presented that the enzyme has a lysyl and/or an arginyl residue as well as a cysteinyl residue that are essential for enzyme activity.[2] Using phosphorothioate analogues of a short oligonucleotide, the stereochemical course of action of this restriction endonuclease has been reported.[3] Note that the recognition sequence is identical to *Pae*R7 I.

[1] T. R. Gingeras, P. A. Myers, J. A. Olson, F. A. Hanberg & R. J. Roberts (1978) *J. Mol. Biol.* **118**, 113.
[2] L. Sun & J. Yin (1990) *Xibei Daxue Xuebao* **20**, 65.
[3] B. J. Moon, S. K. Kim, N. H. Kim & O. S. Kwon (1996) *Bull. Korean Chem. Soc.* **17**, 1031.

Selected entries from *Methods in Enzymology* [vol, page(s)]:
Other molecular properties: recognition sequence, **65**, 11

*Xho*II RESTRICTION ENDONUCLEASE

This Mg^{2+}- or Mn^{2+}-dependent type II restriction endonuclease [EC 3.1.21.4] is obtained from *Xanthomonas holcicola* and recognizes and acts on both strands of DNA at 5′... R∇GATCY...3′[1] where R refers to either A or G and Y refers to either T or C. (The symbol ∇ indicates the cleavage site.)

[1] V. M. Kramarov, A. L. Mazanov & V. V. Smolyaninov (1982) *Bioorg. Khim.* **8**, 220.

Selected entries from *Methods in Enzymology* [vol, page(s)]:
Other molecular properties: recognition sequence, **65**, 11

*Xma*I RESTRICTION ENDONUCLEASE

This type II restriction endonuclease [EC 3.1.21.4] is obtained from *Xanthomonas malvacearum*, is magnesium-ion dependent, and recognizes and acts on both strands of DNA at 5′... C∇CCGGG... 3′, where the symbol ∇ indicates the cleavage site.[1–4] The enzyme will still bind to the recognition site in the absense of Mg^{2+}.[2] The enzyme will also act on DNA if the 5-methylcytosine is present in the CpG position.[3] Binding of the endonuclease induces bending of DNA toward the minor groove.[4] *Xma*I is an isoschizomer of *Sma*I; however, while *Sma*I produces blunt-ended fragments, *Xma*I produces a 5′ extension.

[1] S. A. Endow & R. J. Roberts (1977) *J. Mol. Biol.* **112**, 521.
[2] B. E. Withers & J. C. Dunbar (1995) *NAR* **23**, 3571.
[3] H. Youssoufian & C. Mulder (1981) *J. Mol. Biol.* **150**, 133.
[4] B. E. Withers & J. C. Dunbar (1993) *NAR* **21**, 2571.

*Xma*CI RESTRICTION ENDONUCLEASE

This type II restriction endonuclease [EC 3.1.21.4] is obtained from *Xanthomonas malvacearum* strain C and recognizes and acts on both strands of DNA at 5′ . . . C∇CCGGG . . . 3′. (The symbol ∇ indicates the cleavage site.)

*Xma*JI RESTRICTION ENDONUCLEASE

This type II restriction endonuclease [EC 3.1.21.4] is obtained from *Xanthomonas maltophilia* Jo 85-025 and recognizes and acts on both strands of DNA at 5′ . . . C∇CTAGG . . . 3′. (The symbol ∇ indicates the cleavage site.)

*Xmi*I RESTRICTION ENDONUCLEASE

This type II restriction endonuclease [EC 3.1.21.4] is obtained from *Xanthomonas maltophilia* Jo21-021 and recognizes and acts on both strands of DNA at 5′ . . . GT∇MKAC . . . 3′ where M represents A or C and K represents T or G. (The symbol ∇ indicates the cleavage site.)

*Xmn*I RESTRICTION ENDONUCLEASE

This type II restriction endonuclease [EC 3.1.21.4], obtained from *Xanthomonas manihotis* 7AS1, recognizes and acts on both strands of DNA at 5′ . . . GAANN∇ NNTTC . . . 3′ where N represents any nucleotide.[1,2] Note that *Eco*RI sites which have been cleaved, filled in via DNA polymerase, and blunt-ended ligated will generate *Xmn*I recognition sites. (The symbol ∇ indicates the cleavage site.)

[1] B.-C. Lin, M.-C. Chien & S.-Y. Lou (1980) *NAR* **8**, 6189.
[2] D. O. Nwankwo, J. J. Lynch, L. S. Moran, A. Fomenkov & B. E. Slatko (1996) *Gene* **173**, 121.

*Xor*II RESTRICTION ENDONUCLEASE

This type II restriction endonuclease [EC 3.1.21.4] is obtained from *Xanthomonas oryzae* and recognizes and acts on both strands of DNA at 5′ . . . CGAT∇CG . . . 3′.[1] (The symbol ∇ indicates the cleavage site.)

[1] R. Y.-H. Wang, J. G. Shedlarski, M. B. Farber, D. Kuebbing & M. Ehrlich (1980) *BBA* **606**, 371.

Selected entries from *Methods in Enzymology* [**vol**, page(s)]:
Other molecular properties: recognition sequence, **65**, 11

*Xpa*I RESTRICTION ENDONUCLEASE

This type II restriction endonuclease [EC 3.1.21.4] is obtained from *Xanthomonas papavericola* and recognizes and acts on both strands of DNA at 5′ . . . C∇TCGAG . . . 3′.[1] (The symbol ∇ indicates the cleavage site.)

[1] T. R. Gingeras, P. A. Myers, J. A. Olson, F. A. Hanberg & R. J. Roberts (1978) *J. Mol. Biol.* **118**, 113.

Selected entries from *Methods in Enzymology* [**vol**, page(s)]:
Other molecular properties: recognition sequence, **65**, 11

*Xsp*I RESTRICTION ENDONUCLEASE

This type II restriction endonuclease [EC 3.1.21.4] is obtained from *Xanthomonas* species YK1 and recognizes and acts on both strands of DNA at 5′ . . . C∇TAG . . . 3′.[1] (The symbol ∇ indicates the cleavage site.)

[1] T. Song, B. S. Kang & Y. M. Kim (1998) *Mol. Cells* **8**, 370.

XYLAN ENDO-1,4-β-XYLANASE

This enzyme [EC 3.2.1.8], also known as xylanase, endo-1,4-β-xylanase, and 1,4-β-D-xylan xylanohydrolase, catalyzes the endohydrolysis of 1,4-β-D-xylosidic linkages in xylans.[1-3]

[1] G. Davies, M. L. Sinnott & S. G. Withers (1998) *CBC* **1**, 119.
[2] M. W. Bauer, S. B. Halio & R. M. Kelly (1996) *Adv. Protein Chem.* **48**, 271.
[3] T. Hayashi (1989) *Ann. Rev. Plant Physiol. Plant Mol. Biol.* **40**, 139.

Selected entries from *Methods in Enzymology* [**vol**, page(s)]:
General discussion: **160**, 264, 632, 638, 648, 655, 659, 662, 671
Assay: **160**, 69, 538, 633, 638, 649, 655, 663, 671, 675; nephelometric assay, **160**, 124; type A, **160**, 659

XYLAN ENDO-1,3-β-XYLOSIDASE

This enzyme [EC 3.2.1.32], also known as xylanase and endo-1,3-β-xylanase, catalyzes the random hydrolysis of 1,3-β-D-xylosidic linkages in 1,3-β-D-xylans.[1,2] The enzyme will also act on β-1,3-linked xylotriose, xylotetraose, and xylopentaose.

[1] W. P. Chen, M. Matsuo & T. Yasui (1986) *Agric. Biol. Chem.* **50**, 1183 and 1195.
[2] K. Horikoshi & Y. Atsukawa (1973) *Agric. Biol. Chem.* **37**, 2097.

XYLAN α-1,2-GLUCURONOSIDASE

This glycosidase [EC 3.2.1.131] catalyzes the hydrolysis of α-D-1,2-(4-*O*-methyl)glucuronosyl links in the main chain of hardwood xylans.[1] An alternative substrate is reduced aldobiouronic acid.

[1]M. Ishihara & K. Shimizu (1988) *Mokuzai Gakkaishi* **34**, 58.

1,4-β-D-XYLAN SYNTHASE

This enzyme [EC 2.4.2.24] catalyzes the reaction of UDP-D-xylose with $\{(1,4)\text{-}\beta\text{-}D\text{-xylan}\}_n$ to produce UDP and $\{(1,4)\text{-}\beta\text{-}D\text{-xylan}\}_{n+1}$.[1–3]

[1]M. W. Rodgers & G. P. Bolwell (1992) *BJ* **288**, 817.
[2]R. W. Bailey & W. Z. Hassid (1966) *PNAS* **56**, 1586.
[3]K. Suzuki, E. Ingold, M. Suguyama & A. Komamine (1991) *Plant Cell Physiol.* **32**, 303.

Selected entries from **Methods in Enzymology** [vol, page(s)]:
General Discussion: 8, 407, 409
Assay: 8, 407, 409

XYLAN 1,3-β-XYLOSIDASE

This enzyme [EC 3.2.1.72], also known as exo-1,3-β-xylosidase, catalyzes the hydrolysis of successive xylosyl residues from the non-reducing termini of 1,3-β-D-xylans.[1]

[1]S. Fukui, T. Suzuki, K. Kitahara & T. Miwa (1960) *J. Gen. Appl. Microbiol.* **6**, 270.

XYLAN 1,4-β-XYLOSIDASE

This enzyme [EC 3.2.1.37], also known as β-xylosidase and xylobiase, catalyzes the hydrolysis of 1,4-β-D-xylans resulting in the removal of successive D-xylose residues from the non-reducing termini.[1–4] Xylobiose can also be utilized as a substrate. The sheep liver enzyme has been reported to exhibit other exoglycosidase activities. Other alternative substrates include *p*-nitrophenyl β-D-xylanoside. The enzyme isolated from several sources will act only on xylo-oligosaccharides and not on xylan. **See also** *Xylan Endo-1,4-β-Xylanase*

[1]D. J. Vocadlo, L. F. MacKenzie, S. He, G. J. Zeikus & S. G. Withers (1998) *BJ* **335**, 449.
[2]S. Armand, C. Vieille, C. Gey, A. Heyraud, J. G. Zeikus & B. Henrissat (1996) *EJB* **236**, 706.
[3]S. Saxena, H. P. Fierobe, C. Gaudin, F. Guerlesquin & J. P. Belaich (1995) *Appl. Environ. Microbiol.* **61**, 3509.
[4]H. Kersters-Hilderson, E. Van Doorslaer, M. Lippens & C. K. De Bruyne (1984) *ABB* **234**, 61.

Selected entries from **Methods in Enzymology** [vol, page(s)]:
General discussion: 83, 631; **160**, 633, 662, 663, 675, 679, 684, 696
Assay: 83, 632; **160**, 124, 664, 679, 685, 696

XYLENE MONOOXYGENASE

This FAD- and iron-dependent oxidoreductase catalyzes the hydroxylation of the methyl side chains of toluene and xylenes, serving as a key enzyme in the degradation of these aromatic compounds.[1]

[1]B. Bühler, A. Schmid, B. Hauer & B. Witholt (2000) *JBC* **275**, 10085.

XYLITOL KINASE

This phosphotransferase [EC 2.7.1.122], also known as xylitol phosphotransferase and first isolated from *Streptococcus mutans*, catalyzes the reaction of ATP with xylitol to produce ADP and xylitol 5-phosphate.[1] Note that the product is no longer a *meso* compound.

[1]S. Assev & G. Rölla (1984) *Acta Pathol. Microbiol. Immunol. Scand. Sect. B* **92**, 89.

XYLOGLUCAN 4-GLUCOSYLTRANSFERASE

This glucosyltransferase [EC 2.4.1.168] catalyzes the transfer of a β-D-glucosyl residue from UDP-D-glucose on to a glucose residue in xyloglucan, forming a β-1,4-D-glucosyl-D-glucose linkage; thus, UDP-D-glucose reacts with a (glucosyl)xyloglucan to produce UDP and a (glucosyl-glucosyl)xyloglucan with a new β-1,4-linkage.[1,2] This enzyme works in association with xyloglucan 6-xylosyltransferase [EC 2.4.1.169] to produce xyloglucans. There is a necessity for concurrent transfer of glucosyl and xylosyl residues. This particular enzyme is distinct from that of cellulose synthase [EC 2.4.1.12].

[1]T. Hayashi & K. Matsuda (1981) *JBC* **256**, 11117.
[2]T. Hayashi, T. Koyama & K. Matsuda (1988) *Plant Physiol.* **87**, 341.

XYLOGLUCAN:XYLOGLUCOSYLTRANSFERASE

This transferase [EC 2.4.1.207], also known as xyloglucan endotransglycosylase and endoxyloglucan transferase, catalyzes the hydrolysis of a β-(1 → 4) bond in the backbone of a xyloglucan and transfers the xyloglucanyl segment onto *O*-4 of the non-reducing terminal glucosyl residue of an acceptor (for example, a xyloglucan or an oligosaccharide of xyloglucan).[1] Cello-oligosaccharides cannot act as substrates.

[1]N. M. Steele & S. C. Fry (1999) *BJ* **340**, 207.

XYLOGLUCAN 6-XYLOSYLTRANSFERASE

This transferase [EC 2.4.1.169] catalyzes the reaction of UDP-D-xylose with a (glucosyl)xyloglucan to produce UDP and a (xylosylglucosyl)xyloglucan (with an α-1,6 linkage). Thus, it transfers an α-D-xylosyl residue from UDP-D-xylose to a glucose residue in the formation of a xyloglucan by forming an α-1,6-D-xylosyl-D-glucose linkage. Working in conjunction with xyloglucan 4-glucosyltransferase [EC 2.4.1.168], xyloglucans are synthesized. There is a requirement for concurrent transfers of glucose and xylose residues in this plant biosynthetic pathway.[1–3]

[1] T. Hayashi & K. Matsuda (1981) JBC **256**, 11117.
[2] A. R. White, Y. Xin & V. Pezeshk (1993) BJ **294**, 231.
[3] T. Hayashi, T. Koyama & K. Matsuda (1988) Plant Physiol. **87**, 341.

XYLONATE DEHYDRATASE

This dehydratase [EC 4.2.1.82], also called D-*xylo*-aldonate dehydratase, catalyzes the conversion of D-xylonate to 2-dehydro-3-deoxy-D-xylonate and water.[1]

| D-xylonate | 2-dehydro-3-deoxy-D-xylonate | 2-keto-3-deoxy-D-xylonate |

[1] W. A. Wood (1971) The Enzymes, 3rd ed., **5**, 573.

Selected entries from **Methods in Enzymology** [**vol**, page(s)]:
General discussion: 90, 302
Assay: 90, 302

XYLONO-1,4-LACTONASE

This lactonase [EC 3.1.1.68] catalyzes the hydrolysis of D-xylono-1,4-lactone to produce D-xylonate.[1]

[1] J. Buchert & L. Viikari (1988) Appl. Microbiol. Biotechnol. **27**, 333, and **29**, 375.

D-XYLOSE 1-DEHYDROGENASE

This oxidoreductase [EC 1.1.1.175], also known as NAD^+: D-xylose dehydrogenase, catalyzes the reaction of D-xylose with NAD^+ to yield D-xylonolactone and NADH.

Selected entries from **Methods in Enzymology** [**vol**, page(s)]:
General discussion: 89, 226
Assay: pseudomonad MSU-1, **89**, 226

L-XYLOSE 1-DEHYDROGENASE

This oxidoreductase [EC 1.1.1.113] catalyzes the reaction of L-xylose with $NADP^+$ to produce L-xylono-1,4-lactone and NADPH.[1]

[1] K. Uehara & M. Takeda (1962) JB **52**, 461.

D-XYLOSE 1-DEHYDROGENASE (NADP⁺)

This oxidoreductase [EC 1.1.1.179], also known as D-xylose:$NADP^+$ dehydrogenase, catalyzes the reaction

of D-xylose with $NADP^+$ to produce D-xylono-1,5-lactone and NADPH. Alternative substrates include L-arabinose, 2-deoxy-D-glucose, and D-ribose.[1]

[1] S. Zepeda, O. Monasterio & T. Ureta (1990) BJ **266**, 637.

XYLOSE ISOMERASE

This enzyme [EC 5.3.1.5], which requires magnesium, manganese, or cobalt divalent cations, catalyzes the interconversion of D-xylose and D-xylulose.[1–3] This enzyme from certain sources will also interconvert D-glucose and D-fructose as well as D-ribose and D-ribulose.[1–3]

[1] C. I. F. Watt (1998) CBC **1**, 253.
[2] I. A. Rose (1975) AE **43**, 491.
[3] E. A. Noltmann (1972) The Enzymes, 3rd ed., **6**, 271.

Selected entries from **Methods in Enzymology** [**vol**, page(s)]:
General discussion: 1, 366; 5, 347; 9, 588; 41, 466; 44, 811; 330, 215
Assay: 1, 367, 369; 5, 348; 9, 589; 41, 466, 477
Other molecular properties: glucose isomerase activity (activity, immobilized, 44, 161, 163, 164, 200, 237, 811; coupled to cellulose acetate beads, assay, 135, 287; effect of metal ions, 44, 815, 816; efficiency [effect of pH, 44, 237, 238; of temperature, 44, 238, 239]; immobilized, industrial production and application, 136, 356; immobilization [adsorption, by, 44, 161; collagen, on, 44, 770; controlled-pore alumina, on, 44, 161; cross-linked cell hemogenates, in, 44, 812; DEAE-cellulose, on, 44, 165]; industrial use, 44, 47, 165, 241, 809; kinetics, 44, 769; microbial sources, 44, 770, 811; oxygen sensitivity, 44, 816; pH optimum, 44, 164; productivity, 44, 811; stability, 44, 811; use in reactor, 44, 769); hyperthermophilic, 330, 215; inhibition, 41, 471; 87, 87; *Escherichia coli* (properties, 1, 370; purification, 5, 349; specificity, 1, 370); *Lactobacillus brevis* 41, 466 (activation energy, 9, 592; inhibitors, 41, 471; metal requirement, 9, 592; pH optimum, 9, 591; 41, 470, 471); *Yersinia* (formerly *Pasteurella*) *pestis* (properties, 5, 349; purification, 5, 348); proton isotope exchange, 87, 87

XYLOSE-1-PHOSPHATE URIDYLYLTRANSFERASE

This transferase [EC 2.7.7.11], also known as UTP:xylose-1-phosphate uridylyltransferase, catalyzes the reaction of UTP with α-D-xylose 1-phosphate to produce pyrophosphate, and UDP-D-xylose.[1]

[1] V. Ginsburg, E. F. Neufeld & W. Z. Hassid (1956) PNAS **42**, 333.

D-XYLOSE REDUCTASE

This oxidoreductase catalyzes the reaction of D-xylose with NADPH to produce xylitol and $NADP^+$. The enzyme isolated from several sources also utilizes NADH. D-Xylose reductase has been regarded by some as a variant of aldose reductase (or, aldehyde reductase, EC 1.1.1.21); however, the substrate specificity profile of the latter enzyme is distinctive.[1] **See also** Aldehyde Reductase

[1] B. F. Cooper & F. B. Rudolph (1995) MIE **249**, 188.

Selected entries from *Methods in Enzymology* [vol, page(s)]:
General discussion: 9, 188
Assay: 9, 188
Other molecular properties: *Penicillium chrysogenum*, 9, 188 (kinetics, 9, 190; partial purification, 9, 189; pH optimum, 9, 191; properties, 9, 190; specificity, 9, 190; stability, 9, 191); product inhibition studies, 249, 202

XYLOSYLPROTEIN 4-β-GALACTOSYLTRANSFERASE

This manganese-dependent transferase [EC 2.4.1.133], also known as galactosyltransferase I, catalyzes the reaction of UDP-galactose with an *O*-β-D-xylosylprotein to produce UDP and a 4-β-D-galactosyl-*O*-β-D-xylosylprotein.[1,2] Alternative acceptor substrates include D-xylose and *O*-β-D-xylosyl-L-serine. This enzyme participates in the biosynthesis of chondroitin sulfate.

[1] N. B. Schwartz & L. Roden (1975) *JBC* 250, 5200.
[2] N. B. Schwartz (1976) *JBC* 251, 285.

Selected entries from *Methods in Enzymology* [vol, page(s)]:
General discussion: 28, 649
Assay: 28, 649
Other molecular properties: chondroitin sulfate biosynthesis and, 28, 639, 644; isolation from chick embryo cartilage, 28, 647, 651; properties, 28, 651; reaction, 28, 649; specificity, 28, 642, 651; substrate preparation, 28, 664

D-XYLULOKINASE

This enzyme [EC 2.7.1.17], also known as xylulose kinase, catalyzes the reaction of ATP with D-xylulose to produce ADP and D-xylulose 5-phosphate.[1-3]

[1] W. L. Dills, Jr., P. D. Parsons, C. L. Westgate & N. J. Komplin (1994) *Protein Expr. Purif.* 5, 259.
[2] M. S. Neuberger, B. S. Hartley & J. E. Walker (1981) *BJ* 193, 513.
[3] W. L. Dills, Jr., P. D. Parsons, C. L. Westgate & N. J. Komplin (1994) *Prot. Expr. Purif.* 5, 259.

Selected entries from *Methods in Enzymology* [vol, page(s)]:
General Discussion: 5, 208; 9, 454
Assay: 5, 208; 9, 454
Other molecular properties: *Aerobacter aerogenes* (properties, 9, 458; purification, 9, 455; specificity, 9, 458); L-arabinose isomerase preparation free of, 9, 600; calf liver (properties, 5, 211; purification, 5, 209); sources, 9, 454; D-xylose isomerase, in assay of, 9, 589

L-XYLULOKINASE

This enzyme [EC 2.7.1.53], also known as L-xylulose kinase, catalyzes the reaction of ATP with L-xylulose to produce ADP and L-xylulose 5-phosphate.[1-3]

[1] R. L. Anderson & W. A. Wood (1962) *JBC* 237, 1029.
[2] R. C. Doten & R. P. Mortlock (1985) *J. Bact.* 161, 529.
[3] J. C. Sanchez, R. Gimenez, A. Schneider, W. D. Fessner, L. Baldoma, J. Aguilar & J. Badia (1994) *JBC* 269, 29665.

Selected entries from *Methods in Enzymology* [vol, page(s)]:
General Discussion: 9, 458
Assay: 9, 458

D-XYLULOSE REDUCTASE

This oxidoreductase [EC 1.1.1.9], also known as xylitol dehydrogenase, NAD$^+$:D-*xylo*-dehydrogenase, and NAD$^+$:D-*erythro*-1,2,4-polyalcohol dehydrogenase, catalyzes the reversible reaction of D-xylulose with NADH to produce xylitol (an optically inactive *meso* compound) and NAD$^+$.[1-3] The enzyme will also catalyze the reversible reduction of L-erythrulose, producing erythritol (also a *meso* polyol), and of D-fructose, producing D-sorbitol.

[1] W. B. Jakoby & J. Fredericks (1961) *BBA* 48, 26.
[2] K. H. Do Nascimento & D. D. Davies (1975) *BJ* 149, 553.
[3] R. Lunzer, Y. Mamnun, D. Haltrich, K. D. Kulbe & B. Nidetzky (1998) *BJ* 336, 91.

Selected entries from *Methods in Enzymology* [vol, page(s)]:
General discussion: 5, 317; 9, 163, 174, 186, 191, 193; **18A**, 62
Assay: 5, 318; 9, 163, 186, 188, 191, 193; **18A**, 63
Other molecular properties: *Aerobacter aerogenes* (pH optima, 9, 196; properties, 9, 195; purification, 9, 194; specificity, 9, 195); *Candida utilis*, 9, 163, 165; cytoplasmic, **18A**, 65; properties, 5, 321; 9, 165, 174, 188, 192; purification, 5, 319; 9, 164, 187, 192; specificity, 9, 176; stereospecificity, 87, 103; xylose isomerase, in assay of, 5, 348; 9, 589

L-XYLULOSE REDUCTASE

This oxidoreductase [EC 1.1.1.10], also known as xylitol dehydrogenase (NADP$^+$), catalyzes the reversible reaction of L-xylulose with NADPH to produce xylitol (a *meso* compound) and NADP$^+$.[1-4] The NADPH stereospecificity is B-side. ***See also*** D-*Xylulose Reductase*

[1] C. Arsenis & O. Touster (1969) *JBC* 244, 3895.
[2] A. E. Garrod (1908) *Lancet* II, 214.
[3] R. C. Doten & R. P. Mortlock (1985) *J. Bact.* 161, 529.
[4] A. R. Clarke & T. R. Dafforn (1998) *CBC* 3, 1.

Selected entries from *Methods in Enzymology* [vol, page(s)]:
General discussion: 5, 317; 9, 166, 170; **18A**, 62, 65
Assay: 5, 318; **18A**, 63, 66
Other molecular properties: properties, 5, 321; 9, 174; purification, 5, 319; **18A**, 66; specificity, 9, 176; stereospecificity, 87, 103

Y

YAPSIN 1

This yeast aspartic endopeptidase [EC 3.4.23.41], formerly known as yeast aspartic protease 3, catalyzes the hydrolysis of pro-α-mating factor and prosomatostatin II, always at a basic aminoacyl residue.[1] The synthetic substrate Boc-Arg-Val-Arg-Arg-∇-Mca is also hydrolyzed and Lys9-∇-Lys10ArgArg of ACTH$_{1-39}$ is readily cleaved.[2]

[1]Y. P. Loh & N. X. Cawley (1995) *MIE* **248**, 136.
[2]N. X. Cawley, H.-C. Chen, M. C. Beinfeld & Y. P. Loh (1996) *JBC* **271**, 4168.

Selected entries from *Methods in Enzymology* [vol, page(s)]:
Assay: 248, 138
Other molecular properties: yeast, **248**, 107, 112 (amino acid sequence, **248**, 137; assay, **248**, 138; biological function, **248**, 137; gene, **248**, 137; inhibitor profile, **248**, 142; mammalian homolog, **248**, 138, 144; pH optimum, **248**, 142

YAPSIN 2

This yeast aspartic proteinase, formerly called Mkc7p displays specificity for a lysyl or arginyl residue at P1 and relaxed specificity at P2 (although not a prolyl residue).[1]

[1]H. Komano & R. S. Fuller (1995) *PNAS* **92**, 10752.

YAPSIN 3

This aspartic endopeptidases from *Saccharomyces cerevisiae* is not to be confused with yapsin A (which has also been called yapsin 3).[1] It acts on a number of prohormones.[1]

[1]V. Olsen, N. X. Cawley, J. Brandt, M. Egel-Mitani & Y. P. Loh (1999) *BJ* **339**, 407.

YAPSIN A

This yeast aspartic proteinase [EC 3.4.23.17], formerly called pro-opiomelanocortin converting enzyme and prohormone converting enzyme, catalyzes the hydrolysis of paired basic aminoacyl residues in certain prohormones, either between them, or on the carboxyl side.[1]

[1]Y. P. Loh & N. X. Cawley (1995) *MIE* **248**, 136.

Selected entries from *Methods in Enzymology* [vol, page(s)]:
General discussion: 248, 136
Assay: 248, 138
Other molecular properties: calcium ion effects, **248**, 145; distribution, **248**, 145; inhibitor profile, **248**, 145; pH optimum, **248**, 145; purification, **248**, 141; stability, **248**, 144; subcellular localization, **248**, 145; substrate specificity, **248**, 145

*Yen*I RESTRICTION ENDONUCLEASE

This cold-active type II restriction endonuclease [EC 3.1.21.4] from *Yersinia enterocolitica* acts on both strands of DNA at 5′...CTGCA∇G...3′.[1,2] (The symbol ∇ indicates the cleavage site.)

*Yen*AI, *Yen*BI, *Yen*CI, *Yen*DI, and *Yen*EI restriction endonucleases have been isolated from different strains of *Y. Enterocolitica*, all recognizing 5′...CTGCAG...3′ as well.[1]

[1]M. Miyahara, T. Maruyama, A. Wake & K. Mise (1988) *Appl. Environ. Microbiol.* **54**, 577.
[2]S. A. Kinder, J. L. Badger, G. O. Bryant, J. C. Pepe & V. L. Miller (1993) *Gene* **136**, 271.

Ypt1p PROTEIN

This small yeast G-protein exhibits GTPase activity in the presence of Ypt-GTPase-activating proteins.

Selected entries from *Methods in Enzymology* [vol, page(s)]:
Other molecular properties: yeast enzyme (antibodies, **219**, 383, 385; bacterial expression, **219**, 371; bacterially expressed, purification, **219**, 374; intrinsic GTPase activity, determination, **219**, 381; nucleotide binding, **219**, 378; nucleotide-free preparation, **219**, 379; polyclonal antibodies, **219**, 379, 384)

Z

ZanI RESTRICTION ENDONUCLEASE

This type II restriction endonuclease [EC 3.1.21.4] is obtained from *Zymomonas anaerobia* and acts on both strands of DNA at 5′ ... CC∇WGG ... 3′, where W refers to either A or T.[1]

[1]D. K. Sun & O. J. Yoo (1988) *Korean Biochem. J.* **21**, 419.

β-ZEACAROTENE HYDROXYLASE

This oxidoreductase catalyzes the hydroxylation of β-zeacarotene (or, 7′,8′-dihydro-βΨ-carotene) to produce 3-hydroxy-β-zeacarotene (or, 7′,8′-dihydro-βΨ-caroten-3-ol). It participates in the zeaxanthine biosynthetic pathway in a number of microorganisms, particularly *Flavobacterium*.[1]

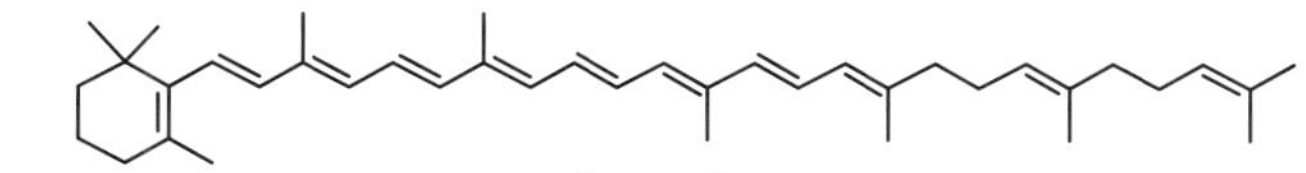

β-zeacarotene

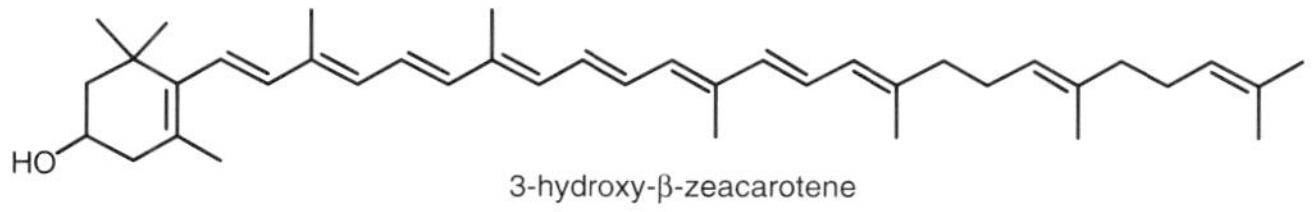

3-hydroxy-β-zeacarotene

[1]J. C. B. McDermott, D. J. Brown, G. Britton & T. W. Goodwin (1974) *BJ* **144**, 231.

ZEATIN O-β-D-GLUCOSYLTRANSFERASE

This enzyme [EC 2.4.1.203], also known as UDPglucose:zeatin O-glucosyltransferase, catalyzes the reaction of

O-β-D-glucosylzeatin

zeatin (*trans*-zeatin) with UDP-D-glucose to produce UDP and O-β-D-glucosylzeatin.[1,2]

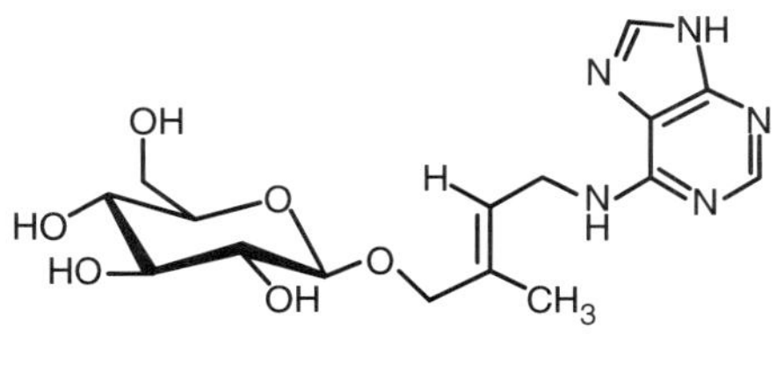

[1]S. C. Dixon, R. C. Martin, M. C. Mok, G. Shaw & D. W. S. Mok (1989) *Plant Physiol.* **90**, 1316.
[2]R. C. Martin, M. C. Mok & D. W. S. Mok (1999) *PNAS* **96**, 284.

ZEATIN REDUCTASE

This enzyme [EC 1.3.1.69; formerly classified as EC 1.1.1.242] catalyzes the reversible reaction of dihydrozeatin with NADP$^+$ to produce zeatin and NADPH.[1]

[1]A. N. Binns (1994) *Ann. Rev. Plant Physiol. Plant Mol. Biol.* **45**, 173.

ZEATIN O-β-D-XYLOSYLTRANSFERASE

This enzyme [EC 2.4.1.204], also known as UDP-xylose:zeatin O-xylosyltransferase, catalyzes the reaction of zeatin (*trans*-zeatin) with UDP-D-xylose to produce UDP and O-β-D-xylosylzeatin.[1–3]

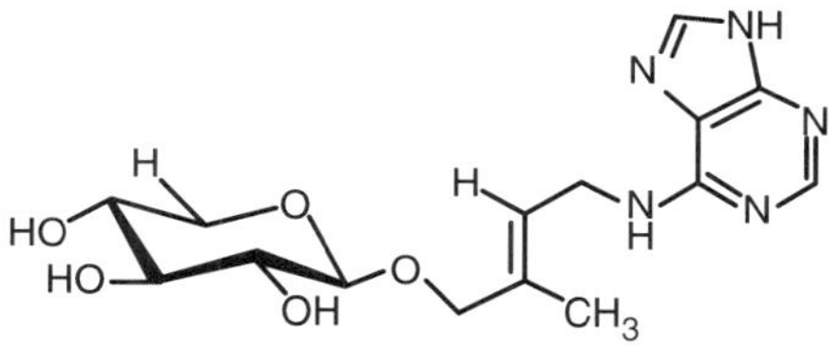

O-β-D-xylosylzeatin
(in the pyranose form)

This xyloside product is important in cytokinin transport, storage, and protection against oxidizing enzymes. The enzyme cannot act on UDP-D-glucose.

[1]J. E. Turner, D. W. S. Mok, M. C. Mok & G. Shaw (1987) *PNAS* **84**, 3714.
[2]R. C. Martin, M. C. Mok & D. W. S. Mok (1993) *PNAS* **90**, 953.
[3]R. C. Martin, M. C. Mok & D. W. Mok (1999) *Plant Physiol.* **120**, 553.

ZEAXANTHINE EPOXIDASE

This epoxidase reportedly catalyzes the NAD(P)H reaction of dioxygen with zeaxanthine (or, β,β-carotene-3,3′-diol) to produce the monoepoxy derivative antheraxanthin, which then undergoes a second epoxidation reaction (this time on the other cyclohexene ring) to generate the diepoxy derivative violaxanthin.[1]

[1]D. Siefermann & H. Y. Yamamoto (1975) *ABB* **171**, 70.

zeaxanthine

antheraxanthin

ZINC D-Ala-D-Ala CARBOXYPEPTIDASE

This zinc enzyme [EC 3.4.17.14], also known as Zn^{2+} G peptidase, KM endopeptidase, and D-alanyl-D-alanine hydrolase, catalyzes the hydrolysis of the D-alanyl-D-alanine peptide bond in bacterial cell wall metabolism.[1–3]

[1] J.-M. Ghuysen, M. Leyh-Bouille, R. Bonaly, M. Nieto, H. R. Perkins, K. H. Schleifer & O. Kandler (1970) *Biochemistry* **9**, 2955.
[2] O. Dideberg, B. Joris, J.-M. Frère, J.-M. Ghuysen, G. Weber, R. Robaye, J.-M. Delbrouck & I. Roelandts (1980) *FEBS Lett.* **117**, 215.
[3] P. Charlier, O. Dideberg, J. C. Jamoulle, J.-M. Frère, J.-M. Ghuysen, G. Dive & J. Lamotte-Brasseur (1984) *BJ* **219**, 763.

Zn^{2+}-EXPORTING ATPase

This so-called P-type ATPase [EC 3.6.3.5] catalyzes the ATP-dependent (ADP- and orthophosphate-producing) transport of Zn^{2+}_{in} to produce Zn^{2+}_{out}.[1,2] Other divalent cations that can be exported by the bacterial and animal proteins include Cd^{2+} and Pb^{2+}. P-type ATPases undergo phosphorylation during the transport cycle.

The idea that a transporter is an enzyme is in keeping with a new definition of enzyme catalysis as the facilitated making/breaking of chemical bonds, not just covalent bonds.[3] This idea builds on Pauling's assertion that any long-lived, chemically distinct interaction (in this case, the persistent location of a solute with respect to the faces of a membrane) can be regarded as a chemical bond. Note also that the equilibrium constant ($K_{eq} = [Ca^{2+}_{out}][ADP][P_i]/[Ca^{2+}_{in}] \times [ATP]$) does not conform to that expected for an ATPase (*i.e.*, $K_{eq} = [ADP][P_i]/[ATP]$). Thus, although the overall reaction yields ADP and orthophosphate, the enzyme is misclassified as a hydrolase, and instead should be regarded as an energase-type reaction. Energases facilitate affinity-modulated reactions by coupling the $\Delta G_{ATP-hydrolysis}$ to a force-generating or work-producing step.[3] In this case, P-O-P bond scission supplies the energy to drive Zn^{2+} transport. *See ATPase; Energase*

[1] P. Oestreicher & R. J. Cousins (1989) *J. Nutr.* **119**, 639.
[2] C. Rensing, Y. Sun, B. Mitra & B. P. Rosen (1998) *JBC* **273**, 32614.
[3] D. L. Purich (2001) *TiBS* **26**, 417.

ZOXAZOLAMINE 6-HYDROXYLASE

zoxazolamine   6-hydroxyzoxazolamine

This cytochrome P450-associated activity catalyzes the reaction of dioxygen, NADPH, and the muscle relaxant zoxazolamine to produce water, $NADP^+$, and 6-hydroxyzoxazolamine.[1] *See also CYP1A1; CYP2E1*

[1] J. E. Tomaszewski, D. M. Jerina, W. Levin & A. H. Conney (1976) *ABB* **176**, 788.

Z-Pro-PROLINAL-INSENSITIVE Z-Gly-Pro-NH-Mec-HYDROLYZING PEPTIDASE

This peptidase, first isolated from bovine serum and also known as ZIP, catalyzes the hydrolysis of Z-Gly-Pro-NH-Mec and is distinct from prolyl endopeptidase. Z-Pro-prolinal, a specific inhibitor of prolyl endopeptidase (at 350 nM), has no effect on this enzyme, even at concentrations 200-fold greater than the reported K_i value for prolyl endopeptidase.[1]

[1] D. F. Cunningham & B. O'Connor (1997) *EJB* **244**, 900.

*Zsp*2I RESTRICTION ENDONUCLEASE

This type II restriction endonuclease [EC 3.1.21.4] is obtained from *Zoogloea* species 2 and acts on both strands of DNA at 5′. . . ATGCA∇T . . . 3′. (The symbol ∇ indicates the cleavage site.)

ZYMASE

This term refers to a collection of enzymes catalyzing alcohol fermentation in yeast extracts. Eduard Büchner first successfully opened yeast cells in 1896 by using fine sand in a mortar and pestle. Upon adding sugars to the resultant cell-free extract, he noticed the evolution of CO_2 bubbles, a sure sign of active metabolism. Büchner's sensational conclusion[1] that biochemical reactions can proceed without any need for intact cells gave birth to investigative enzymology and earned him the 1907 Nobel Prize in Chemistry. Tragically, he was fatally wounded in August of 1917 while serving in a Rumanian hospital near the Eastern front during World War I.[2]

[1] E. Büchner (1897) "Über alkoholische Gärung ohne Hefezellen", *Ber. deut. Chem. Ges.* **30**, 117.
[2] M. Florkin (1972) *Comprehensive Biochemistry*, **30**, 9.

Index

2-Acylglycerol *O*-Acyltransferase, 28
Acylglycerol Kinase, 29
Acylglycerol Lipase, 29
Acylglycerol palmitoyltransferase, 28
1-Acylglycerol-3-Phosphate
   O-Acyltransferase, 29
2-Acylglycerol-3-Phosphate
   O-Acyltransferase, 29
Acylglycerone-Phosphate Reductase, 29
1-Acylglycerophosphocholine
   O-Acyltransferase, 30
2-Acylglycerophosphocholine
   O-Acyltransferase, 30
N-Acylhexosamine Oxidase, 30
Acyl-Lysine Deacylase, 30
N-Acylmannosamine 1-Dehydrogenase,
   30
N-Acylmannosamine Kinase, 30
Acylmuramoyl-Alanine
   Carboxypeptidase, 30
Acylneuraminate Cytidylyltransferase,
   30
N-Acylneuraminate 9-Phosphatase, 31
N-Acylneuraminate-9-Phosphate
   Synthase, 31
Acyloxyacyl hydrolase, 523
N-Acylpeptide hydrolase, 26
Acylphosphatase, 31
Acyl Phosphate:Hexose
   Phosphotransferase, 31
5′-Acylphosphoadenosine Hydrolase, 31
Acyl-protein synthase, 588
Acyl-protein synthetase, 527
Acylpyruvate Hydrolase, 31
Acylsphingosine deacylase, 172
N-Acylsphingosine
   Galactosyltransferase, 32
Acylsphingosine kinase, 172
*Acy*I Restriction Endonuclease, 32
Adamalysin, 32
Adenain, 35
Adenase, 32
Adenine aminase, 32
Adenine aminohydrolase, 32
Adenine Deaminase, 32
Adenine Phosphoribosyltransferase, 32
N^6-Adenine-specific DNA methylase, 775
Adenosinase, 33
Adenosine aminohydrolase, 32
Adenosine Deaminase, 32
Adenosine diphosphatase, 84
Adenosine diphosphate sulfurylase, 796
Adenosine Kinase, 33
Adenosine Monophosphatase, 33
Adenosine Nucleosidase, 33
Adenosine-Phosphate Deaminase, 33
Adenosine 5′-phosphosulfate
   sulfohydrolase, 40
Adenosine Tetraphosphatase, 34

Adenosine Triphosphatase
   (Mg^{2+}-Activated), 34
S-Adenosylhomocysteinase, 34
S-Adenosylhomocysteine Deaminase, 34
Adenosylhomocysteine hydrolase, 34
S-Adenosylhomocysteine Nucleosidase,
   34
Adenosylmethionine:8-amino-7-
   ketononanoate aminotransferase, 34
Adenosylmethionine:8-Amino-7-
   Oxononanoate
   Aminotransferase, 34
S-Adenosylmethionine cleaving enzyme, 35
Adenosylmethionine Cyclotransferase, 35
S-Adenosylmethionine Decarboxylase, 35
S-Adenosylmethionine:erythromycin C
   O-methyltransferase, 307
Adenosylmethionine Hydrolase, 35
S-Adenosylmethionine:ε-*N*-L-lysine
   methyltransferase, 531
S-Adenosyl-L-methionine:precorrin-2
   methyltransferase, 694
S-Adenosylmethionine synthetase, 559
Adenovirus Protease, 35
Adenylate Cyclase, 36
Adenylate deaminase, 77
Adenylate Isopentenyltransferase, 37
Adenylate Kinase, 37
Adenylate:nucleoside phosphotransferase,
   78
Adenylosuccinase, 38
Adenylosuccinate Lyase, 38
Adenylosuccinate Synthetase, 39
Adenylyl cyclase, 37
Adenylyl-[glutamate:ammonia ligase]
   hydrolase, 39
Adenylyl-[Glutamine Synthetase]
   Hydrolase, 39
Adenylylsulfatase, 40
Adenylylsulfate:Ammonia
   Adenylyltransferase, 40
Adenylylsulfate Kinase, 40
Adenylylsulfate Reductase, 40, 41
5′-Adenylylsulfate reductase, 41
Adenylylsulfate Reductase (Glutathione),
   41
Adipate semialdehyde dehydrogenase, 636
Adipsin, 201
ADP-aldose phosphorylase, 54
ADPase, 84
ADP Deaminase, 41
ADP-Dependent Medium-Chain-
   Acyl-CoA Hydrolase, 41
ADP-Dependent Short-Chain-Acyl-CoA
   Hydrolase, 41
ADP-glucose pyrophosphorylase, 369
ADP-glucose-starch glucosyltransferase,
   781
ADP-glucose synthase, 369

ADP-*glyceromanno*-Heptose
   6-Epimerase, 41
ADP-Phosphoglycerate Phosphatase, 41
ADP-ribose-L-arginine cleaving enzyme, 41
ADP-ribose diphosphatase, 41
ADP-ribose phosphohydrolase, 41
ADP-ribose phosphorylase, 753
ADP-Ribose Pyrophosphatase, 41
ADP-Ribosylarginine Hydrolase, 41
ADP-Ribosyl-[Dinitrogen Reductase]
   Hydrolase, 42
ADP-ribosyltransferase (polymerizing), 593
ADP-sugar diphosphatase, 42
ADP-Sugar Pyrophosphatase, 42
ADP-sulfurylase, 796
ADP:Thymidine Kinase, 42
Adrenalin oxidase, 66
[β-Adrenergic Receptor] Kinase, 42
Adrenodoxin Reductase, 42, 319
Aequorin, 743
Aerolysin, 42
Aeromonas proteolytica neutral proteinase,
   860
Aeromonolysin, 860
Aeruginolysin, 42
A-Esterase, 97
*Afe*I Restriction Endonuclease, 43
*Afl*II Restriction Endonuclease, 43
*Afl*III Restriction Endonuclease, 43
Agarase, 43
Agaritine γ-Glutamyltransferase, 43
Agavain, 43
*Age*I Restriction Endonuclease, 43
Ag^+-Exporting ATPase, 43
Agmatinase, 44
Agmatine 4-Coumaroyltransferase, 44
Agmatine Deiminase, 44
Agmatine iminohydrolase, 44
Agmatine Kinase, 44
*Aha*III Restriction Endonuclease, 44
*Ahd*I Restriction Endonuclease, 44
AICAR transformylase, 674
AIR carboxylase, 675
AIR synthetase, 676
L-Alanine-adding enzyme, 845
D-Alanine:D-alanine ligase, 47
D-Alanine:Alanyl-
   Poly(glycerolphosphate) Ligase, 44
β-Alanine aminotransferase, 47
D-Alanine Aminotransferase, 45
L-Alanine Aminotransferase, 45
Alanine Carboxypeptidase, 45
L-Alanine Dehydrogenase, 46
L-Alanine:4,5-dioxovalerate
   aminotransferase, 73
D-Alanine γ-Glutamyltransferase, 46
Alanine:Glyoxylate Aminotransferase, 46
D-Alanine Hydroxymethyltransferase, 46
L-Alanine:α-ketobutyrate aminotransferase,
   250

Butyrate CoA-Transferase, 138
Butyrate Kinase, 138
γ-Butyrobetaine hydroxylase, 138
γ-Butyrobetaine:2-Oxoglutarate
　Dioxygenase, 138
Butyryl-CoA Dehydrogenase, 139
Butyryl-CoA dehydrogenase, 27
Butyryl-CoA Synthetase, 139

C

*Cac*8I Restriction Endonuclease, 140
(+)-δ-Cadinene Synthase, 140
C3 ADP-Ribosyltransferase, 140
Caffeate 3,4-Dioxygenase, 140
Caffeate *O*-Methyltransferase, 140
Caffeoyl-CoA *O*-Methyltransferase, 140
trans-Caffeoyl-CoA 3-*O*-methyltransferase,
　140
Calcidiol-1-monooxygenase, 471
Calcidiol 24-monooxygenase, 471
Calcineurin, 141
Calcium-activated neutral protease, 142
Calcium/Calmodulin-Dependent Protein
　Kinase, 141, 715
Calcium/calmodulin-dependent protein
　kinase type II, 141
Caldecrin, 141
Caldesmon Kinase, 141
Caldesmon-Phosphatase, 141
Callose synthase, 362
Calmodulin-Lysine *N*-Methyltransferase,
　141
Calpain, 142
cAMP-dependent/stimulated protein
　kinase, 712
Camphor ketolactonase I, 142
Camphor 5-*exo*-methylene hydroxylase,
　142
Camphor 1,2-Monooxygenase, 142
Camphor 5-Monooxygenase, 142
cAMP Phosphodiesterase, 209
(*S*)-Canadine Synthase, 143
Canavanase, 90
Cancer Procoagulant, 143
Candidapepsin, 143
Cannizzanase, 328
CANP, 142
Cap methyltransferase, 584
Capreomycin phosphotransferase, 861
Capsular-Polysaccharide
　Endo-1,3-α-Galactosidase, 143
Capsular-Polysaccharide-Transporting
　ATPase, 144
Carbamate Kinase, 144
N-Carbamoyl-D-Amino Acid Hydrolase,
　144
Carbamoylaspartate Decarboxylase, 144
Carbamoylaspartic dehydrase, 262
Carbamoylphosphate Synthetases, 144

Carbamoylputrescine amidase, 145
N-Carbamoylputrescine
　Amidohydrolase, 145
N-Carbamoylsarcosine Amidase, 145
N-Carbamoylsarcosine amidohydrolase,
　145
Carbamoyl-Serine Ammonia-Lyase, 146
O-Carbamoyl-L-serine deaminase, 146
Carbamylaspartotranskinase, 103
Carbamyl phosphokinase, 144
Carbon-dioxide:ammonia ligase, 144
Carbonic Anhydrase, 146
Carbon-monoxide/acetyl-CoA synthase,
　146
Carbon-Monoxide Dehydrogenase, 146
Carbon-Monoxide Oxidase, 147
Carbon Monoxide Oxygenase
　(Cytochrome *b*-561), 147
Carbonyl Reductase (NADPH), 147
Carboxyamidase, 646
Carboxyamidopeptidase, 646
2-Carboxy-D-Arabinitol-1-Phosphatase,
　147
2′-Carboxybenzalpyruvate aldolase, 154
Carboxycathepsin, 645
Carboxycyclohexadienyl Dehydratase,
　147
3-Carboxyethylcatechol 2,3-Dioxygenase,
　147
N^5-(Carboxyethyl)ornithine Synthase,
　148
6-Carboxyhexanoate:CoA ligase, 148
6-Carboxyhexanoyl-CoA Synthetase, 148
3-Carboxy-2-Hydroxyadipate
　Dehydrogenase, 148
4-Carboxy-2-Hydroxymuconate-6-
　Semialdehyde Dehydrogenase, 148
Carboxylate Reductase, 148
Carboxymethylbutenolide lactonase, 498
Carboxymethyl cellulase, 170
S-Carboxymethylcysteine Synthase, 148
Carboxymethylenebutenolidase, 149
Carboxymethylhydantoinase, 149
4-Carboxymethyl-4-Hydroxy-
　isocrotonolactonase, 149
5-Carboxymethyl-2-Hydroxymuconate
　Δ-Isomerase, 149
5-Carboxymethyl-2-Hydroxymuconate-
　Semialdehyde Dehydrogenase, 149
4-Carboxymethyl-4-Methylbutenolide
　Mutase, 149
Carboxymethyloxysuccinate Lyase, 149
Carboxy-*cis*,*cis*-Muconate Cyclase, 149
3-Carboxymuconate cyclase, 149
3-Carboxy-*cis*,*cis*-Muconate
　Cycloisomerase, 150
3-Carboxymuconate lactonizing enzyme,
　150
3-Carboxy-*cis*,*cis*-muconate lactonizing
　enzyme, 150

γ-Carboxymuconolactone
　Decarboxylase, 150
4-Carboxy-2-oxohexenedioate hydratase,
　633
Carboxypeptidase, 150
Carboxypeptidase II, 150
Carboxypeptidase A, 150
Carboxypeptidase A2, 151
Carboxypeptidase B, 151
Carboxypeptidase C, 152
Carboxypeptidase D, 152
Carboxypeptidase D, 556
Carboxypeptidase E, 152
Carboxypeptidase G, G$_1$, and G$_2$, 377
Carboxypeptidase G$_3$, 152
Carboxypeptidase H, 152
Carboxypeptidase LB, 164, 166
Carboxypeptidase M, 153
Carboxypeptidase N, 530
Carboxypeptidase P, 554
Carboxypeptidase R, 153
Carboxypeptidase S, 409
Carboxypeptidase T, 153
Carboxypeptidase Taq, 153
Carboxypeptidase U, 153
Carboxypeptidase IIW, 587
Carboxypeptidase Y, 153
4-(2-Carboxyphenyl)-2-Oxobut-3-Enoate
　Aldolase, 154
Carboxyphosphonoenolpyruvate
　phosphonomutase, 154
Carboxy-terminal procollagen peptidase,
　696
Carboxytripeptidase, 273
Carboxyvinyl-Carboxyphosphonate
　Phosphorylmutase, 154
Caricain, 154
Carnitinamidase, 154
Carnitine acetylase, 154
Carnitine *O*-Acetyltransferase, 154
Carnitine Decarboxylase, 155
L-Carnitine Dehydratase, 155
D-Carnitine dehydrogenase, 154
Carnitine 3-Dehydrogenase, 155
Carnitine *O*-Octanoyltransferase, 155
Carnitine *O*-Palmitoyltransferase, 156
Carnosinase, 156
Carnosinase, 864
Carnosine *N*-Methyltransferase, 156
Carnosine synthase, 156
Carnosine Synthetase, 156
Carotene 7,8-Desaturase, 157
ζ-Carotene desaturase, 157
β-Carotene 15,15′-Dioxygenase, 157
β-Carotene Monooxygenase, 157
Carotene oxidase, 525
Carotene Oxygenase, 157
Carveol Dehydrogenase, 158
Casbene Synthase, 158
Casein Kinase, 158

Glucose-1-Phosphate Guanylyltransferase, 370
Glucose-6-Phosphate Isomerase, 370
Glucose-phosphate kinase, 665
α-Glucose-1-phosphate phosphodiesterase, 371
Glucose-1-Phosphate Phosphodismutase, 371
Glucose-1-Phosphate Thymidylyltransferase, 371
Glucose-1-phosphate transphosphorylase, 371
Glucose-1-phosphate uridylyltransferase, 856
Glucose-1-Phospho-D-Mannosylglycoprotein Phosphodiesterase, 371
Glucose phosphomutase, 665
α-Glucosidase, 371
β-Glucosidase, 372
Glucosidase II, 550
Glucoside 3-Dehydrogenase, 372
β-Glucoside Kinase, 373
Glucosidosucrase, 371
α-Glucosiduronase, 373
Glucosinolate Sulfatase, 373
Glucosulfatase, 406
D-Glucosyl-N-acylsphingosine glucohydrolase, 373
Glucosylceramidase, 373
Glucosylceramide galactosyltransferase, 513
Glucosylceramide synthase, 172
Glucosyl-DNA β-Glucosyltransferase, 373
13-Glucosyloxydocosanoate 2′-β-glucosyltransferase, 449
Glucuronan Lyase, 374
Glucuronate dehydrogenase, 374
Glucuronate Isomerase, 374
Glucuronate-1-Phosphate Uridylyltransferase, 374
Glucuronate Reductase, 374
Glucuronate-2-Sulfatase, 374
α-Glucuronidase, 373
β-Glucuronidase, 374
Glucuronoarabinoxylan Endo-1,4-β-Xylanase, 375
Glucuronokinase, 375
D-Glucuronolactone Dehydrogenase, 375
Glucuronolactone Reductase, 375
Glucuronosyl-Disulfoglucosamine Glucuronidase, 375
Glucuronosyltransferase, 375
Glucuronosyltransferase I, 351
Glucuronoxylan 4-O-Methyltransferase, 376
Glucuronylgalactosylproteoglycan β-1,4-N-Acetylgalactosaminyltransferase, 376

Glucuronyl-N-Acetylgalactosaminyl-proteoglycan β-1,4-N-Acetylgalacto-saminyltransferase, 376
Glu-Glu dipeptidase, 387
Glusulase, 376
Glutaconate CoA-Transferase, 376
Glutaconyl-CoA Decarboxylase, 376
Glutamate Acetyltransferase, 377
Glutamate activating enzyme, 377
D-Glutamate-adding enzyme, 845
Glutamate Adenylyltransferase, 377
[Glutamate:ammonia ligase] adenylyltransferase, 384
D-Glutamate(D-Aspartate) Oxidase, 377
Glutamate Carboxypeptidase, 377
Glutamate Carboxypeptidase II, 378
D-Glutamate Cyclase, 378
Glutamate:cysteine ligase, 386
Glutamate Decarboxylase, 378
Glutamate Dehydrogenases, 378
Glutamate:ethylamine ligase, 809
Glutamate Formimidoyltransferase, 379
Glutamate formiminotransferase, 379
Glutamate formyltransferase, 379
Glutamate 1-Kinase, 379
Glutamate 5-Kinase, 380
Glutamate:Methylamine Ligase, 380
Glutamate mutase, 564
Glutamate:ornithine transaminase, 629
Glutamate:oxaloacetate transaminase, 102
D-Glutamate Oxidase, 380
L-Glutamate Oxidase, 380
Glutamate Racemase, 380
Glutamate-1-Semialdehyde 2,1-Aminomutase, 380
Glutamate-1-semialdehyde aminotransferase, 380
Glutamate-5-Semialdehyde Dehydrogenase, 381
Glutamate-γ-semialdehyde dehydrogenase, 381
Glutamate Synthases, 381
Glutamate:tRNA ligase, 389
L-Glutamic acid activating enzyme, 377
Glutamic–alanine transaminase, 45
Glutamic:aspartic transaminase, 102
Glutamic–pyruvic transaminase, 45
Glutaminase, 381
D-Glutaminase, 382
Glutaminase/Asparaginase, 382
Glutamin-(asparagin-)ase, 382
Glutamine N-Acyltransferase, 382
L-Glutamine amidohydrolase, 381
Glutamine:Fructose-6-Phosphate Aminotransferase (Isomerizing), 382
Glutamine:scyllo-Inosose Aminotransferase, 382
Glutamine:keto-scyllo-inositol aminotransferase, 382
Glutamine:oxo-acid transaminase, 383

Glutamine N-Phenylacetyltransferase, 383
Glutamine:Phenylpyruvate Aminotransferase, 383
Glutamine phosphoribosyl-pyrophosphate amidotransferase, 64
Glutamine:Pyruvate Aminotransferase, 383
Glutamine Synthetase, 383
[Glutamine-Synthetase] Adenylyltransferase, 384
[Glutamine synthetase] deadenylylating enzyme, 39
Glutamine transaminase K, 383
Glutamine transaminase L, 383
Glutamine:tRNA ligase, 385
Glutaminyl cyclase, 384
Glutaminyl-Peptide Cyclotransferase, 385
Glutaminyl-tRNA cyclotransferase, 384
Glutaminyl-tRNA Synthetase, 385
Glutaminyl-tRNA synthetase (glutamine-hydrolyzing), 389
Glutamyl Aminopeptidase, 385
γ-Glutamylcyclotransferase
γ-Glutamylcysteine:glycine ligase, 392
γ-Glutamylcysteine Synthetase, 386
γ-L-Glutamyl-L-cysteinylglycine:spermidine amidase, 392
γ-L-Glutamyl-L-cysteinylglycine:spermidine ligase (ADP forming), 393
γ-D-Glutamyl-L-diamino acid endopeptidase, 274, 386
γ-D-Glutamyl-L-diamino acid endopeptidase I, 386
γ-D-Glutamyl-(L)-meso-Diaminopimelate Peptidase I, 386
Glutamyl Endopeptidase I, 387
Glutamyl Endopeptidase II, 387
γ-Glutamylglutamate Carboxypeptidase, 387
α-Glutamylglutamate Dipeptidase, 387
γ-Glutamylhistamine Synthetase, 387
γ-Glutamyl Hydrolase, 387
γ-Glutamyl kinase, 380
γ-Glutamylmethylamide synthetase, 380
γ-Glutamylphosphate reductase, 381
D-Glutamyltransferase, 387
γ-Glutamyltransferase, 388
D-Glutamyl transpeptidase, 387
γ-Glutamyl Transpeptidase, 388
Glutamyl-tRNA^Gln Amidotransferase, 389
Glutamyl-tRNA Synthetase, 389
Glutarate:CoA ligase, 389
Glutarate Semialdehyde Dehydrogenase, 389
Glutaryl-CoA Dehydrogenase, 389

2-Methylacyl-CoA Dehydrogenase, 562
α-Methylacyl-CoA Racemase (or, Epimerase), 562
3-Methyladenine-DNA glycosylase I, 279
3-Methyladenine-DNA glycosylase II, 279
1-Methyladenosine Nucleosidase, 563
N-Methyl-L-Alanine Dehydrogenase, 563
Methylamine dehydrogenase, 65
Methylamine:Glutamate Methyltransferase, 563
N-Methyl-L-Amino-Acid Oxidase, 563
Methylarsonate Reductase, 563
Methylarsonite Methyltransferase, 563
β-Methylaspartase, 563
β-Methylaspartate ammonia-lyase, 563
Methylaspartate-glutamate mutase, 564
β-Methylaspartate Mutase, 564
Methylated-DNA:[Protein]-Cysteine S-Methyltransferase, 564
2-Methyl-Branched-Chain-Enoyl-CoA Reductase, 564
3-Methylbutanal Reductase, 564
2-Methylcitrate Dehydratase, 565
2-Methylcitrate Synthase, 565
(S)-N-Methylcoclaurine oxidase (C-O phenol-coupling), 122
Methyl-S-Coenzyme-M Methylreductase, 565
β-Methylcrotonyl-CoA Carboxylase, 565
Methylcysteine Synthase, 565
Methylcysteine synthase, 221, 565
5-Methyldeoxycytidine-5′-Phosphate Kinase, 565
Methylenediurea Deaminase, 566
Methylene-Fatty-Acyl-Phospholipid Synthase, 566
4-Methyleneglutamate:ammonia ligase, 566
4-Methyleneglutaminase, 566
4-Methyleneglutamine deamidase, 566
Methyleneglutamine Synthetase, 566
2-Methyleneglutarate Mutase, 566
4-Methyl-3-enelactone methylisomerase, 149
3-Methyleneoxindole Reductase, 566
5,10-Methylenetetrahydrofolate Dehydrogenase, 567
5,10-Methylenetetrahydrofolate Reductase, 567
5,10-Methylenetetrahydrofolate:tRNA (Uracil-5-) Methyltransferase (FADH₂ Oxidizing), 567
Methylenetetrahydromethanopterin Dehydrogenase, 568
3-Methylglutaconyl-CoA Hydratase, 568
Methylglutamate Dehydrogenase, 568
N-Methylglutamate synthase, 563
Methylglyoxalase, 513
Methylglyoxal dehydrogenase, 499
Methylglyoxal reductase, 507
Methylglyoxal Synthase, 568

Methylguanidinase, 568
O-6-Methylguanine-DNA-alkyltransferase, 564
6-O-Methylguanine-DNA methyltransferase, 564
N-Methylhydantoin amidohydrolase, 568
N-Methylhydantoinase (ATP-Hydrolyzing), 568
N-Methylhydrazine Demethylase, 568
4-Methyl-3-Hydroxyanthranilate Adenylyltransferase, 569
4-Methyl-3-hydroxyanthranilate:AMP ligase, 569
2-Methyl-3-hydroxybutyryl-CoA dehydrogenase, 454
Methylhydroxypyridine carboxylate dioxygenase, 456
Methylhydroxypyridinecarboxylate oxidase, 456
2-Methylisocitrate Dehydratase, 569
Methylisocitrate Lyase, 569
Methylitaconate Δ-Isomerase, 569
3-Methyl-2-ketobutanoate dehydrogenase complex, 131
3-Methyl-2-ketobutanoate dehydrogenase (lipoamide), 501
[3-Methyl-2-Ketobutanoate Dehydrogenase (Lipoamide)] Kinase, 569
[3-Methyl-2-Ketobutanoate Dehydrogenase (Lipoamide)]-Phosphatase, 569
3-Methyl-2-Ketobutanoate Hydroxymethyltransferase, 569
N-Methyl-2-Ketoglutaramate Hydrolase, 570
N⁶-Methyl-L-Lysine Oxidase, 570
(R)-2-Methylmalate Dehydratase, 570
(S)-2-Methylmalate Dehydratase, 570
Methylmalonate-Semialdehyde Dehydrogenase (Acylating), 570
Methylmalonyl-CoA Carboxyltransferase, 570
Methylmalonyl-CoA deacylase, 571
Methylmalonyl-CoA Decarboxylase, 571
Methylmalonyl-CoA Epimerase, 571
D-Methylmalonyl-CoA hydrolase, 571
(S)-Methylmalonyl-CoA Hydrolase, 571
Methylmalonyl-CoA Mutase, 571
Methylmalonyl-CoA racemase, 571
Methylmercaptan oxidase, 557
Methylmethionine-sulfonium salt hydrolase, 35
4-Methylmuconolactone methylisomerase, 149
6-O-Methylnorlaudanosoline 5′-O-Methyltransferase, 572
Methyl-ONN-Azoxymethanol Glucosyltransferase, 572
4-Methyloxaloacetate Esterase, 572

3-Methyl-2-oxobutanoate dehydrogenase complex, 131
3-Methyl-2-oxobutanoate dehydrogenase (lipoamide), 501
[3-Methyl-2-oxobutanoate dehydrogenase (lipoamide)] kinase, 569
[3-Methyl-2-oxobutanoate dehydrogenase (lipoamide)]-phosphatase, 569
3-Methyl-2-oxobutanoate hydroxymethyltransferase, 569
Methylphenyltetrahydropyridine N-Monooxygenase, 572
N-Methylphosphoethanolamine Cytidylyltransferase, 572
Methylphosphothioglycerate Phosphatase, 572
Methylquercetagetin 6-O-Methyltransferase, 572
3-Methylquercetin 7-O-Methyltransferase, 572
6-Methylsalicylate Decarboxylase, 573
6-Methylsalicylate Synthase, 573
Methylsterol hydroxylase, 573
Methylsterol Monooxygenase, 573
N⁵-Methyltetrahydrofolate: Corrinoid/Iron-Sulfur Protein Methyltransferase, 573
N⁵-Methyltetrahydrofolate: Homocysteine Methyltransferase, 573
Methyltetrahydroprotoberberine 14-hydroxylase, 574
(S)-cis-N-Methyltetrahydroprotoberberine-14-hydroxylase, 574
Methyltetrahydroprotoberberine 14-Monooxygenase, 574
5-Methyltetrahydropteroyltriglutamate: Homocysteine S-Methyltransferase, 574
Methylthioadenosine Nucleosidase, 574
5′-Methylthioadenosine Phosphorylase, 574
5-Methylthioribose Kinase, 574
5-Methylthioribose-1-Phosphate Isomerase, 575
3-Methyl-1-(trihydroxyphenyl)butan-1-one synthase, 653
Methylumbelliferyl-Acetate Deacetylase, 575
Metridin, 575
Metridium proteinase A, 575
Met-Xaa Dipeptidase, 575
Met-X dipeptidase, 575
Metyrapone Reductase, 575
Mevaldate Reductase, 575
Mevalonate diphosphate decarboxylase, 576
Mevalonate Kinase, 576
Mevalonate-5-Pyrophosphate Decarboxylase, 576
Mexicain, 576

Papaya proteinase II, **189**
Papaya proteinase IV, **407**
Papaya peptidase A, **154**
Papaya peptidase B, **407**
PAPS reductase, **660**
PAPS reductase [thioredoxin-dependent], **660**
PAPS sulfatase, **660**
Paragloboside synthase, **366**
Paraoxonase, **97**
Paraoxon hydrolase, **97**
Parapepsin I, **642**
Parapepsin II, **353**
Particle-bound aminopeptidase, **553**
PC1, **702**
PC2, **702**
PCCase, **702**
*Pci*I Restriction Endonuclease, **639**
PCP, **696**
PCP hydroxylase, **641**
PE II, **296**
Pectate Lyase, 640
Pectinase, **690**
Pectin demethoxylase, **640**
Pectin depolymerase, **690**
Pectinesterase, 640
Pectin Lyase, 640
Pectin methoxylase, **640**
Pectin methylesterase, **640**
Pectin transeliminase, **640**
Penicillin acylase, **640**
Penicillin Amidase, 640
*Penicillium*Nuclease P₁, **641**
Penicillopepsin, 641
Pentachlorophenol dehalogenase, **641**
Pentachlorophenol hydroxylase, **641**
Pentachlorophenol Monooxygenase, 641
Pentalenene Synthase, 641
Pentanamidase, 641
Pentaphosphate guanosine-3′-pyrophosphohydrolase, **415**
trans-Pentaprenyl *trans*Reductase, **642**
Pentokinase, **749**
Pentose-5-phosphate 3-epimerase, **755**
Pentoxifylline Reductase, 642
PEP carboxylase, **663**
PEP carboxykinase, **662**
PEPCK, **662**
PepD, **272**
PepDA, **272**
PepE, **626**
PepF, **626**
PepN, **535**
PepO, **627**
PepP, **864**
Pepsin II, **353**
Pepsin A, 642
Pepsin B, 642
Pepsin C, **353**
Pepsin F, 642

γ-Peptidase, **864**
L-**Peptidase, 528**
Peptidase A (*S. typhimurium* and *E. coli*), **517**
Peptidase D, **864**
Peptidase E, **553**
Peptidase M, **561**
Peptidase N, **535**
Peptidase P, **645, 864**
Peptidase S, **516**
Peptidase T, 643
Peptide-N⁴-(N-Acetyl-β-Glucosaminyl)Asparagine Amidase, 643
Peptide α-N-Acetyltransferase, 643
Peptide Antibiotic Lactonase, 643
Peptide-Aspartate β-Dioxygenase, 643
Peptide hormone inactivating endopeptidase, **231**
Peptide-methionine sulfoxide reductase, **716**
Peptide N-myristoyltransferase, **407**
Peptide N-tetradecanoyltransferase, **407**
Peptide-Transporting ATPase, 644
Peptide-Tryptophan 2,3-Dioxygenase, 644
Peptidoglutaminase I, **646**
Peptidoglycan β-N-Acetylmuramidase, 644
Peptidoglycan Glycosyltransferase, 644
Peptidyl α-amidating enzyme, **646**
Peptidylamidoglycolate Lyase, 644
Peptidyl-D-amino acid hydrolase, **617**
Peptidyl-Asp Metalloendopeptidase, 645
Peptidyl carboxyamidase, **646**
Peptidyl dipeptidase I, **645**
Peptidyl dipeptidase II, **645**
Peptidyl-Dipeptidase A, 645
Peptidyl-Dipeptidase B, 645
Peptidyl-Dipeptidase Dcp, 645
Peptidyl-Dipeptidase (*Streptomyces*), 646
Peptidyl-Glutaminase, 646
Peptidyl-Glycinamidase, 646
Peptidylglycine 2-hydroxylase, **646**
Peptidylglycine Monooxygenase, 646
Peptidyl-Lys Metalloendopeptidase, 646
Peptidylprolyl *cis-trans* isomerase, **647**
Peptidylprolyl Isomerase, 647
Peptidyltransferase, 647
Peptidyl-tRNA hydrolase, **69**
PepX, **865**
Perillyl-Alcohol Dehydrogenase, 647
Periplasmic phosphoglycerotransferase, **554**
Peroxidase, 647
Peroxisomal 3-oxoacyl coenzyme A thiolase, **702**
Peroxisome-Assembly ATPases, 648
Persulfurase, **222**
Pertussis Toxin, 648

PfAPG, **687**
*Pfl*FI Restriction Endonuclease, **648**
PFL activase, **329**
*Pfl*MI Restriction Endonuclease, **648**
PGD₂ 11-ketoreductase, **705**
PGD₂ synthase, **703**
PGE isomerase, **705**
PGH-PGD isomerase, **703**
PGH-PGE isomerase, **705**
PGI₂ synthase, **706**
P-glycoprotein, **867**
PG synthase, **704**
Phage-procoat-leader peptidase, **773**
Phage-T4 UV endonuclease, **275**
Phaseolin, **515**
Phaseollidin Hydratase, 648
Phenolase, **162, 582**
Phenol β-Glucosyltransferase, 648
Phenol hydroxylase, **648**
Phenol o-hydroxylase, **648**
Phenol O-Methyltransferase, 648
Phenol 2-Monooxygenase, 648
Phenol sulfotransferase, **99**
Phenoxazinone Synthase, 649
Phenylacetaldehyde Dehydrogenase, 649
Phenylacetate:CoA ligase, **649**
Phenylacetyl-CoA Hydrolase, 649
Phenylacetyl-CoA Synthetase, 649
Phenylalaninase, **650**
Phenylalanine N-Acetyltransferase, 649
Phenylalanine Adenylyltransferase, 649
Phenylalanine aminotransferase, **650**
Phenylalanine Ammonia-Lyase, 650
Phenylalanine Decarboxylase, 650
Phenylalanine Dehydrogenase, 650
Phenylalanine (Histidine) Aminotransferase, 650
Phenylalanine 4-hydroxylase, **650**
Phenylalanine 2-Monooxygenase, 650
Phenylalanine 4-Monooxygenase, 651
L-Phenylalanine oxidase (deaminating and decarboxylating), **650**
Phenylalanine Racemase, 651
Phenylalanine racemase (ATP-hydrolyzing), **651**
Phenylalanine:tRNA ligase, **651**
Phenylalanyl-tRNA Synthetase, 651
Phenylethanolamine N-Methyltransferase, 652
Phenylglyoxylate Dehydrogenase (Acylating), 652
Phenylpyruvate Decarboxylase, 652
Phenylpyruvate Tautomerase, 653
Phenylserine Aldolase, 653
PHIE, **231**
Phloretin-glucosidase, **407**
Phloretin Hydrolase, 653
Phlorizin hydrolase, **407**
Phloroglucinol Reductase, 653
Phloroisovalerophenone Synthase, 653

The Enzyme Reference: A Comprehensive Guidebook to Enzyme Nomenclature, Reactions, and Methods